# Who Owns the World's Media?

# WHO OWNS THE WORLD'S MEDIA?

## Media Concentration and Ownership around the World

ELI M. NOAM
AND THE INTERNATIONAL MEDIA CONCENTRATION COLLABORATION

OXFORD
UNIVERSITY PRESS

OXFORD
UNIVERSITY PRESS

Oxford University Press is a department of the University of Oxford.
It furthers the University's objective of excellence in research, scholarship,
and education by publishing worldwide.

Oxford New York
Auckland Cape Town Dar es Salaam Hong Kong Karachi
Kuala Lumpur Madrid Melbourne Mexico City Nairobi
New Delhi Shanghai Taipei Toronto

With offices in
Argentina Austria Brazil Chile Czech Republic France Greece
Guatemala Hungary Italy Japan Poland Portugal Singapore
South Korea Switzerland Thailand Turkey Ukraine Vietnam



Published in the United States of America by
Oxford University Press
198 Madison Avenue, New York, NY 10016



Library of Congress Cataloging-in-Publication Data
Noam, Eli M.
Who Owns the World's Media?: media concentration and ownership around the world/ by Eli M. Noam
and contributing authors
p. cm.
Includes bibliographical references and index.
ISBN 978–0–19–998723–8 (hardcover : alk. paper) 1. Mass media—Ownership. 2. Mass media—Economic aspects.
3. Telecommunication—Ownership. 4. Telecommunication—Economic aspects. I. Noam, Eli M.
P96.E25 W58 2015
338.4'730223—dc23
2014010928

9 8 7 6 5 4 3 2

Printed in the United States of America on acid-free paper

The International Media Concentration Collaboration is a group of scholars assembled by Eli M. Noam
who have contributed country chapters to this volume.

# Contents

# Acknowledgments

THIS BOOK is a project that started out as a good and simple idea and then became better but more challenging. At its inception were friends and colleagues who became a mainstay of the project—first, Koichiro Hayashi and Kiyoshi Nakamura of Japan—who then supervised the work on East Asian countries. They were joined by a French team—Patrick Badillo, Dominique Bourgeois, and Jean-Baptiste Lesourd—who recruited many of the European author teams, as well as authored chapter 33, providing a European summary. Together, and with Jason Buckweitz, we formed the project's executive committee. We also organized conferences that brought together the author teams, held in New York, Paris, Brussels, and Yokohama. This project would not exist without their encouragement, contributions, and generosity.

Equally central to this project was Marius Dragomir, senior manager and publications editor of the Open Society Media Program. Jointly with his colleagues Mark Thompson, Gordana Jankovic, and the Open Society Foundation's then president, Arye Neier, OSF supported this project throughout its progress. They warmly encouraged us, yet left us alone in conception, selection, and conclusion.

At Columbia Business School, many people contributed their talent and energy. Jason Buckweitz, Associate Director of the Columbia Institute for Tele-Information (CITI), held this project together with an amazing combination of skills in research, management, technology, and law. He also coauthored chapter 36 on ownership. Essential to the quantitative work was Paul Mutter, the project manager at CITI, who supported both data analysis and editing. He was assisted by teams of summer college interns and research assistants.

Paul succeeded Liwei Wang, who contributed to making the individual-country chapters consistent by providing feedback to their authors and doing other data work, especially on news attention weights.

Many thanks to go to Caitlin Bartow and Corey Spencer, also at CITI, for superb administrative assistance on numerous levels and tasks.

At Oxford University Press, I thank James Cook, editor, for supporting this big project with trust, patience, and good judgment. We also thank Peter Worger, Amy Klopfenstein, and Narmatha Vaithiyanathan for their active contributions.

Most of all, I want to thank my coauthors. Fifty-five of the most talented media researchers from thirty countries came together in this project. They are among the best experts of their countries' media systems and industries. They worked hard in gathering their data, analyzing it, and authoring their country reports. Many of them also waited patiently as chapters on other countries were still being completed and revised. This was followed by my own summaries and interpretations, based on their work and data, which grew larger and broader in order to create a book that is more than a collection of excellent but separate country studies. In the process of this project, we all met on three continents, became friends, and formed a genuine community of scholars. It has been a pleasure and honor to work with them.

Still greater thanks go to my beloved wife Nadine Strossen, who for eighteen years as national president of the American Civil Liberties Union championed media liberty, freedom, and pluralism. She inspires me every day.

# List of Contributors

## MAIN AUTHOR AND EDITOR

**Eli Noam** is Professor of Economics and Finance at the Columbia Business School since 1976, and its Garrett Professor of Public Policy and Business Responsibility. He is also Director of the Columbia Institute for Tele-Information. He served for three years as a Commissioner for Public Services of New York State, and was appointed by the White House to the President's IT Advisory Committee. Noam has published 29 books and over 350 articles in economics journals, law reviews, and interdisciplinary journals, and was a regular columnist for the *Financial Times* online edition. His recent books include *Broadband Networks and Smart Grids (Springer, 2013); Media Ownership and Concentration in America* (Oxford, 2009, "Best Book in Media Economics" of the AJMCR); and *Peer to Peer Video* (Springer, 2008) He is completing a monograph *Tools and Principles for Management in the Information Economy.* He was Chairman of the International Media Management Association (2012–14). Noam has served on boards for the Federal government's telecommunications network, the IRS computer system, the National Commission on the Status of Women in Computing, the Governor's Task Force on New Media, the Nexus Mundi Foundation, the Oxford Internet Institute, and Jones International University (the first accredited online university). He served on advisory boards for the governments of Ireland and Sweden. He received four degrees from Harvard, including a PhD in economics, and honorary doctorates from the University of Munich and the University of Marseilles.

## MEMBERS OF THE PROJECT BOARD AS WELL AS COUNTRY AUTHORS

**Patrick-Yves Badillo** is Professor at Geneva University, and Director- founder of Medi@LAB-Genève, a research institute on media and communications science. He is head of the masters program in "Information Communication et Médias" and is in charge of the advanced studies diplomas in digital communication and in advocacy and external relations.

He is an expert at the European Research Council, member of the "Conseil National des Universités" (2011–2014), the Swiss and the Belgian National Science Foundation, and scientific advisor for the "Chaire d'Economie Numérique Paris Dauphine". He is head of the project "Innovation and Social Networks" supported by the Swiss National Science Foundation, with links to Stanford University. Badillo was professor at Aix-Marseille University. He served as director (1997–2008) of the « Ecole de Journalisme et de Communication de Marseille » (EJCM), and as director of the IRSIC Laboratory and project manager (National Research Agency in France).

He obtained his PhD. in the field of New Media and Innovation (Paris I Panthéon Sorbonne University, and a post master diploma at Paris Dauphine. He first worked at France Télécom - Centre National d'Études des Télécommunications (CNET), as head of studies attached to the General Director of the CNET.

**Dominique Bourgeois** is Professor and Chair in Communication and Media Science, University of Fribourg, Switzerland, since 2009. She is the head of the Department of Communication and Media Research. Formerly, she was Professor in Economics at the University of Angers and then at Aix-Marseille University, France. She obtained an M.S. in Economics and languages from the University of Grenoble, a first PhD. in International Economics and a second PhD. in Economics/Econometrics, Aix-Marseille University. She is head of the Master program in "Business Communication" and of the undergraduate programs in Media and Communication at the University of Fribourg. She is an expert for the Swiss National Science Foundation and the Swiss Association of Communication and Media Research, and member of the advisory board of *Studies in Communication Sciences*. She produced several official reports for various institutions, especially in the field of econometric forecasting, and has published books and journal articles. Her recent research areas include media economics, new media, uses of information and communication technology, e-learning, business communication and the development of *e*-activities.

**Koichiro Hayashi** is Vice President and Professor of Law and Economics at the Institute of Information Security (IISEC), a newly established graduate school specializing in information security. He is also an affiliated researcher with Columbia Institute for Tele-Information (CITI) at the Columbia Business School. He obtained PhD. in economics from Kyoto University (Japan) in 1991 and J.D. from Keio University (Japan) in 2004. After graduating from Tokyo University in 1963, he joined NTT and served in various executive positions including President and CEO of NTT America. In 1997, he changed to academia and engaged in establishing IISEC, after a seven-year professorship at Keio University. His major works are *The Age of Infomunication* (1984) and many articles on interconnection, universal service, copyrights, convergence, media law, digital terrestrial television, and literacy.

**Kiyoshi Nakamura**, Doctor of Science and Professor of Media Economics, is a faculty member of School of International Liberal Studies (SILS) at Waseda University, Tokyo, Japan. His scholarly interests focus on media industrial organization studies centering public broadcasting. He has published books and articles about the policy issues of Japan's broadcasting in a digital environment. He was Visiting Scholar at Harvard University, St. Antony's College, Oxford University, UK

(1996–97), and Columbia University. His major publications in English include co-editorship of *Digital Broadcasting, Policy and Practice in the Americas, Europe and Japan,* (2006), and *Convergence of Telecommunications and Broadcasting in Japan, United Kingdom, and Germany* (2001).

## AUTHORS OF COUNTRY REPORTS (IN THE ORDER OF THEIR REPORTS IN THE BOOK)

**Peggy Valcke** is research professor at KU Leuven in Belgium, part-time professor at the European University Institute in Florence, Italy, and visiting professor at the University of Tilburg in the Netherlands. Her areas of expertise include legal aspects of media innovation, media pluralism, and the interaction between media/telecommunications regulation and competition law.

**Jo Groebel** is the director of the German Digital Institute in Berlin. Before, he was Director General of the European Institute for the Media and professor of Media Psychology at the Utrecht University, amongst other positions. Jo Groebel holds visiting professorships at UCLA, St. Gallen Universtity und Amsterdam University.

**Mikko Grönlund** is Research Manager of the Brahea Centre at the University of Turku. Mr Grönlund has extensive experience in the field development of creative and media industries and businesses, their structure, operations and markets. His studies have been supported by institutions such as the European Commission, Finnish ministries of communication, culture, education and labor, industry federations, and media industry firms. He has authored several publications on media and creative industries.

**Jean-Baptiste Lesourd** is Professor emeritus of Financial and Economic Information at the Université de la Méditerranée, School of Journalism and Communication, Marseille, France. He is currently affiliated with the Research Group on Media, Information and Communication (MédiaSic). His previous position was with the Centre National de la Recherche Scientifique, where he was a Research Professor in Economics and Management.

**Julia Bösch** holds a Master Degree from the University München. She is an entrepreneur, working for the largest incubator of internet start-ups in Germany.

**Max-Josef Meier** holds a Master Degree in Economics from the University of Munich. During his studies he co-founded STYLIGHT (www.stylight.com), an online shopping platform for fashion, where he currently serves as COO & CFO.

**Philipp Rösch-Schlanderer** holds a Master Degree from the University of Munich. During his studies he served as CEO of a real estate investment firm and a beverage distributor with about 30 employees. He is an active member of Munich's entrepreneurship community where he has organized university classes and panel discussions about venture capital in the past. He serves as CEO at eGym (www.eGym.de), the digital fitness revolution.

**Achim Ekkehard Henning Wolf** earned a Bachelor in Economics from London Guildhall University (UK), and master degrees in economics, business research, as well as a PhD in Economics from Ludwig Maximilian University Munich (Germany). His research focuses on innovation and digital media. After working with large telecom providers, internet start ups, and consulting firms, he is now an independent business consultant on subjects of innovation and digital services.

**Dr. Roddy Flynn** is Lecturer at the School of Communications, Dublin City University, where he is Chair of the MA program in Film and Television. He writes extensively on media policy and in particular on questions of media pluralism and media ownership. He is also the author of two forthcoming books on Irish cinema: "The A-Z of Irish Cinema" (with Tony Tracy) and "Cinema and State: Irish Film Policy since 1922".

**Paschal Preston** holds a research professorship in the School of Communications in Dublin City University. He is the founding director of the COMTEC research unit, one of the first social science research centers in the Irish university sector.

**Giuseppe Richeri** is professor emeritus at the University of Lugano (Switzerland), formerly Dean of the Faculty of Communication Sciences and director of the Institute of Media and Journalism. He is currently director of the Observatory of Media and Communications in China that he founded in 2005, and an Invited Professor at the Communication University of China in Beijing. His recent books include *Economia dei Media* (Laterza, Roma 2012) and *Encoding the Olympics* (Routledge, 2010, with L.Qing).

**Benedetta Prario** is senior lecturer at the Faculty of Communication Sciences of the University of Lugano (Switzerland). She is teaching "Media economics and politics" and "Strategies and industries of new media" in the Media Management Master of the University. Her research is focused on new media economics and strategies, in particular on the new forms of television: mobile television, IPTV, over-the-top television and video-on-demand. Author and co-author of books and scientific articles as "Le trasformazioni dell'impresa televisiva verso l'era digitale", "New ways of consumption: the audiences of public service media in Italy and Switzerland" and "The commercialization of television in Switzerland: a matter of neighbours".

**Joost Van Dreunen** is fascinated by games and human behavior. His research explores video games as an entryway to contemporary media culture. He received a Master's degree in Media studies in Amsterdam, and a doctorate from Columbia University for his dissertation on "Social Gaming and Communicative Exchange." Joost is an adjunct professor at NYU Game Center. He is also founder and CEO of the online games research firm SuperData.

**Paulo Faustino** has a PhD in media studies and teaches media management, economics and marketing in Portuguese Catholic University and Polytechnic Leiria Institute. He is President of Media XXI/Formalpress (a consulting, research & publishing company) and director of Media XXI Magazine. Paulo Faustino published 6 books (2 as co-author) and several academic articles. He is editorial board member of the International Journal on Media Management and board member of the European Association of Media Management Education. Previously, he worked 15 years in the media industry and was advisor to government and media companies. He was Vice-President of the Portuguese Press Association, and serves as Chairman of the International Media Management Academic Association.

**Elena Vartanova** is Professor, Dean, and Chair in Media Theory and Media Economics at the Faculty of Journalism, Moscow State University. Her research interests include media systems in Nordic countries, information society, post-Soviet transformation of Russian media, and media economics. Vartanova is an author of three books, an editor of six books, and has published more than 100 research articles in Russian academic magazines.

**Juan P. Artero** is an Assistant Professor of Media Management and Structure of Media Industries at the School of Communication of the University of Navarra. He holds an MSc in Management from City University, London, and a PhD from the University of Navarra. He is research director for the International Center for Media Companies at the IESE Business School, Madrid. He has written several books and academic papers.

**Alfonso Sãnchez-Tabernero** is the Rector (President) of the University of Navarra in Spain, where he also teaches management of media companies. He was Dean of the School of Communications at the University of Navarra from 1996 to 2005. He has published more than fifty articles about media management and media economics in academic journals, and serves on several editorial boards. Recent books include a co-authored work titled *Media Concentration in the European Market: New Trends and Challenges* (2003); *The Content of Media: Quality, Profits and Competition* (2009); and *Innovación en los medios. La ruta del cambio* (with Francisco Pérez-Latre), (2013).

**Robert G. Picard** is fellow of the Royal Society of the Arts, and an affiliated fellow of the Information Society Project at Yale Law School, Yale University. He was formerly Director of Research at the Reuters Institute, University of Oxford, and a research fellow at Green Templeton College, Oxford. He is also on the faculties of the University of Canberra, Australia, the Institute of Media and Entertainment, IESE Business School in New York, the Center for Media Management and Economics at Tsinghua University in China, and the University of Tampere, Finland.

He is the author and editor of 30 books including *Handbook on the Economics of the Media, Media Clusters: Spatial Agglomeration and Content Capabilities (Edward Elgar Publishing), The Economics and Financing of Media Companies* (Fordham University Press), *Media Product Portfolios: Issues in Managing Multiple Products and Services* (Lawrence Erlbaum Publishers), *Media Firms: Structures, Operations and Performance* (Lawrence Erlbaum Publishers), *The Internet and the Mass Media* (Sage), *The Newspaper Publishing Industry* (Allyn & Bacon), *Media Economics: Concepts and Issues* (Sage), *Press Concentration and Monopoly: New Perspectives on Newspaper Ownership and Operation* (Ablex), and *Joint Operating Agreements: The Newspaper Preservation Act and Its Application* (Ablex). He has written hundreds of articles on media issues for scholarly journals and industry publications. Picard has been editor of the *Journal of Media Business Studies* and *The Journal of Media Economics*. He serves on the editorial boards of numerous other journals. Picard received a PhD at the University of Missouri, after completing an MA at California State University, Fullerton, and a BA at La Sierra University. He was a fellow at the Shorenstein Center at the J.F. Kennedy School of Government, Harvard University. and the Reuters Institute in the Department of Politics and International Relations at the University of Oxford. He has been a consultant for government agencies, investment firms, and media companies and media labor organizations. Picard has testified before Congressional and Parliamentary committees and inquiry boards and provided expert testimony in cases involving antitrust and anticartel issues.

**Hüseyin Kemal Bayazit** is a researcher and scholar of innovation strategy, with particular background in communications and media. He has an academic political science and public policy background from Columbia, where he participated in a variety of international studies at CITI. Byazit has been a high level manager in Turkish industry. He has taught at various institutions in Turkey, and his current work focuses on strategies for innovation, both in practice and in conceptual theory.

**Petros Iosifidis** MA and PhD at City University, London and University of Westminster. Taught at the University of North London. Heads the Centre for International Communications & Society in the Department of Sociology.

**Dwayne Winseck** is Professor at the School of Journalism and Communication, Carleton University, Ottawa, Canada. His recent co-authored book, *Communication and Empire: Media, Markets and Globalization, 1860–1930* (2007) won the Canadian Communication Association's award for best book of the year in 2008. A new co-edited book is *Political Economies of the Media: the Transformation of the Global Media Industries.* He lived and taught in Britain, China, Cyprus, and the United States.

**Guillermo Mastrini** is professor of the School of Communication at the National University of Quilmes and at the University of Buenos Aires. He was president of the Argentine Federation of Social Communication Schools. (His publications include: *Globalización y monopolios en la comunicación en América Latina*. He is co-author of *Periodistas y magnates,* (2006); and of *Los dueños de la Palabra,* (2009)

**Ana Bizberge** is a doctoral candidate at the University of Buenos Aires. She holds a MA in Cultural Industries and a BA in

Communication Science. She is a Assistant Professor in a course of Policy and Planning of Communication. Her research centers on communication policies, political economy of media, digitization of audiovisual industries, and convergence. She participates in research projects about policies in communication at the universities of Buenos Aires and Quilmes. She published *Televisión Digital Terrestre: ¿cambio de estatuto de la radiodifusión?* (Prometeo, 2010) and, with Guillermo Mastrini and Diego de Charras, *Las políticas de comunicación en el S.XXI* (La Crujía, 2013). She co-authored with Guillermo Mastrini, Martín Becerra and Fernando Krakowiak several papers about digital television at international journals.

**Martín Becerra** is professor of the School of Communication at the National University of Quilmes and at the University of Buenos Aires. He is also member of the National Council of Scientific and Technical Research and holds a PhD in Communication Sciences from the Autonomous University of Barcelona (Spain). Becerra is author of *Sociedad de la Información: proyecto, convergencia, divergencia*; co-author of *Periodismo de Investigación en la Argentina* (2007); and co-author of *Periodistas y magnate,* (2006), and of *Los dueños de la palabra* (2009).

**Sonia Virginia Moreira** is a journalist, and holds a Master in Journalism from the University of Colorado and Doctor in Communication Sciences from the University of São Paulo. She is Associate Professor at the Rio de Janeiro State University (UERJ) in the area of journalism and media culture. Her work looked at the broadcasting industry in Brazil and regional and international communication. Among her books is *Rádio Nacional, o Brasil em sintonia; O Rádio no Brasil* and others. She was president of the Brazilian Society for Interdisciplinary Studies in Communication (2002–2005).

**Sergio Godoy Etcheverry** is a professor of communications at Pontificia Universidad Católica De Chile. He teaches media policy and management. He received his MA and PhD in Communications from the University of Westminister, and an MBA from the University of Exeter.

**Juan P. Artero** is an Assistant Professor of Media Management and Structure of Media Industries at the School of Communication of University of Navarra. He holds a Masters in Management from City University, London and a PhD in Public Communication from the University of Navarra, Pamplona. He is the research director for the International Center for Media Companies at the IESE Business School, Madrid. He has written the books *Modelos estratégicos de Telecinco* (1990–2005) (2007) and *El mercado de la televisión en España: oligopolio* (2008), and several academic papers. He is also the director of Mediaccion, a medium-sized consulting firm focused on media.

**Rodrigo Gómez Garcia** is Senior Lecturer in Cultural and Communication Policies, and co-coordinates the Observatory of Cultural Industries, Policies and Consumers at the Autónomous Metropolitan University at Cuajimalpa. He was visiting scholar at University of Leeds (2010) and McGill University (2008). He is the vice-chair of Political Economy of the International Association of Media and Communication Research (IAMCR) and former president of the Mexican Association of Communication Research. His recent publications include: Mapping Digital Media: Mexico, Open Society Foundations, 2011.

**Franco Papandrea** is Adjunct Professor in the Faculty of Arts and Design and Member of the News and Media Research Centre at the University of Canberra, Australia. His primary research interests are in communication and media policy and industrial organization of media. As a well-established expert with decades of experience in those fields he has frequently been called to provide research and consultancy services to Australian and International regulatory authorities and other major clients. His appointment to advise the two foremost Australian public inquiries into newspapers (House of Representatives Select Committee on the Print Media – 1991, and the Independent Inquiry into Media and Media Regulation – 2011) twenty years apart has been a highlight in his career.

**Rodney Tiffen** is professor of Government and International Relations at the University of Sydney. He is a former editor of Media International Australia.

**Min Hang** is an associate professor of Media Management and Economics at the School of Journalism and Communication in Tsinghua University, China. She serves as director of the East Asia Institute of Media Management and Transformation Center (MMTC-EA) of Jönköping University, Sweden. She holds business doctorate from Jönköping University, Sweden, and master degrees from Lund University, Sweden, and Hohai University, China. She won the Award of the "Best Paper for Organizational Communication" in the ICA 2006 Convention. Her work on webcasting, with other authors, has won the prize of "Best Book in Media Economics in 2007".

**Anuradha Bhattacharjee** is an alumnus of London School of Economics and an awardee of ICSSR Research Grant to study media ownership concentration in India. She was formerly a fellow at the Centre for Culture, Media and Governance, Jamia Milia Islamia, New Delhi. Prior to this, she was Assistant Professor, Mudra Institute of Communication, Ahmedabad (MICA). A Fellow of the Charles Wallace India Trust (2004), she was a journalist with the Times of India (1991–94) and The Pioneer (2000–02). She was Research Fellow under the Kasturbhai Lalbhai Chair for Social Entrepreneurship at Indian Institute of Management, Ahmedabad (IIMA) in 2006. Her other area of research interest is Polish Refugees in India 1942–48', for which she was awarded the Bendre Prize for Best Dissertation by University of Pune, 2006. The work has been published as "The Second Homeland" (Sage, 2012), translated into Polish and commissioned as a documentary film.

**Tapasya Banergee** has Master Degree in Development Communication from the Gujarat University in India. She is a research trainee at MICORE in the Social and Development.

**Yoshiharu Ichikawa** is Visiting Lecturer at Keio University; Legal Analyst at NHK (Japan Broadcasting Corporation), Specialized in competition law, media law and policy. He holds an LLM from the University of Leicester Contributions in English include: "The Tension Between Competition Policy and State Intervention", *European State Aid Law Quarterly* (2004)

**Hajime Yamada** is a Professor, Faculty of Economics in Toyo University in Japan and is a Member of Engineering Academy of Japan. He is also an Affiliated Fellow of National Institute of Science and Technology Policy.

He received academic degrees from Keio University as well as an MS in Management of Technology from Massachusetts Institute of Technology.

Before being Professor, he was the Executive Manager of Standardization Strategy at NTT. His books include *Management of Technology: Innovating the Future* (in Japanese) (2004). He won the third prize of IEC (International Electrotechnical Commission) Centenary Challenge in 2006 with a paper on Standardization and Patent Pools. He participates in various international, regional and national standardization activities including ITU-T Director's Ad hoc Group on IPR and ISO/IEC JTC1 Special Working Group on Accessibility He was a member of advisory committees in the US and Europe.

**Daeho Kim** is a professor of media and ICT at Inha University in Korea since 1999. Before joining the university, he worked for the regulatory Korea Broadcasting Commission as a senior research fellow. He was also, director of broadcasting policy and convergence policy at KISDI (the Korean Information Strategy Development Institute), a government sponsored think tank on IT and media. He was appointed as an Inha Fellow Professor in 2011. He was selected as a Prestigious Professor from the National Research Foundation of Korea in 2012. He published many papers on the international and Korean academic journals. He serves as a council member of National Economic Advisory Council under the President of Republic of Korea. He serves as a director for the Korea Broadcast Advertising Corporation, Korea Internet and Security Agency, and SBS. He served as a President for the Korea Media Management Association, and Dean for External and International Affairs at the Inha University. He got a PhD in media policy at the University of Birmingham, UK, after graduating from Seoul National University in Korea (BA and MA) in communication studies. He is now researching on media governance and creative economy and innovation.

**Seongcheol Kim** is associate professor in the School of Journalism and Mass Communication at Korea University. He received academic degrees from Seoul National University and Michigan State University. He worked for the SK Group, and led e-government initiatives for the Seoul Metropolitan Government as Information Systems Director. Before joining Korea University, he was the associate dean of the School of Business at Information and Communications University. His research interests include telecommunication management and policy, new media business, mobile media, and convergence. His publications have appeared in major journals.

**Yu-li Liu** is a professor as well as the director of the Master's Program in International Communication Studies at National Chengchi University in Taiwan. She served as one of the Commissioners of the National Communications Commission (NCC) from 2006 to 2008. She earned her PhD degree in telecommunications at Indiana University and was a visiting professor of George Washington University, Fudan University (China), Renmin University (China), Nanyang Technological University (Singapore), Keio University (Japan), and City University of Hong Kong. She has worked on research projects in the areas of convergence, broadband networks, and digital TV in Taiwan.

**Kuo-feng Tseng** is Associate Professor in the Department of Radio & Television, and the associate dean of the Communication College at National Chengchi University, Taiwan. He got a PhD in Telecommunications from Michigan State University, USA in 2001. His research focuses on media economics, innovation and business model, telecommunication policy, and new communication technology.

**Ru-Shou Robert Chen** is Professor at the Department of Radio-TV, National Chengchi University. His publications include *Through a Screen Darkly: Reflections on One Hundred Years of Taiwan Cinema* (2013, in Chinese), *Cinema Empire: Cinema and Cultural Studies* (1995, in Chinese). He also translated *The Death of Cinema: History, Cultural Memory, and the Digital Dark Age* (Paolo Usai, 2001) and *Psychoanalysis and Cinema* (Vicky Lebeau, 2001) into Chinese.

**Ping-Hung Chen** is professor of the Graduate Institute of Mass Communication at National Taiwan Normal University, Taiwan. He received two master degrees from the National Cheng-Chi University of Taiwan and Boston University of the U.S. He worked for newspapers and a cable television conglomerate for six years before going to the US for a doctoral degree in mass communications from the Pennsylvania State University. His research and teaching interests include communications industry analysis, media management and economics, and media literacy. He has published many articles in various academic journals, published the books *Communications Industry Research* (2001) and *(De)constructing Media* (2003).

**Nagla Rizk** is professor of economics and founding director of the Access to Knowledge for Development Center (A2K4D) at the School of Business, The American University in Cairo. She has affiliation with Harvard Law School and Yale Law School. She is member of the Executive Committee of the International Economic Association. She served as Associate Dean at the AUC School of Business, and Chair of its Economics

Department. She taught at Columbia, Yale, and the University of Toronto. She received a PhD in economics from McMaster University in Canada, and an MA and BA in economics from AUC.

**Amit Schejter** is Associate Professor and Head of the department of communication studies at Ben-Gurion University of the Negev, and Visiting Professor and co-director of the Institute for Information Policy at the College of Communications of the Pennsylvania State University. Schejter received an MS in mass communications from Boston University in 1991 and a PhD in communication and information policy from Rutgers in 1995. He served as bureau chief and senior advisor to Israeli ministers of education and culture Yitzhak Navon and Shulamit Aloni. He was director of legal affairs and international relations at the Israel Broadcasting Authority and co-author of the Nakdi Report, the Authority's guidelines for news and current affairs. He then joined the faculty at Tel Aviv University, and in 2000 was appointed vice president for regulatory affairs at Cellcom (Israel). In 2004 he joined Penn State and in 2012 Ben-Gurion University of the Negev. He is the author of numerous articles on communication law and policy, published in four languages. He is co-author (with Akiba Cohen and Dafna Lemish) of *The Wonder Phone in the Land of Miracles: Mobile Telephony in Israel*(2008) and the author of *Muting Israeli Democracy: How Media and Cultural Policy Undermine Freedom of Expression* (2009). In 2014 he was appointed by the government to be the chair of the committee on the future of commercial audiovisual services in Israel.

**Moran Yemini** is a lawyer, whose experience includes representation before the Israeli Supreme Court, international arbitration tribunals and other judicial and quasi-judicial entities, including handling of regulatory matters vis-à-vis governmental entities.

**George Angelopulo** is Professsor at the Department of Communication Science of the University of South Africa. He is also affiliated with Centrum Católica, the business school of the Pontificia Universidad Católica del Perú. Prior to his academic career he worked in media and advertising in Africa.

**Petrus H Potgieter** is Professor in the Department of Decision Sciences at the University of South Africa and a mathematician by training (PhD, University of Pretoria, 1996). He teaches telecommunications regulation for the Department of Communication Science at the same university and is interested in telecommunication economics as well as regulatory issues. He also writes a weblog CyphAfrica.com on general telecommunications and technology issues, with a focus on Africa and South Africa.

**Paul Mutter** holds a BA in History from Rutgers (2010) and an MA in Politics from NYU (2015). He joined CITI in 2012 as a researcher for this book, and then took on the job of project manager. He currently freelances for several online news portals on topics in diplomacy and national security.

**Jason Adam Buckweitz** is the Associate Director of the Columbia Institute for Tele-Information at the Columbia Business School. He holds a Juris Doctorate from New York Law School and degrees in political science from Rutgers University, where he also achieved major technology proficiencies. His work has been published in several publications. At CITI, he has led many research projects and conferences.

# Who Owns the World's Media?

# PART 1

## INTRODUCTION AND OVERVIEW

# 1

# Introduction

MEDIA CONCENTRATION has been an issue around the world. To some observers, the power of large media conglomerates has never been greater. To others, the Internet has brought openness and diversity. Which perspective is correct? The answer has significant policy and business implications.

Large media proprietors and their companies have drawn attention, fear, and ire. The fear is that communications media are increasingly controlled by an ever-shrinking number of firms and that those firms are capable of affecting public opinion, the national agenda, democracy, and global culture. In the United Kingdom, the focus is Rupert Murdoch; in Italy, Silvio Berlusconi; in Mexico, Televisa and Carlos Slim; in Brazil, Globo; in Argentina, Clarin; in Japan, five conglomerates; in France, Vivendi; in Sweden, Bonnier; in Spain, Telefonica; in the United States, Disney, Time Warner, Sumner Redstone, Rupert Murdoch Bill Gates, and Google. Hardly a country, it seems, is without such disputes. Even in tiny Iceland, media concentration created a major governmental crisis in 2004 when the country faced holding its first-ever public referendum on a constitutional dispute arising from the issue.

The debate has become the information-age version of the industrial-age struggle over the control of the means of production. That earlier conflict led, in some countries, to revolution, and in other countries to the socialization of key industries. The United Kingdom, for example, nationalized its heavy industry sector after World War II. The underlying notion was that private ownership of coal and steel companies could be leveraged into control over the economy and society. With hindsight, that fear seems overblown, and this points to the need for a careful analysis of facts and dynamics before launching into policies.

It is necessary to put the debates over media concentration into a wider context. It is the media sector's analog to the debate about inequality in the wider economy and society. The debate over media concentration is part of a much larger mobilization that has been taking place over the control of information resources. This includes advocacy for unimpeded access of content to the Internet ("net neutrality");[1] the "open source," file-sharing,[2] "open innovation,"[3] and "copyleft" movements[4] that have challenged the traditional intellectual property system; the privacy protection advocacy against the use of personal information;[5] the "unlicensed spectrum" initiatives that seek to end the licensed exclusivity of access to airwaves;[6] the push against a "digital divide";[7] the move to municipal and free Wi-Fi connectivity;[8] and more. All of these developments have their particular reasons but also a common thread. They are manifestations of a wider conflict over the extent and nature of control in the information society.

Most observers are familiar with the various flash points but have not always connected the dots and recognized the emerging social movement on the model of environmentalism. For years, information sector companies and governments have touted their activities as the key to the planet's economic and cultural future and the solution to most of its problems. No wonder that control over this sector is being contested by more than business competitors. As the information sector permeates society, society with its internal and international conflicts in turn permeates the information sector.

People have more problems in accepting media power than economic power. They may tolerate the mega-wealth of a Sam Walton, Bernard Arnault, Amancio Ortega, or the Albrecht brothers. But they intensely dislike the media ownerships of Rupert Murdoch, Silvio Berlusconi, or the Marinho and Azcarraga families, even though their wealth, while considerable, is much lower than those at the pinnacle of the pyramid (Murdoch is #83 and Berlusconi #181 in the *Forbes* billionaires list for 2014). And this aversion is no coincidence. People recognize that media is a powerful instrument of influence over their hearts, minds, wallets, and votes. They attribute to media—and therefore to their owners—responsibility for what they dislike in their society. Whether it is violence, gender stereotyping, racism, offensive policy, materialism, hedonism, escapism, low political participation, entrenched status quo, inequality, drug dependency, low literacy—for all these and more, the media system is held partly responsible either as a contributing cause or as an absent remedy. And since so many of society's ills are blamed on media, media reform becomes social reform.

Beyond their influence over mindshare, media also loom big in volume and presence. Their worldwide revenues account for about 7% of the world GDP. As a share of "discretionary income" (what is left after outlays for necessities such as food, shelter, medical services, etc.), it is closer to 20%. And it overlaps with more than half of "discretionary time" (non-work waking time).

With such a large role of media in society and economy, a skewed distribution of their ownership

---

1. Wu, Tim. "Network Neutrality, Broadband Discrimination." *Journal of Telecommunications and High Technology Law* 2 (2003): 141.

2. Noam, Eli, and Lorenzo Pupillo, eds. *Peer to Peer as a Distribution Medium*. New York: Springer, 2008. Benkler, Yochai. *The Wealth of Networks: How Social Production Transforms Markets and Freedom*. New Haven, CT: Yale University Press, 2006.

3. von Hippel, Eric. *Democratizing Innovation*. Cambridge, MA: MIT Press, 2006.

4. Stallman, Richard. "Reevaluating Copyright: The Public Must Prevail." *Oregon Law Review* 75 (Spring 1996): 291–297. Moglen, Eben. *Framing the Debate: Free Expression versus Intellectual Property, the Next Fifty Years*. Barcelona: Universitat Oberta de Catalunya, 2007.

5. Rotenberg, Marc. *Privacy & Human Rights*. Washington, DC: EPIC, 2006.

6. Noam, Eli. "Spectrum Auction: Yesterday's Heresy, Today's Orthodoxy, Tomorrow's Anachronism. Taking the Next Step to Open Spectrum Access." *Journal of Law and Economics* 41, no. 2 (October 1998): 765–790.

7. Mossberger, Karen. 2003. *Virtual Inequality: Beyond the Digital Divide*. Washington, DC: Georgetown University Press, 2003.

8. Lehr, William, and Lee McKnight. "Wireless Internet Access: 3G vs. WiFi?" *Telecommunications Policy* 27, no. 5–6 (June/July 2003): 351–370.

is a major element in the distribution of "information capital" and of wealth more generally. Those who fear an undue influence of media therefore tend to see in each merger or business expansion an affirmation of their dread over mind control. Others are more hopeful. They believe that technology, entrepreneurship, and market forces are self-correcting counter-forces that open media to new voices and pluralism. They point to de-mergers and the periodic failure of once-dominant firms. What makes the debate frustrating is that it is generally devoid of data analysis. Where numbers are used, it is usually as ammunition rather than illumination. This enables people to come to whatever conclusion they desire. There is therefore a need for social science research, done methodically, patiently, and in a multidisciplinary fashion.

What, then, are the facts about the world's media ownership and concentration?

The answer to the empirical question is not as obvious as many sincerely believe. First, although numerous mergers have taken place and have led to large global media companies, the mass media and information sector have also grown rapidly, in volume, attention time, and revenue (even with widespread price declines). And de-mergers have also taken place, such as for Time Warner.

Second, with digital convergence a much discussed tendency, firms have been crossing the lines that once divided the mass media, telecommunications, and computer industries. Software operating system companies are offering phone services, phone companies provide video distribution, and TV set manufacturers become music and game companies.

Third, the internationalization of economies and services means that well-established media firms from countries, typically advanced ones, have gained a presence in other countries, and in the process have grown in size and market power.

New media have been emerging, most notably driven by the Internet, thus creating new distribution channels and content providers. And some internationalization of media also means that some domestic oligopolies have been challenged from entrants from other countries.

Thus, although the fish in the pond may have grown in size, the pond has grown too, and new fish and new, connected ponds have been added.

The same issues led to my earlier book *Media Ownership and Concentration in America*, Oxford University Press, 2009. That book discussed, in 499 pages, ownership and concentration trends for US media industries over a 25-year period. It inspired the present volume.[9]

Why extend the analysis to the world at large? People politicize or personalize media trends—discussing how this government agency or that media mogul are the culprits. But will that identify the nature of the problem? Tolstoy wrote that happy families are all alike, but that each unhappy family is unhappy in its own way. And so the question is, when it comes to media concentration: Is every country unhappy in its own way? Or are there commonalities? Because if we can identify common trends, we can seek the drivers, whether technological, economic, or political. This has policy implications. When drivers are fundamental in nature, it is challenging to deal with them through regulatory policy. But when a country or a region is an outlier, with a media market that is different from similarly positioned countries, a corrective policy might be more effective.

The key question then is whether in media industries the market power achieved by some companies is the result of individual media barons' business acumen, possibly coupled with inept or supportive government regulation; or, alternatively, whether the underlying technology of production and distribution defines economic characteristics that, in turn, lead to market structures of high concentration. In the latter scenario, a few moguls will end up at the top spots, by luck, pluck, skill, stealth, or connections, but their identity is secondary to the fact that the market equilibrium for that medium will be a monopoly or oligopoly. The underlying technology and its economics are the base.

To address this question, this book analyzes the media sector, across countries and across time, and identifies its dynamics, concentration, and ownership trends. This would seem to be a

9. It also inspired this introduction.

straightforward question, but in the policy arena the discussion of media markets is often marked more by heat than by light. Views are strong but numbers are scarce. To many media critics, the sky has been falling for decades. Others believe that the Internet and market forces are overcoming all barriers and that we are in the midst of a flowering of media.

Part of the vehemence of the debate stems from the self-image of its advocates.[10] Opponents of media concentration view themselves as a last line of defense against homogenized news controlled by giant media conglomerates. They point to Italy, where Silvio Berlusconi used his media empire to achieve policy power and public office, and fear for democracy in their country and the world. In contrast, proponents of media deregulation see themselves as removing the shackles of the state from media in the midst of a historic blossoming of information technology. Both sides project themselves as defenders of free speech, either protecting media from the heavy hand of government, or, alternatively, protecting diversity from being choked off by communications empires. Both sides are, to some extent, correct. Yet their principled views are often, consciously or not, colored by a result-oriented pragmatism. In the aforementioned example of Iceland's constitutional crisis over media concentration, it was actually the political left-of-center (normally the home of advocates for ownership limitations) that strenuously defended the dominant media conglomerate—whose editorial positions were supportive of the Left—against the political right-of-center (normally the home of anti-regulation, free market advocates) that tried to break it up.

Given all these perspectives and biases, how then should public policy be determined? At a minimum, it should be based on a solid factual base on the nature of the problem. Principled analyses of issues of public policy are important. But it is also true that a generic overview coupled with factual selectivity can be a labor-saving work method that also enables its practitioners to assume a posture of great moral rectitude. It is essential to proceed beyond selective facts that are culled to suit respective policy preferences. Given the intensity of debate, it should be the role of academia to step back. It is therefore the aim of this book to create a decent fact base, interpret the data, and analyze the underlying dynamics. This process might be more tedious than advocacy and opinion, but it enables informed discussion.

## OVERVIEW OF RESEARCH PERSPECTIVES

The subject of media concentration has, of course, received much attention in academic articles, governmental reports, and opinion pieces, indeed much more than can be covered here. For a partial view of the literature, see the "General References on Media Ownership and Concentration" at the end of the book. That bibliography alone covers 24 pages. Some of it deals with specific media moguls[11] or companies.[12] Other studies deal with industries,[13] countries, and regions[14] and with public regulation, policy, and law.[15] There are also many governmental[16] and NGO reports,[17] economic

10. Noam, Eli. *Media Ownership and Concentration in America*. New York: Oxford University Press, 2009. Part of its introduction is used in the present volume.

11. See, for example, Barry, 2007; Khlebnikov, 2001; Mariscal, 1998; Wolff, 2008.

12. See, for example, Peruško & Popović, 2008; Steinbock, 2001; Chenoweth, 2002; Galambos & Temin, 1987.

13. See, for example, Badillo & Lesourd, 2010; Chan-Olmsted & Litman, 1988; Chen, 2002; Doyle, 2000; Greco, 1999; Iosifidis et al., 2005; Lacy, 1991; Litman, 1988; Richieri, 2004.

14. See, for example, Caspi & Limor, 1999; Chan, 1993; Cornu & Borruat, 2012; Doyle, 2002; Eickelman & Anderson, 2003; Howard, 2008; Kachkaeva, 2012; Sanchez-Tabanero, 1993; Sanchez-Tabanero & Carvajal, 2002; Smith & Tambini, 2012; Vartanova & Smirnov, 2010.

15. See, for example, Bittlingmayer & Hazlett, 2000; Cooper, 2003; Downing, 2011; Noam, 2006, 1992, 1991; Papandrea, 2006; Stockmann, 2012.

16. See, for example, Advisory Panel on Media Diversity, 2004; Brackman et al., 2002; European Commission, 2007; Federal Communication Commission Media Bureau, 2012; Ofcom, 2009; The State Administration of Radio, Film and Television (SARFT), 2009; Ward et al., 2004; UNESCO Institute for Statistics, 2012; Vīķe Freiberga, 2013.

17. See, for example, Commission on Freedom of the Press (Hutchins Commission), 1947; Hans-Bredow-Institut, 2006; Open Society Foundation, 2012.

investigations, and management-oriented analyses.[18] Authors also analyze the impact of media ownership on politics.[19] Only a few examples are provided here, and the reader is referred to the general bibliography and to the bibliographies of its individual country reports.

The most influential works have been of the "big picture" variety,[20] often non-empirical. To simplify considerably, there are two schools of thought, and they can be roughly characterized as those of media pessimists and of media optimists. Media pessimists think of the sky as falling; media optimists see the dawn of a bright new day.

Perhaps the most prominent media pessimist has been Ben Bagdikian. Bagdikian is a Pulitzer Prize-winning journalist and former dean of the journalism school at the University of California, Berkeley. He writes in his widely used book, *The New Media Monopoly*: "Five global-dimension firms, operating with many of the characteristics of a cartel, own most of the newspapers, magazines, book publishers, motion picture studios, and radio and television stations in the United States."[21] These firms are Time Warner, Viacom, Disney, News Corp., and Bertelsmann. He observes, remarkably considering such rival contenders as Hitler, Stalin, and Mao: "This gives each of the five corporations and their leaders more communications power than was exercised by any despot or dictatorship in history" (p. 3).

Bagdikian's mantle has been assumed by Robert McChesney of the University of Illinois, who is much more careful with his numbers than Bagdikian. But Bagdikian's less nuanced numbers have taken on a life of their own. Celebrated film documentary maker Michael Moore picked the number five and globalized it: "By the end of the millennium five men controlled the world's media."[22]

Lawrence Lessig, the noted Stanford law professor (and for a time my friendly co-columnist on new media at the *Financial Times Online*), escalated the number: "Indeed, after the changes that the FCC announced in June 2003, most expect that within a few years, we will live in a world where just three companies control more than 85% of the media."[23]

In academia, the pessimist view has long been the orthodoxy. But there have also been media optimists, who see an abundance of openness and diversity, much of it due to the Internet. (I ignore the self-interested advocacy documents of companies and trade associations.)

Perhaps the most comprehensive expression of the optimist view, for specifically the United States and more generally the world, is Adam Thierer's *Media Myths: Making Sense of the Debate over Media Ownership*.[24] Thierer, an economist at several Washington-based free-market think tanks, concludes: "To the extent that there was ever a 'Golden Age' of media in America, we are living in it today. The media sky has never been brighter and is getting brighter with each passing year. And this is most definitely not a case of looking for silver linings around the clouds; *there are no clouds*."

Much of the literature on media concentration has been stronger in commitment than in empirical evidence, richer in certitude than in numbers, on both sides. One exception is the volume *Who Owns the Media?*[25] a comprehensive data-compiling effort edited by Benjamin Compaine,

---

18. See, for example, Albarran & Dimmick, 1996; Berry & Waldfogel, 2001; Compaine, 1995; Dertouzous & Trautman, 1990; Djankov et al., 2001; Noam, 2005; Owen, 1995; Riordan & Salop, 1995; Vogel, 2013.

19. See, for example, Edwin, 2007.

20. See, for example, Bagdikian, 1997; Baker, 2007; Curran & Seaton, 1997; Eco, 2001; Hallin & Mancini, 2004; Horowitz, 2005; Starr, 2004.

21. Bagdikian, Ben. *The New Media Monopoly*. Boston: Beacon Press, 2004, p. 3.

22. Media reform quotes, at betterworld.net, http://betterworld.net/quotes/media-quotes.htm, accessed on December 28, 2007.

23. Lessig, Lawrence. *Free Culture: How Big Media Uses Technology and the Law to Lock Down Culture and Control Creativity*. New York: Penguin, 2004.

24. Thierer, Adam D. *Media Myths: Making Sense of the Debate over Media Ownership*. Washington, DC: Cato Institute, 2005.

25. I warmly thank Ben Compaine for inspiring the title of the present book and graciously permitting me to use a similar one to his own. Shleifer et al. (2001), was probably similarly inspired by Compaine's title.

Christopher Sterling, Thomas Guback, and J. Kendrick Noble Jr. in 1979, with subsequent newer editions with Douglas Gomery.[26]

Compaine and Gomery, like Thierer, are media optimists. They answer the question of who owns the media: "Thousands of large and small firms and organizations . . . controlled, directly and indirectly, by hundreds of thousands of stockholders, as well as by public opinion" (p. 578). "Looked at as a single industry, there can be little disagreement that there is more competition than ever among media players. The issue could be stopped with a single word, *Internet*. But it goes beyond this development" (p. 574).

Another empirical study, covering the entire world is a 2001 World Bank study by Andrei Shleifer, Caralee McIlesh, Tatiana Nenova, and Simeon Djankov.[27] The study focuses mainly on state ownership of TV and newspapers in developing countries and correlates ownership with factors of political and economic variables. It also identifies the nature of non-state ownership (widespread vs. family owners). They find that newspapers and television media tend to be controlled by families or governments for reasons of political influence or fame. They find that poorer countries have a greater state ownership in media, and that those countries are also characterized by less political freedom.

## EMPIRICAL FINDINGS

It is the nature of a fast-paced, short-attention-span society for readers to seek a single and simple "bumper sticker" answer to complex and divergent set of facts and trends. These people will be disappointed. This book has tried to assemble and sift through vast amounts of information, starting with extensive country reports. That data – several hundred tables – then gets summarized and aggregated in successive stages, first on a country basis at the end of each chapter; then, across countries, media industries, companies, and owners. A large number of results are reported in the chapters that follow, and summarized in chapter 38 and its concentrations, which form the end of the funnel. We report here a few broad findings.

We find that the *trends* generally support the pessimists when it comes to content media but the optimists (for now) in platform media. However, it is the reverse for the *current level* of concentration.

One must distinguish between two categories of media: those creating content and those distributing it over network platforms. (Some media industries combine the two functions.) For both, market concentrations have been quite high. The top four companies in the four major platform media industries which we investigated accounted, by weighted country average, for a dominant 88% of their particular national industry market. For content media industries, the average figure for the main 10 industries was 67%. (These may be different companies in different media industries.) If we pool the industries in each sector into broad categories in order to identify cross-industry conglomeration, we find that the top four firms hold 79% of their country's overall platform media market and 40% of its content market, respectively.

Platform media are and have always been considerably more concentrated than content media (over twice as high), largely due to the much greater capital intensity of network platforms and the higher network externalities. However, the gap has declined. As content companies have become more capital-intensive their market concentration increased. For content media, the average national market share of the top four firms rose by 1.5% per year.[28] The trend of the content industries starts at a much

26. Compaine, Benjamin, Christopher Sterling, Thomas Guback, and J. Kendrick Noble, Jr. *Who Owns the Media?* White Plains: Knowledge Industry, 1979, 1982.

27. Shleifer, Andrei, Caralee Mcliesh, Tatiana Nenova, and Simeon Djankov. "Who Owns the Media?" World Bank, 2001. Also published in the *Journal of Law and Economics* 46 (October 2003): 341–381, as by Djankov, McLiesh, Nenova, and Shleifer.

28. These are the averages weighted by country size. With an arithmetic averaging, the content media rise, on average, by 0.2% p.a., and platform media decline by 0.03%. Using another index for concentration, the so-called HHI, the average annual growth rate rose by 4.2% when countries are weighted by size, and 0.2 % when averaged arithmetically. The HHI is exponential in nature and its growth is roughly the square of that of the C4.

lower level than platforms but rises steadily. In the United States, it rose almost twice as fast as the weighted world average, though from a lower base.

Another way to summarize the trend is to look at the absolute (as opposed to percentage change) in the market share of the top four firms in a content market. Collectively, these firms gained, based on a decade's observations, on average 0.9% per year. When all 10 content industries are pooled, the share of the top four firms rises by 0.5% per year. In the United States, the absolute rise in the market share of the top four firms in the average content industry has been 1.25% per year, over twice the weighted world average, though from a lower base.

Some of these percentages may not look high, but when extended forward beyond the decade measured to another decade or two, it portends a significant rise in content media concentration around the world.

In contrast, the concentration of platform companies has declined since the 1980s, due to the opening to competitive entry, the licensing of multiple mobile operators, and a variety of regulatory measures to increase competition. It declined by an annual −0.6% (and −0.3% for the pooled platform market). Even with these reductions in concentration, the overall concentration platform level is still very high, with the top four firms holding, on average, 88% of the average platform industry. In the United States, the concentration in platforms has declined in the 1980s, driven by regulatory changes, has subsequently risen, and after 1995 declined and risen again with consolidations in mobile and the re-assertion of the wireline incumbents. It has been fluctuating but with an overall upward trend. Thus, the US concentration trend in platform media has been "S-shaped," in an upward-sloping oscillation. Given the similarity of factors around the world, this might be the future trend elsewhere, too, and it would reverse the present downward tendency.

Looking forward then, and considering the fundamental drivers of media discussed above, one must expect further market concentration. Many people have been enticed by the promise of the "long tail," which has indeed enabled niche media creations to reach audiences. But at the same time, digital media also lead to winner-takes-all markets. In their sub-markets, companies like Google, Amazon, or Facebook dominate, and they do so in many countries across the globe. Much of content is accessed thorough them or provided by them. Content production itself moves from the relatively simple text mode with low entry barriers to video creations that are interactive, immersive, and expensive to produce. Print media become integrated with video news and cover many more subjects on a 24/7 basis. As media move online and worldwide, the pressures toward rising concentration are an economic reality. The Internet, once the hoped-for solution, is becoming part of the feared problem.

Thus, the glass is half full. There is now a much greater ability for individuals to express themselves and to reach distant audiences, and for audiences to find highly specialized content. In the sense of options there is a much greater media pluralism than existed in the past. However, if one looks beyond the potential options to the actuality of choices exercised, one finds a greater concentration than before, and often higher than for the traditional offline media.

In this media environment, who then are the world's largest owners of media? Whether by revenues or attention time, it is, by far, the Government of China. Looking for the world's largest *private* media owners, these are not individuals or families but rather the American-based institutional asset management firms State Street Bank, Vanguard, Fidelity, and Capital Group. Of individual or family media owners the largest are Carlos Slim, the two Google founders Larry Page and Sergey Brin, and the Cox family. Rupert Murdoch is #12 and Silvio Berlusconi is #22. Of the major media companies in the world, the top 30 institutional owners hold about 10–20%,[29] the top 30 individual owners hold 8–16%, and the 30 governmental owners

29. The spread is based on different assumptions on asset valuation relative to revenues.

hold 13%. These top 90 owners thus account for 30–50% of the world's major media assets.

## FUNDAMENTAL DRIVERS OF MEDIA CONCENTRATION

The debate over media concentration is usually framed, depending on one's politics, as one of empire-building media moguls undermining either democracy or traditional values. But personalizing the issue of media concentration and putting it in terms of political preferences misses the larger issue. We find and document in this study that media concentration is indeed taking place around the world, and that it is too prevalent and consistent in its trends to be the outcome of national, institutional, and personal particularities. Instead, its drivers are the fundamental economics of media. Various individuals and companies take advantage of these trends and accelerate them but do not cause them.

What are these fundamental economic drivers of media concentration?

### Characteristic #1 of Media: High and growing fixed costs, low and declining marginal costs

Media production usually involves high fixed costs, that is, costs that remain constant independent of the number of units produced. These are mostly the upfront capital investments and development costs. At the same time, the marginal costs—or the incremental costs required to produce and distribute the next unit—are relatively low for media. Media content is typically expensive to produce but cheap to reproduce. Media platforms such as networks are expensive to create but cheap to extend to additional users. The combination of these fixed and marginal costs are average costs. As fixed costs are distributed over more and more units, average costs decline.[30] We can observe these characteristics for films, TV programs, computer software, electronic networks, newspapers, and games. Platform media such as telecommunications networks have always been highly capital-intensive. Products that exhibit these properties are said to have high economies of scale.

There are several business implications. The economies of scale lead to large-sized companies who can produce at lower per-unit cost. This leads to market concentration. There are incentives to reach large size through mergers and to be a first-mover in a product in order to gain scale. In the extreme, one encounters a winner-takes-all near-monopoly.

New media are more capital-intensive than old ones. Their ratio of capital costs to marginal costs is higher than that of traditional media. In consequence, their scale economies are greater and their market concentration is higher. On the whole, the concentration of media industries is the inverse of their age. Book publishing is the oldest of mass media and the least concentrated. It has been followed chronologically by periodicals, newspapers, film, radio, broadcast TV, cable TV, and now the Internet . This chronological order matches pretty much the ranking of the respective industry concentrations.

This move to higher levels of concentration for newer media must be distinguished from the typically exuberant early years of a new medium. But the initial free-for-all gives way as market leaders with resources and scale emerge and end up dominating for a long time. Examples are the early years of film, wireline telecom, radio, cable TV, mobile telecom, and Internet service providers. In each case, thousands of companies entered initially in the United States but only a few dominate today. This is not unique to media. There were, similarly, hundreds of manufacturers of automobiles and of airplanes in the early years of these industries, but they soon gave way to a handful of large producers.

### Characteristic #2 of Media: Network effects

Network effects arise when users benefit each other by sharing a platform or an experience. Individual benefits from media are often

30. Unless marginal costs rise significantly, which they rarely do in media industries.

interdependent with that of other users. Networks have a fundamental economic characteristic: the value of connecting to a network depends on the number of other people already connected. For platform media, the larger the network, the more value it provides to its users and the more valuable it becomes itself. For content media, a major benefit of use is to share experiences with one's peers.

This changes the economics of demand. The more people are on the network or share an experience, the more people are willing to pay for the product. Usually, economic theory implies that price and demand are inversely related—the lower price, the higher the number of customers. But now, at least in some cases, the relationship is the opposite—the higher the number of users, the higher their willingness to pay. Larger networks and content providers thus face a lower price sensitivity by users and might be able to charge higher prices than smaller ones.

There are several formalizations of network effects. Robert Metcalfe, the co-inventor of Ethernet, proposed a "law" according to which the value of a network increases exponentially with the number of nodes.[31]

Network effects have several business implications. As in the case of scale economies, firm size is important. The larger the firm's user base, the more value is provided to users. It also creates advantages to early heavy marketing to create a critical mass. For a film, the attention thus created gains a cumulative advantage.[32] First-movers have an advantage; it is usually easier for a firm to capture market share if it is the first to launch a particular product or service. Interoperability is important, because it enables the users of a new or small product or service to link into a larger one and benefit from its network size. Conversely, firms possessing large network effects will resist the interconnection and interoperation of smaller competitors by proprietary standards, or they will collaborate with each other in an oligopolistic fashion while excluding new challengers.

## Characteristic #3 of Media: Excess supply

Media production increases exponentially at a substantial rate, while media consumption increases linearly and slowly. Given the gap between production (supply) and consumption (demand), excess supply is inevitable. This has consequences for both content style and marketing.[33] Attention is the scarce resource. As Nobelist Herbert Simon observed, "a wealth of information creates a poverty of attention."[34] New media consumption must be mostly supported by substitution from existing media in terms of time or full attention. Inevitably, this leads to competition for "mindshare" and "attention." Compared to 1998, fewer than half as many of the new products make it to the bestsellers lists, reach the top of audience rankings, or win a platinum disc.[35]

In addition to technology, people have been a major driver of the rising supply. There has been a huge increase in the number of information producers. It has been observed, that 90% of all scientists who have ever lived are alive today.[36] This is also true for most or

31. Metcalfe, Robert, and Michael Vizard. "Ontology and Revenge of the System Analyst." *InfoWorld.* Last accessed on June 26, 2007, at http://www.infoworld.com/articles/op/xml/01/12/17/011217opnoise.html. Mathematically, this would be $V = aN^b$. An even stronger relation was proposed by MIT computer scientist David P. Reed. He believes that the value of such a network increases exponentially such that the exponent itself grows exponenetially, $V = a2^n$.

32. Salganik, Michael J., Peter Sheridan Dodds, and Duncan J. Watts. "Experimental Study of Inequality and Unpredictability in an Artificial Cultural Market." Washington, DC: Science, 2006, pp. 854–856. "The Gazillion-Dollar Question." *Economist.* April 20, 2006. Last accessed on August 2, 2012, at http://www.economist.com/node/6794282.

33. "How Much Information." *School of Information Management & Systems, University of California, Berkeley.* 2000. Last accessed on May 14, 2008, at http://www2.sims.berkeley.edu/research/projects/how-much-info/summary.html#consumption.

34. Simon, Herbert. "Designing Organizations for an Information-Rich World." In Martin Greenberger, *Computers, Communication, and the Public Interest.* Baltimore: Johns Hopkins Press, 1971, pp. 37–72.

35. Aris, Annet. Managing Media Companies: Harnessing Creative Value. Hoboken, NJ: Wiley, 2009.

36. Derek John de Solla Price. See Cloud, Wallace. "Science Newsfront." Popular Science 182, no. 3 (March 1963): 17. Last accessed on June 30, 2010, at http://books.google.com/books?id=3iEDAAAAMBAJ&dq=de+solla+price+ninety+percent+scientists&source=gbs_navlinks_s.

all information-based occupations, whether screenwriters, composers, architects, lawyers, engineers, MBAs, and so on. More information workers lead to more information products. In almost any scientific field, more research articles were written just this year alone than in the entire history of human beings before 1900. In the field of chemistry, within a span of 32 years (1907–1938), one million chemistry articles were written and abstracted. In contrast, it took less than one year for a million such articles to be produced in 2010. Every thirty seconds, a new book is published. Every day, fifteen new theatrical feature films and 1,500 television scripted shows are being produced.

The business consequence is greater competition, as well as a greater specialization in content. More product innovation and marketing efforts are necessary. Costs rise per product while audiences fragment.

## Characteristic #4 of Media: Price deflation

When competition occurs the price drops toward marginal cost. In the short term, marginal cost is near-zero and does not cover fixed cost. The result of price competition with low marginal cost has been price deflation in information products and services. This is a good deal for consumers who enjoy a substantial "consumer surplus" in which they must shell out much less than they would be willing to. But it creates a difficult problem for the supplier. And that is indeed what has been happening. Information has become cheaper for many a decade. It is often becoming difficult to charge *anything* for it. Music and online content is increasingly free. Newspaper prices barely cover the cost of paper and distribution; the content is thrown in for free. Price deflation poses a threat to long-term viability, since low prices make it difficult to cover costs and achieve profitability.

Price deflation is one of the fundamental economic trends of our time. The entire competitive part of the information sector—from music to newspapers to telecoms to Internet to semiconductors and anything in-between—has become subject to a gigantic price deflation in slow motion. This price deflation leads to economic pressure, to price wars which squeeze out weaker companies. Therefore, one main strategy for media managers is to avoid such price competition and to focus on product differentiation, price discrimination, consumer lock-in strategies, and competition-limiting mechanisms such as patents and copyrights. It has been observed that "the economics of information requires imperfect markets."[37] And to enable this, perhaps the most important strategy is to reduce competition through industry consolidation.

## Characteristic #5 of Media: High risk distribution of success

The fifth economic characteristic of media is its high risk in the presence of competition. One often observes an "80–20" outcome in which 80% of all media products do not become profitable, 90% of all profits are generated by 10% of the products, and 50% of profits are generated by 1–2% of products.

In consequence, media firms lower their exposure to risk by creating portfolios of products, such as slates of films, multiple labels of music, or various imprints of books, each in turn consisting of multiple products. This transforms a set of highly risky products into a portfolio of moderate risk, thereby enabling such a firm to access more capital at a lower rate than specialized and smaller firms can. With a lower risk profile, a large and diversified media firm holds a long-term advantage over small and specialized firm.

## Characteristic #6 of Media: Convergence of technology—economies of scope

In economic terms, one observes rising "economies of scope," meaning that there are "synergies," where production across several lines of business is cheaper than separate stand-alone production.

37. Evans, Philip, and Thomas S. Wurster. "Information and Things." *Blown to Bits.* Boston: Harvard Business School Press, 2000, pp. 15–21.

Historically, media industries used to be separate from each other. Newspapers, music, TV, telecom, computers, and so on were all realms of their own, each with their own technologies, producers, suppliers, distributors, and cultures.

Starting in the 1970s, integration in the technology began to occur with the increasing overlap of devices, components, and control software. Any content could be encoded as a stream of bits, and then processed, distributed, and displayed in similar ways.[38] In the 1980s, increased integration of technology extended the overlap to consumer electronics and content industries. Today, for example, a smartphone combines the technologies of telecom, computers, radio transceivers, consumer electronics, TV monitors, video game consoles, calculators, cameras, music players, dictaphones, navigation devices, and many more. Similarly, the content streams emanating from major media providers include film, video, text, music, stills, games, peer-to-peer communication, and more.

The implications are that media industries and firms that used to comfortably operate in their separate niches are increasingly facing competition from each other. But it also means that companies can expand more easily to adjoining markets, which leads to vertical and horizontal integration, mergers, and to the emergence of media conglomerates. How these two offsetting trends add up is an empirical question.

### Characteristic #7 of Media: A public good with high government involvement

The seventh economic property of media is that information is often a "public good," that is, a product or service consumed in common, such as national defense, a lighthouse, a scientific discovery, or a TV broadcast. As described by Paul Samuelson, a public good is something "which all enjoy in common in the sense that each individual's consumption in such a good leads to no subtractions from any other individual consumption of that good."

The implication of information being a public good is that it is difficult to charge for information, which leads to an under-production or even non-production of certain categories of such information. This leads to government taking a role in assuring for-profit creation (such as by setting and protecting intellectual property rights), by supporting nonprofit creation (e.g., basic research, funding of the arts, public television, etc.), and by assuring the nationwide distribution of information through various regulatory schemes.

A second factor for government involvement is that they aim to shape media in ways supportive to their policy goals and hold on power. These factors lead them to take a strong role in creating public media and in controlling private ones. This is particularly the case in authoritarian countries. In many developing and emerging nations, a symbiotic relationship of large media organizations and governments leads to the market dominance by a very few media companies or by state-controlled media. In many developed countries, too, public service television organizations, for a variety of policy reasons, held monopoly licenses for a long time and their role is still substantial.

## POLICY IMPLICATIONS

The question is whether these fundamental economic and technological drivers that lead to the concentration trends we identified undermine diversity, pluralism, meritocracy, and the credibility of news. Given these fundamental drivers, it is almost inevitable that the economic equilibrium of media markets, left to themselves, will not be at a level of diversity and pluralism that many people consider necessary. This tension between market equilibrium and social optimum has existed for media for at least a century, and it would be surprising if it were to disappear. Recent decades have led to a reduction of restrictions—"deregulation," "liberalization," and "privatization"— on media ownership and its exercise because of the expectation that

38. Shapiro, Carl, and Hal R. Varian. *Information Rules*. Boston: Harvard Business School Press, 1999, pp. 1–18.

technology and market forces would overcome market power. If this hope is not realized, as the analysis suggests, the pendulum will inevitably swing back to various interventionist approaches of regulation, breakups, ceilings, and subsidization.

But this is becoming more difficult. Government rules on limiting media ownership have worked reasonably well on the operational level when industries were simple and enabled simple tools of control, such as ownership ceilings on TV licenses, or of cross-ownership among media. But outside the ordered broadcast frequency licensing scheme (whose significance keeps shrinking), the government's powers today are much more limited and becoming ever more so. If Google has significant market power in Argentina, how should or could the search engine market there be restructured? If a Korean firm is dominant in interactive games, what then is the Swedish (or the EU's) governmental remedy? If Skype's voice quality declines, who would deal with that, if at all, and how? And these are merely conceptual questions, to which are added those of politics, litigation, international trade, intellectual property rights, and international enforcement. Competition might deal with such problems where it exists, but one must be clear-sighted to recognize that a market equilibrium will often be an oligopoly at best. It is always difficult for laws or regulations to modify fundamental transitions of industries. It is particularly difficult to do so where, as in the case of media, any media policy in a free society needs to be done with a light touch.

It is therefore important for academics, public-policy analysts, NGOs, companies, and governments to think creatively about new approaches to these issues, and to balance the public interest, technological innovation, and investment needs in the emerging environment.

## THE QUESTIONS ADDRESSED

This then is the goal of this book: to answer the question whether, where, and how the world's media are becoming more concentrated. Probably the worst way to approach this question is by anecdote. At any given moment there are journalists, participants, creators, and artists who can tell a compelling story of bias, restriction, and failure of imagination. Yet statistically speaking, most projects would be turned down under any conceivable system, because of limited resources and limited attention relative to the continuously rising information productivity of society. A subjective rejection of worthy projects is inherent to any media system.

The present study was conducted without any preconception as to where the data would lead and without an attempt to prove a point or advocate a position. We did not reach final conclusions until the data for each industry and country had been collected, processed, analyzed, and aggregated, literally up to the last days of work. As the study progressed, we were exhorted to provide policy recommendations. We have resisted doing so here. For the present, the intermingling of normative policy analysis and positive data research would diminish the empirical contribution.

Others have urged us to analyze the societal impacts of media concentration. This, too, is important, but it is not the question that this already long book addresses, and it is a topic that has already received much attention. Instead, this book will engage in the less glamorous task of marshaling the facts systematically.

A main contribution of this book is that it provides a comprehensive data base, across much of the world, across a dozen media industries, and often across two decades. Following often arduous investigation, it assembles the ownerships and market shares of major media firms. This enables the analysis of comparisons, drivers, and trends. The data and the answers might not be complete, but they are more comprehensive than any study that preceded it. National teams of distinguished researchers gathered the data for their respective countries. They followed a common methodology, and we at Columbia tried to make sure that it was followed despite the variations in national data sources and market structures.

The questions that this book addresses follow. They will be answered in the various chapters and summarized in the final chapter, "Findings":

A. How large is the media sector?

B. What are the world's largest national media markets?

C. What are the largest media industries? Is content really "king"?

D. What are the levels and trends of media concentration in different countries and industries?

E. Are the world's media becoming more concentrated?

F. What are the overall concentration trends of Media Industries?

G. Have American media become more concentrated?

H. Does the transition to Internet-based media reduce concentration?

I. What countries have particularly high media concentrations?

J. Where is there a high pluralism of voices, and why?

K. What are the factors for high national media concentration?

L. What countries have particularly high cross-media ownership?

M. What countries have a high foreign ownership of media?

N. What countries are high exporters of media?

O. Do American media companies dominate the world media? Does Hollywood dominate content?

P. Where is there a strong role of public (state) ownership in media concentration?

Q. Are there different market characteristics for media in the countries of the North versus those of the South?

R. Are there trends of convergence among countries in media concentration?

S. What are media industries with particularly high—and low—concentrations? What are the explanations?

T. Is there a convergence in market structure among the various media industries?

U. Who are the world's most dominant media companies?

V. Which companies dominate the attention for news?

W. Who are the media owners, and what do they own?

X. Who are the Institutional Owners?

Y. What is the overall ownership of the Media Sector?

Z. What are the priority problems?

The study proceeds as follows: First, it discusses the project's organization and methodology (chapters 1 and 2). Next, it provides media concentration analysis and narration (chapters 3–31) for each country. It then calculates national, vertical, and ownership concentration trends and aggregates the data into increasingly large segments and sectors. Chapters 32–36 deal with national comparisons (32), European trends (33), industries (34), companies (35), and owners (36). The concluding analysis and findings (chapters 37–38). The reader may consider the structure to be a funnel: from country reports and data, to global summaries of industries, companies, owners, and countries, to conclusions and interpretation. The reader in a hurry may turn to chapter 38.

## Bibliography

For the sources mentioned in this chapter, see the General References on Media Ownership and Concentration at the end of this book.

# 2

# Project Organization and Methodology

## ORGANIZATION

The project covers 30 countries—64% of the world by population and 85% by GDP. Thirty country teams with a total of about 60 reputable researchers from universities around the world contributed their knowledge and analysis. They covered 13 media industries, covered a period of 10 to 25 years, identified the market shares of thousands of companies, assembled their countries' data, and interpreted it. The data are theirs.[1] If this project had done nothing else, generating such work on media concentration in so many countries around the world would be important. We at Columbia University provided a common methodology, feedback, and the overarching summaries. The Columbia Institute for Tele-Information is an academic research center and is not engaged in commercial activities or advocacy. The principal author and editor of this study has not engaged in work for any of the companies discussed. The project was supported by the Media Program of the Open Society Institute.

An executive committee with members from three continents coordinated the first stage of the project. The country researchers worked separately, each an expert on his or her country. They periodically came together for conferences in New York, Paris, Brussels, and Yokohama.

As mentioned, on the subject of media ownership, strong opinions have been more plentiful than a fact base. The function of data analysis—if it was considered at all—has often been to support advocacy rather than enlightenment. In contrast, this study did not start with any preconceived notion or advocacy. Indeed, until the final weeks of a multiyear project, we did not

1. Those who wish to use the various chapters' data extensively need permission from the chapters' authors.

yet know how the numbers added up globally. Our goal was not to make policy recommendations but to provide policymakers with objective measures. One strong point of this project is its consistency across countries, industries, and years, thus making comparisons possible.

Thirteen media industries of three main types were investigated:

1. Content media
   Print media: newspapers, magazines, and books
   Audiovisual media: radio, broadcast TV, video channels, and film
2. Platform media
   Wireline telecom, wireless telecom, and multichannel platforms
   (Cable and satellite TV platforms)
3. Internet media
   ISPs (also in platform media), search engines (also in content media), online news media (also in content media)

The countries studied included the following:

1. Europe
   Belgium, Finland, France, Germany, Ireland, Italy, Netherlands, Poland, Portugal, Spain, Sweden, Switzerland, United Kingdom
2. North America
   Canada, United States
3. Latin America
   Argentina, Brazil,[2] Chile, Mexico
4. BRICS (emerging) Countries
   Brazil, Russia, India, China, South Africa
5. Middle East
   Egypt, Israel, Turkey
6. Asia-Pacific
   Australia, Japan, Korea, Taiwan

To assemble such a fact base has been both exciting and tedious. Writing opinion is much more fun than creating an information-dense book covering 13 information industries, 30 countries, 20 years, and thousands of companies. But the subject does not deserve shortcuts.

Why include 13 industries? Why go beyond mass media? Isn't the key issue the concentration of TV, newspapers, or cable TV? There are several reasons. The first is to recognize that there is no agreement on which media are "important." When people talk about media concentration, they tend to focus on a few examples, often picked based on their particular interest. Those fearing media power in politics are likely to point to television in its various forms. Those believing in media diversity based on the Internet will focus on the market power of internet service providers (ISPs). Those worried about computers and software applications will be concerned about operating systems. Those engaged in cultural studies will be particularly concerned with the film industry and book publishing. Those seeking an efficient national infrastructure to transport the various communications streams will focus on telecom. And so forth.

The importance one attaches to a partial segment of media and communications tends to be a reflection of one's values, pocketbooks, and professional spheres. To disregard the importance of other segments of the information sector can be parochial. And it has no future. Which is the second reason. The underlying technologies of the information sector are growing closer, their basic economics are becoming increasingly similar, and media companies spread their activities across industries.

A third reason is that at any given time, some media will concentrate while others diversify, and the only meaningful way is not to pick and choose examples but to look across the broad sweep of media. Of course, specific problems should not be submerged in a big average, and we therefore also analyze and discuss the data for specific industries, as well as for broader subcategories of industries, and for countries and regions.

To some readers, the selection of industries and companies might not be broad enough and

2. Brazil is listed in two categories.

may cover, in particular, "legacy" media rather than the richness of blogs and user-based publications. We are entirely mindful of those alternatives. But one should not equate newness and attention with economic presence. For now, these activities are, in revenue terms, the "digital pennies" that replace the "analog dollars."

## METHODOLOGY

In order to enable a comparison of countries, industries, companies, and trends, a common basic methodology was required. The question of what firm controls what share of a media industry may seem simple. But it is not. Take the straightforward statement, "Company X has Y percent of media industry Z." Every single element of that statement is potentially contentious.

*Who is X?* Is it part of a conglomerate? Is it a joint venture of several companies? Is it controlled by owners who hold only a minority of shares? What is its nationality?

*What is Y percent?* Is it in terms of revenues? Of circulation? Of audience or readers? What is the geographical market that is counted? What is the language or specialty market that is considered? Do we count broadcast signals across borders as a market participant? How do we know the overall market size, when it includes many small providers that do not show up in any statistics?

*What is media industry Z?* For example, is a weekly paper considered a newspaper or a magazine/periodical? Is a fan magazine in the same market as a news magazine? What exactly is an online news provider, does it include blogs, foreign websites, or information about the weather?

All of these are questions of concept and definitions, and these may well differ from country to country. On top of that are the factual issues of data. Companies are not neatly organized according to the industries and countries we identify. They often do not report their performance at all or they do so in other and nontransparent, nongranular ways. Even if they do, revenues and costs are often shifted for tax and regulatory reasons.

All these questions of concept and fact are complex, contentious, and subjective enough to provide, in the regulatory field, gainful employment for generations of media and antitrust lawyers.

It is therefore inevitable that some experts or interested parties of a particular industry in a particular country will find themselves in disagreement with country authors of this book or with the editor's summaries. We welcome corrections and perspectives. Yet to keep it all in perspective, even medium-sized corrections are unlikely to change the overall findings, given the large number of countries, industries, and firms, their often huge size, and the likelihood that flaws will often cancel each other out.

The important thing to remember is that market definitions and indices are never perfect, and they are most useful if applied consistently over time and across countries. If the thermometer drops it is probably getting colder, whether the scale is Celsius, Fahrenheit, Reaumur, or Kelvin.

None of this is likely to dissuade some people who do not like particular findings—in one direction or its opposite—and who will imagine that the reason must be ideology, obtuseness, or conflicts of interest. The only response we can give, individually and as a team, is that we have tried to identify facts and figures fairly and without preconceptions and that constructive corrections are welcomed.

### Data Sources

"Official" numbers on company revenues and shares are usually not available, except perhaps for some of the regulated telecom operators. Numbers were therefore found by the authors in a variety of ways. Useful source categories were reports by governmental agencies, financial institutions, and consultancies, as well as company annual statements and news articles.

Yet in the end, considerable judgment and detective work were required. Each industry and country had its own sources and institutional peculiarities. Often, reported numbers were not broken down in ways granular enough for the authors' tasks and were instead aggregate for the entire company, across countries and business lines. The effort required from the country chapters' authors to collect and organize highly disparate data in a consistent form was remarkable. For example, industrywide revenues proved difficult to find for industries such as magazine publishing. In several cases we had to estimate and prorate such revenues from those of comparable countries for which they were reported.

For the most recent period, several final updates were done by the editors for a number of industries or countries, especially when there were mergers. The authors are knowledgeable analysts of their country's media industries and are thus also the primary source for the data used in the overarching and summarizing chapters. In a few cases, we have slightly reorganized their data to achieve consistency.

To observe trends over time, we tried to cover a period of about 20 years, with one observation for every four years. It was not always easy to obtain data for past periods or for the most recent years. Due to different data availabilities in the various countries, the most recent available numbers vary. For some countries and industries they might go back to 2009. For others, 2013 data were available. In some cases, the editors updated the numbers to a more recent year.

## Market Definitions

Defining media industries always leaves gray zones. Some industries are intertwined, such as over-the-air TV broadcasters, satellite-delivered program channels, and multichannel video platforms. Another example is the film industry, for which consistency required the inclusion of imports and the exclusion of the film theaters (exhibition) from the market definition of film production and distribution. To deal with such issues, we tried to create consistent definitions, as follows.

- *Newspapers* are daily papers, not including their online versions (news-stand sales, subscriptions, and free distribution).
- *Magazines:* periodicals, mostly consumer oriented.
- *Books:* all books, including textbooks.
- *Radio:* AM, FM, digital terrestrial, and satellite audio broadcasting, both stations and networks.
- *Broadcast TV:* all "free TV" terrestrial video broadcasting by station and networks, as well as the retransmission of such channels over cable and satellite.
- *Video channels:* channels not distributed free over-the-air but for a fee over cable and satellite platform.
- *Film:* production, distribution, and importation of feature-length films. Does not include exhibition (theaters).
- *Multichannel platforms:* cable TV, direct-to-home broadcast satellites, telecom IPTV, and online (OTT) providers. The channels they carry (video channels and TV broadcasters) are not included.
- *Wireline telecom:* telecom companies, cable, and online providers of telecom service. Does not include mobile telecom, ISP service, and IPTV.
- *Wireless telecom:* mobile phone service providers, not including handsets.
- *ISPs*: Internet service access, including broadband and dial-up, using wireline, cable, satellite, or mobile connectivity.
- *Search engines:* major web-based information search systems.
- *Online news media:* online versions of newspapers, magazines, newsletters, and online providers and compilers of regular news. Does not include online blogs.

## Geographic Definitions

The main unit of analysis was *national* rather than local markets. A newspaper company may be small nationally and tiny globally but loom large locally.[3] We first focus on national

3. Several country chapters also discuss local markets.

revenues of a company, and then we aggregate its multinational market shares. In national markets we include all participants, not only those domestically owned or headquartered. That is, we include imports, and do not aim to include exports.

Media activities are typically listed by company, not by title, brand, or corporate division. For example, for Germany, Bertelsmann would be listed as the primary entry, rather than its properties RTL, Random House, or Stern. This makes it easier to identify national and international cross-ownership. Foreign and state ("public") ownerships are identified.

## Concentration Measures

There are several approaches to define media concentration. To many people it is the extent of viewpoint pluralism. To others it is diversity in the type of owners or managers. Still another perspective is the extent of local control versus absentee ownership. These approaches are legitimate, but they are difficult to operationalize on a national level, let alone for transnational research. Instead, this study uses basic tools of market structure definition from the economics of industrial organization. This is not to negate the usefulness of other forms of measuring pluralism, but they are less suitable to cross-industry and cross-country aggregations.

The methodology used had to be fairly simple and straightforward, for several reasons:

1. To facilitate usage by dozens of researchers from around the world who participated in the study and by researchers in the future who wish to deal with different countries, industries, companies, and time periods
2. To create a comfort level in readers and policymakers that the results are based on established ways to measure market concentration rather than on unproven new approaches (That said, several new approaches are also used in this book.)
3. To provide the kind of transparency of methodology, data use, and robustness that more complex econometric specifications with "black box" aspects do not provide

Therefore, instead of constructing a single and potentially complex measure for concentration, this study uses and creates several simple metrics that reflect different aspects of concentration. They are applied separately to subquestions and also put together to identify broad trends and correlations.

Several concentration indices are used, all of which are explained in this chapter:

- The C4 ratio
- The C1 ratio
- The Herfindahl–Hirschman Index (HHI)
- The power index (PI)
- The Net Voice and Net Voice/Capita counts
- The Noam Media Concentration Index (NI) (also known as the media ownership and concentration diversity index (MOCDI)
- The averaged C4, C1, HHI, PI, and NI
- The pooled C4, C1, HHI, PI, and NI

These concentration indices are aggregated across 13 media industries, and across 30 countries, in three ways: arithmetic averages, weighted averages, and pooled. The *weighted average* adjusts, in averaging, for different sizes of industries or countries, which an arithmetic averaging does not do. In contrast, a *pooled* concentration index treats all 13 industries as part of a larger single media sector and looks at market shares and concentration within that larger market.

A pooled measure lowers the concentration numbers, of course, but it does so for all countries, and thus comparisons are still possible. What is gained is a dimension of cross-ownership, insofar as major companies often operate in several industries, and pooling the industries hence shows their overall share in an overall national media market. We also have other ways to identify cross-ownership, discussed later in the chapter.

## The C4 and C1 Ratios as a Measure of Concentration

The C4 index aggregates the market share percentage of the largest four companies in an industry. If the market shares of the top four firms were, for example, 40%, 30%, 10%, and 10%, with five other firms holding 2% each, the C1 would be 40 and the C4 index would be 90. The formula for the C4 ratio is

$$C4_j = \sum_i^4 S_{ij}$$

where $S_i$ = firm's $i$ market share of a given industry $j$ and where firms are ordered by size of market share.

When C4 ranges from 0% to 40%, the industry tends to be competitive if the companies are of roughly equal size. It says that smaller companies serve 60% or more of a market. With a C4 above 40%, the industry is most likely an oligopoly.

The C4 for the film industry in the United States, for example, was 60.7% in 2012: this was the sum of the market shares of Viacom's Paramount, Time Warner's Warner Bros, Disney, and Sony, the firms that had the largest market shares of all film studio-distributors in the United States market that year.

## The HHI Measure of Market Concentration

The C4 index does not account for companies that are below the top four but may still have an important presence. For example, in the US film industry, the share of firms ranked number five and six (Comcast's Universal and 21st Century Fox) is quite similar to those of the top four. Also, the aggregation of the C4 distribution does not account for the distribution of shares within it, which can vary greatly. A C4 of 60% does not show whether competition is relatively high—the top four firms each have about 15% market share—or low (one firm dominates with 55% of the market share in the industry, the remaining three combined account for just 5%).

The Herfindahl–Hirschman Index (HHI) is used to measure concentration in a more informative way than the C4 concentration ratio just described. The HHI rectifies the shortcomings of the C4 by squaring the market shares of the companies in an industry and then adding them up, thus giving extra weight to high market shares. The resulting sum ranges between 0 and 10,000 points (where one company holds 100%, and its squared market share is thus at 10,000). The US government's anti-trust enforcement guidelines hold that an HHI under 1,500 is defined as unconcentrated; moderate concentration occurs in the range of 1,500 to 2,500; and high concentration starts at 2,500. (These numbers had been considerably raised in 2010. Until then, the thresholds were 1,000 and 1,800.)

Although the HHI index is more informative, it lacks the intuitive ease of the C4 index. It is easy to understand that a market share of the top four firms of 80% means that concentration is high. It is less clear what an HHI of 1,900 means.

The formula for this index is

$$\text{HHI} = \sum_{i=1}^{f} S_i^2$$

where $f$ = number of firms participating in an industry,

$S_i$ = each firm's market share,

$i$ = firm in a given industry.

In the earlier example, the HHI would be $\text{HHI} = 40^2 + 30^2 + (2 \times 10^2) + (5 \times 2^2) = 2{,}720$.

## Media Concentration Measured by Voices

One way to look at the diversity of media is through market shares, and this is the approach taken in much of this book. However, there is also another perspective. That perspective says that pluralism is not defined by concentration but rather by the diversity of sources available. Thus, if there are 20 radio stations in the market, their actual audience share is immaterial; what counts is the number of options for a listener, not how popular they are. In this case, there would be 20 "voices." But this needs refinement. The same firm might own several of these stations and thus there would not be 20 truly different voices. It therefore would be more accurate to consolidate voices of the same media organization and calculate "net voices."

A second refinement is that of a limiting principle. Some voices are just too small to be considered an alternative option. Is a college newspaper or a shopping magazine a "voice"? There has to be some threshold for a voice to be a meaningful participant. For reasons of principle and practicality, we chose the threshold that a firm must have at least a 1% share of that media industry's market.[4] A media operation that is tiny does not, in most cases, meaningfully contribute to the pluralism of media choices or shape public opinion. True, tiny publications sometimes can have an impact, but typically only indirectly as a catalyst to larger media operations that pick up the story. On a practical level, it is difficult to define, measure, and count tiny media.

Choosing a threshold of 1% of a market means applying a sliding scale that rises in absolute terms with the size of the market. Comparing countries' voice count on that basis is defensible on the level of pluralism, where voices are relative, not absolute in size.

One may think that picking an absolute rather than relative definition of a voice would be an easy solution, but it is not. The US population is 60 times higher than that of Finland. If the same voice definition based on revenue or circulation definition were applied, and if a newspaper with a 1% penetration in the United States would be counted, the equivalent threshold market share in Finland would have to be 60% of that market. Conversely, a 1% firm in Finland would mean that one would have to count a US newspaper with 0.0167%, a tiny size that would not register many statistics or reports. Neither arrangement is practical. Could one disregard any size definition and just count media? That, too, is impractical. As mentioned, would every college paper, free shopping paper, blogger, and website be counted as a voice? This brings us back to a percentage-based definition.

Counting different voices is one way to measure diversity for user choice. But this does not measure diversity in terms of political perspective or subject matter focus. To measure such differences would be highly subjective to define or measure. Instead, the underlying assumption is that when the number of options rises, different perspectives will emerge. One example is the diversification of TV news perspectives in the United States. After several decades of three fairly centrist national TV news options, the emergence of multichannel cable TV led to the creation of TV news channels with pronounced political perspectives, with Fox News Channel on the right and MSNBC on the left. Thus, more options also meant a wider diversity.

## The Media Ownership and Concentration Diversity Index (MOCDI)

This index (the Noam-index, so named by others, not by the author), takes into account the number of voices available. Market power alone does not reflect media diversity.

The HHI is the sum of the squares of the market shares, or $\Sigma s_i^2$. The problem is that this measure does not cover pluralism well. For example, suppose the radio market consisted, as described earlier, of companies with market shares of 40, 30, 10, 10, and 5 times 2%. It would have an HHI of 2,720.

Most people desire a greater diversity in their media sources than in their computer hardware. They want more choices for themselves and for the political process. Therefore, the question arises of whether the traditional antitrust measure of the HHI is appropriate for media.

The issue is partly whether the concentration threshold for media should be lower, and also whether the HHI methodology itself accounts sufficiently for media pluralism. For example, in the radio example above, if the five smaller stations of 2% share each were replaced by 10 stations, each with 1% of the market, the HHI would decline only trivially, from 2,720 to 2,710. Yet the diversity of the local radio market would clearly be significantly increased by the presence of five additional radio station providers. Ignoring that addition would misrepresent pluralism.

To deal with the pluralism issue, the US government (via its Federal Communication Commission) briefly introduced a "diversity index," which counted each media outlet as a "voice."

4. For radio and book publishing, we use a 2% national market share threshold because a large number of subnational stations or specialty publishers would otherwise dominate country voice counts.

But this approach, which was tainted by its being result oriented as part of a controversial effort to loosen restrictions on ownership, disregarded the size of the various media. It thus equated the mighty *New York Times* with the puny Poughkeepsie Gazette, which happens to be in the same media market. Both were equal voices. The FCC's approach was repudiated by a federal court. This left the question of how to measure concentration and pluralism in media wide open.

Do more voices mean more diversity? The FCC said yes. But the answer is less clear if one takes into account how loud some voices are. With numerous video and Internet avenues existing today, it is easier to speak but harder to be heard. To be heard requires higher volume, which typically means more resources. Thus, one cannot equate a greater number of content providers with greater diversity. The two may work in the same direction but need not. Those who equate the two commit an error of composition: just because it is easier and cheaper for me to reach an audience over the Internet does not mean that it's easier if everybody tries the same. It often becomes much harder and more expensive to get attention, thus necessitating greater marketing resources.

Market power is important. The antitrust HHI is a pretty good litmus test for market power, but it does not make allowance for pluralism. As a radio listener, I am better off with another 20 stations on the dial or another newspaper sold at the news kiosk, even if few people listen to or read them. Their availability provides an option that carries value even if it is unexercised by most readers or authors.

The conclusion is that one should not have to choose between a measure of market power (the HHI) or of pluralism (the number of voices) but ought to incorporate both. The HHI should be replaced with another diversity index. Such an index would take the HHI and divide it by the root of the number of voices.

This can be defined by the equation

$$\text{MOCDI} = \frac{\text{HHI}}{V} = \frac{\sum_{i}^{n} S_i^2}{\sqrt{n}}$$

where $n$ = number of firms participating in an industry,

$S_i$ = each firm's market share,

$i$ = firm in a given industry.

Thus, the less concentrated in market terms and the more numerically diverse a market is, the lower the index.

Using our numeric example, this index would result in

$$\text{MOCDI} = \frac{\text{HHI}}{V} = \frac{2{,}710}{\sqrt{9}} = \sim 903$$

## Aggregations

The study applies various concentration measures to specific industries in specific countries. The measures can be further extended, beyond those limitations. An average C4 for an entire country averages the C4 market measures for the various media industries. Since a straight arithmetic average may overrepresent small industries, it is better to take a weighted average, weighting by the size of that industry. The result represents the average level of concentration prevalent across all media industries of that country, taking their size into account. (It should be understood that the top four companies are not the same in each industry.)

We can also measure the individual companies' share in the world market of that industry. This is found by aggregating a company's revenues in that industry in the countries in which it operates and calculating the percentage of global newspaper revenues that this represents.

These worldwide market shares enable us to calculate worldwide concentration measures such as the worldwide industry C4 (W-C4) and the worldwide HHI (W-HHI) index. Obviously, these measures will usually be much lower than country-specific concentration measures, since the overall market is much larger.

### CROSS-OWNERSHIP OF MEDIA CONCENTRATION

A "pooled" concentration measure such as a pooled C4 looks at the market share of the top four companies in an entire national media market. Unlike the average weighted C4, it does not present a mean of concentration across all 13 industries but describes the companies that dominate the national market overall.

In the extreme, if no firm owns more than one voice, the numerator is zero, and cross-ownership is zero. Where all voices are owned by a single firm, the ratio approaches 1.

### CROSS-OWNERSHIP OF VOICES

We can calculate the percentage of media voices owned by multivoice firms in the overall number of voices.

$$\text{Cross-Ownership of Voices} = \frac{\text{Gross Voices} - \text{Net Voices}}{\text{Gross Voices}}$$

### CONCENTRATION MEASURED BY POWER INDICES

We define a new type of cross-ownership measure, the media power index (MPI). The industry-specific concentration measures (HHI, C4, NI), as well as the averages across industries, are useful measures but they do not capture cross-industry, multimedia market power. A firm may have moderate market shares in multiple media industries but no dominance in any. Looking only at one industry market at a time would understate a company's vertical and horizontal market position. How can one measure such cross-ownership?

The power index introduced here is a cousin to the HHI. Whereas the HHI aggregates the squared market shares of different firms in the same market, the power index aggregates the squared shares of the same firm in different national media markets, adjusting for market size. It is the sum of a company's market shares in the markets it operates, summed up across the various markets in which the company operates, weighted by market size.

There are several variants of the power index. One can aggregate a company's squared market shares across several industries in which it participates in a single nation. Or one can aggregate the company's market shares in a single industry across several nations in which it is active. Or one can do both, aggregating for a given company across industries and nations. Or, in still another variant, one can aggregate in a single nation the aggregations of the several top companies.

We designate the first of these MPIs in which a given company's ($\hat{c}$) activities in a given nation ($\hat{n}$) in the various industries ($i$) are aggregated as

$$\text{Company-National MPI} = \sum_n s^2_{\hat{c},\hat{r},i} \frac{R_{\hat{c},\hat{n},i}}{\sum_i R_{\hat{c},\hat{n},i}}$$

where $R_{n,i}$ is an industry's national revenue.

The second MPI is that for a given company ($\hat{c}$) across all nations ($n$) for a given industry ($\hat{i}$).

$$\text{Company-Industry MPI} = \sum_n s^2_{\hat{c},n,\hat{i}} \frac{R_{\hat{c},n,\hat{i}}}{\sum_n R_{\hat{c},n,\hat{i}}}$$

where $R$ is an industry's revenues in a nation.

This index measures a company's worldwide role in an industry.

The third MPI is a company's overall global market position, across both nations and industries. It measures, in effect, a company's overall role in the overall global media sector.

$$\text{Company-Worldwide MPI} = \sum_n \textit{Company-National MPI}$$

where the company national MPI is given above. The fourth metric—the National MPI—is the overall aggregation of the individual companies' MPI in a given nation.

A fifth measure aggregates these global company power measures to one covering all industries, all companies, and all nations. This is the Media Power Index-Global (MPI-G).

## Foreign Ownership[5]

Additional dimensions of analysis are *foreign ownership* and *public ownership*. Foreign

5. In this section, there is no need to distinguish between China's "integrated" and "segregated" ownership indices because they are the same, so China is rendered as just "China."

ownership means that a company's ultimate controlling owners are operationally headquartered in another country. For example, Telefónica, one of the world's largest telecom companies, is counted as a foreign operator in Argentina or the United Kingdom. Sony, headquartered in Japan, is considered a foreign-owned company in the US market even though its film studio, the former Columbia-TriStar, is one of the traditional Hollywood "Big Six."

Foreign ownership, strictly speaking, is not a measure of concentration, but it is a measure of the openness of a country's media market to the outside and its absence of barriers to such entry, which may raise competitiveness. Such barriers might be legal/regulatory or simply those of difficulty in adapting to a foreign market with its special circumstances.

To calculate the percentage of foreign ownership, identify the imports ($F$) and companies ($c$) that are majority-owned by foreign entities. Their revenues aggregate provide a share of a country's overall media industries' revenue.

$$\text{Foreign Ownership Ratio} = \frac{\sum_c \sum_i F_{c,i}}{\sum_c \sum_i R_{c,i,\hat{n}}}$$

## State Ownership

We also measure public ownership to show how much of a country's media market is controlled by state enterprises. These can be direct government operations, semi-independent organizations like public-service broadcasters, or private-law companies in which the government is a controlling shareholder.

We use the term *public* in the sense of "owned by public authorities," as distinguished from the use of the term in the stock market and investor community, where a "public company" means a firm whose shares are publicly traded in a stock market.

Why is state/public ownership relevant to concentration? The higher the such state ownerships of companies $c$ in an industry $i$ or nation $n$, the less likely the competition is. This can be due to the difficulty of contesting a government operation; it might be legally prohibited; or, conversely, state ownership might indicate a nonviability of competition, for example in the early stages of a new medium.

$$\text{State Ownership Ratio} = \frac{\sum_c \sum_i S_{c,i}}{\sum_c \sum_i R_{c,i,\hat{n}}}$$

## Ownership and Control

The ownership of a media organization is assigned in this study based on actuality of control rather than of legal setup. Berlusconi's Fininvest is structured as a set of over 30 smaller corporations, each holding a small percentage of the whole. Rupert Murdoch's stakes are shares with preferred voting rights, held by a family trust, and they comprise control over the two major firms that were split from each other in 2013, News Corp and 21st Century Fox. Sumner Redstone has a similar setup with Viacom and CBS. Telefonica (Spain) controlled Telecom Italia (Italy) through a stake of about 46% in an investment vehicle Telco, which owned 22.4% of TI. Telefónica's actual share in TI equity was thus only about 10%. In all of these cases we go beyond the legal arrangement that has been created for reasons of tax, corporate law, family succession, control, access to stock markets, and so forth. Similarly, we aggregate media companies that are under common control, regardless of the name of their corporate division or country of operation. Thus, Random House, Gruner + Jahr, and RTL are all counted as part of the same parent company, Bertelsmann. They might operate independently on a day-to-day and even strategic basis, but their top managers are ultimately selected, directed, and coordinated by the same control group, which bears final responsibility.

This would be fairly unexceptional, but the same principle must also be applied to national governments holding multiple state-owned firms. In particular, it is an issue with China. Most Chinese multimedia producers are state owned. However, different parts of the government and different levels of government own certain media properties. So the question is how to view these organizations. Are they a single "firm" with multiple divisions, in the same way that Vivendi's Canal Plus pay-TV operation, its SFR mobile phone operator, and its Universal Music Group are controlled by the same enterprise even if managers get substantial autonomy from interference from the top and from each other? Or, are the various state organizations truly independent of each other? The two perspectives have different adherents. Those inside China often tend to present the latter perspective. But this would make China, by the numbers, the world's most diverse media environment, so this might be pushing an argument a bit too far to be plausible. The other perspective, that of a unified media system where state companies ultimately answer to the same authority or to lower levels who answer to that authority, creates the other extreme: now, China's media is just about the world's most concentrated. To be consistent with the rest of the countries and companies, we tend to the second view. However, to provide the alternative perspective, we often present both sets of numbers, and we use two measures: an *integrated* perspective in which the concentrations are calculated with the state as a single owner, for example ("Government of China"), and a *segmented* perspective in which each state enterprise is deemed an independent entity. Both ownership definitions are used when it is possible, but the former is used in calculating world averages, where one measure must be picked.[6]

There is a large discussion of the issue of media independence in China—hegemony versus diversity. Examples are Chan (1993), Chu (1994), Kennedy (2009), Lee (2006, 2007), Liu (2006), Stockmann and Gallagher (2011), Winfield (2005), Wu (2000), and Zhao (2000). Those who argue for media diversity correctly observe a greater independence by journalists, at least compared to the past, as well as a greater reliance by media on nonstate financial support through advertising and subscriptions. But this diversification of content and business models should not be equated with ownership diversification.

## The Owners of the World's Media

Ultimately, somebody owns media organizations. The question of actual ownership comes in two dimensions: who owns the largest media firms; and whose ownership stakes in media are the largest. We investigate both.

### THE OWNERS OF THE WORLD'S LARGEST MEDIA FIRMS

We select the largest media companies—about 20 platform companies and a similar number of content companies—and investigate their ownership structure by using the ownership records of each of these companies and identified company's owners from companies' annual reports and filings with government agencies, as well as press sources.

We identify the major individual and institutional holders of each company, the number of shares they held, their percentage ownership of the company, and the total value of their shares.

There are several complicating factors. Some shares are not publicly traded and their value has to be estimated. In other cases, companies have several classes of shares.

Still another complicating factor is that for some companies, media activities are only part of what they do. Where companies' activities span media and nonmedia industries, only the former have been used. These issues are discussed in the chapter, "Ownership Analysis."

### THE LARGEST OWNERS OF MEDIA

Who owns the media? There are three kinds of owners: (1) individuals/families, (2) institutions, and (3) governments/public. We identify, measure, and discuss them sequentially. Institutional ownership holdings were determined by examining investment companies' filings to regulatory agencies.[7,8,9] In many cases we also looked at the institutional investor's annual

6. This judgment was taken by the editor, after considerable review of the literature on China's media. It was done independently from the authors of the country report on China and other researchers, and the responsibility is the editor's alone.

7. <http://www.fsa.go.jp/sesc/english/>

8. <http://www.cnbv.gob.mx/en/Paginas/default.aspx>

9. <http://www.finanstilsynet.no/en/>

reports, where available. In the process, we construct an aggregated list of the largest owners of media companies.

We also aggregate the media industries into two subsectors, namely, platform media and content media. *Platform media* refers to the transmission and delivery of media, while *content media* refers to consumed material. A subcategory of content media is *news media,* described further in the next section. Many firms are both providers of content and of platform services. And several industries straddle the platform/content divide. This is particularly true for multichannel platforms. It therefore required some work to segment these firms by activity.

## News Attention

*News media* are the media that provide much of the information for civil and political discourse. In our investigation, they are defined as newspaper, magazines, radio, broadcast TV, online news, and part of multichannel TV. They will be considered based on the time people spend on each medium's news content.

We estimate *news attention shares,* using a number of simplifying assumptions due to a lesser availability of data. The reader should take the results as orders of magnitude rather than as precise metrics.

We can determine the market shares and the concentration indices for these six news media industries. The question is how to add them up in order to get a company's share in the overall news media, both nationally and globally. This really is the question of how important the different news industries are in news terms. Two proxies are used. The first measure is to use an industry's *revenues* to weight the various news industries. The second and better method is to assign weights according to the *attention time* they receive from users as a news source, that is, the time people spent reading, listening to, or watching the medium's news content. Of course, some news attention has more of an impact, such as reading an editorial versus the weather report, and some media may leave more of an impact on a user per time unit. However, trying to gauge and measure subjective dimensions such as importance, value, or impact raises formidable conceptual and methodological barriers.

*Attention weight* of a news medium is calculated by the average time spent by an average user, prorated to the share of that medium's news content in its overall content, and also adjusting for the penetration of that medium in a country's population. The data necessary for usage time were available for some countries, but not for others. Where they were not, we use the average of those that were available. This is a simplifying assumption, but for those countries where the data were available, we did not observe huge differences across countries in the consumption-time among news users of a medium, given access to that medium. What differs more are penetrations, for example, for newspaper readership or of online news access.

News attention for company $c$ in country $n$ for its activities in a given news industry $i$ is

$$A_{\hat{c},\hat{n},\hat{i}} = \text{Population}_{\hat{n}} \times \text{Penetration}_{\hat{n}i} \times \text{Medium's Attention Share}_{\hat{i}} \text{ by users} \times \text{Market share}_{\hat{i}}$$

For all of company $c$'s news activities across industries $i$ in country $n$, news attention would be

$$A_{\hat{c},\hat{n},i} = \sum_{C} A_{\hat{c},\hat{n},i}$$

And the total news attention share of company $c$ in country $n$ and in the world can then be aggregated.

This way, we can calculate the global news media share of companies that are active in several countries, and we can compare the share in worldwide attention of the various media companies.

The top company in news attention has, by definition, the greatest "mindshare" in its society. But there are degrees of dominance. Is media concentration in news associated with high developmental level or with lesser economic development? A regression analysis of the market share of the top firm (the "C1") with the average income in that country might show a possible correlation.

## Factors of Inter-Industry Concentration

When market trends in a single country are considered, the particularities of a particular national situation may obscure a more fundamental explanation. But once 30 countries' figures and trends are analyzed, a fuller picture emerges.

What are factors that can explain concentration levels $K$ that exist in different media industries $i$? To find answers we run regressions over a number of variables $v$. The maturity of an industry, for example, might have provided the time for industry consolidation to take place; the global export intensity of an industry might create scale effects that go beyond a country's borders; or the size of an industry might give more firms an opportunity to flourish, and so forth.

$$K_i = f(v_i)$$

## Factors for Intra-Industry Concentration Differences

How can one explain different industry concentration levels in different countries for the same industry? This is a different question from the one in the previous paragraph. Such an analysis would try to capture country-specific characteristics that affect concentration in that industry. A number of country-specific variables $H_x$ were hypothesized and tested: population size, geographical size, income, education level, per capita spending, regulatory quality (for regulated media industries), years as a democracy since 1900, economic growth, research and development (R&D) spending, and other variables.

$$K_{n,i} = f(H_{x,n})$$

## Identifying National Divergence of Concentration from the Expected

The data that were collected from around the world and subjected to methodological uniformity permit an econometric cross-section analysis. The results show which factor affects concentration levels of different media industries and which factors affect concentration within a media industry.

We then take this analysis a step further and try to extract information from "nonconformity." That is, from a country's industry deviation from the concentration level predicted by the econometric model.

Overall, the models that were estimated provide a prediction: for a given industry and country—using its actual sociodemographic, economic, and geographic values—a prediction for concentration can be readily obtained, based on the explanatory power of the estimated coefficients. But the actually observed concentration levels will tend to deviate from predicted levels. What this means is that we can observe deviations of media from levels that can be explained by poverty, geography, population, and so forth.

These deviations $D$ of the predicted concentration $K_p$ from the one actually observed $K_o$ should be the measure in judging a country's concentration. The difference of actual to predicted values is the residuals of the estimation.

$$D_{n,\hat{i}} = K_{p(n,\hat{i})} - K_{o(n,\hat{i})} = f(H_{x,n}) - K_{o(n,\hat{i})}$$

## Convergence Trends

*Digital convergence* has been discussed for many years, first in terms of similar technology and then of overlapping industries and businesses. We can measure convergence trends of the structure of different media industries with each other. We do so by observing the standard deviation—the measure of the $y$ industries' $I$ divergence from the average $\mu$.

$$\sigma = \sqrt{\frac{1}{y} * \sum_i \left(K_i - \overline{K}\right)^2}$$

## CONCLUSION

In this chapter, we introduced the process and the methodologies of this book. We are now ready to proceed with an analysis of the 30 countries' media markets, followed by five overarching chapters, and two chapters of conclusions.

## Bibliography

For the sources mentioned in this chapter see the General References on Media Ownership and Concentration at the end of this book.

# PART 2

## MEDIA CONCENTRATION AROUND THE WORLD—COUNTRY STUDIES

# A

## Europe

# 3

# Media Ownership and Concentration in Belgium

PEGGY VALCKE, JO GROEBEL, AND MORITZ BITTNER

## INTRODUCTION

Belgium's borders the Netherlands, Germany, Luxembourg, and France which reflects the ethno-linguistic divisions present in its politics, culture, and media markets. In 2008, Belgium had a population of approximately 10.7 million people living in three distinct Communities and three regions.[1] Flanders, which has historical and linguistic ties to the Netherlands, is the heart of the 6.2 million strong Flemish Community that comprises 59% of the population. Three and a half million Belgians (31% of the population) reside in Wallonia, a historically French-speaking area recognized as the French Community of Belgium. There is also a German Community, which is the smallest officially recognized community in Belgium, with fewer than 100,000 inhabitants. The remaining 10% of the population reside in the capital region, Brussels-Capital. Slightly under a million foreign nationals reside in Belgium, of which about a third reside in Brussels-Capital. Therefore, there are four linguistic regions: one for Dutch/Flemish speakers, one for speakers of Belgian French (plus three Walloon dialects, which are related to French), one for German speakers in the "East Cantons," and the multilingual Brussels-Capital Region.[2]

The linguistic, cultural, economic, and ideological differences in the country—especially between Flanders and Wallonia—have led several waves of federalization since 1970, with increased autonomy for regional authorities. As a result, the country is characterized by a

1. Valcke, Peggy, and Lievens, Eva. *Media Law in Belgium*. Alphen aan den Rijn: Kluwer Law International BV, 2011, 21.

2. Flemish, as a language, is very similar to Dutch, but there are a number of differences, most importantly with respect to pronunciations and vocabulary. Belgian French is by far the most common French-derived language in the country, and it is distinct from the dialects of Wallonia that are related to French but like Flemish are fairly distinct in their own right.

complex political system, with a delicate balance among the federal and regional institutions. Between 2010 and 2011, this was underlined by the fact that no national government was able to form for over 500 days following elections, but the regional governments kept functioning, and in doing so demonstrated the reduced relevance of Brussels-Capital's central authority in the federal system. Although recent developments have spurred debates about partitioning the country apart, the movement for partition has not gained sufficient traction to force the question.

Due to these linguistic and cultural differences, the Belgian media landscape is quite unique within Europe. Different media markets can be distinguished that run parallel to the country's officially recognized Communities, that is, the Flemish Community in the north, the French Community in the south, and the much smaller German Community in the east. As a small country with three official languages that achieved statehood relatively recently (the country only came into being in its present form in the 1830s, as a constitutional monarchy), Belgium has long been deeply influenced by the culture and languages of its neighbors. This is especially the case not only for the small German Community, but also for the French-language markets, which are strongly oriented toward France and French television channels.[3] In Flanders, before the introduction of commercial television, audiences relied mainly on channels broadcast from the Netherlands. But since Dutch and Flemish are substantively distinct from one another, the Flemish-language market rapidly turned to Belgian commercial channels as soon as these were introduced.[4]

As part of the federalist compromises with the Communities, authorities at the federal and regional levels share oversight authority for the media sector. During the first state media reforms of the 1970s, around the time Belgium officially became a federal state, regulation of radio and television broadcasting was transferred to the regional level as part of the cultural domain. As a result, commercial broadcasting fell under the exclusive powers of the Flemish, French, and German-language Communities. The Belgian Constitutional Court has ruled that these broadcasting powers cover not only substantive issues like the organization of public broadcasting, licensing, programming requirements, and advertising rules, but also technical aspects such as distribution.[5] It has also ruled that these broadcasting powers do not cover on-demand services offered via digital television platforms, mobile networks, or the Internet. Freedom of the press, journalists' contracts, film, telecom, Internet media, consumer protection, intellectual property rights, and competition law are dealt with at the federal level. The partition of powers over the media sector is a contentious issue that, as a result of 30 years of Belgian Constitutional Court rulings, has led to the formation of a number of legal gray zones in which the overlapping of federal and regional powers remains unresolved.[6]

Each linguistic Community is recognized as a political entity and has a media regulator in charge of supervising the audiovisual sector (radio and television) on its respective territories: the Vlaamse Regulator voor de Media (VRM) for the Flemish-language market, the Conseil Supérieur de l'Audiovisuel (CSA) for the French-language market, and Medienrat for the German-language market. Only the audiovisual sector has Community-based regulatory bodies. The VRM monitors concentration indices and publishes the results in annual reports, while the CSA is not only authorized to monitor

3. Franck, Aline. "Belgique: regards croisés sur deux paysages audiovisuels." *Régulation* 41 (2009), 18.

4. This switchover was also helped by the fact that the legislative framework for commercial broadcasting initially allowed for compulsory shareholding by local press outlets over commercial TV stations. These localized interests help explain why Flanders has some quite successful local channels, while in the French-language market France's leading commercial channels, such as *TF1*, are still very popular.

5. Valcke, Peggy, and Lievens, Eva. *Media Law in Belgium*. Alphen aan den Rijn: Kluwer Law International BV, 2011, 152.

6. The right to reply on radio and television, for instance, is at the same time regulated both in a federal act [the Act of 23 June 1961, On the Right of Reply (*Wet van 23 juni 1961 betreffende het recht tot antwoord*)] and in the Flemish Broadcasting Decree of 2009 (*Decreet van 27 maart 2009 betreffende radio-omroep en televisie*), which has been amended several times since.

these indices, but also to take regulatory action if it determines that the market is becoming too concentrated.[7]

Media groups' operations tend to follow the north-south divide between the Flemish-language market in Flanders and the French-language market in Wallonia, although in recent years some have also developed activities in both Communities.

Linguistic divisions determined media ownership in Belgium for many years.[8] This has changed—resulting in stronger cross-language participations—after 2005. In 2006, the Flemish newspaper publisher Corelio took an interest in, and (eventually, in 2008) bought out, the Walloon media group, Mediabel. Also the country's leading (and formerly independent) financial dailies, *De Tijd* (with a 4.1% market share among Dutch-language newspapers) and *L'echo* (which has a 3.7% market share among French-language newspapers) belong to a single cross-language consortium after their take-over in 2005 by Mediafin, a 50–50% consortium of the Dutch-language De Persgroep and the French-language Rossel.[9] Concentra, another Dutch-language publishing company, and Rossel together publish the free daily *Metro*. Concentra merged in 2013 with Corelio into the firm Mediahuis.

Because media markets in Belgium follow the country's language borders, the markets are not very large compared to those in France, the Netherlands, or Germany. As a result, high concentration levels are inevitable in Belgian media markets. Since the regulatory authorities have accepted this as the norm, there are hardly any ownership restrictions in media legislation, and no takeovers or mergers in the media sector have ever been prohibited by the Competition Authority. The Authority, however, has attached certain conditions and obligations to its approval in recent cases, such as the takeover in Flanders of the only pay-TV provider, the Belgian division of the French Canal + Group, by Telenet in 2003, and the takeover of Wallonia's leading pay-TV provider, *BeTV*, by Tecteo (VOO) in 2008. In those cases, the Competition Council, a body within the Competition Authority, imposed access and nondiscriminatory obligations on the cable operators, prohibiting them from acquiring exclusive broadcasting rights. Even the controversial merger and ownership ruling involving *De Tijd* and *L'Echo* resulted in the Mediafin consortium eventually gaining sole control over these newspapers, although Rossel and De Persgroep are barred from offering bundled products to subscribers and advertisers.[10]

A more recent case is that of the takeover of the commercial broadcaster SBS Belgium's *VT4* and *VijfTV* channels from ProSiebenSat.1 (Germany) by De Vijver Media in 2011, which was a consortium composed of Sanoma Oyj (Finland), the Corelio Group, and Waterman & Waterman (each of which has a 33.3% stake the consortium).[11] The transaction led to a higher concentration of crossownership in the Flemish TV broadcasting sector.[12] Although SBS Belgium was not active in other media markets or segments, its owners had developed activities in TV production (via Waterman & Waterman's

7. CSA. "L'offre de médias et le pluralisme en Fédération Wallonie-Bruxelles." CSA, 2013. 10 Jan. 2013. <http://www.csa.be/pluralisme>; Valcke, Peggy, and Lievens, Eva. *Media Law in Belgium*. Alphen aan den Rijn: Kluwer Law International BV, 2011.

8. Raeymaeckers, Karen, De Bens, Els, Paulussen, Steven, Deprez, Annelore, and Tenret, Yannis, *Media landscape: Belgium*. European Journalism Centre, 8. Nov. 2010. 10 Jan. 2013; <http://ejc.net/media_landscapes/belgium>; Lund, Anker Brink, Raeymaeckers, Karen, and Trappel, Josef, "Newspapers: Adapting and Experimenting." in Trappel, Josef, et al. (eds). *Media in Europe Today*, Bristol/Chicago: Intellect, 2011.

9. *Competition Committee, Directorate for Financial and Enterprise Affairs*. "Annual Report on Competition Policy Developments in Belgium: 2004." OECD, 29 June 2005. 10 Jan. 2013; <http://www.oecd.org/belgium/35110865.pdf>; Valcke, Peggy, and Lievens, Eva. *Media Law in Belgium*. Alphen aan den Rijn: Kluwer Law International BV, 2011, 27–28.

10. Valcke, Peggy, and Lievens, Eva. *Media Law in Belgium*. Alphen aan den Rijn: Kluwer Law International BV, 2011.

11. Briel, Robert. "Sanoma pays €1.225 billion for SBS channels." *Broadband TV News*, 20 Apr. 2011. 9 Jan. 2013. <http://www.broadbandtvnews.com/2011/04/20/sanoma-pays-e1-225-billion-for-sbs-channels/>.

12. Sanoma Media Belgium. "Competition Authorities approve the partnership structure of the SBS TV operations in Belgium." *sanoma.com*, 9 Sep. 2011. 9 Jan. 2013. <http://www.sanoma.com/about-us/sanoma-media-belgium2/news/competition-authorities-approve-the-partnership-structure-of-the-sbs-tv-operations-in-belgium>.

successful TV production house Woestijnvis) and print media (Sanoma and the Corelio Group being leading publishers of newspapers and weekly magazines in Belgium) in previous years. In 2014, Belgian cable operator Telenet bought a 50% stake in De Vijver Media (a.o. through the purchase of Sanoma's shares). The European Commission approved the transaction in February 2015.

## PRINT MEDIA

### Newspapers

The Belgian print media landscape is divided into Flemish- and French-language markets (there is also a German-language market, but it is extremely small and data for it is unavailable). Print companies active in the Flemish north are not necessarily active in the Walloon south and vice versa, although most communitywide and some regional newspapers are available throughout the country. Since the end of World War II (1939–1945), dozens of titles have disappeared in the course of concentration and, more recently, as a result of reduced sales and advertising revenues.[13] The number of independent media firms has shrunk from a high of 34 after the war to just five today: De Persgroep, Mediahuis (Concentra & Corelio), Rossel, IPM, and Tecteo/Editions d'Avenir. Mediahuis is active in both language Communities, while the others are focused on single Community.

De Persgroep, Corelio, and Concentra dominate the paid-subscription Dutch-language newspaper market; Rossel, Tecteo (*Les Editions de l'Avenir*) and IPM dominate the paid-subscription French-language newspaper market. Together with two other print publishers who are active throughout the country—Mediafin, publisher of the financial newspapers *l'Echo* (French-language) and *De Tijd* (Dutch-language), and Mass Transit Media, publisher of the multilingual free daily *Metro*—these publishing groups combined offer eight Dutch/Flemish-language newspapers in Flanders and seven French-language newspapers in Wallonia.

Since 2001, De Persgroep has been the largest newspapers publisher in Flanders (Table 3.1), with a 40% paid-subscription market share for its popular titles *Het Laatste Nieuws* and *De*

Table 3-1. Flanders Daily Newspapers (Market Shares by Circulation), 1999–2010

| | 1999 | 2000 | 2002 | 2004 | 2006 | 2008 | 2010 |
|---|---|---|---|---|---|---|---|
| De Persgroep | 36.9 | 38.5 | 37.2 | 36.2 | 37.0 | 37.5 | 37.0 |
| *VUM* (Corelio Group/Mediahuis) | 41.6 | 40.3 | 35.5 | 36.0 | 33.9 | 33.0 | 33.0 |
| Concentra (including De Vlijt NV), after 2013 part of Mediahuis | 21.4 | 21.2 | 27.3 | 27.9 | 29.1 | 29.6 | 30.0 |
| Total Circulation (1,000) | 1,200 | 1,220 | 1,260 | 1,300 | 1,250 | 1,250 | 1,240 |
| Total Revenue (mil euros) | | | | | 492 | 553 | 525 |
| Total Revenue (mil US$) | | | | | 620 | 813 | 698 |
| C3 | 100 | 100 | 100 | 100 | 100 | 100 | 100 |
| HHI | 2,196 | 3,556 | 3,389 | 3,378 | 3,364 | 3,365 | 3,358 |
| *N* (>1%) | 3 | 3 | 3 | 3 | 3 | 3 | 3 |
| Noam Index | 1,268 | 2,053 | 1,957 | 1,950 | 1,942 | 1,942 | 1,938 |

*Source:* Centrum voor Informatie over de Media (CIM), Studiedienst van de Vlaamse Regering, Vlaamse Regulator voor de Media (VRM), National Bank of Belgium. Concentra and Corelio merged in 2013 into Mediahuis.

13. For example, the popular daily *Het Volk* was acquired by VUM/Corelio in 1994 and integrated in *Het Nieuwsblad/De Gentenaar* in 2008. In 1998, De Persgroep acquired the daily *De Morgen*, and in 1999 VUM/Corelio acquired a majority and then, by 2006, 100% of Mediabel's assets.

*Nieuwe Gazet* (a regional variant of *Het Laatste Nieuws* for the Province of Antwerp) and *De Morgen*, which holds a 6% Flemish-language market share.[14] *Het Laatste Nieuws/De Nieuwe Gazet* is the best-selling title in all of Belgium, with a circulation of 345,000 copies reaching more than one million readers daily as of 2007. Together, these newspapers have a relatively stable market share of 30% for the paid newspaper market, though this falls to and 27.5% when taking into account the free daily *Metro*.

De Persgroep's joint ownership of *De Tijd* and *L'Echo* has helped the company win a greater market share, firmly establishing its leadership in the Flemish newspaper market. That said, the market share of *De Tijd* has fallen in the last ten years from 5% to 4%.

The second largest Flemish-language newspapers publisher is the Corelio Group, which used to be the market leader until 2002, but has now seen its market share continually decline over the past decade. It publishes the daily *De Standaard* and two other popular titles, *Het Nieuwsblad* and *De Gentenaar* (a regional variant of *Het Nieuwsblad* for the Province of Ghent). Together, they form the second largest popular title in Flanders Its main title, *Het Volk*, ceased publication in 2008. In 2006, Corelio was able to acquire the French-language Mediabel. This group publishes ten regional newspapers and several free dailies.

The third Flemish newspaper group, Concentra, publishes two regional titles: *Het Belang van Limburg* and, via its subsidiary De Vlijt NV, the *Gazet van Antwerpen*. Concentra and Rossel together publish *Metro*, in two language versions. The popularity of *Metro* has grown significantly since its launch in 2002. In 2009, it reached a market share of 11% in Flanders. The increase in Concentra's overall market share from 2002 on is primarily due to the success of *Metro*. In 2013, Concentra and Corelio merged and became Merdiahuis.

By far, the largest newspapers publisher in Wallonia is Rossel, which is 100% owned by the Hurbain family (Table 3.2).[15] Rossel controls

## Table 3-2. Wallonia Daily Newspapers (Market Shares by Circulation), 1999–2010

| | 1999 | 2000 | 2002 | 2004 | 2006 | 2008 | 2010 |
|---|---|---|---|---|---|---|---|
| Rossel (Hurbain Family) | 58.5 | 57.7 | 59.1 | 59.4 | 59.9 | 61.0 | 59.4 |
| IPM | 24.8 | 25.4 | 24.4 | 24.4 | 24.3 | 22.2 | 22.8 |
| *Les Editions de l'Avenir* (Corelio Group until 2014, then Tecteo) | 16.7 | 16.8 | 16.5 | 16.2 | 15.8 | 16.8 | 16.6 |
| Others | 0.0 | 0.1 | 0.0 | 0.0 | 0.0 | 0.0 | 1.2 |
| Total Circulation (1,000) | 4,520 | 3,950 | 3,270 | 3,170 | 3,110 | 3,060 | 2,950 |
| Total Revenue (mil euros) | | | 279 | 296 | 307 | 292 | 335 |
| Total Revenue (mil US$) | | | 262 | 367 | 387 | 492 | 446 |
| C3 | 100 | 100 | 100 | 100 | 100 | 100 | 100 |
| HHI[1] | 4,316 | 4,256 | 4,360 | 4,386 | 4,428 | 4,496 | 4,325 |
| *N* (>1%) | 3 | 3 | 3 | 3 | 3 | 3 | 3 |
| Noam Index | 2,492 | 2,458 | 2,517 | 2,532 | 2,557 | 2,596 | 2,497 |

*Source:* Centrum voor Informatie over de Media (CIM), Conseil Supérieur de l'Audiovisuel (CSA), National Bank of Belgium.

[1] HHI in here deviates from the HHI calculated by the CSA for the newspapers sector because the CSA bases its calculations on Rossel and IPM only.

14. Raeymaeckers, Karen, et al. *Media landscape: Belgium*. European Journalism Centre, 8 Nov. 2010. 10 Jan. 2013. <http://ejc.net/media_landscapes/belgium>.

15. Raeymaeckers, Karen, et al. *Media landscape: Belgium*. European Journalism Centre, 8 Nov. 2010. 10 Jan. 2013. <http://ejc.net/media_landscapes/belgium>.

60% of the market with the various titles that it publishes via Rossel & Cie and *Sud Presse,* plus its joint ventures of Mediafin and Mass Transit Media. Rossel & Cie manages *Le Soir,* which is still the leading paid-subscription daily in the French-language market, even though its market share has declined from 18% to 15% in the last ten years. Rossel also publishes popular regional titles under the *Sud Presse* imprint (*La Meuse, La Capitale, La Nouvelle Gazette, La Province, Nord Eclair*), which together represent 22% of the French-language market. It holds 50% of Mediafin, the consortium that controls *L'Echo* in Wallonia and *De Tijd* in Flanders, and 49% of Mass Transit Media, the company that publishes the free daily *Metro* publisher.[16] In Wallonia, *Metro* has done quite well since its launch in 2002, having acquired 18.7% of the French-language market share.

The second, much smaller newspaper group in Wallonia is IPM, which controls less than a quarter of the market.[17] It publishes two titles: *La Libre Belgique,* a conservative daily with a market share of 8.5%, and the popular *La Derniere Heure/Les Sports,* the second largest paid-subscription daily in Wallonia, although it is also facing declining market shares, falling from a market share of 16% in 2002 to 14.3% by 2010.

The regional titles of the Corelio Group, used to fall under the imprint of *Les Editions de L'Avenir. Les Editions's* nine regional titles (including *L'Avenir du Luxembourg*) represent approximately 16% of the French-language newspapers market.[18] In 2013, these papers were acquired by the cable operator Tecteo.

All three groups, IPM, Corelio/Mediahuis, and Rossel, have—via the Audiopresse consortium—a minority share (34%) in RTL Belgium, which controls three television broadcasting channels (*RTL-TVi, Club RTL,* and *Plug RTL*) through the Luxembourg-based RTL subsidiary of CLT-UFA (Table 3.3).

### Table 3-3. Aggregated Daily Newspapers (Market Shares by Circulation), 2006–2010

| | 2006 | 2008 | 2010 |
|---|---|---|---|
| Corelio Group/Mediahuis | 27.6 | 26.8 | 26.6 |
| *VUM* | *21.6* | *21.2* | *20.1* |
| *Les Editions de l'Avenir (after 2013, Tecteo)* | *6* | *5.6* | *6.5* |
| Rossel (Hurbain Family) | 21.4 | 21.4 | 23.1 |
| De Persgroep | 23.6 | 24.1 | 22.6 |
| Concentra (after 2013, Mediahuis) | 18.6 | 19 | 18.3 |
| IPM | 8.8 | 8.7 | 8.9 |
| Others | 0.0 | 0.0 | 0.5 |
| Total Revenue (mil euros) | 771 | 860 | 860 |
| Total Revenue (mil US$) | 972 | 1,264 | 1,144 |
| C4 | 91.2 | 91.3 | 90.6 |
| HHI | 2,200 | 2,194 | 2,166 |
| Corrected HHI[1] | 3,681 | 3,732 | |
| Noam Index | 984 | 981 | 969 |
| *N* (>1%) | 5 | 5 | 5 |

[1]Estimated on the basis that Flanders and Wallonia are two separate markets.

16. It also has a share of the German title, *Grenz Echo,* which has a small daily circulation of 10,000 copies.
17. CSA. "IMP Group." CSA, 2013. 10 Jan. 2013. <www.csa.be/pluralisme/groupes/show/3_groupe_ipm>.
18. CSA. "Group Corelio." CSA, 2013. 10 Jan. 2013. <www.csa.be/pluralisme/groupes/show/7_groupe_corelio>.

## Book Publishing

Despite the small size of the language-specific markets for books the Belgian book publishing industry is very diverse and involves over 150 (though sometimes small) Belgian publishing houses.[19] They are active both in the "nice to have" segment (of general books - including fiction, non-fiction, youth and children's books—and comic books; hereafter: "A-books") and in the "need to have" segment (of educational and school books, scientific books, and professional/informative books; hereafter: "S-books").[20] Few data are available on their activities, so we have been unable to calculate concentration indices for this sector. A scientific report from 2011 estimates the annual turnover in the Flemish book publishing sector between about 0.8 and 1.2 billion euros.[21]

For the so-called A-books, foreign publishers are the market leaders in Belgium, with French firms leading the French-language market and Dutch publishing houses dominating the Flemish market. However, Belgian publishing houses are gaining stronger positions at international level. In 2010, the 100-year old Flemish publishing house Lannoo acquired the renowned Dutch publishing house Meulenhoff, and became the third largest publisher in Flanders.[22] Flemish children's and youth literature is also becoming very popular abroad, as demonstrated by the successes that Flemish authors and illustrators booked at the Bologna Book Fair 2013.[23] ZNU (Zuidnederlandse Uitgeverij), established in the Antwerp region, is the largest publisher of children's books both in Flanders and in the Netherlands. It also owns Standaard Boekhandel, the leader in book distribution, with a 55% market share in a market valued at US$293 million (211 million euros) in 2011.

One area in this sector in which Belgian publishers cooperate with foreign publishers is in comic books/graphic novels, a prominent example of which is Hergé's long-running and internationally syndicated *Tintin* series. The Belgian comic book industry is characterized by cross-ownership among the different communities. For example, the Dargaud-Lombard-Dupuis Group belongs to the Franco-Belgian publisher Média-Participations and produces more than one-third of the company's overall turnover through comic book sales, including the *Asterix* series (which, like *Tintin*, is internationally syndicated). In the Flemish market, WPG Uitgevers België (formerly Standaard Uitgeverij) is the leading book publisher and has a significant stake in the comic book market, with series such as *Luke & Lucy* (in Dutch: *Suske & Wiske;* in French: *Bob & Bobette*), *The Kiekeboes* (in Dutch: *De Kiekeboes*), *FC The Champions* (in Dutch: *FC De Kampioenen*), *The Smurfs* (in Dutch: *De Smurfen*), and the *Red Knight* (in Dutch: *De Rode Ridder*).[24]

Apart from international companies like Amazon's amazon.fr for the French-language market and bol.com for the Dutch/Flemish-language market, the Belgian sites proxis.be and azur.be (both owned by Medio since 2010) dominate the e-book market, though so far e-books have only won a 1.2% audience share in this sector.

## Magazine Publishing

In the Flemish market the three main actors are the Roularta Media Group (*Trends, Knack*), Sanoma Oyj of Finland (*Flair, Humo, Libelle,*

19. For an overview, see <http://www.boekenvak.be/voor-uitgevers/leden> and <http://adeb.be/editors>. 10 Jan. 2013.

20. The publishing sector in Flanders and the Netherlands distinguishes between A-books (fiction, non-fiction, poetry, children's books, comic books), S-books (school books), W-books (scientific books) and O-books (other); <http://www.kvb.nl/feiten-en-cijfers/definities>.

21. A. Guiette, S. Jacobs, A. Schramme and K. Vandenbempt, *Onderzoeksrapport Creatieve Industrieën in Vlaanderen: Mapping en bedrijfseconomische analyse*, Flanders DC and Antwerp Management School. April 2011. 10 Jan. 2013. <http://www.flandersdc.be/sites/default/files/FlandersDC%20-%20Creatieve_industrie_n_in_vlaanderen-_mapping_en_bedrijfseconomische_analyse.pdf>.

22. <http://www.lannoo.be/over-ons>

23. <http://buitenland.vfl.be/en/press/514/bologna-2013-confirms-the-international-reputation-of-flemish-children-s-and-youth-literature.html>

24. <http://foreignrights.wpg.be/content/uploads/STANDAARD-UITGEVERIJ-STRIPS.pdf>

*Story, TeveBlad*), and De Persgroep (*Dag Allemaal, TV Familie, Joepie* and *Blik*). Roularta is also active in Wallonia with *Tendance*, the French-language version of the business magazine *Trends* and the popular *Le Vif/L'Express*.

## AUDIOVISUAL MEDIA

### Radio

As in most of Europe, radio broadcasting was a public monopoly in Belgium until the 1980s. The first commercial FM stations began broadcasting in the early 1980s as "pirate radio," but they were soon legalized in 1982. Since then, the radio sector in the north has developed differently than in the south, because of the divergent policy approaches of the Flemish and French Communities. Whereas the Flemish frequency plan promotes a plurality of small, local FM radios (regulators introduced radio licenses that covered all of Flanders only in 2001), the French Community authorized the creation of communitywide radio networks in 1997. Both VRT, the Flemish public broadcaster, and the French-language public broadcaster RTBF are allowed to carry advertising.

The public service broadcaster VRT is the leading player in the Flemish-language market and offers five radio channels: Radio 1, focusing on the news; Radio 2, a family station; MNM, a pop music channel; Studio Brussel, which offers alternative music programming; and Klara, a classical music and jazz station. An increasing number of local radio channels have been integrated within the larger Flemish networks in the past decade (e.g. Radio Contact, Club FM, Top Radio).

The second main player is the VMMa Group, which acquired one of the two Flanders-wide market licenses for commercial FM radio in 2001 for its Q Music station and bought the second commercial FM license in 2007 for its Joe FM channel (formerly 4FM). VMMa is a joint venture between De Persgroep (50% stake) and the Roularta Media Group (50%).

Other radio stations have a limited geographic scope or only distribute their program via cable networks such as Nostalgie and Be One, which are provincial FM stations, and *Radio* Contact and Top Radio, which are community radio consortiums.

The channels of the public service broadcaster have lost audience shares over the years, especially since the launch of two commercial Flanders FM radio stations in 2002. VRT, however, remains the market leader, holding a stable 60% market share since 2006.

As there are just two main players (VRT and VMMa/Medialaan), it is also useful to calculate the C2 ratio. For the past decade, the C2 and C4 ratios have been nearly equal. The HHI, however, shows that the relation between the two major groups has changed: HHI fell from 7,356 in 2002 to 4,332 in 2010, a drop that can be explained by the growth in VMMa's market shares (Table 3.4).

In contrast to the Flemish radio market, Wallonia's has been characterized by a very strong commercial broadcasting presence, which commands a majority audience share in the French-language sector (Table 3.5). In the last ten years, the public broadcaster RTBF's audience shares stood at one-third of the market, whereas VRT still attracts more than 60% of the Flemish audience share. RTBF's stations include the generalist stations Vivacité (14.5% audience share in 2010) and La Première (7.9%), the rock station Classic 21 (7.4%), the pop and rock station Pure FM (3%), and the classical station Musiq'3 (1.8%).[25]

The RTL Group, a subsidiary of Germany's Bertelsmann AG, is the strongest group in the private radio sector, with audience shares consistently comparable to or slightly exceeding those of RTBF. It controls the two leading commercial stations, Bel RTL and Radio Contact. The RTL Group's Belgian arm, which operates as Radio-Télé Luxembourg (RTL) out of Luxembourg, is the French-language market leader in Belgian radio.

---

25. CSA. "Radio: Audience et parts de marché." CSA, 2013. 10 Jan. 2013. <http://www.csa.be/pluralisme/audience/secteur/2>.

Table 3-4. Flanders Radio Groups (Audience Shares), 2002–2010

| | 2002 | 2004 | 2006 | 2008 | 2010 |
|---|---|---|---|---|---|
| VRT (public) | 85.3 | 72.9 | 63.6 | 62.4 | 61.7 |
| *Radio1* | *11.6* | *11.4* | *7.8* | *8.2* | *8.7* |
| *Radio2* | *35.3* | *32.8* | *32.0* | *30.8* | *30.1* |
| *MNM/Radio Donna* | *29.3* | *18.5* | *15.2* | *12.3* | *9.5* |
| *Studio Brussel* | *6.6* | *7.5* | *7.1* | *9.5* | *11.5* |
| *Klara* | *2.5* | *2.7* | *1.5* | *1.6* | *1.9* |
| VMMa/Medialaan(De Persgroep, Roularta Media Group) | 4.7 | 16.7 | 23.3 | 24.1 | 21.9 |
| *Qmusic* | *3.6* | *11.0* | *16.4* | *18.2* | *14.3* |
| *JoeFM*[1] | *1.1* | *5.7* | *6.9* | *5.9* | *7.6* |
| *Nostalgie* | | | | 1.8 | 5.9 |
| *Be One/ExquiFM* | | | | 0.8 | 0.4 |
| *TopRadio* | 1.1 | 0.6 | 0.8 | 0.9 | 0.8 |
| *Radio Contact* | 1.6 | 1.6 | 0.7 | 0.8 | |
| Others | 7.3 | 8.2 | 11.6 | 9.2 | 9.3 |
| C4 | 90.0 | 89.6 | 86.9 | 89.1 | 90.3 |
| C2 | 90.0 | 89.6 | 86.9 | 86.5 | 83.6 |
| HHI | 7,356 | 5,695 | 4,771 | 4,564 | 4,322 |
| *N* (>1%) | 5 | 4 | 3 | 3 | 3 |
| Noam Index | 3,289 | 2,847 | 2,754 | 2,635 | 2,495 |

*Source:* Centrum voor Informatie over de Media (CIM), Studiedienst van de Vlaamse Regering, Vlaamse Regulator voor de Media (VRM), National Bank of Belgium

[1]Acquired by VMMa in 2007.

The NRJ Group (France) owns 50% of the music station Nostalgie Wallonie (with a 10% audience share), which plays popular music from the 1970s, 1980s, and 1990s, and NRJ (with a 4–5% audience share), which caters to a younger audience. The Belgian firm Mediabel S.A. is the other 50% shareholder in Nostalgie.

Other private stations include Fun Radio (with a 3.8% audience share), Sud Radio (with a 1.1% audience share), and Twizz (with a 0.2% audience share). Concentration indices have remained relatively stable over the past decade (Table 3.5).

## Broadcast Television

Terrestrial broadcasting (first analog, and now digital) has always been limited to the public service broadcasters. Cable is now the most common distribution platform for television programming in Belgium, with 95% of all households connected to a cable network via either Internet Protocol television (IPTV) or traditional cable TV (CATV).

The public service broadcasters have enjoyed a market share of almost 100% in their respective communities for terrestrial broadcasting since their inception, which is why we have not calculated concentration indices for broadcast television in this section.

In Flanders, VRT broadcast on its own analog terrestrial network until the analog switch-off to digital terrestrial television (DTT) was completed in November 2008. Subsequently, the operations of the new digital network were sold off to Norkring Norking (Norway),

## Table 3-5. Wallonia Radio Groups (Audience Shares), 2002–2010

| | 2002 | 2004 | 2006 | 2008 | 2010 |
|---|---|---|---|---|---|
| RTBF (public) | 32.7 | 33.2 | 26.8 | 31.4 | 34.6 |
| RTL Group (Bertelsmann AG, Germany) | 37.5 | 33.6 | 36.6 | 35.9 | 33.8 |
| *INADI (Bel RTL)* | *19.9* | *18.3* | *20.3* | *18.8* | *17.2* |
| *COBELFRA (Radio Contact)* | *14.8* | *12.6* | *15.1* | *16.1* | *16.6* |
| *Joker FM (Mint)* | *2.8* | *2.7* | *1.2* | *1.0* | |
| NRJ Group (France) | 11.1 | 14.2 | 16.1 | 15.0 | 14.5 |
| *Nostalgie* | *6.7* | *8.1* | *9.9* | *10.0* | *9.9* |
| *NRJ Belgique* | *4.4* | *6.1* | *6.2* | *5.0* | *4.6* |
| FM Développement (FUNRadio) | 3.8 | 3.2 | 3.5 | 2.6 | 3.8 |
| RMP (Sud Radio) | 1.0 | 1.2 | 1.5 | 1.1 | 1.1 |
| Groupe IPM (Twizz Radio) | | | 0.8 | 0.2 | |
| Others | 13.9 | 14.6 | 14.7 | 13.8 | 12.2 |
| Total Revenue (mil euros)[1] | | 283 | | 279 | |
| Total Revenue (mil US$) | | 351 | | 410 | |
| C4 | 85.1 | 84.2 | 83 | 84.9 | 86.7 |
| HHI | 2,614 | 2,445 | 2,332 | 2,508 | 2,566 |
| Corrected HHI[2] | | 4,519 | | 3,820 | |
| *N* (>1%) | 5 | 5 | 5 | 5 | 5 |
| Noam Index | 1,255 | 1,188 | 1,139 | 1,206 | 1,213 |

*Source:* Centrum voor Informatie over de Media (CIM), Conseil Supérieur de l'Audiovisuel (CSA); National Bank of Belgium.

[1] This revenue figure is an estimate based on data collected in this study: it is based on average per capita revenues reported for several other countries in this study with similar per capita income.

[2] Estimated on the basis that Flanders and Wallonia are two separate markets.

a subsidiary of the Norwegian state telecom provider Telenor, which acquired the license to develop a digital terrestrial network (DVB-T) in Flanders and, as "common carrier," offers transmission services to package distributors. So far, Norkring has only concluded an agreement with one distributor, Telenet, which launched its Teletenne platform in June 2012 but stopped it again in 2014 due to lack of success. VRT's programming is available to CATV/IPTV users for free via DVB-T. VRT still attracts large audience shares in the audiovisual sector (40% for TV, 70% for radio) and competes fiercely with the two largest commercial broadcasters: VMMa/Medialaan, which offers seven TV channels (VTM, 2BE, Vitaya, Vitaliteit, Jim, Anne, and VTMKzoom), and SBS Belgium (controlled by Telenet), which offers two TV channels (Vier and Vijf, formerly VT4 and VijfTV).

In Wallonia, terrestrial broadcasting has also always been limited to public broadcasting, both in analog and digital modes. Since November 2007, RTBF has been broadcasting its three main TV channels (La Une, La Deux and La Trois), its five radio stations (La Première, Vivacité, Musiqu'3, Pure FM and Classic 21), as well as *Euronews* and the two radio stations run by the German-language public broadcaster BRF, for free on DVB-T. The provisions for digital broadcasting by RTBF stem from its universal service obligation enacted in the 2007–2011 contract it has with the state. Currently, about 95% households in the French Community are covered by DTT. RTBF analog services were terminated in March 2010.[26] The other primary broadcast television channels viewed in Wallonia are offered by leading French media groups: France 2 and France 3 of the public broadcaster France Télévision have a combined 15% market share, and the TF1 Group's commercial TV channels TF1 (with a 17% audience share) and BTV, which is co-owned by the French groups TF1 and AB with a 5% audience share, are also popular.

## Multichannel TV Platforms

Satellite TV is not a significant platform in the Belgian market, so there are no data available to analyze for concentration. As noted earlier, CATV is the near-exclusive distribution channel for TV, with around 95% of Belgian households using it. Cable operators also offer voice-telephony, triple and quadruple play packages, and Internet services. Therefore, there are several market overlaps between cable and wireline telecom providers.

For Internet Protocol television (IPTV) and digital TV services, the state-owned (50% + 1 share), major Belgian telecom company Belgacom faces fierce competition from cable operators, especially from the Internet service provider (ISP) and telecom operator Telenet, which expanded its operations into digital TV in 2005 with Telenet Digital TV. Telenet is the leading CATV operator, with a 2009 market share of 54%. Competitive conditions for TV distribution still differ significantly in Flanders and Wallonia because CATV is offered by different regional cable operators: namely Telenet in Flanders, Tecteo and Brutélé in Wallonia (offering their services under a common brand 'VOO'), and Numericable (formerly Coditel) in Brussels-Capital.

In 2011, the Community media regulators in Belgium, together with the Belgian Institute for Postal Services and Telecommunications (BIPT), adopted a number of resolutions in which they set regionally delineated markets for

**Table 3-6. Multichannel Video Platforms: Cable MSOs, DBS, IPTV (Audience Shares), 2005–2009**

| | 2005 | 2009 |
|---|---|---|
| Telenet (Liberty Global, US) | | 54.6 |
| VOO & Coditel | | 29.3 |
| Belgacom (public) | | 16.2 |
| Total Revenue (mil euros) | 605 | 807 |
| Total Revenue (mil US$) | 720 | 1,122 |
| C3 | 100 | 100 |
| HHI | | 4,094 |
| *N* (>1%) | 3 | 3 |
| Noam Index | | 2,364 |

*Source:* CLEC, 2010.

TV transmission and forced the monopolistic cable providers in the communities to open their networks up to other networks. Cable operators lodged an appeal against the decision, which they lost in November 2014.[27] Mobile operator Mobistar is expected to launch alternative TV packages via the cable in 2015. Its competitor, Base, who had launched an alternative TV package via Belgacom's network under the name *Snow* in February 2013, announced in December 2014 that it will stop its TV service in 2015 and migrate its clients to Belgacom's Scarlet service. CATV/IPTV operators Telenet, Tecteo and Belgacom are also active as pay-TV operators offering premium channels (Table 3.6).

## Video Channels

As noted earlier, CATV is the most common TV distribution platform in Belgium, with over 95% of households connected to the system and subscribing to either CATV or IPTV. Channels included in the basic subscription package of CATV and IPTV providers are "free to air" as a result. The term "pay-TV channel" is reserved for channels that are offered for a premium fee in Belgium, such as sports and film channels. Pay

26. CSA. "Qu'est-ce que le DAB / DAB+?" CSA, 5 Oct. 2012. 10 Jan. 2013. <http://www.csa.be/faqs/9#question_24>.
27. T-Regs. "Belgium: Decisions to mandate resale/access on Cable-TV networks." Last update December 4, 2014. <http://www.t-regs.com/index.php/2010/12/22/belgium-proposals-to-mandate-accessresale-on-cable-tv-networks-2/>.

TV has not developed a substantial market share in Belgium because so much content is available "free to air."

In Flanders, there are three major TV groups. VRT, the public broadcaster, offers three channels: Eén, Canvas, and Ketnet (until 2012 Ketnet was part of Canvas). VMMa/Medialaan offers seven channels: VTM, 2BE, Jim TV, Anne, VTMK-zoom, Vitaya, and Vitaliteit, the two latter of which were formerly owned by Media ad Infinitum.[28] Finally, there is SBS Belgium, a subsidiary of De Vijver Media, which offers two channels, Vier and Vijf. In 2014, Telenet acquired 50% of SBS Belgium, with 25% each for Corelio and W&W.

These major groups offer both linear channels and on-demand services via third party cable and satellite platforms, as well as via the internet and their own mobile application *Stievie*. Besides them, . . . , there are a significant number of small "free to air" digital channels offered by either small independent TV companies, such as Actua TV, or companies owned by large multimedia groups, such as Belgian Business Television, which is owned by Roularta.

Before the launch of commercial television, the Flemish audience could only turn to foreign channels, namely those of the Dutch public broadcaster NPO. However, after the commercial broadcaster VTM was established in 1989, the NPO channels (Ned1, Ned2, and Ned3) started rapidly to lose market share. VTM has also successfully won significant audience shares away from VRT. When VMMa launched a second channel in 1995—in reaction to the establishment of VT4 (which broadcast from London in the United Kingdom to circumvent VTM's Flemish-Community advertising monopoly)—VRT saw its audience shares fall to 23%. Since 2000, however, VRT has gradually regained audience shares, and since 2002 it has regained its place as the largest player. In 2010, its 43% audience share exceeded the combined 38% audience shares of VMMa,/Medialaan SBS Belgium, and Media ad Infinitum.

There are ten regional TV stations in Flanders. First established in 1992, they have a semi-public status and are funded by advertising, as well as by state subsidies. They have been

Table 3-7. Flanders Combined Video Network Industries: Broadcast TV Networks, TV Syndicators, Cable Channels, Pay-TV (Market Shares by Revenue), 1992–2010[1]

| | 1992 | 1995 | 1998 | 2000 | 2002 | 2004 | 2008 | 2010 |
|---|---|---|---|---|---|---|---|---|
| VRT (public) | 31.0 | 23.0 | 30.0 | 31.0 | 36.0 | 39.0 | 40.0 | 43.0 |
| VMMa (Medialaan) | 37.0 | 41.0 | 33.0 | 35.0 | 31.0 | 28.0 | 27.0 | 25.0 |
| SBS Belgium[29] (De Vijver Media) (a) (after 2014, 50% by Telenet, 25% Corelio) | 0.0 | 6.0 | 9.0 | 8.0 | 6.0 | 8.0 | 11.0 | 10.0 |
| Media ad Infinitum | 0.0 | 0.0 | 0.0 | 0.0 | 1.0 | 2.0 | 4.0 | 3.0 |
| NPO (public) (Netherlands) | 9.0 | 6.0 | 5.0 | 5.0 | 5.0 | 4.0 | | |
| Others | 23.0 | 24.0 | 23.0 | 21.0 | 21.0 | 19.0 | 18.0 | 19.0 |
| C4 | 77 | 76 | 77 | 79 | 78 | 79 | 82 | 81 |
| HHI | 2,411 | 2,282 | 2,095 | 2,275 | 2,319 | 2,389 | 2,460 | 2,583 |
| *N* (>1%) | 3 | 4 | 4 | 4 | 5 | 5 | 4 | 4 |
| Noam Index | 2,078 | 1,650 | 1,514 | 1,568 | 1,380 | 1,375 | 1,395 | 1,472 |

*Source:* Centrum voor Informatie over de Media (CIM), Studiedienst van dse Vlaamse Regering, Vlaamse Regulator voor de Media (VRM), National Bank of Belgium.

[1] C4 and HHI are based on the audience shares of the TV companies, not individual channels.

28. Since 2009, VMMa has offered its subscribers access to a digital music channel (Anne) through Belgacom.

29. SBS Belgium after 2014, 50% by Telenet, 25% by Corelio, and 25% by W&W.

Table 3-8. Flanders Combined Video Network Industries: Broadcast TV Networks, TV Syndicators, Cable Channels, Pay-TV (Audience Shares), (Market Shares by Revenue), 2005–2010

| | 2005 | 2006 | 2007 | 2008 | 2009 | 2010 |
|---|---|---|---|---|---|---|
| VRT (public) (including Een/TV1, Canvas, Ketnet and TV2)[1] | 52.2 | 51.4 | 49.5 | 49.4 | 49.7 | 47.9 |
| VMMa (Medialaan) (including *VTM, 2BE* and *KA2*)[2] | 30.0 | 26.4 | 27.0 | 27.3 | 26.7 | 29.2 |
| SBS Belgium (De Vijver Media)[30] (b) (After 2014, 50% Telenet, 25% Corelio) | 9.3 | 10.1 | 10.8 | 10.2 | 9.0 | 10.1 |
| Belgacom Skynet (public) | 3.3 | 6.7 | 6.8 | 7.6 | 8.6 | 8.8 |
| *MTV* (US) | 2.4 | 2.5 | 3.1 | 2.8 | 3.1 | 1.9 |
| Media ad Infinitum | 1.1 | 1.3 | 1.3 | 1.3 | 1.3 | 1.4 |
| Bites Europe | 0.0 | 0.0 | 0.0 | 0.0 | 0.2 | 0.3 |
| Belgian Business Televisie | 0.8 | 0.7 | 0.6 | 0.4 | 0.3 | 0.2 |
| T.Vgas | 0.0 | 0.0 | 0.0 | 0.0 | 0.1 | 0.2 |
| Euro1080 | 0.8 | 0.8 | 0.9 | 1.0 | 1.0 | 0.0 |
| Event TV Vlaanderen | 0.1 | 0.0 | 0.0 | 0.0 | 0.0 | 0.0 |
| Life!TV | 0.0 | 0.1 | 0.1 | 0.0 | 0.0 | 0.0 |
| Total Revenue (mil euros) | 726 | 846 | 868 | 893 | 879 | 893 |
| Total Revenue (mil US$) | 908 | 1,066 | 1,189 | 1,313 | 1,222 | 1,188 |
| C4 | 94.8 | 94.6 | 94.1 | 94.4 | 94.0 | 96.0 |
| HHI | 3,730 | 3,496 | 3,352 | 3,355 | 3,348 | 3,330 |
| *N* (>1%) | 6 | 6 | 6 | 6 | 6 | 6 |
| Noam Index | 1,521 | 1,427 | 1,368 | 1,370 | 1,367 | 1,359 |

*Source*: Centrum voor Informatie over de Media (CIM), Studiedienst van de Vlaamse Regering, Vlaamse Regulator voor de Media (VRM), National Bank of Belgium

[1] Both state subsidies and revenues from advertising and merchandising activities are taken into account for VRT's revenue share.

[2] For VMMa, revenue includes both TV and radio.

successful in attracting viewers, but advertisers and public authorities have been less eager to finance them and, as a result, several of them now face severe financial difficulties.

Filmnet (Richmond-Rupert) was the first pay-TV operator in Flanders, but it never managed to break even in the market. The Canal + Group acquired it in 1996, and it ultimately became part of the CATV operator Telenet's operations in 2003 (Tables 3.7 and 3.8).

Wallonia's public broadcaster RTBF has always had to cope with fierce competition from French channels, both public (France Télévisions) and private (TF1). The largest operator in the French-language market is RTL Belgium, which was the first commercial television broadcaster to obtain a license in Wallonia (in 1987), and it then launched its first channel RTL/TVi two years later. Its owners are the RTL Group's subsidiary CLT-UFA, with a 66% stake, and Audiopresse, a Wallonian print media cooperative, with a 34% stake. In 1995, RTL launched Club RTL, targeting younger demographics, and in 2003, RTL started broadcasting PlugTV, which also focuses on music and youth programming.

30. SBS Belgium after 2014, 50% by Telenet, 25% by Corelio, and 25% by W&W.

Table 3-9. Wallonia Combined Video Network Industries: Broadcast TV Networks, TV Syndicators, Cable Channels, Pay-TV (Audience Shares by %), 2000–2010

| | 2000 | 2002 | 2004 | 2008 | 2010 |
|---|---|---|---|---|---|
| RTBF (public) | 19.7 | 18.9 | 19.6 | 20.1 | 20.4 |
| RLT Group (Bertelsmann AG, Germany) | 23.3 | 24.7 | 25.6 | 26.0 | 27.6 |
| TF1 Group (France) | 16.6 | 15.7 | 17.5 | 17.1 | 16.6 |
| France Télévisions (France) | 17.1 | 14.3 | 14.9 | 15.2 | 15.4 |
| AB Group (France) | 3.7 | 5.9 | 5.7 | 5.3 | 5.6 |
| Tecteo (VOO) | 0.8 | 0.8 | 0.4 | 0.4 | 0.3 |
| Others | 18.8 | 19.7 | 16.3 | 15.9 | 14.1 |
| Total Revenue (mil euros)[1] | | | 1,020 | 1,018 | |
| Total Revenue (mil US$) | | | 1,264 | 1,496 | |
| C4 | 80.4 | 79.5 | 83.3 | 83.7 | 85.6 |
| HHI | 1,866 | 1,841 | 1,866 | 1,884 | 1,920 |
| Corrected HHI[2] | | | 2,190 | 2,255 | |
| *N* (>1%) | 5 | 5 | 5 | 5 | 5 |
| Noam Index | 834 | 823 | 834 | 842 | 859 |

*Source*: Centrum voor Informatie over de Media (CIM; www.cim.be); Conseil Supérieur de l'Audiovisuel (http://www.csa.be/pluralisme/audience/secteur/3); National Bank of Belgium.

[1] This revenue figure is an estimate based on data collected in this study: it is based on average per capita revenues reported for several other countries in this study with similar per capita income.

[2] Estimated on the basis that Flanders and Wallonia are two separate markets.

Until recently, RTL provided its services under a double license: one from the authorities in Wallonia, and another from Luxembourg, where content and investment requirements are less stringent than they are in Belgium. In 2005, RLT decided not to ask for a renewal of its Belgian license and instead opted to continue its operations exclusively under its Luxembourg license taking advantage of the country-of-origin principle outlined in the Audiovisual Media Services Directive.[31]

Wallonia has 11 regional TV stations that are allowed to broadcast advertising. Part of their financing comes from state subsidies, however.

The Canal + Group started a pay-TV channel in 1989, Canal + Belgique. RTBF purchased a 26% stake in the venture, which it sold back to Canal + in 2000 for US$19 million (20.6 million euros). Later, Canal + Belgique became independent and changed its name to BeTV, and the CATV operator Tecteo (VOO) acquired it in October 2008 (Table 3.9). Since September 2014, the video streaming service Netflix is also available in Belgium. Belgacom struck an agreement to offer the Netflix application also via its digital TV decoders.

## Film

The Belgian film industry is dominated by foreign entities, predominantly US-based Hollywood majors like Universal and Warner Brothers. French productions from such studios as Pathé are extremely popular in Wallonia, and, to a lesser extent, Dutch productions are present in the Flemish market. The German Community in the Eastern Cantons gravitates to the German market, though the market is quite small.

31. "Directive 2010/13/Eu of the European Parliament and of the Council of 10 March 2010 on the coordination of certain provisions laid down by law, regulation or administrative action in Member States concerning the provision of audiovisual media services (Audiovisual Media Services Directive)." *Official Journal of the European Union*, 15 Apr. 2012. 10 Jan. 2013. <http://eur-lex.europa.eu/LexUriServ/LexUriServ.do?uri=OJ:L:2010:095:0001:0024:EN:PDF>.

Distribution is also led by international enterprises. Therefore, public subsidies are available to support Flemish and French-language productions in country. Due to the small size of the market, statistically relevant market shares for domestic movie production companies are not kept by public sources.

Belgium was one of the first European countries to adopt the US model of multiplexes, and the country has a total of around 120 cinemas. The Kinepolis Group, with a 50% audience share of 2.2 million theatergoers, dominates the exhibition market by a wide margin. The group is also active in other countries, such as France and Spain.

## TELECOMMUNICATIONS MEDIA

### Wireline Telecom

Unlike the print and audiovisual sectors, the Belgian telecommunications market is nationally operated and regulated. The Belgian Institute for Posts and Telecommunications (BIPT) is the federal telecom regulator and is responsible for regulating bilingual broadcasting activities in Brussels-Capital.

The telecommunications market is structurally similar to most other European countries. Former state monopolists have been privatized but still have leading, if not dominating, market shares. The former public monopolist Belgacom /Proximus still dominates the wireline market, with a market share 71.4% in 2009.[32] The fixed-line telephony market is smaller than the mobile market, although subscriptions for wireline services are still valued at US$5.6 billion (4 billion euros) in 2009, largely because these services are used by ISPs (Table 3.10).

### Wireless Telecom

A majority of Belgian households do not own fixed telephone lines anymore, and the wireless market is characterized by an average

Table 3-10. Wireline Telecom (Market Shares by Revenue), 2001–2009

| | 2001 | 2009 | 2012[1] |
|---|---|---|---|
| Belgacom (public)[33] | 77.0 | 71.4 | 63.1 |
| Telenet (Liberty Global, US) | 23.0 | 16.5 | 29.3 |
| Mobistar (France Telecom) | | 5.5 | 4.3 |
| VOO & Coditel | | 5.5 | |
| KPN Belgium | | 1.1 | |
| BT Belgium//Versatel | | | |
| Others | | | 3.3 |
| Total Revenue (mil euros) | | 4,032 | ~ 4,024[2] |
| Total Revenue (mil US$) | | 5,605 | ~ 5,173 |
| C4 | | 98.9 | 96.7 |
| HHI | | 5,435 | 4,859 |
| *N* (>1%) | | 5 | 3 |
| Noam Index | | 2,431 | 2,805 |

[1] 2012 data added by editors. *Belgium Telecommunications Report - Q3 2014.* London: Business Monitor International, 2014.*ProQuest.*
[2] Industry revenue approximated from Belgacom Financial Reports for FY2012.

32. European Commission, Information Society and Media Directorate-General. "Management Plan 2012." DG INFSO, 13 Dec. 2011. 9 Jan. 2013. <http://ec.europa.eu/atwork/synthesis/amp/doc/infso_mp_en.pdf>.

33. The Belgian state held, in 2014, 56.32% of Belgacom's voting rights.

Table 3-11. Wireless Telecom (Market Shares by Revenue)

| | 2005 | 2006 | 2007 | 2008 | 2009 | 2011[1] |
|---|---|---|---|---|---|---|
| Proximus (public) (Belgacom Mobile) | 51.7 | 49.3 | 48.7 | 47.5 | 46.1 | 41.7 |
| Mobistar (France Telecom, France) | 35.3 | 36.3 | 36.7 | 37.1 | 37.7 | 32.0 |
| Base (KPN, Netherlands) | 13.0 | 14.4 | 14.6 | 15.4 | 16.3 | 19.0 |
| Others | | | | | | 7.3 |
| Total Revenue (mil euros) | 4,178 | 4,289 | 4,163 | 4,112 | 4,162 | 4,800 |
| Total Revenue (mil US$) | 5,223 | 5,404 | 5,703 | 6,045 | 5,785 | 6,700 |
| C3 | 100 | 100 | 100 | 100 | 100 | 92.7 |
| HHI | 4,085 | 3,953 | 3,933 | 3,869 | 3,806 | 3,124 |
| *N* (>1%) | 3 | 3 | 3 | 3 | 3 | 3 |
| Noam Index | 2,358 | 2,282 | 2,270 | 2,234 | 2,197 | 1,804 |

*Source:* National Bank of Belgium.

[1] Data added by editors. Data from Marketline. "Wireless Telecommunication Services Industry Profile: Belgium." (2012). *Wireless Telecommunication Services Industry Profile: Belgium,* 1–30.

of slightly more than one mobile device per capita. With its mobile brand Proximus, Belgacom leads the wireless market with close to 50% of the market share. Its two main competitors are Base, owned by the Dutch operator KPN, and France Telecom's Mobistar. KPN has tried, so far unsuccessfully, to restructure its Belgian properties, having been unable to sell off Base or merge its German mobile brand e-plus with the Spanish telecom major Telefónica. Both Base and Mobistar have recorded slight gains in recent years at the expense of Proximus (Table 3.11).

## INTERNET MEDIA

### Internet Service Providers (ISP)

Wireline telecom networks are now dominated by data and Internet connections. Cable operators, offering triple and quadruple play services, also have significant market shares. Belgacom is the country's largest ISP, with a market share of 47% in 2010, while Telenet holds a growing 36% market share. Belgacom's leading position led the European Commission to express concern over the existence of a *de facto* monopoly. Table 3.12 defines the ISP market without taking into account voice services, which when taken into account would leave Belgacom with a 61% market share, followed distantly by Telenet with just a 14% market share, while smaller players like France Telecom's Mobistar would each hold less than 5%.

In terms of subscriptions, the gap between the DSL-based market leader Belgacom and the cable operator Telenet is closing (Table 3.13).

In 2011, Belgacom had a fixed broadband market share of 46% and a 52% market share for its DSL wireline services. This is well below the European average of 76% in this sector, in part because cable broadband distribution is much higher in Belgium than it is in the rest of the European Union, a situation that is closely related to the near absence of satellite TV distribution that frequently goes along with separate subscriptions for fixed lines. Cable broadband service is not identical with ISP service, although this cross-network situation clearly exists and is cause for concern among Belgian regulators.

### Search Engines

As in most other European countries, Google (US) has a near-monopoly in the search engine market, with a 90% market share. Statistically significant data do not exist for other providers, either foreign or Belgian.

Table 3-12. Internet Service Providers (Market Shares by Revenue), 2005–2010

| | ACCESS | 2005 | 2006 | 2007 | 2008 | 2009 | 2010 |
|---|---|---|---|---|---|---|---|
| Belgacom DSL (public) | DSL | 48.5 | 47.6 | 46.6 | 46.1 | 48.3 | 47.1 |
| Tiscali Belgium (Italy) | DSL | 0.0 | 0.0 | 0.0 | 0.0 | 0.0 | 0.0 |
| Scarlet | DSL | 0.4 | 1.3 | 1.1 | 0.3 | 0.0 | 0.0 |
| Tele2 (Sweden) (including Versatel) | DSL | 2.7 | 4.3 | 0.0 | 0.0 | 0.0 | 0.0 |
| Mobistar (France Telecom, France) | DSL | 0.0 | 0.5 | 0.9 | 0.8 | 1.2 | 1.7 |
| Tele2 (Sweden); Base (KPN, Netherlands) | DSL | 0.0 | 0.0 | 4.7 | 4.9 | 4.8 | 3.2 |
| Telenet (Liberty Global, US) | Cable | 31.0 | 30.8 | 33.1 | 33.8 | 34.4 | 36.0 |
| Chello | Cable | 1.8 | 1.7 | 0.0 | 0.0 | 0.0 | 0.0 |
| Brutele@Home (VOO) | Cable | 2.1 | 2.6 | 2.5 | 2.5 | 2.4 | 2.8 |
| Teledisnet | Cable | 1.5 | 1.5 | 1.5 | 3.8 | 3.6 | 4.7 |
| Others | Cable, DSL | 12.0 | 9.6 | 9.4 | 7.8 | 5.3 | 4.6 |
| Total Revenue (mil euros)[1] | | 551 | | | | 648 | |
| Total Revenue (mil US$) | | 689 | | | | 900 | |
| C4 | | 84.3 | 85.3 | 86.9 | 88.6 | 91.1 | 91.0 |
| HHI | | 3,331 | 3,247 | 3,300 | 3,313 | 3,559 | 3,557 |
| *N* (>1%) | | 6 | 7 | 6 | 5 | 6 | 6 |
| Noam Index | | 1,360 | 1,227 | 1,347 | 1,482 | 1,453 | 1,452 |

[1] This revenue figure is an estimate based on data collected in this study: it is based on average per capita revenues reported for several other countries in this study with similar per capita income.

Table 3-13. Number of Belgian Broadband Subscribers (1,000), 2010–2011

| | 2010 | 2011 |
|---|---|---|
| Belgacom (public) | 1,536 | 1,567 |
| Telenet (Liberty Global, US) | 1,115 | 1,209 |
| Teledisnet | 143 | 165 |
| Tele2 (Sweden) | 154 | 100 |
| Brutele@Home | 79 | 100 |
| Mobistar (France Telecom, France) | 42 | 66 |
| Total Subscribers (1,000) | 3,069 | 3,207 |

## Online News

Like the newspapers market, the online news market is also divided into Walloon and Flemish sectors. The leaders in the Dutch online media are hln.be (*Het Laatste Nieuws*), standaard.be (*De Standaard*), and nieuwsblad.be (*Het Nieuwsblad*). HLN is part of De Persgroep, and the latter two are part of the Corelio Group's *VUM*.

The most successful news site in the French-language market is lesoir.be of Rossel. The website of the public broadcaster RTBF, rtbf.be, is the second most visited news site in the market. The third leading portal is dhnet.be, of IPM's Francophone *La Derniere Heure*.

Table 3-14. Flemish Media Cross-ownership

| | RADIO BROADCASTING | RADIO DISTRIBUTION | TELEVISION CONTENT | TELEVISION BROADCASTING | TELEVISION DISTRIBUTION | NEWSPAPERS | MAGAZINES | ONLINE NEWS | ISPs |
|---|---|---|---|---|---|---|---|---|---|
| Alfacam | X | X | X | | | | | X | |
| Belgacom (public) | | | X | X | X | | | X | X |
| Concentra | X | | X | | | X | X | X | |
| Coreilo | X | | X | | | X | X | X | |
| De Persgroep | X | | X | X | | X | X | X | |
| Roularta | X | | X | X | | | X | X | |
| Sanoma Oyj (Erkko Family, Finland) | | | X | | | | X | X | |
| SBS Belgium (De Vijver Media) | | | X | X | | | | X | |
| Telenet (Liberty Global, US) | | X | X | X | X | | X | X | X |
| VRT (public) | X | | X | X | | | | X | |

## CONCLUSION

The limited number of media groups and high levels of cross-ownership are typical for Belgian media markets. Most media groups have developed a varied portfolio of media products in different markets or segments, but none of them are capable of simultaneously dominating multiple market sectors. The figures in Table 3.14 illustrate this clearly for Flemish media groups.

Although international investments in/by Belgian media groups are a relatively new phenomenon, a growing number of Belgian media groups have developed international ambitions, in search of economies of scale. Examples include the Roularta Group which has expanded its interests in the magazine market toward the French market. De Persgroep has invested in the Dutch newspaper market: after buying the title *Het Parool* from the large Dutch media group PCM in 2003, it acquired a majority share of PCM in 2009 (and controls since then also popular periodicals such as *de Volkskrant, NRC Handelsblad, Trouw,* and *Algemeen Dagblad*). The Dutch media regulator permitted this acquisition on the condition that De Persgroep seek a buyer for NRC Media, publisher of *NRC Handelsblad* and *NRC.next*.[34]

The Belgian situation makes clear that the definition of concentration depends strongly on how the focus is set. If one describes wireline and cable distribution as separate entities, for example, one finds monopolistic trends for Belgacom. If service is generally defined across different technical distribution means, then there is a near duopoly. However, with different services offered by several companies plus the existence of Belgium's three distinct communities, any representation of a specific market will neglect some dimensions of the media landscape.

Belgium, with its strong market divisions along linguistic lines, faces the challenge of small markets: either there is pressure from internationalisation, with foreign groups dominating some sectors, or domestic near-monopolies form because necessary investments in expensive infrastructure for such small audience shares serve as a high entry barrier. However, a number of domestic media players have managed to secure investments in local production by either expanding themselves to neighbouring territories or by striking alliances with international groups. In light of growing European integration, this "heartland of Europe" and its political and administrative center of Brussels-Capital serve as an example of how media pluralism can be achieved in taking on a cross-border perspective.

## Acknowledgement

The authors would like to thank Ingrid Kools and Marthe Van Gorp of the Vlaamse Media Regulator, and Julien Jost of the Conseil Supérieur de l'Audiovisuel, for their help in collecting the data and calculating the concentration indices for the print, radio and television sectors. They also wish to express their gratitude to Evi Werkers (boek.be) for her critical comments on an earlier version of the text. Any errors remaining the responsibilities of the authors.

## Bibliography

### 1. Books and Articles

Flemish Parliament. "Decreet betreffende radio-omroep en televisie van 27 maart 2009." Department of Culture, Youth, Sports and Media, 17 Aug. 2012. 10 Jan. 2013. <http://www.cjsm.vlaanderen.be/media/downloads/mediadecreet_20090327.pdf >.

Briel, Robert. "Sanoma pays €1.225 billion for SBS channels." *Broadband TV News*, 20 Apr. 2011. 9 Jan. 2013. <http://www.broadbandtvnews.com/2011/04/20/sanoma-pays-e1-225-billion-for-sbs-channels/>.

Briel, Robert. "Open Cable for VOO and Numéricable." *Broadband TV News*, 29 Nov. 2012. 10 Jan. 2013. <http://www.broadbandtvnews.com/2012/11/29/open-cable-for-voo-and-numericable/>.

Franck, Aline. "Belgique: regards croisés sur deux paysages audiovisuels." *Régulation* 41, 2009.

34. Raeymaeckers, Karen, et al. *Media landscape: Belgium*. European Journalism Centre, 8 Nov. 2010. 10 Jan. 2013. <http://ejc.net/media_landscapes/belgium>.

Guiette, Alain, Jacobs, Sofie, Schramme, Annick, and Vandenbempt, Koen. *Onderzoeksrapport Creatieve Industrieën in Vlaanderen: Mapping en bedrijfseconomische analyse*, Flanders DC and Antwerp Management School. April 2011. 10 Jan. 2013. <http://www.flandersdc.be/sites/default/files/FlandersDC%20-%20Creatieve_industrie_n_in_vlaanderen-_mapping_en_bedrijfseconomische_analyse.pdf>

Lund, Anker Brink, Raeymaeckers, Karen, and Trappel, Josef, "Newspapers: Adapting and Experimenting." *Media in Europe Today*, Trappel, Josef, et al., eds. Bristol/Chicago: Intellect, 2011.

Raeymaeckers, Karen, De Bens, Els, Paulussen, Steve, Deprez, Annelore, and Tenret, Yannis, *Media landscape: Belgium*. European Journalism Centre, 8 Nov. 2010. 10 Jan. 2013; <http://ejc.net/media_landscapes/belgium>.

Sanoma Media Belgium. "Competition Authorities approve the partnership structure of the SBS TV operations in Belgium." sanoma.com, 9 Sep. 2011. 9 Jan. 2013. <http://www.sanoma.com/about-us/sanoma-media-belgium2/news/competition-authorities-approve-the-partnership-structure-of-the-sbs-tv-operations-in-belgium>.

Valcke, Peggy, and Lievens, Eva. *Media Law in Belgium*. Alphen aan den Rijn: Kluwer Law International BV, 2011.

## 2. Reports

Competition Committee, Directorate for Financial and Enterprise Affairs. "Annual Report on Competition Policy Developments in Belgium: 2004." OECD, 29 June 2005. 10 Jan. 2013. <http://www.oecd.org/belgium/35110865.pdf>.

CSA. "Qu'est-ce que le DAB / DAB+?" CSA, 5 Oct. 2012. 10 Jan. 2013. <http://www.csa.be/faqs/9#question_24>.

CSA. "Corelio Group." CSA, 2013. 10 Jan. 2013. <www.csa.be/pluralisme/groupes/show/7_groupe_corelio>.

CSA. "IMP Group." CSA, 2013. 10 Jan. 2013. <www.csa.be/pluralisme/groupes/show/3_groupe_ipm>.

CSA. "La Capitale." CSA, 2013. 10 Jan. 2013. <http://www.csa.be/pluralisme/offre/service/6>.

CSA. "La Meuse." CSA, 2013. 10 Jan. 2013. <http://www.csa.be/pluralisme/offre/service/9>.

CSA. "La Nouvelle Gazette." CSA, 2013. 10 Jan. 2013. <http://www.csa.be/pluralisme/offre/service/10>.

CSA. "L'offre de médias et le pluralisme en Fédération Wallonie-Bruxelles." CSA, 2013. 10 Jan. 2013. <http://www.csa.be/pluralisme>.

CSA. "La Province." CSA, 2013. 10 Jan. 2013. <http://www.csa.be/pluralisme/offre/service/11>.

CSA. "Nord Eclair." CSA, 2013. 10 Jan. 2013. <http://www.csa.be/pluralisme/offre/service/22>.

CSA. "Presse écrite d'information: Audience et parts de marché." CSA, 2013. 10 Jan. 2013. <http://www.csa.be/pluralisme/audience/secteur/3>.

CSA. "Radio: Audience et parts de marché." CSA, 2013. 10 Jan. 2013. <http://www.csa.be/pluralisme/audience/secteur/2>.

"Directive 2010/13/EU of the European Parliament and of the Council of 10 March 2010 on the coordination of certain provisions laid down by law, regulation or administrative action in Member States concerning the provision of audiovisual media services (Audiovisual Media Services Directive)." *Official Journal of the European Union*, 15 Apr. 2012. 10 Jan. 2013. <http://eur-lex.europa.eu/LexUriServ/LexUriServ.do?uri=OJ:L:2010:095:0001:0024:EN:PDF>.

European Commission, Information Society and Media Directorate-General. "Management Plan 2012." DG INFSO, 13 Dec. 2011. 9 Jan. 2013. <http://ec.europa.eu/atwork/synthesis/amp/doc/infso_mp_en.pdf>.

Ministre pour l'Économie à la Simplification administrative. "LA BELGIQUE Cœur de l'Europe numérique 2010-2015 30 points d'action." *vincentvanquickenborne.be*, 5 Oct. 2009. 10 Jan. 2013. <http://www.vincentvanquickenborne.be/DigitalBelgiumFR.pdf>.

"Wet 23 Juni 1961 betreffende het recht op antwoord (B.S., 8 juli 1961)." Fonds Pascal Decroos voor Bijzondere Journalistiek, n.d. 10 Jan. 2013. <http://www.fondspascaldecroos.org/inhoud/organisatie/wet-23-juni-1961-betreffende-het-recht-op-antwoord-bs-8juli-1961>.

## 3. Websites

*Centrum voor Informatie over de Media/Centre d'Information sur les Médias (CIM)*. cim.be, 2012. n.d. <http://www.cim.be/>.

"Google. be." Alexa, The Web Information Company, 2012. 10 Jan. 2013. <http://www.alexa.com/siteinfo/google.be>.

*National Bank of Belgium*. bbe.be, 2012. n.d. <http://www.nbb.be/>.

*Studiedienst van de Vlaamse Regering*. vlaanderen.be, 2012. n.d. <http://www4.vlaanderen.be/sites/svr/Cijfers/Pages/Excel.aspx>.

*Vlaamse Regulator voor de Media*. vlaamseregulatormedia.be, 2012. n.d. <http://www.vlaamseregulatormedia.be/nl/home.aspx>.

# Belgium—Data Summaries

ELI NOAM AND PAUL MUTTER

BELGIUM IS A MULTILINGUAL country, but so are several other countries in our 30-nation world—Switzerland, Canada, India, China, Spain, Israel, South Africa, Russia, and increasingly the United States (Table 3.15). What is unusual in Belgium is that content media and their regulations are separated along linguistic lines. Of the top five companies, only Belgacom and Liberty Global operate in both the French- and Dutch-speaking regions of Belgium. Belgian media industries have high concentration indices that are caused by the limited number of media groups within each region. On the other hand, the Dutch, German, and French-language markets are relatively small and open to investment and media flows from neighboring countries. The percentage of foreign ownership of Belgian media is 33.0%—a number that would be higher if not for the public company Belgacom's telecom market share, which gives it control of 48.8% of the platform media market (Table 3.17). Belgacom, far and away the largest media company in Belgium by power index and market share, controls nearly half of Belgium's wireless market and almost three-fourths of its wireline, in addition to smaller ISP and multichannel operations. It is also active in the online-based Belgacom-TV. Belgacom's national market share has decreased as Liberty Global's Telenet cable service and Base (owned by the Dutch company KPN) have gradually increased their wireless operations. Mobistar (Owned by Orange) showed no change in percentage of shares of the overall National Media Market.

Belgium has one of the highest rates of multichannel penetration in the European Union. That market is relatively unregulated due to the lack of a national authority. In the multichannel platforms market, Liberty Global dominates with 54.6% market share, after having bought out Telenet in 2012. Liberty Global also operates an ISP service in Belgium, and it has recently moved into Belgian wireline telecom as well, increasing its platform market share by 4.9% from 11.7% in 2004 to 16.6% in 2008 (Tables 3.16 and 3.17).

Table 3-15. National Media Industries Concentration in Belgium

| | 2004/5 | | 2011 OR MOST RECENT | | % CHANGE ANNUAL AVERAGE | |
|---|---|---|---|---|---|---|
| | COMPANY POWER INDEX IN COUNTRY | COMPANY SHARE OF THE OVERALL NATIONAL MEDIA MARKET (%) | COMPANY POWER INDEX IN COUNTRY | COMPANY SHARE OF THE OVERALL NATIONAL MEDIA MARKET (%) | COMPANY POWER INDEX IN COUNTRY | COMPANY SHARE OF THE OVERALL NATIONAL MEDIA MARKET (%) |
| Belgacom—(public) | 2,796 | 45.1 | 1,859 | 36.4 | –6.7 | –1.8 |
| Mobistar (Orange, France) | 421 | 13.6 | 380 | 13.0 | –2.0 | –0.1 |
| Telenet (Liberty Global, US) | 231 | 9.7 | 473 | 13.3 | 20.9 | 0.7 |
| VRT (public)—Flanders | 95 | 3.8 | 87 | 3.8 | –1.6 | 0 |
| KPN (Netherlands) | 56.5 | 4.7 | 132.4 | 7.1 | 26.9 | 0.5 |
| VOO & Coditel | 49.7 | 3.4 | 52.8 | 1.9 | 1.3 | –0.3 |
| Corelio/VUM (with Concerta, now Mediahuis)—Flanders | 46.7 | 1.7 | 44.2 | 1.7 | –1.1 | –0.01 |
| Persgroep (incl. Mediafin)— | 34.1 | 1.4 | 31.9 | 1.4 | –1.3 | –0.01 |
| VMMA/Medialaan—Flanders | 23 | 1.9 | 27.1 | 2.1 | 3.6 | 0.05 |
| Rossel—Wallonia | 28.1 | 1.3 | 33.3 | 1.4 | 3.8 | 0.03 |
| Concentra (now Mediahuis) | 21.2 | 1.1 | 20.9 | 1.1 | –0.3 | 0 |
| Bertelsmann, (Germany)—Wallonia | 36.7 | 2.4 | 41.1 | 2.6 | 2.4 | 0.04 |
| RTBF (public)—Wallonia | 23.9 | 1.9 | 25.7 | 2.1 | 1.5 | 0.02 |
| IPM—Wallonia | 4.7 | 0.5 | 4.9 | 0.6 | 0.8 | 0.003 |
| Bouygues Group (France)—Wallonia | 14.3 | 1.4 | 12.7 | 1.4 | –2.2 | 0.01 |

| | | | | | | |
|---|---|---|---|---|---|---|
| France Télévisions (France) Wallonia | 10.1 | 1.2 | 11.1 | 1.3 | 1.9 | 0.02 |
| Murdoch Group (US) | 1.4 | 0.1 | 0.4 | 0.04 | −15.0 | −0.01 |

| Media Concentration Index | 2004/5 | 2011 or Most Recent | % Change Annual Average |
|---|---|---|---|
| Total Revenue: Nat'l Media Industry (mil US$) | 15,853 | 18,325 | 3.1 |
| Total Voices (*n*) | 36 | 45 | 5.0 |
| Net Voices (*n*) | 32 | 41 | 5.6 |
| Public Ownership (%) | 1.9 | 2.1 | 0.03 |
| Foreign Ownership (%) | 4.9 | 33.0 | 2.0 |
| C4 Average—Weighted | 92 | 89 | −1.0 |
| C1 Average—Weighted | 53 | 45 | −1.0 |
| HHI Average—Weighted | 3,969 | 3,389 | −2.9 |
| Noam Index Average—Weighted | 807 | 1,013 | 5.1 |
| Pooled Overall Sector C4 | 73.0 | 69.8 | −0.6 |
| Pooled Overall Sector HHI | 2,385 | 1,769 | −5.2 |
| Pooled Overall Sector Noam Index | 120 | 91 | −4.8 |
| Market Share of Top Ten Companies: Nat'l Media Industry (%) (Pooled C10) | 88.1 | 84.1 | −0.8 |
| National Media Power Index | 3,929 | 3,393 | −2.7 |

Table 3-16. Top Content Media Companies in Belgium

| | 2004/5 | | 2011 OR MOST RECENT | | % CHANGE ANNUAL AVERAGE | |
|---|---|---|---|---|---|---|
| | COMPANY POWER INDEX IN COUNTRY | COMPANY SHARE OF THE NATIONAL CONTENT MEDIA MARKET (%) | COMPANY POWER INDEX IN COUNTRY | COMPANY SHARE OF THE NATIONAL CONTENT MEDIA MARKET (%) | COMPANY POWER INDEX IN COUNTRY | COMPANY SHARE OF THE NATIONAL CONTENT MEDIA MARKET (%) |
| VRT—Flanders (public) | 389.2 | 15.6 | 318.8 | 13.9 | –3.6 | –0.3 |
| Telenet | 183.5 | 3.4 | 219.9 | 4.0 | 4.0 | 0.1 |
| Corelio—Flanders (now Mediahuis, with Concentra) & Wallonia | 192.1 | 7.0 | 161.5 | 6.1 | –3.2 | –0.2 |
| Persgroup—Flanders | 140.5 | 6.0 | 116.6 | 5.2 | –3.4 | –0.2 |
| VMMA/Medialaan—Flanders | 94.5 | 7.8 | 99.0 | 7.8 | 1.0 | 0.0 |
| Rossel—Wallonia | 115.5 | 5.4 | 121.8 | 5.3 | 1.1 | –0.03 |
| Concentra (now Mediahuis)—Flanders | 87.3 | 4.7 | 76.5 | 4.2 | –2.5 | –0.1 |
| VOO & Coditel | 52.9 | 1.8 | 63.4 | 2.2 | 4.0 | 0.1 |
| Group RTL—Wallonia (Germany) | 150.8 | 9.9 | 150.4 | 9.6 | –0.1 | –0.06 |
| RTBF-Wallonia (public) | 98.3 | 7.9 | 94.1 | 7.5 | –0.9 | –0.1 |

| | | | | | | |
|---|---|---|---|---|---|---|
| Belgacom | 16.1 | 1.0 | 19.3 | 1.2 | 4.0 | 0.04 |
| Murdoch Group (US) | 5.9 | 0.3 | 1.3 | 0.1 | –15.5 | –0.04 |
| **Media Concentration Index** | **2004/5** | | **2011 or Most Recent** | | **% Change Annual Average** | |
| Public Ownership (%) | 7.9 | | 7.5 | | –0.1% | |
| Foreign Ownership (%) | 25.3 | | 23.9 | | –0.3% | |
| C4 Average—Weighted | 73 | | 74 | | 0.3% | |
| HHI Average—Weighted | 1,777 | | 2,098 | | 3.6% | |
| C1 Average—Weighted | 27 | | 30 | | 0.01% | |
| National Power Index | 1,790 | | 2,113 | | 3.6% | |

Table 3-17. Top Platform Media Companies in Belgium

| | 2004/5 | | 2011 OR MOST RECENT | | % CHANGE ANNUAL AVERAGE | |
|---|---|---|---|---|---|---|
| | COMPANY POWER INDEX IN COUNTRY | COMPANY SHARE OF THE NATIONAL PLATFORM MEDIA MARKET (%) | COMPANY POWER INDEX IN COUNTRY | COMPANY SHARE OF THE NATIONAL PLATFORM MEDIA MARKET (%) | COMPANY POWER INDEX IN COUNTRY | COMPANY SHARE OF THE NATIONAL PLATFORM MEDIA MARKET (%) |
| Belgacom | 3,689.1 | 59.2 | 2,511.8 | 48.8 | –6.4% | –2.1 |
| Mobistar (Orange, France) | 542.4 | 15.4 | 704.1 | 18.7 | 6.0 | 0.7 |
| Telenet (Liberty Global, US) | 246.7 | 11.7 | 559.8 | 16.6 | 25.4% | 1.0 |
| Base (KPN) | 74.6 | 6.2 | 179.4 | 9.6 | 28.1% | 0.7 |
| Media Concentration Index | 2004/5 | | 2011 or Most Recent | | % Change Annual Average | |
| Public Ownership (%) | 0 | | 0 | | 0% | |
| Foreign Ownership (%) | 27.6 | | 37.9 | | 2.1% | |
| C4 Average—Weighted | 99 | | 95 | | –0.1% | |
| C1 Average—Weighted | 61 | | 51 | | –2% | |
| HHI Average—Weighted | 4,673 | | 3,815 | | –3.7% | |
| National Power Index | 4,601 | | 4,004 | | –2.6% | |

# 4

# Media Ownership and Concentration in Finland

MIKKO GRÖNLUND

## INTRODUCTION

Finland is a postindustrial, sparsely inhabited Nordic country with a population of approximately 5.4 million. It has two official languages: Finnish, spoken by the vast majority of the population, and Swedish, spoken by 5.4% of Finnish citizens.

Mass media's share of the Finnish gross domestic product (GDP) has declined by a half a percentage point, from 3% to 2.5%, over the past decade. High per capita income, slowed population growth, and the graying of the population, along with a host of other economic, cultural, political, and technological factors, have all affected the country's media landscape. For example, the number of households in Finland is increasing yet the average size of households is declining. In addition, Finland is seeing the proportion of persons 65 or over in the population increasing, and this demographic is projected to rise from 20 % to 26% by 2040.

In Finland, the Ministry of Transport and Communications is responsible for communications policy—including matters relating to information security and data protection, mass media, and postal services—while the Finnish Competition and Consumer Authority (FCCA) monitors business deals and trade practices. Finnish media legislation includes a wide range of regulations and rights, from those embodied in the Constitution to the Communications Market Act of 2003, which in 2009 was amended to enshrine universal Internet access for all Finnish citizens.[1] The Finnish Communications

1. "Communications Market Act." Ministry of Transport and Communications, 2011. 11 Jan. 2013. <http://www.finlex.fi/en/laki/kaannokset/2003/en20030393.pdf>.

Regulatory Authority (FICORA) monitors and promotes communications markets and services in the interests of the general public, business, and industry. Despite the existence of 23 separate media laws, however, there has been no special legislation on media competition, concentration, or ownership. Existing legislation is based on Finnish and EU legislation on general concentration and ownership matters in the economy.

During the 2000s, the growth of the mass media market hinged mostly on audiovisual and online media: the combined turnover of television, radio, and online news sites has doubled since 2001.[2] Print media (newspapers, books, and magazines) turnover in 2011 was only slightly higher than it was in 2001, and the recording industry suffered a decline in revenues that brought it back to pre-2001 levels. As a result of these upward and downward trends, in 2011 the mass media market grew by only 1%, and turnover for all media sectors nearly remained on level with the turnover recorded for 2010, at approximately US$6 billion (4.3 billion euros).[3] The share of print media still accounted for more than half (62%) of the total mass media market's turnover: in 2010 newspapers' market share was 25.6%, while magazines' market share stood at 17%.

## PRINT MEDIA

### Newspapers

Finland's per capita consumption of print media has long been one of the highest in the world, and print editions of newspapers and magazines are still widely consumed. The strength of the newspaper market in the country is often explained by political and cultural factors in the Nordic countries, reflected in the strong position of print media in national advertising expenditures. In 2011, even though total turnover had not increased greatly over previous years, the newspapers' take for advertising revenues stood at 58% of total media advertising spending.

Results of the Finnish National Readership Survey[4] indicate that reading habits change relatively slowly. The readership of print newspapers has remained relatively constant for the past few years, even though reading of online publications has become more common. According to the Finnish National Readership Survey, 96% of the population over 12 years of age have read print newspapers and 60% have read online during an average week. In 2011 the average media usage was 8 hours and 33 minutes per day, and 31 minutes (approximately 6%) of the daily media use fell to newspapers.[5]

In 2011, there were a total of 183[6] paid-for newspapers, of which 48 were dailies (published four to seven times a week) and 135 were nondailies (appearing one to three times a week). The daily newspapers accounted for 71% and nondaily newspapers for 29% of total newspaper circulation.

All major newspaper publishers in Finland have domestic owners and have their roots in the newspaper business. Several newspaper publishers have chosen to meet the challenges

2. Statistics Finland. "Mass media market growing slightly, preliminary data 2010." Official Statistics of Finland (OSF), 26 May 2011. 10 Jan. 2013. <http://www.stat.fi/til/jvie/2010/jvie_2010_2011-2005-26_tie_001_en.html>. & Statistics Finland. "Turnover of electronic media doubled since 2001." Official Statistics of Finland (OSF), 12 Dec. 2012. 10 Jan. 2013. <http://www.stat.fi/til/jvie/2011/01/jvie_2011_01_2012-2012-12_tie_001_en.html>.

3. "The preliminary calculations describe the mass media market at the end user level: for example, the figure on the newspapers market is comprised of retail priced subscription and single copy sales of newspapers, and their revenue from advertising. Overlaps have been eliminated. The figures cover domestic production and imports, but not exports." *See* Statistics Finland. "Mass media market 2010—growth was weak in printed media." Official Statistics of Finland (OSF), 16 Dec. 2011. 10 Jan. 2013. <http://www.stat.fi/til/jvie/2010/jvie_2010_2011-2012-16_tie_001_en.html>.

4. The Finnish National Readership Survey (Kansallinen Mediatutkimus KMT) determines readerships of publications, reading habits, and the structure of the readership. In the survey, data are also collected extensively concerning general media usage, purchasing habits of Finns, attitudes, and use of products and services.

5. In 2010 the method and the target group were changed. Therefore the results are not comparable with previous results.

6. This figure includes members of the Finnish Newspaper Association (FNA) and excludes approximately 20 very small, local, paid-for newspapers that are not members of the FNA.

of the changes in the media market through expansion, either into other media or new territory, so as to strengthen their market position and revenue base. Expansion has been through mergers and acquisitions with other media companies (Grönlund & Björkroth, 2011). The sector is dominated by two firms, which together account for almost half of daily newspaper circulation. The market leader is Sanoma Oyj, the second largest media group in Scandinavia and a leading pan-European publisher with operations in multiple European nations (including Russia).[7, 8] A public limited company (plc) founded in 1999 (although the publishers that it incorporated into its group were first established in the second half of 19th century) and based in Helsinki, Sanoma is a multimedia print and audiovisual group owned by the Jane and Aatos Erkko Foundation, which has a majority 23% stake.[9] The second biggest Finnish player, Alma Media Oyj (plc), was founded in 1998 (although the publishers that it incorporated into its group were first established in the 19th century) and focuses on newspaper publication and online content production (Table 4.1).

## Table 4-1. Daily Newspapers (Market Shares by Circulation), 1996–2010

| | 1996 | 2000 | 2004 | 2008 | 2010 |
|---|---|---|---|---|---|
| Sanoma News (Sanoma Oyj) | 28.8 | 33.7 | 34.0 | 31.0 | 31.0 |
| Alma Media Oyj | 18.1 | 21.4 | 22.1 | 23.8 | 23.0 |
| Keskisuomalainen Oyj | 3.3 | 3.3 | 7.5 | 7.7 | 8.0 |
| TS-Yhtymä Oy | 5.9 | 5.9 | 5.9 | 6.3 | 6.0 |
| Ilkka Oyj | 4.6 | 3.8 | 3.9 | 3.9 | 4.0 |
| Kaleva Oy | 3.6 | 3.6 | 3.6 | 3.8 | 4.0 |
| Esan Kirjapaino Oy | | 3.2 | 3.3 | 3.4 | 3.0 |
| KSF Media Ab | | 3.1 | 2.8 | 3.3 | 3.0 |
| Suomen Lehtiyhtymä Oy[10] | | | 2.6 | 3.2 | 3.0 |
| Pohjois-Karjalan Kirjapaino Oyj | | 2.1 | 2.1 | 2.2 | 2.0 |
| Others | 26.1 | 15.3 | 12.2 | 11.4 | 11.0 |
| Top 10 circulation | 1,750 | 1,973 | 1,970 | 1,885 | 1,763 |
| Average Daily Circulation (1,000) | 2,368 | 2,328 | 2,244 | 2,127 | 1,986 |
| Total daily newspaper revenue (mil euros)[1] | 795 | 965 | 970 | 1,056 | 959 |
| Total Revenue (mil US$) | 930 | 892 | 1,206 | 1,553 | 1,273 |
| C4 | 57.4 | 64.8 | 69.5 | 68.8 | 68 |
| HHI | 1,237 | 1,691 | 1,794 | 1,694 | 1,653 |
| *N* (>1%) | 6 | 9 | 10 | 10 | 10 |
| Noam Index | 505 | 564 | 567 | 536 | 523 |

*Source:* Statistics Finland based on data from The Finnish Audit Bureau of Circulations

[1] This revenue figure (Net sales = Circulation sales + advertising sales in current prices) is from Finnish Newspaper Association

7. Sanoma Oyj. "Sanoma View 2011." Sanoma, 2012. 10 Jan. 2013. <http://www.sanoma.com/content/download/6974/241078/version/4/file/Sanoma_View_2011_EN.pdf>.

8. In 2013, Sanoma Oyj decided to focus on the most important markets in Belgium, Russia, Central Europe, and Eastern Europe. Magazine and digital media companies in Bulgaria, Romania, Serbia, and the Czech Republic were divested.

9. Aatos Erkko was the largest shareholder prior to his death in 2012; he bequeathed his stake in Sanoma Oyj to the Foundation, which as a result found itself in possession of slightly greater than US$1 billion (800 million euros) in investment capital. Some of his relatives also own shares of Sanoma Oyj.

10. Keskisuomalainen Oyj acquired Suomen Lehtiyhtymä Oy in March 2013.

Finland's HHI is relatively low due to the highly regional nature of the newspapers market: for circulation, $HHI_C$[11] is only around 1,000 yet stands at 1,500 for the $HHI_A$.[12, 13] Given the country's historical geography, many towns and their environs are wedded to particular papers, so national monopolies have not formed despite the small size of the country's media market. A national perspective on the newspaper market often overlooks many important nuances of competition, such as the regional character of newspapers. The majority (29 titles) of the 31 seven-day-a-week newspapers are regional newspapers. The regional character of newspaper competition applies to both audience shares and advertising revenue (advertisers need to reach audiences that are often located within specific geographic areas in which their products or services are sold).[14]

Another indicator of strength of the provincial press is that among the 20 biggest newspapers, measured by circulation, 14 are provincial newspapers.[15]

## Book Publishing

The book publishing market is not highly concentrated and has several significant players. Werner Söderström Osakeyhtiö, which traditionally has held the largest market share, publishes business and professional books, dictionaries, and educational materials. Founded in 1878, the company was a subsidiary of Sanoma Oyj, but since 2011 it has operated as a subsidiary of Bonnier AB (Sweden). However, its market shares are slipping: it had a market share of only 17.4% in 2011, down from 25.1% in 2007. Kustannusosakeyhtiö Otava, the number two publisher, also had declining market shares over the same period. Its parent, the Otava Group, operates the country's biggest bookstore chain.[16] Sanoma Pro Oy is the number one educational publisher in the Finnish market. Edita Publishing Oy publishes textbooks and study materials intended for comprehensive school, upper secondary school, vocational training, and on-the-job learning. Edita Publishing Oy is a subsidiary of Edita Oyj, a publicly quoted company wholly owned by the Finnish State. Gummerus Kustannus Oy was founded in 1872, and it has been run as an independent family business ever since (Table 4.2).

## Magazine Publishing

According to the definition of the FPPA (Finnish Periodical Publishers Association), a magazine is a publication that is published regularly at least four times per calendar year; features several articles or editorial material in every issue; is publicly available for subscription or otherwise widely available; and does not primarily feature business bulletins, price lists, announcements, or advertising. It can be anything in terms of size or print paper or it can be exclusively an online publication. Magazines include consumer magazines, trade and organizational magazines, and customer magazines. In 2010, there was a total of 3,056 magazine titles, of which 39 were published weekly, 462 were published once or twice per month, and 2,555 were published four to 11 times a year.[17]

---

11. HHIc represents an HHI calculated from market shares of industry circulation.

12. HHIa represents an HHI calculated from market shares of industry advertising revenues.

13. Grönlund, Mikko, and Björkroth, Tom. "Newspaper market concentration, competitive pressure and financial performance: The case of Finland." *Journal of Media Business Studies* 8:3, 2011. 10 Jan 2013. <http://www.econis.eu/PPNSET?PPN=722188722>.

14. Picard, Robert G. *The Economics* and *Financing of Media Companies*. New York: Fordham University Press, 2002, 53.

15. Grönlund, Mikko, and Björkroth, Tom. "Newspaper market concentration, competitive pressure and financial performance: The case of Finland." *Journal of Media Business Studies* 8:3, 2011. 10 Jan 2013. <http://www.econis.eu/PPNSET?PPN=722188722>.

16. Otava Group. "2011 Annual Report." Otava Group, 2012. 10 Jan. 2012. <http://www.digipaper.fi/otava_vuosikertomus/87387/>.

17. The National Library of Finland, Bibliographical Services.

Table 4-2. Book Publishing (Market Shares by Revenue), 2007–2011

| | 2007 | 2008 | 2009 | 2010 | 2011 |
|---|---|---|---|---|---|
| Bonnier AB (Sweden) | 37.6 | 40.4 | 35.4 | 32.4 | 31.4 |
| Werner Söderström Osakeyhtiö | *25.1* | *26.5* | *21.7* | *19.1* | *17.4* |
| Kustannusosakeyhtiö Tammi | *12.5* | *13.9* | *13.7* | *13.3* | *14.0* |
| Kustannusosakeyhtiö Otava (Otava Oy) | 24.2 | 23.8 | 25.4 | 25.3 | 26.2 |
| Sanoma Pro Oy (Sanoma Oyj) | 16.0 | 15.9 | 20.1 | 23.6 | 25.5 |
| Edita Publishing Oy[1] | 7.1 | 6.7 | 7.2 | 7.2 | 7.2 |
| Gummerus Kustannus Oy | 5.1 | 4.5 | 4.7 | 4.8 | 4.7 |
| Weilin-Göös Oy[2] (Sanoma Oyj) | 4.6 | 4.1 | 4.0 | 3.1 | 1.4 |
| Karisto Oy | 4.1 | 3.3 | 1.9 | 2.2 | 2.1 |
| Schildts & Söderströms Ab | 1.2 | 1.3 | 1.4 | 1.4 | 1.4 |
| Total Revenue (mil euros) | 174 | 159 | 154 | 154 | 143 |
| Total Revenue (mil US$) | 238 | 234 | 214 | 205 | 199 |
| C4 | 84.9 | 86.8 | 88.1 | 88.5 | 90.3 |
| HHI | 2,371 | 2,546 | 2,398 | 2,338 | 2,405 |
| *N* (>1%) | 8 | 8 | 8 | 8 | 8 |
| Noam Index | 838 | 900 | 848 | 827 | 850 |

*Source:* Statistics Finland

[1] Edita Publishing Oy is a subsidiary of Nordic Morning Oyj (Edita Oyj changed it's name to in April 2013). Nordic Morning Oyj is Nordic communications corporation owned by the Finnish State.

[2] Weilin + Göös was closed down in late 2011.

Sanoma Magazines Finland Oy[18] (a subsidiary of Sanoma Oyj), Otavamedia Oy, and A-lehdet Oy are the three leading magazine publishers in Finland. They are followed by Aller Media Oy, Bonnier Publications Oy, and Forma Publishing Group Oy. A number of smaller companies together hold 33% of the market. Otavamedia Oy is a private multimedia company that began as a magazine publisher, and it is now involved in book distribution (through Rautakirja Oyj, whose bookstore chain it acquired in 2011) and e-commerce.

Sanoma Magazines Finland Oy mainly publishes women's magazines, family magazines, and juvenile publications, and it also holds market shares in IT magazines, business directories, and corporate publications. Alma Media focuses on publishing and digital services businesses, and it expanded into the US market in June 2012 when it acquired a 51% stake in the US-based Adalia Media publishing company (Table 4.3).

Market concentration in this sector has declined substantially since 2000.

## AUDIOVISUAL MEDIA

### Radio

Radio broadcasting began in Finland in 1921. A number of small private networks soon developed into the Finnish Amateur Radio League in 1924, which counted a membership of 1,254 private local radio stations. In order to facilitate better reportorial coordination and to establish state control over the market, a radio commission

18. As part of redesign of its consumer media operations, Sanoma merged its Finnish media operations (Sanoma Entertainment Finland, Sanoma Magazines Finland, and Sanoma News) in 2014 into Sanoma Media Finland Oy.

Table 4-3. Magazine Publishing (Market Shares by Revenue), 2007–2011

| | 2007 | 2008 | 2009 | 2010 | 2011 | 2010 NUMBER OF TITLES |
|---|---|---|---|---|---|---|
| Sanoma Magazines Finland Oy (Sanoma Oyj) | 32.9 | 33.0 | 33.7 | 33.3 | 32.5 | 50 |
| Otavamedia Ab (Otava Oy) | 23.2 | 23.9 | 24.8 | 25.0 | 28.4 | 30 |
| A-Lehdet Oy | 13.5 | 12.7 | 13.2 | 12.6 | 12.8 | 12 |
| Aller Media Oy (Aller Holding A/S, Denmark) | 9.2 | 8.9 | 9.1 | 8.3 | 8.3 | 6 |
| Talentum Media Oy (Alma Media Oyj, 32%) | 8.8 | 9.0 | 6.6 | 6.6 | 6.9 | |
| Oy Valitut Palat—Reader's Digest Ab (The Reader's Digest Association, Inc., USA) | 5.0 | 5.1 | 5.2 | 6.9 | 4.3 | 2 |
| Egmont Kustannus Oy AB (Egmont Media Group, Denmark) | 3.5 | 3.4 | 3.5 | 3.3 | 2.9 | 55 |
| Forma Publishing Group Oy (ICA Förlaget, Sweden) | 2.8 | 2.5 | 2.2 | 2.3 | 2.2 | 5 |
| (ICA Förlaget, Sweden) | 0.6 | 0.8 | 1.0 | 1.2 | 1.2 | |
| Bonnier Publications Oy (Bonnier AB, Sweden) | 0.6 | 0.6 | 0.6 | 0.5 | 0.5 | 10 |
| Total Revenue (mil euros) | 415 | 391 | 394 | 421 | 401 | |
| Total Revenue (mil US$) | 568 | 575 | 547 | 560 | 558 | |
| C4 | 78.8 | 78.6 | 80.8 | 79.2 | 82.0 | |
| HHI | 2,010 | 2,030 | 2,096 | 2,070 | 2,175 | |
| *N* (>1%) | 8 | 8 | 8 | 9 | 9 | |
| Noam Index | 711 | 718 | 741 | 690 | 725 | |

Source: Media Statistics based on data from The Finnish Audit Bureau of Circulation, Finland Post Inc./Itella, Ficora, Rautakirja, company annual reports.

was founded by the Helsinki journalists' union that gave the Government of Finland the authority to operate transmission stations. Subsequently, the national broadcasting company Yleisradio, or YLE, was established in 1926.[19] In 1934, the Finnish Parliament legislated that YLE would serve as the national public broadcaster.[20] YLE is still Finland's national public broadcasting company, and its channels continue to dominate the radio market with total market shares of 55.8%.[21] YLE has six nationwide channels, 26 regional windows[22] and two digital

19. YLE is commonly used today to refer to the public broadcaster, though Yleisradio is still officially used as well. YLE became more common after the public broadcaster interested the TV broadcasting market in 1957; for the first 31 years of its existence, Yleisradio was solely a radio broadcaster, hence its name.

20. Noam, Eli. *Television in Europe*. New York: Oxford University Press, 1991.

21. YLE's operations are financed by public broadcasting tax, which replaced TV licenses in 2013. In 2014, Yle tax was 0.68% of employees' earnings and capital incomes up to maximum of 143 euros (140 euros in 2013). Private customers do not pay Yle tax if the amount of tax is less than 51 euros in 2014 (50 million in 2013).

22. Transmitted in windows on nationwide Radio Suomi and Radio Vega channels.

(DVB) radio channels. Private radio broadcasting in Finland started in 1985 when the Council of State granted the first commercial and noncommercial licenses in addition to the existing channels of Yleisradio. The first national commercial radio, Radio Nova, was born in 1997. It was accompanied by a group of seminational network radio stations. As a consequence of changes in the structure of radio broadcasting, the investments started shifting from local radio stations to network radio stations.

As a result of structural changes, the emphasis of advertising has moved from local radio markets to nationwide markets controlled by chain radio stations. At the end of 2010, in addition to YLE's eight public service channels, there were 57 private commercial radio stations.[23]

A handful of big commercial companies—SBS Finland Oy (ProSiebenSat.1 Media AG), MTV Media (Bonnier AB), Nelonen Media (Sanoma Oyj), and NRJ Finland OY (NRJ Group)—along with several other smaller companies, account for the remaining market share. Oy Suomen Uutisradio AB, the second largest company following YLE, has a 11.5% market share. By 2008, Pro Radio Oy had merged into the German broadcaster, ProSiebenSat.1. Likewise, in 2012, Mediasales Finland Oy and Metroradio Finland Oy were absorbed by Nelonen Media, a subsidiary of Sanoma Oyj. All radio broadcasters in Finland exhibit slow growth, and their market share percentages have been stable for at least a decade. The HHI concentration is high (3,435), due to YLE's strong position as public service broadcaster in the market (Tables 4.4, 4.5, and 4.6).

## Broadcast Television

There are three major operators in terrestrial television broadcasting: the public broadcaster YLE and the commercial broadcasters MTV Media (a subsidiary of Bonnier AB ) and Nelonen Media, which is part of Sanoma Oyj's Sanoma Media Finland division. Under the 1957 private broadcasting license, two commercial broadcasters, Tesvisio and Tampere, began broadcasting in 1958 in the cities of Helsinki and Tampere, respectively. YLE also entered the national broadcast market at this time. YLE

### Table 4-4. Radio Group (Market Share of Listening 2010)

| | |
|---|---|
| YLE (public) | 53 % |
| Radio Nova (Bonnier AB, Sweden)[1] | 11.0% |
| SBS-Iskelmäradiot (SBS Broadcasting, Germany)[2] | 7.0% |
| SuomiPop (Communicorp Group,[3] Ireland) | 5.0% |
| Radio Rock (Sanoma Oyj) | 5.0% |
| NRJ (NRJ Group, France) | 4.0% |
| The Voice (SBS Broadcasting) | 3.0% |
| Radio Aalto (Sanoma Oyj) | 1.0% |
| Groove FM (Communicorp Group, Ireland) | 1.0% |

*Source:* Statistics Finland based on data from Ministry of Transport and Communications, Finnpanel, RadioMedia and Statistics Finland

[1] Swedish Modern Times Group MTG AB has a 26% stake in Radio Nova as well. Share of Bonnier AB's is 74%.

[2] SBS Broadcasting was formerly a subsidiary of ProSiebenSat.1 (Germany). Discovery Communications acquired the SBS operations in the Nordic countries ProSiebenSat.1 Media, and in December 2012. SBS Nordic Discovery's operations as well as the previous functions were merged into a new company in the Nordic countries, 9 April 2013, called SBS Discovery Media.

[3] Sanoma Oyj acquired Metroradio Finland's radio channels from Communicorp Group in May 2012.

23. Ten nationwide and 47 regional and local stations of which 16 are chain stations.

## Table 4-5. Radio Group (Market Shares by Revenue), 2004–2010

| | 2000 | 2004 | 2006 | 2008 | 2010 |
|---|---|---|---|---|---|
| YLE (public) | 89.0 | 72.8 | 69.3 | 57.1 | 55.8 |
| Oy Suomen Uutisradio Ab/MTV Oy (Bonnier AB, Sweden) | | 8.2 | 9.8 | 12.4 | 11.5 |
| SBS Finland Oy (ProSiebenSat.1 Media AG, Germany) | 3.0 | 7.9 | 9.2 | 10.7 | 11.9 |
| NRJ Finland Oy Ab (NRJ Group, France) | | 3.0 | 3.4 | 3.9 | 3.5 |
| Nelonen Media (Sanoma Oyj) | | 0.0 | 0.0 | 3.4 | 4.4 |
| Metroradio Finland Oy (Communicorp Group, Ireland) | | 1.4 | 1.2 | 3.0 | 4.0 |
| Others | 5.0 | 6.8 | 7.1 | 9.4 | 8.8 |
| Total Revenue (mil euros) | 405 | 175.8 | 152.2 | 117.8 | 117.6 |
| Total Revenue (mil US$) | 373 | 218.6 | 191.2 | 173.2 | 156.1 |
| C4 | 92 | 93.3 | 91.7 | 84.1 | 83.6 |
| HHI | 7,930 | 5,440 | 4,996 | 3,564 | 3,435 |
| *N* (>1%) | 2 | 5 | 5 | 6 | 6 |
| Noam Index | 5,607 | 2,433 | 2,234 | 1,455 | 1,402 |

*Source:* Statistics Finland based on data from YLE and RadioMedia

## Table 4-6. Market Shares in Radio Advertising, 2004–2010

| | 2004 | 2006 | 2008 | 2010 |
|---|---|---|---|---|
| SBS Finland Oy (ProSiebenSat.1 Media AG, Germany) | 29 | 30 | 25 | 27 |
| MTV Media (Bonnier AB, Sweden) | 30 | 32 | 29 | 26 |
| Nelonen Media (Sanoma Oyj) | – | – | 8 | 11 |
| Metroradio Oy (Communcorp Group, Ireland) | 5 | 4 | 7 | 9 |
| NRJ Finland Oy (NRJ Group, France) | 11 | 11 | 9 | 8 |
| Other channels | 26 | 23 | 22 | 20 |
| Total, € million | 47,8 | 46,8 | 50,5 | 52,0 |

*Source:* RadioMedia

then acquired Tesvisio and Tamvisio's stocks and created TV2 from these channels; YLE also started broadcasting MTV Media's programming on its network for a large fee. By 1982, MTV Media was paying YLE almost two-thirds of its total advertising sales revenue for airtime, and this amounted to almost one-quarter of YLE's annual turnover.[24] Therefore, a public-private duopoly existed throughout the 1980s as MTV Media gained a privileged position in

24. Hujanen, Taisto. "Public Service Media Fee to substitute Television Fee in Finland?" *Central European Journal of Communication* 3, 2009. 10 Jan 2013. <http://ptks.pl/cejc/wp-content/uploads/2012/07/CEJC_Vol3_No1_Hujanen.pdf>.

broadcasting general programming.[25] The presence of a private TV broadcaster, albeit within the confines of a public broadcaster, was at the time highly unusual in Europe.

The market is still dominated by the public-private duopoly of YLE and MTV Media. MTV Media today is the umbrella company of the channels MTV3, SUB TV, AVA, and eight pay-TV channels. Bonnier AB of Sweden acquitted MTV Media in 2005. *MTV3*, launched in 1957 as MTV Media's first national channel, broadcasts a mixture of Finnish programming, as well as syndicated US and British shows. Although YLE's main channels (Yle TV1, Yle TV2, Yle Fem, and Yle Teema) have lost over 20% of their market shares since 2000, the public broadcaster is still the market leader in (digital) broadcast television, and it has expanded into satellite TV[26] as well.

Terrestrially distributed digital television was introduced in 2001, when the number of television channels began to increase. The switch-over to digital was completed in August 2007, when analog terrestrial networks were closed down. In February 2008, the cable networks also switched over to fully digital distribution. The digitalization of terrestrial networks increased the supply of television channels. Analog terrestrial networks had carried only four channels, but with digitalization, households in 2010 had access to an average of 13 channels[27] (Finnpanel Oy, TV-household surveys). All in all, over 30 channels are distributed terrestrially; approximately 15 or so are free-on-air channels. In 2000, the market shares were highly concentrated between the two major companies, YLE and MTV, which

## Table 4-7. TV Broadcasting (Market Shares by Revenue), 2000–2008

| | 2000 | 2004 | 2008 |
|---|---|---|---|
| YLE (public) | 65.0 | 62.0 | 41.0 |
| MTV Media (Bonnier AB, Sweden)[1] | 32.0 | 31.0 | 33.0 |
| Digi TV Plus Oy[2] | | | 6.0 |
| Canal Digital Finland OY (Canal +, France) | 2.0 | 4.0 | 4.0 |
| Oy Viasat Finland Ab | 1.0 | 1.0 | |
| Nelonen Media (Sanoma Oyj) | | | |
| Suomen Urheilutelevisio Oy | | 1.0 | |
| TV5 Finland Oy | | | |
| Others | | 2.0 | |
| Total Revenue (mil euros) | 535 | 590 | 938 |
| Total Revenue (mil US$) | 492 | 732 | 1,379 |
| C4 | 100.0 | 98.0 | 84.0 |
| HHI | 5,254 | 4,823 | 2,822 |
| *N* (>1%) | 4 | 5 | 3 |
| Noam Index | 2,627 | 2,157 | 1,411 |

*Source:* Statistics Finland, 2008

[1] MTV Media has no relation to MTV (US).

[2] Digi TV Plus Oy is a pay-TV operator (started in 2006) for antenna households with approximately 250,000 subscribers and 35% market share. PlusTV is the largest pay-TV operator that packages, markets, sells, and administers program channels that have a permit to broadcast Pay-TV in the terrestrial network on the Finnish market.

25. Noam, Eli. Television *in Europe*. New York: Oxford University Press, 1991.

26. TV Finland is a satellite channel for expatriates. Its programming is a collection of programs on YLE TV1, YLE TV2, YLE Teema, and YLE Fem channels. TV Finland serves Finnish expatriates in Europe.

27. Terrestrial households, 12 channels; and cable/satellite households, 16 channels.

held 65% and 32% market share, respectively. By 2008, the two companies had lost some market shares, bringing the C4 index down from 100% to 84% and distributing shares more evenly. Yet the two companies hold more than half of the TV broadcasting market shares. Nelonen Media has three (Nelonen, Jim, Liv) free-on-air and five (Nelonen Kino, Nelonen Pro 1, Nelonen Pro 2, Ne lonen Perhe, and Nelonen Maailma) pay terrestrial channels (Table 4.7).

## TELECOMMUNICATIONS MEDIA

### Wireline Telecom

The Finnish telecom system has always been different from that of other European countries. After Finland became independent from the Russia Empire in 1917, numerous local telephone companies controlled by local cooperatives of subscribers or small investors formed, so the state enterprise Post and Telecommunications (Finland) served largely rural and remote areas, reaching only about one-third of the population.[28]

The telecommunications sector is characterized by a few companies with activities in the sector that are clearly bigger than others. Moreover, a large number of smaller companies operate in the sector. In the telecom sector, the combined turnover of the ten largest companies encompasses more than 90% of the total turnover in the sector. However, the share of turnover varies from one service to another. Ranked by their turnover from telecom, the three biggest companies are TeliaSonera, Elisa, and DNA. In addition to these, 24 independent, regional telecom operators united under the Finnet Group and a number of other independent telecom operators have activities in Finland (FICORA Market Review, 4/2012—Telecoms operations in Finland).

Elisa, TeliaSonera, DNA, the companies under the Finnet Group, and a few other companies provide telephone sunscriptions to end users. The majority of companies providing services in the fixed network have their own telephone network. Elisa, TeliaSonera, and DNA provide fixed broadband services at the national level. A few other companies offer broadband services in the fixed network, for example, in the biggest cities. The telecom operators within Finnet Group provide these services locally in their operating areas (FICORA Market Review 4/2012—Telecoms operations in Finland).

TeliaSonera was founded in 2003 through the merger of the Swedish telecom major Telia and Finland's Sonera, which was known as Telecom Finland until 1994. In Sweden, Telia was the successor of Televerket, the public monopolist (although with a private law status). TeliaSonera is owned 37% by the Government of Sweden and 13% by Government of Finland. It provides mobile, broadband, fixed voice, and TV broadcasting services.

DNA offers mobile phone service, cable television, and regular telephone service. DNA was originally founded as the mobile phone operator for the Finnet Group until the Helsinki Telephone Association—now Elisa Oyj—left the consortium. Finnet split again when DNA was formed in 2007.

Both the C4 and HHI concentration indices slightly decreased between 2004 and 2010, although both remain high. The basic market structure is still the same as it had been for nearly a century. The major semiprivate operator based in Helsinki, now called TeliaSonera, dominates the market. However, because smaller, locally based independent firms have for the most part consolidated into either DNA, Finnet, or Elisa, concentration has actually increased relative to the market that existed before liberalization began in the 1990s (Table 4.8).[29]

### Wireless Telecom

By 1989, the demand for mobile telephony far exceeded predictions by regulators, and two

28. Noam, Eli. *Telecommunications in Europe*. New York: Oxford University Press, 1992, 212.

29. Edwardsson, Anders. "Liberalization of the telephony markets." TeliaSoneraHistory.com, 2012. 11 Jan. 2013. <http://www.teliasonerahistory.com/building-the-company/article-list/deregulation-of-the-telephony-markets/>.

Table 4-8. Wireline Telecom (Market Shares by Revenue), 2004–2010

| | 2004 | 2007 | 2008 | 2010 |
|---|---|---|---|---|
| Elisa | 38.0 | 33.0 | 36.0 | 38.0 |
| TeliaSonera (public) | 29.0 | 29.0 | 25.0 | 22.0 |
| Finnet Group | 31.0 | 18.0 | 19.0 | 21.0 |
| DNA | | 15.0 | 15.0 | 14.0 |
| Others | 2.0 | 5.0 | 5.0 | 5.0 |
| Total Revenue (mil euros)[1] | 1,835 | | 1,538 | 1,620 |
| Total Revenue (mil US$) | 2,275 | | 2,261 | 2,151 |
| C4 | 98.0 | 95.0 | 95.0 | 95.0 |
| HHI | 3,246 | 2,479 | 2,507 | 2,565 |
| *N* (>1%) | 3 | 4 | 4 | 4 |
| Noam Index | 1,874 | 1,240 | 1,254 | 1,283 |

*Source:* Statistics Finland

[1] This revenue figure is an estimate based on data collected in this study: it is based on average per capita revenues reported for several other countries in this study with similar per capita income.

separate Global System for Mobile Communications (GSM) network operators came into play: the state enterprise Post and Telecommunications (Finland) and the commercial operator Radiolinja, which was originally part of the Finnet Group and one of the world's earliest GSM providers. As in the wireline market, Elisa, TeliaSonera, and DNA now dominate the wireless market; they held nearly even market shares in 2010. The market shares were already very concentrated in 2004 and have become even more concentrated since then. While TeliaSonera held almost half of the total wireless telecommunication market shares in 2004 (36% in 2010), shares since then have become more evenly distributed to Elisa and DNA (Table 4.9).

Table 4-9. Wireless Telecom (Market Shares by Revenue), 2004–2010

| | 2004 | 2005 | 2008 | 2010 |
|---|---|---|---|---|
| Elisa | 27.0 | 28.2 | 37.5 | 39.0 |
| TeliaSonera (public) | 47.0 | 47.1 | 39.5 | 36.0 |
| DNA | 14.0 | 14.0 | 23.1 | 23.0 |
| Others | 12.0 | 10.7 | | 2.0 |
| Total Revenue (mil euros)[1] | 1,459 | | 1,658 | 2,150 |
| Total Revenue (mil US$) | 1,809 | | 2,437 | |
| C4 | 88.0 | 89.3 | 100.0 | 98.0 |
| HHI | 3,134 | 3,210 | 3,500 | 3,346 |
| N (>1%) | 3 | 3 | 3 | 4 |
| Noam Index | 1,809 | 1,853 | 2,021 | 1,932 |

*Source:* Statistics Finland

[1] This revenue figure is an estimate based on data collected in this study: it is based on average per capita revenues reported for several other countries in this study with similar per capita income.

## INTERNET MEDIA

### Internet Service Providers (ISP)

DNA, Elisa, and TeliaSonera provide fixed broadband services at the national level. Also, a few other companies provide broadband services in the fixed network, for example, in the biggest cities. The telecom operators within the Finnet Group and other, smaller telecoms provide these services locally, mainly in their operating areas.

In all, 87% of all households have an Internet connection. Two Internet households out of three (68%) have a fixed broadband connection intended for the home environment. Slightly more than every second (54%) Internet household has a mobile broadband connection, which can also be used outside the home. Mobile connections have continued to grow (FICORA, 2012). As of 2010, the market is in an oligopolistic situation, with Elisa and TeliaSonera each possessing a market share of one-third and DNA holding onto approximately one-fourth of the market. The Finnet Group holds a market share of only 8%. The C4 index has increased significantly in this industry since 2004, because these four companies control the market (Table 4.10).

As early as the late 1990s, traditional media companies started to developed integrated communication products and services. For example, newspaper publishers reacted quickly when classified ads began to move online and they started to offer complementary websites for printed classified ads. Many of the classified advertisement network services have been successful economically. All daily newspapers have had free websites for a long time, and most of them also publish online (paid-for) facsimile editions or optimized editions for different electronic reading devices. In addition to the online services of the printed media, there are two daily online-only newspapers (*Taloussanomat*[30] and *Uusi Suomi*). In 2010, 168 newspapers (49 dailies and 119 nondailies) had homepages that provided at least some news or editorial materials, and the number of magazines online was 251.

Table 4-10. Internet Service Providers (Market Shares by Revenue), 2004–2010

| | 2004 | 2005 | 2008 | 2009 | 2010 |
|---|---|---|---|---|---|
| Elisa | 27.0 | 30.9 | 33.0 | 31.0 | 34.0 |
| TeliaSonera (public) | 32.0 | 30.3 | 30.0 | 32.0 | 32.0 |
| DNA | | | 16.0 | 18.0 | 23.0 |
| Finnet Group | 20.0 | 23.2 | 12.0 | 11.0 | 8.0 |
| Others | 21.0 | 15.6 | 9.0 | 8.0 | 3.0 |
| Total Revenue (mil euros)[1] | 479 | | 552 | | |
| Total Revenue (mil US$) | 594 | | 811 | | |
| C4 | 79.0 | 84.4 | 91.0 | 92.0 | 97.0 |
| HHI | 2,153 | 2,411 | 2,389 | 2,430 | 2,773 |
| *N* (>1%) | 3 | 3 | 4 | 4 | 5 |
| Noam Index | 1,243 | 1,392 | 1,195 | 1,215 | 1,387 |

*Source:* Statistics Finland

[1] This revenue figure is an estimate based on data collected in this study: it is based on average per capita revenues reported for several other countries in this study with similar per capita income.

30. Taloussanomat is a business newspaper that is owned by Sanoma Oyj. The print version of the newspaper was discontinued in December 2007.

## Online News

The role of the Internet as a news source is increasingly important. A total of 38% of all research participants said they mainly use the Internet for following the news in their spare time. For 35% of the respondents, television or teletext was the most important source. For 17% of respondents, printed newspapers were the main source of news, whereas radio was the main news source for 7% of respondents. Those who consider the Internet their main source of news tend to read news in writing from the websites of newspapers, television channels, or online newspapers (FICORA Consumer Survey of Communications Services, 2012).

According to Gallup TNS Metrix, the most popular websites—if search providers and general portals are excluded—are maintained by traditional media companies. The majority of users of the Internet use it for searching for information (95%) or for electronic services (93%). More than four out of five use it for keeping in touch and communication. The majority of users also watch television and videos or listen to music on the Internet (Table 4.11).

In Finland, Internet Protocol television (IPTV) has been provided since approximately

### Table 4-11. Top Twelve Finnish WWW Media Pages 2011 (October Week 40)

| | TYPE OF MEDIA | BROWSERS/ WEEK | SESSIONS/ WEEK | SESSIONS/ BROWSERS | MARKET SHARE |
|---|---|---|---|---|---|
| Iltalehti (Alma Media Oyj) | Tabloid newspaper | 2,442,074 | 17,224,652 | 5.5 | 16.3 |
| Ilta-Sanomat (Sanoma Oyj) | Tabloid newspaper | 2,237,153 | 13,388,412 | 4.8 | 14.9 |
| MTV 3 (Bonnier AB, Sweden) | Television | 1,862,770 | 9,918,915 | 4.6 | 12.4 |
| Helsingin Sanomat (Sanoma Oyj) | Newspaper | 1,399,413 | 7,091,797 | 3.6 | 9.3 |
| YLE (Public broadcaster) | PSB TV and radio | 1,251,891 | 5,902,962 | 3.6 | 8.3 |
| Taloussanomat (Sanoma Oyj) | Online-only business newspaper | 863,664 | 2,100,630 | 2.5 | 5.7 |
| Kauppalehti (Alma Media Oyj) | Business newspaper | 546,072 | 2,460,004 | 3.5 | 3.6 |
| Sub (Bonnier AB, Sweden) | Television | 439,335 | 1,882,803 | 4.0 | 2.9 |
| Nelonen (Sanoma Oyj) | Television | 392,855 | 843,845 | 2.0 | 2.6 |
| Kaksplus (Sanoma Oyj) | Magazine | 308,374 | 549,147 | 1.7 | 2.1 |
| Aamulehti (Alma Media Oyj) | Newspaper | 296,554 | 953,070 | 2.5 | 2.0 |
| Uusi Suomi | Online-only newspaper | 278,138 | 896,277 | 2.6 | 1.9 |
| Others | | | | | 18* |
| C4 | | | | | 52.9 |
| HHI | | | | | 869 |
| *N*(>1%) | | | | | 12 |
| Noam Index | | | | | 251 |

*Source:* TNS Gallup, Netmetrix, Market shares estimated based on average of reporting European countries.

2005. The number of IPTV service users is still relatively small but continues to increase. The biggest servers in the market are KotiTV (Sonera Viihde) and ElisaViihde (Elisa). Netflix and HBO Nordic started their IPTV in late 2012.

## CONCLUSION

Several mass media industries are becoming more and more competitive, such as broadcast television and book publishing. Newspapers and magazines remain highly concentrated despite the regionalism that characterizes their markets. The radio industry is still highly concentrated, with the public broadcaster YLE dominating the market. Wireline and wireless telecom and ISP also remain highly concentrated.

There is a strong and growing role for companies of neighboring Sweden, in particular Bonnier AB and TeliaSonera.[31]

## Bibliography

### 1. Books and Articles


Edwardsson, Anders. "Liberalization of the telephony markets." TeliaSoneraHistory.com, 2012. 11 Jan. 2013. <http://www.teliasonerahistory.com/building-the-company/article-list/deregulation-of-the-telephony-markets/>.

FICORA. "Communications market in Finland—Market review 2011. Helsinki: FICORA, 2012.

FICORA Market Review 4/2012—Telecoms operations in Finland—Telecoms operators' revenues and investments in 2011. Helsinki: FICORA, 2012.

Grönlund, Mikko and Björkroth, Tom. "Newspaper market concentration, competitive pressure and financial performance: The case of Finland." *Journal of Media Business Studies* 8:3, 2011. 10 Jan 2013. <http://www.econis.eu/PPNSET?PPN=722188722>.

Hujanen, Taisto. "Public Service Media Fee to substitute Television Fee in Finland?" *Central European Journal of Communication* 3, 2009. 10 Jan 2013. <http://ptks.pl/cejc/wp-content/uploads/2012/07/CEJC_Vol3_No1_Hujanen.pdf>.

Noam, Eli. *Television in Europe*. New York: Oxford University Press, 1991.

Noam, Eli. *Telecommunications in Europe*. New York: Oxford University Press, 1992.

Nurmela, Sakari. *Consumer Survey on Audiovisual Content Services*. Helsinki: Publications of FICORA, 2012.

Picard, Robert G. *The Economics and Financing of Media Companies*. New York: Fordham University Press, 2002.


### 2. Reports


"Communications Market Act." Ministry of Transport and Communications, 2011. 11 Jan. 2013. <http://www.finlex.fi/en/laki/kaannokset/2003/en20030393.pdf>.

Otava Group. "2011 Annual Report." Otava Group, 2012. 10 Jan. 2012. <http://www.digipaper.fi/otava_vuosikertomus/87387/>.

Sanoma Oyj. "Sanoma View 2011." Sanoma, 2012. 10 Jan. 2013. <http://www.sanoma.com/content/download/6974/241078/version/4/file/Sanoma_View_2011_EN.pdf>.

Statistics Finland. "Mass media market 2001–2006." Official Statistics of Finland (OSF), 27 Feb. 2008. 11 Jan. 2013. <http://www.stat.fi/til/jvie/2006/jvie_2006_2008-2002-27_tie_001_en.html>.

Statistics Finland. "Mass media market 2007." Official Statistics of Finland (OSF), 31 Dec. 2008. 11 Jan. 2013. <http://www.stat.fi/til/jvie/2007/jvie_2007_2008-2012-31_tie_001_en.html>.

Statistics Finland. "Mass media market 2010—growth was weak in printed media." Official Statistics of Finland (OSF), 16 Dec. 2011. 10 Jan. 2013. <http://www.stat.fi/til/jvie/2010/jvie_2010_2011-2012-16_tie_001_en.html>.

Statistics Finland. "Mass media market growing slightly, preliminary data 2010." Official Statistics of Finland (OSF), 26 May 2011. 10 Jan. 2013. <http://www.stat.fi/til/jvie/2010/jvie_2010_2011-2005-26_tie_001_en.html>.

Statistics Finland. "Regular reading of newspapers has diminished." Official Statistics of Finland (OSF), 17 May 2011. 10 Jan 2013. <http://www.stat.fi/til/akay/2009/03/akay_2009_03_2011-2005-17_tie_001_en.html>.

Statistics Finland. "Turnover of electronic media doubled since 2001." Official Statistics of Finland (OSF), 12 Dec. 2012. 10 Jan. 2013. <http://www.stat.fi/til/jvie/2011/01/jvie_2011_01_2012-2012-12_tie_001_en.html>.


31. The State of Finland owns 13.7% of TeliaSonera.

# Finland—Data Summaries

ELI NOAM PAUL MUTTER

FOR A SMALL COUNTRY, Finland has a remarkably low media concentration, though not as low as sometimes portrayed. In overall platforms, the top two providers of Elisa and TeliaSonera hold hefty 22.1% and 17.4% shares, with the third place company Finnet holding 7% (Table 4.12). TeliaSonera is also the dominant telecom in neighboring Sweden, and the governments of the two countries both have partial stakes in the entity (Finland's is 13%, while Sweden's is 50%).

YLE, the public broadcaster, holds 41% of TV broadcasting and 55.8% of radio (one of the highest of any country investigated), making it the largest content producer in Finland with 16.9% of the content market (Table 4.13). Sanoma is Finland's most international media group, with operations in Russia, Belgium, Sweden, and the Netherlands. Domestically, it holds major shares in print media: 31% of daily newspapers, 32.5% of magazines, and 25.5% of book publishing, making it the third largest content media company, after YLE and Bonnier. Bonnier, a major Swedish media firm, controls 33% of TV broadcasting and 31.4% of book publishing, making it runner-up to YLE in content.

Finland's overall concentration measures of average weighted and pooled HHI and C4 are thus not low but intermediate in world comparison. But given the country's small population this is an accomplishment. One can glimpse the problem in looking at the number of voices: in absolute numbers, it is actually the smallest in the 30-country world of this book. It is smaller still when we look at net voices and observe that this figure is one-third lower than the total number, suggesting significant cross-ownership. However, the picture changes dramatically when we look at the number of voices on a per capita basis. Now, Finland is near the top in terms of media options that are independent of each other (Table 4.14).

Table 4-12. National Media Industries Concentration in Finland

| | 2004/2005 | | 2011 OR MOST RECENT | | % CHANGE ANNUAL AVERAGE | |
|---|---|---|---|---|---|---|
| | COMPANY POWER INDEX IN COUNTRY | COMPANY SHARE OF THE OVERALL NATIONAL MEDIA MARKET (%) | COMPANY POWER INDEX IN COUNTRY | COMPANY SHARE OF THE OVERALL NATIONAL MEDIA MARKET (%) | COMPANY POWER INDEX IN COUNTRY | COMPANY SHARE OF THE OVERALL NATIONAL MEDIA MARKET (%) |
| Elisa | 665.6 | 20.0 | 836.3 | 22.1 | 4.3 | 0.343 |
| TeliaSonera (public) (Sweden—50%, Finland—13%) | 861.2 | 22.5 | 542.8 | 17.4 | –6.2 | –0.848 |
| YLE (public) | 526.2 | 8.1 | 302.6 | 7.0 | –7.1 | –0.18 |
| Finnet Group | 320.3 | 10.9 | 107.9 | 5.6 | –11.1 | –0.886 |
| Sanoma | 237.5 | 7.3 | 209.9 | 6.8 | –1.9 | –0.082 |
| Bonnier (Sweden) | 139.6 | 4.4 | 185.5 | 5.8 | 5.5 | 0.225 |
| DNA | 46.9 | 3.3 | 230.9 | 11.3 | 65.5 | 1.3 |
| Otava | 58.8 | 2.5 | 63.3 | 2.3 | 1.3 | –0.038 |
| Alma Media | 68.4 | 3.5 | 75.5 | 3.6 | 1.7 | 0.014 |
| Aller Media (Denmark) | 6.4 | 0.7 | 4.2 | 0.5 | –5.8 | –0.03 |
| A-Lehdet | 13.7 | 1.0 | 9.9 | 0.8 | –4.7 | –0.04 |
| Keskisuomalainen | 7.2 | 1.0 | 8.8 | 1.1 | 3.7 | 0.02 |
| TS-Yhtyma | 4.5 | 0.8 | 5.0 | 0.8 | 1.8 | 0.01 |

| | | | | | | |
|---|---|---|---|---|---|---|
| Vivendi (France) | 1.6 | 0.4 | 2.4 | 0.6 | 8.9 | 0.03 |
| ProSiebenSat.1 (SBS, Germany) | 1.8 | 0.2 | 2.4 | 0.2 | 5.3 | -0.005 |

| Media Concentration Index | 2004/5 | 2011 or Most Recent | % Change Annual Average |
|---|---|---|---|
| Total Revenue: Nat'l Media Industry (mil US$) | 7,568 | 9,267 | 3.7 |
| Total Voices (***n***) | 43 | 44 | 0.39 |
| Net Voices (***n***) | 36 | 36 | 0.0 |
| Public Ownership (%) | 11.0 | 9.3 | -0.44 |
| Foreign Ownership (%) | 16.2 | 21.2 | 0.84 |
| C4 Average—Weighted | 88.0 | 89.1 | 0.19 |
| HHI Average—Weighted | 3,033 | 2,793 | -1.32 |
| C1 Average—Weighted | 42 | 38 | -0.01 |
| Noam Index Average—Weighted | 1,241 | 1,308 | 0.89 |
| Pooled Overall Sector C4 | 61.5 | 57.8 | -0.62 |
| Pooled Overall Sector HHI | 1,196 | 1,124 | -1.01 |
| Pooled Overall Sector Noam Index | 182 | 146 | -3.33 |
| Market Share of Top Ten Companies: Nat'l Media Industry (%) (Pooled C10) | 83.6 | 86.5 | 0.5 |
| National Power Index | 3,018.6 | 2,860.8 | -0.87 |

Table 4-13. Top Content Media Companies in Finland

| | 2004/2005 | | 2011 OR MOST RECENT | | % CHANGE ANNUAL AVERAGE | |
|---|---|---|---|---|---|---|
| | COMPANY POWER INDEX IN COUNTRY | COMPANY SHARE OF THE NATIONAL CONTENT MEDIA MARKET (%) | COMPANY POWER INDEX IN COUNTRY | COMPANY SHARE OF THE NATIONAL CONTENT MEDIA MARKET (%) | COMPANY POWER INDEX IN COUNTRY | COMPANY SHARE OF THE NATIONAL CONTENT MEDIA MARKET (%) |
| YLE (public) | 1,377.9 | 21.3 | 724.9 | 16.9 | -7.9 | -0.733 |
| Sanoma | 621.8 | 19.2 | 502.9 | 16.4 | -3.2 | -0.469 |
| Bonnier (Sweden) | 365.6 | 11.6 | 444.3 | 13.8 | 3.6 | 0.376 |
| Otava | 154.0 | 6.6 | 151.7 | 5.5 | -0.3 | 0.185 |
| Alma Media | 179.2 | 9.2 | 181.0 | 8.6 | 0.2 | -0.097 |
| Keskisuomalainen | 18.9 | 2.5 | 21.1 | 2.6 | 1.9 | 0.02 |
| A-Lehdet | 35.8 | 2.7 | 23.6 | 1.9 | -5.7 | -0.13 |
| Aller Media (Denmark) | 16.6 | 1.8 | 9.9 | 1.2 | -6.7 | -0.10 |
| TS-Yhtymä | 11.7 | 1.9 | 11.8 | 1.9 | 0.2 | 0.00 |
| Vivendi (France) | 4.1 | 1.0 | 5.7 | 1.4 | 6.7 | 0.07 |
| Media Concentration Index | 2004/5 | | 2011 or Most Recent | | % Change Annual Average | |
| Public Ownership (%) | 21.3 | | 16.9 | | -1.10 | |
| Foreign Ownership (%) | 20.6 | | 24.2 | | 0.59 | |
| C4 Average—Weighted | 81.9 | | 78.5 | | -0.56 | |
| HHI Average—Weighted | 2,984 | | 2,576 | | -2.3 | |
| C1 Average—Weighted | 44 | | 38 | | -0.01 | |
| National Power Index | 2,945 | | 2,533 | | -2.3 | |

Table 4-14. Top Platform Media Companies in Finland

| | 2004/2005 | | 2011 OR MOST RECENT | | % CHANGE ANNUAL AVERAGE | |
|---|---|---|---|---|---|---|
| | COMPANY POWER INDEX IN COUNTRY | COMPANY SHARE OF THE NATIONAL PLATFORM MEDIA MARKET (%) | COMPANY POWER INDEX IN COUNTRY | COMPANY SHARE OF THE NATIONAL PLATFORM MEDIA MARKET (%) | COMPANY POWER INDEX IN COUNTRY | COMPANY SHARE OF THE NATIONAL PLATFORM MEDIA MARKET (%) |
| Elisa | 1076.7 | 32.4 | 1435.5 | 37.9 | 5.6 | 0.917 |
| TeliaSonera | 1393.2 | 36.3 | 931.6 | 29.8 | –5.5 | –1.1 |
| Finnet Group | 518.1 | 17.6 | 185.3 | 9.6 | –10.7 | –1.3 |
| Vodafone (UK) | 0 | 0.0 | 146.2 | 8.1 | N/A | 1.4 |
| DNA | 75.8 | 5.4 | 396.3 | 19.4 | 70.5 | 2.3 |
| Media Concentration Index | 2004/5 | | 2011 or Most Recent | | % Change Annual Average | |
| Public Ownership (%) | 4.7 | | 3.9 | | –0.21 | |
| Foreign Ownership (%) | 13.4 | | 19.2 | | 0.95 | |
| C4 Average—Weighted | 91.7 | | 96.7 | | 0.82 | |
| HHI Average—Weighted | 3,064 | | 2,949 | | –0.63 | |
| C1 Average—Weighted | 41 | | 38 | | 0 | |
| National Power Index | 3,064 | | 2,949 | | –0.63 | |

# 5

# Media Ownership and Concentration in France

PATRICK-YVES BADILLO, DOMINIQUE BOURGEOIS, AND JEAN-BAPTISTE LESOURD

## INTRODUCTION

This chapter discusses three key issues regarding media ownership and concentration in France. The first issue is the impact of regulation and privatization in the French media market. In France, traditional media such as the national daily press have been heavily regulated since 1945. D'Armagnac (2005) discusses how this arrangement came to be following World War II:

> [Liberation] gave rise to a number of newspapers which were founded by individuals or families, under a social agreement with the union of press workers—the syndicat du livre (CGT)—on printing and distribution activities, under the control of the French state. This system had for a long time remained at a standstill in its afterwar configuration. But the general picture of the French press is now on the move and it has not reached its final shape.[1]

The agreement with the press workers' union, which affected most national dailies, resulted in a closed-shop arrangement, so printing these dailies became rather expensive over the years. The high costs have also resulted from distribution logistics, which under a 1947 law are carried out by one company, Presstalis (formerly the NMPP). Until 2011, it was controlled by press publishing groups and the Lagardère Group (through its subsidiary Hachette S.A.). In 2011, Lagardère withdrew, and two cooperative firms representing magazines (75% stake)

1. D'Armagnac, Bertrand. "Industriels et familles règnent sur la presse française." *Le Monde*, Jan. 22, 2005 (translated by the author).

and daily newspapers (25% stake) gained ownership of Presstalis, which like all print distributors is facing growing financial difficulties in the 21st century. In contrast, other sectors of French media have been deregulated and their overall financial situation is much better than that of most newspapers.

Privatization in audiovisual media began in the 1980s in France. In 1981 the public sector monopoly in radio ended. Then the broadcasting sector was opened up to commercial operators; previously, there were just a few private radio channels and some TV channels operated by French commercial broadcasters, who transmitted from neighboring countries (specifically, Luxembourg, Germany, Monaco, and Andorra). In the television sector Canal Plus was launched in 1984, and by 1987 the privatization of the state-owned television network TF1 was completed. A further step toward deregulation was the liberalization of the telecom industry that was carried out gradually between 1986 and 1998.

A second key issue for French media is technology: liberalization of the audiovisual and telecom sectors took place just as rapid technological advances were happening in the information and communications technology (ICT) field worldwide. Technological advances, especially in digital technology, have been fundamental for the growth in the telecom and Internet industries. Most French newspapers, as well as most audiovisual media, now have online editions. Some of them have already registered large audience shares; for example, in January 2012, *Le Monde*'s web portal (lemonde.fr) counted 231,000 unique visitors per day and over 21 million unique page views per month.[2] There are now several expanding online titles, including *Rue 89* and *Mediapart,* and the French-language version of AOL's (US) *The Huffington Post.* The present trend among French newspapers and magazines is to publish mostly free online editions, while charging for some specific content.

A third key issue is that of globalization.[3] Before the start of today's globalization wave in the 1980s, the presence of international media groups in France was limited, with the exception of the film industry because Hollywood majors from the United States have had a strong presence in France since the early 1920s. But now, in other media sectors, international media companies have come to hold significant market shares. Conversely, some large French media and communications companies like the Lagardère Group, Vivendi, and France Telecom have developed their own worldwide media activities.

## PRINT MEDIA

### Newspapers

The *Belle Époque,* the period that lasted from the late 19th century to the end of World War I, was a golden age for the French press.[4] But after World War I (1914–1918), national readership fell into decline, a trend that continues to this day: in 1914, there were 322 dailies in France; this had fallen to 179 by 1945, at the end of World War II. By 2011, France counted 66 dailies. The number of *national* dailies stood at 26 in 1945 but by 2011 had fallen to 11. Circulation has also decreased dramatically: in 1914, one of leading national dailies, *Le Petit Parisien,* had a daily circulation of about 3 million copies, while in 1946, even as the country was recovering from World War II, there were 28 national dailies that had a total circulation of about six million copies. In comparison, the two main national dailies publishing today, *Le Monde* and *Le Figaro,* have a combined daily circulation of

2. Alexa, The Web Information Company. "lemonde.fr." 2012. <http://www.alexa.com/siteinfo/lemonde.fr#>. Dec. 31, 2012.

3. For an analysis of globalization, see Badillo, Patrick-Yves, Proulx, Serge. "Mondialisation de la Communication: à la recherche du sens perdu," in *Les essentiels d'Hermès "Sociétés de la Connaissance - Fractures et évolutions."* Paris: CNRS Éditions, 2009, 117–134.

4. For a historical analysis since about 1820, from an economic point of view, see Eveno Patrick. *L'argent de la presse française des années 1820 à nos jours.* Paris: CTHS, 2003; for a more recent analysis, see Charon, Jean-Marie. *La presse quotidienne.* Paris: La Découverte, 2004 and Albert, Pierre. *La presse française.* Paris: La Documentation Française, 2008.

approximately 600,000 copies. The total newspaper market revenues (sales and publicity) stood at US$1.228 billion (1.056 billion euros) in 2008.[5] Our analysis of the French national daily press is based on average daily circulation. We restricted ourselves to the daily national press, with 12 titles, leaving out the 44 regional titles (Table 5.1).[6]

The HHI calculation decreased slightly between 1997 and 2011, falling from 2,366

## Table 5-1. Daily Newspapers (Market Shares by Circulation), 1997–2011

| GROUPS/TITLES | 1997 | 2004 | 2006 | 2008 | 2010 | 2011 |
|---|---|---|---|---|---|---|
| Amaury Group (total) | 31.5 | 38.8 | 39.6 | 39.4 | 37.9 | 37.8 |
| *Aujourd'hui — Le Parisien* | *17.3* | *22.4* | *23.2* | *24.3* | *22.7* | *23.2* |
| *L'Equipe* | *14.2* | *16.4* | *16.4* | *15.0* | *15.2* | *14.6* |
| Socpresse Group (total) | 32.8 | 18.7 | 15.2 | 15.6 | 16.0 | 16.4 |
| *Paris-Turf*[1] | *13.0* | *3.5* | | | | |
| *France-Soir*[2] | *6.2* | | | | | |
| *Le Figaro* | *13.6* | *15.3* | *15.2* | *15.6* | *16.0* | *16.4* |
| Paris Turf | Socpresse | Socpresse | 3.6 | 3.5 | 3.0 | 2.9 |
| France-Soir[3] | Socpresse | 2.8 | 1.2 | 1.1 | 3.7 | 3.7 |
| *Le Monde* (La Vie-Le Monde Group) | 14.3 | 16.6 | 15.9 | 15.8 | 15.4 | 15.0 |
| *La Croix* (Bayard Presse) | 3.5 | 4.3 | 4.6 | 4.8 | 5.1 | 4.8 |
| *Les Echos* (LVMH Group) | 4.9 | 5.3 | 6.2 | 6.4 | 5.8 | 6.1 |
| *La Tribune*[4] | 3.3 | 3.6 | 4.1 | 3.8 | 3.8 | 3.8 |
| *Libération* | 6.3 | 6.5 | 6.1 | 6.1 | 5.7 | 6.1 |
| *L'Humanité* | 2.2 | 2.2 | 2.5 | 2.4 | 2.5 | 2.5 |
| *International Herald Tribune* (New York Times Group, US) | 1.2 | 1.1 | 1.1 | 1.1 | 1.0 | 0.9 |
| Total Circulation (1,000) | 2,739 | 2,236 | 2,233 | 2,155 | 2,070 | 1,956 |
| Total Revenue (mil euros) | | | | 1,056 | | |
| Total Revenue (mil US$) | | | | 1,228 | | |
| C4 | 84.9 | 80.7 | 76.8 | 77.2 | 75.1 | 75.3 |
| HHI | 2,366 | 2,246 | 2,187 | 2,181 | 2,066 | 2,064 |
| *N* (>1%) | 9 | 10 | 11 | 11 | 11 | 9 |
| Noam Index | 789 | 710 | 660 | 658 | 623 | 688 |

*Source*: Our calculations based on data from OJD. <http://www.ojd.com> (3.22.2013 and past years); DGMIC (French Ministry of Culture). <http://www.culturecommunication.gouv.fr/> for total revenue; see also Badillo and Lesourd (2010).

[1] Sold off by the Socpresse Group in 2005.

[2] Sold off by the Socpresse Group in 1999.

[3] *France-Soir* ceased publication as a national print daily in 2012.

[4] *La Tribune* ceased publication as a national print daily in 2012. It was owned by the French investors Alain Weill (20%) and Valérie Decamp (80%) up until then.

5. Source: DGMIC, French Ministry of Culture: DGMIC (Direction Générale des Médias et des Industries Culturelles), Ministère de la Culture et de la Communication. <http://www.culturecommunication.gouv.fr/>. April 2, 2013.

6. The 12 titles include the *International Herald Tribune*, which is published in France and distributed internationally outside of the United States. The data here take into account only the *Tribune*'s paid-subscription rate in France.

in 1997 to 2,064 in 2011. This is still a rather strong concentration. These figures should be seen in the context of the ongoing financial problems French newspapers face. Ownership has not changed significantly—aside from Socpresse's sale of *France-Soir* and *Paris-Turf* in 1999 and 2005, respectively, but the circulation of several titles markedly decreased over the 1997–2008 period, leading to the sale of some of them. For instance, circulation for *France-Soir* fell from 170,499 in 1997 to 71,290 in 2011, and from 355,017 in 1997 to 55,846 in 2011 for *Paris-Turf*. The circulation of *Libération* also declined, from 173,090 in 1997 to 119,205 in 2011. Both *France-Soir* and *La Tribune* ceased publication in 2012 as national print dailies, essentially because they were unprofitable as hardcopy.

## Book Publishing

Concentration indices for book publishing in France since 2005 indicate that the sector is concentrated; the total market shares of the four largest companies remained close to 62% between 2005 and 2011. While for many years the sector consisted of national publishers that were either small or medium-sized enterprises, book publishing is now becoming consolidated into larger publishing houses that have a bigger international presence. Since 2004, Hachette Livre, a branch of the Lagardère Group, has dominated the sector. Hachette Livre is by far the largest French publisher by volume of sales. It is followed by Editis, a fully owned subsidiary of Planeta, the largest book publisher in Spain (Table 5.2).

### Table 5-2. Book Publishing (Market Shares by Revenue), 2005–2011

| | 2005 | 2009 | 2011 |
|---|---|---|---|
| Lagardère Group (Hachette Livre, France) | 32.02 | 38.23 | 36.66 |
| Planeta (Editis, Spain) | 15.56 | 12.63 | 12.70 |
| Lefebvre Sarrut (Editions Lefèvre-Sarrut) | 4.26 | 5.28 | 6.46 |
| Bertelsmann (France Loisirs, Germany) | 7.55 | 6.22 | 6.30 |
| Média Participations (Montagne Family) | 5.82 | 5.37 | 6.12 |
| RCS Media Group (Flammarion Group, Italy) | 4.26 | 4.42 | 4.98 |
| Reed-Elsevier France (UK, Netherlands) | 4.28 | 4.76 | 4.66 |
| La Martinière Group (La Martinière Family) | 4.82 | 4.37 | 4.64 |
| Gallimard Group (Gallimard Family) | 4.77 | 4.09 | 4.55 |
| Panini France (Panini, Italy) | 3.47 | 3.99 | 3.56 |
| De Agostini Group (Editions Atlas, Italy) | 7.32 | 2.94 | 3.14 |
| Wolters Kluwer (Netherlands) | 2.78 | 3.13 | 3.05 |
| Groupe Albin Michel | 2.91 | 2.76 | 1.29 |
| Actes Sud (Nyssen Family, France) | 0.49 | 1.03 | 1.04 |
| Eyrolles (Eyrolles, France) | 0.69 | 0.77 | 0.83 |
| Total Revenue (mil euros) | 5,328 | 5,945 | 5,558 |
| Total Revenue (mil US$) | 6,557 | 8,564 | 7,192 |
| C4 | 61.4 | 62.5 | 62.1 |
| HHI | 1,511 | 2,002 | 1,748 |
| *N* (>1%) | 14 | 15 | 15 |
| Noam Index | 404 | 475 | 451 |

*Source:* Our calculations based on data from <http://www.francelivre.org/Ressources/Le-livre-en-France/Chiffres-cles>.

Book publishers in France fall into several categories. First, there are the large international groups. These include Hachette Livre, valued at US$3.2 billion (2.3 billion euros) in 2011, with a market share of 36.66%. International majors like Planeta (Spain) and Bertelsmann (Germany) compose a second category. Finally, there are a few smaller independent groups from France or Italy that are often family owned.

For the largest publishers, we calculated concentration indices in 2005, 2009, and 2011. The results indicate moderate concentration, which has persisted over that period.

## Magazine Publishing

While newspapers are facing financial problems and declining readership, other print products are steadily growing. This is particularly the case for magazines. There were 754 titles in 1985, increasing to 2,019 in 2010. Trade publications and academic journals constitute a fast-growing magazine market segment, with the number of titles increasing from 1,109 to 1,364 during the same period.

The French magazine industry is highly heterogeneous, and comparable data such as sales or turnover were unavailable; furthermore, there are a number of companies involved in several media, and not only in magazines, thus making it difficult to estimate revenues in magazines only. The figures given in Table 5.3 are, in most cases, only rough estimates. Therefore, concentration indicators calculated from these data are also likely to be only orders of magnitude.

The Lagardère Group is the largest magazine publisher in France, with sales in magazines estimated at US$2.42 billion (1.8 billion euros) in 2011. Most French press groups are much smaller

### Table 5-3. Magazine Publishing (Market Shares by Revenue), 2004–2011

| | 2004 | 2011 | CLASSIFICATION |
|---|---|---|---|
| Lagardère Group (*Paris-Match, Elle*) | 35.90 | 38.61 | G |
| Roularta France (Belgium) (*L'Express, L'Expansion*) | 14.76 | 15.38 | G |
| Prisma (Bertelsmann, Germany) (*Capital, Management, Femme Actuelle, Prima, GEO, Télé Loisirs, Télé 2 Semaines*) | 11.34 | 10.56 | G |
| Mondadori France (Berlusconi Family, Italy) (*Télé poche, Télé star, Auto plus, Auto journal*) | 8.49 | 8.19 | G |
| Bayard Presse (Assumptionists) (*Côté Femme, Notre Temps, Pomme d'Api, Okapi*) | 8.41 | 8.01 | I |
| La Vie-Le Monde (Pierre Bergé, Xavier Niel, Mathieu Pigasse) (*La Vie-Le Monde Diplomatique*) | 12.63 | 7.46 | R |
| Marie Claire (Prouvost Family) (*Marie Claire*) | 2.91 | 5.03 | F |
| Le Nouvel Observateur (*Le Nouvel Observateur, Challenges*) | 2.91 | 3.18 | F |
| Le Point-Artemis (Pinault) (*Le Point*) | 1.20 | 1.75 | G |
| Bauer Editions (Bauer Group, Germany) | 1.17 | 1.27 | F |
| Valmonde (Sud Communication—Pierre Fabre) (*Valeurs Actuelles*) | 0.27 | 0.25 | G |
| Total Revenue (mil euros) | 5,148 | 4,713 | |
| Total Revenue (mil US$) | | | |
| C4 | 74.6 | 72.7 | |
| HHI | 1,957 | 2,070 | |
| *N* (>1%) | 9 | 9 | |
| Noam Index | 652 | 690 | |

*Source*: Our calculations based on data from OJD. <http://www.ojd.com> (3.22.2013 and past years); see also Badillo and Lesourd, op. cit.

than their US and EU counterparts. With a global turnover of US$10.7 billion (7.7 billion euros) in 2011, Lagardère is the only French multimedia group with international subsidiaries. The global turnover of La Vie-Le Monde Group, another French publishing house involved in magazines, stood at just US$863 million (366 million euros) in 2011. The corresponding figures for Pearson (UK) and Bertelsmann (Germany) for 2011 were US$9 billion (6.5 billion euros) and US$21.3 billion (15.3 billion euros), respectively. We can classify the ownership of firms in Table 5.3 into main categories G, F, I, and R:

- G: We find some firms that belong to a large national or international company, which are often publicly traded international groups with activities in the media industry and in other industries as well (the *Grand* category).
- F-I: We find other smaller firms, often private, that are family-owned companies (the *Family* category) or that are private companies owned by independent investors such as not-for-profit foundations and religious organizations (the *Independent* category); for instance, Bayard Presse, which is owned by a religious organization. These two types would be difficult to take over, at least as long as they remain profitable.
- R: We can also distinguish a group whose members had previously been under independent ownership but, after experiencing significant financial difficulties, are classified as "recovering" (the *Recovering* category).

## AUDIOVISUAL MEDIA

### Radio

Radio and TV broadcasting in France were state monopolies between 1945 and 1982. The state company RTF—Radiodiffusion-Télévision de France, later the Office de Radiodiffusion-Télévision de France (ORTF)—operated public service radio and TV channels. However, with respect to both radio and broadcast TV, the public monopoly existed only in principle. Four national commercial radio broadcasters did exist before 1982, despite the public monopoly, operating from nearby extraterritorial jurisdictions. These were *Radio-Luxembourg*, owned by Radio-Télé Luxembourg (RTL) in Luxembourg; *Europe 1*, which broadcast from Germany and is now owned by the Lagardère Group; *Radio Monte-Carlo* (RMC), which broadcast out of the Principality of Monaco; and *Radio-Andorre* in Andorra. Such competition did not exist in the national television market. Additional commercial radio broadcasters soon appeared after the sector was privatized, including many local and community radio stations.

Radio France (public) is the leading player in the market. Broadcasting *France Inter, France Info, France Bleu, France Musique,* and *France Culture,* its 2008 total revenue stood at US$848 million (577 million euros). Next comes the RTL Group of Bertelsmann AG (Germany), which has a 91.6%, stake in *RTL, Fun Radio, and RTL2.* The *NRJ Group,* in which the French businessman Jean-Paul Baudecroux has an 84.9% stake, owns *Chérie FM, Nostalgie, NRJ,* and *Rire et chansons.* Lagardère Active—which is run by a consortium of Lagardère Capital et Management (14%), French private investors (17.5%), foreign investors (56.3%), and other miscellaneous investors—owns *Europe 1, RFM,* and *Virgin Radio. Next Radio TV* (RMC) is owned by Alain Blane Brude (4.4%), WMC-Alain Weill (a 36.3% stake in which Alain Weill retains 49.8% of voting rights), staff (2%), self-detention of shares (0.6%), and outstanding investors (56.7%).

Four of these groups are still the pre-1982 historical radio groups that operated outside of France; Radio France now represents only 22.1% of the national audience share, and the NRJ Group, which only began broadcasting in 1981, is now a major player.

The concentration indices for radio are calculated on the basis of the total audience by group, based on audiences of the 19 largest radio channels (data are not available for smaller channels, which means the data slightly underestimate total audiences for some groups, such as Radio France, whose data do not include figures for its *Le Mouv'* music channel), based on the listeners average for the November-December period each year. Concentration is rather low and has been stable from 2005 to 2011 (Table 5.4).

Table 5-4. Radio Groups (Market Shares by Audience %), 2001–2011

| YEARS | 2001 | 2005 | 2007 | 2009 | 2011 |
|---|---|---|---|---|---|
| Radio France (public) | 21.7 | 20.6 | 20.2 | 21.6 | 21.9 |
| *France Inter* | *10.0* | *8.8* | *8.7* | *9.6* | *10.0* |
| *France Bleu* | *5.4* | *5.4* | *5.5* | *6.3* | *6.2* |
| *France Info* | *5.4* | *4.1* | *4.0* | *3.7* | *3.7* |
| *France Culture* | | *0.8* | *1.0* | *1.1* | *1.1* |
| *France Musique* | *0.9* | *0.9* | *1.0* | *0.9* | *0.9* |
| RTL Group (Bertelsmann, Germany) | 20.0 | 18.1 | 19.5 | 19.2 | 19.2 |
| *RTL* | *13.3* | *11.5* | *13.1* | *12.4* | *11.9* |
| *Fun Radio* | *3.8* | *2.8* | *3.5* | *4.1* | *3.9* |
| *RTL 2* | *2.9* | *3.8* | *2.9* | *2.7* | *2.7* |
| NRJ Group | 22.0 | 18.4 | 16.9 | 15.3 | 14.3 |
| *NRJ* | *7.0* | *7.0* | *6.6* | *5.3* | *6.3* |
| *Nostalgie* | *6.2* | *5.7* | *5.6* | *5.0* | *4.1* |
| *Chérie FM* | *5.2* | *4.2* | *3.4* | *3.2* | *2.3* |
| *Rire et chansons* | *3.6* | *1.5* | *1.3* | *1.8* | *1.6* |
| Lagardère Active | 14.8 | 14.7 | 13.7 | 14.4 | 12.5 |
| *Europe 1* | *8.6* | *8.0* | *7.5* | *8.6* | *7.6* |
| *RFM* | *2.5* | *3.7* | *3.3* | *2.8* | *2.8* |
| *Europe 2/ Virgin Radio* | *3.7* | *3.0* | *2.9* | *3.0* | *2.1* |
| Orbus | 4.5 | 3.9 | 4.1 | 4.5 | 4.3 |
| *Skyrock* | *4.5* | *3.9* | *4.1* | *4.5* | *4.3* |
| Next Radio TV | 1.7 | 3.7 | 4.9 | 5.3 | 6.6 |
| *RMC* | *1.7* | *3.7* | *4.9* | *5.3* | *6.6* |
| Groupe Les Echos | 1.0 | 1.1 | 1.2 | 1.6 | 1.8 |
| *Radio Classique* | *1.0* | *1.1* | *1.2* | *1.6* | *1.8* |
| Espace Group | 1.3 | 1.3 | 1.3 | 0.8 | 1.5 |
| *MFM* | *1.3* | *1.3* | *1.3* | *0.8* | *1.5* |
| Others (Community and local radios . . . ) | 13.0 | 18.2 | 19.0 | 17.3 | 17.9 |
| Total Publicity Revenues (million €) | NA | 3,140 | 3,310 | 3,682 | 4,152 |
| Total Publicity Revenue (million US$) | NA | 3,704 | 4,873 | 5,304 | 5,352 |
| C4 | 78.5 | 71.8 | 68.9 | 70.5 | 67.9 |
| HHI | 1,600 | 1,338 | 1,260 | 1,328 | 1,277 |
| *N* (>1%) | 7 | 8 | 8 | 7 | 8 |
| Noam Index | 605 | 473 | 445 | 502 | 451 |

*Source*: <http://www.mediametrie.fr/radio/>.

## Broadcast Television

The French TV market changed dramatically in the 1980s, with the entry of six broadcast TV channels, including four private channels, due to the privatization of the sector. Though France Télévisions remains the dominant player in this market, it has lost market shares to these new commercial broadcasters. Terrestrial broadcasting is now competing with new distribution technologies such as cable TV (since 1982), satellite TV (since the 1980s), and, from 2000 on, digital terrestrial television (DTT). C4 is almost always equal to 100%, since there have only been four broadcasters present in the sector since 1992, including the state-owned, public-service channels (there were five prior to 1992). HHI is comparatively high and has remained almost constant since 1995 (Table 5.5).

Concentration, however, becomes quite different if we also take into account other television platforms.

## Cable Providers

The first cable networks developed in the 1970s. In 1982 the Plan Câble[7] was launched, with the goal of ensuring the spread of cable. In the first stage, a regulatory ceiling on cable operators to serve no more than 8 million homes led to a multiplicity of operators. Initially, four companies provided service: TDF Cable, which operated since the 1970s, was renamed France Telecom Cable; Lyonnaise Communications; the water utility Générale des Eaux (Compagnie Générale de Vidéocommunication); and CDC (Caisse des Dépôts et Consignations). In 1993, CDC withdrew from cable and sold its networks to France Telecom Cable.

By the end of 1995 there were four main operators:

- France Telecom Cable
- NC Numericable (previously Compagnie Générale de Vidéocommunication) part of Vivendi, and its Canal +
- UPC France, created in 1995, was a subsidiary of United Pan-Europe Communication, a European branch of the American cable and satellite firm, Liberty Media Group
- Lyonnaise Cable became Noos in 2000, and bought networks from France Telecom Cable

### Table 5-5. TV Broadcasting (Market Shares by Audience %), 1995–2011

| GROUPS/CHANNELS | 1995 | 2000 | 2005 | 2007 | 2009 | 2011 |
|---|---|---|---|---|---|---|
| France Télévisions (public) | 43.9 | 42.3 | 39.4 | 37.3 | 33.3 | 29.4 |
| *France 2* | *23.8* | *22.1* | *19.8* | *18.1* | *16.7* | *14.9* |
| *France 3* | *17.6* | *16.8* | *14.7* | *14.1* | *11.8* | *9.7* |
| *France 5* | *1.3* | *1.8* | *3.1* | *3.3* | *3.1* | *3.3* |
| *Arte* | *1.2* | *1.6* | *1.8* | *1.8* | *1.7* | *1.5* |
| Bouygues Group (TF1) | 37.3 | 33.4 | 32.3 | 30.7 | 26.1 | 23.7 |
| Bertelsmann (Germany, M6 Group) | 11.5 | 12.7 | 12.6 | 11.5 | 10.8 | 10.8 |
| Vivendi | 4.4 | 4.1 | 3.6 | 3.4 | 1.7 | 3.3 |
| Others[1] | 2.9 | 7.5 | 12.1 | 17.1 | 28.1 | 32.8 |
| Revenues[2] | No data | 2,865 | 3,094 | 3,257 | 2,571 | 2,762 |
| Revenues[3] | No data | 2,665 | 3,650 | 4,795 | 3,705 | 3,574 |
| C4 | 100.0 | 100.0 | 100.0 | 100.0 | 100.0 | 100.0 |
| HHI | 3,680 | 3,603 | 3,581 | 3,605 | 3,694 | 3,440 |
| *N* (Broadcasting) | 4 | 4 | 4 | 4 | 4 | 4 |
| Noam Index | 1,840 | 1,802 | 1,790 | 1,802.5 | 1,847 | 1,720 |

*Source*: <http://www.mediametrie.fr/television> and <http://www.cnc.fr/web/fr/publications>.

[1] Others include TV channels other than broadcasting channels (cable TV, DTT . . . ).

[2] Advertisement revenues, exclusive of other revenues (pay-TV . . . ), unit: million €.

[3] Advertisement revenues, exclusive of other revenues (pay-TV . . . ), unit: million US $.

7. For a historical perspective about cable, satellite, and telecommunications in France, see Badillo, Patrick-Yves, and Roux, Dominique, eds. *Les 100 mots des télécommunications*. Paris: Presses Universitaires de France, 2009.

In a second stage, new rules were set in 2003, making concentration in the sector possible.

- In 2004, Cinven (a British company that also owns parts of Eutelsat and Amadeus), together with Altice, a Luxembourg-registered cable operator, created the holding company Ypso and bought the cable operations of France Telecom and of NC Numericable.
- UPC Broadband France (Liberty) acquired Noos in 2004.
- In 2006, Liberty Global sold UPC Broadband France, the leader in French cable, for US$1.51 billion to Ypso.
- By 2007/8 Numericable-Completel dominated the entire French cable sector: it included all previous networks—France Telecom Cable, NC Numericable, TDF Câble, UPC, and Noos.

In 2014, Altice bought out Carlyle and Cinven's share in Numericable, raising its stake to a controlling 74.6%. Altice is controlled by Patrick Drahi, son of Moroccan immigrants, and also owns cable operators in Belgium, Israel, and Portugal. In 2014, Altice also acquired telecom operator SFR from Vivendi, subject to approval. In 2009, the turnover of Numericable-Completel was 1.3 billion euros ($1.87 billion). It served 9.4 million households in France (99.6% of cable homes in France), 3.5 million television subscriptions, 1.1 million Internet subscriptions, and 862,000 fixed telephony subscriptions (Table 5.6).

## Satellite TV

French satellite TV became a monopolistic market dominated by Vivendi's Canalsat. There were two operators in the satellite TV industry until 2006: TPS and Canal Satellite (Canalsat), a subsidiary of Canal Plus (Vivendi). TPS was launched in 1996. Initially, the owners were the well-established TV and telecom groups TF1, M6, France Telecom, France Television, CLT, and Lyonnaise des Eaux. By 2002, TPS was owned by the two TV networks TF1 (66%) and M6 (34%). In 2006 the French Minister of Economy, Finance, and Industry approved the acquisition of TPS and Canalsat by Vivendi's Canal+ group.

### Table 5-6. Cable TV Operators (Market Shares by Subscribers), 1998–2012

| CABLE % | YEARS | | | | |
|---|---|---|---|---|---|
| COMPANIES (GROUPS) | 1998 | 2000 | 2004 | 2008 | 2012 |
| Altice/Numericable[1] | 26.0 | 21.5 | 23.3 | ~100 | ~100 |
| Noos (Lyonnaise) | 29.6 | 31.7 | 48.0 | Numericable-Completel | |
| France Telecom Cable (France Telecom, about 27% owned by French State, directly or indirectly) | 37.7 | 25.9 | 24.3 | Numericable-Completel | |
| UPC France (Liberty, USA) | 0.8 | 11.3 | 4.4 | Numericable-Completel | |
| Est Vidéocommunication | 0.0 | 4.2 | Numericable-Completel | | |
| Others | 5.9 | 5.5 | 0.3 | | |
| Total Revenue (million €)* | 416 | 634 | 800 E | 896 | 874 |
| Total Revenue (million US$) | 487 | 586 | 995 | 1,318 | 1,124 |
| HHI | 3,009 | 2,311 | 3,454 | 10,000 | 10,000 |
| Noam | 1,229 | 943 | 1,410 | 10,000 | 10,000 |

*Source*: *ART, ARCEP, and CSA (E estimation based on various documents); data to be interpreted with caution.

[1] Numericable was controlled by the consortium Ypso, owned by Cinven, UK 35%, Carlyle, USA 35%, and Altice, headquartered in Luxembourg 30%. In 2014, Altice acquired the interests of its partners.

A "new" Canalsat was thus created. It was owned by Canal+ France, which was itself owned by the Canal+ Group (65%), Lagardère (20%), TF1 (9.9%), and M6 (5.1%). In 2009, TF1 sold its stake to Canal+ Group, and in 2010 and 2014, M6 and Lagardere did the same. Canal+ Group is a subsidiary of Vivendi and is the leading provider of pay TV in France in 2012. The Autorité de la Concurrence cleared the merger of TPS and Group Canal Plus pay TV services subject to compliance with 33 injunctions. This decision was confirmed by the Conseil d'Etat.

There are two competing satellite platforms: Bis, launched by the AB group in 2007, and Orange, operated by France Telecom (service only available to subscribers whose telephone lines do not permit TV reception via ADSL); but these platforms have achieved only a tiny market share (Tables 5.7 and 5.8).

## Digital Terrestrial Television

In 2010, TF1 bought TMC and NT1, and in 2011, Direct8 and Direct Star were sold by the Bolloré Group to Vivendi's Canal+ Group (Table 5.9).

### Table 5-7. DBS TV (Market Shares by Subscribers), 1996–2011

| | 1996 | 1998 | 2000 | 2004 | 2008 | 2011 |
|---|---|---|---|---|---|---|
| Vivendi (Canalsat) | 100 | 63 | 61 | 64 | 100 | 100 |
| TPS | | 35 | 38 | 36 | | |
| Total Revenue (mil euros) | | 524* | 986* | 1,400 E | 1,600 E | 1,750 E |
| Total Revenue (mil US$) | | 614 | 912 | 1,741 | 2,354 | 2,250 |
| HHI | 10,000 | 5,194 | 5,165 | 5,392 | 10,000 | 10,000 |

*Source*: *ART report; E: authors' estimation based on various documents (to be interpreted with caution).

### Table 5-8. Multichannel TV Platforms: Cable and DBS TV Providers, 1998–2011

| | 1998 | 2000 | 2011 |
|---|---|---|---|
| Altice/Numericable-Completel | 12 | 8 | 34 |
| Noos (Lyonnaise) | 13 | 12 | Numericable |
| France Telecom Cable (France Telecom) | 17 | 10 | Numericable |
| UPC France (Liberty Global, US) | | 4 | Numericable |
| Est Vidéocommunication | | 2 | Numericable |
| Vivendi (Canalsat) | 36 | 37 | 61 |
| *TPS* | 20 | 23 | Vivendi |
| Others | 3 | 3 | 5 |
| Total Revenue (mil euros) | 940 | 1,620 | 2,624 |
| Total Revenue (mil US$) | 1,101 | 1,498 | 3,374 |
| C4 | 85 | 82 | 100 |
| HHI | 2,255 | 2,272 | 4,884 |
| *N* (>1%) | 4 | 6 | 2 |
| Noam Index | 1,128 | 928 | 3,454 |

*Source*: Aggregation of Tables 5.6 and 5.7.

Table 5-9. Digital Video Channels (Market Shares by Audience), 2007–2011[1]

| | 2007 | 2008 | 2009 | 2010 | 2011 |
|---|---|---|---|---|---|
| Free Broadcast TV Channels | 82.9 | 77.0 | 73.3 | 69.6 | 67.0 |
| Vivendi (Canal +) | 3.3 | 3.3 | 3.3 | 3.3 | 3.3 |
| Bolloré Group | 0.6 | 1.2 | 2.1 | 3.0 | 3.5[1] |
| Direct 8 | *0.2* | *0.7* | *1.4* | *2.0* | 2.3 |
| Direct Star | *0.4* | *0.5* | *0.7* | *1.0* | *1.2* |
| Bertelsmann (W9) | 0.9 | 1.8 | 2.5 | 3.0 | 3.4 |
| Bouygues (TF1) | 1.8 | 3.1 | | 4.9 | |
| TMC | *1.2* | *2.1* | *2.6* | *3.3* | *3.5* |
| NT1 | *0.6* | *1.0* | *1.4* | *1.6* | *1.9* |
| NRJ (NRJ12) | 0.4 | 1.0 | 1.5 | 1.9 | 2.3 |
| France Télévisions (France 4) | 0.4 | 0.9 | 1.1 | 1.6 | 2.0 |
| BFMTV (Next Radio TV) | 0.2 | 0.4 | 0.7 | 0.9 | 1.4 |
| i>télé | 0.3 | 0.3 | 0.5 | 0.7 | 0.8 |
| Gulli (Lagardere) | 0.8 | 1.5 | 1.8 | 2.2 | 2.1 |
| Others | 8.4 | 9.5 | 9.2 | 8.9 | 8.8 |
| Total | 100.0 | 100.0 | 100.0 | 100.0 | 100.0 |

*Source*: Our calculations based on data from the CSA *(Guide des chaines thématiques)*: http://www.csa.fr

[1] Vivendi acquired Bolloré's channels in 2012.

## Film

France has historically been the home of major international film production and distribution companies. Pathé Brothers, founded in 1896, soon became a world leader in movie production and distribution. Gaumont et Cie, founded by Léon Gaumont in 1895, also became a global provider. Hollywood majors, however, quickly came to dominate the market in the 1920s: 20th Century Fox, Paramount, Metro-Goldwyn Mayer, RKO, Warner Brothers, Columbia, Universal, and then Disney. Union Générale Cinématographique (UGC), founded in 1970, is a comparatively recent actor in the French film industry. In 2011, the French film sector was dominated by ten large companies, including three French companies: Gaumont, Pathé Brothers, and UGC (Vivendi, Pathé, and Studiocanal, respectively). The Société Nouvelle de Distribution (SND), a subsidiary of the M6 television company founded in 1987, is another French major active today in film distribution. The other majors are mostly subsidiaries of Hollywood giants, and Gaumont has various partnerships. There are also a number of smaller independent French and pan-European film production and distribution companies, such as EuropaCorp and Mars Films.

Due to the lack of data, we have been unable to calculate concentration indices for film production, but we were able to approximate concentration indices for film distribution (Table 5.10).

HHI, as calculated approximately for the top ten film distributers, was 903 in 2000, 880 in 2004, 638 in 2009, and 637 in 2011. The same trend is apparent if we take into account C4, which was 52.1% in 2000, 48.2% in 2004, 42.1% in 2009, and 40.6% in 2011. According to the data, the French film distribution industry is apparently becoming increasingly competitive. This is true in the long term, because in the short

Table 5-10. Film Distribution (Market Shares by Box Office %), 2000–2011

| | 2000 | 2001 | 2002 | 2003 | 2004 | 2005 | 2006 | 2007 | 2008 | 2009 | 2010 | 2011 |
|---|---|---|---|---|---|---|---|---|---|---|---|---|
| Paramount Pictures (Viacom. US) | UIP | | | | | | | 10.2 | 10.0 | 10.0 | 9.4 | 10.9 |
| United International Pictures (UIP)—Universal (Comcast/GE, US) | 12.5 | 13.5 | 6.0 | 11.8 | 11.3 | 12.1 | 7.9 | | | | 6.6 | 7.9 |
| Bac Films | 10.7 | 9.9 | 5.6 | | | | | | | | | |
| Gaumont—BVI | 19.7 | 12.1 | 10.0 | 15.4 | | | | | | | | |
| Vivendi (Pathé, Studiocanal) | 9.2 | 4.5 | 12.2 | 4.3 | 11.1 | 13.5 | 18.0 | 10.6 | 26.8 | 12.8 | 26.8 | 12.8 |
| Sony (Japan)[1] | | | | | | | | 6.5 | 5.6 | 7.7 | 3.9 | 6.5 |
| Gaumont[2] | | | | | 10.4 | 7.6 | 12.5 | | | | | 10.5 |
| Columbia-Tristar (US) | 7.3 | 3.8 | 10.4 | 6.8 | | | | | | | | |
| ARP | 7.0 | | | 5.4 | | | | | | | | |
| Time Warner (Warner Bros., US) | 6.8 | 12.5 | 13.6 | 11.7 | 15.0 | 12.3 | 10.3 | 11.1 | 7.1 | 9.4 | 7.1 | 10.7 |
| Metropolitan FilmExport | 2.7 | 7.4 | 8.7 | 7.5 | 4.1 | 4.4 | 4.7 | 5.9 | 4.8 | 4.9 | 5.0 | 4.2 |
| Pyramide | 2.0 | | | | | | | | | | | |
| Mars Distribution | | 4.9 | 3.6 | 4.5 | 9.4 | | | | | 5.0 | | 5.6 |
| EuropaCorp | | 3.7 | | | 2.7 | | 5.0 | | 2.9 | | 7.4 | |
| Bac Distribution | | | 4.8 | | | | | | | | | |
| TMF Distribution | | | | 3.8 | 3.9 | | | | | | | |
| Century Fox (News Corp, Murdoch Family, US/UK/AUS) | | | | | | 8.3 | 11.1 | 9.2 | 5.0 | 12.2 | 9.7 | 7.8 |
| TF1 Group | | | | | | 8.3 | 5.1 | 6.3 | | 5.7 | | |
| UFD (UGC Fox Distribution) | 9.2 | 14.5 | 14.2 | 8.8 | 7.2 | | | | | | | |
| SND (M6 Group) | | | | | | 3.8 | | | 4.1 | 6.3 | | 4.5 |
| Pan Européenne | | | | | | | | 4.4 | | | | |
| Walt Disney (Buena Vista, US) | | | | | 10.8 | 8.6 | 8.7 | 11.0 | 7.0 | 7.5 | 10.3 | 7.0 |

*continued*

Table 5-10. *continued*

| | 2000 | 2001 | 2002 | 2003 | 2004 | 2005 | 2006 | 2007 | 2008 | 2009 | 2010 | 2011 |
|---|---|---|---|---|---|---|---|---|---|---|---|---|
| Others | 22.1 | 13.2 | 10.9 | 20.0 | 14.1 | 21.1 | 16.7 | 35.0 | 36.7 | 28.9 | 23.9 | 23.8 |
| Total Revenue (million euros) | 367 | 422 | 424 | 407 | 468 | 411 | 452 | 429 | 434 | 469 | 495 | 521 |
| Total Revenue (million US$) | 342 | 372 | 445 | 515 | 637 | 485 | 595 | 631 | 604 | 675 | 662 | 671 |
| C10 | 77.9 | 86.8 | 89.1 | 80.0 | 85.9 | 78.9 | 83.3 | 65.0 | 63.3 | 71.5 | 76.1 | 76.2 |
| C4 | 52.1 | 52.6 | 50.4 | 47.7 | 48.2 | 46.5 | 51.9 | 41.9 | 46.5 | 42.1 | 43.8 | 40.6 |
| HHI | 903 | 920 | 923 | 776 | 880 | 783 | 921 | 577 | 922 | 638 | 713 | 637 |
| *N* (>1%) | 9 | 10 | 10 | 10 | 10 | 9 | 9 | 8 | 8 | 9 | 9 | 10 |
| Noam Index | 301 | 291 | 292 | 245 | 278 | 261 | 307 | 204 | 326 | 213 | 238 | 201 |

*Source*: Our calculations based on data from CNC. <http://www.cnc.fr/>; CNC. *Bilan* 2011 du CNC. May 2012; CNC. *Bilan 2010 du CNC*. May 2011; CNC, *Bilan 2009 du CNC*. May 2010; CNC. *Le marché de la distribution des films*. July 2008.

[1] As of 2007, Sony Pictures Releasing has taken over the French joint distribution activities of Gaumont and Columbia-Tristar (US).

[2] Between 2004 and 2007, Gaumont acted as Columbia-Tristar's (US) French distributor.

term, the market shares of individual studios can increase significantly due to just one very successful film.

The film distribution market is oligopolistic and includes two of the historic French movie majors that are now owned by large French media groups such as Vivendi. However, the independent French distributors SND and Mars, together with the US majors, account for 41.7% of national sales. Smaller companies, however, are numerous: there were 263 smaller companies in 2009 that accounted for a significant share of the sales (28.5%). There were only 13 firms with sales over US$20.9 million (15 million euros) in 2009, while there were 126 smaller companies, with sales under US$20.9 million (15 million euros).

## THE TELECOMMUNICATIONS INDUSTRY

### Wireline Telecom

Since telecom privatization began in the 1980s, the regulatory framework has evolved considerably and competition has increased. Besides France Telecom, the former public monopolist, other major companies now include SFR, which was controlled by Vivendi until it sold out to Altice in 2014; Bouygues Telecom, a subsidiary of the Bouygues Group, which is mainly involved in mobile telecom; and Iliad, a recent entry and Internet service provider (ISP).

Before 1986, the Ministry of Posts and Telecommunications incorporated both a General Directorate of Posts and a General Directorate of Telecommunications. The telecommunications sector was the responsibility of a state monopoly, the *Direction Générale des Télécommunications* (DGT). Since 1986, three other operators were allowed to operate cable networks in the framework of the 1982 Plan Câble. In 1987, the first mobile telephone network licenses were granted. By 1990, the telecommunications sector was privatized, but France Telecom, which has taken over the functions of the DGT, maintained a monopoly on installation and operation for both wireless and wireline services. In other words, a company can enter the market but must use French Telecom's infrastructure. In 1996, France Telecom became a limited liability corporation (SA), and the Telecommunications Regulatory Authority (ART), created in 1997, began to grant licenses to competing operators.

But France Telecom kept a very strong position in the telecommunications sector. Indeed, even though more than 100 licenses were issued by 2000 to other commercial service providers, these licenses were often limited to restricted geographical areas, and France Telecom maintained its monopoly on local and mobile dialing (in these two subsectors, France Telecom kept more than 90% of the market share). However, in the field of long distance and international fixed telephony, new entrants had won 33% of the market by the end of 2000.

In long distance telephony, France Telecom's competitors were Cegetel (which then had 2.7 million customers), Tele2 (with 2.6 million customers), and Neuf Telecom (1.2 million customers). Cegetel was based on SFR, its mobile operator, while Neuf Telecom relied on Bouygues Telecom before the two companies split, and Tele2 built its success on a simple business model of offering a single low price for consumers.

In 2010, production in the information and communication technology sector (which includes computers and the Internet) represented a turnover of US$129 billion (97 billion euros), or 5% of the French GDP.[8] In 2011, electronic communications represented 40.8 billion euros (US$56.8 billion), including all telephony and Internet services, 35 billion euros (US$48.8 billion); advanced services, 2.1 billion euros

8. The ARCEP (Autorité de Régulation des Communications Électroniques et des Postes, previously ART - Autorité de Régulation des Télécommunications) provides no disaggregated data at the level of operators and does not follow the common nomenclature defined by the media concentration consortium. This is the reason that we can use documents of the ARCEP only for a general and aggregated presentation and not for calculating the concentration indicators of the various markets on the basis of the operators.

(US$3 billion); and capacity services, 3.7 billion euros (US$5 billion). Telephony services included fixed wireline services, with 16 billion euros (US$22.4 billion), and mobile telephony, with 19 billion euros (US$26.4 billion).

The two main new wireline players are Iliad (Free) and Altice, which bought out Vivendi's operations in 2014.

Vivendi, as already mentioned, began life as part of the Compagnie Générale des Eaux, originally an industrial conglomerate group formed in the 19th century as a water utility that entered the audiovisual and telecom sectors in the early 1980s. By the early 2000s, Jean-Marie Messier developed the group's media holdings into an international telecom major, but high losses in 2002 meant that after 2003, several of Vivendi's media holdings were sold off. Nowadays, Vivendi controls in particular Universal Music (100%) and the Canal+ Group (100%). It sold off control of Maroc Telecom to Etisalat (UAE), most of SFR to Altice, GVT of Brazil to Telefonica of Spain, and spun off the video game studio Activision Blizzard. The other significant player is Iliad. Xavier Niel, its director, holds about 64% of its shares (Table 5.11).

## Wireless Telecommunication

In the mobile industry, France Telecom maintains a strong position. The second largest French mobile operator is SFR, which was a subsidiary of Vivendi but was sold in 2014 to Altice. SFR was previously a joint venture by Vivendi with Vodafone (UK), the world's second largest mobile operator. Bouygues Telecom, 90% owned by the Bouygues Group, is the third largest French mobile operator. In 2008, mobile virtual network operators (MVNOs) first appeared in

### Table 5-11. Wireline Telecom (Market Shares by Subscribers), 2002–2013[1]

| | 2002 | 2004 | 2008 | 2011 | 2013 |
|---|---|---|---|---|---|
| France Telecom/Orange* | 74.0 | 67.0 | 53.0 | 45.1 | 41.0 |
| Altice (Numericable, SFR) | | | | | 21.1[2] |
| Vivendi (Cegetel/Neuf Telecom/SFR)[3] * | 20.0 | 22.0 | 29.0 | 23.5 | Altice |
| Tele2 (Sweden)* | 2.0 | 4.0 | SFR | | |
| Iliad (Free)* | 4.0 | 7.0 | 18.0 | 22.8 | 22.6 |
| Bouygues Telecom | | | | 5.8 | 8.1 |
| Other (Darty etc.) | | | | 2.8 | 7.6 |
| All telephony and Internet** (US$ million) | 26,073 | 37,724 | 51,522 | 48,769 | 39,787 |
| Fixed services | 14,926 | 19,220 | 24,060 | 22,368 | 19,792 |
| C4 | 100.0 | 100.0 | 100.0 | 97.2 | 92.4 |
| HHI | 5,896 | 5,038 | 3,974 | 3,148 | 2,728 |
| *N* (>1%) | 4 | 4 | 3 | 5 | 5 |
| Noam Index | 2,948 | 2,519 | 2,294 | 1,408 | 1,220 |

*Source*: *Market shares given by Institut de l'Audiovisuel et des Télécommunications en Europe (IDATE) until 2008; ARCEP, 2011.[4]
** ARCEP.

[1] Due to the aggregation of alternative operators, HHI and the Noam Index are only approximate indicators of concentration

[2] In 2014, Altice acquired SFR from Vivendi.

[3] In 2005, Neuf Telecom bought Cegetel, creating Neuf Cegetel. In 2006, Neuf Cegetel bought AOL France (US). In 2007, SFR bought the French activities of Tele2 in the wireline telecommunications sector. In 2008, SFR took the control of Neuf Cegetel.

[4] Official ARCEP numbers are aggregated data with no distinction between operators, and the European Commission keeps disaggregated data confidential.

Table 5-12. Wireless Telecom (Market Shares by Subscribers), 1992–2013

| | 1992 | 1998 | 2000 | 2004 | 2008 | 2011 | 2013 |
|---|---|---|---|---|---|---|---|
| France Telecom/Orange | 74.5 | 50.7 | 48.2 | 47.4 | 43.5 | 39.9 | 38.0 |
| SFR (Vivendi, then Altice) | 25.5 | 38.0 | 34.2 | 35.4 | 33.9 | 31.3 | 29.1 |
| Bouygues Telecom (Bouygues Group) | | 11.3 | 17.6 | 17.0 | 17.6 | 16.7 | 15.8 |
| Iliad (Free) | | | | | | | 12.1 |
| MVNO and others | | | | 0.2 | 5.0 | 12.1 | 5.0 |
| Total Revenue (mil euros) | | 3,949 | 7,890 | 14,862 | 18,669 | 18,957 | 15,056 |
| Total Revenue (mil US$) | | 4,626 | 7,295 | 18,484 | 27,464 | 26,400 | 19,995 |
| C4 | 100.0 | 100.0 | 100.0 | 99.8 | 95.0 | 87.9 | 95.0 |
| HHI | 6,200 | 4,142 | 3,803 | 3,789 | 3,376 | 2,997 | 2,686 |
| *N* (>1%) | 2 | 3 | 3 | 3 | 3 | 3 | 4 |
| Noam Index | 4,384 | 2,391 | 2,195 | 2,188 | 1,949 | 1,730 | 1,343 |

*Source*: ARCEP, INSEE, and various press documents (*Les Echos*); IDATE.

France and now collectively hold 5% of the market (a MVNO is a mobile phone operator that provides services directly to its own customers but does not own key network assets such as licensed frequency allocations or cell towers). Iliad (which owns Free) also entered the mobile market recently, having obtained a mobile telephony license in 2010. In 2014, Iliad made an offer to buy the US mobile operator T-Mobile, owned by Germany's Deutsche Telekom. Market concentration has decreased in recent years with the growth of SFR, Bouygues Telecom, and Iliad. With the development of the "quadruple play" offers (fixed telephony, mobile, television, and Internet), operators are increasingly present in both the fixed and the mobile markets (Table 5.12).

## INTERNET MEDIA

### Internet Service Providers (ISP)

There have been many changes in the French Internet service providers industry during the last decade; in particular, many mergers have occurred. The main operators are now Orange (France Telecom group), Iliad (Iliad Group is 64% owned by its founder, Xavier Niel), and SFR/Altice.

Orange (Wanadoo until 2005) is the leader with a market share of about 45% in 2011. Iliad is second with its main brand, Free; Alice was bought by Free in 2008.

Neuf Cegetel was created by the merger of Neuf Telecom and Cegetel in 2005. Then Neuf Cegetel bought AOL France in 2006 and was bought by SFR in 2008. SFR was a subsidiary in 56% of Vivendi and in 44% of Vodafone, but in 2011 Vivendi bought the 44% of the capital owned by Vodafone, and sold SFR to Altice in 2014. In 2007, Numericable (also owned by Altice) purchased cable TV networks from France Telecom Cable, NC Numericable, TDF Cable, UPC, and Noos, and it became the predominant cable television company and an Internet service provider.

Concentration is high and increased from 2004 to 2008. Bouygues Telecom entered this market by buying the ISP activity of Darty in 2012 (Table 5.13).

### Search Engines

In 2004, Google was already a dominant player of the search engine industry in France, whereas now it is in a quasi-monopoly situation, with a 94% market share in 2012. The market shares of Yahoo! and other search engines such

## Table 5-13. Internet Service Providers (Market Shares by Subscribers), 2004–2014

| | 2004 | 2008 | 2011/2012 | 2013/2014 |
|---|---|---|---|---|
| France Telecom (Orange/Wanadoo) | 42.0 | 47.0 | 45.0 | 41.0 |
| Iliad (Free) | 17.4 | 23.8 | 23.0 | 22.5 |
| Numericable/Altice (cable) | | 5.1 | 27.0 | 25.0 |
| SFR/Neuf Cegetel (Vivendi, then Altice) | | 21.9 | 23.0 | |
| Tele2 (Sweden) | 6.0 | | | |
| AOL France (US) | 12.0 | | | |
| 9 Telecom | 11.0 | | | |
| Tiscali (Italy) | 7.0 | | | |
| Bouygues Telecom (Darty until 2012) | | 1.0 | 6.0 | 7.8 |
| Total Revenue (mil euros) | 2,837 | 6,549 | 9,131 | 9,729 |
| Total Revenue (mil US$) | 3,529 | 9,634 | 12,715 | 12,921 |
| C4 | 82.4 | 92.7 | 97.0 | 96.3 |
| HHI | 2,417 | 3,282 | 3,135 | 2,873 |
| *N* (>1%) | 6 | 5 | 4 | 4 |
| Noam Index | 987 | 1,468 | 1,568 | 1,473 |

*Source*: ARCEP; NB: total revenue includes various Internet sales and broadband sales; market shares are based on data obtained directly from operators.

## Table 5-14. Internet Search Engines (Market Shares by Search Volume), 2004–2012

| | 2004 | 2007 | 2012 |
|---|---|---|---|
| Google (US) | 72.0 | 90.2 | 94.0 |
| Yahoo! (US) | 8.0 | 3.2 | 2.0 |
| Voilà/Orange/France Telecom | 7.0 | 1.7 | 1.0 |
| MSN | 5.0 | 2.3 | |
| AOL France (US) | 3.0 | 1.4 | |
| Lycos (US) | | | |
| Altavista | 0.4 | 0.0 | |
| Iliad (Free) | 0.5 | 0.67 | |
| Alice | 0.5 | Iliad | |
| Bing (Microsoft, US) | | | 2.0 |
| Ask | | | 1.0 |
| Others | 3.6 | 0.5 | 0.0 |
| C4 | 92.0 | 97.4 | 99.0 |
| *N* (>1%) | 5.0 | 5.0 | 5.0 |
| HHI | 5,332 | 8,157 | 8,846 |
| Noam Index | 2,385 | 3,648 | 3,956 |

*Source*: Our calculations based on data from <http://www.atinternet.com/>.

as Ask and Orange quickly decreased in favor of Google. The second-ranked search engines behind Google are Bing (Microsoft), with a market share of 2% in 2012, and Yahoo! (also with a 2% market share). Orange (1%) and Ask (1%) market shares are far behind Google's 94% (Table 5.14).

## Online News

The online news market leaders are those linked to companies in the traditional newspaper industry or to Yahoo and Google. The market is very competitive, with a low concentration. Among the ten main competitors in 2010, six are the online versions of print publications, either newspapers or magazines. As of 2012, Orange and Yahoo have been losing audience shares, and both Google and MSN fell out of the ranks of the top ten. Content enterprises linked to the traditional print sector have the advantage in the French online news market, such as *Le Figaro, Le Monde, Le Parisien-Aujourd'hui*. The HHI remains low, but the C4 has increased from 36% to 48% over the two-year period. The methodology of measurement is not static and will probably change in the future. One could integrate a longer list of online news media: doing so would lead to a lower concentration HHI. Indeed the competition in this field is very high (Table 5.15).

# CONCLUSION

Concentration in French media differs greatly according to the industry. Scale effects favor large groups, like Lagardère Active, Vivendi, and France Telecom/Orange, and Altice. These

### Table 5-15. Online News Media (Market Shares by 1,000 Monthly Unique Visitors), 2010–2012

| | SEPTEMBER/OCTOBER 2010 | | MAY/JUNE 2012 | |
|---|---|---|---|---|
| Lefigaro.fr (Socpresse) | 7,312 | 14% | 9,960 | 15% |
| Lemonde.fr (La Vie-Le Monde) | 5,816 | 11% | 8,850 | 14% |
| Orange News (France Telecom) | 5,546 | 11% | 7,181 | 11% |
| Yahoo! News (US) | 5,034 | 10% | 6,078 | 9% |
| Le Parisien (Amaury) | 4,870 | 9% | 8,580 | 13% |
| 20 Minutes | 4,855 | 9% | 7,602 | 12% |
| L'Express | 4,800 | 9% | 7,840 | 12% |
| Nouvel Obs | 4,524 | 9% | 8,501 | 13% |
| Google News (US) | 4,275 | 8% | | 0% |
| MSN News | 4,245 | 8% | | 0% |
| TF1 News | | | 5,096 | 8% |
| Le Point | | | 4,594 | 7% |
| Others | | | | |
| C4 | | 36 | | 48 |
| HHI | | 1,029 | | 1,048 |
| *N* (>1%) | | 10 | | 10 |
| Noam Index | | 325 | | 331 |

*Source*: Médiamétrie quoted by *Le Figaro* <http://www.lefigaro.fr/medias/2012/12/10/20004-20121210ARTFIG00630-lefigarofr-premier-sitede-presse-enligne.php>.

large groups, whether they belong to the media industry or to the IT sector, can take over smaller groups if these face financial problems. Scope effects are also among the drivers of concentration, since they lead to an "industry meltdown," such as the penetration of telecom and Internet sectors into the media industry.[9] International ownership is also an important concentration-related issue for most if not all media and information industries in France. In the press, inasmuch as several segments of the market are strongly growing (in particular, magazines and trade publications/professional journals), there are many competing international actors on the French market, such as Prisma (Germany), Roularta (Belgium), Mondadori (Italy), and PRISA (Spain). Conversely, several large French press groups are expanding internationally. This is the case with the Lagardère Group, which has subsidiaries in most European countries, the United States, and Canada, as well as Australia and several East Asian nations (South Korea, Singapore, and the People's Republic of China). In telecommunications, the incumbent operator France Telecom is present in many countries, as are Vivendi's wireless and wireline services.

But the situation differs widely among industries. Compared to other countries, the HHI in general is not high.

For newspapers, HHI stood at 2,064 in 2011. Although it had decreased since 1997, it may increase in the future because of the financial difficulties most French national dailies are facing due to overregulation and loss of market shares to Internet media.

For book publishing, the HHI in 2005 was 1,511 and reached 1,748 by 2011. Despite the strengthening of large groups through takeovers and mergers and the growing market penetration of foreign publishing houses like Bertelsmann (Germany), the HHI is stable. The largest actor in this sector is still the French publisher Hachette Livres, a Lagardère Group subsidiary.

The magazine industry in France is in a better state, although we have been unable to calculate long-term concentration indices for the sector. HHI was around 2,000 in 2011.

French radio is becoming less concentrated, decreasing from 1,600 (in 2001) to 1,277 (in 2011). The radio industry shows a low concentration. Since the 1980s this sector has been open to competition, with many new participants competing with the main and incumbent operators.

In broadcast television, we observe a strong concentration phenomenon that is also present in many other Western countries. The HHI for broadcast TV illustrates a stable but highly concentrated industry, as it was 3,680 in 1995 and 3,440 in 2011.

The French film distribution industry is becoming increasingly competitive. As noted previously, the HHI was 903 in 2000 and 637 in 2011.

Cable and satellite TV effectively became quasi monopolies in 2008–2009. Numericable dominated the French cable market and Canalsat controlled the entire satellite market (HHI equals 10,000). If we put together cable and satellite TV, the HHI is lower, of course: 4,884 in 2011.

Due to deregulation in France, the HHI has actually declined both in wireline (5,896 in 2002 and 2,728 in 2013) and wireless telecom (6,200 in 1992 and 2,686 in 2013). While deregulation has induced more competition in wireless and wireline telecommunication industries in France, the concentration indices are still high.

Internet service providers constitute a market with a high HHI (3,135 in 2011). This market is dominated by France Telecom's Orange and Iliad's Free. The search engine market is characterized by the quasi monopoly of Google (HHI was 8,846 in 2012).

We also observe that the online news model of content aggregation is characterized by low concentration indices in France. This is probably due to the fact that it is a new market and existing media outlets are better placed to exploit it, even though they have fewer financial resources to do so than

9. For a more detailed analysis of processes, particularly financial processes, favoring concentration, see Badillo, Patrick-Yves (dir.). *L'écologie des médias*. Bruxelles: Editions Bruylant, 2008, see in particular chapter 6.

telecommunication firms and ISPs normally have. But in this market, the share of technology companies like Google or Microsoft (MSN) has been declining while traditional French media companies are increasing their market shares.

## Bibliography

### 1. Books and Articles


Albert, Pierre. *La presse française*. Paris: La Documentation Française, 2008.

Badillo, Patrick-Yves (Dir.). *L'écologie des médias*. Bruxelles: Editions Bruylant, 2008.

Badillo, Patrick-Yves, and Lesourd, Jean-Baptiste, eds. *The Media Industries and Their Markets: Quantitative Analyses*. London: Palgrave-MacMillan, 2010.

Badillo, Patrick-Yves, and Proulx, Serge. "Mondialisation de la Communication : à la recherche du sens perdu," in *Les essentiels d'Hermès "Sociétés de la Connaissance - Fractures et évolutions."* Paris: CNRS Éditions, 2009, 117–134.

Badillo, Patrick-Yves, and Roux, Dominique, eds. *Les 100 mots des télécommunications*. Paris: Presses Universitaires de France, 2009.

Balle, Francis. *Médias et Sociétés*. 15th ed. Paris: Montchrestien, 2011.

Charon, Jean-Marie. *La presse quotidienne*. Paris: La Découverte, 2004.

Charon, Jean-Marie, Le Floch, Patrick. *La presse en ligne*. Paris: La Découverte, 2011.

d'Armagnac, Bertrand. "Industriels et familles règnent sur la presse française." *Le Monde*, Jan. 22, 2005.

Eveno, Patrick. *La presse*. Paris: Presses Universitaires de France, 2010.

Eveno, Patrick. L'argent de la presse française des années 1820 à nos jours. Paris: CTHS, 2003.

Eveno, Patrick. "Vive la concentration dans les médias! " *Le Monde*, Mar. 29, 2005.

Miège, Bernard. La société conquise par la communication. Grenoble: PUG, 2007.

Noam, Eli. *Media Ownership and Concentration in America*. Oxford: Oxford University Press, 2005.

Noam, Eli. "The Effect of Deregulation on Market Concentration: an Analysis of the Telecom Act of 1996 and the Industry Meltdown." *The Columbia Science and Technology Law Review* 4, 2003, 1–12. Last accessed on Jan. 4, 2013 at <http://www.citi.columbia.edu/elinoam/articles/Effect_of_Deregulation_on_MarketConcentration.pdf>.


### 2. Reports and Websites (Data Sources)


AT Internet. <http://www.atinternet.com/>. 2 April, 2013.

Autorité de régulation des communications électroniques et des postes (ARCEP). <http://www.arcep.fr>. Jan. 4, 2013.

Centre National du Cinéma (CNC). <http://www.cnc.fr/web/fr>. April 2, 2013.

Centre National du Cinéma (CNC). *Le marché de la distribution de films*. Paris: CNC, July 2008, 29.

Conseil Supérieur de l'Audiovisuel (CSA). <http://www.csa.fr/>. April 2, 2013.

DGMIC (Direction Générale des Médias et des Industries Culturelles), Ministère de la Culture et de la Communication. <http://www.culturecommunication.gouv.fr/>. April 2, 2013.

France Livre Book France. <http://www.francelivre.org/Ressources/Le-livre-en-France/Chiffres-cles>. April 2, 2013.

Institut De l'Audiovisuel et des Télécommunications en Europe (IDATE). <http://www.idate.org/en/Home/>. Jan. 4, 2013. Livres Hebdo. <http://www.grouperevuefiduciaire.com/espace_presse/documents/classement_editeurs-LIVRES_HEBDO.pdf>. April 2, 2013.

Médiamétrie. <http://img2.mediasactu.com/radioactu/pdf/sondnovdec2011.pdf>. April 2, 2013.

Médiamétrie. <http://www.mediametrie.com>. April 2, 2013.

Médiamétrie. <http://www.radioactu.com/actualites-radio/4318/mediametrie-enquete-75000-radio-nov-dec-2001/>. April 2, 2013.

OJD (Association pour le Contrôle de la Diffusion des Médias; previously Office de Justification de la diffusion). <http://www.ojd.com>. March 22, 2013.

# France—Data Summaries

ELI NOAM AND PAUL MUTTER

MEDIA CONCENTRATION IN FRANCE has been zigzagging for years (Table 5.16). Even with the full liberalization of telecommunications in France and deregulation of other sectors since 1982, overall concentration is intermediate rather than low by world standards. The number of net voices is high in absolute terms but intermediate on a per capita basis.

Several media industries have seen dramatic increases in concentration: Altice's Numericable-Completel and Vivendi have become duopolists in the multichannel platform market. Numericable-Completel merged with the other three top cable companies, Liberty Global's UPC France, France Telecom Cable, and Noos, to control the cable market. The satellite market was similarly dominated by Vivendi in 2006 when its Canalsat subsidiary merged with its main competitor TPS. Vivendi is one of the largest content producers in France, with a market share of 13% (Table 5.17)—with a rising market shares in multichannel platforms, video channels, and film productions. In 2014, Altice became the second largest platform media company in France (Table 5.18) after Orange/France Telecom, with 25.6%.

Concentration is also high in the newspaper market (Amaury).

The largest media services firm in France is Orange/France Telecom, which is the parent company of the mobile operator Orange: it alone controls 38.4% of the platform market (Table 5.18). Orange has a significant international presence. With over 226 million mobile subscribers, Orange is the sixth-largest wireless provider in the world, and the largest in Egypt (as Mobinil) and the second largest in Belgium (as Mobistar). Orange indirectly controls 45% of the UK wireless market as an equal shareholder with Deutsch Telekom (Germany) in their joint UK venture, EE. The Lagardère

Table 5-16. National Media Industries Concentration in France

| | 2004/5 | | 2011 OR MOST RECENT | | % CHANGE ANNUAL AVERAGE | |
|---|---|---|---|---|---|---|
| | COMPANY POWER INDEX IN COUNTRY | COMPANY SHARE OF THE OVERALL NATIONAL MEDIA MARKET (%) | COMPANY POWER INDEX IN COUNTRY | COMPANY SHARE OF THE OVERALL NATIONAL MEDIA MARKET (%) | COMPANY POWER INDEX IN COUNTRY | COMPANY SHARE OF THE OVERALL NATIONAL MEDIA MARKET (%) |
| France Telecom (32.5% public) | 1,770 | 30.9 | 944 | 23.1 | −6.7 | −1.1 |
| Vivendi | 525 | 16.8 | 387 | 7.4 | −3.8 | −3.8 |
| Bouygues Group | 127 | 6.5 | 111 | 7.7 | −1.8 | 0.2 |
| Lagardère | 116 | 4.9 | 111 | 4.6 | −0.6 | −0.04 |
| Google (US) | 20 | 0.3 | 101 | 1.1 | 56.3 | 0.1 |
| Altice/Numericable-Completel | 2 | 0.3 | 420 | 15.9 | 25.9 | 0.02 |
| France Télévisions (public) | 96 | 3.0 | 64 | 2.8 | −4.8 | −0.03 |
| Vodafone (UK) | 63 | 4.2 | 0 | 0.0 | −14.3 | −0.6 |
| Amaury Group | 65 | 2.1 | 44 | 1.5 | −4.6 | −0.08 |
| Bertelsmann (Germany)[1] | 33 | 3.0 | 39 | 3.1 | 2.4 | 0.03 |
| Iliad | 27 | 2.6 | 212 | 10.5 | 100.1 | 1.1 |
| NRJ Group | 17 | 1.0 | 15 | 1.4 | −1.2 | 0.05 |
| Socpresse Group (Le Figaro) | 16 | 1.1 | 10 | 0.8 | −5.2 | −0.05 |
| La Vie-Le Monde | 17 | 1.4 | 8 | 0.8 | −7.7 | −0.1 |
| Editis (Planéta, Spain) | 14 | 1.1 | 8 | 0.8 | −5.8 | −0.05 |
| Roularta (Belgium) | 8 | 0.7 | 6 | 0.5 | −4.1 | −0.03 |
| Média Participations | 2 | 0.4 | 2 | 0.4 | −0.2 | −0.003 |
| Yahoo (US) | 1 | 0.1 | 0.9 | 0.1 | −1.4 | 0.002 |

*continued*

Table 5-16. *continued*

| | 2004/5 | | 2011 OR MOST RECENT | | % CHANGE ANNUAL AVERAGE | |
|---|---|---|---|---|---|---|
| | COMPANY POWER INDEX IN COUNTRY | COMPANY SHARE OF THE OVERALL NATIONAL MEDIA MARKET (%) | COMPANY POWER INDEX IN COUNTRY | COMPANY SHARE OF THE OVERALL NATIONAL MEDIA MARKET (%) | COMPANY POWER INDEX IN COUNTRY | COMPANY SHARE OF THE OVERALL NATIONAL MEDIA MARKET (%) |
| Media Concentration Index | 2004/5 | | 2011 or Most Recent | | % Change Annual Average | |
| Total Revenue: Nat'l Media Industry (mil US$) | 75,922 | | 93,236 | | 3.3 | |
| Total Voices (*n*) | 84 | | 77 | | –1.2 | |
| Net Voices (*n*) | 60 | | 54 | | –1.4 | |
| Public Ownership (%) | 13.1 | | 10.4 | | –0.4 | |
| Foreign Ownership (%) | 12.1 | | 7.0 | | –0.7 | |
| C4 Average—Weighted | 88 | | 86 | | –0.03 | |
| HHI Average—Weighted | 3,392 | | 2,693 | | –2.9 | |
| C1 Average—Weighted | 45.6 | | 38.7 | | –0.01 | |
| Noam Index Average—Weighted | 841 | | 981 | | 2.4 | |
| Pooled Overall Sector C4 | 59.1 | | 57.2 | | –0.3 | |
| Pooled Overall Sector HHI | 1,358 | | 1,058 | | –3.2 | |
| Pooled Overall Sector Noam Index | 54 | | 65 | | 2.9 | |
| Market Share of Top Ten Companies: Nat'l Media Industry (%) (Pooled C10) | 75.4 | | 78.0 | | 0.4 | |
| National Power Index | 2,950 | | 2,497 | | –2.2 | |

[1] Operating as M6 Group Owned by RTL.

## Table 5-17. Top Content Media Companies in France

| | 2004/5 | | 2011 OR MOST RECENT | | % CHANGE ANNUAL AVERAGE | |
|---|---|---|---|---|---|---|
| | COMPANY POWER INDEX IN COUNTRY | COMPANY SHARE OF THE NATIONAL CONTENT MEDIA MARKET (%) | COMPANY POWER INDEX IN COUNTRY | COMPANY SHARE OF THE NATIONAL CONTENT MEDIA MARKET (%) | COMPANY POWER INDEX IN COUNTRY | COMPANY SHARE OF THE NATIONAL CONTENT MEDIA MARKET (%) |
| Vivendi | 614 | 11.4 | 637 | 13.0 | 0.5 | 0.2 |
| Lagardère | 267 | 11.2 | 276 | 11.4 | 0.5 | 0.03 |
| Google (US) | 47 | 0.7 | 251 | 2.7 | 61.8 | 0.3 |
| France Televisions (public) | 221 | 7.0 | 158 | 7.1 | –4.1 | 0.01 |
| Bouygues Group | 131 | 5.4 | 108 | 6.0 | –2.5 | 0.08 |
| Amaury Group | 150 | 4.8 | 109 | 3.8 | –3.9 | –0.15 |
| Bertelsmann (Germany) | 77 | 6.9 | 97 | | 3.7 | 0.14 |
| Altice/Numericable-Completel | 2 | 0.2 | 50 | 1.5 | 3.9 | 0.002 |
| NRJ Group | 39 | 2.4 | 38 | 3.4 | –0.2 | 0.14 |
| Socpresse Group (Le Figaro) | 37 | 2.5 | 25 | 1.9 | –4.5 | –0.09 |
| La Vie-Le Monde | 40 | 3.3 | | 1.9 | –7.2 | –0.2 |
| Editis (Planéta, Spain) | 31 | 2.5 | 20 | 2.0 | –5.2 | –0.08 |
| Roularta (Belgium) | 19 | 1.6 | 14 | 1.2 | –3.4 | –0.06 |
| Média Participations | 4 | 0.9 | 5 | 0.9 | 0.9 | 0.002 |
| France Telecom (public) | 5 | 0.5 | 3 | 0.3 | –5.8 | –0.03 |
| Yahoo (US) | 2 | 0.2 | 2 | 0.3 | –0.4 | 0.01 |

*continued*

Table 5-17. *continued*

| | 2004/5 | | 2011 OR MOST RECENT | | % CHANGE ANNUAL AVERAGE | |
|---|---|---|---|---|---|---|
| | COMPANY POWER INDEX IN COUNTRY | COMPANY SHARE OF THE NATIONAL CONTENT MEDIA MARKET (%) | COMPANY POWER INDEX IN COUNTRY | COMPANY SHARE OF THE NATIONAL CONTENT MEDIA MARKET (%) | COMPANY POWER INDEX IN COUNTRY | COMPANY SHARE OF THE NATIONAL CONTENT MEDIA MARKET (%) |
| Media Concentration Index | 2004/5 | | 2011 or Most Recent | | % Change Annual Average | |
| Public Ownership (%) | 7.2 | | 7.2 | | 0.0 | |
| Foreign Ownership (%) | 17.0 | | 17.3 | | 0.05 | |
| C4 Average—Weighted | 74 | | 74 | | −0.03 | |
| HHI Average—Weighted | 2,306 | | 2,373 | | 0.4 | |
| C1 Average—Weighted | 33 | | 34 | | 0.0 | |
| National Power Index | 1,744 | | 1,852 | | 0.9 | |

Table 5-18. Top Platform Media Companies in France

| | 2004/5 | | 2011 OR MOST RECENT | | % CHANGE ANNUAL AVERAGE | |
|---|---|---|---|---|---|---|
| | COMPANY POWER INDEX IN COUNTRY | COMPANY SHARE OF THE NATIONAL PLATFORM MEDIA MARKET (%) | COMPANY POWER INDEX IN COUNTRY | COMPANY SHARE OF THE NATIONAL PLATFORM MEDIA MARKET (%) | COMPANY POWER INDEX IN COUNTRY | COMPANY SHARE OF THE NATIONAL PLATFORM MEDIA MARKET (%) |
| France Telecom (public) | 3,117 | 54.1 | 1,576 | 38.4 | –7.1 | –2.2 |
| Altice/Numericable-Completel | 3 | 0.3 | 669 | 25.6 | 34.9 | 3.6 |
| Iliad | 47 | 4.6 | 354 | 17.6 | 94.2 | 1.9 |
| Vivendi | 456 | 20.9 | 218 | 3.6 | –7.5 | –2.5 |
| Bouygues Group | 124 | 7.3 | 113 | 8.8 | –1.3 | 0.2 |
| Vodafone (UK) | 112 | 7.5 | 0 | 0.0 | –14.3 | –1.1 |
| Media Concentration Index | 2004/5 | | 2011 or Most Recent | | % Change Annual Average | |
| Public Ownership (%) | 17.6 | | 12.5 | | –0.7 | |
| Foreign Ownership (%) | 8.5 | | 0 | | –1.2 | |
| C4 Average—Weighted | 98 | | 94 | | –0.7 | |
| HHI Average—Weighted | 4,220 | | 2,909 | | –4.4 | |
| C1 Average—Weighted | 55 | | 42 | | –2.0 | |
| National Power Index | 3,870 | | 2,930 | | –3.5 | |

Group is a conglomerate firm that held stakes in the defense contractor EADS until 2013. Lagardère is a significant player in the book publishing markets of Australia, the United Kingdom, the United States, and Spain, and it also has a large presence in magazines. Iliad and Altice have become major players in platforms. Bouygues has a substantial presence in both platforms and content.

Foreign ownership is low at 7.0% of total national media industries, composed mainly of US and German audiovisual content producers (the six major Hollywood studios and Bertelsmann's RTL Group) and Vodafone.

# 6

# Media Ownership and Concentration in Germany

JULIA BÖSCH, MAX-JOSEF MEIER, PHILIPP RÖSCH-SCHLANDERER, AND ACHIM EKKEHARD HENNING WOLF

## INTRODUCTION

With a population of about 82.2 million people and 35 million households owning at least one TV set, Germany is one of the largest media markets in Europe. When the German-language markets in Switzerland and Austria are included, about 100 million people consume German media products. It is the largest single language space inside the European Union.

Germany has a long history of mass media; several of the world's first newspapers began printing some 400 years ago. However, during the Third Reich (1933–1945), mass media became a tool of the Nazi dictatorship, and subsequently the media experienced a complete reset under the occupying Allied powers. The postwar media system that developed in the zones under French, British, and American control was based on the principle of freedom of the press as stipulated by the Basic Law of 1949.

Until 1990, Germany was a divided country. State and commercial media producers operated freely in the western zones that eventually became the Federal Republic of Germany (FRG) or "West Germany." The media system in the Soviet-administered zone, which was reorganized as the German Democratic Republic (GDR), was highly centralized and operated under the control of the Socialist Unity (Communist) Party. This media system was dismantled during the process of reunification in the 1990s, but patterns of media usage still differ between East and West. The major media production centers are still located within the old West German borders, most eastern newspapers are owned by western firms, and regional broadcasting has been integrated into the system of public and commercial broadcasting developed in West Germany.[1]

Germany has a dual system: in public broadcasting the Länder (states) have a strong role. The Federal Constitution stipulates that the sole responsibility for broadcasting rests with the Länder as part of their cultural sovereignty. As a result, public service broadcasters are created by the Länder and act individually or jointly by treaty agreements among themselves. The exception is the broadcaster Deutsche Welle, which has a federal mandate to provide electronic media services to foreign audiences.

A basic agreement (Rundfunkstaatsvertrag) of all the Länder defines the public and private broadcasting industries. All broadcasting corporations are governed by an independent Broadcasting Council (Rundfunkrat), whose representatives are supposed to reflect the "socially relevant groups" of society, according to a Federal Constitutional Court's ruling; in addition, these broadcasters also have supervisory councils assigned to them.[2] With the advent of commercial broadcasting, all Länder in West Germany drafted new media laws in the 1980s. These laws specifically regulate electronic media outside of the conventional public corporations by handing out commercial radio and TV licenses and deciding what programs may be fed into the cable network. For this purpose, new state supervisory bodies (Landesmedienanstalten) were created, each with a council resembling those of the public broadcasters. There were 14 Landesmedienanstalten as of 2009. Because of Germany's strong federalist system, some TV and most radio broadcasters are regional or even local.[3]

The local and regional newspapers market is strong in Germany and is characterized by a large number of outlets. In 2008, there were 135 independent editorial units, full publishing entities that produce all parts of daily newspapers, and 354 editorial units for weekly newspapers. Including local editions, there are 1,512 newspapers in circulation in Germany. In 2008, total newspaper circulation stood at 20.2 million, most of which is subscription based as opposed to newsstand sales; 95% of the subscription-based press claims to be local, with a circulation of 14.3 million. There are only a small number of national newspapers, and even most of those have a strong regional base: *Bild Zeitung*, *Süddeutsche Zeitung* (SZ), *Frankfurter Allgemeine Zeitung* (FAZ), *Die Welt*, *Frankfurter Rundschau* (FR), and *Tageszeitung* (Taz). In terms of circulation figures, the national newspapers account for approximately 1.7 million daily copies. Another 4.5 million copies are sold in newsstands. The highest-circulation German tabloid paper is *Bild Zeitung*; with a circulation of 3.3 million, it is the top selling paper in the European Union. The most successful and important weekly is *Die Zeit*, with a weekly circulation of 525,000 copies.[4]

The German print press is characterized by a high, albeit decreasing, dependency on advertising income and demonstrates a significant degree of concentration. Despite the profusion of titles, the newspaper market is dominated by a small number of publishers.

The German magazine sector is extremely robust with some 906 general magazines with an annual circulation of 118 million copies and 1,218 specialized periodicals with an annual circulation of 13.6 million. *Der Spiegel* is a weekly news magazine, modeled after the American *Time*, and for long time it held a virtual monopoly in its market with an annual circulation of approximately 1.1 million. With its investigative style of journalism, it is the most influential political publication in Germany.

There are eight relevant news wire agencies. The dominant internationally-active agency is

---

1. Kleinsteuber, Hans J. and Thomass, Barbara. "Media Landscape: Germany." Maastricht: European Journalism Centre. 2012. Aug. 30, 2012, <http://www.ejc.net/media_landscape/article/germany/>.

2. Kleinsteuber, Hans J. and Thomass, Barbara. "Media Landscape: Germany." Maastricht: European Journalism Centre. 2012. Aug. 30, 2012, <http://www.ejc.net/media_landscape/article/germany/>.

3. Kleinsteuber, Hans J. and Thomass, Barbara. "Media Landscape: Germany." Maastricht: European Journalism Centre. 2012. Aug. 30, 2012, <http://www.ejc.net/media_landscape/article/germany/>.

4. "Media Landscape: Germany." European Journalism Centre. 2012. Last accessed Aug. 10, 2012 at http://www.ejc.net/media_landscape/article/germany/

Deutsche Presseagentur (DPA). Because nearly all newspapers subscribe to DPA, it can be regarded as the primary source of news. Other news agencies are complementary sources. Its business model was questioned in 2009 when the biggest regional newspaper, *Westdeutsche Allgemeine Zeitung* (WAZ), canceled its DPA subscription. Other press companies are considering a similar course of action. The Associated Press (AP), Reuters Germany, and Agence France Presse (AFP) rank second, third, and fourth place in the German market, respectively.[5]

Radio is a popular medium in Germany. In 2008, every German citizen listened to 176 minutes every day. Slightly more than a half of these minutes are spent listening to public service broadcasters. They usually offer six regional programs that concentrate on general audiences, as well as special target groups such as culture, news, and youth. In addition, there are two national radio programs based in Berlin (Deutschlandradio Kultur) and Cologne (Deutschlandfunk) with public funding secured by another Länder-level agreement.

Commercial radio is licensed in all Länder; therefore it follows mostly a regional pattern. There are no national broadcasters, but some are active in several Länder, such as NRJ and Klassik Radio. In two southern Länder, local commercial radio is the only option for listeners. In North Rhine-Westphalia, the largest state, 46 local stations have commercial programming but offer local, noncommercial windows of time as well. Noncommercial radio exists in each state. Every state has its own regulations. Some states allow community stations, while others prefer public access, educational stations, campus stations, and so forth. This is also true of television. One Land has no activities at all.[6]

Germans spend about 219 minutes per day watching television, split evenly between public and commercial programs. All regional public broadcasters jointly founded the ARD (Arbeitsgemeinschaft der Rundfunkanstalten Deutschlands) regulatory body, and they contribute financial contributions according to their size to the nationwide TV channel Das Erste, the first and oldest TV program. In addition, they independently organize a regional program (III Programme) that offers regional content, as well as cultural and educational programming.

The Second German Television ZDF (Zweites Deutsches Fernsehen) is based on an agreement of all Länder (ZDF-Staatsvertrag) and is located in Mainz. ARD and ZDF jointly offer a number of specialized programs: Arte (together with France), 3Sat (together with Austria and Switzerland), Kika (for children), and Phoenix (events and documentation).

Today German commercial television is controlled by two media groups that are collectively called Senderfamilien (broadcaster families). One, formerly owned by Leo Kirch, is ProSiebenSat.1 Media AG and consists of the channels Sat.1, ProSieben, N24, Kabel 1, 9live, and others. In 2006, it was acquired by the Anglo-American investment funds Permira and Kohlberg, Kravis & Co. (KKR). It took over the SBS activities of these funds in ten other European countries.

The other is controlled by the German giant Bertelsmann, the largest media company outside of the United States, the largest bookseller in the world, and a global player. RTL Group S.A. owns TV channels in a dozen European countries. In Germany, these include RTL, RTL II, Super RTL, VOX, and N-TV. More programs were introduced in 2009. Some of them are independently owned special interest channels, while others are subsidiaries of international conglomerates such as Viacom, Disney, and NBC Universal. In large cities such as Berlin and Hamburg, regional commercial TV has been established. Over half of German households have cable. A smaller percentage have satellite TV, and a still smaller percentage of households have only terrestrial reception.

---

5. "Media Landscape: Germany." European Journalism Centre. 2012. Last accessed Aug. 10, 2012 at http://www.ejc.net/media_landscape/article/germany/

6. "Media Landscape: Germany." European Journalism Centre. 2012. Last accessed Aug. 10, 2012 at http://www.ejc.net/media_landscape/article/germany/

The only pay-TV company, Premiere, founded by Leo Kirch, went bankrupt. It was taken over by Rupert Murdoch in 2009, renamed Sky, and integrated into Murdoch's European Sky conglomerate. Compared to other European countries, pay TV is not very successful due to the large number of freely accessible channels. In 2009, about 2.4 million viewers subscribed to Sky.[7]

The regulatory body for all matters of telecom, including the non-mass media side of online services, is the Federal Network Agency (Bundesnetzagentur). The largest company in this field is Deutsche Telekom, formerly the state administration for telephony and still partly owned by the federal government. It has entered the market of Internet TV, but so far the resonance is limited: its subsidiary T-Home entertain provided IPTV for about 700,000 subscribers in 2009.[8]

The accountability system for the press was created by the German Press Council (Deutscher Presserat) in 1956. It consists of an equal number of representatives from the journalist and publisher organizations for a total of 20. Members of the general public may appeal directly to the Council. If the Council supports the complaint, the respective newspaper is expected to publish the Council's ruling. The decisions are published on the basis of a Press Codex that is regularly renewed according to recent journalism developments. The effect of this self-regulation is limited, especially in relation to the practices of the yellow press.

The German Advertisement Council (Deutscher Werberat) is a similar organization, consisting of 10 to 12 representatives from the media and advertising agencies. They publish their decisions on complaints in a handbook. Complaints against the public service broadcasters may be brought to the members of the Broadcast Councils. In commercial broadcasting, all television companies are obliged to employ a Commissioner for Youth Protection (Jugendschutzbeauftragter) that reports only to the company. Some media outlets, especially newspapers, have special media sections that provide more transparency for audiences, although they do not criticize the media industry. Journalist organizations such as Netzwerk Recherche (Network for Investigative Journalism) are trying to improve the quality of journalism through more extensive critiques of the industry.[9]

This chapter provides a deeper analysis of the ownership and concentration levels in print, audiovisual, telecom distribution, and online media.

## PRINT MEDIA

### Newspapers

The newspaper market in Germany can be divided into two categories: those mainly sold via subscription and those mainly sold via retail. Readers spend on average of 40 minutes per day reading newspapers sold via subscription and only 20 minutes with those sold via retail; 95% of all newspapers in Germany are sold via subscription.[10] In 2008, Stuttgarter Zeitung acquired the *Süddeutsche Zeitung*, the eighth largest German newspaper with a daily print run of 430,000.[11] This acquisition made Stuttgarter Zeitung the second largest newspaper publisher and the largest publisher of newspapers sold via subscription. This caused the highest increase of concentration since 1989. Axel Springer AG is the largest German newspaper publisher. It is ranked third in the market of subscription newspapers but dominates the retail newspaper market with a market share of 80%.

7. "Media Landscape: Germany." European Journalism Centre. 2012. Last accessed Aug. 10, 2012 at http://www.ejc.net/media_landscape/article/germany/

8. "Media Landscape: Germany." European Journalism Centre. 2012. Last accessed Aug. 10, 2012 at http://www.ejc.net/media_landscape/article/germany/

9. "Media Landscape: Germany." European Journalism Centre. 2012. Last accessed Aug. 10, 2012 at http://www.ejc.net/media_landscape/article/germany/

10. Koschnick Wolfgang J. *Focus Media Lexikon*, München, Fokus Magazin Verlag, 2003.

11. "Süddeutsche geht an Medienholding." June 2007. FAZ. Last accessed on Aug. 10, 2012 at http://www.faz.net/aktuell/wirtschaft/fuer-mehr-als-eine-halbe-milliarde-euro-sueddeutsche-geht-an-medienholding-1488131.html

There are several other prominent newspaper publishers in the market. The privately held WAZ publishing group is the third largest German newspaper publisher and structured similarly to the Stuttgarter Zeitung publishing group with a strong focus on subscription newspapers. DuMont is number four in the German market. While having lost its importance in the subscription market, DuMont is still the third largest publisher for newspapers sold via retail. Ippen publishing group is the fifth largest newspaper publisher and is active in both segments.

Our dataset contains concentration measures from 1989 to 2012. The concentration has been measured by print run per year. From 2006 to 2008, we could observe the largest increase of market concentration in our dataset. In 1989, the market share of the top four players equaled 39.6% while in 2008 it was 40.8%. The market share of the ten largest players grew from 54.8% to 58.5% in the same period.

When looking at the concentration indices, we can observe that HHI and Noam index grow consistently over time. C4 also increased due to acquisition activities. We see differences in the indices between the two newspaper markets. In the subscription market, all three indices increased after 1989. This is driven by the acquisitions during this time period. In 2012, the market share of the five largest players was 33.9%. At first glance, the market for newspapers sold via retail looks different. The market share of the top five groups dropped from 39% in 2000 to 37.4% in 2006 because of Mecom's entry into the German market. In 2005, Mecom acquired some newspapers with its German subsidiary company, BV deutsche Vertragsholding, and became one of the ten largest newspaper publishers for newspapers sold via retail. However, when looking at this market it is more important to focus on the absolute numbers and less on the trends (Table 6.1).

Table 6-1. Market Shares in the German Newspaper Market

| | 1989 | 1993 | 1997 | 2000 | 2004 | 2006 | 2008 | 2012 |
|---|---|---|---|---|---|---|---|---|
| Axel Springer AG | 26.7 | 22.8 | 23.7 | 23.6 | 22.7 | 22.5 | 22.1 | 18.8 |
| Stuttgarter Zeitung | 3.2 | 5.2 | 5.0 | 5.0 | 5.0 | 5.2 | 8.5 | 9.2 |
| Süddeutsche Zeitung | 3.6 | 3.3 | 3.2 | 3.3 | 2.5 | 2.6 | Stuttgarter Zeitung | – |
| Westdeutsche Allgemeine Zeitung (WAZ) (Funke) | 6.0 | 5.6 | 5.9 | 6.0 | 6.0 | 5.6 | 6.0 | 5.7 |
| DuMont Schauberg | 3.3 | 4.5 | 4.0 | 4.4 | 4.0 | 3.9 | 4.2 | 5.5 |
| Ippen-Gruppe | 3.0 | 2.7 | 2.7 | 2.9 | 3.9 | 4.1 | 4.0 | 4.2 |
| Holtzbrinck | | 2.5 | 2.5 | 2.5 | 3.6 | 3.7 | 3.8 | – |
| Frankfurter Allgemeine Zeitung (FAZ) | 2.4 | 3.1 | 3.0 | 3.0 | 3.1 | 3.0 | 3.0 | 2.5 |
| Madsack | 1.9 | 2.1 | 2.3 | 2.4 | 2.5 | 2.5 | 2.5 | 4.9 |
| DDVG | | | | | | 2.2 | 2.4 | 3.1 |
| (Bertelsmann, Gruner + Jahr) | | 3.8 | 3.4 | 2.8 | 2.8 | | | – |
| BV dt. Zeitungsholding | | | | | | | 2.0 | |
| Others | 49.9 | 44.4 | 44.3 | 44.1 | 43.9 | 44.7 | 41.5 | 46.1 |
| Revenue ($US mln) | | | | | 4,163 | | 4,346 | |
| C4 | 39.6 | 38.1 | 38.6 | 39.0 | 37.7 | 37.4 | 40.8 | 39.2 |
| HHI | 801.4 | 651.4 | 687.1 | 685.5 | 650.4 | 637.2 | 669.8 | 558.3 |
| *N* (>1%) | 8 | 10 | 10 | 10 | 10 | 10 | 10 | 10 |
| Noam | 283.3 | 206.0 | 217.3 | 216.8 | 205.7 | 201.5 | 211.8 | 197.4 |

*Source*: Röper (2002, 2004, 2008)

## Book Publishing

With 9 billion € (US $10.8 billion) in revenues, 800 million books, and about 90,000 new titles per year, Germany is the third largest book market in the world.[12] In Germany, as in other European countries, there is a retail price control that forces each publisher to set a fixed price for each title by law. Looking at the different sales channels, we can observe that retail book trade is quite dominant with a market share of 53.6% in 2007; 18% of revenues are generated via direct sale.[13]

Market shares were measured based on the revenues of the top 100 publishers in each year. The figures were taken from the Buchreport Verlagsranking. These 100 publishers were then grouped together by company. Total revenues of the industry were taken from the reports of the German Book Trades Association.[14]

The findings are reported in Table 6.2. In 2007 the three largest book publishers were Springer Science + Business, Klett, and Cornelsen. (Klett offers education books for students and Cornelsen operates in the general education area.)

Among publishers that focus on paperback books, Random House (Bertelsmann) was the largest publisher in 2004, followed by Holtzbrinck and Bonnier of Sweden.[15] Holtzbrinck was restructured in 2009 when long-time CEO and 33% shareholder Dieter von Holtzbrinck sold his shares to his siblings and subsequently acquired a stake in Holtzbrinck Group properties such as *Handelsblatt* and *Zeit*.

In general, there is much change within the top group, as this segment is drastically dependent on best sellers. For example, Carlsen was among the top publishers during the years in which the Harry Potter books were released.

The book market in Germany consolidated after 1989. Many small publishers were merged and sold. At the same time, some of the bigger publishers sold the smaller publishers that they just had acquired a short time before. One example is Axel Springer. Only three years after buying into the book market, Springer sold the book publishers it had acquired. Bertelsmann decided to buy all of Springer's book publishers, but the German antitrust agency intervened and Bertelsmann had to divest parts of

### Table 6-2. Development of Revenues of the Largest (Rank 1—100) Book Publishers (€M)

| YEAR | RANK 1 | RANK 10 | RANK 25 | RANK 50 | RANK 100 |
|---|---|---|---|---|---|
| 2007 | 601 | 150 | 62 | 31 | 7 |
| 2006 | 641 | 164 | 60 | 10 | 7 |
| 2005 | 588 | 144 | 60 | 29 | 7 |
| 2004 | 544 | 185 | 61 | 30 | 7 |
| 2003 | 530 | 161 | 61 | 31 | 6 |
| 2000 | 597 | 135 | 61 | 38 | 7 |
| 1995 | 505 | 83 | 48 | 28 | 9 |
| 1990 | 931 | 56 | 39 | 23 | 7 |

*Source*: *Buchreport.de*. (2008)

12. Eugen Emmerling, "Der deutsche Buchmarkt—Gutenbergs Erben bestehen in der Medienkonkurrenz."
13. "buchreport." *Buchreport.de*. March 27, 2008. Last accessed on August 9, 2012 at http://www.buchreport.de/nachrichten/verlage/verlage_nachricht/datum/2008/03/27/buchreport-ranking-100-groesste-verlage.htm
14. Börsenverein des Deutschen Buchhandels.
15. Dr. Salat, R., "Die Vielfalt des deutschen Buchmarktes." 2005.

the acquisition to Bonnier. Bertelsmann also bought, in 1999, the largest book publisher in Germany, Springer Science, and sold it in 2003 to British private equity investors.

Publishers have expanded their business into a broad range of information and entertainment services, but the structure of the industry is mid-size oriented. The ownership structures are almost constant. Table 6.2 shows publishers' revenues, ranked from 1 to 100 according to revenues from 1990 to 2007.

After 2000, concentration remained constant (Table 6.3). In contrast to the preceding years, the indices remain quite constant. The C4 index shows a very high 23.5% market share for the top four players in 1989. This dropped in 1996 to 12.8% and rose again to 17.7% in 2012. Table 6.3 shows the market shares of the top ten publishers in different years. The common market share of the top ten group runs between 22% and 33%, which is a much smaller concentration compared to the popular magazines. In the magazine segment, the top four own about 60% of circulation.

## Magazines

The most influential magazines in Germany are *Stern*, *Der Spiegel*, and *Focus*. Usually, magazines have a higher circulation compared to newspapers

### Table 6-3. Market Shares of German Publishers

| PUBLISHERS | 1989 | 1992 | 1996 | 2000 | 2004 | 2007 | 2012 |
|---|---|---|---|---|---|---|---|
| Klett | 1.2 | 1.6 | 2.0 | 2.4 | 3.6 | 4.4 | 4.7 |
| Cornelsen | – | 1.3 | 1.5 | 1.6 | 3.7 | 3.8 | 4.4 |
| Medien Union | – | – | – | – | 2.5 | 2.9 | 3.2 |
| Weltbild | – | – | 1.2 | 2.1 | 2.8 | 2.5 | 2.6 |
| Bertelsmann (Random House) | 15.1 | 7.7 | 5.7 | – | 2.3 | 2.5 | 3.6 |
| Bertelsmann Springer Group | – | – | – | 8.8 | – | – | – |
| Springer Science + Business | 3.6 | 2.1 | 1.9 | – | 6.0 | 6.3 | 5.0 |
| Bonnier (Sweden) | – | – | – | – | – | 2.2 | 2.3 |
| Holtzbrinck | 2.7 | 3.2 | 2.9 | 1.8 | 2.1 | 2.1 | 2.2 |
| Weka | 2.0 | 2.8 | 3.1 | 3.0 | 2.1 | 2.1 | 2.0 |
| Langenscheid | 1.5 | 1.8 | 2.1 | 1.6 | – | 1.9 | – |
| Süddeutscher Verlag | – | – | 1.2 | 2.6 | 2.4 | – | – |
| Vogel | – | – | – | 3.0 | 2.3 | – | 1.1 |
| Haufe | – | 0.9 | – | 1.6 | – | – | 2.3 |
| Mairs | – | 1.0 | 1.1 | – | – | – | 1.1 |
| Fink-Kümmerly & Frey | – | 1.0 | – | – | – | – | – |
| Markt + Technik | 1.7 | – | – | – | – | – | – |
| Lübbe | 1.7 | – | – | – | – | – | 0.8 |
| ADAC | 1.2 | – | – | – | – | – | – |
| Thieme | 0.9 | – | – | – | – | – | 1.4 |
| % Top 10 | 31.7 | 23.5 | 22.6 | 28.6 | 29.9 | 30.6 | 32.3 |
| C4 | 23.5 | 15.7 | 12.8 | 17.5 | 16.2 | 17.3 | 17.7 |
| HHI | 265 | 92 | 68 | 124 | 103 | 111 | 121 |
| Noam | 84 | 29 | 22 | 39 | 32 | 35 | 34 |

*Source*: <http://www.boersenverein.de/de/158286> 2012 updates provided by editor and only 40% are sold via subscription. Several readers often share one copy of a magazine. In recent years the number of titles increased while the circulation per title decreased. As economies of scale are important in this industry, these developments led to decreasing margins for publishers. In 1998, there were 39 titles in Germany with a print run higher than 500,000 and 19 of these titles had a print run higher than one million. Six titles crossed a print run of two million.[16] In 2008, only 26 titles had a print run higher than 500,000 and 11 crossed one million. Only two magazines had a print run higher than two million. The number of magazines sold dropped 25.4% from 1998 to 2008, while the average selling price increased by 20.3%.[17] The concentration is measured by print run.[18]

As four publishers make up 60% of the German market, it can be described as oligopolistic.[19] Bauer publishing group is the largest player, with 18.5% market share. With a print run of 8,000,000 magazines, Bauer is the market leader of TV program guides, as well as for magazines for women and teenagers. Burda is the second largest player, and is also focused on TV program guides and women magazines. Furthermore, Burda owns *Focus* and *Bunte*, both among the most popular magazines in Germany. Founded in 1993, *Focus* had become the third largest magazine in Germany by 2008.

Axel Springer AG is the third largest player and specializes on the general interest segment and program guides. General interest magazines like *Bild der Frau, Auto Bild,* and *Sport Bild* refer regularly to *BILD*, the daily newspaper of Axel Springer AG, and vice versa.

Gruner + Jahr (owned by Bertelsmann) is the fourth largest magazine publisher in Germany and focuses on magazines for women and special interest groups. Gruner + Jahr is also the owner of *Stern*, the second largest magazine in Germany.[20]

Since the end of the 1960s, four publishers have dominated the German magazine market. In the 1970s, the C4 index reached its peak with a market share of 67% for the four largest publishers.[21] Within the last 28 years, the C4 index decreased slightly and the four large publishers converged in terms of market shares. HHI and Noam Index also decreased slightly.

Similar to daily newspapers, the market for magazines can be divided into two segments: magazines that are published every 14 days and magazines that are published less frequently. In the first segment, the WAZ Group took over the fourth position from Gruner + Jahr in 2008 and ranked fifth in the total magazine market. This was the first time that the order of the four largest publishers changed over the last 28 years. The largest player, Bauer, lost market share, while Burda, Springer, and WAZ gained market share. C4 increased slightly from 2004 to 2008 while HHI and Noam Index remained almost constant (Table 6.4).

As the revenues are increasing in the segment of magazines that are published less frequently than every 14 days, it is particularly interesting to have a closer look at that segment. From 1980 to 1995, all three indices decreased and reached their lowest points.[22] After 1995, all three indices increased and returned to their 1980 levels by 2008. Burda, Gruner + Jahr, Springer, and Bauer converged in terms of print run over the years. Due to acquisitions, Axel Springer AG grew from a market share of 1% in this segment in 1995 to 6.5% after 2008.

---

16. Informationsgesellschaft zur Feststellung der Verbreitung von Werbeträgern e.V. 2009.

17. Vogel, A. "Publikumszeitschriften Dominanz der Großverlage gestiegen." Media Perspektiven, Sept. 2002, 433.

18. The market shares have been calculated by Andreas Vogel based on IVW Data: Vogel, A. "Publikumszeitschriften Dominanz der Großverlage gestiegen." Media Perspektiven, Sept. 2002, 433; Vogel, A. "Konsolidierte Großkonzerne bereit zu erneutem Wachstum." Media Perspektiven, July 2004, 322; Vogel, A. "Stabile Positionen in schrumpfenden Märkten." Media Perspektiven Sept. 2008, 467.

19. Sjurts, I. *Strategien der Medienbranche.* Wiesbaden: Betriebswirtschaftlicher Verlag Dr. Th. Gabler, 2002, 97.

20. Hans-Bredow-Institut, *Medien von A bis Z.* Wiesbaden: VS Verlag für Sozialwissenschaften, 2006.

21. Vogel, A. "Stabile Positionen in schrumpfenden Märkten." Media Perspektiven Sept. 2008, 467.

22. Vogel, A. "Publikumszeitschriften Dominanz der Großverlage gestiegen." Media Perspektiven, Sept. 2002, 433; Vogel, A. "Konsolidierte Großkonzerne bereit zu erneutem Wachstum." Media Perspektiven, July 2004, 322; Vogel, A. "Stabile Positionen in schrumpfenden Märkten." Media Perspektiven Sept. 2008, 467.

Table 6-4. Market Shares in the German Magazine Publishing Market

| POPULAR MAGAZINES | 1980 | 1985 | 1990 | 1995 | 2000 | 2002 | 2004 | 2006 | 2008 | 2010 | 2012 |
|---|---|---|---|---|---|---|---|---|---|---|---|
| Bauer | 28.5 | 27.9 | 27.6 | 25.6 | 22.3 | 23.4 | 21.1 | 20.7 | 19.5 | 19.0 | 18.5 |
| Burda | 13 | 11.4 | 9.9 | 9.3 | 10.8 | 12.4 | 13.5 | 15.5 | 16.5 | 15.4 | 14.8 |
| Springer | 15.2 | 16.8 | 16.9 | 13.8 | 15.4 | 15.7 | 16.3 | 16.1 | 15.1 | 13.4 | 12.7 |
| Bertelsmann (Gruner + Jahr) | 10.2 | 8.8 | 9.7 | 10.9 | 10.1 | 9.7 | 9.5 | 10.6 | 10.8 | 9.5 | 9.3 |
| Revenue ($US mln) | | | | | | | 5,146 | 5,530 | | | |
| C4 | 66.9 | 64.9 | 64.2 | 59.6 | 58.6 | 61.2 | 60.4 | 62.9 | 61.9 | 57.3 | 55.3 |
| HHI | 1,319 | 1,266 | 1,242 | 1,051 | 951 | 1,041 | 983 | 1,040 | 997 | 868 | 809 |
| *N* (>1%) | 4 | 4 | 4 | 4 | 4 | 4 | 4 | 4 | 4 | 4 | 4 |
| Noam | 659 | 633 | 621 | 526 | 476 | 521 | 492 | 520 | 499 | 434 | 405 |

*Source*: KEK, "Auflagenstärkste Verlagsgruppen" (2009)

## AUDIOVISUAL MEDIA

### Radio

The German radio market is shaped by the regional structure of each German state. National radio stations are rare and of low importance.[23] The number of local stations is high. Of German radio stations, 30% are local, 65% are broadcast statewide, and only 5% are national. Terrestrial radio transmission over VHF, using FM, is dominant, but other types of transmission such as DAB still play a minor role. Because of the regional and local structure of the German radio, the concentration of the market is low. On a nationwide level, no private radio network reaches the critical value of 30% market share, a concentration limit defined by the KEK.[24]

Similar to the German TV market, the radio market is characterized by a dual system of public and private stations. The first private stations were introduced in 1984. In the first wave of licensing, local newspapers publishers played a dominant role.[25]

Following national reunification, in a second phase from 1992 to 1995, new stations were licensed in the former East Germany. Between 1995 and 2000, the radio industry was booming. The number of listeners, as well as advertising revenues, kept growing. At its peak in 2000, statewide private radio stations earned an operating margin of 20 to 30%. After 2001, the economic environment deteriorated and the advertising market decreased. Parallel to that, many statewide private radio stations lost listeners to public radio programs. In 2009, the market share of public radio stations was about 50%, depending on the state.[26]

The number of radio stations grew significantly from 40 public and 100 private channels in 1989 to 59 public and 213 private channels in 2008.[27] In order to measure concentration, the market shares are based on the Media Analyse Radio, a semiannual survey by the Arbeitsgemeinschaft Media-Analyse (AG.MA). All advertising agencies base their media planning on this dataset.

The most important players in radio are media companies that are also engaged in TV. RTL Group (Bertelsmann) holds shares in 19

23. "Hörfunk." KEK. Last accessed on August 10, 2012 at http://www.kek-online.de/Inhalte/hoerfunk.html
24. "Hörfunk." KEK. Last accessed on August 10, 2012 at http://www.kek-online.de/Inhalte/hoerfunk.html
25. Dieter, K. "Praxiswissen Radio." Müller, Esther Raff. 2007, 9.
26. "ALM Jahrbuch 2010".
27. Channels reported by MA Radio.

statewide and regional radio stations.[28] Axel Springer AG is a shareholder of 18 statewide stations and also is invested in Sat.1 and ProSieben. Hubert Burda Media holds interests in various radio stations and RTL II.

Other major players are NRJ, Energy-Holding, Moira Rundfunk, Madsack, Nordwest-Medien-Holding, WAZ (newspaper company), Zeitungsgruppe Ippen (newspaper company), Frank Otto Medienbeteiligung, and Oschmann-Gruppe. The commercial radio market is mainly owned by newspaper publishing companies.[29]

The commercial radio stations with the largest market share are Radio NRW, Antenne Bayern, Hit Radio FFH, ffn, Hit-Radio Antenne, RPR Eins, radio SAW, Hit-Radio Antenne 1 (Stuttgart), R.SH, and Radio PSR.[30] Because the German radio market is mostly state-based, concentration has to be measured on a state level rather than on a nationwide level. We therefore focus on the biggest state, namely Bavaria, to measure the development of market concentration. The revenues shown in Table 6.5 are advertising net revenues.

## Table 6-5. Market Shares of German Radio Stations and Networks in Bavaria (%): 1996–2009

| | 1996 | 2000 | 2003 | 2007 | 2009 |
|---|---|---|---|---|---|
| Public Channels | | | 45.0 | 40.3 | 46.3 |
| Antenne Bayern | | | 19.1 | 28.8 | 27.4 |
| Bayern Funpaket[1] | | | 23.7 | 22.8 | 17.8 |
| Radio Arabella | | | 3.1 | 2.2 | 1.5 |
| Radio Galaxy | | | 1.0 | 1.6 | |
| Radio Gong 96,3 (München) | | | 2.1 | 1.6 | 1.0 |
| Klassik Radio | | | 1.0 | 1.6 | 1.0 |
| Radio-Kombi Ba-Wü | | | | 1.1 | 2.0 |
| Energy München | | | 1.0 | | 1.0 |
| 95.5 Charivari (München) | | | 0.5 | | 0.5 |
| Rock Antenne | | | 1.0 | | 1.5 |
| Radio Melodie | | | 1.0 | | |
| Radio 7 | | | 1.0 | | 0.5 |
| Total Revenues (€ m) | 547 | 742 | 603 | 678 | 680 |
| Total Revenues ($ m) | | 987 | 681 | 929 | 952 |
| C4 | 0 | 0 | 91.0 | 94.0 | 93.0 |
| HHI | 0 | 0 | 2,970 | 2,987 | 3,223 |
| *N* (>1%) | 0 | 0 | 11 | 8 | 9 |
| Noam | | | 895 | 1,056 | 1,074 |
| Sum of Total Market Shares | 0 | 0 | 100 | 100 | 100 |

[1] Bayern Fun Paket is an aggregation of over 50 local radio stations. They are marketed as one package, but they do not have the same ownership structure. Media Analyse Radio reports their market share as an aggregate, and therefore they are treated as one player for the calculation of C4 and HHI. For the calculation of the Noam Index the number of all stations has been taken into account, and there is therefore an enormous difference between the concentration in terms of HHI and Noam Index.

28. "Hörfunk." KEK. Last accessed on January 20, 2009 at http://www.kek-online.de/Inhalte/hoerfunk.html
29. Breunig, C. "Radiomarkt in Deutschland: Entwicklung und Perspektiven." Media Perspektiven, Sept. 2001, 450.
30. VPRT "Hörfunk in Deutschland." 2006, 36.

## Broadcast Television (Over-the-Air)

The terrestrial transmission of TV has a low audience in Germany equaling 5% of the population, or 2,000,000 people.[31] Most people watch TV via cable (50%) or satellite (44%). In 2010, TV started streaming via Internet, capturing an initial 2.5% of viewers.[32] Depending upon location, up to 28 channels can be received via terrestrial TV. Eight public channels, including ZDF and ARD, can be watched across Germany. Terrestrial TV is free. In terms of revenues, terrestrial TV accounts for 5% of all TV viewing in Germany. We therefore assumed that the revenue of terrestrial TV accounts for this percentage of the overall TV revenue in Germany (Table 6.6).

## Multichannel TV Platforms

Cable and satellite platforms are the two major technologies to process TV signals in Germany. From 2005 to 2009, slightly more households received their TV signal by cable than by satellite. Terrestrial broadcasting remains at a relatively low level, while IPTV is gaining ground slowly (Table 6.7).

With a technical reach of up to 28 million households, the German cable TV market is the largest in Europe. The cable network was created by the state-owned Deutsche Bundespost in the 1980s and can be divided into four structural levels. On level 1, the broadcasting company uploads the signal to a satellite. On level 2, the signal is received from the satellite and is sent to central cable backbone nodes. On level 3, the signal is transmitted via coaxial or fiber-optic cable to distribution nodes in the public streets. On level 4, the signal is transmitted via cable on private property to the end consumer.

From 2000 to 2003, Deutsche Telekom, which emerged as the privatized monopolistic provider from Deutsche Bundespost, was forced by EU regulatory officials to sell parts of its cable network in levels 3 and 4 to companies

### Table 6-6. Total Revenues Broadcast TV (Over the Air): 1984–2010

| | 1984 | 1985 | 1988 | 1992 | 1996 | 2000 | 2004 | 2007 | 2010 |
|---|---|---|---|---|---|---|---|---|---|
| Total Revenue (€ m) Terrestrial TV | 44 | 47 | 60 | 164 | 419 | 580 | 562 | 616 | 682 |
| Total Revenue ($ m) Terrestrial TV | | | | | | 540 | 697 | 844 | 907 |

### Table 6-7. Market Shares of TV Signal Distribution Technologies in German Households (%)

| | CABLE | SATELLITE | TERRESTRIAL | IPTV |
|---|---|---|---|---|
| 2005 | 50.4 | 44.0 | 5.6 | 0.0 |
| 2006 | 49.7 | 44.9 | 5.6 | 0.1 |
| 2007 | 48.4 | 45.1 | 5.4 | 0.5 |
| 2008 | 48.8 | 43.0 | 6.1 | 1.4 |
| 2009 | 49.2 | 42.9 | 6.9 | 2.6 |

*Source*: PricewaterhouseCoopers (2010).

31. PricewaterhouseCoopers. "German Entertainment and Media Outlook: 2010–2014—Die Entwicklung des deutschen Unterhaltungs- und Medienmarktes." Frankfurt am Main, 2010, 33.

32. "German Cable Market 2012." *Solon*. Last accessed on 20 January 2009 at http://www.solonstrategy.com/uploads/tx_soloncm003/2008_German_Cable_Market_2012.pdf p. 6.

like Kabel Deutschland and Kabel BW. In 2004, Kabel Deutschland planned to take over several competitors with the potential to gain nearly monopolistic power and therefore was stopped by regulatory offices. In 2005, the two regional cable network providers Ish (Nordrhein-Westfalen) and Iesy (Hessen) attempted a merger that was completed in 2007 under the new name Unitymedia. Through the acquisition of Tele Columbus West in 2007, Unitymedia became the second largest cable network provider after Kabel Deutschland. Kabel BW was third.

Kabel Deutschland offers cable TV, broadband Internet access, and telephone services in 13 of the 16 federal states in Germany, with a customer base of 8.7 million households in 2009. Unitymedia serves in the states of North Rhine-Westphalia and Hesse with a total of 4.5 million households. Kabel BW serves 2.3 million households in Baden-Württemberg. Tele Columbus serves 2.3 million and PrimaCom serves 850,000 households, mainly in eastern Germany.

Besides these major market participants there are several hundred small- to medium-sized companies offering cable TV services on a regional or city level (city-carriers), with customer bases ranging from 2,000 to 300,000. Some of the bigger city carriers are EWETel in Hamburg (Northern Germany), Net-Cologne (Middle Germany), and M-Net in Munich (Southern Germany). Like the large market participants, these companies offer triple play services (TV, Internet access, and telephone services).

German cable providers generated a total revenue of 3.6 billion € (US $5.0 billion) in 2009 (Table 6.8). The main revenue source was TV access with 63% of total revenues. However, since 2007 the provision of broadband Internet access became a larger driving force of business. In 2008, the cable providers gained 710,000 (+ 79%) customers. In 2009, they gained 770,000 (+ 48%) customers in that segment.

The German satellite TV platform provider market can be described as a duopoly between SES ASTRA and Eutelsat S.A. SES ASTRA is a full subsidiary of SES S.A., the worldwide leading provider of satellite communication services. In 2001, SES took over the satellite

**Table 6-8. Revenues and Market Shares of Cable Platform Providers in Germany 2009–2011**

| | COMPANY REVENUES 2009 ($ BILLION) | MARKET SHARE BY REVENUE 2009 | COMPANY REVENUES 2011 ($ BILLION) | MARKET SHARES BY REVENUES 2011 ($ BILLION) |
|---|---|---|---|---|
| Kabel Deutschland | 2.1 | 47 | 2.0 | 45 |
| Unitymedia | 1.3 | 28 | 1.3 | 30 |
| Kabel BW | 0.7 | 16 | 0.8 | 17 |
| Tele Columbia | 0.3 | 6 | 0.3 | 6 |
| Medfort S à r.l (Primacon)[1] | 0.1 | 3 | 0.1* | 2 |
| Total Revenues ($US million) | 4,460 | | 4,480 | |
| C4 | | 97 | | 93 |
| HHI | | 3,294 | | 3,254 |
| *N* (>1%) | | 5 | | 5 |
| Noam | | 1,473 | | 1,455 |

*Source*: "Wirtschaftsfaktor Kabel." Solon Management Consultants. Last accessed on August 10, 2012 at <http://www.anga.de/media/file/10.Solon-Studie_Wirtschaftsfaktor_Kabel_final.pdf>. 2011 updates provided by editors.

[1] Primacon was bought by Medfort S à r l in 2010.

fleet of GE AMERICOM (General Electric). In the process of that acquisition, SES S.A. (SES Global) was founded as a holding company for the two satellite platform providers SES ASTRA and SES WORLD SKIES. The SES S.A. equities can be divided into class A and class B. Class A is owned by private investors. Class B is not publicly traded and is owned by the state of Luxembourg (or companies largely controlled by the state of Luxembourg).

Besides SES ASTRA, Eutelsat S.A. is the largest provider of satellite TV services in Germany and in all of Europe. With a fleet of 30 satellites, more that 4,000 TV channels, and 1,100 radio channels are transmitted to around 190 million households in Europe.[33] The holding company Eutelsat Communications S.A. owns 96.11% of the holding company; 3.89% is owned by other shareholders.[34]

Information about exact market shares of satellite platform providers for DTH reception is not available through Kommission zur Ermittlung der Konzentration (KEK).[35] According to SES ASTRA,[36] the company served up to 17 million German households with satellite signals in 2007. SES ASTRA's market dominance in the DTH segment explains Eutelsat's significantly smaller market share. There are a relatively high number of so-called Dual-Feed-Antennas. These antennas receive TV signals from more than one orbital position. Also, cable providers use various satellite sources simultaneously to feed into their networks.

The German TV landscape is shaped by the dual system of public and private channels. Private TV channels were introduced after the liberalization of the TV market in the 1980s. The concentration in the German TV industry is high, but has declined since the start of the first two private channels, RTL and Sat.1, in 1984. Today, about 40% of the viewing time is spent on public channels and two TV groups dominate the private market: ProSiebenSat.1 Media AG and RTL Group.

The program offered by public stations is mainly financed by tax-like fees, collected by the state. Private stations are primarily financed by the advertising industry and are mostly free. Pay TV only has a small market share of about 8% in Germany. Concentration in the TV industry has been deduced from the concentration of viewing time, measured by the Gesellschaft für Konsumforschung (GFK)/Arbeitsgemeinschaft Fernsehforschung (AGF).[37] Since the advertising spending is allocated according to the viewing time concentration, it is an adequate measurement of the market share. The ownership was assigned according to the annual reports of the Kommission zur Ermittlung der Konzentration (KEK).[38]

The two major public channels in Germany are ARD and ZDF. License fees mainly finance them both, and their program mandate is governed by the Interstate Treaty on Broadcasting (Rundfunkstaatsvertrag).

On January 1, 1984, PKS, which was later renamed Sat.1, became the first private channel in Germany, followed by RTL one day later. This changed the monopoly of the public channels into a public-private mixed broadcasting system. The private channels were broadcast over cable and satellite only.

In the first years after introducing private channels, their market share grew slowly. In 1988, private channels only had a market share of 9.9%. More and more households were connected to cable or satellite, thus making it possible to receive private channels. However, the

---

33. "Investors Information." Eutelsat S.A. 2012. Last accessed on August 10, 2012 at www.eutel-sat.com

34. "Investors Information." Eutelsat S.A. 2012. Last accessed on August 10, 2012 at www.eutel-sat.com

35. "Markt für Übertragungswege." KEK. 2012. Last accessed on August 10, 2012 at http://www.kek-online.de/Inhalte/markt_fuer_uebertragungswege_in_deutschland.html

36. "Investors Information." SES. 2012. Last accessed on August 10, 2012 at http://www.ses.com/3913663/investors

37. "Jährliche Zuschaueranteile seit 1985." KEK. 2009. Last accessed on August 10, 2012 at http://www.kek-online.de/kek/medien/zuschauer/jahr.pdf

38. "Jahresberichte." KEK. Various years. Last accessed on August 10, 2012 at http://www.kek-online.de/Inhalte/jahresberichte.html (The reports of the KEK document the ownership situation since 1996. For the years before 1996, the Media Perspektiven Basisdaten were used.)

quality of the private programming was low and viewers were used to the shows on the public channels.

This situation changed dramatically in the 1990s. In 1992 private channels had a market share of 39.3% and in 1996 a share of 56.2%. The number of channels increased and special interest channels were introduced. Also, new media companies entered the German TV market. While in 1992 all of the private channels belonged either to the Kirch Group (later ProSiebenSat.1) or RTL Group, by 2000 Télévision Francaise, News Corp, and Time Warner owned 2.7% of the German TV market. This number increased to 5.9% in 2007, while the market share of the large TV conglomerates ProSiebenSat.1 and RTL Group decreased to 20.8% and 25.3%, respectively. The market share of the public channels steadily decreased from 1985 until 1996. Since the year 2000, the market share has been stable at around 43%.

RTL Group is part of the worldwide media conglomerate of Bertelsmann AG. Bertelsmann holds 89.8% of the shares of RTL Group.[39] The remaining 10.2% is held by a diverse set of investors. The RTL Group is the largest TV and radio group in Europe, holding shares in 39 TV and 32 radio stations in ten countries, though the highest revenues come from the German stations.[40] RTL Group was founded in 2000 as a merger of CLT-UFA and Pearson TV. Bertelsmann became the major shareholder with 90% of the shares in 2001. German channels belonging to the RTL Group are RTL, RTL II, N-TV, VOX, RTL Living, RTL Crime, RTL Passion, and RTL Shop.

ProSiebenSat.1 Media AG holds shares in 24 free TV, 24 pay TV, and 22 radio networks in 13 European countries.[41] The company was created in 2000 by the merger of Pro Sieben Media AG and Sat.1 Satelliten Fernsehen GmbH. Both were affiliated with the media conglomerate of Leo Kirch, an important figures in German TV. In 2002, Kirch went bankrupt. The US investor Haim Saban took over the majority of the shares of ProSiebenSat.1. In 2005, the publishing house Axel Springer intended to acquire ProSiebenSat.1, but the transaction was blocked by the Bundeskartellamt and KEK due to antitrust concerns. In 2006, Saban sold his stake to the American private equity firms Permira and KKR for about 3 billion €. Since 2008, 100% of the shares of the group have been held by Lavena Holdings 4 GmbH, which is owned by the two private equity companies KKR and Permira. Together with SBS Broadcasting Group, another investment of Permira and KKR, they hold significant investments in European media. German speaking channels of ProSiebenSat.1 Media are Sat.1, ProSieben, Kabel Eins, N24, 9Live, Sat.1 Comedy, and Kabel Eins Classic.

The C4 index decreased from 1985 to 2012. The HHI also strongly decreased from 1985 to 2012. The only increase can be detected from 1996 to 2000, probably caused by a change in regulation. After 1996, companies were allowed to hold a majority stake in more than one TV station, which had previously not been allowed. This facilitated mergers and acquisitions, although the change in the data is not very drastic (Table 6.9).

## Video Portals

Based on the quantity of regular users and the intensity of videos being watched online, video portals represent one of the Internet's most popular features. Video portals sit on the fence between traditional television and the Internet. They allow users to publish their own content, and they provide other users with user-generated content usually at no charge.[42]

Video portals mainly derive their revenues from advertisements, which are either placed on

39. "Elektronische Medien—Entwicklung und Regulierungsbedarf." Holznagel, Bernd/Dörr, Dieter/Hildebrand, Doris. 2008. 1st edition, Munich, 32.

40. "ALM Jahrbuch 2006—Landesmedienanstalten und privater Rundfunk in Deutschland." ALM. 2006. 1st edition, Berlin, 227.

41. "Elektronische Medien—Entwicklung und Regulierungsbedarf." Holznagel, Bernd / Dörr, Dieter / Hildebrand, Doris 2008. 1st edition, Munich, 35.

42. SevenOne Media; Accenture (2008): Videoportale in Deutschland—Im Spannungsfeld zwischen Fernsehen und Internet 3–5.

Table 6-9. National Market Shares of German Satellite and Cable TV Channels (%)[1]

| | 1985 | 1988 | 1992 | 1996 | 2000 | 2004 | 2007 | 2012 |
|---|---|---|---|---|---|---|---|---|
| Public Channels | 96.2 | 84.7 | 52.0 | 40.5 | 43.1 | 44.3 | 43.6 | 37.5 |
| ARD | *43.4* | *37.8* | *22.0* | *14.8* | *14.3* | *13.9* | *13.4* | *12.3* |
| ZDF | *42.6* | *36.2* | *22.0* | *14.4* | *13.3* | *13.6* | *12.9* | *12.6* |
| Other Public (Aggregate Regional) | 10.2 | 10.7 | 8.30 | 11.3 | 15.5 | 16.8 | 17.3 | 12.6 |
| ProSiebenSat.1 Media AG (formerly Kirch Group, now owned KKR/Permira) | | 5.8 | 22.6 | 27.4 | 26.2 | 21.9 | 20.8 | 20.2 |
| Sat.1 | | *5.8* | *13.1* | *13.2* | *10.2* | *10.3* | *9.6* | *9.4* |
| ProSieben | | | *6.5* | *9.5* | *8.2* | *7.0* | *6.5* | *5.9* |
| Kabel 1 | | | | *3.6* | *5.5* | *4.0* | *3.6* | *3.9* |
| DSF (*until 2003) | | | *3.0* | *1.1* | *1.2* | | | |
| N 24 | | | | | | *0.4* | *0.9* | *1.0* |
| Premiere (*1997–2003) | | | | | *1.1* | | | |
| 9Live (*since 2001) | | | | | | *0.2* | *0.2* | - |
| RTL Group (formerly CLT/UFA, now Bertelsmann) | 0.4 | 4.1 | 16.7 | 26.6 | 24.7 | 25.6 | 25.3 | 25.1 |
| RTL | *0.4* | *4.1* | *16.7* | *17* | *14.3* | *13.8* | *12.4* | *12.3* |
| Super RTL, RTL II, VOX | | | | *9.6* | *10.4* | *11.3* | *12.2* | *11.9* |
| n-tv (*since 2002) | | | | | | 0.5 | 0.7 | 0.9 |
| Télévision Francaise (Bouygues, France) | | | | 1.2 | 1.0 | 0.9 | 1.0 | 0.7 |
| Eurosport | | | | *1.2* | *1.0* | *0.9* | *1.0* | *0.7* |
| Viacom (US) | | | | | | 1.4 | 1.8 | 1.5 |
| MTV, MTV2, Viva, Comedy Central, Nick | | | | | | *1.4* | *1.8* | *1.5* |
| EM.Sport Media | | | | | | 1.1 | 1.1 | 0.7 |
| DSF (*since 2005) | | | | | | *1.1* | *1.1* | *0.7* |
| News Corp (US/UK) | | | | | 1.0 | | | |
| 9Live (*2000–2001) | | | | | *1.0* | | | |

*continued*

Table 6-9. *continued*

| | 1985 | 1988 | 1992 | 1996 | 2000 | 2004 | 2007 | 2012 |
|---|---|---|---|---|---|---|---|---|
| Investment Group Permira | | | | | | 2.4 | | |
| Premiere (*2003–2006) | | | | | | *2.4* | | |
| Ufa /Canal Plus France | | | | 0.7 | | | | |
| Premiere (*until 1997) | | | | *0.7* | | | | |
| Time Warner (US) | | | | 0.3 | 0.7 | | | |
| n-tv (*until 2002) | | | | *0.3* | *0.7* | | | |
| Tele-München | | | | | | 0.3 | 0.7 | 1.0 |
| Tele 5 | | | | | | *0.3* | *0.7* | *1.0* |
| Discovery Communications (US) | | | | | | 0.1 | 0.5 | 0.7 |
| DMAX | | | | | | *0.1* | *0.5* | *0.7* |
| MMW Minimovie World (REN Media Group, Lesnevsky Family, Russia) | | | | | | | 0.8 | 0.2 |
| Das Vierte | | | | | | | *0.8* | *0.2* |
| Other | 3.4 | 5.4 | 11.4 | 5.4 | 7.3 | 2 | 4.4 | 12.4 |
| Total Revenues (€ m) | 888 | 1,139 | 3,106 | 7,962 | 11,019 | 11,322 | 11,698 | |
| Total Revenues ($ m) | | | | | 10,137 | 14,048 | 16,027 | 19,974 |
| C4 | 97 | 95 | 92 | 96 | 95 | 94 | 92 | 84.3 |
| HHI | 9,255 | 7,225 | 3,525 | 3,101 | 3,157 | 3,107 | 2,981 | 2,449 |
| *N* (>1%) | 2 | 6 | 8 | 12 | 11 | 14 | 12 | 9 |
| Noam | 6,544 | 4,171 | 2,035 | 1,266 | 1,289 | 1,036 | 994 | 816 |
| Sum of total market shares | 100 | 100 | 100 | 100 | 100 | 100 | 100 | |

[1]All channels with an asterisk * have changed their owner over time.

their website or are integrated specifically into their hosted videos.[43] Hence, the respective market shares have been deducted from the monthly number of videos viewed. Google's video portal comprises YouTube and Google Video. Founded in February 2005, YouTube quickly gained brand awareness and market share. Google acquired YouTube in October 2006, and this acquisition helped Google significantly increase its overall market share. In 2006, Google Video amounted to roughly 8% market share.[44] In the following years, Google's video portals achieved a combined market share of 52.1% in 2008 and 48.3% in 2010. Hence, YouTube and Google Video are by far the biggest players in the German video portals market.

Megavideo had the second biggest market share in 2010. German users viewed over 200,000 videos hosted by this video portal in July 2010, resulting in a market share of less than 3% for Megavideo. ProSiebenSat1, with its video portals such as MyVideo.de, is the third biggest player in the German market. However, those video portals only account for a combined market share of less than 2%.

RTL Group's video portals such as Clipfish.de have an accumulated market share of less than 1% in 2010. Similar to the TV network ProSiebenSat1 mentioned above, the RTL Group publishes user-generated content and its own content on its video portals.

The market includes other competitors such as the video portals from Yahoo and Microsoft. Both corporations entered the market rather late: in 2006 and 2007, respectively.[45]

## Table 6-10. National Market Shares of Video Portals (%)

| | 2008[1] | 2010[2] |
|---|---|---|
| Google Sites (YouTube, Google Video) | 52.1 | 48.3 |
| Megavideo | 1.0 | 2.6 |
| ProSiebenSat1 Sites (MyVideo.de) | 1.5 | 1.2 |
| RTL Group Sites (Clipfish.de) | 1.1 | 0.6 |
| Microsoft Sites | 0.4 | 0.4 |
| Dailymotion.de | 0.2 | 0.3 |
| Yahoo Sites (Yahoo Video) | 0.3 | 0.1 |
| Others (e.g. AOL Sites, T-Online Sites, Vevo, ZDF Mediathek) | 43.4 | 46.6 |
| Sum of Total Market Shares | 100 | 100 |
| Total Revenues (€ m) | n.a. | n.a. |
| Total Revenues ($ m) | n.a. | n.a. |
| C4 | 55.7 | 52.6 |
| HHI | 2,718 | 2,337 |
| Noam | 1,028 | 884 |

[1] ComScore Video Metrix. *Comscore*. October 20, 2008. Last accessed on August 10, 2012 at <http://www.comscore.com/Press_Events/Press_Releases/2008/10/German_YouTube_Viewers>

[2] ComScore Video Metrix. *Comscore*. September 14, 2010. Last accessed on August 10, 2012 at <http://www.comscore.com/Press_Events/Press_Releases/2010/9/Online-Video_Konsum_in_Deutschland_steigt_weiter_stark_an?utm_source=feedburner&utm_medium=feed&utm_campaign=Feed%3A+comscore+%28comScore+Networks%29&utm_content=Google+Feedfetcher>

43. SevenOne Media; Accenture (2008): Videoportale in Deutschland—Im Spannungsfeld zwischen Fernsehen und Internet, 25–27.

44. SevenOne Media; Accenture (2008): Videoportale in Deutschland—Im Spannungsfeld zwischen Fernsehen und Internet, 3–31.

45. "YouTube macht teure Google-Aktie kaum attraktiver." *Faz.net*. October 10, 2006. Last accessed on August 10, 2012 at http://www.faz.net/s/RubF3F7C1F630AE4F8D8326AC2A80BDBBDE/Doc~E70C15174EE9D48D5BB1220F4962150CE~ATpl~Ecommon~Scontent.html

Furthermore, the market is characterized by a multiplicity of small- and medium-sized competitors such as ZDF Mediathek, AOL sites, and Vevo, as well as a high level of illegal and semi-legal video portals. The overall number of video portals in the market has increased continuously and is subject to certain volatility (Table 6.10).

Besides Google's acquisition of YouTube, a major change in ownership did not occur in this comparably young and highly concentrated market.

## Table 6-11. National Market Shares of Film and TV Production and Distribution (in %): 1984–2010

| | 1984[1] | 1989 | 1992 | 1996[2] | 2000 | 2004 | 2008 | 2010[3] |
|---|---|---|---|---|---|---|---|---|
| Warner Bros. Entertainment (Time Warner, US) | 10.0 | 14.1 | 15.2 | 14.7 | 10.0 | 13.5 | 16.6 | 14.8 |
| Paramount Pictures (Viacom, US) | 14.0 | 11.3 | 8.8 | 11.6 | 10.0 | 9.5 | 13.9 | 13.2 |
| DreamWorks SKG | - | - | - | - | 8.0 | 11.1 | - | - |
| Twentieth Century Fox (21st Century Fox/Murdoch, US/UK) | 6.0 | 7.7 | 5.7 | 8.2 | 9.0 | 9.3 | 9.0 | 11.7 |
| Walt Disney Motion Pictures (US) | 9.0 | 11.1 | 11.0 | 14.2 | 13.0 | 9.7 | 10.2 | 11.1 |
| Miramax | 3.0 | 2.4 | 4.1 | - | - | - | - | - |
| Pixar Animation Studios | - | - | - | n.a. | n.a. | 1.6 | - | - |
| Marvel Entertainment | - | - | - | n.a. | 1.0 | 1.2 | 3.1 | - |
| Sony Pictures Entertainment (Japan/US) | - | - | 10.5 | 11.7 | 8.0 | 13.4 | 10.3 | 10.4 |
| Columbia Pictures | 8.0 | 9.8 | - | - | - | - | - | - |
| Tri-Star Pictures | 5.0 | - | - | - | - | - | - | - |
| Universal Pictures (US/France) | 7.0 | 10.5 | 13.3 | 12.1 | 13.0 | 9.1 | 12.8 | 9.7 |
| Other (MGM, Constantin Film, Tobis, etc.) | 38.0 | 33.1 | 31.4 | 27.5 | 28.0 | 21.6 | 24.1 | 29.1 |
| Sum of total market shares | 100 | 100 | 100 | 100 | 100 | 100 | 100 | 100 |
| Total revenues (€ m)[4, 5, 6, 7] | 1,112 | 1,245 | 1,632 | 1,815 | 2,127 | 2,639 | 2,356 | 2,807 |
| Total revenues ($ m) | | | | | 1,957 | 3,272 | 3,487 | 3,733 |
| C4 | 41 | 47 | 50 | 53 | 46 | 48 | 50 | 71 |
| HHI | 560 | 721 | 766 | 903 | 748 | 843 | 933 | 856 |
| *N* (>1%) | 8 | 7 | 7 | 6 | 8 | 9 | 7 | 6 |
| Noam | 198 | 273 | 289,5 | 369 | 265 | 281 | 353 | 349 |

[1]Frank(1993).
[2]Hachmeister and Rager(2003).
[3]PricewaterhouseCoopers(2010).
[4] Uka (2004).
[5] Total revenue in € m for 1988.
[6] Total revenue in € m for 1998.
[7] Total revenue in € m for 2001.

## Film

The market for the production and distribution of film and television content is highly global. The respective German market has been dominated by large American companies. However, several predominantly German players have been able to maintain an aggregated market share of 21% to 38%.

The market shares were obtained from adjacent research that derived the respective market shares from the companies' revenues.[46, 47]

Other market participants include German and foreign companies such as Universum Film AG, Bavaria Film Gruppe, Brainpool TV GmbH, Endemol Gruppe, Constantin Film, Tobis, and MGM.

Besides the industry's major mergers and acquisitions mentioned above, several small market participants have been acquired by the big six German market players for the production and distribution of film and television content. The stagnating HHI between 560 and 856 implies a less concentrated market despite the seemingly dominant handful of big players in the market (Table 6.11).

# TELECOMMUNICATIONS MEDIA

## Wireline Telecom

In 1998, the German market for voice telephony was completely liberalized. The former incumbent Deutsche Telekom was obliged to open its network infrastructure to new entrants and provide access to all customer lines, enabling interconnection, Local Loop Unbundling (LLU), number portability, and carrier preselection.

The calculation of market shares is based on the number of fixed network connections. The German regulator Bundesnetzagentur tracks the ratio between Deutsche Telekom and its competitors.[48] This analysis only includes providers of the direct subscriber lines,[49] not call-by-call or preselection service providers.

Deutsche Telekom AG evolved from the privatization of the Deutsche Bundespost after a reform in the German telecommunication and post system in 1995. With the liberalization of the market in 1998, new players emerged. The main competitors of Deutsche Telekom as providers of direct subscriber lines are Arcor (now Vodafone), and local and regional providers like EWE TEL, HanseNet, ISIS, M-Net, and NetCologne.

Since the liberalization of the fixed network market a lot of merger and acquisition activity could be observed. In 1999, Mannesmann/Arcor acquired the city carrier Isis (Dusseldorf).[50] In 2000, three city carriers (later 14 carriers) merged to become Tropolys.[51] In 2004, Apax became the major shareholder in Tropolys.[52] Apax acquired Versatel in 2005[53] and in 2006 Versatel and Tropolys merged to become Versatel.[54] In 2000, Vodafone acquired Arcor as part of the Mannesmann Corporation.[55] In 2003, Hansenet was acquired by Telecom Italia.[56]

---

46. Sjurts, I. *Strategien der Medienbranche* Wiesbaden: Betriebswirtschaftlicher Verlag Dr. Th. Gabler, 2002.

47. Hickethier, K. 'Film- und Fernsehanalyse. ' 2007.

48. "Tätigkeitsbericht Telekommunikation 2006/2007." Bundesnetzagentur. 2007.

49. German: "Komplettanschluss".

50. "Mannesmann Arcor übernimmt Mehrheit an Düsseldorfer ISIS." *manager-magazin.de*. Last accessed on August 19, 2012 at http://www.manager-magazin.de/geld/artikel/0,2828,58090,00.html

51. "Tropolys: Neue Macht im Ortsbereich?" *Teltarif.de*. Last accessed on August 10, 2012 at http://www.teltarif.de/arch/2000/kw34/s2873.html

52. "TROPOLYS erhält neue Gesellschafterstruktur." *Infosat.de*. Last accessed on Jan. 20, 2009 at http://www.infosat.de/Meldungen/?msgID=10916

53. "Apax kauft Versatel Deutschland." *www.kolloczek.com*. Last accessed on Jan. 20, 2009 at https://www.kolloczek.com/cms/index.php?option=com_content&Itemid=2&task=view&id=15

54. "Fusion vollzogen: Aus Versatel und Tropolys wurde Versatel." *Golem.de*. Last accessed on August 10, 2012 at http://www.golem.de/0601/42967.html

55. "Vodafone und Mannesmann—Ende einer Schlacht." tecchannel.de. 2000. Last accessed on August 10, 2012 at http://www.tecchannel.de/news/themen/business/404428/vodafone_und_mannesmann_ende_einer_schlacht

56. "Kaufpreis soll bei 250 Millionen Euro liegen." Teltarif.de. 2003. Last accessed on August 10, 2012 at http://www.teltarif.de/arch/2003/kw31/s11154.html

## Table 6-12. National Market Shares of German Fixed Network Providers (%)

| | 1984 | 1988 | 1992 | 1996 | 2000 | 2004 | 2007 | 2009 | 2013 |
|---|---|---|---|---|---|---|---|---|---|
| Deutsche Telekom | 100.0 | 100.0 | 100.0 | 100.0 | 98.3 | 92.2 | 82.9 | 70.0 | 56.9 |
| United Internet AG (1&1) | | | | | | | | 4.0 | 10.6 |
| Vodafone, (Arcor, UK) | | | | | 0.2 | 1.2 | 7.4 | 9.3 | 13.2 |
| Telefonica (O2, Spain) | | | | | | | | | 3.8 |
| Telecom Italia (Hansenet, Italy) | | | | | - | 0.6 | 3.4 | 2.8 | |
| Versatel (US) | | | | | - | 0.6 | 1.9 | 2.1 | 1.9 |
| Tropolys | | | | | 0.1 | | | | |
| EWE TEL (EWE AG, Austria) | | | | | - | 0.7 | 1.0 | | |
| NetCologne (GEW Köln AG) | | | | | - | 0.4 | 1.0 | | |
| Mnet (owned by several electricity providers) | | | | | - | 0.3 | 0.5 | | |
| Others | | | | | 1.4 | 4.0 | 1.9 | 11.8 | 13.6 |
| Total revenues (€ m) | - | 19,017 | 19,495 | 22,771 | 34,000 | 37,600 | 37,000 | | |
| Total revenues ($ m) | | 22,210 | 25,312 | 29,375 | 31,280 | 46,624 | 50,690 | | |
| C4 | 100 | 100 | 100 | 100 | 99 | 95 | 93 | 88.2 | 84.5 |
| HHI | 10,000 | 10,000 | 10,000 | 10,000 | 9,663 | 8,520 | 6,948 | 5,015 | 3,542 |
| *N* (>1%) | 1 | 1 | 1 | 1 | 1 | 2 | 6 | 5 | 5 |
| Noam | 10,000 | 10,000 | 10,000 | 10,000 | 9,663 | 6,025 | 2,837 | 2,242 | 1,584 |
| Sum of total market shares | 100 | 100 | 100 | 100 | 100 | 100 | 100 | | |

*Source*: Market shares: Deutsche Telekom, Bundesnetzagentur. Market share competitors: press releases and annual reports of the telco companies listed (2000, 2004), VATM (2007). Total revenue: Annual reports of Deutsche Telekom (numbers for 1988–1996), VATM (numbers for 2000–2007). Sources for 2009 and 2013: Quelle:http://de.statista.com.eaccess.ub.tum.de/statistik/daten/studie/154194/umfrage/marktanteil-der-festnetzanbieter-an-den-festnetzkunden/. 2009 and 2013 updates provided by editors.

Despite the liberalization, the German fixed network market is still dominated by the former monopolist, Deutsche Telekom,[57] although its competitors have gained market share as Deutsche Telekom struggled with poor performance. In 2000, Deutsche Telekom still held a market share of 98.3%. It decreased to 56.9% in 2013. The largest competitor is Arcor, with a market share of 7.4% in 2007. Since 1998, the C4 index has come down from 100 to 84.5. The HHI dropped from its maximum value of 10,000 to 3,542. The Noam index decreased from 10,000 to 1,584 in 2007 (Table 6.12).

## Wireless Telecom

In 2007, the penetration rate of mobile phone contracts had exceeded population. There were 96 million mobile users in a population of 82 million people.[58] Mobile services also started to threaten fixed line revenues. Many German households only started using their mobile

57. "Key Aspects of the German Telecom Market in 2007 as Well as Key Regulatory Developments." Research and Markets. 2007. Last accessed on August 10, 2012 at http://www.redorbit.com/news/technology/1077445/key_aspects_of_the_german_telecom_market_in_2007_as/index.html

58. "Tätigkeitsbericht Telekommunikation 2006/2007." Bundesnetzagentur. 2007, 25.

Table 6-13. National Market Shares of German Mobile Network Operators (%)

| | 1992 | 1996 | 2000 | 2004 | 2008 | 2012 |
|---|---|---|---|---|---|---|
| Deutsche Telekom (T-Mobile, formerly D1) | 38.4 | 43.3 | 39.7 | 38.5 | 36.6 | 32.4 |
| Vodafone (formerly D2, UK) | 61.6 | 46.3 | 40 | 37.8 | 34.1 | 29.9 |
| E-Plus (Netherlands) | | 10.4 | 13.7 | 13.3 | 16.1 | 20.7 |
| Telefonica (Spain; O2 formerly Viag Interkom) | | | 6.6 | 10.4 | 13.2 | 17 |
| Total revenues (€ m) | 115 | 6,000 | 16,200 | 22,108 | 25,600 | 17,088 |
| Total revenues ($ m) | 149 | 7,740 | 14,904 | 26,174 | 37,632 | 27,000 |
| C4 | 100 | 100 | 100 | 100 | 100 | 100 |
| HHI | 5,269 | 4,127 | 3,407 | 3,196 | 2,936 | 2,661 |
| *N* (>1%) | 2 | 3 | 4 | 4 | 4 | 4 |
| Noam | 3,726 | 2,383 | 1,704 | 1,598 | 1,469 | 1,330 |

*Source*: 2012 Minitel GMN Snapshot. 2012 update provided by editors. Mobile networks started on June 30, 1992 in Germany. Sources for market shares: Bundesnetzagentur, Annual reports of the mobile network operators. Sources for total revenue: VATM (number for 2008), Bundesnetzagentur (numbers for 2000, 2004), Solon (1996), estimation by the revenue of D2 (1992). No figure for revenue in 1992, because mobile telephony only started in June 1992 in Germany.

phone connection. The concentration in this market is high, as there are only four mobile network providers.

The market shares were measured according to the number of customers. As mobile service has become a commodity, prices have dropped. This is an appropriate means of measuring market share. The analysis is restricted to mobile network operators. Mobile service providers and mobile virtual network operators are not included.

In 1989 the first private German license for mobile phone service was sold to Arcor D2 (later Vodafone). In 1992, the two providers, D1 (public) and D2 (private), started their service. In 1994 the third provider, E-plus, started its service, and in 1998 O2, formerly Viag Interkom, followed.

There are four players in the market: T-Mobile (formerly D1), Vodafone (formerly Arcor D2), E-Plus (KPN, Netherlands), and O2 [formerly Viag Interkom, now Telefonica (Spain)]. Since their entry, the market shares of the smaller players, E-Plus and O2, increased to 20.7% and 17%, respectively, by 2012.

The second largest player, D2 (later Vodafone), formerly belonged to Mannesmann, which was acquired by Vodafone in a spectacular hostile takeover in 2000. In 2002, the mobile division of Mannesmann was renamed Vodafone D2. With over 17 million customers, E-Plus is the third largest mobile provider in Germany. Since 2000, it has been owned by the Dutch KPN Group. O2 was a part of as Viag Interkom in 1998. Since 2006, it has been owned by the Spanish Telefónica, one of the largest telecommunication companies in the world.

The C4 index has remained constant over time, because there are only four players in the market. The HHI and the Noam index has steadily decreased since 1984. The introduction of E-plus in 1994 and O2 in 1998 lowered the concentration of ownership in the industry (Table 6.13).

## INTERNET MEDIA

### Internet Service Providers (ISP)

Germany has the largest broadband market in Europe. After several years of strong market growth, the country has now entered a phase of consolidation, increased market competition, and saturation of demand for basic broadband access. Major ISPs are currently trying to move broadband services onto a new digital plane that allows service convergence by pushing ahead with cable network upgrades and VDSL deployments.

Until 2003, the market was largely dominated (90.9%) by Deutsche Telekom. Thereafter providers like Arcor (now Vodafone), Freenet, United Internet, or Versatel entered the market and took market shares of up to 49.4% in 2008. Exact data for market shares have not been available for all of the years between 2004 and 2012, but market participants did not change during the period. No major mergers or acquisitions took place (Tables 6.14, 6.15).

## Search Engines

Search Engines classify web-based information and offer their users relevant information according to their search query in a systematic way. Search engines are widely used by individuals as a starting point to navigate through the Internet. The market shares were calculated from the respective search engine's amount of local search queries. Since search engines derive their revenue mainly from selling search advertising space, the amount of search queries is adequate for measuring their market shares.[59, 60]

Google Inc., with its subsidiary Google Germany GmbH, is the dominant player in the German market for search engines. Founded in 1998, Google quickly became the market leader

### Table 6-14. Revenues of Internet Service Providers

| | 2005 | 2006 | 2007 | 2008 | 2009 |
|---|---|---|---|---|---|
| Mobile ($ bn) | 0,39 | 0,83 | 1,74 | 2,83 | 3,00 |
| Fixed ($ bn) | 3,52 | 4,55 | 5,19 | 5,75 | 6,15 |
| Total revenues ($ bn) | 4,79 | 6,57 | 8,85 | 11,28 | 11,62 |

*Source*: PricewaterhouseCoopers (2010)

### Table 6-15. Market Shares Stationary Broadband Internet Access (%)

| | 2001 | 2002 | 2003 | 2004 | 2005 | 2006 | 2007 | 2008 | 2012 |
|---|---|---|---|---|---|---|---|---|---|
| Deutsche Telekom | 97.0 | 92.0 | 90.9 | 82.7 | 61.0 | 49.2 | 48.6 | 50.6 | 45.6 |
| Arcor (Vodafone, UK) | | | | | | | 14.5 | 14.7 | 17.1 |
| 1&1 (United Internet) | | | | | | | 13.5 | 13.8 | 11.8 |
| Alice (Hansenet) | | | | | | | 12.8 | 11.5 | 9.4 |
| Freenet | | | | | | | 7.0 | 4.6 | - |
| Versatel (US) | | | | | | | 3.5 | 3.5 | - |
| Others | 3.0 | 8.0 | 9.1 | 17.3 | 39.0 | 50.8 | 0.1 | 1.3 | 16.1 |
| Total revenue (€ bn) | | | | | 3.52 | 4.55 | 5.19 | 5.75 | |
| Total revenue ($ bn) | | | | | 4.39 | 5.72 | 7.11 | 8.46 | |
| C4 | | | | | | | 88.0 | 92.0 | 83.9 |
| HHI | | | | | | | 2,934 | 3,180 | 2,599 |
| *N* (>1%) | | | | | | | 6 | 6 | 4 |
| Noam | | | | | | | 1,198 | 1,298 | 1,300 |

*Source*: 2012: Statista. 2012, for updates provided by editors. Bundesnetzagentur *Jahresbericht 2011*, 77.

59. Wirtz, Bernd W. (2001). *Electronic Business*, 2nd edition, 243–244.
60. Fredricksen, Clark. "Stat of the Day: Bing Not Making a Dent in Google's Share of Search Revenues." emarketer.com. Last accessed on 20.03.2011 at http://www.emarketer.com/blog/index.php/tat-day-bing-making-dent-googles-share-search-revenues/

in Germany, outpacing well-established search engines like Yahoo or Altavista. Increasing its market share from almost 79% in 2004 to 90.5% in 2013, Google has achieved monopolistic status in the German search engine market (Table 6.16).

Yahoo, with its subsidiary Yahoo Deutschland GmbH, entered the German market in 1996 and established a market share of roughly 20% in 2000. Due to Yahoo's significant loss of market share in the intervening years, the company had a 2.5% share in the German market since 2010.

Microsoft launched its new search engine, Bing, in 2009. MSN Live Search, Microsoft's former search engine, migrated to Bing's search algorithm. Aggregating these search services, Microsoft achieved a 3% market share in 2010. Microsoft has replaced Yahoo as the second biggest player in the market.

The rest of the competition consists of several American companies such as ask.com and AOL, as well as some German search engines. However, these companies only reached a low individual market share in 2010. Furthermore, Google provides the underlying search algorithm for some of these companies.

Aside from the acquisition of Altavista by Yahoo's subsidiary, Overture, in 2003, major changes in ownership did not occur. Originating from a moderately concentrated market with an HHI of 1,028 in 2000, the German market shows a significant increase in concentration during the intervening years. Despite the insolvency of Lycos Europe at the end of 2008[61] and the launch of Microsoft's Bing in 2009,[62] the market remains highly concentrated with an HHI of 8,204 in 2013.

## Online News

According to a survey by the Institut für Demoskopie Allensbach (IFD), since the year 2008 more than 50% of 20–30 years-old respondents consider the Internet their most important source of news.[63] The group gets its news from online news sites such as *Spiegel Online* or *Bild Online*.

The concentration of the news portals was calculated from the amount of visits measured by the organization Informationsgemeinschaft zur Feststellung der Verbreitung von Werbeträgern e.V. (IVW). Since news portals derive their revenue streams mainly from advertisements, which are based on page impressions and therefore on visits, the number of visits of each news portal is an adequate measurement for the market share.

*Bild Online* is the biggest player in the news portal market of Germany, with 22% market share in 2010. As *Bild*'s online spin-off, this news portal is a complete subsidiary of Axel Springer AG. *Bild Online* also manages other Springer-owned websites such as Computerbild.de.

In 2010, *Bild Online* outpaced the long time market leader, *Spiegel Online*, by a small fraction. *Spiegel Online* publishes original content produced by its own editorial team as well as purchased content from news agencies. In addition to that, selected articles from the print edition are available online at no cost. In comparison to *Bild Online, Spiegel Online* focuses on high-quality, serious news.

There are a few additional players. Chip Online, the news portal with the third biggest market share, grew from its offline version *Chip Magazine* and focuses more on tech- and product-related news. Similar to *Chip Online*, *Heise Online* focuses on tech news and was the dominant news portal in 2000. However, *Heise Online* lost significant market share in the following decade and only amounts to roughly 4% market share in 2013. Hence, the common web surfer in the early days of the Internet appeared to be quite tech-savvy.

---

61. Patatlong, Frank. "Der Fall Lycos." *Spiegel Online*. Nov. 26, 2008. Last accessed on August 10, 2012 at http://www.spiegel.de/netzwelt/web/0,1518,592834,00.html

62. Marth, Michale. "Microsoft Bing." *Focus Online*. June 2, 2009. Last accessed on August 10, 2012 at http://www.focus.de/digital/internet/tid-14432/microsoft-bing-was-die-neue-suchmaschine-kann_aid_404459.html

63. Widmann, Britta. "Internet überholt Zeitung als Informationsquelle." *zdnet.de*. Oct. 17, 2008. Last accessed on August 10, 2012 at http://www.zdnet.de/news/tkomm/0,39023151,39197728,00.htm

Table 6-16. National Market Shares of Search Engines (%): 2000–2013

| | 2000[1] | 2004[2] | 2008[3] | 2010[4] | 2013 |
|---|---|---|---|---|---|
| Google (US) | 19.6 | 78.8 | 87.0 | 89.1 | 90.5 |
| Microsoft (US) | 6.2 | 4.3 | 1.8 | 3.3 | 3.2 |
| MSN Live Search | *6.2* | *4.3* | *1.8* | *0.2* | - |
| Bing | - | - | - | *3.1* | *3.2* |
| Yahoo (US) | 19.2 | 5.6 | 3.0 | 2.5 | 1.6 |
| Altavista | 6.7 | - | - | - | - |
| T-Online | - | 1.6 | 2.1 | 2.0 | 1.1 |
| Web.de | 7.6 | 0.3 | 0.5 | 0.5 | 0.7 |
| Lycos | 11.6 | 0.5 | 0.1 | - | - |
| Other | 29.1 | 8.9 | 5.5 | 2.6 | 2.9 |
| Sum of Total Market Shares | 100 | 100 | 100 | 100 | 100 |
| Total Revenues (€ m)[5] | n.a. | 110[6] | 1,476 | 1,867 | 2,277 |
| Total Revenues ($ m) | | 136 | 2,170 | 2,483 | 3,073 |
| C4 | 58.0 | 90.3 | 93.9 | 98.0 | 96.4 |
| HHI | 1,028 | 6,262 | 7,586 | 7,959 | 8,204 |
| *N* (>1%) | 7 | 5 | 5 | 5 | 5 |
| Noam | 389 | 2,800 | 3,393 | 3,559 | 3,669 |

*Source*: 2013 updates provided by editors
[1] Wirtschaftsstandort Österreich (2011).
[2] Luna-Park.de. (2007); Rentaseo.de (2005).
[3] Peters, Berger, and Seitz (2010).
[4] "Marktanteile von Suchmaschinen in Deutschland im April 2011." *Statista.de.*
[5] Bundesministerium für Wirtschaft und Technologie (2010).
[6] "Der Online-Markt in Deutschland." Suchmaschinenkompetenz.de. (2012).

## OWNERSHIP AND CONCENTRATION

In 2000, *Spiegel Online* and *Heise Online* dominated the market of news portals in Germany. However, the market share of *Heise Online* decreased sharply over the years and *Bild Online* became the second major player between 2004 and 2008. By 2010, *Bild Online* had become the market leader, leaving the former dominant news portal, Spiegel Online, behind. Notably, German newspapers such as *Die Welt* or *Süddeutsche Zeitung*, TV news channels such as N-TV, and magazines such as *Focus* entered the market, but were could not gain a dominant position in the news portal market (Table 6.17).

### Internet Portals

An Internet portal aggregates information from diverse sources in a unified way. Apart from the search engine standard, Internet portals offer other services such as e-mail, news, stock prices, educational entertainment, and other features. Internet portals provide a consistent look and feel with access control and procedures for multiple applications, which otherwise would have been different entities altogether (cf. Welfens et al. 2005, 97).

The concentration of Internet portals was deduced from the number of unique visitors. Since Internet portals mainly derive their revenue streams from advertisements that are based on the reach, the amount of unique users

Table 6-17. National Market Shares of News Portals (%): 2000–2012[1]

| | 2000 | 2004 | 2008 | 2010[2] | 2012 |
|---|---|---|---|---|---|
| Axel Springer | 9.5 | 14.9 | 21.5 | 26.6 | 30.2 |
| Bild Online | *9.5* | *14.9* | *16.3* | *22.3* | *25.5* |
| Welt Online | | | *5.2* | *4.3* | *4.7* |
| Bertelsmann | 5.5 | 10.5 | 7.2 | 6.4 | 6.1 |
| N-TV Online | *5.5* | *5.8* | *4.0* | *3.5* | *3.7* |
| Stern.de | | *4.7* | *3.2* | *2.9* | *2.4* |
| Burda | 18.3 | 15.1 | 13.9 | 11.4 | 12.0 |
| Chip Online | *8.4* | *7.9* | *9.4* | *7.4* | *7.6* |
| Focus Online | *9.9* | *7.2* | *4.5* | *4.0* | *5.4* |
| Spiegel Online (incl. Manager Magazin) | 19.9 | 24.8 | 22.5 | 21.6 | 19.1 |
| Kicker Online (Sports, Verlag Nürnberger Presse Druckhaus) | | | 6.6 | 5.0 | 4.9 |
| Süddeutsche.de | 3.3 | 3.1 | 4.0 | 4.2 | 4.0 |
| Heise Online (Heise Medien Gruppe) | 20.8 | 11.5 | 4.9 | 3.9 | 3.7 |
| N24 Online (ProSieben/Sat.1) | | 1.1 | 0.7 | 3.3 | 0.6 |
| FAZ.net | | 2.9 | 3.4 | 3.2 | 2.7 |
| PC-Welt (IDG Communications Media) | 4.8 | 3.9 | 2.9 | 1.6 | 1.4 |
| Others | 18.1 | 12.2 | 12.6 | 12.9 | 15.3 |
| Sum of Total Market Shares | 100 | 100 | 100 | 100 | |
| Total Revenues (€ m) | n.a. | 60[3] | 268[4] | 275[5] | |
| Total Revenues ($ m) | n.a. | 74 | 394 | 366 | |
| C4 | 68.4 | 66.3 | 64.4 | 66.0 | 67.4 |
| HHI | 1,315 | 1,341 | 1,313 | 1,428 | 1,521 |
| *N* (>1%) | 7 | 9 | 10 | 10 | 10 |
| Noam | 497 | 447 | 415 | 452 | 481 |

[1] *Informationsgemeinschaft zur Feststellung der Verbreitung von Werbeträgern e.V.* (2009).
[2] "Besucher deutscher Nachrichtenportale 2010."
[3] Golz (2007).
[4] Nielsen Media Research (2009, 17).
[5] Nielsen Media Research (2011, 24).

per month or per year is an adequate measure for the market share. T-Online, a subsidiary of the Deutsche Telekom AG, is the all-time leader in the German Internet portal market with a market share of 28.3% in 2010. The second biggest player in the German market is *Web.de,* which is a fully owned subsidiary of United Internet. A popular service of Web.de is its free e-mail, FreeMail. In order to enhance the market power of online advertisement, Web.de founded AD Europe in 2004, which is a network of different Internet portals from over 60 countries. Yahoo operates the third largest Internet portal in Germany, with a market share 15% in 2010. Microsoft's Internet portal, *MSN.de,* the fourth biggest player in the German Internet portal market, had a 12.6% market share in 2010.

Table 6-18. National Market Shares of Internet Portals (%): 2000–2010[1]

| | 2000 | 2004 | 2008 | 2010[2] |
|---|---|---|---|---|
| T-Online | 31.0 | 19.5 | 19.6 | 28.3 |
| Web.de (United Internet) | 13.8 | 15.5 | 17.4 | 19.3 |
| Yahoo.de (US) | 17.2 | 12.8 | 15.2 | 15.1 |
| MSN (US) | 2.4 | 13.3 | 14.6 | 12.5 |
| GMX (United Internet) | 4.4 | 11.3 | 11.9 | 12.3 |
| Lycos Europe | 0.7 | 7.9 | 4.6 | |
| Others | 30.6 | 19.8 | 16.7 | 11.9 |
| Sum of total market shares | 100 | 100 | 100 | 100 |
| Total Revenues (€ m) | 300[3] | 385[4] | 655[5] | 1080[6] |
| Total Revenues ($ m) | 276 | 462 | 963 | 1,436 |
| C4 | 66.3 | 61.1 | 66.8 | 75.4 |
| HHI | 1,470 | 1,150 | 1,294 | 1,722 |
| Noam | 600.2 | 469.5 | 528.2 | 770.3 |

[1] "Internet Facts." *Arbeitsgemeinschaft Online Forschung* (2001–2009).
[2] "IVW Online Nutzungsdaten 02–2011." (February 2011).
[3] "Online-Werbeumsatz in Deutschland verdoppelt." *Computerwoche.de.* (2011).
[4] "Online-Report 2007/01." (January 2007, 5).
[5] Nielsen Media Research (2009, 17).
[6] Nielsen Media Research (2011, 24).

While T-Online dominated the market for Internet portals in 2000, it outperformed its competitors only by a narrow margin in 2004 and 2008. United Internet AG, which was an Internet portal in the earlier days of the Internet, bought Web.de in 2005. Today it is the main competitor of T-Online. Therefore, the concentration of Internet portals in Germany decreased from 2000 to 2004. After 2004, the German market for Internet portals indicates a low level of volatility in its concentration, since the biggest players in the market generally maintain their respective market share. The market lacks notable market entries of novel players during the reviewed time frame. Lycos Europe went out of business in 2008, and the remaining players absorbed its market share (Table 6.18).

## CONCLUSION

Freedom of expression is guaranteed in Germany within the Constitution (Grundgesetz, Art.5). Due to the structure of German government, there are a variety of players on different levels. The central players in German audiovisual media policy are the political parties, especially the Länder organizations of the two large parties, the conservative CDU, and the social democratic SPD, which control much of the public broadcasting sector. After years of strong polarization from the 1950s to the 1970s, media policy is now again based on a broad consensus between the Länder. In an agreement between all Länder, the basics of a dual system of broadcasting have been put in place. It includes regulation for media concentration, stating that one company cannot control more than 30% of all TV ratings. The high degree of media concentration, especially the two "Senderfamilien," is causing concern. The update of the Länder agreement, the Rundfunkstaatsvertrag, includes provisions for the EU television directive, especially the provision stating that important events such as the Olympic Games must be broadcast for free. The EU-wide Product Placement Permission (AVMS 2007) was introduced in 2009. In recent times, debates about the future of German public service broadcasting are increasingly influenced by decisions of and challenges to the EU. State subsidies do not exist within the print sector, nor in electronic media. However, a reduced valued-added tax rate and reduced prices for distributing print products via mail serve as a state-generated support for the press.[64]

Besides horizontal concentration inside the specific segments, vertical integration and media convergence rose in the period after 2004. Examples include cable providers entering the market for broadband Internet access (triple play) or the growing acceptance of IPTV offers. These trends will make it harder for future studies to differentiate between content providers and distributor channels.

## Bibliography

### 1. Books and Articles


Dieter, K. Müller and Esther Raff, eds. *Praxiswissen Radio.* VS Verlag fur Sozialwissenschaften. Wiesbaden, 2007, 9.

Frank, B. *Zur Ökonomie der Filmindustrie.* Steuer- und Wirtschaftsverlag, Hamburg 1993, 90–98.

Fredricksen, Clark. "Stat of the Day: Bing Not Making a Dent in Google's Share of Search Revenues." *emarketer.com.* Last accessed on 20.03. 2011 at <http://www.emarketer.com/blog/index.php/stat-day-bing-making-dent-googles-share-search-revenues>

Golz, Sandra. "SPIEGEL Online ist das Maß aller Dinge." May 10, 2007. Last accessed on August 10, 2012 at <http://archiv.medien-mittweida.de/panorama/artikel/1218.html>

Hachmeister, L., and Rager, G. *Wer beherrscht die Medien?* München : Beck, 2003, 70–89.

Hans-Bredow-Institut, *Medien von A bis Z.* Wiesbaden: VS Verlag für Sozialwissenschaften, 2006.

Koschnick, Wolfgang J. *Focus Media Lexikon,* München, Fokus Magazin Verlag, 2003.

Marth, Michale. "Microsoft Bing." *Focus Online.* June 2, 2009. Last accessed on August 10, 2012 at <http://www.focus.de/digital/internet/tid-14432/microsoft-bing-was-die-neue-suchmaschine-kann_aid_404459.html>

"Media Landscape: Germany." European Journalism Centre. 2012. Last accessed August 10, 2012 at <http://www.ejc.net/media_landscape/article/germany>

Patatlong, Frank. "Der Fall Lycos." *Spiegel Online.* November 26, 2008. Last accessed on August 10, 2012 at <http://www.spiegel.de/netzwelt/web/0,1518,592834,00.html>

Peters, A. "Suchmaschinen-Marketing in Direktbanken: Grundlagen, Erfolgsfaktoren und Status quo." 2010. ibi research an der Univ. Regensburg.

Röper, H. "Zeitungsmarkt 2002: Wirtschaftskrise und steigende Konzentration." *Media Perspektiven,* October 2002, 478.

Röper, H. "Bewegung im Zeitungsmarkt 2004." *Media Perspektiven,* June 2004, 268.

Röper, H. "Konzentrationssprung im Markt der Tageszeitungen." *Media Perspektiven,* August 2008, 420.

"Süddeutsche geht an Medienholding." FAZ. June 2007. Last accessed on August 10, 2012 at <http://www.faz.net/aktuell/wirtschaft/fuer-mehr-als-eine-halbe-milliarde-euro-sueddeutsche-geht-an-medienholding-1488131.html>

"Tropolys: Neue Macht im Ortsbereich?" *Teltarif.de.* Last accessed on August 10, 2012 at <http://www.teltarif.de/arch/2000/kw34/s2873.html>

Uka, Walter. Video, in Faulstich, Wilhelm (ed.), *Grundwissen Medien.* 5th Edition, Wilhelm Fink Verlag, 2004, 418–419.

Vogel, A. "Konsolidierte Großkonzerne bereit zu erneutem Wachstum." *Media Perspektiven,* July 2004, 322.

Vogel, A. "Publikumszeitschriften Dominanz der Großverlage gestiegen." *Media Perspektiven,* Sept. 2002, 433.

Vogel, A. "Stabile Positionen in schrumpfenden Märkten." *Media Perspektiven* Sept. 2008, 467.

Widmann, Britta. "Internet überholt Zeitung als Informationsquelle." *zdnet.de.* 17. October 2008. Last accessed on 10 August 2012 at <http://www.zdnet.de/news/tkomm/0,39023151,39197728,00.htm>


### 2. Reports


"Besucher deutscher Nachrichtenportale 2010." *Bitkom.* Last accessed on August 20, 2012 at


64. "Media Landscape: Germany." European Journalism Centre. 2012. Last accessed August 30, 2012 at http://www.ejc.net/media_landscape/article/germany/

<http://www.bitkom.org/files/documents/Datenblatt_Auswertung_2010_Newsgroups.pdf>

Bundesministerium für Wirtschaft und Technologie. *Monitoring-Report Deutschland Digital*, 2010, 59.

"Der Online-Markt in Deutschland." Suchmaschinenkompetenz.de. Last accessed on August 10, 2012 at <http://www.suchmaschinenkompetenz.de/Online-Markt-Suchmaschinenkompetenz.htm>

*Informationsgemeinschaft zur Feststellung der Verbreitung von Werbeträgern e.V.* Last accessed on January 20, 2009 at <http://www.ivwonline.de/ausweisung2/search/ausweisung.php>

"Internet Facts." *Arbeitsgemeinschaft Online Forschung (2001–2009)*. Last accessed on August 10, 2012 at <http://www.agof.de/archiv-pressemitteilungen.1032.de.html>

IVW Online Nutzungsdaten 02–2011. February 2011. Informationsgemeinschaft zur Feststellung der Verbreitung von Werbeträgern e.V. Last accessed on August 10, 2012 at <http://ausweisung.ivw-online.de>

KEK. "Auflagenstärkste Verlagsgruppen." Last accessed on Jan. 20, 2009 at <http://www.kek-online.de/Inhalte/auflagenstaerkste_verlagsgruppen.pdf>

"Marktanteile von Suchmaschinen in Deutschland im April 2011." *Statista.de*. Last accessed March 20, 2011 at <http://de.statista.com/statistik/daten/studie/167841/umfrage/marktanteile-ausgewaehlter-suchmaschinen-in-deutschland/, cited after Webhits.de 2011>.

Nielsen Media Research (2009). Werbemarkt Report, 2008, 17.

Nielsen Media Research (2011). Ad Market Report 2010, 24.

PricewaterhouseCoopers. "German Entertainment and Media Outlook: 2010–2014—Die Entwicklung des deutschen Unterhaltungs- und Medienmarktes." Frankfurt am Main, 2010, 5–119.

Vogel, A. "Stabile Positionen in schrumpfenden Märkten." Media Perspektiven Sept. 2008, 467.

## 3. Websites

*Buchreport.de*. March 27, 2008. Last accessed on Aug. 9, 2012 at <http://www.buchreport.de/nachrichten/verlage/verlage_nachricht/datum/2008/03/27/buchreport-ranking-100-groesste-verlage.htm>

Der Onlinemarkt in Deutschland. 2012. Last Accessed on August 10, 2012 at <http://www.suchmaschinenkompetenz.de/Online-Markt-Suchmaschinenkompetenz.htm>

KEK. "Hörfunk." Last accessed on August 10, 2012 at <http://www.kek-online.de/Inhalte/hoerfunk.html>

Key Aspects of the German Telecom Market in 2007 as Well as Key Regulatory Developments. Research and Markets. 2007. Last accessed on August 10, 2012 at <http://www.redorbit.com/news/technology/1077445/key_aspects_of_the_german_telecom_market_in_2007_as/index.html>

Luna-Park.de. 2007. Last accessed on March 20, 2011 at <http://www.luna-park.de/home/internet-fakten/suchmaschinen-beziehungsgeflecht/2007er-beziehungsgeflecht-suchmaschinen.html>

Marktanteile von Suchmaschinen in Deutschland im April 2011. 2011. Last accessed on August 10, 2012 at <http://de.statista.com/>

Rentaseo.de. 2005. Last accessed on March 20, 2011 at <http://www.rentaseo.de/web-marketing/seo-tips/marktanteile-suchmaschinen-deutschland.html>, cited after webhits.de

Wirtschaftsstandort Österreich (Ed.). Last accessed on March 20, 2011 at <http://www.wsoe.at/index.php?bereich=interessantes\statistiken&language=>

# Germany—Data Summaries

ELI NOAM AND PAUL MUTTER

GERMANY IS ONE OF the largest media markets in Europe. One of its major firms—Bertelsmann—operates in many other countries through the publishing company Random House and the broadcaster RTL Group. Domestically, Bertelsmann holds 25% in both TV broadcasting and video channels. It is still one of the larger companies with 4.0% of the overall national market (Table 6.19). ProSiebenSat.1 (the second big private television network) holds significant market shares in Belgium, Sweden, and the Netherlands. ProSiebenSat.1 controls 30% of TV broadcasting and 20% of video channels. It is no longer owned by German investors but by two private equity firms, Permira (UK) and KKR (US). T-Mobile, the wireless arm of Deutsche Telekom, controls 20% of the wireless market in the United Kingdom directly and controls 34% indirectly as an equal shareholder with Orange (France Telecom) in their joint UK venture, EE. T-Mobile is also a major platform carrier in Central and Eastern Europe, and as T-Mobile USA holds approximately 12% of the large US wireless market.

Germany's first and third channels provided by the regional public ARD broadcasting stations is federated, though not for linguistic reasons as it is in Switzerland and Belgium, but to conform to the constitutional requirement of the states' (Länder) media autonomy. Public broadcasting channels, both national and regional, account for 54.4% of broadcast TV.

ARD and ZDF are Germany's main public broadcasters. In addition to ARD's 25% share in both TV Broadcasting and video channel, the company also holds 36.5% of radio. ZDF holds a smaller, but still formidable 13.5% of radio, which shows the high degree of cross-ownership of content media in Germany.

Concentration is declining overall in Germany because platform media have become a more competitive field: this largely accounts for

Table 6-19. National Media Industries Concentration in Germany

| | 2004/5 | | 2011 OR MOST RECENT | | % CHANGE ANNUAL AVERAGE | |
|---|---|---|---|---|---|---|
| | COMPANY POWER INDEX IN COUNTRY | COMPANY SHARE OF THE OVERALL NATIONAL MEDIA MARKET (%) | COMPANY POWER INDEX IN COUNTRY | COMPANY SHARE OF THE OVERALL NATIONAL MEDIA MARKET (%) | COMPANY POWER INDEX IN COUNTRY | COMPANY SHARE OF THE OVERALL NATIONAL MEDIA MARKET (%) |
| Deutsche Telekom (38% public) | 3,651 | 44.5 | 1,389 | 28.6 | –7.8 | –2.0 |
| Vodafone (UK) | 301 | 8.7 | 236 | 11.2 | 2.7 | 0.3 |
| Telefónica (Spain) | 22.2 | 2.2 | 59.2 | 4.7 | 20.8 | 0.3 |
| ARD (public) | 155.7 | 6 | 161.8 | 5.5 | 0.5 | 0.01 |
| Google (US) | 6.6 | 0.1 | 173.4 | 1.9 | 314.5 | 0.2 |
| ProSiebenSat.1 | 63.4 | 2.9 | 108.5 | 3.8 | 8.9 | 0.1 |
| Bertelsmann | 91.3 | 4.2 | 88.9 | 4.0 | –0.3 | –0.02 |
| Kabel Deutschland | 52.0 | 1.1 | 62.5 | 1.4 | 3.1 | 0.3 |
| E-Plus | 36.3 | 2.7 | 79.7 | 3.9 | 14.9 | 0.02 |
| ZDF (public) | 40.2 | 2.8 | 30.5 | 2.4 | –3.0 | –0.05 |
| Axel Springer | 30.0 | 1.8 | 21.6 | 1.5 | –3.5 | –0.04 |
| Unitymedia | 18.5 | 0.7 | 27.8 | 0.9 | 3.1 | 0.2 |
| Murdoch (US) | 2.7 | 0.3 | 24.4 | 1.1 | 102.2 | 0.1 |
| Bayern Funkpaket | 35.1 | 1.5 | 18.8 | 1.1 | –5.8 | –0.06 |
| Anderer | 0.0 | 0.0 | 16.6 | 2.4 | N/A | 0.3 |
| Bauer | 17.6 | 0.8 | 13.0 | 0.7 | –3.3 | –0.02 |
| Alice | 0.0 | 0.0 | 12.1 | 2.4 | N/A | 0.3 |
| Burda | 7.4 | 0.5 | 8.7 | 0.6 | 2.3 | 0.01 |

| | | | | | | |
|---|---|---|---|---|---|---|
| Versatel (US) | 0.0 | 0.0 | 1.6 | 0.9 | N/A | 0.1 |
| Tele2 | 0.0 | 0.0 | 0.8 | 0.5 | N/A | 0.07 |
| Holtzbrinck | 0.7 | 0.3 | 0.7 | 0.3 | −0.2 | −0.002 |
| Freenet | 0.0 | 0.0 | 0.6 | 0.3 | N/A | 0.03 |
| Keine Angabe | 0.0 | 0.0 | 0.6 | 0.5 | N/A | 0.06 |
| Motorpresse | 0.4 | 0.1 | 0.4 | 0.1 | 0.5 | −0.001 |
| Bonnier (Sweden) | 0.4 | 0.2 | 0.3 | 0.1 | −0.9 | −0.001 |
| Discovery Communications (US) | 0.0 | 0.0 | 0.1 | 0.05 | N/A | 0.01 |
| Bouygues Group | 0.1 | 0.04 | 0.1 | 0.05 | 7.0 | 0.001 |
| Der Spiegel | 0.4 | 0.01 | 0.9 | 0.05 | 19.7 | 0.004 |

| Media Concentration Index | 2004/5 | 2011 or Most Recent | % Change Annual Average |
|---|---|---|---|
| Total Revenue: Nat'l Media Industry (US$ mil) | 127,392 | 5,151 | 1.7 |
| Total Voices (*n*) | 83 | 80 | −0.5 |
| Net Voices (*n*) | 65 | 62 | −0.6 |
| Public Ownership (%) | 25.3 | 18.8 | −0.8 |
| Foreign Ownership (%) | 13.2 | 22.6 | 1.2 |
| C4 Average Weighted | 85.0 | 82.0 | −0.4 |
| HHI Average Weighted | 4,804 | 3,001 | −4.7 |
| C1 Average Weighted | 54 | 40 | −2 |

*continued*

Table 6-19. *continued*

| | 2004/5 | | 2011 OR MOST RECENT | | % CHANGE ANNUAL AVERAGE | |
|---|---|---|---|---|---|---|
| | COMPANY POWER INDEX IN COUNTRY | COMPANY SHARE OF THE OVERALL NATIONAL MEDIA MARKET (%) | COMPANY POWER INDEX IN COUNTRY | COMPANY SHARE OF THE OVERALL NATIONAL MEDIA MARKET (%) | COMPANY POWER INDEX IN COUNTRY | COMPANY SHARE OF THE OVERALL NATIONAL MEDIA MARKET (%) |
| Noam Index Average Weighted | 608 | | 795 | | 3.9 | |
| Pooled Overall Sector C4 | 63.0 | | 49.9 | | –1.6 | |
| Pooled Overall Sector HHI | 2,143 | | 1,091 | | –6.1 | |
| Pooled Overall Sector Noam Index | 60 | | 63 | | 0.5 | |
| Market Share of Top Ten Companies:Nat'l Media Industry (%) (Pooled C10) | 76.9 | | 70.8 | | –0.8 | |
| National Power Index | 4,561 | | 2,620 | | –5.3 | |

Table 6-20. Top Content Media Companies in Germany

| | 2004/5 | | 2011 OR MOST RECENT | | % CHANGE ANNUAL AVERAGE | |
|---|---|---|---|---|---|---|
| | COMPANY POWER INDEX IN COUNTRY | COMPANY SHARE OF THE NATIONAL CONTENT MEDIA MARKET (%) | COMPANY POWER INDEX IN COUN.TRY | COMPANY SHARE OF THE NATIONAL CONTENT MEDIA MARKET (%) | COMPANY POWER INDEX IN COUNTRY | COMPANY SHARE OF THE NATIONAL CONTENT MEDIA MARKET (%) |
| ARD (public) | 412 | 14.7 | 420 | 14.2 | 0.2 | −0.06 |
| Google (US) | 17.5 | 0.2 | 449 | 5.0 | 308.1 | 0.6 |
| ProSiebenSat.1 | 168 | 7.7 | 281 | 9.8 | 8.5 | 0.3 |
| Bertelsmann | 241 | 11.1 | 231 | 10.5 | −0.6 | −0.07 |
| ZDF (public) | 106.4 | 7.5 | 79.1 | 6.3 | −3.2 | 0.2 |
| Axel Springer | 79.4 | 4.8 | 56.1 | 4.0 | −3.7 | −0.1 |
| Kabel Deutschland | 45.4 | 1.0 | 3.5 | 1.2 | 3.1 | 0.3 |
| Bayern Funkpaket | 92.8 | 4.1 | 48.7 | 2.7 | −5.9 | −0.2 |
| Bauer | 46.6 | 2.2 | 33.8 | 1.8 | −3.4 | −0.05 |
| Burda | 19.4 | 1.4 | 22.6 | 1.5 | 2.0 | 0.01 |
| Murdoch (US) | 6.3 | 0.7 | 27.0 | 1.5 | 41.4 | 0.09 |
| Unitymedia | 16.1 | 0.6 | 23.8 | 0.8 | 5.9 | 0.2 |
| Holtzbrinck | 2.0 | 0.7 | 1.9 | 0.7 | −0.4 | −0.01 |
| Motorpresse | 1.1 | 0.3 | 1.0 | 0.3 | −0.7 | 2 |
| Vivendi (France) | 0.0 | 0.0 | 1.0 | 0.3 | N/A | 0.04 |
| Bonnier (Sweden) | 0.9 | 0.4 | 0.9 | 0.4 | −1.1 | −0.01 |
| Bouygues Group (France) | 0.2 | 0.1 | 0.2 | 0.12 | 6.6 | 0.003 |
| Discovery Communications (US) | 0.0 | 0.01 | 0.2 | 0.1 | 1,535.6 | 0.01 |
| Deutsche Telekom | 0.0 | 0.01 | 0.2 | 0.1 | 406.2 | 0.01 |

*continued*

Table 6-20. *continued*

| | 2004/5 | | 2011 OR MOST RECENT | | % CHANGE ANNUAL AVERAGE | |
|---|---|---|---|---|---|---|
| | COMPANY POWER INDEX IN COUNTRY | COMPANY SHARE OF THE NATIONAL CONTENT MEDIA MARKET (%) | COMPANY POWER INDEX IN COUN.TRY | COMPANY SHARE OF THE NATIONAL CONTENT MEDIA MARKET (%) | COMPANY POWER INDEX IN COUNTRY | COMPANY SHARE OF THE NATIONAL CONTENT MEDIA MARKET (%) |
| Media Concentration Index | 2004/5 | | 2011 or Most Recent | | % Change Annual Average | |
| Public Ownership (%) | 22.2 | | 20.5 | | −0.2 | |
| Foreign Ownership (%) | 6.0 | | 12.4 | | 0.8 | |
| C4 Average—Weighted | 67 | | 71 | | 0.4 | |
| HHI Average—Weighted | 1,937 | | 2,257 | | 2.1 | |
| C1 Average—Weighted | 22 | | 27 | | 0.01 | |
| National Power Index | 1,317 | | 1,814 | | 4.7 | |

Table 6-21. Top Platform Media Companies in Germany

| | 2004/5 | | 2011 OR MOST RECENT | | % CHANGE ANNUAL AVERAGE | |
|---|---|---|---|---|---|---|
| | COMPANY POWER INDEX IN COUNTRY | COMPANY SHARE OF THE NATIONAL PLATFORM MEDIA MARKET (%) | COMPANY POWER INDEX IN COUNTRY | COMPANY SHARE OF THE NATIONAL PLATFORM MEDIA MARKET (%) | COMPANY POWER INDEX IN COUNTRY | COMPANY SHARE OF THE NATIONAL PLATFORM MEDIA MARKET (%) |
| Deutsche Telekom | 5,873 | 71.6 | 2,261 | 46.4 | –7.7 | –3.1 |
| Vodafone (UK) | 485 | 14.0 | 384 | 18.2 | 2.6 | 0.5 |
| Telefónica (Spain) | 35.7 | 3.5 | 96.3 | 7.6 | 21.2 | 0.5 |
| E-Plus | 58.5 | 4.4 | 129.8 | 6.3 | 15.3 | 0.02 |
| Kabel Deutschland | 56.1 | 1.2 | 68.2 | 1.5 | 3.1 | 0.3 |
| Unitymedia | 19.9 | 0.7 | 30.3 | 1.0 | 3.1 | 0.2 |
| Anderer | 0.0 | 0.0 | 27.1 | 3.9 | N/A | 0.5 |
| Murdoch (US) | 0.5 | 0.1 | 22.8 | 0.9 | 593.8 | 0.1 |
| Alice | 0.0 | 0.0 | 19.7 | 3.8 | N/A | 0.5 |
| Versatel (US) | 0.0 | 0.0 | 2.7 | 1.4 | N/A | 0.2 |
| Tele2 | 0.0 | 0.0 | 1.3 | 0.9 | N/A | 0.1 |
| Freenet | 0.0 | 0.0 | 1.0 | 0.4 | N/A | 0.06 |
| Keine Angabe | 0.0 | 0.0 | 1.0 | 0.7 | N/A | 0.1 |
| Media Concentration Index | 2004/5 | | 2011 or Most Recent | | % Change Annual Average | |
| Public Ownership (%) | 27.2 | | 17.7 | | –1.2% | |
| Foreign Ownership (%) | 17.6 | | 9.0 | | 1.4% | |
| C4 Average—Weighted | 96 | | 90 | | –8% | |
| HHI Average—Weighted | 6,548 | | 3,468 | | –5.9 | |
| C1 Average—Weighted | 73 | | 48 | | –3% | |
| National Power Index | 6,535 | | 3,126 | | –6.5% | |

the fall in the national power index from 4,561 in 2004 to 2,620 in 2012. In Wireline and ISP platforms Deutsche Telekom has lost market shares to new companies. Before 2000, DT held 100% monopolies in both services, but its wireline share has fallen to 56.9% in 2012. Its ISP market share was at 45.6% and falling. DT's share of the overall platform market declined from 71.6% in 2004, when it was still effectively a fixed line monopolist, to 46.4% in 2012 due to new entrants like Alice, E-Plus, and Keine Angabe, as well as competition from established cable providers like Kabel Deutschland (Tables 6.20, 6.21).

# 7

# Media Ownership and Concentration in Ireland

RODDY FLYNN AND PASCHAL PRESTON

## INTRODUCTION

Irish media markets are dominated by either a handful of relatively large players (e.g., wireline and mobile telephony) or are de facto monopolies (e.g., satellite broadcasting and cable television). This owes much to the small scale of the Irish market and the economies of scale and scope (encouraging the creation of larger media enterprises) that characterize all media industries.

This is not the case in all industries (e.g., broadcast television and print media) because of the presence of international players in Ireland. Political, geographical, and linguistic factors have made it relatively easy to access the Irish media market. This presents difficulties in conducting Irish media market analyses because it is often impossible to identify the "Irish media market" as a discrete entity. Financial reports from Irish media groups do not clearly delineate between domestic and foreign activities. Similarly, the international media groups that often play dominant roles in Irish media sectors rarely present separate accounts of their Irish activities. In film distribution, for example, the Hollywood majors treat the United Kingdom and Ireland as a single market and report combined revenues figures. As a consequence, it is not always possible to produce reliable revenue figures for market shares of individual media firms. In this study we address questions of media concentration in terms of share of circulation, readership, and household penetration.

The openness of the Irish market derives in part from Ireland's former status as a colony of the United Kingdom until it gained independence in 1922. As a consequence, UK-published newspapers competed with indigenous titles from the foundation of the state. In broadcasting, BBC radio broadcasts in Ireland began in

1922 whereas the first Irish radio broadcaster, 2RN, wasn't created until 1926. Irish television began in 1961, although viewers in parts of the eastern seaboard could receive BBC television signals as early as 1936.

In the 1980s and 1990s, competition from UK terrestrial broadcasters was augmented by the arrival of other players, facilitated by new broadcasting technologies (cable, satellite, and their more recent digital variants) and a general European trend toward the liberalization and deregulation of broadcasting markets. Despite this, those Irish media outlets producing content primarily for the local market were largely Irish owned until the close of the 20th century. When private commercial broadcasting was legalized in 1988, the newly established radio and television stations were generally Irish-owned. However, from the late 1990s (and especially between 2000 and 2006) the arrival of foreign capital intensified merger and acquisition activity in the Irish print and broadcast sectors. As a result, an increasing number of Irish media outlets were acquired by international companies.

The presence of overseas players in telecommunications markets such as ISPs and mobile telephony is a more recent phenomenon. In 1998, both players in the Irish mobile phone market were owned by domestic firms but, as of 2011, the sector was dominated by firms headquartered in the United Kingdom, Spain, Singapore, and Hong Kong.

The arrival of privately owned firms into Ireland's media markets led to a reregulation of Irish media markets. Regulation of the various media sectors is spread between a number of institutions. Television and radio stations are licensed by the Broadcasting Authority of Ireland. Fixed and mobile telecommunications services (including cable and satellite television operators) are licensed by the Commission for Communications Regulation (COMREG). There is no specific body for licensing newspapers. However, since 2006 Ireland has had a press council that offers a means of redress for unfair press coverage.

All media outlets in Ireland are also subject to the oversight of the Competition Authority. Until 2002, the Authority treated all business enterprises alike: the approval of the minister for industry and commerce was required whenever one enterprise proposed acquiring 30% or more of another organization. However, the Competition Act of 2002 included a new section dedicated to media ownership that enjoined the minister to consider the impact on plurality and diversity of permitting a media merger to proceed. Thus, the 2002 Act took the ultimate authority for media mergers out of the hands of the Competition Authority and placed it with the minister for enterprise and employment.

The 2002 Competition Act placed the Competition Authority at the center of decisions on corporate mergers (including media mergers). Mergers worth €40m or more must be notified to the Competition Authority for approval. However, if both companies involved in the merger are media companies and at least one operates in Ireland, the Competition Authority must be notified regardless of the merger value.

Once notified of a merger, the Competition Authority decides whether a merger or acquisition will substantially lessen competition. It may to approve a merger (possibly with conditions) or prohibit it. However, in the case of media mergers, even if the Authority decides that a merger raises no competition concerns, Section 23 of the Competition Act empowers the minister for enterprise and employment to block the merger on the basis of what the Competition Act terms "relevant criteria" (Competition Act, 2002, 25).

Between the introduction of the 2002 Competition Act and the spring of 2008 (i.e., the most active period of media mergers and acquisitions for decades), some 89 media mergers were reported to the Competition Authority; 86 of these were cleared after a short (one month) "first phase" review. The Authority has cleared 97% of all mergers notified to it by this process.[1] The remaining three mergers were cleared at

1. Competition Authority. *Strategy Statement 2012–2014*, Dublin: The Competition Authority, 2011, p. 8.

phase two (a longer process taking up to three months) with some conditions attached.[2] In sum, despite a remarkably active period of media mergers in the first half of the 2000s, the state has rarely thoroughly investigated the implications of any of these mergers.

To the extent that it is a public issue in Ireland, debates on media concentration focus on the perception that a small number of companies dominate one or more media markets. "Official" debates around such concentration rarely refer to the potentially negative impact on media pluralism. Such concerns are primarily voiced by Irish academics. For example, from the late 1980s to the late 1990s, "official" discussions of media mergers and concentration in Ireland were dominated by two issues:

1. The introduction of private sector actors and competition in the broadcasting sector in line with neoliberal policy developments in the European Union and international contexts
2. In the print media sector, a concern with a perceived threat to "indigenous" newspapers from external (UK-based) competition[3]

Such debates were informed by economic conceptions of diversity (i.e., competition) as opposed to more political and cultural concerns about plurality. This perspective is still evident in the Competition Authority's approach to evaluating media mergers. The primary concern is "whether or not it will substantially lessen competition in markets for goods or services in the State."[4]

Thus, media sectors in Ireland have been treated more or less identically to less politically sensitive sectors such as manufacturing, despite occasional outbursts of political rhetoric on the subject of media pluralism. In 2009, the state-appointed Advisory Committee on Media Mergers noted this failure to acknowledge the media's political role in the Competition Authority's deliberations: "It is widely recognised that this test while sufficient to protect the public interest in competition, is not in itself always sufficient to protect the public interest in media plurality and diversity."[5]

In the meantime, public concern relating to media mergers and concentration of ownership has focused on the possibility that such media companies may abuse their market dominance for selfish ends. This reflects the perception that the "'amenity potential,' also known as the 'private benefits of control,' arising from owning media outlets is extremely high. In other words, the nonfinancial benefits, such as fame and *influence,* that are obtained by controlling a newspaper or a television station must be considerably higher than those that come from controlling a firm of comparable size in, say, the bottling industry"[6] (italics added).

In Ireland, the perception that media ownership is driven by the desire for influence finds expression in the personification of dominant businesses in their chairmen or chief executives. Two names stand out in this regard. The first is Tony O'Reilly, best known outside Ireland as the former chair of US-based H.J. Heinz, the food and beverages conglomerate and for several decades the richest man in Ireland.

Within Ireland, he is best known as the former chief executive of Independent News and Media (INM), a corporation built around the acquisition of the Independent Group in 1973. INM is the largest print media group in Ireland and has been since the early 1970s.

INM is also one of the largest print media owners in Australia and, until 2012, had significant holdings in India and South Africa. It also owned the London Independent Group for nearly a decade until 2010. O'Reilly's media

2. Advisory Group on Media Mergers. *Report to the Tanaiste and Minister for Enterprise, Trade and Employment.* Dublin: Government Publications, 2008, p. 56.

3. See Horgan, John. *Irish Media: A Critical History Since 1922.* London: Routledge, 2001; and Kenny,Colum." "A double-edged sword: Irish media merger policy in transition." *Journal of Media Business Studies.* 6, 3 ( 2009): 93–108.

4. Horgan, op.cit., p. 50.

5. Horgan, op.cit., p. 51.

6. Djankov, Simon et al. "Who Owns the Media?" *Journal of Law and Economics.* XLVI. 2 ( 2003): 341–381.

interests extend beyond print: INM was a partner in the second largest cable television operator in Ireland from 1992 to 2004, and from 2000 to 2004, O'Reilly was chairman of the Valentia consortium, owner of the former state postal and telecom monopolist, Eircom.

It has been asserted that O'Reilly's strong influence in Irish print media and his cross-media holdings has shaped the editorial content of the INM newspapers. On the day of the 1997 general election, the INM-owned *Irish Independent* called for a vote against the incumbent Fine Gael administration, which *The Independent* had traditionally supported. *The Independent*'s about-face was ascribed to INM dissatisfaction with the failure of the outgoing Fine Gael–led administration to adequately regulate the activities of rural community groups who were piggybacking television distribution systems on the back of legal cable operators. These cable operators included the INM-owned Princes Holdings. Many viewed the editorial as O'Reilly using his media holdings to promote his own interests and those of his corporate empire.

The second figure who exemplifies this focus on the individual is Denis O'Brien. O'Brien secured his first Dublin radio station license in 1989. He subsequently established Esat Telecom as a landline phone provider. In 1995, Esat won the first private mobile phone license in Ireland. In 2000, O'Brien sold his stake in Esat to British Telecom for approximately €320m. His Communicorp investment vehicle subsequently acquired extensive interests in Eastern Europe radio and owns Digicel, now the largest mobile phone operator in the Caribbean. In 2008, he acquired the two Irish national commercial radio stations—Newstalk 106 and Today FM. At around the same time, he began to build a stake in Independent News and Media.

It was not immediately clear why O'Brien would seek to acquire a stake in a newspaper group, at a time when newspapers globally were experiencing falling circulations and ad revenues. In Spring 2011, a state tribunal investigating the circumstances under which Esat Digifone had originally secured its mobile phone license suggested that O'Brien had effectively bribed the Minister for Communications to secure the license. O'Brien denied these assertions. However, different media outlets reported this story in ways that appeared to be influenced by ownership. For example, INM titles, of which O'Brien was the largest shareholder, devoted less space to the story relative to other Irish papers. Furthermore, INM papers focused on the role of the corrupt politician, downplaying O'Brien's involvement. Both the national commercial broadcasters (which are also owned by O'Brien) also adopted a more skeptical attitude to the tribunal's findings than the public service broadcaster.

Consequently, the public formed the opinion that O'Brien's acquisition of dominant stakes in the largest commercial radio stations and print media had been conducted in anticipation of a negative finding from the tribunal. These public perceptions have had a political impact. In October 2011, the Minister for Communications announced new draft legislation exclusively focused on media mergers, stressing "the undesirability of allowing any one individual or undertaking to hold significant interests within a sector or across different sectors of media businesses" (Rabbitte, 2012). However, the Minister also acknowledged the inadequacy of conventional micro-economic competition analysis metrics for assessing media mergers, promising to include criteria relating to pluralism for such mergers.

Both O'Reilly and O'Brien have some involvement in another narrative around Irish media ownership that engendered public skepticism. This involved the 1999 privatization of Eircom, then a state-owned telecommunications operator. In 1996, Eircom (then called Telecom Eireann) was partially privatized when the state sold a 20% share in the company to the Dutch/Swedish KPN/Telia consortium. This initial sell-off was understood to be the first phase of a process that would leave the state with a majority share of 50.1% of the company. However, in 1999, the Minister for Public Service announced that the state would sell its entire stake in Eircom.

The privatization was not a success. By 2001, the market value of Eircom had fallen to €3bn from a peak market capitalization of €10.25bn in 1999. This left Eircom vulnerable to a takeover.

In 2001 Anthony O'Reilly and Denis O'Brien competed to acquire Eircom. O'Reilly's Valentia consortium won the deal. Despite acting as chairman of Valentia, O'Reilly himself owned just 4% of Eircom after 2001, for which he paid €25m. The key institutions behind Valentia then were Providence Equity, a media/entertainment-focused venture capital firm that held a 46.4% stake, the Soros Fund that held a 18.5% stake, and Goldman Sachs, which took a 1.2% holding.[7]

Valentia's acquisition of Eircom was largely funded through debt. In the two years following the acquisition, the consortium asset-stripped Eircom, taking large dividends, selling off existing assets, sweating those that remained, and cutting back on new infrastructure investment. In April 2004, Valentia brought Eircom back to the stock market via a second IPO, making capital gains that nearly doubled their initial investment. However, this came at a cost to the operations of the company: investment in DSL infrastructure was severely curtailed in 2002 and 2003, at precisely the moment that technology was becoming a consumption norm in other OECD countries.

This reluctance to invest in key telecom infrastructures at a time when Ireland sought to brand itself as an information or knowledge economy was compounded in 2006 when Australian venture capitalists Babcock and Brown took over the company, and again concentrated on increasing shareholder value to the detriment of infrastructure investment. Babcock and Brown sold off their shares in September 2009. In 2010 SingTel, a subsidiary of Temasek, the Singaporese government's state holding company, purchased 65% of the company. SingTel inherited a huge debt of €3.8bn that they could not refinance or sell. In March 2012, the company was declared to be legally insolvent. Having previously been the largest privatization in Irish corporate history, the company established a new landmark when it entered the biggest examinership in Irish corporate history. The repeated flipping of Eircom's ownership has undermined public confidence in the virtues or privatization and in the value of free-market competition.

## PRINT MEDIA

### Newspapers

The domestic press in Ireland consists of three or four national dailies, one national evening newspaper, five or six Sunday newspapers and approximately 40 regional/local or "provincial" newspapers. With the exception of the Cork-based *Irish Examiner*, all have a discernible orientation toward Dublin. The has left space for a relatively large number of local and regional papers. There has been a general decline in readership over the past four decades. Between 1971 and 2006, the population increased by 40% while total daily, Sunday, and evening sales declined by 10%. Mirroring international trends, sales have fallen calamitously since 2008. From 2008 to 2011, daily sales fell by 18% and Sunday sales fell by 26%.

UK-based titles are also major players, doubling their share of the Irish daily market since 1970. Their presence creates some ambiguity regarding the number of "Irish" titles. In both the daily and Sunday markets, at least one major "Irish" title is owned or co-owned by a British press group. In the early 1990s, concerns around concentration focused on the dominant position of IMN. The increasing penetration of UK-based media groups has complicated these issues.

INM remains the dominant media group in Ireland. It owns the best-selling daily newspaper—the *Irish Independent*—and has a 50% share (with the UK's Express Group) in the *Irish Daily Star*. It also owns the only truly national evening paper, the *Herald*, and both the best-selling and second best-selling Sunday papers, the *Sunday World* and *Sunday Independent*. INM has also been a major international player with media operations in Australia, New Zealand, South Africa, the United Kingdom, and India. Since 2008, the ownership of Independent News and Media itself has been in flux as cross-media owner Denis O'Brien, whose Communicorp investment vehicle has substantial holdings in the Irish broadcast field (see later in the chapter), increased his personal stake in INM. In 2008,

7. Irish Congress of Trade Unions. *Privatisation: Learning from the Eircom Debacle*. Dublin: ICTU, 2011, p. 7.

O'Brien's stake in INM stake surpassed 25%, thus allowing him to block special resolutions from the INM board. O'Brien's presence was not welcomed by INM. When O'Brien acquired several national and local radio stations in a 2007 deal, INM made submissions to both the Competition Authority and the Irish Government objecting to the deal on the grounds of cross-media ownership. Hostilities between O'Brien and INM persisted. In 2012, O'Brien effectively had INM Chief Executive Gavin O'Reilly removed and replaced by an O'Brien candidate.

Thomas Crosbie Holdings is the second largest newspaper group in Ireland by number of titles (three), but it is significantly smaller than INM.[8] In 2010 the turnover for TCH was €72 million as compared to €630 million for INM. Indeed in 2010, the Irish Times Limited, which only publishes one national daily (albeit with the second largest circulation), recorded higher revenue (€86 million) than TCH. *The Irish Times* is unusual in that it is owned by a trust, a legal status established in 1974 to defend the paper against corporate takeovers. Any profits must be reinvested into the paper. That said, in practice the corporate strategy of the Irish Times Limited differs little from its purely market-driven competitors.

The domestic newspaper market is entirely commercial in its organization. The kind of state subsidies found in Scandinavian countries are unheard of in Ireland: if a newspaper runs into financial difficulties, the state has historically permitted it to collapse. The Irish Press Group, which had successfully operated daily, Sunday, and evening papers since the 1930s went bankrupt in 1995. The Competition Authority prevented Independent News and Media from investing in the ailing Irish Press Group because it gave INM an unacceptably dominant market position.

The position of Independent News and Media appears remarkably dominant, in the domestic market, accounting for nearly 85% of all Sunday paper sales and nearly 50% of daily sales. However, this fails to take account of sales for Irish editions of UK newspapers. The local content of these editions varies. Some publish eight pages of local content around a core that is identical to the UK editions. Others produce editions that would be entirely unrecognizable to UK readers. The "localization" strategy has been very successful. News International's *Irish Sun* and *Irish Sunday Times* both rank fourth (by circulation) in their respective daily and Sunday markets. Before its 2011 closure in the wake of a phone-hacking scandal surrounding the UK edition, the *Irish News of the World* was the third best-selling Sunday edition in Ireland. Similarly, the Daily Mail and General Trust's two Irish titles—the *Irish Mail on Sunday* and the *Irish Daily Mail*—are the third and sixth best-selling papers in the daily and Sunday markets, respectively. As a consequence, News International and the Daily Mail and General Trust have the second and third largest newspaper circulations in Ireland, respectively. As of 2011 the fourth position is occupied by Thomas Crosbie Holdings; this is hotly contested by another a UK-based group, Trinity Mirror, which holds 6.6% of the markets to TCH's 6.7%.

UK-based players are also prominent in the regional press sector. Until the mid-1990s, the regional press was in the hands of myriad small Irish firms save for the 10 titles owned by INM. However, since 1995 there has been a flurry of take-over activity, and since 2010 every regional newspaper with a circulation in excess of 20,000 per week is now in the hands of a media group, the majority of which are UK based. The key groups are the Alpha Newspaper Group (UK), Andersonstown News Group (UK), Dunfermline Press (UK), Independent News and Media, Johnston Press (UK),[9] and Thomas Crosbie Holdings.

When considering concentration number, this study's calculations focus on market reach. Revenue numbers are not reliable. However, there remain questions as to how to define the market for the Irish press sector.

Broadly speaking the Irish press can be divided into national, regional, and local papers.

---

8. In 2014, Thomas Crosbie Holdings exited the market. Many of its former press titles and radio stations were acquired by a new company, Landmark Media.

9. Johnston Press, the largest regional press company at the start of 2014, sold off its Irish holdings to the UK-based Mediaforce.

We must also take into account the three categories of daily paper in Ireland: Irish owned and published, British owned and Irish published, and British owned and British published. These arguably constitute different markets. For the calculation of the HHI index we have measured concentration at two levels:

1. Irish owned national titles only
2. Irish and British titles available in the Irish market

Were we to exclusively focus on Irish-owned titles, INM appears dominant with its daily, evening, and Sunday papers accounting for 80% of the total national newspaper market. This produces an HHI index of 6,603 for 2010 (Table 7.1).

However, this fails to reflect how UK titles constitute a significant proportion of the Irish press market. For every two copies of domestic papers sold, Irish readers buy one "Irish" edition of a UK-published paper. News International publications are particularly prominent in this regard. For a period in 2005, the Irish edition of *The Sun* was second only to the *Irish Independent* in terms of daily sales. In addition, UK groups also sell UK editions of their papers directly into the Irish market. Including all of these categories of newspapers offers the most comprehensive basis for calculating market share within the national newspaper market in Ireland. On this basis the HHI indices for Irish and British newspapers sold in the Irish market in 2000 and 2008 are 3,381 and 3,235, respectively (Table 7.2).

### Table 7-1. Irish-Based Newspaper Groups—%Market Shares by Circulation

| MEDIA GROUP | MARKET SHARE 2010 |
|---|---|
| Independent News and Media | 80 |
| Thomas Crosbie Holdings | 10.6 |
| Irish Times Trust | 9.5 |

*Source:* Authors' extrapolations from Audit Bureau of Circulation figures.

The period from 1995 to 2005 (and especially from 2000 to 2005) saw an extraordinary increase in merger and acquisition activity within the regional press sector. Despite small circulations, such titles were desirable because they attracted reliable advertising revenues. As a consequence the sums offered for small (2–3 title) groups or even individual titles began to soar, peaking between 2003 and 2005 at valuations up to 50 times their pretax earnings. This was largely driven by the arrival of UK-based players with deep pockets like Trinity Mirror, the Dunfermline Press (now the Celtic Media Group), the Johnston Press, and Scottish Radio Holdings. This led "independent" regional papers to become the exception rather than the norm.

Furthermore, the 14-title Johnston Group replaced INM as the largest player by title in the regional market. Despite the increased concentration of ownership, however, the regional press is still relatively competitive. No group in the regional market has a market share in excess of 20%. Thus, this HHI is 938, based on the nine groups singled out in Table 7.2 plus 17 other regional papers.

Although the regional papers are often discussed as a single market, most operate in discrete geographical territories and there is little direct competition between them. This is reflected in Competition Authority policy regarding acquisitions of regional papers. In 2005 the Competition Authority approved Johnston Press's acquisition of the seven-title Leinster Leader Group in 2005. They justified this decision by pointing to the "minimal overlap in circulation" and the "lack of overlap in advertising" between the Irish regional papers Johnston already held and those it was acquiring. Even in those geographical areas where the Authority found there was some circulation overlap between the Johnston Group papers and those of the Leinster Leader group, the Authority found at least three other alternative sources of press advertising.[10]

10. Competition Authority. *Determination of Merger Notification M/05/065—Johnston*. Dublin: The Competition Authority, 2005.

Table 7-2. Irish and UK-Based Newspaper Groups—Irish Market Shares by Circulation (%) 2000–2008

| | 2000 | 2004 | 2008 |
|---|---|---|---|
| Independent News and Media | 53.3 | 50.5 | 52.0 |
| News International (UK) | 17.9 | 18.7 | 17.9 |
| Daily Mail and General Trust (UK) | 1.1 | 8.4 | 9.3 |
| Thomas Crosbie Holdings | 4.5 | 6.5 | 6.7 |
| Trinity Mirror (UK) | 12.3 | 8.6 | 6.6 |
| Irish Times Trust | 5.8 | 5.5 | 5.9 |
| Guardian Media Group (UK) | 0.6 | 0.8 | 0.8 |
| Express Group (UK) | 0.5 | 0.7 | 0.4 |
| Telegraph Group (UK) | 0.4 | 0.3 | 0.3 |
| Independent Print Ltd. (UK) | 0.7 | 0.3 | 0.3 |
| Pearson (UK) | 0.3 | 0.2 | 0.2 |
| Ireland on Sunday | | | |
| Scottish Radio Holdings (UK) | 2.8 | | |
| Sunday Business Post | | | |
| Total Rev (mil €) | 1,000 | 1,600 | 1,500 |
| Total Rev (mil $US) | 930 | 2,100 | 2,140 |
| C4 | 89.3 | 86.2 | 85.9 |
| *N* (>1%) | 7 | 6 | 6 |
| HHI | 3,381 | 3,118 | 3,235 |
| Noam Index | 1,278 | 1,273 | 1,321 |

*Source:* Authors' extrapolations from Audit Bureau of Circulation figures.

These figures for market share create a somewhat misleading impression of the Irish regional press sector. When combined with the figures for national press sales, they offer a fuller picture of the role played by particular media groups in the Irish press sector, especially given that two of national press players, INM and TCH, are also active in the regional market (Table 7.3).

Taking the national and regional markets together, the HHI for 2010 is 2,285, significantly lower than the 3,235 recorded for the national market alone in 2008. Although reliable comparable figures for earlier years are unavailable, the market is unquestionably more concentrated in 2010 than ten years earlier, driven by the increasing market share of UK-based titles and the wave of regional title acquisitions. Again, this is largely driven by UK-based groups. In 2000, the 90 or so titles in the regional market were spread across 75 companies. This was reduced to fewer than 30 by 2010. Despite the more prominent position of overseas parents in the regional market, little concern has been expressed about the potential erosion of local content. This is in large part due to the nature of regional papers: their primary appeal lies in their local-nature. Thus there is little incentive for international parent companies to repackage content produced for other geographical markets and include it in their regional press holdings.

The national market is different. When Daily Mail and General Trust acquired *Ireland on Sunday* it was relaunched as the *Irish Daily Mail,* replacing Irish content with material from the UK paper. However, the extent of this type of

Table 7-3. Irish and UK-based Newspaper Groups—National and Regional Circulation

| GROUP | GROUP CIRCULATION 2010 | GROUP MARKET SHARE 2010 (%) |
|---|---|---|
| INM | 1,049,641 | 43.5 |
| News International (UK) | 331,462 | 13.7 |
| TCH | 199,856 | 8.3 |
| 10 Family-owned regional titles (aggregate) | 126,807 | 5.3 |
| Johnston Press (UK) | 120,126 | 5.0 |
| Trinity Mirror | 118,037 | 4.9 |
| Daily Mirror and General Trust (UK) | 118,037 | 4.9 |
| Irish Times Trust | 105,742 | 4.4 |
| 7 Privately owned regional titles (aggregate) | 63,505 | 2.6 |
| Celtic Media Group (UK) | 51,400 | 2.1 |
| Alpha Newspaper Group (UK) | 31,310 | 1.3 |
| Connacht Tribune | 27,447 | 1.1 |
| Vale/Mallow Star | 16,500 | 0.7 |
| Topic | 14,500 | 0.6 |
| Guardian Media Group (UK) | 11,963 | 0.5 |
| River Media | 9,321 | 0.4 |
| Express Group (UK) | 6,921 | 0.3 |
| Pearson (UK) | 3,680 | 0.2 |
| Telegraph Group (UK) | 5,324 | 0.2 |
| Independent Print Ltd. (UK) | 2,725 | 0.1 |
| Total National and Regional Press Circulation | 2,414,304 | |

*Source:* Authors' extrapolations from Audit Bureau of Circulation figures into larger media groups.

practice has been limited thus far and even *Irish Daily Mail* runs a significant Irish operation in Dublin. The issue is further complicated by the fact that domestic titles have long relied on material sourced from (non-Irish) press agencies and partnerships with UK titles[11] from outside Ireland. In other words, it is not possible to assert that the internationalization of ownership has substantially diluted the domestic content of Irish regional or national papers. Certainly Trinity Mirror's 1997–2002 ownership of *The Sunday Business Post* (before it was acquired by Thomas Crosbie Holdings) did not appear to alter the content of the paper.

The timing of the international and domestic companies who moved into the regional press market in the early 2000s proved to be unfortunate. Though not as badly affected as the national sector in terms of declining circulation and advertising, the local sector has still faced difficulties. The high values placed upon the regional titles at the time of acquisition assumed that conglomeration of production would lead to economies of scale. In practice, many of the regional titles have continued to operate as stand-alone entities and the theoretical benefits of mergers have not been felt. In some cases, most notably with Johnston Press,

11. Thus the *Irish Times* regularly uses material from *The Guardian* and the *Irish Independent* also uses articles from *The London Independent* of London and *The Daily Telegraph*.

the new owners sought to divest their Irish holdings as the economy slid into recession in 2008 and 2009. For the most part, finding buyers has been exceptionally difficult, and Johnston and other groups that expanded their holdings just before the crash seem resigned to retaining their titles until/if the economy rebounds.

### Magazine Publishing

Ireland's print media also includes a periodicals sector, although given the national commitment of the Irish to newspapers (the National Newspapers of Ireland lobby group asserts that 19 out of 20 Irish adults read a paper every day), there are relatively few indigenous magazines. This is largely due to the overwhelming presence in the Irish market of UK-published periodicals. What indigenous titles do exist mirror the categories found in other countries: women's magazines, business journals, and current affairs loom large. Although the Periodical Publishers Association of Ireland lists 59 members, most of these are either very small or engaged in contract publishing (i.e., producing material for corporate clients). Only a handful own more than one "off-the-shelf" title. The largest of these is Harmonia, a company formed in 2004 as a result of a management buy-out of the former Smurfit Publications operation. Harmonia publishes three of the four best-selling women's magazine in Ireland. Others with multiple publicly available titles include Business and Finance, Page 7 Media Limited, and Mediateam, which concentrates on publishing IT-related journals.

## AUDIOVISUAL MEDIA

### Radio

For most of the 20th century in Ireland, radio (and television) broadcasting was the exclusive preserve of state-run or state-owned broadcasters. The Wireless Telegraphy Act of 1926 conferred sole authority to license broadcast operators, private or public, upon the minister for communications. State-run radio broadcasting began in 1926 (although the BBC Belfast transmissions were available in parts of southern Ireland from 1922). In 1988, the Radio and Television Act established an Independent Radio and Television Commission (IRTC) to license and regulate the activities of new privately owned commercial broadcasters. The IRTC subsequently became the Broadcasting Commission of Ireland (BCI) and is now the Broadcasting Authority of Ireland (BAI). As the licensing body, the BAI can determine which individuals and institutions should be permitted to own radio and television stations. The exercise of this power is not limited to new licensees. If an individual/corporation licensed to operate a broadcaster by the BAI seeks to sell that broadcaster to another individual/corporation, the BAI may withhold approval for the transfer of the license. The 2009 Broadcasting Act requires the BAI to have regard for the undesirability of allowing any person/group of persons to have control of/substantial interest in "an undue number" of radio, television, or other communications media when considering awards or transfers of licenses. Since 2001, the BAI has outlined a set of principles on ownership designed to "facilitate the continued development of a viable and diverse broadcasting sector that is characterized by a plurality of ownership."[12]

The dominant player in Irish radio broadcasting is the state-owned broadcaster, RTE, although that dominance has been severely curtailed since the introduction of privately owned competition in 1989. Thus from "officially"[13] holding 100% of the market in 1988, the combined listenership of RTE Radio 1, 2FM and RTE Lyric FM is now 32%.[14] In addition

12. Broadcasting Authority of Ireland. *Ownership and Control Policy 2010*. Dublin: Broadcasting Authority of Ireland, 2010, p. 2.

13. The presence of "pirate" stations in larger urban areas from the late 1970s until the 1988 enactment of legislation providing for private broadcaster meant that RTE *de facto* faced competition prior to 1989. However, such stations rarely published independently audited listenership figures and as a consequence it is not possible to estimate pre-1988 market shares with any confidence.

14. RTE also operates a fourth Irish-language stations called Raidio na Gaeltachta but it's market share is negligible (i.e., sub 1%) and it does not carry advertising.

to RTE there are currently 57 licensed independent "sound broadcasting services" (radio stations) operating in Ireland. These consist of two national independent commercial radio stations, Today FM and Newstalk 106,[15] six regional music stations, 27 local independent commercial stations, and 20 community/community-of-interest stations. The remainder are "institutional" stations often run by hospitals. Local radio usually means a county radio station whose market roughly coincides with county boundaries. Community and community-of-interest stations generally attract small audiences and as a consequence carry little advertising (indeed they are usually expressly prohibited from doing so). In contrast to television broadcasting, UK stations play a minor role in the Irish radio market.

When local radio licenses were initially offered in 1988, the consortia behind individual stations were local in nature, driven by local business people, lawyers, accountants, and so forth. However, after 2001, the BCI relaxed the local radio ownership rules "in the light of the changing broadcasting environment," permitting 100% ownership of broadcasters and more multiple station ownership.[16] This led to a flurry of acquisitions/mergers between 2001 and 2007: by March 2011, only eight of the 35 BAI-licensed commercial radio stations had not been subsumed (Table 7.4).

Within the purely commercial sector, Communicorp dominates. Owned by Denis O'Brien, the businessman who made his fortune in mobile telephony, Communicorp was established with the launch of 98FM in 1989. The group subsequently expanded overseas. By 2012 it owned 42 radio stations in Ireland and Eastern Europe. In July 2007, Communicorp acquired three Irish stations—national broadcaster Today FM, Dublin station FM104, and the Donegal local station Highland Radio from British media group Emap.[17] Since this placed both national commercial broadcasters (Today FM and Newstalk 106) and the two leading Dublin stations in Communicorp's hands, the deal was subject to Competition Authority approval. However, before the Competition Authority could make a recommendation, the Broadcasting Commission of Ireland became involved. Under the BCI's ownership guidelines, any proposal that a single entity own "a number" of radio stations equivalent to more than 15% of the total of those operating in the Irish market requires "careful consideration." Ownership of anything over 25% of all stations is considered completely unacceptable. Consequently the BCI stated that it would not approve a transfer of FM104's license to Communicorp.[18]

This influenced the Competition Authority's assessment of the deal. Although it was ultimately approved, it was subject to the condition that Communicorp divest itself of FM104. Thus in February 2008, Communicorp resold FM104 to UTV, another player actively seeking to acquire more radio stations. UTV grew out of the commercial television franchise for Northern Ireland and, as such, is presently the only "foreign" company to own Irish radio broadcasters. In addition to 12 stations in the United Kingdom, UTV now owns five stations in the Republic of Ireland.

The Communicorp deal prompted questions about the Competition Authority's competence to deal with the media plurality issues thrown up by cross-media ownership. As a result, in March 2008 the minister for enterprise and employment appointed a committee to examine whether competition legislation should take account of diversity of opinion when adjudicating on mergers in the media sector. In April 2008, the Broadcasting Commission of Ireland initiated an investigation into the implications of Denis O'Brien's increased holding in

15. Technically Newstalk is referred to as "quasi-national" since more remote areas of the western seaboard cannot receive its signal.

16. Maguire, Conor. "Setting out new direction for broadcasters." *Irish Times* October 19, 2001, p. 19.

17. Today FM was acquired from its original Irish owners by Scottish Radio Holdings in January 2002. In July 2005, EMAP purchased Scottish Radio Holdings including its Irish properties.

18. "High stakes radio game." *Sunday Business Post*, October 14, 2007, p. 15.

Table 7-4. Radio Station Ownership and Market Share (by Audience) 2008–2011

| NAME OF GROUP/ STATION | NAME OF STATION | % SHARE OF NATIONAL AUDIENCE BY GROUP 2008 | % SHARE OF NATIONAL AUDIENCE BY GROUP 2011 |
|---|---|---|---|
| RTE (Public) | Radio 1 | 35.9 | 32.0 |
| | 2FM | | |
| | Lyric FM | | |
| Thomas Crosbie Holdings | 4FM | 4.6 | 6.0 |
| | Beat FM | | |
| | Red FM | | |
| | WLR | | |
| Communicorp | Newstalk | 26.4 | 25.0 |
| | Today FM | | |
| | 98FM | | |
| | Spin 103.8 | | |
| | Spin Southwest | | |
| UTV (UK) | Cork 96/103FM | 15.5 | 15.0 |
| | FM104 | | |
| | Limerick Live | | |
| | LMFM | | |
| | Q102 | | |
| Vienna Investments | Radio Nova | | 1.0 |
| Wilton Radio | i102–104 | | 4.0 |
| | i103–105 | | |
| Clare FM | KFM | 1.7 | 2.0 |
| | Clare FM | | |
| Radio Kerry Group | Northern Sound/ Shannonside FM | 3.5 | 3.0 |
| | Radio Kerry | | |
| Connaught Tribune | Galway Bay FM | 1.7 | 2.0 |
| Irish Press | Tipp FM | 1.4 | 1.0 |
| Orangold Ltd. | Highland Radio | 1.5 | 1.0 |
| Tindle Group | Midlands 103 | 1.4 | 2.0 |
| River Media | Ocean FM | 0.9 | 1.0 |
| KCLR | KCLR | 0.9 | 1.0 |
| Midwest Radio | Midwest Radio | 1.2 | 1.0 |
| South East Radio | South East Radio | 1 | 1.0 |
| Sunshine 106.8 | Sunshine 106.8 | 0.8 | 1.0 |
| East Coast FM | East Coast FM | 0.8 | 1.0 |
| Phantom | Phantom | 0.8 | 0.0 |
| C4 | | 82.4 | 78.0 |
| HHI | | 2,277 | 1,956 |
| *N* (>1%) | | 12 | 18 |
| Noam Index | | 657 | 461 |

*Source:* Author extrapolations based on JNLR figures.

Independent News and Media for his broadcast radio holdings. That investigation, published in July 2008, concluded that the businessman's increased press holdings should not impact on his right to retain his broadcast interests.

Outside RTE, Communicorp, and UTV, the market shares held by the other radio groups tend to be small; mainly they do not own or operate either national stations or stations aimed at the key Dublin market. Thomas Crosbie Holdings owns four stations but its national market share is only 6% because it only owns non-Dublin local or regional stations. However, the influence of the smaller groups is often greater than their share of the radio market suggests because of their cross-media holdings. Thomas Crosbie Holdings for example owns two national newspaper titles, the *Examiner* and *The Sunday Business Post*. Galway Bay FM is also owned by the Galway-based *Connaught Tribune* newspaper, while River Media, which owns the local Donegal station, also runs several local newspapers in the same market. Perhaps the oddest of these cross-media groups is the Irish Press Ltd., which owns Tipp FM. Established in 1931 the Press group's newspapers entered terminal decline in the late 1980s and were closed in 1995. However, in 1999 the company purchased a stake in the Tipp FM radio station and it remains its main trading activity.

The combined local/national nature of the Irish market makes it difficult to arrive at an objective basis for calculating degrees of concentration. Because the franchise areas of local stations are geographically distinct, competing for different audiences and advertisers, they effectively constitute different markets. To capture overall market share in the various, national, regional, multi-city, and local radio markets we have calculated the market share of individual stations on the basis of absolute numbers of listeners rather than relative share of their respective markets. This allows us to control for the different population sizes within these different markets. On this basis, the HHI figures for 2008 and 2011 (the only years for which nationally comparable figures are available) dropped from 2,277 to 1,956.

As is the case with television, RTE's share of total broadcasting revenues in Ireland includes broadcasting license fees. Thus although RTE only commands 32% of the national radio audience, it takes 53.5% of all broadcasting revenues (Table 7.5).

Again, as is the case with regard to television, the significant discrepancy in the HHI figures based on audience share and revenue share raises doubts about the appropriateness of using the index as a basis for calculating media influence.

## Broadcast Television

The first television broadcasts received in Ireland were spillovers from the United Kingdom, where the BBC and ITV began television broadcasts in 1936 and 1954, respectively. Thus when RTE (the state-owned national broadcaster) began a television service in 1961, it immediately faced external competition. Audience competition did not mean competition for advertising revenues, since neither the BBC nor ITV sold advertising space to Irish advertisers. In one sense then, RTE was not directly competing with either the BBC or ITV. Not until the Radio and Television Act of 1988 was provision made for a purely commercial, privately owned, domestic television broadcaster. Since 1988, the subsequent arrival of cable, satellite, and digital has also hugely expanded the number of international channels available to Irish audiences.

Although the majority of Irish households augment terrestrial channels with cable and satellite channels, 209,000 Irish homes (out of a total of 1.5 million) exclusively rely on terrestrial broadcasts.[19] These homes are effectively limited to five domestic channels—RTE 1, RTE 2, TG4, TV3, and E3. We treat these homes as a separate market. This is a somewhat artificial construction and does not reflect the manner in which either the broadcasters or AGB Nielsen (which audits market share in Ireland) treat the market.

19. Comreg. Irish Communications Market Quarterly Key Data Report Q4 2011. Dublin: Comreg, 2012, p. 68.

Table 7-5. Radio Market Shares (By Revenue) 2008–2011

| | 2008 | 2011 |
|---|---|---|
| Radio Telifis Eireann (Public) | 52.0 | 53.5 |
| Communicorp | 19.8 | 17.3 |
| UTV (UK) | 11.6 | 10.0 |
| Thomas Crosbie Holdings | 3.5 | 3.9 |
| Wilton Radio | | 2.6 |
| Radio Kerry Group | 2.6 | 2.3 |
| Clare FM | 1.3 | 1.2 |
| Vienna Investments | | 0.9 |
| Connaught Tribune | 1.6 | 1.1 |
| Irish Press | 1.1 | 0.8 |
| Orangold Ltd. | 1.1 | 1.0 |
| Tindle Media | 0.7 | 1.1 |
| River Media | 0.7 | 0.6 |
| KCLR | 0.7 | 0.7 |
| Midwest Radio | 0.9 | 1.0 |
| South East Radio | 0.8 | 0.6 |
| Sunshine Radio 106.8 | 0.6 | 0.5 |
| East Coast FM | 0.6 | 0.4 |
| Phantom | 0.6 | 0.3 |
| Total Rev (mil €) | 199.9 | 154.9 |
| Total Rev (mil $US) | 277.8 | 200.6 |
| C4 | 86.9 | 84.7 |
| HHI | 3,261 | 3,297 |
| *N* (>1%) | 17 | 19 |
| Noam | 791 | 756 |

*Source:* Author estimates based on Ipsos/MRBI Joint National Listernship survey and Institute of Advertising Practitioners of Ireland figures.

The two state-owned public service broadcasters—RTE and TG4—currently dominate the over-the-air ("terrestrial") market. RTE 1 Television was the sole television broadcaster from 1961 until 1978 when a second state-owned television channel, RTE 2, began operating. RTE is funded by a combination of advertising and license fees.[20] In addition to the new commercial entrants, the 1990s saw the arrival of a third public service television channel, Teilifís na Gaelige (subsequently renamed TG4), an Irish-language channel. TG4 was originally a wholly owned subsidiary of RTE but in April 2007 was reconstituted as an independent statutory corporation. TG4 also sells advertising, and is subsidized to the tune of €8–10 million per year through RTE's provision of 2–3 hours of daily of Irish-language programming at no cost. However, the bulk of TG4's revenues derive from a direct grant from the Irish parliament.

20. Since 2009, RTE only receives 93% of the license fee, although from 1926 through to 2002 it received all of it. The rest is made available to all broadcasters operating on the island of Ireland (including Northern Ireland) through the Broadcasting Authority of Ireland-administered Sound and Vision Fund.

In the private sector, there are two television stations with a substantial market presence, meaning they are available in 80% or more of Irish homes. Though licensed in 1988, TV3 did not go on-air until 1998 when CanWest Global, the Canadian multimedia giant, took a 45% stake in the company. CanWest was joined by British media group ITV, which took a 45% share in September 2000. The remaining 10% rested with the original Irish investors. The CanWest/ITV investment brought more than foreign capital to TV3. It also brought substantial quantities of (foreign) content to the nascent Irish operation. In 1998, CanWest was the largest non-US buyer of US audiovisual material, while ITV was also a major content producer, second only to the BBC in the United Kingdom. This access to low-cost programming proved crucial to the subsequent success of TV3, which turned a profit with a 10–15% market share. However, the channel also abandoned the programming commitments, specifically with regard to the commissioning of indigenous material, included in its original license bid in 1988. This financial success was reflected in the €265 million price tag attaching to the station when the British private equity group Doughty Hanson bought out the entire company from CanWest, ITV, and the private investors in 2006.

TV3 also owns a second channel, 3E, which began broadcasting in 2006 as Channel 6. Initially supported by private Irish investors through the Kish Media vehicle, the channel struggled for market share from its inception, recording a €10.5 million deficit in 2007. In July 2008, TV3 acquired the channel for approximately €2–3 million. Approving the takeover, the Competition Authority found minimal overlap in the activities of TV3 and Channel 6 and that the newly merged entity would face significant competition for audiences and advertising from other market players.[21] Given that both channels are broad-format entertainment channels, the notion that their activities did not overlap seems perplexing. It is therefore likely that the Authority concluded that Channel 6's small market share meant the merger would not create a significantly larger player.

Treating the terrestrial market as a discrete market one finds there are just three players: RTE, TV3, and TV4, who, as of 2008, split the market 71%, 24%, and 5%, respectively. This translates to a high HHI of 5,642 for 2008, although this is markedly less concentrated than in 1996 (when RTE had a terrestrial monopoly) (Table 7.6).

As the public service broadcaster, RTE receives the lion's share of the broadcasting license fee. Thus RTE's audience share does not reflect its share of broadcasting revenues (i.e., revenues from advertising and the license fee). To calculate overall shares of the terrestrial

### Table 7-6. TV Broadcasting Audience Market Shares (%) 1996—2008

| | 1996 | 2003 | 2004 | 2008 |
|---|---|---|---|---|
| RTE (Public) | 100 | 75 | 75 | 71 |
| TV3 (UK) | | 25 | 25 | 24 |
| TG4 (Public) | | | | 5 |
| C4 | 100 | 100 | 100 | 100 |
| HHI | 10,000 | 6,250 | 6,250 | 5,642 |
| *N* (>1%) | 1 | 2 | 2 | 3 |
| Noam Index | 10,000 | 4,419 | 4,419 | 3,257 |

*Source:* Author extrapolations from Ipsos/MRBI JNLR figures.

21. Competition Authority. *Determination of Merger Notification M/08/021—Tv3/Kish Media.* Dublin: The Competition Authority, 2008.

market in revenue terms we must translate audience share into advertising revenue and then factor in RTE's license fee revenues. In addition, we must take into account the revenues voted directly to TG4. This gives RTE a significantly greater share of the market and, at least until 2007, when TG4 became a separate entity, raised the HHI concomitantly (Table 7.7.).

In theory the level of competition in the terrestrial market was to be augmented by the arrival of digital terrestrial television (DTT). In May 2011, this was launched under the "Saorview" (Freeview) brand, a state-owned service (delivered by RTE), which was immediately available to some 97% of the population. In August 2011, the Broadcasting Authority of Ireland published a report outlining recommendations on new channels to be carried on DTT. For the most part the "new" channels were offered by existing players and offered little prospect of increasing the amount of competition in the Irish market. The only new channels were those jointly proposed by community stations DCTV and NvTv and two other separate proposals to create two new music channels. However, the impact of the 2008 global recession on all the broadcasters connected with these proposals has led to their indefinite deferral. Only the new deferred viewing channels were up and running by October 2012, when the analog signal was switched off permanently.

## Video Channels and Multichannel TV Platforms

The nature of the Irish market renders the introduction of a distinction between the terrestrial market and the "multichannel" market somewhat artificial. Of the 1.5 million Irish homes with television, 1.3 million receive their television via cable or satellite providers. And since the various RTE and TV3 channels have "must carry" status, the vast majority of audiences watch "terrestrial" broadcasters via nonterrestrial means.

The basic packages offered by both the cable monopolist UPC and the satellite monopolist Sky include approximately 200 channels. Their offerings largely overlap, especially with regard to the top 20 channels[22] that constitute the bulk of the competition for the domestic terrestrial broadcasters. The most successful of these come

Table 7-7. TV Broadcasting Market Revenue Shares (%) 2003—2010

| | 2003 | 2004 | 2008 | 2010 |
|---|---|---|---|---|
| RTE (Public) | 85 | 86 | 76 | 74 |
| TV3 (UK) | 15 | 14 | 15 | 15 |
| TG4 (Public) | | | 9 | 10 |
| Revenues €m (US$) | 281.3 (354.4) | 307.6 (415.3) | 413.2 (574.3) | 349 (467.7) |
| C4 | 100 | 100 | 100 | 100 |
| HHI | 7,450 | 7,592 | 6,001 | 5,701 |
| *N* (>1%) | 2 | 2 | 3 | 3 |
| Noam | 5,268 | 5,368 | 3,465 | 3,291 |

*Source:* Author estimates based on Ipsos/MRBI Joint National Listernship survey, Institute of Advertising Practitioners of Ireland figures and RTE Annual Report 2003–2010.

22. This is not always the case, however. UTV, the long-established holder of the independent commercial television licence in Northern Ireland is not officially included on Sky's EPG, apparently because UTV has not been willing to pay the carriage cost charged by Sky. In practice it is possible to manually tune UTV in through the Sky service. In a similar vein, Sky does not offer its Sky Atlantic channel (based on high cost US cable drama content—*Mad Men, Boardwalk Empire,* etc.) through the UPC platform, using it as an incentive for customers to subscribe to the Sky satellite service.

from the United Kingdom and Northern Ireland. The BBC, UTV, Channel 4, and British Sky Broadcasting cumulatively captured about 19% of viewers as of 2011. More recently, US broadcasters have introduced channels repackaging US content: NBC Universal, Viacom, and Discovery treat the United Kingdom and Ireland as a single market in terms of programming, although their cumulative 2011 market share was only 3.1% (Table 7.8).

As most channels have recorded falling market share, especially between 2008 and 2011, the HHI should be falling. That this has not occurred is due to the huge (and increasing) number of channels in the "other" category. Although their average market share is less than 0.2%, these "other" channels now capture one in four viewers (having accounted for just over 10% of the audience in 2003). Those channels that have retained market share or limited the scale of losses have done so through consolidation, for example TV3's 2008 acquisition of Channel 6 or British Sky Broadcasting's acquisition of the Virgin Media Group's Living TV channels in 2010.

Most of the channels with micro-audience shares are UK- and US-based. They often consist of repackaged "classic" television and include a smattering of movie, religious, and sporting content. That said, there are some new domestic cable and satellite-based stations. 3e, in its original incarnation of Channel 6, for example, was limited to cable and satellite distribution. Following its acquisition by TV3 it now has must carry status on the new DTT platform.

Setanta Television, a sports subscription broadcaster, which offers one "free" channel via cable and satellite, also numbers among the new domestic cable channels. Launched in 1990 to acquire pay-per-view rights for Irish sports events marketed to diasporic audiences overseas, it made an aggressive move into the subscription market in mid-2000s, competing with BSkyB for English Premiership soccer

## Table 7-8. TV Channels (By Audience Market Shares (%)) 2003–2011

| | 2003 | 2004 | 2006 | 2008 | 2011 |
|---|---|---|---|---|---|
| RTE (public) | 41 | 42.2 | 42.8 | 40.1 | 35.7 |
| TV3 (UK) | 13.4 | 14 | 12.8 | 12.7 | 14.2 |
| BBC (UK) | 12.1 | 10.9 | 9.9 | 8.8 | 7 |
| SKY (UK) | 7.3 | 6.3 | 4.9 | 5.3 | 5.1 |
| UTV (UK) | 7.7 | 6.7 | 5.6 | 4.5 | 3.5 |
| Channel 4 (UK) | 6.1 | 5.8 | 5.6 | 5.2 | 3.3 |
| NBC Universal (E! Entertainment) (US) | | | | | 0.7 |
| Setanta Ireland | – | – | 0.5 | 0.8 | 0.4 |
| Viacom (Comedy Central, MTV, Nickelodeon, Nick Jr.) (US) | 1.6 | 1.8 | 2 | 3 | 2 |
| Discovery (US) | | | | | 0.4 |
| Virgin Media Group (Living TV) (UK) | | | 0.3 | 1.5 | |
| Channel 6 | | | 0.4 | | |
| Other | 10.6 | 12.3 | 14.9 | 17.9 | 27.5 |
| C4 | | | | | |
| HHI | 2,159 | 2,217 | 2,185 | 1,934 | 1,579 |
| *N* (>1%) | 7 | 7 | 7 | 8 | 7 |
| Noam Index | 816 | 838 | 826 | 684 | 597 |

*Source:* Author extrapolations from AC Nielsen data.

rights. This led the company to the international financial markets, where it raised €400 million between 2005 and 2007. As a result, Benchmark Capital and Goldman Sachs became substantial shareholders, and Doughty Hanson, owners of TV3, paid €100 million for a 20% stake in the company. This valued Setanta at €500 million.

However, in early 2009 the UK arm of the company went bust, owing several hundred million euros. The company was restructured and since July 2009, Setanta has been 40% owned by the holding company of original founders Michael O'Rourke and Leonard Ryan and former Setanta executive Mark O'Meara. The other 60% is owned by Gaiety Investments, the company of Irish concert promoter Denis Desmond.

Another recent—though short-lived—development was the emergence of local, cable-based broadcasters. From 2005 onward, City Channel operated television channels serving three Irish cities. In August 2007, Liberty Global, the US company that also owns the only cable company in Ireland (UPC—formerly NTL and Chorus Communications) took a minority stake in City Channel with a view to acquiring full control. However, City never attracted significant audiences and could not survive the economic downturn leading to its closure in September 2011.

The HHI calculated in the table above is based on audience share. However, because of the specific characteristics of the Irish market, this does not reflect market dominance in terms of revenues. As already noted, the two public service broadcasters receive substantial license fee and state-derived subsidies in excess of their advertising revenues. RTE has only faced competition for advertising revenues since the late 1990s, as UK-based broadcasters such as MTV, Sky, and Nickleodeon began offering Irish advertisers local "opt-out" spots and established sales offices within the Republic of Ireland.[23] As a consequence, audience share again fails to reflect shares of advertising revenues: as Table 7.9 indicates, RTE's 35.7% audience share in 2011 translates into a 55% share of advertising revenue.

Table 7-9. Share of Television Advertising Revenues[1] 2003–2008

| | 2003 | 2004 | 2008 | 2011 |
|---|---|---|---|---|
| RTE | 56 | 58 | 56 | 55 |
| TG4 | | | 4 | 4 |
| TV3 | 18 | 19 | 19 | 22 |
| Setanta | | | 1 | 1 |
| UTV | 11 | 9 | 7 | 6 |
| Channel 4 | 2 | 2 | 2 | 1 |
| Sky | 10 | 9 | 8 | 8 |
| Viacom | 2 | 2 | 4 | 3 |
| Discovery | | | | 1 |

*Source:* Author extrapolations from AC Nielsen and Institute of Advertising Practitioners of Ireland data.

[1] This is an extrapolation calculated by dividing total expenditure on television advertising by the audience share of those channels selling advertising space in Ireland. Given the different audience profiles of different channels this almost certainly overstates the revenues of some channels and understates those of others. However, it is the only consistent basis for making this calculation.

23. Such offices have not necessarily been maintained however. In an instance of market concentration that is hidden to most viewers, Sky Media, which primarily exists to sell advertising time on the various Sky channels to Irish advertisers, also agents (e.g., sales ad space on behalf of) for Viacom's channels and the Discovery Channel. In a similar vein, TV3 agents for NBC Universal.

## Table 7-10. Share of All Television Revenues[1]

| | 2003 | 2004 | 2008 | 2011 |
|---|---|---|---|---|
| RTE (Public) | 75 | 75 | 64 | 64 |
| TG4 | 0 | 0 | 9 | 9 |
| TV3 (UK) | 11 | 12 | 12 | 14 |
| Setanta | 0 | 0 | 1 | 1 |
| UTV | 6 | 5 | 5 | 4 |
| Channel 4 (UK) | 1 | 1 | 1 | 1 |
| Sky (News Corp, UK/US) | 6 | 5 | 5 | 5 |
| Viacom (US) | 1 | 1 | 3 | 2 |
| Discovery (US) | 0 | 0 | 0 | 1 |
| HHI | 5,788 | 5,787 | 4,376 | 4,464 |

*Source:* Author extrapolations from AC Nielsen and Institute of Advertising Practitioners of Ireland data.

[1] This is an extrapolation calculated by dividing total expenditure on television advertising by the audience share of those channels selling advertising space in Ireland. Given the different audience profiles of different channels this almost certainly overstates the revenues of some channels and understates those of others. However, it is the only consistent basis for making this calculation.

We have not calculated an HHI for the multichannel market based on advertising revenues because, as with the terrestrial market, to gain a true understanding of the disposition of all broadcasting revenues we must include license fee revenues (90% plus of which goes straight to RTE) and the revenues voted directly to TG4 by the Irish parliament. As Table 7.10 suggests, this points to a significantly greater degree of concentration in the market than is suggested by calculating the degree of concentration on the basis of audience share alone.

The difference in the two HHI figures for multichannel broadcasting (one based on audience share, the other based on share of revenues) highlights the distinction between two very different approaches to concentration that we might term "competitive" and "pluralistic," respectively. The HHI is primarily an index of the degree of competitiveness or otherwise of a market. However, the competitiveness of a market doesn't necessarily reflect the extent to which it can be regarded as pluralistic. In particular markets it may be the case that the degree of competitiveness in the market mirrors the diversity of content available in that market. That RTE enjoys a position of financial dominance which is significantly greater than its share of the audience reach demonstrates that this is by no means universally true.

The Irish multichannel platform market is based around three different distribution technologies: satellite (DTH), coax cable, and microwave MMDS. It is a highly concentrated market. In 2004, before the NTL/Chorus merger discussed later in the chapter, the HHI for the multichannel sector was 3,643. By 2008, this had risen to 5,008. Furthermore there has been a consistent increase in BSkyB's market share from 28% of all multichannel subscribers in 2002 to 52% in 2008 (Table 7.11).

Historically, there has only been one player in the Irish satellite market, BSkyB. By contrast, the older cable and MMDS markets were initially constituted by a myriad of small-scale players. When MMDS began to emerge as distribution technology in the 1980s, many of the established cable companies, moved into that market too. From a regulatory perspective, however, the satellite and cable/MMDS markets are effectively regarded as one. This found overt expression in 2004 when the Competition Authority considered the merger of cable companies NTL and Chorus, which effectively

Table 7-11. Multichannel Platform Market Shares by Subscriber (%) 2002–2008

| | 2004 | 2008 |
|---|---|---|
| UPC (Liberty, US)[1] | | 48 |
| Chorus (US) | 21 | |
| NTL (UK) | 41 | |
| BSkyB (NewsCorp, US/UK) | 39 | 52 |
| Total Revenues (€m) | 467 | 598 |
| Total Revenues ($USm) | 630 | 831 |
| C4 | 100 | 100 |
| HHI | 3,643 | 5,008 |
| *N* (>1%) | 3 | 2 |
| Noam Index | 2,103 | 3,541 |

[1]BSkyB is 36% owned by News Corporation. UPC is 100% owned by Liberty Global as of 2005.

created an Irish cable monopoly. In giving its approval, the Authority argued that cable and satellite content offerings were substitutable and that the availability of satellite television packages via BSkyB permitted competition in the multichannel market even if there was only one cable/MMDs operator.

As of 2011, the cable market in Ireland is effectively a monopoly controlled by UPC, a subsidiary of Liberty Global. This completes a concentration process that began in the 1970s when the state began to license private cable operators, first in Dublin then in other urban areas, reaching Waterford, Cork, Limerick, and Galway by the mid-1980s. As a consequence cable systems remained geographically concentrated in urban areas, owned by a patchwork of dozens of small-scale operators.

However, in the 1980s and 1990s, two operators began to scale up. RTE Relays, a subsidiary of the broadcaster, had been acquiring operators in the Dublin area since the 1970s. In 1984, it won approval to acquire Dublin Cable Systems from Canadian operator Rogers Cable. The merger attracted the attention of the Restrictive Practices Commission (a precursor of the Competition Authority), but the Commission argued that the availability of programs via terrestrial signals and the emergence of a market in direct-to-home satellite broadcasting constituted sufficient competition in the content delivery market. Two years later in 1986, RTE merged all of its cable subsidiaries into a new company named Cablelink, which, though largely limited to operating in the Greater Dublin area, Galway, and Waterford, was by far the largest operator in terms of subscriber numbers. In 1990, RTE was instructed by the minister for communications to sell its entire stake in Cablelink. The instruction was understood as designed to weaken RTE at a point when the state was seeking to promote competition in all aspects of the Irish broadcasting. However, the RTE Authority refused to make a complete sale and sold only 60% of the company to the successful bidder, Telecom Eireann (later Eircom).

In 1999, the state effectively instructed RTE and Eircom to sell off their respective 40% and 60% stakes in Cablelink. It was suggested that Eircom's stake represented a conflict of interest in that it was not in Eircom's interest as a provider of voice phone services to encourage their cable subsidiary to explore the technical possibilities of offering similar and potentially competitive services over cable. The sale attracted enormous interest from bidders, including Denis O'Brien's Esat Telecom and INM subsidiary Princes Holdings, the other major cable operator in the country. Ultimately, however, the UK-based NTL was successful paying out €680 million to acquire the company.

A second cable company, Princes Holdings, was established in 1992 by Independent News and Media (INM) as a joint venture with John Malone's Tele-Communications International (TCI) and United International Holdings, with the latter two each taking 25% of the stock and INM the remaining 50%. In 1998, TCI bought out UIH's share. Over the course of the 1990s, Princes Holdings acquired a number of cable and MMDS operators outside Dublin before rebranding as Chorus Communications in 2000. The acquisition of Suir-Nore Relays in 2004 was the final major acquisition. In 2004, INM sold its entire stake in Chorus to Liberty Global.

By the start of 2005 there were two cable companies operating in Ireland, but in November 2005 the Competition Authority approved a Liberty Global takeover of NTL. Despite the fact that this created a de facto cable monopoly, the Competition Authority concluded that competition was unaffected because Chorus and NTL had hitherto served distinct geographical markets and were already monopolists within those markets. Furthermore, the satellite packages offered by BSkyB meant Irish consumers retained some choice over their multichannel operator.

However, the Competition Authority did express concern about the degree of cross-ownership between Liberty Global and News Corporation and BSkyB. By 2005, Liberty Global was the largest single shareholder in News Corporation, although the Murdoch family retained a larger number of voting shares. As a consequence the Authority insisted on a management structure designed to maintain a "Chinese wall" between Chorus/NTL (the company was later rebranded as UPC) and BSkyB, especially in negotiations between the two entities for channel space allocation on the UPC cable platform. The cross-ownership issue has arguably become less significant since 2008 when News Corporation and Liberty Global agreed to an asset swap in which Liberty swapped its 16.4% News Corp holding for 38% of DirecTV's shares, held by News Corporation.

## TELECOMMUNICATIONS MEDIA

### Wireline Telecom

Eircom, the former state-owned PTT, is the largest player in the Irish wireline market. However, with the arrival of local loop unbundling and competing infrastructures for voice phone service (most notably cable), the market has become significantly more competitive since 2008.

As in broadcasting, wired telecommunications was directly controlled by the Irish state for most of the 20th century. From 1922 until 1984, phone and mail service were run directly by the Department of Posts and Telegraphs. In 1984, there was a substantial reorganization of the administration of both postal and telecommunications sectors. Two semi-state bodies—An Post and Telecom Eireann—were established to take over day-to-day responsibility for the operation of the postal service and the telephone and telegraph service, respectively. Although telecoms remained a state monopoly, Telecom Eireann immediately adopted a more commercial outlook, anticipating European Commission pressure to end the protected position of European PTTs and to apply a more commercial approach in general.

Ireland's increasingly neoliberal political economic environment combined with Telecom's improving financial fortunes through the 1980s and early 1990s triggered some public debate around the possible privatization of the company. Much of the impetus to privatize derived from the perception that Telecom Eireann would need more flexibility to meet competition down the line. In 1988 chairman Michael Smurfit argued that "new technologies, international trends and European Community requirements" made the liberalization of the Irish market inevitable.[24] Limited competition in terminal equipment provision had been a fact of the Irish market since the 1983 Postal and

24. Telecom Eireann. *Report and Accounts for the year ended 31st March 1988*. Dublin: Telecom Eireann, 1988, p. 10.

Telecommunication Act. To that extent then, Irish telecommunications policy was ahead of the game when the EU's critical Green Paper on Telecommunications Policy appeared in May 1987, stressing the virtues of "liberalizing" European telecommunications markets.

By the early 1990s, competition was in any case increasingly the norm on international routes as Irish companies bypassed Telecom Eireann's network for international calls using sophisticated call-back technologies. By 1992, international competition seriously threatened Telecom Eireann's revenues as the volume of international calls originating in Ireland dropped calamitously. Competition was not limited to foreign companies. In December 1992, the Department of Communications licensed Denis O'Brien's Esat Telecom to lease lines from Telecom Eireann with a view to competing in the long distance business market. Following protracted negotiations on the cost of leasing lines from Telecom, ESAT secured its first customer in October 1993.

Telecom's monopoly of the Irish voice telephony market ended in May 1993 when the EC Council of Telecommunications Ministers approved plans to liberalize the key voice phone market by 1998. In part as a consequence, in 1998 it had become clear that the state intended to privatize Telecom Eireann. Initially it was assumed that the state would retain a "golden share" (i.e., a majority stake) in the company, but it was later decided to privatize the entire company. The 1999 Eircom flotation was (and remains) the largest privatization in Irish corporate history.

Although the launch onto the stock exchange was initially successful (in that the share price grew about 20%), within 12 months the value of the shares had fallen to half their initial value. This led to a 2001 bidding war between two consortia, one led by Denis O'Brien, the other by Tony O'Reilly, to acquire majority shares in the company. The O'Reilly-led Valentia consortium proved successful, and the company was taken private with O'Reilly as executive chairman. Eircom's ownership would shift twice more. In 2006 Australian investment firm Babcock and Brown acquired the company from the Valentia Group. Then in 2009, after Babcock and Brown effectively collapsed, the Singaporean telco SingTel acquired Eircom for €140 million, becoming the fifth owner in a decade. The catastrophic results of this repeated flipping for investment in Irish telecommunications infrastructure and for Eircom itself have already been demonstrated in the earlier discussion.

In the meantime, Esat Telecom had quickly moved into the consumer market, and from 1998 it constituted the first direct competition for Telecom Eireann's domestic consumers. Within three years, however, Esat was purchased by British Telecom, which rebranded the company as BT Ireland. Business and domestic customers were initially slow to switch from Eircom as their fixed line provider. As late as 2006, Eircom accounted for 86% of the domestic landline market. However, this situation has changed rapidly since 2008. As of 2011, Eircom's overall share of the fixed telecom market has fallen to 61.3%, and even among domestic customers it only holds 64% of the market. This is reflected in changes in the HHI index between 2008 and 2011, which fell from 4,775 to (a still concentrated) 4,099 (Table 7.12).

The increased competition is partially due to the exploitation of cable infrastructures to offer voice phone service. After merging Chorus and NTL to create NTL, Liberty Global invested significant capital in upgrading the cable network to facilitate two-way communications. This has allowed UPC to become a significant player in the ISP market (see later in the chapter) and to gain a foothold in the home phone market. This remains small for now, although some surveys suggest that in Dublin up to a third of all home phone subscriptions are now with UPC. There is reason to think that UPC's market share will increase precisely because it can offer voice phone service as part of a bundle of telecom services including broadband Internet and television.

Bundling has also facilitated Vodafone's winning of nearly 20% of the fixed line market. (Vodafone, the largest mobile phone operator in Ireland, offers home phone and broadband Internet bundles.) Vodafone's ascent has also led to acquisitions of smaller fixed lines companies, including Perlico in 2007 and elements of BT's consumer telecom business (as opposed to network/infrastructure) in 2009.

Table 7-12. Wireline Market Shares 2008–2011

| | 2008 | 2011 |
|---|---|---|
| Eircom (Singapore—Singtel) | 67.4 | 61.3 |
| UPC (US—Liberty Global) | | 5.4 |
| Vodafone (inc. BT since 2009) (UK) | | 17.4 |
| BT (UK) | 14.6 | |
| Access/Imagine (Irl) | 2 | 2.3 |
| Cable and Wireless (UK) | | 2 |
| Verizon (US) | 3.3 | |
| Colt (UK) | 2.1 | |
| Others | 10.6 | 11.5 |
| Total Revenues (bil €) | 2.244 | 2 |
| Total Revenues (bil $US) | 3.12 | 2.59 |
| HHI | 4,775 | 4,099 |

*Source:* Commission for Communications Regulation (COMREG).

However, the economic recession has also overcome consumer reluctance to abandon the incumbent operator (Eircom), as subscribers have experienced rapid declines in real incomes. The last quarter of 2010 in particular saw more switching from one fixed line provider to another than in any three-month period since the Irish market was liberalized in December 1998. As the incumbent operator, this disproportionately affected Eircom, which had more customers to lose than any of its competitors.

## Wireless Telecom

There are three dominant players in the Irish mobile phone market: Vodafone Ireland, owned by UK-based Vodafone; O2 Ireland, owned by the Spanish firm Telefonica; and Meteor, owned by Eircom (SingTel). There is also a fourth, smaller player, 3 Ireland, a UMTS-based network owned by Hong Kong-based Hutchinson Whampoa.[25] The mobile phone market was originally a monopoly. The first mobile phone company, Eircell, began operations in 1986 as a wholly owned subsidiary of Eircom (then Telecom Eireann), the state-owned PTT.

In 1995, the state put together an RFP for a second mobile phone license. Several private consortia submitted bids but Esat Digifone, a joint venture between Norwegian PTT Telenor and Irish entrepreneur Denis O'Brien (whose Esat Telecommunications already offered limited fixed line telecom services to the corporate market) won the license and in 1997 started operations. It has subsequently been alleged that the success of the Esat bid was influenced by payments made to then minister for communications, Michael Lowry. The allegation was investigated by public tribunal. The interim findings suggest that payments were made by O'Brien to Lowry and that these materially influenced the decision to award the second telephone license.

Within three years Esat Digifone had established a 40% market share and the HHI fell from 10,000 in 1996 to 5,200 by 2000. By 1999, O'Brien was competing with Telenor to acquire outright ownership of Digifone. The Esat Board treated a Telenor bid for a majority stake as hostile before accepting a friendly bid from British Telecom in 2000. At this point Esat became a wholly owned subsidiary of BT. The following year BT spun off its wireless division (including Digifone) to create an independent company

25. In 2014, Three Ireland acquired O2 Ireland. There are now only three participants in the wireless market.

mmO2 plc. This company traded as a standalone operation until its 2005 acquisition by Spanish telecom company Telefonica.

In meantime, the late 1990s saw Eircom, owner of Eircell, begin a move to sell off the mobile subsidiary. This seemed counterintuitive, since it was the most profitable element of Eircom. However, the context for the decision was the fallout following the privatization of Eircom in 1999. The Eircell sale was an attempt to boost the share price after its postprivatization plummet. In May 2001 the Eircom board (including strategic partners telcom firms Telia and KPN) approved a Vodafone offer of £IR3.2 billion for the company.

The final major player in the Irish mobile market, Meteor, began operations in 2000 after securing the third mobile license in 1998. Meteor, owned by a consortium made up of Western Wireless, RF Communications, and TWG Ireland, was not initially successful. By 2004 its market share was just 7% and the HHI fell only marginally between 2000 and 2004 from 5,200 to 4,458. However, in 2005, Eircom reentered the mobile market by acquiring Meteor for €410 million. Eircom's heavy investment in the branding and marketing of Meteor drove market share to nearly 20% by 2008, causing the HHI to drop to 3,274 the same year.

As of 2011, then, the Irish mobile market was dominated by international players headquartered in the United Kingdom, Spain, Singapore, and Hong Kong. There is evidence that this oligopolistic structure has resulted in unusually high (by international standards) per customer revenues for these operators. In Spring 2005, the Irish telecoms regulator COMREG issued a statement arguing that O2 and Vodafone were at least tacitly colluding in price setting, noting that Vodafone and O2 were "aware of their common interests in ways which are conducive to the creation and maintenance of a position of collective dominance on the market for wholesale mobile access and call origination in Ireland, without the need for the parties to enter into an agreement or to resort to a concerted practice in order to cause prices to align or to deny access."[26] This finding was subsequently upheld by the European Commission, who threatened sanctions if such behavior continued (Table 7.13).

## Table 7-13. Wireless Telecoms Market shares (By Subscriber) 1984–2008

| | 1984 | 1988 | 1992 | 1996 | 2000 | 2004 | 2008 |
|---|---|---|---|---|---|---|---|
| Vodafone (formerly Eircom) (UK) | 100 | 100 | 100 | 100 | 60 | 53 | 42.5 |
| O02 (formerly Esat Digifone) (Telefonica, Spain)[1] | | | | | 40 | 40 | 32.7 |
| Meteor (Eircom subsidiary) (Singapore) | | | | | | 7 | 19.1 |
| 3 Ireland (Hong Kong) | | | | | | | 5.8 |
| Total Revenues (mil €) | | | | | 812 | 1,565 | 1,936 |
| Total Revenues (mil $US) | | | | | 763 | 2,110 | 2,690 |
| HHI | 10,000 | 10,000 | 10,000 | 10,000 | 5,200 | 4,458 | 3,274 |

*Source*: Commission for Communications Regulation (COMREG).

[1]O2 is 100% owned by Telefonica since 2005. In its previous guise as Esat, the company was owned by Denis'O'Brien and Telenor (1997—2000), BT (2000—2001), independent (2001—2005). Meteor is 100% owned by Eircom since its acquisition in 2004, and 3 Ireland is owned by Hutchinson Whampoa.

26. Comreg, Market Analysis—Wholesale Access and Call Origination on Public Mobile Telephony Networks. Dublin: Comreg, 2005, p. 3.

## INTERNET MEDIA

### Internet Service Providers (ISPs)

Relative to other media sectors, the ISP segment has witnessed a significant decline in concentration over the past decade. Although initially the internet service provision market was largely owned by domestic companies, this has changed dramatically in the past decade through the arrival of new ISPs using cable delivery systems and offering mobile broadband.

In the early 1990s, the sector was constituted by a handful of small companies including Ireland On-Line (which was owned by the An Post, the postal operator), Indigo, Internet Ireland, Club Internet, Internet Exchange, and Business Network Ireland. Telecom Eireann (Eircom) also entered the market, establishing TINET as a subsidiary. In 1999, Esat purchased EUnet and Indigo, briefly becoming the largest ISP in the state. Esat would sell its Internet business along with the rest of the company to BT in 2001. As the 2000s progressed, Eircom leapt ahead, accounting for 87% of all subscriptions by 2004. As a consequence the market was highly concentrated with an HHI of 7,642 at this point.

From 2008 onward, UPC began to offer broadband Internet access via their cable network. This coincided with mobile broadband offerings from Vodafone, 02 and later from Meteor. By the end of 2008, although more than the half the 1.2 million broadband connections in Ireland were offered via DSL (a sector dominated by Eircom), a further 104,000 broadband connections (8% of the total) were offered via cable and 309,000 were offered via mobile broadband (i.e. Vodafone and O2).[27] Furthermore a number of new market entrants offering wireless broadband connections also began to emerge from 2005. As a consequence between 2004 and 2008, the number of companies with a 1% or higher market share increased from 4 to 14 and the HHI fell from 7,606 to 3,294.

Inevitably this affected Eircom's dominance. By 2008 Eircom's market share had fallen to 56%. The decline in concentration increased after 2008. As of 2010, Eircom's share of the ISP market was down to 41%, even if one includes the market share of Eircom subsidiary Meteor. Mobile phone companies have been the major beneficiaries of Eircom's decline: 3 Ireland saw its share rise from 1% to 8% in two years, while O2's share rose from 2% to 9%. The biggest single benefactor, however, was Vodafone, whose share of the ISP market leapt from 3% in 2008 to 22% in 2010. This was partly due to the increased take-up of the Vodafone mobile broadband offering. It was also driven by Vodafone's acquisition of two existing ISPs, Perlico, and BT Ireland. As a consequence, the HHI for the Irish ISP market fell to 2,460 in 2010 (Table 7.14).

### Search Engines

Unsurprisingly, the Irish search engine market is dominated by Google, which accounted for 93–95% of all Irish Internet searches between 2008 and 2012 (Table 7.15). According to alexa.com, google.ie and google.com are the first and second most visited websites in Ireland, respectively. This represents a significant shift from the situation a decade ago when Altavista and Yahoo were the dominant players in the market. Although Yahoo subsequently absorbed Altavista, this did nothing to arrest the decline of Yahoo's Irish market share, which fell from 4% to 2% between 2008 and 2009. (Yahoo.com remains the fifth most visited Irish website.) Microsoft's Bing search engine is the only other to present measurable market share figures (typically 2%). The Irish search engine market is highly concentrated with an HHI that consistently approaches 9,000.

Like a number of global tech companies (Facebook, Ebay, etc.), Google has established its European Headquarters in Ireland, where

27. Comreg. Irish Communications Market: Key Data Report—Q4 2008. Dublin: Comreg, 2009, p. 31.

Table 7-14. Internet Service Providers Market Share (%) by Subscriber (2004–2010)

| | 2004 | 2008 | 2010 |
|---|---|---|---|
| Eircom (inc. Meteor) (Singatel, Singapore) | 87 | 56 | 41 |
| Esat BT (UK) | 6 | 6 | |
| UTVnet (UK) | 1 | 1 | |
| UPC (Liberty, US) | | 6 | 12 |
| O2 (Telefonica, Spain) | | 2 | 9 |
| Vodafone (inc. Perlico and BT Ireland from 2009) (UK) | | 3 | 22 |
| 3 (Hong Kong) | | 1 | 8 |
| Access/Imagine/Irish Broadband | | 6 | 2 |
| Smart | | | 1 |
| Digiweb | | 2 | 1 |
| Last Mile | | 1 | |
| Perlico | | 5 | |
| Clearwire (US) | | 2 | |
| Rapid Broadband | | 1 | |
| Other | | | |
| Total Revenues (mil €) | | 403 | 472 |
| Total Revenues (mil $US) | | 560 | 632 |
| C4 | 100 | 74 | 84 |
| HHI | 7,606 | 3,294 | 2,460 |
| *N* (>1%) | 3 | 13 | 8 |
| Noam Index | 4,391 | 914 | 870 |

*Source*: Commission for Communications Regulation (COMREG).

Table 7-15. Search Engines Market Share (%) 2008–2012

| | 2008 | 2010 | 2012 |
|---|---|---|---|
| Google (US) | 93 | 95 | 94 |
| Bing (Microsoft, US) | 0 | 2 | 2 |
| Yahoo (US) | 4 | 2 | 2 |
| Ask Jeeves (US) | 0 | 1 | 0 |
| AVG Search (Czech Republic) | 0 | 0 | 0 |
| Other | 3 | 0 | 1 |
| C4 | 100 | 100 | 100 |
| HHI | 8,665 | 9,034 | 8,844 |
| N (>1%) | 3 | 4 | 4 |
| Noam Index | 5,003 | 4,517 | 4,423 |

*Source*: Alexa.com

it employs 2,000 people. Ireland is an attractive location because of the combination of a low corporation tax rate and ability to engage in transfer pricing (allocating income between units in different countries to avoid taxes). The latter practice on the part of Google attracted the attention of *Bloomberg* in 2010, which suggested that Google had avoided US$3.1 million in taxes between 2007 and 2010. More recently, the Irish economic recession has driven down commercial property rents and labor costs, making Ireland even more attractive for such foreign investors.

Google's dominance of the search engine market both reflects and is driven by the success of the Google Chrome browser. At the time of Chrome's launch in Fall 2008, Microsoft's Internet Explorer and Mozilla's Firefox dominated the market with shares of 63% and 32%, respectively. (Apple's Safari was a distant third with a share of 4%.) By May 2011, Chrome's 25% market share had surpassed that of Firefox (24%), and in April 2012, Chrome overtook Internet Explorer. As of August 2012, the respective market share of the three main browsers are Chrome at 38%, Internet Explorer at 29%, and Firefox at 20%. The increasing popularity of Apple's iPad has increased the share of Safari, which accounted for 10% of all installed browsers in Ireland as of August 2012.

## CONCLUSION

As noted at the outset, the borders of the Irish media market have long been porous but overseas involvement in Irish media, information, and communications industries has intensified over the past two decades. These changes are associated with shifts in the ownership structure of Irish media and changing regulatory practices and structures.

In some media sectors (not discussed above), such as book publishing and film distribution, overseas players have dominated the "Irish" industry since before the foundation of the state. In both cases Ireland was scarcely considered as a market in its own right, and large-scale publishing and film distribution firms in the United Kingdom and United States have regarded Ireland as "icing on the cake," a market in which pricing strategies designed to undercut Irish media producers could be applied.

The same was not historically true of the sectors discussed in this study, such as the national and regional press and the broadcasting sector. Even if UK newspapers and programs were part of the Irish media landscape, they did not dominate it. Nor do they do now. Nonetheless, in the case of both broadcasting and newspapers, the market share enjoyed by overseas newspapers and broadcasters has increased—in some cases dramatically—since the 1980s. In television, for example, the combined market share of the five domestic national broadcasters was 49.9% as of 2011. For the first time since an Irish television service began in 1961, a majority of the audience watches channels originating outside Ireland.

However, not only have overseas channels increased their market share but now they also directly own and run domestic Irish television channels with concomitant implications for local content. Hargreaves Heap et al. note: "The propensity of foreign media business to concentrate more than home ones on foreign themes is a simple consequence of the low marginal costs associated with reusing material that has already been produced for foreign markets. Such material may not be intrinsically especially attractive for the indigenous audience . . . Nevertheless . . . the quality of such foreign outputs . . . can achieve significant audiences."[28]

This describes TV3 scheduling after it was acquired by two international broadcasting giants, CanWestGlobal and ITV. Initial commitments to domestic programming were replaced by wholesale reliance on material produced for the US and UK markets. Even if TV3 couldn't match RTE's audiences, the former

28. Hargreaves Heap, Shaun, et al.*Consultation on the Assessment of Media Mergers under the Competition Act 2002: Final Report.* Dublin: Competition Authority, 2006, p. 50.

could contemplate a lower breakeven point than if it produced or commissioned more expensive domestic content. Although not conceived of as Irish channels, the channels broadcast into the Irish market by BSkyB, Viacom, and NBC Universal adhere to a similar economic logic by repackaging US and UK material for delivery to Irish audiences.

Although the Irish television market has become less concentrated, this has not resulted in the diversity of content boon generally associated with greater competition. The new channels that arrived on Irish screens over the past two decades frequently feature the same US and UK shows that were already available via the domestic broadcasters or those UK broadcasters with a longer history in Ireland. Many of TV3's top-rated shows (*Coronation Street, Emmerdale,* and *The X Factor*) are simulcast on TV3 and ITV in Ireland.

This is not just disappointing for viewers as consumers of television. To the extent that citizens use "media outputs to reflect upon the society in which they live,"[29] the increasing provision and consumption of foreign programming raises questions about the capacity of citizens to reflect upon the conditions of their own country.

As noted above, the same has not been true when UK firms such as the Johnston Press have moved in the regional press sector, where the local focus of the papers is perhaps their key selling point. Therefore there is little evidence of repackaging of material from UK papers belonging to the Johnston stable in Irish papers. Yet in the national market, the increasing market share enjoyed by UK newspaper groups in Ireland has similar implications for Irish citizens and consumers as the increasingly dominant position of international broadcasters in Ireland. The strength of these groups in the Irish market derives from the economies of scale (and scope) they enjoy relative to their Irish competitors. Even if the localization strategies pursued by News International, Trinity Mirror, and the Daily Mail and General Trust diminishes those economies somewhat, they are still able to produce their Irish papers for less than their Irish counterparts. Thus News International's *Irish Sun* retails for €1 while the closest Irish-owned competitor, the *Irish Daily Star* sells for €1.35. Similarly the *Irish Times* sells for €1.90 while the *Irish Daily Mail* also sells for €1. By exploiting economics of scope (re-using material primarily produced for the UK market), these papers become price competitive with Irish newspapers, a factor that contributes to their increasing market share even if the economic recession means that the absolute size of the market is contracting.

As of 2012, there are now a number international media conglomerates operating across a range of media in Ireland. Many of these cross-media players operate "below the radar," working in sectors where the identity of the ultimate owner may not be obvious to the "lay" observer. Thus Viacom's Irish holdings during the 1990s included a film distributor (UIP), cinemas (UCI), and the largest Irish video chain (Xtravision), as well as billboard advertising sites. As of 2012, a variety of regulatory and corporate factors have seen Viacom exit most of these areas (with the exception of advertising). It is doubtful that Irish viewers of the Comedy Central channel will associate it with the billboard sites they pass on their commute home.

Some of these cross-media holdings may not even be that obvious to the parent companies themselves. Did Doughty Hanson, owner of TV3, see synergistic potential in their 2011 acquisition of the UK-based Vue cinema group, which included one Dublin cinema site?

The most visible cross-media players are those engaged in the production of print and broadcast content. News International stands out in this respect, publishing the *Irish Sun, The Times, The Sunday Times,* and the *News of The World* until its closure in 2011. It also owns 36% of BSkyB and thus part-owns a range of free-to-air and subscription channels available in Ireland. In addition to content production firms,

29. Hargreaves Heap, Shaun, et al., p. 51.

it is also a dominant player in broadcasting platforms, effectively monopolizing the satellite television distribution sector in Ireland. Traditionally, this might not have been regarded as a media industry proper because such platform providers have been understood as offering connectivity rather than editorial content. However, that demarcation no longer holds and there is clearly the potential for a conflict between Sky's position as a broadcaster and broadcast platform operator. For example, as a platform operator, Sky's allocation of channels to slots in its electronic program guide (EPG) clearly offers the potential to promote its own channels further up the list than might be justified by audience share alone. However, it is also evident that it is acquiring programming with a view to promote subscriptions to its satellite platform. Sky's channels have been available both to subscribers to the Sky platform and to UPC's cable platform. In other words, there was a separation between Sky's operation as a delivery platform and as a broadcaster. In 2011, however, Sky introduced a new channel—Sky Atlantic—the schedule of which relied upon quality US television drama from HBO, AMC, and Showtime. This channel is offered "free" to any subscriber to the Sky delivery platform but is unavailable (even as a subscription service) to UPC customers.

With technological convergence it is also the case that individual firms are acquiring expanding market shares across a number of communications infrastructure industries. UPC grew out of the cable sector but is now a significant player in both wireline voice phone and Internet service provision. Mobile telecoms players offering mobile broadband—O2, Vodafone, and Meteor—have become Internet service providers.

The arrival of international, cross-media players into the Irish market raises questions about the local regulatory response. That there is a recognition of the special status of media industries—that is, that the commodities they trade in have a political significance—is evident from the special provision for media mergers made in the 2002 Competition Act. However, the efficacy of this regulation is undermined by a number of factors:

- Blurred responsibility for regulation of media ownership
- The overly economistic orientation of the Competition Authority as the body primarily charged with regulating media ownership in Ireland
- Poorly defined criteria for concepts like diversity and pluralism
- The difficulty in accommodating considerations of cross-media ownership
- Jurisdictional limits with regard to international media firms

In terms of blurred responsibility, it is striking that three bodies, the Competition Authority, the Broadcasting Authority of Ireland (BAI), and the Commission for Communications Regulation (Comreg), all have some say over who can own what in terms of the Irish media, information, and communication industries. In practice, Comreg, which licenses wireless, wireline, and broadcast service platform operators, has little engagement with questions of ownership, leaving these to the Competition Authority. In 2005, during a Broadcasting Commission of Ireland (the predecessor to the BAI) consultation on media ownership, Comreg expressed the view that ownership of broadcasting service platforms (such as UPC and Sky) should be excluded from any BCI consideration of mooted media mergers on the grounds that it would create "regulatory confusion" and that in any case Section 23 of the Competition Act already enjoined the Competition Authority to consider broadcasting services platforms in *their* deliberations.[30]

That the BAI sees itself as having a role in regulating ownership has been made clear by its publication of its own ownership policy since 2001. The standard BAI contract with

30. Comreg. *Submission re BCI Ownership and control Policy Consultation*. Dublin: Comreg, 2005.

its licensees requires them to seek BAI consent prior to changing their ownership and control structures. During Communicorp's acquisition of Emap's radio holdings in 2007, both the BAI and the Competition Authority were involved in determining whether and under what conditions the acquisition could proceed. However, this sometimes brings quite different and possibly competing regulatory perspectives to bear on any given media merger or acquisition.

For its part, the Competition Act of 2002 enjoins the minister for enterprise and employment to apply the following "relevant criteria" to any decision relating to a potential media merger:

- The strength and competitiveness of media businesses indigenous to the State
- The extent to which ownership or control of media businesses in the State is spread among individuals and other undertakings
- The extent to which ownership and control of particular types of media business in the State is spread among individuals and other undertakings
- The extent to which the diversity of views prevalent in Irish society is reflected through the activities of the various media businesses in the State
- The share in the market in the State of one or more of the types of business activity falling within the definition of "media business" held by any of the undertakings involved in a given media merger, or by any individual or other undertaking who or which has an interest in such an undertaking[31]

Although ultimate authority as to how to apply these criteria lies with the minister for enterprise and employment, the Competition Act also requires that the Competition Authority form an opinion on how these criteria should be applied. This requires the Authority to make adjudications on matters (most notably the question of what constitutes "plurality") that it has acknowledged fall "outside its area of expertise."[32] The Competition Act assumes that the general orientation of the Competition Authority is to encourage as much competition as possible in all industry sectors because it is assumed that this will benefit consumers. The problem with this point of view was captured by the Report of the Advisory Committee on Media mergers in 2009. Noting that the Authority was "an independent body with expertise in competition matters and the economics of markets," the Group argued that this rendered it "ill-equipped to take ultimate custody of the important public interest issues involved. Its expertise lies in the economics of markets not in issues of plurality or diversity."[33]

In practice, faced with the difficulty of finding objective criteria for concepts like diversity, the Competition Authority has generally applied narrow, economically motivated criteria to determine whether to approve media mergers. This finds expression in a focus on the marketplace implications of a merger to the exclusion of more political considerations. Thus the Authority's discussion of the 2007 Communicorp acquisition of EMAP's radio holdings exclusively focused on the implications for the radio advertising market. So narrow was its focus that the Authority explicitly excluded any consideration of Communicorp owner Denis O'Brien's ownership of Independent News and Media shares on the grounds that the newspaper advertising market was entirely separate to the radio advertising market.

For its part the BAI's viewpoint on ownership derives from the statutory provisions in the various broadcasting acts since 1988. These include the following policy objectives:

31. Competition Act 2002. Dublin: Government Publications.

32. Advisory Group on Media Mergers. *Report to the Tanaiste and Minister for Enterprise, Trade and Employment*. Dublin: Government Publications, 2008, p. 53.

33. Advisory Group on Media Mergers, p. 59.

- To promote open and pluralistic broadcasting services, with particular reference to radio and television services
- To promote diversity in viewpoint, outlet and source, that is, diversity in the opinions expressed, in programming delivery and content, and in the sources of information available to the public
- To contribute to the promotion of diversity in control of the more influential commercial broadcasting services[34]

More political questions of the importance of diversity and pluralism are at the forefront of the BAI's deliberations on media mergers. That said, even the BAI adopts a "by the numbers" approach to regulation of ownership. The 2009 Broadcasting Act requires the BAI to consider the desirability of allowing any person, or group of persons, to have control of/substantial interests in, an "undue number" of radio and television stations. The BAI has interpreted "undue number" as meaning more than 25% of the total number of broadcasters licensed by it. Anything up to 15% is considered an "acceptable" level for any single entity. The BAI states ownership between 15% and 25% would require "careful consideration." What is striking about this is the focus on the *number of stations* rather than more nuanced considerations of market share. Given that there are 37 licensed commercial radio and television stations, a media company would need to own 10 stations before it definitively breached BAI guidelines. Yet, as of 2011, Communicorp holds a 25% combined national and regional audience share on the basis of owning just five stations, primarily because these include two national broadcasters and one Dublin-based station.

Given that Communicorp's Chairman, Denis O'Brien also holds more than 20% of the shares of INM, the question remains as to how the BAI deals with cross-media ownership. The Authority professes to take account of other media holdings when making licensing decisions but it might be argued that its approach to cross-media ownership is unnecessarily narrow. Section 66(2)(i) of the 2009 Broadcasting Act requires the BAI to not merely take account of a media company's ownership of other broadcast media but of "communications media" in general.[35] However the definition of "communications media" appears to exclude ownership and control of websites, film production, and wireless or wireline telecoms.

Furthermore, in assessing whether an individual or entity has more than a reasonable share of the communications media, it adopts a different approach to that used with regard to multiple ownership of radio and television stations. Abandoning the numeric approach the BAI states that it assesses a company's "ability to influence opinion-forming power" (i.e. dominance) by "applying the applicant's audience share of the communications media."[36] Since "communications media" include radio and television stations it is unclear as to which approach takes precedence in practice.

In any case, the focus on audience share alone does not offer a strong basis on which to assess opinion-forming power. This returns us to the thorny question of how more politicized questions relating to diversity and pluralism can be made amenable to even quasi-objective assessment. Ironically, one of the most comprehensive attempts to do this has emerged from the Competition Authority in a bid to address their obligation to offer the Minister for Enterprise and Employment guidance on applying media diversity criteria to the consideration of media mergers. Their 2006 Consultation on the Assessment of Media Mergers acknowledges that a purely numeric approach to diversity is problematic. It describes the "mere counting of numbers of voices" as flawed and "bound to produce anomalies"[37] and rejects the notion that economistic

34. Broadcasting Authority of Ireland. *Ownership and Control Policy 2010*. Dublin: Broadcasting Authority of Ireland, 2010, p. 11.

35. Broadcasting Act 2009. Dublin: Government Publications.

36. Broadcasting Act 2009, p. 16.

37. Hargreaves Heap, Shaun et al. Consultation on the Assessment of Media Mergers under the Competition Act 2002: Final Report. Dublin: Competition Authority, 2006, pp. 8–11.

indices such as the HHI offer a reliable basis for judging diversity. "Contrary to the analysis that supports the use of the HHI, a rise or decrease in the HHI on side of the market does not necessarily indicate consumer or producers welfare will increase or decrease."[38]

As a consequence the Consultation argues that the Competition Authority and/or politicians will have to "exercise judgment when deciding whether any particular merger significantly affects the extent of diversity of view."[39] Crucially, the Consultation stresses that this does not necessarily imply an impossibly subjective approach: "These judgements need not be plucked from the air: they can be informed by relevant data."[40] According to the consultation that data could be captured in the construction of a "media map" combining quantitative and qualitative measures of media power. This would be "a description of the media landscape in Ireland that would minimally contain details of who owns what media businesses, who uses what media outputs, what degree of trust users accord to such outputs, and an understanding of the influence of regulation on these businesses. Ideally, the media map would also provide evidence on the practices of ownership and the orientation of particular media outlets with respect to major political, social and economic issues."[41]

However, the Competition Authority's Consultation fudged the question of who should apply this media map. For its part the Advisory Group on Media Mergers was clear that the role of protecting the public interest in promoting plurality and diversity in the media was an inherently political function. "The Group believes that the identity of the person who discharges the function should ultimately be determined by the nature of the function itself. In this case, the Group sees the function as one essentially of political judgment as to how the public interest is best protected as a result of a particular media merger. A Government Minister is democratically elected, his or her Department is responsible for the relevant functions of Government and he or she is answerable to the Oireachtas (Irish Parliament) for the way in which that statutory power is exercised."[42]

As of spring 2013, these deliberations have not found expression in legislation, although the minister for communications is currently preparing legislation for the Irish cabinet. In any case, it may be that the increasingly international nature of the Irish media landscape will defy local regulation. In 2004, when Ireland held the EU Presidency, the Irish government sought to amend the TV Without Frontiers directive to allow governments to regulate broadcasters providing television services specifically targeting viewers in their jurisdiction. This was resisted by both the European Commissioner Vivianne Reding and several member states. Reding commented that "The "originating country" was responsible for regulation and licensing. That is the general rule of the common market. It is not possible to change that."[43]

In short, supranational and international regulations may trump any national efforts to control the activities of international media players in Ireland.

## BIBLIOGRAPHY

### Books and articles

Djankov, Simon et al. "Who Owns the Media?" *Journal of Law and Economics*. XLVI. 2(2003): 341–381.

"High stakes radio game." *Sunday Business Post*, October 14, 2007, 15.

38. Hargreaves Heap, Shaun et al., p. 25.
39. Hargreaves Heap, Shaun et al., p. 15
40. Hargreaves Heap, Shaun et al., p. 15
41. Hargreaves Heap, Shaun et al., p. 5.
42. Advisory Group on Media Mergers. *Report to the Tanaiste and Minister for Enterprise, Trade and Employment*. Dublin: Government Publications, 2008, p. 59–60.
43. Oliver, Emmet. "Battle for satellite TV control lost." *Irish Times*, March 3, 2004, p. 5.

Horgan, John. *Irish Media: A Critical History Since 1922*. London: Routledge, 2001.
Kenny, Colum. "A double-edged sword: Irish media merger policy in transition." *Journal of Media Business Studies*. 6, 3 (2009): 93–108.
Maguire, Conor. "Setting out new direction for broadcasters." *Irish Times* October 19, 2001, 19.
Oliver, Emmet. "Battle for satellite TV control lost." *Irish Times* March 3, 2004, 5.
Pat Rabbitte. Speech to the Institute of International and European Affairs, March 22, 2012. Government of Ireland Press Office.

## Reports

Advisory Group on Media Mergers. *Report to the Tanaiste and Minister for Enterprise, Trade and Employment*. Dublin: Government Publications, 2008, 53, 56, 59–60.
Broadcasting Authority of Ireland. *Ownership and Control Policy 2010*. Dublin: Broadcasting Authority of Ireland, 2010, 2, 11.
Competition Act 2002 (Dublin: Government Publications)
Competition Authority. *Determination of Merger Notification M/05/065—Johnston*. Dublin: The Competition Authority, 2005.
Competition Authority. *Determination of Merger Notification M/08/021—Tv3/Kish Media*. Dublin: The Competition Authority, 2008.
Competition Authority. *Strategy Statement 2012–2014*. Dublin: The Competition Authority, 2011, 8.
Comreg. *Irish Communications Market: Key Data Report—Q4 2008*. Dublin: Comreg, 2009, 31.
Comreg. *Irish Communications Market Quarterly Key Data Report Q4 2011*. Dublin: Comreg, 2012, 68.
Comreg. *Market Analysis—Wholesale Access and Call Origination on Public Mobile Telephony Networks*. Dublin: Comreg, 2005, 3.
Hargreaves Heap, Shaun, et al. *Consultation on the Assessment of Media Mergers under the Competition Act 2002: Final Report*. Dublin: Competition Authority, 2006, 8–11, 50.
Irish Congress of Trade Unions. *Privatisation: Learning from the Eircom Debacle*. Dublin: ICTU, 2011, 7.
Telecom Eireann. *Report and Accounts for the Year Ended March 31, 1988*. Dublin: Telecom Eireann, 1988, 10.

# Ireland—Data Summaries

ELI NOAM AND PAUL MUTTER

IRELAND HAS AMONG THE highest concentration of foreign media ownership of all 30 countries surveyed at 72.1%, (Table 7.16). It has a relatively small population (under 5 million) and it was part of the United Kingdom until 1922, which partly accounts for the high rate of foreign ownership by UK companies. Aside from British media groups, SingTel (Singapore), owner of Eircom, is the most significant firm present. It is Ireland's largest company in both platform and overall media, by share of the national market and by power index. Its ISP market share has been reduced in recent years, but this is not because of domestic competition: Vodafone (UK), which acquired its rival, UK-based company BT Ireland in 2009, is SingTel's primary competitor.

Though dominated by foreign entities, the market is highly saturated as a result of a high literacy rate and an economic boom in the late 1990s.

One of the most significant impacts on the national media market has been the business rivalry between two media groups controlled by two families. Independent News & Media (INM) is owned by the O'Reilly family, and Communicorp, held by Dennis O'Brien, compete for market share over both print and television properties, and their bidding war over privatization of the public monopolist Eircom saw that firm stripped of assets and resold to an Australian consortium, which went bankrupt after squeezing the firm still further. This then brought in Singapore's telecom incumbent, SingTel, which acquired Eircom in 2010.

Of the content companies, O'Reilly's Independent News & Media (INM) holds the largest share of the overall national market at 10.8%. Within content alone, INM holds a much larger at 29.4%. Liberty Global (Malone, US) has a strong presence in multichannel

Table 7-16. National Media Industries Concentration in Ireland

| | 2004/5 | | 2011 OR MOST RECENT | | % CHANGE ANNUAL AVERAGE | |
|---|---|---|---|---|---|---|
| | COMPANY POWER INDEX IN COUNTRY | COMPANY SHARE OF THE OVERALL NATIONAL MEDIA MARKET (%) | COMPANY POWER INDEX IN COUNTRY | COMPANY SHARE OF THE OVERALL NATIONAL MEDIA MARKET (%) | COMPANY POWER INDEX IN COUNTRY | COMPANY SHARE OF THE OVERALL NATIONAL MEDIA MARKET (%) |
| Eircom (SingTel, Singapore) (public) | 1,896.5 | 28.125 | 1,148.1 | 23.0 | –6.6 | –0.852 |
| Vodafone (UK) | 713.2 | 17.241 | 580.0 | 16.9 | | –0.058 |
| Independent News & Media (INM) (O'Reilly) | 555.8 | 11.006 | 564.2 | 10.8 | | –0.026 |
| RTE (Public) | 456.3 | 6.617 | 360.0 | 5.9 | –3.5 | –0.113 |
| Murdoch Group (US) | 180.6 | 6.828 | 288.1 | 8.5 | 9.9 | 0.274 |
| O2/Telefónica (Spain) | 350.4 | 8.759 | 280.4 | 8.6 | 3.3 | –0.031 |
| Liberty Media (US) (Malone) | 260.8 | 5.802 | 202.9 | 5.9 | –3.7 | 0.032 |
| Google (US) | 18.9 | 0.203 | 127.5 | 1.356 | 96.1 | 0.192 |
| TV3 (UK) | 15.0 | 1.1 | 18.8 | 1.3 | 4.3 | 0.036 |
| Thomas Crosbie Holdings | 4.8 | 1.1 | 9.7 | 1.5 | 17.1 | 0.065 |
| Trinity Mirror (UK) | 16.1 | 1.9 | 9.1 | 1.4 | –7.3 | –0.83 |
| Hutchison Whampoa (3G, Hong Kong) | 0.0 | 0.0 | 8.8 | 1.5 | N/A | 0.254 |
| Communicorp (O'Brien) | 11.3 | 0.571 | 5.9 | 0.339 | –8.0 | –0.039 |
| Irish Times Trust | 6.6 | 1.2 | 7.3 | 1.2 | 1.7 | 0.005 |
| ITV (UK) | 5.4 | 0.559 | 2.5 | 0.344 | –9.0 | –0.036 |
| BBC (UK) (Public) | 4.0 | 0.364 | 2.1 | 0.297 | –7.9 | –0.011 |
| Channel 4 (UK) | 1.1 | 0.194 | 0.4 | 0.127 | –11.0 | –0.011 |

*continued*

Table 7-16. *continued*

| | 2004/5 | | 2011 OR MOST RECENT | | % CHANGE ANNUAL AVERAGE | |
|---|---|---|---|---|---|---|
| | COMPANY POWER INDEX IN COUNTRY | COMPANY SHARE OF THE OVERALL NATIONAL MEDIA MARKET (%) | COMPANY POWER INDEX IN COUNTRY | COMPANY SHARE OF THE OVERALL NATIONAL MEDIA MARKET (%) | COMPANY POWER INDEX IN COUNTRY | COMPANY SHARE OF THE OVERALL NATIONAL MEDIA MARKET (%) |
| Media Concentration Index | 2004/5 | | 2011 or Most Recent | | % Change Annual Average | |
| Total Revenue: Nat'l Media Industry (mil US$) | 9,635 | | 10,257 | | 1.1 | |
| Total Voices (*n*) | 44 | | 47 | | 1.1 | |
| Net Voices (*n*) | 35 | | 38 | | 1.4 | |
| Public Ownership (%) | 6.6 | | 6.4 | | −0.04 | |
| Foreign Ownership (%) | 73.8 | | 72.1 | | −0.43 | |
| C4 Average—Weighted | 91 | | 90 | | −0.16 | |
| HHI Average—Weighted | 4,434 | | 3,956 | | −1.8 | |
| C1 Average—Weighted | 60.0 | | 52.0 | | −0.01 | |
| Noam Index Average—Weighted | 1,745 | | 1,748 | | 0.02 | |
| Pooled Overall Sector C4 | 65.1 | | 59.3 | | −0.97 | |
| Pooled Overall Sector HHI | 1,422 | | 1,165 | | −3.01 | |
| Pooled Overall Sector Noam Index | 260 | | 231 | | −1.9 | |
| Market Share of Top Ten Companies: Nat'l Media Industry (%) (Pooled C10) | 89.3 | | 84.7 | | −0.8 | |
| National Power Index | 4,529 | | 3,650 | | −3.2 | |

Table 7-17. Top Content Media Companies in Ireland

| | 2004/5 | | 2011 OR MOST RECENT | | % CHANGE ANNUAL AVERAGE | |
|---|---|---|---|---|---|---|
| | COMPANY POWER INDEX IN COUNTRY | COMPANY SHARE OF THE NATIONAL CONTENT MEDIA MARKET (%) | COMPANY POWER INDEX IN COUNTRY | COMPANY SHARE OF THE NATIONAL CONTENT MEDIA MARKET (%) | COMPANY POWER INDEX IN COUNTRY | COMPANY SHARE OF THE NATIONAL CONTENT MEDIA MARKET (%) |
| Independent News and Media | 1,550.0 | 30.7 | 1,527.5 | 29.4 | −0.24 | −0.220 |
| RTE (public) | 1,272.5 | 18.5 | 974.7 | 16.1 | −3.9 | −0.396 |
| Murdoch Interests (US) | 317.7 | 14.3 | 382.6 | 15.3 | 3.4 | 0.170 |
| Google (US) | 52.6 | 0.565 | 345.2 | 3.7 | 92.8 | 0.518 |
| Liberty Global (US) (Malone) | 231.3 | 3.7 | 166.8 | 3.5 | −4.7 | −0.043 |
| TV3 (UK) | 41.8 | 3.0 | 51.0 | 3.5 | 3.6 | 0.083 |
| Daily Mail & General Trust (UK) | 42.9 | 5.1 | 48.9 | 5.3 | 2.3 | 0.025 |
| Thomas Crosbie Holdings | 13.3 | 3.0 | 26.2 | 4.0 | 16.1 | 0.163 |
| Trinity Mirror (UK) | 45.0 | 5.2 | 24.6 | 3.7 | −7.5 | −0.250 |
| Communicorp (O'Brien) | 31.5 | 1.6 | 15.9 | 0.918 | −8.3 | −0.113 |
| Irish Times Trust | 18.4 | 3.3 | 19.7 | 3.3 | 1.2 | −0.002 |

*continued*

Table 7-17. *continued*

| | 2004/5 | | 2011 OR MOST RECENT | | % CHANGE ANNUAL AVERAGE | |
|---|---|---|---|---|---|---|
| | COMPANY POWER INDEX IN COUNTRY | COMPANY SHARE OF THE NATIONAL CONTENT MEDIA MARKET (%) | COMPANY POWER INDEX IN COUNTRY | COMPANY SHARE OF THE NATIONAL CONTENT MEDIA MARKET (%) | COMPANY POWER INDEX IN COUNTRY | COMPANY SHARE OF THE NATIONAL CONTENT MEDIA MARKET (%) |
| ITV (UK) | 15.0 | 1.6 | 6.7 | 0.932 | −9.21 | −0.104 |
| BBC (UK) (public) | 11.1 | 1.0 | 5.6 | 0.804 | −8.20 | −0.035 |
| Channel 4 (UK) | 3.1 | 0.541 | 1.0 | 0.344 | −11.17 | −0.033 |
| Media Concentration Index | 2004/5 | | 2011 or Most Recent | | % Change Annual Average | |
| Public Ownership (%) | 18.5 | | 17.3 | | −0.19% | |
| Foreign Ownership (%) | 37.7 | | 39.4 | | 0.44 | |
| C4 Average—Weighted | 87 | | 86 | | −0.23% | |
| HHI Average—Weighted | 3,617 | | 3,736 | | 0.55% | |
| C1 Average—Weighted | 54 | | 53 | | 0.0% | |
| National Power Index | 3,694 | | 3,642 | | −0.24% | |

Table 7-18. Top Platform Media Companies in Ireland

| | 2004/5 | | 2009 OR MOST RECENT | | % CHANGE ANNUAL AVERAGE | |
|---|---|---|---|---|---|---|
| | COMPANY SHARE OF THE OVERALL NATIONAL MEDIA MARKET (%) | COMPANY POWER INDEX IN COUNTRY | COMPANY SHARE OF THE NATIONAL PLATFORM MEDIA MARKET (%) | COMPANY SHARE OF THE NATIONAL PLATFORM MEDIA MARKET (%) | COMPANY POWER INDEX IN COUNTRY | COMPANY SHARE OF THE NATIONAL PLATFORM MEDIA MARKET (%) |
| Eircom (SingTel, Singapore) (public) | 2,956.8 | 43.8 | 1,820.5 | 36.5 | −6.4 | −1.2 |
| Vodafone (UK) | 1,111.9 | 26.9 | 919.6 | 26.8 | −2.9 | −0.015 |
| O2 (Telefónica, Spain) | 546.3 | 13.7 | 444.7 | 13.6 | −3.1 | −0.010 |
| Murdoch Group (US) | 103.9 | 2.7 | 232.7 | 4.5 | 20.7 | 0.302 |
| Liberty Global (US) (Malone) | 277.3 | 7.0 | 224.1 | 7.5 | −3.2 | 0.084 |
| Media Concentration Index | 2004/5 | | 2011 or Most Recent | | % Change Annual Average | |
| Public Ownership (%) | 0 | | 0 | | 0% | |
| Foreign Ownership (%) | 94.0 | | 91.2 | | −0.69 | |
| C4 Average—Weighted | 94 | | 93 | | −0.11 | |
| HHI Average—Weighted | 4,891 | | 4,084 | | −2.75% | |
| C1 Average—Weighted | 64 | | 51 | | −2% | |
| National Power Index | 4,996 | | 3,656 | | −4.47 | |

platforms, but 21st Century Fox's (Murdoch's) share of multichannel platforms (as BSkyB) has risen at its expense. 21st Century Fox also has shares in video channels and film, amounting to a 15.3% of the content media market. RTÉ, Ireland's public broadcasting service, holds 74% of TV broadcasting, as well as 53.5% of radio. Its nearest competitor is TV3, which has only 3.5 % of the content market (Tables 7.17 and 7.18).

# 8

# Media Ownership and Concentration in Italy

GIUSEPPE RICHERI AND BENEDETTA PRARIO

## INTRODUCTION

In 2004, the Italian government passed a new media holdings law expanded the scope of the Italian media market by redefining a number of products and services that were not bound by existing regulatory legislation. Under the new law, no single corporate entity is allowed to own more than 20% of the entire Italian media market. However, the ways in which the Government of Italy has expanded the definition of the media market's size and changed the anti-concentration limits have led to significant debate in Italy—and in the supranational European bodies it has membership in, such as the European Union (EU) and the Organization for Security and Co-operation in Europe (OSCE)—over media concentration and ownership.

The Gasparri Law, named after the then-Minister of Telecommunications Maurizio Gasparri, was enacted in 2004 and renewed in 2011. It defines Italy's media market quite broadly, creating a single economic and regulatory entity—an "integrated system of communications" [sic]—with seven primary divisions: print media, television broadcasting, film, directories, electronic publishing, outdoor advertising, and communication initiatives for products and sponsoring. Following the law's passage, the Italian Communications Regulatory Authority (AGCOM) was tasked with determining the precise size of the media market in order to verify if there were any media holding companies in (retroactive) violation of the new 20% anti-concentration limit. AGCOM's value assessment for 2009, the most recent year for which a full data analysis is available, suggests that the Italian media market is worth approximately US$29 billion.

The three Berlusconi governments mainly pursued reform in the sectors of television broadcasting and telecommunications, where ownership is highly concentrated at the top among a few companies. From the start, the government paid particular attention to television broadcasting, a sector that was (and still is) closely linked to the Berlusconi family's business interests. Fininvest, the holding company in which the Berlusconi family maintains a controlling interest, owns three of Italy's main television channels—Canale 5, Italia 1, and Rete 4—through the commercial broadcaster Mediaset; the leading national publishing house for both books and newspapers, Arnoldo Mondadori Editore; and a host of other Italian media concerns, such as the cinema distributor Medusa.

But within the local and regional media, especially with respect to radio and television, the market is highly fragmented. On average, within one geographic region, there is a range of programming operated by a number of different operators, and the anti-monopoly limitations on ownership concentration are enforced by strict laws that limit the number of licenses one person can get within a single geographic location for most media sources, excluding daily newspapers.

## PRINT MEDIA

### Newspapers

In 2009, Italian newspapers had a net national worth of US$2.730 billion, registering a decrease of 20% over the last three years (2006–2009). Newspapers in Italy can be divided into two categories: national and local. The national papers can further be divided into two subcategories: free or paid.[1] The five major publishers of daily newspapers in Italy are Gruppo Editoriale L'Espresso with 21.9% of the market share, RCS Quotidiani with 33.6%, Il Gruppo 24 ORE with 8.7%, Caltagirone Editore with 10.6%, and Poligrafici Editoriale with 8.2% (Table 8.1).

Revenue for newspapers began shrinking in 2006, and so far the sector has been unable to curb this trend. There are three primary causes. The first reason for declining revenue is sociological: Italians show little inclination to purchase print newspapers anymore, and this has been exacerbated by the global economic downturn that began in 2008. Just 176 newspapers are sold for every 1,000 inhabitants in Italy, making this one of the lowest print-to-consumer ratios among industrialized nations. The second main reason is a reduction in advertising, and this is also due to the economic crisis. The third factor is the impact that "new media" is beginning to have on the print industry, as online services are slowly starting to siphon advertisers and consumers away from hardcopy publications.

According to Accertamenti Diffusione Stampa (ADS), which tracks print circulations in Italy, national newspapers' revenue fell by 19.5% between 2006 and 2009. Sales fell by 11%, advertising by around 20%, and marketing collateral, which includes alternative editorial content such as books and DVDs sold through newspapers, fell by at least 50%.[2]

Antimonopoly laws for radio and television broadcasting that ensure the existence of a large number of regional and local media providers do not apply to daily newspapers: a single corporate entity or individual can control a large number of regional and local newspapers. The main owners are Gruppo Editoriale L'Espresso with *La Repubblica* and its ten local editions, the RCS Media Group with *Corriere della Sera* and its 20 local editions, and Gruppo Caltagirone with five local newspapers. The benefit of this practice is that large portions of many different local newspapers, such as the international section, can be produced by one central editorial office and distributed, thus reducing production costs.

The national distribution network for newspapers is also part of the problem because it does not meet consumer demand in a timely manner. It is also unprofitable: distribution through subscriptions accounts for less than 9% of the total volume of sold print copies, even though point

1. Associazione Italiana Editori (AIE). *Rapporto sullo stato dell'editoria 2009*. Milan: AIE, 2010. <http://culturaincifre.istat.it/sito/libri/sintesiRapporto2009_R.pdf>.

2. Accertamenti Diffusione Stampa (ADS). *Diffusione dei quotidiani 2009*. Milan: ADS, 2010.

Table 8-1. Daily Newspapers (Market Shares by Circulation), 1984–2009

| | 1984 | 1992 | 1996 | 2001 | 2006 | 2009 | 2013[3] |
|---|---|---|---|---|---|---|---|
| Gruppo Editoriale L'Espresso | 27.0 | 27.0 | 27.1 | 27.3 | 27.1 | 29.0 | 21.9 |
| RCS Quotidiani (RCS Media Group) | 27.0 | 26.8 | 27.0 | 27.4 | 26.8 | 26.0 | 33.6 |
| Caltagirone Editore | 8.0 | 8.0 | 7.8 | 8.0 | 10.5 | 9.0 | 10.6 |
| Il Gruppo 24 ORE | 11.5 | 11.4 | 11.4 | 11.6 | 10.6 | 10.0 | 8.65 |
| Poligrafici Editoriale | 9.6 | 9.8 | 9.7 | 9.8 | 8.9 | 9.0 | 8.2 |
| Others | 16.9 | 17.0 | 17.0 | 15.9 | 16.1 | 17.0 | 17.0 |
| Total Revenue (mil euro) | 1,200 | 1,900 | 2,360 | 3,242 | 2,651 | 2,135 | 1,836 |
| Total Revenue (mil US$) | 1,534 | 2,429 | 3,018 | 2,916 | 3,392 | 2,730 | 2,505 |
| C4 | 75.1 | 75.0 | 75.2 | 76.1 | 75.0 | 74.0 | 74.8 |
| HHI | 1,746 | 1,737 | 1,748 | 1,791 | 1,754 | 1,779 | 1,863 |
| *N* (>1%) | 5 | 5 | 5 | 5 | 5 | 5 | 5 |
| Noam Index | 781 | 777 | 782 | 801 | 785 | 796 | 833 |

*Source*: AGCOM, FIEG, FNSI.

of sale distribution costs newspapers a lot more money to maintain. High distribution costs are the result of inefficiencies in the national postal service: more than half the population lives in small- and medium-sized urban areas, so speedy postal distribution is prioritized in those areas, while in geographically "peripheral" towns and villages, newspapers tend to arrive late or not all. In these places, online news sites have stepped in to meet consumers' demands for timely access to information.

According to Audiweb—an organization that tracks the online audience shares of Italian media outlets—on an average day, online news sites count 3.8 million unique visitors and approximately 33 million page hits. Of all the national newspapers with online editions, *La Repubblica*'s Internet portal has achieved the best results, averaging over one million unique visitors and 11.5 million page hits daily. Even so, success in winning a greater audience share online has not substantially increased newspapers' revenue for three reasons: (1) most of the online content is free, (2) because online advertising does not generate as much income as traditional print advertising does, and (3) because paid subscription sites are not very popular.

The concentration indices for this industry are fairly low, and the concentration levels have held constant since 1984. The majority shareholders of the publishers have strong interests in other businesses. Though the newspapers are not significant sources of revenue for them, they are able to still operate them (at a loss) because of the revenue their other business concerns generate. These shareholders hold on to their ownership because having a stake in these newspapers supports their interests in other businesses.

With respect to the reforms for newspapers, the government has sought to reduce subsidies. It has focused not on the dailies, though, but on the newspapers of Italian political parties, which are classified as "political communications" and have up to 70% of the operating costs covered by public funding. It has argued that the cuts are necessary in order to compel parties' newspapers to reduce questionable expenses claimed by the political appointees in these offices.

3. 2013 updates provided by editors.Market shares are generated by using ADS's circulation numbers and calculating the market share for the named companies.Then it was approximated that "Others" remained the same and the left over amount was attributed to the named companies in proportion to their market share.Similarly revenue was estimated by removing 14% of 2009's revenue, the amount in decline sources have stated the Italian market has lost.

### GRUPPO EDITORIALE L'ESPRESSO

Gruppo Editoriale L'Espresso publishes *La Repubblica,* Italy's second most popular national daily newspaper, along with various local editions of the paper and independent local dailies. The shareholding composition of Gruppo L'Editoriale Espresso breaks down into a 53.9% stake held by the members of the Compagnie Industriali Riunite, a 11.7% stake by Eredità Caracciolo di Melito, an entity set up by one of the founders of *La Repubblica,* a 2.7% stake by the bank Fondazione Cassa di Risparmio of Trieste, a 2.4% stake by the investor Giulia Maria Crespi Mozzoni, and a 2.3% stake by the insurance company Assicurazioni Generali S.pA. The government has a 1.9% stake.

### RCS QUOTIDIANI

RCS Quotidiani is a business unit of RCS Media Group, a multimedia-publishing group with interests in daily newspapers, magazines and books, radio broadcasting, new media, and the digital and satellite TV sectors. RCS Media Group is a highly internationalized company with a significant presence in Spain, Portugal, France, the United States, and the People's Republic of China, and it is listed on the Blue Chip segment of the Italian Stock Exchange.

RCS Quotidiani publishes two national daily newspapers: *Corriere della Sera,* the country's highest circulation daily, and *La Gazzetta dello Sport,* Italy's most popular sports magazine. It also publishes various local editions of its flagship paper, *Corriere della Sera.*

The share composition of RCS Media Group is as follows: 65.7% is held by signatories to the RCS Media Group Consultation and Lock-up Shareholders' Agreement, 21.4% is owned by smaller, but similar, entities, and 12.3% consists of government treasury shares.

### IL GRUPPO 24 ORE

Il Gruppo 24 ORE specializes in business, financial, professional, and cultural publishing. Its best-selling media asset is *Il Sole 24 Ore,* which has the highest circulation of any European business daily. The publisher also controls a press agency (*24 Ore Radiocor*), a web portal (www.ilsole24ore.com), and the radio station *Radio 24.*

Confindustria, the Italian manufacturing employers' organization, holds 67.5% of the shares, a 28.8% stake is held by independent shareholders, and the managers of *Il Sole 24 Ore* maintain a 3.7% stake in the group.

### CALTAGIRONE EDITORE

Caltagirone Editore, a family-owned firm, controls *Il Messaggero, Il Mattino,* and *Il Gazzettino,* which are the leading newspapers in the cities of Rome, Naples, and Venice, respectively. It publishes two local newspapers: *Il Corriere Adriatico* in the Marche region, and *Il Nuovo Quotidiano di Puglia* in the Salento region. Incorporated in December 1999, the company has been listed on the Italian Stock Exchange since July 2000; 61% of its shares belong to the head of the family Francesco Gaetano Caltagirone, with a 2% stake held by Gaetano Caltagirone, a relative, and the holding company of the Benetton family, Edizione S.r.l.; 37% of the publisher's shares are publicly floated shares.

### POLIGRAFICI EDITORIALE

Poligrafici Editoriale is a subsidiary of the Riffeser Monti Group, and it owns three local daily newspapers: *Il Resto del Carlino* of Bologna, *La Nazione* of Florence, and *Il Giorno* of Milan. The head of the Group, Maria Luisa Monti Riffeser, owns 64.4% of shares, the U.S. telecom ADV Media owns a 10.3% stake, and the remaining 5.1% stake belongs to the Italian arts foundation Fondazione Cassa Risparmio Trieste.

## Book Publishing

In 2009, the Italian Publishers Association surveyed the industry and found that there were 7,009 publishing entities in Italy—including those that had published as few as one book—of which 1,600 could be considered publishing houses due to their consistent offering of new products in Italian bookstores.

The market value of the book publishing industry in Italy in 2009 was assessed at US$4.3

billion, representing a 4.3% decline since 2008: the decline was primarily a result of a reduction in installment sales. In contrast, over this same period, book sales increased for online vendors by 12%, for mass distributors by 4%, and for bookstores by 2.5%. E-books have registered positive results since 2010, with a turnover of US$1.28 million, although they still represent just 0.03% of the entire book publishing market. Online booksellers have so far had little impact on the book market, although it is possible that they will increase their market share over time.

Although book sales have increased, overall the strength of book publishing industry is weakening because few adult Italians purchase books relative to their counterparts in the wider EU. A recent study found that approximately half of the country's adult population reports that it has never bought books or visited a bookstore.[4] Compared to other industrialized nations, the number of bookstores nationwide is quite low, and many urban areas do not have any bookstores at all, which is uncommon in Western Europe. The result of all this is that even when consumers want to buy books, they find it difficult to do so—which is why online sales are increasing despite the overall market trend. Compared to Spain, Germany, France, and the United Kingdom, Italy ranks second to last in terms of the value of its book market, and its citizens come in at last place in per capita expenditures on books, at US$76.70 per annum.

Assessing the long-term fortunes of the book publishing industry is difficult because the Italian Publishers Association only has been publishing statistics on the sector since 2007. Nonetheless, on the basis of this survey on the shorter term, it can be shown that five publishing houses—Mondadori, RCS Libri, Gruppo Editoriale Mauri Spagnol (GEMS), Feltrinelli, and Giunti—and their subsidiaries together control almost 60% of the market. Of the five, Mondadori, RCS Libri, and GEMS predominate. In 2009, Mondadori registered a turnover of US$1.22 billion, with a market share of 28.4%; RCS Libri, US$562.7 million, with a market share of 12.6%; GEMS, US$396.5 million, with a market share of 9.3%; and Giunti US$262 million, with a market share of 5.8%.

While many smaller publishers represent the remaining 40% market share, none of them individually controls more than 4% of the market (Table 8.2).

Table 8-2. Book Publishing (Market Shares by Revenue), 2007–2009

| | 2007 | 2008 | 2009 |
|---|---|---|---|
| Mondadori (Berlusconi family) | 29.0 | 28.8 | 28.4 |
| RCS Libri | 13.6 | 12.8 | 12.6 |
| GEMS | 8.2 | 8.9 | 9.3 |
| Giunti | 5.4 | 5.5 | 5.8 |
| Feltrinelli | 3.8 | 3.9 | 4.0 |
| Others | 40.0 | 40.1 | 39.9 |
| Total Revenue (mil euro) | 3,670 | 3,553 | 3,407 |
| Total Revenue (mil US$) | 5,028 | 4,868 | 4,668 |
| C4 | 56.2 | 56.0 | 56.1 |
| HHI | 1,137 | 1,118 | 1,101 |
| *N* (>1%) | 5 | 5 | 5 |
| Noam Index | 508 | 500 | 493 |

*Source:* AIE.

4. Associazione Italiana Editori (AIE). *Rapporto sullo stato dell'editoria 2011*. Milan: AIE, 2012. <http://www.aie.it/LinkClick.aspx?fileticket=VhebinUVCfg%3D&tabid=2808>.

### MONDADORI

Mondadori is Italy's largest book publisher and is controlled by Fininvest, the Berlusconi family's financial holding company. The group now has nine separate publishing houses, some of which have more than one imprint, such as Sperling & Kupfer and Piemme. Mondadori also operates in Spain and in the international Spanish-speaking markets through Grupo Editorial Random House Mondadori, a Spanish-language joint venture with Random House, owned since 1998 by Bertelsmann.

### RCS LIBRI

RCS Libri is a division of RCS Media Group. It operates in Italy and France, as well as in Spain, Portugal, the United Kingdom, and the United States. It has expanded its market share in Italy by taking over many smaller publishing houses, such as Rizzoli Editore, Bompiani, Fabbri Editori, Marsilio, and Flammarion.

### GRUPPO EDITORIALE MAURI SPAGNOL

Gruppo Editoriale Mauri Spagnol (GEMS) is run jointly by the Mauri and Spagnol families. It now controls nine publishing houses including Garzanti Libri, Longanesi, and Bollati Boringhieri. The shareholders are the holding company Messaggerie Italiane, with a 73.77% stake; the Spagnol family, with a 23% stake; and Andrea Micheli, with a 3.23% stake.

## AUDIOVISUAL MEDIA

### Radio

Historically, the Italian radio market has been monopolistic. The national public broadcaster Radiotelevisione italiana S.p.A. (RAI), founded in 1954, owns three channels and dominates the airwaves. Although a proposal to privatize some of its assets was considered in the 2000s, the government did not go follow through with it.

Like television broadcasting, the radio sector was liberalized by the government in 1976, opening the market up to competition. Within just a few years, close to 3,000 private broadcasters had emerged to compete with RAI's channels. While some of these run programming interregionally with a consortium of other broadcasters, the majority of them are independent stations in precarious financial situations.

Despite the profusion of new stations, the private sector was not regulated in any way until the early 1990s. As a result of this lack of regulation, long-term data about the number of companies present in the market, annual turnover, and market shares are not available to researchers because the government was not publishing statistics on them. The value of the radio market in Italy in 2009 was US$605 million, and the market share is still highly concentrated in favor of RAI.

The radio industry is still chiefly made up of thousands of local radio broadcasters whose audience shares and turnover are not well documented. For all these reasons the following discussion of the Italian radio market will refer only to the main national radio broadcasters, including RAI (Table 8.3).

Gruppo L'Espresso Editoriale, whose primary business is not in radio, but in magazines and daily newspapers, is the second-largest radio broadcaster in Italy. It owns three national stations: Radio Capital, Radio m20, and Radio Deejay. The Compagnie Industriali Riunite has a 53.9% stake in the group, 11.7% belongs to Eredità Caracciolo di Melito, 2.7% to the Fondazione Cassa di Risparmio of Trieste, 2.4% to Giulia Maria Crespi Mozzoni, 2.3% to Assicurazioni Generali, and the government holds a 1.9% stake.

After Gruppo L'Espresso Editoriale, Gruppo Finelco is the third-largest radio broadcaster in Italy, and it also owns three national stations: Radio Monte Carlo, Radio 105, and Virgin Radio. The group's founder, Alberto Hazan, owns 87% of shares, while the remaining 13% is held by MPS Venture SGR, a company that manages closed-end funds for the Monte dei Paschi di Siena (MPS) Group.

## Table 8-3. Radio Groups (Market Shares, by Revenue) 1995–2009

| | 1995 | 2001 | 2005 | 2009 | 2011 |
|---|---|---|---|---|---|
| RAI (public) | 47.0 | 58.0 | 48.0 | 49.0 | 48.2 |
| Gruppo L'Espresso Editoriale | 13.0 | 8.1 | 12.7 | 11.0 | 13.1 |
| Gruppo Finelco | | 8.5 | 9.5 | 10.0 | 10.1 |
| RTL 102.5[1] | 10.0 | 10.0 | 8.9 | 5.0 | 6.3 |
| Radio Italia Musica | | 4.5 | 4.9 | 4.0 | 4.1 |
| R101 | | 1.0 | 0.5 | 2.0 | 1.1 |
| Radio Dimensione Suono | 11.0 | 7.0 | 6.9 | 8.0 | 7.5 |
| RISMI | 10.0 | | | | n.a |
| Radio Kiss Kiss | | 0.7 | 1.5 | 2.0 | 1.2 |
| Radio 24 | | 1.0 | 1.5 | 2.0 | 2.3 |
| Radio Latte Miele | | 1.0 | 1.5 | | n.a |
| Rin Radio IT.Network | | 0.2 | 1.4 | | n.a |
| Radio Italia 5 | | | | 2.0 | 1.8 |
| CNR | | | 2.7 | 5.0 | 4.3 |
| Total Revenue (mil euro) | 240 | 382 | 481 | 473 | 445 |
| Total Revenue (mil US$) | 329 | 345 | 566 | 648 | 619 |
| C4 | 81.0 | 84.6 | 79.1 | 78.0 | 78.9 |
| HHI | 2,780 | 3,675 | 2,723 | 2,768 | |
| *N* (>1%) | 6 | 9 | 11 | 11 | |
| Noam Index | 1,135 | 1,225 | 821 | 835 | |

*Source*: UPA.

[1] RTL 102.5 does not have any link with the RLT Group.

## Broadcast Television

Italy has only two broadcast television platforms, terrestrial and satellite, since cable TV has never really caught on in Italy as strongly as it has in other countries. The market is still dominated by free channels, with pay channels having gained ground only in the past few years, but since 2000 the Italian television market has changed dramatically.[5]

The first major change occurred in 2004, when the Gasparri Law redefined the limits of market concentration. The second major change emerged with the introduction of digital terrestrial television (DTT), which after 2004 rapidly supplanted traditional analog television (ATV) in Italy; the DTT switch over introduced a wide range of new free and paid channels. The third major change has been the growth of satellite pay TV. And finally, the proliferation of new modes of access, such as Internet Protocol television (IPTV), mobile TV, and web TV, are affecting consumption trends.

Historically, RAI has held a monopoly over broadcast television. In the second half of the 1970s, this monopoly was ended when the government granted private broadcasters the right to transmit local television programming. Following this liberalization, hundreds of local television broadcasters emerged, some of which grouped together into regional or national

5. Richeri, Giuseppe. *Economia dei media*, Rome: Laterza, 2012.

networks by the 1980s. In the early 1980s, competition between the private networks and the public broadcaster increased, as did competition among the private networks. This changed, though, when the Berlusconi family's Fininvest holding company—which already owned one of the biggest private networks, Canale 5, through its subsidiary Mediaset—subsequently acquired two other large private networks, Italia 1 and Rete 4, which until then had been competing with Canale 5. By 1984, the Italian television industry was effectively a duopoly, with 92% of the market held by the public broadcaster RAI and privately owned Fininvest's Mediaset. Local and regional television consists of approximately 400 companies, which focus primarily on the wealthier metropolitan markets, such as Piemonte, Lombardia, and Lazio.

In 2009, the Italian broadcast television was worth US$7.5 billion, representing an increase of 3% from the previous year.[6] The main sources of income were advertising, accounting for 41.2% of revenue (down 9.2% from 2008), subscriptions from pay TV with 33.5% (up 7.4% from 2008), and licensing fees with 19% (up 1.7% from 2008).

Since the C4 index considers all operators in the market, the C2 index was also considered, given the duopolistic nature of the Italian broadcast television market. Combined, these indices indicate that RAI and Mediaset run almost neck and neck in terms of revenue, with RAI's revenue totaling US$3.489 billion and Mediaset's totaling US$3.467 billion within a terrestrial television market valued at approximately US$5.9 billion. The largest broadcaster after the two national giants is Telecom Italia Media (Table 8.4).

## RAI

RAI is the state-owned public radio and television broadcasting service that was founded in 1954. It has many television channels and

**Table 8-4. TV Broadcasting (Market Shares by Revenue), 1984–2012**

| | 1984 | 1992 | 1996 | 1998 | 2002 | 2003 | 2007 | 2009 | 2012 |
|---|---|---|---|---|---|---|---|---|---|
| RAI (public) | 67.0 | 54.0 | 52.7 | 50.9 | 50.1 | 48.7 | 49.0 | 47.1 | 48.3 |
| RTI (Mediaset, Berlusconi family) | 25.0 | 36.0 | 39.1 | 40.2 | 39.1 | 41.3 | 43.0 | 46.0 | 38.9 |
| Telecom Italia Media | | 2.1 | 2.1 | 1.9 | 1.7 | 3.0 | 2.4 | 2.4 | 0.7 |
| Others (national broadcasters) | | 0.5 | | 0.3 | 0.5 | | | | |
| Others (local broadcasters) | 8.0 | 7.4 | 6.1 | 6.7 | 8.6 | 7.0 | 5.6 | 4.5 | 12.1 |
| Total Revenue (mil euro) | 2,500 | 3,640 | 4,397 | 4,223 | 4,918 | 4,781 | 5,616 | 5,883 | 6,375* |
| Total Revenue (mil US$) | 3,425 | 4,987 | 6,024 | 5,786 | 4,844 | 6,412 | 7,182 | 7,524 | 10,072* |
| C4 | 92.0 | 92.1 | 93.9 | 93.0 | 90.9 | 93.0 | 94.4 | 95.5 | 87.9 |
| C2 | 92.0 | 90.0 | 92.0 | 91.0 | 89.2 | 90.0 | 92.0 | 93.1 | 87.2 |
| HHI | 5,114 | 4,216 | 4,311 | 4,210 | 4,042 | 4,086 | 4,256 | 4,340 | 3,847 |
| *N* (>1%) | 2 | 3 | 3 | 3 | 3 | 3 | 3 | 3 | 3 |
| Noam Index | 3,616 | 2,434 | 2,489 | 2,431 | 2,333 | 2,560 | 2,457 | 2,506 | 2,221 |

*Source*: AGCOM; Balance sheets of the players.

*Source for 2012*: Mesco, Mauela. "Italian Telecom and Media Firms Lost Over EUR4B Revenue in 2012." *Wall Street Journal* July 9, 2013. 2012 updates provided by editors. 2012 revenues estimated.

6. Autorità per le Garanzie nelle Comunicazioni (AGCOM). *Relazione annuale 2010*. Rome: AGCOM, 2010. <http://www.agcom.it/Default.aspx?message=downloadpdf&DocID=119>.

radio stations, broadcasts on ATV (until 2012) and, now, DTV, and also offers several satellite and IPTV channels. About half of RAI's revenue comes from license fees, and the other half comes from advertising sales. In 2010, RAI had the highest national audience share, with 48.3%. The shareholders of RAI are the Ministry of Economic Development, which has a 99.56% stake, and the Italian Society of Authors and Publishers (SIAE), which has a 0.44% stake.

### MEDIASET

Mediaset is the largest commercial broadcaster, owned by Silvio Berlusconi, who became one of Italy's richest and most powerful men and subsequently became prime minister. Mediaset started operating in 1978 when it began broadcasting as Telemilano, a local Milan-based broadcaster that became the nationally broadcast Canale 5 two years later. Mediaset soon added its rivals Italia 1, bought from the publishing group Rusconi in 1982, and Rete 4, bought from the publishing group Arnoldo Mondadori Editore in 1984, to its assets. In 1984, Mediaset group became the second largest television broadcaster, in competition with RAI: both had three national television channels on three ATV national networks.

In 1997, Mediaset expanded outside Italy with the acquisition of a 25% stake in the commercial Spanish broadcaster, Telecinco. By 2003, Mediaset became its major shareholder with a 50.1% stake. Its core business is in commercial television, but in recent years it has branched out and expanded into pay TV, web TV, IPTV, and mobile TV. Though it is nearly even with RAI in terms of audience share, Mediaset's share in the broadcast television advertising market is twice as large as RAI's.

The controlling shareholder of the Mediaset Group, with a stake of 39%, is Fininvest, the holding company run by the Berlusconi family founded by former Prime Minister Silvio Berlusconi. The remaining 42% of stakeholders are independent investors.

### TELECOM ITALIA MEDIA

Founded as Telecom Italia Media by SEAT Pagine Gialle in 2003, this commercial broadcaster was taken over by Telecom Italia. Its turnover in 2009 was US$178 million. The main shareholder is Telecom Italia, with a 78% stake, while the remaining 22% stake is traded as free float shares. Telecom Italia's holding company, Telco, is effectively controlled by Telefónica of Spain with a 9.47% stake. The two channels owned by Telecom Italia Media are the generalist channel La7 and the music channel MTV Italia, in joint venture with MTV Networks Europe, a part of Viacom. Though small compared to RAI and Mediaset, Telecom Italia Media has good prospects for expanding its market share due to its ownership by Italy's leading telecommunications provider.

## Multichannel TV Platforms

In 2009, around 780,000 (3%) of Italian households received IPTV. The value of the IPTV industry in 2009 was US$109 million, and its main operators were three telecommunications providers: Telecom Italia with Alice Home TV, Fastweb with La TV di Fastweb, and Wind with Infostrada TV. These operators hold 62%, 31%, and 7% of the market share, respectively.

Cable TV is effectively nonexistent in Italy: the infrastructure has not been developed, and firms display little interest in investing in the sector. If we include in the aggregate the (only) satellite operator, Sky Italia, we can say that the value of the multichannel platform industry is US$3.3 billion. By this reckoning, Sky Italia leads with 97% of the market share, with the three other telecommunications operators hold the remaining 3%.

Until 2002, there were just two satellite pay TV operators in the Italian market: Telepiù, of which 90% was held by French Canal +, and Stream, a joint venture by Rupert Murdoch's News Corporation and Telecom Italia. Neither Telepiù nor Stream ever broke even, and both accrued significant operating losses, so in 2003 Rupert Murdoch took control of Telepiù and created Sky Italia, which is now the only satellite

Table 8-5. Multichannel Video Platforms: Cable MSOS, DBS, IPTV (Market Shares by Revenue), 1992–2012

| | 1992 | 1998 | 2002 | 2003 | 2007 | 2009 | 2012 |
|---|---|---|---|---|---|---|---|
| Sky (News Corporation—UK) | | | | 100.0 | 100.0 | 97.0 | 95.3 |
| Tele + (Canal +—France) | 90.0 | 92.0 | 96.0 | | | | |
| Stream (News Corporation—Australia) | 10.0 | 8.0 | 4.0 | | | | |
| IPTV operators[1] | | | | | | 3.0 | 4.7 |
| Total Revenue (mil euro) | 105.0 | 602.0 | 844.0 | 1,098 | 2,422 | 2,582 | 2,768* |
| Total Revenue (mil US$) | 144.0 | 825.0 | 831.0 | 1,473 | 3,097 | 3,3 | 4,373* |
| C4 | 100.0 | 100.0 | 100.0 | 100.0 | 100.0 | 100.0 | 100.0 |
| C2 | 100.0 | 100.0 | 100.0 | 100.0 | 100.0 | 100.0 | 100.0 |
| HHI | 8,200 | 8,528 | 9,232 | 10,000 | 10,000 | 9,418 | 9,104 |
| *N* (>1%) | 2 | 2 | 2 | 1 | 1 | 1 | 1 |
| Noam Index | 5,798 | 6,030 | 6,528 | 10,000 | 10,000 | 9.418 | 9,104 |

*Source*: AGCOM.

*Source for 2012*: Mesco, Mauela. "Italian Telecom and Media Firms Lost Over EUR4B and TV and on-demand audiovisual services in Italy." 2012, *Wall Street Journal* July 9, 2013. 2012 updates provided by editors. 2012 revenues estimated based on Sky Italia revenue of 2.63 billion euros.

[1] IPTV operators included Fastweb, which had 240,000 subscribers in 2011 but shut down operations in 2012; Wind, which also shut down its service (Infostrada TV) in 2012; and Telecom Italia, which had 242,000 subscribers in 2012.

TV operator in Italy, offering third-party channels like CNN and MTV.[7] In 2009, it had almost 5 million subscribers and a turnover of US$3.2 billion. Satellite pay-TV in Italy has had a very high concentration level, and as of 2003 it has become a News Corporation monopoly. TiVùSat, owned by RAI, Fininvest, and Telecom Italia Media Group, also offers some satellite TV programming, but its primary interests lie elsewhere (Table 8.5).

Sky Italia came into conflict with the Italian government shortly after its inception and, along with the domestic Italian broadcaster Europa 7, brought (in 2005) and won (in 2011) a suit against part of the government's 2004 media reform legislation. Sky Italia came into conflict with the government over the reforms because the government attempted to reduce the amount of advertising shown on Sky Italia's channels by increasing taxes for subscriptions and by barring Sky Italia from entering the new DTT market. Sky Italia asserted this was protectionist because Sky Italia's pay-TV channels had succeeded in taking advertising revenue away from RAI and Mediaset.

The policies the Court found incompatible with Italy's economic obligations as an EU member centered on the fact that as ATV broadcasters prepared to switch over to DTT, the Berlusconi government began providing large incentives for households so that they would start buying digital decoders. This would help cover analog-to-digital conversion costs for not only for RAI, but Mediaset too by ensuring that an audience would exist to receive new DTT frequencies. In the suit, Court of Justice of the EU ruled in 2011: "the Italian subsidies for the purchase of digital terrestrial decoders in 2004 and 2005 constitute State aid which is incompatible with the common market."[8]

7. Prario, Benedetta. Le trasformazioni dell'impresa tv verso l'era digitale, Bern: Peter Lang, 2005.

8. "Judgment in Case C-403/10 P Mediase: *SpA v Commission*." *Press Release No 77/11*. Court of Justice of the European Union. July 28, 2011. Dec 4, 2012. <http://curia.europa.eu/jcms/upload/docs/application/pdf/2011-2007/cp110077en.pdf>.

Sky Italia's other complaint was that while government declared that it sought to increase competition in the multichannel TV sector through the transition, in practice the two broadcasters that already dominated the ATV market, RAI and Mediaset, were favored. The debate over the "digital dividend," that is, deciding how to allocate the new frequencies released by the digital switchover, focused on the existing dominant broadcasters rather than on allowing small or new broadcasters to take advantage of the increasing number of digital

Table 8-6. Film Production/Distribution (Market Shares by Box Office %), 1996–2008

| | 1996 | 2001 | 2006 | 2008 |
|---|---|---|---|---|
| Medusa Film (Fininvest, Berlusconi family) | 11.5 | 19.9 | 13.0 | 16.6 |
| United International Pictures (UIP)—Universal (Comcast/GE, US), Paramount (Viacom, US) | 19.0 | 17.2 | 13.0 | 19.7 |
| Warner Bros. Italia (Time Warner, US) | 11.6 | 11.2 | 7.5 | 9.6 |
| Buena Vista (Walt Disney, US) | 10.0 | 8.6 | 12.0 | 7.7 |
| 20th Century Fox (News Corporation, US) | 8.8 | 5.1 | 6.8 | 6.4 |
| 01 Distribution (RAI Cinema, public) | | 2.5 | 9.5 | 11.0 |
| Sony (Japan/US) | 5.5 | 7.8 | 9.3 | 5.1 |
| Filmauro | 7.8 | 6.8 | 8.2 | 8.0 |
| Eagle Pictures | | 6.2 | 6.0 | 3.9 |
| Bim Distribuzione | | 1.5 | 2.4 | 2.5 |
| Sacher Distribuzione | | 1.3 | 1.3 | 0.2 |
| Mikado | 1.8 | 1.8 | 1.0 | 1.6 |
| Lucky Red Distribution | 3.0 | 2.3 | 0.8 | |
| Cecchi Gori Group | | 1.8 | | |
| Nexo | | 2.7 | 0.6 | |
| Fandango | | 1.2 | 0.4 | 0.4 |
| Key Films | | 1.2 | | |
| Istituto Luce | 1.0 | 1.2 | 0.2 | |
| Lady Film | 0.7 | | | |
| Life International | 0.5 | | | |
| Mediafilm | | | 0.7 | 0.7 |
| Dnc Distribution | | | 0.4 | 0.2 |
| Moviemax | | | | 2.3 |
| Teodora Film | | | | 0.3 |
| Others | 19.7 | 3.7 | 11.4 | 3.8 |
| Total Revenue (mil euro) | 423.0 | 476.0 | 547.0 | 594.0 |
| Total Revenue (mil US$) | 580.0 | 427.8 | 698.4 | 759.5 |
| C4 | 52.1 | 56.9 | 47.5 | 56.9 |
| HHI | 909 | 1,096 | 783 | 1,097 |
| *N* (>1%) | 10 | 18 | 12 | 11 |
| Noam Index | 288 | 258 | 226 | 331 |

*Source:* ANICA, Cinetel.

frequencies to enter the market; hence the government's policies effectively guaranteed that RAI and Mediaset would maintain their existing duopoly. The EU and OCSE delivered several warnings to the Italian government aimed at reducing the favoritism shown to RAI and Mediaset, though as the tables show, the duopolistic structure of broadcast television persists.

## Film

US film distributors have dominated Italy's film distribution market for decades, since both before and after World War II (1939–1945). In 2009, all Italian film distributors together controlled only 24% of the market share. The value of the film distribution industry in Italy in 2009 was US$759 billion, and there was low concentration in this industry. Only in 2001 did the level of concentration increase gradually, due to the take-over of some small players by bigger ones and the failure of several distributors. In 2008, the two biggest players—United International Pictures, which is owned by Comcast/GE, and Fininvest's Medusa Film—owned 16.6% and 19.7% of the market share, respectively. Medusa Film is organized as joint stock company and has been the leading distributor in Italy since the 1990s. Apart from Medusa Film, the main players in the Italian film distribution market are the subsidiaries of US distributors.

In this study we considered the market share of distributors only, and not producers, because the latter segment is highly fragmented, consisting of many small production companies that tend to have very short lives on the market. In addition, for Italy the only useful data available are from 1996 onward (Table 8.6).

# TELECOMMUNICATIONS MEDIA

## Wireline Telecom

The value of the wireline telecom market in Italy is US$20.7 billion and the main operator is Telecom Italia with a 66% market share, followed by Fastweb with 8%, Wind with 14%, and Vodafone Italia, a subsidiary of Vodafone (UK), with 9%. The remaining 3% of the market belongs to a handful of much smaller operators.

In the telecommunications field, the government focused on building broadband networks up to a level of universal service that could contribute to significant economic growth. In pursuing its policies, there arose a prolonged debate between stakeholders in the industry and the government. During the course of the debate, data on the likely total costs were made public, to the government's chagrin, as it had not wanted to publicize this information before working out a network expansion plan with private enterprises. But the main problem with the broadband expansion is the continuing uncertainty about the respective roles of private enterprises and the State. So far, the private enterprises have failed to agree on how to form a consortium to finance, develop, and manage the new networks, and the government has failed to find the formula necessary to implement public funding.

Up until 1998, Telecom Italia ran all of Italy's landline telephone services. After 1998, when the EU legislated extensive liberalization measures for the European telecommunications industry, competition within the Italian market increased and Telecom Italia began to lose some of its share to emerging private operators, although such firms soon encountered difficulties.

The market concentration is expressed here through the more focused C2 index, since the C4 index covers all of the operators (Table 8.7).

### TELECOM ITALIA

Telecom Italia is the largest Italian telecommunications provider. It was established through the merger of several public telephone companies in the 1980s and was privatized in 1998. It operates landline services in Italy, mobile phone services through its subsidiary TIM (Telecom Italia Mobile), Internet and telephony services under the brand Alice, and, as noted earlier, broadcast television through its

Table 8-7. Wireline Telecom (Market Shares by Revenue), 1996–2012

| | 1996 | 2001 | 2006 | 2009 | 2012 |
|---|---|---|---|---|---|
| Telecom Italia (Telco/Telefonica, Spain)[1] | 100.0 | 73.0 | 73.6 | 64.1 | 66 |
| Wind (Vimpelcom, Russia) | | | 6.3 | 7.7 | 14 |
| Albacom, Tele2, Wind-Infostrada | | 17.0 | Wind | | |
| Fastweb (Swisscom, Switzerland) | | | 5.3 | 9.9 | 8 |
| BT Italia (UK) | | | 5.0 | 5.8 | 0 |
| Vodafone Italia (UK) | | | | 4.3 | 9 |
| TeleTU (ex-Tele2) (Sweden) | | | 3.1 | | |
| Tiscali (Soru) | | | 1.2 | 1.7 | 2 |
| Atlanet, Colt, Edisontel, Eplanet | | 10.0 | | | |
| Others | 0.0 | 0.0 | 5.5 | 6.5 | 1 |
| Total Revenue (mil euro) | 13,000 | 15,300 | 17,700 | 16,200 | 11,595 |
| Total Revenue (mil US$) | 17,800 | 13,800 | 22,600 | 20,700 | 18,320 |
| C2 | 100.0 | 90.0 | 79.9 | 74.0 | 80 |
| C4 | 100.0 | 100.0 | 90.2 | 87.5 | 97 |
| HHI | 10,000 | 5,718 | 5,520 | 4,321 | 4,702 |
| *N* (>1%) | 1 | 2 | 6 | 6 | 5 |
| Noam Index | 10,000 | 4,043 | 2,254 | 1,764 | 2,103 |

*Source*: AGCOM.
*Source*: 2012 updates provided by editors.
*Source*: Italy Telecommunications Report. Business Monitor Q1 2013.
[1] Telefónica owns 9.47% of the Telco holding company that controls TI.

subsidiary Telecom Media, which was founded in 2003. The main shareholders of Telecom Italia are Mediobanca, Intesa San Paolo, Generali, and Telefónica, who collectively hold a 22.5% stake, and the remaining 77.5% is split among multiple retailers. Because Telefónica controls 9.47% of Telco, though, it effectively controls Telecom Italia through executive appointments.

### FASTWEB

Fastweb is a broadband provider that specializes in IPTV services. In 2007, the Swiss telecommunications company Swisscom paid US$3.8 billion US for an 82.4% stake in the company, though Fastweb retains its own stock market listing and remains a separate operating entity.

### WIND

Wind was founded in 1997 as a joint venture by Enel, the Italian national power company, France Télécom, and Deutsche Telekom. It is also the third largest mobile phone provider in Italy. Wind is owned by Wind Telecom, which belongs to the Russian telecom VimpelCom Ltd.

## Wireless Telecom

Telecom Italia decided to enter the mobile market when the mobile phone industry in Italy was deregulated in 1994 as Telecom Italia Mobile (TIM). In 2011, the value of the entire wireless telecommunications industry was US$22.6 billion, and TIM is still the main player, holding onto a 33.1% market share. The other players are Vodafone Italia with 30.2%,

## Table 8-8. Wireless Telecom (Market Shares by Revenue), 1996–2012

| | 1996 | 2001 | 2006 | 2009 | 2012 |
|---|---|---|---|---|---|
| TIM/Telecom Italia (controlled by Telco/Telefonica, Spain)[1] | 89.0 | 52.3 | 42.8 | 40.4 | 33.1 |
| Wind (Vimpelcom, Russia) | | | 13.6 | 16.9 | 22.3 |
| Blu-Omintel-ProntoItalia-Wind | 11.0 | 47.7 | Wind | Wind | |
| Vodafone Italia (UK) | | | 35.9 | 35.7 | 30.2 |
| 3 Italy (Hutchison Whampoa, Hong Kong) | | | 7.7 | 7.0 | 9.8 |
| Total Revenue (mil euro) | 3,300 | 12,100 | 18,100 | 17,700 | 13,400 |
| Total Revenue (mil US$) | 4,500 | 10,900 | 23,200 | 22,600 | 21,200 |
| C4 | 100.0 | 100.0 | 100.0 | 100.0 | 95.4 |
| C2 | 100.0 | 100.0 | 79.0 | 76.1 | 63.3 |
| HHI | 8,042 | 5,011 | 3,365 | 3,241 | 2,601 |
| *N* (>1%) | 2 | 2 | 4 | 4 | 4 |
| Noam Index | 5,687 | 3,543 | 1,682 | 1,621 | 1,301 |

*Source*: AGCOM.
*Source for 2012* updates provided by editors. Data provided from "Technology, communications and media: Italy." *Euromonitor.*
[1] Telefónica owns 9.47% of the Telco holding company that controls TI.

Wind with 22.3%, and the Hong Kong-based conglomerate Hutchison Whampoa (operating as 3 Italy) with 9.8%. The high level of competition within the wireless market indicates a substantial reduction in TIM's 1990s monopoly: the firm's market share has fallen from 89% in 1996 to 40.4% in 2009 (Table 8.8).

## INTERNET MEDIA

### Search Engines

The data available for the search engine market were gathered from Audiweb, but unfortunately, the available data only go as far back as 2008. The search engine market is not concentrated, and the largest player in Italy is Google, followed by Ask.com and Virgilio.Ask.com entered the Italian market in 2005, taking over the search engine Excite Italia in a US$7.2 million deal. Virgilio, the leading Italian search provider, was founded in 1996 as a search engine and web directory. It is owned by Matrix, a subsidiary of Telecom Italia (Table 8.9).

### Online News

The main online news media sources are owned by the traditional print publishers, such as Gruppo Editoriale L'Espresso, with *La Repubblica*, and RCS Media Group, with *Corriere della Sera*. The online news media market is less concentrated than the print one. Together, Gruppo Editoriale L'Espresso and RCS Media Group take 40.8% and 16.8% of the total business, respectively.

From an economic point of view, the new media sector is still marginal. For the time being, it has not attracted companies already in the market of "traditional" media to invest heavily in it. Nor does it offer an attractive opportunity for new players wanting to enter the media market because of high start-up costs and the limited penetration of the Internet in Italian households.

The data available for the online news media market were also gathered from Audiweb, though data are only available for 2010 and 2011 (Table 8.10).

## Table 8-9. Internet Search Engines (Market Shares by Search Volume), 2008–2009

| | 2008 | 2009 | 2012 |
|---|---|---|---|
| Google (US) | 49.4 | 50.6 | 47 |
| Ask.com (InterActiveCorp, US) | 10.0 | 8.1 | 5 |
| Virgilio Ricerca | 8.5 | 7.4 | 12 |
| Libero Ricerca | 4.6 | 5.0 | 18 |
| Yahoo! Search (US) | 5.0 | 4.9 | 6 |
| Bing (Microsoft, US) | 5.9 | 4.4 | 9 |
| Trovit | 2.4 | 2.5 | n.a |
| Nasone | 0.5 | 1.5 | n.a |
| Others | 13.7 | 15.6 | 3 |
| Total Access (mil subscribers) | 56,990 | 66,717 | 11,637 |
| C4 | 73.8 | 71.1 | 86 |
| HHI | 2,700 | 2,758 | 2,819 |
| *N* (>1%) | 7 | 8 | 6 |
| Noam Index | 1,020 | 975 | 1,151 |

*Source*: Audiweb.

*Source 2009*: Michael Bonfils, Search Engine Watch March 3, 2010 <http://searchenginewatch.com/article/2066926/Search-Around-the-World-Italy> Update provided by editors.

## Table 8-10. Online News Media (Market Shares by Monthly Unique Visitors), 2010–2011

| | 2010 | 2011 |
|---|---|---|
| Gruppo Editoriale L'Espresso | 44.6 | 40.8 |
| RCS Media Group | 16 | 16.8 |
| Nanopress (Trilud) | 7.9 | 7.2 |
| Liquida (Gruppo Banzai) | 7.5 | 6.4 |
| ANSA | 7.5 | 7.7 |
| La Stampa.it (Editrice La Stampa/FIAT) | 5.6 | 7.6 |
| Il Giornale (Berlusconi family) | 3.7 | 3.3 |
| Adnkronos | 3.1 | 2.8 |
| Quotidiano Libero (Angelucci family) | 2.7 | 2.5 |
| Il Messaggero (Caltagirone family) | 2.5 | 2.5 |
| Others | 24.9 | 19.5 |
| Total monthly unique visitors (1,000) | 44.0 | 46.0 |
| C4 | 76.0 | 72.9 |
| HHI | 2,488 | 2,188 |
| *N* (>1%) | 10 | 10 |
| Noam Index | 445 | 692 |

*Source*: Audiweb.

## CONCLUSION

Any attempt to identify an overarching trend across the main sectors of the Italian media market is impossible. The diversity of legislative and entrepreneurial history, of reference markets, openness to opportunities, and responses to technical innovations has been different for every sector.[9] For example, while for a long time broadcast television seemed to be growing ever more concentrated, for the first time in recent decades the conditions seem to encourage a slow and gradual opening of the market in favor of players other than the historically dominant RAI and Mediaset, such Telecom Italia and Sky Italia. Conversely, print ownership has for decades been less concentrated than broadcast television ownership, but in recent years a decline in advertising investments, caused by decreases in book and magazine sales, has created conditions ripe for mergers and acquisitions, which will serve to consolidate the market.

In terms of wireline and wireless telecommunications, the pan-European deregulation of the 1990s led to a diversification of operators, which reduced Telecom Italia's former monopoly over the sector. However, it should be noted that the Internet, and digital communications in general, have had a limited impact on the Italian market when compared with other European country markers. This is primarily due to low Internet usage in Italy, and because Italian firms have not moved very enthusiastically to embrace online services.

In comparing the Italian media market with other countries, there are three main concerns that raise serious questions about the viability and diversity of Italian media in the coming years.

The first concern is the incomparably high level of ownership concentration for free broadcast TV that has now persisted for over 20 years. As can be understood from the data and commentary in the previous pages, two companies control 87.2% the Italian broadcast television market: the public RAI and the Berlusconi's family's Mediaset. Sky Italia also maintains an effective monopoly over pay satellite TV. In the Italian media market, concentration is most strikingly problematic in the television sector, an acute anomaly when compared to the markets of other developed countries.

The second concern is that the high concentration of advertising investment in the field of television. For several years Italian television has received over half of the total advertising investments made. This exemplifies the anomaly that is the Italian television duopoly, and also indicates how other all other parts of the market—print, online, and radio—must compete for the rest.

The third problem is that of the political-corporate relationship in Italy. A prime example of this was Fininvest founder Silvio Berlusconi, who served as prime minister three times: in 1994, from 2001 to 2006, and again from 2008 to 2011. Berlusconi and his associates in government—a number of whom held direct stakes in his properties—were able to exert considerable influence over RAI and government regulators while in office.

Despite the problems associated with the Gasparri Law, the transition from ATV to DTT still represents an opportunity to finally break apart the RAI-Mediaset duopoly by encouraging the entry of new broadcasters into the market. However, the policies of Berlusconi's last government, which encouraged the consolidation of RAI and Mediaset, has so far limited the potential for change in television broadcast concentration. Thanks to government preference, these two leading broadcasters have obtained a large part of the new DTT complex, and in the process of increasing and diversifying their range of channels have decreased opportunities for other firms. So even though the EU applied pressure on Berlusconi's government to ameliorate this situation, the television duopoly remains and is likely to persist for years to come.

Nonetheless, EU legislation forcing its member states to privatize their public telecom companies to open the market to new competitors

9. Richeri, Giuseppe. *Economia dei media*, Rome: Laterza, 2012.

has still had a positive effect on reducing media monopolization in Italy. Although the regulatory changes should have been implemented with greater transparency, Italy has witnessed an increase in service due to increased competition from both foreign and domestic firms since the regulations were implemented. Such competition has also spurred research and development in the telecommunications sector by creating more incentives for competition. However, the success of these reforms will depend on consistent and timely political decisions affecting media enterprises, decisions that in the past have been difficult to implement due to frequent changes in government. And there is uncertainty within the sector over whether or not there will be sufficient demand to justify investments in infrastructure. In such a situation, the excessive fragmentation of the market and strong competition without proper regulation will discourage investment. The sector is definitely moving to grow high-speed broadband networks and to diversify its offerings, but it is still unclear as to who will undertake the requisite infrastructure investments. A reduction in concentration and the birth of a truly competitive media environment will depend on the market strategies of newcomers in the DTT, IPTV, web, and mobile TV sectors.

## Bibliography

### 1. Books and Articles


Balbi, Gabriele & Prario, Benedetta. "The History of Fininvest/Mediaset's Media Strategy: 30 Years of Politics, the Market, Technology and Italian Society." *Media, Culture and Society* 32.3 (2010): 391–409. <http://mcs.sagepub.com/content/32/3/391.full.pdf>.

Barca, Flavia and Marzulli, Andrea, eds. "L'Industria Della Comunicazione in Italia, XIII Rapporto." Fondazione Rosselli, Istituto di Economia dei Media. Turin: Guerini e Associati, 2011. <http://iem2011.onafets.eu/docs/pres/2011_XIII_rapporto_IEM.pdf>.

Barca, Flavia and Zambardino, Bruno, eds.). "L'Industria della comunicazione in Italia, XIV Rapporto." Fondazione Rosselli, Istituto di Economia dei Media. Turin: Guerini e Associati, 2012. <http://iem.frosselli.org/docs/pres/2012_XIV_rapporto_IEM.pdf>.

"Digital Media in pieno decollo: Video, Social Network, Tablet e Smartphone." Osservatorio New Media e Internet. Milan: Politecnico di Milano, 2012. <http://www.osservatori.net/new_media_e_tv/rapporti/rapporto/journal_content/56_INSTANCE_0HsI/10402/1014735>.

"Judgment in Case C-403/10 P Mediase: SpA v Commission." Press Release No 77/11. Court of Justice of the European Union. July 28, 2011. Dec 4, 2012. <http://curia.europa.eu/jcms/upload/docs/application/pdf/2011-2007/cp110077en.pdf>.

Marta, Cola and Benedetta, Prario. "New Ways of Consumption: The Audiences of Public Service Media in Italy and Switzerland." *Media, Culture and Society* 34:2 (2012): 181–194. <http://mcs.sagepub.com/content/34/2/181.full.pdf+html>.

Prario, Benedetta. *Le trasformazioni dell'impresa tv verso l'era digitale*. Bern: Peter Lang, 2005.

Prario, Benedetta. "Mobile Television in Italy: Value Chains and Business Models of Telecommunications Operators." *Journal of Media Business Studies*, 4, 1 (2007): 1–19.

Richeri, Giuseppe. "Broadcasting and the Market: The Case of Public Television," in Calabrese, Andrew and Sparks, Colin, eds. *Toward a Political Economy of Culture*. Maryland: Rowman & Littlefield Publishers, 2004.

Richeri, Giuseppe. *Economia dei media*. Rome: Laterza, 2012.

"*Televisioni digitali. Le nuove Sofa TV . . . e non solo.*" Osservatorio New Media e Internet. Milan: Politecnico di Milano, 2007. <http://www.osservatori.net/new_media_e_tv/rapporti/rapporto/journal_content/56_INSTANCE_0HsI/10402/98391>.


### 2. Reports


Associazione Italiana Editori (AIE). *Rapporto sullo stato dell'editoria 2011*. Milan: AIE, 2012. Annual. <http://www.aie.it/LinkClick.aspx?fileticket=VhebinUVCfg%3D&tabid=2808>.

Associazione Nazionale Industrie Cinematografiche Audiovisive e Multimediali (ANICA). *Ilcinema italiano in numeri 2009*. Rome: ANICA, 2009. Annual.

Audiweb. *Dati consumo motori di ricerca in Italia 2011*. Rome: Audiweb, 2012.

Autorità per le Garanzie nelle Comunicazioni (AGCOM). *Relazione annuale 2010*. Rome:

AGCOM, 2010. Annual. <http://www.agcom.it/Default.aspx?message=downloadpdf&DocID=119>.

Cinetel. *Dati Distribuzione 2008*. Rome: Cinetel, 2008. Annual.

Federazione Italiana Editori Giornali (FIEG). *La stampa in Italia* 2008–2010. Rome: FIEG, 2011. Annual. <http://www.fieg.it/upload/studi_allegati/La%20Stampa%20in%20Italia%202008-202010.pdf>.

Peresson, Giovanni. *Rapporto sullo stato dell'editoria in Italia 2009*. Milan: Ediser, 2009. Annual. <http://www.aie.it/Portals/21/Files%20allegati/sintesiRapporto2009_R.pdf>.

## 3. Websites

*Auditel*. Auditel, 2012. n.p. <www.auditel.it>.

*Federazione Nazionale Stampa Italiana (FNSI)*. FNSI, 2012. n.p. <www.fnsi.it>.

*Mediaset*. Mediaset, 2012. n.p. <www.mediaset.it>.

*Prima Online*. Prima Online, 2012. n.p. <www.primaonline.it>.

*Radiotelevisione italiana S.p.A. (RAI)*. RAI, 2012. n.p. <www.rai.it>.

*Sky Italia*. Sky Italia, 2012. n.p. <www.sky.it>.

*Telecom Italia*. Telecom Italia, 2012. n.p. <www.telecomitalia.it>.

*Utenti Pubblicità associate (UPA)*. UPA, 2012. n.p. <www.upa.it>.

# Italy—Data Summaries

ELI NOAM AND PAUL MUTTER

FININVEST IS THE MOST diversified multimedia producer in Italy, and it effectively monopolized the commercial part of broadcast TV (Table 8.11). Its owner, Silvio Berlusconi, leveraged this to leader of the Forza Italia party and three terms as prime minister (1994–1995, 2001–2006, and 2008–2011). With Fininvest's control over Mediaset (broadcasting), Medusa (film), Mondadori (books), plus the government's appointees to the governance boards of the public TV broadcaster RAI, Berlusconi held an extraordinary influence over TV. Although Fininvest ranks below the major platform companies nationally in revenue, it holds 23.6% of the content market (Table 8.12), comfortably ahead of its nearest rival 21st Century Fox, which is only active in film and multichannel platforms (its monopoly in multichannel, through the DBS provider Sky Italia, accounts for its 6.2% share of the content market).

Vodafone (UK) and Vimpelcom (Russia) have become significant mobile operators in the country. Telecom Italia's incumbency advantage in wireline (a declining market) is reduced in mobile services, and it is facing increased competition from foreign wireline providers such as BT (UK), 3G (Hong Kong), and Swisscom (Switzerland), so its share of platform media has fallen from 56.8% to 45% (Table 8.13).

Italy's overall media industries show a decline, mostly driven by Telecom Italia's reduced market share. The average HHI declined after 2004 from 4,042 to 3,727, and the national power index declined from 4,040 to 3,654. The combined national market share of the C4 firms of Fininvest, Sky Italia, Vodafone Italia, and Telecom Italia is quite high, 89.8%. If we look at the content sector alone, the picture changes. Of that sector, Fininvest holds 23.6% (though a bit less than five years earlier) (Table 8.12). The public broadcaster RAI (22.9%) follows. In the content media sector, the HHI concentration rose from 2,689 to 3,186 and the power index rose similarly from 2,673 to 3,101 (Table 8.13).

Table 8-11. National Media Industries Concentration in Italy

| | 2004/5 | | 2011 OR MOST RECENT | | % CHANGE ANNUAL AVERAGE | |
|---|---|---|---|---|---|---|
| | COMPANY POWER INDEX IN COUNTRY | COMPANY SHARE OF THE OVERALL NATIONAL MEDIA MARKET (%) | COMPANY POWER INDEX IN COUNTRY | COMPANY SHARE OF THE OVERALL NATIONAL MEDIA MARKET (%) | COMPANY POWER INDEX IN COUNTRY | COMPANY SHARE OF THE OVERALL NATIONAL MEDIA MARKET (%) |
| Telecom Italia1 / Telefonica | 2568.4 | 41.7 | 1581.5 | 29.7 | –5.9 | –1.5 |
| Murdoch Interests (US) | 230.9 | 2.4 | 608.5 | 6.5 | 12.9 | 0.504 |
| Vodafone (Vodafone Italia) (UK) | 472 | 14.5 | 318.7 | 12.3 | –4.8 | –0.269 |
| Fininvest2 (Berlusconi) | 239 | 6.6 | 294.3 | 8.3 | 2.6 | 0.208 |
| RAI (Public) | 258 | 5.4 | 383.1 | 8.0 | 5.1 | 0.331 |
| Vimpelcom (Wind, Russia) | 80.9 | 7.2 | 225.9 | 11.7 | 13.7 | 0.568 |
| Google (US) | 11.2 | 0.221 | 65.2 | 1.4 | 24.7 | 0.146 |
| RCS Media Group | 54.7 | 2.6 | 58.4 | 2.4 | 0.83 | –0.026 |
| Gruppo Espresso | 56.2 | 1.9 | 42.4 | 1.5 | –3.5 | –0.049 |
| Swisscom (Switzerland) | 9.9 | 1.9 | 17.9 | 2.2 | 7.8 | 0.047 |
| Hutchinson Whampoa (3G, Hong Kong) | 21.4 | 2.8 | 31.2 | 3.2 | 4.8 | 0.050 |
| British Telecom (UK) | 8.8 | 1.8 | 0.0 | 0.000 | 100.00 | –0.220 |
| GEMS | 5.3 | 0.651 | 6.2 | 0.665 | 1.8 | 0.002 |
| Il Sole 24ore | 6.0 | 0.573 | 2.9 | 0.353 | –8.4 | –0.028 |
| Caltagirone Editore | 5.9 | 0.575 | 4.4 | 0.440 | –3.6 | –0.017 |
| Poligrafici Editoriale | 4.2 | 0.47 | 2.6 | 0.314 | –5.9 | –0.019 |

| | | | | | | |
|---|---|---|---|---|---|---|
| Giunti | 2.3 | 0.429 | 2.4 | 0.414 | 0.46 | −0.002 |
| Grupo Finelco | 0.8 | 0.09 | 1.0 | 0.1 | 4.9 | 0.003 |

| Media Concentration Index | 2004/5 | 2011 or Most Recent | % Change Annual Average |
|---|---|---|---|
| Total Revenue: Nat'l Media Industry (mil US$) | 64,240 | 65,323 | 0.21% |
| Total Voices (n) | 54 | 52 | −0.46% |
| Net Voices (n) | 41 | 39 | −0.61% |
| Public Ownership (%) | 5.4 | 8.0 | 0.33% |
| Foreign Ownership (%) | | | 0.85% |
| C4 Average—Weighted | 90.1 | 89.8 | 0.0% |
| HHI Average—Weighted | 4042 | 3727 | −0.97% |
| C1 Average—Weighted | 53 | 49 | −1% |
| Noam Index Average—Weighted | 1522 | 1958 | 3.6% |
| Pooled Overall Sector C4 | 69.9 | 62.0 | −0.99% |
| Pooled Overall Sector HHI | 2103 | 1372 | −4.3% |
| Pooled Overall Sector Noam Index | 179 | 194 | 1.03% |
| Market Share of Top Ten Companies: Nat'l Media Industry (%) (Pooled C10) | 86.9 | 85.8 | −0.1% |
| National Power Index | 4040 | 3654 | −1.2% |

[1] Telefonica (Spain) controls the Teleco holding company, which controls TI, with a share of about 9%.
[2] Includes Mediaset (TV and radio), Medusa (film), and Mondadori (publishing).

Table 8-12. Top Content Media Companies in Italy

| | 2004/5 | | 2011 OR MOST RECENT | | % CHANGE ANNUAL AVERAGE | |
|---|---|---|---|---|---|---|
| | COMPANY POWER INDEX IN COUNTRY | COMPANY SHARE OF THE NATIONAL CONTENT MEDIA MARKET (%) | COMPANY POWER INDEX IN COUNTRY | COMPANY SHARE OF THE NATIONAL CONTENT MEDIA MARKET (%) | COMPANY POWER INDEX IN COUNTRY | COMPANY SHARE OF THE NATIONAL CONTENT MEDIA MARKET (%) |
| Fininvest | 879.7 | 24.3 | 840.5 | 23.6 | −0.57 | −0.085 |
| RAI(Public) | 949.7 | 19.8 | 1094.2 | 22.9 | 1.79 | 0.389 |
| Murdoch Group (21st Century Fox, Sky, US) | 284.3 | 3.3 | 574.4 | 6.2 | 9.2 | 0.370 |
| Google (US) | 41.1 | 0.812 | 186.3 | 3.9 | 20.8 | 0.394 |
| RCS Media Group | 201.2 | 9.7 | 166.9 | 6.9 | -2.3 | -0.345 |
| Gruppo Espresso | 206.8 | 7.0 | 121.0 | 4.3 | -6.5 | -0.335 |
| GEMS | 19.7 | 2.4 | 17.7 | 1.9 | -1.3 | -0.062 |
| Caltagirone Editore | 21.6 | 2.1 | 12.5 | 1.3 | -6.6 | -0.107 |
| Poligrafici Editoriale | 15.4 | 1.7 | 7.4 | 0.898 | -8.8 | -0.104 |
| Giunti | 8.5 | 1.6 | 6.9 | 1.2 | -2.7 | -0.049 |
| Media Concentration Index | 2004/5 | | 2011 or Most Recent | | % Change Annual Average | |
| Public Ownership (%) | 19.8 | | 22.9 | | 0.39% | |
| Foreign Ownership (%) | 6.1 | | 14.5 | | 1.20% | |
| C4 Average—Weighted | 75.7 | | 78.3 | | 0.3% | |
| HHI Average—Weighted | 2689 | | 3186 | | 2.31% | |
| C1 Average—Weighted | 39 | | 44 | | 1% | |
| National Power Index | 2673 | | 3101 | | 2.00% | |

Table 8-13. Top Platform Media Companies in Italy

| | 2004/5 | | 2011 OR MOST RECENT | | % CHANGE ANNUAL AVERAGE | |
|---|---|---|---|---|---|---|
| | COMPANY POWER INDEX IN COUNTRY | COMPANY SHARE OF THE NATIONAL PLATFORM MEDIA MARKET (%) | COMPANY POWER INDEX IN COUNTRY | COMPANY SHARE OF THE NATIONAL PLATFORM MEDIA MARKET (%) | COMPANY POWER INDEX IN COUNTRY | COMPANY SHARE OF THE NATIONAL PLATFORM MEDIA MARKET (%) |
| Telecom Italia/Telefónica[1] | 3525 | 56.775 | 2427.1 | 45.0 | –4.6 | –1.5 |
| Vodafone Italia (77% Vodafone, UK & 23% Verizon, US) | 648 | 19.9 | 490.4 | 18.9 | –3.4 | –0.114 |
| Murdoch Interests(US) | 210.9 | 2.109 | 626.8 | 6.578 | | 0.559 |
| VimpelCom (Wind, Russia) | 110.9 | 9.8 | 332.9 | 17.179 | 14.7 | 0.924 |
| Swisscom (Switzerland) | 13.6 | 2.6 | 27.6 | 3.5 | 9.3 | 0.112 |
| Hutchinson Whampoa (3G, Hong Kong) | 29.4 | 3.8 | 48.0 | 4.9 | 6.3 | 0.135 |

| Media Concentration Index | 2004/5 | 2011 or Most Recent | % Change Annual Average |
|---|---|---|---|
| Public Ownership (%) | 0 | 0 | 0% |
| Foreign Ownership (%) | 45.9 | 55.3 | 1.3% |
| C4 Average—Weighted | 95.3 | 96.0 | 0.1% |
| HHI Average—Weighted | 4546 | 4018 | –1.5% |
| C1 Average—Weighted | 59 | 52 | –1.0% |
| National Power Index | 4550 | 3953 | –1.6% |

[1]Telefónica (Spain) controls the Telco holding company, which controls TI.

# 9

# Media Ownership and Concentration in the Netherlands

JOOST VAN DREUNEN

## INTRODUCTION

In the Netherlands, strict anticoncentration regulation exists only on paper, despite the existence of a large number of regulatory agencies.[1] For example, when the commercial broadcaster RTL's RTL 4 channel, inaugurated in 1989, was denied a Dutch commercial operating license, it began transmitting Dutch-language content out of its headquarters in neighboring Luxembourg that proved very popular. Despite the fact that commercial broadcasting licenses were not available in the Netherlands, the government did not respond to RTL 4's actions, though eventually, RTL 4 legally entered the Dutch market when the government decided to grant commercial broadcasting licenses in 1992.[2]

Another instance of limited regulatory enforcement in the Netherlands was the 2007 Temporary Media Concentration Law, which set very high maximum market shares for individual corporate entities in print and audiovisual cross-ownership. After just three years the law was revoked, though, with the understanding

1. These include the Commissariaat voor de Media (an independent organization that oversees media regulation and advises the government on media issues), Stichting Etherreclame (a state agency that oversees commercial broadcasting on public television and radio programs), Stichting Coproductiefonds Binnenlandse Omroep (the agency that subsidizes Dutch film producers), Mediamonitor (the Dutch Media Authority arm that produces concentration surveys for the Ministry of Education, Culture, and Science), the Onafhankelijke Post en Telecommunicatie Autoriteit (the independent post and telecom authority), Nederlandse Omroep Bond (the Dutch broadcasting association), the Institute for Media Auditing, Nationaal Onderzoek Multimedia (a joint industry committee focused on print media), and Stichting KijkOnderzoek (an independent audience survey group).

2. RTL had pursued a similar strategy with French programming broadcast from Luxembourg several decades prior in order to win market shares in the French radio market. As was the case in the Netherlands, the French eventually reversed their restrictive policy.

that Dutch media industries would continue to self-regulate and have Mediamonitor, part of the Dutch Media Authority, produce annual concentration surveys.

One other important factor affecting media concentration in the country is the growing financial insolvency of state-subsidized media producers. While public and commercial producers compete fiercely for audience shares, budget cuts and layoffs are adversely affecting state enterprises. Beginning in 2013, the national media budget will be trimmed by 25% annually—US$64.5 million (50 million euros)—with the cuts gradually increasing to US$258 million (200 million euros) over the next few years.

## PRINT MEDIA

### Newspapers

According to the Commissariaat voor de Media, "[a] pluriform supply of newspapers demands a wide offering of both national and independent regional newspapers."[3] Although the market has historically been regional in nature, a decline in circulation for regional newspapers has led to marked increases in the national dailies' circulation numbers. After a failed attempt in 1988 to institutionalize regulations regarding newspaper ownership, the industry instead chose self-regulation under the condition that no single publisher be allowed to obtain a market share greater than 33%. Like Belgium and Spain, the Netherlands does not have specific legal constraints for newspaper ownership or for cross-ownership of newspapers by media groups active in other market segments.

The newspaper market has become increasingly concentrated, with the top four companies accounting for 95% of the total market shares in 2000. Since then, though, established Dutch publishers such as Telegraaf Media Groep, Wegener, and PCM have been losing market shares to foreign publishers. In 1999 the Swedish publisher Metro began circulating its eponymous free daily, and in 2004 the British private equity firm Apax acquired PCM, and Mecom (UK) became the biggest regional publisher after buying two regional newspapers and Wegener in 2007 (Table 9.1).

Table 9-1. Daily Newspapers (Market Shares by Circulation), 1990–2012

| | 1990 | 1995 | 2000 | 2001 | 2002 | 2003 | 2004 | 2005 | 2009 | 2012 |
|---|---|---|---|---|---|---|---|---|---|---|
| Telegraaf Media Groep | 19.1 | 24.5 | 29.8 | 30.7 | 33.1 | 32.8 | 33.7 | 34.2 | 29.5 | 36.9 |
| Mecom (UK)/ Koninklijke Wegener | | | 28.0 | 24.5 | 24.9 | 25.0 | 24.3 | 22.2 | 21.8 | Telegraaf |
| VNU | 17.0 | 17.8 | Mecom/Koninklijke Wegener | | | | | | | |
| Wegener | 7.3 | 14.8 | | | | | | | | |
| PCM Uitgevers | 12.3 | 30.7 | 30.4 | 30.8 | 24.6 | 24.2 | 23.1 | 19.5 | 19.9 | 22.8 |
| NDU | 17.2 | PCM Uitgevers | | | | | | | | |
| Noordelijke Dagblad Combinatie | 5.5 | 6.8 | 6.8 | 6.9 | 6.0 | 6.1 | 5.9 | 5.8 | 5.2 | 5.5 |
| Reformatorisch Dagblad | 1.2 | 1.2 | 1.3 | 1.4 | 1.2 | 1.3 | 1.3 | 1.3 | 1.2 | 1.3 |

*continued*

3. Brakman, I., van Cuilenburg, J., van der Meulen, L. "A View on Media Concentration: Concentration and Diversity of the Dutch Media in 2001." Hilversum: Commissariaat voor de Media, 2002. Jan 31, 2013. <http://www.mediamonitor.nl/dsresource?objectid=423&type=org>.

Table 9-1. *continued*

| | 1990 | 1995 | 2000 | 2001 | 2002 | 2003 | 2004 | 2005 | 2009 | 2012 |
|---|---|---|---|---|---|---|---|---|---|---|
| Nederlands Dagblad | 0.6 | 0.6 | 0.7 | 0.8 | 0.7 | 0.7 | 0.8 | 0.7 | 0.7 | 0.7 |
| Friesch Dagblad | 0.5 | 0.4 | 0.5 | 0.5 | 0.5 | 0.5 | 0.4 | 0.4 | 0.4 | 0.4 |
| Koninklijke BDU | 0.2 | 0.2 | 0.3 | 0.3 | 0.2 | 0.2 | 0.4 | 0.4 | | 0.3 |
| Others | 19.1 | 3.0 | 2.2 | 4.1 | 8.8 | 9.2 | 10.1 | 15.5 | 21.3 | 1.5 |
| Total Revenue (mil euros) | 667 | 667 | 1,196 | 1,222 | 1,196 | 973 | 968 | 960 | 791 | 660 |
| Total Revenue (mil US$) | 900 | 900 | 1,100 | 1,100 | 1,100 | 1,100 | 1,200 | 1,200 | 1,100 | 900 |
| C4 | 65.6 | 87.8 | 95.0 | 92.9 | 88.6 | 88.1 | 87.0 | 81.7 | 76.4 | 90.3 |
| HHI | 1,187 | 2,127 | 2,645 | 2,542 | 2,359 | 2,326 | 2,297 | 2,079 | 1,771 | 1,914 |
| *N* (>1%) | 7 | 6 | 5 | 5 | 5 | 5 | 5 | 5 | 5 | 4 |
| Noam Index | 456 | 886 | 1,150 | 1,105 | 1,025 | 1,011 | 930 | 903 | 792 | 957 |

*Source:* HOI Online, Mediamonitor, Datamonitor.

Total circulation in 2009 was 1.4 billion annually, down from 1.5 billion in 2001. Free dailies accounted for approximately 21% of the total circulation numbers. The market is only moderately concentrated, with an HHI of 1,914 in 2012, down from a high of 2,645 in 2000 (Figure 9.1).

## Magazine Publishing

Annual circulation in the Netherlands has declined in recent years, from 716 million copies in 2004 to 584 million in 2009.[4] There were 165 different magazine titles in circulation in 2010, slightly down from 182 in 2007. Women's magazines are the most popular titles.

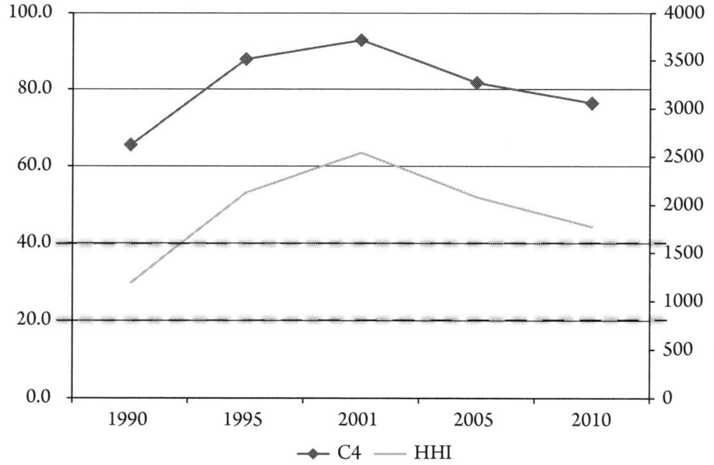


FIGURE 9.1 Daily Newspapers Concentration (HHI and C4), 1990–2010

4. Brakman, I., van Cuilenburg, J., van der Meulen, L. "A View on Media Concentration: Concentration and Diversity of the Dutch Media in 2001." Hilversum: Commissariaat voor de Media, 2002. Jan 31, 2013. <http://www.mediamonitor.nl/dsresource?objectid=423&type=org>.

Of the popular women's magazines, Sanoma Oyj (Finland) has the largest market share, with best-selling titles such as *Libelle* (445,223), *Margriet* (254,957), and *Flair* (111,251). The Telegraaf Media Groep is the second largest magazine publisher, although it publishes the most popular women's periodical, *Vrouw*, which has a weekly circulation of 736,000 (by comparison, the Dutch-language versions of the US magazines *Cosmopolitan* and *Elle* have weekly circulations of 111,415 and 71,451, respectively).

Because broadcasters in the Netherlands retain the rights to publish their programming schedules and offer them as a subscriber service, radio and television magazines are also prevalent in the market. Examples are *Veronica Magazine* (with a weekly circulation of 858,202), *Troskompas* (405,884), *Mikrogids* (400,580), *AVRObode* (302,739), and *NCRV-gids* (233,012) (Table 9.2).

The most noticeable change in terms of ownership was the launch of *Vrouw* by Telegraaf. Its success gave the company a national market

## Table 9-2. Magazine Publishing (Market Shares by Circulation), 2001–2010

| | 2001 | 2004 | 2007 | 2010 |
|---|---|---|---|---|
| Sanoma Uitgevers (Sanoma Oyj, Finland) | 39.5 | 37.3 | 26.6 | 23.8 |
| Programmabladen AKN | 21.5 | 20.8 | 18.2 | 18.0 |
| Telegraaf Media Groep | 0.3 | 0.6 | 13.2 | 14.9 |
| Veronica Uitgeverij (ProSiebenSat.1 Media, Germany) | 12.0 | 12.2 | 13.6 | 12.7 |
| Hilversumse Media Compagnie | 6.8 | 7.2 | 8.5 | 8.4 |
| Audax Publishing | 6.4 | 7.5 | 5.5 | 6.1 |
| VARA Omroepvereniging | 5.3 | 5.2 | 4.8 | 4.2 |
| Omroepvereniging VPRO | 2.6 | 2.7 | 2.8 | 3.0 |
| Evangelische Omroep | 1.7 | 2.4 | 2.1 | 2.1 |
| Reed Business | 1.9 | 1.8 | 1.9 | 1.9 |
| Chellomedia Programming | | | | 1.6 |
| Weekbladpers Tijdschriften | 1.2 | 1.2 | 1.0 | 0.9 |
| Gruner + Jahr (Bertelsmann AG, Germany) | | | 0.7 | 0.9 |
| Hachette Filipacchi Media (France) | 0.4 | 0.4 | 0.6 | 0.6 |
| Vipmedia Publishing en Services | | | | 0.3 |
| De Groene Amsterdammer | 0.2 | 0.2 | 0.2 | 0.3 |
| Pelican Magazines (Hearst Corporation) | | | | 0.2 |
| Uitgeverij Ode | | 0.1 | 0.1 | 0.1 |
| AM van Gaal Media | | 0.1 | | |
| Credits Media | 0.1 | 0.1 | | |
| EB Media/Inspirit Media | 0.0 | 0.1 | 0.1 | |
| Total Revenue (mil euros) | 1,487 | 1,226 | 1,035 | 932 |
| Total Revenue (mil US$) | 1,338 | 1,520 | 1,418 | 1,240 |
| C4 | 79.9 | 77.8 | 71.6 | 69.4 |
| HHI | 2,301 | 2,129 | 1,543 | 1,419 |
| *N* (>1%) | 10 | 10 | 11 | 11 |
| Noam Index | 742 | 686 | 448 | 428 |

*Source:* Mediamonitor.

share of 13.2% by 2007, primarily at the expense of Sanoma Oyj's titles. In addition, the appearance of several smaller magazines in recent years has contributed to an ongoing decline in concentration: HHI stood at 1,419 in 2010, down from 2,301 in 2001 (Figure 9.2).

## AUDIOVISUAL MEDIA

### Radio

The average Dutch citizen spends 200 minutes per day listening to the radio, an increase from 189 minutes per day a decade ago. National public radio stations hold total market shares of 31.6%; if regional public stations are included (12.1%), this figure rises to 43.7%. Commercial stations account for 56.3%. The Telegraaf Media Group has the largest market share of the commercial broadcasters (16.9%) due to its acquisition of Sky Radio Group. RTL Nederland, owned by the German RTL Group (Bertelsmann AG), is the second largest commercial broadcaster due to its ownership of the popular station Radio 538. In 2010, RTL also acquired Radio 10 Gold, and in early 2011 acquired Slam!FM, although the Dutch group Talpa Media Holding now owns all three of these stations (Radio 538, Radio 10 Gold, and Slam!FM).

State-subsidized stations such as Radio 2, Radio 3FM, Radio 4, and Radio 5 increased their audience shares during this period. The market shares of their commercial counterparts also increased, but this is largely the result of these stations acquiring multiple channels from their competitors. Sky Radio Group's market share peaked in 2000 at 14.7%, but then fell to 13.3%; in order to avert this downward trend, Sky acquired the stations Radio Veronica and Classic FM. Four new commercial broadcasters have appeared in the last decade, namely FunX, Radio 6, Q-Music, and 100%NL, whereas the state-owned classical music station Concertzender went off the air.

Regional radio stations in the Netherlands hold substantial market shares. Although they only account for 12.1% of total national market shares, in some provinces these stations have

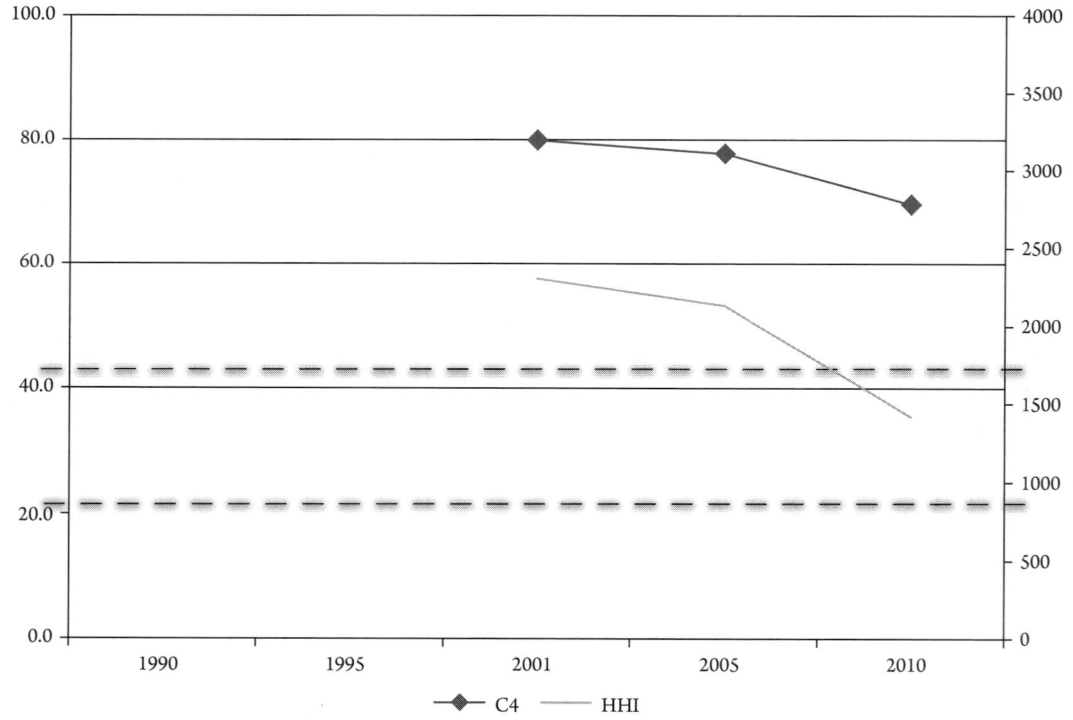


FIGURE 9.2 Magazine Publishers Concentration (HHI and C4), 2001–2010

close to 25% of the regional market shares, especially in the more rural provinces of Groningen (24.2%), Drenthe (22.1%), and Friesland (21.2%).

HHI for the Dutch radio market has historically been high. Despite a decrease from 6,956 in 1990, the concentration index stabilized at 2,428 in 2009 and then dropped to 1,674 in 2012. C4 in 2012 dropped to 71.8%, the lowest it had been in the observed period. (Table 9.3, Figure 9.3).

## Broadcast Television

Until 1992, there were no commercial TV broadcasters licensed to operate in the Netherlands (as noted previously, RTL's stations all transmitted from Luxembourg). Prior to 1992, public channels such as Nederland 1, 2, and 3 held 65% of the total market shares. Due to the entry of commercial operators in the market, this fell to 46.3% in 2012. Nederland 2 and 3 have fared the worst so far in the new market, their

### Table 9-3. Radio Group (Market Shares by Listeners), 1990–2012

| | 1990 | 1995 | 2000 | 2001 | 2002 | 2003 | 2004 | 2005 | 2009 | 2012 |
|---|---|---|---|---|---|---|---|---|---|---|
| Netherlands Public Broadcasting | 83.4 | 57.7 | 46.2 | 45.4 | 45.8 | 45.9 | 44.0 | 42.8 | 43.7 | 32.3 |
| Sky Radio Group (Telegraaf Media Group) | | | | | | | | | 17.5 | 16.9 |
| Sky Radio Ltd. | | 6.9 | 14.7 | 14.1 | 13.3 | 17.1 | 17.4 | 16.5 | | |
| De Persgroep (Belgium) | | | | | | | | 4.5 | 6.9 | 6.6 |
| RadioCorp B. V. | | | | | | | | | 3.8 | 4.4 |
| Talpa Media Holding | | | | | 4.3 | 11.1 | 9.5 | 16.7 | 2.4 | 16.0 |
| RTL Nederland (Bertelsmann AG, Germany) | | 1.3 | 4.7 | 4.3 | | | | | | Talpa |
| SBS Broadcasting (Germany), Telegraaf Media Group, Radio Noordzee (Strengholt) | | 9.0 | 5.1 | 4.8 | Talpa | | | | | |
| Wegener | | 9.2 | 8.1 | 8.1 | 7.3 | Talpa | | | | |
| Radio 538 (RTL Group; Bertelsmann AG, Germany) | | | | | | 10.7 | 11.6 | Talpa | | |
| SKY (Murdoch Family, News Corp., US/UK/Australia) | | 4.6 | 10.0 | 10.9 | 13.0 | Advent International/SKY | | | | |
| Arrow Media Group B. V. | | | 1.3 | 1.8 | 1.7 | 1.9 | 2.3 | 2.5 | 2.1 | 1.7 |
| Slam!FM | | | | | | 0.3 | 0.5 | 1.3 | 2.0 | |
| FD Mediagroep | | | | | | 0.2 | 0.3 | 0.4 | 0.9 | |
| Vereniging Veronica | | 0.9 | | | 0.3 | 0.3 | | 0.1 | 0.3 | |
| TMF Radio VOF | | | | | | | | | 0.2 | |
| 192 Media | | | | | 0.1 | | | | | |
| Bizned | | | 0.7 | 0.6 | 1.8 | 0.5 | | | | |
| Business Nieuws Holding | | | 0.3 | 0.3 | 0.3 | | | | | 4.9 |
| Foreign Commercial Broadcasters | | | | 2.3 | 2.9 | 2.4 | 2.6 | 2.8 | 4.7 | 4.9 |
| Others | 17.0 | 8.3 | 8.8 | 7.2 | 5.7 | 6.0 | 6.5 | 6.7 | 5.4 | 17.2 |
| Total Revenue (mil euros)[1] | | | | | | | 298 | 293 | 278 | 259 |

*continued*

Table 9-3. *continued*

| | 1990 | 1995 | 2000 | 2001 | 2002 | 2003 | 2004 | 2005 | 2009 | 2012 |
|---|---|---|---|---|---|---|---|---|---|---|
| Total Revenue (mil US$) | | | | | | | 369 | 373 | 386 | 355 |
| C4 | 83.4 | 82.8 | 79 | 78.5 | 79.4 | 84.8 | 82.5 | 81.6 | 78.7 | 71.8 |
| HHI | 6,956 | 3,566 | 2,567 | 2,495 | 2,543 | 2,659 | 2,505 | 2,451 | 2,428 | 1,674 |
| *N* (>1%) | 1 | 6 | 7 | 8 | 9 | 7 | 7 | 8 | 9 | 7 |
| Noam Index | 6956 | 1456 | 970 | 882 | 848 | 1005 | 947 | 867 | 809 | 633 |

*Source:* Mediamonitor.
[1] Includes kiosk sales, subscriptions, and freely distributed magazines.

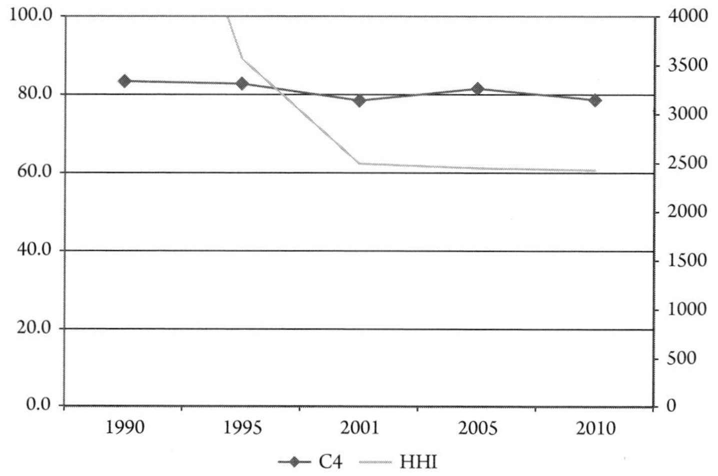


FIGURE 9.3 Radio Group Concentration (HHI and C4), 1990–2010

market shares falling from 25.6% to 9.1% and from 13.9% to 8.6%, respectively. Nederland 1 has done much better, seeing its market shares fall and then rise again (from 25.4% to 28.6%) in the same period. Despite financial problems faced by all public media operators in the Netherlands, Nederland 1 still has the largest single market share.

Among the commercial broadcasters, RTL Nederland, which owns RTL4, RTL5, Yorin, and RTL8, has maintained a market share of 22.1% over the past two decades and has increased it to 32.6%, giving it the second largest market share after Netherlands Public Broadcasting. It has accomplished this increase by introducing new specialized channels, which have offset the fact that its flagship channel, RTL4, saw its market share fall from 22.1% in 1990 to 16.3% in 2009. New entrants during this period, such as SBS Nederland, have managed to win significant market shares: SBS held a 1.8% market share in 1995 to 22% in 2009, largely as a result of SBS 6's popularity and the introduction of NET 5 and V8/Veronica.

Regional broadcasters have also seen their market shares decline since 1992; accounting for 13% of the total market in 1990, these stations accounted for only 6% of the total national market share by 2009. Despite liberalization, international broadcasters—excluding RTL—have only been able to secure a 2.3% national market share combined (Table 9.4).

HHI has declined due to the new commercial entrants, falling from 4,864 in 1990 to 3,559 in 2012. In this period, commercial broadcasters have collectively increased their total market share from 22% in 1990 to 56.3% in 2012. Over the

Table 9-4. TV Broadcasting (Market Shares by Revenue), 1984–2012

| | 1990 | 1995 | 2000 | 2001 | 2002 | 2003 | 2004 | 2005 | 2006 | 2007 | 2008 | 2009 | 2012 |
|---|---|---|---|---|---|---|---|---|---|---|---|---|---|
| Netherlands Public Broadcasting (public) | 64.8 | 47.3 | 40.8 | 40.5 | 40.2 | 39.2 | 42.2 | 40.6 | 41.0 | 39.0 | 41.2 | 40.6 | 46.3 |
| Nederland 1 | *25.4* | *15.7* | *13.3* | *13.3* | *12.4* | *12.8* | *13.0* | *14.5* | *16.4* | *22.2* | *24.9* | *23.9* | *28.6* |
| Nederland 2 | *25.6* | *17.8* | *18.0* | *18.2* | *19.2* | *17.9* | *21.5* | *18.4* | *16.9* | *8.6* | *8.0* | *8.0* | *9.1* |
| Nederland 3 | *13.9* | *13.8* | *9.5* | *9.0* | *8.5* | *8.5* | *7.7* | *7.7* | *7.8* | *8.2* | *8.3* | *8.6* | *8.6* |
| RTL Nederland (Bertelsmann AG, Germany) | 22.1 | 37.1 | 28.3 | 27.8 | 28.3 | 29.8 | 28.2 | 28.6 | 28.9 | 29.3 | 27.9 | 28.9 | 32.6 |
| RTL4 | *22.1* | *25.3* | *15.4* | *16.9* | *17.7* | *18.7* | *18.0* | *17.9* | *17.0* | *16.1* | *15.0* | *16.3* | *19.8* |
| RTL5 | *0.0* | *8.4* | *4.3* | *4.8* | *5.1* | *5.5* | *5.0* | *5.6* | *6.9* | *6.6* | *5.3* | *5.3* | *4.3* |
| Yorin/RTL7 | *0.0* | *3.5* | *8.6* | *6.1* | *5.5* | *5.7* | *5.1* | *5.1* | *5.0* | *5.7* | *5.4* | *5.4* | *6.3* |
| RTL8 | *0.0* | *0.0* | *0.0* | *0.0* | *0.0* | *0.0* | *0.0* | *0.0* | *0.0* | *0.9* | *2.1* | *2.0* | *2.3* |
| SBS Nederland (Sanoma Oyj, Finland) | 0.0 | 1.8 | 17.5 | 17.8 | 18.2 | 19.0 | 19.5 | 20.8 | 21.1 | 22.7 | 22.8 | 22.0 | 18.7 |
| SBS 6 | *0.0* | *1.5* | *10.8* | *11.5* | *10.4* | *10.9* | *11.2* | *11.9* | *12.0* | *12.5* | *12.9* | *13.0* | *10.4* |
| NET 5 | *0.0* | *0.0* | *4.2* | *4.2* | *4.7* | *4.9* | *5.0* | *5.2* | *5.0* | *5.6* | *5.8* | *5.1* | *4.3* |
| V8/Veronica | *0.0* | *0.3* | *2.5* | *2.1* | *3.1* | *3.2* | *3.3* | *3.7* | *4.1* | *4.6* | *4.1* | *3.8* | *4.0* |
| Regional Public Broadcasters | 13.1 | 13.5 | 13.3 | 13.0 | 11.2 | 9.2 | 7.4 | 7.1 | 6.4 | 6.5 | 5.9 | 6.1 | 0.0 |
| AT5 | | | | | | | | 0.2 | | | | 0.2 | 0.1 |
| RNN7 | | | | | | | | 0.1 | | | | | |
| Regional and Foreign TV Broadcasters | 0.0 | 0.3 | 0.0 | 0.9 | 2.1 | 2.7 | 2.7 | 2.6 | 2.6 | 2.5 | 2.2 | 2.2 | 2.3 |
| Total Market Share (Public Broadcasters) | 77.9 | 60.8 | 54.2 | 53.5 | 51.3 | 48.4 | 49.6 | 48.0 | 47.4 | 45.5 | 47.1 | 46.9 | 46.4 |
| Total Market Share (Commercial Broadcasters) | 22.1 | 39.2 | 45.8 | 46.5 | 48.7 | 51.6 | 50.4 | 52.0 | 52.6 | 54.5 | 52.9 | 53.1 | 53.6 |
| Total Revenue (mil euros) | | | | | | | | | | | | 826 | 860 |
| Total Revenue (mil US$) | | | | | | | | | | | | 1,148 | 1,178 |
| C4 | 100.0 | 99.7 | 100.0 | 99.1 | 97.9 | 97.3 | 97.3 | 97.1 | 97.4 | 97.5 | 97.8 | 97.6 | 97.6 |
| HHI | 4,864 | 3,800 | 2,955 | 2,899 | 2,875 | 2,880 | 3,020 | 2,955 | 3,009 | 2,945 | 3,033 | 3,009 | 3,559 |
| *N* (>1%) | 5 | 8 | 10 | 10 | 11 | 11 | 11 | 11 | 11 | 11 | 12 | 12 | 12 |
| Noam Index | 2,175 | 1,344 | 934 | 917 | 867 | 868 | 911 | 891 | 907 | 888 | 876 | 869 | 1,027 |

*Source:* Mediamonitor, Stichting ter Promotie en Optimalisatie van Televisiereclame (SPOT).

same period, public broadcasters' total market share fell from 78% to 46.4% (Figure 9.4).

## Multichannel TV Platforms

The Netherlands has only a handful of cable TV providers providing regional coverage. The largest, Ziggo, emerged from a merger in 2008 of three separate cable companies: Multikabel, Essent Kabelcomm (previously @Home Network), and Casema. The private equity firms Cinven (UK) and Warburg Pincus (US) acquired the three companies and merged them together as Ziggo in 2008, which made the new entity the largest player in the market, with 4.1 million subscribers.

The second largest cable company is UPC, which has 2.3 million subscribers. UPC, too, originated in a merger of smaller cable companies. In 1995, Philips Electronics and United International Holdings (US) acquired regional cable providers and brought them together as UPC in a joint venture. Ziggo and UPC do not actually compete outright because there is no overlap between their regional coverage areas.

The telecommunications provider KPN is the third largest player in the Dutch pay TV market. Offering the most expensive cable package—a monthly subscription to KPN costs US$59.30 (46 euros), compared to UPC US$54.80 (42.50 euros) and Ziggo $54.20 (42 euros)—KPN also has the smallest market share. It entered the cable business by acquiring Caiway and its 158,000 subscribers in 2011 (Table 9.5).

The Dutch cable television market is highly concentrated, with a C2 of 85.6% in 2010. The formation of Ziggo saw HHI increase from 2,554 in 2005 to 3,929 in 2010. Regional cable providers not affiliated with Ziggo or UPC have only a 7.6% combined national market share left (Figure 9.5).

## Video Channels

The video channels market in the Netherlands largely consists of foreign providers. MTV Networks (US), Discovery Communications (US), Eurosport (France), and National Geographic (US) all have significant market shares. However, the overall offering is, compared to US standards, limited. MTV Networks has dominated the market since 2001 and has managed to expand its overall share through a series of acquisitions, such as The Music Factory channel (Table 9.6).

HHI stood at 903 in 2001, and rose to 1,847 in 2012. Similary, C4 increased from 54.3% to 76.6% in that same period (Figure 9.6).

## Film

Historically, major film international distributors like Warner Brothers, Universal Pictures International, and Walt Disney have claimed

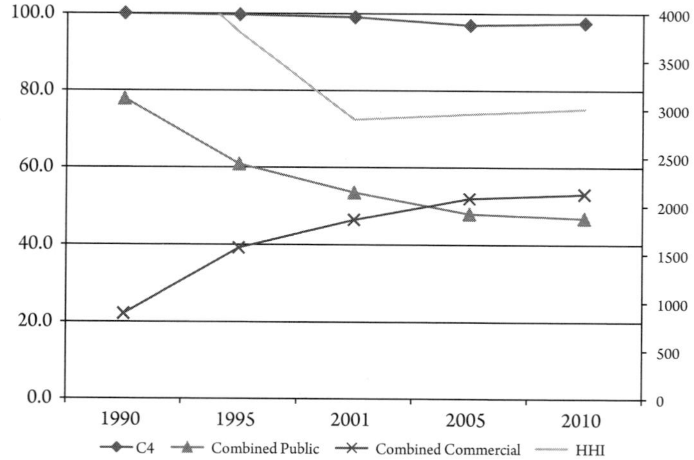


FIGURE 9.4 TV Broadcasting Concentration (HHI and C4), 1990–2010

Table 9-5. Multichannel Video Platforms: Cable MSOS, DBS, IPTV (Market Shares by Revenue), 2001–2010

| | 2001 | 2005 | 2010 |
|---|---|---|---|
| Ziggo (Zesko Holding) | 0.0 | 4.9 | 53.2 |
| Essent Kabelcom (@Home Network) | 27.0 | 28.2 | |
| Casema | 21.7 | 22.7 | Ziggo |
| Multikabel (QuickNet) | 4.9 | Ziggo | |
| UPC (Liberty Global, US) | 38.1 | 34.9 | 32.4 |
| KPN | | | 6.8 |
| Caiway (CAIW Diensten B.V) | 1.0 | 2.2 | KPN |
| Others | 7.2 | 7.2 | 7.6 |
| Total Revenue (mil euros) | 2,859 | 2,074 | 1,852 |
| Total Revenue (mil US$) | 2,573 | 2,592 | 2,463 |
| C4 | 91.8 | 90.6 | 92.4 |
| HHI | 2,682 | 2,554 | 3,929 |
| *N* (>1%) | 5 | 5 | 3 |
| Noam Index | 1,199 | 1,142 | 2,268 |

*Source:* Mediamonitor.[1]

[1]Based on a Radio Advertising Bureau (RAB) survey of a viewing sample consisting of respondents ten years of age and older for daily broadcasting hours. See van der Burg, Miriam; Lauf, Edmond; and Negenborn, Rini. "The Dutch Media." Hilversum: Commissariaat voor Media, 2011. Jan. 31, 2013. <http://www.mediamonitor.nl/dsresource?objectid=11694&type=org>.

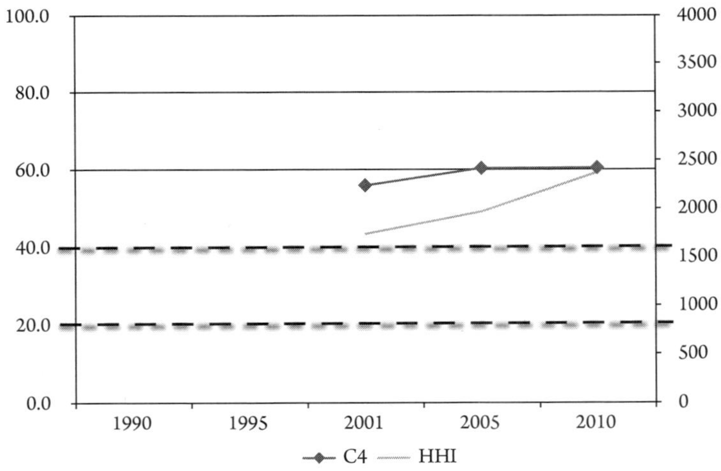


FIGURE 9.5 Multichannel Video Platforms Concentration (HHI and C4), 1990–2010

largest market shares in film distribution. However, state subsidization has led to an increase in the number of independent filmmakers and distributors over the past decade, with these independents holding a collective 14% of total box office receipts in 2008.

There have been three major mergers in the market in the past decade. The first was the global acquisition of Tristar by Sony in 2005; the second, the acquisition of RVC by Entertainment One; and the third was merger of several subsidized independent distributors

Table 9-6. Cable And DBS TV Programming Channels (Market Shares by Revenue), 2001–2012

| | 2001 | 2002 | 2003 | 2004 | 2005 | 2006 | 2007 | 2008 | 2009 | 2012 |
|---|---|---|---|---|---|---|---|---|---|---|
| Viacom International Media Networks | 22.0 | 21.3 | 26.7 | 33.1 | 25.1 | 36.5 | 43.6 | 46.4 | 31.1 | 34.1 |
| *MTV* | 0.4 | 0.5 | 0.5 | 0.8 | 0.7 | 0.7 | 0.8 | 0.7 | 0.9 | 0.7 |
| *TMF* | 0.9 | 0.8 | 0.8 | 0.6 | 0.8 | 0.6 | 0.8 | 0.7 | 0.7 | |
| *Nickelodeon* | 0.7 | 0.6 | 1.4 | 2.4 | 2.2 | 1.8 | 2.1 | 2.2 | 2.1 | 1.8 |
| *Comedy Central* | | | | | | 0.4 | 0.7 | 0.9 | 1.0 | 1.4 |
| *The Box* | | | | 2.1 | 1.7 | | | | | |
| *Kindernet* | | | | | | | | | | 0.2 |
| *TeenNick* | | | | | | | | | | 0.3 |
| Discovery Communications Benelux (Discovery Communications, US) | 14.3 | 15.7 | 15.8 | 16.1 | 15.3 | 26.0 | 23.8 | 27.8 | 19.6 | 19.5 |
| *Discovery Channel* | 1.2 | 1.3 | 1.5 | 1.5 | 1.7 | 1.7 | 1.7 | 1.8 | 1.8 | 1.6 |
| *Animal Planet* | 0.1 | 0.1 | 0.1 | 0.5 | 0.7 | 0.8 | 0.7 | 0.9 | 1.1 | 0.4 |
| Walt Disney Company | 6.6 | 6.7 | 5.0 | 4.8 | 4.5 | 7.3 | 7.9 | 8.2 | 6.1 | 15.6 |
| *Jetix/Disney XD* | 2.0 | 1.7 | 2.1 | 2.1 | 2.2 | 1.9 | 1.8 | 1.7 | 1.6 | 1.3 |
| *Disney Channel* | | | | | | | | | 0.1 | 0.6 |
| *National Geographic Channel* | 0.6 | 0.6 | 0.5 | 0.6 | 0.7 | 0.7 | 0.8 | 0.8 | 0.8 | 1.1 |
| 24Kitchen | | | | | | | | | | 0.3 |
| Eurosport (TF1 Group, France) | 9.9 | 7.9 | 7.9 | 6.5 | 5.7 | 9.4 | 8.9 | 9.3 | 5.4 | 7.0 |
| Het Gesprek | | | | | | | 0.8 | 0.9 | 0.6 | |
| Cartoon Network (Time Warner, US) | 8.1 | 2.8 | | | | | | | | |

| | | | | | | | | | | |
|---|---|---|---|---|---|---|---|---|---|---|
| Others[1] | 34.1 | 44.9 | 44.6 | 39.5 | 35.0 | 20.8 | 14.9 | 7.2 | 37.2 | 23.4 |
| Total Revenue (mil euros)[2] | 2,006 | 2,151 | 2,247 | 2,307 | 2,343 | 2,367 | 2,440 | 2,516 | 2,593 | 2,763 |
| Total Revenue (mil US$) | 2,657 | 2,850 | 2,977 | 3,057 | 3,104 | 3,136 | 3,233 | 3,334 | 3,436 | 3,661 |
| C4 | 54.3 | 51.7 | 55.4 | 60.5 | 50.6 | 79.2 | 84.2 | 91.8 | 62.2 | 76.6 |
| HHI | 903 | 822 | 1,060 | 1,439 | 934 | 2,162 | 2,553 | 3,095 | 1,432 | 1,847 |
| N (>1%) | 6 | 5 | 4 | 4 | 5 | 4 | 5 | 5 | 5 | 4 |
| Noam Index | 857 | 1,271 | 1,519 | 1,490 | 1,055 | 1,291 | 1,264 | 1,402 | 1,251 | 1,203 |

[1]"Others" includes the multisystem operators (MSOs) Zekatel, COGAS/KTMO, Nutsbedrijven Maastricht, REKAM and, Kabe.

[2]This revenue figure is an estimate based on data collected in this study: it is based on average per capita revenues reported for several other countries in this study with similar per capita income.

[3]SBS Nederland was acquired from ProSiebenSat.1 (Germany) in 2011 by Sanoma Oyj of Sweden (67%) and Talpa Media Holding (33%).

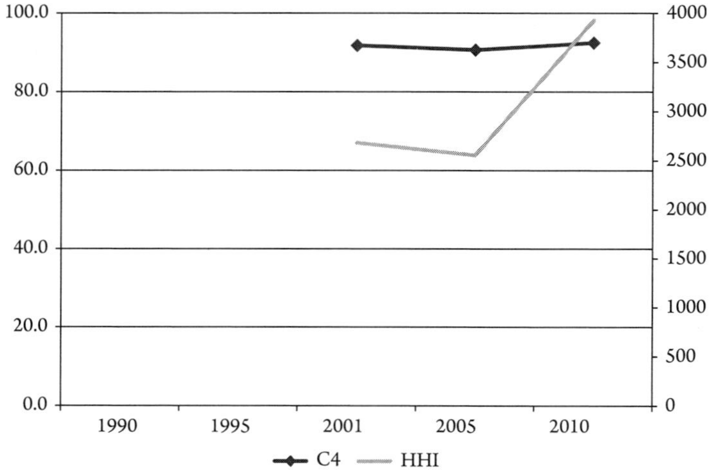


FIGURE 9.6 Video Channels Concentration (HHI and C4), 1990–2010

(Filmmuseum, Holland Film, Filmbank, and the Nederlands Instituut voor Filmeducatie) into Eye Film. These mergers have failed to either significantly reduce or increase overall concentration (Table 9.7).

C4 for the Dutch film production and distribution industry has consistently been over 62% (with the exception of 2006, when it was 55%). Similarly, the HHI indicates moderate (and declining) concentration for the entire period, reaching a high of 1,552 in 2010 and a low of 1,218 in 2005.

In terms of film exhibition, Pathé (France) has historically held the largest market share, (44.9% in 2010). The Dutch exhibitor Jogchem's Theaters ranked second with 12.2%. The Netherlands still has a large number of independent theaters, which collectively accounted for 21.7% of total box office receipts in 2010 (Table 9.8).

The concentration indices for Dutch film exhibitors are increasing, due to market consolidation. HHI rose from 1,702 in 2001 to 2,306 in 2010. C4, however, did not vary significantly during the same period.

## TELECOMMUNICATIONS MEDIA

### Wireline Telecom

Three developments in the Dutch wireline market have worn down the historically dominant market position of the state telecom provider KPN, which was fully privatized in 2006. The first development was heralded by the growth of mobile telephony in the Netherlands. The second challenge to KPN's monopoly came with the arrival of international competitors such as T-Mobile (Deutsche Telekom). Lastly, the transition to digital telephony has proven to be somewhat disadvantageous for KPN (Table 9.9).

Concentration in the Dutch wireline market has declined significantly in recent years, with HHI falling from 3,159 in 2005 to 2,150 in 2011. Concentration is will likely decrease even further as KPN loses market share to other operators.

### Wireless Telecom

Three conglomerates dominate the oligarchical Dutch wireless industry. KPN is the market leader here as well, but it is followed closely by Vodafone and T-Mobile (Table 9.10).

Overall concentration slightly declined between 2005 and 2011, but the entry barriers are high for prospective new providers.

## INTERNET MEDIA

### Internet Service Providers (ISP)

According to a 2009 Eurostat study, the Dutch are among the most active Internet users in Western Europe: 86% of the population between

Table 9-7. Film Production/Distribution (Market Shares by Box Office %), 2001–2010

| | 2001 | 2002 | 2003 | 2004 | 2005 | 2006 | 2007 | 2008 | 2009 | 2010 |
|---|---|---|---|---|---|---|---|---|---|---|
| Warner Bros. Pictures International Holland (Time Warner, US) | 17.4 | 21.4 | 12.4 | 18.1 | 19.8 | 10.2 | 19.6 | 16 | 27.2 | 29 |
| 20th Century Fox Film Netherlands (Murdoch Family, News Corp., US/UK/AUS) | 6.6 | 9.9 | 8.0 | 6.3 | 12.0 | 14.7 | 6.7 | 4.1 | Warner Bros. | |
| Universal Pictures International (Comcast/GE, US) | 27.3 | 15.2 | 20.1 | 21.3 | 22.0 | 16.3 | 22.5 | 26.9 | 16.6 | 17.1 |
| Benelux Film Distributors | | | | | | 3.3 | 3.3 | 5.4 | 7.6 | 11.8 |
| Walt Disney Pictures (US) | 12.3 | 12.4 | 19.4 | 15.2 | 12.1 | 14.8 | 11.2 | 6.5 | 11.4 | 10 |
| Independents | 3.8 | 4.9 | 5.7 | 5.4 | 12.3 | 9.1 | 7.7 | 13.5 | 5.3 | 7.4 |
| Sony Pictures Releasing Holland (Sony, Japan/US) | | | | | 5.1 | 13.6 | 5.2 | 7.8 | 10.4 | 7.4 |
| Columbia Tristar | 4.9 | 11.4 | 8.2 | 7.4 | Sony | | | | | |
| A-Film | 6.7 | 13.0 | 12.5 | 14.6 | 7.4 | 10.6 | 9.8 | 4.8 | 3.9 | 6.0 |
| E1 Entertainment Benelux (E1 Entertainment, US) | | | | | | | | | 8.8 | 5.5 |
| RCV | 13.0 | 6.3 | 9.6 | 4.3 | 3.9 | 4.5 | 8.6 | 7.3 | E1 | |
| Cinéart Nederland | | | | | | | | 1.1 | 3.4 | 1.5 |
| Paradiso Entertainment | 2.6 | 2.7 | 1.5 | 2.0 | 0.9 | 1.0 | 1.9 | 3.1 | 1.6 | 1.0 |
| Cinemien (ABC Theatrical Distribution) | 1.3 | 0.9 | 1.3 | 1.3 | 1.1 | 1.2 | 2.3 | 0.9 | 0.9 | 0.8 |
| Wild Bunch Benelux Distribution (France) | | | | | | | | | 0.8 | 0.7 |
| Eye Film Instituut Distributie | | | | | | | | | 0.4 | 0.3 |
| Filmmuseum Distributie | 0.6 | 0.6 | 0.6 | 0.9 | 0.3 | 0.3 | 0.4 | 0.4 | Eye | |
| Twin Film | | | | | | | | 0 | 0 | 0.1 |
| Moonlight | 1.0 | 0.0 | 0.3 | 0.9 | 1.1 | 0.3 | 0.1 | 0.5 | 0.2 | 0.1 |
| Amstelfilm | | | | | | | | < 0.1 | 0.1 | 0.1 |
| Others | 2.5 | 1.5 | 0.3 | 2.2 | 2 | 0.1 | 0.8 | 1.5 | 1.4 | 1.4 |
| Total Revenue (mil euros) | 150 | 157 | 163 | 154 | 135 | 156 | 160 | 165 | 201 | 219 |
| Total Revenue (mil US$) | 135 | 148 | 184 | 191 | 169 | 197 | 219 | 243 | 279 | 291 |
| C4 | 70.1 | 61.9 | 64.5 | 69.2 | 66.2 | 55.3 | 63.2 | 64.2 | 65.6 | 67.9 |
| HHI | 1,508 | 1,308 | 1,353 | 1,377 | 1,417 | 1,218 | 1,338 | 1,399 | 1,448 | 1,552 |
| *N* (>1%) | 11 | 9 | 10 | 10 | 10 | 11 | 11 | 11 | 10 | 10 |
| Noam Index | 455 | 436 | 428 | 435 | 448 | 367 | 403 | 422 | 458 | 491 |

*Source:* Nederlandse Vereniging van Film Distributeurs.

## Table 9-8. Film Exhibition (Market Shares by Box Office %), 2001–2010

| | 2001 | 2002 | 2003 | 2004 | 2005 | 2006 | 2007 | 2008 | 2009 | 2010 |
|---|---|---|---|---|---|---|---|---|---|---|
| Pathé (France) | 34.6 | 36.3 | 36.2 | 38.3 | 37.8 | 37.6 | 36.7 | 37.2 | 37.1 | 44.9 |
| Jogchem's Theaters | 17.2 | 17.3 | 16.3 | 16.7 | 15.8 | 22.2 | 20.0 | 19.2 | 18.3 | 12.2 |
| Merral | | 1.7 | | | | Jogchem's Theaters | | | | |
| Polyfilm/Utopia | 3.8 | 4.1 | 3.8 | 4.1 | 4.8 | | | | | |
| Minerva | 10.4 | 11.1 | 10.8 | 9.7 | 11.5 | 11.3 | 11.0 | 10.9 | 11.2 | 10.3 |
| Wolff | 9.4 | 9.3 | 8.6 | 8.6 | 8.8 | 9.6 | 9.6 | 10.1 | 9.5 | 5.9 |
| Independents | 21.3 | 20.1 | 20.4 | 17.0 | 16.5 | 15.3 | 18.2 | 17.2 | 18.3 | 21.7 |
| Others[1] | 3.3 | 0.2 | 3.8 | 5.6 | 4.9 | 3.9 | 4.5 | 5.3 | 5.5 | 5.1 |
| Total No. of Screens | 173 | 175 | 164 | 175 | 171 | 163 | 160 | 161 | 164 | 167 |
| Total Revenue (mil euros) | 150 | 151 | 164 | 154 | 135 | 156 | 160 | 165 | 201 | 219 |
| Total Revenue (mil US$) | 163 | 164 | 179 | 168 | 174 | 191 | 214 | 259 | 315 | 293 |
| C4 | 71.5 | 73.9 | 72 | 73.3 | 73.9 | 80.8 | 77.3 | 77.5 | 76.2 | 73.2 |
| HHI | 1,702 | 1,843 | 1,783 | 1,929 | 1,907 | 2,127 | 1,962 | 1,974 | 1,931 | 2,306 |
| *N* (>1%) | 5 | 5 | 5 | 5 | 5 | 5 | 5 | 5 | 5 | 5 |
| Noam Index | 761 | 824 | 797 | 863 | 853 | 951 | 877 | 883 | 864 | 1,031 |

*Source:* Nederlandse Vereniging van Film Distributeurs.

[1] For 2001, 2005, and 2006, the category "Others" was reported separately. "Others" consists of numbers reported by the NFC under "Overigen," and also the reported receipts of the following distributors (where available): Indies, Upstream, C-Film Distributie, and 1 More Fil.

## Table 9-9. Wireline Telecom (Market Shares by Revenue), 2005–2011

| | 2005 | 2008 | 2010 | 2011 |
|---|---|---|---|---|
| KPN | 55.0 | 50.0 | 44.8 | 41.4 |
| Ziggo (Zesko Holding) | 7.0 | 13.0 | 11.0 | 13.8 |
| UPC (Liberty Global, US) | 7.5 | 8.0 | 9.0 | 13.7 |
| Tele2 (Sweden) | 5.0 | 5.0 | 6.5 | 6.9 |
| T-Mobile (Deutsche Telekom, Germany) | 2.0 | 2.5 | 3.1 | 3.4 |
| Others[1] | 23.5 | 21.5 | 25.6 | 20.8 |
| Total Revenue (mil euros)[2] | 5,829 | 5,536 | 5,341 | 5,243 |
| Total Revenue (mil US$) | 7,286 | 7,177 | 7,104 | 7,068 |
| C4 | 75 | 76 | 71 | 76 |
| HHI | 3,159 | 2,764 | 2,263 | 2,150 |
| *N* (>1%) | 5 | 5 | 5 | 5 |
| Noam Index | 1,413 | 1,236 | 1,012 | 962 |

*Source:* Onafhankelijke Post en Telecommunicatie Autoriteit (2011).

[1] For 2001, 2005, and 2006, the category "Others" was reported separately. "Others" consists of numbers reported by the NFC for Three Lines Pictures, Multitone Films, Shooting Star Film Company, Bright Angel Distribution, and European Film Partners BV.

[2] 0.03% market share.

### Table 9-10. Wireless Telecom (Market Shares by Revenue), 2005–2011

| | 2005 | 2008 | 2011 |
|---|---|---|---|
| KPN | 40.0 | 40.0 | 35.0 |
| T-Mobile (Deutsche Telekom, Germany) | 30.0 | 30.0 | 25.0 |
| Vodafone (UK) | 20.0 | 25.0 | 30.0 |
| Others | 10.0 | 5.0 | 10.0 |
| Total Revenue (mil euros) | 4,694 | 6,118 | 7,542 |
| Total Revenue (mil US$) | 5,867 | 7,937 | 10,007 |
| C3 | 90 | 95 | 90 |
| HHI | 2,900 | 3,125 | 2,750 |
| *N* (1%) | 3 | 3 | 3 |
| Noam Index | 1,656 | 1,804 | 1,588 |

*Source:* Onafhankelijke Post en Telecommunicatie Autoriteit.

the ages of 16 and 74 reported using the Internet on a regular basis. Despite high usage rates, Dutch policymakers have so far shied away from assessing market shares among Internet access providers.

Over the past five years, the broadband Internet market in the Netherlands has consolidated. Since 2005, the sector has experienced a series of mergers and acquisitions that have brought smaller providers together, first under XS4ALL, and subsequently, under KPN, and saw Ziggo combine Multikabel, Casema, and Essent Kabelcom into a single operation. The market is now highly concentrated market with an HHI of 2,629 (up from 1,908 five years earlier) (Table 9.11, Figure 9.7).

## Search Engines

As is the case in most EU member states, Google (US) dominates the search engine market. With a virtual monopoly (93% market share in 2010), a host of smaller local and international search engines contend for the remaining 7%, a small share that Google continues to erode in its favor. In 2002, Google only held 32% of the market at the time, and Ilse, a Dutch-language search engine, claimed

### Table 9-11: Internet Service Providers (Market Shares by Revenue), 2001–2010

| | 2001 | 2005 | 2010 |
|---|---|---|---|
| KPN | 38.4 | 34.6 | 42.6 |
| Het Net | | 13.1 | KPN |
| Freeler | | | |
| Tiscali | | 5.3 | KPN |
| XS4ALL | 2.7 | 3.2 | |
| Demon NL | | 1.3 | XS4ALL |
| hcc!net | | 1.0 | |
| CistroN | | 0.1 | |
| World Online | Tiscali | | |

*continued*

Table 9-11. *continued*

| | 2001 | 2005 | 2010 |
|---|---|---|---|
| Ziggo (Zesko Holding) | | | 23.4 |
| Essent Kabelcom @home | 38.4 | 19.2 | Ziggo |
| Casema | 0.8 | 5.8 | |
| Multikabel | | Ziggo | |
| UPC (Liberty Global, US) | 15.3 | 9.2 | 13.8 |
| Tele2 (Kinnevik, Sweden) | | 2.8 | 8.4 |
| Caiway | | | |
| Others | 2.2 | 3.1 | 10.5 |
| Total Revenue (mil euros) | 609 | 1,746 | 1,926 |
| Total Revenue (mil US$) | 548 | 2,183 | 2,561 |
| C4 | 94.8 | 76.2 | 88.3 |
| HHI | 3,186 | 1,908 | 2,629 |
| *N* (>1%) | 4 | 7 | 4 |
| Noam Index | 1,593 | 721 | 1,315 |

*Source:* Mediamonitor.

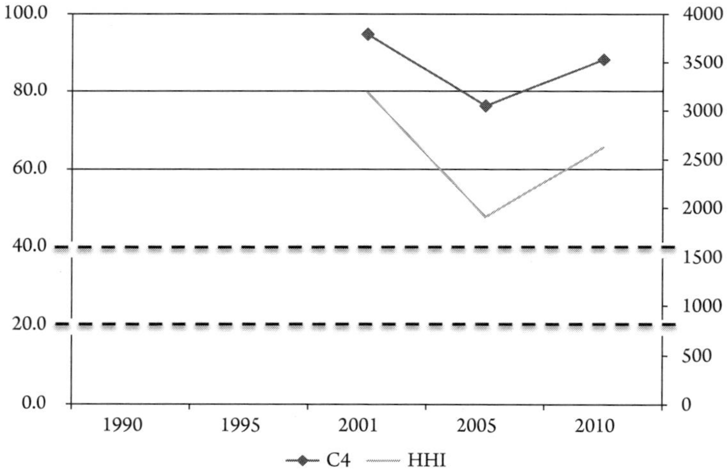


FIGURE 9.7 Internet Service Providers Concentration (HHI and C4), 2001–2010

19%. Other US search engines such as Bing, Yahoo!, and Lycos held market shares of around 8% each. Google, however, then tripled its market share in the Netherlands in just four years as part of its worldwide expansion. The HHI for the Dutch search engine market has greatly increased from 2002 (1,818) to 2010 (8,662) (Table 9.12, Figure 9.8).

## CONCLUSION

There are three main trends in the Dutch media market. The first is that print media is becoming less concentrated. Both newspapers and magazines are losing revenues due to online media growth and a decline in print advertising, and there has been a substantial decline in

Table 9-12. Internet Search Engines (Market Shares by Search Volume), 2002–2010

| | 2002 | 2003 | 2004 | 2005 | 2006 | 2007 | 2008 | 2009 | 2010 |
|---|---|---|---|---|---|---|---|---|---|
| Google (US) | 32.0 | 51.3 | 69.0 | 83.0 | 91.0 | 93.0 | 92.0 | 93.0 | 93.0 |
| MSN/Bing (Microsoft, US) | 18.0 | 16.1 | 19.0 | 8.0 | 5.0 | 3.0 | 3.0 | 3.0 | 3.0 |
| Ilse | 19.0 | 7.5 | 1.0 | 0.5 | n/a | | | | |
| Yahoo! (US) | 8.0 | 7.3 | 3.0 | 2.0 | 1.0 | 1.0 | 1.0 | 1.0 | 1.0 |
| Lycos (US) | 2.0 | 9.0 | 3.0 | 2.0 | 1.0 | 1.0 | 1.0 | 1.0 | 1.0 |
| Vinden.nl | 4.0 | 4.3 | 2.0 | 1.0 | 1.0 | 1.0 | 1.0 | 1.0 | 1.0 |
| Altavista (US) | 5.0 | 4.5 | 2.0 | 1.0 | 1.0 | 1.0 | 1.0 | 1.0 | 1.0 |
| Others | 12.0 | 0.0 | 1.0 | 2.5 | 0.0 | 0.0 | 1.0 | 0.0 | 0.0 |
| Total Revenue (mil euros) | 77 | 115 | 168 | 238 | 390 | 501 | 554 | 599 | 681 |
| Total Revenue (mil US$) | 100 | 150 | 219 | 309 | 507 | 651 | 720 | 779 | 885 |
| C4 | 77.0 | 83.9 | 91.0 | 93.0 | 97.0 | 97.0 | 96.0 | 97.0 | 97.0 |
| HHI | 1,818 | 3.120 | 5,148 | 6,963 | 8,310 | 8,662 | 8,477 | 8,662 | 8,662 |
| *N* (>1%) | 7 | 7 | 6 | 6 | 6 | 6 | 6 | 6 | 6 |
| Noam Index | 687 | 1,179 | 2,102 | 2,843 | 3,393 | 3,536 | 3,461 | 3,536 | 3,536 |

*Source:* Mediamonitor.

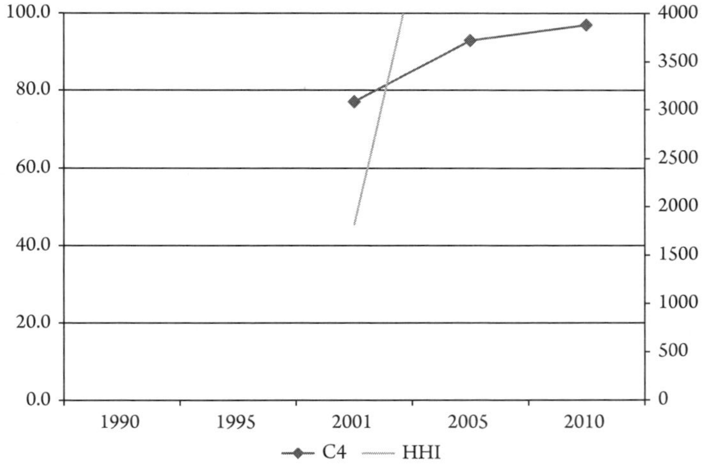


FIGURE 9.8 Search Engines Concentration (HHI and C4), 2002–2010

concentration in the sector, from 2,410 in 2001 to 1,584 in 2010 (Table 9.13).

The second trend is that concentration is increasing in the audiovisual sector. After a period of stagnation between 2001 and 2005, HHI then rose to 3,007 by 2010. The consolidation of cable TV providers has driven this trend, with the growing popularity of video channels further contributing to this upward trend.

The third trend is in online media. Broadband providers and search engines have seen concentration progress along a U-shaped curve between 2001 and 2010. After declining from

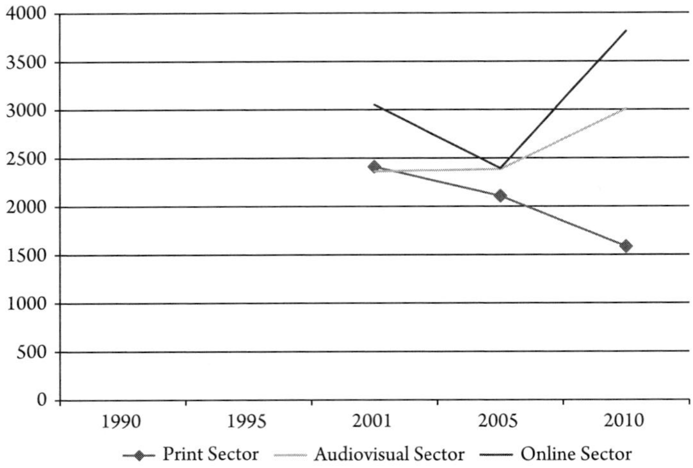


FIGURE 9.9 Aggregate Dutch Media Sectors Concentration (HHI), 2001–2010

## Table 9-13. Aggregate Dutch Media Sectors Concentration (HHI), 2001–2010

| | 2001 | 2005 | 2010 |
|---|---|---|---|
| Print Media | 2,410 | 2,107 | 1,584 |
| Audiovisual Media | 2,362 | 2,384 | 3,007 |
| Online Media | 3,051 | 2,394 | 3,811 |
| Total Revenue (mil euros) | 12,383 | 11,226 | 11,214 |
| Total Revenue (mil US$) | 11,145 | 14,032 | 14,915 |
| **Total HHI** | 2,299 | 2,264 | 2,846 |

3,051 in 2001 to 2,394 in 2005, HHI rose to 3,811 in 2010 (Figure 9.9, Table 9.13).

The Dutch media market is likely to remain highly concentrated. When taken together, the entire industry's HHI grew from 2,264 in 2005 to 2,846 in 2010, and there is a clear trend toward consolidation in most market segments. This has occurred despite the establishment of numerous monitoring agencies, although as noted earlier, the biggest regulatory problem is lack of enforcement.

# Bibliography

## 1. Books and Articles

Noam, Eli. *Media Concentration and Ownership in America*. New York: Oxford University Press, 2009.

Van Os, Michael. "Omzet tv-reclame in tweede kwartaal 2011 naar recordhoogte." *reclamewereld.blog.nl/*, 21 July 2011. Jan 31, 2013. <http://reclamewereld.blog.nl/statistieken/2011/07/21/omzet-tv-reclame-in-tweede-kwartaal-2011-naar-recordhoogte>.

## 2. Reports

Bahlmann, T.; Brakman, I.; and van Cuilenburg, J. "Mediaconcentratie in Beeld: Concentratie en pluriformiteit van de Nederlandse media 2005." Hilversum: Commissariaat voor de Media, 2006. Jan 31, 2013. <http://www.mediamonitor.nl/dsresource?objectid=320&type=org>.

Betzel, Marcel; Lauf, Edmund; Negenborn, Rini; and Vosselman Jan B. "Mediaconcentratie in Beeld: Concentratie en pluriformiteit van de Nederlandse media 2007." Hilversum: Commissariaat voor de Media, 2008. 31 Jan.

2013. <http://www.mediamonitor.nl/dsresource?objectid=8373&type=org>.

Betzel, M.; van der Burg, M.; Lauf, Edmund; and Negenborn, R. "Mediabedrijven en MediaMarkten 2001–2010." Hilversum: Commissariaat voor de Media, 2011. Jan 31, 2013. <http://www.mediamonitor.nl/dsresource?objectid=12148&type=org>.

Brakman, I., van Cuilenburg, J., van der Meulen, L. "A View on Media Concentration: Concentration and Diversity of the Dutch Media in 2001." Hilversum: Commissariaat voor de Media, 2002. Jan 31, 2013. <http://www.mediamonitor.nl/dsresource?objectid=423&type=org>.

Brakman, I.; van Cuilenburg, J.; and van der Meulen, L. "Mediaconcentratie in Beeld: Concentratie en pluriformiteit van de Nederlandse media 2001." Hilversum: Commissariaat voor de Media, 2002. Jan 31, 2013. <http://www.mediamonitor.nl/dsresource?objectid=84&type=org>.

Independent Post and Telecommunications Authority (OPTA). "Jaarverslag 2011." The Hague: OPTA, 2012. Jan 31, 2013. <http://jaarverslag2011.opta.nl/download/OPTA%20Jaarverslag%202011.pdf>.

Lambrechtsen, M.; Panenkoek, J. H.; and Tjooitink, F. "Jaarverslag 2009." Amsterdam: Nederlandse Vereniging van Bioscoopexploitanten & Nederlandse Vereniging van Filmdistributeurs, 2010. Jan 31, 2013. <http://www.filmdistributeurs.nl/files/jaarverslag_nvb_nvf_2009_website.pdf>.

Lambrechtsen, M.; Panenkoek, J. H.; and Tjooitink, F. "Jaarverslag 2010." Amsterdam: Nederlandse Vereniging van Bioscoopexploitanten & Nederlandse Vereniging van Filmdistributeurs, 2011. Jan 31, 2013. <http://www.filmdistributeurs.nl/files/JAARVERSLAG_FINAAL.pdf>.

Nederlandse Vereniging van Filmdistributeurs. "Jaarverslag 2008." Amsterdam: Nederlandse Vereniging van Filmdistributeurs, 2009. Jan 31, 2013. <http://www.filmdistributeurs.nl/files/Jaarverslag_NVF_2008.pdf>.

Stratix Consulting. "Onderzoek FM en AM Etherradio." Hilversum, Stratix Consulting, 2008. Jan 31, 2013. <http://www.stratix.nl/documents/FM_AM_etherradio.pdf>.

van der Burg, Miriam; Lauf, Edmund; and Negenborn, Rini. "The Dutch Media in 2010." Hilversum: Commissariaat voor de Media, 2011. Jan 31, 2013. <http://www.mediamonitor.nl/dsresource?objectid=11694&type=org>.

van der Voort, M. "TV Jaarrapport 2011: Never a Dull Moment." Amstelveen: Stichting ter Promotie en Optimalisatie van Televisiereclame (SPOT), 2011. Jan 31, 2013 <http://www.spot-interactive.nl/downloads/SPOT%20TV%20Jaarrapportag%202011.pdf>.

Ward, David; Fueg, Oliver C.; and D'amo, Alessandro. "A Mapping Study of Media Concentration and Ownership in Ten European Countries." Hilversum: Commissariaat voor de Media, 2004. Jan 31, 2013. <http://www.mediamonitor.nl/dsresource?objectid=435&type=org>.

# Netherlands—Data Summaries

ELI NOAM AND PAUL MUTTER

AT THE NATIONAL LEVEL, concentration has been increasing in the Dutch media market; the national power index rose from 2,378 in 2004 to 2,729 after 2008 (Table 9.14). On the other hand, the entry of free dailies by Swedish and British publishers and the growth of online media have decreased concentration (and caused revenue losses) in both the newspaper and magazine market.

One of the primary drivers for increased concentration has been the multichannel platform market. The largest cable provider, Ziggo, was formed through the merger of three separate cable companies: Multikabel, Essent Kabelcom, and Casema. Its market share in multichannel platforms increased substantially from 4.9% in 2004 to 53.2% in 2010 as a result, making it the third-largest platform media company (11.5 %) in the country and the fifth-largest content firm (4.6%).

Regulatory permission of foreign ownership in television, radio, and newspapers and lenient cross-media ownership restrictions led to an increase in foreign ownership of media companies. Vodafone (UK) and T-Mobile (Germany) have a larger market share in wireless. KPN (the telecom incumbent) is partly (23%) owned by Carlos Slim's America Móvil (Mexico), and that company made a bid in 2013 for the remainder. The RTL Group (Bertelsmann, Germany) established an alternative to the public broadcasters' monopoly, which held until 1992. RTL is now the third largest radio (10.6%) and second largest television broadcaster (32.6%) in the country, giving it control of 4.6% of all Dutch content media. After Ziggo, Liberty Global (US) is the second largest cable provider in the country with 32.4%, while US companies such as Warner Brothers, Disney, and Viacom dominate the relatively small film and video channel markets.

Public broadcasters still lead the radio and broadcast TV markets. The public broadcaster NPB is the largest content media provider with a 7.1% share as of 2013 (Tables 9.15 and 9.16).

Table 9-14. National Media Industries Concentration in the Netherlands

| | 2004/5 | | 2011 OR MOST RECENT | | % CHANGE ANNUAL AVERAGE | |
|---|---|---|---|---|---|---|
| | COMPANY POWER INDEX IN COUNTRY | COMPANY SHARE OF THE OVERALL NATIONAL MEDIA MARKET (%) | COMPANY POWER INDEX IN COUNTRY | COMPANY SHARE OF THE OVERALL NATIONAL MEDIA MARKET (%) | COMPANY POWER INDEX IN COUNTRY | COMPANY SHARE OF THE OVERALL NATIONAL MEDIA MARKET (%) |
| KPN | 1320.9 | 27.6 | 951.8 | 25.1 | –5.59 | –0.5 |
| Ziggo | 16.3 | 2.5 | 317.5 | 9.4 | 370.6 | 1.4 |
| Deutsche Telekom (T-Mobile, Germany) | 206 | 7.4 | 207 | 9.0 | 0.06 | 0.3 |
| Vodafone (UK) | 91.2 | 4.6 | 294.2 | 9.8 | 44.5 | 1.0 |
| Google (US) | 82.7 | 1.0 | 250.1 | 2.7 | 40.5 | 0.338 |
| Liberty Global (US) | 145.8 | 6.4 | 143.7 | 6.9 | –0.28 | 0.1 |
| Netherland Public Broadcasting (public) | 99.8 | 2.4 | 94.6 | 2.2 | –1.1 | –0.054 |
| Telegraaf Media Groep | 54.6 | 1.6 | 52.3 | 1.9 | –0.82 | 0.051 |
| SBS Nederland (Sanoma) (Finland) | 101.5 | 3.1 | 36.4 | 1.7 | –12.8 | –0.290 |
| Viacom (US) | 76.0 | 3.0 | 139.1 | 4.1 | 16.6 | 0.21 |
| Bertelsmann (RTL, Germany) | 36.9 | 1.4 | 42.2 | 1.4 | 2.9 | 0.003 |
| Sanoma (Finland) | 101.50 | 3.132 | 36.40 | 1.684 | –12.83 | –0.290 |
| PCM Uitgevers | 17.7 | 0.909 | 15.3 | 0.670 | –2.8 | –0.048 |
| Mecom (UK) | 23.0 | 1.0 | 0.00 | 0.000 | –20.00 | –0.207 |
| Discovery (US) | 28.2 | 1.9 | 45.5 | 2.3 | 12.2 | 0.10 |
| ProSiebenSat.1 (Germany) | 8.8 | 0.7 | 6.5 | 0.5 | –5.1 | –0.041 |
| Warner Bros. (US) | 2.6 | 0.1 | 8.0 | 0.3 | 42.1 | 0.029 |

*continued*

Table 9-14. *continued*

| | 2004/5 | | 2011 OR MOST RECENT | | % CHANGE ANNUAL AVERAGE | |
|---|---|---|---|---|---|---|
| | COMPANY POWER INDEX IN COUNTRY | COMPANY SHARE OF THE OVERALL NATIONAL MEDIA MARKET (%) | COMPANY POWER INDEX IN COUNTRY | COMPANY SHARE OF THE OVERALL NATIONAL MEDIA MARKET (%) | COMPANY POWER INDEX IN COUNTRY | COMPANY SHARE OF THE OVERALL NATIONAL MEDIA MARKET (%) |
| Comcast (US) | 0.8 | 0.1 | 0.7 | 0.08 | –2.5 | 0.002 |
| Reed Elsevier | 0.2 | 0.1 | 0.2 | 0.1 | –4.7 | 0.01 |
| Media Concentration Index | 2004/5 | | 2011 or Most Recent | | % Change Annual Average | |
| Total Revenue: Nat'l Media Industry (mil US$) | 25,729 | | 30,609 | | 3.8% | |
| Total Voices (*n*) | 60 | | 57 | | –1.0% | |
| Net Voices (*n*) | 54 | | 50 | | –1.5% | |
| Public Ownership (%) | 4.6 | | 4.1 | | –0.09% | |
| Foreign Ownership (%) | 33.7% | | 44.7% | | 2.2% | |
| C4 Average—Weighted | 80 | | 84 | | 0.9% | |
| HHI Average—Weighted | 2732 | | 2754 | | 0.2% | |
| C1 Average—Weighted | 41 | | 40 | | 0.00 | |
| Noam Index Average—Weighted | 1001 | | 1324 | | 6.5 | |
| Pooled Overall Sector C4 | 46.0 | | 53.3 | | 1.5% | |
| Pooled Overall Sector HHI | 927 | | 1001 | | 1.6% | |
| Pooled Overall Sector Noam Index | 49 | | 78 | | 12.0% | |
| Market Share of Top Ten Companies: Nat'l Media Industry (%) (Pooled C10) | 60.5 | | 73.8 | | 2.7% | |
| National Power Index | 2378 | | 2729 | | 2.9% | |

Table 9-15. Top Content Media Companies in the Netherlands

| | 2004/5 | | 2011 OR MOST RECENT | | % CHANGE ANNUAL AVERAGE | |
|---|---|---|---|---|---|---|
| | COMPANY POWER INDEX IN COUNTRY | COMPANY SHARE OF THE NATIONAL CONTENT MEDIA MARKET (%) | COMPANY POWER INDEX IN COUNTRY | COMPANY SHARE OF THE NATIONAL CONTENT MEDIA MARKET (%) | COMPANY POWER INDEX IN COUNTRY | COMPANY SHARE OF THE NATIONAL CONTENT MEDIA MARKET (%) |
| NPB (PUBLIC) | 296.1 | 7.2 | 310.6 | 7.1 | 0.9 | −0.023 |
| Ziggo | 2.4 | 0.5 | 246.7 | 4.6 | 2064.4 | 0.831 |
| Telegraaf Media Group | 161.9 | 4.8 | 171.9 | 6.2 | 1.2 | 0.270 |
| Sanoma (Finland) | 301.1 | 9.3 | 119.5 | 5.5 | -12.1 | -0.752 |
| Viacom (US) | 225.4 | 8.9 | 456.6 | 13.4 | 20.5 | 0.882 |
| Liberty Global (US) | 120.1 | 3.4 | 91.5 | 2.8 | -4.8 | -0.123 |
| PCM Uitgevers | 52.6 | 2.7 | 50.2 | 2.2 | -0.9 | -0.099 |
| Bertelsmann (RTL, Germany) | 109.7 | 4.1 | 138.7 | 4.6 | 5.3 | -0.099 |
| Discovery (US) | 83.8 | 5.5 | 149.3 | 7.7 | 0.2 | 0.4 |
| Liberty Global (US) | 120.1 | 3.4 | 91.5 | 2.8 | -4.8 | -0.123 |
| Mecom (UK) | 68.2 | 3.1 | 0.0 | 0.0 | -20.0 | -0.6 |
| ProSiebenSat.1 (Germany) | 26.1 | 2.1 | 21.5 | 1.7 | -3.6 | -0.09 |
| Reed Elsevier | 0.6 | 0.3 | 0.5 | 0.3 | -0.031 | -0.01 |
| Media Concentration Index | 2004/5 | | 2011 or Most Recent | | % Change Annual Average | |
| Public Ownership (%) | 7.2 | | 7.1 | | -0.2% | |
| Foreign Ownership (%) | 42.4 | | 55.2 | | 2.56% | |
| C4 Average—Weighted | 76 | | 82 | | 1.21% | |
| HHI Average—Weighted | 2527 | | 3044 | | 4.10% | |
| C1 Average—Weighted | 34 | | 42 | | 0.01% | |
| National Power Index | 1868 | | 2842 | | 10.44% | |

Table 9-16. Top Platform Media Companies in the Netherlands

| | 2004/5 | | 2011 OR MOST RECENT | | % CHANGE ANNUAL AVERAGE | |
|---|---|---|---|---|---|---|
| | COMPANY POWER INDEX IN COUNTRY | COMPANY SHARE OF THE NATIONAL PLATFORM MEDIA MARKET (%) | COMPANY POWER INDEX IN COUNTRY | COMPANY SHARE OF THE NATIONAL PLATFORM MEDIA MARKET (%) | COMPANY POWER INDEX IN COUNTRY | COMPANY SHARE OF THE NATIONAL PLATFORM MEDIA MARKET (%) |
| KPN | 1992.8 | 41.6 | 1366.9 | 35.9 | −6.3 | −1.2 |
| T-Mobile (Germany) | 311.3 | 11.2 | 297.7 | 12.9 | −0.9 | 0.3 |
| Ziggo | 23.3 | 3.5 | 348.5 | 11.5 | 278.8 | 1.6 |
| Vodafone (UK) | 137.6 | 6.9 | 423.1 | 14.1 | 41.5 | 1.4 |
| Liberty Global (US) | 158.8 | 7.9 | 166.6 | 8.7 | 0.98 | 0.2 |
| Media Concentration Index | 2004/5 | | 2011 or Most Recent | | % Change Annual Average | |
| Public Ownership (%) | 3.2 | | 2.8 | | −0.1 | |
| Foreign Ownership (%) | 29.3 | | 40.0 | | 2.1 | |
| C4 Average—Weighted | 82 | | 85 | | 0.7 | |
| HHI Average—Weighted | 2837 | | 2628 | | −1.5 | |
| C1 Average—Weighted | 45 | | 39 | | −1% | |
| National Power Index | 2635 | | 2679 | | 0.34 | |

# 10

# Media Ownership and Concentration in Portugal

PAULO FAUSTINO

## INTRODUCTION

Media concentration, and its impact on the pluralism of information, constitutes one of the most important and hotly debated trends of the Portuguese media industry. This report focuses on the major market segments of that sector: newspapers, magazines, television, radio, online publications, and cinema. It aims to objectively quantify the issue of media concentration. For that purpose, diverse information was gathered from accounting data, official statistics, and yearbooks (e.g., APCT and Marktest). Broadly, the information was collected with the purpose of attaining three distinct concentration indicators: the C4 (Concentration Ratio of the four largest companies), the HHI (Herfindhal–Hirschman Index), and the Noam Index. These are applied individually and collectively to provide objective measures of media concentration and contribute to enlighten the debate on the issue of concentration in Portuguese media markets using consistent indicators and analytical data.

The chapter is divided into two main parts. Since we cannot look at the media business without relating it to the telecommunications industry, the indices mentioned were applied to both. The first part of the chapter regards media concentration indicators and general information about media policies and regulation. The second part focuses on the communications sector concentration indicators and provides some data about the regulatory challenges of the relation between media and telecom companies, especially in terms of the convergence trend. The last section features a summary of the main approaches toward the media and communications market structure and some conclusions about the concentration situation and its implications.

First, it is necessary to define the assumptions and the concepts. Next, we must quantify this information, including the market positions of the respective groups. Finally, we will analyze that information with a dynamic perspective to take into account the impact on information and communication technology (ICT) and the market trends. This process requires us to answer the following fundamental questions:

- Is there a concentration trend in the media business of Portugal? If so, to what extent?
- What are the analytical criteria that support this hypothesis?
- What are the types of concentration?
- What are the advantages and disadvantages of this phenomenon?
- Who are the companies or groups that benefit from concentration?
- Which media channels are subject to high levels of concentration?

Many questions remain unanswered or need to be addressed in more detail when debating the media business concentration in Portugal. Generating the data and defining the correct underlying assumptions of their interpretation are the main challenges in answering the questions just outlined.

As is common all over the developed world, Portugal's media outlets have experienced major mergers and business partnerships in recent years. These moves resulted in a small number of conglomerates that control a significant part of the Portuguese media industry. Therefore, discussions about the concentration of media in Portugal are increasingly urgent. In political, legal, academic, and social spheres, as well as among the general public, there are different opinions about the impact of this phenomenon. It is safe to say that all parties believe that there is significant concentration. This discussion is not new and resembles debates about this topic being waged in other countries. Rather than entering a polarized debate on the financial and societal impact of media concentration, it is important to analyze this problem without dogmas and with the understanding and belief that media companies can grow based on a free-market economy as long as certain rules that safeguard the public interest are respected.

The issue of media concentration has frequently been debated by the Portuguese government over the past few years. In 2009, the Socialist Party presented the Law of the Pluralism and Non-Concentration of Media. According to the daily newspaper *Público,* the law intended to "prevent the Government, the regional governments and the autarchies from owning media organs, with the exception of radio and television public service" (Público, 01/23/2009). It was proposed by the government, but rejected by the Assembly of the Republic in October 2008. This law intended to respond to the constitutional provision that it is the state's responsibility to assure media's independence regarding political and economic powers, and to prevent media companies' concentration as well as to disclose media companies' ownership and means of financing.[1] The opposition Social Democratic Party stood against this bill and argued that it did not provide a clear, broad, and strong response to the needs of pluralism in an emerging participatory and deliberative democracy. They criticized this statute for being behind the times because the European Commission was already promoting the establishment of reliable criteria and objective indicators about pluralism. The national president raised this point to justify his veto of the law. Despite the rejection of the proposed Law of the Pluralism and Non-Concentration of Media, the media concentration debate continues. There have been situations that led the regulatory entities involved in the sector to intervene: Autoridade da Concorrência (AdC) focuses on economic market regulation, Autoridade Nacional das Comunicações (ANACOM) focuses on technical regulation of communications, and Entidade Reguladora da Comunicação Social (ERC) focuses on media content regulation. One situation that required regulatory intervention involved the Ongoing Group. It was forbidden from holding simultaneous positions above 20% in two private free-to-air TV channels. In addition, the Media Companies

1. According Article 38, paragraphs 3 and 4, and Article 39, paragraph 1, (b) and (c) of the Portuguese Republic Constitution.

Table 10-1. Advertising Spending by Media 2002–2010 (Thousands €)

| MEDIA | 2002 | 2003 | 2004 | 2005 | 2006 | 2007 | 2008 | 2009 | 2010 | VARIATION 2010–2002 |
|---|---|---|---|---|---|---|---|---|---|---|
| Press | 449,815 | 592,549 | 675,473 | 700,606 | 733,912 | 813,891 | 835,223 | 741,717 | 721,939 | 272,124 |
| Radio | 169,880 | 172,080 | 189,031 | 187,322 | 184,883 | 183,455 | 178,760 | 196,229 | 200,128 | 30,248 |
| TV | 1,528,408 | 1,672,374 | 1,892,813 | 2,483,635 | 2,840,206 | 3,085,068 | 3,330,911 | 3,517,230 | 3,797,464 | 2,269,056 |
| Cinema | n.d. | 9,903 | 13,392 | 13,596 | 14,491 | 21,976 | 23,427 | 22,864 | 23,485 | 13,582 |
| Outdoor | 170,871 | 200,145 | 224,770 | 250,590 | 276,730 | 282,553 | 303,504 | 311,407 | 294,186 | 123,315 |
| Total | 2,318,974 | 2,647,051 | 2,995,479 | 3,635,749 | 4,050,222 | 4,386,943 | 4,671,825 | 4,789,447 | 5,037,202 | 2,718,228 |

*Source*: Media Monitor/Marktest.

Ownership Transparency Initiative, promoted by ERC, forces the industry to provide detailed information about its financial and shareholder structures. This initiative promotes the accountability of media companies.

From the market's perspective, Portugal (with a population of 10 million people) is a relatively small market in terms of media and advertising investment. According to data by Mediamonitor (owned by Marktest), advertising spending has been growing,[2] with television dominating about 72% of the market (Table 10.1).

The main media companies in Portugal are mostly owned by domestic groups and investors, except in the case of Media Capital, which in 2005 was acquired by PRISA Group (Spain) and Bertelsmann (Germany). The latter sold its participation in 2006. Here is a list of media companies with the greatest ownership turnover and whose assets are described throughout the report and market share analysis:

1. ***ZON Multimédia***—Predominantly owned by a group of domestic financial companies (banks like CGD, BES, and BPI, as well as other investors). It is the leading distributor of cable TV.
2. ***Impresa***—Predominantly owned by a group of domestic financial companies. Its main activity areas are television, magazines, newspapers, and distribution.
3. ***Media Capital***—Dominated by PRISA Group (Spain, now UK). Its main activities are television and radio.
4. ***Cofina***—Owned by domestic financial companies and focused on the print segment, particularly newspapers and magazines.
5. ***Controlinveste***—Domestic company whose main assets rely on daily press and relevant stakes in sport channels of pay TV.
6. ***Renascença***—Owned by the Catholic Church; its main assets are three radio channels whose audiences are the most significative.
7. ***Rádio e Televisão de Portugal, SGPS***—Owned by the State. Its main asset is television, in addition to a news agency and radio stations.
8. ***Sonaecom***—Subholding of a domestic firm that operates in mobile telecommunications and the daily press.
9. ***Ongoing***—Owned by financial companies, it has a strong presence in the economic press, including a TV channel that has also invested in the Brazilian market.
10. ***Portugal Telecom/MEO***—The main telecommunications company with investment from large companies with a national presence. It focuses on pay TV.

As in other countries, Portugal has market dynamics that increasingly endanger the market position of the companies that have historically ruled the industry. In the last decade, new players, such as Cofina, Controlinveste, and Ongoing, have solidified their market positions. On the other hand, there is a new trend that illustrates the increasing presence and importance of telecommunication operators, which are becoming more active in the media sector. They have also caused media convergence through content distribution platforms, particularly in pay TV content distribution. Both Portugal Telecom and Sonaecom have demonstrated interest in the content industries, mainly through the development of platforms for paid content distribution. After the government regulators forced it to divest and sell its cable business (particularly the TV Cabo company that now operates as ZON), Portugal Telecom has shown a clear interest and investment in pay TV through the creation of the MEO channel in 2008. Within three years it reached over 1 million subscribers (ANACOM, 2012).

# PRINT MEDIA

## Newspapers

This section examines the degree and the evolution of concentration in the Portuguese newspapers market. For purposes of this study this corresponds to the market of Portuguese

2. According to Marktest Report. The data are presented without commercial discounts, which are usually above 50%.

newspapers with national coverage,[3] including those that focus on general information, sports, and national daily news. To assess the degree and the evolution of concentration in the national daily newspaper market, the variable of circulation was considered. Drawing on these market and share definitions, the list of companies that participate in the Portuguese newspapers market and their respective daily circulation averages are presented in Table 10.2.

The circulation averages of Portuguese daily national newspapers from 2003 to 2010 reveal two interesting trends: the average circulation of paid newspapers (e.g., *Diário de Notícias* or *Jornal de Notícias*) has stagnated or declined while the total circulation (including both paid and free newspapers) has increased significantly in the last decades, thus illustrating the remarkably positive performance of free newspapers in the last decade. However, it should be noted that in 2009, the performance of free newspapers was less impressive, with free newspapers experiencing a small reduction in circulation following the crisis in the advertising markets and the subsequent reduction of free newspapers' revenues. However, the free newspapers recovered in 2010 and increased their market share. Table 10.2 aggregates the market shares of newspapers under the control of the same media group, stressing the relative importance of each media group that participates in the Portuguese newspapers market. In this context, Table 10.2 also reveals the extent of concentration in the industry of national daily Portuguese newspapers (in terms of circulation average), depicting the evolution of the three concentration indicators (C4, HHI and Noam index) for the period of 2003–2010. The concentration indexes described in Table 10.2 reveal two important findings:

1. The market of Portuguese national dailies corresponds to a highly concentrated market.
2. The market grew and diversified (excluding 2008) from 2003 to 2010.

So far, we have evaluated concentration in the Portuguese market of national dailies by examining circulation numbers. But newspapers operate in a two-sided market. Newspapers' core business comprises two distinct but interrelated markets: the circulation market, where newspapers sell content to readers, and the advertising market, where newspapers sell advertising space and contents to advertisers. To evaluate the extent of concentration in the market of Portuguese national dailies, the previous concentration measures of circulation must be complemented with concentration measures of the advertising market. To this end, Table 10.3 depicts data on newspapers' advertising revenues from 2007 to 2010. Table 10.3 also shows the newspapers' market shares in the advertising market by newspaper and by communication group, as well as the concentration indices.

As shown in Table 10.3, Controlinveste has been losing market share in advertising investment. On the contrary, Cofina increased the most in newspaper advertising revenues. In 2009, *i informação*, a new daily owned by a small Portuguese media group (Sojormedia) with an important position in regional and local press, entered the market. It is not shown in Table 10.3 because its advertising revenues are unknown.

The situation of the advertising market seems to mimic the outcomes of the circulation market, which is not surprising given the interrelations between circulation and advertising dollars: advertisers seek advertising space, content relevance, and a media outlet's ability to reach the advertiser's key audience. Consequently, the larger a newspaper's circulation and audience relevant to the advertiser, the more attractive it becomes to advertisers. Given the high degree of concentration of the circulation market, we expect to observe a highly concentrated newspaper industry in terms of advertising revenues. Two groups, Cofina and Controlinveste, dominate about 77% of the market. To confirm the validity of this hypothesis, the concentration

3. For lack of data, the newspaper *Bola* has been excluded from the concentration analysis in terms of average circulation. Despite the recognized importance of this newspaper, data on its circulation are not available as this newspaper is not audited by APCT—Associação Portuguesa Controlo de Tiragem (www.apct.pt). However, this newspaper has a similar audience to that of the newspaper *Record*, according to official data from Marktest. So in this perspective, it was possible to assume a similar circulation performance. In this case, the concentration indexes will decline substantially because it belongs to an independent company.

Table 10-2. Market Share of Daily Average Circulation for Portuguese Newspapers

| | 2003 | | 2005 | | 2007 | | 2009 | | 2010 | |
|---|---|---|---|---|---|---|---|---|---|---|
| | VALUE | % | VALUE | % | VALUE | % | VALUE | % | VALUE | % |
| Controlinveste | 247,537 | 42.9 | 232,688 | 31.0 | 355,149 | 30.6 | 289,240 | 35.6 | 147,126 | 22.3 |
| 24 Horas | *50,824* | *8.8* | *51,181* | *6.8* | *36,923* | *3.2* | *29,262* | *3.6* | | |
| Diário de Notícias | *50,794* | *20.5* | *37,992* | *5.1* | *37,759* | *3.3* | *34,063* | *4.2* | *30,670* | *4.7* |
| Global Notícias | | | | | *150,575* | *13.0* | *104,955* | *12.9* | | |
| O Jogo | *40,677* | *16.4* | *44,878* | *6.0* | *35,976* | *3.1* | *30,068* | *3.7* | *29,945* | *4.5* |
| Jornal de Notícias | *105,242* | *42.5* | *98,637* | *13.1* | *93,916* | *8.1* | *90,892* | *11.2* | *86,511* | *13.1* |
| Cofina | 260,830 | 45.2 | 332,911 | 44.3 | 462,459 | 39.8 | 433,343 | 53.4 | 427,036 | 64.8 |
| Correio da Manhã | *114,643* | *19.9* | *118,254* | *15.8* | *118,859* | *10.2* | *122,264* | *15.1* | *129,080* | *19.6* |
| Jornal Destak | *53,567* | *9.3* | *118,803* | *15.8* | *172,261* | *14.8* | *98,585* | *12.1* | *111,643* | *16.9* |
| Jornal de Negócios | *7,731* | *1.3* | *8,890* | *1.2* | *8,261* | *0.7* | *10,668* | *1.3* | *10,530* | *1.6* |
| Record | *84,889* | *14.7* | *86,964* | *11.6* | *76,582* | *6.6* | *72,570* | *8.9* | *70,721* | *10.7* |
| Meia Hora | | | | | *86,496* | *7.4* | *23,301* | *2.9* | | |
| Metro Portugal | | | | | | | *105,955* | *13.1* | *105,062* | *15.9* |
| Média Capital | | | 121,134 | 16.1 | 176,357 | 15.2 | | | | |
| Metro Portugal | | | *121,134* | *16.1* | *176,357* | *15.2* | | | | |
| Sonae | 56,239 | 9.8 | 50,701 | 6.8 | 44,169 | 3.8 | 38,229 | 4.7 | 35,119 | 5.3 |
| Público | *56,239* | *9.8* | *50,701* | *6.8* | *44,169* | *3.8* | *38,229* | *4.7* | *35,119* | *5.3* |
| IGD | | | | | 87,818 | 7.6 | | | | |
| Diário Desportivo | | | | | *87,818* | *7.6* | | | | |
| Megafin | | | | | 22,935 | 2.0 | 27,103 | 3.3 | 23,217 | 3.5 |
| OJE | | | | | *22,935* | *2.0* | *27,103* | *3.3* | *23,217* | *3.5* |

| | | | | | | | | | | |
|---|---|---|---|---|---|---|---|---|---|---|
| Económico SGPS | 12,299 | 2.1 | 13,557 | 1.8 | 13,534 | 1.2 | 15,222 | 1.9 | 16,088 | 2.4 |
| Diário Económico | *12,299* | *2.1* | *13,557* | *1.8* | *13,534* | *1.2* | *15,222* | *1.9* | *16,088* | *2.4* |
| I Informação | | | | | | | 8,959 | 1.1 | 10,539 | 1.6 |
| Total | 576,905 | 100.0 | 750,991 | 100.0 | 1,162,421 | 100.0 | 812,096 | 100.0 | 659,125 | 100.0 |
| Revenue (USD m) | — | | — | | 347 | | 349 | | 325 | |
| CR4 | 100.0 | | 98.2 | | 93.1 | | 97.0 | | 96.0 | |
| HHI | 3,984.8 | | 3,234.1 | | 2,823.2 | | 4,154.0 | | 4,745.1 | |
| Noam | 1,992.4 | | 1,446.4 | | 1,067.1 | | 1,695.9 | | 1,937.2 | |
| *N* (>1%) | 4 | | 5 | | 7 | | 6 | | 6 | |

*Source*: Elaborated by the author using data from APCT.

## Table 10-3. Newspapers' Advertising Revenues at Official Ad Rates (Thousands €)

| NEWSPAPERS/ MEDIA GROUPS[1] | 2007 | | 2008 | | 2009 | | 2010 | |
|---|---|---|---|---|---|---|---|---|
| | REVENUES | % | REVENUES | % | REVENUES | % | REVENUES | % |
| Controlinveste | 132,961 | 38.4 | 147,211 | 39.9 | 138,839 | 39.8 | 114,243 | 35.2 |
| 24 Horas | *13,434* | *3.9* | *13,604* | *3.7* | *15,574* | *4.5* | *7,114* | *2.2* |
| Diário de Notícias | *39,211* | *11.3* | *40,018* | *10.8* | *35,368* | *10.1* | *32,417* | *10.0* |
| O Jogo | *7,104* | *2.1* | *7,899* | *2.1* | *7,597* | *2.2* | *7,507* | *2.3* |
| Jornal de Notícias | *70,015* | *20.2* | *68,775* | *18.6* | *64,183* | *18.4* | *59,400* | *18.3* |
| Global Notícias | *3,197* | *0.9* | *16,915* | *4.6* | *16,117* | *4.6* | *7,805* | *2.4* |
| Cofina | 120,464 | 34.8 | 129,771 | 35.1 | 146,679 | 42.0 | 137,752 | 42.4 |
| Correio da Manhã | *64,208* | *18.5* | *65,334* | *17.7* | *60,379* | *17.3* | *59,447* | *18.3* |
| Jornal Destak | *32,863* | *9.5* | *40,671* | *11.0* | *31,277* | *9.0* | *27,451* | *8.5* |
| Jornal de Negócios | *8,369* | *2.4* | *9,05* | *2.5* | *11,244* | *3.2* | *13,287* | *4.1* |
| Record | *15,024* | *4.3* | *14,716* | *4.0* | *17,55* | *5.0* | *18,030* | *5.6* |
| Metro Portugal | | | | | *26,229* | *7.5* | *19,537* | *6.0* |
| Sonae | 34,736 | 10.0 | 31,34 | 8.5 | 27,687 | 7.9 | 25,556 | 7.9 |
| Público | *34,736* | *10.0* | *31,34* | *8.5* | *27,687* | *7.9* | *25,556* | *7.9* |
| Media Capital[2] | 29,015 | 8.4 | 28,984 | 7.9 | | | | |
| Metro Portugal | *29,015* | *8.4* | *28,984* | *7.9* | | | | |
| Económica SGPS | 19,712 | 5.7 | 21,177 | 5.7 | 20,503 | 5.9 | 31,274 | 9.6 |
| Diário Económico | *19,712* | *5.7* | *21,177* | *5.7* | *20,503* | *5.9* | *31,274* | *9.6* |
| Socied. Vicra Desportiva, Lda | 9,63 | 2.8 | 10,844 | 2.9 | 15,55 | 4.5 | 15,946 | 4.9 |
| A Bola | *9,63* | *2.8* | *10,844* | *2.9* | *15,55* | *4.5* | *15,946* | *4.9* |
| Total | 346,518 | 100.0 | 369,327 | 100.0 | 349,258 | 100.0 | 324,771 | 100.0 |
| *N* (>1%) | 6 | | 6 | | 5 | | 5 | |
| C4 | 91.5 | | 91.3 | | 95.6 | | 95.1 | |
| HHI | 2,892 | | 2,998 | | 3,461 | | 3,215 | |
| Noam | 1,181 | | 1,224 | | 1,548 | | 1,437 | |

*Source*: Elaborated by the author based on data from Marktest.

[1]The newspapers *OJE* and *Diário Desportivo*, both active in the market since 2007, were excluded for lack of data.

[2]The newspaper *Metro* was sold to Cofina in 2009, which is why Media Capital no longer has a presence in the market of daily newspapers from that year on.

indexes (C4, HHI, and Noam Index) have been computed and reproduced in Table 10.3.

## Book Publishing

According to the study "Book Publishing and Commercialization," prepared by Observatório das Actividades Culturais (OAC), the book publishers have earned 400 million euros, although a GFK study states a turnover of 380 million euros. Given the difficulty of an extensive analysis of the numerous small book companies,[4] this study is based on the eight major companies. However, as the considered earnings stand at 380 million euros, the sample represents about 50% of the book publishing industry. The analysis is based on publishers Porto Editora, Lda. and Leya, S.A., which are mainly responsible for the sector, representing about 50% of the revenue of the total publishing market (Marktest, 2012).

Unlike other media markets, Portuguese book publishers are not integrated into large media groups but consist of autonomous book publishing companies. Foreign ownership has been decreasing (although Bertelsmann has acquired Círculo de Leitores), especially after Porto Editora acquired Bertrand, whose core business is book retailing. Leya is the only Portuguese book publisher partially owned by national and foreign companies with links to Nicolas Berggruen, CEO of Liberty Acquisition Holdings, which has controlled PRISA Group since 2010. The total revenue earned by these seven publishers shows an increase during this period. The publishers Porto Editora, Lda., and Leya S.A. are primarily responsible for this fact, followed by Círculo de Leitores, S.A. (bought by Porto Editora in 2011), and Editora Planeta de Agostini, S.A. (Table 10.4)

Considering a universe of seven publishers, the C4 was 91.3% in 2010, a high value due to

### Table 10-4. Earnings of the Main Publishers (€)

| COMPANY | 2008 | | 2009 | | 2010 | |
|---|---|---|---|---|---|---|
| | REVENUE | %* | REVENUE | %* | REVENUE | %* |
| Editorial Presença | 8,521,591 | 4.4 | 7,735,185 | 4.1 | 8,118,019 | 4.2 |
| Leya SA | 41,878,706 | 21.8 | 51,014,616 | 26.9 | 56,173,063 | 28.9 |
| Porto Editora Lda | 83,023,906 | 43.1 | 87,097,071 | 46.0 | 92,743,376 | 47.8 |
| Coimbra Editora S.A | 4,006,012 | 2.1 | 4,339,461 | 2.3 | 4,242,367 | 2.2 |
| Gradiva-Publicações Lda. | 4,899,600 | 2.5 | 4,234,838 | 2.2 | 4,547,804 | 2.3 |
| Editora Planeta de Agostini, S.A. | 22,695,516 | 11.8 | 16,453,965 | 8.7 | 13,291,869 | 6.9 |
| Círculo de Leitores, S.A (Bertelsmann, Germany) | 27,560,136 | 14.3 | 18,462,749 | 9.8 | 14,953,805 | 7.7 |
| Total | 192,585,467 | 100.0 | 189,337,885 | 100.0 | 194,070,302 | 100.0 |
| Revenues (US $) | 283,363,708 | | 264,027,894 | | 257,606,202 | |
| N (>1%) | 7 | | 7 | | 7 | |
| C4 | 91.0 | | 91.4 | | 91.3 | |
| HHI | 2,705 | | 3,040 | | 3,256 | |
| Noam | 1,023 | | 1,149 | | 1,231 | |

*Source*: Elaborated by the author based on data obtained from annual reports.
*Market share.

4. There are about 250 publishers, according to the Portuguese Association of Publishers and Booksellers.

the significant earnings of one of the companies. Despite the variations from year to year, the C4 index shows a high degree of concentration (C4 > 50%) of the four main publishers in the market. From 2008 to 2010, HHI grew, justified by the sector's increasing concentration. In 2010, HHI was 3,256, while in 2008 it was 2,705 for the same group of seven publishers. The Noam Index was 1,023 in 2008 and 1,231 in 2010 for the same group.

In summary, the C4 in the book publishing market has increased and is expected to continue increasing as a result of very recent movements in the market, which are not yet fully expressed in the available data. In 2008, Leya was formed. It was the first publishing to organize itself as a holding company that incorporated several of the most prestigious Portuguese publishers, including D. Quixote, Asa Oficina do Livro, and Teorema.

The increasing concentration of the book publishing industry is illustrated by two key examples. Babel entered the market with the purchase of Editora Magalhães. In 2011, it was acquired by the Ongoing Group,[5] which spans the publishers Arcádia, Ulisseia, and Verbo, among others.

During the last decade, there was an influx of foreign capital into the Portuguese publishing market for the first time. In 2006, the German group Bertelsmann, which already owned Círculo de Leitores, acquired Bertrand and its associated businesses: publishing, retail trade, and distribution. However, the foreign investment in the national book market is not prevalent today. The main foreign player is Santillana from PRISA Group which operates Planeta Agostini Publishing. Among the analyzed players, Planeta Agostini is the only foreign-owned one. There are other foreign-owned players such as Santillana, though they are excluded from this analysis because they do not have a large enough market share to be considered.

At the end of 2010, Porto Editora, the publishing industry's leader whose base has been scholar books and dictionaries' publishing, was the leader in online sales through the website Wook. It fought for market supremacy and reinforced its leading position by acquiring Bertrand. With this acquisition, Porto Editora added its scholar books business to a library network. With the opening of the Wook store in the North region in 2008, followed by an opening in the South, Porto Editora has innovated the retail book space by capturing more of the value chain. It is a publisher, editor, and distributor. Also, it has created a new and differentiated distribution system with Círculo de Leitores. For further research, it would be interesting to evaluate the impact of this acquisition. The main national editorial groups have been developing businesses in Portuguese-speaking countries, such as Angola, Mozambique, and Brazil. The concentration and consolidation strategies are also part of a business objective to achieve economies of scale in order to reach more competitive positions in these Lusophone markets.

## Magazine Publishing

The magazine industry in Portugal has a dynamic that includes considerable ownership turnover. The competitive set within the newsmagazine segment comprises three general weekly magazines owned by different groups: Visão/Impresa, Sábado/Cofina, and Focus/Impala. They compete for the audience and advertising market. Regarding the question of ownership, the market is concentrated in a small number of players, as shown in Table 10.5.

To calculate the concentration indexes presented in Table 10.5, the options were restricted due to a lack of information. Yet it was possible to estimate the players' market position. Admittedly, the sample presented here (composed of the only players that provided data on advertising investment without commercial discounts, calculated by Marktest) represents about 70% of the market. This estimation is valid, given that the three weekly newsmagazines considered

5. The Ongoing Group did not operate in the media business segment until 2011, when it acquired Babel, which is why it is not included in the analysis.

Table 10-5. Advertising Investment in Magazine Industry (Thousands €)

| | 2005 | | 2007 | | 2009 | | 2010 | |
|---|---|---|---|---|---|---|---|---|
| | REVENUES | % | REVENUES | % | REVENUES | % | REVENUES | % |
| Cofina | 28,091 | 19.5 | 38,427 | 22.5 | 37,596 | 26.3 | 35,915 | 23.5 |
| Media Capital | 12,511 | 8.7 | 14,118 | 8.3 | 12,963 | 9.1 | 12,802 | 8.4 |
| Impala | 39,367 | 27.3 | 49,942 | 29.3 | 31,859 | 22.3 | 36,341 | 23.8 |
| Impresa | 50,070 | 34.7 | 53,176 | 31.2 | 45,405 | 31.8 | 49,261 | 32.3 |
| Motorpress Ibérica (Bertelsmann, Germany) | 5,969 | 4.1 | 6,546 | 3.8 | 5,635 | 3.9 | 2,733 | 1.8 |
| Others | 8,257 | 5.7 | 8,432 | 4.9 | 9,488 | 6.6 | 15,632 | 10.2 |
| Total | 144,265 | 100 | 170,641 | 100 | 142,946 | 100 | 152,684 | 100 |
| Total ($US mil) | 179,808 | | 233,848 | | 199,335 | | 202,671 | |
| *N* (>1%) | 6 | | 6 | | 6 | | 6 | |
| CR4 | 90.14% | | 91.22% | | 89.42% | | 89.42% | |
| HHI | 2,454 | | 2,442 | | 2,339 | | 2,339 | |
| Noam | 1,002 | | 997 | | 955 | | 955 | |

*Source*: Elaborated by the author based on data from Marktest.

here absorb a considerable part of the advertising investments. The main media groups with a significant portfolio of titles are also included, although some titles owned by these groups are missing, due to a lack of official data. Under this assumption, to calculate market shares and concentration levels, the aforementioned magazines are assumed to constitute 100% of the market.

As seen in Table 10.5, Impresa Group has the highest market share in terms of advertisement investment. Over the years, the concentration levels show that the positions of the different players remained relatively constant. The concentration levels tended to decrease slightly. However, three media groups, Impresa, Cofina, and Impala, always retained about 80% of the market, with minor variations over the years. Impresa was the clear leader from 2005 to 2010.

## AUDIOVISUAL MEDIA

### Radio

National radio is dominated by three media groups: Renascença (private/church-owned), Media Capital (private), and RDP (state-owned). The only other relevant radio station is TSF, which is part of another important communication group, Controlinveste, whose presence is greater in the printed press and paid TV. The analysis does not include regional or local radio stations.

As mentioned, this analysis considers the national radio operators. As evident in Table 10.6, RR Group has been losing share. However, it maintained its dominant position. Media Capital had a positive evolution, with Rádio Comercial and Cidade FM primarily responsible for the group's performance. The public RDP Group's share growth was 4%, mostly due to Antena 1's performance. From the data in Table 10.6, RR Group maintained its dominant position in 2009 by introducing the channel Sim. Media Capital Group also evolved positively, introducing Romântica FM and M80 with great success. RDP, on the other hand, lost market share after 2007. TSF maintained a constant market share through the last five years. Drawing on the information presented in Table 10.6, the concentration measure indexes were applied, and one can conclude that the high concentration in the radio segment is a long-standing trend. The

Table 10-6. Radio Broadcasting Audience and Concentration Indexes

| | 2006 | 2007 | 2008 | 2009 | 2010 |
|---|---|---|---|---|---|
| RR Group (Church-owned) | 50.5 | 53.2 | 53.0 | 53.3 | 49.2 |
| RFM | *29.8* | *30.6* | *31.8* | *31.1* | *27.5* |
| Renascença | *18.3* | *19.7* | *18.4* | *17.7* | *15.8* |
| Mega Hits | *2.4* | *2.9* | *2.8* | *2.8* | *3.9* |
| Rádio Sim | | | | *1.8* | *2.0* |
| Media Capital Group (PRISA) | 25.0 | 23.9 | 25.5 | 26.9 | 32.0 |
| Rádio Comercial | *15.3* | *13.8* | *14.1* | *14.9* | *15.4* |
| Cidade FM | *8.5* | *7.8* | *7.6* | *7.1* | *7.1* |
| M80 | | *1.0* | *2.7* | *3.9* | *8.6* |
| Best Rock FM | *1.2* | *1.3* | *1.0* | *1.1* | *0.9* |
| RDP Group (Public) | 17.2 | 16.5 | 14.9 | 13.9 | 13.0 |
| Antena 1 | *9.1* | *8.8* | *7.6* | *7.8* | *7.1* |
| Antena 3 | *7.0* | *6.7* | *6.5* | *5.1* | *5.2* |
| Antena 2 | *1.0* | *0.9* | *0.8* | *0.9* | *0.8* |
| TSF | 7.3 | 6.5 | 6.7 | 5.9 | 5.7 |
| Total | 100 | 100 | 100 | 100 | 100 |
| Revenue Estimate (US $mil) | 232 | | | | 242 |
| *N* (>1%) | 4 | 4 | 4 | 4 | 4 |
| CR4 | 100 | 100 | 100 | 100 | 100 |
| HHI | 3,521 | 3,711 | 3,720 | 3,796 | 3,653 |
| Noam | 1,761 | 1,856 | 1,860 | 1,898 | 913 |

*Source*: Elaborated by the author based on data obtained from Marktest.

market trends have not changed significantly since the turn of the millennium.

This market is highly concentrated and its dynamics are not changing. It was more concentrated in 2009 than in 2003. It is worth noting that the Media Capital Group considerably increased its share in terms of advertising revenues. The RR Group, analyzed through this variable, has experienced a steady reduction in its dominant position, although it owns the radio stations with the largest audiences. TSF lost market share in terms of advertising investment between 2006 and 2010.

According to the study "Caracterização do Sector da Radiodifusão Local" (2009), promoted by Entidade Reguladora da Comunicação Social (ERC), the Portuguese market includes 347 local radio stations, which have, on the whole, large audiences and significant advertising revenue. The same study shows that the advertising revenues earned by these local radio stations in 2007 represented approximately 18% of the audience market and €43.594M in estimated advertising revenues. These indicators show the considerable weight that local radio stations have outside the main metropolitan areas. Even in those areas there are some stations with important market niches mainly due to their close ties and proximity to their local communities. Table 10.7 shows revenues of the national radio market based on undiscounted advertising revenues.

The advertising market has a higher level of concentration than the audience market, as 100% of the advertising investment is concentrated in three groups. Table 10.7 shows that two main groups dominate about 70% of the advertising market. In turn, the market positions remained relatively constant throughout the six

Table 10-7. Radio Advertising Revenues (Thousands €)

| | 2005 | | 2006 | | 2007 | | 2008 | | 2009 | | 2010 | |
|---|---|---|---|---|---|---|---|---|---|---|---|---|
| | VALUE | % | VALUE | % | VALUE | % | VALUE | % | VALUE | % | VALUE | % |
| R/com Group(Church) | 61,294 | 32.7 | 60,821 | 32.9 | 62,599 | 34.1 | 62,608 | 35.0 | 64,323 | 37.1 | 64,821 | 35.7 |
| RFM | *45,076* | *24.1* | *45,761* | *24.8* | *46,121* | *25.1* | *45,843* | *25.6* | *47,427* | *27.4* | *48,846* | *26.9* |
| Renascença | *14,534* | *7.8* | *12,732* | *6.9* | *14,056* | *7.7* | *13,770* | *7.7* | *13,943* | *8.1* | *13,326* | *7.4* |
| Mega Hits | *1,684* | *0.9* | *2,328* | *1.3* | *2,422* | *1.3* | *2,995* | *1.7* | *2,953* | *1.7* | *2,649* | *1.5* |
| Media Capital Group (PRISA) | 72,541 | 38.7 | 66,848 | 36.2 | 65,402 | 35.7 | 63,226 | 35.4 | 61,141 | 35.3 | 63,597 | 35.1 |
| Rádio Comercial | *47,069* | *25.1* | *42,316* | *22.9* | *40,932* | *22.3* | *38,196* | *21.4* | *43,187* | *24.9* | *48,019* | *26.5* |
| Cidade FM | *13,099* | *7.0* | *11,841* | *6.4* | *13,364* | *7.3* | *16,672* | *9.3* | *17,954* | *10.4* | *15,578* | *8.6* |
| RCP | *12,373* | *6.6* | *12,691* | *6.9* | *11,106* | *6.1* | *8,358* | *4.7* | | | | |
| TSF | 53,487 | 28.6 | 57,214 | 31.0 | 55,454 | 30.2 | 52,927 | 29.6 | 47,765 | 27.6 | 52,981 | 29.2 |
| Total | 187,322 | 100.0 | 184,883 | 100.0 | 183,455 | 100.0 | 200,000 | 100.0 | 173,229 | 100.0 | 181,399 | 100.0 |
| Revenues (US$ mil) | 233 | | 232 | | 251 | | 294 | | 242 | | 241 | |
| *N* (>1%) | 3 | | 3 | | 3 | | 3 | | 3 | | 3 | |
| C3 | 100 | | 100 | | 100 | | 100 | | 100 | | 100 | |
| HHI | 3,386 | | 3,347 | | 3,349 | | 3,354 | | 3,385 | | 3,359 | |
| Noam | 1,955 | | 1,933 | | 1,934 | | 1,937 | | 1,954 | | 1,939 | |

*Source*: Elaborated by the author based on data obtained from Marktest.
RDP group cannot be found, since the public radio does not identify advertising investments.

Table 10-8. Radio Market Share of Advertising Revenues 2006–2011 (%)

| | 2006 | 2007 | 2008 | 2009 | 2010 | 2011 |
|---|---|---|---|---|---|---|
| RR Group (Church) | 32.9 | 34.12 | 35.02 | 32.78 | 32.39 | 33.5 |
| RR | 6.89 | 7.66 | 7.7 | 7.11 | 6.66 | 7.91 |
| RFM | 24.75 | 25.14 | 25.64 | 24.17 | 24.41 | 23.91 |
| Mega FM | 1.26 | 1.32 | 1.68 | 1.5 | 1.32 | 1.68 |
| MC Group | 36.16 | 35.65 | 35.37 | 42.88 | 41.14 | 39.91 |
| R Comercial | 22.89 | 22.31 | 21.37 | 22.01 | 23.99 | 24.46 |
| RCP | 6.86 | 6.05 | 4.68 | 3.64 | 1.6 | |
| M80 | | | | 5.1 | 6.09 | 8.34 |
| Romantica FM | | | | 1.59 | 1.02 | |
| Best FM | | | | 1.39 | 0.64 | 0.27 |
| Vodafone FM | | | | | | 0.03 |
| Star FM | | | | | | 0.96 |
| Cidade FM | 6.4 | 7.28 | 9.33 | 9.15 | 7.78 | 5.84 |
| ControlInveste Group | 30.95 | 30.23 | 29.61 | 24.34 | 26.47 | 26.59 |
| TSF | 30.95 | 30.23 | 29.61 | 24.34 | 26.47 | 26.59 |
| CR2 | 69.1 | 69.8 | 70.4 | 76.0 | 74.0 | 73.4 |
| HHI | 3,347.6 | 3,349.0 | 3,354.3 | 3,050.7 | 3,442.4 | 3,421.9 |
| Noam | 1,932.8 | 1,933.5 | 1,936.6 | 2,024.0 | 1,987.5 | 1,975.6 |

*Source*: Elaborated by the author based on data obtained from Marktest.

years analyzed, as shown by the concentration indexes (HHI and Noam), whose variations from year to year are marginal.

Table 10.8 illustrates each radio group's recruitment of advertising investment from 2006–2011.

Until 2009, Media Capital increased its share considerably in terms of advertising revenues, with a slight decrease between 2010 and 2011. The RR Group experienced a steady reduction of its dominant position, although it holds radio stations with large audiences. TSF lost market share in terms of advertising investment in the period between 2006 and 2009, but it recovered in 2010 and 2011.

## Broadcast Television

This section investigates the degree and the evolution of concentration in the Portuguese television market, including free-to-air channels as well as paid/cable channels. Satellite channels were not considered because it is a marginal business and it is promoted by MEO, a recent entrant that merged with Portugal Telecom in 2008, and the data on MEO are insufficient for analysis. According to this market definition, the participants in the Portuguese TV market are described in Table 10.9.

To account for the two-sided nature of TV markets (TV channels constitute platforms between viewers and advertisers), concentration in the Portuguese TV industry was assessed along two dimensions: the advertising market and the market of content diffusion. Therefore, the concentration measures (C4, HHI, and Noam Index) were computed for two variables: audience size and advertising revenues. Each of them assesses the degree of concentration from a different perspective of the Portuguese TV market. To be more precise, the first variable, audience size, assesses concentration from the viewers' perspective, while the second variable, advertising revenues, assesses concentration from the advertisers'

Table 10-9. Free-to-Air TV and Pay-TV Main Players and Channels

| COMPANY—OWNER | CHANNEL—BRAND |
|---|---|
| Media Capital (PRISA Group, Spain, part UK-owned) | • TVI (free to air) |
| Impresa (National Company) | • *SIC* (free to air) |
| | • *SIC* Notícias (cable) |
| | • *SIC* Mulher (cable) |
| | • *SIC* Comédia (cable) |
| | • *SIC* Radical (cable) |
| RTP (public) | • RTP 1 (free to air) |
| | • RTP 2 (free to air) |
| | • RTP Memória (cable) |
| | • RTPN (cable) |
| Other Providers (pay television over cable) | • AXN (Sony Pictures) |
| | • Panda (Chello) |
| | • TV Record (Rede Record) |
| | • MTV PT (MTV Networks) |
| | • GNT (Globosat) |
| | • Fox (Fox Entertainment Group) |
| | • História (Chello) |
| | • Hollywood (Chello) |
| | • People + Arts (BBC) |
| | • MEO (Portugal Telecom) |

*Source*: Elaborated by the author based on data from Marktest.

point of view. Advertising represents the main revenue source for free-to-air TV.

In the case of share of audience, concentration measures were obtained separately for free-to-air and paid television. This division turned out to be necessary given the methodological differences among data available on the audiences of free-to-air channels and paid TV. In the case of free-to-air TV, audience shares correspond to the ratio of the consumers' time spent watching a specific channel and the total time spent by consumers watching TV. In the case of paid TV, audience shares correspond to the number of subscriptions obtained by each group as a fraction of total subscriptions of paid TV. In the paid TV market, the bundling phenomenon prevents consumers from choosing their own portfolio of channels. The free-to-air TV market in Portugal is consists of four TV channels:[6] two public channels (RTP 1 and RTP 2) and two private ones (SIC and TVI) that are owned by powerful media groups (Impresa and Media Capital/Prisa Group, respectively). Table 10.10 presents the audience shares of free-to-air TV channels for 2003–2010.

Considering the ownership structure of the TV channels represented in Table 10.10, it is easy to conclude that C3 is 100% because three operators own the whole market as a consequence of the small number of firms participating in it. The rigorous assessment of the degree of concentration in the Portuguese free-to-air

6. It is important to mention that next year a new free-to-air operator will be chosen. The Portuguese government launched a new license, and there are several candidates for Channel 5.

Table 10-10. Broadcast Television Channel Market Shares 2003–2010 (%)

| CHANNELS | 2003 | 2004 | 2005 | 2006 | 2007 | 2008 | 2009 | 2010 |
|---|---|---|---|---|---|---|---|---|
| RTP1 (public) | 32 | 33 | 34 | 35 | 36 | 35 | 36 | 37 |
| SIC (Impresa) | 35 | 34 | 32 | 30 | 30 | 29 | 29 | 29 |
| TVI (Media Capital) | 33 | 33 | 34 | 35 | 34 | 36 | 35 | 34 |
| Total | 100 | 100 | 100 | 100 | 100 | 100 | 100 | 100 |
| Revenue (US $mil) | | | | 2,743 | 2,935 | 3,168 | 3,338 | 3,548 |
| *N* (>1%) | 3 | 3 | 3 | 3 | 3 | 3 | 3 | 3 |
| Indexes | | | | | | | | |
| C4[1] | 100 | 100 | 100 | 100 | 100 | 100 | 100 | 100 |
| HHI | 3,338 | 3,334 | 3,336 | 3,350 | 3,352 | 3,362 | 3,362 | 3,366 |
| Noam | 1,927 | 1,925 | 1,926 | 1,934 | 1,935 | 1,941 | 1,941 | 1,943 |

*Source*: Elaborated by the author based on data obtained from Marktest.

[1]It includes the two state TV channels—RTP1 and RTP2—for which revenues in 2010 were 15.47% and 0.12%, respectively, as one can see in Table 10.11.

TV market requires the computation of concentration indices, whose values are presented in Table 10.10. Nevertheless, it should be noted that audiences are almost equally distributed among the three media groups, thus creating competitive pressures despite the small number of firms participating in this market. Furthermore, the free-to-air TV market is now substantially more competitive than it was two decades ago after SIC and TVI entered the market in 1991 and 1992, respectively. The degree of competition is compounded by the emergence of other offerings through paid channels. This point will be further explored later in this chapter.

Tables 10.11 and 10.12 emphasize the relative concentration of advertising revenues (without commercial discounts) in a limited number of media groups (RTP, Impresa, Media Capital). The remaining channels (included in the category which gathers almost all pay-TV channels) do not capture a significant share of advertising revenues; their market shares were lower than 7% between 2006 and 2010. Considering a more longitudinal analysis, the progressive growth of pay TV (mainly broadcasted by cable) is noteworthy. There was also a small rise in advertising revenue in the Impresa and Media Capital. RTP Group experienced a slight decrease in advertising revenue. Considering that the core business of RTP, Impresa, and Media Capital is unquestionably the free-to-air segment, the information in Table 10.11 also suggests the existence of an asymmetry between the weight of free-to-air and pay TV in terms of advertising revenues.

Table 10.13 shows multichannel TV platform operators who distribute content, a segment in the Portuguese TV market that has taken root over the last 20 years. Despite the growing weight of paid TV channels, free-to-air TV continues to attract the largest amount of consumers' time. Regarding consumers' subscription choices, the firm ZON (formerly TV Cabo and part of Portugal Telecom) is the market leader. ZON's market leader position is the result of its first-mover advantage. The other major participants are Cabovisão (owned by Canadian company Cogeco until 2012 when it was purchased by ALTICE), PCT, and TVTEL. However, as Table 10.13 shows, the competition in this segment has increased. For example, Portugal Telecom/MEO earned a market share of 30% in 2010, corresponding to about 700,000 subscribers. Another important conclusion is that concentration levels have markedly decreased in the last three years.

Like the radio and print industries, the TV industry is a two-sided market. To get a

Table 10-11. TV Channel and Media Groups Advertising Revenues, 2006–2010 (Thousands €)

| GROUPS | 2006 | | 2007 | | 2008 | | 2009 | | 2010 | |
|---|---|---|---|---|---|---|---|---|---|---|
| | VALUE | % | VALUE | % | VALUE | % | VALUE | % | VALUE | % |
| Media Capital (PRISA) | 1,258,581 | 44.3 | 1,379,080 | 45.0 | 1,612,178 | 48.4 | 1,737,196 | 49.4 | 1,856,915 | 49.1 |
| TVI | *1,258,581* | *44.3* | *1,379,080* | *45.0* | *1,612,178* | *48.4* | *1,737,196* | *49.4* | *1,851,819* | *49.0* |
| TVI24 | — | — | — | — | — | — | — | — | *5,096* | *0.1* |
| Impresa | 1,065,025 | 37.5 | 1,095,441 | 35.7 | 1,062,271 | 32.2 | 1,009,265 | 29.0 | 1,043,488 | 27.6 |
| *SIC* (free to air) | *934,427* | *32.9* | *970,078* | *31.6* | *951,984* | *28.9* | *901,648* | *25.9* | *945,775* | *25.0* |
| *SIC* Notícias (paid) | *90,053* | 3.2 | *90,298* | *3.0* | *79,564* | *2.4* | *74,423* | *2.1* | *83,194* | 2.2 |
| *SIC* Mulher (paid) | *16,686* | *0.6* | *18,786* | *0.6* | *30,723* | *0.9* | *33,194* | *1.0* | *19,510* | *0.1* |
| *SIC* Comédia (paid) | *9,832* | *0.3* | — | — | — | — | — | — | — | — |
| *SIC* Radical (paid) | *14,027* | *0.5* | *16,279* | *0.5* | *11,617* | *0.4* | *11,856* | *0.3* | *12,568* | *0.3* |
| RTP (public) | 419,811 | 14.8 | 460,314 | 15.0 | 493,398 | 14.8 | 591,653 | 16.8 | 647,482 | 17.1 |
| RTP 1 (free to air) | *386,732* | *13.6* | *420,515* | *13.7* | *455,640* | *13.7* | *489,245* | *13.9* | *584,714* | *15.5* |
| RTP 2 (free to air) | *2,197* | *0.1* | *2,127* | *0.1* | *2,791* | *0.1* | *3,623* | *0.1* | *4,566* | *0.1* |
| RTP Memória (paid) | *6,180* | *0.2* | *8,972* | *0.3* | *10,608* | *0.3* | *57,616* | *1.6* | *22,417* | *0.6* |
| RTPN (paid) | *24,702* | *0.9* | *28,700* | *0.9* | *24,359* | *0.7* | *41,169* | *1.2* | *35,785* | *1.0* |
| Others (paid TV) | *96,789* | *3.4* | *130,788* | *4.3* | *151,387* | *4.6* | *160,069* | *4.8* | *232,020* | *6.1* |
| AXN (Sony Pictures) | 28,689 | 1.0 | 52,845 | 1.7 | 53,958 | 1.6 | 57,053 | 1.7 | 66,919 | 1.8 |
| Panda (Chello) | 25,988 | 0.9 | 26,855 | 0.9 | 27,273 | 0.8 | 25,497 | 0.8 | 39,033 | 1.0 |
| Record (Rede Record) | 14,576 | 0.5 | 16,279 | 0.5 | 28,862 | 0.9 | 36,161 | 1.1 | 52,694 | 1.4 |
| MTV PT (Rede MTV) | 10,267 | 0.4 | 13,482 | 0.4 | 9,299 | 0.3 | 5,850 | 0.2 | 7,103 | 0.2 |
| GNT (Globosat) | 8,013 | 0.3 | — | — | — | — | — | — | — | — |

*continued*

Table 10-11. *continued*

| GROUPS | 2006 | | 2007 | | 2008 | | 2009 | | 2010 | |
|---|---|---|---|---|---|---|---|---|---|---|
| | VALUE | % | VALUE | % | VALUE | % | VALUE | % | VALUE | % |
| Fox (Grupo Fox Entertainment) | 6,103 | 0.2 | 17,224 | 0.6 | 27,230 | 0.8 | 34,787 | 1.0 | 50,308 | 1.3 |
| História (Chello) | 1,797 | 0.1 | 816 | 0.0 | 1,507 | 0.1 | 0 | | — | |
| Hollywood (Chello) | 1,356 | 0.1 | 2,667 | 0.1 | 1,391 | 0.0 | 594 | 0.0 | 3,448 | 0.1 |
| People + Arts (BBC) | — | | 620 | 0.0 | 1,867 | 0.1 | 127 | 0.0 | — | — |
| Disney | — | — | — | — | — | — | 4,435 | 0.1 | 10,901 | 0.3 |
| MOV | — | — | — | — | — | — | 2,755 | 0.1 | 1,614 | 0.0 |
| Discovery | — | | — | | 59 | 0.0 | — | | — | |
| TOTAL | 2,840,206 | 100 | 3,065,623 | 100 | 3,330,911 | 100 | 3,517,229 | 100 | 3,797,464 | 100 |
| Revenue (US $mil) | 3,568 | | 4,201 | | 4,901 | | 4,905 | | 5,041 | |
| C4 | 97.6 | | 97.4 | | 97.0 | | 96.9 | | 95.6 | |
| HHI | 3,600 | | 3,544 | | 3,622 | | 3,586 | | 3,507 | |
| Noam | 1,800 | | 1,772 | | 1,811 | | 1,793 | | 1,753 | |

*Source*: Elaborated by the author based on data obtained from Marktest.

Table 10-12. Pay TV-Market Shares by Advertising Revenues 2006–2010 (Thousands €)

| GROUPS | 2006 | | 2007 | | 2008 | | 2009 | | 2010 | |
|---|---|---|---|---|---|---|---|---|---|---|
| | VALUE | % | VALUE | % | VALUE | % | VALUE | % | VALUE | % |
| Media Capital (PRISA) | | | | | | | | | 5,096 | 1.3 |
| TVI24 | | | | | | | | | *5,096* | *1.2* |
| Impresa | 130,598 | 50.6 | 125,363 | 42.7 | 121,994 | 39.5 | 119,473 | 30.9 | 97,713 | 24.9 |
| *SIC* Notícias | *90,053* | *34.9* | *90,298* | *30.7* | *79,564* | *25.8* | *74,423* | *19.3* | *83,194* | *20.3* |
| *SIC* Mulher | *16,686* | *6.5* | *18,786* | *6.4* | *30,723* | *10.0* | *33,194* | *8.6* | *19,510* | *4.8* |
| *SIC* Comédia | *9,832* | *3.8* | | | | | | | | |
| *SIC* Radical | *14,027* | *5.4* | *16,279* | *5.5* | *11,617* | *3.8* | *11,856* | *3.1* | *12,568* | *3.1* |
| RTP (public) | 30,882 | 12.0 | 37,672 | 12.8 | 34,967 | 11.3 | 98,785 | 25.6 | 58,202 | 14.8 |
| RTP Memória | *6,180* | *2.4* | *8,972* | *3.1* | *10,608* | *3.4* | *57,616* | *15.0* | *22,417* | *5.5* |
| RTPN | *24,702* | *9.6* | *28,700* | *9.8* | *24,359* | *7.9* | *41,169* | *10.7* | *35,785* | *8.7* |
| Other (paid TV) | 96,789 | 37.5 | 130,788 | 44.5 | 151,446 | 49.1 | 168,402 | 43.6 | 232,020 | 59.0 |
| AXN (Sony Pictures, Japan) | 28,689 | 11.1 | 52,845 | 18.0 | 53,958 | 17.5 | 57,053 | 14.8 | 66,919 | 16.3 |
| Panda (Chello) | 25,988 | 10.1 | 26,855 | 9.1 | 27,273 | 8.9 | 25,497 | 6.6 | 39,033 | 9.5 |
| Record (Rede Record) | 14,576 | 5.6 | 16,279 | 5.5 | 28,862 | 9.4 | 36,161 | 9.4 | 52,694 | 12.8 |
| MTV PT (Rede MTV) | 10,267 | 4.0 | 13,482 | 4.6 | 9,299 | 3.0 | 5,850 | 1.5 | 7,103 | 1.7 |
| GNT (Globosat) | 8,013 | 3.1 | | | | | | | | |
| Fox (Grupo Fox Entertainment, Newscorp, US) | 6,103 | 2.4 | 17,224 | 5.9 | 27,230 | 8.8 | 34,787 | 9.0 | 50,308 | 12.3 |
| História (Chello) | 1,797 | 0.7 | 816 | 0.3 | 1,507 | 0.5 | | | | |
| Hollywood (Chello) | 1,356 | 0.5 | 2,667 | 0.9 | 1,391 | 0.5 | 594 | 0.2 | 3,448 | 0.8 |
| People + Arts (BBC, UK) | | | 620 | 0.2 | 1,867 | 0.6 | 127 | 0.0 | | |
| Disney (US) | | | | | | | 4,435 | 1.2 | 10,901 | 2.7 |
| MOV | | | | | | | 2,755 | 0.7 | 1,614 | 0.4 |
| Discovery (US) | | | | | 59 | 0.0 | | | | |
| Total | 258,269 | 100.0 | 293,823 | 100.0 | 308,317 | 100.0 | 385,517 | 100.0 | 410,590 | 100.0 |
| Revenue (US $mil) | 324 | | 403 | | 454 | | 538 | | 545 | |
| CR4 | 83.8 | | 82.6 | | 77.7 | | 80.7 | | 68.8 | |
| HHI | 4,104.42 | | 3,966.15 | | 4,104.72 | | 3,504.31 | | 4,324.02 | |
| Noam | 2,369.69 | | 2,289.86 | | 2,369.86 | | 2,023.21 | | 2,496.47 | |

*Source*: Elaborated by the author based on data obtained from Marktest.

### Table 10-13. Multichannel Video Platforms, Market Shares by Subscription, 2007–2010 (%)

| GROUPS | 2007 | 2008 | 2009 | 2010 |
|---|---|---|---|---|
| ZON | 78.4 | 72.3 | 64.4 | 57.9 |
| TVTel[1] | 3 | — | — | — |
| Cabovisão[2] (Cogeco) | 14.9 | 12.4 | 10.2 | 9.4 |
| PT/Meo (Portugal Telecom) | 1 | 13.7 | 23 | 29.9 |
| Optimus | 0.1 | 0.6 | 1 | 1.1 |
| Ar Telecom | 0.6 | 1 | 0.9 | 0.9 |
| Vodafone (UK) | 0 | 0 | 0.3 | 0.7 |
| Others | 2 | 0.1 | 0.1 | 0.1 |
| Total | 100 | 100 | 100 | 100 |
| *N* (>1%) | 4 | 4 | 4 | 4 |
| C4 | 97.3 | 99.4 | 98.6 | 98.3 |
| HHI | 6,379 | 5,571 | 4,785 | 4,342 |
| Noam | 3,190 | 2,786 | 2,763 | 2,171 |

*Source*: Elaborated by the author based on data obtained from ANACOM.
[1]TVTel was bought by ZON on 2008.
[2]Cabovisão was purchased by Altice in 2012.

comprehensive perspective on the TV market configuration (free-to-air and pay TV) and to evaluate the extent of concentration in the Portuguese TV market, the previous analysis must be coordinated with an analysis of concentration concerning advertising revenues. Table 10.13 provides an overview of the situation of the Portuguese TV market in terms of advertising from 2006 to 2010 and an overview of the free-to-air and pay TV broadcasters' concentration, calculated by advertising revenues.

The values shown in Table 10.14 regarding the revenue generated by pay-TV services, published at ANACOM's website in the report on the situation of the communications, were calculated based on total revenues of TV subscription services and on the market shares of the respective operators.

### Table 10-14. Pay-TV Revenues by Operator

| | 2008 | 2009 | 2010 | 2011 |
|---|---|---|---|---|
| ZON/TV Cabo Group | 438,162 | 501,542 | 511,658 | 628,62 |
| PT Group/Meo | 82,712 | 178,805 | 264,130 | 408,236 |
| Cabovisão | 75,380 | 79,633 | 82,921 | 100,164 |
| Optimus (formerly Sonaecom) | 3,333 | 7,706 | 9,272 | 13,993 |
| AR Telecom | 5,757 | 7,239 | 7,948 | — |
| Vodafone | — | 2,413 | 5,828 | 13,410 |
| Other providers | 606 | 1,090 | 1,325 | 1,866 |

*Source*: Elaborated by the author based on data obtained from ANACOM.

Since 2008, the trend is for the strengthening of the most recent operator, Portugal Telecom, and the for Cabovisão and ZON's loss of market share. Revenues have continuously increased in all operators and, despite the prevalence of the mentioned operators, Vodafone reveals a sharply growing trend. This situation can be due to the stabilization of the number of subscribers of the oldest operators, while the most recent ones, with their aggressive policies, are entering into the market to fill the gaps. For the year 2012, despite the introduction of Digital Terrestrial Television (DTT) in Portugal and the global economic and financial crisis, it is expected that the market structure and trends will be stable.

Despite the limited number of free-to-air channels, the market characteristics, and the business model, the free-to-air TV segment captures a much higher market share of the advertising market. This situation, from a business model perspective, reveals a high level of dependence on advertising investment. In contrast, the market share of the pay-TV sector is relatively limited at less than 10% because the main revenues are a product of the number of subscriptions. In any case, this market share will probably evolve positively as the number of pay TV subscribers increases, due to the emergence of new operators in pay-TV via satellite such as MEO. In light of the previous evidence, one expects the Portuguese TV market to exhibit a significant degree of concentration because approximately 90% of advertising revenues are concentrated in two market groups: Impresa/SIC and Media Capital/TVI.

### Table 10-15. TV Segment Market Share of Advertising Revenues 2006–2010 (%)

| SEGMENT | 2006 | 2007 | 2008 | 2009 | 2010 |
|---|---|---|---|---|---|
| Free-to-air TV | 90.91 | 90.42 | 90.74 | 89.01 | 89.60 |
| Pay TV/cable | 9.09 | 9.58 | 9.26 | 10.99 | 10.40 |

*Source*: Elaborated by the author based on data obtained from Marktest.

The results in Table 10.15 corroborate the hypothesis of a highly concentrated market. This result is explained by comparing the advertising revenues of pay TV and free-to-air public TV. A substantial portion of the advertising revenues are concentrated in two media groups, Impresa and Media Capital, thus exerting a considerable level of market power.

## Film

The film industry in Portugal began in 1918, after the restructuring of the production company Invicta Film. During the 1920s, Portuguese film production was mainly dedicated to the transposition of Portuguese literary classics to the screen and the project direction was done by foreign filmmakers. Now the majority of projects are directed by domestic filmmakers. Lusomundo (now part of ZON and Portuguese-owned) entered the market in 1984, and it has been the leader of distribution and exhibition of movies.[7] To get a clearer picture of Lusomundo and other operators' market position, Table 10.16 ranks several operators according to their ability to generate revenue. As the table shows, the highest market shares belong to Lusomundo Audiovisuais, Columbia Tristar Warner (part of US-owned Sony and Time Warner), Castello Lopes Multimédia, and Prisvídeo.

As one can see from Table 10.16, the international film distributors do not operate directly in Portugal, so they are associated with local distributors.

Examining the period between 2008 and 2011, Lusomundo's market share decreased by 16.2%, while Columbia Tristar Warner rose 10.1%. Castello Lopes Multimedia, for its part, suffered a slight decrease of 1.3%. The data show that the market is quite concentrated and that Lusomundo's market share is extremely high compared to other operators. The concentration in distribution in 2010 is particularly evident, according to ICA (Instituto do Cinema e do Audiovisual/Cinema and Audiovisual Institute). Lusomundo had 51.8% market share and the four leading companies—ZON, Prisvídeo,

7. It is not active in the production area.

Table 10-16. Film Studio Market Share by Gross Revenue per 2008–2011 (%)[1]

| | 2008 | | 2009 | | 2010 | | 2011 | |
|---|---|---|---|---|---|---|---|---|
| | VALUE | % | VALUE | % | VALUE | % | VALUE | % |
| ZON—Lusomundo Audiovisuais | 40,919,763.8 | 58.5 | 36,328,113.1 | 49.2 | 425,923,820.3 | 51.8 | 32,323,301.0 | 42.3 |
| Paramount Pictures | | *23.0* | | *19.3* | | *20.3* | *12,687,438.5* | *16.6* |
| Universal Pictures | | *12.6* | | *10.6* | | *11.1* | *6,925,451.3* | *9.1* |
| Walt Disney Studios | | *12.3* | | *10.4* | | *10.9* | *6,831,300.0* | *8.9* |
| Film & TV House | | *6.2* | | *5.2* | | *5.5* | *3,410,348.4* | *4.5* |
| Summit Entertainment | | *4.4* | | *3.7* | | *3.9* | *2,468,762.9* | *3.2* |
| Columbia Tristar | 12,963,606.7 | 18.5 | 19,111,734.9 | 25.9 | 21,245,474.5 | 25.8 | 21,871,029.6 | 28.6 |
| Warner Bros Pictures | | *9.8* | | *13.7* | | *13.6* | *11,540,166.7* | *15.1* |
| Columbia Pictures | | *8.7* | | *12.2* | | *12.2* | *10,330,862.9* | *13.5* |
| Castello Lopes Multimedia | 8,621,321.4 | 12.3 | 12,106,004.0 | 16.4 | 10,932,814.4 | 13.3 | 8,443,531.9 | 11.0 |
| 20th Century Fox | | *12.3* | | *16.4* | | *13.3* | *8,443,531.9* | *11.0* |
| Prisvídeo | 2,371,424.4 | 3.4 | 3,606,666.6 | 4.9 | 5,187,315.8 | 6.3 | 2,229,928.9 | 2.9 |
| Lionsgate | | *3.4* | | *4.9* | | *6.3* | *2,229,928.9* | *2.9* |
| Others | 5,018,515.6 | 7.2 | 2,689,091.2 | 3.6 | 2,285,169.7 | 2.8 | 11,568,171.2 | 15.1 |
| Total | 69,894,631.8 | 100.0 | 73,841,609.8 | 100.0 | 82,243,156.6 | 100.0 | 76,435,962.5 | 100.0 |
| Revenue Estimate (US $ mil) | | | | | | | | |
| *N* (>1%) | 9 | | 9 | | 9 | | 9 | |
| C4 | 60.2 | | 61.6 | | 59.4 | | 56.2 | |
| HHI | 1,231 | | 1,263 | | 1,217 | | 1,008 | |
| Noam | 410 | | 421 | | 406 | | 336 | |

*Source*: Elaborated by the author based on data obtained from ICA.

[1]Table presents market shares of producers/studios, as well as those of the distributors in Portugal who represent several studios. For purposes of concentration measures, we use the producers/studios.

Columbia, and Castello Lopes—had 97.2% audience market share (an average of 92.5% between 2004 and 2010) and 73.3% of the premiere movie market share, a growth of 26.2% when compared to 2004. Between 2004 and 2010, there was an average of 271 feature films per year: 48.1% were from the United States, 35.1% were from Europe, and 17% were coproductions in other regions. Portugal's national production is low. In 2009 and 2010, there were only 22 and 30 films produced in Portugal, respectively. According to a report elaborated by ICA, the market share of domestic films in Portugal is one of the lowest in Europe: "in 2009, the quota was 2.5%, this is the second lowest in the EU-15, occupying the 18th position on the EU's 27 member states ranking" (ICA, 2010).

## TELECOMMUNICATIONS MEDIA

### Wireline Telecom

The 21st century started with Portugal Telecom (PT) leading the telecommunications sector. Nonetheless, there has been a decrease in the penetration rate of the landline phone in Portugal, in part due to the impact of the expansion of mobile communications. Another reason for this trend is related to the expense of landline telephone service in Portugal. In 2006, Portugal had one of the most expensive landline rates, with prices 30% to 40% higher than the EU15 group average. The market for landline telephone service in Portugal has matured and is currently in decline. As Table 10.17 shows, the market share (quantified by consumers with direct access) has steadily decreased from 68.5% in 2007 to 55.3% in 2010.

In the past, the landline communications sector in Portugal was highly concentrated because PT had a very high market share in terms of the number of calls and call duration. The entrance of new competitors has caused PT to lose market share, although it was still the market leader with 68.5% by the end of 2007 and 55.3% by the end of 2010. This segment continues to be highly concentrated, although the HHI decreased from 5,211 in 2007 to 3,653 by 2010.

In Table 10.18, we clearly see Portugal Telecom's leadership between 2007 and 2011.

Table 10-17. Market Share of Fixed Telephone Service (%)

| DIRECT ACCESS CLIENTS | 2007 | 2008 | 2009 | 2010 |
|---|---|---|---|---|
| Portugal Telecom (incumbent) | 68.50 | 63.80 | 57.50 | 55.30 |
| Alternative providers | 31.50 | 36.20 | 42.50 | 44.70 |
| ZON Multimédia (Portugal) | *0.10* | *5.70* | *13.50* | *16.40* |
| Optimus (Portugal) | *21.40* | *19.00* | *16.80* | *16.00* |
| Vodafone (UK) | *1.40* | *3.10* | *4.10* | *4.20* |
| Cabovisão[1] (Cogeco, Canada) | *7.70* | *7.50* | *7.20* | *7.20* |
| ArTelecom (Portugal) | *0.40* | *0.80* | *0.80* | *0.80* |
| Others | *0.50* | *0.10* | *0.10* | *0.10* |
| Total | 100 | 100 | 100 | 100 |
| Revenue Estimate (US $mil) | 1,215 | | | 1,658 |
| C4 | 99 | 96 | 95 | 95 |
| HHI | 5,212 | 4,530 | 3,840 | 3,653 |
| *N* (>1%) | 5 | 5 | 5 | 5 |
| Noam Index | 2,331 | 2,026 | 1,717 | 1,634 |

*Source*: Elaborated by the author from ANACOM data.
[1]Purchased by Altice in 2012.

Table 10-18. Gross Income of Telecommunication Groups (€)

| | 2007 | | 2008 | | 2009 | | 2010 | | 2011 | |
|---|---|---|---|---|---|---|---|---|---|---|
| | REVENUES | % | REVENUES | % | REVENUES | % | REVENUES | % | REVENUES | % |
| Sonaecom | 892,693,691 | 21.5 | 976,200,000 | 23.0 | 949,400,000 | 22.5 | 920,700,000 | 22.3 | 864,000,000 | 21.7 |
| Vodafone Portugal | 1,377,841,059 | 33.1 | 1,493,991,389 | 35.2 | 1,472,745,543 | 34.9 | 1,400,137,007 | 34.0 | 1,315,049,278 | 33.1 |
| Portugal Telecom | 1,891,889,967 | 45.5 | 1,779,319,662 | 41.9 | 1,793,949,224 | 42.6 | 1,800,196,339 | 43.7 | 1,797,273,768 | 45.2 |
| Total | 4,162,424,717 | 100.0 | 4,249,511,051 | 100.0 | 4,216,094,767 | 100.0 | 4,121,033,346 | 100.0 | 3,976,323,046 | 100.0 |
| C4 | 100 | | 100 | | 100 | | 100 | | 100 | |
| HHI | 3,621.5 | | 3,516.9 | | 3,537.8 | | 3,561.7 | | 3,608.9 | |
| *N* (>1%) | 3 | | 3 | | 3 | | 3 | | 3 | |
| Noam Index | 2,090.9 | | 2,030.5 | | 2,042.6 | | 2,056.3 | | 2,083.6 | |

*Source*: Elaborated by the author based on data obtained from Marktest.

Table 10-19. Main Business Metrics of the Mobile Phone Industry and Revenues (Millions €)

| TMN (PORTUGAL TELECOM) | 2002 | | 2003 | | 2004 | | 2005 | | 2006 | | 2007 | | 2008 | | 2009 | |
|---|---|---|---|---|---|---|---|---|---|---|---|---|---|---|---|---|
| | REVENUES | % | REVENUES | % | REVENUES | % | REVENUES | % | REVENUES | % | REVENUES | % | REVENUES | % | REVENUES | % |
| Subscribers | 4,426.0 | 46 | 4,887.0 | 47 | 5,053.0 | 43 | 5,312.0 | 44 | 5,704.0 | 44 | 6,261.0 | 44 | 6,933.0 | 44 | 7,252.0 | 44 |
| Total revenues | 1,474.8 | 47 | 1,522.6 | 46 | 1,606.3 | 45 | 1,557.1 | 44 | 1,502.4 | 43 | 1,542.9 | 42 | 1,593.6 | 43 | 1,517.8 | 43 |
| EBTIDA | 623.2 | 60 | 689.9 | 58 | 746.7 | 56 | 673.5 | 54 | 658.7 | 51 | 679.0 | 50 | 681.9 | 50 | 674.1 | 49 |
| ARPU | 27.1 | - | 25.2 | - | 24.5 | - | 22.8 | - | 21.0 | - | 19.8 | | 18.1 | - | 16.2 | - |
| Vodafone (UK) | Revenues | % | Revenues | % | Revenues | % | Revenues | % | Revenues | % | Revenues | % | Revenues | % | Revenues | % |
| Subscribers | 3,084.7 | 32 | 3,247.9 | 31 | 4,276.3 | 37 | 4,276.3 | 36 | 4,751.0 | 36 | 5,209.1 | 36 | 5,639.0 | 36 | 5,952.0 | 36 |
| Total revenues | 1,063.6 | 34 | 1,153.2 | 35 | 1,325.1 | 37 | 1,325.1 | 38 | 1,393.0 | 40 | 1,511.3 | 41 | 1,495.7 | 40 | 1,424.3 | 40 |
| EBTIDA | 306.8 | 30 | 350.1 | 30 | 412.4 | 31 | 412.4 | 33 | 471.1 | 36 | 533.2 | 39 | 546.3 | 40 | 545.6 | 39 |
| ARPU | 27.4 | | 27.4 | | 24.6 | | 24.6 | | 22.8 | | 22.2 | | 20.0 | | 17.6 | - |
| Optimus (SonaeCom, Portugal) | Revenues | % | Revenues | % | Revenues | % | Revenues | % | Revenues | % | Revenues | % | Revenues | % | Revenues | % |
| Subscribers | 2,119.8 | 22 | 2,305.8 | 22 | 2,350.0 | 20 | 2,350.0 | 20 | 2,601.9 | 20 | 2,893.5 | 20 | 3,191.6 | 20 | 3,432.6 | 21 |
| Total revenues | 607.5 | 19 | 636.2 | 19 | 627.0 | 18 | 627.0 | 18 | 610.4 | 17 | 619.4 | 17 | 629.1 | 17 | 607.0 | 17 |
| EBTIDA | 107.2 | 10 | 146.5 | 12 | 167.0 | 13 | 167.0 | 13 | 169.1 | 13 | 153.7 | 11 | 142.4 | 10 | 166.7 | 12 |
| ARPU | 23.9 | | 22.4 | | 20.6 | | 20.6 | | 19.7 | | 18.2 | | 16.8 | | 14.8 | - |
| Total Mobile Market | | | | | | | | | | | | | | | | |
| Subscribers | 9,631 | | 10,441 | | 11,679 | | 11,938 | | 13,057 | | 14,364 | | 15,764 | | 16,637 | |
| New Clients | 972 | | 810 | | 1,082 | | 1,174 | | 1,114 | | 1,308 | | 1,408 | | 872 | |
| Total Revenues | 3,146 | | 3,312 | | 3,558 | | 3,509 | | 3,506 | | 3,674 | | 3,718 | | 3,549 | |
| EBTIDA | 1,037 | | 1,187 | | 1,326 | | 1,253 | | 1,299 | | 1,366 | | 1,371 | | 1,359 | |
| Revenue (US $mil) | 2,965 | | 3,756 | | 4,427 | | 4,374 | | 4,405 | | 5,035 | | 5,471 | | 4,949 | |
| *N* (>1%) | 3 | | 3 | | 3 | | 3 | | 3 | | 3 | | 3 | | 3 | |
| C3 | 100 | | 100 | | 100 | | 100 | | 100 | | 100 | | 100 | | 100 | |
| HHI | 3,714 | | 3,695 | | 3,735 | | 3,714 | | 3,718 | | 3,741 | | 3,741 | | 3,732 | |
| Noam | 2,144 | | 2,133 | | 2,156 | | 2,144 | | 2,147 | | 2,160 | | 2,160 | | 2,155 | |

*Source:* Elaborated by the author from ANACOM data.

Table 10-20. Mobile Phone Service Operators Market Shares by Gross Income(€)

| | 2007 | | 2008 | | 2009 | | 2010 | | 2011 | |
|---|---|---|---|---|---|---|---|---|---|---|
| | REVENUES | % | REVENUES | % | REVENUES | % | REVENUES | % | REVENUES | % |
| Optimus | 604,384,809 | 17.4 | 629,100,000 | 17.2 | 860,697,911 | 21.9 | 814,969,323 | 21.9 | 795,517,765 | 22.8 |
| Vodafone (UK) | 1,377,841,059 | 39.7 | 1,493,991,389 | 40.8 | 1,472,745,543 | 37.5 | 1,400,137,007 | 37.7 | 1,315,049,278 | 37.7 |
| TMN | 1,492,979,633 | 43.0 | 1,534,831,270 | 42.0 | 1,592,225,805 | 40.6 | 1,499,349,998 | 40.4 | 1,376,495,700 | 39.5 |
| Total | 3,475,205,501 | 100.0 | 3,657,922,659 | 100.0 | 3,925,669,259 | 100.0 | 3,714,456,328 | 100.0 | 3,487,062,743 | 100.0 |
| C4 | 100.0 | | 100.0 | | 100.0 | | 100.0 | | 100.0 | |
| HHI | 3,720.0 | | 3,724.5 | | 3,533.2 | | 3,531.6 | | 3,500.9 | |
| *N* (>1%) | 3 | | 3 | | 3 | | 3 | | 3 | |
| Noam Index | 2,147.8 | | 2,150.3 | | 2,039.9 | | 2,039.0 | | 2,021.2 | |

*Source:* Elaborated by the author based on data obtained from Marktest.

Among the three groups, SonaeCom reached the lowest income during that period.

## Wireless Telecom

Portugal has one of the highest penetration rates of mobile phone service (MPS) in Europe. In 2009, there were 16.5 million subscribers in a population of 10.5 million.

The innovations of 2005 can be considered a turning point in the history of telecommunications. TMN (owned by Portugal Telecom), Vodafone (subsidiary of the UK corporation), and Optimus (owned by the Portuguese Group, Sonaecom) launched new low-cost offers, associated with lower and simpler tariffs and specific brands: Uzo, Vodafone Directo, and Rede 4, respectively. At the same time, companies in this sector began to invest in developing stable relationships with their customers by increasing the number of customer service centers and renovating existing ones, focusing on quality of service, and improving telephone support services for resolving problems. These measures resulted in greater customer engagement, higher customer acquisition rates, and increased revenues. Table 10.19 presents the main business metrics of the mobile phone industry from 2002 to 2009.

Although TMN holds the highest market share in terms of revenues and total number of customers, Vodafone leads in terms of new added clients. By the end of 2004, TMN had added 167,000 new customers, while Vodafone had added 690,000.

The mobile communications market, regardless of the year in question, is highly concentrated. This is mainly due to the high barriers to entry, such as the high investment costs of network infrastructures. TMN has been the market leader since the beginning, followed by Vodafone and then Optimus, which is part of Sonae Group and mainly owned by domestic shareholders. Although there have been some slight oscillations in market shares throughout the years, they have not caused a shift in the rank order of operators during the period analyzed.

As shown in Table 10.20, TMN maintained the market dominance between 2007 and 2011, closely followed by Vodafone. The market has remained relatively stable.

# INTERNET MEDIA

## Internet Service Providers

Internet penetration in Portugal represents one of the most successful examples of the benefits from research and development in the telecom infrastructure. According to ANACOM, broadband penetration has achieved an impressive growth rate since 2002 (Table 10.21). Between 2002 and early 2010, wired Internet users increased by about 88%, mainly because of the increase in ADSL users (ANACOM, 2012).

Despite this scenario, from 2007 onward there was a slower growth rate due to the entrance of mobile service in Portugal. Between 2007 and the quarter of 2010, there was an increase in mobile broadband of about 61% while wired broadband grew about 25%. In comparison with the EU, in 2008 Portugal was the only country, apart from Slovakia, with a high penetration of mobile broadband, reaching 22.4% by the end of that year.

Regarding the number of customers with broadband access, Portugal Telecom is the market leader, with 48.2% market share in 2011. ZON, the leading cable operator, is the second biggest broadband player, with almost 32.9% market share. Together, both operators accounted for more than 80% of the market by the end of 2011, thus enabling us to conclude that broadband is a highly concentrated market (Table 10.22).

One can ascertain that the Internet service sector is highly concentrated both in mobile communications (led by TMN group, followed by Vodafone and Optimus) and in wired communications (led by PT Group).

The values in Table 10.23, showing the revenue generated by Internet access services, were calculated based on total revenues of broadband access services and on the market shares of the respective operators, in the report of the situation of communications published at ANACOM's website.

Overall, total revenues generated have suffered annual increases throughout the period analyzed. However, in individual terms, there are several changes to register. Ar Telecom and

Table 10-21. Broadband Internet Main Access Indicators

| | 2002 | | 2003 | | 2004 | | 2005 | | 2006 | | 2007 | | 2008 | | 2009 | | 2010 | |
|---|---|---|---|---|---|---|---|---|---|---|---|---|---|---|---|---|---|---|
| | (THOU.) | | (THOU.) | | (THOU.) | | (THOU.) | | (THOU.) | | (THOU.) | | (THOU.) | | (THOU.) | | (THOU.) | |
| Users—Total | 261.9 | 100 | 503.0 | 100 | 842.5 | 100 | 1,193.8 | 100 | 1,458.9 | 100 | 1,549.0 | 100 | 1,692.3 | 100 | 1,925.8 | 100 | 2,036.7 | 100 |
| → By ADSL | 52.8 | 20.7 | 184.8 | 36.7 | 424.1 | 50.3 | 700.4 | 58.7 | 916 | 62.8 | 927.7 | 59.9 | 996.5 | 58.9 | 1,108.2 | 57.5 | 1,119.2 | 55 |
| → By cable | 205.4 | 78.4 | 314.6 | 62.5 | 415.1 | 49.3 | 490.1 | 41.05 | 537.6 | 36.84 | 605.7 | 39.1 | 669 | 39.5 | 760,6 | 39.5 | 813.6 | 40 |
| → Others | 3.6 | 1.4 | 3.5 | 0.7 | 3.2 | 0.38 | 3.2 | 0.27 | 5.2 | 0.36 | 15.5 | 1 | 26.6 | 1.57 | 56.5 | 2.9 | 103.8 | 5.1 |
| Penetration of fixed broadband | 2.50 | | 4.80 | | 8.10 | | 11.30 | | 13.80 | | 14.70 | | 15.40 | | 17.50 | | 18.60 | |
| Mobile Access | (thou.) | | (thou.) | | (thou.) | | (thou.) | | (thou.) | | (thou.) | | (thou.) | | (thou.) | | (thou.) | |
| Users—Total | – | | – | | – | | – | | – | | 1,454.5 | 100 | 2,378.8 | 100 | 3,792.5 | 100 | 3,757.1 | 100 |
| Active Users | – | | – | | – | | – | | – | | 659.8 | 45.5 | 1,160.7 | 48.8 | 2,169.8 | 57.2 | 244.4 | 57.74 |
| *Modem* | – | | – | | – | | – | | – | | – | | – | | – | | 1,291.2 | 34.37 |
| Penetration of Mobile Wideband | – | | – | | – | | – | | – | | 13.70 | | 22.40 | | 35.70 | | 35.30 | |
| Operators in activity | 32 | | 25 | | 0 | | 30 | | 28 | | 35 | | 37 | | 35 | | 32 | |

*Source:* Elaborated by the author from ANACOM data.

Table 10-22. Market Share of Internet Access Providers (%)

| | 2008 | 2009 | 2010 | 2011 |
|---|---|---|---|---|
| Portugal Telecom Group* | 41.1 | 43.6 | 46.0 | 48.2 |
| ZON Multimédia* | 27.7 | 31.6 | 32.2 | 32.9 |
| Cabovisão**(Cogeco/Canada) | 9.8 | 7.9 | 7.9 | 7.5 |
| Optimus* | 14.2 | 10.4 | 7.4 | 5.7 |
| TVTEL* | 1.4 | — | — | — |
| Vodafone (UK) | 2.4 | 3.9 | 4.3 | 4.2 |
| Ar Telecom* | 1.5 | 1.5 | 1.3 | 0.8 |
| ONITELECOM* | 0.4 | 0.3 | 0.3 | 0.2 |
| Other Providers* | 1.4 | 0.8 | 0.7 | 0.6 |
| Total | 100 | 100 | 100 | 100 |
| Revenue Estimate (US $mil) | | 1,658 | | |
| C4 | 93 | 94 | 93 | 94 |
| HHI | 2,771 | 3,091 | 3,287 | 3,510 |
| *N* (>1%) | 8 | 6 | 6 | 5 |
| Noam Index | 980 | 1,262 | 1,342 | 1,570 |

*Source:* Elaborated by the author from ANACOM.
*Portuguese companies and investors mainly own all these players.
**Purchased by ALTICE in 2012.

ONITELECOM suffered declines in their revenues from the Internet throughout the whole period analyzed, while the operator Optimus suffered decreases in the years 2009 and 2010. These declines may be caused by the exponential increase in the revenues of access packages, which are not separately identifiable, which include Internet access service, where these operators do not act. The revenues concentrate mainly on the operators PT and ZON, which have also been systematically increasing their market shares.

Table 10-23. Broadband Services Revenues by Operator

| | 2008 | 2009 | 2010 | 2011 |
|---|---|---|---|---|
| PT Group | 254,890 | 283,924 | 317,301 | 518,805 |
| ZON Group | 185,681 | 203,811 | 219,788 | 346,562 |
| Cabovisão | 54,790 | 49,653 | 53,251 | 75,776 |
| Optimus | 74,707 | 58,280 | 44,512 | 56,404 |
| Vodafone | 18,563 | 26,602 | 28,537 | 44,593 |
| Ar Telecom | 9,595 | 9,086 | 7,939 | 8,87 |
| ONITELECOM | 2,340 | 2,088 | 1,810 | 1,127 |
| Other providers | 5,169 | 4,618 | 4,057 | 5,928 |

*Source:* Elaborated by the author from ANACOM.

## Search Engines

Google clearly dominates the Portuguese market, with almost 98% market share, while the other operators are residual (including the Portuguese search engine Sapo). Google's leading position has remained practically unchanged, as is shown in Table 10.24. The extremely concentrated market is illustrated by an HHI of 9,500 for the three years studied.

Despite being able to identify the market share of search engines in Portugal, we could not find official information about the turnover generated by this activity in the domestic market. One of the reasons for this lack of official information is that Google (which, as shown in the table, holds almost 100% of the market), does not own any company legally constituted in Portugal. It operates through a support office, with no telephone contacts available on the Internet. The support provided, especially in matters of institutional communication and relationships with government and other public and private entities, is complemented by Weber & Shandwick Group, an American consultant integrated in the Interpublic Group. The operations in Portugal are also followed by an official representing the Iberian market). We estimate the turnover resulting from the search engines in Portugal, based on the investment in digital advertising in 2011, to be about €32.88M (data obtained by the author directly from advertising central sellers). These advertising experts believe that about 80% of the advertising spending for search engines adds up to €26.31M. Moreover, assuming that about 50% of advertising spending in search engines is not related to the trade advertising centers (on the contrary, they are the result of direct agreements between companies and search engines), we can estimate a total value of about €52.61M.

### Table 10-24. Market Share of Search Engines 2008–2010 (%)

| SEARCH ENGINES | 2008 | 2009 | 2010 |
|---|---|---|---|
| Google (US) | 97.86 | 97.91 | 97.42 |
| Microsoft (US) | | | |
| Windows Live | 0.91 | 0.27 | |
| Bing | | 1.1 | 1.77 |
| Yahoo (US) | 0.49 | 0.48 | 0.42 |
| Sapo | | | 0.24 |
| Ask (US) | 0.07 | 0.12 | 0.11 |
| AltaVista (US) | 0.05 | 0.02 | 0.01 |
| AOL (US) | 0.01 | 0 | 0 |
| Yandex (Russia) | 0 | 0.01 | 0.01 |
| Others | 0.6 | 0.02 | 0.02 |
| Total | 100 | 100 | 100 |
| Indexes | | | |
| C4 | 99.33 | 99.76 | 99.85 |
| HHI | 9,578 | 9,588 | 9,494 |
| *N* (>1%) | 1 | 2 | 2 |
| Noam | 9,578 | 6,780 | 6,713 |

*Source:* StatCounter—GlobalStats.

## Online Publishing

Because of the dynamic nature of online publications, it is important to first analyze the weight of each group in this segment and then consider the extent to which online publications could contribute to the concentration levels of the communication market. The data presented are related to the number of unique users for news and information websites. Due to incomplete information about audiences of each publication in recent years, it is not yet possible to analyze the concentration levels of each publishing segment.

The data presented in Table 10.25 allow us to conclude that generally the main media groups of the offline press are also important in digital publications. In fact, the concentration index C4 was about 50% in 2010, a relatively moderate level. The concentration levels between 2002 and 2010 exhibit a decreasing trend. Media Capital and Impresa are always among the top four media groups. The conclusions are similar to the previous analysis, where moderate degrees of concentration were observed.

There is no official information available regarding the advertising investment on the online media. However, in order to obtain

Table 10-25. Online News Users by Site

| | 2002 | | 2003 | | 2004 | | 2005 | | 2006 | | 2007 | | 2008 | | 2009 | | 2010 | |
|---|---|---|---|---|---|---|---|---|---|---|---|---|---|---|---|---|---|---|
| | USERS | % | USERS | % | USERS | % | USERS | % | USERS | % | USERS | % | USERS | % | USERS | % | USERS | % |
| A Bola | 302 | 6.9 | 409 | 8.1 | 457 | 8.3 | 520 | 8.5 | 1,072 | 8.2 | 1,030 | 7.4 | 1,028 | 7.8 | 1,320 | 7.5 | 1,882 | 7.3 |
| Cofina | 616 | 14.0 | 745 | 14.7 | 806 | 14.7 | 902 | 14.7 | 1,771 | 13.6 | 1,859 | 13.4 | 1,720 | 13.0 | 2,182 | 12.4 | 2,960 | 11.5 |
| Controlinveste | 0 | | 0 | | 0 | | 0 | | 0 | | 1,943 | 14.0 | 1,858 | 14.1 | 2,419 | 13.7 | 3,119 | 12.2 |
| Economica | 104 | 2.4 | 98 | 1.9 | 123 | 2.2 | 150 | 2.5 | 395 | 3.0 | 509 | 3.7 | 507 | 3.8 | 799 | 4.5 | 1,345 | 5.2 |
| Impala | 0 | | 0 | | 0 | | 0 | | 115 | 0.9 | 108 | 0.8 | 100 | 0.8 | 176 | 1.0 | 176 | 0.7 |
| Impresa | 901 | 20.4 | 922 | 18.1 | 966 | 17.6 | 1,029 | 16.8 | 1,991 | 15.3 | 2,039 | 14.7 | 1,842 | 14.0 | 2,508 | 14.2 | 3,240 | 12.6 |
| Lusomundo | 679 | 15.4 | 736 | 14.5 | 793 | 14.5 | 866 | 14.1 | 1,744 | 13.4 | 0 | | 0 | | 0 | | 0 | |
| Media Capital (PRISA) | 767 | 17.4 | 753 | 14.8 | 784 | 14.3 | 817 | 13.3 | 1,874 | 14.4 | 1,921 | 13.9 | 1,690 | 12.8 | 2,345 | 13.3 | 2,984 | 11.6 |
| Motorpress | 66 | 1.5 | 108 | 2.1 | 141 | 2.6 | 183 | 3.0 | 497 | 3.8 | 690 | 5.0 | 984 | 7.5 | 1,264 | 7.2 | 1,654 | 6.5 |
| RDP/RTP (public) | 326 | 7.4 | 464 | 9.1 | 469 | 8.5 | 556 | 9.1 | 1,240 | 9.5 | 1,505 | 10.9 | 1,469 | 11.1 | 1,842 | 10.4 | 2,337 | 9.1 |
| Renascença | 190 | 4.3 | 252 | 5.0 | 290 | 5.3 | 339 | 5.5 | 810 | 6.2 | 802 | 5.8 | 749 | 5.7 | 980 | 5.6 | 1,301 | 5.1 |
| Sonae.com | 457 | 10.4 | 595 | 11.7 | 660 | 12 | 765 | 12.5 | 1,500 | 11.5 | 1,468 | 10.6 | 1,252 | 9.5 | 162 | 0.9 | 2,051 | 8.0 |
| IOnline | 0 | | 0 | | 0 | | 0 | | 0 | | 0 | | 0 | | 444 | 2.5 | 1,252 | 4.9 |
| Sol | 0 | | 0 | | 0 | | 0 | | 0 | | 0 | | 0 | | 1,214 | 6.9 | 1,344 | 5.2 |
| Total Media Groups | 4,408 | 100 | 5,082 | 100 | 5,489 | 100 | 6,127 | 100 | 13,009 | 100 | 13,874 | 100 | 13,199 | 100 | 17,655 | 100 | 25,645 | 100 |
| C4 | 63.6 | | 62.1 | | 61.1 | | 58.9 | | 56.7 | | 56 | | 53.9 | | 53.6 | | 47.9 | |
| HHI | 1,389.1 | | 1,291.3 | | 1,265.4 | | 1,230.8 | | 1,158.4 | | 1,143.8 | | 1,105.8 | | 1,043.5 | | 921.4 | |
| *N* (>1%) | 10 | | 10 | | 10 | | 10 | | 10 | | 10 | | 10 | | 11 | | 12 | |
| Noam | 439.3 | | 408.4 | | 400.2 | | 389.2 | | 366.3 | | 361.7 | | 349.7 | | 314.6 | | 266.0 | |

*Source:* Elaborated by the author based on data obtained from Marktest
*Note:* Top Media groups, according to this variable, are highlighted.

Table 10-26. Advertising Sold by Central Sellers

| SUPPORT | CENTRAL MARKETS | | | | |
|---|---|---|---|---|---|
| | 2008 | 2009 | 2010 | 2011 | 2012 |
| Digital | 20,127,000 | 23,252,000 | 28,908,000 | 32,882,000 | 35,119,000 |

*Source:* Estimative elaborated by the author based on data obtained directly from central advertising sellers

reference value, the author managed to gather information directly from the advertising agencies operating in Portugal, as one can see on Table 10.26.

As mentioned, these data result from the amount exclusively traded by the trade centers, so they do not represent the total advertising market. According to some experts (from the central advertising sales agencies and the Press Association) approached by the author, this value is equivalent to about 50%, so we can estimate that the total online advertising market reaches around €70.238.000(x2).

## Analysis of Cross-Media Concentration by Advertising Revenues

As already mentioned, for the measurement of concentration levels several variables were analyzed for each media segment. For reasons of space, we have chosen only a few of these variables. Regardless of the variable used, the results suggest that the media sector is considerably highly concentrated and that these levels of concentration must be treated with caution and contextualized within the media system in Portugal.

When analyzed in a cross media perspective through the obtained advertising revenue, the concentration levels confirm that the media sector is dominated by two groups: Impresa and Media Capital. They capture advertisement investment of about 300 million euros, while the advertising revenue raised by the other players ranges from 70 to 20 million euros. Table 10.27 shows the advertising revenues from the main media players in Portugal.

According to Robert Picard,[8] media in countries with fewer than 10 million people tends to be concentrated in two or three large groups. In Portugal, the market lies in the hands of four major media groups, which hold 84% of advertisement investment (Table 10.27). The market positions of both groups have also remained relatively stable, as shown by the application of concentration indexes presented at the end of the table. The sample is robust in that it represents more than 70% of the media advertising market in Portugal. There are no precise numbers about the advertising market in Portugal, nor are there statistics on the local and regional media or any smaller players outside of the leading media groups or major newspaper companies. It is estimated that as a whole it may vary between 600 and 800 million euros. Therefore, if we point to an intermediate value (700 million), this sample represents about 70% of the market.

The issue of corporate concentration of media in Portugal has been an integral part of the political agenda and has been a target of analysis and intervention for the regulators. However, the debate has been more focused on a monomedia perspective over a multimedia approach that would allow the measurement of a media group's (and its segments') real potential in the advertising market, as the following table suggests. The target of an in-depth review by regulators was one of the most relevant and controversial situations, and it occurred when Controlinveste purchased Lusomundo Media

8. At a presentation performed in March 2009 in Portugal in a conference organized by the Portuguese Media Confederation.

Table 10-27. Advertising Incomes by Media Group Segment (2006–2010)

| MAIN MEDIA COMPANIES | 2006 | | 2007 | | 2008 | | 2009 | | 2010 | |
|---|---|---|---|---|---|---|---|---|---|---|
| | REVENUES | % SEGMENT INCOME | REVENUES | % SEGMENT INCOME | REVENUES | % SEGMENT INCOME | REVENUES | % SEGMENT INCOME | REVENUES | % SEGMENT INCOME |
| Media Capital (PRISA) | 179.2 | | 167.9 | | 174.4 | | 150.9 | | 151 | |
| Television | *143* | *79.8* | *146* | *86.96* | *153.1* | *87.79* | *135.2* | *89.6* | *134.9* | *89.34* |
| Radio | *12.8* | *7.14* | *12.9* | *7.68* | *13.3* | *7.63* | *12.5* | *8.28* | *12.7* | *8.41* |
| Others (Internet/press) | *6.5* | *3.63* | *9* | *5.36* | *6.7* | *3.84* | *3.2* | *2.12* | *3.3* | *2.19* |
| Productions | | | | | 1.3 | 0.75 | | | | |
| Outdoors | *16.9* | *9.43* | | | | | | | | |
| Impresa | 168 | | 170.8 | | 166.9 | | 143.9 | | 157.1 | |
| Television | *115.8* | *68.93* | *115.8* | *67.8* | *109.2* | *65.43* | *92.7* | *64.42* | *105.9* | *67.41* |
| Newspapers and Magazines | *52.2* | *31.07* | *53.5* | *31.32* | *55.5* | *33.25* | *44.8* | *31.13* | *44.4* | *28.26* |
| Digital | | | *1,5* | *0.88* | *2.2* | *1.32* | *6.4* | *4.45* | *6.8* | *4.33* |
| Controlinveste | 70.8 | | 68.8 | | 65.7 | | 57.3 | | 59.5 | |
| Newspapers and Magazines | *61.2* | *86.44* | *60.1* | *87.35* | *57.5* | *87.52* | *50.2* | *87.61* | *52.1* | *87.56* |
| Radio | *9.6* | *13.56* | *8.7* | *12.65* | *8.2* | *12.48* | *7.2* | *12.57* | *7.4* | *14.2* |
| Cofina | 55.1 | | 61.9 | | 60.2 | | 52.7 | | 54.9 | |
| Newspapers | *40.1* | *72.78* | *44* | *71.08* | *44.8* | *74.42* | *40.1* | *76.09* | *41* | *74.68* |
| Magazines | *15* | *27.22* | *17.9* | *28.92* | *15.4* | *25.58* | *12.6* | *23.91* | *13.9* | *25.32* |
| RTP (Public) | 48.1 | | 51.5 | | 54.2 | | 48.6 | | 49.9 | |
| Radio and Television | *48.1* | *100* | *51.5* | *100* | *54.2* | *100* | *48.6* | *100* | *49.9* | *100* |

*continued*

Table 10-27. *continued*

| MAIN MEDIA COMPANIES | 2006 | | 2007 | | 2008 | | 2009 | | 2010 | |
|---|---|---|---|---|---|---|---|---|---|---|
| | REVENUES | % SEGMENT INCOME | REVENUES | % SEGMENT INCOME | REVENUES | % SEGMENT INCOME | REVENUES | % SEGMENT INCOME | REVENUES | % SEGMENT INCOME |
| Renascença (Church) | 20.3 | | 18.4 | | 17.6 | | 15.4 | | 15.9 | |
| Radio | *20.3* | *100* | *18.4* | *100* | *17.6* | *100* | *15.4* | *100* | *15.9* | *100* |
| Sonaecom | 15.6 | | 14.3 | | 13.5 | | 13.2 | | 12.2 | |
| Newspapers | *15.6* | *100* | *14.3* | *100* | *13.5* | *100* | *13.2* | *100* | *12.2* | *100* |
| Total | 557.1 | | 553.6 | | 552.5 | | 481.9 | | 500.6 | |
| C4 | 85 | | 85 | | 85 | | 84 | | 84 | |
| HHI | 2,299 | | 2,255 | | 2,281 | | 2,252 | | 2,272 | |
| Noam | 869 | | 852 | | 862 | | 851 | | 859 | |

*Source:* Annual accounts of all companies; Renascença and Controlinveste for 2009 and 2010 obtained by extrapolations, applying market's average to the previous year of each company. The data about Controlinveste from 2006, 2007, and 2008 were provided directly from the board company because there are no official data.

*Note:* Media Capital has sold the press and outdoors assets on 2007 to focus on audiovisual business.

from Portugal Telecom in 2005. At that time, this operation was problematic because it could have potentially lead to a dominant position in the relevant market of the daily newspapers. After several studies were conducted, it was proven that the operation would neither increase the concentration, nor lead to predatory market practices. On the other hand, it was also demonstrated that initially the market had not been properly defined and it did not include free papers or daily regional and local journals. Ten months later, the operation was approved.[9] In a more recent situation, regulators have been examining the "complaint" filed by Controlinveste on Cofina's dominant position in the free daily newspapers segment after it acquired *Metro*, which joined *Destak*, the two major dailies in this segment. Specifically, Controlinveste accuses Cofina of having a market position that may lead to predatory practices in the relevant market of free daily advertising.

## CONCLUSIONS

An analysis of the media business concentration measurement indexes leads to the conclusion that it is a sector with considerable concentration levels. Whatever the segment concerned (radio, television, or press), the data show that there are a few operators dominating the market. Several concentration indexes and three models were selected for study: the concentration index C4, the HHI, and the Noam Index. Apart from online publishing, the data reveal a high degree of concentration. However, we do not have enough data to prove that this high degree of concentration has negative impacts related to competition or predatory practices by dominant operators. We cannot conclude that this situation leads to a decrease in information pluralism.

Taking into account the analysis of all the segments, one of the main conclusions is that although there are considerable concentration levels, there has been a trend to reduce those levels, mainly in the press and TV markets. In turn, the moderate concentration level in the online publications that integrate the main communication groups could, from a pluralism perspective, be linked to an increase in the diversity of voices and contribute to a greater balance of market forces. The analysis of information pluralism should be studied from a holistic approach for both offline and online outlets. In this context, the online publications disclose that positive evolution can be perceived due to the growing weight of the digital world in terms of a growth in the size of audiences and the amount of information consumed. This circumstance signifies some moderate optimism related to the existence of a greater plurality of voices in the market.

In Portugal, the property concentration titles issue has been more focused on the daily press. In effect, the high concentration level revealed by the national daily press industry is not surprising given the dominance of Controlinveste and Cofina, which hold a sizeable market share. Despite the existence of high concentration levels, as was proven in this study, one should emphasize that these groups can hardly exploit their dominant position in the market of the national daily press for several reasons.

The barriers to entry by new players into the press industry are relatively limited. In contrast with radio or TV, in which operators need licenses, in the press there is greater freedom of entrance. Licenses are not needed and the dynamics of the market itself allow the creation of alternatives. In this way, one can infer that the legal possibility of free entrance without the need to have a license to work limits the power of existing companies already established in the market. In the case of the national daily press, the entrance mechanism seems to function very well and the distribution boom of free newspapers is proof of this hypothesis, as are the new daily titles such as *i informação*, a property of the Sojormedia/Lena Group.

In Portugal there is an important tradition of weekly newspapers such as *Expresso* (average circulation per edition of 124,143), *SOL* (62,586), and weekly magazines such as *Visão*

9. On this matter, see the book *Concentração, Pluralismo e Regulação dos Media*, Faustino, Paulo 2011. Media XXI Collection.

(116,183) and *Sábado* (77,639). These newspapers have a central role in the Portuguese newspaper market due to the competitive pressure in the national daily press market. On the other hand, there is also an established cultural tradition of reading local and regional weekly newspapers such as *Mirante* (24,877 weekly circulation), *Região de Leiria* (13,091), and *Jornal do Fundão* (15,986) (APCT, 2012). Once again, although these newspapers are not included in the national market of daily press, they end up having a competitive role in that market.

TV and radio markets present high levels of concentration and lack a diversity of voices in national coverage. In the press case, in spite of the absolute levels of concentration revealed in the national daily press market, the increase of the number of voices since 2005 seems to be contributing to slightly reduced concentration levels in this segment. In turn, the emergence of the free newspapers conveys a new dynamic for the industry of national daily press. This trend is quite recent. These newspapers currently represent an important role in the configuration and supply of the national daily press market, holding higher market shares (in terms of circulation and audience) than paid newspapers.

The distribution of free newspapers, as well as the online publications, created alternatives and replaced established patterns. Free newspapers stole share from the paid newspapers, and online publications reinforced their position as a complementary or replacement information product. It is also important to mention that the introduction of the free press has contributed to the expansion of the total size of the market by attracting new readers to the industry, namely young urban consumers. Adding the two effects, in the perspective of information pluralism and diversity of voices, the total impact of digital newspapers and free newspapers seems to create a positive impact on the market, contributing to an increase in the total circulation (at least until 2007). Free newspapers and digital newspapers have stimulated the pluralism and diversity of information by increasing the number of active voices in the industry.

With regard to free television, the market is concentrated but is relatively competitive. There are three major players, one public and two private. In the case of pay TV, ZON has more than 70% market share. However, in the television sector we also need to consider the market dynamics, particularly the transition from analog to digital.[10] It is to be expected that the new technologies (broadband, cable, satellite, DTT, etc.) will contribute to the emergence of new operators, mainly in niche areas. Therefore, as other channels beyond cable and satellite arise (DTT and IPTV), the competition will increase. Yet another aspect that may have great impact on the market will be the possible (partial or full) privatization of RTP (a public channel). In 2012, the coalition government PSD-CDS (of liberal philosophy in economic terms) created a goal to reduce the role of the state in the economy, including the media sector, in order to decrease the public deficit.

Nevertheless, the accomplishment of this eventual privatization has been dividing Portuguese society's opinion: those who favor privatization and those who are against it. In fact, putting aside the discussion about the concept of public service television[11] for further research, it is necessary to take a closer look at the impact on

10. The deadline to switch off was January 2012. There is no doubt that this new space represents an opportunity to respond to a growing supply of communication services, since it assumes the possibility of introducing new broadband applications to reduce the digital division in many countries. However, besides the risks resulting from the transition process, especially in the interior due to literacy levels and regional disparities, it is necessary to have a great coordination capacity, so that the citizens recognize the benefits of DTT implementation. This aspect necessarily requires an increased focus on the model of cooperation between telecommunication companies and media companies, so that the content's quality doesn't decrease due to an excessively instrumental perspective of technology's predominant use as a platform for electronic commerce. Along with this concern, there is another one, already mentioned, which is the crucial importance of identifying a good national strategy to (re)use the free space—digital dividend—preventing it from being occupied by foreigners, particularly Spanish or French, which would have predictable negative impacts on Portugal's cultural affirmation.

11. Many critics believe that these goals have not been met. A privatization would need to deal with the high costs involved, that is, the accumulated deficit of RTP about 800 million euros—the equivalent to the total advertising market in 2010.

the market and on the competition in the media sector. Some experts believe that privatization will cause a negative impact on the market. That is, RTP's privatization may have direct and indirect impacts on the difficult financial balance of the current private operators, information pluralism, and on the general decline of programming quality. They defend the idea that privatization generates excessive competition. However, the market is already too competitive. For example, the net advertising revenues are discounted more than 75% off of the list price. This deep discount is now expected by advertisers and hampers the survival of media outlets and their contribution to the information pluralism.

In the last decade, the communications sector in Portugal has experienced significant changes since the emergence of mobile phones. In the last ten years, new opportunities and innovative services appeared which were aligned with new customer needs. Simultaneously, the marketing and communication strategies led by the players in this market, price reductions of the necessary equipment and services, and the introduction of the prepaid system allowed a rapid expansion of mobile phone use. Despite more than 70% of the market being concentrated in two operators (PT/TMN and Vodafone), it is possible to consider the Portuguese market a competitive one because new entrants could emerge at any time without significant barriers to entry. Regarding landline communications, the market is concentrated in Portugal Telecom, particularly in terms of the residential market. In the business market, there is some competition. Though it is a relatively competitive market, regulators such as the Competition Authority and National Authority of Communications, have felt the need to regulate the market.

Some complaints from competitors were due to the fact that access to the networks is not provided on a predictable timeframe, causing some obstacles for consumers who want to switch to a new provider. Whenever a costumer intends to switch to another provider, there is a long and complicated process with their current provider in order to discourage the client from switching. From a regulatory standpoint, a situation in paid television content distribution that occurred in 2006 had a huge impact on the telecommunications sector. The third operator in the communications ranking, Sonaecom, decided to make a Public Offer of Acquisition. According to Lopes *et al.* (2006), Sonaecom made this offer for several reasons: difficulty of organic growth, cycle of market saturation, and excessive market concentration. Though the Public Offer of Acquisition did not materialize, regulators took advantage of this situation to intervene and compel Portugal Telecom (the market leader) to do a spin-off of PT Multimedia (ZON). This a different ownership structure and generated a more balanced market.

When Portugal Telecom lost their cable business, it was temporarily out of the content distribution market for pay television. In 2008, Portugal Telecom reappeared in the pay television business with the release of MEO, broadcasted by satellite.[12] MEO has had a significant evolution, reaching about 700,000 subscribers in only two years. This is a significant accomplishment given the business cycle's maturity as well as the market's small size. Portugal Telecom's interest in the television sector, including free-to-air channel, has been obvious. In 2009, there was a rumor that that Portugal Telecom would take a significant position in TVI. TVI's owner, PRISA, was having difficulty paying off its debt, a difficulty that this sale would alleviate. In the end, the 30% stake that was for sale was purchased by a local investor supported by a partner who was connected to the Liberty Acquisition Holdings. Portugal Telecom ended up not making the deal, due in part to public opinion. People were against the purchase

12. MEO is Portugal Telecom's commercial service that provides four television services: ADSL, fiber optic network, satellite, and TMN's 3G network. The ADSL and the fiber optic networks offer three services: television, Internet, and telephone. The satellite network only distributes television contents. The television transmission via ADSL works through IPTV (Internet Protocol Television), unlike the WebTV (television via Internet). MEO service came after PT Comunicações and PT Multimédia's separation.

because they were afraid it would reduce competition and increase predatory market practices, including advertising investment.

It is also important to consider that the concentration movements of media companies in Portugal result in part from the need of companies to consolidate the internal market, which is very small, and to gain economies of scale to grow their overseas markets. The international growth of media operations of Portuguese companies has been very slow and cautious, representing one of the most passive economic sectors. In the last several years, due to several factors including the global economic crisis' pressure which has affected internal consumption, media companies have started to be more proactive. Ongoing Group has been investing in the Brazilian market. It launched the newspaper *Brasil Econômico,* bought the newspapers *Dia* and *Correio do Povo,* and has been showing public interest in joining the television market. This group has also created some partnerships in Angola. Cofina launched the free daily newspaper in Brazil about four years ago. In this context, we may even consider that the future of media companies in Portugal is dependent on the acceleration of international growth because of the external market's stagnation.

On the other hand, the book publishers face the same problem in terms of internal growth. For this reason, the international expansion to Brazil and to Portuguese-speaking African countries (especially Angola, Mozambique, and Cape Verde) has been attracting some publishers' attention, particularly Porto Editora and the Leya Group, which have attempted to enter these markets by establishing local subsidiaries.

In the telecommunications industry, the target is to grow internationally. Portugal Telecom has made some steps toward that objective by increasingly assuming the role of major operator for the countries. After selling its 50% stake in Vivo (the Brazilian telecom operator) to Telefónica (the Spanish market leader) for 7.5 billion Euros, it reinvested a significant part of the proceeds into Africa. It also made a large investment of roughly 4 billion Euros (roughly a 25% stake) in Oi, one of the main Brazilian telecom operators. Oi is a leader in landline phones and the 4th largest operator in mobile phones. In association with these types of deals, new opportunities related to pay TV and the convergence between telecommunications and media are apparent.

Technological convergence explains the appearance of new communication services through different networks. According to the Green Paper of 1997, convergence can cause a complete and fast transformation of telecommunications, media, and information technological services. Furthermore, the Green Paper clarifies that "convergence is not only about technologies," but also "about services, new ways of doing business and of interacting with society" (ii). This could cause operators to merge, eliminating the clear differences that existed between them. However, when properly regulated, the concentration risk is minimum. In reality they are cross-platform services and products, and there are financial considerations when entering each sector. Today we have line services combined with television, through web TV or web broadcast news. In the revision of the document in 1999, the Internet (and mainly broadband) is unfairly blamed for media convergence. To access broadband, customers must acquire new equipment, and that takes time. The process is debated in many European institutions and solutions are being created to define and regulate convergence. The Green Paper finds three factors at play when analyzing convergence levels: the availability of the technologies and networks, the particular details of the individual services and markets, and the communication industry as a whole.

What is the true social and political impact of convergence? What are its regulatory and organizational implications for the future? The answers to these questions rest in having the regulation separated from content broadcasting infrastructures. These questions arose when the private market developed and the telecommunications industry was less regulated by government. European institutions have been trying to integrate the cultural and competitive components. The European

Commission, in its communication about the Electronic Commerce, recognized that this fact applied to all the convergence areas. The Green Paper reflects and highlights that the goals of the Member States are not called into question by convergence—indeed, "these objectives are varied and tailored to the specific needs of different sectors, but include national goals such as promoting efficiency, economic welfare, and the public and consumer interest" (18).

When there is a regulatory overlap, the market players can demand a rationalization of the current structures in order to prevent unnecessary administrative barriers. For example, when services can be offered by a single network, the organizations can benefit from having a single regulatory authority as the key contact for questions related to that network regardless of the services offered through it. The current definitions eliminate barriers between the different sector rules and regulatory bodies. The regulation is connected to the definition of activities. Even though the regulation can be neutral towards technologies, as in the broadcasting and telecommunications sector, it can also be associated with technologies used for the service offering, between regulated areas, and in areas where regulations are not currently applied. Convergence does not necessarily imply a unification of different sectors, but instead could refer to new activities and combined with existing services. The resolution of the Green Paper relative to the convergence of the telecommunications and media sectors raised the issue of media concentration and the lack of legislation to defend consumers and media pluralism.

According to European Union guidelines, Portugal has created a regulatory framework in the context of media and telecommunications. As already mentioned, in Portugal there is a Media Regulatory Entity (ERC) for media and a National Communications Authority (ANACOM) for telecommunications. However, these entities have been criticized for their nomination mode (which is the result of an agreement between the two major parties and is accountable to Parliament, in the case of ERC) and the duplication of structures. It is important to reflect on the transformations within the sector, including those resulting from ICT's impact and the disruption of the business models which are increasingly attached to digital distribution models. Progress cannot be reversed, and these innovations represent a decrease of the press's importance as a source of information and opinion.

Although is fundamentally about regulation, it is increasingly important to recognize the regulatory focus on the challenges resulting from convergence, especially in free-to-air and pay television. This need will be reinforced by the transition from an analog to a DTT model (a switch which occurred in January 2012.) This transition will contribute to increased supply of content. It may be appropriate to rethink the regulation model by examining the close relationship between the audiovisual materials' content regulation (ERC) and technical regulation (ANACOM). However, the closeness and cooperation of these entities may materialize in a model that is able to create a permanent mechanism of collaboration where both entities maintain independence, or merge to join the technical and content regulation into the same structure. This merger could generate a multidisciplinary view on some issues and challenges imposed by the convergence or it may create a dominance of technical regulation over content regulation. The main justification of any merger must be to promote pluralism and the diversity of voices as much as possible.

In summary, it makes sense to rethink and promote the discussion about the regulatory model that is most appropriate to Portugal and oriented to the challenges of the *digital dividend*. As described by Faustino (in Albornoz & García Leiva, 2013), the migration from analogue to digital terrestrial television (DTT) enabled a more efficient use of radioelectric spectrum, releasing significant resources—the digital dividend—which have been used for the development of mobile broadband.

The regulation should take into account market logic and contingencies that affect the behavior, programming strategies, and managing practices of public and private media companies. The guarantee of pluralism has its price, which can be justified if regulation contributes to democratic processes' sustainability and development. Regulation, defined as self-regulation and government-imposed regulation, is increasingly necessary. It is possible to identify a model that fits the market characteristics and that corresponds to the specifics of each State and its demands for citizenship and human dignity. With the increasing evolution of media and social values, the regulation must be dynamic, attentive, and adaptable to each new issue that arises. It must try to seek a balance between healthy competition that promotes diversity and pluralism, and regulation that protects the public interest and the freedom of expression.

## Bibliography


Albornoz, L. A. & García Leiva, M. T. (2013), *A Internet e o Futuro da TV.* Lisboa: MediaXXI.

Bogart, Leo. *Commercial Culture, The Media System, and the Public Interest.* New York: Oxford University Press, 1994.

Faustino, P. *Concentração, Pluralismo e Regulação dos Media.* Lisbon: MediaXXI, 2011.

Lopes, C. T., Oliveira, H. S. & Cunha, C. A. S. (2006), O jogo das estratégias nas OPA—caso Sonae/Portugal Telecom. *TOC,* 75, 48–57.

Noam, E. *Media Ownership and Concentration in America.* New York: Oxford Press University Press, 2009.

Público (2009), "PS aprova sozinho lei do pluralismo e não concentração nos meios de comunicação social," Jan. 23, 2009.


### Electronic References


ANACOM—Autoridade Nacional de Comunicações/National Communications Authority. Available at <www.anacom.pt>. Accessed on May 4, 2012.

APCT—Associação Portuguesa Controlo de Tiragem/Portuguese Association for Print Control. Available at <www.apct.pt>. Accessed on May 4, 2012.

APEL—Associação Portuguesa de Editores e Livreiros/Portuguese Association of Publishers and Booksellers. Available at <www apel.pt>. Accessed on May 6, 2012.

ERC—Entidade Reguladora para a Comunicação Social. Available at www.erc.pt. Accessed on May 4, 2012.

European Commission (2007), Green Paper on the Convergence of the Telecommunications, Media and Information Technology Sectors, and the Implications for Regulation. Available at <http://ec.europa.eu/green-papers/>. Accessed on May 14, 2012.

GFK. Available at <www.gfk.com>. Accessed on May 1, 2012.

GMCS—Gabinete para os Meios de Comunicação Social. Available at <www.gmcst.pt>. Accessed on May 12, 2012.

ICA—Instituto do Cinema e Audiovisual/Cinema and Audiovisual Institute. Available at <cwww.ica.pt>. Accessed on May 12, 2012.

Marktest. Available at <www.marktest.pt>. Accessed on May 4, 2012.

Mediamonitor. Available at <www.mediamonitor.pt>. Accessed on May 12, 2012.

OAC—Observatório das Actividades Culturais. Available at www.oac.pt/menuobservatorio.htm. Accessed on May 15, 2012.

Stat Counter—Global Stats. Available at <gs.statcounter.com>. Accessed on May 25, 2012.

# Portugal—Data Summaries

ELI NOAM AND PAUL MUTTER

THE NATIONAL MEDIA INDUSTRY in Portugal is highly concentrated, with a power index of 3,470 in 2010 (Table 10.28). Portugal's C4 companies have a lower market share relative to similar companies in other countries. Overall concentration is high but is slightly declining.

Portugal Telecom (PT) and Vodafone (UK) are the country's largest platform companies, with the former controlling 39.7% of the platform market and the latter 20.8% (Table 10.30). PT's power index declined from 2,166 to 1,811 due to loss of wireline market share to Vodafone and ZON, which is domestically owned.

ZON, a media holding company that owns a satellite and cable operator along with being a movie distributor and a carrier of mobile phone services, with a commanding market share in the multichannel platforms (57.9%) and film distribution markets (42.3%), as well as a growing share in the ISP (32.7%) and wireline markets (16.4%). It is now Portugal's second-largest media group overall. Portugal's other domestic ICT major, Optimus, a GSM/UMTS mobile operator in Portugal, also experienced negative growth during this period.

Portugal's strict television licensing regulations discourage market entrants in broadcast television, though this may change should Portugal decided to privatize its public broadcasting service, RTP. In addition to RTP (37%), two other companies have a share in TV broadcasting—PRISA (34%) and Impresa (29%). There are also a limited number of companies with substantial share in the radio industry. Group RR maintains 35.7% of the market while PRISA owns 35.1% and RTP owns 13%. PRISA remains the second-largest content media producer in the country, controlling 21.2% of the market, just behind RTP with 22.3% (Table 10.29).

The newspaper market, in contrast, is far more open to competition, but the foundation of free dailies has completely upended the business models of traditional print content producers. Cofina, (42.4%) and Controlinveste (35.2%) dominate the daily newspaper industry.

Table 10-28. National Media Industries Concentration in Portugal

| | 2004/5 | | 2011 OR MOST RECENT | | % CHANGE ANNUAL AVERAGE | |
|---|---|---|---|---|---|---|
| | COMPANY POWER INDEX IN COUNTRY | COMPANY SHARE OF THE OVERALL NATIONAL MEDIA MARKET (%) | COMPANY POWER INDEX IN COUNTRY | COMPANY SHARE OF THE OVERALL NATIONAL MEDIA MARKET (%) COUNTRY | COMPANY POWER INDEX IN COUNTRY | COMPANY SHARE OF THE OVERALL NATIONAL MEDIA MARKET (%) |
| Oi + Telemar1 | 1,310 | 26.3 | 1,045.1 | 23.0 | -5.1 | -0. 7 |
| ZON | 653.4 | 10.4 | 506.0 | 118 | -5.6 | 0.3 |
| Vodafone (UK) | 510.9 | 13.8 | 420.7 | 12.0 | -4.4 | -0.4 |
| PRISA Group (Spain) | 304.1 | 9.0 | 293.4 | 9.0 | -0.9 | 0.001 |
| RTP (public) | 281.3 | 8.5 | 330.7 | 9.5 | 4.39 | 0.2 |
| Impresa | 282.5 | 8.5 | 230.3 | 8.1 | -4.6 | -0.07 |
| Google (US) | 38.0 | 0.4 | 212 | 2.2 | 114.7 | 0.4 |
| Sonae.com (Optimus) | 183.8 | 10.3 | 185.8 | 9.3 | -0.3 | -0.2 |
| Controlinveste | 51.6 | 1.3 | 38.6 | 1.2 | -6.3 | -0.03 |
| Cofina | 48.4 | 1.6 | 60.0 | 1.6 | 6.01 | -0.01 |
| Group RR | 20.0 | 0.6 | 20.3 | 0.6 | 0.3 | -0.01 |
| Porto Editora | 42.5 | 1.0 | 38.6 | 0.8 | -2.3 | -0.04 |
| Cabovisão | 34.0 | 3.0 | 20.4 | 2.5 | -10.1 | -0.1 |
| Controlinveste | 51.6 | 1.343 | 38.6 | 1. 2 | -6.3 | -0.03 |
| Media Concentration Index | 2004/5 | | 2011 or Most Recent | | % Change Annual Average | |
| Total Revenue: Nat'l Media Industry (mil US$) | 12,381 | | 15,222 | | 4.6% | |
| Total Voices (*n*) | 57 | | 62 | | 2.2% | |
| Net Voices (*n*) | 41 | | 44 | | 1.8% | |

| | | | |
|---|---|---|---|
| Public Ownership (%) | 10.3 | 11.3 | 0.0% |
| Foreign Ownership (%) | 25.7 | 42.2 | 4.1 |
| C4 Average—Weighted | 98 | 97 | –0.3% |
| HHI Average—Weighted | 3,805 | 3,550 | –1.7% |
| C1 Average—Weighted | 47 | 44 | –0.0 |
| Noam Index Average—Weighted | 1,769 | 1,910 | 2.0 |
| Pooled Overall Sector C4 | 60.8 | 56.2 | –1.2 |
| Pooled Overall Sector HHI | 1,336 | 1,150 | –3.5% |
| Pooled Overall Sector Noam Index | 242 | 212 | –3.1 |
| Market Share of Top Ten Companies: Nat'l Media Industry (%) (Pooled C10)9 | 93.0 | 89.0 | –0.9% |
| National Power Index | 3,824 | 3,470 | –2.3 |

[1]Oi/Telemar of Brazil holds 62% of the merged company.

Table 10-29. Top Content Media Companies in Portugal

| | 2004/5 | | 2011 OR MOST RECENT | | % CHANGE ANNUAL AVERAGE | |
|---|---|---|---|---|---|---|
| | COMPANY POWER INDEX IN COUNTRY | COMPANY SHARE OF THE NATIONAL CONTENT MEDIA MARKET (%) | COMPANY POWER INDEX IN COUNTRY | COMPANY SHARE OF THE NATIONAL CONTENT MEDIA MARKET (%) | COMPANY POWER INDEX IN COUNTRY | COMPANY SHARE OF THE NATIONAL CONTENT MEDIA MARKET (%) |
| PRISA Group | 769.6 | 22.7 | 692.4 | 21.2 | –2.5 | –0.3 |
| RTP (public) | 711.9 | 21.4 | 780.6 | 22.3 | 2.4 | 0.2 |
| Impresa | 715.0 | 21.5 | 543.6 | 19.1 | –6.0 | –0.5 |
| ZON | 602.8 | 8.5 | 331.6 | 6.2 | –11.3 | 6.2 |
| Google (US) | 96.0 | 48.0 | 500.3 | 5.1 | 105.3 | 5.1 |
| Controlinveste | 130.5 | 3.4 | 91.2 | 2.8 | –7.5 | 2.8 |
| Cofina | 122.4 | 3.9 | 141.6 | 2.4 | 3.9 | –0.0 |
| Impala | 22.0 | 0.8 | 12.6 | 0.5 | –0.1 | –0.1 |
| Economica SGPS | 2.9 | 0.5 | 5.7 | 0.6 | 0.2 | 0.0 |
| Cabovisão | 15.3 | 50.3 | 6.7 | 45.9 | –14.1 | –0.1 |
| Scoeid. Vicra Desportiva. Lda | 0.7 | 0.3 | 1.7 | 0.3 | 0.4 | 0.0 |
| Media Concentration Index | 2004/5 | | 2011 or Most Recent | | % Change Annual Average | |
| Public Ownership (%) | 21.4 | | 22.3 | | 0.2 | |
| Foreign Ownership (%) | 30.3 | | 32.1 | | 0.5 | |
| C4 Average—Weighted | 96 | | 95 | | –0.3 | |
| HHI Average—Weighted | 3,305 | | 3,504 | | 1.5 | |
| C1 Average—Weighted | 41 | | 41 | | 0 | |
| National Power Index | 3,496 | | 3,388 | | –0.8 | |

Table 10-30. Top Platform Media Companies in Portugal

| | 2004/5 | | 2011 OR MOST RECENT | | % CHANGE ANNUAL AVERAGE | |
|---|---|---|---|---|---|---|
| | COMPANY POWER INDEX IN COUNTRY | COMPANY SHARE OF THE NATIONAL PLATFORM MEDIA MARKET (%) | COMPANY POWER INDEX IN COUNTRY | COMPANY SHARE OF THE NATIONAL PLATFORM MEDIA MARKET (%) | COMPANY POWER INDEX IN COUNTRY | COMPANY SHARE OF THE NATIONAL PLATFORM MEDIA MARKET (%) |
| Oi + Telemar[1] | 2,166.1 | 43.5 | 1,811.3 | 39.7 | –4.1 | –0.8 |
| Vodafone (UK) | 844.7 | 22.8 | 730.0 | 20.8 | –3.4 | –0.4 |
| ZON | 686.5 | 11.7 | 634.1 | 15.9 | –1.9 | 0.8 |
| Optimus | 296.3 | 16.3 | 319.1 | 15.7 | 1.9 | –0.1 |
| Cabovisão | 45.5 | 4.2 | 30.4 | 3.8 | –8.3 | –0.1 |
| Media Concentration Index | 2004/5 | | 2011 or Most Recent | | % Change Annual Average | |
| Public Ownership (%) | 3 | | 3 | | 0 | |
| Foreign Ownership (%) | 22.8 | | 49.5 | | 6.7 | |
| C4 Average—Weighted | 99 | | 98 | | –0.2 | |
| HHI Average—Weighted | 4,131 | | 3,584 | | –3.3 | |
| C1 Average—Weighted | 51 | | 46 | | –1.0 | |
| National Power Index | 4,039 | | 3,531 | | –3.2 | |

[1]Oi/Telemar of Brazil holds 62% of the merged company; Portugal Telecom shareholders control the other 38%.

# 11

# Media Ownership and Concentration in Russia

ELENA VARTANOVA

## INTRODUCTION

Since 1991, the Russian media market has had to adjust to free market pressures that were unfamiliar to most of the individuals living in the Soviet Union (USSR) at the time of its dissolution. That year, a new mass media law was passed that officially ended state censorship of the media—long a feature of Soviet life—and legalized private ownership of media producers. Market competition, albeit in a distorted form, became the norm during the 1990s, and this period witnessed the rapid growth of private media ownership. Privatization across all economic sectors began in 1993, and wholly new privately owned print and audiovisual media platforms began to emerge. However, it is worth noting that the role of the state in the national economy remained (and still remains) rather influential, and in the media industry the state still holds the crucial position as the owner of many media assets and also provides both formal and informal monetary support to private media companies.

The Soviet media system was characterized by a high degree of centralization and unprofitability, a structure based on the ideological role given to the media by the Communist Party of the Soviet Union (CPSU). State-owned media producers (i.e., all licensed media producers in the USSR) were never concerned about revenues, economy of scale, cost structures, or business strategies. The Soviet media system responded to needs of integrated and heavily centralized political and economic elites—the *nomenklatura*—that were appointed to run media entities. The most salient features of the system were the top-down subordination of editorial content to the ideological control of the CPSU, a complete ban on all private ownership, and heavy investment in state-owned media infrastructure (postal system, terrestrial TV and radio networks, fixed telephone lines, satellite communications).

The media system has changed radically since the early 1990s. In place of radio and newspapers, broadcast TV has emerged as the backbone of the national media market: Russia has nine national terrestrial TV networks, and in the country's 200 largest cities, between 10 and 12 national and regional public broadcasters are available.[1] Since cable and satellite television channels have developed mostly in urban areas, their total penetration remains comparatively low (approximately 25% of the national market). For many Russian families, free over-the-air TV continues to be their primary source of news and entertainment. The federal government has increasingly used broadcast programming to promote Russian integrity and challenge the influence of regional authorities.[2, 3]

In 1990, before the disintegration of the Soviet Union, there were 4,808 newspaper titles in circulation (national, regional, and local), including 43 national daily newspapers.[4] The most characteristic feature of the Soviet press was mass distribution, the near-universal penetration of the national dailies, and the subordination of all daily newspapers to the central (national) newspapers published in Moscow. After privatization began in 1993, the total circulation of the Russian press fell drastically dropped from its 1990 annual circulation rate of 38 billion. Though circulation numbers declined by 85% in the next decade, the overall number of daily national titles increased from 43 to 333 by 2000.[5] Russian publishers have carefully copied the British tabloid model, especially in the case of the popular *Komsomolskaya Pravda* and *Moskovskij Komsomolets* dailies. By combining materials from the central (national) editions with local inserts, these two dailies pioneered a new model for Russian newspapers.

Readership fell drastically during this decade in part because of Russia's financial problems. The crisis in the newspaper market began in early 1990s and was complicated by high prices for postal distribution (home delivery service was still monopolized by the state) and limited retail options. On the contrary, weekly and monthly magazines benefitted from the high prices. With their lower distribution frequency, their readers saved money, and so these magazines managed to gain much more visible financial success for private companies.

Both newspapers and magazines have had to contend with the rapid growth of the Russian Internet. There are over 35 million Russian Internet users (about 30% of the population), and there are some 2,018 online platforms offering varying degrees of news content. Urban centers have the highest rates on Internet penetration, as approximately 73% of the population lacks the necessary infrastructure to access it daily; 6% of the population is believed to have no means of access at all.[6] Distances and obsolescent technology greatly hampers the expansion of a modern telecommunications network in Russia. These problems arise mostly from the low technical level of the national telecommunications infrastructure. Only a small number of Russian Internet users have home access due to low telephone penetration (180 lines per 1,000 inhabitants) and the poor quality of these telephone lines (broadband is limited to the largest industrial centers, fiber optic is practically nonexistent for both commercial and residential purposes). Combined with low per capita incomes and other macroeconomic challenges,

1. Vartanova, Elena L. and Smirnov, Sergey S. "Contemporary Structure of the Russian Media Industry." Rosenholm, Arja; Nordenstreng, Kaarle; and Trubina, Elena, eds. *Russian Mass Media and Changing Values*. London: Routledge: 2010. p 21–40.

2. McNair, Brian. "Power, Profit, Corruption, and Lies: The Russian Media in the 1990s." Curran, James and Park, Myung-Jin, eds. *De-Westernizing Media Studies*. London: Routledge, 2000. p 79–94.

3. Vartanova, Elena L. "The Russian Media Models in the Context of Post-Soviet Dynamics." Hallin, Daniel and Mancini, Paolo, eds. *Comparing Media Systems beyond the Western World*. Cambridge: University Press, 2012. p. 134.

4. Federal Agency on Press and Mass Communication. "Pechat Rossiiskoi Federacii v 2000 godu. Statistichesky Sbornik (Periodical Press of Russian Federation in 2000. Statistical Report)." Moscow: Rossiiskaya Knizhnaya Palata, 2001. p 100.

5. Ibid.

6. Vartanova, Elena L. and Smirnov, Sergey S. "Contemporary Structure of the Russian Media Industry." Rosenholm, Arja; Nordenstreng, Kaarle; and Trubina, Elena, eds. *Russian Mass Media and Changing Values*. London: Routledge: 2010. p 23.

Internet penetration is unlikely to increase significantly in the near future.

The maturation of the advertising market in post-Soviet Russia has become especially important for the media industry as a whole. Advertising became the major source of income for media companies after 1991. The competition for advertising revenue illustrates, however, the continued role of the state in media markets because state-owned media companies secure most of their advertisements from state-owned advertisers (Table 11.1).

In recent years, Russian media studies have paid much attention to concentration issues in the context of the emerging media industry. The political economy perspective is rooted in Soviet media studies, and in this context, media concentration analyses are driven by ideological debates: there is no academic tradition to explore structural issues and the Soviet media system was never regarded as a part of the national economy, since its role was defined primarily by cultural and political priorities.[7] Media studies focused on criticism of foreign media markets using studies translated from English about the corporatization of media in other nations.[8, 9, 10] Though private media in Russia predates the Soviet period, no national newspapers existed because of low per capita income levels, low literacy rates, limited distribution infrastructure, extensive state censorship, a weak advertising market, and the practice of regional authorities heavily subsidizing dailies within their jurisdictions to promote their own political agendas.[11] As many researchers have argued, "Russian mass media appeared as a state enterprise and they existed and still exist as a private state enterprise."[12] Despite this, up until 1917, 74% of all newspapers and magazines were privately owned.[13, 14]

After the establishment of the USSR, nationalization of the national economy occurred in 1917–1918 and by the mid-1920s this process had been completed for licensed media outlets. A new system of political control was established by integrating state and CPSU ownership in order to institute a propaganda apparatus operated by specially appointed officials.[15, 16]

There have been three major stages in the development of the Russia media since 1990.

Leading up to the dissolution of the Soviet Union, the majority of the CPSU, Komsomol (Communist Youth League), and state news services came under the collective ownership of their own staffs, notably the leading CPSU (and Komsomol) dailies of *Pravda, Sovetskaya Rossiya, Komsomolskaya Pravda,* and *Moskovskij Komsomolets.* The legal basis for these practices was established by the Soviet Law on Mass Media (1990), which for the first time in the Soviet Union legalized private ownership in the media, though it left the principles of collective ownership by staff members vague.[17] Following the formal dissolution of the USSR

7. Prokhorov, Yevgeniy P. Osnovy marksistsko-leninskoy teorii zhurnalistiki. (Basics of Marxist-Leninist Journalism Theory). Moscow: Izd-vo Mosk. Un-ta., 1973. p 72.

8. Bagdikian, Ben H. Monopoliya sredstv informacii. (The Media Monopoly). Moscow: Progress, 1987. p 132.

9. Schiller, Herbert *Manipulyatory soznaniem (Mind Managers).* Moscow: Mysl, 1980. p 325.

10. Zassoursky, Yassen N., ed. Na sluzhbe monopoliy. Informacionno-propagandistsky kompleks stran kapitala. (To Serve the Monopolies. The Information and Propaganda Complex of Capitalist Countries). Moscow: Mysl, 1977. p 31.

11. Makhonina, Svetlana Y. Istoriya russkoi zhurnalistiki nachala XX veka. (History of Russian Journalism at the Beginning of the 20th Century). Moscow: Flinta: Nauka. 2009. p 111–120.

12. Trakhtenberg, Anna. "Transformation without Change. Russian Media as a Political Institution and Cultural Phenomenon." Vartanova, Elena L., ed. *Media and Change.* Moscow: MediaMir, 2007. p 124.

13. Makhonina, Svetlana Y. Istoriya russkoi zhurnalistiki nachala XX veka. (History of Russian Journalism at the Beginning of the 20th Century). Moscow: Flinta: Nauka. 2009. p 111–120.

14. Esin, Boris I.; Zassoursky, Yassen N.; and Minaeva, Olga. D., eds. Russkaya zhurnalistika v dokumentakh. Istorija nadzora. (Russian Journalism in Documents. A History of Control). Moscow: Aspekt Press, 2003. p 398.

15. Prokhorov, Yevgeniy P. Osnovy marksistsko-leninskoy teorii zhurnalistiki. (Basics of Marxist-Leninist Journalism Theory). Moscow: Izdatelstvo MGU, 1973. p 72.

16. Ovsepyan, Rafail P. Sovetskaya zhurnalistika i kommunisticheskoye vospitaniye trudyashikhsya. *(Soviet Journalism and Communist Upbringing of Workers).* Moscow: Izd-vo Mosc. Un-ta, 1979. p 212.

17. Zakon RF. "O Sredstvah massovoi informacii." (Law on Mass Media). Moscow. Dec. 27, 1991 no. 2124–1 of Dec. 27, 1991. *Rossiiskaya Gazeta,* 1992.

Table 11-1. Russian Media Advertising (Market Shares by Revenue), 2007–2011

| MEDIA MARKET | 2007 | 2009 | 2010 | 2011 | % GROWTH (2011) |
|---|---|---|---|---|---|
| TV | 4.4 | 2.8 | 3.8 | 4.5 | 18.0 |
| Terrestrial | *4.3* | *2.71* | *3.8* | *4.4* | 18.0 |
| Cable and Satellite | *0.02* | *0.03* | *0.1* | *0.1* | 36.0 |
| Radio | 0.6 | 0.3 | 0.4 | 0.4 | 15.0 |
| Print Media | 2.0 | 1.0 | 1.3 | 1.4 | 6.0 |
| Newspapers | 0.4 | 0.2 | 0.3 | 0.3 | 7.0 |
| Magazines | 0.9 | 0.5 | 0.6 | 0.7 | 8.0 |
| Advertising Publications | 0.7 | 0.3 | 0.4 | 0.4 | 3.0 |
| Outdoor Advertising | 1.6 | 0.7 | 1.0 | 1.2 | 15.0 |
| Internet | 0.5 | 0.5 | 0.9 | 1.4 | 56.0 |
| Media Advertising | 0.2 | 0.2 | 0.4 | 0.5 | 45.0 |
| Contextual Advertising | 0.3 | 0.3 | 0.6 | 0.9 | 63.0 |
| Other Media | 0.1 | 0.1 | 0.1 | 0.1 | 32.0 |
| Indoor Advertising | 0.1 | 0.1 | 0.1 | 0.1 | 35.0 |
| Movie Theaters | 0.01 | 0.01 | 0.02 | 0.03 | 25.0 |
| Total ATL | 9.1 | 5.3 | 7.5 | 9.1 | 21.0 |
| Total BTL | 2.3 | 1.2 | 1.9 | 2.4 | 23.0 |
| Total | 11.3 | 6.6 | 9.4 | 11.4 | 44.0 |

*Source:* AKAR, 2012.

on Dec. 26, 1991, the post-Soviet media landscape in Russia underwent an initial period of commercialization and concentration that lasted through 1996 as almost all media outlets privatized themselves and set up journalists' collectives that had to invent new business models as state financing evaporated.

Due to the cessation of most state revenue streams and a complete lack of managerial experience (both the journalists and the appointed managers they had gotten rid of had no experience with publishing in a market economy), many media producers went bankrupt while the most lucrative properties were taken over by conglomerates dealing in extractive resources and financial services, whose questionable acquisitions of these properties were made with the informal support of state officials. In early 1990s, such influential conglomerates as Onexim, LogoVAZ, Lukoi, and Gazprom established a "media-industrial complex" among themselves, comprising several influential families among the new political and business elites.[18, 19]

Between 1996 and 2000, the process of media concentration along these lines climaxed.[20] Media companies, especially terrestrial TV broadcasters, were acquired or established by "oligarchs"

18. Khlebnikov, Pavel Y. Krestnyi otetz Kremlja Boris Berezovsky (The Kremlin Godfather Boris Berezovsky). Moscow: Detektiv-Press, 2001. p. 381.

19. Puppae, Yakov S. *Oligarchi. Ekonomicheskaya khronika 1992–2000. (Oligarchs. The Economic Chronology 1991–2000).* Moscow: Moscow State University Higher School of Economics, 2000. p. 232.

20. McNair, Brian. "Power, Profit, Corruption, and Lies: The Russian Media in the 1990s." Curran, James and Park, Myung-Jin, eds. *De-Westernizing Media Studies.* London: Routledge, 2000. p. 82.

such as Boris Berezovsky (LogoVAZ), Viktor Chernomyrdin (Gazprom), Vladimir Gousinsky (MOST Group), and Vladimir Potanin and Mikhail Prokhorov (Onexim).

The current third stage in media concentration, which began during the first presidency of Vladimir Putin (2000–2004), has been characterized by the redistribution of media ownership in the context of the reestablishment of direct and indirect state intervention in the media: the number of state-owned media companies has increased, with the government putting renewed emphasis on the national public broadcaster VGTRK and the state news agencies of ITAR-TASS (renamed as TASS in 2014) and RIA Novosti (reorganized into Rossiya Segodnya after restructuring in 2013–2014). Federal investment in Russian media outlets and infrastructure increased in the 2000s; state subsidies totaling US$5.7 billion (174 billion rubles) were disbursed between 2010 and 2012.[21] On the other hand, state intervention has clearly favored the expansion of media companies such as the broadcaster 7TV, the print daily *Kommersant,* and the telecom major MegaFon that belong to politically loyal businessmen. The complexity of the situation has been increased with the rise of mobile telephony and the Internet (in particular broadband ISPs and search engines). Both of these industries have become tightly connected to the Russian media system and its expansion through technological development, and in the 2000s they became the most efficient drivers of the Russian market economy.

## PRINT MEDIA

### Newspapers

Particular features of the modern Russian newspaper industry are its regional character, low revenues compared to other types of media, and declining audience shares. Statistically, the share of national newspapers accounts for 35% of the total circulation, regional papers for 33%, and local newspapers for 32%. Though the total audience for national newspapers does not exceed 20% of the population, non-national newspapers combined account for an additional 20% of the population.[22] Lack of investments in printing and distribution, as well as uneven development of regional advertising markets, means that for non-national papers, regional and local authorities must provide large amounts of both formal and informal financial support. The Russian print media sector still retains many features of the USSR's centrally planned economy and is thus an unattractive investment for most private parties. The regional character of the newspapers market also explains the poor economic status of the sector and the lack of publishers operating at the national level. The national dailies established during the Soviet era still retain the largest market shares, and most private media holding companies active in the market have built themselves up around these decades-old titles: *Komsomolskaya Pravda, Sovetsky Sport, Moskovskij Komsomolets, Trud,* and *Izvestia.* In some cases their publishers are parts of larger national media groups—such as the National Media Group (NMG) in the case of *Izvestia,* formerly owned by Gazprom until 2008, or Media 3 (by 2014, main newspaper asserts were sold and the company stopped its operations at the media industry) in the case of *Trud*; one major exception to this trend is the state-owned *Rossiiskaya Gazeta.* Even the major publishers have great difficulty breaking even, and most do not. Circulation rates show rather stable positions for the established market leaders, albeit with greatly reduced numbers from the Soviet era. Also unique is Pavel Gousev, owner of *Moskovskij Komsomolets,* whose media group is not part of a larger, diversified conglomerate. Industrial groups whose chairmen are close to the Kremlin own NMG (Alexei Mordashov, CEO of Severstal) and *Kommersant* (Alisher Usmanov, CEO of Gazprominvestholding).

---

21. Ivanitsky, Valeriy L. Modernizatsija zhurnaliskiki: metodologicheskii etud (Modernization of Journalism: Methodological Composition). Moscow: Izd-vo Mosk. Un-ta, 2010. p 290.

22. Federal Agency on Press and Mass Communication. "Internet v Rossii. Sostoyanie, tendencii i perspective razvitiya (2011) [The Internet in Russia. Conditions, Tends and Perspectives (2011)]." Moscow: Federal Agency on Press and Mass Communication, 2011. p 82.

In terms of circulation, the leading newspapers publisher in Russia is the Komsomolskaya Pravda group, with the most popular daily *Komsomolskaya Pravda* (700,000 daily copies during the week, 3 million on weekends). Other titles include the former main Soviet sports daily *Sovetsky Sport* and one of the first tabloids set up after 1991, *Ekspress Gazeta*. The total monthly circulation for all newspapers in Russia in 2010 stood at 46.1 million, while total monthly readership was about 30 million copies (weekly, 5.5 million; daily, 2 million). All these newspapers are owned by Komsomolskaya Pravda, which was sold in 2007 by its previous owners, the multimedia group ProfMedia and the Norwegian media group A-pressen to the ECN Group owned by Grigory Berezkin. ECN has a 60% plus-one stake in the publishing house's shares (Table 11.2).

Although until 2015 there were no restrictions on foreign ownership of print media, foreign ownership is rare in the Russian newspaper market (though non-Russian newspapers are available through foreign media groups' Russian distributors). This is in large part because until recently Russian newspapers were wholly dependent the state newsprint monopoly. Another feature of the Russian market that makes it unattractive to investors is the aforementioned high costs of distribution through the postal system. MK and Media 3, for example, have attempted to set up alternative distribution networks in some urban centers to stabilize their circulation numbers.

Though the national dailies are still the most widely circulated newspapers, their former editorial influence during the Soviet era has not been reestablished over the past 20 years under new management due to declining readership and political changes in the country, which partly explains why some of the more financially powerful media and industrial groups are not active in this market, as audiovisual media is by far the more influential shaper of public debates in Russia today.

## Book Publishing

In the last few years, book publishing has been transformed significantly. Nowadays, only one publishing house is still state owned and the rest have been privatized (and, in some cases, have then been removed from the jurisdiction of Russia tax law to off-shore registries). It is difficult to delve deeply into ownership and concentration in this sector, though, because of a lack of transparency: many publishers are privately held companies legally operating from the British Virgin Islands, Cyprus, or Gibraltar (Table 11.3).

## Magazine Publishing

The magazine market is expanding rapidly in Russia: over the past few years its annual growth has exceeded 13%, a rate only surpassed by the magazine markets of India and the People's Republic of China. The total annual circulation of magazines in Russia in 2007 was 1.9 billion copies, 600 million of which were printed abroad. The bulk of the sector's revenue (60%) comes from the Moscow metropolitan area. However, as is the case with newspapers, out of the total number of officially registered magazine titles in the country (31,452 in 2010), only around 50% of them are really being produced and distributed effectively.[23] Again, the high costs and inefficiencies of the national postal service affect the profitability of the sector.

That said, magazine publishing is still more attractive to investors than newspapers are because advertising revenues are higher and distribution is ultimately less expensive because the sector is less dependent on home delivery and involves weekly or monthly print editions. The exposure of Russia to Western consumerism has increased the popularity of publications carrying lifestyle content, which also explains a prevalence of foreign owners in this sector than in any other media platforms.

23. Federal Agency on Press and Mass Communication "Rossiyskaya periodicheskaya pechat. Sostoyanie, tendencii i perspective razvitiya 2011." (The Russian Periodical Press. Conditions, Trends and Perspectives). Moscow: Federal Agency on Press and Mass Communication, 2011. p 84.

Table 11-2. Daily Newspapers (Market Shares by Circulation and Revenue), 2004–2010

| PUBLISHER | PUBLICATIONS | % MARKET SHARE BY CIRCULATION | | % MARKET SHARE BY REVENUE | | | | |
|---|---|---|---|---|---|---|---|---|
| | | 2004 | 2008 | 2006 | 2007 | 2008 | 2009 | 2010 |
| Komsomolskaya Pravda (ECN Group, Grigory Berezkin) | *Komsomolskaya Pravda, Sovetsky Sport* | 29.0 | 18.0 | 36.1 | 40.2 | 38.8 | 46.9 | 39.6 |
| Moskovskij Komsomolets (MK) (Pavel Gousev) | *Moskovskij Komsomolets* | 24.0 | 19.0 | 16.2 | 12.8 | 11.8 | 10.6 | 5.9 |
| Redaktsia Rossiiskaya Gazeta (Federal Public Office) (public) | *Rossiiskaya Gazeta* | 9.0 | 9.0 | | | | 1.0 | 29.6 |
| Sport-Express | *Sport-Express* | 5.0 | 5.0 | | 5.0 | 4.8 | 4.7 | 2.6 |
| Media 3 (until 2014) | *Trud* | 19.0 | 38.0 | | | 2.7 | | |
| National Media Group/NMG (Alexei Mordashov) | *Izvestia* | 7.0 | 5.0 | 10.1 | 7.3 | 7.4 | 8.1 | 4.5 |
| Kommersant (Alisher Usmanov) | *Kommersant* | 3.0 | 2.0 | 37.6 | 34.7 | 34.5 | 28.7 | 17.8 |
| Total Circulation (1,000) | | 3,105 | 4,068 | | | | | |
| Total Revenue (mil RUB) | | | | 4,212 | 6,554 | 7,286 | 6,206 | 12,419 |
| Total Revenue (mil US$) | | | | 156 | 257 | 276 | 214 | 393 |
| C4 | | 83.6 | 87.2 | 100.0 | 59.0 | 58.0 | 52.0 | 57.0 |
| HHI | | 2,054 | 2,409 | 3,028 | 3,057 | 2,917 | 3,227 | 2,824 |
| *N* (>1%) | | 7 | 7 | 4 | 5 | 7 | 6 | 6 |
| Noam Index | | 776 | 911 | 1,514 | 1,367 | 1,191 | 1,317 | 1,153 |

*Sources:* FAPMC, 2011; Integrum-Companies, 2012.

## Table 11-3. Book Publishing (Market Shares by Circulation and Revenue), 2002–2010

| | % MARKET SHARE BY CIRCULATION | | | % MARKET SHARE BY REVENUE | | | | |
|---|---|---|---|---|---|---|---|---|
| | 2002 | 2004 | 2009 | 2006 | 2007 | 2008 | 2009 | 2010 |
| Eksmo (Lorristone Ltd., Cyprus) | 9.6 | 10.7 | 11.3 | 32.6 | 47.2 | 34.2 | 38.1 | 34.2 |
| Prosveschenie (Rosimushchestvo/ Federal Agency for State Property Management) (public) | 8.9 | 5.8 | 6.4 | 23.2 | 23.0 | 19.8 | 21.9 | 22.5 |
| Drofa | 8.8 | 4.8 | 2.3 | 18.0 | 21.2 | 15.7 | | 10.7 |
| AST (Cyprus) | 6.1 | 7.1 | 10.7 | | | 23.5 | 32.9 | 27.1 |
| Olma Press | 2.5 | 1.3 | 1.9 | 17.6 | | | | |
| Flamingo | 2.4 | 4.3 | | 2.3 | 2.2 | 1.7 | 1.6 | 1.2 |
| Rosmen | 2.3 | 1.3 | 1.1 | 0.1 | 0.1 | 0.0 | | 0.2 |
| Balass | 2.3 | 1.7 | | | | | | |
| INFRA-M | 2.0 | | | 6.2 | 6.2 | 5.1 | 5.6 | 4.1 |
| Panorama | 1.4 | | | | | | | |
| Ekzamen XXI | | 3.8 | | | | | | |
| Feniks | | 0.8 | | | | | | |
| Egmont Russia (Egmont, Denmark) | | | 4.5 | | | | | |
| Ripol-klassik | | | 1.1 | | | | | |
| Azbuka-klassik | | | 1.0 | | | | | |
| Others | 53.7 | 58.7 | 58.6 | 0.0 | 0.0 | 0.0 | 0.0 | 0.0 |
| Total Revenue (mil RUB) | | | | 6,696 | 8,033 | 11,484 | 11,803 | 13,881 |
| Total Revenue (mil US$) | | | | 248 | 315 | 435 | 407 | 477 |
| C4 | | | | 91.4 | 97.7 | 93.2 | 98.4 | 94.5 |
| HHI | 315 | 261 | 316 | 2,282 | 3,254 | 2,392 | 3,041 | 2,545 |
| *N* (>1%) | 10 | 9 | 9 | 6 | 5 | 6 | 5 | 6 |
| Noam Index | 99 | 87 | 105 | 932 | 1,455 | 976 | 1,360 | 1,039 |

*Sources:* FAPMC, 2011; Integrum-Companies, 2012.

Hearst-Shkulev Media (formerly Hachette-Filipacchi-Shkulev) is the market leader in magazine publishing, with a total annual circulation of 12 million copies and an estimated readership of 21.6 million people. It has held this spot for 16 years. Originally owned by the Russian media magnate Viktor Shkulev and the Lagardere Group (France), the Lagardere Group sold its share to Hearst (US) in 2011. Nowadays, the company publishes well-established international brands such as *Elle, Elle Girl, Elle Decoration, Elle-Ukraina, Maxim, Maxim-Ukraina, Marie-Clair, Parents, Psychologies,* and *StarHit.* Its most profitable and nationally distributed outlet is the TV guide *Antenna-Telesem.*

Other significant publishers in the market include the Russian-language division of the German publishing house Hubert Burda Media and Independent Media, and the Russian-language division of Sanoma Oyj (Finland). Statistics for the market are skewed due to the returns of unsold copies, which prevent a wholly accurate image of the distribution scale (Table 11.4).

Table 11-4. Magazine (Market Shares by Revenue), 2006–2010

| | 2006 | 2007 | 2008 | 2009 | 2010 |
|---|---|---|---|---|---|
| Hearst-Shkulev Media[1] (49% US owned) | 27.3 | 30.8 | 31.3 | 30.7 | 35.2 |
| Bauer/Logos-Media (Bauer, Germany) | 16.9 | 16.6 | 16.7 | 16.3 | 7.8 |
| Independent Media (Sanoma Oyj, Finland) | 45.7 | 42.7 | 41.3 | 40.1 | 44.5 |
| Edipresse Konliga JSC (Edipresse, Switzerland) | 10.2 | 9.9 | 10.8 | 12.9 | 12.5 |
| Total Revenue (mil RUB)[2] | 4,914 | 6,656 | 7,999 | 5,742 | 5,704 |
| Total Revenue (mil US$) | 182 | 261 | 303 | 198 | 196 |
| C4 | 100.0 | 100.0 | 100.0 | 100.0 | 100.0 |
| HHI | 3,220 | 3,147 | 3,074 | 2,980 | 3,434 |
| *N* (>1%) | 4 | 4 | 4 | 4 | 4 |
| Noam Index | 1,610 | 1,573 | 1,537 | 1,490 | 1,717 |

*Sources:* FAPMC, 2011; Integrum-Companies, 2012; Open sources.

[1] 51%: V. Shkulev, 49%: Hearst (US).

[2] This revenue figure is an estimate based on data collected in this study: it is based on average per capita revenues reported for several other countries in this study with similar per capita income.

## AUDIOVISUAL MEDIA

### Radio

The commercial radio market began to grow in the mid-1990s with the emergence of local FM stations. As a result, traditional "wired broadcasting," a system established in the USSR during the 1920s that relies on public loudspeakers and fixed telephony, has become less popular, due not only to the strong competition in large cities from local FM stations but also to the aging equipment used for wired broadcast services. The key players in the FM radio market are several central (national) stations operating at the national level out of Moscow. Radio broadcasting at the subnational level occurs through local broadcasters joining one or more of the country's 31 existing national networks. The three main networks are the European Media Group (EMG), VPKM (part of ProfMedia), and RMG. Similar to the broadcast TV sector, foreign ownership is almost absent at the market. The exception was EMG, part of Lagardere Group, which in 2011 sold its radio programming to Sibirsky Delovoy Soyuz, for US$162 million (RUB 5.1 billion).

Although the state formally is largely present in ownership structures of the radio industry, this does not guarantee it the leading position at the market. On the contrary, in the radio industry the cost efficiency of privately owned networks and stations gives them an advantage, and their management is less motivated by political concerns than that of TV broadcasters (Table 11.5).

### Broadcast Television

Today, the Russian terrestrial TV market comprises 20 national public broadcasters that can be divided into three major groups. The first group includes the public-private network Channel One Russia, the state broadcaster Rossiya 1, and the commercial provider NTV. The second group comprises the three national public entertainment networks of REN TV, CTC, and TNT. The third group consists of the state broadcasters Rossiya 2 and Kultura, along with the national TVC and 5 Kanal broadcasters and the entertainment channels Domashnyi, MTV and MuzTV, TV 3, Peretz (formerly DTV), 7, Zvezda, and Mir.[24]

24. The major legal form of Russian TV broadcasters, according to Russian legislation, is a JSC (joint stock company) which exists in two types: OJSC (open joint stock company) and CJSC (closed joint stock company).

Table 11-5. Radio Groups (Market Shares by Revenue), 2006–2010

| RADIO GROUP | STATIONS | REVENUE SHARE | | |
|---|---|---|---|---|
| | | 2006 | 2009 | 2010 |
| European Media Group (EMG) | Europe Plus, Retro FM, Radio 7, Fruitcake FM, Fresh Radio | 34.9 | 31.0 | 32.4 |
| VK ProfMedia (ProfMedia) | Stereos, Humor FM, Energy, Radio Alla | 20.7 | 19.3 | 18.2 |
| Russian Media Group (RMG) | Russkoe Radio, Hit FM, Maximum, DFM, Monte Carlo | 23.1 | 11.2 | 12.3 |
| Independent Broadcasters | | | 12.1 | 12.0 |
| Gazprom-Media | Echo of Moscow, Citi-FM, Relax-FM, Radio Child | 6.1 | 9.2 | 10.1 |
| MediaHold | Love Radio, Taxi FM, Radio Dacha | | 2.9 | 3.8 |
| Ob'edinenie Media Company | Business FM, 8 Hits | 0.0 | 3.7 | 3.2 |
| Multimedia Holding (Vitaly Bogdanov, LogoVAZ-Nyus Corporation/News Media Radio Group) | Nashe Radio, Ultra, Best-FM | 5.6 | 2.7 | 3.0 |
| VGTRK (public) | Radio Rossii, Mayak, Yunost, Vesti FM, Kultura | 9.4 | 4.6 | 2.8 |
| Arnold Prize (changed its owner in 2014) | Radio Jazz, Radio Classic | | 2.9 | 2.0 |
| Radio-Centr (Property Department of the City of Moscow) | Govorit Moskva, Russkoe Publichnoe Radio, Radio Sport | | 0.1 | 0.2 |
| Total Revenue (mil RUB) | | 6,858 | 10,527 | 11,788 |
| Total Revenue (mil US$) | | 254 | 363 | 405 |
| C4 | | 88.2 | 73.8 | 74.8 |
| HHI | | 2,342 | 1,758 | 1,822 |
| *N* (>1%) | | 6 | 10 | 10 |
| Noam Index | | 956 | 556 | 576 |

*Source:* TNS, 2011; EMG, 2012; VKPM, 2012.

The number of regional TV companies has been increasing; there are about 700 local stations, both public and private.[25] The development of regional TV is an indicator of the economic situation in Russia: the wealthiest parts of the country have the most commercial regional TV channels. Of all 83 regions, territories, and republics of the Russian Federation, the 10 economically least developed regions receive only public broadcasting. In the 15 most developed regions, including Moscow and St. Petersburg, the number of terrestrial channels available ranges from a maximum of 58 in the capital to approximately 15 in the Tver, Vladimir, and

25. Vartanova, Elena L. and Smirnov, Sergey S. "Contemporary Structure of the Russian Media Industry." Rosenholm, Arja; Nordenstreng, Kaarle; and Trubina, Elena, eds. *Russian Mass Media and Changing Values.* London: Routledge: 2010. p 21–40.

Novosibirsk areas. The overall number of channels available in these regions ranges from 20 to 29.[26]

The combination of state ownership and private funding is crucial for maintaining competition between public and commercial channels, especially at the subnational level. Yet at the same time, state control of terrestrial and satellite transmission networks and licensing puts regional companies in very difficult economic conditions. Local news, the most valuable type of local TV content, has had to adopt rather sensationalist practices in order to increase its ratings. With low salaries and underinvestment in audiovisual technology, production quality is in general quite low.

Attempts by the state to establish technical and structural dominance in the TV market inspired the reorganization of the state-owned VGTRK in 1998. VGTRK now operates two nationally distributed channels, RTR and Kultura. VGTRK also maintains a network of more than 90 state-controlled regional television stations and 100 transmission centers, which was separated into an independent company in 2001. VGTRK's regionally interconnected structure allows it to support the policies of the federal government over the influence of regional authorities.

The dominance of the national channels characterizes the regional TV market. Even the most popular private channels and networks (NTV, CTC, and TNT) have to deal with heavy informal pressure from the state. The political concerns of the Russian political elite are obviously one of the factors in instrumentalizing the role of the national TV channels, but there are also commercial factors, such as advertising, affecting the significant position of TV in the Russian media system. Commercial TV networks (especially since 1999) have formed mutually beneficial alliances of central (national) and regional channels. They will retransmit programs from Moscow with no changes, allow for the partial retransmission of Moscow channels with the addition of regional programs (mostly news and current-affairs programs), and permit the insertion of news and entertainment programming from Moscow on the basis of special agreements. These alliances have increased the probability that regional stations will rely on proven program formats such as talk shows, game shows, and low-cost Russian and foreign soap operas rather than focus on developing new content.

The federal government is the sole owner of VGTRK, through state agencies it is the dominant owner in Channel One Russia, and through a complicated structure of the corporate ownership of the state-owned energy company Gazprom has close links to NTV through Gazprom-Media. The control of ownership of the third group of channels is also kept in the hands of various state agencies, with TVC owned by the Moscow municipal government, Zvezda controlled and financed by the Defense Ministry, and Mir controlled by the governments of the Russian Federation and the Commonwealth of Independent States (CIS). As for private broadcasters, most of them are affiliated with large commercial groups such as NMG and ProfMedia (Table 11.6).

## Multichannel TV Platforms

Cable and satellite TV began operating in Russia after 1991. Although fiber optic cable networks had been installed in many new apartment blocks in the late 1980s, they were intended to inform people in the case of a state of emergency, not to receive foreign cable or satellite programming, which like many foreign radio and terrestrial broadcasts from abroad were officially banned. The fiber optics networks, though, became the first-generation infrastructure for cable TV in many Russian cities, even though from the mid-1990s on, providers were able to offer more technologically sophisticated solutions. Currently, each region in the country has its own cable TV providers (sometimes more; there are at least

26. Vartanova, Elena L. and Kolomiets, Viktor P. eds. "Televidenie v Rossii. Sostoyanie, tendencii i perspective razvitiya. *(Television in Russia. Conditions, Trends and Perspectives)*." Moscow: Report of Federal Agency on Press and Mass Communication, 2011.

## Table 11-6. TV Broadcasting (Market Shares by Viewership and Revenue), 2000–2010

| BROADCASTER | OWNERS | % MARKET SHARE BY VIEWERSHIP | | | % MARKET SHARE BY REVENUE | | | | | | |
|---|---|---|---|---|---|---|---|---|---|---|---|
| | | 2000 | 2004 | 2008 | 2004 | 2005 | 2006 | 2007 | 2008 | 2009 | 2010 |
| Channel One Russia (public) | CJSK ORT-KB: 38%, Goskomimushestvo Rossii: 38.9%, ITAR-TASS: 8.1%, LLC Betas: 11%, TV Technicheskyi Tsentr: 3%[1] | 27.0 | 24.0 | 19.0 | 39.4 | 28.9 | 30.7 | 33.0 | 27.2 | 21.0 | 20.5 |
| VGTRK (public) | VGTRK (public): 100% | 18.0 | 23.0 | 17.0 | | 22.7 | 26.5 | 29.5 | 30.4 | 17.8 | 16.9 |
| NTV (Gazprom-Media) | Gazprom-Media: 35%, LLC PRT-1: 14%, LLC: Aura Media: 51% | 18.0 | 11.0 | 14.0 | 28.4 | 19.0 | 14.8 | 18.3 | 21.7 | 32.6 | 33.5 |
| CTC Media (Modern Times Group, Sweden) | CTC Communications (US): 51%, CTC Media (US): 49% | 5.2 | 10.0 | 8.8 | 32.3 | 29.4 | 25.5 | 17.1 | 19.5 | 20.2 | 21.2 |
| TNT-Teleset | Aura-Media: 51.8%, Gazprombank (public): 30%, Benton Solutions (US): 18.2% | | | | | | | | | 7.6 | 7.3 |
| REN TV (NMG) | REN TV: 100%[2] | 3.9 | 4.9 | 4.9 | | | 2.5 | 2.0 | 1.2 | 0.8 | 0.6 |
| Others | | 28.0 | 27.0 | 36.0 | | | | | | | |
| Total Revenue (mil RUB) | | | | | 32,085 | 45,261 | 67,662 | 66,071 | 80,626 | 104,516 | 119,165 |
| Total Revenue (mil US$) | | | | | 1,150 | 1,605 | 2,506 | 2,951 | 3,054 | 3,604 | 4,095 |
| C4 | | 68.0 | 68.1 | 58.7 | 100.0 | 100.0 | 97.5 | 98.0 | 98.8 | 91.6 | 92.1 |
| HHI | | 1,401 | 1,342 | 944 | 3,396 | 2,575 | 2,521 | 2,596 | 2,516 | 2,288 | 2,331 |
| *N* (>1%) | | 5 | 5 | 5 | 3 | 4 | 5 | 5 | 5 | 5 | 5 |
| Noam Index | | 627 | 600 | 422 | 1,961 | 1,288 | 1,127 | 1,161 | 1,125 | 1,023 | 1,042 |

*Source:* Kommersant, 2011; Integrum-Companies, 2012.

[1] TV Technuchesk Yi Tsentr is the state-owned TV transmission company.

[2] Indirectly, NMG owns ZAO REN TV through a 68% stake in Mediaholding REN TV.

100 subnational cable TV providers in Russia today). All in all, about 18% of Russian viewers subscribe to cable platforms while 12% rely on satellite platforms. The regions most highly developed in this respect are Moscow, St. Petersburg, and the 11 largest industrial centers (with populations greater than a million). At the same time, only 10% of the country's population has access to triple-play services. In 2010, the leading cable platform providers at the national level were the Moscow-based NKS (with 4.7 million subscribers), Comstar (2.4 million), Acado (1.3 million), and the regional providers ER Telecom (880,000) and Moline (870,000).

Direct satellite broadcasting (DBS) is dominated by NTV +, established in 1998 and with a subscriber base of 550,000 in 2010, and Trikolor TV. NTV + was the first company to try experimental high-definition broadcasting (HDTV) in Russia. Trikolor TV, formerly Natsional'naya Sputnikovaya Kompaniya, entered the market in 2005 and offers a package of channels distributed in the European part of Russia (west of the Ural Mountains). In 2011, Tricolor TV had 3.1 million subscribers and became the largest satellite TV platform provider in Russia (and all of Eastern Europe, for that matter). Both cable and satellite TV platforms also provide limited IPTV services (Table 11.7). In total, the Russian market of pay TV in 2010 was valued at US$1.2 billion (RUB34.9 billion).[27]

Table 11-7. Multichannel Video Platforms: Cable MSOS, DBS, IPTV (Market Shares by Revenue), 2006–2010

| | | 2006 | 2007 | 2008 | 2009 | 2010 |
|---|---|---|---|---|---|---|
| Trikolor TV(2010) | Satellite TV | n/a | n/a | n/a | n/a | 9.2 |
| NTK (NMG) | Cable TV | 54.5 | 18.6 | 13.2 | 41.9 | 26.5 |
| AFK Sistema/SMM | IPTV, Cable TV | n/a | 12.5 | n/a | n/a | n/a |
| MTS (before 2010, Multiregion) | Cable TV | 5.1 | 4.2 | n/a | n/a | 1.1 |
| ER Telekom (Perm Financial and Industrial Group) | Cable TV | 6.1 | 4.2 | 7.4 | 23.6 | 19.0 |
| NTV + (Gazprom-Media) | Satellite TV | n/a | 36.5 | 43.1 | n/a | 23.1 |
| Svyazinvest (public) | IPTV, Cable TV | n/a | 3.1 | n/a | n/a | n/a |
| Akado | Cable TV | 10.2 | 8.7 | 22.6 | 22.8 | 14.9 |
| Norilsk Telecom (Kit Finance Investment Bank) | Cable TV | 12.2 | 5.8 | 6.3 | 5.3 | 2.9 |
| TVT | Cable TV | 12.1 | 6.5 | 7.5 | 6.4 | 3.5 |
| Total Revenue (mil RUB) | | 6,264 | 17,631 | 17,398 | 19,778 | 35,822 |
| Total Revenue (mil US$) | | 232 | 653 | 659 | 682 | 1,231 |
| Others | | | | | | |
| C4 | | 88.8 | 76.2 | 86.3 | 94.7 | 83.3 |
| HHI | | 3,425 | 2,027 | 2,693 | 2,904 | 1,918 |
| *N* (>1%) | | 6 | 9 | 6 | 5 | 8 |
| Noam Index | | 1,398 | 6,76 | 1,099 | 1,299 | 678 |

*Sources:* ComNews, 2012; Kommersant, 2011.

27. Wolfgang, Closs. "Rynok televideniya i on-demand servisov v Rossiji (TV Market and On-demand Services in Russia)." Moscow: European Audiovisual Observatory and J'son and Partners Consulting, 2011. p. 69.

## Video Channels

Pay TV operators (cable, satellite, and IPTV) provide subscribers with foreign-language programming translated into Russian (55%), 26% of foreign channels untranslated, and 19% of international Russian channels. The market leaders among translated channels are Animal Planet, The Discovery Channel, and National Geographic (Table 11.8). The breakdown of video channels' popularity in Russia is as follows: entertainment (25% of all cable-delivered channels), educational (20%), movies (13%), sports and music (11% a piece), information (9%), and children's programming (8%).[28]

## Film

Cinema used to be one of the most crucial segments of the Soviet cultural industries and was completely owned and financed by the state. Since 1991, the segment has gone through a

**Table 11-8. Combined Video Network Industries: Broadcast TV Networks, TV Syndicators, Cable Channels, Pay TV (Market Shares by Revenue), 2008–2011**

| | 2008 | 2011 |
|---|---|---|
| VGTRK/Channel One Russia Worldwide (public) | | 16.8 |
| Karusel | | *11.5* |
| Moya Planeta | | 5.3 |
| RBK TV (Onexim Group) | | 8.7 |
| RU.TV (Russian Media Group) | 13.3 | 8.3 |
| MTG Viasat Broadcasting (Sweden) | 34.3 | 20.6 |
| TV 1000 | *20.4* | *8* |
| TV 1000 Russkoe Kino | *13.8* | *7.8* |
| Viasat History | | *4.8* |
| Discovery Communications (US) | 41.2 | 15.2 |
| Discovery Channel | *21.8* | *7.9* |
| Animal Planet | *19.3* | 7.3 |
| Mir | | 6.7 |
| Eurosport (TF1 Group, France) | | 5.4 |
| Disney Channel (Walt Disney, US) | | 5.3 |
| Bridge Media Group | | 4.5 |
| National Geographic (Murdoch Family, News Corp., US/UK/AUS) | 11.3 | 4.4 |
| Dom Kino (Channel 1 Russia) | | 4.2 |
| Total Revenue (mil RUB) | 956 | 5,909 |
| Total Revenue (mil US$) | 36.2 | 187 |
| C4 | 75.4 | 61.3 |
| HHI | 1,763 | 1,242 |
| *N* (>1%) | 4 | 11 |
| Noam Index | 882 | 374 |

*Source:* J'son & Partners Consulting, 2012.

28. Ibid., p 117.

dramatic reorganization that has changed its basic economic structure. Currently, the market comprises film production companies that produce movies and TV serials (mostly private companies), companies providing services (15 state and 12 private studios), 756 theaters, and several independent film distributors. These distributors were formed in the late 1980s and early 1990s in close affiliation with Hollywood majors (MGM, United Artists, Paramount, Universal, 21st/20th Century Fox), as many of the Hollywood majors established joint ventures with privatized Soviet film distributors. However, in response to piracy by unlicensed Russian distributors, the Motion Picture Association of America (MPAA) announced an embargo on US film exports to Russia until 1993, which led to a crisis for the industry due to the dramatic decrease of visits to cinemas (a low of 0.25 per person a year in 1997). This created uncompetitive conditions for Russian film distribution companies and cinema operators who once again dominate the market.

Relying on the long-established popularity of the national cinema, Russian film producers are still highly active. Since 2006, 417 feature films have been produced (on average, 104 films per year). The business model for the segment also involves TV-serial production and distribution. Most film production is based on non-return investments, among which are the investments provided by the Ministry of Culture. From 2006

## Table 11-9. Film Production/Distribution (Market Shares by Box Office %), 2006–2010

| | 2006 | 2007 | 2008 | 2009 | 2010 |
|---|---|---|---|---|---|
| Central Partnership (ProfMedia/Interros) | 6.9 | 15.3 | 7.8 | 16.4 | 27.4 |
| Walt Disney Studios (US)-Sony Pictures (US, Japan) | 0.0 | 14.9 | 14.6 | 27.5 | 19.9 |
| Karo Premier/Karoprokat[1] | 12.5 | 17.5 | 15.7 | 18.7 | 15.2 |
| 21st/20th Century Fox C.I.S. (Murdoch Family, News Corp., US/UK/AUS) | 23.4 | 14.2 | 18.7 | 14.9 | 13.5 |
| United International Pictures (UIP) (Paramount Pictures: Viacom, US) (Universal: Comcast/ GE, US) | 11.8 | 16.8 | 24.6 | 7.6 | 6.1* |
| West | 4.3 | 4.5 | 2.9 | 3.7 | 4.8 |
| Paradise | 5.4 | 6.9 | 4.6 | 3.5 | 3.5 |
| Cascade Film | 20.9 | 3.4 | 2.6 | 0.8 | 2.2 |
| Leopolis | | | | 0.2 | 1.5 |
| Luxor | 0.7 | 1.5 | 1.8 | 1.1 | 1.4 |
| Others | 14.1 | 5 | 6.7 | 5.6 | 4.5 |
| Total Revenue (mil RUB)[2] | 12,349 | 14,436 | 20,589 | 23,387 | 31,861 |
| Total Revenue (mil US$) | 457 | 566 | 790 | 806 | 1,095 |
| C4 | 68.6 | 64.5 | 73.6 | 77.5 | 76.0 |
| HHI | 1,376 | 1,328 | 1,515 | 1,633 | 1,642 |
| *N* (>1%) | 7 | 9 | 9 | 8 | 10 |
| Noam Index | 520 | 443 | 505 | 595 | 519 |

*Source:* J'son & Partners Consulting & European Audiovisual Observatory, 2010.

*Russian distributors.

[1] Warner Brothers (US) distributor.

[2] Because distributors report their revenues in relation to the entire CIS market, the revenue figures for film cited here represent Armenia, Azerbaijan, Belarus, Georgia (until 2008), Kazakhstan, Kyrgyzstan, Moldova, Russia, Tajikistan, Turkmenistan, and Uzbekistan combined.

Table 11-10. Wireline Telecom (Market Shares by Revenue), 2005–2010

| PROVIDERS | OWNERS | 2005 | 2006 | 2009 | 2010 |
|---|---|---|---|---|---|
| Svyazinvest (public) | Russian Federation: 75% + 1, OJSC Rostelecom 25% + 1 | 88.1 | 84.7 | 83.3 | 82.6 |
| TransTeleCom (public) | Russian Railways: 100% | 6.1 | 7.0 | 6.9 | 8.0 |
| Mejregionalnyi TransTeleCom (MegaFon, Alisher Usmanov) | MegaFon: 100% | 2.5 | 5.0 | 4.2 | 3.6 |
| Tattelecom (OJSC Svyazinvestneftekhim) (public) | Government of the Republic of Tatarstan: 100% | 0.4 | 0.6 | 1.6 | 1.7 |
| Bashinformsvyaz (OGSC Rostelecom) (public) | Russian Federation: 29.3%, OGSC Rostelecom: 39.9%, Government of the Republic of Bashkortostan: 18.5% | 1.7 | 1.7 | 1.7 | 1.7 |
| West Koll | N/A | 0.1 | 0.3 | 0.6 | 0.7 |
| Arktel (mostly public) | Russian Federation: 75% + 1, OJSC Mososkovskiy Trastovay Bank Russia: 25% | 0.2 | 0.3 | 0.6 | 0.5 |
| Norilsk Telecom (Kit Finance Investment Bank) | Bank Kit Finance Investment Bank: 100% | 0.2 | 0.3 | 0.5 | 0.5 |
| Start Telecom | N/A | 0.6 | 0.1 | 0.5 | 0.5 |
| Macomnet (Russian Moskovskij Metropolitan Mosteleset) (50% public; 50% US) | Moscow Metro: 50%, Andrew Corporation (US): 50% | 0.0 | 0.0 | 0.3 | 0.3 |
| Total Revenue (mil RUB) | | 240,377 | 255,501 | 32,648 | 328,976 |
| Total Revenue (mil US$) | | 8,524 | 9,463 | 11,258 | 11,305 |
| C4 | | 98.4 | 98.3 | 95.9 | 95.8 |
| HHI | | 7,804 | 7,251 | 7,005 | 6,899 |
| *N* (>1%) | | 4 | 4 | 5 | 5 |
| Noam Index | | 3,902 | 3,626 | 3,133 | 3,450 |

*Sources:* Cnews Analytics, 2011; ComNews, 2011; Integrum-Companies, 2012; Open sources.

on, 279 films have received financial support from the state. Average state support for feature films is US$0.6 million (RUB 19 million) per film, with a limit of US$1 million (RUB29 million) on production and US$0.2 million (RUB5 million) for distribution support. Market players with the largest production budgets—STV, Central Partnership (Paramount Picture's Russian distributor), Art Pictures, CTC Media (Sweden), and VGTRK's Rossiya 1 and 2 channels—received additional state support that amounted to 15% of their production budgets (Table 11.9).

## TELECOMMUNICATIONS MEDIA

### Wireline Telecom

State ownership is still a significant feature of the wireline market. The most powerful player in the market, with a market share of 82.6% in 2010, is the open joint stock company (OJSC) Svyazinvest, established in 1994 through the consolidation of the federal government's shares in 85 separate telecommunications providers. Currently, Svyazinvest includes seven large transregional companies, four open joint-stock telecommunications companies belonging to the Giprosvyaz interregional association, and the leading provider of long distance and international telephony, Rostelecom. In 2011, Rostelecom consolidated the market of fixed telephony by joining seven regional telecommunications companies (Table 11.10).

### Wireless Telecom

Because of the unchallenged state monopoly in the fixed telephony market that prevailed until the early 1990s (and a lack of investment in the sector during the 1990s), mobile telephony experienced a real boom between 1991 and the early 2000s. The number of cellular subscribers in Russia (measured by the number of active SIM cards) in mid-2011 amounted to 225.1 million. The penetration rate of cellular services (number of SIM cards per 100 inhabitants) in Russia at the same time was 155% (209% in Moscow and 212% in St. Petersburg). Even the most underdeveloped regions have an average

Table 11-11. Wireless Telecom (Market Shares by Revenue), 2003–2010

| | 2003 | 2004 | 2005 | 2006 | 2007 | 2008 | 2009 | 2010 |
|---|---|---|---|---|---|---|---|---|
| MTS (Multiregion) | 50.0 | 44.9 | 40.0 | 35.7 | 42.5 | 38.7 | 39.6 | 35.0 |
| Vimpelcom (public) | 31.8 | 31.9 | 33.3 | 34.2 | 32.6 | 38.3 | 35.0 | 30.7 |
| MegaFon (Alisher Usmanov) | 18.2 | 23.2 | 26.7 | 29.5 | 22.6 | 20.7 | 23.1 | 30.7 |
| Sky-Link (mixed state-public Svyazinvest)* | 0.0 | 0.0 | 0.0 | 0.0 | 1.0 | 1.0 | 1.0 | 1.2 |
| Tele2 Russia (Kinnevik, Sweden) | 0.0 | 0.0 | 0.0 | 0.7 | 1.3 | 1.3 | 1.3 | 2.4 |
| Total Revenue (bil RUB) | 129 | 193 | 254 | 348 | 467 | 697 | 788 | 675 |
| Total Revenue (bil US$) | 4.4 | 6.9 | 9.0 | 12.9 | 18.3 | 26.4 | 27.1 | 23.2 |
| C4 | 100.0 | 100.0 | 100.0 | 100.0 | 98.7 | 98.7 | 98.7 | 97.6 |
| HHI | 3,843 | 3,573 | 3,311 | 3,385 | 3,398 | 3,331 | 3,116 | 3,116 |
| *N* (>1%) | 3 | 3 | 3 | 3 | 4 | 5 | 5 | 5 |
| Noam Index | 2,219 | 2,063 | 1,912 | 1,954 | 1,699 | 1,666 | 1,394 | 1,394 |

*Sources:* ComNews, 2011; Cnews Analytics, 2011; AC&M Consulting; RBC, 2011; Integrum-Companies, 2012; Open sources.
*Russian State owns 75% shares minus 1 share, OJSC Rostelekom owns 25% + 1.

penetration rate of 140%.[29] Mobile telephony is also increasingly being used for Internet services (22 million subscribers in 2010) because mobile plans are so widely affordable.

The market is characterized by an oligopoly of MTS (32.9%), MegaFon (24.6%), and Vimpelcom (24.0%). The present strategies of these three Russian mobile operators focus international expansion. In 2010, Vimpelcom merged with the Ukrainian telecommunications provider Kievstar, becoming the largest mobile operator in all of Eastern Europe (Table 11.11).

## INTERNET MEDIA

### Internet Service Providers (ISP)

In 2010, there were 50 million Internet users in Russia (almost 43% of the population), 15.5 million of whom had regular (daily) Internet access. Since 2000, inequities among different regions in terms of access have been going down, and the digital divides among class, age, and gender groups have been lessening. The proportion of women among users is about 40% and more than a third of all users live outside of a major metropolitan area, though a typical user is a male urban professional with a high per capita income between the ages of 25 and 35. Aside from the Moscow and St. Petersburg metropolitan areas and the Urals (which have Internet penetration rates of 51%, 42%, and 38%, respectively), Internet penetration in most other areas barely exceeds one-third of the population.

Around one-third of all connected households (representing 18 million users) have broadband access, while most other home access connections rely on older ASDL lines. The expansion of broadband access will require huge capital investments, and the leading telecom providers—MTS/Komstar, Tsentrtelecom, Vimpelcom, and Akado—have become more active at the broadband segment in Russia in order to take advantage of the market's growth potential (for instance, in 2010 MTS acquired Komstar, merging fixed and mobile operations).

It is extremely difficult to get a detailed picture of the Russian national ISP market due to the profusion of non-broadband access methods and low transparency. The data concerning the broadband segment look quite different compared to market as a whole (Table 11.12).

### Search Engines

In 2010, the search engine Yandex.ru accounted for 47.1% of the market share, followed by Google.ru (14.2%), Mail.ru (34.8%), and Rambler.ru (3.9%). Besides Google, all of these search engines are Russian owned. Yandex.ru was established by Yandex, a Russian IT group that like Google provides a number of other free Internet-based services (email, news, maps, etc.). Yandex.ru is the fifth largest search engine by volume, as it is also popular in the Ukraine, Kazakhstan, Turkey, and Belarus. Another Russian "success story" in the search engine market has been Mail.ru, which also provides free email services to users. Until 2010 it was part of the investment company Digital Sky Technologies (DST), which is now divided into the Mail.ru Group (Russian assets) and DST Global (assets outside of Russia) (Table 11.13).

### Online News

According to Yandex.Novosti in 2011, online media reported at least 50 thousand news items on every working day and at least 15 thousand news items on weekends; 80% of this news flow is produced by 20% leading media outlets. The number of online media with no offline analogues in Russia is limited: among the most popular are rbc.ru, lenta.ru, gazeta.ru, and TV channel Dozhd, which broadcasts on the Internet. Internet products of ria.ru, also attract a lot of users. Still, online versions of traditional media—newspapers *Komsomolskaya Pravda*, *Sport-Express*, *Rossiiskaya Gazeta*, *Argumenty i Fakty*, state broadcaster VGTRK, and radio station Ekho Moskvy are the most popular ones.

29. The high rates stem from the fact that many subscribers have multiple SIM cards from different operators and use different SIM cards in the same device depending on whether the call is work-related or personal.

## Table 11-12. Internet Service Providers (Market Shares by Revenue), 2007–2010

| PROVIDERS | OWNERS | 2007 | 2008 | 2009 | 2010 |
|---|---|---|---|---|---|
| NTK (NMG) | NMG: 51.8%; VTB: 11.6%; Gazprombank (public): 8.3%; Sberbank Capital: 8.3%; Surgutneftegas 10%; Raybrook: 10% | 27.8 | 14.0 | 31.8 | 31.9 |
| Akado (Renova Media Enterprises- Renova Industries) | Viktor Vekselberg: 67%; Yuri Pripachkin: 33% | 46.9 | 50.3 | 34.0 | 31.2 |
| ER Telekom | Perm Financial and Industrial Group: 95%; ER Telekom Management: 5% | 11.5 | 17.8 | 17.9 | 20.1 |
| NetByNet (MegaFon: Alisher Usmanov) | MegaFon: 100% | | | 4.1 | 4.5 |
| Ufanet (Cyprus) | Colswick Holdings Ltd. (Cyprus) 99.9% | 7.0 | 7.0 | 4.7 | 4.2 |
| Pin Telekom (Pin Group) | Pin Group: 100% | 2.2 | 2.0 | 1.5 | 2.4 |
| Gars Telekom | Gars Telekom: 100% | 3.7 | 3.2 | 2.0 | 1.9 |
| Smart Telekom | Promsvyaz: 50%; Tels: 50% | | 3.1 | 2.1 | 1.9 |
| Obit | Obit Ltd.: 100% | 0.9 | 2.5 | 1.9 | 1.9 |
| Total Revenue (mil RUB) | | 11,169 | 16,368 | 26,100 | 33,814 |
| Total Revenue (mil US$) | | 438 | 620 | 900 | 1,162 |
| C4 | | 93.2 | 89.1 | 88.4 | 87.7 |
| HHI | | 3,171 | 3,123 | 2,539 | 2,451 |
| *N* (>1%) | | 6 | 8 | 9 | 9 |
| Noam | | 1,295 | 1,104 | 846 | 817 |

*Sources:* Cnews Analytics, 2011; Integrum-Companies, 2012; Open sources.

## Table 11-13. Internet Search Engines (Market Shares by Search Volume), 2008–2009

| | 2005 | 2006 | 2007 | 2008 | 2009 | 2010 |
|---|---|---|---|---|---|---|
| Yandex.ru (Yandex) | 62.1 | 70.3 | 55.0 | 59.8 | 54.0 | 47.1 |
| Rambler.ru (ProfMedia) | 32.5 | 29.7 | 23.0 | 20.6 | 8.0 | 3.9 |
| Google.ru (Google, US) | 0.0 | 0.0 | 3.3 | 5.6 | 13.5 | 14.2 |
| Mail.ru | 5.4 | *No data* | 18.6 | 14.0 | 24.5 | 34.8 |
| Total Revenue (mil RUB) | 1,607 | 2,781 | 7,650 | 14,071 | 14,935 | 27,179 |
| Total Revenue (mil US$) | 57 | 103 | 300 | 533 | 515 | 934 |
| C4 | 100.0 | 100.0 | 100.0 | 100.0 | 100.0 | 100.0 |
| HHI | 4,943 | 5,828 | 3,915 | 4,229 | 3,761 | 3,642 |
| *N* (>1%) | 3 | 2 | 4 | 4 | 4 | 4 |
| Noam Index | 2,854 | 3,365 | 1,958 | 2,114 | 1,881 | 1,821 |

*Sources:* Integrum-Companies, 2012; Open sources.

Table 11-14. Online News Media (Market Shares by Monthly Unique Visitors), 2005–2011

| | 2005 | 2006 | 2007 | 2008 | 2009 | 2010 | 2011 |
|---|---|---|---|---|---|---|---|
| Rbc.ru | 25.5 | 25.4 | 31.0 | 24.6 | 17.5 | 15.0 | 17.8 |
| Kp.ru (Komsomolskaya Pravda) | 9.2 | 10.9 | 11.6 | 12.2 | 13.4 | 14.5 | 16.4 |
| RIA.ru (RIA Novosti) (public) | 5.0 | 6.7 | 7.2 | 10.0 | 14.4 | 15.0 | 14.5 |
| Vesti.ru (Vesti) | 3.5 | 3.1 | 3.3 | 6.7 | 6.7 | 9.5 | 12.2 |
| Lenta.ru | 14.7 | 16.2 | 13.7 | 15.0 | 13.1 | 9.5 | 9.7 |
| Utro.ru | 15.8 | 14.6 | 12.7 | 10.8 | 6.9 | 8.4 | 8.7 |
| Gazeta.ru | 12.5 | 11.2 | 10.6 | 9.8 | 9.0 | 6.9 | 6.9 |
| Newsru.com | 12.5 | 10.8 | 8.5 | 8.1 | 6.8 | 5.4 | 5.7 |
| Infox.ru | 0.0 | 0.0 | 0.0 | 1.4 | 7.6 | 6.8 | 5.0 |
| Gzt.ru | 1.3 | 1.1 | 1.5 | 1.4 | 4.5 | 9.0 | 3.0 |
| C4 | 68.6 | 67.4 | 68.9 | 62.6 | 58.4 | 54.0 | 61.0 |
| *N* (>1%) | 9 | 9 | 9 | 10 | 10 | 10 | 10 |
| HHI | 1,553 | 1,537 | 1,691 | 1,406 | 1,164 | 1,114 | 1,231 |
| Noam Index | 7 | 7 | 7 | 6 | 6 | 5 | 6 |

*Source:* TNS, 2012.

The number of unique monthly visitors to the most popular Russian online news media reached almost 22 million in 2012, or 61% of total Russian Internet users. Most users search for information on the Internet and read the news. An interesting fact is that 12% of them trust online news media more than other types of media.

Today the main and the most dependable source of income for online media is advertising. According to the report of AKAR (Association of Russian Communication Agencies) in 2012, the Internet experienced remarkable growth in advertising income (+ 35%), reaching 56.3 billion rubles. Almost one-third of Internet advertising (17.9 billion rubles, with growth of 17%) belonged to online news media, while context advertising, with growth of 45%, comprised the remaining two-thirds (38.4 billion rubles).

At the moment there is no clear understanding of the business models functioning in the Russian online media. The main source of income remains advertising, and in many cases advertising in traditional—print or audiovisual—media. However, the revenues for the online activities are not reported separately from those for offline media. In this situation, analyzing the future of paid news content is futile. (Table 11.14).

## CONCLUSION

The main trends in the transformation of media ownership in Russia in the 2000s have been a decrease in commercial media capital and the proportionate increase in state capital and mixed (state and private, non-media) capital. The largest media companies are concentrated (directly or indirectly) in the hands of government-controlled structures (VGTRK, Gazprom-Media) or are private enterprises that are mainly active in non-media sectors like the extractive sector or finance (ProfMedia, NMG).

Both wireline and wireless telecom have aggregated the highest revenues of all media sectors. Of print and audiovisual media, broadcast TV (as the main advertising venue) has showed comparatively good economic results, and magazines have seen their revenue streams stabilize after 2010 (Table 11.15).

From 1991 until the 2008 global financial crisis, US$900 million (RUB23.8 billion) had been spent in the Russian media market on

Table 11-15. Top Russian Media Groups' Annual Revenue (Mil US$), 2008–2011

| MEDIA GROUP | OWNERSHIP DETAILS | PROPERTIES | TOTAL REVENUE (MIL US$) | | |
|---|---|---|---|---|---|
| | | | 2008 | 2009 | 2010 |
| Gazprom-Media | Russian private, Gazprombank | Broadcast TV (NTV, TNT), satellite TV (NTV-Mir, NTV +), book publishing (Sem Dney), newspapers (*Tribuna, Chas Pik*), radio stations (Echo of Moscow, Siti-FM, Popsa, Nekst, Relaks-FM, Detskoye Radio), film production/distribution (NTV-Kino), advertising (NTV-Media), film exhibition (Oktyabr, Kristal-palas) | 1,291 | 1,154 | 1,347 |
| Channel One Russia (public) | Russia, Rosimushestvo—38.9%, Limited liability company "RastrKom-2002" (NMG)—25%, Limited liability Company "Ort-cb"—24%, ITAR—TASS News Agency—9.1%, FGUP "TTC" Ostankino—3% | Broadcast TV (Channel One Russia) | 942 | 756 | 841 |
| VGTRK (public) | Russian, public | Broadcast TV (Rossiya 1, Rossiya 2, Kultura, Vesti 24), cable TV (RTR-Planeta, Sport-Planeta, Bibigon), satellite TV (GTRK), radio (Radio Rossiya, Mayak, Kultura, Vesti-FM, Rossiya), online news (Rossiya 1, Rossiya 2, Kultura, Vesti 24) | 928 | 640 | 693 |
| ProfMedia | Russian private, Interross | Book publishing (Afisha, B2B-Media), radio (Avtoradio, Avtoradio St. Petersburg, Yumor-FM, NRJ, Alla) video channels (TV-3, 2X2, MTV-Rossiya, VH1-Rossiya), film production/distribution (Central Partnership), film exhibition (Cinema Park) | N/A | 556 | 636 |

| | | | | | |
|---|---|---|---|---|---|
| CTC Media (Modern Times Group, Sweden) | Modern Times Group, Sweden | Broadcast TV (CTC, Domashny, Peretz, 31 Kanal, CTC TV Dixi), | 606.5 | 553.5 | 627.7 |
| National Media Group (NMG) | Russian private | Broadcast TV (REN TV), newspapers (*Izvestia*) | N/A | 482.7 | N/A |
| Yandex | 100% of the share capital, registered in the Netherlands Company Yandex NV | Search engines (Yandex.ru, Yandex Dengi, Yandex Probki) | N/A | 311 | 452 |
| Media 3 | Russian private | Newspapers (*Aif, Trud, Ekstra M, Centr Plus*) | N/A | 406.8 | 446.7 |
| Mail.ru Group | 53.2%—Digital Sky Technology (DST) 42.9%—South African holding company Naspers 3.5%—Dmitry Grishin | Search engines (Mail.ru) | 64.0 | 126 | 324 |
| Hearst-Shkulev Media | Russian private—51%, Hearst Corp.—49% | Magazine publishing (*Elle, Elle Girl, Elle Decoration, Maxim, Marie Claire, Parents, Psychologies, Departures, Antenna-Telesem, Va-Bnak, StarHit*) | N/A | 173.5 | 199.3 |
| Eksmo | Russian private | Book publishing | 147.7 | 155.2 | 161.5 |
| Sibirsky Delovoy Soyuz (SDS) | Russian private | Radio (Europa Plus, Retro FM, Radio 7, Keks-FM, Svezhee Radio, Eldoradio) | 66.1 | 112.7 | 133 |
| Google.ru | Larry Page—28.2% of the voting shares, Sergey Brin—27.7%, * Eric Schmidt—10.3%. | Search engines | 29.7 | 69.0 | 132.9 |
| AST | Russian private | Book publishing | 102 | 131 | 129.5 |
| SDS | Russian private | Book publishing | 83.3 | 89.6 | 106.5 |
| Russian Media Group (RMG) | Russian private | Radio (Russkoe Radio, Hit FM, Maximum, DFM, Monte Carlo) | N/A | 40.6 | 50.5 |
| Komsomolskaya Pravda | Russian + Norwegian, media industry | Newspapers (*Komsomolskaya Pravda, Sovetsky Sport*) | 46.9 | N/A | N/A |

*continued*

Table 11-15. *continued*

| MEDIA GROUP | OWNERSHIP DETAILS | PROPERTIES | TOTAL REVENUE (MIL US$) | | |
|---|---|---|---|---|---|
| | | | 2008 | 2009 | 2010 |
| Rambler | Closed Joint-stock Company, "Rambler Media." The largest shareholder Rambler—Vladimir Potanin's holding company, "Prof-Media" (54.8%). About 26% is traded on the alternative market of the London Stock Exchange | Search engines (Rambler.ru) | 109.8 | 41.3 | 36.7 |
| Moskovskij Komsomolets | Russian private | Newspapers (*Moskovskij Komsomolets, MK-Region, MK-Bulvar, MKmobil*), magazine publishing (*Atmosfera, Ochota I Rybalka XXI*) | 14.5 | 22.5 | 23.2 |
| Burda Russia (Hubert Burda Media, Germany) | Hubert Burda Media, Germany | Magazine publishing (*Liza, Priyatnogo Appetita, Otdohni!, Moi Prekrasny Sad, AvtoMir, Dobrye Soveti, Moi Rebenok, Oops!, Chip, Goroskopy, Krossvordy, Lyublu Gotovit*) | N/A | N/A | N/A |
| Sanoma Oyj (Finland) | Finland | Magazine publishing (*Cosmopolitan, Cosmopolitan Beauty, Cosmopolitan Shopping, Harper's Bazaar, Collezioni, Wedding, Grazia, Mamas & Papas, Men's Health, Esquire, Yes!, National Geographic, National Geographic Traveler, Prime Traveller*). Newspapers (*The Moscow Times, The St. Petersburg Times, Vedomosti, Na Rublevke*). | N/A | N/A | N/A |

mergers and acquisitions, and this trend will continue in the short- to medium-term. Transparency is also increasing: since the early 2000s, more firms have begun publishing annual reports online for public viewing. It must be noted, however, that only a few media enterprises have chosen to go through initial public offerings (IPOs) to list themselves on international stock markets (Rambler Media, CTC Media, and RBK-Information Systems).

Some profitable media enterprises were established through the influx of foreign capital into the national media system, but in the 1990s, especially during the period of initial capitalization, the foreign investments were rather unevenly distributed among different media segments. Widescale expansion of foreign media enterprises into the Russian media market only began in the 2000s, in response to increased economic stability. As a result, a number of foreign media groups entered the market: Axel Springer (Germany), Bertelsmann AG (Germany), Hubert Burda Media (Germany), Bonnier AB (Sweden), Modern Times Group (Sweden), Condé Nast (US), the Edipresse Group (Switzerland), Sanoma Oyj (Finland), Trader Classified Media NV (the Netherlands), and others. US online services such as AOL, MSN (Microsoft), and Google have all sought to enter the Russian Internet market, though so far only Google.ru has secured a significant market share (and lags far behind Russian search engines and online news portals). The key development strategy for foreign print and audiovisual media enterprises in Russia is to localize their global brands, an approach that has been most successful in magazine publishing. Notably, few foreign media organizations have decided to open their own offices in Russia; most of them have confined themselves to buying large share-holdings of publishing houses or forming joint enterprises with Russian partners to offer Russian-language editions of their flagships periodicals.

At the subnational level, concentration is increasing, though the process is still in its infancy. Only a few regional companies are comparable with the Moscow metropolitan players in terms of turnover and impact on regional media systems. The most successful are Abak-Press (Altai); the Baltic Media Group and Bonnier AB Grupp Rossiya (St. Petersburg); Korporatsia Fyodorov (Samara); Provintsia (a consortium serving 40 regions); Tomsk Media Group (Tomsk); Chetvyortiy Kanal (Yekaterinburg); and Yuzhniy (Rostov). The most popular reason for the diagonal or cross-concentration of regional media companies is to optimize limited financial resources to maintain as few distribution networks for print and audiovisual media as possible, well as to find new sources of income besides long-time business models based on subscriptions and advertising (though all of the aforementioned regional media companies have become much more active in their respective regional advertising markets). A particular feature of the regional media markets that affects regional media companies and their ownership structures is the informal financial support by regional authorities, especially to local newspapers.

Increased concentration has contributed to the general expansion of the Russian media market and stimulated the development of all its sectors. Large media corporations have been established to accumulate and optimize the distribution of capital necessary for expansion. The Russian model of media concentration as a whole is characterized by centripetal tendencies: privately owned Muscovite and state-owned media enterprises (also based out of Moscow) dominate the national media markets. Given its special administrative status as the capital and a strong economy, Moscow leads the field among other cities in the Russian Federation. It is not surprising, therefore, that Russian media tycoons tended from the very start to develop their businesses in the capital, where conditions were most favorable and the central hubs of Soviet era distribution and infrastructure networks were sited. The growing interest of the larger media companies, especially terrestrial broadcasters, in regional media markets can be accounted for by the regional actors' cultural and physical proximity to the audiences the larger media companies would like to reach.

However, certain peculiarities of the Russian model of media concentration have not changed much over time. The most distinctive characteristic of the Russian media market at the national level is the predominance of a "diagonal" growth strategy (or cross strategy). The fact that this type of concentration prevails in Russia can be accounted for by the way privatization was implemented in the early 1990s. New businessmen who had cut their teeth in a variety of private enterprises in the final years of the USSR's existence used the privatization auctions held by the state to start acquiring any media properties that were available. Mass expansion began in all sectors of the national media industry, first in newspapers, magazines, television, and radio, and later in the film industry and the Internet. Notably, some large media gained control over already existing outlets, while others preferred to start up their own.

Considering the development of the Russian media market during the 1990s in terms of ownership structure, one can observe two important driving forces: state capital (distributed from the state budget) and private/entrepreneurial capital. Of the latter, there are three varieties: that of Russian financial and industrial groups led by a coterie of *nouveux riche* Russian businessmen collectively addressed as the "oligarchs"; that of foreign capital (mostly from pan-European media producers); and that of emerging media groups created solely within the media markets. Although the latter type remains weak in the Russian media industry, it is expanding and creating a separate segment within the industry from foreign-owned media and the outlets run by either oligarchs or the federal government.

Consequently, there exist three main types of media companies in Russia: state, commercial, and hybrids of the two. It is no exaggeration to say that the media industry in Russia has almost without exception developed through total diversification, which manifests itself in the entry of Russian financial, industrial, and financial-industrial groups' capital into the media market. It was the latter type that served as the foundation for most of the still-extant Russian media enterprises.

One of the key reasons why it is difficult to characterize the Russian media is the lack of transparency in the market in general and in the media market in particular. The most obvious problems are the overstated circulation rates of newspapers and magazine and the prevalence of falsified ratings of the broadcasting channels aimed at comforting advertisers. In addition, many of the oligarchs still maintain a high degree of secrecy regarding their holdings, particularly ownership of media companies. In fact, no government agency today possesses comprehensive statistical data on the condition and dynamics of the national media market as a whole. Mapping concentration in the Russian media is difficult due to the lack of public disclosure from media producers, low overall economic transparency, and rapidly changing ownership structures, especially when major media companies owned by industrial groups register their media holding companies offshore or as closed enterprises that have no obligations to make their ownership details public.

Finally, the process of technological convergence is proving beneficial first and foremost to the larger media producers, which are moving to established multimedia newsrooms. Although there is much room for development in this regard, the process can be observed in the establishment of print media titles by RBK Information Systems solely to recycle their existing online content and in the acquisition of Rambler Media by ProfMedia. Online media projects are gradually becoming full-fledged participants in the market. The pioneering companies in this field are the aforementioned RBK Information Systems, and Rambler Media, as well as Yandex, which already offers news content through Yandex.ru. In time, new subsidiaries within the established media groups will emerge to specialize in online media. As media and communication channels are becoming more interwoven, firms such as Gazprom-Media, SMS, and Renova are developing cable and satellite TV platforms in order to pursue new multimedia projects.

Paradoxically, grater commercialization is also contributing to the growth of centripetal trends in the audiovisual market. This is due to two factors. First, Russian television advertising is increasingly being targeted to national, instead of local, audiences. As a consequence, national broadcasters and multisystem operators (MSOs) are the ones mainly benefitting from advertising revenues. Second, the pressures exerted by regional political elites over regional TV companies are much weaker than the pressures exerted by the federal government on national television content producers. Local authorities aim to create favorable economic and regulatory conditions for national TV companies' local affiliates by seeking Moscow-based media moguls' support during election campaigns. This political game is contributing to the centralization of the TV landscape even as moguls undertake efforts to try and maintain autonomy from Moscow.

## Bibliography

### 1. Books and Articles


Bagdikian, Ben H. *Monopoliya sredstv informacii. (The Media Monopoly)*. Moscow: Progress, 1987.

Blinova, Olga N. *Media-imperii Rossii. Na sluzhbe gosudarstva I oligarchii (Media Empires in Russia. In the Service of the State and Oligarchs)*. Moscow: Centr Politicheskoi Informacii, 2001.

Esin, Boris I. *Istoriya russkoi zhurnalistiki (1703—1917). (History of Russian Journalism (1703—1917)*. Moscow: Flinta, Nauka, 2006.

Esin, Boris I.; Zassoursky, Yassen N.; and Minaeva, Olga. D., eds. *Russkaya zhurnalistika v dokumentakh. Istorija nadzora. (Russian Journalism in Documents. A History of Control)*. Moscow: Aspekt Press, 2003.

Garabova, Madina I. *Osobennosti funkcionirovaniya rossiiskih SMI v usloviyah ih monopolizacii I concentracii (Particular Qualities of Russian Media Functioning in Terms of Monopolization and Concentration)*. Diss. Moscow: RUDN, 2004.

Gavruchenkova, Maria G. "Osobennosti processa koncentracii SMI v rossiiskom regione (na primere nizhegorodskoi oblasti) (Particularities of Media Concentration Process in Russian Region (Case of Nizhny Novgorod Area)." *Seriya Mezhdunarodnie Otnosheniya* 1:24, 2003. pp 235–241.

Ilchenko, Sergey N. "Problema koncentracii elektronnyh SMI: peterburgsky opyt (the Issue of Concentration of Electronic Media: The Experience of St. Petersburg)." *Seriya Philologiya* 4:1, 2008. pp 153–156.

Ivanitsky, Valeriy L. *Modernizatsija zhurnalistiki: metodologicheskii etud (Modernization of Journalism: Methodological Composition)*. Moscow: Izd-vo Mosk. Un-ta, 2010.

Kachkaeva, Anna G. *Rossijskie sredstva massovoi informacii, vlast' i capital: k voprosu o koncentracii SMI v Rossii (Russian Mass Media, Power and Capital: On Concentration of Russian Media)*. Moscow: Centre Prava SMI, 2012.

Khlebnikov, Pavel Y. *Krestnyi otetz Kremlja Boris Berezovsky (The Kremlin Godfather Boris Berezovsky)*. Moscow: Detektiv-Press, 2001.

Makhonina, Svetlana Y. *Istoriya russkoi zhurnalistiki nachala XX veka. (History of Russian Journalism at the Beginning of the 20th Century)*. Moscow: Flinta: Nauka. 2009. p 111–120.

McNair, Brian. "Power, Profit, Corruption, and Lies: The Russian Media in the 1990s." Curran, James and Park, Myung-Jin, eds. *De-Westernizing Media Studies*. London: Routledge, 2000. pp 79–94.

Ovsepyan, Rafail P. *Sovetskaya zhurnalistika i kommunisticheskoye vospitaniye trudyashikhsya. (Soviet Journalism and Communist Upbringing of Workers)*. Moscow: Izd-vo Mosc. Un-ta, 1979.

Prokhorov, Yevgeniy P. *Osnovy marksistsko-leninskoy teorii zhurnalistiki. (Basics of Marxist-Leninist Journalism Theory)*. Moscow: Izdatelstvo MGU, 1973.

Puppae, Yakov S. *Oligarchi. Ekonomicheskaya khronika 1992–2000. (Oligarchs. The Economic Chronology 1991–2000)*. Moscow: Moscow State University Higher School of Economics, 2000.

Rantanen, Terhi. *Globalnoe I nacionalnoe. Mass media I communicacii v postcommunisticheskoi Rossii. (The Global and the Local. Mass Media and Communications in Postcommunist Russia)*. Moscow: Izd-vo Mosk. Un-ta, 2005.

Richter, Andrei. *A Landmark for Mass Media in Russia*. Strasburg: European Audiovisual Observatory, 2011.

Schiller, Herbert *Manipulyatory soznaniem (Mind Managers)*. Moscow: Mysl, 1980.

Smirnov, Sergey S. *Koncentraciya sredstv massovoi informacii Rossii v usloviyah transformacii nacionalnoi mediasistemy (1991—2006). (Concentration of Russian Mass Media in the context of national media system transformation (1991—2006)*. Diss. Moscow: Moscow State University, Faculty of Journalism, 2006.

Smirnov, Sergey S. "Koncentraciya SMI v Rossii: puti I trudnisti issledovaniya problemi (Concentration of Mass Media in Russia: Research Methods and Problems)." Vartanova, Elena L., ed. *Media Economics and Management Yearbook 2007*. Moscow: MediaMir, 2007.

Smirnov, Sergey S. "Koncentraciya v industrii SMI (Concentration in the Media Industry)." Vartanova, Elena L., ed. *Osnovy mediabiznesa*. Moscow: Aspekt Press, 2009. pp 152–189.

Trakhtenberg, Anna. "Transformation without Change. Russian Media as a Political Institution and Cultural Phenomenon." Vartanova, Elena L., ed. *Media and Change*. Moscow: MediaMir, 2007.

Vartanova, Elena L. "The Russian Media Models in the Context of Post-Soviet Dynamics." Hallin, Daniel and Mancini, Paolo, eds. *Comparing Media Systems beyond the Western World*. Cambridge: University Press, 2012.

Vartanova, Elena L. and Kolomiets, Viktor P. eds. "Televidenie v Rossii. Sostoyanie, tendencii i perspective razvitiya. *(Television in Russia. Conditions, Trends and Perspectives)*." Moscow: Report of Federal Agency on Press and Mass Communication (FAPMC), 2011.

Vartanova, Elena L. and Smirnov, Sergey S. "SMI Rossii kak industriya razvlecheniy (Russian Media as an Entertainment Industry)." *Medi@ scope* 4:8, 2009. Feb. 5, 2013. <http://www.mediascope.ru/node/446>.

Vartanova, Elena L. and Smirnov, Sergey S. "Contemporary Structure of the Russian Media Industry." Rosenholm, Arja; Nordenstreng, Kaarle; and Trubina, Elena, eds. *Russian Mass Media and Changing Values*. London: Routledge: 2010. pp 21–40.

Zassoursky, Yassen N., ed. *Na sluzhbe monopoliy. Informacionno-propagandistsky kompleks stran kapitala. (To Serve the Monopolies. The Information and Propaganda Complex of Capitalist Countries)*. Moscow: Mysl, 1977.

Zassoursky, Ivan I. *Rekonstrukciya Rossii (Mass media I politica v 90-e) (Reconstruction of Russia (Media and Politics in the 1990s)*. Moscow: Izdatelstvo MGU, 2001.

## 2. Reports

AKAR. "Market Volume of Marketing Communications in Russia in 2011." Moscow: AKAR. 2012. Feb. 8, 2013. <http://www.akarussia.ru/knowledge/market_size/id1865>.

Federal Agency on Press and Mass Communication. "Knizhny rynok v Rossii. Sostoyanie, tendencii i perspective razvitiya (2011) (The Book Publishing Industry in Russia. Conditions, Trends and Perspectives (2011)." Moscow, Federal Agency on Press and Mass Communication, 2011.

Federal Agency on Press and Mass Communication (FAPMC). "Pechat Rossiiskoi Federacii v 2000 godu. Statistichesky Sbornik (The Periodical Press of Russian Federation in 2000. Statistical Report)." Moscow: Rossiiskaya Knizhnaya Palata, 2001.

Federal Agency on Press and Mass Communication (FAPMC). "Pechat Rossiiskoi Federacii v 2004 godu. Statistichesky Sbornik (Periodical Press of Russian Federation in 2004. Statistical Report)." Moscow: Rossiiskaya Knizhnaya Palata, 2005, pp 57–60.

Federal Agency on Press and Mass Communication (FAPMC). "Internet v Rossii. Sostoyanie, tendencii i perspective razvitiya (2011) (The Internet in Russia. Conditions, Tends and Perspectives (2011)." Moscow: Federal Agency on Press and Mass Communication, 2011.

Federal Agency on Press and Mass Communication (FAPMC). "Rossiiskaya Periodicheskaya Pechat. Sostoyanie, tendencii I perspektivi razvitiya (Russian Press. Condition, Trends and Perspectives)" Moscow: Federal Agency on Press and Mass Communication, 2011.

Federal Agency on Press and Mass Communication (FAPMC). "Rossiyskaya poligrafiya. Sostoyanie, tendencii i perspective razvitiya (Printing Industry in Russia. Conditions, Trends and Perspectives)." Moscow: Federal Agency on Press and Mass Communication, 2011.

Federal Agency on Press and Mass Communication (FAPMC). "Radioveschanie v Rossii. Sostoyanie, tendencii i perspective razvitiya (Radio in Russia. Condition, Trends and Perspectives)." Moscow: Federal Agency on Press and Mass Communication, 2011.

Wolfgang, Closs. "Rynok televideniya i on-demand servisov v Rossiji (TV Market and On-demand Services in Russia)." Moscow: European Audiovisual Observatory and J'son and Partners Consulting, 2011.

Zakon RF. "O Sredstvah massovoi informacii." (Law on Mass Media). Moscow. Dec. 27, 1991 no. 2124–2121 of Dec. 27, 1991. *Rossiiskaya Gazeta*, 1992.

## 3. Internet sources

AC&M Consulting http://www.acm-consulting.com/

Cnews Analytics http://cna.cnews.ru/

ComNews http://www.comnews.ru/

EMG http://emg.fm/

Integrum. Companies http://companies.integrum.ru

Kommersant http://www.kommersant.ru/doc/1836614/print

http://www.kommersant.ru/doc/1810554/print

RBC http://www.rbc.ru/

TNS http://www.tns-global.ru/rus/index.wbpъ

VKPM http://www.vkpm.ru/

# Russia—Data Summaries

ELI NOAM AND PAUL MUTTER

RUSSIA IS A HIGHLY concentrated media market, with a Weighted Average HHI of 4,014 in 2010, ranking it seventh internationally. Russia has only possessed independent media outlets since the dissolution of the USSR in 1991, but the number of voices, net and total, has been growing. Declining market shares for wholly state-owned enterprises like Svyazinvest—Russia's largest telecommunications holding company, which is 75% state owned—and VGTRK, a state owned unitary enterprise in broadcasting, also reduced overall concentration. Significantly, much of capital investment for media comes from the state and parastatal organizations, rather than from private investors. The government still controls 26.7% of the national media market (Table 11.16), primarily through the purchase of stakes in ostensibly private companies such as Gazprom Media, which is actually a semi-governmental entity due to its links to its namesake corporate parent.

Gazprom-Media is the most diversified of the top four companies, with a 50% public ownership. The state dominates the content market by owning three of the four national companies. (Channel One Russia and VGTRK are 100% fully state owned.) Together, these networks hold 70.9% of the national broadcast TV market and 40.6% of the entire content market (Table 11.17). The state does not have stakes in the growing online media.

In platform media, state control is declining due to the availability of mobile and fixed line services from private companies. Russia's demographics and geography do not lend the state enterprise Svyazinvest significant advantage, and its market share has nearly halved between 2004 and 2010, from 41.5% to 24.5% (Table 11.18). It lost subscribers to Vimpelcom in mobile and to MTS in cable distribution. MTS became the largest platform media company in Russia with a market share of 23.5%.

Table 11-16. National Media Industries Concentration in Russia

| | 2004/5 | | 2011 OR MOST RECENT | | % CHANGE ANNUAL AVERAGE | |
|---|---|---|---|---|---|---|
| | COMPANY POWER INDEX IN COUNTRY | COMPANY SHARE OF THE OVERALL NATIONAL MEDIA MARKET (%) | COMPANY POWER INDEX IN COUNTRY | COMPANY SHARE OF THE OVERALL NATIONAL MEDIA MARKET (%) | COMPANY POWER INDEX IN COUNTRY | COMPANY SHARE OF THE OVERALL NATIONAL MEDIA MARKET (%) |
| Government of Russia | 3,465 | 40.5 | 1,998 | 26.7 | –7.1 | –2.3 |
| *Svyazinvest (public)* | *3,189* | *36.3* | *1,588* | *19.9* | –8.4 | –2.7 |
| *Gazprom-Media (50% public)* | *45.2* | *1.7* | *108.9* | *3.5* | 23.5 | 0.3 |
| *Channel One Russia (public)* | *86.0* | *2.2* | *35.4* | *1.7* | –9.8 | –0.08 |
| *RIA Novisti* | *0.1* | *0.01* | *4.0* | *0.3* | 958.3 | 0.04 |
| *VGTRK (public)* | *29.7* | *1.4* | *25.2* | *1.5* | –2.5 | –0.02 |
| *Prosveschenie (public)* | *6.4* | *0.3* | *5.0* | *0.2* | –4.2 | –0.01 |
| Multiregion (MTS) | 693 | 17.4 | 671 | 19.2 | –0.5 | 0.3 |
| VimpelCom (39.5% Telenor, Norway) | 480 | 14.4 | 516 | 16.8 | 1.3 | 0.4 |
| MegaFon | 311 | 12.6 | 483 | 16.6 | 9.2 | 0.7 |
| Yandex.ru | 10.6 | 0.2 | 41.8 | 0.9 | 25.7 | 0.1 |
| Mail.ru | 0.1 | 0.02 | 22.8 | 0.7 | 4726.6 | 0.1 |
| Natsionalnaya Media Gruppa (NMG) | 49.8 | 1.2 | 41.5 | 1.5 | –2.8 | 0.04 |
| Eksmo-AST | 0.0 | 0.0 | 36.2 | 0.6 | 0.0 | 0.1 |
| *Eksmo* | 12.7 | 0.4 | 0.0 | 0.0 | –16.7 | –0.07 |
| *AST* | 8.8 | 0.3 | 0.0 | 0.0 | –4.1 | –.05 |
| Akado | 47.5 | 1.1 | 28.4 | 1.1 | –6.7 | 0.0 |
| Komsomolskaya Pravda | 10.0 | 0.3 | 17.5 | 0.6 | 12.5 | 0.06 |
| ER Telekom | 0.4 | 0.1 | 9.0 | 0.5 | 3.4 | 0.1 |

| | | | | | | |
|---|---|---|---|---|---|---|
| Evropeiskaya Media Gruppa (EMG) | 14.9 | 0.4 | 8.6 | 0.3 | -7.1 | -0.03 |
| Sanoma (Finland) | 18.3 | 0.4 | 7.8 | 0.2 | -9.5 | -0.04 |
| Google (US) | 0.0 | 0 | 3.8 | 0.3 | N/A | 0.05 |
| ProfMedia | 9.2 | 0.5 | 19.6 | 0.8 | 18.9 | 0.06 |
| Sony/Disney (Japan, US) | 0.0 | 0.0 | 8.9 | 0.5 | N/A | 0.08 |
| Karo Premier | 0.9 | 0.1 | 1.9 | 0.2 | 20.8 | 0.01 |
| Hearst-Shkulev Media (51% Shkulev, 49% Hearst, US) | 6.5 | 0.2 | 4.9 | 0.1 | -4.1 | -0.02 |
| Kommersant | 10.6 | 0.3 | 2.5 | 0.1 | -12.7 | -0.02 |
| MTG (CTC Media, Sweden) | 2.0 | 0.06 | 1.6 | 0.08 | -3.7 | 0.003 |
| Russkaya Media Gruppa (RMG) | 6.5 | 0.3 | 1.2 | 0.1 | -13.5 | -0.03 |
| Drofa | 3.9 | 0.2 | 1.1 | 0.1 | -11.9 | -0.02 |
| Moskovskij Komsomolets | 2.0 | 0.1 | 0.3 | 0.05 | -14.3 | -0.01 |
| Bauer/Logos-Media (Bauer, Germany) | 2.5 | 0.15 | 0.2 | 0.03 | -15.1 | -0.02 |
| Trud | 0.1 | 0.02 | 0.1 | 0.02 | 1.0 | 0.0 |

| Media Concentration Index | 2004/5 | 2011 or Most Recent | % Change Annual Average |
|---|---|---|---|
| Total Revenue: Nat'l Media Industry (mil US$) | 20,792 | 49,534 | 23.0% |
| Total Voices (*n*) | 52 | 71 | 6.1% |
| Net Voices (*n*) | 43 | 59 | 6.2% |

*continued*

Table 11-16. *continued*

| | 2004/5 | | 2011 OR MOST RECENT | | % CHANGE ANNUAL AVERAGE | |
|---|---|---|---|---|---|---|
| | COMPANY POWER INDEX IN COUNTRY | COMPANY SHARE OF THE OVERALL NATIONAL MEDIA MARKET (%) | COMPANY POWER INDEX IN COUNTRY | COMPANY SHARE OF THE OVERALL NATIONAL MEDIA MARKET (%) | COMPANY POWER INDEX IN COUNTRY | COMPANY SHARE OF THE OVERALL NATIONAL MEDIA MARKET (%) |
| Public Ownership (%) | 40.5 | | 26.7 | | –2.3% | |
| Foreign Ownership (%) | 8.1 | | 8.4 | | 0.05% | |
| C4 Average—Weighted | 98 | | 94.4 | | –0.6% | |
| HHI Average—Weighted | 5,214 | | 4,014 | | –3.8% | |
| C1 Average—Weighted | 61 | | 49 | | –0.02% | |
| Noam Index Average—Weighted | 2,066 | | 1,700 | | –3.0% | |
| Pooled Overall Sector C4 | 74.0 | | 64.4 | | –1.6% | |
| Pooled Overall Sector HHI | 2,160 | | 1,372 | | –6.1% | |
| Pooled Overall Sector Noam Index | 226 | | 245 | | 1.4% | |
| Market Share of Top Ten Companies: Nat'l Media Industry (%) (Pooled C10) | 78.2 | | 70.0 | | –1.4% | |
| National Power Index | 4,927 | | 3,495 | | –4.8% | |

Table 11-17. Top Content Media Companies in Russia

| | 2004/5 | | 2011 OR MOST RECENT | | % CHANGE ANNUAL AVERAGE | |
|---|---|---|---|---|---|---|
| | COMPANY POWER INDEX IN COUNTRY | COMPANY SHARE OF THE NATIONAL CONTENT MEDIA MARKET (%) | COMPANY POWER INDEX IN COUNTRY | COMPANY SHARE OF THE NATIONAL CONTENT MEDIA MARKET (%) | COMPANY POWER INDEX IN COUNTRY | COMPANY SHARE OF THE NATIONAL CONTENT MEDIA MARKET (%) |
| Government of Russia | 2,265 | 34.8 | 2,609 | 40.6 | 2.5 | 1.0 |
| *Gazprom- Media (public, 50%)* | 350.4 | 12.8 | 532.1 | 16.5 | 8.6 | 0.6 |
| *Channel One Russia (public)* | 667 | 16.9 | 189 | 9.2 | –12.0 | –1.3 |
| *RIA Novisti* | 0.5 | 0.1 | 21.5 | 1.5 | 653.3 | 0.23 |
| *VGTRK (public)* | 230 | 10.7 | 134.3 | 8.1 | –6.9 | –0.4 |
| *(a)Prosveshcheniye* | 49.9 | 2.2 | 26.5 | 1.2 | –10.0 | –0.2 |
| Yandex.ru | 84.1 | 1.4 | 253.1 | 5.4 | 20.2 | 0.7 |
| Mail.ru | 0.6 | 0.1 | 138.1 | 4.0 | 3,595.2 | 0.64 |
| Eksmo-AST | — | — | 218.9 | 3.6 | N/A | N/A |
| *Eksmo* | 101 | 3.1 | 0.0 | 0.0 | –16.7 | –0.5 |
| *AST* | 69.6 | 2.6 | 0.0 | 0.0 | 16.7 | –0.4 |
| Komsomolskaya Pravda | 79.5 | 2.4 | 106.0 | 3.8 | 5.5 | 0.2 |
| EMG | 118.2 | 3.4 | 51.9 | 1.6 | –9.4 | –0.3 |
| MTG (CTC Media, Sweden) | 16.3 | 0.5 | 9.7 | 0.5 | –6.7 | –0.001 |
| Sanoma (Finland) | 145.3 | 3.2 | 47.4 | 1.1 | –11.2 | –0.4 |
| Google (US) | 0.0 | 0.0 | 23.0 | 1.6 | N/A | 0.3 |
| ProfMedia | 72.9 | 3.9 | 118.5 | 5.0 | 10.4 | 0.2 |
| Sony/Disney (Japan/US) | 0.0 | 0.0 | 54.1 | 2.8 | N/A | 0.5 |
| Hearst-Shkulev (Russia-US) | 51.6 | 1.9 | 29.7 | 0.8 | –7.1 | –0.2 |
| ER Telekom | 1.1 | 0.2 | 17.9 | 0.9 | 2.6 | 0.1 |
| Kommersant | 84.3 | 2.2 | 15.2 | 0.9 | –13.7 | –0.1 |

*continued*

Table 11-17. *continued*

| | 2004/5 | | 2011 OR MOST RECENT | | % CHANGE ANNUAL AVERAGE | |
|---|---|---|---|---|---|---|
| | COMPANY POWER INDEX IN COUNTRY | COMPANY SHARE OF THE NATIONAL CONTENT MEDIA MARKET (%) | COMPANY POWER INDEX IN COUNTRY | COMPANY SHARE OF THE NATIONAL CONTENT MEDIA MARKET (%) | COMPANY POWER INDEX IN COUNTRY | COMPANY SHARE OF THE NATIONAL CONTENT MEDIA MARKET (%) |
| Akado | 3.0 | 0.3 | 11.0 | 0.7 | 43.7 | 0.07 |
| Karo Premier | 6.8 | 1.1 | 11.7 | 1.3 | 11.9 | 0.03 |
| RMG | 51.8 | 2.2 | 7.5 | 0.6 | –14.2 | –0.3 |
| Drofa | 30.7 | 1.7 | 6.7 | 0.6 | –13.1 | –0.2 |
| Moskovskij Komsomolets | 15.6 | 1.0 | 1.7 | 0.3 | –14.9 | –0.1 |
| Bauer/Logos-Media (Bauer, Germany) | 19.9 | 1.2 | 1.5 | 0.2 | –15.4 | –0.2 |
| Trud | 0.4 | 0.2 | 0.3 | 0.1 | –3.3 | –0.01 |
| Media Concentration Index | 2004/5 | | 2011 or Most Recent | | % Change Annual Average | |
| Public Ownership (%) | 34.8 | | 40.6 | | 1.0% | |
| Foreign Ownership (%) | 19.2 | | 10.9 | | -1.4% | |
| C4 Average—Weighted | 93.4 | | 94.5 | | 0.2% | |
| HHI Average—Weighted | 3,888 | | 3,896 | | 0.04% | |
| C1 Average—Weighted | 51 | | 56 | | 1% | |
| National Power Index | 3,879 | | 4,071 | | 0.8% | |

Table 11-18. Top Platform Media Companies in Russia

| | 2004/5 | | 2011 OR MOST RECENT | | % CHANGE ANNUAL AVERAGE | |
|---|---|---|---|---|---|---|
| | COMPANY POWER INDEX IN COUNTRY | COMPANY SHARE OF THE NATIONAL PLATFORM MEDIA MARKET (%) | COMPANY POWER INDEX IN COUNTRY | COMPANY SHARE OF THE NATIONAL PLATFORM MEDIA MARKET (%) | COMPANY POWER INDEX IN COUNTRY | COMPANY SHARE OF THE NATIONAL PLATFORM MEDIA MARKET (%) |
| Government of Russia (Svyazinvest) | 3,649 | 41.5 | 1,920 | 24.5 | -7.9 | -2.8 |
| MTS | 795 | 19.9 | 822 | 23.5 | 0.6 | 0.6 |
| MegaFon | 357 | 14.4 | 603 | 20.8 | 11.5 | 1.1 |
| Vimpelcom (39.5% Telenor, Norway) | 551 | 16.5 | 632 | 20.6 | 2.5 | 0.7 |
| Akado | 54.1 | 1.2 | 32.5 | 1.2 | -6.6 | -0.003 |
| ER Telekom | 0.3 | 0.1 | 7.4 | 0.4 | 3.7 | 0.06 |
| Media Concentration Index | 2004/5 | | 2011 or Most Recent | | % Change Annual Average | |
| Public Ownership (%) | 41.5 | | 24.5 | | -2.8% | |
| Foreign Ownership (%) | 6.5 | | 8.1 | | 0.3% | |
| C4 Average—Weighted | 99 | | 96.5 | | -0.4% | |
| HHI Average—Weighted | 5,422 | | 4,131 | | -4.0% | |
| C1 Average—Weighted | 63 | | 48 | | -2% | |
| National Power Index | 5,094 | | 3,459 | | -5.4% | |

The Russian-founded mobile provider Vimpelcom is partly owned by Telenor (Norway) and has become the second-largest platform company in Russia, with 20.6% of platforms (Table 11.18). It is headquartered in the Netherlands. Foreign multimedia producers have had limited success in the country since 1991. The exception is magazine publishing. Three of the four major magazine companies—Sanoma (Finland), Hearst Shkulev Media (50% US), Bauer (Germany)—are foreign owned. Even Google (US) has a marginal share of the search engine market here, which is dominated by Yandex.ru and Mail.ru. These platforms provide popular social media and email services, which undercuts Google's appeal for Russian consumers.

# 12

# Media Ownership and Concentration in Spain

JUAN P. ARTERO AND ALFONSO SÁNCHEZ-TABERNERO

## INTRODUCTION

The Spanish national media market has a low circulation rate for newspapers, at 100 newspapers for every 1,000 citizens, a high television consumption figure of 243 minutes per person every day (as of 2013 ), and a moderate use of other media sources, such as magazines, radio, and the Internet. The Spanish national media market's defining characteristics are the weakness of its domestic companies compared to foreign operators, and the existence of a foreign market that shares the same language in Latin America: while Spain has only 46 million inhabitants as of 2010, the Spanish-speaking population in the world numbers 450 million.

The lack of strong national media companies can be explained by late 20th century regulatory developments.[1] In 1985, Spain joined the EU and was viewed as an attractive emerging media market by foreign media producers because it had low entry barriers and a high growth potential. Because of that, foreign companies invested heavily in the television, magazine, and newspaper industries. Spanish media producers were and are not strong enough to protect their own market shares domestically, and there are no legal constraints for foreign direct investment in the Spanish media market. As a result, foreign owners now dominate some Spanish media sectors, such as television broadcasting and book and magazine publishing.

By the end of the 20th century, four national media companies—Telefónica, PRISA, Vocento, and Planeta—were able to win profitable market shares in some media sectors. All of them then

1. De Miguel, Juan Carlos. Los grupos multimedia: estructuras y estrategias de los medios europeos. Barcelona: Bosch, 1993; and Nieto, Alfonso. *Time and the Information Market: The Case of Spain*. Pamplona: Media Market Monographs, 2000.

began to look to Latin American markets. PRISA invested in radio and audiovisual production in several countries, Vocento and Planeta purchased newspapers in Argentina and Colombia, and Telefónica purchased a television company in Argentina.[2]

But the global economic crisis of 2008 hit Spain hard: total advertising expenditures declined by 11% that year and lost another 20% of value in 2009. Some companies were left in deep debt and needed to shed their assets. Vocento sold its stake in Argentine newspapers; and Telefónica decided to focus on its core telecom business by selling off its television operations. Both PRISA and Planeta have curtailed their plans for further internationalization.

When media companies launch strategies of growth and diversification and face risks, they can become more rigid, inflexible, and indebted; they can lose some degree of specialization as well as some innovative spirit; and their internal culture may suffer. But there are also advantages: they can increase in stock value, create synergies and economies of scale, protect their market shares by creating entry barriers, become more attractive for employees, and gain greater bargaining power with suppliers, distributors, and advertisers.

Spanish media enterprises have grappled with five growth strategies since Spain's EU ascension in 1985.[3] The first approach was trading off external for internal growth.[4] The Spanish media groups have grown mainly through internal processes: they have launched new paid and free newspapers, radio stations, websites, and film and television companies. Such decisions have helped them preserve their internal culture and retain skilled employees; on top of that, an internal growth approach requires only moderate capital investments. But companies based in other countries such as the United States or the United Kingdom have been more aggressive in pursuing external growth-driven strategies through mergers and acquisitions, and they have been able to grow faster.

The second strategy centers on the vertical versus horizontal integration debate: should newspapers retain their publication infrastructure or farm it out to concentrate on their core business of newsgathering?[5] During the 1990s, most Spanish audiovisual companies were vertically integrated because they were worried about bottlenecks in the value chain, mainly in broadcasting. However, at the beginning of the 21st century, firms pursued retrenchment strategies to focus more on their core businesses and to raise capital. Media managers discovered that new technologies of communication allowed the development of new channels for content distribution and, as a result, the former bottlenecks started to disappear. Several newspapers, for instance, have sold off their printing plants and distribution operations and are now concentrating on investments in their core business of news gathering.

The third issue is that Spanish companies have been hesitant to look for opportunities in foreign markets.[6] Beyond the investments made by Telefónica, PRISA, Vocento, and Planeta in Latin America during the 1990s, only a few other Spanish media producers have pursued internationalization strategies. The successful gossip magazine *¡Hola!* launched French- and English-language versions; the Recoletos Group bought the Portuguese financial daily *Diario Economico*, invested in Chilean and Argentine print media, and launched some Spanish language newspapers in the United States; the PRISA Group purchased a minority stake in the French daily *Le Monde* and also the Portuguese media group Media Capital;

2. Medina, Mercedes. "Algunas claves de la expansión de los grupos de comunicación españoles en Latinoamérica." *Comunicación y Sociedad* 16(2001): 71–99.

3. Sánchez-Tabernero, Alfonso and Carvajal, Miguel. *Media Concentration in the European Market: New Trends and Challenges*. Pamplona: Universidad de Navarra, 2002; and Kunz, W. M. *Culture Conglomerates: Consolidation in the Motion Picture and Television Industries*. Lanham: Rowman & Littlefield, 2007; and Roger, G. "Media Concentration with Free Entry." *Journal of Media Economics* 22: 3 (2009): 134–163.

4. Woodhull, N. and Snyder, R., eds. *Media Mergers*. New Brunswick: Transaction Publishers, 1998.

5. Gershon, Richard. *Telecommunications Management*. Mahwah: LEA, 2001.

6. Bagdikian, Ben. *The New Media Monopoly*. Boston: Bacon Press, 2004.

and finally, Vocento acquired a minority stake in a French regional newspapers chain. But these are the exceptions to the rule: since 1985, Spanish media companies have not pursued an aggressive internationalization strategy.

On the fourth strategic debate—content specialization versus product variety—homogenous convergence patterns can be identified concerning Spanish media producers' choices.[7] Some companies have built a competitive advantage based on their content specialization. That is the case for Unidad Editorial: even its television, film, and book publishing business units are oriented almost entirely toward current affairs. The same can be said about other media producers focused on news, like Vocento, as well as most regional groups that have based their growth on the success of one or several newspapers. Other companies, like Telecinco and Mediapro, mainly produce and deliver entertainment products. PRISA and Planeta produce both news and entertainment content evenly.

Finally, in terms of platform specialization and multimedia diversification, Spanish media producers are generally platform-skeptic: they are not attached to a given distribution system because they do not have a good sense of consumer preferences here: instead, they launch brand extensions to increase their sources of revenue.[8]

Strategies of growth and diversification of Spanish media companies are limited by a supranational and national legal framework: EU regulation and national laws. EU media regulation is rooted in two directives: the law on free competition and the audiovisual policy, which were first introduced in the 1980s and have been amended several times since. Key documents regulating the policy of competition in the EU are issued in Council Regulations, which authorize the Directorate General for Competition to establish conditions for approving mergers, takeovers, and agreements between companies.[9] The main piece of audiovisual policy in EU media legislation is outlined in the TV Without Frontiers Directive, which contains some restrictions regarding policies on the vertical integration of television broadcasters.[10]

Spanish regulators oversee general legislation on competition to prevent the formation of monopolies, as well as oversee some special legal mechanisms. There is no specific legislation concerning print and online media, although by law no single commercial entity can own more than 27% of the national TV market.[11] Spanish legal framework concerning media concentration has relaxed in the last two decades. Regulators are increasingly taking into account the problems posed by the small size and noncompetitiveness of national companies in Spanish-language markets rather than the existence of dominant positions in the market.

## PRINT MEDIA

### Daily Newspapers

As we can see in Table 12.1, concentration for newspapers has increased during the past quarter century: from 1984 to 2012, the Herfindahl-Hirschman and Noam indices tripled, though they still remain low compared to the indices for newspapers in other EU member states. This trend can be explained by two main factors: first, by the general decline in newspaper readership and advertising income worldwide, which has fostered mergers and acquisitions on newspapers even in Spain; and second, a lack of regulation.[12]

7. Vogel, Harold. *Entertainment Industry Economics: A Guide for Financial Analysis*. 6th ed. New York: Cambridge University Press, 2004.

8. Doyle, Gillian. "The Economics of Monomedia and Cross Media Expansion: A Study of the Case Favouring Deregulation of TV and Newspaper Ownership in the UK." *Journal of Cultural Economics* 24(2000): 1–26.

9. Council of the European Union. Regulation n° 4064/89 on the Control of Concentrations between Undertakings. Brussels: European Union, Dec. 21, 1989.

10. Council of the European Union. *Directive DOCE L 298 of Television without Frontiers*. Brussels: European Union, Oct. 17, 1989. Modified by *Directive 97/36, DOCE L 202*, July 30, 1997.

11. Boletín Oficial del Estado. *Ley 7/2010 de 21 de Marzo, General de Comunicación Audiovisual*. Madrid: Boletín Oficial del Estado, Apr. 1, 2010.

12. Argentesi, E. and Filistrucci, L. "Estimating market power in a two-sided market: The case of newspapers." *Journal of Applied Econometrics* 22(2007): 1247–1266.

Table 12-1. Daily Newspapers (Market Shares by Circulation), 1984–2012

| | 1984 | 1988 | 1992 | 1996 | 2000 | 2004 | 2008 | 2012 |
|---|---|---|---|---|---|---|---|---|
| PRISA (Polanco Family) | 11.3 | 18.6 | 18.6 | 14.1 | 15.8 | 17.7 | 17.5 | 17.6 |
| Godó | 7.7 | 8.6 | 12.2 | 7.3 | 7.1 | 7.6 | 8.2 | 8.5 |
| Vocento | 5.8 | 7.1 | 13.5 | 14 | 13.2 | 18.7 | 18.3 | 18.3 |
| Prensa Española | 6.9 | 9.3 | 10.9 | 7.7 | 7.1 | | | |
| Zeta | 5.8 | 7.6 | 8.7 | 9.9 | 8.9 | 8.6 | 7.5 | 7.0 |
| Diario 16 | 4.3 | 4.6 | 3.5 | | | | | |
| Recoletos | 3.1 | 7.0 | 11.7 | 13.5 | 11.4 | 10.4 | | |
| Unedisa (RCS, Italy) | 0.0 | 0.0 | 6.2 | 6.6 | 7.1 | 7.5 | 16.8 | 15.8 |
| Prensa Ibérica | | | | 7.2 | 7.4 | 7.6 | 7.3 | 7.3 |
| Planeta | | | | | 1.7 | 3.5 | 3.9 | 3.1 |
| Voz | | | | 3.1 | 3.0 | 2.9 | 2.6 | 2.7 |
| Joly | | | | 1.4 | 2.4 | 2.3 | 2.0 | 2.0 |
| Others | 55.1 | 37.2 | 14.7 | 15.2 | 14.9 | 13.2 | 15.9 | 14.6 |
| Total Revenue (mil euro)[1] | | | | | | 4,535 | 4,047 | |
| Total Revenue (mil US$) | | | | | | 5,694 | 5,949 | |
| C4 | 32.0 | 44.0 | 56.0 | 52.0 | 49.0 | 55.0 | 61.0 | 60.0 |
| HHI | 330 | 685 | 1,059 | 895 | 994 | 1,043 | 1,126 | 1,090 |
| *N* (>1%) | 7 | 7 | 8 | 10 | 11 | 10 | 9 | 9 |
| Noam Index | 124 | 258 | 374 | 283 | 300 | 330 | 375 | 363 |

*Sources*: OJD/Introl website, 2012 updates provided by editors using Wikipedia List of Spanish Newspapers by circulation.

[1] This revenue figure is an estimate based on data collected in this study: it is based on average per capita revenues reported for several other countries in this study with similar per capita income.

There are three flagship newspapers in Spain: *El País*, *El Mundo*, and *ABC*, all of which are published in Madrid. *El País* is considered left wing in its political orientation, and it has close ties with the Socialist Workers Party (PSOE). It is the main media offering of the PRISA Group, which is also dominant in radio broadcasting, pay TV, and book publishing. PRISA is partly owned by a Spanish family, the Polancos, but most of its shares are publicly traded. In 2010, the group had fallen into heavy debt, and Liberty Acquisition Holdings, a US investment fund, which acquired 57% of it, although the Polanco family retains most of the shareholding votes, and rescued it.

*El Mundo* is a center-left daily owned by Unidad Editorial, the Spanish subsidiary of the Italian media group RCS. In 2007, Unidad Editorial acquired the Recoletos Group, which publishes the popular sports daily *Marca*, the financial paper *Expansión*, and other popular titles like *Diario Médico* and *Correo farmacéutico*. Unidad Editorial has a weak presence outside of print media, though.

*ABC* is a conservative daily that usually supports the policies of the People's Party (Spain). Formerly published by Prensa Española, in 2000 it merged with a regional chain—Grupo Correo—to create a new venture, Vocento. This new company is partly owned by several Spanish families, and it is also publicly traded on the stock exchange. Vocento is strong in print and online news media, but like Unidad Editorial, it has not really invested in either radio or television broadcasting.

## Book Publishing

The book publishing industry in Spain is not highly concentrated. As of 2000, the leading publisher maintains only a 10% market share, and since records have been kept the C4 concentration index has never risen above 34%. Book publishers in Spain have not suffered the "perfect storm" that occurred in the newspaper market: the number of competitors has not increased substantially since the 1980s, and the rise of the Internet has not had a substantial impact on their business models (Table 12.2).

The Spanish book publishing market presents a paradox: on the one hand, Spain ranks 4th in the world in terms of number of titles published, and Spain accounts for 3.4% of the world publishing market: 110,000 books are published in Spain each year; 22% of them are translations of foreign titles. There are 3,000 publishing companies in Spain, and 15% of their total turnover comes from exports, mainly to Latin America. On the other hand, the number of books sold was 236 million in 2009, which means Spain's per capita book consumption rate per annum is only five. Surveys show that 41% of Spaniards say they are frequent readers of books, 13% are occasional readers, and 45% claim to never or almost never read books.[13]

### Table 12-2. Book Publishing (Market Shares by Revenue), 2000–2008

| | 2000 | 2004 | 2008 |
|---|---|---|---|
| Planeta | 10.7 | 10.6 | 10.2 |
| Anaya | 8.2 | 8.1 | 9.9 |
| Firm 3[1] | 6.8 | 7.1 | 8.0 |
| Firm 4 | 6.7 | 7.0 | 4.8 |
| Firm 5 | 5.5 | 7.0 | 4.5 |
| Firm 6 | 4.3 | 6.2 | 3.5 |
| Firm 7 | 3.8 | 3.8 | 4.0 |
| Firm 8 | 3.3 | 3.4 | 3.6 |
| Firm 9 | 2.5 | 3.4 | 3.1 |
| Firm 10 | 2.3 | 1.8 | 3.1 |
| Others | 45.9 | 41.6 | 45.3 |
| Total Revenue (mil euro)[2] | | 3,153 | 2,794 |
| Total Revenue (mil US$) | | 3,910 | 4,107 |
| C4 | 32.4 | 32.8 | 32.9 |
| HHI | 344 | 391 | 354 |
| *N* (>1%) | 10 | 10 | 10 |
| Noam Index | 109 | 124 | 112 |

*Source:* DBK (data provided specifically by this company to the authors).

[1] In this table, the names of the publishers ranked three through ten nationally are not included because this is a condition set by the source of the data.

[2] This revenue figure is an estimate based on data collected in this study: it is based on average per capita revenues reported for several other countries in this study with similar per capita income.

13. Ministerio de Cultura de España. *La cuenta satélite de la cultura en España*. Ministerio de Cultura: Madrid, 2009.

Sixty-three percent of books in Spain are published in either Madrid or Barcelona. Most of the companies are national, family owned, and midsized. One of the exceptions is Planeta, the biggest book publishing company in Spain: it ranks 7th among international publishing houses and has subsidiaries in France, Portugal, and most Latin American countries. Plaza y Janés is the main foreign-owned publisher, and it is a prime example of recent consolidation in the industry by foreign conglomerates: founded in 1959 as a national company, it was acquired in 2001 by a joint venture between Random House (owned by Bertelsmann) and the Italian publisher Mondadori.

The concentration rate for book publishing is higher if we analyze the book publishing market by categories: for instance, Anaya, second largest publisher in the country, has a dominant position in academic publications, such as reference books and textbooks. Publishers—like the music, movies, and newspapers in the last decade—now face the challenge of adapting to new technologies: the most innovative will learn how to take advantage of e-books and new distribution systems like Kindle. Due to the limited interest so far in online readership and sales of books, it remains to be seen just how those new technologies will influence the book publishing industry's concentration rate.

## Magazine Publishing

Magazine market concentration has decreased slightly over the past decade. The main reason behind this trend is the lack of strong entry barriers in the magazine industry: it is not a very capital-intensive business, and readers are not attached to any particular titles, in contrast with the brand loyalties observed for newspapers.[14] Half of the leading magazine companies in the Spanish market are foreign ones: Hachette Filipacchi (French); Condé Nast (US); Motorpress, G+J, and Springer (German); and Unedisa (Italian). RBA, Zeta, ¡Hola!, PRISA, and Heres are Spanish owned. With the exceptions of G+J, a Bertelsmann's subsidiary, Unedisa, and PRISA, publishing companies operating in Spain have chosen not to diversify their activities into television and radio (Table 12.3).

Table 12-3. Magazine Publishing (Market Shares by Revenue), 2000–2012

| | 2000 | 2004 | 2008 | 2012 |
|---|---|---|---|---|
| RBA | 15.2 | 14.8 | 13.6 | 15.40 |
| Hachette Filipacchi (Lagardere, France) | 14.4 | 14.0 | 12.3 | 17.9 |
| Condé Nast (Advance, US) | 3.2 | 6.1 | 8.7 | 13.5 |
| Motorpress (Motor Presse, Germany) | 4.2 | 6.5 | 8.0 | 9.6 |
| Zeta | 10.0 | 8.2 | 7.1 | 10.9 |
| G+J (Bertelsmann, Germany) | 8.6 | 6.8 | 6.2 | 7.7 |
| Unedisa (RCS, Italy) | 4.2 | 4.4 | 4.7 | 7.1 |
| ¡Hola! | 5.5 | 4.7 | 4.3 | 5.7 |
| PRISA (Polanco Family) | 2.1 | 2.8 | 3.3 | |
| Axel Springer (Germany) | 2.4 | 2.2 | 2.8 | 2.8 |
| Heres | 2.6 | 2.9 | 2.3 | |
| Others | 27.6 | 26.6 | 26.8 | |
| Total Revenue (mil euro)[1] | | 3,203 | 3,049 | |
| Total Revenue (mil US$) | | 3,972 | 4,482 | |

14. Fetschering, M. and Knolmayer, G. "Business Models for Content Delivery: An Empirical Analysis of the Newspaper and Magazine Industry." *International Journal on Media Management* 6(2004): 4–11.

Table 12-3. *continued*

| | 2000 | 2004 | 2008 | 2012 |
|---|---|---|---|---|
| C4 | 48.2 | 43.8 | 42.6 | 53.8 |
| HHI | (a) 705 | 670.5 | (b) 629.4 | 1,100.8 |
| *N* (>1%) | 11 | 11 | 11 | 9 |
| Noam Index | 213 | (c) 202 | 190 | 367 |

*Source:* Infoadex 2011.

[1]This revenue figure is an estimate based on data collected in this study: it is based on average per capita revenues reported for several other countries in this study with similar per capita income.

Until 2012, the leading magazine publisher has been the family-owned RBA. It publishes popular women's weeklies and holds the rights to publish *National Geographic* in Spain. In contrast to many of its smaller domestic competitors, RBA has chosen to focus on external growth: since its founding in 1993, it has bought out several titles, including foreign concerns like the Spanish-language division of the Swiss magazine group Edipresse. The number one in the market is Hearst (formerly Hachette-Filipachi) which has followed the opposite model: it has launched new domestic titles or it has adapted its international brands like *Elle* and *Car and Driver* to the local market.

In 2010, 52% of Spain's adult population read magazines: the number of readers was 21 million, 2.4 million greater than it was in 2005.[15] However, that year, magazines only won 7% of total advertising expenditure on the media, the lowest rate since records started being kept.[16] As a result of declining print advertising revenue, magazine companies are increasingly investing in their online editorial products: in 2010, online readership stood at 2 million. In spite of this shift, the profitability of the magazine industry is still decreasing.

## AUDIOVISUAL MEDIA

### Radio

The Spanish radio broadcasting market has a unique market structure compared with other national radio markets in the European Union. Since its emergence in the 1920s, the radio sector has been dominated by public corporations in most European countries.[17] Deregulation only began in the 1990s.[18] Because of this, the level of concentration in most EU countries is still high and the national public broadcasters dominate the market, though overall concentration has been falling since privatization began.

However, in Spain, the dominant firm is a private company, Unión Radio. It is owned by the PRISA Group, which holds 40% of the entire market with channels such as Cadena Ser and Cadena 40, both leaders in the talk show and music radio genres. This figure—which reached a high of 51.7% in 2008—represents the largest concentration of a radio broadcasting market belonging to a private enterprise in the EU (Table 12.4).

The level of concentration in the radio broadcasting market is stable. C4 has changed very little between 1988 and 2012, and the Herfindahl-Hirschman Index (HHI) has fluctuated between

15. An adult is Spain is defined as anyone over fourteen years of age.

16. Asociación de Revistas de Información. *La Revista*. Madrid: Asociación de Revistas de Información: 2011.

17. European Audiovisual Observatory. *Yearbook: Film and Home Video*. Vol. 3. Strasbourg: European Audiovisual Observatory, 2010.

18. Commission of the European Union. Notice on the Definition of Relevant Market for the Purposes of Community Competition Law. Brussels: European Union. Dec. 9, 1997.

Table 12-4. Radio Group (Market Shares by Revenue) 1988–2008

| | 1988 | 1992 | 1996 | 2000 | 2004 | 2003 | 2012 |
|---|---|---|---|---|---|---|---|
| Unión Radio (PRISA, Polanco Family) | 36.8 | 43.0 | 50.1 | 51.8 | 49.1 | 51.7 | 39.6 |
| Cope | 15.0 | 14.9 | 21.2 | 16.7 | 12.6 | 13.9 | 10.5 |
| RNE (RTVE, public) | 23.0 | 19.3 | 15.6 | 16.8 | 0.0 | 8.7 | 7.6 |
| Antena 3 Radio (Godó)[1] | 10.7 | 13.1 | | | | | |
| Onda Cero (Planeta) | | 11.6 | 12.0 | 11.1 | 10.8 | 12.9 | 13.7 |
| Kiss FM | | | | | 6.2 | 5.6 | 3.6 |
| Punto Radio (Vocento) | | | | | | 2.2 | |
| Others | 14.5 | | 1.1 | 3.6 | 21.3 | 5.0 | |
| Total Revenue (mil euro)[2] | | | | | 1,376 | 1,311 | |
| Total Revenue (mil US$) | | | | | 1,706 | 1,927 | |
| C4 | 84.8 | 90.3 | 98.9 | 96.4 | 78.7 | 87.2 | 71.4 |
| HHI | 2,223 | 2,750 | 3,347 | 3,368 | 2,725 | 3,144 | 1,937 |
| *N* (>1%) | 4 | 5 | 4 | 4 | 4 | 6 | 5 |
| Noam Index | 1,112 | 1,230 | 1,674 | 1,684 | 1,362 | 1,284 | 866 |

*Source:* EGM (Tom Micro media planning software).

[1]Ceased operating in 1994.

[2]This revenue figure is an estimate based on data collected in this study: it is based on average per capita revenues reported for several other countries in this study with similar per capita income.

1,900 and 3,400. In contrast, the Noam Index has decreased, which shows an increase in the number of radio firms, although most of the newcomers still have very small market shares. The continued high level of concentration in the market is mainly due to the lack of legal restrictions on private ownership. The main private operators—Onda Cero, owned by Planeta-Antena 3, and Cope, owned by the Catholic Church of Spain—use their networks of local radio stations to provide several national programs.

Therefore, one of the main competitive advantages comes from distribution. PRISA uses its 440-plus owned or affiliated local radio stations to offer six national programs (one news talk show, five music channels). Cope and Onda Cero have over 200 owned or affiliated local radio stations, and both of them have one news talk show and one music channel apiece. The fourth-place private competitor Kiss FM only has 70 local stations. The national public broadcaster Radio Nacional (RNE) has an 8.7% market share of the news talk show audience, but it does not compete in the advertising market. RNE has three networks that broadcast 24-hour news, music, and cultural programming.

## Broadcast Television

In Spain, there is one public broadcaster, RTVE. Through its two national channels it gets around one-third of the audience share. The remaining two-thirds of the audience share is divided among three private entities: Telecinco-Mediaset, Planeta-Antena 3, and PRISA Televisión. There is one smaller national player, La Sexta, which recently has reached an agreement to be acquired by Planeta, and a variety of regional, local, and niche channels.

The year 2010 was key in the industry because the conversion from analog to digital terrestrial television (DTT) was completed. Going into the DTT transition, the Government of Spain greatly reduced subsidies for the public broadcaster, even though RTVE is no longer allowed to air advertisements (up until 2009, it had captured about 25% of the national advertising market). Then Gestevisión-Telecinco merged with the national channel *Cuatro,* which had a 7% market share, thereby further increasing the Spanish market share of the Italian firm Mediaset, which is owned by the Berlusconi family (Table 12.5).

The launch of new channels has fostered television consumption: the average number of minutes of TV watched per person per day has increased from 210 minutes in 2000 to 243 minutes in 2013.[19] The first private channels began broadcasting in 1990: Telecinco, Antena 3, and Canal+ (which operates in the pay TV market and is French, not Spanish).

The strategies of the leading commercial broadcasters Telecinco and Antena 3 are quite similar: they try to protect their market shares despite the fragmentation of the market. During the late 1990s, they secured 20–25% of the market apiece with their one-channel offerings; now they have to accumulate several channels to hold onto their market share. Telecinco Group, owned by Mediaset, operates with brands such as Telecinco, Cuatro, La Siete, FDF, Boing, and Divinity. In 2010, these channels reached 25.5% of the audience market and won 44% of the advertising expenditures spent on broadcast television. The Planeta Group, owned by the Lara family, operates Antena 3, Nova, Neox, and

## Table 12-5. TV Broadcasting (Market Shares by Revenue), 1984–2013

| | 1984 | 1988 | 1992 | 1996 | 2000 | 2004 | 2008 | 2012 |
|---|---|---|---|---|---|---|---|---|
| TVE (RTVE, public) | 95.0 | 85.0 | 45.5 | 27.9 | 32.4 | 28.2 | 22.0 | 18.9 |
| FORTA (public) | 5.0 | 15.0 | 16.5 | 15.4 | 16.9 | 17.7 | 15.6 | 9.8 |
| Antena 3 (Planeta, Lara Family) | | | 14.7 | 25.0 | 21.5 | 20.8 | 17.1 | 25.8 |
| La Sexta | | | | | | | 5.1 | |
| Canal+/Cuatro (France; PRISA, Polanco Family) | | | 1.7 | 2.2 | 2.1 | 2.1 | 7.8 | 2.1 |
| Telecinco (Mediaset, Berlusconi Family, Italy) | | | 20.8 | 20.2 | 22.3 | 22.1 | 19.0 | 28.1 |
| Others | | | 0.8 | 9.3 | 4.8 | 9.1 | 13.4 | 17.0 |
| Total Revenue (mil euro)[1] | | | | | | 5,290 | 5,292 | |
| Total Revenue (mil US$) | | | | | | 6,560 | 7,779 | |
| C4 | 100.0 | 100.0 | 97.5 | 88.5 | 93.1 | 88.8 | 73.7 | 82.6 |
| HHI | 9,050 | 7,450 | 2,994 | 2,054 | 2,299 | 2,034 | 1,468 | 1,913 |
| *N* (>1%) | 2 | 2 | 5 | 5 | 5 | 5 | 6 | 5 |
| Noam Index | 6,399 | 5,268 | 1,339 | 919 | 1,028 | 910 | 599 | 856 |

*Sources:* Ecotel, Taylor Nelson Sofres (TNS), and Kantar Media websites, 2012 update provided by editors using Barlovento Comunicacion.

[1]This revenue figure is an estimate based on data collected in this study: it is based on average per capita revenues reported for several other countries in this study with similar per capita income.

19. Barlovento Comunicación. *Análisis televisivo del mercado español.* Madrid: Barlovento Comunicación, 2014.

Nitro. Planeta is considered to be center-right in its political orientation. In 2010, it held 16.6% of the audience market and 28% of the advertising television market.[20]

FORTA is another key player in the public service television landscape. But unlike RTVE, it is a confederation of public regional television corporations that remain independent from each other in most decisions.

## Multichannel TV Platforms

As Table 12.6 shows, the multichannel TV market is highly concentrated in Spain. It has evolved from the private monopoly held by Sogecable (of the PRISA Group) in the 1990s toward an oligopoly, albeit one in which the PRISA Group's offerings are still the dominant players. However, the concentration indices have been decreasing each year. More and more of them have had to offer better prices and "triple play" services—TV, telephone, and high-speed Internet access—to compete in the market.

PRISA Televisión (formerly Sogecable) operates through its brand Digital Plus and is still the clear market leader. It distributes 150 channels and has more than 1.8 million subscribers, which in 2010 represented 49% of the pay television market in Spain.[21] PRISA Televisión is the third largest pay-TV operator in the European Union.

## Video Channels

The video and cable network market is very fragmented, yet once again, PRISA Televisión is the leading player here. The industry emerged in Spain in the early 2000s with the launch of new cable and satellite platforms. Both national firms and international

**Table 12-6. Multichannel Video Platforms: Cable MSOS, DBS, IPTV (Market Shares by Revenue), 1996–2008**

| | 1996 | 2000 | 2004 | 2008 | 2012 |
|---|---|---|---|---|---|
| Sogecable (PRISA, Polanco Family) | 100.0 | 77.7 | 83.7 | 73.8 | 42.2 |
| Ono | | | 6.5 | 12.4 | 20.8 |
| Telefónica | | | | 7.5 | 16.7 |
| Orange (France Telecom, France) | | | | 1.3 | 1.7 |
| Auna | | | 5.3 | | |
| Vía Digital | | 20.0 | | | 6.9 |
| Others | | 2.3 | 4.5 | 5.0 | |
| Total Revenue (mil euro)[1] | | | 3,581 | 4,371 | |
| Total Revenue (mil US$) | | | 4,440 | 6,425 | |
| C4 | 100.0 | 97.7 | 95.5 | 95.0 | 81.4 |
| HHI | 10,000 | 6,437 | 7,076 | 5,658 | 2,543 |
| *N* (>1%) | 1 | 2 | 3 | 4 | 5 |
| Noam Index | 10,000 | 4,552 | 4,085 | 2,829 | 1,137 |

*Source:* CMT report (2005 and 2009 editions).

[1]This revenue figure is an estimate based on data collected in this study: it is based on average per capita revenues reported for several other countries in this study with similar per capita income.

20. Infoadex. Estudio de la inversión publicitaria en España. Madrid: Infoadex, 2011.

21. Barlovento Comunicación. *Análisis televisivo del mercado español.* Madrid: Barlovento Comunicación, 2011.

corporations like Sony, Universal, Turner, MTV, Disney, Hearst, Discovery, and Eurosport distribute their brands in Spain. In 2008, PRISA had 14.5% of the market.

The large global distributors operating in Spain have changed their internationalization strategies in the last decade. During the 1980s and 1990s, they sold content to Spanish channels, because broadcasting was fully facilitated by national operators. But after 2000, new technologies reduced these bottlenecks and they discovered that they would make more profit if they moved into content creation for the Spanish national market.[22]

Mediaset and Eurosport are the only two non-Spanish but European content providers that operate significantly in the pay-television-channel market (Table 12.7).

## Table 12-7. Cable And DBS TV Programming Channels (Market Shares by Revenue), 2004–2008

| | 2004 | 2008 |
|---|---|---|
| Sogecable (PRISA, Polanco Family) | 17.1 | 14.5 |
| Factoría de Canales | 0.0 | 6.6 |
| Fox (News Corp., US) | 5.6 | 6.4 |
| Chello (Liberty Global, US/UK) | 7.5 | 6.1 |
| Sony (Japan) | 3.8 | 6.0 |
| Universal (Comcast/GE, US) | 4.8 | 5.5 |
| Turner (Time Warner, US) | 7.5 | 5.2 |
| MTV (Viacom, US) | 3.5 | 4.8 |
| Disney (US) | 3.4 | 4.6 |
| RTVE (public) | 1.8 | 3.5 |
| Discovery (US) | 1.4 | 3.2 |
| Telecinco (Mediaset, Berlusconi Family, Italy) | 5.3 | 2.2 |
| Hearst (US) | 1.0 | 2.1 |
| Eurosport (France) | 2.3 | 1.5 |
| Antena 3 (Planeta) | 0.1 | 0.9 |
| Televisa (Mexico) | 1.1 | 0.4 |
| Others | 33.8 | 26.5 |
| Total Revenue (mil euro)[1] | 2,538 | 2,961 |
| Total Revenue (mil US$) | 3,223 | 4,352 |
| C4 | 37.7 | 33.6 |
| HHI | 482 | 477 |
| *N* (>1%) | 14 | 14 |
| Noam | 121 | 128 |

*Source:* Taylor Nelson Sofres (TNS) website.

[1] This revenue figure is an estimate based on data collected in this study: it is based on average per capita revenues reported for several other countries in this study with similar per capita income.

22. Vogel, Harold. *Entertainment Industry Economics: A Guide for Financial Analysis* 6th ed. New York: Cambridge University Press, 2004.

## Film

Film distribution has a moderate, stable level of concentration. As is the case in most EU member states, the Spanish film distribution industry is dominated by the subsidiaries of US majors, namely Universal, Fox, Warner, Disney, and Sony (through Columbia Pictures). There is a strong vertical integration between production and distribution; exhibitors do not have any real bargaining power in this relationship (Table 12.8).

The movie business in Spain offers one of the most significant examples of the United States' dominant position in that industry on a global scale. The success of the country's Hollywood majors is, of course, in large part due to the talents of screenwriters, directors, actors, producers, managers, and entrepreneurs there.[23] But the position of privilege Hollywood occupies that attracts such talent also comes from the oligopolistic nature of the US film industry.[24] Their products have cost structures that benefit consolidation and the creation of oligopolistic markets: almost all the expenses of producing, marketing, and distributing movies are fixed costs.[25]

But the creative nature of filmmaking makes it very difficult for production studios to control risks and recoup these costs. The Pareto principle is applicable here: in most markets, 20% of the titles account for 80% of the income, and many movies do not reach the break-even point. The best strategy in the film industry to break

### Table 12-8. Film Production/Distribution (Market Shares by Box Office %), 2000–2008

| | 2000 | 2004 | 2008 |
|---|---|---|---|
| Universal (Comcast/GE, US) | | | 22.2 |
| UIP (Universal/Paramount, US) | 15.6 | 16.8 | |
| Fox (News Corp., US) | | 11.9 | 14.0 |
| Sony (Japan) | | | 12.2 |
| Columbia Pictures (US) | 14.5 | | |
| Time Warner (US) | 14.0 | 19.2 | 11.9 |
| Disney (US) | 16.2 | 19.9 | |
| Others | 39.8 | 32.2 | 39.8 |
| Total Revenue (mil euro)[1] | | 893.4 | 831.3 |
| Total Revenue (mil US$) | | 1,114 | 1,222 |
| C4 | 60.2 | 67.8 | 60.2 |
| HHI | 728 | 1,189 | 978 |
| *N* (>1%) | 4 | 4 | 4 |
| Noam Index | 364 | 595 | 489 |

*Source:* ICAA website.

[1]This revenue figure is an estimate based on data collected in this study: it is based on average per capita revenues reported for several other countries in this study with similar per capita income.

23. Balio, T. *The American Film Industry*. Madison: The University of Wisconsin Press, 1985; and Nowell-Smith, G. and Ricci, S. eds. *Hollywood and Europe: Economics, Culture, National Identity 1945-95*. 6th ed. London: BFI Publishing, 1998.

24. Albarran, Alan. *Media Economics: Understanding Markets, Industries and Concepts*. Ames: Iowa State University Press, 2002; and Litman, Barry. *The Motion Picture Mega-Industry*. Boston: Allyn & Bacon, 1988.

25. Bagdikian, Ben. *The New Media Monopoly*. Boston: Bacon Press, 2004.

even is to produce a portfolio of films each year that can guarantee at least one or two blockbusters that will cover the losses of the other titles.

By contrast, in the European Union, linguistic and cultural barriers have resulted in more isolated markets, and national companies have been unable to compete with their US counterparts. As a result, the movies produced in Hollywood account for more than 80% of Spanish box office receipts, while domestic titles account for around 10%.

Spanish distributors have formed alliances with the Hollywood majors. As a consequence, these major groups are distributing not only American blockbusters, but also a large percentage of Spain's most successful films. At the same time, some local producers have been forced to foreclose due to US distributors' use of strategies such as "blind bidding" and "block booking" that restrict exhibitors' freedom.[26]

## TELECOMMUNICATIONS MEDIA

### Wireline Telecom

Spanish media is increasingly branching out into electronic distribution of its products: online versions of newspapers and magazines have video contents, and telecom firms usually control the broadcasting markets. As was the case in most European countries, the Spanish telephone sector was a public monopoly until the late 1990s. Telefónica, a state-owned company founded in 1924, took advantage of that lack of competition to grow and to expand toward Latin America. During the deregulation of the telephone market, Telefónica was privatized and in 1997 the Government of Spain sold its remaining 20.9% stake in the company.

Telefónica's transformation has been a success story for proponents of telecom privatization in the European Union: in 15 years, it has changed from a public monopoly to a competitive multinational corporation. In 2010, it was the second largest Spanish company, after Banco Santander, and one of the largest firms of its kind in the world. It is the leading telecom provider in Argentina, Chile, and Peru, and it has a strong presence in most Latin American countries as well as the United Kingdom, Ireland, Germany, the Czech Republic, and Slovakia. In 2010, it had 270 million subscribers, and 68% of its turnover came from its international operations in these countries.

Telefónica's dominance has not gone unchallenged, though. In 2000, its rivals collectively held less than 1% of the market. Yet 10 years later, by 2010, while the former public monopolist still maintained a dominant position, second-place Ono was able to increase its customer base at a good speed. That year it had 1.7 million subscribers to its telephone services plus 1.4 million to Internet services and 1 million for television services.[27] In addition, international operators such as Orange and Vodafone are slowly increasing their market shares at Telefónica's expense (Table 12.9).

### Wireless Telecom

The wireless telecom market has developed quickly in Spain because it has weaker entry barriers than the wireline telecom market does. Telefónica started offering wireless service in time for the 1982 World Cup soccer tournament, which was held in Spain, but the handsets were large and expensive, so the service was not really in demand. In 1987, a new Telecommunications Act was passed: it recognized the monopoly of Telefónica, but required the liberalization of its emerging handset business.[28]

In 1990, Telefónica only had 54,000 subscribers to its wireless telephone services. Two years later, the EU decreed that European markets should privatize their telecom industries. In order to hold onto its market share, Telefónica

26. Organization for Economic Cooperation and Development. *Competition Policy and Film Distribution*. Paris: Organization for Economic Co-operation and Development, 1996.

27. Comisión del Mercado de las Telecomunicaciones. *Informe Anual 2010*. Madrid: Comisión del Mercado de las Telecomunicaciones, 2011.

28. Boletín Oficial del Estado. *Ley 31/1987 de 18 de Diciembre, de Ordenación de las Telecomunicaciones*. Madrid: Boletín Oficial del Estado, Dec. 19, 1987.

Table 12-9. Wireline Telecom (Market Shares by Revenue), 1996–2008

| | 1996 | 2000 | 2004 | 2008 | 2012 |
|---|---|---|---|---|---|
| Telefónica | 100.0 | 91.6 | 83.3 | 78.6 | 71.9 |
| Ono | | 0.4 | 5.0 | 8.6 | 10.1 |
| Orange (France Telecom, France) | | | | 2.6 | 2.4 |
| Vodafone (UK) | | 0.2 | | 2.0 | 4.6 |
| Euskaltel | | 0.5 | | 1.6 | 1.9 |
| BT (UK) | | 0.3 | | 1.4 | 0.8 |
| R Cable | | | | 1.2 | 1.7 |
| Jazztel (UK) | | 1.1 | | 1.1 | 3.7 |
| Telecable de Asturias | | | | 0.4 | |
| Auna | | 2.9 | 7.2 | | |
| Uni2 | | 1.5 | 3.0 | | |
| Others | | 1.6 | 1.6 | 2.5 | 1.1 |
| Total Revenue (mil euro)[1] | | | 16,306 | 13,647 | 15,335 |
| Total Revenue (mil US$) | | | 20,220 | 20,101 | 19,717 |
| C4 | 100.0 | 97.6 | 98.5 | 91.8 | 90.3 |
| HHI | 10,000 | 8,403 | 7,025 | 6,270 | 5,319 |
| *N* (>1%) | 1 | 5 | 4 | 8 | 7 |
| Noam Index | 10,000 | 3,758 | 3,513 | 2,217 | 2,010 |

*Source:* CMT 2010.

[1]This revenue figure is an estimate based on data collected in this study: it is based on average per capita revenues reported for several other countries in this study with similar per capita income.

launched the new brand Moviline, improved service coverage, and reduced prices. In 1994 had greatly increased its presence to 400,000 subscribers, and it reached 1 million subscribers in 1996.

Spain had 52 million registered cell phones in 2010, a market penetration rate of 111%.[29] Through its brand Movistar, Telefónica is the leading company, but Vodafone has a similar number of subscribers, and the French telecom Orange is not far behind Vodafone. Spanish authorities have been investigating possible price fixing going on among Telefónica, Vodafone, and Orange, but so far they have not found anything that would justify legal action (Table 12.10).

# INTERNET MEDIA

## Internet Service Providers (ISPs)

Telefónica is also the leading Spanish ISP, with a 44.1% market share. Its main competitors are local cable operators like Ono, Euskaltel, R Cable, and Telecable de Asturias and emergent foreign companies like Orange, Jazztel, and Vodafone. The level of concentration is still very high for this sector, but it has started to decrease (Table 12.11).

Competitive advantages in the ISP market are mainly based on six issues: coverage, bandwidth, price, access limits, technological service, and additional services. The ISP

29. Comisión del Mercado de las Telecomunicaciones. *Informe Anual 2010*. Madrid: Comisión del Mercado de las Telecomunicaciones, 2011.

## Table 12-10. Wireless Telecom (Market Shares by Revenue), 1998–2012

| | 1998 | 2000 | 2004 | 2008 | 2012 |
|---|---|---|---|---|---|
| Telefónica | 71.8 | 60.0 | 54.3 | 39.9 | 39.8 |
| Vodafone (UK) | 27.1 | 30.0 | 27.3 | 37.0 | 27.4 |
| Orange (France Telecom, France) | | 10.0 | 18.3 | 20.9 | 21.1 |
| Yoigo | | | | 1.8 | 4.2 |
| Other | | | | | 5.1 |
| Total Revenue (mil euro)[1] | | | 12,971 | 14,738 | |
| Total Revenue (mil US$) | | | 16,084 | 21,665 | |
| C4 | 98.9 | 90.0 | 99.9 | 99.6 | 92.5 |
| HHI | 5,890 | 4,600 | 4,029 | 3,401 | 2,798 |
| *N* (>1%) | 2 | 3 | 3 | 4 | 4 |
| Noam Index | 4,165 | 2,656 | 2,326 | 1,701 | 1,399 |

*Sources:* CMT 2010, 2012 updates provided by editors using data from fiercewireless.com.

[1]This revenue figure is an estimate based on data collected in this study: it is based on average per capita revenues reported for several other countries in this study with similar per capita income.

## Table 12-11. Internet Service Providers (Market Shares by Revenue), 1998–2012

| | 1998 | 2000 | 2004 | 2008 | 2012 |
|---|---|---|---|---|---|
| Telefónica | 92.0 | 31.0 | 66.1 | 59.8 | 44.1 |
| Ono | | 6.0 | 1.4 | 13.8 | 15.1 |
| Orange (Wanadoo/Eresmas, France) | | 21.0 | 20.1 | 10.7 | 11.6 |
| Jazztel (UK) | | 8.0 | 4.6 | 4.6 | 13.0 |
| Vodafone (UK) | | 2.0 | | 3.6 | 5.9 |
| Euskaltel | | | 2.5 | 1.9 | 2.2 |
| R Cable | | | | 1.4 | 1.3 |
| Telecable de Asturias | | | | 1.0 | 1.3 |
| BT (UK) | 4.0 | 5.0 | | | |
| Auna | | 6.0 | 3.1 | | |
| Global One | 3.0 | | | | |
| Others | 1.0 | 21.0 | 2.2 | 3.2 | |
| Total Revenue (mil euro)[1] | | | 4,258 | 4,902 | |
| Total Revenue (mil US$) | | | 5,280 | 7,206 | |
| C4 | 99.0 | 66.0 | 93.9 | 88.9 | 83.8 |
| HHI | 8,489 | 1,567 | 4,812 | 3,922 | 2,519.4 |
| *N* (>1%) | 3 | 7 | 6 | 8 | 8 |
| Noam Index | 4,901 | 592 | 1,965 | 1,387 | 891 |

*Sources:* CMT 2010, 2012 updates provided by editors using data from SNL Kagan & FTTH Council Europe Webinar November 2012.

[1]This revenue figure is an estimate based on data collected in this study: it is based on average per capita revenues reported for several other countries in this study with similar per capita income.

market is oligopolistic by nature because the nature of the market favors large companies: only strong regulation—which does not exist in Spain—can foster more competition among ISPs.

## Search Engines

The level of concentration in the search engine business is even higher than it is in the Spanish ISP market (Table 12.12). In 2011, Google had more than 95% of the Spanish market, while the figure for Europe was 79.2% and for the United States was 65.7%.[30]

The strength of Google in the Spanish search engine market comes from its first-mover advantage and its brand recognition: for most users, it is perceived by users as the only way to find what they are looking for online. This virtuous circle is complemented by the lack of marketing investments by its rivals to compete with Google's established presence.

Traditional media companies are worried about Google's ability to control the online advertising market, and they may eventually ask for government intervention. But so far, Google's advertising income is small and Spanish media groups are paying more attention to their present problems—heavy debt, falling circulation, declining market share, and more competition in their traditional print or broadcasting markets—than to future threats Google's business model poses to them. The Spanish government has shown concerns about concentration in the search engine market, but it has not identified any abuses by Google of its dominant position.

## Online News

This sector does not have strong entry barriers: because of that, C4, HHI, and the Noam Index show moderate level of concentrations. However, during the last few years concentration

### Table 12-12. Internet Search Engines (Market Shares by Revenue), 2004–2012

| | 2004 | 2008 | 2012 |
|---|---|---|---|
| Google (US) | 92.1 | 95.7 | 96.2 |
| Microsoft | 2.4 | 2.1 | |
| Yahoo! (US) | 1.6 | 1.7 | 0.9 |
| Bing | | | 1.7 |
| Ask | | | 0.5 |
| Others | 3.9 | 0.5 | 0.7 |
| Total Revenue (mil euro)[1] | 172.6 | 1,005 | |
| Total Revenue (mil US$) | 214 | 1,478 | |
| C4 | 96.1 | 99.5 | 95.7 |
| HHI | 8,491 | 9,166 | 9,258 |
| *N* (>1%) | 3 | 3 | 4 |
| Noam Index | 4,902 | 5,292 | 4,629 |

*Sources:* StatCounter and Netsuus websites, 2012 updates provided by editors using data from AT Internet Online Intelligence Solutions.

[1]This revenue figure is an estimate based on data collected in this study: it is based on average per capita revenues reported for several other countries in this study with similar per capita income.

30. Comscore. *March 2011 Search Engine Market Share*. Reston, VA: Comscore, 2011.

Table 12-13. Online News Media (Market Shares by Monthly Unique Visitors), 2008–2012

| | 2008 | 2012 |
|---|---|---|
| PRISA (Polanco Family) | 17.5 | 26.1 |
| Unedisa | 7.8 | 18.1 |
| Planeta | 3.6 | 14.6 |
| Mediaset (Berlusconi Family, Italy) | 3.5 | 5.2 |
| Zeta | 3.4 | 5.4 |
| Godó | 3.4 | 5.7 |
| RTVE (public) | 1.8 | 5.3 |
| Yell | 4.6 | 3.2 |
| Schibsted (Sweden) | 1.8 | 3.3 |
| Vocento | 4.3 | 4.7 |
| Others | 47.0 | 35.9 |
| C4 | 34.2 | 64.5 |
| HHI | 462 | 1,382 |
| *N* (>1%) | 8 | 10 |
| Noam Index | 163 | 437 |

*Source:* EGM (Tom Micro media planning software).

has increased, in part because of external growth strategies launched by the three main players in the market: PRISA, Unedisa, and Planeta. All three have acquired offline media holdings—mainly in newspapers and television broadcasting—which produce online content as well (Table 12.13).

## CONCLUSION

Concentration has been stable—with only small measurable changes observed upward or downward—in six sectors: radio broadcasting, film distribution, book publishing, magazine publishing, video-channel and cable broadcasting, and search engines. Concentration has decreased in five sectors, mainly as a result of Spanish deregulation policies: terrestrial television broadcasting, ISPs, multichannel TV, wireline telecom, and wireless telecom. Concentration among newspapers has increased due to the loss of advertising revenue and declining sales.

The most concentrated sectors are for search engines, multichannel TV, radio broadcasting, ISPs, wireless telecom, and wireline telecom, all of which check in at over 3,000 on the HHI. Some of those industries are still dominated by former monopolists that have been able to protect their businesses, thus accounting for the high concentrations.

There is no strong regulatory set of cross-ownership prohibitions in Spain. However, there are only a few cases in which a Spanish or foreign company in Spain has a strong presence across multiple media markets: PRISA is a dominant player in newspapers, radio broadcasting, book publishing, video channels, and cable networks. Planeta ranks among the top four in three sectors: book publishing, terrestrial television, and radio broadcasting. And Telefónica, Vodafone, Ono, and Orange dominate the ISP, wireline, and wireless markets.

Foreign direct investment is substantial because of the market's attraction and the lack

31. Almirón, Nuria. "Grupos privados propietarios de medios de comunicación en España: principales datos estructurales y financieros." *Comunicación y Sociedad* 22: 1 (2009): 243–273.

of legal barriers.[31] International corporations have a strong presence in all Spanish media markets, with the exception of radio broadcasting. Foreign investment comes mainly from the United States for film distribution, ISPs, magazine publishing, video channels, and cable networks; from Germany for book and magazine publishing; from Italy in television broadcasting and newspapers; from France in magazine publishing, ISPs, and wireline and wireless telecom; and from the United Kingdom in ISPs and wireline and wireless telecom.

The analysis of the market as a whole does not allow for identifying monopolistic abuses in regional markets. From a political point of view, this phenomenon is relevant because regional governments have some legislative power in Spain: in some cases, groups with a regional political agenda—by owning a leading print daily as well as a few radio or TV stations—can dominate public opinion in their distribution areas.

Regulators face two main challenges when dealing with concentration issues in Spanish media, entertainment, and telecom at the national level: on the one hand, the legal framework never seems to quite catch up with technological innovations.[32] On the other hand, regulators need to find a balance between (1) excessive levels of concentration that would lead to higher consumer prices due to lack of incentives and (2) the possibility of excessive fragmentation of the industry that would produce small firms unable to compete in international markets.

The level of concentration is high in several industries, but, at the same time, the biggest Spanish media groups have fewer synergies, economies of scale, and brand recognition than their counterparts in other highly industrialized nations. In the last decade, the flow of capital toward media and entertainment industries between Spanish markets and other economies has been unbalanced for a highly industrialized nation like Spain.

Large companies enjoy greater independence and strength to criticize decisions made by political institutions.[33] But in Spain, it seems that the weakening of national firms has produced a shift in bargaining power: nowadays, national and regional governments exert a decisive influence over news media. Governments make decisions about tax and advertising budgets, allocate new broadcasting licenses, and can fine companies for not meeting legal requirements. Government choices here will be critical for the survival of some of the small- and medium-sized media companies.

Regulation of Spanish media concentration began evolving in the late 1990s and in the early 2000s from the "ownership model" into the "market share model" or "total audience model." During the age of scarcity of media outlets, policymakers set limits for ownership of radio and television broadcasting. But the launch of a large variety of new audiovisual products, as well as online content, has made this legal framework outdated. The relevant issue is no longer ownership of a set percentage of a media company but reaching a given market share—adding the audience of different media products owned by the same firm together. So nowadays the controlled market percentage, rather than the ownership percentage of a business, is what generally dominates firms' strategic planning considerations.

The Spanish market is now undergoing a second evolution. The government is trying to make the growth of national companies compatible with the protection of pluralism and free competition. It is demanding more transparency, for starters, and it is paying more attention to external growth; it is more concerned with vertical integrations than before (in order to prevent

---

32. Llorens-Maluquer, Carles. *Concentración de empresas de comunicación y el pluralismo: la acción de la UE.* Doctoral thesis, Universidad Autónoma de Barcelona. Barcelona: TDX, 2001. (Publication No. B.24.148-2002). n.d. <http://www.tdx.cat/handle/10803/4095>.

33. Albarran, Alan. *Media Economics: Understanding Markets, Industries and Concepts.* Ames: Iowa State University Press, 2002; and Edwin, Baker C. *Media Concentration and Democracy: Why Ownership Matters.* Boston: Cambridge University Press, 2007.

bottlenecks that may lead to abuse of dominant positions); and it is increasingly blocking price-fixing arrangements among competitors in the more oligopolistic media sectors.

## Bibliography

### 1. Books and Articles


Albarran, Alan. *Media Economics: Understanding Markets, Industries and Concepts*. Ames: Iowa State University Press, 2002.

Almirón, Nuria. "Grupos privados propietarios de medios de comunicación en España: principales datos estructurales y financieros." *Comunicación y Sociedad* 22:1 (2009): 243–273.

Argentesi, E. and Filistrucci, L. "Estimating market power in a two-sided market: The case of newspapers." *Journal of Applied Econometrics* 22 (2007): 1247–1266.

Bagdikian, Ben. *The New Media Monopoly*. Boston: Bacon Press, 2004.

Balio, T. *The American Film Industry*. Madison: The University of Wisconsin Press, 1985.

De Miguel, Juan Carlos. *Los grupos multimedia: Estructuras y estrategias de los medios Europeos*. Barcelona: Bosch, 1993.

Doyle, Gillian. "The Economics of Monomedia and Cross Media Expansion: A Study of the Case Favouring Deregulation of TV and Newspaper Ownership in the UK." *Journal of Cultural Economics* 24 (2000): 1–26.

Edwin, Baker C. *Media Concentration and Democracy: Why Ownership Matters*. Boston: Cambridge University Press, 2007.

Fetschering, M. and Knolmayer, G. "Business Models for Content Delivery: An Empirical Analysis of the Newspaper and Magazine Industry." *International Journal on Media Management* 6 (2004): 4–11.

Gershon, Richard. *Telecommunications Management*. Mahwah: LEA, 2001.

Kunz, W. M. *Culture Conglomerates: Consolidation in the Motion Picture and Television Industries*. Lanham: Rowman & Littlefield, 2007.

Litman, Barry. *The Motion Picture Mega-Industry*. Boston: Allyn & Bacon, 1988.

Llorens-Maluquer, Carles. *Concentración de empresas de comunicación y el pluralismo: la acción de la UE*. Doctoral thesis, Universidad Autónoma de Barcelona. Barcelona: TDX, 2001. (Publication No. B.24.148-2002). n.d. <http://www.tdx.cat/handle/10803/4095>.

Medina, Mercedes. "Algunas claves de la expansión de los grupos de comunicación españoles en Latinoamérica." Comunicación y Sociedad 16 (2001): 71–99.

Nieto, Alfonso. *Time and the Information Market: The Case of Spain*. Pamplona: Media Market Monographs, 2000.

Nowell-Smith, G. and Ricci, S. eds. *Hollywood and Europe: Economics, Culture, National Identity 1945-95*. 6th ed. London: BFI Publishing, 1998.

Roger, G. "Media Concentration With Free Entry." Journal of Media Economics 22:3 (2009): 134–163.

Sánchez-Tabernero, Alfonso and Carvajal, Miguel. *Media Concentration in the European Market: New Trends and Challenges*. Pamplona: Universidad de Navarra, 2002.

Vogel, Harold. *Entertainment Industry Economics: A Guide for Financial Analysis*. 6th ed. New York: Cambridge University Press, 2004.

Woodhull, N. and Snyder, R., eds. *Media Mergers*. New Brunswick: Transaction Publishers, 1998.


### 2. Reports


Asociación de Revistas de Información. *La Revista*. Madrid: Asociación de Revistas de Información: 2011.

Barlovento Comunicación. *Análisis televisivo del mercado español*. Madrid: Barlovento Comunicación, 2011.

Boletín Oficial del Estado. *Ley 31/1987 de 18 de Diciembre, de Ordenación de las Telecomunicaciones*. Madrid: Boletín Oficial del Estado, Dec. 19, 1987.

Boletín Oficial del Estado. *Ley 7/2010 de 21 de Marzo, General de Comunicación Audiovisual*. Madrid: Boletín Oficial del Estado, Apr. 1, 2010.

Comisión del Mercado de las Telecomunicaciones. *Informe Anual 2010*. Madrid: Comisión del Mercado de las Telecomunicaciones, 2011.

Commission of the European Union. *Notice on the Definition of Relevant Market for the Purposes of Community Competition Law*. Brussels: European Union, Dec. 9, 1997.

Comscore. *March 2011 Search Engine Market Share*. Reston, VA: Comscore, 2011.

Council of the European Union. *Regulation n°. 4064/89 on the Control of Concentrations between Undertakings*. Brussels: European Union, Dec. 21, 1989.

Council of the European Union. *Directive DOCE L 298 of Television without Frontiers*. Brussels: European Union, Oct. 17, 1989. Modified by

*Directive 97/36, DOCE L 202*. Brussels: European Union, July 30, 1997.
European Audiovisual Observatory. *Yearbook: Film and Home Video*. Vol. 3. Strasbourg: European Audiovisual Observatory, 2010.
Infoadex. *Estudio de la inversión publicitaria en España*. Madrid: Infoadex, 2011.
Ministerio de Cultura de España. *La cuenta satélite de la cultura en España*. Ministerio de Cultura: Madrid, 2009.
Organization for Economic Cooperation and Development. *Competition Policy and Film Distribution*. Paris: Organization for Economic Co-operation and Development, 1996.

# Spain—Data Summaries

ELI NOAM AND PAUL MUTTER

CONCENTRATION IN SPANISH MEDIA industries has declined partly because of the entry of foreign media groups into the market (Table 12.14). However, due to a lack of cross-ownership restrictions and the legacies of extensive state ownership of media under the Franco regime, concentration is still moderately high compared to the rest of the EU. Spain's national power index fell from 4,047 in 2004 to 2,860 in 2012, and pooled overall sector HHI fell from 1,823 to 1,195 in the same period. Several markets—such as ISPs and radio—are still heavily concentrated due to the incumbency advantage of former (Telefónica) and present publicly owned companies.

Due to extensive deregulation in the national market, foreign ownership has increased significantly since the 1980s, led by other European media groups. Vodafone (UK) and Orange (France) compete with Telefónica, Spain's leading wireless operator and ISP. At the same time, Spanish multimedia producers are expanding their global reach.

Telefónica is the fifth largest wireless provider in the world by subscribers, with 249 million in 2012, and dominates the mobile markets of Argentina, Chile, and Ireland, while retaining significant company share of the national markets in the United Kingdom, Brazil, and Germany. Within Spain, Telefónica is the largest single platform media company, holding 50.5% of the market (Table 12.16), though this represents a slight loss of market share to its aforementioned competitors, Orange and Vodafone, who are increasing their market shares at Telefónica's expense.

Spain's most diversified multimedia producer—with the largest shares of the radio, multichannel, and video channel markets—is the PRISA Group, which also operates in the Brazilian and Portuguese markets. It controls 11.2% of the content

Table 12-14. National Media Industries Concentration in Spain

| | 2004/5 | | 2011 OR MOST RECENT | | % CHANGE ANNUAL AVERAGE | |
|---|---|---|---|---|---|---|
| | COMPANY POWER INDEX IN COUNTRY | COMPANY SHARE OF THE OVERALL NATIONAL MEDIA MARKET (%) | COMPANY POWER INDEX IN COUNTRY | COMPANY SHARE OF THE OVERALL NATIONAL MEDIA MARKET (%) | COMPANY POWER INDEX IN COUNTRY | COMPANY SHARE OF THE OVERALL NATIONAL MEDIA MARKET (%) |
| Telefónica | 2,899.87 | 40.0 | 1,748.12 | 31.099 | –4.96 | –1.161 |
| PRISA Group | 524 | 8.8 | 204.2 | 6.481 | –7.63 | –0.3 |
| Vodafone (UK) | 165 | 6.0 | 194.68 | 8.356 | 2.28 | 0.291 |
| Google (US) | 24.9 | 0.3 | 157.25 | 1.6 | 66.35 | 0.170 |
| Orange (France Telecom, France) | 105 | 6.3 | 123.56 | 6.886 | 2.17 | 0.073 |
| RTVE (public) | 73.4 | 2.8 | 33.44 | 2 | –6.81 | –0.107 |
| ONO | 9.6 | 1.9 | 73.97 | 5.077 | 84.32 | 0.400 |
| Planeta | 48.7 | 3.0 | 70.95 | 3.462 | 5.68 | 0.059 |
| Fininvest (Italy) (Berlusconi) | 45.3 | 2.2 | 71.07 | 2.663 | 7.11 | 0.052 |
| Vocento | 27.45 | 1.5 | 23.2 | 1.3 | –1.94 | –0.02 |
| FORTA (public) | 28.2 | 1.6 | 8.59 | 0.876 | –8.70 | –0.090 |
| RCS (Italy) | 5.8 | 0.9 | 22.20 | 1.6 | 35.47 | 0.090 |
| RBA | 12.0 | 0.8 | 12.22 | 0.794 | 0.23 | –0.002 |
| Lagardère (France) | 10.7 | 0.8 | 16.51 | 0.922 | 6.83 | 0.020 |
| Disney (US) | 6.6 | 0.5 | 3.9 | 0.4 | –5.02 | –0.003 |
| Zeta | 9.4 | 1.1 | 9.70 | 1.082 | 0.28 | 0.01 |
| Murdoch (US) | 3.6 | 0.4 | 4.8 | 0.5 | 4.42 | 0.01 |
| Godó | 4.6 | 0.6 | 5.19 | 0.625 | 1.67 | 0.002 |
| Anaya | 3.5 | 0.4 | 4.6 | 0.5 | 3.9 | 0.004 |
| Conde Nast (Advance) (US) | 2.0 | 0.3 | 9.39 | 0.696 | 45.71 | 0.020 |

| | | | | | | |
|---|---|---|---|---|---|---|
| Motorpress (Germany) | 2.3 | 0.4 | 4.8 | 0.5 | 13.3 | 0.02 |
| Factoria De Canales | 0.0 | 0.0 | 2.2 | 0.3 | N/A | 0.04 |
| Bertelsmann (Germany) | 2.6 | 0.4 | 3.06 | 0.4 | 2.46 | 0.003 |
| Liberty Media (US) | 2.5 | 0.33 | 1.9 | 0.3 | –3.2 | –0.003 |
| Viacom (US) | 0.5 | 0.16 | 1.2 | 0.240 | 14.07 | 0.011 |
| Bouygues (France) | 0.2 | 0.1 | 0.1 | 0.08 | –6.5 | –0.003 |

| Media Concentration Index | 2004/5 | 2011 or Most Recent | % Change Annual Average |
|---|---|---|---|
| Total Revenue: Nat'l Media Industry (mil US$) | 72,813 | 86,981 | 2.43 |
| Total Voices (*n*) | 72 | 72 | 0 |
| Net Voices (*n*) | 49 | 53 | 1.02 |
| Public Ownership (%) | 4.3 | 2. | -0.2 |
| Foreign Ownership (%) | 21.6 | 29.4 | 1.0 |
| C4 Average—Weighted | 83.5 | 78.5 | –1.26 |
| HHI Average—Weighted | 4,058 | 2,867 | –3.67 |
| C1 Average—Weighted | 53 | 42 | –0.03 |
| Noam Index Average—Weighted | 694 | 683 | –0.2% |
| Pooled Overall Sector C4 | 61.5 | 52.8 | –1.09 |
| Pooled Overall Sector HHI | 1,823 | 1,195 | –4.30 |
| Pooled Overall Sector Noam Index | 57 | 49 | –1.8 |
| Market Share of Top Ten Companies: Nat'l Media Industry (%) (Pooled C10) | 74.5 | 69.5 | –0.6 |
| National Power Index | 4,047 | 2,860 | –3.7 |

Table 12-15. Top Content Media Companies in Spain

| | 2004/5 | | 2011 OR MOST RECENT | | % CHANGE ANNUAL AVERAGE | |
|---|---|---|---|---|---|---|
| | COMPANY POWER INDEX IN COUNTRY | COMPANY SHARE OF THE NATIONAL CONTENT MEDIA MARKET (%) | COMPANY POWER INDEX IN COUNTRY | COMPANY SHARE OF THE NATIONAL CONTENT MEDIA MARKET (%) | COMPANY POWER INDEX IN COUNTRY | COMPANY SHARE OF THE NATIONAL CONTENT MEDIA MARKET (%) |
| PRISA | 612 | 13.9 | 296.1 | 11.2 | −6.45 | −0.3 |
| Google (US) | 64.2 | 0.7 | 401.3 | 4.2 | 65.57 | 0.4 |
| RTVE (public) | 189.3 | 7.1 | 85.3 | 4.8 | −6.9 | 4.8 |
| Planeta | 126 | 7.7 | 181 | 8.8 | 5.5 | 0.1 |
| Fininvest (Italy) (Berlusconi) | 117 | 5.8 | 181.3 | 6.8 | 6.9 | 0.1 |
| Vocento | 71.0 | 3.8 | 59.2 | 3.4 | −2.05 | −0.05 |
| FORTA (public) | 73 | 4.1 | 22 | 2.2 | −8.7 | −0.2 |
| RCS Media Group (Italy) | 15 | 2.2 | 57 | 4.0 | 35 | 0.2 |
| RBA | 31 | 2.1 | 31.2 | 2.0 | 0.11 | −0.007 |
| Zeta | 24.5 | 2.9 | 24.7 | 2.8 | 0.15 | −0.02 |
| Godó | 11.8 | 1.6 | 13.2 | 1.6 | 1.53 | 0.002 |
| Anaya | 9.1 | 1.1 | 11.8 | 1.2 | 3.8 | 0.01 |
| Ono | 2.2 | 0.3 | 27 | 1.3 | 141.0 | 0.1 |
| Motorpress (Germany) | 5.9 | 0.9 | 12.1 | 1.2 | 13.0 | 0.04 |
| Factoria De Canales | 0 | 0 | 5.6 | 0.8 | N/A | 0.1 |
| Liberty Media (US) | 6.4 | 0.9 | 4.8 | 0.8 | −6.5 | −0.02 |

| | | | | | | |
|---|---|---|---|---|---|---|
| Bouygues Group (France) | 0.6 | 0.3 | 0.3 | 0.2 | −6.6 | −0.01 |

| Media Concentration Index | 2004/5 | 2011 or Most Recent | % Change Annual Average |
|---|---|---|---|
| Public Ownership (%) | 11.2 | 7.1 | −0.5 |
| Foreign Ownership (%) | 19.3 | 29.9 | 1.3 |
| C4 Average—Weighted | 60.3 | 61.3 | 0.3 |
| HHI Average—Weighted | 1,532 | 1,606 | 0.6 |
| C1 Average—Weighted | 25 | 25 | 0.0 |
| National Power Index | 1,470 | 1,607 | 1.2 |

Table 12-16. Top Platform Media Companies in Spain

| | 2004/5 | | 2011 OR MOST RECENT | | % CHANGE ANNUAL AVERAGE | |
|---|---|---|---|---|---|---|
| | COMPANY POWER INDEX IN COUNTRY | COMPANY SHARE OF THE NATIONAL PLATFORM MEDIA MARKET (%) | COMPANY POWER INDEX IN COUNTRY | COMPANY SHARE OF THE NATIONAL PLATFORM MEDIA MARKET (%) | COMPANY POWER INDEX IN COUNTRY | COMPANY SHARE OF THE NATIONAL PLATFORM MEDIA MARKET (%) |
| Telefónica | 4,737 | 65.7 | 2,864 | 50.5 | –4.9 | –1.9 |
| Vodafone (UK) | 269 | 9.9 | 320.1 | 13.7 | 2.4 | 0.5 |
| PRISA | 468 | 5.6 | 145 | 3.4 | –8.6 | –0.3 |
| Orange (France) | 172 | 10.3 | 203.1 | 11.3 | 2.26 | 0.1 |
| Ono | 14.2 | 2.9 | 104.3 | 7.5 | 79.2 | 0.6 |
| Media Concentration Index | 2004/5 | | 2011 or Most Recent | | % Change Annual Average | |
| Public Ownership (%) | 0 | | 0 | | 0 | |
| Foreign Ownership (%) | 21.3 | | 28.4 | | 0.9 | |
| C4 Average—Weighted | 98.3 | | 89.6 | | –2.18 | |
| HHI Average—Weighted | 5,659 | | 3,679 | | –4.37 | |
| C1 Average—Weighted | 71 | | 53 | | –5 | |
| National Power Index | 5,666 | | 3,675 | | –4.4 | |

media market in Spain, well ahead of its competitors like the Italian company Fininvest or the public broadcasters RTVE and FORTA (Table 12.15). The PRISA Group is partly owned by the Polanco family, and the rest of it is publicly traded on the national stock exchange. Second to PRISA, with 8.8% of the content market, is Planeta. Like PRISA, it has market shares in many industries: TV broadcasting, video channels, daily newspapers, book publishing, radio, and online news media.

# 13

# Media Ownership and Concentration in Sweden

ROBERT G. PICARD, MART OTS, AND MADISON FORSANDER

## INTRODUCTION

Sweden is the third largest country in Europe by land area but has a population of only 9.5 million people. As an urbanized country with a high standard of living, the Swedish media market benefits from Swedes' high literacy rates, availability of disposable income, and access to an extensive communications infrastructure. Print media, and newspapers in particular, holds a strong position in the Swedish media landscape. Both readership figures and advertising market shares for newspapers are very high by international standards. Sweden is also a global leader in digital television and broadband access.

Though newspapers remain popular, television is now the most widespread consumer medium, followed by newspapers, radio, and the Internet. In terms of advertising spending, newspapers continue to draw nearly 30% of total spending, whereas TV's share of advertising revenues (15%) is declining and will likely lose ground to Internet advertising, which currently has a 14% market share (Table 13.1).

The Government of Sweden has long held that provision of high-quality public broadcasting is essential to the country's political health. It has pursued this in the audiovisual sector by subsidizing a large public service broadcasting network, and in the newspapers sector by providing extensive subsidies. The telecommunications sector was operated by the state and provided universal public service to the country until the 1990s. Sweden was one of the last countries in Europe to privatize the sector.

Because there are often just a few large players in most media sectors in the country, media watchdogs and regulators have long been concerned about concentration in Sweden.

Table 13-1. Media Penetration in Sweden (Audience Shares by %), 2011

| | |
|---|---|
| TV | 85 |
| Newspapers | 77 |
| Radio | 74 |
| Internet | 68 |
| Magazines | 41 |
| Books | 37 |
| CD | 27 |
| Teletext | 25 |
| MP3 | 18 |
| Video/DVD | 12 |
| Cinema | 1 |

*Source:* Nordicom, 2011.

## PRINT MEDIA

### Newspapers

The Swedish newspaper market is divided in two segments. One segment consists of regional and local morning papers, which have a combined circulation of 2.7 million copies, 93% of which are purchased through subscriptions and are home-delivered by the papers themselves. The other segment is shared between two of the largest national daily publishers and their evening-edition tabloids—*Aftonbladet* and *Expressen*, which are sold in grocery stores and news kiosks. Combined, they have a daily circulation of approximately 700,000 copies.

Sweden spends tens of millions of euros a year to subsidize newspapers in order to try to keep concentration low. Press subsidies were introduced by the Swedish Parliament between 1969 and 1975 and consist of three main components—operational support, development support, and joint distribution support. The subsidies were introduced in response to 15 years of growing market concentration in which 40 regional newspaper titles were forced to close. The first response was the institution of the operational support subsidy in 1971. It was aimed at providing financial assistance to the second and third largest newspapers in each region of the country. The papers covered by the operational support law are allowed to spend their subsidies on any aspect of their operations, but in order to qualify for the direct cash subsidy, the papers must have no more than 30% household coverage in their home region, register a circulation exceeding 2,000 paid-for copies, and produce at least 51% of their own original editorial material.[1] As a result, these criteria exclude the evening tabloids from being subsidized. The maximum annual subsidy in 2011 stood at US$2.1 million (1.5 million euros) for provincial papers and US$8.9 million (6.4 million euros) for metropolitan papers. In 2009, 23 dailies received operational support, along with 61 newspapers that do not publish daily editions.

The development support subsidy provides loans to financially weak newspapers to assist them in updating their production technologies. Whereas it encourages collaboration between applicants, it is important that these collaborations are restricted to noneditorial functions in order to preserve a plurality of editorial viewpoints. Lastly, the joint distribution support subsidy is meant to give all newspapers (nominally) equal market access. Since most papers in Sweden are sold by subscriptions and have morning home deliveries, the joint distribution support subsidy was implemented as incentive for the leading dailies to open up their distribution channels to their local competitors: the system subsidizes all participating papers with roughly one US cent per copy.

The total cost of the three press subsidies has remained fairly stable since the 1990s, costing the state US$69.5 million (50 million euros), corresponding to 2.3% of total media turnover, and 15–20% of the receiving newspapers' annual turnover. In addition, outside of the subsidy system, all print publishers benefit from a reduced value-added tax (VAT) for newspapers and books; normal VAT is 25%, but it is only 6% for these printed materials. And finally, parallel to the subsidies, the state has also imposed a 3%

1. Ruth, Evan. "Media Subsidies An ARTICLE 19 Paper." *ARTICLE 19*, 1999. Jan. 4, 2013. <http://www.article19.org/data/files/pdfs/publications/media-subidies.pdf>.

advertising tax on printed media, which generates revenue that more than covers the costs of the press subsidy program. However, since regulators no longer maintain the historical linkage between the advertising tax and the subsidies, most Swedish political parties agree that the advertising tax should be abolished. Even so, there is little political incentive to actually drive through such a change in the laws.

Through the Newspapers Subsidy Council, an agency of the Swedish Ministry of Culture, the state also provides around US$16.7 million (12 million euros) to support the production and distribution of newspapers to the hearing and visually impaired. The funds are distributed to newspaper companies in order to produce audio recordings of their articles and distribute them at normal newspaper subscription prices (Table 13.2).

Despite all of these anticoncentration measures, three companies together hold 69% of the national market share for newspapers. Bonnier AB has consistently led the industry, though its share of the market has fluctuated from 24.8% in 1984 to 35.2% in 1996 and fell to 28.5% in 2008. Stampen and Schibsted, two publicly traded Swedish media groups, compete for second place, together controlling approximately 30% of the market. Mittmedia Fortvaltnings, another privately held Swedish company, has made considerable gains in the sector in recent years, jumping from a 2.6% market share in 1988 to a 7.6% share in 2008. Bonnier's major newspapers are *Dagens Nyheter, Expressen,* and *Sydsvenskan*; Stampen's most widely circulated daily is *Göteborgs-Posten*; and Schibsted's leading dailies are *Aftonbladet* and *Svenska Dagbladet.*

The dominant firm, Bonnier AB, is a family-owned company that was founded in 1804 and is now a highly diversified, multimedia company operating in 17 countries. Within Sweden,

Table 13-2. Swedish Government Press Subsidies

| | OPERATIONAL SUPPORT (MIL EUROS) | DEVELOPMENT SUPPORT (MIL EUROS) | JOINT DISTRIBUTION SUPPORT (MIL EUROS) | TOTAL SUBSIDIES (MIL EUROS) | % CHANGE |
|---|---|---|---|---|---|
| 1990/91 | 42.9 | 2.4 | 8.1 | 53.3 | 16.0 |
| 1991/92 | 42.1 | 3.4 | 7.8 | 53.2 | 0.0 |
| 1992/93 | 41.2 | 3.0 | 7.1 | 51.2 | −3.8 |
| 1993/94 | 41.5 | 1.8 | 7.1 | 50.5 | −1.3 |
| 1994/95 | 40.2 | 0.3 | 7.1 | 47.6 | −5.8 |
| 1995 | 39.7 | 0.1 | 4.2 | 44.0 | −1.0 |
| 1996 | 43.4 | 0.1 | 6.8 | 50.3 | 7.6 |
| 1997 | 45.6 | 0.1 | 6.7 | 52.4 | 4.1 |
| 1998 | 45.9 | | 7.1 | 53.0 | 1.2 |
| 1999 | 43.1 | | 7.1 | 50.3 | −5.2 |
| 2000 | 42.9 | | 7.3 | 50.2 | −0.2 |
| 2001 | 41.1 | | 7.4 | 48.5 | −3.2 |
| 2002 | 40.1 | 7.4 | −0.2 | 47.3 | |
| 2003 | 40.3 | 1.3 | 8.3 | 49.9 | 5.0 |
| 2004 | 41.1 | 1.2 | 8.2 | 50.4 | 1.1 |
| 2005 | 40.8 | 0.4 | 8.1 | 49.3 | −2.2 |
| 2006 | 42.0 | | 8.1 | 50.0 | 1.7 |
| 2007 | 41.9 | | 7.1 | 49.1 | −1.9 |
| 2008 | 42.2 | | 7.2 | 49.4 | 0.6 |

*Source: Dagspressens ekonomi* (2008).

## Table 13-3. Daily Newspapers (Market Shares by Circulation)

| | 1984 | 1988 | 1992 | 1996 | 2000 | 2004 | 2008 | 2011[2] |
|---|---|---|---|---|---|---|---|---|
| Bonnier AB (Bonnier Family) | 24.8 | 24.7 | 23.3 | 35.2 | 30.4 | 26.7 | 28.5 | 41.3 |
| Stampen | 9.0 | 8.7 | 9.8 | 10.6 | 8.8 | 8.2 | 16.5 | 12.4 |
| Schibsted | | | | 9.1 | 14.4 | 14.9 | 14.8 | 10.3 |
| Mittmedia Förvaltnings AB | | 2.6 | | 3.2 | 4.0 | 4.9 | 7.6 | |
| Norrköpings Tidningars Media | | | | | 2.8 | 3.3 | 5.5 | 4.9 |
| Gota Media AB | | | | | | 3.7 | 3.9 | |
| Herenco AB | 2.7 | 4.0 | 3.4 | 3.3 | 3.6 | 3.1 | 2.9 | |
| Metro International (Kinnevik) | | | | | 3.1 | 3.5 | 2.9 | 19.0 |
| Nya Wermlands-Tidningen AB | | 3.7 | 4.3 | 4.0 | 3.6 | | 2.6 | |
| AB Upsala Nya Tidning | | | | | | | 2.3 | |
| VLT AB | 2.8 | | 3.1 | 3.3 | 4.1 | 4.2 | | |
| Centertidningar | 2.7 | | 2.8 | 4.0 | 3.5 | 3.8 | | |
| Sydostpress | 3.2 | 2.7 | 2.8 | 3.0 | 3.3 | | | |
| Svenska Dagbladet Holding | 6.2 | 5.6 | 5.5 | 6.7 | | | | |
| LO | | 7.6 | 7.5 | | | | | |
| Sydsvenska Dagbladet | 6.7 | 6.3 | 6.7 | | | | | |
| A-Pressen | 11.2 | 7.9 | | | | | | |
| Nerikes Allehanda | 2.6 | | | | | | | |
| Others | 28.1 | 26.2 | 30.8 | 17.6 | 18.4 | 23.7 | 12.5 | 12.1 |
| Total Revenue (mil SEK)[1] | 7,206 | 12,481 | 14,225 | 14,818 | 17,954 | 19,762 | 22,051 | |
| Total Revenue (mil US$) | 792 | 1,997 | 2,418 | 2,223 | 1,975 | 2,767 | 3,308 | |
| C4 | 51.8 | 48.9 | 47.3 | 61.6 | 57.7 | 54.7 | 67.3 | 83 |
| HHI | 946 | 922 | 827 | 1,554 | 1,308 | 1,104 | 1,435 | 2,350 |
| *N* (>1%) | 10 | 10 | 10 | 10 | 11 | 10 | 10 | 5 |
| Noam Index | 299 | 292 | 261 | 491 | 394 | 349 | 454 | 1,050 |

[1] Swedish Krona (SEK). Sweden has not switched over to the euro, though some sources give statistics for Swedish media in euros.

2. Source: "Top Daily Newspapers: Sweden 2011". Media Passport. M&M Global. 2011. May 29, 2014. http://mandmglobal.com/media-passport/sweden/market-data

Table 13-4. Book Publishing (Market Shares by Revenue)

| | 1996 | 2000 | 2004 | 2008 |
|---|---|---|---|---|
| Bonnier AB (Bonnier Family) | 26.7 | 23.2 | 28.1 | 22.6 |
| Liber | 29.0 | 18.2 | 15.9 | 13.1 |
| KF/Norstedts Förlagsgrupp | 4.5 | 6.6 | 7.7 | 7.0 |
| Forma Publishing Group | | | | 4.6 |
| B Wahlströms Förlag | 2.6 | 1.9 | 2.0 | |
| Natur och Kultur | 7.4 | | 5.6 | 4.5 |
| Berling Media | 6.0 | 5.4 | 5.0 | 3.9 |
| Studentlitteratur | 6.4 | 5.2 | 4.4 | 3.1 |
| Cydonia | 8.9 | 6.4 | 4.0 | 2.1 |
| Egmont (Denmark) | 6.0 | 5.6 | 5.4 | 2.1 |
| Piratförlaget | | | | 1.3 |
| Sanoma (Finland) | 2.5 | 3.5 | 2.4 | 1.3 |
| Ordfront | | 1.6 | | |
| Others | 0.0 | 22.4 | 19.5 | 34.4 |
| Total Revenue (mil SEK) | 3,935 | 4,900 | 5,619 | 7,000 |
| Total Revenue (mil US$) | 509 | 539 | 787 | 1,050 |
| C4 | 72.0 | 54.4 | 57.3 | 47.3 |
| HHI | 1,835 | 1,059 | 1,234 | 809 |
| *N* (>1%) | 10 | 10 | 10 | 12 |
| Noam Index | 580 | 335 | 390 | 244 |

it has significant market penetration in the print, audiovisual, and film markets.

Overall, HHI values for newspapers have been gradually increasing but have so far stayed within the 1,400–1,500 range (Table 13.3).

## Book Publishing

Sweden has had a highly commercialized book publishing sector since the 19th century, and it is a world leader in the market for translated publications. The two dominant players in the Swedish book publishing industry are Bonnier AB and Liber. Between them, they hold over one-third of the market, though that share has declined from what was once a combined 50% market share. Bonnier AB owns multiple imprints, while Liber is the leading publisher of educational texts in Sweden. From 1996 to 2008, both Bonnier AB and Liber lost shares in the market, as did many of the smaller publishers (Table 13.4).

## Magazine Publishing

Sweden has one of the most active magazine markets in the European Union. It is highly commercialized, and the government has had less influence on this print market than it has on newspapers.

The two major players in the Swedish magazine publication industry are Bonnier AB and Carl Allers Etablissement, which together account for over half of the market. Bonnier, with a 31% Swedish market share in 2008, also publishes magazines in foreign markets: in 2007 it bought the rights to 18 magazines from Time, Inc. (US). Carl Allers Etablissement, a private media corporation with 34 monthly titles, holds 23.5% of the Swedish market share. Smaller publishers include Egmont International Holding and the Forma Publishing Group. Egmont International Holding, founded in 1878, is one of Sweden's oldest media groups, and it is in close competition with Forma Publishing Group, which specializes in book and magazine publishing.

Table 13-5. Magazine Publishing (Market Shares by Revenue)

| | 2004 | 2008 | 2010 |
|---|---|---|---|
| Bonnier AB (Bonnier Family) | 27.7 | 31.0 | 32.9 |
| Carl Allers Etablissement | 21.9 | 23.5 | 24.9 |
| Egmont International Holding (Denmark) | 11.5 | 11.0 | 11.7 |
| Forma Publishing Group | 7.4 | 8.2 | 8.7 |
| LRF Media | 5.3 | 7.4 | 7.8 |
| IDG/International Data Group (US) | 5.0 | 5.1 | 5.4 |
| Talentum | | 3.3 | 3.5 |
| Förlags AB Albinsson & Sjöberg | 3.5 | 2.9 | 3.1 |
| Mentor Online | 2.5 | 2.5 | 2.7 |
| Frida Förlag | | 1.6 | 1.7 |
| Hjemmet Mortensen | 4.3 | | |
| Hachette Filipacchi Sverige (Lagardère Group, France) | 2.6 | | |
| E+T, Ekonomi och Teknik Förlag | 2.5 | | |
| Medströms | | | |
| Others | 5.8 | 3.5 | 5.9 |
| Total Revenue (mil SEK) | 6,200 | 7,200 | 9,520 |
| Total Revenue (mil US$) | 868 | 1,080 | |
| C4 | 68.5 | 73.7 | 78.2 |
| HHI | 1,540 | 1,808 | 2,037 |
| *N* (>1%) | 11 | 10 | 10 |
| Noam Index | 464 | 572 | 644 |

The Swedish magazine market is slightly trending toward higher concentration indices. The C4 jumped from 68.5 in 2004 to 73.7 in 2008, and the HHI increased from 1,540 to 1,808 over the same period. Comparatively, this HHI range is exceptionally high for magazines (Table 13.5).

## AUDIOVISUAL MEDIA

### Radio

Sweden has a strong public service broadcasting tradition that is based off the United Kingdom's BBC model, but audiovisual policy must now be harmonized with supranational EU policies meant to increase the presence of commercial broadcasters and independent production companies. Today, the Swedish public broadcasting network consists of three entities—the TV broadcaster Sveriges Television (SVT), the radio broadcaster Sveriges Radio (SR), and the "educational service" Utbildningsradion (UR). Until the late 1980s commercial broadcasting did not exist in Sweden, thus giving SR's four national channels a monopoly. Since then, increased competition for audience shares has driven public service operations toward a multichannel, multiplatform strategy; today SR includes a portfolio of four national radio channels, 28 local channels, and more than a dozen online stations.

SR has a strong market position as the former public monopolist, in large part due to the licensing fees collected on its behalf. A mandatory US$300 (2,000 SEK) licensing fee for every household with a receiver provides the funding for the public broadcasters. Radiotjänst, a subsidiary, collects the fees. The commercial radio market's combined gross advertising revenue stood at US$1.1 billion (7.3 billion SEK) in 2008; SR took in licensing fees worth US$330 million (2.2 billion SEK) that same year.

SR has argued that the funding is needed to produce content in the absence of commercial alternatives, but commercial broadcasters have argued that there will never be a financial space for them to expand in so long as these licensing funds continue to exist for SR. The commercial networks also must contend with the impact of the state's policy of auctioning off of broadcasting licenses. During the 1990s, when many networks overestimated the market for radio advertising, they overbid on these licenses and suffered heavy losses. The result of this has been two decades of rapid market concentration and cost rationalization among commercial broadcasters. Some officials have proposed that broadcasting permits should now be sold at (relatively low) fixed prices instead of at auction (in addition to public service operations, the state issues broadcasting permits to approximately 900 community radio broadcasters that are often financially supported by local and regional authorities).

The commercial broadcaster Modern Times Group (MTG) occupies a distant second place in the radio market after SR. ProSiebenSat.1 (Germany) is the third main national radio broadcaster, having acquired Scandinavian Broadcasting Group (SBS) in 2007, thus gaining a 10% market share in 2008. Despite SR's decreasing market share, Swedish radio has remained highly concentrated, and it is likely to remain so unless policy toward the commercial broadcasters changes. The HHI reflects this trend, as it stood at 5,840 in 2008 (Table 13.6).

## Broadcast Television

Before the 1990s, terrestrial analog broadcasting in Sweden was dominated by SVT. But since 1991 and the switchover of the country to

Table 13-6. Radio Group (Market Shares by Revenue)

| | 1996 | 2000 | 2004 | 2008 | 2011 |
|---|---|---|---|---|---|
| SR (public) | 84.5 | 71.8 | 75.9 | 75.0 | 74.8 |
| MTG (Kinnevik) | 1.5 | 5.8 | 7.8 | 14.6 | 11.8 |
| ProSiebenSat.1 (Germany) | | | | 1.0 | 9.6 |
| SBS Broadcasting | 1.4 | 4.1 | 5.6 | | |
| NRJ (France) | 1.9 | 6.2 | 2.0 | 1.4 | 1.4 |
| Norrköpings Radio & Co. | 0.6 | | | 0.8 | 0.7 |
| Stampen | 1.6 | | | 0.7 | 0.7 |
| Upsala Nya Tidning | 0.2 | 0.5 | 0.4 | 0.3 | 0.4 |
| Västerbottens-Kuriren | 0.2 | | | 0.3 | 0.4 |
| Mittmedia | 0.4 | 0.8 | 0.6 | 0.2 | 0.1 |
| Fria Media | 1.5 | 5.0 | 3.1 | | |
| Bonnier | 3.4 | 5.4 | 3.0 | | |
| Hallandsposten | 0.2 | | 0.3 | | |
| RTL (Bertelsmann AG, Germany) | | 1.0 | | | |
| VLT | | 0.4 | | | |
| Others | 2.6 | 0.0 | 1.3 | 0.0 | 0.0 |
| Total Revenue (mil SEK) | 2,272 | 2,291 | 2,774 | 2,919 | 3,039 |
| Total Revenue (mil US$) | 341 | 252 | 388 | 438 | 457 |
| C4 | 91.4 | 89.2 | 92.4 | 92.0 | 97.6 |
| HHI | 7,152 | 5,263 | 5,866 | 5,840 | 5,829 |
| *N* (>1%) | 7 | 5 | 4 | 3 | 4 |
| Noam Index | 2,703 | 2,353 | 2,395 | 2,973 | 2,914 |

digital terrestrial television (DTT), a wave of new commercial broadcasters, alongside cable TV operators, have emerged. Four companies dominate the sector: the public broadcaster SVT and the three main commercial broadcasters of TV4 AB, MTG-Viasat Broadcasting, and ProSiebenSat.1.

SVT includes five national channels and is expanding its online broadcasting presence. It has maintained a strong market position thanks to the receipt of substantial state subsidies: in 2008, SVT received US$585 million (3.9 billion SEK) from the government. By way of comparison, the combined advertising revenues for commercial broadcasters stood at US$750 million (5 billion SEK). However, in the long-term, commercial advertising revenues have been growing whereas the licensing fees, when adjusted for inflation, have generally stayed flat since the traditional public broadcasters are not adding many new households (and their US$300.00 installation fees) to their audience shares. Subscription and distribution revenues for commercial channels have also been increasing rapidly over the last decade.

Public opinion favors the public broadcasters, but SVT has also had increasing difficulty in attracting a younger demographic, and its overall market shares are declining. It has also been under government scrutiny for using the licensing fees to compete on digital platforms with on-demand online TV and mobile services. Fortunately, the government recently approved these initiatives as a natural part of public service operations in a digital age, legally classifying public service as a type of content rather than a type of distribution.

SVT's lead has declined substantially since the 1990s, falling from 50% in 1996 to 30% in 2008 (of course, during the 1970s, that market share effectively stood at 100%). Much of this loss can be attributed to the entry into the market of ProSiebenSat.1 (Germany), which expanded its services to Sweden in 2007.[3] The Swedish firms TV4 AB and MTG-Viasat Broadcasting have seen slower but continued growth for their market shares since 1996. TV4 AB, which has been broadcasting since 1990, is a subsidiary of Bonnier AB through Nordic Broadcasting Oy since 2007. MTG-Viasat is a satellite TV platform that operates in Sweden, Norway, Denmark, Finland, Estonia, Latvia, Lithuania, and Ukraine.[4] Its parent, MTG, was initially established and owned by the major Swedish investment company Kinnevik, although it has been spun off in 1997 with share distribution to Kinnevik shareholders and now has a dual listing on the Stockholm Stock Exchange and the NASDAQ in the United States.

Because of the high penetration of cable TV and the decision to shut off analog broadcasting in favor of digital by 2008, distinctions between terrestrial, cable, and satellite channels have largely disappeared in Sweden.

The overall concentration of the broadcasting industry in Sweden remains high, though the C4 decreased from 92% in 1996 to 88% in 2004 due to SVT's decline in market share and the entrance of a number of smaller companies into the market. The HHI and Noam index both register a slow decline from 1996 on, indicating a gradual decrease in concentration in the industry as SVT has gradually lost its monopoly. The industry remains a relatively concentrated one, however, with HHI staying above 2,300 in 2008 and two entities—SVT and Bonnier's TV4 AB—controlling over half of the market shares (Table 13.7).

## Multichannel TV Platforms

For multichannel platforms, by far the largest is the cable operator Com Hem, with 38.4 % of the market, followed by the semi-public telecom Telia (12.7%) and the public Canal Digital Sverige (13%) (Table 13.8). Market concentration dropped in the 1990s but rose again.

3. "ProSievenSat.1 Media AG Company History 2006–2010." ProSievenSat.1 Media AG, 2010. June 12, 2012. <http://en.prosiebensat1.com/en/company/prosiebensat1-media-ag/history/2006-2010>.

4. "MTG.SE brand operations pay-tv Viasat." Modern Times Group, 2010. June 12, 2012. <http://www.mtg.se/en/brands--operations/brands/pay-tv/viasat/>.

Table 13-7. TV Broadcasting (Market Shares by Revenue)

| | 1996 | 2000 | 2004 | 2008 | 2012[5] |
|---|---|---|---|---|---|
| **Terrestrial Broadcasters** | | | | | |
| SVT AB (public) | 47.8 | 40.0 | 39.6 | 29.2 | 26.4 |
| TV4 AB (Bonnier AB, Bonnier Family) | 22.4 | 28.0 | 22.7 | 26.6 | 29.8 |
| MTG—Viasat Broadcasting (Kinnevik) (both broadcaster and satellite) | 15.0 | 16.4 | 17.1 | 23.6 | 21.8 |
| ProSiebenSat.1 (Germany) | | | | 15.7 | 11.3 |
| SBS Broadcasting | 3.2 | 5.8 | 7.7 | | |
| Sveriges Utbildningsradio AB | 4.5 | 3.4 | 3.0 | 2.3 | |
| Lokal-TV Nätverket i Sverige AB | | | 0.1 | | |
| MTV Networks AB (Viacom, US) | | 0.2 | 0.9 | 1.3 | |
| **Satellite-delivered Channels** | | | | | |
| Nonstop Television AB | | | 0.2 | 0.6 | |
| Eurosport Television AB (France) | | 0.5 | 0.6 | 0.6 | |
| Turner Broadcasting System Sweden (Time Warner, US) | | 0.1 | | 0.1 | |
| Axess Publishing | | | | | |
| C MORE[1] (Telenor/TV 4 AB) | 6.9 | 5.2 | 8.1 | | |
| Kunskaps-TV | | 0.5 | | | |
| Others | 0.2 | 0.0 | 0.0 | 0.0 | 10.7 |
| Total Revenue (mil SEK) | 7,200 | 8,947 | 10,250 | 13,550 | |
| Total Revenue (mil US$) | 1,080 | 984 | 1,435 | 2,033 | |
| C4 | 92.1 | 90.2 | 87.6 | 95.1 | 89.3 |
| HHI | 3,090 | 2,729 | 2,514 | 2,369 | 2,188 |
| *N* (>1%) | 6 | 6 | 6 | 6 | 4 |
| Noam Index | 1,261 | 863 | 1,026 | 967 | 1,094 |

[1] Originally Canal+ Television AB, C More was originally established by Vivendi (France). After a series of sales, TV4 AB and Telenor (Norway) now jointly own it, with TV4 AB having a 65% stake and Telenor a 35% stake.

## Video Channels

### SATELLITE TV

The satellite TV market is led by three companies: Teracom, Telenor/Canal Digital Sverige, and MTG/Viasat. Teracom is the market leader, holding nearly half of the total market share in 2008 with 49.8% (though down since 1996, when the company dominated the market at 80%). These three companies together hold 100% of the entire cable TV market. Although the market shares have become more distributed among the three companies since 1996, Teracom still holds half of the total market share.

Teracom benefits greatly from being the Swedish government's terrestrial broadcast transmission service company: it owns all Swedish transmission sites and distributes all terrestrial radio and TV channels—private and public, analog (up until 2008) and digital. Its main product areas are pay TV, radio and television broadcasting, and data capacity services. Telenor, a state-owned Norwegian company, offers mobile and fixed telephony as well as Internet access and content (Table 13.9).

5. http://www.tv4gruppen.se/In-English/About-the-TV4-Group/The-Market/

## Table 13-8. Multichannel Video Platforms: Cable MSOS, DBS, IPTV (Market Shares by Revenue)

| | 1996 | 2000 | 2004 | 2008 | 2012 |
|---|---|---|---|---|---|
| Com Hem (BC Partners, UK) | | | 23.6 | 37.7 | 38.4 |
| Canal Digital Sverige (public) (Telenor, Norway)[1] | 6.8 | 11.3 | 15.7 | 12.5 | 13 |
| Tele2 (Kinnevik) | | 5.4 | 3.0 | | 6.6 |
| UPC Sverige (Liberty Global, US) | | 9.7 | 9.7 | | |
| Telia (public) | 23.0 | 19.5 | | | 12.7 |
| StjärnTVnätet | 9.3 | | | | |
| Sweden on Line (Alte) | 4.6 | | | | |
| Teracom (public) | 34.9 | 23.8 | 17.9 | 20.6 | |
| MTG/Viasat (Kinnevik) | 2.0 | 6.6 | 7.5 | 10.0 | 7.3 |
| Boxer | | | | | 13 |
| Others | 19.5 | 23.8 | 22.6 | 19.2 | 8.9 |
| Total Revenue (mil SEK) | 2,611 | 3,493 | 6,690 | 10,140 | 9,300 |
| Total Revenue (mil US$) | 392 | 384 | 937 | 1,521 | 1,400 |
| C4 | 73.9 | 64.2 | 67.0 | 80.8 | 77.1 |
| HHI | 1,904 | 1,238 | 1,284 | 2,101 | 2,071 |
| *N* (>1%) | 6 | 6 | 6 | 4 | 6 |
| Noam Index | 777 | 506 | 524 | 1,050 | 846 |

[1] Originally a property of Vivendi's (France) Canal+, Canal Digital Sverige has since 2001 been part of Telenor (Norway). The Government of Norway has a 53% stake in the company.

## Table 13-9. DBS TV Programming (Market Shares by Revenue)

| | 1996 | 2000 | 2004 | 2008 |
|---|---|---|---|---|
| Teracom (public) | 80.0 | 62.3 | 46.0 | 49.8 |
| Canal Digital Sverige (public) (Telenor, Norway) | | 20.4 | 34.7 | 26.1 |
| Canal+/Canal Digital Sverige (Vivendi, France) | 15.5 | | | |
| MTG/Viasat (Kinnevik) | 4.5 | 17.3 | 19.3 | 24.1 |
| Others | 0.0 | 0.0 | 0.0 | 0.0 |
| Total Revenue (mil SEK) | 1,471 | 2,161 | 4,090 | 5,940 |
| Total Revenue (mil US$) | 221 | 238 | 573 | 891 |
| C4 | 100 | 100 | 100 | 100 |
| HHI | 6,659 | 4,599 | 3,694 | 3,743 |
| *N* (>1%) | 3 | 3 | 3 | 3 |
| Noam Index | 3,845 | 2,655 | 2,133 | 2,161 |

### CABLE TV

While shares were fairly evenly distributed among three or four companies between 1996 and 2004, shares are now very concentrated between only two (Com Hem and Canal Digital Sverige) as of 2008. Since 2008, Com Hem has dominated the cable TV market, holding a 90.9% market share. Canal Digital Sverige, formerly part of Vivendi (France) but now owned by Telenor (Norway), holds a 4.2% market share. Canal Digital Sverige began as a direct-broadcast satellite service in Sweden, Norway, Denmark, and Finland and has since developed into a cable TV operator in Norway, Sweden, and Denmark; a DTT provider in Finland; and an IPTV operator in Sweden and Denmark.

Com Hem is a private telecommunications firm that was founded in 1983. It was spun off of the former state-owned telecom provider Televerket, which has enabled it to retain its high market share even after privatization. Its low market can also partly be explained by the fact that it has been ranked "as the television brand with the least satisfied customers in Sweden" in 2007, 2008, and 2009, according to the Swedish Quality Index (Svensky Kvalitetsindex), an independent customer survey company."[6] Com Hem was owned by a series of private equity firms, including the Carlyle Group (UK), and in 2011 was acquired by BC Partners (UK), the chief owner of the IntelSat consortium.

The HHI has increased from 3,331 in 1996 to 8,285 in 2008, thus demonstrating an increase of almost 5,000 over the 22-year period; the high C4 is also indicative of the duopolistic nature of the market (Table 13.10).

## Film

Bonnier AB is again one of the major market shareholders, here in the film production and distribution industry, with a market share of 29.5%. Bonnier AB's two main competitors are Schibsted, with a market share of 13.9%, and MTG, with a market share of 10.5%. In 2008, the concentration indices for these three shareholders significantly decreased. In 2008, Bonnier AB went from almost 30% to 22.4%, losing around 8% of its market share. Similarly, Schibsted lost

Table 13-10. Cable TV Programming Channels (Market Shares by Revenue)

| | 1996 | 2000 | 2004 | 2008 |
|---|---|---|---|---|
| Com Hem (BC Partners, UK) | | | 60.7 | 90.9 |
| Telenor/Canal Digital Sverige (public) (Telenor, Norway) | | 9.3 | 5.8 | 4.2 |
| Tele2 (Kinnevik) | | 14.3 | 7.7 | |
| UPC Sverige (Liberty Global, US) | | 25.3 | 25.0 | |
| TeliaSonera (public) | 52.6 | 51.1 | | |
| StjärnTVnätet | 21.3 | | | |
| Sweden on Line (Alte) | 10.6 | | | |
| Others | 15.5 | 0.0 | 0.8 | 4.9 |
| Total Revenue (mil SEK) | 1,140 | 1,332 | 2,600 | 4,200 |
| Total Revenue (mil US$) | 171 | 147 | 364 | 630 |
| C4 | 84.5 | 100.0 | 99.3 | 95.1 |
| HHI | 3,331 | 3,546 | 4,409 | 8,285 |
| *N* (>1%) | 3 | 4 | 4 | 2 |
| Noam Index | 1,923 | 1,773 | 2,204 | 5,858 |

6. "Swedish communications firm Com Hem up for sale." *The Local*, 10 Dec. 2010. Jan. 4, 2013. <http://www.thelocal.se/30744/20101210/#.UOcNkYnjnQY>.

Table 13-11. Film Production/Distribution (Market Shares by Box Office %)

| | 2004 | 2008 |
|---|---|---|
| Bonnier AB (Bonnier Family) | 25.9 | 22.4 |
| Egmont International (Denmark) | 7.1 | 9.6 |
| Schibsted | 13.9 | 7.7 |
| Bonver Videodata AB | 5.2 | 4.9 |
| MPP MediaTec Group | | 4.0 |
| Sony (Japan/US) | | 3.7 |
| De Agostino (Italy) | | 3.6 |
| 20th Century Fox (News Corp., Murdoch Family, US/UK/AUS) | 4.1 | 3.2 |
| Walt Disney Studios (US) | 5.7 | 3.2 |
| MTG (Kinnevik) | 10.5 | 3.2 |
| Universal Pictures Nordic (Comcast/GE, US) | 4.5 | |
| MTV Produktion (Bonnier AD, Bonnier Family) | 2.2 | |
| Kanal 75 | 2.3 | |
| Warner Bros. Entertainment Sweden (Time Warner, US) | 2.3 | |
| Others | 18.6 | 34.5 |
| Total Revenue (mil SEK) | 6,615 | 9,146 |
| Total Revenue (mil US$) | 926 | 1,372 |
| C4 | 57.5 | 44.7 |
| HHI | 1,131 | 751 |
| *N* (>1%) | 10 | 10 |
| Noam Index | 358 | 237 |

around 5%, bringing it down to a 7.7% market share. MTG's shares dipped down to 3.2%.

In 2004, HHI was 1,131 but fell in 2008 to 751. The leading US-based Hollywood majors have a relatively small share of the market (Table 13.11).

# TELECOMMUNICATIONS MEDIA

## Wireline Telecom

Sweden's telecommunications sector was state owned until the 1990s. As a result, several government initiatives have been put in place to support the growth of digital infrastructure, primarily to encourage operating efficiency and sustainable growth.

Expanding broadband infrastructure has been of particular concern, and since the 1990s, successive governments have financed the development of a cable network in remote and rural areas that are not considered economically profitable by commercial operators. The goal is to give every Swedish household the opportunity to receive high-speed Internet services (at a minimum rate of 1MB/s) by the end of 2013; in 2007, 71% of the households met that standard, in part thanks to a tax deduction granted to households that installed broadband between 2001 and 2007. Between 2001 and 2007, US$570 million (3.8 billion SEK) was allocated to expanding high-speed Internet services.

Historically, Televerket had a *de facto* service monopoly over local, long-distance, and international communications (due to its vertical integration) before it became Telia, now the dominant part of TeliaSonera, in 1992.[7] In 2005,

7. Noam, Eli. *Telecommunications in Europe*. New York: Oxford University Press, 1992.

TeliaSonera, Tele2, and Telenor (Norway) controlled 94.6% of the wireline telecom market in Sweden. TeliaSonera dominated the Swedish market with an 80.6% market share in 2005, and is government owned (the Swedish and Finnish governments have a 37% and 13% stake in it, respectively), although the company effectively operates as a private firm despite this. Although by far still the dominant player, by 2010 TeliaSonera's market share had fallen to 62.9% due to increased competition. Tele2 and Telenor (Norway) saw their shares in the market rise: Tele2 increased its share from 9.5% in 2005 to 11.8% in 2010, and Telenor (Norway) increased its share from 4.5% to 7.3% in the same period. However, the driving force reducing concentration in this sector is the growing market share of Com Hem. Com Hem occupied a mere 1.4% of the market in 2005, but by 2010 it commanded 7.5%. Other, smaller companies also made significant strides against TeliaSonera, increasing the collective market shares from 4.1% in 2005 to 10.5% in 2010. Significantly, the HHI dropped from 6,609 in 2005 to 4,205 in 2010 (Table 13.12).

## Wireless Telecom

Sweden has been a leader in mobile communications since the proliferation of 1G technologies in the 1980s, a lead made possible by Nordic Mobile Telephony (NMT). NMT's open specifications minimized the cost of producing hardware and enabled many companies to enter the market.[8] However, by 2010, just four companies (of which NMT was not one) dominated the Swedish wireless telecom market: TeliaSonera, Tele2, Telenor (Norway), and Hi3G. TeliaSonera has dominated the market since 2005, although its grip has slackened slightly, from 43.3% in 2005 to 40.4% in 2010. This is largely due to the entry of Hi3G, a privately owned mobile video communications company, into the market in 2001. Hi3G's share of the market jumped from 3.7% in 2005 to 9% in 2010. Hi3G's entrance has also precipitated losses for Tele2 and Telenor (Norway). Tele2, set up in the 1970s by the Kinnevik investment corporation, is a publicly traded telecommunications operator with services in Croatia, Estonia, France, Germany, Kazakhstan, Latvia, Lithuania, the Netherlands, Norway, Russia, and Sweden.

### Table 13-12. Wireline Telecom (Market Shares by Revenue), 2005–2010

| | 2005 | 2006 | 2007 | 2008 | 2009 | 2010 |
|---|---|---|---|---|---|---|
| TeliaSonera (public) | 80.6 | 74.3 | 70.2 | 67.9 | 65.3 | 62.9 |
| Tele2 (Kinnevik) | 9.5 | 10.9 | 11.6 | 12.4 | 12.4 | 11.8 |
| Telenor (public) (Norway) | 4.5 | 6.3 | 7.0 | 7.3 | 7.5 | 7.3 |
| Com Hem (BC Partners, UK) | 1.4 | 3.2 | 5.2 | 5.9 | 6.7 | 7.5 |
| Others | 4.1 | 5.3 | 5.9 | 6.5 | 8.1 | 10.5 |
| Total Revenue (mil SEK)[1] | 30,328 | | | 26,053 | | |
| Total Revenue (mil US$) | 3,931 | | | 3,908 | | |
| C4 | 96.0 | 94.7 | 94.0 | 93.5 | 91.9 | 89.5 |
| HHI | 6,609 | 5,689 | 5,139 | 4,852 | 4,519 | 4,205 |
| *N* (>1%) | 4 | 4 | 4 | 4 | 4 | 4 |
| Noam Index | 3,305 | 2,845 | 2,570 | 2,426 | 2,259 | 2,103 |

[1] This revenue figure is an estimate based on data collected in this study: it is based on average per capita revenues reported for several other countries in this study with similar per capita income.

8. Noam, Eli. *Telecommunications in Europe*. New York: Oxford University Press, 1992, 210.

Table 13-13. Wireless Telecom (Market Shares by Revenue)

| | 2005 | 2006 | 2007 | 2008 | 2009 | 2010 |
|---|---|---|---|---|---|---|
| TeliaSonera (public) | 43.3 | 43.8 | 43.3 | 42.4 | 41.1 | 40.4 |
| Tele2 (Kinnevik) | 33.8 | 32.3 | 31.8 | 31.8 | 32.0 | 31.6 |
| Telenor (public) (Norway) | 18.2 | 17.4 | 18.3 | 17.4 | 16.8 | 16.5 |
| Hi3G | 3.7 | 5.0 | 5.6 | 7.1 | 8.3 | 9.0 |
| Others | 1.0 | 1.5 | 0.9 | 1.3 | 1.9 | 2.4 |
| Total Revenue (mil SEK)[1] | 24,054 | | | 28,080 | | |
| Total Revenue (mil US$) | 3,127 | | | 4,212 | | |
| C4 | 99.0 | 98.5 | 99.0 | 98.7 | 98.2 | 97.5 |
| HHI | 3,362 | 3,289 | 3,252 | 3,162 | 3,064 | 2,984 |
| *N* (>1%) | 4 | 4 | 4 | 4 | 4 | 4 |
| Noam Index | 1,681 | 1,645 | 1,626 | 1,581 | 1,532 | 1,492 |

[1] This revenue figure is an estimate based on data collected in this study: it is based on average per capita revenues reported for several other countries in this study with similar per capita income.

The overall concentration of the market has remained high. C4 decreased by only 1% between 2005 and 2010, so new entrants were unable to gain any footholds. The firms that compete for the roughly 2% market share not controlled by the four main providers have made some small progress, having seen their 1% market share of 2005 more than double by 2010 (albeit only to 2.4% overall). The HHI's decline between 2005 and 2010 does suggest a trend of deconcentration in the wireless market, but the industry remains highly concentrated and growth has become static (Table 13.13).

## INTERNET MEDIA

### Internet Service Providers (ISP)

The four main providers for Internet services in Sweden are the same market leaders in the telecom market: TeliaSonera, Tele2, Telenor (Norway), and Com Hem. And as in the telecom market, the entry of Hi3G has only slightly chipped away at these four players' dominant market shares; the company has gained a 9.5% market share as of 2010, actually surpassing Com Hem's (9.1%). Though the ISP market has begun to diversify, concentration remains relatively high (Table 13.14).

## CONCLUSION

In Sweden, most media platforms have traditionally been highly concentrated, and in some cases a *de facto* monopoly has existed, notably in the case of the public radio broadcaster Sveriges Radio (SR) and of Televerket (now TeliaSonera), the telecom incumbent. Bonnier AB and TeliaSonera, the two largest Swedish media groups, are likely to retain their commanding places in multiple sectors of the media market. Kinnevik, a private major Swedish media company, has begun to provide a modicum of competition in the wake of Televerket's privatization.

In summation, Swedish media shows declining concentrations in TV broadcasting, book publishing, wireline telecom, film production and distribution, and ISPs. It shows rising concentrations for daily newspapers and magazines, and relatively sustained concentration indices for wireless telecom, satellite and cable TV, and radio broadcasting. Foreign penetration into the Swedish market remained small, and this lack of market penetration is likely to continue given the high entry barriers and small size of the media market compared to those in other EU member states like Germany, Spain, the United Kingdom, and France.

Table 13-14. Internet Service Providers (Market Shares by Revenue)

| | 2005 | 2006 | 2007 | 2008 | 2009 | 2010 |
|---|---|---|---|---|---|---|
| TeliaSonera (public) | 37.9 | 39.0 | 38.4 | 38.7 | 37.9 | 37.2 |
| Tele2 (Kinnevik) | 2.8 | 3.8 | 7.7 | 9.3 | 10.8 | 13.9 |
| Telenor (public) (Norway) | 29.5 | 27.3 | 22.6 | 20.3 | 21.5 | 19.9 |
| Hi3G | | | 2.3 | 4.9 | 6.8 | 9.5 |
| Com Hem (BC Partners, UK) | 16.2 | 16.2 | 15.2 | 13.4 | 11.3 | 9.1 |
| Others | 13.5 | 13.7 | 13.8 | 13.4 | 11.8 | 10.3 |
| Total Revenue (mil SEK)[1] | 7,892 | | | 9,340 | | |
| Total Revenue (mil US$) | 1,026 | | | 1,401 | | |
| C4 | 86.4 | 86.3 | 83.9 | 81.7 | 81.5 | 80.5 |
| HHI | 2,577 | 2,543 | 2,281 | 2,200 | 2,189 | 2,146 |
| *N* (>1%) | 4 | 4 | 5 | 5 | 5 | 5 |
| Noam Index | 1,289 | 1,272 | 1,020 | 984 | 979 | 960 |

[1] This revenue figure is an estimate based on data collected in this study: it is based on average per capita revenues reported for several other countries in this study with similar per capita income.

# Bibliography


## 1. Books and Articles

Noam, Eli. *Television in Europe*. New York: Oxford University Press, 1991.

Noam, Eli. *Telecommunications in Europe*. New York: Oxford University Press, 1992.

Ruth, Evan. "Media Subsidies An ARTICLE 19 Paper." *ARTICLE 19,* 1999. Jan. 4, 2013. <http://www.article19.org/data/files/pdfs/publications/media-subidies.pdf>.

"Swedish communications firm Com Hem up for sale." *The Local,* Dec. 10, 2010. Jan. 4, 2013. <http://www.thelocal.se/30744/20101210/#.UOcNkYnjnQY>.

## 2. Reports

Gustafsson, K. E. and Enström, O. "Dagspressens ekonomi (2008)." Stockholm: Presstödsnämnden, 2009. <http://www.presstodsnamnden.se/wp-content/uploads/2012/01/Dagspressens-ekonomi-2008-webb.pdf>.

"MTG.SE brand operations pay-tv Viasat." Modern Times Group, 2010. June 12, 2012. <http://www.mtg.se/en/brands--operations/brands/pay-tv/viasat/>.

Nordicom. "Daily Media Reach in Sweden 2011." Gothenburg: Nordicom-Sweden, 2011. Jan. 3, 2013. <http://www.nordicom.gu.se/common/stat_xls/2353_10525_English_Mbar_2011.pdf>.

"ProSievenSat.1 Media AG Company History 2006–2010." ProSievenSat.1 Media AG, 2010. June 12, 2012. <http://en.prosiebensat1.com/en/company/prosiebensat1-media-ag/history/2006-2010>.

## 3. Websites

"CANAL+ PROFILE C MORE ENTERTAINMENT AB Company presentation." *CANAL+*, 2013. Jan. 3, 2013. <http://press.canalplus.se/page.asp?guid=2E38E66F-2881-2844CB-A38B-59A0433B6954>.

# Sweden—Data Summaries

ELI NOAM AND PAUL MUTTER

ALTHOUGH SWEDEN'S SMALL POPULATION limits the size of its media groups, the country's concentration indices are among the lowest internationally. Pooled overall sector HHI, for example, decreased from 1,406 in 2004 to 1,232 in 2008 (Table 13.15).

The public broadcaster Sveriges (SVT) is heavily subsidized, but it has been losing market share to commercial operators as they gained an easier licensing system for entry. SVT's market share fell from 40% in 2004 to 26%, with increases in market share for Bonnier, ProSiebenSat.1 (Germany/US), and Kinnevik. SVT now holds 9.7% of the content media market (Table 13.16).

Bonnier is the largest content media producer with a 30% share. It is an international print media group, active in Russia, Finland, the United States, and Israel, and also has investments in film production. In Sweden, Bonnier is large in TV broadcasting (29.8%) and magazines (32.9%).

The publicly owned TeliaSonera—also partly owned by the Government of Finland because it operates across both countries—is the largest company in Sweden by power index and share of the national market. It accounts for 45.9% of the platform market, well ahead of its nearest competitors, Telenor and Kinnevik. Kinnevik, which has holdings in six of the 13 industries surveyed, is the most diversified media group in the country, holding the second largest share of platform media (19.6%) and the third largest in content (13%). ComHem is by far the largest cable TV platform company with a huge market share and with telecom and ISP provision, too. It was acquired in 2011 by BC Partners, a private equity firm in London (Table 13.17).

Table 13-15. National Media Industries Concentration in Sweden

| | 2004/5 | | 2011 OR MOST RECENT | | % CHANGE ANNUAL AVERAGE | |
|---|---|---|---|---|---|---|
| | COMPANY POWER INDEX IN COUNTRY | COMPANY SHARE OF THE OVERALL NATIONAL MEDIA MARKET (%) | COMPANY POWER INDEX IN COUNTRY | COMPANY SHARE OF THE OVERALL NATIONAL MEDIA MARKET (%) | COMPANY POWER INDEX IN COUNTRY | COMPANY SHARE OF THE OVERALL NATIONAL MEDIA MARKET (%) |
| TeliaSonera (public: 37% Sweden, 13% Finland) | 2,141.6 | 32 | 1,256.5 | 25 | –7 | –1.2 |
| Bonnier | 262.5 | 10.1 | 488 | 13.8 | 14.31 | 0.6 |
| Investment AB Kinnevik | 288.5 | 12.2 | 374.7 | 16.4 | 5.0 | 0.7 |
| Sveriges (public) | 292.2 | 5.6 | 204 | 4.5 | –5.0 | –0.2 |
| Com Hem (UK) | 52.1 | 2.9 | 123.1 | 5 | 22.7 | 0.3 |
| Telenor (Norway) | 146 | 7.8 | 110.1 | 7.4 | –4.1 | –0.07 |
| Schibsted (Norway) | 40.1 | 2.7 | 18.2 | 1.8 | –9.2 | –0.2 |
| Teracom (public) | 19.5 | 1.1 | 30.6 | 1.5 | 9.4 | 0.1 |
| ProSiebenSat.1 (Germany, US owned) | 6.3 | 0.9 | 15.5 | 1.4 | 24 | 0.09 |
| Egmont (Denmark) | 9.0 | 0.9 | 10.2 | 1.0 | 2.3 | 0.005 |
| Forma Publishing Group | 3.2 | 0.5 | 6.5 | 0.9 | 17 | 0.06 |
| Media Concentration Index | 2004/5 | | 2011 or Most Recent | | % Change Annual Average | |
| Total Revenue: Nat'l Media Industry (mil US$) | 15,350 | | 19,500 | | 4.5 | |
| Total Voices (*n*) | 62 | | 54 | | –2.2 | |
| Net Voices (*n*) | 38 | | 30 | | –3.5 | |

| | | | |
|---|---|---|---|
| Public Ownership (%) | 18.6 | 15.2 | –0.6 |
| Foreign Ownership (%) | 20.4 | 21.4 | 0.2 |
| C4 Average—Weighted | 82 | 86 | 0.6 |
| HHI Average—Weighted | 3,389 | 2,881 | –2.5 |
| C1 Average—Weighted | 47.4 | 44 | –0.01 |
| Noam Index Average—Weighted | 675 | 1,120 | 11.0 |
| Pooled Overall Sector C4 | 62.0 | 62.6 | 0.09 |
| Pooled Overall Sector HHI | 1,406 | 1,232 | –2.1 |
| Pooled Overall Sector Noam Index | 133 | 223 | 11.2 |
| Market Share of Top Ten Companies: Nat'l Media Industry (%) (Pooled C10) | 80.0 | 83.5 | 0.6 |
| National Power Index | 3,351 | 2,887 | –2.3 |

Table 13-16. Top Content Media Companies in Sweden

| | 2004/5 | | 2011 OR MOST RECENT | | % CHANGE ANNUAL AVERAGE | |
|---|---|---|---|---|---|---|
| | COMPANY POWER INDEX IN COUNTRY | COMPANY SHARE OF THE NATIONAL CONTENT MEDIA MARKET (%) | COMPANY POWER INDEX IN COUNTRY | COMPANY SHARE OF THE NATIONAL CONTENT MEDIA MARKET (%) | COMPANY POWER INDEX IN COUNTRY | COMPANY SHARE OF THE NATIONAL CONTENT MEDIA MARKET (%) |
| Bonnier | 607 | 23.3 | 1,052.3 | 30 | 12.2 | 1.1 |
| Sveriges (Public) | 676 | 13 | 440 | 9.7 | −5.8 | −0.5 |
| Investment AB Kinnevik | 73.4 | 5.9 | 248.3 | 13 | 40 | 1.2 |
| Schibsted (Norway) | 94.4 | 6.3 | 39.2 | 3.8 | −9.7 | −0.4 |
| Com Hem (BC Partners, UK) | 26.0 | 1.1 | 75.4 | 2.0 | 32 | 0.1 |
| ProSiebenSat1 (Germany/US) | 14.6 | 2 | 33.4 | 3.0 | 21.3 | 0.2 |
| Teracom (public) | 15 | 0.8 | 22 | 1.0 | 7.7 | 0.04 |
| Media Concentration Index | 2004/5 | | 2011 or Most Recent | | % Change Annual Average | |
| Public Ownership (%) | 13.9 | | 11 | | −0.5 | |
| Foreign Ownership (%) | 13.8 | | 15.3 | | 0.2 | |
| C4 Average—Weighted | 67 | | 80.2 | | 2.2 | |
| HHI Average—Weighted | 1,819 | | 2,459 | | 5.9 | |
| C1 Average—Weighted | 33 | | 39 | | 0.01 | |
| National Power Index | 1,731 | | 2,427 | | 6.7 | |

Table 13-17. Top Platform Media Companies in Sweden

| | 2004/5 | | 2011 OR MOST RECENT | | % CHANGE ANNUAL AVERAGE | |
|---|---|---|---|---|---|---|
| | COMPANY POWER INDEX IN COUNTRY | COMPANY SHARE OF THE NATIONAL PLATFORM MEDIA MARKET (%) | COMPANY POWER INDEX IN COUNTRY | COMPANY SHARE OF THE NATIONAL PLATFORM MEDIA MARKET (%) | COMPANY POWER INDEX IN COUNTRY | COMPANY SHARE OF THE NATIONAL PLATFORM MEDIA MARKET (%) |
| TeliaSonera (public: 37% Sweden, 13% Finland) | 3,773.5 | 56.4 | 2,335.4 | 46 | –6.4 | –1.7 |
| Investment AB Kinnevik | 452.4 | 17 | 484 | 19.6 | 1.2 | 0.4 |
| Telenor (Norway) (public) | 248.4 | 13.2 | 198 | 13.2 | –3.4 | 0.005 |
| Com Hem (BC Partners, UK) | 72 | 4.2 | 164.4 | 7.5 | 21.4 | 0.5 |
| Media Concentration Index | 2004/5 | | 2011 or Most Recent | | % Change Annual Average | |
| Public Ownership (%) | 22.1 | | 19 | | –0.5 | |
| Foreign Ownership (%) | 25.4 | | 27.0 | | 0.2 | |
| C4 Average—Weighted | 93.9 | | 90.4 | | –0.6 | |
| HHI Average—Weighted | 4,585 | | 3,246 | | –4.9 | |
| C1 Average—Weighted | 58 | | 48 | | –2 | |
| National Power Index | 4,586 | | 3,284 | | –4.73 | |

# 14

# Media Ownership and Concentration in Switzerland

PATRICK-YVES BADILLO AND DOMINIQUE BOURGEOIS

## INTRODUCTION

The Swiss media landscape reflects Switzerland's society. Switzerland is a multilingual and multicultural society with four official languages—German, French, Italian, and Romansch—and about 22% of the population are not Swiss citizens. In consequence, the main broadcaster SRG SSR offers three German-speaking TV channels, two French-speaking TV channels, and two Italian-speaking TV channels, as well as some broadcast programs in the Romansch language. Moreover, many foreign TV channels are accessible in most Swiss regions in their respective languages. As has been observed, Switzerland is characterized by a high penetration of all the types of media, while "on the other hand, the time of use of the media remains moderate."[1]

There are major debates in Switzerland concerning changes in regulation. The public monopoly over radio and TV broadcasting was abolished in the 1980s, but many questions remain regarding the direction that SRG SSR (the main audiovisual group, which is now a non-profit-making association with a public remit) should take and the means to develop commercial broadcasting. Telecom regulation in Switzerland was modified very late compared to other European countries: the telecom market was opened to competition only in 1998 by the Telecommunication Law of April 30, 1997, and there are calls for further revision of that law (which was last revised in 2006).

1. Cornu, Daniel, Borruat, Régis. "Structure des médias suisses," in Cornu, Daniel, et al., *Les Médias en Suisse, Structure et audience*. 7th ed. Lausanne: FCJ/CRFJ, 2012, 9 (translation).

The print press also faces new challenges: in Switzerland the press benefits from preferential postal prices for local and regional newspapers, and from a reduced VAT rate for print publications. This assistance was estimated at US$98.2 million (66.8 million euros) in 2008, which is far less than in other European countries: for example, for 2008 the total amount was estimated to be equal to US$1.9 billion (1.3 billion euros) in France and US$1.8 billion (1.2 billion euros) in Italy.[2] The debate about the role of the state emerged because of the ongoing economic deterioration of the paid-subscription newspapers sector.

And although there are some debates over concentration, media concentration is not considered by many stakeholders to be a major issue in Switzerland: the Federal Council in 2010 rejected further regulation, saying that the sector was sufficiently self-regulated.[3] Moreover, concentration is correlated with the market size: as Switzerland is a small country, high concentration indices should be expected. We will observe that, in general, concentration is relatively low, especially if we take into account this important specificity.

## PRINT MEDIA

Print media is seen as a key pillar of Swiss democracy. Having a wide variety of newspapers has long been viewed as part of political pluralism in such a small country with four official languages. According to the media monitor REMP:

> Switzerland is a press country. The only media database of the Association of Swiss Advertising Companies (ASSP) takes into account 436 newspapers, in the daily, regional weekly and Sunday press category. 347 titles, respectively 79.6%, are published in German, seventy-four (17%) are published in French and thirteen (3%) are written in Italian. Two titles (0.5%) are published in Romansch. This unique supply, in international comparison, is complemented by more than 2,333 general public, financial, economic, thematic and specialized press products.[4]

This can mainly be explained by the high level of print readership in Switzerland: in 2009, REMP reported that 92.4% of the adult Swiss population (14 years and up) regularly read newspapers, and 91.6% regularly read magazines. Switzerland has a high per capita print circulation for newspapers: there were 14 paid-subscription daily newspaper titles per one million inhabitants in 2008, twice as high as the figures for the United States, Germany, France, and Italy.[5]

In the past two decades, a trend for greater concentration has been observed at the ownership level. A few big publishing houses own most of Swiss print titles: Ringier, Tamedia, and NZZ (Neuen Zürcher Zeitung) Medien Gruppe. Nevertheless, there are still a large number of print titles and, from an editorial point of view, a diversity of opinions. That said, "content diversity is not insured by the multiplicity of titles," since several of the smaller newspapers mainly reproduce news agencies' content, and there are editorial collaborations at the national level.[6] On the other hand, various titles belonging to the same company correspond to specific contents for selected audiences.

---

2. Swiss Federal Council, "Garantir la diversité de la presse: Rapport du Conseil fédéral en réponse au postulat Fehr 09.3629 et au postulat de la Commission des institutions politiques du Conseil national (CIP-CN) 09.3980." Bern: OFCOM, June 29, 2011. <http://www.bakom.admin.ch/themen/radio_tv/01153/01156/03479/index.html?lang=fr&download=NHzLpZeg7t,lnp6I0NTU042l2Z6ln1ae2IZn4Z2qZpnO2Yuq2Z6gpJCDeoN3fWym162epYbg2c_JjKbNoKSn6A-->. 23.

3. Ibid., p. 40.

4. Recherches et études des médias publicitaires—Researches and studies on advertising medias (REMP). *Presse écrite*. <http://www.wemf.ch/f/media/presse_ecrite.php>. Dec. 28, 2012 (translation).

5. World Association of Newspapers and News Publishers (WAN-IFRA). *World Press Trends 2010*. WAN-IFRA/ZenithOptimedia, 2010. <http://www.wan-ifra.org/reports/2010/07/23/world-press-trends-2010-edition>. Dec. 28, 2012.

6. Cornu, Daniel, Borruat, Régis, Léchot, Pierre-André & Nerini, Marlène. "Inventaire des médias suisses," in Cornu, Daniel, et al., *Les Médias en Suisse, Structure et audience*. 7th ed. Lausanne: FCJ/CRFJ, 2012. 30.

## Newspapers

We focus here the analysis only on daily newspapers. The number of paid-content newspapers that are published at least four times a week in Switzerland declined by 40% between 1985 and 2006, falling from 124 to 76 titles.[7] At first, this decline was associated with an increase in the circulation of the existing titles: the average circulation of a Swiss daily newspaper was about 10,000 copies in 1985 and 16,000 in 2006. But with the emergence of free dailies since 2000 the situation has deteriorated for paid-content newspapers. Only the Sunday paid-content press has further developed. Concentration will be estimated hereafter only for daily newspapers, Sunday newspapers being excluded.

Tamedia's expanded market share is linked in particular to the development of the free newspapers *20 Minuten* (German) and *20 Minutes* (French) and to various acquisitions: Tamedia bought a 80% share of the Espace Media Group in 2007, it took the control of Zürichsee Presse AG and Zürcher Unterland Medien AG in 2010, and there was the merger between Tamedia and Edipresse (merging their Swiss businesses), which was completed in 2011. The increase in Ringier's market share since 2008 is due to the publication of the free daily *Blick am Abend*. The increased market share of Edipresse in 2008 is correlated with the free newspaper *Le Matin Bleu*, which was published for about a year after October 2005. Concentration of media ownership was very high in 2012 if we consider only the four first companies: the C4 ratio, which was around 55% in 2005, rose to 80% in 2012. The market share of Tamedia alone reached 44.3% in 2012. The HHI and the Noam Index increased as well since 2005: they more than doubled between 2005 and 2012, but in 2012 they stayed at a level that was not extremely high because many small companies were still present in the market in 2012 (Table 14.1).

Concentration in daily newspapers ownership has increased in recent years. The concentration indices are likely to increase further since 2012, because of the economic situation of the daily press. The recent increase in concentration is explained by three correlated main trends: the strong development of free dailies, a decline in advertising revenues,[8] and some important mergers. Tamedia is now the leading publisher of newspapers: the 2011 Tamedia-Edipresse Suisse merger gave birth to a big Swiss publishing conglomerate with an annual turnover of about US$1.2 billion (CHF1.1 billion). It publishes the free daily newspapers *20 Minuten/20 Minutes*—with a print circulation of 700,000 copies in 2011—and 12 paid-content daily newspapers, which also circulate approximately 700,000 copies nationally. Ringier is the second-leading publisher of daily newspapers: it owns the daily tabloid newspaper *Blick*, the free evening sheet *Blick am Abend*, and the leading Sunday edition, *Sonntagsblick*. NZZ Medien Gruppe is an important group mainly in the German-language market of Switzerland. Of course, in such a small country as Switzerland, increasing ownership concentration is inescapable because economic constraints and the growth of online media services have drastically changed the press landscape. But as already stated, the number of titles per capita remains exceptionally high in comparison with other European countries.

## Book Publishing

The Swiss book market is small compared to the press market: the press "budget" of the population was US$1.5 billion (CHF1.9 billion) in 2004, US$1.7 billion (CHF1.8 billion) in 2008, and US$1.6 billion (CHF1.7 billion) in 2010.[9] In contrast, the book "budget" for the Swiss population was US$585 million (CHF722 million)

7. Edi Kradolfer. *La diversité de la* presse *en Suisse, Un aperçu*. Neuchâtel: Office fédéral de la statistique (OFS), 2007. 8. The data here (source: Edi Kradolfer) are more restrictive than those given in the previous page, based on all categories of newspapers (source: REMP).

8. Amez-Droz, Philippe. "L'économie des médias en Suisse," in Cornu, Daniel, and et al., *Les Médias en Suisse, Structure et audience*. 7th ed. Lausanne: FCJ/CRFJ, 2012. 11.

9. Sources: Schweizer Medien. *Medienbudget.ch 2005* and *Medienbudget.ch 2011*. <http://www.schweizermedien.ch/fileadmin/schweizermedien/brancheninfos/medienbudget/2005_Praesentation_Medienbudget.pdf> <http://www.schweizermedien.ch/fileadmin/schweizermedien/brancheninfos/medienbudget/2011_Praesentation_Medienbudget.pdf>. Dec. 28, 2012.

## Table 14-1. Daily Newspapers (Market Shares by Circulation)

| | 2005 | 2008 | 2012 |
|---|---|---|---|
| Tamedia (Coninx Family) | 26.1 | 35.5 | 44.3 |
| Espace Media Groupe | 7.9 | | |
| Edipresse | 9.6 | 13.5 | |
| Ringier (Ringier Family) | 10.0 | 13.6 | 18.2 |
| Neue Zürcher Zeitung (NZZ Medien Gruppe) | 9.0 | 11.5 | 12.3 |
| LZ Medien Gruppe[1] | 4.6 | | |
| AZ Medien Gruppe | 6.6 | 5.9 | 5.7 |
| Südostschweiz Presse | 4.9 | 3.7 | 4.1 |
| Basler Zeitung (BZ Medien Gruppe) | 3.4 | 2.7 | 2.3 |
| Groupe St-Paul | 2.3 | 1.9 | 2.2 |
| Edition Suisse Holding (Philippe Hersant Group) | 1.9 | 1.4 | 2.7 |
| Groupe Rhône Media[2] | 1.5 | 1.3 | |
| Zürichsee Medien[3] | 1.6 | 1.3 | |
| Zürcher Oberland Medien | 1.4 | 1.1 | 1.1 |
| Gassman | 1.4 | 1.1 | 1.1 |
| Fondazione per il Corriere del Ticino | 1.4 | 1.1 | 1.2 |
| Ziegler Druck- und Verlags- | 1.3 | 1.0 | 1.1 |
| Giacomo Salvioni | 1.2 | 0.9 | 1.1 |
| Others | 4.1 | 3.5 | 2.6 |
| Total Revenue (mil CHF)[4] | Ne* | 1,865 | 1,460 |
| Total Revenue (mil US$) | Ne* | 1,727 | 1,558 |
| C4 | 54.6 | 74.0 | 80.5 |
| HHI | 1,137 | 1,824 | 2,520 |
| *N* (>1%) | 18 | 15 | 13 |
| Noam Index | 282 | 471 | 699 |

*Source*: Our calculations[5] based on data from REMP <http://www.remp.ch/> for market shares.

*Ne: not evaluated.

[1]NZZ Medien Group took a majority of the capital in LZ Medien Gruppe in 2007.

[2]In 2010 Edition Suisse Holding (Philippe Hersant Group) took a majority of the capital in Rhône Media, which owns the daily *Le Nouvelliste*.

[3]In 2010 Tamedia took the control of Zürichsee Presse AG, which was previously controlled by Zürichsee Medien AG.

[4]To estimate the value of the market here, we added the advertising revenues (source: Fondation Statistique Suisse en Publicité) and the national "budget," estimated for the population in the country (source: Schweizer Medien).

[5]To calculate the market shares, circulation data for all the companies (source: REMP; sold copies for paid-content newspapers titles) are taken into account, Sunday newspapers being excluded; when a company owns the majority of a publisher's capital, the market share is based on the total circulation of the title. To calculate the HHI and the Noam Index, all the companies with market shares over 1% are taken into account. Data for 2005, 2008, and 2012 correspond to the period from July 2004 to June 2005, from July 2007 to June 2008, and from July 2011 to July 2012, respectively.

in 2004, US$567 million (CHF910 million) in 2008, and US$843 million (CHF878 million) in 2010.[10]

It has not been possible to collect separate data specifically for book publishing in Switzerland, so we can only speak in generalities. First, there are three markets because of the three main languages (Romansch is not widely spoken enough to factor in here). Second, foreign publishing companies dominate the market: it is generally estimated that 80% or 90% of the books circulating in the French-language market are imported from France. The French group Lagardère was present in distribution and bookselling until 2012–2014, but withdrew from these Swiss activities by selling the companies Office du livre de Fribourg (OLF) and Diffulivre, which are the main distribution companies in the French-speaking parts of Switzerland, and the booksellers Payot and Naville.[11]

Swiss publishing companies sell nearly 80% of the books they produce in neighboring countries. In 2008, there were 406 Swiss publishing houses, of which 278 were in German, 121 in French, and seven in Italian.[12] In the German-language market, the publishing houses that play a significant role are the subsidiaries of publishers in Germany, with the exception of Diogenes, Switzerland's biggest independent book publisher.[13]

Between 1998 and 2008, the number of the Swiss publishing houses fell by 20%. The number of bookshops also decreased, from 633 in 1998 to 580 in 2008. The distribution market is highly concentrated in French-speaking Switzerland and has led to high prices, as the four main companies—Office du livre de Fribourg (OLF), Interforum Suisse SA, Diffulivre SA, and Servidis SA—"decide the prices of books sold to bookshops."[14] The Planeta Group (Spain) is the owner of Editis, the French publishing group that runs Interforum. Servidis belongs to Slatkine, a Geneva publishing house, and to the Editions du Seuil (France). In the German-language market of Switzerland, most of the books come from either Germany or Austria; booksellers sometimes order books directly from German or Austrian publishing houses.

There are debates about statistics concerning the book industry in Switzerland. A study by the University of Zürich gave numbers based on different sources. It indicated that the book publishing industry was worth US$785 million (CHF974 million) in 2004, whereas in 1996 it had been worth US$810 million (CHF1 billion).[15] The same study estimated that the overall turnover of all publishing houses in Switzerland in 2004 stood at US$890 million (CHF1.1 billion) and that the overall turnover at the reduced VAT rate for books was US$317 million (CHF393 million) in 2004.[16, 17]

Let us note a debate concerning the regulation of book prices: a regulation in favor of a uniform retail price (a regulation comparable to the French one) was rejected by a popular national referendum that took place in March 2012.

10. Ibid.

11. All these companies are now Swiss companies.

12. FranceLivre BookFrance, le portail international du livre français. <http://www.francelivre.org/>. Apr. 15, 2012.

13. Ibid.

14. Source: Federal Assembly—the Swiss Parliament—Votation populaire du 11 mars 2012, Services du parlement : Argumentaires contre, *04.430, Initiative parlementaire. Maitre Jean-Philippe, Réglementation du prix du livre, Argumentaires contre*, 2012, 1. <http://www.parlament.ch/f/wahlen-abstimmungen/volksabstimmungen/volksabstimmungen-2012/abstimmung-2012-03-11/buchpreisbindung/Documents/04-430-argumentarien-contra-f.pdf>. Dec. 28, 2012 (translation).

15. Trappel, Josef and Uhrmann, Caroline. "Panorama du livre et de la littérature en Suisse sur mandat de l'Office fédéral de la culture." Zurich: University of Zürich, IPMZ (Institut für Publizistikwissenschaft und Medienforschung), June 16, 2006. 6. <http://www.bak.admin.ch/dokumentation/studien/index.html?lang=fr&download=NHzLpZeg7t,lnp6I0NTU042l2Z6ln1ae2IZn4Z2qZpnO2Yuq2Z6gpJCDdn14gGym162epYbg2c_JjKbNoKSn6A-->. Dec. 28, 2012.

16. Ibid., p. 10.

17. Differences between data may be explained by the fact that different sources use different estimation methods. Bookshops that are part of department stores, for example, are not taken into account in company censuses, while their turnovers do appear in VAT statistics. But VAT statistics do not integrate books ordered abroad.

## Magazine Publishing

Here we discuss the *presse grand public,* which does not include regional weekly newspapers, trade journals, or specialized magazines. Market share calculations for magazines include only Swiss consumer magazines and not foreign offerings available in-country.[18] Though many foreign magazines are available on the Swiss market, there is no available data on their circulation. So concentration indices for Swiss magazines should be interpreted in terms of competition among Swiss publishers only. In 2005 and 2008, the C4 ratio was around 75–77% and it reached 87% in 2012, which is very high. The HHI was under 1,800 in 2005 and 2008, which was indicative of a moderately concentrated market, but it climbed to 2,200 in 2012. Concentration has increased, and the sector is now dominated by Ringier, Axel Springer, and Tamedia. Ringier was the market leader for many years, but it is now in competition with Axel Springer and Tamedia (Table 14.2).

### Table 14-2. Magazine Publishing (Market Shares by Circulation)

| (A) | 2005 | 2008 | 2012 |
|---|---|---|---|
| Ringier (Ringier Family) | 27.8 | 23.0 | 26.9 |
| Axel Springer Suisse (Germany) | 1.7 | 26.8 | 29.2 |
| Jean Frey[1] | 18.2 | | |
| Tamedia (Coninx Family) | 10.2 | 9.0 | 23.7 |
| Edipresse | 19.6 | 18.9 | |
| NZZ Medien Gruppe | 6.2 | 6.4 | 7.4 |
| Reader's Digest (US) | 5.4 | 5.2 | 4.2 |
| CT Cinetrade AG | 2.1 | 2.2 | 3.0 |
| Teleradio 7 | 2.7 | 2.8 | 2.6 |
| Weltwoche Verlag (Roger Köppel)[2] | | 2.7 | 2.6 |
| Editions Plus | 2.5 | 1.2 | Data not available |
| Others | 3.5 | 1.8 | 0.3 |
| Total Revenue (mil CHF)[3] | Ne* | 475 E | 581 E |
| Total Revenue (mil US$) | Ne* | 440 E | 620 E |
| C4 | 75.8 | 77.7 | 87.2 |
| HHI | 1,683 | 1,775 | 2,232 |
| *N* (>1%) | 10 | 10 | 8 |
| Noam Index | 532 | 561 | 789 |

*Source*: Our calculations[4] based on data from REMP <http://www.remp.ch/> for market shares.

*Ne: Not evaluated; E: estimate.

[1] Jean Frey was purchased by Axel Springer in 2007.

[2] Until 2006, the magazine *Die Weltwoche* was published by Jean Frey.

[3] To estimate the value of the market here, we added the advertising revenues (source: Fondation Statistique Suisse en Publicité) and the national "budget," estimated for the population in the country (source: Schweizer Medien). Due to some statistical issues it is an estimate (E).

[4] Only Swiss consumer magazines are taken into account. To calculate the market shares, circulation data (source: REMP) for all the companies are taken into account; when a company owns the majority of a publisher's capital, the market share is based on the total circulation of the title.

18. Moreover free magazines distributed by some firms like Migros or Coop are not taken into account.

## AUDIOVISUAL MEDIA

The Swiss audiovisual market is dominated by the broadcaster SRG SSR,[19] which was founded in 1931. In 1991, SRG SSR was restructured as an association with four regional companies: SRG.D (broadcasted in German), RTSR (broadcast in French), Corsi (broadcast in Italian), and SRG.R association (broadcast in Romansch). The regional companies SRG.D and RTSR are each made up of seven member companies. This structure ensures the involvement of many listeners and viewers in each Swiss region. SRG SSR is financed mainly by license fees (accounting for 70%–75% of its income). For a long time SRG SSR was a monopoly, but in 1983 local private and commercial radio channels were authorized. Now private local radio and television companies in Switzerland are entitled to 4% of license fees annually. The revised Federal Radio and Television Act (RTVA), adopted by parliament in 2006 and in force since April 2007, not only has ensured a continued strong public service from the SRG SSR, but also has strengthened commercial radio and television broadcasters. Concerning SRG SSR a convergence process was launched in 2009: now radio and television are amalgamated in each language region.

To analyze concentration in audiovisual media industries, we consider the three main linguistic regions in Switzerland separately. Although SRG SSR is the major Swiss operator in radio and television with 18 radio stations and 7 TV channels, Swiss consumers have access to many European and international channels.

### Radio

The market share of SRG SSR diminished after the end of the public monopoly in the 1980s, but in the past decade, it has grown in all three of Switzerland's language markets, and it is now particularly large in the Italian-language market. The market shares for commercial Swiss radio broadcasters are small, as are the market shares for foreign stations (Table 14.3).

In 2012, the market share for foreign radio channels was only 4.4% in the German-language market, 6.5% in the Italian-language market, and 10.3% in the French-language market. Swiss commercial radio is more developed in the French and German markets (23.8% and 30.2%, respectively) than in the Italian one (11.2%). Each commercial radio channel has a very small market share: for example, in 2012 the market share of Radio 24, which is the largest commercial radio broadcaster in Switzerland and has over 300,000 regular daily listeners, was about 3% in the German-language market.

Concentration is very high in the Italian-language market: in 2012, the C4 ratio was around 95% and the HHI around 6,900. For the same year, the C4 ratios for the other language markets were high but fewer than 80% and the HHI was less than 4,500. In general concentration is high because of SRG SSR's large market share. We provide estimations of the concentration indices at the national level for 2008 and 2012 by taking into account the number of regional listeners (Table 14.4).

### Television

In Switzerland most homes receive radio and television programs by cable. Moreover, terrestrial television has been fully upgraded to digital terrestrial television (DTT). If we look at TV audience shares in the entire television market (combining broadcast TV, cable TV, satellite TV), in the long run the market share of SRG SSR has diminished. If we consider only the last 10 years, though, the situation is less clear: we can say that the market share of SRG SSR tends to be around one-third of the total national audience (slightly higher in the Italian-speaking regions of Switzerland) (Table 14.5).

Concentration is highest in the Italian-language market: in 2012, the C4 ratio was around 80% and the HHI around 2,300, while for the same year the C4 ratios for the other language markets were a little over 60% and the HHI around 1,300. This higher concentration for the Italian-language market is linked to the

19. In French: Société suisse de radiodiffusion et television (SSR); in German: Schweizerische Radio- und Fernsehgesellschaft (SRG); in Italian: Società svizzera di radio-televisione (SSR); in Romansch: Societad svizra da radio e television (SSR).

## Table 14-3. Radio (Market Shares by Audience)[1]

| SWISS LANGUAGE MARKETS (RADIO) | | 2001 | 2004 | 2008 | 2012 |
|---|---|---|---|---|---|
| German-Language Market | SRG SSR | 62.8 | 65.5 | 68.2 | 65.4 |
| | Public foreign stations | 5.4 | 5 | 3.8 | 2.9 |
| | *ARD (Germany)* | *4.3* | *4.2* | *3.4* | *2.7* |
| | *ORF (public) (Austria)* | *1.1* | *0.8* | *0.4* | *0.2* |
| | Private stations in Switzerland | 23.1 | 22.4 | 23.8 | 26.2 |
| | Sepp Trütsch (Eviva) | *0.9* | *1.1* | *1.1* | *1.3* |
| | Belcom AG (Radio 24) | *2.9* | *2.9* | *2.9* | *3* |
| | Ringier AG (Energy Zürich, EnergyBern) | *3.7* | *3* | *2.8* | *3* |
| | Schawinski Roger (Radiol) | _ | _ | *0.9* | *1.2* |
| | BT Holding AG (Argovia) | *2.1* | *2.2* | *1.9* | *2.4* |
| | LZ Medien Holding AG (Pilatus) | *1.8* | *1.8* | *1.7* | *2* |
| | Beelk Holding AG (Sunshine) | *1* | *1.2* | *1* | *1* |
| | Triner Hugo (Central) | *1* | *1.2* | *1* | *1* |
| | St Galler Tagblatt (FM1) | _ | _ | *1.5* | *1.9* |
| | Hagemann Matthias (Basilisk) | *1.4* | *1.2* | *1* | *1.2* |
| | Förderverein Radio BeO (BeO) | *0.9* | *0.8* | *0.8* | *1* |
| | Südostschweiz Presse + Print (Grischa) | *1.4* | *0.6* | *0.6* | *0.6* |
| | Zürichsee Medien AG (Zürisee, Capital FM) | *2.2* | *2.5* | *2.5* | *2.9* |
| | AZ Medien AG (Radio 32) | *1.8* | *1.6* | *1.6* | *1.6* |
| | Heuberger Günter (Top) | *1.5* | *1.4* | *1.6* | *1.1* |
| | Others private | *0.5* | *0.9* | *0.9* | *1* |
| | Others | 8.7 | 7.1 | 4.2 | 5.5 |
| | C4 | 73.7 | 75.6 | 77.3 | 74.1 |
| | HHI | 4,010 | 4,350 | 4,703 | 4,334 |
| | *N* (>1%) | 14 | 13 | 14 | 15 |
| | Noam Index | 1,072 | 1,206 | 1,257 | 1,119 |
| French-Language Market | SRG SSR | 59.2 | 62.6 | 64.7 | 66.2 |
| | Private stations in Switzerland | 21.5 | 21.5 | 21.7 | 21.9 |
| | Sonal FM (Yes FM) | *1.6* | *1.8* | *1.6* | *1.6* |
| | Overshop Holding SA (One FM, LFM) | *5* | *5.5* | *6* | *5.8* |
| | Unicast (Rouge FM) | *2.9* | *2.7* | *2.8* | *3.3* |
| | Association du Chablais (Radio Chablais) | *1.7* | *1.7* | *1.5* | *1.6* |
| | RTN | *1.8* | *2.6* | *2.8* | _ |
| | RJB | *1.7* | *1.1* | *0.9* | _ |
| | RFJ | *2.6* | *1.8* | *1.7* | _ |
| | BNJ Media Holding SA (RTN, RJB, RFJ) | _ | _ | _ | *4.8* |
| | Rhône FM SA | *2.5* | *2.4* | *2.3* | *2.3* |
| | Damien Piller (Fribourg) | *1.7* | *1.9* | *2.1* | *2.5* |
| | Radio France (public) (France) | 5.4 | 5.2 | 4 | 3.7 |
| | NRJ Group (NRJ, Nostalgie) (private) (France) | 5.6 | 4.7 | 3.9 | 3.9 |

*continued*

Table 14-3. *continued*

| SWISS LANGUAGE MARKETS (RADIO) | | 2001 | 2004 | 2008 | 2012 |
|---|---|---|---|---|---|
| | Others | 8.3 | 6 | 5.7 | 4.3 |
| | C4 | 75.2 | 78 | 78.6 | 79.6 |
| | HHI | 3,626 | 4,032 | 4,236 | 4,496 |
| | *N* (>1%) | 12 | 12 | 11 | 10 |
| | Noam Index | 1,047 | 1,164 | 1,292 | 1,422 |
| Italian-Language Market | SRG SSR | 79.8 | 82.9 | 84 | 82.3 |
| | Private stations in Switzerland | 4.2 | 4.5 | 6.7 | 10.4 |
| | Timedia Holding SA (R3iii) | *2.2* | *2.7* | *4.6* | *6.3* |
| | Fondazione UCGBC (Fiume Ticino) | *2* | *1.8* | *2.1* | *4.1* |
| | RAI (public) (Italy) | 2.2 | 1.4 | 1 | 0.6 |
| | RTO (private) (Italy) | 3 | 2.4 | 1.9 | 2.1 |
| | Others | 10.8 | 8.8 | 6.4 | 4.6 |
| | C4 | 87.2 | 89.8 | 91.5 | 94.8 |
| | HHI | 6,391 | 6,891 | 7,086 | 6,855 |
| | *N* (>1%) | 6 | 6 | 6 | 5 |
| | Noam Index | 2,609 | 2,813 | 2,893 | 3,066 |
| Total Revenue (mil CHF)[2] | | | 575 | 602 | 648 |
| Total Revenue (mil US$) | | | 463 | 553 | 691 |

*Source*: Our calculations based on data from Mediapulse AG. <http://www.mediapulse.ch> for market shares.

[1] The market shares are calculated on a listeners' panel sampling (number of minutes of listening by the panel). Comparisons between 2001 and other years have to be made with caution due to changes in the data collection.

[2] This revenue figure is an estimate by adding advertising revenues, including revenues from sponsoring (source: Fondation Statistique Suisse en Publicité) to the radio "budget" estimated for all radio consumers in Switzerland (source: Schweizer Medien, op. cit.; only the budget for "content" is taken into account, not expenses for radio devices).

large market shares of RAI and the channels owned by Mediaset: *Canale 5*, *Italia 1*, and *Rete 4*.

In any case, all the estimated concentration indices are not at high levels. Moreover they tend to diminish between 2008 and 2012. If we consider Switzerland at a whole, the diversity of TV channels is very strong, with public and private Swiss channels and many foreign channels; only SRG SSR has a large market share at the national level. We provide estimations of the concentration indices at the national level for 2008 and 2012 by taking into account the number of regional viewers (Table 14.6).

Concentration appears low at the national level. Of course, the concentration indices are estimated, but it is obvious that the TV market is diversified because of the linguistic and geographic situation of Switzerland. The same conclusion is not applicable for the radio market, because the market share of SRG SSR for radio is large and commercial radio broadcasters have small market shares.

## Film

In the long run, the number of movie viewers has decreased in Switzerland, although for more than two decades, attendance has stayed at a rather high level, between 14 and 19 million per year.

Most of the films distributed in Switzerland are foreign films, in particular films produced in the United States by the Hollywood majors. For the last decade, the market share for Swiss films has fluctuated between 2.3% (in 2001) and 9.5% (in 2006); if we add films that have

Table 14-4. Radio (Estimated Audience Shares by %)

| | GERMAN-LANGUAGE MARKET | | FRENCH-LANGUAGE MARKET | | ITALIAN-LANGUAGE MARKET | | ALL (NATIONAL) | |
|---|---|---|---|---|---|---|---|---|
| | 2008 | 2012 | 2008 | 2012 | 2008 | 2012 | 2008 | 2012 |
| SRG SSR | 68.2 | 65.4 | 64.7 | 66.2 | 84 | 82.3 | 68.08 | 66.35 |
| ARD (Public Germany) | 3.4 | 2.7 | | | | | 2.44 | 1.94 |
| Belcom AG (Radio 24) | 2.9 | 3 | | | | | 2.08 | 2.16 |
| Ringier AG (Energy Zürich, EnergyBern) | 2.8 | 3 | | | | | 2.01 | 2.16 |
| Zürichsee Medien AG (Zürisee, Capital FM) | 2.5 | 2.9 | | | | | 1.79 | 2.09 |
| Overshop Holding SA (One FM, LFM) | | | 6 | 5,8 | | | 1.42 | 1.37 |
| Others | 20.2 | 23.0 | 29.3 | 28.0 | 16.0 | 17.7 | 22.17 | 23.94 |
| Audience (1,000)[1] | 4,505.5 | 4,742.9 | 1,490.3 | 1,552.1 | 281 | 297.2 | 6,276.8 | 6,592.2 |
| C4 | 77.3 | 74.1 | 78.6 | 79.6 | 91.5 | 94.8 | 74.6 | 72.6 |
| HHI | 4,703 | 4,334 | 4,286 | 4,496 | 7,086 | 6,855 | 4,654 | 4,422 |
| *N* (>1%) | 14 | 15 | 11 | 10 | 6 | 5 | 6 | 6 |
| Noam Index | 1,257 | 1,119 | 1,292 | 1,422 | 2,893 | 3,066 | 1,900 | 1,805 |

*Source*: our calculations based on data from Mediapulse AG, <http://www.mediapulse.ch>.

[1] The number of listeners is based on a "universe" estimate, i.e., all Swiss citizens 15 or more years old who live in a home with radio (source: Mediapulse AG).

Table 14-5. Combined Television Network Industries: Broadcast TV Networks, TV Syndicators, Cable Channels, Pay-TV (Estimated Market Shares by Audience)

| SWISS LANGUAGE MARKETS | (B) | 1992 | 1996 | 2000 | 2004 | 2008 | 2012 |
|---|---|---|---|---|---|---|---|
| German-language market | SRG SSR | 31.0 | 34.5 | 33.1 | 35.2 | 34.8 | 30.0 |
| | ARD + ZDF + 3SAT (public) (Germany)[1] | 16.0 | 12.1 | 10.8 | 11.4 | 11.6 | 10.7 |
| | ARD | *8.0* | *6.1* | *5.6* | *5.7* | *5.5* | *4.7* |
| | ZDF | *8.0* | *5.9* | *5.2* | *5.7* | *6.1* | *6.0* |
| | ZDF | *7.0* | *5.0* | *4.3* | *4.8* | *5.0* | *4.7* |
| | 3SAT | *1.0* | *0.9* | *0.9* | *0.9* | *1.1* | *1.3* |
| | ORF1 + ORF2 (public) (Austria) | 8.0 | 7.8 | 6.5 | 6.3 | 4.9 | 3.5 |
| | RTL Group (Bertelsmann AG, Germany) | 14 | 14.6 | 11.9 | 11.7 | 9.5 | 11.5 |
| | ProSiebenSat.1 Media (Germany) | 9.0 | 15.7 | 11.6 | 11.2 | 10.7 | 9.7 |
| | Others | 22.0 | 15.3 | 26.1 | 24.2 | 28.5 | 34.6 |
| | All private Swiss TV (such as TeleZüri) | | | 7.3 | 4.0 | 5.0 | 6.9 |
| | C4 | 70.0 | 76.9 | 67.4 | 69.5 | 66.6 | 61.9 |
| | HHI | 1,558 | 1,857 | 1,531 | 1,671 | 1,574 | 1,253 |
| | *N* (>1%)[2] | | | | | | |
| | Noam Index | 697 | 831 | 685 | 747 | 704 | 560 |
| French- language market | SRG SSR | 36.0 | 35.0 | 35.0 | 32.0 | 32.0 | 29.0 |
| | France télévisions (Public French channels, such as France 2, France 3 . . .) | 20.0 | 21.0 | 17.0 | 18.0 | 17.0 | 13.0 |
| | TF1 (France) | 22.0 | 16.0 | 16.0 | 17.0 | 14.0 | 13.0 |
| | RTL Group (Bertelsmann AG, Germany) | 6.0 | 8.0 | 7.0 | 9.0 | 10.0 | 10.0 |
| | AB Group (France) | | | 5.0 | 4.0 | 3.0 | 2.0 |
| | Others | 16.0 | 20.0 | 20.0 | 20.0 | 24.0 | 33.0 |
| | C4 | 84.0 | 80.0 | 75.0 | 76.0 | 73.0 | 65 |
| | HHI | 2,216 | 1,986 | 1,844 | 1,734 | 1,618 | 1,283 |
| | *N* (>1%)[3] | | | | | | |
| | Noam Index | 1,108 | 993 | 825 | 775 | 724 | 574 |
| Italian-language market | SRG SSR | 33.0 | 33.6 | 34.0 | 37.2 | 35.4 | 37 |
| | RAI (public) (Italy) | 24.0 | 25.1 | 23.7 | 21.0 | 22.0 | 20 |
| | Mediaset (Berlusconi Family, Italy) | 20.0 | 29.3 | 28.4 | 26.8 | 25.6 | 22 |
| | Telecom Italia Media (Italy)[4] | | 2.0 | 1.7 | 1.8 | 1.3 | 2 |

Table 14-5. *continued*

| SWISS LANGUAGE MARKETS | (B) | 1992 | 1996 | 2000 | 2004 | 2008 | 2012 |
|---|---|---|---|---|---|---|---|
| | TicinoNews | | | 0.9 | 1.2 | 1.8 | 1 |
| | Others | 23.0 | 10.0 | 11.3 | 12 | 13.9 | 19 |
| | C4 | 77.0 | 90.0 | 87.8 | 86.8 | 84.8 | 81 |
| | HHI | 2,065 | 2,621 | 2,528 | 2,548 | 2,397 | 2,258 |
| | *N* (>1%)[5] | | | | | | |
| | Noam Index | 1,192 | 1,311 | 1,131 | 1,140 | 1,072 | 1,010 |
| Total Revenue (mil CHF)[6] | | | | | 1,351 | 1,479 | 1,633 |
| Total Revenue (mil US$) | | | | | 1,089 | 1,370 | 1,742 |

*Source*: Our calculations based on data from Mediapulse AG <http://www.mediapulse.ch> for market shares.[7]

[1] 3SAT is a German-language collaboration network that comprises ZDF and ARD (German), ORF (Austria), and Schweizer Fernsehen SF (SRG SSR, Switzerland) that is controlled by ZDF.

[2] The HHI and the Noam Index were calculated for only five firms; *N* (>1%) is not known.

[3] The HHI and the Noam Index were calculated for only four firms for the years 1992 and 1996, and for only five firms for 2000, 2004, 2008, and 2012; *N* (>1%) is not known.

[4] *LA7* is a private channel owned by Telecom Italia since 2001; before 2001, there were other owners and the channel was called *Telemontecarlo*.

[5] The HHI and the Noam Index were calculated for only three firms in 1992, four firms in 1996, and five firms for 2000, 2004, 2008, and 2012; *N* (>1%) is not known.

[6] To estimate the value of the market here, we added advertising revenues, including revenues from sponsoring (source: Fondation Statistique Suisse en Publicité; advertising revenues include the advertising revenues of all the Swiss channels and "foreign advertising windows," i.e., advertising windows operating on foreign TV channels), and the national TV "budget," estimated for all TV viewers in the country (source: Schweizer Medien; only the budget for "content" is taken into account, not expenses for TV devices or expenses for TV access).

[7] The market shares are calculated on a viewers' panel sampling (time of use by the panel. Source: Mediapulse AG). The concentration indices are calculated by taking into account the public channels together. The market shares of the two main German-language channels are indicated in the table, but the concentration indices are based on the sum of all public German-language channels. The HHI and Noam Index are estimated on the basis of the main market shares; *N* (>1%) is not known.

been coproduced with other countries, the share fluctuates between 3.5% (in 2001) and 9.7% (in 2006). There are supportive measures to the Swiss production or coproduction of films.

The main distributors are the Walt Disney Studios, United International Pictures (UIP), Warner Brothers, and 20th Century Fox. These four distributors hold more than 60% of the distribution market. The HHI was around 1,300 in the years 2009–2010, which indicates a moderate concentration, though the C4 ratio was quite high.

There are two main exhibitors in Switzerland: Pathé (UCS Group, France), with a market share (based on the number of admissions) of 25.8% in 2010, and Kitag (Kino-Theater AG, Switzerland), with a market share of 19.8%. Kitag belongs to CT Cinetrade AG; Swisscom, the main Swiss Telecom enterprise, has a 49% stake in CT Cinetrade AG. Other domestic exhibitors include Cinépel SA, Cinérive SA, and Liag-Capitol, though they have smaller market shares. For example, in 2010 Kitag represented 20,363 seats and Pathé 15,893 seats, while Cinépel controlled only 5,558 seats, Liag-Capitol 2,506, and Cinérive 2,288 (Table 14.7).

## TELECOMMUNICATIONS

### Wireline Telecom

The Swiss telecommunications market was opened to competition in 1998 by the Telecommunication Law of April 30, 1997. Swisscom

## Table 14-6. Combined Television Network Industries (Estimated Audience Shares by %)

| | GERMAN-LANGUAGE MARKET | | FRENCH-LANGUAGE MARKET | | ITALIAN-LANGUAGE MARKET | | ALL (NATIONAL) | |
|---|---|---|---|---|---|---|---|---|
| | 2008 | 2012 | 2008 | 2012 | 2008 | 2012 | 2008 | 2012 |
| SRG SSR | 34.8 | 30.0 | 32.0 | 29.0 | 35.4 | 37.0 | 34.2 | 30.1 |
| ARD + ZDF + 3SAT (public) (Germany) | 11.6 | 10.7 | | | | | 8.4 | 7.7 |
| ProSiebenSat.1 Media (Germany) | 10.7 | 9.7 | | | | | 7.7 | 7.0 |
| ORF1 + ORF2 (public) (Austria) | 4.9 | 3.5 | | | | | 3.5 | 2.5 |
| France télévisions (Public French channels, such as France 2, France 3 . . . ) | | | 17.0 | 13.0 | | | 4.0 | 3.1 |
| TF1 (France) | | | 14.0 | 13.0 | | | 3.3 | 3.1 |
| RTL Group (Bertelsmann AG, Germany) | 9.5 | | 10.0 | 10.0 | | | 2.3 | 2.4 |
| Public Italian channels (RAI) | | | | | 22.0 | 20.0 | 0.9 | 0.9 |
| Mediaset (Berlusconi Family, Italy) | | | | | 25.6 | 22.0 | 1.1 | 1.0 |
| Others | 28.5 | 36.6 | 27.0 | 35.0 | 17.0 | 21.0 | 34.5 | 42.3 |
| Audience (1,000)[1] | 5,033 | 5,106.7 | 1,634 | 1,678.5 | 3,080 | 324.6 | 6,975 | 7,109.8 |
| C4 | 66.6 | 61.9 | 73.0 | 65.0 | 84.8 | 81.0 | 54.2 | 47.8 |
| HHI | 1,574 | 1,253 | 1,618 | 1,283 | 2,397 | 2,258 | 1,343 | 1,044 |
| *N* (>1%)[2] | | | | | | | | |
| Noam Index | 704 | 560 | 724 | 574 | 1,072 | 1,010 | 475 | 369 |

*Source*: our calculations are based on data from Mediapulse AG <http://www.mediapulse.ch>.

[1] The number of viewers is based on a "universe" estimate, i.e., all Swiss citizens three or more years old who live in a home with television and phone connections (source: Mediapulse AG). When data are unavailable for some channels for some regions, we treat those market shares as zero.

[2] The HHI and the Noam Index were calculated for only eight firms at the national level; *N* (>1%) is not known.

## Table 14-7. Film Production/Distribution (Market Shares by Box Office %)

| | 2009 | 2010 |
|---|---|---|
| Walt Disney (US) | 21.2 | 18.3 |
| United International Pictures (UIP) (Paramount [Viacom] + NBCUniversal [Comcast/GE], US) | 12.9 | 17.9 |
| Warner Brothers (Time Warner) (US) | 14.3 | 16.3 |
| 20th Century Fox (News Corp., Murdoch Family, US/UK/AUS) | 15.5 | 12.0 |
| Elite-Film (Ascot Elite Entertainment Group) | 11.4 | 11.0 |
| Pathé Film (France) | 8.7 | 5.8 |
| Frenetic Films | 4.4 | 5.5 |
| Filmcoopi Zürich | 2.8 | 2.8 |
| JMH Distributions | 1.6 | 1.5 |
| Xenix Filmdistribution (Switzerland) | 1.0 | 1.5 |
| Total Revenue (mil CHF)[1] | 262 | 262 |
| Total Revenue (mil US$) | 242 | 252 |
| C4 | 63.8 | 64.4 |
| HHI | 1,294 | 1,258 |
| *N* (>1%) | 10 | 10 |
| Noam Index | 409 | 398 |

*Source*: Our calculations based on data from Swiss Federal Statistical Office <http://www.bfs.admin.ch/> for market shares.

[1] To estimate the value of the market here, we added the advertising revenues (source: Fondation Statistique Suisse en Publicité) and the national cinema "budget," estimated for the population in the country (source: Schweizer Medien).

was created in January 1998 by the partition of Swiss Telecom PTT into the two companies Swiss Post and Swisscom Ltd. Swisscom is a public limited company (plc) with a special status under Swiss law; the federal government holds a majority of the capital and voting rights in Swisscom. Although the Telecommunication Law was revised in 2006, Swisscom is today the main telecom and Internet service provider (ISP) in the country.

However, the wireline market has diminished in volume: the 2008 call turnover for fixed telephone lines was 4.5 billion, in comparison to 6.9 billion in 1999, a 35% decrease. Sunrise (UK) is the second main operator, but its market share is much less than that of Swisscom's: in 2008, its market share only reached 10.8%. Sunrise belonged to TDC, a Danish telecom operator, until October 2010, and it is now part of CVC Capital Partners, a British private equity group. The third main operator is Cablecom, a subsidiary of Liberty Global (US), and it posted a 7.8% market share in 2008 and a 8.6% market share in 2011. In 2008 the fourth significant operator was TelCommunication Services (formerly Tele2), which had a 5.8% market share. Sunrise (UK) acquired Tele2 Switzerland in 2009 and continued to offer services under the brand until 2011, when the brand license ran out. In November 2011, Tele2 became Tele4U, a subsidiary of Sunrise (UK). Sunrise (UK) and TelCommunication Services are considered separately until 2008 in Table 14.8. With TelCommunication Services, the market share of Sunrise (UK) for wireline reached 15.1% in 2009, but it declined and stood at just 13.8% in 2010 and 13.6% in 2011. Table 14.8 gives concentration indices on the basis of the number of subscribers. The wireline market is very concentrated, with a C4 ratio over 85% and an HHI over 4,200 in 2011. But concentration has decreased since 2007, because of a slight decrease of the Swisscom market share and the growth of small-size operators.

Table 14-8. Wireline Telecom (Market Shares by Subscribers)

| | 2007 | 2008 | 2009 | 2010 | 2011 PROVISIONAL |
|---|---|---|---|---|---|
| Swisscom (public) | 67.8 | 68.3 | 68.2 | 65.3 | 63.1 |
| Sunrise (CVC Capital Partners, UK) | 11.4 | 10.8 | 15.1 | 13.8 | 13.6 |
| TelCommunication Services (before 2009: Tele2, Sweden; after 2009: Sunrise, UK) | 7.2 | 5.8 | 4.1 | 2.4 | 1.9 |
| Cablecom (Liberty Global, US) | 7.2 | 7.8 | 7.6 | 8.2 | 8.6 |
| mobilezone.com | 0.5 | 0.7 | 0.9 | 1.0 | 1.6 |
| The Phone House/Talk Talk Telecom (Carphone Warehouse, UK) | 1.6 | 1.3 | 1.7 | 1.4 | 1.3 |
| TalkEasy GmbH | 0.0 | 0.0 | 0.3 | 1.5 | 1.3 |
| Orange Network SA (France Telecom, France, until 2012; Apax Partners, UK, since 2012) | 0.0 | 0.1 | 0.2 | 1.2 | 1.2 |
| abalon telecom it ag | 0.6 | 0.6 | 0.6 | 0.6 | 0.5 |
| Backbone Solutions | 0.3 | 0.5 | 0.6 | 0.6 | 0.5 |
| Netstream | 0.3 | 0.4 | 0.4 | 0.6 | 0.5 |
| Finecom Telecommunications | 0.2 | 0.4 | 0.5 | 0.6 | 0.7 |
| Others | 2.8 | 3.3 | 3.9 | 5.3 | 6.9 |
| Total Revenue (mil CHF)[1] | 5,470 | 5,116 | 5,409 | 5,263 | 5,199 |
| Total Revenue (mil US$) | 4,563 | 4,738 | 4,995 | 5,059 | 5,885 |
| C4 | 93.6 | 92.7 | 92.6 | 88.8 | 86.9 |
| HHI | 4,833 | 4,878 | 4,940 | 4,528 | 4,248 |
| *N* (>1%) | 5 | 5 | 4 | 7 | 7 |
| Noam Index | 2,161 | 2,181 | 2,470 | 1,712 | 1,606 |

*Source*: Our calculations based on data from OFCOM <http://www.bakom.admin.ch/> for market shares.
[1] Source OFCOM: Fixed network services.

## Wireless Telecom

The mobile market has grown quickly in Switzerland at the expense of the wireline market. And it is highly concentrated: as of 2011, three operators represented 95.6% of the market. Swisscom, as noted earlier, is the leader here, with a 59.7% market share in 2011. The two other main operators are Sunrise (UK), at 20.8% in 2011, and Orange Suisse, at 15.1% in 2011. In November 2009, Orange and Sunrise (UK) had announced a merger: at that time the resulting joint venture (New Orange) would have represented about 38% of the market, but in April 2010, Comco—the federal competition commission—ruled against the merger on the grounds that it would have reduced wireless telecom competition.[20] But, as explained by the Federal Council in September 2010: "[t]he strong market share held by the incumbent operator [Swisscom] over the years and the balance of power between the various players reflects the

20. OFCOM. "WEKO untersagt Zusammenschluss von Orange und Sunrise." Bern: OFCOM, Apr. 22, 2010. <http://www.news.admin.ch/message/index.html?lang=de&msg-id=32758>. Dec. 28, 2012.

Table 14-9. Wireless Telecom (Market Shares by Subscribers)

| | 2007 | 2008 | 2009 | 2010 | 2014 |
|---|---|---|---|---|---|
| Swisscom (public) | 61.0 | 60.4 | 60.2 | 59.4 | 53.1 |
| Sunrise (CVC Capital Partners, UK) | 18.6 | 19.2 | 19.9 | 20.9 | 25.1 |
| TelCommunication Services (before 2009: Tele2, Sweden; from 2009: Sunrise, US) | 1.0 | 1.1 | 0.5 | Sunrise | Sunrise |
| Salt Niel/Iliad (France) (France Telecom, France, until 2012; Apax Partners, UK, 2012–2014) | 18.4 | 17.3 | 16.7 | 15.5 | 16.7 |
| Lebara (owners: Leon Rasiah, Baskaran Kandiah, Ratheesh Yoganathan, UK) | 0.8 | 1.1 | 1.5 | 2.6 | 2.5 |
| The Phone House AG/Talk Talk (Carphone Warehouse, UK) | 0.2 | 0.1 | 0.2 | 0.1 | 0.1 |
| In&Phone SA | 0.0 | 0.0 | 0.1 | 0.3 | 0.3 |
| Lycamobile AG | 0.0 | 0.8 | 1.3 | 1.0 | 1.0 |
| BEEONE Communications AG | — | — | — | 0.2 | 0.3 |
| Total Revenue (mil CHF)[1] | 4,817 | 5,044 | 4,629 | 4,881 | 10,170 |
| Total Revenue (mil US$) | 4,018 | 4,672 | 4,274 | 4,692 | 10,928 |
| C4 | 99.0 | 98.0 | 97.8 | 98.2 | 98.3 |
| HHI | 4,407 | 4,319 | 4,282 | 4,206 | 3,804 |
| *N* (>1%) | 4 | 5 | 5 | 5 | 5 |
| Noam Index | 2,203 | 1,931 | 1,915 | 1,881 | 1,893 |

*Source*: Our calculations based on data from OFCOM <http://www.bakom.admin.ch/> for market shares. 2014 data for minor providers based on 2011 figures.

[1] Source OFCOM: Mobile network services.

low intensity of competition on the Swiss mobile phone market."[21] In February 2012, Orange Suisse was sold by France Telecom to Matterhorn Mobile, a Swiss company controlled by Apax Partners (UK), a British private equity company. At the end of 2014, Xavier Niel, the owner of the French company Free (Illiad), entered an agreement to buy Orange Suisse, via NJJ Capital, his private holding company. The acquisition is subject to regulatory approvals (Table 14.9).

## INTERNET MEDIA

### Internet Service Providers (ISP)

The number of subscribers to the Internet has increased considerably for the last decade in Switzerland. More than 50% of the whole market. The two other main companies are Cablecom (US), with a 17.1% market share in 2010, and Sunrise (UK), with a 11.7% market share in 2010 (Sunrise and TelCommunication Services are considered separately until 2008 in Table 14.10). Concentration is high, with a C4 ratio of 84.7% and an HHI of 3,300 in 2010.

### Search Engines

Google (US) has a quasi-monopoly in the search engines market, with a market share above 90% in recent years. Nevertheless, the Google's market share has decreased for the last two years (Table 14.11).

21. Swiss Federal Council. "Evaluation du marché des télécommunications: Rapport du Conseil fédéral en réponse au postulat de la CTT-E du 13 janvier 2009 (09.3002)." Bern: OFCOM, Sep. 17, 2010. <http://www.bakom.admin.ch/dokumentation/gesetzgebung/00512/03498/index.html?lang=en>. Dec. 28, 2012 (translation).

Table 14-10. Internet Service Providers (Market Shares by Subscribers)

| | 2007 | 2008 | 2009 | 2010 |
|---|---|---|---|---|
| Swisscom (public) | 48.5 | 52.8 | 53.2 | 53.7 |
| Cablecom (Liberty Global, US) | 19.5 | 17.6 | 17.2 | 17.1 |
| Sunrise (CVC Capital Partners, UK) | 10.5 | 8.6 | 11.9 | 11.7 |
| TelCommunication Services (before 2009: Tele2, Sweden; from 2009: Sunrise, US) | 3.0 | 2.7 | 1.7 | 0.0 |
| green.ch | 1.2 | 2.4 | 2.4 | 2.2 |
| ImproWare | 1.1 | 1.1 | 1.2 | 1.2 |
| Finecom Telecommunications | 0.9 | 1.3 | 1.5 | 1.5 |
| VTX Services | 1.1 | 1.0 | 1.0 | 1.0 |
| Ticinocom | 1.4 | 1.0 | 0.9 | 0.3 |
| WWZ Telekom | 0.5 | 0.6 | 0.6 | 0.7 |
| Netstream AG | 0.4 | 0.3 | 0.5 | 0.7 |
| Others | 11.9 | 10.6 | 9.6 | 9.9 |
| Total Revenue (mil CHF)[1] | 871 | 920 | 964 | 955 |
| Total Revenue (mil US$) | 727 | 852 | 890 | 918 |
| C4 | 81.5 | 81.7 | 83.0 | 84.7 |
| HHI | 2,858 | 3,190 | 3,243 | 3,322 |
| *N* (>1%) | 8 | 9 | 8 | 7 |
| Noam Index | 1,010 | 1,063 | 1,147 | 1,256 |

*Source*: Our calculations based on data from OFCOM <http://www.bakom.admin.ch/> for market shares.

[1] This revenue figure is an estimate based on the "budget" dedicated by the population to Internet access (excluding expenses for content or for devices). Source: Schweizer Medien.

## Online News

Measuring concentration for the online news market in Switzerland is difficult. There is no official definition of this industry in the country, and various criteria have been used to measure online audience shares. Our categories are (1) most popular websites (both Swiss and non-Swiss) and (2) most popular Swiss news websites (Swiss only).

The Ringier Group uses Nielsen reports to rank website popularity in the Swiss market. The most popular sites are those of Google, Facebook, YouTube, and Microsoft. In order to calculate concentration indices, the market may be defined as the sum of all monthly "unique hits" for the main websites (Table 14.12).

The C4, which corresponds to the four main entities, stood at 78% in 2011, which is high, while the HHI stood at 1,885, indicative of a moderately concentrated market. Of course, this approach gives somewhat biased estimates of concentration, since all websites and all unique hits are taken into account, without distinction regarding contents.

For Table 14.13, we used data given by NET-Metrix, which concern only Swiss websites, and we selected the websites that appeared to be the most representative of news. Of course, the estimated indices are debatable, but they give an idea of the evolution of the Swiss online news market. Since foreign firms are not included in the data, the audiences of websites such as Google News, Yahoo! News, LeMonde.fr, and Bild.de are ignored in Table 14.13.

The market shares of public online news sites or news sites that must apply public regulation

## Table 14-11. Internet Search Engines (Market Shares by Search Volume)

| | 2008 | 2009 | 2010 | 2011 | 2012 | 2013 |
|---|---|---|---|---|---|---|
| Google (US) | 96.0 | 96.6 | 97.1 | 91.0 | 91.2 | 89.9 |
| Bing (Microsoft, US) | 0.0 | 1.3 | 1.4 | 3.4 | 3.2 | 3.6 |
| Yahoo! (US) | 2.0 | 1.3 | 0.9 | 3.6 | 3.0 | 3.2 |
| Bluewin (public) (Swisscom) | 0.0 | 0.3 | 0.3 | — | — | — |
| Ask Jeeves (US) | 0.0 | 0.2 | 0.1 | 0.4 | — | — |
| Baidu (China) | — | — | — | — | 0.45 | 0.9 |
| C4 | 98.0 | 99.6 | 99.7 | 98.4 | 97.9 | 97.6 |
| HHI | 9,220 | 9,335 | 9,430 | 8,306 | 8,337 | 8,105 |
| *N* (>1%) | 2 | 3 | 2 | 3 | 3 | 3 |
| Noam Index | 6,520 | 5,390 | 6,668 | 4,795 | 4,813 | 4,680 |

*Source:* Our calculations based on data from StatCounter Global Stats. <http://gs.statcounter.com/>, for months of July except 2013, where it is the month of March.

## Table 14-12. Most Popular Websites in Switzerland (Market Shares by Monthly Unique Visitors), October 2011[1]

| | MARKET PENETRATION (%) | MONTHLY UNIQUE VISITORS (1,000) | ESTIMATED MARKET SHARES (%) |
|---|---|---|---|
| Google (US) | 83.4 | 3,268 | 30.1 |
| YouTube (US) | 47.7 | 1,870 | |
| MSN/Windows Live/Bing (Microsoft, US) | 47.1 | 1,846 | 21.3 |
| Microsoft (US) | 45.4 | 1,777 | |
| Bluewin (public) (Swisscom) | 36.5 | 1,431 | 14.1 |
| Local.ch (Swisscom, public) | 24.9 | 978 | |
| Facebook (US) | 56.1 | 2,198 | 12.9 |
| Wikipedia (US) | 35.9 | 1,405 | 8.2 |
| Apple (US) | 33.6 | 1,317 | 7.7 |
| Search.ch (Tamedia/Swiss Post) | 24.5 | 960 | 5.6 |
| C4 | | | 78.4 |
| HHI | | | 1,885 |
| *N* (>1%)[2] | | | |
| Noam Index | | | 712 |

*Source:* Our calculations based on data from the Nielsen Company (survey in October 2011), in Ringier AG.

[1]Ringier AG. "Abécédaire publicitaire 2012." Zurich: Go4Media, 2012. 27. To calculate the concentration indices, MSN/Windows Live/Bing and Microsoft are considered as one group, as are Google and YouTube; Bluewin and local.ch are also considered to be one group.

[2]The HHI and the Noam Index were calculated for only seven firms; *N* (>1%) is not known.

Table 14-13. Online News Swiss Media (Market Shares by %)[1]

| | 2008 | 2010 | 2012 |
|---|---|---|---|
| Swisscom (public) (www.bluewin.ch/) | 21.1 | 14.2 | 10.3 |
| SRG SSR | 21.6 | 20.3 | 16.5 |
| TV | *15.7* | *15.1* | |
| Radio | *5.9* | *5.3* | |
| Tamedia (Coninx Family) | 19.1 | 21.0 | 26.5 |
| Edipresse | *7.6* | *7.3* | |
| Ringier (Ringier Family) | 12.2 | 13.8 | 12.3 |
| NZZ Medien Gruppe | 6.7 | 6.2 | 5.6 |
| Virtual Network (www.romandie.com/news/) | 3.2 | | |
| Basler Zeitung Medien | 1.8 | 1.7 | |
| Axel Springer Schweiz (Germany) | 1.6 | 6.6 | 6.2 |
| Neue Luzerner Zeitung | 1.3 | | |
| ProSiebenSat.1 Digital (Germany) | | 2.1 | 4.2 |
| Zattoo International | | 1.8 | 1.7 |
| 1&1 Mail & Media GmbH | | | 4.0 |
| Gartenhof Medien | | | 2.1 |
| AZ Medien | | | 2.4 |
| MedienVielfalt Holding (Tito Tettamanti) | | | 1.4 |
| Edition Suisse Holding (Groupe Hersant) | | | 1.2 |
| Others | 12.3 | 1.0 | 5.7 |
| C4 | 69.3 | 62.9 | 65.6 |
| HHI | 1,408 | 1,310 | 1,350 |
| *N* (>1%) | 10 | 12 | 13 |
| Noam Index | 445 | 378 | 374 |

*Source*: Our calculations based on data from NET-Metrix <http://www.net-metrix.ch/>.

[1]The market shares were calculated using NET-Metrix data based on the number of "unique users" per month. They are estimated by an averaging of the data for the period from October 1 to December 31 for every year.

is relatively large: Bluewin, the Swisscom news portal, is a main competitor for traditional print media's news portals in Switzerland, and SRG SSR's online versions of its radio and TV broadcasts have naturally quite large audiences. Nevertheless, we can observe that the main Swiss newspaper companies have a strong online presence and that the market share of Bluewin actually decreased significantly between 2008 and 2012. Tamedia/Edipresse had a market share equal to 26.5% in 2012 and NZZ Medien Gruppe around 6% in 2012. The market share of Ringier is stable around 12–13%, while Axel Springer's market share increased since 2008. There are other companies with quite low market shares. Although the C4 ratio is over 60%, the HHI is quite low because various news websites have developed in recent years. The figures overestimate concentration in the industry because data for foreign websites were not taken into account.

## CONCLUSION

Swiss publishing houses will continue to seek expansion of their cross-platform presence in the coming years. Ringier, for example, is present in print, broadcast, Internet, and mobile telecom applications, as well as in printing, entertainment, and e-commerce as an "integrated multinational media company." In 2012, Michael Ringier, chairman and joint co-proprietor of Ringier AG alongside his sisters Annette Ringier and Evelyn Lingg-Ringier, declared: "We shall always be a press company with the main business based on print media" and "we shall also exercise more businesses which have nothing to do with journalism. In entertainment, in e-commerce, we are going to continue the diversification of our revenue streams to decrease our dependence on economic cycles."[22] In December 2014 Ringier and Axel Springer planned to establish a joint venture in Switzerland. Of course, it will increase the level of concentration in the Swiss press. Tamedia's activities cover newspapers, magazines, printing, and Internet services as well. But every strategy is not necessarily in favor of diversification. Concerning Tamedia, it had expanded its radio and television arms, but since its acquisition of Edipresse, its growth strategy has been refocused on newspapers, magazines, and Internet services. And in the telecom sector, operators offer both wireline and wireless services, as well as Internet services, but they have not entered into the content industries, except for Swisscom's Bluewin.

The presence of foreign operators in Switzerland is likely to grow in the coming years. Switzerland has always been open to media content from other countries, though historically, Swiss companies dominated the domestic media market. With the exception of book publishing and film, the entry of foreign companies is quite recent. But Germany's Axel Springer is now a major company in the magazine industry and Philippe Hersant has entered the French-language newspaper market. In print media, the big Swiss publishing companies, such as Tamedia and Ringier, are mainly involved in newspapers and magazines, but foreign companies dominate the book publishing scene.

In radio and TV broadcasting, SRG SSR is still dominant, but many national and foreign channels are accessible. Concentration is high in radio but quite low in TV. The Swiss commercial broadcasting market is small and has developed more in radio than in TV, particularly for the German- and French-language markets. The TV market reflects the opening of the Swiss market to foreign channels: concentration is low, because many foreign channels in French, Italian, and German are accessible.

In the telecommunications and Internet services sectors, Swisscom is still dominant, but foreign operators are pulling market shares away from it. Swisscom has a special status as a public limited company: its major shareholder, with a stake of 51.2% in December 2013, is the Swiss Confederation (Government of Switzerland), which by law must hold a majority stake in Swisscom's investment capital and voting rights. Sunrise (UK) is an important operator in all sectors, and the mobile operator Orange Suisse, now owned by Xavier Niel, the main holder of Iliad in France (and renamed Salt), is present in wireless telecom, although it suffered a setback when Comco ruled against its merger with Sunrise (UK). Cablecom (US) is also an important operator in wireline telecom and Internet services. The growth of mobile telecom has been a significant trend over the last 10 years.

Concentration is relatively moderate in Switzerland. Although concentration indices appear to be quite high for some industries, the Swiss market is small compared to other European markets. Moreover, most Swiss media sectors are characterized by the presence of some Swiss major groups with large market shares and by a diversity of editorial voices. Concentration is not a simple phenomenon[23]

22. Laugier, Edouard. "La constance du capitalisme familial est un rempart contre la crise." *Le nouvel Economiste*, June 12, 2012. <http://www.lenouveleconomiste.fr/la-constance-du-capitalisme-familial-est-un-rempart-contre-la-crise-15185/#>.UN4r4YnjnQY (translation).

23. See: Badillo, Patrick-Yves. "La concentration des médias : les médias en résilience," in Badillo, Patrick-Yves. *Écologie des médias*. Chapter 6. Bruxelles: Éditions Bruylant, 2008, 115–157. Bourgeois, Dominique, Badillo, Patrick-Yves. *Les médias et les nouveaux médias face aux contraintes économiques*. Online article on the @rchiveSIC portal: <http://archivesic.ccsd.cnrs.fr/docs/00/06/23/92/PDF/sic_00001007.pdf>. Dec. 28, 2012.

and, in particular, ownership concentration does not necessarily mean information concentration. The Swiss major media groups are small compared to big multinational groups. For example, in 2011, Ringier and Tamedia had turnovers around 1.2 billion US$, while the turnover of Bertelsmann was around 21 billion US$ and the turnovers of the main US media companies were over 25 billion US$: Comcast was around 63 billion US$, Walt Disney around 42 billion US$, Google around 41 billion US$, and Time Warner was 29 billion US$ (see Badillo and Proulx about the globalization of the communication sector).[24]

The state of the daily newspaper market illustrates the complexity of the Swiss media landscape. From the standpoint of the C4 ratio, concentration appears high, with C4 equal to 80.5% in 2012, but by observing the HHI, concentration appears less high (HHI standing at 2,520 in 2012). And the Noam Index (in 2012) is only at 699, although it does not take into account the total number of titles, which is very high in Switzerland. However, the future is uncertain for paid-subscription daily newspapers. The merger of Edipresse and Tamedia, for example, has reinforced concentration.

Finally, two contradictory elements must be understood with respect to the Swiss media landscape concerning concentration. First, Switzerland is a small country and, theoretically, high concentration indices should be expected. Some industries are indeed characterized by quite high C4 and HHI; the Swiss market is narrow compared to those of its neighbors, and few new companies can develop. But, as already noted, in comparison to nearby European countries, the diversity of editorial voices is high, particularly in print and audiovisual media. Because of its geographic and linguistic situation, Switzerland offers opportunities to various companies, including foreign companies, and concentration is not as high as it could be for such a small national media market. Of course, many changes can be expected in the coming years: the rise of online media, the development of mobile services, and the race for a share of reduced advertising revenue could accelerate the process of concentration in some industries, especially for print media.

## Acknowledgments

We thank Eli Noam, Philomen Schoenhagen, and Philippe Amez-Droz for their comments. We thank our student assistants as well: Stéphanie Brodard, Angélique Dupuy, Nathalie Hassler, Chiara Marciani, Maude Mocan, Jenny Nunez, Hélène Panchaud, Anne-Florence Pasquier, Tess Sapin, Gabriel Tanner, and Julia Vaudroz.

## Bibliography

### 1. Books and Articles

Amez-Droz, Philippe. "L'économie des médias en Suisse," in Cornu, Daniel et al., *Les Médias en Suisse, Structure et audience.* 7th ed. Lausanne: FCJ/CRFJ, 2012, 115–133.

Badillo, Patrick-Yves. Amez-Droz, Philippe. "Le paradoxe de la presse écrite : un business model introuvable, mais une multiplicité de solutions", *La Revue Européenne des Médias.* n° 29, 2013–2014: 83–89.

Badillo, Patrick-Yves, Amez-Droz, Philippe. "Paywall, versioning, blended media . . . quel avenir pour la presse écrite ?", in Badillo, Patrick-Yves, Roux Dominique, *Le futur est-il e-media?* Paris: Economica, 2014, 182 pages.

Badillo, Patrick-Yves. *Écologie des médias.* Bruxelles: Éditions Bruylant, 2008, 436 pages.

Badillo, Patrick-Yves. "La concentration des médias: les médias en résilience," in Badillo, Patrick-Yves. *Écologie des médias.* Chapter 6. Bruxelles: Éditions Bruylant, 2008, 115–157.

Badillo, Patrick-Yves, Roux Dominique, *Les 100 mots des télécommunications.* Paris: Presses Universitaires de France, Collection Que Sais-je?, 2009, 128 pages.

Badillo, Patrick-Yves, Lesourd, Jean-Baptiste (Eds.). *The Media Industries and Their*

24. Badillo, Patrick-Yves, Proulx Serge. "Mondialisation de la Communication: à la recherche du sens perdu," in *Les essentiels d'Hermès "Sociétés de la Connaissance - Fractures et évolutions."* Paris: CNRS Éditions, 2009, 117–134.

*Markets—Quantitative Analyses*. London: Palgrave Macmillan, 2010, 248 pages.

Badillo, Patrick-Yves, Proulx Serge. "Mondialisation de la Communication: à la recherche du sens perdu," in *Les essentiels d'Hermès "Sociétés de la Connaissance - Fractures et évolutions."* Paris: CNRS Éditions, 2009, 117–134.

Blum, Roger. "Medienstruckturen der Schweiz," in Bentele, Günter, Brosius, Hans-Bernd, Jarren, Otfried, eds. *Öffentliche Kommunikation. Handbuch Kommunikation und Medienwissenschaft*. Wiesbaden: Westdeutscher Verlag, 2003, 366–381.

Bollinger, Ernest. "La peau de chagrin des quotidiens suisses." *Communication et langages*, 114 (1997), 58–64.

Bonfadelli, Heinz, Meier, Werner A., Trappel, Joseph. *Medienkonzentration Schweiz. Formen, Folgen, Regulierung*. Bern, Stuttgart, Wien: Haupt Verlag, 2006.

Bourgeois, Dominique, Badillo, Patrick-Yves. *Les médias et les nouveaux médias face aux contraintes économiques*. Online article on the @rchiveSIC portal: <http://archivesic.ccsd.cnrs.fr/docs/00/06/23/92/PDF/sic_00001007.pdf>. 28 December 2012.

Cornu, Daniel, ed. *Les médias en Suisse. Structures et audience*. 6th ed. Lausanne: FCJ/CRFJ, 2007.

Cornu, Daniel, Borruat, Régis. "Structure des médias suisses," in Cornu, Daniel et al., *Les Médias en Suisse, Structure et audience*. 7th ed. Lausanne: FCJ/CRFJ, 2012, 7–17.

Cornu, Daniel, Borruat, Régis, Léchot, Pierre-André & Nerini, Marlène. "Inventaire des médias suisses," in Cornu, Daniel et al., *Les Médias en Suisse, Structure et audience*. 7th ed. Lausanne: FCJ/CRFJ, 2012, 19–77.

Edi Kradolfer, *La diversité de la presse en Suisse, Un aperçu*. Neuchâtel: Office fédéral de la statistique (OFS), 2007.

Laugier, Edouard. "La constance du capitalisme familial est un rempart contre la crise." *Le nouvel Economiste*, June 12, 2012. <http://www.lenouveleconomiste.fr/la-constance-du-capitalisme-familial-est-un-rempart-contre-la-crise-15185/#.UN4r4YnjnQY>.

Meier, A. Werner. "Das Mediensystem der Schweiz," in Hans-Bredow-Institut für Medienforschung (Hamburg). *Medien: Internationales Handbuch*. Baden-Baden: Nomos, 2009, 592–602.

Meier, A. Werner. "Pluralismus und Vielvalt in Regionalzeitungen – Auswirkungen von Medienkonzentration und Medienkrise auf die Lokalberichterstattung in ausgewählten Regionen in der Schweiz," report to OFCOM. Zurich: Swiss Center for Studies on the Global Information Society, University of Zurich, 2011.

OFCOM. "WEKO untersagt Zusammenschluss von Orange und Sunrise." Bern: OFCOM, Apr. 22, 2010. <http://www.news.admin.ch/message/index.html?lang=de&msg-id=32758>. Dec. 28, 2012.

## 2. Reports

Federal Assembly—the Swiss Parliament—, Votation populaire du 11 mars 2012. "04.430 Initiative parlementaire. Maitre Jean-Philippe, Réglementation du prix du livre, Argumentaires contre." Bern: Assemblée fédérale—le Parlement suisse, 2012. <http://www.parlament.ch/f/wahlen-abstimmungen/volksabstimmungen/volksabstimmungen-2012/abstimmung-2012-03-11/buchpreisbindung/Documents/04-430-argumentarien-contra-f.pdf>. Dec. 28, 2012.

Mediapulse AG. *Rapports annuels*. Bern: Mediapulse AG, various years. <http://www.mediapulse.ch/fr/publications/downloads/rapports-annuels.html>. April 10, 2013.

Schweizer Medien. *Medienbudget.ch 2005*. <http://www.schweizermedien.ch/fileadmin/schweizermedien/brancheninfos/medienbudget/2005_Praesentation_Medienbudget.pdf>. Dec. 28, 2012

Schweizer Medien. *Medienbudget.ch 2011*. <http://www.schweizermedien.ch/fileadmin/schweizermedien/brancheninfos/medienbudget/2011_Praesentation_Medienbudget.pdf>. Dec. 28, 2012.

Recherches et études des médias publicitaires—Researches and studies on advertising medias (REMP). *Presse écrite*. <http://www.wemf.ch/f/media/presse_ecrite.php>. Dec. 28, 2012.

Ringier AG. "Ringier AG Annual Report 2011." Zurich: Ringier AG, 2011. <http://www.ringier.com/sites/default/files/upload/field_publication_file_en_art/low_ringier_jb_2011_inhalt_englisch_21_3.pdf>. Dec. 28, 2012.

Ringier AG. "Abécédaire publicitaire 2012." Zurich: Go4Media, 2012. <http://www.go4media.ch/fr/node/497>.

SRG SSR. "Geschäftsbericht 2011." SRG SSR, 2011. <http://gb.srgssr.ch/index.php?id=745&L=0>. Dec. 28, 2012.

Swisscom. "Swisscom 2011 Annual Report." Bern: Swisscom, 2012. <http://2011.swisscom-report.ch/sites/default/files/Engagiert%20f%C3%BCr%20die%20Schweiz_2011_EN.pdf>. Dec.28, 2012.

Swiss Federal Council. "Evaluation du marché des télécommunications: Rapport du Conseil fédéral en réponse au postulat de la CTT-E du 13 janvier 2009 (09.3002)." Bern: OFCOM, Sep. 17, 2010. <http://www.bakom.admin.ch/dokumentation/gesetzgebung/00512/03498/index.html?lang=en>. 28 Dec. 2012.

Swiss Federal Council, "Garantir la diversité de la presse: Rapport du Conseil fédéral en réponse au postulat Fehr 09.3629 et au postulat de la Commission des institutions politiques du Conseil national (CIP-CN) 09.3980." Bern: OFCOM, June 29, 2011. Dec. 28, 2012. <http://www.bakom.admin.ch/themen/radio_v/01153/0115603479/index.html?lang=fr&download=NHzLpZeg7t,lnp6I0NTU042l2Z6ln1ae2IZn4Z2qZpnO2Yuq2Z6gpJCDeoN3fWym162epYbg2c_JjKbNoKSn6A-->. Dec. 28, 2012.

Tamedia. "Tamedia: Annual Report 2011." Zurich: Tamedia, 2012. <http://www.tamedia.ch/fileadmin/files/documents/finanzberichte/2012/Jahresabschluss_2011_-_Geschaeftsbericht_E.pdf>. Dec. 28, 2012.

Trappel, Josef and Uhrmann, Caroline. "Panorama du livre et de la littérature en Suisse sur mandat de l'Office fédéral de la culture." Zurich: University of Zürich IPMZ (Institut für Publizistikwissenschaft und Medienforschung), June 16, 2006. <http://www.bak.admin.ch/dokumentation/studien/index.html?lang=fr&download=NHzLpZeg7t,lnp6I0NTU042l2Z6ln1ae2IZn4Z2qZpnO2Yuq2Z6gpJCDdn14gGym162epYbg2c_JjKbNoKSn6A-->. Dec. 28, 2012.

World Association of Newspapers and News Publishers (WAN-IFRA). *World Press Trends 2010.* WAN-IFRA/ZenithOptimedia, 2010. <http://www.wan-ifra.org/reports/2010/07/23/world-press-trends-2010-edition>. Dec. 28, 2012.

## 3. Websites

Federal Statistical Office (FSO). <http://www.bfs.admin.ch/bfs/portal/en/index.html>. Dec. 28, 2012.

*FranceLivre BookFrance, le portail international du livre français.* FranceLivre. <http://www.francelivre.org/>. Dec. 28, 2012.

NET-Metrix. <http://www.net-metrix.ch/>. Dec. 28, 2012.

OFCOM (Federal Office of Communications). <http://www.bakom.admin.ch/index.html?lang=en>. Dec. 28, 2012.

*StatCounter Global Stats.* StatCounter Global Stats. <http://gs.statcounter.com/>. Dec. 28, 2012.

Stiftung Werbestatistik Schweiz/Fondation Statistique Suisse en Publicité. <http://www.werbestatistik.ch/index.php?pid=1&l=fr>. 28 Dec. 2012.

WEMF AG für Werbemedienforschung/REMP—Recherches et études des médias publicitaires <http://www.wemf.ch/d/index.php>. Dec. 28, 2012.

# Switzerland—Data Summaries

ELI NOAM AND PAUL MUTTER

OWNERSHIP CONCENTRATION IN SWITZERLAND has been high but stable in recent years—the country's power index stayed 3,633 in 2004 and 3,690 in 2011. C4 is particularly high for newspapers and magazines due to mergers and acquisitions (Table 14.14). Federal restrictions on cross-ownership are relatively limited because the government deemed the national media industries to be "self-regulating." Switzerland's small market size and three language regions predispose it to high levels of internal concentration. Its linguistic diversity has proven conducive to investment by French and German media groups (as well as Italian groups to a lesser extent). Germany's public broadcasters and commercial operator ProSiebenSat.1, along with France Télévisions, hold small market shares in the audiovisual markets. Many Swiss TV viewers tune in to the channels of the neighboring countries. The largest direct foreign operator though, is Liberty Global (US), which controls 74% of the multichannel platforms industry, making it the third-largest platform operator.

The public broadcaster SRG SSR is the largest single entity in the radio and television broadcasting markets in every linguistic group, accounting for high concentration rates in these markets and holding 18% of all content media (Table 14.15).

A lower concentration rate exists in print media, with major firms Tamedia and Ringier with 10% and 8% of the content media market.

Swisscom is the main telecom operator and ISP in Switzerland. It holds a 68.3% market share in wireline while in Europe public incumbents are fast losing market share, and it controlled 57.6% of the

Table 14-14. National Media Industries Concentration in Switzerland

| | 2004/5 | | 2011 OR MOST RECENT | | % CHANGE ANNUAL AVERAGE | |
|---|---|---|---|---|---|---|
| | COMPANY POWER INDEX IN COUNTRY | COMPANY SHARE OF THE OVERALL NATIONAL MEDIA MARKET (%) | COMPANY POWER INDEX IN COUNTRY | COMPANY SHARE OF THE OVERALL NATIONAL MEDIA MARKET (%) | COMPANY POWER INDEX IN COUNTRY | COMPANY SHARE OF THE OVERALL NATIONAL MEDIA MARKET (%) |
| Swisscom (public) | 2,470 | 38.600 | 2,404 | 39.207 | −0.3 | 0.076 |
| Liberty Media (US) | 343 | 7.705 | 372 | 8.304 | 1.05 | 0.075 |
| SRG SSR (public) | 323 | 6.597 | 268 | 5.774 | −2.1 | −0.103 |
| TDC (Sunrise, Denmark) | 98 | 6.955 | 99 | 5.996 | 0.13 | −0.120 |
| Tamedia | 171 | 5.086 | 201 | 4.995 | 2.18 | −0.011 |
| Ringier | 32 | 2.236 | 57 | 2.758 | 9.8 | 0.065 |
| Edipresse | 24 | 1.935 | 29 | 1.935 | 2.6 | 0.000 |
| Axel Springer (Germany) | 24 | 0.910 | 31 | 1.083 | 3.4 | 0.022 |
| ProSiebenSat.1 (Germany/US) | 6 | 0.81 | 5 | 0.73 | −2.4 | −0.01 |
| ARD (Germany) (public) | 2 | 0.54 | 2 | 0.46 | −2.6 | −0.01 |
| ZDF (Germany) (public) | 2 | 0.44 | 1 | 0.39 | −2.4 | −0.01 |

| | | | | | | |
|---|---|---|---|---|---|---|
| France Télévisions (France) (public) | 2 | 0.42 | 1 | 0.31 | –5.3 | –0.01 |

| Media Concentration Index | 2004/5 | 2011 or Most Recent | % Change Annual Average |
|---|---|---|---|
| Total Revenue: Nat'l Media Industry (mil US$) | 13,076 | 17,342 | 4.08% |
| Total Voices (*n*) | 62 | 62 | 0.00 |
| Net Voices (*n*) | 49 | 49 | 0.00 |
| Public Ownership (%) | 28.6 | 28.1 | –0.1 |
| Foreign Ownership (%) | 24.1 | 25 | 0.1 |
| C4 Average—Weighted | 84.0 | 83.4 | –0.08 |
| HHI Average—Weighted | 3,661 | 3,807 | 0.50 |
| C1 Average—Weighted | 55 | 56 | 0.00 |
| Noam Index Average—Weighted | 903 | 1,263 | 4.9 |
| Pooled Overall Sector C4 | 53.3 | 53.6 | 0.03 |
| Pooled Overall Sector HHI | 1,694 | 1,740 | 0.33 |
| Pooled Overall Sector Noam Index | 92 | 103 | 1.5 |
| Market Share of Top Eight Companies: Nat'l Media Industry (%) (Pooled C8) | 76.9 | 79.8 | 0.7% |
| National Power Index | 3,633 | 3,690 | 0.20 |

Table 14-15. Top Content Media Companies in Switzerland

| | 2004/5 | | 2011 OR MOST RECENT | | % CHANGE ANNUAL AVERAGE | |
|---|---|---|---|---|---|---|
| | COMPANY POWER INDEX IN COUNTRY | COMPANY SHARE OF THE NATIONAL CONTENT MEDIA MARKET (%) | COMPANY POWER INDEX IN COUNTRY | COMPANY SHARE OF THE NATIONAL CONTENT MEDIA MARKET (%) | COMPANY POWER INDEX IN COUNTRY | COMPANY SHARE OF THE OVERALL NATIONAL MEDIA MARKET (%) |
| SRG SSR (public) | 903 | 18.464 | 837 | 18.004 | 0.92 | −0.057 |
| Google (US) | 69 | 0.72 | 352 | 3.91 | 0.51 | 0.40 |
| Liberty Global (US) | 281 | 3.795 | 341 | 4.610 | 2.7 | 0.102 |
| Tamedia | 479 | 14.237 | 626 | 15.576 | 3.9 | 0.167 |
| Ringier | 89 | 6.258 | 177 | 8.6202 | 12.3 | 0.293 |
| Edipresse | 68 | 5.416 | 92 | 6.033 | 4.4 | 0.077 |
| Axel Springer Suisse (Germany) | 68 | 2.546 | 96 | 3.378 | 5.2 | 0.104 |
| Neue Zurcher Zeitung | 34 | 4.02 | 49 | 4.38 | 0.05 | 0.04 |
| ProSiebenSat.1 (Germany) | 17 | 2.26 | 16 | 2.28 | −1.2 | 0.00 |
| Media Concentration Index | 2004/5 | | 2011 or Most Recent | | % Change Annual Average | |
| Public Ownership (%) | 18.6 | | 18.1 | | −0.1% / −0.06 | |
| Foreign Ownership (%) | 20.0 | | 23.3 | | 0.4/0.41 | |
| C4 Average—Weighted | 68.2 | | 69.3 | | 0.14 | |
| HHI Average—Weighted | 2,179 | | 2,677 | | 2.9 | |
| C1 Average—Weighted | 40 | | 43 | | 0.00 | |
| National Power Index | 2,115 | | 2,652 | | 3.17% | |

Table 14-16. Top Platform Media Companies in Switzerland

| | 2004/5 | | 2011 OR MOST RECENT | | % CHANGE ANNUAL AVERAGE | |
|---|---|---|---|---|---|---|
| | COMPANY POWER INDEX IN COUNTRY | COMPANY SHARE OF THE NATIONAL PLATFORM MEDIA MARKET (%) | COMPANY POWER INDEX IN COUNTRY | COMPANY SHARE OF THE NATIONAL PLATFORM MEDIA MARKET (%) | COMPANY POWER INDEX IN COUNTRY | COMPANY SHARE OF THE NATIONAL PLATFORM MEDIA MARKET (%) |
| Swisscom (56.9% public) | 3,840 | 59.893 | 3,537 | 57.620 | –0.4 | –0.284 |
| Liberty Global (US) | 378 | 9.879 | 387 | 10.048 | 0.29 | 0.021 |
| Sunrise (TDC, Denmark) | 153 | 10.821 | 146 | 8.827 | –0.55 | –0.249 |
| Orange (France Telecom, France) | 106 | 5.75 | 109 | 6.88 | 0.00 | 0.14 |
| Media Concentration Index | 2004/5 | | 2011 or Most Recent | | % Change Annual Average | |
| Public Ownership (%) | 34.1 | | 32.8 | | –0.2/-016 | |
| Foreign Ownership (%) | 26.5 | | 25.8 | | –0.1% | |
| C4 Average—Weighted | 92.8 | | 90.0 | | –0.35 | |
| HHI Average—Weighted | 4,485 | | 4,341 | | –0.40 | |
| C1 Average—Weighted | 64 | | 62 | | 0.0% | |
| National Power Index | 4,477 | | 4,180 | | –0.83 | |

overall platform media market (Table 14.16). Unlike other incumbents such as neighboring Deutsche Telekom, Swisscom has been able to retain a large share of the mobile market (53.1%) against foreign competitors like Orange (France) (and now Iliad, France) and Sunrise (UK) by offering lower prices. Swisscom is a public limited company, with 56.8% of shares belonging to the Swiss Confederation and the rest publicly traded.

# 15

# Media Ownership and Concentration in Turkey

HÜSEYIN KEMAL BAYAZIT

## INTRODUCTION

Turkish media markets lack overarching cross-ownership rules, despite a profusion of national regulatory agencies and media laws.[1] In practice, politicians, government agencies, the judiciary, and conglomerates all try to game the system for their own partisan and business agendas. This development is the result of a campaign of privatization without liberalization carried out by successive governments beginning in the 1980s. Until the early 1990s, the state controlled all non-print media outlets, and there were no independent regulatory bodies. The privatization of the audiovisual and telecommunications markets led to increased commercialization in the 1990s.[2] From 2003 on, Turkey's efforts to enter the European Union (EU) entailed the adoption of an *acquis communautaire*, an EU outline for rewriting all ownership rules and competition guidelines.[3] Turkish regulators also have obligations to other supranational bodies, such as the OECD, the World Bank, and WTO. On the surface, the *acquis* and ongoing

1. In addition to a basic Competition Law, there are five other significant regulatory measures that are supposed to structure the market: print media is regulated by the Press Law (Law No. 5187), public broadcasting by the Public Broadcasting Law (Law No. 2954), commercial broadcasting by the Private Radio and TV Broadcasting Law (Law No. 6112), wireline telecommunications by the Electronic Communications Law (Law No. 5809), and both the wireless telecom and online media markets by the Mobile Communications and Internet Law (Law No. 5651).

2. Kurban, D. and Sozeri, C. *Case study report: Does media policy promote media freedom and independence? The case of Turkey.* Istanbul: Turkish Economic and Social Studies Foundation, 2011. April 8, 2013. <http://www.tesev.org.tr/Upload/Publication/8a76a788-b00e-4c7c-bf24-6545458100c4/Turkey2.pdf>. p. 8

3. Haurtcourt, Allison. *The European Union and the Regulation of Media Markets.* Manchester: Manchester University Press, 2007, p. 12.

Turkish media reforms are compatible: regulators were initially responsive to EU demands, and media companies behaved similarly to their EU counterparts in making investments in the media markets of Central Asia, Eastern Europe, and the Caucasus.[4] Internally, however, Turkish conglomerates have been successful in finding legal loopholes to bypass regulatory rules, and have been able to veil their actual market shares instead of fully reporting them.[5,6]

Turkey must meet three primary *acquis* obligations: to ensure compliance with EU regulatory laws, prepare the market for full liberalization, and to develop ICT infrastructure. However, these obligations have not been well managed. As a result, most of the initial steps taken in accordance with EU requirements put forward identical provisions, contradictory amendments, and broadly defined responsibilities subject to interpretation.[7] Most significantly, legislators have been unable to promote competition and redress the high level of concentration for the wireline market in particular (with the exceptions of international and long-distance calling, though). The continued dominance of the telecommunications monopolist, Turk Telekom (TT), is one of the reasons that the 2011 progress report for Turkey aired concerns about the implementation of the *acquis* obligations.[8] The state of the wireline market illustrates that merely copying EU rules and standards does not ensure increased efficiency or transparency.

The European Commission's (EC) progress reports on Turkish media summarized several key problems facing regulatory agencies fulfillment of the *acquis*. The Information & Communication Technologies Authority of Turkey (ICTA) is in charge of harmonizing national media laws with EU standards (for spectrum management, licensing, network access, numbering, number portability, rights of way, and tariffs), but Turkey's Electronic Communications Law (ECL) and Mobile Communications and Internet Law are not in compliance with these standards. The ICTA, according to the EU, "confer(s) a margin of discretion on the regulator that could lead to legal unpredictability" in licensing practices, a statement that openly questions the transparency and autonomy of the ICTA.[9]

Legislative policy formulation is a patchwork process involving the prime minister's office, several ministries, parliament, and a number of interest groups and professional organizations. Legislators have generally not been successful in promoting sustainable long-term growth through consistent policymaking or in advancing lower levels of concentration and better cross-ownership controls. From parliament's perspective, the mandates and capabilities necessary for implementing new policies are not present because of a multiplicity of ministries, laws, and national regulatory agencies, as well as the executive branch's insertion of itself into the process. No elected officials want to "own" the regulatory regime due to fears of alienating corporate media and party members, though according to a report of the EC, ambiguity over which Turkish agency is responsible for

---

4. Hallin and Mancini developed three models for the comparison of media systems in Western Europe and North America: the (or Mediterranean) system for France, Greece, and Italy; the Democratic Corporatist (or North / Central European) model for Germany, the Netherlands, and the Scandinavian countries; and the Liberal (or North Atlantic) type for the US, UK, and Canada. *See* Hallin, D. and Mancini, P. *Comparing Media Systems. Three Models of Media and Politics*. London: Cambridge University Press, 2004.

5. Sumer, Burcu. The Impact of Europeanisation on Policy-Making in Turkey: Controversies, Uncertainties and Misfits in Broadcasting Policy (1999–2009). Ankara University, European Research Center, Research Series, No. 35, 2010, p. 131.

6. Papathanassopoulos, S. "The Mediterranean/Polarized Pluralist Media Model Countries." Terzis, George, ed. *European Media Governance: National and Regional Dimensions*. Bristol, UK; Chicago: Intellect, 2007, p. 194.

7. Elmas, Esra and Kurban, Dilek. "Background information report: Media policies and regulatory practices in a selected set of European countries, the EU and the Council of Europe: The case of Turkey." Istanbul: Turkish Economic and Social Studies Foundation, 2010. June 1, 2012. <http://www.mediadem.eliamep.gr/wp-content/uploads/2010/05/Turkey.pdf>.

8. European Commission. "Turkey 2011 Progress Report." Brussels: European Commission, Oct. 12, 2011. April 8, 2013. <http://ec.europa.eu/enlargement/pdf/key_documents/2011/package/tr_rapport_2011_en.pdf>. p. 64.

9. European Commission. "Turkey 2010 Progress Report." Brussels: European Commission, Nov. 9, 2010. April 8, 2013. <http://ec.europa.eu/enlargement/pdf/key_documents/2010/package/tr_rapport_2010_en.pdf>. p. 56.

enforcement is the leading source of confusion here.[10] That ambiguity is the reason that a clear confirmation of policy issues at the ministerial level is rarely observed, and the ex-ante and ex-post regulatory responsibilities are overlapping: ICTA is dedicated to ex-ante regulations, while the Competition Authority (CA) is in charge of the ex-post situation. While their functions seem complementary, in practice both are mandated to perform similar ex-post enforcement duties, and this has led to disputes.[11] While the ECL assigns the ICTA shared responsibility for approving mergers and acquisitions alongside the CA, it has provisions that accept an overriding CA mandate to monitor and audit violations of competition under the Competition Law. It is a common perception in Turkish politics that the state enacts political and economic objectives through regulators. The Radio and TV Supreme Council (RTVSC) is one of the most important media regulators, and one of the most discussed when it comes to transparency and political independence. As is the case with all national regulatory agencies in Turkey, its nine board members are nominated by parliament. Although RTVSC considers itself as an autonomous regulator, the composition and the way boards are appointed belie this assertion.

The relationship between regulatory agencies and the judiciary strains the implementation: the authorization of alternative operators was restricted for five years, from 2004 to 2009, because the ICTA had authorized licenses for three and half years terms, which the High Administrative Court cancelled in 2008. Consequently, the ICTA had to revise the authorization and issue new licenses in 2009. Licensing strictly controls entry to the commercial broadcasting market. The RTVSC and ICTA are mandated to regulate frequency allocation—which was not done for terrestrial TV and radio broadcasting—but whereas RTVSC is responsible for allocating frequencies, ICTA holds a mandate over the planning process. The tasks of RTVSC are explained in the Turkish Constitution, but the duties of ICTA are not clearly specified by any laws. Given that RTVSC and ICTA had similar tasks and responsibilities for frequency allocation and data transfer services, the Ministry of Transport and Communication (MoTC) declared in 2009 that these two agencies would be merged, though no such consolidation has taken place.[12] In any event, the implementation of Turkey's new frequency allocation plan is nontransparent: the goals outlined in the National Spectrum Allocation Plan are contradictory, since they prioritize demands of multiple state agencies yet seek to provide fair competition among commercial and public entities.[13]

There are no cross-ownership restrictions for wireline or wireless telecom, multichannel TV platforms, and Internet service providers (ISPs); capital and revenue caps apply only to broadcasting. In all other markets, firms are allowed to hold assets across different media sectors as long as they follow the Competition Law and existing cross-ownership rules—even though these were initially written for the broadcasting market alone. Turkish conglomerates have easily circumvented these regulations.

The European Competitive Telecommunications Association (ECTA) ranks Turkey—along with the Czech Republic and Bulgaria—in the lowest tier of countries for performance in media regulation and transparency.[14] The

---

10. Katholieke Universiteit Leuven (ICRI), Jönköping International Business School (MMTC), Central European University (CMCS) and Ernst & Young Consultancy Belgium. "Independent Study on Indicators for Media Pluralism in the Member States—Towards a Risk-Based Approach." Leuven: Katholieke Universiteit Leuven, 2009. June 19, 2012. <http://ec.europa.eu/information_society/media_taskforce/doc/pluralism/study/final_report_09.pdf>.

11. Ardıyok, Ş. and Oğuz, F. "Competition law and regulation in the Turkish telecommunications industry: Friends or foes?" *Telecommunications Policy*, 34:4, 2010. April 8, 2013. <http://www.sciencedirect.com/science/article/pii/S030859610900113X>.

12. Yeni Şafak. "Yıldırım: RTÜK ile BTK birleşebilir" (RTVSC and ICTA May Merge). *YeniŞafak*, Dec. 25, 2009. June 21, 2012. <http://yenisafak.com.tr/bilisim/?i=231027>.

13. Sektörel Araştırmave Stratejiler Dairesi Başkanlığı, "Spektrumda Serbestleşme ve Yeni Eğilimler" (Liberalization of Spectrum and Main Trends), Ankara, BilgiTeknolojileri ve İletişim Kurumu, 2010. June 21, 2012. <http://www.tk.gov.tr/kutuphane_ve_veribankasi/raporlar/arastirma_raporlari/dosyalar/Spektrumda_Serbestlesme_Yeni_Egilimler.pdf>.

ICTA's power to enforce rules remains limited, and it lacks the authority to impose fines and sanctions. In addition, it has not imposed either a truly cost-oriented mobile termination rates (MTR) program, or any internal nondiscrimination obligations. The effectiveness of the regulatory regime, for wireline telecom in particular, demands more than *de jure* independence for regulatory bodies.

## THE MAJOR MEDIA INDUSTRY FIRMS

As a result of privatization, the traditional model of media firms as family-owned print publishers was supplanted by conglomerates with cross-ownership holdings in broadcasting and telecommunications. During the second half of the 1990s, wireless, wireless telecom, private broadcasting, and print media became more concentrated through cross-ownership. Similarly, the late 1990s and early 2000s saw commercialization transform into conglomeration. A handful of conglomerates consolidated print media outlets through acquisitions to dominate the industry, and this continued after 2001. In addition, religious and conservative business groups with strong ties to religious sects or communities also entered into market segments through buyouts, thus bringing about further consolidation.

The seven main media groups between 1985 and 1995 were the Dogan Media Group (DMG), controlled by the Dogan Family (with close ties with the Populist Republican Party, and later with the Motherland Party, which is center-right, conservative); the Sabah Group, run by tycoon Dinc Bilgin (with close ties with the True Path Party, center-right, conservative); the Uzan Group of the Uzan family (with close ties to the former one-party regime of the Motherland Party); the Cukurova Group, established by businessman Mehmet Emin Karamehmet; the Aksoy Group (also with close ties to the Motherland Party); the Ihlas Group, led by the head of the Isik religious sect (conservative, pro-religious, and also with close ties to the Motherland Party); and the Feza Group, supported by well-heeled segments of the Sunni Muslim religious community.

Other principal groups with cross-cross-ownership in Turkey include the Albayrak Group (religious and pro-government); the Ciner Group (secular mainstream with strong support of JDP government); the Calik Group's Turkuvaz Medya, whose CEO is the son-in-law of prime minister (mainstream conservative/pro-government); the Dogus Group (secularist, mainstream, reconfigured supporter of JDP government); the Ipek/Koza Group (supporter of the religious community of the Feza Group); and the pro-government, secular STR Group.

After 1995, cross-ownership became even more concentrated across these markets: Dogan alone controls over 60% of total advertising revenues nationwide. The economic crisis of 2001 had a very significant effect on concentration. As a result of it during the first JDP administration, the media assets of the Uzan, Sabah, and Aksoy media groups were all confiscated by the state agency called the Saving Deposit Insurance Fund (SDIF) in 2004, putting the three out of the media business. The Ciner, Dogan, Dogus, and STR media groups acquired their assets. While political parties seek supporters in the press, conglomerates attempt not only to influence policy decisions, but also to maximize their profits through rent-seeking. The creation of the Albayrak, Calik, Ipek/Koza, and STR media groups—which had no previous experience in the media market—to "balance out" against existing media groups affiliated with the JDP's political rivals resulted in increased politicization of media and of clientelism.

All of the major media groups, except for Feza, are large conglomerates and their major business activities are in other sectors such as finance and banking, mining, energy, automotive, tourism, construction, and real estate. The conglomerates use their media outlets to protect these business interests, while there seems

14. Day, Jones and SPC Network. "Regulatory Scorecard 2009." European Communities Trade Mark Association (ECTA), 2010. <http://www.ectaportal.com/en/upload/Scorecards/Regulatory_Scorecard_2009/ECTA%20Regulatory%20Scorecard%20Report%202009.pdf>. p. 2–8.

to be no efficient way to control concentration and cross-ownership. As such, the relationship between ownership and types of financial investment in Turkish media industry has changed significantly in the past two decades. Although technological advancements lower the cost of radio and TV broadcasting, the competition and the required infrastructure demand high investment. Similarly, the profitability of the print media and private broadcasters is directly dependent on deep pockets in other lines of business. The majority of media owners of 1990s and 2000s came from construction and/or banking and finance. The dominance of conglomerates is not only limited to control over the national newspaper market but also other segments of media. Yay-Sat of the Dogan Group and Turkuvaz Dagitim of the Calik Group run a *de facto* duopoly in print media distribution. The only serious recent competition has come from Islamic media outlets, which proliferated in the 2000s, backed by Islamic financiers. Together, media markets have been dominated by conglomerates with core business other than media, and by pro-JDP newcomers subsidized through the preferential awarding of energy and mining licenses, subsidized cheap loans, support by officials, and public works tenders. In sum, the Turkish market's highly differentiated media products have been concentrated through vertical integration and conglomerates accelerated by barriers to entry, economics of scale, high entry costs, inadequate regulations, and absence of cross-ownership rules (except for radio and TV).

## The Conglomerates

The Dogan Group is one of the largest business groups in Turkey, with extensive activities in the media, tourism, energy, real estate, and insurance sectors. Its Dogan Media Group (DMG) controls almost half of Turkey's private media outlets and about 60% of advertising revenues nationwide. It has joint partnerships with a number of well-established international companies, including CNN (owned by Time Warner), Axel Springer, OMV, Universal Music Group, Burda, and Egmont.

DMG has historically been considered a mainstream outlet associated with state-sponsored secularism. This changed in 2009, when the group was fined a massive USD2.5 billion due to the 2006 sale of Dogan TV Holding to the German publisher Axel Springer. In 2011, the group sold its two well-established secularist newspapers—*Milliyet* and *Vatan*. The group's media outlets lowered their secularist tones and became more in line with the conservative and pro-Islamist views of the government.

DMG's portfolio of media firms includes newspapers, magazine and book publishing and distribution, printing, television and radio broadcasting, advertising, online media, an ISP, and fixed line telecom. It operates 8 national newspapers, 24 magazines, and six printing facilities in Turkey and Germany. DMG owns six national dailies *(Hurriyet, Radikal, Posta, Fanatik, Referans,* and *Hurriyet Daily News),* three national TV channels *(Kanal D, Star* and *CNN Turk),* several radio channels *(Radio D, Slow Turk Radio, CNNTürk Radyo, and Radio Moda),* a digital platform (D-Smart), as well as film production and advertising firms (D Produksiyon, ANS, and Hurriyet Yapım), and two distribution firms (DPP and YAYSAT). The group also includes one magazine group Dogan Burda (DB), a joint venture with the German publishing house Burda with 27 titles. Dogan has its own news agency (DHA), two publishing house (Dogan Egmont and Dogan Kitap), one printing house (Dogan Ofset), a music company (DMC), a merchandising company (D&R), Dogan Online, and the telecom division "Smile." In 2005, Dogan acquired 100% of Trader Media East (TME), an online and print classified advertiser active in Central and Eastern European markets. The group also owns one TV station in Romania, Channel Romania D.[15]

The company participates in several joint ventures with foreign firms. Axel Springer owns 44.9% of the shares of Dogan TV Holding, and

15. Dogan Media Holding website, http://www.dyh.com.tr/tr/Hakkimizda.aspx. Accessed on July 12, 2012.

Burda, the German publishing group, holds 44.9% stake of Dogan Yayin Holding. Most of the media assets of the Group are traded on the Istanbul Stock Exchange.

The second largest group is the Turkuvaz Group, which belongs to Calik Holding, owned by Ahmet Calik, a businessman with close ties to the JDP government. As noted earlier, the CEO of TMG is the son-in-law of Prime Minister, and then President, Tayyip Erdogan. The group owns two pro-government national TV channels (*ATV* and *A Haber*), two national newspapers (*Sabah* and *Takvim*), 14 monthly and 5 weekly magazine titles, one printing house, one online news portal, and one distribution firm. All Turkuvaz's media assets are directly owned by Calik and are not publicly traded.[16] In 2007, Calik won control of the defunct Sabah group's media assets. The JDP enabled the process for political ends: it approved credits to the Calik Group from two state banks to finance the buyout of the confiscated media assets of the Sabah Group. TMG broadcasts two pro-JDP TV channels and its management is closely associated with the family of Erdogan.

The Cukurova Group, active since 1977, operated with a secularist inclination up to 2004, but has been pro-government since then. It is owned by Mehmet Emin Karamehmet. It owns two national newspapers (*Akşam and Güneş*), four magazines (*Alem, Platin, Autocar, and Stuff*), two national radio stations (*Alem FM* and *Lig Radyo*), national TV channels *Show TV* and *SkyTürk*, a media marketing firm (Mepas), an advertising marketing (Zedpas), a TV/radio technical infrastructure services firm (Eksen Yayıncilik), a digital satellite TV platform (Digiturk) with the broadcasting rights of Turkish premier football league up to 2014, a leading cellular GSM operator (Turkcell), and the ISP Superonline, Turkey's first ISP.[17]

The Dogus Group has been active since 1999 in the media market, exhibiting a moderate, secularist inclination up to 2005, but it has been pro-JDP since then. It is owned by Ferit Sahenk. The group owns four free news and three thematic channels (*NTV, Star TV, CNBC-E, e2, NTV Spor, NBA TV,* and *Kral TV*). The group also owns seven radio stations (*Kral FM, N101, NTV Radyo, NTV Spor, Radyo 5, Radyo Eksen, Radyo Voyage,* and *Virgin Radyo*) and has license agreements with the American publishers Condé Nast and the National Geographic Society.[18]

Ciner Medya Holding is part of the Ciner Group and is owned by Turgay Ciner. It has three national TV channels (*Haberturk, Kanal1,* and *Bloomberg HT*—the latter is run with Bloomberg L.P.), one newspaper (*Haberturk*), numerous magazine titles like *Marie Claire* and *Maison*, two radio stations (*Haberturk Radyo and Bloomberg HT Radyo*), a news agency (Ajans Haberturk), an online news company (*Haberturk.com*), a production company, and printing facilities. Like the other groups that were characterized as moderate secularist until the mid-2000s, Ciner's media outlets are now very much in line with JDP's politics.[19]

## The Religious Community/Sectarian Media

The only serious competition to these conglomerates has come from the newer religious media groups, which are backed by Islamist associations. After 1990, the proliferation of religious media outlets supported by conservative parties, and the rise of pro-Islamist corporate lobbies, became more visible. Since 2002, the year pro-religious JDP came to power, it has gradually been creating its own media groups. Religious sect and community-based media became powerful, recognizing the abundance of economic and political rents the JDP government could offer them.

The religious and sectarian media is quite diverse, and there are several well-known and visible business groups with close links with the current JDP government. These include

16. Turkuvaz Media Group website, http://www.calik.com/sektorler.aspx?id=24. Accessed on July 19, 2012.
17. Turkmedya website, http://www.turkmedya.com.tr/. Accessed on July 19, 2012.
18. Dogus Group website, http://www.dogusgrubu.com.tr/tr/icerik/47/2/web_sektorler/medya/. Accessed on July 19, 2012.
19. Ciner Yayın Holding website, http://www.parkgroup.com.tr/companies.php?CID=24. Accessed on July 21, 2012.

the Feza Group and Ipek/Koza (community of Nurcu sect), Ihlas Group (Isikcilar sect), Albayrak Group (the Nakshibandia sect), Yeni Asya (Nur sect), and Yeni Mesaj (Icmal sect).

The biggest of these is the Feza Group, controlled by the influential Gulen movement (which makes it a not-for-profit enterprise). It owns four TV channels (*Samanyolu TV, Samanyolu Haber, Mehtap TV, Ebru TV*), the highest circulating newspaper, *Zaman*, two weekly magazines (*Aksiyon* and *Sizinti*), one news agency (Cihan Haber Ajansi), a national radio network (*Burç FM*) and 25 local radio stations, and one printing house. The Ihlas Group, owned by Isikcilar sect leader and pro-JDP businessman Mucahid Oren, is one of the more venerable religious media entities, having been set up in 1970. It owns a newspaper (Turkiye), a TV station (TGRT Haber), a radio station (TGRT FM), a magazine firm, a news agency (Ihlas Haber Ajansi), and a fixed line and Internet services company (Ihlas Net). The family-owned Albayrak Group, composed of 50 firms and with very close personal and familial ties with the Prime Minister Erdogan, owns a newspaper with a moderate Islamist position (*Yeni Safak*) and a TV channel (*TVNET*). The Ipek/Koza Group, with major operations in gold mining and printing industries, is owned by Akin İpek, a pro-Gulen businessman. The group owns a conservative daily (*Bugun*), and two TV outlets (*Kanalturk* and *Bugun TV*).

## Clientelism and Political Parallelism

The regulatory regime has not been successful in limiting high concentration, resulting in augmented "clientelist" relationships between media patrons and the state. Clientelism includes limited development of the mass circulation press caused by an authoritarian political culture, a slow development of democratic and civil society institutions, and a collapse of the "horizontal solidarity" of journalists and the convergence of "journalistic logic" with the logic of party politics and family (i.e., owner) privileges.[20] Moreover, in addition to various forms of protections, Turkish governments, with a strong patrimonial state tradition, support the companies of "eligible" businessmen or families with diverse forms of subsidies, such as credits, incentives, and other advantages. These supports are exclusively discretionary and discriminatory, and the rules and procedures were regularly changed with the political views and the agenda of governments. Consequently, media outlets often reverse their previous positions and roles and become the mouthpieces of the governing political party and of their parent conglomerates.[21]

In Turkey, the combination of market and state power in a patron and client network is a common phenomenon. In fact, the commercialization of the Turkish media sector has not broken the strong linkage between state and media conglomerates, but reversed its direction after 2002, the year JDP came to power. Until the 1990s, the state acted as a patron and media firms were its clients. In that period, the governments exerted enormous power over media content. However, after the deregulation of radio and TV broadcasting segments in early 1990s, the beginning of the period of coalition governments, conglomerates started to put the pressure on politicians and bureaucrats through their media outlets, thus leading to the inversion of the relationship (i.e., the coalition governments became the clients). For the major conglomerates, the main objective behind the involvement in the media business had to do with the indirect benefits of ownership rather than with the desire to develop pluralism and democracy.

In the 1990s, the proliferation of private broadcasting had no effect on the high degree of political parallelism. The industrial conglomerates were able to intervene in political decisions

20. Hallin, D. and Papathanassopoulos, S. (2002) "Political Clientelism and the Media: Southern Europe and Latin America in Comparative Perspective," *Media, Culture & Society* 24: 175–195, p. 186–189.

21. Political parallelism is the extent to which different media outlets reflect distinct political orientations in their news and current affairs reporting, or their entertainment content as well. *See* Hallin, D. and Mancini, P. (2004). *Comparing Media Systems. Three Models of Media and Politics*. London: Cambridge University Press, 2004.

related to their business interests, particularly in the period of coalition governments from 1992 to 2002.

Until the 1990s, the state—through the armed forces—exerted enormous power over media content. After the deregulation of the broadcasting market, the development of coalition governments in place of unitary party or military rule saw politicians and bureaucrats appealing to media owners for favorable media coverage, inverting the media–state relationship. The coalition governments became clients. The proliferation of privately owned broadcasting networks had no effect on the high degree of political parallelism; the instrumentalization of media outlets actually enabled conglomerates to intervene directly in the political process between 1992 and 2002 with select editorializing. Despite the benefits they accrued, though, media groups intervened at their own peril: in 2001, Dinç Bilgin, the owner of the Sabah Group, was arrested and charged with embezzlement. In 2002, the government confiscated all of the Uzan Group's assets when it was found to have not paid back loans to its mobile vendor, Motorola, and the firm was charged with siphoning off deposits in its commercial bank to fund underperforming assets.

Regardless of whether they are diverse or narrowly sectarian, private media groups are still restricted by heavy state censorship. On top of that, the tendency of conglomerates to shrink from controversial reporting, particularly in international coverage, leads to self-censorship. More diverse and vibrant views appear consistently at the local and regional outlets, but these have limited distribution ranges due to restrictions implemented by the courts. There has also been a proliferation of lawsuits brought against journalists, academics, and human rights defenders that have been undermining the freedom of expression. In 2011, watchdog groups reported that Turkey had the dubious distinction of holding "more journalists in prison than any other country" in the EU,[22] in terms of the ratio of journalists to population, and more than any country in the world including China, as stated by a well-known columnist Ugur Dundar.[23] In particular, legislation governing online content and the conditions under which Internet service providers (ISPs) can operate are not in line with EU laws protecting freedom of expression and the right of online access.[24]

Given that the long-awaited directive on media concentration had to be scuttled after lobbying efforts against it, the EU's current directives on media concentration and ownership control lack adequate policy instruments to promote pluralism in Turkish media markets. Creating more pluralist media in Turkey, particularly for new digital platforms, will require much more effective implementation of competition rules and new policy tools to redress cross-ownership.

## PRINT MEDIA

### Newspapers

The history of Turkish national print press starts at the end of the 15th century with the establishment of first print press by Jewish refugees from Spain in two cities of the Ottoman Empire, Istanbul, and Salonika. Much has happened since. Fast forwarding to today: The 1980 *coup d'état* by the Turkish military was a turning point in the structure of print media. Despite the censorship that prevailed under military rule (and the unitary party rule that lasted from 1983 to 1987, when a constitutional referendum was held), the military allowed news outlets to expand into local and regional markets, which greatly expanded the scope of print access. Regional and local print markets are vibrant and diverse, but the national market has

22. European Commission. "Turkey 2011 Progress Report." Brussels: European Commission, p. 65. Oct. 12, 2011. April 8, 2013. <http://ec.europa.eu/enlargement/pdf/key_documents/2011/package/tr_rapport_2011_en.pdf>

23. Ugur Dundar, Sözcü, July 7, 2013, p. 2.

24. "Turkey 'world leader' in imprisoned journalists, IPI report says." *Hürriyet Daily News*, April 8, 2011. April 1, 2013. <http // www.hurriyetdailynews.com/default.aspx?pageid=438&n=ipi-report-declares-turkey-world-leader-of-imprisoned-journalists-2011-2004-08>.

been subject to increasing concentration and political parallelism.

There are 163 national newspapers in Turkey, of which national newspapers have an 80.9% market share by circulation, followed by local papers (2,381) with a 15.3% share and regional dailies (73) with 2.4%. Readership is relatively low given Turkey's economic indicators, literacy rate and population size: the Directorate-General of Press Advertisement has reported a steady decline in circulation numbers, coupled with a fall in the public's level of confidence that the press is more impartial than not.[25] Within the 163 national dailies are seven minority newspapers: the Greek-language *IHO* and *Apoyevmatin*; the Armenian-language *Agos*, *Jamanak*, and *Nor Marmara;* the Ladino-language *Salom;*[26] and the Kurdish-language *Azadiya Welat* (Table 15.1).

Between 1995 and 2005, HHI, C4 the Noam index dropped significantly due to the entrance of other conglomerates in to the market, such as Albayrak, Ipek/Koza, and STR. By 2005, though, 16 out of 21 national newspapers with 1% or more market share were owned by 8 media groups. Meanwhile, DMG doubled its number of titles from three to six by incorporating *Posta, Radikal, Gozcu,* and *Referans,* but its overall market share fell to 37%. In terms of circulation numbers and advertising revenues, the market is highly concentrated.[27, 28]

Concentration indices based on market shares by circulation can be misleading since HHI values based on the market share by revenue are much higher than the HHI values based on market shares by circulation. Between 2006 and 2011, the Turkuvaz Media Group (TMG) of the Calik conglomerate entered the market by acquiring Sabah Group's 15% market share through its two dailies *Sabah* and *Takvim*. In this period, 18 of 23 national newspapers with 1% or more market share fell under the control of nine conglomerates for a combined 89% market share. The market share of the top four conglomerates (C4) slightly decreased from 75% to 71%, due to the doubling of the market share for Feza Group's *Zaman* and the entry of other groups through buyouts. Seven independent newspapers—*Cumhuriyet, Milli Gazete, Sözcü, Sok, Taraf, Yeni Cag,* and *Yeni Akit*—accounted for a combined 13% market share. HHI subsequently declined from 1,996 in 2006 to 1,738 in 2011, while the Noam index (NI) fell from 531 to 424 in the same period, illustrating an increase in the diversity of voices (Table 15.2).[29]

The analysis of the Turkish national newspaper market shows that the NI can be used to assess the hard to measure concept of external pluralism (diversity of voices). The decline in the NI value indicates more diversity of voices, that is, external pluralism. The decline in NI value from 531 to 424 shows the increase in the number of newspapers in the system, rising external pluralism accordingly (Figure 15.1). A recent empirical study of the Turkish newspaper industry supports this observation based on the Noam index by stating that "using micro-individual level survey data it is shown that while internal pluralism within newspaper readership communities is declining, the external pluralism [diversity of voices] is on the rise."[30]

Concentration analysis based on advertising revenues reveals that the concentration assessment based on market share by circulation might be misleading, since HHI values based on the market share by advertising revenues are higher than the HHI values based on circulation data. According to the data on Table 15.2, it is striking to see how the concentration level gets significantly higher when market shares based on advertising

25. Carkoglu, Ali. "Introduction." *Turkish Studies* 11:4, 2010. p. 513.

26. Ladino is a Hebraic-Spanish language used by Turkey's Jewish community.

27. Hallin, D. and Mancini, P. (2004).*Comparing Media Systems. Three Models of Media and Politics.* London: Cambridge University Press, 2004. p. 28.

28. Kaya, Rasit and Cakmur, Baris. "Politics and the Mass Media in Turkey." *Turkish Studies* 11:4, 2010. p. 533.

29. Carkoglu, Ali and Yavuz, Gozde. "Press–party Parallelism in Turkey: An Individual Level Interpretation." *Turkish Studies* 11:4, 2010. p. 613.

30. Carkoglu, Ali and Yavuz, Gozde. *Press–party Parallelism in Turkey: An Individual Level Interpretation,* Turkish Studies, 2010, 11:4, 613–624, p. 613.

Table 15-1. Daily Newspapers (Market Shares by Circulation), 1990–2011

| | 1990 | 1992 | 1996 | 2000 | 2004 | 2008 | 2010 | 2011 |
|---|---|---|---|---|---|---|---|---|
| Dogan Media Group (Dogan Yayin Holding) | 13.8 | 19.0 | 41.3 | 39.7 | 37.8 | 38.2 | 32.7 | 23.7 |
| *Finansal Forum* | | | | *0.6* | *0.3* | *0.5* | *0.4* | *0.0* |
| *Hürriyet Daily News* | | | *17.8* | *15.4* | *12.6* | *12.0* | *11.1* | *10.8* |
| *Milliyet* | *13.8* | *14.7* | *18.7* | *8.9* | *6.5* | *5.3* | *4.4* | *Demirören* |
| *Meydan/Gozcu* | | *4.3* | *4.8* | *3.0* | *3.7* | *0.0* | | |
| *Posta* | | | | *10.6* | *13.6* | *14.6* | *12.2* | *11.3* |
| *Radikal* | | | | *1.2* | *1.1* | *1.0* | *1.2* | *1.6* |
| (a)*Vatan* | | | | | | *4.8* | *3.4* | *Demirören* |
| Demirören | | | | | | | | 6.4 |
| *Milliyet* | | | | | | | | 3.9 |
| *Vatan* | | | | | | | | 2.5 |
| Sabah/Turkuvaz Media Group (Calik Holding) | 21.5 | 28.5 | 18.4 | 20.5 | 17.4 | 14.1 | 11.7 | 11.2 |
| *Sabah* | *21.5* | *28.5* | *18.4* | *15.8* | *10.0* | *9.3* | *8.7* | *8.6* |
| *Takvim* | | | | *4.7* | *7.4* | *4.8* | *3.0* | *2.6* |
| Simavi Family | 30.5 | 23.8 | | Article II. | | | | |
| *Hurriyet* | *17.2* | 23.8 | *Dogan* | | | | | |
| *Gunaydın* | *13.3* | *N/A* | *Dogan* | | | | | |
| Feza Group (*Zaman*) | | 4.7 | 7.7 | 4.5 | 10.8 | 17.5 | 19.6 | 22.7 |
| Ihlas Group (*Turkiye*) | 15.6 | 9.0 | 14.1 | 7.9 | 4.2 | 3.3 | 3.4 | 3.4 |
| Çukurova Group | | | | 9.2 | 9.9 | 8.1 | 6.1 | 5.6 |
| *Aksam* | | | | 4.3 | 4.8 | *4.1* | 3.5 | 3.2 |
| *Gunes* | | | | *4.9* | 3.7 | 3.5 | 2.6 | 2.4 |
| *Halkave Olaylara* | | | | | *1.4* | *0.5* | *0.0* | *0.0* |
| *Tercuman* | | | | | | | | |
| Uzan/STR Group (*Star*) | | | | | 2.3 | 3.3 | 2.6 | 3.8 |
| Albayrak Group (*YeniSafak*) | | | | | 2.9 | 2.5 | 2.5 | 2.5 |

| | | | | | | | | |
|---|---|---|---|---|---|---|---|---|
| Ipek/Koza Group (*Bugun*) | | | | | | 1.2 | 1.7 | 1.7 |
| Habertürk(Ciner Group) | | | | | | | 5.4 | 6.7 |
| Akit/Yeni Akit | | | 1.0 | 2.1 | <1% | <1% | <1% | 1.3 |
| Anadolu'da Vakit | <1% | <1% | <1% | <1% | 1.6 | 1.4 | 1.1 | <1% |
| Bugun | | 7.0 | 4.8 | | | | | |
| Cumhuriyet | 4.1 | 2.1 | 1.3 | 1.1 | 1.3 | 1.8 | 1.3 | 1.3 |
| D. Bugune Tercuman | | | | | 2.8 | <1% | <1% | <1% |
| Gunes | 2.5 | | | | | | | |
| Milli Gazete | <1% | <1% | <1% | <1% | <1% | 1.2 | 1.3 | 1.3 |
| Vatan | | | | | 5.3 | Dogan | | |
| Sok | | | | | | 1.3 | 1.3 | 1.2 |
| Sözcü | | | | | | | 4.6 | 5.4 |
| Tan | 6.2 | | | | | | | |
| Taraf | | | | | | | 1.2 | 1.2 |
| Tercuman | 3.9 | 1.0 | | | | | | |
| Turkiye'de Yeniçag | | <1% | <1% | <1% | 1.1 | 1.3 | 1.1 | 1.3 |
| Yeniyuzyil | | | 6.0 | | | | | |
| Others | 1.9 | 4.9 | 5.4 | 7.9 | 2.8 | 4.8 | 2.4 | |
| C4 | 81.0 | 82.0 | 84.0 | 81.0 | 76.0 | 78.0 | 70.1 | 64.3 |
| HHI | 1,903 | 1,896 | 2,365 | 2,219 | 2,020 | 2,090 | 1,714 | 1,393 |
| *N* (>1%) | 8 | 8 | 8 | 8 | 12 | 13 | 16 | 17 |
| Noam Index | 680 | 677 | 845 | 793 | 577 | 581 | 429 | 338 |

*Source*: Press Advertising Agency.

Table 15-2. Daily Newspapers (Market Shares by Revenue), 2003–2011

| | 2003 | 2004 | 2005 | 2006 | 2007 | 2008 | 2009 | 2010 | 2011 |
|---|---|---|---|---|---|---|---|---|---|
| Dogan Media Group (Dogan Yayin Holding) | 60.0 | 62.0 | 60.1 | 58.7 | 59.3 | 63.0 | 60.9 | 58.3 | 46.4[1] |
| Demirören | | | | | | | | | 13.0 |
| Turkuvaz Media Group (Calik Holding) | 17.0 | 19.0 | 18.0 | 18.5 | 16.5 | 15.3 | 14.2 | 13.4 | 12.5 |
| Cukurova Group | 5.0 | 5.2 | 5.0 | 5.3 | 5.2 | 4.8 | 4.1 | 5.0 | 4.1 |
| STR Group | 3.0 | 2.1 | 2.2 | 1.6 | 1.8 | 2.0 | 2.1 | 2.0 | 2.1 |
| Ihlas Group | 2.0 | 3.1 | 3.7 | 3.5 | 3.2 | 2.7 | 2.3 | 2.5 | 2.2 |
| Feza Group | 4.0 | 4.1 | 4.4 | 5.2 | 6.4 | 7.9 | 8.3 | 9.5 | 10.1 |
| Ciner Group | | | | | | | 2.1 | 2.2 | 2.4 |
| Ipek/Koza Group | | | 1.0 | 1.1 | 1.2 | 1.1 | 1.2 | 1.3 | 1.3 |
| Albayrak Group | 1.2 | 1.1 | 1.2 | 1.4 | 1.3 | 1.2 | 1.1 | 1.2 | 1.4 |
| Vatan | 2.1 | 2.1 | 2.2 | 2.1 | 2.3 | | | | |
| Sözcü | | | | | | | 1.1 | 1.4 | 2.2 |
| Others | 5.7 | 1.3 | 2.2 | 2.6 | 2.8 | 2.0 | 2.6 | 3.2 | 2.7 |
| Total Revenue (mil TRY) | 451 | 645 | 820 | 902 | 1,018 | 986 | 811 | 926 | 997 |
| Total Revenue (mil US$) | 297 | 446 | 607 | 627 | 780 | 764 | 522 | 615 | 596 |
| C4 | 86.0 | 90.0 | 88.0 | 88.0 | 87.0 | 91.0 | 88.0 | 86.0 | 86.0 |
| HHI | 3,949 | 4,268 | 4,006 | 3,825 | 3,879 | 4,302 | 4,014 | 3,714 | 2,637 |
| Noam Index | 1,410 | 1,524 | 1,430 | 1,275 | 1,293 | 1,536 | 1,338 | 1,175 | 839 |
| *N* (>1%) | 8 | 8 | 8 | 9 | 9 | 8 | 9 | 10 | 10 |

*Source*: Author.

[1] 59.0% before sale of two papers to Demirören. Revenue estimated.

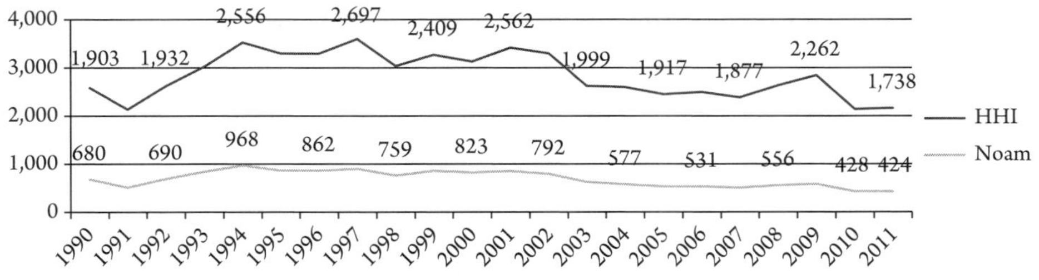


FIGURE 15.1 HHI and MOCDI Index for Market Shares by Circulation

revenues of newspapers rather than circulation are assessed—an HHI of 2,637 versus 1,393.

## Magazine Publishing

While the magazine market is fragmented and pluralistic, it has become more concentrated since the late 2000s. Circulation rates declined and a number of publishers divested themselves from international titles being published under license from US media groups. The distribution network in this market is also duopolistic: the dominance of Dogan Burda—a joint venture between DMG and the German publishing house Hubert Burda Media—and TMG adversely affects smaller outlets' circulation rates.[31] Four major conglomerates (and one smaller publisher, the Mutlu Group) dominate the market: Dogan Burda (27 titles, about half-owned with the German publisher Burda), TMG, the Dogus Group (partnered with Conde Nast), and the Ciner Group. The latter two publish several foreign titles under license, such as *Marie Claire* and *National Geographic,* at a time when many publishers are not renewing licensing agreements with foreign outlets due to declining readership. Despite a decline in HHI due to new entrants since 2006, the market is still highly concentrated (Table 15.3).

## Table 15-3. Magazine Publishing (Market Shares by Circulation), 2006–2011

| | 2006 | 2007 | 2008 | 2009 | 2010 | 2011 |
|---|---|---|---|---|---|---|
| Dogan Burda (Dogan Media Group, Dogan Yayin Holding + Hubert Burda Media, Germany) | 57.5 | 49.9 | 51.9 | 52.0 | 48.8 | 43.9 |
| Dogus Group | | | | | 11.1 | 9.0 |
| Ciner Group | 30.7 | 29.6 | 7.2 | 10.8 | 7.7 | 7.5 |
| *Ciner (GD)* | *8.0* | *7.6* | | | | |
| *Ciner/Sabah Group* | *22.7* | *22.0* | | | | |
| Turkuvaz Media Group (Calik Group) | | | 21.9 | 19.3 | 19.2 | 26.6 |
| Cukurova Group | 3.2 | 9.9 | 9.9 | 6.5 | 3.0 | 3.0 |
| Vatan/Mutlu Group | 8.6 | 10.6 | 9.1 | 11.4 | 10.2 | 10.0 |
| C4 | 100.0 | 100.0 | 93.0 | 94.0 | 89.3 | 89.5 |
| HHI | 4,333 | 3,576 | 3,407 | 3,366 | 3,044 | 2,881 |
| *N* (>1%) | 4 | 4 | 5 | 5 | 6 | 6 |
| Noam Index | 2,167 | 1,788 | 1,521 | 1,503 | 1,242 | 1,176 |

*Source:* Advertisers Association.

31. Sözeri, C. and Güney, Z. "The Political Economy of the Media in Turkey: A Sectoral Analysis." Istanbul: Turkish Economic and Social Studies Foundation Publications, 2011. April 8, 2013. <http://www.tesev.org.tr/Upload/Publication/67e244dd-5c21-24d34-8361-4c7f3d003140/11461ENGmedya2WEB21_09_11.pdf>. p. 68.

Table 15-4. Magazine Publishing (Market Shares by Revenue), 2001–2011

| | 2001 | 2002 | 2003 | 2004 | 2005 | 2006 | 2007 | 2008 | 2009 | 2010 | 2011 |
|---|---|---|---|---|---|---|---|---|---|---|---|
| Dogan Burda (Dogan Media Group, Dogan Yayin Holding + Hubert Burda Media, Germany) | 43.0 | 45.0 | 47.0 | 44.0 | 40.0 | | 41.0 | 44.0 | 49.0 | 48.0 | 52.0 |
| Dogus Group | 4.0 | 4.0 | 4.0 | 3.0 | 4.0 | 5.0 | 7.0 | 8.0 | 7.0 | 9.0 | 8.0 |
| *Ciner Group* | 24.0 | 19.0 | 17.0 | 16.0 | 20.0 | 21.0 | 21.0 | 5.0 | 4.0 | 5.0 | 4.0 |
| *Ciner (GD)* | | | | | | *7.0* | *6.0* | | | | |
| Ciner/Sabah Group | *24.0* | *19.0* | *17.0* | *16.0* | *20.0* | *14.0* | *14.0* | | | | |
| Turkuvaz Media Group (Calik Group) | | | | | | | | 18.0 | 18.0 | 19.0 | 20.0 |
| Cukurova Group | 3.0 | 5.0 | 8.0 | 9.0 | 11.0 | 10.0 | 10.0 | 8.0 | 5.0 | 4.0 | 4.0 |
| Vatan/Mutlu Group | | | | | | 6.0 | 8.0 | 7.0 | 8.0 | 7.0 | 6.0 |
| Others | 26.0 | 27.0 | 24.0 | 28.0 | 25.0 | 16.0 | 13.0 | 10.0 | 9.0 | 8.0 | 6.0 |
| Total Revenue (mil TRY) | 32.0 | 48.3 | 50.0 | 69.0 | 84.0 | 106.0 | 127.0 | 124.0 | 86.0 | 111.0 | 118.0 |
| Total Revenue (mil US$) | 43.2 | 79.2 | 70.4 | 48.0 | *114* | *151* | *165* | *96.0* | *132* | *168* | 71.0 |
| C4 | 74.0 | 73.0 | 76.0 | 72.0 | 75.0 | 84.0 | 87.0 | 90.0 | 91.0 | 92.0 | 94.0 |
| HHI | 2,428 | 2,458 | 2,612 | 2,319 | 2,171 | 2,385 | 2,348 | 2,473 | 2,886 | 2,821 | 3,226 |
| *N* (>1%) | 9 | 9 | 9 | 9 | 9 | 9 | 9 | 9 | 9 | 9 | 9 |
| Noam Index | 809 | 820 | 871 | 773 | 724 | 795 | 783 | 824 | 962 | 940 | 1,075 |

*Source:* Author.

The C4 has increased due to the steady decline of titles that owned by other smaller firms and TMG's acquisition of the Sabah Group's assets from the Ciner Group in 2007.

Table 15.4 reveals interesting results. First, the top two groups' advertising revenue share increased from 60% to 72%, less than circulation-based market share. Second, contrary to the circulation-based market share, the C4 steadily increased from 74% to 94%, due to the decline of titles owned by other smaller firms, as listed on the table, and the entrance of Calik Group's Turkuvaz in 2007 by acquiring Sabah Group's confiscated titles from Ciner Group (Table 15.4).

## AUDIOVISUAL MEDIA

### Radio

From 1927 to 1993, a state-owned company since 1964, the Turkish Radio and Television Corporation (TRT) held a broadcasting monopoly. After 1993, commercial licenses were awarded, and local, regional, and national stations proliferated: in 2011, there were 36 national, 100 regional and 951 local stations in the country. Of those, TRT owned four national and ten regional stations.[32] Commercial stations are subject to the Private Radio and TV Broadcasting Law and operate under the mandate of RTVSC. The heavily state-subsidized and state-owned TRT is subject to the TRT Law (but not to the RTVSC).

While primary revenues sources of private radio stations firms are advertising and sponsorship, due to the differentiation between unit prices there is no correlation between revenues generated by advertisement based on a duration/second measurement and the market share by advertising revenues. For the former, the figures stand as Power Group 18%, Dogus 17%, Spectrum 13%, Dogan 8%, TRT 8%, Cukurova 7%, Saran 4%, Number One 4%, Samanyoluof Feza 3%, Show Radio 2,5%, and Radio Viva 2%, giving a total of 87% of the total market share.[33] Partly as a result of the entry of conservative religious media both the NI and HHI declined between 2001 and 2011. For 2010, the HHI and C4 values for the top 12 station groups are 1,010 and 56, respectively, indicating a moderately concentrated market. The Noam index of 292 is declining, indicating voice diversity. Interestingly, unlike the other traditional media segments, there are two groups (*Power* and *Spectrum*), active only in radio, with a C4 with a combined 31% market share. The big media firms (Dogus, Dogan, and Cukurova) have 32% of the advertising market, meaning that unlike in the other segments, the conglomerates do not dominate radio due to low level of advertising revenues and low level of barriers to entry.

Due to the proliferation of religious media and their consolidation of preexisting stations, the NI and HHI values have declined over the past decade. However, low advertising revenues and lack of entry barriers mean that these new groups have been unable to consolidate their holdings as they have with respect to newspapers (Table 15.5).

### Broadcast Television

With a population of 74 million and nearly 19 million households equipped with TV receivers (of which 17 million are satellite dishes), Turkey is the sixth largest TV market in Europe.[34] Both commercial and public broadcasters are subject to extensive regulation, and as of 2012, commercial broadcasters are increasingly subject to cross-ownership restrictions by the RTVSC. TRT, the national public broadcaster, is funded

32. "Türkiye'de KaçGazete ve TelevizyonVar?" ("How Many Newspapers and TV Channels are there in Turkey?"). *Medyaradar*, Oct. 2, 2008. Sept. 10, 2010. <http://www.medyaradar.com/haber/gundem-21476/turkiyedeki--gazete--televizyon--radyo-vedergi-sayisi-nekadar--peki-kac-iletisim-fakultesi-var--iste-cok-ilginc-rakamlar.html>.

33. ZenithOptimedia data, http://www.bloombergcom/news/2010-2012-06/web-emergingmarkets-to-lead-global-advertisinggrowth-in-011.html

34. "Türkiye'de KaçGazete ve TelevizyonVar?" ("How Many Newspapers and TV Channels are there in Turkey?"). *Medyaradar*, Oct. 2, 2008. Sept. 10, 2010. <http://www.medyaradar.com/haber/gundem-21476/turkiyedeki--gazete--televizyon--radyo-vedergi-sayisi-nekadar--peki-kac-iletisim-fakultesi-var--iste-cok-ilginc-rakamlar.html>.

Table 15-5. Radio Group (Market Shares by Listeners), 2004–2011

| | 2004 | 2005 | 2006 | 2007 | 2008 | 2009 | 2010 | 2011 |
|---|---|---|---|---|---|---|---|---|
| NTV Group (Dogus Group) | 19.6 | 19.5 | 19.5 | 20.3 | 18.2 | 20.3 | 20.4 | 17.9 |
| Power Group | 7.7 | 8.8 | 8.9 | 11.2 | 9.2 | 11.5 | 12.2 | 12.2 |
| Spectrum Medya | 21.5 | 22.8 | 18.3 | 16.1 | 12.0 | 10.7 | 10.9 | 11.0 |
| Dogan Media Group (Dogan Yayin Holding) | 4.2 | 4.5 | 8.7 | 11.7 | 11.8 | 12.2 | 12.0 | 11.0 |
| Number One | 5.6 | 6.7 | 6.2 | 6.8 | 6.7 | 6.8 | 7.0 | 7.2 |
| TRT (public) | 6.8 | 5.5 | 4.5 | 5.1 | 4.8 | 5.0 | 5.2 | 4.2 |
| Cukurova Group | 6.4 | 5.1 | 4.9 | 5.6 | 5.6 | 5.6 | 5.5 | 4.7 |
| Saran Group | 6.3 | 6.1 | 4.5 | 5.2 | 5.1 | 3.6 | 1.7 | 2.5 |
| Turkuvaz Media Group (Calik Holding) | N/A | N/A | N/A | N/A | N/A | 1.4 | 3.2 | 2.3 |
| Ihlas Group | 3.0 | 2.2 | 1.7 | 1.7 | 1.3 | 1.3 | 1.2 | 0.0 |
| Ipek/Koza Group | N/A | N/A | N/A | N/A | N/A | 1.4 | 1.6 | 1.4 |
| Burç FM(Feza Group) | N/A | N/A | 1.2 | 1.5 | 1.8 | N/A | N/A | N/A |
| Radyo Viva | 3.3 | 4.9 | 4.1 | 4.9 | 4.7 | 4.9 | 3.2 | 4.7 |
| Show Radyo | 5.9 | 5.4 | 4.6 | 5.2 | 4.9 | 4.6 | 3.8 | 3.3 |
| Radyo 7 | 3.3 | 3.7 | 3.5 | 3.5 | 3.3 | 3.4 | 1.4 | 2.1 |
| DunyaRadyo | 1.0 | N/A | 1.4 | 1.3 | 1.3 | N/A | N/A | N/A |
| RadyoKlas | N/A | 1.7 | 1.6 | 1.6 | 1.7 | 1.6 | 1.7 | 1.6 |
| Moral FM | 1.6 | 1.2 | 1.3 | 1.5 | 1.5 | 1.5 | 1.2 | 1.1 |
| Total Revenue (mil TRY) | 74.9 | 78.5 | 98.8 | 111.0 | 112.0 | 88.0 | 104.0 | 121.0 |
| Total Revenue (mil US$) | 51.8 | 58.1 | 68.7 | 85.1 | 86.3 | 56.7 | 69.0 | 72.4 |
| C4 | 56.0 | 58.0 | 55.0 | 59.0 | 51.0 | 55.0 | 56.0 | 52.0 |
| HHI | 1,150 | 1,214 | 1,034 | 1,137 | 891 | 987 | 983 | 855 |
| Noam Index | 307 | 325 | 259 | 284 | 223 | 247 | 246 | 214 |
| *N* (>1%) | 14 | 14 | 16 | 16 | 16 | 16 | 16 | 16 |

*Source:* Advertisers' Association and Author.

though an electricity tax that accounts for 70% of its annual budget along with direct state subsidies, some commercial advertising, and licensing fees. TRT broadcasts nine national programs including, in the Kurdish language two regional channels, and four international channels broadcasting in Turkish for the Turkish diaspora, as well as in French, English, and Arabic.

Ending the public broadcasting monopoly in 1993 was aimed to create of private broadcast networks and to bring several established traditional media firms into the national TV market. In 1990, Turkey's first commercial network—then known as Magic Box and now as Star TV (part of DMG)—began broadcasting from the Federal Republic of Germany (West Germany), and its success encouraged other entrants. Between 1990 and 1993, the state made no attempts to regulate the new media groups. Since TRT was directly administered by the state, the RTVSC was formed to serve as a market regulator only in 1994.[35] From 1994 to 2002, ownership regulation as defined by Law No. 3984 was based on

35. Kaya, Raşit. "A Fait Accompli: Transformation of Media Structures in Turkey." METU *Studies in Development*, 21:3, 1994.

capital limits for share-ownership thresholds: a firm was allowed to own only one radio and only one TV station, with the further restriction that shareholders could not hold more than 20% of any outlet's shares. (If the firm owned stakes in multiple outlets, the total share ratio could not exceed 20%).[36] While the law capped cross-ownership between the print and audiovisual markets, it set no limits on the number and variety of media activities in other markets. A new cap based on viewer/listener percentage was set at 20% by a 2002 revision of Law No. 3984, but that cap proved unrealistic and did not act as a barrier against concentration.[37] Instead of preventing concentration, its provisions enhanced the emerging patron–client relationship between media groups and coalition governments, and allowed media groups to rapidly acquire dozens of small, independent networks.[38] Major media companies effectively lobbied against new cross-ownership restrictions imposed. In 2002, Turkey's Constitutional Court ruled that in addressing cross-ownership issues, the law would lead to monopolies.

New broadcasting legislation was enacted in 2011 as Law No. 6112.[39] The new law regarded the share of commercial communication, advertising and sponsorships revenues as the criteria for protecting competition. The new threshold criterion was set at 30% of total market revenue. Although the law does not describe how this share percentage should be calculated, RTVSC officials suggest that the base of calculation is the 3% of total revenues paid to RTVSC on a monthly basis by broadcasters. No entity can hold more than 25% of the shares of one television or radio station. Additionally, the ownership threshold for foreign investors was increased from 25% to 50%, on the condition that no foreign investor can hold stakes in more than two broadcasters. Although the main concern behind the adoption of the new law was to be in harmony with the EU's Audiovisual Media Services Directive (AVMSD), it is not in compliance with AVMSD's content, ownership, and competition rules. Although it largely aligns legislation with the AVMSD, it has failed to comply with the principle of freedom of reception and retransmission.[40] Law No. 6112 expanded commercial entities' investment opportunities, and it has barred political parties, trade unions, religious foundations, and local governments from owning networks (Table 15.6).

Despite an increase in the number of channels since 2002, the market shares by revenue for DMG, TMG, and the Cukurova Group have all increased: combined, the three media groups take in between 55% and 61% of annual advertising revenues (Table 15.7).

## Multichannel TV Platforms

Due to the decline in consumer costs, the proliferation of satellite TV services and the growth of HDTV, there has been a substantial increase in demand for multichannel services since the early 2000s, bringing the total number of registered households to 3.7 million. Free satellite TV broadcasts, however, still dominate the market, reaching 13.3 million households. Offering over 800 channels, their business model is based mostly on advertising revenues, so the only cost incurred by customers comes from installation fees.[41]

36. Besides the shareholder capping rule stating that the capital share of a real or legal entity or a capital group in a company is limited up to 50%, the 2002 Law No. 4756 that amended Law No. 3984 took the 20% audience share threshold as the main criterion to prevent concentration. In addition, Law No. 4756 lifted restrictions on cross-ownership, allowed media companies to bid in public tenders, and trade shares on the Istanbul stock market.

37. İlkiz, F. "İfade Özgürlüğü ve Türk Ceza Hukuku." Istanbul: *Ceza Hukuku Derneği Yayınları* 1, 2003. p. 107.

38. Cankaya, O., and Yamaner, M. B. *Kitle İletişim Özgürlüğü* (Freedom of Mass Communication). Istanbul: Turhan Bookstore, 2006.

39. Özdurdu, Rahime. Medyada Sahiplik ve Yoğunlaşma, Oluşturduğu Sorunlar ve Şeffaflığın Sağlanması. Ankara: Rekabet Kurumu Uzmanlık Tezi, 2011.

40. European Commission. "Turkey 2011 Progress Report." Brussels: European Commission, Oct. 12, 2011. April 8, 2013. <http://ec.europa.eu/enlargement/pdf/key_documents/2011/package/tr_rapport_2011_en.pdf>. p. 65.

41. "17 milyon uydu anteni çöpe gidiyor." *MyNet Finans*, Aug. 10, 2012. Aug. 12, 2012. <http://finanshaber.mynet.com/detay/ekonomi/17-milyon-uydu-anteni-cope-gidiyor/80911>.

## Table 15-6. TV Broadcasting (Market Shares by Viewership), 1999–2011[1]

| | 1999 | 2003 | 2007 | 2009 | 2010 | 2011 |
|---|---|---|---|---|---|---|
| Dogan Media Group (Dogan Yayin Holding) | 16.9 | 16.7 | 23.6 | 22.7 | 24.0 | 21.8 |
| Kanal D | 16.9 | 16.7 | 14.2 | 14.1 | 15.0 | 14.4 |
| Star TV (until 2007) | *0.0* | *0.0* | *9.4* | *8.6* | *8.9* | *7.4* |
| SHOW (Cukurova Group) | 16.5 | 14.2 | 11.9 | 10.7 | 8.4 | 9.4 |
| Ciner Group | | | 10.4 | 1.6 | 0.6 | N/A |
| ATV (until 2002) | | | *10.4* | | | |
| Habertürk TV | | | | 1.6 | 0.6 | |
| ATV (Sabah Group) | 14.3 | 14.2 | | | | |
| Uzan Group[2] | 15.6 | 4.4 | | | | |
| Star TV (until 2011) | *13.4* | *4.4* | | | | |
| Kral TV (until 2008) | 2.2 | | | | | |
| ATV (Turkuvaz Media Group, Calik Holding)[3] | | | | 10.0 | 11.7 | 11.4 |
| TGRT (Ihlas Group) | 8.4 | 7.9 | | | | |
| Samanyolu TV (Feza Group) | Article III. | 2.7 | 5.6 | 5.7 | 4.6 | 4.0 |
| TRT (public) | 5.3 | 8.7 | 3.9 | 3.1 | 0.3 | N/A |
| TRT1 | *4.8* | *7.1* | 3.9 | *3.1* | *0.3* | |
| TRT2 | *0.5* | *1.6* | | | | |
| Kanal 6 | 2.1 | | | | | |
| Kanal 7 | | 5.0 | 6.5 | 4.5 | 3.9 | 3.4 |
| Flash | | 3.2 | 1.8 | 1.8 | 1.7 | 1.9 |
| Fox International Channels (Murdoch Family, News Corp., US/UK/AUS) | | | 4.7 | 8.2 | 7.9 | 8.7 |
| Kanal 1 | | | | | | |
| Cine5[4] | | | | 1.1 | 0.6 | |
| Total Revenue (mil TRY) | 182 | 756 | 1,950 | 1,543 | 2,219 | 2,698 |
| Total Revenue (mil US$) | 441 | 497 | 1,493 | 994 | 1,473 | 1,614 |
| C4 | 61.0 | 53.0 | 46.0 | 43.0 | 44.0 | 51.0 |
| HHI | 1,151 | 882 | 923 | 867 | 885 | 799 |
| Noam Index | 434 | 294 | 327 | 274 | 335 | 303 |
| *N* (>1%) | 7 | 9 | 8 | 10 | 10 | 7 |

*Source:* AGB Nielson, Author.

[1] The data should be taken with grain of salt pending allegations that regulators and broadcasters are colluding to sell Nielsen data from Turkish households in order to rig what households' meters report.

[2] Both Kral and Star TV are now part of the Dogus Group due to the state's seizure of Uzan's assets in 2004.

[3] ATV was first owned by the Sabah Group from its establishment in 1993 until 2002, when it was confiscated by the state depository insurance fund, including assets of Uzan and Aksoy, and their owners imprisoned. The Ciner Group purchased ATV in collaboration with Saban. Ciner then was forced to sell ATV to Çalik Holding in 2007.

[4] Cine5 is now Al Jazeera Turk, having been sold to the Qatar-based network by the Savings Deposit Insurance Fund in 2011.

## Table 15-7. TV Broadcasting (Market Shares by Revenue), 2008–2011

| | 2008 | 2009 | 2010 | 2011 |
|---|---|---|---|---|
| Dogan Media Group (Dogan Yayin Holding) | 31.4 | 30.5 | 32.8 | 27.8 |
| Kanal D | 21.9 | 20.4 | 22.2 | 19.0 |
| Star TV | *8.3* | *8.9* | *9.5* | *7.7* |
| CNN Türk[1] | *1.2* | *1.2* | *1.1* | *1.1* |
| ATV (Turkuvaz Media Group, Calik Holding) | 12.3 | 12.1 | 16.9 | 13.6 |
| Cukurova Group | 11.4 | 12.0 | 8.7 | 13.1 |
| Show TV | *11.4* | *12.0* | *8.7* | *13.1* |
| Fox International Channels (Murdoch Family, News Corp., US/UK/AUS) | 4.2 | 6.1 | 6.8 | 8.1 |
| NTV Group (Dogus Group) | 6.2 | 6.9 | 6.5 | 5.5 |
| NTV | *3.6* | *4.0* | *3.7* | *3.1* |
| CNBC-E[2] | *2.3* | *2.4* | *2.0* | *1.5* |
| NTV SPOR | *0.3* | *0.5* | *0.8* | *0.9* |
| Habertürk TV (Ciner Group) | 0.3 | 1.2 | 1.9 | 1.4 |
| Samanyolu TV (Feza Group) | 2.2 | 2.1 | 1.6 | 1.9 |
| Kanal 7 | 1.8 | 1.8 | 1.8 | 1.6 |
| TRT (public) | 1.4 | 0.7 | 0.9 | 1.3 |
| Total Revenue (mil TRY) | 1,687 | 1,442 | 2,018 | 2,448 |
| Total Revenue (mil US$) | 1,306 | 929 | 1,339 | 1,464 |
| C4 | 61.3 | 61.5 | 67.3 | 63.8 |
| HHI | 1,203 | 1,169 | 1,536 | 1,237 |
| Noam Index | 456 | 443 | 545 | 412 |
| *N* (>1%) | 7 | 7 | 7 | 9 |

*Source:* RTVSC.
[1]Jointly owned by DMG and Turner's CNN (US).
[2]Jointly owned by the NTV Group and CNBC Europe, which are part of Comcast/GE's NBCUniversal (US).

The market is dominated by three platforms because of prohibitions aimed at insulating free terrestrial and the non-fee-based satellite TV networks from competition. In 1998 first state-owned platform Turksat, based on landline cable, was erected. In 2001, the second license was awarded to the Cukurova Group's Digiturk; in 2007, a third license was awarded to the Dogan Group's D-Smart. Although the service structure of D-Smart and DigiTurk is a hybrid of both multichannel video platforms and video channels in European nations, in order to make the segment compatible for international comparison, they are categorized as multichannel TV platforms along with the state-owned landline-based cable TV network TURKSAT Cable TV. The Digiturk and D-Smart operators cannot bundle cable network operators or provide IPTV services due to regulatory restrictions by the RTVSC and ICTA, which distinguishes them from traditional multichannel system operators (MSOs). Like video channels, they do bundle international programming—such as BBC World News, National Geographic, MovieMax, the Discovery Channel, Cartoon Network, CNN International, TV5 Monde, and Al Jazeera International—and distribute them through the video channel transmission monopolist Turksat (both rely entirely on the Turksat 3a satellite for transmission). Currently, there are no major restrictions on the ownership of subscription TV services, but there are considerable entry

Table 15-8. Multichannel TV Platforms (Market Shares by Subscribers), 2005–2011

| 1) | 2005 | 2006 | 2007 | 2008 | 2009 | 2010 | 2011 |
|---|---|---|---|---|---|---|---|
| D-Smart (Dogan Yayan Holding) | 0.0 | 0.0 | 1.1 | 19.0 | 25.1 | 25.3 | 23.2 |
| Digi Turk (Cukurova Group) | 48.6 | 51.0 | 52.6 | 57.7 | 47.4 | 45.5 | 53.7 |
| Turksat Cable TV (State) | 51.4 | 49.0 | 46.3 | 23.3 | 27.5 | 29.2 | 23.1 |
| Total Revenue (mil US$) | 182 | 208 | 289 | 353 | 391 | 425 | 617 |
| C4 | 100.0 | 100.0 | 100.0 | 100.0 | 100.0 | 100.0 | 100.0 |
| HHI | 5,006 | 5,002 | 4,912 | 4,233 | 3,633 | 3,776 | 3,956 |
| *N* (>1%) | 2 | 2 | 3 | 3 | 3 | 3 | 3 |
| Noam Index | 3,550 | 3,547 | 2,839 | 2,447 | 2,100 | 2,183 | 2,287 |

*Source:* Author.

barriers, in relation to both delivery and programming access (Table 15.8).

The state-owned cable TV (landline) operator Turksat had 1.3 million subscribers in 2011, and it provides both infrastructure and broadcasting services for terrestrial networks. Although there are 15 operators authorized to provide cable services as of 2011, Turksat is the only operator that provides nationwide services.

The privatization of TT had a significant impact on the cable TV market. Before 2005, cable providers were usually established as joint ventures between TT and private companies. However, during the course of privatization, TT was required to transfer its cable assets to Turksat by the CA, on the grounds that the separation of these networks was essential for promoting competition. As this was an interim measure, however, there was some ambiguity as to the ownership claims of commercial operators, and this has discouraged further investment infrastructure. A new draft regulation on the privatization of cable network has been presented to the public by RTVSC in order to redress Turksat's monopoly, but it has yet to become law.[42]

Turksat bundles channels, including over the air national TV (and radio) networks as well as international news channels, and it markets packages over its own landline network. Along with D-Smart and DigiTurk, Turksat offers broadband Internet, cable telephony, and cable data services. Of the three, Turksat has the smallest market shares for broadband Internet (6.1% of the total market) and fewer than 100,000 cable telephony and IPTV subscribers each.

## TELECOMMUNICATIONS MEDIA

### Wireline Telecom

The Post, Telegraph, and Telephone (PTT) General Directorate was established as a state enterprise in 1924. As a result of privatization, the PTT was separated from the Ministry of Transport and Communications (MoTC) in 1993 and broken down into two separate entities: the Directorate of Postal Operations and the joint stock company TT, which functions as the owner of all distribution networks, telephone exchanges, and transmission infrastructure. Until 2000, the MoTC and TT were responsible for regulating the market. Law No. 4502 was passed in 2000 to set a deadline for the full liberalization of fixed line voice services not later than the end of 2003. Parliament subsequently passed legislation in 2003—inspired by the EU's 1998 supranational telecommunications

42. "RTVSC's Draft Regulation on Cable Broadcasting." Ankara: RTÜK, 2011. July 21, 2012. <http://www.rtuk.gov.tr/sayfalar/IcerikGoster.aspx?icerik_id=299ea676-661be-494e-9345-33546b785e12>.

regulatory framework—that finally ended TT's monopoly rights. The termination of monopoly rights and the establishment of the ICTA and new regulatory framework did not indicate full liberalization, though, since new entrants were discouraged by the restrictive licensing regime. Subsequently, the 2001 Telecommunication Act capped the stock ratio of domestic and international shareholders to 49%. The Oger Telecom Venture Group (Saudi Arabia) acquired 55% of TT in 2005 (leaving the Turkish state with 30%), thanks to the removal of the upper limit on foreign ownership. In effect, though, privatization occurred without liberalization: the terms and conditions of the sale granted the Oger Group (Saudi Arabia) a 21-year concession, at the end of which all plant and equipment would revert to the state, a move that is incompatible with EU guidelines. Oger itself is part owned by Saudi Telecom, a state-owned enterprise.

Individual licenses are limited to narrowly defined services, and the boundaries of these services are not always clearly defined, leaving significant legal ambiguities. The ICTA did issue such licenses just months after TT's monopoly rights came to an end, but these licenses were not activated for several years because of lengthy interconnection negotiations with TT. Although TT had to complete technical preparation for Type B licenses and Type A licenses by 2004, it was not until 2006 that any interconnection agreements were actually signed. So, unsurprisingly, the market shares of TT for overall fixed voice services in 2008 were 92% by traffic and 81% by revenue. Although local wireline services were opened to competition and number portability opened up in 2009, the Telecom Association interest group contends that TT's monopoly position has not been seriously affected. If wireline tariffs are compared with EU averages, the tariffs of international calls were slightly lower than the EU average, but in terms of purchasing power parity (PPP), local call tariffs were higher.[43,44] Prices charged by TT are higher than those in many OECD and EU countries, and measures recommended by the EU to increase competition at the local level have been delayed (Table 15.9).

The level of competition in fixed broadband services is even lower than it is for fixed local services. Due to pricing and technical issues among ISPs and TT, the local loop has not been unbundled and bitstream access is restricted. In 2008, although there were approximately 80 registered ISPs, TT held a 95% market share in fixed retail broadband connections (in pure resale and the number of bitstream connections, alternative operators' market shares by revenue were 15%).[45] Besides the very slow implementation of fixed services number portability, there remain outstanding issues regarding interconnection rates and pricing tactics among long-distance service providers and the TT, as well as among ISPs and the TT on unbundling the local loop and granting bitstream access.[46]

If cable TV infrastructure were upgraded, it could offer solid competition against TT's monopoly. However, the ill-defined jurisdictional boundaries between the ICTA and CA have given ICTA an area for political maneuvering that has proven detrimental to reducing concentration. While the ICTA was established as an autonomous entity, it has become an important vehicle for political, bureaucratic, and business rent-seekers. The ICTA's stance that it has full authority to develop and enforce regulations is in part responsible for the regulatory

43. Cullen International Inc. Cullen International Report, Namur: Belquim, Cullen International Inc., 2008. Figure 38, 40, 41, 42, and 45.

44. European Commission. "Communication from the Commission to the European Parliament, the Council, the European Economic and Social Committee and the Committee of the Regions Progress Report on the single European electronic communications market 2008 (14th Report)." EUR-Lex, 2009. April 8, 2013. <http://eur-lex.europa.eu/LexUriServ/LexUriServ.do?uri=CELEX:52009DC0140:EN:NOT>.

45. Cullen International Inc. Cullen International Report, Namur: Belquim, Cullen International Inc., 2008. Table A.23 and A.25.

46. European Commission. "Turkey 2009 Progress Report." Brussels: European Commission. June 26, 2011. <http://ec.europa.eu/enlargement/pdf/key_documents/2009/tr_rapport_2009_en.pdf>. p. 52.

## Table 15-9. Telecom Wireline Market Share by Voice Service Revenues (%)

| | 2000 | 2004 | 2005 | 2006 | 2007 | 2008 | 2009 | 2010 | 2011 |
|---|---|---|---|---|---|---|---|---|---|
| Turk Telekom (public until 2005: Oger Group, Saudi Arabia,[1] then 30% public) | 100.0 | 98.0 | 96.0 | 95.0 | 94.0 | 92.0 | 90.0 | 89.0 | 91.0 |
| Superonline | 0.0 | | | 1.3 | 1.3 | 1.7 | 2.2 | 2.2 | 2.1 |
| Millenicom | 0.0 | | | 1.2 | 1.1 | 1.6 | 2.1 | 2.2 | 2.1 |
| Bizfonik | 0.0 | | | | | 1.1 | 1.2 | 1.7 | 1.8 |
| Turknet | 0.0 | | | 1.1 | 1.2 | 1.3 | 1.2 | 1.4 | 1.3 |
| Telsim/Vodafone Turkiye (Vodafone, UK) | 0.0 | | | 1.1 | 1.1 | 1.3 | 1.4 | 1.5 | 1.2 |
| Total Revenue (billion TRY) | 3.12 | 6.61 | 7.42 | 7.11 | 7.53 | 7.74 | 7.55 | 8.58 | 9.21 |
| Total Revenue (mil US$) | 5.02 | 4.63 | 5.49 | 4.97 | 5.78 | 6.00 | 4.87 | 5.70 | 5.87 |
| C4 | 100.0 | 98.0 | 99.0 | 99.0 | 98.0 | 96.0 | 95.0 | 95.0 | 97.0 |
| HHI | 10,000 | 9,604 | 9,216 | 9,031 | 8,842 | 8,474 | 8,114 | 7,938 | 8,296 |
| *N* (>1%) | 1 | 1 | 1 | 5 | 5 | 6 | 6 | 6 | 6 |
| Noam Index | 10,000 | 9,604 | 9,216 | 4,050 | 3,965 | 3,459 | 3,311 | 3,240 | 3,386 |

*Source:* ICTA.

[1] 35% of Oger is owned by Saudi Telecom Co (Saudi government).

environment that shields incumbents from competition. The CA, for its part, is not engaged with politicians or other regulators over the matter. According to experts at the CA, the existing legislation limits the authority's sanctioning powers. Thus, public and civil society support of regulatory capture is needed for the drafting of more effective competition regulations. Although implementing regulations regarding customer rights and quality of services were introduced by the ICTA at the EU's urging, the transparency and accountability of its activities are still in need of substantial improvement.[47] A decree adopted in 2011 authorizes the Ministry of EU to monitor and inspect all kinds of transactions by independent regulatory authorities in line with obligations stemming from the *acquis*. This raises concerns as to the independence of the regulatory authorities in Turkey in carrying out their duties.

## Wireless Telecom

The Turkish mobile market is one of the most concentrated cellular phone markets in Europe. Services began in 1998 with the award of two GSM-900 network licenses to Turkcell and Telsim (the latter then part of the Uzan Group) on the basis of revenue-sharing agreements between them and TT. Later that same year, the government issued the operators licenses, each worth US$500 million (TRY1.6 billion), in a 25-year concession agreement. Until 2001 Turkcell and Telsim were the only mobile operators in the country. In 2001, two GSM-1800 licenses were rewarded to Aycell, a TT subsidiary, and Telecom Italia Mobile's Aria (Italy), in partnership with the public bank Türkiye İş Bankası. These new operators were unable to gain sufficient market shares in the beginning due to the absence of mobile number portability,

47. European Commission. "Turkey 2011 Progress Report." Brussels: European Commission, Oct.12, 2011. April 8, 2013. <http://ec.europa.eu/enlargement/pdf/key_documents/2011/package/tr_rapport_2011_en.pdf>. p. 64.

Table 15-10. Wireless Telecom (Market Shares by Subscribers), 1994–2011

| | 1994 | 1995 | 1997 | 1999 | 2001 | 2003 | 2005 | 2007 | 2009 | 2011 |
|---|---|---|---|---|---|---|---|---|---|---|
| Turkcell (Cukurova Telecom Holdings Ltd, Cukurova Group) | 75.0 | 70.0 | 73.0 | 73.0 | 69.0 | 68.0 | 64.0 | 58.0 | 56.0 | 53.0 |
| Telsim/Vodafone Turkiye (Vodafone, UK) | 25.0 | 30.0 | 27.0 | 27.0 | 30.0 | 21.0 | 22.0 | 26.0 | 25.0 | 28.0 |
| Avea (Turk Telecom; Oger Group, Saudi Arabia) | 0.0 | 0.0 | 0.0 | 0.0 | 2.0 | 10.0 | 14.0 | 16.0 | 19.0 | 19.0 |
| C2/C3 | 100.0 | 100.0 | 100.0 | 100.0 | 100.0 | 100.0 | 100.0 | 100.0 | 100.0 | 100.0 |
| HHI | 6,250 | 5,800 | 6,058 | 6,058 | 5,665 | 5,165 | 4,776 | 4,296 | 4,122 | 3,954 |
| *N* (>1%) | 2 | 2 | 2 | 2 | 3 | 3 | 3 | 3 | 3 | 3 |
| Noam Index | 4,433 | 4,113 | 4,296 | 4,296 | 3,275 | 2,986 | 2,761 | 2,483 | 2,383 | 2,286 |

*Source:* ICTA.

ineffective roaming/interconnection policies, and significant differences between on-net and off-net prices. As a result, Aycell and Avea merged in 2003 to form a new operator, called Avea, in which TT now holds an 81% stake. Telsim was sold to Vodafone (UK) in 2005 after the Uzan Group's assets were seized by the state (Table 15.10, Table 15.11).

Turkey imposes one of the highest value added taxes (VAT) in Europe on mobile operators as a proportion of total cost of ownership. Although the base VAT rate on telecommunications services other than mobile is similar to the rate charged in other countries, Turkey has 45% mobile-specific taxes.[48] Although legislators and experts have recognized that the relationship

Table 15-11. Wireless Telecom (Market Shares by Revenue), 2004–2011

| | 2004 | 2005 | 2006 | 2007 | 2008 | 2009 | 2010 | 2011 |
|---|---|---|---|---|---|---|---|---|
| Turkcell (Cukurova Telecom Holdings Ltd., Cukurova Group) | 74.5 | 70.3 | 7.0 | 63.7 | 63.5 | 61.9 | 57.2 | 52.0 |
| Telsim/Vodafone Turkiye (Vodafone, UK) | 15.2 | 16.9 | 15.3 | 22.2 | 20.9 | 19.6 | 23.9 | 28.1 |
| Avea (Turk Telecom; Oger Group, Saudi Arabia) | 10.3 | 12.8 | 13.7 | 14.1 | 15.6 | 18.6 | 18.9 | 20.0 |
| Total Revenue (mil TRY) | 6,669 | 8,338 | 9,036 | 11,752 | 12,622 | 12,971 | 13,995 | 15,444 |
| Total Revenue (mil US$) | 4,614 | 6,170 | 6,285 | 9,000 | 9,775 | 8,352 | 9,287 | 9,239 |
| C3 | 100.0 | 100.0 | 100.0 | 100.0 | 100.0 | 100.0 | 100.0 | 100.0 |
| HHI | 5,887 | 5,392 | 5,463 | 4,748 | 4,713 | 4,556 | 4,197 | 3,888 |
| *N* (>1%) | 3 | 3 | 3 | 3 | 3 | 3 | 3 | 3 |
| Noam Index | 3,403 | 3,117 | 3,156 | 2,745 | 2,724 | 2,634 | 2,426 | 2,247 |

*Source:* ICTA.

48. Deloitte and GSM Association. "Global Mobile Tax Review 2006–2007." GSMWorld, 2012. June 27, 2012. <http://www.gsma.com/newsroom/wp-content/uploads/2012/03/taxreport1.pdf>.

between the higher taxes and competition has been a shortcoming for more than a decade, the ICTA, MoTC, and Ministry of Finance have only reduced the broadband Internet services tax to redress the matter.

While competitive safeguards have been introduced in mobile number portability and regulatory cost accounting, the rights of new entrants are weak. In wireless markets, competition has been more active, but communication charges imposed on mobile operators are still high. The authorization of mobile virtual network operator services (MVNO) and for regulation authorizing broadband wireless access services operators is pending.

## INTERNET MEDIA

### Internet Service Providers (ISP)

Turkey is the 13th largest Internet market in the world (and sixth in Europe), with 35 million (45% of the population and 41.6% of households) users averaging 32.7 hours per month.[49] With the rapid growth of the market from basic localized Internet access services to large clusters providing a wide range of information services with web connection, TTNet—a TT subsidiary—and a few other providers are well positioned to take advantage of growing market opportunities. They have consolidated their market shares at the expense of small local ISPs, many of which have been forced to shut down. Although there were more than 120 license holders in 2011, only 70 of them are currently active. TTNet's overall market share is 86%, whereas in fixed retail broadband connections (xDSL), its share rises to 95%. Most ISPs are small and serving local markets, often operating as virtual ISPs reselling connectivity.

Between 2006 and 2011 the size of the market increased with the addition of many new households and the rapid adoption of fixed retail broadband connections. By establishing TTNet in 2007, TT extended its telecommunications monopoly to Internet services. Due to the absence of alternative delivery networks, other ISPs have for years been dependent on TT to deliver their services. Although new regulations now guarantee other ISPs access to TT's network, it still benefits from its established relationships with fixed telephony subscribers.

Given the fact that 89% of fixed line Internet access is provided by xDSL, 6% by cable TV network, and 5% by other means, the fixed broadband ISP segment is highly concentrated. The number of bitstream connections was not more than 8,000 in the late 2000s and the combined market shares by revenue for TTNet's competitors only surpassed 15% after 2010.[50] TTNet's monopoly over infrastructure and services endures, since the commercial realization of bitstream access has been significantly delayed by TT's legal challenges. Although ICTA aimed to resolve pure resale and bitstream issues and make bitstream to be readily offered to ISPs, TT has been successful in challenging the ICTA decisions (Table 15.12).

The ICTA agency has not been able to intervene in the retail broadband segment, which is the provenance of the CA. However, regulation has failed to provide a clear delineation of the responsibilities between two agencies. As such, the CA still hesitates to adjudicate cases involving claims of anticompetitive behavior exercised by TT in domains under ex-ante regulation that is enforced by the ICTA, and instead focuses on retail broadband Internet market. Due to continuous advancement in mobile technology, the launch of 3G services in 2010, and the convergence of fixed mobile services, TT does at least face increased competition in wireless services, though it remains to be seen to what extent this will benefit its competitors.

### Search Engines

Google's Turkish-language portal google.com.tr monopolizes the search engine market, reaching 93% of the total online population. Other

49. "Internet Usage in Europe." Internet World Stats, June 30, 2012. April 8, 2013. <www.internerworldstats.com/stats4.htm>.
50. Cullen International Inc. Cullen International Report, Namur: Belquim, Cullen International Inc., 2008.

## Table 15-12. Internet Service Providers (Market Shares by Revenue), 2006–2011

| | 2006 | 2007 | 2008 | 2009 | 2010 | 2011 |
|---|---|---|---|---|---|---|
| TTNet (Turk Telekom; Oger Group, Saudi Arabia) | 92.9 | 93.0 | 93.3 | 90.6 | 85.6 | 82.7 |
| Turksat (public) | 1.0 | 1.0 | 1.1 | 2.3 | 3.8 | 5.3 |
| Others[1] | 6.1 | 6.1 | 5.6 | 7.1 | 10.7 | 12.1 |
| Total Revenue (mil TRY) | | | | | 674.5 | 887.5 |
| Total Revenue (mil US$) | | | | | 447.6 | 530.9 |
| C4 | 97.0 | 97.0 | 96.0 | 97.0 | 94.0 | 93.0 |
| HHI | 8,641 | 8,658 | 8,655 | 8,226 | 7,348 | 6,865 |
| Noam Index | 6,218 | 6,140 | 6,138 | 5,834 | 5,211 | 4,869 |
| *N* (>1%) | 2 | 2 | 2 | 2 | 2 | 2 |

*Source*: ICTA.
[1]Includes ISPs.

popular global search engines have not been able to gain a market share against Google, and their presence is negligible; there are no indigenous search engines competing in the market at this time (Table 15.13).

## Online News

The development of online portals by news agencies and the proliferation of social networking platforms have increased online media awareness.[51] While national daily newspapers, magazines, and broadcasters all produce web content, there are also numerous independent news portals, online magazines, and blogs. Some online news sites—such as MyNet.com, haberler.com, tumgazeteler.com, and ensonhaber.com—host numerous news blogs, exemplifying the potential of independent online journalism. Although audience interest in online media is high, independent online journalism is still relatively underdeveloped. Most

## Table 15-13. Internet Search Engines (Market Shares by Search Volume), 2001–2011

| | 2001 | 2004 | 2006 | 2008 | 2010 | 2011 |
|---|---|---|---|---|---|---|
| Google (US) | 99.0 | 99.0 | 99.0 | 98.0 | 98.0 | 98.0 |
| Others | 1.0 | 1.0 | 1.0 | 2.0 | 2.0 | 2.0 |
| C4 | 99.0 | 99.0 | 99.0 | 99.0 | 99.0 | 99.0 |
| HHI | 9,801 | 9,801 | 9,801 | 9,604 | 9,604 | 9,604 |
| *N* (>1%) | 1 | 1 | 1 | 1 | 1 | 1 |
| Noam Index | 7,001 | 7,001 | 7,001 | 6,860 | 6,860 | 6,860 |

*Source*: comScore, qSearch, Statcounter.

51. Turkey is currently ranked worldwide, sixth for Facebook and eighth for Twitter, with 31 million and 8 million users, respectively. US firms such as Facebook, Windows Live, Twitter, and Netlog.com dominate the social media market. Although no Turkish conglomerates such as DMG and TMG have yet entered the market to offer domestic alternatives to the US networks, independent web portals such as MyNet.com, Gecevip.com, and Kalpkalbe.com have been able to compete with them for market shares, though Facebook is by far the most popular platform.

online news is copied and pasted from traditional media outlets. Only a few online media sites actually employ journalists to do their own reporting. Online news readership has not affected the circulation numbers of daily newspapers and magazines significantly.[52] The Internet users largely gravitate to the websites of established media outlets owned by conglomerates to get their news. Regulatory restrictions on content further impede the development of independent online news sites: Turkey's existing Internet Law limits freedom of expression and restricts citizen's rights of access.[53] While the growth of the Internet has been a very important tool for the development of new digital media and the emergence of some new globally and regionally powerful operators, its influence on concentration of traditional media has been negligible so far. The ICTA's power to ban websites with so-called inappropriate content without recourse to judicial proceedings for the webhosts has limited the growth of the medium, and the regulator is under public (and EU) pressure to relax some of these measures.[54]

Major media conglomerates, including those with no prior investments in media groups, were among the first to invest in online news media. Between 2009 and 2011, the market shares of the top four conglomerates (DMG, Dogus, TMG, and Ciner) increased from 68% to 73%. DMG dominates the market, and the online editions of its dailies *Hurriyet* and *Milliyet* are the fourth and fifth most accessed online news portals in

**Table 15-14. Online News Media (Market Shares By Monthly Unique Visitors), 2009–2011**

| | 2009 | 2010 | 2011 |
|---|---|---|---|
| Dogan Media Group (Dogan Yayin Holding) | 48.0 | 50.2 | 52.0 |
| hurriyet.com.tr | *21.1* | *22.0* | *23.0* |
| milliyet.com.tr | *19.0* | *20.1* | *21.0* |
| E-Kolay News | *7.9* | *8.1* | *8.0* |
| MyNet News | 10.4 | 10.0 | 10.0 |
| haberler.com | 8.2 | 7.1 | 6.4 |
| tumgazeteler.com | 7.5 | 7.0 | 6.5 |
| haberturk.com (Ciner Group) | 7.4 | 7.8 | 7.7 |
| sabah.com.tr (Turkuvaz Media Group, Calik Holding) | 7.1 | 6.1 | 6.0 |
| ntvmsnbc.com (Dogus Group + NBCUniversal, Comcast/GE, US) | 5.8 | 7.0 | 7.2 |
| Ensonhaber.com | 5.6 | 4.8 | 4.2 |
| Total Revenue (mil TRY) | 177 | 230 | 337 |
| Total Revenue (mil US$) | 115 | 152 | 202 |
| C4 | 74.0 | 75.0 | 77.0 |
| HHI | 2,705 | 2,889 | 3,052 |
| Noam Index | 956 | 1,021 | 1,078 |
| *N* (>1%) | 8 | 8 | 8 |

*Source:* ZenithOptimedia, Gemius, comScore Media Metrix.

52. Ipsos-KMG. "İnternet Kullanım Alışkanlıkları Raporu 2010" (Report on Internet Usage Habits 2010). Ipsos-KMG, 2011. Feb. 3, 2011. <http://www.ipsos-kmg.com/node/789>.

53. European Commission. "Turkey 2011 Progress Report." Brussels: European Commission, Oct. 12, 2011. April 8, 2013. <http://ec.europa.eu/enlargement/pdf/key_documents/2011/package/tr_rapport_2011_en.pdf>. p. 27.

54. Tunc, A. and Gorgulu, V. *Mapping Digital Media: Turkey*. Istanbul: Open Society Foundations-Media Program, 2012. April 8, 2013. <http://www.opensocietyfoundations.org/sites/default/files/mapping-digital-media-turkey-20120423.pdf>. p. 8.

Table 15-15. Social Media Market Share by Monthly Unique Visitors (%)

| | 2009 | 2010 | 2011 |
|---|---|---|---|
| Facebook | 45.0 | 49.2 | 52.0 |
| Windows Live Profile | 22.0 | 17.0 | 14.0 |
| MyNet Eksenim | 9.0 | 8.2 | 6.1 |
| Netlog.com | 5.0 | 4.3 | 4.0 |
| Kalpkalbe.com | 5.0 | 4.7 | 4.1 |
| Windows Live People | 4.4 | 4.1 | 4.0 |
| MySpace Sites | 2.6 | 2.2 | 2.1 |
| Gecevip.com | 2.4 | 2.3 | 2.2 |
| Badoo.com | 2.4 | 2.0 | 2.1 |
| Twitter | 2.2 | 6.0 | 9.4 |
| Total Monthly Unique Visitors (million) | 16.9 | 19.2 | 21.5 |
| C4 | 81 | 80 | 82 |
| HHI | 2,683 | 2,884 | 3,087 |
| NOAM | 849 | 913 | 977 |

*Source:* Data compiled from comScore Media Metrix, Socialbakers, StatCounter, and Webrazzi analysis.

Europe, with 9.5 million and 8.8 million unique monthly visitors, respectively (Table 15.14).

Numerous outlets compete a small online advertising revenue pool of just US$191million (TRY318 million). Online news' sites advertising revenues are inadequate to create a self-sufficient and independent online media to be an alternative to the concentrated mainstream media conglomerates.

## Social Media

Table 15.15 shows clearly a highly concentrated market increasingly dominated by *Facebook* among other social media networks. From 2009 to 2011, HHI value increased from 2,683 to 3,087 and C4 fluctuated in between 80% and 82%. The Turkish social media segment is dominated by foreign sites such as Facebook, Windows Live, Twitter, and Netlog.com. Although any major domestic conglomerates with significant market share and investment in online news media are not present in this segment, independent firms such as MyNet, Gecevip.com, and Kalpkalbe.com have been able to manage to stay in competition with global firms (Table 15.15).

# INTERNATIONAL OWNERSHIP (BOTH INBOUND AND OUTBOUND)

International investors have substantial stakes in Turkish media markets. Similarly, Turkish media conglomerates have invested abroad. The first joint venture with a foreign media investor was in private TV broadcasting segment; the CNN-Turk partnership was a joint venture between CNN International (part of AOL-Time Warner) and the Dogan Group in 1998. In 2007, Robert Murdoch's News Corp. ventured with the Ihlas Group in September 2007. News Corporation entered the Turkish TV broadcasting market by acquiring a 50% share of Ihlas Group's TGRT channel, later renamed FOX TV. In 2006, Dogan signed a joint venture with Axel Springer AG of Germany. Axel Springer AG owns 44.9% of Dogan Broadcasting Holding. In 2011, Al-Jazeera (Qatar) bought a 50% share of Cine5 TV.

While there is no foreign direct investment (FDI) in the newspaper segment, it exists in the magazine market. Dogan, as mentioned, has a joint partnership with the German Burda, with a 44.9%

stake, and with Egmont Publishing. Mutlu Magazine Group has partnered with Imako, a Greek publisher. Foreign ownership restrictions up to 50% are valid for all segments, and foreign investment in all media markets are subject to the supervision of the General Directory of Foreign Direct Investment at the Ministry of Economy.

In the wireline market, the Saudi Oger Group has 55% stake at incumbent operator TT that holds 81% of third wireless GSM operator Avea. In wireless segment Vodafone International has 51% stake at the second GSM operator Vodafone Turkey. Although the share and ownership structure is very complex, it can be estimated that Russian Alfa Group and TeliaSonera of Norway have approximately 40% at stake in Turkcell.

Turkish media investments abroad are extensive. The major daily newspapers of the major conglomerates such as *Hurriyet* and *Milliyet* of Dogan, *Sabah* of Calik, and *Zaman* of Feza Group are published and circulated in all Western European countries, the United States, and Canada to reach approximately 4.5 million Turks working abroad. All major public and private national TV channels, by using Turksat 3a satellite, transmit a large footprint covering from Western and Eastern Europe to Central Asia, including Russia, North Africa, and the Arabian Peninsula. Dogan, which owns a TV station in Romania, has been very prolific here. It has also acquired 67% of Trader Media East (TME), a leader in online and print classified advertising with strong local brands serving markets in Russia, Poland, Ukraine, Belarus, Latvia, Croatia, and Kazakhstan, for USD336.5 million. Similar to Dogan, Cukurova's GSM operator Turkcell has several significant investments in GSM services in other countries, 100% stake at the Georgian GSM operator GEOCELL, 100% at MOLDCELL of Moldavia, 100% at KKTCell of Northern Cyprus Turkish Republic, 80% at BEST of Belarus, 55% at Astelit of Ukraine, 51% at Azercell of Azerbaijan, and 51% at Kcell of Kazakhstan, respectively. In addition, TT has 15% stake at ALBtelecom, the incumbent fixed line operator of Albania.

## CONCLUSION

Overall, Turkish media markets are dominated by a handful of vertically integrated conglomerates with cross-ownership in almost every segment, supported by influential religious sects. These interests control more than 90% of advertising revenues. Despite the media market being liberalized on paper, due to political, economic, and regulatory factors it remains highly concentrated, with no real ownership controls in place (again, excluding TV and radio). Turkish media markets show characteristics of monopoly in cable TV, fixed line, ISP, and search engines (the latter is due to Google's worldwide role). There are oligopolies in mobile telecom, the print media, digital platforms, and online news. In most of the segments, particularly wireline, wireless, ISP, cable TV, and digital platforms, the implementation of competition rules have been deliberately delayed.

There are only two unconcentrated media markets, broadcast TV and radio. However, the HHI values based on the market shares of advertising revenues are higher than those based on circulation or subscribers, except for consumer magazines. For example, the *Zaman* daily newspaper is the most circulated newspaper in Turkey with almost one million sold issues/day in 2011. It has 22.7% of market share by circulation, but only 10.1% by advertising revenues.

This private concentration process began when the state relaxed its control over media. Traditional family ownership structures were soon replaced by corporate ownership after the 1980s. As a result of the end of state monopoly in radio and TV markets in 1990, commercial groups began obtaining control of those industries. After 1996, the concentration of ownership further increased by the consolidation of print media and audiovisual media assets by conglomerates.

The absence of overarching cross-ownership rules, weak enforcement of ownership controls, and delayed implementation of competition laws have allowed vested interests to abuse the power of the press. Turkish conglomerates have been successful in finding legal loopholes to bypass regulations, and their complicated ownership structure helps muddy the waters. The largest conglomerates in the larger Turkish economy dominate the media, and there seems to be no efficient way to control the high concentration and cross-media ownership and to promote competition.

Although the regulatory regime concerns the capabilities of vertically integrated media conglomerates to use their power across media segments to influence public opinion on political, social, and economic issues and to twist competition in their favor, the conglomerates are still powerful in almost all highly concentrated media markets.

Besides the legal ambiguities caused by complex legislative process, there has been a multiplicity of ministries, governmental agencies, national regulatory agencies (NRAs) with poor reputations for independence and transparency, overlapping regulatory regimes, slow and ineffective implementation of competition rules and the EU's *acquis*, weak enforcement of ownership limits and control mechanisms, and the absence of cross-ownership rules in most media industries.

The research presented in this chapter reveals six important findings. Some of them may contribute to our understanding of more general comparative media research and theory. First, the findings show that the Noam index is an important tool to measure diversity of voices and external pluralism and may contribute further to comparative media research by identifying the relationship of political parallelism and client networks and clientelism in media systems. Second, it suggests that the regionalization of Turkish media necessitates the development of new comparative media system models. Third, from the perspective of patron-client networks theory, in the 1990s and early 2000s, the relationship between media and state has been inversed. Fourth, Turkish media markets show that there is a fundamental concern about the ability of the EU to raise the level of efficiency (competition) and harmonize the Turkish media market within a single EU market just by arranging the rules. Fifth, the EU's Directives in relation to Turkish regulations illustrate implementation problems of EU policies. Sixth, the EU's current competition provisions contained in the *acquis* are inadequate to address competition and concentration concerns. How to regulate media markets and enforce the rules, particularly of competition, still remain challenging issues.

As we discussed in this chapter, the major regulatory challenge present in Turkey is the domination of media markets by several vertically integrated industrial and financial conglomerates. Also, despite a large number of local, regional, and national news outlets, state censorship is still a major issue, especially for online news. From 2002 to 2011, Article 167 of the Constitution was used to regulate electronic media through ownership restrictions by limiting the number of national broadcasting licenses that could be held by a single individual or entity. But Article 167 never achieved the goal of reducing concentration in that way, since it did not address cross-ownership.

Failure to limit cross-ownership has resulted in augmented clientelist relationships between media patrons and the state. Ruling parties have favored "eligible" businessmen with subsidies, credits, and preferential legal treatment. The proliferation of private broadcasters and the continued concentration of Turkish media markets had no effect on the high degree of political parallelism: the media are polarized, with outlets associated with (or wholly owned by) pro-JDP individuals on one side. The other media are not necessarily opposed to the JDP, but are more concerned with enlarging their profits. Turkish politics and society are highly polarized and deeply divided. The mainstream media, however, are primarily concerned with enlarging their profit margins through higher circulations/ratings. On the other side, conservative, religious community, and sect-based pro-JDP media outlets are principally involved in the dissemination of their perspectives on lifestyles and politics.

With respect to proposed EU reforms for Turkey, there are still concerns regarding the capability of such measures to bring the Turkish media market into line with the European common market, as their competition provisions have so far proven inadequate. Moreover,

Turkish legislators have been unsuccessful in planning for long-term growth and tight ownership controls: a lack of mandates necessary for formulating and implementing policies contributes to ill-defined administrative boundaries among regulatory bodies. The national regulatory agencies' (NRAs) immunity from political interventions, independence, accountability objectivity, and the transparency of NRAs have been disputed since their establishment. Governments express and endorse their political and economic objectives through these authorities. The ICTA has been heavily criticized for exercising discretion that could lead to legal inconsistencies in the discharge of its duties, and other bodies are even more amiss in transparency, accountability, and autonomy.

Although the independence, duties, and powers of the NRAs are well established by law, some aspects of the independence—such as effective structural separation of the regulatory functions from activities associated with ownership or control, appointment and dismissal rules of the board by Cabinet of Ministers and Parliament, independence from state agencies—still need to be ensured. Transparency in the decision-making process also needs to be improved.

In sum, given the convergence of media networks and products, and the emergence of new media, it is evident that problems in Turkish media markets cannot be resolved simply by developing new laws. A complementary and integrated approach should be accepted on all levels to create legislation because enforcement today does not offer fair competition rules for players, but instead protects conglomerates in highly concentrated media markets. Bringing regulatory bodies under the rubric of a single umbrella regulatory agency could be an institutional solution to the multiplicity of responsibilities, overregulation, lack of transparency, slow implementation of reforms, and weak enforcement mechanisms. Given the rapid development of new media outlets, the questions of how to create a new legal framework and the questions of how to regulate the media markets should be brought up into the public policy agenda as soon as possible.

# Bibliography

## 1. Books and Articles


Ardıyok, Ş. and Oğuz, F. "Competition law and regulation in the Turkish telecommunications industry: Friends or foes?" *Telecommunications Policy* 34:4, 2010. April 8, 2013. <http://www.sciencedirect.com/science/article/pii/S030859610900113X>.

Bek, M. G. "Tabloidization of News Media: An Analysis of Television News in Turkey." *European Journal of Communication* 19:3, 2004. April 1, 2013. <http://ejc.sagepub.com/content/19/3/371.full.pdf+html>.

Cankaya, O., and Yamaner, M. B. *Kitle İletişim Özgürlüğü* (Freedom of Mass Communication). Istanbul: Turhan Bookstore, 2006.

Carkoglu, Ali. "Introduction." *Turkish Studies* 11:4, 2010.

Carkoglu, Ali and Yavuz, Gozde. "*Press–party Parallelism in Turkey: An Individual Level Interpretation.*" *Turkish Studies* 11:4, 2010.

Finkel, Andrew. "Who Guards the Turkish Press? A Perspective on Press Corruption in Turkey." *Journal of International Affairs* 54:1, 2000.

Hallin, D. and Mancini, P. *Comparing Media Systems. Three Models of Media and Politics.* London: Cambridge University Press, 2004.

Hallin, D. and Papathanassopoulos, S. "Political Clientelism and the Media: Southern Europe and Latin America in Comparative Perspective." *Media, Culture & Society* 24, 2002. June 21, 2012. <http://www.portalcomunicacion.com/both/opc/client.pdf>.

Haurtcourt, Allison. *The European Union and the Regulation of Media Markets.* Manchester: Manchester University Press, 2007.

İlkiz, F. "İfade Özgürlüğü ve Türk Ceza Hukuku." Istanbul: *Ceza Hukuku Derneği Yayınları* 1, 2003.

Kaya, Rasit. "A Fait Accompli: Transformation of Media Structures in Turkey." *Metu Studies in Development*, 21:3, 1994.

Kaya, Rasit and Cakmur, Baris. "Politics and the Mass Media in Turkey." *Turkish Studies* 11:4, 2010.

Kologlu, Orhan. *Osmanlı'dan 21. Yüzyıla Basın Tarihi* (*A History of the Turkish Press from the Ottomans to the 21st Century*). Istanbul: Pozitif Yayınları, 2006.

Özdurdu, Rahime. *Medyada Sahiplik ve Yoğunlaşma, Oluşturduğu Sorunlar ve Şeffaflığın Sağlanması*. Ankara: Rekabet Kurumu Uzmanlık Tezi, 2011.

Papathanassopoulos, S. "The Mediterranean/ Polarized Pluralist Media Model Countries." Terzis, George, ed. *European Media Governance: National and Regional Dimensions*. Bristol, UK; Chicago: Intellect, 2007.

Safak, Yeni. "Yıldırım: RTÜK ile BTK birleşebilir" (RTVSC and ICTA May Merge). *YeniŞafak*, Dec. 25, 2009. June 21, 2012. <http://yenisafak.com.tr/bilisim/?i=231027>.

Schweizer, Kristen and Rabil, Sarah. "Internet, Emerging Markets to Lead Worldwide Advertising Growth Next Year." *Bloomberg*, Dec. 6, 2010. April 8, 2013. <http://www.bloomberg.com/news/2010-2012-06/web-emerging-markets-to-lead-global-advertising-growth-in-2011.html>.

"17 milyon uydu anteni çöpe gidiyor." *My Net Finans*, Aug. 10, 2012. Aug. 12, 2012 . <http://finanshaber.mynet.com/detay/ekonomi/17-milyonuydu-anteni-cope-gidiyor/80911>.

Sönmez, Mustafa. "Medyada Erozyon ve Ekonomik Nedenler." Gürkan, Nilgün, ed. *Türkiye'de Kitle İletişimi*. Istanbul: Turhan Kitabevi, 2004.

Sumer, Burcu. "The Impact of Europeanisation on Policy-Making in Turkey: Controversies, Uncertainties and Misfits in Broadcasting Policy (1999–2009)." *Ankara University European Research Center Research Series* 35, 2010.

Topuz, Hıfzı. *II. Mahmut'tan Holdinglere Türk Basın Tarihi (History of the Turkish Press from Mahmud II to the Holdings Companies)*. Istanbul: Remzi Kitabevi, 2003.

Tunç, A. "Faustian Acts in Turkish Style: Structural Change in National Newspapers as an Obstacle to Quality Journalism in 1990–2003." O. Spassov, O., ed. *Quality Press in Southeast Europe*. Sofia: SOEMZ/Sofia University, 2003.

Tunc, A. and Gorgulu, V. *Mapping Digital Media: Turkey*. Istanbul: Open Society Foundations-Media Program, 2012. April 8, 2013. <http://www.opensocietyfoundations.org/sites/default/files/mapping-digital-media-turkey-20120423.pdf>.

"Türkiye'de Kaç Gazete ve Televizyon Var?" ("How Many Newspapers and TV Channels are there in Turkey?"). *Medyaradar*, Oct. 2, 2008. Sept. 10, 2010. <http://www.medyaradar.com/haber/gundem-21476/turkiyedeki--gazete--televizyon--radyo-ve-dergi-sayisi-nekadar--peki-kac-iletisim-fakultesi-var--iste-cok-ilgincrakamlar.html>.

"Turkey 'world leader' in imprisoned journalists, IPI report says." *Hürriyet Daily News*, April 8, 2011. April 1, 2013. <http://www.hurriyetdailynews.com/default.aspx?pageid=438&n=ipi-report-declares-turkey-world-leader-of-imprisoned-journalists-2011-2004-08>.

## 2. Reports

Cullen International Inc. Cullen International Report, Namur: Belgium, Cullen International Inc., 2008.

Deloitte and GSM Association. "Global Mobile Tax Review 2006-2007." GSM World, 2012. June 27, 2012. <http://www.gsma.com/newsroom/wp-content/uploads/2012/03/taxreport1.pdf>.

Elmas, Esra and Kurban, Dilek. "Background information report: Media policies and regulatory practices in a selected set of European countries, the EU and the Council of Europe: The case of Turkey." Istanbul: Turkish Economic and Social Studies Foundation Publications, 2010. June 1, 2012. <http://www.mediadem.eliamep.gr/wp-content/uploads/2010/05/Turkey.pdf>.

European Commission. "Communication from the Commission to the European Parliament, the Council, the European Economic and Social Committee and the Committee of the Regions - Progress Report on the single European electronic communications market 2008 (14th Report)." EUR-Lex, 2009. April 8, 2013. <http://eur-lex.europa.eu/LexUriServ/LexUriServ.do?uri=CELEX:52009DC0140:EN:NOT>.

European Commission. "Turkey 2011 Progress Report." Brussels: European Commission, Oct. 12, 2011. April 8, 2013. <http://ec.europa.eu/enlargement/pdf/key_documents/2011/package/tr_rapport_2011_en.pdf>.

"Internet Usage in Europe." Internet World Stats, June 30, 2012. April 8, 2013. <www.internerworldstats.com/stats4.htm>.<http://www.ectaportal.com/en/upload/Scorecards/Regulatory_Scorecard_2009/ECTA%20Regulatory%20Scorecard%20Report%202009.pdf>.

Ipsos-KMG. "İnternet Kullanım Alışkanlıkları Raporu 2010" (Report on Internet Usage Habits 2010). Ipsos-KMG, 2011. Feb. 3, 2011. <http://www.ipsos-kmg.com/node/789>.

Jones Day and SPC Network. "Regulatory Scorecard 2009." European Communities Trade Mark Association (ECTA), 2010.

Katholieke Universiteit Leuven (ICRI), Jönköping International Business School (MMTC), Central European University (CMCS) and Ernst & Young Consultancy Belgium. "Independent Study on Indicators for Media Pluralism in the Member States—Towards a Risk-Based

Approach." Leuven: Katholieke Universiteit Leuven, 2009. June 19, 2012. <http://ec.europa.eu/information_society/media_taskforce/doc/pluralism/study/final_report_09.pdf>.

Kurban, D. and Sozeri, C. "Case study report: Does media policy promote media freedom and independence? The case of Turkey." Istanbul: Turkish Economic and Social Studies Foundation, 2011. April 8, 2013. <http://www.tesev.org.tr/Upload/Publication/8a76a788-b00e-4c7c-bf24-6545458100c4/Turkey2.pdf >.

"RTVSC's Draft Regulation on Cable Broadcasting." Ankara: RTÜK, 2011. July 21, 2012. <http://www.rtuk.gov.tr/sayfalar/IcerikGoster.aspx?icerik_id=299ea676-661be-494e-9345-33546b785e12>.

Sektörel Araştırma ve Stratejiler Dairesi Başkanlığı. "Spektrumda Serbestleşme ve Yeni Eğilimler" (Liberalization of Spectrum and Main Trends). Ankara: Bilgi Teknolojileri ve İletişim Kurumu,2010. June 21, 2012. <http://www.tk.gov.tr/kutuphane_ve_veribankasi/raporlar/arastirma_raporlari/dosyalar/Spektrumda_Serbestlesme_Yeni_Egilimler.pdf>.

Sözeri, C. and Güney, Z. "The Political Economy of the Media in Turkey: A Sectoral Analysis." Istanbul: Turkish Economic and Social Studies Foundation Publications, 2011. April 8, 2013. <http://www.tesev.org.tr/Upload/Publication/67e244dd-5c21-24d34-8361-4c7f3d003140/11461ENGmedya2WEB21_09_11.pdf>.

# Turkey—Data Summaries

ELI NOAM AND PAUL MUTTER

THE TURKISH NATIONAL MEDIA market has been mostly privatized since the 1980s, but competition has arrived at a slower pace (Table 15.16). The existence of several overlapping regulatory regimes has complicated efforts to reduce the market's high overall concentration. Weighted Average HHI declined from 6,967 in 2004 to a still very high 5,012 in 2010. From the chapter's figures, this gives Turkey the third-highest weighted average HHI of all the countries surveyed, after China and South Africa.

The telecommunications market in Turkey is heavily concentrated with the presence of a small number of operators, several of which are owned by foreign media groups such as Vodafone (UK) and the Oger Group (Saudi Arabia). Turk Telecom (TT) is the largest platform company in the country, with 36% of the market. It and Turkcell (with 30%) together control 66% of the platform sector, though both are losing market share to Vodafone and the smaller domestic competitor Avea (which like TT is partly owned by the Oger Group).

The broadcasting industries are competitive due to new market entrants and foreign networks. In contrast, the multichannel platform market is monopolized by the state-owned operator Turksat, which is the only nationwide operator in the market by virtue of its acquisition of TT's cable infrastructure.

Extensive cross-ownership across content media platforms by several conglomerates' media groups is the most outstanding issue in the Turkish market, with a series of mergers and acquisitions that impact on the country's EU ascension process.

The Dogan Group has a large presence in multichannel platforms, daily newspapers, and online news media: it controls 24.9% of the content media market, 2.5 times the share of its nearest competitor, the Cukurova Group (Table 15.17), and it has been increasing its share at a

Table 15-16. National Media Industries Concentration in Turkey

| | 2004/5 | | 2011 OR MOST RECENT | | % CHANGE ANNUAL AVERAGE | |
|---|---|---|---|---|---|---|
| | COMPANY POWER INDEX IN COUNTRY | COMPANY SHARE OF THE OVERALL NATIONAL MEDIA MARKET (%) | COMPANY POWER INDEX IN COUNTRY | COMPANY SHARE OF THE OVERALL NATIONAL MEDIA MARKET (%) | COMPANY POWER INDEX IN COUNTRY | COMPANY SHARE OF THE OVERALL NATIONAL MEDIA MARKET (%) |
| Turk Telekom (55% Oger Telekom, Saudi Arabia; 30% public) | 4,092.1 | 41.8 | 2,676.5 | 29.6 | −4.5 | −1.7 |
| Turkcell (37% TeliaSonera, (Sweden); 13.2% Alfa Group, (Russia); 13.8% Cukurova Group) | 2,287 | 30.7 | 1,279.9 | 24.6 | −6.3 | −0.9 |
| Dogan Group | 248.5 | 4.9 | 170.2 | 4.9 | −4.5 | 0.0 |
| Vodafone(UK) | 95.2 | 6.3 | 374.2 | 13.7 | 41.9 | 1.1 |
| Google (US) | 28.0 | 0.3 | 143.7 | 1.5 | 59.0 | 0.2 |
| Avea (44.6% Oger Telekom, Saudi Arabia; 22% public) | 43.7 | 4.2 | 189.3 | 9.5 | 47.6 | 0.7 |
| Cukurova Group | 49.6 | 1.8 | 101.3 | 2.8 | 14.9 | 0.1 |
| Turksat Cable TV | 47.6 | 0.9 | 17.6 | 0.9 | −9.0 | −0.003 |
| Turkuvaz Group | 22.5 | 1.5 | 18.8 | 1.5 | −2.7 | 0.006 |
| Feza Group | 1.0 | 0.2 | 3.1 | 0.3 | 28.1 | 0.01 |
| 21st Century Fox (US) | 7.0 | 0.6 | 7.8 | 0.9 | 1.8 | 0.1 |
| Dogus | 2.3 | 0.2 | 2.2 | 0.2 | −0.4 | 0.002 |
| Ciner Group | 7.6 | 0.7 | 0.9 | 0.2 | −12.5 | −0.06 |
| Power Group | 0.4 | 0.04 | 0.6 | 0.1 | 7.7 | 0.001 |
| TRT (public) | 3.6 | 0.4 | 1.0 | 0.3 | −10.3 | −0.01 |

| Media Concentration Index | 2004/5 | 2011 or Most Recent | % Change Annual Average |
|---|---|---|---|
| Total Revenue: Nat'l Media Industry (mil US$) | 11,198 | 19,518 | 10.6% |
| Total Voices (*n*) | 57 | 60 | 0.8% |
| Net Voices (*n*) | 40 | 41 | 0.4% |
| Public Ownership (%) | 13.9 | 11.3 | −0.4% |
| Foreign Ownership (%) | 48.9 | 50.1 | 0.2% |
| C4 Average—Weighted | 95.1 | 93.1 | −0.3% |
| HHI Average—Weighted | 6,967 | 5,012 | −4.0% |
| C1 Average—Weighted | 79 | 62 | −0.0% |
| Noam Index Average—Weighted | 1,117 | 1,377 | 3.3% |
| Pooled Overall Sector C4 | 83.7 | 77.4 | −0.0% |
| Pooled Overall Sector HHI | 2,781 | 1,798 | −5.1% |
| Pooled Overall Sector Noam Index | 208 | 139 | −4.8% |
| Market Share of Top Ten Companies: Nat'l Media Industry (%) (Pooled C10) | 93.3 | 89.8 | −0.5% |
| National Power Index | 6,962 | 5,008 | −4.0% |

Table 15-17. Top Content Media Companies in Turkey

| | 2004/5 | | 20011 OR MOST RECENT | | % CHANGE ANNUAL AVERAGE | |
|---|---|---|---|---|---|---|
| | COMPANY POWER INDEX IN COUNTRY | COMPANY SHARE OF THE NATIONAL CONTENT MEDIA MARKET (%) | COMPANY POWER INDEX IN COUNTRY | COMPANY SHARE OF THE NATIONAL CONTENT MEDIA MARKET (%) | COMPANY POWER INDEX IN COUNTRY | COMPANY SHARE OF THE NATIONAL CONTENT MEDIA MARKET (%) |
| Dogan Group | 1,661.8 | 32.7 | 894 | 24.9 | 6.6 | –1.1 |
| Google (US) | 187.3 | 1.9 | 809.3 | 8.3 | 47.5 | 0.9 |
| Cukurova Group | 159.5 | 8.5 | 226.8 | 9.1 | 6.0 | –0.09 |
| Turkuvaz Group | 150.6 | 10.0 | 105.9 | 8.6 | –4.2 | –0.2 |
| Feza Group | 7 | 1.6 | 17.5 | 1.7 | 21.4 | 0.02 |
| Murdoch Group (21st Century Fox, US) | 46.5 | 3.7 | 44.2 | 5.0 | –0.7 | 0.2 |
| Power Group | 2.4 | 0.3 | 3.1 | 0.3 | 4.3 | –0.002 |
| Media Concentration Index | 2004/5 | | 2011 or Most Recent | | % Change Annual Average | |
| Public Ownership (%) | 2.8 | | 1.5 | | –0.2% | |
| Foreign Ownership (%) | 15.8 | | 20.0 | | 0.6% | |
| C4 Average—Weighted | 72.9 | | 67.5 | | –0.8% | |
| HHI Average—Weighted | 2,637 | | 2,290 | | –1.9% | |
| C1 Average—Weighted | 40.0 | | 37.0 | | –0.005% | |
| National Power Index | 2,584 | | 2,271 | | –1.7% | |

Table 15-18. Top Platform Media Companies in Turkey

| | 2004/5 | | 2011 OR MOST RECENT | | % CHANGE ANNUAL AVERAGE | |
|---|---|---|---|---|---|---|
| | COMPANY POWER INDEX IN COUNTRY | COMPANY SHARE OF THE NATIONAL PLATFORM MEDIA MARKET (%) | COMPANY POWER INDEX IN COUNTRY | COMPANY SHARE OF THE NATIONAL PLATFORM MEDIA MARKET (%) | COMPANY POWER INDEX IN COUNTRY | COMPANY SHARE OF THE NATIONAL PLATFORM MEDIA MARKET (%) |
| Turk Telecom (55% Oger Telekom, Saudi Arabia; 30% public) | 4,811.7 | 49.2 | 3,254.2 | 36.0 | –4.6 | –1.9 |
| Turkcell | 2,689.2 | 36.1 | 1,556.2 | 30.0 | –6.0 | –0.9 |
| Vodafone (UK) | 119.9 | 7.4 | 455.0 | 16.6 | 43.8 | 1.3 |
| Avea (44.6% Oger Telekom, Saudi Arabia; 22% public) | 51.4 | 5.0 | 230.2 | 11.5 | 49.7 | 0.9 |
| Dogan Group | 0.0 | 0.0 | 13.9 | 0.6 | N/A | 0.09 |
| Media Concentration Index | 2004/5 | | 2011 or Most Recent | | % Change Annual Average | |
| Public Ownership (%) | 15.9 | | 13.4 | | –0.4% | |
| Foreign Ownership (%) | 54.8 | | 56.6 | | 0.3% | |
| C4 Average—Weighted | 92.6 | | 82.5 | | –0.01% | |
| HHI Average—Weighted | 7,728 | | 5,600 | | –3.9% | |
| C1 Average—Weighted | 86 | | 67 | | –0.03% | |
| National Power Index | 7,732 | | 5,599 | | –3.9% | |

rate of –1.1% per annum. It also has smaller, but still substantial market shares in TV broadcasting and radio. The Turkuvaz Group and Cukurova Group also have operations in TV broadcasting, multichannel platforms, daily newspapers, radio, magazines, and online news, though on a much smaller scale (Table 15.18). Except for Google and 21st Century Fox, the top content producers are all domestically owned groups held by large conglomerates.

# 16

# Media Ownership and Concentration in the United Kingdom

PETROS IOSIFIDIS

## INTRODUCTION

At a time when media companies are becoming major actors in international commerce, and when cable, satellite, and digital platforms are increasingly globalizing media markets, questions about the possible effects of media concentration on the traditional role of the media have increasingly come to the fore.

In the United Kingdom, concentration of media ownership is increasing as evidenced by the reduction of the number of companies in several traditional media sectors, such as print and broadcasting, and the growing number of media firms that are owned by the same parent company. The intensification of concentration can also be viewed as a lever to promote market liberalization that would nurture national champions. Both "New Labor" governments (1997–2010) and the subsequent Conservative-Liberal Democrat coalition government have promoted liberalization.

Excessive media concentration in the United Kingdom is a public concern since it can endanger media pluralism and diversity. For these reasons the United Kingdom has introduced media ownership rules to preserve economic goals (open competition) and social aims (pluralism and diversity). Media regulation in the United Kingdom is governed by the provisions of the Communications Act (2003), which mostly applies to electronic media, and the Enterprise Act (2002), which applies to newspapers. The Office of Communications (Ofcom), the new super-regulatory agency set up in 2003 to replace five existing regulatory bodies—the Independent Television Commission, the Radio Authority, the Office of Telecommunications, the Broadcasting Standards Commission, and the Radiocommunications Agency—is responsible for enforcing this legislation.

British media ownership rules are sector specific: the main objectives for establishing media ownership rules for newspapers, radio, and television broadcasting have been to protect plurality of viewpoints and provide citizens access to a wide range of sources of news and information. The media ownership rules are separate from the merger regime, which applies to all sectors of the economy, including media markets. The primary purpose of the merger regime is to retain and enhance competition, but it may indirectly protect plurality by preventing consolidation in a particular market on competition grounds. The media public interest test, introduced in 2006, allows the Secretary of State for Business, Innovation, and Skills to intervene in newspapers, broadcasting, and cross-media mergers if they raise public interest concerns. The UK regulatory regime for communications law and policy is also determined by EU legislation. Examples include the laws governing the liberalization of UK media and communications law and policy (see below), a process strongly influenced by EU provisions that increasingly foreground competition law as a means of promoting pluralism.

The Communications Act (2003) freed up the communications industry far more than was expected, removing most of the ownership regulations that characterized British broadcasting, as it was thought these deprived companies of the economies of scale and scope required to expand into foreign markets. Rules preventing the joint ownership of television and radio stations were removed; Rupert Murdoch's News International, the UK newspapers arm of News Corporation, was allowed to expand into television broadcasting; and non-European ownership of broadcasting assets was permitted, effectively clearing the field for takeovers by international firms. The decisions to permit the acquisition of *Five* by print publishers and to lift the ban on non-European ownership were particularly controversial, though these provisions proved the Blair government's free market credentials to media owners and received strong support from several major players.[1] This substantial liberalization and reduction of cross-media ownership rules has been accompanied by an increased dependence on competition law, as well as the extension of self-regulation wherever possible.

In the United Kingdom, media ownership is a subject of much political controversy—particularly when it involves media magnate Rupert Murdoch, whose News Corporation has significant market shares in the US, UK, and Australian media markets (and in others as well). The issue of one entity controlling such a large proportion of the United Kingdom's newspapers and broadcasting interests is an area of public concern—particularly when the owner(s) has a close interest in the political agenda of these newspapers.[2] In the 1980s, the Thatcher government allowed News International to take over *The Times* and *Sunday Times* without referring it to the Monopolies and Mergers Commission (MMC), even though the company already controlled *The Sun* and the (now-defunct) *News of the World*. After its defeat in the 1992 elections, the Labor Party realized that it was vital to have Murdoch on board. Murdoch's papers, of course, realized this as well: *The Sun* heralded its role in the Conservative John Major's 1992 victory with the headline "It's The Sun Wot Won It." An example of the allegedly close relationship between Murdoch and former Prime Minister Tony Blair was the former's phone call in 1998 to the Prime Minister enquiring about prospects for further developing News International's interests in the United Kingdom. Despite this, when Murdoch broadcasting giant British Sky Broadcasting (BSkyB) obtained a 17.9% stake in the terrestrial broadcaster ITV in 2008, the Blair government and the Court of Appeal made BSkyB reduce that to a 7.5% stake; BSkyB had to sell off the difference at a loss.

In recent years, emerging issues have been framed in terms of competition considerations.

1. Freedman, Des. *The Politics of Media Policy*. Cambridge: Polity Press, 2008, 119–120; and Tunstall, Jeremy. "The BBC and UK Public Service Broadcasting." *Reinventing Public Service Communication: European Broadcasters and Beyond*. Iosifidis, Petros, ed. Basingstoke: Palgrave Macmillan, 2010, 145–157.

2. Douglas, Torin. "Analysis: Murdoch and Media Ownership in UK." Dec. 22, 2010. July 2, 2012. <http://www.bbc.co.uk/news/uk-12062176>.

One such case in 2011 was the proposed acquisition by News International, a subsidiary of News Corporation, of the remaining 61% of satellite broadcasting networks in the United Kingdom that BSkyB did not already own. The proposal prompted an unprecedented response from nearly all non-News International owned companies in the United Kingdom, both private and public. The companies wrote a letter asking the UK Business Secretary to refer the case to Ofcom on the basis that the takeover would have "serious and far-reaching consequences for media plurality."[3] In the United Kingdom, News International already owns about 34% of the newspaper market, and with the BSkyB deal, it would have controlled two-thirds of pay television. Eventually, in July 2011 the US$16 billion (£10 billion) BSkyB bid was abandoned in the light of the phone hacking scandal—"Hackgate"—at the Murdoch tabloid *News of the World*. Employees of *News of the World* and several other UK papers under the News International banner have been accused of engaging in phone hacking, police bribery, and exercising improper influence in the pursuit of stories. Given that News International also owns a 7.5% stake of the main commercial terrestrial broadcaster ITV plc, this acquisition would have raised competition concerns as well as concerns regarding media pluralism in the UK media.

It is also striking how current debates in the field of broadcasting are focused more than ever on the publicly funded British Broadcasting Corporation (BBC) and its expanding online and digital activities at a time of economic difficulties faced by its competitors. With its strong presence in radio, terrestrial TV, magazines, and digital and online domains, industrialists, some politicians, and academics have strongly criticized the BBC's business practices. Its rival, BSkyB, has run a persistent campaign calling for the government to shrink the scope of public broadcasting. Concerns regarding its expansionist strategies have been partly addressed with the suggestion by the Conservative-Liberal Democrats coalition government to reduce BBC Online's budget by 25% and to freeze BBC license fee income until 2017. However, there are voices that argue that any cut to the BBC license fee raises fundamental questions about the BBC's independence from the government.[4]

## PRINT MEDIA

### Newspapers

There are five national daily "quality" titles—*The Guardian, The Independent, The Financial Times, The Times,* and *The Daily Telegraph*—each owned by a different newspapers group. In addition, there are two national daily "mid-market" titles, the *Daily Mail* and the *Daily Express*. Then come the three national daily "tabloids": *The Sun,* the *Daily Mirror, and the Daily Star. The Sun* belongs to the same group that owns *The Times,* News International. Finally are the 10 Sunday titles—*Sunday Mirror, The People, Sunday Mail, Daily Star Sunday, Mail on Sunday, Sunday Express, The Sunday Times, The Sunday Telegraph, The Observer,* and *The Independent on Sunday*. The Sunday editions follow their eponymous daily papers.

News International is controlled by News Corporation, which is controlled by the Murdoch family and publishes *The Sun, The Times,* and *The Sunday Times*. As already noted, its politics are regarded as being fairly conservative. News Corporation split in 2012 into two separate publicly listed companies—one for newspapers, book publishing, and education (News International); and the other for audiovisual media and entertainment programming (21st Century Fox). The restructure was expected to protect the group's robust film and television assets from the publishing businesses, which

3. Browning, Jonathan. "U.K. Newspapers Join BBC in Opposition to Murdoch BSkyB Deal." *Bloomberg*, Oct. 12, 2010. <http://www.bloomberg.com/news/2010-2010-12/bbc-joins-u-k-newspapers-to-lobby-against-murdoch-bid-for-rest-of-bskyb.html>.

4. Barnett, Steven, and Seaton, Jean. "Why the BBC Matters: Memo to the New Parliament about a Unique British Institution." *The Political Quarterly* 81: 3 (2010): 327–332; and Iosifidis, Petros. *Public Television in the Digital Era: Technological Challenges and New Strategies for Europe*. 2nd ed. Basingstoke: Palgrave Macmillan, 2012.

were undervalued after the phone-hacking scandal's rising legal costs. Given that Rupert Murdoch, chairman and chief executive of News Corp, has served as the chairman of both companies, control still essentially remains in the same hands.

Telegraph Media Group is owned by the British businessmen Sir David Rowat Barclay and Sir Frederick Hugh Barclay, who acquired the business in 2004 for US$1.2 billion (£665 million). It prints *The Daily Telegraph, The Sunday Telegraph,* the weekly magazine *The Spectator, The Scotsman* (plus its Sunday edition, *Scotland on Sunday*), and the *Edinburgh Evening News*. Telegraph offerings are regarded as offering independent political coverage, though they have historically leaned toward the Conservative Party.

The Daily Mail and General Trust is owned by the British media conglomerate Daily Mail and General Trust plc and prints *The Daily Mail, Mail on Sunday, Ireland on Sunday, Mail Today* (a joint venture with the Indian media conglomerate Living Media), and several specialized dailies such as the national urban market daily *Metro*, the classified directory *Loot*, and the *London Lite* "free sheet." Until January 2009, the group also owned the dominant paid-subscription London metro daily *Evening Standard*, but sold the paper to the Russian billionaire Alexander Lebedev. Most of the titles owned by the group are tabloid newspapers that focus on "light news," celebrity lifestyles, and some limited political reporting.

Guardian Media Group (GMG), which is wholly owned by the limited company Scott Trust, owns *The Guardian* and *The Observer; Manchester Evening News* (a regional newspaper); Channel M (a regional TV station); numerous regional radio stations across the United Kingdom under the *Real Radio, Smooth Radio,* and *Rock Radio* brands; and Bauer Radio, a division of the leading international magazine publishing company Bauer. The political perspective of GMG's newspapers is regarded as left leaning.

Independent News and Media's main newspapers are the *Irish Independent* and the *Sunday Independent*, which are Ireland's best-selling newspapers. The O'Reilly family has a controlling stake of 28.2% in the group, with Irish entrepreneur Dennis O'Brien owning the largest stake after that, of 22.5%. Through his company, Communicorp, O'Brien also owns two national radio channels, *Newstalk* and *Today FM*, plus 40 other stations in the EU. In addition, Independent News and Media owns the Belfast Telegraph Group, which publishes the *Belfast Telegraph, Ireland's Saturday Night, Sunday Life,* and *Ads for Free*. These newspapers are regarded as offering independent political coverage. Until 2010, the company published the UK titles *The Independent* and *The Independent on Sunday*, but both newspapers were sold to Alexander Lebedev.

Trinity Mirror plc (managed by businesswoman Sly Bailey) is the result of the takeover of Mirror Group Newspapers by Trinity plc in September 1999. It owns the *Daily Mirror, Sunday Mirror, The People, Daily Record,* and the *Sunday Mail*—in addition to around 120 other (regional) tabloids that run either daily or weekly editions.

Pearson plc, helmed by Dennis Stevenson, is a relatively new entry to the UK media market, having only gotten into the newspaper business in the 1920s. Since the 1990s, it has transformed itself from its roots as an industrial holding company to a major publisher of educational materials and news with three major business groups: Pearson Education, the Penguin Group, and the Financial Times Group, which publishes the *Financial Times* in the global English-language market, *Les Echos* in France, and *Expansion* in Spain. It also owns 50% of The Economist Group (publisher of *The Economist*), plus a host of personal finance magazines.

Finally, Northern & Shell Network is owned by Richard Desmond and publishes the *Daily Express, Sunday Express,* and the *Daily Star*. It also owns the magazines *New!* and *Star* (Table 16.1).

The UK national daily newspapers market is highly concentrated. News International is a clear market leader and has managed to gain and retain very high market shares. In 2012, News International's market share of daily newspapers was more than double compared to that of Trinity Mirror plc's, the second biggest UK media group.

Table 16-1. Daily Newspaper (Market Shares by Circulation) 1997–2012

| | 1997 | 2001 | 2002 | 2008 | 2009 | 2012 |
|---|---|---|---|---|---|---|
| News International (News Corp., Murdoch Family, US/AUS) | 34.4 | 31.8 | 32.2 | 34.8 | 33.8 | 32.5 |
| *The Sun* | *29.7* | *26.6* | *27.3* | *29.2* | *28.1* | *28* |
| *The Times* | *4.7* | *5.2* | *4.9* | *5.6* | *5.7* | *4.5* |
| Trinity Mirror plc | 23.9 | 21.0 | 20.2 | 15.6 | 16.2 | 15.1 |
| *Daily Mirror* | *18.3* | *16.4* | *16.1* | *12.4* | *12.9* | *12* |
| *Daily Record* | *5.6* | *4.6* | *4.1* | *3.2* | *3.3* | *3.1* |
| Northern & Shell (Richard Desmond) | 14.3 | 12.5 | 13.8 | 14.9 | 13.5 | 13 |
| *Daily Star* | *4.9* | *5.6* | *6.5* | *8.1* | *6.9* | *6.7* |
| *Daily Express* | *9.4* | *6.9* | *7.3* | *6.8* | *6.9* | *6.3* |
| Daily Mail & General Trust (*Daily Mail*) | 13.6 | 18.7 | 18.5 | 21.2 | 19.9 | 21.5 |
| Telegraph Group (Barclay Brothers, *Daily Telegraph*) | 7.7 | 7.7 | 7.3 | 7.4 | 7.3 | 6.4 |
| Pearson plc (*Financial Times*) | 1.3 | 3.8 | 3.5 | 4.0 | 4.1 | 3.3 |
| Guardian Media Group (Scott Trust, *The Guardian*) | 2.7 | 3.1 | 3.0 | 3.0 | 3.3 | 2.4 |
| Independent Print Ltd (Alexander Lebedev, *The Independent*) | 2.1 | 1.5 | 1.4 | 1.8 | 1.9 | 1.0 |
| Total Revenue (mil £)[1] | 5,152 | 6,517 | 6,393 | 5,529 | 6,580 | 5,608 |
| Total Revenue (mil US$) | 8,450 | 9,385 | 9,590 | 10,283 | 10,330 | 8,887 |
| C4 | 86.2 | 84 | 84.7 | 86.5 | 83.4 | 82.5 |
| HHI | 2,217 | 2,044 | 2,054 | 2,209 | 2,068 | 1,974 |
| *N* (>1%) | 8 | 8 | 8 | 8 | 8 | 8 |
| Noam Index | 784 | 723 | *726* | *781* | 731 | 698 |

*Source*: Audit Bureau of Circulations, 2010, 2012 http://www.abc.org.uk/.

[1] This revenue figure is an estimate based on data collected in this study: it is based on average per capita revenues reported for several other countries in this study with similar per capita income.

The newspaper is in decline, though, as circulation has been falling steadily over the past four decades. In the period from 1965 to 2005, the cumulative loss of circulation of popular and quality dailies was 29% and 31%, respectively, owing among other factors to the rise of television and the Internet. Since 2008, this has been exacerbated by the global financial crisis. While two of the best-selling newspapers, *The Sun* and the *Daily Mirror,* had about 3.5 million and 2 million readers, respectively, in 2002, this dropped to 2.9 million and 1.3 million, respectively, in 2009. In terms of revenues, for FY2008–2009 newspapers experienced the biggest reduction in advertising revenues among traditional media ever, falling by more than 20%, TV advertising revenues fell by 9%, and radio advertising revenues by 15%.[5]

5. Ofcom. "International Communications Market Report 2010." London: Office of Communications, Dec. 2, 2010. May 30, 2012. <http://stakeholders.ofcom.org.uk/binaries/research/cmr/753567/icmr/ICMR_2010.pdf>.

## Book Publishing

The UK book publishing industry is dominated by four groups: Hachette, Random House, HarperCollins, and Penguin Books. In 2004, the market share of these four was greater than 50%, and although it has been declining slightly since then, the four still account for a considerable overall market share. In the last decade, the top two—Hachette and Random House—lost market shares. Penguin Books saw its market share rising from 10.5% to 11.7%, and HarperCollins began regaining market share in 2011 after several years of decline.

However, the future of book publishing faces a declining economy, a shrinking distribution network, and, above all, a decline in physical book sales. The main problem facing the book publishing industry is the uncertainty surrounding digital publishing. In fact, at the 2011 London Book Fair, a seminar debated the prompt "Authors and readers are all that matter. Publishers will soon be irrelevent."[6] In a similar vein of self-examination, *The Bookseller* magazine that year enquired whether publisher performance based on Nielsen BookScan Total Consumer Market print figures is still a relevant metric in today's increasingly digitalized

### Table 16-2. Book Publishing (Market Shares by Revenue) 2004–2011

| | 2004 | 2008 | 2009 | 2010 | 2011 |
|---|---|---|---|---|---|
| Hachette (Lagardère, France) | 17.0 | 15.9 | 16.4 | 15.2 | 13.9 |
| Random House (Bertelsmann AG, Germany) | 15.5 | 14.8 | 13.7 | 13.8 | 13.2 |
| Penguin Books (UK) | 10.5 | 10 | 9.7 | 11.5 | 11.7 |
| HarperCollins (News Corporation, US/UK/AUS) | 9.0 | 8.3 | 7.6 | 7.1 | 7.7 |
| The Independent Alliance[1] | — | 4.0 | 4.0 | 3.9 | 3.3 |
| Pan Macmillan (Macmillan Group, Holtzbrinck, Germany) | 3.0 | 3.3 | 3.2 | 3.6 | 3.7 |
| Bloomsbury | 2.5 | 2.4 | 2.2 | 2.1 | 2.3 |
| Pearson | 2.2 | 2.4 | 1.9 | 1.8 | 2.4 |
| Oxford University Press | 2.5 | 1.9 | 1.9 | 2.0 | 2.0 |
| John Wiley | 1.7 | 1.5 | 1.5 | 1.5 | 1.5 |
| Simon & Schuster (Viacom, US) | 1.4 | 1.5 | 1.8 | 1.8 | 2.0 |
| Others | 34.7 | 34.0 | 36.1 | 35.7 | 36.3 |
| Total Revenue (mil £)[2] | 3,348 | 4,268 | 3,718 | 3,748 | 3,817 |
| Total Revenue (mil US$) | 6,162 | 6,828 | 5,838 | 5,810 | 6,107 |
| C4 | 52.0 | 49.0 | 47.4 | 47.6 | 46.5 |
| HHI | 752 | 687 | 652 | 649 | 610 |
| *N* (>1%) | 10 | 11 | 11 | 11 | 11 |
| Noam Index | 238 | 207 | 197 | 196 | 184 |

*Source*: Nielsen BookScan TCM 2005, 2009, 2010, 2011, 2012; *The Bookseller*: 2010, 2011.

[1] The *Independent Alliance*, an association of 10 UK publishers, was formed in 2005 and its members are Atlantic Books, Canongate Books, Faber and Faber, Granta Books, Icon Books, Portobello Books, Profile Books, Quercus Publishing, Serpent's Tail, and Short Books.

[2] This revenue figure is an estimate based on data collected in this study: it is based on average per capita revenues reported for several other countries in this study with similar per capita income.

6. Reed, John. "Debate: Will Publishers Soon Be Irrelevant?" *Publishing Talk*, April 18, 2011. Dec. 20, 2012. <http://www.publishingtalk.eu/conferences/debate-will-publishers-soon-be-irrelevant/>.

market. The answer was a qualified yes: digital sales in 2011 may have made up about 10% of the market, perhaps up to 12% for some publishers, but an overwhelming majority of the market is still print-based (Table 16.2).

## Magazine Publishing

The United Kingdom has the world's highest per capita concentration of magazine titles. There are more than 8,000 titles published in Britain, mostly out of London. Consumer magazines make up the bulk of the titles for sale. Uniquely, most UK magazines for consumers—about 90% of them—are sold through newsagents or supermarkets. This is a much higher proportion than in the United States and continental Europe, where subscription services are much more popular. Consumer magazines may be general titles that aim to entertain and inform, or specialist titles aimed at a specific interest or hobby. There are about 2,800 UK consumer magazines.[7]

The biggest UK magazine publishers by sales revenue are Bauer (Germany), which in 2008 took over the then second-largest group, EMAP; IPC Media, owned by Time Inc., the magazine publishing subsidiary of Time Warner in the United States, which in 2001 acquired IPC Media for US$1.7 billion (£1.15 billion) in the biggest magazine deal ever seen in the United Kingdom; Hearst Magazines UK, the product of the US$651 million (£407 million) merger in

### Table 16-3. Magazine Publishing (Market Shares by Revenue) 2004–2011

| | 2004 | 2008 | 2011 | CIRCULATION: 2011 (MIL COPIES) |
|---|---|---|---|---|
| Bauer (Germany) | 18.0 | 20.0 | 26.0 | 5.8 |
| IPC Media/Time Warner (US)[1] | 20.0 | 22.0 | 20.0 | 5.7 |
| Hearst Magazines UK (US/France) | 13.5 | 15.0 | 15.0 | 4.1 |
| BBC Magazines (Public) | 11.5 | 10.0 | 9.5 | 3.0 |
| Condé Nast (Advance Publications, US) | 4.8 | 4.5 | 4.0 | 1.5 |
| Egmont Group (Denmark) | 2.5 | 2.5 | 2.5 | 0.6 |
| Hubert Burda (Germany) | 2.0 | 2.0 | 2.0 | 0.6 |
| Haymarket Media Group (UK) | 1.5 | 1.5 | 1.3 | 0.4 |
| Other | 26.2 | 22.5 | 19.7 | – |
| Total Revenue (mil £)[2] | 3,315 | 3,711 | 4,506 | |
| Total Revenue (mil US$) | 6,067 | 6,903 | 7,210 | |
| C4 | 63.0 | 67.0 | 70.5 | |
| HHI | 1,074 | 1,242 | 1,419 | |
| *N* (>1%) | 8 | 8 | 8 | |
| Noam Index | 380 | 439 | 501 | |

*Source*: Press Gazette Reporters, 2011. http://www.pressgazette.co.uk/magazine.

[1] IPC Media/Time Warner includes *IPC Inspire* (leisure brands), *IPC Connect* (women's weeklies), and *IPC Southbank* (upmarket women's division).

[2] This revenue figure is an estimate based on data collected in this study: it is based on average per capita revenues reported for several other countries in this study with similar per capita income.

7. Magforum. n.d. "Magazine industry sectors." May 22, 2012. <http://www.magforum.com/sectors.htm#key>.

2011 of the National Magazines Company with Hachette Filipacchi Médias, and BBC Magazines. In terms of circulation in 2011, Bauer was the market leader, selling 5.8 million copies with IPC Media/Time Warner a close second, having sold 5.7 million copies. In the same year, Hearst Magazines UK sold about 4.1 million copies, followed by BBC Magazines (2.95 million), Condé Nast (1.54 million), Egmont UK Publishing (610,000 copies), Hubert Burda (550,000 copies), and Haymarket (380,000 copies). Foreign concerns own most UK consumer titles. IPC Media and Condé Nast are owned by the US groups Time Warner and Advance Publications, respectively. The largest British-owned company is BBC Magazines (Table 16.3).

## AUDIOVISUAL MEDIA

### Radio

The public BBC dominates the UK national radio market, while a plethora of private commercial radio stations operate at the local and regional levels. There are 40 local stations in England, and two national services in Wales, Scotland, and Northern Ireland each (altogether there are 10 national BBC stations), including non-English language stations in Wales and Scotland. There are four national commercial broadcasting stations, and approximately 300 local or regional commercial broadcasters, including Global, Bauer, TLRG, UTV, and GMG. There are 146 community radio stations in the United Kingdom outside of the BBC's jurisdiction, and not-for profit community groups with either a geographical or special interest focus run them. The BBC network has a dominant audience share of 47% and the local BBC stations account for an additional 9%. The national commercial broadcasters account for just 10% of the national market share, while the local/regional commercial broadcasters hold 31% of it.[8]

While in 1996 the BBC and commercial radio had an almost equal market share of 50:50, in more recent years, BBC radio stations have surpassed commercial radio: in 2011, BBC's share had risen to 54.5%. One reason is that the BBC national radio retains the best national FM frequencies, while commercial radio's FM frequencies are predominantly local ones and as such have limited range.

Of the commercial groups, the most popular is Global Radio UK, the largest commercial radio company in the country following the acquisitions of Chrysalis Radio and GCap Media in 2008. In June 2012, Total GMG Radio, the United Kingdom's third-largest radio group, reportedly signed a US$111 million (£70 million) deal with market leader Global Radio UK, strengthening the market position of Global Radio and leaving Bauer Radio (owned by foreign interests) in a distant second place. Table 16.4 does not take account of this development, as at the time of writing a regulatory review into the deal was expected to take place to clear it. GMG, owner of the United Kingdom's third-largest radio group, sold its radio business in 2012 to the market leader, Global Radio. Meanwhile, there is speculation that Bauer Radio will be forced to look at further acquisitions in the radio market in order to remain competitive against Global Radio.

Despite the wide availability of other audiovisual platforms, radio remains very popular in the United Kingdom. A survey of RAJAR quarterly listening reports shows that 46.7 million adults tune in each week. The market share of radio via a digital platform stands at 29.1% of all radio listening in 2011, slightly up from 25% in 2010.[9]

8. Ofcom. "Media Ownership Rules Review Consultation Document." London: Office of Communications, July 31, 2009. May 18, 2012, 35–40. <http://stakeholders.ofcom.org.uk/consultations/morr/>.

9. Radio Joint Audience Research. "Quarterly Listening." London: RAJAR. n.d. Dec. 20, 2012. <http://www.rajar.co.uk/listening/quarterly_listening.php>.

Table 16-4. Radio Groups (Market Shares by Revenue), 1996–2011

| | 1996 | 2004 | 2008 | 2011 | LISTENERS: 2011 (MIL PPL) |
|---|---|---|---|---|---|
| BBC | 48.9 | 52.9 | 55.2 | 54.5 | |
| BBC Network Radio[1] | — | *42.0* | *46.7* | *46.6* | |
| BBC Radio One | — | *7.7* | *9.8* | *9.8* | *11.1* |
| BBC Radio Two | — | *16.0* | *16.5* | *16.5* | *13.6* |
| BBC Radio Three | — | *1.4* | *1.1* | *1.1* | |
| BBC Radio Four | — | *11.5* | *12.5* | *19.4* | *10.2* |
| BBC Radio Five Live | — | *4.4* | *4.5* | *4.5* | *6.5* |
| Global Radio UK (UK)[2] | | | 17.0 | 16.3 | |
| Bauer Radio (Germany)[3] | | | 10.9 | 11.7 | |
| GMG Radio (Guardian Media Group)[4] | | | 4.4 | 4.5 | |
| Other | | 11.7 | 12.5 | 13.0 | |
| Total Revenue (mil £)[5] | | 1,559 | 1,581 | 1,729 | |
| Total Revenue (mil US$) | | 2,853 | 2,941 | 2,767 | |
| C4 | | | 87.5 | 87.0 | |
| HHI | | | 3,474 | 3,393 | |
| *N* (>1%) | | | 4 | 4 | |
| Noam Index | | | 1,737 | 1,697 | |

*Source*: RAJAR.

[1] *BBC Radio Three* is a music station and *BBC Radio Four* is the dominant UK news and talk show service, and its early morning news sequence is regarded as an agenda setter in UK national politics. *BBC Radio One* and *Two* are national pop music stations that have some public service elements. *BBC Radio Five Live* offers sports, news, and talk show content.

[2] *Heart Network UK, Classic FM, Capital Network,* and *Gold UK Network*.

[3] *Magic UK* and *Kiss UK*.

[4] *Smooth Radio UK* and *Real Radio UK*.

[5] This revenue figure is an estimate based on data collected in this study: it is based on average per capita revenues reported for several other countries in this study with similar per capita income.

## Broadcast Television

The audience share for terrestrial television broadcasting has been declining over the years, owing to multichannel development, audience fragmentation, and changing audience habits. Commercial broadcasters, faced with fragmentation in audience share and diminishing advertising revenues, have begun to experiment with new business models and portfolio expansion.

For example, in October 2005 Channel 4 launched *More4*, which offered current affairs, drama, and comedy programming to the over 35 demographic, while in November 2005, ITV launched *ITV4*, which was targeted at men and broadcast US sports coverage and Hollywood films (in 2004, it had introduced the entertainment channel *ITV3*). In 2006, the terrestrial commercial channel *Five* launched two new digital services: *Five Life*, providing preschool shows, popular lifestyle shows aimed at women, as well as films and soap operas; and *Five USA*, which offers drama, films, sport, comedy, and children's television content from across the Atlantic. Meanwhile, the BBC has launched numerous over-the-air digital channels like *BBC Three, BBC Four, BBC News 24*, and the children-oriented services *CBeebies* and *CBBC*. But apart from a few exceptions, these ventures have not

## Table 16-5. TV Broadcasting (Market Shares) 1982–2012

| | 1982 | 1986[1] | 1990 | 1994 | 1998 | 2002 | 2004 | 2006 | 2008 | 2010 | 2012[2] |
|---|---|---|---|---|---|---|---|---|---|---|---|
| BBC (One & Two) | 50.0 | 48.0 | 47.0 | 44.8 | 44.5 | 44.3 | 41.4 | 41.6 | 43.6 | 42.6 | 50.1 |
| ITV (GMTV) | 50.0 | 44.0 | 44.0 | 40.7 | 34.5 | 28.4 | 27,2 | 26.0 | 24.1 | 25.4 | 21.8 |
| Channel 4 (S4C) | | 8.0 | 9.0 | 11.5 | 11.3 | 11.8 | 11.6 | 13.0 | 11.1 | 10.0 | 9.8 |
| Five | | | | | 4.7 | 7.4 | 7.9 | 7.4 | 7.5 | 7.0 | 3.3 |
| Others (over-the-air)[3] | | | | 3.0 | 5.0 | 8.0 | 10.0 | 12.0 | 14.0 | 15.0 | 15.0 |
| Total Revenue (mil £)[4] | | | | | 4,446 | | 5,714 | | 7,053 | 8,136 | 10,947 |
| Total Revenue (mil US$) | | | | | 7,381 | | 10,457 | | 13,118 | 12,611 | 16,531 |
| C4 | 100.0 | 100.0 | 100.0 | 97.0 | 95.0 | 91.9 | 88.9 | 88.0 | 86.3 | 85.0 | 95.0 |
| HHI | 5,000 | 4,304 | 4,226 | 3,796 | 3,320 | 2,963 | 2,651 | 2,630 | 2,661 | 2,609 | 3,092 |
| *N* (>1%) | 2 | 3 | 3 | 3 | 4 | 4 | 4 | 4 | 4 | 4 | 4 |
| Noam Index | 3,536 | 2,485 | 2,440 | 2,191 | 1,660 | 1,482 | 1,326 | 1,315 | 1,331 | 1,305 | 1,546 |

*Source*: Broadcasters' Audience Research Board (BARB).

*Source*: 2012 Ofcom Stakeholders report 2012 updates by editors, with "others" estimated.

[1] Shares before 1998 have been rounded to nearest whole number.

[2] Affected by Olympic year in London, carried exclusively by BBC.

[3] "Others (over-the-air)" include over-the-air channels with a viewing share of ≥ 1.0% (i.e., *ITV2, ITV3, ITV4, BBC Three, BBC Four, CBeebies, C4, E4,* etc.).

[4] This revenue figure is an estimate based on data collected in this study: it is based on average per capita revenues reported for several other countries in this study with similar per capita income.

been successful in winning many new viewers for their networks (Table 16.5).

The data show a continuous decline in market shares for the five main terrestrial broadcasters. This can be attributed to the rise of digital cable and satellite subscription channels such as *Sky 1, Sky Sports 1, Sky News, Cartoon Network, Nickelodeon,* and so forth, which in 2010 attracted a cumulative audience share of about 30%. It can also be attributed to the changing habits of the public, with the young population spending more time watching programs online or on mobile platforms. In 2010 a British adult spent on average 242 minutes each day watching TV on a television and this number is increasing, but when people are asked which medium they would miss the most if it were taken away, there are clear differences in response by age-group.[10] Overall, 44% said they would most miss their television set—a decline of 6 percentage points from 2009, while 17% said they would most miss the Internet—more than double the proportion five years ago (8%). For young adults aged 16–24, the picture is quite different: 28% said they would most miss mobile the most, and 26% said the Internet, an increase from just 18% in 2009.

## Multichannel TV Platforms

Tables 16.6 and 16.7 show the continuing growth of cable, satellite, and digital terrestrial television (DTT) platforms. It is interesting to note the prominence of DTT, which from a mere 4% share in 2000 has since 2011 surpassed satellite and cable to reach 55% of the audience share. This is explained by the popularity of the BBC-led Freeview DTT platform, which is appealing to people who do not wish to subscribe to digital TV services. A total of up to 40 channels are available on Freeview through a set-top box priced at less than US$50 (£30). The

10. Ofcom. "Communications Market Report." London: Office of Communications, Aug. 4, 2011. May 18, 2012, <http://stakeholders.ofcom.org.uk/binaries/research/cmr/cmr11/UK_CMR_2011_FINAL.pdf>.

Table 16-6. Multichannel Video Platforms: Cable MSOs, DBS, IPTV (Market Shares by Revenue), 1992–2011

| | 1992 | 1996 | 2000 | 2004[1] | 2008 | 2011 |
|---|---|---|---|---|---|---|
| BSkyB (News Corp., Murdoch Family, US/AUS) | 82.2 | 71.7 | 52.0 | 56.5 | 36.5 | 33.0 |
| Satellite Subscribers (1,000)[2] | *1,893* | *3,542* | *3,963* | *6,946* | *8,860* | *11,012* |
| Virgin Media (US) | 17.8 | 28.3 | 44.0 | 26.6 | 14.0 | 12.0 |
| Cable Subscribers (1,000) | *409* | *1,399* | *3,352* | *3,277* | *3,406* | *3,997* |
| BBC Freeview | | | 4.0 | 16.9 | 49.5 | 55.0 |
| DTT Subscribers (1,000) | | | *303* | *2,075* | *12,017* | *18,376* |
| Total Revenue (mil £)[3] | | | 3,420 | 5,313 | 7,434 | 9,261 |
| Total Revenue (mil US$) | | | 5,164 | 9,722 | 13,827 | 14,818 |
| Total Subscribers (1,000) | 2,302 | 4,941 | 7,618 | 12,298 | 24,283 | 33,385 |
| C4[4] | – | – | – | – | – | – |
| HHI | 7,078 | 5,941 | 4,658 | 4,185 | 3,977 | 4,261 |
| *N* (>1%) | 2 | 2 | 3 | 3 | 3 | 3 |
| Noam Index | 5,004 | 4,200 | 2,689 | 2,416 | 2,296 | 2,460 |

*Source*: Author.

[1] From 2002, figures include homes with two or more reception capabilities.

[2] Satellite and cable figures include homes on both the analog and digital platform.

[3] This revenue figure is an estimate based on data collected in this study: it is based on average per capita revenues reported for several other countries in this study with similar per capita income.

[4] C4 cannot be calculated since the platforms are fewer than four.

service offers a light but diverse line-up of over-the-air services, and it also offers viewers a package including the existing five terrestrial channels, all the BBC's digital channels, and some of BSkyB's channels. In the satellite TV market, BSkyB enjoys a monopoly, having acquired most of the available premium programming for sports and films, thus leaving little room for new market entrants.

The launch of Freeview in 2002 altered the direction of DTT because it meant that the BBC would not seek to compete directly with established satellite and cable offerings. Meanwhile, digital satellite pay TV platforms—particularly Sky TV—continued to expand their market shares with the offering of "premium content" sports and films.

Digital cable, offered by Virgin Media, is a distant third-place platform for UK audiences. Virgin Media operates its own fiber-optic cable network, the only national cable network in the United Kingdom, which in 2010 had a total of approximately 4.8 million cable customers (4 million of them with broadband Internet services), 4.2 million with fixed-line telephone services, and around 3.1 million mobile customers.

News Corporation has the largest market share in the United Kingdom when it comes to satellite-delivered programming. The company retains the lead largely because of the delivery of the premium Sky Sports channels that show live Premier League as well as Champions League football matches, which are very popular in the United Kingdom. However, the audience share of individual satellite and cable channels remains small compared with the share of the traditional terrestrial broadcasting channels, which have also started to build digital portfolios. For example, in 2011 the BBC had the largest market share in multichannel homes at 33.2%, followed by ITV at 23.1%, Channel 4 at

Table 16-7. Combined Video Network Industries: TV Syndicators, Cable Channels, Pay TV (Market Shares by Revenue), 2002–2011

| | 2002 | 2004 | 2006 | 2008 | 2010 | 2011 |
|---|---|---|---|---|---|---|
| News Corp. Ltd. (Murdoch Family, US/AUS)[1] | 34.4 | 34.4 | 32.8 | 32.8 | 32.8 | 38.4 |
| UK Channel Management (UK)[2] | 12.8 | 13.6 | 14.4 | 19.2 | 19.2 | 16.8 |
| Viacom (US)[3] | 15.2 | 12.8 | 16.0 | 10.4 | 8.8 | 7.2 |
| Discovery (US)[4] | 6.4 | 7.2 | 6.8 | 6.8 | 8.0 | 7.2 |
| Turner Broadcasting (Time Warner, US)[5] | 5.6 | 6.4 | 4.0 | 4.0 | 4.0 | 4.0 |
| Walt Disney (US)[6] | 5.6 | 5.6 | 5.6 | 7.2 | 6.8 | 6.4 |
| Other | 20.0 | 20.0 | 20.0 | 20.0 | 20.0 | 20.0 |
| Total Revenue (mil £)[7] | | | | 2,379 | | 3,995 |
| Total Revenue (mil US$) | | | | 4,424 | | 6,392 |
| | | | | | | |
| C4 | 68.4 | 68.0 | 70.0 | 69.6 | 68.8 | 69.6 |
| HHI | 1,675 | 1,656 | 1,633 | 1,667 | 1,648 | 1,917 |
| *N* (>1%) | 6 | 6 | 6 | 6 | 6 | 6 |
| Noam Index | 684 | 676 | 667 | 681 | 673 | 783 |
| Total Non-Broadcasters | 13.8 | 15.0 | 16.5 | 14.8 | 16.0 | 18.3 |
| Total Broadcasters | 86.2 | 85.0 | 83.5 | 85.2 | 84.0 | 72.7 |

*Source*: Broadcasters' Audience Research Board (BARB).

[1] *Sky Sports 1, Sky 1, Sky News, Sky Sports 2, Sky Movies Comedy, Sky Movies Action Thriller*, etc.

[2] *Alibi, Home, Good Food, Watch*, etc.

[3] *Nickelodeon, MTV*, etc.

[4] *Animal Planet, Discovery History, Quest*, etc.

[5] *Cartoon Network, Boomerang*, etc.

[6] *Disney Channel, Disney Cinemagic, Disney Junior*, etc.

[7] This revenue figure is an estimate based on data collected in this study: it is based on average per capita revenues reported for several other countries in this study with similar per capita income.

11.3%, and Channel 5 at 5.9%. These traditional broadcasters have launched numerous digital channels (for example, *BBC3, BBC4, CBBC, ITV2, ITV3, ITV4, E4, Film4, More4, Channel 5 +1, Five USA*, etc.), thus enabling them to get a combined total market share of 73.5% in 2011 for the multichannel market.

## Film

Film distributors' market shares fluctuate from year-to-year. The leading distributor in 2010, Warner Brothers, saw its market share increase thanks to films such as *Harry Potter and the Deathly Hallows: Part 1, Alice in Wonderland*, and *Inception*. The record-breaking *Avatar*, which was released in late 2009 but which took home most of its US$146 million (£94 million) in 2010, made a significant contribution to putting 20th Century Fox in second place in the UK film market. Paramount was the third highest grossing distributor in the United Kingdom that year, with films such as *Shrek Forever After, Iron Man 2*, and *Little Fockers*.

The top 10 distributors had a 94% share of the market in 2010, up from 92% in 2009 but down from 96% in 2004. The remaining

88 distributors handled a total of 379 titles in 2010—62% of the films on release—but gained only a 6% market share. Yet this 6% won by distributors outside of the top 10 in 2010 stands as the second highest share for distributors in this category since 2004. In the last few years, it has ranged from under 3% in 2005 to just below 8% in 2009 (Table 16.8).

## TELECOMMUNICATIONS MEDIA

### Wireline Telecom

The UK telecom market, one of the largest in Europe, is characterized by fierce competition in the mobile and broadband sectors and by an innovative broadcast sector that has pioneered

Table 16-8. Film Production/Distribution (Market Shares by Box Office %)[1] 2004–2010

| | 2004 | 2005 | 2006 | 2007 | 2008 | 2009 | 2010 |
|---|---|---|---|---|---|---|---|
| Warner Bros (Time Warner, US) | 14.7 | 18.2 | 8.2 | 15.6 | 11.0 | 11.2 | 18.3 |
| 20th Century Fox (News Corp., 21st Century Fox, Murdoch Family, US/AUS) | 10.7 | 14.3 | 20.9 | 13.9 | 9.4 | 16.6 | 15.9 |
| Paramount (Viacom, US)[2] | — | — | — | 14.7 | 16.9 | 10.8 | 14.8 |
| Walt Disney (Walt Disney Company, US) | 14.5 | 13.1 | 15.7 | 10.7 | 9.9 | 12.4 | 14.0 |
| Universal (Comcast/GE, US) | — | — | — | 13.9 | 18.5 | 10.5 | 10.2 |
| Sony Pictures (Sony, Japan) | 10.0 | 6.8 | 16.1 | 8.2 | 12.5 | 11.3 | 6.9 |
| eOne Films (Entertainment One Ltd., Canada/US) | — | — | — | — | — | 4.9 | 5.5 |
| Lions Gate (Lions Gate Entertainment Corporation, US) | 1.0 | 0.3 | 2.4 | 2.3 | 2.5 | 2.9 | 3.5 |
| Entertainment (Nigel Green; Entertainment Film Distributors, UK) | 7.9 | 9.4 | 7.9 | 9.5 | 8.0 | 8.6 | 2.5 |
| Optimum (Vivendi, France) | | | | | | | 2.2 |
| Pathé (Pathé Brothers, France) | 2.8 | 3.4 | 3.2 | 1.3 | 2.1 | 2.9 | |
| Momentum (Alliance Films, UK) | 2.2 | 1.9 | 2.3 | 3.4 | 3.5 | | |
| UIP | 29.8 | 29.1 | 18.9 | | | | |
| Top Ten Total[3] | 96.1 | 97.3 | 96.4 | 94.5 | 94.5 | 92.2 | 93.7 |
| Others | 3.9 | 2.7 | 3.6 | 5.5 | 5.5 | 7.8 | 6.3 |
| Total Revenue (mil £)[4] | 902 | | | | 943 | | 1,140 |
| Total Revenue (mil US$) | 1,651 | | | | 1,753 | | 1,767 |
| C4 | 69.7 | 74.7 | 71.6 | 58.1 | 57.8 | 51.5 | 63.0 |
| HHI | 1,605 | 1,704 | 1,451 | 1,136 | 1,178 | 1,024 | 1,208 |
| *N* (>1%) | 9 | 8 | 9 | 10 | 10 | 10 | 10 |
| Noam Index | 535 | 602 | 483 | 359 | 373 | 323 | 382 |

*Source*: Author; BFI Statistical Yearbook 2011.

[1] The British Film Institute (BFI) Statistical Yearbook publishes information on film distributors and exhibitors in the United Kingdom. While the Statistical Yearbook does not publish any data on exhibitors' market shares, in terms of share of box office, it does publish information on distributors' market shares (in terms of total box office receipts).

[2] Until 2006, Paramount and Universal distributed jointly as UIP.

[3] "Top Ten Total" refers to the top 10 distributors of that particular year.

[4] This revenue figure is an estimate based on data collected in this study: it is based on average per capita revenues reported for several other countries in this study with similar per capita income.

business models for digital distribution. As a consequence, mobile and broadband penetration is well ahead of the EU average, while competition in the wireline sector is driving prices down for UK consumers. Mobile communications' share of total telecoms revenue has, as of 2004, been larger than that of fixed lines, and DSL is now almost universal in the country, while Internet coverage (both at work and at home) has reached 79.8% of the population.[11] Digital television is universal as digital switchover occurred in December 2012, and digital radio (DAB) is beginning to make inroads. Consumer prices across the board have fallen steadily, while network capabilities have been greatly expanded following recent investments by British Telecom (BT) and Virgin Media in Next Generation Network (NGN) infrastructure. To comprehend the enormity of the developments one should bear in mind that in the 1990s, most UK households only had access to a basic landline, five TV channels, and their radio service on a set.

British Telecom (BT) dominates wireline services, but it is losing market share. Although historically the telecom sector in the United Kingdom has been driven by growth in basic fixed telephone services, growth has become concentrated more recently in broadband and mobile. The decline in traditional fixed telephone market represents a business challenge to BT, which in the past has relied on those services as its most stable revenue stream. BT's market revenue for all fixed call volumes fell from US$10.7 billion (£5.9 billion) in 2005 to US$6.5 billion (£4.2 billion) in 2010. In the same period, BT's market share fell from 64% to just above 50%, whereas the market shares of Virgin Media and other wireline telecom firms have increased.

Virgin Media provides quadruple services—fixed and mobile telephone, television, and broadband Internet—to UK consumers. In 2006, Virgin Media entered into a licensing agreement with Sir Richard Branson's Virgin Group Limited to license out all the relevant Virgin sub-brands for a term of 30 years, with a 10-year opt-out clause. Branson accepted a mix of shares and cash, making him a 10.7% shareholder of the combined company at the time. However, now that it is listed on NASDAQ and the London Stock Exchange, Sir Richard Branson's Virgin Entertainment Investment Holdings stake has fallen to 5.5% (Table 16.9).[12]

## Wireless Telecom

In contrast to the UK wireline telecom market, which has been characterized by a BT monopoly, the UK's wireless telecom market has been much more competitive. Outside of the new 3G operator, 3UK, before 2010 the other four operators—O2, Vodafone, T-Mobile, and Orange—had roughly equivalent market shares. However, concentration rose when Orange and T-Mobile merged in 2010 under a new company named Everything Everywhere (EE), owned jointly (50:50) by Deutsche Telekom and France Telecom (Table 16.10).

A 2011 report by Ofcom highlights some of the key trends emerging in 2011 regarding both the wireline and wireless sectors.[13] In particular, BT's share of total fixed and mobile voice call volumes fell to 19.4% in 2010, while mobile's share increased to 49.2%, making it likely that more than half of UK voice calls will come from mobile platforms in the next few years. BT's share of fixed voice call volumes also fell to under 40% for the first time in 2011. Meanwhile, total fixed Internet revenues were US$4.7 billion (£3 billion) in 2010, down from US$5 billion (£3.2 billion) in 2009. This decline was primarily the result of falling residential broadband prices. In addition, the decline in the number of fixed lines for businesses accelerated in 2010, falling by 5.2% (Table 16.11).

---

11. Internet World Stats. "United Kingdom: Internet Usage Stats and Market Report—Usage and Population Statistics." Internet World Stats, n.d. March 30, 2012. <http://www.internetworldstats.com/eu/uk.htm>.

12. Virgin Media. "Ownership." *Investor Centre*, n.d. Dec. 20, 2012. <http://investors.virginmedia.com/phoenix.zhtml?c=135485&p=irol-ownershipsummary>.

13. Ofcom. "Communications Market Report." London: Office of Communications, Aug. 4, 2011. May 18, 2012. <http://stakeholders.ofcom.org.uk/binaries/research/cmr/cmr11/UK_CMR_2011_FINAL.pdf>.

## Table 16-9. Wireline Telecom (Market Shares By Revenue)[1] 2005–2012

| | 2005 | 2006 | 2007 | 2008 | 2009 | 2010 | 2012 |
|---|---|---|---|---|---|---|---|
| British Telecom (BT) | 64.0 | 61.0 | 60.0 | 58.0 | 55.0 | 51.0 | 43,8 |
| Virgin Media | 13.0 | 12.0 | 12.0 | 11.0 | 12.0 | 12.0 | 14.6 |
| British Sky Broadcasting Group PLC | | | | | | | 12.7 |
| Others[2] | 23.0 | 27.0 | 28.0 | 31.0 | 33.0 | 37.0 | 29.0 |
| Total Revenue (bil £) | 10,834 | 10,865 | 11,662 | 10,655 | 8,582 | 8,178 | 8,500 |
| Total Revenue (bil US$) | 19,718 | 19,992 | 23,324 | 19,819 | 13,474 | 12,676 | 13,469 |
| C4 | 77.0 | 73.0 | 72.0 | 69.0 | 67.0 | 63.0 | 71.1 |
| HHI | 4,283 | 3,933 | 3,749 | 3,540 | 3,196 | 2,772 | 2,293 |
| *N* (>1%) | 2 | 2 | 2 | 2 | 2 | 2 | 3 |
| Noam Index | 3,029 | 2,781 | 2,651 | 2,503 | 2,260 | 1,960 | 1,324 |

*Source*: Ofcom, 2011
*Source*: Ofcom, 2013 report for 2012 numbers
2012 updates by editors
[1] Includes estimates where Ofcom does not receive data from operators; excludes revenues from NTS voice calls.
[2] "Others" includes Talk-Talk, Sky Talk, O2, Tiscali, Post Office, Orange, Pipex Home Call, Kingston Communications, Tesco Home Phone and Vodafone. Bainbridge, Jane. "Sector Insight: Fixed-line telecoms." *Brand Republic*, May 20, 2011. July 14, 2012. <http://www.brandrepublic.com/analysis/1070132>.
[3] This revenue figure is an estimate based on data collected in this study: it is based on average per capita revenues reported for several other countries in this study with similar per capita income.

## Table 16-10. Wireless Telecom (Market Shares By Revenue), 2005–2012

| | 2005 | 2008 | 2010 | 2012 |
|---|---|---|---|---|
| O2 (Telefónica, Spain) | 22.0 | 25.2 | 28.0 | 29.0 |
| Vodafone | 21.5 | 25.4 | 23.0 | 28.0 |
| Everything Everywhere (EE) | 0.0 | 0.0 | 43.0 | 34.0 |
| T-Mobile (Deutsch Telekom, Germany) | 16.5 | 20.0 | | |
| Orange (France Telecom, France) | 17.0 | 25.0 | | |
| 3UK | 2.3 | 4.5 | 6.0 | 9.0 |
| Others | 20.7 | 0.0 | 0.0 | 0.0 |
| Total Revenue (mil £)[1] | 13,714 | 18,898 | 21,219 | 19,570 |
| Total Revenue (mil US$) | 24,960 | 35,151 | 32,890 | 32,210 |
| C4 | 77.0 | 95.6 | 100 | 100 |
| HHI | 1,513 | 2,325 | 3,198 | 2,862 |
| *N* (>1%) | 5 | 5 | 4 | 4 |
| Noam Index | 677 | 1,040 | 1,599 | 1,431 |

*Source*: Ofcom, 2010; 2011, 2012. 2012 update provided by editors.
[1] This revenue figure is an estimate based on data collected in this study: it is based on average per capita revenues reported for several other countries in this study with similar per capita income.

Table 16-11. Total UK Wireline & Wireless Telecom (Market Shares by %)[1] 2005–2012

| | 2005 | 2006 | 2007 | 2008 | 2009 | 2010 | 2012 |
|---|---|---|---|---|---|---|---|
| 3UK Mobile | 3.5 | 3.6 | 3.7 | 4.1 | 4.3 | 4.9 | 6.3 |
| Everything Everywhere (EE) | 30.0 | 30.9 | 30.5 | 29.9 | 29.8 | 27.2 | 24.0 |
| Orange (France Telecom, France) | 14.8 | 14.7 | 14.5 | 14.8 | 14.8 | | |
| T-Mobile (Deutsch Telekom, Germany) | 15.2 | 16.2 | 16.0 | 15.1 | 15.0 | | |
| O2 Mobile (Telefónica, Spain) | 16.9 | 18.2 | 18.5 | 19.3 | 19.6 | 21.2 | 20.4 |
| Vodafone Mobile | 15.1 | 14.3 | 15.5 | 15.9 | 16.5 | 17.5 | 19.7 |
| Virgin Media (fixed) | 4.9 | 4.5 | 4.2 | 4.3 | 4.2 | 4.3 | 4.3 |
| British Telecom (BT) (fixed) | 26.2 | 23.1 | 21.2 | 19.2 | 16.5 | 14.9 | 13.8 |
| British Sky Broadcasting (fixed) | | | | | | | 3.7 |
| Others (fixed)[2] | 3.6 | 5.4 | 6.4 | 7.3 | 9.2 | 10.0 | 8.6 |
| C4 | 73.4 | 72.2 | 71.2 | 69.5 | 67.6 | 80.8 | 77.9 |
| HHI | 1,699 | 1,610 | 1,570 | 1,530 | 1,493 | 1,860 | 1,851 |
| *N* (>1%) | 8 | 8 | 8 | 8 | 8 | 7 | 7 |
| Noam Index | 601 | 569 | 555 | 541 | 528 | 703 | 702 |

*Source*: Ofcom, 2011.

[1] Includes estimates where Ofcom does not receive data from operators. MVNOs and mobile service provider connections are included within the network operator figures.

[2] "Others (fixed)" includes carrier preselection and wholesale line rental in addition to other licensed operators.

## INTERNET MEDIA

### Internet Service Providers (ISPs)

In the past few decades, ISPs have gone through numerous mergers and acquisitions. However, there are currently a fair number of ISPs on the market, and the competition between them has reduced prices substantially. UK users can now buy high-speed broadband packages (up to 50 MB of broadband) combined with other services for under US$16 (10£) per month (Table 16.12).

### Search Engines

In terms of market share, Google dominates this sector, as it does in most other highly industrialized countries. But in the United Kingdom, Google's market share is even stronger than it is in the United States, where it originated: it controlled 90.7% of all UK web traffic in January 2011.[14] In a very distant second and third place are Bing (formerly MSN) and Yahoo!, both of which are losing market share (Table 16.13).

14. Logan, Stephen. "Search Engine Market Share Statistics—November 2009." *Koozai*, Nov. 18, 2009. June 3, 2012. <http://www.koozai.com/blog/news/industry-news/search-engine-market-share-statistics-november-2009/>; and McDonagh, Kevin. "UK Google Market Share—March 2007." *Intraspin.com*, Apr. 19, 2007. Dec. 20, 2012 <http://www.intraspin.com/webstrategyblog/uk-google-market-share-march-2007>; and Perrin, James. "Search Engine Market Share Statistics—January 2011." *Koozai*, Jan. 11, 2011. June 3, 2012. <http://www.koozai.com/blog/search-engine-news/google/google-still-dominating-uk-and-us-search-markets-007>.

Table 16-12. Internet Service Providers (Market Shares by Revenue) 2004–2012

| | 2004 | 2006 | 2009 | 2012 |
|---|---|---|---|---|
| British Telecom (BT) Broadband | 27.2 | 23.1 | 25.8 | 29.6 |
| Virgin Media | 21.4 | 26.3 | 21.8 | 20.8 |
| Carphone Warehouse (UK)[1] | 18.6 | 22.2 | 25.2 | — |
| Talk-Talk (UK)[2,3] | 7.5 | 10.1 | 10.0 | 17.0 |
| Sky Broadband (News Corp., Murdoch Family, US/UK/AUS)[4] | — | — | 11.0 | 19.4 |
| Orange (France Telecom, France) | 9.3 | 9.2 | 5.75 | 3.0 |
| Cable and Wireless Worldwide (UK)[5] | 1.5 | 1.2 | — | — |
| Others | 22.0 | 18.0 | 10.5 | 8.0 |
| Total Revenue (mil £)[6] | 3,822 | 4,832 | 7,070 | 7,300 |
| Total Revenue (mil US$) | 6,994 | 8,890 | 11,100 | 11,563 |
| C4 | 76.5 | 80.8 | 83.8 | 86.8 |
| HHI | 1,633 | 1,804 | 1,930 | 1,983 |
| *N* (>1%) | 5 | 5 | 5 | 5 |
| Noam Index | 730 | 807 | 863 | 887 |

*Source*: Ofcom, 2010 Source Ofcom, 2013 for 2012. 2012 updates provided by editors.

[1] Includes AOL.

[2] Includes Tiscali.

[3] Talk-Talk was a subsidiary of Carphone Warehouse until 2010 when Carphone demerged their telecom business. http://www.reuters.com/article/2010/03/29/us-best-buy-carphone-idUSTRE62S1OK20100329.

[4] News International operates as an ISP through Sky Broadband, which was founded in 2006. The first available market share was in 2007—which is why Sky's market share in 2004 was 0%, but went up to 11% in 2009.

[5] In 2010, Cable & Wireless PLC became Cable & Wireless Worldwide and in June 2012 Vodafone acquired the UK-based company.

[6] This revenue figure is an estimate based on data collected in this study: it is based on average per capita revenues reported for several other countries in this study with similar per capita income.

### Online News

Established newspapers and television broadcasters dominate the online news media sector, leaving little room for newcomers. Despite the lead the BBC has here, newspapers also have large online market shares. Over the past decade, there has been a fast-growing online market worldwide, attracting even larger audiences from outside the United Kingdom. News websites have seen a decline in domestic readership, but they have compensated by winning new readers abroad. Overall traffic for established news sites run by the BBC, News International, and ITV has shown a slight decline in recent years, owing to competition abroad in the English-language news market (Table 16.14).

## CONCLUSION

In 2009, Ofcom submitted a set of recommendations to the UK Secretary of State concerning local media ownership rules, which were eventually approved.[15]

15. Ofcom. "Media Ownership Rules Review Consultation Document." London: Office of Communications, July 31, 2009. May 18, 2012, 35–40. <http://stakeholders.ofcom.org.uk/binaries/research/cmr/cmr09.pdf>.

Table 16-13. Internet Search Engines (Market Shares by Revenue), 2007–2011

| | 2007 | 2009 | 2011 |
|---|---|---|---|
| Google (US) | 77.0 | 88.7 | 90.7 |
| Yahoo! (US) | 8.0 | 5.5 | 2.9 |
| MSN (US) | 5.0 | — | — |
| Bing (US) | — | 2.9 | 3.9 |
| Ask (US) | 5.0 | — | 1.3 |
| Others | 5.0 | 2.9 | 1.3 |
| Total Revenue (mil £)[1] | 375 | 1,311 | 1,406 |
| Total Revenue (mil US$) | 750 | 2,058 | 2,250 |
| C4 | 95.0 | 97.1 | 98.7 |
| HHI | 6,043 | 7,914 | 8,248 |
| *N* (>1%) | 5 | 4 | 5 |
| Noam Index | 2,703 | 3,956 | 3,689 |

*Source*: Intraspin.com, 2007; Koozai, 2009; 2011.
[1] This revenue figure is an estimate based on data collected in this study: it is based on average per capita revenues reported for several other countries in this study with similar per capita income.

The recommendations consisted mostly of relaxing and simplifying the radio ownership rules and abolishing most of the local rules concerning radio and cross-media ownership. Ofcom's rationale was that given the financial pressures that radio stations face, relaxing regulations may provide opportunities to make them more viable by letting them go under common ownership. Ofcom further justified this position by reporting that a majority of consumers are not opposed to single ownership in radio, so long as they have an alternative source from their local BBC station.

Ofcom also recommended greater liberalization of local cross-media ownership rules, so that the only restriction would be on ownership of three media sources at the same time: local newspapers (with 50% plus local market share), a local radio station, and the regional *Channel 3* license were set as the maximum concentration limits.

Meanwhile, Ofcom advised the government to retain the national cross-media ownership rules. Despite the rise of the Internet and the growth of digital media as a news source, TV remains an important source of news, and newspapers retain an important role in setting the news agenda. Ofcom concluded that there have not yet been such significant changes in the solvency of national media outlets to warrant a relaxation of the anticoncentration measures here.

Ofcom further recommended retaining the restrictions on withholding broadcast licenses in order to continue protecting consumers against undue influence through television and radio by certain owners (i.e., political parties and religious groups).

Ofcom has suggested retaining the appointed news provider rule as well, which aims to ensure that the provision of national and international news to ITV plc is independent of the BBC. This is because *ITV1* remains the most watched source of broadcast news after the BBC, exactly as it was when the rules were liberalized with the Communications Act (2003).[16]

16. "No. 1900 (C. 77) Communications Act 2003." London: The National Archives, July 21, 2003. March 2, 2012. <http://www.legislation.gov.uk/ukpga/2003/21/contents>.

Table 16-14. Online News Media (Market Shares by Revenue), 2007–2011

| | 2007 | 2009 | 2011 |
|---|---|---|---|
| BBC (public) | 35.0 | 34.0 | 32.0 |
| News International (News Corp., Murdoch Family, UK/US/AUS)[1] | 19.2 | 19.0 | 18.0 |
| ITV (ITV plc) | 9.0 | 8.5 | 8.0 |
| Guardian.co.uk (Guardian Media Group) | 7.5 | 7.0 | 7.0 |
| Mail Online (Daily Mail & General Trust) | 6.0 | 6.0 | 6.1 |
| Channel 4 (Channel Four Television Corporation) | 4.0 | 4.0 | 4.0 |
| Mirror Group Digital (Trinity Mirror plc) | 2.2 | 2.0 | 1.8 |
| Telegraph Group (Barclay Brothers) | 1.8 | 1.8 | 1.7 |
| Independent.co.uk (Independent Print Ltd, Alexander Lebedev) | 0.9 | 0.8 | 0.8 |
| Others | 14.4 | 16.9 | 20.6 |
| Total Revenue (mil £)[2] | 375 | 656 | 956 |
| Total Revenue (mil US$) | 750 | 1,030 | 1,530 |
| C4 | 70.7 | 68.5 | 65.0 |
| HHI | 1,829 | 1,736 | 1,559 |
| *N* (>1%) | 8 | 8 | 8 |
| Noam Index | 647 | 614 | 551 |

*Source*: Author, Audit Bureau of Circulations Electronic.

[1] Skynews.com, The Sun Online, and Times Online.

[2] This revenue figure is an estimate based on data collected in this study: it is based on average per capita revenues reported for several other countries in this study with similar per capita income.

With regard to the broadcasting market, the British public has been well served by the mixed economy of a public broadcaster, a regulated private sector, and the arrival of satellite and cable broadcasters.[17] The television market is among the most mature and dynamic in the European Union (EU), consisting of traditional broadcasters with public service obligations that offer high-quality, diverse programs that are universally accessible, as well as digital terrestrial, cable, and satellite channels. Despite the shift toward a more deregulated commercial broadcast market, there is a determination to retain the public service principles that have long shaped British television. This is evidenced by the establishment of the Content Board, a committee of the main Ofcom Board, which will serve as Ofcom's primary forum for the regulation of quality and standards in the broadcasting industry. Support for public service principles is also evidenced by the government's continued support for the BBC in an increasingly crowded marketplace, though it has asked BBC to cut back services and licensing fees. However, independent media critics will continue to argue that technological and regulatory changes have affected the emphasis and character of the programming of both public and private terrestrial

17. Iosifidis, Petros. *Public Television in the Digital Era: Technological Challenges and New Strategies for Europe*. 2nd ed. Basingstoke: Palgrave Macmillan, 2012, 143–163.

channels, which now provide less pluralistic, less distinctive, and less diverse output.

Satellite and cable television channels are also more widely available, but they are not subject to public service obligations. BSkyB, set up in 1990, dominates the satellite TV market with more than 12 million subscribers. The cable TV market developed in the late 1980s with the entry of numerous networks, but in 2006 only two providers had prevailed, NTL and Telewest, who then merged into Virgin Media. A further merger with Virgin Mobile UK in July 2006 created the first "quadruple-play" media company in the United Kingdom, offering TV, Internet, mobile phone, and fixed-line telephone services. Here, Britain stands as an EU leader in the rollout of digital television services. Following the collapse of the pay-DTT consortium ITV Digital in April 2002, the BBC consortium Freeview was launched in late 2002, and in less than a decade it has attracted more than 14 million households by offering free-to-air digital services. At the end of 2012, 100% of the UK households had access to digital TV via terrestrial, satellite, cable, or Digital Subscriber Lines (DSL).

The main point of discussion with regard to UK media concentration will continue to be centered around Rupert Murdoch's media properties, though because the BBC is a powerful player in radio, terrestrial TV, and digital and online services, there will also be many voices (mainly from commercial operators) demanding the government curb its expansion efforts in the online market.

The Communications Act (2003) is the main piece of legislation governing British media. However, the existing law is inefficient in preventing concentration; in fact, it opened the door wide open for more concentration and cross-ownership in the traditional industries like newspapers and broadcasting. This is because it has abolished strict media and cross-media ownership rules, placed greater emphasis on "self-regulation," and made it easier for foreign interests to own UK media companies. As these policies are likely to prevail, the concentration in many industries, especially newspapers, will continue to increase in the coming years, although there are signs that the UK government may adopt a tougher line toward News International in the wake of "Hackgate."

## Bibliography

### 1. Books and Articles

Bainbridge, Jane. "Sector Insight: Fixed-line telecoms." *Brand Republic*, May 20, 2011. July 14, 2012. <http://www.brandrepublic.com/analysis/1070132>.

Barnett, Steven, and Seaton, Jean. "Why the BBC Matters: Memo to the New Parliament about a Unique British Institution." *The Political Quarterly* 81:3 (2010): 327–332.

Browning, Jonathan. "U.K. Newspapers Join BBC in Opposition to Murdoch BSkyB Deal." *Bloomberg*, Oct. 12, 2010. http://www.bloomberg.com/news/2010-2010-12/bbc-joins-u-k-newspapers-to-lobby-against-murdoch-bid-for-rest-of-bskyb.html>.

Freedman, Des. *The Politics of Media Policy*. Cambridge: Polity Press, 2008.

Iosifidis, Petros. *Public Television in the Digital Era: Technological Challenges and New Strategies for Europe*. 2nd ed. Basingstoke: Palgrave Macmillan, 2012.

Logan, Stephen. "Search Engine Market Share Statistics—November 2009." *Koozai*, Nov. 18, 2009. June 3, 2012. <http://www.koozai.com/blog/news/industry-news/search-engine-market-share-statistics-november-2009/>.

McDonagh, Kevin. "UK GOOGLE MARKET SHARE—MARCH 2007." *Intraspin.com*, Apr. 19, 2007. Dec. 20, 2012 <http://www.intraspin.com/webstrategyblog/uk-google-market-share-march-2007>.

Perrin, James. "Search Engine Market Share Statistics—January 2011." *Koozai*, Jan. 11, 2011. June 3, 2012. <http://www.koozai.com/blog/search-engine-news/google/google-still-dominating-uk-and-us-search-markets-007>.

Reed, John. "Debate: Will Publishers Soon Be Irrelevant?" *Publishing Talk*, Apr. 18 2011. Dec. 20, 2012. <http://www.publishingtalk.eu/conferences/debate-will-publishers-soon-be-irrelevant/>.

Stone, Philip. "Book Sales in 2010 Fall by More than 3%." *The Bookseller*, Jan. 5, 2011. July 14, 2012. <http://www.thebookseller.com/news/book-sales-2010-fall-more-3.html>.

Tivnan, Tom and Stone, Philip. "Review of 2011: The publishers." *The Bookseller*, Jan. 1, 2012.

14 July 2012. <http://www.thebookseller.com/user/login?destination=node%2F95921>.

Tunstall, Jeremy. "The BBC and UK Public Service Broadcasting." *Reinventing Public Service Communication: European Broadcasters and Beyond*. Ed. Iosifidis, Petros (Basingstoke: Palgrave Macmillan, 2010). 145–157.

## 2. Reports

The British Film Institute (BFI). *BFI Statistical Yearbook 2011*. London: BFI, 2011. Dec. 20, 2012. <http://www.bfi.org.uk/sites/bfi.org.uk/files/downloads/bfi-statistical-yearbook-2011.pdf.>

"Communications Act 2003." London: The National Archives, July 21, 2003. March 2, 2012. <http://www.legislation.gov.uk/ukpga/2003/21/contents>.

Ofcom. "Media Ownership Rules Review Consultation Document." London: Office of Communications, July 31, 2009. May 18, 2012. 35–40. <http://stakeholders.ofcom.org.uk/binaries/research/cmr/cmr09.pdf>.

Ofcom. "The Communications Market 2009 (August)." London: Office of Communications, 2009. Dec. 20, 2012. <http://stakeholders.ofcom.org.uk/market-data-research/market-data/communications-market-reports/cmr09/>.

Ofcom. "International Communications Market Report 2010." London: Office of Communications, 2010. May 30, 2012. <http://stakeholders.ofcom.org.uk/binaries/research/cmr/753567/icmr/ICMR_2010.pdf>.

Ofcom. "The Communications Market 2010 (August)." London: Office of Communications, 2010. Dec. 20, 2012. <http://stakeholders.ofcom.org.uk/market-data-research/market-data/communications-market-reports/cmr10/>.

Ofcom. "Communications Market Report." London: Office of Communications, Aug 4, 2011. May 18, 2012, <http://stakeholders.ofcom.org.uk/binaries/research/cmr/cmr11/UK_CMR_2011_FINAL.pdf>.

Ofcom. "The Communications Market 2011 (August)." London: Office of Communications, 2011. Dec. 20, 2012. <http://stakeholders.ofcom.org.uk/market-data-research/market-data/communications-market-reports/cmr11/>.

## 3. Websites

Broadcasters' Audience Research Board (BARB). n.d. *Total viewing summary*. Dec. 20, 2012. <http://www.barb.co.uk/viewing/weekly-total-viewing-summary?_s=4>.

Internet World Stats. n.d. "United Kingdom: Internet Usage Stats and Market report—Usage and Population Statistics." Internet World Stats, March 30, 2012. <http://www.internetworldstats.com/eu/uk.htm>.

Magforum.com. n.d. "Magazine Industry Sectors." May 22, 2012. <http://www.magforum.com/sectors.htm#key>.

Point-Topic. "The UK Broadband Market." <http://point-topic.com/brochure/UKP.pdf>.

Radio Joint Audience Research (RAJAR). n.d. "Quarterly Listening." Dec. 20, 2012. <http://www.rajar.co.uk/listening/quarterly_listening.php>.

Virgin Media. n.d. "Ownership." *Investor Centre*, Dec. 20, 2012. <http://investors.virginmedia.com/phoenix.zhtml?c=135485&p=irol-ownershipsummary>.

# United Kingdom—Data Summaries

ELI NOAM AND PAUL MUTTER

MEDIA CONCENTRATION IN THE United Kingdom is among the lowest in the world. Weighted Average HHI, while high in an antitrust sense, were low compared to other countries, at 2,871. However, it has grown in recent years (Table 16.15).

Public ownership of media in the United Kingdom is relatively low, at 9.7%, mostly through the BBC. The BBC accounts for 50% of TV broadcasting, 10% of magazines, 55% of radio, and 32% of online news media, making it one of the most diverse content media companies in the United Kingdom. Of the companies solely involved in content media, it is the largest by overall national market share with 9.7%, and it holds 20.8% of content (Table 16.16).

Since the 1980s, a relaxation on foreign direct investment has led to an increase in foreign ownership of national media industries (52.1% of all properties are foreign owned). Murdoch's 21st Century Fox (US) is the most active and expansionary of these foreign-owned firms. News Corp. is also the major newspaper firm. Its controlling owner, Rupert Murdoch, moved first to London from his native Australia and created a significant print and political presence. He then moved control of the company to New York. Following public controversies, News Corp. split itself in 2013 into News Corp. and 21st Century Fox. These two firms had 8.1% and 11.5% share of content, respectively. 21st Century Fox is the third-largest media company in the United Kingdom based on Sky, which it controls (Table 16.16).

Traditionally, UK media companies have a significant international role: Pearson is a major print publisher; Vodafone (United Kingdom) is the second largest mobile operator in the world, with over 440 million subscribers and large market share of the mobile markets in many countries. Vodafone's global size is vastly greater than its UK presence.

Table 16-15. National Media Industries Concentration in the United Kingdom

| | 2004/5 | | 2011 OR MOST RECENT | | % CHANGE ANNUAL AVERAGE | |
|---|---|---|---|---|---|---|
| | COMPANY POWER INDEX IN COUNTRY | COMPANY SHARE OF THE OVERALL NATIONAL MEDIA MARKET (%) | COMPANY POWER INDEX IN COUNTRY | COMPANY SHARE OF THE OVERALL NATIONAL MEDIA MARKET (%) | COMPANY POWER INDEX IN COUNTRY | COMPANY SHARE OF THE OVERALL NATIONAL MEDIA MARKET (%) |
| Murdoch | 448.6 | 10.5 | 810.6 | 16.4 | 11.5 | 0.9 |
| BT Group | 825.6 | 13.9 | 288.9 | 7.5 | | −0.9 |
| EE (T-Mobile, Germany + Orange) (France)[1] | 0.0 | 0.0 | 299.9 | 9.1 | N/A | 1.3 |
| T-Mobile (Deutsche Telekom, Germany) | 65.3 | 4.0 | *0* | *0* | −14.3 | −0.6 |
| Orange (France Telecom, France) | 75.1 | 4.7 | *0.0* | *0.0* | | −0.7 |
| Telefónica (O2, Spain) | 116.1 | 5.3 | 217.6 | 7.5 | 12.5 | 0.3 |
| BBC (public) | 287.1 | 7.3 | 431.6 | 9.7 | | 0.3 |
| Vodafone | 110.8 | 5.2 | 202.8 | 7.2 | 11.9 | 0.3 |
| Liberty (US) | 0.0 | 0.0 | 142.4 | 6.5 | N/A | 0.9 |
| Google (US) | 42.7 | 0.6 | 148.7 | 1.6 | 35.4 | 0.16 |
| Daily Mail and General Trust | 31.8 | 1.7 | 33.4 | 1.6 | 0.8 | −0.02 |
| Penguin Random House (Germany/UK)[1] | 0.0 | 0.0 | 300.4 | 1.2 | N/A | 0.17 |
| Pearson (UK) | 14.4 | 1.6 | 4.0 | 0.7 | −10.3 | −0.1 |
| Bertelsmann (Germany) | 14.2 | 0.9 | 0.0 | 0.0 | −14.3 | −0.13 |
| ITV | 74.9 | 2.8 | 63.9 | 3.0 | −2.1 | 0.03 |
| Bauer (Germany) | 22.1 | 1.3 | 42.2 | 1.8 | 12.9 | 0.06 |
| Trinity Mirror plc | 37.6 | 1.9 | 16.3 | 1.1 | −8.1 | −0.1 |
| Lagardère (France) | 17.1 | 1.0 | 9.5 | 0.7 | −6.4 | −0.05 |
| Hearst (US) | 10.6 | 0.8 | 13.0 | 0.9 | 3.2 | 0.01 |
| Total Global Radio | 7.9 | 0.5 | 5.9 | 0.4 | −3.6 | −0.02 |
| Channel 4 | 13.6 | 1.2 | 13.0 | 1.4 | −0.7 | −0.02 |
| Viacom (US) | 8.2 | 0.8 | 7.4 | 1.1 | −1.43 | 0.05 |
| Sony (Japan) | 1.6 | 0.2 | 0.68 | 0.1 | −8.2 | −0.01 |

*continued*

Table 16-15. *continued*

| | 2004/5 | | 2011 OR MOST RECENT | | % CHANGE ANNUAL AVERAGE | |
|---|---|---|---|---|---|---|
| | COMPANY POWER INDEX IN COUNTRY | COMPANY SHARE OF THE OVERALL NATIONAL MEDIA MARKET (%) | COMPANY POWER INDEX IN COUNTRY | COMPANY SHARE OF THE OVERALL NATIONAL MEDIA MARKET (%) | COMPANY POWER INDEX IN COUNTRY | COMPANY SHARE OF THE OVERALL NATIONAL MEDIA MARKET (%) |
| Time Warner (US) | 27.4 | 1.5 | 5.6 | 0.5 | –11.4 | –0.16 |
| Disney (US) | 5.5 | 0.5 | 4.9 | 0.5 | | –0.001 |
| The Guardian | 1.8 | 0.5 | 1.5 | 0.4 | –2.4 | –0.01 |
| Condé Nast (Advance, US) | 1.3 | 0.3 | 1.0 | 0.2 | –4.4 | –0.007 |
| Northern & Shell | 17.5 | 1.3 | 12.1 | 1.0 | –4.5 | –0.05 |
| Virgin Media[2] | 129.4 | 6.4 | 0.0 | 0.0 | –14.3 | –0.9 |
| Media Concentration Index | 2004/5 | | 2011 or Most Recent | | % Change Annual Average | |
| Total Revenue: Nat'l Media Industry (mil US$) | 104,098 | | 124,510 | | 2.0% | |
| Total Voices (*n*) | 65 | | 67 | | 0.4% | |
| Net Voices (*n*) | 43 | | 46 | | 1.0% | |
| Public Ownership (%) | 7.3 | | 9.7 | | 0.3% | |
| Foreign Ownership (%) | 33.7 | | 52.1 | | 2.6% | |
| C4 Average—Weighted | 76.7 | | 85.2 | | 1.2% | |
| HHI Average—Weighted | 2,597 | | 2,871 | | 1.5% | |
| C1 Average—Weighted | 37.7 | | 41.6 | | 0.0% | |
| Noam Index Average—Weighted | 1,000 | | 1,242 | | 3.5% | |
| Pooled Overall Sector C4 | 38.2 | | 42.7 | | 0.6% | |
| Pooled Overall Sector HHI | 520 | | 689 | | 4.6% | |
| Pooled Overall Sector Noam Index | 103 | | 123 | | 2.8% | |
| Market Share of Top Ten Companies: Nat'l Media Industry (%) (Pooled C10) | 61.9 | | 71 | | 1.3% | |
| National Power Index | 2,429 | | 2,888 | | 2.7% | |

[1] Book publishing only.
[2] Bought out by Liberty in 2013.

## Table 16-16. Top Content Media Companies in the United Kingdom

| | 2004/5 | | 2011 OR MOST RECENT | | % CHANGE ANNUAL AVERAGE | |
|---|---|---|---|---|---|---|
| | COMPANY POWER INDEX IN COUNTRY | COMPANY SHARE OF THE NATIONAL CONTENT MEDIA MARKET (%) | COMPANY POWER INDEX IN COUNTRY | COMPANY SHARE OF THE NATIONAL CONTENT MEDIA MARKET (%) | COMPANY POWER INDEX IN COUNTRY | COMPANY SHARE OF THE NATIONAL CONTENT MEDIA MARKET (%) |
| Murdoch Interests (US) | 564.3 | 15.7 | 769.5 | 16.7 | 5.2 | 0.14 |
| BBC (public) | 637.3 | 16.7 | 926.4 | 20.8 | 6.5 | 0.6 |
| Google (US) | 96.9 | 1.3 | 319.1 | 3.5 | 32.8 | 0.3 |
| Daily Mail and General Trust | 72.1 | 3.9 | 71.8 | 3.5 | –0.05 | –0.07 |
| ITV | 169.8 | 6.3 | 137.1 | 6.4 | –2.8 | 0.01 |
| Penguin Random House (Germany/UK)[1] | 0.0 | 0.0 | 65.3 | 2.6 | N/A | 0.38 |
| Pearson (UK) | 32.6 | 3.6 | 8.7 | 1.4 | –10.5 | –0.3 |
| Bertelsmann (Germany) | 32.2 | 2.1 | 0.0 | 0.0 | –14.3 | –0.3 |
| Time Inc. (US) | 0.0 | 0.0 | 60.2 | 2.7 | N/A | 0.4 |
| Liberty (US)[2] | 0.0 | 0.0 | 56.1 | 2.1 | N/A | 0.3 |
| Bauer (Germany) | 50.2 | 3.1 | 90.6 | 3.8 | 11.5 | 0.1 |
| Trinity Mirror | 85.3 | 4.3 | 35.0 | 2.4 | –8.4 | –0.3 |
| Northern & Shell | 39.8 | 2.9 | 25.9 | 2.0 | –5.0 | –0.1 |
| Lagardere (France) | 38.8 | 2.3 | 20.3 | 1.5 | –6.8 | –0.1 |
| Total Global Radio | 18.0 | 1.1 | 12.7 | 0.8 | –4.2 | –0.04 |

*continued*

Table 16-16. *continued*

| | 2004/5 | | 2011 OR MOST RECENT | | % CHANGE ANNUAL AVERAGE | |
|---|---|---|---|---|---|---|
| | COMPANY POWER INDEX IN COUNTRY | COMPANY SHARE OF THE NATIONAL CONTENT MEDIA MARKET (%) | COMPANY POWER INDEX IN COUNTRY | COMPANY SHARE OF THE NATIONAL CONTENT MEDIA MARKET (%) | COMPANY POWER INDEX IN COUNTRY | COMPANY SHARE OF THE NATIONAL CONTENT MEDIA MARKET (%) |
| Viacom (US) | 18.7 | 1.7 | 15.9 | 2.4 | –0.02 | 0.01 |
| Virgin Media | 49.8 | 1.9 | 0.0 | 0.0 | –14.3 | –0.3 |
| Media Concentration Index | 2004/5 | | 2011 or Most Recent | | % Change Annual Average | |
| Public Ownership (%) | 16.7 | | 20.8 | | 0.59% | |
| Foreign Ownership (%) | 35.7 | | 40.9 | | 0.74% | |
| C4 Average—Weighted | 74.2 | | 77.7 | | 0.5% | |
| HHI Average—Weighted | 2,201 | | 2,674 | | 3.1% | |
| C1 Average—Weighted | 34 | | 43 | | 1% | |
| National Power Index | 2,061 | | 2,711 | | 4.5% | |

[1] Book publishing only.
[2] Liberty bought out Virgin Media in 2013.

Table 16-17. Top Platform Media Companies in the United Kingdom

| | 2004/5 | | 2009 OR MOST RECENT | | % CHANGE ANNUAL AVERAGE | |
|---|---|---|---|---|---|---|
| | COMPANY POWER INDEX IN COUNTRY | COMPANY SHARE OF THE NATIONAL PLATFORM MEDIA MARKET (%) | COMPANY POWER INDEX IN COUNTRY | COMPANY SHARE OF THE NATIONAL PLATFORM MEDIA MARKET (%) | COMPANY POWER INDEX IN COUNTRY | COMPANY SHARE OF THE NATIONAL PLATFORM MEDIA MARKET (%) |
| BT Group | 1,477 | 25.0 | 540.9 | 14.0 | –9.1 | –1.6 |
| Everything Everywhere (T-Mobile & Orange) | 0.0 | 0.0 | 561.4 | 17.0 | N/A | 2.4 |
| Orange (France Telecom, France) | 134.4 | 8.4 | *0.0* | *0.0* | –14.3 | –1.2 |
| T-Mobile (Deutsche Telekom, Germany) | 116.8 | 7.1 | *N/A* | *0.0* | –14.3 | –1.0 |
| Murdoch Group (Sky, 21st Century Fox, US) | 357.4 | 6.3 | 846.5 | 16.2 | 19.6 | 1.4 |
| Telefónica (O2, Spain) | 207.6 | 9.4 | 407.3 | 14.0 | 13.7 | 0.7 |
| Vodafone | 198.3 | 9.2 | 379.7 | 13.6 | 13.1 | 0.6 |
| Liberty (US) | 0.0 | 0.0 | 217.7 | 10.3 | N/A | 1.5 |
| Hutchison Whampoa (3G, Hong Kong) | 2.3 | 1.0 | 39.2 | 4.4 | 232.7 | 0.5 |
| Virgin Media[1] | 192.1 | 10.0 | 0.0 | 0.0 | | –1.4 |
| Media Concentration Index | 2004/5 | | 2011 or Most Recent | | % Change Annual Average | |
| Public Ownership (%) | 0.0 | | 0.0 | | 0.0% | |
| Foreign Ownership (%) | 32.2 | | 61.8 | | 4.2% | |
| C4 Average—Weighted | 78.6 | | 91.9 | | 1.9% | |
| HHI Average—Weighted | 2,910 | | 3,043 | | 0.7% | |
| C1 Average—Weighted | 41 | | 41 | | 0.0% | |
| National Power Index | 2,722 | | 3,043 | | 1.7% | |

[1] Bought out by Liberty in 2013; Liberty's figures reflect this sale.

At the same time, the UK wireless market has significant foreign ownership through T-Mobile (Germany) and Orange (France), now operating jointly as Everything Everywhere (EE), which holds 9.1% of the national market. Table 16.17 shows what the market shares of both companies would be individually if they were not operating in platforms such as this 50:50 joint venture. The Virgin Media/Liberty deal of 2013 is reflected, with Virgin Media's markets shares transferred to Liberty for the latest year.

# B

## North America

# 17

# Media Ownership and Concentration in Canada

DWAYNE WINSECK

## INTRODUCTION

The media economy in Canada is typically cast as being dwarfed by the world's largest media economy on its doorstep, the United States. In fact, Canada is the eighth largest media economy on the planet. As with most countries, the media economy in Canada has been transformed by extraordinary growth and greater differentiation within the media ecology over the past quarter of a century. New services developed (pay television, wireless telephony, Internet) and, at least during the 1980s and early 1990s, more competition took hold in most sectors of the media economy. However, such trends soon yielded to higher levels of concentration from the late 1990s until the present.

In this chapter, I analyze the evolution of a dozen of the most significant media markets in Canada. I call the totality of these sectors the telecom-media-Internet (TMI) industries, and the analysis is based on a body of data assembled for each sector from 1984 until 2011: wired and wireless telecom services, broadcast television, multichannel pay television, cable, satellite, and IPTV distributors, newspapers, magazines, music, radio, Internet access, search engines, and social network sites. The aim is to create a systematic body of data covering the past two and a half decades where there is currently none. It is also to use standard concentration ratios (CR4) as well as the Herfindhahl-Hirschman Index (HHI) to examine whether media markets have become more or less concentrated over time.

In *Media Ownership and Concentration in America*, Eli Noam[1] laments the lack of a systematic, empirically based portrait of the state

1. Noam, E. *Media Ownership and Concentration in America*. New York: Oxford University, 2009.

of media concentration in the United States and points to two reasons for this: first, the issue is highly politicized and, second, gathering data on complex industries over long spans of time is not easy. Much the same can be said for Canada. Ultimately, the data collected for this element of the International Media Concentration Research project aim to help fill this gap. Furthermore, it will also serve as a stepping stone to a long-term effort to annually update the portrait, broaden and improve the scope and quality of its coverage, and to think long and hard about why we should study this issue in the first place.

## THE HISTORICAL RECORD AND RENEWED INTEREST IN MEDIA CONCENTRATION IN THE 21ST CENTURY

Media ownership and concentration has emerged repeatedly, even if episodically, as a highly contested topic in Canada since the 19th and early 20th centuries. In the early 1900s, the Canadian Pacific Telegraph Company and Great Northwestern Telegraph Company (the latter an arm of the New York–based goliath, Western Union) dominated the telegraph industry west and east of Montreal, respectively. They also shared the exclusive Canadian distribution rights for the Associated Press news service. The two telegraph companies essentially freely gave away the American newswire service to the leading daily newspaper in each city they served as a means to fortify their dominant stake against smaller rivals in the vastly more lucrative telegraph business.

This was a boon to established members of the press and the AP. It helped the telegraph companies stitch up their lock on the telegraph business as well. Such arrangements stifled competition among newspapers and choked the flow of news on the wires. W. F. Maclean, the muckraking journalist, wrote in the *Toronto World*, "Attempts on the part of public service companies [the telegraph companies] to muzzle free expression of opinion by withholding privileges that are of general right cannot be too strongly condemned."[2]

The Winnipeg-based Western Associated Press learned the lessons of concentration and vertical integration the hard way when it established a rival news service in 1907. Its subscribers could not afford to pay two costs—telegraph transmission charges and the news service—and stay in business while the AP made its offering free. In 1910, the Board of Railway Commissioners (BRC), one of the first regulatory bodies in Canada, stepped in to effectively slay the "double-headed news monopoly."[3] In the face of much corporate bluster about the legitimacy of BRC's jurisdiction over this matter, the BRC shot back that the law required it to ensure that rates were just and reasonable because, unless they were, the "telegraph companies could put out of business every newsgathering agency that dared to enter the field of competition with them."[4] Thereafter, the three-fold alliance between the telegraph companies on the one side and Associated Press on the other unraveled with control over the medium forcefully separated from the message.

While there were other moments of concern in the interim, media concentration issues came to a head in the 1970s and early 1980s, when three inquiries were held: (1) the Special Senate Committee on Mass Media, *The Uncertain Mirror* (2 vols.);[5] (2) the Royal Commission on Corporate Concentration prompted partially by Gesca's (Power Corp) bid to acquire the Argus group of newspapers;[6] and (3) the Royal Commission on Newspapers[7] (otherwise known as the Kent Commission), which was called after a spate of newspaper consolidations and closures in 1980.

2. Nichols, M. E. *The Story of the Canadian Press*. Toronto: Ryerson, 1948, p. 41.

3. Blondheim, M., Rehearsal for Media Regulation: Congress Versus the Telegraph-News Monopoly, 1866–1900, *Federal Communications Law Journal* 56 (2004), 299–232.

4. Board of Railway Commissioners (BRC). *The Western Associated Press v. The Canadian Pacific Railway Company's Telegraph and the Great Northwestern Telegraph Company of Canada*. In Sessional papers of the Parliament of Canada. Ottawa: J. De Labroquerie Tache, Printers to the King's Most Excellent Majesty, 1910, p. 275.

5. Canada, Special Senate Committee on the Mass Media. *The Uncertain Mirror* (2 vols.). Ottawa: Queens Printer, 1970.

6. Canada, Royal Commission on Corporate Concentration. *Report*. Ottawa: Minister of Supply and Services Canada, 1978.

7. Canada, Royal Commission on Newspapers, *Report*. Ottawa: Minister of Supply and Services Canada, 1981.

Whatever meager measures did come out of these events soon passed, and there was little official interest in the subject for the following two decades. Instead, the realignment of political, technological, and economic forces evident worldwide took root in Canada. Markets were opened, new media technologies introduced, and regulatory restraints loosened. New media services (pay television, cable and satellite distribution, Internet) meant that the media ecology grew larger and more differentiated. The number of distinct media ownership groups rose in the 1980s and early 1990s, and there was more competition in many sectors but not all (more on this later). The gradual transformation of the media was strictly managed to minimize disrupting the industrial and regulatory status quo. This is mostly because large communication and media companies were seen as national champions with the deep pockets and wherewithal needed to build network infrastructure and fund the high cost of Canadian programming relative to the bargain basement cost of US cultural goods.

Many media players that were once central to the Canadian media system have since disappeared, including Southam, Hollinger, CHUM, Osprey, and many others. Most consolidated into larger groups, although a few such as Canwest Craig Media, and TQS collapsed in bankruptcy and financial disarray. Postmedia, Remstar, and Channel Zero are several new groups that formed in the traditional and emergent media sectors alike. Major players that were big in the 1980s, notably Bell, Rogers, Shaw, and TVA/Videotron (now part of Quebecor Media Inc., QMI), grew considerably and vertically integrated to a degree that is unique in the world.

Bell, Rogers, Shaw, and QMI have roots that reach deep back into the 20th century when they were granted local and regional telephone and cable monopolies. Since the 1990s, they have leveraged their incumbent monopoly status in one area to become competitors with one another. This happens in duopolistic markets for network infrastructure services (telephone, TV distribution, Internet access) and oligopolistic broadcasting markets alongside a few others such as CBC and Astral, and wireless cell phone services who have sizeable, and sometimes even bigger, stakes in such areas. However, there are areas where they do not compete with one another at all. This is also true for newspapers: QMI holds a dominant position alongside a handful of groups that are very strong in newspapers but not much else. This group includes Postmedia (12 dailies and *National Post*), Torstar (*Toronto Star* and Harlequin romance novels), Thomson family (*Globe & Mail*, Thomson Reuters), and Gesca (the La Presse chain of dailies). In areas such as search engines and social networking sites, none of these entities has much sway; Google, Yahoo, Facebook, and LinkedIn rule the roost.

While gradual transformation was the sign of the times in the 1980s and early 1990s, in the late-1990s circumstances shifted abruptly. A slew of mergers and acquisitions set in motion a wave of consolidation in the press, television, telecom, radio, and cable sectors, as well as cross-media ownership on an unprecedented scale (see evidence presented later). These changes thrust the issue of media concentration into the spotlight. While there had been no formal inquiries for more than two decades, three were called between 2003 and 2007: (1) the Standing Committee on Canadian Heritage, *Our Cultural Sovereignty*;[8] (2) the Standing Senate Committee on Transport and Communications, *Final Report on the Canadian News Media*;[9] and (3) the Canadian Radio-Television

8. Canada, the Standing Committee on Canadian Heritage.*Our Cultural Sovereignty: The Second Century of Canadian Broadcasting*. Ottawa: Government of Canada. 2003. <http://www.parl.gc.ca/HousePublications/Publication.aspx?DocId=1032284&Mode=1&Parl=37&Ses=2&Language=E>.

9. Canada, Standing Senate Committee on Transport and Communications. *Final Report on the Canadian News Media*. Ottawa: Government of Canada. Last accessed on August 30, 2011 at <http://www.parl.gc.ca/Content/SEN/Committee/391/TRAN/rep/repfinjun06vol1-e.pdf>. Also see the interim report, which has much more to offer by way of data and evidence. Canada, Standing Senate Committee on Transport and Communications. *Interim Report on the Canadian News Media*. Ottawa: Government of Canada. URL: <http://www.parl.gc.ca/37/3/parlbus/commbus/senate/com-e/tran-e/rep-e/01apr04-e.pdf> [Last accessed, August 30, 2011]

and Telecommunications Commission's *Diversity of Voices* inquiry in 2008.[10]

None of these efforts helped stem the tide. The CRTC's response to its *Diversity of Voices* hearings at least had the merit of adopting, for the first time ever, a formal set of rules and thresholds to guide its decisions in the future about mergers, acquisitions, and consolidations. Its new rules on cross-media ownership also put an end to a situation in which Canada stood unique in the world, except for New Zealand, for having no formal limits on the matter at all. Drawing on lessons learned from Australia, the United Kingdom, and the United States, the CRTC's[11] rules are based on four central elements:

1. **2 out of 3 Rule**: "The Commission, as a general rule, will not approve applications . . . that would result in the ownership or control, by one person, of a radio station, a television station and a newspaper serving the same market."

2. **Ownership Caps**: The CRTC adopted thresholds originally developed by the Competition Bureau for measuring competition in banking services. According to the new rules, any transaction that results in a single ownership group controlling less than 35% of the television broadcasting pay and specialty markets will be seen as not diminishing diversity and approved. Those that fall into the 35–45% range will be considered as *potentially* lessening competition and reviewed. Anything over 45% will be seen as creating excessive concentration and rejected.

3. **Telco/Cable/DTH cross-ownership**: The Commission, as a general rule, will not approve consolidation among the incumbent telephone and cable companies that would result in one company being able to control Internet access and the delivery of programming services in any given local market.

4. **Vertical integration of programming services and producers**: "Access rules" were designed to limit vertically integrated media conglomerates' ability to foreclose access to their broadcast schedules to outside television and film producers. For conventional over-the-air television stations, three quarters of the programming broadcast has to be purchased from outside sources, while for digital specialty and pay services, at least 25% of Canadian programming other than news, sports, and current affairs must be produced and obtained from independent producers.

The results acknowledged that clear guidelines and limits to media concentration are important. They also recognized that concentration in traditional media—newspapers, television, and radio—is still important because traditional media are still the source of the majority of professionally produced media content that cascades across all media, including the Internet. Though the new rules may be better than no rules at all, they are weak. The thresholds adopted import criteria used to review mergers and consolidation in the banking industry, effectively disarming the potential for the regulator to encounter a merger it could not approve because of their extremely high levels. Such a standard also runs roughshod over the qualities that distinguish communications media companies from banks. Instead of striving to maximize the range of diversity and freedom of expression possible, the CRTC's guiding philosophy is the wholly uninspiring goal of achieving as much diversity "as practicable." In other words, it is a technocratic standard rather than a democratic one.

## CANADIAN RESEARCH AND PERSPECTIVES ON MEDIA OWNERSHIP AND CONCENTRATION

The revival of interest in the structure of the media industries stoked a revival in research on the subject. Marc Raboy and David Taras from McGill University and University of Calgary,

10. CRTC. Diversity of Voices. *Broadcasting Public Notice CRTC 2008–4*. 2008. <www.crtc.gc.ca/eng/archive/2008/pb2008-2004.pdf>

11. CRTC. Diversity of Voices. *Broadcasting Public Notice CRTC 2008–4*. 2008. <www.crtc.gc.ca/eng/archive/2008/pb2008-2004.pdf>

respectively, played key roles in the proceedings that led to the *Our Cultural Sovereignty* report.[12] David Skinner and Mike Gasher[13] also published a good overview of the lay of the land historically and as it existed in the late 1990s and early 2000s. Monica Auer[14] has also written thoughtful, empirically rich critiques of television and radio concentration and the CRTC's convoluted reasoning that led to such outcomes. Lastly, Ken Goldstein's[15] intervention on behalf of Canwest during the *Diversity of Voices* hearings, and another study that he wrote on the contemporary state of the press, marshals an impressive body of evidence and argument to the effect that the Internet and proliferation of new distribution platforms have rendered media ownership more dispersed and markets more competitive than ever before. Consequently, the fragmentation of audiences and erosion of the economic base for traditional media as advertising shifts rapidly to the Internet are the real issues, according to Goldstein,[16] rather than anachronistic worries about media concentration.

The periodic attention given to media concentration has generated some important research. This research has tended to cover only short spans of time and is driven by the policy agenda of the day rather than an overarching effort by scholars to create a systematic and coherent portrait of the media industries covering long-term trends. Moreover, the focus has been mainly on the news media at the expense of a larger view of the media ecology and, crucially, the network infrastructure industries.

The CRTC publishes several annual reports, notably the *Communications Monitoring Report*. This and other publications by the regulator are useful sources of information, but they are also hamstrung by several factors. For instance, the CRTC publishes useful data on revenues and market shares for the biggest four or five players in radio, television, cable, and satellite distribution and telecoms, but the data are of relatively recent vintage and do not cover the rest of the players in the field. Moreover, its reports are inconsistent from one year to the next and sometimes appear to be deliberately designed to distract attention from unpleasant truths that might reflect badly on it or the industries it regulates. In addition, most of the information published by the CRTC and Statistics Canada is only done at the aggregate level out of deference to claims of corporate confidentiality and competitive secrets. Pressure on the CRTC by the industries it regulates to minimize disclosure even further for traditional telecoms and broadcasting sectors is growing and intense, and this is especially true for new media such as online video distributors. I have met with CRTC staff numerous times and filed a dozen requests under the Access to Information Act to obtain more detailed information, but to no avail. Just as troubling, the CRTC routinely and randomly discards data more than eight years old. Susan Crawford,[17] a professor at the Cardozo Law School in New York and an expert on media anti-trust issues, reveals the CRTC's minimalist approach to information disclosure overall when she refers to its website as being "truly primitive."

Although debates on media ownership and concentration are commonplace and enduring,

---

12. Canada, the Standing Committee on Canadian Heritage. *Our Cultural Sovereignty: The Second Century of Canadian Broadcasting*. Ottawa: Government of Canada. 2003. <http://www.parl.gc.ca/HousePublications/Publication.aspx?DocId=1032284&Mode=1&Parl=37&Ses=2&Language=E>

13. Skinner, D. & Gasher, M. So Much by So Few: Media Policy and Ownership in Canada. In D. Skinner, J. R. Compton & M. Gasher (eds.). *Converging Media, Diverging Politics: A Political Economy of News Media in the United States and Canada*. Boulder, CA: Westview, 2005, 51–76.

14. Auer, Monica. September 2007. Is Bigger Really Better? TV and Radio Ownership Policy under Review. *Policy Options*, 78–83. URL: <http://www.irpp.org/po/archive/sep07/auer.pdf> [Last accessed August 30, 2011]

15. Goldstein, K. Remarks. Paper Presented at the Ink and Beyond Conference of the Canadian Newspaper Association, Montreal, QC, May 21, 2009. Goldstein, K. 2007. Measuring Media: Ownership and Diversity. Revised Report Prepared for Canwest Mediaworks Inc. Submitted to the CRTC's Diversity of Voices Hearings.

16. Goldstein, K. Remarks. Paper Presented at the Ink and Beyond Conference of the Canadian Newspaper Association, Montreal, QC, May 21, 2009.

17. Crawford, S. June 29, 2011. Say you were Canadian. URL http://scrawford.net/blog/?s=say+you+were+canadian [Last accessed August 30, 2011].

the evidentiary baseline of most of this discussion is rudimentary. The research that does exist is neither systematic nor empirically well grounded. Instead, ideology tends to carry the day, with three basic positions discernable. First, critics often assume a hyperbolic tone, claiming that media concentration is constantly going from bad to worse, robbing Canadians of "democracy's oxygen"[18] or creating "Canada's most dangerous media company," as the subtitle of one book suggests.[19] A second stance mocks concern with media concentration altogether. Writing on the *Final Report on the Canadian News Media,* Chris Dornan[20] set this tone in an article for the *Literary Review of Canada*: "Big Media Bad Thing: How a Senate Committee Wrote a Media Report with Its Head in the Sand." A third view strives to hug the middle ground, quantitatively analyzing reams of media content only to find the evidence "mixed and inconclusive"[21] as if this is the only or main concern and implying that preserving an already constrained status quo might not be a significant problem in its own right.

Ultimately, the episodic nature of policy-driven research agendas, the minimalist stance that the CRTC takes toward information disclosure, and the fact that media ownership and concentration is a highly politicized issue means that we do not have a systematic body of empirical evidence or a coherent portrait of the media industries over time. Philip Savage[22] sums up this state of affairs by stating, "The media ownership debate largely occurs in a vacuum, lacking evidence to ground arguments or potential policy creation either way."[23] Mike Gasher[24] concurs, asking rhetorically, "Who is really trying to measure media concentration and its impact in an empirical way? I honestly . . . cannot think of anyone who does that."[25]

Beyond this, another critically important constraint exists in terms of how the issues are framed. The focus on media content and news media tries to draw the proverbial camel through the eye of the needle but in doing so eclipses broader perspectives on the media ecology, as well as matters related to network infrastructure and the connectivity industries. Tim Wu's *The Master Switch,*[26] in contrast, reaches beyond these limits by illuminating the strong and recurring tendencies toward media consolidation that have existed for more than a century, including today's Internet access, search engines, and online music distribution. He also examines the potential for players to leverage their dominant power in one layer of the network media universe (networks, applications and content, devices) to influence what happens in others.[27] As we saw with the example in regards to the telegraph-news monopoly at the turn of the 20th century discussed earlier, this is not a novel insight, although how such processes manifest themselves in the digital, networked media environment is. Wu also stresses how market power can be transposed into the technological design of communication and media technologies in ways that are intended to buttress and extend market power from one medium to another. The conventional fixation on ideology and content ignores these issues completely.

---

18. Winter, J. *Democracy's Oxygen.* Montreal: Black Rose. 1997.

19. Edge, M. Asper Nation: Canada's Most Dangerous Media Company. Vancouver: New Star Books, 2007.

20. Dornan, C. 2006. "Big Media Bad Thing": How a Senate Committee Wrote a Media Report with Its Head in the Sand. *Literary Review of Canada,* (December). Last accessed, August 30, 2011 at http://findarticles.com/p/articles/mi_7019/is_10_14/ai_n28412445/

21. Soderlund, W. & Hildebrandt, K. *Canadian Newspaper Ownership in the Era of Convergence.* Edmonton, AB: University of Alberta. 2005.

22. Savage, P. & Gasher, M. Gaps in Canadian Media Research. *Canadian Journal of Communication,* 33(2), 291–301. 2008.

23. Savage, P. & Gasher, M. Gaps in Canadian Media Research. *Canadian Journal of Communication,* 33(2), 291–301). 2008, p. 295.

24. Savage, P. & Gasher, M. Gaps in Canadian Media Research. *Canadian Journal of Communication,* 33(2), 291–301. 2008.

25. Savage, P. & Gasher, M. Gaps in Canadian Media Research. *Canadian Journal of Communication,* 33(2), 291–301. 2008.

26. Wu, T. *The Master Switch.* Toronto: Random House. 2009.

27. Noam, E. *Media Ownership and Concentration in America.* New York: Oxford University, 2009.

The extension of market dominance from one medium to another is also problematic because a more concentrated media is a more regulable media. The more regulable media are, the easier it is to require Internet Service Providers (ISPs), search engines (Google), social network sites (Facebook, Twitter), and device manufacturers (RIM, Apple, Hewlett Packard, Sony, etc.) to serve as intermediaries on others' behalf, notably in matters of copyright, sexually oriented content, law enforcement, and national security. The responses taken to consolidation within and across layers of the network media will ultimately shape whether the emergent Internet-centric media ecology turns out to be as open as possible. It also brings into question whether or not it will be guided by notions that human communication by Internet and any other media is a fundamental human right, as a recent UNESCO report affirms,[28] or used as a tool to backstop the interests of the media, entertainment, and software industries, as well as law, order, and national security.

## KEY TRENDS AND DYNAMICS

### The Growth of the Network Media Economy, 1984–2011

The following section sketches the evolution and growth of the network media economy in Canada between 1984 and 2011, pointing to those sectors that have flourished, others that have stayed relatively stable, and the few cases in which they have declined. I then rank order Canada relative to the other large national media markets worldwide. As Figure 17.1 shows, the media economy grew immensely from $37.5 billion in 1984 to $55.4 billion in 2000 and to $69.3 billion in 2011 ("real dollars," CDN.)

The emergence of new media—pay television and wireless cell phone services since the 1980s followed by Internet access and Internet advertising in the latter period—has more than tripled the size of the media economy over the past quarter century. Wireless cell phone service alone is now a $19.3 (2011) billion sector, whereas it was not on the charts in 1984.

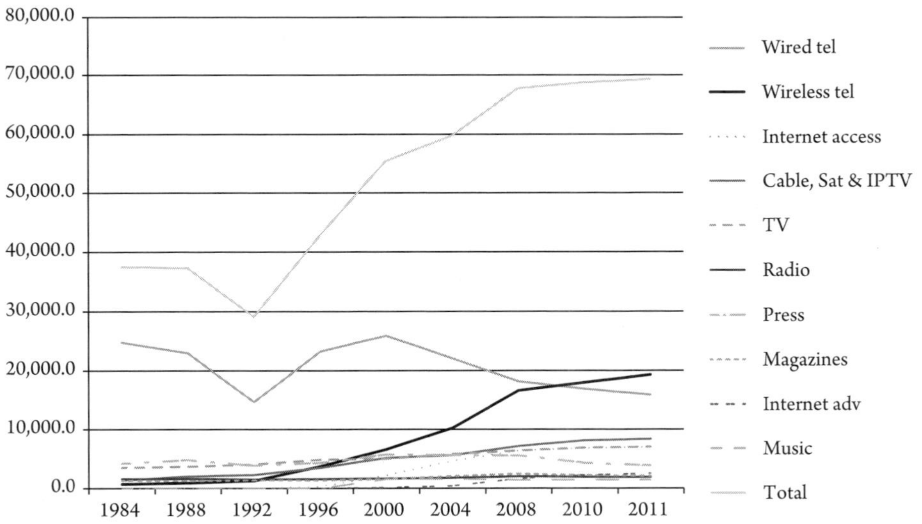


FIGURE 17.1 The Growth of the Network Media Economy, 1984–2011 (Millions, in 2010 Dollars). All graphs in CDN currency unless otherwise stated

28. Dutton, W. H., Dopatka, A., Hills, M., Law, G. & Nash, V. 2011. Freedom *of Connection, Freedom of Expression: The Changing Legal and Regulatory Ecology Shaping the Internet*. UNESCO: Paris. Last accessed on August 30, 2011 at http://unesdoc.unesco.org/images/0019/001915/191594e.pdf

Wireline telecom revenues (excluding Internet access) have fallen sharply by more than a third from $25.9 billion in 2000 to just under $16 billion in 2011, although this has been more than offset by gains from wireless and Internet access. Even with those sharp declines, revenues for the connectivity industries (wireline, wireless, Internet access) have risen sharply from $25.4 billion in 1984 to $34.6 billion in 2000 to just over $42 billion in 2011.

Bracketing wired and wireless telecoms services because they overshadow everything else because of their size in order to focus just on the Internet-centric and traditional media elements in the mix (the network media economy), a similar pattern emerges. Indeed, these 10 sectors grew greatly from $12.1 billion in revenues in 1984 to $23 billion in 2000 to $34.1 billion in 2011 (in real dollars) (Table 17.2).

It is commonplace to hear that the core traditional media—television, music, newspaper, and books—are teetering on the brink of calamity. The slightest whiff of troubles on the horizon, such as "cord cutting," rising Netflix subscriptions, a drop in advertising revenue, or piracy, and the "media in crisis" trope trots out time and time again.

Yet broadcast television revenues—the poster child of a media sector in distress—did drift downward from a high of $3.6 billion in 2006 to $3.4 billion in 2011. Pay and specialty cable channel revenues, in contrast, nearly tripled in the past decade from $1.27 billion in 2000 to over $3.7 billion in 2011 (including the CBC's annual subsidy).

Combine this with the more significant growth in the means of television distribution—cable, satellite, IPTV, and other online video distributors (OVDs)—and the total television universe more than doubled in size between 1984 and 2000 and then grew again to roughly $15.4 billion last year. In short, television remains at the heart of the digital media universe and does not serve as exhibit A for an old medium imperiled by the new. To be sure, there are newcomers on the scene, notably Netflix and other OVDs, but with less than 1% of the television market, their stakes are limited rather than a mortal threat requiring the CRTC and policymakers to revamp the existing rules, as many incumbents are eager to have them do.

The ascent of the Internet has contributed greatly to growth in the size and complexity of the media universe. Internet access rose from $239 million in revenue in 1996 to $7.2 billion in 2011. Internet advertising started from next to nothing in 2000 to $2.5 billion in 2011.

Nonetheless, growth for the network media economy as a whole has stagnated since the economic downturn caused by the global financial crisis (2007 ff). This is typical of a long-term historical tendency for the fortunes of the media economy to hinge on the state of the economy in general, however, as Figure 17.1 and Table 17.1 show with respect to the comparatively milder recession years in the early 1990s when revenues for wireline telecoms and newspapers plummeted. Indeed, total revenues for all media combined in 1992 were down more than 20% from four years earlier.

The impact of the current economic downturn has varied. Revenues largely stagnated between 2008 and 2010, but some media have seen revenues rise substantially (wireless, cable, satellite and other video distribution platforms, Internet access, and Internet advertising) or decline slightly (radio, magazines, and music). When it comes to newspapers and wired telecom services revenues, however, there is no doubt that revenues dropped sharply.

Despite declining circulation and readership since the late-1950s,[29] the newspaper sector's revenues grew until reaching—and then roughly staying steady at—an all-time high between 2000 and 2006. They have fallen greatly since (about 30%). Events in Canada are still not quite nearly as severe as trends in Britain, the United States, and a few other, mostly European, countries. However, now it appears that trends in Canada may simply be drawn out over

29. Goldstein, K. Remarks. Paper Presented at the Ink and Beyond Conference of the Canadian Newspaper Association, Montreal, QC, May 21, 2009.

Table 17-1. The Growth of the Network Media Economy (US$ Million), 1984–2011 (excluding wireline and wireless telecoms)

| | 1984 | 1988 | 1992 | 1996 | 2000 | 2004 | 2008 | 2010 | 2011 |
|---|---|---|---|---|---|---|---|---|---|
| Internet Access | | | | 312 | 2,195 | 4,673 | 6,365 | 6,800 | 6,984.5 |
| Cable, Satellite, IPTV | 1,380 | 2,041 | 2,288 | 3,497 | 5,145 | 5,607 | 7,139 | 8,100 | 8,331 |
| TV | 3,535 | 3,694 | 4,057 | 4,854 | 5,012 | 5,802 | 6,371.4 | 6,865.0 | 7,007.8 |
| Radio | 1,595 | 1,623 | 1,533 | 1,576 | 1,655 | 1,784 | 2,009.2 | 1,916.7 | 1,891.2 |
| Press | 4,219 | 4,803 | 3,858 | 4,324 | 5,731 | 5,600 | 5,544 | 4,300 | 3,856.1 |
| Magazines | 1,369 | 1,272 | 1,401 | 1,362 | 1,585 | 2,055 | 2,458 | 2,202 | 2,071 |
| Internet Advertising | | | | | 134 | 406 | 1,643 | 2,230 | 2,522 |
| | | | | | 1,548 | 1,574 | 1,399 | 1,410 | 1,410.0 |
| Total | 12,098.5 | 13,433.5 | 13,136.4 | 15,925.2 | 23,004.7 | 27,500.2 | 32,928.7 | 33,823.3 | 34,074.2 |

a deferred and slightly longer time frame than in these other countries. In the United States, for example, circulation and advertising revenue in 2010 were down 10 and 40%, respectively, from all-time highs in 2003 (see, for instance, PEJ, 2011; OECD, 2009).[30] The trend in Canada, nonetheless, has been harsh enough to tip the bloated Canwest, once one of the largest media conglomerates in the country, into bankruptcy. It also casts a shadow over prospects for Postmedia, the company that took over the Canwest chain of newspapers in 2010. A standing invitation to its journalists and editorial staff to take early retirement, its convoluted ownership structure designed to meet Canadian cultural policy objectives buried in the income tax code, and a shaky ride in the stock market after its initial IPO surge all raise questions about its long-term viability.

Newspaper Canada says there are 94 dailies across the country, with 66 belonging to four groups: Postmedia (13), Quebecor (36), Power Corp/Gesca (7), and Transcontinental (19). Only about a third actually publish daily. Several have cut their schedules to five or six days per week in the last two years. In addition, many small- to mid-size town dailies have cut back further yet to just one or two days per week, albeit refreshed daily (continuously) with content from regional content factories within these chains. This is significant because the press is the core of the content factory when it comes to news across the media system as a whole. Pound for pound, newspapers pack greater journalistic wallop than other media, so there is good reason for concern.

However, the news is not all bleak in light of the rise in revenue (3.7%) and readership in 2010.[31] The trend has been notable with younger readers who are beginning to obtain news online via social network sites such as Facebook.[32] Lastly, "pure" newspaper publishers such as Torstar and Transcontinental appear to

30. Project for Excellence in Journalism (PEJ). 2010. The state of the news media, 2010. Last accessed on August 30, 2011 at <www.stateofthenewsmedia.org/2010>. OECD. (2010). The Evolution of News and the Internet. Last accessed on August 30, 2011 at <http://www.oecd.org/dataoecd/30/24/45559596.pdf>.

31. Community Media Canada (no date). Read all about it. Last accessed at. <http://www.communitymedia.ca/?s=advertising+revenue+2010>.

32. Canadian Media Research Consortium (April 27, 2011). Social networks transforming how Canadians get the news. Last accessed on August 30, 2011 at <www.cmrcccrm.ca/en/projects/documents/CRMCSocialnewsApril27.pdf>.

Table 17-2. Top 10 Network Media, Entertainment, and Internet Markets by Country (US$ Millions), 1998–2010[1]

| | 1998 | 2000 | 2004 | 2008 | 2009 | 2010 (EST.) | % CHANGE |
|---|---|---|---|---|---|---|---|
| United States | 336,885 | 395,695 | 395,936 | 420,397 | 406,733 | 411,357 | +22% |
| Japan | 94,255 | 100,799 | 114,330 | 141,340 | 156,120 | 157,985 | +68% |
| Germany | 59,919 | 68,981 | 79,877 | 84,635 | 84,100 | 89,905 | +50% |
| China | 23,057 | 27,599 | 32,631 | 66,310 | 72,024 | 81,005 | +247% |
| United Kingdom | 56,738 | 65,319 | 75,637 | 72,346 | 70,478 | 72,605 | +28% |
| France | 39,984 | 46,031 | 53,302 | 63,863 | 58,841 | 59,587 | +49% |
| Italy | 29,626 | 34,107 | 34,494 | 41,528 | 39,890 | 39,924 | +35% |
| Canada | 18,346 | 21,432 | 25,842 | 31,287 | 30,701 | 33,258 | +70% |
| S. Korea | 17,687 | 18,492 | 22,760 | 26,672 | 27,394 | 28,589 | +62% |
| Spain | 19,219 | 22,132 | 25,622 | 28,736 | 27,200 | 27,479 | +43% |
| Total | 695,716 | 797,358 | 860,431 | 977,114 | 973,481 | 999,665 | +44% |

[1] PriceWaterhouseCoopers, *Global Entertainment and Media Outlook*, New York: PWC. 2010 (plus previous editions between 2000 and 2009). IDATE (2009). *DigiWorld Yearbook*. IDATE (2009). *DigiWorld yearbook 2009*. Montpellier, France: IDATE.

be recovering, and there is a chance that Postmedia will make a go of things yet, or that someone else will who can fill its place.

The other media where a slight decline can be seen since 2008 include radio and magazines, while revenues for recorded music and concerts have stayed relatively flat during this period despite widespread cries that the sector has been imperiled by the rise of easy copy and distribution via digital personal media and the Internet. Data for radio and magazines can be found in the tables. However, the declines for these two sectors are of such a modest magnitude and duration that changing the base of measurement from real dollars to current dollars by removing adjustments for inflation turns things around. A plausible argument exists that inflation in the media and ICT sectors has actually pushed costs downward, not upward, and thus it is appropriate to use current dollars rather than just inflation-adjusted real dollars. The rise of the Internet, drawn out instabilities of the global financial crisis, and the rise of a DIY (do-it-yourself) culture of mass self-expression (sharing and piracy) have laid ruin to any of the core elements of the media universe. Of course this does not mean that the demands to adjust to these emergent new realities—technological, economic, cultural, and political—have not been pressing or urgent.

The growth of the network media economy can also be seen relative to Canada's place in the world media economy. Table 17.2 illustrates that the media economy in Canada is actually the eighth largest in the world. In 1998, it accounted for 2.6% of the global media economy; it accounted for over 3% in 2010, all of which casts doubt on the notion that Canada's small media market requires big media to compete.

## THE TWO (OR THREE) WAVES OF MEDIA CONSOLIDATION, 1984–2010

Trends in the telecom and media industries vary considerably over time. Between 1984 and 1996, the advent of new players led to greater diversity in all sectors, except for newspapers, cable, and satellite video distribution, where concentration levels climbed significantly. Concentration levels remained very high in wireline telecoms in the 1980s and early 1990s, too, and in wireless, which was being developed by two companies, Bell and Rogers. As had been the case in many countries, telecom competition moved slowly from the ends of the network into services and then into the network infrastructure. Only in the mid-1990s did competition begin to take hold, although for just a few years before the trend was reversed and concentration levels once again began to climb. Data on wired and wireless telecoms are provided in the tables.

Conventional as well as pay and subscription television channels were already expanding steadily. In terms of ownership it was mostly incumbents and a few newcomers such as Allarcom and Netstar that cultivated the field, with their reach and share of the market growing in tandem with the number of services available and minor shuffles along the way. Most transactions in the 1980s and early 1990s involved players within single sectors, with Conrad Black's takeover of the Southam newspaper chain in 1996 being the poster child of the times. In broadcast television, amalgamation among local and regional ownership groups in the late 1980s and early 1990s produced the large national companies that came to single-handedly own the leading commercial television networks such as CTV, Global, TVA, CHUM, and TQS by the late 1990s. While weighty in their own right, these amalgamations did not have a huge impact on the media as a whole. There were still significant levels of ownership and organizational diversity to be seen. The CBC also remained prominent but was eclipsed by commercial television during this time, as its share of all resources in the television system fell steadily from 46% in 1984 to 26% by 2000 to just above 20% today. Its audience share is under 10%.

This picture of significant, albeit mostly gradual, change shifted dramatically in the late 1990s. Investment poured into mergers and acquisitions, yielding huge media conglomerates with previously unheard of capitalization levels and debts. Figure 17.2 reveals the spike of acquisitions in the telecom and media industries between 1996 and 2000 and again, albeit more modestly, from 2003 until 2007.

We can see a clear trend toward mergers and acquisitions beginning in 1994, mounting steadily to unprecedented levels after that until 2000, a fallback after the collapse of the TMT bubble between 2000 and 2003, and a steady climb again until 2007, before falling off sharply in 2008 with the onset of the global financial crisis. These patterns closely parallel trends in the United States and globally. The processes of consolidation over this period can be seen as occurring in two, possibly three, relatively distinct waves. The results marked a watershed with the circumstances that had prevailed before.

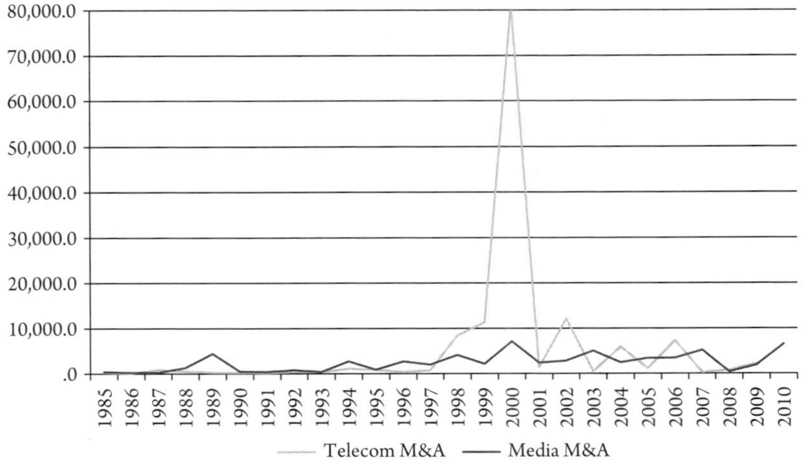


FIGURE 17.2 Financial Investment in Mergers and Acquisitions in Media and Telecoms, 1984–2010 (Millions$)

## Wave 1—1994 to 2000

The first wave of consolidation began in 1994 when the largest cable company, Rogers, took over Maclean-Hunter, a leading magazine publisher that also had interests in broadcasting, cable, and a few other areas. The event inaugurated the rise of major media conglomerates in Canada. Two years later, Conrad Black's Hollinger group took over Southam, the largest and oldest chain of dailies at the time. However, both transactions were still a far cry from what was to follow a few years later.

In 2000, media transactions alone were worth $7.1 billion, more than eight times greater than five years earlier; telecom and Internet acquisitions were more than 10 times what they had been just a few years earlier. Primed by the easy cash of the Telecom-Media-Technology boom, visions of media convergence, and permissive policies of the then Liberal Government, media and telecom companies went on a buying spree. BCE acquired CTV and the *Globe & Mail* for $3.4 billion in 2000, while Quebecor bought the cable company Videotron, the largest French language television network, TVA, as well as the Sun newspaper chain for $7.4 billion between 1998 and 2001. The results were Quebec's biggest media conglomerate and the fourth largest in the country. In 1998, Canwest purchased Western International Communication ($800m) to become the second largest broadcast television network, Global, in Canada followed two years later by its acquisition of the Hollinger (Southam) newspaper chain and the *National Post* ($3.2 billion) from Conrad Black. There was significant consolidation among smaller and larger ISPs, cable companies, and between new wireless rivals and the incumbents.

## Wave 2—2004–2007

As the TMT bubble collapsed in 2000, many of the telecom, media, and Internet firms created during the previous few years went bust or were taken over by the incumbents. As an example, Telus acquired Clearnet (the second largest independent competitor in wireless) in 2000 and PsiNet, the largest independent ISP at the time, the following year. This caused a lull in activity, but by 2003–2004 the processes of consolidation regained steam. Rogers took over the only remaining independent wireless provider, Microcell (Fido), in 2004, which had built up a sizeable (10%) share of the industry.

Canwest was already struggling to bring its debt under control and sold several small dailies to two regional chains, Transcontinental and Osprey (2002–2003). Craig Media, with financing from the US-based private equity fund Providence Equity Partners, expanded its modest A-Channel and created a new station, Toronto One, in 2003. However, the effort failed and Craig was forced into bankruptcy, Toronto One was sold to Quebecor, and Craig's fledgling network of A-Channel stations sold to CHUM in 2004. The latter was the fifth largest broadcaster and owned a handful of television stations that made up the venerable and commercially and culturally innovative CityTV network in a half-dozen major Canadian cities. However, that, too, was short-lived. The debt-laden CHUM was sold after its founder's death to Bell Globemedia in 2006 ($1.6b). Even Bell Globemedia was in disarray, and it abandoned its convergence strategy by scaling back its stake in CTV and the *Globe and Mail* (71–15%) in late 2006.

It also sold its stake in the second French-language commercial network, TQS, the next year. A rebranded CTV Globemedia emerged after the restructuring, with the Thomson family (40%) at the helm and Ontario Teacher's Pension Fund (25%), Torstar (20%), and Bell (15%) all holding minority interests. The last step in this tangled web of affairs occurred when the CRTC allowed CTV Globemedia to keep the A-Channel stations, as well as the 30-odd pay and specialty television services that it had acquired from CHUM, but forced it to sell the CityTV stations. Rogers snapped them up ($375 million) within the year.[33]

33. CRTC. 2006. Change in effective control Bell Globemedia Inc. Broadcasting Decision CRTC 2006–2309. <www.crtc.gc.ca/eng/archive/2006/db2006-2309.htm>

Three other transactions in 2007 marked the high point of the second wave of consolidation. The fifth largest pay and specialty television and radio broadcaster in the country, Astral Media, bought Standard Broadcasting. A small chain of newspapers, Osprey, was sold to Quebecor. Lastly, Canwest and the New York–based investment bank, Goldman Sachs, bought broadcaster and film and television production company Alliance Atlantis for $2.3 billion. Although the fact that Goldman Sachs held two-thirds of the equity in the television services would seem to have put the transaction afoul of Canada's foreign ownership rules, the regulator blessed the transaction based on the fiction that Canwest maintained control of the voting stock and board of directors. The deal gave Canwest ownership majority stakes in 13 specialty and pay television channels (e.g., BBC Canada, HGTV, National Geographic, Showcase, etc.) and a successful film and television production venture.[34] During this second wave of consolidation, media acquisitions neared their dot.com highs, with similar questionable outcomes as debt and capitalization levels soared.

## Wave 3—2010 and Beyond

Whether events since 2010 constitute a third wave of consolidation or just the fallout from the collapse of Canwest is still too early to tell. However, in the past year Canwest's newspapers have been sold to Postmedia ($1.2 billion) and its TV assets sold to Shaw ($2 billion), already the largest video distribution provider and one of the top ISPs in Canada. Shaw also swapped some of the radio stations it owns, via Corus, in Quebec with Cogeco, a mid-size cable company, in return for others in predominantly English-speaking cities. Bell Canada renewed its convergence gambit by buying back CTV ($3.2 billion), giving it a second run to make vertical integration work between its telephone, satellite, and ISP (network infrastructure) businesses. This made it the largest media group in the country with its CTV and A-channel networks, 31 satellite and cable television channels, 28 local television stations, and 33 radio stations. The only real difference between now and 10 years ago when it first tried this strategy is that BCE scaled back its stake in the *Globe & Mail* (15%) and that media conglomerates have fallen out of fashion in most developed capitalist economies (albeit with the obvious exception of the Comcast/NBC–Universal merger approved in the United States in 2011).[35]

The idea that consolidation occurs in waves is important for several reasons. First, it shows that there is a certain periodicity to the run of events rather than a process leading constantly to either greater competition or more consolidation. Second, it shows that there is little that renders the telecom, media, and Internet industries immune to consolidation. In fact, powerful forces push them in the opposite direction. Notably, from the mid- to late 1990s the telecom, media, and Internet sectors have been destinations for capital investment at a rate far out of proportion to their weight in the real economy. In a way, they have been harbingers of the great transformation that has taken place during this time in which the importance of finance and financial markets in the general scheme of the economy as a whole has increased tremendously.

Third, consolidation has given rise to a new kind of entity now at the core of the network media ecology: the media conglomerate. Altogether, 4 massive media conglomerates and 6 large but more specialized companies that are half their size constitute the big 10 media firms in Canada, as outlined in Table 17.3.

---

34. Canadian Radio-television and Telecommunications Commission (2008). *Diversity of Voices* Broadcasting Public Notice CRTC 2008–4. <www.crtc.gc.ca/eng/archive/2008/pb2008-2004.pdf>; Goldstein, K. 2007. Measuring Media: Ownership and Diversity. Revised Report Prepared for Canwest Mediaworks Inc. Submitted to the CRTC's Diversity of Voices Hearings; Communication Energy Paperworkers (CEP). 2007. Canwest MediaWorks Inc on behalf of Alliance Atlantis–Intervention re. Broadcasting notice of Public Hearing CRTC 2007-11. Ottawa: Canada. <www.crtc.gc.ca>

35. In October 2012, the CRTC (2012), in an entirely unprecedented and unexpected decision killed a bid by Bell Canada to acquire the eighth largest media company in Canada, Astral Media. See CRTC. 2012 *BCE Inc on behalf of Astral Media Inc: Broadcasting Decision CRTC 2012-574*. Ottawa: Canada http://crtc.gc.ca/eng/archive/2012/2012-2574.pdf.

Table 17-3. The "Big 10" Media Companies in Canada, 2011 (millions CAN$) (excluding wireline and wireless telecoms)[1]

| 2011 | OWNERSHIP | MARKET SHARE | TOTAL REV. | CABLE & SAT. DIST. | INTERNET ACCESS | TOTAL TV | RADIO | PRESS/ MAGS* |
|---|---|---|---|---|---|---|---|---|
| Bell/CTV | Diversified | 15.3 | 5,373.1 | 1,830.0 | 1,525.0 | 1,857.6 | 160.5 | |
| Shaw | Shaw Family | 15.3 | 5,356.1 | 2,432.7 | 968.6 | 1,759.1 | 195.7 | |
| Rogers | Rogers Family | 11.6 | 4,077.7 | 1,874.0 | 912.0 | 797.0 | 220.8 | 273.9 |
| QMI | Péladeau | 8.5 | 2,986.6 | 1,040.9 | 698.2 | 3,45.2 | | 902.3 |
| CBC | Public | 5.2 | 1,838.8 | | | 1,501.9 | 336.9 | |
| Postmedia | Godfrey, et. al. | 3.3 | *1,168.7* | | | | | *1,168.7* |
| Cogeco | Audet (60%), Rogers (40%) | 3.2 | 1,130.8 | 639.7 | 315.6 | 61.9 | 113.6 | |
| TELUS | Diversified | 3.0 | 1,052.4 | 364.8 | 687.6 | | | |
| Astral | Greenberg | 2.6 | 922.5 | | | 582.2 | 340.3 | |
| Bragg | Privately Held | 1.6 | 558.5 | 326.5 | 232 | | | |
| Others | Diversified | 30.4 | 10,616.6 | | | | | |
| Total | | | 35,081.8 | | | | | |
| NMI $ | | | | | | | | |
| C1 | 15.3 | | | | | | | |
| C4 | 50.7 | | | | | | | |
| C10 | 69.5 | | | | | | | |
| *N* (>1%) | 10 | | | | | | | |
| HHI | 738.5 | | | | | | | |
| Noam Index | 233.5 | | | | | | | |

[1] While I have excluded wired and wireless from the present analysis for reasons explained earlier, I still include Bell and Telus among those on this list because both have very large Internet access providers, which is a topic covered by the present discussion, and the former (Bell) owns CTV.

Table 17.3 shows the sheer size of the four leading TMI conglomerates and six specialized firms in broadcasting and newspapers that follow. It gives us a clear picture of who the key players are, and the media segments in which they operate. It does not, however, indicate whether the media have become more or less concentrated over time.[36] Others also argue that focusing on specific firms and who owns them is misleading because media companies are generally now owned by shareholders and controlled by managers. According to Noam,[37] owner-controlled media firms in the United States fell from 35% to just 20% between 1984 and 2005. Demers and Merskins[38] use such evidence to argue that the managerial revolution has signaled the demise of the media mogul. They argue that this is a good thing because

36. Flew, T. *Understanding Global Media*. London: Palgrave MacMillan. 2007.

37. Noam, E. *Media Ownership and Concentration in America*. New York: Oxford University, 2009, 6.

38. D. Demers & D. Merskin. Corporate News Structure and the Managerial Revolution. *Journal of Media Economics*, 13(2), 2001, 103–121.

media managers do not have ideological axes to grind, although they do have deep pockets and the expertise needed to support better media performance and higher quality journalism than owner-controlled companies.

The upshot from of all this is that the media are more competitive and fragmented than ever. Or are they? The fact that all of the big 10 media firms, except Bell and the CBC, are owner controlled suggests that Demers and Merskins' and Noam's case does not fit the Canadian context. Such arguments usually suggest that the results are part of a steady process of incremental change, but events in Canada suggest that a sharp, dramatic bout of consolidation occurred in the last half of the 1990s and again in the mid-2000s that produced a sharp rise in concentration without altering the structure of media ownership.

The big 10 media firms' share of all revenues (excluding telecom services) reached a high point in 2000 at roughly 68% of all industry revenues—a fairly significant rise from 62% in 1996 and up substantially from 51% in 1992. However, with the collapse of the telecom-media-technology bubble in 2000, the big 10's share of all industry revenue fell to just below 60% in 2006. Thereafter, things reversed course, and by 2011 the big 10 players' share of the TMI ecology had set all-time highs to reach just under 70%. In short, while the TMI ecology has grown much larger and more structurally differentiated over time, the big 10 players' share of it has grown substantially relative to the early 1990s, or modestly if we take 2000 as our base year.

## MEASURING CONCENTRATION TRENDS OVER TIME, 1984–2011

Taken individually, each media sector, except magazines, was highly concentrated in 2011 according to the CR method. HHI scores show similar trends, as Figure 17.3 reveals: wireline and wireless telecoms (3,513.4 and 2,923.7, respectively), conventional television (2,401.4), cable, satellite, and IPTV distribution (1,975.4), pay and specialty TV channels (2,329), and newspapers (1,742.5). Radio (1,218.9) is moderately concentrated, while Internet Access (1,013.4), online news sources (653.2), and the network media as a whole (738.5) are unconcentrated by HHI standards.

In the next few paragraphs, I begin by separating each of the sectors into two categories: one for the network infrastructure industries (wireline and wireless telecom services, ISPs, cable, satellite, and other OVDs) and another for the content industries (newspapers, TV, magazines, radio). At the end, I pool each of these categories together and examine the trends over time using CR and HHI standards.

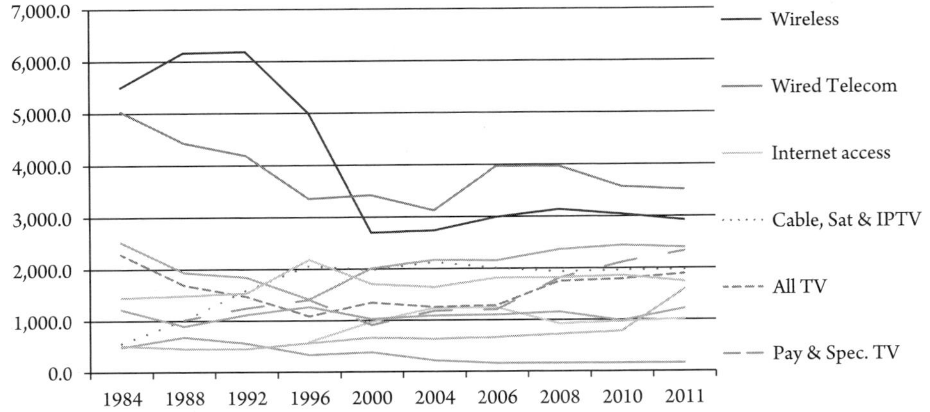


FIGURE 17.3 HHI Scores for the Network Media Industries, 1984–2011

## The Network Infrastructure Industries

Concentration levels have remained very high in wireline telecoms. They slid steadily from 1984 to 1996 as competition was gradually introduced, subscribers were allowed to attach their own equipment to the ends of the network, and competition in information services was opened up. By the mid-1990s, increased competition took hold in services and network infrastructure at the same time that the Telecom-Media-Technology boom was in full swing. HHI scores for wired telecoms dropped from 4,188 in 1992 to a much lower, but still concentrated, 3,123.9 in 2004. Table 17.4 illustrates the trends.

As the telecoms and Internet boom gathered steam new players emerged to become significant competitors in Internet access, with four companies taking more than a third of the ISP market for themselves in 1996: for example, AOL (12.1%), Istar, (7.2%), Hook-Up (7.2%), and Internet Direct (6.2%.) However, the early competitive ISP era yielded to more concentration in the next decade. Although the big four ISPs accounted for a third of all revenues in 1996, by 2000 the number had grown to 54% and a few percentage points higher over the decade. HHI scores also show a significant leap in consolidation between 1996 and 2000, but these are still relatively low levels in comparison to most other sectors. Such an outcome is probably more an indicator of the limits of the HHI method in this particular case, since 94% of high-speed Internet subscribers rely on one or the other of the incumbent cable or telecom companies' ISPs to access the Internet (CRTC, 2011, p. 138).[39] Table 17.5 shows the market share for individual ISPs over time.

The situation with respect to high-speed Internet access reflects the fact that Canada has developed a framework in which it is mostly the incumbent telecom and cable companies that increasingly compete with one another in duopolistic markets, not just for Internet access but also for television (video) distribution. This means that there is some competition in these areas. Yet it is revealing of how limited that competition is once we realize that cable and satellite distribution is one of the only segments assessed for which instead of some early years of falling concentration and more competition, the levels of concentration have risen steadily from low levels in the 1980s (563.9). They peaked in the years between 1996 and 2006 before drifting slightly downward but are still at the top of the concentration scale ever since (high 1900s). Table 17.6 illustrates the trends for multichannel video distributors.

In contrast, greater competition in wireline telecom network infrastructure took hold in the late 1990s and into the early 2000s; the CR4 (81) and HHI (3,123.9) scores reached their lowest level ever in 2004. However, as the aftermath of the collapse of the TMT bubble played out, many of the companies created in the dot.com years collapsed and "ceased to exist" altogether and competition grew feeble as a result.[40] Much the same can be said with respect to wireless services; they have consistently been highly concentrated and continue to be so to this day, despite the advent of three newcomers in the last two years—Mobilicity, Wind Mobile, and Public. Two other competitors, Clearnet and Microcell, managed to garner 13.4% of the market between them, but they were then taken over by Telus and Rogers in 2000 and 2004, respectively. It is still too early to tell whether the newcomers of the last two years will fare any better, but with only 1.6 percent of the market as of 2011, they are a long way from the high tide of competition set a decade ago. Table 17.7 illustrates the trends for wireless telecom services.

Table 17.8 aggregates and presents the data for the CR and HHI scores for each sector of the network infrastructure industries since 1984. Attention then turns to an examination of the data and trends for the content and online segments of the network media industries covered in this study.

39. CRTC, 2011. Communications *Monitoring Report*. Ottawa: Author. <http://www.crtc.gc.ca/eng/publications/reports/policymonitoring/2011/cmr2011.pdf>.

40. CRTC. (2002). Status of Competition in Canadian Telecommunications Markets. Ottawa: Author. 2002, 21. <www.crtc.gc.ca/eng/publications/reports/PolicyMonitoring/2002/gic2002.pdf>.

Table 17-4. Wired Ownership Groups, Revenue, Market Shares, and Concentration Levels from 1984 to 2011

| | 1984 | 1988 | 1992 | 1996 | 2000 | 2004 | 2006 | 2008 | 2010 | 2011 |
|---|---|---|---|---|---|---|---|---|---|---|
| Bell | 69.4 | 64.4 | 62.3 | 55.0 | 54.1 | 51.9 | 59.0 | 58.7 | 54.7 | 54.4 |
| Mar. T&T | *2.2* | *2.6* | *3.7* | *3.3* | *Alliant* | | | | | |
| NB Tel | *1.8* | *2.0* | *2.4* | *2.5* | *Alliant* | | | | | |
| New Tel | *1* | *1.4* | *1.9* | *1.8* | *Alliant* | | | | | |
| Island Tel | *0.1* | *0.1* | *0.1* | *0.2* | *Alliant* | | | | | |
| Telus | | | 8.6 | 9.3 | 20.3 | 18.9 | 20.2 | 20.8 | 22.5 | 21.9 |
| BCTel | *11.6* | *13.7* | *13.5* | *13.7* | *Telus* | | | | | |
| AGT | *7.6* | *7.6* | *Telus* | | | | | | | |
| MTS Allstream | 2.1 | 3.2 | 3.4 | 3.0 | 3.3 | 6.1 | 8.4 | 8.1 | 6.2 | 5.9 |
| SaskTel | 3.0 | 3.7 | 3.7 | 3.4 | 2.7 | 2.5 | 2.2 | 2.0 | 2.9 | 2.9 |
| Rogers | | | | | 1.0 | 2.2 | 1.9 | 2.7 | 2.9 | 3.0 |
| Shaw | | | | 0.04 | 0.1 | 0.2 | 0.4 | 1.4 | 2.3 | 2.7 |
| Quebecor/Videotron | | | | | | 0.05 | 0.6 | 2.0 | 2.4 | 2.7 |
| Videotron | | | | 0.09 | 0.3 Quebecor | | | | | |
| Cogeco | | | | | | | 0.1 | 0.4 | 0.7 | 1.3 |
| Bragg/East-link (7) | | | | | 0.009 | 0.1 | 0.3 | 0.7 | 0.7 | 1.3 |
| AmTel (8) | | | | | | | *0.07* | *Bragg* | | |
| Access Comm | | | | | | 0.05 | 0.05 | 0.08 | 0.08 | 0.8 |
| Cooperative | | | | | | | | | | |
| Small ILECS | 1.2 | 1.2 | 1.2 | 1.2 | 1.3 | 1.9 | 2.0 | 2.4 | 2.4 | 1.7 |
| AT&T (US) | | | | | 6.9 | MTS | | | | |
| Call-Net (Sprint) | | | 0.6 | 4.0 | 3.6 | 4.1 | Rogers | | | |
| 360 Networks | | | | | 1.3 | 1.8 | Bell | | | |
| Group Telecom | | | | | 0.3 | 360 Networks | | | | |

*continued*

Table 17-4. *continued*

| | 1984 | 1988 | 1992 | 1996 | 2000 | 2004 | 2006 | 2008 | 2010 | 2011 |
|---|---|---|---|---|---|---|---|---|---|---|
| FibreLink | | | | 0.04 | 0.1 | GT Telecom (2000) | | | | |
| Axxent (OCI Comm) | | | | | 0.3 | Telus (2002) | | | | |
| Primus | | | | | 0.4 | 1.0 | 0.8 | 0.7 | 0.8 | 0.8 |
| London Telecom | | | | 0.4 | Primus (1999) | | | | | |
| Cannect | | | | | 0.3 | | | | | |
| Futureway/FCI Broadband | | | | | | 0.01 | 0.1 | Rogers (2007) | | |
| Vonage | | | | | | 0.02 | 0.1 | 0.2 | 0.2 | 0.2 |
| Cyberus (VOIP) | | | | | | | 0.003 | 0.01 | 0.01 | 0.01 |
| Skype | | | | | | | | | | |
| Others* | | | | 2.0 | 4.0 | 9.2 | 3.8 | | 1.2 | |
| Total $ | 12,787 | 14,007 | 14,700 | 17,900 | 21,200 | 19,800 | 18,400 | 18,000 | 16,900 | 16,400 |
| Total $US | | | 12,174 | 13,126 | 14,279 | 15,251 | 16,225 | 16,995 | 16,405 | 16,588 |
| C4 | 90.8 | 89.4 | 88.1 | 81.5 | 84.8 | 81.0 | 89.8 | 90.5 | 86.3 | 85.1 |
| *N* (>1%) | 9 | 9 | 9 | 10 | 9 | 9 | 6 | 8 | 8 | 10 |
| HHI | 5,032 | 4,430 | 4,187.6 | 3,357 | 3,422 | 3,123 | 3,973 | 3,968 | 3,572 | 3,513.4 |
| Noam Index | 1,677.5 | 1,477.0 | 1,395.9 | 1,061.8 | 1,140.9 | 1,041.3 | 1,622.1 | 1,402.9 | 1,262.9 | 111.0 |

Table 17-5. ISP Market Shares and Concentration Levels from 1996 to 2011. Internet Service Providers, Market Shares (Based on Revenue) and Concentration Levels, 1996–2011 (1)

| | 1996 | 2000 | 2004 | 2006 | 2008 | 2010 | 2011 |
|---|---|---|---|---|---|---|---|
| Bell (2) | 10.5 | 21.4 | 24.7 | 28.9 | 22.1 | 20.7 | 21.2 |
| Mar. T&T (3) | *0.8* | *Alliant* | | | | | |
| NBTel (4) | *0.8* | *Alliant* | | | | | |
| New Tel (5) | *0.5* | | | | | | |
| Island Tel (6) | *0.05* | *Alliant* | | | | | |
| Shaw | | 12.3 | 10.1 | 11.4 | 11.7 | 13.5 | 13.5 |
| Moffat | | *0.8* | *Shaw (2001)* | | | | |
| Rogers | | 12.4 | 9.0 | 10.5 | 11.2 | 12.4 | 12.7 |
| Quebecor (9) | | | 5.3 | 7.3 | 7.1 | 9.5 | 9.7 |
| Videotron (10) | *3.8* | *4.1* | *Quebecor* | | | | |
| Telus (7) | 2.0 | 8.1 | 10.0 | 9.6 | 9.3 | 9.1 | 9.6 |
| BCTel (8) | *2.7* | *Telus* | | | | | |
| Cogeco | 1.0 | 2.5 | 2.5 | 3.0 | 3.4 | 4.1 | 4.4 |
| Bragg/East-link/Persona | | 0.08 | 0.5 | 1.2 | 2.0 | 2.9 | 3.2 |
| AmTel | | | | 0.9 | Bragg | | |
| SaskTel | 1.1 | 1.7 | 1.6 | 1.7 | 1.6 | 1.7 | 1.6 |
| MTS Allstream | 0.8 | 1.2 | 1.1 | 1.7 | 1.6 | 1.4 | 1.4 |
| AT&T (US) | | *2.6* | *3.1* | *1.7* | | | |
| Access Comm Coop | | | 0.2 | 1.3 | 0.2 | 0.2 | |
| Omineca/Your-link (12) | | | 0.04 | MTS | 0.07 | | |
| AOL (US) | 12.1 | 3.8 | 2.9 | 0.2 | | | |
| Primus (13) | | 1.1 | | 0.04 | | | |
| Call-Net (Sprint) (14) | | 3.3 | 0.3 | Rogers (2004) | | | |
| Look | | 2.2 | 0.06 (15) | 0.2 | 0.1 | | |
| Internet Direct (16) | *6.2* | *Look* | | | | | |
| PsiNet (17) | | 3.4 | | | | | |
| Istar | 7.2 | PsiNet | | | | | |
| Technovision/Uniserve | 0.3 | | | | | | |
| Pathway | | | | | | | |
| Rhythms (18) | | | | | | | |
| Hook-Up | 6.3 | (Netcom 1997) | Pathway (2001) | | | | |
| Craig | | | | 0.02 | 0.01 | | |
| Cybersurf | 0.2 | 0.1 | 0.2 | | | | |
| Others | 43.7 | 18.9 | 28.5 | 20.3 | 29.6 | 24.5 | 22.7 |
| Total Rev. | 239 | 1,800 | 4,200 | 5,000 | 6,200 | 6,800 | 7,200 |
| Total Rev $US | 175 | 1,212 | 3,235 | 4,409 | 5,854 | 6,601 | 7,283 |
| C4 | 33.6 | 54.2 | 53.8 | 54.7 | 54.3 | 56.1 | 57.1 |
| *N* (>1%) | 10 | 14 | 10 | 11 | 9 | 9 | 9 |
| HHI | 591.9 | 974.5 | 1,239.4 | 1,242.5 | 926.0 | 967.7 | 1,013.4 |
| Noam Index | 197.3 | 260.4 | 391.9 | 391.9 | 292.8 | 306.0 | 320.4 |

Table 17-6. Multichannel Video Distribution Platforms—Ownership Groups Market Shares and Concentration Levels, 1984–2011

| | 1984 | 1988 | 1992 | 1996 | 2000 | 2004 | 2006 | 2008 | 2010 | 2011 |
|---|---|---|---|---|---|---|---|---|---|---|
| Shaw (1) | 4.5 | 5.3 | 10.0 | 18.8 | 28.3 | 31.2 | 29.9 | 28.4 | 28.8 | 28.3 |
| Star Choice | | | | *3.4* | *Shaw (1998)* | | | | | |
| Moffat | *4.0* | *2.6* | *1.7* | *1.8* | *3.8* | *Shaw (2001)* | | | | |
| Monarch Cablesystems | *0.2* | *0.2* | *0.4* | *0.4* | *0.7* | *Shaw (2001)* | | | | |
| Okanagan | | | | | | *Monarch (2001)* | | | | |
| Skeena Group | 0.2 | 0.2 | 0.3 | 0.3 | 0.3 | | | | | |
| Dartmouth/ Access Cable | *0.5* | *0.5* | *0.5* | *0.7* | *Shaw (1999)* | | | | | |
| Fundy | | | *2.0* | *2.6* | *Shaw (1999)* | | | | | |
| Cable 2000 | | | *0.7* | *Fundy (1995)* | | | | | | |
| Classic | | | *1.7* | *Fundy (1994)* | | | | | | |
| CUC/Trillium | 0.6 | 4.6 | 5.1 | Shaw (1995) | | | | | | |
| Cablecaster | | | *4.5* | *Shaw (1992/3)* | | | | | | |
| Saskatoon Telecable | | *0.9* | *Shaw (1990)* | | | | | | | |
| Rogers | 21.3 | 22.4 | 30.8 | 35.6 | 26.6 | 24.7 | 22.0 | 23.0 | 21.8 | 21.3 |
| Cable Atlantic/Avalon Cable | *0.6* | *0.9* | *1.0* | *1.0* | *0.1* | *Rogers/Metro Net (2000)* | | | | |
| Maclean-Hunter | *6.6* | *6.7* | *10.6* | | | | | | | |
| Skyline | *2.2* | *1.4* | *Rogers (1990)* | | | | | | | |
| Selkirk | *2.9* | *Maclean-Hunter* | | | | | | | | |

| | | | | | | | | | | |
|---|---|---|---|---|---|---|---|---|---|---|
| Bell TV (DTH) | | | | | 7.2 | 17.0 | 20.0 | 19.8 | 20.7 | 22.0 |
| Quebecor/ Videotron | | | | | | 12.7 | 12.4 | 11.6 | 12.1 | 12.1 |
| Videotron | | *19.7* | *17.7* | *18.8* | *17.9* | *Quebecor (2000)* | | | | |
| CF Cable | *2.5* | *3.0* | *3.0* | *4.9* | *Videotron (1997)* | | | | | |
| Northern Cable | *1.9* | *1.6* | *2.1* | *CFCable (1993)* | | | | | | |
| Telesag | *1.4* | *0.9* | *Videotron (1989)* | | | | | | | |
| Telecable | | | *0.4* | *CFCable (1994)* | | | | | | |
| Cogeco | 0.6 | 2.1 | 6.6 | 5.7 | 8.1 | 9.1 | 8.2 | 7.4 | 7.6 | 7.5 |
| Cableworks/ Western Co-ax | | | *0.6* | *0.9* | *Cogeco* | | | | | |
| Bragg/East-Link/Persona | | | | 1.6 | 2.7 | 2.8 | 2.9 | 4.1 | 3.8 | 3.8 |
| Telus | | | | | | | | 0.01 | 2.7 | 4.2 |
| Regional Cablesystems | *1.9* | *1.7* | *2.1* | *1.5* | *2.7* | *Bragg/East-Link (2001)* | | | | |
| AmTel | | | | | | | *0.1* | *Bragg* | | |
| MTS Allstream | | | | | | 0.2 | 0.5 | 0.7 | 0.7 | 0.8 |
| SaskTel | | | | | | 0.3 | 0.5 | 0.7 | 0.7 | 0.9 |
| Access Comm. Coop | | | | 1.1 | 0.9 | 0.8 | 0.7 | 0.7 | 0.9 | 0.8 |
| Mountain Cable | 0.5 | 0.5 | 0.5 | 0.5 | 0.4 | 0.4 | 0.4 | 0.4 | Shaw | |
| Omineca/Yourk-Link | 0.2 | 0.2 | 0.3 | 0.3 | 0.1 | 0.1 | 0.1 | 0.1 | 0.05 | 0.06 |
| Look (MDS) | | | | | 0.8 | 0.4 | 0.2 | 0.2 | | |

*continued*

Table 17-6. *continued*

| | 1984 | 1988 | 1992 | 1996 | 2000 | 2004 | 2006 | 2008 | 2010 | 2011 |
|---|---|---|---|---|---|---|---|---|---|---|
| Craig | | | | | | | <1 | <1 | <1 | |
| Others | 47.4 | 24.6 | | | | | 2.1 | 2.89 | | |
| Total $ | 716.3 | 1,242.9 | 1,651.4 | 2,677.4 | 4,218.5 | 5,039.4 | 5,791.3 | 6,953.5 | 8,100 | 8,588.3 |
| Total US$ | | | 1,367.6 | 1,963.3 | 2,841.2 | 3,881.5 | 5,106.6 | 6,565.3 | 7,862.7 | 8,686.8 |
| No. of Systems | *18* | *19* | *22* | *18* | *15* | *13* | *14* | *13* | *10* | *10* |
| C4 | 35.3 | 54.1 | 69.1 | 78.9 | 80.9 | 85.6 | 84.3 | 82.8 | 83.4 | 83.7 |
| *N* (>1%) | 10 | 11 | 14 | 12 | 8 | 6 | 6 | 6 | 7 | |
| HHI | 563.9 | 1,014.6 | 1,594.6 | 2,061.2 | 1,977.5 | 2,125.6 | 2,008.6 | 1,935.4 | 1,960.9 | 1,975.4 |
| Noam Index | 178.3 | 305.9 | 426.2 | 595.0 | 699.2 | 867.8 | 820.0 | 790.1 | 741.2 | |

Table 17-7. Wireless Telecommunications Ownership Groups and Market Shares, 1984–2011

| | 1985 | 1988 | 1992 | 1996 | 2000 | 2004 | 2006 | 2008 | 2010 | 2011 |
|---|---|---|---|---|---|---|---|---|---|---|
| Rogers | 34.2 | 25.9 | 25.7 | 50.7 | 30.4 | 29.2 | 36.1 | 39.1 | 38.7 | 37.0 |
| Microcel/Fido | | | | 0.06 | 11.0 | Rogers | | | | |
| Bell | 65.8 | 74.1 | 74.3 | 49.3 | 28.0 | 30.6 | 27.5 | 27.7 | 27.4 | 27.1 |
| Telus | | | | | 29.4 | 30.8 | 30.4 | 28.8 | 28.0 | 28.5 |
| SaskTel | | | | | | | 2.4 | 2.3 | 2.3 | 2.1 |
| MTS Allstream | | | | | | | 1.8 | 1.8 | 1.8 | 1.8 |
| Quebecor/Videtron | | | | | | | 0.1 | 0.3 | 0.3 | 0.6 |
| Clearnet | | | | | 2.4 | Telus | | | | |
| Other (Wind, Mobilicity, Public Mobile) | | | | | | 9.4 | 1.7 | | 0.6 | 1.6 |
| Total $ | 321.0 | 565.2 | 931.0 | 2,175.0 | 5,400.0 | 9,200.0 | 12,696.2 | 16,200.0 | 18,000.0 | 19,300.0 |
| Total $US | | | 771.0 | 1,594.9 | 3,637.0 | 7,086.2 | 11,195.2 | 15,295.6 | 18,206.4 | 19,953.4 |
| (a) No. of Providers | 2 | 2 | 2 | 4 | 5 | 3 | 5 | 5 | 9 | 9 |
| C4 | 100.0 | 100.0 | 100.0 | 100.0 | 98.8 | 100.0 | 96.4 | 97.9 | 96.4 | 94.7 |
| *N* (>1%) | 2 | 2 | 2 | 2 | 5 | 3 | 5 | 5 | 5 | 5 |
| HHI | 5,499.3 | 6,161.6 | 6,181.0 | 5,001.0 | 2,699.3 | 2,737.6 | 2,992.6 | 3,134.2 | 3,041.1 | 2,923.7 |
| Noam Index | 3,888.6 | 4,356.9 | 4,370.6 | 2,500.5 | 1,207.2 | 1,580.6 | 1,338.3 | 1,401.6 | 1,360.0 | 1,307.5 |

Table 17-8. CR and HHI Scores for the Network Infrastructure Industries, 1984–2011

| CR | 1984 | 1988 | 1992 | 1996 | 2000 | 2004 | 2008 | 2010 | 2011 |
|---|---|---|---|---|---|---|---|---|---|
| Wireless | 100.0 | 100.0 | 100.0 | 100.0 | 98.8 | 100.0 | 97.9 | 96.4 | 94.7 |
| Wired | 90.8 | 89.4 | 88.1 | 81.5 | 84.8 | 81.0 | 90.5 | 86.3 | 85.1 |
| Internet Access | | | | 33.6 | 54.2 | 53.8 | 55.3 | 54.3 | 56.1 |
| Cable & Sat. Dist. | 35.3 | 54.1 | 69.1 | 78.9 | 80.9 | 85.6 | 82.8 | 83.4 | 83.7 |
| HHI | 1,984 | 1,988 | 1,992 | 1,996 | 2,000 | 2,004 | 2,008 | 2,010 | 2,011 |
| Wireless | 5,499.3 | 6,161.6 | 6,181.0 | 5,001.0 | 2,699.3 | 2,737.6 | 3,134.2 | 3,041.1 | 2,923.7 |
| Wired Telecom | 5,033.0 | 4,431.0 | 4,187.6 | 3,358.0 | 3,423.0 | 3,124.0 | 3,968.0 | 3,572.0 | 3,513.4 |
| Internet Access | | | | 416.8 | 906.4 | 949.7 | 926.0 | 967.7 | 1,013.4 |
| Cable & Sat. Dist. | 563.9 | 1,014.6 | 1,594.6 | 2,061.2 | 1,977.5 | 2,125.6 | 1,935.4 | 1,960.9 | 1,975.4 |

## The Content Industries

Until the mid-1990s, all aspects of the television industry were moderately concentrated by HHI standards and significantly so by CR measures, with the top four firms, including the CBC, accounting for between 50% and 60% or more of revenues. Competition and diversity made some modest inroads from 1998 to 2004, but the trend abruptly reversed course and levels have climbed steadily since. Shaw's takeover of Canwest's television assets and Bell's buyback of CTV pushed the levels to new extremes by 2011. The four largest commercial television companies—CTV, Global, TVA, and CityTV—as well as more than 100 of the most important cable and satellite channels available are owned by Bell, Shaw, Rogers, and Astral. Together, these four players control 69% of all television revenue (excluding distribution). The CBC accounts for almost 21% and QMI 5% of television revenues, meaning that the top six television groups account for 94% of the entire television industry, with each of its subcomponents (conventional television, pay and specialty channels) each also having high levels of concentration. Tables 17.9 and 17.10 illustrate the trends for conventional television, as well as pay and specialty television services, respectively.

Like the cable industries, there has never been a moment when diversity and competition flourished in the newspaper sector. Consolidation rose steadily from 1984, when the top four players accounted for two-thirds of all revenues, to 1996, when they accounted for nearly three-quarters. It has held fairly steady at three-quarters despite periodic shuffling among the main players at the top. Data on the Newspaper Sector are provided in Table 17.11.

Of all media sectors, magazines are the least concentrated. This has been the case throughout the time frame covered here, and it has become more evident over time with concentration levels falling by one-half on the basis of CR scores and two-thirds for the HHI. Table 17.12 illustrates the trends.

Radio is also among the most diverse and competitive among all of the media sectors on the basis of HHI scores, although the C4 measure suggests that the big four in the sector still dominate. Table 17.13 illustrates the trends.

The trends for each of the content industries are shown in Table 17.14.

## The Network Media as a Whole (Excluding Wired and Wireless Telecoms)

Adding the network infrastructure industries (except wired and wireless telecoms because their revenues tend to overshadow everything else) and the content industries together allows us to chart concentration levels for the pooled network media over time. It is a particularly good indicator of long-term trends. Throughout the 1980s and early 1990s, the HHI for the network media as a whole fell, but by 1996 trends

Table 17-9. Broadcast Television Ownership Groups, Market Shares, and Concentration Levels 1984–2011

| | 1984 | 1988 | 1990 | 1996 | 2000 | 2004 | 2008 (2) | 2010 | 2011 |
|---|---|---|---|---|---|---|---|---|---|
| CBC/Radio Canada (3) | 48.5 | 40.3 | 43.1 | 31.8 | 33.8 | 33.4 | 36.8 | 37.0 | 38.3 |
| Bell | | | | | | | | | 24.0 |
| CTV Globemedia | | | | | | | *24.8* | *27.4* | *Bell* |
| Bell Globemedia (8) | | | | | *0.43* | *21.6* | *CTVgm* | | |
| Baton/CTB (14) | *8.3* | *10.5* | *6.9* | *7.6* | *17.5 BCE* | | | | |
| Electrohome (16) | *1.0* | *2.2* | *2.3* | *2.2* | | | | | |
| Moffat (17) | *0.8* | *0.8* | *0.8* | *0.7* | | | | | |
| Blackburn (22) (CFPL London) | *0.9* | *0.8* | *0.8* | *Baton (1993)* | | | | | |
| MidCanada Comm (23) | *0.9* | *0.8* | *Baton (1989)* | | | | | | |
| Huron Broadcasting | *0.3* | *2* | *Baton (1989)* | | | | | | |
| Yorkton TV | *0.3* | *0.3* | *Baton (1989)* | | | | | | |
| Sunwapta | *0.5* | *0.5 Electrohome* | | | | | | | |
| CHUM | 4.2 | 5.1 | 4.9 | 4.2 | 4.4 | 6.0 | Bellgm (2007) (10) | | |
| Craig (11) | | | | | *1.6* | *1.8 CHUM & QBC (12)* | | | |
| Shaw | | | | | | | | 14.5 | 15 |
| Canwest (5) | 4.9 | 6.7 | 8.9 | 11.9 | 21.2 | 21.3 | 17.3 | Shaw | |
| WIC (18) | 3.3 | 3.4 | 5.8 | 9.7 | | | | | |
| Quebecor TVA (2000Pres) | | | | | 8.9 | 8.6 | 7.3 | 7.4 | 7.5 |

*continued*

Table 17-9. *continued*

| | 1984 | 1988 | 1990 | 1996 | 2000 | 2004 | 2008 (2) | 2010 | 2011 |
|---|---|---|---|---|---|---|---|---|---|
| Videotron (13) | | 6.2 | 6.7 | 7.0 | QMI | | | | |
| TeleMetropole (TVA) (27) | 4.9 | Videotron | | | | | | | |
| Pathonic (24) | 1.0 | 0.9 | Videotron (25) | | | | | | |
| Rogers (7) | 0.9 | 0.9 | 1.9 | 5.3 | 2.7 | 2.4 | 6.1 | 7.3 | 8.5 |
| Maclean Hunter | *1.3* | *4.9* | *5.2 (21)* | | | | | | |
| Selkirk (26) | | 5.8 | | | | | | | |
| Remstar | | | | | | | 1.8 | 1.8 | 1.9 |
| Cogeco TQS (200,108) (40% CTVgm) | 0.4 | 0.9 | 1.8 | 2.5 | 1.3 | 3.6 | Remstar | | |
| TQS (Quebecor 1997–2001) | | | | | 1.8 | | | | |
| TQS (CFCF/Pouliot) (19) | 4.3 | 3.3 | 4.6 | 2.3 | | | | | |
| Radio Nord (9) | 0.4 | 0.5 | 0.6 | 0.6 | 0.8 | 1.0 | 1.0 | 1.2 | 1.2 |
| Other | 7.5 | 1.6 | 6.0 | 4.9 | 9.3 | 6.7 | 3.0 | 2.9 | |
| Conventional TV $ | 1,747.9 | 2,127.4 | 2,378.8 | 2,831.6 | 2,840 | 3,159.9 | 3,381.4 | 3,405.6 | 3,491.9 |
| Conventional TV $US | | | 2,039.1 | 2,076.3 | 1,912.8 | 2,433.9 | 3,192.6 | 3,305.8 | 3,531.9 |
| Conventional TV Owners # | *41* | *34* | *27* | *24* | *20* | *16* | *15* | *13* | |
| C4 | 66.6 | 63.7 | 65.6 | 61.0 | 81.4 | 84.9 | 86.2 | 86.3 | 85.8 |
| HHI | 2,524 | 1,938 | 2,149 | 1,417 | 2,012 | 2,169 | 2,363 | 2,443 | 2,401 |
| *N* (>1%) | 10 | 11 | 10 | 10 | 8 | 8 | 7 | 7 | 7 |
| Noam Index | 798 | 584 | 680 | 448 | 711 | 767 | 893 | 923 | 908 |

Table 17-10. Pay TV and Specialty Ownership Groups, Market Shares, and Concentration Levels from 1984 to 2011

| | 1984 | 1988 | 1992 | 1996 | 2000 | 2004 | 2008 (2) | 2010 | 2011 |
|---|---|---|---|---|---|---|---|---|---|
| Shaw/Corus (4) | | | | 3.7 | 11.3 | 18.7 | 17.5 | 31.7 | 33.9 |
| Canwest | | | | | *1.1* | *2.1* | *16.1* | *Shaw* | |
| WIC | | | *7.4* | *5.1* | *Canwest (1998)* | | | | |
| Alliance Atlantis (9) | | | | *2.7* | *8.0* | *11.7* | *Canwest (2007)* | | |
| Atlantis | | | *2.3* | *5.2* | | | | | |
| Allarcom | | *13.1* | *WIC (1991)* | | | | | | |
| Bell | | | | | | | | | 27.4 |
| CTV Globemedia | | | | | | | 28.4 | 26.3 | Bell |
| Bell Globemedia (8) | | | | | *0* | *15.7* | *CTVgm (2007)* | | |
| CHUM | | *3.5* | *4.0* | *6.5* | *8.8* | *10.5* | *CTVgm/Rogers (2007) (8)* | | |
| Baton/CTV (10) | | | | | *18.9* | | | | |
| Netstar (11) | | *21.8* | *24.1* | *30.6* | *CTV (12)* | | | | |
| Rogers | | 14.1 | 22.5 | 15.4 | 13.5 | 15.8 | 10.9 | 11.4 | 12.3 |
| Astral | | 12.6 | 2.6 | 3.0 | 5.9 | 5.9 | 17.0 | 15.9 | 15.6 |
| CBC/Radio Canada | | | 7.5 | 8.5 | 7.5 | 6.4 | 5.1 | 4.3 | 4.1 |
| Quebecor (5) | | | | | 0.7 | 1.6 | 2.5 | 3.5 | 3.9 |
| Craig | | | | | | *0.4 CHUM & QMI* | | | |
| Videotron | | | | *4.1* | *QMI* | | | | |
| Pelmorex | | | 4.5 | 3.7 | 2.5 | 1.9 | 1.7 | 1.4 | 1.3 |
| Fairchild (Chinavision) | | 1.5 | 0.9 | 2.3 | 1.7 | 1.2 | 1.0 | 0.8 | 0.7 |

*continued*

Table 17-10. *continued*

| | 1984 | 1988 | 1992 | 1996 | 2000 | 2004 | 2008 (2) | 2010 | 2011 |
|---|---|---|---|---|---|---|---|---|---|
| MusicPlus/ MusiqueMax (7) | | | | 0.7 | 0.6 | 0.6 | 0.5 | 0.6 | 0.4 |
| Cogeco (as TQS from 2001–08) | | | | | | | 0.1 (Remstar) | | |
| Other | | 4.1 | 24.2 | 8.5 | 18.5 | 8.1 | 1.7 | 5.0 | |
| Spec and Pay TV $ (14) | 93.8 | 142.4 | 395.2 | 664.5 | 1,270.2 | 2,050.0 | 2,929.9 | 3,459.4 | 3,732.1 |
| Spec and Pay TV in $US | | | 290.6 | 520.2 | 738.3 | 1,509.6 | 2,499.7 | 3,181.6 | 3,858.4 |
| Conventional TV $ | 1,747.9 | 2,127.4 | 2,531.8 | 2,831.6 | 2,840 | 3,159.9 | 3,381.4 | 3,405.6 | 3,491.9 |
| Total TV $ | 1,841.7 | 2,269.8 | 2,927.0 | 3,496.1 | 4,110.2 | 5,209.9 | 6,311.3 | 6,865.0 | 7,224.0 |
| Total TV $US | | | 2,424.0 | 2,563.6 | 2,768.3 | 4,012.8 | 5,959.0 | 6,663.9 | 7,306.8 |
| Owners # (15) | | 8 | 14 | 18 | 18 | 37 | 47 | 39 | |
| C4 | | 61.6 | 61.5 | 61.0 | 52.5 | 61.9 | 72.9 | 85.3 | 89.2 |
| HHI | | 1,019 | 1,247 | 1,407 | 912 | 1,181 | 1,816 | 2,113 | 2,329 |
| *N* (>1%) | | 7 | 9 | 13 | 11 | 12 | 9 | 8 | 8 |
| Noam Index | | 385 | 416 | 390 | 275 | 341 | 605 | 747 | 823 |

Table 17-11. Newspaper Ownership Groups, Revenue, Market Shares, and Concentration Levels from 1984 to 2011

| | 1984 (2) | 1988 | 1992 | 1996 | 2000 | 2004 (3) | 2008 (4) | 2010 | 2011 |
|---|---|---|---|---|---|---|---|---|---|
| Postmedia | | | | | | | | 31 | 29.4 |
| Hollinger | | *4.1* | *4.1* | *41.2 (7)* | *31.2 (5) Hollinger/ Southam* | *28.4* | *27.7* | *Postmedia* | |
| Southam | *27.0* | *27.0* | *27.6* | *Hollinger* | *8.9 (8)* | | | | |
| Quebecor | 8.5 | 8.5 | 8.5 | 8.6 | 20.3 | 21.0 | 25.9 | 23.6 | 22.7 |
| Osprey | | | | | | *5.9* | *Quebecor (2007)* | | |
| Toronto Sun Pub./ Sun Media | *8.8* | *11.0* | *11.1* | *11.2* | *Quebecor (1997)* | | | | |
| Torsta | 9.8 | 9.8 | 9.8 | 9.8 | 13.5 | 13.9 | 13.9 | 11.6 | 11.0 |
| Power Corp/Gesca/ Unimedia | 5.8 | 5.8 | 5.8 | 5.7 | 8.7 | 9.8 | 9.8 | 10.9 | 11.7 |
| Thomson | 20.5 | 20.5 | 20.5 | 12.3 (11) | 3.7 | | | | |
| CTV Globemedia | | | | | | | *7.2* | *7.4 Thomson* | 8.2 |
| Bell Globemedia (BCE/Thomson) | | | | | *6.5* | *6.3* | *CTV Globemedia* | | |
| FP CDN Newspapers | | | | | | 3.1 | 3.5 | 3.9 | 3.4 |
| Transcontinental | | | | | | 3.1 | 3.2 | 2.3 | 2.4 |
| Horizon (now Glacier Cdn. Newspapers) | | | | | 1.2 | 1.9 | 2.9 | 1.5 | 2.2 |
| Halifax Herald | | | | | 2.1 | 2.2 | 2.2 | 2.9 | 2.9 |
| Brunswick News | | | | | 2.3 | 1.9 | 2.1 | 2.1 | 2.2 |

*continued*

Table 17-11. *continued*

| | 1984 (2) | 1988 | 1992 | 1996 | 2000 | 2004 (3) | 2008 (4) | 2010 | 2011 |
|---|---|---|---|---|---|---|---|---|---|
| Irving Group | 2.5 | 2.5 | 2.5 | 2.6 | Brunswick News | | | | |
| Sterling | 0.9 | Hollinger (1986) | | | | | | | |
| St. Catherines Group | 0.9 | 0.9 | 0.9 | 1.0 Southam | | | | | |
| Armadale (Sifton) | 2.4 | 2.4 | 2.4 | Hollinger | | | | | |
| Unimedia | 3.2 | Hollinger (1987) | | | | | | | |
| Trinity (Black Press) | | | | | | 0.4 | 0.3 | 0.4 | 0.6 |
| Annex | | | | | 0.3 | QMI (2003) | | | |
| Nfld. Capital | | | | 0.9 | Hollinger (1997) | | | | |
| Bowes Publ. | 0.3 | Toronto Sun (1987) | | | | | | | |
| Independents | 9.5 | 9.2 | 7.5 | 7.4 | 1.1 | 1.0 | 1.2 | 1.2 | 1.4 |
| # of Independents | 20 | 16 | 13 | 12 | 5 | 4 | 4 | 3 | |
| Avg. Daily Circ. (mill) | 5.6 | 5 | 5.5 | 5 | 5 | 4 | 4.7 | 4.7 | 5.5 |
| Revenues | 2,190.0 | 2,925.0 | 2,790.0 | 3,310.0 | 4,700.0 | 5.033.0 | 5,400.0 | 4,300.0 | 3,975.0 |
| Revenues $US | | | 2,310.5 | 2,427.1 | 3,165.5 | 3,876.6 | 5,098.5 | 4,174.0 | 4,020.6 |
| C4 | 66.1 | 68.3 | 70.3 | 74.3 | 51.4 | 73.1 | 77.3 | 77.1 | 74.8 |
| HHI | 1,451.3 | 1,487.3 | 1,536.9 | 2,183 | 1,710.1 | 1,643.7 | 1,819.3 | 1,861.8 | 1,742.5 |
| *N* (>1%) | 9 | 9 | 9 | 8 | 9 | 11 | 10 | 9 | 10 |
| Noam Index | 484 | 496 | 512 | 772 | 570 | 496 | 575 | 621 | 551 |

Table 17-12. Magazine Publishing Groups, Market Shares, and Concentration Levels 1984–2008

| | 1984 | 1988 | 1992 | 1996 | 2000 | 2004 | 2008 | 2010 EST |
|---|---|---|---|---|---|---|---|---|
| Rogers | | | | 13.0 | 11.7 | 7.6 | 6.3 | 6.3 |
| Maclean Hunter | 13.8 | 17.0 | 11.5 | Rogers 1994 | | | | |
| Transcontinental | 1.7 | 2.5 | 3.3 | 4.3 | 9.4 | 7.0 | 6.2 | 6.2 |
| Telemedia | 14.4 | 16.3 | 17.8 | 1.4 | Transcont. | | | |
| Time Warner (US) | 1.9 | 2.2 | 2.1 | 5.6 | 5.6 | 6.5 | 4.7 | 4.7 |
| American Media (US) | 6.9 | 7.9 | 7.7 | 7.3 | 5.4 | 3.7 | 2.8 | 2.8 |
| Hearst (US) | 1.9 | 2.0 | 1.8 | 2.1 | 3.0 | 3.1 | 2.6 | 2.6 |
| Bauer Media (Germany) | 1.9 | 2.1 | 1.9 | 2.1 | 2.4 | 2.1 | 2.4 | 2.4 |
| Readers Digest (US) | 3.5 | 4.0 | 3.9 | 4.5 | 4.0 | 2.8 | 2.1 | 2.1 |
| National Geographic (US) | 1.7 | 2.0 | 1.9 | 2.0 | 2.2 | 1.9 | 1.7 | 1.7 |
| Quebecor | 3.7 | 4.9 | 3.3 | 4.3 | 3.4 | 1.3 | 1.7 | 1.7 |
| Condé Nast (US) | 1.1 | 1.2 | 1.2 | 1.7 | 1.9 | 2.4 | 1.6 | 1.6 |
| House & Home | 0.2 | 0.2 | 0.3 | 1.2 | 3.8 | 1.7 | 1.5 | 1.5 |
| St. Joseph Media | 0.8 | 0.9 | 0.9 | 2.1 | 1.7 | 1.0 | 1.0 | 1.0 |
| Meredith (US) | 0.6 | 0.7 | 0.7 | 0.9 | 3.8 | 1.3 | 1.0 | 1.0 |
| Wenner (US) | 0.3 | 0.3 | 0.3 | 0.2 | 3.8 | 0.9 | 0.7 | 0.7 |
| Air Canada (En Route) | | | | | | 3.6 | 0.5 | 0.5 |
| Now | | | | | | 0.5 | 0.5 | 0.5 |
| Shaw (Movie Entertainment Magazine) | | | | | | 0.5 | 0.4 | 0.4 |
| CDN Geographic | | | | | | 0.5 | 0.4 | 0.4 |
| Buzz Media (US) | | | | | | 0.2 | 0.3 | 0.3 |
| CTVgm | | | | | | 0.4 | 0.3 | 0.3 |
| Znaimer (Zoomer) | | | | | | | 0.3 | 0.3 |
| Bonnier (Sweden) | | | | 0.5 | | | 0.3 | 0.3 |
| Postmedia (Financial Post) | | | | | | *0.3* | *0.3* | *0.3* |
| LCBO (Food & Drink) | | | | | | 0.3 | 0.3 | 0.3 |
| Torstar (Eye Weekly) | | | | | | 0.3 | 0.2 | 0.2 |
| Alpha (Australia) | | | | | | 0.4 | 0.2 | 0.2 |
| Cottage Life | | | | | | 0.2 | 0.2 | 0.2 |
| Cineplex (Famous) | | | | | | | 0.2 | 0.2 |
| Pearson (UK) | | | | | | 0.2 | 0.2 | 0.2 |
| DecorMag | | | | | | 0.2 | 0.1 | 0.1 |
| Q on Q Media | | | | | | 0.5 | 0.1 | 0.1 |
| Other | 45.6 | 35.8 | 41.4 | 46.8 | 37.9 | 48.6 | 58.9 | 58.9 |
| Total Revenue | 711 | 775 | 994 | 1,011 | 1,300 | 1,847 | 2,394 | 2,394 |
| Total Revenue (USD) | | | 823.1 | 741.5 | 875.6 | 1,422.7 | 2,260.4 | 2,323.9 |
| C4 | 38.8 | 46.0 | 40.9 | 30.4 | 32.1 | 24.8 | 20.0 | 20.0 |
| HHI | 490 | 684 | 563 | 335 | 383 | 217 | 160 | 160 |
| *N* (>1%) | 11 | 11 | 11 | 13 | 14 | 14 | 13 | 13 |
| Noam Index | 148 | 206 | 170 | 93 | 102 | 58 | 44 | 44 |

*Note:* Based on "total revenues": advertising, subscriptions, and news-stand sales.

Table 17-13. Radio Ownership Groups, Market Shares, and Concentration Levels from 1984 to 2011

| | 1984 | 1988 | 1992 | 1996/97 | 2000 | 2004 | 2008 | 2010 | 2011 |
|---|---|---|---|---|---|---|---|---|---|
| Astral Media | | | | | 3.0 | 7.5 | 16.4 | 17.0 | 17.5 |
| Standard | *9.6* | *8.2* | *7.6* | *7.3* | *6.7* | *10.1* | *Astral (2007)* | | |
| Telemedia | *2.7* | *3.3* | *4.2* | *5.5* | *9.2* | *Standard/Astral (2002)* | | | |
| Radio-mutuel | 1.6 | 2.0 | 2.5 | 3.3 | Astral | | | | |
| Rogers | 2.8 | 2.5 | 3.3 | 6.5 | 8.0 | 11.1 | 12.1 | 10.7 | 11.3 |
| Maclean Hunter | 2.3 | 2.6 | 3.2 | *Rogers (1994)* | | | | | |
| Selkirk | *2.9* | *2.6* | *M-H (1989)* | | | | | | |
| CBC/Radio Canada | 32.7 | 27.4 | 30.6 | 32.2 | 24.6 | 23.4 | 19.9 | 21.8 | 17.3 |
| Shaw Corus | 0.6 | 0.4 | 0.5 | 4.0 | 12.3 | 13.5 | 13.4 | 12.6 | 10.0 |
| Metro-media | *0.9* | *1.1* | *1.4* | *1.9* | *2.5* | *Shaw (2001)* | | | |
| Western Int'l Comm | 2.9 | 3.4 | 4.4 | 5.8 | Shaw | | | | |
| Bell | | | | | | | | | 8.2 |
| CTV Globemedia | | | | | | | *8.7* | *8.4* | *Bell* |
| Bell Globemedia | | | | | | | | | |
| CHUM | 3.6 | 5.1 | 6.7 | 5.8 | 7.3 | 7.6 | BGM (2006) | | |
| NewCap | | 0.9 | 1.2 | 1.6 | 2.4 | 3.6 | 4.8 | 5.5 | 5.8 |
| Moffatt | 1.5 | 1.7 | 2.4 Rogers (1991) | | | | | | |
| Cogeco | | | | | | | 1.7 | 2.2 | 5.8 |
| Jim Pattison | | | | | | 2.1 | 2.1 | 2.0 | 2.0 |
| Rawlco | 0.6 | 0.7 | 0.9 | 1.3 | 1.3 | 1.3 | 1.3 | 2.0 | 2.0 |
| Maritime Broadcast | 0.7 | 0.8 | 1.0 | 1.4 | 1.7 | 1.4 | 1.3 | 1.6 | 1.6 |

| | | | | | | | | | |
|---|---|---|---|---|---|---|---|---|---|
| Golden West | 0.6 | 0.8 | 1.0 | 1.3 | 1.6 | 1.7 | 1.7 | 1.5 | 1.5 |
| Okanagen Skeena | 0.3 | 0.4 | 0.4 | 0.4 | | | | | |
| Other | 33.7 | 36.1 | 28.7 | 21.7 | 15.0 | 11.4 | 16.7 | 3.6 | 17.0 |
| All $ | 831.1 | 997.1 | 1,106.2 | 1,201.5 | 1,356.8 | 1,601.6 | 1,990.3 | 1,916.7 | 1,949.5 |
| All $US | | | 916.1 | 881.0 | 913.8 | 1,233.6 | 1,879.2 | 1,860.5 | 1,971.8 |
| C4 | 48.8 | 44.1 | 49.3 | 51.8 | 54.1 | 58.1 | 61.8 | 62.1 | 56.1 |
| *N* (>1%) | 10 | 11 | 13 | 13 | 12 | 11 | 11 | 12 | 11 |
| HHI | 1,218.9 | 897.2 | 1,115.8 | 1,268.5 | 1,028.9 | 1,090.1 | 1,103.2 | 1,156.0 | 980.6 |
| Noam Index | 385.5 | 270.5 | 309.5 | 351.8 | 297.0 | 328.7 | 332.6 | 333.7 | 295.7 |

Table 17-14. CR and HHI Scores for the Content Industries, 1984–2011

| CR | 1984 | 1988 | 1992 | 1996 | 2000 | 2004 | 2008 | 2010 | 2011 |
|---|---|---|---|---|---|---|---|---|---|
| All TV | 64.1 | 59.3 | 57.0 | 52.1 | 66.3 | 62.2 | 75.7 | 79.7 | 81.4 |
| Pay & Spec. TV | | 61.6 | 61.5 | 61.0 | 52.5 | 61.9 | 72.9 | 85.3 | 89.2 |
| Conv. TV | 66.6 | 63.7 | 63.5 | 61.0 | 81.4 | 84.9 | 86.2 | 86.3 | 85.8 |
| Radio | 48.8 | 44.1 | 49.3 | 51.8 | 54.1 | 58.1 | 62.1 | 56.1 | 48.8 |
| Press | 66.1 | 68.3 | 70.3 | 74.3 | 51.4 | 73.1 | 77.3 | 77.1 | |
| Mags | 38.8 | 46.0 | 40.9 | 30.4 | 32.1 | 24.8 | 20.0 | 20.0 | |
| HHI | | | | | | | | | |
| All TV | 2,287.0 | 1,692.4 | 1,474.4 | 1,086.6 | 1,348.6 | 1,258.5 | 1,750.3 | 1,791.7 | 1,890.7 |
| Pay & Spec. TV | | 1,018.9 | 1,247.2 | 1,407.3 | 911.8 | 1,181.3 | 1,816.2 | 2,113.1 | 2,329.0 |
| Conv. TV | 2,523.7 | 1,937.6 | 1,843.6 | 1,416.7 | 2,012.3 | 2,168.7 | 2,363.3 | 2,442.7 | 2,401.4 |
| Radio | 1,218.9 | 897.1 | 1,115.8 | 1,268.5 | 1,028.9 | 1,090.1 | 1,156.0 | 980.6 | 1,218.9 |
| Press | 1,451.3 | 1,487.3 | 1,536.9 | 2,183 | 1,710.1 | 1,643.7 | 1,819.3 | 1,861.8 | 1,742.5 |
| Magazines | 490.0 | 684.0 | 563.0 | 335.0 | 383.0 | 217.0 | 160.0 | 160.0 | 160.0 |

had reversed and levels were even a little bit higher than they had been a dozen years earlier. Thereafter, the number rose steadily to 584 in 2000, where it hovered until dropping to 511.5 in 2008 before once again rising substantially to 600 in 2010 after Shaw took over Global TV from the bankrupt Canwest and an even higher to 693 in 2011 after Bell reacquired CTV (see Figure 17.3). The CR4 standard shows the trend more starkly with the big four media conglomerates, Bell, Shaw, Rogers, and QMI, accounting for more than half of all revenues in 2011, a significant increase in a much larger media universe from the 40% held by the big four media companies in 1984. Over time, some media giants have collapsed (Canwest) and some new players have emerged (Channel Zero, Postmedia, and Remstar), but the long-term trend has been for a substantial rise in concentration levels across the network media as a whole. Figure 17.4 illustrates the trends on the basis of CR4 standards.

This portrait understates media concentration. The national measure used does not fully capture the extent to which QMI dominates the French language media. The shares of media conglomerates in English language markets would be much higher as well if this factor were taken into account.[41] A web of alliances between the big players also blunts the sharp edge of competition. Rogers owns 40% of Cogeco and Bell retains a residual 15% stake in the *Globe and Mail*, while Bell Media, Rogers, QMI, Shaw (Corus), Astral, and Cogeco co-own a dozen or so cable and satellite television channels. There are also several instances of directors from the big 10 sitting on one another's boards, as is the case with Astral and Torstar, as well as Postmedia and Astral.

As we have seen, the Internet neither obviates concerns with concentration within traditional media nor is immune to such tendencies itself. Beyond the network layer represented by ISPs, we can look at search engines, social media, and online news sources. When we do so, the answers are quite clear. Google not only dominates the search engine market, but its dominance is growing. By 2010, it accounted for 81% of searches. Microsoft (6.8%), Yahoo! (5%), and Ask.com (4%) trail far behind, yielding a CR4 of 97% and an HHI of 6,713. The information on market shares with respect to Search Engines is shown in Table 17.15.

41. CRTC, Communications Monitoring Report, Ottawa: Canada, 2011.

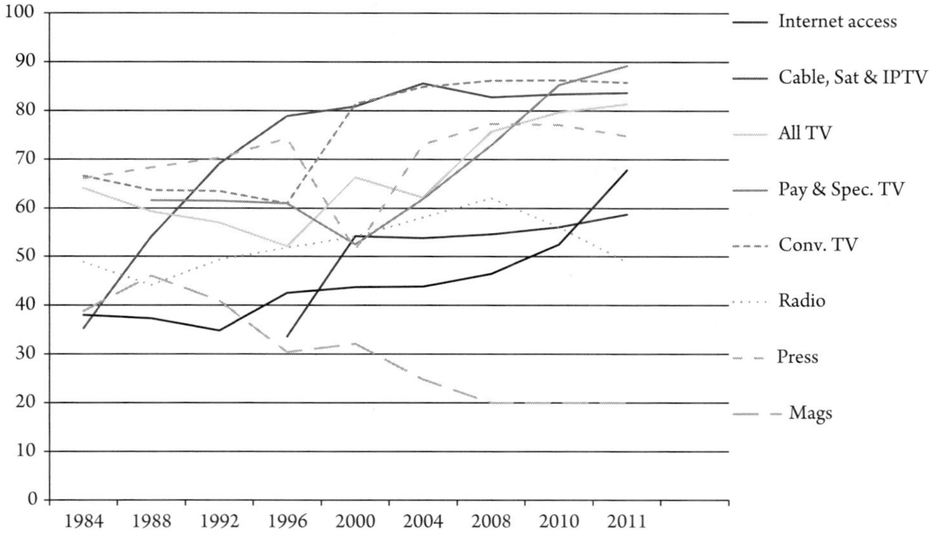


FIGURE 17.4 CR Scores for the Network Media Industries, 1984–2011

Social media sites display a similar trend, with Facebook accounting for 63.2% of time spent on such sites in 2010, trailed by Google's YouTube (20.4%), Microsoft (1.2%), Twitter (0.7%), and News Corp.'s MySpace (0.6%).[42] Again, the CR4 score of 86% and HHI score of 4,426 reveal that social networking sites are highly concentrated.

Similar patterns emerge across the layers of the media ecology, including the interfaces and devices used to access the Internet. The top four web browsers used in Canada, Microsoft's Explorer (52.8%), Google's Chrome (17.7%), Firefox (17.1%), and Apple's Safari (3%), have a combined market share of just over 90%.[43]

## Table 17-15. Search Engines: Market Share of Searches

| | 2004 | 2005 | 2006 | 2008 | 2009 | 2010 |
|---|---|---|---|---|---|---|
| Google (US) | 62.0 | 60.0 | 66.0 | 79.0 | 81.1 | 82.4 |
| Microsoft (US) | 12.0 | 17.0 | 14.0 | 5.0 | 4.4 | 5.9 |
| Yahoo (US) | 15.0 | 16.0 | 13.0 | 4.0 | 3.9 | 4.2 |
| eBay (US) | | | | 2.0 | 2.2 | |
| Ask (US) | 4.0 | 4.0 | 2.0 | 2.0 | 1.9 | |
| Facebook (US) | | | | 4.0 | 1.5 | 1.5 |
| Other | 7.0 | 3.0 | 5.0 | 4.0 | 5.0 | 4.5 |
| CR4 | 93.0 | 97.0 | 93.0 | 93.0 | 91.6 | 98.4 |
| HHI | 4,229 | 4,161 | 4,725 | 6,306 | 6,622 | 6,879 |
| *N* (>1%) | 4 | 4 | 4 | 6 | 6 | 4 |
| Noam Index | 2,115 | 2,081 | 2,363 | 2,574 | 2,703 | 3,440 |

42. Experien Hitwise Canada. 2010. Main data centre: Top 20 Sites & Engines. Last accessed on August 30, 2011 at <www.hitwise.com/ca/datacenter/main/dashboard-10557.html>.

43. Comscore. 2010. *Canada Digital Year in* Review: *Canada*. Last accessed on August 30, 2011 at <http://www.comscore.com/Press_Events/Presentations_Whitepapers/2011/2010_Canada_Digital_Year_in_Review>.

While there are no data available for Canada with respect to smartphone operating systems, US data show that the top four players in 2010 accounted for 93% of all revenues: Google's Android OS (29%), Apple's iOS (27%), RIM (27%), and Microsoft's Windows 7 (10%).[44]

The patterns for websites and online news sources in Canada are somewhat more mixed and ambiguous. With respect to the top 10 websites in Canada, we see a pattern similar to those just described, with the amount of time spent on such sites nearly doubling from 20 to 38% between 2003 and 2008, and with most of the top 15 online news sites belonging to established media companies: cbc.ca, Quebecor, CTV, Globe & Mail, Radio Canada, Toronto Star, Postmedia, and Power Corp. Accounting for almost all of the rest are CNN, BBC, Reuters, MSN, Google, and Yahoo.[45] The picture becomes more mixed and perhaps a bit more ambiguous when we turn to using the CR4 and HHI analyses. On the basis of these measures, we see relatively high levels of concentration with the CR4 method that have stayed mostly constant between 2004 and 2011, while the HHI scores suggest that online news sources have remained relatively diverse and with little change across the period assessed. The trends are illustrated in Table 17.16.

## Table 17-16. Online News Sources, 2004–2011

| NEWS WEBSITE | 2004 (*N* = 1,482) | 2007 (*N* = 1,306) | 2011 (*N* = 1,651) |
|---|---|---|---|
| CBC | 10.6 | 18.3 | 13.8 |
| Google (US) | 5.3 | 9.2 | 10.4 |
| MSN / Sympatico (US) | 18.2 | 11.0 | 14.7 |
| Yahoo (US) | 9.3 | 7.4 | 6.5 |
| CNN (US) | 9.3 | 9.4 | 6.1 |
| CTV | — | 6.2 | 2.9 |
| Canoe | 2.4 | 7.6 | 2.9 |
| Cyberpresse | 3.5 | 3.3 | 3.9 |
| Globe and Mail | 4.1 | 5.9 | 3.6 |
| BBC (UK) | — | 4.9 | 2.8 |
| Toronto Star | 2.6 | 2.4 | 1.5 |
| Global | — | — | 2.0 |
| Other | 32.6 | 14.4 | 31.1 |
| Total (CAD) | 97.9 | 100.0 | 100.2 |
| Total (USD) | 75.0 | 93.0 | 101.0 |
| CR4 | 43.4 | 45.9 | 45.4 |
| HHI | 686 | 855 | 649 |
| *N* (>1%) | 9 | 11 | 12 |
| Noam Index | 229 | 258 | 187 |

*Source:* Table calculated by Fred Fletcher, York University, from the Canadian Internet Project data sets (Charles Zamaria, director). Reports on the 2004 and 2007 surveys are available at www.ciponline.ca.

44. Nielsen. 2011. Who is winning the smartphone battle? Last accessed August 30, 2011 at http://blog.nielsen.com/nielsenwire/online_mobile/who-is-winning-the-u-s-smartphone-battle/.

45. Zamaria, Charles, & Fletcher, Fred. 2008. Canada online! Internet, media and emerging technologies: Uses, attitudes, trends and international comparisons, year two report, 2007. Toronto, ON: Canadian Internet Project. Last accessed August 30, 2011 at <http://www.ciponline.ca/en/docs/2008/CIP07_CANADA_ONLINE-REPORT-FINAL%20.pdf>.

Ultimately, relatively new online media domains appear to be no more impervious to the forces of consolidation than media in the past. As is evident in every one of the online segments just assessed, the CR4 analysis shows generally high levels of concentration. While four giants may compete ferociously among themselves in each sector, four players do not make a competitive market. As Tim Wu[46] shows, new players at different layers in the network media ecology, networks, content and applications, and devices are amassing significant clout and the ability to set the terms of trade for the music industries (Apple), for revenue distribution in the linked economy (Google, an especially important matter for newspapers, television, film, apps, books, and blogs), as well as the conditions for the harvesting and sale of personal information and user-created content (Google, Facebook, and Apple). Noam[47] argues similarly that digitization magnifies the power of economies of scale and that this is leading to a two-tiered digital media system organized around a few large integrator firms, which in turn are surrounded by many smaller, specialist firms. These are the gateways—or better, the sluices—that exist at key junctures across the digital media terrain.

This does not mean that it is time for a massive trust-busting initiative on the digital media frontier, but it is essential to know the empirical trends one way or the other, and to at least keep an open mind, if not altogether vigilant. This is especially true when dominant market power amassed in one layer of the network media ecology is used to lock in users, inhibit information flows, or stifle competition in other layers. This is the crux of debates over network neutrality. It has also been the crux of vertical integration issues since the time of the telegraph and, indeed, time immemorial. As Andrew Odlyzko[48] observes, the crux of the issues revolves around a centuries' old conflict between society's drives for economic efficiency and for fairness that has never been adequately resolved.

> There is no reason to expect that this conflict [between efficiency and fairness] will lessen, and instead there are arguments that suggest it will intensify. Should something like net neutrality prevail, the conflict would likely move to another level. That level might become search neutrality . . . or . . . if 'cloud computing' does become as significant as its enthusiasts claim . . . there might have a push for 'cloud neutrality,' and so on it goes. Obviously, there are gateways and constellations of media power criss-crossing the terrain, but whether and if anything will be done to address those issues will turn on politics and where we place the values of a free press and democracy in the general scheme of things.

## CONCLUDING THOUGHTS

Based on the aims of the International Media Concentration Research Project and my reading of Noam's *Media Ownership and Concentration in America*, four key things from this chapter stand out. First, historically and theoretically informed yet empirically driven research is badly needed. Second, the trajectory of events in Canada is similar to patterns in the United States. Concentration levels declined in the 1980s, rose sharply in the late 1990s until peaking around 2000, when they stayed relatively stable until rising significantly again between 2007 and 2011. All of this took place amidst a major increase in the size of the total network media economy. The strongest players have obtained a bigger piece of a larger pie. Third, media concentration levels in Canada are roughly double those in the United States and high by global standards, at least when compared to most developed capitalist democracies. Fourth, the decades-old vision of media convergence, where a small clutch of large media conglomerates operate across the medium and the message, appears to be far more

46. Wu, T. *The Master Switch*. Toronto: Random House. 2009.

47. Noam, E. *Media* Ownership *and Concentration in America*. New York: Oxford University, 2009, 33–39.

48. Odlyzko, A. "Network Neutrality, Search Neutrality, and the Never-Ending Conflict Between Efficiency and Fairness in Markets." *Review of Network Economics*. 8(1), 2009, 40–60.

prominent in mid-size markets such as Canada, Brazil, and Spain versus the situation in countries that have large media markets such as the United States, Germany, Japan, China, and the United Kingdom.

Changes in media ownership during the past two years have mainly been about shuffling the assets from one large but bankrupt company (Canwest) to others. Canwest's demise magnified concentration levels for newspapers because some of the smaller papers that it had previously sold off (Osprey) were later taken over by some of the largest players (QMI). The Postmedia Group's acquisition of Canwest's newspapers means that there is a new name on the ownership papers, but it does not add to diversity because there has been no net increase in the number of publishers. It does add a new voice to the overall media economy, although in terms of market share, that benefit has been negated by trends elsewhere, notably in television. Whether Postmedia will even survive is an open question, but the standing offer of buyouts to all of its journalists and editors and its precarious stock market valuation do not auger well for the company, the investors, or the future of journalism in Canada.

Concentration in conventional as well as pay and specialty cable and satellite channels has also grown significantly since 2008, mostly because Canwest's television holdings were acquired by a company that was already a big player in the industry, Shaw Media. Bell's takeover of CTV in 2011 pushed the market share of the four biggest television providers (Bell, Shaw, CBC, and Rogers) to a little over 81% in 2011, from 62% just five years earlier. Bell, Shaw, Rogers, and QMI (but not Astral) are all vertically integrated, meaning that they have substantial operations in both distribution networks (telecom, wireless, cable and satellite, and Internet) and broadcasting, and in magazines and newspapers in the case of the latter two, respectively. There is really no debate in Canada about whether these companies use their control over the medium (networks) to influence the flow of messages through them. They do, by throttling the availability of bandwidth to users and hobbling the ability of competitive OVDs such as Netflix to survive, since the bandwidth caps restrict the number of film and television programs that subscribers can download per month (about 10 hours). In contrast, online video services offered by Bell and Shaw are exempt from the bandwidth caps. The only debate in Canada is whether they should be able to act in this way or if there should be a mandatory set of principles and/or structures put into place to separate control over the medium from the message, in the form of network neutrality, common carrier principles, or structural separation. In this one cannot help but have a sense of déjà vu, with the CRTC standing exactly where its predecessor, the BRC, stood a century ago, albeit far less likely to act.

In sum, media concentration has not gone away and, furthermore, this helps explain why ISPs, Google, Facebook, Apple, and (for a time) RIM have become the powerful forces that they are, standing at crucial junctures in the digital media ecology, between traditional and new media, and all along the way. If media history tells us anything, it is that once the structures of a new medium are cemented into place they stay that way for a long time. Indeed, the structure of the industrial media age set down in the late 19th and early 20th centuries has only begun to give way to the network media ecology of the 21st century in the past decade with no small amount of resistance from entrenched interests all along the line.

## Bibliography

### 1. Reports


Bloomberg, United States. "Bloomberg Professional." Bloomberg, United States, 2010.

Canadian Broadcasting Corporation, Canada. "Annual Reports." Canadian Broadcasting Corporation, Canada.

Canadian Newspaper Association (CNA) / Newspaper Canada, Canada. "Newspaper Facts." Canadian Newspaper Association (CNA) / Newspaper Canada, 2000–2008.

Canadian Newspaper Association (CNA), Canada. "Ownership of Canadian Newspapers." Canadian Newspaper Association (CNA), Canada, 2003.

CRTC, Canada. "Communications Monitoring Report 2008–2011." CRTC, Canada, 2011.
CRTC, Canada. "Status of Competition in Canadian Telecommunications Markets 2000–2007." CRTC, Canada, 2007.
CRTC, Canada. "Broadcasting Policy Monitoring Report 2000–2007." CRTC, Canada, 2007.
CRTC, Canada. "Broadcast Monitoring Report and Pay and Specialty Statistical and Financial Summaries." CRTC, Canada, 2007.
Canadian Wireless Telecommunications Association, Canada. "Mobile Wireless Subscribers in Canada." CRTC, Canada.
Financial Post, Canada. "Survey of Industrials." Financial Post, Canada.
FPInfomart, Canada. "Historical Profiles." FPInfomart, Canada.
Friends of Canadian Broadcasting, Canada. "Change in Parliamentary Appropriation." Friends of Canadian Broadcasting, Canada. 2004. Dec. 4, 2008. http://friends.ca/files/PDF/CBCgrant04.pdf>
Internet Advertising Bureau, Canada. "2010 Actual & 2011 Estimated Canadian Online Advertising Revenue Survey." Internet Advertising Bureau, Canada, 2010. Sept. 20, 2010. http://www.iabcanada.com/wp-content/uploads/2011/07/IABCda_2010Act2011Bdg_ONLINEAdRevRpt_FINAL_Eng.pdf>
Parliament of Canada, Senate Committee on Transportation and Communication, Canada. "Interim Report on Canadian News Media." Parliament of Canada, Senate Committee on Transportation and Communication, Canada, 2003.

## 2. Newspapers

PriceWaterhouseCoopers (PWC). (2010). *Global entertainment and media outlook, 2010–14* (plus previous editions between 2000 and 2009). New York: PWC.
Statistics Canada. "Cansim Tables NAICS-defined sectors: 51111 Newspaper Publishers; 51112 Periodical Publishers; 51221-51223 Record Production, Distribution and Music Publishers; 51511 Radio Broadcasting; 51512 Television Broadcasting; 5152 Pay and Specialty Television; 51711 Wired Telecommunications Carriers; 517112 Cable and Other Program Distribution; 5172 Wireless Carriers (except Satellite); 51913 Internet Publishing and Broadcasting, and Web Search Portals."

# Canada—Data Summaries

ELI NOAM AND PAUL MUTTER

CANADA'S INDIVIDUAL MEDIA INDUSTRIES are not particularly highly concentrated if international comparison is the yardstick (Table 17.17). It is the cross-ownership among industries that is high. Four companies—Bell Canada, Telus, Rogers, and Shaw—account for 69.1% of the overall media sector, the highest among the more developed countries in this study. Bell Canada Enterprises/CTV is the largest (28.8%) and most diversified, with holdings in nine media industries, especially in the platform industries of telecom, cable, and ISP. Bell/CTV is both the largest platform company (34.8% of the platform sector) and content company (16.5% of the content sector) in Canada, well ahead of its competitors.

Concentration measures increased in the 1990s as these two major telecom and cable companies consolidated their market positions. Bell Canada's company power index increased to 1,873.5 from 1,231.1, while Rogers' grew from 225 to 465. Rogers displaced Telus as the second largest media provider in Canada in 2004 thanks to its wireless business, but in platform media, Telus is the larger company (20.7% compared to 20.2%) because its wireline business is 10 times larger than Shaw's by revenue (Table 17.19).

Shaw's position increased slightly when it acquired the broadcasting arm of the bankrupt Canwest, which had previously operated Canadian video channels such as the Global Television Network and HGTV Canada. Canwest's newspapers division, which included the *National Post*, was sold to a group of investors who founded Postmedia. Postmedia also owns regional newspapers, including the *Vancouver Sun*, the *Ottawa Citizen*, and the *Gazette* (in Montreal), as well as the online news site Canada.com.

Another major player in the Canadian newspapers industry is Torstar, best known for its flagship publication, the *Toronto Star*.

Table 17-17. National Media Industries Concentration in Canada

| | 2004/5 | | 2011 OR MOST RECENT | | % CHANGE ANNUAL AVERAGE | |
|---|---|---|---|---|---|---|
| | COMPANY POWER INDEX IN COUNTRY | COMPANY SHARE OF THE OVERALL NATIONAL MEDIA MARKET (%) | COMPANY POWER INDEX IN COUNTRY | COMPANY SHARE OF THE OVERALL NATIONAL MEDIA MARKET | COMPANY POWER INDEX IN COUNTRY | COMPANY SHARE OF THE OVERALL NATIONAL MEDIA MARKET (%) |
| Bell Canada Enterprises/CTV | 1,231.1 | 30.1 | 1,837.5 | 28.8 | 7.0 | −0.2 |
| Rogers | 225.0 | 10.4 | 465.0 | 16.6 | 15.2 | 0.9 |
| Telus | 290.1 | 12.5 | 338.9 | 14.1 | 2.4 | 0.2 |
| Shaw | 122.0 | 5.4 | 241.3 | 9.6 | 14.0 | 0.6 |
| Google (US) | 14.0 | 0.2 | 99.2 | 1.2 | 87.2 | 0.1 |
| Canadian Broadcasting Corporation (CBC) (public) | 82.7 | 3.2 | 80.2 | 2.7 | −0.4 | −0.06 |
| Quebecor | 59.0 | 4.0 | 60.7 | 5.2 | 0.4 | 0.2 |
| Postmedia | 72.6 | 2.6 | 47.1 | 1.6 | −5.0 | −0.1 |
| Canwest | 26.0 | 1.4 | 0.0 | 0.0 | −14.3 | −0.2 |
| Astral | 4.9 | 0.8 | 32.3 | 2.0 | 80.8 | 0.2 |
| Torstar | 17.4 | 1.3 | 6.6 | 0.6 | −8.9 | −0.09 |
| Time Warner (US) | 12.2 | 0.7 | 4.8 | 0.2 | −8.6 | −0.06 |
| Murdoch Interests (US) | 10.6 | 0.5 | 2.1 | 0.2 | −0.1 | −0.04 |

*continued*

Table 17-17. *continued*

| | 2004/5 | | 2011 OR MOST RECENT | | % CHANGE ANNUAL AVERAGE | |
|---|---|---|---|---|---|---|
| | COMPANY POWER INDEX IN COUNTRY | COMPANY SHARE OF THE OVERALL NATIONAL MEDIA MARKET (%) | COMPANY POWER INDEX IN COUNTRY | COMPANY SHARE OF THE OVERALL NATIONAL MEDIA MARKET | COMPANY POWER INDEX IN COUNTRY | COMPANY SHARE OF THE OVERALL NATIONAL MEDIA MARKET (%) |
| Microsoft (Bing) (US) | 1.1 | 0.08 | 0.8 | 0.1 | −3.8 | 0.004 |
| Yahoo! (US) | 0.8 | 0.05 | 0.3 | 0.06 | −9.8 | 0.001 |
| Media Concentration Index | 2004/5 | | 2011 or Most Recent | | % Change Annual Average | |
| Total Revenue: Nat'l Media Industry (mil US$) | 43,072 | | 73,733 | | 10.2 | |
| Total Voices (*n*) | 81 | | 78 | | −0.5 | |
| Net Voices (*n*) | 54 | | 50 | | −1.1 | |
| Public Ownership (%) | 3.2 | | 2.7 | | −0.06 | |
| Foreign Ownership (%) | 2.6 | | 2.5 | | −0.03 | |
| C4 Average—Weighted | 77.5 | | 81.7 | | 0.6 | |
| HHI Average—Weighted | 2,283 | | 2,506 | | 1.4 | |
| C1 Average—Weighted | 36 | | 37 | | 0.0 | |
| Noam Index Average—Weighted | 492 | | 767 | | 8.0 | |
| Pooled Overall Sector C4 | 58.4 | | 69.1 | | 1.5 | |
| Pooled Overall Sector HHI | 1,245 | | 1,449 | | 2.4 | |
| Pooled Overall Sector Noam Index | 68 | | 119 | | 10.9 | |
| Market Share of Top 10 Companies: Nat'l Media Industry (%) (Pooled C10) | 73.1 | | 84.4 | | 1.6 | |
| National Power Index | 2,211 | | 3,251 | | 6.7 | |

Table 17-18. Top Content Media Companies in Canada

| | 2004/5 | | 2011 OR MOST RECENT | | % CHANGE ANNUAL AVERAGE | |
|---|---|---|---|---|---|---|
| | COMPANY POWER INDEX IN COUNTRY | COMPANY SHARE OF THE NATIONAL CONTENT MEDIA MARKET (%) | COMPANY POWER INDEX IN COUNTRY | COMPANY SHARE OF THE NATIONAL CONTENT MEDIA MARKET (%) | COMPANY POWER INDEX IN COUNTRY | COMPANY SHARE OF THE NATIONAL CONTENT MEDIA MARKET (%) |
| Bell Canada Enterprises/CTV | 175.6 | 11.1 | 611.9 | 16.5 | 35.5 | 0.8 |
| Google (US) | 39.5 | 0.7 | 303.7 | 3.7 | 95.4 | 0.4 |
| CBC (public) | 234.2 | 9.0 | 245.5 | 8.3 | 0.7 | −0.09 |
| Quebecor | 136.9 | 8.2 | 116.6 | 7.7 | −2.1 | −0.08 |
| Postmedia | 205.5 | 7.3 | 144.3 | 4.9 | −4.3 | −0.3 |
| Shaw | 178.4 | 8.3 | 485.1 | 16.7 | 24.6 | 1.2 |
| Rogers | 126.9 | 8.0 | 124.8 | 9.0 | −0.2 | 0.2 |
| Canwest (until 2010) | 73.7 | 4.0 | 0.0 | 0.0 | −14.3 | −0.6 |
| Astral | 13.7 | 2.2 | 99.0 | 6.2 | 88.5 | 0.6 |
| Torstar | 49.3 | 3.6 | 20.2 | 1.8 | −8.4 | −0.2 |
| Murdoch Interests (US) | 30.1 | 1.3 | 6.4 | 0.5 | −0.1 | −0.1 |
| Media Concentration Index | 2004/5 | | 2011 or Most Recent | | % Change Annual Average | |
| Public Ownership (%) | 9.0 | | 8.3 | | −0.09% | |
| Foreign Ownership (%) | 7.5 | | 97.5 | | 0.01% | |
| C4 Average—Weighted | 67.4 | | 75.4 | | 1.2% | |
| HHI Average—Weighted | 1,475 | | 2,046 | | 5.5% | |
| C1 Average—Weighted | 24 | | 31 | | 1% | |
| National Power Index | 1,363 | | 2,215 | | 8.9% | |

Table 17-19. Top Platform Media Companies in Canada

| | 2004/5 | | 2011 OR MOST RECENT | | % CHANGE ANNUAL AVERAGE | |
|---|---|---|---|---|---|---|
| | COMPANY POWER INDEX IN COUNTRY | COMPANY SHARE OF THE NATIONAL PLATFORM MEDIA MARKET (%) | COMPANY POWER INDEX IN COUNTRY | COMPANY SHARE OF THE NATIONAL PLATFORM MEDIA MARKET (%) | COMPANY SHARE OF THE NATIONAL PLATFORM MEDIA MARKET (%) | COMPANY POWER INDEX IN COUNTRY |
| Bell Canada Enterprises/CTV | 1,807.7 | 40.5% | 2,507.1 | 34.8 | 5.5 | −0.8 |
| Rogers | 278.6 | 11.7 | 630.1 | 20.2 | 18.0 | 1.2 |
| Telus | 448.6 | 19.4 | 502.3 | 20.7 | 1.7 | 0.2 |
| Shaw | 91.1 | 3.8 | 123.1 | 6.2 | 5.0 | 0.3 |
| Quebecor | 16.4 | 1.7 | 33.5 | 4.0 | 14.9 | 0.3 |
| Media Concentration Index | 2004/5 | | 2011 or Most Recent | | % Change Annual Average | |
| Public Ownership (%) | 0.0 | | 0.0 | | 0.0% | |
| Foreign Ownership (%) | 0.0 | | 0.0 | | 0.0% | |
| C4 Average—Weighted | 83.0 | | 84.7 | | 0.2% | |
| HHI Average—Weighted | 2,725.0 | | 2,729 | | 0.02% | |
| C1 Average—Weighted | 42.0 | | 39.0 | | 0.0% | |
| National Power Index | 2,675 | | 3,828 | | 6.2% | |

Through its Metroland Media Group branch, Torstar also publishes regional daily newspapers, including the *Hamilton Spectator*, the *Waterloo Region Record*, and the *Guelph Mercury*. Outside of newspapers, Torstar operates Harlequin Enterprises, a book publisher that specializes in paperback romantic novels. Canadian companies frequently operate across several media as a result of Canadian policy toward concentration and cross-ownership. Canada permits high domestic concentration in order to deal with US firms, which are basically excluded from the market. This double insulation has yielded and shielded large Canadian media firms.

Between 2004 and the most recent period, the average C4 increased from 77.5% to 81.7%, and the weighted HHI increased by 1.4% per annum in the period from 2004 to the most recent period, but the rate of growth is slower than it was in the 1990s.

On the positive side, smaller voices have not been extinguished. Compared to the rest of the countries surveyed, Canada has one of the higher sets of net voices with 50 in 2009 or most recent. And once one looks only at content companies (which include, to remind the reader, one-third of overall cable TV revenues as the share of content, while two-thirds are allocated to platforms), there are two fairly equal-sized large companies, Bell Canada with 16.5%, and Rogers with 16.7%, and there are four fairly equal-sized companies, only one of whom has more than 10% (Table 17.18).

# 18

# Media Concentration in the United States

ELI NOAM

## INTRODUCTION

Recent years have witnessed the expansion of large media firms in the United States. This development has led to fears that American communications media are increasingly controlled by an ever-shrinking number of firms, with global ramifications due to the influence and size of American media firms.

To recapitulate the introduction: On one side of the debate are "media pessimists." Ben Bagdikian, formerly the dean of Journalism at Berkley and Pulitzer Prize award winner, concluded in 2004, "five global-dimension firms, operating with many of the characteristics of a cartel, own most of the newspapers, magazines, book publishers, motion picture studios, and radio and television stations in the United States."[1] These firms were identified as Time Warner, Viacom, Disney, News Corp., and Bertelsmann. Lawrence Lessig, the noted Harvard law professor, further reduced that number of firms: "Indeed, after the changes that the FCC announced in June 2003, most expect that within a few years, we will live in a world where just three companies control more than 85% of the media."[2]

On the other side is the "media optimist" view, whose most comprehensive expression is the Cato Institute's Adam Thierer, who concluded: "To the extent that there was ever a 'Golden Age' of media in America, we are living in it today. The media sky has never been brighter and is getting brighter with each passing year."

Given these diametrically opposed perspectives on the media landscape, one would expect that an empirical analysis would be undertaken to settle the questions. The literature, while

1. Bagdikian, Ben. *The New Media Monopoly.* Boston: Beacon Press, 2004, p. 3.

2. Lessig, Lawrence. *Free Culture: How Big Media Uses Technology and the Law to Lock Down Culture and Control Creativity.* New York: Penguin, 2004.

lively, is surprisingly short of systematic and data-based analysis.[3]

The first American media mogul was Benjamin Franklin, one of the country's founders,[4] who owned seven newspapers, plus magazine and book publishing operations, at the same time that he also functioned for a while as the Postmaster General of his state and later the country. The subsequent history of media followed the pattern, with large firms such as Western Union, AT&T, William Randolph Hearst, Frank Munsey, the Edison Trust, IBM, and others. Thus, media concentration is not a new phenomenon in the United States. On the contrary, it has a long and contentious history, commented upon and fought over at each stage in the country's past. It is one of the fundamental issues of distribution of power and wealth that each generation needs to resolve.

The report that follows aims to provide the data for a reasoned and fact-based discussion based on the concentration trends for 13 media industries. It draws in part on the author's past work on US media concentration[5] and adds several major elements: updates (extending 2005 data to 2012-2014); the addition of several industries (online news and search engines); the addition of several sectoral aggregations such as for telecom, video platforms, and film/TV production; several cross-industry calculations such as the national power index and company power index figures; foreign and public ownership figures; and ownership information. We will begin with the analyses of the newspaper industry.

## PRINT MEDIA

### Newspapers

Newspapers experienced their peak in the early 20th century. The emergence of radio and later television was a major factor in a stagnation of newspaper circulation after World War II. The number of daily newspapers dropped from 2,042 in 1920 to 1,611 in 1990, 1,533 in 1995, 1,437 in 2006, 1,397 in 2009, and 1,382 in 2011.[6] In international comparison, this is a high but declining number. Despite a national population increase of almost one-third, circulation declined from 62.8 million in 1985 to 55 million in 2003, 52 million in 2006, 46 million in 2009,[7] and 41.7 million in 2011,[8] a drop of 33.6%.

The high fixed cost of operations ("first copy cost") in editorial and advertising sales created economies of scale and became the force behind local monopolies and regional chain ownership that replaced most independently owned competitive local daily newspapers. By 2000, 77% of all daily newspapers were owned by chains, though there were 119[9] of these, not a small number. The 10 largest chains owned in 2000 18% of newspaper titles and about 40% of the market by circulation.[10] The circulation share of the 10 largest chains declined somewhat, from 39.4% of circulation in 1984 to 39.9% in 2000 and to 35.4% in 2009.

Newspapers typically operate either locally/regionally or nationally. In contrast to many other countries, and partly due to the size of America, national newspapers did not exist in the United

---

3. Compaine, Benjamin, and Douglas Gomery. *Who Owns the Media?* 2000. Major exceptions are Noam, Eli, *Media Ownership and Concentration in America*. New York: Oxford University Press.

4. Isaacson, Walter. *Benjamin Franklin: An American Life*. New York: Simon & Schuster, 2003, p. 126.

5. Noam, Eli, *Media Ownership* and *Concentration in America*. Oxford University Press, 2009.

6. "Number of Daily Newspapers." *The State of the News Media 2006: An Annual Report on American Journalism*. Journalism.org. Last accessed at http://stateofthemedia.org/2013/newspapers-stabilizing-but-still-threatened/newspapers-by-the-numbers/.

7. "Total Paid Circulation." 2003, 2007. Newspaper Association of America. Last accessed at http://www.naa.org/info/facts04/circulation-daily.html.

8. "A Continued Slide in Paid Circulation." 2012. *Sate of the Media*. Last accessed at http://stateofthemedia.org/files/2012/01/15-Newspaper-A-Continued-Slide-in-Paid-Circulation.png.

9. "Newspaper Ownership." *The State of the News Media 2004: An Annual Report on American Journalism*. Last accessed at http://www.stateofthenewsmedia.org/chartland.asp?id=205&ct=pie&dir=&sort=&col2_box=1.

10. Donald, William H. "Publishing: Industry Profile." Standard & Poor's Industry Surveys (May 3, 2001).

States until recently. Only a few newspapers are distributed nationally, namely *USA Today* (Gannett), *Wall Street Journal* (News Corp.), and *The New York Times*, and none of these titles has a large market share overall. For the most part, the market for newspapers continues to be local and regional. In those markets, a major long-term trend has been the decline of competition. In most US cities, newspapers operate in a near-monopolistic market structure. In 2000, only 20 American cities were served by two or more separately owned, competing full-service dailies. In 2010 that number had declined to 16 (this does not include, however, free but skimpy competing newspapers handed out to commuters).

Table 18.1 provides revenue and national circulation market share figures for the largest newspaper chains. The C4 and C8 index figures reveal a relatively level trend in market concentration of the largest newspaper chains, at about one-quarter and one-third, respectively. In 2013 the national HHI was a low 304, slightly rising after 2000.

The market shares of the biggest firms are modest nationally. Gannett, by far the biggest firm, held a market share of about 9.6% in 2013. It also owns TV and radio stations. The second tier of six companies, having about 3-8% of market circulation, includes McClatchy (5.7% after it acquired 20 of the 32 Knight-Ridder papers in 2006).[11] The Tribune Company (5.9%) was for a while number three after it had bought Times-Mirror in 2000, thus combining the Chicago and Los Angeles markets. However, the company went into financial distress and bankruptcy in 2012 and was broken up. Dow Jones & Co. (8.3% with secondee *The Wall Street Journal*) was bought by Rupert Murdoch's News Corp. in 2007 for $5 billion. Moving into the newspaper business was an unusual move for one of the top media firms. Other large newspaper firms included The New York Times Co (3.5%), which bought the *Boston Globe* in 1993 for US$1.1 billion. That acquisition proved spectacularly unsuccessful, and the New York Times sold it in 2013 for a mere $70 million after losing money for years. Advance Publications (4.8%, owner of the Condé Nast magazine as well as radio and TV stations) and Media News (2.7%, part-owned by Hearst, part-owned by a hedge fund after a 2010 Chapter 11 bankruptcy) are two other large newspaper groups.

The national market share, however, is only one dimension of power. First, some newspapers, such as the *New York Times* and the *Wall Street Journal*, hold much more influence than their firms' circulation shares suggest. Second, and more significant in business terms, a firm's market share might be small nationally but very high locally. As mentioned, the US newspaper industry is mostly a series of parallel local monopolies. We thus report average local concentration too. Table 18.2 shows local concentration based on 30 markets, 10 each for large, medium, and small cities. The share of the top firm (C1) rose from an average 80.3% (60.1% in large cities) to 84.5% (65.8 in large cities and 93.7% in small cities). The HHI measure of local concentration is a huge—7,755 on average (weighted) and higher still in small and medium cities. Even in the big cities it is 5,398.

The city size needed to generally assure a single daily paper in the year 2000 was above 100,000 population, whereas in 1980 it was half of that. Multiple papers on average require a population of more than one million, whereas it was half a million in 1980.

Given the cost and revenue trends, one should expect a slow shift in the United States from local to national newspapers, as it exists in the United Kingdom, France, Italy, and Japan, for example. On the other end of the spectrum of size, there seems to be room for "hyperlocal" papers that provide community news, though not on a daily basis. It is the in-between regional or metropolitan newspapers that are most under pressure.

---

11. McClatchy bought all 32 Knight-Ridder papers and kept 20, including the *Miami Herald* and the *Charlotte Observer*, but sold the *Philadelphia Inquirer* and the *San Jose Mercury News*.

Table 18-1. Daily Newspapers (Market Shares by Circulation)

| – | 1984 | 1988 | 1992 | 1996 | 2000 | 2002 | 2004 | 2006 | 2007 | 2009 | 2013 |
|---|---|---|---|---|---|---|---|---|---|---|---|
| Gannett | 7.6 | 8.6 | 9.5 | 10.5 | 11.0 | 10.3 | 9.9 | 10.2 | 8.9 | 10.0 | 9.6 |
| McClatchy | | | | | 2.0 | 1.9 | 1.9 | 6.5 | 5.4 | 5.9 | 5.7 |
| Knight Ridder | 5.8 | 6.2 | 6.5 | 6.9 | 5.8 | 5.3 | 4.6 | McClatchy | | | |
| Tribune[a] | 4.3 | 3.6 | 2.9 | 2.2 | 5.5 | 4.8 | 4.2 | 4.3 | 4.4 | 3.5 | 5.9 |
| Times Mirror | 4.0 | 4.0 | 4.1 | 4.1 | Tribune | | | | | | |
| Advance | | | | | 4.4 | 3.8 | 3.6 | 3.7 | 3.0 | 3.6 | 4.8 |
| Dow Jones (News Corp)[b] | 3.5 | 3.7 | 3.9 | 4.1 | 3.6 | 3.1 | 3.4 | 3.5 | 5.3 | 3.4 | 8.3 |
| New York Times Co.[c] | 2.8 | 3.2 | 3.6 | 4.0 | 3.6 | 3.1 | 2.9 | 3.0 | 3.0 | 3.0 | 3.5 |
| Media News | 0.0 | 0.8 | 1.7 | 2.5 | 2.7 | 2.2 | 2.6 | 2.6 | 2.5 | 3.8 | 2.7 |
| Hearst | 1.6 | 1.8 | 2.1 | 2.3 | 2.5 | 2.3 | 2.2 | 2.3 | 2.4 | 1.9 | 2.8 |
| Cox | 1.9 | 1.9 | 2.0 | 2.0 | 1.8 | 1.6 | 1.5 | 1.5 | 1.5 | 1.5 | 2.9 |
| Washington Post | 1.8 | 1.8 | 1.7 | 1.6 | 1.6 | 1.1 | 1.0 | 1.0 | 1.3 | 0.9 | 1.8 |
| Others | 66.7 | 66.2 | 63.7 | 59.8 | 55.5 | 60.5 | 62.2 | 61.4 | 62.3 | 62.5 | 49.6 |
| US Total Revenues ($ mil) | 25,170 | 32,280 | 30,639 | 38,075 | 48,670 | 44,100 | 46,700 | 45,690 | 44,289 | 37,848 | 32,109 |
| In 2009 $ (mil) | 51,972 | 58,540 | 46,851 | 52,062 | 60,036 | 52,591 | 53,038 | 48,622 | 45,826 | 37,848 | — |
| US Population (mil) | 235.8 | 244.5 | 255.0 | 265.2 | 281.4 | 290.0 | 294.0 | 300.5 | 303.0 | 310.0 | 313.0 |
| C4 | 21.7 | 22.4 | 24.0 | 25.6 | 26.7 | 24.2 | 22.3 | 24.7 | 24.0 | 23.3 | 29.2 |
| C8 | 31.7 | 33.0 | 34.6 | 36.6 | 39.1 | 34.9 | 33.4 | 36.1 | 34.9 | 35.1 | 45.1 |
| HHI | 155 | 176 | 200 | 230 | 254 | 208 | 188 | 215 | 190 | 202 | 304 |
| *N* (>1%) | 9 | 9 | 10 | 10 | 11 | 11 | 11 | 10 | 10 | 9 | 10 |
| MOCDI Index | 49 | 53[d] | 58 | 67 | 76 | 63 | 57 | 68 | 60 | 67 | 96 |

*Source*: Herrick, Dennis F. *Media Management in the Age of Giants*. Ames: Iowa State Press, 2003, pp. 292–293. Data for 1984–2000. Data for 2000 to 2004 from "The state of the news media, 2006." *The Project for Excellence in Journalism*. Journalism.org. Last accessed on Jan. 8, 2008 at http://www.stateofthenewsmedia.com/2006/contact.asp. Data for 2006 are used to show the effect of the McClatchy/Knight Ridder merger in 2006, using website. http://www.mcclatchy.com/news/2006/story/7528982p-7440749c.html. 2007 revenue from IBISworld.com; circulation figures from The Audit Bureau of Circulation, 2007–8.

[a] In 2007, the Tribune Co. was taken private by real estate investor Sam Zell. In 2008, the company declared chapter 11 bankruptcy.

[b] In 2007, Dow Jones & Co. was acquired by News Corp.

[c] In 2013, The New York Times Co. sold the *Boston Globe*.

[d] 2013 market shares for McClatchy, Gannett, News Corp., The New York Times, and Tribune provided by IBIS world.

Table 18-2. Newspaper Local Concentration

| | 1984 | 1988 | 1992 | 1996 | 2002 | 2006 | 2009 |
|---|---|---|---|---|---|---|---|
| | | | C1 CONCENTRATION | | | | |
| C1 Large Cities | 60.1 | 59.4 | 58.9 | 64.0 | 64.6 | 62.0 | 65.8 |
| C1 Medium Cities | 93.3 | 93.6 | 97.3 | 97.4 | 97.6 | 97.5 | 94.7 |
| C1 Small Cities | 94.8 | 94.9 | 95.0 | 95.2 | 94.4 | 97.3 | 93.7 |
| Weighted Average | 80.3 | 80.2 | 81.1 | 83.3 | 83.4 | 83.1 | 84.5 |
| | | | C4 CONCENTRATION | | | | |
| C4 Large Cities | 98.3 | 97.4 | 95.9 | 96.3 | 96.4 | 96.1 | 89.7 |
| C4 Medium Cities | 100.0 | 100.0 | 100.0 | 100.0 | 100.0 | 100.0 | 100.0 |
| C4 Small Cities | 100.0 | 100.0 | 99.6 | 99.8 | 100.0 | 100.0 | 100.0 |
| Weighted Average | 99.3 | 99.0 | 98.2 | 98.4 | 98.6 | 98.4 | 91.7 |
| | | | HHI CONCENTRATION | | | | |
| HHI Large Cities | 5,047 | 5,081 | 4,996 | 5,571 | 5,562 | 5,464 | 5,398 |
| HHI Medium Cities | 9,064 | 9,083 | 9,588 | 9,602 | 9,629 | 9,622 | 9,679 |
| HHI Small Cities | 8,267 | 8,271 | 8,280 | 8,311 | 8,325 | 8,670 | 8,735 |
| Weighted Average | 7,219 | 7,239 | 7,367 | 7,612 | 7,621 | 7,676 | 7,755 |

*Source:* Local newspaper revenues are based on *Duncan's Radio Market Guide*, 1985, 1989, 1994, 1997, 2002, 2005, and 2008 editions, Gale Broadcasting & Publishing Directory, annual editions. 2009 data: Circulation numbers from Newspaper Association of America, http://www.naa.org/artpage.cfm?AID=1610&SID=1022; Audit Bureau of Circulations, http://abcas3.accessabc.com/ecirc/newstitlesearchus.asp; BurrellesLuce, "Top 100 Daily Newspapers in the U.S. by Circulation," http://www.burrellesluce.com/top100/2005_Top_100List.pdf.

## Book Publishing

Most of the largest publishing companies in the United States were founded in the late 19th century. The book publishing industry evolved into a three-tier industry consisting of a small group of large publishers, a second group of medium-sized houses, and thousands of small publishers. In spite of large acquisition deals in the book industry, the market remained relatively unconcentrated. The C4 concentration level rose to 33.1% in 2009, and the HHI to 365. But these are not especially high numbers.

The number of new titles published each year is one indicator of industry performance. That number remained relatively stable between 1910 and 1950, at around 10,000 titles per year. Output of new titles began to increase enormously after 1950, reaching 51,000 by 1994, 122,000 in 2000, 275,000[12] in 2004,[13] 302,000 in 2009,[14] and 328,259 in 2011.[15]

The German firm Bertelsmann (Random House) was the largest US publisher in 2009 with 10.3% of revenues. Its principal American imprints include Alfred A. Knopf, Anchor,

12. *Bowker New Book Titles and Editions, 2002–2010*. Bowker. Last accessed on August 2, 2012, http://www.bowkerinfo.com/pubtrack/AnnualBookProduction2010/ISBN_Output_2002-2010.pdf.

13. Bowker Annual 2004, p. 508.

14. *Bowker New Book Titles and Editions, 2002–2010* (and earlier editions). Bowker. Last accessed on August 2, 2012, http://www.bowkerinfo.com/pubtrack/AnnualBookProduction2010/ISBN_Output_2002-2010.pdf.

15. "Publishing Market Shows Steady Title Growth in 2011 Fueled Largely by Self-Publishing Sector." Bowker. Last accessed on October 8, 2013, http://www.bowker.com/en-US/aboutus/press_room/2012/pr_06052012.shtml. Bowker, the source for these numbers, changed its cataloguing system in 1999, which resulted in a steep increase in the number of published titles.

Ballantine, Bantam, Crown, Delacorte, Dell, Doubleday, Fawcett, Fodor's, Pantheon, and Random House. Until 2012, when it merged its book publishing with Pearson (7.3%) in a 53:47 joint venture, Bertelsmann had no real presence in the large educational and professional book markets. Penguin Random House, the combined firm, holds 17.6% of the US book market shares. Other major firms include Scholastic (5.4%), one of the world's leading children's book publishers. Reed Elsevier (5%) serves the professional market. News Corp (Harper Collins, 7.3%) also has a strong position.

Market concentration is low, with an HHI of 511 in 2013, rising from 138 over 30 years. During that period the C4 rose from about 21% to 39% (Table 18.3).

There was a strong presence of foreign publishers in the top 10 (Bertelsmann, Pearson, Reed Elsevier, Holtzbrinck, and Lagardére), accounting for 28.5% of US book publishing.

While the publishing of books became global, their retailing went national. The 1960s saw the decline of small local book stores and the emergence of large national bookstore chains, which by the 1990s achieved dominance. The combined market share of the four largest retailers increased from 7% in 1963 to 26% in 1987, 26% in 2004, and 33% in 2011. The face of the bookselling market changed yet again significantly after 1995 as a result of the launch of Amazon.com, the first major electronic book retailer. Amazon became a provider for e-retailing more generally. In the online book retail market, its share was 77% in 2004 and 74% in 2009. In 2011, Barnes & Noble and Amazon were the two largest overall book retailers, with shares of 16% and 27%.[16] That year, the second largest physical bookstore chain, Borders, went bankrupt.

## Magazine Publishing

The magazine industry has been traditionally competitive, with fairly low barriers for entry. In the period between 1954 and 1980, the number of periodicals doubled. The share of the top four and eight publishers also doubled. These firms publish a bouquet of magazines. Group ownership has advantages in printing, physical distribution, promotion to advertisers, and cross-promotion. There are fewer benefits on the content side. But the advantages of scale cannot be very large beyond some point, because the periodical publishing industry historically has shown relatively less concentration of ownership than most other media industries.

The 1980s and 1990s constituted a period of heavy merger-and-acquisition activity. Between 1984 and 2001, at least 18 magazine groups were acquired by their competitors, not always with success.[17]

During the 1980s, foreign firms became active in the acquisition of US magazines. The number of foreign companies with magazine operations in the United States increased from one or two in the late 1970s to 20 or more in the late 1980s. Major foreign owners of US magazines and periodicals were Lagardére/Hachette, Bertelsmann, Reed Elsevier, Holtzbrinck, and Bonnier.

Table 18.4 shows the concentration trends of the magazine industry from 1984 to 2013. Time Inc. (formerly Time Warner) had been the largest magazine company in the United States for a long time with 8.6% in 2012, down from its peak of 14.3% in 2004. In 2007, Time Warner sold 18 of its magazines (out of 130) to the large Swedish publishing firm Bonnier. In 2014, Time Warner spun off of its magazine group Time Inc., the original foundation of the company, after seeking to sell it to another magazine company. This, by itself, did not affect the size of Time Inc.'s magazine operations, but it severed its cross-ownership ties with Time Warner's other media operations. Advance Publications (Condé Nast, owned by the Newhouse family) became the market leader (9.2%), and Hearst third with 7.5%.

16. Milliot, Jim. "As E-books Grow, So Does Amazon." PublishersWeekly.com. Publisher's Weekly, Feb. 8, 2013. Web. Last accessed on July 8, 2013, http://www.publishersweekly.com/pw/by-topic/digital/retailing/article/55888-as-e-books-grow-so-does-amazon.html.

17. Endicott, Craig. "The Advertising Age 100." Advertising Age (August 14, 1995). See also tables 8–16.

Table 18-3. US Book Publishing (Market Shares by Revenues)

| COMPANY | 1984 | 1988 | 1992 | 1996 | 2001 | 2004 | 2006/7 | 2009 | 2013 |
|---|---|---|---|---|---|---|---|---|---|
| Penguin Random House (Bertelsmann, Germany, 58%; Pearson, UK, 47%) | | | | | | | | | 17.6 |
| Bertelsmann (Random House, Bantam, Doubleday) (Germany) | 4.1 | 4.6 | 3.8 | 3.6 | 6.6 | 5.4 | 9.7 | 10.3 | Bertelsmann-Pearson |
| Random House | 4.9 | 5.2 | 4.6 | 4.9 | Bertelsmann | | | | |
| Pearson (Penguin/Addison) (UK) | 3.5 | 3.4 | 5.0 | 9.6 | 6.8 | 5.9 | 4.9 | 7.3 | Bertelsmann-Pearson |
| McGraw Hill | 5.6 | 4.5 | 3.5 | 5.1 | 5.5 | 8.9 | 8.4 | 8.2 | |
| News Corp (Harper Collins) | | 1.6 | 1.8 | 3.3 | 3.6 | 3.1 | 6.7 | 7.3 | |
| Scholastic | 2.5 | 1.6 | 2.0 | 3.4 | 5.1 | 6.0 | 5.0 | 5.4 | |
| Grolier | 0.7 | Hachette | | | Scholastic | | | | |
| Reed Elsevier (Netherlands) | 1.5 | 0.9 | 0.7 | 0.8 | 1.9 | 4.1 | 3.8 | 5.0 | |
| Viacom | | | | 4.4 | 3.0 | 2.5 | 3.0 | 3.3 | |
| Simon and Schuster | 6.3 | 5.7 | 5.8 | Viacom | | | | | |
| Holtzbrinck (Macmillan) (Germany) | | | | | 1.5 | 3.0 | 2.5 | 3.0 | |
| Lagardère/Hachette (Little Brown) (France) | | 0.3 | 0.3 | 0.4 | 0.2 | 1.3 | 3.2 | 2.9 | |
| Time Warner | 0.8 | 0.9 | 0.8 | 1.0 | 1.2 | Lagardere | | | |
| Reader's Digest /(Ripplewood) | 2.1 | 1.8 | 1.9 | 1.8 | 1.4 | 1.3 | 1.2 | 1.5 | |
| Other | 68.0 | 69.5 | 69.8 | 61.7 | 63.2 | 58.5 | 51.6 | 45.8 | |
| Total US Rev. ($ mil) | 8,259 | 12,806 | 17,353 | 19,632 | 21,368 | 26,141 | 26,626 | 23,900 | |
| C4 | 20.9 | 20.0 | 19.2 | 24.0 | 24.0 | 26.2 | 29.7 | 33.1 | 38.6 |
| C8 | 30.5 | 28.4 | 27.2 | 36.1 | 34.0 | 38.9 | 44.7 | 49.8 | 52.7 |
| HHI | 138 | 123 | 119 | 202 | 177 | 224 | 300 | 365 | 511 |
| *N* (>1%) | 8 | 8 | 8 | 9 | 10 | 10 | 10 | 10 | 9 |
| MOCDI Index | 49 | 43 | 42 | 67 | 56 | 71 | 95 | 115 | 170 |

*Source*: Data for 2007 from Book Publishing news.blogspot.com, "Arbor Books Helps Authors Rejected by Top Publishers," Sept. 23, 2007. 2006/7 data estimated from 2004 figures. 2009 data: *Market Share Reporter 2011*, Gale Cengage—September 17, 2010.

Table 18-4. Magazine Publishers (Market Share by Revenue)

| COMPANY | 1984 | 1988 | 1992 | 1996 | 2001 | 2004 | 2007 | 2009 | 2012 |
|---|---|---|---|---|---|---|---|---|---|
| Time Warner[a]/Time Inc | 9.0 | 9.3 | 9.8 | 11.1 | 14.0 | 14.3 | 11.1 | 13.0 | 8.6 |
| Advance Publications[b] | | 3.9 | 4.3 | 4.7 | 5.7 | 7.0 | 11.9 | 12.8 | 9.2 |
| Hearst | 2.2 | 3.2 | 5.2 | 5.2 | 6.1 | 6.0 | 7.2 | 8.0 | 7.5 |
| Source Interlink | | | | | | | 1.6 | 2.2 | 3.0 |
| Primedia[c] | | | 0.9 | 2.5 | 4.5 | 3.4 | Source Interlink | | |
| Reader's Digest/Ripplewood[d] | 1.3 | 3.1 | 3.4 | 3.3 | 2.2 | 2.6 | 2.5 | 2.6 | 1.1 |
| Meredith Corporation[e] | 2.3 | 2.6 | 2.6 | 2.3 | 2.1 | 2.6 | 2.9 | 2.3 | 2.1 |
| Bertelsmann (Gruner & Jahr) (Germany)[f] | | | 3.0 | 3.0 | 2.7 | 2.5 | 1.2 | 1.6 | 0.7 |
| Thomson (Canada) | 3.8 | 3.2 | 3.8 | 3.8 | 2.5 | 2.5 | 2.5 | 2.5 | 2.5 |
| International Data Group | 0.9 | 1.7 | 1.8 | 2.9 | 3.8 | 2.4 | 1.3 | 1.1 | 1.0 |
| McGraw-Hill | 2.1 | 2.2 | 2.2 | 1.9 | 2.7 | 2.1 | 2.8 | 2.2 | 2.0 |
| Reed Elsevier (Netherlands) | | 1.0 | 3.6 | 5.2 | 3.2 | 1.9 | 1.8 | 1.2 | 0.0 |
| Ziff-Davis[g] | 0.5 | 1.2 | 1.5 | 2.0 | 1.6 | 0.9 | 0.2 | 0.2 | 0.2 |
| Lagardère/Hachette Filipacchi (France)[h,i] | | | 1.0 | 1.6 | 1.8 | 1.6 | 1.6 | 2.3 | Hearst |
| Bonnier (Sweden) | | | | | | | 1.0 | 1.0 | 1.0 |
| Macrovision/Gemstar TV Guide[j] | | | | | 2.0 | 1.9 | 0.4 | 0.2 | 0.1 |
| News Corp | | | 0.7 | 0.9 | 1.2 | 1.4 | 1.3 | 1.4 | 1.3 |
| Triangle Publications | 4.0 | News Corp | | | | | | | |
| Others | 73.8 | 68.4 | 56.1 | 49.5 | 43.9 | 46.9 | 49.6 | 46.4 | 60.6 |

*continued*

Table 18-4. *continued*

| COMPANY | 1984 | 1988 | 1992 | 1996 | 2001 | 2004 | 2007 | 2009 | 2012 |
|---|---|---|---|---|---|---|---|---|---|
| Total US Rev. (mil $)[k] | 8,191 | 11,681 | 14,284 | 21,498 | 29,479 | 31,611 | 32,559 | 28,527 | 39,547 |
| C4 | 17.3 | 19.7 | 23.1 | 26.3 | 30.2 | 30.7 | 33.1 | 36.4 | 28.3 |
| C8 | 22.1 | 29.3 | 35.8 | 40.0 | 43.4 | 40.9 | 42.4 | 45.7 | 29.4 |
| HHI | 129 | 150 | 210 | 265 | 351 | 350 | 359 | 438 | 338 |
| *N* (>1%) | 7 | 10 | 12 | 13 | 15 | 14 | 13 | 13 | 10 |
| MOCDI Index | 46 | 48 | 58 | 71 | 91 | 90 | 99 | 126 | 107 |

*Main sources*: Revenues include advertising, subscriptions, and newsstand sales. Source: "100 Leading Media Companies, 2000," at http://www.adage.com/page.cms?pageId=533; "100 Leading Media Companies, 1995," at http://www.adage.com/page.cms?pageId=871; company annual reports and SEC filings. Total market data are from "U.S. Industrial Outlook" for corresponding years. TV Guide was until 1988, when it was purchased by News Corp. Data for 2003 from "100 Leading Media Companies, 2003," owned by Triangle AdAge, August 2004. 2006 data from Seybold online, "Magazine Publishers," May 17, 2007. 2006 revenue from AdAge, "Top 25 Magazine Companies," 2007. 2009 data: "Combined Circulation Revenue for All ABC Magazines 1988–2009." *Magazine Publishers of America*. Last accessed July 7, 2010. http://www.magazine.org/CONSUMER_MARKETING/CIRC_TRENDS/16136.aspx

[a] Time Inc. sold 18 magazines to Bonnier in 2007. It was spun off as a magazine division in 2014, thus severing cross-ownership with Time Warner.

[b] Advance bought Condé Nast in 1959 and Fairchild (from Disney) in 1999.

[c] Primedia, owned by the investment firm KKR, sold most of its magazines to Source Interlink in 2007.

[d] The Reader's Digest Association was acquired in 2006 by Ripplewood Holdings.

[e] Meredith Corp acquired in 2005 Bertelsmann's US women's magazines.

[f] Bertelsmann sold its US magazines in 2005 to Meredith, Morningstar, and Diamandis.

[g] j2 Global bought Ziff-Davis in 2012 and operates it as a subsidiary under the original Ziff-Davis name.

[h] Lagardere has controlled Hachette since 1981 and acquired Diamandis (formerly CBS Magazines) in 1988.

[i] 2006 revenue for Lagardere estimated from 2004 data.

[j] In 1999, Mews Corp sold TV Guide to Gemstar, of which it owned 41%. Gemstar was sold in 2008 to Macrovision.

[k] Total revenue for 1984–2004 of the US magazine industry was estimated by adding advertising to estimated magazine sales revenues. The estimated advertising data source is Publishers Information Bureau. Sunday supplements are excluded from "magazines."

The HHI values of market concentration were low but gradually rising, growing from 129 in 1984 to 351 in 2001, and then remaining fairly flat (338 in 2012). However, the many specialty submarkets of magazines would show a much higher concentration.

## AUDIOVISUAL MEDIA

### Radio

In the United States, no media industry has changed more in its ownership structure than local radio. Until the 1990s, concentration was extremely low because of the severe ceiling on station ownership imposed by government rules. Out of over 10,000 stations, no firm could own more than seven AM and seven FM stations. This was raised to 12 each in 1985, 18 each in 1992, and 20 each in 1994. After 1996, there were no national caps left at all.

In 1984, station ownership was less concentrated nationally than just about any industry in the country and possibly the world. Around that time, consolidations began. On the business end, a major acquisition driver was a leveraged buyout firm, the Dallas-based Hicks Muse Tate and Furst. Capstar, one of its acquisition vehicles, bought about 250 stations in an 18-month period in 1996 and 1997. Several other radio companies emerged and bought a large number of stations. Clear Channel (re-branded iHeartMedia in 2014) became the industry leader. By 2004, the HHI by station ownership had risen to 98, a major increase from near-zero, but still quite unconcentrated. Even industry colossus Clear Channel, with its array of 1,184 stations in 2005, accounted for only 9.2% of stations. The next firms—Cumulus, Citadel/ABC, Viacom, and Hispanic—were in the 0.6–2.4% range.

The market was much more concentrated on a revenue basis than on a station basis. Table 18.5 shows the HHI increasing from 20 in 1984 to 600 by 2011, a large increase in relative terms but still a low number by antitrust standards. The C4 more than quintupled from 8.3% in 1984 to 37.9% in 2004 and to 45.3% in 2011. By 2004 Clear Channel and CBS Viacom had achieved market shares of 18.9% and 11.9%. But that was the peak. Many of the debt-financed acquisitions ran into trouble as the financial markets soured on large and highly leveraged radio companies. In 2007, industry leader Clear Channel agreed to be acquired for $19.5 billion by the private equity firms Bain Capital and Thomas H. Lee. To finance the deal the buyers sold off hundreds of Clear Channel stations. By 2011, the two companies' market shares had declined to 15.1% and 8.5%. Disney left the radio business. The number 3 firm, Citadel, was delisted in 2009 by the New York Stock Exchange when it went into bankruptcy, and it was acquired by Cumulus.

One can also observe a marked decline in the radio content networks. Radio content used to be dominated by a small number of radio networks run by major companies. Until the 1950s, these networks (RCA's NBC, as well as CBS, ABC, and Mutual) were content "wholesalers" to local station "retailers" that consisted of several of their own stations, but mostly of hundreds of affiliated and independently owned local stations. But by the 1960s the networks were a mere shadow of their pre-TV golden age of the 1930s and 1940s. In 2013, of these network companies, only CBS played a role.

One alternative to commercial terrestrial radio is public nonprofit broadcasting, but it has traditionally been a financially weak part of the American media landscape. On the one hand, there are many such stations; to be exact, 2,471, quite a large number. But there are no strong national station groups. Two major public media program networks exist, Public Radio International (PRI) and National Public Radio (NPR).[18]

Local station concentration is low, with an HHI of 600 in 2011 that has barely budged in a decade, after growing in the 1990s. But should one only look at classic broadcast radio? Two

18. The Voice of America is a governmental radio system aimed at foreign countries.

alternatives to over-the-air terrestrial broadcasting, in particular, have been offshoots of video delivery media: cable radio and satellite radio. Three other options are Internet-delivered: online distribution of broadcast stations, non-broadcast music Web sites (such as Pandora, Rhapsody, or Spotify, which over 100 million Americans access per month), and "podcasting." And then there are over-the-air alternatives: low-power FM stations (LPFM), whose weak signals are matched by weak audience counts and economics, and digital radio ("HD-Radio"), which increases the number of channels per broadcaster. Satellite radio has been significant. It is based on a pay-subscription model and offers hundreds of channels on a nationwide basis. Two providers were licensed, Sirius and XM. They competed vigorously, lost money in the process, and then decided that a merger would be more profitable. This was accomplished in 2008, after a regulatory struggle to overcome the government's normally pro-competitive policy. By 2011, this combined firm, Sirius XM had become the nation's largest radio provider by revenues, accounting for 15.4% (see Table 18.5).

The local nature of most radio advertising and news programming makes local ownership concentration a greater concern than national concentration. For that reason, the US government did not remove limits on local ownership entirely when it abolished national ceilings.

Table 18.6 shows local radio concentration trends, based on an analysis of 30 local markets (small, medium, and large). Satellite radio, being national in nature, is not included. Local concentration in 2006 was significantly higher than it was in 1984. The weighted average share of the top firm in a market (C1) grew from about 20% in 1984 to about 34% in 2006 but dropped to 31% in 2010. The top four firms' market share grew from 52.6% to 84.4% and then dropped to 78.1% in 2010 (note that these top four firms may differ in different markets). These are fairly high numbers, and it is not much of a comfort that radio is still among the least concentrated of local media and that concentration has declined after 2002.

## Broadcast Television

Although broadcast television stations have been increasing in number, their total viewership has steadily declined, whether viewed over-the-air (terrestrially), over cable and satellite platforms, or digitally as multicasts. Their audience shares fell from 87.7% of TV audiences in 1984 to 48.6% in 2003, 28.1% in 2007,[19] and 22.8% in 2013.[20] On top of that, the actual viewing of local TV stations by most households is not over the air but of signals retransmitted over cable or satellites. The dominance of the overall TV viewer market by the original three major broadcast networks (ABC, CBS, and NBC) declined considerably, from 64% in 1984 to 21.6% in 2011.[21] But, together with the addition of Fox, they remain by far the largest providers.

In the 1980s, control over the networks passed from the broadcasting industry's founder-owners to more conventional corporate management. The new owners sold off many of the networks' non-TV holdings and expanded their TV-related holdings. A second wave of ownership changes began in 1995 after the FCC's removal of a rule that had limited the major networks' involvement in program production and syndication. This change allowed the networks to vertically integrate upstream into production, but also allowed major content producers (the Hollywood film studios) to do the

19. Dempsey, John. "Cable TV Hits Record Numbers." *Variety* (July 2007).

20. The Nielsen methodology and classification of counting audiences changed somewhat during the period, so that the numbers are not fully comparable. Sources: Diego, Vasquez. "This Week's Broadcast Ratings." *Media Life Magazine*, May 29, 2013, http://www.medialifemagazine.com/this-weeks-broadcast-ratings/; and Diego, Vasquez. "This Week's Cable Ratings." *Media Life Magazine*, May 29, 2013, http://www.medialifemagazine.com/this-weeks-cable-ratings/.

21. Market shares in 2010 were 7.43% for ABC, 8.5% for NBC, and 5.65% for CBS. Guskin, Emily, Tom Rosenstiel, and Paul Moore. "Network: By the Numbers." *State of the News Media 2011*. Pew Research Center. Last accessed on July 17, 2012, http://stateofthemedia.org/2011/network-essay/data-page-5/#audience.

Table 18-5. Radio Group Market Shares, by Revenues

| | 1984 | 1988 | 1992 | 1996 | 2000 | 2001 | 2004 | 2007 | 2008 | 2011 |
|---|---|---|---|---|---|---|---|---|---|---|
| iHeartMedia (Clear Channel) | 0.3 | 0.5 | 1 | 2.8 | 12.4 | 18.7 | 18.9 | 17.2 | 16.9 | 15.1 |
| Jacor | | 0.9 | 0.8 | 1.9 | Clear Channel | | | | | |
| Evergreen | | | 0.6 | 2.5 | Clear Channel | | | | | |
| Chancellor (AM/FM) | | | 0.1 | 2.1 | Clear Channel | | | | | |
| Capstar | | | | 2.1 | Clear Channel | | | | | |
| SFX | | | 0.2 | 2 | Clear Channel | | | | | |
| CBS (III) | | | | | | | | 8.7 | 7.5 | 8.5 |
| Viacom | | | | | 12.5 | 12 | 11.9 | CBS III | | |
| Westinghouse (CBS (II)) | 2.4 | 2.4 | 2.4 | 4.6 | Viacom | | | | | |
| Infinity[a] | 1.3 | 1.3 | 1.8 | Westinghouse | | | | | | |
| CBS (I) | 2.1 | 2 | 2.5 | Westinghouse | | | | | | |
| American Radio Systems[b] | | | | 2.7 | Viacom | | | | | |
| Cumulus | | | | | 1.2 | 1.4 | 1.7 | 1.6 | 1.6 | 6.3 |
| Citadel | | | 0.2 | 0.4 | 1.9 | 1.8 | 4.6 | 3 | 4.4 | Cumulus |
| Disney | | | | 3.4 | 2.2 | 2.3 | Citadel | | | |
| CapCities | | 2.4 | 2.4 | Disney | | | | | | |
| ABC Radio | 2.4 | CapCities | | | | | | | | |
| Cox | 0.8 | 0.8 | 1.2 | 1.6 | 2.4 | 2.5 | 2.6 | 2.2 | 2.1 | 2.6 |

*continued*

Table 18-5. *continued*

| | 1984 | 1988 | 1992 | 1996 | 2000 | 2001 | 2004 | 2007 | 2008 | 2011 |
|---|---|---|---|---|---|---|---|---|---|---|
| Entercom | | | | 0.4 | 1.9 | 1.9 | 2.6 | 2.3 | 2.3 | 2.3 |
| Radio One | | | 0.1 | 0.2 | 1.6 | 1.6 | 2 | 1.6 | 1.6 | 1.7 |
| Emmis | 0.9 | 0.9 | 0.9 | 0.8 | 1.6 | 1.5 | 1.6 | 1.6 | 1.7 | 1.9 |
| Univision | | | | | | | 1.8 | 2.1 | 1 | 2.1 |
| Hispanic | 0 | 0 | 0.2 | 0.6 | 1.3 | 1.4 | Univision | | | |
| Sirius XM | | | | | | | | | 8.5 | 15.4 |
| XM | | | | | | | 1.1 | 5 | Sirius XM | |
| Sirius | | | | | | | 0.4 | 4.6 | Sirius XM | |
| Other | 89.8 | 88.9 | 85.5 | 71.6 | 61 | 54.9 | 50.8 | 50.1 | 52.3 | 44.1 |
| Total Revenue ($ mil) | 5,596 | 7,511 | 8,378 | 11,947 | 18,819 | 17,450 | 18,932 | 20,154 | 19,478 | 16,029 |
| C4 | 8.3 | 7.7 | 8.4 | 13.6 | 29.6 | 35.4 | 37.9 | 35.5 | 37.4 | 45.3 |
| HHI | 20 | 20 | 25 | 75 | 337 | 519 | 546 | 449 | 454 | 600 |
| *N* (>1%) | 4 | 4 | 6 | 10 | 10 | 10 | 10 | 11 | 10 | 9 |
| MOCDI Index | 10 | 10 | 10 | 24 | 107 | 164 | 173 | 135 | 144 | 200 |

*Main sources*: BIA Financial Network, company SEC filings, and Crain Communications. Data for 2004 are from Bear Stearns "*Radio Broadcasting: Fact Book*," April 2005.
[a] The remaining public float of Infinity was merged into Viacom in February 2001 subsequent to Viacom's purchase of CBS in 2000, but it is shown here as a separate entity.
[b] American Radio Systems was created by a three-way merger in 1995.

Table 18-6. Local Radio Concentration (by Market Size)

| | 1984 | 1988 | 1992 | 1996 | 2002 | 2006 | 2010 |
|---|---|---|---|---|---|---|---|
| Local Markets | | | C1 Concentration | | | | |
| C1 Large Cities | 14.1 | 14.6 | 14.4 | 32.3 | 34.3 | 29.9 | 24.1 |
| C1 Medium Cities | 22.3 | 22.2 | 25.0 | 34.8 | 37.8 | 37.5 | 35.0 |
| C1 Small Cities | 25.6 | 24.0 | 23.5 | 35.6 | 31.8 | 35.5 | 35.3 |
| Weighted Average | 19.9 | 19.7 | 20.3 | 34.0 | 34.7 | 33.8 | 31.0 |
| | | | C4 Concentration | | | | |
| C4 Large Cities | 43.0 | 42.2 | 45.8 | 67.7 | 80.3 | 78.8 | 65.8 |
| C4 Medium Cities | 57.2 | 62.6 | 68.0 | 88.8 | 91.8 | 87.1 | 82.6 |
| C4 Small Cities | 61.4 | 60.5 | 64.6 | 77.4 | 82.1 | 87.3 | 85.1 |
| Weighted Average | 52.6 | 53.8 | 58.1 | 77.1 | 84.4 | 83.8 | 78.1 |
| | | | HHI Concentration | | | | |
| HHI Large Cities | 592 | 756 | 816 | 1,733 | 2,139 | 1,989 | 1,356 |
| HHI Medium Cities | 1,094 | 1,298 | 1,572 | 2,457 | 2,798 | 2,569 | 2,471 |
| HHI Small Cities | 1,263 | 1,238 | 1,333 | 2,173 | 2,330 | 2,537 | 2,470 |
| Weighted Average | 939 | 1,062 | 1,200 | 2,085 | 2,400 | 2,139 | 1,526 |

*Source*: Local radio revenues and market shares are based on *Duncan's Radio Market Guide,* 1985, 1989, 1994, 1997, and 2002 editions, respectively. Copyright James H. Duncan, Jr. Local market concentration was established using the *Broadcasting and Cable Yearbook* (1984, 1988, 1992, 1996, and 2001, respectively), published by R.R. Bowker, New Providence, N.J. 2006 and 2010 data are based on Arbitron market ratings.

same downstream into TV distribution networks. A new stage of merger transactions resulted, reflecting the shift of the bottleneck from distribution to the program production side. Content was becoming king.

New TV networks such as Fox gained a foothold, partly via broadcast stations on the less desirable UHF band. Other TV networks served the rapidly growing Spanish language market. Of those, Univision is the largest. In 2013, several station groups consolidated. The Tribune Co. bought the group LocalTV (formerly the TV stations of the *New York Times*) with its 19 stations, making it the number-one local TV broadcaster. Gannett bought Belo and its 20 stations, Sinclair bought 46 stations, and Media General and New Young Broadcasting combined their 30 TV stations.

There is also a public TV broadcasting system, which, like public radio, has been underfunded and over-fragmented into hundreds of individual stations. Its national distribution is through the Public Broadcasting Service (PBS), a network owned and operated by 348 public television stations. The major program producers are the five flagship stations in New York, Boston, Washington, Los Angeles, and Pittsburgh. Public TV's programming added content diversity, but its audience share has hovered around 1%.

TV station ownership is capped by regulation. These ceilings at first limited ownership to 7 stations, and then to 12 stations (14 if the extra stations were controlled by racial minorities or small businesses). Later, a cap was adopted that did not limit the number of stations but rather their total national audience "reach" (the potential audience, not the actual one). Initially the cap limited group ownership to a reach of 25% of the national audience, or 30% if two of the stations were owned by small or minority-controlled businesses.

The rationale for national ownership caps was harder to articulate in a multichannel media environment than for local caps. As cable television and satellite TV increased in penetration and channel capacity, the distributors of video programs could bypass TV stations altogether in accessing viewers on a nationwide scale. The largest of station groups therefore lobbied to remove the caps on their ownership of local TV outlets. Smaller station groups, on the other hand, argued in favor of restrictions on national reach. They feared, in particular, the loss of bargaining power if the national TV network companies would gain local distribution to most American households without having to rely on independently owned local affiliates. This issue split the broadcast industry and reduced its lobbying effectiveness.

Table 18.7 shows concentration of the TV broadcast industry for the revenues of both stations and networks.[22]

In the 1990s, major TV broadcasting firms lost over half of their viewers to cable channel competition but increased their revenue via station acquisitions, made possible by the loosening of restrictions on station ownership. As shown in Table 18.7, the combined revenue share of the top four TV groups was largely flat in the 1980s, from 41.4% in 1984 until 1996, around 41%. It then increased to 61.6% in 2012. The HHI doubled from a low 530 in 1984 to a still low 578 in 2005 to 984 in 2011.

Of course, most video watching is now for cable and satellite channels and online video. However, local TV news is mostly provided by the local stations and their market concentration is hence important. On the local level, the concentration of broadcast TV (i.e., of the local TV stations) declined over 25 years, from a C4 of 90% in 1984 to 75.3% in 2009, and 74.3% in large markets (see Table 18.8). For the market leader, the share dropped on average from 33% to 25% in 2001, and then stabilized at 24.9% in 2009 and 23.7% in large markets. The HHI declined from 2,460 to 1,714, then rose to 1,895 in 2006 and 1,935 in 2009.

## Multichannel TV Platforms

*Cable TV* platform companies are rarely subject to competition from other cable operators within the same franchise area.[23] For a long time, there was no effective competition by other multichannel distribution media. This changed with the entry of high-power digital direct broadcast satellites after 1994.

*DBS*'s share gradually increased to about 27% nationally and even more in rural areas. Cable and DBS are, however, imperfect substitutes. DBS carries a limited set of local TV stations and no local public access and governmental channels. Cable's interactivity enables it also to be a platform for broadband Internet and voice telephony. On the other hand, DBS operations are normally better suited for reaching thin but nationally substantial audiences and for reaching rural, low-density areas.

Besides the two platforms of cable TV and DBS, other multichannel video platforms are:

*Digital terrestrial broadcasting.* Digital Transmission (after 1996) enables multicasting, in which a station can simultaneously broadcast six or more programs and thus create terrestrial multichannel broadcasters. So far, this has not become a competitive factor in the multichannel market.

*Online ("over the top") TV (OTT).* This transmission type enables individualized and interactive video.[24] In primetime viewing periods, online video accounted in 2014 for 67% of all broadband traffic.

*IPTV.* As telecom operators upgraded their networks with fiber, they increasingly

---

22. Combining the two is done to permit a comparison with the data from other countries, where the distinction of stations and networks usually does not exist.

23. Noam, Eli M. ed. *Video Media Competition: Regulation, Economy, and Technology.* 1st ed. New York: Columbia University Press, 1985.

24. Noam, Eli M., Jo Groebel, and Darcy Gerbarg, eds. *Internet Television.* 1st ed. London: Erlbaum, 2004.

Noam, Eli. "Cloud-TV: A National and Global Initiative to Reach the Next Generation of Video Media." Forthcoming.

## Table 18-7. Broadcast TV Stations and Networks, 1984–2009

| | 1984 | 1988 | 1992 | 1996 | 2001 | 2005 | 2009 | 2012 |
|---|---|---|---|---|---|---|---|---|
| 21st Century Fox News Corp (Fox, MNT) | | 3.2 | 8 | 7.9 | 10.9 | 12.6 | 15.5 | 13.3[a] |
| Chris-Craft | 0.3 | 1.1 | 1.5 | 1.3 | Fox | | | |
| CBS (III)[e] | | | | | | | 12.9 | 14.6 |
| Viacom | | | | | 15.7 | 12.5 | CBS | |
| CBS ("I" and "II") | 13.2 | 12.6 | 13.4 | 12 | Viacom | | | |
| UPN | | | | 2.6 | Viacom | | | |
| Comcast | | | | | | | 13.0 | 17.6 |
| GE/NBC, Telemundo | | 12.7 | 14.3 | 10.7 | 12 | 11.6 | 13.0[b] | Comcast |
| RCA (NBC) | 14.3 | GE | | | | | | |
| Disney (ABC) | | | | 11 | 10.7 | 9.7 | 12.8 | 16.1 |
| CapCities | | 11.2 | 10.9 | Disney | | | | |
| ABC | 12 | CapCities | | | | | | |
| Univision (Televisa, Mexico)[c] | | 0.2 | 0.6 | 0.9 | 1.6 | 2.5 | Disney | |
| Gannett | 1.9 | 1.5 | 1.4 | 1.9 | 1.9 | 1.9 | 2.9[d] | 3.2 |
| Belo | 0.9 | 0.7 | 0.8 | 1 | 1.7 | 1.6 | Gannett | |
| Time Warner (CW)[e] | | | | 2 | 2.7 | 2.8 | 1.6 | 1.8 |
| PBS Network | 0.9 | 0.8 | 0.9 | 0.8 | 1 | 1.1 | 0.8 | 0.8 |
| Hearst-Argyle | | | | 0.2 | 1.8 | 1.8 | 1.8 | 1.9 |
| ION Media (Paxson) | | | | | 0.8 | 0.4 | 0.4 | 0.4 |
| Tribune | 1.4 | 1.7 | 1.9 | 2.5 | 3.2 | 3.0 | 3.1 | 1.9 |
| Graham (Washington Post) | | | | | | | 1.3 | 1.3 |
| Advance (Bright House) | | | | | | | 0.9 | 0.9 |
| Others | 55.2 | 54.3 | 46.2 | 45 | 36.1 | 38.5 | 34.5 | 27.8 |
| Total Revenue (mil $) | 18,890 | 23,737 | 25,779 | 33,828 | 35,779 | 42,326 | 40,800 | 43,168 |
| C4 | 41.4 | 39.7 | 46.6 | 41.6 | 49.3 | 46.4 | 54.1 | 61.6 |
| HHI | 530 | 463 | 577 | 468 | 654 | 578 | 753 | 984 |
| *N* | 4 | 7 | 7 | 10 | 11 | 11 | 10 | 8 |
| MOCDI Index | 265 | 175 | 218 | 148 | 197 | 174 | 238 | 348 |

*Main sources*: Station group revenues are from company SEC filings and from "Special Report: Top 25 TV Groups," *Broadcasting and Cable*, p. 48. April 8, 2002. Data for 2004 are from Bear Sterns, *Television Broadcasting: Broadcast Television Fact Book*, December, 2004 and May, 2005. Data for 2006 use 2004 numbers, with only the effect of the Saban acquisition shown. Data from 2008/9 use Bear Sterns, "Television Broadcast Factbook," January 2008. 2013 data repeat 2009 data with assignment to new owners. Market shares calculated from total day shares (Noam 2009, p. 163), ratings for prime time broadcast season (Noam 2009, p. 156), and annual advertising expenditures (Noam 2009, p. 134). Revenue for Viacom consists of combined revenue for CBS and UPN. Total not included in overall market revenue; NBC controlled 32% of Paxson. Revenue from Galavision (Univision) estimated based on the ratio of its share compared to that of PBS and Paxson.

[a] In 2013, News Corp spun off its non-print divisions as 21st Century Fox.

[b] GE sold NBC to Comcast in two stages: 51% in 2011 and 49% in 2013.

[c] In 2007, this was a consortium led by Saban and including the private equity firms Madison Dearborn, Providence Equity, TPG, and T.H. Lee. However, these companies can be seen as placeholders to Mexico's TV giant Televisa, which owns 8% of Univision outright, plus substantial debt that can be converted into stock as soon as US regulations permit foreign broadcast ownership of more than 25%.

[d] In 2013, Gannett acquired Belo.

[e] CW is a joint venture of Time Warner and CBS.

## Table 18-8. Local TV Station Concentration (by Market Size)

| | 1984 | 1988 | 1992 | 1996 | 2002 | 2006 | 2009 |
|---|---|---|---|---|---|---|---|
| Local Market | | | C1 Concentration | | | | |
| C1 Large Cities | 29.1 | 27.9 | 23.6 | 24.5 | 18.4 | 24.1 | 23.7 |
| C1 Medium Cities | 37.1 | 32.6 | 30 | 28.7 | 27.7 | 31.4 | 30.9 |
| C1 Small Cities | 35.9 | 36.3 | 32.9 | 31.3 | 31.2 | 37.6 | 38.7 |
| Weighted Average | 33.5 | 31.8 | 28.2 | 28.0 | 24.9 | 30.2 | 24.9 |
| | | | C4 Concentration | | | | |
| C4 Large Cities | 84.5 | 80.1 | 72.5 | 73.3 | 61.8 | 62 | 74.3 |
| C4 Medium Cities | 95.1 | 90.9 | 87.6 | 88 | 78 | 77.1 | 71.4 |
| C4 Small Cities | 91.4 | 91.4 | 90.1 | 89 | 85.2 | 84.5 | 82.1 |
| Weighted Average | 89.8 | 86.7 | 82.2 | 82.4 | 73.5 | 73.1 | 75.3 |
| | | | HHI Concentration | | | | |
| HHI Large Cities | 2,108 | 1,910 | 1,581 | 1,594 | 1,256 | 1,361 | 1,662 |
| HHI Medium Cities | 2,756 | 2,437 | 2,212 | 2,179 | 1,859 | 1,930 | 2,304 |
| HHI Small Cities | 2,634 | 2,595 | 2,384 | 2,306 | 2,207 | 2,619 | 3,012 |
| Weighted Average | 2,460 | 2,269 | 2,006 | 1,979 | 1,714 | 1,895 | 1,935 |

*Source:* Local television revenues are based on *Duncan's Radio Market Guide,* 1985, 1989, 1994, 1997, and 2002 editions, respectively (copyright James H. Duncan, Jr.). Subsequent concentration values C4 and HHI have been calculated via Nielsen TV Shares. Data used were for July, except in 1996 because the Olympic games held in Atlanta that year were shown during prime time, for which NBC had exclusive rights, thus increasing its market share for that period. In consequence, for 1996 the November period was chosen.

offered a menu of cable-style TV channels. Verizon with its FiOS video service had 5.3 million subscribers by 2014, and AT&T's U-verse video service had 5.7 million.[25]

Concentration for multichannel video program distribution (MVPD) has been very high but declining; the market share of cable as a whole in MVPD declined from 100% in 1992 to 96.7% in 1996, 71.8% in 2009, and 57.4% in 2012. The combined market share of the two DBS providers rose to 27.4%, and IPTV had 8.4% in 2012. OTT was growing rapidly. But the C4, as applied to revenues, increased from 20.7% in 1984 to 42.1% in 1996, 54.2% in 2005, 61.3% in 2009, and 63.2% in 2012, driven by the concentration in the cable industry.

National cable concentration rose considerably from the early days of the cable boom when numerous companies were contesting for franchises. The largest of them, TCI, accounted for only 5.4%. In 2009, in contrast, the market leader Comcast held about 28% of cable (and 22.4% of all multichannel platforms), and Time Warner Cable had about 20% of cable (and 14.3% of multichannel platforms). Three other cable firms (Charter, Cox, and Cablevision) were far behind. The national cable C4 increased from 20.7% in 1984 to 73.2% in 2009. The HHI rose from 162 to 2,011 in 2009. In 2014, Comcast sought the acquisition of Time Warner Cable, thus merging the top two firms. To reduce regulatory concerns, Comcast also entered into an agreement with the number-four cable company, Charter Communications (dominated by John Malone, a major cable magnate in the United States and Europe (Liberty). The two companies would exchange 1.6 million subscribers each, to consolidate their footprints. In addition, Comcast would sell 1.4 million subscribers to Charter,

25. O'Neil, Jim. "Verizon Sees FiOS TV and Internet Subscriber Growth Slow-Again- in Q1" *Ooyala,* 2014, http://www.ooyala.com/videomind/blog/verizon-sees-fios-tv-and-internet-subscriber-growth-slow-again-q1.

and the shareholders of Comcast and Charter would control a new entity into which Comcast would transfer 2.5 million subscribers. Although Comcast shareholders would dominate the new entity directly and through their stake in a new firm called New Charter. At the same time, cable industry revenue rose hugely, from about $8.3 billion in 1984 to $90 billion in 2009.

After 1993, the FCC set limits of 30% for national cable ownership and of 40% for channel occupancy by the MSO's own channels. That ceiling was raised to 36.7% in 1999. These rules were struck down in 2001, and again in 2009, by a federal court.[26]

The DBS industry consists of two providers—EchoStar and DirecTV. DirecTV was owned, in sequence, by Hughes Electronic, a satellite manufacturer and operator controlled by the car maker General Motors, then by Rupert Murdoch's News Corp, as part of the latter's attempt to create a TV footprint that covered the entire world. But in 2006, Murdoch had to swap control to cable TV pioneer John Malone and his Liberty Media, in return for Malone's share in News Corp., which had become a threat. In 2010, due to an FCC condition, Malone had to reduce his control of DirecTV from 24.3% to 3%. Malone then reentered the US cable business in 2013 by buying 27% of the number-five cable company, Charter Communications, and in effect dividing up Time Warner Cable with Comcast. Malone also controls the major European cable TV MSO, UPC, which has been owned by Liberty Global since 2005. In 2013, DirecTV had 20 million subscribers in the United States and 10 million in Latin America. In 2014, the telecom giant AT&T announced its acquisition of DirecTV for $48.5 billion, subsequent to regulatory approvals. This, too, would increase MVPD concentration measures by combining AT&T's U-verse HDTV service with DirecTV's DBS service.

The second US provider is the Dish Network, controlled by Charles Ergen, with 14 million subscribers. Dish also acquired in 2011 the largest home video rental chain, Blockbuster, after Blockbuster's decline and bankruptcy. DirecTV holds 18.9% of the multichannel platform market and Dish has 8.2%. Such concentration, however, is relevant primarily where no cable TV or telecom-based IPTV are available, that is, in low-density rural areas. For the rest of the country and most of its population, there are other choices.

The DBS industry, if seen as a separate product market (e.g., in rural areas), is a highly concentrated duopoly with an HHI of 5,092 in 2009 in revenues. This significant level of concentration is not surprising considering the high barriers to satellite entry in terms of capital and marketing investment required to launch DBS services. At one point, the two firms had signed a merger agreement in order to reduce competition but were blocked by the government. Table 18.9 shows a pooled multichannel distribution platform market (of cable MSOs, DBS, and telecom IPTV). Both the HHI and the C4 measures increased steadily, but to a level still well below the US government's threshold, to 1,233 in 2012. Overall, the industry's revenues rose enormously to over $158 billion. The two major mergers proposed in 2014 would change this number considerably. The AT&T–Comcast deal would raise AT&T's market share from 4.8% to 23.7% and raise the MVPD HHI by 183 points from 1,233 to 1,415. The triangular Comcast–Time Warner Cable–Charter deal would raise Charter's market share to 7.8% and Comcast's to 32.7%. The MVPD HHI would rise by 406 points to 1,639. Together, the two deals would raise the MVPD C4 share by 9.2% of the market to 73%, the C6 by 12.6% (to 82.2%), and the HHI by 588 points, from 1,233 to 1,821. This is an increase of 47.8%.

Table 18.9 presents the multichannel TV market, while Table 18.7 shows the TV broadcast industry. We now put both together as a combined *video retail distribution market*. Table 18.10 shows the shares of companies in a pooled video retail distribution market. It treats TV stations, cable MSOs, DBS, and IPTV as basically interchangeable program delivery systems. (Not included are broadband ISPs that are the

26. *Time Warner Entertainment v. FCC*, 94–1035 D.C. Cir. 2001.

Table 18-9. Multichannel Video Platforms (Cable MSOs, DBS, and IPTV)

| | | 1984 | 1988 | 1992 | 1996 | 2001 | 2005/6 | 2009 | 2012 |
|---|---|---|---|---|---|---|---|---|---|
| Cable TV Operators | Comcast | 1.0 | 3.0 | 3.4 | 5.7 | 10.4 | 23.0 | 22.4 | 23.2 |
| | Jones Intercable | 0.1 | 0.4 | 0.6 | 1.1 | Comcast | | | |
| | Lenfest Communications | 0.9 | 0.8 | 0.9 | 1.4 | Comcast | | | |
| | Storer (50% interest) | | KRR | 1.5 | 1.5 | Comcast | | | |
| | AT&T Broadband | | | | | 18.6 | Comcast | | |
| | MediaOne | | | | 10.0 | AT&T | | | |
| | Continental Cablevision | 3.1 | 5.8 | 5.3 | MediaOne | | | | |
| | TCI | 5.4 | 12.7 | 15.7 | 13.1 | AT&T | | | |
| | Viacom Cable | 2.3 | 2.5 | 1.9 | 1.6 | TCI | | | |
| | Storer (50% interest) | | 1.8 | 1.5 | 1.5 | TCI | | | |
| | Heritage | 1.5 | TCI | | | | | | |
| | Storer Communications | 4.2 | KRR | TCI/Comcast | | | | | |
| | Time Warner Cable[c] | | | | | | | 14.3[a] | 12.9 |
| | Time Warner | | | 9.9 | 13.3 | 9.7 | 14.2 | TWC | |
| | American TV and Communications & Warner Communications | 4.4 | 6.1 | TWE | | | | | |
| | Westinghouse | 6.0 | TWE | | | | | | |
| | Adelphia | | | | | 5 | | | |
| | Century | 9 | 2 | 5 | 3 | Adelphia | | | |
| | Advance (Newhouse Broadcasting/ Bright House) | 1.7 | 1.8 | 2.4 | TWE | | 2.3 | 2.0 | 2.0 |

| | | | | | | | | | |
|---|---|---|---|---|---|---|---|---|---|
| | Charter (27% Liberty) | | | | 1.3 | 6.9 | 5.9 | 4.0 | 4.4 |
| | Marcus Cable | | | 2.1 | 1.5 | Charter | | | |
| | Bresnan | | 0.3 | 0.4 | 0.9 | Charter | | | |
| | Falcon | 0.1 | 0.2 | 0.2 | 0.8 | Charter | | | |
| | Cox Communications | 4.9 | 3.7 | 3.1 | 5.1 | 7.1 | 6.2 | 6.4 | 4.8 |
| | Times Mirror | 3.0 | 2.5 | 2.1 | Cox | | | | |
| | TCA Group | 0.3 | 0.5 | 0.7 | 0.9 | Cox | | | |
| | Cablevision Systems | 1.4 | 3.7 | 2.7 | 1.0 | 1.4 | 2.9 | 4.4 | 3.4 |
| | Mediacom | | | | | 1.0 | 1.4 | 1.2 | 1.1 |
| | Insight | | | | 0.2 | 1.2 | 1.3 | 0.8 | TWC |
| | CableOne | 1.4 | 1.1 | 0.8 | 0.8 | 0.7 | 0.7 | 0.6 | 0.6 |
| | Others | 57.4 | 50.7 | 41.9 | 32.3 | 18.4 | 24.6 | 17.7 | 11.6 |
| | Cable TV Subtotal (market share %): | 100 | 100 | 100 | 96.7 | 80.9 | 82.5 | 71.8 | 61.6 |
| | Cable TV Subtotal ($ mil): | 8,331 | 13,409 | 21,079 | 27,706 | 43,518 | 65,678 | 89,901 | 97,598 |
| DBS Providers | DirecTV[b] | | | | | | | 15.3 | 18.9 |
| | News Corp (DirecTV/Hughes Electronics) | | | | 1.2 | 11.0 | 10.8 | Liberty Media/ DTV | |
| | TCI Satellite Entertainment/Tempo/PrimeStar | | | | 0.8 | DirecTV | | | |
| | US Satellite Broadcasting Corp | | | | 1.0 | DirecTV | | | |
| | Dish Network (Echostar) | | | | 0.2 | 7.4 | 7.0 | 9.3 | 8.2 |
| | Rainbow (Cablevision) | | | | | 0.4 | Echostar | | |
| | Others | | | | 0.1 | 0.3 | 0.3 | 0.3 | 0.3 |
| | DBS Subtotal (market share%): | 0.0 | 0.0 | 0.0 | 3.3 | 19.1 | 18.1 | 24.9 | 27.4 |
| | DBS Subtotal ($ mil): | 0 | 0 | 2 | 943 | 10,270 | 14,494 | 31,185 | 44,010 |

*continued*

Table 18-9. *continued*

| | | 1984 | 1988 | 1992 | 1996 | 2001 | 2005/6 | 2009 | 2012 |
|---|---|---|---|---|---|---|---|---|---|
| IPTV providers | Verizon (FiOS) | | | | | | | 1.8 | 5.0 |
| | AT&T (U-verse) | | | | | | | 1.4 | 4.8 |
| | Others | | | | | | | | 1.0 |
| | IPTV Subtotal ($ mil) | | | | | | | 4,045 | 16,920 |
| | IPTV Subtotal (market share %) | | | | | | | 3.2 | 10.8 |
| | Total US Revenue ($ mil) | 8,331 | 13,409 | 21,081 | 28,649 | 53,788 | 80,172 | 125,131 | 158,528 |
| | C4 | 20.7 | 28.3 | 34.3 | 42.1 | 49.7 | 55.0 | 61.3 | 63.2 |
| | HHI | 162 | 293 | 430 | 532 | 857 | 987 | 1,111 | 1,233 |
| | *N* | 13 | 13 | 14 | 17 | 11 | 10 | 10 | 10 |
| | MOCDI Index | 45 | 81 | 115 | 129 | 258 | 312 | 351 | 390 |

*Main sources*: Revenues from Kagan Research, LLC. Broadband Cable Financial Databook, 2004, as found on http://www.ncta.com/Docs/PageContent.cfm?pageID=309. Starting 1996, 20% of TW's market share has been allocated to MediaOne. Data for 2005 include the Adelphia acquisition by TW and Comcast. TW maintains an ownership state in Bright House after the earlier partnership with Advance Newhouse dissolved in 2003. Also from NCTA.com and company reports of subscriber numbers. Revenues are cable TV only, excluding Internet and telephony revenues. Source: 2009 Cable MSO data, "Comcast Corp.: Financials," Standard & Poor's Net Advantage, Broadcasting and Cable. July 2010. Colony, Sandy. "Insight Announces Third Quarter Results." Insight: Press Release. July 2010. Cable Industry Revenue 1996–2009, National Cable and Telecommunications Association: http://www.ncta.com/Stats/CustomerRevenue.aspx. Last accessed 7/15/10. Amobi, Tuna N., Kolb, Erik B. "S&P 500 Industry Surveys Broadcasting, Cable & Satellite," February 18, 2010. 2009 DBS data: "Domestic Satellite DTH Numbers." SBCA: Satellite Broadcasting and Communications Association, through MediaBiz: Competitive Intelligence. http://www.sbca.com/receiver-network/industry-satellite-facts.htm. Source for system operator customer base: Amobi, Tuna N. and Kolb, Erik B., "S & P 500 Industry Surveys Broadcasting, Cable & Satellite," February 18, 2010. Total revenue figure for 2008 estimated based on DBS subscriber growth, Source: http://www.sbca.com/receiver-network/industry-satellite-facts.htm.

[a] Time Warner Cable was spun off from Time Warner in 2009. In 2010 Liberty reduced voting control to 3%.

[b] In 2008, News Corp. sold DirecTV to Liberty Media. In 2014, AT&T announced its planned acquisition of DirecTV. This would raise AT&T's market share in MVPDs 23.7%.

[c] In 2014 Comcast announced its intention to buy Time Warner Cable. It also entered into a deal with Charter to trade and share millions of subscribers. The deal between Comcast and Charter is complex The deal was abandoned in 2015 after public and regulatory opposition. Subsequently in 2015, Charter announced a complex deal to merge with Time Warner Cable and Bright House.

Table 18-10. Pooled Electronic Video Retail Distribution (TV stations, Cable MSOs, DBS, IPTV)

| | 1984 | 1988 | 1992 | 1996 | 2001 | 2005 | 2009 | 2012 |
|---|---|---|---|---|---|---|---|---|
| Comcast[a] | 0.4 | 1.4 | 2.0 | 3.3 | 8.4 | 22.0 | 25.3 | 29.1 |
| Jones Intercable | 0 | 0.2 | 0.3 | 0.6 | Comcast | | | |
| Lenfest Communications | 0.4 | 0.4 | 0.5 | 0.8 | Comcast | | | |
| Storer (50% interest) | | KRR | 0.9 | 0.9 | Comcast | | | |
| AT&T | | | | | 15 | Comcast | | |
| MediaOne | | | | 5.7 | AT&T | | | |
| Continental Cablevision | 1.4 | 2.8 | 3.0 | MediaOne | | | | |
| TCI | 2.4 | 6.1 | 9.0 | 7.5 | AT&T | | | |
| Viacom Cable | 1.0 | 1.2 | 1.1 | 0.9 | TCI | | | |
| TCI Entertainment/Tempo/Primestar | | | 0 | 0.8 | | | | |
| US Satellite Broadcasting Corp | | | | 0.4 | | | | |
| Storer (50% interest) | | 0.9 | 0.9 | 0.9 | TCI | | | |
| Heritage | 0.7 | TCI | | | | | | |
| Storer Communications | 1.8 | KKR | TCI/Comcast | | | | | |
| GE[c] NBC | | 2.1 | 1.6 | 2.0 | 1.9 | 2.1 | 2.2 | |
| Time Warner Cable (TWC) | | | | | | | 12.2 | 13.4[e] |
| Time Warner | | | 5.7 | 7.7 | 7.8 | 13.6 | TWC | |
| American TV & Warner Communications | 1.9 | 2.9 | TW | | | | | |
| Westinghouse Cable | 2.6 | AOL TW | | | | | | |
| Adelphia | | 0.6 | 0.8 | 0.8 | 4.4 | TW/Comcast | | |
| Century | 0.4 | 0.6 | 0.9 | 0.7 | Adelphia | | | |
| Insight | | | | | | 0.7 | 0.6 | TWC |
| Advance (Newhouse/Bright House) | 0.7 | 0.9 | 1.4 | TW | | 2.2 | 1.0 | 0.3 |

*continued*

Table 18-10. *continued*

| | 1984 | 1988 | 1992 | 1996 | 2001 | 2005 | 2009 | 2012 |
|---|---|---|---|---|---|---|---|---|
| Cox Communications | 2.1 | 1.8 | 1.8 | 3.0 | 5.7 | 6.0 | 5.4 | 5.5 |
| Times Mirror | 1.3 | 1.2 | 1.2 | Cox | | | | |
| TCA Group | 0.1 | 0.2 | 0.4 | 0.5 | Cox | | | |
| Charter[b] | | | | 0.7 | 5.6 | 5.7 | 3.3 | 4.3[d] |
| Marcus Cable | | | 1.2 | 0.9 | Charter | | | |
| Bresnan | | 0.1 | 0.2 | 0.5 | Charter | | | |
| Falcon | 0.4 | 0.1 | 0.1 | 0.4 | Charter | | | |
| Cablevision Systems | 0.6 | 1.8 | 1.6 | 0.6 | 1.1 | 2.8 | 3.6 | 4.3 |
| DirecTV (AT&T)[f] | | | | | | | | 12.9 |
| Liberty Media (DirecTV) | | | | | 6.4 | 7.0 | 9.4 | 0.5 |
| EchoStar (DISH) | | | | 3.0 | 4.5 | 5.0 | 7.6 | 7.6 |
| 21st Century Fox (News Corp)[c] | | | 1.0 | 1.6 | 2.5 | 3.3 | 3.7 | 4.0 |
| Direct TV[d] | | | | 0.7 | Liberty | | | |
| Chris-Craft | 0.3 | 0.9 | 1.0 | 0.9 | Liberty | | | |
| CBS III | | | | | 1.0 | 2.5 | 1.7 | 2.6 |
| Viacom | | | | | 2.1 | CBS | | |
| Group W (CBS II) | 1.8 | 1.5 | 1.4 | 1.6 | Viacom | | | |
| CBS I | 1.9 | 1.4 | 1.4 | Group W | | | | |
| Tribune | 1.4 | 1.4 | 1.4 | 1.7 | 1.7 | 1.5 | 0.6 | 0.9 |
| Disney | | | | 2.2 | 1.4 | 1.4 | 0.8 | 3.6 |
| CapCities | | 2.8 | 2.1 | Disney | | | | |
| ABC | 2.5 | CapCities | | | | 15. | | |
| Mediacom | | | | | | 0.9 | 1.0 | 0.5 |

| | | | | | | | | |
|---|---|---|---|---|---|---|---|---|
| Gannett | 1.9 | 1.3 | 1.0 | 1.3 | 1.0 | 0.9 | 0.4 | 0.5 |
| Belo | 0.9 | 0.6 | 0.5 | 0.7 | 0.9 | 0.5 | 0.4 | |
| Hearst-Argyle | | 0 | 0.1 | 0.1 | 1.0 | 0.9 | 0.5 | 1.4 |
| Netflix | | | | | | | | 2.1 |
| Univision | | 0.1 | 0.3 | 0.4 | 0.5 | 0.5 | 0.4 | 1.3 |
| CableOne | 0.6 | 0.5 | 0.5 | 0.4 | 0.5 | 0.4 | 0.5 | 0.4 |
| ION Media (Paxson) | | | | | | 0.1 | 0.1 | 0.1 |
| Verizon | | | | | | | 1.5 | 1.5 |
| AT&T[f] | | | | | | | 1.2 | 2.5 |
| All Others | 68.7 | 64.2 | 54.7 | 45.8 | 26.6 | 20 | 20.9 | 0.7 |
| Total US Revenue ($ mil) | 19,028 | 27,826 | 36,715 | 49,746 | 66,305 | 83,213 | 150,331 | 177,509 |
| C4 | 9.6 | 14.6 | 19.8 | 24.2 | 37.6 | 48.6 | 54.5 | 61.4 |
| HHI | 54 | 89 | 158 | 205 | 525 | 853 | 855 | 1385 |
| *N* (>1%) | 14 | 14 | 18 | 12 | 17 | 13 | 14 | 13 |
| MOCDI Index | 14 | 24 | 37 | 59 | 127 | 237 | 229 | 384 |

[a] Comcast acquired NBC from GE in 2009 (51%) and 2013 (remaining 49%). Its acquisition of Time Warner Cable was blocked in 2015.
[b] Charter is owned 27% by Liberty after 2012. In 2015, Charter announced complex deals to merge with Bright House and Time Warner Cable.
[c] In 2013, News Corp spun off its non-print divisions as 21st Century Fox.
[d] In 2007, News Corp. sold DirecTV to Liberty Media, which in 2010 ended its control.
[e] Merger with Charter subject to regulatory approvals pending in 2015.
[f] acquisition by AT&T of DirectTV subject to regulatory approvals pending in 2015.

platforms for OTT service. These are almost always cable TV and telecom firms that are already listed.) National concentration in that video retail market increased from a miniscule HHI of 54 in 1984 to 1,385 in 2012–2013, after the full acquisition of NBC by Comcast. The four-firm concentration ratio rose strongly from 9.6% in 1984 to 54.5% in 2009 and 61.4% in 2013. The largest firms in 2013 were Comcast (29.1%), Time Warner Cable (13.4%, potentially merged with Charter and Bright House), DirecTV (12.9%, potentially acquired by AT&T), and Dish (7.6%). These are all cable MSOs or DBS operators, and they dwarfed the local distribution role of the traditional major TV station companies in 2011: CBS (2.6%), Disney/ABC (3.6%), and 21st Century Fox (4.0%).

## Cable and Satellite TV Program Channels

Unlike over-the-air commercial TV networks and stations, which are supported mostly by advertisers, cable and direct broadcast satellite channels are funded additionally by viewer subscriptions, either by direct payments in the case of pay TV or indirectly through payments to the platform providers who pass some of it to the channel providers. Increased cable and satellite penetration sharply raised cable channels' overall viewing market share, both for basic and pay cable, from 12.3% in 1984 to 36.7% in 1996, 51.4% in 2004, 55.2% in 2007, and 59.8% in 2009. However, since that market share was divided among as many as 800 national programming channels in 2011[27] (plus a variety of local access channels), the individual share of each such channel remained small.

But the major cable network companies usually own and offer several such channels, and concentration trends for programming networks must be examined by firm rather than by channel. Also, though no single cable-programming network has consistently attracted even a 2% share of the overall TV viewing audience, some cable programming networks have amassed significant shares in certain submarkets. This is particularly true for Disney's ESPN Sports channels and for Viacom's Nickelodeon children's programming, which have become competitive with the broadcast networks.

In 1984, cable TV was still in its infancy, with four firms accounting for 66.4% (62.4% by just three firms) of the cable-only programming market. As cable penetration increased, the supply for new programming grew, and vertical and horizontal concentration levels fell significantly, with the top four firms' share dropping to a still considerable 53.2% by 1992. New firms entered, such as Discovery Networks and Liberty Media, but the major traditional TV firms also launched new channels and acquired competitors and concentration rose again. Time Warner's acquisition of Turner raised the combined share of the top four firms still further to 57.1% by 1996, 62.1% in 2004, and 64.2% in 2013. The HHI of the cable TV programming industry was intermediate in size. After 1996 it declined somewhat to about 1,200 but then rose to 1,409. Several of the top cable programmers were vertically combined with integrated cable MSOs, which created a gatekeeping power and affected the market access of competitors. Time Warner disintegrated again in 2009 by spinning off Time Warner Cable while Comcast moved in the opposite direction in 2011 (Table 18.11).

We now summarize findings for all video channels—TV broadcast networks, cable and satellite channels, pay TV, and TV-syndicator-supplied programs to TV stations (Table 18.12). The original three TV networks' share of all TV viewing was only 24.9% in 2002, or about one-third of what it used to be; with Fox added, the top four networks' share becomes 35.4%.[28] But this may be a bit simplistic. First, Comcast (NBC) and Viacom/CBS also have interests in other small terrestrial broadcast networks, such as CW and Telemundo. This adds another 3% overall. Second, the four major TV network firms also own many of the cable channels.

27. National Cable & Telecommunications Association. "Other Industry Data." NCTA.org. Last accessed July 17, 2012, http://www.ncta.com/StatsGroup/OtherIndustryData.aspx.

28. Federal Communications Commission, "In the Matter of Broadcast Ownership Rules, Cross-Ownership of Broadcast Stations and Newspapers," MM Docket 01-235: 15, July 2, 2003.

Table 18-11. Cable and DBS TV Programming Channels—Market Shares by Revenues

| | 1984 | 1988 | 1992 | 1996 | 2000 | 2001 | 2004/5 | 2007 | 2008 | 2009 | 2013 |
|---|---|---|---|---|---|---|---|---|---|---|---|
| Time Warner | | | 5.1 | 33.0 | 28.6 | 26.1 | 23.5 | 18.2 | 18.9 | 23.2 | 23.2 |
| TBS | 26.9 | 25.0 | 23.1 | TW | | | | | | | |
| Viacom | 13.8 | 15.1 | 13.8 | 10.9 | 16.6 | 16.1 | 15.8 | 15.9 | 14.2 | 13.6 | 13.6 |
| CBS[a] | | | | | | | Viacom | | 1.3 | 2.3 | |
| Disney | 21.8 | 11.1 | 11.1 | 13.2 | 14.8 | 14.3 | 15.7 | 16.4 | 16.1 | 15.3 | 15.3 |
| Discovery Networks | | 0.2 | 3.0 | 4.3 | 4.3 | 5.3 | 6.4 | 10.3 | 10.4 | 9.9 | 9.9 |
| 21st Century Fox | | | | | | | | | | | 7.7 |
| News Corp[b] | | | | 0.2 | 5.3 | 5.4 | 6.3 | 7.5 | 6.9 | 7.7 | 21st Century Fox |
| Comcast[c] | | | | | 4.3 | 5.3 | 6.4 | 0.8 | 0.8 | 0.7 | 12.1 |
| GE[d] (NBC Universal) | | 2.0 | 2.1 | 2.3 | 2.6 | 2.9 | 7.0 | 15.3 | 15.8 | 11.4 | Comcast |
| RCA (NBC) | 4.0 | GE | | | | | | | | | |
| Liberty Media | | | 2.1 | 1.5 | 3.1 | 3.2 | 4.8 | 1.6 | 1.5 | 1.5 | 1.5 |
| Cablevision Systems | | | | 3.3 | 2.0 | 2.2 | 2.8 | 2.8 | 2.7 | 2.5 | 2.5 |
| Rainbow Media Group | | Cablevision | | | | | | | | | |
| Others | 33.6 | 46.6 | 39.7 | 31.3 | 18.4 | 19.1 | 11.2 | 10.2 | 11.4 | 11.8 | 11.8 |
| Total US Revenue | 2,466 | 4,261 | 6,504 | 10,906 | 20,205 | 22,917 | 26,879 | 26,421 | 25,761 | 27,100 | 28,509 |
| C4 | 66.4 | 51.4 | 53.2 | 57.1 | 65.3 | 61.9 | 62.1 | 65.8 | 65.0 | 63.6 | 64.2 |
| HHI | 1,403 | 980 | 894 | 1,422 | 1,398 | 1,256 | 1,254 | 1,262 | 1,234 | 1,261 | 1,409 |
| *N* (>1%) | 4 | 4 | 7 | 7 | 9 | 9 | 9 | 9 | 9 | 9 | 8 |
| MOCDI Index | 573 | 400 | 298 | 474 | 442 | 397 | 378 | 365 | 357 | 380 | 498 |

*Source*: Company data from Company SEC filings and *Broadcasting and Cable*. Affiliate fee estimates from Veronis Suhler, Communications Industry Forecast, 2000, 1992, 1989. Cable ad revenues from Cable TV Advertising Bureau and Paul Kagan & Associates. Data for 2004 are based on Hoover's company profiles. Other data from http://www.cmcsk.com/phoenix.zhtml?c=118591&p=irol-newsArticle&ID=636294&highlight=. 2007 are data from "Television Broadcasting FactBook," Bear Stearns, Jan. 2008. "Top 10 US TV Station Companies, Ranked by Revenues," *2008 eMarketer Digital Intelligence*. October 2009. "Cable and Premium TV—Summary: 1980 to 2008," *U.S. Census 2008*. http://www.census.gov/compendia/statab/2010/tables/10s1105.pdf. Last accessed 12/15/10.

[a] Viacom acquired CBS and spun it off again.

[b] In 2013, News Corp. spun off its non-print divisions as 21st Century Fox.

[c] Comcast owned 51% of NBC Universal, after 2011, and 100% after 2013.

[d] In 2011/13, Comcast acquired GE's NBC Universal. For the other 2013 data points, 2009 figures are used.

Table 18-12. Combined Video Network Industries (Market Shares of Broadcast TV Networks, Cable Channels, Pay TV)

| | 1984 | 1988 | 1992 | 1996 | 2001 | 2005/6 | 2009 | 2013 |
|---|---|---|---|---|---|---|---|---|
| CBS ("III") | | | | | | 9.9 | 10.3 | 8.5 |
| Viacom[a] | 5.5 | 6.6 | 6.8 | 6.9 | 20.9 | 11.2 | 8.5 | 12.9 |
| CBS ("I" and "II") | 20.0 | 13.7 | 12.8 | 8.8 | Viacom | | | |
| King World | 0.5 | 1.4 | 2.3 | 2.2 | CBS | | | |
| UPN | | | | 2.9 | Viacom | | | |
| Paramount | | | 1.0 | Viacom | | | | |
| Spelling Entertainment | 0.1 | 0.2 | 0.4 | Viacom | | | | |
| Republic | | 0.1 | 0.2 | Viacom | | | | |
| Time Warner | 5.6 | 5.9 | 8.0 | 19.9 | 20.9 | 19.7 | 20.2 | 10.4 |
| Turner | 4.6 | 6.0 | 6.9 | TW | | | | |
| Lorimar Telepic | | 1.0 | TW | | | | | |
| Telepictures | 0.1 | TW | | | | | | |
| Lorimar | 0.3 | TW | | | | | | |
| Newline Cinema | 0.0 | 0.0 | 0.1 | TW | | | | |
| Comcast[b] | | | | 0.0 | 2.6 | 3.2 | 11.6 | 12.6 |
| GE/Universal | | 16.8 | 11.9 | 11.6 | 8.1 | 10.1 | | |
| RCA/NBC | 17.3 | GE | | | | | | |
| Vivendi | | | | | 2.0 | GE | | |
| USA | 0.7 | 0.9 | 1.2 | 1.6 | Vivendi | | | |
| Universal/MCA | 1.1 | 1.0 | Matsushita | Vivendi | | | | |
| Disney | 4.0 | 3.0 | 3.7 | 14.8 | 14.5 | 14.4 | 13.8 | 12.3 |
| Cap Cities | | 14.0 | 11.3 | Disney | | | | |
| ABC | 19.2 | Cap Cities | | | | | | |
| 21st Century Fox | | | | | | | 10.4 | 11.1 |
| News Corp.[c] | | 4.0 | 7.4 | 6.8 | 7.8 | 8.8 | 5.0 | |

| | | | | | | | | |
|---|---|---|---|---|---|---|---|---|
| Discovery Networks | | | 0.8 | 1.5 | 2.6 | 3.2 | 3.4 | 4.0 |
| Liberty Media | | | 1.4 | 0.7 | 1.9 | 2.4 | 1.3 | 1.3 |
| Cablevision | | | | 1.2 | 1.1 | 1.4 | 0.6 | 0.6 |
| PBS | 1.1 | 1.0 | 1.0 | 0.9 | 0.7 | 0.8 | 0.4 | 0.4 |
| Sony | | | 0.4 | 0.4 | 0.3 | 0.4 | | |
| Columbia Pictures | | 0.2 | Sony | | | | | |
| Tristar | 0.1 | Sony | | | | | 0.3 | 0.3 |
| Gannett | | | | 0.5 | 0.3 | 0.3 | 1.7 | 1.9 |
| Multimedia | 0.1 | 0.1 | 0.2 | Gannett | | | 0.1 | 0.1 |
| Paxson/ION | | | | | 0.2 | 0.1 | 0.1 | 0.1 |
| Saban | | | | | | 0.5 | | 1.3 |
| Univision (Galavision, Telefutura) | | | 0 | 0.2 | 0.3 | Saban | | Saban |
| Others | 19.8 | 24.0 | 22.1 | 19.7 | 16.5 | 14.4 | 11.3 | 21.2 |
| Total | 14,767 | 19,927 | 21,700 | 30,552 | 45,773 | 53,589 | 54,900 | 54.900 |
| C4 | 62.1 | 51.1 | 44.0 | 55.1 | 64.3 | 55.3 | 61.2 | 48.9 |
| Pooled HHI | 1,170 | 812 | 672 | 941 | 1,231 | 1,025 | 1,063 | 805 |
| *N* (>1%) | 9 | 11 | 13 | 11 | 10 | 10 | 11 | 11 |
| MOCDI Index | 390 | 245 | 186 | 284 | 389 | 324 | 321 | 243 |

*Source*: For early data, *Gale Broadcasting & Publishing Directory*, 1990, 1996, and 2001 editions, respectively.

[a] Viacom spun off part of its operation to form CBS (III) in 2006. But the two firms remained under control of the same board chairman, Sumner Redstone.

[b] Comcast acquired 51% of NBC Universal in 2011, and 49% in 2013.

[c] In 2013, News Corp spun off its non-print divisions as 21st Century Fox.

If we look at the latter, the data show that the top four traditional broadcasting companies have 44.8% of the cable channel market in 2005 and 50.3% by 2009. Time Warner grew and has a market share of 23% of the cable channel market in 2013. In overall concentration in the overall TV channel program market, the share (by revenues) of the original Big Three network firms (ABC, CBS, NBC), which used to be 56.5% in 1984, declined to about 34.5% in 2013. If we add a fourth firm—21st Century Fox, with its fourth TV network, Fox, which did not exist in 1984—this number rises to 44.9% in 2013. If we add a fifth firm, Time Warner, which has a strong position in cable networks, and ignore the split-up of Viacom/CBS, the share in video content channel of the top five rises to 73.6%. This is a high number. But it does not show much growth. In 2001, the five-firm share in overall networking in its various forms had been 72.2%. Nor does it yet incorporate the emerging challenge from OTT content providers such as Netflix, Amazon, and Google.

## Film

Six Hollywood studio companies historically have dominated the production, financing, distribution, and exhibition of motion pictures. This handful of firms, all located in the Los Angeles area, has exercised an extraordinary dominance over the film medium of the United States and the entire world. But while this market structure has persisted for an extraordinary 90 years or so, it has been far from placid. In 1949, the US government forced the major firms to divest their theater chains. In the 1950s and 1960s, weakened by the emergence of television and rapidly shrinking theater attendance, most studios were acquired by general business conglomerates. Later, they became holdings in the portfolio of communications companies, both American and foreign. Vertical integration increased again as these companies diversified their distribution into TV network ownerships. The major film studio companies themselves became primarily distribution and financing firms that supported and bundled the production of independent production companies.

The firms' market shares fluctuated with box office success. Twentieth Century Fox seesawed from a high of 19.5% in 1970 to a low of 9.4% in 1994 to 13.2% in 2009 and 9.7% in 2013. Disney dropped to a 4.8% market share in 1978 but grew to 21% in 1996, dropped again to 11.6% in 2009, and rose to 15.7% in 2013.

Table 18.13 shows the concentration trend of film distribution. This market has been fairly steady in the aggregate, with six studios accounting for 75% to 85% of the market for decades. "Mini-majors" such as MGM, DreamWorks SKG, Miramax, and Pixar have been absorbed into the six-firm structure. But since no firm accounts for more than 20%, the HHI concentration is flat and in the 1,100–1,300 range, well inside the "moderate" level of US governmental antitrust standards (Table 18.13).

Films are not the only product of film studios. Another major line of business is TV shows. Table 18.14 shows the concentration trends in the production of prime time TV shows. In the 1970s and 1980s, small firms, because of the regulations in place, produced more programs for the major networks. With changes in regulation in the mid-1990s, the Big Three networks soon began to produce more of the shows they broadcasted. More significantly, the Hollywood studios bought two of the three major networks, started a fourth, and added two minor networks. Thus, the major Hollywood studios remained the largest TV show producers. Time Warner, Viacom, Disney, and News Corp. all increased their shares through acquisitions in the 1990s. The C4 increased from 37.0% to 62.7% between 1984 and 2013. The remainder of the market was divided among a handful of relatively small companies. The overall US-only revenue in this industry grew from $2.2 billion in 1984 to $6.8 billion in 2004–2005 to 7.6 million in 2013.

The smaller TV production firms are fragmented and dependent on the graces of the major networks, their own competitors in production. On the other hand, the major production companies also provide programs for rival

Table 18-13. Film Production/Distribution (Market Share by Box Office %)

| | 1970 | 1978 | 1982 | 1984 | 1988 | 1990 | 1992 | 1994 | 1996 | 2000 | 2001 | 2004 | 2006 | 2009 | 2013 |
|---|---|---|---|---|---|---|---|---|---|---|---|---|---|---|---|
| Time Warner (Warner Bros/Warner Comm.) | 5.3 | 13.2 | 10.0 | 11.0 | 12.7 | 13.1 | 14.7 | 16.2 | 15.7 | 11.9 | 21.4 | 17.5 | 14.3 | 19.9 | 17.1 |
| NewLine | | | | | 2.7 | 4.4 | 5.4 | 6.4 | 5.3 | 6.3 | TW | | | | |
| Disney (incl. Miramax and Pixar) | 9.1 | 4.8 | 7.0 | 9.8 | 14.5 | 15.5 | 17.5 | 19.5 | 21.0 | 19.1 | 17.8 | 16.3 | 16.2 | 11.6 | 15.7 |
| Sony (Japan) | | | | | | 13.9 | 11.7 | 9.5 | 10.6 | 8.8 | 10.3 | 16.4 | 20.4 | 13.7 | 10.5 |
| Columbia / Tristar | 19.4 | 11.6 | 10.0 | 10.3 | 11.8 | Sony | | | | | | | | | |
| Tristar | | | | | Columbia | | | | | | | | | | |
| MGM / UA / Pathe | 3.4 | N/A | 11.0 | 8.3 | 3.7 | 2.8 | 2.8 | 2.8 | 5.1 | 1.3 | 5.3 | Sony | | | |
| Orion / Metromedia | | | | | 8.8 | 5.6 | MGM | | | | | | | | |
| 21st Century Fox (ex-News Corp[a]) | | | | | | 13.1 | 11.3 | 9.4 | 12.6 | 9.7 | 10.2 | 9.9 | 15.2 | 13.2 | 9.7 |
| Fox | 19.5 | 13.4 | 14.0 | 11.9 | 10.2 | News Corp | | | | | | | | | |
| Viacom/Paramount | | | | | | | | 14.2 | 12.7 | 11.7 | 10.6 | 6.7 | 10.3 | 13.9 | 8.9 |
| Paramount[b] | 11.8 | 15.8 | 14.0 | 16.3 | 18.1 | 14.9 | 14.6 | Viacom | | | | | | | |
| DreamWorks SKG | | | | | | | | | | 10.3 | 4.8 | 10.0 | Viacom | | 1.8 |
| Comcast[c] | | | | | | | | | | | | | | | 13.1 |
| GE | | | | | | | | | | | | 9.5 | 8.9 | 8.4 | Comcast |
| Vivendi Universal | | | | | | | | | | 14.5 | 11.4 | GE | | | |
| Seagrams | | | | | | | | | 8.4 | Vivendi | | | | | |
| Matsushita | | | | | | | 12.9 | 12.6 | Seagrams | | | | | | |
| Universal / MCA | 13.1 | 16.8 | 19.9 | 16.4 | 11.8 | 13.1 | Matsushita | | | | | | | | |
| Lionsgate | | | | | | | | | | | | 3.2 | 3.6 | 3.8 | 9.8 |
| Summit Entertainment | | | | | | | | | | | | | | 4.6 | Lionsgate |
| Overture Films | | | | | | | | | | | | | | 1.5 | |
| Weinstein Company | | | | | | | | | | | | | 2.5 | 1.9 | 4.5 |

*continued*

Table 18-13. *continued*

| | 1970 | 1978 | 1982 | 1984 | 1988 | 1990 | 1992 | 1994 | 1996 | 2000 | 2001 | 2004 | 2006 | 2009 | 2013 |
|---|---|---|---|---|---|---|---|---|---|---|---|---|---|---|---|
| Others | 18.4 | 24.4 | 14.1 | 16.1 | 5.7 | 3.6 | 9.3 | 9.4 | 8.6 | 6.5 | 8.2 | 10.5 | 8.6 | 7.7 | 8.3 |
| US Market Revenue ($ mil) | 1,430 | 2,644 | 3,453 | 4,031 | 4,458 | 5,022 | 4,871 | 5,396 | 5,912 | 7,661 | 8,413 | 9,406 | 9,407 | 9,817 | 9,480 |
| C4 | 63.8 | 67.2 | 69.0 | 60.7 | 57.4 | 57.5 | 59.5 | 59.4 | 62.0 | 57.2 | 61.2 | 60.1 | 66.1 | 60.7 | 56.4 |
| C6 | 78.2 | 83.6 | 89.0 | 80.8 | 78.7 | 83.6 | 82.5 | 81.4 | 81.0 | 77.2 | 81.7 | 79.6 | 85.3 | 80.7 | 75.9 |
| HHI | 1,190 | 1,043 | 1,158 | 1,067 | 1,179 | 1,229 | 1,198 | 1,231 | 1,245 | 1,171 | 1,278 | 1,184 | 1,319 | 1,265 | 1,182 |
| *N* (>1%) | 7 | 6 | 7 | 7 | 9 | 9 | 8 | 8 | 8 | 9 | 8 | 8 | 8 | 11 | 9 |
| MOCDI Index | 450 | 426 | 438 | 403 | 393 | 410 | 424 | 435 | 440 | 390 | 452 | 419 | 466 | 381 | 394 |

*Main sources*: Entertainment Data Inc. as listed in Screen Digest, Feb. 1, 1997; Wasko, Janet, Movies and Money: Financing the American Film Industry, Ablex Publishing Corporation, 1982, p. 152. Domestic market shares (percentage of US and Canadian rentals) compiled from Variety, Jan. 15, 1975; Feb. 11, 1977; and Jan. 18, 1979. Leisure Time, Standard and Poor Industry Surveys, 1983, 1992, 1996, 2001, 2004, 2013.

[a] In 2013, News Corp spun off its non-publishing divisions as 21st Century Fox.

[b] Viacom's Paramount bought DreamWorks SKG for $1.6 billion in 2006, then sold the DreamWorks film library to private investors for $900 million. DreamWorks Animation remains a separate independent production company, distributed by Paramount. Dreamworks studios was re-created after 2009 with the Reliance conglomerate of India.

[c] GE sold Universal to Comcast (51% in 2011, 49% in 2013).

## Table 18-14. TV Show Production (Market Shares Based on Network Prime Time Program Hours)[a]

| PRODUCER | 1970 | 1975 | 1980 | 1984 | 1988 | 1992 | 1996 | 2001 | 2004/5 | 2010 | 2013 |
|---|---|---|---|---|---|---|---|---|---|---|---|
| 21st Century Fox News Corp. | | | | | | 2.7 | 4.4 | 11.6 | 18.7 | 13.6 | 17.4 |
| 20th Century Fox | 9.2 | 2.5 | 3.8 | 8.0 | 4.6 | NewsCorp | | | | | |
| Stephen J. Cannell | | | 0.6 | 7.0 | 4.9 | NewsCorp | | | | | |
| New World | | | | | 5.2 | NewsCorp | | | | | |
| Disney | 1.9 | 2.5 | 2.9 | 3.0 | 4.3 | 9.8 | 9.5 | 14.1 | 18.7 | 12.6 | 11.8 |
| ABC | | | | | 1.8 | 4.1 | Disney | | | | |
| Viacom (Paramount) | | | 0.3 | 1.3 | 5.4 | 3.8 | 8.9 | 14.7 | CBS | | |
| Spelling | 2.3 | 4.2 | 8.4 | 8.4 | 2.7 | 0.6 | Viacom | | | | |
| Quinn Martin | | 10.3 | 2.8 | Spelling | | | | | | | |
| Paramount | 6.4 | 6.9 | 7.4 | 7.0 | 5.4 | 6.3 | Viacom | | | | |
| CBS (I, II, & III) | 1.9 | 2.5 | 1.1 | 1.0 | 1.0 | 7.8 | 12.2 | Viacom | 14.6 | 18.8 | 16.9 |
| Time Warner (Warner Bros.) | 1.9 | 4.5 | 5.4 | 8.9 | 10.0 | 22.3 | 23.3 | 16.9 | 10.4 | 18.4 | 16.6 |
| Lorimar | | 3.6 | 8.3 | 7.6 | 9.8 | TW | | | | | |
| Comcast (Universal) | | | | | | | | | | 6.2 | 9.4 |
| GE | 1.8 | 2.5 | 5.2 | 0.5 | 0.7 | 1.5 | 6.1 | 8.3 | 10.4 | Comcast | |
| Vivendi (Universal/MCA) | 14.4 | 22.4 | 14.8 | 8.6 | 12.8 | 13.5 | 10.0 | 5.5 | GE | | |
| Endemol (Netherlands) | | | | | | | | 4.0 | 7.5 | 4.4 | News Corp. |
| Sony (Japan) (Columbia) | | | | | 8.1 | 4.7 | 7.8 | 6.2 | 4.2 | 2.6 | 4.2 |
| Columbia | 6.0 | 7.3 | 6.2 | 11.1 | Sony | | | | | | |
| Embassy | | | 2.5 | 4.8 | Sony | | | | | | |
| Tandem (Norman Lear) | | 5.2 | 3.2 | Embassy | | | | | | | |
| MGM | 4.5 | 2.4 | 2.9 | 2.0 | 2.4 | 2.5 | 2.4 | 2.4 | Sony | | |
| MTM | | 5.0 | 6.8 | 8.0 | 2.8 | | | | | | |

*continued*

Table 18-14. *continued*

| PRODUCER | 1970 | 1975 | 1980 | 1984 | 1988 | 1992 | 1996 | 2001 | 2004/5 | 2010 | 2013 |
|---|---|---|---|---|---|---|---|---|---|---|---|
| Alan Landsburg | | | 1.6 | 3.8 | 2.0 | | | | | | |
| Witt-Thomas | | | 1.8 | 1.1 | 2.3 | 2.5 | 3.3 | 3.0 | 2.0 | | |
| Filmways/Orion | 4.4 | N/A | 1.3 | 2.2 | 1.8 | | | | | | |
| Bochco | | | | | | 2.8 | 4.4 | | | | |
| Carsey/Werner | | | | 0.6 | 2.3 | 5.3 | 2.2 | 2.0 | 2.0 | | |
| WGBH (PBS) | | | | | | | | | | 2.0 | 2.0 |
| Others | 45.3 | 18.2 | 12.7 | 6.1 | 9.7 | 9.8 | 5.5 | 11.4 | 11.5 | 23.4 | 20 |
| Total Revenue ($mil) | 696 | 1,056 | 1,602 | 2,236 | 3,122 | 4,358 | 6,083 | 6,400 | 6,776 | 7,249 | 7,559 |
| C4 | 36.0 | 46.9 | 38.9 | 37.0 | 40.7 | 53.4 | 55.0 | 57.2 | 62.4 | 63.4 | 62.7 |
| C6 | 44.9 | 57.1 | 51.9 | 53.0 | 51.5 | 65.0 | 71.7 | 71.7 | 80.3 | 78.6 | 76.3 |
| C8 | 49.1 | 65.8 | 62.5 | 67.6 | 61.6 | 73.8 | 82.2 | 81.2 | 86.5 | 85.6 | 80.1 |
| HHI | 428 | 842 | 619 | 681 | 618 | 988 | 1,119 | 1,005 | 1,211 | 1,181 | 1,116 |
| *N* (>1%) | 11 | 14 | 18 | 16 | 19 | 14 | 12 | 11 | 9 | 8 | 8 |
| MOCDI Index | 129 | 225 | 146 | 170 | 142 | 264 | 323 | 303 | 404 | 418 | 395 |

[a]In 2013, News Corp spun off its non-print divisions as 21st Century Fox. 2013 figures repeat 2012 and add structural changes. Viacom acquired Paramount in 1994, Spelling Entertainment in 1995, and CBS in 1999; Sony bought Columbia and Embassy in 1989; Time Warner acquired Lorimar in 1989; Vivendi acquired Seagram (Universal's parent company) in 2004; film and TV operations were sold to an entity that is 80% owned by GE's NBC; News Corp acquired 20th Century Fox in 1985; Fox acquired New World and Stephen Cannell Productions in 1994; Disney acquired ABC in 1995. TV production revenues were estimated using $1.9 million for ½ hr of programming a prime-time show for 30 weeks of the year and $0.95 million for the remaining 22 weeks with 56 hours of prime time television per week, for all major networks, with 8% increase in cost per year after 1996. 2009 data: TV Guide: www.tvguide.com for shows' shares. 2010 revenue data are based on growth rates from 1996 to 2005 and company 10-K reports.

networks for whose business they compete. These figures should therefore be interpreted cautiously. What the numbers show is that after 1996, the six major Hollywood studios have collectively dominated television production. The major change was vertical integration. By 2004, five of the six major producers, with the exception of Sony, owned TV broadcast networks the traditional three networks plus two to three new ones. This reduced market access for independents as well as for rival network companies. In 2010, ABC chose 9 Disney programs out of 13 regular series, while Fox picked 7 of its own series out of 10. CBS was less vertically integrated, with 3 series.

Combining Tables 18.13 and 18.14 gives us Table 8.15, which shows the aggregated TV and Film production market. The overall concentration in this market went up steadily, peaking around 2001 at a moderately concentrated level of 1,142, before dropping to 956 by 2013. Time Warner has been the top firm throughout the past two decades, with Disney trailing closely, and 21st Century Fox (formerly News Corp.) slowly catching up since 2000. But these numbers must be used with caution. The six "majors" distribute many films produced by other, often by independent producers. Also, cable TV show producers are not included in this table.

Will the Internet change film industry concentration? In the past, improvements in distribution methods have not threatened the production sector in the long term, only the retail distribution. Although the Hollywood production and distribution industry has historically opposed almost any new delivery technology—broadcast television, pay TV, VCRs—it ended up benefiting from them. The Internet, despite its "long tail" of narrowly focused low-budget content production, is not likely to reduce the importance of exquisitely produced, high-budget, popular entertainment with special effects and famous stars.

## TELECOMMUNICATIONS MEDIA

Consumers and organizations use telecommunications services—two-way individualized electronic communication—more than ever before: at home, in the office, on the road, at the beach, when web surfing, chatting with friends, e-mailing, downloading music and video clips, holding a meeting, running a company. The US telecommunications industries generated revenues of $391 billion in 2011 for services and $63 billion for equipment.

### Wireline Telecommunications

The overall concentration trend of the telecom industry has been strongly S-shaped: a huge decline in 1983 with the split-up of AT&T; a gradual decline from 1984 to 1996, a pronounced reconcentration thereafter; and a second decline as cable and online Internet telephony emerged as competitors (Table 18.16).

Before its 1984 divestiture, AT&T accounted for nearly 77% of local telephone revenues nationally. The divestiture decree separated AT&T's 22 local operating companies and reorganized them into seven independent companies. In addition, well over 1,500 independent LECs existed at the time, of which GTE and United Telecommunications were the largest. Before the AT&T divestiture, the national wireline (i.e., overall telecom) HHI was at almost 8,000. The divestiture radically dropped that number to 1,633. The subsequent trend was first greater entry, then a reconsolidation. The major regional Bell companies began to merge into two major groupings, SBC (which was renamed as the new AT&T) and Verizon. Mergers caused the national market share of the top four firms to grow from 48.1% in 1992 to 66.3% in 2012. By the HHI measure, concentration in the national wireline market doubled from moderate (after 1984) to high again after 2009. It then declined again.

In the 2000s, cable companies gained considerable ground in the consumer landline market with the introduction of triple play services. Their collective share increased steadily from 5% in 2002 to almost 20% by 2010. VoIP providers, including Vonage and Skype, also played a role in the wireline market, providing a low-cost alternative to telecom and cable telephony. But their share was below 2%.

Table 18-15. Pooled Film and Network TV Production

| | 1984 | 1988 | 1992 | 1996 | 2001 | 2004/5 | 2009 | 2013 |
|---|---|---|---|---|---|---|---|---|
| Time Warner (Warner Brothers/ Warner Comm.) | 10.4 | 11.6 | 18.3 | 19.6 | 19.8 | 15.0 | 18.7 | 16.9 |
| NewLine | | 1.6 | 2.9 | 2.6 | TW | | | |
| Lorimar | 2.8 | 4.0 | TW | | | | | |
| Disney (incl. Miramax) | 7.5 | 10.3 | 13.9 | 15.2 | 16.5 | 17.9 | 11.7 | 13.8 |
| ABC | | 0.7 | 1.9 | Disney | | | | |
| Sony (Japan) | | 3.3 | 8.4 | 9.2 | 8.7 | 11.7 | 8.7 | 7.8 |
| MGM/UA/Pathe | 6.1 | 3.2 | 2.7 | 3.7 | 4.1 | Sony | | |
| Columbia | 10.8 | 6.9 | Sony | | | | | |
| 21st Century Fox | | | 7.2 | 8.4 | 11.0 | 14.0 | 13.0 | 13.0 |
| News Corp[a] | | | | | | | | |
| Fox | 10.7 | 7.9 | News Corp | | | | | |
| Stephen J. Cannell | 2.5 | 2.0 | News Corp | | | | | |
| New World | | 2.1 | News Corp | | | | | |
| Viacom/Paramount | 0.5 | 2.2 | 1.8 | 10.8 | 12.6 | 4.0 | 7.8 | 5.1 |
| Paramount | 13.2 | 12.9 | 10.7 | Viacom | | | | |
| DreamWorks SKG | | | | | 2.8 | 6.0 | | 1.0 |
| Spelling | 3.0 | 1.1 | 0.3 | Viacom | | | | |
| CBS (I, II, & III) | | 0.4 | 3.7 | 6.2 | Viacom | 6.3 | 7.7 | 7.3 |
| Comcast | | | | | | | 4.6 | 11.5 |
| GE | 0.2 | 0.3 | 0.7 | 3.1 | 3.6 | 10.2 | Comcast | Comcast |
| Vivendi Universal | 3.1 | 5.3 | 6.4 | 5.1 | 9.0 | GE | | |
| Seagrams | | | | 4.1 | Vivendi | | | |
| Matsushita | | | 6.8 | Seagrams | | | | |
| Universal/MCA | 10.7 | 7.0 | Matsushita | | | | | |

| | | | | | | | | |
|---|---|---|---|---|---|---|---|---|
| Lionsgate | | | | | | 1.9 | 2.1 | 5.6 |
| Summit Entertainment | | | | | | | 2.6 | Lionsgate |
| Overture Films | | | | | | | 0.8 | |
| Weinstein Company | | | | | | 1.5 | 1.1 | 2.5 |
| Endemol (Netherlands) | | | | | 1.7 | 3.1 | 1.9 | |
| MTM | 2.9 | 1.2 | | | | | | |
| Alan Landsburg | 1.4 | 0.8 | | | | | | |
| Witt-Thomas | 0.4 | 0.9 | 1.2 | 1.7 | 1.3 | 0.9 | | |
| Filmways/Orion | 0.8 | 5.9 | | | | | | |
| Bochco | | | 1.3 | 2.2 | | | | |
| Carsey/Werner | 0.2 | 0.9 | 2.5 | 1.1 | 0.9 | 0.9 | | |
| Others | 12.7 | 7.4 | 9.5 | 7.0 | 8.0 | 6.7 | 19.4 | 14.7 |
| Total Revenue | 6,160 | 7,580 | 9,229 | 11,995 | 14,813 | 16,417 | 16,616 | 17,030 |
| C4 | 45.4 | 42.6 | 51.2 | 54.7 | 59.8 | 58.5 | 52.0 | 55.2 |
| C6 | 63.3 | 56.6 | 65.3 | 69.4 | 77.6 | 75.1 | 67.6 | 70.3 |
| HHI | 767 | 687 | 894 | 1,004 | 1,142 | 1,090 | 901 | 956 |
| *N* (>1%) | 13 | 17 | 15 | 14 | 11 | 11 | 12 | 10 |
| MOCDI Index | 213 | 167 | 231 | 268 | 344 | 329 | 260 | 302 |

[a] In 2013, spun off from News Corp as 21st Century Fox.

Table 18-16. Wireline Telecom

| | 1983 | 1984 | 1988 | 1992 | 1996 | 2002 | 2006 | 2009 | 2012 |
|---|---|---|---|---|---|---|---|---|---|
| **Telecom Providers (share of overall wireline)** | | | | | | | | | |
| AT&T "II" (SBC) | | 5.8 | 5.7 | 5.2 | 5.8 | 15.8 | 37.7 | 29.6 | 23.1 |
| AT&T "I" | 87.5 | 35.3 | 30.1 | 25.8 | 24.2 | 21.0 | SBC | | |
| Ameritech | | 7.9 | 7.0 | 6.4 | 4.8 | SBC | | | |
| Pacific Telesis | | 5.7 | 6.3 | 5.5 | 4.3 | SBC | | | |
| BellSouth | | 8.7 | 8.5 | 8.0 | 7.1 | 6.9 | SBC | | |
| Verizon | | | | | | 18.2 | 26.4 | 21.3 | 18.6 |
| Worldcom | | | 0.1 | 0.5 | 2.8 | 12.9 | Verizon | | |
| MCI | 1.9 | 1.8 | 4.3 | 7.0 | 10.1 | Worldcom | | | |
| Bell Atlantic | | 8.7 | 7.2 | 7.0 | 11.0 | | | | |
| NYNEX | | 6.6 | 8.4 | 7.2 | 6.6 | Bell Atl | | | |
| GTE (incl. Contel) | 5.0 | 5.0 | 7.5 | 6.8 | 7.2 | Verizon | | | |
| CenturyLink | | | | | | | | 11.4 | 11.5 |
| Embarq[f] | | | | | | | 2.0 | CenturyLink | |
| Qwest | | | 0.1 | 0.3 | 0.8 | 4.8 | 4.5 | | CenturyLink |
| LCI | | | 0.2 | 0.6 | 1.9 | Qwest | | | |
| US West | 0.3 | 5.7 | 6.0 | 5.7 | 5.0 | Qwest | | | |
| Sprint (United)/Softbank (Japan) | | 0.1 | 0.2 | 4.2 | 4.9 | 4.3 | 3.6 | 1.8 | 1.7 |
| Frontier (Citizens Telecom) | | | | | | 1.7 | 2.7 | 3.3 | 2.6 |
| Level 3 | | | | | | | 0.7 | 1.9 | 9.5 |
| Global Crossing | | | | 0.3 | 1.0 | 1.2 | 1.9 | 2.6 | Level 3 |
| Time Warner Telecom | | | | | | 0.1 | 0.1 | 0.2 | Level 3 |
| Other LECs | 5.7 | 8.7 | 8.4 | 9.5 | 8.5 | 8.1 | 9.2 | 9.0 | 3.8 |
| *Total Telecom* | 100.0 | 100.0 | 100.0 | 100.0 | 100.0 | 95.0 | 88.7 | 78.7 | 74.4 |

| | | | | | | | | | |
|---|---|---|---|---|---|---|---|---|---|
| **Cable Providers (share in overall wireline)** | | | | | | | | | |
| Comcast | | | | | | 0.7 | 3.2 | 5.6 | 8.0 |
| Time Warner Cable | | | | | | 1.5 | 3.0 | 5.1 | 4.1 |
| Cablevision | | | | | | 1.5 | 2.5 | 3.0 | 3.3 |
| Cox Communication | | | | | | 1.0 | 1.3 | 2.2 | 2.8 |
| Charter | | | | | | 0.2 | 0.4 | 1.3 | 1.6 |
| Other Cable Providers: | | | | | | 0.1 | 0.4 | 2.4 | 2.7 |
| *Total Cable* | | | | | | 5.0 | 10.8 | 19.6 | 22.5 |
| **VoIP (share of overall wireline–based service)** | | | | | | | | | |
| Vonage | | | | | | | 0.4 | 1.2 | 1.2 |
| Skype | | | | | | | 0.1 | 0.4 | 0.4 |
| Other non- cable VoIP Providers: | | | | | | | 0.0 | 0.0 | 0.0 |
| *Total non-cable VoIP* | | | | | | | 0.5 | 1.7 | 1.7 |
| Overall Wireline Revenue ($ mil) | 92,548 | 104,020 | 122,992 | 150,487 | 186,308 | 247,831 | 243,437 | 214,103 | 174,731 |
| C4 | 94.7 | 60.6 | 54.4 | 48.1 | 52.5 | 67.9 | 72.2 | 64.9 | 66.3 |
| HHI | 7,685 | 1,633 | 1,331 | 1,075 | 1,049 | 1,290 | 2,191 | 1,697 | 1,395 |
| *N* (>1%) | 3 | 10 | 10 | 11 | 12 | 12 | 11 | 14 | 12 |
| MOCDI Index | 4,437 | 516 | 421 | 324 | 303 | 372 | 661 | 454 | 403 |

*Main sources*: Revenue data from annual reports, industry surveys. FCC, Statistics of Common Carriers, 1986, 1990, 1995, 1996, 1997, 1998, 1999, 2000/2001. FCC. Bell Operating Companies aggregated to post-divestiture model for comparability. Figures include Network Access Revenues, mobile backhaul and ISP backbones. Market shares for 1988 and 1992 extrapolated based on adjacent years; revenue from "Acquisition of Teleport Communications" in: CLEC Report 2000, New Paradigm Resources Group, INC. Figures for 2000 for McLead USA, Time Warner Telecom, and NextLink are based on "CLEC Revenue" as listed in CLEC Report 2000, New Paradigm Resources Group. Global Crossing bought Frontier in 1999. Citizens Communications changed its name to Frontier Communications in 2008. Sprint spun off its local wireline business to form Embarq, keeping only its long distance operations. 2006 includes the mergers between AT&T/SBC and Verizon/MCI, assigning 50% of AT&T local share in 2004 to SBC in 2006. Assigned 33% of overall VoIP revenue to local service. 2013 data for Cablevision, Cox, Charter assumed growth of 10%. In 2015, a merger of Charter and Time Warner Cable was pending.

Perhaps the major alternative to the incumbent local telephone companies (ILECs) emerged in the form of cellular mobile wireless. As penetration zoomed and prices dropped, many users opted for wireless service as a substitute for wireline and dropped the latter altogether.

## Mobile Telecommunications

Mobile services, after the sharp decline in concentration level due to the breakup of the original AT&T in 1984, grew tremendously while also consolidating. The concentration trend is shown in Table 18.17. The industry merged after 1984 and again 1995, when it was licensed in a purposefully decentralized fashion on a national level. This led to a second wave of consolidations, in which the telecom firms sought to create national "footprints." The number of such national firms declined after 1996 to six and then four national footprints. A few regional companies still exist, often as partners of the national firms. Deals between SBC (with Ameritech) and Bell South created Cingular, to which AT&T and its name were added in 2004–2005. Verizon Wireless was created out of the cellular operations of Bell Atlantic, NYNEX, GTE, US West, and PacTel (Airtouch), the latter owned by the United Kingdom's mobile phone giant Vodafone. In 1999, Vodafone agreed to merge its considerable US wireless segment into Verizon Wireless, of which it owned 45%. But in 2013, the two companies agreed to end their partnership and Verizon bought out Vodafone. In 2002, Deutsche Telekom acquired Voice-stream and renamed it T-Mobile. In 2004, Sprint acquired the bankrupt Nextel to form Sprint Nextel, which was the third biggest player in the market. By 2005, four companies, all with nationwide footprints, held much of the market (88.4% in 2006 and 93.7% in 2011). The two Bell-successor companies accounted for 50.4% in 2006 and 66.5% in 2013, after the buy-out of Vodafone's share by Verizon (Table 18.17). AT&T tried to acquire T-Mobile from Deutsche Telekom, but the deal was blocked by the US government. T-Mobile then bought Metro PCS, a major discount reseller (MVNO). Sprint, in turn, was acquired by Softbank, a major ISP and mobile network operator in Japan. Other MVNOs are TracFone (America Movil), Cricket (AT&T), Ting, and H20. HHI concentration rose greatly, from 341 in 1984 (in a tiny market of $340 million) to 2,636 in 2013, in a huge market of $160 billion.

## Total Telecom Services

So far, we have looked at the wireline and wireless telecom subindustries in a separate fashion. A third subindustry is ISP service, essential for Internet access by endusers. That industry is discussed further below in the section on the Internet. We can put the three segments together in one table that summarizes the telecommunications sector. This is shown in Table 18.18.

The overall telecommunications services market is dominated by a few key players. AT&T (the old SBC, renamed) and Verizon hold 27.1% and 25% of the market, respectively, in 2012–2013. The HHI plunged from 7,652 in 1983 to 970 in the mid-1990s, largely due to the split-up of the Bell System and competitive entry, but then rose to around 1,948 by 2005. It then dropped again as mobile and cable competition strengthened, (1,491 in 2009) and then reconsolidated slightly. The C4 shows a similar trend, dropping from 95% to 46%, rising again to 72%, and 66% in 2013. The combined market share of non-incumbent new telecom entrants after the AT&T breakup increased, though with numerous consolidations taking place. By 2013, they collectively accounted for about 31% of the telecom market. Cable television companies have gained shares since 2000, with the biggest players, Comcast, Time Warner Cable, and Charter, becoming medium-sized telecom providers. Independent VoIP providers have not yet become a major force in the domestic part of the market.

The national concentration measures overstate the options available to users, since the major local carriers do not usually compete. This leads us to present market shares also in a different way, on the basis of local markets.

Table 18.19 shows the overall telecom market from a *local* perspective. To a residential user, choices available normally include one telecom incumbent (typically either AT&T or Verizon), one cable company, four wireless providers

## Table 18-17. Mobile Cellular Telephone Service

| | 1983 | 1984 | 1988 | 1992 | 1996 | 1998 | 2000 | 2001 | 2004 | 2006 | 2009 | 2013 |
|---|---|---|---|---|---|---|---|---|---|---|---|---|
| AT&T "II" (SBC) | | | | | | | | | | 35.0 | 36.5 | 32.9[a] |
| Cingular[b] | | | | | | | 18.5 | 19.9 | 21.2 | | | |
| SBC | | 5.4 | 6.0 | 7.4 | 7.5 | 11.0 | Cingular | | | | | |
| Ameritech | | 6.5 | 7.5 | 6.5 | 5.5 | SBC | | | | | | |
| BellSouth | | 5.2 | 5.3 | 6.0 | 6.0 | 7.0 | Cingular | | | | | |
| Vanguard / Cellular One | | | | | | | | | | | | |
| AT&T (I) (McCaw) | 44.5 | 0.9 | 2.0 | 10.8 | 11.0 | 12.0 | 16.0 | 14.0 | 14.0 | Cingular | | |
| SNET | | 0.8 | 0.9 | 1.0 | 1.0 | 0.5 | Cingular | | | | | |
| Leap Wireless | | | | | | | | | | | 1.0 | AT&T |
| Verizon Wireless[c] | | | | | | | 13.4 | 15.3 | 15.3 | 15.4 | 17.9 | 33.6 |
| Vodafone (UK) | | | | | | 12.0 | 11.0 | 12.6 | 12.6 | 12.6 | 14.6 | Verizon |
| Bell Atlantic | | 7.6 | 6.3 | 8.9 | 10.0 | 12.0 | Verizon | | | | | |
| Metro Mobile | 2.0 | 3.4 | 3.2 | 2.6 | 2.5 | Bell Atlantic | | | | | | |
| NYNEX[d] | | 6.5 | 6.5 | 6.5 | Bell Atlantic | | | | | | | |
| GTE Mobilenet/Contel[e] | 6.8 | 6.8 | 7.3 | 8.8 | 9.0 | 8.0 | Verizon | | | | | |
| Vodafone (UK) | | | | | | 12.0 | Verizon/Vodafone | | | | | |
| Air Touch (Pac Tel) | 0.0 | 7.0 | 7.1 | 10.3 | 11.0 | Vodafone | | | | | | |
| US West New Vector | 0.0 | 5.5 | 4.1 | 3.9 | Airtouch | | | | | | | |
| MCI | | | | 1.6 | 2.0 | | 0.2 | 0.1 | 0.1 | Verizon | | |
| Alltel[f] | 1.4 | 1.4 | 1.2 | 1.8 | 2.0 | 6.0 | 5.7 | 6.5 | 6.5 | 6.5 | Verizon | |
| Sprint Nextel (Softbank) (Japan)[g] | 1.0 | 1.0 | 1.0 | 4.6 | 5.0 | 6.0 | 7.5 | 11.2 | 11.2 | 18.9 | 16.8 | 16.3 |
| Centel[h] | | | 4.5 | 3.5 | Sprint | | | | | | | |
| Nextel | | | | 3.1 | 3.9 | 4.7 | 5.5 | 7.7 | 7.7 | Sprint | | |

*continued*

Table 18-17. *continued*

| | 1983 | 1984 | 1988 | 1992 | 1996 | 1998 | 2000 | 2001 | 2004 | 2006 | 2009 | 2013 |
|---|---|---|---|---|---|---|---|---|---|---|---|---|
| T-Mobile[i] (Deutsche Telekom, Germany) | | | | | | | 4.3 | 5.9 | 5.9 | 5.9 | 7.9 | 12.4[j] |
| Voicestream Metro PCS[j] | | | | | 0.2 | 2.0 | | | | | | T-Mobile |
| Century | 1.2 | 1.2 | 1.0 | 0.4 | 0.3 | 1.0 | 0.8 | 0.7 | 0.7 | 0.7 | | |
| US Cellular(TDS)[k] | | | | 0.9 | 1.0 | 3.0 | 2.8 | 3.3 | 3.3 | 3.3 | 2.1 | 2.2 |
| Other | 43.1 | 40.8 | 36.1 | 11.4 | 22.1 | 14.8 | 14.3 | 2.8 | 1.5 | 1.7 | 1.0 | 1.0 |
| Total subs | | 0.1 | 2.1 | 16.0 | 44.0 | 69.2 | 109.4 | 128.4 | 182.1 | 194.6 | 285.6 | 302.9 |
| US Total Revenue ($mil) | 200 | 340 | 1,959 | 12,253 | 26,415 | 36,633 | 51,908 | 61,051 | 102,121 | 108,535 | 152,600 | 196,000[l] |
| C4 | 54.7 | 27.9 | 28.4 | 38.8 | 41 | 47 | 58.9 | 61.8 | 63.1 | 81.9 | 85.8 | 95.2 |
| HHI | 2,035 | 341 | 362 | 627 | 602 | 774 | 1,045 | 1,258 | 1,311 | 2,067 | 2,221 | 2,636 |
| *N* (>1%) | 6 | 12 | 14 | 16 | 14 | 12 | 9 | 9 | 9 | 7 | 8 | 5 |
| MOCDI Index | 831 | 98 | 97 | 157 | 161 | 223 | 348 | 419 | 437 | 781 | 785 | 1179 |

*Main sources*: 2009 data: Subscriber counts from companies' websites, annual reports, and press releases. Also *MobileBurn*: Haselton, Todd. "U.S. Cellular reports Q1 2010 figures, Sees Flat Subscriber Growth," May 10, 2010. Last accessed on July 12, 2010 at http://www.mobileburn.com/news.jsp?Id=9408 . The total number of subscribers was found at CTIA.org. 2009 revenue data: *Standard & Poor's Industry Surveys* (Telecommunications—Wireless): http://www.netadvantage.standardpoor.com/NASApp/NetAdvantage/showIndustrySurveyPDF.do?loadIndSurFromMenu=pdf. Last accessed August 23, 2010. CTIA, The Wireless Association: http://www.ctia.org/advocacy/research/index.cfm/aid/10323. Last accessed July 15, 2011.

"Mobile/Cellphone Services – US – Snapshot," *Mintel Group Inc.*, New York, 2013.

[a] Includes 2013 acquisition by AT&T of Leap.

[b] In 2006, SBC acquired Bell South and its part of Cingular (60%).

[c] Vodafone's 45% ownership of Verizon Wireless is reflected in the table for the years of partnership, until full acquisition by Verizon (2013).

[d] Bell Atlantic and NYNEX merged their cellular operations in 1995.

[e] In 1991, GTE Mobilnet merged with Contel Cellular.

[f] Verizon Wireless acquired Alltel in 2008/9 for $28.1 billion, giving it, for 2009, about 33% of the US market. Alltel had bought in 2007 Western Wireless for $6 billion.

[g] In 2012 Sprint was acquired by SoftwareSoftbank (Japan).

[h] Sprint acquired Centel in 1993, then spun it off to create 360 Degrees Communications in 1995.

[i] VoiceStream (1999) was spun off from Western Wireless and acquired Omnipoint Communications and Aerial Communications in 2000. T-Mobile bought the company in 2002.

[j] Includes 2013 acquisition of MetroPCS by T-Mobile.

[k] US Cellular Corp is owned by TDS (80.9%), an LEC that serves about 28 states.

[l] Excluding over half of data which is counted in Internet access.

Table 18-18. Total Telecommunications Services (Wireline, Wireless, and Broadband—Market Shares by Revenues)

| COMPANY (%) | 1983 | 1984 | 1988 | 1992 | 1996 | 2001 | 2005 | 2009 | 2012/2013[a] |
|---|---|---|---|---|---|---|---|---|---|
| **Incumbent Telecom Providers** | | | | | | | | | |
| AT&T (II) (SBC) | | 6.0 | 5.6 | 4.8 | 5.1 | 12.6 | 35.8 | 31.7 | 27.0 |
| Cingular (with BellSouth)* | | | | | | 3.8 | 6.2 | SBC | |
| Ameritech | | 8.1 | 7.0 | 6.4 | 4.9 | SBC | | | |
| PacTel | | 5.9 | 6.3 | 5.9 | 3.8 | SBC | | | |
| SNET | | 0.03 | 0.01 | 0.1 | 0.1 | SBC | | | |
| AT&T (I) | 87.3 | 35.2 | 29.6 | 23.9 | 22.6 | 19.2 | SBC | | |
| McCaw* | 1.0 | 0.03 | 0.03 | 0.8 | AT&T (1) | | | | |
| BellSouth | | 8.8 | 8.5 | 7.9 | 7.0 | 5.5 | SBC | | |
| Leap Wireless* | | | | | | | | 0.4 | AT&T |
| Verizon | | | | | | 17.3 | 22.4 | 17.0 | 24.6 |
| GTE (incl. Contel) | 5.1 | 5.2 | 7.5 | 6.9 | 7.4 | Verizon | | | |
| Bell Atlantic | | 8.7 | 7.2 | 7.1 | 10.9 | 7.9 | | | |
| NYNEX | | 6.6 | 8.4 | 7.1 | 5.8 | Bell Atlantic | | | |
| Metro Mobile* | 0.4 | 0.1 | 0.1 | 0.2 | 0.3 | Bell Atlantic | | | |
| Worldcom (incl. MFS)* | | | 0.1 | 0.5 | 2.5 | 10.2 | Verizon | | |
| MCI* | 1.9 | 1.8 | 4.2 | 6.6 | 9.1 | Worldcom | | | |
| ALLTEL | 0.03 | 0.1 | 0.02 | 0.1 | 0.3 | 1.3 | 1.9 | Verizon | |
| Vodafone (US Mobile)[b]* | | | | | | 3.0 | 3.6 | 5.5 | Verizon |
| Air Touch | | | | | 1.4 | Vodafone | | | |
| US West New Vector* | | 0.2 | 0.1 | 0.3 | Airtouch | | | | |

*continued*

Table 18-18. *continued*

| COMPANY (%) | 1983 | 1984 | 1988 | 1992 | 1996 | 2001 | 2005 | 2009 | 2012/2013[a] |
|---|---|---|---|---|---|---|---|---|---|
| CenturyLink[c] | 0.3 | 0.04 | 0.02 | 0.03 | 0.04 | 0.1 | 0.2 | 6.8 | 5.5 |
| CenturyTel | | | | | | | 0.2 | 0.3 | CenturyLink |
| Embarq (Sprint Wireline) | | | | | | | 1.3 | CenturyTel | |
| Qwest* | | | 0.1 | 0.3 | 0.7 | 3.8 | 3.2 | CenturyLink | |
| LCI* | | | 0.2 | 0.6 | 1.7 | Qwest | | | |
| US West | 0.3 | 5.7 | 5.9 | 5.3 | 4.4 | Qwest | | | |
| Windstream* | | | | | | | | 0.1 | 0.2 |
| Frontier (Citizens Telecom) | | | | | | 1.3 | 1.8 | 1.8 | 1.1 |
| **New Telecom Entrants** | | | | | | | | | |
| Softbank* (Japan) Sprint (incl. Contel and United Telecom)[d] | 0.0 | 0.1 | 0.2 | 4.2 | 4.9 | 5.6 | 7.7 | 7.3 | 8.4 |
| Centel | | | 0.1 | 0.3 | Sprint | | | | |
| Nextel* | | | | 0.2 | 0.5 | 1.5 | 2.3 | Sprint | |
| T-Mobile* (Germany) | | | | | | 1.2 | 1.7 | 3.0 | 5.9 |
| Voicestream MetroPCS* | | | | | 0.02 | | | | |
| McLeod USA* | | | | | | 0.5 | 0.5 | | |
| Electric Lightware* | | | | | | 0.0 | 0.0 | | |
| Level 3* | | | | | | | 0.5 | 1.0 | 4.0 |
| Broadwing* | | | | | | 0.2 | 0.2 | Level 3 | |
| Global Crossing * | | | | 0.3 | 0.9 | 1.0 | 1.3 | 1.4 | Level 3 |
| US Cellular (TDS)* | | | | 0.1 | 0.1 | 0.6 | 1.0 | 0.8 | 1.0 |
| **VoIP (Voice over Internet Protocol)** | | | | | | | | | |
| Microsoft (Skype*) | | | | | | | 0.07 | 0.2 | 0.2 |
| Vonage* | | | | | | | 0.3 | 0.6 | 0.5 |
| Covad (Megapath)* | | | | | | 0.04 | 0.06 | 0.1 | 0.1 |

| | | | | | | | | | |
|---|---|---|---|---|---|---|---|---|---|
| **Cable Telecommunications** | | | | | | | | | |
| Comcast* | | | | | | 0.7 | 3.4 | 5.1 | 5.3 |
| Time Warner Cable* | | | | | | 1.5 | 2.7 | 4.0 | 2.8 |
| Cox Communications* | | | | | | 1.0 | 1.2 | 1.8 | 0.5 |
| Charter* | | | | | | 0.3 | 0.5 | 1.1 | 0.4 |
| Cablevision Systems* | | | | | | 1.3 | 1.8 | 2.0 | 1.7 |
| HughesNET* | | | | | | 0.03 | 0.03 | 0.07 | 0.04 |
| Others: | 3.95 | 9.2 | 8.8 | 11.4 | 5.5 | 7.1 | 0.0 | 7.9 | 10.8 |
| TOTAL US Revenue mil $ | 92,748 | 104,360 | 124,951 | 162,740 | 212,723 | 354714 | 377465 | 403,503 | 416,145 |
| C4 | 95.3 | 60.8 | 54.0 | 46.0 | 50.0 | 59.3 | 72.1 | 62.8 | 65.9 |
| HHI | 7,652 | 1,632 | 1,300 | 972 | 970 | 1,107 | 1,948 | 1,491 | 1,528 |

* Denotes a new telecom entrant after 1978. Does not include divested of merged incumbents.

[a] 2012 figures are 2009 shares with T-Mobile, Sprint, and Vodafone deals. Revenue data are taken from annual reports, industry surveys, and Tables 18.16 and 18.17.

[b] From 1999 onward, Vodafone's share shows its 45% ownership of Verizon Wireless; Verizon numbers show the other 55%, until 2013 when Verizon repurchased Vodafone's share in Verizon.

[c] 2011 data illustrate CenturyLink's acquisition of Qwest, and Level 3's (2011) acquisition of Global Crossing. Ownership of Cingular was split between Bell South (40%) and SBC (60%) until 2006. Cingular revenues have been proportionately allocated.

[d] Sprint data include its wireless and long distance wireline business. Its local wireline business was spun off to form Embarq (later acquired by CenturyLink) in 2006. Sprint was acquired in 2013 by Softbank (Japan).

Table 18-19. Weighted Average of Telecom Connectivity in Local Markets

| SERVICE PROVIDER | 1983 | 1984 | 1988 | 1992 | 1996[a] | 2002 | 2006 | 2009–11 |
|---|---|---|---|---|---|---|---|---|
| LEC Incumbent (plus its mobile service in its wireline franchise territory where applicable) | 100.0 | 100.0 | 100.0 | 100.0 | 100.0 | 86.0 | 73.1 | 59.4 |
| | | | | | | 2.6 | 5.7 | 10.0 |
| Cable TV Provider | | | | | | 4.4 | 7.8 | 12.3 |
| Non-LEC Incumbent Mobile | | | | | | | | |
| AT&T or Verizon in rival's territory | | | | | | 3.0 | 8.0 | 13.3 |
| Sprint (Softbank, Japan) | | | | | | 1.4 | 4.7 | 6.3 |
| T-Mobile (DT, Germany) | | | | | | 0.8 | 1.5 | 3.0 |
| VoIP | | | | | | | 0.3 | 1.0 |
| Weighted Average HHI | 10,000 | 10,000 | 10,000 | 10,000 | 10,000 | 7,428 | 5,496 | 3,907 |
| Weighted Average C4 | 100.0 | 100.0 | 100.0 | 100.0 | 100.0 | 94.8 | 93.6 | 91.3 |

[a] Wireless existed, of course, in 1996 and before, but its price per minute was sufficiently high that it was used mostly for mobile and nomadic applications rather than for home and office usage.

(including that of the local telecom incumbent itself), and several VoIP providers such as Microsoft's Skype. Cable and telecom providers that serve other parts of the country are not part of their choice menu. On that basis, local concentration was 10,000 in the 1980s and 1990s and dropped to 3,907 after 2009, still high but much less so.

## INTERNET MEDIA

### Internet Service Providers

Broadband Internet access grew rapidly and served about 88 million households in the United States by 2013.[29] Cable TV companies used to provide about 55% to 60% of this service, mostly by the five major cable companies, each operating in their franchise territory. About 40% to 45% of the market was offered by the incumbent local telephone companies, mostly three firms, operating each in their own territory. The share of independent DSL providers dropped rapidly as both the dot-com crash and the resistance by ILECs squeezed their business, and technology moved to broadband from dial-up. In rural areas there might be no service option at all, or only that of a costly and slow satellite service.

In addition, mobile-based 4G cellphone service has achieved respectable bit-rates of transmission and is for many people a substitute form of connectivity.

Concentration of broadband Internet providers steadily increased throughout the 2000s, from an HHI of 1,038 in 1999 to 1,458 in 2010. This was largely due to the consolidation in the telecom and cable industries. After 2010, market concentration dropped to 1,390 in 2013. The decrease is due to the emergence of mobile broadband as a medium-speed option. Table 18.20 shows this industry as moderately concentrated on the national level.

The four largest national broadband ISPs are AT&T, Comcast, Verizon, and Time Warner Cable, and they have solidified their position. Initially, the

29. http://www.ntia.doc.gov/blog/2013/household-broadband-adoption-climbs-724-percent.

Table 18-20. Broadband Internet Providers (by Subscribers)

| | 1999 | 2000 | 2001 | 2002 | 2003 | 2004 | 2005 | 2006 | 2010 | 2013 |
|---|---|---|---|---|---|---|---|---|---|---|
| | | | ***Telecom (DSL and Fiber, incl. in-franchise wireless)*** | | | | | | | |
| *Telecom Industry Subtotal Market Share (Share including mobile-only broadband added in parentheses)* | 36.4 | 38.3 | 36.0 | 36.3 | 37.1 | 39.6 | 44.7 | 45.0 | 45.3 | 35.4 (53.0) |
| AT&T (SBC) | 8.5 | 11.7 | 10.9 | 10.2 | 12.5 | 14.9 | 15.5 | 19.9 | 22.2 | 18.2 (25.1) |
| Bell South | 1.3 | 3.2 | 5.0 | 5.7 | 5.2 | 6.1 | 6.5 | AT&T | | |
| Verizon | 3.9 | 8.2 | 9.8 | 9.9 | 8.3 | 10.3 | 11.4 | 12.6 | 13.0 | 9.0 (15.2) |
| CenturyLink | | | | | | | | | 7.3 | 6.0 |
| Qwest | 4.9 | 3.4 | 3.5 | 3.4 | 2.3 | 3.0 | 3.4 | 3.8 | CenturyLink | |
| Windstream | | | | | | | | | 1.4 | 1.2 |
| Covad/Megapath (independent)[a] | 2.5 | 4.2 | 2.8 | 1.9 | 1.8 | 1.8 | 1.3 | 0.9 | 0.9 | 0.8 |
| Other DSL & Fiber Providers: | 15.3 | 7.6 | 4.0 | 5.2 | 7.0 | 3.6 | 6.6 | 7.8 | 0.4 | 0.3 |
| Telecom Industry Subtotal ($ mil): | 423 | 890 | 1,716 | 2,703 | 3,751 | 6,313 | 8,627 | 11,467 | 17,895 | |
| Subscribers (millions): | | | | | | | | | 31.9 | |
| | | | ***Mobile (Net of Wireline Broadband in Franchise Region)*** | | | | | | | |
| AT&T | | | | | | | | | | 6.9 |
| Verizon | | | | | | | | | | 6.2 |
| Sprint | | | | | | | | | | 2.2 |
| T-Mobile | | | | | | | | | | 1.9 |

*continued*

Table 18-20. *continued*

| | 1999 | 2000 | 2001 | 2002 | 2003 | 2004 | 2005 | 2006 | 2010 | 2013 |
|---|---|---|---|---|---|---|---|---|---|---|
| US Cellular | | | | | | | | | | 0.4 |
| Mobile Industry Subtotal ($ mil): | | | | | | | | | | 17.6 |
| Subscribers (millions): | | | | | | | | | | |
| | | | | ***Cable Modem*** | | | | | | |
| *Cable Industry Subtotal (market share %):* | 63.6 | 61.7 | 62.4 | 62.1 | 61.3 | 58.9 | 54.3 | 54.2 | 53.4 | 45.8 |
| Comcast | 6.4 | 6.1 | 7.8 | 7.7 | 18.8 | 19.2 | 19.5 | 20.7 | 22.0 | 18.1 |
| AT&T Broadband | 9.7 | 10.6 | 12.4 | 10.1 | Comcast | | | | | |
| Adelphia | 1.7 | 2.3 | 3.1 | 3.7 | 3.2 | 4.1 | TW/Comcast | | | |
| Time Warner Cable | | | | | | | | | 12.9 | 9.8 |
| Time Warner | 14.8 | 14.5 | 15.7 | 15.4 | 11.5 | 12.4 | 11.2 | 10.7 | TWC | |
| Cox | 8.4 | 7.4 | 7.3 | 7.4 | 7.1 | 7.8 | 6.9 | 5.5 | 5.9 | 4.9 |
| Charter | 4.3 | 4.8 | 5.2 | 5.1 | 5.4 | 6.0 | 4.9 | 4.3 | 4.1 | 3.2 |
| Cablevision | 3.3 | 3.7 | 4.0 | 3.8 | 3.8 | 4.1 | 3.8 | 3.6 | 4.0 | 3.3 |
| Other Cable Providers: | 15.1 | 12.3 | 7.0 | 8.9 | 11.6 | 5.3 | 8.0 | 9.4 | 4.5 | 3.6 |
| Cable Industry Subtotal ($ mil): | 739 | 1,432 | 2,971 | 4,619 | 6,208 | 9,383 | 10,483 | 13,810 | 18,240 | |
| Subscribers (millions): | | | | | | | | | | |
| | | | | | ***Satellite*** | | | | | |
| *Satellite Industry Subtotal (market share %)* | 0.0 | 0.0 | 1.4 | 1.4 | 1.3 | 1.3 | 0.8 | 0.7 | 1.3 | 0.8 |
| Hughes/DirecWay (EchoStar)[b] | | | 1.4 | 1.4 | 1.3 | 1.3 | 0.6 | 0.5 | 0.7 | 0.4 |
| Viasat | | | | | | | | | 0.4 | 0.3 |
| WildBlue | | | | | 0.01 | 0.01 | 0.1 | 0.1 | Viasat | |
| Gilat | | | | | | | 0.1 | 0.1 | 0.1 | 0.1 |
| StarBand | | | | | 0.02 | 0.03 | Gilat | | | |
| Other Satellite Providers: | | | 0.0 | 0.0 | 0.1 | 0.1 | 0.0 | 0.0 | 0.1 | |

| | | | | | | | | | | |
|---|---|---|---|---|---|---|---|---|---|---|
| Satellite Industry Subtotal ($ mil): | 0 | 0 | 75 | 122 | 166 | 223 | 154 | 204 | 478 | |
| Total Subscribers | | | | | | | | 56.2 | 75.1 | 83.6 |
| Total Revenues ($ mil) | 1,162 | 2,322 | 4,762 | 7,444 | 10,125 | 15,930 | 19,291 | 25,493 | 36,800 | 45,414 |
| C4 | 54.9 | 49.1 | 48.8 | 45.6 | 54.4 | 56.8 | 57.6 | 63.9 | 70.1 | 68.6 |
| HHI | 1,038 | 909 | 894 | 865 | 880 | 1,073 | 1,125 | 1,324 | 1,458 | 1,390 |
| *N* (>1%) | 12 | 12 | 13 | 13 | 12 | 12 | 10 | 8 | 9 | 11 |
| MOCDI Index | 300 | 258 | 240 | 254 | 310 | 356 | 468 | 486 | 479 | 419 |

*Source*: CSFB, "RBOC/ILEC Update-Fourth Quarter DSL and Cable Modem Update," April 5, 2002. And also Morgan Stanley, "Broadband Cable Television-Broadband: Grabbing a Bigger Piece of the Pie," June 21, 2002 (Market Share by total DSL Subscribers). 2003 data are based on Wachovia Securities, "North American Broadband Update," June 1, 2004. Total revenues for 2003 are extrapolated from growth percentage in subscriber numbers. 2002 revenue is extrapolated from 2001 and 2003. "Teletruth News Analysis." Teletruth. NewNetworkings.com. November 29, 2007. Last accessed on 8 January 2008 at http://www.newnetworks.com/partwosummary.htm.

Friedland, Jim and Kopelman, Kevin. "Q1:07 Residential Internet Access & VoIP Survey." *Cowen and Company*. May 29, 2007. Hoorigan, John B., and Smith, Aarion. "Home Broadband Adoption 2007, June 2007." Pew Internet & American Life Project. PewInternet.org. June 2007. Last accessed on January 11, 2008, from http://www.pewinternet.org/pdfs/PIP_Broadband%202007.pdf. Goldman, Alex. "Top 21 U.S. ISPs by Subscriber" from http://www.isp-planet.com/researchrankings/2005/usa_history_q42005.html.

[a] After Megapath's acquisition of Covad, market share assumed to have remained the same for 2010 as in 2006.

[b] Hughes was acquired by Echostar in 2011.

C4 shows a strong rise to 70.1% in 2010. With 4G wireless emerging, this declined somewhat to 68.6% by 2013. However, these figures are for the national market. On a local basis, things are different and more concentrated. In a given territory there is no competition among cable providers or among wireline telecom firms, and in the past a duopoly existed. If we report concentration in that fashion, we can observe a much higher and generally increasing concentration level, with an HHI of about 4,670 in 2004 (see Table 18.21). After 2010 local broadband ISP concentration drops to 3,171 in 2013 due to 4G mobile smartphone usage.[30]

The conclusion is therefore, first, that US national broadband market concentration has been steadily rising but is in the moderate territory if viewed nationally. Second, telecom firms in 2011 have taken the lead over cable for the first time, due to their offering of mobile broadband access (Figure 18.1). Nevertheless local broadband ISP concentration is in the range of high concentration, with an HHI of 3,171.

## Search Engines

The search engine industry was initially dominated by Yahoo, Excite, AltaVista, and Infoseek, each in a particular period. Google became popular by using a rank-based search algorithm. By 2001 Infoseek and Excite dropped entirely out of the search market. In 2003 Yahoo acquired Overture, and with it AltaVista and Inktomi. Microsoft's MSN Search, renamed Live Search and then Bing, was number three, followed by IAC's Ask.com.

Table 18-21. Local broadband ISP Concentration (by Subscribers)

| | 1999 | 2000 | 2001 | 2002 | 2003 | 2004 | 2006 | 2010 | 2013 |
|---|---|---|---|---|---|---|---|---|---|
| Cable Provider | 63.6 | 61.7 | 62.4 | 62.1 | 61.3 | 58.9 | 54.2 | 53.4 | 42.8 |
| Wireline Telecom (incl. wireless mobile ISP in own region) | 18.6 | 26.5 | 29.2 | 29.5 | 28.3 | 34.2 | 34.3 | 34.8 | 35.2 |
| Mobile Operators (net of wireline BB from their own in-region telecom) | | | | | | | | | |
| AT&T* | | | | | | | | | 6.9 |
| Verizon | | | | | | | | | 6.2 |
| Sprint | | | | | | | | | 2.2 |
| T-Mobile | | | | | | | | | 1.9 |
| US Cellular | | | | | | | | | 0.4 |
| Independent ISPs | 17.8 | 11.8 | 6.8 | 7.1 | 8.8 | 5.4 | 8.7 | 1.3 | 1.5 |
| Satellite | 0 | 0 | 1.4 | 1.4 | 1.3 | 1.3 | 0.7 | 1.3 | 0.8 |
| HHI | 4,708 | 4,648 | 4,795 | 4,779 | 4,638 | 4,670 | 4,190 | 4,066 | 3,171 |
| C4 | 100 | 100 | 100 | 100 | 100 | 100 | 97.9 | 90.8 | 91.1 |
| *N* (>1%) | 3 | 3 | 4 | 4 | 4 | 4 | 3 | 4 | 7 |
| MOCDI Index | 2,718 | 2,684 | 2,397 | 2,390 | 2,319 | 2,335 | 2,419 | 2,033 | 1,199 |

30. We avoid double-counting AT&T or Verizon customers who get both their wireline broadband and mobile Internet connectivity from the same firm. In 2013, 18% of AT&T or Verizon mobile customers in these firms' wireline franchise territories used smartphones as primary Internet access (one-half of all smartphone use), and the rest used distributed, according to the national breakdown between cable (0.55) and telco broadband (0.45) and among non-wireline-delivered, in-territory mobile providers according to their national mobile shares.

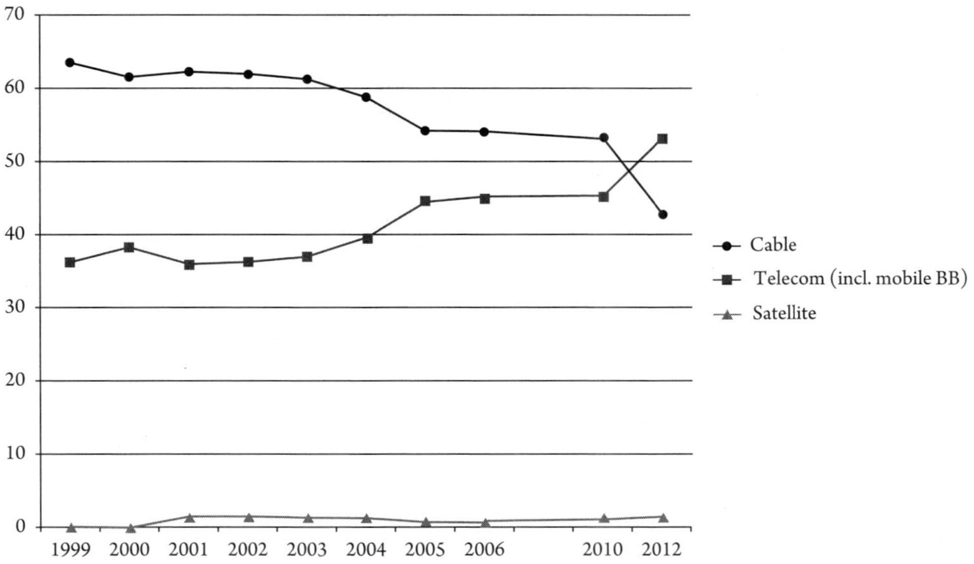


FIGURE 18.1 National Market Shares of Industries Providing Broadband

There are also specialized search engines for jobs, blogs, news, pictures, health, shopping, business, and so forth. And there are metasearch engines that direct a request to multiple search engines. Even so, Google was increasingly dominant, raising its share from 31.5% in 2004 to 67.3% in 2008 and then plateaued at 68.5% in 2010 and 68.4% in 2012.

Concentration is based on the large economies of scale. The cost of creating the database and search algorithms is high, but the usage cost is minimal. Rival search sites compete in the efficiency of the search, or in their specialization. Microsoft unsuccessfully tried to buy Yahoo for 44.6 billion. Google then tried a deal with Yahoo for an advertising partnership, but the deal ran into trouble with Washington antitrust agencies. However, Google reached in 2005 a deal with AOL (owned after 2015 by Verizon) to have AOL Search powered by Google. A similar deal later in 2009 between Microsoft and Yahoo saw Yahoo search fully powered by Bing. Ask.com is powered by Google. These deals added to concentration (Table 18.22).

## Online News

The online news sector has a low and declining concentration, from an HHI of 898 in 2007 to 568 in 2011. This is due in large part to the widening availability of online news providers. The market share of the top four firms also declined from 49.3% to 33.6%, with the share of the market leader Yahoo! News hovering around 11%. The total number of monthly unique visitors has risen steadily over the period studied.

Even so, online news is dominated by traditional offline news providers of print and TV—Time Warner/CNN, Comcast MSNBC, *New York Times*, News Corp (Fox News and *Wall Street Journal*), AP, NPR, and several major newspaper chains. Together, these "news incumbents" had 54.1% in 2011, a fairly high number. The online news HHI, at 658 in 2011, was higher than the national daily newspaper concentration of 304 in 2013. But more important, TV-oriented news and print news now compete head on in online news. Plus the local print newspapers are now available nationwide online and their stories therefore have a wider footprint. There are also "foreign news incumbent" sites, in particular, from the United Kingdom: BBC, Guardian, Telegraph, and Mail. New information sources have emerged that are online-only sites—Yahoo, AOL/Huffington Post (acquired by Verizon), Salon, Drudge Report, Reddit, Digg, and so forth, accounting for 30% by 2011 (Table 18.23).

## Table 18-22. Internet Search Engines (Market Share by Search Volume)

| COMPANY | 1997 | 2001 | 2004 | 2006 | 2008 | 2010 | 2012[c] |
|---|---|---|---|---|---|---|---|
| Google | | 4.5 | 31.5 | 49.6 | 67.3 | 68.5 | 68.4 |
| AOL (Time Warner)[a] | | 7.4 | 3.5 | | | | |
| Microsoft/MSN/Live Search/Bing | | 3.7 | 17.7 | 12.8 | 8.3 | 12 | 15.6 |
| Yahoo[b] | 27.4 | 14.9 | 20.1 | 28.8 | 20.5 | 16 | 13.0 |
| Overture | | 7.4 | Yahoo | | | | |
| FAST | | 0 | Yahoo | | | | |
| AltaVista | 12.1 | 7.4 | Yahoo | | | | |
| Inktomi | 2.3 | 5.3 | Yahoo | | | | |
| IAC Search & Media | | | 4.9 | 5.4 | 3.9 | 3.5 | 3.0 |
| Ask.com (previously Ask Jeeves) | | 3.6 | IAC | | | | |
| Excite | 20.3 | 7.4 | IAC | | | | |
| Go.com/Infoseek (Disney) | 13.2 | 7.4 | | | | | |
| LookSmart | | 5.2 | | | | | |
| FindWhat (MIVA) | | 1.5 | | | | | |
| Lycos (Terra/Telefonica/Duam) | 4.5 | 4.1 | | | | | |
| HotBot | | 0 | | | | | |
| Other | 20 | 20.1 | 22.3 | 3.4 | | | |
| Revenues (in $ mil) | 193 | 1,000 | 3,900 | 6,800 | 10,546 | 11,700 | 14,800 |
| C4 | 73.1 | 37.2 | 74.2 | 96.6 | 100 | 100 | 100 |
| HHI | 1,514 | 618 | 1,746 | 3,483 | 5,034 | 5,105 | 5,506 |
| *N* (>1%) | 6 | 13 | 5 | 4 | 4 | 4 | 4 |
| MOCDI Index | 618 | 171 | 781 | 1,741 | 2,517 | 2,553 | 2,753 |

*Source*: Revenues include the following categories: banner ads, buttons, and sponsorships on search pages; license fees; maintenance fees; and paid placement. 1997 "total revenue" is the sum of the 10K revenues (minus non-US revenues) of the following companies: Excite, Infoseek, Lycos, Yahoo, Search.com, and Inkitomi. Calculation based on Alta Vista's revenues of 20 million. 2001 total revenue and market shares from "The Internet Search Market," Salomon Smith Barney, October 1, 2002. In 2002 Infoseek was part of Disney's Go. Con, Excite used other search services for its portal, and Lycos was part of terra.com. 2004 data are compiled from Nielsen Netratings and SearchEngineWatch.com, 2005 Market shares from Nielsen methodology, which measures "time spent." Market shares for 2008–2012 are taken from the Comscore search engine rankings for December 2008, December 2010, and June 2012. Revenues for 2010 and 2011 were taken from the Interactive Advertising Bureau's "Internet Advertising Revenue Report" for 2011.

"US Search Engine Market Share Data," *AccuraCast*; http://www.accuracast.com/search-daily-news/seo-7471/us-search-engine-market-share-data-jan-2009/. "Bing Passes Yahoo to Become No. 2 US Search Engine" http://blog.nielsen.com/nielsen-wire/online_mobile/bing-overtakes-yahoo-as-the-2-search-engine/; *Nielsen*. "Internet Ad Revenues Reach Record Quarterly High of $6.3 Billion in Q4 '09," *iab.net*. http://www.iab.net/about_the_iab/recent_press_releases/press_release_archive/press_release/pr-040710. Last accessed January 11, 2011.

[a]AOL was part of AOL-Time Warner from 2000 to 2009. Its market shares from 2006 onward are added to the Google shares, due to the deal between Google and AOL in 2005 that saw AOL Search powered by Google. In 2015, Verizon acquired AOL.

[b]Yahoo's 2012 share of 13% can be added to Bing, since Microsoft powers the Yahoo search. Similarly, Google powers the Ask.com search. (http://news.bbc.co.uk/2/hi/business/8174763.stm).

[c] Revenue is for 2011.

Table 18-23. Online News Media (Market share by Monthly Unique Visitors)[a]

| | 2007 | 2008 | 2009 | 2010 | 2011 |
|---|---|---|---|---|---|
| Yahoo! News | 15.2 | 11.1 | 11.4 | 11.2 | 11.0 |
| Associated Content | | | 2.8 | 4.0 | |
| Time Warner (CNN) | 11.5 | 12.6 | 11.7 | 9.9 | 8.8 |
| Comcast/GE (MSNBC) | 13.0 | 12.4 | 12.6 | 8.9 | 7.8 |
| New York Times | 6.1 | 6.4 | 6.1 | 4.4 | 4.7 |
| AOL News/Verizon | 8.2 | 7.9 | 6.5 | 5.8 | 8.4 |
| Huffington Post | 1.5 | 1.8 | 2.1 | 3.2 | |
| News Corp | 5.5 | 6.1 | 7.4 | 7.2 | 7.3 |
| Fox News | 3.5 | 3.9 | 4.4 | 4.3 | 4.0 |
| WSJ Online | 2.0 | 2.2 | 3.0 | 2.9 | 3.4 |
| Disney (ABCNews) | 5.2 | 4.0 | 3.6 | 3.7 | 3.5 |
| WashingtonPost | 3.8 | 5.2 | 4.7 | 4.0 | 4.4 |
| WashingtonPost.com | 3.8 | 3.0 | 3.1 | 2.8 | 3.5 |
| Slate | | 2.1 | 1.6 | 1.2 | 1.0 |
| Tribune Newspapers | 5.4 | 4.8 | 5.0 | 3.2 | 4.0 |
| LA Times | | | | 2.3 | 3.5 |
| BBC (UK, public) | 3.5 | 1.9 | 2.8 | 1.8 | 2.3 |
| CBS | 4.2 | 3.5 | 3.7 | 2.7 | 3.4 |
| Gannett | 9.6 | 8.1 | 6.8 | 6.2 | 6.2 |
| Gannett Newspapers | 5.4 | 4.4 | 3.9 | 3.7 | 3.0 |
| USA Today | 4.2 | 3.7 | 3.0 | 2.5 | 3.2 |
| AP | 4.5 | 1.9 | 1.2 | 0.9 | 0.7 |
| Advance | | 2.4 | 2.2 | 2.8 | 3.0 |
| DrudgeReport | | 1.2 | 1.0 | 3.7 | 3.1 |
| McClatchy | 4.2 | 3.5 | 2.7 | 2.5 | 3.1 |
| Mail Online (UK) | | | | 2.6 | 3.5 |
| Topix | | 1.7 | 2.0 | 1.2 | 1.5 |
| Telegraph (UK) | | 1.0 | 1.4 | 1.1 | 1.1 |
| Guardian (UK) | | 1.2 | 1.1 | 1.1 | 1.6 |
| NPR (Public) | | 1.3 | 1.4 | 1.1 | 1.0 |
| Salon | | | | | 0.7 |
| Reddit | | | | 2.3 | 3.3 |
| Digg | | | | 2.3 | 1.7 |
| Others | 0.0 | 1.8 | 4.7 | 9.4 | 3.9 |
| Total Monthly Unique Visitors (in thousands) | 213,695 | 294,525 | 356,485 | 344,903 | 393,022 |
| Revenue (US$ mil) | 7,915 | 7,773 | 6,858 | 7,605 | 8,366 |

*continued*

Table 18-23. *continued*

| | 2007 | 2008 | 2009 | 2010 | 2011 |
|---|---|---|---|---|---|
| C4 | 49.3 | 44.2 | 39.4 | 37.2 | 33.6 |
| HHI | 977 | 827 | 810 | 723 | 658 |
| N (>1%) | 20 | 28 | 29 | 32 | 30 |
| MOCDI Index | 204 | 156 | 143 | 126 | 116 |

*Main sources*: Nielsen Top News Sites lists (pooled by Cyberjournalist.net) used as main source: 2007 data are from http://www.cyberjournalist.net/top-news-sites-for-february-2007/; 2009 data are from http://www.cyberjournalist.net/top-news-sites-for-january-5/; 2011 data are from http://blog.nielsen.com/nielsenwire/online_mobile/january-2011-top-u-s-web-brands-and-news-sites/; http://www.ebizmba.com/articles/news-websites; http://www.poynter.org/latest-news/mediawire/131916/yahoo-grew-unique-visitors-in-april-while-new-york-times-others-declined/. Revenue numbers are estimated based on advertising revenue produced by NAA (Newspaper Association of America).

[a] Google News and Bing News are not included because they are pure news aggregators without content creation (such as Yahoo) or selection and screening (such as Digg).

Tribune Company filed for bankruptcy in 2012. Its properties, including the LA Times, were transferred to the company's senior debt holders including Oaktree Capital Management, JPMorgan Chase, and Angelo, Gordon & Co.

# SUMMARIES

## Industry Trends

Having analyzed 13 industries, we can now show the concentration trend of several media sectors (Table 18.24 and Figure 18.2). By weighted average, print media are the least concentrated sector, though they have slightly consolidated over three decades. Average concentration of the audiovisual media industries is higher, but in the low range of the antitrust guidelines for concentration. Telecom media are considerably higher and trending up. Even higher and faster rising are concentrations for online media (ISPs, search engines, online news).

The trend for telecom media resembles an S-shape, with rapid increase especially after 1996, and a decline after 2009; for platform media more generally, it is an upward sloping S-curve with some decline after 2005. For content media, concentration rose from 280 in 1984 to 1,042 in 2013, still unconcentrated but less than before. For content media, the trend has been one of steady increase, first rising up to about 2,000, then slowing down in growth somewhat, and then rising again. For audiovisual media, it is an inverted U-shape peaking in 2004 at 943; print and content media generally have a gradually increasing trend; and for online media a greatly rising trend. The overall concentration of all media is progressing in a gradually rising S-shaped trend and reached 1,565 in 2013, up from 940 in 1984 (post-AT&T divestiture) and 717–719 between 1992 and 1996.[31]

Table 18.25 presents industry concentration organized along two dimensions: the magnitude of concentration and its growth trend. The combination that deserves most concern in terms of pluralism and competition are those industries that are characterized by both a high and growing concentration (the northwest corner). Of the 13 industries analyzed, these are platform media (telecom, cable, and ISP) generally, mostly driven by mobile telecom. Second are search engines, which also drive online media to be a high-concentration, high-growth sector.

Also with high concentration, but stable in trend, are TV stations, newspapers, and broadband ISPs, all locally. Those with high but declining concentration are telecom, radio, and multichannel platforms, all locally. Generally, we observe very high concentrations for local media, but at least they are stable or declining.

Concentration is growing (though it is still moderate) for TV production and broadband ISPs. And it is growing, though still at a low

31. The trend for all 13 industries taken together more or less resembled the trend of the two main telecom media, due to their great weight in the overall media sector.

Table 18-24. US Industry Weighted Average HHI Sector Averages (Weighed by Industry Revenues), 1984–2013

| INDUSTRY | 1984 | 1988 | 1992 | 1996 | 2000/1 | 2004/5 | 2008/9 | 2011 | 2013 |
|---|---|---|---|---|---|---|---|---|---|
| Print Media (Daily Newspapers, Magazines, Book Publishing) | 150 | 164 | 184 | 206 | 265 | 393 | 339 | 399 | 668 |
| Audiovisual Media (TV Broadcasting, Video Channels, Multichannel Platforms, Radio, Film) | 489 | 459 | 537 | 601 | 900 | 943 | 873 | 967 | 907 |
| Telecom Media (Wireless and Wireline Telecommunications) | 2,145 | 1,385 | 932 | 857 | 1,060 | 2,143 | 1,915 | 2,174 | 2,160 |
| Online Media (ISPs, Search Engines, Online News Media) | | | | | 840 | 1,321 | 2,073 | 2,072 | 2,036 |
| All Media Industries | 940 | 775 | 714 | 719 | 972 | 1,518 | 1,399 | 1,579 | 1,565 |
| All Content Media | 280 | 280 | 337 | 401 | 577 | 778 | 826 | 934 | 1,042 |
| All Platform Media | 1,555 | 1,248 | 993 | 955 | 1,218 | 1,875 | 1,702 | 1,897 | 1,817 |

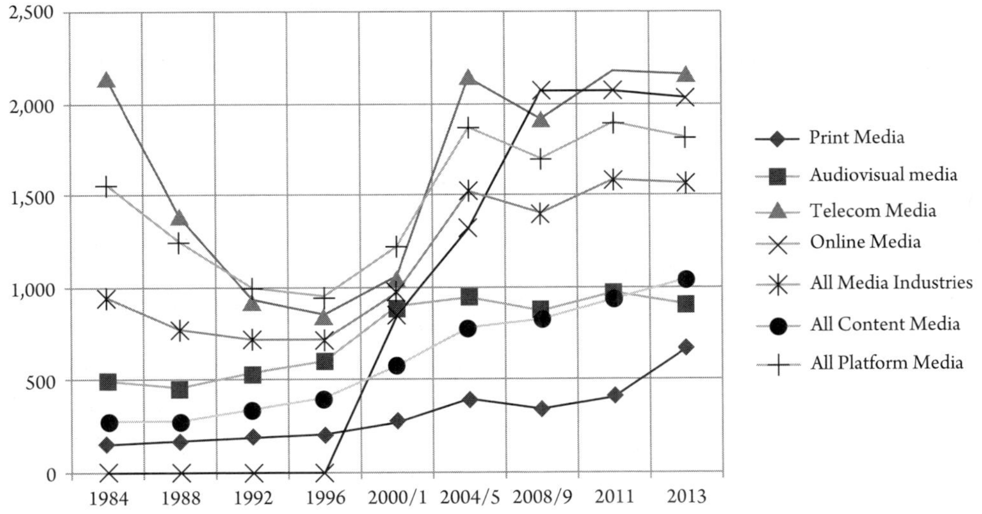


FIGURE 18.2 Average Market Concentration in Main Sectors of U.S. Media (Weighed by Industry Revenues), 1984–2013

level, for radio, TV broadcasting, multichannel TV platforms, and the various print media.

Looking at the broad categories, content media have been steadily growing in concentration and are now in the intermediate range. Platform Media are high and growing in concentration, after a period of decline in the 1980s and early 1990s.

We thus observe a generally rising media concentration. No industry sector has a low and declining concentration, which might offset some of the others. However, concentration is not at the levels that media critics often imagine. Most industries are at levels that are defined by the US government as "low concentration" (HHI < 1,500) or "intermediate concentration"

Table 18-25. Concentration Levels and Trends of US Media Industries

| | HIGH | INTERMEDIATE | LOW |
|---|---|---|---|
| Growing | Platform media Mobile Search engines Online media | All media Content media TV production Wireline telecom Broadband ISPs | Print media Audiovisual media All electronic video distribution Radio Broadcast TV Multichannel TV Platforms Print media Books Magazines |
| Stable | TV stations locally Newspapers locally Broadband ISPs locally Wireline Telecom | All video channels Cable channels Film | Newspapers Online news |
| Declining | Telecom locally Radio locally Multichannel video-platforms locally | | |

(1,500 < HHI < 2,500). A threshold of 1.800 was the previous definition. Where high media concentration exists, it is on the local level, less so on the national level. High national concentration does exist for platform media and search engines. Platform media have been a regulatory battlefield for over a century, and search engines, that is, Google, have received much critical attention recently.

A few companies increased their role in overall media even if their primary industries might not have risen in concentration. This is due to expansion into other industries and will now be discussed.

## Cross-ownership

So far, we have looked at industries separately. Many companies, however, have a presence in several industries. Tracking their market share in one industry alone would not show their overall position in the media sector. To do so, one might add up overall revenues. But that would not take into account major market shares in one or two industries, as opposed to small shares in a large number of industries. Large shares in an industry provide special power, which is why the market shares are squared for purposes of an HHI index. We therefore create a media power index for companies (MPI–C) and report it in Table 18.25. The index takes a company's market shares in each industry the company operates in, squares them in the same way as in an HHI calculation, and adds them across industries for the company, weighted by industry size. This measure represents a company's overall "power" in the overall media sector. The findings are shown in Figure 18.3 and Table 18.31.

Telecom and cable companies—that is, platform companies—dominate the top of the chart due to their high market shares in high-revenue industries and their cross-media involvements. By far the highest are AT&T and Verizon, and their power index figures have risen. Among the top eight, the only non-platform exception is Google, ranked fourth. Its share in its industry is very high, but that industry's revenues are intermediate only. Comcast is a mixed-sector company, both in platform and content, and also has a high and rising power index score. Large content firms take up the next three spots. Time Warner (and News Corp.) rank in the middle, and both have segmented themselves and thus

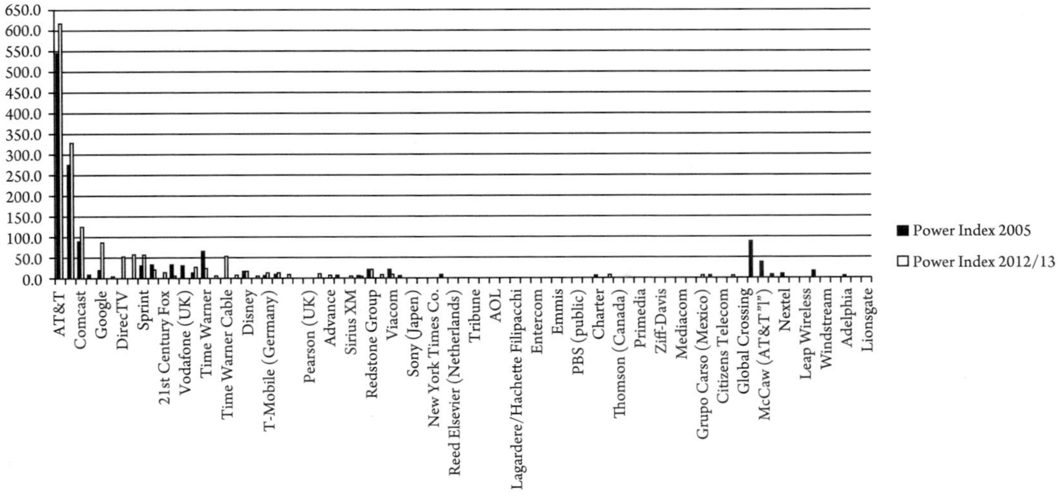


FIGURE 18.3 Company Power Indices

downsized and specialized, as did Viacom before. Disney ranks in the middle of the table.

The most significant changes in company power index were the rise of Google, the rise of Comcast, the rise of Liberty, and the split of News Corp.; and the mergers in the mobile and telecom sectors by AT&T and Verizon. Since 2004, Time Warner's PI has fallen by over half due to the formation of Time Warner Cable as an independent entity and the spin-offs of TW Telecom, AOL, TW Cable, and Time Inc. The Viacom-CBS split more than halved Viacom's 2004 PI by 2008, and it continues to decline. Comcast, after finalizing its acquisition of GE in 2013, has seen its power index increase from 77 in 2004 to 166 in 2013. The growth of Liberty Media in this period is also notable. The merger of Bertelsmann and Pearson's book publishing arms into Penguin Random House significantly lifted the two companies' share of that market.

Table 18.26 shows the sums of the top Power Index companies in the United States. The power index score of the single company with the highest power index peaked in 2008–2009 and decreased. For the top four and ten companies, the aggregate power index rose and then plateaued.

### Table 18-26. Top Company Power Indices in the US (2004–2013)

| | 2004/05 | 2008/09 | 2012/13 |
|---|---|---|---|
| P1 | 542 | 763 | 616 |
| P4 | 994 | 1,182 | 1,157 |
| P10 | 1,216 | 1,369 | 1,395 |
| P20 | 1,342 | 1,431 | 1,511 |

## Who Owns American Media?

Because of limited space, we do not discuss the ownership structure of the major media firms in this chapter but refer the reader to the analysis in Chapter 36, "Media Ownership."

It is necessary to clarify who is being measured as "US media." These are media companies that are either US-headquartered (including their non-US operations) or US-based divisions of foreign firms, such as Sprint, Random House, or Sony Pictures.

Equity ownership and control are not identical. For Google, for example, the founders own about 15% but together hold more than 50% of the voting power. Comcast is controlled by the Roberts family which holds roughly 33% of the voting power but very little of dividend-entitled shares. Some of the owners are "insiders": individuals, such as the founders, their heirs, or senior executives of a company. Others are financial institutions such as mutual funds, private equity funds, and banks. Some large funds such as BlackRock, Vanguard, and State Street own a small percentage of many top media companies.

The ownership by "insiders" declined significantly for most of the largest media firms. Even for content media firms, where such ownership is higher than for platforms, it fell from 35% to 20%. At the same time, institutional ownership rose from 40% to 57% for the content sector and from 49% to 61% for the platform sector.

The largest institutional owners hold only a few percent (at most) of any single media company but have stakes in many firms. Thus, the view that the information industries are owned by a small group of media moguls is not an accurate one. A better description is one of a number of institutions owning many small slices of a big pie.

Table 18.27 lists major owners, both individual and institutional, and aggregates their media holdings in top US and international media companies. More detail is provided in Chapter 36 on media ownership.

Vanguard is the largest owner of US media companies, with $62 billion of holdings. It has over $1.5 billion of overseas media holdings. Vanguard has approximately 3.8% of the total estimated value of the US media sector. State Street is the second largest owner of top US media companies, with over $61 billion, plus $4 billion in foreign media companies. Of individual owners, the largest are Larry Page and Sergey Brin, the Google founders, each with $26 billion, Michael Bloomberg ($24 billion), the Cox Family with $24 billion, Mark Zuckerberg (Facebook), and the Roberts family, which holds the majority voting of Comcast. Its stake is valued at $18.5 billion. Large foreign owners are the Government Pension Fund of Norway ($5.6 billion); Baillie Gifford & Co, (a Scottish investment firm with $5.4 billion); Masayoshi Son (the Japanese founder of Softbank, which now owns Sprint, $4.75 billion); and AXA of France (an insurance company, $1.5 billion).

The total equity value of the top 35 owners of media (excluding the US public (governmental) entities) is almost $600 billion, and over $44 billion internationally.

The overall value of the US media sector is estimated based on its revenue multiple. (For more detail, see Chapter 36.) This multiple is estimated to be 2.07. It results in the estimates of company value shown in Table 18.28.

Based on these sectoral values, the top 10 owners of US media own about $350 billion worth of US media holdings, accounting for 21.2% of all US media. The top 20 owners of US media hold $503 billion worth of US media holdings, accounting for 30.5% of all US media assets. The top 35 hold almost $598 billion, amounting to 36.2% of all US media assets. If a more conservative multiple of revenues is used to estimate asset values, such as a multiple of 1, then the shares just listed would approximately double.

### Foreign Ownership of US Media Companies

Table 18.29 shows the percentage of US media industries owned by foreign companies. The book publishing industry in 2013 was over one-third owned by five foreign companies, with a considerable rise since 2004–2005. About one-fourth of wireless telecom was also owned by foreign companies, T-Mobile (since 2001), Softbank (since 2013), and Vodafone (until it sold its 45% share of Verizon Wireless in 2013). About 15% of the US film industry is owned by Sony. (That percentage used to be higher when Matsushita (Japan) and then Vivendi (France) controlled Universal.) On the whole, foreign ownership of media in America is low.

## CONCLUSIONS

Did concentration in the US media sector rise in recent years? Yes. Looking at Figure 18.2 and Table 18.24, weighted average HHI for all media industries rose from 940 in 1984 (unconcentrated) to about 1,565 in 2013 (moderately concentrated).

Table 18-27. Top US Media Owners as of September 2013

| RANK | NAME OF OWNER | MAJOR MEDIA STAKES (>$1 BIL) | AGGREGATE HOLDINGS OF US MEDIA COMPANIES (IN $ MIL) | AGGREGATE HOLDINGS OF NON-US MEDIA COMPANIES (IN $ MIL) |
|---|---|---|---|---|
| 1 | Vanguard | Google<br>Comcast<br>Disney<br>Time Warner<br>21st Century Fox<br>Time Warner Cable<br>Yahoo<br>CBS | $62,277 | $1,522 |
| 2 | State Street | Microsoft<br>Google<br>AT&T<br>Verizon<br>Disney<br>Amazon<br>Comcast<br>21st Century Fox<br>Time Warner<br>DirecTV<br>CBS<br>Time Warner Cable<br>Yahoo | $60,845 | $3,999 |
| 3 | Fidelity | Google<br>Disney<br>Comcast<br>Yahoo<br>Time Warner<br>21st Century Fox<br>Comcast | $46,285 | $216.16 |
| 4 | Capital Group | Google<br>Comcast<br>Time Warner<br>21st Century Fox<br>Time Warner Cable<br>CBS | $32,455 | $2,748 |
| 5 | Larry Page | Google | $26,791 | |
| 6 | Sergey Brin | Google | $26,050 | |

*continued*

Table 18-27. *continued*

| RANK | NAME OF OWNER | MAJOR MEDIA STAKES (>$1 BIL) | AGGREGATE HOLDINGS OF US MEDIA COMPANIES (IN $ MIL) | AGGREGATE HOLDINGS OF NON-US MEDIA COMPANIES (IN $ MIL) |
|---|---|---|---|---|
| 7 | Cox Family (Chambers, Kennedy, Parry-Sheden, Anthony) | Cox Communications | $24,000 | |
| 8 | Michael Bloomberg | Bloomberg | $24,000 | |
| 9 | BlackRock | Microsoft | $23,470 | $850 |
| 10 | T. Rowe Price Inc. | Google<br>AT&T<br>Time Warner<br>Disney<br>Comcast<br>Netflix | $23,300 | $2,874 |
| 11 | Mark Zuckerberg | Facebook | $19,000 | |
| 12 | Dodge & Cox | Comcast<br>Time Warner<br>21st Century Fox<br>Time Warner Cable<br>News Corp.<br>Google | $18,689 | $1,390 |
| 13 | Massachusetts Financial Services | Disney<br>SBA Communications<br>Google<br>Viacom<br>Comcast | $18,512 | $284 |
| 14 | Brian Roberts Family | Comcast | $18,500 | |
| 15 | Newhouse Family | Advance Publications | $17,100 | |
| 16 | Wellington Management | Verizon<br>Comcast<br>Time Warner<br>Disney<br>Google<br>Yahoo | $13,906 | $192 |
| 17 | Janus | Google<br>21st Century Fox<br>Time Warner Cable | $13,539 | $347 |

Table 18-27. *continued*

| RANK | NAME OF OWNER | MAJOR MEDIA STAKES (>$1 BIL) | AGGREGATE HOLDINGS OF US MEDIA COMPANIES (IN $ MIL) | AGGREGATE HOLDINGS OF NON-US MEDIA COMPANIES (IN $ MIL) |
|---|---|---|---|---|
| 18 | Goldman Sachs | Google | $11,724 | $625 |
| 19 | Murdoch Family Trust | 21st Century Fox<br>News Corp. | $11,606 | |
| 20 | JP Morgan Chase | Google | $11,123 | $6,865 |
| 21 | ClearBridge | Comcast | $10,808 | $171 |
| 22 | Charles Ergen | Dish Network | $10,600 | |
| 23 | Steve Jobs Trust | Disney | $8,371 | |
| 24 | Eric Schmidt | Google | $8,200 | |
| 25 | Hearst Family | Hearst Corporation | $7,000 | |
| 26 | US Public Entities | Public Broadcasting | $5,700 | |
| 27 | Government Pension Fund of Norway | Microsoft<br>Google<br>AT&T | $5,652 | $9,586 |
| 28 | David Geffen | Dreamworks | $5,600 | |
| 29 | John Malone | Liberty Global/Liberty Media | $5,600 | |
| 30 | Baillie Gifford & Co | Google<br>Facebook | $5,434 | $2,138 |
| 31 | Masayoshi Son | Sprint | $4,752 | $3,900 |
| 32 | Coatue | CBS | $4,451 | $161 |
| 33 | Oppenheimer | Google | $4,307 | $2,491 |
| 34 | State Farm | Disney<br>AT&T | $4,197 | $322 |
| 35 | Berkshire Hathaway | DirecTV | $4,044 | |
| 36 | Thornburg | Liberty Global | $1,920 | $4,361 |
| 37 | AXA Financial (France) | Google | $1,521 | $101 |
| | Total Aggregate Media Holdings | | $601,329 | $45,145 |

Table 18-28. US Media—Total Revenues and Company Values

| | 2013 TOTAL REVENUE (MIL US$) | 2013 COMPANY VALUE (MIL US$) |
|---|---|---|
| US Content Media | 258,367 | 534,820 |
| US Platform Media | 537,527 | 1,112,681 |
| US All Media | 795,890 | 1,647,492 |

Table 18-29. Percentages of Foreign Ownership for Media Industries (2004/5, 2009)

| INDUSTRY | % FOREIGN OWNED | |
|---|---|---|
| | 2004/5 | 2013 |
| Book publishing | 19.7% | 35.8% |
| Film | 16.4% | 13.7% |
| Wireless | 18.5% | 22.5% |
| Magazines | 8.5% | 7.6% |
| Other 8 industries | 0.0% | 0.0% |

We observe an S-shaped concentration trend in platform media, declining until 1996, increasing until 2005, and slightly declining thereafter. In contrast, the content media sector followed a gradual and steady pattern of growing concentration, rising from 280 to 1,042 in 2013. Its average concentration is at a much lower level than the platform sector, and it is fairly low by the standards of the US government's Antitrust Merger Guidelines. But the gap in concentration between content and platforms has narrowed. Whereas in 1984 content media HHI concentration was only 13.3% of that of the platform sector, it was 70% in 2012. This is a big change. By revenue, platform media accounted for 51.9% of the total US media market in 1984, and this increased to 67.5% by 2013. Content accounted for 45.1% in 1984 and dropped to 32.5% by 2013. Thus, overall concentration is weighted more heavily to platform media, which increased in concentration more slowly after 2004–2005, and this partly obscures the rise in content concentration.

What are the reasons for this change? Partly, it is the general relaxation of ownership ceilings and antitrust enforcement. And partly, it is the fundamental shift in the technology. Content media have become more technological, digital, and capital-intensive. Generally, the more electronic and "digital" a media subsector is, the more highly it is concentrated, and more general economic dynamics of the information sector (see also Chapter 38 on findings).

The Internet sector exemplifies these dynamics. If anything, its greater rate of change drives it faster to concentration, and with it media more generally. This pours cold water over the hope that the Internet will solve the American media concentration issue.

## Bibliography

For a listing of the literature, see the General References on Media Ownership and Concentration at the end of the book.

# United States of America—Data Summaries

ELI NOAM AND PAUL MUTTER

CONCENTRATION IS RELATIVELY LOW in the United States compared to other countries, partly due to the country's size in terms of population, geography, and GDP. The content industry with the highest level of concentration is that of search engines (Table 18.30). This is not unexpected, as Google and Microsoft dominate the industry worldwide. Concentration varies greatly. It is highest in the telecommunications and Internet media industries and lowest in print. Among the top four companies, two are telecommunications providers. AT&T and Verizon together control domestically 41.7% of the wireline, 66.5% of the wireless markets, and 44.4% of the overall platform market. (Table 18.32). Also among the top four is the multimedia provider Comcast, accounting for 9.7% of the national platform market (and 9.4% of content; seeTable 18.31).

Comcast is the most diverse large media company in the United States. It raised its share in TV broadcasting, video channels, and film through the acquisition of NBC Universal. It maintains high shares in multichannel platforms (its core activity), along with wireline telecom and ISPs. In 2014, Comcast sought to acquire Time Warner Cable. Time Warner, too, is active in several industries, but it has spun off its cable and Broadband ISP platforms (Time Warner Cable), enterprise telecom (TW Telecom), music activities (WMG), and publishing (Time, Inc.).

Public (governmental) ownership of media is extremely low in the United States, with less than 1% (0.4%) of the national market by total share (and 0.5% in platforms). Foreign ownership of media is on the lower end of the international spectrum, with foreign firms controlling 10.1% of the national market—primarily in mobile

Table 18-30. National Media Industries Concentration in the United States

| | 2004/05 | | 2011 OR MOST RECENT | | % CHANGE ANNUAL AVERAGE | |
|---|---|---|---|---|---|---|
| | COMPANY POWER INDEX IN COUNTRY | COMPANY SHARE OF THE OVERALL NATIONAL MEDIA MARKET (%) | COMPANY POWER INDEX IN COUNTRY | COMPANY SHARE OF THE OVERALL NATIONAL MEDIA MARKET (%) | COMPANY POWER INDEX IN COUNTRY | COMPANY SHARE OF THE OVERALL NATIONAL MEDIA MARKET (%) |
| AT&T | 604 | 17.6 | 449 | 16.1 | −2.8 | −0.2 |
| Verizon | 298 | 12.4 | 366 | 14.0 | 2.5 | 0.2 |
| Comcast | 77 | 4.7 | 166 | 9.9 | 12.7 | 0.5 |
| GE | 11 | 1.1 | | | −11.1 | −0.12 |
| Google | 26 | 0.5 | 88 | 1.3 | 26.9 | 0.09 |
| Liberty Media | 1 | 0.4 | 5 | 0.6 | 33.2 | 0.02 |
| DirecTV | 0.0 | 0.0 | 72 | 3.8 | N/A | 0.4 |
| Softbank (Japan, Sprint) | 25 | 3.1 | 67 | 4.4 | 19.1 | 0.1 |
| Murdoch | 29 | 3.0 | 19 | 1.9 | −3.9 | −0.1 |
| *21st Century Fox* | 0.0 | 0.0 | 14 | 1.2 | N/A | 0.2 |
| *News Corp.* | 30 | 3.0 | 5 | 0.7 | −11.8 | −0.3 |
| Vodafone (UK) | 25 | 2.0 | 0.0 | 0.0 | −11.1 | −0.2 |
| EchoStar (DISH) | 11 | 1.1 | 14 | 1.6 | 3.0 | 0.06 |
| Time Warner | 67 | 4.1 | 24 | 1.2 | −7.2 | −0.3 |
| Time Inc. | 0.0 | 0.0 | 4 | 0.4 | N/A | 0.05 |
| Time Warner Cable | 3 | 1.1 | 43 | 4.1 | 129.9 | 0.3 |
| TW Telecom | 0.0 | 2.0 | 0.0 | 0.0 | −11.1 | −0.2 |
| Disney | 20 | 1.6 | 26 | 1.7 | 2.9 | 0.01 |
| Microsoft | 2 | 0.1 | 5 | 0.3 | 20.1 | 0.02 |
| T-Mobile (Germany) | 5 | 0.9 | 38 | 3.1 | 66.6 | 0.24 |
| Cox | 7 | 1.5 | 8 | 2.0 | 3 | 0.06 |

| | | | | | | |
|---|---|---|---|---|---|---|
| Penguin-Random House (Germany/UK) | 0.0 | 0.0 | 9 | 0.5 | | 0.06 |
| *Bertelsmann (Germany)* | 2 | 0.3 | 3 | 0.3 | 17.1 | 0.001 |
| *Pearson (UK)* | 1 | 0.2 | 0.0 | 0.0 | –11.1 | –0.01 |
| CenturyLink | 0.0 | 0.0 | 31 | 2.9 | | 0.32 |
| Advance | 3 | 0.6 | 5 | 0.7 | 6.4 | 0.01 |
| Gannett | 8 | 0.9 | 5 | 0.6 | –4.9 | –0.04 |
| Sirius XM | 1 | 0.4 | 5 | 0.6 | 33.2 | 0.02 |
| Clear Channel | 10 | 0.5 | 5 | 0.3 | –6.2 | –0.03 |
| Redstone Group | 26 | 1.4 | 21.2 | 1.7 | –1.9 | 0.03 |
| *CBS (ex-Viacom)* | 0.0 | 0.0 | 25 | 1.4 | | 0.2 |
| *Viacom* | 26 | 2.1 | 8 | 0.6 | –10.0 | –0.21 |
| Hearst | 2 | 0.6 | 6 | 0.7 | –17.0 | 0.02 |
| Sony (Japan) | 4 | 0.2 | 1 | 0.1 | –7.3 | –0.01 |
| McGraw Hill | 3 | 0.5 | 2 | 0.3 | –3.8 | –0.01 |
| New York Times | 1 | 0.3 | 1 | 0.3 | 3.9 | –0.001 |
| Yahoo | 11 | 0.5 | 4 | 0.4 | –6.8 | –0.01 |
| Reed Elsevier (Netherlands) | 0.8 | 0.3 | 0.8 | 0.2 | –1.2 | –0.01 |
| Cumulus | 0.1 | 0.05 | 0.8 | 0.1 | 95.5 | 0.01 |
| Tribune | 2 | 0.6 | 2 | 0.4 | –2.1 | –0.02 |
| IAC | 0.3 | 0.06 | 0.2 | 0.06 | –5.0 | 0.0 |
| AOL | 0.1 | 0.04 | 0.7 | 0.09 | 53.9 | 0.01 |
| McClatchy | 0.5 | 0.2 | 1 | 0.3 | 22.3 | 0.01 |
| Lagardere/Hachette Filipacchi (France) | 0.2 | 0.1 | 0.3 | 0.09 | 3.6 | –0.01 |
| Meredith Corporation | 0.3 | 0.1 | 0.2 | 0.1 | –3.6 | –0.002 |
| Entercom | 0.2 | 0.08 | 0.1 | 0.05 | –5.0 | –0.003 |
| Radio One | 0.1 | 0.06 | 0.1 | 0.03 | –5.5 | –0.003 |
| Emmis | 0.1 | 0.05 | 0.1 | 0.04 | –0.2 | –0.001 |
| Univision | 0.5 | 0.2 | 0.1 | 0.04 | –9.1 | –0.02 |

*continued*

Table 18-30. *continued*

| | 2004/05 | | 2011 OR MOST RECENT | | % CHANGE ANNUAL AVERAGE | |
|---|---|---|---|---|---|---|
| | COMPANY POWER INDEX IN COUNTRY | COMPANY SHARE OF THE OVERALL NATIONAL MEDIA MARKET (%) | COMPANY POWER INDEX IN COUNTRY | COMPANY SHARE OF THE OVERALL NATIONAL MEDIA MARKET (%) | COMPANY POWER INDEX IN COUNTRY | COMPANY SHARE OF THE OVERALL NATIONAL MEDIA MARKET (%) |
| PBS (public) | 0.1 | 0.07 | 0.0 | 0.04 | –6.2 | –0.003 |
| Scholastic | 1 | 0.2 | 0.9 | 0.2 | –4 | –0.01 |
| Charter | 3 | 0.8 | 3 | 1.2 | 1.3 | 0.05 |
| Cablevision | 4 | 1.5 | 6 | 1.7 | 4.6 | 0.03 |
| Thomson (Canada) | 0.3 | 0.1 | 0.3 | 0.1 | 0.4 | 0.0 |
| Reader's Digest | 0.4 | 0.2 | 0.1 | 0.1 | –7.5 | –0.01 |
| Primedia | 0.6 | 0.2 | 0.0 | 0.0 | –11.1 | –0.02 |
| Source Interlink | 0.0 | 0.0 | 0.5 | 0.15 | N/A | 0.02 |
| Ziff-Davis | 0.0 | 0.04 | 0.0 | 0.01 | –10.5 | –0.004 |
| Citadel | 0.6 | 0.1 | 0.0 | 0.0 | –11.1 | –0.02 |
| Mediacom | 0.2 | 0.17 | 0.2 | 0.22 | 0.08 | 0.01 |
| Netflix | 0.0 | 0.0 | 1 | 0.5 | N/A | 0.05 |
| Grupo Carso (Mexico) | 0.0 | 0.0 | 10 | 1.5 | N/A | 0.2 |
| Qwest | 8 | 1.7 | 0.0 | 0.0 | –11.1 | –0.2 |
| Citizens Telecom | 3 | 1 | 2 | 0.6 | –5.0 | –0.05 |
| Level 3 | 0.2 | 0.3 | 20 | 2.1 | 1201.4 | 0.2 |
| Global Crossing | 1 | 0.7 | 0.0 | 0.0 | –11.1 | –0.08 |
| Cingular | 70 | 3.3 | 0.0 | 0.0 | –14.3 | –0.5 |
| McCaw (AT&T "I") | 31 | 2.2 | 0.0 | 0.0 | –11.1 | –0.2 |
| Alltel | 7 | 1.0 | 0.0 | 0.0 | –11.1 | –0.1 |
| Nextel | 9 | 1.2 | 0.0 | 0.0 | –11.1 | –0.1 |

| | | | | | | |
|---|---|---|---|---|---|---|
| US Cellular | 2 | 0.5 | 1 | 0.6 | −3.2 | −0.01 |
| Leap Wireless | 0.0 | 0.00 | 0.2 | 0.2 | N/A | 0.02 |
| Bell South | 18 | 2.7 | 0.0 | 0.0 | −11.1 | −0.3 |
| Windstream | 0.0 | 0.02 | 0.1 | 0.08 | 24.4 | 0.01 |
| Covad | 0.0 | 0.02 | 0.0 | 0.05 | 4.5 | 0.003 |
| Adelphia | 3 | 0.7 | 0.0 | 0.0 | −11.1 | −0.1 |
| Knight Ridder | 2 | 0.3 | 0.0 | 0.0 | −11.1 | −0.04 |
| Lionsgate | 0.1 | 0.05 | 1 | 0.1 | 75.6 | 0.01 |

| Media Concentration Index | 2004/05 | 2011 or Most Recent | % Change Annual Average |
|---|---|---|---|
| Total Revenue: Nat'l Media Industry (mil US$) | 652,565 | 790,581 | 3.0 |
| Total Voices (*n*) | 101 | 99 | −0.2 |
| Net Voices (*n*) | 56 | 59 | 0.6 |
| Public Ownership (%) | 0.4 | 0.4 | 0.0 |
| Foreign Ownership (%) | 4.0 | 10.1 | 0.7 |
| C4 Average—Weighted | 58 | 68 | 1.2 |
| HHI Average—Weighted | 1,366 | 1,574 | 1.7 |
| C1 Average—Weighted | 26 | 25 | 0.1 |
| Noam Index Average—Weighted | 227 | 409 | 8.9 |
| Pooled Overall Sector C4 | 39 | 44 | 0.5 |
| Pooled Overall Sector HHI | 564 | 638 | 1.5 |
| Pooled Overall Sector Noam Index | 24 | 33 | 4.0 |
| Market Share of Top Ten Companies: Nat'l Media Industry (%) (Pooled C10) | 54 | 61 | 0.9 |
| National Power Index | 1,386 | 1,540 | 1.2 |

Table 18-31. Top Content Media Companies in the United States

| | 2004/05 | | 2011 OR MOST RECENT | | % CHANGE ANNUAL AVERAGE | |
|---|---|---|---|---|---|---|
| | COMPANY POWER INDEX IN COUNTRY | COMPANY SHARE OF THE NATIONAL CONTENT MEDIA MARKET (%) | COMPANY POWER INDEX IN COUNTRY | COMPANY SHARE OF THE NATIONAL CONTENT MEDIA MARKET (%) | COMPANY POWER INDEX IN COUNTRY | COMPANY SHARE OF THE NATIONAL CONTENT MEDIA MARKET (%) |
| Google | 69 | 1.4 | 258 | 3.8 | 30.6 | 0.3 |
| Comcast | 65 | 3.6 | 178 | 9.4 | 19.7 | 0.6 |
| Disney | 55 | 4.2 | 76 | 4.9 | 4.2 | 0.07 |
| Time Warner | 135 | 7.8 | 71 | 3.6 | –5.3 | –0.5 |
| DirecTV | 0 | 0 | 70 | 3.7 | | 0.4 |
| Redstone Group | 69 | 3.8 | 62 | 5 | –1.0 | 0.1 |
| Murdoch | 53 | 5.8 | 56 | 5.5 | 0.6 | –0.03 |
| Time Warner Cable | 0 | 0 | 33 | 2.5 | | 0.3 |
| Penguin-Random House (Germany/UK) | 0 | 0 | 28 | 1.6 | | 0.2 |
| *Bertelsmann (Germany)* | 4 | 0.9 | 10 | 1 | 20.1 | 0.02 |
| *Pearson (UK)* | 4 | 0.6 | 5 | 0.7 | 3.8 | 0.002 |
| Hearst | 6 | 1.5 | 17 | 2.2 | 19.8 | 0.02 |
| Advance | 9 | 1.7 | 16 | 2.2 | 8.1 | 0.06 |
| Liberty Media | 3 | 0.8 | 15 | 1.3 | 44.8 | 0.06 |
| Sirius XM | 3 | 0.8 | 15 | 1.3 | 8.1 | 0.06 |
| Microsoft | 5 | 0.4 | 14 | 0.9 | 23.2 | 0.06 |
| Gannett | 23 | 2.6 | 14 | 1.9 | –4.2 | –0.08 |
| Clear Channel | 28 | 1.5 | 14 | 0.9 | –5.7 | –0.06 |
| EchoStar (DISH) | 9 | 1 | 13 | 1.6 | 4.4 | 0.07 |
| Yahoo | 31 | 1.3 | 13 | 1.1 | –6.4 | –0.03 |

| | | | | | | |
|---|---|---|---|---|---|---|
| Time Inc. | 0 | 0 | 11 | 1.3 | | 0.1 |
| McGraw Hill | 9 | 1.2 | 7 | 1 | –3.1 | –0.02 |
| Cox | 5 | 1 | 6 | 1.3 | 2.1 | 0.04 |
| AT&T | 0 | 0 | 5 | 1 | | 0.1 |
| Verizon | 0 | 0 | 5 | 1 | | 0.1 |
| Tribune | 6 | 1.5 | 5 | 1.1 | –1.2 | –0.04 |
| Sony (Japan) | 10 | 0.6 | 4 | 0.4 | –6.9 | –0.03 |
| New York Times | 3 | 0.8 | 4 | 0.8 | 5.3 | 0.004 |
| McClatchy | 1 | 0.5 | 4 | 0.8 | 25.6 | 0.03 |
| Scholastic | 4 | 0.6 | 3 | 0.5 | –3.7 | –0.02 |
| Cablevision | 2 | 0.6 | 3 | 0.9 | 7.1 | 0.03 |
| Lionsgate | 0.4 | 0.1 | 3 | 0.3 | 84.1 | 0.03 |
| Reed Elsevier (Netherlands) | 2 | 0.7 | 2 | 0.4 | –0.2 | –0.03 |
| Cumulus | 0.2 | 0.1 | 2 | 0.4 | 106 | 0.03 |
| AOL | 0.3 | 0.1 | 2 | 0.3 | 60.3 | 0.02 |
| Charter | 2 | 0.5 | 2 | 0.6 | –0.03 | 0.02 |
| Source Interlink | 0 | 0 | 1.3 | 0.4 | | 0.05 |
| Thomson (Canada) | 0.8 | 0.3 | 0.9 | 0.4 | 1.5 | 0.01 |
| Lagardere/Hachette Filipacchi (France) | 0.5 | 0.3 | 0.7 | 0.3 | 5.1 | –0.01 |
| Meredith Corporation | 0.9 | 0.3 | 0.7 | 0.3 | –2.9 | –0.003 |
| IAC | 0.8 | 0.2 | 0.5 | 0.2 | –4.3 | 0.002 |
| Reader's Digest | 1 | 0.5 | 0.4 | 0.3 | –7.1 | –0.02 |

*continued*

Table 18-31. *continued*

| | 2004/05 | | 2011 OR MOST RECENT | | % CHANGE ANNUAL AVERAGE | |
|---|---|---|---|---|---|---|
| | COMPANY POWER INDEX IN COUNTRY | COMPANY SHARE OF THE NATIONAL CONTENT MEDIA MARKET (%) | COMPANY POWER INDEX IN COUNTRY | COMPANY SHARE OF THE NATIONAL CONTENT MEDIA MARKET (%) | COMPANY POWER INDEX IN COUNTRY | COMPANY SHARE OF THE NATIONAL CONTENT MEDIA MARKET (%) |
| Entercom | 0.5 | 0.2 | 0.3 | 0.1 | –4.4 | –0.01 |
| Univision | 1 | 0.6 | 0.3 | 0.1 | –8.9 | –0.05 |
| Radio One | 0.3 | 0.2 | 0.2 | 0.1 | –5.0 | –0.01 |
| Emmis | 0.2 | 0.1 | 0.2 | 0.1 | 0.9 | –0.001 |
| Mediacom | 0.2 | 0.2 | 0.2 | 0.2 | 1.2 | 0.01 |
| PBS (public) | 0.2 | 0.2 | 0.1 | 0.1 | –5.7 | –0.01 |
| GE | 30 | 3 | 0 | 0 | –11.1 | –0.3 |
| Primedia | 2 | 0.4 | 0 | 0 | –11.1 | –0.05 |
| Ziff-Davis | 0.1 | 0.1 | 0 | 0.03 | –10.5 | –0.01 |
| Citadel | 2 | 0.4 | 0 | 0 | –11.1 | –0.04 |
| Adelphia | 3 | 0.5 | 0 | 0 | –11.1 | –0.06 |
| Knight Ridder | 4 | 0.9 | 0 | 0 | –11.1 | –0.1 |
| Media Concentration Index | 2004/05 | | 2011 or Most Recent | | % Change Annual Average | |
| Public Ownership (%) | 1.1 | | 1.0 | | 0.0 | |
| Foreign Ownership (%) | 2.9 | | 3.1 | | 0.02 | |
| C4 Average—Weighted | 42 | | 52 | | 1.1 | |
| HHI Average—Weighted | 666 | | 1,080 | | 6.9 | |
| C1 Average—Weighted | 16 | | 20 | | 0.4 | |
| National Power Index | 664 | | 1,035 | | 6.2 | |

Table 18-32. Top Platform Media Companies in the United States

| | 2004/05 | | 2011 OR MOST RECENT | | % CHANGE ANNUAL AVERAGE | |
|---|---|---|---|---|---|---|
| | COMPANY POWER INDEX IN COUNTRY | COMPANY SHARE OF THE NATIONAL PLATFORM MEDIA MARKET (%) | COMPANY POWER INDEX IN COUNTRY | COMPANY SHARE OF THE NATIONAL PLATFORM MEDIA MARKET (%) | COMPANY POWER INDEX IN COUNTRY | COMPANY SHARE OF THE NATIONAL PLATFORM MEDIA MARKET (%) |
| AT&T | 963 | 28.1 | 678 | 23.8 | −3.3 | −0.5 |
| Verizon | 475 | 19.8 | 551 | 20.6 | 1.8 | 0.1 |
| Comcast | 85 | 5.4 | 160 | 9.0 | 9.7 | 0.4 |
| Liberty Media | 0.3 | 0.2 | 0.3 | 0.2 | −1.5 | 0.004 |
| DirecTV | 0.0 | 0.0 | 73 | 2.3 | | 0.3 |
| Softbank (Japan, Sprint) | 39 | 4.9 | 101 | 6.7 | 17.6 | 0.2 |
| Murdoch | 15 | 1.4 | 0.0 | 0.0 | −11.1 | −0.2 |
| Vodafone (UK) | 40 | 3.1 | 0.0 | 0.0 | −11.1 | −0.3 |
| EchoStar (DISH) | 11 | 1.2 | 14 | 1.7 | 2.3 | 0.05 |
| Time Warner | 27 | 1.9 | 0.0 | 0.0 | −11.1 | −0.2 |
| Time Warner Cable | 5 | 1.8 | 48 | 4.8 | 88.2 | 0.3 |
| TW Telecom | 0.0 | 3.1 | 0.0 | 0.0 | −11.1 | −0.01 |
| T-Mobile (Germany) | 9 | 1.5 | 58 | 4.7 | 62.7 | 0.4 |
| Cox | 8 | 1.8 | 9 | 2.3 | 2.7 | 0.1 |
| CenturyLink | 0.0 | 0.0 | 47 | 4.4 | | 0.5 |
| The Washington Post | 0.1 | 0.1 | 0.1 | 0.1 | 1.5 | 0.003 |
| Charter | 3 | 1.0 | 4 | 1.5 | 1.4 | 0.06 |
| Cablevision | 5 | 2.0 | 7 | 2.1 | 3.6 | 0.01 |
| Mediacom | 0.3 | 0.2 | 0.2 | 0.2 | −0.5 | 0.004 |
| Grupo Carso (Mexico) | 0.0 | 0.0 | 14 | 2.3 | | 0.3 |
| Qwest | 12 | 2.8 | 0.0 | 0.0 | −11.1 | −0.3 |

*continued*

Table 18-32. *continued*

| | 2004/05 | | 2011 OR MOST RECENT | | % CHANGE ANNUAL AVERAGE | |
|---|---|---|---|---|---|---|
| | COMPANY POWER INDEX IN COUNTRY | COMPANY SHARE OF THE NATIONAL PLATFORM MEDIA MARKET (%) | COMPANY POWER INDEX IN COUNTRY | COMPANY SHARE OF THE NATIONAL PLATFORM MEDIA MARKET (%) | COMPANY POWER INDEX IN COUNTRY | COMPANY SHARE OF THE NATIONAL PLATFORM MEDIA MARKET (%) |
| Citizens Telecom | 4 | 1.6 | 2 | 0.9 | −5.3 | −0.1 |
| Level 3 | 0.3 | 0.4 | 30 | 3.2 | 1140.1 | 0.3 |
| Global Crossing | 2 | 1.1 | 0.0 | 0.0 | −11.1 | −0.1 |
| Cingular | 111 | 5.3 | 0.0 | 0.0 | −11.1 | −0.8 |
| McCaw (AT&T "I") | 49 | 3.5 | 0.0 | 0.0 | −11.1 | −0.4 |
| Alltel | 11 | 1.6 | 0.0 | 0.0 | −11.1 | −0.2 |
| Nextel | 15 | 1.9 | 0.0 | 0.0 | −11.1 | −0.2 |
| US Cellular | 3 | 0.8 | 2 | 0.9 | −3.6 | 0.004 |
| Leap Wireless | 0.0 | 0.0 | 0.4 | 0.4 | | 0.04 |
| Bell South | 29 | 4.3 | 0.0 | 0.0 | −11.1 | −0.5 |
| Windstream | 0.0 | 0.04 | 0.1 | 0.1 | 22.6 | 0.01 |
| Covad | 0.0 | 0.03 | 0.1 | 0.1 | 3.68 | 0.004 |
| Adelphia | 4 | 0.8 | 0.0 | 0.0 | −11.1 | −0.1 |

| Media Concentration Index | 2004/05 | 2011 or Most Recent | % Change Annual Average |
|---|---|---|---|
| Public Ownership (%) | 0.7 | 0.5 | 0.0 |
| Foreign Ownership (%) | 4.6 | 13.7 | 1.0 |
| C4 Average—Weighted | 67 | 77 | 1.1 |
| HHI Average—Weighted | 1,781 | 1,827 | 0.3 |
| C1 Average—Weighted | 31 | 28 | −0.3 |
| National Power Index | 1,815 | 1,799 | −0.1 |

communications due to participation by several European firms like Vodafone (UK), which used to have a 45% stake in Verizon Wireless, as well as Deutsche Telekom (Germany, T-Mobile), and Softbank (Japan) with Sprint. The percentage of US industries that are foreign-owned is high in book publishing (25.5%) and film (13.7%)—the latter due primarily to Sony (Japan). It is moderate in magazines (7.4%).

# C

## Latin America

# 19

# Media Ownership and Concentration in Argentina

GUILLERMO MASTRINI, ANA BIZBERGE, AND MARTÍN BECERRA

## INTRODUCTION

Argentina is one of the most developed Spanish-language media markets in South America due to its high literacy rates and well-developed communications infrastructure. State intervention has characterized the market during the financial difficulties of the early 2000s. New subsidy policies were implemented in 2002 and 2003, and since 2004 the government has passed legislation that exempts media firms from a national bankruptcy law that would allow foreign creditors to purchase their assets, extended their licensing arrangements, and granted tax exemptions. According to figures from the Cultural Information System of Argentina (SINCA), media production represented 3.5% of the GDP in 2009, up from 2.4% in 2004.[1]

Argentine newspapers are confronted with declining subscriptions and advertising revenues, while digitization has opened up many new possibilities for audiovisual producers.[2] Traditional print media companies are looking for ways to retain traditional customers by applying new business models: the daily *La Nación*, for example, has an agreement with other companies to give its subscribers special discounts for a variety of retail and food purchases, as well as club memberships.

The government has promoted the expansion of digital TV platforms in the country, in part because it is increasing its own programming in

1. SINCA. "Cuenta Satélite de Cultura." sinca.cultura.gov.ar, 2012. July 13, 2012 <http://sinca.cultura.gov.ar/sic/estadisticas/csc/index.php>.

2. Albornoz, Luis. Periodismo digital. Los grandes diarios en la red. Buenos Aires: La Crujia, 2007.

this medium. The state has subsidized the construction of antenna sites, offered free decoders to low-income households, and appropriated funding for content production on new government-sponsored channels.[3] The analog-to-digital conversion process was undertaken for cable TV in 2007. By 2009, terrestrial broadcasters had adopted the Japanese ISDB-T (SBTVD) standard that was introduced to Latin America by Brazil. Of Argentine households, 98% have access to a television set, and this sector is the main recipient of advertising investments.

Digitization has also spurred the entry telecommunications and online actors in the audiovisual sector. Since the 1990s, telecom companies and Internet service providers (ISPs) have wanted to offer video content directly to customers. However, regulations have not kept pace with these developments, so those regulations in place have actually contributed to the creation of vertically integrated multimedia groups in the supply chain, resulting in greater concentration.

Two milestones for concentration in Argentina took place between 1989 and 2012 and are the focus of this chapter. The first was the privatization of the audiovisual and telecom sectors in the 1990s through deregulation and the legalization of foreign direct investment in media production. This process culminated in the consolidation of two leading commercial media groups: the Buenos Aires–based Grupo Clarín, and Telefónica Argentina, a subsidiary of Telefónica (Spain). With the privatization of the state telecom operator, the mobile telecom and ISP markets have become increasingly concentrated.

The second milestone was a series of laws enacted under the Néstor Kirchner Administration (2003–2007) that greatly favored established media groups. During that administration's time in office, the government aimed at preserving the status quo of national media groups that faced heavy debts from an economic crisis that saw a huge devaluation of the country's currency.[4] The most significant regulation passed between 2003 and 2007 was the Law 25.750 for the Preservation of Cultural Goods and Property (2003), which placed a 30% cap on foreign direct investment in Argentine media companies. Complementary measures included Decree No. 527/2005—which suspended existing TV licensing agreements for a decade—and the state's 2007 authorization of Grupo Clarín's acquisition of Cablevisión and Multicanal, the country's two main cable TV operators.[5]

Increased state focus on the sector since 2001 through ownership reforms has had a mixed impact on concentration, with the earlier policies tending to help increase concentration and newer policies working to undo this legacy. In fact, due to Grupo Clarín's dominance over the market—and, its supporters contend, the critical editorial line many of its properties have taken against the Cristina Fernandez de Kirchner administrations—significant tensions have emerged between this media giant and the government since 2008. At the start of her administration, a proposal to raise the agricultural export tax led to nationwide strikes and demonstrations. This conflict eventually boiled over into an open dispute between the government and Grupo Clarín. As a result, the ruling Judicialist Party moved to amend Broadcasting Law No. 22.285, which dated back to the military dictatorship era (1976–1983), in order to break up Grupo Clarín's hold over the cable TV market.[6] Grupo Clarín strongly opposed this policy, contending that the government was

3. Bizberge, Ana. Mastrini, Guillermo; and Becerra, Martin. "La Television Digital Terrestre en Argentina: entre la geopolitica regional y la iniciativa estatal." *La transicion a la television digital terrestre en Iberoamerica: diagnostic y prospective.* Badillo, Angel and Sierra, Francisco, eds. Quito: CIESPAL, 2011.

4. In economic terms the 1:1 exchange rate between the ARS and US$ set forth in the 1990s was broken, and as a result, the economic crisis led to the succession of five presidents in just two years, followed by Néstor Kirchner taking office.

5. Romero, Simon and Schmall, Emily. "Battle Between Argentine Media Empire and President Heats Up Over a Law." *The New York Times,* Nov. 30, 2012. Jan. 11, 2013. <http://www.nytimes.com/2012/12/01/world/americas/media-law-ratchets-up-battle-between-kirchner-and-clarin-in-argentina.html?pagewanted=all>.

6. Bizberge, Ana. Mastrini, Guillermo; and Becerra, Martin. "La Television Digital Terrestre en Argentina: entre la geopolitica regional y la iniciativa estatal." *La transicion a la television digital terrestre en Iberoamerica: diagnostic y prospective.* Badillo, Angel and Sierra, Francisco, eds. Quito: CIESPAL, 2011.

motivated only by a desire to shut down criticism of its policies. A new regulatory framework was introduced for audiovisual services in October 2009, the Audiovisual Communication Services Law (Ley de Servicios de Comunicación Audiovisual No. 26.522), including among its key objectives the deconcentration of the industry and the launching of new state-run media initiatives.

There are few quantitative studies on the concentration processes in Argentine media, and media producers do not often disclose their annual reports. This situation has partially improved because public companies must now publish annual financial statements, and the state has pressed for financial disclosures by all media producers.

Margarita Graziano conducted a sociological analysis of terrestrial broadcasters in 1974 based on the composition of their boards of directors. She found that the owners were deeply involved in the economic system as a whole, most noticeably in agribusiness and commercial advertising. Military officers and clergymen were also represented among the boards of directors for media providers, reflecting the role of the military and the Catholic Church as part of the ruling elite. Graziano also demonstrated how audiovisual content production was concentrated in Buenos Aires while in the rest of the country most private channels purchased TV programs from the capital: for political reasons, in Argentina there are no audiovisual providers with full national coverage (except for those operated by the state, which do not have large audience shares).[7] Also in 1974, Heriberto Muraro published a series of articles in the political magazine *Crisis* that closely analyzed the business models of Argentine media producers.

During the military dictatorship from 1976 to 1983, studies dealing with the social sciences, and with communications in particular, were paralyzed by state censorship and the harassment of writers. It was not until the 1990s that interest in the economic dynamics of the media could resume in academia. A key author was Octavio Getino. Writing in 1995 as liberalization, concentration, and foreign direct investment were becoming the norm, his work did not focus on concentration in particular. In 2003, the Ministry of Culture published a study analyzing the structure of the national media market and discussed in detail how concentration increased during the late 1990s and how many smaller enterprises went bankrupt after 2001 due to the currency crisis.[8]

## PRINT MEDIA

### Newspapers

Our 2006 research examining concentration in multiple Latin American markets was the first published study in Argentina that included an analysis of concentration in the telecom sector in the country.[9] We found that in all of the Latin American counties surveyed, the concentration levels of traditional print and audiovisual media producers were lower than those of the telecom providers, but that Argentine print concentration was comparatively high because two publishers control over 50% of the market. In Argentina, there are approximately 180 daily newspapers, which are defined as those that publish more than four times a week. Of those, only 37 publish seven days a week and few dailies have nationwide coverage. Except for the two leading dailies, *Clarín* and *La Nación*, whose sales outside of the capital's metropolitan area account for approximately 20% of their audience shares, sales for other dailies are concentrated in the cities where they are published.

7. Rivals of the ruling Judicialist Party, which was formed in 1947 by Juan Perón, have long pursued such regionalism in order to keep the "Peronists" from controlling all of the country's broadcasting stations. It is not dissimilar from efforts by Brazilian media owners to maintain control over local and regional broadcasting networks at the expense of the federal government.

8. Becerra, Martin et al. "La concentracion de las industrias culturales." *Industrias culturales: Mercado y politicas publicas en Argentina*. Buenos Aires: Ciccus, 2003.

9. Mastrini, Guillermo and Becerra, Martin. Periodistas y Magnates. Estructura y concentracion de las industrias culturales en America Latina. Buenos Aires: Prometeo, 2006.

In 2010, *Clarín* had a national market share of nearly 30% and a daily circulation of approximately 315,000 copies. In comparison, *Clarín*'s closest competitor, the family-owned daily *La Nación,* had a circulation slightly less than half of that (155,000). Though it remains the market leader, since 2000 *Clarín*'s market share has declined: in 1996 it reached a national market share of 45.4%, and this had fallen by 2010 to 28.9%. On top of that, overall industry sales fell due to the growth of online news (Table 19.1).

Grupo Clarín has expanded its print operations through capital reinvestment in its newspapers.[10] In addition to *Clarín,* Grupo Clarín also owns the Cimeco publishing house, which controls several regional newspapers, including the fourth best-selling paper in the country, *La Voz del Interior.*[11] Grupo Clarín also controls the majority shareholding of the main free daily, *La Razón. Clarín*'s dominant position in this sector is augmented by its effective monopoly over traditional classified advertisements (with a 41.5% share of the national market). Since the late 1970s, it has also held a significant stake in Papel Prensa S.A., the country's main newsprint producer alongside the owners of *La Nación,* further enhancing its dominance in the print sector.

In 2008, the entire daily newspaper market in Argentina accounted for 31.6% of total national advertising spending, worth US$649 million

**Table 19-1. Daily Newspapers (Market Shares by Circulation), 1984–2010**

| | 1984 | 1988 | 1992 | 1996 | 2000 | 2004 | 2008 | 2010 |
|---|---|---|---|---|---|---|---|---|
| Grupo Clarín[1] | 43.9 | 38.8 | 46.9 | 50.3 | 30.3 | 34.7 | 34.9 | 28.9 |
| *Clarín* | 38.5 | 38.8 | 41.3 | 45.4 | 30.3 | 34.7 | 30.3 | 14.2 |
| *La Voz del Interior* | 5.4 | | 5.6 | 4.9 | | | 4.7 | 8.3 |
| La Nación (La Nación, Mitre Family) | 14.1 | 17.3 | 13.4 | 13.0 | 11.1 | 13.4 | 12.9 | 4.9 |
| El Día *(Diario Popular)* | 8.2 | 7.2 | 7.5 | | | 6.2 | 7.4 | |
| Crónica | | | | | 14.9 | 9.7 | | |
| Ámbito Financiero | | | | | 7.3 | | | |
| Others | 33.8 | 36.6 | 32.3 | 36.7 | 36.4 | 36.0 | 44.8 | 43.7 |
| Total Daily Circulation (1,000) | 1,432 | 1,403 | 1,549 | 1,361 | 1,507 | 1,160 | 1,196 | 1,090 |
| Total Revenue (mil ARS)[2] | | | | | | 1,229 | 1,422 | |
| Total Revenue (mil US$) | | | | | | 418 | 455 | |
| C4 | 66.2 | 63.4 | 67.7 | 63.3 | 63.6 | 64.0 | 55.2 | 56.3 |
| HHI | 2,195 | 1,861 | 2,431 | 2,702 | 1,315 | 1,514 | 1,441 | 1,130 |
| *N* (>1%) | 3 | 3 | 3 | 2 | 4 | 4 | 3 | 4 |
| Noam Index | 1,267 | 1,074 | 1,404 | 1,911 | 657 | 757 | 832 | 565 |

*Source*: IVC.

[1]Our figures do not include all of Grupo Clarín newspapers, such as those put out by Cimeco, only *Clarín* and *La Voz del Interior.*

[2] This revenue figure is an estimate based on data collected in this study: it is based on average per capita revenues reported for several other countries in this study with similar per capita income.

10. Grupo Clarín is a family-owned enterprise. The Noble and Magnetto families, represented by Director Ernestina Herrera de Noble and CEO Héctor Magnetto, along with the minority shareholders José Aranda and Lucio Plagiaro, own Grupo Clarín with a combined 70.99% stake.

11. "Interior" refers to the towns and cities that are not part of the metropolitan area of Buenos Aires.

(Argentinian Peso [ARS]2 billion). Of that, newspapers based in the capital's metropolitan area accounted for 85% of the share, while the rest of the country combined accounted for the remaining 15%.

Concentration has been gradually declining in the newspaper market. Since 2004, five new newspapers have been launched in Buenos Aires in competition with the four mainstream papers. Though their financial position is still unstable, they have managed to achieve a combined market share of approximately 20%. The decline in overall C4 is mainly the result of these new dailies competing with *Clarín* at the regional level.

## Book Publishing

Until the 1970s, Argentine book publishers were the market leaders in the Latin America. The domestic market was not excessively concentrated, and there were very few foreign-owned publishing houses. But with the military coup in 1976, publication and readership plunged due to state censorship. Book sales began to approach precoup levels after civilian government returned in 1989, but the most noticeable event for the sector during the 1990s was the concentration of ownership in production and distribution. Several publishing houses based in Spain acquired the main Argentine publishing companies. The industry was also affected by the government's decision to set the ARS-US$ exchange at a 1:1 ratio during the 1990s, such that book production was moved abroad to reduce overhead.

Following the delinking of the ARS and US$ in 2001, the Argentine publishing industry began to regain its international competitiveness. Though the market is still characterized by concentrated production, distribution, and marketing, publishing and sales for both the domestic and foreign markets have increased considerably, as has the traditional role of native publishers. During the 2000s, most firms operating in the Argentine market fell under the umbrella of two groups: Cámara Argentina del Libro, which accounts for most of the small- and medium-sized domestic publishing houses, and Cámara Argentina de Publicaciones, which comprises the larger, foreign-owned publishers (Table 19.2).

The book publishing market is not very transparent when it comes to disclosing sales

Table 19-2. Book Publishing (Circulation), 1996–2007

| | EDITED TITLES | AVERAGE PRINTED COPIES PER TITLE | TOTAL ANNUAL CIRCULATION |
|---|---|---|---|
| 1996 | 8,835 | 4,787 | 42,296,878 |
| 1997 | 10,374 | 5,137 | 53,289,819 |
| 1998 | 11,535 | 4,724 | 54,490,652 |
| 1999 | 12,291 | 5,851 | 71,914,910 |
| 2000 | 11,717 | 6,338 | 74,264,135 |
| 2001 | 12,378 | 4,444 | 55,005,271 |
| 2002 | 9,537 | 3,534 | 33,708,268 |
| 2003 | 14,375 | 2,650 | 38,096,586 |
| 2004 | 18,828 | 2,974 | 55,985,843 |
| 2005 | 19,636 | 3,430 | 67,346,651 |
| 2006 | 22,598 | 3,761 | 84,998,863 |
| 2007 | 26,320 | 3,525 | 92,768,996 |

*Source*: SINCA, 2007.

data. Cámara Argentina del Libro does issue an annual report on its members' output, but this does not include information on sales or comparative market shares. Therefore, we have been unable to calculate concentration indices for this sector.

We were able to calculate market shares by type of publisher, of which Brazil has three, each defined by their total annual turnover. Smaller publishers—those reporting annual revenues less than or equal to US$1 million (ARS3.1 million)—accounted for 42% of all published annual titles. Medium-sized publishers, defined as those with annual turnovers in the range of US$1 to 10 million (ARS3.1 to 31.2 million) published 48% of all titles in circulation. Finally, the large publishing companies, those with an annual income greater than US$10 million (ARS31.2 million) only published 10% of all titles. It is clear, however, that this 10% accounts probably for a significantly higher percentage of income from sales.

### Magazine Publishing

Before 1976, the magazine market was highly concentrated: four publishing houses—Julio Korn, Atlántida, the Abril Group (Brazil), and Dante Quinterno—dominated the sector. Circulation declined during the period of military rule, however, and the sector only began to regain its former vitality after the early 1990s. By 2000, the old market structure had undergone several significant changes. Two of the main publishers fell away from their former dominant positions: Atlántida did not recover a large market share, and the Abril Group chose to refocus on its Brazilian operations (one of its former holdings became the independent publisher Editorial Publiexpress). Grupo Clarín has since entered the market through its print publishing arm AGEA. Another newspaper publisher—Editorial Perfil, the publisher of the *Perfil* tabloid—has also entered the book publishing market. Group Clarín and Editorial Perfil entered the market partly to expand their advertising revenues through the magazines that are sold with the Sunday editions of their papers, in particular Grupo Clarín's *Viva* magazine. The Mexican media conglomerate Grupo Televisa has also entered the market to sell magazines promoting its audiovisual content (Table 19.3).

The main source of income for the sector is advertising, valued at US$110 million (ARS344 million) in 2008, representing 5% of total national advertising spending. Due to the C4 index fluctuating between 40% and 50% since 1988, it is not possible to determine concentration trends for the sector.

One phenomenon that is not measured in these data but deserves mention is that of magazines that are distributed by cable TV providers. Subscribers to these magazines pay lower prices than the rest of the market, and given that cable TV penetration is very high in Argentina (over 70% of households), magazines linked with TV distributors are among those with the highest readership.

## AUDIOVISUAL MEDIA

### Radio Broadcasting

Radio broadcasting is the smallest media sector in the country measured in market shares from advertising revenue, and while there are a large number of independent private broadcasters in total nationally, the market is still strongly concentrated because of regionalism. Argentina has one of the highest rates of radio ownership in Latin America, with 99% of households capable of receiving broadcasts, but only the national public broadcaster, Radio Nacional, covers most of Argentina. In the 1990s, the state planned to promote chain stations in order to increase the range of regional actors, but because the economic downturn began just as the plan went into effect, advertising investment in radio collapsed, scuttling the initiative. The Audiovisual Communication Services Law (26522/09) in 2009 imposed new restrictions on the development of a national commercial network: under the law, only 30% of all broadcast content can be retransmitted nationally, thus limiting the reach of powerful commercial entities, like Grupo Clarín, in radio.

The radio broadcaster with the largest audience share is the AM station Radio 10, which

Table 19-3. Magazine Publishing (Market Shares by Circulation), 1984–2013

| | 1984 | 1988 | 1992 | 1996 | 2000 | 2004 | 2008 | 2013 |
|---|---|---|---|---|---|---|---|---|
| Editorial Atlátida (Vigil Family) | 24.6 | 19.4 | 22.8 | 16.7 | 9.3 | 11.6 | 14.6 | 28.7 |
| Editorial Perfil | | 11,2 | 15,5 | 26,5 | 14,5 | 11,3 | 13,6 | 14.4 |
| *Editorial Perfil* | | *9.7* | *15.5* | *23.7* | *13.2* | *9.9* | *13.6* | |
| *Editorial Primavera S.A.* | | *1.5* | | *2.8* | *1.3* | *1.4* | | |
| Artegrafico Editorial Argentina (Grupo Clarín) | | | | 1.6 | 6.1 | 8.8 | | |
| Editorial Publiexpress (Grupo Publiexpress) | | | | 4.4 | 5.3 | 7.3 | 12.0 | 13.1 |
| Editorial Televisa Argentina (Azcárraga Family, Grupo Televisa, Mexico) | | 11.0 | 10.0 | 2.4 | 2.4 | 2.6 | 8.2 | 10.7 |
| Others | 75.4 | 58.4 | 51.7 | 48.5 | 62.5 | 58.5 | 51.6 | 33.1 |
| Total Annual Circulation (1,000) | 1,446 | 1,135 | 1,367 | 2,068 | 2,495 | 2,065 | 9,991 | 8,620 |
| Total Revenue (mil ARS)[1] | | | | | | 1,000 | 1,438 | |
| Total Revenue (mil US$) | | | | | | 340 | 460 | |
| C4 | 24.6 | 41.6 | 48.3 | 50.0 | 35,2 | 39.0 | 48.4 | 66.9 |
| HHI | 606 | 593 | 860 | 874 | 334 | 371 | 609 | 1,317 |
| *N* (>1%) | 1 | 2 | 2 | 6 | 6 | 6 | 4 | 4 |
| Noam Index | 606 | 419 | 608 | 357 | 136 | 151 | 305 | 659 |

*Source*: IVC.

[1] This revenue figure is an estimate based on data collected in this study: it is based on average per capita revenues reported for several other countries in this study with similar per capita income.

belongs to Grupo Infobae, a media enterprise led by the entrepreneur Daniel Hadad that also includes the FM radio stations FM Mega, Pop, Vale, and TKM. In 2012 Grupo Infobae sold Radio 10, Vale, TKM, Mega and Pop, as well as C5N to Grupo Indalo. In the AM radio market, Radio Mitre belongs to Grupo Clarín, and Telefónica held AM Continental until it sold it to its rival, the PRISA Group (Spain). Even though the law does not allow foreign investors to hold more than a 30% stake in a radio broadcaster, PRISA is allowed to circumvent this legal limitation with the acquiescence of the state. There are also several FM stations with national audience shares, such as Rock&Pop, recently acquired by Grupo Montena, and PRISA's FM 105.5 Hit (Table 19.4).

According to the Federal Audiovisual Communication Authority (AFSCA), in 2008 there were 1,940 registered private radio stations in the country, of which only 438 were licensed commercial radio stations. Religious entities, indigenous tribes, frontier schools, and public universities own 9% of these 438 stations, and 67% were commercial radio stations with preliminary permits, which since the 1980s have been granted to "pirate radio" stations to legalize their operations.

Due to low advertising revenues, there is little content production in the sector outside of the networks headquartered in the Buenos Aires metropolitan area. In 2008, advertising revenues barely exceeded US$60 million (ARS188 million), representing less than 4% of total national advertising expenditures. Advertising income for the most part is concentrated in the Buenos Aires metropolitan area, and to a

## Table 19-4. Radio Group (Audience Shares by Ratings), 2000–2010[1]

| | 2000 | 2004 | 2008 | 2010 |
|---|---|---|---|---|
| Grupo Infobae | | | | |
| *Radio 10* | 11.3 | 34.3 | 36.2 | 15.6 |
| *FM Mega* | 7.4 | | | |
| PRISA Group (Polanco Family, Spain) | 8.3 | 12.3 | 8.9 | 5.5 |
| *AM Continental* | | *12.3* | *8.9* | |
| *FM 105.5 Hit* | *8.3* | | | |
| *AM Mitre* (Grupo Clarín) | 8.2 | 23.5 | 16.0 | 13.5 |
| *La Red* (Grupo América) | | 10.4 | 8.0 | |
| *Rock&Pop* (Grupo Moneta) | | | 9.8 | 12.3 |
| *Del Plata* (Grupo Electroingeniería) | | | 9.6 | |
| Others | 65.0 | 19.5 | 11.5 | 53.1 |
| Total Revenue (mil ARS) | | | 703 | |
| Total Revenue (mil US$) | | | 225 | |
| C4 | 35.0 | 80.5 | 64.5 | 46.9 |
| HHI | 316 | 1,989 | 1,897 | 607 |
| *N* (>1%) | 4 | 4 | 6 | 4 |
| Noam Index | 158 | 995 | 775 | 304 |

*Source:* IBOPE.

[1] Audience shares are only comprehensively calculated for the Buenos Aires metropolitan area. Though national chains have been restricted over long periods, broadcasters outside of the capital region do, in fact, often retransmit programming from Buenos Aires, so real audience concentration indicators are much higher than the data in Table 19.4 indicate.

lesser degree, in the provincial cities of Córdoba and Rosario.

In revenue terms, the sector is highly concentrated: the top four radio stations capture more than 51% of advertising market share, leaving the remaining 49% to be split among hundreds of other broadcasters. For most radio stations outside of Buenos Aires, their business consists of selling airtime to advocacy groups or political activists to promote their own content. Argentine law formally prohibits this practice, so it is impossible to track the amounts involved because statistics are not kept (even though in practice, the government turns a blind eye to this behavior).

## Broadcast Television

The number of terrestrial broadcasting stations is low in Argentina (44), and *N* (>1%) has remained steady since the 1970s. From the 1990s on, ownership has become more concentrated because of the Buenos Aires–based channels buying out channels in the rest of the country.[12] As with radio, there are no commercial stations that provide full national coverage, so the majority of provincial stations simply rebroadcast programming content from the leading Buenos Aires–based channels.[13] The most popular networks are Telefónica's Telefe (formerly Channel 11) and Grupo Clarín's El Trece (also known as Channel 13), both of

12. Mastrini, Guillermo and Becerra, Martin. Periodistas y Magnates. Estructura y concentracion de las industrias culturales en America Latina. Buenos Aires: Prometeo, 2006.

13. TV stations in Argentina are local and do not belong to any one national network. However, the local stations of the country purchase their programming to the Buenos Aires-based stations, creating an informal nationwide network transmission system.

which began as commercial broadcasters in the 1960s, were nationalized under the military dictatorship, and then reprivatized in the 1990s. In the 1990s, the government favored privatization and capital centralization for TV broadcasters. The enactment of the State Reform Law 23.696 in 1989, authorizing the privatization of utilities, amended Broadcasting Law 22.285/80 to legalize the privatization of broadcast television as well.[14]

State Reform Law 23.696 (1989) eliminated the legal limit of allowing no more than three TV licenses per legal entity. Cross-ownership was authorized with the elimination of the ban on print media from participating in license bidding. The period from 1995 to 2000 was characterized by foreign direct investment.[15] Telefónica and its then-partner from the United States, Citicorp Equity Investments (CEI), gained control of Cointel (now the holding company for its Telefónica Argentina division), Cablevisión TCI (the first cable TV operator in the country), and Torneos y Competencias (a popular sports channel). This partnership eventually came to an end in favor of Telefónica, thus expanding the role of the media branch of Telefónica International, Admira, which in 1999 acquired the TV broadcaster Telefe and increased its share in Canal 9 (then Azul Televisión), along with three other provincial channels, thereby becoming a major TV provider. Grupo Clarín also saw increased foreign investment in its operations: the US investment bank Goldman & Sachs bought a share of 18% of Grupo Clarín in 1999.[16] Because of the US financial crisis in 2008, though, Goldman Sachs reduced its equity stake here to 9.1% and in 2012 sold it off entirely to the CEO of Fontinalis Partners (US), Ralph Booth.

In the 2001 economic crisis, the broadcasting sector was greatly impacted by the devaluation of the ARS, as the companies held their debts in US$ and required state bailouts to avoid bankruptcy. The administrations of Fernando De La Rúa (2000–2001), Eduardo Duhalde (2002–2003), and Néstor Kirchner (2003–2007) bolstered the economic recovery of the media system while expanding the scope of government involvement in the sector.[17] To prevent Grupo Clarín from being acquired by international investors, Néstor Kirchner saw to it that the Cultural Goods Act (Law 25.750) was passed in 2003 to cap foreign ownership in Argentine media groups at 30% (Table 19.5).[18]

There are two other broadcasters with significant market shares after the two leading networks: Canal 9, which is one of the 30 Latin American channels owned by the Mexican media tycoon Ángel González's Albavision network, and América TV (Channel 2), which is owned by Grupo Ávila in partnership with the politician Francisco De Narváez (despite the existence of laws that prohibit legislators from having shareholdings in audiovisual media entities).

In 2000, the four main Buenos Aires–based commercial channels of *Telefe*, *El* Trece, Canal 9, and América TV were the undisputed market leaders, with 96% of the aggregate national audience shares and 46.6% of the aggregate national

---

14. During privatization Editorial Atlántida—owned by the Vigil Family—acquired Telefé. By 1997, the station was bought by the spanish telecom operator Telefónica. Rossi, Diego. "La radiodifusion entre 1990–1995: exacerbacion del modelo privado-comercial." *Mucho ruido, pocas leyes. Economia politica de comunicacion en la Argentina (1920–2004).* 2nd ed. Mastrini, Guillermo ed. Buenos Aires: La Crujia, 2008.

15. Mutual Investment Promotion and Protection Agreement (1991); Amendment to the National Constitution (1994); Decree 1005/99. *See* Mastrini, Guillermo, ed. *Mucho Ruido y pocas leyes. Economia y politicas de comunicacion en la Argentina (1920–2004).* 2nd ed. Buenos Aires: La Crujia, 2008.

16. Albornoz, Luis and Hernández, Pablo. "La radiodifusion entre 1995–1999: concentracion, denacionalizacion y ausencia del control public." *Mucho ruido, pocas leyes. Economia y politicas de comunicacion en la Argentina (1920–2004).* 2nd ed. Mastrini, Guillermo, ed. Buenos Aires: La Crujia, 2008.

17. Becerra, Martin and Mastrini, Guillermo. Los duenos de la palabra. Acceso, estructura y concentracion de los medios en la America latina del Siglo XXI. Buenos Aires: Prometeo, 2009.

18. Mastrini, Guillermo and Becerra, Martin. Periodistas y Magnates. Estructura y concentracion de las industrias culturales en America Latina. Buenos Aires: Prometeo, 2006.

Table 19-5. TV Broadcasting (Market Shares by Revenue), 2000–2010

| | 2000 | 2004 | 2008 | 2010 |
|---|---|---|---|---|
| *Telefe* (Telefónica, Spain) | 24.8 | 28.2 | 21.5 | 13.8 |
| *Canal 9* (Ángel González, Mexico) | 5.9 | 15.4 | 19.4 | |
| *América TV/Channel 2* (Grupo Ávila) | | 13.1 | 18.3 | |
| *El Trece* (Grupo Clarín) | 15.8 | 25.5 | 24.5 | 23.5 |
| Others | 53.4 | 17.9 | 15.4 | 62.7 |
| Total Revenue (mil ARS) | 1,067 | 1,150 | 2,669 | 4,708 |
| Total Revenue (mil US$) | 1,067 | 391 | 854 | 1,224 |
| C4 | 46.6 | 82.1 | 84.6 | 37.3 (C2)[1] |
| HHI | 903 | 1,848 | 1,817 | 743 |
| *N* (>1%) | 3 | 4 | 4 | 2 |
| Noam Index | 521 | 924 | 909 | 525 |

*Source*: Getino (1995), IBOPE.

[1] Figure 19.1 in this text shows the evolution of C2, i.e., the market shares of *Telefe* and *El Trece*. It was not possible to measure C4 in 2010 due to the unavailability of *Canal 9* and *América TV* (*Channel 2*) data. The C4 levels of the four main channels, which increased from 46% in 2000 to 82% in 2004 to 89% in 2008, are remarkably high.

market shares for advertising revenue.[19] *Telefe* and *El Trece* together broadcast 79% of all nationally syndicated programming.[20] In 2004, supported by the overall economic recovery, revenues for broadcast TV advertising rose to pre-2001 levels, which enabled broadcasters to increase content production (Figure 19.1).[21]

*Telefe* and *El Trece* together accounted for 40.7% of national advertising income in 2000, with market shares of 24.8% and 15.8%, respectively. By 2008, *El Trece* had pulled ahead of *Telefe*, with a total market share of 25.4% compared to *Telefe*'s 21.5%. Since then, *Telefe*'s market share has dropped substantially, to 13.8% in 2010, against a smaller decline for *El Trece*, which fell from 25.4% to 23.5% in 2010. The explanation for this overall decline can be attributed to the fact that advertising space for prime-time programming has decreased.[22] Advertising costs have also increased due to increased concentration.[23]

It is still too early to analyze The impact on of the 2009 Audiovisual Communication Services Law (26.522) in terrestrial broadcasting is limited. The new regulatory framework ensures nonprofit players such as community media, service cooperatives, and NGOs a combined 33% national audience share, has set quotas for

19. The national public broadcaster *TV Pública* has only a 4% aggregate national audience share. *See* Mastrini, Guillermo and Becerra, Martin. *Periodistas y Magnates. Estructura y Concentracion de las industrias culturales en America Latina*. Buenos Aires: Prometeo, 2006.

20. AFSCA. "11 informe Contenidos de la Television Abierta Argentina." Autoridad Federal de Servicios de Comunicación Audiovisual (AFSCA), 2011. November 20, 2011. <http://www.afsca.gob.ar/web/Varios/informes/Informe_TV_Abierta_11_marz_a_may11.pdf>.

21. Becerra, Martin and Mastrini, Guillermo. Los duenos de la palabra. Acceso, estructura y concentracion de los medios en la America latina del Siglo XXI. Buenos Aires: Prometeo, 2009.

22. AAP. "Argentine Official Advertising Spend Report 2010." Asociación Argentina de Publicidad (AAP), 2010. Nov. 20, 2011. <http://190.220.150.93/webaaap/wp-content/inversion_publicitaria/2010/informe_inversion_publicitaria_2010.pdf>.

23. Crettaz, J. "La Televisión abierta argentina no es rentable." *La Nación* newspaper, Jan. 9, 2011. Nov. 20, 2011. <http://www.lanacion.com.ar/1339818-la-television-abierta-argentina-no-es-rentable>. & Del Río, J. and Quiroga, C. "Queremos ganar la batalla del rating pero no a cualquier precio." *Apertura*, July 29, 2011. Nov. 20, 2011. <http://www.apertura.com/notas/259680-queremos-ganar-la-batalla-del-rating-per-no-cualquier-precio>.

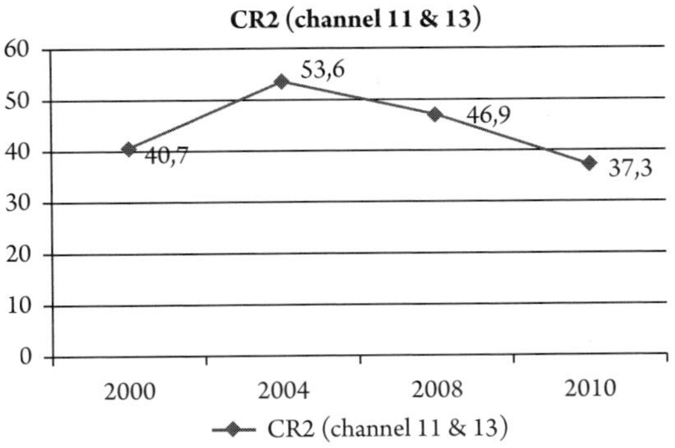


FIGURE 19.1 C2 (Telefe and El Trece), 2000–2010

national (70%) and local (30%) programming content, and restricts commercial airtime to 12 minutes per broadcast hour. However, after almost five years of its enactment the nonprofit players are still awaiting the call for bids to hold an audiovisual communication service license. The implementation of the new regulatory framework is complex. Almost immediately approved, Clarín Group interposed legal measures challenging the implementation of the Law. In October 2013, after a four-year legal battle, the Supreme Court declared the constitutionality of the articles questioned by Clarín.

## Multichannel TV Platforms

In 2000, Argentina had the largest number of multisystem operator (MSO) subscribers in Latin America: 5.3 million cable TV subscribers and 150,000 satellite TV subscribers. Of 800 operators, four accounted for 65.6% of total national subscriptions and 55.7% of the total revenues in a market valued at US$2 billion (ARS2 billion).[24] Increased monetary stability from the mid-2000s to the present has further increased the potential subscriber base in the country.[25] As was the case with broadcast TV, the 1990s were marked by renewed domestic investment and an influx of foreign capital (Table 19.6).

From 1989 to 1994, the large MSOs were established through the acquisition of provincial cable TV providers. It is estimated that in the mid-1990s, there were around 1,600 such providers, while by 2000, at least half of those had been absorbed by their larger competitors.[26] In 1994, foreign direct investment in the sector increased, due to the ratification of Mutual Investment Promotion and Protection Agreements with other countries.[27] Grupo Clarín's pay TV channels include Todo Noticias, Magazine 24, Volver, Nickelodeon, and TyC Sport (with a 50% stake). In 2004, the main players were Grupo Clarín's Cablevisión, Grupo Vila-Manzano's Multicanal, and Supercanal, the latter owned jointly by Grupo Vila-Manzano (80%) and Grupo Clarín (20%). Cablevisión and Multicanal held a combined 55.7% of total market revenues. This trend became even stronger in 2007, when Grupo Clarín, in partnership with the US investment fund Fintech, acquired both Cablevisión and Multicanal.[28]

In parallel, the satellite TV market has also become more concentrated. In 1996, Televisión Satelital Directa al Hogar (TDH) was established as a partnership with the Argentine firms Grupo Vila-Manzano (70%), the Argentine firm Tevycom (7%), and the Brazilian investor Laser (23%).[29] Until 1998, TDH was the only

24. Mastrini, Guillermo and Becerra, Martin. Periodistas y Magnates. Estructura y Concentracion de las industrias culturales en America Latina. Buenos Aires: Prometeo, 2006.

25 Albornoz, Luis. Al fin solos . . . la nueva television del Mercosur. Buenos Aires: La Crujia, 2000.

26. Mastrini, Guillermo and Becerra, Martin. Periodistas y Magnates. Estructura y Concentracion de las industrias culturales en America Latina. Buenos Aires: Prometeo, 2006.

27. Albornoz, Luis and Hernández, Pablo. "La radiodifusion entre 1995–1999: concentracion, denacionalizacion y ausencia del control public." *Mucho ruido, pocas leyes. Economia y politicas de comunicacion en la Argentina (1920–2004)*. 2nd ed. Guillermo, Mastrini, ed. Buenos Aires: La Crujia, 2008.

28. On December 7, 2007, the Anti-Trust Agency (Comisión Nacional de Defensa de la Competencia, CNDC) approved the joint Cablevisión-Multicanal transaction. Grupo Clarín took over 60% of Cablevisión while Fintech (US) kept the remaining 40%. In addition, Cablevisión kept a 98.5% share in Multicanal and 100% of Teledigital (a digital provider in the Interior) and Fibertel. Through Multicanal, it also acquired a 97% stake in Prima. *See* Becerra, Martin and Mastrini, Guillermo. *Los duenos de la palabra. Acceso, estructura y concentracion de los medios en la America latina del Siglo XXI*. Buenos Aires: Prometeo, 2009.

29. Albornoz, Luis, et al. "'Al fin solos': el nuevo escenario de las comunicaciones en la Argentina." *Comunicación y Sociedad* 35, 1999. Jan. 14, 2013. <http://catedras.fsoc.uba.ar/gpost/material/006.doc>.

### Table 19-6. Multichannel Video Platforms: Cable MSOs, DBS, IPTV (Market Shares by Revenue), 2000–2008

| | 2000 | 2004 | 2008 |
|---|---|---|---|
| Cablevisión (Grupo Clarín) | 24.4 | 30.9 | 55.1 |
| Multicanal (Grupo Clarín) sAME | 23.8 | 23.2 | |
| Red Intercable (cooperatives) | 0.0 | 0.0 | 0.0 |
| Telecentro (Alberto Pierri) | 0.0 | 0.0 | 0.0 |
| VCC (Samuel Liberman) | 0.0 | 0.0 | 0.0 |
| Supercanal (Grupo Vila-Manzano) | 7.5 | | 5.8 |
| DirecTV Latin America (DirecTV, US), owned by Liberty | 0.0 | 0.0 | 19.9 |
| Others | 44.3 | 45.9 | 19.2 |
| Total Revenue (mil ARS) | 2,000 | 2,353 | 5,097 |
| Total Revenue (mil US$) | 2,000 | 800 | 1,631 |
| C4 | 55.7 | 54.1 | 80.8 |
| HHI | 1,218 | 1,494 | 3,466 |
| *N* (>1%)[1] | 6 | 6 | 6 |
| Noam Index | 497 | 610 | 1,415 |

*Source*: Getino (1995), LAMAC, Mastrini and Becerra (2006, 2009), Convergencia.

[1] In addition to the companies presented in the tables, we found several other companies that have market shares greater than 1%.

satellite TV provider operating in Argentina. After 1998, TDH's monopoly was broken as a result of a mutual satellite-broadcasting agreement signed between Argentina and the United States, enabling Grupo Clarín, in partnership with the US satellite TV provider Galaxy Latin America (GLA)—now DirecTV Latin America (US)—to become the country's second largest satellite provider by purchasing a 51% equity stake in DirecTV Argentina. SKY Television, owned by Rupert Murdoch's News Corporation, briefly participated in the market but suspended its operations in the country in 2002.[30] In 2002, Grupo Clarín had to sell off its stake in GLA due to the group's rising debts. DirecTV (US) subsequently acquired the stake, and GLA's operations were rebranded as DirecTV Latin America.

Between 2004 and 2008, DirecTV became the second largest pay TV operator after Cablevisión. In 2008, it counted 750,000 subscribers, and in 2010 it reached 1.1 million subscribers.[31] By 2008, the four companies that controlled the market (Cablevisión, Red Intercable, Telecentro, and DirecTV) collectively controlled 71.7% of the total subscriber base on the market, with Cablevisión holding onto a 44.7% audience share (Table 19.7).[32]

In 2008, there was a slight decline in C4 that positioned the four leading companies,

30. Albornoz, Luis and Hernández, Pablo. "La radiodifusion entre 1995–1999: concentracion, desnacionalizacion y ausencia del control public." *Mucho ruido, pocas leyes. Economia y politicas de comunicacion en la Argentina (1920–2004)*. 2nd ed. Mastrini, Guillermo, ed. Buenos Aires: La Crujia, 2008.

31. "DirecTV cruzo la barrera del millon de abonados en la Argentina." iProfesional.com, Dec. 10, 2010. July 16, 2012. <http://tecnologia.iprofesional.com/notas/108569-DirecTV-cruz-la-barrera-del-milln-de-abonados-en-la-Argentina>.

32. Run as a cooperative, Red Intercable is composed of 200 small- and medium-sized cable TV companies operating in 19 provinces.

## Table 19-7. Multichannel Video Platforms: Cable MSOs, DBS, IPTV (Audience Shares by Subscribers), 1984–2013

| | 1984 | 1988 | 1992 | 1996 | 2000 | 2004 | 2008 | 2010 | 2013 |
|---|---|---|---|---|---|---|---|---|---|
| Cablevisión (Grupo Clarín) | 0.3 | 0.6 | 4.0 | 11.8 | 26.7 | 30.3 | 44.7 | 51.4 | 42.7 |
| VCC (Samuel Liberman) | 0.3 | 0.6 | 2.7 | 14.4 | | | | | |
| Multicanal (Grupo Clarín) | | | 0.1 | 18.2 | 26.7 | 27.8 | | | |
| Red Intercable (cooperatives) | | | | 13.6 | 5.5 | 6.3 | 4.6 | 6.7 | |
| Telecentro (Alberto Pierri) | | | | | | 4.2 | 4.3 | 4.3 | 6.8 |
| Supercanal (Grupo Vila-Manzano) | Article I. | | | | 9.1 | 8.5 | 6.6 | 8.0 | |
| DirecTV Latin America (DirecTV, US) | | | | | | | 11.5 | 17.9 | 25.7 |
| Others | 99.5 | 98.8 | 93.3 | 42.0 | 32.0 | 22.9 | 28.3 | 11.6 | 17.3 |
| Total Revenue (mil ARS) | | | | | 2,000 | 2,353 | 5,097 | | |
| Total Revenue (mil US$) | | | | | 2,000 | 800 | 1,631 | | |
| C4 | 0.6 | 1.2 | 6.7 | 58.0 | 68.0 | 77.1 | 71.7 | 88.3 | 75.2 |
| HHI | 0 | 1 | 23 | 863 | 1,536 | 1,821 | 2,217 | 3,090 | 2,530 |
| *N* (>1%) | 0 | 0 | 2 | 4 | 4 | 5 | 5 | 5 | 3 |
| Noam Index | | | 16 | 432 | 768 | 815 | 991 | 967 | 1,461 |

*Source*: Getino (1995), LAMAC, Mastrini and Becerra (2006, 2009), Convergencia.

and a new rebound was seen in 2010, when they reached 88.3% (Table 19.7).

As mentioned earlier for broadcast TV, the 2009 Audiovisual Communication Services Law was meant to reduce concentration in this sector as well. In spite of the new legislation, though, no significant changes have yet occurred in the pay TV market. Cablevisión held onto a 51.4% audience share, followed by DirecTV with 17.9%, Supercanal with 8%, Red Intercable 6.7%, and Telecentro with 4.3%.

## Film

The state encourages domestic film production and distribution through subsidy payments. Since 1994, with the passing of Law 24.377/94 for National Movie-Making Promotion and Regulation (Ley de Fomento y Regulación de la Actividad Cinematográfica Nacional), the state has granted substantial funding for national production and has established screening quotas for the distribution of movies produced in-country (Table 19.8).

Thanks to state subsidies, the number of Argentine-made movies premiered per year has increased significantly, though the revenue share of domestically produced films has increased only moderately. In other words, the increase in offerings has not yet been matched by consumption from the Argentine audience, in part because Argentine movies are generally withdrawn from theaters sooner than foreign films are (Table 19.9).

Argentina is ranked number one in annual film production in Latin America, producing 70–90 pictures a year, a number it is able to

Table 19-8. Film Production/Distribution (Market Shares by Box Office %), 2000–2010

| | 2000 | 2004 | 2008 | 2010 |
|---|---|---|---|---|
| Walt Disney Pictures (US) | 26.4 | 23.7 | 22.9 | 25.3 |
| Sony (Japan/US) | 16.1 | 8.1 | | |
| Distribution Comp. | 7.2 | 8.2 | 9.6 | |
| United International Pictures (UIP)—Universal (Comcast/GE, US) | | | 25.2 | 22.6 |
| Warner Bros. (Time Warner, US) | | | 13.3 | 16.4 |
| 20th Century Fox (News Corp., Murdoch Family, US/UK/AUS) | | | | 11.5 |
| Primer Plano | 3.4 | 0.7 | | |
| Others | 47.5 | 59.3 | 30.4 | 25.1 |
| Total Revenue (mil ARS) | 165 | 270 | 359 | 662 |
| Total Revenue (mil US$) | 165 | 91.8 | 115 | 172 |
| C4 | 53.05 | 40.76 | 71.0 | 75.8 |
| HHI | 1,019 | 696 | 1,427 | 1,552 |
| *N* (>1%) | 4 | 4 | 4 | 4 |
| Noam Index | 510 | 348 | 714 | 776 |

*Source*: INCAA.

Table 19-9. Domestic vs. Foreign Films (Market and Audience Shares by Box Office %), 2000–2010

| | AUDIENCE SHARES | | MARKET SHARES | |
|---|---|---|---|---|
| | DOMESTIC FILMS | FOREIGN FILMS | DOMESTIC FILMS | FOREIGN FILMS |
| 2000 | 17.7 | 82.3 | 10.9 | 89.1 |
| 2004 | 22.8 | 77.2 | 14.4 | 85.7 |
| 2008 | 27.2 | 72.8 | 25.0 | 75.0 |
| 2010 | 30.2 | 69.9 | 18.8 | 81.2 |

*Source:* INCAA.

sustain in part because of state subsidies. State support has had less of an impact on film distribution, and the bulk of it is in the hands of US-based Hollywood majors. The main Argentine distribution company, Primer Plano, has a very low market share at home. Though not shown in Tables 19.8 and 19.9, Grupo Clarín also has a small stake in this market through the production company Patagonik Films, which consists of Grupo Clarín, Walt Disney Productions (US), and Cinecolor.

## TELECOMMUNICATIONS MEDIA

### Wireline Telecom

The state-owned telecommunications company, Empresa Nacional de Telecomunicaciones

(ENTel), monopolized fixed telephony services between 1950 and 1990, leaving audience shares of just 6% for Ericsson (Sweden) and 4% combined for several independent provincial cooperatives.[33]

This changed when, in August 1989, the Carlos Menem Administration introduced the Economic Emergency Law (Ley de Emergencia Económica) and the State Reform Law (Ley de Reforma del Estado 23.696), which granted the president unprecedented powers over the disbursement of state properties.[34] In January 1990, an international public bid took place to privatize ENTel. The winning bidders were Telefónica (Spain) and Telecom Argentina, founded as a joint venture between Telecom Italia and France Telecom.[35] As a result, the fixed telephony market in Argentina became a geographically defined duopoly, with Telefónica Argentina dominating the southern part of the country and Telecom Argentina dominating the northern part. The sector became an effective a monopoly, though, because Telefónica (Spain) controlled in Europe the holding company, Telco, that in turn controlled Telecom Italia. However, a measure of duopolistic competition has returned to the market because Telecom Italia's stake in Telecom Argentina was acquired in 2013 for $960 million by the Mexican investment fund Fintech, run by David Martinez, who has had good relations with the Kirchners. That transaction is still subject to governmental approvals. (Tables 19.10 and 19.11).

Telefónica has expanded largely through acquisitions, starting in the 1990s with its acquisition of 50% of the former state monopolist ENTel's wireline telecom infrastructure, both with its own capital and in partnership with other companies. Telefónica Argentina (and Telecom Argentina) also holds the monopoly over long-distance calling and has benefited from the telecom regulatory framework that allows it to take part in competitive market segments, including cellular telephony and personal communications (PCS), which has significantly increased concentration.[36] Even though the basic services became free in 2000, the market is still dominated by Telefónica.[37]

Small telephone or cable TV companies offer commercial triple-play services through business partnerships.[38] In Buenos Aires, the company Telecentro, founded by former politician and entrepreneur Alberto Pierri, pioneered household triple-play services in 2008.[39] Nevertheless, Telecentro has, in terms of market share, had more of an impact on pay TV than on fixed telephony, given that many users have acquired their Telecentro packages with IP telephony but run their lines through the ADSL networks offered by Telefónica subsidiaries. Ultimately, all new entrants like Telecentro end up following this route, thus giving the duopolists a distinct advantage.

---

33. Bizberge, Ana. "Audiencias publicas por la renegociacion de contratos de telefonia." X Jornadas Nacionales de Investigadores en Comunicación: Una década de encuentros para (re) pensar los intercambios y consolidar la red. Oct. 19–21, 2006, National University of San Juan. REDCOM, San Juan: REDCOM, 2006. <http://www.redcomunicacion.org/memorias/p_jornadas_p.php?id=568&idj=5 >.

34. The Economic Emergency Law allows the Executive Branch to legislate without regard for the National Congress. State Reform Law 23.696 authorized the privatization of state-owned companies.

35. In 2003, the Werthein Group bought half of the shares that Telecom France owned through Nortel Inversora in Telecom Argentina. In 2010, Telefónica took control over Telecom Italia's operations. In Argentina, the Competency Defense National Commission (CDNC, by its acronym in Spanish) decided that the merger would not influence local companies operations because Telefónica cannot cast votes on Telecom Argentina's board of directors.

36. For more information on the privatization of ENTel, see Abeles, Forcinito, and Schorr (2001) and López A. and Felder, R. (1997, 1999).

37. Mastrini, Guillermo and Becerra, Martin. Periodistas y Magnates. Estructura y Concentracion de las industrias culturales en America Latina. Aires: Prometeo, 2006.

38. The triple-play offer is not always the result of network convergence: it is sometimes due to the establishment of business partnerships between companies providing services through their own networks, but with a unified billing system.

39. Bizberge, Ana. "La carrera por la TV digital y el Triple Play en la TV paga en Argentina." Seventh ENACOM National Meeting of Communication Programs. Dec. 10–12, 2009, Universidad Nacional de Comahue. General Roca, Rio Negro, 2009. <https://www.zotero.org/groups/lecotec/items/itemKey/TX8VBHKM>.

Table 19-10. Wireline Telecom (Audience Shares by Subscribers), 2000–2010

| | 2000 | | 2004 | | 2008 | | 2010 | |
|---|---|---|---|---|---|---|---|---|
| | TOTAL SUBSCRIBERS (1,000) | % | TOTAL SUBSCRIBERS (1,000) | % | TOTAL SUBSCRIBERS (1,000) | % | TOTAL SUBSCRIBERS (1,000) | % |
| Telefónica Argentina (Telefónica, Spain) | 4,738 | 56.1 | 4,180 | 53.4 | 4,656 | 52.0 | 4,622 | 52.9 |
| Telecom Argentina (Telecom Italia, Italy/Fintech, Mexico)[1] | 3,713 | 43.9 | 3,650 | 46.6 | 4,300 | 48.0 | 4,107 | 47.1 |
| Total Subscribers (1,000) | 8,451 | | 7,830 | | 8,956 | | 8,729 | |
| C4 | | 100.0 | | 100.0 | | 100.0 | | 100.0 |
| HHI | | 5,074 | | 5,023 | | 5,008 | | 5,017 |
| *N* (>1%) | | 2 | | 2 | | 2 | | 2 |
| Noam Index | | 3,588 | | 3,552 | | 3,541 | | 3,548 |

*Source*: CNC, 2008.

[1] As of 2013, Fintech now controls Telecom Argentina instead of TI/Telefónica.

Table 19-11. Wireline Telecom (Market Shares by Revenue), 2000–2008

| | 2000 | 2004 | 2008 |
|---|---|---|---|
| Telefónica Argentina (Telefónica, Spain) | 53.0 | 48.3 | 61.1 |
| Telecom Argentina (Telecom Italia, Italy/Fintech, Mexico) | 47.0 | 51.7 | 38.9 |
| Total Revenue (mil ARS) | 4,404 | 6,182 | 7,688 |
| Total Revenue (mil US$) | 4,404 | 2,102 | 2,460 |
| C4 | 100.0 | 100.0 | 100.0 |
| HHI | 5,019 | 5,006 | 5,246 |
| *N* (>1%) | 2 | 2 | 2 |
| Noam Index | 3,549 | 3,540 | 3,710 |

*Source*: CNC, 2008.

## Wireless Telecom

Movicom, a subsidiary of Bellsouth (US), was the first wireless provider to enter the Argentine market. Miniphone, a joint venture by Telefónica Argentina and Telecom Argentina, then entered the market in 1990. Competition between the two companies led to a significant expansion of services in Argentina. In 1992, CTI was set up to offer a national service alternative to Movicom and Miniphone. CTI's majority shareholder was Verizon (US), until it was bought out in 2003 by America Móvil (owned by Mexican Carlos Slim. In the late 1990s, Miniphone was dissolved by the government and split into two companies: Personal (Telecom Italia) and Unifón (Telefónica Argentina).[40] In this way, the same companies that held the basic fixed telephony monopoly became the main mobile service providers, competing with each other's own assets (Tables 19.12 and 19.13).[41]

In 2005, Telefónica Móviles, part of Telefónica Argentina, acquired 100% of Movicom and merged it with Unifón, forming a new company, Movistar Argentina. Movistar was launched in the Latin American market in April 2005 and became the Argentine market leader.[42] América Móvil was the first to launch 3G services in late 2007. In 2008, the company introduced a regional brand change and started operating as Claro throughout Latin America.

# INTERNET MEDIA

## Internet Service Providers (ISP)

Under Argentine law, the Internet is considered a "value-added service."[43] Its development in Argentina is strongly linked with telephone carriers' licenses. In 1995, Startel, which was jointly owned by Telefónica Argentina and

40. Decree 266/98 was widely criticized because it was favorable to the operators that had a dominant position in the industry, leading to a considerable market concentration, unlike declared intentions that indicated the need to encourage the development of national wireless communication networks to ensure a competitive landscape.

41. Mastrini, Guillermo and Becerra, Martin. Periodistas y Magnates. Estructura y concentracion de las industrias culturales en America Latina. Buenos Aires: Prometeo, 2006.

42. Becerra, Martin and Mastrini, Guillermo. Los duenos de la palabra. Acceso, estructura y concentracion de los medios en la America Latina del siglo XXI. Buenos Aires: Prometeo, 2009.

43. CNT Resolution 1083/95 defines Value-Added Telecommunication Services as "those services which, using networks, links and/or telecommunication systems for support, offer features differentiating them from the basic service, applying processes that make information available, act on Information or even enable subscribers to interact with information." *See* CNC. "Resolución CNT 1083/1995." Comision Nacional de Comunicaciones (CNC), 1995. July 16, 2012. <http://www.enre.gov.ar/web/bibliotd.nsf/($IDWeb)/943B4F96BE336D9A03257134006E13AA>.

Table 19-12. Wireless Telecom (Audience Shares by Subscribers), 2000–2010

| | 2000 | | 2004 | | 2008 | | 2010 | |
|---|---|---|---|---|---|---|---|---|
| | TOTAL SUBSCRIBERS (1,000) | % | TOTAL SUBSCRIBERS (1,000) | % | TOTAL SUBSCRIBERS (1,000) | % | TOTAL SUBSCRIBERS (1,000) | % |
| Movistar (Telefónica, Spain) | | | 3,632 | 28.0 | 15,228 | 35.9 | 16,149 | 33.5 |
| Movicom (Telefónica, Spain; Telecom Italia, Italy) | 1,672 | 25.7 | | | | | | |
| Unifón (Telefónica, Spain) | 1,746 | 26.8 | | | | | | |
| Personal (Telecom Italia, Italy/ Fintech, Mexico)[1] | 2,040 | 31.3 | 3,835 | 29.6 | 11,844 | 28.0 | 16,333 | 31.1 |
| CTI/Claro (América Móvil, México) | 1,052 | 16.2 | 3,350 | 25.8 | 14,339 | 33.8 | 18,812 | 32.4 |
| Nextel (Bellsouth/NII Holding, US)[2] | | | 2,150 | 16.6 | 967 | 2.3 | | 3.0 |
| Total Subscribers (1,000) | 6,512 | | 12,967 | | 42,378 | | 36,190 | |
| C4 | | 100.0 | | 100.0 | | 100.0 | | 100.0 |
| HHI | | 2,622 | | 2,602 | | 3,223 | | 3,148 |
| *N* (>1%) | | 4 | | 4 | | 4 | | 4 |
| Noam Index | | 1,311 | | 1,301 | | 1,611 | | 1,574 |

[1] Fintech controls Personal as of 2013, along with TI's other Argentina assets.
[2] Nextel 2010 marketshare supplied by editors.

Table 19-13. Wireless Telecom (Market Shares by Revenue), 2000–2010

| | 2000 | 2004 | 2008 | 2010 |
|---|---|---|---|---|
| Movistar (Telefónica, Spain) | | 27.7 | 25.5 | 30.4 |
| Movicom (Telefónica, Spain; Telecom Italia, Italy) | 34.9 | | | |
| Unifón (Telefónica Spain) | 14.3 | | | |
| Personal (Telecom Italia, Italy) | 31.4 | 31.6 | 25.5 | 31.2 |
| CTI/Claro (América Móvil, México)[1] | 19.3 | 28.9 | 25.2 | 35.4 |
| Nextel (Bellsouth/NII Holding, US)[2] | | 11.8 | 23.8 | 3 |
| Total Revenue (mil ARS) | 2,477 | 4,876 | 6,671 | 7,448 |
| Total Revenue (mil US$) | 2,477 | 1,658 | 2,135 | 1,900 |
| C4 | 100.0 | 100.0 | 100.0 | 100.0 |
| HHI | 2,783 | 2,740 | 2,501 | 3,160 |
| *N* (>1%) | 4 | 4 | 4 | 4 |
| Noam Index | 1,392 | 1,370 | 1,251 | 1,580 |

[1] In 2003, América Móvil, which is owned by Mexican tycoon Carlos Slim, acquired Verizon's (US) shares in CTI.
[2] Nextel's 2010 marketshare supplied by editors.

Telecom Argentina, first entered the market, at which point access grew rapidly. According to the regulatory agency, the National Communications Commission (CNC), Argentina went from having 1.7 million Internet users in 2003 to 3.75 million in 2008, a 116% increase. By 2008, broadband access had doubled, reaching 3.7 million connections, or 28% of all households. Household access accounted for 91% of all Internet usage in Argentina that year (94% of it broadband). The most widely used broadband platform is ADSL (68%), followed by cable modems (30%) and satellite (2%).

Telecom Argentina and Telefónica Argentina employ ENTel's landline network, which makes ADSL the preferred platform for most users. After privatization, the two companies won a 10-year exclusivity contract to use ENTel's network. In contrast, MSOs such as Grupo Clarín that were consolidated during the 1990s had to build their own private networks from the ground up with cable modem technology to offer Internet and pay TV.

Internet access is concentrated in the Buenos Aires metropolitan area, where more than 65% of total connections are located. The provinces of Santa Fe and Córdoba account for 16% of Internet connections, while the rest of the provinces account for the remaining 17%.[44] By 2000, Grupo Clarín's, Telefónica Argentina, and Telecom Argentina together controlled 75.6% of the total market revenues, valued at approximately US$115 million (ARS115 million) (Table 19.14).[45]

Telefónica Argentina has pressed, so far unsuccessfully, to win a license to provide triple- and quadruple-play services. The group instead offers audiovisual services through a business partnership with DTH operator DirecTV (US). And through the Speedy broadband service, it provides audiovisual content under a subscription and rental model.

44. CNC. "Evolucion y penetracion del servicio." Comision Nacional de Comunicaciones (CNC), 2009. July 16, 2012. <http://www.cnc.gov.ar/ciudadanos/internet/evolucion.asp#iconsumo>.

45. Mastrini, Guillermo and Becerra, Martin. Periodistas y Magnates. Estructura y concentracion de las industrias culturales en America Latina. Buenos Aires: Prometeo, 2006.

## Table 19-14. Internet Service Providers (Audience Shares by Subscribers), 2004–2010

| | 2004 | | 2008 | | 2010 | |
|---|---|---|---|---|---|---|
| | TOTAL SUSCRIBERS (1,000) | % | TOTAL SUSCRIBERS (1,000) | % | TOTAL SUSCRIBERS (1,000) | % |
| Telecom Argentina (Telecom Italia, Italy/Fintech, Mexico)[1] | 591 | 36.9 | 1,042 | 27.7 | 1,380 | 29.9 |
| Prima (Grupo Clarín) | | | | | | |
| Advance/Speedy (Telefónica, Spain) | 409 | 25.6 | 1,079 | 28.7 | 1,439 | 31.2 |
| Fibertel (Grupo Clarín) | | | 939 | 25.0 | 1,128 | 24.5 |
| Ciudad Internet (Grupo Clarín) | 280 | 17.5 | | | | |
| Uol Sinectis (Brazil) | 220 | 13.8 | | | | |
| Others | 100 | 6.3 | 697 | 18.6 | 662 | 14.4 |
| Total Subscribers (1,000) | 1,600 | | 3,757 | | 4,609 | |
| Total Revenue (mil ARS) | | | | | | |
| Total Revenue (mil US$) | | | | | | |
| C4 | | 93.7 | | 81.4 | | 85.6 |
| HHI | | 2,552 | | 2,563 | | 2,676 |
| *N* (>1%) | | 5 | | 4 | | 4 |
| Noam Index | | 1,141 | | 1,281 | | 1,338 |

*Source*: INDEC, CNC, Barometro CISCO.

[1] Fintech obtained full control of Telecom Argentina in 2013.

## Search Engines

In ranking the top 20 websites in Argentina by total traffic, the top search engines were Google (US) and its local platform, Google Argentina. Some other search engines have slowly made their way into the rankings in recent years, including Yahoo!, Bing, and Google Espana (Table 19.15).

## Online News

The emergence of online news platforms have not only changed the relationship between journalists and their audiences, but also have challenged traditional media business models.[46] Thirteen of the top 100 websites belong to information and news media, and most of them are the online editions of the leading daily newspapers (Table 19.16). As we detailed in our 2012 report "Mapping Digital Media: Argentina":

> [T]he news media website with the highest visitorship is www.clarin.com, Grupo Clarín's portal, which posts the hard-copy edition of the newspaper together with the latest news updates and other services, and is listed 13th on the overall Argentinean website ranking. The second most highly visited is the portal of the sports paper Ole (also owned by Grupo Clarin) and ranks 14th on the general list. It is followed by www.lanacion.com and www.infobae.com which are ranked 15th

### Table 19-15. Search Engines (Alexa.com Rankings), 2010–2011

| RANKING | 2010 | 2011 |
|---|---|---|
| 1 | Facebook (US) | Google Argentina (US) |
| 2 | Google Argentina (US) | Facebook (US) |
| 3 | Google (US) | YouTube (US) |
| 4 | YouTube (US) | Google (US) |
| 5 | Windows Live (Microsoft, US) | Windows Live (Microsoft, US) |
| 6 | Yahoo! (US) | Blogger.com (US) |
| 7 | Blogger.com (US) | Yahoo! (US) |
| 8 | Taringa! | MercadoLibre.com |
| 9 | MercadoLibre.com (MercadoLibre) | Taringa! |
| 10 | MSN (Microsoft, US) | Wikipedia (US) |
| 11 | Wikipedia (US) | Twitter (US) |
| 12 | Twitter (US) | *Clarin.com* (Grupo Clarín) |
| 13 | *Clarin.com* (Grupo Clarín) | MSN (Microsoft, US) |
| 14 | *Ole* (Grupo Clarín) | *lanacion.com* (Mitre Family, La Nación) |
| 15 | *lanacion.com* (Mitre Family, La Nación) | *infobae.com* (Grupo Infobae) |
| 16 | Poringa! | LinkedIn (US) |
| 17 | *infobae.com* (Grupo Infobae) | *Ole* (Grupo Clarín) |
| 18 | Megaupload (Hong Kong) | Wordpress (US) |
| 19 | Bing (US) | xvideos.com (US) |
| 20 | Google Espana (US) | t.co (Twitter, US) |

*Source*: Alexa.com.

46. Albornoz, Luis. Periodismo Digital. Los Grandes Diarios en la Red. Buenos Aires: La Crujia, 2007.

Table 19-16. Online News (Alexa.com Rankings), 2010–2011

| WEBSITE | OVERALL MEDIA SITE RANKING | ALEXA RANKING |
|---|---|---|
| Clarin | 1 | 13 |
| Ole | 2 | 14 |
| La Nacion | 3 | 15 |
| Infobae | 4 | 17 |
| Perfil | 5 | 42 |
| Pagina 12 | 6 | 46 |
| Primicias ya | 7 | 53 |
| Minuto 1 | 8 | 58 |
| El Argentino | 9 | 59 |
| La voz del interior | 10 | 63 |
| Cancha Ilena | 11 | 68 |
| Todo Noticias | 12 | 70 |
| Telefe | 13 | 76 |

*Source*: Becerra, Marino, and Mastrini (2012).

and 17th respectively on the general list. Finally, www.perfil.com is the fifth most visited news site but is ranked 42nd in the general list. Online-only news sites are ranked considerably lower and outside the top [50] most visited sites in Argentina. The leading sites in this sub-category are www.primiciasya.com (ranked 53rd overall) and www.minutouno.com (ranked 58th). The online version of a provincial newspaper with the highest visitorship is www.lavozdelinterior.com from Córdoba (owned by Grupo Clarín), listed [at] 63rd. Grupo Clarín's channel TN ranks 70th of online broadcasters, and the terrestrial channel with the highest visitorship is Telefé (www.telefe.com), which ranks 76th. Overall, 13 of the top 100 websites belong to information and news media, and most of them are the online versions of major newspapers.[47]

## CONCLUSION

The high levels of concentration in Argentine media are in large part the result of deregulation in the 1990s, when restrictions on media cross-ownership were removed. As a result, small media producers have largely disappeared, and independent media producers in the provinces have been acquired by the media majors based in Buenos Aires. One area where concentration has not increased, though, has been in content production for the audiovisual market, particularly in cable television programming. This increased production capacity is reflected in an increase of new content in the country and the range of programs now produced in Argentina to be broadcast in other Spanish-language media markets.

Concentration has stabilized in the telecommunications sector, albeit at the maximum possible level for both wireless and wireline. In the print and audiovisual sectors there is

47. Becerra, Martin, Marino, Santiago, and Mastrini Guillermo. "Mapping Digital Media: Argentina." Open Society Foundation (OSI), Feb. 29, 2012. July 16, 2012. <http://www.soros.org/sites/default/files/mapping-digital-media-argentina-20120404.pdf>.

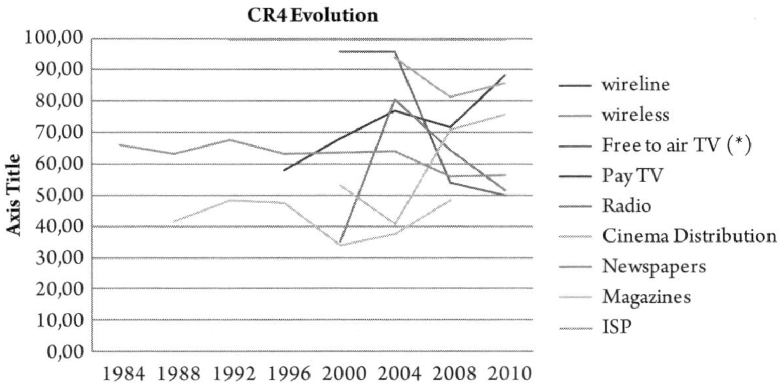


FIGURE 19.2 C4 Indices (All Media), 1984–2010

considerable concentration, ranging from 50% to 85%, in favor of the largest ISPs, MSOs, radio broadcasters, film distributors, and publishers of newspapers. Only in three markets is C4 below 50%: film production, magazine publication, and broadcast television[48] (Figure 19.2).

While the telecommunications sector remains stable, the audiovisual and print sectors show significant variations. In the pay TV, film distribution, magazine publications, and ISP markets, there has been a steady increase in concentration. In the case of broadcast TV and daily newspapers, concentration indicators have decreased, although the decrease in concentration for broadcast TV is due to change in the measurement criteria made by the organizations that track it. Downward trends in the broadcast TV market and sales of newspapers can be attributed to market transformations and the impact of digitization rather than to governmental action.

The two dominant media groups in Argentina will continue to be Grupo Clarín and Telefónica. Grupo Clarín is the more diversified in its media holdings, and it has achieved a dominant position that has turned it into a privileged stakeholder with large advertising revenues. The group has only a few assets in telecommunications and Internet media (ISPs). Conversely, Telefónica boasts control over the landline and mobile telephony and ISP market, in addition to a significant audiovisual presence through Telefe, but it has disposed of other assets in the cultural industries, such as the sale of AM Continental to the PRISA Group.

The link between media ownership concentration and pluralism and diversity is a complex one. Particularly in Argentina, the issue of diversity and pluralism is one of the drivers of the confrontation between the government and Grupo Clarín. Reforms would redress past government policies of deregulation that have had adverse effects on media consumers. When the outgoing Nestor Kirchner Administration approved the merger of Cablevisión and Multicanal under Grupo Clarín, for instance, service costs soon began to increase substantially and many local signals stopped broadcasting.

Aside from the declarations and policies of the national government aimed at promoting wider diversity in the media market, no fundamental changes have yet been observed in the structure of media ownership. Despite the determination of the state, the largest media groups have so far retained most of their media assets, including Grupo Clarín's much-debated cable TV merger, even though the government has ruled against the merger and demanded it sell off these properties. High concentration will remain present without major structural reforms.

48. The C4 of free-to-air-TV has been calculated over the total TV audiences, including pay TV, because there are no separate measures. In a country with a high penetration of cable TV, this distorts the results.

# Bibliography

## 1. Books and Articles


Abeles, Martin, Forcinito, Karinal, and Schorr, Martin. *El oligopolio telefónico Argentino frente a la liberalización del Mercado*. Buenos Aires: Basualdo. Universidad Nacional de Quilmes, 2001.

Albornoz, Luis, et al. "'Al fin solos': el nuevo escenario de las comunicaciones en la Argentina." *Comunicación y Sociedad* 35, 1999. Jan. 14, 2013. <http://catedras.fsoc.uba.ar/gpost/material/006.doc>.

Albornoz, Luis. *Al fin solos . . . la nueva television del Mercosur*. Buenos Aires: La Crujia, 2000.

Albornoz, Luis. *Periodismo digital. Los grandes diarios en la red*. Buenos Aires: La Crujia, 2007.

Albornoz, Luis and *Hernández, Pablo*. "La radiodifusion entre 1995–1999: concentracion, denacionalizacion y ausencia del control public." *Mucho ruido, pocas leyes. Economia y politicas de comunicacion en la Argentina (1920–2004)*. 2nd ed. Mastrini, Guillermo, ed. Buenos Aires: La Crujia, 2008.

Becerra, Martin et al. "La concentracion de las industrias culturales." *Industrias culturales: Mercado y politicas publicas en Argentina*. Buenos Aires: Ciccus, 2003.

Becerra, Martin and Mastrini, Guillermo. *Los duenos de la palabra. Acceso, estructura y concentracion de los medios en la America latina del Siglo XXI*. Buenos Aires: Prometeo, 2009.

Becerra, Martin, Marino, Santiago, and Mastrini Guillermo. "Mapping Digital Media: Argentina." Open Society Foundation (OSI), Feb. 29, 2012. July 16, 2012. <http://www.soros.org/sites/default/files/mapping-digital-media-argentina-20120404.pdf>.

Bizberge, Ana. "Audiencias publicas por la renegociacion de contratos de telefonia." X Jornadas Nacionales de Investigadores en Comunicación: Una década de encuentros para (re) pensar los intercambios y consolidar la red. Oct. 19–21, 2006, National University of San Juan. REDCOM, San Juan: REDCOM, 2006. <http://www.redcomunicacion.org/memorias/p_jornadas_p.php?id=568&idj=5 >.

Bizberge, Ana. "La carrera por la TV digital y el Triple Play en la TV paga en Argentina." Seventh ENACOM National Meeting of Communication Programs. Dec. 10–12, 2009, Universidad Nacional de Comahue. General Roca, Rio Negro, 2009. <https://www.zotero.org/groups/lecotec/items/itemKey/TX8VBHKM>.

Bizberge, Ana, Mastrini, Guillermo, and Becerra, Martin. "La Television Digital Terrestre en Argentina: entre la geopolitica regional y la iniciativa estatal." *La transicion a la television digital terrestre en Iberoamerica: diagnostic y prospective*. Badillo, Angel and Sierra, Francisco, eds. Quito: CIESPAL, 2011.

Crettaz, J. "La Televisión abierta argentina no es rentable." *La Nación* newspaper, Jan. 9, 2011. Nov. 20, 2011. <http://www.lanacion.com. ar/1339818-la-television-abierta-argentina-no-es-rentable>.

Del Río, J. and Quiroga, C. "Queremos ganar la batalla del rating pero no a cualquier precio." *Apertura*, July 29, 2011. Nov. 20, 2011. <http://www.apertura.com/notas/259680-queremos-ganar-la-batalla-del-rating-per-no-cualquier-precio>.

"DirecTV cruzo la barrera del millon de abonados en la Argentina." <iProfesional.com, Dec. 10, 2010. July 16, 2012. http://tecnologia.iprofesional.com/notas/108569-DirecTV-cruz-la-barrera-del-milln-de-abonados-en-la-Argentina>.

Getino, O. *Las industrias culturales en la Argentina: Dimensión económica y políticas públicas*. Buenos Aires: Colihue, 1995.

Graziano, M. "Los dueños de la televisión argentina." *Comunicación y Cultura* 3, 1974.

López, A. and Felder, R. "Nuevas relaciones entre el Estado y los usuarios de Servicios públicos en la postprivatización." Buenos Aires: INAP, 1997. Jan. 16, 2013. <http://www.sgp.gov.ar/contenidos/inap/publicaciones/docs/estado/partusu.pdf>.

López, A. and Felder, R. "Regulación de las Telecomunicaciones en Argentina." Buenos Aires: INAP, 1999. Jan. 16, 2013. < http://www.sgp.gov.ar/contenidos/inap/publicaciones/docs/modernizacion/totalcnc.pdf>.

Mastrini, Guillermo, ed. *Mucho Ruido y pocas leyes. Economia y politicas de comunicacion en la Argentina (1920–2004)*. 2nd ed. Buenos Aires: La Crujia, 2008.

Mastrini, Guillermo and Becerra, Martin. *Periodistas y Magnates. Estructura y concentracion de las industrias culturales en America Latina*. Buenos Aires: Prometeo, 2006.

Muraro, H. "La Manija." *Crisis* 1–4, 1974.

Romero, Simon and Schmall, Emily. "Battle Between Argentine Media Empire and President Heats Up Over a Law." *The New York Times*, Nov. 30, 2012. Jan. 11, 2013. <http://www.nytimes.com/2012/12/01/world/americas/media-law-ratchets-up-battle-between-kirchner-and-clarin-in-argentina.html?pagewanted=all>.

Rossi, Diego. "La radiodifusion entre 1990–1995: exacerbacion del modelo privado-comercial." *Mucho ruido, pocas leyes. Economia politica de comunicacion en la Argentina (1920–2004)*. 2nd ed. Mastrini, Guillermo ed. Buenos Aires: La Crujia, 2008.

## 2. Reports

AAP. "Argentine Official Advertising Spend Report 2010." Asociación Argentina de Publicidad (AAP), 2010. Nov. 20, 2011. <http://190.220.150.93/webaaap/wp-content/inversion_publicitaria/2010/informe_inversion_publicitaria_2010.pdf>.

AFSCA. "11 informe Contenidos de la Television Abierta Argentina." Autoridad Federal de Servicios de Comunicación Audiovisual (AFSCA), 2011. November 20, 2011. <http://www.afsca.gob.ar/web/Varios/informes/Informe_TV_Abierta_11_marz_a_may11.pdf>.

CNC. "Resolución CNT 1083/1995." Comision Nacional de Comunicaciones (CNC), 1995. July 16, 2012. < http://www.enre.gov.ar/web/bibliotd.nsf/($IDWeb)/943B4F96BE336D9A03257134006E13AA>.

CNC. "Balance General AL 31/12/08." Comision Nacional de Comunicaciones (CNC), 2008. Jan. 15, 2013. <http://www.cnc.gov.ar/institucional/archivos/BALANCE%20AL%-2031-2012-2008.pdf>.

CNC. "Evolucion y penetracion del servicio." Comision Nacional de Comunicaciones (CNC), 2009. July 16, 2012. < http://www.cnc.gov.ar/ciudadanos/internet/evolucion.asp#iconsumo>.

LAMAC. "TV Paga en Argentina 2012." Latin American Multichannel Advertising Council (LAMAC), 2012. Jan. 16, 2012. <http://www.lamac.org/files/factbook/FactBook2012.pdf>.

SINCA. "Cuenta Satélite de Cultura." sinca.cultura.gov.ar, 2012. July 13, 2012 < http://sinca.cultura.gov.ar/sic/estadisticas/csc/index.php>.

## 3. Websites

"Argentina." Alexa, The Web Information Company, 2011. Nov. 27, 2011. <http://www.alexa.com/search?q=argentina&r=home_home&p=bigtop>.

*Conectar Igualdad*. conectarigualdad.gob.ar, 2011. Nov. 27, 2011. <www.conectarigualdad.gob.ar>.

"El Plan 'Más Cerca Verano' de TDA, Argentina Conectada e Igualdad Cultural inicia su recorrido." Ministry of Federal Planning, Public Investment and Services, 2011. Nov. 27, 2011. <http://www.argentinaconectada.gob.ar/notas/3717-el-plan-ms-cerca-verano-tda-argentina-conectada-e-igualdad-cultural-inicia-su-recorrido>.

*Grupo IBOPE*. ibope.com.ar, 2011. Nov. 27, 2011. <http://www.ibope.com.ar/ibope/wp/>.

*Instituto Verificador de Circulaciones*. ivc.org.ar, 2011. <http://www.ivc.org.ar/>. Nov. 27, 2011.

*Instituto Nacional de Cine y Artes Audiovisuales*. incaa.gov.ar, 2011. Nov. 27, 2011. <http://www.incaa.gov.ar/>.

*Televisión Digital Abierta*. tda.gob.ar, 2011. Nov. 27, 2011. <www.tda.gob.ar>.

Nov. 7, 2011. <http://www.ivc.org.ar/>.

# Argentina—Data Summaries

ELI NOAM AND PAUL MUTTER

ARGENTINA HAS THE SIXTH-HIGHEST level of foreign media ownership of all the countries surveyed. Three of Argentina's top four media service providers are foreign groups: Telefónica (Spain), Telecom Italia, and América Móvil (Mexico). The country's liberalized ownership laws attracted large investments in the telecom field, and its high rate of cable television penetration means there is a market for foreign video industries. The top four national media companies—Telefónica, Fintech/Telecom Italia (Fintech is in the process of acquiring Telecom Argentina, from Telecom Italia), Grupo Clarín, and America Móvil—have a high combined market share and power index value relative to other countries: Argentina ranked seventh overall for the Pooled Overall Sector C4 in 2008.

Telefónica (Spain) increased its company power index in Argentina from 1,078 to 1,275, and it is by far the largest media company in the country with 26.1% of the national market (Table 19.17). But it also has a big TV presence, giving it 5.2% of content (Table 19.18), and its platform market share is 37.8%, the highest of all platform companies (Table 19.19). What used to be the second largest platform enterprise, Telecom Italia, has seen its national market share decline, and as of 2013 it has been in the process of exiting from country, having its 22.7% stake Telecom Argentina bought out by the investment company Fintech.

Of the content producers, Grupo Clarín holds by far the largest share at 24.3%. It has significant market share in the TV broadcasting, multichannel platforms, and daily newspapers. Its market share of the multichannel industry increased from 54.1% in 2004 to 55.1% in 2008, and its shares in TV broadcast and daily newspapers were stable at 23.5% and 45.1%, respectively, while its

Table 19-17. National Media Industries Concentration in Argentina

| | 2004/5 | | 2011 OR MOST RECENT | | % CHANGE ANNUAL AVERAGE | |
|---|---|---|---|---|---|---|
| | COMPANY POWER INDEX IN COUNTRY | COMPANY SHARE OF THE OVERALL NATIONAL MEDIA MARKET (%) | COMPANY POWER INDEX IN COUNTRY | COMPANY SHARE OF THE OVERALL NATIONAL MEDIA MARKET (%) | COMPANY POWER INDEX IN COUNTRY | COMPANY SHARE OF THE OVERALL NATIONAL MEDIA MARKET (%) |
| Telefónica (Spain) | 1,078 | 26.5 | 1,275 | 26.1 | 3.1 | –0.07 |
| Grupo Clarín | 593 | 13.4 | 722 | 16.3 | 3.6 | 0.5 |
| Fintech (Mexico) | 0.0 | 0.0 | 651 | 18.3 | N/A | 3.1 |
| Google (US) | 26 | 0.3 | 160 | 1.7 | 86 | 0.2 |
| América Móvil (Grupo Carso, Mexico) | 227 | 7.9 | 264 | 7.5 | 2.7 | –0.07 |
| NII Holdings (Mexico/US)[1] | 38 | 3.2 | 2 | 0.6 | –15.8 | –0.4 |
| DirecTV (US) | 0 | 0.0 | 72 | 3.6 | N/A | 0.6 |
| Canal 9 (Mexico) | 15 | 1.0 | 51 | 2.6 | 39.2 | 0.3 |
| Radio 10 (Grupo Indalo) | 30 | 0.9 | 6 | 0.4 | –13.5 | –0.08 |
| América 2 (Grupo Ávila) | 14 | 1.1 | 47 | 2.7 | 40.4 | 0.3 |
| Editorial Atlántida | 13 | 0.8 | 42 | 1.5 | 38.5 | 0.1 |
| Editorial Perfil | 12 | 0.8 | 11 | 0.7 | –2.0 | –0.01 |
| La Nación | 201 | 1.2 | 2 | 0.3 | –15.0 | –0.15 |
| Grupo Publicexpress | 5 | 0.5 | 9 | 0.7 | 12.4 | 0.02 |
| Disney (US) | 8 | 0.4 | 12 | 0.5 | 7.58 | 0.02 |
| Supercanal | 7 | 1.0 | 6 | 1.0 | –2.9 | 0.01 |
| Grupo Televisa (Mexico) | 1 | 0.2 | 6 | 0.5 | 136.0 | 0.06 |
| El Dia | 5 | 0.6 | 5 | 0.5 | 0.8 | –0.01 |
| Grupo Monera | 0 | 0.0 | 4 | 0.3 | N/A | 0.05 |
| Del Plata | 0 | 0.0 | 2 | 0.24 | N/A | 0.04 |
| PRISA (Spain) | 4 | 0.3 | 1 | 0.14 | –13.4 | –0.03 |

*continued*

Table 19-17. *continued*

| | 2004/5 | | 2011 OR MOST RECENT | | % CHANGE ANNUAL AVERAGE | |
|---|---|---|---|---|---|---|
| | COMPANY POWER INDEX IN COUNTRY | COMPANY SHARE OF THE OVERALL NATIONAL MEDIA MARKET (%) | COMPANY POWER INDEX IN COUNTRY | COMPANY SHARE OF THE OVERALL NATIONAL MEDIA MARKET (%) | COMPANY POWER INDEX IN COUNTRY | COMPANY SHARE OF THE OVERALL NATIONAL MEDIA MARKET (%) |
| Murdoch Interests (US) | 5 | 0.3 | 3 | 0.2 | –7.6 | –0.01 |
| Telecom Italia | 1,221 | 27.2 | 0 | 0.0 | –16.7 | –4.5 |
| Media Concentration Index | 2004/5 | | 2011 or Most Recent | | % Change Annual Average | |
| Total Revenue: Nat'l Media Industry (mil US$) | 6,091 | | 9,018 | | 8.0% | |
| Total Voices (*n*) | 52 | | 52 | | 0% | |
| Net Voices (*n*) | 47 | | 48 | | 0.4% | |
| Public Ownership (%) | 0.0 | | 0.0 | | 0% | |
| Foreign Ownership (%) | 68.1 | | 62.8 | | –0.9% | |
| C4 Average—Weighted | 89.5 | | 85.2 | | –0.7% | |
| HHI Average—Weighted | 3,182 | | 3,151 | | –0.2% | |
| Noam Index Average—Weighted | 910 | | 1,171 | | 4.8% | |
| C1 Average—Weighted | 42 | | 44 | | 0.0% | |
| Pooled Overall Sector C4 | 75 | | 68 | | –1.1 | |
| Pooled Overall Sector HHI | 1,702 | | 1,374 | | –3.2% | |
| Pooled Overall Sector Noam Index | 148 | | 105 | | –4.8% | |
| Market Share of Top 10 Companies: Nat'l Media Industry (%) (Pooled C10) | 83.3 | | 81.3 | | –0.3% | |
| National Media Power Index | 3,341 | | 3,366 | | 0.1% | |

[1] Formerly Sprint-Nextel; now separate from Softbank/Sprint.

Table 19-18. Top Content Media Companies

| | 2004/5 | | 2009 OR MOST RECENT | | % CHANGE ANNUAL AVERAGE | |
|---|---|---|---|---|---|---|
| | COMPANY POWER INDEX IN COUNTRY CONTENT SECTOR | COMPANY SHARE OF THE NATIONAL CONTENT MEDIA MARKET (%) | COMPANY POWER INDEX IN COUNTRY CONTENT SECTOR | COMPANY SHARE OF THE NATIONAL CONTENT MEDIA MARKET (%) | COMPANY POWER INDEX IN COUNTRY CONTENT SECTOR (% HHI) | COMPANY SHARE OF THE NATIONAL CONTENT MEDIA MARKET ("DELTA" SHARE) |
| Grupo Clarín | 1,198 | 30.2 | 927 | 24.3 | –3.8 | –1.0 |
| Telefónica (Spain) | 186 | 6.6 | 72 | 5.2 | –10.2 | –0.2 |
| Canal 9 (Mexico) | 55 | 3.6 | 143 | 7.4 | 26.3 | 0.6 |
| Radio 10 (Grupo Haddad) | 108 | 3.2 | 17 | 1.1 | –14.1 | –0.3 |
| América 2 | 50 | 4.0 | 131 | 7.5 | 27 | 0.6 |
| DirecTV (US) | 0 | 0 | 66 | 3.3 | N/A | 0.6 |
| Editorial Perfil | 44 | 3.0 | 30 | 2.1 | –5.4 | –0.2 |
| Grupo Publicexpress | 18 | 1.9 | 24 | 1.9 | 5.6 | –0.01 |
| Grupo Televisa (Mexico) | 2 | 0.7 | 16 | 1.5 | 100.6 | 0.14 |
| Grupo Monera | 0 | 0 | 11 | 0.86 | N/A | 0.14 |
| Del Plata | 0 | 0 | 6 | 0.7 | N/A | 0.1 |
| PRISA (Spain) | 14 | 1.1 | 2 | 0.4 | –14.1 | –0.1 |
| Murdoch Interests (US) | 17 | 1.0 | 7 | 0.6 | –9.7 | –0.06 |
| Media Concentration Index | 2004/5 | | 2011 or Most Recent | | % Change Annual Average | |
| Public Ownership (%) | 0 | | 0 | | 0% | |
| Foreign Ownership (%) | 18.6 | | 27.1 | | 1.4% | |
| C4 Average—Weighted | 69.5 | | 69.9 | | 0.1% | |
| HHI Average—Weighted | 1,768 | | 1,914 | | 1.4% | |
| C1 Average—Weighted | 35 | | 35 | | 0% | |
| National Power Index | 2,035 | | 2,114 | | 0.65% | |

## Table 19-19. Top Platform Media Companies

| | 2004/5 | | 2011 OR MOST RECENT | | % CHANGE ANNUAL AVERAGE | |
|---|---|---|---|---|---|---|
| | COMPANY POWER INDEX IN COUNTRY | COMPANY SHARE OF THE OVERALL NATIONAL MEDIA MARKET (%) | COMPANY POWER INDEX IN COUNTRY | COMPANY SHARE OF THE OVERALL NATIONAL MEDIA MARKET (%) | COMPANY POWER INDEX IN COUNTRY | COMPANY SHARE OF THE OVERALL NATIONAL MEDIA MARKET (%) |
| Telefónica | 1,416 | 34.1 | 1,946 | 37.8 | 6.2 | 0.6 |
| Fintech (Mexico) | 0 | 0.0 | 1,015 | 28.5 | N/A | 4.8 |
| Grupo Clarín | 363 | 7.0 | 608 | 11.8 | 11.2 | 0.8 |
| América Móvil (Mexico) | 313 | 10.9 | 441 | 11.6 | 5.2 | 0.1 |
| NII Holdings (Mexico/US)[1] | 52 | 4.4 | 3 | 1.0 | –15.7 | –0.6 |
| DirecTV (US) | 0 | 0.0 | 75 | 3.8 | N/A | 0.6 |
| Telecom Italia (Italy) | 1,684 | 37.5 | 0 | 0 | –16.7 | –6.2 |
| Media Concentration Index | 2004/5 | | 2011 or Most Recent | | % Change Annual Average | |
| Public Ownership (%) | 0 | | 0 | | 0% | |
| Foreign Ownership (%) | 86.8 | | 82.6 | | –0.7% | |
| C4 Average—Weighted | 97 | | 93.8 | | –0.5% | |
| HHI Average—Weighted | 3,718 | | 3,840 | | 0.55% | |
| C1 Average—Weighted | 44.0 | | 50.0 | | 1% | |
| National Power Index | 3,836 | | 4,046 | | 1% | |

[1] Formerly Sprint-Nextel; now separate from Softbank/Sprint.

radio market share fell to 13.5%. On some level, Clarín has been beneficiary of an Argentine climate increasingly restrictive against foreign direct investment in media. However, as part of a growing conflict between Clarín and the Peronist government, new laws were passed that would force the company to reduce cross-ownership by divesting itself of some of its subsidiaries. The regulations, enacted in 2009, aimed to restrict the number of station licenses a single company can own, decreasing the number from 24 to 10, in addition to other restrictions. Clarín appealed the decision, but lost its appeal before the Supreme Court in 2013.

# 20

# Media Ownership and Concentration in Brazil

SONIA VIRGINIA MOREIRA

## INTRODUCTION

For many years, the relationship in Brazil among media producers, consumers, and regulators has not been particularly close. But over the past decade, they have come together in deliberating media legislation regarding shareholding and concentration issues. The proliferation of Internet platforms for traditional media sources like radio, TV broadcasting, and print publications has altered circulation patterns and consumer demands. This is mainly due to the significant growth of a new stratum of consumers, the "C class," Brazil's newly emergent middle class that has grown substantially mainly from 2006 to 2012.[1]

Public and corporate administrators have moved to demark a new legal environment for communications, which is inevitable for political reasons—the necessity of reorganizing the sector and instituting new concession rules—as well as for business reasons, in particular by bringing traditional media platforms into the national broadband system. A brief timeline of the government's actions between January 2009 and January 2011 provides a few clues about the recent course of these issues:

- In 2010, the federal government renationalized Telebras to lead the implementation of the National Broadband Plan. First established in 1972, Telebras was the core of the national telecom market for over three decades. It was responsible for more than 95% of public telecom services in the country until

1. Secretariat for Social Communication (SECOM). "Additional Facts on Brazil's Middle Class." The Federal Government of Brazil, Aug. 8, 2011. Dec. 26, 2012. <http://www.brasil.gov.br/para/press/press-releases/august-1/additional-facts-on-brazils-middle-class/br_model1?set_language=en>.

it was privatized in July 1998. One month earlier, in June 1998, the Telebras system marked the installation of 18.2 million fixed terminals and 4.6 million phones distributed over 22.9 million localities.[2] In November 2010, six months after it was renationalized, the federal government announced that Telebras's fiber optic network would be used to broadcast public digital terrestrial television (DTT), which in the original plan was predicted to have required a net investment of US$1.4 billion (R$1.5 billion).[3]

- In November 2010, the President of the Senate reinstated the Social Communication Council, and requested that unions and media companies nominate representatives. As an auxiliary organ of Congress, anticipated in the 1988 Constitution and formally set up in 2002, the Council's role had been gradually reduced: in 2007 only one meeting was held, and after 2008 no new members joined the Council.[4]
- In December 2010, a draft version of the federal legislation for media was circulated unofficially as the "General Law for Social Communication," which among other issues anticipated the creation of a National Communications Agency (NCA) as a regulator.
- In June 2010, a bill on "audiovisual communication with conditional access" was submitted to a vote in the Senate. In September 2011, it was issued as Law No. 12.485.[5]

Expanded information collection of invoicing, composition, and mergers in Brazilian media markets demonstrates that the investigation and organization of the (not always) available information has been deemed necessary by regulators. The task still to be accomplished will provide legislators, governments, society, and fellow researchers with reliable data to evaluate these government interventions in the communications sphere of the economy.

The challenge for Brazilian research on ownership and concentration in the media and telecommunications industry is the lack of comprehension of strategic matters. Government policies and corporate agreements very often ground adopted decisions in restricted spheres. But because these decisions affect all consumers and producers indirectly, it is imperative to advance beyond circumstantial studies that do not map out continuities and trends. This means investing in academic studies that are impartial and independent.

When it comes to media concentration and regulation in Brazil, the literature on political economy of communications constitutes the most frequent theoretical reference. Lima (2001), Bolaño (2007), Jambeiro (1998), Brittos and Haussen (2009), Possebon (2007), Santos and Silveira (2007), and Dantas (2002) have authored work that deals with issues in monopoly, concentration, and national public policy within the broader perspective of the political economy of information access and Brazilian culture. There are practically no grounding studies focused on the economy of communications, the object of which would be to study in detail the forces and constraints that regulate media market structures internally and externally, and to examine consumer demand. Likewise, there are no analyses, as already identified by Noam and Wolfson (1997), that consider the assumption that "the telecommunications industry, long organized . . . along geographic and product lines that were both a shield and a weapon, is being transformed by contradictory forces: on the one hand, the trend toward global expansion by carriers, and on the other hand, fragmentation and entry in local communications," in a period of reduced competition and of greater

2. Telebras. "Conheça a Telebras—Histórico." Telebras, Jan. 5, 2012. Dec. 26, 2012. <http://www.telebras.com.br/inst/?page_id=41>.

3. "Publicado decreto: I Conferência Nacional de Comunicação é official." *Pulsar Brasil*, Apr. 17, 2009. Dec. 29, 2010. <http://www.brasil.agenciapulsar.org/nota.php?id=4408>.

4. Costa, Rosa. "Conselho de Comunicação vai ser revigorado." *O Estado de S. Paulo*, Nov. 9, 2010. Dec. 28, 2010. <http://www.observatoriodaimprensa.com.br/artigos.asp?cod=615IPB004>.

5. Presidência da República. "Lei Nº 12.485, de 12 de setembro de 2011." *Dispõe sobre a comunicação audiovisual de acesso condicionado*. Sept. 13, 2011. Jan. 22, 2013. <http://www.planalto.gov.br/ccivil_03/_Ato2011-2014/2011/Lei/L12485.htm>.

enterprise integration, generating a complex market web.[6] For example, media conglomerates have grown over the past decade in Brazil. Telecom service contractors, for example, reached a market value of US$73 billion (R$79.3 billion) in shares negotiated in the stock market at the end of September 2010, according to the Brazilian Stock Exchange.[7]

This large volume of investment is the result of growth in the market that began in 1999, following the 1998 privatization of regional wireline telecoms that had been part of Telebras. Brazil went from having 25 million wireline and 15 million mobile subscribers in 1999 to having 41.8 million and 191.5 million, respectively, by 2010.[8] In this same decade, telephone operators contributed to the evolution of broadband services (which were practically nonexistent in 1998); as of 2010, there are 12.8 million broadband subscribers in Brazil. Cable TV companies have also benefitted from these circumstances: in 2010, they had 9.1 million subscribers, though the slower rhythm of expansion in this period was partly due to economic reasons (limited popularity of the paid service) and to infrastructure (lack of cables and retransmission towers for wireless cable systems).

Cable's slow growth was also due to flawed political decisions, such as the privatization of the Telebras system in 1998, which included the selling of Embratel, a state company that also operated the commercial BrasilSat satellites, to the US telecom major Verizon (America Movil, owned by the Mexican telecom tycoon Carlos Slim, subsequently acquired Embratel). This decision delayed the dissemination of the direct-to-home (DTH) and satellite cable TV systems.[9]

Almost simultaneously to the privatization of telephony, the Law of Community Broadcasting—Law 9.612 of 19 February 1998—was instituted in Brazil.[10] As the first specific legislation for radio and TV broadcasting to be enacted in around 30 years, the law allowed hundreds of local clandestine broadcasting stations, popularly known as *rádios piratas* (pirate radio), to transmit from 2001 on with the consent of the Ministry of Communications, after they had been operating without licenses for almost 20 years in nearly all Brazilian states.

Despite criticism regarding the kinds of financing permitted and limited antenna range, communitarian low-frequency radios registered their largest increase ever in the first decade of the 21st century, growing from 980 channels in 2001 to 3,897 by 2009.[11] This shows the interest of Brazilians, independent of social background, in alternative media forms, and it is one of the elements that must be considered in order to understand the dynamics of online social networking in Brazil.

Another issue is the revenue generated by federal taxes on commercial telecoms. According to the National Agency of Telecommunications (ANATEL) and the National Treasury, between 2001 and the first trimester of 2010, US$20.3 billion (R$22.1 billion) was collected—of which US$14.6 billion (R$15.9 billion) went to supply the Fiscal Telecommunications Fund

6. Noam, Eli and Wolfson, A. J. eds. "The End of Territoriality in Communications." *Globalism and Localism in Telecommunications.* New York: Elsevier, 1997. Dec. 26, 2012. <http://www.citi.columbia.edu/elinoam/articles/TheEndofTerritorialityinCommunications.pdf>.

7. Moreira, Sonia Virgínia. "Geografias do público e do privado na configuração de um marco legal para as comunicações." *Intercom—Sociedade Brasileira de Estudos Interdisciplinares da Comunicação,* Sep. 2–6, 2011. Dec. 26, 2012. <http://www.geografias.net.br/papers/2011/R6-0804-1.pdf>.

8. Moreira, "Geografias do público e do privado. . . ."

9. "[A]ll the Brazilian information that passes through space—be it military, governmental, or from national private companies—[passes] through private satellites controlled by only one company, Star One, owned by the Mexican billionaire Carlos Slim. Actually, Brazil is a simple lessor of a spatial retransmitter that has as its main purpose the generation of profit for its owner." Sequeira, C. D. "Brasil devassado." *IstoÉ* 2166, May 13, 2011. July 16, 2012 at <http://www.istoe.com.br/reportagens/137133_BRASIL+DEVASSADO?pathImagens=&path=&actualArea=internalPage>.

10. Cardoso, F. H. and Motta, Sergio. "LEI Nº 9.612, DE 19 DE FEVEREIRO DE 1998: Institui o Serviço de Radiodifusão Comunitária e dá outras providências." Presidência da República Casa Civil Subchefia para Assuntos Jurídicos, Feb. 19, 1998. Jan. 15, 2011. <http://www.planalto.gov.br/ccivil_03/Leis/L9612.htm>.

11. Moreira, Sonia Virgínia. "Geografias do público e do privado na configuração de um marco legal para as comunicações." *Intercom—Sociedade Brasileira de Estudos Interdisciplinares da Comunicação,* Sept. 2–6, 2011. Dec. 26, 2012. <http://www.geografias.net.br/papers/2011/R6-0804-1.pdf>.

(FISTEL), US$4.3 billion (R$4.7 billion) went to for the Universalization of Funds of the Telecommunications Services (FUST), and US$1.4 billion (R$1.5 billion) went to the Fund for the Technological Development of Telecommunications (FUNTTEL).[12] However, FUST, created to universalize telecom services and facilitate access for low-income and rural consumers, has not seen any of these programs yet implemented.

Some contextual data may contribute to a better understanding of the economic and cultural context for Brazilian media over the last 50 years, starting with the demographic numbers. From a predominantly rural country in 1960, Brazil has evolved into a markedly urban nation—with 84% of the population of 191 million inhabitants living in cities by 2010.[13] Internal migratory processes motivated urban expansion, especially by the movement of the inhabitants of the Northeast region, toward the metropolitan areas of São Paulo and Rio de Janeiro.

Families who dominated the production of print newspapers and magazines for decades gradually added to their publishing business radio and television broadcasting, news agencies, and book publishing companies. Until the late 1990s, 10 families controlled most of the media industry in Brazil: the Marinho (news, print publishing, broadcasting) and Brito (news and radio broadcasting) families in Rio de Janeiro; the Frias (newspapers), Mesquita (newspapers and radio broadcasting), Civita (print publishing), and Saad (broadcasting) families in São Paulo; the Sirotsky family (newspapers and broadcasting) in the southern states of Rio Grande do Sul and Santa Catarina; the Câmara family (newspapers and broadcasting) in the midwestern states of Goiás and Tapajós; the Maiorana family (newspapers and radio broadcasting) in the northern state of Pará; and the Abravanel family (TV broadcasting) in São Paulo.[14] A commonality among these families is their interest in the mass media industry; just a few families ventured into other businesses, such as finance.[15]

Since the first half of the 1990s, these families expanded their actions to add content distribution structures, online news portals, and Internet service providers (ISPs). Brazilian media conglomerates such as Folhapar (originated in the Frias family), Globopar (originated in the Marinho family), the Abril Group (originated in the Civita family), and the Estado Group (originated in the Mesquita family) are examples of the changes that took place in media production and ownership.

In the case of telecommunications, the main market and legal changes occurred after 1998, with the approval of the new General Telecommunications Law. It broke up the state monopoly of the sector, dividing Telebras into 12 holdings of wireline and mobile telephone services, privatizing Embratel (up until then a state company that managed satellite broadcasting service) and creating the National Agency of Telecommunications (ANATEL). The largest telecom company that emerged was Tele Norte Leste Participações S.A., which administered 16 state wireline operators that covered 64% of the country.

Telemar (created as Consórcio Telemar, later renamed Telemar Participações) has its origin in national groups that reorganized and diversified their investments to dedicate themselves to the exploration of this sector. They included the Andrade Gutierrez Group (which created AG Telecom Participações S.A.); the electronics manufacturer Inepar S.A. (which in 1999 created the business division Inepar S.A. Indústria e Construções and started a joint venture with Lucent Technologies); and La Fonte, a textile and metallurgy manufacturer (created La

12. Moreira, "Geografias do público e do privado . . ."

13. *Censo Demográfico 2010*. Rio de Janeiro: IBGE, 2011. May 25, 2011. <http://www.ibge.gov.br/home/estatistica/populacao/censo2010/default.shtm>.

14. *Moreira, S. V. and Helal, C. L. R.* "Notes on Media, Journalism Education and News Organizations in Brazil." *Journalism* 10, no. 1 (2009): 91–107. <http://jou.sagepub.com/content/10/1/91.full.pdf>.

15. One of the families that did invest outside of the media sector was the Abravanels, who invested in the Panamericano Bank that went bankrupt in 2011, an event that became a public scandal when accounting fraud worth US$2.5 billion (R$2.5 billion) was discovered.

Fonte Telecom S.A.). The federal government would also maintain a stake in Telemar Participações by means of the Companhia de Seguros Aliança do Brasil, created in 1997 as a participant of the state conglomerate Banco do Brasil; and of the BNDESPAR, an investment branch of the Banco Nacional de Desenvolvimento Econômico e Social (BNDES). Control of most of the wireless telecom firms, and of the wireline sector, changed into the hands of foreign groups, mainly European and North American. These include the companies Telecom Italia/TIM (Italy); Telefónica (Spain); Telmex/America Móvil (Mexico); Bell South, Sprint, and NII Holdings, Inc. (United States); Portugal Telecom (Portugal); France Telecom and Vivendi (France).

## PRINT MEDIA

### Newspapers

Four groups dominate the daily newspapers market, with a combined 89% market share: Folhapar, Infoglobo, S.A. O Estado de S. Paulo, and the RBS Group, the latter of which operates in the southern states of Santa Catarina and Rio Grande do Sul. All of these dailies were the starting point for what later became Brazil's first family-owned broadcasting and telecom conglomerates. There was no foreign participation in this industry until April 2010, when the Ongoing Group of Portugal acquired the Arca Group, owner of the *O Dia* daily (Table 20.1).

Table 20-1. Daily Newspapers (Market Shares by Circulation), 2000–2010

| | 2000 | 2004 | 2008 | 2009 | 2010 |
|---|---|---|---|---|---|
| Folhapar S.A.[1] (Frias Family) (*Folha de S. Paulo, Agora SP, Expresso, Valor Econômico*) | 30.7 | 34.5 | | 27.3 | 27.9 |
| Infopar (Marinho Family) Infoglobo Comunicação e Participações S.A. | 20.8 | | 30.1 | 27.2 | 28.4 |
| Estado Group (*O Estado de S. Paulo, Jornal da Tarde*) | 27.7 | 25.3 | 21.7 | 21.1 | 25.2 |
| RBS Group (Sirotsky Family) (*Diário Gaúcho, Zero Hora*) | 6.4 | 10.3 | 9.7 | 6.9 | 6.7 |
| Diários Associados Group | | 7.0 | 8.4 | | |
| Arca Group (Ongoing, Portugal) (*Meia Hora*)[2] | 10.0 | 8.6 | 6.2 | | 7.5 |
| Jornal do Brasil | | 6.9 | 7.6 | | |
| Sada Group (Vittorio Medioli) (*O Tempo, Super Notícia*) | | 2.1[3] | 3.5 | | |
| Caldas Junior Group (Record Group) (*Correio do Povo*) | | | 2.9 | | |
| Areté Editorial (*Lance!*) | 1.0 | | | | |
| Others | 3.4 | 5.3 | 9.9 | 17.5 | 4.3 |
| Total Revenue (bil R$)[4] | 2.0 | 2.4 | 3.4 | 3.1 | 3.0 |
| Total Revenue (bil US$) | 1.1 | 1.2 | 1.9 | 1.6 | 1.7 |
| Average Daily Circulation ANJ (mil) | 7.8 | 6.5 | 8.4 | 8.1 | 8.1 |
| Average Daily Circulation IVC (mil)[5] | 3.9 | 3.3 | 4.3 | 4.1 | 4.3 |
| C4 | 89.2 | 78.7 | 69.9 | 82.5 | 89 |

Table 20-1. *continued*

| | 2000 | 2004 | 2008 | 2009 | 2010 |
|---|---|---|---|---|---|
| HHI | 2,419 | 2,111 | 1,658 | 1,978 | 2,321 |
| *N* (>1%) | 7 | 7 | 8 | 4 | 5 |
| Noam Index | 855 | 798 | 586 | 989 | 1,038 |

*Source:* National Association of Newspapers (ANJ), 2006–2011; Circulation Audit Institute (IVC), 2011; *InterMeios* Project, 2001–2011.

[1] *Valor Econômico* was launched in May 2000 as a joint venture of Infoglobo and Folhapar, each with 50% of the shares.

[2] The Carvalho family controlled the Arca Group until April 2010, when the Ongoing Group, based in Portugal, bought it.

[3] In 2006.

[4] This revenue figure is an estimate based on data collected in this study: it is based on average per capita revenues reported for several other countries in this study with similar per capita income.

[5] IVC has the most reliable (certified) data on circulation, but not all newspapers are affiliated with IVC, so ANJ's daily circulation estimates are included here as well.

## Book Publishing

The federal government is the largest consumer of books in Brazil: it accounted for 40% of all book sales in the past decade, primarily textbooks. Mainly national companies controlled the industry until 1999, when France's Vivendi became the first foreign investor through the acquisition of a percentage of Editora Abril's book publishing branch. The book publishing sector has begun to register ample movement in terms of mergers and acquisitions in the last decade, beginning in 1999 with Vivendi's investment in the Abril Group's Editora Abril. The most significant shift in this segment occurred with the heavy investment of Spanish firms, especially the PRISA Group, which between 2001 and 2005 acquired Editora Moderna, the third largest educational book publisher and, in 2008, had taken over the leadership of the book publishing market. PRISA controls 13.9% of the market share, with the Abril Group's imprints holding 11.5% and the Saraiva and Ediouro Publicações groups with 9.3% and 5.1%, respectively. In 2003, Planeta Editora of Spain started a local venture and succeeded in acquiring the rights to bestselling Brazilian authors like Paulo Coelho. And in 2004, the Spanish firm Edições SM entered the Brazilian market as Edições SM Brasil with a US$14.8 million (R$20 million) investment. Other important international book publishers with a presence in Brazil include the Darby Overseas Investment Fund, which held a 35% stake in the Siciliano Group from 1998 to 2008. Almost 15% of the Saraiva Group belongs to foreign investors, including the International Financial Corporation, a World Bank division, that in 1999 acquired 2.6% of the book publisher. In 2008, the German publishing company Bertelsmann AG was interested in the Rio de Janeiro-based Record Editorial Group, but there was no acquisition. (Table 20.2).

The first major sign of an emerging e-book market in the country was the acquisition by Penguin, a division of the Pearson Group, of a 45% share of Companhia das Letras, part of Schwarcz Editors Ltd. Another sign in this area was the arrival of Amazon.com from the United States, which will almost certainly affect the market for electronic books and the business model of the major national bookstore chains after starting its e-book operations in Brazil in December 2012.[16] According to the first "book census" of Brazil using data collected between November 2010 and April 2011, there are 750 active publishers in the country, one-third of which do not match the criteria of UNESCO for firms to publish at least five titles and 5,000 copies per year. Of the

16. "Com livros digitais, Amazon inicia operação no Brasil" *Folha de S. Paulo*, Dec. 6, 2012. Jan. 22, 2013 <http://www1.folha.uol.com.br/mercado/1196810-com-livros-digitais-amazon-inicia-operacao-no-brasil.shtml>.

## Table 20-2. Book Publishing (Market Shares by Revenue), 1996–2010[1]

| | 1996 | 2000 | 2004 | 2006 | 2008 | 2009 | 2010 |
|---|---|---|---|---|---|---|---|
| PRISA (Polanco Family, Spain) | | | | | 15.2 | 15.9 | 13.9 |
| Moderna | | 4.4 | 6.6 | 6.8 | 9.5 | | |
| Objetiva | | | | | | | |
| Abril Group[2] (Civita Family; Naspers, South Africa) | 13.7 | 12.7 | 16.1 | 11.9 | 11.6 | 15.1 | 11.5 |
| Saraiva Group (Saraiva Family) | 4.6 | 8.6 | 9.5 | 8.5 | 10.8 | 10.1 | 9.3 |
| Siciliano | *0.7* | *0.7* | *0.5* | *0.3* | | | |
| Ediouro Publicações[3] | 0.5 | 2.8 | 5.4 | 6.3 | 7.9 | 7.6 | 5.1 |
| Editora FTD[4] | 7.4 | 8.3 | 9.3 | 10.2 | 9.9 | 9.9 | 3.6 |
| Sextante[5] | | 1.5 | 1.0 | 1.1 | 2.7 | 3.6 | 3.6 |
| Record Editorial Group[6] | 2.7 | 2.4 | 5.4 | 4.2 | 4.5 | 3.0 | 2.1 |
| Others | 65.5 | 54.6 | 42.1 | 45.5 | 34.2 | 34.8 | 53.0 |
| Total Revenue (mil R$)[7] | | | 1,847 | | 2,521 | | |
| Total Revenue (mil US$) | | | 942 | | 1,412 | | |
| C4 | 30.1 | 36.2 | 41.7 | 40.1 | 47.5 | 51.0 | 39.8 |
| HHI | 292 | 364 | 542 | 466 | 670 | 761 | 467 |
| *N* (>1%) | 5 | 7 | 7 | 7 | 7 | 7 | 7 |
| Noam Index | 103 | 121 | 181 | 165 | 253 | 287 | 176 |

*Source:* National Association of Magazine Editors (ANER), 2007–2010; Circulation Audit Institute (IVC), 2010–2011; InterMeios Project, 2001–2010.

[1] There are two other important book publishing houses operating in this sector, but their market shares are not tracked. They are São Paulo's Schwarcz Editors Ltd. (which has seven imprints, including Penguin Book's Brazilian arm); and Rocco Editors of Rio de Janeiro.

[2] Ática/Scipione. Naspers, the South African media group, has a 30% stake in Abril.

[3] Agir, Nova Fronteira/Nova Aguilar, Thomas Nelson Brazil (joint). The US publishing house Thomas Nelson launched the venture in 2007 and has a 50% stake in the joint venture, which primarily publishes religious (Christian) literature.

[4] Quinteto.

[5] Intrinseca.

[6] Editora Record, BCD União de Editoras, Harlequin Enterprises (joint). The Canada-based publisher Harlequin inaugurated a Brazilian arm in 2005 with the Record Editorial Group, specializing in romantic fiction.

[7] This revenue figure is an estimate based on data collected in this study: it is based on average per capita revenues reported for several other countries in this study with similar per capita income.

498 publishers that met this specification, only 16 demonstrated incomes of more than US$16.4 million (R$28 million) annually. Almost half of them earn less than US$0.4 million (R$0.6 million) annually.[17]

## Magazines

Two families dominate the magazine industry: the Marinhos, of Globopar, and the Civitas, of the Abril Group. Globopar and Abril control the magazine sector with approximately 51% combined of the national market in 2009. The Abril Group, though, is the dominant player, with its publishing strategy focused on both weekly and monthly editions. The Abril Group is the most traditional magazine printing company in the country, founded in the 1950s by Victor Civita, an American publisher of Italian origin, to print Portuguese editions of Walt Disney comic books.

In 2004, the Abril Group became the first Brazilian media conglomerate to attract foreign

17. Freitas, G. "Censo reavalia dimensões do mercado editorial brasileiro." *O Globo*, Aug. 16, 2011. Jan. 27, 2012 <http://oglobo.globo.com/blogs/prosa/posts/2011/08/16/censo-reavalia-dimensoes-do-mercado-editorial-brasileiro-399149.asp>.

direct investment (permitted on the constitutional amendment approved two years previously by Congress). The Capital Group, one of the largest US hedge fund managers, acquired a 13.8% stake in Abril. By that time, Editora Abril was the magazine industry's market leader, with over 200 publications and 26 million readers nationwide. In addition, it held TVA, the first pay TV service launched in Brazil, and owned 70% of Viacom's local MTV Channel, MTV Brasil. The uncertainty regarding the successors of its CEO Roberto Civita, though, led the editorial group to invest in education, which was viewed as more promising than the magazine industry, by creating the Abril Educação Company in 2009, which would have no relation to the rest of Abril Media. However, with changes in the publishing industry, together with a fragmentation of audiences interests, Abril Media has either closed down companies (such as the record company Abril Music, the music website *Usina do Som,* and the pay TV channels Fiz TV and Idea TV) or sold its shareholdings, as was the case with the BOL Internet service provider; its licenses for HBO Brazil, DirecTV Latin America, ESPN Brazil, and Eurochannel; and the direct-to-home (DTH) satellite service TVA. Presently, the Abril Group is focused on its print publishing ventures, as well as on its print distribution and logistics arm DBG (which is a monopolist in print distribution and logistics), the digital-out-of-home (DOOH) advertiser Elemidia, and MTV Brasil (Table 20.3).[18]

## AUDIOVISUAL MEDIA

### Radio

In Brazil, domestic producers dominate the radio market, with no foreign participation, though foreign players did help shape the sector when it was first emerging. During the 1940s, the presence in the country of the Office of the Coordinator of Inter-American Affairs led by Nelson Rockefeller shaped the local radio business to the American model.[19] The concentration of the Brazilian radio industry in urban centers is clear, but it is also evident that current regional networks are sources of economic and political power, mostly owned by local family groups with economic power in these places. One example is the Edson Queiroz Group, based in Fortaleza (the capital of the northeastern state of Ceará), which started out in the gas distribution business in 1951. Less than a decade later, it had acquired a radio broadcasting station, then received a TV broadcasting license, and in the 1980s entered newspaper publishing.

At present, the radio industry is a fragmented sector with more than 9,000 stations—4,988 AM, FM, short wave (SW), and tropical wave (TW) stations plus 4,150 community radios—according to ANATEL. Politics is a key issue to consider when analyzing the radio business: besides the substantial participation of politicians, or rather, *because* of it, the government has struggled since 2008 to introduce a public broadcasting model.

For a commercial media industry, it is characterized by a considerable imbalance in terms of ownership: it is estimated that politicians (deputies, senators, governors, and a few mayors) own around 30% of these concessions, even though the Constitution prohibits this practice. In other words, a public broadcasting system at the national-federal level would lessen the politicians' control and they have little reason to support it. Data compiled in 2008 by the Institute for Studies and Research in Communication (Epcom) revealed that 271 Brazilian politicians, in contravention to the Constitution's Article No. 54, were partners or directors in 348 private radio and TV broadcasting stations. Of these, 54% were mayors, 21% were state

18. Costa, L. M. "Para onde vai a Editora Abril?" *Observatório da Imprensa* 659, Sep. 14, 2011. Jan. 23, 2012. <http://www.observatoriodaimprensa.com.br/news/view/para-onde-vai-a-editora-abril>. & Gindre, Gustavo. "O fracasso do Grupo Abril." *Observatório da Imprensa,* 686, Mar. 15, 2012. May 25, 2012. <http://www.observatoriodaimprensa.com.br/news/view/_ed686_o_fracasso_do_grupo_abril>.

19. The Office was established to carry out the "Good Neighbor Policy," a United States diplomatic initiative in Latin America during the 1930s and 1940s.

Table 20-3. Magazine Publishing (Market Shares by Circulation), 2000–2010

| | 2000 | 2004 | 2006 | 2008 | 2009 | 2010 |
|---|---|---|---|---|---|---|
| Abril Group (Civita Family; Naspers, South Africa) | | | | | | |
| Editora Abril | 55.3 | 52.5 | 51.2 | 34.6 | 35.9 | 37.3 |
| Editora Globo (Globopar S.A., Marinho Family) | 19.6 | 15.3 | 18.4 | 13.7 | 15.1 | |
| Editora Escala | | 7.1 | | | 2.9 | 5.7 |
| Ediouro Group | | | | 0.5 | 1.2 | |
| Others | 25.1 | 25.1 | 30.4 | 51.2 | 46.2 | 42.0 |
| Total Revenue (bil R$)[1] | 2.0 | 2.7 | 3.7 | 5.5 | 5.6 | 6.0 |
| Total Revenue (bil US$) | 1.1 | 1.4 | 1.7 | 3.1 | 2.9 | 3.4 |
| C4 | 74.9 | 74.9 | 69.6 | 48.8 | 55.1 | 58.0 |
| HHI | 3,442 | 3,041 | 2,960 | 1,385 | 1,526 | 1,652 |
| *N* (>1%) | 2 | 3 | 2 | 2 | 4 | 2 |
| Noam Index | 2,434 | 1,756 | 2,093 | 800 | 763 | 954 |

*Source:* National Association of Magazine Editors (ANER), 2007–2010; Circulation Audit Institute (IVC), 2010–2011; InterMeios Project, 2001–2010.

[1] This revenue figure is an estimate based on data collected in this study: it is based on average per capita revenues reported for several other countries in this study with similar per capita income.

representatives, 18% were deputies, 7% were senators, and one of them was a governor.[20]

Religious groups also control radio concessions, in particular those linked to Neo-Pentecostal churches.[21] More recently, this imbalance spread to community broadcasting stations, which have grown the most since their regulation by the government with the 1998 Law of Community Broadcasting. In addition, a new expedient was discovered by political groups to cover up their participation in the radio broadcasting sector: the purchase of licenses to operate educational radio channels. A similar phenomenon occurs in television broadcasting, but it is less prevalent because the licensing fees in TV cost more than radio (Table 20.4).

The high share for "Others" is another consequence of the participation of politicians in the radio broadcasting industry. Besides ownership, it is difficult to obtain data on financial reports or advertising shares of radio stations and/or networks belonging to politicians.

## Broadcast Television

Globopar, with 52.4% of the market share, is by far the predominant group in terrestrial broadcasting. Regional groups hold a significant stake of 24%, and the Universal Church Group holds 14.7% of the market. Three commercial groups (Bandeirantes, Silvio Santos, and Rede TV!) and the state-owned Empresa Brasil de Comunicação (EBC) control most of the rest of industry.[22] Globopar is the only company, among private groups, that releases public information about its ownership structure. The group operates in the communications and entertainment sectors and in online infrastructure.

20. Marini, A. R. "Cresce o número de políticos donos de meios de comunicação." Fórum Nacional para a democratização da Comunicação (FNDC), 2008. Nov. 14, 2010. <http://www.fndc.org.br/internas.php?p=noticias&cont_key=236850>.

21. Specifically, the Universal Church of the Kingdom of God (Igreja Universal do Reino de Deus), a Neo-Pentecostal group in Brazil that purchased the Record Television Network in the 1980s to broadcast televangelical programming.

22. 2010 data.

### Table 20-4. Radio Group (Market Shares by Revenue), 2000–2010

| | 2000 | 2004 | 2008 | 2009 | 2010 |
|---|---|---|---|---|---|
| Globopar S.A. (Globo Radio System, Marinho Family)[1] | 12.4 | 11.2 | 8.8 | 9.1 | 9.8 |
| Bandeirantes Communication Group (Bandeirantes Radio System, Saad Family) | 4.0 | 3.2 | | 3.7 | 4.7 |
| RBS Group (Sirotsky Family) (*Rede Gaúcha de Rádio*) | 5.9 | 5.4 | 4.1 | 4.0 | 4.2 |
| EBC (public) MEC AM and FM, Radiobras Radio Network | | | 2.7 | 2.8 | 4.2 |
| Estado Group (Mesquita Family and The Walt Disney Company—2007 *Estadão ESPN (AM & FM Radio)* | | 1.5 | 2.6 | 3.8 | 4.0 |
| Radio Panamericana (*Jovem Pan Sat*) | | | 2.7 | 2.5 | 1.8 |
| Alfa Group *Rede Transamerica de Rádio*[2] | | 1.9 | 2.1 | 2.5 | |
| Diários Associados Group | | 1.8 | 1.6 | | |
| GRPCOM—Grupo Paranaense de Comunicação (Pereira and Lemansky families) | | 1.2 | 0.9 | 1.0 | 1.5 |
| Jaime Câmara Organization (Câmara Family) (Câmara Family[3] | 0.9 | 0.9 | 0.7 | 0.7 | |
| Others | 76.8 | 72.9 | 72.6 | 69.9 | 69.8 |
| Total Revenue (mil R$)[4] | 643 | 1,000 | 1,700 | 2,000 | 2,200 |
| Total Revenue (mil US$) | 351 | 505 | 946 | 1,031 | 1,251 |
| C4 | 23.3 | 21.3 | 18.8 | 19.7 | 22.9 |
| HHI | 207 | 172 | 119 | 121 | 136 |
| *N* (>1%) | 3 | 6 | 7 | 4 | 5 |
| Noam Index | 103 | 65 | 40 | 61 | 61 |

*Source:* National Agency of Telecommunications (ANATEL), 2007–2010; Tudo Radio 2010 <http://tudoradio.com/noticias>, InterMeios Project, 2001–2010.[5]

[1] Includes *Radios Globo* and *CBN.*

[2] Includes *Transamerica Pop, Transamerica Light,* and *Rede Transamerica Hits.*

[3] Includes *Anhanguera* and *Araguaia Radio.*

[4] This revenue figure is an estimate based on data collected in this study: it is based on average per capita revenues reported for several other countries in this study with similar per capita income.

[5] Figures regarding radio revenues are not kept by ANATEL. The InterMeios Project, a private organization that collects the annual reports provided by media companies under a non-identification agreement, has published the ones cited here. The "Total Revenue" given in Table 20.4 represents just a small portion of the real advertising income for radio.

Globopar's dominance dates from the period of military dictatorship (1964–1985). TV Globo was founded in 1965 under that regime, which assisted Globopar's rapid growth as the main TV network in the country by allowing the US firm Time-Life Inc. to participate in the broadcasting company. In the early 1960s, as Time-Life searched a way to enter the Brazilian market, Roberto Marinho's TV broadcasting project needed investments and knowledge. In 1962, three years before the launching of TV Globo's programming, two contracts with Time-Life were signed in New York: the generic aspects of the agreement were defined in the main contract, and the technical aspects were presented in the technical assistance agreement.

Time-Life was responsible for providing technical information; training the TV Globo's professionals; assisting in planning, construction, and operation of the studios and equipment; and guiding the acquisition of films and programs produced abroad. So, "[t]hese were the ingredients for a perfect plot, written by a local baron and funded by a multinational, under the gaze of a government which conveniently ignored the unconstitutional nature of the agreement."[23]

In relation to TV broadcasting, the most significant movement happened within the public sector, with the launch in 2007 of EBC, officially created to "fill in a gap in the radio broadcasting system and with the objective of implanting and administrating the public channels, those that, for their independent publishing distinguish themselves from the state or government channels."[24] The company was formed from the incorporation of two state television channels: of the TVE Brasil (Brazil Educational TV), then under the administration of the Roquette Pinto Association of Educational Communication (ACERP), and RADIOBRAS, a company that from its creation in 1975 was linked to three state organs: the Ministry of Communications, the Ministry of Justice, and the Office of the President (Table 20.5).

## Multichannel TV Platforms

Until February 2012, Globopar and America Móvil (Mexico) shared control of the cable TV industry as partners of the cable TV provider *NET* Cable. Today this is the largest multiservice cable provider in Latin America and is present in over 100 Brazilian cities, offering cable TV, ISP, and wireline telecom services. In March 2012, the *Comissão de Valores Imobiliários—CVM*—the Brazilian equivalent of the US Securities and Exchange Commission—announced that America Móvil had assumed control as the major stockholder of the cable provider, with 92.2% of stocks acquired by its subsidiaries Embratel and Embrapar. The participation of the Marinho family was reduced to 6% of the assets.[25] Besides being one of the major moves in the cable and satellite providers' sector, it also strengthened the ever-growing presence of Carlos Slim, the Mexican media tycoon, in Brazilian telecoms. It also confirmed the goal set by Slim's America Móvil[26] in 2010, aimed at the acquisition of 100% of the preferred shares of the cable TV operator as a strategy to maintain control over his companies in Brazil, thus creating better conditions to integrate these operations.[27]

In addition to cable TV, television providers also transmit channels via DTH and multichannel multipoint distribution service (MMDS). As of September 2011, the pay-TV market has been regulated by Act 12.485/11, which opened the market to national and foreign telecommunications companies.[28] The new law unified the regulation on transmission via satellite (DTH), cable, or microwave (MMDS). Even though foreign companies are now permitted to enter the cable TV distribution market (in the previous rule, their participation was limited to no more than a 49% stake in any one provider), they have stayed out of content production but are able to offer combined services of pay-TV, telephone, and broadband services. The law remained in limbo for five years before its approval in the Senate. The delay was due, to some extent, to

23. Sousa, Helena. "Time-Life/Globo/SIC: um caso de reexportação do modelo americano de televisão?" Presented at IAMCR—*International Association for Mass Communication Research*, Glasgow, 1998, July 25–30. June 23, 2012 <http://bocc.ubi.pt/pag/sousa-helena-time-life-sopcom.html>.

24. *Conheça a Empresa*. EBC. Aug. 20, 2010. <www.ebc.com.br/empresa>.

25. Drska, M. "Embratel assume controle da Net por R$ 6,4 milhões." *Valor* Econômico, Mar. 7, 2012.

26. In 2000, Slim divided its businesses in fixed and mobile, creating América Móvil and Telmex. For nearly four years, Telmex spun off in Telmex Mexico and Telmex Internacional—the latter having a stake in Embratel and Net, and data assets and pay TV in Latin America. In 2012, America Movil acquired Carso, the holding company that controlled Telmex Mexico and Telmex International.

27. Afonso, R. "Com oferta, Telmex antecipa plano de assumer controle da NET." *Computerworld*, Aug. 5, 2010. Nov. 14, 2010. <http://computerworld.uol.com.br/telecom/2010/08/05/com-opa-telmex-antecipa-estrategia-de-assumir-controle-da-net/>.

28. Rousseff, Dima et al. "**LEI Nº 12.485, DE 12 DE SETEMBRO DE 2011**: Dispõe sobre a comunicação audiovisual de acesso condicionado." Presidência da República Casa Civil Subchefia para Assuntos Jurídicos, Sep. 11, 2011. Nov. 23, 2011. <http://www.planalto.gov.br/ccivil_03/_Ato2011-2014/2011/Lei/L12485.htm>.

## Table 20-5. TV Broadcasting (Market Shares by Revenue), 1996–2010

| | 1996 | 2000 | 2004 | 2006 | 2008 | 2009 | 2010 |
|---|---|---|---|---|---|---|---|
| Globo TV Network (Globopar S.A., Marinho Family) | 31.0 | 42.7 | 43.7 | 45.4 | 53.9 | 61.8 | 52.4 |
| Record TV Network (Universal Church Group) | 1.0 | 4.8 | 6.0 | 9.1 | 9.5 | 16.5 | 14.7 |
| Bandeirantes Group | | 4.6 | 3.0[1] | 5.7 | 6.1[2] | 7.3 | |
| SBT TV Network (Silvio Santos Group) | 11.7 | 8.6[3] | 7.0 | 6.3 | 5.0 | 5.1 | 4.2 |
| Manchete TV Network (Bloch Group) | 2.4 | 1.0 | 2.3 | 2.4 | 2.2 | 2.2 | |
| EBC Brasil (public) | | | | | 2.8 | 2.7 | 2.3 |
| ACERP/TVE Brasil | | 0.7 | 1.7 | 1.7 | EBC Brasil | | |
| Radiobras | | 1.9 | 1.3 | 1.4 | EBC Brasil | | |
| Fundação Padre Anchieta (State of São Paulo) (public) | | | | | | | |
| Cultura TV Network | | 0.5 | 1.2 | 1.4 | 1.6 | | 1.3 |
| Others | 34.7 | 34.6 | 26.8 | 17.5 | 19.0 | 4.5 | 25.2 |
| Total Revenues (mil R$)[4] | 5,542 | 8,223 | 10,354 | 12,605 | 13,600 | 14,400 | 18,332 |
| Total Revenues (mil US$) | 5,535 | 4,494 | 5,233 | 5,800 | 7,568 | 7,422 | 10,425 |
| C4 | 46.1 | 60.7 | 59.7 | 66.5 | 74.5 | 90.7 | 73.6 |
| HHI | 1,105 | 1,947 | 2,015 | 2,228 | 3,073 | 4,183 | 2,983 |
| *N* (>1%) | 4 | 6 | 8 | 8 | 7 | 6 | 5 |
| Noam Index | 552 | 688 | 712 | 788 | 1,161 | 1,708 | 1,334 |

*Source:* National Agency of Telecommunications (ANATEL), 2011; InterMeios Project, 2001–2010; CADE, 2004.

[1] In 2003.

[2] In 2007.

[3] In 1999.

[4] This revenue figure is an estimate based on data collected in this study: it is based on average per capita revenues reported for several other countries in this study with similar per capita income.

the inclusion of national quotas as compulsory in the programming. The law establishes a quota of three hours a week of content produced in Brazil—half of it created by companies with no links to traditional broadcasting groups. Also, 30% of the resources of the Audiovisual Sector Fund (FSA), introduced by Act 12.2485/11, will be allocated to productions originating in the north, northeast, and midwest regions. Managed by the National Agency of Cinema (ANCINE), the Fund supervises the interdependent development of the entire audiovisual production chain in Brazil (Table 20.6).

In August 2006, Globopar and DirecTV (US) announced their association in the business of satellite communications operated by SKY Brasil, a property of Rupert Murdoch's News Corp., and DirecTV Brazil. DirecTV acquired the shares of News Corp. and Liberty Media in SKY Brasil. The groups did not report the value of the transaction, but according to the announcement, the DirecTV Group owned approximately 74% of SKY Brasil and Globo owned approximately 26%. This operation completed a move initiated in October 2004 involving News Corp. (US/UK/AUS),

## Table 20-6. Multichannel Video Platforms: Cable MSOs, DBS, IPTV (Market Shares by Revenue), 2000–2010

| | 2000 | 2004 | 2006 | 2008 | 2009 | 2010 |
|---|---|---|---|---|---|---|
| Globopar/America Móvil (NET Cable) (Mexico) | 52.3 | 37.5 | 34.5 | 40.6 | 43.0 | 52.3 |
| Globopar (Liberty Media, US; News Corp., Murdoch Family, US/UK/AUS) (SKY TV) | 19.0 | 12.1 | 12.7 | 15.3 | 14.0 | 19.0 |
| Telefónica (Spain)[1] | | 7.5 | | 4.7 | 4.0 | 3.8 |
| America Móvil (*Claro TV*) (Mexico) | | | | 7.2[2] | 3.0 | 6.4 |
| Embratel TV | | | 6.6[3] | | | |
| Vivax | | 5.2 | Globopar/ America Móvil | | | |
| Telemar Participações S.A. (Oi TV) (Portugal Telecom, Portugal)[4] | | | | | 0.1 | 0.3 |
| Infovias/Cemig *WayBrasil* | 12.8 | 1.0 | 1.0 | Telemar (Oi TV) | | |
| Others | 9.3 | 23.7 | 30.9 | 32.2 | 34.6 | 32.4 |
| Total Revenue (mil R$)[5] | 1,070 | 4,000 | 5,500 | 9,300 | 10,700 | 12,220 |
| Total Revenues (mil US$) | 584 | 2,020 | 2,530 | 5,170 | 5,510 | 6,950 |
| C4 | 84.1 | 62.3 | 54.8 | 67.8 | 64.0 | 67.5 |
| HHI | 3,260 | 1,637 | 1,396 | 1,954 | 2,070 | 2,179 |
| *N* (>1%) | 3 | 5 | 4 | 4 | 4 | 4 |
| Noam Index | 1,882 | 732 | 698 | 977 | 1,035 | 1,090 |

*Source:* National Agency of Telecommunications (ANATEL), 2008; InterMeios Project, 2001–2010.
[1] Ownership of Brazil's TVA network transferred from the Abril Group to Telefónica in 2007.
[2] In 2007.
[3] In 2005.
[4] Portugal Telecom holds a 22.4% stake in Telemar Participações S.A.
[5] This revenue figure is an estimate based on data collected in this study: it is based on average per capita revenues reported for several other countries in this study with similar per capita income.

Globopar (Brazil), Grupo Televisa S.A. (Mexico), and Liberty Media International (US) to join two DTH platforms—DirecTV Latin America and SKY Latin America—into a single platform operating in the main areas of the region. As a result, the DirecTV Group and its affiliate, SKY Mexico (in which the DirecTV Group had a 41% share) started broadcasting to approximately 4 million subscribers in Latin America by that time.

In the last quarter of 2011, the cable provider NET, wireless provider Claro, and the long-distance services provider Embratel (all now controlled by America Móvil[29]) announced a

29. In 2000, Slim divided its businesses in fixed and mobile, creating América Móvil and Telmex. For nearly four years, Telmex spun off in Telmex Mexico and Telmex Internacional. The latter having a stake in Embratel and Net, and data assets and pay TV in Latin America. In 2010, America Movil acquired Carso, the holding company that controlled Telmex Mexico and Telmex International.

commercial venture that enabled the three companies to create packages of products and services. The companies started integrating their networks and prepared the 2012 launch of cable, fixed, and mobile services. The Law of Conditional Access Services (2011) for the pay TV sector also benefited the four dominant groups operating in Brazil: Spain's Telefónica, Mexico's America Móvil, Telecom Italia (Italy), and Brazil's Telemar with Portugal Telecom.

### Film

International conglomerates dominate the Brazilian film sector, especially the US majors in Hollywood. Today, the market includes Sony, the Walt Disney Company, Viacom, Time Warner, News Corporation, and Comcast/GE's Universal. In 2010, Sony and Disney held 10.7% of the market share each; Viacom (through its subsidiary Paramount), 19.2%; 20th Century Fox (News Corp.), 11.1%; and Time Warner (Warner Bros.), 12.6% of the film distribution market. The four largest distributors control 53.6% of the very lucrative Brazilian market: among Latin American countries, Brazil stands as the second largest market in revenues, following Mexico's, and it was also the second largest producer of feature films in the region in 2008, with 82 local films released, behind Argentina (85) and ahead of Mexico (70) (Table 20.7).[30]

## TELECOMMUNICATIONS MEDIA

### Wireline Telecom

In 1998, the Brazilian government privatized the national telecom sector. State wireline and wireless companies were sold in a public auction by a sum of US$15.3 billion (R$18.3 billion). Today, the Telemar Group dominates the wireline market with 42.1% of the market share.[31] The position of the Telemar group in 2010 was a result of previous mergers, especially the one managed by Brazil's Opportunity Group. Together with Telecom Italia, it acquired the Tele Centro Sul carrier for US$1.5 billion in 1998 (R$1.8 billion) during the privatization process of Telebras. Soon after that, the Tele Centro Sul was transformed into Brasil Telecom, which was acquired by Telemar in 2008. Now operating under the brand Oi, 26% of its shares were acquired in 2010 by Portugal Telecom.

Telefónica (Spain) ranks second in the sector, with 27% of the market share, and America Móvil ranks third, with a 17% share through Embratel's 17%. The United States' MCI (formerly WorldCom) was in control of the main Brazilian telecom company until July 2004, when Telmex International (now America Móvil) acquired it in a US$204 million deal (R$400 million). Vivendi entered the Brazilian wireline market in 2009 by acquiring the Brazilian provider GVT after a long dispute with Telefónica. Sercomtel, a joint venture of the Londrina City Hall (a medium size city in the southern state of Paraná) and the local electricity company Copel, is one of the very few regional groups to stand among the "giants" of media conglomerates. Intelig, a joint venture of Telecom Italia and Docas Investments of Brazil, and was fully acquired by Telecom Italia in 2010, completes the wireline telecommunication industries operating in the country (Table 20.8).

### Wireless Telecom

Telefónica (Spain) ranks first in the Brazilian wireless market. After almost four years of combined business, Portugal Telecom sold its share of 30% in Vivo and simultaneously announced the acquisition of 25.6% of Oi, Telemar's telecom division, until then the only Brazilian telecommunications group in the sector. In three years, all movements within the sector had come to concentrate among the same established players: Telemar's acquisition of the wireless carrier Brasil Telecom in 2008, Vivendi's

30. Marché du film. *FOCUS 2012—World Film Market Trends.* Cannes: Marché du film, 2011. July 10, 2011. <http://www.obs.coe.int/oea_publ/market/focus.html>.

31. The biggest firm in the wireline telecommunication sector, controlled by five groups of diversified fields gathered under the Telemar umbrella.

Table 20-7. Film Production/Distribution (Market Shares by Box Office %), 1996–2010

| | 1996 | 2000 | 2002 | 2005 | 2007 | 2008 | 2009 | 2010[1] |
|---|---|---|---|---|---|---|---|---|
| Sony (US) | | | | 16.2 | | | | 10.7 |
| Columbia | 19.4 | 21.2 | 23.6 | | | | | |
| Walt Disney Studios (US) | 19.2 | 10.8 | 11.1 | 14.0 | 9.9 | 9.3 | 9.2 | 10.7 |
| Viacom (Paramount) | 9.6 | 2.3 | 9.8 | 13.3 | 18.5 | 7.1 | 19.6 | 19.2 |
| Comcast/GE (Universal/ UIP) (US) | 9.6 | 7.6 | 9.8 | 6.8 | 9.1 | 5.8 | | |
| Twentieth Century Fox (News Corp., Murdoch Family, US/UK/AUS) | 11.1 | 13.2 | 18.4 | 14.8 | 13.5 | 26.6 | 18.7 | 11.1 |
| Warner Brothers (Time Warner) (US) | 12.6 | 25.1 | 20.6 | 14.6 | 13.4 | 11.7 | 12.9 | 12.6 |
| Paris Films | 2.5 | | | 1.5 | 4.5 | 7.3 | 6.7 | 2.5 |
| PlayArte | 3.0 | 3.9 | 5.7 | 2.1 | 5.6 | 4.8 | 3.3 | 3.0 |
| Imagem Filmes | | | 0.1 | 0.1 | 7.2 | 6.3 | 3.2 | |
| Rio Filmes S.A. | 0.5 | 0.2 | 0.1 | 1.3 | 1.4 | 1.7 | 1.9 | 0.5 |
| Others | 10.7 | 13.3 | 8.2 | 15.7 | 6.5 | 3.6 | 12.3 | 10.7 |
| Total Revenue (mil R$)[2] | 183.0 | 150.4 | 282.5 | 401.0 | 423.4 | 557.7 | 756.9 | 183.0 |
| Total Revenue (mil US$) | 72.0 | 82.1 | 101.0 | 264.1 | 217.5 | 310.3 | 390.1 | 104.1 |
| C4 | 62.3 | 70.3 | 73.7 | 59.2 | 55.8 | 54.9 | 60.4 | 53.6 |
| HHI | 1,227 | 1,449 | 1,668 | 1,109 | 1,098 | 1,134 | 1,055 | 895 |
| *N* (>1%) | 8 | 7 | 7 | 9 | 9 | 9 | 8 | 7 |
| Noam Index | 434 | 548 | 590 | 553 | 351 | 378 | 373 | 338 |

*Source:* National Agency of Cinema (ANCINE); InterMeios Project, 2001–2010; Focus, 2009, 2011; Filme B.

[1] For 2010, it was only possible to find the aggregated totals for Sony-Columbia/Disney and Paramount/Universal-UIP. We estimate that an equal share is held between Sony-Colombia and Disney, and also for Paramount and Universal-UIP.

[2] This revenue figure is an estimate based on data collected in this study: it is based on average per capita revenues reported for several other countries in this study with similar per capita income.

acquisition of GVT in 2009, and Portugal Telecom's 2010 entrance into the Telemar Group. Telecom Italia, second in the market, made just one move in the wireless market in the past decade: together with Opportunity Group and Citigroup (US), it held Brasil Telecom from 2004 to 2008.

America Móvil, controller of Claro, on the contrary, entered the market in 2002 as Telmex, with the strategy of acquiring small regional companies—at first ATL, then Claro Digital, Americel, Tess, BSE, and BCP (in a dispute with Telefónica and Embratel). Those steps created the basis for its most valuable asset, the Latin American division of AT&T, which wanted to withdraw from the market. That acquisition, valued at US$68.3 million (R$207 million) in 2003, opened markets for America Móvil in Colombia, Chile, Peru, Brazil, and Argentina. In 2004, Telefónica launched a counterattack by purchasing 100% of Bell South's Latin American operations.[32] With that move, Telefónica entered Venezuela, Colombia, Ecuador, and Uruguay, thus

32. BellSouth Brazil held until 2003 a participation in BCP Northeast, one of the six companies now part of the Telmex holding, and until 2004 it had shares of TCO (Tele Centro Oeste Celular), now part of Telefónica.

Table 20-8. Wireline Telecom (Market Shares by Revenue), 2006–2012

| | 2006 | 2007 | 2008 | 2009 | 2010 | 2012 |
|---|---|---|---|---|---|---|
| Telemar Participações S.A. (Telemar/Oi) (Portugal Telecom, Portugal) | 58.9 | 56.6 | 53.7 | 51.51 | 47.6 | 42.1 |
| Brasil Telecom | | | Telemar Participações S.A. | | | |
| Telecom Italia (Italy)/Opportunity Group[1] | 0.1 | 0.1 | 0.2 | 0.1 | 0.1 | 0.1 |
| Telefónica (Spain) | 31.2 | 30.4 | 28.4 | 27.1 | 26.9 | 23.8 |
| America Móvil (Mexico)[2] | 5.7 | 8.5 | 12.3 | 14.4 | 16.8 | 27.9 |
| Embratel | | | | | | |
| Bell South (US) | | | | | | |
| Vivendi (France) | | | | 3.5 | 5.0 | 8.2 |
| GVT (Global Village Telecom) | 1.7 | 2.0 | 2.5 | Vivendi | | |
| Algar S.A. | 1.7 | 2.0 | 2.5 | 3.5 | 5.0 | 2.0 |
| CTBC | | | | | | |
| Sprint/France Telecom | 0.1 | 0.1 | 0.7 | | | |
| Intelig Telecom (Docas Investments/ Telecom Italia, Italy) | | | | 1.1 | 1.5 | 1.1 |
| Sercomtel | 0.2 | 0.2 | 0.2 | *0.2* | *0.4* | *0.4* |
| Others | 0.5 | 0.5 | 0.5 | 0.5 | 0.3 | 0.5 |
| Total Revenue (bil R$)[3] | 70.0 | 70.0 | 76.0 | 90.0 | 91.8 | |
| Total Revenue (bil US$) | 32.2 | 36.0 | 42.3 | 46.4 | 52.2 | |
| C4 | 97.5 | 97.5 | 96.9 | 96.5 | 96.2 | 102 |
| HHI | 4,481 | 4,208 | 3,855 | 3,620 | 3,321 | 3,190 |
| *N* (>1%) | 5 | 5 | 5 | 6 | 6 | 6 |
| Noam Index | 2,004 | 1,882 | 1,457 | 1,368 | 1,255 | 1,302 |

*Source:* ANATEL 2011 Annual Report—National Agency of Telecommunications; Prêmio Melhores do Brasil, *Brasil Econômico*, June 2012; Anuário Telecom 2010, Plano Editorial, August 2011; Atlas Brasileiro de Telecomunicações 2011, Converge Comunicações, May 2012.

[1] Telefónica has a controlling 9.47% stake in Telecom Italia through the Telco consortium as of 2007. Telecom Italia data by communication from company and Mintel

[2] In 2004, Mexico's Telmex acquired Embratel, formerly part of WorldCom.

[3] This revenue figure is an estimate based on data collected in this study: it is based on average per capita revenues reported for several other countries in this study with similar per capita income.

consolidating its position in South and Central America.[33] In 2007, Telefónica also paid $3.12 billion for a 42.3% share of Telco, the company that was born after the sale of the Pirelli Group's Telco stake (Table 20.9).

The wireless industry, one of the most profitable sectors of local telecommunications, comprised 242.2 million users at the end of 2011, representing a growth of 19.4% over the previous year.

33. "Telmex and Telefonica Step on the Gas Pedal in Latin America." Universia Knowledge@Wharton, Oct. 6, 2004. Jan. 11, 2011. <http://www.wharton.universia.net/index.cfm?fa=viewArticle&id=845&language=english>.

Table 20-9. Wireless Telecom (Market Shares by Revenue), 1999–2013

| | 1999 | 2000 | 2002 | 2004 | 2005 | 2006 | 2007 | 2008 | 2009 | 2010 | 2013 |
|---|---|---|---|---|---|---|---|---|---|---|---|
| Telefónica (Spain) | 45.8 | 20.8 | 35.3 | 27.5 | 30.4 | 30.8 | 38.8 | 32.9 | 30.9 | 29.5 | 28.6 |
| Vivo[1] | | | | | | | | | | | |
| Telemig Celular | 9.9 | 5.8 | 6.4 | 4.3 | 5.0 | 5.0 | [2] | | | | |
| Telecom Italia (Italy)[3] | 23.2 | 16.2 | 12.3 | 9.6 | 9.2 | 28.5 | 31.3 | 27.0 | 24.7 | 25.8 | 27.2 |
| TIM Participações (Italy) | | | | | | | | | | | |
| America Móvil (Mexico) | | | | 18.8 | 20.6 | 23.7 | 25.2 | 24.0 | 22.7 | 25.5 | 25.1 |
| Claro | | | | | | | | | | | |
| Bell South (US) | 23.9 | 10.2 | 15.3 | | | | | | | | |
| BCP[4] | | | | | | | | | | | |
| Telemar Participações S.A. (Portugal Telecom, Portugal) | | | 7.9 | 7.9 | 8.7 | 8.2 | 9.4 | 14.8 | 17.1 | 18.9 | 18.6 |
| Oi Móvel | | | | | | | | | | | |
| Amazonia Celular | | 7.2 | 3.6 | 2.7 | 1.9 | 1.4 | 1.2 | [5] | [6] | | |
| Brasil Telecom Móvel | | | | | | | | [7] | | | |
| Telecom Italia (Italy)/opportunity Group | | | | 0.3 | 2.2 | 3.4 | 4.3 | 4.2 | | | |
| NII Holdings Inc. (US) | | 4.6 | 2.2 | 1.6 | 1.3 | 1.6 | 2 | 2.7 | 3.3 | 0.6 | |
| Nextel Comunicacões Ltda. | | | | | | | | | | | |

| | | | | | | | | | | | |
|---|---|---|---|---|---|---|---|---|---|---|---|
| Algar Telecom | 1.9 | 1.0 | 1.1 | 0.5 | 0.7 | 0.6 | 0.7 | 0.6 | 0.4 | 0.3 | 0.4 |
| CTBC Celular | | | | | | | | | | | |
| Sercomtel Celular | 0.0 | 0.3 | 0.3 | 0.2 | 0.2 | 0.1 | 0.1 | 0.3 | 0.1 | 0.04 | 0.1 |
| Total Revenue (bil R$)[8] | 12.0 | 13.0 | 17.0 | 27.0 | 32.0 | 36.0 | 39.0 | 48.0 | 53.0 | 57.0 | 102 |
| Total Revenue (bil US$) | 6.7 | 7.1 | 6.1 | 13.6 | 21.1 | 16.6 | 20.0 | 26.7 | 27.3 | 32.4 | 42.8 |
| C4 | 102.8 | 53.0 | 70.8 | 63.8 | 68.9 | 91.1 | 104.8 | 98.7 | 95.4 | 99.7 | 99.5 |
| HHI | 3,365 | 866 | 1,748 | 1,288 | 1,542 | 2,427 | 3,233 | 2,632 | 2,384 | 2,543 | 2,534 |
| *N* (>1%) | 4 | 6 | 6 | 5 | 6 | 6 | 6 | 6 | 5 | 4 | 4 |
| Noam Index | 1,272 | 306 | 583 | 408 | 488 | 768 | 1,143 | 931 | 901 | 961 | 1,267 |

*Source:* ANATEL 2011 Annual Report, National Agency of Telecommunications; Prêmio Melhores do Brasil, *Brasil Econômico*, June 2012; Anuário Telecom 2010, Plano Editorial, August 2011; Atlas Brasileiro de Telecomunicações 2011, Converge Comunicações, May 2012. *2013 updates provided by editors.*

[1] In 2010, Telefónica bought out Portugal Telecom's stake in Vivo's wireless services, becoming the sole controller.

[2] Telefónica/Portugal Telecom (Portugal).

[3] Telefonica controlled Telecom Italia from 2007-2014, when they separated again.

[4] Purchased by America Móvil in 2003.

[5] Telefónica/ Portugal Telecom (Portugal).

[6] Telemar Participações S.A.

[7] Telemar Participações S.A.

[8] This revenue figure is an estimate based on data collected in this study: it is based on average per capita revenues reported for several other countries in this study with similar per capita income.

## INTERNET MEDIA

### Internet Service Providers (ISP)

The early years of the commercial Internet in Brazil began with the launching of the BBS Mandic ISP in 1993 (Mandic Internet, by 1995), the first ISP to have an investment group—Garantia Partners Investimentos—as associate. In 1998, Mandic Internet's revenue stood around US$10.8 million (R$13 million), and in 1999 the Argentinean group El Sitio acquired it. A few ISPs initiated their commercial activities in 1996, such as UOL (Folhapar) and BOL (Abril Group), both properties of print publishers. Both Folhapar and the Abril Group soon decided to form a joint venture, which lasted from 1996 to 2005, when Portugal Telecom became a partner in UOL. Aleksandar Mandic, the founder and first owner of Mandic, was also part of the founding team of iG, which became the largest Brazilian free Internet service provider in 2000. In those early years, foreign ISPs began entering the market in strength: the Canadian Primus Telecommunications Group acquired 51% of Matrix in 1998, and PSI Net (US) acquired Openlink in 1999.

The year 1999 witnessed many moves in the ISP sector with the arrival of bigger foreign companies in Brazil, who were attracted by the telecom privatization. Many regional small providers from the southern and southeastern regions also launched ISPs, such as Telefónica's Terra Networks Brasil, which was based in the infrastructure of two small Internet providers, ZAZ and Nutecnet. American Online (AOL) also ventured into the sector with a joint venture business backed by AOL USA, the Cisneros Group from Venezuela, and Brazil's Itaú Bank. Its poor performance in the market facilitated its acquisition by Telefónica in 2006.

In general terms, the most recent transactions were those of Brasil Telecom, acquired by Telemar in 2008, the acquisition of GVT by Vivendi in 2009, the acquisition of 22.4% of Oi Internet by Portugal Telecom; the entrance of the Hypermarcas (a Brazilian medicinal, cosmetics, food and cleaning service company) in UOL-Folhapar with a 29% stake and (in 2010) the Net Virtua transfer of 91.7% of its shares to Telmex. In 2010, there were two major moves in the sector: Portugal Telecom left the Folha/UOL Group and acquired 22.4% of Telemar Participações and Hypermarcas acquired 29% of the Folha/UOL Group (Table 20.10).

The last two years have demonstrated great growth of Internet users nationwide. Of the total accesses, about 50% were related to e-commerce, while entertainment (audiovisual and film) accounted for more than 60% of households requests.[34]

### Search Engines

As of 2012, Google is the leading search engine in Brazil, holding 90.2% of the market share, followed distantly by Microsoft's Bing, with 1.3%, and the Yahoo/Cadê search system, which a market share less than 1%.[35] Google Brasil established its first regional office in 2005 in São Paulo. One year later, it inaugurated in Belo Horizonte, the state capital of Minas Gerais, the Center for Research and Development in Brazil to train personnel. The center was established after Google acquired Akwan, a company owned by local Internet R&D groups.

Two national search engines, both created in 1995, attempted to wrest dominance from Google. Inspired by Yahoo, Cadê—a Brazilian Portuguese colloquialism meaning "where is it?"—used similar interface and categorization hierarchies, but also invested in subsections for more specific searches. In 1999, starMedia (Mexico), which entered the national market in its expansion to Latin American countries, acquired Cadê, but the venture did not last long. When the "dot.com bubble" collapsed in 2000, Cadê's market share fell fast and Yahoo Brasil finally acquired it in 2002. AltaVista was

34. IBOPE Nielsen Online. "Mais internautas em casa." *IBOPE*, May 16, 2011. Dec. 12, 2011. <http://www.ibope.com.br/pt-br/noticias/Paginas/Mais%20internautas%20em%20casa.aspx>.

35. "Segundo Pesquisa, Google Brasil é o pior buscador e Bing, o melhor." *IDGNOW!* Mar. 15, 2012—http://idgnow.uol.com.br/internet/2012/02/16/segundo-pesquisa-google-brasil-e-pior-buscador-e-bing-o-melhor/

Table 20-10. Internet Service Providers (Market Shares by Revenue), 2000–2012

| | 2000 | 2004 | 2007 | 2008 | 2009 | 2010 | 2012 |
|---|---|---|---|---|---|---|---|
| Globopar/Telmex (Mexico)[1] | 1.6 | 7.8 | 22.9 | 20.5 | 21.2 | 22.4 | 25.9 |
| NET Virtua | Globo/ Telecom Italia | | Telmex | | | Telmex | |
| Folhapar S.A. | 30.8 | 32.4 | 24.6 | 18.2 | 16.2 | | |
| Universo Online (UOL) | Folha de S. Paulo/Abril Group | Folha-UOL S.A./Portugal Telecom (Portugal) | | | | Folha-UOL S.A./ Hypermarcas | |
| Telefónica (Spain) | 19.1 | 29.2 | 26.4 | 15.1 | 14.2 | 12.5 | 27.7 |
| Terra Networks Brasil S.A. | | | | | | | |
| RBS Group | | | | | | | |
| ZAZ/Nutecnet | Telefónica | | | | | | |
| AOL Latin America Inc. | 2.0 | | | | | | |
| AOL Brasil Ltda. | | | Telefónica | | | | |
| Telemar Participações S.A. | | | 7.2 | 15.1 | 15.8 | 15.4 | 30.3 |
| Oi Internet/Oi Velox | | | | | | Portugal Telecom (Portugal) | |
| Brasil Telecom | 6.75 | 4.8 | 4.4 | | | | |
| BRT Serviços de | | | | Telemar Part. S.A. | | | |
| Internet Group Ltd. | 8.2 | 16.4 | | | | | |
| iG | | Brasil Telecom | | | | | |

*continued*

Table 20-10. *continued*

| | 2000 | 2004 | 2007 | 2008 | 2009 | 2010 | 2012 |
|---|---|---|---|---|---|---|---|
| Vivendi (France) | | | | | 9.7 | 10.2 | 10.6 |
| GVT S.A. | | 7.5 | 10.1 | 8.4 | Vivendi | | |
| Pop | | | | | | | |
| Others | 31.6 | 1.9 | 4.4 | 22.7 | 22.9 | 23.8 | 7.0 |
| Total Revenue (mil R$)[2] | | 1,400 | 2,900 | 4,400 | 6,000 | 7,100 | |
| Total Revenue (mil US$) | | 708 | 1,489 | 2,450 | 3,090 | 4,040 | |
| C4 | 64.9 | 85.8 | 84.0 | 68.9 | 67.4 | 66.0 | 94.5 |
| HHI | 1,435 | 2,310 | 1,998 | 1,277 | 1,256 | 1,246 | 2.469 |
| *N* (>1%) | 6 | 6 | 5 | 5 | 5 | 4 | 4 |
| Noam Index | 586 | 943 | 816 | 571 | 562 | 557 | 1,235 |

*Source:* The Brazilian Association of Internet Providers (ABRANET), Ibope/NetRatings 2010, <http://www.abranet.org.br/index.php/ibopenetratings/category/16-ibopenetratings-dados-de-2010>; *ICT Households and Enterprises 2011,* CETIC.br <http://op.ceptro.br/cgi-bin/cetic/tic-domicilios-e-empresas-2011.pdf>; *Acesso à Internet e posse de telefone móvel celular para uso pessoal 2005,* The Brazilian Institute of Geography and Statistics (IBGE) 2006, <www.ibge.gov.br/home/estatistica/populacao/acessoainternet/comentarios.pdf>; *Anuário Telecom 2011,* Plano Editorial <www.anuariotelecom.com.br/anutel/2011/servicos.shtml>; *Atlas Brasileiro de Telecomunicações 2011,* Converge Comunicações, May 2012.

[1] Telmex's stake stood at 34% in 2005, 49% in 2007, and 91.9% in 2010.

[2] This revenue figure is an estimate based on data collected in this study: it is based on average per capita revenues reported for several other countries in this study with similar per capita income.

the other national search engine, and it allowed advanced research in a multimedia search for photos, music, and videos, but it has also only held a very small market share, less than 1% (Table 20.11).

### Online News

Four large groups currently operating in the telecommunications and media sectors in Brazil concentrate the delivery of online news services. Altogether, they control 70% of the market: Folhapar, through its news portal *UOL*, accounts for 31.9% of the market; *Terra Notícias* (Telefónica) accounts for 21.2%; *Globo.com* (Globopar), for 13.3%; and *iG* (Telemar), for 12% of the market share.

Foreign participation in the sector is limited mainly to Telefónica's *Terra Notícias*; Portugal's Ongoing Group, which maintains online sites of the daily newspapers *O Dia* and *Brasil Econômico*; and the South African Naspers' stake in *Abril.com*. A former president of the portal recalls that *iG* had a controversial start offering free access to the Internet, "but it created the largest portfolio of electronic addresses of Brazil and then turned itself into a major portal for information and entertainment" (Table 20.12).[36]

## CONCLUSION

The demand for broadband represents a significant challenge to Brazilian providers. Broadband integration implicates the complex online web of networking, products, and markets. Today, Brazil's access to sites in Colombia, Ecuador, and Peru, for instance, is done via the United States, through a submarine cable, and returns through another cable via the Pacific. The countries of UNASUR (Union of South American Nations) ratified a plan in 2012 to integrate the broadband networks of neighboring countries. The optical ring at the center of the plan will allow online traffic to stay in South America, with the added economy of time and cost. In the case of Brazil, the optical ring will also reduce the percentage of the Internet data routed to the United States (35% of Brazilian internal traffic).[37]

### Table 20-11. Internet Search Engines (Market Shares by Monthly Search Volume), 2008–2012

| | 2008 | 2009 | 2010 | 2012 |
|---|---|---|---|---|
| Google (US) | 97.8 | 97.4 | 97.5 | 96.2 |
| Microsoft Bing (US) | 0.0 | 0.8 | 1.8 | 1.3 |
| Yahoo/ Cadê (US) | 1.0 | 1.5 | 0.7 | 0.5 |
| Altavista | 0.9 | 0.2 | 0.1 | |
| Others | 0.3 | 0.1 | 0.0 | 1.6 |
| Monthly Search Volume (mil) | 31.5 | 34.0 | 39.0 | |
| C4 | 99.7 | 99.9 | 100.0 | 98 |
| HHI | 9,560 | 9,480 | 9,506 | 9,256 |
| *N* (>1%) | 2 | 2 | 2 | 2 |
| Noam Index | 5,520 | 5,473 | 5,488 | 6,545 |

*Source:* Stat Counter, Global Stats, "Top 5 Search Engine in Brazil, Dec 2008 to Dec 2010" <http://gs.statcounter.com/#search_engine-BR-monthly-200812-201012>; *Anuário Telecom 2011*, Plano Editorial <www.anuariotelecom.com.br/anutel/2011/servicos.shtml>; "Market share de buscadores, sistemas operacionais e navegadores," Webartz, December 2010 <http://www.webartz.com.br/tecnologia-da-informacao/market-share-de-buscadores-sistemas-operacionais-e-navegadores/>.

36. Costa, Rosa. "Conselho de Comunicação vai ser revigorado." *O Estado de S. Paulo*, Nov. 9, 2010. Dec. 28, 2010. <http://www.observatoriodaimprensa.com.br/artigos.asp?cod=615IPB004>.

37. Santanna, R. "Anel óptico dará à internet brasileira mais autonomia." *Folha de S. Paulo*, Mar. 9, 2012. n.d.

Table 20-12. Online News Media (Market Shares by Revenue), 2005–2010

| | 2005 | 2006 | 2007 | 2008 | 2009 | 2010 |
|---|---|---|---|---|---|---|
| Folhapar S.A.[1] | 29.5 | 25.7 | 43.0 | 44.4 | 44.8 | 31.9 |
| *Folha Online* | | | | | | 5.0 |
| *UOL* | | | | | | 26.9 |
| Telefónica (Spain) | 25.3 | 35.1 | 37.0 | 30.8 | 35.6 | 21.2 |
| *Terra Notícias* | | | | | | |
| Globopar S.A. | | | | | | 13.3 |
| *Globo.com* | | | | | | |
| Infoglobo S.A. | | | | | | |
| *O Globo Online* | | | | | | |
| Telemar Participações S.A./Portugal Telecom (Portugal) | | | | | 11.3 | 12.1 |
| *iG* | | | | | | |
| O Estado Group | | | | | 8.56 | 4.9 |
| *Estadao.com.br* | | | | | | |
| *AE, Agência Estado* | | | | | | |
| Abril Group (Civita Family/Naspers, South Africa) | | | | | | 6.4 |
| *Abril.com* | | | | | | |
| Universal Church Group | | | | | | 4.9 |
| *R7.com* | | | | | | |
| Thomson Reuters (UK) | | | | | | |
| *Reuters Brasil* | | | | | | |

| | | | | | | |
|---|---|---|---|---|---|---|
| Arca Group | | | | | | Ongoing Group (Portugal) |
| *O Dia Online* | | | | | | |
| Ongoing Group (Portugal) | | | | | | |
| *O Dia Online* | | | | | | |
| *Brasil Econômico Online* | | | | | | |
| *Jornal do Brasil Online* | | | | | | |
| Others | 45.2 | 39.2 | 20.0 | 24.8 | 0 | 25.4 |
| Total Revenues (mil R$)[2] | 531.2 | 722.0 | 1,207 | 1,300 | 1,600 | 3,040 |
| Total Revenues (mil US$) | 349.9 | 332.2 | 620.0 | 723.4 | 824.7 | 1,729 |
| C4 | 89.3 | 58.8 | 80.0 | 75.1 | 100.0 | 70.0 |
| HHI | 2,699 | 1,791 | 3,215 | 2,916 | 3,472 | 1,875 |
| *N* (>1%) | 2 | 2 | 2 | 2 | 4 | 7 |
| Noam Index | 1,558 | 1,266 | 2,273 | 2,062 | 1,736 | 625 |

*Source:* InterMeios Project, 2010, www.projetointermeios.com.br/relatoriosInvestimento.aspx; "Most-visited News Websites in Brazil," *ComScore Data Mine*, Nov. 2010, <http://www.comscoredatamine.com/2010/11/most-visited-news-websites-in-brazil>; Top Sites in Brazil, Alexa <http://www.alexa.com/topsites/countries/BR>; "Pesquisa Anuário Meio&Mensagem 2010," IVC Brasil, Auditoria Web, Feb., 2011, <http://www.ivcbrasil.org.br/conteudos/pesquisas_estudos/Pesquisa%20Anu%C3%A1rio%20MM.pdf>.

[1] Data regarding the Folhapar Group are included here in the totals concerning *Folha Online* (the online edition of the *Folha de S. Paulo* daily).

[2] This revenue figure is an estimate based on data collected in this study: it is based on average per capita revenues reported for several other countries in this study with similar per capita income.

Internally, the Brazilian government created the National Broadband Program (PNBL) in 2010 to "promote digital inclusion, expanding of e-gov services, and also to increase the country's competitiveness and technological autonomy."[38] The program was officially inaugurated by decree No. 7175 on May 12, 2010, and aims to greatly expand broadband access. 85% of Brazilian wireline connections are in cities that meet 53% of the population, and account for 71% of GDP. The cities of lower income represent only 2% of broadband connections.[39]

The new regulations for audiovisual and telecom services boost the expansion of the paid TV market, since the rules facilitate the licensing and authorize the services' expansion into areas not currently served. Analysts believe that, in this circumstance, Telefónica should see a rapid expansion of its service because it is the largest mobile operator in the country. America Móvil—a major player in the wireline and wireless industry as the controller of the country's main cable TV operator (and satellite transmitters)—also benefited from the process of reinforcing connections within the group companies. Data released by the International Telecommunication Union (ITU) in 2011 pointed out that among the five largest mobile phone markets in the world, four were BRICS (Brazil-Russia-India-China-South Africa) countries: China (900 million), India (812 million), Russia (220 million), and (in fifth place) Brazil (211 million).[40]

In the context of the multichannel industry, the most significant growth is expected in the satellite and cable services. Data released in 2012 by the National Telecommunications Agency (ANATEL) indicated that the pay TV market grew 2.4% in 2011—with services available in about 13.3 million households. Also according to ANATEL, pay TV services reached 43 million views nationwide, and in January 2012 alone, the sector added 309,800 new subscribers.[41] The share of satellite services (DTH) reached 55.5% of the subscriber base, while the services provided via cable accounted for 42.7%. Adding up to the new regulation for the cable TV industry, ANATEL also reopened the issuing of licenses for new cable TV services, which had been closed since 2003.[42]

Overall, at the beginning of the second decade of the 21st century, Brazil was experiencing a special moment of growth in media and telecom services. The difficulties in growth are mainly attributable to outdated technology (when compared to infrastructure in other highly industrialized nations) and the slow process of improving regulation, a battle that has been ongoing for 50 years now, which has kept Brazil tethered to an archaic legislation that does not correspond to the nation's actual stage of media development.

Evaluating these shortcomings precisely is difficult because although most companies are responsible for publishing their balance sheets each year, these publications are not always easy to find. These publications include detailed reports and quantitative/qualitative documents with added numbers per sector produced by consultancy firms that charge companies for the material if it becomes public. Primary sources—official documents, legal texts, technical reports, annual balance sheets, and internal documents—constitute rare official references. Market projections and audience records published by independent institutions, like the Grupo de Mídia São Paulo and Projeto Inter-Meios, are more readily available. In Brazil, there is still a lack of systematic organization of information that could provide media producers, consumers, and regulators with accurate figures concerning corporations' assessments

38. Programa Nacional de Banda Larga. "Brasil Conectado: Programa Nacional de Banda Larga." Secretaria-Executiva do Comitê Gestor do Programa de Inclusão Digital, 2010. Jan. 14, 2011. <http://www4.planalto.gov.br/brasilconectado/forum-brasil-conectado/documentos/3o-fbc/documento-base-do-programa-nacional-de-banda-larga>.

39. Currently, around 700 cities have broadband Internet service of with a speed of 1Mbps offered at a monthly rate of US$10 (R$20). The wireline telephony companies Oi, Telefónica, Algar Telecom and Sercomtel provide this service.

40. The US ranked 3rd (312 million). ITU. *Trends in Telecommunication Reform 2010–2011: Enabling Tomorrow's Digital World.* 11th ed. Geneva: ITU, 2011. Dec. 26, 2012. <http://www.itu.int/dms_pub/itu-d/opb/reg/D-REG-TTR.12-2010-SUM-PDF-E.pdf>.

41. Reuters. "TV paga atinge 13 milhões de assinantes em janeiro." *Veja*, Feb. 28, 2012. Feb. 7, 2012 at <http://veja.abril.com.br/noticia/economia/tv-paga-atinge-13-milhoes-de-assinantes-em-janeiro>.

42. Tavares, M. "Anatel aprova regulamentação para nova lei de TV por assinatura." *O Globo*, Mar. 23, 2012.

of their business interests. This would improve the quality and relevance of wide-ranging public policies. The collection of information followed by a comparative reading of the industries could assist in such an important task.

## Bibliography

### 1. Books and Articles


Afonso, R. "Com oferta, Telmex antecipa plano de assumer controle da NET." *Computerworld*, Aug. 5, 2010. Nov. 14, 2010. <http://computerworld.uol.com.br/telecom/2010/08/05/com-opa-telmex-antecipa-estrategia-de-assumir-controle-da-net/>.

Bolaño, C. R. S. *Qual a lógica das políticas de comunicação no Brasil*? São Paulo: Paulus, 2007.

Brittos, V. C. and Haussen, D. F., eds. *Economia política, comunicação e cultura*. Porto Alegre: EdiPUCRS, 2007.

Cardoso, F. H. and Motta, Sergio. "LEI Nº 9.612, DE 19 DE FEVEREIRO DE 1998: Institui o Serviço de Radiodifusão Comunitária e dá outras providências." Presidência da República Casa Civil Subchefia para Assuntos Jurídicos, Feb. 19, 1998. Jan. 15, 2011. <http://www.planalto.gov.br/ccivil_03/Leis/L9612.htm>.

Costa, C. T. "Criadores e ex-presidentes do iG comentam 10 anos do portal." *iG 10 Anos*, Jan. 29, 2010. Mar. 25, 2010. <http://aniversario10anos.ig.com.br/criadores+ex+presidentes+comentam+aniversario+de+10+anos+do+portal.html>.

Costa, Rosa. "Conselho de Comunicação vai ser revigorado." *O Estado de S. Paulo*, Nov. 9, 2010. Dec. 28, 2010. <http://www.observatoriodaimprensa.com.br/artigos.asp?cod=615IPB004>.

Costa, L. M. "Para onde vai a Editora Abril?" *Observatório da Imprensa* 659, Sep. 14, 2011. Jan. 23, 2012. <http://www.observatoriodaimprensa.com.br/news/view/para-onde-vai-a-editora-abril>.

Cruz, R. "Quase 90% das cidades brasileiras não têm competição na telefonia fixa." *O Estado de S. Paulo*, Feb. 10, 2011.

Damasco, C., Rabello J. B., and Gandour, R. "Governo vai propor concessão única para todas as mídias." *O Estado de S. Paulo*, Jan. 27, 2011, page A4.

Dantas, J. G. "O comportamento dos telespectadores diante da programação televisiva neopentecostal." *Protestantismo em Revista* 16 (May/Aug., 2008): 7–24.

Dantas, M. *A lógica do capital informação*, 2nd ed. Rio de Janeiro: Contraponto, 2002.

Freitas, G. "Censo reavalia dimensões do mercado editorial brasileiro." *O Globo*, Aug. 16, 2011. Jan. 27, 2012 <http://oglobo.globo.com/blogs/prosa/posts/2011/08/16/censo-reavalia-dimensoes-do-mercado-editorial-brasileiro-399149.asp>.

Gindre, Gustavo. "O fracasso do Grupo Abril." *Observatório da Imprensa*, 686, Mar. 15, 2012. May 25, 2012. <http://www.observatoriodaimprensa.com.br/news/view/_ed686_o_fracasso_do_grupo_abril>.

Laudratti, L. "Convergência tecnológica." *Conjuntura Econômica* 64, nº 9 (2011): 18–26.

Lee, P. and Stewart, D. *Technology, Media & Telecommunications Prediction*. London: Deloitte Touche Tohmatsu Limited, 2011. Jun. 10, 2011. <http://www.deloitte.com/view/en_GX/global/industries/technology-media-telecommunications/tmt-predictions-2011/index.htm>.

Lima, V. A. de. *Mídia: teoria e política*. São Paulo: Editora Fundação Perseu Abramo, 2001.

Marini, A. R. "Cresce o numero de políticos donos de meios de comunicação." Fórum Nacional para a democratização da Comunicação (FNDC), 2008. Nov. 14, 2010. <http://www.fndc.org.br/internas.php?p=noticias&cont_key=236850>.

Miron, W. "Vaca leiteira." *Teletime*, 15, no. 151 (Jan./Feb., 2012): 31–33.

Monte, Fabiana. "Vivendi supera Telefónica e compra GVT." *Computerworld*, Nov. 13, 2009. Nov. 13, 2010. <http://computerworld.uol.com.br/telecom/2009/11/13/vivendi-surpreende-mercado-e-compra-gvt/>.

Moreira, Daniela. "Google inaugura novo centro de pesquisas em BH e busca engenheiros." *IDG Now!*, Nov. 10, 2006. Mar. 28, 2010. <http://idgnow.uol.com.br/mercado/2006/11/10/idgnoticia.2006-11-10.1445065867/>.

Moreira, Sonia Virgínia. "Geografias do público e do privado na configuração de um marco legal para as comunicações." *Intercom—Sociedade Brasileira de Estudos Interdisciplinares da Comunicação*, Sept. 2–6, 2011. Dec. 26, 2012. <http://www.geografias.net.br/papers/2011/R6-0804-1.pdf>.

Moreira, S. V. "Para além dos clichês: o Brasil e o contexto internacional da radiodifusão digital." In: Magnoni, A. F. e Carvalho, J. M. de. *O novo rádio, cenários da radiodifusão na era digital*. São Paulo: Editora SENAC SP, 2010, 177–201.

Moreira, S. V. and Helal, C. L. R. "Notes on Media, Journalism Education and News Organizations in Brazil." *Journalism* 10, no. 1 (2009): 91–107. <http://jou.sagepub.com/content/10/1/91.full.pdf>.

Moura, G. *Tio Sam chega ao Brasil—A penetração cultural americana*. São Paulo: Brasiliense, 1985.

Noam, Eli and Wolfson, A. J. eds. "The End of Territoriality in Communications." *Globalism and Localism in Telecommunications*. New York: Elsevier, 1997. Dec. 26, 2012. <http://www.citi.columbia.edu/elinoam/articles/TheEnd ofTerritorialityinCommunications.pdf >.

Possebon, S. "O mercado de comunicações – um retrato até 2006." *Políticas de comunicação, buscas teóricas e práticas*. Ramos, M. and Santos, S. eds. São Paulo: Paulus, 2007, 277–304.

Reuters. "TV paga atinge 13 milhões de assinantes em janeiro." *Veja*, Feb. 28, 2012. Feb. 7, 2012 at <http://veja.abril.com.br/noticia/economia/tv-paga-atinge-13-milhoes-de-assinantes-em-janeiro>.

Rousseff, Dima et al. "LEI Nº 12.485, DE 12 DE SETEMBRO DE 2011: Dispõe sobre a comunicação audiovisual de acesso condicionado." Presidência da República Casa Civil Subchefia para Assuntos Jurídicos, Sep. 11, 2011. Nov. 23, 2011. <http://www.planalto.gov.br/ccivil_03/_Ato2011-2014/2011/Lei/L12485.htm>.

Santanna, R. "Anel óptico dará à internet brasileira mais autonomia." *Folha de S. Paulo*, Mar. 9, 2012. n.d.

Santos, S. and Silveira, E. "Serviço público e interesse público nas comunicações" *Políticas de comunicação, buscas teóricas e práticas*. Ramos, M. and Santos, S. eds. São Paulo: Paulus, 2007, 49–82.

Sequeira, C. D. "Brasil devassado." *IstoÉ* 2166, May 13, 2011. July 16, 2012 at <http://www.istoe.com.br/reportagens/137133_BRASIL+DEVASSADO?pathImagens=&path=&actual Area=internalPage>.

Sousa, Helena. "Time-Life/Globo/SIC: um caso de reexportação do modelo americano de televisão?" Presented at IAMCR—*International Association for Mass Communication Research*, Glasgow, 1998, July 25–30. June 23, 2012 <http://bocc.ubi.pt/pag/sousa-helena-time-life-sopcom.html>.

Tavares, M. "Anatel aprova regulamentação para nova lei de TV por assinatura." *O Globo*, Mar. 23, 2012.

Ursini, N., Figueiredo, O. "O rádio em números." *Meio & Mensagem*, Sep. 12, 2012. Jan. 25, 2013. <http://www.meioemensagem.com.br/home/midia/em_perspectiva/2012/05/04/O-radio-em-numeros.html#.UTStqzeFPcw>

Wallsten, Scott. "Understanding International Broadband Comparisons." Washington, DC: Technology Policy Institute, 2009. Jan. 18, 2011. <http://www.techpolicyinstitute.org/files/understanding%20international%20broadband%20comparisons%202009%20update%207-209.pdf>.

Wharton School. "Telmex and Telefonica Step on the Gas Pedal in Latin America." *Universia Knowledge@Wharton*, Oct. 6, 2004. Jan. 11, 2011. <http://www.wharton.universia.net/index.cfm?fa=viewArticle&id=845&language=english>.

## 2. Reports

*ANATEL* 2011. *Annual Report*—National Agency of Telecommunications of Brazil. Feb. 28, 2013. http://www.anatel.gov.br/Portal/verifica Documentos/documento.asp?numero Publicacao=278637&pub=original&filtro=1&documentoPath=278637.pdf

*Censo Demográfico* 2010. Rio de Janeiro: IBGE, 2011. May 25, 2011. <http://www.ibge.gov.br/home/estatistica/populacao/censo2010/default.shtm>.

Congresso Nacional, "Concentração da Mídia, debates no Conselho de Comunicação Social." Brasília: Gráfica do Senado, 2004.

Glasberg Assessoria, Consultoria e Representações S/A. "Atlas Brasileiro de Telecomunicações." São Paulo, Converge Comunicações, 2011.

Grupo de Mídia. *Mídia Dados Brasil 2010*. São Paulo: Grupo de Mídia, 2010. Last accessed on Jul. 17, 2012. <http://midiadados.digitalpages.com.br/home.aspx?edicao=4>.

IAB Brasil 2012. Indicadores de Mercado. Last accessed on Jan. 27, 2013. <http://www.iabbrasil.org.br/indicadores/Indicadores-de-Mercado-IAB-Brasil.pdf>

IBGE 2006. *Acesso à Internet e posse de telefone móvel celular para uso pessoal 2005*, The Brazilian Institute of Geography and Statistics 2006. <www.ibge.gov.br/home/estatistica/populacao/acessoainternet/comentarios.pdf>

IBOPE Nielsen Online. "Mais internautas em casa." *IBOPE*, May 16, 2011. Dec. 12, 2011. <http://www.ibope.com.br/pt-br/noticias/Paginas/Mais%20internautas%20em%20casa.aspx>.

ITU. *Trends in Telecommunication Reform 2010–2011: Enabling Tomorrow's Digital World*. 11th ed. Geneva: ITU, 2011. Dec. 26, 2012.

<http://www.itu.int/dms_pub/itu-d/opb/reg/D-REG-TTR.12-2010-SUM-PDF-E.pdf>.

IVC Brasil 2010. "Pesquisa Anuário Meio& Mensagem 2010," Auditoria Web, Feb. 2011. <http://www.ivcbrasil.org.br/conteudos/pesquisas_estudos/Pesquisa%20Anu%C3%A1rio%20MM.pdf>

Marché du film. *FOCUS 2012—World Film Market Trends*. Cannes: Marché du film, 2011. July 10, 2011. <http://www.obs.coe.int/oea_publ/market/focus.html>.

Ministério das Comunicações. "Brasil em Alta Velocidade, um Plano Nacional para Banda Larga." Brasília, MiniCom, 2009.

Ministério das Comunicações/Secretaria de Comunicação Social da Presidência da República. "Caderno 1ª Conferência Nacional de Comunicação." Brasília, MiniCom, 2010.

Programa Nacional de Banda Larga. "Brasil Conectado: Programa Nacional de Banda Larga." Secretaria-Executiva do Comitê Gestor do Programa de Inclusão Digital, 2010. Jan. 14, 2011. <http://www4.planalto.gov.br/brasilconectado/forum-brasil-conectado/documentos/3o-fbc/documento-base-do-programa-nacional-de-banda-larga>.

"Publicado decreto: I Conferência Nacional de Comunicação é official." *Pulsar Brasil*, Apr. 17, 2009. Dec. 29, 2010. <http://www.brasil.agenciapulsar.org/nota.php?id=4408>.

Senado Federal. "PLC - Projeto de lei da Câmara, Nº 116 de 2010." *Projetos e Materias Legislativas*, Feb. 15, 2011. Dec. 26, 2012. <http://www.senado.gov.br/atividade/materia/detalhes.asp?p_cod_mate=97352>.

Secretariat for Social Communication (SECOM). "Additional Facts on Brazil's Middle Class." The Federal Government of Brazil, Aug. 8, 2011. Dec. 26, 2012. <http://www.brasil.gov.br/para/press/press-releases/august-1/additional-facts-on-brazils-middle-class/br_model1?set_language=en>.

Telebras. "Conheça a Telebras—Histórico." Telebras, Jan. 5, 2012. Dec. 26, 2012. <http://www.telebras.com.br/inst/?page_id=41>.

## 3. Websites

*ABRANET*. Rio de Janeiro, 2011. Aug. 2, 2011 <http://www.abranet.org.br/>.

ANATEL. "National Telecommunications Agency: 2007–2010 Indicators." Brasilia, 2010. Jun. 8, 2011. <http://www.anatel.gov.br/Portal/exibirPortalInternet.do>.

ANCINE 2012. Orçamento Global Anual. Feb. 16, 2013 <http://fsa.ancine.gov.br/o-que-e-fsa/orcamento-global-anual>

"Anuário Telecom 2004–2010." São Paulo, Plano Editorial, 2011. Jan. 20, 2012. <http://www.anuariotelecom.com.br/anutel/2010/index.html>.

*BM&FBovespa*. São Paulo, 2011. Jan. 11, 2011. <http://www.bmfbovespa.com.br/pt-br/mercados/acoes.aspx?idioma=pt-br>.

CADE 2012. Conselho Administrativo de Defesa Econômica, Pesquisa Processual, Feb. 23, 2013. <http://www.cade.gov.br/Default.aspx?a8889b6caa60b241d345d069fc>

comScore 2010. "Most-visited News Websites in Brazil," *ComScore Data Mine*, Nov., 2010—<http://www.comscoredatamine.com/2010/11/most-visited-news-websites-in-brazil>

*Conheça a Empresa*. EBC, Aug. 20, 2010. <www.ebc.com.br/empresa>.

CADE. "Conselho Administrativo de Defesa Econômica." Brasília, 2011. Nov. 3, 2011. <http://www.cade.gov.br/>.

CVM. "Comissão de Valores Mobiliários." Rio de Janeiro, 2010. Dec. 21, 2010. <http://www.cvm.gov.br/page.asp>.

Filme B. "Portal sobre o mercado de cinema no Brasil." n.p. Jan. 9, 2011 <http://www.filmeb.com.br/portal/html/portal.php>.

IDGNOW! "Segundo Pesquisa, Google Brasil é o pior buscador e Bing, o melhor." Mar. 15, 2012 http://idgnow.uol.com.br/internet/2012/02/16/segundo-pesquisa-google-brasil-e-pior-buscador-e-bing-o-melhor/

Lei de Radiodifusão Comunitária Brasília, 1999. Last accessed on Jan. 15, 2011. <http://www.planalto.gov.br/ccivil_03/leis/l9612.htm>.

Rousseff, Dima et al. "LEI Nº 12.485, DE 12 DE SETEMBRO DE 2011: Projeto InterMeios. "Relatórios de Investimentos 2011." n.p. Feb. 28, 2011. <http://www.projetointermeios.com.br/relatorios-de-investimento>.

Stat Counter 2010. Global Stats—"Top 5 Search Engine in Brazil," Dec. 2008 to Dec. 2010. <http://gs.statcounter.com/#search_engine-BR-monthly-200812-201012>

# Brazil—Data Summaries

ELI NOAM AND PAUL MUTTER

BRAZIL IS RANKED 13TH-HIGHEST in levels of Pooled HHI concentration, meaning that the top firms are big overall, not just in their core industries (Table 20.13). The national power index declined from 3,570 in 2004–2005 to 2,868 in 2009 or most recent. Foreign ownership of Brazil's platform media is very high: the three top platform companies out of four listed in Table 20.3 are all foreign-owned (Telefónica, Telmex, Telecom Italia). In addition, the company with the largest market share, Telemar, with its Oi operator, was 22.4% owned by the Portuguese company, Telecom Portugal, until they merged. Telemar holds the largest share of the overall national market, at 24.7%, and about one-third of platform media market share. Partly in response to concerns about foreign ownership in relation to the implementation of national digitization, Brazil renationalized the former public telecom monopolist Telebras in 2010 from Carlos Slim's America Móvil (Mexico).

Beyond telecom, Globo Group is the dominant player in Brazil's content media industries, controlling an extraordinary 35.1% of the content market (Table 20.14). Globo Group controls 52.4% of the Brazilian TV broadcasting market, 53.6% of the multichannel platforms market, 28.4% of the daily newspapers market, and 1.7% of the ISP market. It also has significant holdings in magazines (19.6%) and radio (9.8%).

Along with Globo Grupo, Folhapar and Estado de São Paulo also hold significant shares of daily newspapers, at 27.9% and 25.2% respectively. Folhapar also owns 31.9% of online news media. Another important content company is Abril Group, which owns 48.5% of magazines and 11.5% of book publishing.

Radio has been an important political and economic entity at the sub-national level, due to Brazil's geography and underdeveloped ICT infrastructure. Liberal ownership controls meant that politicians are able to own radio licenses, and they often use them for electoral purposes (Table 20.15).

Table 20-13. National Media Industries Concentration in Brazil

| | 2004/5 | | 2011 OR MOST RECENT | | % CHANGE ANNUAL AVERAGE | |
|---|---|---|---|---|---|---|
| | COMPANY POWER INDEX IN COUNTRY | COMPANY SHARE OF THE OVERALL NATIONAL MEDIA MARKET (%) | COMPANY POWER INDEX IN COUNTRY | COMPANY SHARE OF THE OVERALL NATIONAL MEDIA MARKET (%) | COMPANY POWER INDEX IN COUNTRY | COMPANY SHARE OF THE OVERALL NATIONAL MEDIA MARKET (%) |
| Oi/Telemar Participações | 1,627 | 32.4 | 876 | 24.7 | –7.7% | –1.3 |
| Telefónica (Spain) | 716 | 24.8 | 542 | 20.8 | –4.1 | –0.7 |
| Globo Group | 199 | 5.1 | 407 | 8.5 | 17.5 | 0.6 |
| Telmex (Grupo Carso, Mexico) | 550 | 12.6 | 552 | 21.0 | 0.06 | 1.4 |
| Telecom Italia (Italy) | 313 | 15.2 | 149 | 9.2 | –3.4 | –1.0 |
| Google (US) | 10 | 0.1 | 55 | 0.6 | 73.8 | 0.08 |
| Abril Group | 90 | 1.5 | 65 | 1.5 | –4.7 | 0.01 |
| Folhaper | 33 | 1.0 | 33 | 1.3 | 0.04 | 0.06 |
| DirecTV (US) | 0 | 0 | 17 | 1.0 | N/A | 0.2 |
| Liberty Global (US) | 0.76 | 0.14 | 0 | 0 | –16.7 | 0.02 |
| Vivendi (France) | 0 | 0 | 27 | 3.4 | N/A | 0.56 |
| Estado de São Paulo | 10 | 0.42 | 9 | 0.44 | –1.9 | 0.004 |
| Universal Church Group | 3 | 0.4 | 18 | 1.3 | 104.8 | 0.14 |
| PRISA Group (Spain) | 0.5 | 0.08 | 2 | 0.15 | 49.7 | 0.01 |
| Bandeirantes Group | 0.69 | 0.2 | 5 | 0.6 | 94.6 | 0.07 |
| Murdoch Interests (US) | 1 | 0.1 | 0.4 | 0.03 | –10.7 | –0.02 |
| Silvio Santos Group | 5 | 0.7 | 2 | 0.4 | –14.2 | –0.1 |

*continued*

Table 20-13. *continued*

| | 2004/5 | | 2011 OR MOST RECENT | | % CHANGE ANNUAL AVERAGE | |
|---|---|---|---|---|---|---|
| | COMPANY POWER INDEX IN COUNTRY | COMPANY SHARE OF THE OVERALL NATIONAL MEDIA MARKET (%) | COMPANY POWER INDEX IN COUNTRY | COMPANY SHARE OF THE OVERALL NATIONAL MEDIA MARKET (%) | COMPANY POWER INDEX IN COUNTRY | COMPANY SHARE OF THE OVERALL NATIONAL MEDIA MARKET (%) |
| Saravia | 3 | 0.5 | 1 | 0.3 | –9.5 | –0.02 |
| Sony/Disney (Japan & US) | 3 | 0.1 | 1 | 0.07 | –9.3 | –0.01 |
| Editora FTD | 1 | 0.12 | 0.14 | 0.04 | –0.14 | –0.01 |
| Diarios Associados Group | 0.8 | 0.1 | 1 | 0.1 | 3.7 | 0.001 |
| Jornal do Brasil | 0.75 | 0.1 | 0.77 | 0.1 | 0.5 | –0.001 |
| RBS Group | 2 | 0.2 | 0.8 | 0.1 | –9.8 | –0.01 |
| EBC (public) | 0.67 | 0.2 | 0.61 | 0.2 | –1.5 | 0.001 |
| Ongoing | 1 | 0.13 | 0.75 | 0.10 | –5.9 | –0.01 |
| Bloch Group | 0.36 | 0.16 | 0.4 | 0.18 | 1.5 | 0.004 |
| Sada Group | 0.1 | 0.03 | 0.2 | 0.05 | 22.6 | 0.002 |
| Media Concentration Index | 2004/5 | | 2011 or Most Recent | % Change Annual Average | | |
| Total Revenue: Nat'l Media Industry (mil US$) | 76,023 | | 127,050 | 11.2 | | |
| Total Voices (*n*) | 50 | | 44 | –2.0 | | |
| Net Voices (*n*) | 36 | | 26 | 1.6 | | |
| Public Ownership (%) | 0.2 | | 0.2 | 0.0 | | |

| | | | |
|---|---|---|---|
| Foreign Ownership (%) | 60.7 | 65.8 | 0.8 |
| C4 Average—Weighted | 83.7 | 91.7 | 1.3 |
| HHI Average—Weighted | 2,903 | 2,773 | −0.7 |
| C1 Average—Weighted | 45.0 | 38 | −1 |
| Noam Index Average—Weighted | 856 | 1,235 | 7.4 |
| Pooled Overall Sector C4 | 85.0 | 75.7 | −1.6 |
| Pooled Overall Sector HHI | 2,084 | 1,660 | −3.4 |
| Pooled Overall Sector Noam Index | 182 | 273 | 8.3 |
| Market Share of Top 10 Companies: Nat'l Media Industry (%) (Pooled C10) | 93.8 | 92.6 | −0.2 |
| National Media Power Index | 3,570 | 2,868 | −3.3 |

Table 20-14. Top Content Media Companies in Brazil

| | 2004/5 | | 2011 OR MOST RECENT | | % CHANGE ANNUAL AVERAGE | |
|---|---|---|---|---|---|---|
| | COMPANY POWER INDEX IN COUNTRY | COMPANY SHARE OF THE NATIONAL CONTENT MEDIA MARKET (%) | COMPANY POWER INDEX IN COUNTRY | COMPANY SHARE OF THE NATIONAL CONTENT MEDIA MARKET (%) | COMPANY POWER INDEX IN COUNTRY | COMPANY SHARE OF THE NATIONAL CONTENT MEDIA MARKET (%) |
| Globo Group | 1,203 | 30.6 | 1,641 | 35.1 | 6.1 | 0.8 |
| Google (US) | 72 | 0.7 | 298 | 3.1 | 52.3 | 0.4 |
| Abril Group | 643 | 10.5 | 353 | 8.2 | –7.5 | –0.4 |
| Folhapar | 163 | 4.9 | 132 | 4.4 | –3.1 | –0.08 |
| Universal Church Group | 18 | 3.0 | 98 | 6.9 | 76 | 0.7 |
| Bandeiranted Group | 5 | 1.62 | 25 | 3.5 | 68.2 | 0.3 |
| Murdoch Interests (US) | 6 | 0.6 | 2 | 0.2 | –11.1 | –0.1 |
| Silvio Santos Group | 24 | 3.4 | 8 | 1.9 | –11.2 | –0.3 |
| Jornal do Brasil | 5 | 0.8 | 4 | 0.6 | –3.6 | –0.04 |
| EBC (Public) | 5 | 1.6 | 3 | 1.3 | –5.1 | –0.06 |
| Ongoing (Arca Group) (Portugal) | 8 | 1.0 | 4 | 0.5 | –8.5 | –0.07 |

| | | | | | | |
|---|---|---|---|---|---|---|
| Bloch Group | 3 | 1.1 | 2 | 1.0 | −2.8 | −0.02 |
| Sada Group | 0.5 | 0.2 | 0.9 | 0.3 | 13.3 | 0.003 |

| Media Concentration Index | 2004/5 | 2011 or Most Recent | % Change Annual Average |
|---|---|---|---|
| Public Ownership (%) | 1.6 | 1.3 | −0.1% |
| Foreign Ownership (%) | 6.2 | 10.4 | 0.7% |
| C4 Average—Weighted | 62 | 68 | 0.9% |
| HHI Average—Weighted | 1,987 | 2,453 | 4.0% |
| C1 Average—Weighted | 41 | 45 | 1% |
| National Power Index | 2,326 | 2,746 | 3% |

Table 20-15. Top Platform Media Companies in Brazil

| | 2004/5 | | 2011 OR MOST RECENT | | % CHANGE ANNUAL AVERAGE | |
|---|---|---|---|---|---|---|
| | COMPANY POWER INDEX IN COUNTRY | COMPANY SHARE OF THE NATIONAL PLATFORM MEDIA MARKET (%) | COMPANY POWER INDEX IN COUNTRY | COMPANY SHARE OF THE NATIONAL PLATFORM MEDIA MARKET (%) | COMPANY POWER INDEX IN COUNTRY | COMPANY SHARE OF THE NATIONAL PLATFORM MEDIA MARKET (%) |
| Oi/Telemar Participações | 1,892 | 37.7 | 1,071 | 30.0 | –7.2 | –1.3 |
| Telefonica (Spain) | 829 | 28.6 | 657 | 25.0 | –3.5 | –0.6 |
| Telmex (Mexico) | 640 | 14.6 | 676 | 25.6 | 0.9 | 1.8 |
| Telecom Italia (Italy) | 364 | 17.7 | 305 | 11.2 | –2.7 | –1.1 |
| Globo Group | 35 | 0.9 | 129 | 2.5 | 45.1 | 0.3 |
| Media Concentration Index | 2004/5 | | 2011 or Most Recent | | % Change Annual Average | |
| Public Ownership (%) | 0 | | 0 | | 0% | |
| Foreign Ownership (%) | 69.6 | | 78.2 | | 1.4% | |
| C4 Average—Weighted | 87.2 | | 97.1 | | 1.7% | |
| HHI Average—Weighted | 3,052 | | 2,846 | | –1.13% | |
| C1 Average—Weighted | 46 | | 37 | | –2% | |
| National Power Index | 3,772 | | 2,896 | | –3.9% | |

# 21

# Media Ownership and Concentration in Chile

SERGIO GODOY E.
SCHOOL OF COMMUNICATIONS, PONTIFICIA UNIVERSIDAD CATÓLICA DE CHILE

## INTRODUCTION

Chile is a middle-income country of 17 million inhabitants that ranks atop most indicators for welfare, technological development, and the economy within Latin America.[1,2] It also shows high levels of free expression[3] and low corruption,[4] which are important when considering institutional issues related to media ownership and the rule of the law. Although its media sector is relatively small in comparison to Brazil, Mexico, and Argentina, it is particularly dynamic, predominantly market-driven, and mostly open to foreign investors. It is also highly centralized in the capital city, Santiago, and shows increasing levels of ownership concentration, especially in newspapers (which are a duopoly), pay-television, telecommunications, cinema distribution, and, in recent years, radio. An additional problem to consider is the country's unequal income distribution.[5]

As in most of Latin America, broadcast television is still the predominant medium

1. The author wishes to acknowledge journalist Carla Pia Ruiz for her assistance in the preparation of this study.

2. Chile and Argentina lead the way as the countries with the highest Human Development Index (HDI) in the region, with 0.783 and 0.775 HDI scores respectively. See UNDP (2010): *Human Development Report 2010. The Real Wealth of Nations: Pathways to Human Development*, UNDP, New York, <http://www.beta.undp.org/content/dam/undp/library/corporate/HDR/HDR_2010_EN_Complete_reprint-1.pdf> (accessed on Sept. 20, 2011).

3. According to Reporters Without Borders, Costa Rica, Chile, and Uruguay lead the ranking of press freedom in the region, and ahead of Spain, France, and Italy. See Reporters Without Borders: *Press Freedom Index 2010*, <http://en.rsf.org/press-freedom-index-2010,1034.html> (accessed on Sept. 15, 2011).

4. According to Transparency International, Chile ranks 21st in the world list of less corrupt nations, just below the United Kingdom and ahead of Belgium, France, and the USA. The other leading Latin American countries are Uruguay (24th) and Costa Rica (41st). See Transparency International: *Corruption Perceptions Index 2010 Results*, <http://www.transparency.org/policy_research/surveys_indices/cpi/2010/results> (accessed on Sept. 2011).

5. Castro, C. "Chile" in *Industrias de Contenidos en Latinoamérica*. Documento de Grupo de Trabajo eLAC2007. CEPAL/ECLAC, Santiago, 2008,56–60.

Table 21-1. Aggregate Advertising Expenditure in Chilean Media, 2004–2008

| | 2004 NOMINAL $M[1] | MARKET SHARE % | 2008 NOMINAL $MIL | MARKET SHARE % |
|---|---|---|---|---|
| Broadcast Television | 283.5 | 48.8 | 281.2 | 48.4 |
| Newspapers | 169.6 | 29.2 | 166.7 | 28.7 |
| Radio | 47.6 | 8.2 | 41.8 | 7.2 |
| Magazines | 19.8 | 3.4 | 16.8 | 2.9 |
| Outdoor | 44.2 | 7.6 | 50.5 | 8.7 |
| Cinema | 1.7 | 0.3 | 1.2 | 0.2 |
| Cable TV | 9.3 | 1.6 | 9.9 | 1.7 |
| Online | 5.2 | 0.9 | 13.4 | 2.3 |
| Total | 580.9 | 100 | 794.1 | 100 |

*Source:* ACHAP.

[1]Nominal values at each year's rates at 636 pesos per dollar. Calculations based on peso figures.

in audience and in advertising income (see Table 21.1), but the industry is fragmented because the law forbids control of more than one franchise in the same area, a principle likely to remain for digital broadcasting. In addition, a strong public broadcaster, Television Nacional de Chile (TVN), competes effectively against six private networks.

Radio is universally available, with a high per capita exposure, yet individual stations are considerably less influential than the networks. Only four radio networks dedicate themselves to broadcast hard news, while the others are oriented to entertainment and music. Deregulation of the industry, plus the regulator's enthusiasm for allocating as much spectrum as possible for free and with a minimum of restrictions, has generated 1,780 radio franchises operating in 2009,[6] with an average of 8,988 listeners per station. In contrast, there are only 6 national television networks out of 125 licenses, 68 are only available through pay-television platforms, and 113 only have local coverage.[7]

Forty percent of households have some form of subscription television, though income by subscription is similar to the advertising expenditure shared by all media types (see Table 21.1). In contrast to traditional media, large international companies control this sector and the telecommunications market with a triple offer of subscription -based television, Internet, and phone service. By 2011, half of all Chileans were Internet users, and nearly 100% had access to mobile phones.

The press is more conventional in ownership. It is a duopoly of two family holdings, El Mercurio and COPESA, with a third international player (Sweden's MTG) playing a secondary role. COPESA, bought at the turn of the century by an emerging businessman with a highly diversified portfolio, has become a multimedia conglomerate by expanding into radio, broadcast television, and cable.

Due to the country's small population and poor reading habits (despite high literacy rates), book publishing is fairly small (5,170 new titles were published in 2010),[8] and the share of

6. Instituto Nacional De Estadísticas: *Cultura y tiempo libre, Informe Anual 2009.* Instituto Nacional de Estadísticas, Santiago, 2010. <http://www.ine.cl/canales/menu/publicaciones/calendario_de_publicaciones/pdf/291110/cult09_29112010.pdf> [accessed on Sept. 17, 2011].

7. Castro, op. cit.

8. Cámara Chilena Del Libro: *Informe estadístico 2010,* Agencia chilena I.S.B.N., Santiago, Chile, 2011. <http://www.camaradellibro.cl/archivos/estadisticas/isbn2010.pdf> [accessed on Sept. 15, 2011].

magazine advertising expenditure has fallen by 50% since 2000 to $20.3 million (see Table 21.1). Although film is experiencing a relative boom by Chilean standards (14 full-length Chilean features were screened in 2009, a figure similar to the previous decade's average),[9] film production is a labor-intensive craft by small, isolated producers who have little business acumen. The average Chilean movie costs between $500,000 and $1 million. Despite many improvements in the recent two decades, they are still poorly marketed.[10] Theater owners, in contrast, are an oligopoly of multinational companies that run multiscreen cinema houses, often placed within popular shopping malls.

The main issues about media concentration and ownership in Chile are similar to those in other countries. Many authors, activists, and specialized bodies have pointed out that media ownership concentration is Chile's main problem and a considerable threat to freedom of expression.[11] Additional issues include

- The conflicts between "old" and "new" media, in terms of business models
- Digital convergence and its ability (or inability) to enhance choice, freedom of expression, and ownership de-concentration
- Ownership of digital infrastructure by a few powerful, large telecom companies
- The dilemma between a business view of media and a broader "public interest" perspective[12]

This chapter comments on these issues throughout the various media and related communication sectors following the structure used in the rest of this book. So it starts with print media—that is, newspapers (the most relevant), book publishing, and magazines. Then the audiovisual media follow: radio and broadcast television (which are universally accessed), multichannel TV platforms (to which 40% of the population has access), and film (small and fragmented in terms of production, but consolidated and growing in exhibition). The fourth part deals with the telecommunications sector (both wireline and wireless), which is closely related to the fifth part devoted to Internet media (especially ISPs) because these areas are controlled by bigger, transnational companies that are becoming the backbone of digital convergence and so are gradually displacing the traditional media in content distribution and processing (texts, videos, static images, and audio).

There are several points to note here before proceeding. The first relates to the chapter's structure. While each sector has many very distinct characteristics and dynamics (which are explained in detail), the boundaries between media platforms and industries are blurring due to digital convergence, as well as corporate mergers and acquisitions. Thus, studying ownership concentration separately without an overall, convergent perspective is risky.

The second issue relates to the quantitative data presented here and their interpretation. We mostly used publicly available data and tried to compare at least two points in time (the latest figure available, plus at least one previous date) so as to have an idea of the market dynamics. Yet their quality is patchy, to say the least. Regulated sectors, notably broadcast television

9. CAEM: *El Cine en Chile en el 2009*, Cámara de Exhibidores Multisalas de Chile A.G., Santiago, March 2010, <http://www.apct.cl/descargas/El.cine.en.Chile.en.el.2009-Marzo2010.pdf> (accessed on April 1, 2011).

10. Matías Lira, cineasta y socio de BF Distribution, plantea un nuevo modelo de negocios: La fórmula para tener nuestro "Chilewood," El Mercurio, Sept. 25, 2011, B5, <http://diario.elmercurio.com/2011/09/25/economia_y_negocios/enfoques/noticias/60ddb94c-1add-45c7-b7bc-bd71f61ff2a8.htm> (accessed on September 26, 2011).

11. A. Cerda & P. Peña: chile, in A. Finlay (ed.):*Global Information Society Watch 2009,* Uruguay: APC/Hivos 104–105; Godoy, S. & Gronemeyer, M.: S. Godoy and M. Gronemeyer: *Mapping Digital Media: Country Report—Chile.* A report by the Open Society Foundations, London, Oct. 23, 2012. http://www.opensocietyfoundations.org/reports/mapping-digital-media-chile.

12. P. Iosifidis, J. Steemers & M. Wheeler: *European Television Industries.* London: BFI, 2005; D. McQuail: *Mass Communication Theory.* 5th Ed., London & Thousands Oaks, 2005; C. Nissen: *Public Service Broadcasting in the Information Society.* Strasbourg: Council of Europe, Directorate General of Human Rights, Media Division, 2006. Available at <http://www.coe.int/t/dghl/standardsetting/media/Doc/H-Inf(2006)003_en.pdf> (accessed on Dec. 10, 2011); J. Curran J. & J. Seaton: *Power without responsibility. The Press and Broadcasting in Britain* (5th Ed.), London & New York: Routledge, 1997.

and telecommunications, have the most accurate numbers. In the other media platforms, data quality and accuracy vary considerably—especially in terms of revenues, which are very hard to get, in contrast to audience indicators. In addition, there are problems of comparability between the different media types: for example, the consumption of television is measured in terms of rating points, radio in terms of audience reach,[13] newspapers in average weekly readership (we discarded circulation), websites in unique visitors, and telecommunications in percentage of lines or connections. Last but not least, there are inconsistencies and discontinuities of these data over time. Although great effort was put into producing the most reliable figures possible within the constraints of this study, quantifying media concentration in Chile is not an easy exercise and should be undertaken with care, matching it against the overall background information provided in the following pages.

Taking all this into account, the conclusions at the end of the chapter confront the evidence examined with the Noam Index. Despite the shortcomings mentioned, there are many interesting findings.

## PRINT MEDIA

### Newspapers

Newspapers in Chile are similar to those in other Latin American countries in terms of market and ownership structure: most have been owned by politically powerful families defending—until recently at least—a conservative ideology. General Augusto Pinochet's 1973 military coup meant closure for the historically strong Chilean partisan press. Two family-owned, anti-communist holdings were left alone: the broadsheet *El Mercurio,* controlled by the Edwards family since the 1830s, and *La Tercera,* a tabloid owned by the Picó-Cañas family. Both groups owned additional, smaller papers. They were active supporters of the right-wing dictatorship and its pioneering neoliberal economic agenda. They formed a powerful duopoly.[14] In the early 2000s, La Tercera became the COPESA consortium and was bought by businessman Alvaro Saieh.[15] The El Mercurio and COPESA publishing groups still dominate the industry.

Table 21.2 depicts the main newspaper owners in Chile. Table 21.3 shows the share of weekly readership in 2010 and 2011 by the most influential newspapers of the dominant duopoly: *El Mercurio,* the populist tabloid *Las Ultimas Noticias* (owned by El Mercurio group of the Edwards family), COPESA's main stream *La Tercera,* and the populist *La Cuarta.*

Although readership figures are evenly split between the two main newspaper holdings, revenues are more concentrated in the outlets owned by the El Mercurio Group (targeted to higher income groups in contrast to those from COPESA), as revealed by Table 21.4. This table is based on estimations by Megatime, an agency that measures advertising effectiveness. These figures should be taken with care because they are based on published prices of advertising

---

13. Rating points are the percentage of the public who is exposed to a certain channel or program *at a certain moment,* with respect to the total universe considered. This indicator does not distinguish if in the next measurements there is a completely different group of people exposed to the content. In contrast, reach is the total percentage of the public who *has been exposed* to a channel or program at least in two different moments. For example, a station that has an average rating of 5 points but a reach of 10% means that it there was always a 5 % of the population exposed to it each time the measurement was taken. But as the public was changing, it finally reached 10% of the population when all the different moments of measurement were considered.

14. See for instance K. Dermota: *Chile Inédito, periodismo en democracia,* Ediciones B, Barcelona, 2002; M. O. Monckeberg: *Los Magnates de la Prensa en Chile. Concentración de los Medios de Comunicación en Chile,* Random House Mondadori, Santiago, 2009. Author and journalist Maria Olivia Monckeberg has said her book has been deliberately ignored by the mainstream media (El Ciudadano: María Olivia Mönckeberg: *"En los medios no tengo la posibilidad de hacer periodismo,"* Nov. 12, 2009, <http://www.elciudadano.cl/2009/11/12/maria-olivia-monckeberg-%E2%80%9Cen-los-medios-no-tengo-la-posibilidad-de-hacer-periodismo%E2%80%9D/>, accessed on Mar. 1, 2011).

15. La Nación Domingo: *La apuesta del empresario por ampliar su poder mediático: Todos Los Caminos Conducen A Saieh,* Sept. 12, 2004, quoted in <http://miguelpaz.blogspot.com/2004/09/reportaje-alvaro-saieh-siete7.html> (accessed on Mar. 1, 2011); La Nación: *Álvaro Saieh Bendeck Escribe A La Nación. La historia según Copesa,* July 31, 2005, <http://www.lanacion.cl/prontus_noticias/site/artic/20050730/pags/20050730194208.html> (accessed on Feb. 1, 2011).

Table 21-2. Main Newspaper Titles and Their Owners

| OWNER | CITY/REGION OF DISTRIBUTION | TITLES |
|---|---|---|
| Grupo Edwards / El Mercurio (Chile) | Santiago | *El Mercurio de Santiago** |
| | | *Las últimas Noticias** |
| | | *La Segunda** |
| | | *El líder de Melipilla* |
| | Valparaiso and San Antonio | *El Mercurio de Valparaíso* |
| | | *La Estrella de Valparaíso* |
| | | *El Líder de San Antonio* |
| | North of Chile | *La Estrella de Arica* |
| | | *La Estrella de Iquique* |
| | | *La Prensa de Tocopilla* |
| | | *El Mercurio de Calama* |
| | | *La Estrella del Loa* |
| | | *La Estrella del Norte* |
| | | *El Mercurio de Antofagasta* |
| | South of Chile except Concepcion | *El Austral de Temuco* |
| | | *Diario Austral (Valdivia)* |
| | | *El Austral de Osorno* |
| | | *La Estrella de Chiloé* |
| | | *El Llanquihue* |
| | Concepcion | *El Sur (Concepción)* |
| | | *La Estrella (Concepción)* |
| | | *Crónica Chillán* |
| Alvaro Saieh / COPESA (Chile) | Santiago | *La Tercera** |
| | | *La Hora* |
| | | *La Cuarta** |
| | | *Pulso* |
| | Concepcion | *El Diario de Concepción* |
| Metro International (Sweden) | Santiago | *Publimetro* |
| La Nacion S.A. (Chile) | National | *La Nacion (closed in 2012)* |
| | | *El Nortino* |
| | | *Diario Oficial (Official Gazette)* |

*Source:* Based on data from Asociacion Nacional de la Prensa, ANP; Publimetro.cl; Vera (2005).
*Note*: asterisk (*) indicates national reach.

multiplied by the number of advertisements printed by each newspaper. The frequent discounts to clients were not considered. Therefore, the aggregated income of newspapers in Table 21.4 ($650.4 million in 2008) is almost four times as much as the more realistic advertising expenditures shown in Table 21.1 ($166.7 million the same year). Yet specialists roughly consider shares of income accurate. There are no other publicly available sources, and newspaper companies are not required by law to disclose their financial information because they are neither publicly listed in the stock exchange nor do they belong to a regulated market such as telecommunications and broadcast television.

Many attempts to reduce the dominance of El Mercurio and COPESA failed. The most dramatic attempt was by *La Epoca,* a center-left

Table 21-3. Share of Main Newspaper Weekly Readership Rates, 2010 vs. 2011

| MEDIA HOLDING[1] | 2010 | % | 2011 | % |
|---|---|---|---|---|
| El Mercurio SAP Newspapers (Edwards family, Chile): | 1,106.4 | 34.4 | 765.7 | 16.1 |
| *El Mercurio* (daily broadsheet) | *551.9* | | *765.7* | |
| *Las Ultimas Noticias* (tabloid) | *473.3* | | *1,006.1* | |
| *La Segunda* (evening paper, weekdays only) | *81.2* | | *162.3* | |
| COPESA Newspapers (A. Saieh, Chile): | 1,623.3 | 50.5 | 2,098.3 | 44.0 |
| *La Tercera* (daily) | *410.2* | | *666.5* | |
| *La Cuarta* (tabloid) | *789.6* | | *795.9* | |
| *La Hora* (free evening paper, weekdays only) | *423.5* | | *635.9* | |
| Others: | | | | |
| *Publimetro* (free paper, MTG Sweden) | 464.9 | 14.5 | 732.3 | 15.4 |
| *La Nacion* (State-owned, paper version ceased in 2011) | 22.3 | 0.7 | — | — |
| Average Weekly Readership, 000s | 3,216.9 | 100.0 | 4,764.7 | 100.0 |
| C4 | | 100.0 | | 75.5 |
| HHI | | 3,939 | | 2,434 |
| Number of Firms with Market Share >1% (*n*) | | 4 | | 3 |
| Noam Index | | 1,969 | | 1,405 |

*Source:* ANP.

[1]Only the titles depicted were measured. *Las Ultimas Noticias* added online readership figures to its paper-only ones in 2012.

publication linked to Christian-Democrat investors created at the end of the Pinochet regime, and *El Metropolitano,* a commercial venture backed by the Hites family. Both went bankrupt in the late 1990s.The following decade, two new ideologically motivated projects also failed: *Diario Siete* and *Le Monde Diplomatique,* which is now a weekly publication of limited impact.

The only successful challenger so far has been a big multinational with a novel business model, Sweden's Metro International Group (MTG), which introduced the free daily *Publimetro* in the early 2000s.[16] Despite their usual stance in favor of the free market, competition, and free speech (at least in Communist countries such as Cuba), the incumbents tried to block the newcomer by suing Santiago's Metro system to prevent distribution of *Publimetro* inside its facilities.[17] The new paper is now distributed at the Metro's station entrances. Its impact is considerable, given the exponential expansion of the Metro network in the last decade and its role as the backbone of Santiago's public transportation system. *Publimetro* is more a complement than a threat to El Mercurio and COPESA, since its contents are mostly drawn from press agencies and are designed for fast reading during an average Metro ride. Therefore, it is no real match to the duopoly's own distinctive, original news content.

In 2010, the owner of COPESA, Alvaro Saieh, bought a 20% stake of the powerful cable operator VTR for $340 million, as well as the small analog Channel 22 UHF for an additional $2.5 million.[18] COPESA became the sole traditional

16. http://www.publimetro.cl/.

17. Larenas, Ramos & Siegel (2002): *El fenómeno de los diarios gratuitos, una nueva modalidad de hacer periodismo.* Tesis para optar al grado de Licenciado en Comunicación Social. Santiago: Escuela de Periodismo, Universidad Diego Portales. Available at <http://www.comunicacionyletras.udp.cl/files/Larenas_Ramos_y_Siegel.pdf> (accessed on Sept. 15, 2011).

18. El Mostrador: *Compra señal 22 en US$2.5 millones: Álvaro Saieh tiene "mentalidad televisiva,"* July 6, 2010, <http://www.elmostrador.cl/noticias/negocios/2010/07/06/alvaro-saieh-tiene-mentalidad-televisiva/> (accessed on Mar. 1, 2011).

Table 21-4. Estimated Share of Advertising Revenues per Newspaper Group, 2007–2008[1]

| MEDIA HOLDING/TITLE | 2007 NOMINAL $M | % | 2008 NOMINAL $M | % |
|---|---|---|---|---|
| El Mercurio SAP Newspapers (Edwards family, Chile) | 381.6 | 56.6 | 357 | 54.9 |
| *El Mercurio* (daily broadsheet) | *316* | | *293.5* | |
| *Las Ultimas Noticias* (tabloid) | *49.0* | | *48.4* | |
| *La Segunda* (evening paper, weekdays only) | *16.6* | | *15.1* | |
| COPESA Newspapers (A. Saieh, Chile) | 166.5 | 24.7 | 165.9 | 25.5 |
| *La Tercera* (daily) | *130.6* | | *125.2* | |
| *La Cuarta* (tabloid) | *19.6* | | *20.2* | |
| *La Hora* (free evening paper, weekdays only) | *16.3* | | *20.5* | |
| Others: | | | | |
| *Publimetro* (free paper, MTG, Sweden) | 45.7 | 6.8 | 51.3 | 7.9 |
| *Estrategia* (business paper, Ed. Gestion, Chile) | 41.2 | 6.1 | 37.1 | 5.7 |
| *Diario Financiero* (business paper, Claro group, Chile) | 35.3 | 5.2 | 34.7 | 5.3 |
| *La Nacion* (public)[2] | 4.2 | 0.6 | 4.5 | 0.7 |
| Total | 674.5 | 100 | 650.4 | 100 |
| C4 | | 99.4 | | 99.3 |
| HHI | | 3,921 | | 3,786 |
| Number of Firms with Market Share >1% (*n*) | | 5 | | 5 |
| Noam Index | | 1,960 | | 1,693 |

*Source:* Megatime.

[1] Values are based on published price of advertising by number of advertisements. Exchange rate of 636 pesos per dollar (2008).

[2] *La Nacion* is not considered in the calculations.

media consortium with a stake—although a minority one—in the far more profitable telecommunications industry.

Until its final disappearance at the end of 2012, *La Nacion* was another marginal player in the newspaper sector. Originally it was a private broadsheet, nationalized in the 1920s. In contrast to the public broadcaster TVN, it was always openly controlled by the government and had relatively low readership rates, which became marginal after the 1990s. However, it was able to survive for decades due to cross-subsidization from the highly profitable *Diario Oficial*. After adverse coverage of right-wing president Sebastian Piñera's 2009 campaign, the print version of *La Nacion* was closed in late 2010 by the new Piñera government.[19] The online-only version suffered the same fate two years later.

## Book Publishing

The small size of the Chilean market, the lack of book readership habits among the population, and high taxation (VAT is 19%, and books are not exempted) create an adverse environment for book production despite the country's high literacy rates. In 2010, 5,107 titles were published.

19. <http://www.lanacion.cl/> . See also El Mostrador: Lo que está detrás del fin de la edición en papel del matutino: La estrategia del gobierno para sacar a los accionistas privados de La Nación. Oct. 18, 2010. <http://www.elmostrador.cl/noticias/pais/2010/10/18/la-estrategia-del-gobierno-para-sacar-a-los-accionistas-privados-de-la-nacion/> (accessed on Dec. 15, 2010).

Only a few of these printed more than a 1,000 copies, but there are no updated public figures available. Nearly 90% of all published titles are concentrated among three main players, Spain's Santillana being the most important followed by the newspaper consortium COPESA.

## Magazines

The magazine market in Chile is shrinking in terms of advertising income (down by almost a quarter during the 2000s according to Table 21.1), readership, and political influence. Of the historical titles that once shaped political discussion from the 1960s up to the 1980s, only *Que Pasa* remains. All of the others have disappeared or are no longer relevant, as the sector has shifted to entertainment, lifestyle, and sports.

Originally a right-wing publication that opposed the socialist government of Salvador Allende and applauded the military coup that ousted him in 1973, *Que Pasa* is now a relatively

Table 21-5. Weekly and Bi-Weekly Magazines: Average Total Readers in Santiago 2009–2011

| OWNER | TITLE | CONTENT | 2009[1] 000S | SHARE % | 2011[2] 000S | SHARE % |
|---|---|---|---|---|---|---|
| COPESA (Chile) | *Paula* | Female | *126.8* | | *119.1* | |
| (A. Saieh) | *Que Pasa* | News & business | *67.8* | | *45.4* | |
| Total COPESA | | | 194.6 | 32 | 164.5 | 24 |
| Ed. Tiempo Presente (Chile) | *Cosas* | General/Glamour | *86.5* | | *102.1* | |
| Total Tiempo Presente | | | 86.5 | 14 | 102.1 | 15 |
| El Mercurio (Chile) | *Dato Avisos* | Advertising | *69.3* | | *76.9* | |
| Total El Mercurio | | | 69.3 | 12 | 76.9 | 11 |
| Luksic Group (Chile) | *Capital* | News & business | *21.4* | | *38.3* | |
| Total Luksic | Same | | 21.4 | 4 | 38.3 | 6 |
| Publimetro (MTG, | *El Gráfico* | Sports | | | *21.4* | |
| Sweden) | *Golf Digest* | Sports | | | *1.2* | |
| Total Publimetro | | | | 0 | 22.6 | 3 |
| Televisa (Mexico) | *Caras* | General/Glamour | *87.1* | | *105* | 15 |
| | *Vanidades* | Female | *28.5* | | *28.7* | 4 |
| Total Televisa | | | 115.6 | 19 | 133.7 | 19 |
| The Clinic (Chile) | *The Clinic* | Political satire | *115.1* | | *139.8* | |
| Total, The Clinic | | | 115.1 | 19 | 139.8 | 21 |
| | | | 602.5 | 100 | 677.9 | 100 |
| Revenue Estimate ($US mil) | | | | 163 | | |
| C4 | | | | 100 | | 84 |
| HHI | | | | 2,102 | | 1,649 |
| Number of Firms with Market Share >1% (*N*) | | | | 6 | | 7 |
| Noam Index | | | | 858 | | 794 |

*Source:* ANP/Ipsos Search.
[1]Oct 2008/Nov 2009.
[2]Jan/Jun 2011.

harmless, business-oriented, and elegantly designed weekly controlled by COPESA, the second largest newspaper consortium and holder of cable TV interests. Most other titles that rank at the top of current readership figures (see Table 21.5) also belong to large consortia (including El Mercurio), and have relatively inoffensive, business savvy content that complements the premium advertising of expensive cars, financial products, clothing, and the like. The exception to the rule, with high political impact, non-corporate ownership, and sharp and incisive contents, is the semi-satirical *The Clinic,* named after the place in which general Augusto Pinochet was arrested in London in 1998 on grounds of human rights abuses. *The Clinic* began as a satirical and intellectual broadsheet published every two weeks (hence its classification as a magazine), but it has gradually increased its news content.

## AUDIOVISUAL MEDIA

### Radio

Similar to the newspaper industry, radio has historically been a private venture in Chile and predominantly featured entertainment and music, as well as information. The Pinochet dictatorship (1973–1990) created Radio Nacional AM out of confiscated left-wing outlets, but this government-controlled news and talk network was privatized after the return to democracy in the 1990s. To distinguish itself from its Communist counterparts, the Pinochet regime tolerated some dissent in radio, as this medium was deemed less influential than TV and the press.[20] Radio's subsequent diversity of ownership and editorial views in difficult times—including the 2010 earthquake and tsunami, when the Internet and mobile networks collapsed—has consistently placed it among the most credible and reliable social institutions in the country. This explains its universal penetration and high rates of per capita listenership (Table 21.6).

Radio is highly deregulated and therefore experienced extensive ownership concentration and internationalization after 1990. To appease increasingly concerned local broadcasters, a reciprocity clause was included in article 9 of the Press Law N°19,733 of 2001 by which foreign investors can own more than a 10% stake of a radio station provided Chileans are entitled to the same condition in the investor's home country. This clause came too late to revert the well-entrenched internationalization of the industry, but it blocked further expansion of foreign conglomerates. Local broadcasters invoked this clause in 2010 to prevent the renewal of some regional radio franchises controlled by Iberoamericana Radio Chile, the local arm of the PRISA Group (Spain).[21] The issue was still unsolved at the time of this writing.[22]

Iberoamericana now controls the 12 most successful networks of national reach—60% of audience share and 37% of radio's advertising income—since it bought four networks from Colombia's CRC in 2007 for $75 million.[23] This merger caused considerable concern, at least among the local broadcasters, which may explain why Iberoamericana created the country's one

---

20. M. Lasagni, P. Edwards, and Bonnefoy: *La Radio en Chile*, 3° ed., CENECA, Santiago, 1988; E. Tironi and G. Sunckel: "Modernizacion De Las Comunicaciones Y Democratizacion De La Politica. Los Medios En La Transicion A La Democracia En Chile," *Estudios Públicos*, 52 (primavera), 1993, 215–246. Secretaría de Comunicación y Cultura (SECC): *Reseña de Medios N°32: Perspectivas de la Radio en Chile*, SECC, Santiago, 1996.

21. El Mostrador: *La pelea que calienta el dial*, Feb. 1, 2011, <http://www.elmostrador.cl/noticias/pais/2011/02/01/la-pelea-que-calienta-el-dial/> (accessed on Feb. 3, 2011); Luis Pardo, President of the Chilean Association of Radio Broadcasters ARCHI: *Reciprocidad y Subtel*, Sección Cartas al Director, El Mercurio, Jan. 8, 2011, B2.

22. El Mercurio: *Resurge debate sobre concesiones radiales de grupo PRISA*, Jan. 12, 2011, available at. <http://www.mercurio.cl/2011/01/12/espectaculos/mas/noticias/46B6E44A-2C2D-4ED1-8546-7FEC9913AC08.htm?id={46B6E44A-2C2D-4ED1-8546-7FEC9913AC08}> (accessed on Jan. 2011).

23. Extranoticias.cl: *Preocupa fusión de radios Iberoamerican con CRC*, Sept. 2007, <http://www.extranoticias.cl/preocupa-fusion-de-radios-iberoamerican-con-crc/> (accessed on Feb. 1, 2011); Radio Tierra: Grupo español PRISA concentra mercado radiofónico, July 31, 2007, <http://www.radiotierra.cl/node/526> (accessed on Feb. 1, 2011); El Mercurio Online: *Corte Suprema autoriza fusión que crea al mayor grupo radial en historia de Chile*, Nov. 22, 2007, <http://www.emol.com/noticias/nacional/detalle/detallenoticias.asp?idnoticia=282784> (accessed on Feb. 1, 2011).

## Table 21-6. Owners of FM Radio Stations in Santiago[1]

| | |
|---|---|
| Iberoamericana Radio Chile S.A. (Grupo Prisa, Spain) | FM Dos 98.5 FM |
| | Futuro 88.9 FM |
| | Imagina 88.1 FM |
| | Concierto 88.5 FM |
| | Pudahuel 90.5 FM |
| | ADN Radio 91.7 FM* |
| | Radio Activa 92.5 FM |
| | Rock & Pop 94.1 FM |
| | Radio Uno 97.1 FM |
| | Corazón 101.3 FM |
| | 40 Principales 101.7 FM |
| COPESA (Alvaro Saieh, Chile) | Duna 89.7 FM |
| | Carolina 99.3 FM |
| | Beethoven 96.5 FM |
| | Zero 97.7 FM |
| | Paula 106.9 FM |
| | Disney 104.9 FM |
| Bio Bio Comunicaciones S.A. (Mosciatti family, Chile) | Bio-Bio 99.7 FM* |
| Canal 13 (Luksic Group and Universidad Catolica de Chile, Chile) | Sonar 105.7 FM |
| | Play 100.9 FM |
| | Maria 89.3 FM |
| | Oasis 100.1 FM |
| | Horizonte 103.3 FM |
| Compañía Chilena de Comunicaciones S.A. (Carlos Figueroa and others, Chile) | Cooperativa 93.3 FM* |
| | Universo 93.7 FM |
| Sociedad Nacional de Agricultura, SNA (Chile) | Agricultura 92.1 FM* |
| César Molfino (Chile) | El Conquistador 91.3 FM |

*Source:* Own elaboration based on the information provided by the respective companies.
[1]News/information stations are marked with an asterisk (*).

and only purely Chilean musical station, UNO 97.1 FM in Santiago,[24] and kept ADN 91.7 FM as one of Chile's four existing news and talk national radio networks along with Cooperativa, Bio Bio, and Agricultura.[25] Despite this trend toward ownership concentration, radio is still highly valued by the public and widely considered as pluralist and varied (Table 21.7).

Most stations are dedicated to music and light entertainment. Although there are only four radio networks of national coverage dedicated to news—Cooperativa, Bio Bio, Agricultura, and ADN Radio—they are respected sources of information though very different from one another. Cooperativa belongs to a group of Christian-Democrat, center-left investors (Compañia Chilena de Comunicaciones). Bio Bio belongs to the fiercely independent, non-Catholic Mosciatti family from Concepción. Agricultura belongs to SNA, the National

24. <http://www.radiounochile.cl/>

25. <http://www.adnradio.cl/>. The previous news outlet owned by CRC was called W Radio and was changed into ADN in 2008.

Table 21-7. Aggregated Audience Reach of FM Radio per Operator in Santiago

| | 2004 | 2008 |
|---|---|---|
| Cia Chilena Comunicaciones Stations[1] | 9.2 | 8.5 |
| COPESA Stations | 8.7 | 7.2 |
| Iberoamericana Stations (Spain) | 41.8 | 43.8 |
| Mosciatti Family Station | 0.9 | 2 |
| SNA Station | 2.2 | 2.2 |
| UC/Luksic Group Stations | 2.3 | 3.3 |
| Other Stations | 37.3 | 25.8 |
| Total Revenue ($US mil) | 80.1[2] | 95.7 |
| C4 | 62.0 | 62.8 |
| HHI | 1,919 | 2,062 |
| Number of Firms with Market Share >1% (*n*) | 5 | 6 |
| NoamIndex | 858 | 842 |

*Source:* Based on IPSOS/Search audience measurements.
[1]Stations grouped for referential purposes only according to station ownership of March 2012.
[2]Estimated number.

Agricultural Association, traditionally associated with the so-called landed aristocracy. ADN Radio is owned by the Spanish held Iberoamericana Radio Chile. This diversity has helped radio stay ahead of most social institutions in terms of credibility in various surveys since the 1980s.[26]

Yet radio is not a very transparent market. There are no reliable figures about how income is spread among the existing operators depicted in Table 21.8. The National Institute of Statistics, INE, calculated that the aggregate income of all stations was $95.7 million in 2008.

## Broadcast Television

Until 1990, when private operators were authorized, television was a not-for-profit duopoly of

Table 21-8. Aggregate Income of the Chilean Radio Industry 2005–2008 According to Source of Funds, US$mil (Exchange Rate at 560 Pesos per Dollar)

| | ADVERTISING | LEASING OF AIRTIME | OTHER SOURCES | TOTAL |
|---|---|---|---|---|
| 2005 | 67.2 | 2.6 | 6.2 | 82.2 |
| 2006 | 80.0 | 3.4 | 5.6 | 94.5 |
| 2007 | 78.4 | 3.4 | 5.6 | 93.0 |
| 2008 | 82.1 | 4.0 | 4.8 | 95.7 |
| Total 2005–2008 | 307.6 | 13.4 | 22.2 | 365.4 |
| % of Total | 84.0 | 4.0 | 6.0 | 100.0 |

*Source:* Instituto Nacional de Estadísticas INE's Cultura y Tiempo Libre Annual Reports 2006 to 2010.

26. S. Godoy & M. Gronemeyer: Mapping Digital Media: Country Report -Chile. A report by the Open Society Foundations. London, Oct. 23, 2012. <http://www.opensocietyfoundations.org/reports/mapping-digital-media-chile>.

state and university broadcasting. Lack of public funding and absolute reliance on advertising after general Pinochet liberalized the economy in 1975 generated a "trout and parrot mixture" of a system in which elitist, public-service principles of European origin coexisted with often conflicting mass appeal and market rules.[27]

The Latin American norm in media ownership consists of private monopolies controlled by family conglomerates. In broadcast television, that was the case of Brazil's Globo, Mexico's Televisa, and Venezuela's Venevision and RCTV[28] (before the socialist government of Hugo Chavez did not renew RCTV's licenses in 2007).[29] In Chile, ownership concentration in this sector has been prevented by law. The law stated that a broadcaster could only operate one television channel in the same geographical area, a feature kept by the recent Law of Digital TV N°20,750 of March 25, 2014 (see below). Although just one university broadcaster remains, the small but profitable UCV from Universidad Catolica de Valparaiso, the market is still led by TVN, an unusual public broadcaster due to its combination of sustained mass appeal, economic success, and an absolute prohibition to receive public funding. In 2009, it attracted 33% of advertising expenditures in television,[30] a dominant position that gave it political independence from government (Table 21.9).

Table 21-9. TV Broadcasting: Terrestrial Channels 2008 vs. 2004, by Revenues (Nominal US$mil)[1]

| BROADCASTER/CURRENT OWNER[2] | 2004 | % | 2008 | % |
|---|---|---|---|---|
| 2 Telecanal (Guillermo Cañedo, Mexico) | — | 0 | 2.63 | 1 |
| 4 Red (Angel Gonzalez, Mexico) | 26.79 | 11 | 65.26 | 18 |
| 5 UCV (Universidad Catolica Valparaiso, Chile) | N/A | | N/A | |
| 7 TVN (Public) | 83.8 | 33 | 106.29 | 30 |
| 9 MEGA (Bethia Group, Chile) | 47.2 | 19 | 59.47 | 17 |
| 11 CHV (Time Warner, US) | 13.08 | 5 | 15.62 | 4 |
| 13 UCTV (Universidad Catolica/ Luksic group, Chile) | 81.71 | 32 | 105.51 | 30 |
| Total | 252.58 | 100 | 354.78 | 100 |
| C4 | | 95.0 | | 95.0 |
| HHI | | 2,620 | | 2,430 |
| Noam Index | | 1,172 | | 992 |
| Number of Firms with Market Share >1% (*n*) | | 5 | | 6 |

*Source:* SVS.
[1]Exchange rate 2004 and 2008: 636 pesos per dollar. Nominal values for each year, not corrected.
[2]Ownership in March 2012.

27. S. Godoy: An Unlikely Mixture. An Evaluation of Chile's Market Oriented Model of Public Television. Lambert Academic Publishing, Köln, 2009.
28. Fox, E.: *Latin American Broadcasting, from tango to telenovela*, John Libbey/University of Luton, Luton, Hertforshire, 1997.
29. Ministerio del Poder Popular para la Comunicación e Información de Venezuela (2007): Libro blanco sobre RCTV, Caracas, Venezuela. Available at <http://www.rnv.gov.ve/noticias/docs/libro_blanco_RCTV-Web.pdf>.
30. Source: SVS, La Tercera.

In 2012, TVN stood alone as the only successful not-for-profit operator after the sale in 2010 of a 67% stake of its traditional rival Universidad Catolica de Chile's Canal 13 (UCTV) to the Chilean businessman Andronico Luksic.[31] Until then, UCTV and TVN dominated the broadcasting industry with a highly unusual combination of market behavior and a public-service remit. High costs, corporate confusion, and increasing competition against less prudish players (which were not afraid of vulgarity or sensationalism) left UCTV almost bankrupt. Almost simultaneously to the partial sale of UCTV, Time Warner acquired another former university network, Chilevision (CHV), from the newly elected national president Sebastián Piñera, a businessman who promised to sell most of his companies if elected. Piñera bought CHV in 2005 for $40 million from Venezuela's Venevision and sold it five years later to the US-based holding for $140 million (Table 21.10).[32]

The sector faced uncertainly as Chile's Congress debated two legal drafts proposed by the Executive in 2008 as a response to digitization: one reforming the law of TVN and the other governing television.[33] Despite many advantages, the proposals also had serious deficiencies that delayed their approval. They were considered mere upgrades of outdated legislation: television was treated as if neither pay-TV nor convergence existed, isolated regulatory bodies were kept intact, and incumbent broadcasters

Table 21-10. TV Broadcasting Market Shares by Audience (Household Rating Points, Daily Average)

| | 2005 | 2006 | 2007 | 2008 | 2009 | 2010 |
|---|---|---|---|---|---|---|
| 2 Telecanal (Guillermo Cañedo, Mexico) | 0.4 | 3.2 | 2.8 | 2.5 | 1.6 | 1.7 |
| 4 Red (Angel Gonzalez, Mexico) | 10.5 | 8.8 | 8.3 | 6.3 | 4.4 | 4.8 |
| 5 UCV (Universidad Catolica Valparaiso, Chile) | 1.1 | 1.5 | 1.5 | 1.4 | 1.5 | 1.6 |
| 7 TVN (Public, Chile) | 20.7 | 19.9 | 20.6 | 18.9 | 19.1 | 18.4 |
| 9 MEGA (Bethia Group, Chile) | 19.3 | 17.6 | 18.5 | 20.7 | 18.3 | 17.6 |
| 11 CHV (Time Warner, US) | 12.6 | 14.8 | 14.6 | 16.7 | 16.5 | 17.1 |
| 13 UCTV (Universidad Catolica/Luksic group, Chile) | 21 | 17.1 | 17.6 | 18.1 | 17.1 | 13 |
| Total for Broadcast Channels | 85.6 | 82.9 | 83.9 | 84.5 | 78.5 | 74.2 |
| Others: All Pay-Television Channels | 14.4 | 17.1 | 16.1 | 15.5 | 21.5 | 25.8 |
| C4 | 73.6 | 69.4 | 71.3 | 74.4 | 71 | 66.1 |
| HHI | 1,512 | 1,307 | 1,369 | 1,440 | 1,289 | 1,138 |
| Noam Index | 617 | 494 | 517 | 544 | 487 | 430 |
| Number of Firms with Market Share >1% (*n*) | 6 | 7 | 7 | 7 | 7 | 7 |

*Source:* Time IBOPE.

31. La Tercera: La televisión chilena ad portas de su mayor cambio en las últimas tres décadas, suplemento Negocios, August 15, 2010, 10–11.

32. La Nación: *Chilevisión ya es oficialmente de Sebastián Piñera*, April 15, 2005, <http://www.lanacion.cl/prontus_noticias/site/artic/20050415/pags/20050415162651.html> (accessed on Feb., 2011); El Mundo: *Time Warner compra Chilevisión al presidente Sebastián Piñera*, Aug. 25, 2010, <http://www.elmundo.es/elmundo/2010/08/25/comunicacion/1282748887.html> (accessed on Feb. 1, 2011).

33. Presidencia de la República: Mensaje de S.E. la Presidenta de la Republica con el que inicia un proyecto de ley que modifica la ley N° 19.132 de Televisión Nacional de Chile. Mensaje N° 943–356, Santiago, Oct. 27, 2008. // Presidencia de la República: Mensaje de S.E. la Presidenta de la Republica, con el que inicia un proyecto de ley que permite la introducción de la televisión digital terrestre (Law of the National Television Council N°18.838). Mensaje N° 942–356, Santiago, Oct. 24, 2008.

(including TVN) received unexplained preferential treatment (each of them would get an automatic, free-of-charge 6MHz digital multiplex in the almost vacant UHF band, which they wanted to be perpetual).[34] At the same time, the prohibition to control more than one franchise in the same zone was maintained. Therefore, ownership concentration looked more difficult to attain in this sector than elsewhere.

Complications also existed for the former university channels. According to law, the terrestrial franchise allocated to UCTV and CHV's original operators, Universidad Catolica de Chile and Universidad de Chile, respectively, could not be transferred. Thus, the broadcasters UCTV and CHV were merely enjoying a temporary lease from their actual university owners. The new draft for digital terrestrial television stated that current franchisees will automatically be awarded a 6MHz digital multiplex channel. It also introduced further confusion as it was not clear whether the universities or the buyers of UCTV and CHV were entitled to 6MHz digital multiplex channels for free. These channels are valuable slices of spectrum and therefore highly coveted properties.

## Multichannel TV Platform

Pay-television in Chile is an oligopoly, like most markets worldwide: the main players are large multinational telecommunication firms that also dominate phone and broadband (landlines and mobile.) They are Spain's Movistar/Telefonica, Mexico's Claro (Telmex in its home country), United States's VTR Globalcom, and Italy's Entel (these last two have minority stakes of Chilean investors though). Their yearly estimated receipts from both cable and satellite TV range from \$803.9 to \$921.6 million (there are no publicly available official figures), an amount similar to the advertising expenditure in all Chilean media channels.[35] This income is generated from 1.9 million subscriptions.[36] Internet users in Chile are estimated at half of the population, while 38.8% of households are subscribed to either cable or satellite television. Yet these companies are not completely comparable to traditional media firms, as they are more focused on providing connectivity rather than producing content.

Until the middle of the 2000s, pay-television was predominantly provided by cable, and it was dominated by VTR Globalcom, a firm resulting from the merger of preexisting cable operators VTR and Metropolis/Intercom. Until 2009, VTR only participated in telecommunications, providing pay-television, landline phone service, and Internet connections. That year it was granted a 3G license to operate mobile phone services as well. VTR started operating mobile phone services in 2012. Eighty percent of the company belongs to the US-based Liberty Media (John Malone), and the remaining 20% belongs to Alvaro Saieh,[37] a Chilean businessman who also owns the newspaper consortium COPESA, plus six radio networks and Channel 22 UHF in Santiago (Table 21.11).[38]

In the second half of the decade, satellite television quickly grew while cable was relatively stalled. According to the latest available statistics from SUBTEL, the regulator of telecoms, more than 80% of pay-television subscriptions were controlled by three international holdings: VTR (USA plus Saieh), Mexico's Claro (an affiliate of Telmex), and Spain's Telefonica.[39]

34. Godoy, S: "Televisión digital en Chile: ¿es posible más y mejor TV para los chilenos?," *Temas de la Agenda Pública*, Año 4, N°31, Dirección de Asuntos Públicos UC, Santiago, Octubre 2009.

35. Source: own calculation based on rates advertised on each provider's website.

36. Source: SUBTEL, <http://www.subtel.cl/prontus_subtel/site/artic/20070212/pags/200702> (accessed on April 1, 2011).

37. Source: <http://www.bnamericas.com/company-profile/es/VTR_Globalcom_S,A,-VTR_Globalcom> (accessed on March 30, 2011).

38. Terra.cl: *Álvaro Saieh compra señal de televisión Más Canal 22*, July 6, 2010, <http://economia.terra.cl/noticias/noticia.aspx?idNoticia=201007062249_INV_79140440> (accessed on Feb. 1, 2011).

39. Source: SUBTEL, ibid.

Table 21-11. Share of Subscriptions of Pay-Television per Operator, 2006–2011

| OPERATOR / ORIGIN | 2006 | 2007 | 2008 | 2009 | 2010 | 2011[1] |
|---|---|---|---|---|---|---|
| Liberty Global (US) | 89.0 | 80.0 | 71.0 | 64.2 | 56.6 | 56.0 |
| VTR (US/Chile) | *82.0* | *72.0* | *63.0* | *53.9* | *46.3* | *45.0* |
| DirecTV (US) | *7.0* | *8.0* | *8.0* | *10.3* | *10.3* | *11.0* |
| Telmex/Claro (Mexico) | 4.0 | 4.0 | 10.0 | 18.3 | 19.2 | 20.0 |
| Telefonica (Spain) | 4.0 | 15.0 | 18.0 | 16.9 | 17.0 | 18.0 |
| Others | 2.0 | 2.0 | 1.0 | 0.6 | 7.2 | 6.0 |
| Estimated income, $ mil | | | | | | 921.6 |
| C4 | 97.0 | 99.0 | 99.0 | 99.4 | 92.8 | 94.0 |
| HHI | 6,805 | 5,489 | 4,457 | 3,632 | 2,907 | 2,870 |
| Noam Index | 3,929 | 3,169 | 2,573 | 2,097 | 1,679 | 1,657 |
| Number of Firms with Market Share >1% (*n*) | 3 | 3 | 3 | 3 | 3 | 3 |

*Source:* SUBTEL, own estimations of income based on published prices of subscriptions.
[1]Income estimate of 2011 considers the whole year, subscriptions only cover the period January–June.

Pay-television operators, and VTR in particular, refused to pay any fees for carrying the networks' contents, and these networks have sued VTR for this. Their perspective was summarized by TVN's CEO, who compared cable/satellite operators to "an imaginary discotheque owner who has grown rich by charging customers dearly while playing music from radio stations without paying them any copyright fees for the music that attracts the customers."[40] Until 2014, the courts rejected these claims given that terrestrial broadcasts were freely available. In addition, incumbents did not object to cable carrying their broadcasts for free before the 1990s. The chances of a monopoly seemed greater because pay-television's subscription-based income (which does not vary according to audience ratings) is long ago bigger than terrestrial television's. And subscribers have been steadily growing.[41]

The struggle of payment did not turned into open war because VTR distributed TVN's news channel, TVN 24 Horas, and UCTV's highbrow Canal 13 Cable for free. As legislation forbade terrestrial franchisees to operate more than one channel, cable was the only alternative to distribute these not-for-profit yet commercially operated outlets. In addition, VTR fostered local channels in the main regional cities where it operates[42]—a strategy that helps bolster local subscriptions as well as public sympathy for the distributor. Finally, the Law of Digital TV of 2014 established that pay operators must compensate the terrestrial networks, and also follow must-carry rules to favor local channels operating a terrestrial franchise.

Cable and satellite-only channels share three characteristics. First, taken individually, each of them has very low ratings. All of them put together attracted almost 25% of the audience share in 2010 (see Table 21.11); the rest went to the terrestrial broadcasters (audience measurements via People Meter include homes that have

40. Source: Mauro Valdes, CEO of TVN, Seminar "La Transformación de la Televisión," Circulo De Marketing Icare, Casapiedra, Santiago de Chile, Apr. 5, 2011. Valdes created the analogy of the discotheque and also pointed out that most of the audience of pay-television in Chile concentrates around eight channels, most of them the terrestrial ones.

41. S Godoy. "Televisión digital en Chile: aspectos regulatorios y modelo de negocio," *Cuadernos de Información* N°21, 2007–2, 74–81; Godoy, S: "Televisión digital en Chile: ¿es posible más y mejor TV para los chilenos?," *Temas de la Agenda Pública*, Año 4, N°31, Dirección de Asuntos Públicos UC, Santiago, Octubre 2009.

42. F. Fernández, S. Godoy, D. Grassau, P. Julio, S. Pellegrini, S. Puente, and J. Soto: *Ventanas y Espejos; televisión local en red*, El Mercurio/Aguilar, Santiago, 2009.

subscribed to pay-television as well as those without a subscription). Second, the most watched channels are predominantly dedicated to children and films or series. Finally, most belong to international holdings from the United States.

Table 21.12 depicts the 10 most watched cable and satellite-only channels in 2009 and 2011. The most watched channel in 2009 had 1.03 rating points (Discovery Kids) and 1.20 in 2011 (Disney Channel). Most of these outlets belong to US-based transnational holdings: Fox (Nickelodeon, Cinecanal), Time Warner (Cartoon Network, TNT, HBO, CNN), MTV (Nickelodeon, MTV), and the Discovery Channel.

## Film

As mentioned, filmmaking in Chile is still a relatively precarious craft made by small, isolated producers who are becoming increasingly more sophisticated and market-driven. This shift comes after decades of a militant, socially conscious perspective with a strong influence of the *cinema d'auteur* as a contrast to commercially

**Table 21-12. Audience Share of the Top 10 Most Viewed Cable/Satellite TV Channels, 2009–2011[1]**

| | 2009 | | | 2011 | | |
|---|---|---|---|---|---|---|
| *N°* | CHANNEL | RATINGS | % | CHANNEL | RATINGS | % |
| 1 | Discovery Kids (Children, USA) | 1.03 | 7.5 | Disney Channel (Children, USA) | 1.20 | 6.8 |
| 2 | Disney Channel (Children, USA) | 0.89 | 6.5 | Discovery Kids (Children, USA) | 1.03 | 6.6 |
| 3 | Cartoon Network (Children, USA) | 0.56 | 4.1 | Cartoon Network (Children, USA) | 0.72 | 6.3 |
| 4 | Disney XD (Children, USA) | 0.38 | 2.8 | Disney XD (Children, USA) | 0.61 | 6.2 |
| 5 | Canal Fox (Film, USA) | 0.31 | 2.2 | Canal Fox (Film, USA) | 0.43 | 6.0 |
| 6 | Nickelodeon (Children, USA) | 0.29 | 2.1 | Nickelodeon (Children, USA) | 0.39 | 6.0 |
| 7 | TNT (Film, USA) | 0.29 | 2.1 | Play House Disney (Children, USA) | 0.37 | 6.0 |
| 8 | HBO (Film, USA) | 0.23 | 1.7 | TNT (Film, USA) | 0.35 | 6.0 |
| 9 | Cinecanal (Film, USA) | 0.23 | 1.7 | Cinecanal (Film, USA) | 0.29 | 5.9 |
| 10 | CNN Chile (News, USA) | 0.22 | 1.6 | HBO (Film, USA) | 0.27 | 5.9 |
| | Remaining Channels (197) | 9.35 | 67.9 | Remaining channels (197) | 12.16 | 17.8 |
| | Total | 13.78 | 100.0 | | 17.82 | 100.0 |
| C4 | | | 20.8% | | | 26.0% |
| HHI | | | 144 | | | 383 |
| Noam Index | | | 59 | | | 191 |
| Number of Firms with Market Share >1% (*n*) | | | 6 | | | 4 |

*Source:* Time Ibobe, Department of Studies TVN.

[1]Due to the extremely low viewership figures of cable/satellite channels and high margin of error of audience measurements (+ /– 2.5%), only the top 10 channels were considered for the calculations, for referential purposes.

Table 21-13. Attendance to Cinemas in Chile, 2000–2010

| YEAR | ATTENDANCE TO CINEMAS (TICKETS SOLD) | VARIATION FROM PREVIOUS YEAR | % CHANGE |
|---|---|---|---|
| 2000 | 9,340,476 | –292,797 | –3.04% |
| 2001 | 11,064,343 | 1,723,867 | 18.46% |
| 2002 | 11,454,115 | 389,772 | 3.52% |
| 2003 | 11,442,377 | –11,738 | –0.10% |
| 2004 | 12,658,778 | 1,216,401 | 10.63% |
| 2005 | 10,722,860 | –1,935,918 | –15.29% |
| 2006 | 10,524,251 | –198,609 | –1.85% |
| 2007 | 11,455,550 | 931,299 | 8.85% |
| 2008 | 11,886,801 | 431,251 | 3.76% |
| 2009 | 14,442,596 | 2,555,795 | 21.50% |

*Source:* Chilean Chamber of Multiplex Cinema Exhibitors.

oriented movies from Hollywood. After the end of the military dictatorship in 1990, the combination of increased freedom of speech, growing expertise in producing commercials, targeted money incentives from the State for preproduction, and cheap digital cameras, Chilean filmmakers manage to screen nearly a dozen long features per year, reaching 14 new productions in 2009, an historically high figure.[43] Digitalization has cut costs of long feature films from an average of $500,000 to $1 million: for these very modest productions, by European or North American standards, film stock and cameras are a very high proportion of overall costs.[44] Yet attendance to cinemas has stayed relatively flat throughout the 2000s, as shown in Table 21.13.

In contrast, exhibition is mostly in hands of foreign cinema chains: Cinemark and Cine Hoyts from the United States. Together, they own 86% of all cinema venues in Chile, most of which are located in highly popular shopping malls.[45]

## TELECOMMUNICATIONS MEDIA

In contrast to mass media, where local firms still play an important role, telecoms are an oligopoly dominated by four big multinationals, three foreign (Telefonica, Claro, and VTR) and the Italian/Chilean Entel. As a regulated sector, their revenues are publicly available at SVS, the Exchange and Securities' regulator (*Superintendencia de Valores y Seguros*), and are shown in Table 21.14. These figures do not include income generated by their pay-television operations as Chilean (preconvergent) legislation considers this to be different from pay-television and there is no legal obligation to disclose any financial details. These anachronistic distinctions and subsequent lack of reliable data prevent a complete view of phenomena such as ownership concentration in the convergent scenario where communications and media are taking place.

In each subsector there is a market leader among these big four holdings. The following

43. CAEM: *El Cine en Chile en el 2009*, Cámara de Exhibidores Multisalas de Chile A.G., Santiago, March 2010, <http://www.apct.cl/descargas/El.cine.en.Chile.en.el.2009-Marzo2010.pdf> (accessed on Apr. 1, 2011).

44. Fernando Acuña, former Executive Producer of Nueva Imagen Producciones and former Production Manager of Canal 13 UCTV, interviewed on Apr. 1, 2011.

45. CAEM, op. cit.

Table 21-14. Income of Main Telecom Companies 2004 vs. 2008, US$mil[1]

| | 2004 NOMINAL US$MIL | % | 2008 NOMINAL US$MIL | % |
|---|---|---|---|---|
| Telmex Claro (Mexico) | 72.1 | 3.2 | 126.9 | 4.4 |
| Entel (Italy/Chile) | 1,088.80 | 48 | 1,677.50 | 58 |
| Telefónica (Spain) | 1,105.10 | 48.7 | 1,077.50 | 37.3 |
| Liberty VTR[2] (USA/Chile) | 2.5 | 0.1 | 9.8 | 0.3 |
| Total | 2,268.50 | 100 | 2,891.60 | 100 |
| C4 | | 100.0% | | 100.0% |
| HHI | | 4,686 | | 4,775 |
| Noam Index | | 2,705 | | 2,757 |
| Number of Firms with Market Share >1% (*n*) | | 3 | | 3 |

*Source:* SVS.
[1]Exchange rate 2004 and 2008: 636 pesos per dollar. Nominal values for each year, not corrected.
[2]Liberty's VTR figures show only its fixed telephone business, not pay-TV.

tables show their respective share of subscriptions in landline phone service (Table 21.15), mobile phone service (Table 21.16), pay-television systems (Table 21.11), and fixed Internet connections (Table 21.17). When each of these subsectors is examined individually, smaller providers emerge (such as Nextel, a niche mobile company, and Telsur, a regional telecom provider in southern Chile), but their weight is still negligible compared to Telefonica, Entel, Claro, and VTR. The spectrum and telecoms regulator, SUBTEL, does not provide information about revenue.

## Wireline Telecom

Table 21-15. Fixed Telephone Lines per Operator 2008 vs. 2010

| | 2008 | % | 2010 | % |
|---|---|---|---|---|
| Entel (Italy/Chile) | 178,515 | 5 | 195,972 | 5 |
| Telefonica (Spain) | 2,120,974 | 58 | 1,944,739 | 53 |
| Telmex's Claro (Mexico) | 113,508 | 3 | 188,965 | 5 |
| Telsur (Chile) | 201,306 | 6 | 180,759 | 5 |
| Liberty's VTR (US/Chile) | 613,604 | 17 | 638,176 | 18 |
| Other | 3,208 | 8 | 308,895 | 8 |
| Total | 1,310 | 100 | 3,646,471 | 100 |
| Revenue Estimate ($US) | 1,045 | | | |
| C4 | | 86 | | 81 |
| HHI | | 3,723 | 3,546 | |
| Noam Index | | 1,665 | 1,448 | |
| Number of Firms with Market Share >1% (*n*) | | 5 | | 5 |

*Source:* SUBTEL.

## Table 21-16. Total Mobile Subscribers per Operator

| | 2008 | SHARE | 2010 | SHARE |
|---|---|---|---|---|
| Telmex Claro (Mexico) | 2,749.21 | 19 | 4,465.52 | 22 |
| Entel (Italy/Chile) | 5,741.09 | 39 | 7,229.29 | 36 |
| Telefonica (Spain) | 6,299.09 | 43 | 8,129.25 | 41 |
| Total | 14,796.59 | 100 | 19,851.62 | 100 |
| C4 | | 100 | | 99 |
| HHI | | 3,731 | | 3,461 |
| Noam Index | | 2,154 | | 1,998 |
| Number of Firms with Market Share >1% (*n*) | | 3 | | 3 |

*Source:* SUBTEL. Total estimated revenue in 2008: $1.59 billion.

## Table 21-17. Fixed-Line Internet Connections per Operator

| | 2008 | % | 2010 | % |
|---|---|---|---|---|
| Telefonica (Spain) | 662,113 | 46 | 788,996 | 43 |
| Telmex (Mexico) | — | 0 | 127,368 | 7 |
| Telsur (Chile) | 74,919 | 5 | 92,771 | 5 |
| Liberty's VTR (US/Chile) | 590,074 | 41 | 698,146 | 38 |
| Entel (Chile/Italy) | 32,257 | 2 | 21,632 | 1 |
| Other | 79,646 | 6 | 89,886 | 5 |
| Total | 1,439,009 | 100 | 1,818,799 | 100 |
| C4 | | 98.0 | | 93.0 |
| HHI | | 3,862 | | 3,393 |
| Noam Index | | 1,931 | | 1,517 |
| Number of Firms with Market Share >1% (*n*) | | 4 | | 5 |

*Source:* SUBTEL. Total estimated revenue in 2008: $276 million.

## Wireless Telecom

Mobile phone service has expanded quickly in Chile: subscriptions exceeded the local population in 2010 when mobiles reached 19.8 million, twice as many as in 2005. Only three of the four large landline providers are strong: VTR, the main player in pay-television, started offering 3G mobile in 2012 and struggles to dominate the so-called triple-play of fixed/mobile phone service, Internet connections, and pay-television (where the other companies were weaker than VTR).[46]

46. S Godoy. "Televisión digital en Chile: aspectos regulatorios y modelo de negocio," *Cuadernos de Información* N°21, 2007–2, 74–81.

## INTERNET MEDIA

### Internet Service Providers

The telecom companies (Claro, Entel, Telefonica, and VTR) provide Internet connections (Tables 21.17 and 21.18) and also offer fixed and mobile phone service (Tables 21.15 and 21.16), as well as pay-television. Mobile connections are relatively recent but are growing fast: they added up to 5.8 million by the end of 2010.

### Online News

Concerning news sites online, the only publicly available comparative figures for audience impact are depicted in Table 21.19. As these

Table 21-18. Mobile Line Internet Connections per Operator, 000s

| | 2009 | | | | 2010 | | | |
|---|---|---|---|---|---|---|---|---|
| | 2G | 3G | 2G + 3G | % | 2G | 3G | 2G + 3G | % |
| Telefonica (Spain) | 2,058.0 | 266.8 | 2,324.8 | 32.1 | 2,643.90 | 345.7 | 2,989.6 | 51.2 |
| Telmex Claro (Mexico) | 694.6 | 171.8 | 866.4 | 12 | 801.9 | 275.8 | 1,077.7 | 18.5 |
| Entel (Italy/Chile) | 3,848.2 | 191.8 | 4,040.0 | 55.9 | 1,152.0 | 617.1 | 1,769.1 | 30.3 |
| Total | 6,600.8 | 630.4 | 7,231.2 | 100 | 4,597.8 | 1,238.6 | 5,836.4 | 100 |
| C4 | | | | 100 | | | | 100 |
| HHI | | | | 4,298 | | | | 3,884 |
| Noam Index | | | | 2,482 | | | | 2,242 |
| Number of firms with market share >1% (*n*) | | | | 3 | | | | 3 |

*Source:* SUBTEL.

Table 21-19. Unique Visitors of Online News Sites, 000s, 2011–2012[1]

| | 2011 (DECEMBER) | | 2012(MAY) | |
|---|---|---|---|---|
| COMPANY | UNIQUE VISITORS (000S) | MARKET SHARE % | UNIQUE VISITORS (000S) | MARKET SHARE % |
| El Mercurio (Chile) | 7,204.1 | 57.8 | 6,030.5 | 54.2 |
| Copesa (Chile) | 3,333.1 | 26.8 | 3,293.1 | 29.6 |
| Cía. Chilena Comunicaciones (Chile) | 1,021.9 | 8.2 | 891.5 | 8 |
| La Nacion (Public) | 557.3 | 4.5 | 630.9 | 5.7 |
| Iberoamericana (Spain) | 311.5 | 2.5 | 279.9 | 2.5 |
| Total | 12,427.9 | 100 | 11,125.8 | 100 |
| C4 | | 74.0 | | 72.1 |
| HHI | | 1,656 | | 1,651 |
| Noam Index | | 740 | | 738 |
| Number of Firms with Market Share >1% (*n*) | | 5 | | 5 |

*Source:* IAB/Comscore.

[1]Only IAB affiliates are considered.

Table 21-20. Top 10 Most Popular Websites in Chile, 2010

| | TOTAL UNIQUE VISITORS, 000S | % |
|---|---|---|
| 1. Google Sites (includes YouTube) | 6,797 | 17.21 |
| 2. Facebook.Com | 6,219 | 15.75 |
| 3. Microsoft Sites (includes MSN) | 6,069 | 15.37 |
| 4. Terra—Telefonica | 4,12 | 10.43 |
| 5. Yahoo Sites | 3,397 | 8.60 |
| 6. Wikimedia Foundation Sites | 3,215 | 8.14 |
| 7. Empresa El Mercurio S.A.P. | 2,888 | 7.31 |
| 8. Taringa.net | 2,495 | 6.32 |
| 9. Grupo La Tercera | 2,175 | 5.51 |
| 10. Hi-Media Group (includes Fotolog) | 2,114 | 5.35 |
| Total | 39,489 | 100.00 |
| C4 | | 58,8 |
| HHI | | 1,182 |
| Number of Firms with Market Share >1% (*n*) | | 10 |
| Noam Index | | 374 |

*Source:* ComScore Media Metrix.

numbers only include the sites affiliated with the International Advertising Bureau (IAB), they should be taken with care. Those related to El Mercurio group represent 57.8% of unique visitors quantified by IAB in 2011 and 54.2% in 2012. This is followed by COPESA, the other dominant newspaper holding, and by two radio consortia: Compañia Chilena de Comunicaciones (owner of Radio Cooperativa, one of the four main radio news networks) and the Spanish consortium Iberoamericana (owner of ADN radio, another radio news network). The table includes the state-owned La Nacion that disappeared in 2012.

## Search Engines

There are no significant local search engines to consider. Most Chilean portals refer to sites such as Google and Yahoo—which have local versions with the suffix "cl"—for these purposes. Table 21.20 shows the 10 most popular sites in Chile in terms of unique visitors measured by ComScore. In 2010, the different Google sites attracted 17.2% of visits compared to 8.6% on Yahoo.

# CONCLUSION

## Qualitative Considerations: Concentration and Political Diversity

How has consolidation of ownership affected the pluralism of media and the diversity of voices? Opinions in this respect are widely divided in Chile. There is a segment of activists, academics, politicians, NGOs, and professionals who are convinced that media concentration is on the rise and that media diversity and pluralism are jeopardized as a result. Many of them belong to the political left, but not all. They are generally deeply suspicious of the supposedly conservative ideology of Chilean media owners—especially of the newspaper duopoly of COPESA and El Mercurio, owned by Alvaro Saieh and the Edwards family, respectively.[47] Both conglomerates thrived during the Pinochet years (1973–1990) when all other newspapers were shut down. The market was left to these two active supporters of the regime and promoters of its open-market economic policies within a politically repressive environment.[48] Although Saieh bought COPESA from the Picó Cañas family in 2000 and did not previously participate in the group's newspapers (*La Tercera* and *La Cuarta*), he held important posts during the military regime and became very wealthy.[49]

After the influence of General Pinochet faded away (he died in 2006), preoccupation about media ownership shifted from the old pro-dictatorship versus pro-democracy divide to the more subtle debate between commercial and non-commercial media. Those sharing this latter view highlight an apparently irresistible expansion of private, for-profit, and occasionally multinational media holdings combined with the decline of university television, the increase of commercialization of the public broadcaster TVN, and the difficulties faced by community media, notably radio. Commercially oriented media, say some of these critics, are not interested in genuine information and news, but in entertainment and advertising.[50]

Most media owners and executives have a completely different view of media concentration and ownership. They are grouped in associations such as ARCHI (radio), ANATEL (terrestrial television, including the public broadcaster TVN), ANP (newspapers), and ANDA (advertisers), as well as industry consultants and some government officials from the technical branches of government such as SUBTEL or the Ministry of Economic Development (the one in charge of promoting the Internet for economic growth.) This group, more influential and wealthy than the one described previously, has a market-focused view of the situation and stresses the fiercely competitive nature of the Chilean media landscape. They are proud of the relative technical quality, content diversity, and pluralism available to the Chilean public, especially when contrasted to other Latin American countries, where the political and economic power of conglomerates such as Brazil's Globo, Mexico's Televisa, or Argentina's Clarin is considerably higher than that of the newspaper duopoly of El Mercurio and COPESA (which are irreconcilable market rivals).[51]

In addition, terrestrial television, the most powerful medium in terms of news impact, is fragmented among public, university, private, national, and multinational operators, none of which has (at least until now) monopoly power of its own. The new entrants, Time Warner and the Luksic family, who bought CHV and Canal 13 UCTV, respectively, in 2010, are not politically conservative either.[52] In radio, despite the recent ownership concentration described earlier, there are literally more than 1,500 radio stations available nationwide. The medium is consistently

---

47. See for instance K. Dermota: *Chile Inédito, periodismo en democracia*, Ediciones B, Barcelona, 2002; M. O. Monckeberg: *Los Magnates de la Prensa en Chile. Concentración de los Medios de Comunicación en Chile*, Random House Mondadori, Santiago, 2009. Author and journalist Maria Olivia Monckeberg has said her book has been deliberately ignored by the mainstream media (El Ciudadano: María Olivia Mönckeberg: *"En los medios no tengo la posibilidad de hacer periodismo,"* Nov. 12, 2009, <http://www.elciudadano.cl/2009/11/12/maria-olivia-monckeberg-%E2%80%9Cen-los-medios-no-tengo-la-posibilidad-de-hacer-periodismo%E2%80%9D/>, accessed on Mar. 1, 2011).

48. E. Tironi & G. Sunckel: "Modernización de las comunicaciones y democratización de la política. Los medios en la transición a la democracia en Chile," *Estudios Públicos* N° 52, primavera 1993, 215–246.

49. La Nación Domingo: *La apuesta del empresario por ampliar su poder mediático: Todos Los Caminos Conducen A Saieh*, 12 September 2004, quoted in <http://miguelpaz.blogspot.com/2004/09/reportaje-alvaro-saieh-siete7.html> (accessed on Mar. 1, 2011); La Nación: *Álvaro Saieh Bendeck Escribe A La Nación. La historia según Copesa*, July 31, 2005, <http://www.lanacion.cl/prontus_noticias/site/artic/20050730/pags/20050730194208.html> (accessed on Feb. 1, 2011).

50. E. Carmona: *Implicaciones éticas de la concentración de medios y sus vinculaciones políticas en América Latina*, Presentation at the seminar "Periodismo, Ética y Democracia" (Journalism, Ethics, and Democracy), CIESPAL, Quito, Ecuador, Oct. 20–22, 2009, linked at Fucatel.cl: <http://www.observatoriofucatel.cl/wp-content/uploads/2010/04/Ponencia-Ernesto-Carmona-en-Ciespal.pdf> (accessed on March 1, 2011).

51. Eduardo Arriagada, expert on media business, Universidad Catolica de Chile's School of Communication, interviewed on April 14, 2011.

52. Eduardo Arriagada, expert on media business, Universidad Catolica de Chile's School of Communication, interviewed on April 14, 2011.

ranked among the most credible institutions in the country, together with the Fire Brigade and the *Carabineros*, the uniformed police force.[53]

Pay-television is highly concentrated, and customer satisfaction with the service is weak due to mediocre customer service. But there are no reports about this industry restricting freedom of expression and/or the availability of news. On the contrary, cable distributors facilitated the creation in 2009 of the news-only channels CNN Chile and TVN 24 Horas (broadcast television operators were forbidden by law to operate more than one channel).[54]

What about media ownership being transparent? According to Chilean corporate law, as in most countries, only companies that trade shares on the stock exchange need to regularly inform the public about their ownership composition, financial situation, and similar information. Media companies do not have such a structure and therefore are not obligated to provide information. The exception is terrestrial television: franchisees must be organizations based in Chile led by Chilean executives, even if the investment capital comes from other countries as with CHV, La Red, and Channel 2 Telecanal. Furthermore, the industry's regulator, CNTV, is entitled to request from the franchisees "all the information necessary to accomplish its (regulatory) functions; those requested are obliged to present the requested information."[55]

Nevertheless, there are no big mysteries concerning who owns the main telecommunication companies, terrestrial channels, pay-television operators, newspapers, and radio stations: Chilean media outlets are relatively small companies that are far less structurally complex than their peers in larger industrialized countries.[56] It is much more difficult to know the ownership composition of the many hundreds of smaller media outlets—especially radio stations. But they are not a concern in these matters.

## Qualitative Considerations: International Ownership and Alternative Media

Chile's communication and media markets are a very open to the world. Telecommunications are mostly owned by foreigners. Being a small country, there are no examples of exports like Mexican and Brazilian *telenovelas* or Hollywood films. Chilean-only media companies are relatively small, and despite their cultural affinity with local audiences, they face tough international competition. A significant example is the Spanish radio holding Iberoamericana Radio Chile, who concentrates the largest audience shares and revenues with its menu of attractive and complementary music networks (including the only station fully dedicated to Chilean music). More recently, Time Warner bought Chilevision, one of the most successful broadcast television networks, from the current President Sebastian Piñera after he took office in 2010.

Within the pro-market, liberal vision that prevails in Chile—in contrast to Argentina, Venezuela, Ecuador, and Bolivia—the only serious legal effort to prevent foreign operators was the reciprocity clause introduced in 2003, by which foreign investors seeking a radio license must demonstrate that a Chilean can get one in their own countries, too.[57] Yet this clause applies only to radio, and it came too late to prevent the entrenchment of companies such as Iberoamericana Radio Chile.[58]

Although largely ignored by mainstream media, there are also concerns about the excessive centralization of the media and the funding sources in Santiago, notably advertising and its related audience measurement methods. This

53. P. Gómez, S. Godoy, W. Pullen: "El nuevo consumidor chileno exige empresas más amigables y responsables," *Qué Pasa*, N° 2073, Dec. 31, 2010, 50–67, <http://www.quepasa.cl/articulo/16_4210_9.html>.

54. S. Godoy & M. Gronemeyer: Mapping Digital Media: Country Report -Chile. A report by the Open Society Foundations. London, Oct. 23, 2012. http://www.opensocietyfoundations.org/reports/mapping-digital-media-chile.

55. Law of the National Television Council N°18,838 of 1989, article 12-d.

56. Eduardo Arriagada, expert on media business, Universidad Catolica de Chile's School of Communication, interviewed on April 14, 2011.

57. Law on the Press N°19,733 of 2001, article 9.

58. Godoy & Gronemeyer, op. cit.

has harmed regional and local outlets, most of which have been absorbed by companies of national reach based in the capital. Many others have simply disappeared.

Finally, there are also concerns among grassroots and community organizations about the limited audience impact and resources of community citizen media as alternatives to market-driven, professional mainstream media outlets. In Chile there are no special safeguards or incentives for these types of media. In fact, legislation severely restricts the conditions of community radio to operate: they only can apply to a single channel at the high end of the FM band; licenses last three years; operators can neither sell advertising, make political propaganda, nor form networks; and their transmitting power is limited to 1 watt, which is even lower than a light bulb or a cell phone. A proposal to end these discriminatory conditions is being discussed in Congress at the time of this writing; among other things it proposes increasing transmission power up to 25 watts in urban areas; however, the outcome is not yet clear.[59]

## Quantifying Ownership Concentration in Chile

In summary, after most restrictions from the Pinochet years have gradually disappeared, many experts, NGOs, and citizen organizations consider ownership concentration as one of the top threats to freedom of expression in this deeply market-oriented, capitalistic nation, in contrast to countries where the main concern is terrorism, organized crime, or government intervention in the media.

Yet it has been a rather imprecise debate. There are almost no examples of quantification, such as those performed for this study, aside from rough shares of audience, revenue, or broadcasting licenses.[60] Besides, numbers are incomplete and inconsistent. Their quality is varied, ranging from good (as in terrestrial television and telecommunications) to dubious (newspapers and radio) to even nonexistent (radio again, as well income figures, or audience measurements outside the largest cities). Furthermore, data are isolated between different media types and markets—an important defect in times of growing convergence.

Anyway, an effort to quantify the phenomenon by using the Noam Index largely confirms the qualitative picture previously described. And despite the inaccuracies of the existing information, there are many valuable findings.

The first finding is the lopsided availability of audience and revenue share figures needed to quantify concentration. While audience measurements are in fair supply, only broadcast television and telecommunications show reliable income figures, as they are regulated sectors. For the rest of the communications markets there are only broad estimates available (Table 21.21).

When analyzing the concentration of income shares alone in two moments selected for this chapter (see Table 21.20), it is interesting to note that telecommunications (excluding pay-television) show the highest Noam Index value (2,757), followed by broadcast television (2,430) and newspapers (1,893). Yet concentration (i.e., the index) dropped in broadcast television and newspapers compared to the previous values calculated for both sectors (Table 21.22).

Audience share concentration (Table 21.21) also shows some unusual aspects within many predictable traits. Perhaps the most important are the comparatively higher levels of Noam concentration indexes in the growing sectors of mobile Internet and telephony (2,242 and 1,998,

59. Ibid. See also: Saez, C: *TV Digital en Chile. Politicas Publicas y Democracia*. Santiago: Editorial Universitaria, 2014.

60. An interesting exception was a report from FNE, the National Economic Prosecutor Office (Fiscalía Nacional Económica) warning about the purchase of a rival holding by Iberoamericana Radio Chile, owned by the Spanish group Prisa in 2007. FNE used the Herfindahl-Hischmann Index to existing audience and revenue estimations in radio, and concluded that Iberoamericana's subsequent market power could be harmful. This recommendation was ignored by the Free Competition Court (Tribunal de la Libre Competencia) given higher concentration levels in other (non-media) markets. See "Informe a TDLC sobre compra de acciones de Iberoamerican Radio Chile por GLR Chile" (Report to the Free Competition Court concerning the purchase of Iberoamerican Radio Chile's shares by GLR Chile), FNE, March 10, 2007; and "Informe a TDLC en compra de acciones de Iberoamerican Radio Chile por GLR" (Report to the Free Competition Court concerning the purchase of Iberoamerican Radio Chile's shares by GLR), FNE, Apr. 4, 2007. Available at <http://www.fne.gob.cl/?s=informe+iberoamerican+radio+chile> (accessed on March 1, 2013).

## Table 21-21. Noam Indexes of Concentration in Different Chilean Media and Communication Industries: Share of Revenues

| | LATEST NOAM INDEX | RANK | PREVIOUS NOAM INDEX | GROWTH | RANK |
|---|---|---|---|---|---|
| Telecommunications | 2,757 | 1 | 2,705 | 51 | 1 |
| Broadcast TV | 2,430 | 2 | 2,620 | –190 | 3 |
| Newspapers | 1,893 | 3 | 1,960 | –68 | 2 |
| Magazines | | | | | |
| Radio | | | | | |
| Pay-television operators | | | | | |
| Pay-television channels | | | | | |
| Cinema | | | | | |
| Landlines | | | | | |
| Mobile phones | | | | | |
| Fixed Internet | | | | | |
| Mobile Internet | | | | | |
| Online news sites | | | | | |
| Websites | | | | | |

*Source:* Own calculations based on previous tables. Blank spaces show no reliable data available.

## Table 21-22. Noam Indexes of Concentration in Different Chilean Media and Communication Industries: Audience Share

| AUDIENCE SHARE | LATEST NOAM INDEX | RANK | PREVIOUS NOAM INDEX | GROWTH | RANK |
|---|---|---|---|---|---|
| Mobile Internet | 2,242 | 1 | 2,482 | –240 | 7 |
| Mobile phones | 1,998 | 2 | 2,154 | –156 | 5 |
| Pay-television operators | 1,657 | 3 | 2,097 | –440 | 9 |
| Landlines | 1,448 | 4 | 1,616 | –169 | 6 |
| Magazines | 1,408 | 5 | 2,102 | –694 | 11 |
| Newspapers | 1,405 | 6 | 1,969 | –564 | 10 |
| Fixed Internet | 1,385 | 7 | 1,727 | –342 | 8 |
| Radio | 842 | 8 | 858 | –16 | 3 |
| Online news sites | 738 | 9 | 740 | | 2 |
| Broadcast TV | 430 | 10 | 544 | –114 | 4 |
| Websites | 374 | 11 | | | |
| Pay-television channels | 191 | 12 | 59 | 133 | 1 |
| Cinema | | | | | |
| Telecommunications | | | | | |

*Source:* Own calculations based on previous tables. Blank spaces show no reliable data available.

respectively) and pay-television operators (not the channels per se). Landlines, magazines (an increasingly weak actor though), newspapers, and fixed Internet come next, ranging between 1,448 and 1,385. The rest of the communication sectors fall below 1,000 points. Radio (842) appears even more concentrated than broadcast television (430), which is usually a highly consolidated market in the rest of Latin America. Again, when comparing the latest concentration values with those from one or two years before, concentration falls in all sectors except the already non-concentrated sector of pay-television channels (in which great diversity is available through the handful of available distributors).

In summary, and noting the fact that figures in Chile are problematic and inconsistent, there is evidence of higher concentration in telecommunications (and for pay-television distributors, who are the same companies) than among the traditional media (where newspapers are confirmed as the less diverse). Also, concentration is higher in terms of revenue than in terms of audience share. This is relevant because income cannot be shared in the way that media contents are shared by audiences: unlike money, contents are non-excludable and non-rival in consumption, the classic attributes of public goods. So analyzing audience and income concentration separately is misguiding in terms of real market power—a combination of both would be better, as recognized in the literature long ago.[61] The same occurs when different media platforms are analyzed in isolation. Technological convergence and corporate mergers are making this distinction gradually risky and even dangerous, because they may give a false impression of diversity (or concentration) depending on how the discussion is framed.

## Bibliography

### 1. Books and Articles


A. Cerda and P. Peña: Chile, in A. Finlay (Ed.):*Global Information Society Watch* 2009, Uruguay: APC/Hivos,. 104–105.

J. Curran J. and J. Seaton: *Power without Responsibility. The Press and Broadcasting in Britain,* 5th ed., London & New York: Routledge, 1997.

K. Dermota: *Chile Inédito, Periodismo en Democracia,* Barcelona: Ediciones B, 2002.

F. Fernández, S. Godoy, D. Grassau, P. Julio, S. Pellegrini, S. Puente, and J. Soto: *Ventanas y Espejos; televisión local en red,* Santiago: El Mercurio/Aguilar, 2009.

E. Fox: *Latin American Broadcasting, from Tango to Telenovela,* Luton, Hertforshire: John Libbey/ University of Luton, 1997.

S. Godoy. "Televisión digital en Chile: aspectos regulatorios y modelo de negocio," *Cuadernos de Información*N°21, 2007–2002, 74–81.

S. Godoy: *An Unlikely Mixture. An Evaluation of Chile's Market Oriented Model of Public Television,* Köln: Lambert Academic Publishing, 2009.

S. Godoy: "Televisión digital en Chile: ¿es posible más y mejor TV para los chilenos?," *Temas de la Agenda Pública,* Año 4, N°31, Dirección de Asuntos Públicos UC, Santiago, Octubre 2009.

P. Iosifidis: "Cómo medir la concentración medial," *Cuadernos de Información* N°12, 1997, 94–111.

P. Iosifidis, J. Steemers, and M. Wheeler: *European Television Industries,* London: BFI, 2005.

M. Lasagni, P. Edwards, and Bonnefoy: *La Radio en Chile,* 3rd ed., Santiago: CENECA, 1988.

D. McQuail: *Mass Communication Theory.* 5th ed., London & Thousands Oaks: Sage, 2005.

M. O. Monckeberg: *Los Magnates de la Prensa en Chile. Concentración de los Medios de Comunicación en Chile,* Santiago: Random House Mondadori, 2009.

Saez, C: *TV Digital en Chile. Politicas Publicas y Democracia.* Santiago: Editorial Universitaria, 2014.

E. Tironi and G. Sunckel: "Modernización de las comunicaciones y democratización de la política. Los medios en la transición a la democracia en Chile," *Estudios Públicos*N° 52, primavera 1993, 215–246.


### 2. Reports and Other Documents


CAEM: *El Cine en Chile en el 2009,* Cámara de Exhibidores Multisalas de Chile A.G., Santiago,


61. P. Iosifides: "Cómo medir la concentración medial," *Cuadernos de Información* N° 12, 1997, 94–111.

March 2010, <http://www.apct.cl/descargas/El.cine.en.Chile.en.el.2009-Marzo2010.pdf> (accessed on April 1, 2011).

Cámara Chilena Del Libro: *Informe estadístico 2010*, Agencia chilena I.S.B.N., Santiago, Chile, 2011. <http://www.camaradellibro.cl/archivos/estadisticas/isbn2010.pdf> (accessed on Sept. 15, 2011).

C. Castro. "Chile," in *Industrias de Contenidos en Latinoamérica*. Documento de Grupo de Trabajo eLAC2007, Santiago: CEPAL/ECLAC, 2008, 56–60.

S. Godoy and M. Gronemeyer: *Mapping Digital Media: Country Report—Chile*. A report by the Open Society Foundations, London, Oct. 23, 2012. <http://www.opensocietyfoundations.org/reports/mapping-digital-media-chile>.

Instituto Nacional De Estadísticas: *Cultura y tiempo libre, Informe Anual 2009*. Instituto Nacional de Estadísticas, Santiago, 2010. <http://www.ine.cl/canales/menu/publicaciones/calendario_de_publicaciones/pdf/291110/cult09_29112010.pdf> (accessed on Sept. 17, 2011).

M. Larenas, C. Ramos &M. Siegel (2002): *El fenómeno de los diarios gratuitos, una nueva modalidad de hacer periodismo*. Tesis para optar al grado de Licenciado en Comunicación Social. Santiago: Escuela de Periodismo, Universidad Diego Portales. Available at <http://www.comunicacionyletras.udp.cl/files/Larenas_Ramos_y_Siegel.pdf> (accessed on Sept. 15, 2011).

Ministerio del Poder Popular para la Comunicación e Información de Venezuela (2007): Libro blanco sobre RCTV, Caracas, Venezuela. Available at <http://www.rnv.gov.ve/noticias/docs/libro_blanco_RCTV-Web.pdf>.

C. Nissen: *Public Service Broadcasting in the Information Society*. Strasbourg: Council of Europe, Directorate General of Human Rights, Media Division, 2006. Available at <http://www.coe.int/t/dghl/standardsetting/media/Doc/H-Inf(2006)003_en.pdf> (accessed on Dec. 10, 2011).

Reporters Without Borders: *Press Freedom Index* 2010, <http://en.rsf.org/press-freedom-index-2010,1034.html> (accessed on Sept. 15, 2011).

Secretaría de Comunicación y Cultura (SECC): *Reseña de Medios N°32: Perspectivas de la Radio en Chile*, Santiago: SECC, 1996.

Transparency International: *Corruption Perceptions Index* 2010 *Results*, <http://www.transparency.org/policy_research/surveys_indices/cpi/2010/results> (accessed on Sept. 15, 2011).

UNDP (2010): *Human Development Report 2010. The Real Wealth of Nations: Pathways to Human Development*, New York: UNDP, <http://www.beta.undp.org/content/dam/undp/library/corporate/HDR/HDR_2010_EN_Complete_reprint-1.pdf> (accessed on Sept. 20, 2011).

## 3. Websites

http://www.adnradio.cl/.
http://www.anp.cl/.
http://www.camaradellibro.cl
http://www.cntv.cl
http://www.elmostrador.cl/
http://www.emol.com
http://es.rsf.org/
http://www.fne.gob.cl/
http://www.lanacion.cl/
http://www.latercera.com/
http://www.megatime.cl
http://www.mercurio.cl
http://www.observatoriofucatel.cl/
http://www.opensocietyfoundations.org
http://www.publimetro.cl/
http://www.quepasa.cl
http://www.radiotierra.cl
http://www.radiounochile.cl/
http://www.subtel.cl
http://www.transparency.org/
http://www.tvn.cl/
http://www.undp.org

## 4. Interviews

Fernando Acuña, former executive producer of Nueva Imagen Producciones and former production manager of Canal 13 UCTV, interviewed on April 1, 2011.

Prof. Eduardo Arriagada, expert on media business, Universidad Catolica de Chile's School of Communication, interviewed on April 14, 2011.

Prof. Claudia Lagos, coordinator of the Freedom of Expression Program, Institute of Communication and Image, Universidad de Chile, interviewed on Dec., 2010.

## 5. Newspaper Articles and Presentations

E. Carmona: *Implicaciones éticas de la concentración de medios y sus vinculaciones políticas en América Latina*, Presentation at the seminar "Periodismo, Ética y Democracia" (Journalism, Ethics, and Democracy), CIESPAL, Quito, Ecuador, Oct. 20–22, 2009, linked at Fucatel.cl: <http://www.observatoriofucatel.cl/wp-content/uploads/

2010/04/Ponencia-Ernesto-Carmona-en-Ciespal.pdf> (accessed on Mar. 1, 2011).
El Mercurio: *Matías Lira, cineasta y socio de BF Distribution, plantea un nuevo modelo de negocios: La fórmula para tener nuestro "Chilewood,"* Sept. 25, 2011, B5, <http://diario.elmercurio.com/2011/09/25/economia_y_negocios/enfoques/noticias/60ddb94c-1add-45c7-b7bc-bd71f61ff2a8.htm> (accessed on Sept., 2011).
El Mercurio: *Resurge debate sobre concesiones radiales de grupo PRISA,* Jan. 12, 2011, available at <http://www.mercurio.cl/2011/01/12/espectaculos/mas/noticias/46B6E44A-2C2D-4ED1-8546-7FEC9913AC08.htm?id={46B6E44A-2C2D-4ED1-8546-7FEC9913AC08}> (accessed on Jan. 15, 2011).
El Mercurio Online: *Corte Suprema autoriza fusión que crea al mayor grupo radial en historia de Chile,* Nov. 22, 2007, <http://www.emol.com/noticias/nacional/detalle/detallenoticias.asp?idnoticia=282784> (accessed on Feb. 1, 2011).
El Mostrador: *Compra señal 22 en US$2.5 millones: Álvaro Saieh tiene "mentalidad televisiva,"* July 6, 2010, <http://www.elmostrador.cl/noticias/negocios/2010/07/06/alvaro-saieh-tiene-mentalidad-televisiva/> (accessed on Mar. 1, 2011).
El Mostrador: *La pelea que calienta el dial,* Feb. 1, 2011, <http://www.elmostrador.cl/noticias/pais/2011/02/01/la-pelea-que-calienta-el-dial/> (accessed on Feb. 3, 2011).
El Mostrador: *Lo que está detrás del fin de la edición en papel del matutino: La estrategia del gobierno para sacar a los accionistas privados de La Nación.* Oct. 18, 2010. <http://www.elmostrador.cl/noticias/pais/2010/10/18/la-estrategia-del-gobierno-para-sacar-a-los-accionistas-privados-de-la-nacion/> (accessed on Dec. 15, 2010).
El Mundo: *Time Warner compra Chilevisión al presidente Sebastián Piñera,* Aug. 25, 2010, <http://www.elmundo.es/elmundo/2010/08/25/comunicacion/1282748887.html> (accessed on Feb. 1, 2011).
Extranoticias.cl: *Preocupa fusión de radios Iberoamerican con CRC,* Sept. 7, 2007, <http://www.extranoticias.cl/preocupa-fusion-de-radios-iberoamerican-con-crc/> (accessed on Feb. 1, 2011).
P. Gómez, S. Godoy, W. Pullen: "El nuevo consumidor chileno exige empresas más amigables y responsables," *Qué Pasa,* N° 2073, Dec. 31, 2010, 50–67, <http://www.quepasa.cl/articulo/16_4210_9.html>.
La Nación: *Álvaro Saieh Bendeck Escribe A La Nación. La historia según Copesa,* July 31, 2005, <http://www.lanacion.cl/prontus_noticias/site/artic/20050730/pags/20050730194208.html> (accessed on Feb. 1, 2011).
La Nación: *Chilevisión ya es oficialmente de Sebastián Piñera,* Apr. 15, 2005, <http://www.lanacion.cl/prontus_noticias/site/artic/20050415/pags/20050415162651.html> (accessed on Feb. 1, 2011).
La Nación Domingo: *La apuesta del empresario por ampliar su poder mediático: Todos Los Caminos Conducen A Saieh,* Sept. 12, 2004, quoted in <http://miguelpaz.blogspot.com/2004/09/reportaje-alvaro-saieh-siete7.html> (accessed on March 1, 2011).
La Tercera: *La televisión chilena ad portas de su mayor cambio en las últimas tres décadas,* suplemento Negocios, August 15, 2010, 10–11.
Luis Pardo, President of the Chilean Association of Radio Broadcasters ARCHI: *Reciprocidad y Subtel,* Sección Cartas al Director, El Mercurio, Jan. 8, 2011, B2.
Radio Tierra: Grupo español PRISA concentra mercado radiofónico, July 31, 2007, <http://www.radiotierra.cl/node/526> (accessed on Feb. 1, 2011).
Terra.cl: *Álvaro Saieh compra señal de televisión Más Canal 22,* July 6, 2010, <http://economia.terra.cl/noticias/noticia.aspx?idNoticia=201007062249_INV_79140440> (accessed on Feb. 1, 2011).
Mauro Valdes, CEO of TVN, Seminar "La Transformación de la Televisión," Circulo De Marketing Icare, Casapiedra, Santiago de Chile, April 5, 2011.

## 6. Legislation

General Law of Telecommunications No. 18,168 of 1982.
Law of Digital TV N°20,750 of 2014.
Law of Network Neutrality N°20,453 of 2010.
Law of Television Nacional de Chile N°19.132 of 1992.
Law of the National Television Council N°18,838 of 1989, modified by Law N°19.131 of 1992.
Law of the Press N°19.733 of 2001.
Law of Transparency N°20,285 of 2008.
Presidencia de la República: *Mensaje de S.E. la Presidenta de la Republica con el que inicia un proyecto de ley que modifica la ley N° 19.132 de Televisión Nacional de Chile.* Mensaje N° 943–356, Santiago, 27 October 2008.
Presidencia de la República: *Mensaje de S.E. la Presidenta de la Republica, con el que inicia un proyecto de ley que permite la introducción de la televisión digital terrestre* (Law of the National Television Council N° 18.838). Mensaje N° 942–356, Santiago, October 24, 2008.

# Chile—Data Summaries

ELI NOAM AND PAUL MUTTER

CHILE HAS THE HIGHEST percentage of foreign media ownership internationally, at 62.4% (Table 21.23). Telefónica (Spain), one of the top ICT providers in Latin America, is the largest platform media company in Chile, and it also has a 2.6% share in content media through its multichannel video platforms (Table 21.24). DirecTV (US) is the second largest of Chile's C4 companies, dominating the multichannel platform market. In content, in addition to Telefonica and America Movil, traditional content producers such as El Mercurio and COPESA are larger (Table 21.24), as well as the incumbent multichannel provider DirecTV.

America Móvil (Carlos Slim's Telmex/Grupo Carso) moved into the Chilean telecom market in 2004 when it bought the Latin American division of AT&T. Between 2004 and 2010, its market share of platforms, under the brand name Claro Chile, nearly doubled from 9.2% to 15.4%.

El Mercurio is the largest content company in the country. The company operates a daily newspaper by that name. El Mercurio, together with COPESA, form a duopoly in the newspaper market. COPESA, however, has also expanded into the radio, broadcast TV, and cable industries. In contrast, Bethia Group started as a holding corporation involved in other industries before it moved into the media industry in 2012 with its purchase of Radio Candela and the TV networks MEGA and ETC TV (Table 21.25).

Table 21-23. National Media Industries Concentration in Chile

| | 2004/5 | | 2011 OR MOST RECENT | | % CHANGE ANNUAL AVERAGE | |
|---|---|---|---|---|---|---|
| | COMPANY POWER INDEX IN COUNTRY | COMPANY SHARE OF THE OVERALL NATIONAL MEDIA MARKET (%) | COMPANY POWER INDEX IN COUNTRY | COMPANY SHARE OF THE OVERALL NATIONAL MEDIA MARKET (%) | COMPANY POWER INDEX IN COUNTRY | COMPANY SHARE OF THE OVERALL NATIONAL MEDIA MARKET (%) |
| Telefónica (Spain) | 1,894 | 31.0 | 1,134 | 26.3 | –6.7 | –0.78 |
| DirecTV (US) | 811 | 11.8 | 641.2 | 14.3 | –3.5 | 0.42 |
| Entel PCS | 322 | 10.2 | 369 | 11.1 | 2.4 | 0.14 |
| El Mercurio | 548 | 10.0 | 374 | 7.1 | –5.3 | –0.49 |
| America Movil, (Telmex) (Mexico) | 78 | 5.5 | 208 | 10.7 | 27.4 | 0.88 |
| COPESA | 142 | 5.5 | 99 | 4.0 | –5 | –0.25 |
| Universidad Catolica/ Luksic Group | 64 | 2.2 | 58 | 2.1 | –1.6 | –0.01 |
| TVN (public) | 67 | 2.0 | 56 | 1.9 | –2.6 | –0.03 |
| Iberoamericana (Spain) | 34 | 0.8 | 33 | 0.76 | –0.78 | –0.01 |
| Albavision (Mexico) | 8 | 0.68 | 20 | 1.1 | 28.8 | 0.08 |
| Bethia Group | 22 | 1.2 | 18 | 1.1 | –3.1 | –0.02 |
| Bio Bio Comunicaniones | 0 | 0.02 | 0.1 | 0.03 | 82.6 | 0.004 |
| Media Concentration Index | 2004/5 | | 2011 or Most Recent | | % Change Annual Average | |
| Total Revenue: Nat'l Media Industry (US$ mil) | 4,102 | | 5,657 | | 6.3 | |
| Total Voices (*n*) | 43 | | 45 | | 0.78 | |
| Net Voices (*n*) | 31 | | 32 | | 0.54 | |

| | | | |
|---|---|---|---|
| Public Ownership (%) | 2.0 | 1.9 | −0.03 |
| Foreign Ownership (%) | 59.4 | 62.4 | 0.5 |
| C4 Average—Weighted | 95.6 | 93.2 | −0.41 |
| HHI Average—Weighted | 4,277 | 3,607 | −2.6 |
| C1 Average—Weighted | 56 | 46 | −2 |
| Noam Index Average—Weighted | 1,531 | 1,641 | 1.2 |
| Pooled Overall Sector C4 | 53.1 | 52 | −21.5 |
| Pooled Overall Sector HHI | | 1,230 | −2.0 |
| Pooled Overall Sector Noam Index | 171 | 124 | −4.6 |
| Market Share of Top 10 Companies: Nat'l Media Industry (%) (Pooled C10) | 83.2 | 82.1 | −0.2 |
| National Power Index | 4,216 | 3,319 | −3.6 |

Table 21-24. Top Content Media Companies in Chile

| | 2004/5 | | 2011 OR MOST RECENT | | % CHANGE ANNUAL AVERAGE | |
|---|---|---|---|---|---|---|
| | COMPANY POWER INDEX IN COUNTRY | COMPANY SHARE OF THE NATIONAL CONTENT MEDIA MARKET (%) | COMPANY POWER INDEX IN COUNTRY | COMPANY SHARE OF THE NATIONAL CONTENT MEDIA MARKET (%) | COMPANY SHARE OF THE NATIONAL CONTENT MEDIA MARKET (%) | COMPANY POWER INDEX IN COUNTRY |
| El Mercurio | 1,307 | 23.9 | 993 | 18.8 | −4 | −0.85 |
| DirecTV | 584 | 6.6 | 448 | 8.0 | −3.9 | 0.24 |
| COPESA | 338 | 13.1 | 263 | 10.6 | −3.7 | −0.42 |
| Telmex/Claro (Mexico) | 1 | 0.3 | 57 | 2.9 | 4,167.4 | 1.2 |
| Universidad Catolica/ Luksic Group | 152 | 5.2 | 153 | 5.6 | 0.13 | 0.08 |
| TVN (public) | 160 | 4.8 | 150 | 5 | −1.0 | 0.03 |
| Albavision(Mexico) | 18 | 1.6 | 54 | 3.0 | 34.0 | 0.23 |
| Bethia Group | 53 | 2.8 | 48 | 2.8 | −1.5 | 0.01 |
| Telefonica (Spain) | 1 | 0.3 | 46 | 2.6 | 637.6 | 0.38 |
| Bio Bio Comunicaciones | 0 | 0.04 | 0.2 | 0.09 | 95.7 | 0.01 |
| Media Concentration Index | 2004/5 | | 2011 or Most Recent | | % Change Annual Average | |
| Public Ownership (%) | 4.8 | | 5.0 | | 0.2% | |
| Foreign Ownership (%) | 32.7 | | 41.6 | | 1.5 | |
| C4 Average—Weighted | 94 | | 92.8 | | −0.20 | |
| HHI Average—Weighted | 3,369 | | 3,318 | | −0.25 | |
| C1 Average—Weighted | 46 | | 45 | | 0 | |
| National Power Index | 3,228 | | 3,113 | | −0.59 | |

Table 21-25. Top Platform Media Companies in Chile

| | 2004/5 | | 2011 OR MOST RECENT | | % CHANGE ANNUAL AVERAGE | |
|---|---|---|---|---|---|---|
| | COMPANY POWER INDEX IN COUNTRY | COMPANY SHARE OF THE NATIONAL PLATFORM MEDIA MARKET (%) | COMPANY POWER INDEX IN COUNTRY | COMPANY SHARE OF THE NATIONAL PLATFORM MEDIA MARKET (%) | COMPANY SHARE OF THE NATIONAL PLATFORM MEDIA MARKET (%) | COMPANY POWER INDEX IN COUNTRY |
| Telefónica (Spain) | 3,261 | 53.2 | 1,791 | 40.7 | –7.5 | –2.1 |
| DirecTV (US) | 974 | 15.5 | 758 | 18.1 | –3.7 | 0.43 |
| Entel PCS | 554 | 17.6 | 592 | 17.8 | 1.1 | 0.02 |
| Telmex/Claro (Mexico) | 134 | 9.2 | 299 | 15.4 | 20.7 | 1.0 |
| Media Concentration Index | 2004/5 | | 2011 or Most Recent | | % Change Annual Average | |
| Public Ownership (%) | 0 | | 0 | | 0% | |
| Foreign Ownership (%) | 77.9 | | 74.2 | | –0.6 | |
| C4 Average—Weighted | 96.8 | | 93.4 | | –0.57 | |
| HHI Average—Weighted | 4,932 | | 3,781 | | –3.9 | |
| C1 Average—Weighted | 62 | | | | –3 | |
| National Power Index | 4,939 | | | | –5.0 | |

# 22

# Media Ownership and Concentration in Mexico

JUAN ENRIQUE HUERTA WONG AND RODRIGO GÓMEZ GARCÍA

## INTRODUCTION

The Government of Mexico promotes economic competitiveness through regulations and actions that enforce a level playing field for the economic agents involved. It is also responsible for creating the necessary conditions for a pluralist media system in tune with the diversity of the population. Pundits have often presented their arguments for either more or less federal regulation without much data about the industry's actual concentrated nature.[1] Debates take place in an environment where an audiovisual duopoly and a telecommunications monopoly face the public interest.[2]

An example of this is the debate between the two private owners whose media holdings dominate the audiovisual and telecommunications sectors: Emilio Azcárraga Jean and Carlos Slim. Azcárraga is the CEO of Grupo Televisa, the largest Spanish-language media group in the world, while Slim is the CEO of the national telecommunications major Telmex and its mobile arm, Telcel. On March 23, 2012, Azcárraga published a broadside in *The Wall Street Journal* titled "Carlos Slim Needs More Competition."[3]

1. For instance, on October 10, 2004, President Vicente Fox issued a decree that abrogated the Federal Law for Radio and Television regarding state-sponsored advertising content.

2. An example of commercial operators and the government's collaboration in this regard was the passage of the 2006 Radio, Television, and Telecommunications Law, which was called the "Televisa Law" due to that media group's influence in devising the legislation, as well as in its unanimous approval in just five minutes by the Deputies Chamber without debate.

3. Azcárraga, Emilio. "Carlos Slim Needs More Competition." *The Wall Street Journal*. 22 Mar. 2012. Jan. 16, 2013. <http://online.wsj.com/article/SB10001424052702304724404577293070405826732.html>.

Azcárraga took to the influential US business daily to announce that Grupo Televisa is open to a third, fourth, or even a national broadcaster entering the Mexican market—albeit under certain conditions set by Grupo Televisa. First, if these new broadcast competitors were to be US-based companies, then they ought to be governed by the same mechanism that in the US prohibits foreign investors "from owning more than 25% of any television station."[4] Second, Azcárraga asserted that if Slim moved to invest in broadcast television, then the telecommunications sector should also be opened to greater competition, by citing an OECD study that states:

> Mexico loses 2.2% of its gross domestic product each year because of astronomically high cellphone rates, low Internet penetration, and mediocre connectivity. Mexico has 10% as many wireless Internet subscribers per 100 inhabitants as Turkey. Its cellular phone rates are by far the most expensive in the OECD. Relative to other OECD countries, Mexico is ranked last in terms of investment in telecommunications per capita; but, says the study, "profit margins of the incumbent nearly double the OECD average."

Slim struck back against Azcárraga's editorial by arguing that Mexico's high pay-TV subscription rates are not the result of his Grupo Carso's monopoly in telecommunications—the group's holdings account for 70% of Mexico's mobile phone market, 74% of its fixed broadband services, and 80% of the country's landline telephony market—but due to the lack of competitiveness in the audiovisual sector. In summary, Azcárraga claims that Grupo Carso's domination of the national telecom market accounts for one of the highest rates of fixed and cell phone fees in the world, and correspondingly, Slim claims that Grupo Televisa's domination in broadcast and cable TV is the cause of Mexico's extremely high-paid TV subscription rates.

We have prepared this chapter in order to illustrate how concentration indices are often overlooked in the public debate over media concentration and ownership. They would better inform the discussion of initiatives and public policies that are proposed to reorganize the Mexican media market. First, we provide information about the dimensions of the Mexican media market in order to contextualize the data and the results that are to follow. The three indices we use (C4, HHI, and the Noam Index) to measure concentration emphasize the governing role of the main corporations and total concentration in the media market. The US approach to measuring competitiveness is not followed here as a regulatory model *for* Mexico per se, but as an international standard of comparison. A review of the main media and telecommunication companies in Mexico will demonstrate that while these media majors act globally, they think locally when it comes to asset ownership.

## PRINT MEDIA

### Newspapers

It is not possible to calculate daily readership for newspapers in Mexico accurately because publishers inflate the print runs beyond projected demand, despite the fact that Mexico has a certification authority set up to prohibit this practice. We can offer some general observations based on other socioeconomic data: specifically, that despite an increase in literacy (the national rate now stands at over 90%, up from 85% in the early 1990s), readership does not seem to be rising accordingly. In part, this is because of Mexico's 28.8 million registered households, 42% of have access to a multichannel platform and 90% own a radio, and

4. This is a sore point for Azcárraga because he would like to significantly expand Grupo Televisa's 5% stake in the Spanish-language Univisión network (it is the fifth largest terrestrial broadcaster in the United States by audience share), but US law prohibits him from doing so.

as a result the national media culture is much more audiovisually inclined.[5] In addition, the reading public regards political commentary and reporting in Mexican dailies as hopelessly partisan due to the role of party politics in the industry.

In the 1960s and 1970s, for example, the daily *El Excélsior* was the leading daily, running highly influential editorials in 250,000-copy daily print runs. But *El Excélsior* began to suffer financially when state-sponsored advertising was withdrawn in the late 1970s due to the paper's growing criticism of the long-ruling party, the Institutional Revolutionary Party (PRI). Despite the departure of the "offending" editorial staff in 1976, allegedly under intense government pressure, political advertising did not rebound and the daily's economic slide continued. This suggests that it had been the state-sponsored political advertising, and not the daily's high audience shares, that made its market leadership possible during the 1960s and 1970s.[6] Our research suggests that competitiveness in the industry only truly began when new political parties finally challenged the ruling party's political monopoly in the 1990s. The newspaper market in Mexico was stabile for most of the period the PRI held power, with few new dailies opening after the late 1960s, but when the opposition National Action Party (PAN) won the presidency for the first time in 2000, a number of new titles emerged under new publishers. Of course, because new dailies and editorial voices have become associated with PAN and other opposition parties, readers are still generally cynical about Mexican newspapers' political coverage.

The leading newspaper publisher in Mexico today is the Organizacion Editorial Mexicana (OEM), which expanded under PRI rule when it acquired many of the independent media group Garcia Valseca's titles. Today, OEM owns *Esto*, a sports daily with the highest print circulation numbers nationwide, and a network of regional newspapers under the brand name *El Sol* (except for the state of Monterrey, where *El Sol*, originally a Garcia Valseca title, belongs to the publishing company Grupo Reforma). OEM's audience shares have begun to fall due to the rise of new regional players, since Mexican newspapers operate on a regional marketing and distribution model pioneered by Garcia Valseca. Increased competition for audience shares and advertising revenues takes place mainly in Mexico City and other metropolitan areas with a population of at least one million. While there is more competition among newspapers in these cities today than there has ever been before, the audience shares in these urban markets have not added a significant number of new readers. In fact, the dailies with the highest print circulations in Mexico have shown a relatively significant decrease over the past few years. These newspapers include the two leading sports dailies, *Esto* and *Ovaciones*; the crime blotter *La Prensa*; and the popular regional daily *El Norte*. As noted earlier, newspapers focusing on political coverage are not print circulation leaders due to the public's disdain for their coverage (Table 22.1).

When disaggregating publishing groups by titles, concentration indices show a notable decrease as a result of increased regional competition. But the cost of advertising for the dailies focusing on political content is not at all correlated with these dailies' circulation numbers, suggesting that print runs have more to do with advertising revenue than audience shares. Those dailies specializing in political coverage, in fact, have the highest advertising costs. For example, in 2010, *El Excélsior* had print runs that were consistently 80% less than *Esto*'s totals in the same period, and a full-page advertisement in *Esto* cost only one-third of what it would cost to run the same advertisement in *El Excélsior* (Table 22.2).

5. Gómez, Rodrigo; Sosa, Gabriel; Bravo, Jorge; and Tellez, Primavera. *Mapping Digital Media: México*. Open Society Foundations–Media Program, London: Open Society Foundations–Media Program, 2011. June 13, 2011. <http://www.soros.org/initiatives/media/articles_publications/publications/mapping-digital-media-mexico-20110613>.

6. Werbowski, Michael. "'El Excelsior': The Rise and Fall of a Great Mexican Newspaper." *Worldpress.org*, Feb. 18, 2007. Jan. 17, 2013. <http://www.worldpress.org/Americas/2678.cfm#down>.

## Table 22-1. Daily Newspapers (Market Shares by Circulation), 2004–2010

| | 2004 | 2008 | 2010 |
|---|---|---|---|
| Organizacion Editorial Mexicana (OEM) | 54.9 | 56.7 | 59.4 |
| Grupo Reforma | 12.6 | 11.7 | 9.9 |
| El Universal | 12.6 | 9.0 | 5.4 |
| Demos/La Jornada | 4.5 | 6.3 | 4.5 |
| Grupo Milenio (Grupo Multimedios) | 5.4 | 6.3 | 4.5 |
| Editorial Ovaciones | | 8.1 | 6.3 |
| Others[a] | 10.0 | 10.0 | 10.0 |
| Total Annual Circulation (1,000) | 2,794 | 3,279 | 4,320 |
| Total Revenue (mil MXN)[1] | 1,015 | 1,822 | 2,400 |
| Total Revenue (mil US$) | 132 | 164 | 213 |
| C4 | 86 | 86 | 81 |
| HHI | 3,381 | 3,578 | 3,736 |
| *N* (>1%) | 5 | 6 | 6 |
| Noam Index | 1,512 | 1,461 | 1,525 |

[1] Estimated.

## Table 22-2. Newspapers Advertising Costs, 2010[1]

| | TOTAL COST (MXN) | TOTAL COST (US$) |
|---|---|---|
| *Esto* | 38,275 | 3,062 |
| *La Prensa* | 30,213 | 2,417 |
| *El Excélsior* | 115,000 | 9,200 |
| *El Universal* | 157,088 | 12,567 |
| *El Financiero* | 116,250 | 9,300 |
| *Ovaciones* | 44,500 | 3,560 |
| *Reforma* | 132,325 | 10,586 |
| *La Jornada* | 108,750 | 8,700 |
| *El Norte* | 79,875 | 6,390 |
| *Ovaciones 2a Ed* | 44,500 | 3,560 |
| *Milenio Diario* | 129,325 | 10,346 |

[1] Based on the cost of a full-page ad, the most expensive type of print ad buy-in, for each paper's lead section (the US equivalent would be "Section A" in a daily such as *The New York Times* or *The Washington Post*).

The public mistrust of political coverage partly explains why sports and crime newspapers have larger audience shares than political dailies, and these audience shares help them keep their advertising costs lower. In contrast, the low readership of the other dailies compels them to raise advertising costs in order to break even. Interestingly, most of a daily's revenue from political advertising comes directly from the political parties, so these ad placements are not based on any measure of their effectiveness in influencing polling numbers, but have more to do with the parties wanting to maintain a positive relationship with certain newspapers to secure their editorial endorsements.

## AUDIOVISUAL MEDIA

### Radio

There are 1,594 radio stations in Mexico. Of that number, 1,053 have commercial licenses and 541 are nonprivate broadcasters that are not allowed to feature commercial advertising. There are also two national public networks, IMER and Radio Educación, and several regional networks that belong to universities, local government offices, and indigenous tribes. Due to the plurality of state and nonstate broadcasters in the radio market, it manages to show greater social diversity and inclusion than television broadcasting, though social and community networks are technically considered "illegal" stations (they do not have commercial licenses or nonprivate permissions). Corporate sponsors are gradually increasing their presence in this sector: the percentage of overall advertising spending on radio increased from 8% of the national total in 2004 to 10% by 2010, amounting to US$400 million (MXN 5 billion).[7] Pop and folk music stations are ranked as the most popular types of programming.[8]

The available data for this sector refer only to the number of stations. This allows us to identify the groups' size by stations, but not by audiences. The indices' results suggest a low concentration in the Mexican radio market, though not all radio groups operate at the national level (Table 22.3).

### Broadcast TV

Broadcast television in Mexico has always been a highly concentrated market and has a monopolistic structure that is dominated by Grupo Televisa. Today, its three national networks and one regional broadcaster hold the greatest audience shares nationwide. Its news shows are the most popular in the country and are regarded as agenda setters in political debates.[9] Grupo Televisa also has four multichannel platforms offering triple-play services, an interest in film production and distribution (Videocine), the radio market (Televisa Radio), and in wireless telecom (Iusacell).[10] Its international holdings include a stake in the Spanish television broadcaster *laSexta* and a 5% stake in the Spanish–language television network Univisión that operates in the United States. Grupo Televisa has now become the main audiovisual content producer and distributor in the international Spanish-language markets.[11]

Between 1972 and 1993, Grupo Televisa's Channel 2 (Canal de las Estrellas), Channel 5 (Canal 5), Channel 9 (formerly Channel 8), and regional Channel 4 were the only commercial broadcasters in the country. Its only other "competitor" was the public broadcaster Imevisión, organized in 1982, with its three main channels: Channel 7, Channel 13, and Channel 22.[12] During this period, Grupo Televisa's executives had a tacit understanding with PRI leaders that as long as its channels were not overly critical of the PRI, political roadblocks would not be thrown up to prevent Grupo Televisa from expanding its operations.[13] This allowed Grupo Televisa to implement a successful vertical integration

7. Cofetel. "Documento de referencia a la consulta pública para la licitación de frecuencias para televisión abierta." Cofetel, 2011. Nov. 9, 2011. <http://www.cft.gob.mx:8080/portal/wp-content/uploads/2012/06/Documento_de_referencia_Consulta_TV.pdf>. 4.

8. ARVM Staff. "Registran números positivos en Radio." Asociación de Radio del Valle de México (ARVM), Oct. 15, 2011. Jan. 17, 2013. <http://laradioenmexico.mx/registran-numeros-positivos-en-radio/>.

9. ARVM Staff. "Registran números positivos en Radio."

10. Grupo Salinas and Grupo Televisa both own a 50% stake in Iusacell.

11. Sánchez–Ruiz, E. "El audiovisual Mexicano: ¿Concentrar para competir?" *Global Media Journal Edición Iberoamericana* 1:2, 2004. Feb. 8, 2012 <http://gmje.mty.itesm.mx/articulos2/sanchezruiz_OT04.html>.

12. Trejo, R., ed. *Televisa quinto poder.* Mexico City: Claves Latinoamericanas, 1985.

13. Orozco, G. *Historias de La televisión en América Latina.* Barcelona: Gedisa, 2002; and Trejo, R., ed. *Televisa quinto poder.* Mexico City: Claves Latinoamericanas, 1985.

Table 22-3. Radio Group (Number of Stations), 2011

| | # STATIONS | % STATIONS |
|---|---|---|
| Grupo ACIR | 169 | 16.0 |
| Grupo Radiorama (PRISA Group, Spain) | 116 | 11.0 |
| Azcárraga Family | 116 | 11.0 |
| Televisa Radio Group | *64* | *6.0* |
| Grupo Fórmula (Radio Fórmula, Azcárraga Family) | *52* | *5.0* |
| Grupo CMR[1] | 94 | 9.0 |
| Grupo Ramsa | 74 | 7.0 |
| Grupo Radio Centro (GRC) | 68 | 6.5 |
| Corporadio | 52 | 5.0 |
| MVS Radio (MVS Comunicaciones) | 52 | 5.0 |
| Grupo Radio México (GRM) | 27 | 2.5 |
| Grupo Multimedios | 21 | 2.0 |
| Cadena Rasa | 21 | 2.0 |
| Firmesa | 21 | 2.0 |
| ABC Radio | 16 | 1.5 |
| Grupo Imagen | 16 | 1.5 |
| Others | 190 | 18.0 |
| Totals | 1,053 | 100.0 |
| Total Revenue (mil MXN) | 5,000 | |
| Total Revenue (mil US$) | 400 | |
| C4 | 47 | |
| HHI | 743 | |
| *N* (>1%) | 14 | |
| Noam Index | 199 | |

*Source*: PAPIIT Project, UNAM IN301111.[2]

[1] Both GRC and GRM's board of directors are chaired by the same man, Francisco Aguirre Gomez.

[2] Vega Montiel, Aimée, coordinator. "II Seminario Internacional Acceso y participación de las mujeres en las industrias de comunicación: de los diagnósticos a las experiencias." Ciudad Universitaria. Torre II de Humanidade, Mexico City, Mexico. Nov. 8, 2012. <http://www.ceiich.unam.mx/0/60EncFic.php?tblEncuentrosAcad_id=1066>.

strategy in distribution and content production and to accumulate capital to invest in cable and satellite TV.

Grupo Televisa's monopoly was challenged after 1993, when Imevisión was privatized. At that time, the Mexican conglomerate Grupo Salinas entered the market through its subsidiary TV Azteca (now usually referred to simply as "Azteca"), becoming Grupo Televisa's primary competitor during the 1990s. Azteca offers three primary channels—Channel 7, Channel 13 (Azteca Trece), and Channel 40 (Proyecto 40)—along with an international channel based out of the United States, Azteca América. The competition between the two media groups reached a climax in 1997, when Azteca won the highest ratings of any broadcast TV network in Mexico to date.[14] The market shares of both companies have varied little since then, although Grupo Televisa's four domestic channels retain

14. Alva, A. "Entre la crisis financiera y la guerra de los *ratings* televisivos." *Revista Mexicana de Comunicación*, 1999). 7–10.

a combined market share three times that of Azteca's three. Since Azteca primarily broadcasts at a regional level, it has not been able to surpass its 1997 high-water mark, and from the 2000s to the present it has retained on average 28–30% of the national market share.

Since 2009, Mexican broadcast television consumed the lion's share of national advertising spending, which differentiates it from other Latin American countries such as Argentina and Brazil. In 2009, Argentine broadcast television consumed a 42% share of its national advertising spending, while the Brazilian broadcasting market took in 40% of its respective national advertising spending.[15] Mexican broadcast television, though, has held between 65% and 58% of national advertising revenue in recent years, and in the 1990s and 2000s it often accounted for up to 70% of national advertising spending.[16] In 2010, the sector's advertising revenue was valued at US$4.2 billion (MXN 52 billion), or 58% of all national advertising spending in Mexico.[17] Though still higher than the advertising revenues of other countries' broadcast TV markets, considering the market share of the sector as a whole, concentration indices have been falling due to new market entries and greatly reduced viewership in terrestrial television. In 2005, for example, the average Mexican viewer saw up to 527 minutes of broadcast television programming per day, but by 2009 this number had fallen to an average of 273 minutes per day.[18] The decline is the result of increased growth among multichannel platforms (Tables 22.4, 22.5).

## Table 22-4. TV Broadcasting (Audience Shares by Prime Time Viewership) 1995–2010

| | 1995 | 2000 | 2005 | 2010 |
|---|---|---|---|---|
| Grupo Televisa (Azcárraga Family)[1] | 91.0 | 70.4 | 68.6 | 68.3 |
| Azteca (Grupo Salinas)[2] | 7.0 | 26.8 | 30.0 | 28.2 |
| Others | 2.0 | 2.8 | 2.5 | 3.5 |
| C4 | 98.0 | 97.2 | 98.6 | 96.5 |
| HHI | 8,334 | 5,682 | 5,612 | 5,472 |
| *N* (>1%) | 2 | 2 | 2 | 2 |
| Noam Index | 5,893 | 4,018 | 3,968 | 3,869 |

*Source*: Observatory with the companies' data (F-20 Reports and Annual reports presented to the BMV) and IBOPE.

[1] Data for *Channel 2, Channel 4, Channel 5,* and *Channel 9.*

[2] Data for *Channel 7, Channel 13,* and from 2007 on, *Channel 40.*

15. AAAP. "Informe oficial Latinoamericano de inversión publicitaria 2009." Buenos Aires: Asociación Argentina de Agencias de Publicidad (AAAP), 2010. Dec. 1, 2011. <http://www.aapublicidad.org.ar/wp-content/inversion_publicitaria/2009/argentina/informe_inversion_publicitaria_2009.pdf>.

16. Cofetel. "Documento de referencia a la consulta pública para la licitación de frecuencias para televisión abierta." Cofetel, 2011. Nov. 9, 2011. <http://www.cft.gob.mx:8080/portal/wp–content/uploads/2012/06/Documento_de_referencia_Consulta_TV.pdf>; and Gómez, R. *El impacto del Tratado de Libre Comercio de América del Norte (TLCAN) en la industria audiovisual mexicana (1994–2004)*. Diss. Universidad Autónoma de Barcelona, 2006. Bellaterra: UAB, 2007. <http://www.tdx.cat/bitstream/handle/10803/4197/rgg1de1.pdf;jsessionid=F8D3D1A7F49F60B1D7D2B046E0C3AEA5.tdx2?sequence=1>.

17. Duarte, Enrique. "TV digital acecha dominio de televisoras." CNNExpansión.com, Mar. 28, 2011. Jan. 17, 2013. <http://www.cnnexpansion.com/negocios/2011/03/23/televisa-y-tv-azteca-compartiran-mercado>.

18. Gómez, Rodrigo; Sosa, Gabriel; Bravo, Jorge; and Tellez, Primavera. *Mapping Digital Media: México*. Open Society Foundations-Media Program, London: Open Society Foundations-Media Program, 2011. June 13, 2011. <http://www.soros.org/initiatives/media/articles_publications/publications/mapping-digital-media-mexico-20110613>. 17.

Table 22-5. TV Broadcasting (Market Shares by Revenue), 2006–2010

| | 2006 | 2007 | 2010 |
|---|---|---|---|
| Grupo Televisa (Azcárraga Family) | 67.0 | 63.0 | 61.0 |
| Azteca (Grupo Salinas) | 29.0 | 27.0 | 30.0 |
| Other | 3.0 | 10.0 | 9.0 |
| Total Revenue (mil MXN) | 29,333 | 31,578 | 35,175 |
| Total Revenue (mil US$) | 2,640 | 2,842 | 2,814 |
| C4 | 96 | 89 | 91 |
| HHI | 5,346 | 4,673 | 4,746 |
| *N* (>1%) | 3 | 3 | 3 |
| Noam Index | 3,123 | 2,701 | 2,772 |

*Source*: Observatory of industries, policies and consuming culture with data from Cofetel (2011), the companies' annual reports (F-20) and AAAP (2007, 2009).

## Multichannel TV Platforms

Cable and direct-to-home (DTH) satellite are the primary platforms in use for Mexican multichannel distribution. According to the Federal Commission of Telecommunications (Cofetel), the Mexican pay-TV market was valued at US$1.2 billion (MXN 8 billion) in 2005 and US$2.6 billion (MXN 32.5 billion) in 2010. Since 2007, all cable TV companies have been offering triple-play packages that include Internet, television, and telephony services. In fact, investment in the sector (in 2006, it stood at US$418 million (MXN 4.6 billion) and reached US$1.1 billion (MXN billion) by 2010) represents 18.8% of total telecommunications investment in the country.

DTH is the fastest growing market segment in the paid TV sector, with its audience shares rapidly increasing.[19] The spectacular growth of satellite TV is the result of the market entry of a new provider, Dish-México (MVS Comunicaciones), in 2008. Its entry ended the monopoly that SKY México had held over the market since 2005.[20] Satellite providers also have the advantage of being able to broadcast nationally, while cable companies do not have their own nationwide infrastructure to transmit. In 2010, the increased competition among multichannel platforms resulted in a 1.1% decrease in service costs (Figure 22.1).[21]

Grupo Televisa also dominates the pay-TV market in Mexico. It has rapidly increased its audience shares in recent years, leading to an oligopolistic structure. Grupo Televisa's properties include 15 pay-TV channels, the regional operator TVI-MTY (in Monterrey, the country's third most populous metropolitan area) and Cablemás, which operates in 43 cities and 16 states. By the end of 2011, Grupo Televisa held four multichannel platforms, three cable companies, and one satellite TV provider, representing a 50% multichannel market share.[22] The Mexican Federal Commission

19. "DTH grew an 83.8% in Mexico." *Dataxis*, Mar. 17, 2011. Jan. 17, 2013. <http://nextvlatam.com/index.php/1-cable-dth/dth-grew-an-83-88-in-mexico/>.

20. Dish México was formed in 2008 by Grupo MVS-Multivisión (51%) and Charles Ergen (49%), the owner of Dish Network in the US. SKY México is registered with the name Innova and it is part of DirecTV Latin America. Its stock share is divided as follows: Grupo Televisa (58.7%) and DirecTV. (41.3%).

21. Cofetel. "Documento de referencia a la consulta pública para la licitación de frecuencias para televisión abierta." Cofetel, 2011. Nov. 9, 2011. <http://www.cft.gob.mx:8080/portal/wp-content/uploads/2012/06/Documento_de_referencia_Consulta_TV.pdf>.

22. Huerta-Wong, J. and Gómez, R. "Concentración y diversidad de los medios de comunicación y las telecomunicaciones en México." *Comunicación y Sociedad* 19, 2013. Jan. 17, 2013. <http://www.comunicacionysociedad.cucsh.udg.mx/sites/default/files/a5.pdf>. 113–152.

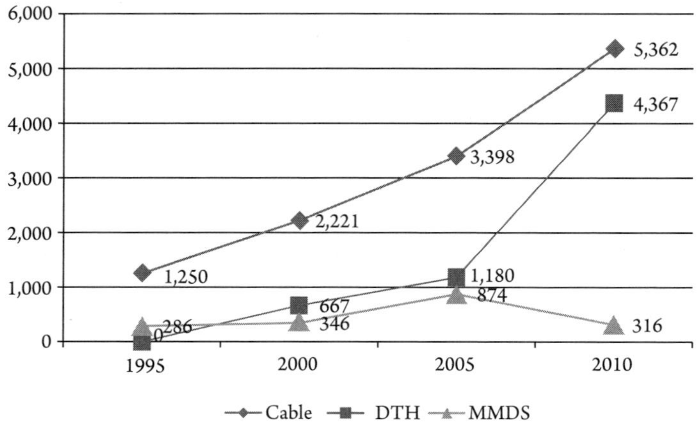


FIGURE 22.1 Multichannel Platforms (Audience Shares by Subscription), 1995–2010

for Economic Competition (CFC) authorized the acquisitions of TVI–MTY and Cablemás on the grounds that Grupo Televisa was "the only firm that can balance [out] Telmex" in the market.[23] The CFC also imposed must-offer and must-carry conditions on Grupo Televisa, mandating that it offer its broadcast and paid TV channels to all multichannel providers.[24] These obligations drew criticism from Grupo Televisa because they were not applied to Grupo Televisa's main DTH competitor, Dish-México, when it entered the market (Table 22.6).

## Film

The film industry has not yet recovered from the financial crisis it suffered between 1994 and the early 2000s when the Mexican peso experienced severe devaluation (the impact of the currency devaluation was so severe that in 1997 and 1998, there were only 9 and 11 feature films, respectively, produced in-country).[25] In 2010, Mexican productions accounted for 6% of audience shares and 5.6% of box office sales. In contrast, US productions from the Hollywood majors accounted for 90% of audience shares and 91% of box office sales.[26] This was despite a history of quotas and restrictions on American imports, including a long prohibition against dubbing into Spanish from a foreign language—a severe handicap in a country where illiteracy used to be high. Mexico has been among the top 20 countries worldwide in box office receipts since 2007, and it is ranked fifth in number of sold tickets worldwide per annum.[27] Ticket sales have increased steadily since 2007, with a per capita rating of 1.7 tickets per annum, and so the number of theaters in the country has increased in response from 3,958 in 2006 to 4,946 in 2010.

In film distribution, the US-based Hollywood majors have completely dominated the market. The two main Mexican distribution companies, Gussi and Corazón, combined hold only a 10.1% market share (Table 22.7).[28]

23. Gómez, R. and Sosa, G. "La concentración en el mercado de la televisión restringida en México." *Comunicación y Sociedad* 14, 2010. Jan. 17, 2013. <http://redalyc.uaemex.mx/src/inicio/ArtPdfRed.jsp?iCve=34615372005>. 109–142.

24. Gómez, R. and Sosa, G. "La concentración en el mercado. . ."

25. Sánchez–Ruiz, E. "Globalization, Cultural Industries and Free Trade: The Mexican Audiovisual Sector in the NAFTA Age." Mosco, V. and Schiller, D., eds. *Continental Order? Integrating North America for Cybercapitalism*. Lanham: Rowman & Littlefield, 2001.

26. IMCINE. "Anuario estadístico de cine mexicano 2010." Mexico City: IMCINE, 2010. Jan. 17, 2013. <http://www.imcine.gob.mx/temporales/ANUARIO_IMCINE_2010.pdf>.

27. CANACINE. "Asistencia al Cine." CANACINE, 2010. Oct. 22, 2011. <http://www.canacine.org.mx/index.php/resultados-2010/asistencia-al-cine.html>.

28. CANACINE. "Participación de mercado por distribuidora." CANACINE, 2010. Oct. 22, 2011. <http://www.canacine.org.mx/index.php/resultados-2010/participacion-de-mercado-por-distribuidora.html>.

Table 22-6. Multichannel Video Platforms: Cable MSOS, DTH, MMDS (Audience Shares by Subscribers), 2007–2010

| | 2007 | | 2008 | | 2009 | | 2010 | |
|---|---|---|---|---|---|---|---|---|
| | TOTAL SUBSCRIBERS (1,000) | % | TOTAL SUBSCRIBERS (1,000) | % | TOTAL SUBSCRIBERS (1,000) | % | TOTAL SUBSCRIBERS (1,000) | % |
| Megacable | 1,298 | 20.0 | 1,483 | 20.0 | 1,592 | 20.0 | 1,757 | 17.0 |
| Telecable (Grupo Hevi) | 150 | 2.0 | 183 | 2.0 | 220 | 3.0 | 300 | 3.0 |
| Cablecom (Slim Carlos, Grupo Carso) | 357 | 5.0 | 610 | 8.0 | 747 | 9.0 | 780 | 8.0 |
| Cablemás, TVI-MTY, Cablevisión, SKY México (Azcárraga Family, Grupo Televisa) | 3,098 | 48.0 | 3,393 | 46.0 | 3,742 | 47.0 | 5,012 | 50.0 |
| Dish México (MVS Comunicaciones) | 500 | 8.0 | 422 | 6.0 | 730 | 9.0 | 1,324 | 13.0 |
| Others | 1,113 | 17.0 | 1,341 | 18.0 | 1,003 | 13.0 | 874 | 9.0 |
| Total Subscribers (1,000) | 6,517 | | 7,431 | | 8,035 | | 10,046 | |
| Total Revenue (mil MXN) | | 1,333 | | | | | | 32,500 |
| Total Revenue (mil US$) | | 1,200 | | | | | | 2,600 |
| C4 | | 81 | | 80 | | 85 | | 88 |
| HHI | | 2,797 | | 2,620 | | 2,771 | | 3,022 |
| *N* (>1%) | | 5 | | 5 | | 5 | | 5 |
| Noam Index | | 1,250 | | 1,171 | | 1,239 | | 1,351 |

*Source*: Observations of industries, policies, and consuming culture from Cofetel (2011) and the companies' annual reports from 2008 to 2011.

## Table 22-7. Film Production/Distribution (Market Shares by Box Office %), 2010

| | 2010 |
|---|---|
| Walt Disney Pictures (US) | 20.1 |
| Paramount (Viacom, US) | 14.9 |
| Warner Bros. (Time Warner, US) | 16.8 |
| 20th Century Fox (News Corp., Murdoch Family, US/UK/AUS) | 12.6 |
| Universal (GE/Comcast, US) | 7.4 |
| Sony-Columbia (Japan/US) | 9.1 |
| Corazón | 5.2 |
| Gussi | 4.9 |
| Videocine (Azcárraga Family, Grupo Televisa) | 2.8 |
| Others | 6.2 |
| Total Revenue (mil MXN)[1] | 10,925 |
| Total Revenue (mil US$) | 874 |
| C4 | 65 |
| HHI | 1,263 |
| *N* (>1%) | 9 |
| Noam Index | 421 |

*Source*: CANACINE, 2010.

[1] This revenue figure is an estimate based on data collected in this study: it is based on average per capita revenues reported for several other countries in this study with similar per capita income.

## Table 22-8. Film Exhibition (Market Shares by Box Office %), 2010

| | TOTAL # OF THEATERS | BOX OFFICE % |
|---|---|---|
| Cinépolis | 2,305 | 61.0 |
| Cinemex-MMC | 1,655 | 26.0 |
| Cinemark (US) | 312 | 6.0 |
| Lumiere | 140 | 2.0 |
| Others | 536 | 5.0 |
| Total | 4,946 | 100.0 |
| C4 | 89.0 | 92.0 |
| HHI | 3,452 | 4,437 |
| *N* (>1%) | 5 | 5 |
| Noam Index | 1,543 | 2,219 |

*Source*: CANACINE (2011), IMCINE (2011).

Unlike the distribution market, the exhibition market is led by Mexican enterprises and is highly concentrated as a result of a merger between Cinemex and Multimedios Cinemas. Today, almost 90% of the exhibition market is divided among just two companies, Cinépolis and Cinemex-Multimedios. Cinépolis also operates in several Central American countries (Table 22.8).

# TELECOMMUNICATIONS MEDIA

## Wireline Telecom

The Mexican telecommunications market has grown substantially since 1998. That year, the entire sector generated US$8.9 billion (MXN 80.9 billion) in revenue. By 2010, its annual turnover was US$31 billion (MXN 388 billion). With growth, though, have come high concentration indices: the wireline (and wireless) sector is characterized by oligopoly.[29]

The start of this process can be dated back to 1990, when the Mexican government decided to privatize the state monopolist Telmex. The government considered breaking Telmex up into several regional companies, divvying up Telmex's assets by the type of service they provided and selling the entire entity to a private owner. The government chose the latter option in 1996 based on the view that consumer costs would fall and the availability of fixed telephony would increase, and the media tycoon Carlos Slim's conglomerate Grupo Carso gained control of all of Telmex's assets.[30] Twenty years later, Mexico now faces a national debate over the future of Telmex because, due to the absence of competitors in the market, Mexican consumers are charged some of the highest per capita fixed telephony rates in the world.

Grupo Carso's Telmex has held a monopoly over landline telephony in Mexico since 1990, when it started joint operations with Southwestern Bell (AT&T, United States) and France Telecom (France). In 2010, Telmex had 15.6 million landline subscribers and 7.4 million broadband subscribers. In 2010, the company's landline segment reported US$9.1 billion (MXN 114 billion) in revenue.[31]

It is estimated that the total number of wireline subscribers in Mexico stands at is 19.4 million, including those operated by the cable companies (providers do not make detailed figures available for independent research). The data that are available do suggest, however, that privatization has not had the desired effects of lowering prices and increasing the number of lines in service. There are 17.3 telephone lines per 100 people in Mexico, a rate much lower than in the United States and Canada in 1980.[32] Telmex's emphasis on protecting its local and long-distance call markets has been accompanied by government regulations favoring vertical and horizontal concentration plans. Because Telmex charges such high rates as a carrier, competitors for local services have focused on commercial line rather than residential line expansion. Axtel/Avantel has become Telmex's only real competitor in the wireline market, and it offers satellite services rather than traditional terrestrial connections. Some cable companies—such as Maxcom, Megacable, and Cablevisión—have started to offer IP telephony, but this has not won an appreciable audience share yet (Table 22.9).

As with local telephony, Telmex is the main long-distance telephone service carrier and provider. Telmex has offered its services to end users for a rate below the one legally established for other long-distance providers.

---

29. Mariscal, J. and Rivera, E. "Regulación y competencias en las telecomunicaciones mexicanas." Mexico City: United Nations (UN) CEPAL, 2007. Jan 17, 2013. <http://www.eclac.cl/publicaciones/xml/7/29037/Serie%2083.pdf>.

30. Mariscal, Judith. *The Mexican Telecommunications Reform: A Political Economy Approach*. Diss. The University of Texas at Austin, 1998. Austin: UT, 1998.

31. Telmex. "Informe Annual 2010: 20 Anos de desarrollo tecnológico." Telmex, 2011. Jan. 17, 2013. <http://www.telmex.com/mx/corporativo/pdf/pt_descarga.jsp?a=01TRIM12n.pdf>.

32. Mariscal, Judith. *The Mexican Telecommunications Reform: A Political Economy Approach*. Diss. The University of Texas at Austin, 1998. Austin: UT, 1998.

Table 22-9. Wireline Telecom (Market Shares by Revenue), 1988–2010

| | 1988 | 1992 | 1996 | 2008 | 2010 |
|---|---|---|---|---|---|
| Telmex (Carlos Slim, Grupo Carso) | 100.0 | 100.0 | 100.0 | 85.0 | 85.0 |
| Axtel/Avantel | | | | 5.0 | 5.2 |
| Alestra | | | | 1.0 | 3.5 |
| Maxcom | | | | 2.0 | 2.0 |
| Megacable | | | | 1.0 | 2.3 |
| Cablevisión, Cablemás (Azcárraga Family, Grupo Televisa) | | | | | 1.7 |
| Others | | | | 6.0 | 7.0 |
| Total Subscriptions (1,000) | 4,381 | 6,754 | 8,826 | 18,937 | 19,891 |
| Total Revenue (mil MXN)[1] | | | | 34,000 | |
| Total Revenue (mil US$) | | | | 3,060 | |
| C4 | 100 | 100 | 100 | 93 | 96 |
| HHI | 10,000 | 10,000 | 10,000 | 7,292 | 7,324 |
| *N* (>1%) | 1 | 1 | 1 | 5 | 6 |
| Noam Index | 10,000 | 10,000 | 10,000 | 3,261 | 3,002 |

*Source*: Authors' table using data from The Competitive Intelligence Unit (2011) and the companies.

[1] This revenue figure is an estimate based on data collected in this study: it is based on average per capita revenues reported for several other countries in this study with similar per capita income.

The CFC therefore accused Telmex of monopolistic behavior, but no government action has yet been brought against Telmex. Whether smaller competitors will be able to continue doing business given Telmex's reduced long-distance rates is an open question (Table 22.10).

## Wireless Telecom

Grupo Carso also dominates the mobile telephony market. At the beginning of the 1990s, Telcel, Carso's subsidiary specializing in mobile phones, became the main competitor for Iusacell, the then-dominant firm owned jointly by Grupo Televisa and Azteca's parent, Grupo Salinas. As a result of financial difficulties in the mid-1990s, Iusacell decided to focus mostly on post–payment telephone lines for urban markets. In response, Telcel focused on offering prepaid service contracts to low and median income populations outside of the major metropolitan centers. Telcel's campaign was extremely successful (the demographic targeted represented no less than 50% of the entire population and was underserved by Iusacell), and Telcel emerged as the clear market leader by 1995. Mobile telephony is the most popular personal communication medium in Mexico, with 84% of the public having coverage.[33] Each month, 700,000 new mobile lines are added to the market—mostly by low-income consumers and predominantly by Telcel—and the mobile market is the main driver of the telecommunications sector's overall growth.[34]

In 2010, Telmex reorganized Telcel into América Móvil, which offers wireless products and directory services in Mexico and is now a major wireless provider in Latin American

33. Flores-Roux, Ernesto and Mariscal, Judith. *Política de Generación de Infraestructura de Telecomunicaciones en México*: Una Crítica. Mexico City: Centro de Investigación y Docencia Económicas, 2010. <http://telecomcide.org/docs/publicaciones/DTAP-246.pdf>.

34. Mariscal, Judith; Bonina, Carla; and Luna, Julio. *Los nuevos escenarios del mercado en América Latina. Documento de Trabajo.* Lima: International Development Research Centre, 2007. <http://www.dirsi.net/sites/default/files/dirsi_07_PD03_es.pdf>.

Table 22-10. Wireline Telecom—Long-Distance Telephony (Audience Shares by Subscribers), 1997–2010

| | 1997 | 1998–2000 | 2000–2004 | 2010 |
|---|---|---|---|---|
| Telmex (Carlos Slim, Grupo Carso) | 85.2 | 71.9 | 73.4 | 97.0 |
| Axtel/Avantel | 6.0 | 13.6 | 9.9 | 2.1 |
| Alestra (AT&T, US) | 4.9 | 9.3 | 8.8 | |
| Iusacell (Grupo Salines/Grupo Televisa) | 0.6 | 1.55 | 1.5 | |
| Marcatel | 0.2 | 0.5 | 1.0 | |
| Protel | 0.1 | 0.4 | 0.2 | |
| Telnor (Telmex) (Carlos Slim, Grupo Carso) | 3.0 | 2.3 | 2.7 | |
| Bestel (Carlos Slim, Grupo Carso) | | 0.2 | 1.1 | |
| Maxcom | | | <0.1 | |
| TeleReunión[1] | | | 0.9 | |
| Axtel | | | 0.6 | |
| GTM | | | <0.1 | |
| Others | | | | 0.9 |
| Total Revenue (mil MXN) | | | | |
| Total Revenue (mil US$) | | | | |
| C4 | 99.0 | 97.1 | 94.7 | 100.0 |
| HHI | 7,329 | 5,447 | 5,568 | 9,433 |
| *N* (>1%) | 7 | 8 | 12 | 12 |
| Noam Index | 3,665 | 2,436 | 2,236 | 6,690 |

*Source*: Mariscal and Rivera (2007); Telmex (2011).
[1] TeleReunión merged with Maxcom in 2012.

markets through its international operator, Claro. At the end of 2010, Telcel held a market share of 79.5% in Mexico (reflecting the strategy to win lower income customers, 58.7 million people, representing 90.1% of all Telcel subscribers, held prepaid plans). In 2010, the revenues generated by Telcel's wireless services in Mexico were US$11.7 billion (MXN 146 billion), or 26% of its consolidated income that year and Telcel users accounted for 28.5% of the total number of mobile phone users in Mexico. Telcel's main domestic competitors are Iusacell and Telefónica (Spain), although it also competes with Nextel Mexico, owned by NII Holdings (US), in certain areas of the country.[35]

Internationally, Grupo Carso and Spain's Telefónica are the main competitors in most of the world's Spanish-language markets. In Mexico, Telcel holds a much higher audience and market share than Telefónica's Movistar. Concentration is high for wireless, but not as high as it is for wireline (Table 22.11).

## INTERNET MEDIA

### Internet Service Providers (ISP)

Since 2000, home Internet access rates have risen from 5% at the start of the decade to 61% by 2010, though the digital gap has grown wider between metropolitan and rural areas. The ISP

35. Nextel Mexico retains the Nextel brand name even though Nextel (US) proper is now part of Sprint (US).

Table 22-11. Wireless Telecom (Market and Audience Shares by Revenue and Subscribers), 1997–2009

| | | 1997 | 2000 | 2004 | 2008 | 2009 | 2012 |
|---|---|---|---|---|---|---|---|
| Telcel (Carlos Slim, Grupo Carso) | % Subscribers | | | | 72.0 | 71.0 | 69.9 |
| | % Revenue | 63.7 | 73.1 | 77.2 | 69.0 | 69.0 | 70.8 |
| Movistar México (Telefonica, Spain) | % Subscribers | | | | 20.0 | 21.0 | 19.1 |
| | % Revenue | 11.4 | 9.0 | 15.1 | 14.0 | 14.0 | 22 |
| Pegaso[1] | % Subscribers | | | 4.0 | Movistar (2002) | | |
| | % Revenue | | | | | | |
| Iusacell (Grupo Salinas/Grupo Televisa) | % Subscribers | | | | 4.0 | 4.0 | 7.1 |
| | % Revenue | 22.9 | 11.7 | 3.9 | 6.0 | 5.0 | 5.7 |
| Unefon[2] | % Subscribers | | | | | Iusacell (2008) | |
| | % Revenue | | 1.1 | 3.8 | | | |
| Portatel | % Subscribers | | | | | | |
| | % Revenue | 1.9 | 1.1 | | | | |
| Nextel (NII Holdings, US) | Subscribers | | | | 3.0 | 4.0 | 3.9 |
| | % Revenue | | | | 12.0 | 12.0 | 4 |
| Total Subscribers (1,000) | | | | | 77,395 | 83,150 | 98,909 |
| Total Revenue (bil MXN) | | | | | 163 | 230 | 265.1 |
| Total Revenue (bil US$) | | | | | 14.6 | 16.1 | 20.1 |
| *N* (>1%) | | 4 | 6 | 4 | 4 | 4 | 4 |
| C4 (Audience Shares) | | | | | 99.0 | 100.0 | 100 |
| HHI (Audience Shares) | | | | | 5,609 | 5,514 | 5,316 |
| Noam Index (Audience Shares) | | | | | 2,805 | 2,757 | 2,658 |
| C4 (Market Shares) | | 99.9 | 97.8 | 100.0 | 100.0 | 100.0 | 100 |
| HHI (Market Shares) | | 4,716 | 5,580 | 6,218 | 5,137 | 5,126 | 5,545 |
| Noam Index (Market Shares) | | 2,358 | 2,278 | 3,109 | 2,569 | 2,563 | 2,773 |

[1] Movistar México acquired Pegaso in 2002.
[2] Iusacell purchased Unefon in 2008.

## Table 22-12. Internet Service Providers (Market Shares by Revenue), 2007–2011

| | 2007 | 2008 | 2009 | 2010 | 2011 |
|---|---|---|---|---|---|
| Prodigy (Telmex) (Carlos Slim, Grupo Carso) | 74.2 | 78.2 | 83.9 | 74.0 | 76.9 |
| Grupo Televisa (Azcárraga Family) | 14.0 | 11.5 | 8.9 | 12.0 | 6.8 |
| Cablevisión | 14.0 | 11.5 | 8.9 | 8.0 | 6.8 |
| Cablemás | | | | 3.0 | |
| TVI-SKY (News Corp., Murdoch Family, US/UK/AUS) | | | | 1.0 | |
| Megacable | 8.6 | 8.2 | 7.2 | 6.0 | 5.9 |
| Axtel/Avantel | 2.8 | 2.1 | | 7.0 | 10.4 |
| Maxcom | 0.4 | | | 1.0 | |
| Total Revenue (mil MXN) | 4,311 | | | 9,988 | |
| Total Revenue (mil US$)[1] | 388 | | | 799 | |
| C4 | 99.6 | 100.0 | 100.0 | 99.0 | 100.0 |
| HHI | 5,784 | 6,319 | 7,170 | 5,706 | 6,103 |
| *N* (>1%) | 5 | 4 | 3 | 6 | 6 |
| Noam Index | 2,892 | 3,160 | 4,140 | 2,552 | 3,051 |

[1] This revenue figure is an estimate based on data collected in this study: it is based on average per capita revenues reported for several other countries in this study with similar per capita income.

## Table 22-13. Online News Media (Market Shares by Monthly Unique Visitors), 2009–2010

| | 2009 | 2011 |
|---|---|---|
| *El Universal de México* | 9.5 | 15.8 |
| *MSN* (Microsoft, US) | 10.9 | |
| *elmundo.es* (Unidad Editorial S.A., Spain) | 6.0 | 3.8 |
| *Yahoo! News* (US) | 5.5 | |
| *oem.com.mx* (OEM) | 4.7 | 4.3 |
| *elpais.com* (PRISA Group, Spain) | 4.3 | 2.8 |
| *mediosmasivos.com.mx* (Medios Masivos Mexicanos S.A.) | | 11.6 |
| *aee.com.mx* (AEE Periodicos Network) | | 14.4 |
| *milenio.com* (Grupo Multimedios) | 3.1 | 7.0 |
| *20minutos.es* (Multiprensa y Más S.L., Spain) | | 4.9 |
| *excelsior.com.mx (El Excélsior)* | | 2.3 |
| *CNN.com* (Time Warner, US) | 2.8 | |
| Total Revenue (mil MXN) | Estimate based on visitor number | |
| Total Revenue (mil US$) | | |
| C4 | 32 | 49 |
| HHI | 341 | 711 |
| *N* (>1%) | 8 | 9 |
| Noam Index | 121 | 237 |

*Source*: comScore.com.

Table 22-14. Sector Concentration (Noam Index), 1988–2010

| | 1988 | 1992 | 1995–1996 | 1997 | 2000 | 2004–2005 | 2007 | 2008 | 2009 | 2010 |
|---|---|---|---|---|---|---|---|---|---|---|
| Broadcast TV (by Audience Shares) | | | 5,910 | | 4,029 | 3,980 | | | | 3,880 |
| Broadcast TV (by Market Shares) | | | | | | 3,791 | 3,314 | | | 3,366 |
| Multichannel TV Platforms (by Audience Shares) | | | | | | | 1,250 | 1,171 | 1,239 | 1,352 |
| Radio Group (by # of Stations) | | | | | | | | | | 176 |
| Film (by Box Office %) | | | | | | | | | | 2,219 |
| Wireline Telecom (by Market Shares) | 10,000 | 10,000 | 10,000 | | | 3,261 | | 3,002 | | |
| Wireline Telecom—Long-Distance Telephony (by Market Shares) | | | | 3,665 | 2,436 | 2,236 | | | | 6,690 |
| Wireless Telecom (by Market Shares) | | | | 2,358 | 2,278 | 3,108 | | 2,569 | 2,563 | |
| ISPs (by Total # of Subscribers) | | | | | | | 2,892 | 3,160 | 4,144 | 2,263 |

market is highly concentrated: Telmex's Prodigy has a distinct advantage over its competitors as both a bidder and a carrier and as a result dominates the market (Table 22.12).

### Online News

In contrast to the print sector, online political coverage is the most popular web content, and the most successful sites are those that provide free content, though this means that existing outlets' online editions actually dominate the sector rather than new online-only outlets. Paid online subscription service is not popular and has not caught on (Table 22.13).

## CONCLUSION

Mexico has one of the most concentrated audiovisual and telecommunication markets in the world, and this is unlikely to change in the near term. Looking forward, we believe discussion of concentration studies in Mexico needs to move toward more descriptive analyses. While new technologies such as multichannel platforms have introduced some competitiveness to the market, concentration indicators in general have either stabilized or increased since 2000. Moreover, both Grupo Televisa and Azteca are now further expanding their interests in telecommunications through triple-play service offers, which would only accentuate concentration in paid TV, Internet access, and telephony (Table 22.14).

Though the media market is highly concentrated, political rivalry will make it difficult for reforms to take place. In the 2012 presidential election, Grupo Televisa was accused of influencing its audiences to benefit Enrique Peña Nieto, the PRI candidate. In turn, Grupo Carso was accused of helping Nieto's rival, Andres Manuel Lopez-Obrador, leader of the opposition Party of the Democratic Revolution (PRD). Given that Nieto won his 2012 reelection campaign, increased regulatory scrutiny is more likely to be seen in the telecommunications sector rather than in the audiovisual sector, as Grupo Televisa's executives have made it quite clear to Mexican politicians that reforms in the audiovisual sector should primarily come at the expense of Grupo Carso's monopolistic practices, rather than Grupo Televisa's.

The ongoing dispute between Emilio Azcárraga Jean, Carlos Slim, and their respective political partisans over the opening up of the audiovisual and telecom markets to greater competition will shape the future of the Mexican media market for years to come.

## Bibliography

### 1. Books and Articles


Alva, A. "Entre la crisis financiera y la guerra de los *ratings* televisivos." *Revista Mexicana de Comunicación,* 1999, no. 57, 7–10.

Azcárraga, Emilio. "Carlos Slim Needs More Competition." *The Wall Street Journal.* 22 Mar. 2012. Jan. 16, 2013. <http://online.wsj.com/article/SB10001424052702304724404577293070405826732.html>.

Castañeda, R. *Prensa Vendida.* Mexico: Grijalbo, 1993.

"DTH grew an 83.8% in Mexico." *Dataxis,* Mar. 17, 2011. Jan. 17, 2013. <http://nextvlatam.com/index.php/1-cable-dth/dth-grew-an-83-88-in-mexico/>.

Duarte, Enrique. "TV digital acecha dominio de televisoras." CNNExpansión.com, Mar. 28, 2011. Jan. 17, 2013. <http://www.cnnexpansion.com/negocios/2011/03/23/televisa-y-tv-azteca-compartiran-mercado>.

Fernández, F. *Los medios de difusión masiva en México.* Mexico City: Editorial Juan Pablos, 1982.

Flores-Roux, Ernesto and Mariscal, Judith. *Política de Generación de Infraestructura de Telecomunicaciones en México: Una Crítica.* Mexico City: Centro de Investigación y Docencia Económicas, 2010. <http://telecomcide.org/docs/publicaciones/DTAP-246.pdf>.

García, Julio S. *Estos Años.* Mexico: Océano, 1995.

Gómez, R. *El impacto del Tratado de Libre Comercio de América del Norte (TLCAN) en la industria audiovisual mexicana (1994–2004).* Diss. Universidad Autónoma de Barcelona, 2006. Bellaterra: UAB, 2007. <http://www.tdx.cat/bitstream/handle/10803/4197/rgg1de1.pdf;jsessionid=F8D3D1A7F49F60B1D7D2B046E0C3AEA5.tdx2?sequence=1>.

Gómez, R. "TV Azteca y la industria televisiva mexicana en tiempos de integración regional (TLCAN) y desregulación económica." *Comunicación y Sociedad* 1, 2004. Jan. 18, 2013. <http://redalyc.uaemex.mx/src/inicio/ArtPdfRed.jsp?iCve=34600104>. 52–90.

Gómez, R. and Sosa, G. "La concentración en el mercado de la televisión restringida en México." *Comunicación y Sociedad* 14, 2010. Jan. 17, 2013. <http://redalyc.uaemex.mx/src/inicio/ArtPdfRed.jsp?iCve=34615372005>. 109–142.

Gómez, Rodrigo, Sosa, Gabriel, Bravo, Jorge, and Tellez, Primavera. *Mapping Digital Media: México.* Open Society Foundations–Media Program, London: Open Society Foundations–Media Program, 2011. June 13, 2011. <http://www.soros.org/initiatives/media/articles_publications/publications/mapping-digital-media-mexico-20110613>.

Huerta-Wong, J. and Gómez, R."Concentración y diversidad de los medios de comunicación y las telecomunicaciones en México." *Comunicación y Sociedad* 19, 2013. Jan. 17, 2013. <http://www.comunicacionysociedad.cucsh.udg.mx/sites/default/files/a5.pdf>. 113–152.

Huerta-Wong, J. and Lozano, J. "La programación televisiva mexicana desde la perspectiva de la diversidad." Cerdán, J. and Aceves, F., eds. *Anuario de investigación de la Comunicación VIII.* Mexico: Consejo Nacional de Escuelas e Investigadores de las Ciencias de la Comunicación, 2001.

Lozano, J. "Public policies and research on cultural diversity and television in Mexico." *Journal of Broadcasting & Electronic Media* 50:3, 2006. 467–481. Jan. 18, 2013. <http://www.tandfonline.com/doi/abs/10.1207/s15506878jobem5003_7>.

Marchant, Ivan. "Futuro Digital Mexico." comScore, Aug. 7, 2011. Jan 18, 2013. <http://www.comscore.com/Insights/Presentations_and_Whitepapers/2012/Futuro_Digital_Mexico>.

Mariscal, Judith. *The Mexican Telecommunications Reform: A Political Economy Approach.* Diss. The University of Texas at Austin, 1998. Austin: UT, 1998.

Mariscal, Judith, Bonina, Carla, and Luna, Julio. *Los nuevos escenarios del mercado en América Latina. Documento de Trabajo.* Lima: International Development Research Centre, 2007. <http://www.dirsi.net/sites/default/files/dirsi_07_PD03_es.pdf>.

Mariscal, J. and Rivera, E. "Regulación y competencias en las telecomunicaciones mexicanas." Mexico City: United Nations (UN) CEPAL, 2007. Jan 17, 2013. <http://www.eclac.cl/publicaciones/xml/7/29037/Serie%2083.pdf>.

Noam, Eli. *Media Ownership and Concentration in America.* New York: Oxford University Press, 2009.

Orozco, G. *Historias de La televisión en América Latina.* Barcelona: Gedisa, 2002.

Paxman, A. and Saragoza, A. "Globalization and Latin Media Powers: The Case of Mexico's Televisa." Mosco, V. and Schiller, D., eds. *Continental Order? Integrating North America for Cybercapitalism.* Lanham: Rowman & Littlefield, 2001.

Sánchez-Ruiz, E. "El audiovisual Mexicano: ¿Concentrar para competir?" *Global Media Journal Edición Iberoamericana* 1:2, 2004. Feb. 8, 2012 <http://gmje.mty.itesm.mx/articulos2/sanchezruiz_OT04.html>.

Sánchez-Ruiz, E. "Globalization, Cultural Industries and Free Trade: The Mexican Audiovisual Sector in the NAFTA Age." Mosco, V. and Schiller, D., eds. *Continental Order? Integrating North America for Cybercapitalism.* Lanham: Rowman & Littlefield, 2001.

Sinclair, J. *Latin American Television: A Global View.* New York: Oxford University Press, 1999.

Toussaint, F., ed. *¿Televisión pública en México?* Mexico City: Consejo Nacional para la Cultura y las Artes, 1993.

Trejo, R. "Muchos medios en pocas manos: concentración televisiva y democracia en América Latina." *Intercom—Revista Brasileira de Ciências da Comunicação* 33:1, 2010. Jan. 18, 2013. <http://portcom.intercom.org.br/revistas/index.php/revistaintercom/article/download/146/139>. 17–51.

Trejo, R., ed. *Las redes de Televisa.* Mexico City: Claves Latinoamericanas, 1988.

Trejo, R., ed. *Televisa quinto poder.* Mexico City: Claves Latinoamericanas, 1985.

Vega Montiel, Aimée, coordinator. "II Seminario Internacional Acceso y participación de las mujeres en las industrias de comunicación: de los diagnósticos a las experiencias." Ciudad Universitaria. Torre II de Humanidade, Mexico City, Mexico. Nov. 8, 2012. <http://www.ceiich.unam.mx/0/60EncFic.php?tblEncuentrosAcad_id=1066>.

Werbowski, Michael. "'El Excelsior': The Rise and Fall of a Great Mexican Newspaper." *Worldpress.org*, Feb. 18, 2007. Jan. 17, 2013. <http://www.worldpress.org/Americas/2678.cfm#down>.

## 2. Reports

AAAP. "Informe oficial Latinoamericano de inversión publicitaria 2009." Buenos Aires:

Asociación Argentina de Agencias de Publicidad (AAAP), 2010. Dec. 1, 2011. <http://www.aapublicidad.org.ar/wp-content/inversion_publicitaria/2009/argentina/informe_inversion_publicitaria_2009.pdf>.

ARVM Staff. "Registran números positivos en Radio." Asociación de Radio del Valle de México (ARVM), Oct. 15, 2011. Jan. 17, 2013. <http://laradioenmexico.mx/registran-numeros-positivos-en-radio/>.

CANACINE. "Asistencia al Cine." CANACINE, 2010. Oct. 22, 2011. <http://www.canacine.org.mx/index.php/resultados-2010/asistencia-al-cine.html>.

CANACINE. "Participación de mercado por distribuidora." CANACINE, 2010. Oct. 22, 2011. <http://www.canacine.org.mx/index.php/resultados-2010/participacion-de-mercado-por-distribuidora.html>.

Cofetel. "Documento de referencia a la consulta pública para la licitación de frecuencias para televisión abierta." Cofetel, 2011. Nov. 9, 2011. <http://www.cft.gob.mx:8080/portal/wp-content/uploads/2012/06/Documento_de_referencia_Consulta_TV.pdf>.

IMCINE. "Anuario estadístico de cine mexicano 2010." Mexico City: IMCINE, 2010. Jan. 17, 2013. <http://www.imcine.gob.mx/temporales/ANUARIO_IMCINE_2010.pdf>.

IMCINE. "Anuario estadístico de cine mexicano 2011." Mexico City: IMCINE, 2011. Jan. 17, 2013. <http://www.imcine.gob.mx/media/26138.pdf>.

Telmex. "Informe Annual 2010: 20 Anos de desarrollo tecnológico." Telmex, 2011. Jan. 17, 2013. <http://www.telmex.com/mx/corporativo/pdf/pt_descarga.jsp?a=01TRIM12n.pdf>.

# Mexico—Data Summaries

ELI NOAM AND PAUL MUTTER

MEXICO HAS THE FIFTH-HIGHEST concentration rate of the countries surveyed, with a weighted average C4 of 97% in 2010 (Table 22.15). This is due to the hold over electronic media and telecommunications that two companies possess. Carlos Slim's Grupo Carso, through its América Móvil branch, effectively controls the wireless, wireline, and ISP markets. In platform media, it has a 67% market share (Table 22.17). Emilio Azcárraga's family controls Grupo Televisa with its 69% market share in television broadcasting, which translates into a 46% market share in content media alone (Table 22.16). In newspapers, the largest Mexican print media group (OEM) held 59.4% of the market in 2012. Such control has spurred Mexican legislators to pass laws that create a new regulatory agency for electronic media, IFT, with considerable independence and professionalism in comparison to the past regulator, to reduce market power. Slim's wireline Telmex and wireless Telcel subsidiaries were said to cost the Mexican economy each year $25 billion due to the absence of real competition. America Móvil was such a giant that its revenue was larger than that of the next five largest companies in Mexico combined.

In order to encourage deconcentration, Mexico's foreign direct investment rules for telecommunications and broadcast television have been relaxed to allow an ownership up to 100% and 49% in these industries, respectively, doubling the pre–existing limits for both. These limits accounted for the fact that of all the Spanish-language markets surveyed, Mexico is the only one without a presence by Telefónica (Spain)—or any foreign participation in its wireline markets. In contrast, América Móvil, often operating under the name of "Claro", is the fourth-largest wireless provider internationally with 252 million subscribers.

Mexico's newspaper market is unique in that much of its advertising revenue comes from

Table 22-15. National Media Industries Concentration in Mexico

| | 2004/5 | | 2011 OR MOST RECENT | | % CHANGE ANNUAL AVERAGE | |
|---|---|---|---|---|---|---|
| | COMPANY POWER INDEX IN COUNTRY | COMPANY SHARE OF THE OVERALL NATIONAL MEDIA MARKET (%) | COMPANY POWER INDEX IN COUNTRY | COMPANY POWER INDEX IN COUNTRY | COMPANY POWER INDEX IN COUNTRY | COMPANY SHARE OF THE OVERALL NATIONAL MEDIA MARKET (%) |
| Grupo Carso | 4,095 | 57 | 4,185 | 59 | 0.4 | 0.1 |
| Televisa | 700 | 15.1 | 580 | 14.2 | –3 | 0.2 |
| TV Azteca | 102 | 4 | 83 | 3 | –3.2 | –0.1 |
| NII Holdings (Mexico/US)[1] | 97 | 8 | 11 | 3 | –15 | –1 |
| Megacable | 23.4 | 1.4 | 26 | 2 | 1.88 | 0.08 |
| Organizacion Editorial Mexicana (OEM) | 18 | 0.3 | 19 | 0.3 | 0.6 | –0.002 |
| Google (US) | 5.2 | 0 | 17 | 0.2 | 38.04 | 0.03 |
| MVS Comunicaciones | 2 | 0.3 | 14 | 1.1 | 104 | 0.13 |
| Disney (US) | 6.3 | 0.3 | 12 | 0.6 | 14 | 0.04 |
| Time Warner (US) | 4 | 0.3 | 8 | 0.5 | 14 | 0.04 |
| Murdoch Interests (US) | 5 | 0.3 | 5 | 0.4 | –1 | 0.01 |
| El Universal | 1.0 | 0.1 | 0.6 | 0.06 | –7 | –0.003 |
| Grupo Reforma | 1.0 | 0.1 | 0.5 | 0.05 | –7.6 | –0.004 |
| La Jornada | 0.1 | 0.03 | 0.1 | 0.02 | –2 | –0.001 |
| Grupo Multimedios | 0.2 | 0.03 | 0.1 | 0.02 | –6.4 | –0.001 |
| Medios Masivos Mexicanos | 0.1 | 0.007 | 0.2 | 0.02 | 33.5 | 0.002 |

*continued*

Table 22-15. *continued*

| | 2004/5 | | 2011 OR MOST RECENT | | % CHANGE ANNUAL AVERAGE | |
|---|---|---|---|---|---|---|
| | COMPANY POWER INDEX IN COUNTRY | COMPANY SHARE OF THE OVERALL NATIONAL MEDIA MARKET (%) | COMPANY POWER INDEX IN COUNTRY | COMPANY POWER INDEX IN COUNTRY | COMPANY POWER INDEX IN COUNTRY | COMPANY SHARE OF THE OVERALL NATIONAL MEDIA MARKET (%) |
| Media Concentration Index | 2004/5 | | 2011 or Most Recent | | % Change Annual Average | |
| Total Revenue: Nat'l Media Industry (mil US$) | 21,710 | | 30,526 | | 7 | |
| Total Voices (*n*) | 31 | | 33 | | 1.1 | |
| Net Voices (*n*) | 28 | | 30 | | 1.2 | |
| Public Ownership (%) | 0.0 | | 0.0 | | 0.0% | |
| Foreign Ownership (%) | 19 | | 20 | | 0.2 | |
| C4 Average—Weighted | 97 | | 97 | | −0.1 | |
| HHI Average—Weighted | 5,936 | | 5,315 | | −2 | |
| C1 Average—Weighted | 69 | | 68 | | 0 | |
| Noam Index Average—Weighted | 2,576 | | 2,078 | | −3.22 | |
| Pooled Overall Sector C4 | 83.7 | | 75.7 | | −2.0% | |
| Pooled Overall Sector HHI | 3,685 | | 3,777 | | 0.4 | |
| Pooled Overall Sector Noam Index | 679 | | 447 | | −6 | |
| Market Share of Top Ten Companies: Nat'l Media Industry (%) (Pooled C10) | 96.4 | | 97 | | 0.1 | |
| National Power Index | 5,204 | | 5,298 | | 0.3 | |

[1] Formerly Nextel; it is no longer part of Sprint (and not acquired by Softbank).`

Table 22-16. Top Content Media Companies in Mexico

| | 2004/5 | | 2011 OR MOST RECENT | | % CHANGE ANNUAL AVERAGE | |
|---|---|---|---|---|---|---|
| | COMPANY POWER INDEX IN COUNTRY | COMPANY SHARE OF THE NATIONAL CONTENT MEDIA MARKET (%) | COMPANY POWER INDEX IN COUNTRY | COMPANY SHARE OF THE NATIONAL CONTENT MEDIA MARKET (%) | COMPANY POWER INDEX IN COUNTRY | COMPANY SHARE OF THE NATIONAL CONTENT MEDIA MARKET (%) |
| Grupo Televisa | 3,613 | 56 | 2,623 | 46.0 | –5 | –2 |
| TV Azteca | 628 | 21.7 | 528 | 17.6 | –4.0 | –1.0 |
| OEM | 139 | 2.3 | 136 | 2.2 | –0.6 | –0.03 |
| Megacable | 44.8 | 2.2 | 51.7 | 3.0 | 3.8 | 0.2 |
| Disney (US) | 39 | 2 | 73 | 4 | 15 | 0.3 |
| MVS Comunicaciones | 4.0 | 0.7 | 30 | 2.3 | 108 | 0.3 |
| Time Warner (US) | 27.1 | 1.6 | 51 | 3 | 15 | 0.24 |
| Murdoch Interests (US) | 30 | 1.7 | 29 | 2.3 | –0.5 | 0.01 |
| Grupo Carso | 3 | 0.6 | 11 | 1.4 | 51 | 0.14 |
| Google (US) | 32.1 | 0 | 108 | 1.2 | 40 | 0.2 |
| Grupo Reforma | 6 | 0.5 | 3.3 | 0.34 | –7 | –0.02 |
| Medios Masivos Mexicanos | 0.5 | 0.04 | 1.5 | 0.1 | 35 | 0.02 |
| El Universal | 6.3 | 0.5 | 4 | 0.4 | –6.5 | –0.02 |
| Grupo Multimedios | 1.1 | 0.2 | 0.7 | 0.2 | –6.1 | –0.01 |
| La Jornada | 0.8 | 0.17 | 0.7 | 0.12 | –1.5 | –0.0 |
| Media Concentration Index | 2004/5 | | 2011 or Most Recent | | % Change Annual Average | |
| Public Ownership (%) | 0.0 | | 0.0 | | 0.0% | |
| Foreign Ownership (%) | 9.0 | | 15.8 | | 1.14 | |
| C4 Average—Weighted | 91 | | 85 | | –1 | |
| HHI Average—Weighted | 4,589 | | 3,778 | | –3 | |
| C1 Average—Weighted | 60 | | 51 | | –1 | |
| National Power Index | 4,598 | | 3,699 | | –3.3 | |

## Table 22-17. Top Platform Media Companies in Mexico

| | 2004/5 | | 2011 OR MOST RECENT | | % CHANGE ANNUAL AVERAGE | |
|---|---|---|---|---|---|---|
| | COMPANY POWER INDEX IN COUNTRY | COMPANY SHARE OF THE NATIONAL PLATFORM MEDIA MARKET (%) | COMPANY POWER INDEX IN COUNTRY | COMPANY SHARE OF THE NATIONAL PLATFROMMEDIA MARKET (%) | COMPANY POWER INDEX IN COUNTRY | COMPANY SHARE OF THE NATIONAL PLATFORM MEDIA MARKET (%) |
| Grupo Carso | 5,867 | 75 | 4,778 | 67 | –3.1 | –1.3 |
| Grupo Televisa | 134 | 7.2 | 197 | 8.3 | 7.9 | 0.18 |
| NII Holdings (Mexico/US)[1] | 116 | 10 | 13 | 3.2 | –15 | –1.1 |
| Megacable | 19.3 | 1.2 | 21.3 | 1.6 | 1.7 | 0.1 |
| MVS | 1.6 | 0.3 | 12 | 1 | 103 | 0.1 |
| TV Azteca | 0 | 0 | 0 | 0 | 0 | 0 |
| Media Concentration Index | 2004/5 | | 2011 or Most Recent | | % Change Annual Average | |
| Public Ownership (%) | 0.0 | | 0.0 | | 0.0% | |
| Foreign Ownership (%) | 21 | | 21 | | –0.04 | |
| C4 Average—Weighted | 98 | | 98 | | 0.1 | |
| HHI Average—Weighted | 6,198 | | 5,603 | | –2 | |
| C1 Average—Weighted | 77 | | 70 | | –1 | |
| National Power Index | 6,298 | | 5,407 | | –2.4 | |

[1] Formerly Nextel; it is no longer part of Sprint (and not acquired by Softbank).

the country's political parties rather than commercial enterprises. At least 58% of total commercial advertising spending in Mexico per annum goes to broadcast television—and, by extension, Grupo Televisa, which also has a 50% market share of the multichannel platform market through its Cablemás, second only to the domestic operator Megacable. Because so much news is consumed through audiovisual media rather than print, even the largest Mexican print media group (OEM), which held 59.4% of newspaper market share in 2012 (down from an even higher 63% in 2009), accounts for less than 3% of overall content media revenues.

# D

# Asia-Pacific

# 23

# Media Ownership and Concentration in Australia

FRANCO PAPANDREA AND RODNEY TIFFEN

## INTRODUCTION

Australian history has been shaped by the "tyranny of distance," not only internationally, but also domestically.[1] At the time of federation in 1901—when the continent's six British colonies entered a political union—media markets were primarily local and thus reflected parochial content and advertising. The national media market is a relatively recent development. Australian media have tended toward high levels of concentration due to a variety of factors that differ between certain industries. The moderate size of the market, combined with economies of size and scope in the print and audiovisual segment, has always constrained market entrants. In some industries, most notably electronic media, regulatory interventions have sought to limit the impact of market forces on ownership concentration. In other industries, such as telecommunications, the former state monopoly has been supplanted by open competition, but ownership concentration nonetheless remains relatively high in the hands of the former state-owned monopolist Telstra. A high rate of technological change in recent decades has been a catalyst for the development of new industries and in forcing traditional producers to alter their business models. While the Internet has been instrumental in the growth of "new media" and the emergence of new global players such as Google, at the local level, its transformative effects on concentration have been limited. Usage of new media is widespread. In 2012, 55% of the population

1. Blainey, Geoffrey. *The Tyranny of Distance: How Distance Shaped Australia's History*. Melbourne and London: Macmillan, 1968.

held Internet subscriptions.[2] Yet domestic news consumption is still dominated by traditional media sources, which have come to control some of the most popular domestic online news portals.

Rules to safeguard media diversity were first applied to radio broadcasting in 1935 when ownership of multiple licences was restricted at the local, state and federal levels. The primary aim of the rules was to constrain dominance of commercial radio licences by newspaper interests.[3] Similar rules preventing any one individual or entity from acquiring more than one commercial license in the same locality, and more than two nationally were applied when television was introduced in 1956. Despite these measures, newspapers' owners dominated both commercial radio and commercial TV broadcasting.

Notwithstanding this long-held concern over cross-ownership by newspapers, specific restrictions were not introduced until 1987. The federal government argued that they were needed to "curb major expansion in television by existing newspapers and radio interests that already have considerable influence over the formation of public opinion."[4] These changes had immediate and far-ranging effects: between 1986 and 1987, 13 of 19 metropolitan newspapers changed hands and 11 of 17 metropolitan commercial TV stations changed owners. Cross-ownership between newspapers and broadcasting was almost completely eliminated, but concentration within both industries increased.

The interests surrounding this change were complex: the Labor Government at the time was inclined to introduce aggregation into rural areas, giving consumers more products to choose from and undermining highly profitable local monopolies. With support for the change by the metropolitan networks, the roll-out began in 1989 after an intense internal debate (the other main parties, the Nationals and Liberals, were also divided despite officially opposing Labor's initiative). Kerry Packer held TV stations through his Nine Network but owned no newspapers at the time, and his operations were thus unaffected. Rupert Murdoch had become a US citizen in order to buy 20th Century Fox, giving up his Australian citizenship. Since he had to divest himself from his Australian TV holdings (but not his News Ltd.'s newspapers), the regulations came at a propitious time for him. Conversely, the other two major media groups—Fairfax Media and The Herald and Weekly Times (HWT)—owned both TV stations and daily newspapers, so the new restrictions made it almost impossible for either to expand its businesses.

Subsequent to the change, the widely diffused ownership structure of the commercial television industry was transformed into one where three urban networks covered the major metropolitan areas and two regional networks dominated the rest of the country. Almost a decade after their introduction, the cross-ownership rules became the focus of a major political debate between a new Conservative government intent on their removal and the parties in opposition. After several years of debate, legislation abolishing the restrictions was passed in 2006 that introduced the concept of "independent voices" in local media markets: at least five independent media voices in metropolitan areas and at least four in other areas. Given the existence of restrictions on cross-ownership of radio and broadcast TV licenses in local markets, the 2006 media diversity rules were largely redundant and proved ineffective in decreasing concentration.[5] For radio, the strict limits on cross-ownership introduced in 1935 were removed with the passage of the 1992 Broadcasting Services Act. Since then, the only remaining restriction is a limit of no more than two stations in common

---

2. ABS (Australian Bureau of Statistics). "8153.0—Internet Activity, Australia, Dec 2012." Canberra: ABS, 2012. Accessed May 27, 2013. <http://www.abs.gov.au/ausstats/abs@.nsf/Products/8153.0~December+2012~Chapter~Sector?OpenDocument>.

3. Joint Parliamentary Committee on Wireless Broadcasting (W. Gibson, Chair). *Report.* Canberra: Government Printing Office, 1942.

4. Duffy, M. (MP, the Hon). "Second Reading Speech." *House of Representatives Debates*, Apr. 29, 1987. 2,194.

5. Papandrea, Franco. "Media Diversity and Cross Media Regulation." *Prometheus* 24: 3, 2006. 301–322.

ownership in the same licensing market: there are no limits on the number of licenses that can be held in different local markets. The removal of the cross-ownership restrictions entrenched the already relatively concentrated media markets and opened the doors to greater concentration. The ownership changes that followed led to the reemergence of cross-ownership of radio stations and newspapers (APN News and Media) and broadcast radio and TV (Southern Cross Media Group). Today, most cities and towns are served by three commercial television services, several commercial radio services (at least two for every locale), and the television and radio services of the publicly-owned Australian Broadcasting Corporation (ABC) and Special Broadcasting Service (SBS).

The new regulations were introduced partly to counterbalance the planned replacement of existing constraints on cross-ownership of more than two broadcasting stations. The new, broader rules provided for common ownership of any number of commercial television stations, provided their aggregate reach did not exceed a 60% share of the population. Subsequent to the change, the widely diffused ownership structure of the commercial broadcasting industry was transformed into one where three urban networks covered the major metropolitan areas and two regional networks dominated the rest of the country. The removal of the cross-ownership restrictions not only entrenched the already relatively concentrated market, but it also opened the doors to greater concentration in other media industries.

Other media industries are not subject to specific ownership regulation. Starting in 1991, the telecommunications sector underwent reforms that fully privatized and liberalized the market. The transformation began with the establishment of a second fully private carrier, Optus—fully owned by the Singaporean state-owned wireless provider SingTel—providing both wireline and wireless services in competition with Telstra. A third domestic mobile operator, Vodafone, was licensed and entered the market in 1993. The 1997 Telecommunications Act introduced a fully competitive market, and Telstra was fully privatized. Telstra's former monopolist status and ownership of the wireline consumer access network has given it a distinct advantage over its competitors. Telstra has been able to use its position as Australia's dominant wireline carrier and largest wireless carrier to establish itself as the largest Internet service provider (ISP) in the country as well.

Widespread access and use of the Internet have brought many changes with major implications for media industries, but hoped-for increases in news sources and the mitigation of concentration in traditional media has only partially materialized. In the domestic arena, little has changed since the Productivity Commission reported in 2000 that the majority of Australians obtaining their news from the Internet were accessing the sites of the established traditional media organizations.[6] So far, *Crikey.com* is the only significant independent alternative Australian news portal.

In the following sections, we observe that in recent years there has been a significant increase in the presence of cross-ownership by conglomerates.

## PRINT MEDIA

### Newspapers

The zenith for Australian newspapers' diversity came in 1923, when there were 26 metropolitan dailies and 21 independent owners.[7] Since then, the trend has been toward a reduction in both the number of titles and owners. The worldwide Great Depression of the 1930s led to several closures and weakened other titles. During this time, *The Herald and Weekly Times* (HWT), led (but not owned) by Sir Keith Murdoch, started acquiring titles. After further consolidation in the years after WWII, three owners now dominate the national market: News Corp.'s News Ltd., with six titles accounting for 65% of metropolitan and daily circulation; Fairfax Media

6. Productivity Commission. "Broadcasting Report no. 11." Canberra: Ausinfo, 2000.

7. Mayer, Henry. *The Press in Australia*. Melbourne: Lansdowne Press, 1964.

with four titles accounting for 25%; and Seven West Media Ltd.'s sole title, with a 10% market share. The market, both at the national and subnational levels, is entirely commercial. Sydney and Melbourne are the only cities in the country with competing, locally produced daily newspapers. Print media is highly concentrated when compared against world standards.

Australia also has a well-established provincial daily press consisting in 2011 of 37 newspapers of varying size and quality, but with little impact or newsgathering capacity at the national level. Nearly all these began as locally owned enterprises, but by 2008, only two remained as such: all of the others have been absorbed into larger conglomerates.

The number of newspapers in circulation declined from 56 to 48 between 1984 and 2010. However, seven of the eight closures were of metropolitan afternoon newspapers, all of which closed between 1987 and 1993. This was in part because of changing commuting, advertising, and reading habits, but the 1980s ownership changes in all Australian traditional media accelerated their demise. Apart from these seven afternoon newspapers, there has been only one other closure in the period under review, that of a small provincial daily. Australian newspapers' overall good health paints a contrasting picture to the state of affairs facing counterparts in other English-language speaking markets: compared with 1992 levels, the 2008 circulation rate was around 90%. That said, the substantial population growth Australia has experienced since the 1980s means that the ratio of sales to population has significantly declined (Table 23.1).

In 2011, the third biggest print group was Seven West Media, which owns two titles in Western Australia, most importantly the morning daily *West Australian*, formerly part of HWT. Its circulation has been slightly worse than average in recent years, hence its declining market share. APN, owned by the Irish businessman Tony O'Reilly's Independent News & Media (INM), owns several provincial daily newspapers until it began selling its shares off in 2015. It began a substantial Australian presence during the 1987 shake-up, and has acquired more titles since. Rural Press was formed following internal divisions in the Fairfax family in the late 1980s and built a stable of provincial dailies, which in 2007 became part of Fairfax's operations when the two companies merged, boosting the share of Fairfax papers in the total national circulation. Before the late 1980s disruptions, Fairfax had accounted for about one-quarter of national daily circulation. By far, the largest newspapers' publisher in the country is Rupert Murdoch's News Ltd., whose titles account for nearly 6 out of 10 daily sales.

Australian newspapers were unusually one-sided in their partisan allegiances for many years.[8, 9] Beginning in the 1960s, the situation began to improve, as many newspapers allowed reporters more editorial independence and there was greater pluralism of perspectives in commentary. However, such pluralism is precarious, and there have been continuing accusations of bias, especially against News Ltd. (Figure 23.1).

## Book Publishing

For most of Australian history, British publishing houses have dominated the book market, but a local industry grew quickly in the last quarter of the twentieth century. The share of Australian book publishing revenue accounted for by Australian books is considerably smaller than it is for domestic newspapers and magazines. In 2010, it stood at 48%, up from 44% in 2001. Although book publishing, like most other content creation industries, has been greatly affected by the digital revolution, e-books of various types have only become financially successful in recent years.[10] Total revenue in the sector has been growing slowly, up from a total of US$1.21 billion

8. Select Committee on the Print Media (M. Lee, Chair), House of Representatives. "Report." *News & Fair Facts.* Canberra: Australian Government Publishing Service, 1992.

9. Souter, Gavin. *Heralds and Angels: The House of Fairfax, 1841–1990.* Melbourne: Melbourne University Press, 1991.

10. Lee, Jenny. "Digital Technologies in Australia's Book Industry." Canberra: Book Industry Strategy Group, 2010. Accessed May 27, 2013. <http://www.innovation.gov.au/Industry/BooksandPrinting/BookIndustryStrategyGroup/Documents/DigitalTechnologiesInAustraliasBookIndustry.pdf>.

Table 23-1. Daily Newspapers (Market Shares by Circulation)

| | 1984 | 1988 | 1992 | 1996 | 2000 | 2004 | 2008 | 2011 |
|---|---|---|---|---|---|---|---|---|
| News Ltd. (News Corp., Murdoch, US/UK/AUS) | 25.0 | 56.2 | 58.1 | 56.6 | 57.2 | 57.7 | 57.8 | 57.5 |
| The Herald and Weekly Times (HWT) | 46.4 | News Ltd. | | | | | | |
| Fairfax Media | 23.7 | 18.9 | 20.2 | 21.5 | 22.0 | 21.4 | 28.6 | 28.5 |
| Rural | 1.5 | 1.9 | 2.2 | 2.7 | 3.9 | 5.3 | Fairfax Media | |
| Seven West Media (Kerry Stokes) | | | 8.3 | 8.6 | 8.1 | 7.8 | 7.7 | 8.1 |
| APN News & Media (Independent News & Media) (Ireland) | 0.8 | 4.5 | 5.5 | 5.5 | 5.2 | 5.3 | 5.3 | 5.3 |
| Others | 2.7 | 18.5 | 5.6 | 5.1 | 3.5 | 2.4 | 0.6 | 0.6 |
| # of Titles in Circulation | 56 | 53 | 49 | 48 | 48 | 48 | 48 | 48 |
| Total Circulation (mil copies) | 4.5 | 4.0 | 3.3 | 3.2 | 3.1 | 3.1 | 3.0 | 2.8 |
| Total Revenue (mil US$) | 1,459 | 2,104 | 1,567 | 2,784 | 2,901 | 3,427 | 5,447 | 4,900 |
| Total Revenue (mil A$) | 1,694 | 2,648 | 3,096 | 3,544 | 4,858 | 4,929 | 5,697 | 4,653 |
| C4 | 96.6 | 81.5 | 92.1 | 91.7 | 92.5 | 92.2 | 99.4 | 99.4 |
| HHI | 3,343 | 3,530 | 3,888 | 3,777 | 3,864 | 3,904 | 4,246 | 4,212 |
| *N* (>1%) | 4 | 4 | 5 | 5 | 5 | 5 | 4 | 4 |
| Noam Index | 1,672 | 1,765 | 1,739 | 1,689 | 1,728 | 1,746 | 2,123 | 2,106 |

*Source*: Audit Bureau of Circulation for circulation data; Australian Press Council for ownership data.

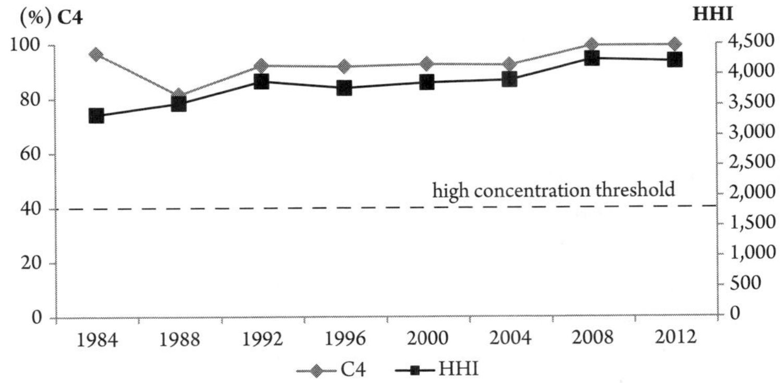


FIGURE 23.1 Daily Newspapers Concentration Indices (1984–2008)

(A$1.5 billion) in 2004 to US$1.1 billion (A$1.7 billion) million in 2008.[11] In 2008, there were approximately 4,000 book publishers in Australia, but almost 70% published only one title that year, while 295 (8%) published five titles or more, and 23 (fewer than 1%) published over 100 titles.[12] The market is the least concentrated of the three print markets under review here.

11. The figures in Table 23.2 suggest a higher growth rate, but that is more because Nielsen has been extending its methodology, rather than due to increased sales.

12. Hanover, Wilton Hanford (ed.). Book Industry Strategy Group. "Final Report to Government." Canberra: Book Industry Strategy Group, 2011. Accessed May 27, 2013. <http://www.innovation.gov.au/Industry/BooksandPrinting/BookIndustryStrategyGroup/Documents/BISGFinalReport.pdf>.

Table 23-2. Book Publishing (Market Shares by Revenue)

| | 2004 | 2006 | 2008 |
|---|---|---|---|
| Penguin/Pearson (UK) | 13.6 | 13.3 | 14.3 |
| Hachette (Lagardere) (France) | | 11.6 | 13.0 |
| Random House (Bertelsmann AG) (Germany) | 12.4 | 9.9 | 10.1 |
| HarperCollins (News Corp.) Murdoch Family | 11.8 | 9.4 | 9.3 |
| Pan Macmillan (UK) | 6.3 | | |
| Others | 55.9 | 55.8 | 53.3 |
| Total Revenue (mil US$) | 669 | 749 | 1,160 |
| Total Revenue (mil A$) | 870 | 974 | 1,740 |
| C4 | 44.1 | 44.2 | 46.7 |
| HHI | 518 | 498 | 562 |
| *N* (>1%) | 4 | 4 | 4 |
| Noam Index | 259 | 249 | 281 |

*Source*: Nielsen BookScan.

The available data are not comprehensive and are only available for recent years. The leading industry monitor is Nielsen BookScan, which supplied the data in the table below. Their figures do not include educational publishing, which comprises around one-third of the total market volume. However, there has been a high turnover rate of small and medium-size publishers and, as the table suggests, one means of survival is having a large parent company.

Hachette, whose parent company is Lagardere (France), was the second ranking company in 2006 and 2008, having grown by acquiring such publishers as Hodder Headline (UK) and Little, Brown and Company (US). Apart from these acquisitions, the market share of the top four companies has been stable since 1984 (Table 23.2).

## Magazine Publishing

Only after WWII did magazine publishing became a major industry. Three large Australian companies and some international publishers, such as Time Warner and Reader's Digest, dominate the magazine market. A plethora of small circulation periodicals also exist, some of which are fairly short-lived. The most reliable data, however, concern the larger magazines. Table 23.3 gives the data for the top 50 circulating magazines in each of the years selected.

The revenue figures in Table 23.3 do not reflect circulation numbers: total magazine circulation fell by 12.5% between 2000 and 2008 despite an increase in revenue (Figure 23.2).

# AUDIOVISUAL MEDIA

## Radio

The Australian commercial radio industry was reconfigured in the 1930s as a dual public–private system. The public service broadcaster Australian Broadcasting Corporation (ABC) which operates as a mixture of national and regional/local networks. Public broadcasting was augmented with the establishment of the Special Broadcasting Service (SBS) in 1978, which was expanded from the Sydney and Melbourne metropolitan areas to provide nationwide coverage alongside the ABC. Apart from some relatively minor income generated from its ancillary business activities, the ABC is funded entirely by the public. The overwhelming proportion of the SBS's funding also comes from

Table 23-3. Magazine Publishing (Market Shares by Circulation)

| | 2000 | 2004 | 2008 |
|---|---|---|---|
| Publishing and Broadcasting Limited (PBL) (Packer) | 46.0 | 46.0 | 44.0 |
| Pacific Magazines (Seven West Media) (Kerry Stokes) | 19.0 | 22.0 | 27.0 |
| News Ltd. (News Corp., Murdoch, US/UK/AUS) | 5.0 | 1.0 | 14.0 |
| FPC (The Independent Print Media Group) | 7.0 | 11.0 | News Ltd. |
| Reader's Digest (US) | 6.0 | 4.0 | 6.0 |
| Horwitz[1] | 3.0 | 2.0 | N/A |
| Time Warner (US) | 5.0 | 4.0 | 1.0 |
| Others | 8.0 | 9.0 | 7.0 |
| Total Circulation (mil copies) | 8.1 | 7.6 | 7.1 |
| Total Revenue (mil US$) | 1,027 | 1,343 | 2,210 |
| Total Revenue (mil A$) | 1,720 | 1,932 | 2,311 |
| C4 | 78.0 | 83.0 | 93.0 |
| HHI | 2,621 | 2,758 | 2,898 |
| *N* (>1%) | 7 | 7 | 5 |
| Noam Index | 991 | 1,042 | 1,296 |

*Source*: Audit Bureau of Circulation for circulation data; Australian Press Council for ownership data.
[1]Horwitz dropped out of the leading 50 magazines in this period.

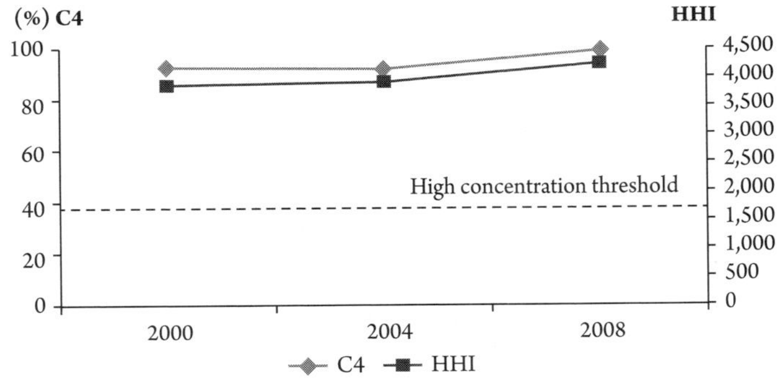


FIGURE 23.2 Magazines Concentration Indices

the public, though unlike the ABC, the SBS can sell advertising.

Prior to 1987, strict regulation of commercial radio prohibited the ownership of more than one station in a local broadcasting area, more than four in any one state, and more than eight nationally. These limits were relaxed in 1987 to allow common ownership of up to 50% of the stations in any one state, and up to 16 stations nationally. As for commercial television, the introduction of regulation prohibiting cross-ownership of radio, television, and newspapers in the same market that year led to the divestiture of many media assets by their owners, including radio stations. Some instances of cross-ownership have reemerged since 2006. Another major factor that influenced the structure of the radio market was the introduction of FM radio in the 1970s. The new entrants in FM commercial radio were able to carve out large shares of their local markets, and the ABC also expanded with the development of additional FM services.

Restructuring was given further impetus by the promulgation of the 1992 Broadcasting Services Act, which removed restrictions on foreign ownership. The limits on ownership of radio licences were also largely removed. The only restriction retained was the limitation that no single individual or entity could own more than two commercial radio stations in any local broadcasting area. The new legislation also provided for the licensing of Aboriginal radio services, and community radio stations.

Since 1992, there have been significant changes in the ownership of several major networks. Some new entities have emerged from mergers and acquisition of existing stations, and some foreign print publishers entered the market as well: Independent News & Media (Ireland) purchased a string of existing stations to form the Australian Radio Network (ARN), while Trinity Mirror plc (UK) acquired several regional stations in 1996 and subsequently acquired several new licenses to establish FM services in urban markets.

In 2011, the Austereo Group merged with Southern Cross Media (previously acquired by the Macquarie Bank Group) to form the Southern Cross Austereo Group with interests in several state capital city radio stations, two regional radio networks, and a regional TV network (Table 23.4).

While many networks have changed hands since 1992, the effect of the market restructuring on ownership concentration of ownership has been insignificant. HHI in particular has remained reasonably steady and shows moderate concentration. The ABC has retained its status as the largest single operator, although its share of market revenue has declined significantly. The Austereo Radio Network, founded in 1980, has become the largest commercial operator and the ABC's main competitor. In 2011, Austereo and the Southern Cross Media Group merged to form Southern Cross Austereo with interests in both radio and broadcast TV (Figure 23.3).[13]

## Broadcast Television

Broadcast TV in Australia was established in 1956 as a dual public–private system in which the government strictly controlled market entry and the size of the industry. Both the ABC and SBS operate nationwide public TV networks. The commercial sector developed under strict licensing provisions: owners were prohibited from holding a controlling interest—defined as 15% or more of the stock in the entity that held the station's license—in more than one television station in any one region and two stations nationally. As previously mentioned, in 1987 the national two-station ownership restriction was replaced by limiting ownership to an aggregate population reach of 60%. The introduction of the cross-ownership restrictions prohibited owners from having a controlling interest in more than one print daily, commercial television station, or commercial radio station in the same local market.

The number of commercial operators in the market was also strictly regulated. Initially, two commercial television licenses were allocated in the country's "mainland" capital cities.[14] In all other areas, only a single commercial television license was issued. A third commercial operator was progressively licensed in each of these capitals from 1965 to 1988. Finally, after 1988, commercial operators in adjoining states and territories were permitted to extend distribution into their neighboring markets. Foreign nationals and entities were prohibited from holding a controlling interest in a television station: these restrictions were lifted in 2006.

Until 1987, excessive concentration within the industry was held in check by the two-station ownership rule. Nonetheless, because newspapers were major players in commercial

13. Previously, in late 2007, Southern Cross Media sold several capital city radio stations to Fairfax Media.

14. Canberra, Melbourne, Perth, Adelaide, Brisbane, Sydney, and Darwin are the state and territorial capitals of "mainland" Australia. Hobart, in Tasmania, is not on the mainland.

## Table 23-4. Radio Group (Market Shares by Revenue)

| | 1992 | 1996 | 2000 | 2004 | 2007 | 2012 |
|---|---|---|---|---|---|---|
| Australian Broadcasting Corporation (ABC) (public) | 33.2 | 29.1 | 27.3 | 24.4 | 23.7 | 27.2 |
| Southern Cross Austereo | | | | | | 23.2 |
| Austereo Radio Network (Village Roadshow) (US) | 7.9 | 21.5 | 16.1 | 15.8 | 21.5 | Southern Cross |
| Hoyts Media | 9.9 | Austereo | | | | |
| Australian Media Network (ARN) (Independent News & Media) (Ireland) + (Clear Channel) (US) | 5.2 | 10.0 | 13.1 | 11.8 | 10.4 | 9.6 |
| Wesgo | 5.4 | | | | | |
| SEA FM | | 2.1 | R G Capital | | | |
| R G Capital | | | 3.7 | 4.9 | Regional Media) | |
| Rural Press (Fairfax Media) | | 3.5 | (Trinity Mirror plc) | | | |
| Broadcast Media Group | | 1.8 | (Trinity Mirror plc) | | | |
| Regional Media (Macquarie Bank Group) | | | | | 9.8 | Southern Cross |
| Trinity Mirror plc (UK) | | | 6.9 | 13.8 | 8.8 | 9.0 |
| Tricom/Southern Cross Media Group/Fairfax Media | 1.4 | 2.5 | 3.9 | 6.5 | 7.5 | 6.7 |
| Radio Superhighway/Macquarie Network | | 1.4 | 2.8 | 3.6 | 3.1 | 4.4 |
| Broadcast Operations | | | 3.4 | 3.0 | 3.0 | |
| Special Broadcasting Service (SBS) (public) | 3.0 | 2.7 | 2.7 | 2.9 | 2.4 | 2.6 |
| Others | 34.0 | 29.8 | 14.7 | 13.0 | 15.4 | 17.2 |
| Total Revenue (mill US$) | 516 | 639 | 629 | 815 | 1,181 | 1,488.6 |
| Total Revenue (mil A$) | 722 | 830 | 1,195 | 1,060 | 1,299 | 1,454.3 |
| C4 | 56.6 | 59.8 | 68.8 | 66.1 | 65.4 | 69.1 |
| HHI | 1,330 | 1,445 | 1,279 | 1,271 | 1,386 | 1,522 |
| *N* (>1%) | 7 | 9 | 9 | 9 | 9 | 7 |
| Noam Index | 503 | 482 | 426 | 424 | 462 | 575 |

*Source*: Estimates by authors based on original data from: Australian Communication and Media Authority, Broadcasting Financial Results (various years); Australian Broadcasting Corporation, Annual Reports; Special Broadcasting Service, Annual Reports; Australian Broadcasting Tribunal, Annual Reports; and Communications Law Centre, Update, "Media Ownership Update" (various years).

broadcasting, the aim of the cross-ownership prohibitions introduced that year was to break apart existing cross-holdings over time. Pre-existing breaches were sanctioned ("grandfathered") for as long as the ownership of a station remained unchanged. By 1992 all such breaches had been resolved by changes in the ownership of major commercial broadcasters.

Aggregation of regional television markets began in 1989 and gradually led to the

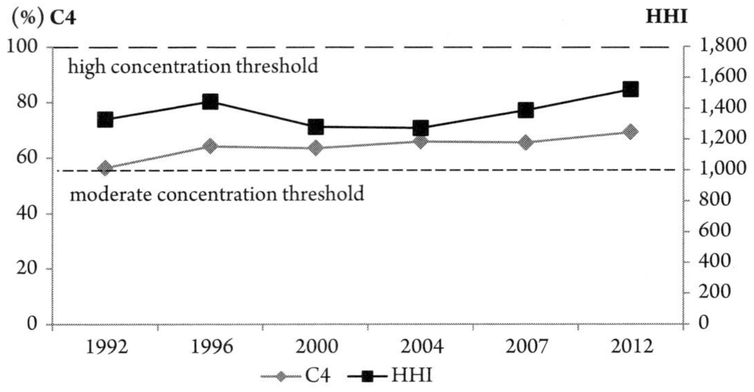


FIGURE 23.3 Radio Concentration Indices

formation of new regional television networks. The post-1987 regulatory system helped major commercial networks gain control of individual stations in each of the mainland's capital cities, reaching 75% of the national population—legislation was enacted to increase the existing 60% cap to permit this. The population reach restrictions, combined with implementation of market aggregation policies, contributed significantly to the formation of regional commercial networks. The resulting concentration of commercial broadcasting in the hands of a small number of major players significantly altered the Australian TV market. Commercial ownership is now almost entirely concentrated in three major metropolitan network operators—Seven Network, Nine Network, and Ten Network—and three smaller regional networks. The remaining independent operators are economically small entities in control of small individual licenses of some regional services—and folded into "Others" in Table 23.5. In terms of programming, the three aforementioned regional networks—Prime Network, Southern Cross Media Group, and Win Television—are affiliated with their metropolitan counterparts and retransmit that programming with only minor local inputs.

Because of the cross-ownership restrictions in place from 1987 to 2006 and the ban on foreign ownership that applied until 2006, foreign and domestic conglomerates had a limited presence in the market. Except for a brief period from 1987 to 1990, Packer's Publishing and Broadcasting Limited (PBL) controlled the Nine Network before its sale to private equity investors CVC Capital Partners (UK) in 2006. News Ltd. sold its ownership of Ten Network when the cross-ownership rules were introduced in 1987. A consortium of interests headed by Canwest (Canada) acquired Ten Network in 1992 and held it until 2009. Although Canwest had a 57.5% stake in the network, it held only 15% of the stock to comply with restrictions on foreign ownership. Kerry Stokes acquired 20% of Seven Network in 1995, making him its largest shareholder. The network has since expanded its interests into other media sectors including print.

In 1998, the government implemented a digital transmission conversion plan for over-the-air broadcasting. Although each of the incumbent operators broadcasts up to three channels each, the preexisting industry structure and concentration levels influenced the digitization policy. Available spectrum was allocated to established operators and bans were placed on the use of the digital spectrum to start new services in competition with established operators.

C4 and HHI for the industry indicate a moderate level of concentration. The top four operators (the three main commercial networks and the ABC) account for 75% of the total market revenue. The individual shares of revenue are not highly dissimilar for the three commercial operators, each in the range of 18–25% in 2007.

## Table 23-5. TV Broadcasting (Market Shares by Revenue)

| | 1988 | 1992 | 1996 | 2000 | 2004 | 2007 | 2012 |
|---|---|---|---|---|---|---|---|
| Seven Network (Seven West Media) (Kerry Stokes) | 17.0 | 21.2 | 22.8 | 26.2 | 22.1 | 24.9 | 26.4 |
| Australian Broadcasting Corporation (ABC) (public) | 14.6 | 16.1 | 12.5 | 11.6 | 10.2 | 12.4 | 13.6 |
| Nine Network (Nine Entertainment) (CVC Capital Partners) (UK)[1] | 23.6 | 24.4 | 25.7 | 27.9 | 25.8 | 21.5 | 21.6 |
| Ten Network (Canwest) (Canada) (until 2009)[2] | 17.6 | 11.5 | 15.4 | 13.6 | 17.9 | 18.2 | 15.7 |
| Southern Cross Media Group | | 7.6 | 2.7 | 3.5 | 7.6 | 7.4 | |
| Win Television | | 4.2 | 4.0 | 4.2 | 4.7 | 4.6 | |
| Prime Network | | 2.9 | 4.4 | 4.1 | 4.6 | 4.5 | |
| Special Broadcasting Service (SBS) (public) | 2.5 | 2.5 | 3.1 | 3.0 | 3.5 | 3.0 | 4.0 |
| Others | 24.7 | 9.7 | 9.3 | 6.0 | 3.6 | 3.5 | 17.3 |
| Total Revenue (mil US$) | 1,266 | 1,706 | 2,289 | 2,286 | 3,001 | 3,982 | 4,903 |
| Total Revenue (mil A$) | 1,407 | 2,305 | 2,861 | 4,114 | 3.901 | 4,380 | 4,790 |
| C4 | 72.9 | 73.2 | 76.5 | 79.2 | 76.0 | 77.0 | 77.2 |
| HHI | 1,378 | 1,525 | 1,631 | 1,836 | 1,691 | 1,671 | 1,611 |
| *N* (>1%) | 5 | 8 | 8 | 8 | 8 | 8 | 5 |
| Noam Index | 616 | 539 | 577 | 649 | 598 | 591 | 721 |

*Source*: Estimates by authors based on original data from: Australian Communication and Media Authority, Broadcasting Financial Results (various years); Australian Broadcasting Corporation, Annual Reports; Special Broadcasting Service, Annual Reports; Australian Broadcasting Tribunal, Annual Reports; and Communications Law Centre, Update, "Media Ownership Update" (various years).

[1] Owned by Publishing and Broadcasting Limited (PBL) since 1990, the Packer family sold Nine Entertainment to private equity investors CVC Capital Partners (UK) in 2006.

[2] Canwest sold its stock in TEN in 2009.

In the same year, the ABC's share of industry revenue was 12.4% (Figure 23.4).

The Australian TV production industry is the beneficiary of substantial direct and indirect financial assistance from federal and state-support agencies. The sector also benefits from domestic content regulations imposed on free-to-air and subscription television operators. Given the prevailing programming agreements and affiliations among the metropolitan and regional commercial operators, which have few means of obtaining popular programming from other sources or producing their own, local content at the regional level is very low.

## Multichannel TV Platforms

Prior to the passage of the 1992 Broadcasting Services Act, the broadcast licensing regime prohibited the establishment of multisystem operators (MSOs) likely to compete with broadcast free-to-air television. The Act sanctioned a limited form of satellite delivery of multichannel services, and two licenses were subsequently

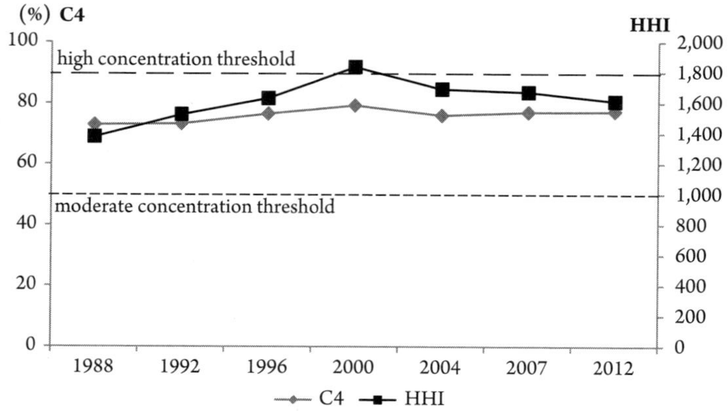


FIGURE 23.4 Broadcast Television Concentration Indices

auctioned. Because of technical and policy delays, subscription TV channels commenced operating only in 1995 on satellite, terrestrial multipoint distribution, and cable platforms.[15] The first MSO licenses were issued in 1997. There are no major restrictions on the ownership of subscription television services. However, there are considerable barriers to entry into the industry, both in relation to delivery platforms and access to programming.

MSOs are relative latecomers in Australia largely because of regulatory prohibitions aimed at insulating commercial free-to-air TV from competition. According to the Australian Subscription Television and Radio Association, only 34% of the population had subscription TV services in 2011.[16] Current regulatory provisions constrain access of subscription providers to broadcasting rights for popular sporting events. Control of major sources of content by established media became a potent barrier to entry into the industry, as well as a major constraint to competitiveness. Control of the output of several major Hollywood producers (not available to competitors) enabled Foxtel—a DBS, cable, and IPTV provider founded in 1995 as a 50:50 joint venture between News Ltd. and Telstra—to establish a dominant position from the start and to continue consolidating that position ever since. Under a territorial program distribution arrangement between the two operators, Foxtel's market was confined to five state capitals on the mainland and the Gold Coast metropolitan area near Brisbane, while Austar, founded in 1994 and acquired by Foxtel in 2012 from Liberty Global (US), distributed the same programming throughout the rest of the country. The Singaporean-owned ICT provider Optus attempted to enter the market but faced considerable difficulties in growing its customer base due to having a limited cable distribution network at its disposal and failing to secure access to the US-produced content controlled and distributed by Foxtel. In 2002, after reaching a program-sharing agreement with Foxtel, Optus withdrew from the market and became a reseller of Foxtel programming content to its cable customers. A very small number of other players operate localized cable distribution systems in regional areas and limited satellite service.

Drama and sporting programs on subscription television are affected by regulation. Subscription television providers, including their drama channel package suppliers, are subject to the Australian content provisions of the 1992 Broadcasting Services Act that require them to

15. Tiffen, Rodney. "From Technological Abundance to Commercial Monopoly in Australian Pay TV: Key Relationships in Institutionalising Subscription Television." Kenyon, Andrew T. (ed.). *TV Futures: Digital Television Policy in Australia*. Melbourne: Melbourne University Press, 2007.

16. ASTRA (Australian Subscription Television and Radio Association). 2011. May 31, 2011. <http://astra.org.au/pages/facts-figures>.

invest at least 10% of their total drama channels' program expenditure on domestically produced shows. Access to transmission rights of major sporting events is constrained by anti-siphoning rules. Australia has one of the world's most extensive mandated lists of sporting events that are to be made available for general viewing on free-to-air TV. Subscription television licensees can acquire the rights to broadcast events on the anti-siphoning list only if those rights are not acquired by over-the-air broadcasters (Table 23.6, Figure 23.5).

## Film

The Australian film and video production and post-production services industry is made up of more than 2,000 establishments, most of which are very small in size. According to the Australian Bureau of Statistics, only 18 of these establishments had more than 100 employees in 2007.[17] These 18 studios accounted for 39% of industry employment and 31.3% of its total revenue. The main producers are Village Roadshow, with 31.6% of total production expenditure in 2009, Endemol Southern Star with 9.8%, GTV Holdings with 4.3%, and Beyond International with 1.3%. Of these, Endemol (Netherlands), GTV, and Beyond International are primarily engaged in TV production.

In 2007, productions made primarily for television accounted for more than 72% of all production expenditures. An additional 13% of production expenditure was devoted to production of commercials, station, and program promotions. Only 14.5% was spent on non-television productions, the main components of which were US$167 million (A$184 million) for feature films and US$65.2 million (A$71.7 million) for corporate marketing and training media. The aggregate production expenditure of the top four producers was 47% of the total; the largest producer alone accounted for approximately 32% of the total.[18]

**Table 23-6. Multichannel Video Platforms: Cable MSOs, DBS, IPTV (Market Shares by Revenue)**

| | 1996 | 2000 | 2004 | 2008 | 2012 |
|---|---|---|---|---|---|
| Australis Media | 14.0 | (in liquidation) | | | |
| Foxtel (Telstra: 50% News Ltd., 50%)[1] | 30.7 | 54.4 | 56.6 | 67.8 | 95.0 |
| Austar (Liberty Global) (US) (until 2012) | 19.5 | 27.9 | 26.7 | 27.2 | Foxtel |
| Optus (SingTel) (Singapore) (public) | 34.9 | 16.4 | 12.1 | Foxtel reseller | |
| Others | 0.9 | 1.3 | 4.5 | 4.0 | 5.0 |
| Total Revenue (mil US$) | 169 | 601 | 1,010 | 2,153 | 3,292 |
| Total Revenue (mil A$) | 211 | 1,082 | 1,313 | 3,230 | 3,230 |
| C3/C4 | 99.0 | 99.0 | 95.5 | 96.5 | 90.0 |
| HHI | 2,736 | 4,009 | 4,068 | 5,337 | 9,025 |
| *N* (>1%) | 4 | 3 | 3 | 2 | 1 |
| Noam Index | 1,368 | 2,314 | 2,348 | 3,773.6 | 9,025 |

*Source:* Authors.

[1] The original ownership breakdown was Telstra, 50%; News Ltd., 25%; and PBL, 25%. News Ltd. acquired PBL's 25% stake—managed through Consolidated Media Holdings—in 2012.

---

17. ABS (Australian Bureau of Statistics). "8679.0—Television, Film and Video Production and Post-Production Services, Australia, 2006–07." Canberra: ABS, 2008. Accessed May 27, 2013. <http://www.abs.gov.au/AUSSTATS/abs@.nsf/Lookup/8679.0Explanatory%20Notes12006-12007?OpenDocument>.

18. Screen Australia. "Research: Production Industry, Drama." screenaustralia.gov, May 31, 2011. Accessed May 27, 2013. <http://www.screenaustralia.gov.au/research/statistics/mpfeaturesspending.aspx>.

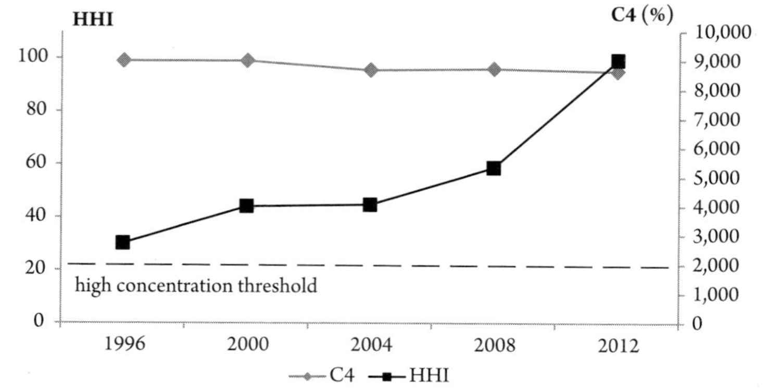


FIGURE 23.5 Multichannel TV Platforms: Concentration Indices

Approximately 30 feature films are produced domestically each year, with approximately half of total production expenditure coming from a small number of high-budget foreign features filmed in Australia. Foreign feature films take in the lion's share of box office revenue. Overseas producers significantly influence the level of activity in feature film production, even though they produce on average only two to three features a year. In 2010, total production expenditure amounted to US$435 million (A$435 million), of which US$69 million (A$69 million) related to foreign productions. From 2001 to 2010, the average annual proportion of production activity funded by foreign producers was 44%, with a peak of 74% between 2004 and 2005.[19]

The major Hollywood studios dominate film distribution. In addition to the main international distributors, 30 small, specialized distributors are active in the industry and account for an aggregate market share of 7–15% in the period covered by the research data. Among the international distributors, the Australian company Village Roadshow, in association with Warner Brothers (US), has the largest market share (Table 23.7, Figure 23.6).

The cinema theater industry was initially characterized by widespread localized ownership but eventually saw the emergence of major chains that now control over 60% of box office revenue. Village Roadshow (AUS) has grown into an international media conglomerate. Its early expansion involved the acquisition of other drive-in and cinema houses. Then, it expanded into film distribution in the 1960s and film production in the 1970s. In 1988, it further broadened its interests with the acquisition of the De Laurentis Entertainment Group (US). In the 1990s, it expanded its cinema exhibition operations internationally, acquiring cinemas in 20 countries, which were subsequently scaled back to just the Australian and Singaporean markets. It also further broadened its media interests with the purchase of the Australian Triple M radio network in 1993, which was later transformed into the Austereo Group, the largest commercial radio network in Australia in terms of market share. In 2008, its Village Roadshow Pictures interests merged with Concord Music Group to form the Village Roadshow Entertainment Group.

The decline in the industry concentration indices during the period under review is largely a reflection of the combined effect of United International Pictures' split into Paramount and Universal in 2006 and increased market share by small, specialized independent distributors. Concentration in the Australian cinema exhibition industry has changed very little during the review period. The share of the market held by each of the three major cinema chains remained relatively stable. While there appears to be some ongoing slow rationalization among the smaller

19. Screen Australia. "Research: Production Industry, Drama."

## Tabe 23-7. Film Production/Distribution (Market Shares by Box Office %)

| | 2003 | 2004 | 2006 | 2008 | 2009 |
|---|---|---|---|---|---|
| Village Roadshow/Warner Bros (AUS/US) | 22.0 | 23.0 | 20.0 | 23.0 | 21.0 |
| 20th Century Fox (Murdoch Family, News Corp., US/UK/AUS) | 11.0 | 11.0 | 20.0 | 15.0 | 16.0 |
| United International Pictures (UIP) (Universal + Paramount)[1] | 21.0 | 27.0 | 19.0 | (dissolved) | |
| Paramount (US) | UIP | | | 18.0 | 13.0 |
| Universal (US) | | | | 14.0 | 11.0 |
| Walt Disney [Buena Vista International (2003–2006)] (US) | 25.0 | 16.0 | 15.0 | 12.0 | 13.0 |
| Sony (Columbia TriStar) (Japan) | 14.0 | 13.0 | 15.0 | 8.0 | 12.0 |
| Others | 7.0 | 11.0 | 11 | 11 | 15 |
| Total Revenue (mil US$) | 581 | 631 | 643 | 904 | 875 |
| Total Revenue (mil A$) | 813 | 820 | 817 | 1,356 | 963 |
| C4 | 82.0 | 79.0 | 74.0 | 70.0 | 63.0 |
| HHI | 1,867 | 1,804 | 1,611 | 1,482 | 1,300 |
| *N* (>1%) | 5 | 5 | 5 | 6 | 6 |
| Noam Index | 835 | 807 | 720 | 605 | 531 |

*Source*: Screen Australia.
[1] United International Pictures split into Paramount and Universal after 2006.

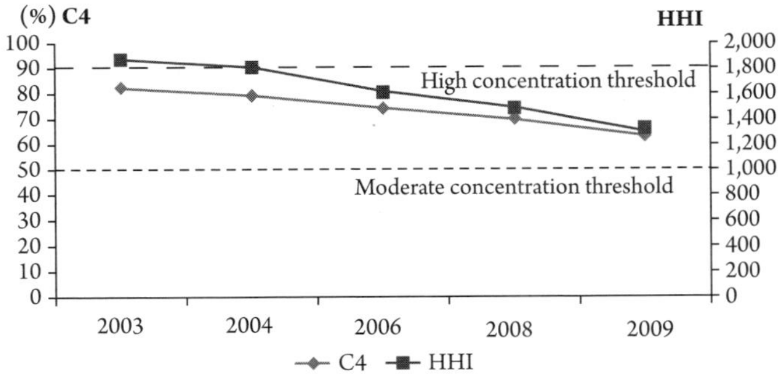


FIGURE 23.6 Film Distribution Concentration Indices

operators in the industry, the resultant minor gains in market share are being made by mid-sized operators.

The Australian cinema exhibition industry is made up of three large chain operators—Greater Union, Hoyts, and Village Roadshow—and some 140 smaller exhibitors. Prior to 2000, the number of operators in the industry was declining at a rate of approximately 3% per annum.[20] In 2000, there were 173 cinema exhibitors in operation, with 97 of them earning less than US$0.6 million (A$1 million) annually. While the slow

20. ABS (Australian Bureau of Statistics). "8654.0—Motion Picture Exhibition, Australia, 1999–2000." Canberra: ABS, 2001. Accessed May 27, 2013. <http://www.abs.gov.au/ausstats/abs@.nsf/mf/8654.0/>.

decline in the number of operators has continued over the past decade, the industry itself has grown both in terms of cinema screens and audience size.

In 2009, independent operators controlled 31% of total screens. The three major chains controlled over 1,000 cinema screens (or over 50% of the total), the rest being controlled by several other smaller chains. Overall, there is low to medium concentration in the industry, with little change recorded over the period from 1998 to 2009 (Table 23.8, Figure 23.7).

Table 23-8. Film Exhibition (Market Shares by Screen Ownership)

| | 1998 | 2000 | 2002 | 2004 | 2006 | 2008 | 2009 |
|---|---|---|---|---|---|---|---|
| Greater Union | 24.2 | 24.2 | 24.9 | 23.1 | 24.2 | 23.8 | 24.6 |
| Hoyts | 17.6 | 19.2 | 18.5 | 18.9 | 17.7 | 17.2 | 16.7 |
| Village Roadshow (US) | 13.5 | 12.5 | 12.6 | 11.8 | 11.5 | 11.3 | 11.3 |
| Reading | | 4.6 | 5.1 | 6.8 | 6.9 | 7.8 | 7.5 |
| Palace | | | 3.1 | 2.6 | 3.0 | 3.8 | 3.8 |
| Grand | | | | | | 1.9 | 1.9 |
| Dendy | | | | | | 1.3 | 1.3 |
| Wallis Theatres | 1.4 | 1.1 | 1.0 | 1.7 | 1.4 | 1.4 | 1.2 |
| Others | 43.3 | 38.5 | 34.7 | 35.1 | 35.3 | 31.5 | 31.7 |
| Total Revenue (mil US$) | 391 | 412 | 476 | 631 | 643 | 904 | 876 |
| Total Revenue (mil A$) | 434 | 742 | 857 | 820 | 836 | 1,356 | 964 |
| Domestic Films: Total Revenue Share (mil US$) | 15.9 | 32.4 | 23.5 | 8.3 | 29.7 | 33.9 | 44.1 |
| Domestic Films: Total Revenue Share (mil A$) | 19.9 | 58.3 | 42.3 | 10.8 | 38.6 | 50.9 | 48.5 |
| Domestic Films: Market Share (%) | 4.0 | 7.9 | 4.9 | 1.3 | 4.6 | 3.8 | 5.0 |
| C4 | 56.7 | 60.5 | 61.2 | 60.6 | 60.3 | 60.1 | 60.1 |
| HHI | 1,080 | 1,133 | 1,161 | 1,084 | 1,088 | 1,073 | 1,087 |
| *N* (>1%) | 4 | 5 | 6 | 6 | 6 | 8 | 8 |
| Noam Index | 540 | 507 | 474 | 443 | 444 | 380 | 384 |

*Source*: Screen Australia.

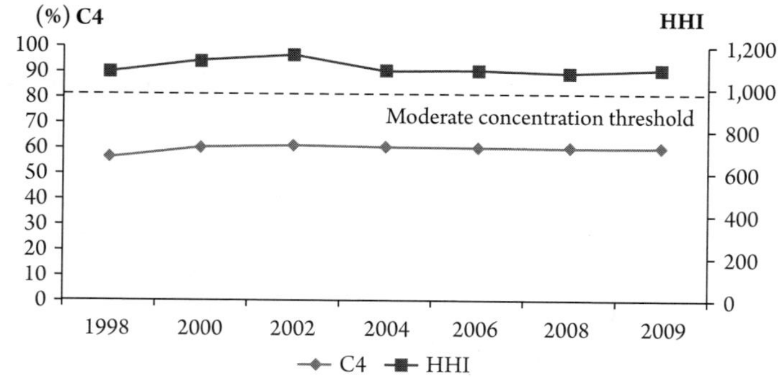


FIGURE 23.7 Cinemas Concentration Indices

## TELECOMMUNICATIONS MEDIA

### Wireline Telecom

Australia embarked on a two-stage program of full-scale liberalization of its telecommunications sector. The first stage lasted from 1991 to 1997 and involved the licensing of a full-line carrier, Optus, to provide wireline and wireless services in competition with the former state monopolist, Telstra. Initially established as a government-owned enterprise from legacy public telecommunications monopoly operators, it was progressively privatized starting in 1997 and became a fully private company in 2007. The liberalization arrangements mandated interconnection: Telstra had to grant private carriers access to its wireline customer access network (CAN).

The second stage began with the 1997 Telecommunications Act, which inaugurated a fully competitive market. Telstra was transformed into a private sector corporation listed on the stock exchange, and its stock was progressively sold to a wide spectrum of institutional and individual stockholders, though the state still retains a small stake (less than 5%). Service provider or carrier licenses were easily secured and by the end of that year, the Australian Communications Authority had issued 77 carrier licenses and many more service provider licenses to prospective operators in the liberalized regime. Many, however, had very small market shares, and some never pursued their licenses or had to close down shortly after opening.[21] The end of the 1990s economic boom in the global ICT industry in 2000 led to a shakeout, causing many more companies to withdraw from the sector: in 2001, there were only 54 carriers and 88 service providers listed by the Telecommunications Industry Ombudsman.[22]

The weaker market outlook of the early 2000s impacted all carriers, with many scaling back their investment plans. Nonetheless, competition intensified as the new entrants sought to establish themselves in the market. Telstra's former monopoly status and ownership of the wireline CAN (notwithstanding the interconnection requirements established in the 1990s) provided it with an ongoing advantage over its competitors, though its market share declined rapidly, from 92.9% in 1997 to 78.8% in 2003. The erosion slowed down after 2003 but continued throughout the period covered by the analysis (see Table 23.9 for details.) By the end of the period, Telstra was still the dominant entity in the market, controlling more than 70% of industry revenue. This erosion of the market is particularly evident in the HHI ratio but less so in the C4, which is based on the aggregate share of the largest four operators (see Figure 23.8 for details.)

### Wireless Telecom

When Australia embarked on liberalization of its telecommunications, the wireless sector was still in its infancy, but growing rapidly. Beginning with Telstra and Optus, a third mobile-only carrier entered the market in 1993, Vodafone plc (UK), to compete with the two original carriers in the wireless telephony market. Telstra's incumbent position did not grant it the same level of market dominance it held in wireline. Optus and Vodafone moved quickly to establish their own wireless networks and began to compete strongly with Telstra. After 1997, a fourth wireless carrier, Hutchison Telecommunications International (Hong Kong), established its own network. Several other independent service providers, relying on the established networks to carry their services, also entered the market.

The largest of the new independent service providers, OneTel, began operating in 1999. Its major stockholders included James Packer of PBL and Lachlan Murdoch, son of Rupert, and previously associated with News Ltd. It quickly established itself with an aggressive marketing strategy, but then failed when it got

---

21. The basic difference between carriers and service providers is that the former own or operate a network for the carriage of their services and those of other service providers.

22. Productivity Commission. "Telecommunications Competition Regulation Inquiry Report." Canberra: Productivity Commission, 2001. Accessed May 31, 2011. <http://www.pc.gov.au/projects/inquiry/telecommunications/docs/finalreport>.

## Table 23-9. Wireline Telecom (Market Shares by Revenue), 1997–2009

| | 1997 | 1999 | 2001 | 2003 | 2005 | 2007 | 2009 | 2012 |
|---|---|---|---|---|---|---|---|---|
| Telstra | 92.9 | 90.0 | 84.1 | 78.8 | 77.0 | 73.6 | 71.8 | 71.4 |
| Optus (SingTel) (Singapore) (public) | 6.4 | 7.5 | 10.9 | 13.3 | 13.5 | 12.0 | 11.6 | 15.0 |
| Telecom NZ (incl. AAPT) (New Zealand) | 0.7 | 2.1 | 2.5 | 3.7 | 3.5 | 3.6 | 2.9 | 1.6 |
| Powertel | 0.0 | 0.1 | 0.1 | 0.3 | 0.5 | Telecom NZ | | |
| Primus (Primus Telecom) (US) | 0.0 | 0.4 | 0.9 | 1.3 | 1.5 | 1.1 | 0.8 | 1.6 |
| Agile Communications | 0.0 | 0.0 | 0.0 | 0.0 | 0.1 | 0.2 | 0.3 | 0.9 |
| Soul Pattinson | 0.0 | 0.0 | 0.0 | 0.0 | 0.1 | 1.4 | 1.2 | 2.3 |
| Chime | 0.0 | 0.0 | 0.1 | 0.2 | 0.5 | 0.6 | 1.1 | 1.3 |
| Uecomm Operations | 0.0 | 0.0 | 0.1 | 0.1 | 0.4 | 0.6 | 0.9 | Optus |
| Macquarie Telecom | 0.0 | 0.0 | 0.1 | 0.4 | 0.5 | 0.6 | 0.5 | 0.7 |
| Reach Networks Australia | 0.0 | 0.0 | 0.3 | 0.3 | 0.0 | 0.0 | 0.0 | 0.0 |
| Others | 0.0 | 0.1 | 0.8 | 1.4 | 2.3 | 6.2 | 8.9 | 5.3 |
| Total Revenue (mil US$) | 9,600 | 9,000 | 8,100 | 10,900 | 11,700 | 13,400 | 13,900 | 12,994 |
| Total Revenue (mil A$) | 12,727 | 13,632 | 15,851 | 16,265 | 15,367 | 15,800 | 17,222 | 12,694 |
| C4 | 100.0 | 99.9 | 98.5 | 97.2 | 95.5 | 89.7 | 86.4 | 90.3 |
| HHI | 8,673 | 8,160 | 7,202 | 6,410 | 6,126 | 5,575 | 5,302 | 5,336 |
| *N* (>1%) | 2 | 3 | 3 | 4 | 4 | 5 | 5 | 6 |
| Noam Index | 6,133 | 4,711 | 4,158 | 3,205 | 3,063 | 2,495 | 2,371 | 2,178 |

*Source*: Australian Communication and Media Authority "Carriers Eligible Revenue Returns Data," Australian Competition and Consumer Commission (various published reports); Companies' annual reports; Productivity Commission (2001), Telecommunications Competition Regulation Inquiry Report; and authors' estimates.

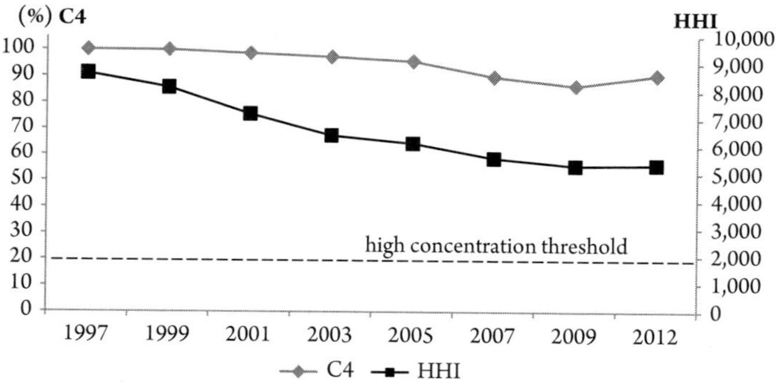


FIGURE 23.8 Wireline Telecom Concentration Indices

itself into credit difficulties and its major stockholders withdrew from the enterprise.

Telstra gradually lost market share as the other major carriers established themselves. Its losses were nominally greatest in the early years of market liberalization and the introduction of full competition. However, the loss of market share did not translate into reductions of market revenue. With market growth, its wireless revenue continued to grow in absolute terms, but at a slower rate than its competitors. Telstra remains the largest operator in the market, with its market share in recent years stabilizing around 43%.

The smallest of the major carriers, Hutchison, entered the market in 2000 with its 3G network. It gradually increased its market share but remained relatively small compared to the three previously established wireless carriers. In 2009, it entered into a 50:50 joint venture with Vodafone, combining the wireless operations of the two carriers in Australia.

The changes in industry concentration during the period under observation are reflected in the HHI. The C4 index did not change significantly due to the fact that only in 2001 were there more than four carriers present, excluding resellers (Table 23.10, Figure 23.9).

## Table 23-10. Wireless Telecom (Market Shares by Revenue)

| | 1997 | 1999 | 2001 | 2003 | 2005 | 2007 | 2009 | 2012 |
|---|---|---|---|---|---|---|---|---|
| Telstra | 67.0 | 57.0 | 48.0 | 46.0 | 43.0 | 43.0 | 42 | 58 |
| Optus (SingTel) Singapore) (public) | 22.0 | 26.0 | 29.0 | 32.0 | 34.0 | 32.0 | 33 | 27 |
| Vodafone-Hutchison (UK + Hong Kong) | | | | | | | 24 | 15 |
| Vodafone Australia (Vodafone plc) (UK) | 6.0 | 12.0 | 15.0 | 17.0 | 16.5 | 18.0 | Vodafone-Hutchison | |
| Hutchison Telecommunications International (Hutchinson Whampoa) (Hong Kong) | | | 3.0 | 3.0 | 5.0 | 6.0 | | |
| OneTel | | 0.9 | 2.8 | (in liquidation) | | | | |
| Others (Resellers) | 5.0 | 4.0 | 2.0 | 2.0 | 1.0 | 1.0 | 1.0 | <1.0 |
| Total Revenue (mil US$) | 2,232 | 2,612 | 2,994 | 4,332 | 7,281 | 8,715 | 9,638 | 15,361 |
| Total Revenue (mil A$) | 2,973 | 4,000 | 5,700 | 5,805 | 9,490 | 9,570 | 10,560 | 15,007 |
| C4 | 95.0 | 96.0 | 95.2 | 98.0 | 99.0 | 99.0 | 99.0 | 99.0 |
| HHI | 5,009 | 4,070 | 3,387 | 3,438 | 3,302 | 3,233 | 3,429 | 4,318 |
| *N* (>1%) uderesellers) | 3 | 4 | 5 | 4 | 4 | 4 | 3 | 3 |
| Noam Index | 2,892 | 2,035 | 1,515 | 1,719 | 1,651 | 1,617 | 1,979 | 2,493 |

*Source*: Australian Communication and Media Authority, "Carriers Eligible Revenue Returns Data," Australian Competition and Consumer Commission (various published reports); Companies' annual reports; Productivity Commission (2001), and authors' estimates.

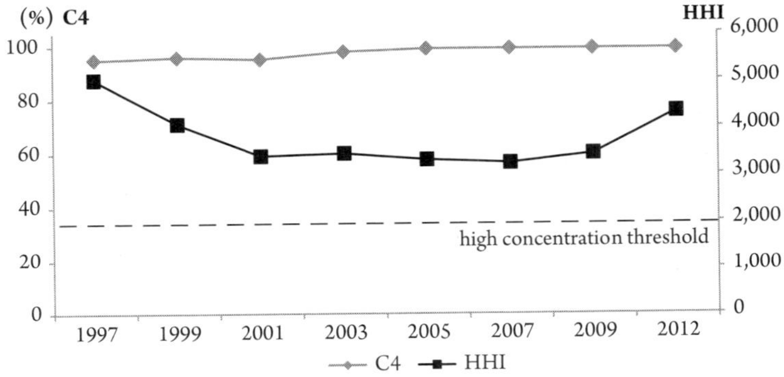


FIGURE 23.9 Wireless Telecom Concentration Indices

## INTERNET MEDIA

### Internet Service Providers (ISPs)

In 2000, the number of ISPs operating in the industry was almost 700 and remained around that level for several years. After 2006, the number rapidly declined, contracting to just below 400 by the end of 2008 with many of the smaller operators disappearing. Thereafter the number of small and medium ISPs continued to decline.[23] The vast majority of those left are small (fewer than 1,000 subscribers), with many servicing localized geographic markets and often operating as "virtual" ISPs reselling connectivity provided by major operators. Fewer than 10% of ISPs have a subscriber base of more than 10,000, with only 11 having more than 100,000 subscribers. The larger ISPs are mainly associated with telecommunications carriers.

Internet connectivity by households increased rapidly between 1998 to 2008. After 2003, there was rapid adoption of broadband. In 1998, only 16% of Australian households had access to the Internet. By 2008, the proportion of households with Internet access had increased to 72%.[24] Incumbent telecom carriers extended their established positions in the telephony market to the provision of Internet services, giving them an advantage through vertical integration. As alternative delivery platforms were not available, other ISPs were reliant on access to the networks of incumbent carriers to deliver their services. While access was guaranteed by regulation, the incumbent telecommunications carriers had the advantage of established relationships with customers. This was particularly so for Telstra, with its established national network presence. The ISP industry transitioned from basic localized access services to large hubs providing a wide range of ICT services as the telecom carriers consolidated their market shares at the expense of the small local providers. The consolidation is clearly reflected in the transition of HHI from low concentration levels pre-2000 to moderate concentration levels in the first half of the following decade and to high concentration levels by 2010.

The aggregate market share of the top four ISPs more than doubled between 1998 and 2008, with C4 increasing from 31.3% in 1998 to 65.5% in 2008. Much of the increase in C4 is due to the growth in the market share of the largest ISP, Bigpond (owned by Telstra), which increased from 13% to 39.8% in the same period. Optus also experienced a high rate of growth in its market share, increasing from 1.7% to 11.1%. Of particular note is the decline in the market share of OzEmail, one of the successful early entrants in the ISP market that had built up a sizable dial-up Internet service. OzEmail was eventually taken over by iiNet, transforming the latter from a small regional operator to the third largest ISP in Australia.

It should be recalled that the overall market grew rapidly in the period under review. Consequently, the growth in the size of the individual ISPs was much greater than would be implied by a simple comparison of the market shares held in 2000 and 2010 (Table 23.11, Figure 23.10).

### Search Engines

The Internet search engine market in Australia is dominated by the main global search engines, specifically Google (US), whose market share now exceeds 90%. Yahoo (US) and AltaVista (US) have both lost considerable market share to Google. In the period under review, Yahoo!'s market share declined from 33.3% in 2001 (the largest in the market) to a share of only 2.6% in 2010. In the same period, AltaVista's share declined from 15.1% (second largest in the market) to 0.1% in 2010.

While the global search engines dominate the market, there are several well-established Australian search engines available to users.

---

23. ABS (Australian Bureau of Statistics). "8153.0—Internet Activity, Australia, Dec 2012." Canberra: ABS, 2012. Accessed May 27, 2013. <http://www.abs.gov.au/ausstats/abs@.nsf/Products/8153.0~December+2012~Chapter~Sector?OpenDocument>.

24. ABS (Australian Bureau of Statistics). "8153.0."

## Table 23-11. Internet Service Providers (Market Shares by Revenue)

| | 2000 | 2004 | 2008 | 2010 |
|---|---|---|---|---|
| Bigpond (Telstra) | 24.4 | 31.6 | 39.8 | 42.2 |
| Optus (SingTel) (Singapore) (public) | 7.0 | 9.8 | 11.1 | 11.8 |
| iiNet | 0.8 | 3.3 | 8.7 | 9.6 |
| OzEmail (until 2005) | 12.8 | 8.2 | iiNet (from 2005) | |
| iPrimus (Primus Telecom) (US) | 4.6 | 8.3 | 5.9 | 5.3 |
| TPG/SP Telemedia | 3.4 | 5.7 | 5.2 | 5.1 |
| Chariot (until 2008) | 1.3 | 1.6 | TGP (from 2008) | |
| Others | 45.7 | 31.5 | 29.3 | 26.0 |
| Total Revenue (mil US$) | 836 | 1,947 | 3,729 | 3,562 |
| Total Revenue (mil A$) | 1,400 | 2,800 | 3,900 | 4,200 |
| C4 | 48.8 | 57.9 | 65.5 | 68.9 |
| HHI | 843 | 1,277 | 1,845 | 2,066 |
| *N* (>1%) | 6 | 7 | 6 | 5 |
| Noam Index | 344 | 483 | 753 | 924 |

*Source*: Actual and estimates by authors based on data from a variety of sources, including Australian Bureau of Statistics, 8153.0—Internet Activity, Australia (various years), companies annual reports, media reports and other sources.

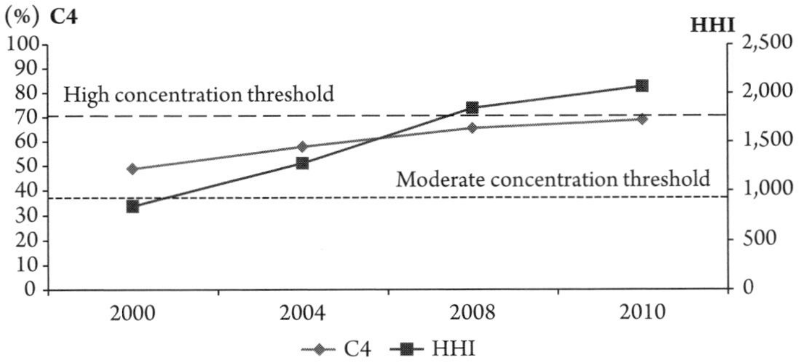


FIGURE 23.10 ISP Concentration Indices

Most of them, however, have limited coverage and depth relative to what is offered on the global engines. Sensis benefited considerably from its ownership by Telstra and its relationship with Bigpond: Sensis maintained its own database and sponsored listings, and in 2001 had 7.5% of the Australian market. Its linkages to Telstra appear to have insulated it somewhat from Google's market takeover, and its share of the market remained relatively stable at about 8%. In 2008, however, it entered into a commercial arrangement with Google, including adoption of the Google search engine for its operations. Other Australian search engines include Ansearch, Anzwers, Howzat, WebWombat, and WotBox. Some tend to operate specialized services, and their market share is very small.

The concentration indices indicate that the search engine market has always been highly concentrated. The C4 values indicate that the four largest search engines in aggregate controlled over 90% of the market throughout the period under review, rising to over 98% at the end of the period. The rise of Google at the expense of other previously popular global

Table 23-12. Internet Search Engines (Market Shares by Revenue)

| | 2001 | 2004 | 2006 | 2008 | 2010 |
|---|---|---|---|---|---|
| Google (US) | 12.3 | 54.0 | 60.2 | 87.8 | 92.5 |
| (ii) Sensis (Telstra) | 7.5 | 7.1 | 8.3 | Google | |
| Yahoo (US) | 33.3 | 21.4 | 16.8 | 3.9 | 2.6 |
| MSN (ninemsn) (US)/ Microsoft (Bing) (US) | 13.0 | 14.2 | 12.8 | 6.7 | 3.4 |
| AltaVista (US) | 15.1 | Yahoo | | | |
| Ask (US) | 0.0 | 1.1 | 0.8 | 0.4 | 0.4 |
| Others | 18.8 | 2.2 | 1.2 | 1.2 | 1.1 |
| Total Revenue (mil US$) | 10.2 | 89.0 | 297 | 793 | 957 |
| Total Revenue (mil A$) | 20 | 128 | 400 | 829 | 1,128 |
| C4 | 73.7 | 96.7 | 98.0 | 98.8 | 98.9 |
| HHI | 1,714 | 3,623 | 4,134 | 7,769 | 8,575 |
| *N (>1%)* | 5 | 5 | 4 | 3 | 3 |
| Noam Index | 766 | 1,620 | 2,067 | 4,485 | 4,951 |

*Source*: Based on Roy Morgan Survey Data (Press Releases 2006/453 and 2006/496), IAB Australia, and various other sources including Hitwise and http://statcounter.com/.

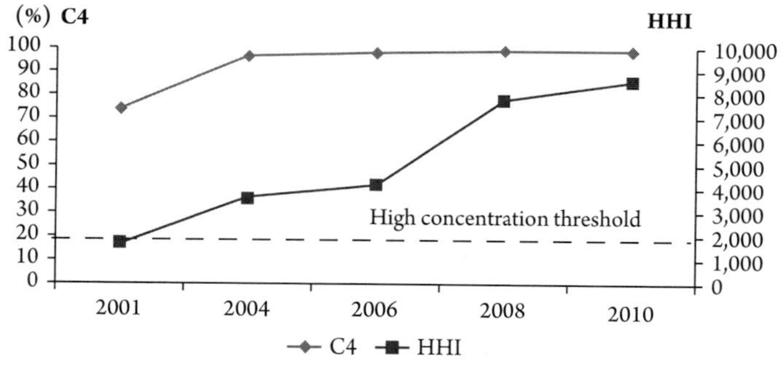


FIGURE 23.11 Search Engines Concentration Indices

engines has resulted in dramatic increases in the HHI (Table 23.12, Figure 23.11).[25]

## CONCLUSION

Four key corporations historically dominated Australian newspapers, magazines, television, and radio, and had various other media interests. The first was The Herald and Weekly Times (HWT), which until 1986 was the largest publisher of newspapers. Its early driving force was Sir Keith Murdoch, father of Rupert Murdoch, though he did not have a controlling ownership (after his death, collective management was effected). It was the first publisher to extend its holdings into several state capitals, and it had

25. Hitwise Australia. "Weekly Online Trends." Experian, Sept. 13, 2010. Accessed May 27, 2013. <http://www.hitwise.com/au/datacentre/main/dashboard-1706.html>; and Roy Morgan Research (RMR). "Google Overwhelms Search Competitors in Australia." RMR, Feb. 1, 2006; and Roy Morgan Research (RMR). "It's Google Then the Rest When Searching Online In Australia." RMR, May 18, 2006.

the highest circulation rates in every state capital, except Sydney, where it had no presence. In the aftermath of the 1986–87 media ownership policy changes Rupert Murdoch moved decisively to gain control of HWT. The introduction of those laws was widely seen as the last chance to buy into TV, and there was a frenzy of takeovers, with stations commanding prices that would have seemed impossible a few years prior.[26]

Apart from HWT, the other three companies were unusual in the modern corporate world in that a central proprietor dominated them. The flavor of media concentration issues as they related to newspapers and broadcasting was thus shaped by the unusual degree to which major corporations were associated with dominant individuals and families.

The most famous and controversial was Rupert Murdoch's News Ltd., which he began with a relatively small inheritance from his father—the afternoon daily *The News*. Murdoch built this into a global company, News Corp. (now headquartered in the US) with holdings in the United Kingdom, the United States, and several other countries. Murdoch has been controversial because he has been seen as driving his media companies downmarket in the quest for greater audience share and because his political influence on his outlets is much greater than is the case with many other media proprietors.[27]

The Packer organization (PBL Media)—led by Sir Frank Packer from the 1930s to his death in the 1970s, then by Kerry until his death in 2005, and since by his son James—owned only one daily, the Sydney *Daily Telegraph*. While PBL Media sold that newspaper to Rupert Murdoch in 1972, it was also, the largest magazine group in the country, and crucially set up the first TV network to own a strategic Sydney-Melbourne metropolitan area channel package. Kerry controlled the largest TV operator in the country until 1987, when he sold to Alan Bond, though a few years later, he regained control of the network. Later, PBL Media acquired a 25% stake in Foxtel, in partnership with News Ltd. and Telstra. Like Rupert Murdoch, Kerry was internally very controlling, with little regard for editorial independence, and had strong politically conservative views.[28] After Kerry's death, his son James sold his interests in free-to-air TV and magazine publishing (although initially keeping his interest in the pay-TV operator, Foxtel) and concentrated instead on investing in casinos.

The most patrician of the Australian media families, and one that traced its ownership back even further than the Murdochs and Packers, was the Fairfax family, which has been managing print properties in the country since 1841. The late 1980s brought a disastrous internal family split, which cut the company's size and ended family ownership.[29] It remained strong in newspapers, but no longer had a presence in broadcasting.

The mixture of concentrated ownership and powerful personalities has meant that media policy has sometimes been a contentious issue in Australia, although the influence of the public controversies on policy content is more difficult to discern. One controversy arose in the early 1980s over the so-called Murdoch amendments, when the government changed the requirement for TV licensees from being an Australian resident to being an Australian citizen, a move that was seen as designed to accommodate Murdoch. The largest public controversy came in 1991, when it appeared that a consortium involving Kerry Packer might acquire Fairfax Media, which had gone into liquidation. For a period the Canadian mogul, Conrad Black, became the largest shareholder in Fairfax.

An opportunity to reduce media concentration came in the late 1990s with the conversion from analog to DTT transmission, but the government chose to give the extra available

---

26. Tiffen, R. "Media Policy." Brett, J., J. Gillespie and M. Goot (eds.). *Developments in Australian Politics*. Macmillan, Melbourne, 1994.

27. Chenoweth, Neil. *Virtual Murdoch: Reality Wars on the Information Superhighway*. London: Vintage, 2002; and Wolff, Michael. *The Man Who Owns the News: Inside the Secret World of Rupert Murdoch*. New York, Knopf, 2008.

28. Barry, Paul. *The Rise and Rise of Kerry Packer Uncut*. Sydney: Bantam, 2007.

29. Souter, Gavin. *Heralds and Angels: The House of Fairfax, 1841–1990*. Melbourne: Melbourne University Press, 1991.

frequencies to the existing networks. Again there was some public controversy, but it had no impact on the policy.

Much of the policy concern in Australia in relation to media diversity has been focused on the capacity of media conglomerates to use their power across media to influence public opinion and to skew competition in their favor. This has been a concern from the early days of the development of audiovisual media, starting with radio in the 1920s and 1930s and reinforced by the introduction of television in 1956.

The Australian press has traditionally enjoyed full freedom of expression, and participation in the industry has never been subject to regulation. Indeed, the Australian federal constitution contains no powers for regulation or control of the press. In contrast, the constitutional powers in relation to electronic transmissions (initially telegraph) are used to justify extensive regulation of electronic media. According to the Joint Parliamentary Committee on Wireless Broadcasting in 1942, the purpose of the ownership limits was to constrain the extension of influence of newspapers into the new medium, as there was "little multiple ownership of broadcasting licenses by other interests."[30] The concern with media influence on public opinion was also cited as the justification for restrictions on foreign ownership and control of media. In this regard, as stressed in 1951 by the then Prime Minister Robert Menzies, the issue was:

> whether the Government should permit or even encourage a state of affairs in which the most intimate form of propaganda known to modern science that is being conducted in this country, one that is going into every home and is reaching every man, woman and child in this country, should be in the hands of people who do not belong to this country.[31]

While these restrictions remained in place and ensured some limits on concentration within each broadcasting medium, they did not prevent a degree of cross-ownership from arising. The 1987 changes were designed to trade off a deeper influence within electronic media by allowing the formation of networks with a prohibition of control in more than one established traditional media outlet in the same local market. These regulations remained in force until 2006: foreign ownership restrictions were abolished in 1992 for radio, and for television in 2006. It should be noted, however, that foreign investment in major traditional media remains under the purview of the Foreign Investment Review Board. The application of ownership and control (including foreign control) limits and the cross-ownership regulations was a major constraining influence on the formation of media conglomerates with interests in electronic media. The impact of these factors is evident in the main media interests of the various conglomerates summarized in Table 23.13.

Table 23.13 reflects the situation in 2011, four years after the abolition of the 1987 cross-ownership restrictions in 2007. Although there have been some changes in the ownership of traditional media assets since then, they have not been extensive. Combinations of media assets that would have been prohibited by the old cross-ownership rules include

- The acquisition by Fairfax Media of several radio stations owned by the Southern Cross Media Group
- The acquisition of the *Western Australian* daily by Seven Network Limited
- The acquisition of the Southern Cross TV regional network by the former Macquarie Media Group (owner of a regional radio network with a similar coverage area and its subsequent 2011 merger with Austereo to form the Southern Cross Austereo Group)

30. Joint Parliamentary Committee on Wireless Broadcasting (W. Gibson, Chair). *Report.* Canberra: Government Printing Office, 1942.

31. House of Representatives. *Debates, 215:28. November 1951.* Canberra: Commonwealth of Australia Parliament, 1951. 2,926.

Table 23-13. Australian Media Crossownership Holdings, 2011

| | TELSTRA | OPTUS (SINGTEL, SINGAPORE) | NEWS CORPORATION (MURDOCH, US-AUSTRALIA) | NINE ENTERTAINMENT CO (FORMERLY PBL MEDIA, PACKER) | SEVEN NETWORK LTD (STOKES) | SOUTHERN CROSS MEDIA GROUP (FORMERLY MACQUARIE MEDIA GROUP) | APN NEWS & MEDIA (O'REILLY, IRELAND) | VILLAGE ROADSHOW |
|---|---|---|---|---|---|---|---|---|
| Wireline Telecommunication | Telstra (72% market share) | Optus (12% market share) | | | | | | |
| Wireless Telecommunication | Telstra (42% market share) | Optus (33% market share) | | | | | | |
| TV Broadcasting | | | | Nine Network | Seven Network | Southern Cross TV (regional network) | | |
| Radio | | | | | | Ownership of several radio stations | Australian Radio Network | Austereo |
| Pay TV | 50% owner of Foxtel (63% market share) | OptusVision (Foxtel reseller) | 25% owner of Foxtel (63% market share) | 25% owner of Foxtel (63% market share) | | | | |
| Film | | | Fox Studios (Sydney) | | | | | Village Roadshow |
| Movie Theatres | | | | | | | | Village Roadshow |

*continued*

Table 23-13. *continued*

| | TELSTRA | OPTUS (SINGTEL, SINGAPORE) | NEWS CORPORATION (MURDOCH, US-AUSTRALIA) | NINE ENTERTAINMENT CO (FORMERLY PBL MEDIA, PACKER) | SEVEN NETWORK LTD (STOKES) | SOUTHERN CROSS MEDIA GROUP (FORMERLY MACQUARIE MEDIA GROUP) | APN NEWS & MEDIA (O'REILLY, IRELAND) | VILLAGE ROADSHOW |
|---|---|---|---|---|---|---|---|---|
| ISPs | Bigpond (largest ISP (40% market share) | Optus (second ranked ISP 12% Market share) | | | | | | |
| Search Engines | Sensis | | | 50% ninemsn | Yahoo!7 (joint venture with Yahoo) | | | |
| Magazines | | | 14% of market | ACP Magazines 44% market share | Pacific Magazines | | | |
| Daily Newspapers | | | newspapers with combined circulation of 57% of market | | *Western Australian* daily | | several regional newspapers | |
| Books | | | HarperCollins (incl. Angus & Robertson) | | | | | |

- APN News and Media's acquisition of radio stations in regional markets where it published newspapers
- Other common ownership of regional radio and regional newspapers in the same market

All of the other combinations illustrated in Table 23.13 would have been permitted by the previous regulations, as they did not involve common ownership of daily newspapers, TV, and radio entities with overlapping coverage areas.

The two most prominent cross-media conglomerates have been News Corporation, which has relocated its headquarters from Australia to the United States, and Nine Entertainment Company, formerly PBL Media. While News Corporation's assets include newspapers, magazines, book publishing (HarperCollins), subscription television (Foxtel), and film production studios, they do not extend to broadcast media. Although Murdoch did own some TV stations before the introduction of the 1987 cross-ownership restrictions, he sold those assets on becoming a US citizen to avoid conflict with the then-prevailing prohibitions of foreign ownership. The timing of the 1987 changes was advantageous for Murdoch as his TV holdings sold for much more than they would have done if he had been forced to sell immediately after renouncing his Australian citizenship.

After selling and then regaining Nine Network, Kerry Packer sought to broaden his interests in television and magazines to daily newspapers but was frustrated by the cross-media rules. Except for a brief period in the late 1980s, PBL Media was the owner of Nine Network, then Australia's consistently highest-rated network. PBL was consequently barred from owning daily newspapers or radio services in any major metropolitan areas, and its assets were concentrated in broadcast TV, magazine publishing, and multichannel TV. In 2006, PBL Media's television and magazine interests were sold off to the Nine Entertainment Company, with James Packer selling a 50% share of the new entity to a private equity company, CVC Capital Partners (UK). Packer retained an interest in Foxtel. Consequently, CVC acquired all of the capital in Nine Entertainment.

Kerry Stokes controls the Seven Network, which has recently acquired a print daily, the *Western Australian*, an acquisition that would not have been possible under the former regulatory regimen. Before acquiring a strategic interest in the Seven Network in 1996, Stokes owned several broadcast and newspaper assets, including a television station and a print daily in the federal capital, Canberra. He sold those assets with the onset of the 1987 regulations. The Seven Network also owns Pacific Magazines (the second largest Australian magazine publisher) and Yahoo!7 (a joint venture combining Yahoo search and online capabilities with the Seven Network's content), as well as interests in wireless broadband and voice-over Internet protocol (VOIP) telephony.

International owners have substantial stakes in Australian media. Indeed, Australia is probably unique not only in having one company controlling 65% of the circulation of daily newspapers, but also in that the company is now headquartered abroad, in the United States.

The acquisition of a controlling interest in the broadcasting sector was specifically prohibited until 1992 in relation to radio and 2006 in relation to TV. Since these years, foreign investment in radio and TV is governed by the provisions of general foreign investment policy that is broadly encouraging of foreign investment consistent with the national interest. Newspapers, broadcast TV, and radio are defined as sensitive industries within that policy and all foreign investment proposals involving 5% or more of a company's stock must obtain prior approval.[32] In most recent cases, approval has been granted subject to conditions regarding Australian residency of the CEO, retention of Australian incorporation and headquarters, and the inclusion of a majority of Australian directors on the board of directors. Foreign investment

32. Treasurer "Australia's Foreign Investment Policy." Canberra: Foreign Investment Review Board, 2011. Accessed May 31, 2011. <http://www.firb.gov.au/content/_downloads/AFIP_2013.pdf>.

proposals in other media industries, except for the special case applied to privatization of Telstra, are subject only to the broader thresholds of non-sensitive foreign investment. In the case of Telstra, "aggregate foreign ownership . . . is limited to 35% of the privatized equity and individual foreign investors are only allowed to own up to 5%."[33]

Until recently, two of the three main commercial broadcasting networks were controlled by foreign entities. Canada's Canwest acquired 50.1% of Ten Network in 1992 when the network was in receivership: to get around the then ruling prohibition of foreign control, the network was formally acquired by a corporation in which Canwest owned marginally less than the 15% share of the prescribed foreign control threshold. It held the remainder of this firm's economic interest in the form of subordinated debentures and convertible debentures, which were not captured by the legislated definition of "company interests." Following abolition of the restrictions on foreign ownership of television in 2006, the government approved Canwest's conversion of all its debentures into stock, thus giving the company formal control of the network.[34] Foreign control of Ten Network ended in 2009 when Canwest, faced with financial difficulties in Canada, sold its stock to Australian institutional investors. In 2007, the government approved the purchase of the Nine Network by CVC Capital Partners from PBL. The transaction also included the purchase of ACP Magazines, the largest magazine publisher in Australia, which did not require government approval.

The removal of foreign investment controls on radio in 1992 was quickly followed by the entry of foreign investors: Australian Provincial Press (APN), owned by Independent News & Media (Ireland) and already established in the Australian market as publishers of more than a dozen regional daily newspapers and other periodicals, became the first foreign entrant into the radio industry. Forming a joint venture with Clear Channel (US), it gradually purchased several metropolitan radio stations to avoid breaching the prevailing cross-ownership prohibition against owning newspapers and radio licenses in the same region. Another major foreign investor in radio was the Trinity Mirror plc (UK), which entered the market in 1996 and acquired one metropolitan and over 50 regional stations. By 2004, it had expanded its metropolitan stations to nine and began to focus exclusively on the metropolitan market. Soon after, it sold all its regional stations to the Macquarie Media Group.

In telecommunications, the special foreign investment rules relating to Telstra have ensured its status as an Australian controlled carrier. The three other major carriers operating in Australia, however, are all foreign controlled. Optus, the second major carrier operating both wireline and wireless networks in competition with Telstra, is currently a subsidiary of SingTel (Singapore), and was formerly owned by Cable and Wireless (UK). The other two major wireless carriers are Vodafone Australia, owned by Vodafone plc (UK), and Hutchison Telecommunications International, a subsidiary of Hutchison Whampoa (Hong Kong), now operating in a joint venture. Other foreign investors in the industry include Telecom New Zealand (New Zealand), Primus Telecom (US), and Reach Networks (Hong Kong).

Telstra has several significant investments in telecommunications services in other countries. These include TelstraClear, the third-ranked telecommunications carrier in New Zealand, and CSL New World Mobility Group, Hong Kong's largest wireless provider. Another significant Australian-related venture overseas is the US-based Village Roadshow Entertainment Group, formed from a merger of the Australian Village

33. Costello, P. "Foreign Investment Proposal: Red Earth Holdings b.v.—Acquisition of an additional interest in PBL Media Holdings Pty Limited and the PBL Media Holdings Trust." Media Statement No. 088, Sept. 4, 2007. Accessed May 31, 2011. <http://ministers.treasury.gov.au/DisplayDocs.aspx?doc=pressreleases/2007/088.htm&pageID=003&min=phc&Year=&DocType=0>.

34. Costello, P. "Foreign Investment: Canwest Global Communications Corp—Acquisition of Interest in Ten Network Holdings Limited." Media Statement No. 087, Aug. 20, 2007. Accessed May 31, 2011. <http://ministers.treasury.gov.au/DisplayDocs.aspx?doc=pressreleases/2007/078.htm&pageID=003&min=phc&Year=&DocType=0>.

Roadshow Pictures with Concord Music Group (US). Besides Village Roadshow and Telstra, no other Australian media groups have significant overseas investments.

Before 1987, prohibition of ownership of more than two television stations ensured a low level of concentration. While the two station rule operated, major networking leverage was attained by owning the Sydney and Melbourne stations. Packer owned the two Nine stations, Murdock the two Ten stations, while the two Seven stations were split between HWT and Fairfax. The removal of the two stations' limit was quickly followed by changes in ownership of the three major commercial operators, with the new owners moving promptly to acquire properties in other metropolitan markets. This is reflected by the gradual rise in both the C4 and HHI from 1988 to 2000. Although concentration increased, the impact was attenuated by other ongoing ownership controls preventing ownership of more than one television station in any local market and prohibitions of entry into the industry by new players. Concentration peaked in 2000 when one of the three main networks was temporarily in excess of the 75% population reach limit following several acquisitions. That year was also a time of considerable turmoil in the industry, with all three major commercial networks experiencing financial difficulties that forced them into receivership. The slight subsequent decline in the concentration ratio is due to the mandated divestiture of assets to ensure that new owners remained within the population reach limit.

The situation in broadcast radio is much more fluid. Like broadcast television, broadcast radio had been subject to considerable change in ownership controls immediately prior to the period under review. The entry of new players in the radio market, although strictly controlled by licensing, was not prohibited, as was the case for television. During the decade preceding 1988, several new FM radio broadcasters established themselves in the main metropolitan and regional markets. These, combined with the changes to the preexisting controls of ownership, brought about extensive changes in the industry. The combined effect of the two-station ownership limit in any one local market and the licensing of up to 12 new large FM stations in the main metropolitan areas, in addition to the market share for public broadcasting, seem to have reduced concentration. Both C4 and HHI indicate low to moderate concentration in the radio market.

While the initial establishment of multicast subscription television in Australia was delayed by government regulation, subsequent development of the industry has not been subject to regulatory restrictions of ownership. However, entry into the industry faces considerable barriers: while some independent content aggregators are available to all MSOs, there is a significant level of exclusivity in access to Hollywood productions and popular sports programs. Foxtel was able to exploit a degree of vertical integration with its partial owners—Telstra for cable distribution in major cities and News Corp. for access to 20th Century Fox content—to become the dominant player. Apart from a couple of very small operators, the Australian subscription TV market was until recently subdivided into two exclusive submarkets, each served by a single supplier of the same programming: Foxtel in major metropolitan areas and Austar in regional areas. Foxtel's purchase of Austar in 2012 has created a virtual national monopoly. The high levels of both the C4 and HHI reflect the high level of concentration in the market.

Liberalization of the telecommunications sector has had a major impact on industry concentration. With competition from new carriers and many smaller service providers, the market share of the incumbent carrier has been declining gradually over the years. However, the incumbent enjoys substantial economic advantages over its competitors, not the least of which is its ownership of the nationwide wireline CAN. After almost two decades of competition, Telstra still controls more than 70% of industry revenue. The position is unlikely to change significantly in the foreseeable future. However, the government has recently embarked on the building of a national broadband fiber-optic network including the CAN (fiber-to-the-home) with the intention of providing access to all carriers on an equal basis. When completed, it will eliminate

Telstra's monopoly control of the CAN and is likely to intensify competition in the market, with consequential reductions in concentration.

In the wireless sector, liberalization has produced more intense market competition. The wireless sector was still in an early development stage when liberalization was put in place, giving the new entrant carriers a more balanced basis to compete with the incumbent. While Telstra's market share has declined significantly, it remains the largest operator. However, as the market has grown considerably, the current market shares of Telstra and its competitors relate to a much larger pie than was the case in the early 1990s.

The nature of the Australian film and video production industry is not conducive to a high degree of concentration. The industry is largely made up of medium and small enterprises, with many very small establishments engaged in the provision of freelance services to larger production houses. This structure lends itself to the "project nature" of much of the industry's main production activity for which specialist skills and services are typically brought together for the duration of a project. A large proportion of the industry output is made-to-order for television internally by the networks themselves or commissioned from other producers. More than two-thirds of all establishments are engaged in production of outputs for television. In contrast, the major international distributors dominate film distribution. During the period under review, the HHI ratio for film distribution declined from just above the high concentration threshold to lower levels in the moderate concentration band. The C4 ratio shows a similar decline. While the doubling of the share of the market held by some 30 independent specialist distributors also made a small contribution to the decline, it is unlikely that their market share will grow sufficiently to pose a threat to the majors in the foreseeable future.

In the period from 1998 to 2008, concentration in the Australian ISP market increased considerably, with the transformation of a relatively unconcentrated market to a highly concentrated one. The period coincides with extensive growth in household Internet access.[35] During the same period there was also a transformation of household Internet connections from telephone dial-up to largely ADSL broadband. The growth of the household market provided a significant advantage to incumbent telecommunications carriers over whose network the services were delivered. Telstra was able to exploit this advantage to secure the lion's share of the market. Other carriers, despite deregulation of the telecommunications industry, were largely dependent on access to Telstra's network to deliver their services. Consequently, both their appeal to customers and market growth is relatively constrained.

The analysis indicates a tendency for increased market concentration in all electronic media markets. While regulation has constrained high levels of concentration in mass media markets, such as broadcast TV and radio, the inherent economies of scale provide incentives for greater concentration. In television, for example, there are major economies of scale in both programming and the supply of national advertising: in both of these areas, major broadcasters were able to increase their market share through commercial agreements.

The tendency toward concentration is also evident in the new media. As for all information services, once content is created the cost of making it available to larger numbers of users is very small. Increased popularity of a service enables providers to exploit a virtuous circle by increasing investment in content and thus increasing its appeal to users. Less popular services face a vicious circle fed by the loss of users and reduced capacity to invest in improvements. Bigpond in the ISP market and Google in the search engine market display some of these characteristics. Bigpond, for example, is one of the most popular websites in Australia and offers its ISP subscribers unmetered access to its own

35. ABS (Australian Bureau of Statistics). "8146.0—Household Use of Information Technology, Australia, 2008–09." Canberra: ABS, 2010. Accessed May 27, 2013. <http://www.abs.gov.au/AUSSTATS/abs@.nsf/allprimarymainfeatures/5B15663305C2B5DCCA25796600153023?opendocument>.

content. Subscribers to other ISPs incur a usage charge to download the same content via their service providers, which lack the capacity to provide comparable, unmetered content for their subscribers. While there are several specialized Australian search engines, they command only very small market shares, so the rise of Google at the expense of other engines has resulted in dramatic increases in concentration ratios.

The measures of concentration used in the research discussed in this chapter are based on traditional industry definitions. While convergence brought about by the development and rapid growth of online information services may have eroded traditional industry boundaries, the resultant impact on industry concentration does not appear to be significant. Traditional media not only continue to dominate in the domestic market via traditional distribution platforms, but they have also extended their presence into the online world. The most popular online news services are associated with traditional media, including daily newspapers and broadcasters. Among the "thousands of voices" accessible online, very few have the capacity to challenge the influence of traditional media on public opinion.

## Bibliography

### 1. Books and Articles


Barry, Paul. *The Rise and Rise of Kerry Packer Uncut.* Sydney: Bantam, 2007.

Blainey, Geoffrey. *The Tyranny of Distance: How Distance Shaped Australia's History.* Melbourne and London: Macmillan, 1968.

Chenoweth, Neil. *Virtual Murdoch: Reality Wars on the Information Superhighway.* London: Vintage, 2002.

Duffy, M. (MP, the Hon). "Second Reading Speech." *House of Representatives Debates,* Apr. 29, 1987.

House of Representatives. *Debates, 215:28. November 1951.* Canberra: Commonwealth of Australia Parliament, 1951.

Mayer, Henry. *The Press in Australia.* Melbourne: Lansdowne Press, 1964.

Papandrea, Franco. "Media Diversity and Cross Media Regulation." *Prometheus* 24:3, 2006. 301–322.

Souter, Gavin. *Heralds and Angels: The House of Fairfax, 1841–1990.* Melbourne: Melbourne University Press, 1991.

Tiffen, R. "Media Policy." Brett, J., J. Gillespie and M. Goot (eds.). *Developments in Australian Politics.* Macmillan, Melbourne, 1994.

Tiffen, Rodney. "From Technological Abundance to Commercial Monopoly in Australian Pay TV: Key Relationships in Institutionalising Subscription Television." Kenyon, Andrew T. (ed.). *TV Futures: Digital Television Policy in Australia.* Melbourne: Melbourne University Press, 2007.

Tiffen, Rodney and Gittins, Ross. *How Australia Compares.* 2nd ed. Melbourne: Cambridge University Press, 2009.

Wolff, Michael. *The Man Who Owns the News: Inside the Secret World of Rupert Murdoch.* New York: Knopf, 2008.


### 2. Reports


ASTRA (Australian Subscription Television and Radio Association). 2011. May 31, 2011. <http://astra.org.au/pages/facts-figures>.

Australian Broadcasting Corporation (ABC). *Annual Report.* Various years.

Australian Broadcasting Tribunal (ABT). *Annual Report.* Various years.

ABS (Australian Bureau of Statistics). "8146.0—Household Use of Information Technology, Australia, 2008–09." Canberra: ABS, 2010. Accessed May 27, 2013. <http://www.abs.gov.au/AUSSTATS/abs@.nsf/allprimarymainfeatures/5B15663305C2B5DCCA25796600153023?opendocument>.

ABS (Australian Bureau of Statistics). "8153.0—Internet Activity, Australia, Dec 2012." Canberra: ABS, 2012. Accessed May 27, 2013. <http://www.abs.gov.au/ausstats/abs@.nsf/Products/8153.0~December+2012~Chapter~Sector?OpenDocument>.

ABS (Australian Bureau of Statistics). "8654.0—Motion Picture Exhibition, Australia, 1999–2000." Canberra: ABS, 2001. Accessed 27 May 2013. <http://www.abs.gov.au/ausstats/abs@.nsf/mf/8654.0/>.

ABS (Australian Bureau of Statistics). "8679.0—Television, Film and Video Production and Post-Production Services, Australia, 2006–07." Canberra: ABS, 2008. Accessed May 27, 2013. <http://www.abs.gov.au/AUSSTATS/abs@.nsf/Lookup/8679.0Explanatory%20Notes12006-12007?OpenDocument>.

Australian Communication and Media Authority (ACMA). *Broadcasting Financial Results.* Various years.

Communications Law Centre. *Update*, "Media Ownership Update" special issues. Various years.

Costello, P. "Foreign Investment: Canwest Global Communications Corp—Acquisition of Interest in Ten Network Holdings Limited." Media Statement No. 087, Aug. 20, 2007. Accessed May 31, 2011. <http://ministers.treasury.gov.au/DisplayDocs.aspx?doc=pressreleases/2007/078.htm&pageID=003&min=phc&Year=&DocType=0>.

Costello, P. "Foreign Investment Proposal: Red Earth Holdings b.v. —Acquisition of an Additional Interest in PBL Media Holdings Pty Limited and the PBL Media Holdings Trust." Media Statement No. 088, Sept. 4, 2007. Accessed May 31, 2011. <http://ministers.treasury.gov.au/DisplayDocs.aspx?doc=pressreleases/2007/088.htm&pageID=003&min=phc&Year=&DocType=0>.

Federal Communications Commission (FCC). "2002 Biennial Regulatory Review: Report and Order and Notice of Proposed Rulemaking." FCC: Washington, DC, 2003.

Hanover, Wilton Hanford (ed.). Book Industry Strategy Group. "Final Report to Government." Canberra: Book Industry Strategy Group, 2011. Accessed May 27, 2013. <http://www.innovation.gov.au/Industry/BooksandPrinting/BookIndustryStrategyGroup/Documents/BISGFinalReport.pdf>.

Hitwise Australia. "Weekly Online Trends." Experian, Sept. 13, 2010. Accessed May 27, 2013. <http://www.hitwise.com/au/datacentre/main/dashboard-1706.html>.

Joint Parliamentary Committee on Wireless Broadcasting (W. Gibson, Chair). *Report*. Canberra, Government Printing Office, 1942.

Lee, Jenny. "Digital Technologies in Australia's Book Industry." Canberra: Book Industry Strategy Group, 2010. Accessed May 27, 2013. <http://www.innovation.gov.au/Industry/BooksandPrinting/BookIndustryStrategyGroup/Documents/DigitalTechnologiesInAustraliasBookIndustry.pdf>.

Productivity Commission. "Broadcasting Report no. 11." Canberra: Ausinfo, 2000.

Productivity Commission. "Telecommunications Competition Regulation Inquiry Report." Canberra: Productivity Commission, 2001. Accessed May 31, 2011. <http://www.pc.gov.au/projects/inquiry/telecommunications/docs/finalreport>.

Roy Morgan Research (RMR). "Google Overwhelms Search Competitors in Australia." RMR, Feb. 1, 2006.

Roy Morgan Research (RMR). "It's Google Then The Rest When Searching Online In Australia." RMR, May 18, 2006.

Screen Australia. "Research: Production Industry, Drama." screenaustralia.gov, May 31, 2011. Accessed May 27, 2013. <http://www.screenaustralia.gov.au/research/statistics/mpfeaturesspending.aspx>.

Select Committee on the Print Media (M. Lee, Chair), House of Representatives, Report, *News & Fair Facts*. Canberra: Australian Government Publishing Service, 1992.

Special Broadcasting Service (SBS). *Annual Report*. Various years.

Treasurer. "Australia's Foreign Investment Policy." Canberra: Foreign Investment Review Board, 2011. Accessed May 31, 2011. <http://www.firb.gov.au/content/_downloads/AFIP_2013.pdf>.

# Australia—Data Summaries

ELI NOAM AND PAUL MUTTER

THERE HAS BEEN FIERCE competition among traditional media families in the newspapers market—the Murdochs, who own News Corp., News Ltd., and 21st Century Fox; the Fairfaxes, who own Southern Cross and formerly owned Seven Network; and the Packers, formerly owners of PBL Media (Table 23.14).

News Corp.'s News Ltd. print media group in Australia has political clout equal to its economic weight: it is the second largest media company overall (with a PI of 433), and as a leading content producer holds 20.8% of the content market (Table 23.15). Due to citizenship laws passed in the 1980s alongside new cross-ownership regulations, News Corp.'s Australian properties are concentrated in print (newspapers and book publishing), in contrast to its substantial electronic media holdings in the United Kingdom and the United States. In newspapers, News Corp. dominates, with a 57.5% industry market share, and it has made substantial advances in magazines as well.

Australia's rank in foreign ownership of its media is the result of Optus (SingTel) and News Corp.'s high market shares in their respective industries—even though within the platform industry, Telstra controls 61.1% of the market (Table 23.16), well ahead of the next largest companies, SingTel and Vodafone (UK). Also, the numbers assume that the Murdochs are non-Australian foreigners, which is true only in a legalistic sense.

The Average-Weighted C4 of Australia's overall media increased from 89.1% in 2004 to 89.5% in 2009 or most recent, putting it in the top third internationally for this measure. The combined national market share for the top four firms (C4) in Australia increased from 66.5% in 2004 to 67.8% in 2012 or most recent, of which 47.6% are in telecom; of the country's top four multimedia producers, only News Corp. (US) saw its power index increase, by 2.1% per year, between 2004 and 2012.

Table 23-14. National Media Industries Concentration in Australia

| | 2004/5 | | 2011 OR MOST RECENT | | % CHANGE ANNUAL AVERAGE | |
|---|---|---|---|---|---|---|
| | COMPANY POWER INDEX IN COUNTRY | COMPANY SHARE OF THE OVERALL NATIONAL MEDIA MARKET (%) | COMPANY POWER INDEX IN COUNTRY | COMPANY SHARE OF THE OVERALL NATIONAL MEDIA MARKET (%) | COMPANY POWER INDEX IN COUNTRY | COMPANY SHARE OF THE OVERALL NATIONAL MEDIA MARKET (%) |
| Telstra | 2,683 | 40.9 | 2,577 | 41.4 | −0.5 | 0.06 |
| Murdoch Interests (US) | 367 | 7.1 | 433 | 9.0 | 2.1 | 0.2 |
| SingTel (Optus) (Singapore) (public) | 342 | 13.7 | 289 | 12.8 | −2.1 | −0.12 |
| Nine Entertainment | 158 | 4.8 | 131 | 4.1 | −2.3 | −0.1 |
| Google (US) | 8 | 0.2 | 163 | 1.8 | 45.5 | 0.2 |
| Seven Network | 73 | 3.9 | 106 | 4.5 | 4.8 | 0.08 |
| Fairfax Media | 49 | 2.3 | 70 | 2.5 | 4.6 | 0.02 |
| Liberty Media (US) | 23 | 0.8 | 0.0 | 0.0 | −100.0 | −0.1 |
| Australian Broadcasting Corporation (ABC) (public) | 25 | 1.6 | 40 | 2.1 | 6.1 | 0.07 |
| Village Roadshow (US) | 9 | 0.6 | 2 | 0.2 | −17.5 | −0.06 |
| Pearson (UK) | 4 | 0.3 | 0.0 | 0.0 | −100.0 | −0.04 |
| Lagardère (France) | 3 | 0.2 | 4 | 0.3 | 4.2 | 0.01 |
| Bertelsmann (Germany) | 3 | 0.3 | 0.0 | 0.0 | −100.0 | −0.03 |
| Southern Cross Media Group | 7 | 0.9 | 23 | 1.4 | 16.9 | 0.07 |
| Yahoo (US) | 0.3 | 0.03 | 0.0 | 0.03 | −24.9 | −0.001 |
| Microsoft (Bing) (US) | 0.6 | 0.04 | 0.2 | 0.07 | −11.0 | 0.003 |

| Media Concentration Index | 2004/5 | 2011 or Most Recent | % Change Annual Average |
|---|---|---|---|
| Total Revenue: Nat'l. Media Industry (mil US$) | 31,908 | 50,130 | 7.1% |
| Total Voices (*n*) | 47 | 35 | −3.2% |
| Net Voices (*n*) | 32 | 23 | −3.5% |
| Public Ownership (%) | 4.0 | 4.6 | 0.1% |
| Foreign Ownership (%) | 32.8 | 32.4 | −0.06% |
| C4 Average—Weighted | 89.1 | 89.5 | 0.0% |
| HHI Average—Weighted | 3,989 | 4,238 | 0.8% |
| C1 Average—Weighted | 53.0 | 54.0 | 0.0% |
| Noam Index Average—Weighted | 1,068 | 1,806 | 8.6% |
| Pooled Overall Sector C4 | 66.5 | 67.8 | 0.2% |
| Pooled Overall Sector HHI | 1,981 | 2,018 | 0.36% |
| Pooled Overall Sector Noam Index | 155 | 188 | 2.6% |
| Market Share of Top Ten Companies: Nat'l. Media Industry (%) (Pooled C10) | 81.0 | 84.3 | 0.4% |
| National Media Power Index | 3,898 | 3,984 | 0.31% |

Table 23-15. Top Content Media Companies in Australia

| | 2004/5 | | 2011 OR MOST RECENT | | % CHANGE ANNUAL AVERAGE | |
|---|---|---|---|---|---|---|
| | COMPANY POWER INDEX IN COUNTRY | COMPANY SHARE OF THE NATIONAL CONTENT MEDIA MARKET (%) | COMPANY POWER INDEX IN COUNTRY | COMPANY SHARE OF THE NATIONAL CONTENT MEDIA MARKET (%) | COMPANY POWER INDEX IN COUNTRY | COMPANY SHARE OF THE NATIONAL CONTENT MEDIA MARKET (%) |
| Murdoch Interests (US) | 1,125 | 21.1 | 1,013 | 20.8 | –1.3 | –0.05 |
| Nine Entertainment | 477 | 14.0 | 387 | 12.0 | –2.6 | –0.25 |
| Google (US) | 25 | 0.5 | 482 | 5.2 | 44.6 | 0.6 |
| Seven Network Ltd. | 227 | 12.0 | 313 | 13.3 | 4.1 | 0.16 |
| Fairfax Media | 152 | 7.1 | 206 | 7.2 | 3.9 | 0.02 |
| Australian Broadcasting Corporation (Public) | 78 | 4.9 | 118 | 6.3 | 5.4 | 0.2 |

| Media Concentration Index | 2004/5 | 2011 or Most Recent | % Change Annual Average |
|---|---|---|---|
| Public Ownership (%) | 6.0 | 7.6 | 0.2% |
| Foreign Ownership (%) | 38.1 | 35.2 | –0.4% |
| C4 Average—Weighted | 80.4 | 83.4 | 0.4% |
| HHI Average—Weighted | 2,554 | 3,150 | 2.9% |
| C1 Average—Weighted | 38.0 | 41 | 0% |
| National Power Index | 2,314 | 2,930 | 3.3% |

Table 23-16. Top Platform Media Companies in Australia

| | 2004/5 | | 2011 OR MOST RECENT | | % CHANGE ANNUAL AVERAGE | |
|---|---|---|---|---|---|---|
| | COMPANY POWER INDEX IN COUNTRY | COMPANY SHARE OF THE NATIONAL PLATFORM MEDIA MARKET (%) | COMPANY POWER INDEX IN COUNTRY | COMPANY SHARE OF THE NATIONAL PLATFORM MEDIA MARKET (%) | COMPANY POWER INDEX IN COUNTRY | COMPANY SHARE OF THE NATIONAL PLATFORM MEDIA MARKET (%) |
| Telstra | 3,946 | 59.9 | 3,832 | 61.1 | –0.4 | 0.16 |
| SingTel Optus Pty Limited (Singapore) | 502 | 20.0 | 438 | 19.3 | –1.7 | –0.1 |
| Vodafone (UK) | 92 | 5.6 | 104 | 7.0 | 1.6 | 0.17 |
| Media Concentration Index | 2004/5 | | 2011 or Most Recent | | % Change Annual Average | |
| Public Ownership (%) | 3.0 | | 3.1 | | 0.01% | |
| Foreign Ownership (%) | 30.3 | | 30.9 | | 0.1% | |
| C4 Average—Weighted | 93.3 | | 92.7 | | –0.1% | |
| HHI Average—Weighted | 4,672 | | 4,796 | | 0.3% | |
| C1 Average—Weighted | 60% | | 61% | | 0.0% | |
| National Power Index | 4,593 | | 4,524 | | –0.2% | |

# 24

# Media Ownership and Concentration in the People's Republic of China

MIN HANG AND THE CHINA MEDIA CONCENTRATION RESEARCH GROUP

## INTRODUCTION

Media ownership and concentration has not been a hotly debated topic in China since 1949, when state control of media became the norm after the People's Republic of China was established. Since the economic reforms of the 1980s—"socialism with Chinese characteristics"—most citizens have accepted the continued role of the Communist Party of China (CPC) in the country's media and government in exchange for greater economic opportunities, especially in the wake of the post-2008 global economic slowdown. However, the emergence and rapid development of new media have rekindled interest in media pluralism, and the wave of market deregulation has allowed some degree of media commercialization.[1] Some media critics have begun to argue that high media concentration may hinder the development of professional journalism, and that a more diversified media system would enable people in the country to have a clearer picture of both the world and the country as they really are.

To a certain extent, developments in media ownership and concentration in China have always been shaped by two conflicting forces: a desire for media pluralism, on the one hand, and the desire for "national champions." Though state and party control play a major part in the latter vision, it is also a strongly felt nationalistic desire among Chinese leaders. In Chinese academic circles, for example, contrary to

1. Li, Ziwei. "Media Transformation in China in the New Media Era." *Contemporary Manager* 23: 6, 2005. 182–183.

conventional wisdom in the West, low (de facto) levels of concentration in some sectors (especially print media) are not seen as an indication for healthy market competition but as a sign of market fragmentation, a lack of competitiveness of major firms (especially in an international context), inefficient resource allocation, and poor economies of scale. Underlying this is a deep regionalism among Chinese provinces promoted by local governments to create greater economic autonomy from Beijing.

The Chinese government regulates and administers media industries, particularly the audiovisual media and telecommunications sectors, through a number of regulatory bodies. The State Administration of Radio, Film, and Television (SARFT) is responsible for the content side of the audiovisual sector, as well as its related online content. As a ministerial-rank government agency directly under the State Council, it owns and directly controls the big national media outlets: China Central Television (CCTV), China National Radio (CNR), China Radio International (CRI), and the China Film Group (CFG). The Ministry of Industry and Information Technology (MIIT) sets policies for telecommunications and online media and is also responsible for technological standards pertaining to the audiovisual sectors, which SARFT then implements.

The State-owned Assets Supervision and Administration Commission (SASAC) is the nominal owner of the state-owned enterprises (SOEs) responsible for overseeing a wide range of media market activities, including mergers, stock sales, executive appointments, legislation, and legal issues. In the context of media industries, it owns and controls the primary telecommunications providers of China Telecom, China Mobile, and China Unicom.

The General Administration of Press and Publication (GAPP) is the regulator for print media. Unlike SARFT, it does not claim ownership for companies in these industries: major outlets, such as the *People's Daily* and Xinhua News Agency's *Reference News*, were established before the founding of the People's Republic. Due to their importance to the CPC, they remain controlled by its executive today. This practice has continued for newspapers formed since 1949. Magazine and book publishing industries, in contrast, are more "liberal" in terms of ownership. Owners of companies in these two industries include government ministries (in the areas of their responsibility), institutional organizations, commercial entities, and institutional investors.

The Publicity Department (PD) of the Communist Party assumes a special role in China's media ecosystem. In principal, it oversees the content side of Chinese media. The editors-in-chiefs or directors of major media outlets are required to report to the PD weekly for instructions on how and to what extent certain stories can be covered. The PD works closely with government regulators and agencies, including SARFT, GAPP, the State Council's Information Office, and in some cases, the Public Security Bureau.

## PRINT MEDIA

### Newspapers

For most of the history of the People's Republic, many cities had a single authorized state-run daily; local markets were virtual monopolies with little competition. After economic reforms were introduced in the 1970s and 1980s, the distribution side was reformed, while the operational and editorial sides have largely adhered to the preexisting model.[2] As the government gradually reduced funds allocated to newspapers in the 1990s, publishers had to commercialize and adopt commercial practices. The rapid development of the Internet in the 2000s further pressured the newspaper industry to transform its operating practices.

2. Zhang, Dianyuan. "Social Motivation of System Innovation of Chinese Newspaper." *Science Economy Society* 28: 4, 2010. 146–150.

Table 24-1. Daily Newspapers: National (Market Shares by Circulation), 1992–2007

| | 1992 | 1996 | 2000 | 2004 | 2007 |
|---|---|---|---|---|---|
| C4 | 81.2 | 72.3 | 66.0 | 60.4 | 60.3 |
| HHI | 2,104 | 1,734 | 1,595 | 1,408 | 1,393 |
| *N* (>1%) | 9 | 8 | 9 | 9 | 8 |
| Noam Index | 391 | 285 | 262 | 207 | 273 |

From 1992 on, many newspapers—though not all—had to be financially independent in accordance with the new market reforms.[3] Operations at the provincial and municipal levels were consolidated to form media groups. Notable examples include the People's Daily Group at the national level and the Guangzhou Daily Group at the municipal level. In the national newspaper market, concentration is declining, with the HHI falling from 2,104 in 1992 to 1,393 in 2007. All companies listed here—and all state-level newspapers—are publicly owned (Table 24.1).

The *People's Daily*—the most influential newspaper in China—is the flagship newspaper of the People's Daily Group. The newspaper was established as an organ of the CPC in 1948, before the rest of the country's party administered media system came into existence. It has since remained a key organization directly under the control of the Communist Party's Central Committee. Its editor-in-chief holds ministerial rank and is appointed by the Organization Department of the Communist Party. In addition to *People's Daily,* the group also publishes *Global Times,* an English-language daily with a relatively liberal political stance. Other activities of the group include a web portal, *People's Daily Online,* and a search engine Jike (launched in 2011), with the aim to compete with privately owned search engines such as Baidu.

*Reference News,* published by Xinhua News Agency, has the largest daily circulation numbers in China. It was initially available only to party cadres and government organizations but was made available to the general public in the 1980s, rapidly increasing its circulation numbers.[4] The head of the Xinhua News Agency, similar to People's Daily Group, also holds a ministerial rank and is subordinate to the State Council (Cabinet). Before the 1980s, it was the only official news outlet that carried reporting from abroad for its audience.

The declining circulation figures for national newspapers are indicative of the ascent of local newspapers: since 2000, local newspaper markets have become increasingly competitive and adopted market practices. A host of evening newspapers came into being in their respective cities and have attracted ever-larger readership and advertising fees, dwarfing the traditional national dailies in those cities. Print media groups became a natural recourse for consolidating resources and promoting operational efficiency, particularly from the standpoint of local governments owning these various newspapers. Consequently, local governments have taken an active role in the consolidation process, both financially and operationally (Table 24.2).

The market is extremely fragmented—among the least concentrated in countries investigated

3. Song, Zifang. "Reconsideration of Commercialization in China's Newspaper Industry and Its Outcomes." *Journal of Journalism and Communication* 3: 1, 1997. 49.

4. Ma, Shengrong. "A Discussion on Media Development and Strategy in China." *International Communications* 7: 6, 2011. 27–23.

Table 24-2. Daily Newspapers: National and Provincial (Market Shares by Circulation), 2003–2012[1]

| | 2003 | 2005 | 2008 | 2012 |
|---|---|---|---|---|
| People's Daily Group (*People's Daily, Global Times*) (Communist Party of China) (public) | 4.4 | 4.8 | 5.4 | 11.1 |
| Xinhua News Agency (*Reference News*) (State Council) (public) | 4.7 | 4.1 | 4.5 | 6.4 |
| China Youth League (*China Youth Daily*) (public) | 1.5 | 1.0 | 0.8 | 1.0* |
| All-China Federation of Trade Unions (*Worker's Newspaper*) (public) | 1.4 | 0.9 | 0.7 | 2.1 |
| Central Military Commission (*People's Liberation Army Daily*) (public) | 0.9 | 0.8 | 0.7 | 1.7* |
| Guangdong Provincial Government (*Guangzhou Daily, Yangcheng Evening News, Southern Daily*) (public) | 10.5 | 14.5 | 13.7 | 7.9 |
| Shanghai Municipal Government (*Xinmin Evening News*) (public) | 6.5 | 6.4 | 6.1 | 1.9 |
| Beijing Municipal Government (*Beijing Evening News, Beijing Youth*) (public) | 5.9 | 5.7 | 5.4 | 3.1 |
| Jiangsu Provincial Government (*Yangtse Evening Post*) (public) | 3.0 | 2.7 | 2.6 | 3.2 |
| Zhejiang Provincial Government (*Qianjiang Evening News*) (public) | 2.0 | 2.3 | 2.7 | 2.1 |
| Hubei Provincial Government (*Chutian Metro Daily, Wuhan Evening News*) | 2.3 | 2.7 | 2.9 | 2.6 |
| Shandong Provincial Government (*Qilu Evening News, Peninsula City News*) (public) | 2.7 | 2.6 | 2.7 | 5.7 |
| Tianjin Provincial Government (*Today Evening News*) (public) | 1.8 | 2.0 | 1.6 | 2.1 |
| Hebei Provincial Government (*Yanzhao Metro Daily*) (public) | 0.8 | 1.1 | 1.4 | 2.1 |
| Others | 51.0 | 47.8 | 48.3 | 47.0 |
| Total Daily Circulation (1,000) | 82,000 | 96,600 | 107,000 | 128,000 |
| C4 | 27.6 | 31.4 | 30.6 | 31.2 |
| HHI | 263 | 357 | 339 | 312 |
| *N* (>1%) | 12 | 12 | 11 | 14 |
| Noam Index | 76.0 | 103 | 102 | 83 |

*Source*: World Association of Newspapers, 2012 updates provided by editors, using Chinadaily.com.cn; * indicates estimate.

[1] Market Shares and Index calculations by author based on assumption that publically owned organization are segmented in ownership.

in this project. But if we take all levels of government as a single "public" entity, the HHI would be in effect 10,000. However, operational independence has become increasingly pronounced—compared to nominal ownership—in recent years. This can be attributed not only to provincial autonomy, but also to the economic management approaches adopted by local governments, which shun cross-regional consolidation. Although nominally these print media groups are all owned by the government, they are in fact controlled and operated by distinct government entities, that is, different central, provincial, and local governments. More pluralist reporting is now permitted by the authorities, except when it concerns particularly sensitive topics such as China's foreign policy. A few local print groups have taken a distinctively more "liberal" approach in reporting, largely reflecting the respective local governments' liberal approach in economic management. One can make a strong case that in the southern province of Guangdong, due to its distance from Beijing and history as the starting point for the ongoing economic reforms, the newspaper market there has been particularly independent-minded. In the case of Southern Daily Group—owned by the Guangdong provincial government—in addition to two popular titles/groups (*Southern Daily* and *Southern Metropolis Daily*), the group also publishes *Southern Weekend*, often described in Western media as "China's most influential liberal newspaper."

## Book Publishing

Until recently, most book publishers operated as units of their respective administrative agencies, be they government ministries, institutional organizations such as unions and universities, or local governments. Beginning 2000, a series of mergers and consolidations were undertaken that have increasingly corporatized the market, though ownership ultimately remains with the aforementioned public entities.[5] Market concentration is still low, though it has increased significantly since the 2000s. The competitiveness of major publishing houses has been characterized by the government as "weak and scattered," especially when gauged against international peers.[6] China's largest publishing company in terms of revenue, Jiangsu Phoenix Publishing & Media, for example, posted revenue of US$996 million (RMB6.8 billion) in 2009, equivalent to 11% of the total annual revenue of Pearson plc. (UK), despite the much greater potential market size for Jiangsu Phoenix's publications.

There have been major efforts by governments at various levels, particularly local ones, to consolidate different publishing houses within their administrative regions. As a result, local publishing houses controlled by local governments have amassed substantial market shares. Another characteristic of China's publishing industry is that the profits of publishing firms rely heavily on textbooks and other teaching materials. The top 10 publishing companies in the People's Republic all rely on main and supplementary teaching materials to break even and turn a profit.[7] Other types of books published by these firms account for only a small proportion of their profits (Table 24.3).

Most publishing houses in China are, in effect, owned by public entities: China Science Publishing and Media Group, one of the three top national publishers, is managed by the Chinese Academy of Science through its National Asset Supervision division. The Chinese Academy of Science is, in turn, under the auspices of the State Council (Cabinet). University publishers are in general controlled by their respective universities, most of which are state-run.

China Education (Publishing and Media Holdings) is the country's largest publication house in terms of volume sold: strictly speaking, it was established in 2011 through the merger

5. Wei Yushan. "Strategies for China's Publishing Industry." *Publishing Research* 8: 5, 2010. 5–6.

6. General Administration of Press and Publication, "National Press and Publication Statistics Network." Beijing: General Administration of Press and Publication: 2006.

7. Su, Lei. "A Study on China's Teaching Material Publishing." *Science-Technology & Publication* 6: 10, 2011. 25.

Table 24-3. Book Publishing (Market Shares by Revenue), 2003–2012

| | 2003 | 2004 | 2005 | 2006 | 2007 | 2008 | 2009 | 2012 |
|---|---|---|---|---|---|---|---|---|
| China Education Publishing and Media (Ministry of Education) (public) | 6.5 | 4.7 | 4.7 | 4.7 | 3.8 | 4.5 | 7.2 | 6.1 |
| China Publishing Group (public) | 4.4 | 4.6 | 4.7 | 4.6 | 5.1 | 5.7 | 5.4 | 9.6 |
| China Science Publishing and Media (Chinese Academy of Science) (public) | 3.7 | 4.0 | 3.6 | 3.9 | 4.0 | 4.3 | 4.5 | 6.0* |
| Jiangsu Phoenix Publishing and Media (public) | 2.3 | 2.1 | 1.3 | 1.4 | 2.6 | 6.8 | 12.8 | 9.4 |
| China South Media/ Hunan Publishing Investment Holding (public) | 1.1 | 1.2 | 1.0 | 1.3 | 1.5 | 4.7 | 8.1 | 9.1 |
| Beijing Publishing Group (public) | 3.7 | 5.2 | 4.2 | 3.2 | 4.9 | 7.7 | 8.0 | 11.6 |
| Zhejiang Publishing United Group (public) | 3.5 | 4.3 | 3.9 | 4.1 | 5.9 | 7.4 | 10.3 | 15.3* |
| Shanghai Century Publishing Group (public) | 2.1 | 1.1 | 0.5 | 1.5 | 0.6 | 1.7 | 2.4 | 5.4* |
| People's Medication Publishing (Ministry of Health) (public) | 0.8 | 1.5 | 0.8 | 1.0 | 1.3 | 2.2 | 2.0 | 3.4* |
| Peking University Press (public) | 0.0 | 0.8 | 1.4 | 1.1 | 1.0 | 0.9 | 1.0 | 1.4* |
| Renmin University Press (public) | 0.9 | 1.2 | 1.1 | 1.0 | 1.4 | 1.1 | 1.1 | 1.2* |
| Tsinghua University Press (public) | 1.0 | 1.2 | 1.7 | 1.4 | 1.6 | 1.2 | 1.3 | 1.4* |
| FLTRP: Foreign Language Teaching & Research Press (Beijing Foreign Studies University) (public) | 1.3 | 1.6 | 1.1 | 1.3 | 1.3 | 1.1 | 1.3 | 1.1* |
| Others | 68.6 | 66.6 | 69.8 | 68.5 | 65.1 | 50.7 | 35.5 | 19 |
| Total Revenue (mil RMB) | 24,130 | 26,030 | 26,940 | 27,260 | 29,640 | 35,110 | 46,740 | 72,500 |
| Total Revenue (mil US$) | 2,915 | 3,144 | 3,289 | 3,419 | 3,898 | 5,144 | 6,842 | 11,486 |
| C4 | 18.3 | 18.8 | 17.6 | 17.2 | 19.9 | 27.6 | 39.2 | 45.9 |
| HHI | 113 | 114 | 94 | 92 | 126 | 261 | 510 | 762 |
| *N* (>1%) | 8 | 9 | 7 | 9 | 8 | 9 | 9 | 13 |
| Noam Index | 40 | 38 | 35 | 31 | 44 | 87 | 170 | 214 |

*Source*: 2012 updates provided by editors, using Chinadaily.com.cn; * indicates estimate.

of Higher Education Press and People's Education Press, as well as another smaller, state-owned publishing house and two other publishing companies. Although before 2011, the two educational publishing houses were separate corporate entities, the Ministry of Education ultimately controlled both entities. The Higher Education Press was traditionally the publishing arm of the Chinese Ministry of Education, and People's Education Press was exclusively authorized by the Ministry of Education to design the curriculum for the government's Nine-Year Compulsory Education plans.

Other large publishing houses include Beijing Publishing House and Shanghai Century Publishing. They are all majority owned by the respective local governments. As a case in point, Jiangsu Phoenix Publishing & Media is 72.1% owned by the Jiangsu provincial government through a state holding company. Other significant shareholders of the company include Hony Capital, with a 5.9% stake, and the National Social Security Fund (NSSF), with a 2.0% stake. Hony Capital, founded in 2003, is a leading China-focused private equity firm controlled by Legend Holdings, a holding company of the Chinese Academy of Sciences. The National Council for the Social Security Fund, an organization subordinate to the State Council (Cabinet), manages the NSSF, established in 2000. It is important to note that although private capital is still entering the sector, private investors only hold minority stakes in the larger publishing groups.

## Magazine Publishing

China's periodical industry has undergone a tremendous transformation since the 1970s.[8] Unlike the newspaper industry, which is tightly controlled by the government, magazines have attracted significant private sector investment and a number of independent operators. Compared with book publishing, which has been well established since the early 20th century, magazine publishing only started in the 1950s—only really coming into its own in 1970s—and except for literary and current affairs magazines, much of its content follows the patterns set by Western magazines. The dramatic expansion of the industry is reflected in advertising growth: total advertisement sales were valued at US$446 million (RMB3.1 billion) in 2008, a 286% increase from the total in 1978. Many foreign titles have entered the market via joint ventures or cooperative publishing in the past three decades. After China joined the World Trade Organization (WTO) in 2001, Chinese magazine publishers faced greater international competition. However, due to language barriers and economic and ideological protectionism, demand for Western media has been channeled into cooperative publishing efforts between Chinese and foreign media groups (Table 24.4).

If all publishers with public affiliations were defined as a single state-run entity, the HHI would be well over 2,500, though not as close to 10,000, as the newspapers or book publishing markets would be. This is due to the existence of publishing companies managed by non-publicly affiliated industry associations, such as Rayli and Caijing, and privately owned publishers such as Caixin Media, established in 2010 by Hu Shuli, the founder and former editor-in-chief of *Caijing*. The increase in HHI measured since 2003 is mainly due to the ascendance of consumer titles and the network effect created by certain media groups publishing Chinese editions of international titles.

Trends Media Group, the largest magazine publisher in China, is a case in point. Established in 1993, it was controlled by the National Tourism Administration, a government agency under the State Council. It now publishes 17 magazines, over half of which are Chinese editions of foreign titles such as *Cosmopolitan, Esquire, Good Housekeeping,* and *Harper's Bazaar,* all printed in Mandarin through licensing cooperation with Hearst (US) and IDG (US). Trends

8. Huang, Wanwu. "Market Economy and Transformation of Periodical Industry in China." *Chinese Journal of Scientific and Technical Periodicals* 4: 3, 1993. 27–28.

## Table 24-4. Magazine Publishing (Market Shares by Circulation), 2003–2012[1]

| | 2003 | 2004 | 2005 | 2006 | 2007 | 2012 |
|---|---|---|---|---|---|---|
| Trends Media Group (National Tourism Association) (public) | 8.3 | 11.1 | 11.2 | 15.5 | 15.3 | 16.1* |
| Rayli (China National Light Industry Council) | 2.6 | 3.8 | 5.0 | 7.5 | 7.9 | 23.6 |
| Elle (Shanghai Century Publishing Group, under license from Hearst, US) (public) | 4.3 | 4.2 | 4.5 | 5.2 | 5.7 | 10.7 |
| Vogue (China Pictorial, under license from Condé Nast, US) (China International Publishing Group) (public) | 0.0 | 0.0 | N/A | 1.9 | 4.1 | 7.8 |
| DuZhe (Gansu People's Press) (public) | 5.6 | 5.3 | 5.2 | 4.6 | 4.8 | 2.7 |
| China Science Publishing and Media (Chinese Academy of Sciences) (public) | 3.9 | 4.1 | 4.0 | 4.2 | 4.4 | 4.6* |
| Ban Yue Tan (Xinhua News Agency) (public) | 5.3 | 4.9 | 4.3 | 3.9 | 3.5 | 2.6 |
| Marie Claire (China Sports Publications, under license from Hearst US, prev. Lagardere) (General Administration of Sports) (public) | 1.0 | 1.2 | 1.3 | 1.5 | 2.0 | 4.5 |
| Caijing (Stock Exchange Executive Council) | 1.2 | 1.4 | 1.5 | 1.9 | 1.8 | 0.4 |
| Fortune (CCI Asia-Pacific, Hong Kong) | 2.2 | 1.8 | 2.4 | 1.6 | 1.9 | 0.1 |
| Gu Shi Hui (Shanghai Literature and Arts Publishing) (public) | 1.8 | 1.6 | 1.7 | 1.3 | 1.2 | 1.8 |
| Zhiyin | 1.3 | 1.4 | 1.3 | 1.2 | 1.2 | 2.2 |
| Sanlian Life Weekly (China Publishing Group) (public) | 1.0 | 1.2 | 1.2 | 1.4 | 1.3 | 1.8* |
| Car Magazine Press | 0.6 | 0.6 | 0.7 | 0.9 | 1.2 | 1.7* |
| Shanghai Fashion | 1.3 | 1.1 | | | | |
| World Executive's Digest (Global Sources, Hong Kong) | | 1.2 | 1.3 | | | |
| CinMedia (Cinderella Media Group, Hong Kong) | | | | | 2.2 | 2.2* |
| Others | 59.6 | 55.1 | 54.4 | 47.4 | 41.5 | 16.8 |
| Total Revenue (mil RMB) | 3,800 | 4,900 | 5,500 | 5,900 | 10,900 | 34,860 |
| Total Revenue (mil US$) | 459 | 592 | 672 | 740 | 1,433 | 5,522 |
| C4 | 23.5 | 25.5 | 25.9 | 32.8 | 33.7 | 58.2 |
| HHI | 184 | 240 | 250 | 395 | 422 | 1,066 |
| *N* (>1%) | 13 | 14 | 13 | 13 | 15 | 13 |
| Noam Index | 51.0 | 64.0 | 69.0 | 110 | 109 | 296 |

*Source*: 2012 updates provided by editors.

[1] Market Shares and Index calculations by author based on assumption that publically owned organization are segmented in ownership.

Media Group also has licensing agreements with National Geographic, Condé Nast, Rodale, and EMAP. *Elle China*, the first international fashion magazine launched in China (in 1988), was published by Shanghai Translation Publishing, which is now a subsidiary of Shanghai Century Publishing, under license, originally from the French publisher Lagardere. In 2011, Hearst purchased *Elle China* from Lagardere. The transaction was part of a global deal that saw Hearst acquiring most of Largardere's foreign titles in 14 countries.[9]

Non-fashion-related magazines, and domestically published magazines in general, typically fall short of attracting big advertising spending compared to the international glossies like *Elle China*. Thus, titles such as *Du Zhe*, *Ban Yue Tan*, and *Gu Shi Hui*, which are more popular among younger or older audiences, are disproportionally reflected in Table 24.4 in terms of readership while their market shares by revenue are actually much lower.

Foreign titles exist within a tightly (though not consistently) enforced regulatory environment. With the exception of Bloomberg's (US) *Businessweek*, which entered China in the 1980s, no other foreign titles are allowed to publish Chinese editions in Mainland China. There are two ways for foreign titles to get around this restriction: (1) registering in Hong Kong or Macao to enter the Mainland market through subscription-only services or (2) seeking local licensing partners, which has been the approach adopted by most fashion and entertainment titles. There is some restriction as to how much of the content can be translated directly from the original foreign version: up to 50% for fashion-related magazines, and up to 60% for technology- and science-related ones. In most cases, the foreign publishers license out use of the title. But they are generally restricted from exerting editorial influence or making capital investments.

## AUDIOVISUAL MEDIA

### Radio

National organizations such as China National Radio (CNR) and China Radio International (CRI) have traditionally dominated the market. During the past two decades, more and more publicly owned stations have sprung up at the provincial, municipal, and county levels. Since 2000, local radio broadcasting markets have become increasingly competitive, and local stations outperform national stations in many of the local markets. Beijing, Shanghai, and Guangzhou are urban markets where local stations ratings are the strongest (Table 24.5).[10]

Most non-national stations are only active in local and, in some cases, neighboring regions. For instance, Shanghai Oriental Radio is widely consumed in the Yangzi River Delta region that surrounds Shanghai. This low level of concentration is likely to continue as new digital technology makes it possible for radio stations to meet the demand of niche markets, thus channeling audience and advertising revenues into smaller, local stations. This contributed to the increased popularity of local stations, which integrate real-time local traffic information into their programming.

Regarding ownership, national players CNR and CRI are both directly subordinate to the State Administration of Radio, Film, and Television (SARFT). Stations at the local level are subordinate to the respective local administrations of Radio, Film, and Television.

China National Radio (CNR), the traditional national broadcasting station, was founded in 1940 by the CPC. Formerly known as the Yan'an Xinhua Broadcasting Station, the radio station assumed its current name in 1949, shortly after the founding of the People's Republic. It expanded into six channels in 1984,

9. Hearst Corporation. "Hearst Corporation Completes the Acquisition of the Majority of Hachette China Operations." Hearst.com, Dec. 6, 2011. Feb 28, 2013. <http://www.hearst.com/press-room/pr-20111206a.php>.

10. Lv, Jie. "Rise of China's Local Radio Media." *China Newspaper Industry* 14: 24, 2011. 70–71.

Table 24-5. Radio Group (Market Shares by Revenue), 1992–2008

| | 1992 | 1996 | 1999 | 2004 | 2008 |
|---|---|---|---|---|---|
| China Radio International (public) | Radio Peking | 6.0 | 9.7 | 11.9 | 9.3 |
| Radio Peking | 3.5 | China Radio International | | | |
| China National Radio (public) | 13.1 | 6.5 | 7.0 | 5.9 | 4.6 |
| Southern Media Corporation (Radio Guangdong) (public) | 8.0 | 4.9 | 5.4 | 6.3 | 4.4 |
| Shanghai Media Group (Shanghai Media & Entertainment Group) (public) | 6.5 | 8.8 | 8.9 | 5.8 | 4.6 |
| Radio Shanghai | | | | | |
| Shanghai Oriental Radio | | | | | |
| Radio Tianjin (public) | 1.5 | | 2.4 | | 4.0 |
| Radio Jiangsu (public) | 2.5 | 1.8 | | 3.3 | 3.3 |
| Radio Shandong (public) | 2.0 | | 2.6 | 3.2 | 2.4 |
| Radio Zhejiang (public) | | | 3.2 | 4.3 | 2.2 |
| Radio Liaoning (public) | | | | 2.9 | 2.1 |
| Radio Foshan (public) | 3.0 | 1.4 | | | |
| Radio Shenzhen (public) | 2.5 | 2.4 | | 5.5 | |
| Radio Guangzhou (public) | 4.0 | 2.4 | | | |
| Radio Nanjing (public) | | 2.0 | | | |
| Radio Heilongjiang (public) | | | 2.6 | | 1.9 |
| Radio Qingdao (public) | | | 2.3 | | |
| Radio Henan (public) | | | | 3.1 | |
| Others | 53.3 | 63.8 | 56.0 | 47.8 | 61.3 |
| Total Revenue (mil RMB) | 199 | 873 | 1,252 | 3,123 | 7,257 |
| Total Revenue (mil US$) | 37 | 105 | 151 | 377 | 1,063 |
| C4 | 31.7 | 26.2 | 30.9 | 29.9 | 22.9 |
| HHI | 335 | 201 | 284 | 338 | 194 |
| *N* (>1%) | 10 | 9 | 9 | 10 | 10 |
| Noam Index | 106 | 67.0 | 95.0 | 107 | 61.0 |

and currently broadcasts 10 channels. CRI was also founded one year after CNR in 1941. Its English-language service was first launched in 1947.[11] With its more than 30 overseas bureaus, it currently has the most comprehensive foreign media service in East Asia. Despite intensified competition in the market, CNR and CRI still have the largest overall market shares. Among the top 10 stations with the largest number of listeners in cities across the country, CNR's *Sounds of China, Music Radio,* and *Business Radio* are, respectively, the first, second, and third most popular, collectively accounting for 10% of the total national market share. Chinese radio broadcasting firms began to integrate digital audio broadcasting (DAB)

11. China Radio International. "History % Milestones: CRI English Service." cri.cn, 2004. Feb. 28, 2013. <http://english.cri.cn/about/history.htm>.

with their traditional broadcasting methods, following SARFT's China Multimedia Mobile Broadcasting (CMMB) standard.[12] Advertising in the radio market began in the 1980s when stations were permitted to transmit advertisements and receive advertising fees. By the 1990s, due to the defunding of state subsidies for media, radio stations had come to rely more heavily on advertising fees. By 2000, national radio advertising revenue reached US$181 million (RMB1.5 billion), increasing to US$364 million (RMB3 billion) in 2004, and surpassing US$1 billion (RMB7 billion) in 2008.[13]

## Broadcast Television

Terrestrial broadcasting in China is unique in that it is primarily delivered via cable. Unlike in many Western countries, where private entities played major roles in the introduction of multichannel alternatives to terrestrial platforms, in the People's Republic, the government rolled out a national cable system in the 1990s. As a result, cable is the predominant transmission medium, and often the only delivery option in urban centers. As of 2010, there were 188 million households with cable access, up from 100 million in 2000.[14, 15] Though 30% of all households with a television set still rely on traditional terrestrial TV, advertising revenue is concentrated on cable platforms due to its greater penetration of urban markets over rural markets, where on average only five channels can be accessed. Since the 1990s, cable-delivery TV has provided its viewers with 20 to 40 channels, among them 13 CCTV channels, 15 provincial channels, and 5 local channels.

In China, the public broadcasters took over the newly created cable TV system when it was introduced, merely moving from over-the-air delivery to cable delivery. The pick-up rate of pay-TV channels remains low due to the wide range of content available on the "free" channels for which viewers pay utility fees for access. For these reasons, we take into account only the "free" channels in aggregating the broadcast television market (Table 24.6).

There were 354 TV stations in China in 2001, broadcasting a total of 1,206 channels. In 2007, the number of stations consolidated to 296, though the total number of channels increased to 2,984.[16] Concentration has been declining since 2000, largely owing to the growth of several provincial stations, such as Shanghai Oriental TV and Hunan TV.

Launched in 1958, CCTV remains the most popular audiovisual outlet in China, and it is expanding its international presence (with the launch of CCTV America in 2012). CCTV's household penetration rate stood at 96% at the end of 2008. One of its most important programs, *The CCTV New Year's Gala*, a Chinese Lunar New Year's special, routinely attracts over 700 million viewers annually. In recent years, though, it has lost market share among younger demographics due to a perceived decline in program quality compared with new programming from provincial broadcasters. Market shares by revenue are shifting from CCTV to local stations, though CCTV's overall dominant position remains unchallenged. Since its founding, CCTV has been considered one of the "big three" media outlets in China, along with *People's Daily* and Xinhua News Agency. It is both owned and regulated by the SARFT, and its chief is subordinate to the minister in charge of the SARFT.

Shanghai Media Group (SMG) is the second largest broadcaster in the People's Republic.

---

12. Li, Zhanjun. "A Look at the Radio Industry in the New Media Era." *China Broadcasts* 23: 4, 2011. 76–77.

13. The State Administration of Radio, Film and Television (SARFT). "China Radio, Film, and TV Development." Beijing: SARFT: 2009.

14. Zhang, Wenjun. "A New Definition of Wireless Broadcast Television in China." Yonsei University. BMSB 2012, Seoul, South Korea. June 28, 2012. Feb. 28, 2013. <http://www.ieee-bmsb2012.org/images/program/BMSB2012_Keynote_5_A_New_Definition_of_Wireless_Broadcasting_China_WenjunZhang.pdf>.

15. Buddle, Paul. "China's cable TV market in transition." BuddeBlog.com, July 6, 2012. Feb. 28, 2013. <http://www.buddeblog.com.au/frompaulsdesk/chinas-cable-tv-market-in-transition/>.

16. National Bureau of Statistics of China. *China Statistical Yearbook*. Beijing: National Bureau of Statistics of China: 2007.

## Table 24-6. TV Broadcasting (Market Shares by Revenue), 2001–2012[1]

| | 2001 | 2002 | 2003 | 2004 | 2005 | 2006 | 2007 | 2012 |
|---|---|---|---|---|---|---|---|---|
| CCTV (public) | 65.7 | 64.4 | 53.4 | 61.3 | 62.4 | 60.7 | 60.4 | 59.1 |
| Shanghai Media Group (Shanghai Media & Entertainment Group) (public) | 18.8 | 19.5 | 21.0 | 20.3 | 17.2 | 19.2 | 18.3 | 12.7 |
| Hunan Media Group (public) | 3.8 | 6.7 | 8.2 | 11.1 | 9.8 | 9.1 | 10.3 | 8.1 |
| Jiangsu Media Group (public) | 1.2 | 1.4 | 1.5 | 1.7 | 1.9 | 1.6 | 1.4 | 6.3 |
| Zhejiang Media Group (public) | 0.8 | 1.0 | 0.8 | 1.3 | 0.6 | 0.3 | 0.4 | 0.9 |
| Southern Media Corporation (public) | 0.3 | 0.3 | 0.4 | 0.3 | 0.5 | 0.4 | 0.6 | 1.1 |
| Others | 9.4 | 6.8 | 14.8 | 4.1 | 7.7 | 8.8 | 8.6 | 11.8 |
| Total Revenue (bil RMB) | 19.4 | 21.0 | 30.0 | 58.4 | 78.1 | 110 | 161 | 206 |
| Total Revenue (bil US$) | 2.3 | 2.5 | 3.6 | 7.1 | 9.5 | 13.8 | 21.2 | 33 |
| C4 | 89.5 | 92.0 | 82.6 | 92.7 | 91.3 | 91.6 | 90.4 | 86.2 |
| HHI | 4,899 | 4,832 | 4,796 | 4,837 | 4,269 | 3,964 | 3,876 | 3,761 |
| *N* (> 1%) | 4 | 4 | 4 | 5 | 4 | 4 | 4 | 5 |
| Noam Index | 2,450 | 2,416 | 2,398 | 2,163 | 2,135 | 1,982 | 1,938 | 1,682 |

*Source*: 2012 data supplied by editors.

[1] Market Shares and Index calculations by author based on assumption that publically owned organization are segmented in ownership.

Currently a subsidiary of Shanghai Media & Entertainment Group (SMEG), a local conglomerate owned by the Shanghai municipal government, SMG was formed in 2001 through the merger of Shanghai Television Station, Oriental Television Station, the People's Radio Station of Shanghai, and Oriental Radio Shanghai. Although SMEG's nominal owner, Shanghai Municipal Culture, Radio, Film, and Television Administration, ultimately reports to the SARFT, SMEG retains a considerable level of operational autonomy.[17] SMG broadcasts 11 traditional analog channels, as well as 90 paid TV digital cable TV channels, an IPTV service, and 29 radio channels. The group also operates and owns five sports centers and 14 cultural arts centers, which are connected to its sports media and entertainment interests. With its widely diversified media business, SMG has been attracting high advertisement revenue in recent years. SMG has traditionally been geared toward local audience in the Shanghai metropolitan area, but it is now also trying to better cater to national audiences.

Trailing SMG are other provincial media groups, such as Hunan Media Group and Jiangsu Media Group. In the past 10 years, the flagship channels of these stations have become increasingly popular due to their development of new entertainment programming, including reality TV, drama series, and dating shows. For example, Hunan Media Group has focused on its "Happy China" brand, with a wide range of hugely popular entertainment shows and talent contests.

CCTV has traditionally dominated the advertising market.[18] Its annual ad auction is considered a major brand promotion

17. Xie, Liya. "New Business Model of Shanghai Media Group." *Ceocio China* 26: 5, 2010. 41–44.
18. Zheng, Baoyu. "Business Model for China's TV Broadcasting Industry." *Modern Audio-Video Arts* 13: 1, 2009. 119.

opportunity for foreign and domestic companies alike: in 2011, a total of US$2 billion (RMB12.7 billion) was committed toward CCTV ads, and this increased to US$2.3 (RMB14.3 billion) in 2012.[19] Provincial TV companies, in contrast, are attracting an ever-larger proportion of total advertising revenue in the past five years, thanks to their lower base and popular programming on their flagship channels.

The government has set out to fully transition the television sector from analog to digital by 2015. However, given the predominance of cable delivery in the country, this for the most part means rolling out a digital cable TV system nationwide, except for far-flung rural areas. This will allow most households to overcome the 40-channel limit in the current cable system and make premium content more accessible. Along with the platform conversion, several programming providers, most notably CCTV and SMEG, will offer more paid TV channels. In 2011, the platform conversion rate stood at 47% for the 188 million cable-accessible households, but the pick-up rate of paid TV packages was much lower.[20] This is because programming available on "free" channels is already wide-ranging and the demand for additional specialized content (such as sports) is low.

To most households in Mainland China, channels from the special administrative region of Hong Kong, Taiwan, and other countries are generally only available in places with a substantial number of foreign expatriates (such as luxury hotels), or otherwise received through illegal or exempt-under-special-circumstance satellite dishes. In the province of Guangdong, a few Hong Kong–based channels are accessible to the public. One trend affecting the TV market in recent years is the spread of Internet-based usage of TV programs. The development of webcasting makes it possible for audiences to view TV programs online, which propels the current trend for TV broadcasters to move their content onto their websites. In the meantime, IPTV adoption is not low in international comparison, forecasted to reach 8% in 2015.[21]

## Film

The film industry is arguably the most dynamic sector of the national Chinese media market. Before the 1990s, film companies were all state-owned studios, of which Beijing Film Studio and Shanghai Film Studio were the largest. At the turn of the century, these traditional film studios were all transformed and merged into larger film groups, and many privately owned studios entered the market. The current major players are the state-owned China Film Group and Huaxia Film Distribution, and the privately owned Huayi Brothers and Poly Bona Film studios.[22] On the production side, 70% of content making is completed or participated in by private production companies. On the distribution side, 90% of the approximately 300 distributors in operation are privately owned. But the market is still relatively concentrated, with the top 10 enterprises holding over 80% of the combined market share.[23] Publicly owned companies took up over half of the overall share (Table 24.7).

Despite box-offices revenue surpassing that of Japan in 2012, making China the second largest film market in the world after India, the

19. Burkitt, Laurie. "CCTV's Ad-Sales Auction Draws Record Bids." *The Wall Street Journal*, Nov. 9, 2011. Feb. 28, 2013. <http://online.wsj.com/article/SB10001424052970204190704577025770417699372.html>.

20. "China to become Asia's Top Pay-TV Economy in 2011 with $7.5bn in Revenue." *VOD Professional*, June 22, 2011. Feb. 28, 2013. <http://www.vodprofessional.com/research/china-to-become-asia%E2%80%99s-top-pay-tv-economy-in-2011-with-$75bn-in-revenue/>.

21. Zhang, Wenjun. "A New Definition of Wireless Broadcast Television in China." Yonsei University. BMSB 2012, Seoul, South Korea. June 28, 2012. Feb. 28, 2013. <http://www.ieee-bmsb2012.org/images/program/BMSB2012_Keynote_5_A_New_Definition_of_Wireless_Broadcasting_China_WenjunZhang.pdf>.

22. Wei, Ting. "Solutions to Marketing Issues in China's Film Industry." *Inquiry into Economic Issues* 9: 6, 2008. 101–105.

23. "Research and Markets: China's Film Industry Report, 2011 Confirms China As the Fastest Growing Region In the Global IMAX Business." *BusinessWire*, Sept. 29, 2011. Feb. 28, 2013. <http://www.businesswire.com/news/home/20110929005867/en/Research-Markets-Chinas-Film-Industry-Report-2011>.

## Table 24-7. Film Production/Distribution (Market Shares by Box Office %), 1992–2012

| | 1992 | 2002 | 2003 | 2004 | 2005 | 2008 | 2010 | 2012 |
|---|---|---|---|---|---|---|---|---|
| China Film Group (public) | | 4.0 | 12.0 | 18.0 | 33.0 | 45.0 | 36.2 | 38.2 |
| Beijing Film Studio | 21.0 | China Film Group | | | | | | |
| Huaxia Film Distribution (public) | | | 20.0 | 9.0 | 14.0 | 12.7 | 21.1 | 23.5 |
| Huayi Brothers | 4.1 | 17.6 | 17.8 | 23.0 | 30.8 | 16.0 | 8.6 | 10.2 |
| Poly Bona Film | | | | | | 9.8 | 4.1 | 4.5 |
| Shanghai Film Group (Shanghai Media & Entertainment Group) (public) | 13.9 | | | | | | 2.8 | 2.0* |
| Media Asia | | | 21.0 | 9.0 | 12.3 | | | 3.4 |
| Beijing New Picture Company | | 59.0 | 18.0 | 26.0 | | | | 8.4* |
| BFA Youth Film Corporation | | 8.5 | | | | | | 2.1* |
| Century Hero Film | | 4.0 | | | | | | 1.0* |
| Ningxia Film Studio (public) | | | | | | 5.3 | | 1.3* |
| Changchun Film Studio (public) | 14.0 | | | | | | | 3.4* |
| Emei Film Studio (public) | 8.0 | | | | | | | 1.9* |
| Others | 39.0 | 6.9 | 11.2 | 15.0 | 9.9 | 27.2 | 27.2 | 27.2* |
| Total Revenue (mil RMB) | 17.0 | 141 | 271 | 520 | 783 | 2,110 | 7,260 | |
| Total Revenue (mil US$) | 3.0 | 17.0 | 33.0 | 64.0 | 96.0 | 304 | 1,073 | |
| C4 | 56.9 | 89.1 | 76.8 | 76.0 | 90.1 | 72.8 | 70.0 | 76.5 |
| HHI | 911 | 3,895 | 1,626 | 1,691 | 2,385 | 2,310 | 1,854 | 2,248 |
| HHI (public) | 1,306 | 3,879 | 2,106 | 2,015 | 3,309 | 4,065 | 3,703 | |
| *N* (>1%) | 5 | 5 | 5 | 5 | 4 | 5 | 5 | 5 |
| Noam Index | 407 | 1,742 | 727 | 756 | 1,192 | 1,033 | 829 | 1,005 |

*Source*: 2012 updates provided by editors;* indicates estimate.

number of films produced each year in China is still a fraction of that in India or the United States.[24] Thus, performance of a handful of films determines the market shares for each year. In terms of overall concentration level, the industry is highly concentrated due to China Film Group and Huaxia Film Distribution capturing around half of the market share (the other public entities operate at the provincial level). Despite the increasing successes of private filmmakers and the loosening of regulation in recent years, concentration is still high.

China Film Group (CFG) is the largest and most influential film enterprise in China. Its predecessor, China Film Corporation, was established in 1949. Subsequently renamed China Film Studio, by 1999 it had become a conglomerate involved in almost all aspects of the film business. Owned and supervised by the SARFT, it serves as the "gatekeeper" of China's film industry, with roles in film importation and exportation and management of international co-productions, in addition to its primary production and distribution functions.

24. "Production: World Cinema Stats (1)." *Screenville*, June 15, 2009. Feb. 28, 2013. <http://screenville.blogspot.com/2009/06/production-world-cinema-stats-1.html>.

Huaxia Film Distribution was established in 2003, with the aim of breaking the decade-long monopoly foreign film distribution by CFG. It is 20% owned by China Broadcasting Networks, a holding company of SARFT. Other major shareholders include CFG (itself), Shanghai Film Studio, and Changchun Film Studio, each with a 10% stake.[25] Huayi Brothers Media, founded in 1994, is the country's most rapidly expanding, privately owned entertainment company. Aside from its main film production and distribution business, it has become one of the biggest TV production companies in China. Its main shareholders are its two founders, Wang Zhongjun and Wang Zhonglei, with 26% and 8% ownership stakes, respectively; other major shareholders include Ma Yun (Jack Ma), owner of China's biggest e-commerce group, Alibaba, with a 10% stake, and Tencent Holdings, one of China's top online media groups, with a 5% stake.[26] In 2011, Huayi Brothers announced its plans to build a studio complex on the outskirts of Shanghai, East Asia's largest.

One issue that has garnered increased attention is foreign film importation into China, particularly films from the US Hollywood majors. In the first half of 2012, box-office revenue of foreign films reached US$852 million (RMB5.4 billion), accounting for two-thirds of the total box-office receipts in China up to that point.[27] This was achieved despite the strict rules imposed on foreign film importation, as CFG was, for a long time, the only licensed distributor of foreign films (though the establishment of Huaxia Film Distribution merely created a duopoly). However, films co-produced by Chinese and foreign studios are treated as domestic films and do not count toward import quotas.[28] This has been a major factor propelling industry incumbents in Mainland China and overseas to pursue co-productions and joint ventures with foreign studios in recent years.

This restrictive import model faced growing international pressure, particularly from US film producers who seek greater access to the Chinese market. In 2007, the United States initiated a WTO dispute against China, arguing that the Chinese film industry should no longer be protected as "developing," but had in fact reached the size of many other "developed" markets. The WTO ruled in 2009 that a number of Chinese film import restrictions were inconsistent with WTO obligations, and to comply with WTO regulations China had to lower its import restrictions in 2012. First, the import quota for films under a revenue-sharing model was raised to 34 per annum (up from 20). The 14 extra slots are required to be used for "enhanced" productions, namely IMAX, 3D, and animated films. This move is likely to lead to increased concentration in the market, given the duopoly on foreign film distribution and the fact that domestic studios lack the capacity to produce this many enhanced films. Second, the revenue share that foreign rights owners' are to receive will increase from the original 13–17% to 25%.[29] This measure is also likely to diminish the market power of the two licensed national film importers and create a more competitive market environment for both private Chinese players and foreign stakeholders.

To overcome some of these restrictions, a whole slew of cross-border corporate activities have taken place since 2009. In 2012, News Corporation acquired a 19.9% stake in the Bona Film Group in order to gain a stronger foothold

25. "Film sector monopoly broken up." *China Daily*, June 10, 2003. Feb. 28, 2013. <http://www.chinadaily.com.cn/en/doc/2003-2006/10/content_244133.htm>.

26. Hexun. "Huayi Brother Stock Distribution Analysis." Hexun.com, 2009. July 26, 2012. <http://stockdata.stock.hexun.com/2009_sdgd_300027.shtml>.

27. Burkitt, Laurie. "Foreign Films Dominate China Box Office." *The Wall Street Journal*, July 19, 2012. Feb. 28, 2013. <http://online.wsj.com/article/SB10000872396390444097904577536820360717562.html>.

28. Italian Trade Commission. "Overview of China Film Market." Shanghai: Italian Trade Commission, June 14, 2011. Feb. 28, 2013. <http://www.ice.gov.it/paesi/asia/cina/upload/174/Final%20Report%20%20Market%20Report%20of%20China%20Film%20Market%202011.pdf>.

29. Waxman, Sharon. "White House Gets China to Open Market to U.S. Movies." *Reuters*, Feb. 18, 2012. July 18, 2012. <http://www.reuters.com/article/2012/02/18/idUS420251887620120218>.

Table 24-8. Film Exhibition (Market Shares by Box Office %), 2003–2010

| | 2003 | 2004 | 2005 | 2008 | 2010[1] |
|---|---|---|---|---|---|
| Wanda Cinema Circuit | | | | 11.8 | 16.0 |
| China Film Stellar Chain | 9.0 | 8.0 | 9.0 | 12.4 | 13.8 |
| Shanghai Allied Circuit | 15.3 | 13.0 | 12.0 | 11.3 | 12.3 |
| China Film South Cinema Circuit | | | 6.5 | 9.7 | 10.9 |
| Beijing New Video Circuit | 9.0 | 9.0 | 10.5 | | 9.1 |
| Guangdong North Cinema Circuit | | 5.0 | | | |
| Zhejiang Cinema Circuit | 6.0 | | | | |
| Others | 60.7 | 65.0 | 62.0 | 54.8 | 37.9 |
| Total Revenue (mil RMB) | 950 | 1,580 | 2,000 | 4,300 | 10,250 |
| Total Revenue (mil US$) | 117 | 195 | 245 | 630 | 1,514 |
| C4 | 39.3 | 35.0 | 38.0 | 45.2 | 53.0 |
| HHI | 432 | 339 | 378 | 515 | 799 |
| *N* (>1%) | 4 | 4 | 4 | 4 | 5 |
| Noam Index | 216 | 170 | 189 | 257 | 357 |

[1] Italian Trade Commission. "Overview of China Film Market." Shanghai: Italian Trade Commission, June 14, 2011. Feb. 28, 2013. <http://www.ice.gov.it/paesi/asia/cina/upload/174/Final%20Report%20%20Market%20Report%20of%20China%20Film%20Market%202011.pdf>.

in China. DreamWorks Animation (DWA) SKG (US) announced a plan to establish a studio in Shanghai, alongside three Chinese firms: China Media Capital, a state-backed private equity firm, Shanghai Media Group, and Shanghai Alliance Investment. The new studio, named Oriental DreamWorks, will be 45% owned by DWA and 55% by the Chinese firms, and is slated to open in 2016. Walt Disney (US) has announced its plan to establish a joint venture with Tencent Holdings and the animation arm of China's Ministry of Culture, aiming to revive the market for animated films.

In the other direction, Chinese film companies have started to establish relationships with their foreign counterparts, seeking to increase their distribution channels overseas. In 2011, Bona Film Group and Huayi Brothers each acquired a 20% stake in China Lion Film Distribution (US), which has an exclusive partnership with the US theater chain AMC to introduce Chinese films to the North America market.[30] On the exhibition side, Wanda Group acquired AMC Entertainment Holdings in 2012 in a high-profile US$2.6 billion (RMB16.4 billion) takeover, creating in effect the world's largest cinema chain (Table 24.8).[31]

In 2007 there were 34 cinema chains in China.[32] Compared to the moderately to highly concentrated distribution market, the exhibition market is rather competitive, with HHI staying under 1,000 in the period surveyed.

## TELECOMMUNICATIONS MEDIA

### Wireline Telecom

Deemed a "strategically important" industry, the telecommunications industry has been carefully managed by the government for the past

30. "Chinese Studios Huayi Brothers, Bona take Stakes in U.S.-Based China Lion." *The Hollywood Reporter.* Oct. 31, 2011. July 18, 2012. <www.hollywoodreporter.com/news/chinese-studios-huayi-brothers-bona-255242>.

31. Mider, Zachary. " China's Wanda to Buy AMC Cinema Chain for $2.6 Billion." *Bloomberg,* May 21, 2011. July 18, 2012. <http://www.bloomberg.com/news/2012-2005-21/china-s-wanda-group-to-buy-amc-cinema-chain-for-2-6-billion.html>.

32. China Film Press. *China Film Yearbook.* Beijing: China Film Press, 2007.

20 years. The three largest telecommunications providers are all majority-owned by the state: China Telecom (83%), China Mobile (74%), and China Unicom (58%).

China Telecom came into existence in 1998 with the split of the China Postal Telecommunication Co. into China Post and China Telecom. China Unicom traces its origin to 1994, when the State Council (Cabinet) founded China United Telecommunications with a focus on code division multiple access (CDMA) technology, and China Mobile originated in 1994 when China Postal Telecommunication Co. launched a mobile communication office. In 1999, the Ministry of Information Industry (the MIIT's predecessor) reorganized the sector: the original China Telecom was split into China Telecom, China Mobile, and ChinaSat; its paging services were merged into China Unicom.

In 2001, the sector was reorganized again: China Telecom was divided into northern and southern sections; together with China Netcom, China Mobile, China Unicom, China Railcom, and ChinaSat, a "5 + 1" structure took shape. The reorganized China Telecommunication Co. consisted of the original 21 southern regions of China Telecom, and primarily managed the fixed network services of the region. The reorganized China Netcom consisted of the original 10 northern regions of China Telecom and the original China Netcom's service areas. China Jitong mainly oversaw the fixed network services of the northern region. China Mobile focused on GSM mobile service. China Unicom operated both CDMA and GSM mobile services. China Railcom concentrated on railway communication service, and China Satcom worked on satellite communication. There was little overlap between the firms.

In 2008, the MIIT reorganized the industry again, combining all of the aforementioned entities into just three: China Telecom, China Unicom, and China Mobile (Table 24.9).

## Wireless Telecom

Following the MIIT's 2008 reorganization of the telecommunications industry, China Telecom took over Unicom's CDMA mobile service, as well as China Satcom's satellite service. China Unicom acquired China Netcom and was issued the operating license of 3G CDMA2000 and 3G wideband–code division multiple access (W-CDMA) mobile services. China Railcom

Table 24-9. Wireline Telecom (Market Shares by Revenue), 2000–2012[1]

| | 2000 | 2004 | 2008 | 2009 | 2010 | 2011 | 2012 |
|---|---|---|---|---|---|---|---|
| China Mobile (public) | 0.0 | 0.0 | 5.1 | 7.1 | 7.7 | 10.7 | 11.1 |
| China Telecom (public) | 82.4 | 63.3 | 61.1 | 60.1 | 59.5 | 59.5 | 58.6 |
| China Unicom (public) | | | 33.8 | | | | |
| China Netcom | 17.2 | 32.6 | China Unicom | 32.8 | 32.8 | 29.8 | 30.3 |
| China Railcom | 0.0 | 4.2 | 0.0 | | | | |
| China Jitongcom | 0.4 | 0.0 | 0.0 | | | | |
| Total Revenue (bil RMB) | 201 | 170 | 158 | | | | 165 |
| Total Revenue (bil US$) | 24.3 | 20.5 | 23.1 | | | | 26.4 |
| C4 | 100.0 | 100.0 | 100.0 | 100.0 | 100.0 | 100.0 | 100.0 |
| HHI | 7,093 | 5,078 | 4,899 | 4,738 | 4,675 | 4,542 | 4,476 |
| *N* (>1%) | 2 | 3 | 3 | 3 | 3 | 3 | 3 |
| Noam Index | 4,095 | 2,932 | 2,829 | 2,735 | 2,699 | 2,622 | 2,584 |

*Source*: 2012 updates provided by editors.

[1] Market Shares and Index calculations by author based on assumption that publically owned organization are segmented in ownership.

was merged into China Mobile, aiming to help the latter research and develop time division–synchronous code division multiple access (TD-SCDMA) 3G technology, a standard being promoted by the People's Republic as an alternative to W-CDMA. The reorganization resulted in overlapping services among the three large operators, partly to promote competition in the market.[33]

The wireless market in China is almost the most concentrated market among all 30 countries surveyed for this book. Though not as concentrated as the wireline sector, the market structures of the two industries in China are similar: one player dominates (in wireline's case, China Telecom, and in wireless's case, China Mobile) and another player follows—in both cases China Unicom. For the wireless market, though, competition is increasing and market shares are evening out among the major players. China Mobile's share of new mobile subscribers is now down to 49%, the first time that it has fallen below 50%. China Telecom has been making strides in the mobile market since 2008, with revenue of its mobile segment exceeding its fixed broadband revenues by 2011 China Telecom and China Unicom were the two companies to sell the Apple iPhone in China, and as a result, China Mobile's average revenue per user (ARPU) has become less competitive, declining to US$11 (RMB71) in 2012.[34]

The ownership structure of China Telecom was formed in the early 2000s. In 2002, China Telecom went public, with the government retaining majority control (70.9%) through China Telecommunications Corporation (now fully owned by the SASAC). A 12% stake is split among several provincial governments' investment firms, including Guangdong Rising Assets Management, Zhejiang Financial Development Company (6.94%),[35] and Zhejiang Financial Development Company (2.64%). The remaining 17.15% of shares are free-floaters on the Hong Kong and New York Stock Exchanges.[36] In 2003, when SASAC was established, the assets and ownership of China Telecommunications Corporation, the government holding company, were transferred to the SASAC.

The state, through the SASAC, also majority owns China Mobile, the country's largest mobile operator, through a holding company, China Mobile Communications Corporation (CMCC). Listed on both the New York and Hong Kong Stock Exchanges, China Mobile is 74% owned by CMCC. Institutional and general investors own the remaining 26%.

The ownership structure of China Unicom was established in 2008–9, when China Unicom sold its CDMA business to China Telecom and merged the remainder of the company with China Netcom, a past state-owned operator in the ISP and wireline businesses. As of 2012, it is 57.97% owned by the SASAC.[37] Telefónica, the Spanish operator, was also one of the company's major shareholders, with 9.57% share. But this number decreased to 5% because of the Spanish operator's financial shortfalls in recent years.[38] General investors hold the remaining shares.

All three operators, though ultimately owned by the SASAC, compete fiercely with each other in the wireline, wireless, and ISP markets. This is purposefully established by the MIIT in order

---

33. I-Research. "China Online Search Annual Report." Beijing: I-Research, 2009.

34. Wang, Jane. "China Mobile's dominance eroded as competitors improve mobile performance." *Ovum*, Mar. 29, 2012. Feb. 28, 2013. <http://ovum.com/2012/03/29/china-mobiles-dominance-eroded-as-competitors-improve-mobile-performance/>.

35. Guangdong Rising Assets Management Co., Ltd is one of the biggest state-owned enterprises regulated directly by the Assets Supervision and Administration Commission of Guangdong Province. The total asset of Guangdong Rising Assets Management Co., Ltd is US$11.5 billion (RMB73.2 billion).

36. China Telecom Corporation Limited. "Share Information." China Telecom Corporation Limited, 2012. June 15, 2012. <http://www.chinatelecom-h.com/eng/ir/shareinfo.html>.

37. China Unicom (Hong Kong) Limited. "Company Profile." China Unicom (Hong Kong) Limited, 2012. July 17, 2012 at <http://www.chinaunicom.com.hk/en/aboutus/about_profile.html>.

38. Mark, Scott. "Telefonica to Sell Shares Back to China Unicom for $1.4 Billion." *The New York Times*, June 11, 2012. Feb. 28, 2013. <http://dealbook.nytimes.com/2012/06/11/telefonica-to-sell-stake-in-china-unicom-for-1-4-billion/>.

Table 24-10. Wireless Telecom (Market Shares by Revenue), 2000–2012[1]

| | 2000 | 2004 | 2008 | 2012 |
|---|---|---|---|---|
| China Mobile (public) | 81.5 | 71.2 | 80.1 | 63.9 |
| China Telecom (public) | 0.0 | 0.0 | 1.1 | 14.4 |
| China Unicom (public) | 18.3 | 28.4 | 18.7 | 21.5 |
| China Satcom | 0.2 | 0.4 | 0.0 | 0.2 |
| Total Revenue (bil RMB) | 140 | 215 | 349 | 772.37 |
| Total Revenue (bil US$) | 16.9 | 26.0 | 51.1 | 126.49 |
| C4 | 100.0 | 100.0 | 100.0 | 100.0 |
| HHI | 6,973 | 5,873 | 6,772 | 4,752 |
| *N* (>1%) | 2 | 2 | 3 | 3 |
| Noam Index | 4,026 | 4,153 | 3,910 | 2,744 |

*Source*: 2012 updates provided by editors, GSMA.

[1] Market Shares and Index calculations by author based on assumption that publically owned organization are segmented in ownership.

to foster competition. Despite state-orchestration, the series of reorganizations has provided telecom users in China with lower fees and broad access—an outcome mirrored those produced by market forces in Western countries (Table 24.10).[39]

## INTERNET MEDIA

### Internet Service Providers (ISP)

There are 10 primary backbone networks in China: CHINANET, CSTNET, CERNET, CHINAGBN, UNINET, CNCNET, CMNET, CIETNET, CGWNET, and CSNET. The access ISPs that compete in this market are the three primary telecom operators: China Telecom (CHINANET), China Mobile (CMNET), and China Unicom (UNINET) (Table 24.11).[40]

The significant increase in HHI between 2004 and 2008 is the result of the merger of China Netcom into China Unicom.

### Search Engines

Baidu became China's top search engine in 2006, displacing Yahoo! (US). Despite the hype surrounding Google, the US search company, it was never a dominate player or even came close in terms of actual market share. As the second largest search engine, Google (US) has a much smaller presence; the 32.3% market share it held in 2009 was half that of Baidu's. Aside from censorship issues, language barriers and poor localization are also major factors for Google's weak market position in China. Google's departure from China in 2010 due to declining market share and concerns about censorship did not make a huge impact from a market concentration point of view. The impact was also mitigated by the fact that Google is still accessible in Mainland China through google.hk.com (Table 24.12).

The overall concentration level, albeit high in absolute terms and still rising, is relatively low compared to other countries'. The Chinese market is one of the few major markets

39. "Rewired." *The Economist*, May 29, 2008. Feb. 28, 2013. <http://www.economist.com/node/11465595>.

40. China Internet Network Information Center. "Statistical Report on China's Internet Use and Current Status." Beijing: China Internet Network Information Center, 2006. 2012 updates provided by editors, Chinadail.com.cn, China Internet Watch.

## Table 24-11. Internet Service Providers (Market Shares by Revenue), 1998–2012[1]

| | 2004 | 2008 | 2012 |
|---|---|---|---|
| China Mobile (public) | 21.6 | 21.6 | 4.0 |
| China Telecom (public) | 45.2 | 53.8 | 51.4 |
| China Unicom (public) | 10.1 | 24.6 | 36.5 |
| China Netcom | 19.9 | | |
| China Railcom (Ministry of Railways) (public) | 3.0 | | 4.1 |
| China Satcom (public) | 0.3 | | 4.0 |
| Total Revenue (bil RMB) | 31.2 | 72.5 | 268.1 |
| Total Revenue (bil US$) | 3.8 | 10.6 | 42.9 |
| C4 | 96.8 | 100.0 | 100.0 |
| HHI | 3,016 | 3,963 | 4,023 |
| *N* (>1%) | 5 | 3 | 5 |
| Noam Index | 1,349 | 2,288 | 1,799 |

[1] Market Shares and Index calculations by author based on assumption that publically owned organization are segmented in ownership.

## Table 24-12. Internet Search Engines (Market Shares by Search Volume), 2003–2012

| | 2003 | 2004 | 2005 | 2006 | 2007 | 2008 | 2009 | 2010 | 2011 | 2012 |
|---|---|---|---|---|---|---|---|---|---|---|
| Baidu | 11.2 | 18.7 | 26.0 | 55.2 | 60.4 | 63.6 | 63.9 | 71.0 | 77.3 | 78.6 |
| Google (US) | 2.1 | 5.6 | 14.4 | 21.7 | 21.2 | 27.2 | 32.3 | 26.8 | 19.3 | 15.6 |
| Yahoo! (US) | 43.7 | 34.4 | 26.9 | 7.2 | 13.8 | 6.2 | 1.0 | | | |
| Sougou (Sohu) | 12.5 | 14.5 | 10.6 | 6.5 | 1.8 | 1.0 | 1.1 | 1.0 | 1.7 | 3.1 |
| Soso (Tencent Holdings) | | | | | | | | 0.6 | 1.3 | 1.4 |
| Aiwen (Sina) | 16.7 | 8.9 | 6.7 | 2.9 | 0.3 | 0.3 | 0.3 | | | |
| Others | 13.8 | 17.9 | 15.4 | 6.5 | 2.5 | 1.7 | 1.4 | 0.6 | 0.4 | 1.3 |
| Total Revenue (mil RMB) | 500 | 1,250 | 1,040 | 1,390 | 2,900 | 5,000 | 6,960 | 10,900 | 18,860 | 22,120 |
| Total Revenue (mil US$) | 60.0 | 151 | 127 | 174 | 381 | 733 | 1,019 | 1,610 | 2,908 | 3,508 |
| C4 | 86.7 | 85.5 | 77.9 | 90.6 | 97.2 | 98 | 98.3 | 99.4 | 99.6 | 98.7 |
| HHI | 2,470 | 1,823 | 3,326 | 3,612 | 4,291 | 4,824 | 5,129 | 5,641 | 6,273 | 6,433 |
| *N* (>1%) | 5 | 5 | 5 | 5 | 4 | 4 | 4 | 3 | 4 | 4 |
| Noam Index | 1,105 | 815 | 1,487 | 1,615 | 2,146 | 2,412 | 2,565 | 3,256 | 3,136 | 3,216 |

*Source*: 2012 updates provided by editors, Chinadaily.com.cn, China Internet Watch.

around the globe where Google does not have a monopoly.

Regarding ownership, there are few public operators. The only publicly owned search engine is Jike, created by the People's Daily Group in 2011. Currently, its market share is still negligible and is therefore not included in Table 24.12.

## Online News

China's online news market is dominated by four companies: Sina, Sohu, NetEase (163.com), and Tencent Holdings.[41] Sina was founded in 1998 and listed on NASDAQ in 2000. As a website that mainly caters to the global Chinese Diaspora, sina.com has about 230 million registered users and records three billion page views daily. Sohu, Sina's main competitor, was founded the same year as Sina and is also listed on NASDAQ; NetEase and Tencent were both founded in 1997. NetEase's email service has more than 200 million users.

Tencent originated from its instant messaging service, OICQ, and is the social networking platform with the largest online community, beating out both Weibo and Renren, China's Twitter and Facebook equivalents, respectively (Table 24.13).

All four leading companies are privately owned. Sina is owned by Western institutional investors. Together, the "Big Four" account for more than half of the total traffic of online news sites. Publicly owned Internet news sites are in most cases the online portals of the national public media groups, such as renmin.com (The People's Daily Group) and xinhua.net (Xinhua News Agency). Although more and more online news sites are being launched, the market has been rather concentrated in absolute terms and by international comparison, particularly with respect to market shares by advertising revenue.

Table 24-13. Online News Media (Market Shares by Monthly Unique Visitors), 2004–2009

| | 2004 | 2005 | 2006 | 2007 | 2008 | 2009 |
|---|---|---|---|---|---|---|
| Sina.com | 39.9 | 42.5 | 40.9 | 38.2 | 39.0 | 38.9 |
| Sohu.com | 33.2 | 33.0 | 31.3 | 31.4 | 27.5 | 32.4 |
| 163.com (NetEase) | 6.4 | 4.7 | 11.6 | 17.2 | 15.5 | 19.1 |
| Tencent.com (Tencent Holdings) | 19.0 | 14.2 | 12.3 | 8.5 | 9.5 | 7.8 |
| Tom.com | 0.8 | 0.8 | 0.8 | 0.9 | 1.3 | 0.6 |
| China.com | 0.3 | 0.2 | 0.4 | 0.2 | 0.3 | 0.1 |
| Ynet.com | 0.02 | 0.01 | 0.01 | 0.02 | 0.01 | 0.01 |
| 21CN | 0.04 | 0.02 | 0.02 | 0.03 | 0.02 | 0.1 |
| MSN.cn (MSN, Microsoft, US) | 0.1 | 0.1 | 0.2 | 0.01 | 0.2 | 0.2 |
| Dayoo.com | 0.01 | 0.03 | 0.02 | 0.02 | 0.01 | 0.02 |
| Total Revenue (mil RMB) | 1,220 | 1,690 | 2,350 | 3,120 | 4,770 | 6,320 |
| Total Revenue (mil US$) | 147 | 206 | 295 | 410 | 699 | 925 |
| C4 | 98.5 | 94.3 | 96.0 | 94.3 | 91.3 | 93.2 |
| HHI | 3,097 | 3,113 | 2,934 | 2,813 | 2,608 | 2,989 |
| *N* (> 1%) | 4 | 4 | 4 | 4 | 5 | 4 |
| Noam Index | 1,549 | 1,557 | 1,467 | 1,407 | 1,166 | 1,495 |

41. I-Research. "Annual Report on China's Online News Media Development." Beijing: I-Research, 2009.

The fall in HHI in 2008 was largely due to the 2008 Beijing Olympic Games pushing a large stream of revenue to sites with strong sports sections; after 2008, the concentration level immediately rebounded to pre-Olympics levels.

## CONCLUSION

Conglomeration or cross-ownership in the media industry in China is not as widespread as in other country markets. One notable exception to this trend is SMEG: chiefly operating in TV broadcasting, pay-TV, radio, and film—and to a lesser extent in online and print news media—it is regarded as having good brand recognition, consistent program quality, and objective reporting. Most telecommunications operators and local TV and radio broadcasters also exercise some cross-ownership, as do a few print media groups, but unlike SMEG, this mainly reflects how the industries are grouped in the regulatory framework, rather than any underlying commercial drive for cross-ownership.

Despite their market power, there is still a significant gap between these media corporations in Mainland China and their Western counterparts. Even with China's larger potential market size and state intervention, Chinese media groups generally have fewer resources to pursue international ventures and are tightly constrained by government policies from doing so. Except for SMEG, CCTV, China Film Group, and Baidu, most Chinese media producers have yet to reap the benefit of economies of scale. From a regulatory perspective, while regulators in other countries, particularly advanced economies, are vigilant about merger and acquisition attempts, regulators in China generally encourage consolidation in the media and entertainment industries, and actively promote an atmosphere of cultural exuberance, according to Michael Tung, chief investment officer of China Media Capital, a fund set up to help develop media groups.[42]

Foreign ownership, on the whole, is very rare in the Chinese media market. Due to ideological reasons, the government has traditionally been protective of media industries, especially news sources. As for telecommunications, the government is concerned about penetration by more advanced foreign players for both security and economic reasons. Therefore, foreign ownership almost exclusively exists in the form of joint ventures or foreign companies' minority stakes in Chinese companies. As in other economic sectors, the government's industrial policy has set upper limits on how much share foreign entities can own in these domestic operations—especially in the media sector than other sectors. However, the government is extremely concerned over the indirect influence that foreign companies can have on these domestic operations. In magazine publishing, for instance, Chinese companies are not allowed to cede editorial control to their foreign partners.

Outbound foreign ownership, though increasingly encouraged by the government, is still rare. Given the huge and quickly expanding domestic market, the incentive to acquire foreign companies is mainly to establish better brand recognition overseas and obtain niche market shares. The success of Wanda Cinema Chain in acquiring AMC and the difficulties the telecommunications operator Huawei has faced abroad due to other countries' national security reservations over granting the company access to sensitive domestic systems are isolated cases that shed little light on the prospect of foreign ownership by Chinese firms.

Under the CPC, a significant proportion of media enterprises ultimately report to the party or the state, whatever their degree of operational independence. As the market reforms deepen, it is only natural that they will become increasingly market-driven, with government holding majority or minority stakes depending on sector and circumstance, barring any major political upheavals. Companies such as SMEG

42. Barboza, David. "China Yearns to Form Its Own Media Empires." *The New York Times*, Oct. 5, 2010. Feb. 28, 2013. <http://www.nytimes.com/2009/10/05/business/global/05yuan.html>.

and Lenovo will eventually be, or in Lenovo's case, have already become, widely recognized commercial entities subject to very limited state influence, despite the government's majority stake (the Chinese Academy of Science still holds a 36% majority stake in Lenovo).

One issue unique to China is regional market segmentation. In the early stages of the market reforms, regional focus propelled companies' growth at the provincial and municipal levels. But as local markets became saturated, such regionalism inhibited interregional franchising options. Established media companies and brands are constrained by their market reach, and uncompetitive entities continue to survive despite low profit margins and must be subsidized by local governments despite the state's efforts to cut subsidies for media producers. But many local governments have their own agendas, other than those of central government, and these are very often more commercially—rather than ideologically—oriented. This is particularly true in TV broadcasting and newspaper industries. Although stances of these media outlets are highly homogenized on national issues, their editorial style is much more diversified.

In recent years, even the state has recognized the importance of media pluralism and has gradually loosened control on the day-to-day management of many publicly owned media outlets. In order to head off social discontent over wealth disparities and corruption, it has given much more leeway to the publicly owned media organizations and the media system in general to report on these issues. There are virtually no state-owned players online, thanks to its late development and the industries' fragmented structure. In industries such as film and magazine publishing, more and more private enterprises have sprung up and have grown into major players. The government did not seek to curb the growth of companies such as Baidu, Sina, and Huayi Brothers, but instead sought to utilize them, intentionally or unintentionally, as an effective tool to push forward its incremental reform agenda.

Concentration of print media is generally low. There is strong regional segmentation in the newspaper industry. If we only consider national dailies, the HHI has stayed below 1,800 over the past 20 years. Counting both national and local newspapers, the HHI is even lower, around 300. In contrast, book and magazine publishers are more concentrated. Recent consolidations have yet to lead to the formation of companies with significant market power in the book and magazine publishing markets, and concentration is increasing for these two print media sectors, though it remains relatively low. Serious market segmentation also exists in radio and TV broadcasting industries. The former better resembles the newspaper market, with national players losing market shares at the expense of ascending local players. TV broadcasting, in contrast, is heavily concentrated—with HHI hovering around 4,000—due to CCTV's dominance, but as with radio, the growth of provincial stations has reduced CCTV's market share. Among mass media industries, the film industry most resembles a fully competitive market because private firms play a significant role alongside the state-run studios. Despite various advantages enjoyed by the China Film Group, it is by no means a monopolist.

The telecommunications industry remains dominated by the state, though private investors have been allowed to purchase 20–40% of the major firms' total shares. The HHI for wireline, though it has fallen significantly in recent years, still stands around 5,000. Concentration for wireless is even higher, with HHI near 6,700, in large part due to the dominance of China Mobile. The ISP industry is less concentrated, but in fact it is increasing in concentration due to the merger of China Unicom and China Netcom. Internet media is not as heavily regulated as other media sectors, especially from a distribution point of view. The state lets private players operate relatively freely in that regard, though this is not present to the same degree on the content side of things.

Taken together, mass media industries with significant up-front content production costs, such as TV broadcasting and film, have very high concentration indices. In contrast, mass media industries with significant local barriers, including newspapers and radio broadcasting, have extremely low concentration indices, as do the fragmented book and magazine publishing markets. Platform media industries, including wireline telecom, wireless telecom search engines, and ISPs, are highly concentrated due to their inherent network effect and the state intervention. The online news sector, freed from the geographical barriers within the newspaper market, is more highly concentrated.

China is unique in the level of state control exercised over the market, which maintains a high level of government control but also allows a significant degree of operational freedom in media companies' day-to-day business in telecommunications, film, and TV broadcasting. In addition, China's sheer size (in population and geography) contributes to regionalism in terms of economic management by local governments and how the market is formed in the newspapers, radio, and TV broadcasting markets. Underlying this is the increased impact of commercial and competitive forces—though not necessarily in the form of private ownership.

## Bibliography

### 1. Books and Articles


Barboza, David. "China Yearns to Form Its Own Media Empires." *The New York Times*, Oct. 5, 2010. Feb. 28, 2013. <http://www.nytimes.com/2009/10/05/business/global/05yuan.html>.

Buddle, Paul. "China's cable TV market in transition." BuddeBlog.com, July 6, 2012. Feb. 28, 2013. <http://www.buddeblog.com.au/frompaulsdesk/chinas-cable-tv-market-in-transition/>.

Burkitt, Laurie. "CCTV's Ad-Sales Auction Draws Record Bids." *The Wall Street Journal*, Nov. 9, 2011. Feb. 28 2013. <http://online.wsj.com/article/SB10001424052970204190704577025770417699372.html>.

Burkitt, Laurie. "Foreign Films Dominate China Box Office." *The Wall Street Journal*, July 19, 2012. Feb. 28, 2013. <http://online.wsj.com/article/SB10000872396390444097904577536820360717562.html>.

China Film Press. *China Film Yearbook*. Beijing: China Film Press, 2007.

China Radio International. "History % Milestones: CRI English Service." cri.cn, 2004. Feb. 28, 2013. <http://english.cri.cn/about/history.htm>.

China Telecom Corporation Limited. "Share Information." China Telecom Corporation Limited, 2012. June 15, 2012. <http://www.chinatelecom-h.com/eng/ir/shareinfo.html>.

"China to Become Asia's Top Pay-TV Economy in 2011 with $7.5bn in Revenue." *VOD Professional*, June 22, 2011. Feb. 28, 2013. <http://www.vodprofessional.com/research/china-to-become-asia%E2%80%99s-top-pay-tv-economy-in-2011-with-$75bn-in-revenue/>.

China Unicom (Hong Kong) Limited. "Company Profile." China Unicom (Hong Kong) Limited, 2012. July 17, 2012 at <http://www.chinaunicom.com.hk/en/aboutus/about_profile.html>.

"Chinese Studios Huayi Brothers, Bona take Stakes in U.S.-Based China Lion." *The Hollywood Reporter*. Oct. 31, 2011. July 18, 2012 <www.hollywoodreporter.com/news/chinese-studios-huayi-brothers-bona-255242>.

"Film sector monopoly broken up." *China Daily*, June 10, 2003. Feb. 28, 2013. <http://www.chinadaily.com.cn/en/doc/2003-06/10/content_244133.htm>.

Hearst Corporation. "Hearst Corporation Completes the Acquisition of the Majority of Hachette China Operations." Hearst.com, Dec. 6, 2011. Feb. 28, 2013. <http://www.hearst.com/press-room/pr-20111206a.php>.

Hexun. "Huayi Brother Stock Distribution Analysis." Hexun.com, 2009. July 26, 2012. <http://stockdata.stock.hexun.com/2009_sdgd_300027.shtml>.

Huang, Wanwu. "Market Economy and Transformation of Periodical Industry in China." *Chinese Journal of Scientific and Technical Periodicals* 4:3, 1993.

Li, Zhanjun. "A Look at the Radio Industry in the New Media Era." *China Broadcasts* 23:4, 2011.

Li, Ziwei. "Media Transformation in China in the New Media Era." *Contemporary Manager* 23:6, 2005.

Lv, Jie. "Rise of China's Local Radio Media." *China Newspaper Industry* 14:24, 2011.

Ma, Shengrong. "A Discussion on Media Development and Strategy in China." *International Communications* 7:6, 2011.

Mark, Scott. "Telefonica to Sell Shares Back to China Unicom for $1.4 Billion." *The New York Times*, June 11, 2012. Feb. 28, 2013. <http://dealbook.nytimes.com/2012/06/11/telefonica-to-sell-stake-in-china-unicom-for-1-4-billion/>.

Mider, Zachary. "China's Wanda to Buy AMC Cinema Chain for $2.6 Billion." *Bloomberg*, May 21, 2011. July 18, 2012. <http://www.bloomberg.com/news/2012-05-21/china-s-wanda-group-to-buy-amc-cinema-chain-for-2-6-billion.html>.

National Bureau of Statistics of China. *China Statistical Yearbook*. Beijing: National Bureau of Statistics of China: 2007.

"Production - World Cinema Stats (1)." *Screenville*, June 15, 2009. Feb. 28, 2013. <http://screenville.blogspot.com/2009/06/production-world-cinema-stats-1.html>.

"Research and Markets: China's Film Industry Report, 2011 Confirms China As the Fastest Growing Region In the Global IMAX Business." *BusinessWire*, Sept. 29, 2011. Feb. 28, 2013. <http://www.businesswire.com/news/home/20110929005867/en/Research-Markets-Chinas-Film-Industry-Report-2011>.

"Rewired." *The Economist*, May 29, 2008. Feb. 28, 2013. <http://www.economist.com/node/11465595>.

Song, Zifang. "Reconsideration of Commercialization in China's Newspaper Industry and Its Outcomes." *Journal of Journalism and Communication* 3:1, 1997.

Su, Lei. "A Study on China's Teaching Material Publishing." *Science-Technology & Publication* 6:10, 2011.

Wang, Jane. "China Mobile's dominance eroded as competitors improve mobile performance." *Ovum*, Mar. 29, 2012. Feb. 28, 2013. <http://ovum.com/2012/03/29/china-mobiles-dominance-eroded-as-competitors-improve-mobile-performance/>.

Waxman, Sharon. "White House Gets China to Open Market to U.S. Movies." *Reuters*, Feb. 18, 2012. July 18, 2012. <http://www.reuters.com/article/2012/02/18/idUS420251887620120218>.

Wei, Ting. "Solutions to Marketing Issues in China's Film Industry." *Inquiry into Economic Issues* 9:6, 2008.

Xie, Liya. "New Business Model of Shanghai Media Group." *Ceocio China* 26:5, 2010.

Wei Yushan. "Strategies for China's Publishing Industry." *Publishing Research* 8:5, 2010.

Zhang, Dianyuan. "Social Motivation of System Innovation of Chinese Newspaper." *Science Economy Society* 28:4, 2010.

Zhang, Wenjun. "A New Definition of Wireless Broadcast Television in China." Yonsei University. BMSB 2012, Seoul, South Korea. June 28, 2012. Feb. 28, 2013. <http://www.ieee-bmsb2012.org/images/program/BMSB2012_Keynote_5_A_New_Definition_of_Wireless_Broadcasting_China_WenjunZhang.pdf>.

Zheng, Baoyu. "Business Model for China's TV Broadcasting Industry." *Modern Audio-Video Arts* 13:1, 2009.

## 2. Reports

China Internet Network Information Center. "Statistical Report on China's Internet Use and Current Status." Beijing: China Internet Network Information Center, 2006.

General Administration of Press and Publication, "National Press and Publication Statistics Network." Beijing: General Administration of Press and Publication: 2006.

I-Research. "Annual Report on China's Online News Media Development." Beijing: I-Research, 2009.

I-Research. "China Online Search Annual Report." Beijing: I-Research, 2009.

Italian Trade Commission. "Overview of China Film Market." Shanghai: Italian Trade Commission, June 14, 2011. Feb. 28, 2013. <http://www.ice.gov.it/paesi/asia/cina/upload/174/Final%20Report%20%20Market%20Report%20of%20China%20Film%20Market%202011.pdf>.

The State Administration of Radio, Film and Television (SARFT). "China Radio, Film, and TV Development." Beijing: SARFT: 2009.

# China—Data Summaries

ELI NOAM AND PAUL MUTTER

BY POPULATION, THE PEOPLE'S Republic of China is the largest single national media market in the world, though it also has one of the lowest per capita revenues for media industries of the countries surveyed. Most Chinese multimedia producers are state-owned, the highest percentage of all countries surveyed: as a single entity, the Chinese government has a 98.6% market share of all media. However, different parts of the government and different levels of government own certain media properties. So the fundamental question is how to view the overall ownership system of these organizations. Is it a single "firm" with multiple divisions, in the same way that Vivendi's Canal+ pay-TV operation, its former SFR mobile phone operator, and its Universal Music Group are controlled by the same enterprise even if managers get substantial autonomy from interference from the top and from each other? Or, are the various state organizations truly independent of each other directly or through common control? The two perspectives have different adherents. Those inside China tend to present the latter perspective. But this would make China, by the numbers, the world's most diverse media environment, so this might be pushing an argument a bit too far. The other perspective, that of a unified media system where state companies ultimately answer to the same authority or to those who answer to that authority, creates the other extreme: now, China's media is just about the world's most concentrated, more or less on par with Egypt, also a country without a tradition of democratic pluralism. We need not resolve this question one way or the other, and often present both sets of numbers. The editors, using their judgment without involving the authors, use several measures: "integrated" perspective where the concentrations are calculated with the state as a single owner ("Government of China" in Tables 24.14, 24.15,

Table 24-14. National Media Industries Concentration in China

| | 2004/5 | | 2011 OR MOST RECENT | | % CHANGE ANNUAL AVERAGE | |
|---|---|---|---|---|---|---|
| | COMPANY POWER INDEX IN COUNTRY | COMPANY SHARE OF THE OVERALL NATIONAL MEDIA MARKET (%) | COMPANY POWER INDEX IN COUNTRY | COMPANY SHARE OF THE OVERALL NATIONAL MEDIA MARKET (%) | COMPANY POWER INDEX IN COUNTRY | COMPANY SHARE OF THE OVERALL NATIONAL MEDIA MARKET (%) |
| Government of China | 9,835.2 | 98.6 | 9,571.9 | 96.5 | –0.3 | –0.3 |
| *China Mobile (public)* | 2,266.6 | 32.8 | 2,491.9 | 40.5 | 1.2 | 1.0 |
| *China Telecom (public)* | 1,524.4 | 24.9 | 820.1 | 21.2 | –5.8 | –0.5 |
| *China Unicom (public)* | 362.6 | 13.2 | 527.4 | 20.5 | 5.7 | 0.9 |
| *CCTV (public)* | 223.3 | 3.6 | 187.4 | 3.2 | –2.0 | –0.06 |
| *Jiangsu Phoenix Publishing and Media (public)* | 0.1 | 0.2 | 2.5 | 0.2 | 324.6 | 0.004 |
| *Shanghai Media Group (SMG) (public)* | 24.7 | 1.2 | 8.8 | 0.7 | –8.1 | –0.07 |
| *China Film Group (public)* | 0.4 | 0.02 | 7.5 | 0.2 | 253.9 | 0.02 |
| *Xinhua News Agency (public)* | 0.4 | 0.08 | 0.3 | 0.08 | –2.8 | –0.001 |
| *Trends Media Group (public)* | 1.2 | 0.1 | 6.9 | 0.4 | 56.7 | 0.04 |
| *Hunan Media Group (public)* | 7.3 | 0.7 | 3.5 | 0.4 | –6.5 | –0.03 |
| *People's Daily (public)* | 0.2 | 0.03 | 0.3 | 0.03 | 11.1 | –0.001 |
| *China Science Publishing and Media Group (public)* | 0.7 | 0.2 | 1.0 | 0.2 | 5.1 | 0.004 |
| *China National Radio (public)* | 0.9 | 0.08 | 0.4 | 0.05 | –6.4 | –0.004 |
| Baidu | 9.0 | 0.2 | 103.8 | 1.3 | 131.7 | 0.15 |
| Sina.com | 6.3 | 0.2 | 6.7 | 0.2 | 0.7 | 0.002 |
| Google (US) | 1.4 | 0.1 | 4.1 | 0.3 | 24.3 | 0.03 |
| Sohu.com | 4.4 | 0.16 | 4.8 | 0.2 | 1.1 | 0.01 |
| 163.com | 0.1 | 0.02 | 1.6 | 0.09 | 249.1 | 0.01 |

| | | | | | | |
|---|---|---|---|---|---|---|
| Huayi Brothers | 0.6 | 0.03 | 0.5 | 0.05 | −0.9 | 0.003 |
| Caijing | 0.0 | 0.01 | 0.0 | 0.01 | −9.8 | 0.0 |
| Yahoo! (US) | 0.2 | 0.02 | 0.0 | 0.0 | −12.5 | −0.003 |

| Media Concentration Index | 2004/5 | 2011 or Most Recent | % Change Annual Average |
|---|---|---|---|
| Total Revenue: Nat'l Media Industry (mil US$) | 58,898 | 208,702 | 31.8% |
| Total Voices (*n*) (Integrated) | 57 | 65 | 1.8% |
| Total Voices (*n*) (Segmented) | 62 | 70 | 1.6% |
| Net Voices (*n*) (Integrated) | 16 | 17 | 0.8% |
| Net Voices (*n*) (Segmented) | 55 | 62 | 1.6% |
| Public Ownership (%) | 98.6 | 96.5 | −0.3 |
| Foreign Ownership (%) | 0.1 | 0.5 | 0.05% |
| C4 Average (Integrated)—Weighted | 99.3 | 99.3 | −0.01% |
| C4 Average (Segmented)—Weighted | 93.0 | 94.5 | 0.2% |
| HHI Average (Integrated)—Weighted | 9,859 | 9,700 | −0.2% |
| HHI Average (Segmented)—Weighted | 4,877 | 3,960 | −0.02 |
| C1 Average—Weighted (Integrated) | 99 | 98 | 0.0% |
| C1 Average—Weighted (Segmented) | 61 | 58 | 0.0% |
| Noam Index Average (Integrated)—Weighted | 8,902 | 7,491 | −2.0% |

*continued*

Table 24-14. *continued*

| | 2004/5 | | 2011 OR MOST RECENT | | % CHANGE ANNUAL AVERAGE | |
|---|---|---|---|---|---|---|
| | COMPANY POWER INDEX IN COUNTRY | COMPANY SHARE OF THE OVERALL NATIONAL MEDIA MARKET (%) | COMPANY POWER INDEX IN COUNTRY | COMPANY SHARE OF THE OVERALL NATIONAL MEDIA MARKET (%) | COMPANY POWER INDEX IN COUNTRY | COMPANY SHARE OF THE OVERALL NATIONAL MEDIA MARKET (%) |
| Noam Index Average (Segmented)—Weighted | 984 | | 1,050 | | 0.01% | |
| Pooled Overall Sector C4 (Integrated) | 99.1 | | 98.5 | | −0.1% | |
| Pooled Overall Sector C4 (Segmented) | 83.5 | | 85.4 | | 0.2% | |
| Pooled Overall Sector HHI (Integrated) | 9,720 | | 9,311 | | −0.5% | |
| Pooled Overall Sector HHI (Segmented) | 2,046 | | 2,526 | | 2.9% | |
| Pooled Overall Sector Noam Index (Integrated) | 2,042 | | 1,539 | | −3.1% | |
| Pooled Overall Sector Noam Index (Segmented) | 99.2 | | 68.7 | | −3.8% | |
| Market Share of Top 10 Companies: Nat'l Media Industry (%) (Pooled C10) | 99.3 | | 99.2 | | 0.0% | |
| National Power Index (Integrated) | 9,858 | | 9,702 | | −0.2% | |
| National Power Index (Segmented) | 4,836 | | 4,237 | | −0.02% | |

Table 24-15. Top Content Media Companies in China

| | 2004/5 | | 2011 OR MOST RECENT | | % CHANGE ANNUAL AVERAGE | |
|---|---|---|---|---|---|---|
| | COMPANY POWER INDEX IN COUNTRY | COMPANY SHARE OF THE NATIONAL CONTENT MEDIA MARKET (%) | COMPANY POWER INDEX IN COUNTRY | COMPANY SHARE OF THE NATIONAL CONTENT MEDIA MARKET (%) | COMPANY POWER INDEX IN COUNTRY | COMPANY SHARE OF THE NATIONAL CONTENT MEDIA MARKET (%) |
| Government of China | 8,871 | 90.4 | 7,467 | 79.2 | –1.4 | –1.4 |
| *CCTV (public)* | 1,530 | 25.0 | 1,109 | 18.8 | –3.4 | –0.8 |
| *Jiangsu Phoenix Publishing and Media (public)* | 0.6 | 0.5 | 15 | 2.2 | 278.7 | 0.2 |
| *Shanghai Media Group (public)* | 169 | 8.5 | 51 | 4.2 | –8.7 | –0.5 |
| Baidu | 62 | 1.1 | 614 | 7.8 | 112.1 | 0.8 |
| Sina.com | 43.4 | 1.1 | 40 | 1.0 | –1.1 | –0.01 |
| Google (US) | 10 | 0.44 | 24 | 1.6 | 19.3 | 0.14 |
| 163.com | 0.5 | 0.1 | 10 | 0.5 | 213.5 | 0.05 |
| Caijing | 0.1 | 0.1 | 0.0 | 0.06 | –10.2 | –0.004 |
| Media Concentration Index | 2004/5 | | 2011 or Most Recent | | % Change Annual Average | |
| Public Ownership (%) | 90.4 | | 79.2 | | –1.4% | |
| Foreign Ownership (%) | 0.4 | | 3.0 | | 0.3% | |
| C4 Average (Integrated)—Weighted | 95.4 | | 95.8 | | 0.04% | |
| C4 Average (Segmented)—Weighted | 53.5 | | 67.2 | | 1.7% | |
| HHI Average (Integrated)—Weighted | 9,032 | | 8,224 | | –1.1% | |
| HHI Average (Segmented)—Weighted | 2,213 | | 2,406 | | 0.01% | |
| C1 Average—Weighted (Integrated) | 92 | | 88 | | –1% | |
| C1 Average—Weighted (Segmented) | 31 | | 38 | | 1.0 | |
| National Power Index (Integrated) | 9,028 | | 8,234 | | –1.1% | |
| National Power Index (Segmented) | 1,974 | | 2,333 | | 0.02% | |

Table 24-16. Top Platform Media Companies in China

| | 2004/5 | | 2011 OR MOST RECENT | | % CHANGE ANNUAL AVERAGE | |
|---|---|---|---|---|---|---|
| | COMPANY POWER INDEX IN COUNTRY | COMPANY SHARE OF THE NATIONAL PLATFORM MEDIA MARKET (%) | COMPANY POWER INDEX IN COUNTRY | COMPANY SHARE OF THE NATIONAL PLATFORM MEDIA MARKET (%) | COMPANY POWER INDEX IN COUNTRY | COMPANY SHARE OF THE NATIONAL PLATFORM MEDIA MARKET (%) |
| Government of China | 10,000 | 100 | 10,000 | 100 | 0.0 | 0.0 |
| *China Mobile (public)* | 2,654 | 98.2 | 2,999 | 99.9 | 1.6 | 0.2 |
| *China Telecom* | 1,785 | 29.2 | 987 | 25.5 | –5.6 | –0.5 |
| *China Unicom* | 425 | 15.4 | 635 | 24.6 | 6.2 | 1.1 |
| Media Concentration Index | 2004/5 | | 2011 or Most Recent | | % Change Annual Average | |
| Public Ownership (%) | 100 | | 100 | | 0.0% | |
| Foreign Ownership (%) | 0.0 | | 0.0 | | 0.0% | |
| C4 Average (Integrated)—Weighted | 100 | | 100 | | 0.0% | |
| C4 Average (Segmented)—Weighted | 99.8 | | 100 | | 0.03% | |
| HHI Average (Integrated)—Weighted | 10,000 | | 10,000 | | 0.0% | |
| HHI Average (Segmented)—Weighted | 5,333 | | 4,276 | | –0.02% | |
| C1 Average—Weighted (Integrated) | 100.0 | | 100.0 | | 0.0% | |
| C1 Average—Weighted (Segmented) | 66 | | 62 | | –1% | |
| National Power Index (Integrated) | 10,000 | | 10,000 | | 0.0% | |
| National Power Index (Segmented) | 5,325 | | 4,624 | | –0.02% | |

and 24.16),[43] and the "segmented" perspective. Both ownership definitions are shown, and the former is used in calculating world averages.

Reflecting the high level of state ownership, China's Weighted Average HHI ranks first internationally at 9,859 when state ownership is treated as a single enterprise, and 4,877 when state enterprises are treated as individual entities. China's power index is 9,858 when the unified measurement is used (and 4,836 when the non-unified measure treats state enterprises individually), the highest of all the countries surveyed. Telecommunications constitute the largest segments of China's national media industries: by revenue, both wireless and wireline exceed all other countries markets except for the US's, despite their lower prices. Three public telecoms —China Unicom, China Telecom, and China Mobile—are by far the largest media groups in China. The fourth largest company is the state TV organization CCTV.

The structure of the system partially explains why, despite the size of the worldwide Chinese diaspora, the international presence of Chinese media groups is small. Limits on foreign direct investment between China and other countries limit Chinese investment in media abroad and vice versa. Since most media are state owned, and the government controls entry into media and infrastructure industries, foreign firms have almost no presence in the Chinese national media industries. In content media, state ownership is slightly lower, at 79.2%; down from 90.4% in 2005, because private firms are permitted in magazines, film, search engines, and online news (Table 24.15). The massive rise of online media, though, means that private companies like Baidu (7.8% of the content market in 2008) are gaining against state enterprises—though the cumulative state total dwarfs private ownership.

Restrictions toward foreign media ownership in China are partly due to political and economic protectionism. Critics, to be fair, need to acknowledge that the United States, Canada, and other countries also have several foreign ownership restrictions based on similar factors.

Concentration is also high in several of the industries open to private enterprise. In search engines, where Google, Microsoft, and Yahoo have been rebuffed, the domestic company Baidu holds 78.6% of the market. Two of the four companies—Sina.com and Sohu.com—control over 71% of online news media.

43. In these three tables, the state is presented first as a single owner under the label "integrated." We also present "segmented" data, which treat each state enterprise as an independent entity.

# 25

# Media Ownership and Concentration in India

ANURADHA BHATTACHARJEE, LIWEI WANG, AND TAPASYA BANERJEE

## INTRODUCTION

The Indian media market is greatly affected by the country's geographical size, the seventh largest in the world by land area, and demographics. India's 28 states and seven union territories were originally created on a linguistic basis, and there are 22 "recognized" languages in India, including the official national languages of English and Hindi. Most Indians are fluent in at least two languages, including English, and have a preferred vernacular language for media consumption. Such vernacular media outlets play a key role in the politics of the states and union territories.[1]

Historically, most Indian mass media outlets have been state-owned, with the exception of newspapers, magazines, and film studios. Liberalization of the media market really only began in 2001, when private enterprises were finally licensed to operate in the audiovisual and telecommunications sectors.[2] Substantial restrictions on foreign direct investment (FDI) remain in place, however: 100% FDI is allowed only for the film industry and online media, while there are caps for telecommunications and print media (74%), cable and satellite television (49%), and radio broadcasting (20%).[3] The

1. ETV has traditionally dominated Telegu-language programming in South India, the ABP Group is strongest in the Bengali-language print market, and Zee Bangla and ETV Bangla dominate the Bengali-language audiovisual market. Nationally, NDTV and Network18 offer programming in English and Hindi, while Zee offers programming primarily in Hindi but also adds 12 regional languages as well as English through a joint venture with Time Warner (US).

2. Ipsos. "India Pulse." Mumbai: Ipsos, 2012. Feb. 18, 2013. <http://www.ipsos.com/sites/ipsos.com/files/India%20Pulse%20-%20October%202012.pdf>.

3. "Industry and Services: Media and Entertainment." New Delhi: Ministry of External Affairs, Government of India Investment & Technology Promotion (ITP) Division, 2009. May 10, 2012. <http://www.indiainbusiness.nic.in/industry-infrastructure/service-sectors/media-entertainment.htm>

most restrictive measures affect the radio and terrestrial TV markets: the state-run All India Radio (AIR) monopolizes news broadcasting. Though privately owned Indian television networks are allowed to broadcast news content, foreign investment in such networks is highly restricted with a 26% cap.

Due to the legacy of state control over many media producers, these enterprises indulge in rent-seeking behavior due to the importance of political and familial patronage in politics.[4] It is estimated that one-third of all news channels and 60% of all cable distributors are owned by politicians and political activists (or their relatives).[5, 6] Politicians who own cable operators or newspapers—either directly, or indirectly through patronage networks—use undeclared revenue from these operations to fund their election campaigns.[7] Industrial conglomerates also practice rent-seeking, though they do so not to raise funds, as their other holdings are used to subsidize their often unprofitable media properties.[8] Instead, they hope to use the media properties to pressure politicians and head off negative reporting concerning their business practices. What this entrails was illustrated in a dispute between Jindal Steel and Power Limited (JSPL) and the Zee media group. JSPL accused Zee of extortion: the network was videotaped as it allegedly proposed an US$2 million (INR100 million) advertising deal to JSPL in exchange for downplaying reports about JSPL obtaining favors from state officials.[9]

Further demonstrating the extent to which official collusion underwrites such dealings is the fact that the chairman of JSPL, Navin Jindal, was a Member of Parliament in the long-dominant Indian National Congress (INC) party. Concerns that political favoritism was undermining regulatory measures led the Telecom Regulatory Authority of India (TRAI) to recommend in 2012 that provincial governments should not be allowed to enter the cable or satellite broadcasting and distribution because clientelism and patronage already characterized media outlets affiliated with politicians' families.[10, 11] TRAI recommended that the functions of the national public broadcaster, Prasar Bharti, and the Ministry of Information an Broadcasting (MIB) be further segregated from one another in order to reduce the possibility of political pressure on media producers.

Regulators are also concerned about increased concentration through mergers and acquisitions spearheaded by major industrial groups entering the media market. In 2012, for example, the Indian petrochemicals major Reliance Industries financed the takeover of Ushodaya Enterprises' ETV by Network18 Media & Investments Limited in exchange for a stake in Network18. The Competition Commission of India (CCI) ruled in 2012 that Reliance now controlled Network18 indirectly due to this deal.[12] In 1996, Tara Teleservices services was formed to spearhead the Tata Group's foray into telecommunications after it had helped establish the state-owned Videsh

---

4. Khanna, T. and Palepu, K. "The Evolution of Concentrated Ownership in India: Broad Patterns and a History of the Indian Software Industry." Cambridge: National Bureau of Economic Research, 2004. June 14, 2012. <http://www.nber.org/papers/w10613>.

5. Kohli-Khandekar, Vanita. "When Politicians Own the Media." *Business Standard*, July 3, 2012. Feb. 20, 2013. <http://business-standard.com/india/news/vanita-kohli-khandekar-when-politicians-ownmedia/479159/>.

6. Khanna, T. and Palepu, K. "The Evolution of Concentrated Ownership in India: Broad patterns and a History of the Indian Software Industry." Cambridge: National Bureau of Economic Research, 2004. June 14, 2012. <http://www.nber.org/papers/w10613>.

7. Kohli-Khandekar, Vanita. "When Politicians Own the Media." *Business Standard*, July 3, 2012. Feb. 18, 2013. <http://business-standard.com/india/news/vanita-kohli-khandekar-when-politicians-ownmedia/479159/>.

8. Khanna, T. and Palepu, K. "The Evolution of Concentrated Ownership in India: Broad Patterns and a History of the Indian Software Industry." Cambridge: National Bureau of Economic Research, 2004. June 14, 2012. <http://www.nber.org/papers/w10613>.

9. Shukla, Archana. "With FIR, Jindal gives police tapes of 'meetings' with Zee." *Indian Express*, Oct. 12, 2012. Oct. 25, 2012. <http://www.indianexpress.com/news/with-fir-jindal-gives-police-tapes-of--meetings--with-zee/1015622>.

10. Singh, Prabhakar; Regional Head, Centre for Monitoring Indian Economy. Personal interview with Anuradha Bhattacharjee. May 24, 2012.

11. Manoj, C G. "TRAI against govt entering broadcast, distribution." *The Indian Express*, Dec. 29, 2012. Dec. 29, 2012. <http://www.indianexpress.com/news/trai-against-govt-entering-broadcast-distribution/1051593/>.

12. Subramanian, N. Sundaresha. "RIL Controls Network18 Group Indirectly, says CCI." *Business Standard*, June 23, 2012. June 23, 2012. <http://www.business-standard.com/india/news/ril-controls-network18-group-indirectly-says-cci/478250/>.

Sanchar Nigam Limited (VSNL), India's first Internet service provider (ISP) in 1986. VSNL and two of its subsidiaries were then merged to form Tata Communications in 2008.[13] The Essel Group has expanded into the pay-TV market as Wire and Wireless India Limited (WWIL). And Reliance Communications, the media subsidiary of the industrial giant Reliance Industries, is one of the largest mobile operators in the world, with over 150 million subscribers.[14, 15]

## PRINT MEDIA

### Newspapers

The first Indian-language newspapers appeared in 1822, 42 years after the first English-language newspapers began publishing.[16] The vernacular press was strongly nationalistic and pro-independence, while the English-language ones reflected the British Indian establishment and were anti-independence. As a result, the authorities often shut down the non-English-language dailies. Kinship networks were important for maintaining these newspapers' finances and distribution networks during this period, so after 1947, when India achieved independence from Great Britain, the value of the print market was extremely small: the First Press Commission (FPC) reported in 1956 that the print media sector was valued at only US$2 million (INR110 million) for the 127 newspapers surveyed, a rate of return of less than 1%. The FPC reported that newspapers continued to remain in business during the first decade of independence in spite of this low rate of return for two reasons: the overall effects of the economic boom caused by WWII and rent-seeking behavior by Indian political elites, which continues to this day.[17, 18] Then and now, the Government of India financially subsidizes many small- and medium-sized newspapers. The increased availability of capital from the 1950s on, however, has been significant for publishers: by 1980, commercial advertising revenues were valued at US$27.4 million (INR1.5 billion) and rose to US$2 billion (INR110 billion) over the next 25 years.[19] Even so, newspapers are still capital-intensive and characterized by low rates of return. Only Kasturi & Sons (publisher of *The Hindu*) and Indian Express Newspapers (publisher of *The Indian Express*) are unique in that they are exclusively invested in media properties and have no interests in other businesses, unlike most of their competitors (Table 25.1).

The Times Group (BCCL) publishes *The Times of India* and *The Economic Times*, which have a combined daily circulation of 4.3 million copies; *The Times of India* is the largest English-language daily in circulation worldwide.[20] The Times Group publishes 13 dailies in total, along with 11 magazines—eight of which are produced in a joint venture with the BBC (UK)—and owns a number of radio stations. The Times Group's main competitor, HT Media, was formed from the Hindustan Times Group, which had offered an IPO (one of the largest ever for an Asian media group) in response to the loss of revenue it incurred in a price war with the Times Group.[21] Its properties also include newspapers,

13. *Tata Communications: About*. tatacommunications.com, 2012. June 28, 2012. <http://www.tatacommunications.com/about/network.asp>.

14. "Reliance Communications Annual Report 2010-11." Hyderabad: Reliance Communications, 2011. Jan. 16, 2012. <http://www.bseindia.com/bseplus/AnnualReport/532712/5327120311.pdf>.

15. After Dhirubhai Ambani, the founder of Reliance Industries, passed away on July 6, 2002, his sons, Mukesh Ambani and Anil Ambani, split Reliance Industries. Anil Ambani became head of Reliance Infocomm, Reliance Energy, Reliance Capital and RNRL under the brand of Reliance ADAG. In December 2011, Reliance ADAG rejoined Reliance Industries.

16. Beck, Sanderson. "British India 1818–1875." *India & Southeast Asia to 1875*. Santa Barbara: World Peace Communications, 2004. <http://wiki.phalkefactory.net/images/f/fc/British_India_1818-1875.pdf>.

17. Kohli-Khandekar, Vanita. *The Indian Media Business*. 2nd ed. New Delhi: SAGE Publications, 2006. 24.

18. Khanna, T. and Palepu, K. "The Evolution of Concentrated Ownership in India: Broad Patterns and a History of the Indian Software Industry." Cambridge: National Bureau of Economic Research, 2004. June 14, 2012. <http://www.nber.org/papers/w10613>.

19. Kohli-Khandekar, Vanita. *The Indian Media Business*. 2nd ed. New Delhi: SAGE Publications, 2006. 24–25.

20. "NRS 2006—Key Findings." *The Hindu*, 2006. Feb. 21, 2013. <http://www.hindu.com/nic/nrs.htm>.

21. Mitra, Chandan; Managing Editor, *The Pioneer*. Personal interview with Anuradha Bhattacharjee. May 24, 2007.

## Table 25-1. Daily Newspapers (Market Shares by Revenue), 2000–2010

| | 2000 | 2004 | 2008 | 2010 |
|---|---|---|---|---|
| The Times Group (BCCL) | 26.9 | 28.3 | 31.4 | 29.2 |
| HT Media | N/A | 8.2 | 9.7 | 9.7 |
| Bhaskar Group | N/A | N/A | 6.7 | 7.9 |
| Deccan Chronicle Holdings Ltd. (DCHL) | N/A | 1.9 | 6.4 | 7.1 |
| Jagran Prakashan Ltd. | 0.7 | 1.8 | 5.6 | 6.8 |
| Kasturi & Sons | 8.8 | 7.4 | 6.8 | 6.1 |
| ABP Group | 5.4 | 5.0 | 5.0 | 5.1 |
| Manorama Group | 5.2 | 5.0 | 4.0 | 4.3 |
| Amar Ujala Publications | N/A | N/A | 2.7 | 3.2 |
| Lok Prakashan | 3.9 | 3.8 | 2.9 | 2.6 |
| Mathrubhumi Printing and Publishing Company Ltd. | 2.7 | 2.8 | 2.2 | 2.6 |
| Sakal Papers | N/A | 2.3 | 2.3 | 2.4 |
| Express Publications | 3.0 | 2.0 | 1.5 | 1.6 |
| Sandesh | 2.8 | 2.1 | 1.2 | 1.3 |
| Diligent Media Corp. | N/A | N/A | 0.7 | 1.2 |
| Ushodaya Enterprises | 7.2 | 7.0 | 0.9 | 0.8 |
| Rajasthan Patrika | 1.8 | 2.0 | 0.5 | 0.7 |
| Others | 51.8 | 51.0 | 45.5 | 46.1 |
| Total Revenue (mil INR) | 4,250 | 6,080 | 1,228 | 1,255 |
| Total Revenue (mil US$) | 91.0 | 139.0 | 25.0 | 25.3 |
| C4 | 48.2 | 48.9 | 54.5 | 53.9 |
| HHI | 960 | 1,074 | 1,313 | 1,228 |
| *N* (>1%) | 12 | 15 | 15 | 16 |
| Noam Index | 277 | 277 | 339 | 307 |

*Source*: Center for Monitoring Indian Economy (CMIE).

magazines, and radio stations. According to the Indian Readership Survey (IRS), the *Hindustan Times* had a daily readership of 3.5 million in 2010.[22] In the Hindi-language market, *Dainik Jagran*, published by Jagran Prakashan Ltd., and *Dainik Bhaskar*, published by the Bhaskar Group, are the top two titles, both with daily circulations of 16.4 million.[23] The third-highest circulation Hindi-language daily (12.2 million copies), *Hindustan Dainik*, was formerly part of HT Media until 2009 when it became a separate entity.

The English-language market is concentrated in the metropolitan areas of New Delhi, Mumbai, Kolkata, and Chennai, which are home to most of India's white collar, English-speaking population. Though the vernacular press has higher absolute circulation figures, English-language publications receive far more advertising revenue. In 2008, for instance, The Times Group's titles collectively held a 3.7% market share measured by circulation, but 31.4% measured by advertising revenue.

22. Indian Readership Survey (IRS). *IRS Q4 2010 Topline Findings*. Mumbai: Media Research Users Council, 2012. Feb. 18, 2012. <http://www.pdfone.com/download/11_meri-saheli-hindi/irs-2010-q4-topline-findings-hansa-research-market-research.html>.

23. Indian Readership Survey (IRS). *IRS Q2 2011 Topline Findings*. Mumbai: Media Research Users Council, 2012. Feb. 21, 2013. <http://mruc.net/irs2011q2-topline-finding.pdf>.

Several vernacular language media groups also hold significant market shares. Malayala Manorama Co., which publishes *Malayala Manorama* in the Malayalam language, has nearly 10 million daily readers. Ushodaya Enterprises, with its Telegu-language daily *Eenadu* has a total daily readership of 6 million.[24] Ushodaya Enterprises also has interests in film production and distribution, and until 2010, it was heavily invested in the pay-TV market through its subsidiary ETV. DB Corp, which publishes the Hindi-language *Dainik Bhaskar* (founded in 1958), has a daily circulation of over 2 million copies daily. Jagran Prakashan Limited (JPL), publisher of the Hindi-language daily *Dainik Jagran,* has a daily circulation of 2.7 million copies and a total readership of 16.4 million as of 2010.[25, 26] The ABP Group publishes the Bangla-language *Ananda Bazar Patrika,* with a daily circulation of approximately 6 million copies. The group also publishes eight magazines and owns three pay-TV channels (Table 25.2).[27]

## Book Publishing

There are approximately 16,000 publishing houses in India, with a combined annual output of 80,000 titles. Of these, 25% are published in Hindi, with English accounting for slightly less (20%), and the remaining 55% falls across the vernacular languages.[28] The market's fortunes have been closely connected to government textbook orders since the 1940s: increased literacy—up from 20% of the population in 1948 to 64.8% in 2006—is in large part the

Table 25-2. Daily Newspapers (Market Shares by Circulation), 1984–2008[1]

| | 1984 | 1988 | 1992 | 1996 | 2000 | 2004 | 2008 |
|---|---|---|---|---|---|---|---|
| Indian Express Group (*Indian Express*) | 5.4 | 5.4 | 3.2 | N/A | N/A | N/A | N/A |
| Bhaskar Group (*Dainik Bhaskar*) | N/A | N/A | N/A | N/A | 3.9 | 5.3 | 5.6 |
| Manorama Group (*Malayalam Manorama*) | 5.2 | 5.4 | 4.1 | 4.5 | 4.7 | 4.1 | 4.2 |
| The Times Group (BCCL) (*The Times of India*) | 4.5 | 4.7 | 3.9 | 5.8 | 5.5 | 5.0 | 3.7 |
| Mathrubhumi Printing and Publishing Company Ltd. (*Mathrubhumi Malayalam*) | 3.7 | N/A | N/A | 3.4 | N/A | N/A | N/A |
| Kasturi & Sons (*The Hindu*) | N/A | 3.6 | N/A | N/A | N/A | N/A | 3.3 |
| ABP Group (*Anand Bazar Patrika*) | 3.6 | 3.4 | N/A | N/A | N/A | N/A | N/A |
| Punjab Kesri Group (*Punjab Kesari*) | N/A | N/A | 3.5 | 3.9 | N/A | N/A | N/A |
| Lok Prakashan Ltd. (*Gujarat Samachar, Gujarati*) | N/A | N/A | 3.3 | 3.9 | N/A | N/A | N/A |
| Jagran Prakashan Ltd. (*Dainik Jagran*) | N/A | N/A | N/A | N/A | 3.6 | 6.3 | 6.3 |
| HT Media (*Hindustan Times*) | N/A | N/A | N/A | N/A | N/A | 3.6 | N/A |

*Source*: Audit Bureau of Circulation, India.

[1] Measuring market shares by circulation is complicated by the fact that several print publishers do not subscribe to the Audit Bureau of Circulation (ABC) and instead publish their own circulation figures.

24. Indian Readership Survey (IRS). *IRS Q1 2012 Topline Findings*. Mumbai: Media Research Users Council, 2012. Feb. 18, 2013. <http://mruc.net/irs2012q1-topline-findings.pdf>.

25. Audit Bureau of Circulation. "January-June 2010." Mumbai: Audit Bureau of Circulation, 2010. & Audit Bureau of Circulation. "July-December 2010." Mumbai: Audit Bureau of Circulation, 2010.

26. Indian Readership Survey (IRS). *IRS Q2 2011 Topline Findings*. Mumbai: Media Research Users Council, 2012. Feb. 21, 2013. <http://mruc.net/irs2011q2-topline-finding.pdf>.

27. *ABP: Mjoy*. abp.in, 2011. Aug. 2, 2011. <http://www.abp.in/index.php?show=22002>.

28. Malhotra, Dina N. "The Panorama of Indian Book Publishing." *60 Years of* Book *Publishing in India, 1947–2007*. Malhotra, Dina N., ed. New Delhi: Federation of Indian Book Publishers, 2006. 14.

result of government literacy campaigns.[29] The textbook market, worth at least US$17.9 million (INR981 million), was nationalized in the mid-1950s by the state. The Ministry of Human Resources Development oversees textbook publishing in India, and the National Council for Educational Research and Training (NCERT) was set up to publish and distribute textbooks. The affiliated state bodies publish books in vernacular languages.[30] Until the 1990s, the commercial publishing industry was marginalized and several foreign firms left the market due to the nationalization campaign.

The largest companies in the textbook sector today are Oxford University Press, Macmillan, S. Chand and Co., Rupa Sterling, and Vikas.[31] After market reforms were enacted in 1991, restrictions on FDI were relaxed: the 1999 Foreign Exchange Management Act (FEMA) replaced the draconian Foreign Exchange Regulation Act (FERA), so several international publishers that had left the country in the 1950s returned and set up subsidiaries. These now include SAGE Publications, Picador, Random House, Routledge, Pearson, Butterworth–Heinemann, the Penguin Group, HarperCollins, Scholastic, and Cambridge University Press.[32] According to a 2006 report, India's publishing industry is worth US$20 billion (INR882 billion) and generates US$1.5 billion (INR66.2 billion) in revenue annually.[33]

## Magazine Publishing

Magazine circulation has risen from 11 million copies in 1984 to 75 million by 2005 (Table 25.3).[34] As with newspapers, advertising expenditures are skewed in favor of English-language titles with smaller (but higher-income) audiences. The largest magazine publisher is the India Today Group (ITG), which entered the market in 1975 with *India Today,* a biweekly magazine styled after the US *TIME* magazine, that is now published weekly in five languages. In 2000, ITG took in 3.5% of the total advertising revenue in the print media market, though by 2010 the group's national market share had fallen to 1.7%.[35]

In 2012 the Aditya Birla Group (ABG), a retail conglomerate that also owns the wireless provider Idea Cellular purchased 27.5 % stake in Living Media India, a subsidiary of ITG for US$ 63.8 million (INR 3.5 billion).[36]

# AUDIOVISUAL MEDIA

## Radio

Commercial radio broadcasting began in India in 1925 with the Indian Broadcasting Company (IBC), which then became the national public broadcaster All India Radio (AIR) in 1936 due to financial difficulties.[37] Radio news content is as of this writing still monopolized by AIR, and it took more a decade for AIR to cede its monopoly on entertainment programing. In 1993, AIR began selling FM channels to commercial entities, but in 1998 all of these channels were pulled off the air when their contracts expired because the government decided that

29. Malhotra, Dina N. "The Panorama of Indian Book Publishing." *60 Years of Book Publishing in India, 1947–2007.* Malhotra, Dina N., ed. New Delhi: Federation of Indian Book Publishers, 2006. 10–11.

30. Kaul H. K. "Role of Books in National Development." *60 Years of Book Publishing in India, 1947–2007.* Malhotra, Dina N., ed. New Delhi: Federation of Indian Publishers, 2006. 202.

31. Gupta, R. K. "College Level Publishing." *50 Years of Book Publishing in India Since Independence.* Malhotra, Dina N., ed. New Delhi: Federation of Indian Publishers, 1998. 210.

32. Singh, T. "Publishing in English." *60 Years of Book Publishing in India, 1947–2007.* Malhotra, Dina N., ed. New Delhi: Federation of Indian Publishers, 2006. 57.

33. Bhattacharjee, Anuradha and Banerjee, Tapasya. "Media Concentration in India: 2nd Interim Report." New York: Columbia Business School, 2011. 20 Jan. 2011. <http://www4.gsb.columbia.edu/filemgr?file_id=738014>.

34. Audit Bureau of Circulation. "January-June 2010." Mumbai: Audit Bureau of Circulation, 2010. & Audit Bureau of Circulation. "July-December 2010." Mumbai: Audit Bureau of Circulation, 2010.

35. *Centre for Monitoring of Indian Economy (CMIE).* cmie.com, 2012. July 8, 2012. <http://www.cmie.com/>.

36. May, Mumbai. "Aditya Birla Group Acquires Stake in Living Media India." *Business Standard,* 19 May 2012. May 19, 2012. <http://business-standard.com/india/news/aditya-birla-group-acquires-stake-in-living-media-india/474783/>.

37. Telecom Regulatory Authority of India (TRAI). "Recommendations on Issues Relating to Entry of Certain Entities into Broadcasting and Distribution Activities." New Delhi: TRAI, 2008. Feb. 19, 2013. <http://www.cccindia.co/corecentre/Database/Docs/DocFiles/trai1.pdf>.

Table 25-3. Magazine Publishing (Market Shares by Revenue), 1996–2008

| | 1996 | 2000 | 2004 | 2008 |
|---|---|---|---|---|
| Living Media Group (ITG) | 28.0 | 17.0 | 11.0 | 24.0 |
| Outlook Group (Rajan Raheja Group) | N/A | 7.0 | 6.0 | 5.0 |
| Business India Group | 9.0 | 5.0 | 5.0 | N/A |
| Citra Lekha Group | 6.0 | 5.0 | 4.0 | 3.0 |
| The Times Group (BCCL) | N/A | 5.0 | N/A | 4.0 |
| Others | 57.0 | 66.0 | 74.0 | 64.0 |
| Total Circulation (1,000) | 11,018 | 12,972 | 16,356 | 17,381 |
| Total Revenue (mil INR) | N/A | N/A | 5,300 | 13,700 |
| Total Revenue (mil US$) | N/A | N/A | 121 | 279 |
| C4 | 43.0 | 34.0 | 26.0 | 36.0 |
| HHI | 901 | 388 | 198 | 626 |
| *N* (>1%) | 3 | 5 | 4 | 4 |
| Noam Index | 520 | 174 | 99 | 313 |

*Source*: Audit Bureau of Circulation, India; AdEx India.

instead, it would auction off 108 radio licenses in geographical areas known as "circles" beginning in 2000. The next attempt at privatizing the FM market began in 1999, and the "circle" licenses were issued in 2001. Under the new licenses, broadcasters would pay an annual licensing fee worth 15% of their annual turnover. Licensing costs were exorbitant: ten licenses in Mumbai alone cost their bidders a combined US$20.9 million (INR975 million), whereas the advertising income for the transmission circle itself was valued at only US$4.3 million (INR200 million). Overbidding took place on national scale: the license fees for the 108 circles cost their buyers at least US$83.5 million (INR3.9 billion), while the market was valued at no more than US$24.1 million (INR1 billion). Losses continue to plague radio broadcasters: in 2004, 20 national broadcasters that had taken part in the 2001 auctions collectively posted total revenues of US$27.5 million (INR1.2 billion), but their expenses totaled US$54.9 million (INR2.4 billion), including US$ million US$25.2 million (INR1.1 billion) in licensing fees.

In 2005, the government announced an expansion of the number of licenses to 250 in 91 cities, valued at US$249 million (INR11 billion). The continued prevalence of high annual licensing fees, however, led commercial broadcasters to bring suits against the state. As a result, the state has since adopted a revenue-sharing model more favorable to the broadcasters for a third licensing phase (FM-3). In this phase, the state will auction off 338 new frequencies beginning in 2013, adding 700 stations to the market. The FM-3 auction will also, for the first time, permit commercial broadcasters to air news programming, breaking AIR's decades-long monopoly. TRAI has recommended that the government increase the FDI for the sector from 20% to 26%.

Due to the development of FM channels offering entertainment content, the market share of AIR has fallen from an absolute monopoly in 2000 to a 30.6% market share in 2008. Its closest competitor is BCCL, through its subsidiary ENIL, with a second-place 28% market share (Table 25.4).

## Broadcast Television

AIR began broadcasting television programming in 1959, and continued to do so until 1982, when Doordarshan (DD) was established by the state. Up until 1975, only seven Indian cities had access to the national terrestrial TV network.

Table 25-4. Radio Group (Market Shares, by Revenue), 2000–2008

| | 2000 | 2004 | 2008 |
|---|---|---|---|
| All India Radio (AIR) (Prasar Bharti) (public) | 100.0 | 49.2 | 30.6 |
| ENIL (BCCL) | 0.0 | 23.8 | 28.1 |
| Music Broadcast Pvt. Ltd. (MBIL) | 0.0 | 0.0 | 12.3 |
| HT Media | 0.0 | 0.0 | 3.3 |
| Reliance Industries | 0.0 | 0.0 | 3.1 |
| MiD-Day Multimedia-Jagaran Prakashan (Jagaran Prakashan) + BBC Worldwide (BBC, UK) (public) | 0.0 | 1.7 | 2.4 |
| Others | 0.0 | 25.4 | 20.4 |
| Total Revenue (mil INR) | N/A | 2,400 | 8,000 |
| Total Revenue (mil US$) | N/A | 54.9 | 163 |
| C4 | 100.0 | 74.6 | 74.1 |
| HHI | 10,000 | 2,984 | 1,902 |
| *N* (>1%) | 1 | 3 | 6 |
| Noam Index | 10,000 | 1,723 | 776 |

*Source*: IndiaStat.com, ENIL, Indiantelevision.com, HT Media, MiD-Day Multimedia-Jagaran Prakashan, FICCI-PWC.

Since then, access has grown exponentially: the number of households with television sets has increased from 2.8 million in 1984 to 138 million in 2010.[38, 39] All homes connected to cable and satellite TV service receive DD (Table 25.5).

## Multichannel TV Platforms

Cable TV operators emerged when color transmission became available in 1982. The market was established by unlicensed providers, who would set up master antennae for apartment buildings and then sell VCRs to individual units for a monthly fee. Multisystem operators (MSOs) remain, then and now, largely confined within wealthier urban neighborhoods, whereas local cable operators (LCOs) mainly operate in rural areas and poorer urban neighborhoods. In 1991, satellite TV channels became available, and in response, international broadcasters began encrypting their signals to prevent piracy, which spurred the government to begin regulating the market. Only in 1995 were cable operators retroactively legalized by the Cable Television Networks (Regulation) Act. Following the law's passage, an operators' consortium was formed that brought the total number of cable operators down from 60,000 to 30,000.[40] Indian MSOs are largely unregulated because ownership in the industry, classified as "small-scale," is highly politicized: politicians (and their families) often own LCOs. The Sun TV Network and its distribution arm, Sumangali Cable Vision, for instance, belong to the parliamentarian Dayanidhi Maran. Sun TV contains some of the most watched channels in the southern state of Tamil Nadu. The company posted revenues of US$398 million (INR19.2 billion) in 2008 from its television operations. This does not include (undisclosed) revenue from Sumangali Cable Vision's

38. Singhal, Arvind; Doshi, J. K.; Rogers, Everett M.; and S. Adnan, Rahman. "Diffusion of Television in India." *Media Asia* 15:4, 1988. Aug. 5, 2012. <http://utminers.utep.edu/asinghal/Articles%20and%20Chapters/Singhal-Rogers-Doshi-Rahman-1988-Diffusion%20of%20television.pdf>. 222–229.

39. KPMG and FICCI. "Hitting the High Notes: FICCI-KPMG Indian Media and Entertainment Industry Report 2011." kmpg.com, 2011. Feb. 18, 2013. <http://www.kpmg.com/IN/en/IssuesAndInsights/ThoughtLeadership/FICCI-KPMG-Report-2011.pdf>. 16.

40. Kohli-Khandekar, Vanita. *The Indian Media Business*. 3rd ed. New Delhi: SAGE Publications, 2010. 80.

Table 25-5. TV Broadcasting (Market Shares by Revenue), 1984–2008

| | 1984 | 1988 | 1992 | 1996 | 2000 | 2004 | 2008 | 2012 |
|---|---|---|---|---|---|---|---|---|
| Doordarshan (DD) (Prasar Bharti) (public) | 100.0 | 100.0 | 100.0 | 100.0 | 100.0 | 100.0 | 100.0 | 100.0 |
| Total Revenue (mil INR) | 3,100 | 16,100 | 36,000 | 57,300 | 53,800 | 66,500 | 81,800 | 60,000 |
| Total Revenue (mil US$) | 250 | 1,073 | 2,057 | 1,769 | 1,152 | 1,522 | 1,687 | 1,030 |
| C4 | 100.0 | 100.0 | 100.0 | 100.0 | 100.0 | 100.0 | 100.0 | 100.0 |
| HHI | 10,000 | 10,000 | 10,000 | 10,000 | 10,000 | 10,000 | 10,000 | 10,000 |
| *N* (>1%) | 1 | 1 | 1 | 1 | 1 | 1 | 1 | 1 |
| Noam Index | 10,000 | 10,000 | 10,000 | 10,000 | 10,000 | 10,000 | 10,000 | 10,000 |

*Source*: Ministry of Information and Broadcasting (MIB).

operations, since Sumangali Cable Vision is a wholly owned subsidiary of Sun TV Network.

Lack of infrastructure remains a major challenge for cable and satellite operators: with more than 500 available channels and a carriage capacity of only 100, some MSOs have imposed a "placement fee" in addition to carriage fees.[41] As a result, channel providers have pursued backward integration, setting up companies individually or as joint ventures, and undertaken a lobbying campaign to mandate digitization. The later effort has been successful, and so digitization is now being carried out in four phases, the first of which was completed in the major metropolitan areas of New Delhi, Mumbai, Kolkata, and Chennai by the end of 2012.[42] All other major cities (those with a population of one million) are to complete the second phase of the campaign by the end of 2013. Phase three will be implemented shortly thereafter for all other municipalities, while the rest of the country is supposed to complete the process by the end of 2014.[43] The high cost of the digitization campaign has led many companies to opt for direct-to-home (DTH) platforms. This programming has reached 13% of the estimated 134 million homes in India with television access, and the market was valued at US$618 million (INR30 billion) in 2008, but most DTH providers still operate at a loss. There is still room for growth, however, as the estimated market value is US$3.1 billion (INR149 billion) (Tables 25.6 and 25.7).[44]

## Video Channels

Hoteliers and their guests drove the initial demand for video channels in India during the 1990s so as to bring Indian hotels up to international standards; that is, upgrading channel access to include 24/7 programming from CNN (US) and other networks. In 1990, Hutchison Whampoa (Hong Kong) launched Satellite Television Asia Region (STAR) TV after purchasing a satellite that could transmit into both India and China. STAR TV then sought for an Indian partner to lease transponders to provide Hindi-language content and partnered with Zee to launch Asia Today. Launched in 1992, it became India's first video channel, and for several years operated as a joint venture with Zee. Later, it added MTV (US) and BBC (UK) to its list, though these were dropped when Rupert

41. LCOs operate as monopolists or duopolists within a small geographical region, usually not much larger than a major metropolitan area, which partly explains the highly fragmented nature of the industry.

42. PricewaterhouseCoopers (PwC). "PwC India Entertainment and Media Outlook 2011." New Delhi: PwC India, 2011. Feb. 18, 2013. <http://www.pwc.in/en_IN/in/assets/pdfs/publications-2011/India_Entertainment_and_Media_Outlook_2011.pdf>.

43. "Time schedule for digitization of cable system." *India Telecom Online*, Apr. 17, 2011, Dec. 25, 2012. <http://www.indiatelecomonline.com/time-schedule-for-digitization-of-cable-system/>.

44. Kohli-Khandekar, Vanita. *The Indian Media Business*. 3rd ed. New Delhi: SAGE Publications, 2010. 80.

## Table 25-6. Multichannel Video Platforms: Cable MSOS (Market Shares by Revenue), 2004–2008

| | 2004 | 2008 | 2013 |
|---|---|---|---|
| Hathway Bhawani Cabletel & Datacom[1] | 11.5 | 12.8 | 23.5 |
| Zee (Essel Group) | 6.0 | 11.4 | 11 |
| DEN Networks[2] | N/A | 3.0 | 18.5 |
| Asianet Satellite (Asianet Communications: STAR TV; Jupiter Entertainment, US)[3] | 0.1 | DEN Networks | DEN Networks |
| Others | 82.4 | 40.8 | 47 |
| Total Revenue (bil INR) | 99.0 | 167 | |
| Total Revenue (bil US$) | 2.3 | 3.4 | |
| C4 | 17.6 | 59.0 | 53 |
| HHI | 6,924 | 1,666 | 1,016 |
| *N* (>1%) | 3 | 3 | 3 |
| Noam Index | 3,998 | 962 | 587 |

*Source*: Kohli-Khandekar (2010).

[1] Rupert Murdoch's News Corp. owns a 17% stake in Hathway Bhawani Cabletel & Datacom.

[2] Includes STAR TV, a 50:50 joint venture between DEN Networks and with News Corp. that carries 25 channels.

[3] STAR TV is a 50:50 joint venture between DEN Networks and with News Corp.

## Table 25-7. Multichannel Video Platforms: DTH (Market Shares by Revenue), 2004–2008

| | 2004 | 2008 | 2012[45] |
|---|---|---|---|
| Dish TV (Zee, Essel Group) | 100.0 | 43.0 | 27.3 |
| Tata Sky (Tata Group; Murdoch Family, News Corp., US/UK/AUS) | 0.0 | 37.2 | 19.1 |
| Sun Direct (Sun TV Network, Sun Group) | 0.0 | 19.2 | 15.3 |
| Bharti Airtel | 0.0 | 0.0 | 16.5 |
| Reliance Communications (Reliance Industries) | 0.0 | 0.6 | |
| Total Revenue (mil INR) | N/A | 3,050 | 91,883 |
| Total Revenue (mil US$) | N/A | 7.0 | 1,500 |
| C4 | 100.0 | 99.9 | 78.2 |
| HHI | 10,000 | 3,601 | 1,616 |
| *N* (>1%) | 1 | 3 | 4 |
| Noam Index | 41,943 | 2,079 | 808 |

*Source*: Kohli-Khandekar (2010).

45. 2012 update provided by editors. Data from Media Partners Asia 2013 DTH industry study and forecast report.

Murdoch's News Corp. purchased Hutchison Whampoa's 63.6% stake in STAR TV in 1993. The BBC and MTV slots were soon filled by Sun Network TV, originally a Telegu-language channel in South India that has since expanded to offer 20 additional channels.[46] Domestic content providers, such as New Delhi Television (NDTV), began moving to STAR TV over DD because of its more favorable advertising contract terms. Further consolidation can be expected in the foreseeable future, and concentration will continue to increase as more mergers and acquisitions take place.

At present Zee's operations consist of 38 television channels, a film production house, and two newspapers.[47] Zee's pay channels are currently distributed by a joint venture called Media Pro Enterprise India Pvt. Ltd. Zee has a 74% stake in it, and the US company Turner International (Time Warner) holds the remaining 26%.

Sun TV Network has 32 TV channels in several languages. Its parent, Sun Group, owns 45 radio stations, 2 daily newspapers, and 4 magazine titles, the film production company Sun Pictures, Sun Direct (DTH), and Sumangali Cable Vision (SCV).[48]

The New Delhi Television Group (NDTV) began operating in the 1990s as a content provider for DD, but then launched a 24-hour news channel in 2003 and now has seven pay-TV channels.[49] NDTV has content-sharing arrangements with several print media partners, especially the newspapers the *Indian Express* and *The Hindu*.

The Network18 Group consists of TV 18 and Network18. TV 18 is a joint venture of Network18 and NBCUniversal, owned by GE/Comcast (US). In India, Network 18 owns 13 television channels, a film production company, and interests in magazine publishing and printing as well. In a joint venture with Sun TV Network, it distributes several vernacular language channels in South India. In FY2011–12, the company reported revenues of US$49.2 million (INR2.7 billion).[50]

One of the most significant acquisitions in the Indian audiovisual market was the 2012 deal between Reliance Communications and Network18. The Network18 Group had a total debt of about US$49.2 million (INR4.7 billion), and it was in the process of acquiring the assets of Ushodaya Enterprises' Eenadu TV (ETV).[51] The deal, estimated to be worth US$273 million (INR15 billion), was structured to help Network18 pay off its debt and simultaneously acquire ETVs assets.[52] Besides securing indirect control of Network18, Reliance also gained exclusive video channel content in several vernacular languages for 4G mobile distribution.

Other important companies in the video channel market are ESPN (US)—which has been able to take advantage of the popularity of cricket in India to carve out a niche—BBCL, the India Today Group (ITD), and, until 2012, the STAR TV-ABP joint venture Media Content & Communications Services (MCCS) (Table 25.8).[53]

## Film

In India, distinctions among film producers, distributors, and exhibitors are blurred: most studios in this sector manage at least two of these activities directly, most notably the old family-owned studios such as Rajshri Productions, RK Films, Yashraj Films, and Ushodaya Enterprises. These studios were able to hang on through difficult times after independence. From the 1940s

46. *Sun TV Channel.* Sun Group, 2009. July 5, 2012. <http://www.sun.in/television.aspx>.

47. "Zee Entertainment Enterprises Limited Annual Report 2011-2012." Mumbai: Zee, 2012. Dec. 27, 2012. <http://vidz.zeecdn.com/zeetele/pdfs/zeear2011-2121-e72d9565b5c91b6.pdf>.

48. "Sun TV Network Net Profit Declines 12.42% in the June 2012 Quarter." *IIFL*, Aug. 3, 2012. Dec. 26, 2012. <http://www.indiainfoline.com/Markets/News/Sun-TV-Network-net-profit-declines-12.42-percent-in-the-June-2012-quarter/4441514693>.

49. *NDTV.* New Delhi: NDTV Convergence Limited, 2010. Oct. 25, 2010. <http://www.ndtv.com/convergence/ndtv/corporatepage/index.aspx>.

50. Network18. "Annual Report: 2011–2012." New Delhi: Network18, 2012. Dec. 27, 2012. <http://www.network18online.com/reports/AnnualReportFinal210812.pdf>.

51. Gokhale, N. "Reliance-TV 18 Deal Fires Up Media Stocks by Up to 15%." *The Times of India*, May 24, 2012. May 24, 2012. <http://lite.epaper.timesofindia.com/mobile.aspx?article=yes&pageid=12&edlabel=ETBG&mydateHid=04-01-2012&pubname=&edname=&articleid=Ar01205&format=&publabel=ET>.

52. Gokhale, N. "Reliance-TV 18 Deal Fires Up Media Stocks..."

53. MCCS was dissolved in 2012.

Table 25-8. Cable And DBS TV Programming Channels (Market Shares by Revenue), 2000–2010

| | 2000 | 2004 | 2008 | 2010 |
|---|---|---|---|---|
| Sun TV Network (Sun Group) | 9.5 | 14.4 | 18.5 | 22.3 |
| Zee (Essel Group) | 47 | 31.1 | 30.1 | 30.2 |
| ESPN India (Walt Disney, US) | 9.4 | 13.5 | 16.9 | 15.2 |
| New Delhi Television (NDTV) | 6.8 | 3.9 | 6.6 | 5.3 |
| India Today Group (ITG) | N/A | 9.1 | 5.0 | 4.8 |
| Network18 Group | N/A | 2.8 | 9.0 | 7.9 |
| Asianet Communications (STAR TV; Jupiter Entertainment, US)[1] | N/A | 3.3 | 3.3 | 3.9 |
| Sahara India Commercial Corporation | N/A | 2.3 | 3.6 | 3.1 |
| Times Global Broadcasting Co. (BCCL) | N/A | N/A | 2.2 | 3.5 |
| Malayalam Communications | N/A | 1.2 | 0.7 | 0.8 |
| Raj Television Network | 3.1 | 2.3 | 1.2 | 0.8 |
| Mavis Satcom | 0.3 | 2.6 | 0.7 | 0.7 |
| MCCS India (ABP; STAR TV) | N/A | 5.6 | 0.1 | 0.1 |
| Others | 27.4 | 32.1 | 25.5 | 24.4 |
| Total Revenue (mil INR) | 9,900 | 15,400 | 46,600 | 59,200 |
| Total Revenue (mil US$) | 212 | 352 | 961 | 1,290 |
| C4 | 72.6 | 68.0 | 74.5 | 75.6 |
| HHI | 2,442 | 1,517 | 1,712 | 1,793 |
| *N* (>1%) | 5 | 12 | 10 | 9 |
| Noam Index | 1,092 | 438 | 541 | 598 |

*Source*: Center for Monitoring Indian Economy (CMIE).
[1] STAR TV is a 50:50 joint venture between DEN Networks and with News Corp.

on, studios were increasingly hard-pressed to secure capital, and so a number of other established studios had to shut down. Financing films became a risky business during this period, and the prevailing model was for production companies to come together for individual projects and then disband upon the conclusion of one. This trend continued through the 1980s and 1990s, when the number of moviegoers dropped from 12.2 million in 1986 to 5 million in 1992. By late 1990s, the industry began to turn its fortunes around by expanding its operations to include the sale of music and television rights, growing overseas distribution networks to cater to the Indian Diaspora, and successfully producing several blockbusters that reinvigorated national audiences' interest in domestic productions.

By 2008, market value had grown exponentially, by 350%, partly thanks to the transition to digital projectors and the opening of multiplexes after 1997 when PVR Cinemas opened in Delhi. As box office receipts increased, financial enterprises began purchasing assets in film production and distribution, thus substantially increasing the availability of capital. In addition, in 1998, the government decided to allow up to 100% FDI in the industry.[54] Domestic film production has increased to a global high of over 3,000 in recent years as a result of these liberalization policies: 1,000 domestic movies grossed

54. Kohli-Khandekar, Vanita. "The filmi ecosystem." *Business Standard*, Oct. 9, 2012. Feb. 19, 2013. <http://www.business-standard.com/india/news/vanita-kohli-khandekarfilmi-ecosystem-/488947/>.

about US$2 billion (INR97 billion) from the sale of 3.3 billion tickets nationally in 2008.[55] The market has expanded at a compounded annual rate of growth (CAGR) of 17.7% between 2006 and 2009, and CAGR is expected to increase at a rate of 9.1% through 2016 before plateauing.[56]

The more organized production and distribution networks that existed in the 1940s are reemerging. Although concentration remains low, the number of companies in production and distribution fell from 59 in 2002 to 49 in 2004 to 36 in 2008 (Table 25.9).

Table 25-9. Film Production/Distribution (Market Shares by Box Office %), 2000–2008

| | 2004 | 2008 | 2010 |
|---|---|---|---|
| Eros Entertainment | 4.3 | 3.8 | 12.9 |
| Yash Raj Films | 7.2 | 5.7 | 0.0 |
| Sun Pictures (Sun Group) | N/A | 0.9 | 9.1 |
| Baba Arts | 0.7 | 1.1 | 6.5 |
| Mukta Arts | 2.8 | 5.1 | 4.7 |
| Shree Ashtavinayak Cine Vision | 1.4 | 3.8 | 3.4 |
| Reliance Mediaworks (Reliance Industries) | N/A | 4.9 | 5.3 |
| Viacom 18 Media (Network18, Reliance ADAG) | N/A | 1.4 | 3.3 |
| UTV (UTV; Walt Disney, US) | 3.4 | 5.6 | 9.3 |
| Fox Star Studio (20th Century Fox, Murdoch Family, News Corp., US/UK/AUS) | N/A | 2.9 | 4.1 |
| Warner Bros. (Time Warner, US) | N/A | 4.2 | 1.3 |
| Columbia TriStar (Sony, Japan/US) | 2.5 | 6.5 | 0.8 |
| Universal (GE/Comcast, US) | N/A | 5.8 | 0.9 |
| Tips Industries | 5.4 | 1.1 | 1.2 |
| K Sera Sera | 2.0 | 2.0 | 1.5 |
| Inox | 0.6 | 0.0 | 0.0 |
| PVR Cinemas | N/A | 1.1 | 1.0 |
| Mediaone Global Entertainment | N/A | 0.6 | 0.7 |
| Fame India | 0.6 | 0.5 | 0.2 |
| Midvalley | N/A | 1.4 | 0.8 |
| Shringar | 0.5 | 0.5 | N/A |
| National Film Development Corp. | N/A | 0.7 | N/A |
| Others | 68.5 | 40.3 | 33.0 |
| Total Revenue (mil INR) | 18,600 | 32,500 | 24,400 |
| Total Revenue (mil US$) | 426 | 670 | 535 |
| C4 | 20.3 | 23.6 | 37.9 |
| HHI | 132 | 259 | 476 |
| *N* (>1%) | 8 | 16 | 13 |
| Noam Index | 47.0 | 65.0 | 132.0 |

*Source:* Center for Monitoring Indian Economy (CMIE).

55. Kohli-Khandekar, Vanita. *The Indian Media Business*. 3rd ed. New Delhi: SAGE Publications, 2010. 117.

56. KPMG and FICCI. "In the interval, but ready for the next act: FICCI-KPMG Media and Entertainment Industry Report 2009." kmpg.com, 2009. Feb. 18, 2013. <http://www.kpmg.com/IN/en/ThoughtLedership/In%20the%20interval-%20But%20ready%20for%20the%20next%20act.pdf>.

## TELECOMMUNICATIONS MEDIA

### Wireline Telecom

Until the 1990s, handsets were considered a luxury good in India, and call rates were prohibitively expensive. In 1991, the long-distance calling market was partially privatized. TRAI was set up in 1997 to oversee the market as it was liberalized. Increased per capita incomes in India have driven (and reinforced) the growth of the market. Until 2000, the Department of Telecommunications (DoT) was the sole landline provider, after which point it lost significant market share because of the poor quality of its services.

Bharti Airtel, the fifth largest telecom operator in the world, enjoyed a greater market share than its nearest rival, Tata Communications, in the early 2000s as a result of an aggressive marketing campaign.[57] By 2008, Tata Communications and Bharti Airtel were close competitors, though DoT still dominated the market, and a fourth market entrant, Quadrant Televentures, had been reduced to a marginal entity. In 2010, Tata Communications finally overtook Bharti Airtel's market share by pursuing a strategy of building new infrastructure in underserved rural and urban areas (Table 25.10).

The industry has remained concentrated due to the limited number of licenses granted and a lack of capital investment in infrastructure. These conditions will persist, since wireless telecom is an established, more affordable option with a high household penetration rate.

### Wireless Telecom

There were 4 million registered mobile users in India in 2004 and 874 million in 2011.[58]

Table 25-10. Wireline Telecom[1]

| | 2000 | 2004 | 2009 | 2011 |
|---|---|---|---|---|
| Department of Telecommunications (public) (Govt. of India) | 98.1 | 88.3 | 86.6 | 81.0 |
| BSNL (Bharat Sanchar Nigam Ltd.) (public) | | | *77.1* | *70.4* |
| MTNL (Mahangar Telephone Nigam Ltd.) (public) | | | *9.5* | *10.6* |
| Tata Teleservices (Tata Group) | 0.5 | 3.0 | 2.5 | 4.3 |
| Bharti Airtel | 0.3 | 8.0 | 7.3 | 10.2 |
| Reliance Communications (Reliance Industries) | | | 3.0 | 3.9 |
| Quadrant Televentures | 0.0 | 0.7 | 0.5 | 0.6 |
| Total Revenue (bil INR) | 50.0 | 262 | | |
| Total Revenue (bil US$) | 1.1 | 6.0 | | |
| C4 | 100.0 | 100.0 | 99.7 | 99.3 |
| HHI | 9,630 | 7,865 | 7,568 | 6,699 |
| *N* (>1%) | 1 | 3 | 4 | 4 |
| Noam Index | 9,630 | 4,541 | 3,784 | 3,350 |

*Source*: Center for Monitoring Indian Economy (CMIE); Bharti Airtel.

[1] 2000–2004 by revenue, 2009–2011 by subscribers. Tables consolidated by editor.

57. "Quarterly Report on the Results for the Third Quarter and Nine Months Ended December 31, 2011." New Delhi: Bharti Airtel, 2012. July 12, 2012. <http://www.airtel.in/QuarterlyResult/Quarterly_report_Q3_11.pdf>.

58. Telecom Regulatory Authority of India (TRAI). "Point of Interconnection (POIs) Congestion report of the Cellular Mobile Telephone Service Providers (CMTS) and Basic Telephone Service (Wireline) Providers for the months of October, November & December 2011." TRAI, 2012. Feb. 12, 2012. <http://www.trai.gov.in/WriteReadData/PIRReport/Documents/NW-report-December11.pdf>. 4.

The market opened up in 1992 when the DoT invited bids for mobile services in the four largest metropolitan areas—New Delhi, Mumbai, Kolkata, and Chennai—which led to significant overbidding by prospective licensees; mobile services were then introduced commercially two years later. By 1998, eight mobile operators (and some landline operators that had also entered the market) had defaulted on their license fees. Only Bharti Airtel (then Bharti Tele-ventures) reported a profit for that year, which came to a mere US$0.6 million (INR25 million). In response, a new national telecommunications policy was implemented in 1999, and a market boom began after 2002; three years in, the number of mobile phones overtook landlines. Since 2000, the availability of mobile services though wireless has increased greatly, particularly in rural areas, where there is still little wireline infrastructure. The market continues to grow rapidly: 130 million new subscribers were added in 2008, up from 95 million new subscribers in 2007. In 2008, India's wireless phone subscriber base stood at 386 million individuals.[59]

Reliance Communications took a very different route into this market from other firms, namely Bharti Airtel, by offering a low-end wireless in local loop (WiLL) technology to low-income customers. This means Reliance does not have to bid for a telecommunications license like its competitors do. It has acquired a controversial reputation among other providers because after securing low-income customers, Reliance's focus shifted to imposing higher-spectrum fees and marketing more expensive full-service platforms (Table 25.11).

Concentration has increased in recent years, and one event in particular stands out because the increase in concentration represented the fallout of an anti-corruption investigation by regulators: in 2012, Bharti Airtel, Vodafone India and Idea Cellular together gained 6.9 million subscribers when the Supreme Court of India ordered the cancellation of 122 licenses, including ones for international groups such as Telenor (Norway), Sistema (Russia) and Etisalat (UAE) that had been awarded in 2008.[60] TRAI stated it sought to lower concentration in 2008 by awarding these 122 licenses, but the Supreme Court ordered their annulment due to corruption charges that had been brought against the then-minister of telecommunications. The investigation had been started because of significant discrepancies between the licensing fees that were estimated and the sums actually collected. Some US$190 billion (INR1.8 trillion) was lost as a result of the now-defunct licenses, according to the Comptroller and Auditor General of India (CAG). Other state agencies have estimated the losses to be lower, though at a still massive US$32.8 billion (INR310 billion) deficit.

## INTERNET MEDIA

### Internet Service Providers (ISP)

Initially, when the Internet became available in India 1995, access was limited to the major cities under the aegis of the public monopolist Videsh Sanchar Nigam Limited (VSNL), which is now part of Tata Communications. By 2008, there were 12.9 million Internet subscribers in India and by 2011, the market penetration rate reached 10.2%, or 121 million active users.[61, 62] The ISP market is highly integrated with the wireless market, so much so that the two are effectively undifferentiated by regulators.

There is no equity cap to foreign investment in ISPs that do not provide gateways or infrastructure. However, foreign-owned Indian subsidiaries firms must divest 26% of their equity

59. "Wireless Subscribers Base at 386 mn + Airtel Losing to Reliance + Broadband Growth Slows Down." *goBroadband,* Apr. 23, 2009. July 2, 2012. <http://convergence.in/blog/2009/04/23/wireless-subscribers-base-at-386-mn-airtel-losing-to-reliance-broadband-growth-slowsdown/>.

60. Taneja, Mansi. "Top Three Players Consolidate Position in Telecom Market." *Business Standard,* Apr. 22, 2012. <http://www.business-standard.com/india/news/top-three-players-consolidate-position-in-telecom-market/472221/>.

61. "How Cable TV Began and Spread in India." indiancabletv.net, Feb. 4, 2010. July 9, 2012. <http://indiancabletv.net/cabletvhistory.htm>.

62. Sengupta, Debjoy. "India Achieves a 10.2% Internet Penetration Rate." *The Economic Times,* Apr. 10, 2012. Apr. 10, 2012. <http://articles.economictimes.indiatimes.com/2012-2004-10/news/31318824_1_internet-users-penetration-number-of-facebook-users>.

Table 25-11. Wireless Telecom (Market Shares by Revenue), 2000–2010

| | 2000 | 2004 | 2008 | 2010 | 2013[63] |
|---|---|---|---|---|---|
| Bharti Airtel | 20.9 | 25.2 | 29.1 | 34.1 | 30.9 |
| Reliance Communications (Reliance Industries) | 2.8 | 2.6 | 21.0 | 15.0 | 7.6 |
| Vodafone India (Vodafone, UK) | 35.0 | 24.0 | 19.3 | 23.0 | 23.4 |
| Idea Cellular (ABG Group) | 10.8 | 14.7 | 9.9 | 11.2 | 16.2 |
| BSNL/Department of Telecommunications (public) | N/A | 23.3 | 15.0 | 10.4 | 6.6 |
| Tata Communications (Tata Group) + DoCoMo (Japan) | 2.7 | 4.9 | 2.2 | 1.9 | 7.5 |
| Dishnet Wireless (Siva Group) | N/A | N/A | 0.7 | 1.5 | |
| Sun Network (Sun Group) | 2.9 | 2.9 | 2.0 | 2.1 | |
| Loop Mobile | 14.4 | 3.2 | 0.8 | 0.6 | |
| Sistema Shyam Teleservices | N/A | 0.7 | 0.2 | 0.2 | |
| Others | 18.8 | 12.7 | 20.9 | 16.6 | 7.7 |
| Total Revenue (bil INR) | 23,7 | 140 | 778 | 1,056 | 1,489 |
| Total Revenue (bil US$) | 0.5 | 3.2 | 15.8 | 23.2 | |
| C4 | 81.2 | 87.3 | 79.1 | 83.4 | 78.1 |
| HHI | 2,012 | 2,019 | 1,991 | 2,164 | 1,922 |
| *N* (>1%) | 7 | 8 | 7 | 8 | 6 |
| Noam Index | 760 | 713 | 753 | 765 | 785 |

*Source*: Center for Monitoring of Indian Economy (CMIE).

in favor of Indian owners within five years as a condition for entering the market, and their services are subject to stringent licensing and data security requirements (Table 25.12).[64]

## Search Engines

Google (US) and Yahoo! (US) dominate the market. Consumers were quick to adopt Google and Yahoo! because English is seen as the language of the upwardly mobile young Indian professional and also because English-speaking members of the Indian Diaspora had heavily invested in developing ICT platforms for the Indian market. Both Google and Yahoo! are deepening their presence in India by offering more Hindi- and vernacular language content. However, the non-English market is relatively underserved when compared with the depth of the average search volume for English-language results (Table 25.13).

## Online News

One of the first Indian Internet portals was IndiaWorld.com, launched in 1995. The site initially charged a subscription fee but eventually

63. 2013 updates provided by editors using data from "Top Indian Telecoms Based on Service Revenue Market Share." *Telecom-Lead*. N.p., Nov. 18, 2013. Web. May 12, 2014. <http://www.telecomlead.com/telecom-services/top-indian-telecoms-based-service-revenue-market-share-72248/>.

Khare, Shobhit. "India Telecom: Consolidation Calling. . .– Detailed Report." *Motilal Oswal*. N.p., June 2013. Web. May 12, 2014. <http://www.motilaloswal.com/site/rreports/HTML/635076879701331004/index.htm>.

64. "Wireless Subscribers Base at 386 mn + Airtel Losing to Reliance + Broadband Growth Slows Down." *goBroadband*, Apr. 23, 2009. July 2, 2012. <http://convergence.in/blog/2009/04/23/wireless-subscribers-base-at-386-mn-airtel-losing-to-reliance-broadband-growth-slowsdown/>.

Table 25-12. Internet Service Providers (Market Shares by Revenue), 1996–2008[1]

| | 1996 | 2000 | 2004 | 2008 | 2013 |
|---|---|---|---|---|---|
| Department of Telecommunications (public) | 100.0 | | 0.9 | 0.3 | 69.1 |
| Reliance Communications (Reliance Industries) | 0.0 | | 34.3 | 48.3 | 11.9 |
| Sify | 0.0 | 33.9 | 14.4 | 8.1 | |
| Exlservice.com | 0.0 | | 7.9 | 8.1 | |
| Hathway Cable & Datacom | 0.0 | | 6.0 | 5.4 | 1.7 |
| Bharti Airtel | 0.0 | | | 5.7 | 6.5 |
| Dishnet (Zee) | 0.0 | 5.3 | 7.0 | 0.8 | |
| rediff.com | 0.0 | 4.9 | | | |
| Ortel Communications | 0.0 | 0.1 | | | |
| Others | 0.0 | 55.8 | 36.4 | 30.1 | 10.8 |
| Total Revenue (mil INR) | N/A | 1,700 | 21,600 | 68,100 | 135,904 |
| Total Revenue (mil US$) | N/A | 36.4 | 494 | 1,384 | 2,200 |
| C4 | 100.0 | 44.2 | 63.6 | 70.1 | 89.2 |
| HHI | 10,000 | 1,201 | 1,532 | 2,526 | 4,962 |
| *N* (>1%) | 1 | 3 | 5 | 5 | 4 |
| Noam Index | 10,000 | 693 | 685 | 1,130 | 2,481 |

*Source*: Telecom Regulatory Authority (TRAI).

[1] Tata Communications and VSNL are not included in Table 25.12 because they operate in the gateways and overseas business.

Table 25-13. Internet Search Engines (Market Shares by Revenue), 2008

| | 2008 |
|---|---|
| Google | 81.4 |
| Yahoo! | 9.4 |
| Ask.com | 1.9 |
| MSN (Microsoft) | 1.7 |
| Rediff.com | 1.5 |
| Facebook | 0.8 |
| C4 | 94.4 |
| HHI | 6,724 |
| *N* (>1%) | 5 |
| Noam Index | 3,015 |

*Source*: ComScore, TH@T! Company.

shifted to a free-content model. Reflecting the limited growth of the market in its early days, in 1997, the site generated 95% of its traffic from outside India. By 1997, though, Indian traffic numbers rose so much that the site's owners created specialized sports, news, and lifestyle portals. In 1999, IndiaWorld.com was bought by Satyam Infoway for US$115 million (IN$;5 billion) and was registering 13 million page views a month.[65] Another early major Indian portal was rediff.com, also started in 1995. In 2010, it had 95 million registered users and had revenues of US$21.8 million (INR992 billion).[66, 67] BCCL's The Times Group launched Indiatimes.com in 1999, and the group's sites now collectively receive over 400 million monthly page views. The Times Group's sites represent the country's third largest online news portal in terms of traffic, just behind rediff.com and Yahoo! (Table 25.14).[68]

The *Indian Express* and *Hindustan Times* have both successfully used their online presence to compensate for their limited print distribution network outside of the major metropolitan areas. Despite the growing potential of the market, most Indian online news outlets are unconnected to a print publication.

## CONCLUSION

The most significant area of growth for Indian media markets has been in the provision of vernacular language content across all media. Concentration is increasing for the rapidly expanding telecommunications and pay-TV markets, but not the print media market, which

Table 25-14. Online News Media (Market Shares by Monthly Unique Visitors), 2010–2011

| | 2010 | 2011 |
|---|---|---|
| NDTV.com (NDTV) | 4.2 | 4.9 |
| The New York Times Online (The New York Times Group, US) | 12.0 | 11.9 |
| CNN.com (Time Warner, US) | 2.0 | 3.9 |
| Yahoo! News Network (US) | 8.3 | 32.7 |
| Oneindia.in | 0.8 | 8.0 |
| Times of India Online (The Times Group, BCCL) | | 23.7 |
| HT Media | | 8.2 |
| BBC.co.uk (UK) (public) | | 5.7 |
| The Indian Express Group | | 4.5 |
| Others | | 0.0 |
| C4 | 26.4 | 76.5 |
| HHI | | 1,996 |
| *N* (>1%) | | 9 |
| Noam Index | | 665 |

*Source*: ComScore.

65. Kohli-Khandekar, Vanita. *The Indian Media Business*. 2nd ed. New Delhi: SAGE Publications, 2006. 211.

66. "Investors." rediff.com, 2012. July 4, 2012. <http://investor.rediff.com/>.

67. Singh, Anubhuti. "Rediff Records 15 Percent Growth in Registered User Base." *Media Newsline*, Oct. 28, 2010. July 4, 2012. <http://www.medianewsline.com/news/122/ARTICLE/6876/2010-2010-28.html>.

68. "Case Study—eCommerce Sites: Indiatimes.com." Mumbai: Human Factors International, 2009. July 4, 2012. <http://www.humanfactors.com/about/Indiatimes.asp>.

is also growing significantly, in contrast to the downward trends in most of the other countries surveyed in this study (the newspaper market will continue to grow at a 17% CAGR between 2013 and 2015).[69] In addition, the 2011 Indian census suggests that while 26% of the Indian population in still illiterate, this represents a marked improvement from 35% in 2001.[70] Since consumption of print media is closely linked to literacy rates, this is good news for the market. The audiovisual market has grown faster than the print market, since it is not directly linked to literacy, and the telecommunications market has grown the fastest of all.

Indian media analysts are particularly concerned about the strong ties between content producers and political elites. Their ownership of local news outlets is closely linked to the problem of objective news coverage in industrial disputes and in tracking campaign finances. Disclosure requirements for media producers are met with a bare minimum of effort. The recommendations by TRAI that public operators stay out of the satellite TV market is causing a political uproar because politicians and their patronage networks wish to take advantage of this growing market.

TRAI's ability to contest such a policy is very limited. It lacks the authority of a body such as the US's Federal Communications Commission (FCC) or the UK's Ofcom, and is only able to initiate review proposals for market-entry regulations or guidelines on mergers and acquisitions. The absence of a true umbrella media regulator means that advertising content regulations are quite lax. There is a preference among media producers for advertiser-oriented supplements and meeting the demands of advertisers. BCCL has actually institutionalized these practices with its "MediaNet" placement system, which provides firms with paid advertorials in The Times Group's titles. The phenomenon of large industrial conglomerates investing heavily in media companies—either through direct ownership or indirect actions such as offering or withholding advertising revenue—is still largely undiscussed in the Indian context.

TRAI would either have to have its mandate greatly expanded to address these issues, or be subsumed into a wholly new agency that would actually have the authority of an FCC or Ofcom. Such an agency would not be well received by many politicians or by media conglomerates, especially those in the pay-TV market, who benefit from the gaps in the existing regulatory structures.

## Bibliography

### 1. Books and Articles


Beck, Sanderson. "British India 1818–1875." *India & Southeast Asia to 1875*. Santa Barbara: World Peace Communications, 2004. <http://wiki.phalkefactory.net/images/f/fc/British_India_1818-1875.pdf>.

Biwalkar, Meghana. "Why Go FM Became Radio One." rediff.com, July 18, 2006. June 10, 2012. <http://inhome.rediff.com/money/2006/jul/18radio.htm>.

"Case Study—eCommerce Sites: Indiatimes.com." Mumbai: Human Factors International, 2009. July 4, 2012. <http://www.humanfactors.com/about/Indiatimes.asp>.

Demers, David P. *Global Media: Menace or Messiah?* Cresskill: Hampton Press, 1999.

Doshi, J. K. et al. "Diffusion of Television in India." *Media Asia* 15:4, 1988. Feb 21, 2013. <http://utminers.utep.edu/asinghal/Articles%20and%20Chapters/Singhal-Rogers-Doshi-Rahman-1988-Diffusion%20of%20television.pdf>.

Gokhale, N. "Reliance-TV 18 Deal Fires Up Media Stocks by Up to 15%." *The Times of India*, May 24, 2012. May 24, 2012. <http://lite.epaper.timesofindia.com/mobile.aspx?article=yes&


69. "Print Media to Grow at 17% CAGR during FY13-15: Report." *Business Standard*, Dec. 26, 2012. <http://www.business-standard.com/india/news/print-media-to-grow-at-17-cagr-during-fy13-15-report/496802/>/

70. "Literacy Rate in Indian State: Census 2011." *TheOnlineGK*, Apr. 2, 2011, Dec. 26, 2012. <http://theonlinegk.wordpress.com/2011/04/02/literacy-rate-in-indian-state-census-2011/>.

pageid=12&edlabel=ETBG&mydateHid=04-01-2012&pubname=&edname=&articleid=Ar01205&format=&publabel=ET>.
"How Cable TV Began and Spread in India." indiancabletv.net, Feb. 4, 2010. July 9, 2012. <http://indiancabletv.net/cabletvhistory.htm>.
Indiantelevision.com Team. "Radio City FY'09 Revenue Pegged at Rs 1.15 bn." indiatelevision.com, May 6, 2009. Feb. 21, 2013. <http://www.indiantelevision.com/headlines/y2k9/may/may46.php>.
Indiantelevision.comTeam. "White PAPER on CAS and Cable Industry." indiatelevision.com, Jan. 2, 2004. Feb. 21, 2013. <http://www.indiantelevision.com/headlines/y2k4/jan/jan6.htm>.
"Investors." rediff.com, 2012. July 4, 2012. <http://investor.rediff.com/>.
Jagannathan, N. S. *Independence and the Indian Press—Heirs to a Great Traition.* 1st ed. New Delhi: Konark Publishers, 1999.
Karkhanis, Sharad. *Indian Politics and the Press.* New Delhi: Vikas, 1980.
Kathuria, Raiat; Singh, H. V.; and Soni, Anita. "Telecom Policy Reform in India." Washington, DC: The World Bank, 2005. Feb. 21, 2013. <http://siteresources.worldbank.org/INTRANETTRADE/Resources/Singh.pdf>.
Kaul, H. K. *Books and National Development.* Apr. 27–29, 1968, Seoul, South Korea. Seoul: Academy House, 1968.
Khanna, T. and Palepu, K. "The Evolution of Concentrated Ownership in India: Broad Patterns and a History of the Indian Software Industry." Cambridge: National Bureau of Economic Research, 2004. June 14, 2012. <http://www.nber.org/papers/w10613>.
Kohli-Khandekar, Vanita. "The Filmi Ecosystem." *Business Standard,* Oct. 9, 2012. Feb. 19, 2013. <http://www.business-standard.com/india/news/vanita-kohli-khandekarfilmi-ecosystem-/488947/>.
Kohli-Khandekar, Vanita. *The Indian Media Business.* 3rd ed. New Delhi: SAGE Publications, 2010.
Kohli-Khandekar, Vanita. "When Politicians Own the Media." *Business Standard,* July 3, 2012. Feb. 18, 2013. <http://business-standard.com/india/news/vanita-kohli-khandekar-when-politicians-ownmedia/479159/>.
"Literacy Rate in Indian State: Census 2011." *The Online GK,* Apr. 2, 2011, Dec. 26, 2012. <http://theonlinegk.wordpress.com/2011/04/02/literacy-rate-in-indian-state-census-2011/>.
Malhotra, Dina N., ed. *50 Years of Book Publishing in India Since Independence.* New Delhi: Federation of Indian Publishers, 1998.
Malhotra, Dina N., ed. *60 Years of Book Publishing in India, 1947–2007.* New Delhi: Federation of Indian Publishers, 2006.
Manoj, C. G. "TRAI Against Govt Entering Broadcast, Distribution." *The Indian Express,* 29 Dec. 2012. Dec. 29, 2012. <http://www.indianexpress.com/news/trai-against-govt-entering-broadcast-distribution/1051593/>.
May, Mumbai. "Aditya Birla Group Acquires Stake in Living Media India." *Business Standard,* May 19, 2012. May 19, 2012. <http://business-standard.com/india/news/aditya-birla-group-acquires-stake-in-living-media-india/474783/>.
Piramal, G. *Business Maharajas.* New Delhi: Penguin, 1996.
"Print media to grow at 17% CAGR during FY13-15: Report." *Business Standard,* Dec. 26, 2012. Dec. 26, 2012. <http://www.business-standard.com/india/news/print-media-to-grow-at-17-cagr-during-fy13-15-report/496802/>/
Rathore, Sumantha. "The Best Way Out?" *afaqs! Reporter,* Aug. 16–31, 2009. Feb. 21, 2013. <http://www.afaqsreporter.com/perl/tbr/story.html?id=3966>.
Robin, Jeffrey. *India's Newspaper Revolution: Capitalism, Technology and the Indian-Language Press.* 3rd ed. New Delhi: Oxford, 2010.
"Search Engine Market Share: Why Is Search Engine Market Share So Important?" Leesburg: TH@T! Company, 2007. Feb. 21, 2013. <http://www.buildtelligence.com/s-e-market-share.htm>.
Sengupta, Debjoy. "India Achieves a 10.2% Internet Penetration Rate." *The Economic Times,* Apr. 10, 2012. Apr. 10, 2012. <http://articles.economictimes.indiatimes.com/2012-2004-10/news/31318824_1_internet-users-penetration-number-of-facebook-users>.
Sheth, Jagdish and Sisodia, Rajendra. *Rule of Three: Surviving and Thriving in Competitive Markets.* New York: Free Press, 2002.
Shukla, Archana. "With FIR, Jindal Gives Police Tapes of 'Meetings' with Zee." *Indian Express,* Oct. 12, 2012. Oct. 25, 2012. <http://www.indianexpress.com/news/with-fir-jindal-gives-police-tapes-of--meetings--with-zee/1015622>.
Singh, Anubhuti. "Rediff Records 15 Percent Growth in Registered User Base." *Media*

*Newsline*, Oct. 28, 2010. July 4, 2012. <http://www.medianewsline.com/news/122/ARTICLE/6876/2010-10-28.html>.

Singhal, Arvind; Doshi, J. K.; Rogers, Everett M.; and S. Adnan, Rahman. "Diffusion of Television in India." *Media Asia* 15:4, 1988. Aug. 5, 2012. <http://utminers.utep.edu/asinghal/Articles%20and%20Chapters/Singhal-Rogers-Doshi-Rahman-1988-Diffusion%20of%20television.pdf>. 222–229.

Subramanian, N. Sundaresha. "RIL Controls Network18 Group Indirectly, Says CCI." *Business Standard*, June 23, 2012. June 23, 2012. <http://www.business-standard.com/india/news/ril-controls-network18-group-indirectly-says-cci/478250/>.

"Sun TV Network Net Profit Declines 12.42% in the June 2012 Quarter." *IIFL*, Aug. 3, 2012. Dec. 26, 2012. <http://www.indiainfoline.com/Markets/News/Sun-TV-Network-net-profit-declines-12.42-percent-in-the-June-2012-quarter/4441514693>.

Taneja, Mansi. "Top Three Players Consolidate Position in Telecom Market." *Business Standard*, Apr. 22, 2012. <http://www.business-standard.com/india/news/top-three-players-consolidate-position-in-telecom-market/472221/>.

Thierer, Adam. "The Rule Of Three: The Nature of Competition in the Digital Economy." *Forbes*, June 29, 2012. Dec. 25, 2012. <http://www.forbes.com/sites/adamthierer/2012/06/29/the-rule-of-three-the-nature-of-competition-in-the-digital-economy/>.

"Time Schedule for Digitization of Cable System." *India Telecom Online*, Apr. 17, 2011, Dec. 25, 2012. <http://www.indiatelecomonline.com/time-schedule-for-digitization-of-cable-system/>.

Trikha, Jatin; Garg, Manish; and Singh, Supreet. "Market Sizing and Future Growth Potential of Telecom in ABS Segment." Presentation, July 30, 2009. Feb. 21, 2013. <http://www.scribd.com/doc/17855326/Airtel-Ppt>.

"Wireless Subscribers Base at 386 mn + Airtel Losing to Reliance + Broadband Growth Slows Down." *goBroadband*, Apr. 23, 2009. July 2, 2012. <http://convergence.in/blog/2009/04/23/wireless-subscribers-base-at-386-mn-airtel-losing-to-reliance-broadband-growth-slowsdown/>.

## 2. Reports

Audit Bureau of Circulation. "January-June 2010." Mumbai: Audit Bureau of Circulation, 2010. & Audit Bureau of Circulation. "July-December 2010." Mumbai: Audit Bureau of Circulation, 2010.

Broadcasting Corporation of India (Prasar Bharati). "Annual Report 2007–08." New Delhi: Prasar Bharati, 2008. <http://prasarbharati.gov.in/NR/rdonlyres/5A7A0DC9-B6B6-4E7A-9C38-021E363E0367/3599/PBFINALREPORT.doc>.

Broadcasting Corporation of India (Prasar Bharati). "Annual Report 2008–09." New Delhi: Prasar Bharati, 2009. Feb. 21, 2013. <http://prasarbharati.gov.in/NR/rdonlyres/5A7A0DC9-B6B6-4E7A-9C38-021E363E0367/3634/EnglishPDF.pdf>.

Doordarshan and Annual reports of Ministry of I&B. Reports for 1984, 1988, 1992, 1996, 2000 and 2004 <http://prasarbharati.gov.in/NR/rdonlyres/5A7A0DC9-B6B6-4E7A-9C38-021E363E0367/3634/EnglishPDF.pdf>

Emerging Markets Information Service (EMIS), ISI Emerging Markets: A Euromoney Institutional Investor Company, India Reports.

Entertainment Network (India) Limited (ENIL). "Annual Report 2007–08." Mumbai: ENIL, 2008. Feb. 21, 2013. <http://www.enil.co.in/pdf/annual_reports/enil_annual_report_2007_2008.pdf>.

Ernst & Young. "The Animation and Gaming Industry in India: A Report by Ernst & Young." New Delhi: NASSCOM, 2009. July 12, 2012 <http://www.nasscom.org/animation-and-gaming-industry-indiaa-report-ernst-young>.

Federation of Indian Chamber of Commerce and Industry (FICCI), Media and Entertainment Division Report, "FICCI Frames" of March 2000, 2002, 2004, 2005, 2009.

Government of India. "Report of the Press Commission." New Delhi: Government of India Press, 1954.

Government of India, Telecom Regulatory Authority of India (TRAI) Annual reports of 1997–98, 1999–2000, 2000–01, 2003–04

HT Media. "Annual Report 2007–08." New Delhi: HT Media, 2008. Feb. 21, 2013. <http://www.htmedia.in/HTMLCorpImages/HTMediaCorpSite/pdf/Reports/Annual_Report_FY_2007-2008.pdf>.

Indian Readership Survey (IRS). *IRS Q1 2012 Topline Findings*. Media Research Users Council, 2012. Feb. 20, 2013. <http://mruc.net/irs2012q1-topline-findings.pdf>.

Indian Readership Survey (IRS). *IRS Q2 2011 Topline Findings*. Media Research Users Council,

2012. Feb. 20, 2013. <http://mruc.net/irs2011q2-topline-finding.pdf>.

Indian Readership Survey (IRS). *IRS Q4 2010 Topline Findings*. Mumbai: Media Research Users Council, 2012. Feb. 18, 2012. <http://www.pdfone.com/download/11_meri-saheli-hindi/irs-2010-q4-topline-findings-hansa-research-market-research.html>.

"Industry and Services: Media and Entertainment." New Delhi: Ministry of External Affairs, Government of India Investment & Technology Promotion (ITP) Division, 2009. May 10, 2012. <http://www.indiainbusiness.nic.in/industry-infrastructure/service-sectors/media-entertainment.htm>

"Investing in India, Foreign Direct Investment, Policy and Procedures of 2006." New Delhi: Department of Policy and Promotion, Ministry of Commerce and Industry, Government of India, 2007. Dec. 25, 2012. <http://dipp.nic.in/English/Archive/FDI_Manual/FDI_Manual_text_Latest.pdf>.

Ipsos. "India Pulse." Mumbai: Ipsos, 2012. Feb. 18, 2013. <http://www.ipsos.com/sites/ipsos.com/files/India%20Pulse%20-%20October%202012.pdf>.

KPMG and FICCI. "Hitting the High Notes: FICCI-KPMG Indian Media and Entertainment Industry Report 2011." kmpg.com, 2011. Feb. 18, 2013. <http://www.kpmg.com/IN/en/IssuesAndInsights/ThoughtLeadership/FICCI-KPMG-Report-2011.pdf>.

KPMG and FICCI. "In the interval, but ready for the next act: FICCI-KPMG Media and Entertainment Industry Report 2009." kmpg.com, 2009. Feb. 18, 2013. <http://www.kpmg.com/IN/en/ThoughtLedership/In%20the%20interval-%20But%20ready%20for%20the%20next%20act.pdf>.

"Media Guide India." New Delhi: Lowe Lintas and Partners, 2001.

MiD-Day Multimedia. "23rd Annual Report 2003-2004." Mumbai: Next MediaWorks Ltd., 2004.

MiD-Day Multimedia. "27th Annual Report 2008-2008." Mumbai: Next MediaWorks Ltd., 2008. Feb. 21, 2013. <http://nextmediaworks.com/annual-reports/annualreport07-08.pdf>.

Ministry of Information and Broadcasting (MIB). "Annual Report 2010–2011." New Delhi: Additional Director General (In-charge), Publications Division, Ministry of Information and Broadcasting, Government of India, 2011. Feb. 21, 2013. <http://www.mib.nic.in/writereaddata/html_en_files/document/docu_other/ar/AnnualReport2010-11-English.pdf>.

Ministry of Information and Broadcasting (MIB). "Annual Report 2011–2012." New Delhi: Additional Director General (In-charge), Publications Division, Ministry of Information and Broadcasting, Government of India, 2012. Feb. 21, 2013. <http://www.mib.nic.in/writereaddata/html_en_files/document/docu_other/ar/AnnualReport2011-12-English.pdf>.

NDTV. "Highlights for the Quarter ended 30 June 2012." New Delhi: NDTV, 2012. Dec. 27, 2012. <http://www.ndtv.com/convergence/ndtv/corporatepage/images/PressRelease Q1FY2013_2.pdf>.

Network18. "Annual Report: 2011–2012." New Delhi: Network18, 2012. Dec. 27, 2012. <http://www.network18online.com/reports/AnnualReportFinal210812.pdf>.

"NRS 2006—Key Findings." *The Hindu*, 2006. Feb. 20, 2013. <http://www.hindu.com/nic/nrs.htm>.

PricewaterhouseCoopers (PwC). "The Indian Entertainment & Media Industry: Sustaining Growth, Report 2008." New Delhi: PwC India, 2008. Feb. 18, 2013. <http://industry.bfi.org.uk/media/pdf/9/e/The_Indian_Entertainment___Media_Industry.pdf>.

PricewaterhouseCoopers (PwC). "PwC India Entertainment and Media Outlook 2011." New Delhi: PwC India, 2011. Feb. 18, 2013. <http://www.pwc.in/en_IN/in/assets/pdfs/publications-2011/India_Entertainment_and_Media_Outlook_2011.pdf>.

"Quarterly report on the results for the third quarter and nine month ended December 31, 2011." New Delhi: Bharti Airtel, 2012. July 12, 2012. <http://www.airtel.in/QuarterlyResult/Quarterly_report_Q3_11.pdf>.

"Reliance Communications Annual Report 2010-11." Hyderabad: Reliance Communications, 2011. Jan. 16, 2012. <http://www.bseindia.com/bseplus/AnnualReport/532712/5327120311.pdf>.

RK Swamy/BBDO. "Media Market Guide India 1997-98." TV Navigator, Media Division, cited in p 24, Revenue of All India Radio for years 1991–92, 1995–96 and 2003–04.

Telecom Regulatory Authority of India (TRAI). "Annual Report 2004–05." New Delhi: TRAI, 2006. Feb. 20, 2013. <http://www.trai.gov.in/WriteReadData/UserFiles/Documents/AnuualReports/ar_04_05.pdf>.

Telecom Regulatory Authority of India (TRAI). "Annual Report 2006–07." New Delhi: TRAI, 2008. Feb. 20, 2013. <http://www.trai.gov.in/

WriteReadData/UserFiles/Documents/AnuualReports/ar_06_07.pdf>.
Telecom Regulatory Authority of India (TRAI). "Annual Report 2007–08." New Delhi: TRAI, 2009. Feb. 20, 2013. <http://www.trai.gov.in/WriteReadData/UserFiles/Documents/AnuualReports/ar_07_08.pdf>.
Telecom Regulatory Authority of India (TRAI). "Point of Interconnection (POIs) Congestion report of the Cellular Mobile Telephone Service Providers (CMTS) and Basic Telephone Service (Wireline) Providers for the months of October, November & December 2011." TRAI, 2012. Feb. 12, 2012. <http://www.trai.gov.in/WriteReadData/PIRReport/Documents/NW-report-December11.pdf>.
Telecom Regulatory Authority of India (TRAI). "Recommendations on Issues relating to entry of certain entities into Broadcasting and Distribution activities." New Delhi: TRAI, 2008. Feb. 19, 2013. <http://www.cccindia.co/corecentre/Database/Docs/DocFiles/trai1.pdf>.
Zee Entertainment Enterprises Limited. "Annual Report 1999–2000." Mumbai: Zee Entertainment Enterprises Limited, 2000. Dec. 27, 2012. <http://vidz.zeecdn.com/zeetele/pdfs/zeeannualreport1999-bba65e04c3a412b.pdf>.
Zee Entertainment Enterprises Limited. "Annual Report 2000–2001." Mumbai: Zee Entertainment Enterprises Limited, 2001. Dec. 27, 2012. <http://vidz.zeecdn.com/zeetele/pdfs/zeeannualreport2001-2563b8804483360.pdf>.
Zee Entertainment Enterprises Limited. "Annual Report 2002–2003." Mumbai: Zee Entertainment Enterprises Limited, 2003. Dec. 27, 2012. <http://vidz.zeecdn.com/zeetele/pdfs/fullannualreports-2bcee2059a01293.pdf>.
Zee Entertainment Enterprises Limited. "Annual Report 2004–2005." Mumbai: Zee Entertainment Enterprises Limited, 2005. Dec. 27, 2012. <http://vidz.zeecdn.com/zeetele/pdfs/annualreport2005-92c4498514b21a1.pdf>.
Zee Entertainment Enterprises Limited. "Annual Report 2006–2007." Mumbai: Zee Entertainment Enterprises Limited, 2012. Dec. 27, 2012. <http://vidz.zeecdn.com/zeetele/pdfs/annualreport2007-9a87a39a200cbcb.pdf>.
Zee Entertainment Enterprises Limited. "Annual Report 2008–2009." Mumbai: Zee Entertainment Enterprises Limited, 2009. Dec. 27, 2012. <http://vidz.zeecdn.com/zeetele/pdfs/annualreportfulldownload-55e7cd63e7273ab.pdf>.
Zee Entertainment Enterprises Limited. "Annual Report 2010–2011." Mumbai: Zee Entertainment Enterprises Limited, 2011. Dec. 27, 2012. <http://vidz.zeecdn.com/zeetele/pdfs/annualreport2011-e6876aa5f3f67f3.pdf>.
Zee Entertainment Enterprises Limited. "Annual Report 2011–2012." Mumbai: Zee Entertainment Enterprises Limited, 2012. Dec. 27, 2012. <http://vidz.zeecdn.com/zeetele/pdfs/zeear2011-121-e72d9565b5c91b6.pdf>.

## 3. Websites

*ABP: Mjoy.* abp.in, 2011. Aug. 2, 2011. <http://www.abp.in/index.php?show=22002>.
*AdEx India.* Television Audience Measurement Media Research (TAM), 2012. <http://www.tamindia.com/tamindia/>.
*AIR Radio (1990–2013).* indiastate.com, 2013. Feb. 21, 2013. <http://www.indiastat.com/media/21/radiofm/236/airradio19902013/450129/stats.aspx>.
*Centre for Monitoring of Indian Economy (CMIE).* cmie.com, 2012. July 8, 2012. <http://www.cmie.com/>.
*NDTV.* New Delhi: NDTV Convergence Limited, 2010. Oct. 25, 2010. <http://www.ndtv.com/convergence/ndtv/corporatepage/index.aspx>.
*Sun TV Channel.* Sun Group, 2009. July 5, 2012. <http://www.sun.in/television.aspx>.
*Tata Communications: About.* tatacommunications.com, 2012. June 28, 2012. <http://www.tatacommunications.com/about/network.asp>.
*Worldspace Indian Limited.* franchiseindia.com, 2012. Oct. 28, 2012. <http://www.franchiseindia.com/categoryDetails.php?catid=13&id=FIHL4110692&pageID=1>.

## 4. Interviews

Mitra, Chandan; Managing Editor, *The Pioneer.* Personal interview with Anuradha Bhattacharjee. May 24, 2007.
Singh, Prabhakar; Regional Head, Centre for Monitoring Indian Economy. Personal interview with Anuradha Bhattacharjee. May 24, 2012.

# India—Data Summaries

ELI NOAM AND PAUL MUTTER

INDIA HAS THE SECOND-LARGEST domestic media market in the world by population size, but per capita spending is low (Table 25.15). India's overall national power index declined from 4,311 to 2,923 due to privatization in the market. The company that had the highest company power index—the state telecom monopolist BSNL—dropped from 2,679 to 1,205.33 due to competition from privately owned mobile platform companies such as Reliance and Bharti Enterprises, the latter of which became the top platform company in India after 2004 (Table 25.17). BSNL service prices and reliability were such that many customers left when alternatives became available after 2000. BSNL's market share as a platform provider declined by 4.18% per annum while Bharti Enterprises' increased by 2.23% and Reliance's by 1.5%.

Public ownership fell substantially, from 44.2% to 22.7%. The state companies BSNL and the Broadcasting Corporation of India, also known as Prasar Bharati, had been the two top firms, bot moved down to the number two and five spots. The Indian government is particularly concerned about retaining control over the production of broadcast news content on both radio and television, which ensures that Prasar has held onto its market leadership even as it fell from 31.3% in content in 2004 to 12.3% in 2009. Substantive restrictions for foreign direct investment are present in both content and platform media. A total of 17.1% of media industries are foreign-owned, largely due to mobile provider Vodafone's presence (UK).

The Times of India Group is the most diverse content company, with 29.2% of daily newspapers, 2.18% of radio, and 23.7% of online news media, and smaller shares in video channels and magazines. It controls 11.5% of the content media market (Table 25.16), while in the overall media market the presence of the telecom

Table 25-15. National Media Industries Concentration in India

| | 2004/5 | | 2011 OR MOST RECENT | | % CHANGE ANNUAL AVERAGE | |
|---|---|---|---|---|---|---|
| | COMPANY POWER INDEX IN COUNTRY | COMPANY SHARE OF THE OVERALL NATIONAL MEDIA MARKET (%) | COMPANY POWER INDEX IN COUNTRY | COMPANY SHARE OF THE OVERALL NATIONAL MEDIA MARKET (%) | COMPANY POWER INDEX IN COUNTRY | COMPANY SHARE OF THE OVERALL NATIONAL MEDIA MARKET (%) |
| BSNL (Public) | 2,679 | 34.0 | 1,205 | 20.2 | –9.2 | –2.3 |
| Bharti Enterprises | 156 | 8.1 | 640 | 20.5 | 51.8 | 2.1 |
| Prasar Bharti (Public) | 1011 | 10.2 | 241 | 2.5 | –12.7 | –1.3 |
| Reliance Industries (Anil Dhirubhai Ambani Group) | 40 | 1.7 | 129 | 9.2 | 37.8 | 1.3 |
| Vodafone (UK) | 122 | 5.1 | 282 | 12.3 | 22.0 | 1.2 |
| Zee Entertainment | 34 | 1.9 | 70 | 3.1 | 17.6 | 0.2 |
| Google (US) | 78 | 0.71 | 96 | 1.2 | 4.0 | 0.08 |
| BCCL (The Times of India Group) | 83 | 3.1 | 62 | 2.3 | –4.2 | –0.13 |
| Tata Group | 7 | 1.9 | 7 | 2.1 | 0.42 | 0.04 |
| Sun Group | 7 | 0.97 | 18 | 1.9 | 29.1 | 0.15 |
| Disney (US) | 5 | 0.41 | 8 | 0.57 | 12.4 | 0.03 |
| Living Media Group | 0.96 | 0.09 | 4 | 0.15 | 47.3 | 0.01 |
| Eros Entertainment | 0.52 | 0.12 | 2 | 0.16 | 49.2 | 0.01 |
| Network18 Group | 0.18 | 0.06 | 2 | 0.28 | 165.6 | 0.04 |
| Murdoch Interests (US) | 0.30 | 0.12 | 11 | 1.2 | 588.1 | 0.18 |

| | | | | | | |
|---|---|---|---|---|---|---|
| DEN Networks (50% Murdoch Interests, US) | 0.00 | 0.13 | 5 | 0.78 | N/A | 0.11 |

| Media Concentration Index | 2004/5 | 2011 or Most Recent | % Change Annual Average |
|---|---|---|---|
| Total Revenue: Nat'l Media Industry (mil US$) | 15,192 | 43,457 | 31.0% |
| Total Voices (*n*) | 59 | 70 | 3.1% |
| Net Voices (*n*) | 46 | 54 | 2.9% |
| Public Ownership (%) | 44.2 | 22.7 | –3.6% |
| Foreign Ownership (%) | 8.0 | 17.1 | 1.5% |
| C4 Average—Weighted | 75 | 80 | 0.80% |
| HHI Average—Weighted | 4,906 | 2,935 | –6.7% |
| C1 Average—Weighted | 52 | 42 | –2% |
| Noam Index Average—Weighted | 3,330 | 1,710 | –8.1% |
| Pooled Overall Sector C4 | 57.4 | 62.2 | 0.80% |
| Pooled Overall Sector HHI | 1,387 | 1,140 | –3.0% |
| Pooled Overall Sector Noam Index | 160 | 58 | –10.6 |
| Market Share of Top Ten Companies: Nat'l Media Industry (%) (Pooled C10) | 70.9 | 80.2 | 1.5% |
| National Power Index | 4,311 | 2,923 | –5.4% |

Table 25-16. Top Content Media Companies in India

| | 2004/5 | | 2011 OR MOST RECENT | | % CHANGE ANNUAL AVERAGE | |
|---|---|---|---|---|---|---|
| | COMPANY POWER INDEX IN COUNTRY | COMPANY SHARE OF THE NATIONAL CONTENT MEDIA MARKET (%) | COMPANY POWER INDEX IN COUNTRY | COMPANY SHARE OF THE NATIONAL CONTENT MEDIA MARKET (%) | COMPANY POWER INDEX IN COUNTRY | COMPANY SHARE OF THE NATIONAL CONTENT MEDIA MARKET (%) |
| Prasar Bharati (public) | 3,097 | 31.3 | 1,191 | 12.3 | –10.3% | –3.2 |
| Google (US) | 78 | 2.2 | 476 | 5.8 | 85.7% | 0.6 |
| Zee | 34 | 3.1 | 204 | 8.0 | 83.7 | 0.82 |
| BCCL (The Times of India Group) | 254 | 9.4 | 308 | 11.5 | 3.5% | 0.3 |
| Tata Group | 0.0 | 0.0 | 4 | 0.87 | N/A | 0.14 |
| Sun Group (Sun Pictures, India) | 15 | 1.1 | 78 | 3.8 | 0.71% | 0.46% |
| Disney (US) | 14 | 1.3 | 39 | 2.8 | 0.30 | 0.26 |
| Living Media Group | 3 | 0.27 | 18 | 0.76 | 86.7 | 0.08 |
| Murdoch Interests (US) | 1 | 0.38 | 22 | 2.6 | 3.8 | 0.37 |
| Eros Entertainment | 2 | 0.37 | 10 | 0.79 | 0.90 | 0.07 |
| Hathway | | 1.8 | 49 | 3.0 | 0.24 | 0.21 |
| DEN Networks | 0.0 | 0.12 | 8 | 1.5 | N/A | 0.23 |
| Media Concentration Index | 2004/5 | | 2011 or Most Recent | | % Change Annual Average | |
| Public Ownership (%) | 31.2 | | 12.3 | | –3.2% | |
| Foreign Ownership (%) | 9.0 | | 20.4 | | 1.9% | |
| C4 Average—Weighted | 61 | | 66 | | 0.76% | |
| HHI Average—Weighted | 3,814 | | 2,673 | | –5.0% | |
| C1 Average—Weighted | 47 | | 39 | | –1% | |
| National Power Index | 3,594 | | 2,677 | | –4.3% | |

Table 25-17. Top Platform Media Companies in India

| | 2004/5 | | 2011 OR MOST RECENT | | % CHANGE ANNUAL AVERAGE | |
|---|---|---|---|---|---|---|
| | COMPANY POWER INDEX IN COUNTRY | COMPANY SHARE OF THE NATIONAL PLATFORM MEDIA MARKET (%) | COMPANY POWER INDEX IN COUNTRY | COMPANY SHARE OF THE NATIONAL PLATFORM MEDIA MARKET (%) | COMPANY POWER INDEX IN COUNTRY | COMPANY SHARE OF THE NATIONAL PLATFORM MEDIA MARKET (%) |
| BSNL | 3,977 | 50.5 | 1,510 | 25.4 | –10.3 | –4.2 |
| Bharti Enterprises | 231 | 12.1 | 801 | 25.5 | 41.1 | 2.2 |
| Reliance Anil Dhirubhai Ambani Group | 59 | 2.5 | 162 | 11.5 | 29.3 | 1.5 |
| Vodafone (UK) | 181 | 7.5 | 354 | 15.4 | 16.0 | 1.3 |
| Zee | 34 | 1.2 | 36 | 1.9 | 0.92 | 0.11 |
| Tata Group | 10 | 2.8 | 8 | 2.5 | –3.9 | .06 |
| Hathway | 22 | 2.0 | 25 | 1.7 | 0.03 | –0.06 |
| Den Networks | 0.0 | 0.00 | 4 | 0.6 | N/A | 0.10 |
| Media Concentration Index | 2004/5 | | 2011 or Most Recent | | % Change Annual Average | |
| Public Ownership (%) | 50.5 | | 25.4 | | –4.2% | |
| Foreign Ownership (%) | 7.5 | | 16.3 | | 1.5% | |
| C4 Average—Weighted | 82 | | 83 | | 0.27% | |
| HHI Average—Weighted | 5,435 | | 3,001 | | –7.5% | |
| C1 Average—Weighted | 54 | | 43 | | –2% | |
| National Power Index | 4,682 | | 3,020 | | –5.9% | |

companies reduces its share to 2.3%. Zee Entertainment holds significant shares in video channels (30%) and multichannel platforms (19.4%), which amount to 8.0% of the overall national market and make it the third largest content producer in India. Several content producers had high growth rates based on expansion and entry from some of India's largest industrial conglomerates, such as the Tata Group, Bharti Enterprises, and Reliance Industries, that have been buying into the mobile telecom sector in particular.

# 26

# Media Ownership and Concentration in Japan

KIYOSHI NAKAMURA, TERUAKI ASARI, YOSHIHARU ICHIKAWA, KOICHIRO HAYASHI, HAJIME YAMADA, AND SHO YAMAGUCHI

## INTRODUCTION

The Internet has revolutionized the ways in which Japanese audiences consume traditional media sources, such as print, radio, and television. Although it is true that new media heavily rely on traditional media as sourcing for its products, new media growth is compelling firms to reinvent their business models as established revenue sources decline and concentrations change in the market's individual sectors. Historically, media concentration in Japan has changed only gradually, due to traditional business practices and consistent regulatory policies. However, since the 1980s, there have been substantial changes in the makeup of the telecommunications sector.

## PRINT MEDIA

### Newspapers

Japan's four main national daily newspapers are the *Asahi Shimbun, Mainichi Shimbun, Yomiuri Shimbun,* and *Nikkei Shimbun*. There are also several regional papers, such as the Tokyo-based *Sankei Shimbun,* the Nagoya-based *Chunichi Shimbun,* the Hokkaido-based *Hokkaido Shimbun*, and the Fukuoka-based *Nishi-Nippon Shimbun,* as well as a number of local dailies.

The national newspapers' circulation accounts for about one-half of the dailies' print circulation in Japan. No new papers have entered the newspaper industry in over a half-century, and the industry's circulation rates have stabilized since 1984. There has been only a small measurable increase in the Herfindahl–Hirschman Index (HHI), which was observed during the 1990s, most likely because the financial newspaper, *Nikkei Shimbun,* expanded its market share during the rapid expansion of the "bubble economy" in the late 1980s and early 1990s.

One particular feature of Japan's national newspapers is that they are the major stockholders among television broadcasters; though only in the case of Fuji Media Holdings may broadcasters exercise the right to subsidize newspapers to reduce their operating losses (Table 26.1).

Japan has the highest per capita circulation rate for newspapers in the industrialized world. However, Japanese newspapers now face a formidable challenge from the Internet, because younger readers prefer online editions to the paper ones. At the same time, older readers are cancelling their subscriptions in order to conserve their pension money. Circulation and advertising revenue are in decline. In spite of the papers' unprofitability, Japanese ownership rules shield newspaper companies from shareholder pressure.

## Book Publishing

There are four major publishing houses in Japan: Kodansha, Shogakkan, Shueisha, and Kadokawa Holdings. Since these publishers, except for Kadokawa, are not publicly listed on

### Table 26-1. Daily Newspapers (Market Shares by Circulation)

| | 1984 | 1988 | 1992 | 1996 | 2000 | 2004 | 2008 | 2012 |
|---|---|---|---|---|---|---|---|---|
| Yomirui Shimbun | 18.4 | 18.9 | 19.6 | 19.6 | 19.6 | 19.6 | 19.5 | 19.7 |
| Asahi Shimbun | 15.7 | 15.7 | 16.6 | 16.2 | 15.9 | 16.1 | 15.6 | 15.5 |
| Maimichi Shimbun | 8.9 | 8.2 | 8.1 | 7.7 | 7.7 | 7.7 | 7.5 | 6.6 |
| Nihon Keizai Shimbun | 4.3 | 5.0 | 5.9 | 5.7 | 5.4 | 5.9 | 5.9 | 6.7 |
| Sankei Shimbun | 4.1 | 4.0 | 3.9 | 3.8 | 3.8 | 4.1 | 4.3 | 3.1 |
| Chunichi Shimbun | 6.1 | 5.9 | 6.2 | 6.3 | 6.9 | 6.8 | 5.3 | 6.0 |
| Hokkaido Shimbun | | | 2.3 | 2.3 | 2.4 | 2.4 | 2.3 | |
| Nishi-nippon Shimbun | | | 1.6 | 1.6 | 1.6 | 1.6 | 1.7 | |
| Tokyo Shimbun | | | 1.5 | 1.4 | 1.3 | 1.2 | 1.1 | |
| Shizouka Shimbun | | | 1.4 | 1.4 | 1.4 | 1.4 | 1.4 | |
| Chugoka Shimbun | | | 1.4 | 1.4 | 1.4 | 1.4 | 1.4 | |
| Kobe Shimbun | | | 1.0 | 1.0 | 1.1 | 1.1 | 1.1 | |
| Kyoto Shimbun | | | 1.0 | 1.0 | 1.0 | 1.0 | 1.0 | |
| Kahoku Shimbun | | | 0.9 | 1.0 | 1.0 | 1.0 | 1.0 | |
| Niigata Nippo | | | 0.9 | 0.9 | 1.0 | 1.0 | 1.0 | |
| Others | 42.5 | 42.3 | 28.0 | 28.8 | 28.8 | 27.8 | 29.9 | 42.4 |
| Circulation (1,000) | 47,515 | 50,598 | 49,948 | 51,571 | 51,890 | 51,408 | 51,491 | |
| Total Revenue (100 mil yen)[1] | | | | | | 16,641 | 17,435 | 19,154 |
| Total Revenue (mil US$) | | | | | | 15,403 | 16,893 | 23,077 |
| C4 | 49.1 | 48.7 | 50.5 | 49.8 | 50.1 | 50.2 | 49.2 | 48.5 |
| HHI | 752 | 761 | 829 | 814 | 805 | 818 | 770 | 762 |
| *N* (>1%) | | | 13 | 14 | 15 | 15 | 15 | 15 |
| Noam Index | | | 230 | 218 | 208 | 211 | 199 | 197 |

*Source*: Nihon Shinmun Kyokai, Nihon ABC Kyokai 2012 updates provided by editors, data from Mintel Group report.

[1] This revenue figure is an estimate based on data collected in this study: it is based on average per capita revenues reported for several other countries in this study with similar per capita income.

Table 26-2. Book and Magazine Publishing (Market Shares by Revenue), 1996–2008

| | 1996 | 2000 | 2004 | 2008 |
|---|---|---|---|---|
| 1–5 | 23.3 | 23.3 | 22.3 | 20.9 |
| 6–50 | 28.3 | 30.9 | 29.6 | 29.2 |
| 51–100 | 12.0 | 9.9 | 11.3 | 12.5 |
| 101–200 | 10.5 | 10.9 | 11.6 | 12.2 |
| 201–500 | 13.1 | 12.7 | 13.2 | 13.2 |
| 501–2,000 | 12.8 | 11.6 | 11.5 | 11.6 |
| 2,001– | 0.0 | 0.8 | 0.6 | 0.4 |
| Total Revenue (100 mil yen) | 34,669 | 33,452 | 29,149 | 24,626 |
| Total Revenue (mil US$) | 31,892 | 31,043 | 26,980 | 23,860 |
| C5 | 23.3 | 23.3 | 22.3 | 20.9 |
| HHI | 131 | 134 | 123 | 112 |
| *N* (>1%) | 5 | 5 | 5 | 5 |
| Noam Index | 59 | 60 | 55 | 50 |

*Source*: Nippon Shuppan Hanbai.

the Japanese stock exchange and have diversified holdings in other media sectors like film, it is difficult to estimate their sales revenue for print products in addition to revenues earned from reselling copyrights (Table 26.2).[1]

There are over 4,260 publishing entities in Japan. Superficially, the market structure is very competitive. Though the HHI has been stable since 1996, the distribution of books and magazines is much more concentrated. Two distributors, Tohan Corporation and Nippon Shuppan Hanbai, Inc., account for 70% of the total volume. Revenue from book and magazine sales has been declining in recent years due to a shift in consumer preference for audiovisual media.

Books and magazines are sold on commission in Japan, so bookstores can return unsold books to the publisher. Prices for Japanese books, magazines, and newspapers—in addition to CDs and records—are set using resale price maintenance (RPM) that is exempted by the Antimonopoly Act (anti-trust law). The pretext it to maintain a high content quality, but in practice it prevents price competition among retailers. However, emerging online bookstores such as Amazon.com, used booksellers such as Book-Off, and the growing popularity of e-books are forcing Japanese publishers to reevaluate their business models.

## AUDIOVISUAL MEDIA

### Radio

All of Japan's 47 prefectures have their own radio stations, except for major cities such as Tokyo, Osaka, Nagoya, Sapporo, and Fukuoka. In Tokyo there are three commercial radio stations: Nippon Cultural Broadcasting, Nippon Broadcasting System, and TBS Radio and Tokyo Radio

1. Since it is impossible to precisely calculate the HHI in this sector, we estimated it. First, we divided the total revenue of the industry by five to determine the average revenue of each publishing house. The average revenue was then divided by the total sales registered by the five main publishing houses to arrive at an average market share estimate for the top five companies. We calculated that 4.2% was the market share for the top five publishers. We applied the same method to estimate the market share for the publishing houses ranked from 6 to 50 on the national scale.

Table 26-3. Radio Group (Market Shares, by Revenue)

| | 1984 | 1988 | 1992 | 1996 | 2000 | 2004 | 2008 | 2012 |
|---|---|---|---|---|---|---|---|---|
| TBS Radio & Communications (TBS R&C) | 9.8 | 8.2 | 7.4 | 6.1 | 6.0 | 7.0 | 6.9 | 7.5 |
| Nippon Cultural Broadcasting (NCB) | 9.0 | 9.0 | 7.6 | 6.8 | 4.3 | 4.8 | 4.7 | 4.3 |
| Nippon Broadcasting System (NBS) | 14.5 | 14.2 | 14.9 | 14.3 | 14.5 | 12.9 | 11.5 | 10.7 |
| Mainichi Broadcasting System (MBS) | 3.9 | 3.8 | 3.7 | 3.6 | 3.9 | 2.5 | 2.3 | 3.3 |
| Tokai Radio Broadcasting | 4.0 | 4.0 | 4.2 | 4.3 | 2.0 | 1.9 | 1.8 | 1.9 |
| Tokyo FM Broadcasting (FM) | 7.1 | 8.7 | 7.1 | 9.2 | 11.9 | 12.1 | 7.7 | 8.1 |
| FM 802 | 0.0 | 1.2 | 3.0 | 2.2 | 2.0 | 2.8 | 3.2 | 1.9 |
| Others | 51.7 | 50.9 | 52.1 | 53.5 | 55.4 | 56.0 | 61.9 | 62.6 |
| Total Revenue (100 mil yen) | 1,772 | 2,299 | 2,774 | 2,812 | 2,630 | 2,262 | 1,953 | 1,475 |
| Total Revenue (mil US$) | 746 | 1,793 | 2,190 | 2,587 | 2,441 | 2,094 | 1,892 | 1,777 |
| C4 | 40.4 | 40.1 | 47.0 | 36.4 | 36.7 | 36.8 | 30.8 | 31.0 |
| HHI | 543 | 520 | 474 | 451 | 468 | 438 | 318 | 273 |
| *N* (>1%) | 22 | 23 | 23 | 21 | 18 | 16 | 20 | 20 |
| Noam Index | 116 | 108 | 99 | 98 | 110 | 109 | 71 | 61 |

*Source:* NAB Japan Research Institute 2012 updates provided by editors, based on data from Mintel.

Communications. The market share of the Tokyo-based radio stations has been expanding because local radio stations are losing audience share.

There has been a sharp decline in radio advertising revenues, from 152 billion yen in 1984 to 116 billion yen in 2006, which translated into a 25% revenue loss in a 20 year-period, largely due to the proliferation of Internet radio stations (Table 26.3).

Most of the Japanese AM radio stations, especially in local areas, have strong financial relationships with television broadcasters. Nippon Broadcasting System (NBS), for example, was once the parent company of Fuji TV, but was bought out by Fuji TV in 2005. There are two main AM Radio networks: JRN (Japan Radio Network), centering around two Tokyo-based key stations, Nippon Radio Station and Bunka Radio Station, as well as NRN (National Radio Network), which is organized by a Tokyo-based key station, TBS Radio Station. Some local AM stations belong to both networks.

In contrast with AM Radio, the market size of FM Radio has been increasing steadily in recent years, tripling in size between 1984 and 2006. Since the Ministry of Internal Affairs and Communications (MIC) allows multiple FM Radio stations to broadcast from individual cities, there have been many new market entries, particularly from community radio stations. Every prefecture of Japan has at least one FM station, matching the existing 47 AM stations.

Concentration is also decreasing in the FM sector: FM Tokyo, the leading FM station in the country, now holds no more than 30% of the market share, while new FM stations in Tokyo and Osaka have increased their market shares at its expense. Nevertheless, when the "bubble economy" burst in 2001, FM advertising revenues began declining.

## Broadcast Television

Since 1953, television broadcasting in Japan has adhered to a dual structure, made up of one nationwide public broadcaster, Nippon Hoso Kyokai (NHK), and multiple commercial broadcasters, which are supported mainly by their advertising revenues. When the television broadcasting network began in Japan in the 1950s, the government planned for a wide distribution of

Table 26-4. TV Broadcasting (Market Shares by Revenue)

| | 1984 | 1988 | 1992 | 1996 | 2000 | 2004 | 2008 | 2012 |
|---|---|---|---|---|---|---|---|---|
| Nippon Television Network (NTV) | 18.5 | 18.7 | 17.9 | 19.5 | 20.5 | 19.2 | 18.4 | 19.1 |
| Tokyo Broadcasting System Television (TBS) | 19.4 | 19.6 | 17.5 | 17.4 | 16.8 | 17.0 | 17.6 | 15.9 |
| Fuji Television | 19.1 | 21.4 | 21.9 | 21.2 | 21.3 | 22.1 | 21.5 | 17.1 |
| TV Asahi Corporation | 14.1 | 14.3 | 14.2 | 14.9 | 14.8 | 14.9 | 15.5 | 18.8 |
| TV Tokyo Corporation | 3.0 | 3.7 | 4.4 | 4.6 | 4.8 | 5.0 | 5.3 | 7.0 |
| Japan Broadcasting Corporation (NHK) (public) | 22.9 | 19.5 | 21.2 | 19.3 | 19.0 | 19.2 | 19.1 | 16.1 |
| Others | 2.6 | 2.6 | 2.6 | 2.7 | 2.5 | 2.3 | 2.5 | 6.0 |
| Total Revenue (100 mil yen) | 14,695 | 18,317 | 23,363 | 26,703 | 28,912 | 29,288 | 27,640 | |
| Total Revenue (mil US$) | 6,187 | 14,287 | 18,445 | 24,564 | 26,830 | 27,109 | 26,780 | |
| C4 | 80.0 | 79.2 | 78.5 | 77.5 | 77.6 | 77.6 | 76.6 | 71.1 |
| HHI | 1,827 | 1,797 | 1,784 | 1,761 | 1,766 | 1,770 | 1,749 | 1,572 |
| *N* (>1%) | 6 | 6 | 6 | 6 | 6 | 6 | 6 | 6 |
| Noam Index | 746 | 734 | 728 | 719 | 721 | 723 | 714 | 642 |

*Source*: NAB Japan Research Institute, NHK, 2012 updates provided by editors. Source: <http://www.Stat.go.jp>.

broadcasting stations throughout Japan in order to advance localism, and set a target for making either three or four channels available in every prefecture. At the same time, the Ministry of Post and Telecommunications (now MIC) allowed stations to form affiliations so they could broadcast programs nationwide (Table 26.4).

Five Tokyo-based commercial broadcasters carry most of the country's television programming (over 70% of the market share). The public broadcaster, NHK maintains a market share of approximately 20%. There were 25 new entries to the market in the 1990s, but these are still minor players compared to the six stations named in Table 26.4 (above).

One of the major reasons these five commercial broadcasters dominate the market is that these broadcasters, often referred to as "the key stations", are capable of producing programs and set up common broadcast networks with affiliated local broadcasters. Such a "program network" or "news network" reduce stations' operating costs. NNN is organized by NTV with 30 local stations, JNN by TBS with 28 local stations, FNN by Fuji TV with 28 local stations, ANN by TV Asahi Corporation with 26 local stations, and TXN by TV Tokyo, the smallest of five key stations, with 6 local stations as of 2013.

Commercial broadcasters are prohibited from owning another broadcaster in the same region. In addition, cross-ownership of three different media sources is usually prohibited unless it can be demonstrated that a monopoly will not result. Despite anti-trust regulations, there are in fact strong financial relationships among the five Tokyo-based key stations and the main newspapers through cross-shareholdings: Asahi Shimbun, the second-highest circulation newspaper, holds 24.7% of TV Asahi; Nikkei Shimbun, the fourth-highest circulation newspaper, owns 33.3% of TV Tokyo; Yomiuri Shimbun, the highest-circulation newspaper, owns 15.2% of Nippon TV; and Fuji TV owns 39.9% of *Sankei Shimbun,* the fifth-highest circulation newspaper. Although shareholding of a television broadcaster by a newspaper company is limited to one-third of the broadcaster's total shares, both institutions are strongly linked to each other because they share executives at the corporate level and journalists at the operational level.

The most important revision of the 1950 Broadcast Law made in 2008 was to allow the formation of "broadcasting holding companies." Fuji TV, TBS, and TV Tokyo changed over to a holding company system in order to facilitate the rescue of local TV stations that had been suffering from the heavy financial burden of switching over to digital terrestrial television (DTT). The MIC plans on revising the law because its provisions have been weakened due to the advent of the Internet.

## Multichannel TV Platforms

In Japan, communication satellites were not originally meant to broadcast television signals, so "broadcast" and "communications" satellite systems, or BS and CS, were legally separate transmission platforms up until 2009, when Japanese regulatory law caught up with the reality that there is no meaningful technological difference between BS and CS anymore. However, since the legal distinction was maintained well past its utility as a technical marker, the entities that fell under BS and CS have evolved differently from one another.

Those that have historically been described as broadcast satellite systems formed an oligopoly in which a handful of broadcasters dominate, and this oligopoly still dominates the satellite television market: the public broadcaster NHK, five commercial channels owned by the five main commercial terrestrial broadcasters, and pay television channels such as WOWOW and Star Channel, which specialize in cinema. If viewers pay their subscription fees to NHK terrestrial broadcasting services, they can enjoy NHK satellite programming by paying an additional charge. All the programs offered by the five commercial broadcasters, in contrast, are free. Since new entries into the market are regulated by the MIC, the HHI has been stable for more than a decade, hovering around 4,000. The growth of satellite TV can be attributed to three main factors. First, the demand for flat-panel TVs equipped with satellite receivers has grown sharply, due to DTT conversion, and this is helped along by government subsidies. Second, there has been an increase in advertising revenues because of the rising popularity of new types of programs, such as travel shows targeting an elderly audience. Third, TV shopping channels have led to an increase in revenue for the networks.

In the past, communications satellites (CS) were used exclusively for non-broadcast transmissions, but the legal changes meant that new entities began forming in the 1990s that used the former CS-only platforms. One was SkyPerfecTV! founded in 1998 after the merging of PerfecTV! with JSkyB, a former subsidiary of Rupert Murdoch's News Corp. The other significant new entry was the US satellite TV provider DirecTV, which began offering services in 1997. After battling with SkyPerfecTV! for two years, DirecTV withdrew from the Japanese market in 2000, and its subscribers were transferred to SkyPerfecTV!. SkyPerfecTV! offered more than 300 channels to 3.7 million subscribers as of 2011 (Table 26.5).

## Video Channels

Cable television (CATV) in Japan began when local broadcasters started enhancing terrestrial television broadcasting for communal reception facilities, such as apartment buildings and hotels, and for remote, rural areas. When satellite TV started broadcasting in 1989, CATV providers began to retransmit their programming and expanded over time. Since then, regulations on foreign capital investment (1993), mergers and acquisitions (1994), and market entry (2002) have been gradually relaxed to allow the market to grow. Japanese CATV, like the counterparts in other countries, provides triple-play services—for example, high-speed Internet access, television, and telephone subscriptions.

There were also around 11 million subscribers for both multichannel TV services through CATV and satellite broadcasters in Japan as of 2010, constituting about 20% of total households. However, neither multichannel nor cable programming have yet proven more popular than DTT. So although over 40 million households are able to access cable programming, so far only around 7.3 million have subscribed. Multiple system operators (MSOs) were

Table 26-5. Multichannel Video Platforms: Cable MSOs, DBS, IPTV (Market Shares by Subscribers)

| | 1992 | 1996 | 2000 | 2004 | 2008 |
|---|---|---|---|---|---|
| Tokyo Cable Network (TCN) | 4.9 | 1.6 | 0.7 | | |
| SkyPerfecTV! | | 55.5 | 44.3 | 39.0 | 43.6 |
| SkyPort | 4.5 | 21.6 | | | |
| Jupiter Telecom (J:COM) | | 2.4 | 10.9 | 18.1 | 22.1 |
| Titus Communications | | | 1.8 | | |
| Sapporo Cable TV | 2.8 | | | | |
| Cable West | | | | 2.6 | |
| Mediatti Broadband Communications | | | | | 1.9 |
| JCN | | | | 3.5 | 5.4 |
| Bay Communications | | | | | 1.4 |
| Information Network | 3.8 | 1.8 | | | |
| Reinan Cable Network | 3.5 | | | | |
| Kintetsu Cable Network | 2.8 | | | | |
| iTSCOM (Tokyu Cable Television) | 2.4 | 2.6 | | 1.3 | 1.4 |
| K-CAT | | | | | 1.3 |
| TEPCO Cable TV | | 1.7 | | | 1.0 |
| Osaka Central Cable Network | | | 0.9 | | |
| Fukuoka Cable Network | | | | 1.0 | |
| STARCAT Cable | | | | 0.8 | |
| Lakecity Cable Vision | 2.2 | | | | |
| Obihiro Cable TV | 2.2 | | | | |
| Chubu Cable Network | | | | 0.8 | |
| Himawari Network | | 1.7 | 0.8 | | |
| Nagoya Cable Network | | 1.6 | 0.7 | | |
| Hanshin City Cable | | | | 0.8 | |
| ZTV | | | | | 1.0 |
| Others | 70.9 | 9.5 | 37.6 | 32.1 | 20.9 |
| Total Revenue (100 mil yen) | 530 | 1,481 | 3,624 | 5,879 | 7,561 |
| Total Revenue (mil US$) | 418 | 1,362 | 3,385 | 5,442 | 7,326 |
| Subscribers (1,000) | 425 | 1,552.1 | 5,010 | 8,380 | 10,408 |
| C4 | 16.7 | 82.1 | 58.3 | 63.2 | 73.0 |
| HHI | 171 | 220 | 2,032 | 1,876 | 1,773 |
| *N* (>1%) | 33 | 25 | 5 | 6 | 8 |
| Noam Index | 30 | 44 | 909 | 766 | 627 |

*Source*: Monthly Broadcasting Journal (Gekkan Hoso Journal), Japan Satellite Broadcasting Association.

Table 26-6. Cable and DBS TV Programming Channels (Market Shares by Revenue)

| | 1992 | 1996 | 2000 | 2004 | 2008 |
|---|---|---|---|---|---|
| NHK-BS (public) | 56.9 | 57.6 | 60.2 | 60.3 | 57.7 |
| WOWOW | 43.1 | 42.4 | 34.1 | 30.8 | 27.2 |
| BS Nippon Corporation | | | 1.0 | 1.2 | 2.6 |
| BS-TBS | | | 1.3 | 2.4 | 3.5 |
| Fuji Satellite Broadcasting (BS-FUJI) | | | 1.1 | 1.6 | 2.5 |
| Asahi Satellite Broadcasting (BS Asahi) | | | 1.1 | 1.7 | 3.1 |
| BS Japan Corporation | | | 1.2 | 2.1 | 2.4 |
| Nippon BS Broadcasting Corporation (BS11 Digital) | | | | | 1.0 |
| Total Revenue (100 mil yen) | 802 | 1,390 | 1,787 | 2,031 | 2,311 |
| Total Revenue (mil US$) | 633 | 1,279 | 1,658 | 1,880 | 2,239 |
| C4 | 100.0 | 100.0 | 100.0 | 96.9 | 91.6 |
| HHI | 5,094 | 5,114 | 4,797 | 4,596 | 4,112 |
| *N* (>1%) | 2 | 2 | 7 | 7 | 8 |
| Noam Index | 3,602 | 3,616 | 1,813 | 1,737 | 1,454 |

*Source*: NAB Japan Research Institute, Radio & Television Yearbook (NHK Nenkan).

formed for the first time in 1995, and the HHI of CATV has been increasing ever since then due to mergers and acquisitions by MSOs, up from 99 in 1996 to 1,288 in 2008. The largest MSO in Japan is J:COM, which merged with Titus Communications (2000), then with Cable West (2008), and most recently with Mediatti Communications (2009). Its market share has increased sharply, to 37.5% of total CATV revenues, as a result.

While CATV providers maintain a "local" monopoly status in each area, they face challenges from various competitors. Free video-sharing sites such as YouTube, Nikonikodouga (a largest Japanese video-sharing site), and Facebook undercut their dominance. Telecom providers such as NTT and KDDI also now offer similar Internet Protocol television (IPTV) multichannel packages (Tables 26.6, 26.7, and 26.8).

Table 26-7. Cable and DBS TV Programming Channels (Market Shares by Subscribers)

| | 1992 | 1996 | 2000 | 2004 | 2008 |
|---|---|---|---|---|---|
| SkyPerfecTV! | 0.0 | 72.0 | 100.0 | 100.0 | 100.0 |
| SkyPort | 100.0 | 28.0 | 0.0 | 0.0 | 0.0 |
| Total Revenue (100 mil yen) | – | 71 | 1,161 | 2,346 | 2,894 |
| Total Revenue (mil US$) | – | 65 | 1,078 | 2,172 | 2,804 |
| Subscribers (1,000) | 19 | 328 | 2,220 | 3,149 | 2,737 |
| C4 | 100.0 | 100.0 | 100.0 | 100.0 | 100.0 |
| HHI | 10,000 | 5,964 | 10,000 | 10,000 | 10,000 |
| *N* (>1%) | 1 | 2 | 1 | 1 | 1 |
| Noam Index | 19 | 328 | 2,220 | 3,310 | 3,567 |

*Source*: Japan Satellite Broadcasting Association.

Table 26-8. Cable and DBS TV Programming Channels (Market Shares by Subscribers)

| | 1992 | 1996 | 2000 | 2004 | 2008 | 2012 |
|---|---|---|---|---|---|---|
| Jupiter Telecom (J:COM) | | 2.8 | 19.6 | 29.6 | 34.1 | 39.0 |
| Titus Communications | | 0.0 | 3.3 | | | |
| Sapporo Cable TV | 3.0 | 1.4 | | | | |
| Cable West | | | | 4.3 | | |
| Mediatti Broadband Communications | | | | | 2.9 | |
| JCN | | | 2.0 | 5.8 | 8.8 | 11.0 |
| iTSCOM (Tokyu Cable Television) | 5.3 | 3.2 | 2.4 | 2.1 | 2.2 | |
| Bay Communications | | | | 2.2 | 2.2 | |
| K-CAT | | | | | 2.0 | |
| Bic TOKAI | | | | 1.0 | 1.4 | |
| Information Network | 4.1 | 2.2 | 1.2 | 0.5 | 0.4 | |
| Reinan Cable Network | 3.7 | 1.6 | 0.8 | 0.4 | 0.3 | |
| Himawari Network | 0.8 | 2.0 | 1.4 | 1.0 | 0.9 | |
| Catch Network | | 1.4 | 1.2 | 1.0 | 0.9 | |
| Hanshin City Cable | | | 1.8 | 1.2 | | |
| Others | 83.9 | 89.8 | 70.3 | 55.0 | 46.4 | |
| Total Revenue (100 mil yen) | 530 | 1,410 | 2,463 | 3,533 | 4,667 | 5,512 |
| Total Revenue (mil US$) | 418 | 1,297 | 2,287 | 3,270 | 4,522 | 6,640 |
| C4 | 16.1 | 10.2 | 27.1 | 41.9 | 48.0 | 56 |
| HHI | 167 | 99 | 448 | 1,066 | 1,288 | 1,642 |
| *N* (>1%) | 33 | 23 | 15 | 14 | 12 | 12 |
| Noam Index | 29 | 21 | 116 | 285 | 372 | 474 |

*Source*: Monthly Broadcasting Journal (Gekkan Hoso).

## Film

Japan's film industry enjoyed its golden age in the 1950s, but the number of Japanese theatergoers declined sharply by the 1990s, to one tenth of the market size due to the emergence of TV and home video. However, there has been an increase in the number of theatergoers in the 2000s, mainly due to the emergence of multiplexes and the popular revival of Japanese films.

There are three major components of the film industry: production, distribution, and exhibition. The three major players in the Japanese film market—Toho, Toei, and Shochiku—dominate not only the production aspects, but also in distribution and exhibition. These three studios accounted for 82.8% of total film industry revenue (including exhibition) in 2009. They have been reducing their own production companies and investing in independent film producers through consortiums.

Japanese filmmakers used a "block booking" system, which limits distribution flexibility by fixing the schedule of running films about a year ahead, but they voluntarily abandoned this system in 2000. For foreign movies, Toho and Shochiku control most distribution and exhibition in Japan, although some US studios and distributors have high market shares in Japan. Generally speaking, the ratio of foreign movies to Japanese movies in the Japanese market has been 50–50, but this has been increasing in favor of the Japanese studios in recent years (Table 26.9).

Table 26-9. Film Production/Distribution (Market Shares by Box Office %)

| | 1992 | 1996 | 2000 | 2004 | 2008 |
|---|---|---|---|---|---|
| Shochiku | 11.3 | 7.6 | 4.1 | 9.7 | 8.2 |
| Toho | 20.6 | 16.9 | 11.5 | 25.7 | 37.9 |
| Toei | 10.6 | 7.5 | 4.6 | 4.5 | 6.1 |
| Toho-Towa | 5.2 | 3.3 | 3.9 | 0.9 | 7.2 |
| Nippon Herald | 9.6 | 7.8 | 4.5 | 3.6 | |
| Herald Ace | 0.7 | | | | |
| Ace Pictures | | 0.5 | | | |
| ASMIK-ACE | | | | 0.4 | 3.1 |
| Shochiku Fuji | 1.3 | 4.3 | | | |
| Gaga Humax | 0.4 | 5.9 | 8.7 | 5.0 | |
| Gaga Communications | | | | | 4.0 |
| ASCII Pictures | 1.1 | | | | |
| Kadokawa Pictures | | | | | 2.0 |
| United International Pictures (UIP) Japan | 9.0 | 18.4 | 13.5 | 5.7 | |
| Paramount Pictures Japan | | | | | 4.2 |
| Warner Entertainment Japan | 12.0 | 6.1 | 6.3 | 16.1 | 8.4 |
| Buena Vista International (Japan) | | 6.8 | 12.7 | 9.9 | |
| Walt Disney Studio Japan | | | | | 5.6 |
| Sony Pictures International (Japan) | 8.4 | 4.0 | 10.1 | 6.9 | 3.5 |
| Twentieth Century Fox Film (Japan) | 5.6 | 3.7 | 7.4 | 4.1 | 4.0 |
| Others | 4.2 | 7.2 | 12.7 | 7.5 | 5.8 |
| Total Revenue (100 mil yen)[1] | 624 | 633 | 855 | 1,055 | 974 |
| Total Revenue (mil US$) | 493 | 582 | 793 | 977 | 944 |
| C4 | 54.5 | 50.7 | 47.8 | 61.4 | 61.8 |
| HHI | 1,113 | 997 | 821 | 1,389 | 1,771 |
| *N* (>1%) | 12 | 12 | 11 | 10 | 12 |
| Noam Index | 321 | 282 | 248 | 439 | 511 |

*Source:* Film Industry Yearbook (Eiga Nenkan).

[1] After 2000, the Motion Picture Producers Association (MPAA) began releasing the numbers for total box office revenues, instead of just the distribution revenue, which is estimated to be about half that of the box office revenue. The figures for 2000, 2004 and 2008 are half of the totals released by the MPAA.

One of the reasons behind Toho's dominance is that Toho owns 61 theaters, the largest single ownership share in Japan. Toho owns these theaters through its subsidiary, Toho Cinemas Co., and through film producers, in particular TV stations, as a part of its diversification strategy.

## Music

Japan has the second-largest music market in the world after the United States. However, the Japanese music industry, as elsewhere, faces significant challenges due to new digital methods of distribution that have adversely affected CD

Table 26-10. Music Distributors (Market Shares by Revenue)

| | 1996 | 2000[1] | 2004 | 2008 |
|---|---|---|---|---|
| Columbia Music Entertainment | | | 4.4 | 2.7 |
| Nippon Columbia | 5.4 | 4.2 | | |
| Victor Entertainment | 11.6 | 19.4 | 6.8 | 5.9 |
| King Records (Japan) | 4.2 | 2.9 | 3.6 | 4.5 |
| Teichiku | 1.8 | 1.4 | | |
| Universal Music (Japan)[2] | | | 8.9 | 9.8 |
| PolyGram | 11.4 | | | |
| EMI Music Japan[3] | 11.4 | 11.8 | 8.5 | 4.9 |
| Nippon Crown | 8.5 | 1.5 | 1.2 | 0.8 |
| Tokuma Japan Communications | 1.9 | 1.3 | 0.7 | 1.3 |
| Sony Music Entertainment (Japan) | 13.9 | 16.5 | 26.0 | 31.1 |
| Pony Canyon | 8.1 | 9.8 | 10.9 | 10.2 |
| Warner Music Japan | 4.3 | 6.8 | 2.9 | 3.1 |
| Bandai Music Entertainment | 1.0 | | | |
| VAP | 4.3 | 4.1 | 7.2 | 5.6 |
| BMG Japan[4] | 4.9 | | | |
| BMG Funhouse | | 3.0 | 3.5 | |
| Pioneer LDC[5] | 4.0 | 3.6 | | |
| Geneon Entertainment | | | 5.3 | |
| Bing | | | | 2.0 |
| J DISK | 3.3 | 3.8 | | |
| J DISK Bing | | | 1.1 | |
| Avex | | 9.8 | 7.6 | |
| Avex Marketing | | | | 16.7 |
| Total Revenue (100 mil yen) | 7,402 | 6,418 | 5,954 | 5,372 |
| Total Revenue (mil US$) | 6,809 | 5,956 | 5,511 | 5,205 |
| C4 | 48.3 | 57.5 | 54.3 | 67.8 |
| HHI | 868 | 1,110 | 1,189 | 1,581 |
| *N* (>1%) | 16 | 15 | 15 | 13 |
| Noam Index | 217 | 287 | 307 | 438 |

*Source*: Record Tokushin Shuppansha Y.K.

[1] After 2000, the Motion Picture Producers Association (MPAA) began releasing the numbers for total box office revenues, instead of just the distribution revenue, which is estimated to be about half that of the box office revenue. The figures for 2000, 2004, and 2008 are half of the totals released by the MPAA.

[2] France's Vivendi Universal Entertainment Company acquired Universal Music in 2004. Though it retains the Universal name, it is no longer affiliated with Universal Pictures (US).

[3] Formerly Toshiba EMI: Toshiba sold its stake to EMI Group in 2007. EMI Group's music assets, including EMI Music Japan, became part of Universal Music Group in 2012.

[4] BMG Japan was reorganized into Sony Music Japan in 2009 after its purchase from Bertelsmann AG of Germany.

[5] Pioneer LDC became Geneon Entertainment in 2003: in 2009, Geneon became part of Universal Pictures Japan and is now known as Geneon Universal Entertainment.

sales while Internet-based music download services have expanded in recent years. Although the primary sources of revenue are derived from the sale of CDs and copyright royalties, paid music downloads' revenue share has also grown, accounting for 12% of the total revenue in 2008.

There are three main music distributors in Japan: Sony, Universal, and Avex. EMI, once the fourth ranked, was acquired by Universal. BMG was sold to Sony by Bertelsmann in 2008 and as a result, Sony's market share increased to more than 30%. Market share is highly volatile, with a distributor's fortunes depending on which label produces the most hit songs or albums in a fiscal year.

As noted above, the music distribution system has changed rapidly from direct CD sales to Internet-based music download services. One of the unique features in the Japanese music market that has been observed over the past several years is the expansion of mobile content, such as ringtones and playback ringtones downloaded for mobile phones, particularly for "J-pop" properties, a genre unique to Japan that dominates the domestic music scene (Table 26.10).

Due to new digital distribution technologies and the advent of digital recording technologies, independent record labels have gradually gained market shares and are able to use the Internet more effectively for marketing purposes. As a result, musicians or singers have acquired greater negotiating powers over traditional record labels. Coupled with the burgeoning popularity of smart-phones, these factors might alter the market structure of the Japanese music industry in the near future.

## TELECOMMUNICATIONS MEDIA

### Wireline Telecom

The public corporation Nippon Telegraph and Telephone (NTT) was privatized in 1985—though the Government of Japan still holds a controlling stake in it—and in 1999 it became a holding company. The NTT Group comprises two regional companies, NTT EAST and NTT WEST; a wireless company, NTT DOCOMO; a data processing and system integration company, NTT DATA; and a company specializing in inter-prefecture and international telecom, NTT Communications. There are several other smaller subsidiaries as well. NTT monopolized the telephone market for over 30 years prior to its privatization, so NTT EAST and NTT WEST still dominate the national wireline telecom market, and they accounted for 78% of the total market shares in 2008. In spite of NTT's dominance, though, the HHI for wireline telecom has begun declining, from 9,546 in 2004 to a still very high 7,749 in 2008, due to new market entries. And on the technological end, part of this decline can be attributed to the decreasing popularity of landline telephones. NTT EAST and WEST have lost subscribers since 2005, when the wireline subscription decline became apparent, with subscriptions falling from 58 million in 2004 to 35 million in 2010. Internet Protocol (IP) phone services have been increasing significantly, but are offered by Internet-service providers (ISP) other than NTT (Table 26.11).

KDD was the international carrier before the liberalization in 1985. Subsequently, DDI, IDO, Japan Telecom, TTNet, and others entered the market.

NTT has been keen on deploying NGN (Next Generation Networks) to the existing fiber-optic access service area and constructing IP network infrastructure in order to give NTT a competitive lead. How to set up policies that would promote a level playing field and provide a balance between infrastructure competition and service competition has been a subject of lively debate.

### Wireless Telecom

Wireless telecom is Japan is highly competitive. NTT DOCOMO, as a major subsidiary company of the former monopolist, NTT, has been losing market shares since 1992 due to stiff competition from new entries. au—KDDI and SoftBank are NTT DOCOMO's main competitors.

There used to be more firms in the sector, such as the IDO, Cellar, and TuKa Groups, Digital Phone, and J-PHONE, which operated

Table 26-11. Wireline Telecom (Market Shares by Revenue)

| | 1984 (PUBLIC) | 1988 | 1992 | 1996 | 2001 | 2004 | 2008 | 2012 |
|---|---|---|---|---|---|---|---|---|
| Nippon Telegraph and Telephone (NTT) | 95.6 | 92.8 | 87.8 | 82.1 | 84.3 | 85.3 | 72.7 | 72.3 |
| KDDI | | | | | 8.9 | 9.3 | 10.7 | 9.9 |
| KDD | 4.4 | 4.6 | 3.7 | 3.6 | | | | |
| DDI | | 0.7 | 3.6 | 5.6 | | | | |
| Teleway Japan | | 0.2 | 0.6 | 1.2 | | | | |
| SoftBank Telecom | | | | | | 4.1 | 10.3 | 6.2 |
| Japan Telecom | | 0.6 | 3.2 | 4.5 | 5.6 | | | |
| K-Opticom | | | | | | | 0.9 | 2.2 |
| Others | | | 0.1 | 3.0 | 1.2 | 1.2 | 6.3 | 11.6 |
| Total Revenue (100 mil yen)[1] | | | | | | 49,371 | 46,579 | 64,560 |
| Total Revenue (mil US$) | | | | | | 45,698 | 45,130 | 7,778 |
| C4 | 100.0 | 98.9 | 100.0 | 95.8 | 98.8 | 98.7 | 94.6 | 90.6 |
| HHI | 9,154 | 8,634 | 7.748 | 6,806 | 7,217 | 7,379 | 5,507 | 5,369 |
| *N* (>1%) | 2 | 2 | 4 | 5 | 3 | 3 | 3 | 4 |
| Noam Index | 6,473 | 6,105 | 3,874 | 3,043 | 4,167 | 4,260 | 3,179 | 2,685 |

*Source*: MIC 2012 updates provided by editor, data from Technology, Communications and Media: Japan Country Briefing *Euromonitor* December 20, 2013.

[1] This revenue figure is an estimate based on data collected in this study: it is based on average per capita revenues reported for several other countries in this study with similar per capita income.

during the 1990s and into the early 2000s. But these companies were consolidated in a series of mergers: au—KDDI was formed in 2000 by the merging of DDI Corp., KDD Corp., and IDO Corp., and later it incorporated both Cellar and TuKa Group into its operations.

SoftBank's acquisitions were as follows: J-PHONE was formed in 1999 from Digital Phone, which was founded in 1991 as the mobile phone division of Japan Telecom. British Vodafone increased its share in Japan Telecom, but after 2003, Vodafone decided to pull out from the Japanese market entirely and sold its interests in Vodafone Japan to SoftBank, which created SoftBank Mobile Corp. in 2006 and now holds J-PHONE's former assets as well. SoftBank's success in expanding its market shares stems largely from its contract with Apple to release the iPhone in Japan in 2008 (Table 26.12).

In 2008, SoftBank Mobile started to sell Apple's iPhone, which became a smash hit, opening the smart-phone market in Japan. The number of SoftBank Mobile subscribers has been increasing ever since. In order to better compete, KDDI's au started to offer new services in partnership with Skype in 2010. Though three companies compete fiercely, NTT DOCOMO continues to hold onto around half of the total market shares.

The government has introduced various pro-competition policies to enhance competition in the mobile phone market, which is arguably the world's most technologically advanced. For example, the MIC introduced mobile number portability (MNP) among carriers in 2006. However, this did not have as much impact on market shares as the MIC had hoped. The MIC also required network operators to unlock the subscriber identity module (SIM) cards that are expected to expand

Table 26-12. Wireless Telecomm (Market Shares by Revenue)

| | 1984 | 1988 | 1992 | 1996 | 2000 | 2004 | 2008 | 2012 |
|---|---|---|---|---|---|---|---|---|
| Cellular | | | | | | | | |
| NTT (DOCOMO) | 100.0 | 100.0 | 61.4 | 48.4 | 51.6 | 53.2 | 49.7 | 45.8 |
| KDDI[1]—au | | | | | | 19.5 | 28.3 | 27.4 |
| IDO Group | | | 38.6 | 11.1 | 6.9 | | | |
| Cellar Group | | | | 24.8 | 11.0 | | | |
| TuKa Group | | | | | 6.1 | 4.2 | | |
| SoftBank Mobile | | | | | | | | 26.8 |
| Digital Phone Group[2] | | | | 15.8 | | | | |
| J-Phone Group[3] | | | | | 14.4 | | | |
| Vodafone[4] | | | | | | 17.2 | 17.3 | |
| E-Mobile | | | | | | | 0.4 | |
| PHS[5] | | | | | | | | |
| DDI Pocket | | | | | 5.8 | 3.3 | | |
| NTT DOCOMO (NTT Personal) | | | | | 2.5 | 1.8 | | |
| Astel | | | | | 1.7 | 0.7 | | |
| Willcom[6] | | | | | | | 4.3 | |
| Total Revenue (100 mil yen)[7] | | | | | | 53,487 | 66,361 | 63,228 |
| Total Revenue (mil US$) | | | | | | 49,508 | 64,297 | 76,178 |
| Subscribers (mil) | 0.3 | 0.15 | 1.38 | 10.20 | 56.85 | 87.06 | 107.34 | 134.3 |
| C4 | 100.0 | 100.0 | 100.0 | 100.0 | 100.0 | 100.0 | 99.6 | 100.0 |
| HHI | 10,000 | 10,000 | 4,515 | 3,325 | 3,509 | 3,865 | 3,591 | 3,567 |
| *N* (>1%) | 1 | 1 | 2 | 4 | 8 | 6 | 4 | 4 |
| Noam Index | 10,000 | 10,000 | 3,193 | 1,663 | 1,241 | 1,578 | 1,796 | 1,784 |

*Source*: Telecommunications Carriers Association (TCA), MIC. 2012 data provided by editors, data from MIC and Mintel.

[1] IDO, Cellar, and TuKa Group all ceased operating as distinct entities within au by 2008.

[2] SoftBank Mobile's brand name up to 1994.

[3] SoftBank Mobile's brand name from 2003 to 2006.

[4] SoftBank Mobile's brand name from 2003 to 2006; SoftBank Mobile was adopted in 2006.

[5] "Personal Handy-phone System," which was created in Japan as an alternative to 2G.

[6] Willcom is a wholly owned subsidiary of SoftBank.

[7] This revenue figure is an estimate based on data collected in this study: it is based on average per capita revenues reported for several other countries in this study with similar per capita income.

options for customers in selecting their carrier. In consequence, NTT DOCOMO began to offer SIM-free handsets in 2011.

## Internet Service Providers (ISP)

Approximately 30 million people use broadband services in Japan, and 49.5% of total subscribers have a fiber optic service line at home. Internet phone and television services are available over these fiber optic lines as well.

The so-called Big Five ISPs are Yahoo! Japan, Biglobe, OCN, Nifty, and So-net. These companies have different backgrounds. SoftBank owns Yahoo! Japan. The other telecom-owned ISP is the Open Computer Network (OCN), which is the largest nationwide ISP, with over seven million subscribers and is owned by NTT Communications Co., a part of the NTT Group. The three other major ISPs are owned by major hardware manufactures: Biglobe is a subsidiary of NEC, Nifty is a subsidiary of Fujitsu, and So-net is a subsidiary of Sony.[2] There are thousands of small- and medium-sized ISPs as well.

Since there are no consistent data available, the estimates of HHI only suggest that the concentration has been decreasing in recent years (Table 26.13).

## Search Engines

Japan's search engine market is duopolistic: Yahoo! and Google dominate. In contrast to the United States, where Yahoo! has lost market share to Google, Yahoo! sites in Japan have surged above 50% of the market share in 2009, whereas Google's share has slipped. Yahoo! Japan's major shareholders are SoftBank and Yahoo! Inc. In 2010, Yahoo! Japan also became a partner of Taobao Marketplace, a subsidiary of China's largest e-commerce company, the Alibaba Group.

In 2010, however, Yahoo! Japan decided to adopt Google's search engine and advertising platform technology in place of Microsoft's Bing. Moreover, in 2011, the popular online mall Rakuten also switched its portal site search engine to Google. Both companies seemed to reach these decisions in order to concentrate their human resources on applications. These shifts may affect the market structure of online businesses.

### Table 26-13. Internet Service Providers (Market Shares by Revenue)

| | 1996 | 2000 | 2004 | 2008 | 2012 |
|---|---|---|---|---|---|
| Nifty (Fujitsa) | 33.7 | | | – | – |
| Infoweb[1] | 3.7 | 16.8 | 14.1 | | |
| NEC (BIGLOBE) | 33.0 | 13.2 | 11.2 | 7.8 | 8.4 |
| KDDI (au one net) | | 6.6 | 7.6 | 7.7 | 11.3 |
| Softbank (Yahoo! BB) | | | 12.8 | 15.9 | 13.4 |
| Japan Telecom[2] | | 5.5 | | | |
| NTT | | 8.2 | 12.0 | 28.8 | 33.1 |
| Asahi Net/People | 10.2 | | | | |
| MSN Japan | 3.6 | | | | |
| Others | 16.4 | 49.7 | 42.3 | 39.8 | 33.8 |
| Total Revenue (100 mil yen)[3] | | | 15,431 | 20,118 | 19,273 |
| Total Revenue (mil US$) | | | 14,283 | 19,492 | 23,220 |
| C4 | 80.5 | 44.8 | 50.1 | 52.7 | 66.2 |
| HHI | 2,351 | 598 | 690 | 1,202 | 1,473 |
| *N* (>1%) | 5 | 5 | 5 | 4 | 4 |
| Noam Index | 1,051 | 267 | 309 | 601 | 737 |

*Source*: Nikkei Business Daily (Nikkei Sangyo Shimbun). 2012 data added by editors, calculated from company annual reports.

[1] Infoweb was purchased from Fujitsu Limited in 1999. @nifty left the ISP market in 2006.

[2] SoftBank purchased Japan Telecom in 2006.

[3] This revenue figure is an estimate based on data collected in this study: it is based on average per capita revenues reported for several other countries in this study with similar per capita income.

2. The data from 1996 to 2004 include all services. The data in 2008 include broadband services such as FTTH, digital subscriber lines (xDSL), CATV, and fixed wireless access (FWA), but exclude dial-up and Integrated Services Digital Network (ISDN) services.

Table 26-14. Internet Search Engines (Market Shares by Search Volume)

| | 2008 | 2009 | 2012 |
|---|---|---|---|
| Yahoo! Japan | 49.8 | 51.3 | 55.1 |
| Google Japan | 39.9 | 38.2 | 34.9 |
| Rakuten | 1.8 | 2.3 | 2.1 |
| Microsoft Japan (Bing, MSN etc.) | 1.9 | 1.7 | 1.0 |
| NTT (public) (goo etc.) | 1.5 | 1.5 | 1.3 |
| Amazon Japan | 1.3 | 1.0 | |
| Excite Japan | 0.9 | 0.8 | |
| NEC (BIGLOBE) | 0.6 | 0.6 | 1.2 |
| Mixi Inc. | 0.3 | 0.3 | |
| Nifty Corporation (@nifty) | 0.2 | 0.3 | |
| Others | 1.8 | 1.9 | 4.4 |
| Search Queries (mil/month) | 6,291 | 6,792 | |
| Total Revenue (100 mil yen)[1] | 4,127 | 1,803 | |
| Total Revenue (mil US$) | 3,999 | 1,929 | 4,176 |
| C4 | 93.4 | 93.5 | 93.4 |
| HHI | 4,084 | 4,108 | 4,262 |
| *N* (>1%) | 6 | 6 | 6 |
| Noam Index | 1,667 | 1,677 | 1,740 |

*Source*: comScore. 2012 updates by editors from <http://www.slideshare.net/infocubic/english-presentaion2013ny>. Revenues estimated.
[1]This revenue figure is an estimate based on data collected in this study: it is based on average per capita revenues reported for several other countries in this study with similar per capita income.

Bing, a Microsoft-owned portal site of Microsoft and goo, which is a product of NTT Communications Co.'s NTT Resonant, are also major players in the Japanese search engine market.

There are several Japanese portal sites competing in this sector as well: Rakuten is one of Japan's leading e-commerce companies, with over 50 million registered users. It covers many services, from e-commerce to travel services (Table 26.14).

## Online News

Yahoo! News is the frontrunner in Japanese online news markets, and is owned by SoftBank. Online sites like Yahoo! News have become one of the primary news sources for Japanese audiences. Since more advertising expenditures go to online outlets rather than to newspapers, this shift may prove to be a fatal blow to newspaper companies (Table 26.15).

## CONCLUSION

The pace of change in Japanese media concentration has been more gradual when compared with that of the United States. Media concentration and ownership in Japan have not changed drastically in years for several reasons. First, the Japanese government has kept a friendly relationship with the media, which created vested interests in terms of allocation of privileges, such as broadcasting licenses and spectrum allocation. Second, Japan tends to prefer evolution to revolution, making the incremental approach preferable, even in IT services. Third, Japanese media companies, particularly traditional media such as newspapers and broadcasting, have avoided making IPOs: although some broadcasters have been listed on the stock exchange in recent years, private family ownership or cross-ownership is much more common in Japan. Fourth, Japanese media, as well as

Table 26-15. Online News Media (Market Shares by Monthly Unique Visitors), 2010

| | DOMAIN | PAGE VIEWS (MIL/MONTH) | MARKET SHARE |
|---|---|---|---|
| Yahoo! News | (headlines.yahoo.co.jp, news,yahoo.co.jp) | 12,759,641 | 42.7 |
| Yomiuri Online | (yomiuri.co,jp) | 1,265,398 | 4.2 |
| BIGLOBE News | (news.biglobe.ne.jp) | 1,204,092 | 4.0 |
| MSN Sankei News | (sankei.jp.msn.com) | 1,124,069 | 3.8 |
| Fuji TV | (fujitv.co.jp) | 1,072,909 | 3.6 |
| NHK (public) | (nkh.or.jp) | 939,778 | 3.1 |
| Yahoo! Sports | (sports.yahoo.co.jp) | 920,926 | 3.1 |
| asahi.com | (asahi.com) | 806,319 | 2.7 |
| Infoseek News | (news.infoseek.co.jp) | 789,972 | 2.6 |
| AFP BB News | (afpbb.com) | 764,674 | 2.6 |
| Total Revenue (100 mil yen)[1] | 1,689 | | |
| Total Revenue (mil US$) | 877 | | |
| C4 | 54.7 | | |
| HHI | 1,965 | | |
| *N* (>1%) | 23 | | |
| Noam Index | 410 | | |

*Source*: Video Research Interactive.

[1] This revenue figure is an estimate based on data collected in this study: it is based on average per capita revenues reported for several other countries in this study with similar per capita income.

Japanese industry as a whole, exhibit little in the way of drive for mergers or acquisitions in order to expand into new business arenas.

However, the technological environment surrounding Japanese media has changed rapidly due to the growth of the Internet. It is now quite easy for media consumers to access online services. These innovations have impacted how traditional media sources do business, leading to a sharp reduction in advertising revenue. As a result, traditional media are now being forced to enter nontraditional markets. ISPs, search engines, video-sharing sites, and social media forums are creating new ways for people to communicate, congregate, and share information. Media innovation forces changes in communication policies, which alternatively promote the development of new software and hardware in communications technology. Japan's geography also gives rise to new demands of online services: on March 11, 2011, a devastating earthquake struck Japan, triggering a massive tsunami and a nuclear crisis as well. This unprecedented disaster made the Japanese realize how important it is to have access to reliable media sources in emergencies. In this case, social media platforms like Twitter provided a platform for airing urgent pleas for help and for passing on information on the location of emergency shelters and open gas stations.

When we discuss media concentration issues, the interactive relationships between media and society should not be ignored. Japan faces many social and economic challenges from a rapidly aging population and a sharp decline in the birth rate, which are changing consumer demographics. The graying of Japan's population particularly disfavors traditional media sources like print newspapers, while the low birth rate does not bode well for Internet-oriented companies who rely on digital natives to purchase more of their products. There is a great need to bridge the divide between digital natives and digital immigrants. The existence of this digital divide widens the gap of opportunities for consumers' access to communications technologies and information relevant to a wide variety of socio-economic activities.

We should also keep in mind that the assessment of the HHI is only the first step in achieving a balanced view of media diversity and pluralism. Regarding the use of the HHI as a measurement of media concentration, we should consider determining the relevant media markets to analyze. Since media companies are attempting to expand into new venues like smart-phones and broadband access, the market boundaries among traditional media fields have been blurred. It is also important that not only the supply-side economics for media be considered, but the demand-side as well. A media concentration index should account for the power of influence by the media in terms of the number of people reached compared to the amount of time and money needed to acquire the information that the consumer receives.

## Bibliography

### 1. Books and Articles


Hoso Journal (Broadcasting Journal), "Nihon no Cable Television (CATV Industry in Japan)." Gekkan Hoso Journal (Monthly Broadcasting Journal) from July 1992, July 1996, July 2000, July 2004 and July 2008.

JiJi Eiga Tsushinsha, "Eiga Nenkan (Film Industry Yearbook)" from 1993, 1997, 2001, 2005 and 2009

NAB Japan Research Institute, "Management Data Book 2009."

Nihon ABC Kyokai (Japan Audit Bureau of Circulation), "Shimbun Report and Hanki Hakkosha Report" from January—June 2008.

Nihon Hoso Kyokai (Japan Broadcasting Corporation), "NHK Nenkan (Radio & Television Yearbook)" from 1985, 1989, 1993, 1997, 2001, 2005, and 2009.

Nihon Shimbun Kyokai (The Japan Newspaper Publishers & Editors Association), "Zenkoku Shimbun Guide" from 1993, 1997, 2001 and 2005.

Nippon Shuppan Hanbai (Japan Publication Sales), "Shoten Keiei Seminar Kaiho Tokushugo (Bookstore Bulletin, Special Issue)" from 1996, 2000, 2004 and 2008.

Nikkei Sangyo Shimbun, "Nikkei Share Chosa (Nikkei Market Share Research)" from 1997, 2001, 2005, 2009.

Record Tokushin Shuppansha, "Nikkan Record Tokushin Shimo-ki Kessan Tokushu-go" from 1997, 2001, 2005 and 2009.


### 2. Reports


Ministry of Internal Affairs and Communications, Japan. "2009 White Paper: Information and Communications in Japan." Ministry of Internal Affairs and Communications, Japan. 2009. <http://www.soumu.go.jp/johotsusintokei/whitepaper/ja/h21/pdf/index.html>.

Ministry of Internal Affairs and Communications, Japan. "2001 White Paper: Information and Communications in Japan." Tokyo: Ministry of Internal Affairs and Communications, Japan, 2001. <http://www.soumu.go.jp/johotsusintokei/whitepaper/ja/h13/pdf/index.htm>.

Ministry of Internal Affairs and Communications, Japan. "2005 White Paper: Information and Communications in Japan." Tokyo: Ministry of Internal Affairs and Communications, Japan, 2005. <http://www.soumu.go.jp/johotsusintokei/whitepaper/ja/h17/pdf/index.html>.

Ministry of Posts and Telecommunications, Japan. "1985 White Paper: Communications in Japan (Status Report on Communications)." Ministry of Posts and Telecommunications, Japan. 1985. <http://www.soumu.go.jp/johotsusintokei/whitepaper/ja/s60/pdf/index.html>.

Ministry of Posts and Telecommunications, Japan. "1989 White Paper: Communications in Japan (Status Report on Communications)." Tokyo: Ministry of Posts and Telecommunications, Japan, 1989. <http://www.soumu.go.jp/johotsusintokei/whitepaper/ja/h01/pdf/index.html>.

Ministry of Posts and Telecommunications, Japan. "1993 White Paper: Communications in Japan (Status Report on Communications)." Tokyo: Ministry of Posts and Telecommunications, Japan, 1993. <http://www.soumu.go.jp/johotsusintokei/whitepaper/h05.html>.

Ministry of Posts and Telecommunications, Japan. "1997 White Paper: Communications in Japan (Status Report on Communications)." Tokyo: Ministry of Posts and Telecommunications, Japan, 1997. <http://www.soumu.go.jp/

johotsusintokei/whitepaper/ja/h09/pdf/index.html>.

## 3. Websites

"Number of Subscribers." TCA.org. n.d. Telecommunications Carriers Association, n.d. July 1, 2012. <http://www.tca.or.jp/english/database/index.html>.

Shicho-Setaisuu (Subscribers). Japan Satellite Broadcasting Association, n.d. July 1, 2012. <http://www.eiseihoso.org/data/>.

"Yahoo! Attracts More than Half of All Searches Conducted in Japan in January 2009 – comScore, Inc." comScore, Mar. 10, 2009. <http://www.comscore.com/Insights/Press_Releases/2009/3/Japan_Search_Engine_Rankings>.

# Japan—Data Summaries

ELI NOAM AND PAUL MUTTER

MOST SECTORS WITHIN JAPAN'S national media industries are the second-largest in the world by revenue, after the United States. But advertising revenues have declined for traditional media groups (Table 26.16). Overall concentration is declining due to the entry of several new multimedia producers, with the national power index declining from 3,585 in 2004 to 3,274 after 2008.

Japan has the twelfth-highest percentage of public ownership of media internationally among 30 countries. This is due to the public telecom provider NTT (which is 33.7% state-owned) and to the NHK television network. NHK's share ownership in TV broadcasting and video channels has remained steady at 16.1% and 25.6%, respectively.

The largest content media companies by market shares are the Asahi Shimbun group (11.1%), NHK (9.0%), and Fujisankei (8.6%). After those are NTV (8.0%), the Tokyo Broadcasting System (7.0%), SkyPerfecTV! (6.9%), and the Yomiuri Shimbun (6.0%).

Overall, concentration is lower in content than in platforms. The Japanese broadcaster Fujisankei (NBS) holds substantial but not overwhelming market shares in TV broadcasting, video channels, daily newspapers, radio, and online news. Similarly, the Asahi Shimbun group maintains intermediate market shares in multichannel, video channels, daily newspapers, and online news. Although most companies exhibit a high level of cross ownership, SkyPerfecTV!, Toho, and J:COM maintain strong ownership in only one industry; SkyPerfecTV! owns 55.6% of video channels, and Toho controls approximately 37.9% of the film market. J:COM and SkyPerfecTV! have market shares of 22.1% and 43.6% in multichannel platforms, respectively. Because of the size of the larger

Table 26-16. National Media Industries Concentration in Japan

| | 2004/5 | | 2011 OR MOST RECENT | | % CHANGE ANNUAL AVERAGE | |
|---|---|---|---|---|---|---|
| | COMPANY POWER INDEX IN COUNTRY | COMPANY SHARE OF THE OVERALL NATIONAL MEDIA MARKET (%) | COMPANY POWER INDEX IN COUNTRY | COMPANY SHARE OF THE OVERALL NATIONAL MEDIA MARKET (%) | COMPANY POWER INDEX IN COUNTRY | COMPANY SHARE OF THE OVERALL NATIONAL MEDIA MARKET (%) |
| NTT (33.7% public) | 2,730.1 | 38.0 | 2,223.0 | 38.0 | -2.6 | −0.07 |
| KDDI | 136.0 | 8.6 | 290.0 | 13.0 | 16.0 | 0.6 |
| Sky Perfect JSAT | 145.6 | 3.2 | 152.0 | 3.1 | 0.6 | −0.01 |
| SoftBank (Japan Telecom) | 111.0 | 7.5 | 276.0 | 12.4 | 21.3 | 0.7 |
| NHK (public) | 77.3 | 3.7 | 44.0 | 2.4 | −6.15 | −0.19 |
| Fujisankei | 79.7 | 4.0 | 35.0 | 2.3 | −8.01 | −0.24 |
| Asahi Shimbun | 57.6 | 3.8 | 59.4 | 3.4 | 0.43 | −0.05 |
| NTV | 57.5 | 3.0 | 41.4 | 2.1 | −3.99 | −0.12 |
| J:COM | 15.8 | 9.0 | 21.6 | 1 | 5.22 | 0.015 |
| Tokyo Broadcasting System | 45.7 | 2.8 | 29.2 | 2 | −5.1 | −0.1 |
| Yomiuri Shimbun | 34.1 | 1.7 | 34.1 | 1.6 | −1.3 | −0.02 |
| Google (US) | 36.6 | 0.92 | 21.6 | 0.62 | −5.9 | −0.04 |
| Toho | 7.4 | 0.3 | 11.0 | 0.3 | 6.9 | 0.000 |
| Yahoo (US) | 8.4 | 0.5 | 7.3 | 0.4 | −2.0 | −0.01 |
| Mainichi Shimbun | 5.3 | 0.7 | 3.6 | 0.5 | −4.7 | −0.02 |
| WOWOW | 5.1 | 0.3 | 3.2 | 0.2 | −5.3 | −0.01 |
| Nihon Keizai Shimbun | 7.0 | 1.3 | 9.2 | 1.3 | 4.4 | 0.004 |
| Chunichi Shimbun | 4.1 | 0.6 | 2.9 | 0.45 | −4.3 | −0.02 |

*continued*

Table 26-16. *continued*

| | 2004/5 | | 2011 OR MOST RECENT | | % CHANGE ANNUAL AVERAGE | |
|---|---|---|---|---|---|---|
| | COMPANY POWER INDEX IN COUNTRY | COMPANY SHARE OF THE OVERALL NATIONAL MEDIA MARKET (%) | COMPANY POWER INDEX IN COUNTRY | COMPANY SHARE OF THE OVERALL NATIONAL MEDIA MARKET (%) | COMPANY POWER INDEX IN COUNTRY | COMPANY SHARE OF THE OVERALL NATIONAL MEDIA MARKET (%) |
| Rakuten | 0.0 | 0.0 | 0.1 | 0.04 | N/A | 0.005 |
| Sony | 0.5 | 0.07 | 0.1 | 0.03 | –11.8 | –0.007 |

| Media Concentration Index | 2004/5 | 2011 or Most Recent | % Change Annual Average |
|---|---|---|---|
| Total Revenue: Nat'l Media Industry (mil US$) | 173,874 | 235,803 | 5.01% |
| Total Voices (*n*) | 64 | 72 | 1.8 |
| Net Voices (*n*) | 45 | 53 | 2.5 |
| Public Ownership (%) | 16.7 | 15.2 | –0.2 |
| Foreign Ownership (%) | 2.5 | 1.8 | –0.1 |
| C4 Average—Weighted | 84.2 | 84.4 | 0.02 |
| HHI Average—Weighted | 3,758 | 3,328 | –1.6 |
| C1 Average—Weighted | 48 | 46 | 0 |
| Noam Index Average—Weighted | 713 | 635 | –1.58 |
| Pooled Overall Sector C4 | 58.7 | 66.8 | 1.15 |
| Pooled Overall Sector HHI | 1,700 | 1,818 | 1.0 |
| Pooled Overall Sector Noam Index | 92 | 73 | –3 |
| Market Share of Top Ten Companies: Nat'l Media Industry (%) (Pooled C10) | 77.0 | 80.4 | 0.5 |
| National Power Index | 3,585 | 3,274 | –1.24 |

Table 26-17. Top Content Media Companies in Japan

| | 2004/5 | | 2011 OR MOST RECENT | | % CHANGE ANNUAL AVERAGE | |
|---|---|---|---|---|---|---|
| | COMPANY POWER INDEX IN COUNTRY | COMPANY SHARE OF THE NATIONAL CONTENT MEDIA MARKET (%) | COMPANY POWER INDEX IN COUNTRY | COMPANY SHARE OF THE NATIONAL CONTENT MEDIA MARKET (%) | COMPANY POWER INDEX IN COUNTRY | COMPANY SHARE OF THE NATIONAL CONTENT MEDIA MARKET (%) |
| Sky Perfect JSAT | 285.5 | 5.8 | 354.2 | 6.9 | 3.4 | 0.15 |
| NHK (public) | 228.6 | 11.0 | 162.4 | 9.0 | –4.1 | –3 |
| Fujisankei | 236.0 | 11.8 | 129.3 | 8.6 | –6.5 | –0.5 |
| Asahi Shimbun | 170.5 | 11.2 | 219.2 | 12.6 | 4.1 | 0.2 |
| NTV | 170.0 | 9.0 | 153.0 | 8.01 | –1.43 | –0.12 |
| Tokyo Broadcasting System | 135.1 | 8.2 | 108.0 | 7.0 | –3.0 | –0.16 |
| Yomiuri Shimbun | 101.0 | 5.3 | 114.7 | 6.0 | 1.9 | 0.11 |
| Google (US) | 108.3 | 2.7 | 79.6 | 2.3 | –3.8 | –0.06 |
| J:COM | 15.4 | 0.85 | 26.3 | 1.2 | 10.1 | 0.05 |
| Toho | 22.0 | 0.85 | 40.6 | 1.07 | 12.1 | 0.03 |
| Yahoo (US) | 8.0 | 0.5 | 27.2 | 1.6 | 59.7 | 0.3 |
| Mainichi Shimbun | 15.8 | 2.02 | 13.1 | 1.9 | –2.4 | –0.02 |
| WOWOW | 15.2 | 1.07 | 11.9 | 0.9 | –3.1 | –0.01 |
| Nihon Keizai Shimbun | 20.7 | 1.5 | 33.8 | 1.5 | 9.0 | –0,01 |
| Chunichi Shimbun | 12.1 | 1.8 | 10.6 | 1.3 | –1.8 | –0.06 |
| Rakuten | 0.0 | 0.0 | 0.3 | 0.13 | N/A | 0.02 |

*continued*

Table 26-17. *continued*

| | 2004/5 | | 2011 OR MOST RECENT | | % CHANGE ANNUAL AVERAGE | |
|---|---|---|---|---|---|---|
| | COMPANY POWER INDEX IN COUNTRY | COMPANY SHARE OF THE NATIONAL CONTENT MEDIA MARKET (%) | COMPANY POWER INDEX IN COUNTRY | COMPANY SHARE OF THE NATIONAL CONTENT MEDIA MARKET (%) | COMPANY POWER INDEX IN COUNTRY | COMPANY SHARE OF THE NATIONAL CONTENT MEDIA MARKET (%) |
| Softbank (Japan Telecom) | 8.4 | 1.6 | 28.3 | 1.6 | 59.7 | 0.01 |
| Sony | 1.6 | 0.2 | 0.3 | 0.1 | –11.2 | –0.02 |
| Media Concentration Index | 2004/5 | | 2011 or Most Recent | | % Change Annual Average | |
| Public Ownership (%) | 11.0 | | 9.0 | | –0.3 | |
| Foreign Ownership (%) | 6.9 | | 5.8 | | –0.2 | |
| C4 Average—Weighted | 69.9 | | 66.3 | | –0.5 | |
| HHI Average—Weighted | 1,781 | | 1,725 | | –0.5 | |
| C1 Average—Weighted | 25.5 | | 24.8 | | 0.0 | |
| National Power Index | 1,619 | | 1,531 | | –0.8 | |

Table 26-18. Top Platform Media Companies in Japan

| | 2004/5 | | 2011 OR MOST RECENT | | % CHANGE ANNUAL AVERAGE | |
|---|---|---|---|---|---|---|
| | COMPANY POWER INDEX IN COUNTRY | COMPANY SHARE OF THE NATIONAL PLATFORM MEDIA MARKET (%) | COMPANY POWER INDEX IN COUNTRY | COMPANY SHARE OF THE NATIONAL PLATFORM MEDIA MARKET (%) | COMPANY POWER INDEX IN COUNTRY | COMPANY SHARE OF THE NATIONAL PLATFORM MEDIA MARKET (%) |
| NTT (33.7% public) | 4,124 | 58.2 | 3,049 | 52.2 | –3.7 | –0.9 |
| KDDI | 205.1 | 13.0 | 398.0 | 17.7 | 13.4 | 0.67 |
| Softbank (Japan Telecom) | 148.0 | 9.0 | 180.0 | 11.3 | 5.5 | 0.6 |
| JSAT (SkyPerfecTV!) | 74.2 | 2.0 | 77.1 | 1.8 | 0.5 | –0.02 |
| J:COM | 16.0 | 1.0 | 20.0 | 1.0 | 3.4 | 0.002 |
| Media Concentration Index | 2004/5 | | 2011 or Most Recent | | % Change Annual Average | |
| Public Ownership (%) | 19.6% | | 17.6% | | –0.3% | |
| Foreign Ownership (%) | 0.5% | | 0.4% | | –0.02% | |
| C4 Average—Weighted | 91.6 | | 91.1 | | –0.07 | |
| HHI Average—Weighted | 4,768 | | 3,924 | | –2.5 | |
| C1 Average—Weighted | 60.2 | | 54.0 | | –1.0 | |
| National Power Index | 4,589 | | 3,922 | | –2.1 | |

multimedia groups like Yomirui Shimbun and Fujisankei, though, even in content their market shares are relatively low (Table 26.17) despite higher power indices.

NTT has the largest company power index in the overall national media industries and as a platform company (Table 26.18), but it is losing market share as a result of a decline in demand for wireline services and competition in the wireless and ISP markets from fully commercial operators like SoftBank and KDDI. NTT's share of the platform market declined from 58.2% in 2004 to 52.2% in 2012 (Table 26.18).

# 27

# Media Ownership and Concentration in South Korea

DAEHO KIM AND SEONGCHEOL KIM

## INTRODUCTION

The Korean media industry has changed from a monopolistic or oligopolistic structure to a more competitive market structure. Before 1987, South Korean media were characterized by extensive state intervention. In terrestrial broadcasting, three state-funded providers constituted the entirety of the sector; print publishing was restricted by censorship and licensing; and as in a number of other countries, the telecommunications industry was government-run as well. But after free elections were held in 1987 and a new government came into office, new democratic politics and increased economic liberalization led to the privatization of the country's media markets. As private media ownership grew, the National Assembly passed both a Periodicals Law and a Broadcasting Law in 1987 that together deregulated print and audiovisual media ownership. Print licensing became less restrictive, commercial television networks were launched, cable TV providers entered the market, and the privatization of the state telecom monopolist was announced. Technological innovations spurred the introduction of new market segments in the 2000s: satellite broadcasting was licensed in 2002, mobile TV in 2005, and IPTV in 2008. Finally, after a lobbying campaign lasting several years, print publishers managed to convince the government to legalize greater media cross-ownership in the audiovisual sector (2001 legislation had already removed some anti-concentration measures in the pay-TV market). In 2010, the Broadcasting Act was amended to remove anti-concentration limits on terrestrial broadcasting ownership. The debate over the deregulation of media ownership focused on the issue of whether deregulatory policy would lead to the development of media conglomerates or see concentration indices fall further. Critics of the legislation contended that deregulation would lead to the formation of vertically integrated media groups that would dominate multiple platforms. In fact, the departure of traditional conglomerates, such as Samsung and Hyundai, from media markets in the 1990s did open these markets—particularly audiovisual markets—to such trends. Though the formation of multimedia groups is a relatively new phenomenon in South Korea, it is likely that vertical integration will increase among South Korean media producers, especially in the audiovisual sector.

## PRINT MEDIA

### Newspapers

In the 1980s, South Korean consumers had a choice of six national daily newspapers. Then in 1987, the Periodicals Law did away with government licensing requirements for daily newspapers. With the industry deregulated, several new commercial publishers entered the market and concentration fell dramatically: the number of daily newspapers increased rapidly; by 1993, South Koreans had 70 national and local dailies to choose from. The leading commercial operators enjoyed high market shares because South Korea's advertising industries expanded during this period, thanks to a booming economy. By 2009, 288 daily newspapers were available in South Korea, 10 of them national dailies.

In 2010, three major newspapers, *The Chosun Daily* (24.3%), *JoongAng Daily* (21.8%), and *The Dong-A Ilbo* (18.3%), accounted for 64.4% of total market shares. These three dailies are collectively referred to as the "Chojoongdong" due to their market dominance and shared political conservatism. All belong to private, family-owned media groups: JoongAng Media Network (JMnet) owns *JoongAng Daily*; *The Chosun Daily* and *The Dong-A Ilbo* are also family-owned newspapers. Due to the influence of evangelical Christianity in South Korean culture, church ownership of national dailies is not uncommon: *Kookmin Ilbo* is owned by Yoido Full Gospel Church, and *Segye Times* is owned by the Unification Church.

The daily newspaper market has been moderately concentrated since 1980, and concentration has decreased from 1,774 in 1985 to 1,467 in 2008, though concentration increased slightly in 2009.

Before 2010, cross-ownership of newspapers and commercial broadcasters was prohibited. In 2010, even though there were questions about print publishers' ability to survive in the broadcasting industry due to their declining revenues, regulators permitted JMnet and other print publishers to purchase stakes in commercial broadcasters. The Chojoongdong, along with the Maeil Business Newspaper & MK Inc., publisher of the largest financial daily in South Korea, received broadcasting licenses (Table 27.1).

### Book Publishing

The book publishing industry is mostly composed of small- and medium-sized firms. As a result, the sector is highly deconcentrated. Even the "largest" publishers have relatively small market shares. The Woongjin Group's Woongjin ThinkBig, which specializes in textbook publishing and is the largest publishing house in the country, posted a total revenue of US$60 million (KRW70 billion) in 2008, a sum that represented just 2% of total book publishing revenue in the country. Other major book publishers, such as Sigongsa and Gimmyoung Publishers, are independently owned, as are most of the 2,691 publishing houses operating in South Korea, and they are much smaller than Woongjin ThinkBig. The average revenue for a publishing house in 2009, for example, was only US$1.2 million (KRW1.5 billion) (Table 27.2).

### Magazine Publishing

South Korea's magazine publishing market is structured much like the book publishing market. Even though the value of the magazine market had shrunk in recent years due to declining advertising revenue, the number of magazine titles in circulation has substantially increased. Most of the large publishers of books and newspapers, such as Seoul Media Group, JMnet, Chosun Media, and Woongjin

## Table 27-1. Daily Newspapers (Market Shares by Circulation), 1985–2010

| | 1985 | 1990 | 1995 | 2000 | 2004 | 2008 | 2009 | 2010 |
|---|---|---|---|---|---|---|---|---|
| *The Chosun Ilbo* (Bang Family, Chosun Media) | 18.9 | 22.9 | 22.9 | 25.1 | 24.9 | 24.0 | 25.1 | 24.3 |
| *JoongAng Ilbo* (Hong Family, JMnet-Samsung Group) | 22.8 | 23.8 | 23.6 | 20.3 | 21.6 | 19.7 | 20.4 | 21.8 |
| *The Dong-A Ilbo* (Kim Family, Dong-A Media Group) | 19.3 | 14.1 | 18.7 | 18.9 | 18.6 | 17.1 | 19.1 | 18.3 |
| *Kyunghyang Shinmun* (cooperative) | 10.1 | | | 4.3 | 4.7 | 8.8 | 5.2 | 4.8 |
| *Seoul Newspaper* (public) | 12.7 | 14.1 | 9.4 | 5.5 | 5.3 | 6.2 | 5.8 | 6.3 |
| Hankook Ilbo Media Group | 16.2 | 17.1 | 15.0 | 14.0 | 6.8 | 5.5 | 5.8 | 5.4 |
| Hankyoreh (cooperative)[1] | | 2.6 | 2.8 | 4.5 | 5.0 | 4.9 | 4.2 | 5.3 |
| Munhwa Ilbo | 0.0 | | 3.2 | 3.2 | 3.9 | 4.5 | 4.8 | 4.6 |
| Kukmin Ilbo | | 2.2 | 1.8 | 2.2 | 2.6 | 3.6 | 3.9 | 3.6 |
| Maeil Business Newspaper & MK Inc. | | | | | 2.1 | 3.4 | 3.4 | 3.4 |
| Segye Ilbo | | 3.2 | 2.6 | 2.1 | 4.6 | 2.2 | 2.3 | 2.2 |
| Total Revenue (bil KRW) | 227 | 391 | 1,149 | 1,832 | 1,071 | 1,498 | 1,505 | 1,472 |
| Total Revenue (mil US$) | 250 | 548 | 1,493 | 1,649 | 1,392 | 1,348 | 1,204 | 1,325 |
| C4 | 77.2 | 78.0 | 80.2 | 78.3 | 71.8 | 69.7 | 69.9 | 69.3 |
| HHI | 1,774 | 1,804 | 1,772 | 1,683 | 1,594 | 1,467 | 1,569 | 1,564 |
| *N* (>1%) | 6 | 8 | 9 | 10 | 11 | 11 | 11 | 11 |
| Noam Index | 724 | 638 | 591 | 532 | 481 | 442 | 473 | 472 |

*Source:* Korea Press Foundation.

[1] 65% of Hankyoreh's shares are publicly traded and its employees own the remaining 35%, making them the largest single shareholding group.

## Table 27-2. Book Publishing (Market Shares by Revenue), 2000–2009

| | 2000 | 2003 | 2004 | 2005 | 2006 | 2007 | 2008 | 2009 |
|---|---|---|---|---|---|---|---|---|
| Total Revenue (bil KRW) | 2,996 | 3,538 | 2,439 | 2,041 | 3,158 | 3,152 | 3,444 | 4,110 |
| Total Revenue (mil US$) | 2,696 | 2,830 | 3,170 | 3,266 | 3,158 | 3,476 | 3,110 | 3,288 |
| *N* (>1%) | 1,718 | 1,517 | 1,709 | 2,261 | 2,168 | 2,757 | 2,761 | 2,691 |

*Source:* Ministry of Culture, Sports and Tourism.

ThinkBig, are active in magazine publishing, but just as Woongjin ThinkBig does not dominate the book market, it and these other publishers do not dominate the magazine market. The most popular titles are Korean-language monthlies produced in the country and foreign newsweeklies translated into Korean (Table 27.3).

## AUDIOVISUAL MEDIA

### Radio

Three groups constitute radio broadcasting market in Korea (Table 27.4). First, terrestrial broadcasters have dominated the radio

Table 27-3. Magazine Publishing (Market Shares by Revenue), 2002–2009

| | 2002 | 2003 | 2004 | 2005 | 2006 | 2007 | 2008 | 2009 |
|---|---|---|---|---|---|---|---|---|
| Total # of Titles in Circulation | 4,030 | 3,795 | 3,929 | 4,289 | 4,661 | 5,008 | 4,993 | 7,956 |
| Total Revenue (bil KRW) | | 723 | 558 | 478 | 812 | 918 | 1,741 | 1,094 |
| Total Revenue (mil US$) | | 578 | 725 | 765 | 812 | 1,010 | 1,567 | 875 |

*Source:* Ministry of Culture, Sports and Tourism.

broadcasting market. The radio market has long been dominated by MBC, which began the first radio service in Korea. Second, religious radio broadcasting has also been strong. CBS (Christian Broadcasting System) has a special position in that this Protestant channel led Korea's radio broadcasting and still maintains a high degree of popularity. PBC (Pyunghwa Broadcasting Company) is owned by the Catholic Church. BBS (Buddhist Broadcasting System) is operated by a Buddhist group. FEBC (Far East Broadcasting Company) is owned by another Protestant church. WBS (Won-Buddhist Broadcasting System) is operated by Won-Buddhism, another Buddhism group. Third, since 2000, regional broadcasting increased its market share. K-FM (Kyunggi FM), which covers the largest Kyunggi province, led regional broadcasting. KNN (Korea New Network), which covers the second largest city, followed.

The radio broadcasting market has been highly concentrated. However, its concentration ratio has decreased because of the increasing popularity of regional radio broadcasting. Podcasts also tend to make the radio broadcasting market less concentrated.

Table 27-4. Radio Group (Market Shares by Revenue), 1990–2011

| | 1990 | 1994 | 1998 | 2002 | 2006 | 2010 | 2011 |
|---|---|---|---|---|---|---|---|
| KBS (public) | 18.6 | 17.1 | 9.8 | 11.1 | 11.6 | 9.7 | 8.2 |
| MBC (public) | 57.6 | 54.9 | 63.9 | 48.9 | 49.6 | 50.6 | 47.6 |
| SBS (SBS Media Holdings) | — | 6.9 | 6.1 | 7.4 | 7.8 | 9.3 | 10.4 |
| CBS | 11.9 | 12.1 | 10.6 | 15.8 | 12.9 | 13.1 | 13.9 |
| PBC | 2.8 | 3.9 | 2.9 | 4.4 | 3.9 | 2.9 | 3.2 |
| BBS | 2.9 | 3.9 | 3.1 | 4.8 | 4.3 | 3.4 | 3.8 |
| FEBC | — | 1.0 | 1.0 | 1.3 | 1.7 | 1.4 | 1.4 |
| K-FM | — | — | — | 1.6 | 2.0 | 1.6 | 1.8 |
| WBS | — | — | — | — | 1.7 | 1.5 | 1.8 |
| KNN | — | — | — | — | — | 1.1 | 1.3 |
| Others | 6.2 | 0.2 | 2.6 | 5.2 | 4.5 | 5.4 | 6.6 |
| Total Revenue (bil KRW) | 95.3 | 148.6 | 137.2 | 282.2 | 279.9 | 256.5 | 260.4 |
| Total Revenue (mil US$) | 136.1 | 165.1 | 91.5 | 235.2 | 243.4 | 233.2 | 236.7 |
| C4 | 91 | 91 | 90.4 | 83.2 | 81.9 | 82.7 | 80.1 |
| *N* (>1%) | 5 | 7 | 7 | 9 | 10 | 10 | 13 |
| HHI | 3,822 | 3,485 | 4,348 | 2,865 | 2,865 | 2,940 | 2,669 |
| Noam Index | 1,709 | 1,318 | 1,644 | 955 | 906 | 929 | 740 |

*Source:* Korea Broadcast Advertising Corporation.

## Broadcast TV

Before 1991, the public service broadcasters KBS, EBS, and MBC were the only available networks in the country. KBS was the primary public service broadcaster, followed by MBC (which KBS held a majority stake in); EBS served as educational public service broadcaster. KBS and EBS are still fully owned and funded by the Government of South Korea, but MBC now finances itself through commercial advertising revenues, and it is administered by two non-profit organizations, the Foundation for Broadcasting Culture (70%) and the Jeongsu Scholarship Foundation (30%). The National Assembly established the Foundation in 1988 to reduce the level of political control exercised by the state through KBS over MBC's programming.

In 1991, the first commercial broadcaster, Seoul Broadcasting System (SBS), entered the market. A financial consortium led by SBS Media Holdings Group, SBS immediately began to win market shares at the expense of the public broadcasters. In 1996, four regional commercial channels began broadcasting: TJB, KBC, TBC, and PSB. However, since their channels all had network affiliations with SBS, competition did not significantly increase in real terms among the four national terrestrial broadcasters. Even though concentration indices have decreased over the past two decades due to an increase in the number of commercial broadcasters, the Noam index was still high at 2,660 in 2010. As of 2012, analog service has been phased out in favor of digital terrestrial television (DTT) (Table 27.5).

## Multichannel TV Platforms

Cable TV service became available in South Korea during the 1960s because terrestrial TV reception was usually of such low quality that local cable operators relayed the state broadcasting channels to households. Only in 1995, though, was the first multichannel license awarded by the state, and the first cable channels were introduced with strict licensing restrictions to prevent vertical and horizontal integration. Multisystem operators (MSOs) did not enter the national market until 2000.

As a result of these regulations, concentration has remained low, but since the lifting of them, concentration has been increasing due to mergers and acquisitions carried out by the MSOs Tboard and CJ Hellovision between 2003 and 2009. As a result, Tboard and CJ Hellovision are now the two largest pay-TV providers in the

### Table 27-5. TV Broadcasting (Market Shares by Revenue), 1980–2010

| | 1980 | 1985 | 1990 | 1996 | 2000 | 2004 | 2007 | 2008 | 2009 | 2010 |
|---|---|---|---|---|---|---|---|---|---|---|
| KBS (public) | 41.0 | 73.0 | 68.0 | 53.0 | 49.0 | 46.0 | 45.0 | 36.6 | 39.1 | 37.3 |
| MBC (public) | 59.0 | 27.0 | 32.0 | 24.0 | 26.0 | 26.0 | 27.0 | 32.8 | 30.4 | 30.4 |
| SBS (SBS Media Holdings) | 0.0 | 0.0 | 0.0 | 20.0 | 22.0 | 22.0 | 22.0 | 16.9 | 16.5 | 18.3 |
| EBS (public) | 0.0 | 0.0 | 0.0 | 3.0 | 3.0 | 6.0 | 6.0 | 2.6 | 2.9 | 3.1 |
| Others | 0.0 | 0.0 | 0.0 | 0.0 | 0.0 | 0.0 | 0.0 | 11.1 | 11.1 | 10.9 |
| Total Revenue (bil KRW) | | | | | 2,871 | 2,129 | 2,973 | 3,307 | 3,551 | 3,540 |
| Total Revenue (mil US$) | | | | | 2,584 | 2,768 | 3,270 | 2,967 | 2,841 | 3,186 |
| C4 | C2[1] | C2 | C2 | 100.0 | 100.0 | 100.0 | 100.0 | 88.9 | 88.9 | 89.1 |
| *N* (>1%) | 2 | 2 | 2 | 4 | 4 | 4 | 4 | 4 | 4 | 4 |
| HHI | 5,162 | 6,058 | 5,648 | 3,794 | 3,570 | 3,312 | 3,274 | 2,708 | 2,734 | 2,660 |
| Noam Index | 3,650 | 4,284 | 3,994 | 1,897 | 1,785 | 1,656 | 1,637 | 1,354 | 1,367 | 1,330 |

*Source:* Korea Communications Commission.

[1] C2 = 100.0.

country. The Government of South Korea waved anti-concentration measures for pay-TV in 2001 to build up the sector so that cable MSOs could compete with the existing terrestrial broadcasters and with the IPTV platforms offered by telecommunication providers.

In 2010, the Chojoongdong "big three" newspapers' publishers and Maeil Business Newspaper & MK Inc. were each licensed by the Korean Communications Commission to provide their own cable TV channels. In amendments to the Broadcasting Act that outline the new policy, these networks are considered to be the equivalent of a terrestrial broadcasting channel. Comprehensive cable channels such as these will attract greater audience shares and increase competition in the audiovisual sector but they will need time to do so (Table 27.6).

The combined market shares of the Tbroad, CJ Hellovision, and C&M reached 66% in 2010. The Taekwang Group, which is one of big business conglomerates in Korea, owns Tboard. Taekwang Group also owns t.cast, a video channel, and film distributor. CJ Hellovision is a part of the CJ Group, which focuses on media and entertainment content. C&M is an independent MSO. Finally, HCN, the fourth largest MSO in South Korea, is owned by Hyundai Department Store Group. C&M and CMB are independently owned. These top five cable MSOs all provide triple-play services (Figure 27.1).

## Table 27-6. Multichannel Video Platforms: Cable MSOs (Market Shares by Revenue), 2001–2010

| | 2001 | 2004 | 2007 | 2008 | 2009 | 2010 |
|---|---|---|---|---|---|---|
| Tbroad (Taekwang Group) | 7.3 | 20.8 | 21.2 | 25.9 | 26.2 | 25.6 |
| Curix (as of 2009) | 6.8 | 5.5 | 5.3 | | | |
| Hanbit (as of 2003) | 16.2 | | | | | |
| CJ Hellovision (CJ Group) | 7.3 | 9.0 | 18.1 | 18.6 | 22.3 | 22.2 |
| On Media (as of 2009) | 6.2 | 3.7 | 2.6 | 2.9 | | |
| Eujin (as of 2006) | 7.8 | | | | | |
| C&M | 17.6 | 17.4 | 17.9 | 18.3 | 18.4 | 18.2 |
| HCN (Hyundai Department Store Group) | 10.2 | 5.5 | 7.1 | 8.3 | 8.6 | 8.6 |
| CMB | 0.0 | 8.6 | 5.0 | 5.6 | 4.6 | 5.4 |
| GS | 0.0 | 2.3 | 4.1 | 3.8 | 3.7 | 3.5 |
| KT (IPTV + Skylife) | 0.0 | 21.0 | 18.7 | 14.3 | 12.6 | 14.5 |
| SK Broadband | 0.0 | 0.0 | 0.0 | 0.0 | 0.0 | 0.0 |
| LG U + | 0.0 | 0.0 | 0.0 | 0.0 | 0.0 | 0.0 |
| Others | 20.6 | 6.2 | 0.0 | 2.3 | 3.6 | 2.0 |
| Total Subscribers (1,000) | 9,900 | 12,952 | 16,852 | 17,338 | 19,931 | 21,272 |
| Total Revenue (KRW bil) | 798 | 1,217 | 2,075 | 2,708 | 3,142 | 2,979 |
| Total Revenue (mil US$) | 638 | 1,523 | 2,283 | 2,437 | 2,514 | 2,681 |
| C4 | 52.0 | 68.0 | 76.0 | 77.0 | 80.0 | 81.0 |
| *N* (>1%) | 8 | 9 | 9 | 8 | 7 | 7 |
| HHI | 927 | 1,411 | 1,574 | 1,679 | 1,790 | 1,805 |
| Noam Index | 328 | 343 | 433 | 557 | 666 | 651 |

*Source:* Korean Communications Commission.

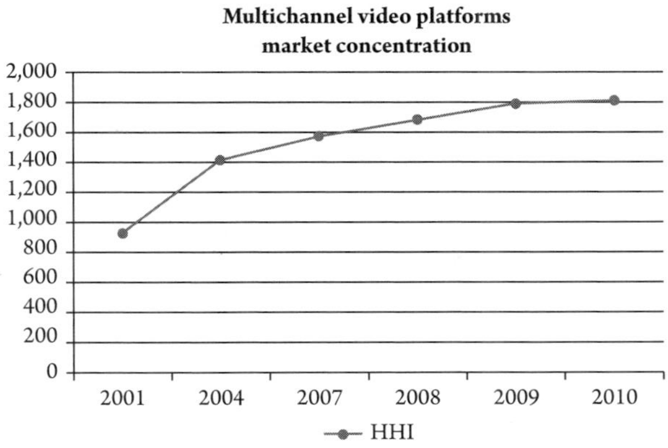


FIGURE 27.1 Multichannel Video Platforms Market Concentration (HHI), 2001–2010

Cable TV is by far the largest pay-TV platform in South Korea, with over 15 million subscribers. Household penetration rates stood at 83.7% in 2008, higher than every other country's rates except for Japan and Switzerland. As of 2012, there were three IPTV services—KT Olleh TV, BTV, and MyLG—operated by the telecom providers Korea Telecom (KT), SK Broadband (SK), and LG U + (LG), respectively. Together they constitute the second-largest pay-TV platform market segment, with a combined subscriber base of 7 million in 2013. So far, though, these telecoms' IPTV services have failed to break even (their actual revenues have not been disclosed yet). After IPTV, direct broadcast satellite (DBS) constitutes the third largest MSO. Korea Telecom's KT Skylife monopolizes satellite TV in South Korea, though with just 3.5 million subscribers it is the smallest MSO platform in the country.

## Film

Although the South Korean film market is not highly concentrated, this has increased slightly in recent years with the expansion of film theater multiplexes. CJ Entertainment, Showbox, and Lotte Entertainment are the three South Korean market leaders. CJ Entertainment, part of CJ E&M, is vertically integrated and owns the country's largest multiplex chain (CJ CGV); its market share stood at 35.5% in 2011. Lotte Entertainment is a relatively new operator, but like CJ Entertainment, its Lotte Cinema has been growing at Showbox's expense. The country's third largest multiplex chain is Megabox, with a market share of 9.2%. It is owned by JMNet. (Table 27.7).

In the film market, domestic distributors have benefitted from a state-mandated screen quota system. Every theater had to show Korean-made productions on 146 days, or 40%, of the year. This was subsequently lowered to 73 days.

# TELECOMMUNICATIONS MEDIA

## Wireline Telecom

South Korea's wireline communications market was a public monopoly until 2002, when the former state operator Korea Telecom (KT) was fully privatized 15 years after the process was announced in 1987. The state was reluctant to implement the process rapidly, but South Korea's 1995 World Trade Organization (WTO) ascension, followed by the 1997 East Asian financial crisis, forced the state to privatize the company at a faster pace. Liberalization policies included licensing the ISPs LG Dacom and LG Hanaro to handle international services, and then domestic long distance and local telephony services as well (Table 27.8).

The wireline telecommunications market is highly concentrated. KT is the clear market leader, with a 72.8% market share in 2009, while LG U + held just 16.6% and SK Broadband 9.9% (Figure 27.2).

## Table 27-7. Film Production/Distribution (Market Shares by Box Office %), 2000–2011

| | 2000 | 2004 | 2006 | 2008 | 2009 | 2010 | 2011 |
|---|---|---|---|---|---|---|---|
| CJ Entertainment (CJ E&M, CJ Group) | 23.3 | 24.4 | 23.2 | 30.1 | 29.2 | 28.0 | 35.5 |
| Showbox | 0.0 | 18.0 | 20.1 | 10.2 | 15.2 | 6.5 | 9.2 |
| Lotte Entertainment (Lotte Group, Japan/Korea) | 0.0 | 0.0 | 0.0 | 10.0 | 11.3 | 9.8 | 15.0 |
| Sony Pictures (Japan/US) | 12.7 | 4.9 | 10.0 | 8.3 | 9.4 | 9.9 | 9.4 |
| 20th Century Fox Korea (Murdoch Family, News Corp., US/UK/AUS) | 4.8 | 6.6 | 5.9 | 6.8 | 8.3 | 12.3 | 8.2 |
| NEW | 0.0 | 0.0 | 0.0 | 6.1 | 5.8 | 6.7 | 8.9 |
| Warner Brothers Korea (Time Warner, US) | 5.4 | 8.0 | 5.8 | 5.1 | 5.7 | 8.9 | 4.9 |
| Cinergy | — | — | — | 4.7 | 3.9 | 3.3 | 1.6 |
| UIP Korea (Universal-Paramout, US) | 7.5 | 4.2 | 7.6 | 4.1 | 2.1 | 2.5 | 1.7 |
| SK Telecom (SK) | — | — | — | 3.6 | 2.1 | 0.0 | — |
| Cinema Service | 15.6 | 17.2 | 11.0 | — | — | — | — |
| Others | 30.7 | 16.7 | 16.4 | 11.0 | 7.0 | 12.1 | 5.6 |
| Total Revenue (bil KRW) | 500 | 569 | 805 | 926 | 1,174 | 1,122 | 1,194 |
| Total Revenue (mil US$) | 455 | 739 | 805 | 833 | 939 | 1,001 | 1,075 |
| C4 | 59.0 | 68.0 | 64.0 | 59.0 | 65.0 | 60.0 | 69.0 |
| HHI | 1,056 | 1,364 | 1,289 | 1,340 | 1,459 | 1,313 | 1,834 |
| *N* (>1%) | 6 | 7 | 7 | 10 | 10 | 9 | 9 |
| Noam Index | 431 | 516 | 487 | 424 | 461 | 438 | 611 |

*Source:* Korean Film Council.

## Table 27-8. Wireline Telecom (Market Shares by Revenue), 2004–2009

| | 2004 | 2005 | 2006 | 2007 | 2008 | 2009 | 2011[1] |
|---|---|---|---|---|---|---|---|
| Korea Telecom (KT) | 83.6 | 81.7 | 79.9 | 79.0 | 75.7 | 72.8 | 72.5 |
| SK Broadband (SK) | 5.7 | 6.7 | 7.4 | 7.7 | 9.3 | 9.9 | 8.5 |
| LG U + (LG Corporation) | 9.6 | 10.7 | 11.9 | 12.6 | 14.3 | 16.6 | 10.9 |
| Onse Telecom | 1.1 | 0.9 | 0.8 | 0.7 | 0.7 | 0.7 | |
| Others | | | | | | | 8.1 |
| Total Revenue (bil KRW) | 4,930 | 3,954 | 6,173 | 5,492 | 6,452 | 6,424 | 9,836 |
| Total Revenue (mil US$) | 6,409 | 6,326 | 6,173 | 6,041 | 5,807 | 5,139 | 9,479 |
| C4 | 100.0 | 100.0 | 100.0 | 100.0 | 100.0 | 100.0 | 91.9 |
| HHI | 7,115 | 6,835 | 6,581 | 6,459 | 6,021 | 5,673 | 5,447 |
| *N* (>1%) | 4 | 3 | 3 | 3 | 3 | 3 | 3 |
| Noam Index | 3,558 | 3,946 | 3,800 | 3,729 | 3,476 | 3,275 | 3,115 |

*Source:* Korea Communications Commission, Korea Association for ICT Promotion.

[1] 2011 data provided by editors using Resource: Korea Communications Commission: Telecommunication Market Competition Assessment(2012) Retrieved from:http://www.kcc.go.kr/user.do?mode=view&page=A02050101&dc=K02050101&boardId=1022&cp=1&boardSeq=35585

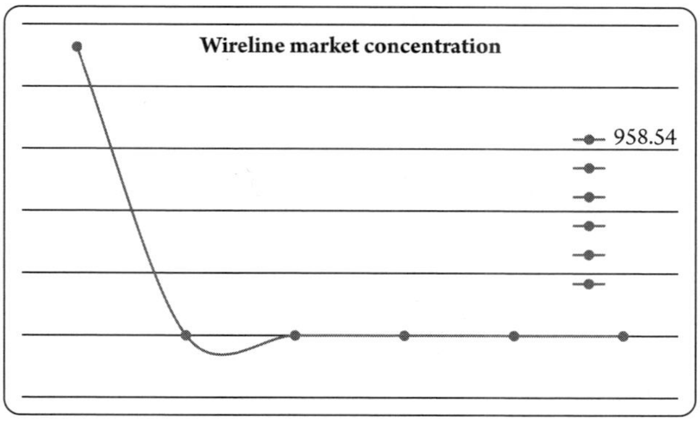


FIGURE 27.2 Wireline Market Concentration (HHI), 2004–2009

## Wireless Telecom

South Korea is a world leader in mobile phone technology: it was one of the first countries where both 3G and 4G services, such as Long Term Evolution (LTE) and mobile WiMAX(WiBro), were available. The market has also evolved considerably: in 1984, Korea Mobile Telecom (KMT) was established by the state as an arm of KT to provide mobile service, and for the first few years the sector was, like wireline, a KT monopoly. Then, during the privatization drive of the early 1990s, KMT was acquired by the SK Group and was renamed SK Telecom. At the same time, the state awarded a second national carrier license to Shinsegyi Telecom, creating a duopoly that persisted until 1997. That year, three new mobile service providers, including Korea Telecom Freetel (KTF), Hansol M.com, and LG Telecom (LGT, later LG U +) were awarded mobile licenses. In 2000, KTF was fully absorbed by KT and Hansol M.Com became part of KT as well, leaving the market with just three operators: KT, SK Telecom (SKT), and LG U + (Table 27.9).

Although these three providers have dominated the market for a decade, the introduction of mobile virtual network operator (MVNO) services may alter the market structure in favor of reduced concentration. (Figure 27.3).

### Table 27-9. Wireless Telecom (Market Shares by Revenue), 1996–2011

| | 1996 | 2000 | 2004 | 2008 | 2009 | 2010 | 2011 |
|---|---|---|---|---|---|---|---|
| SKT (SK) | 90.9 | 40.8 | 51.3 | 50.5 | 50.6 | 50.6 | 50.5 |
| Shinsegyi (as of 2000) | 9.1 | 13.1 | SKT | | | | |
| KT (formerly KTF) | 0.0 | 19.7 | 32.1 | 31.5 | 31.3 | 31.6 | 31.6 |
| Hansol M.com (as of 2001) | 0.0 | 11.7 | KT | | | | |
| LG U + | 0.0 | 14.7 | 16.6 | 18.0 | 18.1 | 17.8 | 17.9 |
| Total Revenue (bil KRW) | 2,676 | 12,446 | 14,098 | 17,739 | 20,745 | 19,604 | 19,749 |
| Total Revenue (mil US$) | 3,345 | 11,203 | 12,689 | 15,965 | 16,596 | 17,664 | 17,774 |
| C4 | 100.0 | 100.0 | 100.0 | 100.0 | 100.0 | 100.0 | 100.0 |
| HHI | 8,346 | 2,577 | 3,938 | 3,867 | 3,868 | 3,876 | 3,869 |
| *N* (>1%) | 2 | 5 | 3 | 3 | 3 | 3 | 3 |
| Noam Index | 5,902 | 1,152 | 2,274 | 2,233 | 2,233 | 2,238 | 2,234 |

*Source:* Korea Communications Commission, Korea Association for ICT Promotion.

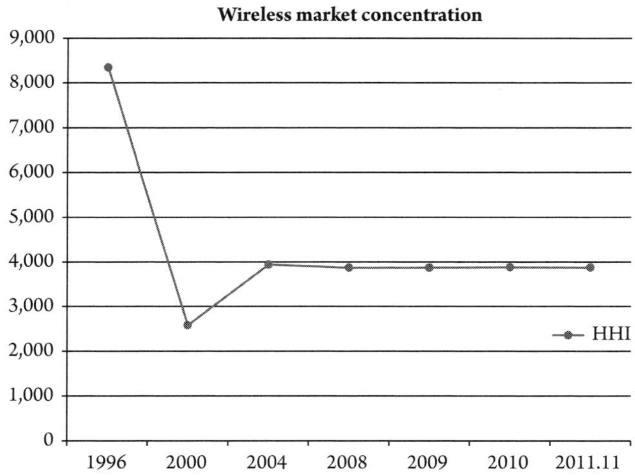


FIGURE 27.3 Wireless Telecom Market Concentration (HHI), 1996–2011

## INTERNET MEDIA

### Internet Service Providers (ISP)

The state has played a significant role in increasing broadband access in the country. It launched a series of broadband infrastructure initiatives beginning in 1999 under the aegis of "Cyber Korea 21." Since then, several other capacity-building initiatives have followed (e-Korea, IT839, and Cyber Infrastructure) that have spent trillions of won over the past 10 years to improve infrastructure so that now South Korea possess the world's most accessible broadband Internet network. Following KT's privatization, several new licenses were granted, leading to the formation of Dacom and Hanaro and then the entry of SKT to the broadband industry through its acquisition of LG Hanaro (Table 27.10).

In late 1990s, telecoms such as KT and LG Hanaro began offering consumers broadband access. In the 2000s, cable TV operators also entered the market, gaining market shares by offering access at significantly lower prices. Due to the competition between telecom providers and cable TV operators, concentration indices have decreased, from 4,407 in 2005 to 2,895 in 2009, though in 2011 the HHI was still high at 2,954 and the C4 stood near 100 (Figure 27.4).

### Search Engines

The search engine market is highly concentrated, and the level of concentration is increasing—but in contrast to the experience of several other countries, this is not being driven by the entry of Google (US) into the market. South Korean search engines dominate the market. Up until 2003, the search engine Daum, owned by Daum Communications, was the market leader; the NHN Corporation's Naver search engine then supplanted Daum in 2004. NHN also maintains Knowledge iN, a knowledge exchange service similar to Wikipedia (US), with 200 million entries by 15 million users. NHN's platforms integrate news, e-commerce, and online gaming—such as its own highly popular Hangame portal. Thanks to a business strategy centered on localization, simplifying user interfaces, and greater content integration, Naver held a 68.8% market share in 2011 while Daum held only 22.1% (Table 27.11).

US search engines like Google, Yahoo!, and Microsoft's MSN have not been able to compete with Naver and Daum in South Korea. In particular, Google and MS have an extremely weak presence. Although Google Korea began service in the mid-2000s, its market share has not surpassed 5%. However, in mobile Internet searches, Google has a 10% market share thanks to its mobile operating system (OS) Android.

Table 27-10. Internet Service Providers (Market Shares by Revenue), 2004–2011

| | 2004 | 2005 | 2006 | 2007 | 2008 | 2009 | 2010 | 2011 |
|---|---|---|---|---|---|---|---|---|
| KT | 58.3 | 61.1 | 52.1 | 48.4 | 47.6 | 42.5 | 43.1 | 43.5 |
| SK Broadband (SK) | 25.5 | 24.5 | 26.3 | 25.3 | 23.2 | 23.5 | 23.2 | 23.4 |
| Thrunet (as of 2008) | 9.1 | 7.5 | | | | | | |
| LG U + (LG) (as of 2010) | | | | | | | 16.1 | 16.0 |
| LG Powercom (until 2009) | 0.0 | 0.4 | 5.8 | 11.6 | 14.1 | 15.4 | LG U + | |
| LG Dacom (until 2009) | 2.4 | 2.6 | 1.1 | 0.8 | 0.3 | 0.1 | | |
| Cable TV MSOs(CJ Hellovision, T-Broad, C&M, HCN) | 0.0 | 0.0 | 11.1 | 12.6 | 13.7 | 17.2 | 16.4 | 16.1 |
| Onse Telecom (until 2007) | 3.6 | 3.1 | 2.1 | 0.3 | 0.0 | 0.0 | 0.0 | 0.0 |
| Dreamline (until 2006) | 1.1 | 0.9 | 0.5 | 0.0 | 0.0 | 0.0 | 0.0 | 0.0 |
| Others | 0.0 | 0.0 | 0.9 | 1.0 | 1.1 | 1.3 | 1.2 | 1.1 |
| Total Revenue (bil KRW) | 2,284 | 1,858 | 3,435 | 3,209 | 3,957 | 4,458 | 4,002 | 4,090 |
| Total Revenue (mil US$) | 2,969 | 2,972 | 3,435 | 3,530 | 3,561 | 3,566 | 3,602 | 3,681 |
| C4 | 96.5 | 96.2 | 95.3 | 97.9 | 98.6 | 98.6 | 98.9 | 98.9 |
| HHI | 4,152 | 4,407 | 3,570 | 3,278 | 3,192 | 2,895 | 2,928 | 2,954 |
| *N* (>1%) | 6 | 5 | 6 | 4 | 4 | 4 | 4 | 4 |
| Noam Index | 1,695 | 1,971 | 1,457 | 1,639 | 1,596 | 1,448 | 1,464 | 1,477 |

*Source:* Korea Communications Commission, Korea Association for ICT Promotion.

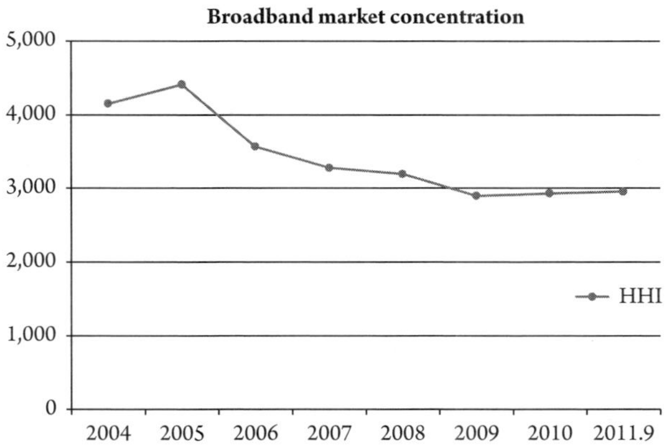


FIGURE 27.4 Broadband ISP Market Concentration (HHI), 2004–2011

Aside from this niche segment, joint ventures with South Korean web portals are necessary to win greater market shares: in 2010, MSN entered into a short-lived alliance with JMnet to launch Joinsmsn.com. At the end of 2012, Yahoo! stopped offering Korean-language service and shut down its Korea subsidiary (Figure 27.5).

## CONCLUSION

Before the 2000 Broadcasting Law, massed media ownership was strictly regulated in Korea and cross-ownership of print and audiovisual properties was prohibited. However, faced with revenue losses due to a decline in advertising

## Table 27-11. Internet Search Engines (Market Shares by Search Volume), 2000–2011

| | 2000 | 2003 | 2004 | 2007 | 2008 | 2009 | 2010 | 2011 |
|---|---|---|---|---|---|---|---|---|
| Naver (NHN Corporation) | 13.1 | 6.0 | 40.4 | 63.3 | 67.0 | 66.4 | 63.6 | 68.8 |
| Daum (Daum Communications) | 42.4 | 52.2 | 32.3 | 14.7 | 13.0 | 20.4 | 23.4 | 22.1 |
| SK | 6.6 | 9.6 | 5.5 | 13.6 | 12.2 | 2.2 | 4.3 | 2.4 |
| KT | 0.0 | 13.5 | 10.8 | 8.4 | 7.7 | 1.2 | 1.0 | 0.1 |
| Yahoo! Korea (Yahoo!, US) | 27.8 | 18.6 | 10.9 | 0.0 | 0.0 | 2.9 | 2.4 | 1.2 |
| Dacom (LG) | 10.0 | 0.0 | 0.0 | 0.0 | 0.0 | 0.0 | 0.8 | 0.4 |
| Others | 0.0 | 0.0 | 0.0 | 0.0 | 0.0 | 1.9 | 4.4 | 0.0 |
| Total Revenue (bil KRW)[1] | 0 | 0 | 161 | 612 | 750 | 825 | 1,044 | 1,244 |
| Total Revenue (mil US$) | 0 | 0 | 209 | 673 | 675 | 660 | 940 | 1,120 |
| C4 | 93.3 | 93.9 | 94.4 | 100.0 | 99.9 | 91.9 | 93.8 | 94.6 |
| HHI | 2,886 | 3,381 | 2,941 | 4,479 | 4,866 | 4,845 | 4,621 | 5,230 |
| *N* (>1%) | 5 | 5 | 5 | 4 | 4 | 5 | 5 | 4 |
| Noam Index | 1,291 | 1,512 | 1,315 | 2,240 | 2,433 | 2,167 | 2,067 | 2,615 |

*Source:* Korea Communications Commission, Korea Association for ICT Promotion.

[1] This revenue figure is an estimate based on data collected in this study: it is based on average per capita revenues reported for several other countries in this study with similar per capita income.

revenue, print publishers led the way in pushing for greater deregulation of cross-ownership, and they finally succeeded at this effort in 2010 (Table 27.12).

Print publishers (newspapers) are now allowed to hold up to a 20% stake in a terrestrial broadcaster, and publishers can bid for commercial broadcasting licenses. Although several civil society groups contested the decision, the Korean Communications Commission (KCC) stated that it had authorized these measures to promote the development of larger media groups. Regulations governing horizontal and vertical ownership in the cable TV market have also been modified: a cable TV company may now own up to a 33% stake in the national MSO market. The government hopes

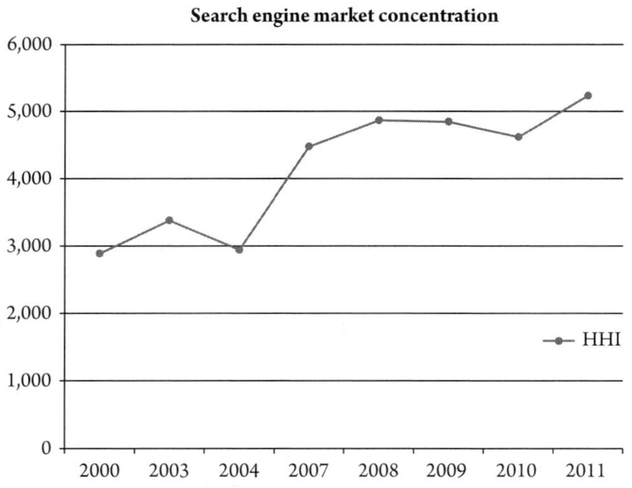


FIGURE 27.5 Search Engine Market Concentration (HHI), 2000–2011

Table 27-12. Media Ownership Restrictions in South Korea (Market Shares by %), 2010

| | SOUTH KOREAN CONGLOMERATES | FOREIGN OWNERSHIP | SOUTH KOREAN PRINT PUBLISHERS (NEWSPAPERS) |
|---|---|---|---|
| DTT | 0–20% | 0% | 0–20% |
| General Interest Channels | 0–30% | 0–20% | 0–30% |
| News Channels | 0–49% | 0–20% | 0–49% |
| Satellite TV | 49–100% | 33–49% | 33–49% |
| Cable TV | | 49% | 33–49% |
| IPTV | | 49% | 49% |

this will enable the emergence of a larger cable service provider. Foreign ownership and investment rules were also (begrudgingly) relaxed for every sector except terrestrial broadcasting in order to meet the terms of the South Korea–US Free Trade Agreement (FTA) that went into effect in 2012. Finally, the 2010 regulatory changes now require the KCC to review all media ownership laws every two years and propose amendments to them. The KCC has established a Media Diversity Committee whose responsibility is promoting media diversity as well as market competitiveness in order to monitor the effects of mergers and acquisitions of independent media producers.

The more closely regulated audiovisual and telecommunications industries—in which the state still grants licenses—are still more concentrated than the less regulated print and online media sectors. At the same time, the less regulated media sectors are becoming more concentrated overall due to the incentives deregulation has established for them to lead mergers and acquisitions. Though the cable TV market is not as concentrated as broadcast TV, concentration is increasing as cable MSOs move to take advantage of changes in media ownership laws. In the film industry, where there are no legal ownership restrictions, the market shares of the three largest distributors are increasing. And in the similarly unregulated search engine market, NHN's increased dominance has seen concentration rise as well. However, traditional print media markets do not display high concentration levels despite their lack of strict regulations. Both book and magazine publishing are highly fragmented, and concentration in the daily newspaper market is decreasing as more and more titles are establishing themselves (Table 27.13).

Media groups formed through mergers and acquisitions among media producers and their holding companies are likely to be the trend in South Korean media markets. Business conglomerates have mostly withdrawn from media ownership: in 1997, Samsung closed its media division during the East Asian financial crisis, and the Hyundai and Daewoo conglomerates soon followed Samsung's lead. Small- and medium-sized enterprises replaced them. The CJ Group expanded its media business through a series of mergers and acquisitions over a 10-year period, and by 2011 the CJ Group's E&M had emerged as a leading media group with stakes in film (CJ CGV), several video channels, and a successful music publishing arm (Mnet Media).

This trend will be helped along by the state, since regulators are focused on deregulation to increase competitiveness. Successive administrations have generally encouraged increased concentration in the media industry in order to increase the global competitiveness of South Korean media producers. In 2011, the KCC and the Ministry of Culture, Tourism, and Sports jointly announced that they would support increased conglomeration by relaxing ownership regulations.

Table 27-13. Media Concentration and Regulation, 2009–2011

| | STRONG REGULATION | WEAK OR NO REGULATION |
|---|---|---|
| Concentration Increased (HHI > 1,800) | Multichannel TV platforms (Cable TV) | Film, search engines |
| Concentration Decreased (HHI < 1,800) | Broadcast TV, radio, wireline telecom, wireless telecom | Newspapers, books, magazines |

## Bibliography

### 1. Books and Articles


Curran, J. "Rethinking Media and Democracy." *Mass Media and Society*. Curran, James and Gurevitch, Michael, eds. 3rd ed. London: Arnold, 2000. 120–154.

Jin, D. "Political and Economic Processes in the Privatization of the Korea Telecommunications Industry: A Case Study of Korea Telecom, 1987–2003." *Telecommunications Policy* 30, 2006. 3–13.

Noam, Eli. "Deregulation and Market Concentration: An Analysis of Post-1996 Consolidations." *Federal Communications Law Journal* 58:3, 2006. Jan. 29, 2013. <http://www.law.indiana.edu/fclj/pubs/v58/no3/l-noam.pdf>. 539–549.

Organization of Economic Co-operation and Development (OECD). "Total Broadband Subscriptions in the OECD Area." *OECD Communication Outlook 2011*. Paris: OECD, 2011.

Yonhap News Agency. "(New Focus): Entry of new cable broadcasters heralds media big bang." *Yonhap News Agency, 31* Dec. 2010. Sept. 5, 2012. <http://english.yonhapnews.co.kr/national/2010/12/31/41/0302000000AEN20101231005700315F.HTML>.


### 2. Reports


International Telecommunications Union (ITU). "Measuring the Information Society 2012." Geneva: IUT, 2012.

Korea Association for ICT Promotion, *Statistics of Broadcasting and Telecommunications Industry*. 2011 <http://www.kait.or.kr/index.jsp>

Korea Broadcast Advertising Corporation, *Statistics of Advertising Industry*. 2010–2012. <http://adstat.kobaco.co.kr/>

Korea Communications Commission. *Annual Report on Broadcasting Industry*, 2003–2011. <http://www.kisdi.re.kr/kisdi/fp/kr/board/listSingleBoard.do?cmd=listSingleBoard&sBoardId=BCAST_DB2>

Korea Press Foundation. *Statistics of Newspaper Industry*. 2011 <http://www.kpf.or.kr/journal/mdata_result_view.jsp?ctg=%C1%B6%BB%E7%BA%D0%BC%AE%BC%AD&bd_seq=7372&pg=1&url=4>

Korean Film Council. *Korean Cinema 2011*. Seoul: Dong Kwang Printing. 2011.

Ministry of Culture, Sports and Tourism of Korean government. *Annual Report of Contents Industry*. 2008–2011. <http://www.mcst.go.kr/web/dataCourt/research/researchView.jsp>

Waldman, Stephen and the Working Group on Information of Communities. "The Information Needs of Communities: The Changing Media Landscape in a Broadband Age." Washington, DC: FCC, 2011. Jan. 29, 2013. <http://www.fcc.gov/info-needs-communities>.

# South Korea—Data Summaries

ELI NOAM AND PAUL MUTTER

SOUTH KOREA HAS MADE the transition from substantially state-owned media industries in the 1980s to an almost entirely commercialized market by the 2010s; only broadcast television is still led by public entities, KBS with 37.3% and MBC with 18.3% (Table 27.14). The South Korean national market also has one of the lowest percentages of foreign ownership or participation internationally, partly due to legal restrictions. In telecommunications, the domestic providers KT and the SK and LG industrial groups (the incumbent) dominate the market.

Market concentration in South Korea is quite high. The pooled overall sector C4 is 79.9% and the Weighted C4 Average is 94%. The four largest national media companies control three quarters of the market: KT, SK Group, LG Group, and KBS. Of these, KT, SK, and LG are the three largest platform media companies, controlling 43.8%, 32.5%, and 14.7% of the platform market, respectively (Table 27.16).

KBS holds 15.4% of the content market, making it the largest media producer in the country, followed by the other two TV broadcasters, MBC (a second public broadcaster, with 7.8% share of content media) and the commercial SBS (also with 7.8%) (Table 27.15).

Table 27-14. National Media Industries Concentration in South Korea

| | 2004/5 | | 2011 OR MOST RECENT | | % CHANGE ANNUAL AVERAGE | |
|---|---|---|---|---|---|---|
| | COMPANY POWER INDEX IN COUNTRY | COMPANY SHARE OF THE OVERALL NATIONAL MEDIA MARKET (%) | COMPANY POWER INDEX IN COUNTRY | COMPANY SHARE OF THE OVERALL NATIONAL MEDIA MARKET (%) | COMPANY POWER INDEX IN COUNTRY | COMPANY SHARE OF THE OVERALL NATIONAL MEDIA MARKET (%) |
| KT | 2,372 | 39.8 | 1,851 | 35.7 | –3.1 | –0.6 |
| SK Group | 1,228 | 26.4 | 1,184 | 26.3 | –0.5 | –0.02 |
| Government of Korea | 614 | 8.2 | 413 | 6.1 | –4.7 | –0.3 |
| *KBS (public)* | 204 | 4.5 | 110 | 3.0 | –6.6 | –0.2 |
| *MBC (public)* | 47 | 2.2 | 27 | 1.5 | –6.06 | –0.1 |
| *EBS (public)* | 3 | 0.6 | 0.8 | 0.2 | –11.15 | –0.05 |
| LG Group | 142 | 9.7 | 191 | 11.9 | 4.95 | 0.3 |
| NHN | 12 | 0.3 | 131 | 1.9 | 144.1 | 0.2 |
| CJ Group | 20 | 1.1 | 66 | 2.4 | 34.1 | 0.2 |
| Tbroad | 23 | 1.1 | 43 | 1.7 | 12.9 | 0.08 |
| SBS | 47 | 2.2 | 27 | 1.5 | –6.06 | –0.1 |
| Chosun Ilbo | 30 | 1.2 | 19 | 0.8 | –5.05 | –0.06 |
| C&M | 16 | 0.9 | 22 | 1.2 | 5.3 | 0.04 |
| Jmnet (JoongAng Ilbo) | 22 | 1.0 | 16 | 0.7 | –4.4 | –0.05 |
| Dong-A-Ilbo | 17 | 0.9 | 11 | 0.6 | –4.9 | –0.04 |
| Daum | 8 | 0.2 | 14 | 0.6 | 11.3 | 0.05 |
| Showbox | 8 | 0.5 | 2 | 0.2 | –10.4 | –0.03 |

| Media Concentration Index | 2004/5 | 2011 or Most Recent | % Change Annual Average |
|---|---|---|---|
| Total Revenue: Nat'l Media Industry (mil US$) | 28,933 | 40,558 | 5.74% |
| Total Voices (*n*) | 45 | 48 | 0.95% |
| Net Voices (*n*) | 40 | 43 | 1.07% |
| Public Ownership (%) | 8.2 | 6.1 | −0.35% |
| Foreign Ownership (%) | 0.9 | 0.9 | 0.0% |
| C4 Average—Weighted | 95 | 94 | −0.17% |
| HHI Average—Weighted | 4,588 | 4,036 | −2% |
| C1 Average—Weighted | 58 | 54 | −1% |
| Noam Index Average—Weighted | 1,861 | 1,726 | −1.03% |
| Pooled Overall Sector C4 | 84.1 | 79.9 | −0.6% |
| Pooled Overall Sector HHI | 2,452 | 2,162 | −1.69% |
| Pooled Overall Sector Noam Index | 282 | 185 | −4.92% |
| Market Share of Top Ten Companies: Nat'l Media Industry (%) (Pooled C10) | 91.6 | 89.4 | −0.3% |
| National Power Index | 4,565 | 3,997 | −1.78% |

Table 27-15. Top Content Media Companies in South Korea

| | 2004/5 | | 2011 OR MOST RECENT | | % CHANGE ANNUAL AVERAGE | |
|---|---|---|---|---|---|---|
| | COMPANY POWER INDEX IN COUNTRY | COMPANY SHARE OF THE NATIONAL CONTENT MEDIA MARKET (%) | COMPANY POWER INDEX IN COUNTRY | COMPANY SHARE OF THE NATIONAL CONTENT MEDIA MARKET (%) | COMPANY POWER INDEX IN COUNTRY | COMPANY SHARE OF THE NATIONAL CONTENT MEDIA MARKET (%) |
| Government of Korea | 3,032 | 30.4 | 2,135 | 30.5 | –4.2 | –1.3 |
| *KBS (public)* | 1,007 | 22.2 | 568 | 15.4 | –6.2 | –0.97 |
| *MBC (public)* | 231 | 10.7 | 140 | 7.8 | -5.7 | –0.4 |
| *EBS (public)* | 17 | 2.8 | 4 | 1.3 | –11.01 | –0.2 |
| CJ Group | 82 | 3.9 | 229 | 7.4 | 25.5 | 0.5 |
| NHN | 58 | 1.4 | 677 | 9.8 | 151.5 | 1.2 |
| SBS | 232 | 10.7 | 140 | 7.8 | -5.7 | –0.4 |
| Chosun Ilbo | 148 | 5.9 | 100 | 4.1 | –4.6 | –0.3 |
| Jmnet (JoongAng Ilbo) | 111 | 5.1 | 80 | 3.7 | -3.9 | –0.2 |
| Tbroad | 37 | 1.8 | 74 | 2.9 | 14.2 | 0.2 |
| Dong–A Ilbo | 82 | 4.4 | 57 | 3.1 | –4.5 | –0.2 |
| C&M | 26 | 1.5 | 37 | 2.1 | 6.3 | 0.08 |
| DAUM | 37 | 1.2 | 70 | 3.2 | –12.5 | 0.3 |

| | | | | | | |
|---|---|---|---|---|---|---|
| SK Group | 1 | 0.2 | 0.8 | 0.3 | –3.4 | 0.02 |
| Showbox | 41 | 2.3 | 12 | 1.3 | –10.2 | –0.15 |

| Media Concentration Index | 2004/5 | 2011 or Most Recent | % Change Annual Average |
|---|---|---|---|
| Public Ownership (%) | 39.4 | 30.5 | –1.48% |
| Foreign Ownership (%) | 4.5 | 4.9 | 0.05% |
| C4 Average—Weighted | 85 | 83 | –0.3% |
| HHI Average—Weighted | 4,004 | 3,726 | –0.99% |
| C1 Average—Weighted | 39.4 | 30.5 | –1.3% |
| National Power Index | 3,960 | 3,668 | –1.05% |

Table 27-16. Top Platform Media Companies in South Korea

| | 2004/5 | | 2011 OR MOST RECENT | | % CHANGE ANNUAL AVERAGE | |
|---|---|---|---|---|---|---|
| ARTICLE II. | COMPANY POWER INDEX IN COUNTRY | COMPANY SHARE OF THE NATIONAL PLATFORM MEDIA MARKET (%) | COMPANY POWER INDEX IN COUNTRY | COMPANY SHARE OF THE NATIONAL PLATFORM MEDIA MARKET (%) | COMPANY POWER INDEX IN COUNTRY | COMPANY SHARE OF THE NATIONAL PLATFORM MEDIA MARKET (%) |
| KT | 2,963 | 49.3 | 2,288 | 43.8 | –3.25 | –0.78 |
| SK Group | 1,539 | 33.1 | 1,467 | 32.5 | –0.7 | –0.08 |
| LG Group | 178 | 12.2 | 237 | 14.7 | 4.7 | 0.36 |
| Tbroad | 19 | 0.9 | 36 | 1.4 | 12.6 | 0.07 |
| CJ Group | 4 | 0.4 | 27 | 1.2 | 93.5 | 0.12 |
| Media Concentration Index | 2004/5 | | 2011 or Most Recent | | % Change Annual Average | |
| Public Ownership (%) | 0 | | 0 | | 0% | |
| Foreign Ownership (%) | 0 | | 0 | | 0% | |
| C4 Average— Weighted | 98 | | 96 | | –0.16% | |
| HHI Average—Weighted | 4,736 | | 4,110 | | –2% | |
| C1 Average—Weighted | 60 | | 54 | | –1% | |
| National Power Index | 4,716 | | 4,073 | | –1.95% | |

# 28

# Media Ownership and Concentration in Taiwan

YU-LI LIU, YUNTSAI CHOU, KUO-FENG TSENG, RU-SHOU ROBERT CHEN, YI-HSUAN CHIANG, AND PING-HUNG CHEN

## INTRODUCTION

Ownership and control are two important factors affecting the operation and performance of the media in Taiwan.[1] The owners of media outlets influence the media industry's operations. The influence can come from personal, commercial, or political interests. It is generally believed that highly concentrated media tend to be detrimental to the diverse content and opinions of the public. However, the appropriate limit of the media's market shares and concentration in a democratic country is a hotly debated topic. In Taiwan, the newspapers are largely commercial, but most of them are inclined toward specific political parties or political figures. Neither the government nor political parties are allowed to own any electronic media except for several radio stations that have been owned and operated by the government since December 2003. Taiwan only has clear laws and regulations for cable TV's horizontal and vertical integration. The Publication Law was abolished in 1999, so there is no regulation for print media. The multiple ownership rule is vague and controversial. It is difficult to calculate the market concentration when media are not publicly listed in the stock market or if there is no legitimate system for auditing circulation. Some media outlets oppose an auditing system because they do not want to make their circulation numbers transparent to the public. It is even more difficult to calculate the market power of converged media. In this

1. Liu, Yu-li. Scarcity of Spectrum, Political Access, and Broadcast Autonomy: The Open Broadcasting Channel Debate in Taiwan. Ph.D. dissertation, Indiana University. 1992.

chapter, every valid method has been taken to calculate the market concentration of the different media outlets and to interpret the trend of media development and media concentration.

There is no clear regulation of cross-media ownership in Taiwan. It is only indicated in the Enforcement Rules of the Radio and TV Act.[2] In 2002, *China Times* acquired a satellite channel company, CTi, which owns three channels carried by cable operators. In 2006, *China Times'* owner, Albert Yu, wanted to buy the biggest radio station, the Broadcasting Corporation of China (BCC), and a terrestrial TV company, China Television Company (CTV). This acquisition case received a great deal of attention from the media, the political parties, and the public because BCC and CTV were both owned by the Kuomintang (KMT).[3] The NCC approved Yu's application because there was no law restricting multiple ownership. However, it was criticized by the ruling Democratic Progressive Party (DPP) and several media critics and scholars. With this new acquisition, Yu owned newspapers, magazines, terrestrial TV stations, satellite channels, and radio stations. Quickly following this deal, Yu sold BCC to Jaw Shaw-kong[4] whose wife had to reduce her shares from 34% to less than 10% in another radio station, UFO. In 2008, Yu sold *China Times*, CTV, and CTi to Want Want's owner Tsai Eng-meng,[5] who has a large business interest in China. The NCC held hearings and finally passed the case by giving several orders to Want Want's owner. However, since Want Want could not fulfill the NCC's special orders, the case was still pending.

In December 2003, the Legislative Yuan (Congress) revised the Broadcasting Law, Cable Radio and TV Law, and Satellite Radio and TV Law to require that all the political parties, government, and military to withdraw their shares from the electronic media within two years.[6] The Legislative Yuan wanted to prevent biased media reporting, especially during elections. Since then, the Broadcasting Corporation of China and China Television Company have sold their shares. In 2006, Chunghwa Telecom could not provide IPTV services because the government owned a 41% stake in the company at the time. It was instructed by the regulator to change its non-transparent business model to an open platform model in order to comply with the laws. Meanwhile, some companies were fined because the government held indirect ownership over them. The industry urged the government to revise three broadcasting related laws and to relax the restriction on government shares to 10% of indirect shares in order to avoid unknown investors in the stock market.[7] Another controversial case concerned horizontal integration and vertical integration. The Cable Radio and TV Act set up a cap for horizontal integration and vertical integration. No multiple system operators (MSO) can own more than one-third of cable systems or have more than one-third of the subscribers. Within the same franchise area, the MSO cannot own more than half of the systems.

With regard to vertical regulation, the cap is one-fourth. It means programs provided by system operators and their affiliated enterprises shall not exceed 25% of the usable channels. In 2010, Taiwan Mobile (Taiwan Fixed Network's biggest shareholder) wanted to acquire the MSO, Kbro. At that time, Taiwan Fixed Network owned six cable systems. If it acquired Kbro, it would become the biggest MSO in Taiwan, and

2. Article 18 and Article 19 of the Enforcement Rules of the Radio and TV Act.

3. Kuomintang (KMT) is also called the Nationalist Party. It lost to the Communist Party the ruling of Mainland China during the Internal War. The then-KMT leader, Generalissimo Chiang Kai-shek, fled to Taiwan with his military groups and hundred thousands of Mainlanders. It has been the ruling party most of the time except 2000–2008.

4. A former politician with links to Pro-New Party. <http://en.wikipedia.org/wiki/Broadcasting_Corporation_of_China>.

5. Want Want is a food manufacturer from Taiwan. Its owner Tsai Eng-Meng is recognized for his entrepreneurial success in China's snack food industry. He is No. 1 on Forbes 2012 Taiwan Rich List. Flannery, Russel. "Taiwan Rich List's New No. 1 Tsai Eng-Meng Answers Critics: 'What's Wrong With Peace?'" May 23, 2012. <http://www.forbes.com>.

6. Liu, Y. L. et al. (2012). White Paper of Digital TV and New Media Platform Policy, National Chengchi University, Taipei, Taiwan.

7. In the stock market, people can buy and sell stocks every day. There is no record showing who owns the shares in a short period of time.

its total subscribers would be almost one-third of the market. Taiwan Mobile also had a shopping channel and a children's channel. Therefore, the NCC held a hearing.

The owner of Taiwan Mobile, the Tsai family, knew that it would be difficult to acquire Kbro, because they owned an MSO. Therefore, they decided to use a new company, Da-Fu, to acquire 80% of Kbro, a transaction valued at US$1.2 billion. The Fair Trade Commission approved the acquisition with several conditions, as did the NCC. Since the proposed acquisition did not violate the Cable Radio and TV Act, it was difficult for the NCC to deny the application.[8] Both the Fair Trade Commission and the NCC granted the application but set a number of conditions. For example, after the acquisition, Kbro and Taiwan Fixed Network could not have the same board members and supervisors. Table 28.1 shows the market share of major cable MSOs in 2014. (Table 28.1).

In Taiwan, cable MSOs, satellite channels, advertising companies, telephone companies, music companies, and film companies all have some foreign share ownership. The Telecommunications Act and the Cable Radio and TV Act both set a cap on direct and indirect ownership for Type I telephone companies and cable system operators.[9] Foreign ownership is not allowed in terrestrial radio and TV. The three biggest cable MSOs were all acquired by international investors Carlyle Group (US), MBK (Korea), and Macquarie (Australia) between 2007 and 2008. The two biggest MSOs were

Table 28-1. The Market Share of Major Cable TV MSOs (2014)

| MSO | MAJOR SHAREHOLDER | SYSTEM OPERATOR | SUBSCRIBERS | SUBSCRIBERS IN THE CABLE TV MARKET (%) |
|---|---|---|---|---|
| Kbro | Da Fu Media (Richard Tsai, Taiwan) | 12 | 1,049,028 | 21.03 |
| CNS | MBK | 10 | 1,187,053 | 23.80 |
| TBC | Macquarie (Australia) | 4 | 698,624 | 14.00 |
| Taiwan Fixed Network | Taiwan Mobile (Richard Tsai, Taiwan) | 5 | 495,866 | 9.94 |
| Taiwan Optical Platform | Taiwan Technology (Rocky Chien, Taiwan) | 4 | 296,718 | 5.95 |
| Independent | | 24 | 1,261,286 | 25.28 |
| Total | | 59 | 4,988,575 | 100 |

Total revenue in March, 2014 (estimated on a per household basis): 1.072 billion US dollars (4,988,575 households x NT $537 x12 month)

8. Chih-chuan Tai (2012). Regulatory Conditions for Media Mergers and Their Oversight-A Case Study of Dafu Media Acquiring Kbro Cable. MA Thesis. Taiwan University.

9. According to the Telecommunications Act, a Type I telecommunications enterprise means an enterprise that installs telecommunication line facilities and equipment in order to provide telecommunication services. The aforementioned telecommunication line facilities and equipment refer to network transmission facilities connecting the sending and receiving terminals, the switching facilities installed to be integrated with the network transmission facilities, and the auxiliary facilities of both.

acquired by large domestic groups in 2011. The public and academics seem to be more tolerant of foreign participation in telephone companies. However, they are not comfortable with foreign participation in MSOs because they view cable systems as a conduit for free speech and expression.

# PRINT MEDIA

## Newspapers

Taiwan's print publishing industry has a long history. In the 1950s, the government introduced a ban on establishing new newspapers because it wanted to control the media. There were about 30 newspapers at that time. The magazine industry was also affected by the ban, although to a lesser degree. In the 1970s, the *China Times* and *United Daily News* were the major newspapers. They were owned and controlled by the Yu and Wang families. In 1980, *Liberty Times* joined the market. After martial law was lifted in 1987 and the government's ban on new newspapers ended in 1988, the number of magazines grew by 50%, and the number of newspapers increased sharply.[10] The Asian financial crisis in the late 1990s hit the industry, and with the rising cost of paper and the increasing importance of satellite and Internet services, the newspaper and magazine markets have been shrinking since 2000.

Newspaper readership has also declined. *China Times* and *United Daily* have suffered the greatest losses in terms of advertising revenue. Frequent changes in ownership have also led to a rapidly changing media market. In 2003, the *Apple Daily* entered the Taiwanese market. Its sensational reporting has helped it secure the number 1 market position in terms of circulation. Currently, *Apple Daily, Liberty Times, United Daily News,* and *China Times* are the four major newspapers in the market.

Total advertising volume in Taiwanese newspapers reached historically high levels in 1997. This was followed by the popularization of cable TV, the introduction of satellite TV news, falling readership among young readers, and the popularity of the Internet. All of these forces combined led to the shrinking advertising volume of the newspaper industry. By 2002, advertising had dropped to NT$12.19 billion (US$406 million), half of the 1997 volume.[11] The entrance of *Apple Daily* into the market in 2003 popularized the use of template-style pages. Although this resulted in a slight increase in advertising volume, it was not enough to offset the overall downward trend. In addition, cable TV stole market share from newspapers, dedicated TV news channels were established, and Internet news websites began to appear. These trends contributed to the drop in newspaper readership and advertising volume.

Due to the unavailability of newspaper sales figures, the concentration ratio must be calculated based on overall advertising volume. However, such information comes from a variety of sources. *Brain Magazine* and *Advertising Yearbook Taiwan* both provide information about total newspaper advertising revenue. Unfortunately, they derive their information from a variety of sources, so the respective numbers for any given year often vary. Some numbers are calculated based on published volume, while others are based on effective advertising volume. A few of these numbers exclude commission, while others include it. This means that data sometimes differ widely. To establish an authoritative standard, this study uses the advertising volume per newspaper group as given in the *Advertising Yearbook of Taiwan* as its basic standard. However, we have excluded newspaper sales revenue from our calculations, thereby affecting the accuracy of the CR4 and HHI values.

In the early newspaper market, the three big newspaper groups were China Times, United Daily News, and Liberty Times. After Apple Daily entered the market, it became one of the big four. In the newspaper market, CR4 exceeds 93%, and HHI remains above 2,326 between

10. Tien Bin Wang (2003). The History of Taiwan's Newspaper. Taipei: Asiapac Books.

11. Tsung-Heng Lee (2008). The key success factors of Apple Daily marketing strategy. MA Thesis. Ming-chuan University, Taiwan.

Table 28-2. Market Share, Revenue, and the Degree of Concentration of Daily Newspapers[1]

| % | 2004 | 2006 | 2008 | 2010 | 2012 |
|---|---|---|---|---|---|
| China Times Group | 26.0 | 23.1 | 19.9 | 20.3 | 18.0 |
| United Daily News Group | 32.3 | 26.9 | 21.1 | 21.2 | 19.3 |
| Liberty Times News Group | 22.6 | 29.2 | 28.5 | 30.4 | 31.4 |
| Apple Daily (UK) | 12.4 | 17.8 | 25.5 | 23.1 | 24.7 |
| Others | 6.7 | 3.1 | 5.1 | 5.1 | 6.6 |
| Total Advertising Revenues (NT mil) | 18,097 | 14,771 | 11,079 | 11,956 | 9,522 |
| Total Advertising Revenues (USD mil) | 603.2 | 492.4 | 369.3 | 398.5 | 317.4 |
| Total Revenues (NT mil) | 28,738 | 22,944 | 16,673 | 18,174 | 15,059 |
| Total Revenues (USD mil) | 957.9 | 764.8 | 555.8 | 605.8 | 502.0 |
| *N* (>1%) | 4 | 4 | 4 | 4 | 4 |
| CR4 | 93.2 | 97.0 | 94.9 | 94.9 | 93.4 |
| HHI | 2,428 | 2,433 | 2,326 | 2,341 | 2,293 |
| Noam Index | 1,214 | 1,088 | 1,163 | 1,047 | 1,147 |

China Times Revenues: Advertisements (60%); subscriptions (34%); newsstand sales (6%)

Apple Daily Revenues: Advertisements (97%); subscription and newsstand sales (3%). The revenue figures in Table 28.2 were estimated based on China Times and Apple Daily revenue percentages.

[1] All figures were calculated based on advertising revenues only.

2004 and 2010. Although HHI declined to 2,293 in 2012, CR4 still exceeds 93% (Table 28.2). Therefore the newspaper market as a whole still remains highly concentrated. The following table provides these details.

## Book Publishing

Book publishing is much less concentrated than the newspaper and magazine markets. The biggest book publishing group is Cité Media Holding Group, which owns five publishing groups:

- Cité Publishing Group publishes diverse genres
- Business Weekly Media Group focuses on business and economics
- PC Home Publication Group focuses on technology, how-to, and lifestyle
- Sharp Point's focuses on leisure and popular culture
- Nong Nong Intermedia Group publishes Taiwan's leading lifestyle, fashion, and infant-care guides for women

Altogether, there are 40 publishing companies in Media Holding Group, and its market share of the industry is about 10% (textbooks not included). Eurasian Publishing Company and Crown account for about 4% each. These are followed by Yuan-Liou Publishing Company and CommonWealth. The book publishers are all local companies except for Cité, of which 49% of the shares were acquired by Tom Group Limited (a joint venture between Hutchison Whampoa, Cheung Kong [Holdings] Limited, and other strategic investors in Hong Kong).[12]

The revenue from book publishing remained stable between 2007 and 2011. In 2007, the total revenue from book publishing was US$1.81 billion, but in 2008 it declined to US$1.7 billion.

12. The imported books are mainly from the United States, Japan, and China. In 2009, the total revenue of the imported books accounted for US$8,646,000 (according to the statistics provided by Taiwan's Customs).

In 2008, 63.8% of the book publishers did not belong to any publishing group.[13] As of March 2012, the revenues from digital publishing are still insignificant.

There are no available data on the revenues of individual publishing groups. The only available data are of the circulation of books distributed by Book.Com.Co., Ltd. Since the price of each book is different, in Table 28.3 we reveal only the books sold by Book.Com.Co., Ltd., which is one of the three major book distributors. The other two are Eslite and Kingstone.

## Magazine Publishing

In 1945, after the end of 50 years of Japanese rule, there were about 126 magazines in Taiwan. In 1947, because of the 228 Incident (also known as the 228 Massacre, an anti-government uprising), only 51 magazines were left. In 1952, this number increased to 220. In 1970, some non–ruling party magazines appeared, such as *Formosa*. In the meantime, China Times and United Daily News both published their own magazines, *China Times Weekly*

### Table 28-3. Market Share, Revenue, and the Degree of Concentration of Book Publishing[1]

| CIRCULATION % | 2008 | 2009 | 2010 | 2011 |
|---|---|---|---|---|
| Cité (HK) | 6.6 | 12.0 | 8.8 | 7.8 |
| Tongli | 2.5 | 3.5 | 3.3 | 3.3 |
| Eurasian | 2.3 | 1.8 | 1.5 | 1.5 |
| Crown | 2.2 | 2.8 | 2.1 | 1.7 |
| Yuan-Liou | 2.1 | 2.0 | 2.2 | 1.7 |
| China Times | 2.9 | 2.9 | 2.9 | 2.5 |
| CommonWealth | 2.7 | 2.1 | 2.1 | 2.5 |
| Locus | NA | 1.1 | 1.4 | 1.2 |
| Sun color | NA | 1.9 | 2.4 | 2.1 |
| Taiwan Kadokawa | 2.1 | 2.2 | 2.3 | 2.0 |
| Linking | NA | 0.8 | 1.1 | 1.0 |
| Global Group Holdings | NA | 1.0 | 0.9 | 0.8 |
| Spring | 1.7 | 1.0 | 0.8 | 1.0 |
| Others | 74.9 | 65.1 | 68.2 | 71.0 |
| Total (circulation) | 7,228,893 | 8,890,000 | 11,320,000 | 14,000,000 |
| Revenue (US$ bn) | 1.7 | 1.045 | 1.225 | NA |
| *N* (>1%) | 9 | 12 | 11 | 10 |
| CR4 | 14.7 | 21.1 | 17.4 | 16.1 |
| CR8 | 23.5 | 29.2 | 26.1 | 23.5 |
| HHI | NA | 192 | 123 | 99 |
| Noam Index | NA | 58 | 34 | 27 |

[1] The revenue figures were released by the Ministry of Finance.

13. See note 12.

and *United Monthly*, respectively. In 1981, Commonwealth was established. In 1986, the Democratic Progressive Party created *Asia*. After martial law was lifted in 1987, many more magazines were established. In 1996, there were 5,493 magazine companies, and by 2001 this number had grown to 7,236. However, it declined drastically to 3,909 the next year because of the influence of the Internet. Thereafter, it increased steadily and reached 7,544 in 2010. Currently, Cité Media Holding Group is not only the biggest book publisher, but also the biggest magazine company. One of its magazines, *Business Weekly*, is a best-selling magazine in Taiwan.

The advertising volume in the magazine industry has seen a constant increase since the 1990s. Annual advertising volume has hovered between $70 and $80 billion NT (US$2.66 billion), and the market for magazine advertising in Taiwan is now saturated. As brand competition between magazines has intensified, advertising volume has been concentrated in the more well-known magazines. This has led to a slight, insignificant increase in concentration.

## Table 28-4. Market Share, Revenue, and the Degree of Concentration of Magazines and Periodicals[1]

| % | 1985 | 1988 | 1992 | 1996 | 2000 | 2004 | 2008 |
|---|---|---|---|---|---|---|---|
| Cité (HK) | 1.4 | 1.6 | 2.9 | 7.4 | 4.0 | 9.7 | 10.2 |
| CommonWealth Magazine Group | 6 | 5 | 3.5 | 4.6 | 2.6 | 3.3 | 4.3 |
| Business Today | 0 | 2 | 2.1 | 2.5 | 1.4 | 1.9 | 4.0 |
| Commonwealth Publishing Group | 0 | 0 | 0 | 0 | 1.8 | 1.4 | 1.8 |
| Next Media (Next Magazine, HK) | 0 | 0 | 0 | 0 | 0 | 7 | 10.4 |
| China Times (China Times Weekly) | 14.5 | 12.6 | 7.8 | 11.3 | 5.7 | 5.4 | 3.1 |
| TVBS (TVBS Weekly) | 0 | 0 | 0 | 0 | 3.9 | 2.4 | 1.9 |
| Condé Nast—VOGUE (USA) | 0 | 0 | 0 | 0 | 1.5 | 2.5 | 1.8 |
| Hearst- ELLE (US) | 0 | 0 | 0 | 2.4 | 1.6 | 2.5 | 1.6 |
| Reader's Digest (US) | 3.6 | 3.8 | 1.5 | 2.5 | 0 | 0 | 0 |
| Mee-Hwa Report | 3.4 | 2.7 | 1.5 | 0 | 0 | 0 | 0 |
| TV Guide | 2.2 | 2.0 | 0 | 0 | 0 | 0 | 0 |
| Families | 16.5 | 1.9 | 0 | 0 | 0 | 0 | 0 |
| U.S. News & World Report (US) | 0 | 3.0 | 0 | 0 | 0 | 0 | 0 |
| Scoop Weekly | 0 | 2.6 | 2.8 | 0 | 0 | 0 | 0 |
| Others | 52.4 | 62.8 | 77.9 | 72.0 | 81.5 | 62.9 | 58.4 |
| Total Advertising Revenues (NT mil) | 1,125 | 2,100 | 4,360 | 3,987 | 8,264 | 8,313 | 6,723 |
| Total Advertising Revenues (USD mil) | 35 | 66 | 136 | 125 | 258 | 260 | 210 |
| Total Revenue (NT mil) | | | | | 9,800 | 22,100 | 22,100 |
| Total Revenue (USD mil) | | | | | 326.7 | 736.7 | 736.7 |
| *N* (>1%) | 7 | 10 | 7 | 6 | 8 | 9 | 9 |
| CR4 | 40.6 | 24.4 | 17.0 | 25.8 | 16.2 | 25.4 | 28.9 |
| CR8 | 47.6 | 33.7 | 22.1 | 30.7 | 22.5 | 34.7 | 37.5 |
| HHI | 550 | 235 | 98 | 222 | 81 | 207 | 269 |
| Noam Index | 208 | 74 | 37 | 91 | 29 | 69 | 90 |

[1] According to a survey of 157 magazine companies, on average the revenues of the magazine companies come from advertisements (44.1%), sales (54.6%), and others (1.3%). However, advertising revenue accounts for 40% and 55% of the two big magazine companies.

Having calculated a concentration benchmark for the magazine market using CR4 and CR8, a compilation of statistics has shown the magazine industry to be a low concentration industry. CR4 hovers between 16.2 and 40.6%. With the exception of the figures for 1985, when the CR8 value was 47.6%, CR8 hovered between 22.1 and 47.6%. Table 28.4 provides more details.[14] An overall view of Taiwan's magazine industry shows that for a long time there has been a large number of competing magazines and that advertising budgets have been concentrated among those with high circulation numbers. The four largest magazines were exclusively local magazines, but as multinational companies bought up local magazines, changes occurred among the top four and the top eight magazines. Now the top four and top eight are national news and fashion magazines. The fact that CR4 and CR8 point to low concentration shows that the diversity in Taiwan's magazine industry has not diminished as larger company groups have formed.

## AUDIOVISUAL MEDIA

### Radio

Taiwan's broadcasting is unique in that its radio system has a mixed ownership that includes the government, political parties, military, and private sector. In 1988, there were 192 radio stations owned by 33 companies: 28.6% of the stations were privately owned, 30.7% owned by the government, 24.4% owned by the ruling Kuomintang (KMT) party, and 16.1% owned by the military.[15] General Secretary Chang Jun-hsiung of the Democratic Progressive Party (DPP), the then-opposition party, launched an unauthorized Underground Democratic Radio in 1992, which directly challenged the rule of limited participation in the media industry set by the KMT-led government. The broadcast market was opened in 1993 when the Legislative Yuan (Congress) passed the amendments to the Broadcasting TV Act at the end of the KMT ruling era. Since then, 143 new radio stations (66 medium-power and 77 low-power[16]) and 29 existing stations were licensed. A total of 171 licensed radio stations were in operation by the end of 2010.[17] However, the actual number of radio stations is far higher, because numerous underground radio stations broadcast in central and southern parts of Taiwan without licenses.

Regrettably, the liberalization approach did not deter the growth of underground radio. At the end of 1994, nearly half of the radio stations in Taipei were still broadcast underground. They were all divided by political allegiance. These underground radio stations mostly broadcast through low-power stations over the white space set between medium-power and low-power stations that prevent interference. The low-power radio equipment is cheap, making the cost of illegal broadcasting low. These stations earn profits by advertising and selling uncertified medications and nutrition products. Government efforts at cracking down on the underground radio stations have not been very effective, since they can be easily re-installed.

As shown in Table 28.5, the total annual revenue of the radio broadcasting industry amounted to about $119.6 million US dollars in 2012, down from $126.8 million US dollars in 2011. Compared with other media outlets, the radio industry is much smaller in terms of its annual revenue. It generates only one-third of the annual revenue of terrestrial TV and one-tenth of that of cable TV.

The top radio broadcaster in the industry, the Broadcasting Corporation of China (BCC), dominates the market with a 20% share. BCC holds five nationwide networks and three regional channels. Two of its nationwide

---

14. All figures in Table 28.4 were calculated by advertising revenues only, because the total revenue figures were incomplete. The Ministry of Finance only released the magazine revenue figures from 1999 to 2008.

15. Peng, Shin-yi, Yuchi Wang, and Yuntsai Chou. The auction and recall planning and management of spectrum frequencies. Taipei: NCC research report PG9601–0077, 2007.

16. Medium-power stations refer to those whose signals are transmitted within the city range, while low-power ones refer to those which transmit signals within town-based areas.

17. Peng, Shin-yi et al. The auction and recall planning and management of spectrum frequencies. Taipei: NCC research report PG9601–0077, 2007.

Table 28-5. Market Share and Concentration in the Radio Broadcasting Industry, 2001–2007[1]

| % | 2001 | 2002 | 2003 | 2004 | 2005 | 2006 | 2007 | 2008 | 2009 | 2010 | 2011 | 2012 |
|---|---|---|---|---|---|---|---|---|---|---|---|---|
| BCC | 26.9 | 26.8 | 25.4 | 23.6 | 21.2 | 19.3 | 18.8 | 19.8 | 22.0 | 21.8 | 20.9 | 20.9 |
| Chengsheng | | | | | | 9.5 | 8.9 | 8.9 | 9.4 | 8.5 | 8.2 | 8.5 |
| UFO | 8.9 | 9.2 | 9.6 | 9.4 | 9.7 | 4.7 | 4.1 | 4.0 | 3.2 | 3.3 | 3.6 | 2.8 |
| News 98 | | | | | | 3.7 | 3.7 | 3.8 | 3.2 | 3.4 | 2.9 | 2.5 |
| Voice of Taipei | 8.0 | 5.9 | 5.2 | 4.8 | 5.0 | 2.8 | 2.6 | 2.0 | 2.3 | 2.1 | 2.7 | 2.8 |
| Kiss | | | | | | 3.6 | 3.5 | 3.3 | 2.7 | 2.2 | 1.6 | 1.8 |
| Others | 51.8 | 53.9 | 56.1 | 58.4 | 60.4 | 56.4 | 58.4 | 58.2 | 57.2 | 58.7 | 60.1 | 60.7 |
| Total Revenue (USD mil) | 111.0 | 108.0 | 110.0 | 114.0 | 110.0 | 121.1 | 121.8 | 119.2 | 108.6 | 123.6 | 126.8 | 119.6 |
| N (>1%) | 4 | 4 | 4 | 4 | 4 | 8 | 9 | 9 | 9 | 9 | 9 | 9 |
| C4 | 48.2 | 46.1 | 44.0 | 41.6 | 39.6 | 37.2 | 35.4 | 36.4 | 37.8 | 37.0 | 35.6 | 36.1 |
| HHI | 888 | 855 | 778 | 683 | 582 | 533 | 499 | 533 | 621 | 594 | 550 | 548 |
| Noam Index | 397 | 383 | 348 | 342 | 260 | 188 | 166 | 176 | 207 | 198 | 183 | 183 |

*Source*: NCC website, Ministry of Finance.

[1] The revenues of seven state-owned stations and Radio Taiwan International are not recorded, because their budgets come from the government.

networks, Pop Network (i like radio) and Music Network (i radio), are dedicated to twenty-four-hour live music broadcasting, so it appeals to a very large audience. Other state-owned nationwide broadcast networks, such as Police Commercial Radio (PCR), Education Radio (ER), and Fushing Radio (FSR), in comparison command a less loyal audience. UFO and the Voice of Taipei are commercial stations that are ranked as the third largest ones according to their annual market revenues. ICRT is the only nationwide network broadcast in a foreign language (English). The remaining radio networks are either medium-power or low-power ones, and they consequently cannot reach a wide range of audience beyond their communities. Table 28.5 records the market share of the radio broadcasting industry.

In 2012, the top five radio stations collectively comprised about 37.5% of the market. The HHI index in radio industry was consistently below 1,000 (499-621). These two indices combined show that radio broadcasting is less concentrated than other media in Taiwan.

## Broadcast Television

There are five terrestrial television stations in Taiwan. Before 1997, three TV stations (TTV, CTV, and CTS) were in operation. The fourth commercial station, FTV, was established in 1997. The fifth station, established in 1998, is a public TV channel. TTV was established in 1962. It was owned by three major groups: 48.95% by the Taiwan Provincial Government (through six Taiwan province-owned banks), 40% by a Japanese consortium (Fuji, Hitachi, NEC, and Toshiba), and 11.05% by domestic private investors. CTV was established in 1968 and ended TTV's six-year monopoly. The Broadcasting Corporation of China (BCC), the largest radio station and owned by the KMT party, acquired 50% of CTV; 28% of the remaining shares went to private radio stations and 22% to other private investors.[18]

18. Liu (1992).

CTS was established in 1971. The Ministry of National Defense owned 51% and the Ministry of Education owned 49%. However, in the 1980s the shareholder structure of CTS was changed to make CTS look more diversified. In reality, the Ministry of National Defense still had great influence. Therefore, although the three TV stations claimed that they were all private enterprises that rely on profits from advertising, the operation of the stations was private in name only.[19]

FTV, the fourth-ranked network, was owned by two companies. Most of the shareholders were considered pro-DPP. In 2003, the Legislature revised three broadcasting-related laws and required the government, military, and political parties to withdraw their shares from broadcast, cable TV, and satellite TV within two years. TTV, CTV, and CTS changed their shareholders' profiles significantly.

The ratings and advertising revenues of the four commercial TV stations have declined drastically. Starting in 2001, the total advertising revenues of cable channels were greater than those of terrestrial TV. While TTV used to lead the TV news market, its market share dropped by almost 10% from 1998 to 2008. CTV's strength used to be its drama series. It became number two in 2008 because its competitor, FTV, produced a lot of local content that appealed to the public. In 2006, the Legislature passed a law requesting that CTS change its role from that of a commercial station to a semi-public TV station. This change established some limitations on its operations, such as restrictions on commercials for children's programs and airing election commercials.[20] Before CTS became a semi-public TV station, there were a number of debates around this topic and a good deal of speculation about its future.[21] As a result, its advertising revenues and total revenues declined and became the lowest among the terrestrial stations in 2009 and 2012.

FTV has a solid positioning strategy and focuses on locally produced programs. It has been number one in the market in terms of its revenue since 2006. Public Television Service's budget mainly comes from government appropriation, which amounts to 900 million NT per year. It also has other income such as audience donations, sales, and the production of programs for Hakka TV and Taiwan Indigenous TV. The HHI of the broadcast TV has been above 2,158. It is obvious that the broadcast TV market is like an oligopoly (Table 28.6).

## Multichannel TV Platforms

Before the Cable TV Act was passed in 1993, cable TV was considered illegal and was nicknamed the "fourth channel" because there were only three terrestrial TV stations at that time.[22]After quite a few turbulent years battling the unlicensed "fourth channel," Taiwan entered the cable TV era. However, cable TV operators were criticized as being associated with local gangs and offering low picture quality due to poor network installation. Furthermore, the operators were busy driving one another out of service zones to monopolize the market, which was usually done at the cost of consumer viewing rights, such as blocking unaffiliated channels even if they were very popular.[23]

By the year 2000, MSOs such as ETTV (now Kbro) and Hoshin (now CNS) emerged after a series of hostile takeovers. The Cable TV Act allows for a maximum of five system operators providing service in each service zone, while in

---

19. Liu (1992).

20. The Law asked the government to acquire all the private shares of CTS and donate them to the Public Television Service Foundation. The chairman of PTS is also chairman of CTS, which can continue to carry commercials but cannot air commercials on children's program. Also, CTS has obligations to provide programs serving the public interest.

21. Before CTS became a semi-public station, it did not know how to do the programming after its role was changed. If it wanted to have more advertising revenues, its programs needed to have high ratings. But to get high ratings it had difficulty to provide too many public TV -like programs

22. Shih, Jun-ji, Kungchung Liu, Shu Yeh, and Yuntsai Chou (2003). *The market survey of cable TV and the competition policy analysis.* Taipei: GIO research report.

23. Shih, Jun-ji, et al. (2003). *The market survey of cable TV and the competition policy analysis,* Taipei: GIO research report.

Table 28-6. Market Share, Revenues, and Concentration of Broadcast TV[1]

| % | 1998 | 2000 | 2002 | 2004 | 2006 | 2007 | 2008 | 2009 | 2010 | 2011 | 2012 |
|---|---|---|---|---|---|---|---|---|---|---|---|
| Taiwan TV Enterprise (public) | 29.2 | 23.3 | 17.4 | 24.2 | 17.1 | 17.0 | 19.3 | 16.8 | 18.4 | 17.0 | 17.7 |
| Chinese Television System | 27.3 | 24.0 | 21.4 | 19.0 | 16.6 | 17.2 | 17.2 | 14.9 | 16.4 | 15.8 | 15.8 |
| China Television Corp (China Times) | 30.8 | 26.5 | 25.0 | 20.1 | 17.7 | 19.0 | 19.2 | 16.2 | 16.4 | 16.3 | 17.0 |
| Formosa TV (DPP-affiliated) | 9.8 | 18.4 | 24.3 | 23.7 | 31.9 | 31.1 | 32.4 | 30.1 | 35.4 | 36.9 | 33.0 |
| Public Television Service (public) | 2.8 | 7.9 | 11.9 | 13.0 | 16.7 | 15.7 | 11.8* | 22.1 | 13.4 | 13.9 | 16.6 |
| Total revenue (NT million) | 21,205 | 14,597 | 11,024 | 10,820 | 8,969 | 8,778 | 9,074 | 9,528 | 9,517 | 9,405 | 8,556 |
| Total revenue (USD million) | 707 | 487 | 367 | 361 | 299 | 293 | 303 | 318 | 318 | 314 | 285 |
| *N* (>1%) | 5 | 5 | 5 | 5 | 5 | 5 | 5 | 5 | 5 | 5 | 5 |
| CR4 | 97.2 | 92.1 | 88.1 | 87.0 | 83.3 | 84.3 | 88.2 | 85.2 | 86.6 | 86.1 | 84.2 |
| HHI | 2,654 | 2,219 | 2,118 | 2,082 | 2,177 | 2,158 | 2,230 | 2,158 | 2,308 | 2,364 | 2,214 |
| Noam Index | 1,327 | 993 | 947 | 931 | 974 | 965 | 997 | 965 | 1,032 | 1,057 | 990 |

*Source*: Rainmaker. XKM International Corp. from Adm Magazine; Annual report of PTS; NCC website.

*In 2008, the Legislative Yuan (Congress) froze the execution of 50% of PTS' budget. It was unfrozen in 2009. Therefore, PTS' budget in 2009 looked high.

[1] Only total revenues for terrestrial TV were available. PTS revenues are publicly available online. The four commercial TV revenues were calculated individually based on the advertising revenues estimated by multiple sources. They were given weights based on the total revenues.

practice only one operator evolved in each zone through the MSO consolidation. There are 47 effective service zones at present, 33 of which are monopolies and the remaining 14 duopolies. At the local level, the industry has become monopolized and consumers lack choice in subscribing to pay-TV services.[24]

The regulated tariff structure makes the cable TV business in Taiwan highly profitable. Its EBITDA (earnings before interest, taxes, depreciation, and amortization) was once said to be about 40%, which lured foreign capital into the business. The Carlyle Group thus acquired local independence in the north-middle parts of Taiwan in 2000 to constitute Taiwan Broadband Communications (TBC) that later became the third largest MSO.[25] Since then, foreign private equity funds have been targeted at the two major MSOs due to the high expected profit returns from a US$20 flat charge per month set by the government. Subsequently, the top three MSOs, Kbro (formerly ETTV), China Network Systems (CNS, formerly Hoshin), and TBC, have been held by the Carlyle Group, MBK Partners LP[26], and Macquarie International Infrastructure Fund of Australia, respectively, even though there is a 60% cap on indirect foreign investment in cable TV systems.

24. Tsai, Zhi-hong, Ying-chin Shieh, Chinghe Chen, Nian-chong Tsai, Yuntsai Chou, Shin-hong Chen, and Chun Li. *The white paper on digital convergence policy*. Taipei: III research report, 2009.

25. Shih, Jun-ji, et al. (2003). *The market survey of cable TV and the competition policy analysis,* Taipei: GIO research report.

26. MBK Partners LP is a Korea-registered fund established in March 2005, one of the largest independent private equity firms in the North-Asia region, with over 3.7 billion in capital under management.

The statistics of the National Communication Commission (NCC) show that 4.98 million households subscribed to cable TV service in 2014, which accounts for a 60.1% penetration rate. However, the actual penetration rate is higher since numerous families just bypass their local cable operators and connect wires to the nearby cable TV households without paying the operators.[27] It is estimated that 81.2% households in Taiwan have access to cable TV programming.[28] Only 15.9% of households receive TV via broadcast signals. Cable TV is the predominant platform of video entertainment in Taiwan.

Table 28.7 shows the total revenue of cable TV from 2002 to 2012. The value of the total revenue has grown from US$632 million to US$1,281.2 million over a period of eleven years, equivalent to a growth of 100%. The compound annual growth rate (CAGR) in 2002–2009 was 4.6%. Unlike radio broadcasting and terrestrial TV, cable TV exhibits a trend of increasing revenue streams. This is the main reason that foreign private equity funds that focus on controlled investment are interested in investing in the cable TV industry.

Table 28.8 shows the market shares of Taiwan's cable TV operators. As the top five are MSOs, they comprise a total of nearly 70% of the market. Although the independents in total account for over a 25% market share, they are smaller on an individual basis and some of them serve in franchise areas with less than 20,000 households. This results in fewer economies of scale.

The largest MSO in Taiwan, Kbro, owns 12 cable TV systems. The second largest MSO, CNS, owns 10. The third largest MSO, TBC, owns 4. The fourth largest MSO, TFN, owns 5. Finally, the fifth largest, Taiwan Optical Platform (TOP), has 4 systems.[29] The major stakeholder of TFN, Taiwan Mobile, sought to merge with Kbro in 2009. In order to avoid the one-third market cap limit, Mr. Tsai, the chairman of Taiwan Mobile, established Da-Fu Media Technology Co. to merge with Kbro. The merger case was finally approved by the Fair Trade Commission in October 2010 and NCC in November 2010, albeit with several conditions.[30] This merger makes the second largest mobile service provider in Taiwan the largest MSO as well. Its market share after the merger rose to 33.19%, reaching the threshold (one-third) for horizontal integration.

It is imperative to notice that the market shares of each of the top five MSOs do not fluctuate dramatically through time. Due to the horizontal integration restriction (a MSO cannot own the systems whose household subscription in total exceeds a one-third market share), the MSOs cannot raise their penetration rate indefinitely. All MSOs are reluctant to fiercely compete with the existing companies and avoid overbuilding in a given service zone. Therefore, competition is limited.

## Table 28-7. Total Revenue of the Cable TV Industry in Taiwan, 2002–2010

| | 2002 | 2003 | 2004 | 2005 | 2006 | 2007 | 2008 | 2009 | 2010 |
|---|---|---|---|---|---|---|---|---|---|
| Total Revenue (USD million) | 632.0 | 711.0 | 739.0 | 773.0 | 813.0 | 842.0 | 896.0 | 917.0 | 947.0 |

*Source*: MSOs' annual financial reports.

27. Tsai et al., op. cit.

28. NCC, "TV digitization and convergence," Internal document delivered to the Executive Yuan, 2011.

29. Tsai et al., *The white paper on digital convergence policy* Taipei: III research report, 2009.

30. The conditions are (1) the company cannot rent head facilities from other MSOs; (2) the company cannot produce extra programming neither resell more channels beyond the current standing; (3) the company cannot discriminate other MSOs or competitive platform aggregators in licensing its channels; (4) the company must achieve the degree of digitization as designated by the Executive Yuan; (5) the company should devote resources in producing HD programming in order to develop the cultural creative industry.

Table 28-8. Market Share and Concentration in the Cable TV Industry, 2001–2012

| % | 2001 | 2002 | 2004 | 2006 | 2008 | 2010 | 2012 |
|---|---|---|---|---|---|---|---|
| Kbro | 25.6 | 24.2 | 24.1 | 24.0 | 23.5 | 22.5 | 20.8 |
| CNS (Korea) | 20.6 | 22.1 | 23.0 | 23.3 | 23.1 | 21.1 | 21.6 |
| TBC (Australia) | 13.9 | 14.5 | 13.6 | 14.6 | 14.6 | 13.5 | 14.9 |
| TFN | 10.7 | 10.0 | 10.2 | 9.8 | 11.0 | 10.9 | 10.2 |
| TOP | 5.2 | 5.1 | 6.9 | 6.7 | 6.6 | 5.9 | 6.5 |
| Others (independents) | 24.0 | 24.0 | 22.3 | 21.7 | 21.3 | 26.2 | 25.9 |
| Total Revenue (USD Mil) | | 632.0 | 739.0 | 813.0 | 896.0 | 947.0 | 1,281.2 |
| *N* (>1%) | 5 | 5 | 5 | 5 | 5 | 5 | 5 |
| CR4 | 71.0 | 71.0 | 71.0 | 72.0 | 72.0 | 68.0 | 67.5 |
| HHI | 1,416 | 1,413 | 1,444 | 1,471 | 1,462 | 1,284 | 1,268 |
| Noam Index | 578 | 577 | 646 | 600 | 654 | 574 | 161 |

*Source*: NCC website.

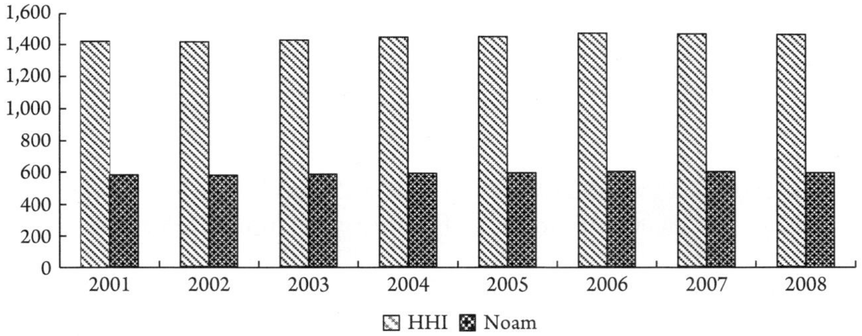


FIGURE 28.1 The HHI and Noam Indices of Taiwan's Cable TV Industry

With respect to industry concentration, Table 28.8 shows that the top four MSOs together make up more than 68-70% of the market. Figure 28.1 also illustrates the scores for the HHI and Noam indices. Compared with CR4, the medium-score for the HHI index reflects a moderately concentrated market. The reason for this lies in the fact that a high number of independents deflate the value of the HHI index. In this sense, the high score of the Noam index undoubtedly measures the concentrated reality of the cable TV industry more appropriately. We can conclude that the cable TV industry in Taiwan is unevenly concentrated in the case of the big four, while being scarcely diverse in the case of the independents.

## Video Channels

Taiwanese audiences have watched the cable- and satellite-delivered TV channels since the end of the 1980s. Many of those programs were the Hong Kong and USA dramas or movies which infringed upon international copyright laws and were retransmitted by VHS cassettes in most local cable systems. In 1997, the first legal channel license was issued, and after that various types of programming were booming in the satellite TV business. By the end of 2013, there were more than 240 satellite TV channels licensed in Taiwan, and around 150 domestic and 90 overseas channels. There were over eight 24-hour domestic news channels, fifteen movie channels, thirteen shopping channels, and many Japanese or Korean drama channels. Due to the limited capacity of the analog cable TV system, the average number of available channels is less than 100.[31] The Taiwan government hoped to increase the percentage of digital cable system household penetration to 100% by the end of 2014. At that time, The capacity will increase to over 1,000 channels. The penetration of digital cable in 2014 year was over 70%.

For most satellite channels, there are two major sources of revenue. One is the programming fees paid by the cable system based on their ratings, and the other is advertising revenue. Cable systems usually expend US$8 monthly per household to pay for the programming fees that are shared by the carried channels. The total annual programming fees from the cable systems are around US$480 million (5 million subscribers × $8/month × 12 months). Data on the revenues of the individual satellite TV companies are not available at this time but might be calculated from the audience share. Some satellite TV companies receive more revenues from advertisements, while some receive more from programming fees. For companies like CTI and SET, advertising revenues amount to about three-fourths of the total revenue. However, some companies rely more on programming fees or airtime rental fees. The average percentage of programming fees for satellite TV companies was 34% in 2003, but declined to 27% in 2005. The average percentage of advertising revenue was 37% in 2003, and increased to 40% in 2005.

Because we could not obtain data on the revenue for each satellite TV company, this study collected the advertising market shares in order to analyze the concentration and market power of satellite TV companies (Table 28.9). The yearly advertising data for each channel from 1996 to 2004 were collected from the Adm magazines provided by Rainmaker XKM International Corp. This study compiled each channel into the conglomerate ownerships. However, Rainmaker Corp after 2004 stopped releasing the advertising information for individual channels to the Adm magazines. From 2005 to 2008, this study compiled a similar data set from Brain magazine. The 2010 and 2012 data were transformed from the rating share of AGB Nielsen but the total advertising revenues are significantly lower than for the previous years.

Figure 28.2 shows that from 1996 to 2012, the advertising revenues of the satellite TV channels increased from US$201 million to US$857 million. Most of the increased revenues came from the decreases in the broadcasting TV advertising revenues. The total advertising revenues of broadcast and satellite TV did not change very much after 1998. For the

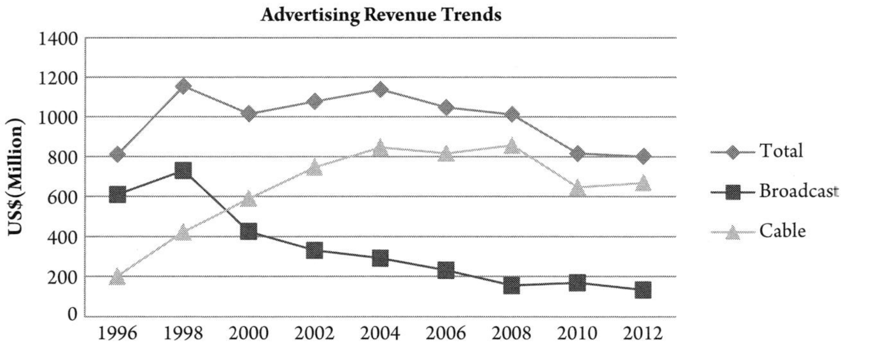


FIGURE 28.2 Advertising Revenue Trends of Broadcasting and Cable TV

*Source:* Rainmaker XKM International Corp. from Adm Magazine and Brain Magazine.

31. Chen, Ping-Hung. "Who Owns Cable Television? Media Ownership Concentration in Taiwan." Journal of Media Economics, 15, 1, (2002): 44–55. Chen, Ping-Hung. "Transnational Cable Channels in the Taiwanese Market." Gazette, 66, 2, (2004): 167–183.

Table 28-9. Market Shares of Taiwan Satellite TV Channels (%): 1996–2012[1]

| | 1996 | 1998 | 2000 | 2002 | 2004 | 2006 | 2008 | 2010 | 2012 |
|---|---|---|---|---|---|---|---|---|---|
| TVBS (TVB, HK) | 30.8 | 17.2 | 15.4 | 17.9 | 15.2 | 12.6 | 11.5 | 9.6 | 9.1 |
| ERA | | | | | | 5.3 | 9.3 | 6.5 | 7.1 |
| Star TV (News Corp, UK) | 12.0 | 4.9 | 7.2 | 8.6 | 10.3 | 11.9 | 11.0 | 9.9 | 10.2 |
| SET | 9.7 | 10.4 | 11.0 | 13.4 | 12.3 | 16.5 | 16.6 | 19.3 | 19.4 |
| VL | 9.4 | 8.5 | 6.4 | 8.5 | 12.9 | 11.3 | 11.5 | 8.0 | 8.9 |
| GTV | | 4.4 | 7.8 | 8.5 | 6.8 | 10.9 | 10.0 | 10.5 | 8.6 |
| ET | 3.6 | 12.9 | 15.9 | 16.7 | 19.6 | 19.3 | 16.3 | 17.8 | 19.2 |
| CTI (China Times) | 3.8 | 4.8 | 3.6 | 4.5 | 6.2 | 9.6 | 10.9 | 7.3 | 7.1 |
| Unique | 0.9 | 1.2 | 1.8 | 2.9 | 2.4 | 2.7 | 2.9 | 3.2 | 2.9 |
| Other Individuals | 29.8 | 35.7 | 30.9 | 19.0 | 14.3 | 0.0 | 0.0 | 7.8 | 7.5 |
| Revenues($ mil) | 200 | 423 | 590 | 747 | 847 | 817 | 857 | 647 | 669 |
| *N* (>1%) | 6 | 8 | 8 | 8 | 8 | 9 | 9 | 9 | 9 |
| CR4 | 61.9 | 49.0 | 50.0 | 56.7 | 59.9 | 60.3 | 55.9 | 57.6 | 57.9 |
| HHI | 1,302 | 710 | 780 | 1,030 | 1,127 | 1,319 | 1,241 | 1,161 | 1,193 |
| Noam Index | 460 | 251 | 276 | 364 | 398 | 467 | 414 | 387 | 398 |

*Source*: Data of the 1996 to 2008 years from the Rainmaker XKM International Corp. published in Adm and Brain Magazine and 2010 & 2012 from the Nielsen database.

[1] TVBS and ERA were part of the same conglomerate before 2005. The percentages for 2005–2008 did not include the other individual channels.

last 10 years, with the slowdown in economic growth in Taiwan, the advertising revenues of the satellite TV channels have remained fixed at US$750–850 million.

TVBS is owned by TVB (Television Broadcast Limited of Hong Kong) and Star TV is owned by News Corp. Except for these two conglomerates, others are owned and operated by the local media conglomerates. Among them, ET and VL import many overseas movies channels, and VL dominates sports programming. SET and GTV have produced many dramas and recently they sold those programs to China and other Asian countries. CTI was owned by the Want Want China Times Group, which also owns the *China Times* (newspaper) and China Broadcasting Company.

In 1996, the top firm TVBS/ERA had the largest market share with 30.8% and CR4 held 61.9%. However, in 1997 the Taiwanese government began to issue many new licenses to the satellite TV channels, and the number of channels by the other conglomerates quickly increased. CR4's market share dropped to less than 50% in 1998 and then increased to 57.9% in 2012. In 2005, although the ownership of ERA and TVBS had separated, CR4's share did not significantly decrease. However, this is explained by examining the data closely. The data set covering the period from 2005 to 2008 did not include other individual channels and the percentages were a little inflated. In 2008, CR4's share was approximately 55.9% and satellite TV remained a highly concentrated market.

In Figure 28.3, most of the CR4s were over 50% and represented a concentrated market. However, the HHI between 1998 and 2000 was below 1,000 (indicating a competitive market) because the market shares of the top four firms were even.

## Film

Taiwan's film production had been monopolized by the Central Motion Picture Company, owned by the ruling KMT party since the 1950s. Starting in the 1980s, Taiwan opened

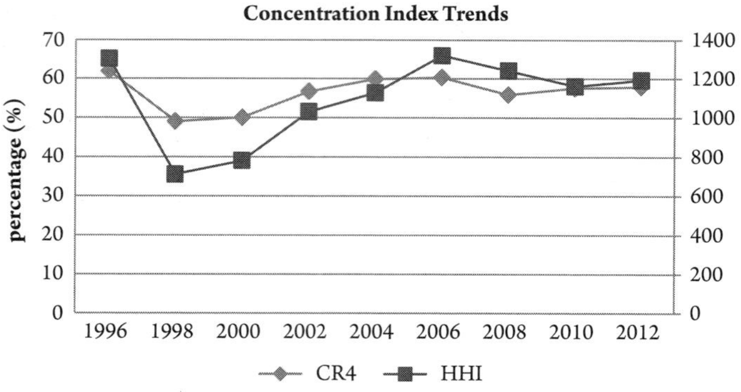


FIGURE 28.3 The CR4 and HHI Indices

its film market to the outside world, mainly to Hollywood. This led to the closing of domestic film studios. Domestic production has dropped to 20–30 films a year for the past 30 years. Film distribution chains are tightly controlled by Hollywood branch companies, such as Buena Vista (Disney), Warner Bros. (Time Warner), and Fox (News Corp.). They signed long-term contracts with theaters to secure exhibition rights. In 1998, the international theater chain, the Warner Village Roadshow, built a 17-screen Cineplex in Taipei. It drove many traditional theaters out of business.

Before 2007, Taiwanese films were in no position to fight Hollywood movies in terms of box office earnings. Hollywood films usually took away more than 90% of the yearly gross, while Taiwanese films usually had less than 2%. However, Taiwanese cinema has rebounded since 2007. Films like *Lust, Caution*, and *Secret* brought local audiences back to the theater. In 2008, *Cape No. 7* succeeded in becoming the highest box office earner, $530 million NT (US$16 million), in the history of Taiwanese cinema. Box office earnings for Taiwanese films have begun to pile up since then. Films such as *Seedy Bale, Mon-Ga, You Are the Apple of My Eye*, and *Black and White* all earned $200 million NT box office records or higher.

However, distribution is still controlled by foreign companies. Table 28.10 illustrates the film distribution from 1998–2011. Figure 28.4, and Figure 28.5 illustrate the top four distributors' concentration ratios in terms of box office revenue and the numbers of films distributed. All of the top four distributors are Hollywood agents. Taiwan's market is dominated by Hollywood productions. As the high concentration ratios reveal, CR4 reached its peak from 2005 to 2007 and took over almost the entire market. Between 70% and 80% of distributors operate and function each year; however, the top four monopolize the market. It should also be noted that three of the top four also distribute domestic films, including *Cape No. 7*.

In Figure 28.6, we can see that the film production industry's concentration ratio is extraordinarily high. From 2005 to 2009, the accumulated box office income of Yen Ping, Serenity, and 3rd Vision could not even compete with that of Central Motion Picture Company, a KMT-owned company operated continuously for more than half a century.

In 2005, four major distributors equally divided the market. However, the share of Buena Vista rose after 2006 to have a higher market share (Figure 28.7). This increase in market share may have been caused by the decision to distribute Taiwanese films. Fox followed suit and, with the aid of the box office success of *Avatar*, became the biggest distributor in 2009.

Since Warner Village Roadshow opened its first multiplex in 1998, independent theaters in Taiwan have been forced to either close down or merge with theater chains. Fewer than 100 out of 319 administrative districts have movie theaters today and most of the theaters are in metropolitan areas. A handful of companies own major theater chains. From the exhibition industry's box office allocation, we see that Vie Show has the largest market share (Figure 28.8).

Table 28-10. Film Distribution

| % | 1998 | 2000 | 2004 | 2007 | 2008 | 2009 | 2010 | 2011 |
|---|---|---|---|---|---|---|---|---|
| Buena Vista (Disney, US) | 15.0 | 33.1 | 28.5 | 35.57 | 29.05 | 25.62 | 19.46 | 17.04 |
| UIP (Paramount-Viacom & Universal-Comcast/GE, US) | 24.5 | 18.9 | 18.5 | 16.41 | 22.73 | 18.15 | 12.28 | 26.19 |
| Warner (Time Warner, US) | 13.2 | 11.2 | 17.5 | 23.3 | 10.17 | 10.05 | 20.07 | 12.06 |
| FOX (News Corp, US) | 17.8 | 10.5 | 15.9 | 8.0 | 8.08 | 14.31 | 17.23 | 14.24 |
| Others | 29.6 | 26.3 | 19.6 | 16.72 | 29.97 | 31.87 | 30.96 | 30.47 |
| Total Revenues (NT mil) | 2,987 | 2,559 | 2,585 | 2,992 | 2,340 | 2,268 | 2,538 | 2,831 |
| Total Revenues (USD mil) | 93 | 80 | 81 | 100 | 78 | 76 | 85 | 94 |
| *N* (>1%) | 4 | 4 | 4 | 4 | 4 | 4 | 4 | 4 |
| CR4 | 70.5 | 73.7 | 80.4 | 83.3 | 70.0 | 68.1 | 69.0 | 69.5 |
| HHI | 1,316 | 1,689 | 1,714 | 2,141 | 1,529 | 1,292 | 1,229 | 1,325 |
| Noam Index | 658 | 845 | 857 | 1,071 | 765 | 646 | 614 | 663 |

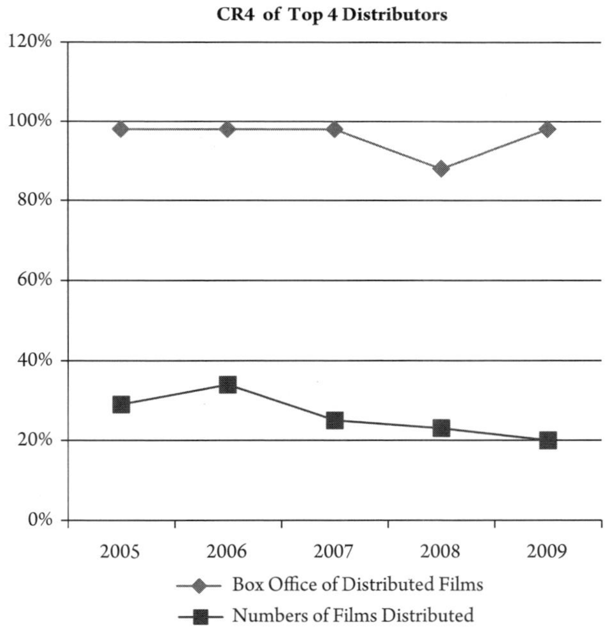


FIGURE 28.4 Distributors' Concentration Ratio: 2005–2009

The CR4 and HHI charts (Figure 28.9) show that both production and distribution are highly concentrated. The HHI numbers are all higher than 1,800, which indicates a high concentration.

The huge popularity of *Cape No. 7*, *Mon Ga*, *Seedy Bale*, and *Black and White* prove that local audiences still love domestic products. Although both distribution and exhibition are highly concentrated and tightly controlled by foreign

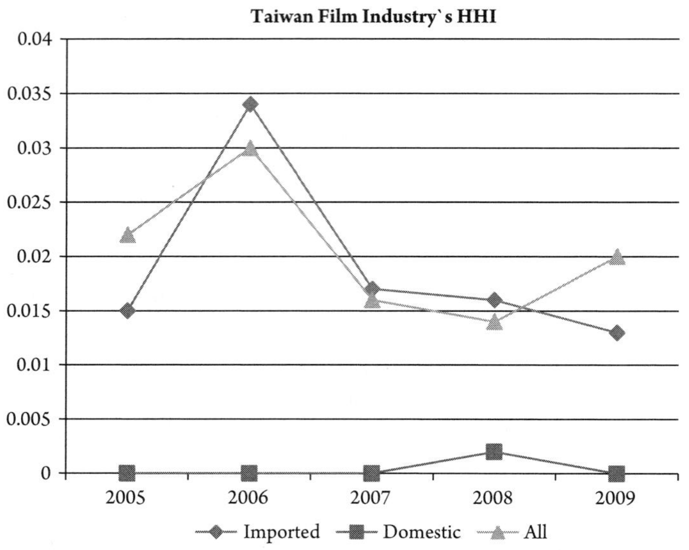


FIGURE 28.5 Taiwan Film Industry's HHI Index

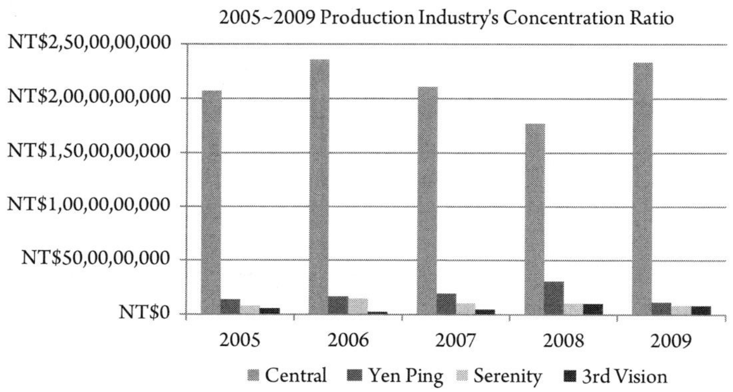


FIGURE 28.6 Film Production Industry's Concentration Ratio (2005–2009)

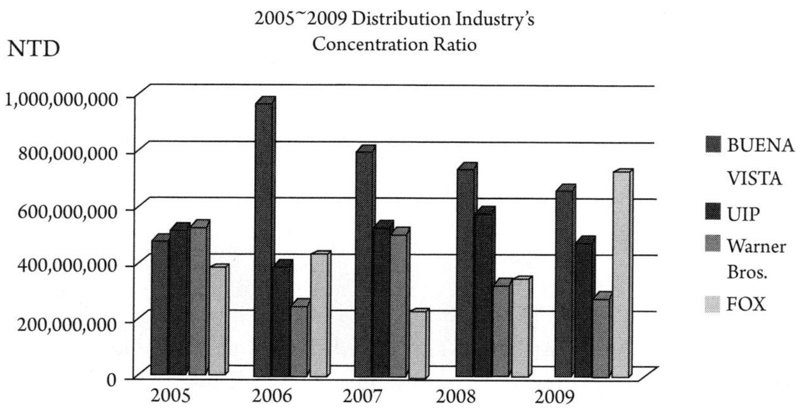


FIGURE 28.7 Film Distribution Industry's Concentration Ratio (2005–2009)

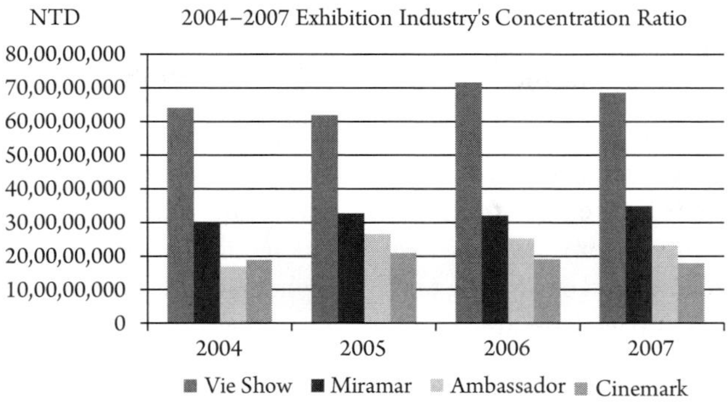


FIGURE 28.8 Exhibition Industry's Concentration Ratio (2004–2007)

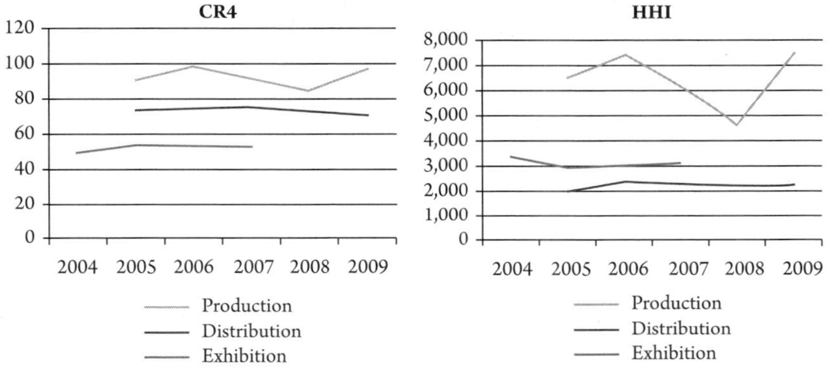


FIGURE 28.9 Film Industry's Concentration Ratio Graph 10 Film Industry's HHI

companies, it remains to be seen whether the recent success of Taiwanese films can reverse this trend.

## Music Industry

Over the past few years, the revenue of Taiwan's music market dropped from over US$400 million in 1997 to around US$43 million in 2012. This is mainly caused by the free P2P downloading of MP3.[32] In order to survive in the continuously shrinking music market, many local music firms merged into transnational music corporations such as Warner, Universal, or Sony to take advantage of lower production costs or more marketing channels. On the other hand, following the introduction of cultural diversity and freedom in the creative music industry, more than 70% of popular Chinese music CDs have been produced in Taiwan. They were then sold in China, Hong Kong, Singapore, or Malaysia. Therefore, the purpose of the mergers between transnational music corporations and Taiwanese local music companies was not simply to focus on the local market, but also on the other Pan-Chinese cultural markets.

Although the online music market has continuously grown in recent years, its 2013 market share is around 1/3 of the CD market, but no official sales data have been released. Therefore,

32. Alejandro, Zentner. "File Sharing and International Sales of Copyrighted Music: An Empirical Analysis with a Panel of Countries." Topics in Economic Analysis & Policy 5, Iss. 1, (2005): Article 21. Last accessed on July 17, 2012 at <http://www.bepress.com/bejeap/topics/vol5/iss1/art21>; Survey of Taiwan Music Market (2010). Last accessed on July 17, 2012 at <http://info.gio.gov.tw/public/Attachment/05209533371.doc>.

we have focused on the traditional CD market. The data were collected from two online websites, GMusicBillboard and CDtop. The two websites provide the weekly market shares of the top 20 music production companies. The data for GMusicBillboard were collected from the major consumer chain stores and the two top music stores that comprised 65% of the total market. The data for CDtop were obtained from the other large music chain stores accounting for 15% of the overall market. The individual stores accounted for the remaining 20% share, but there are no specific revenue data for each music company. Therefore, this study only includes the market shares for GMusicBillboard and CDtop. These two websites provide weekly CD market shares, this study compiled and weighted to calculate annual market shares.

From 2005 to 2012, CR4 amounted to over 50% and the music industry and can be considered a highly concentrated market (Table 28.11). In addition, most of the top companies are transnational music corporations, such as Avex, Gold Typhoon (EMI), Sony, Universal, and Warner Music. While their individual market shares might change every year, together they account for nearly 70% of the market. The music market share is heavily dependent on celebrities. If the contract of a major music celebrity in Taiwan is transferred to another transnational music corporation, the rankings in terms of market share will change, too. Although there are still many

Table 28-11. National Market Shares of Taiwan Music Production and Distribution (%): 2005–2008[1]

| % | 2005 | 2006 | 2007 | 2008 | 2010 | 2012 |
|---|---|---|---|---|---|---|
| ALFA Music | 2.7 | 2.9 | 0.6 | – | 0.1 | 0.1 |
| Avex (Japan) | 11.8 | 13.1 | 15.8 | 20.2 | 13.7 | 9.0 |
| B`in Music | – | 1.5 | 3.1 | 4.5 | 5.0 | 10.8 |
| Forward Music | 2.5 | 3.5 | 2.4 | 3.2 | 1.3 | 0.6 |
| Gold Typhoon (EMI) | 16.8 | 10.8 | 14.2 | 8.8 | 17.7 | 16.2 |
| Jingo | 0.6 | 1.4 | 0.8 | 2.3 | 0.0 | 0.0 |
| HiGH NoTE | 1.8 | 0.5 | 1.5 | 1.9 | 0.0 | 0.6 |
| HIM | 2.8 | 3.3 | 4.9 | 2.4 | 6.0 | 7.0 |
| LINFAIR Records | 1.8 | 1.6 | 2.7 | 3.0 | 0.7 | 2.5 |
| Rock Records | 3.1 | 5.7 | 3.8 | 1.7 | 1.0 | 5.3 |
| SEED |  | 0.0 | 1.5 | 1.9 | 7.8 | 3.0 |
| Sony Music (Japan) | 19.3 | 14.8 | 12.2 | 12.5 | 13.8 | 9.3 |
| Universal Music Taiwan (Vivendi, France) | 18.9 | 15.4 | 15.1 | 15.0 | 7.8 | 9.2 |
| Warner Music Taiwan (US) | 9.5 | 11.3 | 9.2 | 11.7 | 16.0 | 19.5 |
| Other | 8.4 | 14.2 | 12.2 | 10.9 | 9.1 | 7.1 |
| Total Revenues ($ mil) | 150 | 78 | 71 | 58 | 59 | 43 |
| *N* (>1%) | 11 | 12 | 12 | 13 | 11 | 10 |
| CR4 | 66.8 | 54.6 | 57.2 | 59.5 | 61.2 | 55.8 |
| HHI | 1,278.4 | 943.9 | 976.5 | 1,065.6 | 1,216.1 | 1,154.9 |
| Noam Index | 369.1 | 252.3 | 261.0 | 295.5 | 366.6 | 365.2 |

*Source*: compiled from online data by the authors.

[1] GMusicBillboard website: <http://www.g-music.com.tw/GMusicBillBoard0.aspx>. CDtop website: <http://www.5music.com.tw/cdtop.asp>

local music companies, most of their market shares are very small.

Since the market shares of the top four companies are fairly even, the HHI index is only slightly above 1,000 from 2005 to 2012. This can be considered moderately concentrated. In 2006 and 2007, the HHI index was below 1,000 and the market can be considered competitive. The Noam Index, which is calculated by dividing the HHI index by the square root of the number of music companies, significantly increased in 2010 and 2012.

## TELECOMMUNICATIONS MEDIA

### Wireline Telecom

There are four fixed-line networks in Taiwan. Chunghwa Telecom (CHT), the oldest carrier, used to be owned by the Ministry of Transportation and Communication (MOTC). Before 1996, it was a business unit of the Directorate General of Telecommunications (DGT) under the MOTC. Even though CHT was separated from the DGT in 1996, its chairman is still nominated by the government. The telecommunication liberalization started in 1994. CHT became a privatized company in 2005. However, the MOTC still owns 35% of CHT.

In 2000, the government granted three new fixed-line licenses to Taiwan Fixed Network (TFN), New Century Infocomm Tech Co (NCIT), and Asia Pacific Broadband Telecom (APBT). CHT has remained the largest carrier in terms of market share. Because CHT controls the last mile, the other three new fixed-line networks cannot compete with it in the local phone market. They prefer to promote wireline services to small and medium businesses and commercial buildings. In the local phone market, the market share of the three new carriers dropped from 15% in 2006 to 13.8% in 2008, then rose to 15.6% in 2010, and even rose to 19.3% in 2012. From 2006–2008, the HHI was between 7,291 and 7,502, showing a high concentration, but then declining from 2008 to 2012 to 6,654 (Table 28.12).

### Wireless Telecom

The mobile phone industry is composed of six 2G operators belonging to three telecom groups—Chunghwa Telecom, Taiwan Mobile,[33] and FarEasTone Telecommunications Co[34]—, five 3G services, and six 4G operators (four 4G

**Table 28-12. Market Shares of Taiwan Telecom Companies (Wireline), 2006–2008**

| % | 2006 | 2007 | 2008 | 2010 | 2012 |
|---|---|---|---|---|---|
| CHT (35% public) | 84.9 | 85.7 | 86.2 | 84.4 | 80.71 |
| APBT | 5.3 | 3.9 | 2.2 | 3.42 | 4.22 |
| TFN | 5.1 | 5.2 | 6.3 | 5.97 | 5.46 |
| NCIT | 4.6 | 5.1 | 5.3 | 6.22 | 9.62 |
| Total Revenues (NT mil) | 104,440 | 102,032 | 99,566 | 95,487 | 103,504 |
| Total Revenues (USD mil) | 3,264 | 3,189 | 3,111 | 2,984 | 3,235 |
| C4 | 100.0 | 100.0 | 100.0 | 100.0 | 100.0 |
| *N* (>1%) | 4 | 4 | 4 | 4 | 4 |
| HHI | 7,291 | 7,419 | 7,502 | 7,209 | 6,654 |
| Noam Index | 3,646 | 3,709 | 3,751 | 3,605 | 3,327 |

33. Taiwan Mobile is owned by Mr. Richard Tsai who also owns Taiwan Fixed Network, a cable MSO Kbro, and Fubon Financial Holding company.

34. FarEasTone is owned by Mr. Douglas Hsu who also owns a fixed network New Century Infocomm Tech Co (NCIT) and Far Eastern Group.

players and two new entrants). Vibo did not bid for 4G spectrum and was acquired by Taiwan Star Cellular Co., a new entrant of 4G. Asia Pacific Telecom was acquired by Ambit Microsystems Corp (backed by Hon Hai Group), another new entrant in 4G. Until August 2014, the two new entrants have not started to provide 4G services; therefore they are not listed in Table 28.13. The first 3G service started in July 2003, while the other four operators started two years later.[35] Since the three big 3G operators also provide 2G services, they have actively worked to migrate their subscribers from 2G to 3G or 4G.

At the end of 1997, when the private mobile operators joined the market, the penetration was only 6.88%. In 2003 the penetration was 111.33%.[36] There were three regional and three national 2G operators. KG Telecom was established in 1997. It served the northern part of Taiwan. It was acquired by FarEasTone in 2004, and was completely merged into FarEasTone in 2010. In the central part of Taiwan, MobiTai Communications Co., Ltd, the 2G operator, was also established in 1997. It was acquired completely by Taiwan Mobile in 2006. In the southern part of Taiwan, TransAsia provided mobile phone service. It was established in 1988 and acquired by Taiwan Mobile in 2002. Both MobiTai and TransAsia's brand names were changed to Taiwan Mobile in 2008.

Asia Pacific Telecom launched its 3G service in 2003. In order to be the first 3G operator, it adopted CDMA2000 which was different from the W-CDMA adopted by the other four operators. In 1998, CHT had a 53.7% market share. However, its market share began to decline in 1999. In 2000, its market share was

Table 28-13. Market Shares of Telecom Companies (Wireless), 1998–2012

| % | 1998 | 2000 | 2002 | 2004 | 2006 | 2008 | 2010 | 2012 |
|---|---|---|---|---|---|---|---|---|
| CHT (35% public) | 53.7 | 30.7 | 34.8 | 35.4 | 34.4 | 34.8 | 34.59 | 32.53 |
| Taiwan Mobile | 22.4 | 29.8 | 31.3 | 30.7 | 28.4 | 26.2 | 27.76 | 29.12 |
| TransAsia | 3.2 | 5.1 | TM | | | | | |
| MobiTai | 3.9 | 3.4 | 2.7 | TM | | | | |
| FarEasTone | 11.1 | 21.1 | 19.3 | 34.0 | 32.1 | 31.2 | 24.82 | 26.57 |
| KG | 5.7 | 9.9 | 11.9 | FarEasTone | | | | |
| Asia Pacific Telecom | | | | | 3.4 | 5.6 | 9.19 | 8.64 |
| Vibo | | | | | 1.8 | 2.2 | 3.64 | 3.14 |
| Total Revenues (NT mil) | 54,804 | 151,559 | 179,418 | 198,171 | 212,240 | 207,552 | 210,941 | 222,443 |
| Total Revenues (USD mil) | 1,713 | 4,736 | 5,607 | 6,193 | 6,633 | 6,486 | 6,687 | 7,496 |
| *N* (>1%) | 6 | 6 | 5 | 3 | 5 | 5 | 5 | 5 |
| CR4 | 93 | 92 | 92 | 92 | 89 | 92 | 96 | 97 |
| HHI | 3,566 | 2,413 | 2,437 | 2,404 | 2,273 | 2,583 | 2,681 | 2,697 |
| Noam Index | 1,456 | 985 | 995 | 1,388 | 804 | 1,155 | 1,199 | 1,206 |

35. The first 3G service was provided by Asia Pacific Telecom. The other four 3G operators including the three existing mobile operators and one new player Vibo. Liu and Wang (2010).

36. Kao (2004).

less than one third of the mobile phone market. In 2002, Taiwan mobile acquired TransAsia. In 2004, the added market share of FarEasTone and KG Telecom was 34%, which was only 1.4% behind CHT's market share. From 1998 to 2004, the revenue of Taiwan Mobile made it number two in the market. Because of the merger effect from 2004 to 2008, FarEasTone became number two in the market if the revenues of its acquired company, KG Telecom, are added to the total. In order to compete with CHT and FarEasTone, Taiwan Mobile acquired 67% of MobiTai in 2004 and the rest of the shares in 2006. Since the three regional mobile phone companies were all acquired by the three big players, there were three big and two small mobile phone operators before the government opened up 4G. Between 2002 and 2010, CHT continues to account for at least one third of the market, with Taiwan Mobile and FarEasTone right behind it. Between 2006–2008, FarEastone held the second highest market share and Taiwan Mobile came in third. However, this changed after 2010. Taiwan Mobile took second place while FarEasTone fell to third. Meanwhile, the combined market share of Asia Pacific Telecom and Vibo held less than 8% between 2006–2008, but exceeded 10% between 2010–2012. (Table 28.13).

## INTERNET MEDIA

### Internet Service Providers (ISP)

Broadband services rely on the last mile of lines. Since CHT controls the last mile, the other three fixed-line networks, TFN, NCIT, and APBT, cannot compete with it in the broadband market. CHT's subsidiary company, HiNet, has dominated the broadband market.[37] This also explains why TFN, NCIT, and APBT only focus on commercial buildings and metropolitan areas for their broadband services. Since 2008, cable operators have also become active in providing cable modem services.

In Taiwan, there are more DSL/FTTx subscribers than cable modem (CM) subscribers. From 2003 to 2008, HiNet accounted for about 80% of the broadband market. No fixed network could compete with it. HiNet tried to urge its subscribers to change to fiber-optic service since 2009. In December 2012, the market share for CHT's HiNet was 67.9% for the wireline broadband. Meanwhile, other fixed-line networks and other DSL service providers occupy 12.7% of the market share. As for the cable operators, they occupy 19.3% of the market share. Therefore, the ISP market is highly concentrated.

It seems that the telecommunications liberalization policy was not effective. HiNet has been dominant in the DSL and FTTx markets since 2000. The HHI was 5,467 in 2007 but declined to 4,737.1 in 2012. CR4 was 91.41 in 2007 but fell to 83.6 in 2012. Therefore, the ISP market is highly concentrated even though HiNet's market share slightly declined from 2010 to 2012 (Table 28.14).

### Search Engines

Between 2007 and 2010, the top four search engine companies were Yahoo-Kimo, Google, Baidu, and Microsoft. Yahoo-Kimo Search and Google Search always occupied the top two positions from 2005 to 2010. Furthermore, Microsoft's Bing (MSN) maintained its third place until 2009, and in 2010 Baidu Search in Taiwan surpassed Bing to become number three. Baidu, the biggest search engine in China, which controls 63% of China's market share, set up a local search service especially for Taiwanese Internet users.[38]

Another three search engines in the top ten were from China, including Soso, Gougou, and Sogou. Soso is mainly produced in a simplified form of Chinese, and is owned by Tencent, which also owns QQ, the most popular instant messenger service in China. Xunlei's Gougou is a search engine that focused on movies, music, games, and software search because its parent company's two other related businesses operate

37. Liu, Y. L. (2008). "Broadband Policy, Market Competition, and User Adoption in Taiwan," in Dwivedi, Y. K., et al. eds. 47–59. *Handbook of Research on Global Diffusion of Broadband Data Transmission*, New York: IGI Global.

38. Barboza, David. "Baidu's Gain from Departure Could Be China's Loss." January 13, 2010. The New York Times. Last accessed on July 17, 2012 at <http://www.nytimes.com/2010/01/14/technology/companies/14baidu.html>.

Table 28-14. Wireline Broadband Subscribers, 2007–2010[1]

| % | 2007 | 2008 | 2009 | 2010 | 2011 | 2012 |
|---|---|---|---|---|---|---|
| HiNet (CHT, 35% public) | 72.7 | 70.7 | 71.4 | 66.2 | 67.6 | 67.9 |
| Sparq (NCIT) | 1.0 | 0.8 | 5.6 | 5.5 | 5.2 | 4.4 |
| TFN | 3.4 | 4.1 | 3.6 | 2.9 | 2.6 | 2.0 |
| APBT | 0.7 | 0.6 | 0.6 | 0.5 | 0.4 | 0.3 |
| Other DSL | 12.1 | 10.8 | 6.8 | 7.8 | 5.6 | 6.0 |
| CNS (cable MSO) | 1.4 | 2.4 | 4.4 | 4.5 | 4.4 | 4.4 |
| Kbro (cable MSO) | 3.3 | 3.6 | 4.2 | 4.6 | 4.9 | 5.3 |
| Taiwan Mobile | 1.3 | 1.8 | 2.3 | 2.6 | 2.8 | 3.1 |
| TBC (cable MSO) | 2.3 | 2.6 | 2.8 | 2.9 | 3.0 | 3.1 |
| Taiwan Optical Platform (cable MSO) | 1.3 | 1.4 | 1.6 | 1.6 | 1.2 | 1.1 |
| Other Cable Operators | 1.1 | 1.4 | 1.1 | 1.4 | 2.2 | 2.3 |
| Total Subscriber | 4,741,231 | 4,968,120 | 4,943,665 | 5,265,026 | 5,464,082 | 5,561,711 |
| Total Revenues (NT mil) | 20,591.47 | 24,975.97 | 24,285.51 | 27,486.4 | | |
| Total Revenues (USD mil) | 686.38 | 832.53 | 809.52 | 916.21 | | |
| *N* (>1%) | 10 | 10 | 10 | 10 | 10 | 10 |
| CR4 | 91.4 | 89.2 | 88.1 | 84.0 | 83.3 | 83.6 |
| HHI | 5,467 | 5,171 | 5,239 | 4,538 | 4,695 | 4,737 |
| Noam Index | 1,933 | 1,829 | 1,852 | 1,605 | 1,660 | 1,675 |

[1] Calculated based on subscriber accounts.

in the areas of video players and download managers. Gougou is a resource index tool, providing search and download links for films, software, songs, books, and other contents. Sogou is a search engine of Sohu, the second largest Chinese portal site.

In terms of total unique visitors, the second biggest search engine, Google Search, has had significant growth in the last six years. The number one search engine, Yahoo-Kimo, remains the same. MSN Search, the third biggest search engine, gained a significant increase from 16.9% to 26.8% mainly between 2007 and 2008.[39]. Nevertheless, when the number of unique visitors of several portal websites such as Yam Search or PChome Online Search began to decline, Q&A search engines (e.g., Sina's iASK) and specialized search engines (e.g., FindPrice) became more popular.

From 2005 to 2010, this industry's HHI was between 2,274 and 2,814. The values indicate that Taiwan's search engine market was deemed to be moderately concentrated (see Table 28.15 for details). According to IAMA, total revenue in the Taiwan Search Ads market reached $2.664 billion NT in 2010, up 432.8% compared to $0.5 billion NT in 2005. The market share of paid search by revenue continued to increase from 16.4% to 31.16% of total online advertising.[40]

39. InsightXplorer (2010). *Access Rating Online, ARO* [Database]. Retrieved from <http://aro.insightxplorer.com/index.jsp>.

40. Internet Advertising and Media Association in Taipei (IAMA). Search Ads included paid search and content match, which is charged based on online users' clicks.

## Table 28-15. Search Engines, 2005–2010

| MARKET SHARES (% OF TOP 10 RANKED SITES' VISITORS) | 2005 | 2006 | 2007 | 2008 | 2009 | 2010 |
|---|---|---|---|---|---|---|
| tw.search.yahoo.com (US) | 38.4 | 45.0 | 42.2 | 40.0 | 38.7 | 39.7 |
| google.com(.tw&hk.cn) (US) | 21.8 | 23.9 | 29.7 | 29.9 | 30.8 | 34.4 |
| baidu.com (China) | 3.4 | 4.1 | 7.0 | 8.6 | 8.6 | 11.5 |
| bing.com/MSN (Microsoft, US)[1] | 15.2 | 9.6 | 8.3 | 12.4 | 11.5 | 5.5 |
| soso.com (China) | | | | | | 2.6 |
| gougou.com (China) | | | | 2.1 | 1.8 | 1.8 |
| sogou.com (China) | 5.9 | | | | | 1.7 |
| findprice.com.tw | | | | | | 1.4 |
| google.pchome.com.tw | | | | 2.3 | 5.3 | 1.3 |
| search.yam.com | 3.9 | 4.2 | 3.3 | 2.2 | 1.6 | |
| iask.com (China) | | | | | 1.0 | |
| ask.com (US) | | | | | 0.8 | |
| information.com (US) | 2.2 | 2.0 | 2.2 | 1.4 | | |
| seekmo.com (US) | | | | 1.2 | | |
| uk.search.yahoo.com (US) | | 2.7 | 2.5 | | | |
| adbaaz.com (US) | | | 2.2 | | | |
| megaclick.com (US) | | | 1.3 | | | |
| openfind.com.tw | 3.4 | 3.9 | 1.2 | | | |
| hisearch.hinet.net | 2.8 | 2.6 | | | | |
| t2t2.com (China) | | 2.0 | | | | |
| infoseek.co.jp (Japan) | 3.1 | | | | | |
| *N* (>1%) | 10 | 10 | 10 | 10 | 10 | 10 |
| C4 | 81.3 | 82.7 | 87.3 | 89.3 | 87.7 | 88.4 |
| HHI | 2,274 | 2,763 | 2,814 | 2,631 | 2,494 | 2,646 |
| Noam Index[2] | 804 | 874 | 890 | 832 | 882 | 837 |

*Source*: Estimates by authors based on original data from InsightXplorer's Access Rating Online (ARO). <http://www.insightxplorer.com/en/product_aro.html>.

[1] MSN data is aggregated from search.msn.com.tw, search.msn.com, search.live.com.

[2] Number of voices for Noam index defined as 10.

## Online News

The online news sector's penetration in Taiwan is about 60% of Taiwan's general Internet market.[41] Five major online news websites are from traditional newspaper groups: the United Daily News (udn.com), Apple Daily (nextmedia.com), China Times, and Liberty Times.[42] Furthermore, NOWnews, previously known as the Eastern Broadcasting Group's ETToday, was sold to Chung Hwa United Group in 2008, and is now operated by its telecom company. In addition, there are a number of news websites with lower reach rates, such as Epoch Times and CNA (Central News Agency).

41. InsightXplorer (2010). "ARO Watch: Overview of Taiwan Online News Usage Overview," accessed December 15, 2011, <http://news.ixresearch.com/?p=1847>.

42. The ranking here excludes news aggregators and television news channel websites, such as tw.news.yahoo.com, news.google.com and news.ftv.com.tw.

As of December 2010, the Epoch Times, which is available in eighteen languages on the Internet, has reached 8.92% of total Internet users in Taiwan.[43] CNA, Taiwan's national press agency, has been in operation longer than any other media outlet in the country. However, it is not the major online news platform in Taiwan, and its low reach rate barely made the top ten in 2006, 2007, 2009, and 2010. All other news providers such as Chinese media companies People.com.cn, China.com.cn, and Xinhuanet.com have reach rates lower than 3%.

According to ARO, NOWnews was ranked as the second most popular online news content provider from 2008 to 2010. This has contributed to the increase in the HHI from 1,689 in 2008 to 1,834 in 2009. The HHI value has been stable ever since. (See Table 28.16 for details.) Newtalk.tw, a new entrant in the top ten online news outlets in 2010, was founded by

## Table 28-16. Online News, 2005–2010

| MARKET SHARES (% TOP 10 RANKED SITES' VISITORS) | 2005 | 2006 | 2007 | 2008 | 2009 | 2010 |
|---|---|---|---|---|---|---|
| udn.com (United Daily News) | 27.5 | 23.7 | 28.8 | 30.4 | 31.3 | 29.1 |
| nownews.com | | | | 14.3 | 19.8 | 20.4 |
| nextmedia.com.tw (Apple Daily)[1] | 8.3 | 9.3 | 13.0 | 13.4 | 12.1 | 16.6 |
| chinatimes.com | 15.6 | 13.6 | 14.8 | 13.0 | 13.2 | 12.4 |
| libertytimes.com.tw | 6.0 | 6.9 | 10.6 | 11.6 | 9.7 | 8.0 |
| epochtimes.com | 7.3 | 9.6 | 8.1 | 7.3 | 5.1 | 5.5 |
| people.com.cn | 2.6 | | 2.3 | | 1.5 | 3.4 |
| cna.com.tw (The Central News Agency) | | 3.7 | 1.9 | | 2.9 | 1.9 |
| newtalk.tw | | | | | | 1.5 |
| china.com.cn | | | | | | 1.3 |
| xinhuanet.com | 3.4 | 7.2 | 3.4 | 3.5 | 3.1 | |
| nextmedia.com.hk | | | | | 1.3 | |
| epochtimes.com.tw | 4.7 | 2.6 | 2.7 | 2.6 | | |
| ettoday.com | 20.4 | 21.4 | 14.3 | 2.3 | | |
| cri.cn (China Radio International) | | | | 1.8 | | |
| huaxia.com | | 2.1 | | | | |
| cdn.com.tw (Central Daily News) | 4.2 | | | | | |
| *N* (>1%) | 10 | 10 | 10 | 10 | 10 | 10 |
| C4 | 71.8 | 68.4 | 71.0 | 71.0 | 76.4 | 78.5 |
| HHI | 1,632 | 1,508 | 1,630 | 1,689 | 1,834 | 1,803 |
| Noam Index[2] | 516 | 477 | 515 | 534 | 580 | 570 |

*Source*: Estimates by authors based on original data from InsightXplorer's Access Rating Online (ARO). <http://www.insightxplorer.com/en/product_aro.html>.

[1] Apple Daily changed its domain name from appledaily.com.tw to 1-apple.com.tw, and then to nextmedia.com.tw.

[2] Number of voices for Noam index defined as 10.

43. InsightXplorer (2010). *Access Rating Online, ARO* [Database]. Retrieved from < http://aro.insightxplorer.com/index.jsp>.

independent journalists and media professionals. It has positioned itself in social business, and it reached 2.35% of Taiwanese Internet users in December 2010.

## CONCLUSION

In Taiwan, newspapers, terrestrial TV, film production, and distribution, wireline, wireless, ISP, and search engine companies are highly concentrated. Satellite TV, music, and online news companies are moderately concentrated. Cable system operators used to be moderately concentrated. After Da-Fu Media merged with Kbro MSO in 2010, the cable TV market has become highly concentrated. Even though magazine, book, and radio companies are not concentrated, there are still big players in each market. International players have important roles in the cable and film industries. Taiwan is very open to international satellite channels. Even though terrestrial TV has to broadcast 70% domestic content, Japanese drama and Korean drama can be shown on prime time.

Before 2003, people in Taiwan were more concerned with the media's political affiliation than with media concentration. It was believed that most of the news media outlets were biased. There were many campaigns to urge the government and political parties to withdraw their influence from the media. The Legislative Yuan revised three broadcasting-related laws and asked the government and political parties to withdraw their shares from the electronic media.

In 2006, people became concerned about the media concentration issue and the foreign ownership issue. In the fixed network and broadband market, CHT is the dominant player. In the mobile phone market, there are three major players: Chunghwa Telecom, Taiwan Mobile, and FarEasTone Telecommunications Co. CHT is still leading but does not significantly surpass its competitors. In the telecommunications industry, the important competition issues include the last mile, local loop unbundling, and the wholesale price. It seems that people do not worry too much about the degree of concentration within the telecommunications industry as much as the issue of media concentration itself.

There are horizontal and vertical integration caps for cable operators. However, the major MSOs can still control the channel line-up and the programming prices. The subscribers are

Table 28-17. Degree of Concentration in Telecommunications and Media Industries

| INDUSTRY | CONCENTRATION DEGREE | NOTE |
|---|---|---|
| Newspaper | Highly concentrated | |
| Magazine | Unconcentrated | |
| Book | Unconcentrated | |
| Terrestrial TV | Highly concentrated | |
| Radio | Unconcentrated | |
| Cable operators | Moderately concentrated | Since November 2010, highly concentrated |
| Satellite TV companies | Moderately concentrated | |
| Film production and distribution | Highly concentrated | |
| Music | Moderately concentrated | |
| Wireline | Highly concentrated | |
| Wireless (mobile) | Highly concentrated | |
| ISP | Highly concentrated | |
| Online news | Moderately concentrated | |
| Search engine | Highly concentrated | |

forced to pay for the channels they do not watch because the system is not tiered. In recent years, there have been more mergers. However, the regulators have been facing a dilemma because there are no clear legal restrictions on cross-media and multimedia ownership. In order to compete with CHT's IPTV service, cable operators lobbied the government to lift the one-third ban in relation to the horizontal integration. CHT also lobbied the government to lift government ownership restrictions.

How should market power be measured, especially in the age of convergence? Should the regulator set a cap for each individual medium or its aggregated market power? How should the magic formula for media concentration be determined?

When the government faces ownership and control issues, it is important to obtain all the data regarding media concentration. However, in Taiwan, if a media company is not listed publicly it usually protects its revenue data and that causes difficulties for the researchers collecting data. If there is no data available for revenues, collected information on advertising dollars, subscription numbers, circulation numbers, or ratings is used as a proxy. Therefore, the authors urge the Taiwan government to collect relevant data and to open up information such as revenues, advertising dollars, subscription numbers, or circulation numbers to the public. This will help anyone interested or concerned to monitor concentration of telecom and media industries in Taiwan (Table 28.17).

## Bibliography

Alejandro, Zentner. "File Sharing and International Sales of Copyrighted Music: An Empirical Analysis with a Panel of Countries." *Topics in Economic Analysis & Policy* 5, Iss1, (2005): Article 21. Last accessed on July 17, 2012 at <http://www.bepress.com/bejeap/topics/vol5/iss1/art21>

Barboza, David. "Baidu's Gain from Departure Could Be China's Loss." January 13, 2010. The New York Times. Last accessed on July 17, 2012 at <http://www.nytimes.com/2010/01/14/technology/companies/14baidu.html>

Chen, Ping-Hung. "Transnational Cable Channels in the Taiwanese Market." Gazette, 66, 2, (2004): 167–183.

Chen, Ping-Hung. "Who Owns Cable Television? Media Ownership Concentration in Taiwan." Journal of Media Economics, 15, 1, (2002): 44–55.

GIO. Taiwan Publishers Survey, 2008.

InsightXplorer(2010). "ARO Watch: Overview of Taiwan Online News Usage Overview," accessed December 15, 2011, <http://news.ixresearch.com/?p=1847>.

Liu, Y. L. (2008). "Broadband Policy, Market Competition, and User Adoption in Taiwan," in Dwivedi, Y. K., et al. eds. 47–59. *Handbook of Research on Global Diffusion of Broadband Data Transmission*, New York: IGI Global.

NCC, "TV digitization and convergence," Internal document delivered to Executive Yuan, 2011.

Peng, Shin-yi, Yuchi Wang, and Yuntsai Chou. *The auction and recall planning and management of spectrum frequencies*. Taipei: NCC research report PG9601–0077, 2007.

Shih, Jun-ji, Kungchung Liu, Shu Yeh, and Yuntsai Chou. *The market survey of cable TV and the competition policy analysis*. Taipei: GIO research report, 2003.

Survey of Taiwan Music Market (2010). Last accessed on July 17, 2012 at <http://info.gio.gov.tw/public/Attachment/05209533371.doc>.

Tsai, Zhi-hong, Ying-chin Shieh, Chinghe Chen, Nian-chong Tsai, Yuntsai Chou, Shin-hong Chen, and Chun Li. *The white paper on digital convergence policy*. Taipei: III research report, 2009.

# Taiwan—Data Summaries

ELI NOAM AND PAUL MUTTER

TAIWAN'S AVERAGE HHI STOOD at 3, 375 after 2008. Taiwan's national power index of 3,408 in 2008 is also moderately high (Table 28.18). There has been a major increase in the number of licensed print and audiovisual outlets since the rescinding of martial law decrees in 1987. But content media are still concentrated due to delayed market entry and frequent changes in ownership among media properties (Table 28.19). Platform media are still more concentrated, given the predominance of Chunghwa Telecom in the wireline, wireless, and ISP markets: the operator holds 34.8% of wireless, 86.2% of wireline, and 70.7% of ISPs, ahead of its primary competitors FarEasTone and Taiwan Mobile, who do not have shares in the non-mobile telecom services (Table 28.20).

Foreign ownership overall is relatively low due to strict limits on overseas investment in Taiwan's national media industries. It is most pronounced in the multichannel platform market, where the Carlyle Group (US), MBK (South Korea), and Macquarie (Australia) are the leading companies.

Among content companies, the China Times News Group is the largest and most diverse, holding 8.4% of the content media (the state holds 9%). It is active in TV broadcasting, video channels, multichannel platforms, daily newspapers, magazines, book publishing, and online news media. Liberty Times News Group, and United Daily News Group both have formidable shares in daily newspapers and online news, but they are about half the size of China Times by shares in their content media (Table 28.19): 5.1% and 3.6%, respectively. US (21st Century Fox, as Star TV) and Hong Kong (TVBS and Next Media) have similar shares of the content market as well, where investment is somewhat more liberalized than in telecommunications.

Table 28-18. National Media Industries Concentration in Taiwan

| | 2004/5 | | 2011 OR MOST RECENT | | % CHANGE ANNUAL AVERAGE | |
|---|---|---|---|---|---|---|
| | COMPANY POWER INDEX IN COUNTRY | COMPANY SHARE OF THE OVERALL NATIONAL MEDIA MARKET (%) | COMPANY POWER INDEX IN COUNTRY | COMPANY SHARE OF THE OVERALL NATIONAL MEDIA MARKET (%) | COMPANY POWER INDEX IN COUNTRY | COMPANY SHARE OF THE OVERALL NATIONAL MEDIA MARKET (%) |
| Government of Taiwan | 2,647 | 41.0 | 2,343 | 38.3 | –1.9 | –0.4 |
| *Chunghwa Telecom (CHT) (35.3% public)* | *2,479.8* | *38.8* | *2,190.0* | *36.4* | *–2.0* | *–0.4* |
| *Formosa TV* | *14.4* | *0.6* | *29.1* | *0.8* | *17.0* | *0.03* |
| *Taiwan TV* | *15.0* | *0.6* | *6.2* | *0.4* | *–9.8* | *–0.04* |
| *BCC* | *7.8* | *0.5* | *7.7* | *0.5* | *–0.3* | *–0.005* |
| FarEasTone (FET) | 508.5 | 15.0 | 312.7 | 11.8 | –6.4 | –0.5 |
| Taiwan Mobile | 427.7 | 15.5 | 390.7 | 15.3 | –1.4 | –0.04 |
| China Times News Group | 62.2 | 2.9 | 25.3 | 1.8 | –9.7 | –0.19 |
| Kbro | 31.0 | 1.4 | 39.5 | 2.1 | 4.6 | 0.1 |
| CNS (South Korea) | 27.8 | 1.3 | 41.9 | 2.2 | 8.4 | 0.15 |
| Liberty Times News Group | 34.9 | 1.6 | 34.0 | 1.1 | –0.42 | –0.07 |
| Next Media (Hong Kong) | 12.1 | 1.0 | 24.7 | 1.1 | 17.3 | 0.01 |
| United Daily News Group | 72.4 | 2.2 | 16.4 | 0.8 | –12.9 | –0.24 |
| ET | 23.1 | 1.2 | 16.0 | 1.0 | –5.5 | –0.04 |
| TBC (Australia) | 10.0 | 0.8 | 20.0 | 1.5 | 16.5 | 0.1 |
| Yahoo! (US) | 0.9 | 0.03 | 9.1 | 0.2 | 144.7 | 0.03 |
| TVBS (TVB, HK) | 13.9 | 0.9 | 7.7 | 0.7 | –7.4 | –0.04 |
| Murdoch Group (21st Century Fox, US) | 6.9 | 0.7 | 8.2 | 0.7 | 3.2 | 0.01 |
| Google (US) | 0.3 | 0.01 | 7.4 | 0.21 | 389.8 | 0.03 |
| Cité Media Holding Group (Hong Kong) | 4.6 | 0.49 | 4.7 | 0.48 | 0.14 | –0.004 |
| Time Warner (US) | 1.8 | 0.10 | 0.8 | 0.07 | –8.6 | –0.005 |

| | | | | | | |
|---|---|---|---|---|---|---|
| Disney (US) | 4.7 | 0.2 | 1.7 | 0.1 | –10.7 | –0.01 |
| Microsoft (Bing, US) | 0.1 | 0.01 | 0.2 | 0.03 | 3.2 | 0.004 |
| Comcast (US) | 0.5 | 0.05 | 1.0 | 0.08 | 0.2 | 0.00 |

| Media Concentration Index | 2004/5 | 2011 or Most Recent | % Change Annual Average |
|---|---|---|---|
| Total Revenue: Nat'l Media Industry (mil US$) | 14,079 | 14,677 | 0.7% |
| Total Voices (*n*) | 54 | 60 | 1.9% |
| Net Voices (*n*) | 43 | 49 | 2.3% |
| Public Ownership (%) | 41.0 | 38.4 | –0.4% |
| Foreign Ownership (%) | 6.7 | 6.4 | –0.05% |
| C4 Average—Weighted | 86.3 | 84.9 | –0.23% |
| HHI Average—Weighted | 3,537 | 3,375 | –0.76% |
| C1 Average—Weighted | 46 | 49 | 0.4% |
| Noam Index Average—Weighted | 1,060 | 980 | –1.3% |
| Pooled Overall Sector C4 | 74.4 | 71.6 | –0.5% |
| Pooled Overall Sector HHI | 2,217 | 1,904 | –2.1% |
| Pooled Overall Sector Noam Index | 63 | 41 | –6.0% |
| Market Share of Top Ten Companies: Nat'l. Media Industry (%) (Pooled C10) | 83.3 | 81.4 | –0.3% |
| National Power Index | 3,929 | 3,408 | –2.2% |

Table 28-19. Top Content Media Companies in Taiwan

| | 2004/5 | | 2011 OR MOST RECENT | | % CHANGE ANNUAL AVERAGE | |
|---|---|---|---|---|---|---|
| | COMPANY POWER INDEX IN COUNTRY | COMPANY SHARE OF THE NATIONAL CONTENT MEDIA MARKET (%) | COMPANY POWER INDEX IN COUNTRY | COMPANY SHARE OF THE NATIONAL CONTENT MEDIA MARKET (%) | COMPANY POWER INDEX IN COUNTRY | COMPANY SHARE OF THE NATIONAL CONTENT MEDIA MARKET (%) |
| Government of Taiwan | 684 | 9.1 | 706 | 9.0 | 0.5 | −0.003 |
| China Times News Group | 246.5 | 12.0 | 116.8 | 8.4 | −8.8 | −0.6 |
| Next Media (Hong Kong) | 49.6 | 4.2 | 113.8 | 5.1 | 21.6 | 0.15 |
| Liberty Times News Group | 142.7 | 6.3 | 156.9 | 5.1 | 1.65 | −0.21 |
| United Daily News Group | 296 | 9.2 | 75.5 | 3.6 | −12.4 | −0.9 |
| ET | 94.6 | 4.8 | 71.6 | 4.4 | −4.1 | −0.07 |
| TVBS(TVB, Hong Kong) | 56.9 | 3.7 | 35.6 | 3.1 | −6.2 | −0.1 |
| Murdoch Group (21st Century Fox, US) | 28 | 2.8 | 38 | 3.3 | 0.06 | 0.1 |
| Cite Media Holding Group(HK) | 19.0 | 2.0 | 21.6 | 2.2 | 2.3 | 0.03 |
| Best | 2.2 | 0.3 | 3.9 | 0.5 | 18.2 | 0.05 |
| UFO | 3.0 | 0.3 | 0.3 | 0.1 | −0.15 | −0.03 |
| Voice of Taipei | 0.8 | 0.16 | 0.3 | 0.1 | −0.1 | −0.01 |
| Baidu (China) | 0.03 | 0.01 | 3.5 | 0.3 | 19.3 | 0.05 |

| Media Concentration Index | 2004/5 | 2011 or Most Recent | % Change Annual Average |
|---|---|---|---|
| Public Ownership (%) | 3.20 | 3.19 | 0.0% |
| Foreign Ownership (%) | 19.6 | 22.2 | 0.5% |
| C4 Average—Weighted | 64 | 60 | −0.7 |
| HHI Average—Weighted | 1,930 | 1,830 | −0.9 |
| C1 Average—Weighted | 27.6 | 25.6 | −0.3% |
| National Power Index | 1,864 | 1,789 | −0.7 |

Table 28-20. Top Platform Media Companies in Taiwan

| ARTICLE II. | 2004/2005 | | 2011 OR MOST RECENT | | % CHANGE ANNUAL AVERAGE | |
|---|---|---|---|---|---|---|
| | COMPANY POWER INDEX IN COUNTRY | COMPANY SHARE OF THE NATIONAL PLATFORM MEDIA MARKET (%) | COMPANY POWER INDEX IN COUNTRY | COMPANY SHARE OF THE NATIONAL PLATFORM MEDIA MARKET (%) | COMPANY POWER INDEX IN COUNTRY | COMPANY SHARE OF THE NATIONAL PLATFORM MEDIA MARKET (%) |
| STATE | 3,282 | 51.3 | 2,796 | 46.5 | -2.5 | -0.8 |
| FarEasTone | 673.0 | 19.8 | 399.0 | 15.0 | -6.8 | -0.8 |
| Taiwan Mobile | 563.6 | 20.3 | 495.0 | 19.1 | -2.0 | -0.2 |
| Media Concentration Index | 2004/5 | | 2011 or Most Recent | | % Change Annual Average | |
| Public Ownership (%) | 18.1 | | 16.4 | | -0.25% | |
| Foreign Ownership (%) | 2.5 | | 2.0 | | -0.1% | |
| C4 Average—Weighted | 93.4 | | 91.8 | | -0.3% | |
| HHI Average—Weighted | 4056 | | 3,802 | | -1% | |
| C1 Average—Weighted | 52.5 | | 48.1 | | -0.7% | |
| National Power Index | 4,597 | | 3,856 | | -2.7% | |

# E

## Middle East and Africa

# 29

# Media Ownership and Concentration in Egypt

NAGLA RIZK

## INTRODUCTION

The purpose of this chapter is to shed light on media concentration and diversity in Egypt.[1,2] Of particular interest is the interplay between economic and political liberalization, and their impact on the concentration and diversity of the various media industries. The study also examines the impact of media concentration practices on access to content, tools, and platforms.

This chapter covers 10 media industries. Since published data on this subject are not readily available, the figures were collected expressly for this study. The chapter includes an analysis of quantitative and qualitative data gathered through interviews with key stakeholders, industry reports, and marketing surveys by private research agencies. In Egypt, there is a legacy of scarce statistics that reflect media market dynamics, market shares, distribution, and advertising. This is coupled with the lack of transparency of ownership dynamics. This scarcity of information poses a serious research challenge and the lack of transparency and openness raises questions about the control and regulation of media in Egypt.

The chapter starts by introducing the main issues surrounding media concentration in Egypt. This is followed by in-depth coverage of the 10 media industries: print (daily newspapers), audiovisual media (broadcast television, radio, multichannel platforms cable and satellite, and video

1. Research for this chapter was undertaken during the period July 2011-July 2012. As such, the analysis covers the period following the Mubarak regime up to the rule of the Supreme Council of Armed Forces. The chapter does not cover the period under the presidency of Mohammed Mursi or Abdel Fattah El Sisi, save for brief mention of the 2014 Constitution.

2. I would like to thank Lina Attalah and Mariam El Abd for their invaluable contribution to this research. I am also grateful to the research support of Nagham El Houssamy, Youssef El Shazli and Nadine Weheba. Thanks are also due to all our interviewees. I am particularly indebted to Hisham Kassem for sharing his knowledge and for helping with fact-checking this work.

channels), telecommunications distribution media (wireline and wireless), and online media (Internet service providers, search engines, and to a lesser extent, online media). Sectors are covered in varying degrees of detail depending on data availability.

Media in Egypt has been the subject of the interplay of political and economic powers. Since the military takeover in 1952, Egypt has had a history of controlling and manipulating the public sphere through government-owned and government-controlled media. Gamal Abdel Nasser, president from 1954 to 1970, completely nationalized the media. His successor, Anwar Sadat, president from 1970 to 1981, maintained a tight grip on the media, though he did allow political parties to establish newspapers to promote their ideologies and candidates in 1978.

President Hosni Mubarak, who served from 1981 to 2011, oversaw the liberalization of the media in the 2000s as a product of the liberal economic policies of the 1990s that were created by the Economic Reform and Structural Adjustment Program (ERSAP). This program integrated Egypt into the global economy. Foreign and local pressures were placed on Mubarak to execute political reform. This pressure caused some loosening of the restrictions on media, although the regime's focus was largely on economic growth as "a precondition to democratic change."[3]

Chronologically, media privatization was led by the telecommunications sector in line with the development of Egypt's data backbone. Private participation in telecommunications can be traced back to 1998 with the entry of cellular operators Mobinil and Vodafone.[4] Audiovisual and print media broke away from state monopoly status soon after in 2001. Dream TV became the first private satellite broadcast.[5] Subsequently, the radio industry witnessed its first privately owned station, Nile Radio Production, in 2003.[6] As for daily newspapers, the founding of *Al Masry Al Youm* in 2004 marks the return of the first privately owned and licensed Egyptian newspaper since Nasser's complete nationalization of the press in 1961.[7]

While Mubarak's policies of the early 2000s eased the political barriers to entry and ended state monopoly over most media sectors, they also introduced a new form of concentration. Given the high cost of initial operation, media ownership was limited to a small number of elite businessmen who could afford to invest in such financially burdensome ventures. Therefore, a segment of the media became owned by a small number of "anchor investors"[8] who have a vested interest in the media because they can exert their influence in a way that favors the political environment necessary for their businesses.[9]

While the privatization of the media meant freedom from the state's monopoly over content, freedom of expression remained largely limited by an arsenal of laws. Despite the fact that Article 48 of the 1971 Constitution, now abrogated, guaranteed "freedom of the press, printing, publication, and mass media," the regime used a host of constitutional clauses, laws, and regulations to wield an iron fist over the media. Chapter 3 in Part V of the Constitution granted the president executive powers to issue or stifle legislative laws and to generally rule by decree.[10] Article 48

---

3. Sharp, Jeremy M. "Egypt, Congress and U.S. Relations." Congressional Research Service. Order Code RL33003. Aug. 12, 2008. Last accessed on June 27, 2012 at <http://fpc.state.gov/documents/organization/109518.pdf>.

4. National Telecom Regulatory Authority (NTRA). "Consumer Protection: 'Know Your Rights' Awareness Campaign." National Telecom Regulatory Authority. Apr. 20, 2010. Last accessed on June 27, 2012 at <http://www.tra.gov.eg/english/dpages_dpagesdetails.asp?ID=469>.

5. Sakr, Naomi. *Arab Television Today*. New York: I. B. Tauris, 2007.

6. Good News Group. "Good News for Radio and TV." Good News Group. 2006. Last accessed on Jan. 16, 2012 at <http://www.goodnewsgroup.com/gng9.swf>.

7. Al Masry Al Youm. "About Al Masry Al Youm." Al Masry Al Youm. Last accessed on Jan. 16, 2012 at <http://www.almasryalyoum.com/>.

8. Attalah, Lina and Rizk, Nagla. "Egypt's Evolving Media Landscape: Access, Public Interest and Control," in *A New Frontier, An Old Landscape*. Report for the Ford Foundation, November 2011, 118–136. Last accessed on January 16, 2012 at <http://participatorymedia.lab.asu.edu/files/NewFrontierOldLandscape.pdf>.

9. Ibid.

10. "Constitutional Declaration." Cabinet. Last accessed on Mar. 7, 2012 at <http://www.cabinet.gov.eg/AboutEgypt/Egyptian_constitution.aspx>.

of the 1971 Constitution dictated that freedom of the media "shall be guaranteed. . . . However, in case of a declared state of emergency," censorship may be imposed "in matters related to public safety or for purposes of national security."[11] The spirit of media control continued to prevail throughout different constitutional documents that were passed in Egypt since 2011, including the recently passed Constitution in 2014. This constitution includes some relative guarantees for freedom of expression such as the protection of freedom of information and media freedoms and independence, that have not yet translated into an opening up of the media ecosystem.[12]

Additionally, Article 178 of the Penal Code confers authorities the right to imprison and fine anyone who distributes or publishes photos, manuscripts, or any form of material "violating public morals."[13] Indeed, Article 179 of the Penal Code calls for imprisonment for the defamation of the president of the republic.[14] Furthermore, Article 80 describes the fines that can be imposed upon "whoever carries out an activity which influences national interests of the country."[15] More informally, Mubarak's regime continuously interrogated, detained, fined, and even jailed journalists and bloggers. Control of content continued through pressure on TV anchors and newspaper editors.

It is therefore not surprising that Egypt continued to score poorly on the press freedom index of Reporters Without Borders (*Reporters Sans Frontières*, RSF) and has been described as a "difficult situation" at least since 2007.[16] While scores improved slightly between 2007 and 2010, Egypt's performance deteriorated and plummeted after 2011.

As SCAF led the post-Mubarak Egypt in transition, political repression continued. Journalists were attacked, a number of television licenses were revoked,[17] and digital activists were subjected to military trials. Noteworthy here is Law 313 of 1956 amended as Law 14 of 1967, as it prohibits publication and broadcast of information about the Armed Forces, including public knowledge about the military, without prior consent from the director of military intelligence.[18] While SCAF pledged to allow any new station to operate, it repealed this policy by closing Al Jazeera Mubasher Misr and sending warning letters to stations threatening to revoke licenses for not adhering to content as stated in its licenses.[19]

With continuing political pressure on the media before the revolution and during the post-revolution interim military rule, the broad use of this new technology by activists provided alternative outlets for expression. During the Mubarak era, improvements in the country's data backbone helped to substantially increase Internet usage. In 2003, online forums began to emerge as a channel of open expression and public organization. They were replaced in 2006 by social media channels such as blogs, Facebook, and Twitter that allowed great participation in the public discourse and proved to be efficient methods for organizing and mobilizing the public.

---

11. "Constitutional Declaration." Cabinet. Last accessed on Mar. 7, 2012 at <http://www.cabinet.gov.eg/AboutEgypt/Egyptian_constitution.aspx>.

12. Reporters Without Borders. "Country File- Egypt." March 2014. Last accessed on 26 Jan. 2015, at <http://en.rsf.org/report-egypt,149.html>.

13. Penal Code No. 58 of 1937 [Egypt]. August 1937. Last accessed on Mar. 7, 2012 at <http://www.unhcr.org/refworld/docid/3f827fc44.html>.

14. "Egypt: Article 19 and the Justice Initiative Submission to the UN Universal Periodic Review" in the Seventh Session of the UPR Working Group. 2010. Last accessed on Jan. 16, 2012 at <http://www.article19.org/data/files/pdfs/submissions/egypt-article-19-and-the-justice-initiative-submission-to-the-un-universAl p.pdf>.

15. "Urgent Communication to the Egyptian Public Prosecutor a Call for Arresting the Egyptian Minister of Mass Media and Information & Prosecuting Him for Stimulating Hatred and Murder." *The Arab Center for the Independence of the Judiciary and the Legal Profession (ACIJLP)*. 25 Feb. 2011. Last accessed on Jan. 16, 2012 at <http://www.acijlp.org/?browser=view&ID=24'>.

16. "Press Freedom Index%20 Images%20 of%20 Press%20 Freedom%20 Index%20 2007–2012." Explow News. Last accessed on June 4, 2012 at <http://explow.com/Press_Freedom_Index images of press freedom index 2007–2012>.

17. "Egypt Authorities Freeze New Satellite TV Station Licenses." *X Index*. Sept. 9, 2011. Last accessed on Jan. 16, 2012 at <http://www.indexoncensorship.org/2011/09/egypt-authorities-freeze-new-satellite-tv-station-licenses/>.

18. "Presidential Decree No. 313/1956, Amended by Law 14/1967." *United Group: Attorneys at Law, Legal Researcher and Human Rights Advocates*. Last accessed on Jan. 16, 2012 at <http://ug-law.com/index.php?option=com_content&view=article&id=424%3ApresidentiAl decree-no-3131956-amended-by-law-no-141967-&catid=64%3Amedia-laws&Itemid=89&lang=en>.

19. Kassem, Hisham. Personal interview. Feb. 21, 2012.

Bloggers and social media sites became an increasingly important source of alternative news and views as online audiences continued to grow. In response, mainstream media, including the national press, continued to integrate technologies into its expanding platforms.

Crowdsourced news and political discourse using digital platforms such as Twitter and Facebook gained ground over traditional news media.[20] These online communities created a more interactive audience through producing consumer-generated diversified content.

In March 2012, 11.86% of the Egyptian population (9.9 million people) had Facebook accounts, twice as many people who had accounts in December 2010.[21] By June 2012, that number reached 11.34 million people, 13.56% of the population.[22] Between January and March 2011, the number of active Twitter users in Egypt reached 131,204.[23]

Although the government led developments in Egypt's Internet backbone with the objective of realizing economic gains from technological improvement, it ended up giving rise to citizen media by providing a vent for tabooed political criticism.[24] This is an example of the disconnect between the economic and political agendas typical of Egypt's regime. In this case, they strengthened the technological backbone to promote investments and growth under the economic freedoms agenda, while simultaneously restricting online freedoms of political expression that flowed through these technological pipes. Lifting technological and economic barriers to entry was sufficient to trigger political dissent, even in the presence of the political repression of activists.

In addition to the above, understanding Egypt's media landscape necessitates a consideration of distinctive audience habits of the Egyptian population. As of 2008, 22% of Egypt's population lived below the national poverty line.[25] As of 2006, only 66% were literate.[26] In a country of 84 million people,[27] widespread poverty and illiteracy and the intricate relationship between these two characteristics strongly shape media consumption both in choice of medium and pattern of consumption. Low literacy rates mean consumption is most popular in its visual and audio forms, television and radio. Because of the increased interactivity available on the Internet, online usage is rapidly increasing, even among illiterates.[28]

Also, shared access is an interesting development in the media landscape of Egypt. This ranges from residents of apartment buildings sharing access to an ADSL line, to using cheap satellite receivers to view satellite channels without subscriptions. Also, in less affluent areas, it is a cultural norm and a common economic activity for locals to gather at cafes to watch satellite TV with friends and family.[29] This is done

20. Attalah, Lina and Rizk, Nagla. "Egypt's Evolving Media Landscape: Access, Public Interest and Control," in A New Frontier, An Old Landscape. Report for the Ford Foundation, November 2011, 118–136. Last accessed on January 16, 2012 at <http://participatorymedia.lab.asu.edu/files/NewFrontierOldLandscape.pdf>.

21. "Facebook Usage Factors and Analysis," in Arab Social Media Report 1, no. 1. Dubai School of Government, 2011, 6. Last accessed on Mar. 5, 2012 at <http://unpan1.un.org/intradoc/groups/public/documents/dsg/unpan044212.pdf.p.6>.

22. "Egypt Facebook Statistics." *Social Bakers*. Last accessed on Mar. 5, 2012 at <http://www.socialbakers.com/facebook-statistics/egypt>.

23. "Civil movements: The impact of Facebook and Twitter," in *Arab Social Media Report* (Vol 1, No. 2). Dubai School of Government, 2011, 16. Last accessed on Mar. 5, 2012 at <http://www.dsg.ae/en/Publication/Pdf_En/DSG_Arab_Social_Media_Report_No_2.pdf. 16>.

24. Attalah, Lina and Rizk, Nagla. "Egypt's Evolving Media Landscape: Access, Public Interest and Control," in A New Frontier, An Old Landscape. Report for the Ford Foundation, November 2011, 118–136. Last accessed on January 16, 2012 at <http://participatorymedia.lab.asu.edu/files/NewFrontierOldLandscape.pdf>.

25. World Bank. "Egypt, Arab Rep." World Bank Development Indicators. Last accessed on Dec. 28, 2011 at <http://data.worldbank.org/country/egypt-arab-republic>.

26. Egypt: The National Report on Literacy and Adult Education. United Nations Educational, Scientific and Cultural Organization, 2009. Last accessed on Dec. 19, 2011 at <www.unesco.org/fileadmin/MULTIMEDIA/.../UIL/.../Egypt.pdf>.

27. "Egypt." The World Factbook. Last accessed on Mar. 7, 2012 at <https://www.cia.gov/library/publications/the-world-factbook/geos/eg.html>.

28. "Media Usage." Insights MENA. 2010. Last accessed on Jan. 16, 2012 at <http://www.insightsmena.com/en/#!place=category&cat=Media+Usage>.

29. Howard, P.N. "Ownership Diversity in Muslim Media Systems." World Information Access Project (WIA). Oct. 6, 2008. Last accessed on Jan. 16, 2012 at <http://www.wiaproject.org/index.php/68/ownership-diveristy-in-muslim-media-systems>.

to reduce costs, but is also an outcome of a culture that favors collective and informal communal activities.

Finally, vertical integration in Egypt is a privilege exercised by the state. The government monopolizes infrastructure and private companies are either politically or economically prohibited from entering the market. The infrastructural aspects of telecom, terrestrial broadcast, and print media are examples of this imbalance.

## PRINT MEDIA

### Newspapers

#### INFRASTRUCTURE AND REGULATION

The constitution in Egypt allows newspapers to be owned by public or private legal entities and political parties.[30] The press functions under the auspices of the Ministry of Information,[31] where print media operates under the Press Authority Law no. 96 of 1996 and the Publications Law no. 20 of 1936. In compliance with these laws, a publishing license can only be obtained through notification to the Supreme Press Council,[32] subject to the latter's discretion to refuse publishing. Newspapers can only be licensed in the form of joint-stock companies after incurring substantial minimum capitalization fees.[33] Press legislation limits ownership by a single individual to a maximum of a 10% share in any newspaper.[34] However, this law was applied unevenly and those who wished to circumvent it simply appropriated shares to their families and relatives.[35]

Generally, the Supreme Press Council is regarded as the main vehicle for government influence in the sector. It is important to note that the Shura Council (the upper house of parliament) legally owns national press outlets.[36] In commonly fraudulent elections prior to the revolution on January 25, 2011, the formerly ruling National Democratic Party used to win majority seats in the Shura Council. The head of the Shura Council has also been the de facto head of the Supreme Press Council, thereby exposing the latter to significant government control at a time when election results were highly controlled by the ruling regime.[37] Moreover, the Supreme Press Council retains the executive rights to issue licenses to private newspapers.[38] It is yet to be seen how this process will be affected by the 2014 Constitution which rules the setting up of newspaper only by notification and not by authorization.[39]

In order to circumvent opaque licensing criteria, expensive start-up costs, and restrictions on content, over 200 Egyptian-owned publications (including newspapers) registered their titles abroad[40] in countries such as

30. Rugh, William. A. *Arab Mass Media: Newspapers, Radio & Television in Arab Politics*. Westport, CT: Praeger, 2004. 122.

31. Sakr, Naomi. *Satellite Realms: Transnational Television, Globalization and The Middle East*. New York: I. B. Tauris, 2001. 30.

32. "Egypt: Article 19 and the Justice Initiative Submission to the UN Universal Periodic Review" in the Seventh Session of the UPR Working Group. 2010. Last accessed on Jan. 16, 2012 at <http://www.article19.org/data/files/pdfs/submissions/egypt-article-19-and-the-justice-initiative-submission-to-the-un-universAl p.pdf>.

33. Ibid.

34. Bureau of Democracy, Human Rights and Labor. 2010 Human Rights Report: Egypt. U.S. Department of State, 2011. Last accessed on June 27, 2012 at <http://www.state.gov/documents/organization/160456.pdf>.

35. Kassem, Hisham. Personal interview. Feb. 21, 2012.

36. Black, Jeffrey. "Egypt Press: More Free, Still Fettered." Arab Media and Society. 2008. Last accessed on Jan. 16, 2012 at <http://www.arabmediasociety.com/?article=572>.

37. "Egypt: Article 19 and the Justice Initiative Submission to the UN Universal Periodic Review" in the Seventh Session of the UPR Working Group. 2010. Last accessed on Jan. 16, 2012 at <http://www.article19.org/data/files/pdfs/submissions/egypt-article-19-and-the-justice-initiative-submission-to-the-un-universAl p.pdf>.

38. Black, Jeffrey. "Egypt Press: More Free, Still Fettered." Arab Media and Society. 2008. Last accessed on Jan. 16, 2012 at <http://www.arabmediasociety.com/?article=572>.

39. Arab Republic of Egypt. "Constitution of the Arab Republic of Egypt." 18 January 2014. Last accessed on 26 Jan. 2015, at <http://www.sis.gov.eg/Newvr/Dustor-en001.pdf>.

40. "News Media in Egypt." Press Reference. Last accessed on Jan. 8, 2012 at <http://www.pressreference.com/Co-Fa/Egypt.html>.

Cyprus. However, these publications printed their publications locally.[41] Shortly after, a new bureau was established at the Ministry of Information called "censorship on foreign publications" to block this loophole created by independent publishers. The justification for creating this bureau was to protect national security and the morality of the state from foreign conspiracies. In 2001, most licenses were revoked in Cyprus since most publishers did not pay taxes, nor submit annual legal paperwork there. Very few investors now resort to foreign licenses.[42] More private ventures around print media with local licenses emerged after *Al Masry Al Youm* daily was established in 2003, supported by the capital of powerful businessmen.

State-run distribution firms distribute all privately owned newspapers, including independent or partisan titles. *Al Ahram* and *Al Akhbar* are the largest two printing and distribution houses in Egypt. *Al Ahram* has an estimated 7,500 distribution points within the nation.[43] Given the private sector's point of view, securing a large fleet that can cater to Egypt's national needs[44]could result in higher costs as the sheer size of *Al Ahram* allows it to benefit from economies of scale in distribution.[45] A distribution arm may be set up at the rough cost of LE12 million (US$2 million), half of which would cover equipment and vehicles. The rest would cover operational costs. Private media operations have not shown interest in investing in this activity.[46]

The state also monopolizes the print industry for newspapers and newsprint.[47] With the exception of *Al Masry Al Youm*, Egyptian newspapers use state-owned print houses for both printing and distribution.[48] The government uses this monopoly in printing and distribution to influence market competition, and as a vehicle for censorship because newspapers' printing and distribution could be delayed if the content is deemed controversial.[49]

## OWNERSHIP

The government owns controlling stakes in *Al Ahram* (established in 1875),[50] *Al Akhbar* (established in 1952),[51] and *Al Gomhouriya* (established in 1953).[52] As national press organizations, the law sets ownership dynamics whereby the Shura Council holds 51% of establishment capital and affiliates, with the remaining 49% held by the establishment employees. Accordingly, 50% of the net profits are to be allocated to establishment employees, while the other 50% is to be allocated for "expansion and renovation projects."[53] Indeed, Article 22 of the Law on Press Authority states, "National press establishments shall be considered the property of the state and the Shura Council shall exercise ownership rights over national press establishments."[54] Furthermore, past editors-in-chief

41. "Egypt: Article 19 and the Justice Initiative Submission to the UN Universal Periodic Review" in the Seventh Session of the UPR Working Group. 2010. Last accessed on Jan. 16, 2012 at <http://www.article19.org/data/files/pdfs/submissions/egypt-article-19-and-the-justice-initiative-submission-to-the-un-universAl p.pdf>.

42. Kassem, Hisham. Personal interview. Feb. 21, 2012.

43. El Hefnawy, Magdi. Personal interview. June 30, 2011.

44. Ibid.

45. Ibid.

46. Kassem, Hisham. Personal interview. Feb. 21, 2012.

47. Rugh, William. A. *Arab Mass Media: Newspapers, Radio & Television in Arab Politics.* Westport, CT: Praeger, 2004.

48. El Galad, Magdy. Interview. Al Nas TV. Last accessed on Jan. 16, 2012 at <http://www.youtube.com/watch?v=TSnLjXi2rdI>.

49. The Cairo Institute for Human Rights Studies. *Media in North Africa: Obstacles and challenges.* United Nations Economic Commission for Africa, 2007, 4. Last accessed on Jan. 16, 2012 at <http://www.uneca.org/africanmedia/documents/NorthAfrica-eng-report.pdf. 4>.

50. Al Ahram. "About Us." Al Ahram. Last accessed on May 29, 2012 at <http://english.ahram.org.eg/UI/Front/Aboutus.aspx>.

51. Ibid.

52. Al Gomhouria Newspaper, Last accessed on July 5, 2012 at <http://www.algomhuria.net.eg/algomhuria/today/pdf/>.

53. "Freedom of Expression Project: The Press in Egyptian Legislation" in Middle East and Islamic Studies Collection. Cornell University Library. Last accessed on Jan. 16, 2012 at <http://www.library.cornell.edu/colldev/mideast/egpress.htm>.

54. Ibid.

of all three papers have been appointees by the president[55] and were commonly selected from members of the formerly ruling National Democratic Party.[56] Since these newspapers have been run by a succession of sycophantic managers who steered clear of red lines surrounding political criticism, this administrational appointment system provided for self-censorship.[57] After the revolution on January 25, 2011, the Shura Council set a series of guidelines for the positions of editors-in-chief, while it opened a nomination door for candidates.

*Al Ahram* is a giant in Egypt's print industry. Apart from issuing its daily newspaper, its flagship product, the company is the largest printing, publishing, and distribution house in Egypt.[58] *Al Ahram* officially quotes its daily circulation at an estimated 750,000 in 2009, claiming approximately 80% of daily newspaper circulation in Egypt and declaring itself one of the most widely circulated in the Arab world.[59, 60] This is a political overstatement and comes in sharp contrast to available statistics. Our own estimate of the market share of *Al Ahram* in 2009 is 29%.[61] *Al Ahram* issues other publications and magazines.[62] According to the officially quoted figures and available statistics in 2011, *Al Akhbar* ranked second in the market and *Al Gomhouria* third,[63] both of which follow the same structure and operate in the same industries as *Al Ahram*, but on a relatively smaller scale.[64]

An alternative voice to the national press existed in the form of partisan newspapers, funded by the individual parties[65] and government subsidies.[66] The press law allowed political parties to establish their own papers that were generally referred to as opposition papers.[67] Due to their reliance on government printing facilities and subsidies, as well as their need for government approval to legally establish a political party, news coverage by partisan papers tends to adopt a cautious attitude toward political criticism.[68] Moreover, often co-opted party leaders themselves assume the papers' editorial.[69] An example is *Al Wafd*, established in 1984 by the economically and politically liberal *Wafd* Party,[70] and managed by the chairman of the Wafd Party, Al-Sayyid Al-Badawi.[71] Of all partisan newspapers, *Al Wafd* enjoyed the largest circulation,[72] estimated at 60,000 in 2003.[73]

---

55. Karlekar, Karen D. and Cook, Sarah G., ed. *Freedom of the Press 2008: A Global Survey of Media Independence.* New York: Rowman and Littlefield Publishers, Inc. 2008, 121

56. Amin, Hussein. *Strengthening the Rule of Law and Integrity in the Arab World: Report on the State of the Media in Egypt.* The Arab Center for the Development of the Rule of Law and Integrity, 2009, 6. Last accessed on June 10, 2012 at <http://www.arabruleoflaw.org/Files/PDF/Media/English/P2/Egypt_MediaReportP2_En.pdf. 6>.

57. Ibid.

58. El Hefnawy, Magdi. Personal interview. June 30, 2011.

59. Amin, Hussein. *Strengthening the Rule of Law and Integrity in the Arab World: Report on the State of the Media in Egypt.* The Arab Center for the Development of the Rule of Law and Integrity, 2009, 6. Last accessed on June 10, 2012 at <http://www.arabruleoflaw.org/Files/PDF/Media/English/P2/Egypt_MediaReportP2_En.pdfp.6>.

60. A view from the field offers a different perspective: publisher Hisham Kassem argues that by 2009 Al Ahram were distributing only 400,000 closer to 40%, and at present revolve round 200,000 (Kassem, Hisham. Personal interview. Feb. 21, 2012).

61. See table for details. Source: Newspaper readership study 2002–2010. Pan Arab Research Center (PARC), 2011.

62. "Al Ahram Establishment Websites." Al Ahram International North America Inc. Limited. Last accessed on Jan. 16, 2012 at <http://aina.ahram.org.eg/Ahram USA%5COnline Ads%5CAl-Ahram Websites.pdf>.

63. El Hefnawy, Magdi. Personal interview. June 30, 2011.

64. Ibid.

65. Allam, Rasha. "Media Landscape: Egypt." European Journalism Centre. Nov. 8, 2010. Last accessed on Dec. 29, 2011 at <http://www.ejc.net/media_landscape/article/egypt/>.

66. Rugh, William. A. *Arab Mass Media: Newspapers, Radio & Television in Arab Politics.* Westport, CT: Praeger, 2004. 122.

67. Ibid.

68. Rugh, William. A. *Arab Mass Media: Newspapers, Radio & Television in Arab Politics.* Westport, CT: Praeger, 2004. 122. 127

69. Kassem, Hisham. Personal interview. Feb. 21, 2012.

70. Al Wafd. Official Website. Last accessed on June 27, 2012 at <http://www.alwafd.org/digital/>

71. Ibid.

72. Allam, Rasha. "Media Landscape: Egypt." European Journalism Centre. Nov. 8, 2010. Last accessed on Dec. 29, 2011 at <http://www.ejc.net/media_landscape/article/egypt/>.

73. Rugh, William. A. *Arab Mass Media: Newspapers, Radio & Television in Arab Politics.* Westport, CT: Praeger, 2004.

The first private Egyptian-owned newspaper to be registered locally was *Al Masry Al Youm. Al Masry Al Youm* was founded in 2003[74] by Salah Diab, a prominent member of *Al Wafd Party*.[75] Controlling stakes in the paper are held by a group of businessmen including Diab, Ahmed Bahgat (owner of satellite channel Dream TV), and Naguib Sawiris (owner of majority stakes in Orascom Telecom).[76] It is the only newspaper in Egypt with its own independent print firm.[77] *Al Masry Al Youm* enjoys the highest circulation figures among privately owned newspapers,[78] and many consider it a rival to the dominant state press. Indeed, its circulation figures peaked at 200,000 in 2011[79] since its first issue in 2004,[80, 81] but these figures are generally in decline. This drop is partially due to expanding competition.

Until 2004, when the first license for a privately owned Egyptian daily for *Al Masry Al Youm* was issued, private sector aspirations exploited narrow leeway to get their newspapers in print. Privately owned Egyptian newspapers licensed their titles abroad, but operated within Egypt. Among the first newspapers to acquire an offshore private license was *Al Alam Al Youm*, founded in 1991 and headquartered in London.[82] Printing business and financial news, the newspaper gained popularity with Egypt's business community. Good News Group, a media and entertainment company, owns *Al Alam Al Youm*.[83]

Following the same model, *Al Dostour* was founded in 1995 and registered its permission to publish in Cyprus.[84] It began to locally print and distribute its newspaper using government facilities.[85] *Al Dostour* began as a weekly paper and later became a daily. In 1998, circulation for the paper was estimated at 50,000.[86] In the same year, due to an article on Islamist groups, the government closed down the paper under accusations of inciting sectarian controversy in Egypt.[87] In 2005, it acquired an Egyptian license for the first time to resume its weekly publication[88]and went daily in 2007.[89] In 2010, prior to the parliamentary elections, *Al Dostour* was acquired from the original founder by *Al Wafd's* Badawi and Reda Edward for LE16 million (US$3.2 million).[90] This deal was done within Badawi's efforts to expand his investments to

74. Al Masry Al Youm. "About Al Masry Al Youm." Al Masry Al Youm. Last accessed on Jan. 16, 2012 at <http://www.almasryalyoum.com/en/node/54>.

75. El Galad, Magdy. Interview. Al Nas TV. Last accessed on Jan. 16, 2012 at <http://www.youtube.com/watch?v=TSnLjXi2rdI>.

76. Ibid.

77. Ibid.

78. Al Masry Al Youm. "About Al Masry Al Youm." Al Masry Al Youm. Last accessed on Jan. 16, 2012 at <http://www.almasryalyoum.com/en/node/54>.

79. "Al Masry Al Youm: Transforming the Egyptian Press." Tavaana. Media Monitor. Jan. 28, 2011. Last accessed on Jan. 16, 2012 at <http://www.tavaana.org/archive.jsp?restrictids=nu_repeatitemid&restrictvalues=2071502000341295552773910&lang=en>.

80. Al Masry Al Youm. "About Al Masry Al Youm." Al Masry Al Youm. Last accessed on Jan. 16, 2012 at <http://www.almasryalyoum.com/en/node/54>.

81. A different view on the numbers is offered by veteran publisher Hisham Kassem, who argues that "The circulation of Al Masry Al Youm in the first quarter of 2011 reached 550,000, but is down to 170,000" (Kassem, Hisham. Personal interview. Feb. 21, 2012).

82. GoodNews4Me. "Information." GoodNews4Me. Al Alam Al Youm. Last accessed on Jan. 16, 2012 at <http://www.gn4me.com/alalamalyoum/info.jsp?edition_id=3238>.

83. Good News Group. "Companies" Good News Group. Last accessed on Jan. 16, 2012 at <http://www.goodnewsgroup.com/.

84. Black, Jeffrey. "Egypt Press: More Free, Still Fettered." Arab Media and Society. 2008. Last accessed on Jan. 16, 2012 at <http://www.arabmediasociety.com/?article=572>.

85. Rugh, William. A. *Arab Mass Media: Newspapers, Radio & Television in Arab Politics*. Westport, CT: Praeger, 2004. 127.

86. "Egypt." Press Freedom and Media Development. Last accessed on Jan. 16, 2012 at <http://www.wan-press.org/pfreedom/articles.php?id=4404>.

87. "Al Dustour." The Media Briefing. Last accessed on Jan. 16, 2012 at <http://www.themediabriefing.com/products/Al dustour/>.

88. "Egypt." Press Freedom and Media Development. Last accessed on Jan. 16, 2012 at <http://www.wan-press.org/pfreedom/articles.php?id=4404>.

89. Newspaper readership study 2002–2010. Pan Arab Research Center (PARC), 2011.

90. Nasrawi, Saif. "Wafd Party President Buys Al Dustour Newspaper." Al Masry Al Youm. Aug. 23, 2010. Last accessed on Jan. 16, 2012 at <http://www.almasryalyoum.com/en/node/70155>.

Table 29-1. Daily Newspaper Concentration

DAILY NEWSPAPERS

| MARKET SHARES BY READERSHIP (%) | 2002 AND 2003 | 2004 AND 2005 | 2006 AND 2007 | 2008 AND 2009 | 2010 |
|---|---|---|---|---|---|
| Al Ahram (Public) | 42.0 | 38.3 | 38.5 | 29.4 | 28.7 |
| Al Akhbar (Public) | 30.2 | 36.1 | 31.7 | 28.9 | 23.5 |
| Al Gomhouriya (Public) | 21.8 | 20.1 | 21.5 | 21.6 | 20.0 |
| Al Wafd (Private/Partisan, Egypt) | 3.9 | 3.6 | 2.8 | 2.4 | 1.9 |
| Al Ahrar (Private/Partisan, Egypt) | 0.6 | 0.5 | 0.7 | 0.7 | 0.7 |
| Al Masry Al Youm (Private/Independent, Egypt) | N/A | 0.5 | 3.5 | 8.1 | 15.4 |
| Al Dostoor (Private, Egypt) | N/A | N/A | N/A | 3.8 | 5.2 |
| Al Alam Al Youm | 0.0 | 0.2 | 0.2 | 0.3 | 0.1 |
| Other | 1.4 | 0.6 | 1.1 | 4.8 | 4.6 |
| Revenue Estimates (mil $) | | 123.0 | | 148.0 | |
| C4 | 97.9 | 98.1 | 95.2 | 88.0 | 87.6 |
| HHI | 3,168 | 3,188 | 2,972 | 2,278 | 2,065 |
| Number of Players ($n$ with % >1%) | 4 | 4 | 5 | 6 | 6 |
| Noam Index | 1,584 | 1,594 | 1,329 | 930 | 843 |

*Source*: PARC Newspaper Readership Study 2002–2010 (we bought the study results).
*Note*: The Noam index only includes all voices that have a market share larger than 1%.
*Note 2*: N/A means not applicable because the company had not started up yet.
*Note 3*: Revenue estimates are for the years 2004 and 2008.

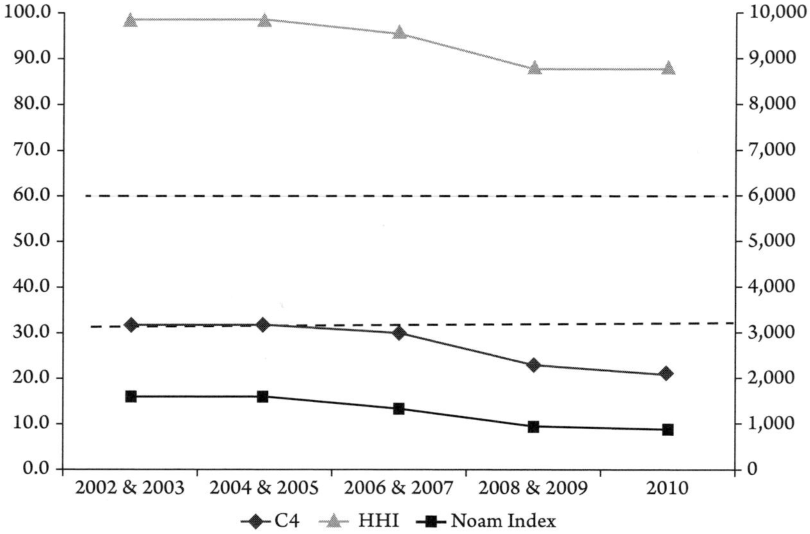


GRAPH 29.1 Daily Newspaper Concentration

include media in print and television (Al Hayat TV Satellite channel).[91] Much to the dismay of the public, Badawi expelled *Al Dostour's* editor-in-chief, Ibrahim Eissa, a famous Egyptian journalist who was popular for his criticism of the Mubarak regime and *Al Dostour's* editorial team.[92] In 2011, Al Dostour ranked fifth in circulation of daily newspapers.[93]

### CONCLUSION

The newspaper industry in Egypt has historically been under a host of direct and indirect controls by the government. Since 2004, the entry of powerful businessmen to the scene has meant that business interests have also come to influence print media, albeit in different forms than government censorship. As ownership diversity increased, concentration figures declined in the industry. The appropriation of significant market share for new entrants is reflected in the HHI's steep fall starting in 2004. Although the market is still constrained by a number of economic barriers and political agendas, a greater number of voices in the market and a positive step toward media plurality are reflected in the gap between the HHI and the Noam Index (see Table 29.1 and Graph 29.1). In general, the expansion and diversity of print media ownership are positive from an access to knowledge perspective, despite the fact that structural impediments, political and economic, continue to limit the ability to start a newsprint operation.

## AUDIOVISUAL MEDIA

Television is Egypt's most popular mass medium: 90% of Egyptians own a television set and 70% own a satellite receiver.[94] Radio ranks

91. "Al Sayed Al Badawy." Ahram Online. 19 Nov. 2011. Last accessed on Jan. 16, 2012 at <http://english.ahram.org.eg/NewsContent/33/102/26705/Elections-/Whos-who/AlSayed-AlBadawi.aspx>.

92. "Al Sayed Al Badawy." Ahram Online. 19 Nov. 2011. Last accessed on Jan. 16, 2012 at <http://english.ahram.org.eg/NewsContent/33/102/26705/Elections-/Whos-who/AlSayed-AlBadawi.aspx>.

93. Nasrawi, Saif. "Wafd Party President Buys Al Dustour Newspaper." Al Masry Al Youm. Aug. 23, 2010. Last accessed on Jan. 16, 2012 at <http://www.almasryalyoum.com/en/node/70155>.

94. Mekay, Emad. "TV Stations Multiply as Egyptian Censorship Falls." *The New York Times*. July 13, 2011. Last accessed on Jan. 16, 2012 at <http://www.nytimes.com/2011/07/14/world/middleeast/14iht-M14B-EGYPT-MEDIA.html?_r=2&pagewanted=all>.

second with approximately 80% of households in 2007 owning a radio set.[95] The high level of exposure offered by audiovisual media gives it strategic political importance. Subsequently, it attracts acute interest from the ruling regime.

Until 2012, Egypt's authorities kept an iron grip over audiovisual media, including television, radio, and satellite TV. Control is maintained in three ways: regulatory measures, maintenance of a monopoly over the industry or industry infrastructure, and informal censorship. The latter is practiced by tacitly ensuring that key players and gatekeepers in privately owned outlets are political allies to the ruling regime.

Thus far, the state has retained full discretion over the granting of licenses to media corporations. These licenses are issued without a clear set of rules or procedures and in "a highly selective manner."[96] On the ownership side, the Egyptian government owns major media players in order to directly compete in the market. Additionally, it monopolizes the infrastructural backbone for media corporations to limit competition to an approved set of players. Although subject to state influence, Egypt has witnessed unprecedented successful private initiatives in the post-millennium era and particularly after the revolution of January 25, 2011.

## Radio

Radio is the second most popular form of media in Egypt.[97] Due to its strategic political importance, the radio industry in Egypt is almost completely dominated by the government through the state-run Egyptian Radio and Television Union (ERTU) (Graph 29.2). ERTU has remained a monopoly, with very limited participation from international companies' Arabized

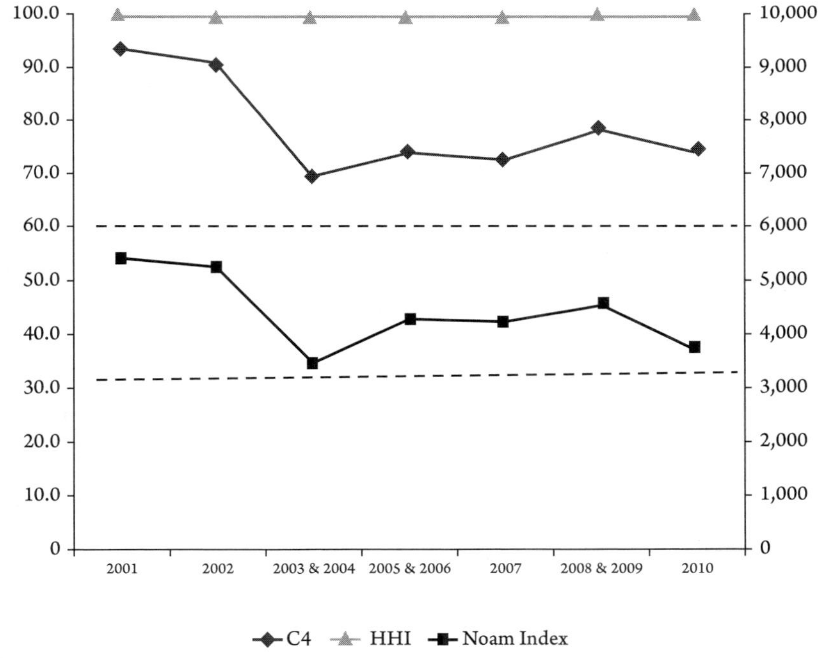


GRAPH 29.2 Radio Concentration

95. "Target 8." International Telecommunications Union. 2010. Last accessed on July 8, 2012 at <http://www.itu.int/ITU-D/ict/publications/wtdr_10/material/WTDR2010_Target8_e.pdf. 158>.

96. The Cairo Institute for Human Rights Studies. *Media in North Africa: Obstacles and challenges.* United Nations Economic Commission for Africa, 2007, 4. Last accessed on Jan. 16, 2012 at <http://www.uneca.org/africanmedia/documents/NorthAfrica-eng-report.pdf>.

97. Allam, Rasha. "Media Landscape: Egypt." European Journalism Centre. Nov. 8, 2010. Last accessed on Dec. 29, 2011 at <http://www.ejc.net/media_landscape/article/egypt/>.

stations, including BBC Arabic radio service, Voice of America Arabic service, and Monte Carlo Arabic service. In 2003, Nile Radio Productions was the first private Egyptian company to enter the radio market.

In 1981, ERTU's radio networks were divided into seven divisions according to content, ranging from commercial to religious (Qur'an), educational, and cultural stations. The local radio network expanded to cover local cities and administrative districts including Alexandria, Upper Egypt, Sinai, and overseas.[98] Qur'an channel has the largest audience and accounts for the greatest part of ERTU's radio market share.[99]

Nile Radio Productions (NRP) was established in 2003 and consists of two music stations: Nile FM broadcasts in English and Nogoom FM broadcasts in Arabic. Good News Group, an entertainment and media conglomerate in the Middle East and North Africa, owns a 25% stake in NRP.[100] In contrast to ERTU's archaic content that includes outdated music and radio presenters broadcasting in classic Arabic, these two stations provide current pop culture entertainment. Meanwhile, NRP was prohibited by the government to air any newscasts; content was restricted to entertainment only.[101] Even though NRP's coverage is geographically limited to Cairo, its modern content and savvy advertising explain why it gained ground quickly. Its success is evident in the sharp dip in HHI and NI in 2003 (see Table 29.2). Its recent web launch helped it reach listeners in Alexandria, Egypt's second largest urban city, and also earned it an expatriate and international audience abroad. Amr Adeeb, managing director of NRP, head of the entertainment division at the Good News Group, reported advertising revenues in the first year of operation to be a record-breaking $3.2 million.[102]

Emad Eddin Adeeb, the owner, was close to the Mubarak regime, and this is believed to have aided him in investing in a private radio venture.[103] In fact, after the launch of the private radio channels, the newspaper *Al Alam Al Youm* provided positive commentary on Mubarak's pledge to run the first multi-candidate elections in 2005. Moreover, the general manager of the radio channels, Amr Adeeb, provided critical commentary on the president's opponents in his nightly talk show broadcasted via satellite.[104]

NRP was a game changer because it commercialized the long neglected radio market in Egypt. In response to NRP's emergence as a new player, ERTU altered its content strategy and in 2009 established its own media agency, *Sawt Al Qahira* (Voice of Egypt), responsible for commercializing ERTU's media activities.[105] In the same year, ERTU launched the Egyptian Radio Network (ERN) with four new FM stations with nationwide coverage,[106] including Radio Misr in 2009,[107] a private–public radio station, with presenters broadcasting in Egyptian dialect and providing news bulletins interspersed between contemporary music.[108] Many are detecting the changing nature of Radio Misr's content, which became more politicized in the aftermath of the January 25 revolution when politics became an inevitable topic for conversation in all media outlets.

---

98. Allam, Rasha. "Media Landscape: Egypt." European Journalism Centre. Nov. 8, 2010. Last accessed on Dec. 29, 2011 at <http://www.ejc.net/media_landscape/article/egypt/>.

99. Newspaper readership study 2002–2010. Pan Arab Research Center (PARC), 2011.

100. Good News Group. "Companies" Good News Group. Last accessed on Jan. 16, 2012 at <http://www.goodnewsgroup.com/>.

101. Allam, Rasha. "Media Landscape: Egypt." European Journalism Centre. Nov. 8, 2010. Last accessed on Dec. 29, 2011 at <http://www.ejc.net/media_landscape/article/egypt/>.

102. Shahine, Alaa. "New Media Flourish in Egypt." UCLA Center for Communications and Community. Apr. 27, 2004. Last accessed on Jan. 17, 2012 at <http://www.c3.ucla.edu/newsstand/global/new-media-flourish-in-egypt/>.

103. Sakr, Naomi. *Arab Television Today*. New York: I. B. Tauris, 2007. 31.

104. Ibid., 30–31.

105. ERTU. "Sectors." ERTU. Last accessed on Dec. 15, 2011 at <http://ertu.org/1/sectors.asp>.

106. Sout Al Qahira. "About Sout Al Qahira." Sout Al Qahira. Last accessed on Jan. 17, 2012 at <http://www.soutalqahira-eg.com/about-ern.aspx>.

107. "Radio Masr." Facebook Page. Last accessed on Jan. 17, 2012 at <http://www.facebook.com/RadioMasr?sk=info>.

108. Allam, Rasha. "Media Landscape: Egypt." European Journalism Centre. Nov. 8, 2010. Last accessed on Dec. 29, 2011 at <http://www.ejc.net/media_landscape/article/egypt/>.

Table 29-2. Radio Concentration

RADIO

| MARKET SHARES BY LISTENERSHIP (%) | 2001 | 2002 | 2003 AND 2004 | 2005 AND 2006 | 2007 | 2008 AND 2009 | 2010 |
|---|---|---|---|---|---|---|---|
| Egyptian Radio and Television Union (ERTU) (public) | 96.6 | 95.2 | 82.2 | 85.1 | 84.1 | 88.0 | 85.4 |
| Nile Radio Productions (NRP) (Local, Private) | N/A | N/A | 12.7 | 11.7 | 13.0 | 10.0 | 10.4 |
| Eza'at Masr | N/A | N/A | N/A | N/A | N/A | N/A | 2.2 |
| BBC (UK) | 1.3 | 0.6 | 1.0 | 0.6 | 0.5 | 0.3 | 0.2 |
| Monte Carlo (France) | 1.3 | 1.0 | 0.7 | 0.5 | 0.1 | 0.1 | 0.1 |
| Voice of America (US) | 0.4 | 2.6 | 3.0 | 1.8 | 1.8 | 1.1 | 1.3 |
| Radio France (France) | N/A | 0.0 | 0.0 | 0.0 | 0.0 | 0.0 | 0.0 |
| Deutsche Welle (DW) (Germany) | N/A | N/A | N/A | N/A | N/A | N/A | 0.0 |
| Others | 0.4 | 0.6 | 0.4 | 0.3 | 0.4 | 0.4 | 0.4 |
| Revenue Estimates (mil $) | | | 13.0 | | | 39.0 | |
| C4 | 99.6 | 99.4 | 98.9 | 99.2 | 99.4 | 99.4 | 99.3 |
| HHI | 9,335 | 9,071 | 6,929 | 7,383 | 7,245 | 7,845 | 7,408 |
| Number of players (*n*) | 3 | 3 | 4 | 3 | 3 | 3 | 4 |
| Noam Index | 5,390 | 5,237 | 3,464 | 4,262 | 4,183 | 4,530 | 3,704 |

*Source*: PARC Radio Reach survey 2001–2010.

*Note*: For the years 2003–2006, the sampling was conducted over a two-year period and so values represent two years' work.

N/A: No data available.

Many groups flirting with the idea of setting up a radio station have kicked off online channels, which also shows the difficulty of appropriating the airwaves. In turn, the government's Draft Broadcast Law in 2008 included "computer networks" and "digital media" in its scope of audiovisual content to bring these outlets into the legal constraints placed on other media outlets.[109]

## Broadcast Television (Over-the-Air)

Broadcast television is the monopoly of the Ministry of Information, which owns and manages state television and radio through the ERTU.[110] State monopoly over audiovisual terrestrial broadcasts through the ERTU is legally granted by Law no. 13 of 1979, and its amendment by Law no. 223 of 1989.[111] By 2008, new laws officially suspended the state's monopoly by allowing for the allocation of terrestrial broadcast licensing[112] but maintained restrictions on broadcasting content.[113] Although hopeful entrepreneurs are in line applying for these terrestrial licenses, by spring 2013 none had been issued (Graph 29.3).[114]

ERTU reports to the Ministry of Information,[115] who in turn delivers "political 'guidance' on an annual basis."[116] It dominates the audiovisual market, owning Egyptian Television Network (ETN), the only terrestrial broadcaster, which includes eight local channels. Owned and run by the state since its establishment in 1960,[117] terrestrial television has been a mouthpiece for each governing regime. Anger and discontent with the state media's biased coverage of key events has made the Maspero building, which houses the ERTU, a site for major protests since the January 25 revolution.

ERTU also owns a major stake in Nile Television Network (NTN), a main satellite and channel provider in the Arab world. It owns 50% of Egyptian Media Production City (EMPC), established in 2000[118] as a main provider of equipment, studios, and technical support for audiovisual media content. It benefits from tax exemptions and other legal incentives,[119] thus attracting private players to license Egyptian-owned satellite channels.

ERTU also owns a 40% stake in the joint stock company Nilesat, the only provider of satellite broadcasting services in Egypt and a major provider to the Arab World, airing more than 600 channels, 76% of which are free (Table 29.3).[120] These include Egyptian TV (ETV), which sells its specially produced programs and soap operas to the Arab World, Nile TV, Nile News, and seven other specialized channels.[121]

109. Full Text: Draft Egyptian Broadcast Law. Arab and Media Society, 2008, Article 1, Clause 4. Last accessed June 27, 2012 at <http://www.arabmediasociety.com/articles/downloads/20080729105525_AMS6_draft_media_law_english.pdf>.

110. Saleh, Ibrahim. "The Crossing Roads of Egypt: An Avenue of Participation Crucifies the Stagnation Development." *Global Media Journal* 2, no. 4 (2004). Last accessed on June 27, 2012 at <http://lass.calumet.purdue.edu/cca/gmj/OldSiteBackup/SubmittedDocuments/archivedpapers/fall2004/refereed/saleh.htm>.

111. "Egypt: Article 19 and the Justice Initiative Submission to the UN Universal Periodic Review" in the Seventh Session of the UPR Working Group. 2010. Last accessed on Jan. 16, 2012 at <http://www.article19.org/data/files/pdfs/submissions/egypt-article-19-and-the-justice-initiative-submission-to-the-un-universAl p.pdf>.

112. Sakr, Naomi. *Arab Television Today*. New York: I. B. Tauris, 2007. 45.

113. "Egypt: Article 19 and the Justice Initiative Submission to the UN Universal Periodic Review" in the Seventh Session of the UPR Working Group. 2010. Last accessed on Jan. 16 2012 at <http://www.article19.org/data/files/pdfs/submissions/egypt-article-19-and-the-justice-initiative-submission-to-the-un-universAl p.pdf>.

114. Sakr, Naomi. *Arab Television Today*. New York: I. B. Tauris, 2007. 45.

115. Kotb, Ahmed. "Media Control." Al Ahram Weekly Online, 1056, July 14–20, 2011. Last accessed on Jan. 17, 2012 at <http://weekly.ahram.org.eg/2011/1056/eg692.htm>.

116. Sakr, Naomi. *Satellite Realms: Transnational Television, Globalization and The Middle East*. New York: I. B. Tauris, 2001. 30.

117. Egyptian Television Network. "About Us." Egyptian Television Network (ETN). Last accessed on Jan. 17, 2012 at <http://www.egytv.net/aboutus.aspx>.

118. "The Media Free Zone." Egyptian Media Production City (EMPC). Last accessed on Dec. 29, 2011 at <http://www.empc.com.eg/English/MediaFreeZone.html>.

119. Ibid.

120. NileSat. "Nilesat Company Profile." NileSat. Last accessed on June 27, 2012 at <http://nilesat.com.eg/banners/AboutUs/CompanyProfile.aspx>.

121. Egyptian Television Network. "About Us." Egyptian Television Network (ETN). Last accessed on Jan. 17, 2012 at <http://www.egytv.net/aboutus.aspx>.

Table 29-3. Broadcast TV Concentration

TV BROADCASTING

| MARKET SHARES BY VIEWERSHIP (%) | 1984 | 1988 | 1992 | 1996 | 2000 | 2003 AND 2004 | 2008 AND 2009 | 2010 |
|---|---|---|---|---|---|---|---|---|
| Egyptian Radio and Television Union (ERTU) (Public) | 100.0 | 100.0 | 100.0 | 100.0 | 100.0 | 100.0 | 100.0 | 100.0 |
| Revenue Estimates (mil $) | | | | | | 162.0 | 243.0 | |
| C1 | 100.0 | 100.0 | 100.0 | 100.0 | 100.0 | 100.0 | 100.0 | 100.0 |
| HHI | 10,000 | 10,000 | 10,000 | 10,000 | 10,000 | 10,000 | 10,000 | 10,000 |
| Noam Index | 10,000 | 10,000 | 10,000 | 10,000 | 10,000 | 10,000 | 10,000 | 10,000 |
| Number of players (*n*) | 1 | 1 | 1 | 1 | 1 | 1 | 1 | 1 |

*Source:* ERTU is the only company allowed to broadcast terrestrial television.

*Note:* Revenue estimates are for the years 2004 and 2008 based on the figures in other developing countries in this project provided by the editor.

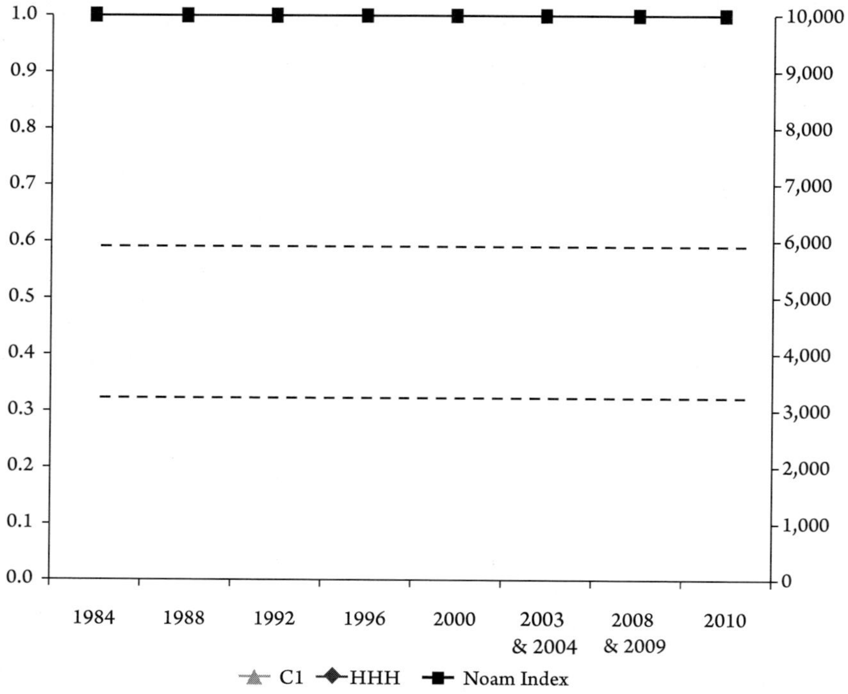


GRAPH 29.3 Broadcast TV Concentration

The launch of the first private satellite channels, Dream TV in 2001 and Mehwar in 2002,[122] was "a face-saving procedure" in response to growing political pressures on state-led liberalization.[123] Egyptian state television—both terrestrial and satellite—has been exposed to competition by private satellite broadcasters in entertainment media. However, Decree 411 of 2000 placed restrictions on broadcast content and genre.[124] Since private broadcasters are not permitted to supply their own newscasts,[125] limited competition ensures that the government maintains a tight grasp on political censorship. Even though ERTU owns a share in both Dream TV and Mehwar, neither of these channels air newscasts.[126] To circumvent restrictions on news content, private channels produced talk shows with cautious commentary on issues in the news. This became the only permissible form of political expression. Nevertheless, these talk shows gained popularity and professional television journalism offered an alternative news narrative to that of ERTU. This was especially true of the coverage of the January 25 revolution and the ensuing events.

122. Allam, Rasha. "Media Landscape: Egypt." European Journalism Centre. Nov. 8, 2010. Last accessed on Dec. 29, 2011 at <http://www.ejc.net/media_landscape/article/egypt/>.

123. The Cairo Institute for Human Rights Studies. *Media in North Africa: Obstacles and challenges*. United Nations Economic Commission for Africa, 2007, 4. Last accessed on Jan. 16, 2012 at <http://www.uneca.org/africanmedia/documents/NorthAfrica-eng-report.pdf>.

124. Sakr, Naomi. *Arab Television Today*. New York: I. B. Tauris, 2007. 30.

125. IREX. "Media sustainability Index 2005: Middle East and North Africa." IREX. Last accessed on May 24, 2012 at <http://www.irex.org/system/files/MSI05_MENA_EG.pdf. 26>.

126. Sakr, Naomi. *Arab Television Today*. New York: I. B. Tauris, 2007. 30.

## Multichannel TV Platforms

While cable television was introduced in other parts of the world in the 1970s and 1980s, this was not the case in Egypt because it was not a priority for the state at that time.[127]

In 1990, ERTU, along with private shareholders, received approval from the government to establish Cable News Egypt (CNE) as "the first Pay-TV project in the Arab world."[128] While carrying the "cable" title, CNE did not use cable technology, but actually used terrestrial broadcasting technology for "rebroadcasting."[129] CNE was established within a cooperative agreement with CNN for 25 years, with the main purpose of retransmitting CNN International in Egypt. The chair of ERTU also chaired CNE. Additionally, ERTU retained control over "CNE's ultra high frequency channel (UHF)."[130]

Later in the 1990s, Egypt "bypassed cable straight into satellite television."[131] Three main multichannel platform providers cater to Egypt and the Arab region with a combined bundle of channels: Arab Radio Television (ART), Orbit, and Showtime Arabia, established in 1993, 1994, and 1996, respectively (Graph 29.4).[132] ART was founded by Saudi Arabia's Sheikh Saleh Kamel through his Arab Digital Distribution Company (ADD) and is based in the United Arab Emirates.[133,134] Apart from operations in the Middle East and North Africa, ADD owns Pehla, another satellite platform in Asia,[135] and Firstnet, which broadcasts to Europe and the Americas.[136] Kamel is also a partner in the prominent channel provider, the Middle East Broadcasting Center (MBC),[137] and ART also owns major film production studios in the region.[138]

The Orbit network is a subsidiary of the Saudi investment company, Mawarid Group.[139] Mawarid Group also owns subsidiaries in telecommunications, satellite television, and radio networks.[140] The Orbit network targets its content toward a more affluent customer profile and was based in Rome until 2005, when it moved to Bahrain.[141] Showtime Arabia is the only one of three pay-TV networks that is not Saudi-owned. The Kuwait Company Projects Holding (KIPCO) owns controlling shares in Showtime Arabia, and the remaining 21% is owned by the CBS Corporation.[142] Showtime and Orbit

127. Hamdy, Naila. Personal interview. June 10, 2012.

128. Nile Engineering Group. Last accessed on June 10, 2012 at <http://www.nileengineering.com/English/Products--and--Services/NileTel-Stores>.

129. Hamdy, Naila. Personal interview. June 10, 2012.

130. Amin, Hussein. *Strengthening the Rule of Law* and *Integrity in the Arab World: Report on the State of the Media in Egypt.* The Arab Center for the Development of the Rule of Law and Integrity, 2009, 6. Last accessed on June 10, 2012 at <http://www.arabruleoflaw.org/Files/PDF/Media/English/P2/Egypt_MediaReportP2_En.pdf>.

131. Hamdy, Naila. Personal interview. June 10, 2012.

132. "The Arab World's Major Pay TV Operators Provide A Choice of 140 Pay TV Channels." AMEinfo. Jan. 15, 2008. Last accessed on Jan. 17, 2012 at <http://www.ameinfo.com/143724.html>.

133. Attia, Tarek. "No More Free TV." Al Ahram Weekly, no. 526. Mar. 22–28, 2001. Last accessed on May 14, 2012 at <http://weekly.ahram.org.eg/2001/526/li3.htm>.

134. In December 2009, six ART sports channels were sold to AlJazeera for $2.75 billion. Source: "Al Jazeera in $2.75bn ART channels deal" Zawya.com (<http://www.zawya.com/story/ZAWYA20091105051037/>).

135. "Press Releases." Pehla: Premium Asian Entertainment. Last accessed on Jan. 17, 2012 at <http://www.pehlatv.net/pehla/PressReleases/54/ADD-launches-Hallmark-channel-on-its-platform>.

136. "TBS Feature Interview: Sheikh Saleh Kamel." Transnational Broadcasting Studies (TBS) Archives. May 1998, 1 (Fall 1998). Last accessed on Jan. 17, 2012 at <http://www.tbsjournal.com/Archives/Fall98/Interviews1/Sheikh_Saleh/sheikh_saleh.html>.

137. Ibid.

138. "Arab Radio and Television Network." Wikipedia. Sept. 21, 2011. Last accessed on Dec. 19, 2012 at <http://en.wikipedia.org/wiki/Arab_Radio_and_Television_Network>.

139. Cochrane, Paul. "Saudi Arabia's Media Influence." Arab Media and Society. 2007. Last accessed on Jan. 17, 2012 at <http://www.arabmediasociety.com/?article=421> and <http://www.insightsmena.com/en/#!place=category&cat=Media+Usage>.

140. "Industrial Conglomerates: Mawarid Group Company LLC." Bloomberg Businessweek. Last accessed on Jan. 17, 2012 at <http://investing.businessweek.com/research/stocks/private/people.asp?privcapId=22595361>.

141. Sakr, Naomi. *Arab Television Today.* New York: I. B. Tauris, 2007. 5, 196.

142. "Saudi Arabia: Media Market Description" in Worlds Association of Newspapers: World Press Trends 2008. 2008. Last accessed on Jan. 17, 2012 at <http://www.ihudaif.com/wp-content/uploads/2010/01/WORLD-ASSOCIATION-OF-NEWSPAPERS—WORLD-PRESS-TRENDS-2008.pdf>.

merged in 2009.[143] This merger came under the Panther Media Group; a company registered in the Dubai International Financial Center and is jointly owned by KIPCO and the Orbit Group.[144]

Table 29.4 only addresses the pay-TV market for platform providers for which data were available. It is worth noting that pay-TV has captured a minimal share of the satellite market. Estimates reveal that only 4% of the total population in 2009 subscribed to a network.[145] In 2005, Booz and Co (previously Booz and Allen) confirmed these figures in a study that reported only 1 million subscribers in the Arab region for the three main platform providers: ART, Orbit, and Showtime.[146] Pay networks are in direct competition with free-to-air satellite, including state-owned Nilesat,[147] Saudi Sheikh Walid Al Ibrahim's MBC,[148] the international Lebanese Broadcasting Corporation (LBC) based in the Cayman Islands,[149] and Dubai Media Incorporated (DMI), which is partly owned by ART's Kamel and the government.[150]

Additionally, pay-TV network subscriptions are hampered by receivers that use sharing servers to illegally decode transmissions by pay-TV platforms. These decoders offer the same array of channels with no subscription fees. Hence, Egyptians could view the same content as pay-TV platform subscribers with only an investment in a receiver and DSL connection. To combat illegal

## Table 29-4. Multichannel Video Platform Concentration

MULTICHANNEL PLATFORMS (CABLE, SATELLITE)

| MARKET SHARES BY VIEWERSHIP (%) | 2005 | 2006 | 2007 AND 2008 | 2009 |
|---|---|---|---|---|
| ART (Saudi) | 69.50 | 80.10 | 88.10 | 91.50 |
| Orbit (Mawarid Group, Saudi Arabia*) | 20.67 | 9.30 | 7.07 | 6.86 |
| Showtime (KIPCO Kuwait 79%, CBS USA 21%) | 9.83 | 10.58 | 4.84 | 1.62 |
| Star Select | | | | |
| Revenue Estimates (mil $) | 75.0 | | 164.0 | |
| C4 | 100.0 | 100.0 | 100.0 | 100.0 |
| HHI | 5,354 | 6,614 | 7,835 | 8,422 |
| Noam Index | 3,091 | 3,819 | 4,524 | 4,862 |
| Number of players (*n*) | 3 | 3 | 3 | 3 |

*Source:* PARC PayTV Reports 2005–2009.

* Orbit and Showtime merged in July 2009. Data for this year refer to each company separately.

143. "Showtime Arabia and Orbit Group Merge Pay TV Operations." AMEinfo.com. July 12, 2009. Last accessed on Jan. 17, 2012 at <http://www.ameinfo.com/203504.html>.

144. "Group Companies: Orbit Showtime Network." KIPCO: Kuwait Projects Company (Holding). Last accessed on Jan. 17, 2012 at <http://www.kipco.com/GroupCompanies.asp?q_pageid=56>.

145. Newspaper readership study 2002–2010. Pan Arab Research Center (PARC), 2011.

146. Sakr, Naomi. *Arab Television Today*. New York: I. B. Tauris, 2007. 189.

147. Nile Television Network (NTN). Last accessed on June 27, 2012 at <http://www.ntntv.net/>.

148. Sakr, Naomi. "Satellite Television and Development in the Middle East." Middle East Research and Information Project 2010. Last accessed on May 29, 2012 at <http://www.merip.org/mer/mer210/satellite-television-development-middle-east>.

149. Jaafar, A. "Mixed Results for LBC – Rotana: Arab Conglomerate Tries to Compete with MBC." Variety. Oct. 2, 2008. Last accessed on Jan. 17, 2012 at <http://www.variety.com/article/VR1117993363>.

150. Sakr, Naomi. *Arab Television Today*. New York: I. B. Tauris, 2007.

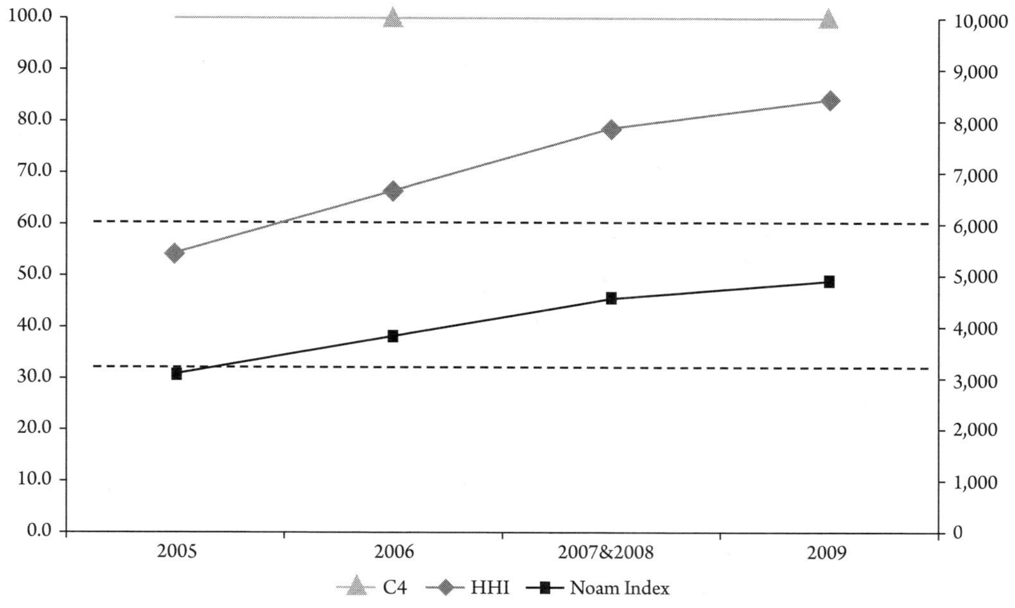


GRAPH 29.4 Multichannel Platform Concentration

viewing in December 2010, the Orbit Showtime Network successfully distributed set-top boxes that use a conditional access system (CAS) to all its subscribers. These boxes bar unofficial reception of its channels by other receivers. The network has also lobbied the government to enforce new regulatory measures that would cease illegal viewership.[151]

## Video Channels

Satellite channels have brought a regional dimension to the market and eliminated redundant local media.[152] Table 29.5 and Graph 29.5 for video channel providers illustrate one of the lowest HHI levels across the spectrum of industries studied in this chapter. Egypt has traditionally been a pioneer for entertainment media.[153] Although figures reveal that the industry remains concentrated, ownership shows substantial participation from regional players, most notably Gulf States. Within the last decade, oil-rich Arab states with ample investment power have imported Egyptian ownership and regulation policies. Initially, this allowed Egyptian professionals to program content and create their own competing entertainment industry.[154] Today, with two of the largest media empires owned by Saudi Arabia, the country has become a major contender in satellite airwaves with free-to-air television. This includes Western channels such as Newscorp's Fox International that have been dubbed for Arab viewers.

## INFRASTRUCTURE AND REGULATION

The Arab Satellite and Communications Organization (Arabsat), headquartered in Riyadh, Saudi Arabia, was established in 1976. Jointly owned by the Arab League of Nations, Arabsat shares are concentrated in the hands of oil-rich nations including Saudi Arabia, Kuwait,

151. Williams, Sean. "Walking the Plank." Arabian Business. Dec. 15, 2010. Last accessed on Jan. 17, 2012 at <http://www.arabianbusiness.com/walking-plank-366838.html>.

152. Fakhredinne, Jihad. "Pan-Arab Satellite Television: Now the Survival Part." Allied Media Corporation. Last accessed on Jan. 17, 2012 at <http://www.allied-media.com/ARABTV/pan-arab-sat.htm>.

153. Fandy, Mamoun. *(Un)Civil War of Words: Media and Politics in the Arab World*. Westport, CT: Praeger, 2007. 20.

154. Ibid.

Table 29-5. Video Channel Concentration

VIDEO CHANNELS

| MARKET SHARES BASED ON VIEWERSHIP (%) | 2001 | 2002 | 2003 AND 2004 | 2005 AND 2006 | 2007 | 2008 AND 2009 | 2010 |
|---|---|---|---|---|---|---|---|
| ERTU (Public) | | | | | | | |
| Egypt Satellite Channel (ESC) (Public) | 6.6 | 7.4 | 10.9 | 3.0 | 2.0 | 2.1 | 1.8 |
| NTN (Public) | 53.8 | 40.0 | 23.0 | 16.7 | 18.3 | 14.7 | 10.0 |
| Al Jazeera (Qatar Government) | 8.7 | 6.4 | 6.2 | 6.3 | 5.6 | 3.8 | 2.1 |
| MBC (Saudi Arabia) | 4.7 | 3.2 | 10.3 | 11.9 | 12.7 | 11.3 | 9.2 |
| Future TV Network (Private, Lebanon) | 1.8 | 1.4 | 1.1 | 0.2 | 0.2 | 0.2 | 0.0 |
| ART (Saudi Arabia) | 1.1 | 6.0 | 7.1 | 7.5 | 2.0 | 2.4 | 1.4 |
| DMI (Public, Dubai) | 6.2 | 3.2 | 3.5 | 2.8 | 2.3 | 1.7 | 0.3 |
| BBC (Pubic, UK) | 0.4 | 0.4 | 0.1 | 0.0 | 0.0 | 0.1 | 0.0 |
| CNN (Time Warner, USA) | 3.3 | 1.0 | 0.1 | 0.0 | 0.0 | 0.1 | 0.0 |
| Dream TV Network (Private) | N/A | 17.4 | 15.7 | 12.3 | 13.3 | 9.8 | 6.4 |
| Al Mehwar (83% private, 17% public) | N/A | 4.4 | 4.7 | 3.4 | 4.3 | 4.9 | 4.2 |
| Melody Entertainment Holding (Private) | N/A | 1.6 | 2.7 | 4.0 | 2.3 | 6.2 | 5.7 |
| Saudi TV (Saudi Arabia) | N/A | 0.8 | 0.4 | 0.1 | 0.1 | 0.1 | 0.1 |
| Rotana (Private, Saudi Arabia) | N/A | N/A | 4.1 | 21.7 | 20.4 | 15.4 | 9.7 |
| LBC (merged with Rotana in 2007) | 4.7 | 3.2 | 2.7 | 0.6 | | N/A | N/A |
| Mazzika (Private) | N/A | N/A | 4.4 | 2.3 | 1.8 | 1.2 | 1.1 |
| Al Majd TV (Private, Saudi Arabia) | N/A | N/A | 0.3 | 5.0 | 2.7 | 0.9 | 2.4 |
| Spacetoon (Private, Headquarters in Dubai) | N/A | N/A | 0.6 | 0.3 | 0.3 | 0.4 | 0.1 |
| Al Rai TV (Private, Kuwait) | N/A | N/A | N/A | 0.2 | 0.0 | 0.1 | 0.1 |
| OTV (Private) | N/A | N/A | N/A | N/A | 0.1 | 0.3 | 0.3 |
| Hayat (Private) | N/A | N/A | N/A | N/A | 0.0 | 2.1 | 13.3 |
| Al Baraheen International (Saudi Arabia) | N/A | N/A | N/A | N/A | 9.5 | 9.6 | 6.1 |

| | | | | | | | |
|---|---|---|---|---|---|---|---|
| Modern TV | N/A | N/A | N/A | N/A | N/A | 2.0 | 3.9 |
| Panorama | N/A | N/A | N/A | N/A | N/A | 7.0 | 12.2 |
| Al Rahma TV (Private, Sheikh Saleh) | N/A | N/A | N/A | N/A | N/A | N/A | 2.9 |
| Moga Comedy (Private) | N/A | N/A | N/A | N/A | N/A | N/A | 3.0 |
| Others | 8.7 | 3.6 | 2.26 | 1.8 | 2.1 | 3.8 | 4.0 |
| Revenue Estimates (mil $) | | | 333.0 | | | 510.0 | |
| C4 | 75.3 | 71.2 | 59.8 | 62.6 | 64.6 | 51.2 | 45.2 |
| HHI | 3,114 | 2,090 | 1,172 | 1,214 | 1,256 | 919 | 782 |
| Number of Players (*n*) | 9 | 12 | 13 | 12 | 13 | 15 | 17 |
| Noam Index | 1,038 | 603 | 325 | 351 | 348 | 237 | 190 |

*Source:* PARC Household Viewership Survey 2001–2010.
N/A: Data not available.

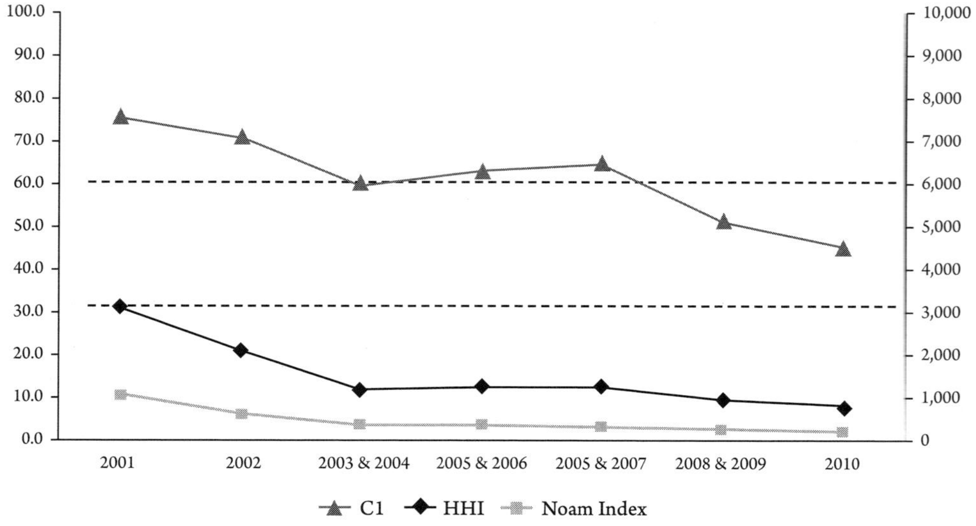


GRAPH 29.5 Video Channel Concentration

Libya, and Qatar, which ranked as the four top shareholders.[155]

The second satellite company in the region is Egypt's Nilesat, partly owned by the Egyptian government through its media vehicle, ERTU.[156] Egypt first notified the International Telecommunications Union of its intention to obtain an orbital slot after being expelled from the Arab League, and consequently Arabsat, in 1979 because of its peace treaty with Israel. However, its allocation was not assigned until 1995.[157] With one satellite headquartered in Saudi Arabia and the other in Egypt, both countries were able to launch rival attempts at controlling regional satellite broadcasting.[158] A legal case unfolded after SCAF announced in September 2011 that it would ban the issuing of new satellite licenses. Saudi Arabia announced the transfer of all of its channels from Nilesat to Arabsat, a tactical move that caused the Egyptian television industry to sustain losses.[159]

While these satellites have allowed for transnational television to proliferate, the country that owns the satellite platform also controls content. For example, news sources revealed that Nilesat suspended Al Jazeera's signal on January 30, 2011, during the toppled regime's battle to survive the revolutionary ascent of 2011.[160] Moreover, self-censorship imposed by nations broadcasting on Nilesat and Arabsat exposed transnational television to a level of state-control similar to local terrestrial channels.[161]

In Egypt, the Ministry of Information is responsible for licensing satellite channels and requires already existing satellite channels to acquire permission for live coverage of events

155. Arab Satellite Communications Organization. "About Us. Corporate Governance." Arab Satellite Communications Organization. Last accessed on June 27, 2012 at <http://www.arabsat.com/pages/ArabLeague.aspx>.

156. NileSat. "Nilesat Company Profile." NileSat. Last accessed on June 27, 2012 at <http://nilesat.com.eg/banners/AboutUs/CompanyProfile.aspx>.

157. Sakr, Naomi. *Satellite Realms: Transnational Television, Globalization and The Middle East.* New York: I. B. Tauris, 2001.

158. Ibid., 161.

159. "Saudi Arabia Moves its Satellite Channel to Arabsat." Al Masry Al Youm. Sept. 8, 2011. Last accessed on Mar. 12, 2012 at <http://www.almasryalyoum.com/en/node/493638>.

160. Sennitt, Andy. "Al Jazeera Kicked out of Egypt, Nilesat Signal Cut." Media Network. Jan. 30, 2011. Last accessed on June 27, 2012 at <http://blogs.rnw.nl/medianetwork/al-jazeera-kicked-out-of-egypt-nilesat-signal-cut>.

161. Sakr, Naomi. *Satellite Realms: Transnational Television, Globalization and The Middle East.* New York: I. B. Tauris, 2001. 163

and the distribution of news to other channels.[162] Despite this, there is no specialized regulatory body governing the broadcast market.[163] New channels wishing to start up in Egypt had to abide by a minimum capitalization of US$4.5 million and vow to maintain the same genre of content.[164] Decree 411 of 2000 is concerned with television licensing and restricts companies from broadcasting content of their choice, especially with regard to news content.[165]

## OWNERSHIP

ERTU was the first company to use Arabsat by launching the Egyptian Space Channel (ESC) in December 1990.[166] The ESC became so popular that Kuwaiti and Bahraini governments arranged to distribute the channel nationally through terrestrial networks.[167] Also owned by ERTU,[168] NTN started in 1993 under the name Nile TV International.[169] It offers nine specialized channels, all of which became free-to-air as of 2005.[170]

Privately owned Dream TV, Egypt's first satellite contender, launched in 2001 with start-up capital of US$5 million,[171] a miniscule figure relative to Gulf-financed channel providers. *Al Mehwar* followed Dream TV as the second private satellite channel and launched in early 2002. Well-known businessmen own 83% of the channel. The government owns the remaining 17% through ERTU, EMPC, and Nilesat.[172] While the content aired on both channels is theoretically considered "editorially independent,"[173] both channels are restricted from broadcasting news despite ERTU's ownership stake in these private companies.[174]

The launch of the first private channels, owned largely by business moguls with close ties to Mubarak's regime, initiated a trend in Egypt where businessmen have shown a vested interest in the power of media. This also includes *Al Hayat* Channel, owned by Badawi,[175] and Sawiris' ONTV. After the revolution of January 25, 2011, additional channels have acquired licenses[176] and are run by business magnates[177]

162. Hamzawy, Amr and Dunne, Michele. "Media Freedom Restricted as Egyptian Parliamentary Elections Approach." *Carnegie Endowment for International Peace.* Last accessed on Jan. 17, 2012 at <http://egyptelections.carnegieendowment.org/2010/10/25/media-freedom-restricted-as-egyptian-parliamentary-elections-approach>.

163. Allam, Rasha. "Media Landscape: Egypt." European Journalism Centre. Nov. 8, 2010. Last accessed on Dec. 29, 2011 at <http://www.ejc.net/media_landscape/article/egypt/>.

164. Mekay, Emad. "TV Stations Multiply as Egyptian Censorship Falls." *The New York Times.* July 3, 2011. Last accessed on Jan. 16, 2012 at <http://www.nytimes.com/2011/07/14/world/middleeast/14iht-M14B-EGYPT-MEDIA.html?_r=2&pagewanted=all>.

165. Sakr, Naomi. *Arab Television Today.* New York: I. B. Tauris, 2007. 30.

166. Allam, Rasha. "Media Landscape: Egypt." European Journalism Centre. Nov. 8, 2010. Last accessed on Dec. 29, 2011 at <http://www.ejc.net/media_landscape/article/egypt/>.

167. Sakr, Naomi. *Satellite Realms: Transnational Television, Globalization and The Middle East.* New York: I. B. Tauris, 2001. 32.

168. *Nile Television Network (NTN).* Last accessed on 27 June 2012 at <http://www.ntntv.net/>.

169. State Information Service (SIS). Egypt Yearbook 2007. 2007. Last accessed on July 3, 2012 at <http://www.sis.gov.eg/En/ListTemplate.aspx?Category_ID=870>.

170. State Information Service (SIS). Egypt Yearbook 2007. 2007. Last accessed on July 3, 2012 at <http://www.sis.gov.eg/En/ListTemplate.aspx?Category_ID=870>.

171. Hamdy, Naila. "A Dream TV Come True." Transnational Broadcasting Studies (TBS), No. 8, (Spring 2002). Last accessed on Jan. 17, 2012 at <http://www.tbsjournal.com/Archives/Spring02/sirhan.html>.

172. Hamdy, Naila. "El Mehwar the Mercurial." Transnational Broadcasting Studies (TBS). No. 9, (Fall/Winter 2002). Last accessed on January 17, 2012 at <http://www.tbsjournal.com/Archives/Fall02/Mehwar.html>.

173. Hamdy, Naila. "El Mehwar the Mercurial." *Transnational Broadcasting Studies* (TBS*). No.* 9, (Fall/Winter 2002). Last accessed on January 17, 2012 at <http://www.tbsjournal.com/Archives/Fall02/Mehwar.html>.

174. Sakr, Naomi. *Arab Television Today.* New York: I. B. Tauris, 2007.

175. "Al Sayed Al Badawy." Ahram Online. Nov. 19, 2011. Last accessed on Jan. 16, 2012 at <http://english.ahram.org.eg/NewsContent/33/102/26705/Elections-/Whos-who/AlSayed-AlBadawi.aspx>.

176. Mekay, Emad. "TV Stations Multiply as Egyptian Censorship Falls." *The New York Times.* July 13, 2011. Last accessed on Jan. 16, 2012 at <http://www.nytimes.com/2011/07/14/world/middleeast/14iht-M14B-EGYPT-MEDIA.html?_r=2&pagewanted=all>.

177. Attalah, Lina and Rizk, Nagla. "Egypt's Evolving Media Landscape: Access, Public Interest and Control," in *A New Frontier, An Old Landscape.* Report for the Ford Foundation, November 2011, 118–136. Last accessed on January 16, 2012 at <http://participatorymedia.lab.asu.edu/files/NewFrontierOldLandscape.pdf>.

such as Mohammed Al Amin's CBC.[178] Amin also owns considerable shares in *Al Nahar* satellite channel, *Al Youm Al Sabae* newspaper,[179] and *Al Watan* daily.

In the regional landscape, the early 1990s saw a boost from Saudi-owned satellite channels including MBC and ART.[180] MBC was first launched in 1991 from London, and has been headquartered in Dubai Media City[181] since 2002. MBC is a market leader among Arab media conglomerates with strong viewership in the Pan-Arab region. MBC owns 10 television stations in addition to two radio stations, online interfaces, and documentary producers.[182]

Saudi control of the market is furthered by Rotana Group, which is considered one of the largest media companies in the region. Owned by Al Walid Bin Talal and part of the Kingdom Holding Company,[183] Rotana brings strong representation from the Gulf across a number of entertainment media outlets. News sources in early 2010 announced plans from management to list Rotana on the London and Dubai stock exchange with reported valuations for Rotana Group amounting to just over US$770 million.[184] Rotana's satellite division hosts 11 channels featuring music and film entertainment. A new general entertainment channel targeted specifically at the Egyptian market[185] has recently been announced. Apart from satellite television, Rotana owns the largest record label in the Arab World,[186] prints its own magazine, and is a major distributor of Arabic movies, with a library exceeding 1,600 films.[187]

As early as 2003, when Prince Talal fully acquired Rotana,[188] it became clear that his strategy was to work toward consolidating a media empire. That same year he reduced his previous stake in ART to 5% in addition to creating the Rotana music channel, based on ART's previously existing music channel.[189] Prince Talal increased his stake in LBC to 85% in July 2008, and merged it with Rotana.[190] LBC was already a conglomerate in its own right, privately owned by Lebanese businessmen. Rotana's cross-ownership in the Arab world also extends to global partners. Talal owns a 7% stake in News Corp[191] and conversely, Rupert Murdoch has come to own a 9% stake in Rotana as of 2010.[192] This partnership between two burgeoning media empires is expected to grow as News Corp has the option to extend its stake in Rotana. Talal also owns sizeable shares in Time Warner and Euro Disney.[193] Rotana's

178. Ibid.

179. Solayman, Hanan. "New TV Channels Hit Egypt's Airwaves." Daily News Egypt. July 17, 2011. Last accessed on June 27, 2012 at <http://www.thedailynewsegypt.com/new-tv-channels-hit-egypts-airwaves.html>.

180. Sakr, Naomi. *Satellite Realms: Transnational Television, Globalization and The Middle East.* New York: I. B. Tauris, 2001. 33.

181. Kawach, Nadim. "Saudi cleric slams MBC over TV series." Emirates 24/7. Aug. 18, 2010. Last accessed on May 29, 2012 at <http://www.emirates247.com/entertainment/films-music/saudi-cleric-slams-mbc-over-tv-series-2010-2008-18-11.280816>.

182. "MBC, Largest Media Group in the Middle East Chooses Magnolia as Strategic Platform." Magnolia. Mar. 9, 2011. Last accessed on May 29, 2012 at <http://www.magnolia-cms.com/our-company/news/press-releases/mbc-usa.html>.

183. "Rotana Launches Egyptian General Entertainment Channel." June 1, 2011. Arab News. Last accessed on July 5, 2012 at <http://mediame.com/news/broadcasting_news/rotana_launches_egyptian_general_entertainment_channel>.

184. "Murdoch Buys into Rotana." Arab News. 23 Feb. 2010. Last accessed on Dec. 19, 2011 at <http://arabnews.com/saudiarabia/article21517.ece>.

185. "Rotana Launches Egyptian General Entertainment Channel." June 1, 2011. Arab News. Last accessed on July 5, 2012 at <http://mediame.com/news/broadcasting_news/rotana_launches_egyptian_general_entertainment_channel>.

186. "Murdoch Buys into Rotana." Arab News. 23 Feb. 2010. Last accessed on Dec. 19, 2011 at <http://arabnews.com/saudiarabia/article21517.ece>.

187. "Rotana Launches Egyptian General Entertainment Channel." June 1, 2011. Arab News. Last accessed on July 5, 2012 at <http://mediame.com/news/broadcasting_news/rotana_launches_egyptian_general_entertainment_channel>.

188. Sakr, Naomi. *Arab Television Today.* New York: I.B. Tauris, 2007. 122

189. Ibid.

190. Jaafar, A. "Mixed Results for LBC – Rotana: Arab Conglomerate Tries to Compete with MBC." Variety. 2 Oct. 2008. Last accessed on Jan. 17, 2012 at <http://www.variety.com/article/VR1117993363>.

191. "Murdoch Buys into Rotana." Arab News. Feb. 23, 2010. Last accessed on Dec. 19, 2011 at <http://arabnews.com/saudiarabia/article21517.ece>.

192. Ibid.

193. Ibid.

merger with LBC and expansion into other media forms is often considered part of a strategy designed to take on MBC.[194]

Another important player is Al Jazeera. Owned and financed by the government of Qatar, Al Jazeera Satellite Channel (JSC) was launched in November 1996 as the first Arab channel to offer 24-hour news.[195] In 2006, with a US$150 million grant from the Emir of Qatar,[196] Al Jazeera International was deployed, earning it attention from Western audiences.[197] After Mubarak stepped down in February 2011,[198] Al Jazeera Mubasher Misr was launched without a license to cover live events in Egypt. It was shut down in September 2011.[199] Al Jazeera has been the site of tensions with the Egyptian and Saudi regimes, exemplifying regional media battles. After ousting his father in 1995, the new Crown Prince of Qatar, Sheikh Hamad Bin Khalifa Al Thani, developed a rift with Saudi Arabia and Egypt. Initially, Al Jazeera attacked Egypt and Saudi Arabia because of their negative newspaper coverage of this bloodless coup and because they housed the former Qatari regime's supporters.[200] Owned by the Emir himself, Al Jazeera quickly earned a reputation for uncensored criticism of Egyptian and Saudi rule.

The market has included other regional conglomerates. For example, Future Television Network is a private Lebanese channel founded in 1993 by assassinated Lebanese Prime Minister Rafik Al Hariri and is presently owned by Lebanese businessmen. *Al Rai* is a Kuwaiti-based conglomerate, which owns a TV channel (broadcast on Nilesat and Arabsat), radio stations, and the newspaper *Al Rai*. In 2008–2009, the conglomerate expanded to include a printing press, *Al Nasher*, and it now owns the entire production chain from producing content to publishing it. It also produces TV drama series in Egypt and in Syria.[201]

Although the video channels sector remains concentrated, a number of smaller noteworthy players exist. Especially important is the surge in the number of private channels since the onset of the January 25 revolution. Whether or not these channels have obtained licenses to broadcast, and why some have been shut down, remains unclear.[202] While some channels are completely new and are either owned by businessmen or represent certain political parties, some preexisting channels have launched live news coverage channels. None have been able to obtain an actual license to do so.[203]

An increasingly important local player adding to the voices on the scene is Sawiris' ONTV. Another channel owned by Sawiris is OTV, which began broadcasting in 2007 as an entertainment channel.[204] The quest to broadcast live by non-state-owned channels an issue of contention. For example, ONTV's

---

194. Hagey, Keach. "News Corp Eyes Stake in Saudi Firm Rotana Media." The National. Sept. 2, 2009. Last accessed on Jan. 17, 2012 at <http://www.thenational.ae/business/media/news-corp-eyes-stake-in-saudi-firm-rotana-media>.

195. "Human Resources Division." Al Jazeera Network. Last accessed on June 27, 2012 at <http://recruitment.aljazeera.net/english/>.

196. Khalil, Joe. "News Television in the Arabian Gulf... Period of Transitions." Global Media Journal 5, no. 8 (2004).

197. Al Jazeera. "Corporate Profile." Al Jazeera. Last accessed Dec. 29, 2011 at <http://www.aljazeera.com/aboutus/2006/11/2008525185555444449.html>.

198. "Minister: Al Jazeera Mubasher Misr Shut Down for Disregarding Law." Al Masry Al Youm. Sept. 12, 2011. Last accessed on Jan. 17, 2012 at <http://www.almasryalyoum.com/en/node/494878>.

199. "Al Jazeera Closed for Operating Unlicensed, Sawiris Considers Qatar-based Channel." Egypt Independent. Sept. 17, 2011. Al Mary Al Youm. Last accessed on Jan. 17, 2012 at <http://www.almasryalyoum.com/en/node/496619>.

200. Fandy, Mamoun. *(Un)Civil War of Words: Media and Politics in the Arab World.* Westport, CT: Praeger, 2007. 46.

201. Al Rai Media Group. "Al Rai Media Group." Al Rai Media Group. Last accessed on June 27, 2012 at <http://www.alraimedia.com/Alrai/MediaGroup.aspx>.

202. "Al Jazeera Closed for Operating Unlicensed, Sawiris Considers Qatar-based Channel." Egypt Independent. Sept. 17, 2011. Al Mary Al Youm. Last accessed on Jan. 17, 2012 at <http://www.almasryalyoum.com/en/node/496619>.

203. Mekay, Emad. "TV Stations Multiply as Egyptian Censorship Falls." *The New York Times.* July 13, 2011. Last accessed on Jan. 16, 2012 at <http://www.nytimes.com/2011/07/14/world/middleeast/14iht-M14B-EGYPT-MEDIA.html?_r=2&pagewanted=all>.

204. Fahim, Joseph. "OTV Hits the Airwaves." Daily News Egypt. Jan. 30, 2007. Last accessed on Dec. 26, 2011 at <http://www.dailystaregypt.com/article.aspx?ArticleID=5308>.

24-hour news station was launched in 2011. Yet, it took a few months before a live coverage channel could launch.[205]

Other popular local Egyptian players include Melody Entertainment Holding and *Mazzika*. Melody was established in 2000 and is engaged in film and music production, radio, mobile entertainment, artist management, and advertising.[206] *Mazzika* is an entertainment channel that is part of the larger *Alam Al Fan* media conglomerate, owns another TV channel, Zoom, and is a large record label that rivals Saudi-based Rotana.[207]

Religious TV programming also has considerable viewership in Egypt. In 2007, Mohamed Hassan, an influential Salafi preacher in Egypt, launched *Al Rahma* TV.[208] At the time, Salafi channels' content focused on religious doctrine and advice and refrained from criticizing the government.[209] The former regime allowed channels to exist with the goal of positioning them as rivals to the Muslim Brotherhood, which was not allowed to operate a TV channel of its own. These channels include *Al Nass,* and *Al Hafith,* which are owned by Saudi conglomerate *Al Baraheen*. They have faced shutdowns by the government based on claims that they were advertising unlicensed medical products.[210]

## CONCLUSION

Although recent law amendments alleviated the official monopoly of the state over terrestrial television, control over the industry has not been relinquished. No terrestrial licenses have been issued and complete concentration resides in the hands of government.

In the arena of satellite television, there is a clear dichotomy in the approach between entertainment and news media. The Egyptian Center for News, incorporating ERTU's and NTN's news channels, broadcasts news bulletins. Alternatively, even though the government holds a stake in privately owned *Al Mehwar* and Dream TV, they are prohibited from airing news and can only host shows with censored political commentary. Meanwhile, the boom in satellite television channels that followed the Mubarak-era control over televised political content is still marred by intermittent clampdowns.

The major privately owned Arab media empires are mostly owned by members of the ruling regimes in the Gulf, making this ownership quasi-private in nature. This showcases the typical marriage between media business and politics in the Arab world. This marriage allows for similar content regulation in different countries and provides for self-censorship within each country within the Arab world.

The quasi-private nature of this industry also brings regional politics into the satellite sphere. Media battles between ERTU's space channels and Al Jazeera are a case in point. Another example is Saudi and Egyptian ownership of satellites Arabsat and Nilesat, which polarizes infrastructural power between these two countries. Although the table and graph for video channel providers exhibit the lowest HHI levels and NI levels in audiovisual media, in reality the figures do not reveal the concealed concentration presented by the ownership dynamics in the hands of politically prominent individuals. Nonetheless, 2001 marked significant private entry into the Egyptian market as an accessible medium across the country to fill a critical

---

205. "Al Jazeera Closed for Operating Unlicensed, Sawiris Considers Qatar-based Channel." Egypt Independent. Sept. 17, 2011. Al Mary Al Youm. Last accessed on Jan. 17, 2012 at <http://www.almasryalyoum.com/en/node/496619>.

206. "Spotlight." Melody Holding. Last accessed on June 27, 2012 at <http://www.melodyholding.com/>.

207. Rizk, Nagla. "Stories from Egypt's Music Industry: De Facto Commons as Alternative to Copyright" in Rizk, Nagla and Shaver, Lea, eds. *Access to Knowledge: New Research on Intellectual Property and Development*. Bloomsbury Academic, 2010, 103. Last accessed on May 29, 2012 at <http://www.aucegypt.edu/Business/A2K4D/Documents/A2KEgypt.pdf>.

208. Field, Nathan and Hamam, Ahmed. "Salafi Satellite TV in Egypt." Arab Media and Society. 2009. Last accessed on Jan. 8, 2012 at <http://www.arabmediasociety.com/?article=712>.

209. Ibid.

210. Mattar, Eman. "El Hefaz and El Nas TV Channels Halted on TV and El Faraeen Warned." Business Investment Banking News. Dec. 12, 2010. Last accessed on Jan. 18, 2012 at <http://www.bibnews.com/en/news_details.php?id=5012>.

informational gap, Egypt's authoritarian regime was wary of opening it up for private ownership. Hence, the barring of any political content from private channels falls outside the scope of the concentration figures shown in Table 29.5. Again, political concentration becomes evident as the regime tightened its reign on radio content and ownership. Media activists see the opening up of radio licensing to nongovernmental enterprises as a key tool to fill the information gap and allow access to knowledge, particularly in the country's peripheries.

Meanwhile, players in the entertainment industry, including satellite channel providers, satellite platforms, and print media have exhibited substantial horizontal integration. A point in case is Rotana, which is backed by the financial clout of the Saudi royal family and has used its brand name to leverage its entry into radio, film production, satellite broadcasting, and magazines.

## TELECOMMUNICATIONS MEDIA

The Egyptian telecom sector is characterized by substantial growth and activity. Telecom revenues experienced some of the highest growth rates in the Middle East and Africa, amounting to $6.4 billion in 2010.[211] Its constant evolution, including deregulation by state authorities, mergers, acquisitions, and high-profile ownership disputes since the inception of the first cellular operator license in 1998 lends the industry a reputation for dynamism. Yet, an examination of the ICT market in Egypt as a whole indicates that growth rates differ among the different segments. In general, wireless is growing faster than wireline for both phone and Internet services.

As evident in the Table 29.6, although mobile phone penetration is experiencing exponential growth and wireless broadband is following suit, fixed broadband is growing at a slower rate, whereas the wireline phone market is steadily shrinking.

The growth, or lack thereof, of each market can be attributed to unique factors within the Egyptian context. To illustrate, high poverty rates have made shared access to ADSL and wirelines common. This matches the trend of shared access of other media referred to earlier in this chapter. Accordingly, official indicators may reveal a spuriously low rate for the fixed market in both Internet and phone. Indeed, statistics reveal, "on average, for every DSL line there are 2.7 households."[212] Generally, lagging demand in the Internet market relative to the cellular phone is due to the 66% literacy rate.[213] Penetration

Table 29-6. Growth of Telecom Sector

| INDUSTRY INDICATOR | 2009 | 2010 | 2011 |
|---|---|---|---|
| Mobile telephone Penetration (Individual) (%) Subscribers per 100 inhabitants | 72% | 90% | 100%[1] |
| Mobile flash drive Internet Subscriptions (millions) | 0.35 | 1.43 | 2.18 |
| Fixed line penetration (%) | 13.4% | 12.2% | 11.29%[2] |
| ADSL subscriptions (millions) | 1.3% | 1.9% | 2.1% |

*Source:* Egypt ICT Indicators Portal, MCIT.

[1] As of October 2011.

[2] "Egypt Major African Mobile Markets: Future Growth Prospects." Africa and Middle East Telecom Week. Last accessed Jan. 18, 2012 at <http://www.africantelecomsnews.com/resources/AfricaOpp_Egypt.shtml>.

211. Rao, Madanmohan. Mobile Africa Report 2011: Regional Hubs of Excellence and Innovation. MobileMonday, 2011. Last accessed on Jan. 18, 2012 at <http://www.mobilemonday.net/reports/MobileAfrica_2011.pdf>.

212. El Azhary, Emad. Personal interview. Nov. 24, 2011.

213. *Egypt: The National Report on Literacy and Adult Education. United Nations Educational, Scientific and Cultural Organization, 2009. Last accessed on Dec. 19, 2011 at* <www.unesco.org/fileadmin/MULTIMEDIA/.../UIL/.../Egypt.pdf>.

rates for the mobile market greatly exceed that of wireline despite higher costs because of the extensive network coverage of wireless lines, as well as the historical bureaucratic difficulties of getting a main landline. Furthermore, competition in the mobile phone market has created an array of affordable schemes and subscriptions to address the less affluent segments of society.

From 1957 to 1998, the telecommunications industry was embodied in the Arab Republic of Egypt National Telecom Organization (ARENTO) because of the legacy of the Nasserist regime. Functioning under the auspices of the Ministry of Transportation and Communication, ARENTO was a state-owned company that encompassed all telecom regulatory and operational functions and was also responsible for road networks (such as highways) and postal services.[214] In 1998, ARENTO was transformed into Telecom Egypt (TE), a joint stock company.[215] This marks the beginning of significant restructuring impetus at the public policy level.[216] Noteworthy, however, is the fact that Law no. 19 of 1998 requires that the Egyptian government maintain a majority stake in TE.[217] In 1999, the Ministry of Communications and Information Technology (MCIT) was inaugurated by presidential decree with the mandate to regulate and develop Egypt's information and communications technology sector, separating regulatory functions from TE's operational role in the market.[218] These functions were then relayed to the National Telecom Regulatory Authority (NTRA), established in 2003.[219]

NTRA's mandate encompasses the liberalization of the telecom sector, expanding the broadband network capacity, developing technological foundations for increased data services and reforming telecom regulations in order to promote Egypt's information sector and e-economy.[220] In 2009, NTRA launched two triple-play licenses[221] that allow cable television, Internet, and phone over a single-access broadband connection. This has created an increased potential for media convergence, as the three wireless operators integrate a host of services into one subscription. The implications of a triple play license and the potential benefit to these companies remain vague since the auctioned license is geographically limited and is in its experimental phase.

Today, the structure of the telecom industry in Egypt resembles the one prevalent in a number of other countries (e.g., France and the UAE). The state retains the monopoly over the infrastructural backbone of the industry and offers wholesale services to licensed operators through its joint-stock company, TE. The state has allowed private and foreign participation in retail markets through offering bandwidth, hosting other infrastructural services at economically viable prices.[222]

## Wireline

TE continues to be Egypt's sole wireline provider (Table 29.7; Graph 29.6). The Ministry of Transport and Communications was in charge of setting the overall strategy for the industry

214. Saleh, Nivien. Third World Citizens and the Information Technology Revolution. New York: Palgrave Macmillan, 2010.

215. Telecom Egypt. "Basic Information." Telecom Egypt. Last accessed Nov. 27, 2011 at <http://ir.telecomegypt.com.eg/Basic Information.asp#>.

216. Telecom Egypt. "Company Profile." Telecom Egypt. Last accessed on Nov. 27, 2011 at <http://ir.telecomegypt.com.eg/Basic Information.asp#>.

217. Ibid.

218. MCIT. Ministry of Communications and Information Technology eProtal. Last accessed on Nov. 29, 2011 at <http://www.mcit.gov.eg/TeleCommunications/Historical_Prespective>.

219. National Telecom Regulatory Authority (NTRA). "About Us." National Telecom Regulatory Authority (NTRA). Last accessed on June 27, 2012 at <http://www.tra.gov.eg/english/DPages_DPagesDetails.asp?ID=180&Menu=5>.

220. National Telecom Regulatory Authority (NTRA). "Media Convergence Issues in Egypt and the Region and Prospects for the Future." Scribd. 2009. Last accessed Nov. 27, 2011 at <www.scribd.com/doc/38482708/Doc9-NTRA>.

221. Telecom Egypt, "Annual Report 2009: Equipped For A New Era," Giza, 2009, 24. Last accessed on Nov. 27, 2011 at <http://ir.telecomegypt.com.eg/admin/uploads/annual-report-2009/FINAL %20 ANNUAL%20web.pdf. 24>.

222. Telecom Egypt. "Wholesale." Telecom Egypt. 2008. Last accessed on Jan. 3, 2012 at <http://www.telecomegypt.com.eg/English/wholesale_MPLS_IP-VPN.asp#>.

## Table 29-7. Wireline Telecommunications Concentration

WIRELINE TELECOMMUNICATIONS

| MARKET SHARES (%) | 1984 | 1988 | 1992 | 1996 | 2000 | 2004 | 2006 | 2007 | 2008 | 2009 | 2010 |
|---|---|---|---|---|---|---|---|---|---|---|---|
| Telecom Egypt (Public, Egypt) | 100.0 | 100.0 | 100.0 | 100.0 | 100.0 | 100.0 | 100.0 | 100.0 | 100.0 | 100.0 | 100.0 |
| Revenue from wireline ONLY (mil EGP)[1] | | | | | | | 5,361.4 | 4,999.7 | 3,014.9 | 2,618.3 | |
| Total Rev,[2] including services (mil EGP) | | | | | 3,181.8 | 7,802.2 | 9,488.4 | 9,993.1 | 10,116.9 | 9,960.3 | 10,318.0 |
| Revenue (mil $)[3] | | | | | | 1260.5 | 1647.9 | 1754.2 | 1897.0 | 1779.8 | 1811.7 |
| C1 | 100.0 | 100.0 | 100.0 | 100.0 | 100.0 | 100.0 | 100.0 | 100.0 | 100.0 | 100.0 | 100.0 |
| HHI | 10,000 | 10,000 | 10,000 | 10,000 | 10,000 | 10,000 | 10,000 | 10,000 | 10,000 | 10,000 | 10,000 |
| Noam Index | 10,000 | 10,000 | 10,000 | 10,000 | 10,000 | 10,000 | 10,000 | 10,000 | 10,000 | 10,000 | 10,000 |
| Number of players (*n*) | 1 | 1 | 1 | 1 | 1 | 1 | 1 | 1 | 1 | 1 | 1 |

*Source:* Telecom Egypt Annual Reports 2006–2009 at www.telecomegypt.com.eg.

[1] Source: Emad El Azhary, Email Interview 29 November 2011.

[2] Including revenues from other sources e.g. leased optical fiber cables and presence in wireless telephony and fixed broadband.

[3] Converted from original Total Rev in EGP using respective year exchange rate from the Central Bank of Egypt's Annual Reports.

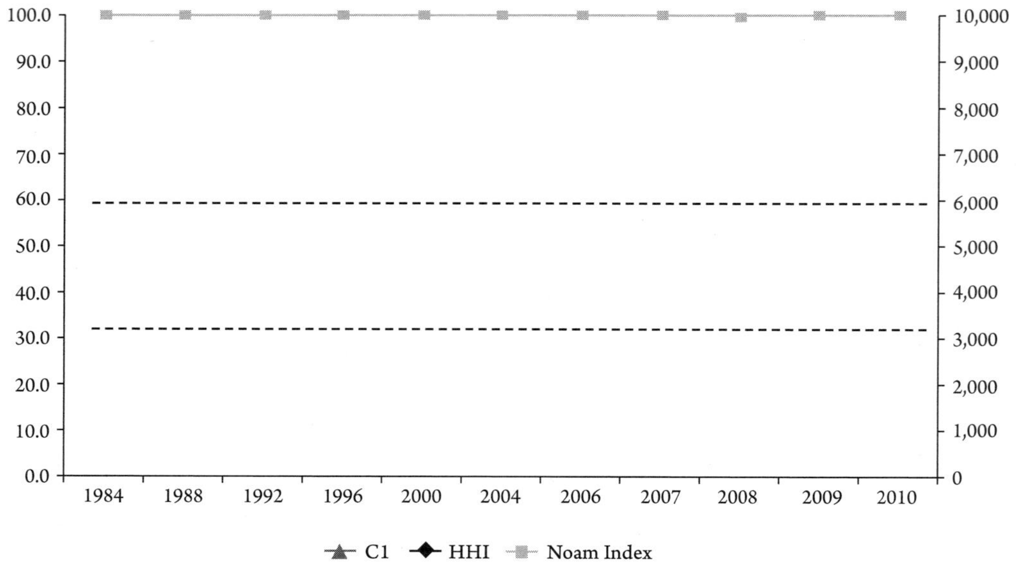


GRAPH 29.6 Wireline Telecommunications Concentration

while ARENTO strictly held responsibilities for operational functions.[223] The Ministry lacked a clear and calculated vision for telecom development in Egypt and liberally reallocated ARENTO's revenues toward other infrastructural projects, a common practice in most less developed countries.[224] ARENTO also had to dedicate the sum of US$17 million of its annual revenues toward the Ministry of Transport and Communications, further debilitating its ability to expand its network coverage.[225]

The concentration of telecom in the hands of one state company bore negative implications on the supply and accessibility of landlines. ARENTO's customers suffered from long waiting lists, creating a backlog of 1.2 million lines in 1994 due to a lack of infrastructural capacity.[226] Additionally, poor network coverage limited to cities, as well as substandard customer service and failure to collect delinquent revenues, gave the company its notorious reputation for inefficiency.[227]

In 1998, the liberalization of the telecommunication sector was kick-started with ARENTO becoming TE. At that point, it became legally equipped to play a strategic role in other telecom-related industries through mergers and acquisitions.[228] The same year saw the issuance of licenses for two payphone operators, Menatel and Nile Telecom. In 1998, a consortium of private and public corporations, including Orascom, France Telecom, and *Al Ahram* Press, bought out TE's 70% share in Egyptian Company for Mobile Services (ECMS).[229] Additionally, Vodafone acquired the second operating license and entered the market as Misrfone. The introduction of these players proved a success in terms of enhanced service and quality standards.[230]

223. Telecom Egypt, "Annual Report 2009: Equipped For A New Era," Giza, 2009, 24. Last accessed on Nov. 27, 2011 at <http://ir.telecomegypt.com.eg/admin/uploads/annual-report-2009/FINAL%20 ANNUAL%20 web.pdfp.24>.

224. Saleh, Nivien. *Third World Citizens and the Information Technology Revolution*. New York: Palgrave Macmillan, 2010. 147.

225. Saleh, Nivien. *Third World Citizens and the Information Technology Revolution*. New York: Palgrave Macmillan, 2010. 148–149.

226. K&M Engineering and Consulting Corporation. "ARENTO Cost of Service Study." USAID. National Telecommunications Organization & the U.S. Agency for International Development, 2011, 105. Last accessed on Dec. 19, 2011 at <http://pdf.usaid.gov/pdf_docs/PNABS218.pdf>.

227. Ibid.

228. El Azhary, Emad. Personal interview. Nov. 24, 2011.

229. Saleh, Nivien. *Third World Citizens and the Information Technology Revolution*. New York: Palgrave Macmillan, 2010.

230. Ibid.

## OWNERSHIP

In 2005, 20% of the TE's total shares were made available in an initial public offering on the Cairo and London Stock Exchanges,[231] while the Egyptian government retained the remaining 80%.[232] TE offers retail and wholesale services. Its main retail service is fixed line telephony. Its cross-ownership in wireless telecom is through its 45% ownership stake in Vodafone Egypt.[233] TE also owns TEData,[234] the leading Internet service provider for fixed broadband. TE's presence in these different segments ensures that the company maintains its status as a strong player in Egypt's ICT sector and gives the government a firm footing in the telecom industry through indirect ownership.

Although TE has been experiencing dwindling retail revenues and subscriptions in the fixed wireline market, these are countered by growing wholesale revenues from leased optical fiber cables that run across the entire country, coupled with its presence in wireless telephony as well as wireless and fixed broadband.[235] While TE is the sole provider of wireline services in Egypt, its diversified, tactical, and low-risk investment strategy and infrastructural assets established it as a key constituent in the ICT sector. In addition to local and international wireline services, economic barriers to entry for new players mean that TE monopolizes the nation's optic fiber backbone and provides a range of wholesale services that include the leasing of telephony capacity and data services to wireless operators and ISPs.[236]

In 2008, NTRA announced the bid through an auction[237] for a second fixed-line license,[238] heralding the end of TE's monopoly. Nevertheless, it postponed the bid several times on grounds of the then deteriorating economic conditions, which would have allegedly affected the success of the auction.[239] The license would span for a period of 20 years and offer triple play functions, with the exception of mobile services.[240]

The decision to offer the second license was an attempt to increase competition in the wireline market. Strategically, NTRA prohibited any of the three wireless operators, Mobinil, Vodafone, or Etisalat, to participate in the auction to protect the "welfare of the telecom user."[241] Allowing these operators who were already strongly present in wireless and ISP segments to bid for a fixed line license reinforced an even stronger oligopoly. The same ruling, however, did not apply to the shareholders of the mobile operators.[242] TE's Senior Vice President for Commercial Affairs attributed the failure of the bid for a second fixed line license to the fact that

231. Telecom Egypt. "Ownership Structure." Telecom Egypt. Last accessed on June 27, 2012 at <http://telecom.tedata.net/uipages/Ownership_Structure.aspx>.

232. Telecom Egypt. "Company Milestones." Telecom Egypt. Last accessed on June 27, 2012 at <http://ir.telecomegypt.com.eg/Company%20 Milestones.asp>.

233. Telecom Egypt. "Company Fact Sheet." Telecom Egypt. Last accessed on June 27, 2012 at <http://ir.telecomegypt.com.eg/Company%20 Fact%20 Sheets.asp>.

234. "Subsidiaries and Investments." Telecom Egypt. Last accessed on June 27, 2012 at <http://ir.telecomegypt.com.eg/Subsidiaries.asp>.

235. El Azhary, Emad. Personal interview. Nov. 24, 2011.

236. Telecom Egypt. "Company Fact Sheet." Telecom Egypt. Last accessed on June 27, 2012 at <http://ir.telecomegypt.com.eg/Company%20 Fact%20 Sheets.asp>.

237. Reuters. "Egypt to Auction 2nd Fixed-line License on June 19." Balancing Act Africa, no. 394. Last accessed on Jan. 4, 2012 at <http://www.balancingact-africa.com/news/en/issue-no-394/telecoms/egypt-to-auction-2nd/en>.

238. National Telecom Regulatory Authority (NRTA). "The NTRA Offering the Second Fixed-Line License." National Telecommunications Regulatory Authority. Last accessed on Dec. 19, 2011 at <http://www.tra.gov.eg/english/news_newsdetails.asp?PID=36&ID=129>.

239. Middle East Rating and Investors Service. "Egyptian Company for Mobile Services S.A.E (Mobinil)." Merisratings.com. November 2009. Last accessed Jan. 10, 2011 at <http://www.mobinil.com/Investors/FinancialReports/doc/CreditRating-ReportNov092ndIssue.pdf>.

240. Ibid.

241. National Telecom Regulatory Authority (NRTA). "The NTRA Offering the Second Fixed-Line License." National Telecommunications Regulatory Authority. Last accessed on Dec. 19, 2011 at <http://www.tra.gov.eg/english/news_newsdetails.asp?PID=36&ID=129>.

242. Ibid.

barriers to entry to this market were no longer political, but economic.[243] Rival fixed-line telephony is difficult due to the shrinking market and major upfront investments. In contrast, the retail revenue in wireless markets attracts more attention from investors because it is an easier option.

TE's outbound operations span across a number of IT industries and include three companies: Xceed, Centra, and MERC. Xceed, a leading regional call center, began operations in 2003. TE also owns Centra, an IT solutions provider established in 2002. In addition to providing complete IT solutions, the company produces top-quality PCs, notebooks, and servers. MERC is a joint stock company established in 2001, which builds, operates, and manages wireless communication stations.[244]

## Wireless

Although the mobile market is relatively in its nascence, with the first private telecom operator founded in 1998, the industry has been evolving at a brisk pace. ARENTO founded the Egyptian Company for Mobile Services (ECMS) in 1996, and a year later opened the bid for cellular operating licenses to the private sector.[245] In May 1998, Mobinil acquired ECMS. Also in 1998, Mobinil acquired Egypt's first GSM license.[246] In 2001, ARENTO granted the second GSM license to Click GSM, renamed Vodafone Egypt.[247]

Mobinil and Vodafone's duopoly lasted from 1998 until 2007. In 2007, the government licensed a third operator, Etisalat, to increase competition and improve network coverage of Egypt's growing wireless market.[248] Despite the presence of a solid duopoly, Etisalat's entry heralded record growth rates in Egypt's mobile market.[249] Although Vodafone and Mobinil minimized their prices and removed administrative fees in an effort to retain customers,[250] Etisalat successfully countered these attempts. It was the first operator to provide 3.5G technologies introducing mobile Internet among other new data services,[251] as well as the cheapest call rates.[252] These services and rates helped it earn 1 million subscribers within the first 50 days of operating.[253]

From the onset, Mobinil and Vodafone have had almost an equal 50% market share. Using subscribers as a proxy puts Mobinil in the lead with slightly over 51% market share, whereas using ARPU would rank Vodafone ahead of Mobinil. The market for mobile telephony has been growing steadily. Upon entry into the market, Etisalat was able to capture 8.6% of the market, while Mobinil and Vodafone remain in the same relative positions, with 48.7% and 42.7% market shares, respectively, in 2008.

---

243. El Azhary, Emad. Personal interview. Nov. 24, 2011.

244. "Subsidiaries and Investments." Telecom Egypt. Last accessed on June 27, 2012 at <http://ir.telecomegypt.com.eg/Subsidiaries.asp>.

245. Hassanein, Leila. "Access to Infrastructure: Egypt." GISWatch. 2008. Last accessed Dec. 22, 2010 at <http://www.giswatch.org/sites/default/files/Egypt_0.pdf>.

246. Saleh, Nivien. *Third World Citizens and the Information Technology Revolution*. New York: Palgrave Macmillan, 2010.

247. Al Bawaba. "Click to Become IPO in Egypt." Al Bawaba News. May 2, 2001. Last accessed on 18 Dec. 2011 at <http://www.albawaba.com/news/click-gsm-become-ipo-egypt>.

248. Hassanein, Leila. "Access to Infrastructure: Egypt." GISWatch. 2008. Last accessed Dec. 22, 2010 at <http://www.giswatch.org/sites/default/files/Egypt_0.pdf>.

249. Kotler, P and Armstrong, Gray, eds. "Pricing: Understanding and Capturing Customer Value." In *Principles of Marketing*. New York: Pearson Prentice Hall, 2012. Last accessed on June 27, 2012 at <http://catalogue.pearsoned.co.uk/assets/hip/gb/hip_gb_pearsonhighered/samplechapter/0273752502.pdf>.

250. Ibid.

251. Etisalat. "Corporate Information." *Etisalat*. Last accessed June 27, 2012 at <http://etisalat.com.eg/etisalat/Etisalat_Portal_En/about/corporate.htm?_pageid=42,1&_dad=portal&_schema=PORTAL&siteAlias=etisalat&sitePath=Etisalat_Portal_En&kpAlias=about&pageAlias=corporate>.

252. Kotler, P and Armstrong, Gray, eds. "Pricing: Understanding and Capturing Customer Value." In *Principles of Marketing*. New York: Pearson Prentice Hall, 2012. Last accessed on June 27, 2012 at <http://catalogue.pearsoned.co.uk/assets/hip/gb/hip_gb_pearsonhighered/samplechapter/0273752502.pdf>.

253. Etisalat. "Corporate Information." *Etisalat*. Last accessed June 27, 2012 at <http://etisalat.com.eg/etisalat/Etisalat_Portal_En/about/corporate.htm?_pageid=42,1&_dad=portal&_schema=PORTAL&siteAlias=etisalat&sitePath=Etisalat_Portal_En&kpAlias=about&pageAlias=corporate>.

Early in May 2011, NTRA announced it would start taking tenders for a fourth mobile operator using a "mobile virtual network license," which uses the infrastructure and network of other existing operators.[254] In this case, the eventual license holder would resell mobile services that the company buys from Mobinil, Vodafone Egypt, or Etisalat. TE has announced interest in acquiring the mobile virtual network license, which the mobile operators claim is "an unfair advantage in the marketplace," given that TE already has a monopoly over wireline.[255]

The wireless telecommunications sector in Egypt includes significant private sector participation and foreign influence compared to other media industries. All operators are joint stock companies. While Mobinil is headquartered in Egypt, Vodafone is headquartered in the United Kingdom and Etisalat Egypt is located in the UAE. The three companies operate within the jurisdiction of NTRA. Although technically the market is dominated by an oligopoly, in reality, it is highly competitive. Prices have been declining and mobile penetration rates in Egypt have been soaring. Official figures indicate that penetration has reached 100 subscribers per 100 inhabitants as of October 2011.[256]

### OWNERSHIP

The ownership structure of each of the three private operators bears interesting insight. The Egyptian Company for Mobile Services (Mobinil), a local private firm listed on the Egyptian stock exchange, offers wireless phone and data services through USBs and mobile Internet. Although first in the wireless market, Mobinil was the last of the three companies to enter the fixed broadband market in 2010 by acquiring Egypt's second largest DSL provider, LinkdotNet. Until early 2011, Orascom Telecom Holding (OTH) and France Telecom owned 35% and 36% of the shares, respectively, in Mobinil. The remaining 29% was publicly traded.[257] In February 2012, France Telecom acquired 71% of the shares in Mobinil, with the remaining 29% owned by OTH.[258] In June 2012, France Telecom acquired 94% of Mobinil.[259]

OTH is one of Egypt's largest conglomerates, owned by business mogul Sawiris and family. Apart from Mobinil in Egypt, Orascom Telecom owns significant stakes in foreign telecom enterprises in over 10 countries in the Middle East and Africa, ranging from wireless and wireline telecom, ISPs, and IT hardware.[260] France Telecom is one of the largest telecom networks worldwide, with global operations in a range of telecom-related industries including ISP and cable television (Orascom Group, 2011). In 2008, Mobinil filed a legal suit against TE on the back of TE lowering its fixed-to-mobile rates despite a legal agreement between the two (interconnect agreement). On September 3, 2008, NTRA supported TE in its position, ruling in favor of TE. Consequently, Mobinil filed a lawsuit against the regulator in November 2008 as its decision contradicts the contracts between TE and the operator. In September 2009, Mobinil also filed

254. "Telecom Watchdog Decision Clears Way for 4th Egyptian Mobile Operator." Ahram Online. May 7, 2012. Last accessed on June 4, 2012 at <http://english.ahram.org.eg/NewsContent/3/12/40995/Business/Economy/Telecom-watchdog-decision-clears-way-for-th-Egypti.aspx>.

255. "Telecom Watchdog Decision Clears Way for 4th Egyptian Mobile Operator." Ahram Online. May 7, 2012. Last accessed on June 4, 2012 at <http://english.ahram.org.eg/NewsContent/3/12/40995/Business/Economy/Telecom-watchdog-decision-clears-way-for-th-Egypti.aspx>.

256. Ministry of Communications and Information Technology (MCIT). "ICT Infrastructure and Access." Egypt ICT Indicators. Last accessed on Jan. 5, 2012 at <http://www.new.egyptictindicators.gov.eg/en/Indicators/_layouts/KeyIndicatorsViewer.aspx>.

257. Mobinil. "Annual Report 2010: Egyptian Company for Mobile Services (ECMS)." Mobinil. Last accessed on June 27, 2012 at <http://www.mobinil.com/investors/AnnualReport2010.aspx>.

258. Sahota, Dawinderpal. "France Telecom Increases Stake in Egypt's Mobinil." Telecoms.com, Feb. 13, 2012. Last accessed on July 27, 2012 at <http://www.telecoms.com/tag/mobinil/>.

259. "An Overly French Mobinil." Egypt Independent, no. 4, June 7, 2012, 11. Last accessed on June 27, 2012 at <http://www.scribd.com/doc/96300554/Egypt-Independent-Jun-07>.

260. Orascom Telecom. "Our History." Orascom Telecom. Last accessed on June 27, 2012 at <http://www.orascomtelecom.com/about/Contents/default.aspx?ID=395>.

for arbitration against TE. At the end of 2009, NTRA issued a decree changing the interconnect tariffs in favor of TE. Mobinil continues its legal effort against this decision.[261]

What began as a dispute over an agreement signed in 2001 between Orascom Telecom and France Telecom (partners in Mobinil's holding company), evolved into a case of international arbitration between the two[262] in 2009. An international court ruled that France Telecom was obliged to buy Orascom Telecom's shares in Mobinil for US$45.5 per share by April 2009.[263] The Egyptian Capital Markets Authority (CMA) and Orascom Telecom argued that the arbitration results meant that France Telecom was to make an offer for the entire company, including minority shareholders' shares. While France Telecom argued that this should be a voluntary offer, it nevertheless made several offers which were rejected on the grounds of the price being lower than that of the other shares.[264] In late 2009, the NTRA confirmed that in the case of Orascom Telecom selling its shares to France Telecom, the presence of an Egyptian partner in Mobinil was mandatory.[265] In April 2010, Orascom Telecom and France Telecom ended all legal disputes between them with regards to their ownership of Mobinil. The signed agreement does not change voting rights and gives Orascom Telecom "put option rights" to sell all its Mobinil and ECMS stake to France Telecom at certain periods up to November 2013.[266]

The other two wireless carriers involve some foreign ownership: 66% of Etisalat Egypt is owned by Etisalat UAE, with Egypt Post, National Bank of Egypt, and Commercial International Bank owning the remaining 34%.[267] Etisalat Egypt is a subsidiary of Etisalat UAE, the largest carrier and network provider in the Middle East, operating in Asia, the Middle East, and Africa.[268] The Emirati government owns a 60% share of Etisalat UAE and the remaining 40% is traded in the Abu Dhabi Securities Exchange (ADX).[269] In 2008, Etisalat Egypt acquired EgyNet and Nile Online, two of the top five contenders in the fixed broadband market.

Vodafone Egypt is a subsidiary of Vodafone Group Plc. and is listed on the London Stock Exchange. It operates in six continents worldwide. Vodafone Egypt is jointly owned by TE (45%) and Britain's Vodafone.[270] Apart from a host of data services including wireless telephony and broadband, Vodafone Egypt was first to enter the fixed line broadband market in 2006 by acquiring Raya, one of the top five DSL providers in Egypt.

---

261. Mobinil. "Annual Report 2010: Egyptian Company for Mobile Services (ECMS)." Mobinil. Last accessed on June 27, 2012 at <http://www.mobinil.com/investors/AnnualReport2010.aspx>.

262. "The Dispute over the Ownership of Egypt's Mobinil." *International Telecommunications Union Regulation Toolkit.* Dec. 22, 2011. International Telecommunications Union. Last accessed on July 27, 2011 at <http://www.ictregulationtoolkit.org/en/PracticeNote.3264.html>.

263. Middle East Rating and Investors Service. "Egyptian Company for Mobile Services S.A.E (Mobinil)." Merisratings.com. November 2009. Last accessed Jan. 10, 2011 at <http://www.mobinil.com/Investors/FinancialReports/doc/CreditRatingReportNov092ndIssue.pdf>.

264. "The Dispute over the Ownership of Egypt's Mobinil." *International Telecommunications Union Regulation Toolkit.* Dec. 22, 2011. International Telecommunications Union. Last accessed on July 27, 2011 at <http://www.ictregulationtoolkit.org/en/PracticeNote.3264.html>.

265. "NTRA Says Finding Local Partner in Mobinil Mandatory." Arab Finance Brokerage Company. Dec. 16, 2009. Last accessed Jan. 10, 2012 at <http://www.arabfinance.com/news/newsdetails.aspx?id=157214>.

266. Middle East Rating and Investors Service. "Egyptian Company for Mobile Services S.A.E (Mobinil)." Merisratings.com. November 2009. Last accessed Jan. 10, 2011 at <http://www.mobinil.com/Investors/FinancialReports/doc/CreditRatingReportNov092ndIssue.pdf>.

267. Etisalat. "Our History." Etisalat. Last accessed on June 27, 2012 at <http://etisalat.com/html/history.html> and "Etisalat Led Group Signs $2.88b Deal." *MENAFN.* Aug. 23, 2006. Last accessed on 4 Jan. 2011 at <https://www.menafn.com/qn_news_story_s.asp?StoryId=1093124586>.

268. Etisalat. "About Us." Etisalat. Last accessed on June 27, 2012 at <http://etisalat.com.eg/etisalat/Etisalat_Portal_En/about/corporate.htm?_pageid=42,1&_dad=portal&_schema=PORTAL&siteAlias=etisalat&sitePath=Etisalat_Portal_En&kpAlias=about&pageAlias=corporat>.

269. Etisalat. "Our History." Etisalat. Last accessed on June 27, 2012 at <http://etisalat.com/html/history.html>.

270. Telecom Egypt. "Company Milestones." Telecom Egypt. Last accessed on June 27, 2012 at <http://ir.telecomegypt.com.eg/Company%20 Milestones.asp>.

State ownership is evident in all three companies. There is a French presence in Mobinil. As of December 2010, the French government owns about 27% of France Telecom. The Egyptian government owns almost half of Vodafone Egypt and the UAE government controls Etisalat. These operators offer a wide range of mobile telephony and Internet services and are not allowed to bid for fixed licenses to prevent monopolistic or oligopolistic structures.[271]

Teletech, an Egyptian consortium led by Vodafone, was the first to receive a triple play license from MCIT. Link followed with its consortium, which includes Orascom Telecom and Mobinil.[272] Mobinil and Vodafone's acquisition of these licenses is in line with a trend toward business models with integrated services, indicative of strong potential for media convergence in the future of the Egyptian market.

## ANALYSIS

Although HHI figures are high (see Table 29.8), indicating a concentrated market, the decline from almost 7,000 to 4,300 between 1998 and 2008 shows greater diversity and competition (Graph 29.7). As evident in the table, C3 remains at 100% throughout the timeline and the HHI and NI take a significant dip at the entry of each new firm. This denotes the successful absorption of sufficient market share despite monopolistic barriers. The ability of new entrants to overcome these barriers exists for several reasons.

First, the mobile market in Egypt is a growing market that was for the most part unsaturated, allowing each entrant a share of a growing pie. Second, all three companies are publicly listed companies and enjoy the backing of large multinational corporations with significant telecom-related expertise and financial resources, allowing them to undergo onerous costs of entry and enter the market on strong footing in the face of fierce price wars.

In turn, the general decline in prices helped enlarge the wireless market even further by making mobile subscriptions more affordable to poorer segments of society, including rural areas that were cut off from main landlines. In Egypt as in many developing countries, mobile phones are easier to acquire than landlines.[273] Even though the latter may be cheaper, we see falling subscriptions[274] and revenues per line for fixed lines.[275] Price wars heightened with the entry of Etisalat further saturating the market, and penetration levels grew at an annual rate of 18.31% in 2010 and peaked at 100% in October 2011.[276] With penetration levels this high, companies focused their efforts on providing superior customer and data services to gain an edge over their competitors, a level of competition that is ideal for advancing consumer welfare despite the oligopolistic state of the industry.

Third, the role of the Egyptian government as an indirect market player in providing the infrastructure and regulating the market helped level the playing field for Etisalat's entrance, thereby encouraging companies to revert to price wars to gain advantage, all for the benefit of the consumer. For example, optical fiber leases from TE for all three operators are non-preferential,[277] thereby

---

271. Middle East Rating and Investors Service. "Egyptian Company for Mobile Services S.A.E (Mobinil)." Merisratings.com. November 2009. Last accessed Jan. 10, 2011 at <http://www.mobinil.com/Investors/FinancialReports/doc/CreditRatingReportNov092ndIssue.pdf>.

272. Ramadan, Amr. "Egypt to Sign Triple-play Licenses, Says Kamel at Euromoney." *Daily News Egypt*. Sept. 30, 2010. Last accessed on July 5, 2012 at <http://www.thefreelibrary.com/Egypt+to+sign+triple-play+licenses,+says+Kamel+at+Euromoney.-a0238670369>.

273. Kotler, P and Armstrong, Gray, eds. "Pricing: Understanding and Capturing Customer Value." In *Principles of Marketing*. New York: Pearson Prentice Hall, 2012. Last accessed on June 27, 2012 at <http://catalogue.pearsoned.co.uk/assets/hip/gb/hip_gb_pearsonhighered/samplechapter/0273752502.pdf>.

274. Ministry of Communications and Information Technologies (MCIT). Dataset. Last accessed on July 5, 2012 at <http://www.mcit.gov.eg/Indicators/Ind_Communications.aspx>.

275. Telecom Egypt. "Annual Reports 2006–2009." Telecom Egypt. Last accessed on May 14, 2012 at <http://ir.telecomegypt.com.eg/Annual%20 Reports.asp>.

276. Ministry of Communications and Information Technology (MCIT). "ICT Infrastructure and Access." Egypt ICT Indicators. Last accessed on Jan. 5, 2012 at <http://www.new.egyptictindicators.gov.eg/en/Indicators/_layouts/KeyIndicatorsViewer.aspx>.

277. El Azhary, Emad. Personal interview. Nov. 24, 2011.

Table 29-8. Wireless Telecommunications Concentration

WIRELESS TELECOMMUNICATIONS

| MARKET SHARES BY SUBSCRIBERS (%) | 1998 | 2000 | 2002 | 2004 | 2006 | 2008 | 2013 |
|---|---|---|---|---|---|---|---|
| Mobinil (Private, Egypt) (94% France Telecom, 5% Orascom Telecom, 1% free float) | 81.0 | 54.6 | 52.3 | 53.3 | 51.5 | 48.7 | 35.1 |
| Vodafone Group Plc. (Private, UK) | | | | | | | |
| Vodafone Egypt (55% Britain's Vodafone, 45% Telecom Egypt, Public) | 19.0 | 45.4 | 47.7 | 46.7 | 48.5 | 42.7 | 40.6 |
| Etisalat (Public/Private, UAE) | | | | | | | |
| EtisalatMisr (66% Etisalat UAE, 34% Egypt Post, National Bank of Egypt and Commercial International Bank) | 0.0 | 0.0 | 0.0 | 0.0 | 0.0 | 8.6 | 24.3 |
| Total Rev (mil EGP) | | | 5,204.5 | 10,056.5 | 15,754.1 | 23,030.4 | |
| Revenue[1] (mil $) | | | 1,156.6 | 1,624.7 | 2,736.1 | 4,318.4 | |
| C3 | 100.0 | 100.0 | 100.0 | 100.0 | 100.0 | 100.0 | 100.0 |
| HHI | 6,922 | 5,042 | 5,011 | 5,022 | 5,005 | 4,269 | 3,471 |
| Noam Index | 4,895 | 3,565 | 3,543 | 3,551 | 3,539 | 2,465 | 2,004 |
| Number of players ($n$) | 2 | 2 | 2 | 2 | 2 | 3 | 3 |

*Source:* Figures obtained from the Ministry of Communications and Information Technology (MCIT).
*Note:* If we are to calculate market shares by revenue we expect almost the same situation, but with Vodafone Egypt ranking first instead of Mobinil since Vodafone's ARPU is higher.
[1] Converted from original in EGP using respective year exchange rate from the Central Bank of Egypt's Annual Reports

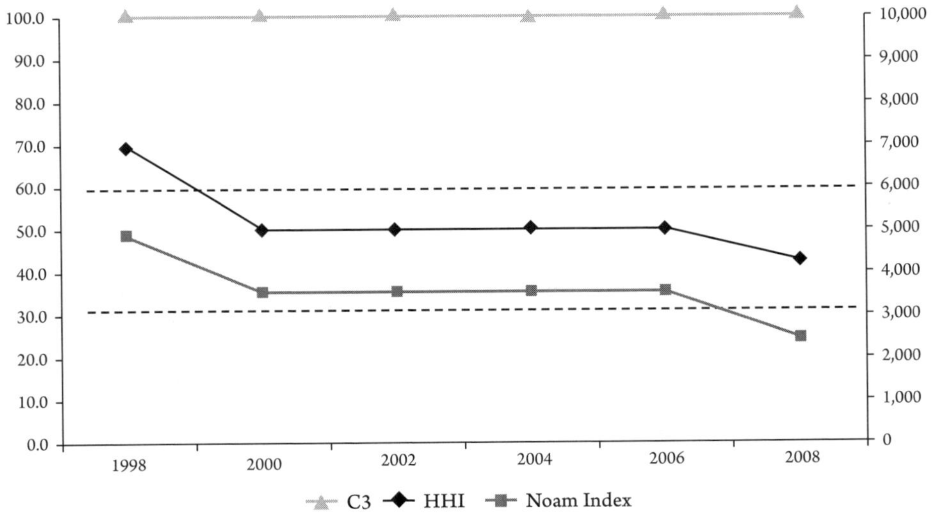


GRAPH 29.7 Wireless Telecommunications Concentration

maintaining a similar cost structure for older and newer companies alike. No laws prevent private wireless companies from establishing their own fiber lines, but it is more cost-effective for them to lease out TE's infrastructure.[278] The wholesale price is in effect manipulated by TE to ensure it is always cheaper for operators to lease infrastructure from TE rather than build their own new infrastructure.

In 2008, NTRA launched mobile number portability services (MNP),[279] which made it worthwhile for customers to bear the switching costs of moving between operators without having to change their phone numbers.

## CONCLUSION

While the Egyptian wireline industry is totally concentrated, the wireless industry is open for a strong private sector presence. The wireline state monopoly is a case that rests purely on economic motivation. This is contrary to ERTU in the case of terrestrial broadcasting and radio, where state monopoly emanates from a political motivation intended to curb freedom of expression.

While the wireless industry is less concentrated, the government retains some authority. The transformation of ARENTO into TE meant that the government had relinquished some of its overbearing role in a pertinent and growing industry. The extent to which it let go of the reins in the telecom market is dubious. Apart from the fact that TE gained firm footing in the wireless and fixed broadband market through indirect ownership, economic barriers to entry mean that TE monopolizes the nation's optic fiber backbone for wireless and fixed telecom and leases out this infrastructure to all three companies, Mobinil, Vodafone, and Etisalat. Dwindling revenues and subscription numbers in the fixed wireline market are countered by growing revenues from leased optical fiber cables that run across the entire country. This, coupled with a presence in wireless telephony, as well as wireless and fixed broadband, help the company maintain its status as a strong player in Egypt's ICT sector.

Meanwhile, increasing media convergence has meant that field players have delved into a cross-industrial endeavor. A case in point is private cellular operators bundling wireless

278. Ibid.

279. National Telecom Regulatory Authority (NTRA). "Mobile Number Portability Guidelines." National Telecom Regulatory Authority. 2008. Last accessed on Dec. 15, 2011 at <http://www.tra.gov.eg/Presentations/NumberPotabilityConsumersGuide24032008_En.pdf>.

Internet with their services and acquiring broadband companies. Given the flexibility of connectivity and the ease of use, wireless telephony has expanded widely in Egypt, as is the case in other developing countries. Mobile phones have a developmental angle as they empower and facilitate businesses, but they are also used for political mobilization. Indeed, mobile phones have promoted political participation, which surely was unwelcome by the regime. Eventually, the regime cut off telephone connectivity and text messaging during the revolution of January 25, 2011, generating another political victory of the state over the wireless industry. This backfired against the regime because it fueled further political dissent in addition to the collateral damage manifested in economic losses. This is another instance where, contrary to what the regime in Egypt had planned, the political aspects of media use could not be separated from economic facets.

## INTERNET MEDIA

### Internet Service Providers (ISP)

TE leases out wholesale services to ISPs as well as wireline and wireless companies.[280] The Egyptian ISP market is open to the private sector, with the number of ISPs amounting to 220[281] companies in 2010. Nonetheless, Internet penetration rates are markedly low (Table 29.9). Low Internet consumption in Egypt, especially relative to telecom-related industries, may be due to high poverty rates, low literacy levels, and inadequate infrastructure in certain geographical areas.[282]

ISPs in Egypt are divided into class A, B, and C licenses. Class A licenses have the right to own their own infrastructural equipment as well as rights to international gateways through TE. They can thus provide wholesale services to class B and C ISPs.[283] Class B ISPs retain the right to infrastructure equipment, but gain access to international gateways through class A members. Class C ISPs can only partake in retail Internet services and lease out infrastructure from class A and B organizations. However, all ISPs, lease out the fiber optic cables from TE. Only four ISPs possess class A licenses: TEData, Nile Online, Linkdotegypt and Egynet[284]; another four retain class B licenses: Raya, YallaMisr, Noor, and Menanet.[285] For the purpose of this chapter the analysis addresses the retail market for class A and B ISPs. Market shares are calculated by the total number of subscribers per ISP (including indirect subscribers from class C ISPs, eliminating any duplicates).

The data set for this analysis is only for fixed-line broadband (i.e., DSL services). Cable television based on broadband or FTTX does not exist commercially. However, trials have been made. The reason for the absence of broadband-based cable TV is a lack of competition by cable operators to pursue this sector. Triple play licenses were issued by NTRA in 2009 that allow cable television, Internet, and telephone over a single-access broadband connection. LinkdotEgypt and Tivotech obtained the licenses; however, both licenses have been dormant.[286]

280. Telecom Egypt. "Company Fact Sheet." Telecom Egypt. Last accessed on June 27, 2012 at <http://ir.telecomegypt.com.eg/Company%20 Fact%20 Sheets.asp>.

281. Mobinil. "Annual Report 2010: Egyptian Company for Mobile Services (ECMS)." Mobinil. Last accessed on June 27, 2012 at <http://www.mobinil.com/investors/AnnualReport2010.aspx>.

282. National Telecom Regulatory Authority (NTRA). "Broadband: The Speedy Access." National Telecom Regulatory Authority. Last accessed on Dec. 15, 2011 at <http://www.tra.gov.eg/presentations/broadband_En.pdf>.

283. Monsef, Olfat A. "Pioneering Liberalization & Regulation of Telecommunications in The Arab Region." *International Telecommunications Union*. November 2003. Last accessed on Mar. 12, 2012 at <www.ituarabic.org/PreviousEvents/2003/reg/Doc7.ppt>.

284. Abdel Hafez, Fatma H.M., and Wahba, Khaled (2004). "Assessment of the 'Free Internet Access' Project on the Internet Market in Egypt: System Dynamics View." Presented at *Systems Dynamics Society Conference*. Oxford, England, July 25–29, 2004. Last accessed Jan. 5, 2011 at <http://www.systemdynamics.org/conferences/2004/SDS_2004/PAPERS/276WAHBA.pdf>.

285. Ministry of Communications and Information Technologies (MCIT). Dataset. Last accessed on July 5, 2012 at <http://www.mcit.gov.eg/Indicators/Ind_Communications.aspx>.

286. El Azhary, Emad. Personal interview. Nov. 24, 2011.

### Table 29-9. Internet Industry Indicators

| INDUSTRY INDICATOR | 2009 | 2010 | 2011 (OCT) |
|---|---|---|---|
| Internet penetration (Users per 100 Inhabitants) | 22% | 29% | 34% |
| Proportion of households using a computer | 44% | 44% | 48% |
| Proportion of individuals using a computer from any location in last 12 months | 24% | 26% | 32% |

*Source:* Egypt ICT Indicators: Infrastructure and Access Portal, Ministry of Communications and Information Technology (MCIT). "Key Indicators." Egypt ICT Indicators. May 2010. Last accessed on Dec. 15, 2011 at <http://www.new.egyptictindicators.gov.eg/en/Indicators/_layouts/KeyIndicatorsViewer.aspx>.

### Table 29-10. TE DATA Home ADSL Price List (EGP) for Unlimited Access

| SPEED (KBPS) | 1 MONTH | 3 MONTHS | 6 MONTHS | 12 MONTHS |
|---|---|---|---|---|
| 512 | 95 | 280 | 555 | 1,090 |
| 1,024 | 140 | 415 | 820 | 1,610 |
| 2,048 | 220 | 650 | 1,290 | 2,530 |
| 4,096 | 380 | 1,125 | 2,230 | 4,380 |
| 8,192 | 695 | 2,055 | 4,020 | 8,000 |
| 16,384 | 1,355 | 4,010 | 7,845 | 15,600 |
| 24,576 | 1,950 | 5,770 | 11,310 | 22,500 |

*Source:* TE Data. "Home ADSL Price List." TE DATA 2012. Last accessed Jan. 5, 2012 at <http://www.tedata.net/eg/en/Home-ADSL/Home-ADSL-Prices-List>.

The license does not include wireless broadband USB subscriptions or mobile Internet. Mobinil, Vodafone, and Etisalat are the only providers of wireless broadband services. Nonetheless, their recent acquisition in the fixed line market is addressed in the chapter, as it bears implications on ownership and concentration of ISPs in Egypt.

## HISTORY AND REGULATION

The Internet was introduced into Egypt in 1993 by the Information Decision Support Center that granted exclusive services to Egyptian universities to connect with the European Academic Research Network for academic purposes.[287] Ever since, the government and the MCIT have taken quick and concrete steps to propel Egypt into the digital scene. In 1996, the government had further developed the infrastructural backbone for the Internet in order to officially open the data services door to private investors.[288] After 1998, broadband services gained ground and DSL connections were later introduced in 2002.[289]

The NTRA announced the Free Internet Initiative in 2002 in an effort to increase penetration levels through allowing Internet users

287. MCIT. "Telecom Reform Milestones." Ministry of Communication and Information Technology. Last accessed on Nov. 29, 2011 at <http://www.mcit.gov.eg/TeleCommunications/Telecom_Reform_Milestones>.

288. Ibid.

289. Ibid.

access priced at the rate of a regular telephone call instead of charging subscription fees (Table 29.10).[290] ISPs abided by a revenue-sharing model in which TE, the sole telephone landline provider, received a 30% share of revenues from "Internet calls" and the licensed ISP retained the remaining 70%.[291] This initiative succeeded in increasing penetration levels. This was followed by the broadband initiative in 2004, with a goal of transforming Egypt into an "information society" under the umbrella of the e-access program.[292]

## OWNERSHIP

Only class A ISPs are licensed to build their own infrastructure. TEData is the leading ISP in Egypt and is a subsidiary of TE. It held 70% of the infrastructure capacity in Egypt and 61% of the DSL market share in 2009.[293] TEData is also active in Jordan.[294] Vodafone gradually acquired Raya; a class B licensed ISP ranked in the top five largest ISPs by market share, raising its stake in Raya from 51% in 2006 to 97% in 2008.[295] Etisalat Egypt closely followed suit and acquired all shares of Egynet and Nile Online (both previously partly owned by TE).[296]

Mobinil was last to enter the DSL market and acquired Linkdotnet from Orascom Telecom Holding, which had the second largest market share after TEData, in 2010 (Table 29.11). Linkdotnet was established in 1995 and is owned by Orascom Telecom Holding.[297] It operates in the Middle East and Asia with offices in the UAE, Saudi Arabia, Algeria, and Pakistan.[298]

The remaining three companies are YallaMisr, Noor, and Menanet. YallaMisr is owned by a Kuwaiti conglomerate, Al Kharafi Group.[299] Even after diligent research, we have not reached any conclusions on Noor's ownership. Menanet was originally owned by Gamal Marwan, chief executive officer of Menanet, when it was established in 1999. In 2001, it was acquired by Africaonline (headquartered in London and Nairobi)[300] and resold to Marwan in 2004.[301]

Egypt's three private wireless companies, Mobinil, Vodafone, and Etisalat, propelled themselves into the DSL market through acquisition of four out of the eight main class A and B ISPs in Egypt. Although their operations remain under separate entities, this is expected to change with the advent of bundling and triple play licenses on

---

290. Monsef, Olfat A. "Pioneering Liberalization & Regulation of Telecommunications in The Arab Region." *International Telecommunications Union*. November 2003. Last accessed on Mar. 12, 2012 at www.ituarabic.org/PreviousEvents/2003/reg/Doc7.ppt. Abdel Hafez, Fatma H.M., and Wahba, Khaled (2004). "Assessment of the 'Free Internet Access' Project on the Internet Market in Egypt: System Dynamics View." Presented at *Systems Dynamics Society Conference*. Oxford, England, July 25–29, 2004. Last accessed Jan. 5, 2011 at <http://www.systemdynamics.org/conferences/2004/SDS_2004/PAPERS/276WAHBA.pdf>.

291. Monsef, Olfat A. "Pioneering Liberalization & Regulation of Telecommunications in The Arab Region." *International Telecommunications Union*. November 2003. Last accessed on Mar. 12, 2012 at www.ituarabic.org/PreviousEvents/2003/reg/Doc7.ppt. Abdel Hafez, Fatma H.M., and Wahba, Khaled (2004). "Assessment of the 'Free Internet Access' Project on the Internet Market in Egypt: System Dynamics View." Presented at *Systems Dynamics Society Conference*. Oxford, England, July 25–29, 2004. Last accessed Jan. 5, 2011 at <http://www.systemdynamics.org/conferences/2004/SDS_2004/PAPERS/276WAHBA.pdf>.

292. National Telecom Regulatory Authority (NTRA). "Broadband: The Speedy Access." National Telecom Regulatory Authority. Last accessed on Dec. 15, 2011 at <http://www.tra.gov.eg/presentations/broadband_En.pdf>.

293. Telecom Egypt, "Annual Report 2009: Equipped For A New Era," Giza, 2009, 24. Last accessed on Nov. 27, 2011 at <http://ir.telecomegypt.com.eg/admin/uploads/annual-report-2009/FINAL%20 ANNUAL%20 web.pdf>.58.

294. "Telecom Egypt Announces the Acquisition of Remaining TE Data Stake." Telecom Egypt. Jan. 5, 2010. Last accessed on Mar. 12, 2012 at <http://ir.telecomegypt.com.eg/press%20 releases/press %20 releases/TE%20 Data%20 PR%20 -%20 Final.pdf>.

295. Hot Telecom. Country profile: Egypt 2008. Apr. 30, 2008. Last accessed on Jan. 8, 2012 at <http://www.hottelecoms.com/Egypt.pdf>.30.

296. "EtisalatMisr Acquires NOL and EgyNet." Arab Finance. Oct. 27, 2008. Last accessed Jan. 8, 2012 at <https://www.arabfinance.com/News/newsdetails.aspx?Id=122348>.

297. LinkdotNet. "News and Events." LinkdotNet. Last accessed on June 27, 2012 at <http://www.link.net/English/Linkcorp/News+and+Events/OTH-Logo.htm>.

298. LinkdotNet. "News and Events." LinkdotNet. Last accessed on June 27, 2012 at <http://www.link.net/English/Linkcorp/News+and+Events/OTH-Logo.htm>.

299. YallaMisr. "About Us." YallaMisr. Last accessed on June 27, 2012 at <http://www.yallamisr.com/about.htm>.

300. Africa Telecom. Information Gatekeepers Inc., 2:3, March 2001.

301. "Gamal Marwan Buys 100 Per Cent of MenaNet Communications SAE (MenaNet) from The African Lakes Corporation plc (ALC)." *AME Info*. May 24, 2004. Last accessed on Mar. 12, 2012 at <http://www.ameinfo.com/40072.html>.

Table 29-11. Internet Service Provider Concentration

INTERNET SERVICE PROVIDERS

| MARKET SHARES BY SUBSCRIBERS (%) | 2004 | 2005 | 2006 | 2007 | 2008 | 2009 | 2010 | 2011 |
|---|---|---|---|---|---|---|---|---|
| Telecom Egypt/TE (Public) | | | | | | | | |
| TEData | | 30.0 | 44.3 | 52.0 | 59.0 | 60.7 | 62.9 | 63.0 |
| LinkdotNet (Private, Orascom Telecom Holding OTH, Egypt) | | | | | 30.2 | 28.0 | N/A | N/A |
| EgyNet (Private, Egypt) | | | | | N/A | N/A | N/A | N/A |
| Nile Online (also known as the Egyptian Company for Internet and Digital Infrastructure) (Private, Egypt) | | | | | N/A | N/A | N/A | N/A |
| YallaMisr (Kuwait, Al Kharafi) | | | | | 0.7 | 0.2 | 0.3 | 0.3 |
| Noor | | | | | 0.5 | 0.4 | 0.5 | 0.5 |
| Mena Net (Private, Egypt, Gamal Marwan) | | | | | 0.1 | 0.0 | 0.0 | 0.0 |
| Vodafone Egypt (UK) | | | | | | | | |
| Vodafone (previously Raya)(Private, UK) | | | | | 1.9 | 3.8 | 6.0 | 8.0 |
| Etisalat U.A.E. (Public/Private, U.A.E.) | | | | | | | | |
| Etisalat Misr (Public/Private, UAE) | | | | | 7.7 | 6.7 | 5.7 | 5.8 |
| Mobinil (Orascom, France Telecom, France) | | | | | N/A | N/A | 24.4 | 22.5 |
| Others | | 70 | 55.7 | 48 | 0 | 0.2 | 0.2 | 0 |
| Revenue Estimates (mil $) | 280.0 | | | | 591.0 | | | |
| C4 | | | | | 98.7 | 99.2 | 99.0 | 99.2 |
| HHI | | | | | 4,453 | 4,532 | 4,620 | 4,570 |
| Noam Index | | | | | 2,226 | 2,266 | 2,310 | 2,285 |
| Number of players (*n*) | | | | | 4 | 4 | 3 | 4 |

*Source:* Figures obtained from Ministry of Communications and Information Technology (MCIT).

*Note:* This is only for ADSL (fixed line) service providers and does not include wireless broadband. On the wireless broadband front, there is only Vodafone, Mobinil and Etisalat, TE Data are not in that market.

*Note 2:* For GSM (wireless broadband) Wael Fakharny said that out of 1.5 m lines Etisalat, Vodafone, and Mobinil have 47%, 40%, and 13%, respectively for 2011.

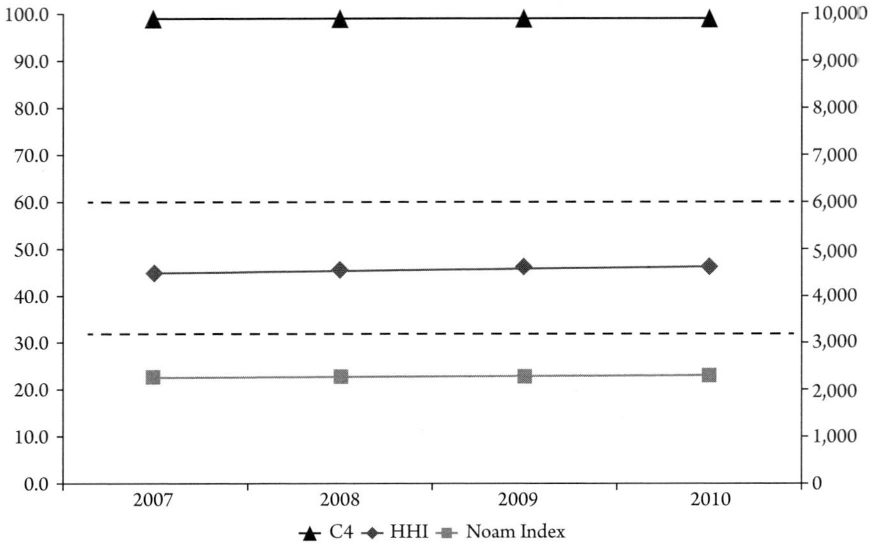


GRAPH 29.8 Internet Service Provider Concentration

offer by the NTRA. This adds another dimension to media convergence (Graph 29.8).

## CONCLUSION

Similar to broadcast, the indirect control of the ISP industry by the government comes through infrastructure ownership, despite opening up the market for investors to capitalize on this infrastructure. It is through this arrangement that the government owns the figurative "kill switch." As events culminated in the January 25 revolution, the government shut down the Internet it has been promoting and developing over the past decade. This was mainly done through the State Security apparatus pressuring large ISPs to physically switch off all domestic Internet supply. Again, this is a case where the government exerted political pressure. Unfortunately, this was done in compliance with Article 67 of the Telecommunications Law, which grants the government the right to "subject to their administration all telecommunication services and networks of any operator or service provider," when deemed necessary for national security. This law reflects the heavy hand of the regime on communication and media in general.[302]

### Search Engines

International players dominate the search engine market in Egypt with Google leading the market with over 90% market share from 2009 to 2011, and all other search engines (Bing, Yahoo, Ask Jeeves, etc.) competing for the remaining traffic. This includes Arabized search engines, such as Yahoo's Maktoob (Table 29.12; Graph 29.9).[303]

### Online News Media Summary

Online interfaces for news have been widely proliferating in Egypt as part of the Internet boom and an increased appetite for news consumption after the revolution of January 25, 2011. Two main types of news websites have been sprawling. One is solely based on an online presence through fully fledged portals that include text and multimedia content, as well as interactive content built around engagement with online social media users. The other is based on portal

302. National Telecom Regulatory Authority (NTRA). "Telecom Law." National Telecom Regulatory Authority. Last accessed on June 4, 2012 at <http://www.tra.gov.eg/uploads/law/law_en.pdf>.

303. Fakharany, Wael. Email interview. Sept. 14, 2011.

Table 29-12. Search Engine Concentration

| SEARCH ENGINES | | | | | |
|---|---|---|---|---|---|
| MARKET SHARES (%) | 2004 | 2008 | 2009 | 2010 | 2011 |
| Google (Private, USA) | | | 93.0 | 95.0 | 97.0 |
| Others (Yahoo, Bing, maktoob, ask.com) | | | 7.0 | 5.0 | 3.0 |
| Monthly Search Volume (million access) | | | | | |
| Revenue Estimates (mil $) | 9.0 | 45.0 | | | |
| C1 | | | 93.0 | 95.0 | 97.0 |
| HHI | | | 8,649 | 9,025 | 9,409 |
| Noam Index | | | 8,649 | 9,025 | 9,409 |
| Number of players (*n*) | | | 1 | 1 | 1 |

*Source*: Figures obtained from Wael Fakharany, Regional Head for Google Middle East.

projects embedded in bigger mainstream journalism outlets. The latter are usually created by mainstream media because of the dire need to invest in web-based projects that provide more current information at a lower price than printed news.

Islam Online belongs to the first category as it was one of the first news websites, funded by the government of Qatar and with no activities other than online. Similarly, *AlYoum7* started exclusively as a digital project. Only recently did it move to printing a daily paper. *Masrawy* is a third example of a purely web-based news website with high traffic. This model helped news organizations to create a culture of a digital newsroom and circumvent problems in the print industry such as distributing and printing through Al Ahram.

The second group includes websites affiliated with major media outputs, particularly print outlets. In this arena, both state-run newspapers have invested in fully fledged website projects such as *Al Ahram's* portal, as well as private enterprises such as *Al Masry Al Youm*. Television channels

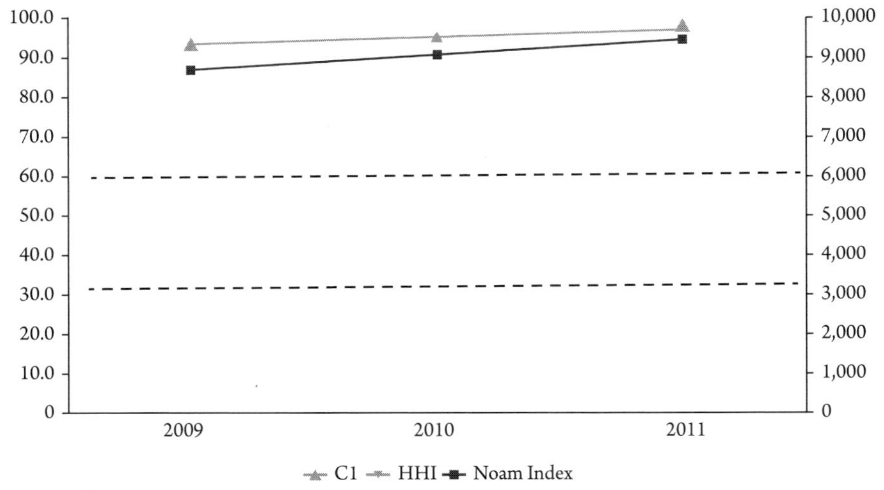


GRAPH 29.9 Search Engine Concentration

also make use of digital platforms, particularly YouTube. Poor Internet connections get in the way of high-quality streaming, which undermines the proliferation of online news videos. ONTV's YouTube channel is very successful because its content is spread virally and solicits feedback.

The biggest obstacle for news websites is that advertising and commercial revenue is very limited. Most online news media starts as a complement to an actual physical project, and online news advertisements become part of the package of advertising in both print and online. In advertisement contracts, there are two quotations for print and online. Therefore, the traditional media outlets that monopolize advertising revenues in print will get to monopolize online revenues on websites. The cost of online advertising is based on impressions. After an ad reaches the target impressions online, it gets removed. Many agree that a business model of a pure digital media enterprise is yet to be found.[304]

The term "digital media" includes websites and mobile phone applications as well as SMS services that are becoming a large part of the industry. Digital development strategies are relatively new in media organizations. While some organizations invest in creating digital development positions, the norm is that online news media efforts are sporadic and not systematically, nor sequentially, planned.[305]

Information about ratings, traffic, and ad revenues from these different websites are difficult to find. There is an issue with transparency in this industry, as no independent institution exists that is in charge of and able to collect and audit market indicators for the different online news sites. State and private companies alike fail to provide information on their operations. Real circulation and traffic figures are concealed in order to amass greater advertising revenues.[306]

## CONCLUSION

The evolution of regulations and power politics to engineer ownership underlies the configuration of the media ecosystem in Egypt. Although the contemporary media landscape in Egypt has become more dynamic within the last decade, government involvement is a cornerstone to the development, or lack thereof, of each sector. The extent to which the state inhibits or encourages private sector participation in media is contingent upon the power of each medium to mobilize public opinion. Accordingly, industries such as telecommunications were allowed to become more market oriented and open to private sector participation. On the other hand, tight reins were kept on audiovisual media, with private sector entry prohibited until the inauguration of the free media zone in 2000. Even with private participation in mass media, ownership trends and content are restricted to a select few.

As in other Arab countries, Egypt's policies demonstrate a clear divide between economic and political freedoms, which in turn cause economic and political concentration. In a typical Arab regime manner, Egypt focused on encouraging economic freedoms in the strictest neoclassical sense while simultaneously continuing to harshly stifle political freedoms. Despite a heralded move toward market liberalization and private participation, contradictory policies have been noted to stifle these economic currents.

This was reflected in the fast pace of Internet and telephone connectivity within policies encouraged to promote economic freedoms and foreign investments. Ironically, such digital connectivity gave way to a generation of savvy political activists that eventually played a strong role in toppling the Mubarak regime. They remain actively engaged in fighting the current military rule. Nonetheless, despite accomplishments in facilitating the spread of information and communications technologies, government surveillance and censorship on ICT activities have increased.

Government surveillance was not only inflicted on individuals. Among the most disconcerting evidence of political repression is the government's iron clad rule of private enterprise. Vodafone revealed in 2009 that Egyptian

304. Abou Hatab, Fathy. Personal interview. May 16, 2012.
305. Ibid.
306. Ibid.

authorities had obtained communications data from their organization after the April 2008 bread riots and convicted 22 people by December in connection with these protests.[307] Egypt's black out during the January 25 revolution was an affirmation of the force of politics on economic, organizational, and individual freedoms. It is not surprising that Egypt fared relatively well on indices of business success, but it performed dismally on democracy and freedom indices. Whenever freedom of expression was in question, political reign overruled economic expansion, bringing in political concentration as an additional dimension of media concentration.

In addition to direct surveillance and censorship, the state indirectly influenced media diversity and content. Of particular interest has been the role of businessmen as an additional source of control over media, especially in the case of newspapers and satellite TV. These businessmen were generally aligned with the ruling regime of Egypt, or were themselves members of the royal families of Saudi Arabia and Qatar. This impacts content in a manner that enhances the political concentration of the regimes. Indeed, with Egypt going through the growing pains of a hopeful democracy, both Saudi Arabia and Qatar have been trying to play pivotal roles in the region's politics. Such change in the geopolitical scene in the Arab world will have considerable ramifications on the state of the media in the region and in Egypt.

The months following Egypt's revolution of January 25, 2011, witnessed a series of turbulent actions as Egypt stood at the crossroads of shedding the old regime and moving on to a new era. There were some initial indicators of a loosening of the reins on freedom of expression and the promotion of a political openness to match the consistent drive toward economic openness. Matching political freedom to economic freedom was never on Mubarak's agenda, be considered one of the tests for a successful transition to democracy.

## Works Cited


Abdel Fattah, Amira. *Press Freedom in Egypt.* MA dissertation, The American University in Cairo, 2008.

Abdel Hafez, Fatma H.M., and Wahba, Khaled (2004). "Assessment of the 'Free Internet Access' Project on the Internet Market in Egypt: System Dynamics View." Presented at *Systems Dynamics Society Conference.* Oxford, England, July 25–29, 2004. Last accessed Jan. 5, 2011 at <http://www.systemdynamics.org/conferences/2004/SDS_2004/PAPERS/276WAHBA.pdf>.

Abou Hatab, Fathy. Personal interview. May 16, 2012.

Africa Telecom. *Information Gatekeepers Inc.*, 2:3, March 2001.

Al Ahram. "About Us." *Al Ahram.* Last accessed on May 29, 2012 at <http://english.ahram.org.eg/UI/Front/Aboutus.aspx>.

"Al Ahram Establishment Websites." *Al Ahram International North America Inc. Limited.* Last accessed on Jan. 16, 2012 at <http://aina.ahram.org.eg/Ahram %20USA%5COnline%20 Ads%5CAl-Ahram%20 Websites.pdf>.

Al Bawaba. "Click to Become IPO in Egypt." *Al Bawaba News.* May 2, 2001. Last accessed on Dec. 18, 2011 at <http://www.albawaba.com/news/click-gsm-become-ipo-egypt>.

"Al Dustour." *The Media Briefing.* Last accessed on Jan. 16, 2012 at <http://www.themediabriefing.com/products/Al dustour/>.

Al Gomhouriyya Newspaper, Last accessed on July 5, 2012 at <http://www.algomhuria.net.eg/algomhuria/today/pdf/>.

"Al Jazeera Closed for Operating Unlicensed, Sawiris Considers Qatar-based Channel." *Egypt Independent.* July 5, 2012. Al Mary Al Youm. Last accessed on Jan. 17, 2012 at <http://www.almasryalyoum.com/en/node/496619>.

Al Jazeera. "Corporate Profile." *Al Jazeera.* Last accessed Dec. 29, 2011 at <http://www.aljazeera.com/aboutus/2006/11/2008525185555444449.html>.

Allam, Rasha. "Media Landscape: Egypt." *European Journalism Centre.* Nov. 8, 2010. Last accessed on Dec. 29, 2011 at <http://www.ejc.net/media_landscape/article/egypt/>.

Al Masry Al Youm. "About Al Masry Al Youm." *Al Masry Al Youm.* Last accessed on Jan. 16, 2012 at <http://www.almasryalyoum.com/en/node/54>.


307. "Internet Filtering in Egypt." OpenNet Initiative. Aug. 6, 2009. Last accessed on Feb. 13, 2012 at <http://opennet.net/sites/opennet.net/files/ONI_Egypt_2009.pdf>.

"Al Masry Al Youm: Transforming the Egyptian Press." *Tavaana*. Media Monitor. 28 Jan. 2011. Last accessed on Jan. 16, 2012 at <http://www.tavaana.org/archive.jsp?restrictids=nu_repeatitemid&restrictvalues=207150200034129555277 3910&lang=en>.

Al Rai Media Group. "Al Rai Media Group." *Al Rai Media Group*. Last accessed on June 27, 2012 at <http://www.alraimedia.com/Alrai/MediaGroup.aspx>.

"Al Sayed Al Badawy." *Ahram Online*. 19 Nov. 2011. Last accessed on Jan. 16, 2012 at <http://english.ahram.org.eg/NewsContent/33/102/26705/Elections-/Whos-who/AlSayed-AlBadawi.aspx>.

Al Wafd. *Official Website*. Last accessed on June 27, 2012 at <http://www.alwafd.org/digital/>.

Amin, Hussein. *Strengthening the Rule of Law and Integrity in the Arab World: Report on the State of the Media in Egypt*. The Arab Center for the Development of the Rule of Law and Integrity, 2009, 6. Last accessed on June 10, 2012 at <http://www.arabruleoflaw.org/Files/PDF/Media/English/P2/Egypt_MediaReportP2_En.pdf>.

"An Overly French Mobinil." *Egypt Independent*, no. 4, June 7, 2012, 11. Last accessed on 27 June 2012 at <http://www.scribd.com/doc/96300554/Egypt-Independent-Jun-07>.

"Arab Radio and Television Network." *Wikipedia*. Sept. 21, 2011. Last accessed on Dec. 19, 2012 at <http://en.wikipedia.org/wiki/Arab_Radio_and_Television_Network>.

Arab Satellite Communications Organization. "About Us. Corporate Governance." *Arab Satellite Communications Organization*. Last accessed on June 27, 2012 at <http://www.arabsat.com/pages/ArabLeague.aspx>.

Attalah, Lina and Rizk, Nagla. "Egypt's Evolving Media Landscape: Access, Public Interest and Control," in *A New Frontier, An Old Landscape*. Report for the Ford Foundation, November 2011, 118–136. Last accessed on January 16, 2012 at <http://participatorymedia.lab.asu.edu/files/NewFrontierOldLandscape.pdf>.

Attia, Tarek. "No More Free TV." *Al Ahram Weekly*, no. 526. Mar. 22–28, 2001. Last accessed on May 14, 2012 at <http://weekly.ahram.org.eg/2001/526/li3.htm>.

Black, Jeffrey. "Egypt Press: More Free, Still Fettered." *Arab Media and Society*. 2008. Last accessed on Jan. 16, 2012 at <http://www.arabmediasociety.com/?article=572>.

Bureau of Democracy, Human Rights and Labor. *2010 Human Rights Report: Egypt*. U.S. Department of State, 2011. Last accessed on June 27, 2012 at <http://www.state.gov/documents/organization/160456.pdf>.

"Civil Movements: The Impact of Facebook and Twitter," in *Arab Social Media Report* (Vol 1, No. 2). Dubai School of Government, 2011, 16. Last accessed on Mar. 5, 2012 at <http://www.dsg.ae/en/Publication/Pdf_En/DSG_Arab_Social_Media_Report_No_2.pdf>.

Cochrane, Paul. "Saudi Arabia's Media Influence." *Arab Media and Society*. 2007. Last accessed on Jan. 17, 2012 at <http://www.arabmediasociety.com/?article=421> and <http://www.insightsmena.com/en/#!place=category&cat=Media+Usage>.

"Constitutional Declaration." *Cabinet*. Last accessed on Mar. 7, 2012 at <http://www.cabinet.gov.eg/AboutEgypt/Egyptian_constitution.aspx>.

"Control" in Hawtin, Dixie and Puddephatt, Andrew, eds. *A New Frontier, an Old Landscape*. New York: Ford Foundation, 2011, 118–136. Last accessed on Jan. 16, 2012 at <http://participatorymedia.lab.asu.edu/files/NewFrontierOldLandscape.pdf>.

Dubai Media Incorporated. "About Dubai Media Incorporated." *Dubai Media Incorporated*. Last accessed on June 27 2012 at <http://dmi.gov.ae/about_En.asp>.

"Egypt: Article 19 and the Justice Initiative Submission to the UN Universal Periodic Review" in the *Seventh Session of the UPR Working Group*. 2010. Last accessed on Jan. 16, 2012 at <http://www.article19.org/data/files/pdfs/submissions/egypt-article-19-and-the-justice-initiative-submission-to-the-un-universAl p.pdf>.

"Egypt Authorities Freeze New Satellite TV Station Licenses." *X Index*. 9 Sept. 2011. Last accessed on Jan. 16, 2012 at <http://www.indexoncensorship.org/2011/09/egypt-authorities-freeze-new-satellite-tv-station-licenses/>.

"Egypt Facebook Statistics." *Social Bakers*. Last accessed on Mar. 5, 2012 at <http://www.socialbakers.com/facebook-statistics/egypt>.

"Egypt Major African Mobile Markets: Future Growth Prospects." *Africa and Middle East Telecom Week*. Last accessed Jan. 18, 2012 at <http://www.africantelecomsnews.com/resources/AfricaOpp_Egypt.shtml>.

"Egypt." *Press Freedom and Media Development*. Last accessed on Jan. 16, 2012 at <http://wwwwan-press.org/pfreedom/articles.php?id=4404>.

*Egypt: The National Report on Literacy and Adult Education*. United Nations Educational, Scientific and Cultural Organization, 2009. Last accessed on Dec. 19, 2011 at www.unesco.org/fileadmin/MULTIMEDIA/.../UIL/.../Egypt.pdf.

"Egypt." *The World Factbook*. Last accessed on Mar. 7, 2012 at <https://www.cia.gov/library/publications/the-world-factbook/geos/eg.html>.

Egyptian Television Network. "About Us." *Egyptian Television Network (ETN)*. Last accessed on Jan. 17, 2012 at <http://www.egytv.net/aboutus.aspx>.

El Azhary, Emad. Personal interview. Nov. 24, 2011.

El Galad, Magdy. Interview. Al Nas TV. Last accessed on Jan. 16, 2012 at <http://www.youtube.com/watch?v=TSnLjXi2rdI>.

El Hefnawy, Magdi. Personal interview. June 30, 2011.

ERTU. "Sectors." *ERTU*. Last accessed on Dec. 15, 2011 at <http://ertu.org/1/sectors.asp>.

Etisalat. "About Us." *Etisalat*. Last accessed on June 27, 2012 at <http://etisalat.com.eg/etisalat/Etisalat_Portal_En/about/corporate.htm?_pageid=42,1&_dad=portal&_schema=PORTAL&siteAlias=etisalat&sitePath=Etisalat_Portal_En&kpAlias=about&pageAlias=corporate>.

Etisalat. "Corporate Information." *Etisalat*. Last accessed June 27, 2012 at <http://etisalat.com.eg/etisalat/Etisalat_Portal_En/about/corporate.htm?_pageid=42,1&_dad=portal&_schema=PORTAL&siteAlias=etisalat&sitePath=Etisalat_Portal_En&kpAlias=about&pageAlias=corporate>.

"Etisalat Led Group Signs $2.88b Deal." *MENAFN*. 23 Aug. 2006. Last accessed on Jan. 4, 2011 at <https://www.menafn.com/qn_news_story_s.asp?StoryId=1093124586>.

"EtisalatMisr Acquires NOL and EgyNet." *Arab Finance*. Oct. 27, 2008. Last accessed 8 Jan. 2012 at <https://www.arabfinance.com/News/newsdetails.aspx?Id=122348>.

Etisalat. "Our History." *Etisalat*. Last accessed on June 27, 2012 at <http://etisalat.com/html/history.html>.

"Facebook Usage Factors and Analysis," in *Arab Social Media Report* 1, no. 1. Dubai School of Government, 2011, 6. Last accessed on Mar. 5, 2012 at <http://unpan1.un.org/intradoc/groups/public/documents/dsg/unpan044212.pdf>.

Fahim, Joseph. "OTV Hits the Airwaves." *Daily News Egypt*. Jan. 30, 2007. Last accessed on Dec. 26, 2011 at <http://www.dailystaregypt.com/article.aspx?ArticleID=5308>.

Fakharany, Wael. Email interview. Sept. 14, 2011.

Fakhredinne, Jihad. "Pan-Arab Satellite Television: Now the Survival Part." *Allied Media Corporation*. Last accessed on Jan. 17, 2012 at <http://www.allied-media.com/ARABTV/pan-arab-sat.htm>.

"False Freedom: Online Censorship in the Middle East and North Africa." *UNHCR: Human Rights Watch*. Nov. 15, 2005. Last accessed on Feb. 13, 2012 at <http://www.unhcr.org/refworld/docid/45cc445f2.html>.

Fandy, Mamoun. *(Un)Civil War of Words: Media and Politics in the Arab World*. Westport, CT: Praeger, 2007.

Field, Nathan and Hamam, Ahmed. "Salafi Satellite TV in Egypt." *Arab Media and Society*. 2009. Last accessed on Jan. 18, 2012 at <http://www.arabmediasociety.com/?article=712>.

"Freedom of Expression Project: The Press in Egyptian Legislation" in *Middle East and Islamic Studies Collection*. Cornell University Library. Last accessed on Jan. 16, 2012 at <http://www.library.cornell.edu/colldev/mideast/egpress.htm>.

*Full Text: Draft Egyptian Broadcast Law*. Arab and Media Society, 2008, Article 1, Clause 4. Last accessed June 27, 2012 at <http://www.arabmediasociety.com/articles/downloads/20080729105525_AMS6_draft_media_law_english.pdf>.

"Gamal Marwan Buys 100 Per Cent of MenaNet Communications SAE (MenaNet) from The African Lakes Corporation plc (ALC)." *AME Info*. May 24, 2004. Last accessed on 12 Mar. 2012 at <http://www.ameinfo.com/40072.html>.

Good News Group. "Companies" *Good News Group*. Last accessed on Jan. 16, 2012 at <http://www.goodnewsgroup.com/>.

Good News Group. "Good News for Radio and TV." *Good News Group*. 2006. Last accessed on Jan. 16, 2012 at <http://www.goodnewsgroup.com/gng9.swf>.

GoodNews4Me. "Information." *GoodNews4Me*. Al Alam Al Youm. Last accessed on Jan. 16, 2012 at <http://www.gn4me.com/alalamalyoum/info.jsp?edition_id=3238>.

"Group Companies: Orbit Showtime Network." *KIPCO: Kuwait Projects Company (Holding)*. Last accessed on Jan. 17, 2012 at <http://www.kipco.com/GroupCompanies.asp?q_pageid=56>.

Hagey, Keach. "News Corp Eyes Stake in Saudi Firm Rotana Media." *The National*. 2 Sept. 2009. Last accessed on Jan. 17, 2012 at <http://www.

thenational.ae/business/media/news-corp-eyes-stake-in-saudi-firm-rotana-media>.
Hamdy, Naila. "A Dream TV Come True." *Transnational Broadcasting Studies (TBS)*, No. 8, (Spring 2002). Last accessed on Jan. 17, 2012 at <http://www.tbsjournal.com/Archives/Spring02/sirhan.html>.
Hamdy, Naila. " El Mehwar the Mercurial." *Transnational Broadcasting Studies (TBS)*. No. 9, (Fall/Winter 2002). Last accessed on Jan. 17, 2012 at <http://www.tbsjournal.com/Archives/Fall02/Mehwar.html>.
Hamdy, Naila. Personal interview. June 10, 2012.
Hamzawy, Amr and Dunne, Michele. "Media Freedom Restricted as Egyptian Parliamentary Elections Approach." *Carnegie Endowment for International Peace*. Last accessed on Jan. 17, 2012 at <http://egyptelections.carnegieendowment.org/2010/10/25/media-freedom-restricted-as-egyptian-parliamentary-elections-approach>.
Hassanein, Leila. "Access to Infrastructure: Egypt." *GISWatch*. 2008. Last accessed Dec. 22, 2010 at <http://www.giswatch.org/sites/default/files/Egypt_0.pdf>.
Hot Telecom. *Country profile: Egypt 2008*. Apr. 30, 2008. Last accessed on Jan. 8, 2012 at <http://www.hottelecoms.com/Egypt.pdf>.
Howard, P.N. "Ownership Diversity in Muslim Media Systems." *World Information Access Project (WIA)*. Oct. 6, 2008. Last accessed on Jan. 16, 2012 at <http://www.wiaproject.org/index.php/68/ownership-diveristy-in-muslim-media-systems>.
"Human Resources Division." *Al Jazeera Network*. Last accessed on June 27, 2012 at <http://recruitment.aljazeera.net/english/>.
"Industrial Conglomerates: Mawarid Group Company LLC." *Bloomberg Businessweek*. Last accessed on Jan. 17, 2012 at <http://investing.businessweek.com/research/stocks/private/people.asp?privcapId=22595361>.
"Internet Filtering in Egypt." *OpenNet Initiative*. Aug. 6, 2009. Last accessed on Feb. 13, 2012 at <http://opennet.net/sites/opennet.net/files/ONI_Egypt_2009.pdf>.
IREX. "Media sustainability Index 2005: Middle East and North Africa." *IREX*. Last accessed on May 24, 2012 at <http://www.irex.org/system/files/MSI05_MENA_EG.pdf 26>.
Jaafar, A. "Mixed Results for LBC – Rotana: Arab Conglomerate Tries to Compete with MBC." *Variety*. Oct. 2, 2008. Last accessed on Jan. 17, 2012 at <http://www.variety.com/article/VR1117993363>.
Karlekar, Karen D. and Cook, Sarah G., ed. *Freedom of the Press 2008: A Global Survey of Media Independence*. New York: Rowman and Littlefield Publishers, Inc. 2008, 121.
Kassem, Hisham. Personal interview. 21 Feb. 2012.
Kawach, Nadim. "Saudi Cleric Slams MBC over TV Series." *Emirates 24/7*. Aug. 18, 2010. Last accessed on May 29, 2012 at <http://www.emirates247.com/entertainment/films-music/saudi-cleric-slams-mbc-over-tv-series-2010 2008-18-11.280816>.
Khalil, Joe. "News Television in the Arabian Gulf... Period of Transitions." *Global Media Journal* 5, no. 8 (2004).
K&M Engineering and Consulting Corporation. "ARENTO Cost of Service Study." *USAID*. National Telecommunications Organization & the U.S. Agency for International Development, 2011, 105. Last accessed on Dec. 19, 2011 at <http://pdf.usaid.gov/pdf_docs/PNABS218.pdf>.
Kotb, Ahmed. "Media Control." *Al Ahram Weekly Online*, 1056, July 14–20, 2011. Last accessed on Jan. 17, 2012 at <http://weekly.ahram.org.eg/2011/1056/eg692.htm>.
Kotler, P and Armstrong, Gray, eds. "Pricing: Understanding and Capturing Customer Value." In *Principles of Marketing*. New York: Pearson Prentice Hall, 2012. Last accessed on June 27, 2012 at <http://catalogue.pearsoned.co.uk/assets/hip/gb/hip_gb_pearsonhighered/samplechapter/0273752502.pdf>.
LinkdotNet. "News and Events." *LinkdotNet*. Last accessed on June 27, 2012 at <http://www.link.net/English/Linkcorp/News+and+Events/OTH-Logo.htm>.
Madany, Sherine. "Egypt Postpones Second Fixed Line Auction for a Year." *Daily News Egypt*. Sept. 8, 2008. Last accessed on Dec. 19, 2011 at <http://www.dailystaregypt.com/article.aspx?ArticleID=16317>.
Mattar, Eman. "El Hefaz and El Nas TV Channels Halted on TV and El Faraeen Warned." *Business Investment Banking News*. Dec. 12, 2010. Last accessed on Jan. 18, 2012 at <http://www.bibnews.com/en/news_details.php?id=5012>.
"MBC, Largest Media Group in the Middle East Chooses Magnolia as Strategic Platform." *Magnolia*. Mar. 9, 2011. Last accessed on May 29, 2012 at <http://www.magnolia-cms.com/our-company/news/press-releases/mbc-usa.html>.
MCIT. *Ministry of Communications and Information Technology eProtal*. Last accessed on Nov. 29, 2011 at <http://www.mcit.gov.eg/TeleCommunications/Historical_Prespective>.

MCIT. "Telecom Reform Milestones." *Ministry of Communication and Information Technology*. Last accessed on Nov. 29, 2011 at <http://www.mcit.gov.eg/TeleCommunications/Telecom_Reform_Milestones>.

"Media Usage." *Insights MENA*. 2010. Last accessed on Jan. 16, 2012 at <http://www.insightsmena.com/en/#!place=category&cat=Media+Usage>.

Mekay, Emad. "TV Stations Multiply as Egyptian Censorship Falls." *The New York Times*. July 13, 2011. Last accessed on Jan. 16, 2012 at <http://www.nytimes.com/2011/07/14/world/middleeast/14iht-M14B-EGYPT-MEDIA.html?_r=2&pagewanted=all>.

Middle East Rating and Investors Service. "Egyptian Company for Mobile Services S.A.E (Mobinil)." Merisratings.com. November 2009. Last accessed Jan. 10, 2011 at <http://www.mobinil.com/Investors/FinancialReports/doc/CreditRatingReportNov092ndIssue.pdf>.

"Minister: Al Jazeera Mubasher Misr Shut Down for Disregarding Law." *Al Masry Al Youm*. Sept. 12, 2011. Last accessed on Jan. 17, 2012 at <http://www.almasryalyoum.com/en/node/494878>.

Ministry of Communications and Information Technologies (MCIT). *Dataset*. Last accessed on July 5, 2012 at <http://www.mcit.gov.eg/Indicators/Ind_Communications.aspx>.

Ministry of Communications and Information Technology (MCIT). "ICT Infrastructure and Access." *Egypt ICT Indicators*. Last accessed on Jan. 5, 2012 at <http://www.new.egyptictindicators.gov.eg/en/Indicators/_layouts/KeyIndicatorsViewer.aspx>.

Ministry of Communications and Information Technology (MCIT). "Key Indicators." *Egypt ICT Indicators*. May 2010. Last accessed on Dec. 15, 2011 at <http://www.new.egyptictindicators.gov.eg/en/Indicators/_layouts/KeyIndicatorsViewer.aspx>.

Mobinil. "Annual Report 2010: Egyptian Company for Mobile Services (ECMS)." *Mobinil*. Last accessed on June 27, 2012 at <http://www.vmobinil.com/investors/AnnualReport2010.aspx>.

Monsef, Olfat A. "Pioneering Liberalization & Regulation of Telecommunications in The Arab Region." *International Telecommunications Union*. November 2003. Last accessed on Mar. 12, 2012 at www.ituarabic.org/PreviousEvents/2003/reg/Doc7.ppt.

"Murdoch Buys into Rotana." *Arab News*. Feb. 23, 2010. Last accessed on Dec. 19, 2011 at <http://arabnews.com/saudiarabia/article21517.ece>.

"Muslim Brotherhood to Launch Satellite TV Channel in April." *Al Masry Al Youm*. Mar. 12, 2011. <http://www.almasryalyoum.com/en/node/349787>.

Nasrawi, Saif. "Wafd Party President Buys Al Dustour Newspaper." *Al Masry Al Youm*. Aug. 23, 2010. Last accessed on Jan. 16, 2012 at <http://www.almasryalyoum.com/en/node/70155>.

National Telecom Regulatory Authority (NTRA). "About Us." *National Telecom Regulatory Authority*. Last accessed on Nov. 30, 2011 at <http://www.tra.gov.eg/english/DPages_DPagesDetails.asp?ID=180&Menu=5>.

National Telecom Regulatory Authority (NTRA). "About Us." *National Telecom Regulatory Authority (NTRA)*. Last accessed on June 27, 2012 at <http://www.tra.gov.eg/english/DPages_DPagesDetails.asp?ID=180&Menu=5>.

National Telecom Regulatory Authority (NTRA). "Broadband: The Speedy Access." *National Telecom Regulatory Authority*. Last accessed on Dec. 15, 2011 at <http://www.tra.gov.eg/presentations/broadband_En.pdf>.

National Telecom Regulatory Authority (NTRA). "Consumer Protection: 'Know Your Rights' Awareness Campaign." *National Telecom Regulatory Authority*. Apr. 20, 2010. Last accessed on June 27, 2012 at <http://www.tra.gov.eg/english/dpages_dpagesdetails.asp?ID=469>.

National Telecom Regulatory Authority (NTRA). "Media Convergence Issues in Egypt and the Region and Prospects for the Future." *Scribd*. 2009. Last accessed Nov. 27, 2011 at www.scribd.com/doc/38482708/Doc9-NTRA.

National Telecom Regulatory Authority (NTRA). "Mobile Number Portability Guidelines." *National Telecom Regulatory Authority*. 2008. Last accessed on Dec. 15, 2011 at <http://www.tra.gov.eg/Presentations/NumberPotabilityConsumersGuide24032008_En.pdf>.

National Telecom Regulatory Authority (NTRA). "Telecom Law." *National Telecom Regulatory Authority*. Last accessed on June 4, 2012 at <http://www.tra.gov.eg/uploads/law/law_en.pdf>.

National Telecom Regulatory Authority (NRTA)."The NTRA Offering the Second Fixed-Line License." *National Telecommunications Regulatory Authority*. Last accessed on Dec. 19, 2011 at <http://www.tra.gov.eg/english/news_newsdetails.asp?PID=36&ID=129>.

"News Media in Egypt." *Press Reference*. Last accessed on Jan. 8, 2012 at <http://www.pressreference.com/Co-Fa/Egypt.html>.

*Newspaper readership study 2002–2010*. Pan Arab Research Center (PARC), 2011.
*Nile Engineering Group*. Last accessed on June 10, 2012 at <http://www.nileengineering.com/English/Products--and--Services/NileTel-Stores>.
NileSat. "Nilesat Company Profile." *NileSat*. Last accessed on June 27, 2012 at <http://nilesat.com.eg/banners/AboutUs/CompanyProfile.aspx>.
*Nile Television Network (NTN)*. Last accessed on June 27, 2012 at <http://www.ntntv.net/>.
"NTRA Says Finding Local Partner in Mobinil Mandatory." *Arab Finance Brokerage Company*. Dec. 16, 2009. Last accessed Jan. 10, 2012 at <http://www.arabfinance.com/news/newsdetails.aspx?id=157214>.
Office of the United Nations High Commissioner for Human Rights. "International Covenant on Civil and Political Rights." *United Nations Human Rights*. Last accessed on Jan. 17, 2012 at <http://www.ohchr.org/EN/Pages/WelcomePage.aspx>.
Orascom Telecom. "Our History." *Orascom Telecom*. Last accessed on June 27, 2012 at <http://www.orascomtelecom.com/about/Contents/default.aspx?ID=395>.
*Penal Code No. 58 of 1937 [Egypt]*. August 1937. Last accessed on Mar. 7, 2012 at <http://www.unhcr.org/refworld/docid/3f827fc44.html>.
"Presidential Decree No. 313/1956, Amended by Law 14/1967." *United Group: Attorneys at Law, Legal Researcher and Human Rights Advocates*. Last accessed on Jan. 16, 2012 at <http://ug-law.com/index.php?option=com_content&view=article&id=424%3ApresidentiAl decree-no-3131956-amended-by-law-no-141967-&catid = 64%3Amedia-laws&Itemid = 89&lang = en>.
"Press Freedom Index 2011–2012." *Reporters Without Borders*. Last accessed on June 4, 2012 at <http://en.rsf.org/spip.php?page=classement&id_rubrique=1043>.
"Press Freedom Index Images of Press Freedom Index 2007–2012." *Explow News*. Last accessed on June 4, 2012 at <http://explow.com/Press_Freedom_Index%20images%20of%20press%20freedom%20index%202007-202012>.
"Press Releases." *Pehla: Premium Asian Entertainment*. Last accessed on Jan. 17, 2012 at <http://www.pehlatv.net/pehla/PressReleases/54/ADD-launches-Hallmark-channel-on-its-platform>.
"Radio Masr." *Facebook Page*. Last accessed on Jan. 17, 2012 at <http://www.facebook.com/RadioMasr?sk=info>.
Ramadan, Amr. "Egypt to Sign Triple-play Licenses, Says Kamel at Euromoney." *Daily News Egypt*. Sept. 30, 2010. Last accessed on July 5, 2012 at <http://www.thefreelibrary.com/Egypt+to+sign+triple-play+licenses,+says+Kamel+at+Euromoney.-a0238670369>.
Rao, Madanmohan. *Mobile Africa Report 2011: Regional Hubs of Excellence and Innovation*. MobileMonday, 2011. Last accessed on Jan. 18, 2012 at <http://www.mobilemonday.net/reports/MobileAfrica_2011.pdf>.
Reuters. "Egypt to Auction 2nd Fixed-line License on June 19." *Balancing Act Africa*, no. 394. Last accessed on Jan. 4, 2012 at <http://www.balancingact-africa.com/news/en/issue-no-394/telecoms/egypt-to-auction-2nd/en>.
Rizk, Nagla. "Stories from Egypt's Music Industry: De Facto Commons as Alternative to Copyright" in Rizk, Nagla and Shaver, Lea, eds. *Access to Knowledge: New Research on Intellectual Property and Development*. Bloomsbury Academic, 2010, 103. Last accessed on May 29, 2012 at <http://www.aucegypt.edu/Business/A2K4D/Documents/A2KEgypt.pdf>.
"Rotana Launches Egyptian General Entertainment Channel." June 1, 2011. *Arab News*. Last accessed on July 5, 2012 at <http://mediame.com/news/broadcasting_news/rotana_launches_egyptian_general_entertainment_channel>.
Rugh, William. A. *Arab Mass Media: Newspapers, Radio & Television in Arab Politics*. Westport, CT: Praeger, 2004.
Sahota, Dawinderpal. "France Telecom Increases Stake in Egypt's Mobinil." Telecoms.com, Feb. 13, 2012. Last accessed on July 27, 2012 at <http://www.telecoms.com/tag/mobinil/>.
Sakr, Naomi. *Arab Television Today*. New York: I. B. Tauris, 2007.
Sakr, Naomi. *Satellite Realms: Transnational Television, Globalization and The Middle East*. New York: I.B. Tauris, 2001.
Sakr, Naomi. "Satellite Television and Development in the Middle East." *Middle East Research and Information Project 2010*. Last accessed on May 29, 2012 at <http://www.merip.org/mer/mer210/satellite-television-development-middle-east>.
Saleh, Ibrahim. "The Crossing Roads of Egypt: An Avenue of Participation Crucifies the Stagnation Development." *Global Media Journal* 2, no. 4 (2004). Last accessed on June 27, 2012 at <http://lass.calumet.purdue.edu/cca/gmj/OldSiteBackup/SubmittedDocuments/archivedpapers/fall2004/refereed/saleh.htm>.
Saleh, Nivien. *Third World Citizens and the Information Technology Revolution*. New York: Palgrave Macmillan, 2010.

"Saudi Arabia: Media Market Description" in *Worlds Association of Newspapers: World Press Trends 2008*. 2008. Last accessed on Jan. 17, 2012 at <http://www.ihudaif.com/wp-content/uploads/2010/01/WORLD-ASSOCIATION-OF-NEWSPAPERS---WORLD-PRESS-TRENDS-2008.pdf>.

"Saudi Arabia Moves Its Satellite Channel to Arabsat." *Al Masry Al Youm*. Sept. 8, 2011. Last accessed on Mar. 12, 2012 at <http://www.almasryalyoum.com/en/node/493638>.

Sennitt, Andy. "Al Jazeera Kicked out of Egypt, Nilesat Signal Cut." *Media Network*. Jan. 30, 2011. Last accessed on June 27, 2012 at <http://blogs.rnw.nl/medianetwork/al-jazeera-kicked-out-of-egypt-nilesat-signal-cut>.

Shahine, Alaa. "New Media Flourish in Egypt." *UCLA Center for Communications and Community*. Apr. 27, 2004. Last accessed on Jan. 17, 2012 at <http://www.c3.ucla.edu/newsstand/global/new-media-flourish-in-egypt/>.

Sharp, Jeremy M. "Egypt, Congress and U.S. Relations." *Congressional Research Service*. Order Code RL33003. Aug. 12, 2008. Last accessed on June 27, 2012 at <http://fpc.state.gov/documents/organization/109518.pdf>.

"Showtime Arabia and Orbit Group Merge Pay TV Operations." AMEinfo.com. July 12, 2009. Last accessed on Jan. 17, 2012 at <http://www.ameinfo.com/203504.html>.

"Shut Down Gadaffi State TV." *Change.org*. July 13, 2011. Last accessed on Mar. 12, 2012 at <http://www.change.org/petitions/shut-down-gaddafi-state-tv>.

Solayman, Hanan. "New TV Channels Hit Egypt's Airwaves." *Daily News Egypt*. July 17, 2011. Last accessed on June 27, 2012 at <http://www.thedailynewsegypt.com/new-tv-channels-hit-egypts-airwaves.html>.

Sout Al Qahira. "About Sout Al Qahira." *Sout Al Qahira*. Last accessed on Jan. 17, 2012 at <http://www.soutalqahira-eg.com/about-ern.aspx>.

"Spotlight." *Melody Holding*. Last accessed on June 27, 2012 at <http://www.melodyholding.com/>

State Information Service (SIS). *Egypt Yearbook 2007*. 2007. Last accessed on July 3, 2012 at <http://www.sis.gov.eg/En/ListTemplate.aspx?Category_ID=870>.

"Subsidiaries and Investments." Telecom Egypt. Last accessed on June 27, 2012 at <http://ir.telecomegypt.com.eg/Subsidiaries.asp>.

"Target 8." *International Telecommunications Union*. 2010. Last accessed on July 8, 2012 at <http://www.itu.int/ITU-D/ict/publications/wtdr_10/material/WTDR2010_Target8_e.pdf>.

"TBS Feature Interview: Sheikh Saleh Kamel." *Transnational Broadcasting Studies (TBS) Archives*. May 1, 1998 (Fall 1998). Last accessed on Jan. 17, 2012 at <http://www.tbsjournal.com/Archives/Fall98/Interviews1/Sheikh_Saleh/sheikh_saleh.html>.

TE Data. "Home ADSL Price List." *TE DATA 2012*. Last accessed Jan. 5, 2012 at <http://www.tedata.net/eg/en/Home-ADSL/Home-ADSL-Prices-List>.

"Telecom Egypt Announces the Acquisition of Remaining TE Data Stake." *Telecom Egypt*. Jan. 5, 2010. Last accessed on Mar. 12, 2012 at <http://ir.telecomegypt.com.eg/press%20releases/press%20releases/TE%20Data%20PR%20-%20Final.pdf>.

Telecom Egypt, "Annual Report 2009: Equipped For A New Era," Giza, 2009, 24. Last accessed on Nov. 27, 2011 at <http://ir.telecomegypt.com.eg/admin/uploads/annual-report-2009/FINAL%20ANNUAL%20web.pdf>.

Telecom Egypt. "Annual Reports 2006–2009." *Telecom Egypt*. Last accessed on May 14, 2012 at <http://ir.telecomegypt.com.eg/Annual%20Reports.asp>.

Telecom Egypt. "Basic Information." *Telecom Egypt*. Last accessed Nov. 27, 2011 at <http://ir.telecomegypt.com.eg/Basic%20Information.asp#>.

Telecom Egypt. "Company Fact Sheet." *Telecom Egypt*. Last accessed on June 27, 2012 at <http://ir.telecomegypt.com.eg/Company%20Fact%20Sheets.asp>.

Telecom Egypt. "Company Milestones." *Telecom Egypt*. Last accessed on June 27, 2012 at <http://ir.telecomegypt.com.eg/Company%20Milestones.asp>.

Telecom Egypt. "Company Profile." *Telecom Egypt*. Last accessed on Nov. 27, 2011 at <http://ir.telecomegypt.com.eg/Basic%20Information.asp#>.

Telecom Egypt. "Ownership Structure." *Telecom Egypt*. Last accessed on June 27, 2012 at <http://telecom.tedata.net/uipages/Ownership_Structure.aspx>.

Telecom Egypt. "Wholesale." *Telecom Egypt*. 2008. Last accessed on Jan. 3, 2012 at <http://www.telecomegypt.com.eg/English/wholesale_MPLS_IP-VPN.asp#>.

"Telecom Watchdog Decision Clears Way for 4th Egyptian Mobile Operator." *Ahram Online*. May 7, 2012. Last accessed on June 4, 2012 at <http://english.ahram.org.eg/NewsContent/

3/12/40995/Business/Economy/Telecom-watchdog-decision-clears-way-for-th-Egypti.aspx>.

"The Arab World's Major Pay TV Operators Provide A Choice of 140 Pay TV Channels." *Arab Advisors Group*. Jan. 14, 2008. Last accessed on July 26, 2014 at <http://www.arabadvisors.com/Pressers/presser-140108.htm-0>.

The Cairo Institute for Human Rights Studies. *Media in North Africa: Obstacles and Challenges*. United Nations Economic Commission for Africa, 2007, 4. Last accessed on Jan. 16, 2012 at <http://www.uneca.org/africanmedia/documents/NorthAfrica-eng-report.pdf>.

"The Dispute over the Ownership of Egypt's Mobinil." *International Telecommunications Union Regulation Toolkit*. Dec. 22, 2011. International Telecommunications Union. Last accessed on July 27, 2011 at <http://www.ictregulationtoolkit.org/en/PracticeNote.3264.html>.

"The Emergency Law in Egypt." *Fidh: Worldwide Human Rights Movement*. Nov. 17, 2001. Last accessed on Jan. 16, 2012 at <http://www.fidh.org/THE-EMERGENCY-LAW-IN-EGYPT>.

"The Media Free Zone." *Egyptian Media Production City (EMPC)*. Last accessed on Dec. 29, 2011 at <http://www.empc.com.eg/English/MediaFreeZone.html>.

UNICJ. *ICJ Submission to the Universal Periodic Review of Egypt*. International Commission of Jurists, 2009. Last accessed on Jan. 16, 2012 at <http://www.icj.org/IMG/UPRSubmission-Egypt.pdf>.

United Nations. "Article 19." *The United Nations Declaration of Human Rights*. Last accessed on June 27, 2012 at <http://www.un.org/en/documents/udhr/index.shtml#a19>.

"Urgent Communication to the Egyptian Public Prosecutor a Call for Arresting the Egyptian Minister of Mass Media and Information & Prosecuting Him for Stimulating Hatred and Murder." *The Arab Center for the Independence of the Judiciary and the Legal Profession (ACIJLP)*. Feb. 25, 2011. Last accessed on Jan. 16, 2012 at <http://www.acijlp.org/?browser=view&ID=24'>.

Williams, Sean. "Walking the Plank." *Arabian Business*. Dec. 15, 2010. Last accessed on Jan. 17, 2012 at <http://www.arabianbusiness.com/walking-plank-366838.html>.

World Bank. "Egypt, Arab Rep." *World Bank Development Indicators*. Last accessed on Dec. 28, 2011 at <http://data.worldbank.org/country/egypt-arab-republic>.

YallaMisr. "About Us." *YallaMisr*. Last accessed on June 27, 2012 at <http://www.yallamisr.com/about.htm>.

# Egypt—Data Summaries

ELI NOAM AND PAUL MUTTER

Egypt's media industries are in transition following the several changes in government since 2011. Due to the limited liberalization of the national media industries prior to 2011, Egypt has the fourth highest weighted average HHI of all the countries surveyed at 5,041 (PI is ranked #5, at 5,071) (Table 29.13). Egypt's newspapers and video channels markets exhibit one of the highest concentrations of state ownership outside of China, a situation that the several regime changes have not altered (Table 29.14). Despite commercialization in the 1990s and 2000s, state regulators and censors wield significant influence over all broadcast and print content.

Egypt has the second highest percentage of state ownership of media after China, led in particular through the Egyptian Radio and Television Union (ERTU) and the main newspaper groups Al Ahram, Al Akhbar, and Al Gomhouria. ERTU has a legal monopoly over both TV broadcasting and radio, while the three newspaper groups collectively account for nearly 60% of daily newspapers.

In stark contrast to China, however, Egypt is ranked seventh for foreign ownership of media. This is due to mobile communications, where the two main foreign firms present are Vodafone (UK) and Orange (French Telecom), which owns the country's largest media enterprise, Mobinil, together with the domestic company Orascom (Table 29.15).

Where Egypt's media is diverse is in film production. Egypt has traditionally been the center for Arab filmmaking. This role declined under the Nasser regime when the industry was nationalized, but it re-emerged when private film production was liberalized. The role as a regional center has its drawbacks, however: to be commercially successful, a film has to be

Table 29-13. National Media Industries Concentration in Egypt (Subsidiary organizations denoted by italics)

| | 2004/5 | | 2011 OR MOST RECENT | | % CHANGE ANNUAL AVERAGE | |
|---|---|---|---|---|---|---|
| | COMPANY POWER INDEX IN COUNTRY | COMPANY SHARE OF THE OVERALL NATIONAL MEDIA MARKET (%) | COMPANY POWER INDEX IN COUNTRY | COMPANY SHARE OF THE OVERALL NATIONAL MEDIA MARKET (%) | COMPANY POWER INDEX IN COUNTRY | COMPANY SHARE OF THE OVERALL NATIONAL MEDIA MARKET (%) |
| Government of Egypt | 4,111 | 51.2 | 3,311 | 45.5 | -2.8 | -0.81 |
| *Telecom Egypt (public)* | *3,329* | *41.4* | *2,746* | *37.3* | *-2.5* | *-0.58* |
| *Egyptian Radio and Television Union (public)* | *428* | *5.1* | *340* | *3.6* | *-2.9* | *-0.22* |
| *Al Ahram (public)* | *44* | *1.2* | *10* | *0.44* | *-11.0* | *-0.1* |
| *Al Akhbar (public)* | 39 | *1.1* | *15* | *0.5* | *-8.7* | *-0.08* |
| *Nile Television Network (public)* | 44 | *1.9* | 7 | *0.7* | *-12.0* | *-0.18* |
| *Al Gomhouria (public)* | *12* | *0.61* | 7 | *0.4* | *-5.6* | *-0.03* |
| Mobinil (Orange, France) | 1,132 | 21.2 | 704 | 20.7 | -5.4 | -0.08 |
| Vodafone (UK) | 263 | 10.2 | 275 | 12.7 | 0.6 | 0.35 |
| Etisalat (UAE) | 0 | 0.0 | 322 | 13.6 | N/A | 1.9 |
| LinkDotNet | 63 | 2.1 | 58 | 2.1 | -1.0 | 0.0 |
| Hayat | 0 | 0.0 | 11 | 0.85 | N/A | 0.1 |
| Google (US) | 19 | 0.2 | 53 | 0.6 | 25.4 | 0.05 |
| Panorama Action TV | 0 | 0.0 | 10 | 0.8 | N/A | 0.11 |
| ART (Saudi Arabia) | 4 | 0.6 | 0.2 | 0.09 | -13.4 | -0.07 |
| Rotana (Saudi Arabia) | 1 | 0.3 | 6 | 0.6 | 48.3 | 0.04 |
| MBC (Saudi Arabia) | 9 | 0.8 | 5 | 0.6 | -5.4 | -0.04 |
| Melody Entertainment | 0.6 | 0.2 | 2 | 0.4 | 35.6 | 0.02 |
| Dream TV | 20 | 1.3 | 3 | 0.4 | -12.4 | -0.1 |

| | | | | | | |
|---|---|---|---|---|---|---|
| Al Baraheen (Saudi Arabia) | 0 | 0.0 | 2 | 0.4 | N/A | 0.06 |
| Al Masry Al Youm | 0.0 | 0.02 | 4 | 0.3 | 8,321.9 | 0.04 |
| Al Mehwar | 1.2 | 0.3 | 0.8 | 0.2 | –5.4 | –0.01 |
| Al Jazeera (Qatar) | 3 | 0.5 | 0.3 | 0.1 | –13.0 | –0.05 |

| Media Concentration Index | 2004/5 | 2011 or Most Recent | % Change Annual Average |
|---|---|---|---|
| Total Revenue: Nat'l Media Industry (mil US$) | 4,079 | 7,979 | 13.7 |
| Total Voices (*n*) | 29 | 35 | 3.0 |
| Net Voices (*n*) | 26 | 32 | 3.3 |
| Public Ownership (%) | 51.3 | 48.3 | –0.42 |
| Foreign Ownership (%) | 37.1 | 50.3 | 1.9 |
| C4 Average—Weighted | 92.2 | 93.6 | 0.2 |
| HHI Average—Weighted | 6,269 | 5,041 | –2.8 |
| C1 Average—Weighted | 65 | 53 | –2 |
| Noam Index Average—Weighted | 2,729 | 2,741 | 0.06 |
| Pooled Overall Sector C4 | 84.7 | 92.5 | 1.1 |
| Pooled Overall Sector HHI | 3,186 | 2,853 | –1.5 |
| Pooled Overall Sector Noam Index | 385 | 227 | –5.8 |
| Market Share of Top 10 Companies: Nat'l Media Industry (%) (Pooled C10) | 89.3 | 97.9 | 1.2 |
| National Power Index | 5,653 | 4,774 | –2.2 |

Table 29-14. Top Content Media Companies in Egypt

| | 2004/5 | | 2011 OR MOST RECENT | | % CHANGE ANNUAL AVERAGE | |
|---|---|---|---|---|---|---|
| | COMPANY POWER INDEX IN COUNTRY | COMPANY SHARE OF THE NATIONAL CONTENT MEDIA MARKET (%) | COMPANY POWER INDEX IN COUNTRY | COMPANY SHARE OF THE NATIONAL CONTENT MEDIA MARKET (%) | COMPANY POWER INDEX IN COUNTRY | COMPANY SHARE OF THE NATIONAL CONTENT MEDIA MARKET (%) |
| Government of Egypt | 3,493 | 44.0 | 2,828 | 35.2 | –2.7 | –1.3 |
| *Egyptian Radio and Television Union (Public)* | *1,914* | *22.9* | *2,159* | *22.7* | *1.8* | *–0.03* |
| *Al Ahram (public)* | *198* | *5.2* | *65* | *2.8* | *–9.6* | *–9.6* |
| *Al Akhbar (Public)* | *176* | *4.9* | *97* | *3.4* | *–6.4* | *–0.2* |
| *Nile Television Network (NTN) (Public)* | *195* | *8.6* | *44* | *4.4* | *–11.1* | *–0.6* |
| *Al Gomhouria (Public)* | *54* | *2.7* | *47* | *2.4* | *–1.9* | *–0.05* |
| Google (US) | 85 | 0.9 | 337 | 3.5 | 42.1 | 0.4 |
| Hayat | 0.0 | 0.0 | 72 | 5.4 | N/A | 0.8 |
| Al Baraheen (Saudi Arabia) | 0.0 | 0.0 | 15 | 2.5 | N/A | 0.4 |
| Rotana (Saudi Arabia) | 6 | 13.7 | 38 | 49.5 | 0.7 | 0.4 |
| MBC (Saudi Arabia) | 39 | 3.8 | 34 | 3.7 | –0.02 | 0.0 |
| Al Masry Al Youm | 0.03 | 0.07 | 28 | 22.8 | 118.2 | 0.3 |
| Dream TV | 90 | 5.7 | 17 | 2.6 | –0.1 | –0.5 |
| Panorama Entertainment TV | 0 | 0.0 | 60 | 5.0 | N/A | 0.7 |
| Melody Entertainment | 3 | 1.0 | 13 | 2.3 | 0.6 | 0.2 |
| Al Mehwar (public) | 6 | 1.4 | 5 | 1.4 | –0.02 | 0.0 |

| | | | | | | |
|---|---|---|---|---|---|---|
| Nile Radio Productions | 2 | 0.2 | 3 | 0.3 | 0.07 | 0.02 |
| ART (Saudi Arabia) | 18 | 2.6 | 1 | 0.6 | –13.4 | –0.3 |
| Al Jazeera (Qatar) | 14 | 2.3 | 2 | 0.9 | –0.1 | –0.2 |

| Media Concentration Index | 2004/5 | 2011 or Most Recent | % Change Annual Average |
|---|---|---|---|
| Public Ownership (%) | 44.3 | 35.5 | –1.3% |
| Foreign Ownership (%) | 25.2 | 21.7 | –0.5% |
| C4 Average—Weighted | 65.6 | 59.6 | –0.9% |
| HHI Average—Weighted | 3,893 | 3,510 | –1.4% |
| C1 Average—Weighted | 48 | 41 | –1% |
| National Power Index | 3,874 | 3,488 | –1.4% |

Table 29-15. Top Platform Media Companies in Egypt

| | 2004/5 | | 2011 OR MOST RECENT | | % CHANGE ANNUAL AVERAGE | |
|---|---|---|---|---|---|---|
| | COMPANY POWER INDEX IN COUNTRY | COMPANY SHARE OF THE NATIONAL PLATFORM MEDIA MARKET (%) | COMPANY POWER INDEX IN COUNTRY | COMPANY SHARE OF THE NATIONAL PLATFORM MEDIA MARKET (%) | COMPANY POWER INDEX IN COUNTRY | COMPANY SHARE OF THE NATIONAL PLATFORM MEDIA MARKET (%) |
| Government of Egypt (Telecom Egypt) | 4,289 | 53.3 | 3,402 | 47.5 | –3.0 | –0.8 |
| Mobinil (Orange, France) | 1,458 | 27.4 | 836 | 22.6 | –6.1 | –0.7 |
| Vodafone (UK) | 339 | 13.2 | 326 | 15.1 | –0.5 | 0.3 |
| Etisalat (UAE) | 0.0 | 0.0 | 382 | 16.1 | N/A | 2.3 |
| LinkDotNet | 81 | 2.7 | 69 | 2.5 | –2.1 | –0.03 |
| Media Concentration Index | 2004/5 | | 2011 or Most Recent | | % Change Annual Average | |
| Public Ownership (%) | 42.5 | | 46.3 | | 0.9% | |
| Foreign Ownership (%) | 51.5 | | 50.3 | | –0.3% | |
| C4 Average—Weighted | 100 | | 100 | | 0% | |
| HHI Average—Weighted | 6,954 | | 5,882 | | –3.9% | |
| C1 Average—Weighted | 69.9 | | 63.9 | | –2.1% | |
| National Power Index | 6,721 | | 5,331 | | –5.2% | |

acceptable in many countries with widely different levels of political or social standards, and this often results in bland creations.

Because of Egypt's large population and history as a center content production and distribution in the Middle East, the Nilesat network of communications satellites—owned by the ERTU—are important electronic media distribution systems for several neighboring Arabic-language markets.

# 30

# Media Ownership and Concentration in Israel

AMIT M. SCHEJTER AND MORAN YEMINI

## INTRODUCTION

The first Hebrew-language daily in what became the State of Israel appeared in 1886, when the territory was part of the Ottoman Empire. The British Mandate of Palestine, which was formed out of these territories and became the State of Israel in 1948, already had a vibrant print sector on the eve of independence in 1948. A multilingual market existed (and still does) as a consequence of the presence of a Palestinian Arab population and of successive waves of Jewish immigration from all over the world.[1] The economy in which these outlets operated was characterized until the early 1980s by central planning. Following the downfall of the ruling coalition led by the Labor Party and the rise of the Likud Party in 1977, Israel experienced a dramatic transition from "a socialist inspired mixed, highly centralized, highly planned state-centered, protectionist economy to a much more decentralized and internationally oriented neo liberal one."[2] Nowhere has this change been more evident than in the media and telecommunications industries.

Political and state control in the media and telecommunications sectors prevailed until the mid-1960s. Newspapers owned by various political parties such as the now defunct *Davar*, *Herut*, and *Lamerhav* operated alongside commercial

1. Assaf, M. "The Arabic Press in Israel." *International Communication Gazette* 7, 1961. pp. 23–25; Masalha, O. A. "Arabic Press in Israel.", Aug 2, 2009. *available at* http://www.reader.co.il/article/48627העיתונות-הערבית-בישראל/ (accessed Jan 25, 2013).

2. Aronoff, M. "Radical Change in Israel: A Review Essay." *Political Science Quarterly* 116: 3, 2001. pp. 447–453.

newspapers (such as *Haaretz* and *Yediot Aharonot*, both of which are still published daily) since the pre-State years. Some party papers still exist as well, among them the Hebrew-language *Hamodia* (owned by the religious party Agudath Israel) and the Arabic-language *Al-Ittihad* (owned by the Israeli Communist Party), but for the most part private media groups now own and publish the leading Hebrew-, Russian-, and English-language dailies.[3] The government-controlled radio broadcasting from independence, until 1965, when the Israel Broadcasting Authority (IBA) was established as a quasi-public service broadcaster. The government also launched general broadcast television in 1968,[4] and in 1969 assigned it to be managed under the auspices of the IBA. When Likud came to power in 1977, there was only a single black and white terrestrial channel in Israel operated by the IBA, while fixed telephony was run as a state monopoly so "woefully inadequate" that in the 1980s, intercity calling meant consumers had to employ a complicated and (very often) unreliable process.[5]

The first two Likud-led government coalitions (1977–1981 and 1981–1984) established three public committees to evaluate the possibility of establishing commercial broadcasting, to auction licenses for "regional" (sub-national) cable television concessions, and to establish a national telecom provider.[6] A telecommunications law was first passed in 1982 and led to the founding of the public monopolist Bezeq in 1984 (Bezeq was subsequently fully privatized in 2005). Following this, an economic stabilization program was launched in 1985 to reign in hyperinflation.[7] The 1982 Telecommunications Law[8] was subsequently amended in 1986,[9] forming the Cable Broadcasting Council (CBC), which in 1989 began awarding concessions to cable service operators. In 1990, the Second Authority for Radio and Television was established to oversee the introduction of commercial broadcast television.[10] The first commercial terrestrial broadcasting channel, Channel 2, was launched in 1993, and commercial radio broadcasting began shortly thereafter. The second Yitzhak Rabin government (1992–1995) that pursued these measures also began the process of awarding licenses for competing mobile telephony operators to the monopoly until then held by a Bezeq subsidiary. The early 2000s saw additional changes in the media landscape, with the development and licensing of additional commercial media outlets. In 2000 direct-to-home (DTH) satellite service was introduced, and in 2002, a second commercial terrestrial channel was established (Channel 10). Dial-up Internet service, introduced in the 1990s, transitioned to broadband in the mid-2000s.

The growth in Israeli media should also be seen within the context of the rapid adoption of new media technology by Israeli consumers. In 1970, already, 53% of households had a television set (today, household penetration is universal).[11] By 1991, almost half of Israeli households owned VCRs, which was double the number of those owning other expensive consumer goods at the time such as dishwashers, microwave ovens, and PCs. Israelis rapidly adopted cable television once it became available: legal cable TV franchises that were awarded as of 1989 started operating in 1991 (in response to the popularity of pirate services), and by the

---

3. The only daily in Arabic is *Al-Ittihad*.

4. Educational broadcasting was launched as early as 1966.

5. Aharoni, Y. *The Israeli Economy: Dreams and Realities*. London: Routledge, 1991, p. 232; Gandal, N. "Compatibility, Standardization, & Network Effects: Some Policy Implications." *Oxford Review of Economic Policy* 18, 2002. pp 80–91.

6. Schejter, A. *Muting Israeli Democracy: How Media and Cultural Policy Undermine Free Expression*. Urbana, IL: University of Illinois Press, 2009.

7. Ben-Bassat, A. "The Obstacle Course to a Market Economy in Israel." *in The Israeli Economy, 1985–1998: From Government Intervention to Market Economics*. Ben-Bassat, A., ed. Cambridge: MIT Press, 2002.

8. S.H. 1960, 218 (Aug 22, 1982) ("Telecommunications Law").

9. Telecommunications Law (Amendment No. 4), 1986, S.H. 1192, 224 (Aug 13, 1986).

10. Second Authority for Radio and Television Law, 1990, S.H. 1304, 58 (Feb 13, 1990) ("Second Authority Law").

11. Schejter, A. *Muting Israeli Democracy: How Media and Cultural Policy Undermine Free Expression*. Urbana, IL: University of Illinois Press, 2009, p. 25.

summer of 1992, more than 50% of all households were subscribed to a cable service, which rose to 70% by 1996.[12]

Similar trends were observed with regard to contemporary media platforms. In 2001, only 2.7% of all households used DSL broadband. However, in 2002 the government finally licensed cable companies to provide Internet access and by the end of 2004, household Internet access rates reached 43%, a figure matched at the time internationally only by South Korea.[13] Mobile phones were also adopted rapidly, even more so than Internet service, though not as quickly as cable TV had.[14] In 1994, mobile phone market penetration reached just 2.5% of the population. Within three years that number had risen to 29%, then to 74% by 2002 and by the end of 2009, digital mobile phones were present in 91.8% of households in Israel, three-quarters of which possessed at least two mobile lines.[15]

Media concentration and ownership concerns were significant issues in the transition that Israeli media and telecommunication providers were making from a mostly public sector to a mostly private market. In fact, such anxieties were already present in the draft legislation for the Second Authority for Radio and Television Law as it made its way through the Knesset (Israeli parliament) in the mid-1980s. The original bill focused on the effects that the new Channel 2 might have on the advertising revenue of newspapers. A compensation mechanism was proposed to offset such losses. Two High Court decisions, HCJ 757/84[16] and HCJ 3424/90,[17] limited the IBA from increasing its TV advertisement space beyond PSAs and underwriting credits. In the course of further Knesset debate over the Second Authority Law, a number of academics and senior Ministry of Justice officials protested a provision allowing newspapers-broadcast TV cross-ownership, a protest that led to a legal limitation on such practice. The final version of the Second Authority Law was passed in 1990 and limited the holdings of newspapers publishers in a commercial broadcaster to a 24% stake. No similar restriction was created with regard to cable TV, however. As a result, Israel's two leading print publishers, the Yediot Aharonot Group and the Ma'ariv Group, ended up with significant holdings in both broadcast and cable TV. Further concerns over cross-ownership raised in the early 2000s overtook the original concerns over the financial health of the newspapers industry. In the process of overhauling the Telecommunications Law and the Second Authority Law in light of the introduction of broadband Internet and DTH satellite, the formation of the new broadcast channel—Channel 10—and the pending expiration of the first round of radio and television broadcasting licenses, stricter cross-ownership rules were established. Under the new rules, cross-ownership of newspapers, cable, and broadcast television was prohibited. The amendment to the Telecommunications Law that set this policy down has few precedents or comparisons in other countries' regulatory systems.

Until liberalization increased in the 1980s, antitrust regulation in Israel for all sectors was minimal and did not exist in the audiovisual and telecommunications markets since these were state-operated. According to the requirements of the Israeli antitrust laws, monopoly status is

12. See Schejter, *id.*, at 25–26; Lehman-Wilzig, S. and Schejter, A. "Mass Media in Israel." *in Mass Media in the Middle East.* Kamalipour, Y. and Mowlana, H., eds. Westport, CT: Greenwood Press, 1994. pp. 109–125.

13. Schejter, A. and Lee, S. "The Evolution of Cable Regulatory Policies and Their Consequences: Comparing South Korea and Israel." *Journal of Media Economics* 20:1, 2007. pp. 1–28.

14. Cohen, A.; Lemish, D.; and Schejter, A. *The Wonder Phone in the Land of Miracles: Mobile Telephony in Israel.* Cresskill: Hampton Press, 2008.

15. Schejter, A. and Cohen, A. "Israel: Chutzpah and Chatter in the Holy Land." *in Perpetual Contact.* Katz, J. and Aakhus, M., eds. Cambridge: Cambridge University Press, 2002. pp. 30–42; Central Bureau of Statistics. *Israeli Population Demographic Characteristics, 1990–2009,* 2010. *available at* <http://www.cbs.gov.il/www/statistical/isr_pop_heb.pdf> (accessed Jan 25, 2013).

16. HCJ 757/84 *The Daily Newspaper Union in Israel v. The Minister of Education and Culture,* PD 41(4) 337 (1987).

17. HCJ 3424/90 *The Daily Newspaper Union in Israel v. The Minister of Education and Culture,* PD 45(2) 24 (1991).

conferred on companies that have market shares greater than 50% in one sector of the economy. In 1995, the national public telecom operator Bezeq was declared a monopoly, though in 2001 this designation was removed from its ISP subsidiary, Bezeq International.[18] At the same time, Bezeq was declared a monopoly in the provision of broadband Internet, a decision that was ratified in 2004; in 2009, the antitrust court found that Bezeq had also abused its monopoly status.[19] The Yediot Aharonot Group was labeled a monopolist in the print daily newspaper sector during this period, first in 1995 and then again in 1999, though both declarations were rescinded in 2010 as the group's print properties were by then rapidly losing readers and revenue.[20]

The challenge of enforcing these rulings was made clear in a 1994 ruling on a cable TV cartel formed in 1989 when Israel's five cable TV providers of the time teamed up to create Israel Cable Programming Ltd. (ICP) in order to jointly purchase all programming and "package" the four cable channels (movies, sports, children, and "family") each were required to provide their subscribers. Responding to complaints that ICP's role in the video channel market amounted to cartelization, the antitrust commissioner brought a suit against the operators that resulted in ICP being declared an illegal cartel that must dissolve immediately as there was no legal basis for the existence of a national cable TV provider.[21] Before the ruling could be fully enforced, however, in 1996 the Knesset (under pressure from ICP) amended the Telecommunications Law to legalize the cartel.[22] Content and platform ownership among multichannel operators became one and the same, until DTH satellite was introduced.[23]

So despite several antitrust actions meant to limit cross-ownership, concentration has increased in several sectors. When cable TV companies were licensed to provide Internet access in 2001, their licenses and Bezeq's own license required them to maintain structural separation between their ISP and infrastructure segments, and in the case of Bezeq, regarding their long distance and mobile telephony service as well. In 2006, however, the Israeli conglomerate IDB Holdings purchased controlling shares in the mobile provider Cellcom adding to its involvement in two of Israel's main ISPs of the time, 013 Barak and Netvision. After 013 Barak and Netvision merged to form 013 Netvision in January 2007, IDB Holdings then merged the new enterprise with Cellcom, all with the approval of Israeli regulators. A similar process took place when the cable provider HOT (which in the early 2000s became a legal national cable infrastructure) merged with the small mobile operator MIRS (rebranded as HOT Mobile) and received a license covering both its ISP and mobile assets. While a government committee recommended in 2011 ending the structural separation of Bezeq's assets as well, that development has yet to take place, though in practice, Bezeq's holdings include already a mobile operator (Pelephone), the ISP Bezeq International, and the DTH provider Yes. The 2011 tendering of a fifth national mobile operating license ("Golan Telecom") alongside a handful of virtual mobile operators (among them "Rami Levi," "You," and "Home") and the passage of new digital terrestrial television (DTT)

---

18. Tokatly, O. *Communication Policy in Israel*. Tel Aviv: Open University, 2000.

19. A.T. 801/08, *Bezeq the Israeli Communications Corporation v. the Antitrust Commissioner* (published in *Nevo*, Dec 16, 2009). *See also* Tal, Y. and Ivry-Omer, D. *Regulation of Electronic Communications Services in Israel: The Need to Establish a Communications Authority*. Jerusalem: Israel Democracy Institute, 2009.

20. Limor, Y. "Mass Media in Israel." *in Trends in Israeli Society*. Ya'ar, E. and Shavit, Z., eds. Tel Aviv: Open University, 2003; Ezrahi, Y., Goshen, Z. and Leshem, S. *Cross Ownership—Control and Competition in the Israeli Media: Economic and Legal Aspects and Their Impact on the Israeli Democracy*. Jerusalem: Israel Democracy Institute, 2003; Calcalist. "In Accordance with the Newspaper's Request: Yediot Aharonot Not a Monopoly" calcalist.co.il, Jan 3, 2010. <http://www.calcalist.co.il/local/articles/0,7340,L-3384283,00.html> (in Hebrew) (accessed Jan 25, 2013).

21. A.T. 455, *I. C. S. Israel Cable Systems, Ltd. v. The Antitrust Commissioner* Jan 10, 1994 (on file with the authors).

22. Telecommunications Law (Amendment No. 13) (1996), S.H. 1590, 302.

23. In 2000, the Cable and Satellite Broadcasting Council forced the cartel to partially separate MSO content from platform ownership by making them offer their channels to DTH satellite providers at normal market rates.

legislation in 2012, were all aimed at increasing competition, and indeed prices in the mobile market dropped dramatically, but it remains to be seen if the trend toward vertical integration among distributors can be offset.

Studying concentration among the different Israeli media industries requires some methodological creativity. The available data do not conform to the traditional definitions of market shares as the "share of the actual sales (either in quantity sold or dollar volume) for a product in a given period and in a given geographical area."[24] Adhering to such a definition is impossible in the case of Israel because with only a few exceptions in radio and television broadcasting, all media outlets in Israel are commercial, but not publicly traded enterprises, which have no public obligation to disclose their market shares as measured by revenue. There is no verifiable data, therefore, with regard to all of their sales. In some cases, we were able to acquire data that allowed us to present a number we believe provides a verifiable measure of the audience shares; in other cases we use regulatory or private sources that provide estimates of market shares based on total consumers[25] (Table 30.1).

The second challenge in measuring concentration has to do with survey methodology. Over the twenty-nine year period under consideration (1984–2013), there was no standardized system for the regulators and firms measuring concentration. Even data acquired from the same sources do not display consistent survey methodologies: the Israeli Marketing Union (formerly the Israeli Advertisers Union) conducted industry-funded, semi-annual phone surveys among the public, questioning its media consumption preferences, but over the years different subcontractors have been used to conduct this survey so there is little methodological consistency. Since 1998 such surveys have been handled somewhat more consistently by Single Source Research Ltd. ("Single Source"), a subsidiary of the Israeli market research company Teleseker, which holds the license for the TGI brand in Israel, operated in cooperation with the UK research agency, BMRB International. While TGI changed in 2011 as it was merged into the international research firm Kantar Media, the methodology for the media consumption study remained the same. Single Source's sister company, Tele-Gal, which holds the industry contract for TV audience measurement in Israel using People Meter technology, was also folded into Kantar Media in 2011. Additionally, due to the complexity of conducting studies in more than one language and the fact that non-native Hebrew-language speakers in Israel often fall into lower income brackets, many of the studies conducted by the Israeli advertising industry reflect media consumption only among native Hebrew-language speakers.

Similar methodological challenges exist in measuring the size of the market, which we have determined by assessing the national expenditure on advertising in media markets using multiple sources; and investments in communications infrastructure (infrequently) reported by the Israeli Ministry of Communications. This proves problematic for evaluating market shares, though, as revenue alone cannot explain market shares (Table 30.2). Some of the market players, for example, are conglomerates that subsidized their properties to promote certain services at different times; others, especially in the broadcast sector, are public entities, which are not profit-oriented, and therefore revenue figures cannot reflect their audience share.

## PRINT MEDIA

### Daily Newspapers

Several media groups still in existence today predate the State of Israel's founding in 1948. Most

24. Cooper, L. G. and Nakanishi, M. *Market-Share analysis: Evaluating Competitive Marketing Effectiveness*. Norwell: Kluwer Academic Publishing, 1998.

25. In the telecommunications sector some reliable data may be obtained from the Ministry of Communications. In the media sector most data regarding audience shares can be derived from periodical exposure surveys conducted for commercial purposes. In the case of the latter, the market may include consumers who consumed more than one brand of a particular product. In these instances, we normalized the market shares to reflect a market in which their sum equaled 100%.

Table 30-1. Advertising Spending in Israel (Mil US$[1]), 1984–2013

| | 1984 | 1988 | 1992 | 1994 | 1998[2] | 2000[3] | 2004[4] | 2008[5] | 2012[6] | 2013[7] |
|---|---|---|---|---|---|---|---|---|---|---|
| Television (Broadcast) | 2.0 | 14.0 | 18.0 | 305.0 | 237.0 | 261.0 | 274.0[8] | 448.0[9] | 371.0[10] | 434.0[11] |
| Daily Newspapers | 99.0 | 228.0 | 354.0 | 452.0 | 447.0 | 603.0 | 391.0 | 440.0 | 260.7 | 260.0 |
| Internet | 0.0 | 0.0 | | | | 3.0 | 23.5 | 134.0 | 146.0 | 182.3 |
| Billboards/Outdoors | | | | 66.4 | 65.0 | 78.0 | 31.4 | 62.7 | 44.3 | 51.4 |
| Radio | | | | 30.8 | 68.0 | 98.0 | 54.8 | 58.5 | 57.8 | 67.7 |
| Film | | | | 1.3 | 1.0 | | 7.8 | 9.9 | 10.2 | 11.9 |

*Source*: Tokatly O. (1984–1998 data); Ynet (2000 data) Knesset Research and Information Center (2004–2008 data); Israel Marketing Union (2012–2103 data).

[1] Revenue figures originally reported in Israeli Shekels (ILS), except for the year 2000. Conversion to US dollars has been made based on the Bank of Israel's official currency exchange rate at the end of June of each relevant calendar year.

[2] Data for 1984–1998 derived from Tokatly, O. *Communication Policy in Israel*. Tel Aviv: Open University, 2000.

[3] See "Growth in Newspaper Advertising and Standstill in Television in 2000," *Ynet*, June 10, 2001 (in Hebrew), *available at* <http://www.ynet.co.il/articles/1,7340,L-808150,00.html> (relying on data released by the Advertisers Union) (accessed Jan 25, 2013).

[4] Data derived from Agmon T., and Tsadik, A. *Analysis of the Economic Consequences of Concentration and Cross-Ownership in the Media*, Jerusalem: Knesset Research and Information Center, Nov 2, 2011,18 (in Hebrew), *available at* <http://www.knesset.gov.il/mmm/data/pdf/m02952.pdf> (accessed Jan 25, 2013).

[5] Data derived from Goldsmith, R. and Schwartz O., *Advertising by Law in the Printed Press and on the Internet: Cost, Availability and Effectiveness*, Jerusalem: Knesset Research and Information Center, Mar 24, 2009, 13 (in Hebrew), *available at* <http://knesset.gov.il/mmm/data/pdf/m02206.pdf> (accessed Jan 25, 2013).

[6] Israel Marketing Union and Ifat Advertising Monitoring. *An Increase of 4% in Advertising Expenditure in the Year 2013, available at* http://www.ishivuk.co.il/message/1111 (accessed Jul 14, 2014).

[7] Ibid.

[8] 4.49 ILS per 1 US$.

[9] 3.35 ILS per 1 US$.

[10] 3.92 ILS per 1 US$.

[11] 3.62 ILS per 1 US$.

dailies are published in Hebrew, though historically, multiple publications in a plethora of languages held significant audience shares in the first few decades of the state's existence. More recent market entrants primarily serve Jewish immigrants from the former Soviet Union, such as the Russian-language *Vesti*, and some papers run English-language editions, either online or in print as well. Now that Hebrew fluency is more widespread, a number of the older multilingual dailies have closed down, such as the pre-1948 *Uy Kelet* (Hungarian language), *Israel Nachrichten* (German language), *Letzte Nayes* (Yiddish language), and *Nasha Strana* (Russian language) newspapers.

Just as the economy moved away from its socialist foundations toward more neoliberal policies, press ownership liberalized as well and party-affiliated dailies ceased publication in favor of privately owned publishers, though weekly local newspapers—which have proliferated since the 1980s—in the Hebrew- and Arabic-language markets have generally been exempted from such changes. With the exception of the free daily *Israel HaYom* (and its English-language edition *Israel Today*), which was set up in 2007 by American investors, the other papers' circulation rankings remained nearly constant between 1984 and 2007. Some smaller circulation newspapers, however, were established during this period, and most of them have now shut down, usually as a result of their failure to establish a significant following in a more commercial market.

Until *Israel HaYom* displaced it in 2012, the Yediot Aharonot Group's flagship daily *Yediot Aharonot* was the national leader in terms of market shares from the 1980s to

Table 30-2. Telecommunications Revenue in Israel (Mil US$[1]), 1996–2013

| | 1996[2] | 2000[3] | 2004[4] | 2008[5] | 2012[6] | 2013[7] |
|---|---|---|---|---|---|---|
| Fixed Line Telephony | 1,889[8] | 1,627[9] | 1,265[10] | 1,241[11] | 1,004[12] | 1,032[13] |
| Internet Access (Infrastructure) (Dial-up (exclusively until circa 2000), ADSL and Cable) | Included in Fixed Line Telephony | Included in Fixed Line Telephony | 151 | 341 | 445 | 527 |
| Multichannel TV Platforms | 569 | 443 | 670 | 1,104 | 1,000 | 1,081 |
| International Telephony | 896 | 623 | 335 | 482 | 332 | 317 |
| Mobile Telephony | 1,130 | 2,448 | 2,977 | 4,759 | 3,401 | 3,014 |
| ISPs | Included in International Telephony | Included in International Telephony | 248 | 510 | 534 | 489 |
| Others | 62 | 107 | 101 | N/A | N/A | N/A |
| Total Revenue (mil US$) | 4,547 | 5,250[14] | 5,749 | 8,440 | 6,705 | 6,461 |

*Source*: Ministry of Communications.

[1] Revenue figures originally reported in Israeli Shekels (ILS). Conversion to US dollars has been made based on the Bank of Israel's official currency exchange rate at the end of June of each relevant calendar year.

[2] Data for the year 2000 obtained from The Report of the Committee for Detailed Recommendations Regarding Policy and Competition Rules in the Telecommunications Industry in Israel 2008 (Mar 9, 2008) (in Hebrew), 45 (the "Gronau Report"), *available at* <http://www.moc.gov.il/new/documents/gronau2008/gronau1.pdf>. An executive summary of the Gronau Report in English may be found at <http://www.moc.gov.il/sip_storage/FILES/5/1355.pdf>.

[3] Data for the year 2000 obtained from The Gronau Report, *id.*

[4] Data for the year 2004 obtained from Ministry of Communications, *Summary of Revenues of the Telecommunications Market for the Year 2006*, Oct 23, 2007 (in Hebrew), *available at* <http://www.moc.gov.il/sip_storage/FILES/5/1245.pdf >.

[5] Data for the year 2008 obtained from Ministry of Communications, *Summary of Revenues of the Telecommunications Market for the Year 2011*, Nov 28, 2012 (in Hebrew), *available at* <http://www.moc.gov.il/sip_storage/FILES/6/3086.pdf>.

[6] Data for the years for the year 2012 obtained from Ministry of Communications, *Summary of Revenues of the Telecommunications Market for the Year 2013*, 10 July, 2014 (in Hebrew) (on file with the authors).

[7] Data for the year 2013 obtained from the 2013 Revenue Summary, *id.*

[8] 3.20 ILS per 1 US$.

[9] 4.08 ILS per 1 US$.

[10] 4.49 ILS per 1 US$.

[11] 3.35 ILS per 1 US$.

[12] 3.92 ILS per 1 US$.

[13] 3.62 ILS per 1 US$.

[14] Fluctuations in the ILS-US$ exchange rate over the years may obscure the trend of growth in the Israeli telecommunications sector's revenue figures. For example, the total revenue, in ILS, for the year 2000 was nearly 50% higher than that for the year 1996 (ILS 21,420 million and ILS 14,553 million, respectively), but the growth rate seems much lower when presented in US dollars due the significant change in the ILS-US$ exchange rate from 1996 to 2000.

the late 2000s. Established in 1936 by the Mozes family, the Yediot Aharonot Group runs several other dailies as well and is also involved in book and magazine publishing, and online news production. In addition to *Yediot*, the group has also launched the Russian language daily *Vesti*, the financial news daily *Calcalist*, and the (defunct) free daily *24 Dakot*. Yediot is also active in the local weeklies market.

Prior to 1984, *Ma'ariv* was Israel's bestselling daily. *Ma'ariv* was founded in 1948 by a group of former *Yediot Aharonot* staff and became part of the journalist-owned cooperative known as the Ma'ariv Group, which distributed *Ma'ariv* and several affiliated magazines and weeklies. In the early 1990s, the British media mogul Robert Maxwell acquired control of the Ma'ariv Group.[26] Until then a broadsheet, *Ma'ariv*'s declining readership by this time led to its reorganization as a tabloid. Following Maxwell's death in 1991, Ma'ariv was sold to the Nimrodi family and became the first Israeli daily to be purchased by a publicly traded conglomerate, Hakhsharat Hayeshuv. During the 1990s, Ma'ariv's owners, like Yediot's, invested in broadcast television, cable TV, local weeklies, an online news portal, and book publishing, and were the owners of Israel's largest music publisher, Hed Arzi. For a short time the group also published a Russian-language daily, *Vremya*. Ultimately though, after losing its audiovisual properties and facing declining readership, the Nimrodi family sold their stake in the Ma'ariv Group in 2011 to another Israeli conglomerate, IDB Holdings. However, after a series of financial reversals for IDB's owners (and in response to *Ma'ariv*'s further loss of audiences shares, mostly to *Israel HaYom*), Ma'ariv Group went into receivership and was run by a court-appointed trusteeship until Makor Rishon Hameuchad (HaTzofe), owned by the publisher Shlomo Ben-Zvi, purchased the paper, consolidated Ma'ariv Group's titles and laid off most of *Ma'ariv*'s staff. The laid off staff then went on to publish a new weekly, *SofHashavua*, ("the weekend") in early 2013 in partnership with the *Mirkaei Tikshoret* group, publisher of the *Jerusalem Post* (see below). In an ironic twist of events, *Mirkaei Tikshoret* took over *Ma'ariv* in 2014 when it verged liquidation again, and merged it with the new weekly to create *Ma'ariv-Hashavua* (Ma'ariv This Week), as a daily newspaper and *Ma'ariv-SofHashavua* (Ma'ariv This Weekend) as its weekend edition.

The Ha'aretz Group is Israel's oldest publisher of newspapers, having been founded in 1918. The Schocken family, which fled Germany in 1934, purchased Ha'aretz in 1937 and has owned it since, pursued a different business model from the Mozes family or *Ma'ariv*'s various owners after their own attempts to acquire broadcast television licenses in the 1990s failed. The Ha'aretz Group initially responded by setting up a tabloid, *Hadashot*, which only ran from 1984 to 1993. The Ha'aretz Group was more successful in establishing its presence on the local weeklies market segment in the 1980s, and was one of the first media groups to create online news portals.[27] The Haa'retz Group also launched in 1997 an English-language edition of its flagship daily *Ha'aretz* in partnership with the *International Herald Tribune*, the international edition of *The New York Times*, which in 2013 became the *International New York Times*. Though it has not held market shares as high as *Yediot* or *Ma'ariv* historically, *Ha'aretz* is still highly influential due to its reputation as the paper of the Israeli upper middle class. Its owners are the Schocken family (60%), the family of the German publisher-aristocrat Alfred Neven Dumont (20%), and the Israelu-Russion billionaire Leonid Nevzlin.

Other important newspapers include the financial daily *Globes*, owned by the Globes Group, and The Jerusalem Post Group, which prints *The Jerusalem Post*. The Globes Group also has interests in the cable TV market and the Yediot Aharonot Group. *The Jerusalem Post*, which is published in English, only entered the Hebrew-language market in 2007, with the free daily *Israel Post*.[28] Originally owned by the national labor union (the "Histadrut"), the

26. Historical Jewish Press. "Ma'ariv." <http://web.nli.org.il/sites/JPress/Hebrew/Pages/maariv.aspx>

27. Haaretz launched the first Israeli news portal, *IOL*, (Israel Online) as early as 1995. In 2001, IOL was merged into the highly-popular portal *Walla.co.il* (which the Haaretz Group managed in partnership with Bezeq until 2010). Also in 2001, Haaretz launched *Haaretz Online*, an online edition of Haaretz's print edition (which later started to offer original content a well) and shortly thereafter the financial news website themarker.com, which later went on to become both a print daily and a magazine. *See* http://www.haaretz.co.il/misc/about-haaretz.

28. *The Jerusalem Post* was known as *The Palestine Post* from 1932 to 1950.

Post Group was acquired by the Hollinger Group (Canada) in 1989 and then sold to the Mirkaei Tikshoret Group in 2004, whose other holdings include several radio networks, national pay-TV channels, as of 2013, the weekly *SofHashavua* ("weekend"), and as of 2014 *Ma'ariv*.[29]

Israeli newspapers face a new challenge in the twenty-first century due to the market entry of the free daily *Israel HaYom*. It is owned by the US billionaire and casino owner Sheldon Adelson, a prominent donor to conservative political causes in America. In the data for 2008, it was already the second most read daily in the country (Table 30.3). Statistics published in 2012 have now rated its market share as higher than that of *Yediot Aharonot*'s. In 2008, the Yediot Aharonot Group tried to compete in this market with *24 Dakot*, but were unable to absorb the losses of distributing it for free and so discontinued its publication in 2009. In 2011, Yediot started instead to distribute a condensed version of its news and sports pages free to riders of the Israeli railway system and on university campuses.

Until the appearance of *Israel HaYom*, the period from 1988 to 2010 reflected a stable picture in which *Yediot Aharonot*, *Ma'ariv*, and *Haaretz* were read by more than 80% of Israeli newspaper consumers, with *Yediot*'s lead and *Haaretz*'s share staying constant (though the 1988 and 1992 data for the Haaretz Group include *Hadashot*, and from 2008 it includes the print edition of *TheMarker*).

There are no Arabic-language editions of the major dailies currently in circulation for the Palestinian Israeli minority, who account for approximately 20% of the population. Phone surveys conducted in 2008 and 2009 of Palestinian Israeli citizens, and obtained by the authors, show significant differences between the two years, differences that raise doubts over the validity of the reported market shares. The Israeli Arabic-language papers with the highest circulation rates are all privately owned weeklies. *Panorama* was read by 25.6% of the sample in 2008 and 34% in 2009; *Kul el Arab*, 22.9% in 2008 and 30.8% in 2009; and *Al-Sinara* 17.4% and 24.8%, respectively. The Israeli Communist Party's Arabic-language daily *Al-Ittihad* was read in 2009 by 9.2% of those surveyed.[30]

## Book Publishing

Data for the book publishing industry were received on condition that the names of the publishing houses not be revealed, so they are ranked numerically. The data are also not based on actual sales by the publishers since the publishing houses choose not to release that data in the public domain. Concentration indices for the sector were calculated using comparative data from the Israeli Book Publishers Association regarding the amount of space rented by each publisher for the display and sale of books during the country's annual Hebrew Book Week. According to the Association, this serves as a reliable indicator of market shares and trends.

We can say that the major players in the newspapers and magazine industry—the Yediot Aharonot and Ma'ariv groups—are also significant players in the book publishing industry. In addition, the book publishing industry's national book fair numbers obscure the fact that most sales are made through retail purchases. The retail distribution segment is in fact a duopoly consisting of the booksellers Steimatzky and Tzomet Sfarim, both of which also own stakes in publishing houses (Table 30.4). The Israeli Book Publishers Association filed complaints with the antitrust authorities against both chains for alleged anticompetitive behavior.[31] In response, the Knesset passed in 2013 a provisional order, to be intact for 3–4 years, which limits price reductions by booksellers due to the negative effect their price war was having on authors' royalty payments.[32]

---

29. However, this merger is not reflected in this study as it took place shortly before this study went to press.

30. See Single Source/TGI, *Sector Media Index: Exposure to Newspapers among the Israeli Palestinian Population, 2008–2009* (on file with authors).

31. Levy, R. "Books on sale - but at what cost?" *Haaretz*, June 8, 2012. <http://www.haaretz.com/business/books-on-sale-but-at-what-cost-1.435170> (accessed Jul 14, 2014).

32. *See* The Protection of Literature and Writers in Israel (Provisional Order) Law, 2013, S.H 2407, 208 (Aug 8, 2013).

Table 30-3. Daily Newspapers (Market Shares by Audience Reported Exposure), 1984–2013[1]

| | 1984[2] | 1988[3] | 1992[4] | 1996[5] | 2000 | 2004 | 2008 | 2012 | 2013[6] |
|---|---|---|---|---|---|---|---|---|---|
| Yediot Aharonot Group (Mozes Family) | 55.4 | 50.0 | 56.0 | 58.3 | 53.2 | 52.6 | 43.4[7] | 36.4 | 39.0 |
| Ma'ariv Group (various owners) | 21.6 | 19.9 | 15.8 | 23.7 | 28.4 | 28.9 | 14.2 | 9.0 | 4.2 |
| Haaretz Group (Schocken Family)[8] | 9.4 | 17.9 | 15.5 | 7.3 | 10.0 | 8.9 | 12.6 | 10.1 | 9.5 |
| The Jerusalem Post Group (various owners) | 2.7 | 2.9 | 2.3 | 1.0 | 1.1 | 0.5 | 4.1 | 7.3 | 7.6 |
| Globes Group | 0.0 | 1.9 | 2.7 | 2.6 | 3.5 | 3.7 | 3.2 | 2.9 | 3.5 |
| *Telegraph* | — | — | — | 0.5 | 0.0 | 0.0 | 0.0 | 0.0 | 0.0 |
| Makor Rishon Hameuchad (HaTzofe) (various owners)[9] | No Data | No Data | 0.9 | 0.9 | 1.4 | 1.0 | 1.5 | 1.3 | 1.5 |
| *Al Hamishmar* (Hashomer Hatzair) | No Data | No Data | 0.5 | 0.0 | 0.0 | 0.0 | 0.0 | 0.0 | 0.0 |
| *Davar* (Histadrut) | 4.0 | 4.2 | 2.3 | 1.1 | 0.0 | 0.0 | 0.0 | 0.0 | 0.0 |
| *Israel HaYom* (Sheldon Adelson) | — | — | — | — | — | — | 21.0 | 32.9 | 34.7 |
| *Hamodi'a* (Agudat Yisrael World Federation) | No Data | No Data | No Data | No Data | 2.3 | 2.1 | No Data | No Data | No Data |
| *Yated Neeman* (Yetedot Publishing) | No Data | No Data | No Data | No Data | No Data | 2.1 | No Data | No Data | No Data |

*continued*

Table 30-3. *continued*

| | 1984[2] | 1988[3] | 1992[4] | 1996[5] | 2000 | 2004 | 2008 | 2012 | 2013[6] |
|---|---|---|---|---|---|---|---|---|---|
| Others[10] | 6.7 | 3.2 | 3.9 | 4.6 | No Data | No Data | No Data | No Data | No Data |
| Total Advertising Revenue (mil US$)[11] | 99.0 | 228.0 | 354.0 | 452.0[12] | 603.0 | 392.0 | 410.0 | 260.7 | 260.0 |
| C4 | 90.4 | 92.0 | 90.0 | 91.9 | 95.1 | 94.1 | 91.2 | 88.4 | 90.8 |
| HHI | 3,647 | 3,230 | 3,660 | 4,023 | 3,757 | 3,704 | 2,714 | 2,654 | 2,905 |
| *N* (>1%) | 4 | 5 | 5 | 5 | 6 | 6 | 4 | 7 | 7 |
| Noam Index | 1,102 | 936 | 1,158 | 1,273 | 1,332 | 1,313 | 904 | 1,005 | 1,100 |

*Source*: Israeli Marketing Union (1984–1996); Single Source/TGI/Kantar Media (2000–2013).

[1] Data relate to audience consumption of daily newspapers on a regular weekday (as opposed to weekends and holidays). Since some people may be exposed to more than one newspaper, the numbers usually add up to more than 100%. Figures have therefore been normalized to add up to 100%, while maintaining the relative share of each entity.

[2] Data for the year 1984 obtained from Ya'acov Solan, "How Do the Israeli Dailies Market Themselves (I)", 69 *Otot 85*, 1985, pp. 30–31, relying on data of the Israeli Advertisers Union (now the Israeli Marketing Union). Notably, the research methodology employed in the 1984 survey allowed respondents to indicate only one newspaper, which they considered as their primary print news source. This limitation was revoked in subsequent surveys. *Id.*

[3] Data for the year 1988 obtained from Israeli Advertisers Union, *Exposure to the Media—Telephonic Survey Conducted by the Teleseker Institute*, 1988 (on file with authors).

[4] Data for the year 1992 obtained from Israeli Advertisers Union, *Exposure to the Media—Survey Conducted by Gallup Israel*, 1992 (on file with authors).

[5] Data for the year 1996 obtained from Israeli Advertisers Union, *Exposure to the Media—Survey Conducted by the Geocartography*, 1996 (on file with authors).

[6] Data for the years 2000–2013 obtained from Single Source/TGI/Kantar Media (on file with authors).

[7] Includes *24 Dakot*, which ceased publication in 2009, and *Calcalist*, which began publishing in 2008. All other years are for *Yediot Aharonot's* audience exposure only.

[8] From 1984 to 1992, *the tabloid Hadashot* is included with *Haaretz*. For 1996, only *Haaretz* is included. From 2000 to 2008, *The International Herald Tribune* and *Haaretz* are included. From 2008, *The International Herald Tribune*, *Haaretz*, and *TheMarker* are all included.

[9] *Hatzofe* ("the Observer") was established in 1937 as the newspaper of the Zionist-religious movement (*Hamizrachi*) and later functioned as the mouthpiece of the Israeli National-Religious Party (*Mafdal*), until it was transferred to private hands in 2003. In 2007 *Hatzofe* was merged into the struggling weekly *Makor Rishon* (established a decade earlier). *Makor Rishon* ("First Source") has since been circulated as a daily (from 2009 to subscribers only), with *HaTzofe* as a weekly supplement. Data for 1992–2004 refer to *HaTzofe*; Data for 2008 refer to *Makor Rishon- HaTzofe.*

[10] The category of "Others" includes newspapers for which there were only accumulated data. The category is considered to be one daily for the purpose of HHI calculation.

[11] Assuming the expenditure on advertising, as specified in Table 30.1 above, equals newspapers' revenue from advertising.

[12] Amount of spending on advertising in 1994.

Table 30-4. Book Publishing (Market Shares by Space Rented In Hebrew Book Week), 1997–2008[1]

| | 1997[2] | 2000 | 2004 | 2008 | | 1997[2] | 2000 | 2004 | 2008 |
|---|---|---|---|---|---|---|---|---|---|
| 1 | 0.1 | 0.5 | | | 36 | 0.9 | 0.4 | 1.0 | 0.9 |
| 2 | 0.8 | 1.9 | 0.9 | 1.2 | 37 | 0.9 | 0.8 | 0.5 | 0.9 |
| 3 | 4.7 | 4.1 | 4.0 | 0.0 | 38 | 1.5 | 1.4 | 1.2 | 0.2 |
| 4 | 1.5 | 1.5 | 0.9 | 0.0 | 39 | 0.2 | 0.3 | 0.8 | 0.6 |
| 5 | 0.1 | 0.1 | 0.1 | 0.0 | 40 | 0.5 | 0.4 | 0.6 | 0.2 |
| 6 | 0.1 | 0.0 | 0.0 | 0.0 | 41 | 0.6 | 0.6 | 0.6 | 0.9 |
| 7 | 2.9 | 0.0 | 0.0 | 0.0 | 42 | 0.1 | 0.1 | 0.2 | 0.2 |
| 8 | 0.1 | 0.0 | 0.0 | 0.0 | 43 | 4.6 | 4.0 | 5.5 | 4.6 |
| 9 | 0.5 | 0.1 | 0.0 | 0.0 | 44 | 2.1 | 2.1 | 1.1 | 1.1 |
| 10 | 9.2 | 7.7 | 3.0 | 1.0 | 45 | 1.5 | 1.0 | 0.6 | 0.0 |
| 11 | 0.3 | 0.3 | 0.4 | 0.5 | 46 | 1.0 | 1.2 | 0.0 | 0.0 |
| 12 | 4.6 | 4.8 | 7.1 | 4.3 | 47 | 0.6 | 0.0 | 0.0 | 0.0 |
| 13 | 2.1 | 1.2 | 1.2 | 0.9 | 48 | 0.1 | 0.1 | 0.0 | 0.0 |
| 14 | 3.8 | 6.4 | 8.7 | 10.1 | 49 | 0.4 | 0.3 | 0.0 | 0.0 |
| 15 | 0.5 | 0.5 | 0.5 | 0.6 | 50 | 0.1 | 0.1 | | 0.1 |
| 16 | 2.3 | 2.0 | 1.7 | 0.0 | 51 | 4.3 | 4.3 | 5.6 | 3.7 |
| 17 | 0.1 | 0.2 | 0.0 | 0.0 | 52 | 0.7 | 0.5 | 0.0 | 0.0 |
| 18 | 0.9 | 0.8 | 1.1 | 1.3 | 53 | 0.8 | 0.6 | 0.8 | 0.5 |
| 19 | 7.0 | 8.4 | 8.9 | 12.2 | 54 | — | 0.5 | 1.2 | 0.6 |
| 20 | 0.5 | 0.1 | 0.0 | 0.0 | 55 | — | 1.2 | 1.5 | 0.0 |
| 21 | 4.8 | 5.1 | 8.7 | 10.1 | 56 | — | 0.1 | 0.1 | 0.0 |
| 22 | 1.3 | 1.0 | 1.3 | 1.1 | 57 | — | 0.4 | 0.3 | 0.3 |
| 23 | 0.2 | 0.4 | 0.4 | 1.2 | 58 | — | 0.6 | 0.0 | 0.0 |
| 24 | 7.2 | 5.8 | 6.3 | 9.8 | 59 | — | 2.0 | 0.9 | 2.6 |
| 25 | 1.5 | 1.4 | 0.5 | 0.0 | 60 | — | — | 0.9 | 0.0 |
| 26 | 1.4 | 1.1 | 0.0 | 0.0 | 61 | — | 1.0 | 0.4 | 1.5 |
| 27 | 0.5 | 0.5 | 0.9 | 1.1 | 62 | — | 0.3 | 0.0 | 0.0 |
| 28 | 12.4 | 12.1 | 8.5 | 10.7 | 63 | — | 0.8 | 0.6 | 0.0 |
| 29 | 1.1 | 0.7 | 1.9 | 1.5 | 64 | — | — | 2.5 | 2.3 |
| 30 | 1.8 | 1.1 | 1.2 | 1.0 | 65 | — | — | — | 1.4 |
| 31 | 0.3 | 0.5 | 0.0 | 0.0 | 66 | — | — | — | 0.9 |
| 32 | 2.5 | 2.5 | 2.8 | 6.1 | C4 | 35.9 | 34.6 | 34.8 | 43.1 |
| 33 | 0.5 | 0.3 | 0.4 | 0.4 | HHI | 516 | 492 | 537 | 674 |
| 34 | 0.7 | 0.7 | 0.0 | 0.0 | N (>1%) | 24 | 25 | 23 | 21 |
| 35 | 0.9 | 1.0 | 1.6 | 1.7 | Noam Index | 105 | 98 | 111 | 147 |

*Source*: Israeli Book Publishers Association (on file with authors).

[1] Data originally provided in units of meters of space rented. Figures have been converted to percentage and normalized to add up to 100%, while maintaining the relative share of each entity.

[2] No data available for 1996.

Table 30-5. Magazine Publishing (Market Shares by Audience Reported Exposure), 1986–2012[1]

| | 1984(6)[2] | 1988[3] | 1992[4] | 1996[5] | 2000 | 2004 | 2008 | 2012[6] |
|---|---|---|---|---|---|---|---|---|
| Yediot Aharonot Group (Mozes Family) | 32.0 | 26.4 | 38.3 | 55.4 | 33.5 | 31.0 | 29.4 | 33.5 |
| Olam Ha'Isha | 9.0 | 7.0 | | | | | | |
| Ma'ariv Group (various owners) | 16.0 | 14.2 | 14.7 | 16.0 | 18.6 | 21.7 | 21.6 | 14.3 |
| Bamahane (government)[7] | 12.0 | 13.8 | 11.7 | No Data | 2.2 | 2.5 | 3.1 | 4[8] |
| Koteret Rashit | 3.0 | 3.5 | 0.0 | 0.0 | 0.0 | 0.0 | 0.0 | 0.0 |
| Lahiton | 2.0 | 1.5 | 0.0 | 0.0 | 0.0 | 0.0 | 0.0 | 0.0 |
| Globes-Monitin (Globes Group) | 7.0 | 8.6 | 1.0 | 0.0 | 0.0 | 0.0 | 1.6 | 2.2 |
| TURBO Magazine | 4.0 | 3.9 | 0.0 | 0.0 | 0.0 | 0.0 | 0.0 | 0.0 |
| HaOlam HaZeh | 3.0 | 2.8 | 2.7 | 0.0 | 0.0 | 0.0 | 0.0 | 0.0 |
| Moto Media Group | – | No Data | 1.9 | 2.3 | No Data | No Data | 2.4 | 23.1[9] |
| SBC | – | – | 6.7 | 8.7 | 17.0 | 19.6 | 19.0 | |
| Hadashot Sport & Toto[10] | No Data | No Data | 2.8 | 6.4 | 0.0 | 0.0 | 0.0 | 0.0 |
| Auto (Auto Group) | – | 4.0 | 3.8 | 3.8 | 3.6 | 3.2 | 2.9 | 3.3 |
| Anashim uMachshevim | – | 2.1 | 2.3 | 3.2 | 1.0 | 0.9 | 0.8 | 0.5 |
| Eretz VeTeva | – | – | 1.9 | 2.5 | 4.7 | 2.1 | 1.7 | 2.3 |
| Mada Populari (Popular Science Israel) | – | – | – | 1.7 | 1.8 | 1.0 | 0.6 | 0.4 |
| TIME (Time Warner, US) | – | 4.1 | 4.3 | 0.0 | 0.0 | 0.0 | 0.0 | 0.0 |
| Newsweek(Tina Brown, US) | – | 3.1 | 0.0 | 0.0 | 0.0 | 0.0 | 0.0 | 0.0 |
| Histadrut | No Data | No Data | 2.9 | No Data | No Data | No Data | No Data | No Data |
| Mashehu | No Data | No Data | 0.8 | 0.0 | 0.0 | 0.0 | 0.0 | 0.0 |
| Anashim | – | – | – | – | 5.3 | 3.2 | 0.4 | 0.5 |
| Teva Hadvarim | – | – | – | – | 3.5 | 2.6 | 1.8 | 1.7 |
| Binyan Vediur | – | – | – | – | 8.8 | 5.5 | 2.9 | 2.3 |
| Atmosphere (El Al) | | | | | No Data | 0.7 | 0.5 | 0.5 |

| | | | | | | | | |
|---|---|---|---|---|---|---|---|---|
| Bayit veNoy | – | – | No Data | No Data | No Data | 2.1 | 1.5 | 1.4 |
| Haaretz Group (Schocken Family) | | | | No Data | No Data | 2.4 | 5.6 | 3.4 |
| Time Out Israel | – | – | – | – | – | 1.1 | 0.9 | 1.0 |
| Cigar | – | – | – | – | – | 0.2 | 0.3 | 0.3 |
| Lilac | – | – | – | – | – | 0.1 | 0.1 | No Data |
| Yerushalmi | – | – | – | – | – | – | 0.2 | No Data |
| Eretz Acheret | – | – | – | – | – | No Data | 0.5 | 0.8 |
| Yayin veGourmet | – | – | – | – | No Data | No Data | 1.0 | 1.0 |
| Al Hashulchan | – | – | No Data | No Data | No Data | No Data | 1.2 | 1.8 |
| Makor Rishon Group | – | – | – | – | – | – | No Data | 3.0 |
| Others[11] | 12.0 | 5.8 | 4.2 | No Data | No Data | No Data | No Data | 3.8 |
| C4 | 69.0 | 63.0 | 73.3 | 88.8 | 77.9 | 81.1 | 75.6 | 74.3 |
| HHI | 1,736 | 1,330 | 1,952 | 3,481 | 1,907 | 1,898 | 1,771 | 1,949 |
| *N* (>1%) | 9 | 13 | 13 | 9 | 11 | 13 | 14 | 13 |
| Noam Index | 550 | 356 | 522 | 1,160 | 576 | 461 | 370 | 415 |

*Source*: Israeli Marketing Union (1986–1996); Single Source/TGI/Kantar Media (2000–2012).

[1] Figures have been normalized to add up to 100%, while maintaining the relative share of each entity.

[2] No data available for 1984. Data for 1986 obtained from Israeli Advertisers Union, *Exposure to the Media Survey 1986*, 1986 (on file with authors).

[3] Data for 1988 obtained from Israeli Advertisers Union, *Exposure to the Media—Telephonic Survey Conducted by the Teleseker Institute*, 1988 (on file with the authors).

[4] Data for 1992 obtained from Israeli Advertisers Union, *Exposure to the Media—Survey Conducted by Gallup Israel*, 1992 (on file with the authors).

[5] Data for 1996 obtained from Israeli Advertisers Union, *Exposure to the Media—Survey Conducted by the Geocartography*, 1996 (on file with authors).

[6] Data for 2000–2012 obtained from Single Source/TGI/Kantar Media (on file with authors).

[7] Published by the Israeli Defense Forces (IDF).

[8] Includes data for the Air Force Magazine.

[9] Following the acquisition of SBC by Moto Media Group.

[10] The magazine was published for 18 years (1980–1998), the last 8 of which under a contract and in cooperation with the Israeli Council for the Arrangement of Sports Betting. *See* C.A. (Tel-Aviv District Court) 1749/99 *Shvakim Hadashim Publishing Ltd. v. The Council for the Arrangement of Sports Betting* (published in *Nevo,* Mar 15, 2009). A sports newspaper by a similar name, *Hadashot Hasport* (the Sports News) operated for 31 years (1954–1985), 26 of which (from 1959) as a daily. *See* http://he.wikipedia.org/wiki/%D7%97%D7%93%D7%A9%D7%95%D7%AA_%D7%94%D7%A1%D7%A4%D7%95%D7%A8%D7%98 (in Hebrew).

[11] The category of "Others" is considered to be one magazine for the purpose of HHI calculation.

## Magazine Publications

The two largest magazine publishers in Israel are the Yediot Aharonot and the TimeOut Israel groups; the latter after purchasing the magazines associated with the Ma'ariv, SBC and Moto Media groups in previous years. Since these acquisitions took place late in 2013 and in mid 2014, they are not reflected in the data.[33] All other publishers, including the Ha'aretz Group, have much smaller market shares. The magazine industry appears to be highly competitive, but most magazines have niche audience shares, so competitiveness within this market is not among equal producers with overlapping national markets (Table 30.5). Had we conducted the study with regards to more specific interests, in particular magazines targeting women, the magnitude of both Yediot Aharonot and Ma'ariv's (prior to its sale to Time-Out) market shares would have been even more apparent. Still, the diversity in interest areas and audiences addressed by this industry is a demonstration of a robust consumer market.

According to the Israeli Central Bureau of Statistics (ICBS) data, there were 285 magazine titles circulating in 1985, 284 in 1990, and 368 in 1997.[34] 1997 is the last year for which official data exist on the total number of magazine titles in circulation, so for the Noam Index we used the number of magazines for which the Israeli Marketing Union (and Subsequently Single Source/TGI/Kantar Media) tracked market shares, based on readership surveys.

# AUDIOVISUAL MEDIA

## Radio

The Israel Broadcasting Authority (IBA), a public service broadcaster (PSB) that purportedly adheres to the European PSB model, was founded in 1965, continuing seamlessly the state's radio service, which in itself was inherited from the British colonial government. From a single radio station, IBA's radio network grew over the years, until by the 1990s it consisted of eight thematic stations and one short wave service targeting audiences in the region and throughout the Jewish world. Its "competition" in the years leading to the launch of commercial stations in the 1990s was a state-owned operator, the Israeli Defense Forces' (IDF) Radio, which broadcast news and entertainment programming. Only IBA's radio, however, officially sold advertising. Even so, two major unlicensed stations operated nationally and sold their own advertising: The Voice of Peace, which broadcast from the sea and ceased broadcasting in 1993, and *Arutz Sheva* ("Channel Seven"), which broadcast from a Jewish settlement in the occupied territories until shut down by the courts in 2003. In the mid-1990s, a network of commercially based regional radio stations was established. In order to regulate the commercial radio and television industry, the Knesset enacted the Second Authority Law.

As the commercial stations do not broadcast nationally, while the IBA and IDF Radio do, the audience shares for each of the commercial stations, even after more than a decade of being on air, are relatively small. Their cumulative share, however, is significant, and when combined is equivalent to the total shares for IBA's eight domestic stations—Reshet Aleph (culture and talk radio), Reshet Bet (news and current affairs), Reshet Gimel (Israeli music), Reshet Dalet (Arabic), REQA (Russian and Amhari service to new immigrants), 88FM (traffic), Kol Ha Musica (classical music), and Reshet Moreshet (religious)—and IDF Radio's two (Galei Tzahal and Galgalatz).[35] While IDF Radio had become as popular as IBA's radio by

33. *See* Orbach, L. "Compromise: The Moto Group to be divided up between TimeOut and Tchelet," *Globes*, Sep 3, 2013, *available at http://www.globes.co.il/news/article.aspx?did=1000877411* (accessed Jul 14, 2014); Orbach, L. "After 47 years: "'At"' is separated from 'Ma'ariv' – sold to Yuval Sigler," *Globes*, June 2, 2014, *available at http://www.globes.co.il/news/article.aspx?did=1000943708* (accessed Jul 14, 2014).

34. Central Bureau of Statistics, "Daily Newspapers and Periodicals Published in Israel, by Frequency, Language, Type and Subject," *in 1998 Statistical Abstract of Israel*, 1998, pp. 26–29, *available at* <http://www.cbs.gov.il/archive/shnaton49/st26-04_h.shtml>.

35. IBA's ninth radio service, Reshet Heh ("E" Network), originally broadcast over one dozen non-Hebrew language channels abroad, but due to cost-cutting measures now only transmits some Persian-language programming into Iran.

Table 30-6. Radio Group (Market Shares by Audience Reported Exposure), 1984–2013[1]

| | 1984 | 1988 | 1992 | 1996 | 2000 | 2004 | 2008 | 2012 | 2013 |
|---|---|---|---|---|---|---|---|---|---|
| IBA (public) | No Data | 60.1 | 56.9 | 59.4 | 42.8 | 36.8 | 35.4 | 33.4 | 31.9 |
| IDF Radio (government)[2] | No Data | 36.5 | 34.9 | 31.2 | 26.7 | 30.9 | 33.1 | 32.2 | 32.2 |
| Arutz Sheva[3] | – | No Data | 5.8 | 5.5 | 2.9 | 1.0 | 0.0 | 0.0 | 0.0 |
| The Voice of Peace | No Data | 4.7 | 2.8 | 0.0 | 0.0 | 0.0 | 0.0 | 0.0 | 0.0 |
| Lelo Hafsaka Radio Group (David Weisman)[4] | – | – | – | Included in "Others" | 8.1 | 7.6 | 7.7 | 11.1 | 11.5 |
| 99 FM | – | – | – | – | 2.3 | 3.7 | 2.7 | | |
| Radius Group (David Ben-Basat)[5] | – | – | – | Included in "Others" | 4.0 | 6.2 | 6.7 | 7.1 | 7.0 |
| Darom Radio Group (Joseph Saban)[6] | – | – | – | – | 5.4 | 4.8 | 4.1 | 2.4 | 2.6 |
| Tel Aviv Radio | – | – | – | – | Included in "Darom Radio Group" | Included in "Darom Radio Group" | Included in "Darom Radio Group" | 1.6 | 1.8 |
| Haifa Radio[7] | – | – | – | Included in "Others" | 2.8 | 2.6 | 1.9 | 2.1 | 2.3 |

*continued*

Table 30-6. *continued*

| | 1984 | 1988 | 1992 | 1996 | 2000 | 2004 | 2008 | 2012 | 2013 |
|---|---|---|---|---|---|---|---|---|---|
| Jerusalem Radio | – | – | – | Included in "Others" | 1.2 | 1.2 | 1.1 | 1.3 | 1.2 |
| Red Sea Radio | – | – | – | Included in "Others" | 0.4 | 0.4 | 0.4 | 0.3 | 0.3 |
| Kol Hai | – | – | – | – | 1.0 | 2.0 | 3.7 | 3.0 | 3.0 |
| Kol Rega | – | – | – | – | 1.2 | 1.4 | 1.4 | 1.4 | 1.3 |
| Emtza Haderech | – | – | – | – | 1.2 | 1.2 | 1.6 | 1.1 | 1.1 |
| Kol Barama | – | – | – | – | – | – | – | 3.0 | 3.7 |
| Others | No Data | No Data | No Data | 8.6 | No Data | No Data | No Data[8] | No Data | No Data |
| Advertising Revenue (mil US$) | | | | 30.8[9] | 98 | 55.8 | 58.5 | 57.8 | 67.7 |
| C4 | | 100 | 100 | N/A | 83.0 | 81.5 | 82.9 | 83.8 | 82.6 |
| HHI | | 4,966 | 4,497 | 4,606 | 2,683 | 2,458 | 2,500 | 2,356 | 2,278 |
| *N* (>1%) | | 3 | 4 | 3 | 11 | 11 | 11 | 12 | 12 |
| Noam Index | | 2,921 | 2,248 | 1,633 | 745.3 | 682.8 | 722.5 | 654.4 | 632.8 |

*Source*: Israeli Marketing Union (1988–1996), Single Source/TGI/Kantar Media (2000–2013)

[1] Figures have been normalized to add up to 100%, while maintaining the relative share of each entity.

[2] Broadcast by the Israeli Defense Forces (IDF).

[3] Unlicensed. Though Arutz Sheva temporarily received government authorization to operate in 1999, the High Court of Justice overruled the government in 2002.

[4] Until 2009 Lelo Hafsaka (non-stop) Radio Group included the Tel-Aviv based Lelo Hafsaka Radio (103FM) and Tzafon (north) Radio (104.5FM), whose franchise covers northern areas of Israel. In 2009 the Group acquired an additional radio station, 99FM, whose franchise covers highly populated areas at the center of Israel.

[5] Includes Radius 100FM and Lev Hamedina (heart of the land) Radio 91FM.

[6] IData for 2000–2008 includes Darom (south) Radio (97FM) only and Tel-Aviv Radio (102FM). Data for 2012–2013 includes Darom Radio and Galey Israel Radio (102.5FM), which commenced broadcasting in February 2010 and covers the areas of Judea and Samaria. As of 2009, Tel-Aviv Radio is no longer fully owned by the Darom Radio Group, with 50% of its shares being acquired by the owners of Haifa Radio. The data relating to Tel-Aviv Radio for the years 2012–2103 has therefore been separated from that of the Darom Radio Group.

[7] As of 2009 Haifa Radio owns 50% of Tel Aviv Radio as well.

[8] Besides the highly-popular Voice of Peace and Channel 7, dozens of other pirate radio stations have operated in Israel over the years, alongside legal stations. *See generally* Limor, Y. and Naveh, H., *The Pirate Radio in Israel*, Haifa: Pardes Publishing, 2007.

[9] Amount of spending in 1994, as specified in Table 30.1 above.

the 2000s, the shares of the regional commercial stations have not changed significantly since they were launched and reflect a steady standing in the market.

## Broadcast Television

Until 1992, Channel 1 of the Israel Broadcasting Authority was the only licensed national terrestrial station in Israel. In 1986, the state began experimenting with a new channel, Channel 2, in part due to concerns that neighboring countries would take control of available broadcast frequencies not in use.[36] These tentative broadcasts proved to be quite popular, as can be seen in the 1988 numbers in Table 30.6, which suggest that Israeli audiences wanted a choice of broadcasters. As the data for 1998 illustrate, when Channel 2 was formally launched in 1993, the shift in consumption was dramatic, with the new broadcaster winning a 62.5% audience share by 1998. Although the IBA's monopoly over news programming and educational content was already being offset in the late 1980s by the market entry of cable TV providers, it was not until Channel 2 was launched in 1993 that the IBA's monopoly truly began to crumble.[37]

Israel's second commercial broadcaster, Channel 10, launched in 2002. Buoyed by the perception that allowing Channel 2 to employ three franchisees—Keshet, Reshet, and Telad—had been a successful experiment in limiting concentration in the television industry, Israeli officials stipulated that Channel 10 would be permitted to operate with two franchisees, realizing the limitations of the Israeli advertising market to sustain more than two more commercial television operations. However, an economic downturn caused by the collapse of the "dot-com bubble" worldwide and the violence of the Second Palestinian Uprising, or Intifada (2002–2005) within Israel itself resulted in Channel 10 consolidating its operations into one franchise, known as "Israel 10." Further changes in legislation led to Channel 2 being reduced to two franchises (Keshet and Reshet) as of 2005 and Knesset legislation passed in 2011 was aimed at allowing Keshet and Reshet to develop independent operations as sole licensees of new channels, delivered over DTT, cable, and satellite, once Channel 2's license expires in 2015.

Channel 10 was not technically a terrestrial network when it began broadcasting: due to spectrum constraints, it was first delivered over an unencrypted satellite feed (the IBA employed a similar system for its Channel 33, a satellite network serving the Palestinian-Israeli minority). Most Channel 10 viewers received the network via cable or DTH. In 2009, however, digital terrestrial television (DTT) was launched in Israel and analog broadcasting was phased out by the end of 2010. The new DTT service now covers IBA's Channel 1 and Channel 33, the commercial broadcasters Channel 2 and Channel 10, and another public network, Channel 99 which is similar to the US's C-SPAN as it broadcasts live debates from the Knesset, as well as Knesset-produced political programming.[38]

2005 saw the exit of several enterprises that had entered the broadcasting sector in the 1990s. The Yediot Aharonot Group had held a strategic share in one of Channel 2's franchises (Reshet) up until this time, as had the Ma'ariv Group (Telad). Two prominent Israeli banks, Bank Leumi and Bank Hapoalim, were also invested in these franchise holders. However, once cross ownership was banned in 2001 and banks were required to divest from non-financial holdings like television, ownership of the franchise holders transferred to a series of investment firms and other media groups. For a brief period in the mid-2000s, both the Channel 2 franchise holder Keshet and the telecom provider Bezeq were owned by Haim Saban, an Israeli-American media mogul. As noted, Channel 10, which was originally meant to be

36. Tokatly, *supra* note 18, at —.

37. Each franchise was allotted two days of broadcasting, with a rotation of broadcasts on the Sabbath (a peculiarity of the Israeli media market, due to Friday night into Saturday being the weekly holy day of obligation in Judaism).

38. Davidson, R. and Schejter, A. "'Their Deeds are the Deeds of Zimri; but They Expect a Reward Like Phineas': Neoliberal and Multicultural Discourses in the Development of Israeli DTT Policy." *Communication, Culture and Critique* 4, 2011. pp. 1–22.

## Table 30-7. TV Broadcasting (Audience Shares by Ratings), 1984–2013[1]

| | 1984 | 1988 | 1992 | 1996(8) | 2000 | 2004 | 2008 | 2012 | 2013 |
|---|---|---|---|---|---|---|---|---|---|
| IBA (public)[2] | 100.0 | 83.7 | No Data | 37.5 | 38.7 | 29.5 | 15.7 | 11.3 | 10.0 |
| Channel 2 | — | 16.3 | No Data | 62.5 | 61.3 | 52.0 | 53.8 | 56.9 | 60.6 |
| Channel 10 | — | — | — | — | | 18.5 | 30.5 | 30.2 | 27.4 |
| Channel 23 | — | — | — | — | — | — | — | 1.6 | 1.9 |
| Advertising Revenue (mil US$) | 2 | 14 | 18 | 305[3] | 261 | 274 | 448 | 371 | 434 |
| C4 | 100 | 100 | | 100 | 100 | 100 | 100 | 100 | 100 |
| HHI | 10,000 | 7,271 | | 5,312 | 5,255 | 3,916 | 4,071 | 4,280 | 4,527 |
| *N* (>1%) | 1 | 2 | | 2 | 2 | 3 | 3 | 4 | 4 |
| Noam Index | 10,000 | 5,157 | | 3,794 | 3,754 | 2,303 | 2,395 | 2,140 | 2,263 |

*Source*: Israeli Marketing Union (1988); Tele-Gal/Kantar Media (1998–2013).

[1] Figures for 1988 have been obtained from Israeli Advertisers Union, *Exposure to the Media—Telephonic Survey Conducted by the Teleseker Institute*, 1988 (on file with authors). Figures for 1998–2013 relate to the average whole-day share of each entity during the relevant calendar year, as measured by Peoplemeter technology. Figures have been normalized to add up to 100%, while excluding rating measures of other, non-broadcast channels. Data provided by Telegal/Kantar Media on file with authors.

[2] Includes both *Channel 1* and *Channel 33*.

[3] Amount of spending for 1994, as specified in Table 30.1 above.

operated by two franchise holders, consolidated its operations into one due to financial difficulties. More recent partners in Israel 10 include Rupert Murdoch's News Corporation International, the Israeli-American Hollywood producer Arnon Milchan, and the US cosmetics magnate Ron Lauder. The network's single largest shareholder since its inception has been the Israeli billionaire Yossi Maiman, through the private holding company Merhav. As the transition to a new licensing system was looming toward the end of 2014 it was not clear whether channel 10 would in fact continue to broadcast. At the same time, the Knesset passed legislation in 2014 liquidating IBA, and establishing a new public broadcaster to start operations in 2015.

The reestablishment of the public broadcaster was a direct consequence of what had been perceived as the irrelevancy of its television broadcast. By 2010, the audience share for IBA's channels had fallen to just 10%, Channel 2's total audience share reached more than 60% and Channel 10 stood with 27.4%, gained primarily at the expense of Channel 1 (Table 30.7). Still, the numbers for broadcast TV should be seen in the general proportion for the consumption of audiovisual fare. According to data for 2013, the audience shares of all three over the air channels were equivalent to approximately one-third of total television viewing in Israel when cable and satellite are included.

### MULTICHANNEL TV PLATFORMS

Cable TV became (legally) available in Israel in 1989 and was first envisioned as a sub-national service, though the local franchises were each a monopoly in their designated territory. As a result, we describe the market shares in 1992 and 1996 as 100% (Table 30.8). Even though the five cable providers present in Israel at the time—Tevel, Idan (ICS), Gvanim, Golden Channels, and Matav—were not operating at the national level, their programming offering was identical. Each company was providing service in a different district of Israel, so there was no direct competition among them until they eventually merged to form HOT. The legislation that authorized this was enacted concomitantly with legislation banning cross-ownership of cable and terrestrial broadcasters, which saw Yediot and Ma'ariv divest from Golden Channels and Matav, respectively.

Table 30-8. Multichannel Video Platforms: Cable MSOS, DBS, IPTV (Audience Shares by Subscribers), 1984–2013

| | 1984 | 1988 | 1992 | 1996 | 2000[1] | 2004[2] | 2008[3] | 2012[4] | 2013[5] |
|---|---|---|---|---|---|---|---|---|---|
| HOT (cable) | — | — | 100.0 | 100.0 | 93.0 | 67.0 | 62.0 | 60.7 | 57.3 |
| Yes (Bezeq) (satellite) | — | — | — | — | 7.0 | 33.0 | 38.0 | 39.3 | 33.4 |
| Idan Plus (DTT) | — | — | — | — | — | — | — | No Data | 9.3 |
| Total Market Revenue (mil ILS)[6] | | | | | | 3,010 | 3,701 | 3,922 | 3,914 |
| Total Market Revenue (mil US$) | 0.0 | 0.0 | | | | 670[7] | 1,104[8] | 1,000[9] | 1,081[10] |
| C4 | 0.0 | 0.0 | 100.0 | 100.0 | 100.0 | 100.0 | 100.0 | 100.0 | 100.0 |
| HHI | 0.0 | 0.0 | 10,000 | 10,000 | 8,698 | 5,578 | 5,288 | 5,229 | 4,485 |
| *N* (>1%) | 0.0 | 0.0 | 1 | 1 | 2 | 2 | 2 | 2 | 3 |
| Noam Index | 0.0 | 0.0 | 10,000 | 10,000 | 6,168 | 3,956 | 3,750 | 3,708 | 2,592 |

*Source*: Council for Cable and Satellite Broadcasting (CSBC) (until 2012); Israel Audience Research Board (2013).

[1] Data obtained from Council for Cable and Satellite Broadcasting, *2000 Annual Report*, 2001, pp. 11–12, *available at* <http://www.moc.gov.il/sip_storage/FILES/3/1033.pdf>.

[2] Data obtained from Council for Cable and Satellite Broadcasting, *2004 Annual Report*, 2005, pp. 20–21, *available at* <http://www.moc.gov.il/sip_storage/FILES/9/1029.pdf>.

[3] Data obtained from Council for Cable and Satellite Broadcasting, *2008 Annual Report*, 2010, pp. 13–14, *available at* <http://www.moc.gov.il/new/documents/doch2008.pdf>.

[4] Data obtained from Council for Cable and Satellite Broadcasting, *2012 Annual Report*, 2013, pp. 14–15, *available at* http://www.moc.gov.il/sip_storage/FILES/5/3435.pdf.

[5] Data obtained from study of the Israel Audience Research Board, on file with authors. Data presented does not take into account households whose access to multichannel television is obtained through a private satellite dish.

[6] Data for 2004 derived from Ministry of Communications, *Summary of Revenues of the Telecommunications Market for the Year 2006*, Oct 23, 2007 (in Hebrew); data for 2008 derived from Ministry of Communications, *Summary of Revenues of the Telecommunications Market for the Year 2011*, Nov 28, 2012 (in Hebrew); data for the years 2102 and 2013 derived from *Summary of Revenues of the Telecommunications Market for the Year 2013*, 10 July 2014 (in Hebrew).

[7] 4.49 ILS per 1 US$.

[8] 3.35 ILS per 1 US$.

[9] 3.92 ILS per 1 US$.

[10] 3.62 ILS per 1 US$.

Since the introduction of Yes's DTH service in 2000, cable subscribers began moving to DTH, though this trend has slowed somewhat since 2004. Since Bezeq has a controlling stake in Yes, it serves as Bezeq's primary platform and allows Bezeq to compete indirectly with HOT. The introduction of digital terrestrial broadcasting contributed to a slightly more competitive landscape, however its modest 6-channel offering does not make it a direct rival to multichannel cable and satellite, which offer hundreds of channels as well as video-on-demand services. HOT, after 2009, has been controlled by the French entrepreneur Patrick Drahi. Drahi owns Altice, which controls France's major cable company Numericable and its #2 mobile operator SFR, the major Portuguese telecom firm, and other media properties. In Israel, he also owns the #4 mobile operator MIRS.

## Video Channels

A complex regulatory history[39] led to the emergence of different regulatory regimes based on the ownership structure of the channels offered over the multichannel platforms. Each of the platforms was allowed some proprietary channels as the CSBC was interested in promoting channels that included original Israeli productions. Must-carry designations were applied to the terrestrial Channels 1 and 2, to Israeli Educational Television's Channel 23, the Knesset's Channel 99, and the then-satellite based Channels 10 and 33, plus several other networks considered "designated channels." These government-licensed "designated channels" are privately owned and include the Russian-language Channel 9 (Israel Plus), the Israeli music network Channel 24, the Arabic language channel Hala-TV, and the home shopping network, owned jointly by Globus Group Productions (controlled by movie producer Yoram Globus) and by the Dahaf Group, one of Israel's first advertising firms. The rest of the extremely diverse offering consists of locally owned and packaged channels devoted to a wide range of cultural, educational, and religious programming as well as syndicated international content, some of which is translated into Hebrew (e.g., National Geographic) while the rest is retransmitted in its original format, particularly programming from networks like the BBC, CNN, France24, Fox News, and their like.

The cable operator HOT and the RGE Group (in which one of the partners, Aviv Giladi, also has, as of the date of this research, a 24% stake in Reshet) were the major players in the video channels sector until the late 2000s, when HOT's position was taken by the emerging content aggregator Dori Media Group. HOT's main assets were the cable operators' original content channels, notably Channel 3 (family programming and repacked US entertainment programming), and the movie channel, (Channel 4). In recent years however, these channels were transferred to Dori Media and re-packed as HOT 3 and HOT Family, respectively. The RGE Group's assets include the traditional sport channel group known as Sport 5, which operates about half a dozen basic and premium sport services, the traditional children's channel, known as channel 6, and the science-oriented Channel 8. By 2008, as the Dori Media Group started operating HOT 3 and HOT Movies (Channel 4), the HOT channels' audience shares had fallen by more than 50% (Table 30.9). Keshet's increased audience share among video channels reflects its ownership of the government-licensed and designated music channel known as Channel 24, which has been growing in popularity and of the comedy channel Bip (which has since been replaced by HOT Comedy Central). Telad has grown by running more repackaged foreign programming from US cable networks like E! and The History Channel.

The video channel market is similar in its audience shares' breakdown to the magazines market, with two leading media groups—HOT and RGE in 2004, RGE and Dori Media since 2008—competing at a national level and a number of niche providers not in direct competition with one another constituting the rest of the market. The video channel market therefore has the lowest HHI of all Israeli media markets surveyed here.

39. *See* Schejter, A. "From a Tool for National Cohesion, to a Manifestation of National Conflict: The Evolution of Cable Television Policy in Israel, 1986–1998." *Communication Law and Policy* 4:2, 1999. pp. 177–200 and Schejter, A. and Lee, S. "The Evolution of Cable Regulatory Policies and Their Consequences: Comparing South Korea and Israel." *Journal of Media Economics* 20:1, 2007. pp. 1–28.

## Film

Though the film industry emerged in the 1920s, for decades it was composed primarily of independent producers with little capital to their names. This changed when the 1999 Motion Pictures Law was passed,[40] which created a fund to support Israeli movie production by levying license funds from commercial broadcasters and cable TV operators. The quadrupling of state

**Table 30-9. Cable and DBS TV Programming Channels (Audience Shares by Audience Reported Exposure), 2004–2013[1]**

| | 2004 | 2008 | 2012 | 2013 |
|---|---|---|---|---|
| HOT | 20.1 | 9.8 | 6.6 | 5.9 |
| RGE Group | 23.6 | 22.9 | 22.5 | 16.8 |
| Telad | 5.5 | 4.5 | 2.7 | 2.7 |
| Dori Media Group | 5.2 | 16.0 | 17.4 | 20.0 |
| Aneney Tikshoret | 4.4 | 3.2 | 6.6 | 7.3 |
| Keshet | 2.0 | 2.6 | 6.2 | 4.3 |
| Channel 24[2] | 5.8 | 4.0 | | |
| National Geographic (US) | 6.8 | 5.6 | 6.5 | 7.2 |
| Yes (Bezeq) | 4.0 | 7.6 | 10.3 | 9.0 |
| AXN (Sony, Japan/US) | 3.8 | 2.2 | 0.0 | 0.0 |
| Russian-language channels excluding Channel 9 (multiple owners) | 2.7 | 1.5 | 1.7 | 1.4 |
| CNN (Time Warner, US) | 2.5 | 0.9 | 0.6 | 0.5 |
| Discovery Channel (Discovery Communications, US) | 1.8 | 1.4 | 1.8 | 2.9 |
| Channel 23 (government educational channel) | 1.6 | 2.0 | –[3] | – |
| Fashion TV | 1.4 | 0.6 | 0.4 | 0.8 |
| Shopping Network | 1.3 | 0.9 | 0.9 | 1.0 |
| BBC World (BBC, UK) (public) | 1.1 | 1.0 | 0.6 | 0.9 |
| TV5 Monde (TV5MONDE S.A., France) | 0.7 | 0.6 | 0.7 | 0.7 |
| Sky News (Murdoch Family, News Corp., US/UK/AUS) | 0.5 | 1.0 | 0.6 | 0.6 |
| METV (UK) | 0.3 | 0.3 | 0.2 | 0.3 |
| MTV Networks (US) | 4.7 | 2.6 | 2.6 | 2.9 |
| The Hallmark Channel (Crown Media Holdings, US)[4] | | 4.5 | 0.0 | 0.0 |
| Talit Communications | | 1.0 | 1.5 | 1.0 |
| Hop! | | 1.0 | 3.1 | 2.6 |
| ISG Media (India) | | 0.5 | 0.5 | 1.0 |
| Channel 9 (Israel Plus) | | 1.0 | 1.5 | 1.5 |
| Kabala Channel | | | 0.7 | 0.7 |
| Charlton Group | | | 3.9 | 3.7 |
| Slutzky Communication Channels | | | | 4.0 |
| C4 | 56.3 | 56.3 | 56.8 | 53.1 |

*continued*

40. S.H. 1698, 53 (Jan 10, 1999).

Table 30-9. *continued*

| | 2004 | 2008 | 2012 | 2013 |
|---|---|---|---|---|
| HHI | 1,199 | 1,068 | 1,132 | 992 |
| *N* (>1%) | 18 | 20 | 15 | 15 |
| Noam Index | 262 | 210 | 231 | 198 |

*Source*: Single Source/TGI/Kantar Media (2004–2013).

[1] When cable service was originally launched in 1989, it consisted of the joint channels offered by the ICP cartel plus a wide range of satellite channels from abroad. Unfortunately, no data are available for the period between 1989 and 2004 with respect to audience shares. Since the ratings of most channels delivered over the multichannel platforms are not measured by People Meter, data are based on surveys examining viewing habits, performed by Single Source/TGI/Kantar Media. The numbers have been normalized to add up to 100%, while maintaining the relative share of each entity. Original figures provided by Single Source/TGI/Kantar Media on file with authors.

[2] Keshet currently holds approximately 90% of the shares of Channel 24, half of which it purchased in September 2008 and the remaining half in 2011. After the first round of acquisition the channel was re-branded, shut down and re-launched in March 2009. *See, e.g.,* Bar-Zohar, O. "Meimad Forces Keshet to Buy the Rest of Its Shares in Channel 24," *TheMarker*, June 17, 2011, *available at* http://www.themarker.com/advertising/1.657025 (accessed Jan 25, 2013). Channel 24's figure for 2008 remains separate from those of Keshet since the latter purchased its shares in Channel 24 (which do not amount to control) only towards the end of that year, and the effects of that step took shape only in the following year.

[3] Data for 2012 onwards has been included in the section relating to broadcast channels.

[4] The Hallmark Channel ceased broadcasting in Israel in December 2009.

subsidies in film production led to a doubling of film production in four years' time: now approximately 10% of all films distributed in Israel are produced in the country, a number of which have gone in recent years to compete in international film festivals (Table 30.10). Close to 70% of films distributed in Israel are of US production.[41] According to the Israel Film Fund in each of the years 2008–2009, approximately 9.5 million tickets were sold in the box offices (with the number of cinemas, screens, and seats estimated at 120, 380, and 80,000, respectively).[42] These figures seem highly exaggerated, however, as according to other estimates, the total number of tickets sold between 2005 and 2009 reached 9.8 million.[43] In the following years ticket sales have increased, with sales between 2010 and mid-2012 estimated at 12.2 million, while the number of cinemas has decreased (with the market pushing toward fewer, multi-screen complexes).[44] The Film Fund further notes that 10 film distribution companies operate in Israel (as of 2009),[45] though in practice 3 groups (Globus Group, Israel Theaters Group, and United King Group) dominate, through vertical integration, both the distribution and the screening markets.[46] In recent years, the United King Group, owned by the brothers Leon and Moshe Edery, has emerged as by far the most dominant player in the production and distribution of Israeli films.[47]

41. See Israel Film Fund, *Film and Television in Israel—Facts and Figures*, 2009 *available at* <http://filmfund.org.il/userfiles%5Cfile/%D7%A7%D7%95%D7%9C%D7%A0%D7%95%D7%A2%20%D7%95%D7%98%D7%9C%D7%95%D7%95%D7%99%D7%96%D7%99%D7%94%20%D7%91%D7%99%D7%A9%D7%A8%D7%90%D7%9C.pdf>.

42. *Id.*

43. See, e.g., Pery, T. "The Film Industry is Alive and Kicking: 12.2 Million Tickets Sold in the Last Year and a Half," *Globes*, June 17, 2012, *available at* <http://www.globes.co.il/news/article.aspx?did=1000756070>.

44. *Id.*

45. See Israel Film Fund, *supra* note 54.

46. All three groups operate as major film distributors and run major chains of movie theaters. Globus Group, controlled by Yoram Globus, runs the Globus Max chain; The Israel Theaters Group, controlled by Moshe (Mooky) Geridinger, runs the Rav-Chen and Yes Planet chains (and is also the third largest cinema operator in Europe); and the United King Group, controlled by the Edery brothers, runs the Cinema City chain (not to be confused with the Cinema City International brand, under which Israel Theaters operates in Europe).

47. See, e.g., Kurtz, A. "Light, Camera, Edery!," *Calcalist*, Feb 23, 2012, *available at* <http://www.calcalist.co.il/local/articles/0,7340,L-3563000,00.html>.

## Table 30-10. Film Production/Distribution, 1985–2012

| | TOTAL FILM DISTRIBUTION[1] | NUMBER OF ISRAELI PRODUCTIONS[2] | GOVERNMENT SUBSIDIES (MIL US$)[3] |
|---|---|---|---|
| 1985 | 210 | 14 | N/A |
| 1990 | 229 | 12 | N/A |
| 1996 | 186 | 9 | ~5.0[4] |
| 2000 | 180 | 9 | 5.2[5] |
| 2004 | 228 | 27 | 12.2[6] |
| 2008 | 240 | 28 | 17.3[7] |
| 2012 | 265 | 39 | 17.1[8] |

*Source*: ICBS & Haaretz (1985–1996); ICBS & The Knesset Research and Information Center (2000–2012).

[1] For the 1985, 1990 and 1996 figures regarding film distribution in Israel *see* Central Bureau of Statistics, "Full Length Films—Production in Israel and Imports," *in Statistical Abstract of Israel 1997*, 1998 *available at* <http://www.cbs.gov.il/archive/shnaton48/st26-05.gif> . For the 2000 and 2004 figures *see* Central Bureau of Statistics, "Full Length Films, by Continent and Country of Production," *in Statistical Abstract of Israel 2005*, 2006 p. 380, *available at* <http://www.cbs.gov.il/reader/shnaton/templ_shnaton.html?num_tab=st09_03&CYear=2005>. For the 2008 figure *see* Central Bureau of Statistics "Full Length Films Screened in Movie Theaters in Israel, by Continent and Country of Production," *in Statistical Abstract of Israel 2009*, 2010 p. 462, *available at* <http://www.cbs.gov.il/reader/shnaton/templ_shnaton.html?num_tab=st09_09&CYear=2009>. For the 2012 figure *see* Central Bureau of Statistics, "Full Length Films, by Continent and Country of Production," *in Statistical Abstract of Israel 2013*, p. 497 *available at* <http://www.cbs.gov.il/shnaton64/shnaton64_all.pdf>.

[2] *Id.*

[3] For an estimate of government subsidies provided in 1996 *see* Anderman, N. "A Decade for the Motion Picture Law: What Have We Gained from It?" *Haaretz*, Jan 11, 2011 (in Hebrew), *available at* <http://www.haaretz.co.il/gallery/1.1156419>. For the accurate amounts of government subsidies provided in 2000, shortly before the Motion Picture Law came into effect, and in 2004–2008, following its implementation see Pensirer, U. *Governmental Support of the Israeli Film Industry*, Jerusalem: Knesset Research and Information Center, 2011, 7, *available at* <http://www.knesset.gov.il/mmm/data/pdf/m02827.pdf>. It should be noted that the figures in the table reflect the entire budget dedicated to the subsidization of the film industry, including feature films, documentaries, student films, film festivals and so on. However, under the Motion Picture Law, approximately 60% of the budget should be invested in supporting feature films. *Id.*

[4] 3.2 ILS per 1 US$.

[5] 4.08 ILS per 1 US$.

[6] 4.49 ILS per 1 US$.

[7] 3.35 ILS per 1 US$. The amount of subsidies in ILS in 2004 and 2008 was, in fact, identical (58 million).

[8] 3.92 ILS per 1US$. In ILS terms government subsidies have in fact increased from ILS 58 million in 2008 to ILS 67 million in 2012. This increase does not reflect in the figures presented in US$ terms due to fluctuations in the ILS-US$ exchange rate.

# TELECOMMUNICATIONS MEDIA

## Wireline Telecom

From a national monopoly that as late as the 1980s had difficulties managing intercity calling, the Israeli telecom industry has grown substantially over the past twenty years to provide consumers with near-universal access to digital telephony and broadband Internet. Government telephony was corporatized in 1984 as Bezeq, now a publicly traded corporation, in which the public owns close to 70%, and Shaul Elovitch, who purchased it from the Israeli-American billionaire Haim Saban in 2010 owns the strategic share of 30%. Since fixed telephony service was universal by the time the sector was fully privatized, the state allowed new entrants to target specific segments of the wireline market, without making licensing contingent on providing universal service to consumers. HOT entered the wireline market in November 2004, offering services over its own cable infrastructure. Five additional providers entered the market since 2008—Bezeq Group's Bezeq International, IDB Group's Cellcom and 013-Netvision, Partner and 012. Some (e.g., Cellcom) chose to develop their own proprietary infrastructure for that purpose, while others (e.g., 013-Netvision) preferred to transfer calls over existing infrastructures using Voice over Broadband technology.[48] At the same time, Bezeq was not required to unbundle its own network and is still the dominant player. A consolidation process that took place in the Israeli telecommunications market during 2010 and 2011 has left three potential competitors to the dominant Bezeq—HOT, the Partner Group (which acquired 012-Smile), and the IDB Group (which merged 013-Netvision into Cellcom)[49] (Table 30.11).

Bezeq was licensed ahead of HOT to provide broadband infrastructure, though the market has since become a duopoly. As a result, DSL entered the market first, though cable access has quickly gained ground and is close to reaching parity (Table 30.12).

## Table 30-11. Wireline Telecom (Market Shares by Revenue), 1984–2013

| | 1984 | 1988 | 1992 | 1996 | 2000 | 2004 | 2009[1] | 2013 |
|---|---|---|---|---|---|---|---|---|
| Bezeq | 100.0 | 100.0 | 100.0 | 100.0 | 100.0 | 99.0 | 82.0 | 68.0 |
| HOT | — | — | — | — | — | X[2] | 14.0 | 20.0 |
| Partner Group | — | — | — | — | — | — | 3.0 | |
| IDB | — | 0.0 | 0.0 | 0.0 | 0.0 | 0.0 | 1.0 | |
| Total Market Revenue (mil ILS)[3] | | | | | | 5,680 | 4,160 | 3,735 |
| Total Market Revenue (mil US$) | | | | 1,889 | 1,627 | 1,265[4] | 1,241[5] | 1,032 |
| C4 | 100.0 | 100.0 | 100.0 | 100.0 | 100.0 | 100.0 | 100.0 | ~99.0 |
| HHI | 10,000 | 10,000 | 10,000 | 10,000 | 10,000 | 10,000 | 6,930 | 5,168 |
| *N* (>1%) | 1 | 1 | 1 | 1 | 1 | 2 | 4 | ~4 |
| Noam Index | 10,000 | 10,000 | 10,000 | 10,000 | 10,000 | 7,092[6] | 3,465 | 2,317 |

*Source*: Ministry of Communications (until 2009); Ministry of Communications, Ministry of Finance & Hot's Financial Reports (2013).

[1] Figures obtained from the Report of the Committee for the Examination of Bezeq's Tariff Structure and for the Setting of Tariffs for Wholesale Services in the Wireline Sector, Oct 4, 2011 ("Hayak Committee Report"), p. 48, *available at* <http://www.moc.gov.il/sip_storage/FILES/2/2642.pdf>. It should be noted that while figures for 2009 are presented in the Hayak Committee Report, which was published in 2011, as if the mergers in the wireline market have already occurred, they in fact relate to a period shortly prior to those mergers. If we take 2009 as a year in which seven competitors still operated in the market, the Noam Index would be significantly lower (2,625). The HHI would not be significantly affected, however, in light of the very small market shares of all players, except for Bezeq and HOT. See also Table 30-2.

[2] Insignificant market share (HOT provided wireline services for less than two months in 2004).

[3] Data for 2004 derived from Ministry of Communications, *Summary of Revenues of the Telecommunications Market for the Year 2006*, Oct 23, 2007 (in Hebrew); data for 2008 derived from Ministry of Communications, *Summary of Revenues of the Telecommunications Market for the Year 2011*, 28 November, 2012 (in Hebrew); data for the years 2012 and 2013 derived from from Ministry of Communications, *Summary of Revenues of the Telecommunications Market for the Year 2013*, 10 July, 2014 (in Hebrew).

[4] 4.49 ILS per 1US$.

[5] 3.35 ILS per 1US$.

[6] HOT's market share at the end of 2004 was too small to be considered for the calculation of the HHI. We have considered HOT's market entrance for the purpose of calculating the Noam Index in order to reflect the latter's emphasis on the number of "voices," or market players, in addition to each player's market share.

48. See Report of the Committee for the Examination of Bezeq's Tariff Structure and for the Setting of Tariffs for Wholesale Services in the Wireline Sector, Oct 4, 2011, pp. 47, 50, *available at* <http://www.moc.gov.il/sip_storage/FILES/2/2642.pdf>.

49. *Id.*, at 47–48.

Table 30-12. Broadband (Market Shares by Subscribers), 2004–2013[1]

| | 2004 | 2008 | 2012 | 2013 |
|---|---|---|---|---|
| Bezeq | 66.6 | 59.7 | 60.0 | 63.0 |
| HOT | 33.3 | 40.3 | 40.0 | 37.0 |
| Total Market Revenue (mil ILS)[2] | 679 | 1,144 | 1,744 | 1,907 |
| Total Market Revenue (mil US$) | 151[3] | 341[4] | 445[5] | 527[6] |
| C4 | 100 | 100 | 100 | 100 |
| HHI | 5,544 | 5,188 | 5,200 | 5,338 |
| *N* (>1%) | 2 | 2 | 2 | 2 |
| Noam Index | 3,920 | 3,679 | 3,687 | 3,785 |

*Source*: Ministry of Communications (2004–2008); Bezeq financial reports (2012–2013)

[1] Figures for 2004 and 2008 based on number of subscribers reported in the Report of the Committee for the Examination of Bezeq's Tariff Structure and for the Setting of Tariffs for Wholesale Services in the Wireline Sector, Oct 4, 2011 (In Hebrew); figures for 2012 and 2013 derived from Bezeq the Israeli Communications Company Ltd., 2013 Financial Report, 5 March, 2014, p. 40, *available at* <http://maya.tase.co.il/bursa/report.asp?report_cd=882660-00&Type=Pdf> ("Bezeq's 2013 Financial Report").

[2] Data for 2004 derived from Ministry of Communications, *Summary of Revenues of the Telecommunications Market for the Year 2006,* 23 October, 2007 (in Hebrew); data for 2008 derived from Ministry of Communications, *Summary of Revenues of the Telecommunications Market for the Year 2011,* 28 November, 2012 (in Hebrew); data for 2012 and 2013 derived from Ministry of Communications, *Summary of Revenues of the Telecommunications Market for the Year 2013,* 10 July, 2014 (in Hebrew).

[3] 4.49 ILS per 1 US$.

[4] 3.35 ILS per 1 US$.

[5] 3.92 ILS per 1 US$.

[6] 3.62 ILS per 1 US$.

## Wireless Telecom

The first public tender for a mobile license was issued in Israel in 1987. Motorola Israel, part of Motorola (US), established the country's first domestic mobile operator, Pelephone, together with the wireline monopolist Bezeq.[50] In 1993, the government issued a second tender, which was won by Cellcom, a consortium owned at the time by BellSouth (US), the Safra Group (Brazil), and the Israeli investment firm IDB. The launch of Cellcom services significantly reduced consumer costs: Cellcom charged subscribers no connection fees for the first five years, compared with the US$150 (ILS420) charged by Pelephone, and also did not charge monthly subscription fees for two years, as compared with the US$25 (ILS70) charged by Pelephone. Israel soon emerged, as a result, to be a world leader in market penetration and usage rates.[51]

In 1997, a third operator entered the market: Partner Communications, whose major shareholder at the time was Hutchison Whampoa (Hong Kong). Within three years, the penetration rate of cell phones in the Israeli market rose from 44.5% to 73.8%, with the share of households with at least two cell phones expanding from 9.2% to 37.7%.[52] In 2001, a fourth license was awarded to MIRS, then also owned by Motorola (US). While penetration rates continued to rise between 2004 and 2008, the data demonstrate that competition has stalled and the market has stabilized among three primary operators and one smaller operator (MIRS). This development did not go unnoticed by the government, which authorized the formation of

50. Schejter, A. "Israeli cellular telecommunications policy." *Telecommunications Policy*, 30(1):14–28, 2006.

51. See Schejter and Cohen, *supra* note 15.

52. Schejter, *supra* note 86.

## Table 30-13. Wireless Telecom (Market Shares by Subscribers), 1988–2013

| | 1988 | 1992 | 1996[1] | 2000[2] | 2004(3)[3] | 2008[4] | 2012[5] | 2013[6] |
|---|---|---|---|---|---|---|---|---|
| Pelephone (Bezeq) | 100.0 | 100.0 | 48.4% | 35.6 | 29.4 | 29.0 | 28.2 | 26.8 |
| Cellcom (IDB) | — | — | 51.6% | 44.5 | 35 | 35.0 | 32.1 | 31.5 |
| Partner Communications | — | — | — | 19.8 | 31.6 | 32.0 | 29.9 | 29.5 |
| MIRS/HOT Mobile | — | — | — | | 4.0 | 5.0 | 7.7 | 7.7 |
| Golan Telecom | — | — | — | — | — | — | 1.3 | 3.1 |
| MVNOs (multiple operators) | — | — | — | — | — | — | 0.8 | 1.4 |
| Total Market Revenue (mil ILS)[7] | | | | | 13,368 | 15,945 | 13,332 | 10,910 |
| Total Market Revenue (mil US$) | | | 1,130 | 2,448 | 2,977[8] | 4,759[9] | 3,401 | 3,014 |
| C4 | 100 | 100 | 100 | 100 | 100 | 100 | 13,332 | 10,910 |
| HHI | 10,000 | 10,000 | 5,005 | 3,640 | 3,103 | 3,115 | 3,401 | 3,014 |
| *N* (>1%) | 1 | 1 | 2 | 3 | 4 | 4 | 97.9 | 95.5 |
| Noam Index | 10,000 | 10,000 | 3,549 | 2,104 | 1,551 | 1,557 | 2,781 | 2,651 |

*Source*: Ministry of Communications (1996–2008); Bezeq Financial Reports (2102–2103).

[1] Data derived from Ministry of Communications, *Connectivity Tariffs with Regard to the Completion of Calls and SMS to the Mobile Network*, Letter from Dr. Assaf Cohen, Deputy Director-General for Budget and Economic Issues to the Minister of Communications, July 28, 2004, p. 3 ("Letter Regarding Connectivity Tariffs"), *available at* <http://www.moc.gov.il/new/documents/about/asaf_10.2.05.pdf>. See also Table 30-2.

[2] Data derived from Letter Regarding Connectivity Tariffs. *Id.*

[3] Data presented are for the end of 2003 and derived from Letter Regarding Connectivity Tariffs. *Id.*

[4] Data derived from Ministry of Communications, *MVNOs in the Israeli Mobile Market*, Presented before the Knesset Economic Committee by Dr. Assaf Cohen, Deputy Director-General for Budget and Economic Issues, June 2009 (in Hebrew), *available at* <http://www.moc.gov.il/sip_storage/FILES/4/1714.pdf>.

[5] Data derived from Bezeq the Israeli Communications Company Ltd., *2013 Financial Report*, 5 March 2014, p. 71, available at http://maya.tase.co.il/bursa/report.asp?report_cd=882660-00&Type=Pdf.

[6] *Id.*

[7] Data for 2004 derived from Ministry of Communications, *Summary of Revenues of the Telecommunications Market for the Year 2006*, 23 October, 2007 (in Hebrew); Data for 2008 derived from Ministry of Communications, *Summary of Revenues of the Telecommunications Market for the Year 2011*, Nov 28, 2012 (in Hebrew); data for 2012 and 2013 derived from from Ministry of Communications, *Summary of Revenues of the Telecommunications Market for the Year 2013*, 10 July, 2014 (in Hebrew).

[8] 4.49 ILS per 1 US$.

[9] 3.35 ILS per 1 US$.

mobile virtual network operators (MVNOs) and awarded a fifth mobile license in 2011 to Golan Telecom, owned by the French entrepreneurs Xavier Niel, Patrick and Gerard Pariente and Michael Golan. MIRS itself was purchased in 2011 by HOT, which was taken over by another French investor, Patrick Drahi, and rebranded as HOT Mobile. The result of these changes is already apparent in the 2012 and 2013 figures (Table 30.13).

## INTERNET MEDIA

### Internet Service Providers (ISPs)

Israeli Internet access infrastructure is controlled by Bezeq and HOT, so regulatory law prohibits them from providing bit stream access and as a result our data for ISPs are presented separately from those describing Israeli Internet access infrastructure (consumers are required to purchase infrastructure and Internet service independently). This regulatory landscape changed in 2011 when HOT was allowed to provide bit stream access as well through a subsidiary, Hot-Net. A recommendation to allow Bezeq the same is still pending regulatory approval (the Bezeq Group provides bit stream access through the subsidiary Bezeq International, but is required to maintain structural separation between the latter and the Group's other businesses).

## Table 30-14. Internet Service Providers (Market Shares by Subscribers), 2000–2013[1]

| | 2000[2] | 2004[3] | 2008[4] | 2012[5] | 2013[6] |
|---|---|---|---|---|---|
| Bezeq International (Bezeq) | 23.0 | 35.0 | 36.0 | 37.0 | 36.0 |
| 012 Smile | N/A | N/A | 28.0 | 26.0 | 27.0 |
| 012 Golden Lines | 3.2 | 14 .0 | 012 Smile | | |
| Internet Gold | 28.0 | 17.0 | | | |
| 013 Netvision (IDB) | N/A | N/A | 36.0 | 31.0 | 27.0 |
| Netvision | 32.0 | 23.0 | 013 Netvision (IDB) | | |
| 013 Barak | 4.5 | 11.0 | | | |
| Surfree | 7.0 | 0.0 | 0.0 | 0.0 | 0.0 |
| Nonstop | 4.6 | 0.0 | 0.0 | 0.0 | 0.0 |
| HOT Net | – | – | – | 5.0 | 9.0 |
| Others | 7.7 | 0.0 | 0.0 | 1.0 | 1.0 |
| Total Revenue (mil ILS)[7] | | 1,115 | 1,710 | 2,083 | 1,770 |
| Total Revenue (mil US$) | | 248[8] | 510[9] | 534[10] | 489[11] |
| C4 | 90 | 89 | 100 | 99 | 99 |
| HHI | 2,439 | 2,360 | 3,376 | 3,032 | 2,836 |
| *N* (>1%) | 7 | 5 | 3 | 4 | 4 |
| Noam Index | 445 | 1,055 | 1,951 | 1,360 | 1,272 |

*Source*: Globes (2000); Companies' Financial Reports (2004–2008) ); Ministry of Communications & Financial Reports (2012–2013)

[1] The data for 2000 reflect only dial-up, while the data from 2004 on reflects broadband Internet.

[2] *See* Hirsch, S. "The TNS/Teleseker Survey: Walla-The Most Visited Site; Netvision-The Leading Provider," *Globes*, Jan 31, 2001, *available at* http://www.globes.co.il/news/article.aspx?did=467380 (accessed Feb 8, 2012).

[3] *See* Bezeq the Israeli Communications Company Ltd., *2004 Financial Report*, Mar 3, 2005, p. 83, *available at* http://maya.tase.co.il/bursa/report.asp?report_cd=129531.

[4] *See* Bezeq the Israeli Communications Company Ltd., *2008 Financial Report*, Mar 23, 2009, p. 95, *available at* http://maya.tase.co.il/bursa/report.asp?report_cd=432655; Netvision Ltd., *2008 Financial Report*, Mar 9, 2009, p. 15, *available at* http://maya.tase.co.il/bursa/report.asp?report_cd=430210.

[5] Data for 2012 derived from Ministry of Communications, *Summary Report of Public's Complaints for the Year 2012* (2013), *available* at http://www.moc.gov.il/sip_storage/FILES/0/3170;pdf; and Scailex Corporation Ltd., *2012 Financial Report, available at* http://maya.tase.co.il/bursa/report.asp?report_cd=805862-00&Type=Pdf. Scailex was, at the relevant time, a grandparent company of 012 Smile.

[6] Data for 2013 derived from Ministry of Communications, *Summary Report of Public's Complaints for the Year 2013* (2014), *available* at http://www.moc.gov.il/sip_storage/FILES/2/3532.pdf; and Scailex Corporation Ltd., *2013 Financial Report, available at* http://maya.tase.co.il/bursa/report.asp?report_cd=888778-00&Type=Pdf. Scailex was, at the relevant time, a grandparent company of 012 Smile.

[7] Data for the year 2004 derived from Ministry of Communications, *Summary of Revenues of the Telecommunications Market for the Year 2006*, Oct 23, 2007 (in Hebrew); data for the year 2008 derived from Ministry of Communications, *Summary of Revenues of the Telecommunications Market for the Year 2011*, Nov 28, 2012 (in Hebrew); data for the years 2012 and 2013 derived from Ministry of Communications, *Summary of Revenues of the Telecommunications Market for the Year 2013*, Jul 10, 2014 (in Hebrew).

[8] 4.49 ILS per 1 US$.

[9] 3.35 ILS per 1 US$.

[10] 3.92 ILS per 1 US$.

[11] 3.62 ILS per 1 US$.

The original companies that acquired international communications and ISP licenses—012 Golden Lines, Internet Gold, 013 Barak, and Netvision—led the consolidation in the industry and concentration has been trending upward ever since. There were over 30 ISPs operating in Israel in 2000, and as of 2008 there were only 3: Bezeq International, 012 Smile, and 013 Netvision (Table 30.14). As noted above, the most recent round of consolidation took place after 2008: 012 Smile has been bought out by the mobile provider Partner Communications, which is a publicly traded corporation in which the strategic share has been held since January 2013 by Haim Saban, following his divestment of Bezeq, and 013 Netvision was merged into Cellcom by IDB.

## Search Engines

The data for search engines are based on exposure to core search engine sites and do not take into account all properties where search activity may be observed. Most Israeli online news portals, which will be discussed in the following section, are for search metric purposes affiliated with Google.

The search engine market is not an Israeli market. Concentration is increasing due to the rise of Google's market shares, but this had no substantial internal effect on Israeli businesses, since there are no domestic search engine players (Table 30.15).

## Online News

The major print publishing groups—Yediot Aharonot, Ma'ariv, and Haaretz—have all developed popular websites providing news. Ynet.co.il is owned and operated by the Yediot Aharonot Group, though it is managed by a separate editorial staff and generally does not share content with the print daily *Yediot Aharonot*. It currently has an English-language version (ynetnews.com) companion site, and from 2002 to 2005, it had an Arabic-language version as well (arabynet.com).[53] In addition, the Yediot Group operates calcalist.co.il, a financial news website which shares content with the print daily, *Calcalist*. *Ma'ariv*'s affiliated website nrg.co.il maintained a separate editorial staff for a few years as well, though as a result of Ma'ariv's financial difficulties and its eventual changing of hands it now shares substantial content with the print daily and since 2014, with the liquidation of *Ma'ariv* and the transfer of its paper edition to the Jerusalem Post group, nrg.co.il was purchased by *Israel Hayom*. *Haaretz* operates two sites—in Hebrew and English (haaretz.co.il and haaretz.com, respectively)—both of which originally relied on print edition materials. In 2012 and 2013 Haaretz became the first Israeli news site to launch a pay wall, first on its English language service—for which it now produces original content available only online and provided over mobile applications for smartphones and tablets—and subsequently for the Hebrew edition. For about a decade the Haaretz Group partnered with Bezeq International to manage *Walla!* (walla.co.il), Israel's most popular online news site, a partnership that ended in 2010.[54] The Haaretz Group also operates *themarker.com*, which was established as an online business news site, but eventually also became an economic daily and a monthly print magazine known as *The Marker*. In addition to the three traditionally leading newspapers, others such as *Globes* (globes.co.il), *Israel HaYom* (israelhayom.com) and The Jerusalem Post Group (jpost.com) also offer online content which is derived from their print editions.[55]

53. Levy, R. "Books on sale—but at what cost?" *Haaretz*, June 8, 2012. <http://www.haaretz.com/business/books-on-sale-but-at-what-cost-1.435170> (accessed Jan 25, 2013).

54. Kornfeld, T. "Ha'aretz Departs from Walla Website" *Bizportal*, May 1, 2010, *available* at <http://www1.bizportal.co.il/article/233845> (in Hebrew) (accessed Jan 25, 2013).

55. globes co.il offers original content, alongside content derived from the print edition.

Table 30-15. Internet Search Engines (MARKET SHARES BY REPORTED AUDIENCE EAPOSURE (2000–2008[1]); BY SAMPLE OF PAGE VIEWS 2012–2013[2])

| | 2000[3] | 2004[4] | 2008[5] | 2012[6] | 2013[7] |
|---|---|---|---|---|---|
| Yahoo! (US) | 34.7 | 17.5 | 10.8 | 0.5 | 1.1 |
| Microsoft (MSN, US)[8] | 34.0 | 31.5 | 18.5 | 1.4 | 1.5 |
| Altavista (US) | 19.2 | 0.0 | 0.0 | 0.0 | 0.0 |
| Lycos (US) | 4.2 | 0.0 | 0.0 | 0.0 | 0.0 |
| Excite (US) | 4.2 | 0.0 | 0.0 | 0.0 | 0.0 |
| Infoseek (US) | 3.6 | 0.0 | 0.0 | 0.0 | 0.0 |
| Google (US) | 0.0 | 51.0 | 70.7 | 97.8 | 96.7 |
| Ask | No data | No data | No data | 0.3 | 0.6 |
| C4 | 77.1 | 100 | 100 | 100 | 100 |
| HHI | 2,777 | 3,900 | 5,457 | 9,567 | 9,356 |
| *N* (>1%) | 6 | 2 | 2 | 2 | 3 |
| Noam Index | 1,138 | 2,254 | 3,150 | 4,784 | 4,678 |

*Source*: Teleseker Internet Monitor (TIM) (2000–2008); StatCounter Global Stats (2012–2013)
[1] Figures have been normalized to add up to 100%, while maintaining the relative share of each entity.
[2] Figures have been normalized to add up to 100%, while maintaining the relative share of each entity.
[3] Data derived from the January 2001 Telesker Internet Monitor (TIM) survey (on file with authors).
[4] Data derived from the November 2004 TIM survey (on file with authors).
[5] Data derived from the December 2008 TIM survey (on file with authors).
[6] Data derived from StatCounter Global Stats (http://gs.statcounter.com)
[7] Data derived from StatCounter Global Stats (http://gs.statcounter.com)
[8] Both msn.com/co.il and hotmail.com/co.il are included in this data from 2004 on.

Audiovisual content providers are also building an online presence. Channel 2's franchise Keshet launched mako.co.il in 2008 while Reshet, experimented with a few formats, among them a domain within the Ynet website until launching reshet.tv. The online news site nana10.co.il, which started as Netvision's content arm in 1999, when Netvision was a major dial-up ISP, became a partnership between 013 Netvision and Channel 10 in 2007. RGE's Sport 5 channel launched a popular sports website (sport5.co.il), while its online competitor One (one.co.il) has gone the opposite way, launching in 2011 its own cable sports channel. It is partially owned by the Yediot group. Earlier independent online news portals such as Netking and Start have had to close down due their inability to compete with these media groups (Table 30.16).

## CONCLUSION

When describing media concentration in Israel, an additional perspective that needs to be considered is that of cross-ownership, which has been a concern of the legal system since the launch of the neoliberal economic policies first begun in the late 1970s by the Likud Party. As noted above, cross-ownership was initially characterized by the involvement of the Yediot Aharonot and Ma'ariv groups in Channel 2's franchise holders and two cable TV outlets. Yediot benefited the most from what the system allowed at the time: the group held more than 50% of the newspaper market, as well as stakes in Channel 2 and the cable franchise Golden Channels before DTH and Channel 10 came into the picture. The restructuring of the regulatory landscape that followed the introduction of broadband Internet service gave lawmakers

## Table 30-16. Online News Media (Market Share by Audience Reported Exposure), 2000–2012[1]

| | 2000 | 2004 | 2008 | 2012 |
|---|---|---|---|---|
| *Walla!* | 18.2 | 20.2 | 19.7 | 19.8 |
| nana.co.il (013 Netvision-Channel 10) | 13.3 | 11.0 | 8.6 | 6.8 |
| MSN (Microsoft, US) | 10.9 | 14.1 | 7.0 | 2.9 |
| hotmail.co.il | 11.7 | | | |
| Haaretz Group (Schocken Family) | 13.5 | 10.4 | 7.4 | 7.1 |
| mako.co.il (Keshet) | 0.0 | 0.0 | 8.4 | 12.2 |
| Tapuz | 6.7 | 7.6 | 6.2 | 6.0 |
| nrg.col.il (Ma'ariv) | 5.3 | 9.7 | 7.4 | 5.2 |
| Start | 5.3 | 6.2 | 0.0 | 0.0 |
| Yediot Aharonot Group (ynet.co.il; calcalist.co.il) (Mozes Family) | 7.8 | 16.3 | 18.8 | 24.3[2] |
| Netking | 4.3 | 0.0 | 0.0 | 0.0 |
| globes.co.il (Globes Group) | 3.0 | 4.4 | 4.6 | 4.0 |
| sport5.co.il (Sport 5, RGE) | | | 3.4 | 4.2 |
| One | | | 4.1 | 4.0 |
| Reshet | | | 4.5 | 3.4 |
| C4 | 56.7 | 61.6 | 55.5 | 63.4 |
| HHI | 1,135 | 1,311 | 1,152 | 1,361 |
| *N* (>1%) | 11 | 12 | 12 | 11 |
| Noam Index | 342 | 378 | 333 | 411 |

*Source*: Teleseker Internet Monitor (TIM).

[1] Data derived from TIM surveys for January 2001, November 2004, December 2008 and November 2012 (on file with authors). Figures have been normalized to add up to 100%, while maintaining the relative share of each entity.

[2] Calcalist.co.il was launched in 2008 and reflected in data for 2012.

an opportunity to reduce Yediot and Ma'ariv's cross-ownership. Both groups had to pull out of Channel 2's franchises and Ma'ariv left the cable business entirely, first. Although the Yediot Aharonot and the Globes Groups were left with minority shares in HOT, these shares were eventually sold to HOT's new lead investor Patrick Drahi in 2011.

However, even as the print publishers left the audiovisual market, a new cross-ownership structure emerged for Internet and pay-TV services. Yes, the DTH satellite operator, is controlled by Bezeq, which thus became a direct competitor with HOT on three fronts: broadband Internet access, wireline telephony, and multichannel video content. In 2014, a merger was approved between Bezeq and Yes.

Bezeq has always been involved—through its subsidiaries—in mobile services (Pelephone) and ISPs (Bezeq International), though it is required to maintain structural separation from them. HOT is required to structurally separate its broadband operations from its video channels, but is allowed, like Bezeq, to provide basic telephony. And while Bezeq itself cannot offer an ISP to consumers, its subsidiary Bezeq International can. This policy may be on the verge of changing. In 2011, a committee appointed by the Minister of Communications (The Hayak Committee) recommended the elimination of the structural separation of the Bezeq group by allowing a merger of the Internet access and the infrastructure operations.[56] Such a move has already occurred with regard to Bezeq's competitors: Cellcom, the largest mobile operator took over 013 Netvision, the second largest Israeli ISP, and Partner Communications, the second largest mobile operator, purchased 012 Smile, which is both the third largest Israeli ISP and a growing player in the wireline market. Indeed, in the years following the latest data provided in our tables, there is a growing level of cross-ownership in the telecom distribution industries.

Further concentration of content-producing industries is a trend for Israel as well. The most notable example of this in recent years was the purchase of *Ma'ariv* by IDB, and then the merger of Cellcom and 013 Netvision. Following *Ma'ariv's* acquisition by the IDB Group, there were suggestions that a merger between 013 Netvision's nana.co.il (which had partnered with the financially troubled Channel 10) and *Ma'ariv*'s nrg.co.il online news portal was next.[57] This has not come to pass, however, and in the meantime IDB divested itself from *Ma'ariv* in 2012 because of its own unrelated financial difficulties. So even though the online content merger did not occur—Channel 10 has since accepted the terms of an Israeli government bailout to remain afloat and independent—IDB has now sold off *Ma'ariv* to Shlomo Ben-Zvi, publisher of *Makor Rishon*, who has himself since sold *Ma'ariv* to the *Mirkaei Tikshoret* group.[58] Such mergers are indicative of the concentration trend in the national media market.

The overall trend that emerges from this data is that traditional print concentration in Israel is declining, while contemporary digital media is formulating into a vertically integrated, highly concentrated distribution system. Magazines, book publishing, pay TV and online news portals demonstrate, on the face of it, healthy competition while concentration for newspapers and broadcast television concentration seems to be declining. However, assessing that this trend is a positive development for de-concentration and that these industries are truly competitive would be a superficial statement that does not take into account cross-ownership, and the lack of competition in the distribution markets (particularly broadband infrastructure and ISPs). As we have seen from the description of the news market over the Internet, that market is but a reflection of the existing players in the traditional media markets. The nature and impact of the purportedly growing competitiveness in the newspaper market is yet to be seen.

Audiovisual media is Israel is a closed and highly controlled system when it comes to television news.[59] Most programming falls into the entertainment category (where there is more competition). Therefore, the following examples of the interference of powerful individuals who control major media assets of their own demonstrate the weakness of the data in describing concentration trends in Israel.

In 2005, an investigative piece about IDB was pulled off the air due to the conglomerate's pressure on Keshet.[60] Keshet's chairman justified the decision to pull the piece in a radio interview by arguing that an investigative program does not belong in his company's lineup, which should "focus on entertainment" instead.[61] Then, in 2008, all the commercial television franchises refused to air a documentary that was critical of the politically influential Ofer family's activities

---

56. Horesh, H. "The Interim Recommendations of the Hayak Committee: To Eliminate the Separation between Internet Access Provision and Infrastructure Operations" *NRG*, Mar 3, 2011, *available* at <http://www.nrg.co.il/online/16/ART2/218/376.html> (in Hebrew) (accessed Jan 25, 2013).

57. Bar-Zohar, O. and Levin, R. "Nana Website in Negotiations towards a Merger with Nochi Dankner's NRG-Ma'ariv" *TheMarker*, July 13, 2012, *available* at <http://www.themarker.com/advertising/1.668188> (in Hebrew) (accessed Jan 25, 2013).

58. Associated Press. "Israel's Channel 10 agrees to financial rescue deal, but future still unclear." *Haaretz*, Dec 2, 2012, *available at* <http://www.haaretz.com/news/national/israel-s-channel-10-agrees-to-financial-rescue-deal-but-future-still-unclear-1.481936> (accessed Jan 25, 2013).

59. Schejter, supra note 6.

60. Aberbach, L. "The Facts You Did Not Know" *Walla!*, Dec 28, 2005, *available at* <http://b.walla.co.il/?w=//832026> (in Hebrew) (accessed Jan 25, 2013).

61. Zach, Y. "Chairman of Keshet's Board of Directors: We Made a Mistake when We Took 'Uvda' over to Us" *NRG*, Dec 23, 2005, *available at* <http://www.nrg.co.il/online/1/ART1/024/288.html> (in Hebrew) (accessed Jan 25, 2013).

(even Yes, which co-produced the documentary, refused to broadcast it in the end).[62] Only the IBA agreed to air the controversial production. Finally, in 2011, *Israel HaYom*'s owner Sheldon Adelson reportedly used his personal connections with a major shareholder in Channel 10 to force the network to air an unprecedented apology over what he regarded as an unflattering portrayal in one of their stories. Three of the channel's top personnel resigned in protest.

Media regulation and control can be described using four parallel narratives: technological, economic, democratic, and cultural.[63] Liberal democracies and market economies that pursue liberalization without effective regulatory oversight end up allowing market players to dictate programming based on their corporate interests.[64] Based on this assessment, we can conclude that the Israeli media market is embracing a model that will less effectively serve democratic ideals. While some academics believe that "in small economies social goals should be given little or no independent weight in formulating competition policy,"[65] such an approach is not applicable in media markets, whose regulation traditionally employs a combination of social and economic considerations. Napoli identified three such fundamental differences between communications regulation and the regulation of other industries: the unique potential for social and political impact, the ambiguity of classifying decisions along economic lines and the potential overlap and interaction between economic and social concerns within individual decisions.[66] Thus, while the Israeli media market can be characterized by shortage of resources and small audience shares, it is not the smallness of the Israeli production segments that is alarming, as some comparative cases suggest,[67] but rather the threat increasingly concentrated distribution poses to the health of Israeli democracy.

It is possible that greater public investment in media production and distribution may help to redress these imbalances. As the data regarding the film industry demonstrate, Israeli policymakers have found a way to enhance local film production. Not only are there more Israeli movies produced each year, they have also become internationally recognized for their quality. In addition, the fact that broadcasters and video channels partner in the production of films provides these movies with a "second life" on the television screen. Similar (though arguably, not as quality-oriented) is the situation in the television industry, where strict production quotas for local fare apply to Channels 1, 2, and 10, and to the cable and satellite sectors, quotas that help maintain a vibrant industry relative to the size of the economy.

Developing and maintaining a competitive market in a small economy is a difficult challenge. Israel is not only small in land and population size, but unlike countries of comparable size—such as Austria, Belgium, and Switzerland—that are also part of a supranational market (the EU), Israel's media market (and its economy in general) is not very integrated with the markets of its neighbors due to their history of interstate conflict since 1948. So, on the one hand, the relative openness of the Israeli economy since the mid-1980s and its size have made it a susceptible candidate for fast diffusion of advanced telecommunications, an opportunity it has undoubtedly seized.[68] On the other hand, since Israel has a small

---

62. Yudelovich, M. "The Ofer Brothers: War Movie" *YNet*, Aug 19, 2008, *available at* <http://www.ynet.co.il/articles/0,7340,L-3583905,00.html> (in Hebrew) (accessed Jan 25, 2013).

63. Schejter, A. and Han, S. "Regulating the Media: Four Perspectives". in *Politics of Regulation Handbook*. Levi-Faur, D., ed. London: Edward Elgar, 2011. pp. 243–253.

64. Curran, J. and Seaton, J. *Power Without Responsibility: The Press, Broadcasting and New Media in Britain*. 6th ed. London: Routledge, 2003.

65. Gal, M. *Competition Policy for Small Market Economies*. Cambridge, MA: Harvard University Press, 2003. p 50.

66. Napoli, P. "The Marketplace of Ideas Metaphor in Communications Regulation." *Journal of Communication* 49, 1999. pp. 151–169.

67. Puppis, M. "Media Regulation in Small States." *International Communication Gazette*, 71:2, 2009. pp. 7–17.

68. Antonelli, C. *The Diffusion of Advanced Telecommunications in Developing Countries*. Part 56. Paris: Organization for Economic Co-operation and Development (OECD) Development Centre, 1991.

economy, it is extremely difficult to create within it truly competitive markets. The threat the current structure of the Israeli market dictates is more in the cross-ownership of channels providing news than in outlets providing entertainment, and if the trend continues in distribution, this will have a negative influence on outlets' ability to freely distribute content.

The Internet does not seem to offer much promise in reducing concentration because ISPs and online news portals are so heavily concentrated in favor of established groups and further mergers. The resolution must lie in the strengthening of noncommercial sources for news and information. Indeed, low audience shares of IBA's outlets and the fact that its Internet presence is near-nonexistent attests that not much is being done in this direction as well. The overwhelming growth in the size of local production industries as a result of direct regulatory involvement is further proof that in conditions of market failure regulatory support is crucial and can be effective, but it cannot be limited simply to supporting more entertainment programming and preferential licensing agreements. The bottom line is that high levels of concentration, growing levels of vertical integration in distribution, and the weakness of the IBA all point to further increasing media concentration in Israel.

# Bibliography

## 1. Legislation


Motion Picture Law, 1999, S.H. 1698, 53 (Jan 10, 1999).

Protection of Literature and Writers in Israel (Provisional Order) Law, 2013, S.H 2407, 208 (Aug 8, 2013).

Second Authority for Radio and Television Law, 1990, S.H. 1304, 58 (Feb 13, 1990).

Telecommunications Law (Amendment No. 4), 1986, S.H. 1192, 224. (Aug 13, 1986).

Telecommunications Law (Amendment No. 13), 1996, S.H. 1590, 302. (May 10, 1996).

Telecommunications Law (Amendment No. 25), 2001, S.H. 1807, 530. (Aug 7, 2001).


## 2. Court Decisions


A.T. 455, *I. C. S. Israel Cable Systems, Ltd. v. The Antitrust Commissioner* (Jan 10, 1994) (on file with the authors).

A.T. 801/08, *Bezeq the Israeli Communications Corporation v. the Antitrust Commissioner* (published in *Nevo*, Dec 16, 2009).

C.A. (Tel-Aviv District Court) 1749/99 *Shvakim Hadashim Publishing Ltd. v. The Council for the Arrangement of Sports Betting* (published in *Nevo*, Mar 15, 2009).

HCJ 757/84 *The Daily Newspaper Union v. The Minister of Education and Culture*, PD 41(4), 337 (1987).

HCJ 3424/90 *The Daily Newspaper Union v. The Minister of Education and Culture*, PD 45(2), 24 (1991).


## 3. Surveys and Reports


Agmon, T. and Tsadik, A., *Analysis of the Economic Consequences of Concentration and Cross-Ownership in the Media*, Jerusalem: Knesset Research and Information Center, Nov 2, 2011 (in Hebrew), *available at* <http://www.knesset.gov.il/mmm/data/pdf/m02952.pdf>.

Bezeq the Israeli Communications Company Ltd., *2004 Financial Report*, Mar 3, 2005, *available at* <http://maya.tase.co.il/bursa/report.asp?report_cd=129531>.

Bezeq the Israeli Communications Company Ltd., *2008 Financial Report*, Mar 23, 2009, available at <http://maya.tase.co.il/bursa/report.asp?report_cd=432655>.

Bezeq the Israeli Communications Company Ltd., *2013 Financial Report*, Mar 5, 2014, *available at* <http://maya.tase.co.il/bursa/report.asp?report_cd=882660-00&Type=Pdf>.

Book Publishers Association of Israel, *Space Rented in Hebrew Book Week by Publisher, 1997–2008* (data provided by request of the authors and on file with the authors).

Central Bureau of Statistics, "Daily Newspapers and Periodicals Published in Israel, by Frequency, Language, Type and Subject," *in 1998 Statistical Abstract of Israel*, 1999, pp. 26–29, available at <http://www.cbs.gov.il/archive/shnaton49/st26-04_h.shtml>.

Central Bureau of Statistics, "Full Length Films-Production in Israel and Imports," *in* Statistical *Abstract of Israel 1997*, 1998, *available at* <http://www.cbs.gov.il/archive/shnaton48/st26-05.gif>.

Central Bureau of Statistics, "Full Length Films, by Continent and Country of Production," *in Statistical Abstract of Israel 2005*, 2006, p. 380, *available at* <http://www.cbs.gov.il/reader/shnaton/templ_shnaton.html?num_tab=st09_03&CYear=2005>.

Central Bureau of Statistics, "Full Length Films Screened in Movie Theaters in Israel, by Continent and Country of Production," *in Statistical Abstract of Israel 2009*, 2010, p. 462, *available at* <http://www.cbs.gov.il/reader/shnaton/templ_shnaton.html?num_tab=st09_09&CYear=2009>.

Central Bureau of Statistics, "Full Length Films, by Continent and Country of Production," *in Statistical Abstract of Israel 2013*, 2014 p. 497, *available at* <http://www.cbs.gov.il/shnaton64/shnaton64_all.pdf>.

Central Bureau of Statistics, *Israeli Population: Demographic Characteristics, 1990–2009*, 2010, *available at* <http://www.cbs.gov.il/www/statistical/isr_pop_heb.pdf>.

Council for Cable and Satellite Broadcasting, *2000 Annual Report*, 2001, *available at* <http://www.moc.gov.il/sip_storage/FILES/3/1033.pdf>.

Council for Cable and Satellite Broadcasting, *2004 Annual Report*, 2005, *available at* <http://www.moc.gov.il/sip_storage/FILES/9/1029.pdf>.

Council for Cable and Satellite Broadcasting, *2008 Annual Report*, 2010, *available at* <http://www.moc.gov.il/new/documents/doch2008.pdf>.

Council for Cable and Satellite Broadcasting, *2012 Annual Report*, 2013, *available at* <http://www.moc.gov.il/sip_storage/FILES/5/3435.pdf>.

Goldsmith, R. and Schwartz, O., *Advertising by Law in the Printed Press and on the Internet: Cost, Availability and Effectiveness*, Jerusalem: Knesset Research and Information Center, Mar 24, 2009 (in Hebrew), *available at* <http://knesset.gov.il/mmm/data/pdf/m02206.pdf>.

Israeli Advertisers Union/Marketing Union, *Exposure to the Media Survey 1986*, 1986 (on file with the authors).

Israeli Advertisers Union/Marketing Union, *Exposure to the Media - Telephonic Survey Conducted by the Teleseker Institute*, 1988 (on file with the authors).

Israeli Advertisers Union/Marketing Union, *Exposure to the Media - Survey Conducted by Gallup Israel*, 1992 (on file with the authors).

Israeli Advertisers Union/Marketing Union, *Exposure to the Media - Survey Conducted by the Geocartography*, 1996 (on file with the authors).

Kantar Media/Telegal, *Daypart Analysis of Television Ratings 2012–2013* (data provided by request of the authors and on file with the authors).

Kantar Media/TGI, *Audience Exposure to Daily Newspapers, Radio Stations, Magazines and Television Channels 2012–2013*, (data provided by request of the authors and on file with the authors).

Ministry of Communications, *Connectivity Tariffs with Regard to the Completion of Calls and SMS to the Mobile Network*, Letter from Dr. Assaf Cohen, Deputy Director-General for Budget and Economic Issues to the Minister of Communications, July 28, 2004, *available at* http://www.moc.gov.il/new/documents/about/asaf_10.2.05.pdf.

Ministry of Communications, *MVNOs in the Israeli Mobile Market, Presented before the Knesset Economic Committee by Dr. Assaf Cohen, Deputy Director-General for Budget and Economic Issues*, June 2009 (in Hebrew), *available at* <http://www.moc.gov.il/sip_storage/FILES/4/1714.pdf>.

Ministry of Communications, *Summary Report of Public's Complaints for the Year 2012*, 2013, *available at* <http://www.moc.gov.il/sip_storage/FILES/0/3170.pdf>.

Ministry of Communications, Summary Report of Public's Complaints for the Year 2013, 2014, *available at* <http://www.moc.gov.il/sip_storage/FILES/2/3532.pdf>.

Ministry of Communications, *Report of the Committee for Detailed Recommendations Regarding Policy and Competition Rules in the Telecommunications Industry in Israel 2008*, Mar 9, 2008 (in Hebrew), *available at* <http://www.moc.gov.il/new/documents/gronau2008/gronau1.pdf>.

Ministry of Communications, *Report of the Committee for the Examination of Bezeq's Tariff Structure and for the Setting of Tariffs for Wholesale Services in the Wireline Sector*, Oct 4, 2011, *available at* <http://www.moc.gov.il/sip_storage/FILES/2/2642.pdf>.

Ministry of Communications, *Summary of Revenues of the Telecommunications Market for the Year 2006*, Oct 23, 2007 (in Hebrew), *available at* <http://www.moc.gov.il/sip_storage/FILES/5/1245.pdf>.

Ministry of Communications, *Summary of Revenues of the Telecommunications Market for the Year 2011*, Nov 28, 2012 (in Hebrew), *available at* <http://www.moc.gov.il/sip_storage/FILES/6/3086.pdf>.

Ministry of Communications, *Summary of Revenues of the Telecommunications Market for the Year 2013*, 10 July, 2014 (in Hebrew) (on file with authors).

Netvision Ltd., *2008 Financial Report*, Mar 9, 2009, *available at* <http://maya.tase.co.il/bursa/report.asp?report_cd=430210>.

Pensirer, U., *Governmental Support of the Israeli Film Industry*, Jerusalem: Knesset Research and Information Center, 2011, *available at* <http://www.knesset.gov.il/mmm/data/pdf/m02827.pdf>.
Single Source/TGI, *Sector Media Index: Exposure to Newspapers among the Israeli Palestinian Population, 2008–2009* (data provided by request of the authors and on file with the authors).
Single Source/TGI, *Audience Exposure to Daily Newspapers and Radio Stations – TGI Survey for the Years 1998, 2000, 2004 and 2008* (data provided by request of the authors and on file with the authors).
Single Source/TGI, *Audience Exposure to Magazines and Television Channels – TGI Survey for the Years 2000, 2004 and 2008* (data provided by request of the authors and on file with the authors).
Scailex Corporation Ltd., 2012 *Financial Report*, 2013, *available at* <http://maya.tase.co.il/bursa/report.asp?report_cd=805862-00&Type=Pdf>.
Scailex Corporation Ltd., 2013 *Financial Report*, 2014, *available at* <http://maya.tase.co.il/bursa/report.asp?report_cd=888778-00&Type=Pdf>.
TNS/Telegal (Israeli Audience Research Board), *Day-part Analysis of Television Ratings 1998–2008* (data provided by request of the authors and on file with the authors).
TNS/Teleseker Internet Monitor, *TIM Internet Survey - January 2001* (on file with the authors).
TNS/Teleseker Internet Monitor, *TIM Internet Survey - November 2004* (on file with the authors).
TNS/Teleseker Internet Monitor, *TIM Internet Survey - December 2008* (on file with the authors).
TNS/Teleseker Internet Monitor, *TIM Internet Survey - November 2012* (on file with the authors).

## 4. Books and Articles

Aberbach, L. "The facts you didn't know" *Walla!*, Dec 28, 2005, *available at*, <http://b.walla.co.il/?w=//832026> (accessed Jan 25, 2013).
Aharoni, Y. *The Israeli Economy: Dreams and Realities*. London: Routledge, 1991.
Anderman, N. "A Decade for the Motion Picture Law: What Have We Gained from It?" *Haaretz*, Jan 11, 2011 (in Hebrew), *available at* http://www.haaretz.co.il/gallery/1.1156419 (accessed Jan 25, 2013).
Antonelli, C. *The Diffusion of Advanced Telecommunications in Developing Countries. Part 56*. Paris: Organization for Economic Co-operation and Development (OECD) Development Centre, 1991.
Aronoff, M. "Radical Change in Israel: A Review Essay." *Political Science Quarterly* 116:3, 2001, pp. 447–453.
Assaf, M. "The Arabic Press in Israel." *International Communication Gazette* 7, 1961, pp. 23–25.
Associated Press. 'Israel's Channel 10 agrees to financial rescue deal, but future still unclear." *Haaretz*, Dec 2, 2012, *available at* http://www.haaretz.com/news/national/israel-s-channel-10-agrees-to-financial-rescue-deal-but-future-stillunclear-1.481936 (accessed Jan 25, 2013).
Bar-Zohar, O. "Meimad Forces Keshet to Buy the Rest of Its Shares in Channel 24." *TheMarker*, June 17, 2011 (in Hebrew), *available at* <http://www.themarker.com/advertising/1.657025>.
Bar-Zohar, O. and Levin, R. "Nana Website in Negotiations towards a Merger with Nochi Dankner's NRG-Ma'ariv." *TheMarker*, July 13, 2012 (in Hebrew), *available at* http://www.themarker.com/advertising/1.668188 (accessed Jan 25, 2013).
Ben-Bassat, A. 'The Obstacle Course to a Market Economy in Israel." *in The Israeli Economy, 1985–1998: From Government Intervention to Market Economics*. Ben-Bassat, A., ed. Cambridge: MIT Press, 2002.
Calcalist. "In Accordance with the Newspaper's Request: Yediot Aharonot Not a Monopoly." *Calcalist*, Jan 3, 2010 (in Hebrew), *available at* http://www.calcalist.co.il/local/articles/0,7340,L-3384283,00.html (accessed Jan 25, 2013).
Cohen, A., Lemish, D., and Schejter, A. *The Wonder Phone in the Land of Miracles: Mobile Telephony in Israel*. Cresskill: Hampton Press, 2008.
Cooper, L. and Nakanishi, M. *Market-Share Analysis: Evaluating Competitive Marketing Effectiveness*. Norwell: Kluwer Academic Publishing, 1998.
Curran, J. and Seaton, J. *Power Without Responsibility: The Press, Broadcasting and New Media in Britain*. 6th ed. London: Routledge, 2003.
Davidson, R. and Schejter, A. "'Their Deeds are the Deeds of Zimri; but They Expect a Reward Like Phineas': Neoliberal and Multicultural Discourses in the Development of Israeli DTT Policy." *Communication, Culture and Critique* 4, 2011, pp. 1–22.
Ezrahi, Y., Goshen, Z., and Leshem, S. *Cross Ownership - Control and Competition in the Israeli Media: Economic and Legal Aspects and Their Impact on the Israeli Democracy*. Jerusalem: Israel Democracy Institute, 2003.

Gal, M. *Competition Policy for Small Market Economies*. Cambridge, MA: Harvard University Press, 2003.
Gandal, N. "Compatibility, Standardization, & Network Effects: Some Policy Implications." *Oxford Review of Economic Policy* 18, 2002, pp. 80–91.
Historical Jewish Press. *Ma'ariv, available at* <http://web.nli.org.il/sites/JPress/Hebrew/Pages/maariv.aspx> (accessed Jan 25, 2013).
Haaretz. *About Us*, available at http://www.haaretz.co.il/misc/about-haaretz (accessed Jan 25, 2013).
Hirsch, S. "The TNS/Teleseker Survey: Walla-The Most Visited Site; Netvision-The Leading Provider." *Globes*, Jan 31, 2001 (in Hebrew), *available at* http://www.globes.co.il/news/article.aspx?did=467380 (accessed Feb 8, 2012).
Horesh, H. "The Interim Recommendations of the Hayak Committee: To Eliminate the Separation between Internet Access Provision and Infrastructure Operations." *NRG*, Mar 3, 2011 (in Hebrew), *available at* http://www.nrg.co.il/online/16/ART2/218/376.html (accessed Jan 25, 2013).
Israel Film Fund, *Film and Television in Israel—Facts and Figures*, 2009, *available at* http://filmfund.org.il/userfiles%5Cfile/%D7%A7%D7%95%D7%9C%D7%A0%D7%95%D7%A2%20%D7%95%D7%98%D7%9C%D7%95%D7%95%D7%99%D7%96%D7%99%D7%94%20%D7%91%D7%99%D7%A9%D7%A8%D7%90%D7%9C.pdf (accessed Jan 25, 2013).
Israel Marketing Union and Ifat Advertising Monitoring. *An Increase of 4% in Advertising Expenditure in the Year 2013*, 2013, *available at* http://www.ishivuk.co.il/message/1111 (accessed Jul 14, 2014).
Kornfeld, T. "Ha'aretz Departs from Walla Website." *Bizportal*, May 1, 2010 (in Hebrew), *available at* http://www1.bizportal.co.il/article/233845 (accessed Jan 25, 2013).
Kurtz, A. "Light, Camera, Edery!" *Calcalist*, Feb 23, 2012 (in Hebrew), *available at* http://www.calcalist.co.il/local/articles/0,7340,L-3563000,00.html (accessed Jan 25, 2013).
Lehman-Wilzig, S. and Schejter, A. "Mass Media in Israel." *In Mass Media in the Middle East*. Kamalipour, Y. and Mowlana, H., eds. Westport, CT: Greenwood Press, 1994, pp. 109–125.
Levy, R. "Books on sale - but at what cost?" Haaretz, June 8, 2012, *available at* http://www.haaretz.com/business/books-on-sale-but-at-what-cost-1.435170 (accessed Jan 25, 2013).
Limor, Y. "Mass Media in Israel." *in Trends in Israeli Society*. Ya'ar, E. and Shavit, Z., eds. Tel Aviv: Open University, 2003.
Limor, Y. and Naveh, H. *The Pirate Radio in Israel*. Haifa: Pardes Publishing, 2007.
Masalha, O. A. "The Arabic Press in Israel", 2009 (in Hebrew), *available at* http://www.reader.co.il/article/48627/העיתונות-הערבית-בישראל
Napoli, P. "The Marketplace of Ideas Metaphor in Communications Regulation." *Journal of Communication* 49, 1999, pp. 151–169.
Orbach, L. "Compromise: The Moto Group to be divided up between TimeOut and Tchelet." *Globes*, Sep 3, 2013 (in Hebrew), *available at* http://www.globes.co.il/news/article.aspx?did=1000877411 (accessed Jul 14, 2014).
Orbach, L. "After 47 years: 'At' is separated from 'Ma'ariv' – sold to Yuval Sigler." *Globes*, June 2, 2014 (in Hebrew), *available at* http://www.globes.co.il/news/article.aspx?did=1000943708 (accessed Jul 14, 2014).
Pery, T. "The Film Industry is Alive and Kicking: 12.2 Million Tickets Sold in the Last Year and a Half." *Globes*, June 17, 2012 (in Hebrew), *available at* http://www.globes.co.il/news/article.aspx?did=1000756070 (accessed Jan 25, 2013).
Puppis, M. "Media Regulation in Small States." *International Communication Gazette*, 71:2, 2009, pp. 7–17.
Schejter, A. "From a Tool for National Cohesion, to a Manifestation of National Conflict: The Evolution of Cable Television Policy in Israel, 1986–1998." *Communication Law and Policy* 4:2, 1999, pp. 177–200.
Schejter, A. "Israeli cellular telecommunications policy." *Telecommunications Policy*, 30(1), 2006, pp. 14–28.
Schejter, A. *Muting Israeli Democracy: How Media and Cultural Policy Undermine Free Expression*. Urbana, IL: University of Illinois Press, 2009.
Schejter, A. and Cohen, A. (2002). "Israel: Chutzpah and Chatter in the Holy Land." Perpetual Contact. Katz, J. and Aakhus, M., eds. Cambridge: Cambridge University Press, 30–42.
Schejter, A. and Lee, S. "The Evolution of Cable Regulatory Policies and Their Consequences: Comparing South Korea and Israel." *Journal of Media Economics* 20:1, 2007. pp. 1–28.
Solan, Y. "How Do the Israeli Dailies Market Themselves (I)." *Otot* 85 69, 1985, pp. 30–31.
Tal, Y. and Ivry-Omer, D. *Regulation of Electronic Communications Services in Israel: The Need to*

*Establish a Communications Authority*. Jerusalem: Israel Democracy Institute, 2009.
Tokatly, O. (2000). *Communication Policy in Israel.* Tel Aviv: Open University, 2000.
Ynet. "Growth in Newspaper Advertising and Standstill in Television in 2000." *Ynet*, Jun 10, 2001 (in Hebrew), *available at* http://www.ynet.co.il/articles/1,7340,L-808150,00.html (accessed Jan 25, 2013).
Yudelovich, M. "The Ofer Brothers: A War Movie." *Ynet*, Aug 19, 2008 (in Hebrew), *available at* http://www.ynet.co.il/articles/0,7340,L-3583905,00.html (accessed Jan 25, 2013).
Zach, Y. "Chairman of Keshet's Board of Directors: We made a Mistake by Bringing Uvda over to Us." *NRG*, Dec 23, 2005, *available at* http://www.nrg.co.il/online/1/ART1/024/288.html (accessed Jan 25, 2013).

# Israel—Data Summaries

ELI NOAM AND PAUL MUTTER

SEVERAL THINGS STAND OUT in Israel's media concentration scores. Foremost is the country's high number of media voices. Even in absolute numbers, they are fairly high in international comparison. And on a per capita basis, given the small population size, the number of voices per capita is fourth highest in the world, and even higher if one considers that for the other top-ranked countries include media that spill over from across their borders. A second observation is that in 2004, Israel had the seventh-highest public ownership of media after China, India, Egypt, Russia, and Poland, and Switzerland. This was because 70% of Bezeq—with its 31.4% of the Israeli market—was publicly owned, along with the public broadcaster IBA. However, Bezeq has since then has been fully privatized, putting public ownership of Israeli media overall at just 1.9%, which is now low internationally.

Media concentration is high in Israel. Weighted Average C4 rose to 96% after 2008. This is due to heavy concentration in the telecommunications and ISP markets. The telecom provider Bezeq dominates the national media market, with a wireline dominance of 81.2%. It is also the third-largest content media company because of its multichannel and online news media holdings with 38% and 20% shares of those industries, respectively (Table 30.18). Bezeq's share of platforms has been declining, from 51.7% to 42.3% between 2004 and 2010, because it now faces greater challenge from its near competitors listed in Table 30.19, the wireless provider Cellcom Israel (owned by the industrial group IDB, with 17.3% of the platform media market) and the wireless/ISP company Partner Communications (with 19.3%).

Table 30-17. National Media Industries Concentration in Israel

| | 2004/5 | | 2011 OR MOST RECENT | | % CHANGE ANNUAL AVERAGE | |
|---|---|---|---|---|---|---|
| | COMPANY POWER INDEX IN COUNTRY | COMPANY SHARE OF THE OVERALL NATIONAL MEDIA MARKET (%) | COMPANY POWER INDEX IN COUNTRY | COMPANY SHARE OF THE OVERALL NATIONAL MEDIA MARKET (%) | COMPANY POWER INDEX IN COUNTRY | COMPANY SHARE OF THE OVERALL NATIONAL MEDIA MARKET (%) |
| Bezeq | 2,512 | 38.44 | 1,670 | 32.570 | −3.7 | −0.652 |
| Partner Communications | 466 | 14.756 | 403 | 14.246 | −1.5 | −0.057 |
| IDB (Cellcom) | 572 | 16.344 | 451 | 15.938 | −2.4 | −.0451 |
| 013 Netvision | 21 | 0.914 | IDB | | N/A | N/A |
| HOT | 500 | 10.116 | 543 | 14.833 | 0.95 | 0.524 |
| Channel 2 | 119 | 2.611 | 217 | 3.939 | 9.2 | 0.148 |
| Yediot Aharonot Group | 181 | 23.401 | 70 | 1.981 | −6.8 | −0.188 |
| Channel 10 | 15 | 0.795 | 44 | 1.601 | 22.0 | 0.090 |
| Google (US) | 6 | 0.120 | 72 | 0.743 | 120.0 | 0.069 |
| RGE Group | 28 | 1.2 | 21 | 1.276 | −2.7 | 0.010 |
| Israeli Broadcasting Authority (IBA) (public) | 49 | 1.6 | 15 | 0.875 | −7.7 | −0.079 |
| Ma'ariv Holdings | 56 | 2.0 | 3 | 0.331 | −2.4 | −0.185 |
| Haaretz | 5 | 0.588 | 4 | 0.405 | −3.4 | −0.020 |
| Altice | 8 | 1.868 | 10 | 2.028 | 3.97 | 0.018 |
| IDF (Public) | 8 | 0.294 | 10 | 0.338 | 1.6 | 0.005 |
| Israel HaYom | 0 | 0.0 | 5 | 0.4 | N/A | 0.05 |
| Kinneret Zemora | 2 | 0.14 | 2 | 0.15 | 0.01 | 0.0 |
| Globes | 1 | 0.282 | 0.8 | 0.223 | −2.6 | −0.007 |
| Time Warner (US) | 2 | 0.181 | 0.7 | 0.079 | −6.4 | −0.011 |

*continued*

Table 30-17. *continued*

| | 2004/5 | | 2011 OR MOST RECENT | | % CHANGE ANNUAL AVERAGE | |
|---|---|---|---|---|---|---|
| | COMPANY POWER INDEX IN COUNTRY | COMPANY SHARE OF THE OVERALL NATIONAL MEDIA MARKET (%) | COMPANY POWER INDEX IN COUNTRY | COMPANY SHARE OF THE OVERALL NATIONAL MEDIA MARKET (%) | COMPANY POWER INDEX IN COUNTRY | COMPANY SHARE OF THE OVERALL NATIONAL MEDIA MARKET (%) |
| Microsoft (Bing, US) | 3 | 0.098 | 0.1 | 0.026 | –10.9 | –0.008 |
| Yahoo (US) | 0.7 | 0.041 | 0 | 0.008 | –10.97 | –0.004 |
| Media Concentration Index | 2004/5 | | 2011 or Most Recent | | % Change Annual Average | |
| Total Revenue: Nat'l Media Industry (mil US$) | 6,375 | | 7,429 | | 1.84 | |
| Total Voices (*n*) | 64 | | 65 | | 0.26 | |
| Net Voices (*n*) | 45 | | 42 | | –0.62 | |
| Public Ownership (%) | 28.8 | | 1.2 | | –3.06 | |
| Foreign Ownership (%) | 1.7 | | 2.2 | | 0.06 | |
| C4 Average—Weighted | 96 | | 92 | | –0.43 | |
| HHI Average—Weighted | 2,100 | | 1,691 | | –2.2 | |
| C1 Average—Weighted | 52 | | 46 | | –1 | |
| Noam Index Average—Weighted | 1,478 | | 1,554 | | 1.3% | |
| Pooled Overall Sector C4 | 79.7 | | 74.4 | | –0.58 | |
| Pooled Overall Sector HHI | 2,100 | | 1,691 | | –2.2 | |
| Pooled Overall Sector Noam Index | 138 | | 143 | | 1% | |
| Market Share of Top Ten Companies: Nat'l Media Industry (%) (Pooled C10) | 92.6 | | 88.6 | | –0.4 | |
| National Power Index | 4,688 | | 3,601 | | –2.39 | |

Table 30-18. Top Content Media Companies in Israel

| | 2004/5 | | 2011 OR MOST RECENT | | % CHANGE ANNUAL AVERAGE | |
|---|---|---|---|---|---|---|
| | COMPANY POWER INDEX IN COUNTRY | COMPANY SHARE OF THE NATIONAL CONTENT MEDIA MARKET (%) | COMPANY POWER INDEX IN COUNTRY | COMPANY SHARE OF THE NATIONAL CONTENT MEDIA MARKET (%) | COMPANY POWER INDEX IN COUNTRY | COMPANY SHARE OF THE NATIONAL CONTENT MEDIA MARKET (%) |
| Channel 2 | 528.0 | 11.6 | 823.0 | 15.0 | 6.2 | 0.37 |
| Yediot Aharonot Group | 803.0 | 16.3 | 265.0 | 7.5 | –7.5 | –0.976 |
| HOT | 781.0 | 14.791 | 607.0 | 12.1 | –2.5 | –0.300 |
| Channel 10 | 65.0 | 3.53 | 166 | 6.064 | 17.2 | 0.282 |
| Google (US) | 27.0 | 0.531 | 272.0 | 2.815 | 100.7 | 0.254 |
| Bezeq | 174.0 | 6.125 | 233.0 | 8.996 | 3.8 | 0.319 |
| RGE | 124.0 | 5.258 | 80.0 | 4.834 | –3.9 | –0.047 |
| Israel Broadcasting Authority (public) | 219.0 | 7.057 | 57.0 | 3.314 | –8.2 | –0.416 |
| Ma'ariv Holdings | 249.0 | 8.880 | 11.0 | 1.255 | –2.4 | –0.847 |
| Haaretz (Shocken) | 23.0 | 2.61 | 13.0 | 1.536 | –4.6 | –0.119 |
| IDF (Public) | 37.0 | 1.306 | 37.0 | 1.279 | –0.27 | –0.003 |
| Israel HaYom | 0.0 | 0.0 | 17.8 | 1.5 | N/A | 0.17 |
| KinneretZemora | 7.0 | 0.6 | 7.0 | 0.6 | –0.7 | 0.0 |
| Globes (Fishman) | 5.0 | 1.250 | 3.0 | 0.845 | –3.9 | –0.045 |
| Jerusalem Post | 0.07 | 0.14 | 8.0 | 1.0 | 1235.9 | 0.1 |
| Microsoft (US) | 12.0 | 0.436 | 0.2 | 0.098 | –10.9 | –0.037 |
| Media Concentration Index | 2004/5 | | 2011 or Most Recent | | % Change Annual Average | |
| Public Ownership (%) | 12.7 | | 4.6 | | –0.9 | |
| Foreign Ownership (%) | 7.5 | | 8.5 | | 0.11 | |
| C4 Average—Weighted | 83 | | 76 | | –0.75 | |
| HHI Average—Weighted | 3,147 | | 2,842 | | 1.1 | |
| C1 Average—Weighted | 44 | | 41 | | 0 | |
| National Power Index | 3,134 | | 2,800 | | –1.18 | |

Table 30-19. Top Platform Media Companies in Israel

| | 2004/5 | | 2011 OR MOST RECENT | | % CHANGE ANNUAL AVERAGE | |
|---|---|---|---|---|---|---|
| | COMPANY POWER INDEX IN COUNTRY | COMPANY SHARE OF THE NATIONAL PLATFORM MEDIA MARKET (%) | COMPANY POWER INDEX IN COUNTRY | COMPANY SHARE OF THE NATIONAL PLATFORM MEDIA MARKET (%) | COMPANY POWER INDEX IN COUNTRY | COMPANY SHARE OF THE NATIONAL PLATFORM MEDIA MARKET (%) |
| Bezeq | 3,192 | 47.835 | 2,186 | 41.024 | -3.5 | -0.757 |
| Partner Communications | 601 | 19.047 | 547 | 19.355 | -1 | 0.034 |
| Cellcom Israel | 738 | 21.097 | 547 | 17.362 | -2.9 | -0.415 |
| HOT | 418 | 8.757 | 519 | 15.816 | 2.7 | 0.784 |
| 013 Netvision | 27 | 1.155 | 65 | 2.414 | 16.2 | 0.140 |
| Media Concentration Index | 2004/5 | | 2011 or Most Recent | | % Change Annual Average | |
| Public Ownership (%) | 33.5 | | 0.0 | | -3.72 | |
| Foreign Ownership (%) | 0.0 | | 0.0 | | 0% | |
| C4 Average—Weighted | 99 | | 97 | | -0.22 | |
| HHI Average—Weighted | 3,180 | | 2,623 | | -1.95 | |
| C1 Average—Weighted | 54 | | 47 | | -1 | |
| National Power Index | 5,011 | | 3,889 | | -2.49 | |

Several smaller cable providers have been gradually eroding IDB/Cellcom, Partner, and Bezeq's combined market share, their growth driving the 4.9% annual decline of Israel's national power index because concentration is decreasing in the telecom sector.

Although Bezeq's large overall national market share of 32% (Table 30.17) results from its strength in telecom platforms, it also has a strong presence in multichannel platforms (the DBS provider Yes) and online news media, with 38% and 20%, respectively. The cable firm HOT (a merger of three regional firms) controls all of cable, which accounts for the remaining 68% of multichannel. It is now controlled by the French investor and owner of Altice, Patrick Drahi. Ma'ariv Holdings and Yedioth Aharonot Group (Mozes) are two other content companies that have substantial shares in daily newspapers, multichannel, and online news media. Channel 2 and Channel 10 collectively control 84% of TV broadcasting.

Foreign ownership is low (0.6%) but rising. Revenue losses for newspapers and television broadcasters led to an increase in outside investment since 2007. The entry of ideologically inspired investors from the United States such as Ron Lauder and Sheldon Adelson from the right, and from Germany (DuMont Schauberg 20%) from the left is noteworthy because Israel has had (and still does have) a very low rate of foreign ownership of its media industries. Lauder's acquisition of Channel 10 gives him the largest stake in the broadcast television market in Israel, and Adelson's establishment of the *Israel HaYom* free daily newspaper in 2007 redrew the map of the newspapers industry, severely undercutting the traditional business models of paid dailies like *Haaretz* and *Yediot Aharonot*, which are associated with family ownership and are fierce rivals themselves. In contrast, HOT's Patrick Drahi keeps a low profile.

# 31

# Media Ownership and Concentration in South Africa

GEORGE ANGELOPULO AND PETRUS H. POTGIETER

## INTRODUCTION

This chapter covers a significant period and arguably the most interesting time in the evolution of South African media. During the initial period of assessment (1984–94), South African media operated under a system of pervasive state intervention and racial segregation known as apartheid. These restrictive policies were most evident in the electronic and broadcasting sectors, but these practices were to some degree manifested in all media markets with respect to ownership and state censorship. Following the first full democratic elections held in 1994, extensive economic liberalization and editorial diversification occurred, but state interventionism and ideological influences from the apartheid era remain factors in the structure, composition, and diversity of the South African media landscape.

There is a substantial, if incomplete, body of industry data and publicly accessible information that has made this study possible. But there are a number of challenges in producing an assessment of media concentration in South Africa: a lack of accessible market information in some media industries and a lack of comparable information across some sectors. Some of the data used in the study give an incomplete view of a particular media sector. For example, advertising revenue does not represent all revenue sources for newspapers or broadcasters, while unique browsers per month only hint at the revenue generation of online news portals through other means. This chapter's application of the C4, HHI, and Noam Index ratios across South Africa's media landscape does, however, establish a baseline measurement of concentration and diversity and serves as a benchmark for further studies.

Concentration in South African media markets has attracted wide research attention, but the focus of most studies is on issues peripheral to concentration per se, its measurement or economic specification. Media concentration has, for example, been considered in relation to freedom of expression (Bussiek, 2006), marketing (Hunter, 2006), individual media sectors and companies (Bauer, 2009; Hadland and Thorne, 2004; Hlengani et al., 2006), the flow of information (Band, 1992), democratization (Botma, 2011; Tomaselli, 2003), employment demographics, management, and ownership (Leshin, 2010), and government relations and control (Wasserman, 2005). The study described in this chapter therefore addresses a significant blind spot in the current view of South Africa's media industry.

Horwitz and Currie (2007) have described how, subsequent to the first democratic elections in 1994, liberalization was essentially abandoned in favor of an aggressive policy of privatization that has benefited entities and individuals close to the incumbent African National Congress (ANC) political party.[1] Less pronounced, but similar, developments have occurred in broadcasting and other areas, where private companies that were influential before 1994 maintained their position in return for creating some space for companies that were either owned by or very close to members of the government or ruling party. This, in practice, reflects the national economic policy and is evident in the position of the company MultiChoice, which is part of Naspers, South Africa's leading media group. MultiChoice, initially licensed during the apartheid era by the now-defunct National Party, maintained a pay-TV monopoly until very recently. The company is still keen on maintaining close political ties with members of the government today though: MultiChoice itself funded the sale in 2006 of a minority stake to a group led by a former Independent Communications Authority of South Africa (ICASA) chief regulator whose brother was Chief Justice of South Africa at the time and was an active member of the ANC.

The state has always been strongly involved in media, especially in radio and television broadcasting services accessed by lower income groups, in the ISP market and in telecommunications in general. The regulatory authority for the broadcasting and telecommunications industries in South Africa is currently ICASA, which was established in 2000 in a merger between the South African Telecommunications Regulatory Authority (SATRA) and the Independent Broadcasting Authority (IBA), the body responsible for broadcasting regulation following the first democratic elections, with the Postal Regulator also incorporated in 2005. Prior to 1994, broadcasting fell under the direct authority of the Department of Home Affairs, and the later changes arose in order to ensure independence in the governance and regulation of the broadcasting and telecommunications industries.[2] These changes were required, as in many other countries, by the World Trade Organization (WTO), which South Africa joined in 1995. South Africa's Competition Commission, operating under the Competitions Act of 1998, was established after the country's WTO ascension, but the body has so far not been active in the media sector except for minor forays into telecommunications. The trends in the degree of concentration within the telecommunications and broadcasting industries described below have to a large extent been determined by the policies and their implementation by ICASA and the institutions that preceded it.

Although media consumption is mainly in English, there is a lively Afrikaans-language market for all kinds of media. Other languages have a much smaller market share, except for radio and music in general. Zulu is the only other language with a print daily (the paid tabloid *Isolezwe*) that has a modest circulation. There is so far very little online content in languages other than Afrikaans and English.

South Africa now has a program of Black Economic Empowerment (BEE) that seeks to promote ownership and control in the mainstream economy by individuals who were

1. Horwitz, Robert B. and Currie, W. "Another instance where privatization trumped liberalization: The politics of telecommunications reform in South Africa—A ten-year retrospective." *Telecommunications Policy* 31: 8–9, 2007. 445–462.

2. "The Independent Communications Authority of South Africa." *ICASA*, 2011. July 4, 2012. <http://www.icasa.org.za/>; "National Association of Broadcasters." *NAB*, 2011. July 4, 2012. <http://www.nab.org.za>.

disenfranchised before 1994, their descendants, as well as the descendants of immigrants who would have been disenfranchised before 1994.[3] The ICASA has been free to impose its own additional rules in this regard. Although the enforcement powers of the state are limited when it comes to the operation of small private companies, they are quite stringent in any operation for which a license is required. Research in other industries has found that the main effect of these rules has been to entrench the interests of established media producers (those predating the end of apartheid in 1994) and to restrict new market entrants.[4] It is often the case that licenses are granted to all applicants who qualify, but the qualification process can be quite problematic and requires resources that are only available to large, often incumbent operators.

## PRINT MEDIA

### Newspapers

Four media groups dominate the South African newspaper industry: Media24, Caxton/CTP, Independent News & Media, and Avusa. While ownership of these groups has changed somewhat since 1984, the companies and their constituent newspapers have on the whole remained dominant in the South African market.

Media24 publishes over 50 newspaper titles, including the *Daily Sun* tabloid, South Africa's highest selling daily. The company's majority shareholder is Naspers, a media group that was formed in 1915. Media24 was established as an independent company in 2000 when Naspers's newspapers and other print media interests were divested from their original holding company, Nasmedia.[5]

Caxton/CTP publishes over eighty titles, a large number of which are free community newspapers that derive the bulk of their revenue from advertising. The company originated in 1902 and only entered the newspapers business in 1961. It went public in the 1940s, ownership changed twice in the 1960s, and it acquired a newsprint firm in the 1980s. It has since formed a partnership with the National Empowerment Consortium (NEC), one of the first companies formed to advance the BEE campaign, and merged with Perskor, another print publisher, in the 1990s. Caxton/CTP is a diversified print media company that also draws revenue from printing, book publishing, stationery, and packaging enterprises.[6]

Independent News & Media is the largest publisher of newspapers in South Africa and is part of the Independent News & Media (INM) Group founded by the Irish media magnate Tony O'Reilly.[7] It accounts for a third of all newspapers sold in the country and more than half of the newspapers published in English. Its South African operations were originally held by the Anglo-American Argus Group, which INM acquired in the 1990s, first with the purchase of a third of the shares and later with the acquisition of the remaining balance.

Avusa is a media and entertainment company with significant newspaper holdings in South Africa. The company originated with two newspapers: *The Rand Daily Mail* in 1902 and *The Sunday Times*, in 1906 (only the *Times* is still in circulation). Avusa was formed from Johnnic Communications, a company formed with the purchase of Times Media by the NEC from an Anglo-American company in the 1990s.

English-language newspapers evolved in parallel with the mining industry and were largely funded by dominant mining interests. The original newspapers in what is now the INM stable were, for example, funded by the diamond magnate Cecil Rhodes, and today's Avusa Group evolved from a number of newspapers originally owned by the mining financier Abe Bailey.

---

3. Schneiderman, D. "Promoting equality, Black economic empowerment, and the future of investment rules." *South African Journal on Human Rights* 25:2, 2009. 246–279.
4. Ponte, S. and van Sittert, L. "The chimera of redistribution in post-apartheid South Africa: 'Black Economic Empowerment' (BEE) in industrial fisheries." *African Affairs* 106:424, 2007. 437–462.
5. "Media24." *Media24*, 2011. July 4, 2012. <http://www.media24.com/>.
6. "The Caxton and CTP Group." *Caxton*, 2011. June 15, 2012. <http://www.caxton.co.za>.
7. "Independent News & Media PLC." *INM*, 2011. July 4, 2012. <http://www.inmplc.com/>.

The Afrikaans-language press developed independently as an extension of Afrikaner nationalism and as a counter to British hegemony in politics and the economy.

While the companies mentioned above may be seen as the dominant drivers in the development of South Africa's newspaper industry, three other significant sections of the newspaper industry should be noted. The indigenous press for black Africans originated in the early 1800s but suffered throughout the years leading up to democracy due to a lack of resources and capital. Second, a comparatively small but politically important group of newspapers—sometimes closely associated with the first group—began publication from the 1950s onward with the intention of giving voice to anti-apartheid sentiments. Of these, what is now The Mail & Guardian, a newspaper owned by M&G Media, is one example, along with *The Sowetan* daily, which is now part of Avusa. Third, as mentioned above, the South African newspaper landscape is characterized by a large number of free community newspapers such as those published by Caxton/CTP.

South Africa's newspaper industry has always reflected the prevailing economic, political, and social conditions of the day, and this is as true today as it has been in the past. While there have always been publications regarded as government outlets, newspapers have consistently given voice to the incumbent government's opposition, both pre- and post-1994, and even during the period of censorship that prevailed before 1994. Prior to 1994, the press was subjected to varying degrees of government control and censorship—*The Rand Daily Mail* was sometimes forced to withdraw reports that reflected poorly on the National Party, and *The World*, a paper for black South Africans, was shut down in the late 1970s—but the press has since 1994 subsequently enjoyed the greatest period of freedom in its history, as press freedom is now guaranteed in the South African Constitution. From 2010, however, there had been sustained attempts by government to introduce legislation to govern access to, and dissemination of, information, with the first such bill being legislated in November 2011. Many believe that such legislation will significantly retard media freedom in South Africa, and indeed the demand for newspapers.

The market has shown some stagnation in the growth of traditional newspapers but significant growth for tabloids and local newspapers. Overall, growth in the market is supported by rapid population growth, increased literacy and access to newspapers, and greater disposable income. Despite the fundamental changes that have occurred in the ownership of newspapers since 1984, the underlying structure of the market has not changed significantly. In terms of C4, the newspapers market in South Africa has remained consistently concentrated since 1984, and at a relatively high level. The HHI and Noam Index measures also indicate high levels of concentration and moderate diversity as shown in Table 31.1.

## Book Publishing

By the end of the nineteenth century, a number of small publishers had become established in the then capital of the Cape Colony (South Africa), Cape Town. By the early 1900s, other, mainly British, publishers had also established operations in what had by then become the Union of South Africa. The 1930s and 1940s saw the rise in primarily Afrikaans-language publishers, and later, between 1948 and 1994, several anti-apartheid publishers entered the market. Since 1994, there has been a significant change in local publishing companies' ownership and control, largely as a reflection of the BEE campaign and the entry of international publishers.[8]

The trend identified for book publishing in South Africa in this study is for the relatively short period of 2004–2008, for which very limited industry information is available. In the absence of primary data, industry concentration is assessed with secondary data, and the findings certainly contain some margin of error. The findings have, however, been included in order to offer a provisional insight into the sector.

8. DACST. "The South African Publishing Industry." Pretoria: Department of Arts, Culture, Science and Technology, 1998.

Table 31-1. Daily Newspapers (Market Shares by Circulation), 1997–2011

| | 1997[1] | 2000 | 2004[2] | 2008 | 2011 |
|---|---|---|---|---|---|
| Media24 (Naspers) | 24.4 | 26.7 | 31.4 | 36.1 | 35.0 |
| Independent News and Media Group (O'Reilly Family, Ireland) | 35.5 | 36.1 | 31.2 | 27.5 | 26.5 |
| Avusa | 21.1 | 20.5 | 16.8 | 17.2 | 17.7 |
| Caxton/CTL | 15.0 | 13.5 | 15.1 | 15.0 | 14.3 |
| Community Dailies | 1.2 | 0.9 | 1.6 | 2.2 | 1.7 |
| M&G Media Ltd. | 0.5 | 0.4 | 1.3 | 0.9 | 1.0 |
| Capital Media | 0.2 | 0.2 | 1.0 | 0.8 | 1.7 |
| TNA Media | | | | | 1.5 |
| Others | 2.2 | 1.6 | 1.7 | 0.5 | 0.7 |
| Total Revenue (mil ZAR) | 1,817 | 2,392 | 3,687 | 6,441 | 7,479 |
| Total Revenue (mil US$) | 423 | 345 | 572 | 781 | 1,031 |
| C4 | 96 | 97 | 94 | 96 | 93 |
| HHI | 2,524 | 2,622 | 2,477 | 2,584 | 2,450 |
| *N* (> 1%) | 5 | 4 | 6 | 5 | 7 |
| Noam Index | 1,130 | 1,311 | 1,011 | 1,156 | 926 |

*Source:* AIS/The Nielsen Company, South African Advertising Research Foundation.

[1] For newspapers, 1997 is assessed instead of 1996 because comprehensive data was not available prior to this date.

[2] Revenue from inserts in newspapers is only included from 2004 onward.

Book publishing in South Africa covers a wide spectrum that includes trade, education, academic, and mixed publishing sectors, and retail sales account for most of the revenue. Information on the size and composition of the South African book industry is not directly accessible as published industry research expressly excludes the competitive position of individual companies. By 2011, the largest publishing houses were Pearson in the academic sector, Maskew Miller Longman and Heinemann in the textbook market, LexisNexis in legal and accounting publications, and Random House Struik in trade publications.

Between 2002 and 2005, about a third of the country's large publishers were locally owned, a third internationally owned, and a third had mixed local/international ownership, while sizeable foreign ownership was also evident among small and medium-sized publishers.[9] By the end of the period of assessment the size of the sector by net turnover was upward of US$512 million (ZAR3.2 billion) and growing, but at a slower rate than either GDP growth or inflation.[10]

The findings suggest that by all three measures used in this study there was a reduction in media concentration in the publishing industry between 2004 and 2008, based on the number of titles registered by publishers with South Africa's Legal Deposit Library. All three indices suggest a significant reduction in media concentration and an increase in diversity over this period. Data collection was changed in the final year of assessment making it incomparable with

9. Galloway, F., Venter, R. M. R., and Bothma, T. P. "ASA industry surveys. Executive summary: Broad trends over four years (2002–2005)." Pretoria: School of Information Technology, University of Pretoria, 2006.

10. Calculated using the turnover of 95% of large companies, 63% of medium companies, and 29% of small companies in the industry. *See* Galloway, F. and Struik, W. "Annual book publishing industry survey report 2008." Pretoria: School of Information Technology, University of Pretoria, 2009.

Table 31-2. Book Publishing (Market Shares by Revenue), 2007–2009

| | 1984 | 1988 | 1992 | 1996 | 2000 | 2004 | 2008 | 2011 |
|---|---|---|---|---|---|---|---|---|
| Total Revenue (mil ZAR) | | | | | | 1,756 | 3,206 | |
| Total Revenue (mil US$) | | | | | | 272 | 389 | |
| C4 | | | | | | 55 | 31 | |
| HHI | | | | | | 903 | 460 | |
| *N* (> 1%) | | | | | | 17 | 22 | |
| Noam Index | | | | | | 219 | 98 | |

*Source:* South African Legal Deposit Library, Publishers' Association of South Africa.

the 2004 and 2008 data. For this reason, only these two years are included in Table 31.2.

## Magazine Publishing

South African magazine publishers are mostly self-owned. The largest media owner by number of titles and revenue is Naspers's Media24, which publishes over 60 magazine titles. With its 58% stake in New Media Publishing, the second largest publisher by titles, Media24 controls over 80 titles.[11] Other major companies in the magazine industry are Avusa and Caxton/CTP. Avusa's magazine division includes popular consumer titles and trade publications for a range of industries. Caxton/CTP publishes 10 monthlies and 3 weeklies and has a 30% stake in another (though smaller) magazine publisher, RamsayMedia.[12] Highbury Safika Media, which was established in 2005 by a consortium of Highbury Monarch Communications, Safika Investments, and RMB Corvest, publishes approximately 30 titles.

Magazines operate in a relatively open, competitive market in South Africa. Roughly half the market is serviced by the larger magazine companies, and these account for over 60% of advertising spend. Media24, the largest company in the sector, alone receives between 33% and 43% of the entire sector's total advertising expenditure.

The magazine market, if assessed by individual titles, may be described as unconcentrated, with great diversity throughout the period of assessment and a trend toward lower concentration. By ownership, however, it becomes evident in Table 31.3 that concentration is higher.

# AUDIOVISUAL MEDIA

## Radio

Regular national broadcasting began in 1924, and the Schlesinger Corporation—which was founded as a cinema group—established African Broadcasting Corporation (ABC) in 1927 as the single national commercial broadcaster. In 1936, the government purchased the ABC and renamed it the South African Broadcasting Corporation (SABC), modeling it on the UK's BBC services.[13] Radio stations were developed for the individual language groups in the country, and the first new commercial station was launched in 1950, after which commercial regional music stations were introduced in the following decade.[14]

11. "Media24." *Media24*, 2011. July 4, 2012. <http://www.media24.com/>; "New Media Publishing." *New Media Publishing*, 2011. July 4, 2012. <http://www.newmediapub.co.za/Home>.

12. MDDA. "Trends of Ownership and Control of Media in South Africa." Johannesburg: Media Development and Diversity Agency, 2009; "RamsayMedia." *RamsayMedia*, 2011. July 4, 2012. <http://www.ramsaymedia.co.za>.

13. Rosenthal, E. *You Have Been Listening . . . The Early History of Radio in South Africa*. Cape Town: Purnell, 1974.

14. "IBIS Media Manager. Media history." *IBIS*, 2011. July 25, 2012. <http://www.mediamanager.co.za/medialibrary/mediahistory/sector/broadcast>.

Table 31-3. Magazine Publishing (Market Shares by Revenue), 1997–2011

| | 1997[1] | 2000 | 2004 | 2008 | 2011 |
|---|---|---|---|---|---|
| Media24 (Naspers) | 33.2 | 37.4 | 42.8 | 38.9 | 39.5 |
| Avusa | 13.3 | 12.2 | 10.5 | 11.1 | 8.7 |
| Caxton/CTP | 6.3 | 5.0 | 6.3 | 7.1 | 7.2 |
| RamsayMedia | 6.4 | 7.5 | 7.8 | 6.7 | 5.6 |
| Associated Magazines | 3.6 | 4.5 | 4.7 | 4.4 | 4.8 |
| Condé Nast (Advance Publications, US) | 0.0 | 2.5 | 2.9 | 3.3 | 3.1 |
| Primedia[2] | 0.8 | 0.8 | 2.0 | 1.9 | 1.1 |
| Mafube Pub | 2.4 | 1.7 | 1.8 | 1.9 | 1.6 |
| New Media Publishing | 0.8 | 0.2 | 0.8 | 1.7 | 2.4 |
| Martin Creamer | 1.9 | 3.6 | 1.5 | 1.3 | 1.2 |
| Overdrive Publishing | | | | 1.1 | 1.3 |
| The Publishing Partnership | | | | 1.1 | 1.0 |
| Bike SA | 0.5 | 0.9 | 0.9 | 1.0 | 0.8 |
| Others | 30.8 | 23.9 | 17.9 | 18.6 | 21.8 |
| Total Revenue (ZAR mil ZAR) | 921 | 1,194 | 1,738 | 2,680 | 2,728 |
| Total Revenue (mil US$) | 214 | 172 | 269 | 325 | 376 |
| C4 | 59 | 62 | 67 | 64 | 61 |
| HHI | 1,435 | 1,698 | 2,091 | 1,787 | 1,780 |
| *N* (> 1%) | 7 | 8 | 9 | 13 | 12 |
| Noam Index | 542 | 600 | 697 | 496 | 514 |

*Source:* AIS/The Nielsen Company, South African Advertising Research Foundation.

[1] For magazines, 1997 is assessed instead of 1996 because comprehensive data was not available prior to this date.

[2] Primedia is a South African company and unrelated to companies of the same name operating elsewhere in the world.

The SABC maintained its monopoly on radio transmitted within the borders of South Africa. It was, however, exposed to some competition that originated beyond its borders. A number of primarily South African-financed private stations were launched from neighboring countries such as Mozambique and Swaziland with stations that included LM Radio and Swazi Music Radio, and from the Bantustans of Transkei and Bophuthatswana within its borders, with private stations such as Capital Radio and Radio 702.[15] Inadequate funding, poor signal strength and reliance on transmission systems with limited ranges prevented these stations from competing on an equal footing with the SABC network.

Competition was also provided by the public broadcasters of the "black African homelands" set up as segregated cantonments for the indigenous population of South Africa: Bop Radio (Bophuthatswana), Radio Thohoyandou (Venda), and Radio Ciskei (Ciskei), among others.[16] In 1985, the SABC undertook a competitive strategy to improve its market shares. It closed Springbok Radio, its first commercial radio station,

15. "Radio in South Africa." *MediaClubSouthAfrica*. 2011. July 4, 2011. <http://www.mediaclubsouthafrica.com/index.php?option=com_content&view=article&id=118%3ARadio+in+South+Africa&catid=36%3Amedia_bg&Itemid=54>.

16. "IBIS Media Manager. Media history." *IBIS*, 2011. July 25, 2012. <http://www.mediamanager.co.za/medialibrary/mediahistory/sector/broadcast>; "Radio in South Africa." *MediaClubSouthAfrica*. 2011. July 4, 2012. <http://www.mediaclubsouthafrica.com/index.php?option=com_content&view=article&id=118%3ARadio+in+South+Africa&catid=36%3Amedia_bg&Itemid=54>.

launched Radio Metro to compete with the "homeland" stations, and restructured a number of its older radio stations.[17] After 1994, the radio industry underwent a fundamental change. The Independent Broadcasting Authority (IBA), created to regulate the broadcasting industry independently of government, introduced a three-tiered broadcasting system that allowed for commercial, public, and community radio.[18] Under the IBA's new regulations, SABC was to remain a public broadcaster. In practice, as the national public broadcaster, it had still not relinquished many of its commercial operations and continued in part as a commercial broadcaster. This anomaly was addressed in the relicensing of the SABC in 2005, when its operation was reorganized into commercial and public divisions. In 1995, the first community radio stations were launched. These are not-for-profit radio stations that are based in communities and owned by those communities, with "community" defined by geographical area or common interest such as language or religion. As part of the market's restructuring by the IBA, six of its stations were privatized in 1996 and in 1997 the IBA awarded eight new commercial radio licenses. In 2000, the IBA merged with the Telecommunications Authority to become the Independent Communications Authority of South Africa (ICASA), the body that currently regulates South African broadcasting.[19]

Up until 1996, radio was concentrated in the hands of the South African Broadcasting Corporation (SABC), and while the SABC is still the market leader, two new groups have emerged: Primedia and Kagiso Media. From its origins as the owner of Radio 702, Primedia now has holdings in four radio stations and has evolved into a multimedia group owned by its management, the Mineworkers Investment Company, the Kirsh Consortium, and Brait SA.[20] Kagiso Media was founded in 1997 and has developed into a listed investment holding company with interests in a number of radio stations in which it has varying degrees of control.[21] Other companies with significant but smaller holdings in radio are African Media Entertainment and Hosken Consolidated Investments Ltd.

With minor exceptions, both HHI and the Noam Index ratios in Table 31.4 reflect an incremental reduction in concentration over the period of the study, and C4 has also only declined slightly since 1996.

## Broadcast Television

South Africa's terrestrial broadcasting industry was from 1984 to 1998 largely monopolized by the national public broadcaster, the South African Broadcasting Corporation (SABC). The SABC's television service was first launched in 1976 with SABC 1. It was originally funded through license fees, but advertising was added to its revenue stream in 1978, a situation that remains unchanged. In 1982 a second SABC channel, SABC 2, was added, and a third in 1998 (SABC 3).[22] The first alternative station to the SABC's was Bop TV, which started out as the public television station of Bophuthatswana. Although it was the first challenger to SABC's monopoly, Bop TV never had a major impact on South African broadcast television in terms of market or audience shares. Launched in 1984, Bop TV was eventually centralized under the SABC's authority in 1997 and shut down in 2003.[23]

The state's overall television monopoly in South Africa, though, ended with the launch of the M-Net subscription television channel in

17. "IBIS Media Manager. Media history." *IBIS*, 2011. July 25, 2012. <http://www.mediamanager.co.za/medialibrary/mediahistory/sector/broadcast>.

18. "Radio in South Africa." *MediaClubSouthAfrica*. 2011. July 4, 2012. <http://www.mediaclubsouthafrica.com/index.php?option=com_content&view=article&id=118%3ARadio+in+South+Africa&catid=36%3Amedia_bg&Itemid=54>.

19. "National Association of Broadcasters." *NAB*, 2011. July 4, 2012. <http://www.nab.org.za>.

20. "Primedia." *Primedia*, 2011. Last accessed on July 4, 2012. <http://www.primedia.co.za>.

21. "Kagiso Media." *Kagiso Media*, 2011. July 4, 2012. <http://www.kagisomedia.co.za/kagisomedia/content/en/about-us>.

22. SABC. "About the SABC: Mandate." SABC, 2011. July 4, 2012. <http://www.sabc.co.za/wps/portal/SABC/SABCMANDATE>.

23. "IBIS Media Manager. Media history." IBIS, 2011. July 25, 2012. <http://www.mediamanager.co.za/medialibrary/media-history/sector/broadcast>; "Media Statement." info.gov.za: South African Government Information, 2003. July 25, 2012. <http://www.info.gov.za/speeches/2003/03071116111001.htm>.

Table 31-4. Radio Group (Market Shares by Revenue), 1984–2011

| | 1984 | 1988 | 1992 | 1996 | 2000 | 2004 | 2008 | 2011 |
|---|---|---|---|---|---|---|---|---|
| SABC (public) | 84.4 | 45.3 | 61.6 | 64.1 | 55.0 | 45.0 | 45.2 | 31.6 |
| Kagiso Media | 0.0 | 0.0 | 0.0 | 0.0 | 16.5 | 24.6 | 28.5 | 21.2 |
| Primedia | 15.6 | 32.7 | 25.7 | 34.6 | 19.1 | 22.7 | 18.1 | 31.1 |
| African Media Entertainment | | | | | 3.4 | 2.5 | 3.8 | 11.2 |
| Hosken Consolidated Investments Ltd. | 0.0 | 0.0 | 0.0 | 0.0 | 2.6 | 2.2 | 3.0 | 2.3 |
| Community Stations[1] | | | | | | 0.6 | 0.7 | 2.0 |
| Classic FM | | | | | 1.2 | 2.4 | 0.6 | 0.6 |
| Bophuthatswana Broadcasting Corp. (public) | | 8.7 | 6.3 | 1.3 | 0.0 | 0.0 | 0.0 | 0.0 |
| Punt Radio | | | | | 2.1 | | | |
| Independent Stations | | 13.3 | 6.4 | | | | | |
| Total Revenue (mil ZAR) | 68.4 | 61.6 | 178 | 323 | 1,226 | 1,930 | 3,345 | 4,528 |
| Total Revenue (mil US$) | 46.4 | 27.1 | 62.4 | 75.1 | 178 | 299 | 405 | 624 |
| *N* (> 1%) | 2 | 3 | 3 | 3 | 7 | 6 | 5 | 6 |
| C4 | 100.0 | 86.7 | 93.6 | 100.0 | 94.0 | 95.0 | 96.0 | 95.0 |
| HHI | 7,368 | 3,197 | 4,495 | 5,313 | 3,690 | 3,161 | 3,209 | 2,553 |
| Noam Index | 5,210 | 1,846 | 2,595 | 3,068 | 1,395 | 1,290 | 1,435 | 1,142 |

*Source:* AIS/The Nielsen Company, South African Advertising Research Foundation.

[1] For 2000 and earlier, there were no individual statistics for community radio stations, a group of nonprofit, cooperative local stations.

1986, when it was permitted a free-to-air time slot between 17:00 and 19:00 daily, a practice which continued until 2007 when this free-to-air window was discontinued entirely.[24] During this time, M-Net is best described as a subscription service with an encoded channel, despite the small opportunity that it had in free-to-air (for this reason it is not reflected in Table 31.5 data, but is included in Tables 31.6 and 31.7). The conversion of analog to digital terrestrial television (DTT) was to have begun in 2013.

Following initial steps toward full liberalization in the broadcast industry in the mid-1990s, the new MIDI Television network was licensed to operate under the name e.tv, becoming the first significant over-the-air competitor to the SABC. e.tv was launched in 1998 and became the second largest terrestrial broadcaster by 2001, a position that it has held ever since.[25] Hosken Consolidated Investments Ltd. has a majority stake (63%) in e.tv.[26]

## Multichannel TV Platforms

In 1986, a consortium of print publishers were awarded a license to operate the pay-TV service M-Net, which was not permitted to broadcast news programming (as this was retained as the sole domain of the SABC for political purposes) at first. In 1993, M-Net's customer service divisions were split apart to form MultiChoice, and then, in 1996 MIH Holdings. MultiChoice

24. "Icasa calls time on M-Net's free window." *IOL*, 19 Mar. 2005. July 25, 2012. <http://www.iol.co.za/news/south-africa/icasa-calls-time-on-m-net-s-free-window-1.236807>; "M-Net: History." M-Net, 2011. July 4, 2012 <http://www.mnetcorporate.co.za/ArticleDetail.aspx?Id=82>.

25. "e.tv profile." e.tv, 2011. July 4, 2012. <http://www.etv.co.za/pages/about_us>.

26. "Hosken Consolidated Investments Ltd." *HCI*, 2011. July 4, 2011. <http://www.hci.co.za/corp_profile.htm>.

## Table 31-5. TV Broadcasting (Market Shares by Revenue), 1984–2011

| | 1988 | 1992 | 1996 | 2000 | 2004 | 2008 | 2011 |
|---|---|---|---|---|---|---|---|
| SABC (public) | 99.0 | 98.7 | 99.6 | 89.0 | 74.8 | 69.2 | 67.6 |
| e.tv (Hosken Consolidated Investment Ltd.) | 0.0 | 0.0 | 0.0 | 10.9 | 25.2 | 30.9 | 32.4 |
| Bophuthatswana Broadcasting Corporation (public) | 1.0 | 1.0 | 1.3 | 0.4 | 0.0 | 0.0 | 0.0 |
| Total Revenue (mil ZAR) | 345 | 771 | 1,017 | 2,250 | 4,466 | 6,784 | 8,888 |
| Total Revenue (mil US$) | 152 | 270 | 237 | 324 | 692 | 822 | 1,182 |
| C4 | 100.0 | 100.0 | 100.0 | 100.0 | 100.0 | 100.0 | 100.0 |
| HHI | 9,805 | 9,735 | 9,917 | 8,045 | 6,230 | 5,734 | 5,618 |
| *N* (> 1%) | 2 | 2 | 2 | 2 | 2 | 2 | 2 |
| Noam Index | 6,933 | 6,884 | 7,012 | 4,645 | 4,405 | 4,054 | 3,973 |

*Source:* AIS/The Nielsen Company, South African Advertising Research Foundation.

## Table 31-6. Multichannel Video Platforms: DBS (Market Shares by Revenue), 1984–2011

| | 1984 | 1988 | 1992 | 1996 | 2000 | 2004 | 2008 | 2011 |
|---|---|---|---|---|---|---|---|---|
| DStv (MultiChoice Africa-Naspers) | 0.0 | 100.0 | 100.0 | 100.0 | 100.0 | 100.0 | 100.0 | 100.0 |
| Total Revenue (mil ZAR) | 0.0 | 51 | 274 | 633 | 756 | 1,111 | 3,180 | 5,796 |
| Total Revenue (mil US$) | 0.0 | 23 | 96 | 147 | 109 | 172 | 385 | 799 |
| C4 | 0.0 | 100.0 | 100.0 | 100.0 | 100.0 | 100.0 | 100.0 | 100.0 |
| HHI | 0.0 | 10,000 | 10,000 | 10,000 | 10,000 | 10,000 | 10,000 | 10,000 |
| *N* (> 1%) | 0 | 1 | 1 | 1 | 1 | 1 | 1 | 1 |
| Noam Index | 0.0 | 10,000 | 10,000 | 10,000 | 10,000 | 10,000 | 10,000 | 10,000 |

*Source:* AIS/The Nielsen Company, South African Advertising Research Foundation.

began broadcasting the satellite network DStv in 1995. DStv now transmits multiple channels on satellite, including news, entertainment, children's, sport, business, education, and religious programming, as well as SABC's terrestrial programming and foreign content from channels such as RAI (Italy) and Deutsche Welle (Germany).[27] The majority shareholder of M-Net and DStv is Naspers, but the nature of the group's shareholding has changed since 1994, with BEE transactions resulting in the transfer of 20% of Naspers's shares in MultiChoice to Phuthuma Nathi Investments.[28] Until 2008, M-Net/MultiChoice's channels were the only satellite broadcasting channels available, South Africa does not have any cable TV providers; cable services are a monopoly reserved for the wireline operator Telkom, which has so far

27. "DStv Channels." *DStv*, 2011. July 4, 2012. <http://www.dstv.com/Pages/Channels/>.

28. "The media in South Africa." *MediaClubSouthAfrica*. 2011. July 4, 2012. <http://www.mediaclubsouthafrica.com/index.php?option=com_content&view=article&id=110:the-media-in-south-africa&catid=36:media_bg%20>; "MultiChoice Investment Centre." *MultiChoice*, 2011. July 4, 2011. <http://www.multichoice.co.za/multichoice/view/multichoice/en/page54>.

Table 31-7. Combined Video Network Industries: Broadcast TV Networks, TV Syndicators, Pay-TV (Market Shares by Revenue) 1984–2011

| | 1984 | 1988 | 1992 | 1996 | 2000 | 2004 | 2008 | 2011 |
|---|---|---|---|---|---|---|---|---|
| SABC (public) | | 86.0 | 73.0 | 61.0 | 67.0 | 60.0 | 47.0 | 41.0 |
| DStv (MultiChoice Africa-Naspers) | 0.0 | 13.0 | 26.0 | 38.0 | 25.0 | 20.0 | 31.0 | 39.0 |
| e.tv (Hosken Consolidated Investment Ltd.) | 0.0 | 0.0 | 0.0 | 0.0 | 8.0 | 20.0 | 22.0 | 20.0 |
| Bophuthatswana Broadcasting Corporation (public) | | 1.0 | 1.0 | 0.0 | 0.0 | 0.0 | 0.0 | 0.0 |
| Total Revenue (mil ZAR) | 209 | 397 | 1,045 | 1,650 | 3,006 | 5,577 | 9,965 | 14, 683 |
| Total Revenue (mil US$) | 142 | 175 | 366 | 384 | 433 | 865 | 1,208 | 2,024 |
| C4 | | 100.0 | 100.0 | 100.0 | 100.0 | 100.0 | 100.0 | 100.0 |
| HHI | | 7,598 | 5,990 | 5,240 | 5,140 | 4,392 | 3,659 | 3,616 |
| *N* (> 1%) | | 3 | 3 | 2 | 3 | 3 | 3 | 3 |
| Noam Index | | 4,387 | 3,459 | 3,705 | 2,968 | 2,536 | 2,112 | 2,088 |

*Source:* AIS/The Nielsen Company, South African Advertising Research Foundation.

chosen not to develop the network for television programming.

The only significant competition DStv currently faces is from On Digital Media's TopTV. The SABC attempted to enter the pay-TV arena with two channels, AstraPlus and AstraSport, in 1997, but both channels ceased broadcasting due to inadequate levels of funding. In 2007, four new broadcast licenses were awarded to Walking on Water, On Digital Media, a satellite service of e.tv called e-Sat and Telkom Media, a division of Telkom, the national telecommunications operator. On Digital Media's TopTV now competes directly against DStv. Telkom's Super 5 Media has not yet launched- Telkom sold its 75% stake to Shenzen Media, and Super 5 Media has obtained a number of launch date extensions.[29] e-Sat opted to operate as a content provider instead of an MSO, and Walking on Water has also not launched at the time of writing.

## Combined Video Network Industries

In the South African context, it is best to assess concentration in the television industry as a whole. The reason is that both broadcast and DBS services are so highly concentrated. The dynamics between these sectors, however, is particularly interesting, and more indicative of the general trend in the concentration of the television industry in South Africa.

Television was introduced comparatively late in South Africa because it was widely seen by the government as a threat to the apartheid system. The Minister of Posts and Telegraphs in the early 1960s denounced television as "a miniature bioscope over which parents would have no control . . . showing race-mixing" and as "the devil's own box, for disseminating communism and immorality."[30] The government passed legislation for the introduction of television only in 1971, and five more years passed until SABC 1 was launched. SABC remained the sole legal terrestrial broadcaster until 1986, when M-Net was launched (MultiChoice's DStv entered service in 1995). In 1998, e.tv, the first private terrestrial television channel was launched. Although four new broadcasting licenses were awarded in 2007, only one of these has so far gone on air, doing so in 2011. The consolidated market has evolved over the period

29. Editor. "Super 5 Media wants more time." *TechCentral*, Feb 15, 2011. July 4, 2012. <http://www.techcentral.co.za/super-5-media-wants-more-time/21116.

30. Editor. "The story behind Sentech's iconic tower." *TechCentral*, Oct 27, 2010. July 25, 2012. <http://www.techcentral.co.za/the-story-behind-sentechs-iconic-tower/18447>.

of assessment from very highly concentrated to a condition of low but increasing diversity.

## Film

South Africa's first cinema houses (Africa's Amalgamated Theatres and the Empire Theatres Company) were bought out by the Schlesinger Corporation, which dominated film importation and distribution in South Africa until the late 1950s. In 1931, Schlesinger merged its interests with Kinemas, establishing African Consolidated Films and African Consolidated Theatres under the African Film Productions (AFP) banner. A decade later 20th Century Fox bought out AFP, renaming it South African Screen Productions. In 1969, the insurance company Sanlam, which owned the film producer/distributor Ster Theatres and Films, bought out 20th Century Fox's South African division and formed Suid Afrikaanse Teaterbelange Beperk (Satbel), which operated as Kinekor due to regulatory concerns. When television was introduced in 1976, cinema attendance dropped significantly and the government permitted the merger of the two entities as Ster-Kinekor. Ster-Kinekor was subsequently acquired by Interleisure in the late 1980s and finally, by Primedia in 1997. Ster-Kinekor Theatres is now the largest exhibitor in South Africa.

South Africa's second largest exhibitor is the NuMetro Cinema Group, which was established in 1987 with the acquisition of United International Pictures' (UIP) South African distributor, Metro Group.[31] The new group became part of CNA Gallo, which listed on the stock exchange as MEGA.[32] Its interests were fully acquired by Avusa's predecessor, Johnnic, in 1998, and the company is now a subsidiary of Avusa. NuMetro operates 24 multiplexes across South Africa with 196 screens. Avusa also controls NuMetro Home Entertainment, which markets film content for home use and represents a number of international studios and brands such as Warner Brothers Home Video and the BBC.

HHI and the Noam Index demonstrate that from 1984, the market has been characterized by a period of consistency with slightly increasing concentration as shown by Table 31.8.

Since the 1980s, Ster-Kinekor, NuMetro, and UIP have been the primary film distributors in the country. Ster-Kinekor Entertainment's market share has declined from 50% in 1984 to 34% in 2011, UIP's market share has fallen from 50% in 1984 to 29% in 2011 and NuMetro's market share

### Table 31-8. Film Exhibition (Market Shares by Box Office %), 1996–2011

| | 1996 | 1996 | 2004 | 2008 | 2011 |
|---|---|---|---|---|---|
| Ster-Kinekor Theatres (Primedia) | 69.0 | 58.0 | 57.0 | 60.3 | 61.0 |
| Nu Metro Theatres (Avusa) | 23.0 | 27.0 | 33.0 | 26.3 | 27.0 |
| Independent Exhibitors | 8.0 | 15.0 | 10.0 | 13.3 | 12.0 |
| Total Revenue (mil ZAR) | 330 | 445 | 656 | 617 | 764 |
| Total Revenue (mil US$) | 77 | 64 | 102 | 75 | 105 |
| C4 | 100.0 | 100.0 | 100.0 | 100.0 | 100.0 |
| HHI | 5,355 | 4,318 | 4,437 | 4,511 | 4,594 |
| *N* (> 1%) | 3 | 3 | 3 | 3 | 3 |
| Noam Index | 3,092 | 2,493 | 2,562 | 2,604 | 2,652 |

*Source:* Ster-Kinekor Entertainment.

31. DeBarros, L. "An assessment of emerging independent filmmaking in the Gauteng region: Relating to difficulties in training, funding and distribution." safilm.org: SA Film, 1995. July 25, 2012. <http://www.safilm.org.za/reading/thesis.html>.

32. DACST. "The South African Film and Television Industry Report." Pretoria: Department of Arts, Culture, Science and Technology, 1998.

Table 31-9. Film Distribution (Market Shares by Box Office %), 1996–2011

| | 1996 | 2000 | 2004 | 2008 | 2011 |
|---|---|---|---|---|---|
| Ster-Kinekor Entertainment (Primedia) | 50.0 | 57.0 | 37.0 | 31.0 | 34.0 |
| United International Pictures (UIP) (Paramount Pictures: Viacom, US) (Universal: Comcast/GE, US) | 50.0 | 21.0 | 22.0 | 29.0 | 29.0 |
| Nu Metro Film Distribution (Avusa) | 0.0 | 22.0 | 41.0 | 40.0 | 37.0 |
| Total Revenue (mil ZAR) | 330 | 445 | 656 | 617 | 617 |
| Total Revenue (mil US$) | 77 | 64 | 102 | 75 | 85 |
| C4 | 100.0 | 100.0 | 100.0 | 100.0 | 100.0 |
| HHI | 5,000 | 4,174 | 3,534 | 3,402 | 3,366 |
| *N* (> 1%) | 3 | 3 | 3 | 3 | 3 |
| Noam Index | 2,887 | 2,410 | 2,040 | 1,964 | 1,943 |

*Source:* Ster-Kinekor Entertainment.

has risen to 37%, as seen in Table 31.9. The three distributors represent the major international film distributors in South Africa. Ster-Kinekor, for example, distributes Walt Disney and Sony Pictures productions, while NuMetro distributes 20th Century Fox and Warner Brothers' productions.

From its inception the local industry has produced films that have enjoyed varying degrees of success at the South African box office, but these generally contribute only a small percentage to total box office takings in the South African market, the bulk of the revenue being generated by imported films, mainly from the Hollywood majors in the United States.

## TELECOMMUNICATIONS MEDIA

### Wireline Telecom

By the mid-1980s, the inefficiencies of the state-run system set up in the 1940s and 1950s had become abundantly clear, with a tremendous backlog and long waiting times even for residential and business lines in the major urban centers. As part of a general liberalization of the economy, a fully state-owned company, Telkom, was split off from the South African Post Office—which had managed fixed telephony operations since 1958—in 1991. As in many other countries, long-distance call rates had previously subsidized line rental and local calls, and Telkom had to address this imbalance, especially as callback services for international destinations made inroads in the 1990s.

In 1997, the government sold 30% of its stake in Telkom to SBC Communications, an American "Baby Bell" company that would eventually become AT&T, and Malaysia Telekom, and that year Telkom was granted a license under the Telecommunications Act by the then telecom regulator, SATRA. It was argued that Telkom required an extended period of protection from competition in order to expand telephone services, especially in previously disenfranchised areas. Despite this protection, expansion did not occur to any significant extent.[33] In 2002 Telkom's exclusivity license expired but it retained its monopoly and political influence. In 2005, for example, 15.1% of Telkom was acquired by the Elephant Consortium, led by the outgoing Director General of the Department of Communications, a former SBC/Telkom executive and a senior government official. By 2007,

33. Gillwald, A. "Good intentions, poor outcomes: Telecommunications reform in South Africa." *Telecommunications Policy* 29: 7, 2005. 469–491; Horwitz, Robert B., and Currie, W. "Another instance where privatization trumped liberalization: The politics of telecommunications reform in South Africa—A ten-year retrospective." *Telecommunications Policy* 31: 8–9, 2007. 445–462.

SBC Communications and Malaysia Telekom had completely sold their shares in Telkom and the government currently retains a large (but still a minority) stake in the company.

The ANC's telecommunications' policies since 1994 have been marred by incoherence and the failure to introduce competition to Telkom, "in part because the ANC leadership has been loath to trust democratic structures outside of its immediate control," according to Horwitz and Currie.[34] The total number of wireline connections declined by 2011 to pre-1997 levels, with just over 4 million lines in service, giving a wireline tele-density of well below 10%.[35] Wireless substitution has no doubt played a role, and for many consumers there is little advantage in adopting a wireline service with fixed monthly fees and small savings on most calls. The network externality that is realized in advanced economies from expansive wireline deployment remains unrealized.

A Second National Operator (SNO), intended to be Telkom's competitor, was finally licensed in 2005. Neotel (controlled by the Indian telecom major Tata Communications) has so far had a very limited impact on the consumer market. It has a CDMA wireless network on which it provides "fixed" wireless services with geographic numbers and without roaming between area codes. It provides wireline service to business premises as well as servicing a niche Internet access market. The wireline telephony market in South Africa remains in practice a complete monopoly, with the attendant inefficiencies, high prices, and a highly politicized business model heavily influenced by ANC politics.

The wireline telephony market, although being challenged in a small way by VoIP services (running mainly on Telkom's ADSL network), is still fully concentrated in the incumbent, Telkom, giving the highest possible score on all measures of concentration in Table 31.10.

## Wireless Telecom

Prior to 1994, Telkom had run an analog car phone network that was decommissioned when the GSM networks were launched. Just prior to the 1994 elections, the outgoing National Party government licensed two GSM cellular networks (MTN and Vodacom). The two GSM networks rapidly expanded consumer access to voice services, multiplying the number of people with access to telephones several fold. The success of mobile wireless

### Table 31-10. Wireline Telecom (Market Shares by Subscribers), 1984–2011

| | 1984 | 1988 | 1992 | 1996 | 2000 | 2004 | 2008 | 2011 |
|---|---|---|---|---|---|---|---|---|
| Telkom[1] | 100.0 | 100.0 | 100.0 | 100.0 | 100.0 | 100.0 | 100.0 | 100.0 |
| Total Subscribers (1,000) | 3,890 | 4,743 | 3,524 | 4,259 | 4,962 | 4,726 | 4,451 | 3,995 |
| Total Revenue[2] (mil ZAR) | 1,629 | 3,426 | 10,600 | 13,320 | 26,700 | 40,500 | 56,300 | 33,079 |
| Total Revenue (mil US$) | 1,104 | 1,508 | 3,717 | 3,100 | 3,850 | 6,279 | 6,823 | 4,563 |
| C4 | 100.0 | 100.0 | 100.0 | 100.0 | 100.0 | 100.0 | 100.0 | 100.0 |
| HHI | 10,000 | 10,000 | 10,000 | 10,000 | 10,000 | 10,000 | 10,000 | 10,000 |
| *N* (> 1%) | 1 | 1 | 1 | 1 | 1 | 1 | 1 | 1 |
| Noam Index | 10,000 | 10,000 | 10,000 | 10,000 | 10,000 | 10,000 | 10,000 | 10,000 |

*Source:* Office of the Postmaster General, Telkom.

[1] Neotel and VoIP service have small market shares.

[2] Total revenue is given as total Telkom revenue from all sources. Up to 2008, this included income from Vodacom.

34. Ibid., 445.

35. Muller, R. "Telkom's copper line battle." June 13, 2011. *MyBroadband*, July 13, 2011. July 25, 2012. <http://mybroadband.co.za/news/telecoms/26477-telkom%E2%80%99s-copper-line-battle.html>.

telephony in South Africa has been spectacular. Vast areas of the country that had previously been without service of any kind are now fully covered by mobile networks. Vodacom and MTN were licensed in spite of opposition from the ANC.

Telkom initially had a 50% share in Vodacom, which was very profitable for the wireline monopolist until it divested from Vodacom in 2008–2009. Release of a new frequency (1,800 MHz) and new number ranges were supposed to be delayed until the Third Mobile Operator (TMO) had been licensed, and that license was finally awarded to Cell C (controlled by Oger Telecom, a division of Saudi Oger Ltd.) in a highly politicized process which was directly connected to an arms transaction involving the Government of Saudi Arabia.[36]

The profitability of the Vodacom/MTN duopoly was supported from the start by some of the highest mobile interconnection rates in the world, though these rates have become reduced slightly. Nevertheless, large unsatisfied demand among the majority of the population and the lackluster performance of Telkom meant that the high interconnection rates did not prevent spectacular growth in access to wireless telephony services, although the number of actual monthly minutes per consumer remains fairly low. All of the large telecommunications companies have significant involvement from either the state, state-owned

## Table 31-11. Wireless Telecom (Market Shares by Revenue), 1988–2011

| | 1988 | 1992 | 1996 | 2000 | 2004 | 2008 | 2011 |
|---|---|---|---|---|---|---|---|
| Vodacom (Vodafone, UK) | 0.0 | 0.0 | 58.0 | 56.9 | 50.8 | 53.7 | 50.9 |
| Subscribers (1,000) | 0 | 0 | 553 | 5,108 | 1,284 | 2,763 | 3,170 |
| MTN (MTN Group) | 0.0 | 0.0 | 42.0 | 43.1 | 40.5 | 33.4 | 33.7 |
| Subscribers (1,000) | 0 | 0 | 400 | 3,877 | 10,235 | 17,169 | 20,968 |
| Cell C (Oger Telecom-Saudi Oger Ltd., Saudi Arabia) | 0.0 | 0.0 | 0.0 | 0.0 | 8.7 | 12.4 | 13.2 |
| Subscribers (1,000) | 0 | 0 | 0 | 0 | 2,197 | 6,400 | 8,200 |
| Virgin Mobile (Richard Branson, Virgin Group, UK)[1] | 0.0 | 0.0 | 0.0 | 0.0 | 0.0 | 0.5 | 0.5 |
| Subscribers (1,000) | 0 | 0 | 0 | 0 | 0 | 250 | 300 |
| Telkom/8ta[2] (Telkom) | 100.0 | 100.0 | 0.0 | 0.0 | 0.0 | <0.1% | 1.8 |
| Subscribers (1,000) | 3 | 13 | 0 | 0 | 0 | 5 | 11,403 |
| Total Subscribers (1,000) | 3 | 13 | 953 | 8,985 | 2,527 | 5,145 | 6,231 |
| Total Revenue (mil ZAR) | | | | | 45,380 | 87,410 | |
| Total Revenue (mil US$) | | | | | 7,040 | 10,590 | |
| C4 | 100.0 | 100.0 | 100.0 | 100.0 | 100.0 | 100.0 | 99.5 |
| HHI | 10,000 | 10,000 | 5,129 | 5,094 | 4,297 | 4,152 | 3,898 |
| *N* (> 1%) | 1 | 1 | 2 | 2 | 3 | 3 | 4 |
| Noam Index | 10,000 | 10,000 | 3,627 | 3,602 | 2,481 | 2,397 | 1,949 |

*Source:* MTN, MyBroadband, Office of the Postmaster General, Telkom, Frost & Sullivan.

[1] Virgin Mobile is an MVNO JV with Cell C: both companies are privately held and data were compiled from press and other reports.

[2] From 2008 on, the new Telkom network had a much smaller footprint than the rest and was in fact dropped in 2011 for an entirely new network called 8ta, which has a roaming agreement with MTN.

36. Feinstein, A. *After the Party: Corruption, the ANC and South Africa's Uncertain Future*. 2nd ed. London: Verso, 2010.

companies, or through organizations and individuals with strong links to the ANC (with the exception of Neotel).

Concentration remained high from 1993 to 2001, when its level was reduced with the launch of Cell C, the third operator in the market. By 2011 there were two national GSM networks (MTN and Vodacom) and three other infrastructure networks with coverage in the metropolitan areas (Telkom/8ta, Cell C, and Neotel) plus roaming agreements with the national operators (in the case of Telkom/8ta and Cell C) in rural areas. In the case of Cell C, the regulator mandated the roaming agreement but Telkom's 8ta had actually started out by searching for a roaming partner on a commercial basis. Neotel provides a completely mobile wireless Internet access service on its CDMA network, but otherwise provides fixed wireless telephony services only.

HHI indicates steadily decreasing degrees of concentration, with the most significant period of reduction occurring between 1992 and 1996 due to the market entry of MTN and Vodacom. The next significant, but smaller drop occurred with the launch of Cell C in 2001, and this trend continued to 2008. The same, but slightly more accentuated pattern is evident in the Noam Index in Table 31.11, which fell from a level of 10,000 in 1992, to 3,627 in 1996, and then to 1,857 by 2008.

## INTERNET MEDIA

### Internet Service Providers (ISP)

For South African consumers, Internet access is almost universally bundled with (wireline or wireless) telephony, or is provided as a stand-alone product on a wireless telephony network. Consumer Internet platforms are in practice no different from the telecom distribution networks (fiber optic lines are not widely distributed either for residential or commercial use). The City of Johannesburg is planning to allow some consumer access to its new fiber optic network through resellers, but efforts driven by local authorities have in the past not proved to be very successful. There is some commercial provision of wireless access over WiMax using high-powered WiFi, but this practice takes place in a legal limbo that has discouraged further expansion. Very high access charges for ADSL service by Telkom have spurred the growth of community WiFi networks, notably for gamers in the area of Pretoria and Johannesburg.

South Africa has only the fourth largest number of Internet users in Africa, despite having the largest economy on the continent.[37] Although the numbers are rather imprecise, South Africa's position behind Nigeria, Morocco, and Egypt suggests that the country's policy of "managed liberalization" and its emphasis on expanding access beyond traditionally privileged communities have not been wholly successful. Nevertheless, the impact of smart phones has been considerable and wireless networks are responsible for the majority of private connections. As a result of the economics of wireless networks (shared spectrum in a given locality), the total Internet traffic per user is particularly low for South Africa, as per-megabyte pricing on UMTS networks remains relatively high due to the scarcity of spectrum(s).

In principle, South Africa has a converged and liberal licensing framework. However, issues that at the time of writing hang in the air include the finalization of a WiMax auction and the proper allocation of spectrum in those bands (since 2008), and the unbundling of the local loop, a process that has been going on since the late 1990s. Lack of movement in these areas has undermined the impact of the Altech court case of 2008. The case proceeded from a newly instituted licensing system and concerned the conversion of existing licenses to the new framework: the ICASA intended to make 600 licenses available but some of

37. "Africa Internet Usage and Population Statistics." *Internet World Stats*, 2011. July 25, 2012. <http://www.internetworldstats.com/stats1.htm#africa>.

the license holders wished to receive a license that would allow them to build and operate their own national or local networks. These applicants won, and the ICASA was forced to convert the licenses to these broader "ECNS" licenses. This resulted in a burst of activity in the rollout of trunk network capacity, which has allowed the national network to expand significantly.

Nevertheless, consumer Internet service provision remains mainly the business of the dominant wireless providers and the sole wireline provider, Telkom. However, there is a market in a customized bitstream access environment, where ISPs provide a data stream to clients of Telkom's ADSL service. The client pays a monthly fee to Telkom and an additional charge for a mandatory wireline voice service.[38] ISPs then provide a range of access products on top of that, ranging from per gigabyte (GB) pricing to all-you-can-eat data plans (as of 2010), although the latter have their traffic aggressively managed, sometimes even through "rolling caps" which will restrict users to 20 gigabytes of data for any 10-day period. By December 2010, only 731,500 ADSL connections were active in the country, with the majority connecting at only 384 kbps.[39] From a population of some 50 million, this represents a very low level of wireline broadband penetration since there is next to no fiber or cable access in the country. South Africa's high national crime rates also impact ISPs: only gated communities usually offer fiber optic service to their residents, and the high-capacity market for households is restricted to the Telkom ADSL service, which is affected by frequent theft of copper wiring.

The Internet Service Providers Association (ISPA) of South Africa has expressed grave concern about the Regulation of Interception of Communications Act (RICA) that has been implemented in South Africa, which requires Internet service providers to verify the identity of every customer in-person.[40] Only very large ISPs can comply with RICA by establishing branch offices to receive customers. All but the largest ISPs offer online signup and instantly activated accounts on a national basis, which makes the market for data access as competitive as the market for line rental is a monopoly. RICA also effectively requires Internet cafés to do the same, and outlaws casual WiFi use. Should full implementation of RICA go ahead, it would be a tremendous setback for competition in the ISP industry and expanding Internet access.

The provision of Internet service by ADSL, cable, and fiber was completely monopolized by Telkom's ADSL service during the period of the study. Three significant groups of players in the broader industry have been omitted from Table 31.12 however:

i. Dial-up Internet service providers (especially in the 1990s);

ii. Providers of Internet content to ADSL subscribers of the Telkom network (most of which have grown out of the first category); and

iii. Wireless service providers, including the mobile networks.

All three categories play a significant role. The first two categories are very similar (the second having in the main replaced the first during the period of the study) and the third is fast becoming the dominant way to access the Internet, even on traditional computers and at home. However, subscriber data for wireless services are difficult to distinguish from casual users of very modest data services on their cellular handsets, which many networks report as their "data users."

38. "Telkom SA Limited—DSL." Telkom SA, 2011. July 4, 2012. <http://www.telkom.co.za/products_services/dsl/cost_dsl_cost.html>.

39. Muller, R. "Telkom releases latest ADSL subscriber numbers." *MyBroadband*, Dec 22, 2010. July 25, 2012. <http://mybroadband.co.za/news/adsl/17431-Telkom-releases-latest-ADSL-subscriber-numbers.html>.

40. Editor. "ISPs want rethink on Rica." *TechCentral*. June 13, 2011. July 4, 2012. <http://www.techcentral.co.za/isps-want-rethink-on-rica/23753/>.

## Table 31-12. Internet Service Providers: ADSL, Cable, Fiber Optic (Market Shares by Subscribers), 2004–2011

| | 2004 | 2008 | 2011 |
|---|---|---|---|
| Telkom | 100.0 | 100.0 | 100.0 |
| Total Subscribers (1,000) | 58 | 548 | 827 |
| Total Revenue (mil ZAR) | N/A | N/A | N/A |
| Total Revenue (mil US$) | N/A | N/A | N/A |
| C4 | 100.0 | 100.0 | 100.0 |
| HHI | 10,000 | 10,000 | 10,000 |
| *N* (> 1%) | 1 | 1 | 1 |
| Noam Index | 10,000 | 10,000 | 10,000 |

*Source:* Telkom.

Note: Residential infrastructure wired Internet service providers. There has been no unbundling of the local loop in South Africa and there is an insignificant amount of optical fiber and no cable provision.

## Table 31-13. Internet Search Engines (Market Shares by Search Volume), 1996–2011

| | 2009 | 2011 |
|---|---|---|
| Google (US) | 93.4 | 91.8 |
| Bing (Microsoft, US) | 3.9 | 3.3 |
| Yahoo! (US) | 2.1 | 3.0 |
| Ask (US) | 0.3 | 0.4 |
| AltaVista (US) | 0.1 | |
| MSN (Microsoft, US) | 0.0 | |
| AOL (US) | 0.0 | |
| Others | 0.1 | 1.6 |
| Total Revenue (mil ZAR) | N/A | N/A |
| Total Revenue (mil US$) | N/A | N/A |
| C4 | 99.7 | 100.0 |
| HHI | 8,751 | 8,438 |
| *N* (> 1%) | 3 | 3 |
| Noam Index | 5,052 | 4,872 |

*Source:* NetMarketShare.

## Search Engines

During the early days of commercial consumer Internet, South African search engines such as Aardvark and Ananzi held significant market shares, especially as US providers such as Yahoo! and MSN did not initially localize their search results. Since then, however, Google has come to dominate the search market and provides its main interface in several South African languages. The three most popular websites in South Africa are Google.co.za, Facebook, and Google.com (Table 31.13).[41]

41. "Alexa—Top Sites in South Africa." *Alexa*. 2011. July 4, 2012. <http://www.alexa.com/topsites/countries/ZA>.

## Online News

The provision of online news is dominated by the traditional news publishers, Naspers (News24) and the Independent News & Media Group (INM) in Table 31.14. The bilingual Naspers news portals News24/Nuus24 (English and Afrikaans, respectively) are the market leaders but the group also has websites for individual titles like *Beeld* and *Die Burger*, a pattern also followed by the INM with iol.co.za as general news portal and individual sites for the *Pretoria News*, *The Star*, and other papers.

# CONCLUSION

South Africa's media landscape is characterized by great diversity when viewed as a whole, with a myriad of individual titles, publishers, stations, or sites available to the media consumer. If one considers the media by category, however, it is evident that most are dominated by between one and four companies. In general, print media is characterized by the lowest degree of concentration and the highest degree of diversity, although a number of companies are dominant within the industry. Media24, INM, Avusa, and Caxton/CTP account for the largest share

Table 31-14. Online News Media (Market Shares by Monthly Unique Visitors), 2009–2011

| | 2009 | 2011 |
|---|---|---|
| Media24 (Naspers) | 27.8 | 32.6 |
| MSN (Microsoft, US)[1] | 15.0 | 24.5 |
| iol.co.za (O'Reilly Family, INM, Ireland) | 13.6 | 21.3 |
| Avusa/TIMESlive | 11.1 | 10.2 |
| TEAMtalk Media | 7.8 | |
| MyBroadband | 5.5 | 11.4 |
| Primedia Online | 4.3 | |
| Mail & Guardian Online (UK) | 4.0 | |
| Moneyweb Holdings | 3.1 | |
| SABC (public) | 2.7 | |
| Creamer Media (Pty) Ltd. | 2.3 | |
| ITWeb Limited | 1.6 | |
| Business News Online | 0.4 | |
| South Africa: The Good News | 0.4 | |
| The Daily Maverick | 0.2 | |
| Cape Media | 0.1 | |
| PRAAG | 0.1 | |
| Total industry revenue (mil ZAR) | N/A | N/A |
| Total industry revenue (mil US) | N/A | N/A |
| C4 | 67.6 | 89.8 |
| HHI | 1,458 | 2,352 |
| *N* (> 1%) | 12 | 5 |
| Noam Index | 421 | 1,052 |

*Source:* Digital Media and Marketing Association (DMMA), Nielsen Online Ranking Report (2009).

[1] We have included the MSN portal in the study since it carries local news as well as wire services like Reuters and the AFP.

of the market; Media24 and Avusa hold significant shares in magazine publishing as well; and in book publishing, it is assumed (but cannot be confirmed by this study) that a small number of publishers hold strong positions in the sectors of the book publishing industry in which they operate.

In audiovisual media, there is a high degree of concentration across all sectors. In broadcast television, the SABC and e.tv are still the only broadcasters. Pay-TV (DBS) in South Africa is a monopoly dominated by Naspers's MultiChoice's DStv. Film has two dominant players in Ster-Kinekor and Avusa's Nu Metro. It is only in radio that a diversity of stations exists, but the greatest share of market still resides within three media groups: the SABC, Kagiso Media, and Primedia.

Telkom, the original wireline service provider in which the state holds a 39.8% minority stake, is still the wireline monopolist but now faces increased competition in wireless from Vodacom and MTN.[42] MTN began operating beyond South Africa's borders at an early stage, and it now operates successfully across the rest of Africa and the Middle East. Telkom and Vodacom have also operated within South Africa and beyond its borders during the period of this assessment. Vodacom has a wider subscriber base than MTN in South Africa, but MTN's international operations are more extensive than those of Vodacom, which had limited international exposure due to agreements undertaken with Vodafone (UK), its major shareholder. Online media is dominated by the print media giants, especially Naspers's Media24, although there are smaller independent news sites, including MyBroadband and The Daily Maverick.

One characteristic of the media that is evident in this study is that growth has continued in all sectors since 1994, except for fixed line telephony. With some exceptions, the entrenched patterns of concentration remain little altered over the period of assessment, particularly in industries that are relatively free from regulation and where the cost of entry is relatively high, such as newspapers, film, and magazines. It is only in sectors that had monopoly status, such as consolidated television and wireless telecom, where significant changes have become evident when competition has been introduced. In these sectors it appears that the degree of concentration will settle into a fixed range unless the sector is again structurally altered with the introduction of new competition, or significant changes in regulatory policies. This does not, however, appear to be the case where a new entrant appears to be a comparatively weaker alternative to the incumbents, as appears to have been the case with the Cell C and Virgin Mobile entries to the mobile phone market.

While it is difficult to draw any conclusions from the broad digital sphere that includes mobile telephony and online media because of the paucity of data on the latter, the findings do suggest a pattern that may ensue with the introduction of new providers and media producers. The initial incumbents appear to consolidate their market shares very swiftly in markets that are relatively restricted (MTN and Vodacom in mobile telephony); while in completely open markets, the field remains highly competitive for a period, with a small number of competitors then moving to the fore and establishing highly concentrated conditions (for example, the film industry prior to the late 1950s, and search engines and even news portals in the present).

The impact of the Internet on media competition has been relatively muted. The Internet has certainly affected the scale and nature of media consumption in South Africa very deeply, but its effects have essentially been limited to the most economically active sections of the South African population, leaving a large majority of the population untouched. Further, existing media companies entered the Internet arena in the areas of content, bandwidth, and infrastructure provision, although it must be said that a parallel industry has developed in support of these companies with the provision of technology.

42. "Telkom investor relations overview." Telkom SA, 2011. July 4, 2012. <https://secure1.telkom.co.za/ir/overview/shareholding.html>.

In contrast to the Internet, mobile telephony has had an unprecedented impact on the population as a whole. Certainly more so than in other developed countries, the introduction of mobile telephony has had a significant positive effect on the quality of life of almost every South African citizen. It has enabled a degree of personal and business interaction that would not have been possible by any other medium, and it is no exaggeration to say that mobile telephony has had a significantly positive effect on economic conditions in the country as well.

South Africa's small middle class is the primary consumer of online media, though there is a fair amount of strictly mobile content consumption on relatively basic cellular handsets, the diffusion of which has been encouraged by a market in which high handset subsidies have always prevailed (even for prepaid customers). At the higher income end, traditional print and audiovisual media face increasing competition from online media, but this is mitigated by two factors:

- Robust population growth and steady economic growth support a growing market that generates new titles in daily newspapers and magazines especially aimed at the lower end,
- Naspers's Media24 is very strongly placed in the local content market especially with its news portals News24.com (and its Afrikaans equivalent, Nuus24.com).

Readership of foreign newspapers online is widespread among the urban educated classes who might even prefer such publications to local ones.

The South African consumer is faced with a broad, modern range of media. Since the main language of commerce and inter-ethnic communication is English, access to the products of the main international markets is common. There is a small but vibrant market for films produced in India and Nigeria, which cater to niche markets in the country, such as South Africa's South Asian diaspora community. Newspapers and magazines reprint a lot of foreign content directly (or in Afrikaans), and the shelves are well stocked with local editions of foreign publications. Foreign print media consists mainly of UK and US publications, both imported and localized. There is also a print market serving the large Zimbabwean expatriate community. Popular television series and films are mainly imported from the United States. Local content rules apply to radio and television and in the case of radio these have been very supportive of the local music industry, whereas the rebroadcasting of old material often satisfies these criteria.

Throughout the period covered in this study, many of the issues have centered on state control and censorship. The advent of a liberal constitution in 1994, and, in fact, the period since Nelson Mandela's release from jail in 1990, did considerably improve the diversity and freedom of media and information in South Africa. Nevertheless, high levels of state ownership and intervention are still significant concerns.

The five main issues that suggest a trend of greater media concentration in South Africa are the continued dominance SABC and Telkom; the incipient attempts to muzzle the media through new legislation for the "protection of state information"; the anticompetitive aspects of the ownership redistribution policy through Black Economic Empowerment (BEE) Legislation; the high barriers to entry in most classes of media; and the regulatory agency ICASA's competence in exercising its mandate more efficiently and with less political interference.

## Bibliography

### 1. Books and Articles

Band, D. "Media Ownership and the Flow of Information." *Nieman Reports* 46:4, 1992.

Bauer, T. "A democratized market? Development of South Africa's daily newspapers 1990–2006." *Global Media Journal—African Edition* 3:1, 2009. 1–17.

Botma, G. J. "Going back to the crossroads: Visions of a democratic media future at the dawn of the new South Africa." *Ecquid Novi: African Journalism Studies* 32:2, 2011. 75–89.

Bussiek, H. "African Media Barometer: South Africa." *Rhodes Journalism Review* 6:26, 2006. 31–40.

"Cell C subscribers up 32% to 2.9m in FY 2005." *MyBroadband*, Apr 25, 2006. Feb 5, 2013. <http://mybroadband.co.za/nephp/2590.html>.

DeBarros, L. "An assessment of emerging independent filmmaking in the Gauteng region: Relating to difficulties in training, funding and distribution." safilm.org: SA Film, 1995. July 25, 2012. <http://www.safilm.org.za/reading/thesis.html>.

DMMA. "Top websites in South Africa for June 2011." dmma.co.za, July 12, 2011. Feb 8, 2013. <http://www.dmma.co.za/news/top-websites-in-south-africa-for-june-2011/>.

Editor. "ISPs want rethink on Rica." *TechCentral*. June 13, 2011. July 4, 2012. <http://www.techcentral.co.za/isps-want-rethink-on-rica/23753/>.

Editor. "Super 5 Media wants more time." *TechCentral*, Feb 15, 2011. July 4, 2012. <http://www.techcentral.co.za/super-5-media-wants-more-time/21116.

Editor. "The story behind Sentech's iconic tower." *TechCentral*, Oct 27, 2010. July 25, 2012. <http://www.techcentral.co.za/the-story-behind-sentechs-iconic-tower/18447>.

e.tv. "e.tv profile." e.tv, 2011. July 4, 2012. <http://www.etv.co.za/pages/about_us>.

Feinstein, A. *After the Party: Corruption, the ANC and South Africa's Uncertain Future*. 2nd ed. London: Verso, 2010.

Gillwald, A. "Good intentions, poor outcomes: Telecommunications reform in South Africa." *Telecommunications Policy* 29:7, 2005. 469–491.

Hadland, A., and Thorne, K. *The People's Voice: The Development and Current State of the South African Small Media Sector*. Cape Town: HSRC Publishers, 2004.

Hlengani, T., Ryan, B., and Fife, I. "Whole sector is printing money." *Financial Mail* 185: 9, 2006.

Horwitz, Robert B. and Currie, W. "Another instance where privatization trumped liberalization: The politics of telecommunications reform in South Africa—A ten-year retrospective." *Telecommunications Policy* 31:8–9, 2007. 445–462.

Hunter, C. "The ever-increasing cost of media." *Journal of Marketing*, 2006. 42–43.

Leshin, M. "Writing change: affirmative action as a tool for newspaper transformation at the Durban Post." *Global Media Journal*—African *Edition* 4:1, 2010. 51–72.

"Media Statement." info.gov.za: South African Government Information, 2003. July 25, 2012. <http://www.info.gov.za/speeches/2003/03071116111001.htm>.

M-Net. "M-Net: History." M-Net, 2011. July 4, 2012. <http://www.mnetcorporate.co.za/ArticleDetail.aspx?Id=82>.

Muller, R. "Telkom releases latest ADSL subscriber numbers." *MyBroadband*, Dec 22, 2010. July 25, 2012. <http://mybroadband.co.za/news/adsl/17431-Telkom-releases-latest-ADSL-subscriber-numbers.html>.

Muller, R. "Telkom's copper line battle." June 13, 2011. *MyBroadband*, July 13, 2011. July 25. 2012. <http://mybroadband.co.za/news/telecoms/26477-telkom%E2%80%99s-copper-line-battle.html>.

Ponte, S. and van Sittert, L. "The chimera of redistribution in post-apartheid South Africa: 'Black Economic Empowerment' (BEE) in industrial fisheries." *African Affairs* 106:424, 2007. 437–462.

"Radio in South Africa." *MediaClubSouthAfri*ca. 2011. July 4, 2011. <http://www.mediaclubsouthafrica.com/index.php?option=com_content&view=article&id=118%3ARadio+in+South+Africa&catid=36%3Amedia_bg&Itemid=54>.

Rosenthal, E. *You have been listening . . . The early history of radio in South Afri*ca. Cape Town: Purnell, 1974.

Schneiderman, D. "Promoting equality, Black economic empowerment, and the future of investment rules." *South African Journal on Human Rights* 25:2, 2009. 246–279.

SABC. "About the SABC: Mandate." *SABC*, 2011. July 4, 2012. <http://www.sabc.co.za/wps/portal/SABC/SABCMANDATE>.

Staff. "Icasa calls time on M-Net's free window." *IOL*, Mar 19, 2005. July 25, 2012. <http://www.iol.co.za/news/south-africa/icasa-calls-time-on-m-net-s-free-window-1.236807>.

"Telkom investor relations overview." Telkom SA, 2011. July 4, 2012. <https://secure1.telkom.co.za/ir/overview/shareholding.html>.

"The media in South Africa." *MediaClubSouthAfri*ca. 2011. July 4, 2012. <http://www.mediaclubsouthafrica.com/index.php?option=com_content&view=article&id=110:the-media-in-south-africa&catid=36:media_bg%20>.

Tomaselli, K. "Media ownership and democratization: the political economy of the South African print media," in: Hyden G., Leslie M., and Ogundimu F. F., eds. *Media and Democracy in Afri*ca. London: Transaction Books, 2003.

Wasserman, H. "Talking of change: constructing social identities in South African media debates." *Social Identities* 11:1, 2005. 75–85.

## 2. Reports


"Africa Internet Usage and Population Statistics." *Internet World Stats*, 2011. July 25, 2012. <http://www.internetworldstats.com/stats1.htm#africa>.

"Alexa—Top Sites in South Africa." *Alexa*. 2011. July 4, 2012. <http://www.alexa.com/topsites/countries/ZA>.

DACST. "The South African Film and Television Industry Report." Pretoria: Department of Arts, Culture, Science and Technology, 1998.

DACST. "The South African Publishing Industry." Pretoria: Department of Arts, Culture, Science and Technology, 1998.

Galloway, F. and Struik, W. "Annual book publishing industry survey report 2008." Pretoria: School of Information Technology, University of Pretoria, 2009.

Galloway, F., Venter, R. M. R., and Bothma, T. P. "ASA industry surveys. Executive summary: Broad trends over four years (2002–2005)." Pretoria: School of Information Technology, University of Pretoria, 2006.

MDDA. "Trends of Ownership and Control of Media in South Africa." Johannesburg: Media Development and Diversity Agency, 2009.

MTN Group Limited. "Integrated Business Report—for the year ended December 31, 2011." mtn-investor.com, 2012. Feb 8, 2013. <http://mtn-investor.com/mtn_ar2011/pdf/full.pdf>. Annual.

Office of the Postmaster General. "Annual Report of the Postmaster General 1989–1990." Pretoria: Telkom SA, n.d. Feb 8, 2013. <https://secure1.telkom.co.za/apps_static/ir/pdf/financial/history/republic_of_south_africa/1989-1990.pdf>.

Office of the Postmaster General. "Annual Report of the Postmaster General 1990–1991." Pretoria: Telkom SA, n.d. Feb 8, 2013. <https://secure1.telkom.co.za/apps_static/ir/pdf/financial/history/republic_of_south_africa/1990-1991.pdf>.

Snymon, Soekie, van der Merwe, Cloirwyn, and Maree, Carolina, eds. "Telkom: The first annual report 1991-10-01 to 1993-03-31." Pretoria: Telkom SA, 1993. Feb 8, 2013. <https://secure1.telkom.co.za/apps_static/ir/pdf/financial/history/telkom_annual_report/1991-1993.pdf>.

Telkom SA. "Telkom Group Annual Report 2011." Pretoria: Telkom SA, 2011. Feb 8, 2013. <http://telkom.investoreports.com/telkom_ar_2011/>.


## 3. Websites


"DStv channels." *DStv*, 2011. July 4, 2012. <http://www.dstv.com/Pages/Channels/>.

"Hosken Consolidated Investments Ltd." *HCI*, 2011. July 4, 2011. <http://www.hci.co.za/corp_profile.htm>.

"IBIS Media Manager. Media history." *IBIS*, 2011. July 25, 2012. <http://www.mediamanager.co.za/medialibrary/mediahistory/sector/broadcast>.

"Independent News & Media PLC." *INM*, 2011. July 4, 2012. <http://www.inmplc.com/>.

"Kagiso Media." *Kagiso Media*, 2011. July 4, 2012. <http://www.kagisomedia.co.za/kagisomedia/content/en/about-us>.

"Media24." *Media24*, 2011. July 4, 2012. <http://www.media24.com/>.

"MultiChoice Investment Centre." *MultiChoice*, 2011. July 4, 2011. <http://www.multichoice.co.za/multichoice/view/multichoice/en/page54>. "National Association of Broadcasters." *NAB*, 2011. July 4, 2012. <http://www.nab.org.za>.

"Primedia." *Primedia*, 2011. Last accessed on July 4, 2012. <http://www.primedia.co.za>.

"RamsayMedia." *RamsayMedia*, 2011. July 4, 2012. <http://www.ramsaymedia.co.za>.

"Telkom SA Limited – DSL." Telkom SA, 2011. July 4, 2012. <http://www.telkom.co.za/products_services/dsl/cost_dsl_cost.html>.

"The Caxton and CTP Group." *Caxton*, 2011. June 15, 2012. <http://www.caxton.co.za>.

"The Independent Communications Authority of South Africa." *ICASA*, 2011. July 4, 2012. <http://www.icasa.org.za/>.

# South Africa—Data Summaries

ELI NOAM AND PAUL MUTTER

SOUTH AFRICA'S RELATIVELY LOW number of voices (*n*) in Table 31.15 is the result of the country's media industries not being liberalized until after the end of apartheid in 1994. Concentration is especially pronounced due to Telkom's monopoly in the ISP and wireline markets, the multimedia group Naspers' control of the multichannel market, and the wireless market being a duopoly between Vodacom and the domestic operator MTN. Telkom dominates the platform market. The company has a legal monopoly in the ISP and wireline markets.

In content media, Naspers controls the multichannel market because it offers satellite service, the only multichannel distribution platform available to consumers. Naspers also has a strong presence in daily newspapers—its origin—with 36%, as well as in magazines with 39%. It accounts for 17.9% of the content market as a result, a slight increase from 17.5% in 2004 (Table 31.16).

Still larger in the content sector is the publicly owned company, SABC with 26.7% of the content market. SABC maintains a monopoly over TV broadcasting with 69%, as well as 45% of the radio market.

The largest media organization, however, is the public incumbent Telkom. It has the highest company power index and share of the national market in South Africa, followed by Vodacom, which is owned by Vodafone (UK) and controls 34.1% of the platform market (Table 31.17). Attempts to license new operators have so far had no effect on the incumbent's monopoly, as the government is reluctant to fully liberalize the industry after partially privatizing Telkom. Thus, for both the content and the platform sectors of media, the two largest organizations are both state owned and run.

Table 31-15. National Media Industries Concentration in South Africa

| | 2004/5 | | 2009 OR MOST RECENT | | % CHANGE ANNUAL AVERAGE | |
|---|---|---|---|---|---|---|
| | COMPANY POWER INDEX IN COUNTRY | COMPANY SHARE OF THE OVERALL NATIONAL MEDIA MARKET (%) | COMPANY POWER INDEX IN COUNTRY | COMPANY SHARE OF THE OVERALL NATIONAL MEDIA MARKET (%) | COMPANY POWER INDEX IN COUNTRY | COMPANY SHARE OF THE OVERALL NATIONAL MEDIA MARKET (%) |
| Telkom (39.8% public) | 4,110 | 41.1 | 2,553 | 26.5 | –5.4 | –2.1 |
| Vodacom (65% Vodafone, UK + 35% Telkom) | 1,163 | 23 | 1,405 | 27.6 | 3.0 | 0.7 |
| MTN | 739 | 18.3 | 616 | 18.3 | –2.4 | 0.003 |
| Naspers Group | 166.4 | 3.0 | 276.2 | 4.6 | 9.4 | 0.2 |
| SABC (public) | 286.7 | 4.2 | 308.6 | 5.1 | 1.1 | 0.13 |
| Google (US) | 11.2 | 0.1 | 78.1 | 0.9 | 85.6 | 0.1 |
| Hosken Consolidated Investment | 28.2 | 1.2 | 63.7 | 2.0 | 18.0 | 0.13 |
| Independent News & Media (INM) (Ireland) (O'Reilly) | 36.2 | 1.2 | 40.2 | 1.5 | 1.6 | 0.05 |
| Kagiso Media | 11.6 | 0.5 | 14.4 | 0.7 | 3.4 | 0.03 |
| Avusa | 12.7 | 0.8 | 18.7 | 1.2 | 6.7 | 0.05 |
| Caxton | 9.0 | 0.7 | 11.8 | 0.9 | 4.4 | 0.03 |
| Primedia | 10.0 | 0.5 | 32.5 | 1.1 | 30.7 | 0.08 |
| Nu Metro Film Distribution | 11.0 | 0.3 | 6.0 | 0.2 | –6.5 | –0.02 |

| | | | | | | |
|---|---|---|---|---|---|---|
| Ster Kinekor Entertainment | 8.9 | 0.2 | 5.0 | 0.1 | –6.2 | –0.01 |
| Ramsay Media | 1.0 | 0.1 | 0.6 | 0.1 | –6.1 | –0.004 |

| Media Concentration Index | 2004/5 | 2011 or Most Recent | % Change Annual Average |
|---|---|---|---|
| Total Revenue: Nat'l Media Industry (mil US$) | 15,619 | 19,525 | 3.6% |
| Total Voices ($n$) | 42 | 46 | 1.4% |
| Net Voices ($n$) | 31 | 35 | 1.8% |
| Public Ownership (%) | 28.6 | 25.3 | –0.5% |
| Foreign Ownership (%) | 20.4 | 27.9 | 1.1% |
| C4 Average—Weighted | 99.0 | 98.4 | –0.1% |
| HHI Average—Weighted | 6,646 | 5,535 | –2.39% |
| C1 Average—Weighted | 72 | 64 | –1% |
| Noam Index Average—Weighted | 2,442 | 2,351 | –0.5% |
| Pooled Overall Sector C4 | 86.4 | 79.6 | –1.1% |
| Pooled Overall Sector HHI | 2,594 | 1,909 | –0.04% |
| Pooled Overall Sector Noam Index | 282 | 226 | –2.8% |
| Market Share of Top 10 Companies: Nat'l Media Industry (%) (Pooled C10) | 97.2 | 95.0 | –0.3% |
| National Power Index | 6,644 | 5,536 | –2.4% |

Table 31-16. Top Content Media Companies in South Africa

| | 2004/5 | | 2011 OR MOST RECENT | | % CHANGE ANNUAL AVERAGE | |
|---|---|---|---|---|---|---|
| | COMPANY POWER INDEX IN COUNTRY | COMPANY SHARE OF THE NATIONAL CONTENT MEDIA MARKET (%) | COMPANY POWER INDEX IN COUNTRY | COMPANY SHARE OF THE NATIONAL CONTENT MEDIA MARKET (%) | COMPANY POWER INDEX IN COUNTRY | COMPANY SHARE OF THE NATIONAL CONTENT MEDIA MARKET (%) |
| SABC (Public) | 2,175 | 31.7 | 1,617 | 26.7 | −3.7 | −0.7 |
| Naspers | 775 | 17.5 | 836 | 18 | 1.1 | 0.06 |
| Google (US) | 85 | 0.9 | 409.5 | 4.5 | 54.7 | 0.5 |
| Hosken Consolidated Investments | 214 | 8.8 | 334 | 10.7 | 8.0 | 0.3 |
| INM (Ireland) | 275 | 9.0 | 210 | 8.1 | −3.34 | −0.13 |
| Kagiso Media | 88 | 3.6 | 75.3 | 3.6 | −2.1 | −0.003 |
| Avusa | 96.5 | 6.3 | 98.1 | 6.1 | 0.2 | −0.03 |
| Caxton | 68.5 | 5.0 | 61.8 | 4.7 | −1.4 | −0.05 |
| Primedia | 76 | 3.7 | 165 | 5.5 | 16.8 | 0.3 |
| Nu Metro | 83.3 | 2.0 | 31.2 | 0.8 | −8.9 | −0.2 |
| Ster Kinekor | 67.8 | 1.8 | 26.4 | 0.78 | −8.7 | −0.15 |
| Ramsay Media | 8.0 | 1.0 | 3.2 | 0.6 | −8.6 | −0.1 |
| Media Concentration Index | 2004/5 | | 2011 or Most Recent | | % Change Annual Average | |
| Public Ownership (%) | 31.7 | | 26.8 | | −0.7 | |
| Foreign Ownership (%) | 11.9 | | 14.7 | | 0.4 | |
| C4 Average—Weighted | 92.4 | | 92.9 | | 0.07 | |
| HHI Average—Weighted | 4,055 | | 3,945 | | −0.4 | |
| C1 Average—Weighted | 52 | | 50 | | 0.0 | |
| National Power Index | 4,045 | | 3,932 | | −0.4 | |

Table 31-17. Top Platform Media Companies in South Africa

| | 2004/5 | | 2011 OR MOST RECENT | | % CHANGE ANNUAL AVERAGE | |
|---|---|---|---|---|---|---|
| | COMPANY POWER INDEX IN COUNTRY | COMPANY SHARE OF THE NATIONAL PLATFORM MEDIA MARKET (%) | COMPANY POWER INDEX IN COUNTRY | COMPANY SHARE OF THE NATIONAL PLATFORM MEDIA MARKET (%) | COMPANY POWER INDEX IN COUNTRY | COMPANY SHARE OF THE NATIONAL PLATFORM MEDIA MARKET (%) |
| Telkom (39.8% public) | 4,734 | 47.3 | 3,511 | 32.8 | –4.8 | –2.0 |
| Naspers | 74.1 | 0.7 | 144.2 | 1.4 | 13.5 | 0.1 |
| Vodacom (65% Vodafone, UK and 35% Telkom, South Africa) | 1,340 | 26.4 | 1,737 | 34.1 | 4.2 | 1.1 |
| MTN | 852 | 21.0 | 761 | 22.6 | –1.5 | 0.2 |
| Media Concentration Index | 2004/5 | | 2011 or Most Recent | | % Change Annual Average | |
| Public Ownership (%) | 28.1 | | 25.0 | | –0.4 | |
| Foreign Ownership (%) | 21.7 | | 31.0 | | 1.3 | |
| C4 Average—Weighted | 100 | | 99.7 | | –0.1 | |
| HHI Average—Weighted | 7,039 | | 5,910 | | –2.3 | |
| C1 Average—Weighted | 74 | | 67 | | –0.01 | |
| National Power Index | 7,039 | | 5,914 | | –2.3 | |

# PART 3

## SUMMARIES AND ANALYSES

# 32

# National Media Concentrations Compared

## INTRODUCTION

This chapter examines and summarizes country data across all of the 30 nations surveyed in this study. To remind the reader: these countries were chosen on the basis of their importance in the media world, their GDP, and population size. Combined, they account for 90% of media revenues, 82% of the world's GDP, and 63% of the world's population. Twenty-two of the countries are OECD members. The study covers 95.2% of overall OECD GDP and 94.8% of its population.

Tables 32.1 and 32.2 show the size of the media sectors in the various countries, first for content media, then for platform media. They show the large share of US market (36.5% and 31.4%). It is followed by Japan (8.7% and 10.4%), China (4.8% and 10.5%), the United Kingdom (7.9% and 4%), Germany (7.6% and 5.4%), France (5.1% and 4.2%), and Spain (4.7% and 3.2%). Together, these seven countries account for 75.3% of the 30-country world in content and 69% in platforms. In platforms, Canada (3.3% and 3%), Italy (3.1% and 2.6%), and Brazil (3.2% and 6.3%) join these seven countries to control 80.8% of the sector (84.8% for content).

Table 32.3 shows the worldwide revenues for the 13 media industries in the 30 countries: the total value of all media activities in 2013 was US$2.4 trillion, up from US$1.7 trillion in 2005.[1] The new ICT industries—wireless and ISPs—grew very rapidly, while the older wireline industry has seen a low growth per annum (0.2%) due to the expansion of wireless services at its expense. The print newspapers industry has been shrinking by –1.2% per annum. The most pronounced change in this period of time has been in search engines (a 23.7% per annum increase). In absolute terms, by far the largest increase was that of mobile wireless, which grew, on average, by $40 billion per year, and became the largest media industry.

1. Revenues are listed in US dollars and at the prevailing exchange rate in the years cited.

Table 32-1. Content Media Revenue, 30 Countries (Mil US$)

| | 2004/5 | 2011/3 | PERCENTAGE (%) |
|---|---|---|---|
| United States | 243,167 | 268,222 | 36.5% |
| Japan | 58,770 | 63,861 | 8.7% |
| United Kingdom | 45,912 | 58,004 | 7.9% |
| Germany | 48,194 | 55,999 | 7.6% |
| France | 32,856 | 37,462 | 5.1% |
| China | 8,598 | 35,272 | 4.8% |
| Spain | 28,254 | 34,088 | 4.6% |
| Canada | 15,217 | 24,089 | 3.3% |
| Brazil | 10,662 | 23,354 | 3.2% |
| Italy | 17,453 | 22,873 | 3.1% |
| Australia | 10,284 | 16,998 | 2.3% |
| Netherlands | 8,674 | 9,323 | 1.3% |
| Russia | 2,674 | 9,122 | 1.2% |
| Sweden | 6,638 | 9,041 | 1.2% |
| India | 4,957 | 8,774 | 1.2% |
| South Korea | 5,846 | 7,827 | 1.1% |
| Portugal | 4,893 | 6,450 | 0.9% |
| Switzerland | 4,672 | 5,562 | 0.8% |
| Belgium | 3,854 | 5,011 | 0.7% |
| Mexico | 3,533 | 4,825 | 0.7% |
| Finland | 2,890 | 3,868 | 0.5% |
| Ireland | 3,455 | 3,788 | 0.5% |
| South Africa | 2,059 | 3,725 | 0.5% |
| Turkey | 1,675 | 3,465 | 0.5% |
| Argentina | 1,675 | 3,229 | 0.4% |
| Taiwan | 3,441 | 3,181 | 0.4% |
| Chile | 1,719 | 2,129 | 0.3% |
| Poland | 1,802 | 1,988 | 0.3% |
| Israel | 1,436 | 1,961 | 0.3% |
| World Total Content Media | 586,173 | 734,749 | 100.00% |
| World GDP | 52,000,000 | 72,000,000 | |
| Content Media as a % of World GDP | 1.1% | 1.0% | |

ISPs grew by $12 billion a year, and search engines by $3.5 billion.

Four concentration indices form the basis for country-by-country comparisons: the C4 and C1 indices, the Herfindahl-Hirschman Index (HHI), the MOCDI (Noam) Media Concentration Index (NI), and Power Index (PI). Each is explained below in its respective section, which details the following: (1) what the index is, (2) what the index represents, (3) what the data show, (4) the world averages of the index, (5) the top and bottom countries relative to that world average, and (6) the US concentration numbers, as the largest of media markets.

The concentration indices described below are used across the 13 media industries in several different ways: weighted average, unweighted (arithmetic average), and pooled. The "weighted

Table 32-2. Platform Media Revenue, 30 Countries (Mil US$)

| | 2004/5 | 2011/3 | PERCENTAGE (%) |
|---|---|---|---|
| United States | 409,398 | 521,350 | 31.4% |
| China | 50,300 | 173,430 | 10.5% |
| Japan | 115,104 | 171,942 | 10.4% |
| Brazil | 65,361 | 103,697 | 6.3% |
| Germany | 79,198 | 89,152 | 5.4% |
| France | 43,066 | 68,955 | 4.2% |
| United Kingdom | 58,186 | 66,506 | 4.0% |
| Spain | 44,559 | 52,893 | 3.2% |
| Canada | 27,855 | 49,644 | 3.0% |
| Italy | 46,787 | 42,450 | 2.6% |
| Russia | 18,117 | 36,492 | 2.2% |
| India | 10,235 | 34,683 | 2.1% |
| Australia | 21,624 | 33,132 | 2.0% |
| South Korea | 23,087 | 32,730 | 2.0% |
| Mexico | 18,177 | 25,701 | 1.6% |
| Netherlands | 17,055 | 21,286 | 1.3% |
| Turkey | 9,523 | 16,053 | 1.0% |
| South Africa | 13,561 | 15,800 | 1.0% |
| Belgium | 11,999 | 13,525 | 0.8% |
| Poland | 9,617 | 12,691 | 0.8% |
| Switzerland | 8,404 | 11,781 | 0.7% |
| Taiwan | 10,638 | 11,495 | 0.7% |
| Sweden | 8,712 | 10,459 | 0.6% |
| Portugal | 7,489 | 8,772 | 0.5% |
| Egypt | 3,166 | 6,721 | 0.4% |
| Ireland | 6,180 | 6,469 | 0.4% |
| Argentina | 4,416 | 5,789 | 0.3% |
| Israel | 4,939 | 5,468 | 0.3% |
| Finland | 4,678 | 5,399 | 0.3% |
| Chile | 2,382 | 3,528 | 0.2% |
| World Total Platform Media | 1,153,813 | 1,657,993 | 100.00% |
| World GDP | 52,000,000 | 72,000,000 | |
| Platform Media as a % of World GDP | 2.2% | 2.3% | |

average" adjusts, in averaging, for different sizes of industries. In contrast, a "pooled" concentration index treats all 13 industries as part of a larger single media sector and looks at market shares and concentration within that larger market.

A pooled measure lowers the concentration numbers, of course, but it does so for all countries, and thus comparisons are still possible. What is gained is a dimension of cross-ownership, insofar as major companies often operate in several industries, and pooling the industries hence shows their overall share in an overall national media market. We have also another way to identify cross-ownership, through the power index, to be discussed below.

Table 32-3. Industry Revenues (30 Countries)

| | 2004/05 (MIL US$) | 2011/3 (MIL US$) | ANNUAL GROWTH TREND (8-YRS) |
|---|---|---|---|
| Wireless | 416,773 | 740,038 | 9.7% |
| Wireline | 581,810 | 589,986 | 0.2% |
| Multichannel Platforms | 131,234 | 247,666 | 11.1% |
| Broadcast TV | 146,227 | 184,379 | 3.3% |
| ISP | 67,303 | 162,033 | 17.6% |
| Newspapers | 116,295 | 106,734 | –1.0% |
| Book Publishing | 64,021 | 72,224 | 1.6% |
| Magazines | 60,830 | 79,590 | 3.9% |
| Video Channels | 56,855 | 70,636 | 3.0% |
| Radio | 45,231 | 47,687 | 0.7% |
| Search Engines | 14,242 | 41,300 | 23.7% |
| Film | 26,607 | 30,710 | 1.9% |
| Online News | 12,558 | 19,760 | 7.2% |
| Total 13 Media Industries | 1,739,986 | 2,392,742 | 4.7% |

Another distinction is that of the two subsectors of the overall media industry, namely "platform media" and "content media." Platform media refers to the transmission and delivery of media, while content refers to consumed material. Of course, many firms are both providers of content and of platform services. And several industries straddle the platform/content divide. This is particularly true for multichannel platforms. A cable television provider offers both transmission of media content over its distribution plant, similar to the telecom's provision of network connectivity, and it also provides a bundle of channels to subscribers. The subscriber typically pays for both bundled together, but that is a billing and economic convention. When the same cable TV operator offers Internet connectivity, and the user also subscribes to the video services of a content provider such as Netflix, the distinction becomes clearer. In traditional cable TV service, the operator secures rights to various channels of content and pays their providers a fee, which becomes part of the cost that is then covered by the subscription. (To make things still more complicated, some of those channels might be created by the cable company itself.) To allocate revenues among content and platform services of a multichannel provider, we used their cost share in overall costs. This resulted in an allocation one-third of multichannel revenues toward content and the other two-thirds toward platforms.

Another way to look at media concentration diversity is to count "voices" rather than market shares. The basic idea is that regardless of a low popularity of a media outlet, its mere existence provides a valuable option to the public.

Additional dimensions of analysis are "Foreign Ownership" and "Public Ownership." The latter shows how much of a country's media market is controlled by state enterprises. These can be direct state operations, or semi-independent organizations like public-service broadcasters, or private-law companies in which the government is a controlling shareholder. "Foreign Ownership" means that a company's ultimate controlling owners are operationally headquartered in another country. For example, Telefónica, one of the world's largest telecom companies, is counted as a foreign operator in Argentina or the United Kingdom. Sony, headquartered in Japan, is considered a foreign-owned company in the US market even though its film studio, the former Columbia-TriStar, is one of the traditional Hollywood "Big Six."

A discussion of the primary trends in concentration at the national level follows the presentation of each table. The text also briefly notes specific issues of competitiveness and diversity in the countries, although that is discussed more thoroughly in the country chapters by their respective authors.

These authors are knowledgeable analysts of their country's media industries, and are thus the main source for the data used in this chapter. In a few cases, we have slightly recast their data to achieve consistency in market definitions. We have also updated numbers to account for subsequent mergers and spin-offs.[2]

## CONCENTRATION MEASURED BY VOICES

We start with media voices. Pluralism requires the vigorous interplay of media "voices"—newspapers, magazines, TV networks, radio stations, and so on. This is expressed, in Graph 32.1, in terms of the total number of voices ($n$), which represents how many entities are present. But, not literally all. To include every college newspaper or specialty magazine as a "voice" would be an impossible task in data terms and would also clutter up the big picture. We therefore consider only those entities that achieve, within their content-producing industry, at least 1% in terms of market share and are thus on a society's "radar screen."[3] This total number of voices, however, contains duplication when a company owns, for example, a TV network as well as a radio network. To account for this, in Graph 32.2 we subtract the number of "duplicate" owners from the total $n$, arriving at a "net $n$." For example, in Italy the holding company Fininvest owns a book publisher (Mondadori), three broadcasting networks (Mediaset), and a film producer/distributor (Medusa). They are counted as one "voice" because they are

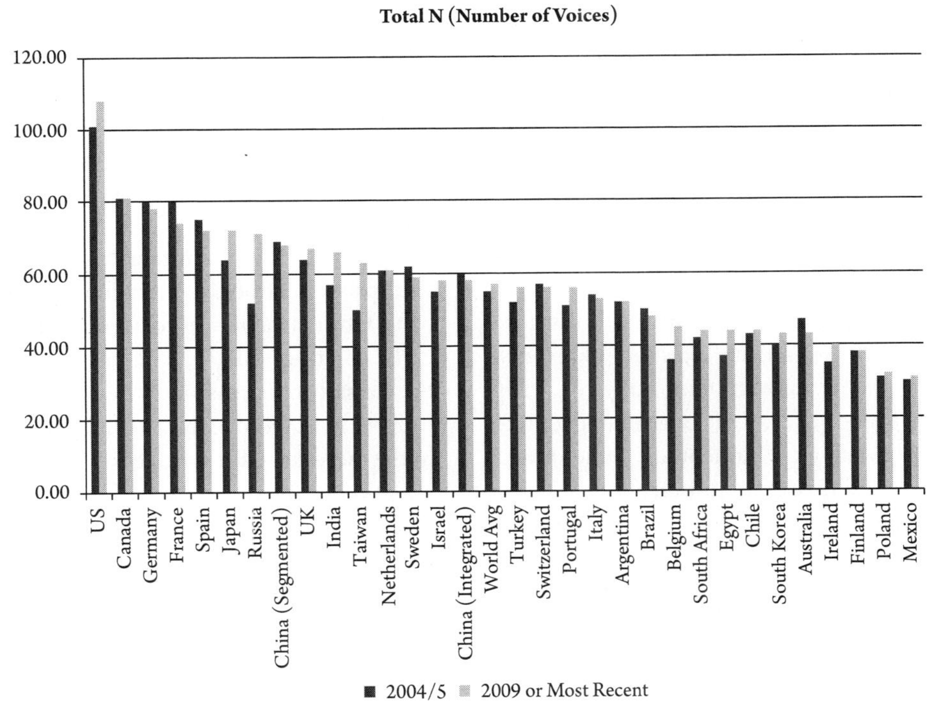


GRAPH 32.1 Countries' Total $N$ (Number of Voices), Companies with 1% of Market Share

2. Due to different data availabilities in the various countries, the most recent available numbers vary. For some countries and industries, they might go back to 2011. For others, 2013 data are available.

3. For book publishing, with its large number of small firms, the cut-off is $N$ (>2%) instead of the standard $N$ (>1%) used for the other industries.

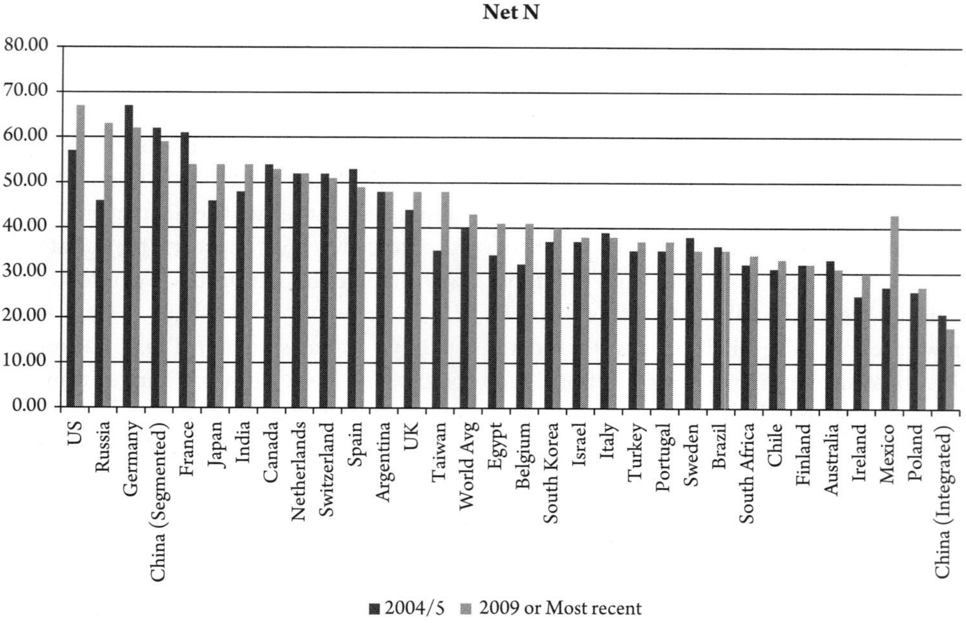


GRAPH 32.2 Countries' Net *N* (Number of Voices), (Companies with >1% of Market Share)

all owned by Silvio Berlusconi and his family. When these and other cross-holdings are taken into account, net *n* in Italy stands at 39, whereas total *n* equals 52. Similarly, state-owned entities, such as Russia's various state-run broadcast networks (Gazprom Media, First Channel, and VGTRK) or South Korea's public broadcasting companies all count as one "voice."

Within China, we display voices both as "Integrated"—that is, all state-owned enterprises count as a single voice in an industry, regardless of which agency or level of government controls them—as well as "Segmented," which treats each state enterprises as an independent entity. As elsewhere, only the "Integrated" number is included in world averages.

To properly assess this, one needs to recognize that the definition of a voice, being tied to a market share (1%), varies across countries, and this must be taken into account. This is further discussed in the chapter "Analyses."

The findings show:

- Since 2004, the world average of net voices per country slightly increased from 41 to 42, meaning that, on average, a country typically has 42 independent content media producers active in its national media market, possessing at least 1% in one of the media markets.
- China has the lowest net *n* due to the extent of state control: with the state as a single owner, the country has merely 17 net voices whose size is above 1% in their industry. If each state enterprise is counted as an independent company, China's net *n* rises to 62, almost the highest in the world.
- Russia had an increase in its net *n* of voices from 42 in 2004 to 58 by 2012. This was due to the increase in the number of new media outlets entering the market.
- The United States has the largest number of total voices (99) and net voices (59). It is followed by Russia (58 net), France and India (both 54), Japan and Spain (each with 53), Germany (52), and Canada and the Netherlands (each with 50).

However, things look very different when one takes into account the size of a country in terms of population (Graph 32.3).

Per capita voices are calculated by dividing net *n* over the population of each country.

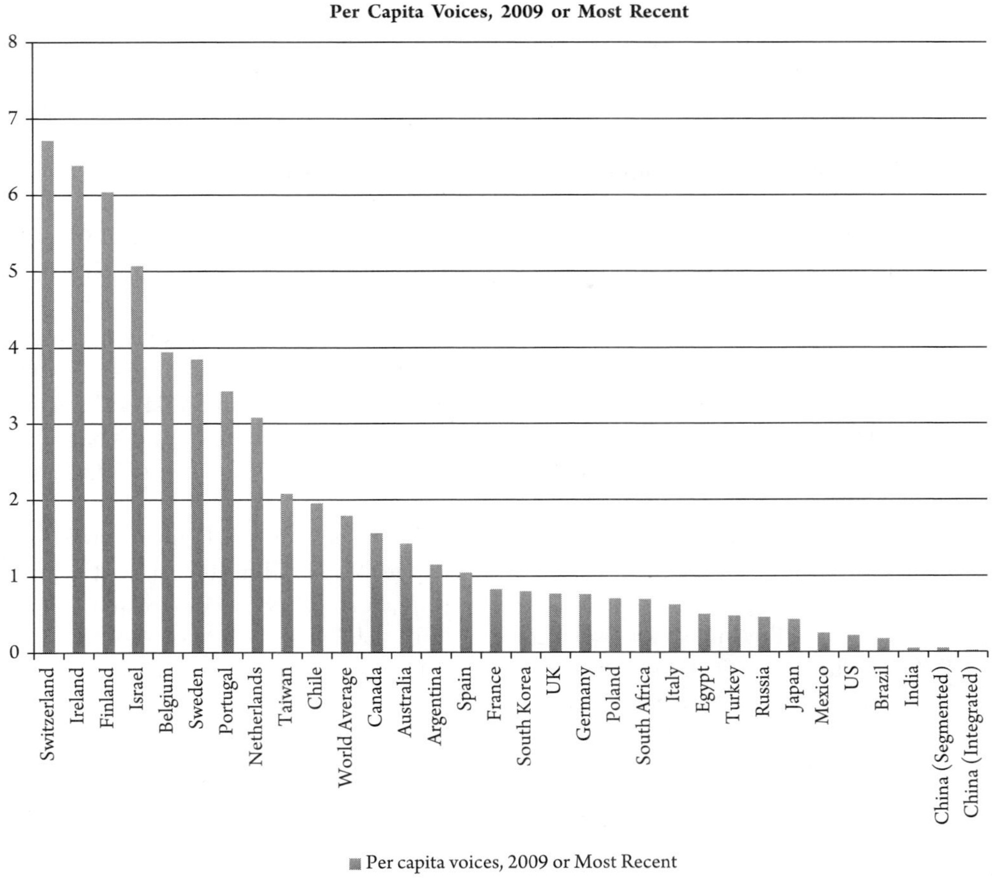


GRAPH 32.3 Countries' Per Capita *N* (Number of Voices), Companies with >1% of Market Share (2011 or Most Recent)

- Worldwide, while the United States has the highest number of voices, on a per capita basis it is actually quite low (0.2 per one million people), due to its population size (and higher threshold for a voice definition.)
- Brazil and India, whose net *n* values are 26 and 54 to China's 17, are also close to the bottom. In China, it is a combination of state ownership and population that lands it in last place for voices per capita. Even with the average net *n* that separates out state enterprises, China would still be at the bottom.
- Countries with a high count of voices per capita tend to be relatively small but with active politics and culture. Their cut-off point for voices is also lower (see Chapter 37 "Analyses"). Europe has many such countries. The European countries' average is over twice as high as the North American and just about any region.
- The highest number of per capita voices are found in Ireland and Switzerland. Both countries have many media offerings spilling over from neighboring countries, which inflates the number. In the next tier are Israel, as well as Finland, Sweden, and Belgium (the latter three also counting major cross-border spill-ins).
- On the other extreme are countries that are large, as well as poor and sometimes nondemocratic, and there the number of voices per capita is the smallest—China, India, Egypt, Mexico. Another finding is that the number of media "voices" in richer countries is two times as large as in emerging countries, on average.

## CONCENTRATION MEASURED BY THE MARKET SHARE OF THE TOP FOUR FIRMS

"C4" is a concentration measure that presents the combined market share of the top four companies in any media industry (Graph 32.4). When C4 ranges from 0% to 40%, the industry tends to be competitive if the companies are of roughly equal size. It says that smaller companies serve 60% or more of a market. With a C4 above 40%, the industry is most likely an oligopoly.

The C4 for the film industry in the United States, for example, was 56.4% in 2013: this was the sum of the market shares of Viacom's Paramount, Time Warner's Warner Bros, Disney, and Sony, which had the largest market shares of all film studio-distributors in the US market that year.

C4 represents the film industry in isolation, not as 1 of 13 industries in the US's $790 billion national media market. A weighted average C4 for an industry weights these market shares by the revenue in that industry; for film, this would mean the weight is total industry revenue of $9.5 billion. When this is done for all 13 industries, and an average is then taken, the result represents the average level of concentration prevalent across all media industries, taking their size into account. It should be understood, moreover, that the top four companies are not the same in each industry. Later we will look at top companies across all media.

What do the findings show?

- The world average for unweighted C4 was very high 80.6% in 2004 and has slightly increased to 80.7% in 2013. For weighted C4, the world average has increased marginally, from an extraordinarily high 88.2% in 2004 to 88.6% in 2013. Weighted C4 is so high primarily because of the large revenues of platform media (i.e., a high weight on top of a high C4 concentration).
- The countries with the highest unweighted average C4s are China (97%), South Africa (94%), Portugal (91%), and Ireland (87%). Chile, Poland, South Korea, Mexico, and Finland are all at 86% while the United Kingdom has 85%. For weighted average, the highest are China (99%), South Africa (98%), Portugal (97%), South Korea (94%), Russia (94%), Poland (96%), Mexico (97%), and Egypt (94%).
- The countries with the lowest weighted average C4s are Canada and Germany (both

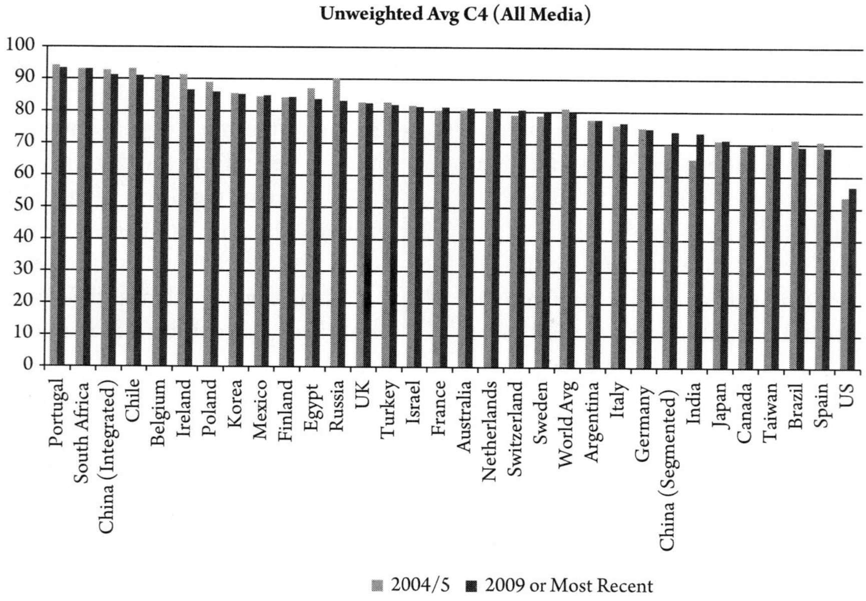


GRAPH 32.4 Countries' Unweighted Average C4 – All Media

82%), Spain (79%), India (80%), and the United States (67%). "Lowest" is a relative term when the numbers are so high, but they are lower than the world average (88.5%). Unweighted, the least concentrated countries are Japan and Canada (both 71%), Taiwan (70%), Spain (69%), and the United States (58%).

- Platform average C4 is higher than content, with a world average of 93.3% weighted, and 87.9% unweighted. C4 concentration is remarkably similar—and high—across countries. Only the United States is below 75% for both weighted and unweighted platform C4, and this is partly explained by the regional nature of wireline telecom in a geographically large country.
- Content media's C4 stood at 64.2% unweighted and 76.5% weighted.
- C4 was lowest in the United States for the measures all media, platform media, and content media, both unweighted and weighted. However, C4 has increased across all industries in this period for the United States (See Chapter 38).
- Other significant increases in C4 were seen in India and Sweden, and somewhat so in the Netherlands, the United Kingdom, Brazil, South Africa, Egypt, Poland, and China.

## CONCENTRATION MEASURED BY THE MARKET SHARE OF THE TOP FIRM IN EACH INDUSTRY (C1)

"C1" presents the market share of the top company in each of the 13 media industries (Graph 32.5). (This is not the top company in the overall national 'pooled' media market, which describes the number 1 company in the whole of a country, discussed later in this chapter).

The C1 for the broadcast TV industry in the United States, for example, was 17.6% in 2013: this was Comcast's share in this industry. The average of this and the other number 1 companies in each industry (AT&T for wireline, Google for search engines, etc.) is the unweighted C1. We also calculate a weighted average C1, using industry size (revenues).

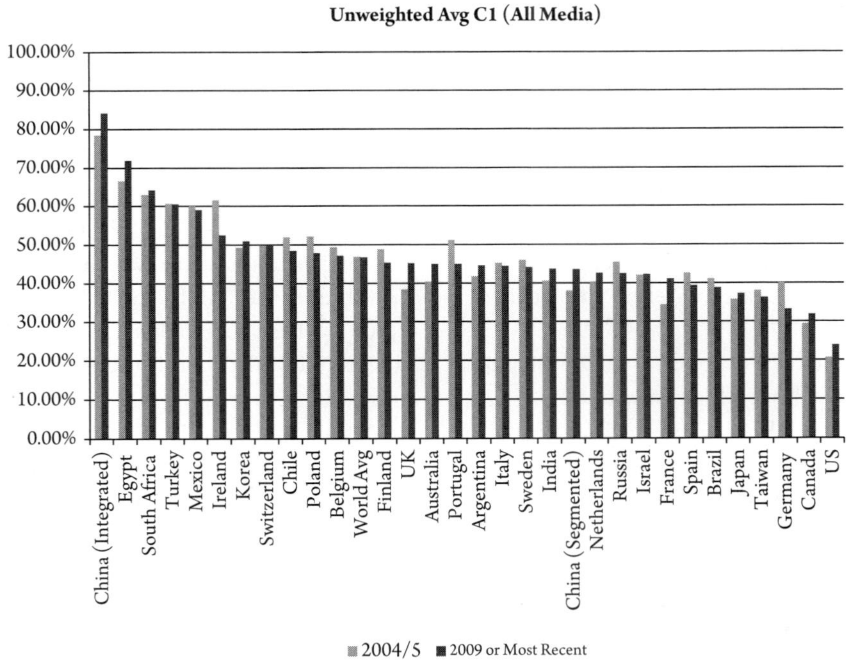


GRAPH 32.5 Countries' Unweighted Average C1 – All Media

- The world average unweighted C1 was 46.7% in 2004 and has dropped slightly to 45.6%. It is high in particular due to platform media, where incumbent telecom providers dominate telecoms and public broadcasters are strong in Europe and Asia. The search engines industry is dominated by Google.
- The countries with the highest unweighted average C1s are China (86%), Egypt (64%), South Africa (662%), Turkey (53%), Mexico (59%), Ireland (52%), South Korea and Switzerland (both 50%).
- The countries with the lowest weighted average C1s are the Netherlands, Germany, the United Kingdom, and Spain (each around 40–42%), Finland (38%), Canada (37%), and as the lowest, the United States (24%). Here, too, "lowest" is a relative term when the numbers are so high, the world average being 48.3%. There are at least 17 countries with C1s in the 40–50% range. For unweighted C1, the least concentrated nations are Canada (32%), Spain (36%), Brazil and France (both 38%), and the United States (22%).
- For platform media, weighted C1 concentration is high across countries at 52.9% (though declining from 60.6% in 2004). Only the United States (26%), Canada (39%), Brazil (37%), and Finland (38%) are below 40%. China is at 100% due to state monopoly of telecommunications; the next most concentrated countries are Mexico (70%), South Africa and Turkey (both 67%), though these figures represent a decline relative to 2004 numbers. Platform concentration is are high but falling, as the mobile market expands and the still-concentrated wireline market loses weight to mobile.
- Content media average C1 is 40.9% weighted and rising. China is highest with 88%. Egypt, Mexico, South Korea, South Africa, Brazil, Italy, Ireland, Australia, the Netherlands, Switzerland, and the United Kingdom all surpass 40%.
- C1 was lowest in the United States for all media, both platforms and content, and both unweighted and weighted. However, as with C4, the C1 has increased across all industries in this period for the United States. Significant decreases occurred in Russia, Turkey, and Germany — driven by the decline of public incumbents.

## CONCENTRATION MEASURED BY HHI

The Herfindahl–Hirschman Index (HHI) measures concentration in a more informative (though less intuitive) way than the C4 concentration ratio described above. C4 does not account for companies that are below the top four but may still have an important presence. For example, in the US film industry, the shares of firms ranked number five and six in 2013 (Lionsgate Films and Sony Entertainment) are quite similar to those of the top four. Also, the aggregation of the C4 distribution does not account for the distribution of shares within it, which can vary greatly. A C4 of 60% does not show whether competition is relatively high—the top four firms each have about 15% market share—or low (one firm holds 55% of the market share in the industry, the remaining three combined account for just 5%).

The HHI rectifies this shortcoming of the C4 by squaring the market shares of the companies in an industry and then adding them up, thus giving extra weight to high market shares (Graph 32.6). The resulting sum ranges between 0 and 10,000 points (where one company holds 100%, and its squared market share is thus at 10,000). The US government's anti-trust enforcement guidelines hold that HHIs under 1,500 are defined as "unconcentrated"; "moderate concentration" occurs in the range of 1,500–2,500, and the "high" concentration starts at 2,500. (These thresholds had been raised considerably in 2010. Until then, they were 1,000 and 1,800 respectively).

Although the HHI index is more informative, it lacks the intuitive simplicity of the C4 index. It is easy to assess that a market share of the top four firms of 80% is high. It is less clear what an HHI of 2,600 means.

Findings:

- Looking at concentration around the world, our first finding is that it is high everywhere.

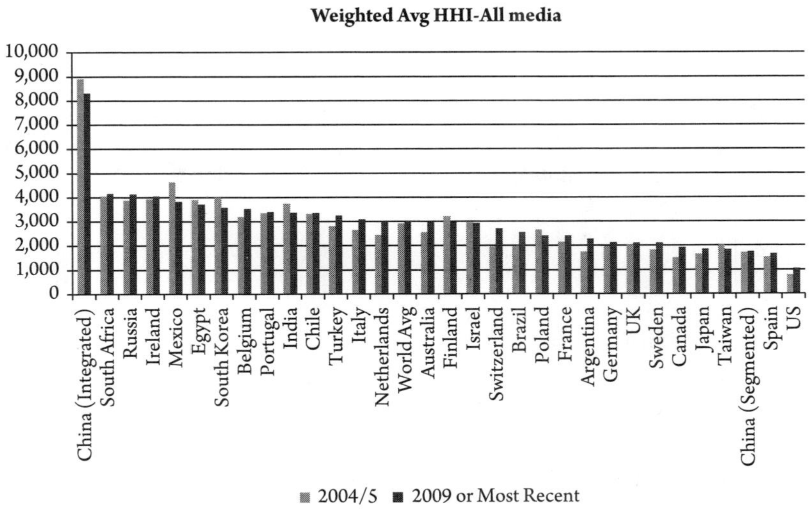


GRAPH 32.6 Countries' Weighted Average HHI – All Media

The average media industry in the average country has an unweighted HHI of 3,448 — well inside the range of "highly concentrated" of antitrust.

- World average weighted national HHI shows a high level of concentration that is declining, from 4,306 to 3,729, i.e., by –1.9% p.a.
- The countries with the highest weighted average HHIs are China (integrated, 9,700), South Africa (5,535), Egypt (5,041), Turkey (5,012), and Mexico (5,315). The first three have high HHIs due to state ownership of media, the second due to privately owned conglomerates favored by the state. In South Africa, the state operators SABC in broadcasting and Telkom in ICT infrastructure dominate, while the private company Naspers is strong in the print and video channels industries. Mexico is dominated by two media groups, Carso (Telmex) and Televisa.
- The countries with the lowest weighted average HHIs are Brazil (2,773), France (2,694), the Netherlands (2,754), Canada (2,506), and the United States (1,389). India is relatively low (2,935 in 2012) and declining (it was 4,906 in 2004) because the historic government monopolies in telecom and broadcasting are offset by media diversity in other sectors.
- In the United States, only search engines and wireless have HHIs over 2,500 (highly concentrated), while for audiovisual media, the average HHI is under 1,500. To visualize that number, an average HHI of 1,171 would be achieved by six equal-sized firms holding 14% apiece. Or, it is reached by five firms with shares of 25%, 17%, 13%, 8%, and 5%.

Graphs 32.6A and 32.6B separate out content and platforms' weighted HHIs.

- Average national content media HHI is very slightly increasing—from 2,871 to 2,906 — while platform is decreasing (from a very high 4,872 to a still high 4,026). This reflects the overall global media trends in which content ownership remains consolidated while in platforms the mobile providers are becoming more competitive at the expense of the more monopolistic wireline industry.
- The United States is the least concentrated national market for both content (1,080) and platform media (1,742). China has the highest HHIs in both sectors. Even when state ownership is segmented, its platform concentration is well above the world average.
- Of all the countries surveyed, India's content and platform sectors show the biggest divergence: in platforms, the country is low, at (3,001) in international comparison

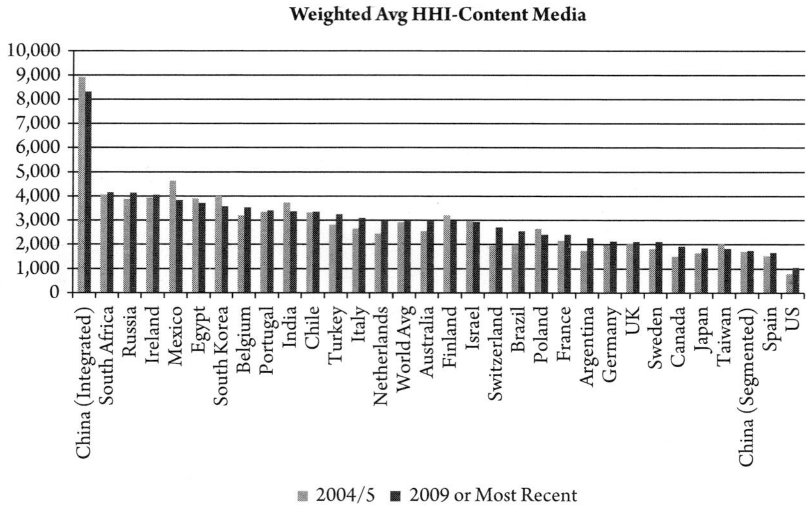


GRAPH 32.6A Countries' Weighted Average HHI – Content Media

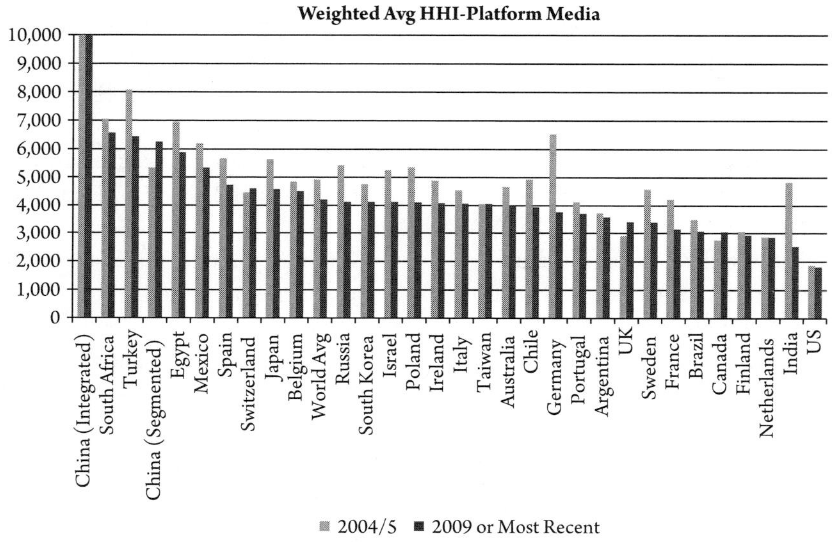


GRAPH 32.6B Countries' Weighted Average HHI – Platform Media

(4,027), but is above average concentrated in content media (2,673) by international comparison (2,907 world average) due to the legacy of state ownership in broadcasting.

- Content HHI concentration rose in the United States, Spain, Canada, Germany, Sweden, the United Kingdom, Argentina, France, Brazil, Switzerland, Australia, Netherlands, Italy, Belgium, Portugal, Ireland, and Russia.
- It declined in China, Mexico, South Korea, India, Finland, Israel, Egypt, Turkey, Poland, South Africa, Taiwan, and Russia.
- For platform media, HHI concentration dropped in most countries, in particular India, Sweden, France, Germany, Chile, Poland, Israel, South Korea, Russia, Japan, Mexico, and Turkey. It rose in the United Kingdom, Australia, and Argentina, and stayed flat in both China (10,000 when integrated) and Canada.

## CONCENTRATION MEASURED BY NOAM INDEX

The Noam Index (NI) (so named by others, not by the author, and kindly accepted) is a diversity index that uses the net *n* voice measures for industries and countries to account for media pluralism (Graph 32.7). It uses market shares as the HHI does. It also incorporates the number of voices present to calculate how pluralistic a market is in terms of number of outlets (voices). The NI is calculated by taking an industry HHI and dividing it by the square root of the number of voices present in that industry. It is a content media measure and thus not applied to platform industries.

- The world average of the Weighted Noam Index has been slightly decreasing: from 1,627 in 2005 to 1,621 in 2013, which means content diversity has increased overall worldwide. There is a unique reason for this, even as content concentration increases overall by other measures:
- Emerging markets have been driving this slightly downward trend, such as China (down from a hugely high 8,902 to 7,491), South Africa (from 2,442 to 2,351) Russia (from 2,066 to 1,700), South Korea (from 1,874 to 1,716), Mexico (from 2,576 to 2,078), India (from 3,330 to 1,710), and Taiwan (from 1,060 to 980). New market entrants in content media increased the number of net voices. Of such emerging economies, only Brazil is an exception, with the index rising to 1,338 in 2013 from 1,157 due to market consolidation.
- If these emerging markets are excluded, the weighted average Noam Index for the world rises substantially: from 1,154 in 2004 to 1,326 by 2013.
- In the United States, the Noam Index increased from 227 in 2004 to 409 in 2013, but still the lowest measure of all 30 countries. The second lowest is Japan, at 635 and declining.

## CONCENTRATION MEASURED BY POWER INDEX

The industry-specific concentration measures (HHI, C4, NI), as well as the averages across industries are useful measures, but they do not capture cross-industry, multi-media market power. A firm may have moderate market shares in multiple media industries but no dominance in any. Looking only at one industry market at a time would understate a company's vertical and horizontal market position.

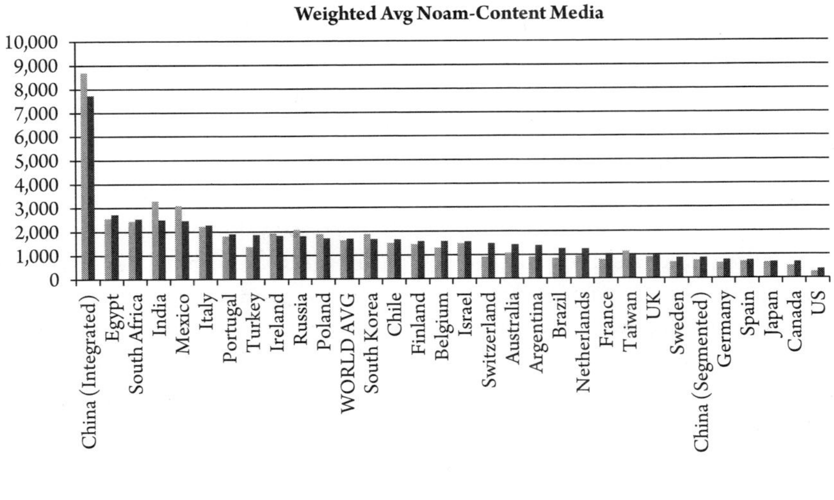


GRAPH 32.7 Countries' Weighted Average Noam Index – Content Media

How can one measure such cross-ownership? We propose a separate index for this: the Media Power Index (MPI).

There are several variations on the Power Index. They reflect the three dimensions of aggregation: across companies, across nations, and across industries. Of the eight possible combinations, five are of interest:

The *Media Power Index* for a single company in a single nation (MPI-CN) expresses a company's cross-ownership across the several media industries, but in a single country.

Second, the Media Power Index for each company in a single country can be aggregated for that nation's Media Power Index (MPI-N).

Third, a Media Power Index can also be calculated for a company across the world in a single industry. This is the MPI-CI.

Fourth, a Media Power Index showing the global power of a company can be calculated across both the various industries and the world's nations. This is the MPI-CW.

A fifth measure aggregates these global company power measures to one covering all industries, all companies, and all nations. This is the Media Power Index Global (MPI-G).

In the first case, a firm can be compared to other companies in the national market. The basic formula starts with the market share of a company in an industry, and squares that share, just as for calculating HHI. The result is then multiplied by the share of total industry revenue in that country in the total value of all the country's 13 media industries. This is done for all of the industries where the company is active in a particular country, and added together. For example, Telstra in Australia is active in the following industries: Multichannel 45%; Wireless 58%; Wireline 71.4 %; and ISPs 42.2%. It has a company PI of 2,577 in the latest year. The sum of all of Australia's media industry revenues is $50.1 billion, including the nine industries in which Telstra does not have a presence. Telstra, then, has a MPI-C of:

(((45%)^2*$3.3 billion) + ((58%)^2*$15.4 billion) + ((71.4%)^2*$13 billion) + ((42.2%)^2*$2.7 billion))/$50.1 billion = 2,577.

This represents the "media power" the company has in the Australian national market because it accounts for market power across several industries. Companies with the highest PIs are those with large market shares in high-value industries, which is why, for example, Google has a Company Power Index in the United States of 147 compared to AT&T's 385: Google in 2012 was only active in 1 of the 13 industries (search engines), and this industry produces much less revenue than the platform industries in which AT&T is active. (On the other hand, Google has a strong presence almost everywhere in the world and has thus a much higher Global Power Index in content media. More on that in the "Companies" chapter.)

A *National* Media Power Index is the sum of all company-PIs in a country, so for Australia that would be Telstra's 2,577 plus Rupert Murdoch's enterprises combined CPI of 433 plus Nine Entertainment's 131 plus SingTel's 289 and so on—which are all calculated using the same method as Telstra. When all companies are added together Australia's National Media Power Index (MPI-N) comes to 3,984. A high national PI relative to other countries means that there are significantly large companies that have large market shares in high-revenue media industries.

The world average National Media Power Index, for all media, stands at 3,701; down from 2004's 4,229.

In Graph 32.8 we can see the countries with the highest National Media Power Indices are China (9,702), South Africa (5,536), Mexico (5,298), Turkey (5,143), Egypt (4,774), Russia (4,092), and South Korea (3,997). These are, arguably, the countries with the greatest overall media concentration. Their high National Media PIs, though decreasing from the early 2000s due to liberalization and greater competition, stem from the larger cross-holdings that individual media groups (including the state as an owner, in China, South Africa, Egypt, Russia, and South Korea) hold in their national markets, or from one company holding a majority of the shares in a single high-value industry. For South Africa, these are Telkom (both a monopolist and cross-holder) and Naspers. In Mexico, they are the company power indices of America Móvil and Grupo Televisa. Turkey shows a similar

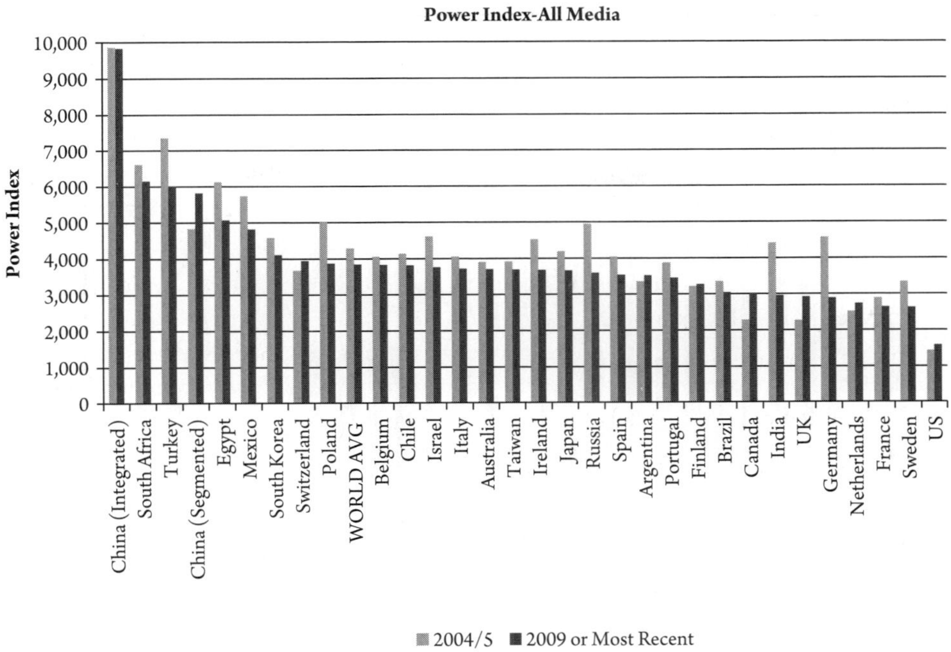


GRAPH 32.8 Country Media Power Indices – All Media

trend. Cross-ownership is a major cause of the high concentration, with few companies controlling most industries. For example, Dogan holds some of the highest market shares across industries including newspapers, magazines, radio, TV broadcasting, and online news media. The high National Media PI, though, is primarily the result of Turkcell and TT's domination of the telecom industries. South Korea is the result of a highly concentrated ICT sector and a large state role in audiovisual news media. Russia's stems from high telecom concentration and the government's television shares.

The countries with the lowest National Media Power Index are Germany (2,603), the United Kingdom (2,888), the Netherlands (2,726), Sweden (2,887), France (2,524), and the United States (1,482). Although major global companies are present in these markets, their market shares are rarely dominant enough to generate high national media Power Index scores. The US MPI-CN reflects the moderately oligopolistic nature of its media industries: though the largest firms have significant market share, near-monopolies are not present in high-revenue media industries on a national basis, except in search engines. But it is also rising, from 1,386 in 2005 to 1,482 in 2013.

## CONCENTRATION MEASURED BY POOLED C4

We move to the next set of concentration measures. A "pooled" concentration measure such as a pooled C4 looks at the market share of the top four companies in an entire national media market (Graph 32.9). Unlike the average weighted C4, it does not present a mean of concentration across all 13 industries, but describes the companies that dominate the national media market overall.

On average, the pooled overall sector C4 for the world is 66.8%, which means that, on average, four companies control 66.8% of each country's national media market. This is a high percentage primarily because it includes the inherently less competitive platform industries. We will therefore separate at platform media and content media:

- Pooled Overall Sector C4 is lowest in Germany (49.9%), in the United Kingdom, and in the United States (both 42.7%).

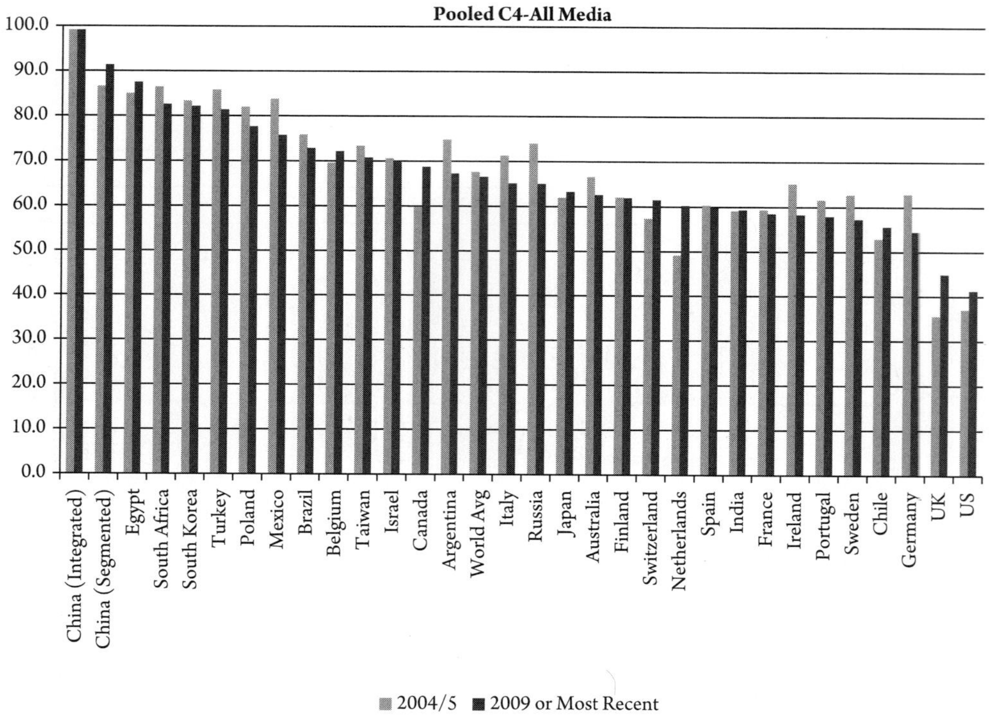


GRAPH 32.9 Countries' Pooled Overall Sector C4 – All Media

- If we separate pooled C4 for content and platform, the average concentration diverges. For content, the world average is 49.9%, much lower than it is for platforms, where it is a huge 86.7%. In other words, four firms (at most) account for almost all platform revenues in each country.
- When content alone is measured, China (91%), Mexico (69.9%), Portugal (68.9%), Turkey (67.2%) Ireland (66%), South Africa (63.4%), Italy (59.7%), and Russia (55.3%) have the highest pooled C4s. There are not many companies per industry in these countries' high-revenue industries. In Italy, Berlusconi's Fininvest and the public broadcast RAI dominate the content market.

## CONCENTRATION MEASURED BY POOLED C1

We move to the next set of concentration measures. A "pooled" concentration measure such as a pooled C1 looks at the market share of the top company of an entire national media market (Graph 32.10). It does not present a mean of concentration across all 13 industries like pooled C4, but describes the companies that dominate their total national market by revenue.

On average, the pooled overall sector C1 for the world is 32.2%, which means that, on average, one firm controls 32.2% of each country's national media market. Pooled Overall Sector C1 is lowest in Ireland (Eircom, 23%), Portugal (Oi Telemar, 23%), Finland (Elisa, 22.1%), India (Prasar Bharati, 20.5%), the United States (AT&T, 15.8%), and the United Kingdom (Murdoch, 16.4%). It is generally the case that pooled platform C1s are the same as pooled "all media" C1s globally, since the lead firm is usually a telecom firm.

If we separate pooled C1 for content and platform, average concentration diverges. For content, the world average is 23.6%, much lower than it is for platforms (42.6%). Only Mexico (45.7%) and China (79.2%) exceed 40% for pooled content C1.

When content alone is measured, China (state enterprises cumulative: 79.2%, of which CCTV has 18.8%), Mexico (Televisa, 45.7%),

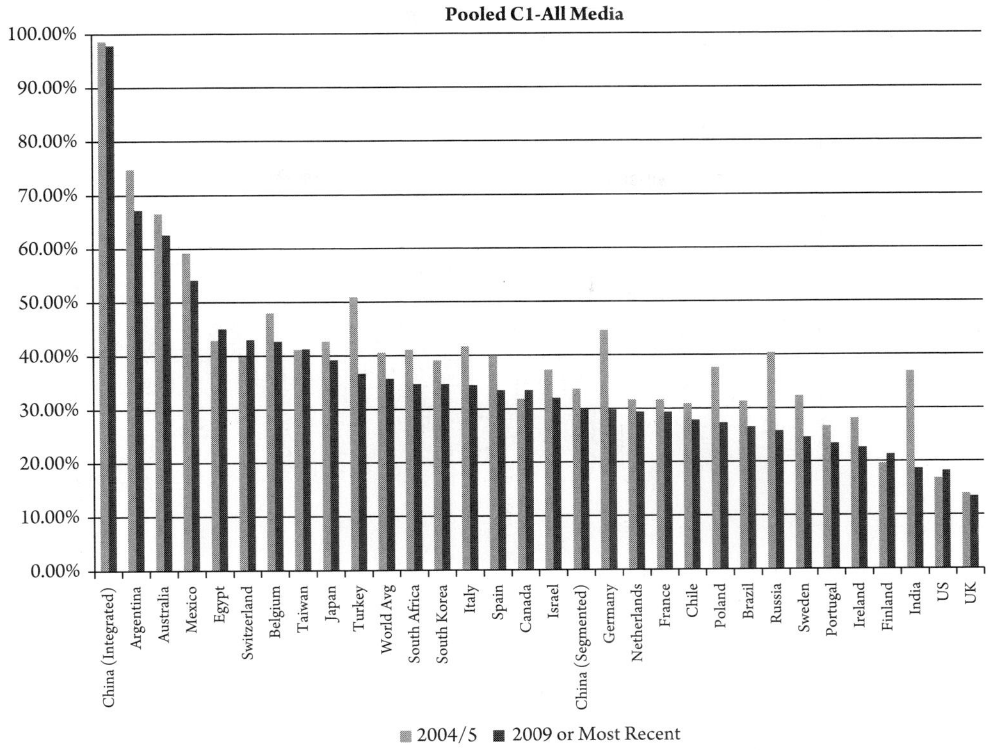


GRAPH 32.10 Countries' Pooled Overall Sector C1 – All Media

Egypt (state enterprises cumulative, 35.2%), Russia (state enterprises cumulative, 36.4%), and Turkey (Dogan, 33%) have high pooled C1s. Brazil and South Korea are also over 30% (Globo and the public broadcasters, respectively). Argentina is also high (Clarín, 24.3%), along with Italy (Fininvest, 23.6%). Ireland (Independent News and Media, 29.4%) and South Africa (Naspers, 26.7%) are also above average. The United States' top firm, Comcast, scores 9.3%.

## POOLED HHI

The Pooled HHI shows what the C4 index does not reveal: the distribution of market shares of the very top companies. The difference here is that the pooled measures use the major companies' shares of the national market as a whole, instead of treating each industry separately.

Graph 32.11 shows the world average for pooled overall HHI stands at 1,824 — down from 2,116 in 2004. This is, of course, much lower than the average sector HHI on an average weighted industry basis: the world average weighted HHI is 3,729. For content, it is 1,132 and for platforms, it is 2,908.

The countries with the highest pooled overall sector HHIs are China (9,311), Mexico (3,777), Egypt (2,853), South Korea (2,162), and Australia (2,038). These countries had large top-company shares in the national markets held by a few firms—America Móvil and Grupo Televisa in Mexico, Telstra in Australia, Mobinil and the Government of Egypt in Egypt, Liberty (Telenet) and Belgacom in Belgium, and the Government of China. South Africa (1,909) and Taiwan (1,904) also have high HHIs.

The countries with the lowest pooled overall sector HHIs are India (1,140), Finland (1,124), Germany (1,085), The Netherlands (1,001), the United Kingdom (689), and the United States (616). These lower-than-average pooled HHIs indicate that no company in these countries is particularly big in an overall sense, even if it might be large in a particular industry, such as Deutsche Telekom and AT&T in the German

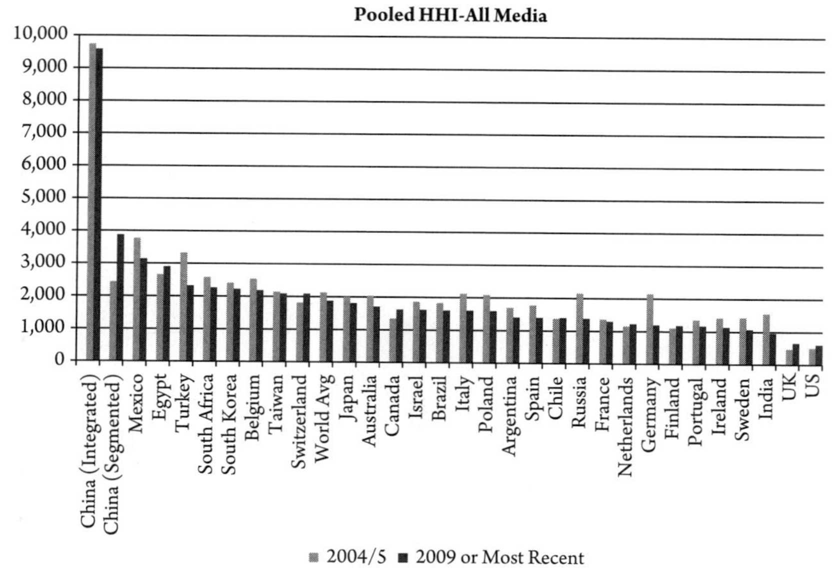


GRAPH 32.11 Countries' Pooled Overall Sector HHI – All Media

and American wireline markets (respectively), or the state in Indian broadcasting. The US pooled overall sector HHI of 564 reflects an intermediate share (relative to other countries) by the very top companies like AT&T and Verizon.

In both content and platform pooled HHI, China and Mexico stand near the top due to heavy concentration in their audiovisual and telecommunications industries. Russia's relatively high pooled content HHI of 1,522 indicates the extent to which the government has invested in media ownership (Gazprom, First Channel, and VGTRK are controlled by the Kremlin, and are the largest broadcasters outside of the privately owned enterprise CTC). Germany's significant platform media decline (from 5,354 in 2004 to 2,663 in 2013) is the result of Deutsche Telekom's market losses to new entrants in wireline, wireless, and ISPs.

## FOREIGN OWNERSHIP

We now move to another type of measure, that of foreign ownership.[4] This is, strictly speaking, not a measure of concentration, but it is nevertheless an indicator of the openness of a country's media market to the outside and the absence of barriers to such entry, which tends to raise competitiveness. Such barriers might be legal/regulatory, or simply those of difficulty in adapting to a foreign market with its special circumstances.

Foreign ownership of media has been increasing, rising from a world average of 24.5% in 2004 to 28.2% in 2013 for all media industries. The growth has been driven primarily by internationalization in telecom, especially mobile, and in the non-broadcast audiovisual media, enabled by usually more relaxed laws favoring domestic ownership.

To calculate the percentage of foreign ownership, we ignore small stock ownership by investment funds seeking diversification by purchasing stakes in foreign companies. The companies considered in Graph 32.12 are controlled by foreign entities. Their revenues are added together and calculated as a share of a country's overall media industries' revenue.

The countries with the highest percentage of market shares owned by foreign principal owners are Ireland (72.1%), Brazil (65.8%),

4. In this section, there is no need to distinguish between China's "integrated" and "segmented" ownership indices because they are the same, so China is rendered as just "China."

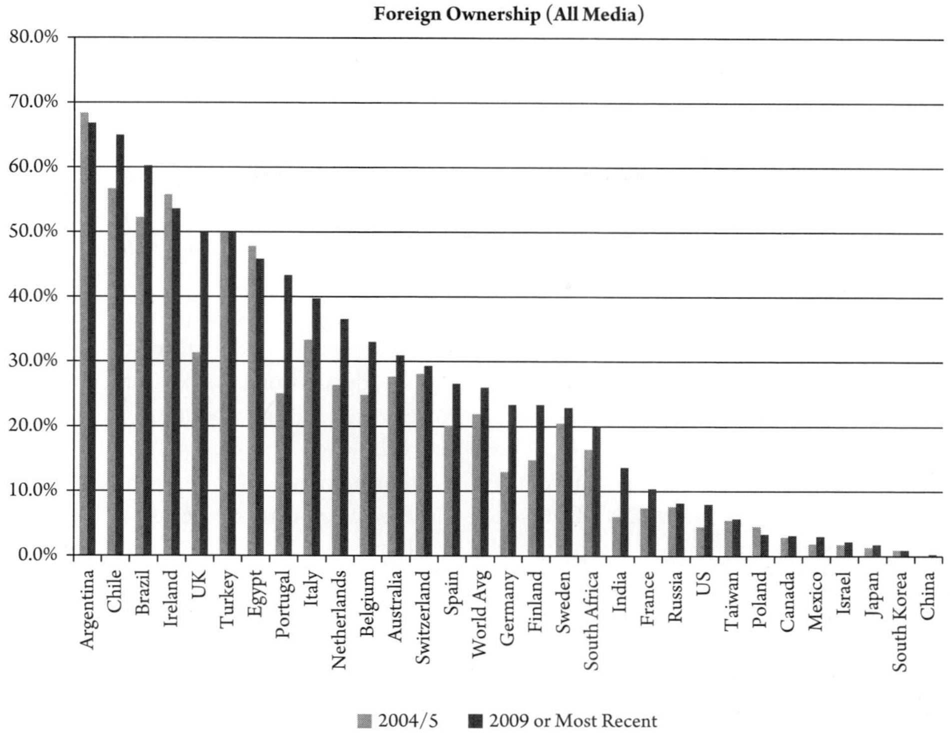


GRAPH 32.12 % Foreign Ownership in Country Markets – All Media

Argentina (62.8%), Chile (62.4%), and the United Kingdom (52.1%).

Spain's Telefónica is responsible for the high concentrations of foreign ownership in the Latin American countries. US-based 21st Century Fox and News Corp., as well as foreign telecoms, account for the UK's share, especially with the formation of EE, a joint T-Mobile and Orange mobile and ISP service. Ireland has historically been a place for UK and US investors, now joined by Asian ones; major companies include Liberty (US), mobile provider Eircom (owned by SingTel of Singapore), Rupert Murdoch's companies, and also Telefónica.

In Chile, foreign investors from Mexico, Spain, and Argentina control most of the media market encouraged by proximity of geography, culture, or language. Argentina's own high foreign ownership percentage stems from the domination of its telecom industries by three international groups: Telefónica, Telecom Italia (and now by the Mexican investment company Fintech), and Grupo Carso's Telmex. This followed the sale of the monopoly state telecom operator in order to reduce the country's debt burden.

A similar trend is apparent in Brazil, where Telefónica, Telmex, and Telecom Italia (dominated by Telefónica) have high shares. Foreign ownership in Turkey is also high (49.3%), solely due to the privatization and sale of Turk Telecom, the former public incumbent, to the Oger Group of Saudi Arabia and Lebanon. Effectively, Turkey traded a high public ownership percentage for a high foreign one.

The countries with the lowest foreign ownership are countries with major regulatory restrictions of such ownership: Japan (1.8%), South Korea (0.9%), and China (0.5%) almost entirely in licensed magazines and Hollywood film imports. In South Korea, there is a state-mandated screen quota system that requires every movie theater to show Korean-made films 40% of the year. The Chinese government also has many restrictions on media ownership and import for all industries. South Korea's low foreign ownership rates are the legacy of security concerns. In Israel, foreign ownership has been growing (to 17.5%) because ownership rules are being relaxed. American owners have been permitted to buy into Israeli news media (TV stations and newspapers), and there is significant French ownership in cable platforms (Altice-HOT). Mexico used to have high barriers, but Telefónica's entry changed this. Foreign ownership in Mexico was 20% in 2012: still much lower than it is in Chile (62.4%), Argentina (62.8%), or Brazil (65.8%).

Between 2004 and 2013, the biggest increases were in India, Portugal, the Netherlands, Germany, Spain, and the United Kingdom. The Netherlands and Spain experienced more US companies' entry, and in Germany' Deutsch Telekom lost its near-monopoly on platform media. In Portugal Oi of in Brazil obtained a majority (62%) control over Portugal Telecom (which retains 38% control of their venture, after Oi's 2013 buyout). In India, Google (US) remains the most dominant foreign company with an 81.4% share in the search engine industry. Telecoms constitute the bulk of non-domestic ownership Vodafone (UK) has a 12.3% share of the national market in India, for example.

The United States has a foreign ownership percentage of 8.6%, mostly due to the mobile carriers Sprint (now owned by Softbank of Japan) and T-Mobile (Deutsche Telekom, Germany). Vodafone (UK) used to have a 45% stake in Verizon Wireless, but resold it to its partner Verizon in 2013. The United States has restrictive foreign direct investment regulations on the ownership of broadcasting assets, which has retarded efforts of Grupo Televisa of Mexico, a country with even more restrictive media ownership laws, to expand its presence to the U.S. Spanish-language broadcasting market. Televisa, however, has significant convertible debt in Univision, and large blocks of shares are held in place by several US private equity funds, ready for conversion into stock when the law permits.

Like public ownership of media described below, ownership by foreign entities is a double-edged sword. As part of a global market power it reduces competition, as in the case of the Hollywood film studios. On the other hand, it can bring in new players with deep pockets to contest national incumbents, as in the case of wireline telecom. We find that content media industries are significantly domestically owned, but that foreign ownership has been increasing overall in the world (Table 32.4). The major exception is mobile telecom where, where foreign ownership is high. It is highest in Latin America, the Middle East, and in Europe.

Foreign ownership within content and platform industries varies greatly, averaging 21% worldwide for content, an increase from 18.3% in 2004. Within national media markets, platforms usually account for most of the non-domestic ownership through mobile operators. Graph 32.12A shows foreign ownership in content media, which is highest in the Netherlands (55.2%) due to the growth of American and German companies there, and Ireland (39.4%), due to the presence of British outlets in newspapers, video channels, multichannel

Table 32-4. Foreign Ownership (All Media): By Region

| | 2004/05 | 2011 OR MOST RECENT |
|---|---|---|
| Latin America | 51.8% | 52.7% |
| Middle East | 32.5% | 39.0% |
| Europe | 26.4% | 31.8% |
| BRICS | 19.4% | 23.9% |
| Asia-Pacific | 10.7% | 10.4% |
| North America | 3.3% | 5.5% |

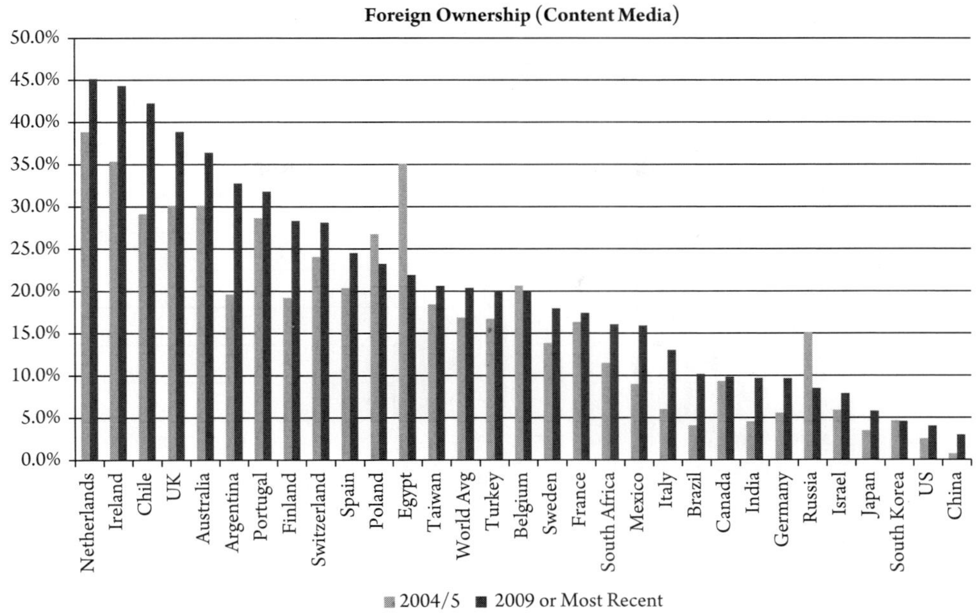


GRAPH 32.12A % Foreign Ownership in Country Markets – Content Media

platforms, and broadcast TV. It is also high in the United Kingdom (for European mobile providers in telecom) and in Chile, where telecom is dominated by foreign firms. Foreign ownership of content media is lowest in Russia (9.8%), Japan (5.8%), the United States (3.1%), South Korea (4.9%), and China (3%). Table 32.5 shows the regional breakdown: content media ownership abroad is highest in Europe at 26.9%, due to the cross-border presence of Murdoch's News Corp. and 21st Century Fox, Liberty, ProSiebenSat.1, Sanoma, Fininvest, Bouygues, PRISA, and Bertelsmann. However, overall transnational holdings are low.

In platform media, however, foreign ownership is much higher: at an average worldwide of 31.5%, up from 26.6% in 2004/05 (Graph 32.12B). It is highest, but static, in Latin America (64% regionally) due to the presence of companies such as America Móvil and Telefónica (Table 32.6). It is lowest in North America (5.7%). Within Europe, Italy and Ireland account for the largest non-domestic platform ownership. In Italy it is accounted by VimpelCom's Wind in mobile telecom and by Telefónica's temporary control over Telecom Italia's holding company, Telco. In Ireland it is caused by SingTel of Singapore (controlled by

## Table 32-5. Foreign Ownership (Content Media): Region

| | 2004/05 | 2011 OR MOST RECENT |
|---|---|---|
| Europe | 23.1% | 26.9% |
| Latin America | 16.6% | 23.7% |
| Middle East | 20.5% | 19.7% |
| Asia-Pacific | 17.3% | 17.0% |
| BRICS | 9.3% | 11.6% |
| North America | 5.2% | 5.3% |

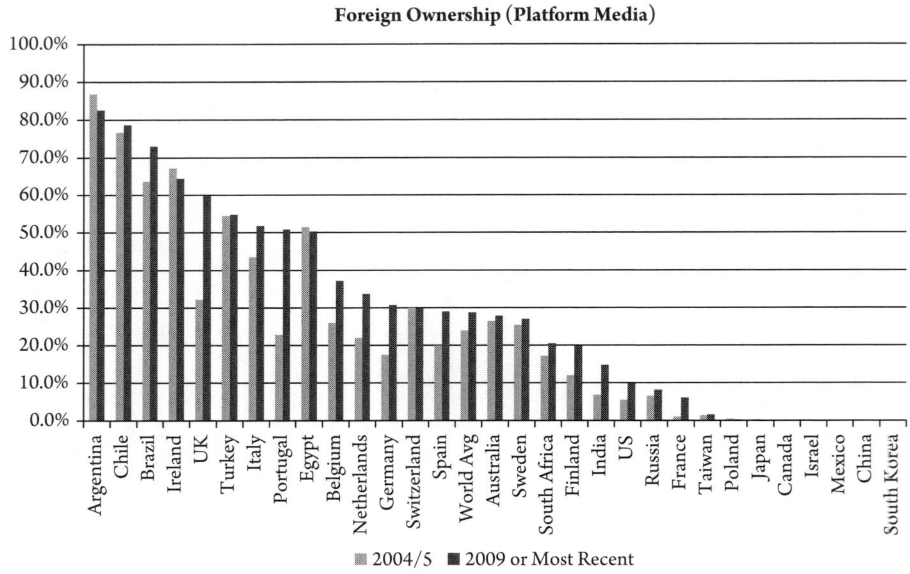


GRAPH 32.12B % Foreign Ownership in Country Markets – Platform Media

the Singapore government's Temasek Holdings) and its control of the former Irish state incumbent, Eircom. In Portugal foreign ownership is high due to Brazil's Oi taking control in a merged Oi-Portugal Telecom (Oi Telemar). The Saudi-owned Oger Group obtained control in Turkey over Turk Telecom, with a 54.8% stake acquired during the privatization of the former state incumbent. Domestic firms dominate Japan, Spain, China, Mexico, France, Israel, South Korea, Mexico, Germany, and Poland in platforms. Of these countries, Mexico, Germany, Spain, and France's platform companies have the largest international export presence (America Móvil, Telefónica, Deutsche Telekom, and Orange).

## PUBLIC OWNERSHIP

The final measure for analysis is public (state or parastatal) ownership (Graph 32.13). We use the term "public" in the sense of "owned by public authorities," as distinguished from the use of the term in the stock market and investor community, where a "public company" means a firm whose shares are publicly traded in a stock market. Why is state/public ownership relevant to concentration? The higher such ownership in an industry or country, the less likely competition is. This can be due to the difficulty of contesting a government operation; or it might be legally prohibited; or, conversely, state ownership might indicate a phase of non-viability of

Table 32-6. Foreign Ownership (Platform Media): Region

| | 2004/05 | 2011 OR MOST RECENT |
|---|---|---|
| Latin America | 63.8% | 64.0% |
| Middle East | 34.7% | 42.0% |
| Europe | 28.0% | 35.7% |
| BRICS | 21.1% | 26.6% |
| Asia-Pacific | 8.3% | 8.3% |
| North America | 2.3% | 5.7% |

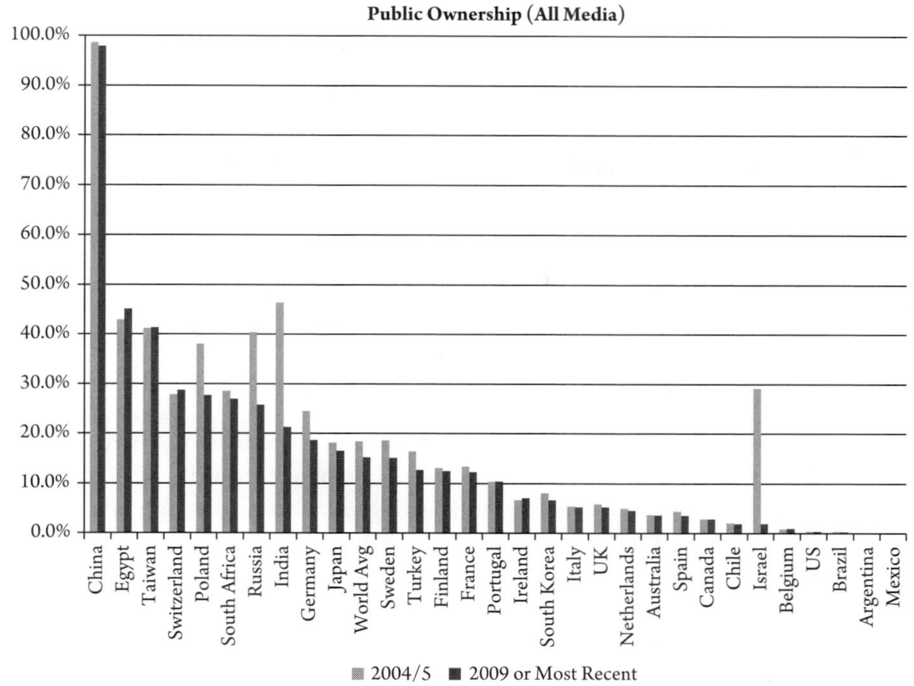


GRAPH 32.13 % Public Ownership in Country Markets – All Media

competition, for example, in the early stages of a new medium. The early days of the Internet in the United States are an example. Public ownership used to be very high in many countries, with both telecom and TV entirely state-owned into the 1980s. Then came liberalization and privatization, which reduced that share.

Public ownership is decreasing internationally: for all media, it declined from 18.6% in 2005 to 15.3% in 2013. Ownership is flat in Latin America for content media (1.7% in both 2004 and 2013); in all other content and platform market regions, it continues to decline (Table 32.7, Table 32.8). It is falling fastest in the Middle East for both content (20.2% to 14.3%) and platforms (34.2% to 21.4%). It is highest in countries where public incumbents, usually in platform industries (wireline, wireless, ISPs, multichannel), control the national cable infrastructure. China, of course, is the largest market with public ownership of all media at 79.2% and of that, 100% in platforms.[5]

## Table 32-7. Public Ownership (Content Media): Region

| | 2004/05 | 2011 OR MOST RECENT |
|---|---|---|
| BRICS | 37.8% | 31.2% |
| Europe | 16.2% | 15.4% |
| Middle East | 20.2% | 14.3% |
| Asia-Pacific | 15.2% | 12.8% |
| North America | 5.1% | 4.7% |
| Latin America | 5.0% | 4.6% |

5. In this section, there is no need to distinguish between China's averaged and state ownership indices because they are the same.

Table 32-8. Public Ownership (Platform Media): Region

| | 2004/05 | 2011 OR MOST RECENT |
|---|---|---|
| BRICS | 44.0% | 35.5% |
| Middle East | 34.2% | 21.4% |
| Asia-Pacific | 10.2% | 9.3% |
| Europe | 11.7% | 9.2% |
| North America | 0.3% | 0.3% |
| Latin America | 0.0% | 0.0% |

Public ownership in content media is highest in China (79.2%), Egypt (35.5%). In Russia (36.4%), and South Korea (31.6%). In Russia and South Korea this is due to state-owned broadcasters, while in Egypt it is a combination of public broadcasting and state control of the main daily newspapers. Underfunding of public broadcasting in the United States (1% in 2013) and Latin America (4.6% regionally) accounts for their tiny numbers.

Several European countries have high public ownership rates, mainly because of telecom and broadcasting. But EU governments do not directly control broadcast TV and radio companies such as the BBC (UK), RTÉ (Ireland), Radio France, or ARD and ZDF (Germany). State ownership is very high in Egypt, South Korea, Taiwan, India, China, and Russia because there, governments directly administer content and platform media companies.

China, not surprisingly for a People's Republic, holds the largest percentage of public-owned market share. China's state ownership of media is the highest by far, as the state effectively holds complete monopolies in all industries except for film, magazines, online news, and search engines. In almost every media industry, the top four companies are owned by a government agency, either at the federal or provincial level. Major public companies in China are China Mobile, China Central Television, Shanghai Media Group, China Film Group, Hunan Media Group, and China Telecom.

The Egyptian government dominates its country's television broadcasting and wireline telecom industries, contributing to its 48.3% market share nationally. Likewise, it formerly held a complete majority over newspaper publishing (94.5%), but saw its market share decline to 72.2% after liberalizing the industry. Other industries in which the Egyptian government is active include radio broadcasting, video channels, ISPs, and wireless telecom. However, ERTU (the state-owned electronic media firm) and the state-owned Nile Television Network (NTN) lost market shares to privately owned video channels like Al Jazeera once these were given permission to enter the market.

The Russian government held a 28.9% market share in all media, down from 40.5% in 2004. The main industries it operates in are broadcast television and wireline telecommunications. The large drop in Russia's market share can be attributed to the large increase in revenues for wireless telecom companies that the state does not control directly. Unlike in China and Egypt, in Russia significant state ownership of content involves holding controlling stakes in corporations through a complex system of holding companies.

In 2004, India had the third largest percentage of public ownership for all media after China and Egypt at 44.2%, due to the monopoly on news content and wireline that public companies have held. However, since then this number has decreased quite dramatically to 22.7%, the result of greater competition emerging in the radio and wireless industries. Privatization gave customers more options, and consumers switched away from the state monopolists in these sectors, the state telecom Bharat Sanchar Nigam and the public broadcaster Prasar

Bharati. Platform public ownership halved from 50.5% in 2004 to 25.4% in 2013.

The countries with the lowest percentage of publicly owned market share are Argentina, Brazil, Mexico, and the United States, all of which had less than 1% of their national media markets held by state-owned media groups (for platforms, it is less than 1% in these countries). In Israel, the state and parastatal share dropped dramatically, from one of the world's highest to one of the lowest, with a major step being the privatization of the telecom incumbent Bezeq. What separates the United States from most other countries in public ownership is that it virtually never had a major state-owned telecom incumbent. AT&T effectively filled that role as a private company for several decades, until its breakup in 1984.

## Bibliography

For the sources mentioned in this chapter, see the general bibliography at the end of this book.

# 33

# European Trends

PATRICK-YVES BADILLO, DOMINIQUE BOURGEOIS, AND JEAN-BAPTISTE LESOURD

## INTRODUCTION

This chapter compares the concentration processes in media and information-related industries in the European countries (Russia and Turkey are included in the analysis). As far as possible, we used five data for five years to envision the long-term dynamics of concentration: 1996, 2000, 2004, 2008, and 2012.[1] The graphs that follow use the HHI calculated in the preceding country chapters, taken as a quantitative indicator of concentration. In a few cases, we estimated HHI by interpolation of actual data.

The result found for several countries, and especially for France, is that industries with a higher technological content are more concentrated than the traditional media industries, which are more labor-intensive. We will test that working hypothesis on an industry per industry basis.

## THE INDUSTRY OF NATIONAL DAILY NEWSPAPERS

In many countries, such as the United Kingdom, the industry of regional newspapers is inexistent or small; in other countries, such as France or the United States, regional newspapers are more important, but highly concentrated in each region, with a monopoly or quasi-monopoly situation. For France, the HHI was calculated only for national newspapers.

In Graph 33.1, we observe two groups of countries: a small group with high concentration indices, around 3,000, and a larger group with lower concentration indices.

1. We have sometimes used data from other proximate years where it was missing data, especially for 2012.

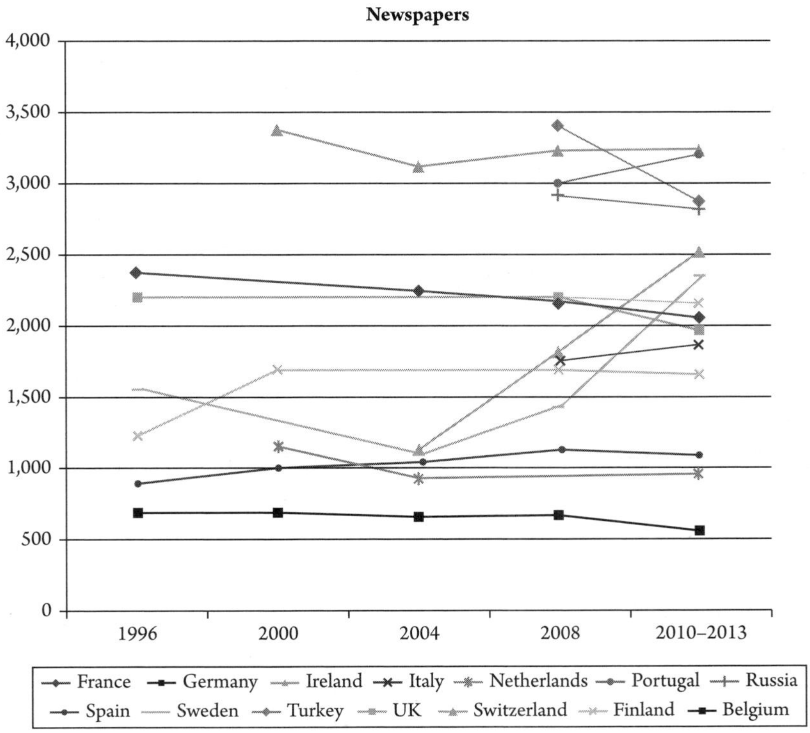


GRAPH 33.1 Daily Newspapers HHI for European Countries

The first group includes Portugal, Ireland, Turkey and Russia. Ireland and Portugal are comparatively smaller countries, for which a high concentration is logical from an economic point of view. Switzerland, a small country, has a lower concentration index, although it increased considerably on the considered period. Turkey and Russia are specific countries with specific media contexts.

Concerning the second group, a majority of countries has an HHI under 2,000.

The main results for the European newspapers industry is a high variability of the HHI.

## BOOK PUBLISHING

Graph 33.2 shows that the European book industry is characterized by a majority of countries with HHI indices under 1,200. Nevertheless three countries have HHI indices around or over 2,500: Portugal, Russia and Finland. With the exception of these three countries the HHI in the book industry is stable and low.

## MAGAZINES

In Graph 33.3 we observe three groups of countries over the recent years:

- Countries with high levels of the HHI, above 3,000: Russia, Portugal and Turkey;
- Countries with a concentration around 2,000: Finland, France, Switzerland and Sweden;
- Countries with moderate or low concentration: Netherlands (which has had a strong decline of its HHI), United Kingdom, Spain and Germany.

## RADIO

One can classify European countries into four groups (see Graph 33.4):

(1) two countries with comparatively low concentration, the HHI being under 1,500: France and Turkey;

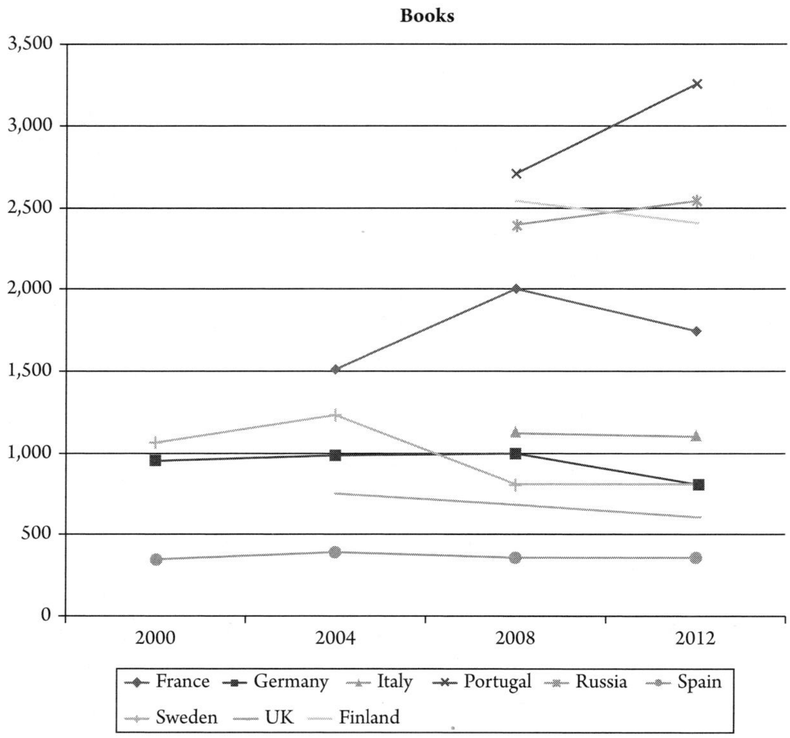


GRAPH 33.2 Book Publishing HHI for European Countries

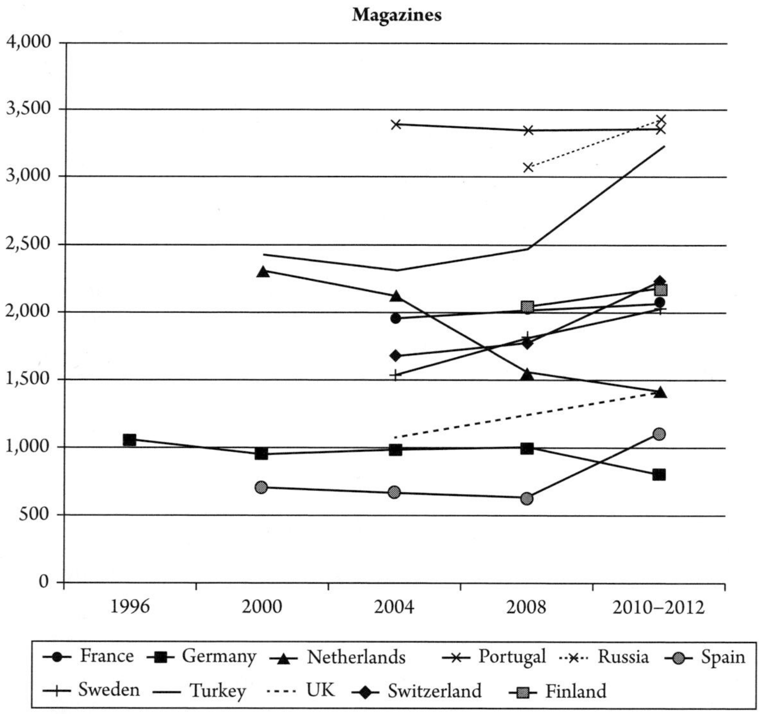


GRAPH 33.3 Magazine Publishing HHI for European Countries

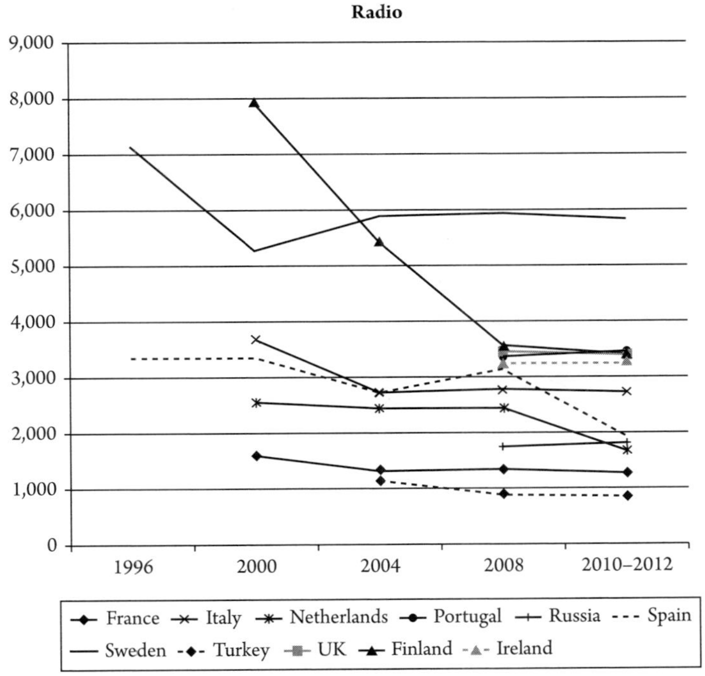


GRAPH 33.4 Radio HHI for European Countries

(2) some countries with the HHI under 2,000: Russia, Netherlands and Spain;

(3) most countries with high concentration, with HHI above 2,800 in the recent years: Italy, Ireland, Finland, Portugal, Germany and United Kingdom; let us underline a strong decrease of the HHI for Finland;

(4) for Sweden, the HHI is close to 6,000.

In general the radio industry has a high level of concentration in Europe, with two exceptions.

## FILM INDUSTRY

The European countries under study constitute a rather homogeneous group with respect to concentration in the film industry. The HHI indicator is below 1,700 over the period 2000–2011 (see Graph 33.5). The film industry is moderately or low concentrated in Europe.

## BROADCAST TV

Graph 33.6 provides the HHI indicators for broadcast television. The data are not fully comparable since the HHI can be calculated on various bases, such as turnovers or audiences.

Moreover the distinction between the terrestrial market and the "video channel" satellite market may be sometimes artificial.

The Turkey, with its low concentration level, can be considered as an exception: its HHI indicator is just above 1,000 in 2012.

Three countries are characterized by a level of concentration around 2,000: Spain, Russia and Sweden.

For most European countries the level of concentration in the broadcast TV industry is high: the HHI is above 3,000. Ireland is the country with the highest concentration, although the HHI decreased over the 2004–2012 period. One can note also a decreasing

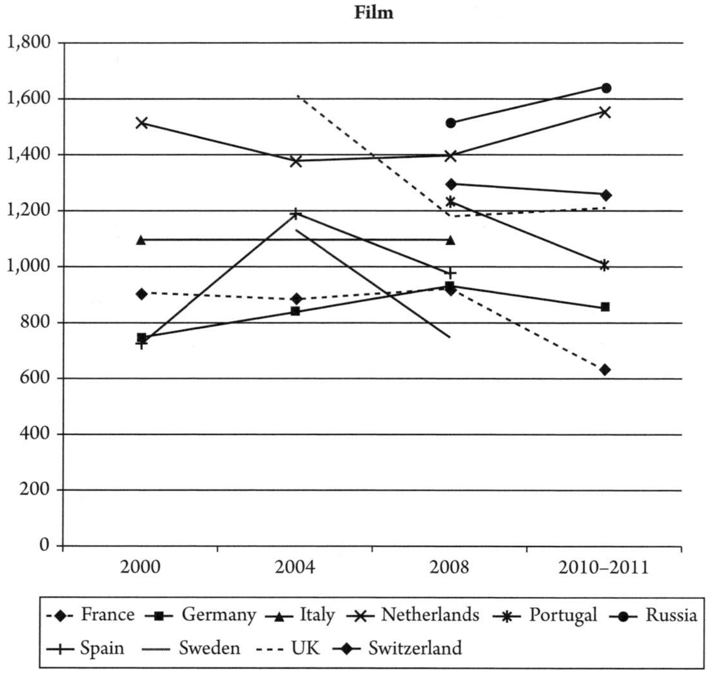


GRAPH 33.5 Film Production HHI for European Countries

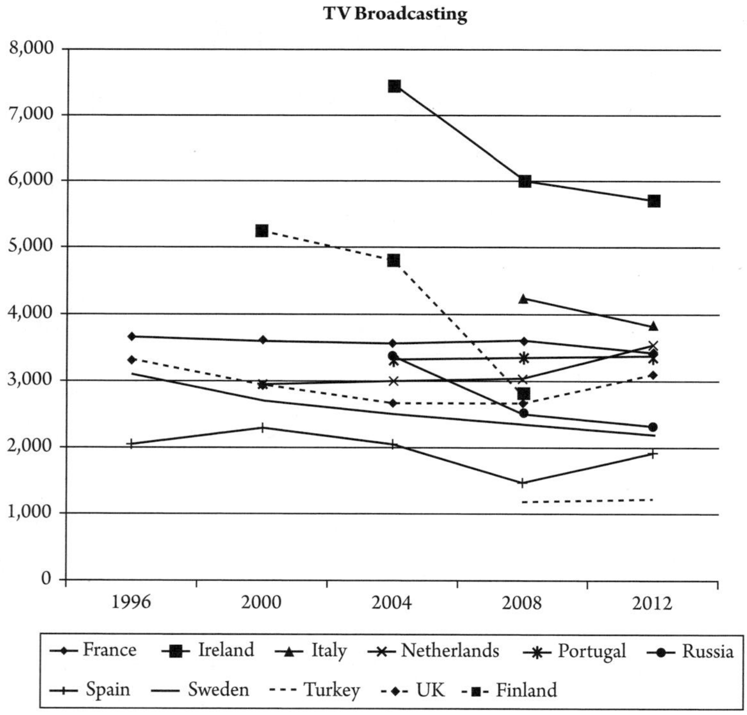


GRAPH 33.6 Broadcast TV HHI for European Countries

trend in concentration for Finland, Sweden, France and Italy for the recent years.

Although situations vary greatly across European countries, the concentration level of broadcast TV is generally high. This conclusion is, however, open to discussion for several reasons: the present situation is very different from that 40 years ago since public monopolies have disappeared, and since the number of available channels is considerably higher than it used to be; moreover, today's analyses that are pertinent for the TV industry must take into account all available distribution technologies, and not only broadcast TV.

## COMBINED VIDEO NETWORK INDUSTRIES (MULTICHANNEL TV)

It has not been possible to make an assessment of concentration in the combined video network industries for European countries. Little data is available on the market shares of cable and satellite TV channels and, when there is data, it is not comparable. Moreover, by taking into account all video network industries (including broadcast TV, cable and satellite channels), concentration may appear quite moderate in some countries, although they may have few companies in cable or/and satellite TV.

In the case of Italy, there is now a quasi-monopoly for satellite TV as the market is controlled by News Corporation (Cable TV is quite nonexistent in Italy). News Corporation has also the largest market share in the United Kingdom for satellite delivered programming. But, in the case of United Kingdom, as noted by Iosifidis[2]: "the audience share of individual satellite (and cable) channels remains small compared with the share of the traditional channels which have developed a digital portfolio of channels. For example, in 2011 the BBC had the largest market share in multichannel homes at 33.2%, followed by ITV at 23.1%, Channel 4 at 11.3% and Channel 5 at 5.9%. The traditional broadcasters have launched numerous digital channels . . . enabling them to get a combined total market share of 73.5% in 2011 for the multichannel market." In the Netherlands, the cable and DBS TV programming channels market largely consists of foreign providers and is dominated by MTV networks (Viacom, US) with a market share of 34.1%. In Germany, the terrestrial transmission of TV has a low audience; when considering satellite and cable TV channels, 37.5% of the viewing time was spent on public channels in 2012 and two TV groups dominate the private market: ProSiebenSat.1 Media AG (KKR/Permira) and RTL Group (Bertelsmann). In Spain, Sogecable (PRISA) is the main company on the market of cable and DBS TV programming channels, but concentration is low because the market is fragmented.

In small countries such as Switzerland, Belgium, and Ireland, if we consider all TV networks, the market share of public channels is still important but people can access many other channels, including foreign channels. In Switzerland, although SRG SSR (public) had about 30% of the market in 2012, concentration for TV networks is indeed low at the national level, because there are many accessible TV channels in various languages. In Ireland there are also many accessible channels, but RTE (public) is dominant with 35.7% audience share and 55% share of advertising revenue in 2011. In Flanders (Belgium), VRT (public) dominates the market of combined video network industries (with an audience market share equal to 43% in 2010), followed by VMMa (with an audience market share equal to 25% in 2010). In Wallonia, the RTL group is the main group with 27.6% of the market (combined video network industries) in 2010, followed by RTBF (public), with 20.4% of the market.

In Europe, the multichannel market has changed considerably since the 1980s because of deregulation and development of cable TV, satellite TV, and digital terrestrial television (DTT). Although public companies remain important

2. See the Iosifidis chapter on the United Kingdom, in this volume.

players in many countries, they have lost market shares to commercial companies, including foreign groups. But there are various landscapes both for distribution technologies and for concentration on the multichannel market.

## MULTICHANNEL VIDEO PLATFORMS: CABLE MSOS, DBS, IPTV PROVIDERS

Graph 33.7 shows concentration for multichannel video platforms. We are dealing with new industries (cable and satellite) that are highly concentrated.

In Italy the (only) satellite operator, Sky Italia, holds 97% of the market share (Cable TV is quite nonexistent in Italy). In the other European countries, the level of concentration generally remains at a high level. France ended up with two quasi-monopolies, one for cable and one for satellite. In Ireland, similarly, the cable market is a monopoly controlled by UPC and the satellite market is controlled by BSkyB. In the United Kingdom, one observes a high concentration (for cable + satellite + DTT). The Dutch cable market is highly concentrated (The Netherlands have only cable TV providers providing regional coverage). One can note that in Portugal, the concentration level decreased between 2007 and 2010 and that concentration is less high in Sweden and Russia.

## WIRELINE TELECOMMUNICATIONS

The industry of fixed telecommunications has been characterized by a deregulation process since the 1980s in Europe. Graph 33.8 below shows the evolution since the year 2,000. The industry remains highly concentrated, with HHI indicators above 2,000 for all European countries; however, a general trend of decreasing concentration can be observed. Two countries still have a very high HHI in recent years: Turkey with an HHI above 8,000 and Russia with an HHI near 7,000. Spain is at the third concentration rank with an HHI above 5,000. For other European countries, the HHI

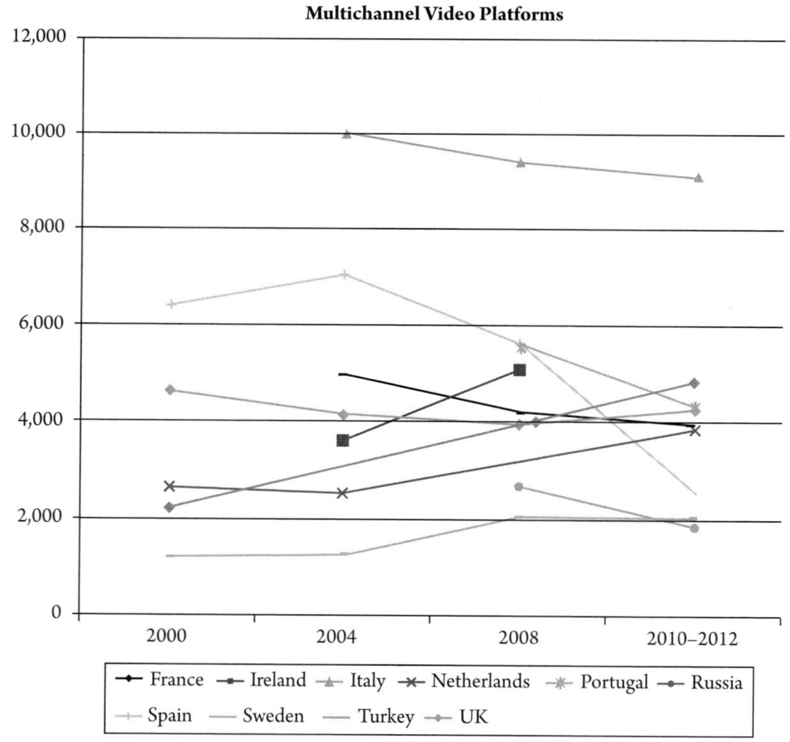


GRAPH 33.7 Multichannel Video Platforms HHI for European Countries

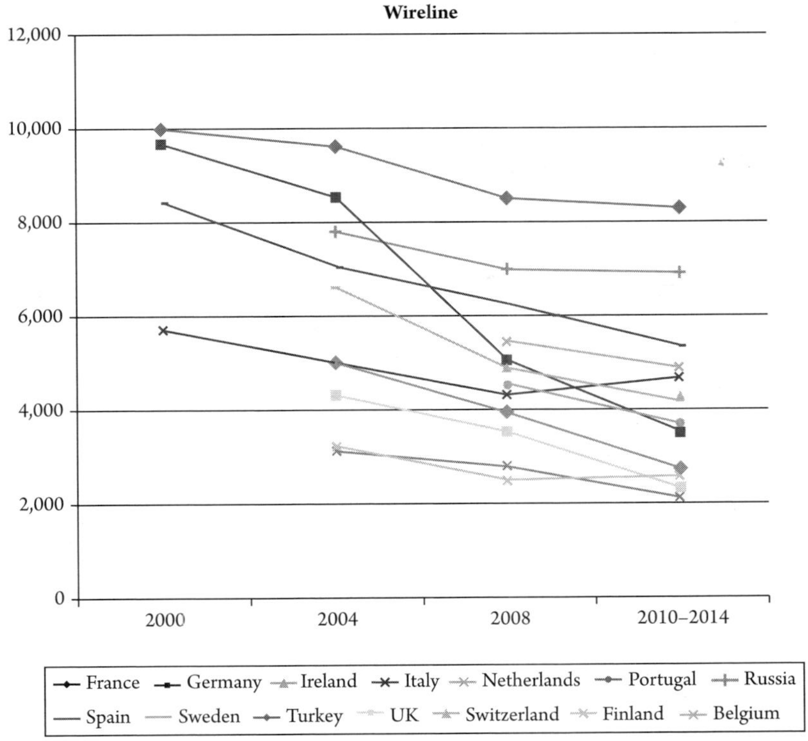


GRAPH 33.8 Wireline HHI for European Countries

indicators remain very high in 2010–2014, with HHI indicators between 2,000 and 5,000 and with decreasing trends.

The incumbent operators still keep significant market shares; the C4 indicators are mostly above 80%, and even equal to 100% for some countries. However, the general trend in industry concentration is downwards, with the mean HHI about 4,200 in 2010–2012, versus more than 8,000 in 2000.

Fixed telephony is declining, because of the growth of the mobile and of the Internet industries. Graph 33.8 describes only partly the evolution of the telecommunication industry.

## WIRELESS TELECOMMUNICATION

Graph 33.9 shows the evolution of concentration for the wireless telecommunication industry. It shows a decreasing trend in concentration (except for the United Kingdom between 2004 and 2010) and, moreover, a homogenization process, since the HHI for various European countries get closer and closer during the 2000–2012 period.

The industry is quite concentrated but with a decreasing average trend, from about 4,400 in 2000 to 3,100 in 2010–2014. With the exceptions of Turkey and Switzerland, all the European countries have HHI indicators under 3,500 in 2010–2012.

## ISP

For Internet Service Providers, graph 33.10 shows that concentration varies among European countries: the HHI indicators are situated between about 2,000 and 3,500 in 2010–2014, except for Turkey which has an HHI around 7,000.

For certain countries one observes strong fluctuations of the HHI indicator: in the case of Spain, the HHI considerably increased between 2000 and 2004, and decreased later; in the case of Ireland, the HHI strongly decreased (from around 7,600 in 2004 to about 2,500 in 2010).

The European ISP industry is highly concentrated (mean value of the HHI above 3,000

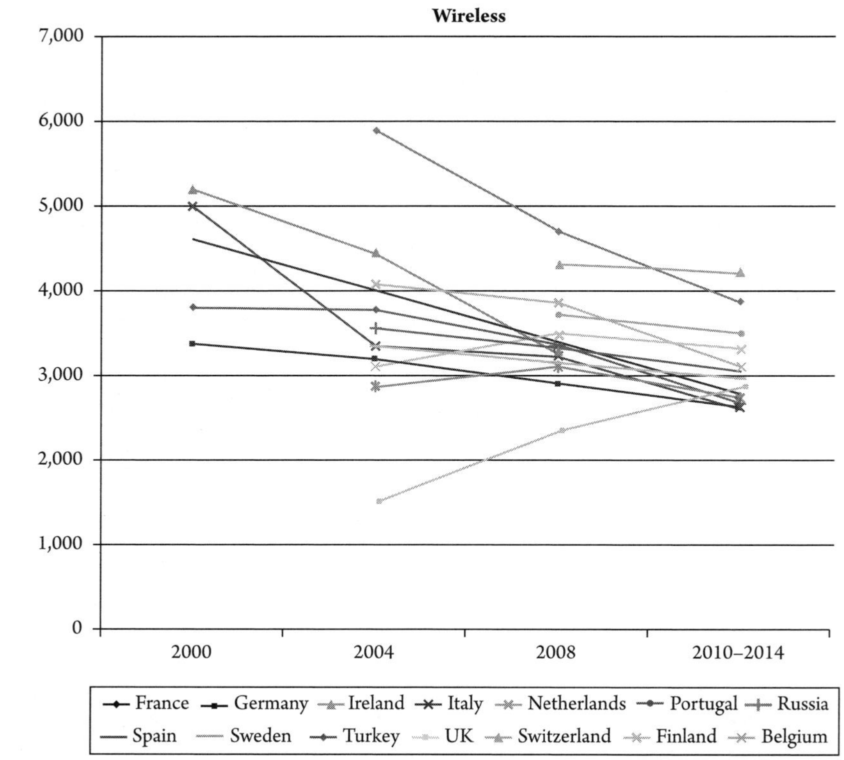


GRAPH 33.9 Wireless HHI for European Countries

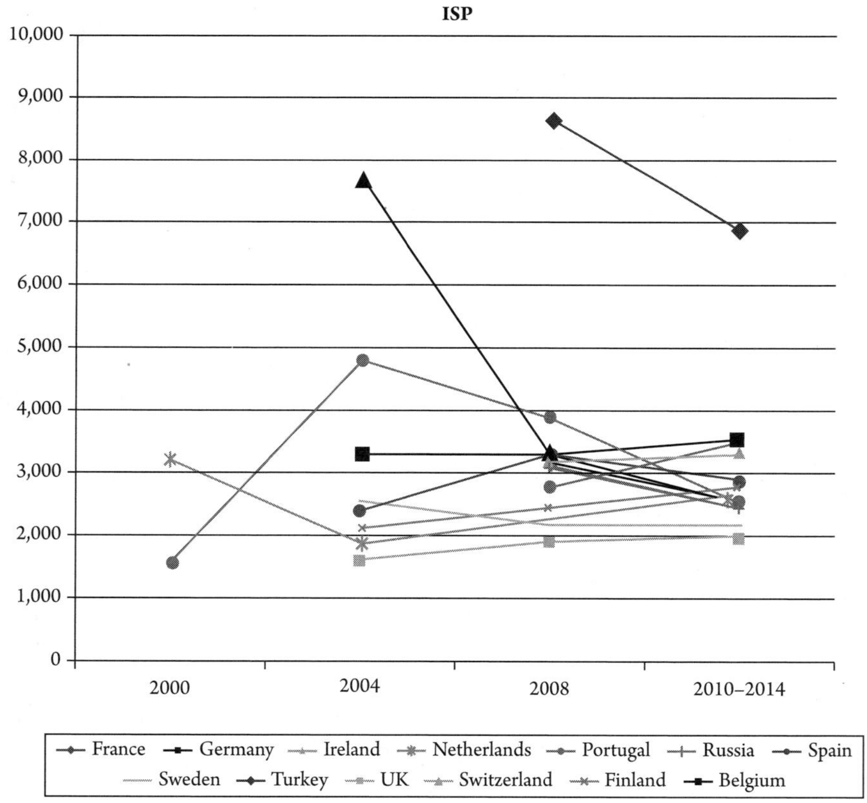


GRAPH 33.10 Internet Service Providers HHI for European Countries

in 2010–2014). Concentration differs among various countries; however, one can observe a trend of homogenization with HHI indicators generally being in the 2,000–3,500 interval in 2008–2014. This level of concentration is lower than that observed in the wireline telecommunication industry, and very near the level observed in the wireless telecommunication industry.

## SEARCH ENGINES

Generally speaking, the search engines industry has been subject to strongly increasing concentration during the 2000s. As shown by Graph 33.11, the cases of France, Germany and Netherlands are very illustrative of this trend. Moreover, in 2010–2013, for European countries, with the exception of Russia, the concentration indicators were very high. Of course, this reflects the fact that generally Google is in a near-monopoly situation, and other search engines having only small market shares today. In Russia there are Russian-owned search engines, and, in particular, the search engine Yandex.ru accounted for 65% of the market share in 2010.

## ONLINE NEWS

The industry of online information is a rather recent and emerging industry, and we do not have data for all European countries. However, on the basis of available figures, one can observe, as a rule, a limited concentration in Europe, since, for most countries, the HHI indicator was below 1,800 between 2000 and 2012. Turkey and Italy are the only countries with HHI over 2,000 in 2012. Graph 33.12 includes data for 2010 and 2012 in order to observe recent trends. In the case of Italy, the HHI decreased between 2010 and 2012. For the United Kingdom, the HHI decreased between 2008 and 2012. In the cases of other countries, we observe increasing concentration indicators between 2010 and 2012. In particular, there was a strong increase of concentration in the case of Spain between 2008 and 2012.

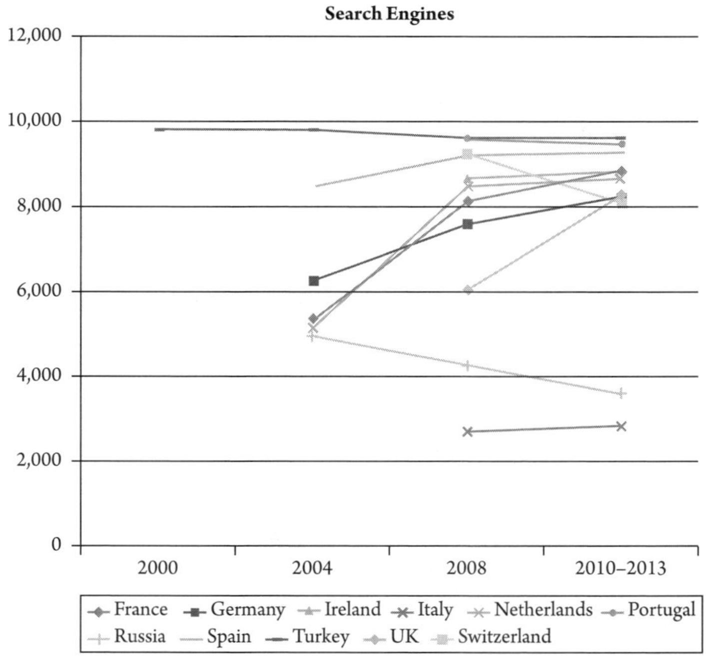


GRAPH 33.11 Search Engine HHI for European Countries

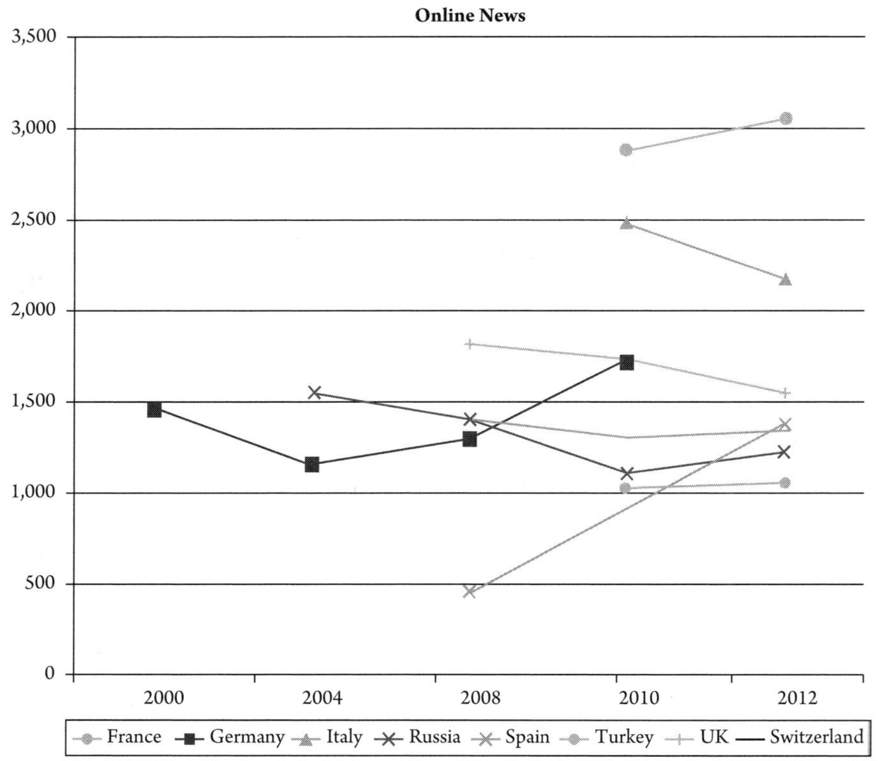


GRAPH 33.12 Online News HHI for European Countries

The HHI for Switzerland was under 1,400 on the period 2008–2012. However, one can consider that concentration is smaller than that for the online information industry in Switzerland: our data only concern Swiss firms and do not include the market shares of enterprises such as Google or Yahoo. For a small country such as Switzerland, concentration in the online news industry is low.

As a whole, in 2010–2012, the industry of European online information was at a low concentration level.

## CONCLUSION

In the beginning of this chapter we considered the hypothesis whereby the industries with a higher technological content are more concentrated than the traditional, more labor-intensive, media industries. Table 33.1 aims to sum up our essential results on the basis of the 2008 and 2010–2012 figures for European countries and Graph 33.13 gives the average values of the HHI for the various industries in 2010–2012. One can see that concentration indices are comparatively small (mean values of the HHI below or around 2,000 in 2010–2012) for most content industries, that is for daily newspapers, books, magazines, cinema, online news, and that concentration indices are comparatively high (mean value of the HHI above 2,000) for the telecommunication industry (wireline and wireless), the cable and satellite providers, Internet Service Providers, and the search engines industry. However, concentration is also important for content industries such as radio and broadcast TV.

Thus, the above hypothesis must be slightly reformulated: while it is true that high technology industries are, as a rule, strongly concentrated in Europe, concentration also seems high in some content industries: this is, more precisely, true of the radio and TV industries. Finally, the traditional industries that produce written contents such as daily newspapers, books, magazines, just as the new industry of online information, are comparatively less

Table 33-1. Mean HHI values in 2008 and 2010–2012 for European countries

| INDUSTRY | MEAN HHI VALUE IN 2008 | MEAN HHI VALUE IN 2010–2012 |
|---|---|---|
| Daily newspapers | 2,152 | 2,008 |
| Books | 1,451 | 1,896 |
| Magazines | 1,931 | 2,116 |
| Radio | 3,375 | 2,968 |
| Film | 1,112 | 1,166 |
| Broadcast TV | 2,826 | 3,067 |
| Wireline telecommunications | 4,668 | 4,264 |
| Wireless telecommunications | 3,466 | 3,119 |
| Multichannel video Platforms (Cable and satellite providers) | very high HHI* | very high HHI* |
| ISP | 3,579 | 3,056 |
| Search engines | 7,285 | 7,391 |
| On line news | 1,143 | 1,692 |

*The average was not estimated because of the variety of situations and the small amount of data available for both the cable and satellite providers industries.

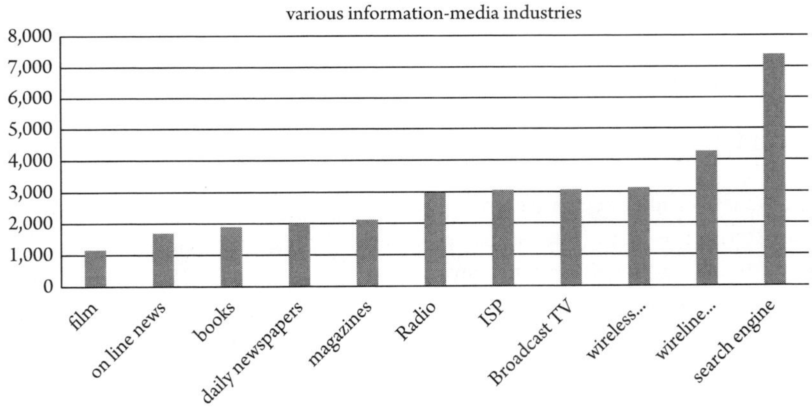


GRAPH 33.13 Mean HHI values in 2010–2012 for European countries

concentrated in Europe. The same is true of the film industry, which comparatively can be considered as a labor-intensive industry.

Finally, one must mention that quite large variation of concentration indices can be observed across European countries for some industries, while a homogenization process with decreasing concentration has characterized the telecommunication industry (wireline and wireless) since the 1990s.

# Bibliography

For the sources mentioned in this chapter, see the general bibliography at the end of this book.

# 34

# Media Industries in International Comparison

## INTRODUCTION

This chapter discusses the concentration of the 13 media industries the countries surveyed. National data for an industry are measured against the other 30 countries. We also compare the major regions of the world by averaging the countries for six regions: North America, Latin America, the emerging BRICS countries, Europe, the Middle East, and Asia-Pacific.[1]

Average HHI is measured in two ways. First, by the arithmetic mean of the industry concentrations in the various countries, which is referred to simply as the average. The second measure is a weighted HHI average, with the weight being the national media industries' revenues as a share in global industry revenues as proxies for size by population and wealth.

Our "world" consists of the 30 countries of the study. We also compare the average per capita spending on the various media among countries, and the share of a particular medium in overall national consumption.

Content media[2] account for a combined $735 billion, 1% of the world's GDP (really the "GWP"), and platform media[3] accounted for $1.7 trillion, 2.3% of world GDP. Together then, media spending—both direct consumer spending and media advertising—add up to 3.3% of world GDP, or $2.4 trillion.

1. North America: Canada and the United States. Latin America: Argentina, Brazil, Chile, and Mexico. BRICS: Brazil, Russia, India, China, and South Africa. Europe: Belgium, Finland, France, Germany, Ireland, Italy, The Netherlands, Poland, Portugal, Spain, Sweden, Switzerland, and the United Kingdom. Middle East: Egypt, Israel, and Turkey. Asia-Pacific: Australia, Japan, South Korea, and Taiwan.

2. "Content media" includes newspapers, magazines, book publishing, film, radio, broadcast television, video channels, search engines, online news media, and one-third of the multi-channel industry's total revenues.

3. "Platform media" are defined as wireless and wireline telecom, ISPs, and two-thirds of multi-channel industry revenues. The allocation for the multi-channel industry is based on its dual role of both platform and content aggregator, and approximates the cost shares.

## DAILY NEWSPAPER INDUSTRY CONCENTRATION ACROSS THE WORLD

Daily newspapers are a concentrated industry by standard definitions that peg the line to "highly concentrated" at an HHI of 2,500 (formerly 1,800). The 2013 world average arithmetic HHI was 2,484 (Graph 34.1). However, the weighted average was significantly lower, at 1,328, suggesting that smaller and poorer countries have a higher concentration. Despite these numbers, newspapers were one of the less concentrated media industries. But once one looks at specific countries one can often observe strong shares of the top firms. Table 34.1 shows the market share of a country's top company. In Mexico, OEM holds 59.4%; in Turkey, Dogan 46.4%; in Australia, News Corp., 57.5%, and in Chile, El Mercurio, 54.9%. Other highly concentrated markets exist—in France, with the Amaury Group at 30.2%; in Ireland, Independent News and Media (INM) with 52%; in Russia, Komsomolskaya Pravda with 39.6%; in Switzerland, Tamedia with 44.3%; in the United Kingdom, Murdoch's News Corp. with 32.5%; and in Portugal, Cofina with 42.4%. The countries with the lowest concentrations are Germany and the United States (the latter with an HHI of 304). Gannett held a 9.6% national market share across the US. The low concentration is in part due to the local character of the US and German newspapers.

Australia's high HHI of 4,212 stems from the high market shares of three main media groups, including News Corp.'s (US) News Ltd.—Rupert Murdoch's original home base—which consolidated its ownership over regional dailies. Ireland's HHI of 3,235 in 2012 was greatly affected by increased consolidation in the Irish market, with two multimedia groups: Independent News and Media and Communicorp, gaining control of most of the market through buyouts. Several newspapers have a much higher share in revenue than in readers, for example in Mexico, where there is a wide discrepancy between share of advertising revenues and readership. This is true for India, too, because advertisers are focused on the urban middle class demographic. India's concentration is relatively low (1,228) due to the abundance of regional and vernacular language newspapers, as well as due to government

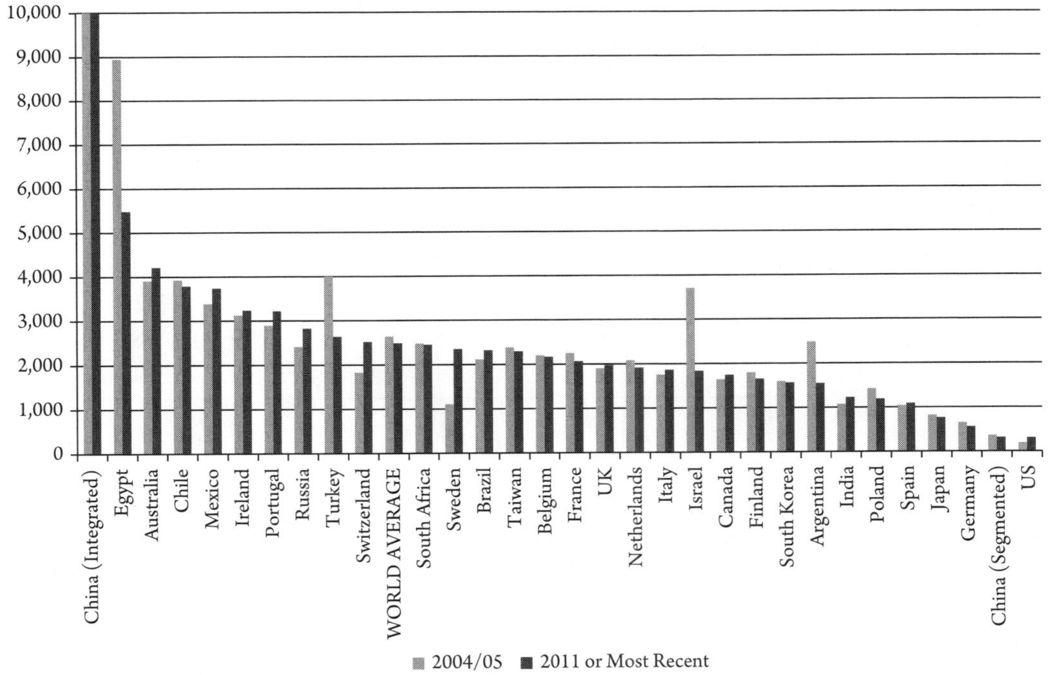


GRAPH 34.1 Daily Newspapers Concentration by Country

Table 34-1. Countries' Top Daily Newspaper Companies, 2011 or More Recent

| COUNTRY | TOP COMPANY | MARKET SHARE OF TOP COMPANY (%) | NATIONAL HHI |
|---|---|---|---|
| China | Government of China (public) | 100% | 10,000 |
| | *People's Daily (public)* | *11.1%* | |
| Egypt | Government of Egypt (public) | 72.2% | 5,481 |
| | *Al Akhbar (public)* | *28.7%* | |
| Australia | Murdoch (US) | 57.5% | 4,212 |
| Chile | El Mercurio SAP | 54.9% | 3,786 |
| Mexico | OEM | 59.4% | 3,736 |
| Ireland | Independent News and Media | 52.0% | 3,235 |
| Portugal | Cofina | 42.4% | 3,215 |
| Russia | Komsomolskaya Pravda | 39.6% | 2,824 |
| Turkey | Dogan Group | 46.4% | 2,637 |
| Switzerland | Tamedia | 44.3% | 2,520 |
| South Africa | Naspers Group | 35.0% | 2,450 |
| Sweden | Bonnier | 41.3% | 2,350 |
| Brazil | Globo Group | 28.4% | 2,321 |
| Taiwan | Liberty Times News Group | 31.4% | 2,293 |
| Belgium | Corelio | 26.6% | 2,166 |
| France | Amaury Group | 30.2% | 2,064 |
| UK | Murdoch Group (News Corp., US) | 32.5% | 1,974 |
| Netherlands | Telegraaf Media Groep | 36.9% | 1,914 |
| Italy | RCS Media Group | 33.6% | 1,863 |
| Israel | Yedioth Ahronoth Group | 39.0% | 1,835 |
| Canada | Postmedia Network Canada | 29.4% | 1,743 |
| Finland | Sanoma Oyj | 31.0% | 1,653 |
| South Korea | Chosun Ilbo | 24.3% | 1,564 |
| Argentina | Grupo Clarín | 37.6% | 1,547 |
| India | BCCL (The Times of India Group) | 29.2% | 1,228 |
| Poland | Agora | 24.1% | 1,188 |
| Spain | Vocento | 18.3% | 1,090 |
| Japan | Yomiuri Shimbun | 19.7% | 762 |
| Germany | Axel Springer AG | 18.8% | 558 |
| US | Gannett | 9.6% | 304 |

subsidies. In Mexico, OEM has control of nearly two-thirds of newspapers' advertising revenues, giving the market an HHI of 3,736.

In China, there are many newspaper organizations controlled by various ministries or administrative levels of the governmental and party structure. Many people therefore consider Chinese media to be merely in the nature of rival divisions of the same enterprise, in the same way that the different parts of Time Warner or Vivendi may differ and contest each other at times, but in the end they are part of the same company, and their top managers are hired and fired by the same top leadership. In contrast, other observers argue that the various news media are separate and autonomous competitors. Who is right? If we adopt the "integrated" perspective of unified control, then China's newspaper industry, given that no private newspapers are allowed, is just about the most concentrated in the world. (HHI = 10,000). But if we take the view that publications controlled by the Ministry of Defense are not part of the same control grouping as those controlled by the Foreign Ministry or the Province of Hebei, then China's media are just about the most unconcentrated in the world (HHI = 312). The truth is probably somewhere in-between. We present therefore both perspectives: the "integrated" ownership of 10,000 and the "segregated" ownership of 312. For averaging international concentration figures, we use the "integrated" numbers (in this case, HHI = 10,000). The "separated" data are shown for comparative purposes.

Aggregating China's state operations does not single out that country. On the governmental level, we do the same for other countries where the state owns several media. We do not show the "segregated" data for these countries because there is much less state control overall, so the numbers for media industries in South Korea, Taiwan, Russia, and Egypt are also "integrated" data. The same type of aggregation is also done for private companies. Rupert Murdoch's various newspapers, for example, are added together, even though they might be entirely independent of each other, and even compete amongst themselves. Murdoch owns several major companies—News Corp and 21st Century Fox—and these are aggregated into a "Murdoch Group." We repeat this process for two other American media owners: John Malone (who controls the "Liberty"-branded assets) and Sumner Redstone, who controls both Viacom and CBS.

Of the countries surveyed, about half experienced an increase in newspapers concentration after 2004, in particular, Mexico, Australia, Chile, Russia, and Switzerland, while about half experienced a decrease, in particular, Portugal, Poland, Israel, and Egypt. Unweighted HHI declined (to 2,484 from 2,636) – driven by Turkey, Israel, and Egypt – but weighted HHI rose from 1,114 to 1,328 because by share of the world market, Turkey, Egypt, and Israel are small.

Concentration in EU countries is below the world average but considerably higher than for North America. Concentration is significantly higher in the BRICS countries, Latin America, and the Middle East. In all regions except the Mideast and Latin America, concentration is increasing (Graph 34.2).

Of course, media concentration is affected by the size of a country's market. A large market can sustain more competitors, in the same way that a large city can accommodate more movie theaters than a small town. Hence, one should look at the market size to assess the industry concentration in a country. We do this with more variables and details in the chapter "Country

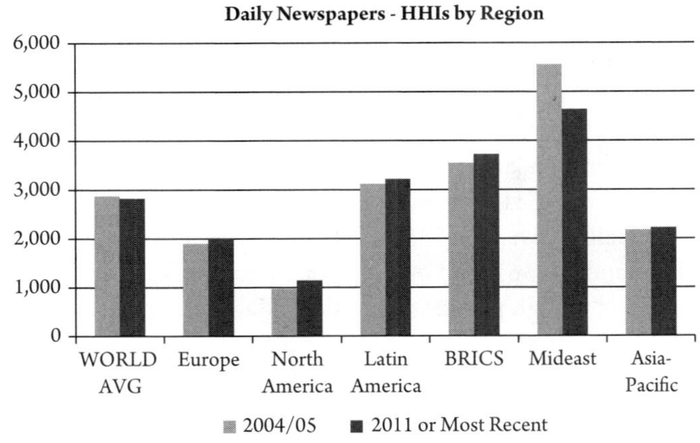


GRAPH 34.2 Daily Newspapers Regional Concentration

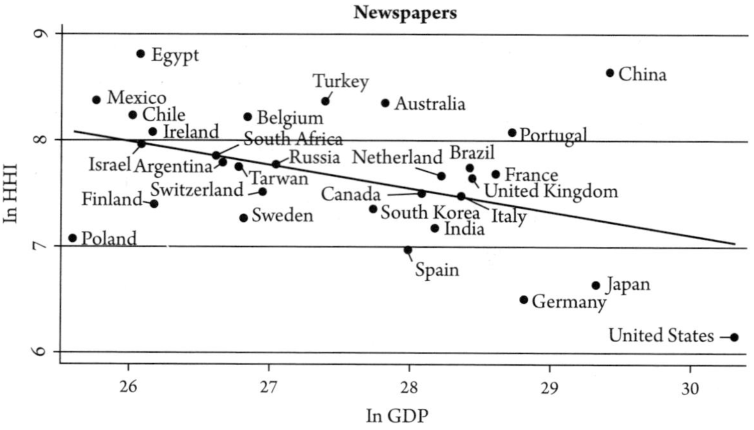


GRAPH 34.3 Newspapers Industry Concentration and GDP

Comparisons." Here, we present the correlation of GDP and concentration in the newspaper industry. The GDP is used because it incorporates both the income level of a country and its population. A large market may exist because its people are richer, or because there are more of them. Graph 34.3 is a log-log presentation of this relation. What it shows is:

1. A negative correlation of GDP size and HHI concentration.
2. "Outlier" countries: several high-concentration countries are well above the regression line. They includes Egypt, Turkey, Australia, Portugal, and China. Low concentration countries relative to their GDP include Germany, Japan, and the United States, allowing for their market size.

Total global (i.e., 30 country) revenue for the newspaper industry is $107 billion. We also compare the per capita revenues (directly through purchase and indirectly through advertising). The world average for the world is $114. It is highest in Ireland ($605 annually), Sweden ($458), Finland ($307), Switzerland ($256), and Australia ($247). This reflects the high penetration in these countries due to the popularity of print content and advertising. On top of that, Ireland and Australia have some of the highest industry concentrations, which might lead to high prices. Sweden subsidizes its newspapers. Per capita spending was lowest in China ($0.50), Mexico ($1.90), Egypt ($2.30), and Russia ($3.60). The US per capita spending was $132, slightly above the world average. Newspaper spending, as a share of national consumption, is relatively small in the US economy as a whole (0.21%). It is highest in Ireland (0.88%), Sweden (0.74%), Australia (0.44%), and the United Kingdom (0.37%). The total newspaper industry share of world GDP was 0.24% in 2013.

The US newspaper industry has the largest share of the total world industry at 30.1%. Japan accounts for 17.6% and the United Kingdom has 8.3%. Together, these three countries account for 56% of the 30-country world newspaper revenue (Table 34.2).

## MAGAZINE INDUSTRY GLOBAL CONCENTRATION

The magazine industry is fairly unconcentrated, with its arithmetic average HHI stable at 1,761 in 2013 (Graph 34.4) (Table 34.3). Weighted average HHI was even lower, at 1,031—which means that the larger magazine markets by income and population size have below-average concentration. Magazines have the fourth lowest industry concentration after film, online news media, and book publishing.

Brazil's magazine industry HHI was extremely high in 2004, but declined significantly thereafter, as Abril's high national market share

Table 34-2. Daily Newspapers in National Markets, 2011 or Most Recent

| COUNTRY | COUNTRY SHARE IN WORLD NEWSPAPER INDUSTRY TOTAL (%) | DAILY NEWSPAPERS SHARE OF NATIONAL GDP (%) | PER CAPITA REVENUE OF DAILY NEWSPAPERS ($) |
|---|---|---|---|
| Argentina | 0.4% | 0.14% | 14.6 |
| Australia | 4.0% | 0.44% | 247.0 |
| Belgium | 1.1% | 0.23% | 135.2 |
| Brazil | 1.6% | 0.10% | 11.2 |
| Canada | 3.8% | 0.28% | 151.5 |
| Chile | 0.6% | 0.37% | 48.8 |
| China | 0.5% | 0.01% | 0.5 |
| Egypt | 0.1% | 0.08% | 2.3 |
| Finland | 1.2% | 0.50% | 306.5 |
| France | 4.0% | 0.16% | 88.5 |
| Germany | 4.1% | 0.13% | 69.1 |
| India | 2.4% | 0.20% | 2.6 |
| Ireland | 2.0% | 0.88% | 604.8 |
| Israel | 0.2% | 0.13% | 42.8 |
| Italy | 2.3% | 0.11% | 53.6 |
| Japan | 17.6% | 0.38% | 190.5 |
| Mexico | 0.2% | 0.02% | 1.9 |
| Netherlands | 0.8% | 0.11% | 70.1 |
| Poland | 0.5% | 0.10% | 17.0 |
| Portugal | 0.4% | 0.19% | 54.8 |
| Russia | 0.4% | 0.03% | 3.6 |
| South Africa | 1.0% | 0.37% | 26.5 |
| South Korea | 1.2% | 0.15% | 35.2 |
| Spain | 5.6% | 0.39% | 167.4 |
| Sweden | 3.1% | 0.74% | 457.5 |
| Switzerland | 1.5% | 0.30% | 256.4 |
| Taiwan | 0.5% | 0.13% | 28.1 |
| Turkey | 0.6% | 0.09% | 10.5 |
| UK | 8.3% | 0.37% | 184.3 |
| US | 30.1% | 0.23% | 132.1 |
| Average Newspaper Industry Share of GDP (%) | | 0.21 | |
| Total Newspaper Industry Share of World GDP (%) | | 0.24 | |
| Total Newspaper Revenue (mil $) | | 106,734 | |
| Average Per Capita Spending ($) on Newspapers (incl. advertising) | | | 113.8 |

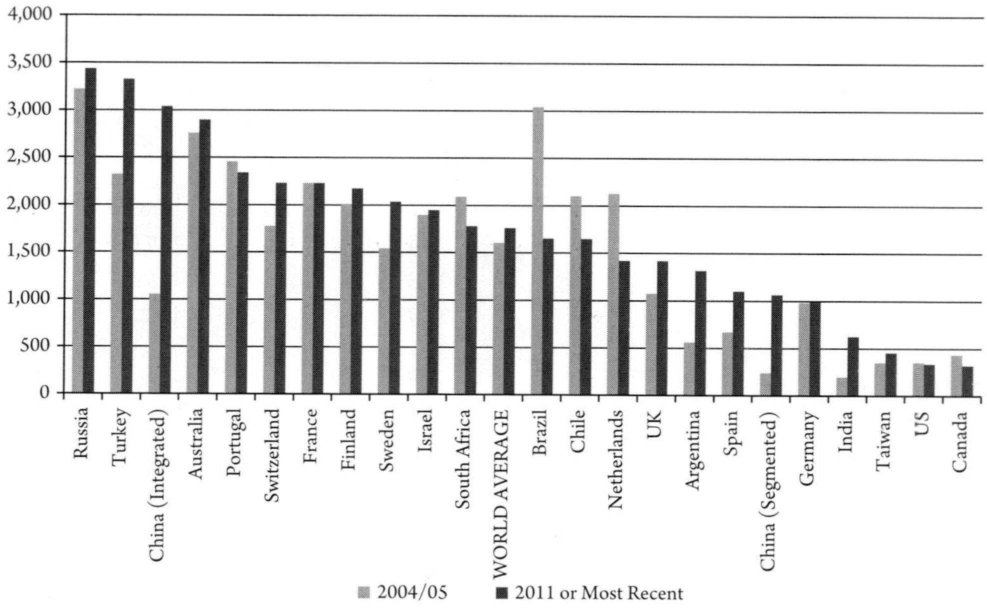


GRAPH 34.4 Magazine Concentration by Country

decreased from 68.3% to 48.5% due to the rise of specialty titles. New weeklies, such as *Caras* increased the aggregate market share of small companies (below 1% market share) from a very low figure before 2000—when the market was a duopoly between Abril and the Globo Group—to 42% by 2010. Low concentration rates were in the United States (157), India (626), Taiwan (455), and Canada (320). India used to rank lowest of all countries in 2004 (198), but has increased to 626 as a result of the rise in the Living Media Group's national market share (to 24%) due to its partial acquisition by the ABG Group.

Concentration is highest in Russia (3,434): in 2012, the combined share of Sanoma, Hearst Shkulev Media-InterMedia Group, and Bauer, was 87.5% of the national market. Concentration in France (Lagardère, 29.4% market share) and South Africa (Naspers, 39.5% market share) are also high. Dogan's share in Turkey (in a joint venture with Burda of Germany) rose to 52% from 44% due to several rounds of mergers and acquisitions in which foreign publishers participated heavily.

Chinese state ownership is lower in magazines than in book publishing or newspapers because some private ownership exists, especially in the form of licensed foreign publications like Condè Nast (7.8%) and Caijing (0.4%), thus the HHI is 3,039, less than that of Turkey (3,326), Australia (2,898), and Russia (3,434).

Magazine markets expanded (and HHIs fell correspondingly) in Latin America as overall economic growth in these regions created new consumer cultures, higher literacy rates, and household with incomes high enough to afford magazines and the products advertised in them (Graph 34.5). Overall, however, the trends for magazines are not favorable. The growth of digital media has made it harder for publishers to retain advertisers and paid subscribers in the hardcopy format and to maintain revenues.

The United States holds the largest world market share in this industry, with 49.7%. Magazines account for 0.28% of the US national GDP, one of the highest percentages of all the countries surveyed (alongside the United Kingdom and Sweden), the result of the strong magazine diversity and a very active advertising market that targets specific consumer segments. Per capita spending is lowest in India ($0.30), Turkey ($2.50), and Russia ($1.80). The United States, together with Germany (6.9%) and the United Kingdom (9.1%), accounts for 65.7% of the world market. The US's per capita spending

Table 34-3. Countries' Top Magazine Publishers, 2011 or Most Recent

| COUNTRY | TOP COMPANY | MARKET SHARE OF TOP COMPANY (%) | NATIONAL HHI |
|---|---|---|---|
| Russia | Sanoma Oyj (Finland) | 44.5% | 3,434 |
| Turkey | Dogan Group | 52.0% | 3,326 |
| China | Government of China | 53.2% | 3,039 |
| | *Rayli (public)* | 23.6% | |
| Australia | Nine Entertainment | 44.0% | 2,898 |
| Portugal | Impresa | 32.3% | 2,339 |
| Switzerland | Axel Springer (Germany) | 29.2% | 2,232 |
| France | Lagardère | 29.4% | 2,229 |
| Finland | Sanoma Oyj | 32.5% | 2,175 |
| Sweden | Bonnier | 32.9% | 2,037 |
| Israel | Yedioth Ahronoth Group | 33.5% | 1,949 |
| South Africa | Naspers Group | 39.5% | 1,780 |
| Brazil | Abril Group | 48.5% | 1,652 |
| Chile | COPESA | 24.0% | 1,649 |
| Netherlands | SBS Nederland (Sanoma Oyj, Finland) | 23.8% | 1,419 |
| UK | Bauer (Germany) | 26.0% | 1,419 |
| Argentina | Editorial Atlátida | 28.7% | 1,317 |
| Spain | Hachette (Lagardere, France) | 17.9% | 1,101 |
| Germany | Bauer | 18.5% | 997 |
| India | Living Media Group | 24.0% | 626 |
| Taiwan | Next Media (Hong Kong) | 13.5% | 455 |
| US | Hearst | 10.3% | 338 |
| Canada | Rogers | 6.3% | 320 |

annually on magazines is $163; higher than that of Germany ($87.9) and France ($72.2), though not of Sweden ($191). The world average per capita spending on magazines is $65.8. Magazine consumption, as a share of overall consumption, averages 0.15%. It is high in the United States (0.28%), Spain (0.29%), the United Kingdom (0.3%), and Sweden (0.31%). It is lowest in India (0.02%) and Russia (0.01%) (Table 34.4) .

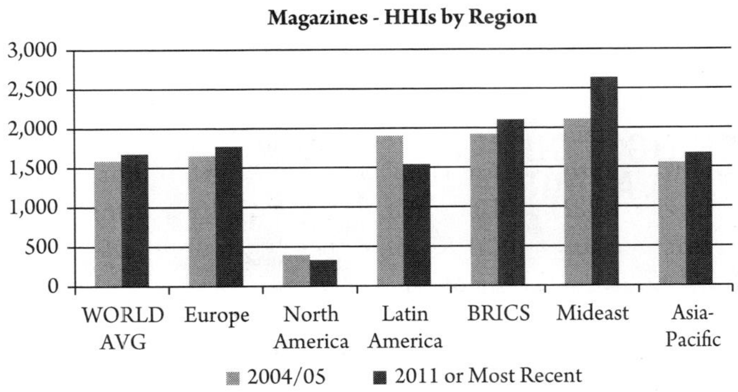


GRAPH 34.5 Magazines Regional Concentration

Table 34-4. Magazine Publishers in National Markets, 2011 or Most Recent

| COUNTRY | COUNTRY SHARE IN WORLD MAGAZINE INDUSTRY TOTAL (%) | MAGAZINE PUBLISHERS SHARE OF NATIONAL GDP (%) | PER CAPITA REVENUE OF MAGAZINES ($) |
|---|---|---|---|
| Argentina | 0.6% | 0.15% | 14.8 |
| Australia | 2.8% | 0.22% | 126.6 |
| Brazil | 4.3% | 0.21% | 22.5 |
| Canada | 2.9% | 0.16% | 87.6 |
| Chile | 0.2% | 0.09% | 12.2 |
| China | 6.9% | 0.12% | 5.3 |
| Finland | 0.7% | 0.22% | 134.3 |
| France | 4.4% | 0.13% | 72.2 |
| Germany | 6.9% | 0.16% | 87.9 |
| India | 0.4% | 0.02% | 0.3 |
| Israel | 0.1% | 0.04% | 13.5 |
| Netherlands | 1.6% | 0.15% | 96.5 |
| Portugal | 0.2% | 0.06% | 17.4 |
| Russia | 0.2% | 0.01% | 1.8 |
| South Africa | 0.5% | 0.14% | 9.7 |
| Spain | 5.6% | 0.29% | 126.1 |
| Sweden | 1.7% | 0.31% | 190.7 |
| Switzerland | 0.8% | 0.12% | 102.0 |
| Taiwan | 0.3% | 0.05% | 11.8 |
| Turkey | 0.2% | 0.02% | 2.5 |
| UK | 9.1% | 0.30% | 149.5 |
| US | 49.7% | 0.28% | 162.7 |
| Average Magazine Industry Share of GDP (%) | | 0.15 | |
| Total Magazine Industry Share of World GDP (%) | | 0.20 | |
| Total Magazine Revenue (mil US$) | | 79,590 | |
| Average Per Capita Spending ($) on Magazines | | | 65.8 |

## BOOK PUBLISHING INDUSTRY CONCENTRATION ACROSS THE WORLD

The book publishing industry is unconcentrated in almost every country (Graph 34.6). By unweighted HHI, it ranks as the third-least concentrated media industry, and it remained essentially flat with an unweighted average global HHI of 1,690 in 2013 (1,633 in 2004). When weighted, average world industry HHI rose substantially from 1,022 to 2,160 due to the increase in China's share of the industry (to 15.9%): when China is excluded, book publishing for the rest of the world remained unconcentrated (below 1,000) and static from 2004 to 2013. China's industry concentration is highest at 10,000 due to the

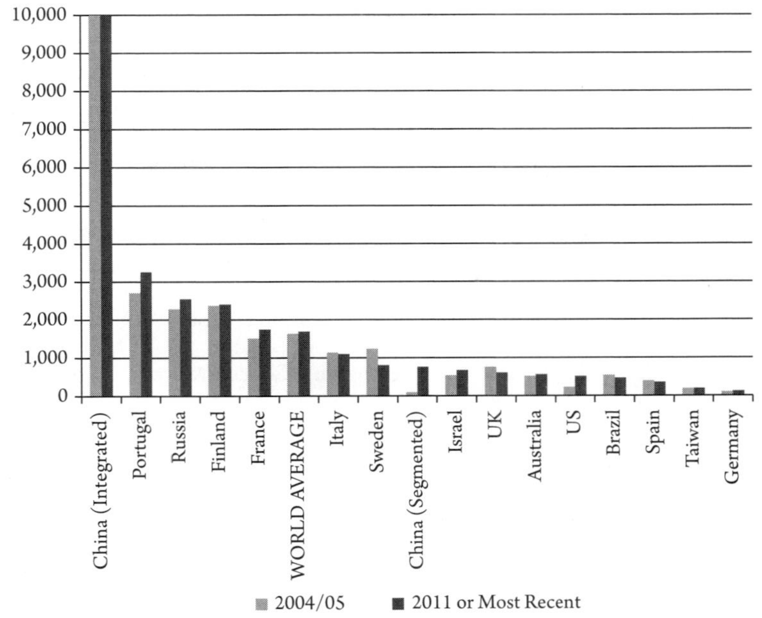


GRAPH 34.6 Book Publishing Concentration by Country

## Table 34-5. Countries' Top Book Publishers, 2011 or Most Recent

| COUNTRY | TOP COMPANY | MARKET SHARE OF TOP COMPANY (%) | NATIONAL HHI |
|---|---|---|---|
| China | Government of China[1] | 100% | 10,000 |
| | *Zhejiang Publishing United (Public)* | 15.3% | |
| Portugal | Porto Editora | 47.8% | 3,256 |
| Russia | Eksmo-AST | 61.3% | 2,545 |
| Finland | Bonnier (Sweden) | 31.4% | 2,405 |
| France | Lagardère | 29.4% | 1,748 |
| Italy | Fininvest (Mondadori) | 28.4% | 1,101 |
| Sweden | Carl Allers | 24.9% | 809 |
| Israel | Kinneret Zemora | 12.2% | 674 |
| UK | Bertelsmann-Pearson (Germany/UK)[2] | 24.9% | 610 |
| Australia | Bertelsmann-Pearson (Germany/UK) | 24.4% | 562 |
| US | Bertelsmann-Pearson (Germany/UK) | 17.6% | 511 |
| Brazil | PRISA (Spain) | 13.9% | 467 |
| Spain | Planeta | 10.2% | 354 |
| Taiwan | Cité Media Holding Group (Hong Kong) | 7.8% | 192 |
| Germany | Axel Springer | 6.3% | 121 |

[1]The Government of China owns all book publishing companies, for a 100% of the market. If each publisher is considered separately, industry HHI = 762.

[2]In book publishing, Bertelsmann (Germany) and Pearson (UK) operate as Penguin Random House internationally as of 2013. The combined market share for this joint venture in the US is 17.6%–Bertelsmann with 10.3% and Pearson with 7.3%. It is 24.4% in Australia–Bertelsmann with 10.1% and Pearson with 14.3%.

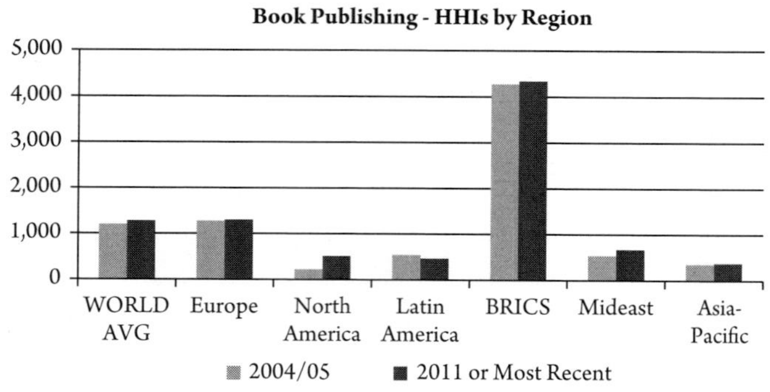


GRAPH 34.7 Book Publishing Regional Concentration

## Table 34-6. Book Publishers in National Markets, 2011 or Most Recent

| COUNTRY | COUNTRY SHARE IN WORLD BOOK PUBLISHING INDUSTRY TOTAL (%) | BOOK PUBLISHERS SHARE OF NATIONAL GDP (%) | PER CAPITA SPENDING ON BOOKS ($) |
|---|---|---|---|
| Australia | 1.6% | 0.12% | 102.2 |
| Brazil | 2.0% | 0.09% | 14.4 |
| China | 15.9% | 0.24% | 17.0 |
| Finland | 0.3% | 0.08% | 73.7 |
| France | 10.0% | 0.26% | 227.2 |
| Germany | 13.3% | 0.28% | 234.6 |
| Israel | 0.1% | 0.05% | 23.1 |
| Italy | 6.5% | 0.21% | 153.6 |
| Portugal | 0.4% | 0.11% | 48.0 |
| Russia | 0.7% | 0.03% | 6.7 |
| Spain | 5.7% | 0.27% | 177.8 |
| Sweden | 1.5% | 0.24% | 223.4 |
| Taiwan | 0.7% | 0.13% | 45.0 |
| UK | 8.5% | 0.25% | 194.8 |
| US | 33.1% | 0.17% | 151.3 |
| Average Book Publishing Industry Share of GDP (%) | | 0.17 | |
| Total Book Publishing Industry Share of World GDP (%) | | 0.20 | |
| Total Book Publishing Revenue (mil $) | | 72,224 | |
| Average Per Capita Spending ($) on Book Publishing | | | 112.8 |

monopoly on book publishing by state enterprises (in particular, textbooks). Concentration in the Russian market is also high (2,545), rising by 4% per annum—only a handful of publishing houses are present, including a state enterprise, and the two major publishers AST and Eksmo merged in 2012. France's high industry HHI (1,745) is the result of increasing consolidation of the market by international publishing houses, led by the Lagardère Group with its extraordinarily high share of 29.4%, at the expense of smaller French publishers. In the United States, concentration stood at 511, up from 224: this was in large part due to the merger of Bertelsmann and Pearson's book publishing divisions into Penguin-Random House.

Taiwan has a very low HHI of 192. The German book publishing is the least concentrated of all the book industries at 121 and the third largest in the world by revenue (at $9.6 billion) with substantial niche content. These factors lead its top 10 publishing houses to have only single-digit shares of the national market, the largest being 6.3% for Axel Springer (see Table 34.5).

Concentration rose worldwide. It is highest in the BRICS countries, largely due to the expansion of the Chinese market (Graph 34.7). It is lowest in North America, Asia-Pacific, and Latin America, though it is rising in all three of these regions as publishers consolidate.

Total global revenue for the industry is $72.2 billion. Per capita per annum revenue averages $112.8 worldwide. It is highest in Germany ($235, partly due to the high prices that accompany retail price maintenance) and France ($227). It is $151 in the United States and is lowest in Russia ($6.70). The United States accounts for 33.1% of the world industry total, with China (15.9%), Germany (13.3%) and France (10%). Average book industry share as a percentage of GDP is 0.17%. It is highest in France (0.26%) and Germany (0.28%), followed by the United Kingdom (0.25%) (Table 34.6).

## RADIO INDUSTRY CONCENTRATION ACROSS THE WORLD

Radio broadcasting is highly concentrated, with a global unweighted average HHI of 2,627 (weighted, HHI is 1,854) (Graph 34.8). Because in many markets most of the stations operate

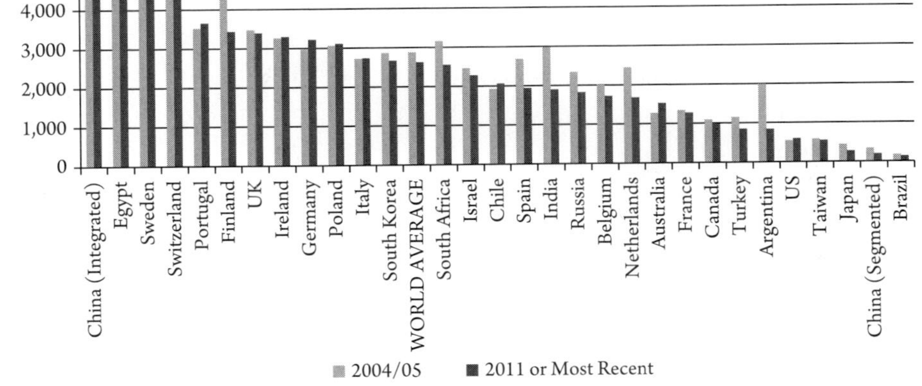


GRAPH 34.8 Radio Concentration by Country

at the regional level, actual concentration is still higher. Its concentration puts it in seventh place for industry concentration – the median of our 13 industries. The most significant change took place in India, where concentration fell from 2,984 to under 1,901 once the state's legal monopoly was abolished. In the United States, industry HHI rose to 600 in 2013. Although the US radio sector is unconcentrated by international standards—due to geography, a history of regulation favoring localism, and the fragmentation of public broadcasting—its post-1995 figures represented

## Table 34-7. Countries' Top Radio Broadcasters, 2011 or More Recent

| COUNTRY | TOP COMPANY | MARKET SHARE OF TOP COMPANY (%) | NATIONAL HHI |
|---|---|---|---|
| China | Government of China[1] | 100% | 10,000 |
| | *China Radio International (public)* | 9.3% | |
| Egypt | Government of Egypt (ERTU, public) | 85.4% | 7,408 |
| Sweden | Sveriges (public) | 74.8% | 5,829 |
| Switzerland | SRG SSR (public) | 66.4% | 4,422 |
| Portugal | Group RR | 35.7% | 3,653 |
| Finland | Yle (public) | 55.8% | 3,435 |
| UK | BBC (public) | 54.5% | 3,393 |
| Ireland | RTE (public) | 53.5% | 3,297 |
| Germany | ARD (public) | 36.5% | 3,223 |
| Poland | Radio RMF FM | 36.4% | 3,115 |
| Italy | RAI (public) | 48.2% | 2,739 |
| South Korea | Government of South Korea | 55.8% | 2,669 |
| | *SBS* | 47.6% | |
| South Africa | SABC (public) | 31.6% | 2,553 |
| Israel | IDF (public) | 32.2% | 2,278 |
| Chile | Iberoamericana (Spain) | 43.8% | 2,062 |
| Spain | PRISA | 39.6% | 1,937 |
| India | Prasar Bharati (public) | 30.6% | 1,901 |
| Russia | EMG | 32.4% | 1,822 |
| Belgium | VRT | 30.9% | 1,721 |
| Netherlands | Netherland Public Broadcasting (public) | 32.3% | 1,674 |
| Australia | Australian Broadcasting Corporation (public) | 27.2% | 1,526 |
| France | France Télévisions (Radio France, public) | 21.9% | 1,277 |
| Canada | Astral | 17.5% | 981 |
| Turkey | Dogus | 17.9% | 855 |
| Argentina | Radio 10 | 15.6% | 843 |
| US | Liberty (Sirius XM) | 15.4% | 600 |
| Taiwan | BCC (public) | 20.9% | 548 |
| Japan | Fujisankei (NBS) | 10.7% | 273 |
| Brazil | Globo Group | 9.8% | 136 |

[1]Most Chinese radio is state-owned at the subnational level. No single station has a very large share (segmented HHI = 194), but government entities control all stations.

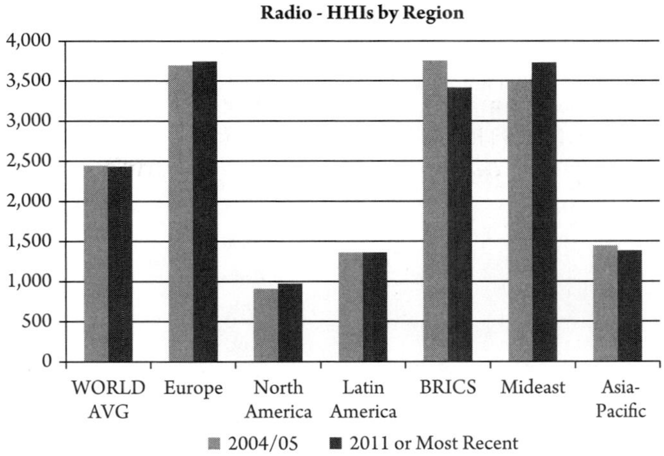


GRAPH 34.9 Radio Regional Concentration

a significant increase in ownership from the 1980s, when industry HHI, measured for both station ownership and market revenue, was below 100. Since then, large station groups (iHeartRadio and CBS) as well as the satellite radio operator Sirius XM have emerged.

Concentration levels are lowest in the United States, Mexico, Turkey, Brazil, and Japan (Table 34.7). Concentration is highest in China (10,000) and Egypt (7,408) because radio broadcasting is under government control, though private ownership is permitted in Egypt. It is also very high in several European countries, including Finland, Sweden, and Switzerland—dominated by the national public broadcasters. Here, there is no legal monopoly, yet the incumbency advantage of national public broadcasters is strong—credibility as well as funding through user fees or special taxes. In several countries, where the national public broadcaster's legal monopoly was abolished, concentration declined sharply, as it has in India.

Brazil's HHI concentration is the lowest of all countries surveyed — at 136 — partly because of the absence of a national public broadcaster, which is due to political opposition by regional and local authorities toward the formation of such an entity. Competing political parties and religious denominations control a significant number of stations at the regional level (like in the United States, Brazil's geographic size has reduced the formation of national radio stations).

Worldwide radio concentration is stable. Regionally, compared to the world average, Europe, the BRICS countries, and the Middle East have high concentrations (Graph 34.9). It is lowest in North America and Latin America.

The world average per capita industry revenue for radio is \$28.3. It is highest in Germany (\$105) and France (\$84.5), and lowest in India (\$0.10) and Egypt (\$0.50) (Table 34.8). The United States holds by far the largest market share of the industry in the world with 33.6% and has a high per capita revenue per annum rate of \$50.7. In contrast, per capita spending in Russia, Turkey, and China is \$2.80, \$1.00, and \$0.80, respectively. The average share of GDP is 0.08%. It is highest in Germany (0.3%) and France (0.2%), followed by South Africa, the United States, Spain, Australia, the United Kingdom, Ireland, and Canada, all around 0.1%. It is lowest in India and Turkey (0.01%).

## TV BROADCASTING INDUSTRY CONCENTRATION ACROSS THE WORLD

This section deals with traditional free, over-the-air, terrestrial TV broadcasting. Other sections deal with pay-TV platforms and with channels distributed over satellite and cable platforms. These three segments are partly overlapping. The industry system varies by country, and for

Table 34-8. Radio Broadcasters in National Markets, 2011 or Most Recent

| COUNTRY | COUNTRY SHARE IN WORLD RADIO INDUSTRY TOTAL (%) | RADIO BROADCASTERS SHARE OF NATIONAL GDP (%) | PER CAPITA SPENDING ON RADIO (US$) |
|---|---|---|---|
| Argentina | 0.5% | 0.07% | 5.6 |
| Australia | 3.1% | 0.15% | 65.6 |
| Belgium | 14.8% | 0.08% | 37.3 |
| Brazil | 7.8% | 0.08% | 6.4 |
| Canada | 4.1% | 0.14% | 57.2 |
| Chile | 0.2% | 0.05% | 5.5 |
| China | 2.2% | 0.02% | 0.8 |
| Egypt | 0.1% | 0.02% | 0.5 |
| Finland | 0.3% | 0.06% | 28.9 |
| France | 11.2% | 0.20% | 84.5 |
| Germany | 18.0% | 0.25% | 105.1 |
| India | 0.3% | 0.01% | 0.1 |
| Ireland | 0.4% | 0.08% | 43.7 |
| Israel | 0.1% | 0.03% | 8.6 |
| Italy | 1.3% | 0.03% | 10.2 |
| Japan | 3.9% | 0.04% | 14.6 |
| Netherlands | 0.7% | 0.04% | 21.3 |
| Poland | 0.4% | 0.04% | 4.8 |
| Portugal | 0.5% | 0.10% | 22.6 |
| Russia | 0.8% | 0.03% | 2.8 |
| South Africa | 1.3% | 0.22% | 12.4 |
| South Korea | 0.5% | 0.03% | 4.8 |
| Spain | 4.0% | 0.13% | 41.7 |
| Sweden | 1.0% | 0.10% | 48.6 |
| Switzerland | 1.4% | 0.14% | 87.5 |
| Taiwan | 0.3% | 0.03% | 5.2 |
| Turkey | 0.2% | 0.01% | 1.0 |
| UK | 5.8% | 0.11% | 44.1 |
| US | 33.6% | 0.11% | 50.7 |
| Average Radio Industry Share of GDP (%) | | 0.08 | |
| Total Radio Industry Share of World GDP (%) | | 0.10 | |
| Total Radio Revenue (mil $) | | 47,687 | |
| Average Per Capita Spending ($) on Radio | | | 28.3 |

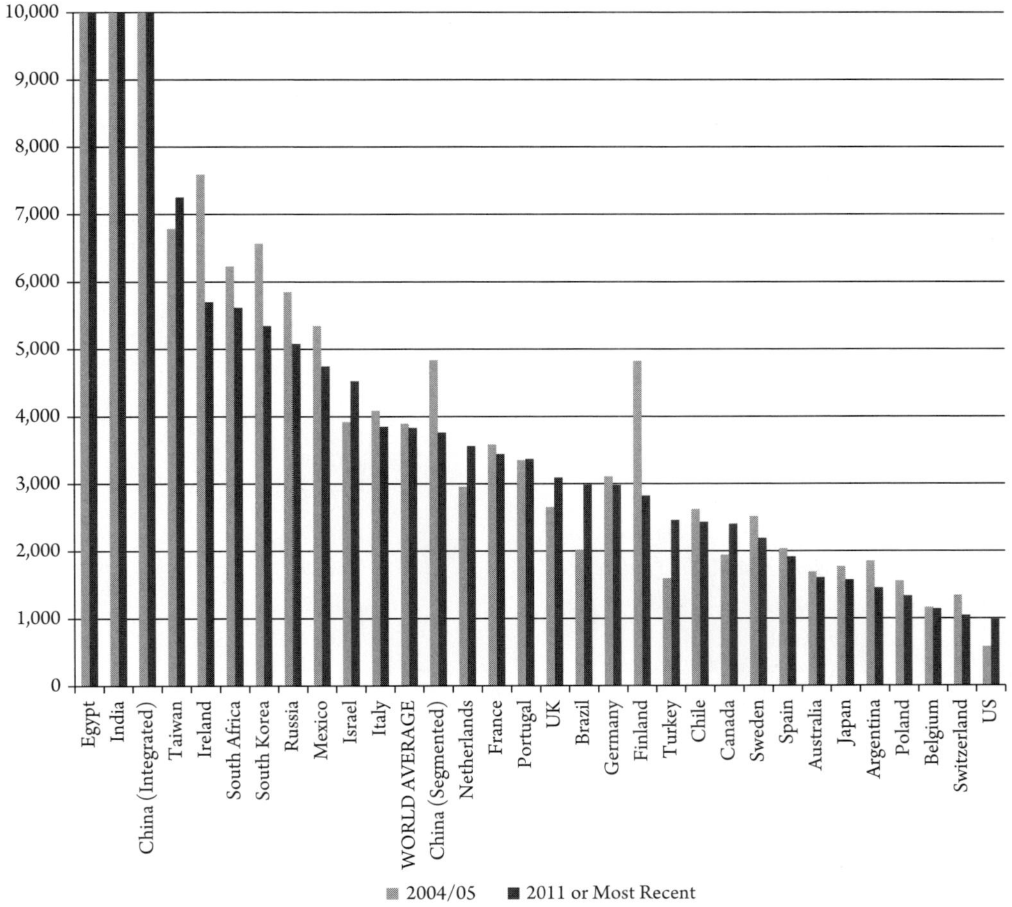


GRAPH 34.10 TV Broadcasting Concentration by Country

an apples-to-apples comparison it needs to be disaggregated. Subsequent aggregations are possible, either along the dimensions of all content channels, or of free channels/platforms vs. pay-channels/platforms, or for all channels.

The broadcast TV industry, with a world average unweighted HHI of 3,830 is highly concentrated (Graph 34.10). Even so, it holds an intermediate — sixth — place in concentration among the 13 media industries. The weighted average HHI is lower, but still high, at 2,839 in 2013, up from 2,367 in 2004. The difference indicates that the relatively large markets are less concentrated. Few countries have more than four TV networks or station groups. Japan has six major ones (NHK, NBS, TV Asahi Shimbun, Tokyo Broadcasting System, TV Tokyo, and Fuji TV). The United States has four major networks—ABC, CBS, NBC, and Fox—plus five small networks and several independent station groups.

Concentration is highest in China, Egypt, India, and Taiwan, while it is lowest in the United States, Switzerland, and Turkey (Table 34.9). (It is low in Switzerland due to the prevalence of cross-border TV stations from Germany, France, Italy, and Austria.) The high-concentration countries have strong state systems: India, Egypt, and China have state broadcast monopolies. In Taiwan, of the five terrestrial broadcasters, only one (China Times News Group, held by the Want-Want conglomerate) is not owned by the state or affiliated with the ruling political party. The US's concentration in this industry is the lowest internationally (984)

## Table 34-9. Countries' Top TV Broadcasters, 2011 or Most Recent

| COUNTRY | TOP COMPANY | MARKET SHARE OF TOP COMPANY (%) | NATIONAL HHI |
|---|---|---|---|
| China | Government of China | 100% | 10,000 |
| | *China Central Television (CCTV, public)* | 59.1% | |
| Egypt | Government of Egypt (ERTU, public) | 100.0% | 10,000 |
| India | Prasar Bharati, Doordarshan (public) | 100.0% | 10,000 |
| Taiwan | Government of Taiwan | 83.6% | 7,255 |
| | *Formosa TV (public)* | 36.9% | |
| Ireland | RTE (public) | 74.0% | 5,701 |
| South Africa | SABC (public) | 67.6% | 5,618 |
| South Korea | Government of South Korea | 70.8% | 5,348 |
| | *KBS (public)* | 37.3% | |
| Russia | Government of Russia | 70.9% | 5,081 |
| | *Gazprom Media (public)* | 33.5% | |
| Mexico | Televisa | 61.0% | 4,746 |
| Israel | Channel 2 | 60.6% | 4,527 |
| Italy | RAI (public) | 48.3% | 3,847 |
| Netherlands | Netherland Public Broadcasting (public) | 46.3% | 3,559 |
| France | France Televisions (public) | 29.4% | 3,440 |
| Portugal | RTP (public) | 37.0% | 3,366 |
| UK | BBC (public) | 50.1% | 3,092 |
| Brazil | Globo Group | 52.4% | 2,983 |
| Germany | ProSiebenSat.1 | 30.0% | 2,981 |
| Finland | Yle (public) | 41.0% | 2,822 |
| Chile | TVN (public) | 30.0% | 2,430 |
| Turkey | Dogan Group | 39.2% | 2,458 |
| Canada | Canadian Broadcasting Corporation (public) | 38.3% | 2,401 |
| Sweden | Bonnier | 29.8% | 2,188 |
| Spain | Finivest (Mediaset, Italy) | 28.1% | 1,913 |
| Australia | Seven Network | 26.4% | 1,606 |
| Japan | NTV | 19.1% | 1,572 |
| Argentina | Grupo Clarín | 23.5% | 1,454 |
| Poland | TP (public) | 44.3% | 1,334 |
| Belgium | VRT | 21.2% | 1,142 |
| Switzerland | SRG SSR (public) | 30.1% | 1,044 |
| US | Comcast (NBC) | 17.6% | 984 |

for reasons that also apply to the radio market: a high degree of regionalism, a long history of commercial players in the market, a fragmentation of public TV, and residual restrictions on ownership: the top network in 2013 was Comcast's NBC with 17.6%.

In Egypt and India there are legal monopolies over TV news programming content in favor of public broadcasters ERTU and Prasar Bharati's Doordarshan service. Ireland's two national public broadcasters, RTÉ and TG4, hold a near duopoly due to their subsidy packages, along

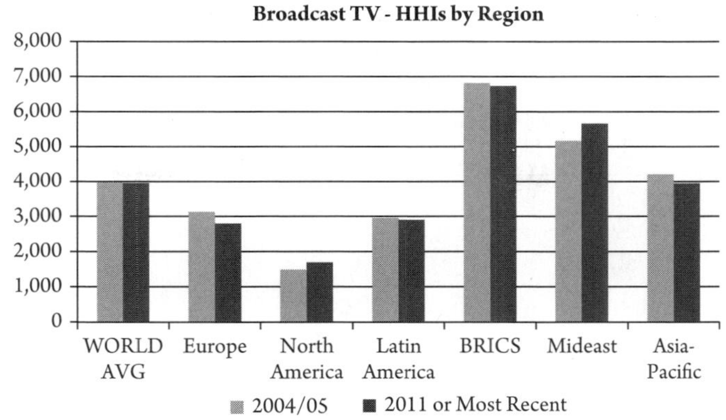


GRAPH 34.11 Broadcast TV Regional Concentration

with advertising and licensing fees. The South African market's public incumbent, SABC, benefits from a combination of state subsidies, licensing and public sector advertising. Finland's HHI was high (for Europe) in 2004, but declined from 4,823 to 2,822. The Government of China controls 100% of the broadcast market in the country, of which CCTV accounts for a market share of 59.1%.

The United Kingdom has an increasing TV industry concentration, but historically lower than in the 1980s due to the increasingly even distribution of market share among commercial broadcasters. In that decade, the market was effectively a duopoly between BBC and the ITV collaborative commercial network arrangement. Turkey's low HHI (799) is maintained by regulations against commercial crossownership. In Mexico, the Televisa Law of 2006 assigned digital spectrum to the two national television networks in the country: to the dominant (and politically well-connected) Televisa, and to TV Azteca. In Italy, industry HHI is well above the world average because Berlusconi's Mediaset is the largest private network, with 38.9% of the market, and the public broadcaster RAI controls the largest market share (48.3%).

The worldwide TV broadcasting industry concentration showed a slight decrease, driven by the European and Asia-Pacific countries (Graph 34.11). The HHI of North America was the lowest of all regions at 1,693 in 2013. The BRICS countries have the highest concentration, at 6,736.

The largest national TV markets are the United States (23.4%), Japan (14.5%), Germany (8.7%), and the United Kingdom (9%) (Table 34.10). TV revenues (both from use charges and through advertising) average $110.8. They are highest in Portugal ($332), Finland ($255), the United Kingdom ($264), Japan ($209), and Germany ($196). It is intermediate in the United States ($137). It is lowest in India ($0.80), China ($8.30), and Egypt ($2.90).

The percentage of TV spending as a share of national GDP is 0.38%. It is highest in Portugal, the United States, Finland, Mexico, Germany, Japan, Sweden, and Spain. It is lowest in India and Egypt.

## MULTI-CHANNEL PLATFORMS INDUSTRY CONCENTRATION ACROSS THE WORLD

Multi-channel platforms include cable TV distribution networks offered by Multiple Systems Operators (MSOs); direct broadcast satellite to the home (DBS or DTH); IPTV service by telecom companies over their networks: micro wave transmission to the home (MMDS); digital terrestrial broadcasting for pay bundles; and online video-on-demand cloud providers.

Media concentration in the multi-channel platforms industry across the 30 countries is high and remaining so, being flat in trend. The arithmetic average HHI is 3,929. Weighted it is 2,148 (Graph 34.12). Multi-channel platforms are the most concentrated of audiovisual

Table 34-10. TV Broadcasters in National Markets, 2011 or Most Recent

| COUNTRY | COUNTRY SHARE IN WORLD INDUSTRY TOTAL (%) | TV BROADCASTERS SHARE OF NATIONAL GDP (%) | PER CAPITA SPENDING ON BROADCAST TV (US$) |
|---|---|---|---|
| Argentina | 0.7% | 0.39% | 30.2 |
| Australia | 2.7% | 0.50% | 216.0 |
| Belgium | 1.5% | 0.55% | 244.0 |
| Brazil | 5.7% | 0.64% | 53.0 |
| Canada | 1.9% | 0.25% | 102.4 |
| Chile | 0.2% | 0.20% | 20.5 |
| China | 6.1% | 0.24% | 8.3 |
| Egypt | 0.1% | 0.14% | 2.9 |
| Finland | 0.7% | 0.54% | 255.4 |
| France | 1.9% | 0.13% | 56.5 |
| Germany | 8.7% | 0.46% | 195.9 |
| India | 0.6% | 0.08% | 0.8 |
| Ireland | 0.3% | 0.19% | 101.7 |
| Israel | 0.2% | 0.22% | 54.9 |
| Italy | 5.5% | 0.46% | 165.7 |
| Japan | 14.5% | 0.54% | 209.1 |
| Mexico | 1.5% | 0.29% | 24.5 |
| Netherlands | 0.6% | 0.14% | 70.5 |
| Poland | 0.5% | 0.17% | 22.5 |
| Portugal | 1.9% | 1.46% | 331.6 |
| Russia | 2.2% | 0.28% | 28.7 |
| South Africa | 0.6% | 0.42% | 23.4 |
| South Korea | 1.7% | 0.36% | 65.0 |
| Spain | 4.2% | 0.51% | 168.4 |
| Sweden | 1.1% | 0.46% | 216.3 |
| Switzerland | 0.9% | 0.34% | 220.5 |
| Taiwan | 0.2% | 0.08% | 13.5 |
| Turkey | 0.9% | 0.24% | 21.8 |
| UK | 9.0% | 0.68% | 263.7 |
| US | 23.4% | 0.31% | 136.6 |
| Average TV Broadcasting Industry Share of GDP (%) | | 0.38 | |
| Total TV Broadcasting Industry Share of World GDP (%) | | 0.37 | |
| Total TV Broadcasting Revenue (mil $) | | 184,379 | |
| Average Per Capita Spending ($) on TV Broadcasting | | | 110.8 |

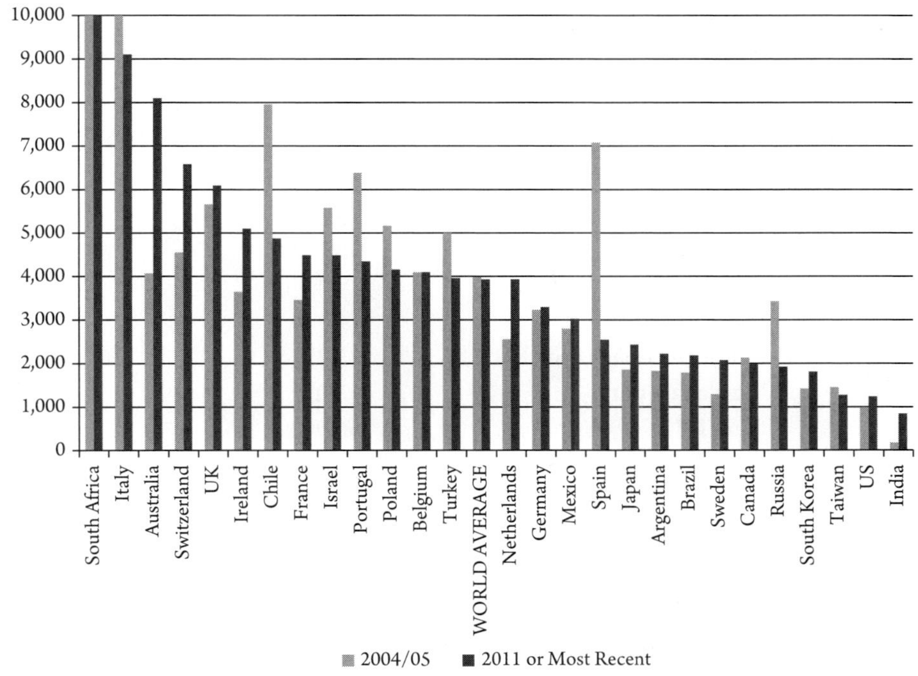


GRAPH 34.12 Multichannel Platforms Concentration by Country

industries and the third among all 13 media industries. The highest concentrations exist in Italy and Switzerland (Table 34.11). In these countries, there is little or no cable TV, and satellite TV is dominated by a single firm (Naspers in South Africa with 100%, 21st Century Fox's Sky Italia in Italy with 95.3%, and Liberty in Switzerland with 74%). India is the least concentrated market (842): this is due to the fragmentation of cable TV, many of whose operators function as unregistered/unlicensed entities. In the United States, which is the second-least concentrated market (1,233), there are very loose ownership ceilings imposed on cable operators. In Taiwan, even though there are only five major cable operators, ownership ceilings mean that each of them has a nearly even market share of 20% in cable optic platforms. In other countries (Chile, Turkey, and Portugal) the HHI fell from extremely high (over 6,000) to merely very high 5,000 and under level, as a result of new commercial operators entering the market.

The greatest declines took place in Chile, Portugal, and Turkey, while the largest increases occurred in India and Spain (though India remains unconcentrated with industry HHI of less than 1,800).

Liberty (as UPC) and 21st Century Fox (including both Sky Broadcasting and Star TV) are the two largest multinational companies in this industry. In Italy, Sky Italia, part of Rupert Murdoch's 21st Century Fox, dominates. IPTV platforms collectively hold less than 5% of the total market. Concentration increased most in Switzerland, the Netherlands, and Ireland.

Duopolies exist in several markets. Concentration is high in Ireland — with an HHI of 5,102 — because both cable and satellite broadcasting are effectively monopolies controlled by companies that are also active in the multi-channel industry: Liberty Media (US) which bought up Virgin Media's cable operation, and 21st Century Fox (Sky Broadcasting) in satellite TV. The United Kingdom is also a duopoly with a high industry HHI of 6,089 due to the presence of the aforementioned two American companies. In France, the duopoly of Altice/Numericable in cable and Vivendi in DBS results in an industry HHI of 4,488 based on a series of mergers and acquisitions.

## Table 34-11. Countries' Top Multichannel Platforms, 2011 or Most Recent

| COUNTRY | TOP COMPANY | MARKET SHARE OF TOP COMPANY (%) | NATIONAL HHI |
|---|---|---|---|
| South Africa | Naspers Group | 100.0% | 10,000 |
| Italy | Murdoch Group | 95.3% | 9,104 |
| Australia | Foxtel[1] (50% Murdoch, US + 50% Telstra, Australia) | 90.0% | 8,100 |
| Switzerland | Liberty (US) | 74.0% | 6,577 |
| UK | Murdoch Group (BSkyB, US) | 73.3% | 6,089 |
| Ireland | Murdoch Group | 52.0% | 5,102 |
| Chile | DirecTV (US) | 56.0% | 4,875 |
| France | Vivendi | 61.0% | 4,488 |
| Israel | Altice (France)/HOT | 57.3% | 4,485 |
| Portugal | ZON | 57.9% | 4,342 |
| Poland | Polsat | 56.7% | 4,156 |
| Belgium | Telenet (Liberty Media, US) | 54.5% | 4,094 |
| Turkey | Cukurova Group | 53.7% | 3,956 |
| Netherlands | Ziggo | 53.2% | 3,929 |
| Germany | Kabel Deutschland | 45.0% | 3,294 |
| Mexico | Televisa | 50.0% | 3,022 |
| Spain | PRISA | 42.2% | 2,543 |
| Japan | JSAT (SkyPerfecTV) | 43.6% | 2,430 |
| Argentina | Grupo Clarín | 55.1% | 2,217 |
| Brazil | Globo Group | 53.6% | 2,179 |
| Sweden | Com Hem (BC Partners, UK) | 38.4% | 2,071 |
| Canada | Shaw | 28.3% | 1,975 |
| Russia | NMG | 26.5% | 1,918 |
| South Korea | Tbroad | 25.6% | 1,805 |
| Taiwan | CNS (South Korea) | 21.6% | 1,268 |
| US | Comcast | 23.2% | 1,233 |
| India | Zee Entertainment Enterprises | 19.4% | 842 |

[1] 21st Century Fox runs Foxtel as a joint venture with Telstra, so their 50:50 split in the venture would give Telstra 45% market share, and Fox 45%.

Many cable companies operate at the regional level. National concentration numbers can therefore be deceptive from a consumer's choice perspective. In each area, there is typically only one cable firm and maybe two DBS (DTH) providers. Occasionally (but increasingly), there may be a wireline telecom firm offering IPTV. Online broadband TV might change that to a more open system on the content level, but on the transmission level such online TV would use the broadband facilities of the ISP cable and telecom firms. This kind of TV is in its early development. This will change rapidly.

Concentration in multi-channel platforms is high in Europe as foreign companies (mostly American) consolidated their presence, while declining slightly in Latin America as more foreign companies entered new markets (Graph 34.13). It was lowest in North

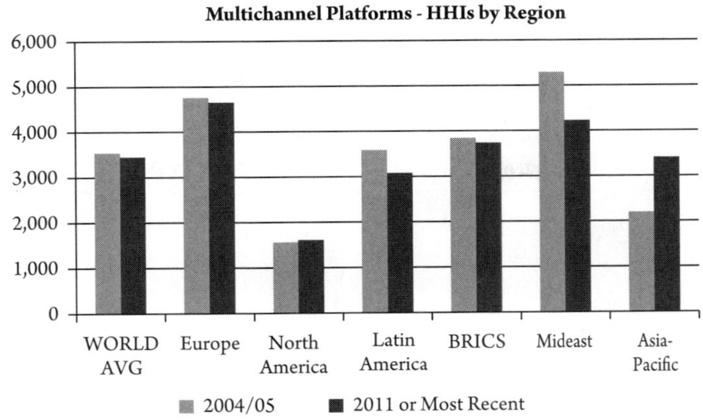


GRAPH 34.13 Multichannel Platforms Regional Concentration

America, but rising. The biggest decline was in the Mideast, driven by increasing competition in Turkey and Israel for growing pay-TV demand.

The United States has an extraordinary high share of this industry, at 64%, and not counting the presence of US firms such as Liberty Media, 21st Century Fox, DirecTV, and Dish internationally. Per capita spending was highest in the United States ($502), reflective of the extent of subscribership, and also of high prices. In most other heavily cabled countries—Canada, Switzerland, The Netherlands, United Kingdom, Ireland, Finland, and Sweden—the spending was less than half of that. The world average per capita spending stands at $104.6. As a share of the national GDP (1.13%), US spending is the largest of all countries, while it is lowest in Germany and South Africa (Table 34.12).

## VIDEO CHANNEL INDUSTRY CONCENTRATION ACROSS THE WORLD

We turn to "video channels"—non-terrestrial and typically payment-based channels, as delivered over subscription cable, DBS, IPTV, and online. Also included are encrypted conditional access pay-channels over terrestrial broadcasting. Globally, the video channel industry is highly concentrated, with a world average arithmetic HHI of 1,957 unweighted and 1,948 weighted (Graph 34.14). Even so, among the audiovisual industries it is the second least concentrated audiovisual media industry (only film is less concentrated). Among reporting countries, concentration is highest in Japan (3,902), France (2,928) and Chile (2,584) while it is lowest in Israel (992), Egypt (818), and Spain (477). The US's industry HHI is moderately concentrated at 1,409 and rising. (Table 34.13).

As in the multi-channel platforms industry, concentration has been relatively static. It declined slightly in all regions except Latin America, where it rose to 2,584 (Graph 34.15).

Of the reporting countries, the Netherlands ($219) have the highest per capita spending, while it is extremely low in India at just $1 (Table 34.14). This is in part due to the relatively small size of the market compared to broadcasting in India: the industry's share of the national GDP in European countries is on average three times what it is in India, where the market is worth only over $1.3 billion and has a large number of participants. The United States has a 40.4% share of the industry's total world revenue of $70.6 billion.

## FILM INDUSTRY GLOBAL CONCENTRATION

Six US-based Hollywood major "studios" dominate the industry worldwide (Table 34.15). They are Universal (Comcast), Paramount (Viacom), 20th Century Fox (owned by 21st Century Fox, spun off from Rupert Murdoch's News Corp. in

Table 34-12. Multichannel Platforms in National Markets, 2011 or Most Recent

| COUNTRY | COUNTRY SHARE IN WORLD MULTICHANNEL INDUSTRY TOTAL (%) | MULTICHANNEL PLATFORMS SHARE OF NATIONAL GDP (%) | PER CAPITA SPENDING ON MULTICHANNEL PLATFORMS (US$) |
|---|---|---|---|
| Argentina | 0.7% | 0.51% | 40.3 |
| Australia | 1.3% | 0.33% | 145.6 |
| Belgium | 0.5% | 0.23% | 102.0 |
| Brazil | 2.8% | 0.42% | 35.3 |
| Canada | 3.5% | 0.61% | 251.8 |
| Chile | 0.4% | 0.52% | 53.3 |
| France | 2.0% | 0.18% | 77.2 |
| Germany | 1.8% | 0.13% | 54.8 |
| India | 2.0% | 0.38% | 3.9 |
| Ireland | 0.3% | 0.34% | 180.7 |
| Israel | 0.4% | 0.55% | 136.8 |
| Italy | 1.8% | 0.20% | 71.9 |
| Japan | 4.2% | 0.21% | 81.2 |
| Mexico | 1.0% | 0.26% | 22.6 |
| Netherlands | 1.0% | 0.30% | 147.5 |
| Poland | 0.1% | 0.03% | 4.4 |
| Portugal | 0.6% | 0.61% | 138.2 |
| Russia | 0.5% | 0.09% | 8.6 |
| South Africa | 0.1% | 0.12% | 6.7 |
| South Korea | 1.1% | 0.30% | 54.7 |
| Spain | 2.6% | 0.42% | 139.1 |
| Sweden | 0.6% | 0.31% | 148.9 |
| Switzerland | 0.4% | 0.21% | 132.9 |
| Taiwan | 0.5% | 0.32% | 55.2 |
| Turkey | 0.2% | 0.09% | 8.3 |
| UK | 5.6% | 0.57% | 220.5 |
| US | 64.0% | 1.13% | 501.7 |
| Average Multichannel Industry Share of GDP (%) | | 0.35 | |
| Total Multichannel Industry Share of World GDP (%) | | 0.55 | |
| Total Multichannel Revenue (mil $) | | 247,666 | |
| Average Per Capita Spending ($) on Multichannel | | | 104.6 |

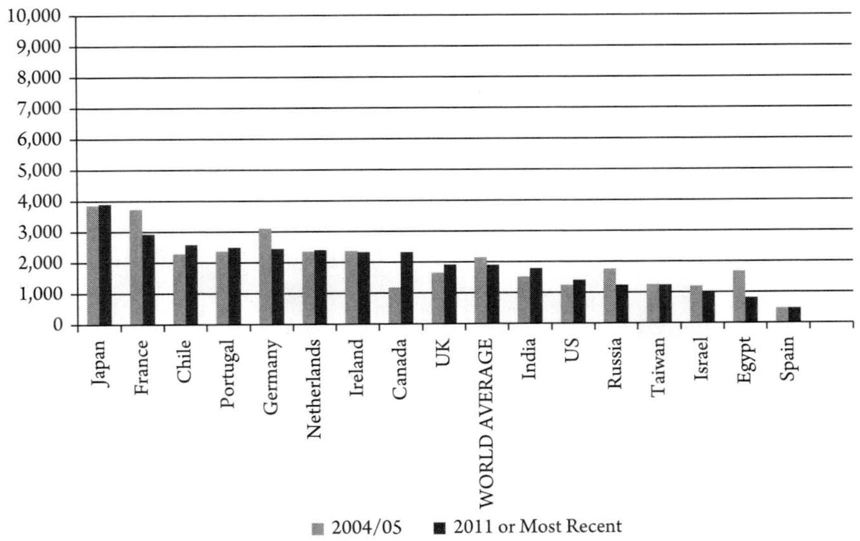


GRAPH 34.14 Video Channel Concentration by Country

## Table 34-13. Countries' Top Non-Broadcast Video and Pay TV Channel Provider (Non-Broadcast Channels), 2011 or Most Recent

| COUNTRY | TOP COMPANY | MARKET SHARE OF TOP COMPANY (%) | NATIONAL HHI |
|---|---|---|---|
| Japan | JSAT (SkyPerfecTV) | 55.6% | 3,902 |
| France | Vivendi | 49.5% | 2,928 |
| Chile | Discovery (US) | 38.5% | 2,584 |
| Portugal | Impresa | 24.9% | 2,497 |
| Germany | Bertelsmann | 25.1% | 2,449 |
| Netherlands | Viacom (US) | 34.1% | 2,406 |
| Ireland | RTE (public) | 35.7% | 2,335 |
| Canada | Shaw | 33.9% | 2,329 |
| UK | Murdoch (BSkyB, US) | 38.4% | 1,917 |
| India | Zee Entertainment Enterprises | 30.2% | 1,793 |
| US | Time Warner | 23.2% | 1,409 |
| Russia | MTG (Sweden) | 20.6% | 1,242 |
| Taiwan | SET | 16.6% | 1,241 |
| Israel | Dori Media Group | 20.0% | 992 |
| Egypt | Hayat | 13.3% | 818 |
| Spain | PRISA | 14.5% | 477 |

2013), Disney, Warner Brothers (Time Warner), and the Japanese-owned Sony Pictures.

Surprisingly, given Hollywood's global dominance, the film industry has a moderate overall concentration in most countries. The worldwide arithmetic average HHI was only 1,403 (weighted 1,294) (Graph 34.16). This made film production and distribution the least concentrated media industry overall by this measure. The explanation is that while these

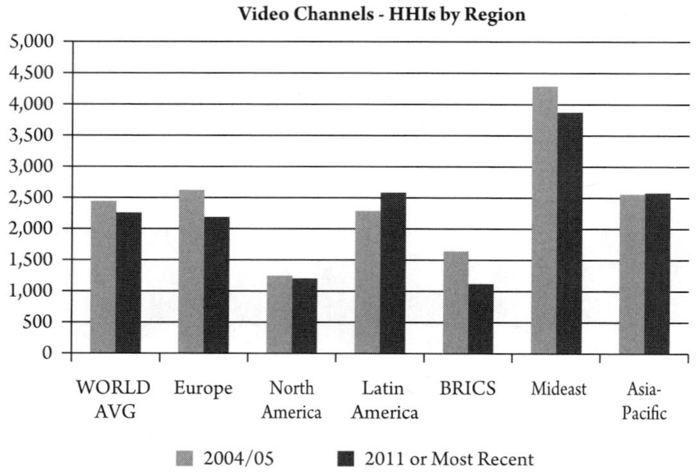


GRAPH 34.15 Video Channel Regional Concentration

**Table 34-14. Non-Broadcast Video Channels in National Markets, 2011 or Most Recent**

| COUNTRY | COUNTRY SHARE IN WORLD VIDEO CHANNEL INDUSTRY TOTAL (%) | VIDEO CHANNELS SHARE OF NATIONAL GDP (%) | PER CAPITA SPENDING ON VIDEO CHANNELS (US$) |
|---|---|---|---|
| Canada | 10.3% | 0.51% | 211.8 |
| Chile | 0.5% | 0.22% | 22.0 |
| Egypt | 0.7% | 0.20% | 94.4 |
| France | 10.2% | 0.26% | 113.7 |
| Germany | 4.6% | 0.09% | 39.7 |
| India | 1.8% | 0.10% | 1.0 |
| Ireland | 0.6% | 0.18% | 94.6 |
| Israel | 0.8% | 0.28% | 70.3 |
| Japan | 7.4% | 0.11% | 40.6 |
| Netherlands | 5.2% | 0.44% | 219.2 |
| Portugal | 0.8% | 0.22% | 50.9 |
| Russia | 0.3% | 0.01% | 1.3 |
| Spain | 6.2% | 0.29% | 94.2 |
| Taiwan | 1.2% | 0.22% | 36.9 |
| UK | 9.0% | 0.26% | 101.9 |
| US | 40.4% | 0.20% | 90.2 |
| Average Video Channel Industry Share of GDP (%) | | 0.22 | |
| Total Video Channel Industry Share of World GDP (%) | | 0.20 | |
| Total Video Channel Revenue (mil $) | | 70,636 | |
| Average Per Capita Spending ($) on Video Channels | | | 80.2 |

Table 34-15. Countries' Top Film Studios/Distributors, 2011 or Most Recent

| COUNTRY | TOP COMPANY | MARKET SHARES OF TOP COMPANY (%) | NATIONAL HHI |
|---|---|---|---|
| China | Government of China | 63.0 | 5,154 |
| | *China Film Group (public)* | 45.0 | |
| South Africa[1] | Sony | 16.5 | 3,366 |
| | Disney | 16.5 | |
| Taiwan | Sony | 16.6 | 2,257 |
| Chile | Disney | 13.7 | 2,062 |
| South Korea | CJ Group | 30.1 | 1,834 |
| Japan | Toho | 37.9 | 1,771 |
| Russia | Warner Brothers/Karo Premier | 18.7 | 1,642 |
| Netherlands | Warner Brothers | 27.2 | 1,522 |
| Argentina | Disney | 22.9 | 1,522 |
| Brazil | Paramount | 19.6 | 1,484 |
| Australia | Village Roadshow/Warner Brothers | 21.0 | 1,300 |
| Mexico | Grupo Televisa | 6.9 | 1,263 |
| Switzerland | Sony | 16.6 | 1,258 |
| UK | Universal | 18.5 | 1,208 |
| US | Warner Brothers | 19.9 | 1,182 |
| Ireland | Universal | 18.5 | 1,178 |
| Canada | Warner Brothers | 19.9 | 1,103 |
| Italy | Fininvest (Medusa) | 16.6 | 1,097 |
| Israel | Sony[3] | 16.6 | 1,027 |
| Poland | Sony | 16.6 | 1,027 |
| Belgium | Sony | 16.6 | 1,027 |
| Finland | Sony | 16.6 | 1,027 |
| Portugal | Disney[2] | 15.3 | 1,008 |
| Spain | Universal | 22.2 | 978 |
| Turkey | Sony | 16.6 | 909 |
| Germany | Warner Brothers | 16.6 | 856 |
| France | Vivendi | 26.8 | 637 |
| Sweden | Sony | 16.6 | 751 |
| India | Eros Entertainment | 12.9 | 476 |
| Egypt | Sony | 6.6 | 165 |

[1] Ster-Kinekor Entertainment, owned by Primedia, is a South African distributor for both Sony and Disney, whose market shares tie for the highest in the country.

[2] ZON, the Portuguese distributor for Paramount, Universal, and Walt Disney in this market, has a combined share of 42.3%. Of this, Disney's was the single largest, at 15.3%, in 2011/12, based on global shares.

[3] Sony data for several countries estimated based on global share.

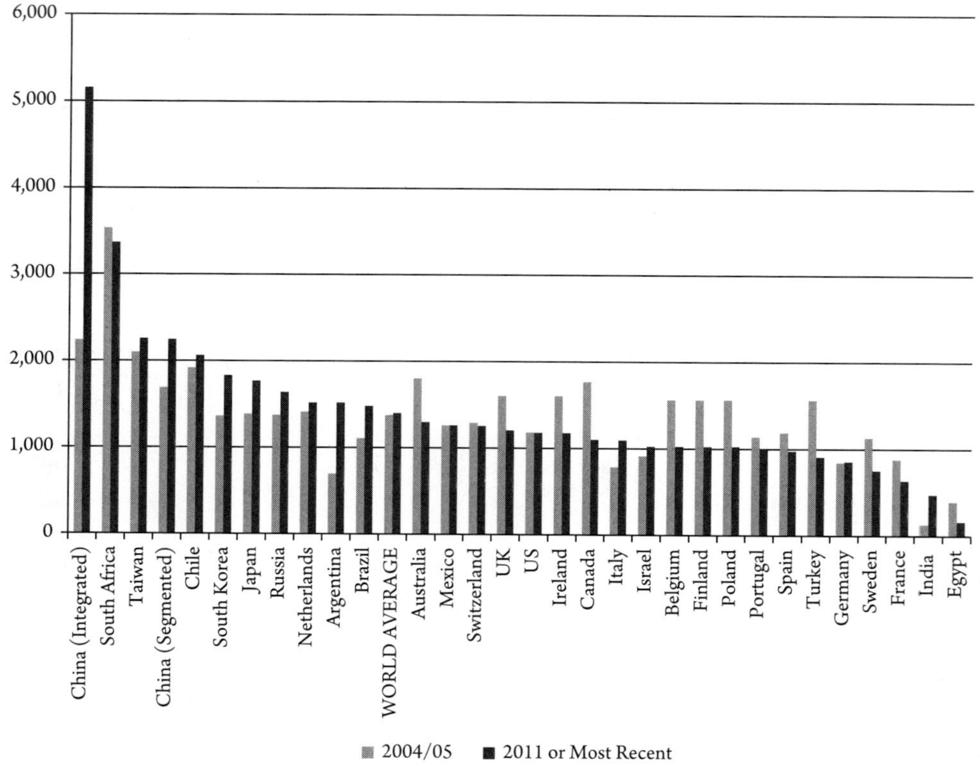


GRAPH 34.16 Film Industry Concentration by Country

six firms dominate in the aggregate, with a 76.4% combined world average market share, none dominates internationally, and none dominates a particular country. Market shares fluctuate somewhat from year to year based on the box office performance of particular attractive films but rarely exceed 20% or drop below 10%. This kind of oligopoly has a lower concentration measure than those of many other media, where often one or two firms stand out. What makes film unusual is that the same firms, for all their intermediate market shares, have such presence in just about every country in the world.

National market concentration is actually highest in protectionist markets where there is a domestic industry, often state supported. Concentration is high and increasing in China, due to the consolidation of smaller publicly owned studios which account for more than half of production, into two larger state-owned studios. One of them is the China Film Group, with 45% of the domestic market. The import of foreign films is limited to 34 per year. There is also a private studio, Huayi Brothers. Japan's high HHI is due to the dominance of Toho as a domestic distributor and exhibitor of foreign films and longtime producer of popular genre content. It has 37.9% of the Japanese market. In South Korea, domestic films must be shown at least 40% of the time, as a way to give domestic producers an opening. The CJ Group, a large media firm, has a share of 30.1%. In France, Vivendi, through its near-monopoly pay-TV service Canal Plus and its production/distribution arm Studio Canal (which can supply funds and distribution), has 26.8%. On the other hand, concentration is lowest in the two countries with substantial homegrown film industries where protectionism and state support are not major factors but cultural difference is, leading to a role as regional production centers. This is the case for Egypt and India.

India has an enormously active and fragmented domestic film industry. Hollywood's share in India is fairly low at 16.1%, of which

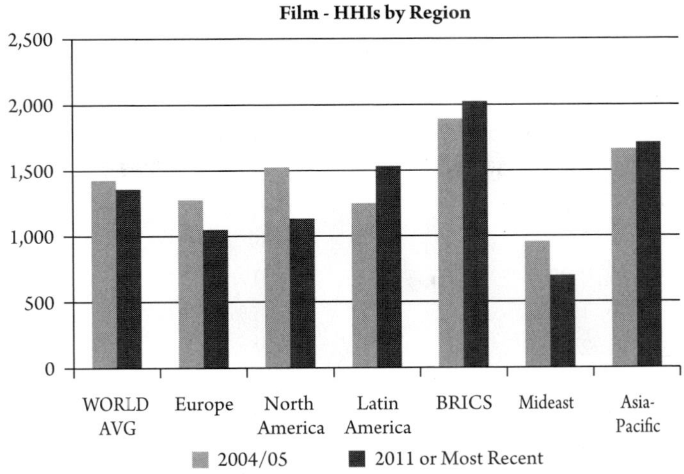


GRAPH 34.17 Film Industry Regional Concentration

21st Century Fox accounted for 4.1% for 2012. In Egypt, Hollywood has only 20% combined, of which Sony held 6.6% in 2012. Overall, World C6 is 76.4%—the sum of the averages market shares the Hollywood "Big Six" (including Sony).

The average HHI for film has been stable on a global scale, but the trends vary by region. Between 2004 and 2012, the industry became less concentrated in the European Union (Graph 34.17). Similarly, film industry concentration in North America slightly declined. It remains low in the Middle East primarily due to the strong indigenous cinema production houses in Egypt and to a lesser extent, Turkey.

Per capita film consumption revenues average a fairly low $31.6. It is highest in Australia ($79.6), France ($85.3), Germany ($91.3), Switzerland ($63.8), and the United States ($60). (Table 34.16). It is low in Latin America, and even lower in China ($1.6). Worldwide industry revenues are $30.7 billion, of which the Hollywood companies take in approximately $22.9 billion. As a sector of national economies (with a world average of 0.06%), film production and consumption in Egypt is the highest at 0.16% of GDP, more than twice that in the United Kingdom or United States (0.07% each). Cairo is the film capital of the Arab world.

## WIRELINE TELECOM INDUSTRY CONCENTRATION ACROSS THE WORLD

"Wireline" used to be fairly synonymous with "telecommunications." More recently, however, mobile wireless telecom eclipsed it in volume and penetration, especially in the developing world. However, it would be superficial to view wireline as a relic like the telegraph. Business communications, long distance calls, international communications, and Internet data traffic travel over wires of increasing power such as fiber. Also, most mobile communications use wireless only for the tail end close to the end-users, with everything in between travelling over wires. Wireline and wireless are overlapping markets though they have different regulatory and global ownership arrangements.

The wireline telecommunications sector is the second most highly concentrated media industry, with an average HHI of 5,208, significantly more so than wireless' arithmetic mean HHI of 3,493 (Graph 34.18). Concentration is highest in China, Egypt, Turkey, and South Africa and lowest in the United States, The Netherlands, and Finland. The United States market has followed a unique trajectory. In the 1980s, 80% of the market was controlled by

Table 34-16. Film in National Markets, 2011 or Most Recent

| COUNTRY | COUNTRY SHARE IN WORLD FILM INDUSTRY TOTAL (%) | FILMS SHARE OF NATIONAL GDP (%) | PER CAPITA SPENDING ON FILMS (US$) |
|---|---|---|---|
| Argentina | 0.6% | 0.05% | 8.5 |
| Australia | 2.8% | 0.09% | 77.1 |
| Belgium | 0.2% | 0.01% | 13.3 |
| Brazil | 1.3% | 0.02% | 4.0 |
| Canada | 2.9% | 0.06% | 51.5 |
| Chile | 0.2% | 0.04% | 8.4 |
| China | 3.5% | 0.02% | 1.6 |
| Egypt | 0.9% | 0.16% | 6.6 |
| Finland | 0.4% | 0.05% | 50.7 |
| France | 8.8% | 0.10% | 85.3 |
| Germany | 12.2% | 0.11% | 91.3 |
| India | 1.7% | 0.04% | 0.9 |
| Ireland | 0.4% | 0.05% | 53.0 |
| Israel | 0.1% | 0.01% | 5.1 |
| Italy | 2.5% | 0.03% | 25.0 |
| Japan | 5.9% | 0.04% | 28.1 |
| Mexico | 2.8% | 0.09% | 15.2 |
| Netherlands | 0.9% | 0.03% | 34.9 |
| Poland | 0.5% | 0.03% | 8.7 |
| Portugal | 0.9% | 0.12% | 52.5 |
| Russia | 3.6% | 0.08% | 15.3 |
| South Africa | 0.3% | 0.03% | 3.4 |
| South Korea | 3.5% | 0.12% | 43.9 |
| Spain | 4.0% | 0.08% | 52.9 |
| Sweden | 0.2% | 0.01% | 13.8 |
| Switzerland | 0.8% | 0.05% | 63.8 |
| Taiwan | 0.3% | 0.02% | 7.3 |
| Turkey | 1.1% | 0.05% | 9.3 |
| UK | 5.8% | 0.07% | 56.4 |
| US | 30.9% | 0.07% | 60.0 |
| Average Film Industry Share of GDP (%) | | 0.06% | |
| Total Film Industry Share of World GDP (%) | | 0.06% | |
| Total Film Industry Revenue (mil $) | | 30,710 | |
| Average Per Capita Spending ($) on Films | | | 31.6 |

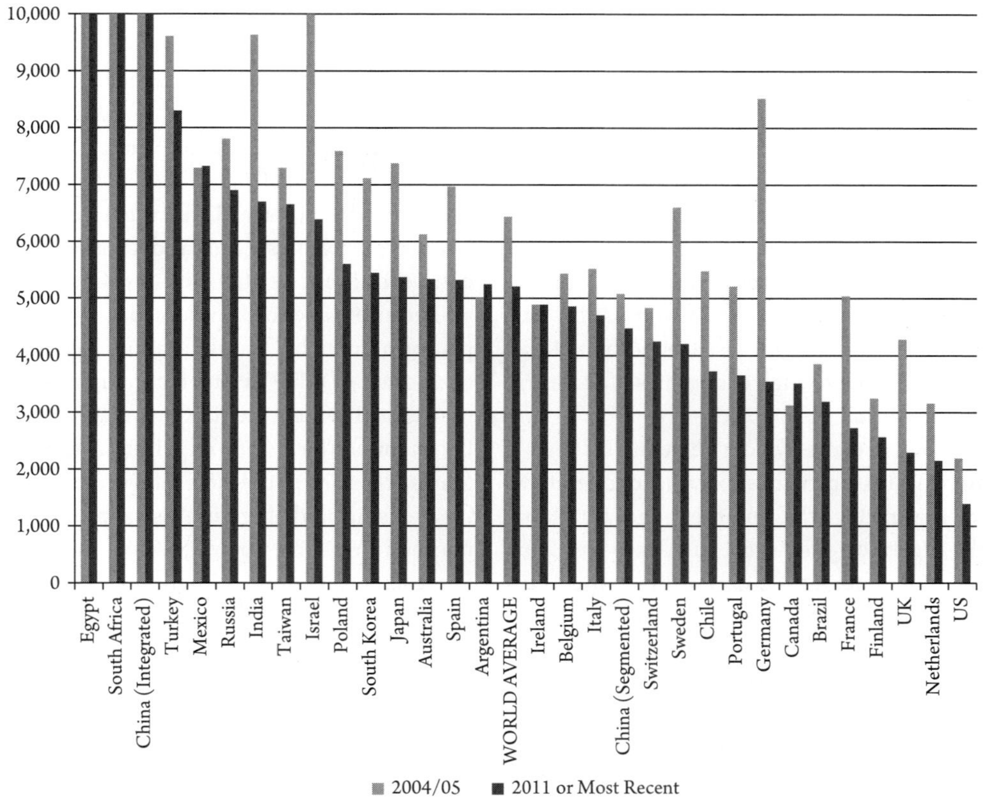


GRAPH 34.18 National Wireline Telecom Industries by Country

one firm, AT&T. Following the government's breakup of that provider into eight parts, concentration declined enormously. The regional providers subsequently reconverged into three entities, Verizon, a new AT&T, and a smaller firm called CenturyLink. Even so, other providers entered or remained, and industry HHI is low in international comparison. For the United States, the industry used to be historically near-monopolistic, with an HHI of about 7,685, but became much less concentrated thereafter, with the industry HHI falling instantly to 1,633 in 1984 and to 1,049 in 1996. Since then, it rose again due to consolidations to 2,191 by 2006 but then fell to 1,395 in 2013 as cable TV companies' telecom business grew.

But one must recognize that in the United States, as well as in Finland, and Canada, the major wireline providers operate in separate territories for their local service, and from the consumer's perspective there is less competition than the national HHI figures suggest.

South Africa and Egypt have industry HHIs of 10,000 because the incumbent wireline telecom operators (Telkom and Telecom Egypt, respectively) have a legal monopoly in their respective markets. China's industry is also state-owned, but separated into two regional companies, so the segmented HHI (4,476) is much smaller than the integrated measure (10,000).

In countries with high industry HHIs in 2004—Turkey, Japan, Germany, Israel,[4] Sweden,

4. Wireline market shares dropped significantly for Bezeq (B Communication), which had 99% in 2004 and 82% in 2013. Altice-HOT holds 14% of market shares and Partner Communications gained 3%, both much smaller companies than Bezeq.

and India[5]—concentration fell significantly. In all of these countries, formerly state-owned incumbents lost market shares to new entrants. HHI remains high in many countries despite the transfer of state ownership of the public incumbent to a commercial operator ("privatization") and an opening to new entrants ("liberalization"), which often includes an opening to foreign firms. New operators entered the market but the formerly public incumbents still dominate the top companies' list (see Table 34.17). Partly this is due to high economies of scale and network effects.

Finland's relatively low telecom HHI is a legacy of its history, with subscriber cooperatives operating in the urban markets and a national public operator providing rural service. This system goes back to the czarist days when Finland was a reluctant Russian province. Although the cooperatives have disappeared, rather than merge with the state-owned incumbent they joined into different commercial operators.

Graph 34.19 with its scattergram shows the relation of wireline industry concentration and GDP. Outliers are, for high concentration, not only China, South Africa, and Egypt—usually among the most concentrated—but also Japan (Graph 34.19). Among low concentration countries, adjusted for GDP, are the United States, the Netherlands, Finland, and the United Kingdom.

Worldwide average HHI has been decreasing. Every region shows a similar decreasing trend in concentration across the industry (Graph 34.20).

Because wireline services are an essential platform industry within national economies, their share of GDP as a media industry is higher than that of content media. On average, the industry accounts for 1.17% of GDP. In the United States, it is 1.5% of the US's total economic activity and is similarly high in Australia (1.3%), Germany (1.5%), and Spain (1.3%) (Table 34.18). Per capita spending is highest in Switzerland ($745) and the United States at $664 annually. It is lowest in countries such as India, Mexico, China, and Egypt because of lower usage and much lower penetration (i.e., the total revenue is generated by fewer users than total national population). The world average per capita spending stands at $288.

## Table 34-17. Countries' Top Wireline Telecom Providers, 2011 or Most Recent

| COUNTRY | TOP COMPANY | MARKET SHARE OF TOP COMPANY (%) | NATIONAL HHI |
|---|---|---|---|
| China | Government of China | 100% | 10,000 |
| | China Telecom (public) | 58.6% | |
| South Africa | Telkom (public) | 100.0% | 10,000 |
| Egypt | Government of Egypt (Telecom Egypt, public) | 100.0% | 10,000 |
| Turkey | Turk Telekom (Oger Group, Saudi Arabia/Turkey, public) | 91.0% | 8,296 |
| Mexico | Grupo Carso | 85.0% | 7,324 |
| Russia | Government of Russia (Svyazinvest, public) | 82.6% | 6,899 |
| India | BSNL (public) | 81.0% | 6,699 |
| Taiwan | Government of Taiwan (CHT, public) | 80.7% | 6,654 |

5. Wireline revenue decreased substantially from $9.6 billion in 2004 to $6.7 billion in 2013.

Table 34-17. *continued*

| COUNTRY | TOP COMPANY | MARKET SHARE OF TOP COMPANY (%) | NATIONAL HHI |
|---|---|---|---|
| Israel | Bezeq | 82.0% | 6,390 |
| Poland | TP (public) | 73.3% | 5,600 |
| South Korea | KT | 72.5% | 5,447 |
| Japan | NTT (33% public) | 72.3% | 5,369 |
| Australia | Telstra | 71.4% | 5,336 |
| Spain | Telefónica | 71.9% | 5,319 |
| Argentina | Telefónica (Spain) | 61.1% | 5,246 |
| Ireland | SingTel (Singapore) | 61.3% | 4,888 |
| Belgium | Belgacom | 63.1% | 4,859 |
| Italy | Telecom Italia | 66.0% | 4,702 |
| Switzerland | Swisscom (public) | 63.1% | 4,248 |
| Sweden | TeliaSonera (Sweden/ Finland) (50% public) | 62.9% | 4,205 |
| Chile | Telefónica (Spain) | 53.0% | 3,723 |
| Portugal | Oi (Brazil)/Portugal Telecom | 55.3% | 3,653 |
| Germany | Deutsche Telekom (38% public) | 56.9% | 3,542 |
| Canada | Bell Canada Enterprise/CTV | 54.4% (regional) | 3,513 |
| Brazil | Oi (Telemar Participações)/ Portugal Telecom | 42.1% (regional) | 3,190 |
| France | France Telecom (27% public) | 41.0% | 2,728 |
| Finland | Elisa | 38.0% (regional) | 2,565 |
| UK | BT | 43.8% | 2,293 |
| Netherlands | KPN | 41.4% | 2,150 |
| US | AT&T | 26.7% (regional) | 1,395 |

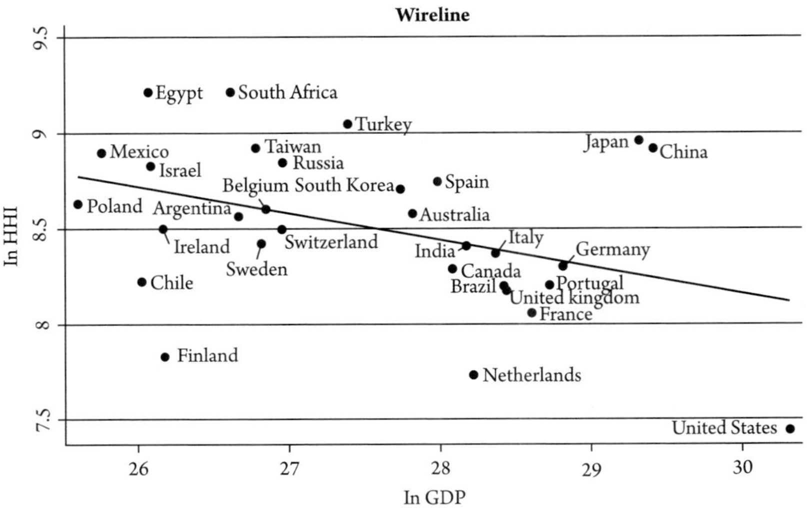


GRAPH 34.19 Wireline Industry Concentration and GDP

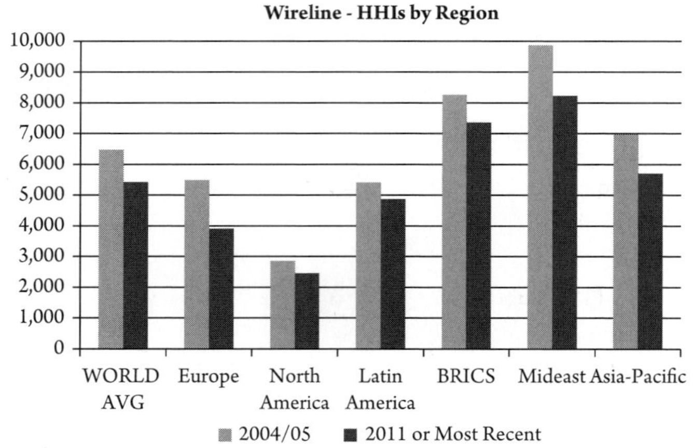


GRAPH 34.20 Wireline Telecom Regional Concentration

## Table 34-18. Wireline Telecoms in National Markets, 2011 or Most Recent

| COUNTRY | COUNTRY SHARE IN WORLD WIRELINE INDUSTRY TOTAL (%) | WIRELINE SHARE OF NATIONAL GDP (%) | PER CAPITA SPENDING ON WIRELINE (US$) |
|---|---|---|---|
| Argentina | 0.4% | 0.78% | 60.7 |
| Australia | 2.2% | 1.32% | 572.4 |
| Belgium | 0.9% | 1.05% | 470.3 |
| Brazil | 8.8% | 3.19% | 265.4 |
| Canada | 2.8% | 1.17% | 480.8 |
| Chile | 0.2% | 0.59% | 60.4 |
| China | 4.5% | 0.56% | 19.5 |
| Egypt | 0.3% | 1.03% | 21.9 |
| Finland | 0.4% | 0.84% | 398.3 |
| France | 3.8% | 0.82% | 353.4 |
| Germany | 8.6% | 1.46% | 619.7 |
| India | 1.0% | 0.46% | 4.8 |
| Ireland | 0.4% | 1.06% | 563.0 |
| Israel | 0.2% | 0.63% | 157.1 |
| Italy | 3.1% | 0.83% | 301.3 |
| Japan | 10.8% | 1.29% | 499.3 |
| Mexico | 0.5% | 0.31% | 26.7 |
| Netherlands | 1.2% | 0.85% | 423.2 |
| Poland | 0.8% | 0.96% | 129.1 |
| Portugal | 0.3% | 0.68% | 155.0 |
| Russia | 1.9% | 0.78% | 79.2 |
| South Africa | 0.8% | 1.64% | 90.4 |
| South Korea | 1.6% | 1.07% | 193.4 |
| Spain | 3.3% | 1.29% | 426.8 |
| Sweden | 0.7% | 0.88% | 415.7 |
| Switzerland | 1.0% | 1.15% | 744.9 |

Table 34-18. *continued*

| COUNTRY | COUNTRY SHARE IN WORLD WIRELINE INDUSTRY TOTAL (%) | WIRELINE SHARE OF NATIONAL GDP (%) | PER CAPITA SPENDING ON WIRELINE (US$) |
|---|---|---|---|
| Taiwan | 0.5% | 0.81% | 139.4 |
| Turkey | 1.0% | 0.87% | 79.3 |
| UK | 2.3% | 0.56% | 214.8 |
| US | 35.6% | 1.49% | 664.0 |
| Average Wireline Industry Share of GDP (%) | | 1.01 | |
| Total Wireline Industry Share of World GDP (%) | | 1.17 | |
| Total Wireline Industry Revenue (mil $) | | 589,986 | |
| Average Per Capita Spending ($) on Wireline | | | 287.7 |

## WIRELESS TELECOMMUNICATIONS INDUSTRY CONCENTRATION ACROSS THE WORLD

The international wireless market displays a high but stable average HHI (unweighted), which fell from 3,691 to 3,493, still making it the fourth most highly concentrated media industry (Graph 34.21). The emerging market structure for mobile wireless is three to four national footprints, typically two large firms and one or two smaller players, plus a few resellers (MVNOs). Among the national operators, the incumbent traditional wireline operators, formerly the monopoly providers, typically have the highest market share, for example, Orange in France, NTT in Japan, Deutsche Telekom (T-Mobile) in Germany, Telmex in Mexico, Telefónica in Spain and Argentina, and so on (Table 34.19) The number 2–4 firms are often foreign companies, and often the national incumbents of other countries (e.g., Orange, Telefónica, Deutsche Telekom, etc.). These firms then can leverage their domestic experience, resources, and economies of scale (and the needs of developing markets) into increasingly global footprints. This process is still taking place. At present, there are over a dozen of such firms. Consolidation is likely, probably through international partnerships.

Overall, the market is less concentrated than the wireline market. Incumbents, too, had to build their business, and had fewer advantages over new entrants than for wireline. And the entrants were often large telecom incumbents themselves that moved into other countries' markets rather than small new entrants. The most concentrated markets are China, Mexico, Turkey, and Switzerland. The least concentrated markets are India, the United States, and Taiwan. India's low concentration shows that low income in a country is not a fundamental factor for concentration. This is true for the other BRICS countries too, except for China.

In the United Kingdom, concentration has risen sharply from a low HHI of 1,513 to a high 2,862 because of the merger between T-Mobile and Orange in 2010, operating under the name Everything Everywhere (EE). Concentration also increased considerably in the United States

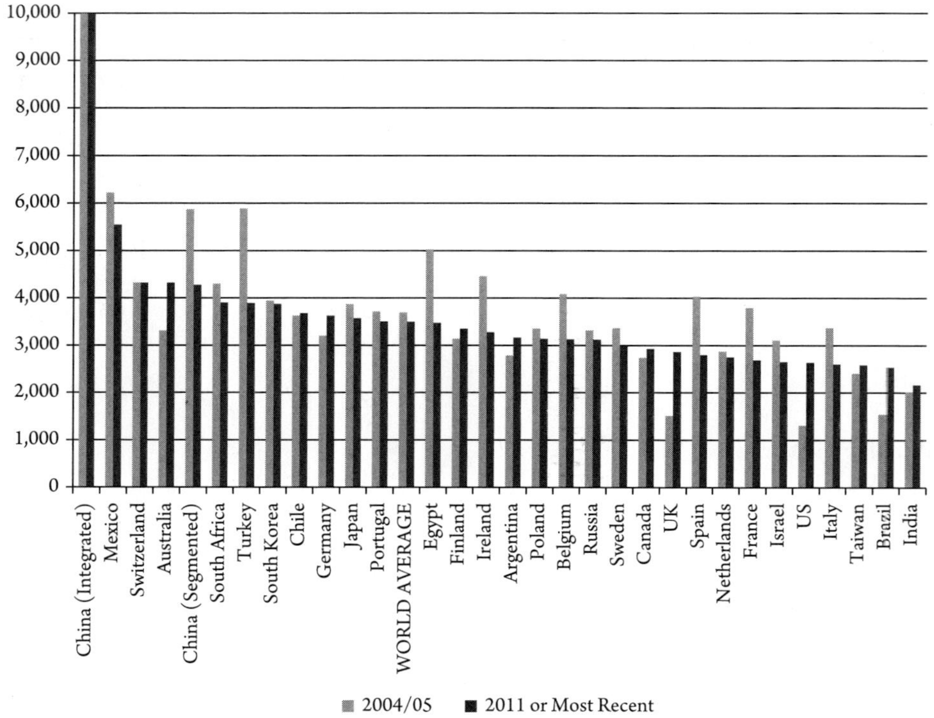


GRAPH 34.21 Wireless Telecom Concentration by Country

## Table 34-19. Countries' Top Wireless Telecoms, 2011 or Most Recent

| COUNTRY | TOP COMPANY | MARKET SHARE OF TOP COMPANY (%) | NATIONAL HHI |
|---|---|---|---|
| China | Government of China | 100% | 10,000 |
| | *China Mobile (public)* | 63.9% | |
| Mexico | Grupo Carso | 70.8% | 5,545 |
| Switzerland | Swisscom (public) | 60.4% | 4,319 |
| Australia | Telstra | 58.0% | 4,318 |
| South Africa | Vodacom[1] | 50.9% | 3,898 |
| Turkey | Turkcell[2] | 52.0% | 3,888 |
| South Korea | SK Group | 50.5% | 3,869 |
| Germany | Deutsche Telekom (38% public) | 32.4% | 3,622 |
| Japan | NTT (33% public) | 45.8% | 3,567 |
| Portugal | Oi (Brazil)/Portugal Telecom | 39.5% | 3,501 |
| Egypt | Mobinil (Orange, France Telecom) | 35.1% | 3,471 |
| Chile | Telefónica (Spain) | 41.0% | 3,461 |
| Finland | Elisa | 39.0% | 3,346 |
| Ireland | Vodafone (UK) | 42.5% | 3,274 |
| Argentina | América Móvil (Mexico) | 35.4% | 3,160 |
| Poland | PTTK Centertel | 33.4% | 3,136 |
| Belgium | Belgacom | 41.7% | 3,124 |

Table 34-19. *continued*

| COUNTRY | TOP COMPANY | MARKET SHARE OF TOP COMPANY (%) | NATIONAL HHI |
|---|---|---|---|
| Russia | MTS | 35.0% | 3,116 |
| France | France Telecom (27% public) | 38.0% | 2,686 |
| Sweden | TeliaSonera (Sweden/Finland) (50% public) | 40.4% | 2,984 |
| Canada | Rogers | 37.0% | 2,924 |
| UK | Everything Everywhere (Deutsche Telekom & Orange) | 34.0% | 2,862 |
| | *Deutsche Telekom (T-Mobile, Germany)* | 17% | |
| | *Orange (France Telecom, France)* | 17% | |
| Spain | Telefónica | 39.8% | 2,798 |
| Netherlands | KPN | 35.0% | 2,750 |
| Israel | Cellcom Israel | 31.5% | 2,651 |
| US | Verizon | 33.6% | 2,636 |
| Italy | Telecom Italia | 33.1% | 2,601 |
| Taiwan | Government of Taiwan (CHT, public) | 32.5% | 2,583 |
| Brazil | Telefónica (Spain) | 28.6% | 2,534 |
| India | Bharti Enterprises | 34.1% | 2,164 |

[1] 65% Vodafone, UK. 35% Telkom South Africa.
[2] 37% TeliaSonera, Sweden; 13.2% Alfa Group, Russia; 13.8% Cukurova Group.

(from 1,311 to 2,636) due to mergers. Even so, both of these countries are relatively unconcentrated in international comparison. In the United States, AT&T holds a share of 32.1% of the industry, while Verizon holds a share of 33.6%. The American increase in industry HHI is a result of consolidation of several companies by several operators (AT&T, Verizon, Sprint, T-Mobile) to reach a structure of four national firms.

In China, three state-owned companies, into which the government has consolidated the assets of smaller operators, dominate the industry. China Mobile holds 63.9% of the market, followed by China Unicom (21.5%) and China Telecom (14.4%). In Egypt, the market is effectively a duopoly with two foreign-owned incumbents dominating the market: Vodafone and Orange's Mobinil. Brazil's high HHI increase (from 1,542 to 2,534) stems from the aggressive mergers and acquisitions policy pursued by the market entrant America Móvil (Mexico), a subsidiary of Carlos Slim's operations. America Móvil bought several operators in South America, including AT&T's Latin American operations, Argentina's Techtel, Mexico's Metrored, and Verizon's Caribbean operations.

Regionally, concentration in North America, the BRICS countries, and Latin America rose, and declined in the rest of the world on average (Graph 34.22). It used to be significantly lower in North America before 2004, but not anymore.

Per capita spending in the United States is high—at $506 about $42 per month—and accounts for 21.6% of the total global revenue in this sector. China (17.1%) and Japan (10.7%) follow. As a share of GDP, wireless spending averages a hefty 1.53%, the highest worldwide. It is highest in South Africa (3.8%) (Table 34.20). This is due to a combination of high penetration, high prices due to the high market share of the state-owned Telkom. Other countries with high per capita spending are Israel, the United Kingdom, Switzerland, the Netherlands, and Japan, all of which are higher than the United States. Per capita spending is much lower in Argentina, China, Egypt, and India. This is mostly due to lower prices and partly due to a lower penetration rate.

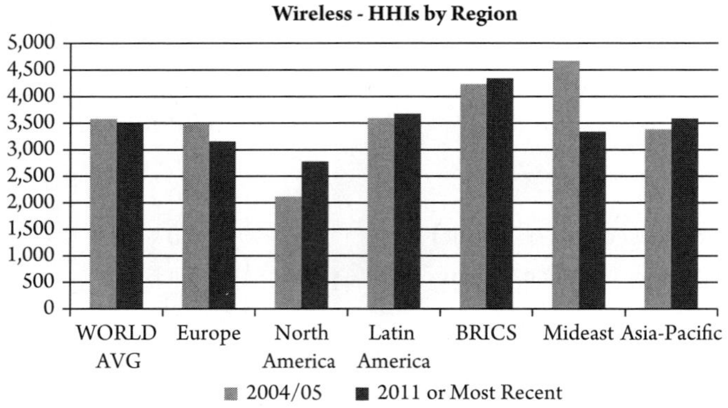


GRAPH 34.22 Wireless Telecom Regional Concentration

## Table 34-20. Wireless Telecom in National Markets, 2011 or Most Recent

| COUNTRY | COUNTRY SHARE IN WORLD WIRELESS INDUSTRY TOTAL (%) | MOBILE SHARE OF NATIONAL GDP (%) | PER CAPITA SPENDING ON MOBILE (US$) |
|---|---|---|---|
| Argentina | 0.3% | 0.60% | 46.9 |
| Australia | 2.1% | 1.56% | 676.7 |
| Belgium | 0.9% | 1.37% | 609.1 |
| Brazil | 5.8% | 2.61% | 217.6 |
| Canada | 2.7% | 1.41% | 578.3 |
| Chile | 0.2% | 0.90% | 91.9 |
| China | 17.1% | 2.66% | 93.5 |
| Egypt | 0.6% | 2.45% | 52.3 |
| Finland | 0.3% | 0.95% | 451.3 |
| France | 4.1% | 1.12% | 483.4 |
| Germany | 3.6% | 0.78% | 330.1 |
| India | 3.1% | 1.79% | 18.6 |
| Ireland | 0.4% | 1.11% | 584.3 |
| Israel | 0.4% | 1.52% | 381.5 |
| Italy | 2.9% | 0.96% | 348.7 |
| Japan | 10.7% | 1.60% | 617.0 |
| Mexico | 2.7% | 2.04% | 175.1 |
| Netherlands | 1.4% | 1.20% | 599.2 |
| Poland | 1.0% | 1.48% | 198.9 |
| Portugal | 0.6% | 1.84% | 417.3 |
| Russia | 3.1% | 1.61% | 162.5 |
| South Africa | 1.4% | 3.80% | 209.7 |
| South Korea | 2.4% | 2.01% | 362.7 |
| Spain | 2.9% | 1.42% | 468.9 |
| Sweden | 0.6% | 0.94% | 448.1 |
| Switzerland | 0.6% | 0.84% | 541.0 |
| Taiwan | 0.9% | 1.63% | 279.6 |
| Turkey | 1.2% | 1.37% | 124.9 |

Table 34-20. *continued*

| COUNTRY | COUNTRY SHARE IN WORLD WIRELESS INDUSTRY TOTAL (%) | MOBILE SHARE OF NATIONAL GDP (%) | PER CAPITA SPENDING ON MOBILE (US$) |
|---|---|---|---|
| UK | 4.4% | 1.33% | 513.7 |
| US | 21.6% | 1.14% | 506.0 |
| Average Wireless Industry Share of GDP (%) | | 1.53 | |
| Total Wireless Industry Share of World GDP (%) | | 1.47 | |
| Total Wireless Industry Revenue (mil $) | | 740,038 | |
| Average Per Capita Spending ($) on Wireless | | | 353.0 |

## INTERNET SERVICE PROVIDER INDUSTRY CONCENTRATION ACROSS THE WORLD

There are basically five types of Internet service providers for broadband connectivity (ignoring the remaining dial-up ISPs): (1) telecom companies; (2) cable TV companies; (3) other technology platforms such as satellite and terrestrial microwave, with small shares due to cost and low speed; (4) mobile wireless operators, whose networks increasingly provide speeds that approach those of wireless-based broadband. (Of course, the speed of the wireline networks also keeps rising rapidly, and with it the needs of applications and expectations of users); lastly, (5) there are ISPs without a network facilities platform of their own, but riding on top of telecom networks. Such independent providers are important, in particular, in Japan and several European countries. To some extent, they become possible through the regulation of "unbundling" the telecom incumbents' system into network elements, and through a price regulation of the terms under which they must be offered. The result can be a more competitive ISP market. Negative consequences include lower incentive to invest in network upgrades, both by incumbents and new entrants.

The ISP market is highly concentrated, worldwide, with the arithmetic average HHI standing at 3,616, down from 3,829 in 2005 and ranked fifth out of all industries for concentration (weighted HHI is 3,085) (Graph 34.23). The highest concentrations exist in South Africa, China, Turkey, and Mexico. The largest ISPs, by total subscribers, are China Telecom, which holds 51.4% of the Chinese market, followed by China Unicom, also state-owned, with 36.5%. The lowest concentrations are in the United States, Brazil, Japan, and Canada. In the United States, the regionalism of the telecom and cable TV markets results in the second lowest HHI of the surveyed countries, with national concentration at 1,390 (local ISP HHI is 4,790). A similar regional situation exists in Canada. Ireland had one of the most concentrated markets in 2004, but HHI fell from 7,642 to 3,343 in 2011 due to Eircom's losing market share to new entrants.[6]

6. Between 2004 and 2008, the incumbent Eircom's market share fell by half, from 87% to 41%, and two new companies entered the market: Liberty Global and Vodafone.

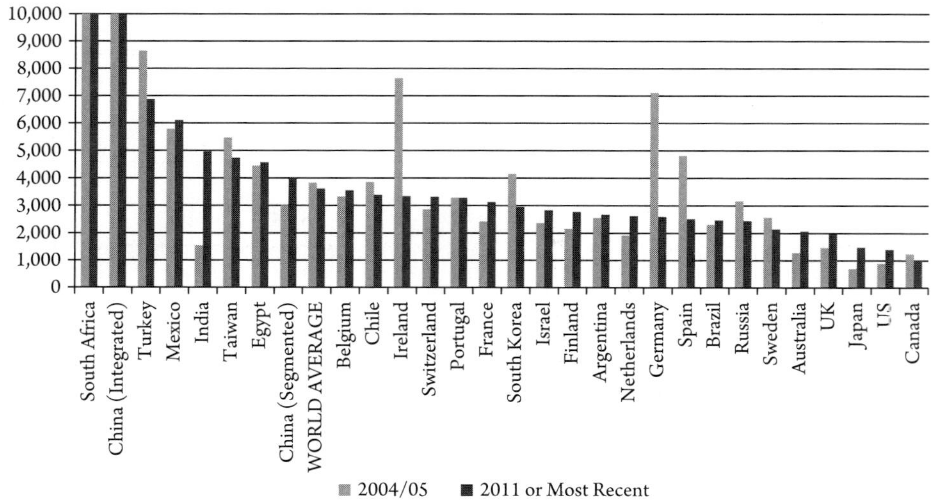


GRAPH 34.23 ISP Concentration by Country

The biggest international ISP provider outside of its home country is a US cable company, Liberty Global, which owns cable distribution systems in several European countries. Liberty Global has high shares of the ISP market in Switzerland, Ireland, The Netherlands, Belgium, and Chile. In 2013, Liberty Global completed the acquisition of Virgin Media in the United Kingdom, expanding its total reach to 25 million broadband subscribers globally (and 19.4% of the British market).

Telefónica is the leading ISP in Spain, with a 59.8% market share in 2013, and is able to use its incumbency as a wireline and wireless operator. Spain has been unbundling regulations, though, so Telefónica has seen its share decline from 66.1% in 2004. Telefónica is also the biggest ISP in Chile and Argentina, with market shares of 46% and 28.7%, respectively, and number two in Brazil (15.1%). Telefónica has more ISP customers outside of Spain than Liberty Global has outside of the United States. It operates in 22 countries.

The main ISP operator in France is Orange (France Telecom), which led the ISP sector with a 45% market share there in 2013. France has unbundling regulations. Orange also has a mobile ISP presence in Germany, Spain, and the United Kingdom, where it operates in a venture with Germany's T-Mobile. The ISP sector in France has gone through several changes, including many mergers that have caused increasing concentration since 2008.

Although the US's industry HHI of 1,390 after 2008 was low in international comparison, it marked an increase from 880 in 2004. This in crease in concentration is the result of market consolidation by incumbent local telephone companies.

South Africa has the highest HHI index in the industry next to China due to the monopoly maintained by the incumbent public wireline monopolist, Telekom (Table 34.21). Turkey's HHI has gone down slightly due to new market entrants, but the market share of the incumbent is high. Turk Telecom (TT) was protected from competition for a decade after it was privatized by a restrictive licensing system. There is little by way of cable TV platform options.

The Mexican market is nearly monopolized by the wireline incumbent Telmex, which as in South Africa and Turkey controls most of the wireline infrastructure in the country, allowing it to set rates. In Switzerland, the public operator Swisscom holds 53.7% and in Germany, incumbent Deutsch Telekom has 45.6%. Canada's geography, which has produced a regionalized telecom sector, also affects the scope of the ISP market, as in the United States and Finland.

Concentration is relatively low internationally due to the presence of many regional providers. Concentration is also lower in countries

Table 34-21. Countries' Top ISPs, 2011 or Most Recent

| COUNTRY | TOP COMPANY | MARKET SHARE OF TOP COMPANY (%) | NATIONAL HHI |
|---|---|---|---|
| China | Government of China | 100% | 10,000 |
| | *China Telecom (public)* | 51.4% | |
| South Africa | Telkom (public) | 100.0% | 10,000 |
| Turkey | Turk Telekom (Oger Telekom, Saudi Arabia/ Turkey, public) | 82.7% | 6,865 |
| Mexico | Grupo Carso | 76.9% | 6,103 |
| India | BSNL (public) | 69.1% | 4,962 |
| Taiwan | Government of Taiwan (CHT, public) | 67.9% | 4,737 |
| Egypt | Government of Egypt (Telecom Egypt, public) | 64.2% | 4,570 |
| Belgium | Belgacom | 47.1% | 3,557 |
| Chile | Telefónica (Spain) | 43.0% | 3,393 |
| Ireland | SingTel (Singapore) | 41.0% | 3,343 |
| Switzerland | Swisscom (public) | 53.7% | 3,322 |
| Portugal | Oi (Brazil)/Portugal Telecom | 48.2% | 3,287 |
| France | France Telecom (27% public) | 45.0% | 3,135 |
| South Korea | KT | 43.5% | 2,954 |
| Israel | Bezeq | 36.0% | 2,836 |
| Finland | Elisa | 34.0% | 2,773 |
| Argentina | Telefónica (Spain) | 31.2% | 2,676 |
| Netherlands | KPN | 42.6% | 2,629 |
| Germany | Deutsche Telekom (38% public) | 45.6% | 2,599 |
| Spain | Telefónica | 44.1% | 2,519 |
| Brazil | Oi (Telemar Participações)/Portugal Telecom | 30.3% | 2,469 |
| Russia | NMG | 31.9% | 2,451 |
| Sweden | TeliaSonera (Sweden/Finland) (50% public) | 37.2% | 2,146 |
| Australia | Telstra | 42.2% | 2,066 |
| UK | BT | 29.6% | 1,983 |
| Japan | NTT (33% public) | 33.1% | 1,473 |
| US | AT&T | 18.2% | 1,390 |
| Canada | Bell Canada Enterprise/CTV | 21.2% | 1,013 |

with cable TV platforms, such as to the United States, Canada, Argentina, Australia, Sweden, Finland, the Netherlands, and Switzerland.

The ISP market concentration has slightly declined worldwide since 2004 (Graph 34.24). Decreases can be observed in all regions except for North America and the BRICS countries.

The highest per capita spending on ISPs is in Canada ($211), France ($201), the United Kingdom ($184), Japan ($172), the Netherlands ($153), Portugal ($155), Finland ($150), and Australia ($113) (Table 34.22). The United States has the largest share internationally of 28%. It is still high, but slightly lower, in the United States at $144. The worldwide average is $89.4. Revenues of ISPs as a share of GDP average at 0.29%. It is highest in Portugal (0.68%) and Spain (0.47%), and lowest in Turkey, Mexico, and Russia (0.08% each).

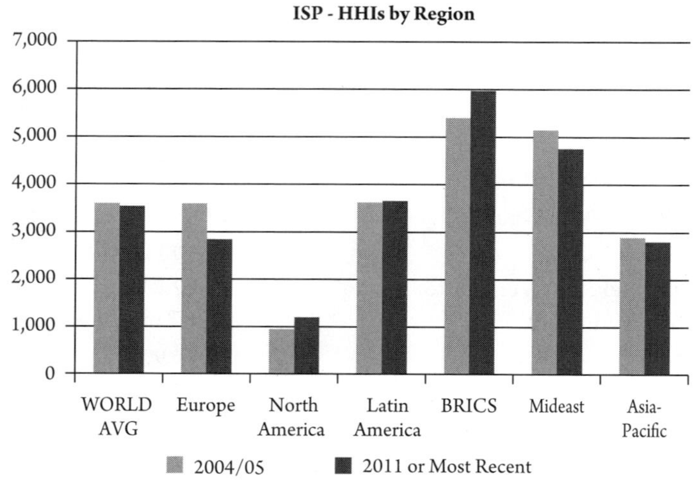


GRAPH 34.24 ISPs Regional Concentration

## Table 34-22. ISP in National Markets, 2011 or Most Recent

| COUNTRY | COUNTRY SHARE IN WORLD ISP INDUSTRY TOTAL (%) | ISPS SHARE OF NATIONAL GDP (%) | PER CAPITA SPENDING ON ISPS (US$) |
|---|---|---|---|
| Argentina | 0.2% | 0.11% | 8.3 |
| Australia | 1.6% | 0.26% | 112.9 |
| Belgium | 0.6% | 0.18% | 81.8 |
| Brazil | 2.5% | 0.25% | 20.5 |
| Canada | 4.5% | 0.51% | 211.1 |
| Chile | 0.2% | 0.16% | 16.0 |
| China | 12.7% | 0.43% | 15.2 |
| Egypt | 0.4% | 0.34% | 7.2 |
| Finland | 0.5% | 0.32% | 150.2 |
| France | 7.8% | 0.47% | 200.9 |
| Germany | 5.2% | 0.24% | 103.4 |
| India | 1.4% | 0.17% | 1.8 |
| Ireland | 0.4% | 0.26% | 137.4 |
| Israel | 0.3% | 0.25% | 61.9 |
| Japan | 13.6% | 0.44% | 171.5 |
| Mexico | 0.5% | 0.08% | 7.0 |
| Netherlands | 1.6% | 0.31% | 153.4 |
| Portugal | 1.0% | 0.68% | 155.0 |
| Russia | 0.7% | 0.08% | 8.1 |
| South Africa | 0.3% | 0.15% | 8.3 |
| South Korea | 2.3% | 0.42% | 75.1 |
| Spain | 4.4% | 0.47% | 156.0 |
| Sweden | 0.9% | 0.31% | 149.0 |
| Switzerland | 0.6% | 0.18% | 116.2 |

Table 34-22. *continued*

| COUNTRY | COUNTRY SHARE IN WORLD ISP INDUSTRY TOTAL (%) | ISPS SHARE OF NATIONAL GDP (%) | PER CAPITA SPENDING ON ISPS (US$) |
|---|---|---|---|
| Taiwan | 0.6% | 0.23% | 39.5 |
| Turkey | 0.3% | 0.08% | 7.2 |
| UK | 7.1% | 0.48% | 184.4 |
| US | 28.0% | 0.32% | 143.7 |
| Average ISP Industry Share of GDP (%) | | 0.29 | |
| Total ISP Industry Share of World GDP (%) | | 0.34 | |
| Total ISP Industry Revenue (mil $) | | 162,033 | |
| Average Per Capita Spending ($) on ISPs | | | 89.4 |

## SEARCH ENGINE INDUSTRY CONCENTRATION ACROSS THE WORLD

The search engines industry is the most concentrated media industry of all 13 surveyed here, by both arithmetic and weighted indexing. The average HHI concentration was an extraordinarily high 7,760 in 2013 (weighted: 6,333), up from 6,809 in 2004 (weighted 4,278) (Graph 34.25). The $41.3 billion industry is concentrated as a result of Google's dominance. The countries with the highest concentrations include Finland, Turkey, Chile, and Portugal. The countries with the lowest levels of concentration are Taiwan, Italy, South Korea, Russia, Japan, China, and the United States. The United States has a somewhat lower industry HHI of 5,506, reflecting the presence of several smaller companies.

Industry HHI is over 8,000 in 21 of the 30 countries surveyed. Countries where industry HHI is under 6,000 (South Korea, the US, Japan, Russia, Italy, Taiwan) have particular characteristics that inhibit a Google monopoly (Table 34.23). Italy has a large number of specialized, indigenous search engines that reduce Google's share of the national market. In Japan, the domestic operation of the US-based Yahoo is majority owned by the entrepreneurial Japanese company, Softbank, and it had an incumbency advantage of an early presence. In South Korea, a series of corporate lawsuits brought about by domestically owned search engines and suggested by the government led to a regulatory regime that bars Google from indexing enough Korean-language content. In China, Google is no longer officially present due to a dispute with the state over censorship (Google is accessible, but there is no Google China), and the market is dominated by the commercial platform Baidu, one of the few privately owned large companies operating in the Chinese national media market. In Russia, the search engine Yandex was established prior to Google's entry to the market, and its analytics seem to be better optimized for Russian-language content than Google's are.

US companies dominate. Google is far ahead. Yahoo and Microsoft tend to be number 2 and 3 in the various national markets where there are not strong local entities, such as Mail.ru and Yandex.ru in Russia. Google's success is attributed to its rank-based search algorithm, its applications and productivity tools, and acquisition of other Internet media properties, such as YouTube and DoubleClick.

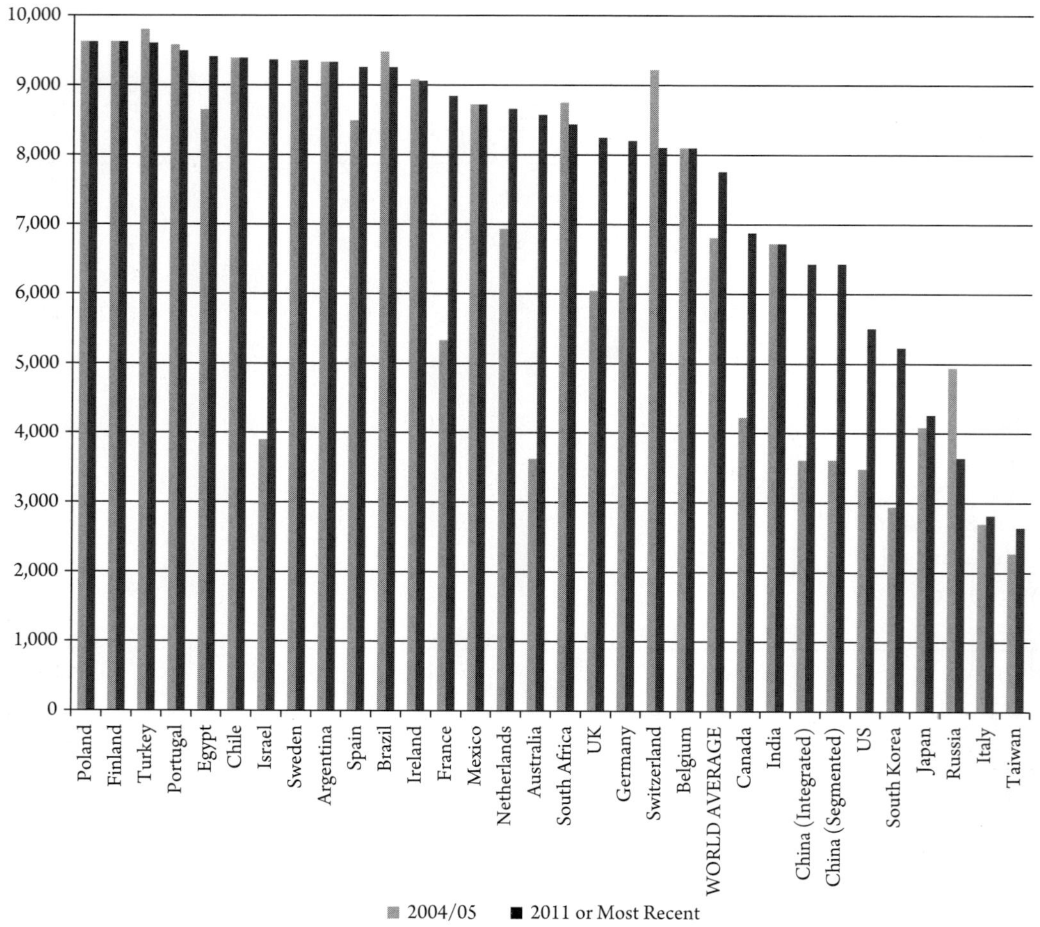


GRAPH 34.25 Search Engine Concentration by Country

## Table 34-23. Countries' Top Search Engines, 2011 or Most Recent

| COUNTRY | TOP COMPANY | MARKET SHARE OF TOP COMPANY (%) | NATIONAL HHI |
|---|---|---|---|
| Poland | Google (US) | 98.1% | 9,624 |
| Finland | Google (US) | 98.1% | 9,624 |
| Turkey | Google (US) | 98.0% | 9,604 |
| Portugal | Google (US) | 97.4% | 9,494 |
| Egypt | Google (US) | 97.0% | 9,409 |
| Chile | Google (US) | 96.6% | 9,390 |
| Israel | Google (US) | 96.7% | 9,365 |
| Sweden | Google (US) | 96.7% | 9,356 |
| Argentina | Google (US) | 96.6% | 9,332 |
| Spain | Google (US) | 96.2% | 9,258 |
| Brazil | Google (US) | 96.2% | 9,256 |
| Ireland | Google (US) | 94.0% | 9,061 |
| France | Google (US) | 94.0% | 8,846 |

Table 34-23. *continued*

| COUNTRY | TOP COMPANY | MARKET SHARE OF TOP COMPANY (%) | NATIONAL HHI |
|---|---|---|---|
| Mexico | Google (US) | 93.4% | 8,724 |
| Netherlands | Google (US) | 93.0% | 8,662 |
| Australia | Google (US) | 92.5% | 8,575 |
| South Africa | Google (US) | 91.8% | 8,438 |
| UK | Google (US) | 90.7% | 8,248 |
| Germany | Google (US) | 90.5% | 8,204 |
| Switzerland | Google (US) | 89.9% | 8,105 |
| Belgium | Google (US) | 90.0% | 8,100 |
| Canada | Google (US) | 82.4% | 6,879 |
| India | Google (US) | 81.4% | 6,724 |
| China | Baidu | 78.6% | 6,433 |
| US | Google | 68.4% | 5,506 |
| South Korea | NHN | 68.8% | 5,230 |
| Japan | Google (US) | 34.9% | 4,262 |
| Russia | Yandex.ru | 47.1% | 3,642 |
| Italy | Google (US) | 47.0% | 2,819 |
| Taiwan | Yahoo (US) | 39.7% | 2,646 |

Yahoo is best known for its web search engine, Yahoo Search, and has the third largest global search directory after Google. Outside of the United States, where it has lost significant market share to Google, it leads in Taiwan, with a market share of 39.7%. Between 2001 and 2004, Yahoo Search was powered using Google's search engine, but as of 2009 is powered by Microsoft's Bing.

Google's top competitor in the United States is Microsoft's Bing. While it still has only a fraction of the international market, Bing experienced rapid growth and is challenging Google in the United States, accounting for 16% of the US market, and international markets such as Australia, Taiwan, and Israel.

In the handful of countries where Google is not the dominant search engine, the largest search engine by volume tends to be one based in that country. Baidu is the largest search engine in China, holding 78.6% of the Chinese market share. Baidu is also the world's second largest search engine by number of searches. Baidu is a privately owned content media producer in a country where most platform and content media producers are state-owned. Google occupies a much smaller share of the Chinese market with 15.6% in 2012. In 2005, Google entered the Chinese market with the launch of Google China, but had to deal with government censorship and cyber-attacks on its servers. In 2009, the company decided to end its self-censorship, resulting in Google China being banned in 2010, though its services were still available to users who access the search engine using a Taiwanese domain name. Other factors affecting Google's low market share in China are language barriers and a poor localization of content.

Yahoo and Google dominate the Japanese market, creating a duopoly. Yahoo Japan's decision to adopt Google's technology to power its Internet search engine and search advertising platform ultimately helped Google's market share from diminishing further in

Japan. In 2012, Google's share of the market was 34.9%%. Despite its name, Yahoo is not the majority owner of Yahoo Japan: Softbank (Japan) is. In this sense, Yahoo Japan is a Japanese-owned company. In contrast, Google is seen as a foreign company, and it does not appear to overtake Yahoo Japan.

The case of search engine usage in South Korea parallels the situation in China, as Google was unable to acquire a large share of the market. The top search engines in South Korea are NHN, with 68.8% of the market in 2012, followed by Daum, with 22.1%. An important factor in explaining Google's small market share in the country is that the government blocks access to its websites to Google, a practice it does not extend to South Korean-based search engines. Thus, government websites are not indexed by Google and cannot be accessed through its search engine. With so much content blocked, it would actually be burdensome for many South Koreans to use Google over other search engines. Google lobbied the South Korean government to allow it to index Korean-language governmental content.

Yandex is the largest search engine in Russia, followed by Google, Mail.ru, and Prof-Media. Yandex controlled 47.1% of the market in 2010. Unlike the cases in several East Asian countries where Google cannot compete due to government roadblocks, the main reason for Yandex's success at the expense of Google is the popularity of its user interface among Russian consumers, and its compatibility with the linguistic and grammatical rules of the Russian language. Yandex's search algorithm caters to the difficulty of the inflections specific to the Russian language.

Given the fears that Google is leveraging its search market power into the advertising market, competitors and various governments have brought antitrust lawsuits against the company.

Average concentration rose after 2004 worldwide and in every region (Graph 34.26). Where it was already highest (Latin America), it rose least, and where it was lowest (North America and Asia-Pacific), it rose most.

Japan accounts for a 10.1% share of the world industry total, second to the US's 35.8%. Search engines constitute a low percentage of countries' total economic activity as a media industry, and only 0.07% of GDP. Per capita spending in this industry by consumers and advertisers is, on average, $20.5. They are highest in the United States ($46.8), The Netherlands ($53), and the United Kingdom ($35.9) (Table 34.24). They are lowest in Egypt, India, and Mexico ($0.50).

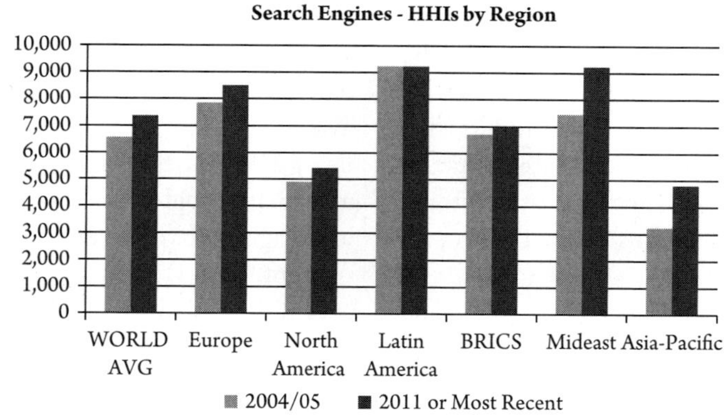


GRAPH 34.26 Search Engine Regional Concentration

Table 34-24. Search Engines in National Markets, 2011 or Most Recent

| COUNTRY | COUNTRY SHARE IN WORLD SEARCH ENGINE INDUSTRY TOTAL (%) | SEARCH ENGINES SHARE OF NATIONAL GDP (%) | PER CAPITA SPENDING ON SEARCH ENGINES (US$) |
|---|---|---|---|
| Argentina | 0.4% | 0.05% | 3.8 |
| Australia | 2.3% | 0.10% | 42.2 |
| Belgium | 0.8% | 0.07% | 30.0 |
| Brazil | 1.8% | 0.05% | 3.8 |
| Canada | 2.6% | 0.08% | 31.2 |
| Chile | 0.2% | 0.04% | 3.6 |
| China | 8.5% | 0.07% | 2.6 |
| Egypt | 0.1% | 0.03% | 0.5 |
| Finland | 0.4% | 0.06% | 30.7 |
| France | 2.6% | 0.04% | 16.8 |
| Germany | 7.4% | 0.09% | 37.6 |
| India | 1.5% | 0.05% | 0.5 |
| Ireland | 0.4% | 0.06% | 32.2 |
| Israel | 0.1% | 0.03% | 7.2 |
| Italy | 4.7% | 0.09% | 31.7 |
| Japan | 10.1% | 0.08% | 32.6 |
| Mexico | 0.1% | 0.01% | 0.5 |
| Netherlands | 2.1% | 0.11% | 53.0 |
| Poland | 0.5% | 0.04% | 5.6 |
| Portugal | 0.8% | 0.14% | 31.8 |
| Russia | 2.3% | 0.06% | 6.5 |
| South Africa | 0.4% | 0.07% | 3.6 |
| South Korea | 2.7% | 0.13% | 22.9 |
| Spain | 3.6% | 0.10% | 32.0 |
| Sweden | 0.7% | 0.06% | 30.5 |
| Switzerland | 0.6% | 0.05% | 30.6 |
| Taiwan | 0.2% | 0.02% | 3.7 |
| Turkey | 0.7% | 0.04% | 3.9 |
| UK | 5.4% | 0.09% | 35.9 |
| US | 35.8% | 0.11% | 46.8 |
| Average Search Engine Industry Share of GDP (%) | | 0.07 | |
| Total Search Engine Industry Share of World GDP (%) | | 0.08 | |
| Total Search Engine Industry Revenue (mil US$) | | 41,300 | |
| Average Per Capita Spending ($) on Search Engines | | | 20.5 |

## ONLINE NEWS INDUSTRY CONCENTRATION ACROSS THE WORLD

Online news media are only moderately concentrated, with an average arithmetic HHI of 1,621 (Graph 34.27). The industry ranks tenth out of the 13 industries for such concentration. By weighted average HHI, the industry is also unconcentrated, at 1,279 in 2013. Online news media is the least concentrated of the three Internet media industries analyzed. It is also less concentrated than newspapers by both measures, though not in all countries. The countries with the highest concentrations are Brazil, China, Switzerland, and Turkey (Table 34.25). Turkey's online news rose strongly after 2004, because the Dogan Media Group, the largest newspaper owner in Turkey, bought out other newspaper companies' and dominated the online news industry. In Brazil, four media groups—Folhapar, Globo, Telefónica, and Telemar—dominate the industry and collectively hold 70% of the national market share.

The countries with the lowest levels of online news concentration are Canada, the United States, Spain, and Mexico. The US's industry HHI of 569 reflects its relatively low concentration in the broadcast TV and daily newspapers market at the national level, plus several independent news-style sites. That said, the incumbency advantage of the major multimedia producers here means this sector is almost twice as concentrated in the United States as the daily newspapers industry, though less concentrated than for broadcast TV news.

Brazil's rising online news industry HHI is the result of the increasing market shares held by the country's largest audiovisual and telecommunications media groups. In China, online news media is one of the national media industries not dominated by state-owned companies (the other being search engines). But the online news market has become increasingly concentrated because new entrants such as Caijing have not been able to attract significant advertising revenues away from the more established firms such as China.org.cn.

Mexico's low HHI of 711 illustrates the popularity of many online editions of print and audiovisual content, especially the Spanish-language editions of US content producers.

Many of the world's major online news companies were successful due to their strong brands in print news, coupled with economies of scope in extending traditional news operations to the online mode. Examples are News Corp., Dogan Group (Turkey), Folhapar (Brazil), Axel Springer (Germany), Gruppo

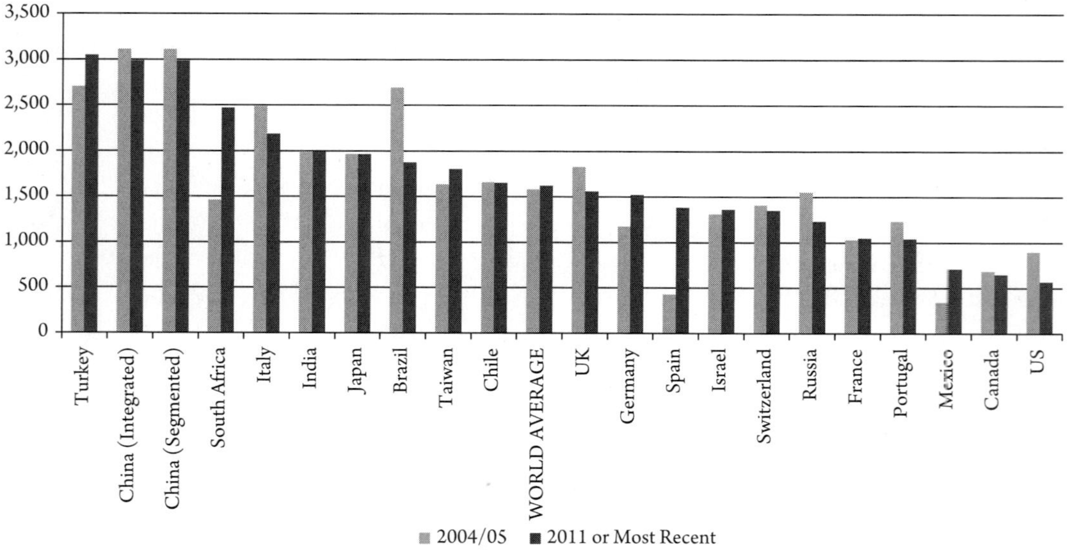


GRAPH 34.27 Online News Concentration by Country

## Table 34-25. Countries' Top Online News Media Outlets, 2011 or Most Recent

| COUNTRY | TOP COMPANY | MARKET SHARE OF TOP COMPANY (%) | NATIONAL HHI |
|---|---|---|---|
| Turkey | Dogan Group | 52.0% | 3,052 |
| China | Sina.com | 38.9% | 2,989 |
| South Africa | Naspers Group | 32.6% | 2,471 |
| Italy | Gruppo Espresso | 40.8% | 2,188 |
| India | Yahoo | 32.7% | 1,996 |
| Japan | Yahoo (US)[1] | 15.9% | 1,965 |
| Brazil | Folhapar | 31.9% | 1,875 |
| Taiwan | United Daily News Group | 29.1% | 1,803 |
| Chile | El Mercurio | 54.2% | 1,651 |
| UK | BBC (public) | 32.0% | 1,559 |
| Germany | Axel Springer | 30.2% | 1,521 |
| Spain | PRISA | 26.1% | 1,382 |
| Israel | Yedioth Ahronoth Group | 24.3% | 1,361 |
| Switzerland | Tamedia | 26.5% | 1,350 |
| Russia | RBC.ru | 17.8% | 1,231 |
| France | Socpresse Group | 15.0% | 1,048 |
| Portugal | Impresa | 14.2% | 1,041 |
| Mexico | El Universal | 15.8% | 711 |
| Canada | Bing (Microsoft, US) | 14.7% | 649 |
| US | Yahoo | 11.0% | 569 |

[1] Shares here indicate its 34.74% share in Yahoo! Japan.

Espresso (Italy), the Guardian (United States, United Kingdom, Australia), and The Times of India Group. Similar advantages accrue to TV and cable news providers including BBC, Time Warner (CNN), and Comcast (MSNBC). Other companies started their business exclusively online, such as Sina.com, Sohu.com, and 163.com in China, Yandex.ru and Mail.ru in Russia, and Yahoo.

The global online news media industry is moderately concentrated. Concentration is declining in North America, where it fell from 774 to 609 and, in the BRICS countries, and Europe (Graph 34.28).

Revenues in online news—both directly through charges on users, and indirectly through advertising—are low, averaging $8.90 per capita worldwide. In comparison, average per capita revenue for newspapers is $113.8, about 13 times higher, and for magazines it is $65.8. Online news has revenues of $19.8 billion, a figure that is rising rapidly (from $12.6 billion in 2004), but still only a fraction of print news at $107 billion. Similarly, as a percentage of GDP, online news on average is 0.04%. It is highest in Brazil (0.11%) (Table 34.26). It is 0.06% in the United States, and it is lowest in Mexico at 0.01%.

Table 34.27 compares the concentration of online news with that of print newspapers. Online news has a fairly unconcentrated HHI of 1,621; often lower than of print newspapers, which have a worldwide average concentration of HHI of 2,484. But in certain countries, online news media HHIs are actually higher than print dailies' HHIs. This is the case in the United States (569 for online news, 304 for newspapers), Italy (2,188 for online news, 1,863 for newspapers), Germany (1,521 for online news,

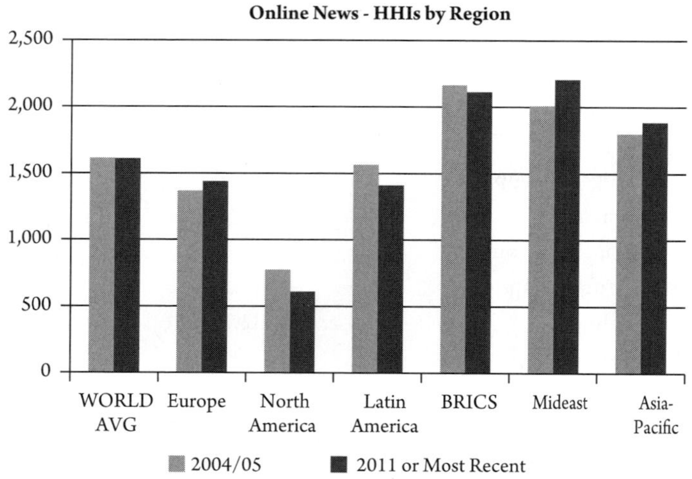


GRAPH 34.28 Online News Regional Concentration

## Table 34-26. Online News Media in National Markets, 2011 or Most Recent

| COUNTRY | COUNTRY SHARE IN WORLD NEWS MEDIA INDUSTRY TOTAL (%) | ONLINE NEWS MEDIA SHARE OF NATIONAL GDP (%) | PER CAPITA SPENDING ON ONLINE NEWS MEDIA (US$) |
|---|---|---|---|
| Brazil | 8.7% | 0.11% | 8.8 |
| Canada | 0.5% | 0.01% | 2.9 |
| Chile | 0.2% | 0.03% | 2.7 |
| China | 4.7% | 0.02% | 0.7 |
| France | 4.8% | 0.03% | 14.8 |
| Germany | 1.9% | 0.01% | 4.5 |
| India | 3.5% | 0.05% | 0.6 |
| Israel | 0.2% | 0.02% | 4.7 |
| Italy | 4.4% | 0.04% | 14.4 |
| Japan | 9.2% | 0.04% | 14.2 |
| Mexico | 0.3% | 0.01% | 0.5 |
| Portugal | 0.8% | 0.06% | 14.5 |
| Russia | 4.7% | 0.06% | 6.5 |
| South Africa | 0.7% | 0.05% | 2.7 |
| Spain | 3.4% | 0.04% | 14.5 |
| Switzerland | 0.6% | 0.02% | 13.9 |
| Taiwan | 0.3% | 0.02% | 2.7 |
| Turkey | 1.0% | 0.03% | 2.7 |
| UK | 7.7% | 0.06% | 24.4 |
| US | 42.3% | 0.06% | 26.5 |
| Average News Media Industry Share of GDP (%) | | 0.04 | |

Table 34-26. *continued*

| COUNTRY | COUNTRY SHARE IN WORLD NEWS MEDIA INDUSTRY TOTAL (%) | ONLINE NEWS MEDIA SHARE OF NATIONAL GDP (%) | PER CAPITA SPENDING ON ONLINE NEWS MEDIA (US$) |
|---|---|---|---|
| Total News Media Industry Share of World GDP (%) | | 0.04 | |
| Total News media Industry Revenue (mil $) | | 19,760 | |
| Average Per Capita Spending ($) on News Media | | | 8.9 |

Table 34-27. HHI of Print Newspapers vs. Online News Media

| | DAILY NEWSPAPERS | ONLINE NEWS MEDIA |
|---|---|---|
| Argentina | 1,547 | N/A |
| Australia | 4,212 | N/A |
| Belgium | 2,166 | N/A |
| Brazil | 2,321 | 1,875 |
| Canada | 1,743 | 649 |
| Chile | 3,786 | 1,651 |
| China (Integrated) | 10,000 | 2,989 |
| Egypt | 5,481 | N/A |
| Finland | 1,653 | N/A |
| France | 2,064 | 1,048 |
| Germany | 558 | 1,521 |
| India | 1,228 | 1,996 |
| Ireland | 3,235 | N/A |
| Israel | 1,835 | 1,361 |
| Italy | 1,863 | 2,188 |
| Japan | 762 | 1,965 |
| Mexico | 3,736 | 711 |
| Netherlands | 1,914 | N/A |
| Poland | 1,188 | N/A |
| Portugal | 3,215 | 1,041 |
| Russia | 2,824 | 1,231 |
| South Africa | 2,450 | 2,471 |
| South Korea | 1,564 | N/A |
| Spain | 1,090 | 1,382 |
| Sweden | 2,350 | N/A |
| Switzerland | 2,520 | 1,350 |
| Taiwan | 2,293 | 1,803 |
| Turkey | 2,637 | 3,052 |
| UK | 1,974 | 1,559 |
| US | 304 | 569 |
| World Average | 2,484 | 1,621 |

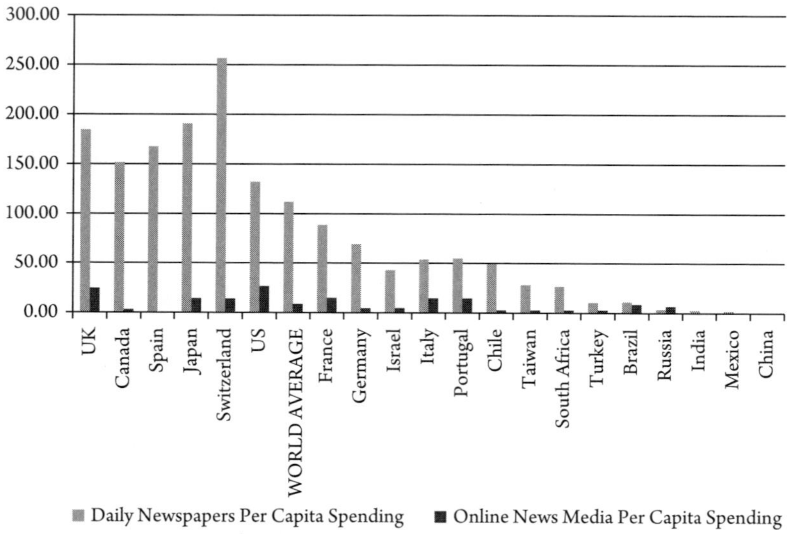


GRAPH 34.29 Daily Newspapers vs. Online News Per Capita Spending, 2011 or Most Recent

558 for newspapers), Japan (1,965 for online news, 762 for newspapers), Turkey (3,052 for online news, 2,637 for newspapers), India (1,996 for online news, 1,228 for newspapers), Spain (1,382 for online news, 1,090 for newspapers), South Africa (2,471 for online news, 2,450 for newspapers), and Brazil (2,916 for online news, 2,321 for newspapers).

The lower concentration for online news is positive in comparison to print newspapers in terms of diversity. On the negative side of the ledge is that revenues per capita are dismal in comparison (see Graph 34.29).

## THE MUSIC INDUSTRY

Although the music industry is outside the scope of this study, Table 34.28 provides, as of 2013, the regional market shares for the top three companies: Sony, Vivendi, and

Table 34-28. Global and Regional Market Shares of the Music Group Majors (2013)[1]

| | SONY (JAPAN) | VIVENDI (FRANCE) | WARNER MUSIC (US) | OTHERS | HHI |
|---|---|---|---|---|---|
| North America | 23.6 | 34.9 | 17.6 | 23.9 | 2,085 |
| Europe | 22.2 | 38.4 | 17.3 | 22.1 | 2,267 |
| Asia (excluding Japan) | 18.5 | 22.8 | 13.4 | 45.3 | 1,042 |
| Japan[2] | 21.3 | 17.4 | 10.0 | 23.3 | 1,162 |
| Latin America | 30.0 | 23.7 | 8.9 | 37.4 | 1,541 |
| Australasia | 24.4 | 31.6 | 17.8 | 26.2 | 1,911 |
| Africa | 24.2 | 36.0 | 7.9 | 31.9 | 1,944 |
| World | 22.9 | 32.2 | 16.1 | 28.8 | 1,820 |

[1] Based on market shares before sale of BMG (Germany) music labels to Sony in 2008, and sale of EMI on market shares before sale of BMG (Germany) music labels t to Warner (1/3) in 2013. Warner Music Group is independent of Time Warner.

[2] Japanese HHI is calculated using Avex (14.7%), King (7.4%) and J-Stormtorm (14.7%), King (7.4%)

(Time) Warner Music. The high market shares around the world of the surviving three major music groups, especially of France's Vivendi (which owns Universal Music Group), are remarkable.

## SUMMARIES

The largest media industries, by far and away, are mobile and wireline telecom ($740 billion, 30.9%, and $590 billion, 24.7%) followed by multi-channel platforms (10.4%). Of purely content media, broadcast TV is largest (7.7%) and ranked fourth in size. It is followed by ISPs (6.8%), newspapers (4.5%), magazines (3.3%), book publishing (3%), video channels (3%), radio (2%), search engines (1.7%), film (1.3%), and online news (0.8%). If one looks only at content media, these percentages would scale up by about three times, with broadcast TV largest at 25.9%.

Content media[7] account for a combined $735 billion, 1% of the world's (30-country) GDP, and platform media[8] account for $1.7 trillion, 2.3% of world GDP. Together then, media spending—receipts by the providers from consumers directly and by advertisers, and from government sources—add up to $2.4 trillion, or 3.3% of world GDP. To that we can add other media activities or related businesses. Music sales accounted for $16.5 billion worldwide[9] in 2012. Video game revenue was $67 billion in 2011. Consumer electronics devices used in the consumption of media account for the following global revenues: mobile phone handsets $210 billion; TV sets $100 billion; consumer microcomputers $160 billion; and music players $30 billion. If these are included, the consumption of media content, platforms, and devices is almost $2.9 trillion, 5.9% of the 30-country gross world product of $61.2 trillion USD.

These are revenues on the platform/vendor/producer end. The price to end-users/consumers rises by the retail and wholesale markups for print media, film, music, games, and most consumer electronics. If we make a fairly realistic assumption that the wholesale and retail distribution chain doubles the price of content and devices to consumers, the share in GDP rises to $3.6 trillion, or 7.3%. (There is, however, some double counting involved, due to a vertical value chain, as is often the case in GDP and industry aggregations.)

The content market is largest, by far, in the United States ($268 billion), followed by Japan ($63.9 billion), the United Kingdom ($58 billion), Germany ($56 billion), France ($37.5 billion), and Spain ($34.1 billion). Although China is perhaps the largest country by audience size, its content is sixth internationally, at $35.3 billion.

The largest platform media markets are the United States ($521 billion), China ($173 billion), Japan ($172 billion), Brazil ($104 billion), Germany ($89.2 billion), France ($69 billion), and the United Kingdom ($66.5 billion). China and India's ($34.7 billion), growth have been especially significant, with the market value nearly tripling in the period under review.

We can now compare the concentrations of the global average HHI of the 13 industries (Table 34.29).

Due to the worldwide market power of Google, the search engine industry is the most concentrated (7,760) and experienced the highest increase in its already high concentration (1.7%) during the observed period (Graph 34.30). In comparison, the worldwide average HHI of the wireline industry, the second least concentrated, has been decreasing by 2.4% annually. However, the presence of formerly or currently incumbent public telecom monopoly providers makes the wireline telecommunication

7. "Content media" include newspapers, magazines, book publishing, film, radio, broadcast television, video channels, search engines, online news media, and one-third of the multi-channel industry's total revenues. The allocation for the multi-channel industry is based on its dual role as both platform and content aggregator, and approximates the cost shares.

8. "Platform media" are defined as wireless and wireline telecom, ISPs, and two-thirds of multi-channel industry revenues.

9. According to the International Federation of the Phonographic Industry Report 2012, as reported by Billboard here: <http://www.billboard.com/biz/articles/news/digital-and-mobile/1549915/ifpi-digital-music-report-2013-global-recorded-music>.

## Table 34-29. Average HHI Industry Concentration (Arithmetic Averages of Companies)

| | 2004/05 | 2011 OR MOST RECENT | % CHANGE PER ANNUM (8-YRS) |
|---|---|---|---|
| Search Engines | 6,809 | 7,760 | 1.7% |
| Wireline | 6,440 | 5,208 | –2.4% |
| Multichannel Platforms | 3,982 | 3,929 | –0.2% |
| Broadcast TV | 3,984 | 3,830 | –0.5% |
| ISP | 3,829 | 3,616 | –0.7% |
| Wireless | 3,691 | 3,493 | –0.7% |
| Radio | 2,874 | 2,627 | –1.1% |
| Daily Newspapers | 2,636 | 2,484 | –0.7% |
| Video Channels | 2,004 | 1,957 | –0.3% |
| Magazines | 1,602 | 1,761 | 1.2% |
| Book Publishing | 1,633 | 1,690 | 0.4% |
| Online News | 1,580 | 1,621 | 0.3% |
| Film | 1,378 | 1,403 | 0.2% |

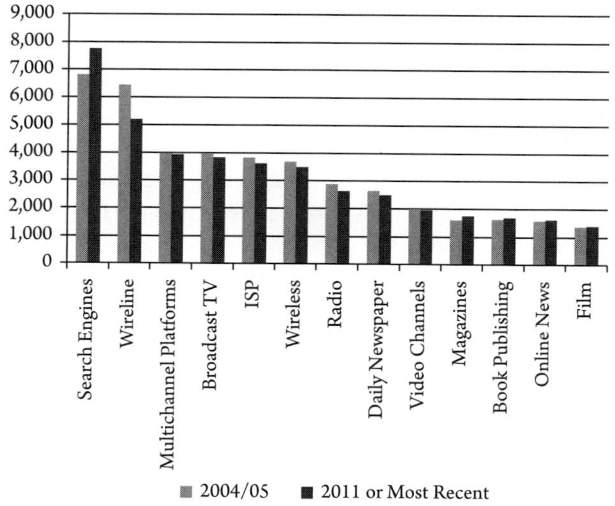


GRAPH 34.30 Average HHI Industry Concentration (Arithmetic Averages of Companies)

sector still one of the most highly concentrated. At the lower end are the print media industries such as daily newspapers, magazines, and book publishing industries. Surprisingly, the film industry has the lowest average HHI of all media industries (1,403), due to the absence of a single dominating firm. That industry, however, is unusual insofar as the same six firms dominate in almost all countries.

## APPENDIX:CHINA MEDIA CONTROL: INTEGRATED, SEGMENTED, OR AVERAGED

We present the two approaches for Chinese media in the tables and graphs to calculate the concentration figures in the chapter, so that the reader may utilize different perspectives. (Table 34.30).

Table 34-30. China HHI Data 2004/05 vs. 2011 or Most Recent

| | 2004/2005 | | | 2011 OR MOST RECENT | | |
|---|---|---|---|---|---|---|
| | SEGMENTED HHI | INTEGRATED HHI | AVERAGED HHI | SEGMENTED HHI | INTEGRATED HHI | AVERAGED HHI |
| Newspapers | 357 | 10,000 | 5,179 | 312 | 10,000 | 5,156 |
| Magazines | 240 | 1,053 | 647 | 1,066 | 3,039 | 2,053 |
| Books | 94 | 10,000 | 5,047 | 762 | 10,000 | 5,381 |
| Radio | 338 | 10,000 | 10,000 | 194 | 10,000 | 10,000 |
| Broadcast TV | 4,837 | 10,000 | 7,419 | 3,761 | 10,000 | 6,881 |
| Multichannel Platforms | N/A | N/A | N/A | N/A | N/A | N/A |
| Video Channels | N/A | N/A | N/A | N/A | N/A | N/A |
| Film | 1,691 | 2,240 | 1,966 | 2,248 | 5,154 | 3,701 |
| Wireline | 5,078 | 10,000 | 7,539 | 4,476 | 10,000 | 7,238 |
| Wireless | 5,873 | 10,000 | 7,936 | 4,275 | 10,000 | 7,138 |
| ISP | 3,016 | 10,000 | 6,508 | 4,023 | 10,000 | 7,012 |
| Search Engines | 3,612 | 3,612 | 10,000 | 6,433 | 6,433 | 10,000 |
| Online News | 3,113 | 3,113 | 3,113 | 2,989 | 2,989 | 2,989 |

“Segmented” HHI treats each of the different state-owned companies as separate firms. The second approach is to assume that the different state enterprises are in the nature of divisions of the same overall organization. This is the “Integrated” definition. The third approach is to combine both of these two metrics in an “averaged” concentration measure.

## Bibliography

For the sources mentioned in this chapter, see the general bibliography at the end of this book.

# 35

# The World's Major Media Companies

## INTRODUCTION

This chapter presents the world's top media companies in 13 industries across 30 countries. We measure them by revenue as well as by a measure we call the "power index." At the end of the chapter we aggregate the findings across all industries to obtain the world's largest media firms, both by revenues and by power index.

Due to the sheer number of companies worldwide and in a particular industry, inclusion in the analysis of this section requires firms to meet one of three criteria. They must be ranked in the world's top 30 firms in terms of revenue; or be the number 1 company in their country market; or have a share of their national market that is greater than 20%. Due to the second criterion, there will be at least one company represented from each country. For China, we list companies and organizations separately, as well as in the aggregate where they are owned by the state.

To explain the nature of the information that is being developed, look at Table 35.1. That table lists the major newspaper companies worldwide. It provides five different measures for their presence in the newspaper industry. The right-most column shows the individual companies' share in the 30-country world. For example, in Table 35.1, the Murdoch Group (News Corp., specifically) has a world newspapers market share of 7.9%. This is found by aggregating the company's newspaper revenues in the countries in which it operates newspapers and calculating the percentage of global (30-country) newspaper revenues that this represents. In this case, News Corp.'s total newspaper revenues are \$8.4 billion (third column from the right); and worldwide newspaper revenues are \$107 billion, from the bottom row. News Corp.'s global market share is therefore 7.9%.

These worldwide market shares enable us to calculate worldwide concentration measures such as the Worldwide Industry C4 and the worldwide HHI index. At the bottom of Table 35.1, they are given as W-C4 = 17%, and

Table 35-1. The World's Major Daily Newspaper Companies (First In Their Countries; Or, >20%; Or, in Top Thirty by Revenue)

| COMPANIES | COMPANY-POWER INDEX IN WORLD INDUSTRY | 2011 OR MOST RECENT REVENUE (MIL $) | MARKET SHARES BY COUNTRY BY REVENUE (%) | MARKET SHARE OF WORLD INDUSTRY TOTAL (%) |
|---|---|---|---|---|
| Murdoch Group (News Corp., US) | 249 | 8,416 | Australia (57.5%), Ireland (17.9%), UK (32.5%), US (8.3%) | 7.90% |
| Yomiuri Shimbun (Japan) | 68.2 | 3,698 | Japan (19.7%) | 3.50% |
| Independent News & Media (Ireland) | 62.1 | 1,615 | Australia (5.3%), Ireland (52%), South Africa (26.5%) | 1.50% |
| Bonnier (Sweden) | 52.9 | 1,366 | Sweden (41.3%) | 1.30% |
| Government of China | 46.4 | 495 | China (100%) | 0.50% |
| *Xinhua (China)* | *0.2* | 32 | *China (6.4%)* | *0.03%* |
| *People's Daily (China)* | *0.6* | 55 | *China (11.1%)* | *0.10%* |
| Asahi Shimbun (Japan) | 42.2 | 2,909 | Japan (15.5%) | 2.70% |
| RCS (Italy) | 40.4 | 1,782 | Italy (33.6%), Spain (15.8%) | 1.70% |
| Daily Mail & General Trust (UK) | 40.2 | 2,110 | Ireland (9.3%), UK (21.5%) | 2.00% |
| Amaury Group (France) | 36.8 | 1,302 | France (30.2%) | 1.20% |
| Fairfax (Australia) | 32.8 | 1,229 | Australia (28.5%) | 1.20% |
| Postmedia (Canada) | 32.6 | 1,182 | Canada (29.4%) | 1.10% |
| Tamedia (Switzerland) | 28.6 | 690 | Switzerland (44.3%) | 0.60% |
| Gannett (US) | 27.7 | 3,082 | US (9.6%) | 2.90% |
| BCCL (The Times of India Group) (India) | 20.2 | 7,39 | India (9.2%) | 0.70% |
| Trinity Mirror (UK) | 19.9 | 1,483 | Ireland (6.6%), UK (15.1%) | 1.40% |
| Quebecor (Canada) | 19.4 | 913 | Canada (22.7%) | 0.90% |
| Vocento (Spain) | 18.7 | 1,089 | Spain (18.3%) | 1.00% |
| El Mercurio (Chile) | 18.4 | 357 | Chile (54.9%) | 0.30% |

| | | | | |
|---|---|---|---|---|
| PRISA (Spain) | 17.6 | 1,083 | Spain (17.6%), Portugal (7.9%) | 1.00% |
| Axel Springer (Germany) | 16.5 | 923 | Germany (18.8%), Poland (21.2%) | 0.90% |
| Northern & Shell (UK) | 14.1 | 1,155 | UK (13%) | 1.10% |
| Globo Group (Brazil) | 12.8 | 483 | Brazil (28.4%) | 0.50% |
| Folhapar (Brazil) | 12.4 | 474 | Brazil (27.9%) | 0.40% |
| Dogan Group (Turkey) | 12 | 277 | Turkey (46.4%) | 0.30% |
| Naspers (South Africa) | 11.8 | 361 | South Africa (35%) | 0.30% |
| Telegraaf Media Group (Netherlands) | 11.5 | 332 | Netherlands (36.9%) | 0.30% |
| Sanoma (Finland) | 11.5 | 395 | Finland (31%) | 0.40% |
| Grupo Espresso (Italy) | 11.3 | 549 | Italy (21.9%) | 0.50% |
| Tribune (US) | 10.5 | 1,894 | US (5.9%) | 1.80% |
| Estado Group (Brazil) | 10.1 | 428 | Brazil (25.2%) | 0.40% |
| The New York Times (US) | 8.8 | 1,734 | US (5.4%) | 1.60% |
| Nihon Keizai Shimbun (Japan) | 7.9 | 1,258 | Japan (6.7%) | 1.20% |
| Mainichi Shimbun (Japan) | 7.7 | 1,239 | Japan (6.6%) | 1.20% |
| Corelio (Belgium) | 7.6 | 304 | Belgium (26.6%) | 0.30% |
| Cofina (Portugal) | 7.6 | 191 | Portugal (42.4%) | 0.20% |
| Chosun Ilbo (South Korea) | 7.3 | 322 | South Korea (24.3%) | 0.30% |
| Government of Egypt | 7.2 | 107 | Egypt (72.2%) | 0.10% |
| *Al Ahram (Egypt) (public)* | 0.8 | 34.8 | Egypt (23.5%) | 0.00% |
| *Al Akhbar (public)* | 1.1 | 42.5 | Egypt (28.7%) | 0.00% |
| *Al Gomhouria (public)* | 0.6 | 29.6 | Egypt (20%) | 0.00% |
| Advance (US) | 6.9 | 1,541 | US (4.8%) | 1.40% |
| Grupo Clarín (Argentina) | 6 | 171 | Argentina (37.6%) | 0.20% |
| Komsomolskaya Pravda (Russia) | 5.8 | 156 | Russia (39.6%) | 0.10% |
| Rossel (Belgium) | 5.7 | 264 | Belgium (23.1%) | 0.20% |
| De Persgroep (Belgium) | 5.5 | 259 | Belgium (22.6%) | 0.20% |

*continued*

Table 35-1. *continued*

| COMPANIES | COMPANY-POWER INDEX IN WORLD INDUSTRY | 2011 OR MOST RECENT REVENUE (MIL $) | MARKET SHARES BY COUNTRY BY REVENUE (%) | MARKET SHARE OF WORLD INDUSTRY TOTAL (%) |
|---|---|---|---|---|
| OEM (Mexico) | 5.4 | 97.4 | Mexico (59.4%) | 0.10% |
| Controlinveste (Portugal) | 5.2 | 159 | Portugal (35.2%) | 0.10% |
| Torstar (Canada) | 4.6 | 442 | Canada (11%) | 0.40% |
| Liberty Times News Group (Taiwan) | 4.6 | 158 | Taiwan (31.4) | 0.10% |
| Yediot Aharonot Group (Israel) | 3.7 | 101 | Israel (29%) | 0.10% |
| Next Media (Hong Kong) | 2.9 | 124 | Taiwan (24.7%) | 0.10% |
| Agora (Poland) | 2.7 | 121 | Poland (24.1%) | 0.10% |
| United Daily News Group (Taiwan) | 1.8 | 96.9 | Taiwan (19.3%) | 0.10% |
| The Washington Post (US) | 1 | 578 | US (1.8%) | 0.50% |
| Pearson (UK) | 0.9 | 293 | UK (3.3%) | 0.30% |
| Worldwide Industry HHI Concentration (W-HHI) | | | | 126 |
| Worldwide Industry C4 (30-countries) concentration | | | | 17% |
| Total Global Newspapers Revenue (mil US$) | | | | 106,734 |

W-HHI = 126 for 2013. Obviously, these measures will be much lower than country-specific concentration measures, since the fish now swim in a larger pond. Different countries are different markets. But that is truer for newspapers than it is for film or search engines.

The second column from the right identifies the market share of News Corp. in the newspaper market in the several countries where it is active: 57.5% in Australia, 32.5% in the United Kingdom, 17.9% in Ireland, and 8.3% in the United States.

The third column from the right shows the revenues of a company in its newspaper business, excluding other media industries across the 30 countries. For News Corp., this was, as mentioned above, $8.4 billion in 2013.

The left-most column is the "Company Power Index in World Industry" of a company for that particular industry, and it requires explanation. This measure represents a company's overall "power" in an industry. The power index is a cousin to the Hirschman–Herfindahl Index (HHI). Whereas the HHI aggregates the squared market shares of different firms in the same market, the power index aggregates the squared shares of the *same* firm in *different* national media markets, adjusting for market size. It is the sum of a company's squared market shares in the markets it operates, and summed up across the various markets in which the company operates, weighted by market size.

There are several variants of the power index.

One can aggregate a company's ($c$) squared market shares ($s$) across several industries ($i$) in which it participates in a single nation ($n$). Or, one can aggregate the company's squared market shares in a single industry across several nations in which it is active. Or, one can do both, aggregating for a given company across industries and nations. Or, in still another variant, one can aggregate for a single nation the individual aggregations of its several top companies.

We designate the first of these media power indices (MPI), in which a given company's ($\hat{c}$) activities in a given nation ($\hat{n}$) in the various industries ($i$) are aggregated as:

$$\text{Company-National-MPI} = \sum_{n} s^2_{\hat{c},\hat{n},i} \frac{R_{\hat{c},\hat{n},i}}{\sum_i R_{\hat{c},\hat{n},i}}$$

where $R_{n,i}$ is an industry's national revenue.

The second MPI is that for a given company ($\hat{c}$) across all nations ($n$) for a given industry ($\hat{i}$):

$$\text{Company-Industry-MPI} = \sum_{n} s^2_{\hat{c},\hat{n},i} \frac{R_{\hat{c},\hat{n},i}}{\sum_n R_{\hat{c},\hat{n},i}}$$

where $R$ is an industry's revenues in a nation. This index measures a company's worldwide role in an industry.

The third MPI is a company's overall global market position, across both nations and industries. It measures, in effect, a company's overall role in the overall global media sector:

$$\text{Company-Worldwide-MPI} = \sum_{n}(\text{Company-National-MPI}).$$

The fourth metric—the National MPI—is the overall aggregation of the individual companies' MPI in a given nation. In doing so, it is a measure similar to that of a "Pooled" HHI, where each company's revenue share in the overall industry is squared. But there is a difference. Suppose companies A and B both have revenues of $1 billion in each of the same Nations 1 and 2. In Nation 1, each of them dominates (50%) two industries. In Nation 2, their revenues are distributed over 10 industries, for 10% shares each. The National MPI of Nation 1 will be much higher than that of Nation 2, even if the companies are of the same size.

As an example, we look at News Corp. Its overall global market share in the newspaper industry is 7.9%, but this would not account for its very high market shares in several countries—particularly Australia (57.5%) and the United Kingdom (32.5%). The reason why an HHI calculation squares the market shares is to account for an exponential increase in market power due to size. We now do the same across countries. This gives us the relative market power of a company within an industry. News Corp.'s Company-Industry MPI-WI for newspapers is the sum of its power index in four countries where it operates newspapers, the United Kingdom, United States, Australia, and Ireland, weighted by the size of these markets.

News Corp.'s US market share is 8.3%. The US daily newspaper market earned $32.1 billion in revenue in 2012, 30% of the global newspaper market ($107 billion). The US component of News Corp. company worldwide newspaper power index is thus $8.3^2 \times 0.3 = 20.7$.

Similar calculations are done for the three other countries in which News Corp. operates newspapers. The share of Ireland's newspapers in the world is 2%, the United Kingdom's 8%, and Australia's 4%. Multiplied by the square of the company's market shares, weighted, and summed up, the result is 249, as listed in Table 35.1.

News Corp.'s Company-Industry MPI score of 249 is the highest for newspapers. The company also has the highest worldwide market share. It is thus the leader on both scores. On the other hand, the Yomiuri Shimbun is only the third highest worldwide market share (3.5%), but its very high market share in a high-revenue country (Japan) propels it to the number 2 spot by power index (68.2). The companies in the tables that follow are ordered by the size of their power index.

## TOP COMPANIES: DAILY NEWSPAPERS

As mentioned, total revenue for the newspapers industry of the 30-country world is $107 billion. The top company by power index is News Corp. (United States). Its newspaper revenue is $8.4 billion, giving it a 7.9% share of the world newspapers' market by revenue and a global PI of 249 in this industry. The other largest firms in the order of their power index are Amaury Group (France, $1.3 billion in revenue), Independent News and Media (Ireland, $1.6 billion), Yomiuri Shimbun (Japan, $3.7 billion), Daily Mail & General Trust (United Kingdom, $2.1 billion), Fairfax (Australia, $1.2 billion), RCS Media Group (Italy, $1.7 billion), the Asahi Shimbun (Japan, $2.9 billion), and Gannett (United States, $3.1 billion). All of these are headquartered in English-speaking countries, Western Europe, or Japan.

Companies that achieve a large share of the US market, besides Gannett and News Corp., are Advance and Tribune. Gannett has the largest market share inside the United States and a power index of 27.7 globally. Its best-known daily is the national *USA Today*, and it also owns several local and regional titles.

Gannett is fairly typical for newspaper firms in that it does not operate internationally. Not many newspaper companies do, in contrast to magazine and book publishing. Exceptions are News Corp., which operates in Australia, Ireland, United Kingdom, United States, and Papua New Guinea. The Ireland-based Independent News & Media (INM), owned by Tony O'Reilly, also operates internationally in Australia and in South Africa. The RCS Media Group operates in both France and Spain. British papers, like most British companies, such as the Daily Mail & General Trust or Trinity Mirror operate in both the United Kingdom and Ireland.

The People's Daily is the largest of several state-owned print media groups in China. However, its revenues were small, at $54.9 million, and number two by market share in the country (6.4%) Xinhua/Reference News even smaller, at $31.7 million, due to limitations placed on advertising (the entire industry in China is worth just $495 million) and the low newsstand prices set in order to generate mass circulation. *Xinhua*'s editor-in-chief holds a ministerial rank in the government. China's newspaper industry is relatively revenue-poor but power-index-rich. For the Japanese firms, it is the other way around. In France, the newspaper group owned by Philippe Amaury has a power index of 36.8 and a 30.2% market share. It owns the national daily, *Aujourd'hui en France*. In Mexico, the Organización Editorial Mexicana (OEM), founded by Mario Vazquez Raña, is the largest publisher of newspapers. OEM owns *Esto* and *El Sol* and holds an astonishing 59.4% share of the market by revenue, though its share by circulation is lower, in the 30–40% range, because it heavily competes with the more popular sports and crime tabloids that have high circulation, but attract less advertising revenue.

High market shares exist in many of the countries surveyed here. (Country-by-country numbers are provided in the chapter on "Industries.") Dogan Media Group (46.4% by advertising revenue in Turkey), Komsomolskaya Pravda (39.6% by revenue in Russia), and El Mercurio (54.9% by revenue in Chile) predominate for different reasons. Dogan doubled its number of titles by buying out several other papers in Turkey, and as part of the Dogan Holding industrial conglomerate, it is well placed to attract advertising revenue. Komsomolskaya Pravda (owned by the energy company ECN, closely linked to Gazprom Bank) has an incumbency advantage as a leading daily in the USSR, and El Mercurio dominates the Santiago market in Chile, where most of the population and advertisers are present.

In India, The Times of India Group, owned by the BCCL conglomerate, publishes the largest English-language daily in circulation (9.2% domestic market share) and 13 other dailies. However, national concentration levels are low due to the abundance of regional and vernacular-language dailies, which have higher absolute circulation rates but much lower incomes than the urban, nationwide English-language papers like *The Times of India*. Such papers are the primary focus of advertisers because the English-speaking market segments consist of white-collar workers who have a larger share of disposable income.

Within the world market (defined as the 30 countries analyzed), the worldwide industry HHI concentration—that is, the squared market share of the companies in the world market—is 126. Since internationalization is low, this W-HHI is low. It might rise in the future. As newspapers become increasingly unprofitable due to free online content, international and domestic consolidation is likely, which will raise the W-HHI measure.

## TOP COMPANIES: MAGAZINE PUBLISHING

By power index, the world's most significant magazine firms are: Hearst (110), Abril (100), Time Inc. (81.1), Bauer (85), Advance Publications (46.3), and Nine Entertainment (53.8) (Table 35.2). By revenue, however, the industry is dominated by US firms: Hearst leads with 8.6%, followed by Time Inc., (part of Time Warner from 1990 to 2014) with 6.4% of the global magazines market in dollar terms, and then Advance, with 5.1%. Abril of Brazil and Nine Entertainment in Australia have high PIs due to the fact that the each control nearly half of their national markets, but just 2.1% and 1.2% of global revenues. Bertelsmann's Gruner + Jahr, although having a PI of 12.8 worldwide due to its modest market shares, is one of the largest by revenue (1.8%).

Despite the seemingly low entry barriers and large number of titles, the magazine industry is often dominated by a single firm or oligopoly. In Germany, Bauer, Burda, and Bertelsmann account for 62.9% of the industry. In Finland, Sanoma holds a 32.5% market share (as well as 44.5% in Russia). In Turkey, Dogan-Burda, a Turkish-German venture, controls 52% of the market. In Australia, Nine Entertainment has 44%; in South Africa, Naspers 39.5%; in Portugal, Impresa 32.3%; in Chile, Copesa 24%; and in Israel, Yedioth Aharonot 33.5%.

In the United States, the top three companies increased their market shares in recent years. Advance's Condé Nast division has "lifestyle" magazines with popular titles such as *The New Yorker*, *Vogue*, and *Vanity Fair*. It was expanded by acquisition of Disney's magazine group Fairchild Publications. Time Inc. increased its market share in the United States through acquisitions. It sold several of its magazine titles in 2007 but still operates 90 titles and has a US market share of 8.6%.

Publishers in the magazine industry are fairly internationalized. The major US publishers Time Inc., Hearst, and Advance have high market shares in several European countries. European firms such as Bauer primarily compete in other European countries. Several companies, such as Bertelsmann, Lagardère (France), and Reed Elsevier (the Netherlands), own magazines and periodicals in the United States as well. In addition, well-known magazine brands get franchised to other countries, where various domestic publishers create local-language variations, either independently or in a joint venture. Examples are *Playboy Magazine*, *Fortune Magazine*, *Cosmopolitan*, or *Vogue*.

Some publishers' market shares in foreign markets are higher than at home. Time Inc. holds 22% of the British market through IPC, compared to 8.6% in the United States. Time

Table 35-2. The World's Major Magazine Publishers (First In Their Countries; Or, >20%; Or, in Top Thirty by Revenue)

| COMPANIES | COMPANY POWER INDEX IN WORLD INDUSTRY | 2011 OR MOST RECENT REVENUE (MIL $) | MARKET SHARES (%) | MARKET SHARE OF WORLD INDUSTRY TOTAL REVENUE (%) |
|---|---|---|---|---|
| Government of China | 190 | 2,888 | China (52.3%) | 3.60% |
| *Rayli (public)* | *38.6* | *1,303* | *China (23.6%)* | 1.60% |
| *Trend Media Group (public)* | *18* | *889* | *China (16.1%)* | 1.10% |
| Hearst (US) | *110* | *6,869* | Spain (17.9%), China (15.2%), UK (15%), US (10.3%), Taiwan (1.6%), Russia (35.2%) | 8.60% |
| Abril (Brazil) | 100 | 1,649 | Brazil (48.5%) | 2.10% |
| Bauer (Germany) | 85.2 | 2,913 | UK (26%), Germany (18.5%), Russia (7.8%) | 3.70% |
| Time Inc (US) [1] | 81.1 | 5,096 | UK (22%), US (8.6%), Canada (4.7%) | 6.40% |
| Nine Entertainment (Australia) | 53.8 | 972 | Australia (44%) | 1.20% |
| Advance Publications (Condé Nast, US) | 46.3 | 4,086 | US (9.2%), South Africa (3.1%), Taiwan (2.3%), Spain (13.5%), China (7.8%) | 5.10% |
| Lagardère (France)[2] | 38.3 | 1,035 | France (29.4%) | 1.30% |
| Sanoma (Finland) | 21.1 | 563 | Finland (32.5%), the Netherlands (23.8%), Russia (44.5%) | 0.70% |
| Seven Network (Australia) | 20.2 | 596 | Australia (27%) | 0.70% |
| Bonnier (Sweden) | 18.8 | 453 | Sweden (32.9%) | 0.60% |
| Axel Springer (Germany) | 18.3 | 1,009 | Switzerland (29.2%), Germany (12.7%), Spain (2.8%) | 1.30% |
| Globo Group (Brazil) | 16.5 | 667 | Brazil (19.63%) | 0.80% |

| | | | | |
|---|---|---|---|---|
| Burda (Germany) | 15.6 | 962 | Germany (14.8%), UK (2%) | 1.20% |
| RBA (Spain) | 13.4 | 690 | Spain (15.4%) | 0.90% |
| Bertelsmann (Gruner + Jahr, Germany) | 12.8 | 1,440 | Germany (9.3%), France (8.6%), Portugal (1.8%), Spain (7.7%), US (0.7%), and 18 other countries | 1.80% |
| BBC (UK, public) | 8.2 | 684 | UK (9.5%) | 0.90% |
| Naspers (South Africa) | 7.4 | 148 | South Africa (39.5%) | 0.20% |
| Roularta (Belgium) | 6.7 | 433 | France (12.3%), Belgium (38%) | 0.50% |
| Murdoch Group (US)[3] | 6.2 | 823 | Australia (14%), US (1.3%) | 1.00% |
| Editorial Atlátida (Argentina) | 6 | 198 | Argentina (43.1%) | 0.20% |
| Motorpress (Germany) | 5.9 | 607 | Spain (9.6%), Germany (3.2%) | 0.80% |
| Otava (Finland) | 5.7 | 158 | Finland (28.4%) | 0.20% |
| Ringier (Switzerland) | 5.6 | 166 | Switzerland (26.9%) and 10 other countries, such as Romania and Hungary | 0.20% |
| Pearson (UK) | 5.1 | 541 | UK (7.5%) | 0.70% |
| Dogan Group (Turkey) | 4.8 | 73 | Turkey (52%) | 0.10% |
| Telegraaf Media Groep (Netherlands) | 3.5 | 184 | Netherlands (14.9%) | 0.20% |
| Egmont (Denmark) | 2.9 | 341 | Sweden (11.7%), UK (2.5%) | 0.40% |
| Living Media Group (India) | 2 | 66 | India (24%) | 0.10% |
| Impresa (Portugal) | 1.9 | 46 | Portugal (32.3%) | 0.10% |
| La Vie-Le Monde (France) | 1.6 | 211 | France (6%) | 0.30% |
| Forma Publishing Group (Sweden) | 1.3 | 119 | Sweden (8.7%) | 0.20% |
| COPESA (Chile) | 1.2 | 39 | Chile (24%) | 0.05% |
| Yedioth Ahronoth Group (Israel) | 1.2 | 27 | Israel (33.5%) | 0.03% |
| Rogers (Canada) | 1.2 | 146 | Canada (6.3%) | 0.20% |
| TC Transcontinental (Canada) | 1.1 | 144 | Canada (6.2%) | 0.20% |

*continued*

Table 35-2. *continued*

| COMPANIES | COMPANY POWER INDEX IN WORLD INDUSTRY | 2011 OR MOST RECENT REVENUE (MIL $) | MARKET SHARES (%) | MARKET SHARE OF WORLD INDUSTRY TOTAL REVENUE (%) |
|---|---|---|---|---|
| Cofina (Portugal) | 1 | 33 | Portugal (23.5%) | 0.04% |
| The Clinic (Chile) | 0.9 | 34 | Chile (21%) | 0.04% |
| Moto Media Group (Israel) | 0.5 | 17 | Israel (21.3%) | 0.02% |
| Next Media (Hong Kong) | 0.5 | 28 | Taiwan (13.52%) | 0.04% |
| Ma'ariv Modi'in (Israel) | 0.2 | 11 | Israel (14.3%) | 0.01% |
| Worldwide Industry HHI Concentration (W-HHI) | | | | 189 |
| Worldwide Industry C4 (30-countries) Concentration | | | | 23.8% |
| Total Global Magazines Revenue (mil US$) | | | | 79,590 |

Warner had acquired IPC Media in 2001 for $1.7 billion in the biggest magazine deal in the United Kingdom, and renamed it Time Inc. UK.

Hearst holds 35.2% in Russia and 15% in the United Kingdom, while having 10.3% in the United States, and through licensed versions of its main international titles 15.2% of the Chinese market. Hearst attained this share of the Russian market by buying out the market leader Shkulev-Lagardère joint venture in 2011. Advance Publications has a substantial share of the market in Spain (13.5%) plus shares in Taiwan, China (through licensed titles as well), and South Africa. Hearst and Advance have benefitted from the substantial growth for the industry in China, worth $5.5 billion in 2013.

German publishers are also active worldwide. Axel Springer changed the market landscape of Switzerland with its purchase of the Jean Frey Group in 2007, holding 29.2% of that market. It has joint ventures in Eastern Europe with the Swiss firm Ringier. Bauer raised its share in the British market to 26% when it purchased the second leading publisher, EMAP, in 2008.

With high market share levels in the large French market (29.4%), Lagardère has the sixth largest company power index in the industry. In 2011, Lagardère sold off its international titles to Hearst.

Other international magazine activities include Bertelsmann, which has far-flung operations through its Gruner + Jahr subsidiary in Belgium, Croatia, Denmark, France, Germany, Greece, Hungary, Italy, Luxembourg, The Netherlands, Russia, the United Kingdom, Spain, and the United States. Bertelsmann is the third leading magazine publisher in France with 8.6%. Ringier of Switzerland is present in the magazine markets of China, Croatia, the Czech Republic, Germany, Hungary, Romania, Serbia, Slovakia, Switzerland, and Vietnam.

The global revenue of the industry is $79.6 billion. The world industry concentration, calculated using the market shares of the top companies within the global market, is W-HHI =189. As with all print media, this relatively low number is due to the large number of publishers in most countries. The worldwide C4 is about 24%.

## TOP COMPANIES: BOOK PUBLISHING

The largest book publisher by power index is the Chinese state, with a PI of 1,590 and a share of the world market at 15.9% (Table 35.3). The next largest firms are European companies: the German/UK collaboration of Bertelsmann and Pearson with their joint book publishing firm Penguin Random House (168), the French company Lagardère (Hachette) (108), and Italy's Fininvest (Mondadori) (52.1). These, plus Murdoch's HarperCollins ($2.3 billion in revenue) and McGraw-Hill ($2 billion), are also among the largest companies by revenue and global market share. The merger of Bertelsmann and Pearson's book divisions in 2013 gives the joint company revenues of $6.6 billion and a 9.2% share of the global book publishing market; Lagardère ranks second, with a 5.3% share and $3.8 billion. Lagardère's Hachette Livre division has 29.4% of the large market in France.

In the United States, by revenue and power index, Bertelsmann-Pearson is the leading book publisher with 17.6%. The company also has high shares in the United Kingdom (24.9%), Australia (24.4%), Portugal (7.7%), France (5%), and Germany (2.5%). Its main imprints are Random House, Dell, and Pantheon. In 2006, Hachette Livre purchased Time Warner's book publishing company in the United States.

In Russia, two publishers, Eksmo-AST and Prosveshcheniye (state-owned), comprise 61.3% of the market. Concentration in Russia has risen since 2012 with the merger of these two leading publishers.

The top ten privately owned companies by share of revenue are Bertelsmann-Pearson (9.2%), and Lagardère (5.3%) as mentioned, plus Murdoch-News Corp. (3.2%), Fininvest (1.8%), Planeta (1.6%), Redstone-CBS (1.1%), RCS (1.2%), GEMS (0.6%), Anaya (0.6%), and Axel

Table 35-3. The World's Major Book Publishers (First In Their Countries; N (>20%); Or, Top Thirty by Revenue)

| COMPANIES | COMPANY POWER INDEX IN WORLD INDUSTRY | 2011 OR MOST RECENT REVENUE (MIL $) | MARKET SHARES (%) | MARKET SHARE OF WORLD INDUSTRY TOTAL (%) |
|---|---|---|---|---|
| Government of China | 1,590 | 11,486 | China (100%) | 15.9% |
| *Beijing Publishing Group (public)* | *21.4* | *1,332* | *China (11.6%)* | *1.8%* |
| *China Publishing Group (public)* | *15.0* | *1,114* | *China (9.7%)* | *1.5%* |
| *China Science & Technology Publishing Group (public)* | *2.9* | *494* | *China (4.3%)* | *0.7%* |
| *Chinese Academy of Science (public)* | *5.7* | *689* | *China (6%)* | *1.0%* |
| *Zhejiang Publishing United Group (public)* | *37.2* | *1,757* | *China (15.3%)* | *2.4%* |
| *Jiangsu Phoenix Publishing and Publishing and Media (China) (public)* | *7.4* | *781* | *China (6.8%)* | *1.1%* |
| Bertelsmann-Pearson (Penguin Random House) | 168 | 6,629 | Portugal (7.7%), France (5.0%), Germany (2.5%), US (17.6%), Australia (24.4%), UK (24.9%) | 9.2% |
| *Bertelsmann (Germany, 53% of PRH)* | *54.9* | *4,004* | *Portugal (7.7%), France (5%), Germany (2.5%), US (10.3%), Australia (10.1%), UK (13.2%)* | *5.5%* |
| *Pearson (UK, 47% of PRH)* | *32.5* | *2,625* | *US (7.3%), Australia (14.3%) UK (11.7%)* | *3.6%* |
| Lagardère (Hachette, France) | 108 | 3,804 | France (29.4%), US (2.9%), Australia (13%) UK (13.9%) | 5.3% |
| Fininvest (Mondadori, Italy) | 52.1 | 1,326 | Italy (28.4%) | 1.8% |
| Planeta (Spain) | 16.4 | 1,167 | Spain (10.2%), Portugal (6.9%), France (10.15%) | 1.6% |
| Murdoch Group (Harper Collins, US) | 24.0 | 2,323 | US (7.3%), UK (7.7%), Australia (9.3%) | 3.2% |

| | | | | |
|---|---|---|---|---|
| Eksmo-AST (Russia) | 24.8 | 292 | Russia (61.3%) | 0.4% |
| McGraw Hill (US) | 22.3 | 1,960 | US (8.2%) | 2.7% |
| RCS (Rizzoli, Italy) | 13.2 | 880 | Italy (12.6%), Spain (7.1%) | 1.2% |
| Bonnier (Sweden) | 10.1 | 300 | Sweden (22.6%), Finland (31.4%) | 0.4% |
| Porto Editora (Portugal) | 8.1 | 123 | Portugal (47.8%) | 0.2% |
| PRISA (Moderna, Spain) | 3.7 | 196 | Spain (13.9%) | 0.3% |
| GEMS (Italy) | 5.5 | 434 | Italy (9.3%) | 0.6% |
| Anaya (Spain) | 5.5 | 407 | Spain (9.9%) | 0.6% |
| Axel Springer (Germany) | 5.2 | 603 | Germany (6.3%) | 0.8% |
| Média Participations (France) | 2.3 | 351 | France (4.9%) | 0.5% |
| Redstone Group (CBS, US) | 3.6 | 789 | US (3.3%) | 1.1% |
| Prosveshcheniye (Russia) (public) | 3.3 | 107 | Russia (22.5%) | 0.1% |
| Abril (Brazil) | 2.6 | 162 | Brazil (11.5%) | 0.2% |
| Saravia (Brazil) | 1.7 | 131 | Brazil (9.5%) | 0.2% |
| Editora FTD (Brazil) | 0.3 | 50.8 | Brazil (3.6%) | 0.1% |
| Liber (Sweden) | 2.5 | 138 | Sweden (13.1%) | 0.2% |
| Giunti (Italy) | 2.2 | 271 | Italy (5.8%) | 0.4% |
| Otava (Finland) | 1.9 | 52.1 | Finland (26.2%) | 0.1% |
| Leya (Portugal) | 3.0 | 74.3 | Portugal (28.9%) | 0.1% |
| Sanoma (Finland) | 1.8 | 50.7 | Finland (25.5%) | 0.1% |
| Cité Media (Hong Kong) | 0.4 | 40.8 | Taiwan (7.8%) | 0.1% |
| Holtzbrinck (Germany) | 0.6 | 205 | Germany (2.1%) | 0.3% |
| Kinneret Zemora (Israel) | 0.2 | 11.1 | Israel (12.2%) | 0.02% |
| Worldwide Industry HHI Concentration (W-HHI) | | | | 394 |
| Worldwide Industry C4 (30-countries) Concentration | | | | 33.6% |
| Total Global Books Revenue (mil US$) | | | | 72,224 |

Springer (0.8%). Together the share of these 10 large-scale multimedia companies in overall book publishing is only 27.6% (though if China is counted as a single entity, C10 rises to 60.4%). W-HHI is the highest of the print media industries, at 394.

## TOP COMPANIES: RADIO

The two leading companies by power index are public broadcasters: Germany's ARD (240) and the UK's BBC (172) (Table 35.4). Their market shares in their home countries are quite high: 36.5% and 54.5%, respectively. There are other public broadcasters with very high market shares in Egypt (ERTU, 85.4%), Finland (Yleisradio, 55.8%), Sweden (Sveriges Radio, 74.8%), Switzerland (SSR, 66.4%), Ireland (RTÉ, 53.4%), South Africa (SABC, 31.6%), Chile (Iberoamericana, 43.8%), Israel (IBA, 31.9%), Italy (RAI, 48.2%), the Netherlands (NPB, 32.3%), and Belgium (VRT, 30.9%). The Government of China has a PI of 223, but individual companies within that figure are not significant on a global scale since so many are regional or municipally owned and transmitted.

Smaller but substantial market shares for public radio exist also in France (Radio France, 22.1%), India (Prasar Bharati, 30.6%), Australia (ABC, 27.2%), Canada (CBC, 17.3%), and Taiwan (BCC, 20.9%). The large national shares of public radio broadcasters are remarkable insofar as they are usually not based on a formal monopoly: rather, public radio prevails in the face of private competition in almost every country except for the United States, Brazil, Argentina, Portugal, Russia, and Spain. Of the privately owned radio companies, highest by power index are Spain's PRISA (69.8), Bertelsmann's RTL (43.3), and the US-based firms Sirius XM (79.7 satellite from pay-radio) and Clear Channel (iHeartRadio, 76.6). By revenue, the largest radio broadcasters are ARD ($3.2 billion). Sirius XM ($2.5 billion) – owned by Liberty Media – and Clear Channel ($2.4 billion).

In the United States, Clear Channel has benefited tremendously from Telecommunications Act of 1996, which repealed the ceiling on national station ownership and enabled the company to purchase over 1,200 stations throughout the country. However, Clear Channel's market share peaked at 18.9% in 2004. Subsequently its market shares declined (to 15.1% in 2013) as it has sold off hundreds of stations due to the financial pressures of over-expansion. It was bought by a consortium of private equity firms (Bain Capital and Thomas H. Lee Partners) and renamed iHeartRadio. By advertising revenues, however, it is the world's largest radio company.

The Redstone Group's CBS is the second largest commercial terrestrial radio broadcaster by revenue, with hundreds of stations in local US markets. It too has sold off many stations due to increased competition from satellite and Internet radio, decreasing its market share from 11.9% in 2004 to 8.5% in 2013.

Other large commercial radio stations include Union Radio in Spain (owned by the PRISA Group), which holds 39.6% of the Spanish market with over 400 stations. Since 2009, PRISA has also been the principal owner across the border of Portugal's largest broadcaster, Media Capital, which has a 35.1% market share in Portugal. Other large radio groups are EMG in Russia (32.4%), Radio 10 in Argentina (13.5%), Bertelsmann (16.9% in Belgium, 19.2% in France and 10.6% in the Netherlands), Total Global Radio in the United Kingdom (16.3%), NBS in Japan (10.7%), and NRJ (14.5% in Belgium and 14.3% in France), and Lagardère in France (12.5%).

The leading radio broadcaster by power index, ARD, captures 36.5% of the German market. ARD is actually a collective of a dozen independent regional broadcasters, and their individual power indices and revenues are much lower when examined separately. In contrast to public radio broadcasters in several other countries, it offers limited advertisements. It also provides several national channels by its Deutschland Radio affiliate. ZDF, the second German public radio operator, has a national market share of 13.5%. Radio in Finland, Sweden, and Egypt are dominated by public broadcasters.

Private radio station ownership is often localized by city or region; often, only public

Table 35-4. The World's Major Radio Companies (First In Their Countries; Or, >20%; Or, in Top Thirty by Revenue)

| COMPANIES | COMPANY POWER INDEX IN WORLD INDUSTRY | 2011 OR MOST RECENT REVENUE (MIL $) | MARKET SHARES (%) | MARKET SHARE OF WORLD INDUSTRY TOTAL REVENUE (%) |
|---|---|---|---|---|
| ARD (Germany) (public) | 240 | 3,152 | Switzerland (1.9%), Germany (36.5%) | 6.60% |
| Government of China (China) | 223 | 1,063 | China (100%) | 2.20% |
| *China Radio International (public)* | *2* | *99* | *China (9.3%)* | |
| BBC (UK) (public) | 172 | 1,508 | UK (54.5%) | 3.20% |
| Liberty Group (SiriusXM, US) | 79.7 | 2,468 | US (15.4%) | 5.20% |
| Clear Channel/IHeartRadio (US) | 76.6 | 2,420 | US (15.1%) | 5.10% |
| PRISA (Spain) | 69.8 | 860 | Spain (39.6%), Argentina (5.5%), Portugal (35.1%) | 1.80% |
| SRG SSR (Switzerland) (public) | 63.9 | 459 | Switzerland (66.4%) | 1.00% |
| France Televisions/Radio France (France) (public) | 53.8 | 1,172 | France (21.9%) | 2.50% |
| Sveriges Radio (Sweden) (public) | 53.6 | 342 | Sweden (74.8%) | 0.70% |
| Bertelsmann (RTL, Germany) | 45.8 | 1340.91 | Germany (2.4%), France (19.2%), Netherlands (10.6%), Belgium (16.9%) | 2.80% |
| ZDF (Germany) (public) | 32.9 | 1,161 | Germany (13.5%) | 2.40% |
| RAI (Italy) (public) | 30.2 | 298 | Italy (48.2%) | 0.60% |
| NRJ Group (France) | 24.8 | 830 | Belgium (14.5%) Finland (4.5%) France (14.3%) | 1.70% |
| Redstone Group (CBS, US) | 24.3 | 1,362 | US (8.5%) | 2.90% |

*continued*

Table 35-4. *continued*

| COMPANIES | COMPANY POWER INDEX IN WORLD INDUSTRY | 2011 OR MOST RECENT REVENUE (MIL $) | MARKET SHARES (%) | MARKET SHARE OF WORLD INDUSTRY TOTAL REVENUE (%) |
|---|---|---|---|---|
| ABC (Australia) (public) | 23.1 | 405 | Australia (27.2%) | 0.80% |
| Lagardère (France) | 17.5 | 669 | France (12.5%) | 1.40% |
| Total Global Radio (UK) | 15.4 | 451 | UK (16.3%) | 0.90% |
| SABC (South Africa) (public) | 13.1 | 197 | South Africa (31.6%) | 0.40% |
| CBC (Canada) (public) | 12.4 | 341 | Canada (17.3%) | 0.70% |
| RTÉ (Ireland) (public) | 12.1 | 108 | Ireland (53.5%) | 0.20% |
| Yle (Finland) (public) | 10.2 | 87 | Finland (55.8%) | 0.20% |
| EMG (Russia) | 8.9 | 131 | Russia (32.4%) | 0.30% |
| VRT (Belgium) (Public) | 8.2 | 127 | Belgium (30.9%) | 0.30% |
| Bauer (Germany) | 7.9 | 324 | Germany (11.7%) | 0.70% |
| NPB (Netherlands) (public) | 7.8 | 115 | Netherlands (32.3%) | 0.20% |
| Planeta (Spain) | 7.6 | 264 | Spain (13.7%) | 0.60% |
| Group RR (Portugal) | 6.5 | 86.4 | Portugal (35.7%) | 0.20% |
| Government of Egypt (ERTU) (public) | 6 | 33 | Egypt (85.4%) | 0.10% |
| Kagiso Media (South Africa) | 5.9 | 132 | South Africa (21.2%) | 0.30% |
| Rogers (Canada) | 5.3 | 223 | Canada (11.3%) | 0.50% |
| Radio RMF FM | 5.1 | 66.9 | Poland (36.4%) | 0.10% |
| NBS/Fujisankei (Japan) | 4.5 | 200 | Japan (10.7%) | 0.40% |
| Shaw (Canada) | 4.1 | 197 | Canada (10%) | 0.40% |
| Iberoamericana (Chile) (public) | 3.9 | 41.9 | Chile (43.8%) | 0.10% |
| Radio ZET | 3.5 | 55.1 | Poland (30%) | 0.10% |

| | | | | |
|---|---|---|---|---|
| Polskie Radio (public) | 3.4 | 54.6 | Poland (29.7%) | 0.10% |
| Prasar Bharati (AIR, India) (public) | 3.2 | 49.9 | India (30.6%) | 0.10% |
| RTVE (Spain) (public) | 3.1 | 168 | Spain (8.7%) | 0.40% |
| BCCL (The Times of India Group) (India) | 2.7 | 45.8 | India (28.1%) | 0.10% |
| Globo Group (Brazil) | 2.5 | 123 | Brazil (9.8%) | 0.30% |
| Tokyo FM Broadcasting (Japan) | 2.2 | 140 | Japan (7.5%) | 0.30% |
| IBA (Israel) (public) | 1.4 | 21.6 | Israel (31.9%) | 0.00% |
| Government of Taiwan | 1.1 | 25.1 | Taiwan (20.9%) | 0.10% |
| *BCC (Taiwan) (public)* | 1.1 | 25.1 | Taiwan (20.9%) | 0.10% |
| Grupo Clarín (Radio 10) (Argentina) | 0.9 | 30 | Argentina (13.5%) | 0.10% |
| Worldwide Industry HHI Concentration (W-HHI) | | | | 154 |
| Worldwide Industry C4 (30-countries) Concentration | | | | 20% |
| Total Global Radio Revenue (mil US$) | | | | 47,687 |

broadcasters are national in scope. Not many companies own interests in foreign countries' radio markets. An exception is Bertelsmann's RTL Group, which has stakes in the Belgian, Dutch, French, and German radio markets. Indeed, RTL originated in the transborder oriented Radio Luxembourg. Public broadcasters tend to be confined to their countries' markets, except through broadcasting operations aimed at representing their country, such as by the Voice of America, BBC World Service, Radio Vatican, or Deutsche Welle.

A third category of radio operators is that of satellite broadcasting. Sirius XM is America's leading satellite radio broadcaster, and since 2011 is the US's and the world's largest radio broadcaster by revenue. The company is a result of a $13 billion merger in 2008 of two separate providers, Sirius and XM. Pre-merger, the two companies had tremendous content and marketing costs as they tried to gain the upper hand. Once merged and not facing satellite competition, the company quickly reported the first profits for either company in 2009. (The merger, eliminating satellite radio competition, had been supported by several public interest organizations after the companies promised to set aside 8% of its channels for minority and public interest programming, and also agreed to other conditions.)

Sirius XM is 49% owned by John Malone's Liberty Group, which also controlled at one time the largest satellite TV provider, DirecTV. In 2011, Sirius XM overtook Clear Channel to become America's leading radio broadcaster with 25 million subscribers, revenues of $2.5 billion, and a US market share of 15.1%. The company also owns 25% of Sirius XM Canada, which has about 2.3 million subscribers.

Brazil's large number of companies with small market shares stems from the country's geographical vastness and the fact that politicking over licenses favors local companies.

Looking at the world as a single market, the worldwide industry concentration (W-HHI) is a low 154, reflecting the national nature of the radio medium and its ownership.

## TOP COMPANIES: BROADCAST TV

In this section, we look at terrestrial over-the-air TV broadcast organizations, also called "free TV." This includes national networks as well as local and regional stations, which in some countries are well-funded entities. In contrast, and discussed further below, are channels that are "pay-TV." These include channels that are delivered by satellite directly to the home (DTH or DBS) or to cable distribution, and which require subscriptions. Terrestrial pay-TV, such as Sky in the United Kingdom, is also included in that category. All those offerings are described in the section "Non-broadcast Video Channels." Also not part of "Broadcast TV" is a third category—the physical distribution platforms of cable TV (mostly using coax and fiber plants), and satellite broadcast platforms that deliver channels directly to homes or to cable distributors. These are described separately in the section "Multichannel Platforms."

There are overlaps in these categories and line-drawing problems exist. For example, most "free" (i.e., broadcast) channels are also delivered over satellite or cable. Another example is that platforms providers usually bundle channels into the subscription, and some of these channels might be their own and others are those from alternative providers. Whichever definition of markets one uses, line-drawing and definitional problems arise. Our separation is that of "platforms" versus "content," and within the latter, we differentiate between "Free" and "Pay" channels. While these distinctions do not always match the industry structure of some countries, we have tried to be consistent across countries. We also note that after isolating the two categories of video channels, they can be readily merged, as is done in the chapter on US media.

The world's television broadcasters with the highest power indices are the Government of China with 607 (mostly CCTV, plus provincial TV organizations), the BBC (UK) with 225, Globo Group (Brazil) with 155, RAI (Italy) with 127, Fininvest (Italy) with 116, Comcast (US) with 72.5. and Bertelsmann

(RTL), operating networks in several European countries, with 68.2 (Table 35.5). The global market share of the top four firms is only 19.5% of this $184 billion international industry, as most terrestrial broadcasters limit their activities to their home countries. This is part of the legacy of the industry where national public service broadcasters held exclusive national licenses. In part it is due to the ownership restrictions on foreign ownership, as exists for example in the United States, Canada, Mexico, Japan, South Korea, and others.

The largest companies by revenue are the BBC in the UK ($8.3 billion, 4.5%), Comcast/NBC ($7.6 billion, 4.1%) and Disney/ABC ($7 billion, 3.8%) in the US. Other major firms are Fininvest (Italy, $6.1 billion, 3.3%), 21st Century Fox (US, $5.7 billion, 3.1%), Globo (Brazil, $5.5 billion, 3%), RAI (Italy, $4.9 billion, 2.6%). The top German firms are Bertelsmann ($5.3 billion, 2.7%) and ProSiebenSat.1 ($4.9 billion, 2.7%). The top Japanese broadcasters are Fuji TV ($4.6 billion, 2.5%), Tokyo Broadcasting System ($4.3 billion, 2.3%), NTV ($5.2 billion, 2.8%), TV Asahi ($5 billion, 2.7%).

Many of the largest television broadcasters are public companies. Prasar Bharati is India's public TV broadcaster, with a legal monopoly over broadcasting television. In China, CCTV, with its several channels in multiple languages, has 59.1% of the national market there, and a company PI of 212. This is the result of the Chinese market rapidly growing from $2 billion in total revenues in 2004 to $11.2 billion in 2013. CCTV also has the largest television audience in the world measured by total viewers. Given the growth of the Chinese market, the rise of CCTV to the top of the world's broadcast organizations, by both revenue and audience share, seems likely.

The BBC is the largest publicly owned TV broadcaster outside of China by revenue ($8.3 billion), followed by RAI in Italy with $4.9 billion and a global share of 2.6%, NHK of Japan ($4.3 billion, 2.3%), ARD of Germany ($4.1 billion, 2.2%), ZDF of Germany ($2 billion, 1.1%), and Prasar Bharati in India ($1 billion, 0.6%). These public broadcasters are, in terms of revenue, almost as large as major private TV broadcasters. Some are purely financed by a viewer license fee (i.e., an earmarked tax): the BBC is an example. Others also run commercial advertisements, for example, ARD.

The largest broadcast television market by revenues is that of the United States, with total industry revenues of $43.2 billion. Total viewership decreased for the four major US TV broadcast networks as a result of multi-channel options on cable and satellite. In response, they expanded their local station ownership and induced the regulators to substantially relax ownership ceilings on station ownership. In consequence, US broadcast TV industry concentration increased but in a declining market. Even so, due to the legacy of ownership restrictions, as well as due to the absence of a nationwide and well-funded public service TV broadcaster, the TV industry's concentration in the United States, though rising, is relatively low by antitrust standards at 984 and especially in comparison to other countries, when the world average HHI for the industry is 3,830 unweighted and 2,839 weighted (See Chapter 34).

France Télévisions was formed as an in 2000 when the government consolidated several of its public television channels under a single holding company. It broadcasts six channels, of which Channel 2 is the second most watched channel in the country. It used to receive a large share of its revenue from advertising, but in 2009 the government banned advertising on public television. The company became mostly funded by viewers' license payments.

In Germany, there are two main public systems: ZDF, a nationwide institution, and ARD, an association of 10 independent regional public broadcasters. Together, ARD and ZDF account for approximately 40% of the German broadcast TV market through their shares of license fees and advertising. They compete with the main commercial broadcasters, ProSiebenSat.1, and Bertelsmann's RTL Group, within the German market.

Italy's highly concentrated market is split between public and private terrestrial television providers. Silvio Berlusconi's Mediaset (owned by the holding company Fininvest) is Italy's largest commercial broadcaster, with

Table 35-5. The World's Major TV Broadcasters (First In Their Countries; Or, >20%; Or, in Top Thirty by Revenue)

| COMPANIES | COMPANY POWER INDEX IN WORLD INDUSTRY | 2011 OR MOST RECENT REVENUE (MIL $) | MARKET SHARES (%) | MARKET SHARE OF WORLD INDUSTRY TOTAL REVENUE (%) |
|---|---|---|---|---|
| Government of China | 607 | 11,200 | China (100%) | 6.10% |
| *CCTV* | *212* | *6,619* | *China (59.1%)* | 3.60% |
| *Shanghai Media Group (public)* | *9.8* | *1,422* | *China (12.7%)* | 0.80% |
| *Hunan Media Group (public)* | *4* | *907* | *China (8.1%)* | 0.50% |
| *Jiangsu Media Group (public)* | *2.4* | *706* | *China (6.3%)* | 0.40% |
| BBC (UK) (public) | 225 | 8,282 | UK (50.1%) | 4.50% |
| Globo Group (Brazil) | 155 | 5,463 | Brazil (52.4%) | 3.00% |
| RAI (Italy) (public) | 127.4 | 4,865 | Italy (48.3%) | 2.60% |
| Fininvest (Mediaset, Italy) | 116 | 6,121 | Italy (38.9%), Spain (28.1%), Switzerland (1%) | 3.30% |
| Government of Russia | 112 | 2,903 | Russia (70.9%) | 1.60% |
| *Gazprom Media (public)* | *24.9* | *1,372* | *Russia (33.5%)* | 0.70% |
| *First Channel (public)* | *9.3* | *840* | *Russia (20.5%)* | 0.50% |
| *VGTRK (public)* | *6.3* | *692* | *Russia (16.9%)* | 0.40% |
| Government of South Korea | 86.9 | 2,256 | South Korea (70.8%) | 1.20% |
| *KBS (public)* | *24* | *1,188* | *South Korea (37.3%)* | 0.60% |
| *MBC (public)* | *16* | *969* | *South Korea (30.4%)* | 0.50% |
| *EBS (public)* | *0.2* | *99* | *South Korea (3.1%)* | 0.10% |
| ProSiebenSat.1 (Germany) | 78.7 | 4,930 | Germany (30%), Switzerland (7%) | 2.70% |
| Comcast (NBC, US) | 72.5 | 7,598 | US (17.6%) | 4.10% |

| | | | | |
|---|---|---|---|---|
| Bertelsmann (RTL, Germany) | 68.2 | 5,292 | Germany (25.3%), Netherlands (32.6%), France (10.8%), Switzerland (2.4%), Belgium (15.4%) | 2.90% |
| Disney (ABC, US) | 60.7 | 6,950 | US (16.1%) | 3.80% |
| Televisa (Mexico) | 56.8 | 1,717 | Mexico (61%) | 0.90% |
| Prasar Bharati (DD, India) (public) | 55.9 | 1,030 | India (100%) | 0.60% |
| ARD (Germany) (public) | 54.5 | 4,075 | Germany (25%), Switzerland (3.9%) | 2.20% |
| NTV (Japan) | 53 | 5,115 | Japan (19.1%) | 2.80% |
| TV Asahi Corporation (Japan) | 51.3 | 5,035 | Japan (18.8%) | 2.70% |
| Redstone Group (CBS, US) | 49.9 | 6,303 | US (14.6%) | 3.40% |
| ITV (UK) | 42.6 | 3,604 | UK (21.8%) | 2.00% |
| NBS/Fujisankei (Japan) | 42.5 | 4,579 | Japan (17.1%) | 2.50% |
| Murdoch Group (US) | 41.4 | 5,741 | US (13.3%) | 3.10% |
| NHK (Japan) (public) | 37.6 | 4,312 | Japan (16.1%) | 2.30% |
| Tokyo Broadcasting Service (Japan) | 36.7 | 4,258 | Japan (15.9%) | 2.30% |
| SABC (South Africa) (public) | 29.3 | 799 | South Africa (67.6%) | 0.40% |
| CBC (Canada) (public) | 28.1 | 1,353 | Canada (38.3%) | 0.70% |
| Planeta (Spain) | 28.1 | 2,007 | Spain (25.8%) | 1.10% |
| RTP (Portugal) (public) | 26.3 | 1,313 | Portugal (37%) | 0.70% |
| PRISA (Spain) | 22.4 | 1,370 | Portugal (34%), Spain (2.1%) | 0.70% |
| Seven Network (Australia) | 18.5 | 1,294 | Australia (26.4%) | 0.70% |

*continued*

Table 35-5. *continued*

| COMPANIES | COMPANY POWER INDEX IN WORLD INDUSTRY | 2011 OR MOST RECENT REVENUE (MIL $) | MARKET SHARES (%) | MARKET SHARE OF WORLD INDUSTRY TOTAL REVENUE (%) |
|---|---|---|---|---|
| Bonnier (Sweden) | 17.9 | 1,061 | Finland (33%), Sweden (29.8%) | 0.60% |
| France Télévisions (France) (public) | 16.9 | 1,105 | France (29.4%), Switzerland (3.1%) | 0.60% |
| Impresa (Portugal) | 16.2 | 1,029 | Portugal (29%) | 0.60% |
| RTVE (Spain) (public) | 15.1 | 1,470 | Spain (18.9%) | 0.80% |
| RTÉ (Ireland) (public) | 13.9 | 346 | Ireland (74%) | 0.20% |
| TV Azteca (Mexico) | 13.7 | 844 | Mexico (30%) | 0.50% |
| NPB (Netherlands) (public) | 13.7 | 545 | Netherlands (46.3%) | 0.30% |
| Dogan Group (Turkey) | 13.5 | 633 | Turkey (32.9%) | 0.30% |
| Government of Egypt (ERTU) (public) | 13.2 | 243 | Egypt (100%) | 0.10% |
| ZDF (Germany) (public) | 12.9 | 2,007 | Germany (12.1%), Switzerland (3.9%) | 1.10% |
| Yle (Finland) (public) | 12.6 | 565 | Finland (41%) | 0.30% |
| Nine Entertainment (Australia) | 12.4 | 1,059 | Australia (21.6%) | 0.60% |
| Government of Taiwan | 11.9 | 263 | Taiwan (83.6%) | 0.10% |
| *Formosa TV (public)* | *2.3* | *116* | *Taiwan (36.9%)* | 0.10% |
| *BCC (public)* | *0.3* | *44* | *Taiwan (13.9%)* | 0.00% |
| *China TV Corp. (public)* | *0.4* | *50* | *Taiwan (15.8%)* | 0.00% |
| *Taiwan TV (public)* | *0.5* | *53* | *Taiwan (17%)* | 0.00% |
| Bouygues Group (TF1, France) | 11 | 901 | France (23.7%), Switzerland (3.1%) | 0.50% |

| | | | | |
|---|---|---|---|---|
| Bell Canada Enterprises/CTV (Canada) | 11 | 848 | Canada (24%) | 0.50% |
| CTC (Russia) | 10 | 868 | Russia (21.2%) | 0.50% |
| TP (Poland) (public) | 9.2 | 383 | Poland (44.3%) | 0.20% |
| SRG SSR (Switzerland) (public) | 8.6 | 524 | Switzerland (30.1%) | 0.30% |
| Sveriges TV (Sweden) (public) | 7.7 | 537 | Sweden (26.4%) | 0.30% |
| Hosken (South Africa) | 6.7 | 383 | South Africa (32.4%) | 0.20% |
| VRT (Belgium) (public) | 6.5 | 569 | Belgium (21.2%) | 0.30% |
| Investment AB Kinnevik (Sweden) | 5.2 | 443 | Sweden (21.8%) | 0.20% |
| Shaw (Canada) | 4.3 | 530 | Canada (15%) | 0.30% |
| Grupo Clarín (Argentina) | 3.7 | 288 | Argentina (23.5%) | 0.20% |
| Sanoma (SBS Nederland, Finland) | 2.2 | 220 | Netherlands (18.7%) | 0.10% |
| Grupa TVN (Poland) | 2.1 | 181 | Poland (21%) | 0.10% |
| TVN (Chile) (public) | 2.1 | 181 | Poland (21%) | 0.10% |
| Channel 10 (Israel) | 1.8 | 119 | Israel (27.4%) | 0.10% |
| Universidad Catholica (Chile) | 1.7 | 106 | Chile (30%) | 0.10% |
| Telefónica (Spain) | 1.3 | 169 | Argentina (13.8%) | 0.10% |
| Channel 2 (Israel) | 0 | 8 | Israel (1.9%) | 0.00% |
| Worldwide Industry HHI Concentration (W-HHI) | | | | 206 |
| Worldwide Industry C4 (30-countries) Concentration | | | | 18.5% |
| Total Global Broadcast TV Revenue (mil US$) | | | | 184,379 |

38.9% of the market in 2013 for its three national networks. The public broadcaster, RAI, has 48.3%.

Many companies' high power indices can be attributed to politics as well as programming and marketing savvy. Brazil's Globopar (part of the Globo Group) rose to dominance partly due to its licensing agreements made under the country's military dictatorship from 1964 to 1985. Grupo Televisa leads the Mexican market partially due to the company's close political ties to PRI, the ruling party of the 1980s and 1990s, which allowed it to greatly expand its market reach.

Internationalization of terrestrial television broadcasting is fairly limited. Sometimes this can be explained by legal restrictions against foreign ownership. In the United States, Rupert Murdoch had to trade his Australian passport for an American one before being able to buy enough stations to launch Fox (US networks are not licensed and are not limited to ownership by citizens; but local stations are, though exceptions are possible). As in radio, one of the few TV broadcasters that operates internationally is the RTL Group, Bertelsmann's TV subsidiary. With market shares of 25.3% in Germany, 32.6% in the Netherlands, 10.8% in France, 2.4% in Switzerland, and 15.4% in Belgium, RTL is the second largest privately owned broadcaster in Europe. Italy's Mediaset, in addition to its 38.9% of the Italian market, also holds 28.1% of the Spanish market under its broadcaster Telecinco.

The worldwide concentration for the TV broadcasting industry was 206. Concentration is lower than in the other video industries because broadcasting has been subject to licensing, which enabled governments to limit the entry by multinational TV firms and the creation of multinational footprints.

## TOP COMPANIES: MULTI-CHANNEL PLATFORMS

Of the world's top 45 multi-channel platform firms, 5 are satellite-platform providers such as 21st Century Fox's Sky, DirecTV, or Dish, and 6 firms are IPTV providers, mostly incumbent telecom firms. The other 29 firms are cable TV platform providers. Of these, five are mixed cable/DBS or cable/IPTV providers.

In the multi-channel platform industry, the top companies by power index are 21st Century Fox (PI of 510, and a world market share by revenue of 7.2%), which often operates as the satellite operator Sky; Comcast (United States, 345, 14.9%); DirecTV (252, 12.9%) and Time Warner Cable (United States, 107, 8.3%) (Table 35.6). The American market is by far the largest multi-channel platform market (64% of a total world industry valued at $248 billion in 2013). By revenue, the top companies are the American firms Comcast ($36.8 billion), DirecTV ($32 billion), Time Warner Cable ($20.5 billion), 21st Century Fox ($17.8 billion), Dish Network ($13 billion), and Liberty ($7.6 billion).

The multi-channel platform industry has become highly internationalized by large companies like 21st Century Fox, DirecTV, Liberty (UPC), Telmex, and Telefónica. 21st Century Fox holds market shares across the world in Brazil, Germany, Ireland, India, Italy, and United Kingdom. Liberty operates cable services in Australia, Austria, the Czech Republic, Germany, Hungary, Ireland, the Netherlands, Poland, Romania, Slovakia, and Switzerland. It re-entered the US cable market through acquiring a large share of Charter Communications in 2013, and through negotiations with other cable firms. It also had control of DBS provider DirecTV but reduced it subsequently. Telmex expanded into Latin America (Brazil, Chile, and Mexico). Telefónica has large market shares in both Spain and Chile.

In Italy, 21st Century Fox (as Sky Italia) has a 95.3% market share due to buyouts of other satellite providers and since no cable TV option exists as an alternative except for a few IPTV providers. In France, a duopoly of Vivendi (DBS) and Altice (Numericable) emerged after a series of acquisitions that took place over the 2000s.

The industry has a W-HHI of 599 for 2013, which is much higher than that of video channels (431) or broadcast TV (206).

Table 35-6. The World's Major Multichannel Platform Companies (First In Their Countries; Or, >20%; Or, in Top Thirty by Revenue)

| COMPANIES | COMPANY POWER INDEX IN WORLD INDUSTRY | 2011 OR MOST RECENT REVENUE (MIL $) | MARKET SHARES (%) | MARKET SHARE OF WORLD INDUSTRY TOTAL REVENUE (%) |
|---|---|---|---|---|
| Murdoch Group (US)[a, b] | 510 | 17,843 | Australia (45%), Germany (26%), India (9.3%), Ireland (52%), Italy (95.3%), UK (73.3%) | 7.20% |
| Comcast (US) | 345 | 36,778 | US (23.2%) | 14.90% |
| DirecTV (US) | 252 | 32,033 | Argentina (19.9%), Brazil (17.7%), Chile (56%), US (18.9%) | 12.90% |
| Time Warner Cable (US)[c] | 107 | 20,450 | US (12.9%) | 8.30% |
| Liberty (US)[d] | 82.7 | 7,617 | Poland (28.7%), Ireland (48%), US (1.2%), UK (26.7%), Switzerland (74%), Netherland (32.4%) | 3.10% |
| Globo (Brazil) | 80.6 | 3,725 | Brazil (53.6%) | 1.50% |
| SkyPerfecTV (Japan) | 79.9 | 4,538 | Japan (43.6%) | 1.80% |
| Vivendi (France) | 73.4 | 2,979 | France (61%) | 1.20% |
| PRISA (Spain) | 46.2 | 2,711 | Spain (42.2%) | 1.10% |
| Dish Network (US) | 43 | 12,999 | US (8.2%) | 5.20% |
| Altice (France) | 37.1 | 2,280 | France (34%), Israel (57.3%) | 0.90% |
| Kabel Deutschland (Germany) | 36.6 | 2,016 | Germany (45%) | 0.80% |
| Ziggo (Netherlands) | 28.1 | 1,310 | Netherlands (53.2%) | 0.50% |
| Shaw (Canada) | 28.1 | 2,458 | Canada (28.3%) | 1.00% |
| Telstra (Australia) | 27 | 1,488 | Australia (45%) | 0.60% |
| Televisa (Mexico) | 26.2 | 1,300 | Mexico (50%) | 0.50% |

*continued*

Table 35-6. *continued*

| COMPANIES | COMPANY POWER INDEX IN WORLD INDUSTRY | 2011 OR MOST RECENT REVENUE (MIL $) | MARKET SHARES (%) | MARKET SHARE OF WORLD INDUSTRY TOTAL REVENUE (%) |
|---|---|---|---|---|
| J:COM (Japan)[e] | 20.5 | 2,300 | Japan (22.1%) | 0.90% |
| Grupo Clarín (Argentina) | 20 | 899 | Argentina (55.1%) | 0.40% |
| ZON (Portugal) | 20 | 856 | Portugal (57.9%) | 0.30% |
| Bell Canada Enterprises/CTV (Canada) | 17 | 1,911 | Canada (22%) | 0.80% |
| Unitymedia (Germany) | 16.3 | 1,344 | Germany (30%) | 0.50% |
| Verizon (US) | 16 | 7,926 | US (5%) | 3.20% |
| Rogers (Canada) | 15.9 | 1,850 | Canada (21.3%) | 0.70% |
| Cox (US) | 14.7 | 7,609 | US (4.8%) | 3.10% |
| AT&T (US) | 14.7 | 7,609 | US (4.8%) | 3.10% |
| Naspers (Multichoice, South Africa) | 13.7 | 340 | South Africa (100%) | 0.10% |
| Telenet (Belgium) | 13.5 | 611 | Belgium (54.5%) | 0.20% |
| ONO (Spain) | 11.2 | 1,336 | Spain (20.8%) | 0.50% |
| Telefónica (Spain) | 8.9 | 1,503 | Spain (16.7%), Chile (18%), Brazil (3.8%) | 0.60% |
| Com Hem (Baring Capital Partners, UK) | 8.3 | 538 | Sweden (38.4%) | 0.20% |
| Zee (India) | 7.4 | 951 | India (19.4%) | 0.40% |
| Cablevision (US) | 7.4 | 5,390 | US (3.4%) | 2.20% |
| Cukurova Group (Turkey) | 7.2 | 331 | Turkey (53.7%) | 0.10% |
| Tbroad (South Korea) | 7.1 | 686 | South Korea (25.6%) | 0.30% |
| Charter (US)[d] | 6.6 | 5,073 | US (3.2%) | 2.00% |
| Quebecor (Videotron, Canada) | 5.1 | 1,051 | Canada (12.1%) | 0.40% |
| Bezeq (Israel) | 4.9 | 361 | Israel (33.4%) | 0.10% |

| | | | | |
|---|---|---|---|---|
| VOO & Coditel (Belgium) | 3.9 | 329 | Belgium (29.3%) | 0.10% |
| NMG (Russia) | 3.5 | 326 | Russia (26.5%) | 0.10% |
| Telmex (Grupo Carso, Mexico) | 3.3 | 837 | Brazil (6.4%), Chile (20%), Mexico (8%) | 0.30% |
| Teracom (Sweden) (public) | 2.4 | 288 | Sweden (20.6%) | 0.10% |
| CNS (South Korea) | 2.4 | 277 | Taiwan (21.6%) | 0.10% |
| Kbro (Taiwan) | 2.2 | 266 | Taiwan (20.8%) | 0.10% |
| Polsat (Poland) | 2.2 | 95.8 | Poland (56.7%) | 0.04% |
| ER Telekom (Russia) | 1.8 | 234 | Russia (19%) | 0.10% |
| Dogan Group (Turkey) | 1.3 | 143 | Turkey (23.2%) | 0.10% |
| Akado (Russia) | 1.1 | 183 | Russia (14.9%) | 0.10% |
| Tata Group | 0.4 | 230 | India (4.7%) | 0.10% |
| Worldwide Industry HHI Concentration (W-HHI) | | | | 599 |
| Worldwide Industry C4 (30-countries) Concentration | | | | 43% |
| Total Global Multichannel Revenue (mil US$) | | | | 247,666 |

[a] Murdoch and Telstra have a joint venture for MSOs in Australia called Foxtel; the share allocated reflects Fox/Sky's partial ownership.
[b] 50/50 split with Tata Group for DEN Network.
[c] Separated from Time Warner in 2009.
[d] US market share includes Liberty's 23% stake in Charter.
[e] J:COM is a joint venture between KDDI and the Sumimoto Corporation.

## TOP COMPANIES: NON-BROADCAST VIDEO CHANNELS

The most significant video channel companies, ordered by their power index are Vivendi of France (PI of 247, world revenue share of 5.1%, revenues of $3.6 billion), Time Warner (222, 9.4%, $6.6 billion), 21st Century Fox (164, 7.4%. $5.2 billion), Viacom (142, 8.4%), Disney (118, 8.3%, $5.8 billion), and the Japanese network SkyPerfecTV (228, 4.1%, $2.9 billion). Other large providers are Discovery Communications (72.8, 6%, $4.3 billion) Comcast (59, 4.9%, $3.5 billion), Shaw Canada (119, 3.5%, $2.5 billion), Bertelsmann (39.6, 2.2%, $1.6 billion) of Germany, and Bouygues (29.7, 2.2%, $1.6 billion) (Table 35.7).

The $70.6 billion industry is highly internationalized, mostly due to the satellite form of delivery to cable firms with a coverage footprint that easily covers several countries. Time Warner's position as the third-largest firm in the world stems from its strong role in the American market, where it holds a 23.3% market share. Its acquisition of Turner Broadcasting in 1996 secured its position as the market leader, by adding the news channel CNN as well as TNT and TBS. It also owns HBO, America's leading pay-cable channel. Disney owns the ABC family of channels and the hugely profitable sports channel ESPN. Discovery Communications offers several documentary-style channels. Discovery Communications is a market leader in Russia and operates in many other countries as well. Viacom owns many channels focusing on young viewers, in particular by channels such as MTV, Comedy Central, and Nickelodeon.

Vivendi's first place ranking in the market by company PI derives from its ownership of the Canal + Group, which has a 50% share (and significant revenues) in France through the Canal + network of television channels, including also Sport +, Extreme Sports, Planète, and CinéCinéma. France Télévisions (Bouygues) owns several video channels, including Gulli and Mezzo (co-owned with Lagardère); Planète Thalassa and Planète Justice (co-owned with Vivendi); France2 +, TV5Monde, Euro News, and Arte, all in collaboration with other public broadcasters. TF1's Eurosport channel, offered by Bouygues in conjunction with Discovery Communications (United States), is available across Europe, Australia, and parts of Asia.

In India, 30.2% of the market is held by Zee TV, which operates channels such as Zee Cinema HD, Zee TV HD, Zee Action, Zee Salaam, and Zee Business. SkyPerfecTV and WOWOW dominate the Japanese market. In Canada, Bell Canada/CTV owns TSN, RDS, The Movie Network, and Super Écran.

The 30-country worldwide concentration in the video channels industry is W-HHI = 431, higher than the figure for broadcast TV (218) due to its much greater internationalization, but lower than for multichannel platforms (701). Table 35.8 names the most important channels owned by the top 11 video channel providers in the world market.

The line between broadcast television and non-broadcast video channels is not always well defined. The industries are overlapping. To reflect this, in Table 35.9, we present them together as one combined TV channel industry valued at $255 billion. Table 35.9 includes several companies that were excluded in preceding tables because they did not meet the size criteria for those industry tables. The results show China's CCTV, the BBC, Fininvest, Globo, RAI, Disney, the Government of Russia, Viacom, Comcast, Murdoch, Bertelsmann, ProSiebenSat.1, Vivendi, SkyPerfecTV, Time Warner, and ARD leading the combined video industries market by Power Index. By revenues, the top firms are Disney ($12.8 billion), Redstone ($12.3 billion), the Chinese government as a whole ($11.2 billion), Comcast ($11.1 billion), the Murdoch Group ($11 billion), the BBC ($9.4 billion), Time Warner ($7.4 billion), and Bertelsmann ($6.8 billion). Of the several Chinese entities, CCTV as a standalone is still higher than Italy's Fininvest ($6.2 billion), the tenth largest video industries company by revenue. The C10 is 36.3%.

Table 35-7. The World's Major Non-Broadcast Video Channel Companies (First In Their Countries; Or, >20%; Or, in Top Thirty by Revenue)

| COMPANIES | COMPANY POWER INDEX IN WORLD INDUSTRY | 2011 OR MOST RECENT REVENUE (MIL $) | MARKET SHARES (%) | MARKET SHARE OF WORLD INDUSTRY TOTAL REVENUE(%) |
|---|---|---|---|---|
| Vivendi (France) | 247 | 3,633 | France (49.2%), Germany (2.8%) | 5.10% |
| SkyPerfecTV (Japan) | 227.8 | 2,894 | Japan (55.6%) | 4.10% |
| Time Warner (US) | 222 | 6,614 | Chile (23.7%), US (23.3%), UK (4%) | 9.40% |
| Murdoch Group (21st Century Fox/Sky, US) | 164 | 5,236 | Portugal (12.3%), Chile (12.7%), Ireland (5.1%), Taiwan (11%), Russia (4.4%), India (1.9%), UK (38.4%), US (7.7%), Spain (6.4%), Poland (6.6%), Israel (7.8%) | 7.40% |
| Redstone Group (Viacom, US) | 142 | 5,956 | Poland (2.6%), Ireland (2%), Israel (2.9%), UK (7.2%), Chile (6.9%), Germany (3.4%), Spain (4.8%), US (13.6%), Netherland (34.1%) | 8.40% |
| Shaw (Canada) | 119 | 2,477 | Canada (33.9%) | 3.50% |
| Disney (US) | 118 | 5,832 | US (15.3%), India (15.2%), Netherlands (15.6%), Portugal (2.7%), UK (6.4%), Spain (4.6%), Russia (5.3%) | 8.30% |
| Bell Canada Enterprises/CTV (Canada) | 78 | 2,002 | Canada (27.4%) | 2.80% |
| Discovery Communications (US) | 73 | 4,252 | Chile (38.5%), Israel (2.9%), US (9.9%), Netherlands (19.5%), Russia (15.2%), UK (7.2%), Germany (2%) | 6.00% |
| Comcast (US) | 59.1 | 3,450 | US (12.1%) | 4.90% |
| NHK (Japan, public) | 48.3 | 1,333 | Japan (26.5%) | 1.90% |
| Bertelsmann (RTL, Germany; M6, France) | 39.6 | 1,550 | France (10.2%), Germany (25.1%) | 2.20% |
| WOWOW (Japan) | 33 | 1,098 | Japan (21.1%) | 1.60% |

*continued*

Table 35-7. *continued*

| COMPANIES | COMPANY POWER INDEX IN WORLD INDUSTRY | 2011 OR MOST RECENT REVENUE (MIL $) | MARKET SHARES (%) | MARKET SHARE OF WORLD INDUSTRY TOTAL REVENUE(%) |
|---|---|---|---|---|
| Bouygues Group (TF1, France) | 29.7 | 1,563 | France (16.2%), Netherlands (7%), Russia (5.4%), Germany (2%), Spain (1.5%) | 2.20% |
| ARD (Germany) (public) | 28.5 | 809 | Germany (24.9%) | 1.10% |
| Astral (Canada) | 25.2 | 1,140 | Canada (15.6%) | 1.60% |
| ProSiebenSat.1 (Germany) | 18.8 | 657 | Germany (20.2%) | 0.90% |
| Zee (India) | 16.7 | 390 | India (30.2%) | 0.60% |
| PRISA (Spain) | 13 | 631 | Spain (14.5%) | 0.90% |
| Sun Group (India) | 9.1 | 288 | India (22.3%) | 0.40% |
| ZDF (Germany) (public) | 7.3 | 410 | Germany (12.6%) | 0.60% |
| Impresa (Portugal) | 4.8 | 136 | Portugal (24.9%) | 0.20% |
| Lagardère (France) | 4 | 454 | France (6.3%) | 0.60% |
| France Télévisions (France) (public) | 3.7 | 432 | France (6%) | 0.60% |
| SET (Taiwan) | 3.3 | 142 | Taiwan (16.6%) | 0.20% |
| Liberty (US | 3.2 | 693 | Section 1.01 US (1.5%), Spain (6.1%) | 1.00% |
| Dori Media Group (Israel) | 3.1 | 111 | Israel (20%) | 0.20% |
| Factoria de Canales (Spain) | 2.7 | 287 | Spain (6.6%) | 0.40% |
| Cablevision (US) | 2.5 | 713 | US (2.5%) | 1.00% |
| RGE (Israel) | 2.2 | 93.2 | Israel (16.8%) | 0.10% |
| TVB (TVBS, Hong Kong) | 1.6 | 98.6 | Taiwan (11.5%) | 0.10% |
| China Times News Group (Taiwan) | 1.4 | 93.4 | Taiwan (10.9%) | 0.10% |

| | | | | |
|---|---|---|---|---|
| Hayat (Egypt) | 1.3 | 67.8 | Egypt (13.3%) | 0.10% |
| Panorama (Egypt) | 1.1 | 62.2 | Egypt (12.2%) | 0.10% |
| Kinnevik (Sweden) | 1.1 | 38.4 | Sweden (20.6%) | 0.10% |
| Network18 Group (India) | 1.1 | 102 | India (7.9%) | 0.10% |
| Government of Egypt (NTN) (public) | 1 | 60.2 | Egypt (11.8%) | 0.10% |
| Rotana (Saudi Arabia) | 0.7 | 49.5 | Egypt (9.7%) | 0.10% |
| Altice (France) | 0.3 | 32.7 | Israel (5.9%) | 0.00% |
| Worldwide Industry HHI Concentration (W-HHI) | | | | 431 |
| Worldwide Industry C4 (30 Countries) Concentration | | | | 33.5% |
| Total Global Video Channel Revenue (mil US$) | | | | 70,636 |

Table 35-8. The World's Top Video Channels owned by Non-Broadcast Video Channel Companies by Power Index

| TOP COMPANIES | MAIN CHANNELS |
|---|---|
| Time Warner (US) | CNN, TNT, TBS, Cartoon Network, HBO, Cinemax |
| Redstone Group (US) | MTV, VH1, Nickelodeon, Comedy Central, CMT, TV Land, Showtime |
| Disney (US) | ESPN, ABC Family, Disney XD, A&E, Bio, Lifetime, History |
| Murdoch Group (US) | Fox News, Fox Sports, FX, National Geographic |
| Discovery Communications (US) | Discovery Channel, TLC, Animal Planet, OWN |
| Vivendi (France) | Canal+, Canal+ Sport, Canal+ Cinema, D8, D17, i>télé, Planète |
| Comcast (US) | MSNBC, CNBC, Bravo, E!, SyFy, USA |
| SKY PerfecTV (Japan) | Channel 267, Channel 89, 57 SD |
| Shaw (Canada) | Slice, Showcase, HGTV Canada |
| Bell Canada Enterprises/CTV (Canada) | TSN, RDS, The Movie Network, Super Écran |
| WOWOW (Japan) | WOW Prime, WOW Live, WOW Cinema |
| France Télévisions (France) (public) | Gulli, Mezzo, Planète Thalassa, Planète Justice |
| Zee Entertainment (India) | Zee Next, TEN Sports, Zing |

## TOP COMPANIES: FILM

Film is a relatively static market by revenues, which totaled $30.7 billion in 2013. The companies with the largest CPI-I are the "big six" Hollywood studios: Disney (219), Time Warner (180), Sony-owned Columbia-TriStar (151), 21st Century Fox (139), Comcast's Universal Studios (110), and Viacom (102). Also in the top are the China Film Group (176) and the Japanese studio Toho (84) due to their high domestic shares. By revenue, by far the largest firms are Warner Bros ($4.4 billion), Sony ($3.3 billion), Paramount ($3.3 billion), Disney ($3.3 billion), 21st Century Fox ($3.2 billion), and Universal ($2.5 billion). The next tier of film companies in revenue terms are the China Film Group with revenues of $1.2 billion and Toho with $683 million. In Europe, Vivendi (StudioCanal Productions) is strong in France, Germany, and the United Kingdom. Vivendi used to own the Hollywood major studio Universal before it was bought by GE and then Comcast.

Major East Asian countries are among the few markets where the six American studios (including Japan-owned Sony) are not the dominant firms (Table 35.10). The state-owned China Film Group maintains its high market share partly due to severe governmental restrictions against foreign films. Only 34 foreign films may be imported into China each year (an increase from 20 in 2008). But since the national market is still fairly low income, the China Film Group achieves only relatively small revenues.

The South Korean government, too, restricts the foreign films that can enter the country's film market. Regulations state that a theater must show domestic films 40% of the year. In Japan, Toho has considerable influence over the fairly sizable domestic market. Toho is Japan's largest movie studio, known for its science fiction and animated films. It is also the leading distributor of American films within the Japanese market and has a strong hold on theaters in the country.

In other Asian countries, the film industry is less concentrated. A strong domestic film industry in India has kept foreign companies from dominating, but languages, cultural divisions, and entrepreneurial production firms have kept

Table 35-9. The World's Major Television Channel Companies (Broadcast and Non-broadcast Channels), 2011 or Most Recent

| COMPANY | 2011 OR MOST RECENT POWER INDEX | REVENUE (MIL $) | MARKET SHARE IN BROADCAST AND OTHER VIDEO CHANNELS |
|---|---|---|---|
| Disney (US) | 76.6 | 12,782 | 5.0% |
| Redstone Group (US) | 75.4 | 12,259 | 4.8% |
| Government of China | 439 | 11,200 | 4.4% |
| *CCTV (China) (public)* | *153* | *6,619* | *2.6%* |
| *Shanghai Media Group (China) (public)* | *7.1* | *1,422* | *0.6%* |
| *Hunan Media Group (China) (public)* | *2.9* | *907* | *0.4%* |
| *Jiangsu Media Group (China) (public)* | *1.7* | *706* | *0.3%* |
| Comcast (US) | 68.8 | 11,047 | 4.3% |
| Murdoch Group (US) | 75.4 | 10,977 | 4.3% |
| BBC (UK) (public) | 170 | 9,386 | 3.7% |
| Time Warner (US) | 62.0 | 7,391 | 2.9% |
| Bertelsmann (Germany) | 60.3 | 6,843 | 2.7% |
| Fininvest (Italy) | 83.9 | 6,217 | 2.4% |
| NHK (Japan) (public) | 40.6 | 5,644 | 2.2% |
| ProSiebenSat.1 (Germany) | 62.1 | 5,587 | 2.2% |
| Globo Group (Brazil) | 112 | 5,463 | 2.1% |
| NTV (Japan) | 38.3 | 5,115 | 2.0% |
| Asahi Shimbun (Japan) | 37.1 | 5,108 | 2.0% |
| ARD (Germany) (public) | 47.3 | 4,884 | 1.9% |
| RAI (Italy) (public) | 92.1 | 4,865 | 1.9% |
| NBS (Japan) | 30.7 | 4,637 | 1.8% |
| Tokyo Broadcasting Service (Japan) | 26.6 | 4,341 | 1.7% |
| Discovery Communications (US) | 20.2 | 4,252 | 1.7% |
| Vivendi (France) | 68.4 | 3,633 | 1.4% |
| ITV (UK) | 30.8 | 3,604 | 1.4% |
| Shaw (Canada) | 36.0 | 3,007 | 1.2% |
| Government of Russia | 80.9 | 2,935 | 1.2% |
| *Gazprom Media (Russia) (public)* | *18.0* | *1,372* | *0.5%* |
| *First Channel (Russia) (public)* | *6.7* | *840* | *0.3%* |
| *VGTRK (Russia) (public)* | *4.6* | *692* | *0.3%* |
| SkyPerfecTV (Japan) | 63.1 | 2,894 | 1.1% |
| Bell Canada Enterprises/CTV (Canada) | 29.5 | 2,850 | 1.1% |
| Bouygues (France) | 16.2 | 2,464 | 1.0% |
| ZDF (Germany) (public) | 11.3 | 2,417 | 0.9% |
| Government of South Korea | 86.9 | 2,256 | 0.9% |
| *KBS (South Korea) (public)* | *24.0* | *1,188* | *0.5%* |
| *MBC (South Korea) (public)* | *16.0* | *969* | *0.4%* |
| *EBC (South Korea) (public)* | *0.2* | *99* | *0.0%* |

*continued*

Table 35-9. *continued*

| COMPANY | 2011 OR MOST RECENT POWER INDEX | REVENUE (MIL $) | MARKET SHARE IN BROADCAST AND OTHER VIDEO CHANNELS |
|---|---|---|---|
| Planeta (Spain) | 20.3 | 2,046 | 0.8% |
| PRISA (Spain) | 19.8 | 2,001 | 0.8% |
| Nihon Keizai Shimbun (Japan) | 5.2 | 1,932 | 0.8% |
| Televisa (Mexico) | 41.1 | 1,717 | 0.7% |
| CBC (Canada) (public) | 20.8 | 1,652 | 0.6% |
| Channel 4 (UK) | 6.3 | 1,633 | 0.6% |
| France Télévisions (France) (public) | 13.2 | 1,537 | 0.6% |
| Universal Church Group (Brazil) | 8.8 | 1,533 | 0.6% |
| RTVE (Spain) (public) | 10.9 | 1,470 | 0.6% |
| RTP (Portugal) (public) | 19.5 | 1,393 | 0.5% |
| Gannett (US) | 1.7 | 1,381 | 0.5% |
| Seven Network (Australia) | 13.4 | 1,294 | 0.5% |
| Rogers (Canada) | 5.4 | 1,199 | 0.5% |
| Impresa (Portugal) | 13.0 | 1,165 | 0.5% |
| Astral (Canada) | 7.0 | 1,140 | 0.4% |
| WOWOW (Japan) | 9.1 | 1,098 | 0.4% |
| Bonnier (Sweden) | 13.0 | 1,061 | 0.4% |
| Nine Entertainment (Australia) | 9.0 | 1,059 | 0.4% |
| Prasar Bharati (India) (public) | 40.4 | 1,030 | 0.4% |
| CTC (Russia) | 7.2 | 868 | 0.3% |
| TV Azteca (Mexico) | 9.9 | 844 | 0.3% |
| Hearst (US) | 0.6 | 820 | 0.3% |
| Tribune (US) | 0.6 | 820 | 0.3% |
| SABC (South Africa) (public) | 21.2 | 799 | 0.3% |
| Ten Network (Australia) | 4.8 | 770 | 0.3% |
| FORTA (Spain) (public) | 3.0 | 762 | 0.3% |
| Bandeirantes Group (Brazil) | 2.2 | 761 | 0.3% |
| Cablevision (US) | 0.7 | 713 | 0.3% |
| Liberty (US) | 0.9 | 693 | 0.3% |
| Dogan Group (Turkey) | 9.8 | 633 | 0.2% |
| SBS (South Korea) | 4.2 | 583 | 0.2% |
| VRT (Belgium) (public) | 4.7 | 569 | 0.2% |
| Yle (Finland) (public) | 9.1 | 565 | 0.2% |
| The Washington Post (US) | 0.3 | 561 | 0.2% |
| Quebecor (Videotron, Canada) | 1.2 | 550 | 0.2% |
| NPB (Netherlands) (public) | 9.9 | 545 | 0.2% |
| Sveriges (Sweden) (public) | 5.6 | 537 | 0.2% |
| SRG SSR (Switzerland) (public) | 6.2 | 524 | 0.2% |
| RTÉ (Ireland) (public) | 12.2 | 502 | 0.2% |

Table 35-9. *continued*

| COMPANY | 2011 OR MOST RECENT POWER INDEX | REVENUE (MIL $) | MARKET SHARE IN BROADCAST AND OTHER VIDEO CHANNELS |
|---|---|---|---|
| NRJ Group (France) | 1.4 | 497 | 0.2% |
| Kinnevik (Sweden) | 4.1 | 482 | 0.2% |
| Lagardère (France) | 1.1 | 454 | 0.2% |
| Silvio Santos Group (Brazil) | 0.7 | 438 | 0.2% |
| Zee (India) | 4.6 | 390 | 0.2% |
| Advance (US) | 0.1 | 389 | 0.2% |
| Hosken (South Africa) | 4.9 | 383 | 0.2% |
| TP (Poland) (public) | 6.7 | 383 | 0.2% |
| Southern Cross Media (Australia) | 13.2 | 363 | 0.1% |
| VMMA (Belgium) | 1.7 | 346 | 0.1% |
| Turkuvaz-Kalyone Group (Turkey) | 2.7 | 331 | 0.1% |
| Cukurova Group (Turkey) | 2.4 | 312 | 0.1% |
| RTBF (Belgium) (public) | 1.4 | 306 | 0.1% |
| Government of Egypt | 9.8 | 303 | 0.1% |
| *ERTU (public)* | *9.5* | *243* | *0.1%* |
| *NTN (public)* | *0.2* | *60* | *0.02%* |
| Next Radio TV (France) | 0.5 | 302 | 0.1% |
| Sun Group (India) | 2.5 | 288 | 0.1% |
| Grupo Clarín (Argentina) | 2.7 | 288 | 0.1% |
| Factoria de Canales (Spain) | 0.7 | 287 | 0.1% |
| Government of Taiwan | 8.6 | 263 | 0.1% |
| *Formosa TV (Taiwan)* | *1.7* | *116* | *0.05%* |
| *Taiwan TV (Taiwan)* | *0.4* | *53* | *0.02%* |
| *China TV Corp. (Taiwan)* | *0.3* | *50* | *0.02%* |
| *BCC (Taiwan) (public)* | *0.2* | *44* | *0.02%* |
| EBC (Brazil) (public) | 0.2 | 240 | 0.1% |
| Canal 9 (Mexico) | 1.8 | 238 | 0.1% |
| Bloch Group (Brazil) | 0.2 | 229 | 0.1% |
| America 2 (Argentina) | 1.6 | 224 | 0.1% |
| Sanoma Oyj (Finland) | 1.6 | 220 | 0.1% |
| TVN (Chile) (public) | 1.5 | 181 | 0.1% |
| Grupa TVN (Poland) | 1.5 | 181 | 0.1% |
| Telefónica (Spain) | 0.9 | 169 | 0.1% |
| China Times News Group (Taiwan) | 0.8 | 147 | 0.1% |
| SET (Taiwan) | 0.9 | 142 | 0.1% |
| Channel 10 (Israel) | 1.3 | 119 | 0.05% |
| Dori Media Group (Israel) | 0.9 | 111 | 0.04% |
| Universidad Catholica (Chile) | 1.3 | 106 | 0.04% |
| Network18 Group (India) | 0.3 | 102 | 0.04% |

*continued*

Table 35-9. *continued*

| COMPANY | 2011 OR MOST RECENT POWER INDEX | REVENUE (MIL $) | MARKET SHARE IN BROADCAST AND OTHER VIDEO CHANNELS |
|---|---|---|---|
| TVB (TVBS, Hong Kong) | 0.4 | 98.6 | 0.04% |
| RGE (Israel) | 0.6 | 93.2 | 0.04% |
| Hayat (Egypt) | 0.4 | 67.8 | 0.03% |
| Yomiuri Shimbun (Japan) | 0.0 | 62.5 | 0.02% |
| Panorama (Egypt) | 0.3 | 62.2 | 0.02% |
| Bezeq (Israel) | 0.2 | 50.0 | 0.02% |
| Rotana (Egypt) | 0.2 | 49.5 | 0.02% |
| Altice (France) | 0.08 | 32.7 | 0.01% |
| Channel 2 (Israel) | 0.03 | 32.1 | 0.01% |
| Worldwide Industry HHI Concentration (W-HHI) | | | 199 |
| Worldwide Industry C4 (30-countries) concentration | | | 18.5% |
| Worldwide Industry C10 (30-countries) concentration | | | 36.8% |
| Total Combined Video Industries Revenue (mil US$) | | | 255,015 |

the domestic industry fragmented, though less than in the past.

By revenue as well as company-industry power index, the six leading global companies are American-based, though Sony Pictures is Japanese-owned. Disney, one of the strongest companies, has made numerous high-profile acquisitions, including Miramax (until 2010), Marvel Entertainment, LucasFilm, and Pixar. Universal has repeatedly changed hands across several countries and industries within its recent history; it was owned by the Japanese consumer electronics giant Matsushita, then by the Canadian firm Seagram, and then, between 2000 and 2004, by the French company Vivendi. In 2004, 80% of the company was transferred to GE to form part of NBC Universal, which Comcast then acquired in 2011 and fully controlled by 2013.

The film industry is one of the most internationalized media industries in this study, with the six Hollywood majors operating in every country surveyed. Table 35.11 breaks down the market share of the "big six" in these countries. It must be understood, however, that only part of the majors' revenues and activities are those of production. Most of their distribution is films produced by others, often in collaboration.

For 2012, this is the average market share of the big six in each country: Sony averaged 16.6%, Warner Brothers 15.4%, Disney 14.3%, Universal 12.2%, 21st Century Fox 9.5%, and Paramount 8.4%. The numbers are significantly lower in highly regulated markets (China and South Korea), or ones with strong domestic film industries (Egypt and India).

Historically, Universal and Paramount have often distributed together through their UIP joint venture. In many counties the American firms distribute directly. In others, they go through local distributors. In Portugal, ZON Lusomundo distributes most of the major American studios' films, accounting for 58.5% of the nation's box office in 2009. Even in India, the Hollywood Six have made inroads. Disney partnered with a domestic Indian company, UTV, chosen because it had a higher capitalization

Table 35-10. The World's Major Film Studio/Distributors (First In Their Countries; Or, >20%; Or, in Top Thirty by Revenue)

| COMPANIES | COMPANY POWER INDEX IN WORLD INDUSTRY | 2011 OR MOST RECENT REVENUE (MIL $) | MARKET SHARES BY COUNTRY BY REVENUE (%) | MARKET SHARE OF WORLD INDUSTRY TOTAL REVENUES (%) |
|---|---|---|---|---|
| Disney (US) | 219 | 3,343 | See Table 35.11 | 14.3 |
| Time Warner (Warner Bros., US) | 180 | 4,421 | See Table 35.11 | 15.4 |
| Government of China (China) (China Film Group) | 176 | 1,215 | China (45.0) | 3.9 |
| Sony (Columbia-TriStar, Japan) | 151 | 3,275 | See Table 35.11 | 16.6 |
| 21st Century Fox (Murdoch, US)[a] | 139 | 3,193 | See Table 35.11 | 9.5 |
| Comcast (Universal, US) | 110 | 2,470 | See Table 35.11 | 12.2 |
| Redstone Group (Viacom, Paramount, US) | 102 | 3,346 | See Table 35.11 | 8.4 |
| Toho (Japan) | 84 | 683 | Japan (37.9) | 2.2 |
| CJ Group (South Korea) | 24 | 251 | South Korea (30.1) | 0.8 |
| Bonnier (Sweden) | 23 | 315 | Sweden (22.4) | 1.0 |
| Huayi Brothers (China) | 22 | 549 | China (16.0) | 1.4 |
| Vivendi (StudioCanal, France) | 16 | 724 | France (26.8), Germany (2.8), UK (0.8) | 2.3 |
| Karo Premier (Russia)[b] | 8 | 137 | Russia (18.7) | 0.44 |
| Fininvest (Medusa, Italy) | 7 | 126 | Italy (16.6) | 0.41 |
| ProfMedia (Russia) | 6 | 120 | Russia (16.4) | 0.39 |
| Eros Entertainment (India) | 4 | 83 | India (12.9) | 0.27 |
| Egmont (Denmark) | 4 | 135 | Sweden (9.6) | 0.44 |
| Showbox (South Korea) | 3 | 85 | South Korea (10.2) | 0.27 |
| RAI (Italy) | 3 | 84 | Italy (11.0) | 0.27 |
| Schibsted (Norway) | 3 | 108 | Sweden (7.7) | 0.35 |
| Sun Group (Sun Pictures, India) | 2 | 61 | India (9.1) | 0.2 |
| Baba Arts (India) | 1 | 44 | India (6.5) | 0.14 |
| Grupo Televisa (Mexico) | 1 | 62 | Mexico (6.9) | 0.08 |
| Investment AB Kinnevik (Sweden) | 1 | 45 | Sweden (3.2) | 0.15 |
| Worldwide Industry HHI Concentration (W-HHI) | | | | 1,057 |
| Worldwide Industry (30-Country) C4 Concentration | | | | 47.0 |
| Worldwide Industry (30-Country) C6 Concentration | | | | 76.4 |
| Total Global Film Revenue | | | | $30,710 |

[a] 21st Century Fox spun off from News Corp. in 2013.

[b] Distributes Warner Bros. in Russia, in addition to producing and exporting its own content within the markets of Russia and other former Soviet republics.

Table 35-11. Market Shares For Hollywood Distributors (% Market Share By Box Office Revenues, 2012)[a]

| | WARNER BROS. | PARAMOUNT | 21st CENTURY FOX | DISNEY | SONY | UNIVERSAL[b] | TOTAL % 6 MAJOR PRODUCTIONS |
|---|---|---|---|---|---|---|---|
| Argentina | 13.3 | 10.2 | 8.9 | 22.9 | 6.3 | 10.2 | 71.8 |
| Australia | 21.0 | 13.0 | 16 | 13 | 12 | 11 | 86.0 |
| Belgium | 15.4 | 8.4 | 9.5 | 14.3 | 16.6 | 12.2 | 76.4 |
| Brazil | 12.9 | 19.6 | 18.7 | 10.7 | 10.7 | 5.8 | 78.4 |
| Canada | 19.9 | 13.9 | 13.2 | 11.6 | 13.7 | 4.2 | 76.5 |
| Chile | 12.3 | 9.1 | 8.9 | 13.7 | 6 | 8.1 | 58.1 |
| China | 0.6 | 2.3 | 0.1 | 0.9 | 0.9 | 1.5 | 6.3 |
| Egypt[c] | 6.2 | 3.4 | 3.8 | 5.8 | 6.6 | 4.8 | 30.6 |
| Finland | 15.4 | 8.4 | 9.5 | 14.3 | 16.6 | 12.2 | 76.4 |
| France | 9.4 | 10 | 12.2 | 7.5 | 7.7 | 1.6 | 48.4 |
| Germany | 16.6 | 13.9 | 9.0 | 10.2 | 10.3 | 12.8 | 72.8 |
| India | 3.5 | 2.6 | 4.1 | 0.2 | 3.1 | 2.6 | 16.1 |
| Ireland | 11.0 | 16.9 | 9.4 | 9.9 | 12.5 | 18.5 | 78.2 |
| Israel | 15.4 | 8.4 | 9.5 | 14.3 | 16.6 | 12.2 | 76.4 |
| Italy | 9.6 | 9.9 | 6.4 | 7.7 | 5.1 | 9.9 | 48.6 |
| Japan | 8.4 | 8.4 | 4 | 14.3 | 3.5 | 12.2 | 50.8 |
| Mexico | 16.8 | 14.9 | 12.6 | 20.1 | 9.1 | 7.4 | 80.9 |
| Netherlands | 27.2 | 8.4 | 4.1 | 11.4 | 10.4 | 16.6 | 78.1 |
| Portugal | 11.9 | 11.1 | 14.0 | 14.3 | 12.2 | 11.1 | 74.6 |

| | | | | | | | |
|---|---|---|---|---|---|---|---|
| Russia | 15.2 | 3.1 | 13.5 | 10.0 | 10.0 | 3.1 | 54.9 |
| South Africa | 18.5 | 14.5 | 18.5 | 16.5 | 16.5 | 14.5 | 99.0 |
| South Korea | 5.7 | 2.1 | 8.3 | 8.3 | 9.4 | 2.1 | 35.9 |
| Spain | 11.9 | 8.4 | 14 | 19.9 | 12.2 | 22.2 | 88.6 |
| Sweden | 15.4 | 8.4 | 9.5 | 14.3 | 16.6 | 12.2 | 76.4 |
| Switzerland | 15.4 | 8.4 | 9.5 | 14.3 | 16.6 | 12.2 | 76.4 |
| Taiwan | 15.4 | 8.4 | 9.5 | 14.3 | 16.6 | 12.2 | 76.4 |
| Turkey | 9.6 | 11.1 | 9.5 | 14.3 | 16.6 | 11.1 | 72.2 |
| UK | 11 | 16.9 | 9.4 | 9.9 | 12.5 | 18.5 | 78.2 |
| US | 19.9 | 13.9 | 13.2 | 11.6 | 13.7 | 4.2 | 76.5 |
| World Averages | 15.4 | 8.4 | 9.5 | 14.3 | 16.6 | 12.2 | 76.4 |

[a] In some cases, where a company's national share was unavailable, it was estimated as the global share of the company.

[b] For Argentina, India, Italy, South Africa, South Korea, and Turkey, the market shares for Universal and Paramount is allocated to reflect a 50:50 split of the combined UIP number reported.

[c] For Egypt, the total percentage of Egyptian firms is subtracted from the overall total, and 10% allocated to non-Hollywood imports. The rest is allocated among the Hollywood "Big Six" according to the international breakdown.

than the next two largest indigenous production houses, Sun Pictures and Baba Arts.

The worldwide industry concentration in the film industry is highest of all the audiovisual industries, at 1,057. Even though the Hollywood majors average up to 76.4% market share within individual country markets, none dominates, and rarely has any firm more than 20% or less than 10%.

## TOP COMPANIES: WIRELINE TELECOM

Worldwide industry concentration (W-HHI), measured by the market shares of the top companies, is 308. National wireline concentration is usually high, and often even higher than the numbers show because the incumbents in several countries operate on a regional basis and do not compete in other regions.

National companies have a domestic incumbency advantages because many of them were former (or current) state wireline operators. But they lack many of these incumbency advantages when they enter other countries. Wireline operators that are more internationalized typically possess historical and linguistic ties with the new markets. In wireless, in contrast, the incumbents typically did not receive a monopoly when that industry emerged in the 1980s.

Internationalization of the wireline industry is more common in several of the smaller markets, where established European providers have been able to move in after the privatizations of the 1990s. Deutsche Telekom has a stake in the Netherlands; Telefónica is the largest provider in Latin America, where it dominates the Argentine, Chilean, and Brazilian markets. It is the fourth largest company by revenue and by power index in the world.

AT&T holds a 26.7% market share in the United States and is one of the top five companies by power index (254) and the largest company by revenue ($56.1 billion) and worldwide market share (9.5%) (Table 35.12).[1] This incarnation of the AT&T brand is a consolidation of the regional "Baby Bell" firms Southwestern Bell, Ameritech, Pacific Telesis, and Bell South, as well as the large remnant of the original AT&T that was broken apart in 1984. Verizon holds a 18.6% market share in the United States and revenue of $39 billion (6.6% global). The company formed with the merger of the "Baby Bells" Nynex and Bell Atlantic, as well as of GTE and MCI.

Wireline telecom providers achieve some of the highest revenues and power indices of any companies in this study. Other top companies by power index, revenue, and market share of the wireline global market are NTT (567, $46.2 billion, 7.8%), Deutsche Telekom (278, $29.1 billion, 4.9%), and Telefónica (245, $30.6 billion, 5.2%). The Chinese government has a PI of 448 and revenues of $26.4 billion for 4.5% global wireline market share. In every country, the former monopolist—which usually had been previously state-owned—is still number one as a privatized company in a liberalized market.

Japan's NTT, accounted for 72.3% of Japan's wireline market in 2012, giving it the second highest power index and third highest yearly revenue of any wireline provider. As a former public monopolist (the Japanese government still holds 32% of its stock), NTT maintained an edge over wireline competitors following its privatization partly due to its heavy investment in infrastructure upgrade.

Public ownership still exists in several major markets. Russia's Svyazinvest is the result of the merger of several public companies in 1994, giving it a clear majority over the country's wireline market. China has two regional state-owned firms, China Telecom and China Unicom. India's Bharat Sanchar Nigam has 81% of the market. While its inefficient service has made it an unpopular option, its market share has not changed significantly. What *has* changed is the value of the market: Indian wireline fell from $9.6 billion in 2004 to $6.7 billion in 2011 while mobile revenues surged.

1. AT&T is lower today than in previous years because of its declining share in the United States due to rival entry. In 2004, it had a high company power index of 595, ahead of all other wireline providers in the world. At that time it had a 37.7% U.S. market share.

TeliaSonera is the result of the 2003 merger between the Swedish and Finnish public incumbents. Grupo Carso's Telmex, Mexico's largest provider, has expanded its activities into other Latin American countries such as Brazil and Chile, though it is still far behind Telefónica.

## TOP COMPANIES: WIRELESS TELECOM

Worldwide Industry Concentration (W-HHI) in wireless telecom was 655 in 2013. Because the largest mobile operators have acquired licenses in many countries, their global market shares are much higher than in the wireline industry. On the other hand, their national market shares tend to be lower than those of wireline incumbents.

Generally, the emerging market structure in the wireless mobile telecom industry is that of three or four nationwide companies, sometimes with a few resellers (MVNOs—Mobile Virtual Network Operator). Among the national operators, the incumbent traditional wireline operators, formerly a monopoly, typically have the highest market share, for example, Orange in France, NTT in Japan, Deutsche Telekom in Germany, America Móvil/Claro/Telmex in Mexico, Telefónica in Spain and Argentina, and so on. The number 2 to number 4 firms are often foreign companies, and those are often the national incumbents of other countries, for example, Orange, Telefónica, Deutsche Telekom, and so on. These firms then can leverage their domestic experience, resources, and economies of scale (and the needs of developing markets) into increasingly global footprints. This process is still taking place. At present, there are over a dozen of such firms. Consolidation of the major firms is likely, probably through international partnerships.

By power index, revenue, and share of the industry, by far the largest operator is the Government of China (1,709; $127 billion, 17.1%), of which China Mobile (698, $80.8 billion, 10.9%) would still be the largest single mobile provider by PI and revenue in the world when treated as a separate entity. Following it come several firms with similar size: Verizon (244, $53.7 billion, 7.4%), (AT&T (234, $52.6 billion, 7.1%), NTT (224, $36.2 billion, 4.9%), Vodafone (197, $50.4 billion, 6.8%), Softbank (135, $48.1 billion, 6.5%), Deutsche Telekom/T-Mobile (92.6, $35.7 billion, 4.9%), France Telecom/Orange (104, $26.3 billion, 3.5%), America Móvil/Claro (185, $35.9 billion, 4.9%), and Telefónica (164, $41.3 billion, 5.6%) (Table 35.13).

The wireless telecom industry has been thoroughly internationalized. Vodafone, Deutsche Telekom/T-Mobile, Telefónica, Orange (France Telecom), Telecom Italia, América Móvil/Claro (part of Carlos Slim's conglomerate), KPN, Vimpelcom, and TeliaSonera all operate across borders. Vodafone operates networks in over 30 countries and has partner networks in an additional 40.[2] Telefónica, which has the largest private company power index, provides service in 20 countries,[3] and Deutsche Telekom in over 30 countries.[4] France Telecom/Orange operates in 38 countries,[5] and controls Orascom, which is active in 12 countries across Africa and South Asia.

The industry is also highly concentrated in Mexico, where Grupo Carso's América Móvil/Claro holds more than two-thirds of the market, and in Turkey (via Turkcell). In Mexico, political opposition to America Movil's (Telmex) dominance has been forcing reform on the wireless industry. América Móvil/Claro has market shares in Latin American countries such as Argentina, Mexico, Brazil, and Chile—where it competes primarily with Telefónica.

2. Its markets include the Democratic Republic of Congo, Egypt, Ghana, South Africa, the United States, the United Kingdom, Australia, India, and New Zealand.

3. This includes the Czech Republic, Germany, Ireland, Italy, Slovakia, Spain, the United Kingdom, Puerto Rico, Venezuela, China, Argentina, Brazil, Chile, Colombia, Costa Rica, the Dominican Republic, Ecuador, Guatemala, Panama, and Peru.

4. This includes the United States, Canada, Mexico, Brazil, Argentina, Germany, France, Spain, India, and China.

5. This includes the United States, Russia, India, Vietnam, France, Spain, and the United Kingdom. VimpelCom (Russia) owns Wind (Italy).

Table 35-12. The World's Major Wireline Telecom Companies (First In Their Countries; Or, >20%; Or, in Top Thirty by Revenue)

| COMPANIES | COMPANY-POWER INDEX IN INDUSTRY | 2011 OR MOST RECENT REVENUE (MIL $) | MARKET SHARES (%) | MARKET SHARE OF WORLD TOTAL REVENUES (%) |
|---|---|---|---|---|
| NTT (Japan) (33% public) | 567 | 46,247 | Japan (72.3%) | 7.80% |
| Government of China | 448 | 26,400 | China (100%) | 4.50% |
| *China Telecom (public)* | *154* | *15,470* | *China (58.6%)* | 2.60% |
| *China Unicom (public)* | *41.1* | *7,999* | *China (30.3%)* | 1.40% |
| *China Mobile (public)* | *5.5* | *2,930* | *China (11.1%)* | 0.50% |
| Deutsche Telekom (Germany, 38% public)[a] | 278 | 29,083 | Germany (56.9), Hungary (40%), Netherlands (3.4%) | 4.90% |
| AT&T (US) | 254 | 56,130 | US (26.7%), Mexico (3.5%) | 9.50% |
| Telefónica (Spain)[b] | 245 | 30,583 | Spain (71.9%), Chile (53%), Brazil (23.8%), Germany (3.8%), Argentina (61.1%) | 5.20% |
| Oi + Telemar Participações (Brazil)[c] | 165 | 22,893 | Brazil (42.1%), Portugal (55.3%) | 3.90% |
| Telecom Italia (Italy) | 135 | 12,091 | Italy (66%) | 2.00% |
| Government of Russia (Svyazinvest) (public) | 131 | 9,338 | Russia (82.6%) | 1.60% |
| Verizon (US) | *123* | *39,027* | US (18.6%) | 6.60% |
| Telstra (Australia) | 112 | 9,278 | Australia (71.4%) | 1.60% |
| Telmex (Grupo Carso, Mexico) | 106 | 17,217 | Brazil (27.9%), Chile (5%), Mexico (85%) | 2.90% |
| Korea Telecom (South Korea) | 84.4 | 6,872 | South Korea (72.5%) | 1.20% |
| Bell Canada Enterprises/CTV (Canada) | 83.2 | 9,024 | Canada (54.4%) | 1.50% |
| Turk Telecom (Oger Group, Saudi Arabia) | 82.4 | 5,342 | Turkey (91%) | 0.90% |
| Telkom (South Africa) (public) | 77.3 | 4,563 | South Africa (100%) | 0.80% |
| Bharat Sanchar Nigam (India) (public) | 66.7 | 4,860 | India (81%) | 0.80% |
| France Telecom (France) (27% public) | 63.9 | 9,644 | France (41%), Spain (2.4%) | 1.60% |
| CenturyLink (US) | *47* | *24,130* | US (11.5%) | 47 |

| | | | | |
|---|---|---|---|---|
| TP (public) | 45.1 | 3,628 | Poland (73.3%) | 2.90% |
| BT (UK) | 43.8 | 5,899 | UK (43.8%) | 1.00% |
| Swisscom (Switzerland) (public) | 41.7 | 3,825 | Switzerland (63.1%), Italy (8%) | 0.60% |
| Government of Taiwan (CHT) (public) | 35.7 | 2,611 | Taiwan (80.7%) | 0.40% |
| Belgacom (Belgium) | 34.9 | 3,264 | Belgium (63.1%) | 0.60% |
| Level 3 (US) | *32.1* | *19,933* | US (9.5%) | 3.40% |
| Telecom Egypt (Egypt) (public) | 30.7 | 1,812 | Egypt (100%) | 0.30% |
| TeliaSonera (Sweden/Finland) (50% public) | 28 | 2,931 | Sweden (62.9%), Finland (22%) | 0.50% |
| Altice (France)[c] | 24.8 | 9,397 | Brazil (8.2%), France (22.1%), Israel (14%), Portugal | 1.60% |
| Comcast (US) | *22.8* | *16,786* | US (8%) | *22.8* |
| SingTel (Singapore) (public) | 21.5 | 3,537 | Australia (15%) Ireland (61.3%) | 0.60% |
| KPN (Netherlands) | 20.5 | 2,926 | Netherlands (41.4%) | 0.50% |
| Vodafone (UK) | 19.5 | 9,836 | Germany (13.2%), Ireland (17.4%), Turkey (1.2%), Spain (4.6%), Portugal (4.1%), Italy (9%) | 1.70% |
| Iliad (France) | 19.4 | 5,055 | France (22.6%) | 0.90% |
| Liberty Media (US) | 15.4 | 5,097 | UK (14.6%), Belgium (29.3%), Ireland (5.4%), Switzerland (8.6%), Netherlands (13.7%) | 0.90% |
| Bezeq (Israel) | 14.1 | 1,018 | Israel (82%) | 0.20% |
| Telus (Canada) | 13.5 | 3,633 | Canada (21.9%) | 0.60% |
| United Internet (Germany) | 9.7 | 5,373 | Germany (10.6%) | 0.90% |
| Fintech Group (Mexico) | 6.3 | 957 | Argentina (38.9%) | 0.20% |
| Time Warner Cable (US) | *6* | *8,603* | US (4.1%) | 1.50% |

*continued*

Table 35-12. *continued*

| COMPANIES | COMPANY-POWER INDEX IN INDUSTRY | 2011 OR MOST RECENT REVENUE (MIL $) | MARKET SHARES (%) | MARKET SHARE OF WORLD TOTAL REVENUES (%) |
|---|---|---|---|---|
| Elisa (Finland) | 5.3 | 817 | Finland (38%) | 0.10% |
| Cablevision (US) | 3.9 | 6,924 | US (3.3%) | 1.20% |
| Murdoch (US) | 3.7 | 1,711 | UK (12.7%) | 0.30% |
| Cox (US) | 2.8 | 5,875 | US (2.8%) | 1.00% |
| Bouygues Telecom (France) | 2.5 | 1,812 | France (8.1%) | 0.30% |
| Citizens Telecom (US) | 2.4 | 5,455 | US (2.6%) | 0.90% |
| Sprint-Nextel (Softbank, Japan) | 1 | 3,567 | US (1.7%) | 0.60% |
| MegaFon (Russia) | 0.2 | 407 | Russia (3.6%) | 0.10% |
| Worldwide Industry HHI Concentration (W-HHI) | | | | 308 |
| Worldwide Industry C4 (30-countries) Concentration | | | | 26% |
| Total Global Wireline Revenue (mil US$) | | | | 589,986 |

[a] Operates as Magyar Telekom in Hungary.

[b] Telefónica controlled Telecom Italia through 9.12 ownership in an intermediary consortium. In 2014, these arrangements were being dissolved or in transition.

[c] In 2007 Telemar Participações changed its name to Oi. Portugal Telecom bought 23.4% of Oi's shares in 2010. The remaining shares are split between construction firm Andrade Gutierrez (AG) and telecom firm LF Tel, which each own 4.5% of Oi. Additionally, AG and LF began a joint venture, also called Telemar, which now owns 17.7% stake in Oi. Oi and Telemar merged in 2013 with PT (Portugal) controlling 38% of the venture and Oi (Brazil) the other 62%. However, when PT suffered investment reverses, Oi sold PT's Portugal operations to France's Altice in 2015. TP remained Oi's largest shareholder.

Table 35-13. The World's Major Wireless Telecom Companies (First In Their Countries; Or, >20%; Or, in Top Thirty by Revenue)

| COMPANIES | COMPANY POWER INDEX IN INDUSTRY | 2011 OR MOST RECENT REVENUE (MIL $) | MARKET SHARES (%) | MARKET SHARE OF WORLD TOTAL (%) |
|---|---|---|---|---|
| Government of China | 1,709 | 126,490 | China (100%) | 17.10% |
| *China Mobile (China) (public)* | *698* | *80,827* | *China (63.9%)* | *10.90%* |
| *China Telecom (83% Public)* | *35.4* | *18,215* | *China (14.4%)* | *2.50%* |
| *China Unicom (China) (public)* | *79* | *27,195* | *China (21.5%)* | *3.70%* |
| Verizon (US)[a] | 244 | 53,726 | US (33.6%) | 7.30% |
| AT&T (US) | 234 | 52,607 | US (32.9%) | 7.10% |
| NTT (Japan) (33% public) | 224 | 36,198 | Japan (45.8%) | 4.90% |
| Vodafone (UK)[a] | 197 | 50,402 | South Africa (33.1%), UK (28%), Germany (29.9%), India (23%), Netherlands (30%), Portugal (37.7%), Spain (27.4%), Turkey (28.1%), Australia (15%), Egypt (22.3%), Finland (18%), Ireland (42.5%), Italy (30.2%) | 6.80% |
| América Móvil/Claro (Telmex, Mexico) | 185 | 35,910 | Argentina (35.2%), US (6.2%), Mexico (70.8%), Chile (22%), Brazil (25.1%) | 4.90% |
| Telefónica (Spain)[c, d] | 164 | 41,326 | Spain (39.8%), UK (29%), Chile (41%), Argentinia (30.4%), Brazil (28.6%), Germany (17%), Mexico (22%), Ireland (32.7%) | 5.60% |
| Softbank (Japan) | 135 | 48,106 | US (16.3%), Argentina (3%), Japan (26.8%), Mexico (4%) | 6.50% |
| *Sprint-Nextel* | *57.8* | *26,925* | *US (16.3%), Argentina (3%), Mexico (4%)* | *3.60%* |
| *Softbank Mobile* | *76.7* | *21,181* | *Japan (26.8%)* | *2.90%* |
| Orange (France Telecom, France)[b] (27% Public) | 104 | 26,226 | Switzerland (17.3%), Portugal (3.4%), Egypt (35.1%), UK (17%), Spain (21.1%), Belgium (32%), France (38%) | 3.50% |

*continued*

Table 35-13. *continued*

| COMPANIES | COMPANY POWER INDEX IN INDUSTRY | 2011 OR MOST RECENT REVENUE (MIL $) | MARKET SHARES (%) | MARKET SHARE OF WORLD TOTAL (%) |
|---|---|---|---|---|
| Deutsche Telekom/T-Mobile (Germany) (38% public) | 92.6 | 36,553 | Germany (32.4%), Netherlands (25%), US (12.4%), UK (17%) | 4.90% |
| KDDI (Japan) | 80.2 | 21,656 | Japan (27.4%) | 2.90% |
| Telecom Italia (Italy)[d] | 74.2 | 18,659 | Italy (33.1%), Brazil (27.2%) | 2.50% |
| Telstra (Australia) | 69.8 | 8,909 | Telstra (58%) | 1.20% |
| SK Group (South Korea) | 61.3 | 8,976 | South Korea (50.5%) | 1.20% |
| VimpelCom (Wind, Russia) | 44.9 | 11,850 | Italy (22.3%), Russia (30.7%) | 1.60% |
| MTS (Russia) | 38.4 | 8,120 | Russia (35%) | 1.10% |
| Rogers (Canada) | 36.9 | 7,383 | Canada (37%) | 1.00% |
| Bharti Enterprises (India) | 36.5 | 7,911 | India (34.1%) | 1.10% |
| Altice (France)[f] | 35.2 | 9,137 | France (29.1%), Israel (7.7%) | 1.20% |
| Turkcell (TeliaSonera, Sweden/ Finland (public) + Alfa Group, Russia + Cukurova Group, Turkey) | 33.8 | 4,804 | Turkey (52%) | 0.60% |
| MegaFon (Russia) | 29.5 | 7,122 | Russia (30.7%) | 1.00% |
| Oi + Telemar Participações (Brazil)[f] | 29.4 | 9,724 | Brazil (18.6%), Portugal (39.5%) | 1.30% |
| Korea Telecom (South Korea) | 24 | 5,617 | South Korea (31.6%) | 0.80% |
| Telus (Canada) | 21.9 | 5,687 | Canada (28.5%) | 0.80% |
| Swisscom (public) | 21.1 | 2,582 | Switzerland (60.4%) | 0.30% |
| Bell Canada Enterprises/CTV (Canada) | 19.8 | 5,407 | Canada (27.1%) | 0.70% |

| | | | | |
|---|---|---|---|---|
| KPN (Netherlands) | 16.6 | 3,502 | Netherlands (35%) | 0.50% |
| SingTel (Singapore) (public) | 16.4 | 4,147 | Australia (27%), Ireland (19.1%) | 0.60% |
| MTN (South Africa) | 16.3 | 3,569 | South Africa (33.7%) | 0.50% |
| Belgacom (Belgium) | 15.7 | 2,794 | Belgium (41.7%) | 0.40% |
| E-Plus (Germany) | 15.6 | 5,589 | Germany (20.7%) | 0.80% |
| TeliaSonera (Sweden/Finland) (50% public) | 13.6 | 2,579 | Sweden (40.4%), Finland (36%) | 0.30% |
| PTTK Centertel | 11.5 | 2,548 | Poland (33.4%) | 0.30% |
| Polkomtel | 11.1 | 2,502 | Poland (32.8%) | 0.30% |
| Bouygues Telecom (France) | 10.3 | 4,835 | France (15.8%) | 0.70% |
| Polska Telefonia Cyfrowa | 9.7 | 2,334 | Poland (30.6%) | 0.30% |
| CHT (Taiwan) (public) | 9.3 | 2,108 | Taiwan (32.5%) | 0.30% |
| Oger (Saudi Arabia) | 7.5 | 3,246 | Turkey (20%), South Africa (13.2%) | 0.40% |
| Taiwan Mobile (Taiwan) | 7.4 | 1,887 | Taiwan (29.1%) | 0.30% |
| Hutchison Whampoa (3G, Hong Kong) | 6.4 | 5,133 | UK (9%), Ireland (5.8%), Italy (9.8%) | 0.70% |
| FarEasTone (Taiwan) | 6.2 | 1,725 | Taiwan (26.6%) | 0.20% |
| Iliad (France) | 6.1 | 3,703 | France (12.1%) | 0.50% |
| Kinnevik (Sweden) | 5.7 | 1,331 | Sweden (31.6%) | 0.20% |
| Elisa (Finland) | 5 | 950 | Finland (39%) | 0.10% |
| Cellcom (Israel) | 4 | 949 | Israel (31.5%) | 0.10% |
| Idea Cellular (India) | 3.9 | 2,598 | India (11.2%) | 0.40% |
| Partner Communications (Israel) | 3.5 | 889 | Israel (29.5%) | 0.10% |

*continued*

Table 35-13. *continued*

| COMPANIES | COMPANY POWER INDEX IN INDUSTRY | 2011 OR MOST RECENT REVENUE (MIL $) | MARKET SHARES (%) | MARKET SHARE OF WORLD TOTAL (%) |
|---|---|---|---|---|
| Telecom Egypt (public) (Egypt) | 3.1 | 997 | Egypt (23.1%) | 0.10% |
| Bezeq (Israel) | 2.9 | 808 | Israel (26.8%) | 0.10% |
| Entel (Chile) | 2.8 | 572 | Chile (36%) | 0.10% |
| Sunrise (Denmark) | 2.1 | 821 | Switzerland (19.2%) | 0.10% |
| Televisa (Mexico) | 0.9 | 1,146 | Mexico (5.7%) | 0.20% |
| Worldwide Industry HHI Concentration (W-HHI) | | | | 655 |
| Worldwide Industry C4 (30-countries) Concentration | | | | 38.3% |
| Total Global Wireless Revenue | | | | 740,038 |

[a] Market share and revenues reflect the 2013 sale of Vodafone's 45% stake in Verizon Wireless to Verizon Communications (US).
[b] In 2013 Orange/France Telecom sold off its 20% interest in Sonae, the parent company of Optimus in Portugal.
[c] Telefónica acquired E-Plus in Germany.
[d] Telefónica controlled Telecom Italia through an intermediary consortium until 2014.
[e] Vimpelcom, owned by the Alfa Group of Russia, owns "Wind" in Italy.
[f] Oi and Portugal Telecom merged in 2013, but sold in 2015 PT's Portugal operations to France's Altice.

In the United States, wireless telecom has consolidated from a highly fragmented regional patchwork to four providers with national footprints; industry C4 is 95.2%. Verizon Wireless is the result of mergers bringing together the wireless operations of Bell Atlantic, NYNEX, GTE, US West, PacTel, and Alltel. Vodafone used to own 45% of Verizon Wireless, with the rest held by Verizon. In 2013, Verizon Communications bought out this stake for $130 billion in one of the largest corporate transactions in history. Deutsche Telekom owns T-Mobile in the United States. Softbank bought Vodafone's Japan operations. Its acquisition of Sprint-Nextel's wireless services in 2013 made it the third largest mobile operator in the United States, with 16.3% of the market.

## TOP COMPANIES: INTERNET SERVICE PROVIDERS

By company power index, the top ISP companies are China Telecom (335), Deutsche Telekom (109), Orange (167), and Telefónica (111) (Table 35.14). By revenue, the top ISP companies are AT&T ($8.3 billion), Comcast ($8.2 billion), China Telecom ($10.6 billion), NTT ($7.3 billion), and Telefónica ($4.8 billion). Except for Telefónica, these four operate as ISPs only in their home market. The Government of China is the largest single owner, its control of China Telecom, China Unicom, and China Mobile gives it a huge PI of 1,268 and 12.7% of the ISP market.[6]

In the United States, the major Internet service providers are AT&T, Comcast, Verizon, and Time Warner Cable (TWC). Comcast, in 2015, was poised to acquire TWC. Cable MSOs held a 42.8% share of the US market in 2013, followed by wireline telecoms with 35.4% and mobile providers that had an 18.6% share, net of the local wireline telecom in the same territory. Due to the consolidation in the telecom and cable industries, and the move from dial-up to broadband, the concentration of Internet service providers has increased since 1999.

The Internet service provider industry has been internationalized through companies such as America Movil (Carlos Slim, Mexico), Telefónica (Spain), Orange (France Telecom), Vodafone (United Kingdom), Liberty (United States), and SingTel (Singapore). A legal monopoly exists in South Africa. In Turkey, Turk Telekom is a near-monopoly with a market share of 82.7% due to licensing privileges.

Worldwide Industry Concentration (W-HHI) is 307, only slightly less concentrated than wireline – reflecting the fact that many operators are either cable or telecommunications providers within individual national markets, where their incumbency advantage ensures large shares of the subscriber base.

We also calculate the largest telecom companies in the world by aggregating their wireline, mobile, and ISPs. As with Table 35.9, this includes companies that were too small to make the top firms' industry tables (35.12 for wireline, 35.13 for wireless, and 35.14 for ISPs). The world's combined telecom results are presented in Table 35.15.

The overall size of the telecom sector, (wireline, wireless, and ISP), broadly defined, is a huge $1.5 trillion. China collectively commands 11.6% of the market, of which the largest share, is due to China Mobile (4.9%). The largest private telecom firms by revenues are AT&T (7.8%), Verizon (6.5%), NTT (6%), Telefónica (5.1%), Deutsche Telekom/T-Mobile (4.7%), Vodafone (4.2%), Softbank and Grupo Carso (both 3.7%), and Orange (2.9%).

The results of Table 35.15 mirror those of the company rankings in the wireline and wireless industries. It is AT&T's sizable share in platforms industries in the US market that puts it well ahead of China Mobile in revenues, and accounts for Verizon and NTT's places as well (NTT is dominant in Japan by a wide margin). The more internationalized companies (Softbank, Deutsche Telekom, Orange, Telefónica, Oi + Telemar, SingTel, Telecom Italia, Vivendi, and Claro/Móvil) also rank highly.

6. Minor ISP shares are held by two smaller state enterprises as well: China Railcom and China Satcom.

Table 35-14. The World's Major Internet Service Providers (First In Their Countries; Or, >20%; Or, in Top Thirty by Revenue)

| COMPANIES | COMPANY POWER INDEX IN WORLD INDUSTRY | 2011 OR MOST RECENT REVENUE (MIL $) | MARKET SHARES BY COUNTRY BY REVENUE (%) | MARKET SHARE OF WORLD INDUSTRY TOTAL REVENUES (%) |
|---|---|---|---|---|
| Government of China | 1,268 | 20,540 | China (100%) | 12.70% |
| *China Telecom (China) (public)* | *335* | *10,558* | *China (51.4%)* | *6.50%* |
| *China Unicom (China) (public)* | *169* | *7,497* | *China (36.5%)* | *4.60%* |
| *China Mobile (China) (public)* | *2* | *822* | *China (4%)* | *0.50%* |
| Orange (France Telecom, France) (27% public) | 167 | 7,038 | France (45%), Spain (6%), Egypt (22.5%), UK (3%) | 4.30% |
| NTT (Japan) (33% public) | 149 | 7,272 | Japan (33.1%) | 4.50% |
| Telefónica (Spain) | 111 | 4,808 | Spain (43%), Chile (43%), Argentina (31.2%), Germany (3.4%), Brazil (27.7%) | 3.00% |
| Deutsche Telekom (Germany) (38% public) | 109 | 3,858 | Germany (45.6%) | 2.40% |
| AT&T (US) | 92.8 | 8,265 | US (18.2%) | 5.10% |
| Comcast (US) | 91.8 | 8,220 | US (18.1%) | 5.10% |
| British Telecom (UK) | 62.5 | 3,423 | UK (29.6%) | 2.10% |
| Altice (France)[a] | 57.4 | 3,477 | France (27%), Israel (9%) | 2.10% |
| Telmex (Grupo Carso, Mexico) | 43.4 | 1,595 | Mexico (29.2%), Brazil (23.8%), Chile (7%) | 1.00% |
| Korea Telecom (South Korea) | 43 | 1,601 | South Korea (43.5%) | 1.00% |
| Iliad (France) | 41.5 | 2,925 | France (23%) | 1.80% |
| SingTel (Singapore) (public) | 34.8 | 1,345 | Australia (11.8%), Ireland (41%) | 0.80% |
| KPN (Netherlands) | 28.7 | 1,091 | Netherlands (42.6%), Belgium (3.2%) | 0.70% |
| Telstra (Australia) | 28.2 | 1,081 | Australia (42.2%) | 0.70% |
| Time Warner Cable (US) | 26.9 | 4,451 | US (9.8%) | 2.70% |
| Murdoch Group (US) | 26.9 | 2,243 | UK (19.4%) | 1.40% |

| | | | | |
|---|---|---|---|---|
| CHT (Taiwan) (public) | 26.1 | 622 | Taiwan (67.9%) | 0.40% |
| Telkom (public) (South Africa) | 25.9 | 419 | South Africa (100%) | 0.30% |
| Japan Telecom (Softbank, Japan) | 24.3 | 2,944 | Japan (13.4%) | 1.80% |
| Oi-Telemar (Brazil/Portugal)a | 23.8 | 799 | Portugal (48.2%) | 0.50% |
| Verizon (US) | 22.7 | 4,087 | US (9%) | 2.50% |
| Turk Telekom (Oger Group, Saudi Arabia) | 22.4 | 439 | Turkey (82.7%) | 0.30% |
| Talk-Talk (UK) | 20.6 | 1,966 | UK (17%) | 1.20% |
| Bell Canada Enterprises/CTV (Canada) | 20.2 | 1,544 | Canada (21.2%) | 1.00% |
| Vodafone (UK) | 19.1 | 2,128 | Germany (17.1%), Egypt (8%), Portugal (4.2%), Spain (5.9%), Ireland (22%) | 1.30% |
| TeliaSonera (Sweden/Finland) (50% public) | 17.1 | 781 | Sweden (37.2%), Finland (32%) | 0.50% |
| Swisscom (Switzerland) (public) | 16.3 | 493 | Switzerland (53.7%) | 0.30% |
| Telecom Egypt (public) (Egypt) | 15 | 379 | Egypt (64.2%) | 0.20% |
| Belgacom (Belgium) | 12.3 | 424 | Belgium (47.1%) | 0.30% |
| SK Group (South Korea) | 12.3 | 858 | South Korea (23.3%) | 0.50% |
| ZON (Portugal) | 11.1 | 545 | Portugal (32.9%) | 0.30% |
| ONO (Spain) | 10.1 | 1,088 | Spain (15.1%) | 0.70% |
| Ziggo (Netherlands) | 8.7 | 599 | Netherlands (23.4%) | 0.40% |
| NMG (Russia) | 7.3 | 371 | Russia (31.9%) | 0.20% |
| Akado (Russia) | 7 | 363 | Russia (31.2%) | 0.20% |
| Elisa (Finland) | 5.8 | 276 | Finland (34%) | 0.20% |
| Liberty (US) | 5.2 | 586 | Netherlands (13.8%), Switzerland (17.1%), Ireland (12%) | 0.40% |
| Bezeq (Israel) | 3.9 | 176 | Israel (36%) | 0.10% |

*continued*

Table 35-14. *continued*

| COMPANIES | COMPANY POWER INDEX IN WORLD INDUSTRY | 2011 OR MOST RECENT REVENUE (MIL $) | MARKET SHARES BY COUNTRY BY REVENUE (%) | MARKET SHARE OF WORLD INDUSTRY TOTAL REVENUES (%) |
|---|---|---|---|---|
| Telenor (Norway) | 3.4 | 279 | Sweden (19.9%) | 0.20% |
| LinkDotNet (Egypt) | 2.9 | 165 | Egypt (28%) | 0.10% |
| 013 Netvision (Israel) | 2.2 | 132 | Israel (27%) | 0.10% |
| Partner Communications (Israel) | 2.2 | 132 | Israel (27%) | 0.10% |
| Reliance (India) | 1.9 | 262 | India (11.9%) | 0.20% |
| Fintech (Mexico) | 1.9 | 100 | Argentina (29.9%) | 0.10% |
| Grupo Clarín (Argentina) | 1.2 | 82.3 | Argentina (24.5%) | 0.10% |
| Globo Group (Brazil) | 0.1 | 67.1 | Brazil (1.7%) | 0.04% |
| MegaFon (Russia) | 0.1 | 52.3 | Russia (4.5%) | 0.00% |
| Worldwide Industry HHI Concentration (W-HHI) | | | | 443 |
| Worldwide Industry C4 (30-countries) Concentration | | | | 34.5% |
| Total Global ISP Revenue | | | | 162,033 |

[a] In 2007 Telemar Participações changed its name to Oi. Portugal Telecom bought 23.4% of Oi's shares in 2010. The remaining shares are split between construction firm Andrade Gutierrez (AG) and telecom firm LF Tel, which each own 4.5% of Oi. Additionally, AG and LF began a joint venture, also called Telemar, which now owns 17.7% stake in Oi. Oi and Telemar merged in 2013, with PT (Portugal) controlling 38% of the venture and Oi (Brazil) the other 62%. In 2015, Oi sold PT's Portugal operations to Altice.

Table 35-15. Market Share of The World's Major Telecom Companies (Wireline, Mobile, and ISP)

| COMPANY | TOTAL COMBINED TELECOM POWER INDEX | TOTAL COMBINED TELECOM REVENUE | % MARKET SHARE OF TELECOM INDUSTRIES (30 COUNTRIES) BY REVENUE |
|---|---|---|---|
| Government of China | 1,162 | 173,430 | 11.6% |
| *China Mobile (China) (public)* | *346* | *81,649* | *5.5%* |
| *China Telecom (China) (public)* | *115* | *44,243* | *3.0%* |
| *China Unicom (China) (public)* | *73.8* | *42,692* | *2.9%* |
| AT&T (US) | 226 | 117,002 | 7.8% |
| Verizon (US) | 172 | 96,841 | 6.5% |
| NTT (Japan) (33% public) | 351 | 89,716 | 6.0% |
| Telefónica (Spain)[1] | 190 | 76,717 | 5.1% |
| Deutsche Telekom (Germany) (38% public) | 168 | 69,494 | 4.7% |
| Vodafone (UK) | 107 | 62,366 | 4.2% |
| Grupo Carso (Mexico) | 139 | 54,722 | 3.7% |
| Softbank (Japan) | 67.1 | 54,617 | 3.7% |
| Orange (France Telecom, France, 27% public) | 94.8 | 42,907 | 2.9% |
| Oi Telemar (Brazil/Portugal) | 79.8 | 32,618 | 2.2% |
| Telecom Italia (Italy) | 90.3 | 30,750 | 2.1% |
| KDDI (Japan) | 45.8 | 30,471 | 2.0% |
| Comcast (US) | 19.0 | 25,006 | 1.7% |
| CenturyLink (US) | 18.6 | 24,130 | 1.6% |
| Altice (France) | 33.5 | 22,011 | 1.5% |
| Level 3 (US) | 12.7 | 19,933 | 1.3% |
| Telstra (Australia) | 82.1 | 19,268 | 1.3% |
| Bell Canada Enterprises/CTV (Canada) | 44.9 | 15,975 | 1.1% |
| Vimpelcom (Russia) | 24.5 | 14,415 | 1.0% |
| Korea Telecom (South Korea) | 49.9 | 14,090 | 0.9% |
| Time Warner Cable (US) | 5.3 | 13,053 | 0.9% |
| Iliad (France) | 15.2 | 11,682 | 0.8% |
| Telus (Canada) | 16.6 | 10,019 | 0.7% |
| SK Group (South Korea) | 31.7 | 9,834 | 0.7% |
| Government of Russia (Svyazinvest) | 51.7 | 9,647 | 0.6% |
| SingTel (Singapore) (public) | 20.4 | 9,543 | 0.6% |
| BT (UK) | 24.1 | 9,322 | 0.6% |
| Oger Group (Saudi Arabia) | 38.7 | 9,027 | 0.6% |
| Rogers (Canada) | 19.2 | 8,805 | 0.6% |
| Bharat Sanchar Nigam (India) (public) | 35.1 | 8,793 | 0.6% |
| Swisscom (Switzerland) (public) | 28.7 | 8,253 | 0.6% |
| MTS (Russia) | 19.0 | 8,120 | 0.5% |
| MegaFon (Russia) | 14.7 | 7,582 | 0.5% |

*continued*

Table 35-15. *continued*

| COMPANY | TOTAL COMBINED TELECOM POWER INDEX | TOTAL COMBINED TELECOM REVENUE | % MARKET SHARE OF TELECOM INDUSTRIES (30 COUNTRIES) BY REVENUE |
|---|---|---|---|
| KPN (Netherlands) | 19.5 | 7,520 | 0.5% |
| Cablevision (US) | 1.5 | 6,924 | 0.5% |
| Bouygues (France) | 6.1 | 6,774 | 0.5% |
| Belgacom (Belgium) | 23.1 | 6,482 | 0.4% |
| United Internet 1&1 (Germany) | 4.6 | 6,371 | 0.4% |
| TeliaSonera (Sweden/Finland) (50% public) | 20.7 | 6,291 | 0.4% |
| Cox (US) | 1.1 | 5,875 | 0.4% |
| Liberty (US) | 6.7 | 5,682 | 0.4% |
| E-Plus (Germany) | 7.7 | 5,589 | 0.4% |
| Citizens Telecom (US) | 0.9 | 5,455 | 0.4% |
| Hutchison Whampoa (Hong Kong) | 3.2 | 5,133 | 0.3% |
| Telkom (South Africa) (public) | 33.4 | 4,982 | 0.3% |
| Turkcell (TeliaSonera, Sweden/Finland (public) + Alfa Group, Russia + Cukurova Group, Turkey) | 16.8 | 4,804 | 0.3% |
| Reliance (India) | 3.8 | 3,976 | 0.3% |
| Murdoch Group (US) | 4.4 | 3,954 | 0.3% |
| TP (Poland) (public) | 17.8 | 3,628 | 0.2% |
| MTN (South Africa) | 8.1 | 3,569 | 0.2% |
| US Cellular (US) | 0.5 | 3,518 | 0.2% |
| Kinnevik (Tele2, Sweden) | 3.8 | 3,450 | 0.2% |
| Alice (Germany) | 1.4 | 3,431 | 0.2% |
| Government of Egypt (Telecom Egypt) | 15.3 | 3,189 | 0.2% |
| ONO (Spain) | 2.4 | 3,080 | 0.2% |
| Charter (US) | 0.3 | 2,938 | 0.2% |
| Government of Taiwan (CHT) | 7.4 | 2,730 | 0.2% |
| Idea Cellular (India) | 1.9 | 2,598 | 0.2% |
| PTTK Centertel (Poland) | 5.7 | 2,548 | 0.2% |
| Vonage (US) | 0.2 | 2,518 | 0.2% |
| Polkomtel (Poland) | 5.5 | 2,502 | 0.2% |
| Polska Telefonia Cyfrowa (Poland) | 4.8 | 2,334 | 0.2% |
| Taiwan Mobile (Taiwan) | 3.8 | 2,112 | 0.1% |
| Elisa (Finland) | 5.2 | 2,044 | 0.1% |
| Bezeq (Israel) | 7.4 | 2,001 | 0.1% |
| Talk-Talk (UK) | 2.2 | 1,966 | 0.1% |
| BIGLOBE (Japan) | 1.0 | 1,845 | 0.1% |
| FarEasTone (Taiwan) | 3.1 | 1,725 | 0.1% |
| JazzTel (UK) | 1.0 | 1,666 | 0.1% |

Table 35-15. *continued*

| COMPANY | TOTAL COMBINED TELECOM POWER INDEX | TOTAL COMBINED TELECOM REVENUE | % MARKET SHARE OF TELECOM INDUSTRIES (30 COUNTRIES) BY REVENUE |
|---|---|---|---|
| Fintech (Mexico) | 3.9 | 1,650 | 0.1% |
| Ziggo (Netherlands) | 1.8 | 1,575 | 0.1% |
| MTS Allstream (Canada) | 0.5 | 1,440 | 0.1% |
| Shaw (Canada) | 1.0 | 1,431 | 0.1% |
| K-Opticom (Japan) | 0.2 | 1,407 | 0.1% |
| Optimus (Portugal) | 1.8 | 1,378 | 0.1% |
| Quebecor (Videotron, Canada) | 0.5 | 1,274 | 0.1% |
| Versatel (US) | 0.2 | 1,259 | 0.1% |
| Televisa (Mexico) | 0.5 | 1,252 | 0.1% |
| Partner Communications (Israel) | 2.0 | 1,058 | 0.1% |
| DNA (Finland) | 1.4 | 1,048 | 0.1% |
| SaskTel (Canada) | 0.1 | 1,017 | 0.1% |
| Cellcom (Israel) | 2.0 | 949 | 0.1% |
| Yogio (Spain) | 0.2 | 910 | 0.1% |
| NCIT (Taiwan) | 0.5 | 909 | 0.1% |
| Skype (US) | 0.0 | 839 | 0.1% |
| Sunrise (Denmark) | 1.0 | 821 | 0.1% |
| ZON (Portugal) | 1.5 | 817 | 0.1% |
| NII (US-Mexico) | 0.2 | 804 | 0.1% |
| Portugal Telecom (Portugal) | 2.6 | 799 | 0.1% |
| Bharti Enterprises (India) | 18.6 | 755 | 0.1% |
| Tata Group (India) | 0.1 | 699 | 0.05% |
| Folhapar (Brazil) | 0.7 | 655 | 0.04% |
| Entel (Chile) | 1.4 | 628 | 0.04% |
| Windsteam (US) | 0.1 | 590 | 0.04% |
| Cogeco (Canada) | 0.1 | 536 | 0.04% |
| Euskaltel (Spain) | 0.1 | 533 | 0.04% |
| Hi3G (Sweden) | 0.3 | 512 | 0.03% |
| Essent Kabelcom (Netherlands) | 0.6 | 492 | 0.03% |
| Bragg (Canada) | 0.1 | 471 | 0.03% |
| Mediacom (US) | 0.0 | 452 | 0.03% |
| R Cable (Spain) | 0.1 | 429 | 0.03% |
| FreeNet (Germany) | 0.1 | 389 | 0.03% |
| NMG (Russia) | 0.8 | 371 | 0.02% |
| Covad (US) | 0.0 | 363 | 0.02% |
| Akado (Russia) | 0.8 | 363 | 0.02% |
| Primus (US) | 0.1 | 344 | 0.02% |
| Soul Pattinson (Australia) | 0.0 | 299 | 0.02% |

*continued*

Table 35-15. *continued*

| COMPANY | TOTAL COMBINED TELECOM POWER INDEX | TOTAL COMBINED TELECOM REVENUE | % MARKET SHARE OF TELECOM INDUSTRIES (30 COUNTRIES) BY REVENUE |
|---|---|---|---|
| DirecTV (US) | 0.5 | 293 | 0.02% |
| Telenor (Norway) | 0.4 | 279 | 0.02% |
| iiNet (Australia) | 0.2 | 246 | 0.02% |
| Cabovisão (Portugal) | 0.1 | 244 | 0.02% |
| Telecom NZ (New Zealand) | 0.0 | 208 | 0.01% |
| LinkDotNet (Egypt) | 0.3 | 165 | 0.01% |
| 013 Netvision (Israel) | 0.2 | 132 | 0.01% |
| Grupo Clarín (Argentina) | 0.1 | 82.3 | 0.01% |
| Globo Group (Brazil) | 0.0 | 67.1 | 0.004% |
| Finnet (Finland) | 0.7 | 64.9 | 0.004% |
| Worldwide Industry HHI Concentration (W-HHI) | | | 276 |
| Worldwide Industry C4 (30-countries) concentration | | | 32.0% |
| Worldwide Industry C10 (30-countries) concentration | | | 56.2% |
| Total Combined Telecom Revenue (mil US$) | | | 1,492,057 |

[1]Includes Telefónica's 9.12% stake in Telecom Italia's holding company.

## TOP COMPANIES: SEARCH ENGINES

The search engines industry is highly concentrated. This is due to the ubiquitously large market shares of Google (United States), which has a staggeringly high company power index of 4,943 and controls 64.8% of the entire industry worldwide (Table 35.16). The next three top companies are Baidu (China) with a power index of 525, Yahoo (United States) with 98 and Bing (Microsoft, US) with 92.8. The top companies by search engine revenue include Google's $26.8 billion, Yahoo, a distant second with $3 billion, Baidu with $2.8 billion, and Microsoft with $2.5 billion. Yahoo's share of the world market is declining. The share of Bing, Microsoft's successor to MSN, is increasing.

Google holds a major market share, often over 90%, in almost every country surveyed here. In the United States, Yahoo, Excite, Altavista, and Infoseek originally led the search engine industry, each for a while, but Google, using a rank-based search algorithm, emerged as the most popular and successful search engine. Although there are metasearch and specialized engines for jobs, news, pictures, and so on, Google dominates the search engine industry because of its comprehensive indexing system. From 2006 to 2012, Google raised its US market share from 49.6% to 68.4%. Microsoft unsuccessfully tried to buy Yahoo for $44.6 billion in 2008, but was rebuffed by Yahoo's then-CEO and cofounder, Jerry Yang, and the deal was withdrawn in 2009. Since 2005, AOL has had its Search powered by Google.

Table 35-16. The World's Major Search Engines (First In Their Countries; Or, >20%; Or, in Top Thirty by Revenue)

| COMPANIES | COMPANY POWER INDEX IN WORLD INDUSTRY | 2011 OR MOST RECENT REVENUE (MIL $) | MARKET SHARES (%) | MARKET SHARE OF WORLD INDUSTRY TOTAL REVENUES (%) |
|---|---|---|---|---|
| Google (US) | 4,943 | 26,747 | Chile (96.6%), Finland (98.1%), Mexico (93.4%), Poland (98.1%), US (68.4%), Argentina (96.6%), Belgium (90%), Brazil (96.2%), Canada (82.4%), Egypt (97%), Ireland (94%), Israel (96.7%), Netherlands (93%), Portugal (97.4%), South Africa (91.8%), Spain (96.2%), Sweden (96.7%), Switzerland (89.9%), Turkey (98%), UK (90.7%), China (15.6%), France (94%), Germany (90.5%), Japan (34.9%), Australia (92.5%), Italy (47%), South Korea (1.7%), Taiwan (35.7%), India (81.4%), Russia (14.2%) | 64.80% |
| Baidu (China) | 525 | 2,767 | China (78.6%) Taiwan (11.5%) | 6.70% |
| NHN (South Korea) | 128 | 770 | South Korea (68.8%) | 1.90% |
| Yahoo (US) [a] | 98.3 | 3,035 | India (9.4%), UK (2.9%), South Africa (3%), US (13%), Italy (6%), Brazil (1.5%), Canada (4.2%), France (2.5%), Israel (1.1%), Taiwan (39.7%), Australia (1.3%), South Korea (1.2%), Japan (17.3%) | 7.30% |
| Microsoft (Bing, US) | 92.8 | 2,484 | Argentina (6.7%), India (1.7%), Israel (1.5%), South Africa (3.3%), US (16%), UK (3.9%) | 6.00% |
| Yandex.ru (Russia) | 50.2 | 440 | 47.1 | 1.10% |
| Softbank (Japan) [a] | 38.4 | 814 | Japan (19.5%) | 2.00% |
| Mail.ru (Russia) | 27.4 | 325 | Russia (34.8%) | 0.80% |
| Daum (South Korea) | 13.2 | 247 | South Korea (22.1%) | 0.60% |
| Sohu.com (China) | 0.8 | 109 | China (3.1%) | 0.30% |

*continued*

Table 35-16. *continued*

| COMPANIES | COMPANY POWER INDEX IN WORLD INDUSTRY | 2011 OR MOST RECENT REVENUE (MIL $) | MARKET SHARES (%) | MARKET SHARE OF WORLD INDUSTRY TOTAL REVENUES (%) |
|---|---|---|---|---|
| Rakuten (Japan) | 0.4 | 88 | Japan (2.1%) | 0.20% |
| ProfMedia (Russia) | 0.3 | 36 | Russia (3.9%) | 0.10% |
| Deutsche Telekom (Germany) | 0.3 | 64 | Germany (2.1%) | 0.20% |
| SK Group (South Korea) | 0.2 | 27 | South Korea (2.4%) | 0.10% |
| Nine Entertainment (Australia) | 0.1 | 16 | Australia (1.7%) | 0.00% |
| AOL (US) | 0.1 | 15 | US (1.4%) | 0.00% |
| Worldwide Industry HHI Concentration (W-HHI) | | | | 4,339 |
| Worldwide Industry C4 (30-countries) Concentration | | | | 84.8% |
| Total Global Search Engine Revenue (mil US$) | | | | 41,300 |

[a] Yahoo Japan is owned by Softbank and Yahoo. Its revenue and market shares have been proportionally allocated to both companies.

In 2009, a similar deal took place, with Yahoo search powered by Microsoft's Bing. Bing has slightly increased its share of the US market—as well as the much larger Chinese market.

Of non-US engines, only Baidu and the Russian engines have significant holdings across several countries: Baidu in Taiwan, and the Russian companies in most of the former Soviet republics. Google is not dominant in several East Asian countries largely due to government policies. In China, Google content can only be accessed indirectly because of state censorship regulations. In South Korea, lawsuits by domestic companies such as NHN with governmental encouragement, and the denial of access to governmental information have restricted Google.

In Russia, Yandex.ru holds the largest market share because its algorithms, it is claimed, are better suited to indexing Russian-language content. In several European countries, new and established firms have attempted to compete with Google and Bing. In Portugal, Aeiou (aeiou.pt); in France, Voila (voila.fr); in Germany, Fireball (fireball.de); in The Netherlands and Belgium, Vinden (vinden.be/vinden.nl). However, it is difficult for these companies to reach the level of Google or Bing because of lack of resources and technology. Economies of scale had an impact.

## TOP COMPANIES: ONLINE NEWS MEDIA

The online news industry is fairly new, fast moving, and lacking in reliable data. Even the definition of what constitutes this industry is unclear. Hard news is intermingled with commentary, gossip, personal observations, and commercial messages. Some is provided with a profit motive, and much is offered to be part of a community, to influence or simply get attention.

Established news firms typically dominate national markets. Daily newspapers and broadcast companies are the main providers. There are also other, emerging operations that are not extensions of established news organizations.

The companies with the highest power indices in this industry, based on the reports of the various country teams of this study, is Yahoo (116), followed by Folhaper (Brazil) with a PI of 89, BBC with 82.7, and Grupo Espresso in Italy with 73.9 (Table 35.17). The companies with the largest revenue are the American firms Yahoo with $1.2 billion, the Murdoch Group with $894 million, Time Warner with $742 million, AOL with $703 million, and Comcast with $653 million. Yahoo accounts for 7.7% of the world online news market, giving it the largest share of any company, by revenues.

Yahoo News is partnered with Disney's ABC network. It has a 15.9% share of the Japanese market, due to Softbank's partial ownership of Yahoo Japan. News Corp. holds significant shares of American and British markets, partly a result of its presence in print media in both countries. The BBC and the Guardian have international followings, as does CNN (Time Warner).

AOL owns both AOL News and The Huffington Post. AOL itself was acquired in 2015 by Verizon. Comcast's share comes partly from several NBC sites, including MSNBC's online content. Other major US websites include *USA Today* (Gannett). *The Wall Street Journal, New York Times,* and *Financial Times* (Pearson) are subscription-based. The main US broadcasters (ABC, NBC, CBS, and Fox) all have free online news portals as well.

Sina and Sohu are two Chinese companies with high market shares of 38.9% and 32.4%, respectively, in a large national market. Sina has more than three billion page views daily. Globo leads the expanding Brazilian online news market (with $230 million in the industry). In the United Kingdom, the BBC is a publicly owned company that dominates the domestic market (32%) with a power index of 82.7 and annual revenue of $722 million. The Daily Mail and The Guardian have low power indices (8.1 and 4.9, respectively), but relatively high revenues of $386 million and $251 million, including from the United States. Dogan Media Group (DMG) dominates the Turkish market with a market share of 52%. Regulatory restrictions on content impede the development of DMG's competitors.

There is a high interconnectedness between online news and print media in several countries. In the United Kingdom, Murdoch has the highest market share in online news media partially due to its strength in print and video media. It is

Table 35-17. The World's Major Online News Companies (First In Their Countries; Or, >20%; Or, in Top Thirty by Revenue)

| COMPANIES | COMPANY POWER INDEX IN WORLD INDUSTRY | 2011 OR MOST RECENT REVENUE (MIL $) | MARKET SHARES (%) | MARKET SHARE OF WORLD TOTAL INDUSTRY (%) |
|---|---|---|---|---|
| Yahoo (US) [a] | 116 | 1,523 | India (32.7%), Japan (15.9%), US (11%), France (9%) | 7.70% |
| Folhapar (Brazil) | 89 | 552 | Brazil (31.9%) | 2.80% |
| BBC (public, UK) | 82.7 | 722 | UK (32%), India (5.7%), US (2.3%) | 3.70% |
| Gruppo Espresso (Italy) | 73.9 | 358 | Italy (40.8%) | 1.80% |
| Sina.com (China) | 70.8 | 360 | China (38.9%) | 1.80% |
| Sohu.com (China) | 49.1 | 300 | China (32.4%) | 1.50% |
| Murdoch Group (US) | 48.3 | 894 | UK (18%), US (7.4%) | 4.50% |
| Telefónica (Spain) | 39.3 | 367 | Brazil (21.2%) | 1.90% |
| Time Warner (CNN) (US) | 33 | 742 | US (8.8%), Canada (6.1%) | 3.80% |
| AOL/Verizon (US) | 29.9 | 703 | US (8.4%) | 3.60% |
| Dogan Group (Turkey) | 27.6 | 105 | Turkey (52%) | 0.50% |
| Comcast (US) | 25.8 | 653 | US (7.8%) | 3.30% |
| RCS (Italy) (Rizzoli) | 23.7 | 269 | Spain (18.1%), Italy (16.8%) | 1.40% |
| PRISA (Spain) | 23.2 | 175 | Spain (26.1%) | 0.90% |
| Softbank (Japan) | 21.1 | 275 | Japan (15.1%) | 1.40% |
| BCCL (The Times of India Group) (India) | 19.9 | 165.9 | India (23.7%) | 0.80% |
| 163.com (China) | 17.1 | 177 | China (19.1%) | 0.90% |
| Axel Springer (Germany) | 17.1 | 117 | Germany (30.2%), Switzerland (6.2%) | 0.60% |
| Gannett (US) | 16.3 | 519 | US (6.2%) | 2.60% |
| Globo (Brazil) | 15.5 | 230 | Brazil (13.3%) | 1.20% |
| RCB.ru (Russia) | 15 | 166 | 17.8 | 0.80% |
| The New York Times (US) | 14.4 | 477 | US (4.7%), India (11.9%) | 2.40% |
| Telemar Participações (with Oi) (Brazil) | 12.8 | 209 | Brazil (12.1%) | 1.10% |

| | | | | |
|---|---|---|---|---|
| Socpresse (France) | 10.7 | 141 | France (15%) | 0.70% |
| The Washington Post (US) | 8.2 | 368 | US (4.4%) | 1.90% |
| Daily Mail and General Trust (UK) | 8.1 | 386 | UK (), US (3.5%) | 2.00% |
| Naspers (South Africa) | 7.2 | 43.7 | South Africa (32.6%) | 0.20% |
| Planeta (Spain) | 7.2 | 98.1 | Spain (14.6%) | 0.50% |
| El Mercurio (Chile) | 6.8 | 24.9 | Chile (54.2%) | 0.10% |
| Tribune (US) | 6.8 | 335 | US (4%) | 1.70% |
| Disney (US) | 5.2 | 293 | US (3.5%) | 1.50% |
| ITV (UK) | 5 | 122 | UK (8%) | 0.60% |
| Guardian Media Group (UK) | 4.9 | 241 | UK (7%), US (1.6%) | 1.20% |
| Microsoft (MSN, US) | 4.1 | 33.9 | South Africa (24.5%), Israel (2.9%) | 0.20% |
| McClatchy (US) | 4.1 | 259 | US (3.1%) | 1.30% |
| Tamedia (Switzerland) | 3.9 | 29.2 | Switzerland (26.5%) | 0.10% |
| Pearson (UK) | 3.8 | 107 | UK (7%) | 0.50% |
| Advance (US) | 3.8 | 251 | US (3%) | 1.30% |
| Burda (Germany) | 2.7 | 43.9 | Germany (12%) | 0.20% |
| United Daily News Group (Taiwan) | 2.7 | 18.3 | Taiwan (29.1%) | 0.10% |
| HT Media (India) | 2.4 | 57.4 | India (8.8%) | 0.30% |
| COPESA (Chile) | 2 | 13.6 | Chile (29.6%) | 0.10% |
| Yomiuri Shimbun (Japan) | 1.6 | 76.4 | Japan (4.2%) | 0.40% |
| Impresa (Portugal) | 1.6 | 22 | Portugal (14.21%) | 0.10% |
| Controlinveste (Portugal) | 1.5 | 21.2 | Portugal (13.7%) | 0.10% |
| Fininvest (Italy) | 1.4 | 63.9 | Spain (5.2%), Italy (3.3%) | 0.30% |

*continued*

Table 35-17. *continued*

| COMPANIES | COMPANY POWER INDEX IN WORLD INDUSTRY | 2011 OR MOST RECENT REVENUE (MIL $) | MARKET SHARES (%) | MARKET SHARE OF WORLD TOTAL INDUSTRY (%) |
|---|---|---|---|---|
| Now News (Taiwan) | 1.3 | 12.9 | Taiwan (20.4%) | 0.10% |
| NBS (Japan) | 1.2 | 65.5 | Japan (3.6%) | 0.30% |
| Next Media (Hong Kong) | 0.9 | 10.5 | Taiwan (16.6%) | 0.10% |
| Ringier (Switzerland) | 0.8 | 13.5 | Switzerland (12.3%) | 0.10% |
| Bertelsmann (Germany) | 0.7 | 22.3 | Germany (6.1%) | 0.10% |
| Bezeq (Walla, Israel) | 0.7 | 7.3 | Israel (19.8%) | 0.00% |
| El Universal (Mexico) | 0.7 | 8.7 | Mexico (15.8%) | 0.00% |
| Ciner Group (Turkey) | 0.6 | 15.6 | Turkey (7.7%) | 0.10% |
| Worldwide Industry HHI Concentration (W-HHI) | | | | 193 |
| Worldwide Industry C4 (30 Countries) Concentration | | | | 19.6% |
| Total Global Online News Revenue | | | | 19,760 |

[a] Yahoo Japan is equally owned by Softbank and Yahoo. Its revenue and market shares have been allocated to both companies in equal measure.

the same in Germany where Axel Springer AG controls 18.8% of the newspaper market, 12.7% of the magazine market, and 30.2% of the online media market. India's Times of India Group is the market leader in print media with 9.2%, and also has a substantial share of the online market with 23.7%. Reflecting its dominant position in the print sector (54.9%), *El Mercurio* in Chile holds a majority share over online news with 54.2%.

The low W-HHI of 193 reflects the diversity of companies within the sector: though media groups that own print daily newspapers have an incumbency advantage, the fact that search engines and broadcasters also have online holdings means that no single type of company dominates the global market.

## THE WORLD'S LARGEST MEDIA FIRMS

After exploring the largest companies for each industry, we can now identify the world's largest companies (and state organizations) across all media industries. We discuss this in greater detail in Chapter 37 "Analysis of Media Concentration." But we will conclude here with such a summary.

These results are based on our authors' country data. We also calculate and report the overall worldwide, power index for the companies across platform industries. The power index, as described before, is the sum of market shares of a company, squared, across all countries and industries where it operates, and weighted by the revenue of that country and industry's overall media market.

Graph 35.1 shows the top 12 global overall media companies ordered by revenues, including the Chinese government as a single entity. These revenues account for their activities across the 13 media industries. Some of these firms are active in non-media industries too, or in media not covered here, such as music. These operations are not included.

We find that all of the top 12 companies are primarily telecommunications platform providers. This is not surprising, given the analysis of the preceding sections: these are concentrated industries globally and producing higher revenues than most content media industries. The top five ownership organizations are the Government of China (the combination of China Mobile, China Unicom, China Telecom, CCTV, and many other media activities),

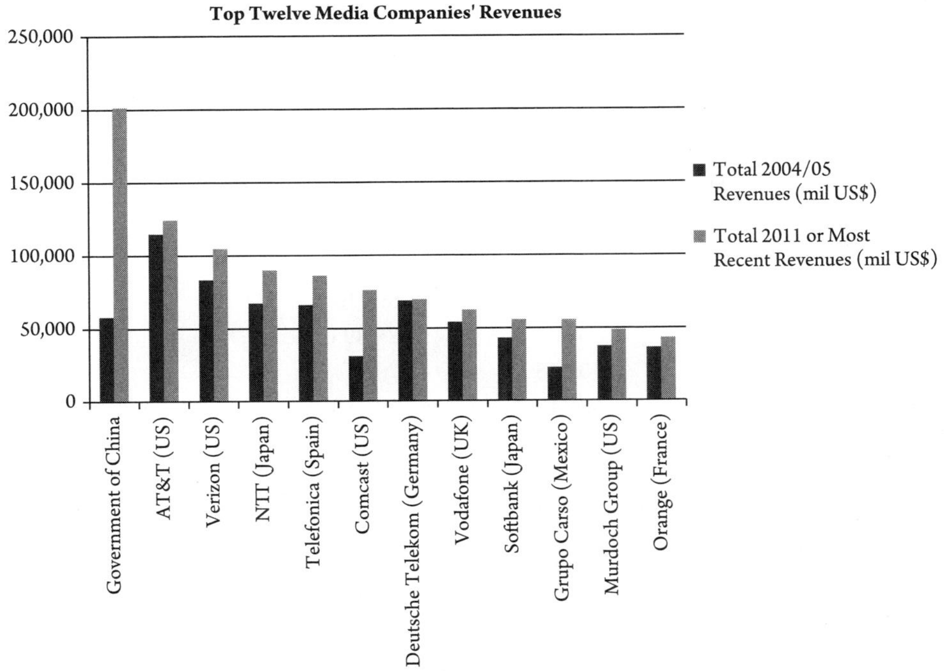


GRAPH 35.1 Top 12 media companies by revenue

AT&T (United States), Telefónica (Spain), NTT (Japan), and Verizon (United States). Even separately, the three main telecoms would be among the world's largest overall media organizations by revenues.

In the top 12, only the Murdoch Group, Comcast (a United States platform cable company) and Softbank (with mobile, ISP, search, and other online activities in Japan and the United States) are private-sector firms with a significant content presence, too (Carso has some audiovisual media holdings) (Table 35.18). Telefónica and Grupo Carso's rise since 2004 is primarily due to the great expansion of the Latin American wireless market.

We next turn to the platform sector and look at the top 100 platform companies ordered by revenues (this also means separating the revenues of companies that are both platform and content providers) (Table 35.19).

The integrated entity "Government of China" is at the top by a wide margin with total revenues of US$173 billion dollars, and a huge power index of 1,046.

Table 35-18. Top 12 Media Companies by Revenue

| COMPANY | TOTAL 2004/05 REVENUES (MIL US$) | TOTAL 2011 OR MOST RECENT REVENUES (MIL US$) | % CHANGE PER ANNUM (8 YRS) |
|---|---|---|---|
| Government of China[a] | 58,069 | 201,366 | 30.8% |
| *China Mobile (public)* | *19,326* | *84,579* | *42.2%* |
| *China Telecom (public)* | *14,685* | *44,243* | *25.2%* |
| *China Unicom (public)* | *7,768* | *42,692* | *56.2%* |
| *CCTV (public)* | *2,146* | *6,619* | *26.1%* |
| *Shanghai Media Group (public)* | *732* | *1,493* | *13.0%* |
| *Zhejiang Publishing (public)* | *128* | *1,757* | *159%* |
| *Beijing Publishing Group (public)* | *138* | *1,332* | *108%* |
| *Rayli (public)* | *23* | *1,303* | *712%* |
| *Hunan Media Group (public)* | *389* | *907* | *16.7%* |
| *Trend Media Group (public)* | *66* | *889* | *157%* |
| *China Satcom (public)* | *93* | *874* | *105%* |
| *China Railcom (public)* | *114* | *842* | *79.8%* |
| *Jiangsu Media Group (public)* | *60* | *706* | *136%* |
| AT&T (US) | 114,958 | 124,611 | 1.0% |
| Verizon (US) | 83,074 | 104,767 | 3.3% |
| NTT (Japan) (33% public) | 67,093 | 89,716 | 4.2% |
| Telefónica (Spain)[b] | 66,054 | 85,895 | 3.8% |
| Comcast (US) | 30,815 | 75,954 | 18.3% |
| Deutsche Telekom (Germany) (38% public) | 68,746 | 69,558 | 0.1% |
| Vodafone (UK) | 54,055 | 62,366 | 1.9% |
| Softbank (Japan) | 43,109 | 55,707 | 3.7% |
| Grupo Carso (Mexico) | 22,700 | 55,559 | 18.1% |
| Murdoch Group (US) | 37,320 | 48,425 | 3.7% |
| Orange (France) (27% public) | 36,459 | 43,138 | 2.3% |

[a] Not all state enterprises are listed separately; revenue totals for GOC reflect all enterprise revenues.
[b] Includes the share of Telecom Italia (Italy) revenues for 2013.

Table 35-19. Top 100 Platform Media Companies, 2013 Top 100 Platform Media Companies (Sorted by Revenue)

| COMPANY/ORGANIZATION | TOTAL GLOBAL POWER INDEX | TOTAL PLATFORM REVENUES (MIL US$) |
|---|---|---|
| Government of China | 1,046 | 173,430 |
| *China Mobile (China) (public)* | *312* | *81,649* |
| *China Telecom (China) (public)* | *103* | *44,243* |
| *China Unicom (China) (public)* | *66.4* | *42,692* |
| AT&T (US) | 205 | 122,100 |
| Verizon (US) | 156 | 102,151 |
| NTT (Japan) (32% public) | 316 | 89,716 |
| Telefónica (Spain)[1] | 172 | 77,724 |
| Deutsche Telekom (Germany) (38% public) | 151 | 69,494 |
| Vodafone (UK) | 96.7 | 62,366 |
| Grupo Carso (Mexico) | 125 | 55,283 |
| Softbank (Japan) | 60.4 | 54,617 |
| Comcast (US) | 51.6 | 49,647 |
| Orange (France Telecom, France, 27% public) | 85.3 | 42,981 |
| Oi Telemar (Brazil/Portugal)[2] | 71.8 | 32,618 |
| Telecom Italia (Italy) | 81.3 | 30,750 |
| KDDI (Japan) | 41.3 | 30,471 |
| Time Warner Cable (US) | 15.4 | 26,755 |
| CenturyLink (US) | 16.7 | 24,130 |
| Altice (France) | 33.9 | 23,539 |
| DirecTV (US) | 25.7 | 21,755 |
| Telstra (Australia) | 76.6 | 20,265 |
| Level 3 (US) | 11.4 | 19,933 |
| Bell Canada Enterprises/CTV (Canada) | 42.1 | 17,256 |
| Murdoch Group (US) | 55.0 | 15,909 |
| Vimpelcom (Russia) | 22.0 | 14,415 |
| Korea Telecom (South Korea) | 45.2 | 14,351 |
| Iliad (France) | 13.7 | 11,682 |
| Cox (US) | 2.5 | 10,973 |
| Liberty (US) | 14.3 | 10,785 |
| Cablevision (US) | 2.1 | 10,535 |
| Telus (Canada) | 15.0 | 10,263 |
| Rogers (Canada) | 18.9 | 10,045 |
| Government of Russia | 46.8 | 9,838 |
| *Svyazinvest (Russia) (public)* | *46.5* | *9,647* |
| *Gazprom Media (Russia) (public)* | *0.3* | *191* |
| SK Group (South Korea) | 28.6 | 9,834 |
| SingTel (Singapore) (public) | 18.4 | 9,543 |
| BT (UK) | 21.7 | 9,322 |

*continued*

Table 35-19. *continued*

| COMPANY/ORGANIZATION | TOTAL GLOBAL POWER INDEX | TOTAL PLATFORM REVENUES (MIL US$) |
|---|---|---|
| Oger Group (Saudi Arabia) | 34.9 | 9,027 |
| Bharat Sanchar Nigam (India) (public) | 31.6 | 8,793 |
| Dish Network (US) | 4.3 | 8,709 |
| Swisscom (Switzerland) (public) | 25.9 | 8,253 |
| MTS (Russia) | 17.1 | 8,120 |
| KPN (Netherlands) | 17.6 | 7,632 |
| MegaFon (Russia) | 13.2 | 7,582 |
| Bouygues (France) | 5.5 | 6,774 |
| Belgacom (Belgium) | 20.9 | 6,604 |
| United Internet 1&1 (Germany) | 4.2 | 6,371 |
| Charter (US) | 0.9 | 6,336 |
| TeliaSonera (Sweden/Finland) (50% public) | 18.6 | 6,291 |
| E-Plus (Germany) | 7.0 | 5,589 |
| Citizens Telecom (US) | 0.9 | 5,455 |
| Hutchison Whampoa (Hong Kong) | 2.9 | 5,133 |
| Telkom (South Africa) (public) | 30.0 | 4,982 |
| Turkcell (TeliaSonera, Sweden/Finland (public) + Alfa Group, Russia + Cukurova Group, Turkey) | 15.1 | 4,804 |
| Reliance (India) | 3.4 | 3,976 |
| ONO (Spain) | 3.3 | 3,975 |
| TP (Poland) (public) | 16.0 | 3,628 |
| MTN (South Africa) | 7.3 | 3,569 |
| US Cellular (US) | 0.4 | 3,518 |
| Kinnevik (Tele2, Sweden) | 3.4 | 3,512 |
| Alice (Germany) | 1.3 | 3,431 |
| Government of Egypt (Telecom Egypt) | 13.8 | 3,189 |
| Shaw (Canada) | 3.7 | 3,078 |
| SkyPerfecTV (Japan) | 8.0 | 3,040 |
| Government of Taiwain (CHT) | 6.7 | 2,730 |
| Idea Cellular (India) | 1.7 | 2,598 |
| PTTK Centertel (Poland) | 5.1 | 2,548 |
| Vonage (US) | 0.2 | 2,518 |
| Polkomtel (Poland) | 5.0 | 2,502 |
| Ziggo (Netherlands) | 4.5 | 2,453 |
| Polska Telefonia Cyfrowa (Poland) | 4.3 | 2,334 |
| Bezeq (Israel) | 7.2 | 2,243 |
| Taiwan Mobile (Taiwan) | 3.4 | 2,200 |
| Televisa (Mexico) | 3.0 | 2,123 |
| Elisa (Finland) | 4.7 | 2,044 |
| Vivendi (France) | 7.3 | 1,996 |
| Quebecor (Videotron, Canada) | 1.0 | 1,978 |

Table 35-19. *continued*

| COMPANY/ORGANIZATION | TOTAL GLOBAL POWER INDEX | TOTAL PLATFORM REVENUES (MIL US$) |
|---|---|---|
| Talk-Talk (UK) | 2.0 | 1,966 |
| BIGLOBE (Japan) | 0.9 | 1,845 |
| PRISA (Spain) | 4.6 | 1,817 |
| FarEasTone (Taiwan) | 2.8 | 1,725 |
| JazzTel (UK) | 0.9 | 1,666 |
| Fintech (Mexico) | 3.5 | 1,650 |
| Mediacom (US) | 0.1 | 1,620 |
| J:COM (Japan) | 2.1 | 1,541 |
| MTS Allstream (Canada) | 0.4 | 1,440 |
| K-Opticom (Japan) | 0.2 | 1,407 |
| ZON (Portugal) | 3.4 | 1,391 |
| Optimus (Portugal) | 1.7 | 1,378 |
| Kabel Deutschland (Germany) | 3.7 | 1,351 |
| Versatel (US) | 0.2 | 1,259 |
| Partner Communications (Israel) | 1.8 | 1,058 |
| DNA (Finland) | 1.3 | 1,048 |
| SaskTel (Canada) | 0.1 | 1,017 |
| Cogeco (Canada) | 0.3 | 973 |
| Cellcom (Israel) | 1.8 | 949 |
| Yogio (Spain) | 0.2 | 910 |
| NCIT (Taiwan) | 0.5 | 909 |
| Unitymedia (Germany) | 1.6 | 900 |
| Tata Group (India) | 0.2 | 853 |
| Skype (US) | 0.0 | 839 |
| Sunrise (Denmark) | 0.9 | 821 |
| NII (US-Mexico) | 0.2 | 804 |

[1] This figure includes both Telefónica and a 9.12% stake in Telecom Italia.

[2] Oi and Portugal Telecom (PT) merged in 2013, creating a new company in which Oi's shareowners will be the majority stakeholders

The combined state holdings of China are largest in revenue terms, and it is far ahead of the second group of very substantial telecom firms: AT&T, Telefónica, Verizon, NTT, Deutsche Telekom, Vodafone, Grupo Carso, and Orange. A third tier then adds somewhat different types of platform companies—Softbank, Comcast, TWC, Altice, the Murdoch Group, Bell Canada, and DirecTV with their significant TV and Internet media activities. Also in that size league are Telecom Italia, BT, Telstra, KT, SingTel, Vimpelcom, and Oi Telemar.

Then follows a fourth tier, comprising several companies that are typically their countries' number 2 telecom incumbent (or for the United States, number 3) or the primary new entrant—KDDI, Century Link, MTN, the SK Group, Bharat Sanchar Nigam, MTS, Rogers, TeliaSonera, and Telus.

By power index, China's number one position is greater still, given its population and the dominance of the triumvirate of state firms. Several firms' PI ranking is stronger than their revenue ranking, which is the case where a company dominates a platform in a medium-sized or poor country—BT (United Kingdom), KT (South Korea), Svayazinvest (Russia), Oger and Turkcell (Turkey), and Telkom (South Africa). Conversely, firms are lower ranked in PI terms than their revenues would suggest if they operate in either large or rich markets as a medium-sized market share company—Century Link and Dish Network in the United States; Liberty in Europe; Bharati, Reliance, and BSN in India, and MTN (South Africa).

We next look at *content* companies. Graph 35.2 shows the top 15 content media companies by revenue in 2004/05 and 2011 or most recently. Included in those revenues are the revenues made in the industries of film, TV, multi-channel (one-third of multi-channel revenues), search engine, newspapers, book publishing, magazines, video channel, and online news.

In 2004/05, the content company with the highest revenues was Murdoch's News Corp./21st Century Fox, with $26.5 billion, and it retained this spot in 2013 with $32.5 billion (Table 35.20). Time Warner was second largest in 2004/5, with $22.8 billion. However, after several spin-offs, its revenues had dropped to $12.6 billion. Google, with $7.4 billion in 2004/05, became the third largest firm in 2013, with $26.8 billion and had the highest growth rate since it quadrupled its revenues At such a rate, Google is already or shortly will be the world's largest content company.

The largest content entity is the Government of China, with a PI of 359 for all its activities in content (the largest component of which is CCTV, with a PI of 53.2). By revenues, the combined state ownership of China makes it the second largest globally after Murdoch's properties. The commercial company with the highest power index for content is Google (US), with a measure of 278. Third are Rupert Murdoch's holdings, coming in at 129. The BBC (73.2) is fourth, and Comcast ranks fifth in content (without platforms) (67.5).

Several firms rank high in revenue but much lower in power index—Disney, Sumner

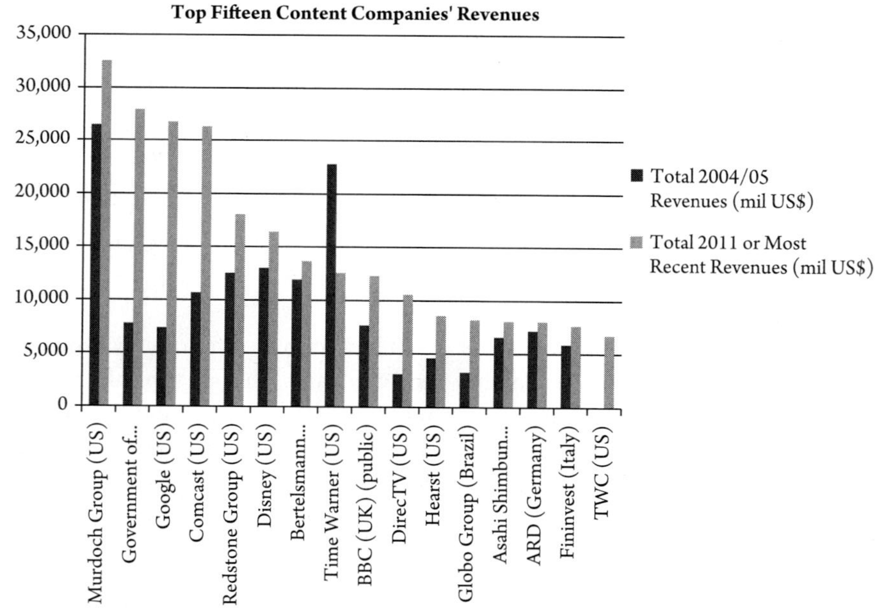


GRAPH 35.2 Top 15 Content Media Companies by Revenue[1]

[1] Time Warner Cable did not exist independently in 2004/05. It is shown here as "#16" in the world to visualize the original parent Time Warner's drop in the rankings.

Table 35-20. Top 100 Content Media Companies, 2013 Top 100 Content Media Companies (Sorted by Revenue)

| COMPANY/ORGANIZATION | TOTAL GLOBAL POWER INDEX | TOTAL CONTENT REVENUES (MIL US$) |
|---|---|---|
| Murdoch Group (US) | 129 | 32,516 |
| Government of China | 359 | 27,936 |
| *CCTV (China) (public)* | *53.2* | *6,619* |
| Google (US) | 278 | 26,747 |
| Comcast (US) | 67.5 | 26,307 |
| Redstone Group (US) | 32.5 | 18,041 |
| Disney (US) | 35.9 | 16,418 |
| Bertelsmann (Germany) | 30.7 | 13,650 |
| Time Warner (US) | 29.9 | 12,554 |
| BBC (UK) (public) | 73.2 | 12,301 |
| DirecTV (US) | 28.0 | 10,571 |
| Hearst (US) | 12.5 | 8,588 |
| Globo Group (Brazil) | 52.2 | 8,195 |
| ARD (Germany) (public) | 32.0 | 8,036 |
| Asahi Shimbun (Japan) | 19.0 | 8,017 |
| Fininvest (Italy) | 34.6 | 7,732 |
| Time Warner Cable (US) | 11.9 | 6,749 |
| Advance (US) | 6.2 | 6,267 |
| Lagardère (France) | 16.3 | 5,962 |
| NHK (Japan) (public) | 14.1 | 5,701 |
| Liberty (US) | 14.7 | 5,675 |
| ProSiebenSat.1 (Germany) | 21.6 | 5,587 |
| NBS (Japan) | 11.2 | 5,485 |
| Vivendi (France) | 32.6 | 5,341 |
| RAI (Italy) (public) | 34.1 | 5,247 |
| PRISA (Spain) | 20.1 | 5,210 |
| NTV (Japan) | 13.3 | 5,115 |
| Time Inc. (US) | 8.8 | 5,096 |
| Gannett (US) | 5.1 | 4,983 |
| Yahoo (US) | 8.7 | 4,559 |
| Tokyo Broadcasting Service (Japan) | 9.4 | 4,482 |
| SkyPerfecTV (Japan) | 30.8 | 4,391 |
| Dish Network (US) | 4.8 | 4,290 |
| Discovery Communications (US) | 7.0 | 4,252 |
| Shaw (Canada) | 15.9 | 4,015 |
| Bell Canada Enterprises/CTV (Canada) | 12.7 | 3,982 |
| Yomiuri Shimbun (Japan) | 10.0 | 3,837 |
| ITV (UK) | 10.8 | 3,726 |
| ZDF (Germany) (public) | 6.1 | 3,578 |

*continued*

Table 35-20. *continued*

| COMPANY/ORGANIZATION | TOTAL GLOBAL POWER INDEX | TOTAL CONTENT REVENUES (MIL US$) |
|---|---|---|
| Planeta (Spain) | 9.3 | 3,576 |
| Pearson (Penguin, UK) | 4.0 | 3,566 |
| Bonnier (Sweden) | 16.2 | 3,496 |
| Cox (US) | 2.0 | 3,442 |
| Government of Russia | 29.1 | 3,324 |
| *Gazprom Media (Russia) (public)* | *6.6* | *1,507* |
| Sony (Japan) | 6.3 | 3,275 |
| Bauer (Germany) | 9.7 | 3,237 |
| Nihon Keizai Shimbun (Japan) | 2.9 | 3,189 |
| Universal Church Group (Brazil) | 3.1 | 3,150 |
| Tribune (US) | 1.9 | 3,049 |
| RCS (Italy) | 7.8 | 2,930 |
| Baidu (China) | 29.5 | 2,767 |
| McGraw Hill (US) | 2.4 | 2,751 |
| France Télévisions (France) (public) | 8.1 | 2,709 |
| Axel Springer (Germany) | 5.4 | 2,652 |
| Verizon (US) | 1.8 | 2,616 |
| Bouygues (France) | 5.7 | 2,539 |
| AT&T (US) | 1.6 | 2,511 |
| Daily Mail & General Trust (UK) | 6.1 | 2,496 |
| Cablevision (US) | 1.1 | 2,491 |
| Government of South Korea | 23.1 | 2,472 |
| *KBS (South Korea) (public)* | *6.0* | *1,208* |
| Clear Channel (US) | 5.0 | 2,420 |
| Seven Network (Australia) | 7.2 | 2,240 |
| Televisa (Mexico) | 17.3 | 2,239 |
| The New York Times (US) | 1.7 | 2,210 |
| Rogers (Canada) | 4.1 | 2,179 |
| McClatchy (US) | 1.5 | 2,090 |
| Nine Entertainment (Australia) | 8.9 | 2,048 |
| CBC (Canada) (public) | 8.0 | 1,994 |
| Abril (Brazil) | 11.2 | 1,922 |
| Quebecor (Videotron, Canada) | 3.8 | 1,849 |
| The Washington Post (US) | 0.5 | 1,821 |
| Independent News and Media (Ireland) | 9.2 | 1,758 |
| Trinity Mirror (UK) | 3.3 | 1,716 |
| Charter (US) | 0.7 | 1,674 |
| RTVE (Spain) (public) | 4.0 | 1,673 |
| Channel 4 (UK) | 2.2 | 1,633 |
| Astral (Canada) | 3.2 | 1,485 |

Table 35-20. *continued*

| COMPANY/ORGANIZATION | TOTAL GLOBAL POWER INDEX | TOTAL CONTENT REVENUES (MIL US$) |
|---|---|---|
| RTP (Portugal) (public) | 6.9 | 1,441 |
| Amaury Group (France) | 5.6 | 1,424 |
| NRJ Group (France) | 2.1 | 1,327 |
| Mainichi Shimbun (Japan) | 1.1 | 1,301 |
| Scholastic (US) | 0.9 | 1,291 |
| Kinnevik (Sweden) | 3.2 | 1,240 |
| Impresa (Portugal) | 4.8 | 1,233 |
| Sanoma Oyj (Finland) | 4.7 | 1,229 |
| Fairfax (Australia) | 4.8 | 1,229 |
| Reed Elsevier (Netherlands) | 0.8 | 1,219 |
| Postmedia (Canada) | 4.7 | 1,189 |
| Source Interlink (US) | 0.5 | 1,186 |
| Vocento (Spain) | 2.8 | 1,163 |
| Northern & Shell (UK) | 2.0 | 1,155 |
| Dogan Group (Turkey) | 6.6 | 1,135 |
| Chunichi Shimbun (Japan) | 0.9 | 1,126 |
| WOWOW (Japan) | 3.2 | 1,098 |
| Softbank (Japan) | 2.7 | 1,090 |
| Prasar Bharati (AIR, DD, India) (public) | 14.2 | 1,080 |
| Telefónica (Spain) | 2.4 | 1,031 |
| Folhapar (Brazil) | 4.2 | 1,026 |
| Burda (Germany) | 1.8 | 1,007 |
| SRG SSR (Switzerland) (public) | 6.3 | 1,001 |
| SABC (South Africa) (public) | 8.2 | 996 |

Redstone's group (CBS-Viacom), Bertelsmann, Time Warner, Fininvest, ARD, and the BBC. These firms have typically a strong but non-dominant market share in medium or large markets.

A third tier are medium-sized firms in medium-sized industries (often print or video)—Advance, Lagardère, Bouygues, PRISA, TWC, and a set of Japanese firms: NBS/Fujisankei, NHK, NTV, TBS, and the Asahi Shimbun.

## Bibliography

For the sources mentioned in this chapter see the general bibliography at the end of this book.

# 36

# The Owners of the World's Media

There are few industries whose ownership and control are more important than those of media, given their central role in social, cultural, economic, and political life.[1] To provide the base for a factual analysis, this chapter examines who owns the world's major media companies. In a second step, we identify the major owners of media properties around the world.

There are various types of ownership. First, there are *non-commercial* arrangements. These include direct governmental ownership (e.g., Voice of America for foreign broadcasts); semi-independent public corporations, such as the BBC; or media owned by charitable foundations set up by the original owners. Then there are several types of *private* ownership—by individuals, families, partnerships, shareholders, institutional investors, private equity funds, and venture capitalists.

Stock companies come in several types. Some are owned and controlled by individuals and families through "privately held" (that is, largely non-traded) shares or by holding special voting stock. Other media firms rely on a "public" ownership of shares, in order to access outside financing. This, in turn, has led to a significant ownership role by institutional investors that buy stakes in companies on behalf of smaller investors.

Decades ago, private media ownership was primarily in the hands of a small group of individuals and families. But by the late twentieth century, new technology and new media had created a massive need for capital, requiring many sources of investment and hence a more widespread ownership. This has been most evident for platform companies whose networks are capital-intensive. It has also been true for the startup Internet media industries where relatively young companies need to rely on outside equity capital—initially often by venture capital investors—to finance their future growth since they have only limited access to debt capital or to internal funds.

1. The extensive data that follow were identified and selected, and tables constructed by and under the supervision of Jason Buckweitz, to whom goes much of the credit for the data work.

In contrast, in content-oriented media companies the founders have usually relinquished ownership control to a much lesser extent. They were able to self-finance or expand more gradually, to use debt rather than equity, and to require a lesser capital investment. But they, too, have widened ownership where vertical, horizontal, or global expansion beckoned.

## OWNERSHIP OF THE LARGEST MEDIA FIRMS

To proceed, we select the largest media companies—about 20 platform companies and a similar number of content companies—and investigate their ownership structure. We examined the ownership records of each of these companies and identified the company's owners. In addition to the companies' annual reports, information was culled from Securities and Exchange Commission filings for American companies,[2] and comparable agencies in other countries, where such reports exist and are available.

We also identified the top 10 institutional holders of each company,[3] the number of shares they held, their percent ownership of the company, and the total value of their shares. This results in Table 36.1 (Short Version) for the Top 20 Platform Companies and Table 36.2 (Short Version) for the Top 20 Private Content Companies. For space reasons, these tables are listed in an abbreviated form, identifying only the top two owners. The more detailed information is provided in the Appendix to this chapter as Table 36.1 (Full Version) and Table 36.2 (Full Version).

There are several complicating factors. The value of publicly traded companies can be observed where the share price is reported. But other shares are not publicly traded and their value has to be estimated. In other cases, companies have several classes of shares. One class might have superior rights of voting, others have preferred dividends, and possibly no voting rights. In those cases, estimates were usually done by reference to the value of other shares or companies. For example, the Roberts family controls the giant cable TV company Comcast. The asset value in its untraded Class B stock was calculated by taking the market cap value of Comcast's largest competitor, Time Warner Cable, which has only a single category of shares, scaling it up to Comcast's size proportionately to subscribers, and adding the value of NBC Universal when it was acquired by Comcast. The resultant extrapolation of the company's value was prorated to the Roberts family according to its one-third undiluted control.

In identifying the ultimate shareholder in an ownership chain and attribution, other work has used a test of at least 20% of voting rights (not cash flow rights) at every link of the ownership chain (LaPorta et al. 1999).

Another complicating factor is that for some companies, media activities are only part of what they do. Where companies' activities span media and non-media industries, only the former have been used.

To the best of our knowledge, such an ownership analysis has not been done before for media companies.

### Observations:

- Many of the wireline companies, even after the corporatizations and privatizations (fully or partly) of the 1980s, have major majority ownership stakes by their governments. This includes NTT in Japan (33%), Deutsche Telekom in Germany (32%), Orange in France (27%), China Telecom (78%), Svyazinvest in Russia (53%), and Telkom in South Africa (50.7%). These governmental ownership stakes are particularly high given the fragmentation of the private stock holdings in platform companies among numerous small investors who have traditionally considered telecom shares as safe "widows-and-orphans" stock.

2. <http://www.sec.gov/edgar.shtml>.
3. As listed by Morningstar Financial: <http://investors.morningstar.com>.

Table 36-1. Ownership of the World's Top 20 Platform Companies (Short Version) (as of Sept. 2013)

| | LARGEST OWNERS | NUMBER OF SHARES OWNED | PERCENT OF COMPANY OWNED | PERCENT OF VOTING POWER | VALUE OF SHARES (MIL $), PRORATED TO PLATFORMS |
|---|---|---|---|---|---|
| NTT (Japan) | Government of Japan | ~436,816,470 | 33 | 33 | 27,672 |
| | Japan Trustee Services Bank | 67,508,000 | 5.1 | 5.1 | 3,421 |
| AT&T | Vanguard | 437,604,939 | 8.2 | 8.2 | 15,092 |
| | State Street | 233,415,579 | 4.4 | 4.4 | 8,262 |
| Deutsche Telekom | Government of Germany | 1,395,738,354 | 32 | 32 | 22,227 |
| | BlackRock(US) | 131,286,639 | 3.0 | 3.0 | 1,672 |
| China Mobile (China) | Government of China | ~2,947,010,345 | 74.1 | 74.1 | ~192,051 |
| | Lazard (US) | 25,062,318 | 0.63 | 0.63 | 1,306 |
| Telefónica (Spain) | Banco Bilbao Vizcaya Argentaria | 261,798,437 | 5.8 | 5.8 | 3,659 |
| | Caja de Ahorros y Pensiones de Barcelona, "la Caixa" | 254,697,815 | 5.6 | 5.6 | 3,560 |
| Vodafone (UK) | Capital Research and Management (US) | 2,472,830,993 | 5.1 | 5.1 | 7,814 |
| | Legal and General (UK) | 2,178,073,425 | 4.4 | 4.4 | 6,882 |
| Orange (France) | Government of France | 701,986,766 | 27 | 27 | 8,924 |
| | Franklin Templeton Investment (US) | 54,960,745 | 2.1 | 2.1 | 558 |
| Telecom Italia (Italy) | Telefonica (Spain) (42.3% of Telco S.p.A, which owned 22.4% of Telecom Italia in 2013) | ~1,835,254,130 | 9.5 | 9.5 | 1,491 |
| | BlackRock | | 10.1 | 10.1 | |

| | | | | | |
|---|---|---|---|---|---|
| Verizon (US) | Vanguard | 251,846,095 | 8.8 | 8.8 | 12,524 |
| | State Street | 120,621,754 | 4.2 | 4.2 | 6,072 |
| América Móvil (Mexico) | Carlos Slim | 12,376,000,000(AA) | 52.6 | 52.6 | 29,217 |
| | AT&T (US) | 5,739,000,000 (AA) | 23.8 | 23.8 | 13,198 |
| Telemar (Oi, Brazil & Portugal Telecom) | AG Telecom (Brazil) (telecom holding of Andrade Guiterrez conglomorate) | | 11.3 | 11.3 | 442 |
| | Bindes (Brazil) | | 8.1 | 8.1 | 291 |
| China Telecom | Government of China | 62,977,197,275 | 78 | 78 | 40,147 |
| | Colonial First State (Australia) | 2,200,924,795 | 2.7 | 2.7 | 1,122 |
| Telstra (Australia) | Capital Research and Management (US) | 283,496,500 | 2.3 | 2.3 | 1,224 |
| | BlackRock (US) | 78,225,739 | 0.63 | 0.63 | 337 |
| Svyazinvest (Russia) | Government of Russia | 1,564,341,770 | 53.2 | 53.2 | 5,881 |
| | Marshall Capital (Private Equity, Russia) | 309,042,118 | 10.5 | 10.5 | 1,161 |
| 21st Century Fox (US) | Murdoch Family | 311,103,499 (B) | 13.5 | 39.0 | 9,965 |
| | Vanguard | 152,697,912 (A) | 6.6 | 0.00 | 4,658 |

Table 36-2. Ownership of World's Top Private Content Companies (Short Version) (as of September 2013)

| | | LARGEST OWNERS | NUMBER OF SHARES OWNED | PERCENT OF COMPANY OWNED | VOTING PERCENT | VALUE OF SHARES (MIL $) |
|---|---|---|---|---|---|---|
| Google (US) | | Larry Page | 24,355,252 (B) | 7.3 | 28.2 | 26,717 |
| | | Sergey Brin | 23,764,586 (B) | 7.2 | 27.5 | 26,047 |
| Murdoch Group | News Corp. (US) | Murdoch Family | 77,855,793 (B) | 13.5 | 39.0 | 1,640 |
| | | Managers Investment Group | 72,291,747 (A) | 12.5 | 0.0 | 1,102 |
| | 21st Century Fox (US) | Murdoch Family | 311,423,171 (B) | 13.5 | 39.0 | 9,965 |
| | | Vanguard | 152,697,912 (A) | 6.6 | 0.0 | 4,658 |
| Comcast (US) | | Roberts Family | 6,856,323 (A) | 0.26 | 0.22 | 368 |
| | | | 9,444,375 (B) | 0.36 | 33.4 | 18,500 |
| | | Vanguard | 227,152,526 (A) | 8.6 | 7.1 | 9,690 |
| | | | 25,104,652 (K) | 0.95 | 0.0 | 1,025 |
| Globo Group (Brazil) | | Roberto Marinho | | 33.3 | 33.3 | 6,666 |
| | | João Marinho | | 33.3 | 33.3 | 6,666 |
| | | José Marinho | | 33.3 | 33.3 | 6,666 |
| Lagardère (France) | | Arnaud Lagardère | 12,196,950 | 9.3 | 14.9 | 467 |
| | | DNCA Finance (France) | 3,210,000 | 2.5 | 2.5 | 77 |
| Disney (US) | | Steve Jobs Trust | 138,000,007 | 7.7 | 7.8 | 8,371 |
| | | Vanguard | 148,460,516 | 7.4 | 8.4 | 8,044 |
| Bertelsmann (Germany) | | 19.1% directly owned by Mohn families. 80.9% owned by three foundations whose voting is controlled by the Mohn families. | | 100 | 100 | 4.431 |

| | | | | | | |
|---|---|---|---|---|---|---|
| Berlusconi Group | Mondadori | Fininvest (Berlusconi) | 130,623,800 | 53 | 53 | 241 |
| | | Silchester (UK) | 32,832,269 | 13.3 | 13.3 | 65 |
| | Mediaset | Fininvest (Berlusconi) | 483,800,000 | 41 | 41 | 2,896 |
| | | MacKenzie Cundill Invest (Canada) | 31,743,385 | 2.8 | 2.8 | 126 |
| Malone Group (US) | Liberty Global (Malone) | John Malone | 8,751,613 (B)<br>2,730,878 (C) | 3.4<br>1.1 | 35.9<br>0 | 868<br>204 |
| | | Vanguard | 11,144,802 | 4.3 | 4.6 | 868 |
| | Liberty Media | John Malone | 8,496,800 (B) | 7.0 | 40.4 | 1,226 |
| | | Berkshire Hathaway (Warren Buffet) | 5,622,340 | 4.6 | 2.7 | 808 |
| ProsiebenSat1 (Germany) | KKR (US) | | 35, 663, 944 | 16.3 | 16.3 | 1,444 |
| | Permia (US) | | 35, 663, 944 | 16.3 | 16.3 | 1,444 |
| | Telegraaf Media (Netherlands) | | 13, 127, 832 | 6.0 | 6.0 | 531 |
| Televisa (Mexico) | Emilio Azcárraga Jean | | 53,275,414,049 | 14.7 | 23.7 | 535 |
| | BlackRock (US) | | 32,073,554,097 | 8.9 | 7.8 | 572 |
| Fuji Television (Japan) | Shikanai Family | | | 100 | 100 | 7,000 |
| Yahoo (US) | Vanguard | | 87,549,068 | 7.9 | 7.9 | 2,157 |
| | Fidelity | | 76,079,094 | 6.8 | 6.8 | 1,835 |
| Bouygues (France) | SCDM (Bouygues Family) (France) | | 67,317,686 | 20.5 | 29.2 | 1,653 |
| | Bouygues Employees (France) | | 77,825,813 | 23.7 | 28.7 | 1,911 |

*continued*

Table 36-2. *continued*

| | LARGEST OWNERS | | NUMBER OF SHARES OWNED | PERCENT OF COMPANY OWNED | VOTING PERCENT | VALUE OF SHARES (MIL $) |
|---|---|---|---|---|---|---|
| Redstone Group | CBS (US) | National Amusements (Sumner Redstone) | 33,742,639 (A) | 5.4 | 79.0 | 1,828 |
| | | Gamco (US) | 2,338,784 (A) | 0.37 | 5.5 | 121 |
| | Viacom | National Amusement (Sumner Redstone) | 40,409,751 (A) | 8.5 | 79.0 | 4.863 |
| | | Vanguard | 40, 908,747 (B) | 8.5 | 0.0 | 5.280 |
| Time Warner (US) | Vanguard | | 84,839,919 | 9.1 | 9.1 | 4,906 |
| | Capital Group | | 71,699,627 | 7.7 | 7.7 | 4,131 |
| Time Warner Cable (US) | Dodge & Cox | | 28,144,881 | 9.5 | 9.5 | 2,897 |
| | Vanguard | | 26,529,901 | 9.0 | 9.0 | 2,547 |
| Baidu (China) | Robin Yanhong Li (China) | | 5,581,324 | 16.0 | 16.0 | 7,534 |
| | Handsome Reward Limited(UK) | | 5,490,000 | 15.7 | 15.7 | 7,411 |
| Softbank (Japan) | Masayoshi Son | | 240,567,906 | 20.2 | 20.2 | 18,900 |
| | Japan Trustee Services Bank | | 102,420,000 | 8.6 | 8.6 | 6,463 |
| Yomiuri Shimbun Holdings (NTV) (Japan) | Shoriki Family | | 69,803,000 | 27 | 27 | 1,279 |
| | CBNY-Orbis Funds (Japan) | | 19,037,000 | 7.2 | 7.2 | 348 |

- Few of the top 20 platform companies have major individual owners. The main exception is America Movil, with Carlos Slim holding 52%. Several other large platform firms have high individual ownership stakes but originated as a media or Internet company with a presence in platforms. They are 21st Century Fox (Sky TV) with the Murdoch family, Softbank in Japan owned by Masayoshi Son, and Comcast owned by the Roberts family. Most of these companies have a dual stock structure. This is designed to allow an individual owner to maintain control while accessing outside capital investors. The other major platform companies are majority owned by large institutional investors or governments, and none of them has a dual stock structure.
- US institutional investors have good-sized ownership stakes in many of the top 20 platform companies around the world, though rarely over 5%.
- The largest institutional investor in the top 20 platform companies is the Vanguard Group based in the United States. Vanguard owns assets in 10 of the top 20 platform companies, with a total value of $47.5 billion. (See Table 36.1 (Full Version) and Table 36.3.)
- The largest non-US institutional investor in the top 20 platform companies by asset value is Legal & General, a multinational financial service company based in London. Its total major platform assets add up to over $3.7 billion.

We now turn to the ownership of the largest content media companies (Table 36.2, Short Version).

### Observations:

- Content firms tend to be privately owned. Most of the top content companies have major individual owners—Globo Group (Marinho family, Brazil); Lagardère (Lagardère family, France); Bertelsmann (Mohn family, Germany); Fininvest (former Prime Minister Berlusconi, Italy); Softbank (Masayoshi Son, Japan); Fuji Television (Shikanai family, Japan); TF1 (Bouygues family, France); Fox/News Corp (Murdoch family); Google (Larry Page and Sergey Brin); Comcast (Roberts family); Liberty (John Malone); CBS/Viacom (Sumner Redstone); Baidu; Yaomuri; and Disney (Steve Jobs family with 7.8%). The exceptions are Time Warner/Time Warner Cable; Yahoo; and ProSiebenSat1.[4]
- Great discrepancies of equity percentage and voting percentage exist for several of the content media: Google, NewsCorp, 21st Century Fox, Comcast, Lagardère, Bertelsmann, Liberty, Televisa, Bouygues, CBS, and Viacom. In each case, a founder or his heirs protect their control through complex stock structures.
- Public service TV and radio companies are owned by the state in a variety of legal structures. This includes, in Germany, ARD, a cooperative of the several regional states' public broadcasters; BBC, which is a crown corporation; Prasar Bharati, a public corporation owned by the state of India; France Télévisions; NHK (Japan); and RAI (Italy), all public service broadcasters.
- In China and Egypt, state ownership in content media goes beyond the national public service broadcaster and also includes newspapers, book publishing, and magazines.

## THE LARGEST OWNERS OF MEDIA

Who owns the media? There are three kinds of owners: (1) individuals/families, (2) institutions, and (3) governments/public. We will discuss them sequentially.

We use Tables 36.1 and 36.2 (Long Versions, in the Appendix) to construct an aggregated list of the largest owners of media companies. Holdings were determined by examining investment companies' 13-F filings to the Securities and

4. The latter is the only non-US media firm in the top 20 that is not controlled by an individual family, after having been owned by a group lead by Haim Saban.

## Table 36-3. Media Billionaires: The Largest Individual Owners of Media (as of September 2013)

| OWNER | VALUE OF SHARES ($ BILLION) |
|---|---|
| Carlos Slim (America Movil, Mexico)[a] | 29.2 |
| Larry Page (Google, US) | 26.7 |
| Sergey Brin (Google, US) | 26.0 |
| Cox family (Chambers, Kennedy, Parry-Sheden, Anthony) (Cox Communications, US) | 24.0 |
| Michael Bloomberg (Bloomberg LP, US) | 24.0 |
| David Thompson family (Thompson Reuters, Canada) | 20.3 |
| Marinho family (Globo, Brazil) | 20.0 |
| Mark Zuckerberg (Facebook, US) | 19.0 |
| Brian Roberts family (Comcast, US) | 18.5 |
| Newhouse family (Advance Publications, US) | 17.1 |
| Sawiris Family (Orascom Group, Egypt) | 12.0 |
| Rupert Murdoch family (News Corp, 21st Century Fox; US) | 11.6 |
| Charles Ergen (Dish Network, US) | 10.6 |
| Jobs family (Disney, US) | 8.7 |
| Masayoshi Son (Softbank, Japan) | 8.6 |
| Eric Schmidt (Google, US) | 8.2 |
| Edward Rogers family (Rogers cable & telecom, Canada) | 7.6 |
| Robin Yanhong Li (Baidu, China)[b] | 7.53 |
| Hearst family (Hearst publishing, broadcasting, cable, US) | 7.0 |
| Xavier Niel (Iliad, France) | 6.6 |
| Ambani family (Reliance, India)[a] | 6.2 |
| Silvio Berlusoni (Fininvest, Italy)[a] | 6.2 |
| David Geffen (film and music, US) | 6.0 |
| Avande Krishnan (Moxis mobile, Malaysia) | 5.85 |
| John Malone (Liberty, US) | 5.6 |
| Denis O'Brien (Communicorp Radio, Ireland) | 5.2 |
| Roberto Civita (Abril Publishing, Brazil) | 4.9 |
| Patrick McGovern (IDG Publishing Media, US) | 4.7 |
| Sumner Redstone family (CBS & Viacom, US) | 4.7 |
| Gustavo Cisneros (TV, Venezuela) | 4.4 |
| Mohn family (Bertelsmann, Germany) | 4.3 |
| Sunil Mittal (Bharti Airtel, India)[a] | 4.2 |
| Arnon Milchan (film, Israel) | 4.2 |
| Dustin Moskovitz (Facebook, US) | 3.8 |
| Francis Bouygues family (Bouygues, France)[a] | 3.7 |
| Kalanithi Maran (Sun Media, India) | 3.3 |
| Haim Saban (TV, US) | 3.1 |

Table 36-3. *continued*

| OWNER | VALUE OF SHARES ($ BILLION) |
|---|---|
| Friede Springer (Axel Springer SE, Germany) | 3.0 |
| William Ding (online games, China) | 3.0 |
| Arnaud Lagardere (Lagardere, France) | 2.4 |
| Mike Adenuga (Globecom Mobile, Nigeria) | 2.35 |
| Bill Gates (Microsoft, US) | 2.3 |
| Shikanai Family (Fuji Fujisankei, Japan) | 2.3 |
| Eduardo Saverin (Facebook, US) | 2.2 |
| Oprah Winfrey (Harpo TV production) | 2.18 |
| Bonnier Family ( Bonnier, Sweden) | 2.1 |
| Hiroshi Yamauchi family (Nintendo, Japan) | 2.1 |
| Taksin Shinawatra family (Mobiletelecom, Thailand) | 1.7 |
| Chey family (SK Telecom, Korea) | 1.56 |
| Zygmunt Solorz-Zak (Polsal TV & mobile, Poland) | 1.45 |
| Elisabeth Badinter (Publicis Advertising, France) | 1.4 |
| Shoriki family (Yomiuri Shinibun, Japan) | 1.28 |
| de Noble family (Clarin, Argentina) | 1.0 |
| Aydiu Dogan (Dogan Media, Turkey) | 1.0 |
| Barry Diller (IAC, US) | 1.0 |
| Ferit Sakenk (Dogus Media, Turkey) | 1.0 |
| Lee Jay-Hyan (CJ, Korea) | 0.85 |
| Tony O'Reilly (INM, newspapers, Ireland) | 0.7 |
| Hajime Satomi (Sega, Japan) | 0.57 |
| Azcarraga family (Televisa, Mexico) | 0.54 |

[a] Media activities only
[b] Includes beneficial ownership in behalf of non-Chinese investors

Exchange Commission[5] in the United States and similar filings in other countries.[6] In many cases we also looked at the institutional investor's annual reports, where available.

## Individual Owners

We begin with individual owners, including families and heirs of original company founders.

The table of the largest individual media owners was created by using a variety of sources and approaches. In most cases, the shareholding and reported share prices were the basis of calculation. In other cases, we started with reported net worth from sources such as the business magazines *Forbes* or *Bloomberg* and in Wikipedia, and so on, and proceeded by subtracting major non-media holdings. Our list may differ from lists of the "Richest People" for several reasons:

5. <http://www.sec.gov/edgar.shtml>.

6. <http://www.fsa.go.jp/sesc/english/>; <http://www.cnbv.gob.mx/en/Paginas/default.aspx>; <http://www.finanstilsynet.no/en/>.

1. Owners often hold substantial assets in non-media companies and these are not included.

2. Media companies often are engaged in non-media activities.

3. Major owners may, in fact, hold only moderate ownership percentages in "their" company.

For example, Bill Gates is often reported to be worth over $65 billion, and that might be true. But not in media. First, Gates has substantially reduced his stake in Microsoft to 4.5% by late 2013. Second, Microsoft's media activities are only a modest part of its activities. Media-type revenues are calculated to be 12.66% of its overall revenues. This includes the search engine Bing and game software, both of which are content media, and estimated to account together for 10% of Microsoft's revenues. Microsoft's Skype is a platform activity and accounts for about 2.66% of Microsoft's value. Thus, Gates's media assets would be a percentage (4.5%) of a percentage (12.66%) = 0.574% of Microsoft's total market value of about $320 billion (i.e., about $1.82 billion). Most of Gates's net worth is attributable to MS's non-media activities such as operating software, etc. Third, other Gates assets are in different companies. For example, he owns $481 million in America Movil stock and this would be added to his media holdings, for a total of $2.3 billion. Finally, part of the wealth attributed to him has been donated to the Bill and Melinda Gates Foundation.

Table 36.3 identifies the world's largest individual owners of media. Sources are Tables 36.1 and 36.2, *Forbes*,[7] *Bloomberg*,[8] Wikipedia, as well as information providers and other search tools.

- There are 56 media billionaires as individuals or families. There are other billionaires who own media properties but at a level below $1 billion.
- Among traditional media, significant wealth is encountered in TV (19), newspapers (11), magazines (8), film (6), and even books (3), but not in radio (1) and telecom (1).
- Of the billionaires, two-thirds substantially created their media properties themselves (37) rather than inheriting established firms (19).
- The "new media billionaires" were mostly active in Internet (12), mobile telecom (11), and information services (2).
- A significant share of the 56 billionaires is US citizens (22); 4 each from Japan and France; 3 from India; 2 each from Germany, Canada, Turkey, and China; 5 from Latin America; and 2 from Africa.
- Of the 17 media billionaires in developing countries and BRICS, half originated in "new" media—mobile telecom (7) and Internet (2). The other half is based in classic old media of TV (7) and publishing (1), run by very established families.
- The top 10 individual owners hold, in aggregate, $225 billion worth of media companies. The top 20 individual owners hold $313 billion. And the top 50 had $422 billion.

For 2013, the media sector's global aggregate (i.e., 30-countries) revenues were about $2.25 trillion dollars. If we estimate a multiple of enterprise value (EV) to revenue of 2.07 (see discussion below), this would give a rough estimate for valuation of $4.7 trillion. Thus, the share of the top 10 individual owners in the world's media companies is about 4.8%, the top 20 owners about 6.7%, and the top 50 had 9.0%. This is a low percentage. But even if we reduced the multiple to a very conservative value of 1, these percentages would only double, and the top 50 media billionaires had "only" 18% of the world's media by value.

## Institutional Owners

One thing to note is the prevalence of institutional investors in media ownership. While large individual and family ownership stakes

7. <http://www.forbes.com/billionaires/list/>.

8. <http://www.bloomberg.com/billionaires/>.

exist, of course, the largest are not personal but institutional.

Institutional owners control the shares they hold in two ways. First, they own shares outright in their own account, partly to earn a dividend return and often for the potential gain in value. In some cases they might have been part of an investment bank consortium that created and marketed the public shares in an initial or secondary public offering (IPO or SPO), and they may have kept shares for gradual sale.

Larger in volume is the second way in which institutional owners hold shares as asset managers. They manage other people's money through various forms of investment funds which they control, but which are indirectly benefiting their funds' investors. In practice, the two forms of holdings are intermingled. And it is conceptually unclear what the difference is of managing on behalf of a fund's investors and limited partners or on behalf of their own shareholders and partners.

Institutional ownership is not a recent phenomenon, but it has increased with the growth of mutual funds and pension funds. Mutual funds are companies that seek and manage the money of investors and invest it in a portfolio of stocks, bonds, and other assets. They attempt to optimize return for a given risk level, or category of investment. In some countries, government rules aimed at protecting investors from imprudent risk-taking limit fund investment in any single company to no more than, for example, 5% of assets in any one company, and to no more than 10% of any company's outstanding shares.[9] This limits the capacity of any individual fund to exercise much control over a firm.

The term "institutional investors" formally applies only to entities that hold themselves out as primarily engaged in investing, reinvesting, or trading securities. Accordingly, actual institutional ownership of media companies may be understated by most data sources. For example, if Company A is partly owned by a Company B, which does not fall into the category of "institutional investor," then B's investment will not be counted in the overall institutional ownership figures for Company A, even though B may be largely owned by institutions. Moreover, the SEC does not require disclosure of institutional investors that own less than 5% of a company. However, many companies list such ownerships anyway.

The trend toward institutional ownership accelerated with the emergence of private equity (PE) funds as acquirers of stock market traded "public" media and communications firms. PE funds pool the financial resources of large investors, which are often financial institutions such as pension funds. They then buy up companies, withdraw their shares from public trading, reorganize them, and eventually may sell them back to the wider investor public.

In contrast to the public institutional funds with their numerous small investors, private equity funds are limited by law and strategy to deep-pocket investors whose identities are not disclosed.

After 2005, large US-based private equity firms such as Bain, Blackstone, Carlyle, KKR, Providence, or Texas Pacific—and their equivalents elsewhere, especially in London—acquired major media and communications companies. These include, in the United States, Clear Channel, MGM, Univision, Primedia, PanAmSat, in The Netherlands VNU (which owns Nielsen Media Research in the United States). Other PE-held firms include ProSieben-Sat1 in Germany, TDC in Denmark, Eircom in Ireland (for two ill-fated rounds), SBS in Luxembourg, EMI in the United Kingdom, and PRISA in Spain. Still other firms were taken

9. Institutional investors are required to file disclosure information when they own 5% or more of a publicly traded company. They are limited in their ability to profit on shares held for less than six months once they reach the 10% ownership threshold. If a mutual fund wants to promote itself as "diversified" and gain pass-through tax benefits, then the regulated 75% of the fund cannot own more than 10% of a company in its portfolio. Also, a "diversified" fund cannot have more than 5% of its total assets invested in a single company (Investment Company Act of 1940). Therefore, only 25% of a fund, the unregulated portion, can be concentrated in a single stock. Pension funds are generally less regulated than mutual funds and fall under the Employee Retirement Income Security Act of 1974 (ERISA), which require each fund to diversify. But under ERISA, pension fund managers are given some leeway to avoid diversification if it is "clearly prudent" not to do so.

fully private by their own majority shareholders, such as Bertelsmann and Cox.

These institutional owners hold a great number of media stocks, with stakes valued at often many billions of dollars. Table 36.4 provides an example, using the Vanguard Group, a major American investment fund with significant media holdings. Only stakes whose value was above $20 million are listed.

- Vanguard, based in the United States, is the largest institutional investor in the top 20 content companies, including companies from overseas. Vanguard owns assets in 12 of the top 20 content companies. Vanguard, given its origin as an index fund, seems invested in almost every major media company—in the United States, the five major network and content providers, the three major traded cable TV companies, and two major search engines; in Europe, three major TV companies; in Canada, Singapore, France, and Germany, major telecoms.
- More interesting than the components of Vanguard's portfolio is their magnitude, and what is missing. Not included are US telecom firms such as Verizon and AT&T. Also less represented is Asia. Hugely present is Google ($20 billion) and Comcast ($11 billion), Disney ($10 billion), and Time Warner and 21st Century Fox ($5 billion each), not counting another $3 billion for the TWC spin-off.
- Vanguard's portfolio is diversified, but has large holdings in media companies. Its stake in Google is almost as high as those of the founders Brin and Page (though without the voting power). It holds more shares in Comcast than the Roberts family (again, without the votes). It is by far the largest shareholder in Time Warner, Liberty, and Disney (except for Steve Jobs' widow). And so on. Thus, on any objective measure, it is a huge media owner. And yet, hardly anybody has heard of its CEO, F. William McNabb III, or its headquarters location, Malvern, Pennsylvania, even though his company's investments in media qualify him to be an über-media-mogul, albeit as a manager. McNabb does not even have a Wikipedia entry.

Table 36.5 (Short Version) is a summary of the larger Table 36.5 (Full Version) of the Appendix. It compiles the media holdings of large financial institutions. For institutional owners in several countries the data might not be complete since reporting is partial only.

The institutional investors chosen for this table are those that appeared as large stakeholders for one or several of the top media companies, as listed in Table 36.4. In the alternative, they are among the largest asset managers in the world and in their country.

## Observations:

- The total media assets of the top 10 institutional owners add up to $332.5 billion; the top 20 have $423.4 billion and the top 30 have $449 billion.
- Of overall media value, estimated as $4.7 trillion before including state-owned media, the top 10 institutional owners hold 6.1%, and the top 30 hold 9.6%. We use again a multiple of enterprise value to revenue of 2.07. A more conservative multiple of 1 would about double those percentages.
- Of the headquarters of those top 30 asset management companies, 73% are based in the United States (22 companies).
- US institutional owners dominate this list. There are several reasons:
  - US-based asset management firms are particularly active and operate in other major financial centers. Overall, North American firms account for 50% of the asset management industry's funds under management. UK firms have about 10% and about 5% each for Switzerland, Japan, Germany, and France.[10] Total US-wide managed assets

10. "The World's 500 Largest Asset Managers." Towers Watson, 2013. <http://www.towerswatson.com/en/Insights/IC-Types/Survey-Research-Results/2013/11/The-Worlds-500-Largest-Asset-Managers-Year-end-2012>.

Table 36-4. The Media Holdings of the Vanguard Group (as of Sept 2013)

| | COMPANIES OWNED | NUMBER OF SHARES OWNED | % OF COMPANY OWNED | % OF VOTING POWER | VALUE OF SHARES ($ MIL) |
|---|---|---|---|---|---|
| The Vanguard Group (US) | Google (US) A | 23,255,893 | 6.9 | 2.7 | 20,616 |
| | Comcast (US) A | 227,152,526 | 8.6 | 7.1 | 9,981 |
| | Comcast (US) K | 25,104,652 | 0.95 | 0 | 1,006 |
| | Disney (US) | 148,460,516 | 7.4 | 8.4 | 9,612 |
| | Time Warner (US) | 84,839,919 | 9.1 | 9.1 | 5,383 |
| | 21st Century Fox (US) A | 152,697,912 | 6.6 | 0 | 4,999 |
| | Time Warner Cable (US) | 26,529,901 | 8.9 | 8.9 | 2,986 |
| | Yahoo (US) | 87,549,068 | 7.9 | 7.9 | 2,649 |
| | CBS (US) B | 46,523,377 | 7.4 | 0 | 2,601 |
| | Liberty Global (Malone) (US) K | 12,282,027 | 4.8 | 0 | 914 |
| | Liberty Global (Malone) (US) A | 11,144,802 | 4.3 | 4.58 | 873 |
| | Liberty Media (Malone) (US) A | 4,461,937 | 3.7 | 2.12 | 652 |
| | China Mobile (China) | 29,434,000 | 1.5 | 1.46 | 334 |
| | Singapore Telecommunications (Singapore) | 7,802,432 | 0.05 | 0.05 | 179 |
| | Vodafone Group (UK) | 50,117,938 | 1.0 | 1.03 | 168 |
| | Rogers Communications (Canada) | 1,624,590 | 0.31 | 0.31 | 76 |
| | Telefonica (Spain) | 4,180,947 | 0.09 | 0.09 | 64 |
| | Bouygues (France) | 1,649,698 | 0.5 | 0.56 | 59 |
| | Lagardère (France) | 1,706,578 | 1.3 | 1.17 | 54 |
| | Shaw Communications (Canada) | 1,862,797 | 0.41 | 0 | 45 |
| | Deutsche Telekom (Germany) | 3,211,193 | 0.08 | 0.08 | 45 |
| | TELUS (Canada) | 1,053,142 | 0.17 | 0.17 | 36 |
| | China Telecom (China) | 68,850,000 | 0.09 | 0.09 | 36 |
| | Vivendi (France) | 1,409,341 | 0.11 | 0.11 | 33 |
| | TDC (Denmark) | 3,214,884 | 0.40 | 0.40 | 27 |
| | Sony Corp. (Japan) | 1,061,478 | 0.10 | 0.10 | 23 |
| | France Telecom (France) | 1,921,686 | 0.07 | 0.07 | 23 |
| | ProSiebenSat.1 (Germany) | 505,387 | 0.23 | 0.23 | 21 |
| | Total, including stakes <15M | | | | 63,800 |

Table 36-5. Largest Institutional Owners of Media (as of Sept 2013) (Short Version)

| COMPANY | TOTAL MEDIA HOLDINGS ($ BIL) |
|---|---|
| State Street Bank (US) | 64.8 |
| Vanguard Group (US) | 63.8 |
| Fidelity (US) | 46.5 |
| Capital Group (US) | 35.2 |
| T. Rowe Price Assoc. (US) | 26.1 |
| BlackRock (US) | 24.3 |
| Dodge & Cox (US) | 20.0 |
| Massachusetts Finance (US) | 18.7 |
| JP Morgan Chase (US) | 17.9 |
| Government Pension Fund of Norway | 15.2 |
| Wellington Management (US) | 14.0 |
| Janus Group (US) | 13.8 |
| Goldman Sachs (US) | 12.3 |
| ClearBridge (US) | 10.9 |
| Legal & General (UK) | 7.7 |
| Baille Gifford (UK) | 7.5 |
| Handsome Reward Ltd (UK) | 7.4 |
| Oppenheimer Funds (US) | 6.7 |
| Thornburg Invest. Mgmt. (US) | 6.2 |
| Coatue Management (US) | 4.6 |
| State Farm (US) | 4.5 |
| Berkshire Hathaway (US) | 4.0 |
| Deutsche Bank (Germany) | 3.3 |
| Viking Global Investors (US) | 3.2 |
| Lazard (US) | 3.0 |
| Horizon Asset Management (US) | 1.9 |
| AXA Financial (France) | 1.6 |
| Commerzbank (Germany) | 1.4 |
| Scout Capital Management (US) | 1.0 |
| BNP Paribas (France) | 1.0 |
| Children's Investment Fund (UK) | 0.8 |
| HSBC (UK) | 0.6 |
| Cascade Investment (US) | 0.6 |
| Lloyds Banking Group (UK) | 0.5 |
| MedioBanca (Italy) | 0.2 |

were, in 2012, $35.9 trillion in North America, $24 trillion in Europe, and $4.8 trillion in Japan.
- The old-age pension system in the United States is based on individual investment accounts (such as "401(k)" plans) rather than on a tax-based "pay-as-you-go" governmental pension plan prevalent in Europe.
- In the United States, there are many huge endowments of private universities, museums, and so on, that are run by asset management firms.
- There exists a greater willingness and ability to invest pension funds and other forms of savings in equities (stocks) rather than bonds, which are safer in producing income but do not provide ownership rights.

For these and other reasons, the stock investment portfolios of US-based institutional investors are quite large and are spread around the world. They include—but are not—media assets. Given the size and expertise of the asset management firms, they also attract money to manage from investors around the world, not just the United States.

## State Ownership

The third category of ownership is that by government, in various forms—direct operations, public corporations, stock holdings, nonprofits under state or parliamentary control, and so on. Such ownership is quite prevalent. Most countries have established public service television networks. Many have a governmental role in wireline and mobile telecom, going back one century of full state monopoly over telegraphs and telephones. In the 1980s, they were privatized, but frequently large share holdings remained with the state.[11]

How much are these and other government holdings worth? It is straightforward to value government shares in a company where those shares are traded in a stock exchange and thus have a publicly available price. For example, Table 36.1 shows that the value of the Chinese State's holdings in China Telecom is $40 billion, based on the market price for the stock. Similarly, Japan's share in NTT is worth $27.6 billion. But it is harder to value public service TV or radio broadcasting, which are fully owned by state entities. Thus, any methodology will be a rough order-of-magnitudes only. In stock valuation, a frequent measure where profits are unknown or nonexistent is to estimate enterprise value (EV) as a multiple of sales revenues. Such multiples vary by industry and country. One report by New York University,[12] shows the multiples, for the United States in 2014, as:

| | |
|---|---|
| Telecom (Wireline) | 1.91 |
| Telecom (Wireless) | 2.68 |
| Publishing | 1.32 |
| Newspapers | 1.57 |
| Cable TV | 2.37 |
| Information Services | 3.16 |
| Entertainment | 2.34 |
| Advertising | 1.22 |

The average value is 2.07. We then make the assumption that this media multiple also applies to the media sector in other countries.[13] This results in estimates of the value of state holdings (Table 36.6).

11. Noam, Eli. *Television in Europe*. New York: Oxford University Press, 1991; Noam, Eli. *Telecommunications in Europe*. New York: Oxford University Press, 1992.

12. "Revenues Multiples by Sector." < http://people.stern.nyu.edu/adamodar/New_Home_Page/datafile/psdata.html>

13. It could be argued that the United States has a well-developed stock market, and thus US multiples are higher. But it is equally possible to argue the opposite, that the US media market, by being somewhat more competitive, constrains profitability, and that the growth potential in emerging countries is also much higher. Given that both conflicting scenarios are plausible, we assume that they cancel each other out.

Table 36-6. Public (State) Ownership Revenues and Asset Value, 2009 or Most Recent

| | PUBLICLY OWNED MEDIA REVENUE (MIL US$) | PUBLICLY OWNED MEDIA ASSETS (BIL US$) |
|---|---|---|
| China | 153,239 | 317.2 |
| Japan | 32,508 | 67.2 |
| Germany[a] | 14,650 | 29.9 |
| France | 12,775 | 26.4 |
| Russia | 12,302 | 25.4 |
| India | 9,605 | 19.8 |
| Norway[b] | | 16.3 |
| UK | 6,717 | 13.9 |
| Taiwan | 5,870 | 12.1 |
| South Africa | 5,625 | 11.6 |
| Switzerland | 4,418 | 9.1 |
| Poland | 4,065 | 8.4 |
| Egypt | 3,634 | 7.5 |
| Italy | 3,463 | 7.1 |
| Spain | 3,108 | 6.4 |
| Sweden | 2,905 | 6.0 |
| Turkey | 2,825 | 5.8 |
| US | 2,770 | 5.7 |
| South Korea | 2,233 | 4.6 |
| Australia | 1,617 | 3.3 |
| Canada | 1,599 | 3.3 |
| Portugal | 1,592 | 3.2 |
| Netherlands | 1,286 | 2.7 |
| Finland | 1,169 | 2.4 |
| Ireland | 736 | 1.5 |
| Brazil | 252 | 0.5 |
| Belgium | 145 | 0.3 |
| Israel | 142 | 0.2 |
| Chile | 106 | 0.2 |
| Argentina | 0.0 | 0.0 |
| Mexico | 0.0 | 0.0 |

[a] excludes ARD share of the Länder.

[b] Includes the Government Pension Fund of Norway which invests worldwide, with media assets of $15.2 billion.

## Observations:

- The media holdings of the Chinese state dwarf those of all other governments.
- Even countries that are market-oriented, such as Japan and Germany, have substantial state ownership in the media sector.
- The total state holding of the 30 countries is $631.3 billion. This accounts for 13.4% of total estimated 30-world media values.[14]

## The World's Largest Media Owners

We can now integrate the three major categories of owners into a single Table 36.7 that shows the largest media owners, across all three categories.

- We can summarize sectoral ownership of the 13 media industries observed. (The lower numbers provided are those for a multiple of enterprise value/revenue of 2.0, the higher one is for a multiple of 1.0):
  - *Top 30 individual owners: 7.9–15.8%*
  - *Top 30 institutional owners: 9.6–19.2%*
  - *Top 30 government holdings: 13.4%*
- These 90 owners, then, account for 30.9–38.4% of the world's media assets.
- With the government holdings mostly in declining industries (broadcast TV and wireline telecom), state holdings are likely to decrease in their share (depending on the revenue-enterprise value multiple, which may be in the range of 1–2).
- As long as innovation favors startups, the share of media assets held by entrepreneurs will increase at the expense of established players.
- Since individual small investors are better prepared to invest in established firms, technological innovation favors the delegation of investment decisions to professional asset managers (i.e., institutional investors).
- The overall upshot is that institutional investors will grow their ownership share, while state ownerships and individual non-entrepreneurial ownership will likely decline.

Public attention has centered on highly visible media moguls such as Murdoch, Redstone, or Berlusconi. But that kind of personalized portrayal has a certain antiquated quality to it, given the prevalence of institutional investors, dot-com startup founders and national governments. We will discuss state ownership first.

### STATE OWNERSHIP

The extent of state ownership of media organization is often underestimated. In some countries, public service television organizations are seen as independent from the government, but this may be more in terms of culture and perception rather than law. Even the BBC—the quintessential independent public service broadcaster—is funded by a government-set earmarked tax, its Board of Governors is appointed by the cabinet, and its world service was funded, until 2014, by the government budget.

More generally, the extent of state media ownership around the world is huge. Perhaps the best qualification is that of Shleifer et al.[15] in a World Bank Study of 97 countries.[16] In 21 of these countries, the government had monopolies in daily newspapers and in 43 countries—almost half—there was a government monopoly of TV (with local news). Sixty percent of the major TV stations around the world were state-owned and 29% of major newspapers. This is probably an understatement.

Distinguishing state and private ownership is not always clear-cut. Sometimes media are technically not owned by the state but by a ruling party. In other cases, political leaders or their families own media. This is the case in Saudi Arabia (two of five major papers by the

---

14. A different Enterprise Value multiple would not alter that number, since the change would modify both numerator and denominator and cancel out.

15. Shleifer, Andrei, Caralee Mcliesh, Tatiana Nenova, and Simeon Djankov. "Who Owns the Media?" World Bank 2001. Also Published in the *Journal of Law and Economics*, 46 (October 2003).

16. For the remaining countries there were no data, which suggest less developed media systems.

Table 36-7. Top Media Owners Worldwide (as of Sept. 2013)

| OWNER OR ASSET MANAGER | VALUE OF MEDIA HOLDINGS ($ BILLIONS) |
|---|---|
| Government of China | 317.2 |
| Government of Japan | 67.2 |
| State Street (US) | 64.8 |
| Vanguard (US) | 63.8 |
| Fidelity (US) | 46.5 |
| Capital Group (US) | 35.2 |
| Government of Germany | 29.9 |
| Carlos Slim (America Movil, Mexico) | 29.2 |
| Larry Page (Google, US) | 26.7 |
| Government of France | 26.4 |
| T. Rowe Price Assoc. (US) | 26.1 |
| Sergey Brin (Google, US) | 26.0 |
| Government of Russia | 25.4 |
| BlackRock (US) | 24.3 |
| Cox family (Chambers, Kennedy, Parry-Sheden, Anthony) (Cox Communications (US) | 24.0 |
| Michael Bloomberg (Bloomberg LP, US) | 24.0 |
| David Thompson family (Thompson Reuters, Canada) | 20.3 |
| Marinho family (Globo, Brazil) | 20.0 |
| Dodge & Cox (US) | 20.0 |
| Government of India | 19.8 |
| Mark Zuckerberg (Facebook, US) | 19.0 |
| Massachusetts Finance (US) | 18.7 |
| Brian Roberts family (Comcast, US) | 18.5 |
| JP Morgan Chase (US) | 17.9 |
| Newhouse family (Advance Publications, US) | 17.1 |
| Government of Norway[17] | 16.2 |
| Wellington Management (US) | 14.0 |
| Government of the UK | 13.9 |
| Janus Group (US) | 13.8 |
| Goldman Sachs (US) | 12.3 |
| Government of Taiwan | 12.1 |
| Sawiris Family (Orascom, Egypt) | 12.0 |
| Murdoch family (News Corp./21st Century Fox, US) | 11.6 |
| Government of South Africa | 11.6 |
| ClearBridge (US) | 10.9 |
| Charles Ergen (Dish Network, US) | 10.6 |
| Government of Switzerland | 9.1 |
| Jobs Family (Disney, US) | 8.7 |

17. Includes the Government Pension Fund of Norway which invests worldwide, with media assets of $15.2 billion.

Table 36-7. *continued*

| OWNER OR ASSET MANAGER | VALUE OF MEDIA HOLDINGS ($ BILLIONS) |
|---|---|
| Masayoshi Son (Softbank, Japan) | 8.6 |
| Government of Poland | 8.4 |
| Eric Schmidt (Google, US) | 8.2 |
| Legal & General (UK) | 7.7 |
| Edward Rogers Family (Rogers Cable, Canada) | 7.6 |
| Robin Yanhong Li (Baidu, China) | 7.53 |
| Baille Gifford (UK) | 7.5 |
| Government of Egypt | 7.5 |
| Government of Italy | 7.1 |
| Hearst Family (Hearst Publishing and TV, US) | 7.0 |
| Oppenheimer Funds (US) | 6.7 |
| Xavier Nile (France) | 6.6 |
| Government of Spain | 6.4 |
| Thornburg Invest. Mgmt. (US) | 6.2 |
| Ambani family (Reliance, India)[a] | 6.2 |
| Silvio Berlusconi (Fininvest, Italy)[a] | 6.2 |
| Government of Sweden | 6.0 |
| David Geffen (film and music, US) | 5.6 |
| Avande Krishnan (Moxis mobile, Malaysia) | 5.9 |
| Government of Turkey | 5.8 |
| Government of the United States | 5.7 |
| John Malone (Liberty, US) | 5.6 |
| Denis O'Brien (Communicorp Radio, Ireland) | 5.2 |

[a] Media activities only

Royal family; the appointment of editors-in-chief of newspapers is subject to government approval). In Singapore, the politically dominant Lee family controls companies that control the Singapore Press Holdings, which owns all five major newspapers.

According to the World Bank Shleifer et al. (2001) study, state ownership is more widespread in countries that are poorer, or have lower educational levels or autocratic regimes. The study shows that state ownership in a country is associated with indicators such as journalists jailed, Internet censorship, and lower civil liberties. The authors do not find empirical evidence of any benefits of state ownership.

This is in contrast to the public interest explanation of state ownership of media, according to which the public benefit externalities of media, coupled with the inability to exclude listeners and viewers, leads to an underinvestment by private firms, thus requiring a governmental role.

## INSTITUTIONAL OWNERSHIP

Our study finds that the total ownership stakes held by large institutions in various media sectors are significant. And even though the ownership percentages of institutions like State Street Bank or Vanguard in any single firm are usually intermediate or small in percentage terms, these institutional investors are often among the largest owners of the companies. Institutional investors rarely have interests greater than 5% in companies, since they seek diversification of their portfolios and are subject to a variety of legal limitations and reporting requirements.

Generally, the stake of institutional investors is much larger than those of individuals. In 2013, State Street had $65 billion invested in major media companies. Rupert Murdoch, in comparison, had "only" $11.6 billion. Dodge & Cox, with $20 billion, had more money tied up in media than Berlusconi, Malone, Redstone, and Lagardère combined. Yet hardly anyone outside the financial community has heard of Dodge & Cox, whereas the other four are well known.

One reason is that while individual holdings tend to be more concentrated in one firm, institutional investors' assets are scattered. The average number of media sector companies in the portfolio of the top 20 institutional investors in 2013 was 50. And the largest of institutional investors in media and communications held only between 4% and 30% of their total assets invested in the largest of information industry companies, which include IT. In contrast, large individual owners are much more heavily focused in their ownership. For that reason, they are usually much more involved in the day-to-day running of the firm and often control "their" companies.

Institutional investors are usually viewed as primarily concerned with short- or medium-term gain, gauging corporate performance solely according to stock price and earnings. But institutional investors cannot easily liquidate large stakes and often are in for the long haul. They have the capability to intervene, and top management knows that. In 1997, institutional investors became dissatisfied with the composition of Walt Disney's board of directors, which *Business Week* had named the "worst board in America." It included individuals with close ties to CEO Michael Eisner, such as his personal attorney and his architect. Eisner was forced to make changes in response to the institutional investor criticism, but his troubles with institutional and pension funds continued, and they led, eventually, to his ouster.

In 2003, US mutual fund Tweedy Browne, which held 18% of the newspaper holding firm Hollinger International shares, initiated an investigation that uncovered misspending at the newspaper chain (*Chicago Sun-Times*, *Daily Telegraph*, and several other papers). The discovery led to the resignation of Lord Conrad Black from his position as CEO, the sale of the company, and to Black's criminal conviction.

In 2006, several "activist" institutional shareholders, led by Carl Icahn, challenged Time Warner's conglomerate structure, advocating a breakup of the company. They argued that the sum of the parts was more valuable than the whole. Time Warner's management opposed the shareholder resolution and prevailed in a formal sense. But within a few years it sold or spun off these parts of the company: Warner Music Group, Time Warner Cable, AOL, TW Telecom, Time Books, and Time Inc. magazines.

In theory, fund managers might be tempted to oppose content that would negatively affect other holdings of their portfolio.[18] If Fidelity holds large ownership positions in tobacco companies and in Disney, it is possible that Disney's ABC TV network management might pull its punches in producing programs about the addictiveness of nicotine. Since direct intervention by institutional owners would often not remain confidential and backfire, either such instances are rare or they are implicit and require no direct communication. In contrast, for individual ownership there is ample evidence for direct intervention into content matters by the major individual owners, including on just that tobacco issue.

Generally, institutional investors will prefer safe mainstream content rather than controversial one that may make some of their investors unhappy. Similar incentives for safe mainstream

18. Soloski, John, and Robert Picard. "The New Media Lords: Why Institutional Investors Call the Shots." *Columbia Journalism Review* (September 19, 1996): 11.

content exist also for corporate media management. It also exists for individual owners, unless their personal politics and commercial interests are intermingled. Institutional ownership might affect content quality through greater pressures for short-term profitability. Yet it may also shield managers from control by erratic principal owners.

As this chapter shows, the popular belief that technological convergence in the media sector, plus worldwide mergers, has resulted in a small group of media moguls is not an accurate one. They do exist, of course, as Table 36.3 shows. But there are also a large number of fund managers who control, on behalf of their fund investors, narrow and big slices of a very big pie. While they do not micro-manage the companies' leadership, often in contrast to large individual owners, they nevertheless put performance pressure on them through their buy and sell decisions.

Thus, the actual ownership of firms has shown two seemingly contradictory trends. On the other hand, it has become more fragmented—a large number of institutional stakeholders with small-percentage stakes. On the other hand, some of these institutional owners have stakes in many media and information firms. If they acted in concert, they could exercise industry-wide power. But there is, so far, little evidence for such joint activity.

# APPENDIX:THE WORLD'S OWNERSHIP OF MEDIA—JASON BUCKWEITZ

Table 36-1. Full Version Ownership of the World's Top 20 Platform Companies (as of Sept. 2013)

| | MAJOR OWNERS | NUMBER OF SHARES OWNED | PERCENT OF COMPANY OWNED | PERCENT OF VOTING POWER | VALUE OF SHARES (MIL $) |
|---|---|---|---|---|---|
| NTT (Japan) | Under Japanese law no person or institution of non-Japanese Nationality may own more than 33% of NTTs voting shares. | | | | |
| | Government of Japan (Public) | ~436,816,470 | 33 | 33 | 27,672 |
| | Japan Trustee Services Bank (Japan) | 67,508,000 | 5.1 | 5.1 | 3,421 |
| | Master Trust Bank of Japan (Japan) | 37,138,000 | 2.81 | 2.81 | 1,882 |
| | State Street (US) | 32,488,000 | 2.45 | 2.45 | 1,646 |
| | JP Morgan Chase (US) | 27,676,000 | 2.1 | 2.1 | 1,402 |
| | NTT Employee Share-Holding Association (Japan) | 10,273,000 | 0.78 | 0.78 | 520 |
| AT&T | Vanguard (US) | 437,604,939 | 8.24 | 8.24 | 15,092 |
| | State Street (US) | 233,415,579 | 4.39 | 4.39 | 8,262 |
| | Evercore (US) | 232,820,447 | 4.38 | 4.38 | 8,241 |
| | BlackRock (US) | 204,170,442 | 3.84 | 3.84 | 6,870 |
| | Capital Research (US) | 143,148,651 | 2.7 | 2.7 | 4,816 |
| | Northern Trust (US) | 77,328,866 | 1.46 | 1.46 | 2,737 |
| | Bank of New York Mellon (US) | 75,778,353 | 1.43 | 1.43 | 2,682 |
| | Price (T.Rowe) Associates (US) | 72,256,871 | 1.36 | 1.36 | 2,651 |
| Deutsche Telekom | Government of Germany (Public) | 1,395,738,354 | 32 | 32 | 22,227 |
| | Government of Germany (direct) | 654,252,353 | 15 | 15 | 10,418 |
| | KfW Bankengruppe (State Controlled Investment Vehicle) (Germany) | 741,486,001 | 17 | 17 | 11,808 |
| | BlackRock (US) | 131,286,639 | 3.01 | 3.01 | 1,672 |

| | | | | | |
|---|---|---|---|---|---|
| | Government Pension Fund of Norway (Public) | 82,117,602 | 1.91 | 1.91 | 1,046 |
| | Vanguard Group (US) | 50,356,048 | 1.14 | 1.14 | 641 |
| | Legal and General (UK) | 24,829,037 | 0.57 | 0.57 | 316 |
| | AEGON Ireland (Ireland) | 22,201,117 | 0.52 | 0.52 | 282 |
| | Deka Investment (Germany) | 12,634,462 | 0.29 | 0.29 | 160 |
| | Union Investment (Germany) | 11,428,679 | 0.26 | 0.26 | 145 |
| | Pioneer Investment (US) | 11,257,391 | 0.26 | 0.26 | 143 |
| China Mobile (China) | Government of China (Public) | ~2,947,010,345 | 74.08 | 74.08 | ~192,051 |
| | Lazard (US) | 25,062,318 | 0.63 | 0.63 | 1,306 |
| | Matthews International Capital (US) | 15,722,343 | 0.39 | 0.39 | 823 |
| | Fisher Asset Management (US) | 3,827,916 | 0.1 | 0.1 | 198 |
| | Neuberger Berman Group(US) | 3,527,862 | 0.09 | 0.09 | 182 |
| | Dimensional Fund Advisors(US) | 4,722,004 | 0.11 | 0.11 | 253 |
| | Rockefeller Financial Services(US) | 1,559,556 | 0.04 | 0.04 | 80 |
| | Bank of New York Mellon (US) | 1,606,962 | 0.04 | 0.04 | 85 |
| | Arrowstreet Capital (US) | 1,745,523 | 0.04 | 0.04 | 90 |
| | Harris (Everett) & Company (US) | 1,781,661 | 0.04 | 0.04 | 92 |
| | Macquarie Group Limited (Australia) | 1,507,075 | 0.04 | 0.04 | 80 |
| Telefónica (Spain) | Banco Bilbao Vizcaya Argentaria (Spain) | 261,798,437 | 5.753 | 5.753 | 3,659 |
| | Caja de Ahorros y Pensiones de Barcelona, "la Caixa" (Spain) | 254,697,815 | 5.596 | 5.596 | 3,560 |
| | BlackRock (US) | 177,257,649 | 3.895 | 3.895 | 2,478 |
| | Société Generale (France) | | 2.92 | 2.92 | |
| | Government Pension Fund of Norway (Public) | | 2.15 | 2.15 | |
| | Vanguard Group (US) | | 1.12 | 1.12 | |
| | Franklin Mutual Advisers (US) | | 0.74 | 0.74 | |
| Vodafone (UK) | Capital Research and Management (US) | 2,472,830,993 | 5.11 | 5.11 | 7,814 |
| | Legal and General (UK) | 2,178,073,425 | 4.39 | 4.39 | 6,882 |
| | Franklin Mutual Advisers (US) | 851,430,263 | 1.75 | 1.75 | 2,690 |

*continued*

Table 36-1. *continued*

| | MAJOR OWNERS | NUMBER OF SHARES OWNED | PERCENT OF COMPANY OWNED | PERCENT OF VOTING POWER | VALUE OF SHARES (MIL $) |
|---|---|---|---|---|---|
| Vodafone (UK) (*cont.*) | Vanguard Group (US) | 731,196,044 | 1.51 | 1.51 | 2,310 |
| | Dodge & Cox (US) | 619,468,400 | 1.28 | 1.28 | 1,957 |
| | Scottish Widows Inv Partnership (Ireland) | 437,785,921 | 0.89 | 0.89 | 1,383 |
| | Insight Investment Management (Glb) (UK) | 384,819,424 | 0.79 | 0.79 | 1,216 |
| | AXA Life Europe (France) | 344,054,837 | 0.71 | 0.71 | 1,087 |
| | BlackRock (US) | 303,833,502 | 0.62 | 0.62 | 960 |
| Orange (France) | Government of France (Public) | 701,986,766 | 27 | 27 | 8,924 |
| | French Government (France) | 350,342,189 | 13.45 | 13.45 | 4,453 |
| | Fond Stratégique d'Investissement (Government Investment Fund)(France) | 351,644,577 | 13.5 | 13.5 | 4,470 |
| | Société Generale (France) | | 4.99 | 4.99 | |
| | Barclay's (UK) | | 4.57 | 4.57 | |
| | Franklin Templeton Investment (US) | 54,960,745 | 2.11 | 2.11 | 558 |
| | Government Pension Fund of Norway (Public) | 31,086,505 | 1.17 | 1.17 | 316 |
| | BlackRock (US) | 27,386,104 | 1.06 | 1.06 | 278 |
| | Amundi (Capital Investment) (France) | 20,807,729 | 0.79 | 0.79 | 211 |
| | Vanguard Group (US) | 18,375,443 | 0.7 | 0.7 | 186 |
| | DNCA Finance (France) | 17,367,533 | 0.66 | 0.66 | 176 |
| | Legal and General (UK) | 15,227,679 | 0.57 | 0.57 | 154 |
| | Natixis Asset Management (France) | 10,385,480 | 0.39 | 0.39 | 105 |
| Telecom Italia (Italy) | Telco S.p.A | ~4,338,525,259 | 22.387 | 22.387 | ~3,525 |
| | Telefonica (42.3% of Telco ) | ~1,835,254,130 | 9.47 | 9.47 | ~1,491 |
| | Generali Group (28.1% of Telco) | ~1,218,980,832 | 6.29 | 6.29 | ~990 |

| | | | | | |
|---|---|---|---|---|---|
| | Intesa San Paolo S.p.A (10.6% owner of Telco S.p.A) | ~459,298,024 | 2.37 | 2.37 | ~298 |
| | Mediobanca S.p.A (10.6% owner of Telco S.p.A) | ~459,298,024 | 2.37 | 2.37 | ~298 |
| | Sintonia S.A (8.4$ of Telco S.p.A) | ~361,605,036 | 1.88 | 1.88 | ~235 |
| | Findim Group | ~966,270,020 | 4.986 | 4.986 | ~628 |
| | BlackRock | | 10.1 | 10.1 | |
| | Dodge & Cox (US) | 403,097,000 | 2.08 | 2.08 | 262 |
| | Government Pension Fund of Norway (Public) | 357,758,174 | 1.86 | 1.86 | 232 |
| | Legal and General (UK) | 228,018,407 | 1.17 | 1.17 | 148 |
| | Vanguard Group (US) | 161,299,317 | 0.84 | 0.84 | 104 |
| | Pictet Asset Management (Switzerland) | 117,312,266 | 0.61 | 0.61 | 76 |
| | Grantham, Mayo, Van Otterloo (US) | 105,706,212 | 0.54 | 0.54 | 68 |
| | Majedie Asset Management (UK) | 56,915,587 | 0.29 | 0.29 | 36 |
| | RBC (Canada) | 53,082,034 | 0.27 | 0.27 | 34 |
| | Exane Asset Management (France) | 42,638,329 | 0.22 | 0.22 | 27 |
| Verizon (US) | Vanguard (US) | 251,846,095 | 8.8 | 8.8 | 12,524 |
| | State Street (US) | 120,621,754 | 4.22 | 4.22 | 6,072 |
| | Capital Research (US) | 97,935,900 | 3.42 | 3.42 | 4,813 |
| | BlackRock (US) | 78,147,989 | 2.73 | 2.73 | 3,933 |
| | Capital World Investors (US) | 74,336,675 | 2.6 | 2.6 | 3,653 |
| | Fidelity (US) | 44,917,104 | 1.57 | 1.57 | 2,207 |
| | Northern Trust (US) | 42,896,884 | 1.5 | 1.5 | 2,159 |
| | Wellington (US) | 41,331,439 | 1.44 | 1.44 | 2,031 |
| | Bank of New York Mellon (US) | 40,672,222 | 1.42 | 1.42 | 2,047 |
| | JP Morgan Chase | 32,796,020 | 1.15 | 1.15 | 1,650 |
| América Móvil (Mexico) | Three share classes of stock. L shares have no voting rights but receive a preferred dividend. AA and A shares have full voting rights. | | | | |
| | Carlos Slim Family Trust (AA) | 12,376,000,000 | 52.623 | 52.623 | ~29,217 |

*continued*

Table 36-1. *continued*

| | MAJOR OWNERS | NUMBER OF SHARES OWNED | PERCENT OF COMPANY OWNED | PERCENT OF VOTING POWER | VALUE OF SHARES (MIL $) |
|---|---|---|---|---|---|
| Telemar Participações (Oi, Brazil & Portugal Telecom) (merges 2014) | AT&T (US) (AA) | 5,739,000,000 | 23.765 | 23.765 | ~13,198 |
| | Telemar is owned by a conglomerate of 8 companies. It owns 62.8% of the merged company with Portugal Telecom. | | | | |
| | AG Telecom (Brazil) (telecom holding of Andrade Guiterrez conglomorate) | | 11.28 | 11.28 | 442 |
| | LF Tel (Brazil) | | 7.4 | 7.4 | 291 |
| | Bindes Partiipacoes (Brazil) (Investment Bank) | | 8.1 | 8.1 | 197 |
| | Bratel Brasil (Brazil) (Wholly owned subsidiary of Portugal Telecom.) | | 7.5 | 7.5 | 181 |
| | Fundacao Atantico- FATL (Brazil) (Pension fund) | | 7.13 | 7.13 | 7.13 |
| | PREVI (Brazil) (Banco do Brazil Employee pension fund) | | 6.01 | 6.01 | 146 |
| | FUNCEF (Brazil) (Federal Savings Foundation pension fund) | | 4.64 | 4.64 | 112 |
| | Petros (Brazil) (Investment Fund) | | 4.64 | 4.64 | 112 |
| | Esporito Santo Group (Portugal) | | 3.84 | 3.84 | 3.85 |
| | RS Holding (Rocha dos Santos de Ameida) (Portugal) | | 3.8 | 3.8 | 3.82 |
| | Telemar Norta Lesta (Oi) (Brazil) | | 3.8 | 3.8 | 3.18 |
| | Government Pension Fund of Norway (Public) | | 1.9 | 1.9 | 1.8 |
| | UBS (Switzerland) | | 1.78 | 1.78 | 1.78 |
| | Visabeira Group (Portugal) | | 1.0 | 1.0 | 1.0 |
| | BlackRock (US) | | 0.9 | 0.9 | 0.9 |

| | | | | | |
|---|---|---|---|---|---|
| China Telecom | Government of China | 62,977,197,275 | 78 | 78 | 40,147 |
| | China Telecom Group (Public) (Government owned) (China) | 57,361,602,470 | 70.89 | 70.89 | 36,568 |
| | Guangdong Province (Public) | 5,615,594,805 | 6.94 | 6.94 | 3,579 |
| | Colonial First State (Australia) | 2,200,924,795 | 2.72 | 2.72 | 1,122 |
| | BlackRock (US) | 2,142,209,109 | 2.66 | 2.66 | 1,092 |
| | Vanguard (US) | 454,607,051 | 0.56 | 0.56 | 231 |
| | Government Pension Fund of Norway (Norway) | 377,657,000 | 0.47 | 0.47 | 192 |
| | AXA (France) | 446,122,000 | 0.55 | 0.55 | 227 |
| | China Citic Bank (China) | 301,662,000 | 0.37 | 0.37 | 153 |
| | Capital Research and Management (US) | 230,138,000 | 0.28 | 0.28 | 117 |
| Telstra (Australia) | Telstra has 1.4 million individual shareholders. | | | | |
| | Capital Research and Management (US) | 283,496,500 | 2.28 | 2.28 | 1,224 |
| | BlackRock (US) | 78,225,739 | 0.63 | 0.63 | 337 |
| | MLC Investments (Australia) | 78,641,722 | 0.63 | 0.63 | 339 |
| | Thornburg (US) | 119,273,356 | 0.96 | 0.96 | 515 |
| | Australian Investment Co Services (Australia) | 55,687,000 | 0.45 | 0.45 | 240 |
| | Newton Investment (UK) | 50,000,000 | 0.4 | 0.4 | 216 |
| | Blair William & Co (US) | 46,746,861 | 0.38 | 0.38 | 201 |
| | BlackRock (US) | 45,895,971 | 0.36 | 0.36 | 198 |
| | Vanguard (US) | 64,407,404 | 0.52 | 0.52 | 278 |
| | Argo Investments (Australia) | 38,954,800 | 0.31 | 0.31 | 168 |

*continued*

Table 36-1. *continued*

| | MAJOR OWNERS | NUMBER OF SHARES OWNED | PERCENT OF COMPANY OWNED | PERCENT OF VOTING POWER | VALUE OF SHARES (MIL $) |
|---|---|---|---|---|---|
| Svyazinvest[19] (Russia) | Government of Russia (Public) | 1,564,341,770 | 53.15 | 53.15 | 5,881 |
| | MarshallCapital (Private Equity Group, Russia) | 309,042,118 | 10.5 | 10.5 | 1,161 |
| | Mobitel (Slovenian Government) (Public) | 123,911,173 | 4.21 | 4.21 | 465 |
| | Vnesheconombank (Government of Russia) | 72,109,827 | 2.45 | 2.45 | 271 |
| 21st Century Fox (US) | There are two classes of shares, Class A (Equity, Non-Voting) and Class B (Voting). | | | | |
| | Murdoch Family | 311,103,499 (B) | 13.48 | 39.0 | 9,965 |
| | Vanguard | 152,697,912 (A) | 6.61 | 0.00 | 4,658 |
| | Capital Group (US) | 94,872,222 (A) | 4.11 | 0.00 | 2,894 |
| | Dodge & Cox (US) | 93,300,035 (A) | 4.04 | 0.00 | 2,904 |
| | State Street Corporation (US) | 73,301,178 (A) | 3.17 | 0.00 | 2,236 |
| | Janus Group (US) | 71,520,919 (A) | 3.10 | 0.00 | 2,215 |
| | Fidelity (US) | 42,536,129 (A) | 1.84 | 0.00 | 1,297 |
| | | 7,291,358 (B) | 0.32 | 0.91 | 230 |
| | BlackRock (US) | 53,051,483 (A) | 2.30 | 0.00 | 1,618 |
| | Invesco (US) | 16,831,497(B) | 0.73 | 2.11 | 530 |
| | Bank of New York Mellon (US) | 28,811,418 (A) | 1.25 | 0.00 | 879 |
| | Children's Investment Fund (UK) | 25,629,967 (A) | 1.11 | 0.00 | 781 |
| | JP Morgan Chase (US) | 10,551,414 (B) | 0.46 | 1.32 | 346 |
| | Elliott Management (US) | 10,095,822 (B) | 0.44 | 1.26 | 310 |

19. In July 2013 Svyazinvest, previously pWrimarily owned by the Russian Government, was merged with Rostelecom and became partly private.

| | | | | | |
|---|---|---|---|---|---|
| | Waddell & Reed (US) | 9,826,425 (B) | 0.43 | 1.23 | 308 |
| | Deutsche Bank (Germany) | 4,898,516 (B) | 0.21 | 0.61 | 150 |
| | D.E. Shaw & Co. (US) | 4,750,070 (B) | 0.21 | 0.59 | 146 |
| | Government Pension Fund of Norway (public) | 4,331,050 (B) | 0.19 | 0.54 | 113 |
| | CQS (UK) | 4,184,246 (B) | 0.18 | 0.52 | 128 |
| | UBS (Switzerland) | 4,024,316 (B) | 0.17 | 0.50 | 123 |
| Oger Telecom | Harriri family (Lebanon) owns 55% of Oger Telecom, which owns 55% of Turk Telecom, and other telecoms. | | 65 | 65 | 4,754 |
| | Saudi Telecom Co | | 35 | 35 | 2,560 |
| British Telecom (UK) | AXA (France) | 550,700,679 | 6.9 | 6.9 | 2,825 |
| | Legal and General (UK) | 303,166,335 | 3.72 | 3.72 | 1,555 |
| | Government Pension Fund of Norway (Public) | 142,964,029 | 1.75 | 1.75 | 733 |
| | Vanguard (US) | 118,768,177 | 1.51 | 1.51 | 609 |
| | OFI Global Asset Management (US) | 111,273,097 | 1.41 | 1.41 | 570 |
| | AEGON Ireland (Ireland) | 75,047,170 | 0.95 | 0.95 | 384 |
| | Scottish Widows Inv Partnership (Scotland) | 71,618,522 | 0.91 | 0.91 | 367 |
| | Aviva Life International (UK) | 66,273,571 | 0.84 | 0.84 | 339 |
| | Artemis (UK) | 64,493,936 | 0.81 | 0.81 | 330 |
| | Invesco (US) | 60,802,392 | 0.77 | 0.77 | 311 |
| Telkom (South Africa) | Government of South Africa (Public) | 260,800,668 | 50.7 | 50.7 | 689 |
| | Allan Gray Unit Trust Mgmt (South Africa) | 15,894,952 | 3.09 | 3.09 | 34 |
| | Vanguard Group (US) | 15,715,693 | 3.08 | 3.08 | 34 |
| | Grantham, Mayo, Van Otterloo & Co (US) | 14,960,788 | 2.92 | 2.92 | 32 |
| | Dimensional Fund Advisors(US) | 12,196,340 | 2.38 | 2.38 | 26 |
| | BNY Mellon Fund Advisers (US) | 5,063,058 | 1 | 1 | 11 |
| | Dreyfus Corp. (US) | 4,219,924 | 0.82 | 0.82 | 9 |

*continued*

Table 36-1. *continued*

| | MAJOR OWNERS | NUMBER OF SHARES OWNED | PERCENT OF COMPANY OWNED | PERCENT OF VOTING POWER | VALUE OF SHARES (MIL $) |
|---|---|---|---|---|---|
| Telkom (South Africa) (*cont.*) | Momentum Collective Investments (South Africa) | 3,641,490 | 0.71 | 0.71 | 7 |
| | Old Mutual Unit Trust Mgrs (South Africa) | 3,242,695 | 0.63 | 0.63 | 7 |
| | STANLIB Collective Investments (South Africa) | 2,216,679 | 0.43 | 0.43 | 4 |
| | MET Collective Investments Limited (South Africa) | 1,398,936 | 0.27 | 0.27 | 3 |
| Softbank (Japan) | Masayoshi Son (Japan) | 240,567,906 | 20.2 | 20.2 | 18,900 |
| | Japan Trustee Services Bank (Trust Account) (Japan) | 102,420,000 | 8.6 | 8.6 | 6,463 |
| | JP Morgan Chase (US) | 91,902,000 | 7.7 | 7.7 | 5,799 |
| | State Street Company (US) | 60,571,000 | 5.1 | 5.1 | 3,822 |
| | The Master Trust Bank of Japan(Trust Account) (Japan) | 54,367,000 | 4.6 | 4.6 | 3,430 |
| | SSBT OD05 OMNIBUS ACCOUNT - TREATY CLIENTS (Chinese Government) | 21,125,000 | 1.8 | 1.8 | 1,333 |
| | Trust & Custody Services Bank (Trust Account) (Japan) | 20,360,000 | 1.7 | 1.7 | 1,284 |
| | JP Morgan Chase (US) | 16,897,000 | 1.4 | 1.4 | 1,066 |
| | Bank of New York Mellon (US) | 9,558,000 | 0.8 | 0.8 | 603 |

| | | | | | |
|---|---|---|---|---|---|
| Comcast (US) | Comcast has three classes of stock. Class A has 0.1333 vote per share. Class B share has 15 votes per share. Class K has no vote, just equity. | | | | |
| | BRCC (Roberts family trust) | 6,856,323 (A) | 0.26 | 0.22 | 368 |
| | | 9,444,375 (B) | 0.36 | 33.37 | 18,500 |
| | Vanguard (US) | 227,152,526 (A) | 8.61 | 7.13 | 9,690 |
| | | 25,104,652 (K) | 0.95 | 0.00 | 1,025 |
| | Fidelity (US) | 147,574,158 (A) | 5.59 | 4.63 | 6,294 |
| | | 28,914,184 (K) | 1.10 | 0.00 | 1,181 |
| | Capital Group (US) | 131,850,346 (A) | 5.00 | 4.14 | 5,623 |
| | | 22,000,000 (K) | 0.84 | 0.00 | 898 |
| | BlackRock (US) | 80,641,798 (A) | 3.06 | 2.53 | 3,439 |
| | | 35,293,832 (K) | 1.34 | 0.00 | 1,441 |
| | State Street Corporation (US) | 100,479,208 (A) | 3.81 | 3.15 | 4,284 |
| | Dodge & Cox (US) | 98,812,470 (A) | 3.74 | 3.10 | 4,213 |
| | Wellington Management (US) | 46,398,200 (A) | 1.76 | 1.46 | 1,978 |
| | ClearBridge (US) | 39,933,826 (K) | 1.51 | 0.00 | 1,631 |
| | MFS Investment Management (US) | 40,014,544 (K) | 1.52 | 0.00 | 1,634 |
| | Northern Trust (US) | 34,261,885 (A) | 1.31 | 1.08 | 1,461 |
| | JP Morgan Chase (US) | 30,669,310 (A) | 1.17 | 0.96 | 1,308 |
| | Arnhold & S. Bleichroeder (US) | 24,398,308 (K) | 0.93 | 0.00 | 996 |
| | Government Pension Fund of Norway (Norway) (public) | 25,228,035 (K) | 0.93 | 0.00 | 1,030 |
| | Eagle Capital (US) | 14,516,974 (K) | 0.55 | 0.00 | 593 |
| | Templeton (US) | 10,270,387 (K) | 0.39 | 0.00 | 419 |

Table 36-2. Ownership of the World's Top 20 Content Companies (Full Version) (as of Sept. 2013)

| | MAJOR OWNERS | NUMBER OF SHARES OWNED | PERCENT OF COMPANY OWNED | VOTING PERCENT | VALUE OF SHARES ($ MIL) |
|---|---|---|---|---|---|
| Google (US) | Google has a dual-stock system, Class A shares (1 vote) and Class B shares (10 votes) Class B shares are not publicly traded and are held by Google insiders, mostly Sergey Brin, Larry Page, and Eric Schmidt. | | | | |
| | Larry Page | 24,355,252 (B) | 7.3 | 28.16 | 26,717 |
| | Sergey Brin | 23,764,586 (B) | 7.2 | 27.48 | 26,047 |
| | Eric Schmidt | 5,205,664 (B) | 1.6 | 6.02 | 5,720 |
| | Fidelity (US) | 26,934,595 | 8.09 | 3.11 | 22,151 |
| | Vanguard (US) | 23,255,893 | 6.98 | 2.69 | 19,574 |
| | Capital Group (US) | 15,853,849 | 4.76 | 1.83 | 13,213 |
| | T. Rowe Price Associates (US) | 11,604,607 | 3.48 | 1.34 | 9,410 |
| | State Street Corp.(US) | 11,100,106 | 3.33 | 1.28 | 9,772 |
| | BlackRock (US) | 7,222,271 | 2.17 | 0.84 | 6,358 |
| | JP Morgan Chase (US) | 6,822,933 | 2.05 | 0.79 | 6,006 |
| | Bank of New York Mellon (US) | 4,861,862 | 1.46 | 0.56 | 4,280 |
| | Northern Trust (US) | 3,962,235 | 1.19 | 0.46 | 3,488 |
| News Corp. (US) | Two classes of stock, Class A (non-voting) and Class B (voting). | | | | |
| | Murdoch Family | 77,855,793 (B) | 13.45 | 39 | 1,640 |
| | Managers Investment Group (US) | 72,291,747 (A) | 12.5 | 0.0 | 1,102 |
| | Rhumbline Advisers (US) | 3,670,563 (A) | 0.6 | 0.0 | 55 |
| | | 26,700 (B) | 0.01 | 0.01 | 0.77 |
| | Advisory Research, Inc. (US) | 3,263,065 (A) | 0.6 | 0.0 | 49 |
| | Parametric Portfolio Associates (US) | 2,678,093 (A) | 0.5 | 0.0 | 40 |
| | | 364,981 (B) | 0.06 | 0.18 | 5 |

| | | | | | |
|---|---|---|---|---|---|
| | Corsair Capital (US) | 1,574,000 (A) | 0.3 | 0.0 | 24 |
| | Gruss & Co. (US) | 985,325 (A) | 0.2 | 0.0 | 15 |
| | 40 North Industries (US) | 585,860 (A) | 0.1 | 0.0 | 8 |
| | Mar Vista Investment Partners (US) | 499,287 (A) | 0.1 | 0.0 | 7 |
| | Nisa Investment Advisors (US) | 244,113 (A) | 0.04 | 0.0 | 3 |
| | Hudson Bay Capital (US) | 218,300 (A) | 0.04 | 0.0 | 3 |
| | | 50,000 (B) | 0.01 | 0.03 | 0.77 |
| | Baker Ellis Asset Management | 23,300 (B) | 0.01 | 0.01 | 0.36 |
| | S & Co (US) | 29,700 (B) | 0.01 | 0.02 | 0 .47 |
| | Mirae Asset Global Investments (South Korea) | 55,822 (B) | 0.01 | 0.03 | 0.86 |
| | Reynolds Capital Management (US) | 18,000 (B) | 0.01 | 0.01 | 0.28 |
| | Axiom International Investors (US) | 18,862 (B) | 0.01 | 0.01 | 0.26 |
| | Riverpoint Capital Management (US) | 7,565 (B) | 0.01 | 0.01 | 0.12 |
| | Ancora Advisors (US) | 7,450 (B) | 0.01 | 0.00 | 0.11 |
| 21st Century Fox (US) | There are two classes of shares, Class A (Equity, Non-Voting) and Class B (Voting) | | | | |
| | Murdoch Family | 311,423,171 (B) | 13.48 | 39.00 | 9,965 |
| | Vanguard (US) | 152,697,912 (A) | 6.61 | 0.00 | 4,658 |
| | Capital Group (US) | 94,872,222 (A) | 4.11 | 0.00 | 2,894 |
| | Dodge & Cox (US) | 93,300,035 (A) | 4.04 | 0.00 | 2,904 |
| | State Street (US) | 73,301,178 (A) | 3.17 | 0.00 | 2,236 |
| | Janus (US) | 71,520,919 (A) | 3.10 | 0.00 | 2,215 |
| | Fidelity (US) | 42,536,129 (A) | 1.84 | 0.00 | 1,297 |
| | | 7,291,358 (B) | 0.32 | 0.91 | 230 |
| | BlackRock (US) | 53,051,483 (A) | 2.30 | 0.00 | 1,618 |
| | Invesco (US) | 16,831,497 (B) | 0.73 | 2.11 | 530 |
| | Bank of New York Mellon (US) | 28,811,418 (A) | 1.25 | 0.00 | 879 |
| | Children's Investment Fund (UK)(The) | 25,629,967 (A) | 1.11 | 0.00 | 781 |
| | JP Morgan Chase (US) | 10,551,414 (B) | 0.46 | 1.32 | 346 |

*continued*

Table 36-2. *continued*

| | MAJOR OWNERS | NUMBER OF SHARES OWNED | PERCENT OF COMPANY OWNED | PERCENT OF VOTING POWER | VALUE OF SHARES (MIL $) |
|---|---|---|---|---|---|
| 21st Century Fox (US) (*cont.*) | Elliott Management (US) | 10,095,822 (B) | 0.44 | 1.26 | 310 |
| | Waddell & Reed (US) | 9,826,425 (B) | 0.43 | 1.23 | 308 |
| | Deutsche Bank (Germany) | 4,898,516 (B) | 0.21 | 0.61 | 150 |
| | D.E. Shaw & Co. (US) | 4,750,070 (B) | 0.21 | 0.59 | 146 |
| | Government Pension Fund of Norway (Norway) (public) | 4,331,050 (B) | 0.19 | 0.54 | 113 |
| | CQS (UK) | 4,184,246 (B) | 0.18 | 0.52 | 128 |
| | UBS (Switzerland) | 4,024,316 (B) | 0.17 | 0.50 | 123 |
| Comcast (US) | Comcast has three stock classes. Class A are voting shares with 0.1333 votes per share, Class B has 15 votes per share. Class K has no votes, and is equity only. Class B shares hold an undilutable 33 1/3% of voting power in the company. | | | | |
| | BRCC Holding (Roberts Family Holding Group) (USA) | 6,856,323 (A) | 0.26 | 0.22 | 368 |
| | | 9,444,375 (B) | 0.36 | 33.37 | 18,500 |
| | Vanguard (US) | 227,152,526 (A) | 8.61 | 7.13 | 9,690 |
| | | 25,104,652 (K) | 0.95 | 0.00 | 1,025 |
| | Fidelity (US) | 147,574,158 (A) | 5.59 | 4.63 | 6,294 |
| | | 28,914,184 (K) | 1.10 | 0.00 | 1,181 |
| | Capital Group (US) | 131,850,346 (A) | 5.00 | 4.14 | 5,623 |
| | | 22,000,000 (K) | 0.84 | 0.00 | 898 |
| | BlackRock (US) | 80,641,798 (A) | 3.06 | 2.53 | 3,439 |
| | | 35,293,832 (K) | 1.34 | 0.00 | 1,441 |
| | State Street (US) | 100,479,208 (A) | 3.81 | 3.15 | 4,284 |

| | | | | | |
|---|---|---|---|---|---|
| Comcast (US) | Dodge & Cox (US) | 98,812,470 (A) | 3.74 | 3.10 | 4,213 |
| | Wellington Management (US) | 46,398,200 (A) | 1.76 | 1.46 | 1,978 |
| | ClearBridge Investments (US) | 39,933,826 (K) | 1.51 | 0.00 | 1,631 |
| | MFS Investment Management (US) | 40,014,544 (K) | 1.52 | 0.00 | 1,634 |
| | Northern Trust (US) | 34,261,885 (A) | 1.31 | 1.08 | 1,461 |
| | JP Morgan Chase. (US) | 30,669,310 (A) | 1.17 | 0.96 | 1,308 |
| | Arnhold & S. Bleichroeder (US) | 24,398,308 (K) | 0.93 | 0.00 | 996 |
| | Government Pension Fund of Norway (Norway) (public) | 25,228,035 (K) | 0.93 | 0.00 | 1,030 |
| | Eagle Capital Management (US) | 14,516,974 (K) | 0.55 | 0.00 | 593 |
| | Templeton (US) | 10,270,387 (K) | 0.39 | 0.00 | 419 |
| Globo Group (Brazil) | Privately held company run by brothers Roberto, João, and José Marinho, grandsons of founder Irineu Marinho. | | | | |
| | Roberto Marinho | | 33.3 | 33.3 | 6,666 |
| | João Marinho | | 33.3 | 33.3 | 6,666 |
| | José Marinho | | 33.3 | 33.3 | 6,666 |
| Lagardère (France) | Arnaud Lagardère, heir of Guy Lagardere, is largest shareholder with 9.3% of all shares and 15% of voting rights. | | | | |
| | Arnaud Lagardère | 12,196,950 | 9.30 | 14.94 | 467 |
| | DNCA Finance (France) | 3,210,000 | 2.45 | 2.51 | 77 |
| | International Value Advisers (US) | 2,634,121 | 2.01 | 1.80 | 63 |
| | Amundi Capital (France) | 2,235,116 | 1.70 | 1.75 | 53 |
| | Edmond de Rothschild Asset Management (France) | 1,767,588 | 1.35 | 1.38 | 42 |
| | The Vanguard Group (US) | 1,706,578 | 1.30 | 1.17 | 41 |
| | BlackRock (US) | 1,361,666 | 1.04 | 0.93 | 32 |

*continued*

Table 36-2. *continued*

| | MAJOR OWNERS | NUMBER OF SHARES OWNED | PERCENT OF COMPANY OWNED | PERCENT OF VOTING POWER | VALUE OF SHARES (MIL $) |
|---|---|---|---|---|---|
| Disney (US) | Steve Jobs Trust | 138,000,007 | 7.7 | 7.79 | 8,371 |
| | Vanguard (US) | 148,460,516 | 7.4 | 8.38 | 8,044 |
| | Fidelity (US) | 103,334,320 | 5.8 | 5.83 | 6,138 |
| | State Street Corporation (US) | 77,161,033 | 4.3 | 4.35 | 4,872 |
| | Massachusetts Financial Services (US) | 57,066,534 | 3.2 | 3.22 | 3,241 |
| | BlackRock (US) | 48,613,370 | 2.7 | 2.74 | 3,069 |
| | State Farm (US) | 42,206,018 | 2.4 | 2.38 | 2,665 |
| | Bank of New York Mellon (US) | 30,293,150 | 1.7 | 1.71 | 1,913 |
| | T. Rowe Price Associates (US) | 29,167,554 | 1.6 | 1.65 | 1,656 |
| | Wellington Management (US) | 28,484,413 | 1.6 | 1.61 | 1,617 |
| | Northern Trust (US) | 28,465,082 | 1.6 | 1.61 | 1,797 |
| | Roy Disney's 4 Children (Tim Disney, Roy Disney, Abigail Disney, and Susan Disney Lord) | 4,466,196 | 0.3 | 0.25 | 272 |
| | Michael Eisner | 14,024,352 | 0.8 | 0.79 | 857 |
| Bertelsmann (Germany) | 19.1% shares directly owned by Mohn families. 80.9% owned by foundations associated with the Mohn families and whose voting is controlled by them. | | 19.1 | 100 | 4,431 |

| | | | | | | |
|---|---|---|---|---|---|---|
| Berlusconi Group | Fininvest is controlled by founder and former Prime Minister of Italy Silvio Berlusconi and his family, including his daughter Marina Berlusconi, the company Chairwoman. Berlusconi indirectly controls 96% of Fininvest through 38 privately held companies, all called Holding Italiana, numbers 1-38. Fininvest's media subsidiaries include Mediaset (TV); Medusa (film production, a 100% subsidiary of Mediaset), Mondadori (print). These subsidiaries are traded individually on the Italian Stock Exchange. | | | | | |
| | Mondadori | Fininvest (Berlusconi) | 130,623,800 | 53 | 53 | 241,821,391 |
| | | Silchester | 32,832,869 | 13.32 | 13.32 | 66,000 |
| | Mediaset | Fininvest (Berlusconi) | 483,800,000 | 41 | 41 | 2,896,450,125 |
| | | McKenzie Cundill Inv. (Canada) | 31,743,385 | 2.8 | 2.8 | |
| Malone Group (US) | Liberty is split into two holding companies: Liberty Global operates wireline and cable TV services, and Liberty Media operates satellite and other pay-TV services. Liberty Global has three classes of stock, Class A (1 Vote per Share), Class B (10 Votes per Share), and Class C (Equity only). Liberty Media has three classes of stock, Class A (1 vote per share) and Class B (10 votes per share). John Malone, founder, controls both companies. | | | | | |
| | Liberty Global (US) | John Malone | 8,751,613 (B) | 3.41 | 35.98 | 868 |
| | | | 2,730,878 (C) | 1.06 | 0 | 204 |

continued

Table 36-2. *continued*

| MAJOR OWNERS | | NUMBER OF SHARES OWNED | PERCENT OF COMPANY OWNED | PERCENT OF VOTING POWER | VALUE OF SHARES (MIL $) |
|---|---|---|---|---|---|
| Malone Group (US) (*Cont.*) | Coatue Management | 9,146,979 | 3.56 | 3.76 | 712 |
| | Vanguard | 11,144,802 | 4.34 | 4.58 | 868 |
| | Capital World Investors | 7,583,872 | 2.95 | 3.12 | 591 |
| | Cascade Investment | 8,736,009 | 3.40 | 3.59 | 680 |
| | Thornburg Investment Management | 10,816,837 | 4.21 | 4.45 | 842 |
| | T. Rowe Price Associates | 5,668,594 | 2.21 | 2.33 | 441 |
| | Goldman Sachs | 5,482,340 | 2.13 | 2.25 | 427 |
| | State Street | 5,244,201 | 2.04 | 2.16 | 408 |
| | Fidelity | 8,327,101 | 3.24 | 3.42 | 648 |
| | Blue Ridge Cap Holdings | 4,330,000 | 1.69 | 1.78 | 337 |
| | Spo Advisory Corp | 15,426,265 | 6.01 | 0 | 1,142 |
| | Eagle Capital Management | 12,849,099 | 5.00 | 0 | 951 |
| | Vanguard | 12,282,027 | 4.78 | 0 | 909 |
| | Davis Selected Advisers | 5,827,348 | 2.27 | 0 | 431 |
| | Tiger Technology Management | 4,300,000 | 1.67 | 0 | 318 |
| | Capital World Investors | 3,672,815 | 1.43 | 0 | 272 |
| | Jennison Associates | 3,656,120 | 1.42 | 0 | 270 |
| | Dodge & Cox | 7,175,642 | 2.79 | 0 | 531 |
| | GE Asset Management | 3,564,225 | 1.39 | 0 | 264 |
| | Eton Park Capital Management | 3,483,700 | 1.36 | 0 | 258 |

| | | | | | | |
|---|---|---|---|---|---|---|
| | Liberty Media (Malone) (US) | John Malone | (Class B) 8,496,800 | 7.01 | 40.4 | 1,226 |
| | | Berkshire Hathaway (Warren Buffet) | 5,622,340 | 4.64 | 2.7 | 808 |
| | | Capital World Investors | 4,144,651 | 3.42 | 2.0 | 596 |
| | | Tiger Technology Management | 3,845,000 | 3.17 | 1.8 | 553 |
| | | ClearBridge Advisors | 3,770,631 | 3.11 | 1.8 | 542 |
| | | Fidelity | 6,261,744 | 5.17 | 3.0 | 900 |
| | | Vanguard | 4,461,937 | 3.68 | 2.1 | 641 |
| | | BlackRock | 3,064,177 | 2.53 | 1.5 | 440 |
| | | Horizon Asset Management | 2,687,072 | 2.22 | 1.3 | 386 |
| | | Principal Global Investors | 2,520,820 | 2.08 | 1.2 | 362 |
| | | D. E. Shaw | 2,184,176 | 1.80 | 1.0 | 314 |
| ProsiebenSat 1 (Germany) | Lavena Holding, a Joint venture between KKR, an American private equity firm and Permia, a UK based equity firm, own 44% ($1,997,069,092)of shares and are the largest owner. Other Institutional owners include: | | | | | |
| | Lavena Holding | | 71,327,887 | 32.6 | 32.6 | 2,888 |
| | KKR (US) | | 35,663,944 | 16.3 | 16.3 | 1,444 |
| | Permia (US) | | 35,663,944 | 16.3 | 16.3 | 1,444 |
| | Telegraaf Media Group (Netherlands) | | 13,127,832 | 6.0 | 6.0 | 531 |
| Televisa (Mexico) | Televisa has 4 classes of stock. Class A elects 11 of 20 board members, votes on dividends and only Mexican citizens can vote these shares, Class B elects 5 board members, Class D preferred dividends and elects 2 board members, and Class L has limited Voting rights and elects 2 board members. A, B, and D must be owned by a Mexican citizens or must be held in trust in Mexico. | | | | | |

*continued*

Table 36-2. *continued*

| | MAJOR OWNERS | NUMBER OF SHARES OWNED | PERCENT OF COMPANY OWNED | PERCENT OF VOTING POWER | VALUE OF SHARES (MIL $) |
|---|---|---|---|---|---|
| Televisa (Mexico) (*cont.*) | Azcárraga Trust (Emilio Azcárraga Jean) | 53,275,414,049 | 14.70 | 23.70 | 535 |
| | A Shares | 52,991,825,705 | 42.99 | 23.64 | 529 |
| | B Shares | 67,814,604 | 0.11 | 0.029 | 1 |
| | D Shares | 107,886,870 | 0.12 | 0.012 | 2 |
| | L Shares | 107,886,870 | 0.12 | 0.012 | 2 |
| | BlackRock | 32,073,554,097 | 8.85 | 7.77 | 572 |
| | A Shares | 6,853,323,525 | 5.56 | 3.06 | 68 |
| | B Shares | 6,030,924,702 | 10.22 | 2.56 | 120 |
| | D Shares | 9,594,652,935 | 10.65 | 1.07 | 191 |
| | L Shares | 9,594,652,935 | 10.65 | 1.07 | 191 |
| | William H. Gates III | 26,954,635,500 | 7.43 | 6.51 | 481 |
| | A Shares | 5,759,537,500 | 4.67 | 2.57 | 57 |
| | B Shares | 5,068,393,000 | 8.59 | 2.15 | 101 |
| | D Shares | 8,063,352,500 | 8.95 | 0.90 | 161 |
| | L Shares | 8,063,352,500 | 8.95 | 0.90 | 161 |
| | First Eagle Investment Management | 16,630,725,735 | 4.59 | 4.01 | 297 |
| | A Shares | 3,553,573,875 | 2.88 | 1.59 | 35 |
| | B Shares | 3,127,145,010 | 5.30 | 1.32 | 62 |
| | D Shares | 4,975,003,425 | 5.52 | 0.55 | 99 |
| | L Shares | 4,975,003,425 | 5.52 | 0.55 | 99 |
| Fuji Television (Japan) | Fuji Television is a wholly owned subsidiary of Fuji Media Holdings, which is 100% owned by the Fujisankei Communications Group, a Japanese Keiretsu. The Shikanai family maintains control of Fujisankei Communications Group. | | | | |

| | | | | | |
|---|---|---|---|---|---|
| | Shikanai Family | | 100 | 100 | 7,000 |
| Yahoo (US) | Jerry Yang, co-founder, owns about 1% of the company. CEO Marissa Mayer, owns 1%. | | | | |
| | Vanguard (US) | 87,549,068 | 7.9 | 7.9 | 2,157 |
| | Fidelity (US) | 76,079,094 | 6.8 | 6.8 | 1,835 |
| | Third Point (US) | 62,000,000 | 5.6 | 5.6 | 1,458 |
| | Wellington Management (US) | 44,189,301 | 4.0 | 4.0 | 1,039 |
| | State Street (US) | 40,063,743 | 3.6 | 3.6 | 1,006 |
| | Invesco (US) | 29,335,139 | 2.6 | 2.6 | 690 |
| | BlackRock (US) | 25,522,396 | 2.3 | 2.3 | 641 |
| | Capital Group (US) | 19,067,798 | 1.7 | 1.7 | 448 |
| | Bank of New York Mellon (US) | 17,846,935 | 1.6 | 1.6 | 448 |
| | Deutsche Bank (Germany) | 13,815,580 | 1.2 | 1.2 | 325 |
| Bouygues (France) | SCDM (Bouygues Family) (France) | 67,317,686 | 20.5 | 29.2 | 1,653 |
| | Bouygues Employees (France) | 77,825,813 | 23.7 | 28.7 | 1,911 |
| | Arnhold & S. Bleichroeder (US) | 16,418,948 | 5.0 | 5.5 | 403 |
| | DNCA Finance (France) | 5,226,200 | 1.6 | 1.8 | 128 |
| | Government Pension Fund of Norway (Norway) | 4,044,091 | 1.3 | 1.4 | 99 |
| | BlackRock (US) | 3,508,205 | 1.1 | 1.2 | 86 |
| | New Jersey Division of Pensions and Benefits (US) | 2,400,000 | 0.7 | 0.8 | 58 |
| | Vanguard (US) | 1,649,698 | 0.5 | 0.6 | 40 |
| | Legal and General (UK) | 1,566,210 | 0.5 | 0.5 | 38 |
| | Natixis (France) | 1,372,242 | 0.4 | 0.5 | 33 |
| | Amundi Capital (France) | 1,246,830 | 0.4 | 0.4 | 30 |

*continued*

Table 36-2. *continued*

| | | MAJOR OWNERS | NUMBER OF SHARES OWNED | PERCENT OF COMPANY OWNED | PERCENT OF VOTING POWER | VALUE OF SHARES (MIL $) |
|---|---|---|---|---|---|---|
| Redstone Group | CBS(US) | CBS has two classes of stock. Class A (Voting) and Class B ( non-voting) National Amusements, a holding corporation 80% owned by Sumner Redstone and 20% by daughter Sheri Redstone, is the controlling shareholder. | | | | |
| | | National Amusements (Sumner Redstone) | 33,742,639 (A) | 5.36 | 79.0 | 1,828 |
| | | Gamco Investors(US) | 2,338,784 (A) | 0.37 | 5.48 | 121 |
| | | Gabelli Funds (US) | 1,906,000 (A) | 0.30 | 4.46 | 98 |
| | | Gruss Asset Mgmt (US) | 1,500,000 (A) | 0.24 | 3.51 | 77 |
| | | Pacific Heights Asset Mgmt (US) | 700,000 (A) | 0.11 | 1.64 | 36 |
| | | Renaissance Technologies (US) | 265,600 (A) | 0.04 | 0.62 | 13 |
| | | Dimensional Fund Advisors (US) | 120,498 (A) | 0.019 | 0.28 | 6 |
| | | The Vanguard Group (US) | 46,523,377 (B) | 7.39 | 0.00 | 2,231 |
| | | Waddell & Reed (US) | 33,510,724 (B) | 5.33 | 0.00 | 1,564 |
| | | Capital Group (US) | 23,747,500 (B) | 3.77 | 0.00 | 1,108 |
| | | State Street Corporation (US) | 23,192,294 (B) | 3.69 | 0.00 | 1,133 |
| | | Janus Group (US) | 22,238,540 (B) | 3.53 | 0.00 | 1,047 |
| | | BlackRock (US) | 15,545,215 (B) | 2.47 | 0.00 | 759 |
| | | Coatue Management (US) | 13,976,475 (B) | 2.22 | 0.00 | 652 |
| | | PointState Capital (US) | 9,570,660 (B) | 1.52 | 0.00 | 467 |
| | | Fidelity (US) | 8,988,794 (B) | 1.43 | 0.00 | 419 |
| | | Goldman Sachs (US) | 8,712,523 (B) | 1.38 | 0.00 | 406 |
| | Viacom | Vanguard | 40,908,747 (B) | 8.58 | 0 | 5.28 |
| | | National Amusement (Sumner Redstone) | 40,409,751 (A) | 8.47 | 79.0 | 4.563 |

| | | | | | |
|---|---|---|---|---|---|
| Time Warner (US) | Vanguard (US) | 84,839,919 | 9.1 | 9.1 | 4,906 |
| | Capital Group (US) | 71,699,627 | 7.7 | 7.7 | 4,131 |
| | | 59,961,376 | 6.4 | 6.4 | 3,466 |
| | Dodge & Cox (US) | | | | |
| | JP Morgan Chase (US) | 49,460,496 | 5.3 | 5.3 | 2,859 |
| | State Street (US) | 36,937,372 | 4.0 | 4.0 | 2,135 |
| | Wellington Management (US) | 31,421,367 | 3.4 | 3.4 | 1,810 |
| | Fidelity (US) | 28,114,135 | 3.0 | 3.0 | 1,619 |
| | BlackRock (US) | 25,795,977 | 2.8 | 2.8 | 1,491 |
| | Viking Global Investors (US) | 22,364,094 | 2.4 | 2.4 | 1,288 |
| Time Warner Cable (US) | Dodge & Cox (US) | 28,144,881 | 9.5 | 9.5 | 2,897 |
| | Vanguard (US) | 26,529,901 | 9.0 | 9.0 | 2,547 |
| | Capital Group (US) | 18,738,817 | 6.3 | 6.3 | 1,800 |
| | State Street (US) | 11,765,247 | 4.0 | 4.0 | 1,130 |
| | Franklin Resources (US) | 11,435,571 | 3.9 | 3.9 | 1,098 |
| | Janus (US) | 9,719,402 | 3.3 | 3.3 | 933 |
| | Invesco (US) | 9,468,413 | 3.2 | 3.2 | 909 |
| | Capital Group (US) | 9,050,000 | 3.1 | 3.1 | 869 |
| | BlackRock (US) | 8,181,412 | 2.8 | 2.8 | 785 |
| | T. Rowe Price Associates (US) | 7,396,015 | 2.5 | 2.5 | 710 |
| Baidu (China) | Baidu is controlled by non-Chinese investors. To comply with Chinese law Baidu has Variable Interest Entities owned by Robin Yanhong Li and Eric Yong Xu who technically own Baidu. The beneficial owners are non-Chinese investment firms, which provide personal loans to the nominal shareholders in exchange for the shares as collateral and an assignment of their voting rights to a person of the companies choosing. owners include: | | | | |

*continued*

Table 36-2. *continued*

| | MAJOR OWNERS | NUMBER OF SHARES OWNED | PERCENT OF COMPANY OWNED | PERCENT OF VOTING POWER | VALUE OF SHARES (MIL $) |
|---|---|---|---|---|---|
| Baidu (China) (*cont.*) | Robin Yanhong Li (China) | 91,324 | 0.003 | 0.0009 | 156.6 |
| | Melissa Ma (China) (Robin Li's Wife) | 1,676,667 | 0.04 | 16 | 2,875 |
| | Handsome Reward Ltd(UK) (Owned by Robin Li) | 5,490,000 | 15.7 | 52.4 | 9,415 |
| | Baillie Gifford & Co (UK) | 3,036,275 | 8.7 | 8.7 | 4,098 |
| | T. Rowe Price Associates (US) | 1,766,548 | 5.1 | 5.1 | 2,384 |
| | OppenheimerFunds | 2,980,070 | 8.5 | 8.5 | 4,023 |
| | Capital Research | 2,023,354 | 5.8 | 5.8 | 2,731 |
| | Sands Capital Management | 924,349 | 2.6 | 2.6 | 1,247 |
| | Artisan Partners | 876,270 | 2.5 | 2.5 | 1,182 |
| | Lazard Asset Management | 847,253 | 2.4 | 2.4 | 1,143 |
| | Thornburg | 1,206,805 | 3.5 | 3.5 | 1,629 |
| Softbank (Japan) | Masayoshi Son (Japan) | 240,567,906 | 20.2 | 20.19 | 18,900 |
| | Japan Trustee Services Bank (Trust Account) (Japan) | 102,420,000 | 8.6 | 8.60 | 6,463 |
| | JP Morgan Chase (US) | 91,902,000 | 7.7 | 7.71 | 5,799 |
| | State Street Company (US) | 60,571,000 | 5.1 | 5.08 | 3,822 |
| | Master Trust Bank of Japan (Trust Account) (Japan) | 54,367,000 | 4.6 | 4.56 | 3,430 |

| | | | | | |
|---|---|---|---|---|---|
| | SSBT OD05 OMNIBUS ACCOUNT - TREATY CLIENTS (Probably Chinese Government) | 21,125,000 | 1.8 | 1.77 | 1,333 |
| | Trust & Custody Services Bank (Trust Account) (Japan) | 20,360,000 | 1.7 | 1.71 | 1,284 |
| | JP Morgan Chase (US) | 16,897,000 | 1.4 | 1.4 | 1,066 |
| | Bank of New York Mellon (US) | 9,558,000 | 0.8 | 0.8 | 603 |
| Yomiuri Shimbun Holdings (NTV) (Japan) | Controlled by the Shoriki Family, through Yomiuri Shimbun Holdings (Japan) | 69,803,000 | 27 | 27 | 1,279 |
| | CBNY-Orbis Funds (Japan) | 19,037,000 | 7.2 | 7.2 | 348 |
| | Japan Trustee Services Bank (Japan) | 9,820,000 | 3.8 | 3.8 | 180 |
| | Teikyo University (Japan) | 9,553,000 | 3.7 | 3.7 | 175 |
| | State Street Bank and Trust Company (US) | 7,963,000 | 3 | 3 | 145 |

Table 36-4. (Full Version) - Ownership in Media Companies by Major Institutional Owners (as of September 2013)

| OWNERS | COMPANIES | SHARES | % EQUITY | % VOTES | VALUE |
|---|---|---|---|---|---|
| State Street (US) | Microsoft (US) | 328,873,653 | 3.93 | 3.93 | $10,744,302,244 |
| | Google (US) A | 11,100,106 | 3.33 | 1.28 | $9,772,200,319 |
| | AT&T Inc. (US) | 233,415,579 | 4.43 | 4.43 | $8,024,827,606 |
| | Verizon Communications Inc (US) | 120,621,754 | 4.21 | 421 | $5,712,646,269 |
| | Disney (US) | 77,161,033 | 4.32 | 4.35 | $4,872,719,233 |
| | Amazon (US) | 14,350,930 | 3.14 | 3.14 | $4,288,918,940 |
| | Comcast (US) A | 100,479,208 | 3.81 | 3.15 | $4,284,433,429 |
| | Softbank (Japan) | 60,571,000 | 5.09 | 5.08 | $3,822,314,178 |
| | 21st Century Fox (US) A | 73,301,178 | 3.17 | 0 | $2,236,418,940 |
| | Time Warner (US) | 36,937,372 | 3.96 | 3.96 | $2,135,718,849 |
| | DirecTV (US) | 27,371,015 | 5.21 | 5.21 | $1,678,938,060 |
| | CBS (US) B | 23,192,294 | 3.69 | 0 | $1,133,407,407 |
| | Time Warner Cable (US) | 11,765,247 | 3.98 | 3.98 | $1,130,169,626 |
| | Yahoo (US) | 40,063,743 | 3.59 | 3.59 | $1,006,801,861 |
| | Discovery Communications | 8,862,035 | 3.61 | 4.21 | $691,593,211 |
| | Netflix Inc | 2,218,239 | 3.74 | 3.74 | $668,599,417 |
| | Liberty Global (Malone) (US) A | 5,244,201 | 2.04 | 2.15 | $408,680,584 |
| | Electronic Arts Inc. (US) | 11,682,102 | 3.78 | 3.78 | $308,290,672 |
| | Scripps Networks Interactive | 3,495,299 | 2.35 | 0 | $258,896,797 |
| | Sirius XM Radio | 55,105,456 | 1.05 | 0.84 | $208,298,624 |
| | Dish Network Corp (US) | 3,490,533 | 0.77 | 0.13 | $174,387,029 |
| | Yomiuri Shimbun Holdings (NTV) (Japan) | 7,963,000 | 3.00 | 3.00 | $145,961,790 |
| | Charter Communications | 1,124,215 | 1.11 | 1.11 | $144,214,300 |
| | Thomson-Reuters Corp | 3,599,932 | 0.44 | 0.44 | $124,521,648 |
| | TW Telecom Inc. | 4,094,590 | 2.69 | 2.69 | $121,691,215 |
| | Total, including stakes <100M | | | | $64,845,405,540 |

| | | | | | |
|---|---|---|---|---|---|
| The Vanguard Group (US) | Google (US) A | 23,255,893 | 6.98 | 2.69 | $20,616,349,145 |
| | Comcast (US) A | 227,152,526 | 8.61 | 7.13 | $9,981,081,992 |
| | Disney (US) | 148,460,516 | 7.36 | 8.38 | $9,612,818,411 |
| | Time Warner (US) | 84,839,919 | 9.10 | 9.10 | $5,383,941,260 |
| | 21st Century Fox (US) A | 152,697,912 | 6.61 | 0 | $4,999,329,639 |
| | Time Warner Cable (US) | 26,529,901 | 8.98 | 8.98 | $2,986,736,255 |
| | Yahoo (US) | 87,549,068 | 7.85 | 7.85 | $2,649,234,798 |
| | CBS (US) B | 46,523,377 | 7.39 | 0 | $2,601,587,242 |
| | Comcast (US) K | 25,104,652 | 0.95 | 0 | $1,006,194,452 |
| | Liberty Global (Malone) (US) K | 12,282,027 | 4.78 | 0 | $914,028,449 |
| | Liberty Global (Malone) (US) A | 11,144,802 | 4.34 | 4.58 | $873,752,477 |
| | Liberty Media (Malone) (US) A | 4,461,937 | 3.68 | 2.12 | $652,156,712 |
| | China Mobile (CN) | 29,434,000 | | | $334,958,920 |
| | Singapore Telecommunications (SG) | 7,802,432 | | | $179,924,082 |
| | Vodafone Group (UK) | 50,117,938 | | | $168,396,272 |
| | Total, including stakes <100M | | | | $63,800,023,915 |
| Fidelity (US) | Google (US) A | 26,934,595 | 8.09 | 3.11 | $23,877,518,468 |
| | Disney (US) | 103,334,320 | 5.8 | 5.83 | $6,690,897,220 |
| | Comcast (US) A | 147,574,158 | 5.59 | 4.63 | $6,484,408,503 |
| | Yahoo (US) | 76,079,094 | 6.8 | 6.8 | $2,331,063,440 |
| | Time Warner (US) | 28,114,135 | 3.0 | 3.0 | $1,784,123,007 |
| | 21st Century Fox (US) A | 42,536,129 | 1.84 | 0.00 | $1,392,632,863 |
| | Comcast (US) K | 28,914,184 | 1.10 | 0 | $1,158,880,495 |
| | Liberty Media (Malone) (US) A | 6,261,744 | 5.17 | 2.98 | $915,216,503 |
| | Liberty Global (Malone) (US) A | 8,327,101 | 3.24 | 3.42 | $652,844,718 |
| | CBS (US) B | 8,988,794 | 1.43 | 0.00 | $502,653,360 |
| | 21st Century Fox (US) B | 7,291,358 | 0.32 | 0.91 | $238,646,147 |
| | Nippon Telegraph & Telephone (Japan) | 75,600 | | | $123,757,200 |
| | Total, including stakes <100M | | | | $46,501,836,820 |

*continued*

Table 36-4. *continued*

| OWNERS | COMPANIES | SHARES | % EQUITY | % VOTES | VALUE |
|---|---|---|---|---|---|
| Capital Group (US) | Google (US) A | 15,853,849.00 | 4.76 | 1.83 | $13,213,383,650 |
| | Comcast (US) A | 131,850,346.00 | 5.0 | 4.14 | $5,623,417,256 |
| | Time Warner (US) | 71,699,627.00 | 7.69 | 7.69 | $4,131,332,507 |
| | 21st Century Fox (US) A | 94,872,222.00 | 4.11 | 0 | $2,894,551,493 |
| | Baidu (China) | 2,023,354.40 | 5.79 | 5.79 | $2,731,528,440 |
| | Time Warner Cable (US) | 18,738,817.00 | 6.34 | 6.34 | $1,800,050,761 |
| | CBS (US) B | 23,747,500.00 | 3.77 | 0 | $1,108,770,775 |
| | Comcast (US) K | 22,000,000.00 | 0.84 | 0 | $898,700,000 |
| | Time Warner Cable (US) | 9,050,000.00 | 3.06 | 3.06 | $869,343,000 |
| | Liberty Media (Malone) (US) A | 4,144,651.00 | 3.42 | 1.97 | $596,332,386 |
| | Liberty Global (Malone) (US) A | 7,583,872.00 | 2.95 | 3.12 | $591,011,145 |
| | Yahoo (US) | 19,067,798.00 | 1.71 | 1.71 | $448,665,286 |
| | Liberty Global (Malone) (US) K | 3,672,815.00 | 1.43 | 0 | $272,045,407 |
| | Total, including stakes <100M | | | | $35,203,110,732 |
| T. Rowe Price Associates (US) | Google (US) A | 11,604,607 | 3.48 | 1.34 | $9,410,598,733 |
| | AT&T | 90,744,940 | | | $3,126,163,183 |
| | Baidu (China) | 1,766,548 | 5.1 | 5.1 | $2,384,839,665 |
| | Time Warner Inc | 34,095,865 | | | $2,132,014,438 |
| | Disney (US) | 29,167,554 | 1.6 | 1.65 | $1,656,717,067 |
| | Comcast | 34,397,486 | 2.86 | 5.75 | $1,524,496,580 |
| | Netflix | 4,968,383 | | | $1,503,929,534 |
| | Time Warner Cable (US) | 7,396,015 | 2.5 | 2.5 | $710,461,200 |
| | McGraw-Hill Companies | 8,621,759 | | | $542,136,206 |
| | Liberty Global (Malone) (US) A | 5,668,594 | 2.21 | 2.33 | $441,753,530 |
| | Vodafone Group | 11,276,833 | | | $381,269,724 |
| | Cablevision Systems | 18,898,845 | | | $333,375,626 |

| | | | | | |
|---|---|---|---|---|---|
| | Sprint Nextel Corporation | 46,913,325 | | | $310,566,212 |
| | Yahoo | 6,976,870 | | | $207,352,576 |
| | DirecTV | 3,381,820 | | | $159,148,449 |
| | Washington Post Company | 258,830 | | | $150,512,233 |
| | Dreamworks Animation | 4,829,500 | | | $140,635,040 |
| | Dish Network Corp | 2,759,446 | | | $129,859,529 |
| | Electronic Arts | 4,640,290 | | | $125,380,636 |
| | New York Times Company | 10,630,700 | | | $120,764,752 |
| | Total, including stakes <100M | | | | $26,175,685,790 |
| BlackRock (US) | Google (US) A | 7,222,271 | 2.17 | 0.84 | $7,484,006,101 |
| | Comcast (US) A | 80,641,798 | 3.06 | 2.53 | $3,847,420,183 |
| | Disney (US) | 48,613,370 | 2.7 | 2.74 | $3,350,433,460 |
| | 21st Century Fox (US) A | 53,051,483 | 2.30 | 0.00 | $1,847,252,638 |
| | Time Warner (US) | 25,795,977 | 2.8 | 2.8 | $1,807,524,108 |
| | Comcast (US) K | 35,293,832 | 1.34 | 0 | $1,638,692,620 |
| | Time Warner Cable (US) | 8,181,412 | 2.8 | 2.8 | $966,552,014 |
| | CBS (US) B | 15,545,215 | 2.47 | 0.00 | $924,473,936 |
| | Yahoo (US) | 25,522,396 | 2.3 | 2.3 | $846,577,875 |
| | AT&T Inc (US) | 7,831,419 | | | $284,045,567 |
| | Amazon Com Inc (US) | 734,284 | | | $266,324,807 |
| | Televisa (Mexico) D | 9,594,652,935 | | | $191,893,059 |
| | Televisa (Mexico) L | 9,594,652,935 | | | $191,893,059 |
| | Bouygues (France) | 3,508,205 | 1.1 | 1.2 | $139,556,395 |
| | Televisa (Mexico) A | 6,853,323,525 | | | $137,066,471 |
| | Televisa (Mexico) B | 6,030,924,702 | | | $120,618,494 |
| | Total, including stakes <100M | | | | $24,320,774,848 |

*continued*

Table 36-4. *continued*

| OWNERS | COMPANIES | SHARES | % EQUITY | % VOTES | VALUE |
|---|---|---|---|---|---|
| Dodge & Cox (US) | Comcast (US) A | 99,057,632 | | 3.10 | $4,213,363,720 |
| | Time Warner (US) | 105,719,400 | | 6.43 | $3,466,966,760 |
| | 21st Century Fox (US) A | 93,300,035 | 4.04 | 0 | $2,904,278,333 |
| | Time Warner Cable (US) | 28,144,881 | 9.52 | 9.52 | $2,897,208,656 |
| | News Corp | 96,431,826 | | | $1,659,591,725 |
| | Google | 1,303,140 | | | $1,152,040,917 |
| | Vodafone Group | 23,430,538 | | | $791,952,184 |
| | Sony Corp | 31,729,550 | | | $688,213,940 |
| | Liberty Interactive | 26,063,108 | | | $617,956,291 |
| | Grupo Televisa | 21,242,992 | | | $598,415,085 |
| | DISH Network Corp | 11,672,634 | | | $552,232,315 |
| | Liberty Global (Malone) (US) K | 7,175,642 | 2.79 | 0 | $531,499,803 |
| | Total, including stakes <100M | | | | $20,079,994,470 |
| Massachusetts Financial Services (US) | Walt Disney Company | 55,643,840 | | | $3,646,897,274 |
| | SBA Communications Corporation | 26,347,000 | | | $2,037,676,980 |
| | Disney (US) | 57,066,534 | 3.2 | 3.22 | $1,653,322,388 |
| | Google | 1,647,168 | | | $1,483,159,482 |
| | Viacom | 16,177,979 | | | $1,359,112,016 |
| | Comcast Corporation (CMCSK) | 29,075,772 | | | $1,254,619,562 |
| | AT&T | 27,127,456 | | | $933,455,761 |
| | Time Warner | 14,025,107 | | | $903,357,142 |
| | Time Warner Cable | 7,304,738 | | | $830,183,474 |
| | Verizon Communications | 12,889,332 | | | $617,914,576 |
| | Autodesk, Inc. | 10,536,390 | | | $431,359,807 |
| | Qualcomm Incorporated | 5,070,550 | | | $353,772,274 |
| | eBay | 6,037,891 | | | $332,687,794 |

| | | | | | |
|---|---|---|---|---|---|
| | Cisco Systems | 12,978,570 | | | $319,791,965 |
| | Check Point Software Technologies | 5,233,733 | | | $301,829,382 |
| | Amazon | 849,622 | | | $265,133,041 |
| | Discovery Communications - A | 2,973,381 | | | $245,184,997 |
| | News Corp - A | 14,408,770 | | | $240,770,547 |
| | Comcast Corporation (CMCSA) | 4,482,824 | | | $199,082,214 |
| | Yahoo! | 5,461,590 | | | $168,544,667 |
| | Mobile Telesystems | 7,525,695 | | | $168,199,283 |
| | Facebook | 3,237,737 | | | $149,939,600 |
| | Vodafone Group Plc | 4,055,383 | | | $135,814,777 |
| | Century Link | 4,015,408 | | | $130,741,684 |
| | Microsoft Corp | 3,397,380 | | | $111,909,697 |
| | Total, including stakes <100M | | | | $18,797,231,139 |
| JP Morgan Chase (US) | Google (US) A | 6,822,933 | 2.05 | 0.79 | $6,006,705,525 |
| | Softbank (Japan) | 91,902,000 | 7.7 | 7.71 | $5,799,447,220 |
| | Time Warner (US) | 49,460,496 | 5.3 | 5.3 | $2,859,805,878 |
| | Comcast (US) A | 30,669,310 | 1.17 | 0.96 | $1,308,046,071 |
| | Softbank (Japan) | 16,897,000 | 1.4 | 1.42 | $1,066,279,947 |
| | 21st Century Fox (US) B | 10,551,414 | 0.46 | 1.32 | $346,297,407 |
| | DISH Network | 4,973,000 | | | $232,985,050 |
| | CBS Corp | 2,640,000 | | | $148,500,000 |
| | Total, including stakes <100M | | | | $17,989,328,776 |
| Government Pension Fund of Norway | Vodafone Group PLC | | 2.28% | 2.28% | $2,657,767,846 |
| | Telenet Group Holding NV | | 4.26% | 4.26% | $2,145,767,427 |
| | Samsung Electronics Co Ltd | | 0.84% | 0.84% | $1,668,189,152 |
| | Microsoft | | 0.74% | 0.74% | $1,542,833,077 |
| | Google | | 0.60% | 0.22% | $1,306,228,878 |
| | Telefonica SA | | 2.15% | 2.15% | $1,243,289,943 |
| | AT&T Inc | | 0.67% | 0.67% | $1,205,766,852 |

*continued*

Table 36-4. *continued*

| OWNERS | COMPANIES | SHARES | % EQUITY | % VOTES | VALUE |
|---|---|---|---|---|---|
| Government Pension Fund of Norway (*cont.*) | Comcast Corp | | 0.99% | 0.69% | $923,735,032 |
| | Verizon | | 0.69% | 0.69% | $788,794,895 |
| | Deutsche Telekom AG | | 1.59% | 1.59% | $734,911,644 |
| | Telefonaktiebolaget LM Ericsson | | 2.16% | 1.26% | $675,933,852 |
| | Walt Disney | | 0.79% | 0.79% | $646,563,303 |
| | BT Group PLC | | 1.86% | 1.86% | $519,463,340 |
| | China Mobile | | 0.19% | 0.19% | $415,811,657 |
| | Softbank Corp | | 1.03% | 1.03% | $394,421,166 |
| | America Movil | | 0.47% | 1.48% | $388,120,995 |
| | Kabel Deutschland Holding | | 5.99% | 5.99% | $373,983,136 |
| | France Telecom | | 1.35% | 1.35% | $371,846,195 |
| | British Sky Broadcasting Group PLC | | 1.75% | 1.75% | $334,322,213 |
| | MTN Group | | 0.88% | 0.88% | $326,586,105 |
| | Virgin Media | | 3.50% | 3.50% | $321,791,407 |
| | Viacom | | 1.15% | 0.00% | $291,166,857 |
| | Telecom Italia | | 1.81% | 1.57% | $285,010,513 |
| | Yahoo | | 1.17% | 1.17% | $260,193,897 |
| | Pearson | | 1.71% | 1.71% | $254,957,955 |
| | Naspers | | 0.96% | 0.35% | $242,111,485 |
| | China Unicom | | 0.67% | 0.67% | $237,617,624 |
| | Ziggo | | 3.25% | 3.25% | $201,014,054 |
| | TeliaSonera | | 0.72% | 0.72% | $200,484,371 |
| | Axiata Group | | 1.15% | 1.15% | $198,967,261 |
| | Directv | | 0.64% | 0.64% | $181,024,150 |
| | Telefonica Brazil | | 0.68% | 0.00% | $174,350,436 |
| | Portugal Telecom | | 4.15% | 4.15% | $174,061,012 |

| | | | | | |
|---|---|---|---|---|---|
| | China Telecom | | 0.37% | 0.37% | $155,594,282 |
| | Singapore Telecommunications | | 0.38% | 0.38% | $154,462,767 |
| | ProSiebenSat.1 Media | | 2.50% | 0.00% | $145,457,019 |
| | NTT DOCOM | | 0.24% | 0.24% | $141,132,688 |
| | Tele2 | | 1.86% | 1.30% | $140,795,227 |
| | Turkcell Iletisim Hizmetleri | | 0.99% | 0.99% | $133,332,067 |
| | Koninklijke KPN | | 2.00% | 2.00% | $133,021,166 |
| | Focus Media Holdings | | 3.99% | 3.99% | $125,346,631 |
| | Grupo Televisa | | 0.87% | 0.87% | $125,039,366 |
| | Publicis Groupe | | 0.99% | 0.86% | $117,208,223 |
| | Chunghwa Telecom Co | | 0.47% | 0.47% | $113,191,129 |
| | Telstra Corp | | 0.21% | 0.21% | $112,391,864 |
| | CBS Corp | | 0.74% | 0.74% | $104,118,990 |
| | Telekomunikasi Indonesia Persero Tbk PT | | 0.57% | 0.57% | $101,444,309 |
| | Telekom Austria | | 3.18% | 3.18% | $100,787,861 |
| | Total, including stakes <100M | | | | $15,239,297,623 |
| Wellington Management (US) | Verizon Communications (US) | 45,222,499 | | | $2,169,775,502 |
| | Comcast (US) A | 46,398,200 | 1.76 | 1.46 | $1,978,883,230 |
| | Time Warner (US) | 31,421,367 | 3.4 | 3.4 | $1,810,499,166 |
| | Disney (US) | 28,484,413 | 1.6 | 1.60 | $1,617,914,658 |
| | Google Inc | 1,795,921 | | | $1,589,946,821 |
| | Walt Disney Co | 24,292,691 | | | $1,572,951,742 |
| | Yahoo (US) | 44,189,301 | 4.0 | 4.0 | $1,039,774,252 |
| | Facebook Inc | 11,010,828 | | | $519,490,865 |
| | Digitalglobe Inc | 10,419,564 | | | $337,072,895 |
| | Netflix Inc | 773,052 | | | $233,848,230 |
| | Dreamworks Animation SKG | 7,243,404 | | | $203,539,652 |
| | Yandex N V | 4,489,488 | | | $164,719,315 |
| | Liberty Interactive | 5,304,291 | | | $127,568,199 |

*continued*

Table 36-4. *continued*

| OWNERS | COMPANIES | SHARES | % EQUITY | % VOTES | VALUE |
|---|---|---|---|---|---|
| Wellington Management (US) (*cont.*) | SK Telecom | 608,299 | | | $120,035,642 |
| | TIVO | 9,706,899 | | | $118,424,168 |
| | CBS Corp (Class B) | 2,108,381 | | | $117,900,666 |
| | Total, including stakes <100M | | | | $14,098,776,035 |
| Janus Group (US) | Google | 2,127,610 | | | $1,910,742,713 |
| | 21st Century Fox (US) A | 71,520,919 | 3.10 | 0.00 | $1,733,667,077 |
| | Time Warner Cable (US) | 9,719,402 | 3.3 | 3.3 | $1,109,566,932 |
| | CBS (US) B | 17,276,832 | 3.53 | 0.00 | $982,878,972 |
| | Time Warner Cable | 8,539,271 | | | $976,038,675 |
| | Total News Corp | | | | $887,455,187 |
| | Total Comcast | | | | $856,707,665 |
| | Microsoft | 24,522,747 | | | $810,722,016 |
| | AT&T | 14,108,461 | | | $487,306,243 |
| | Total Discovery Communications, Inc. | | | | $481,938,324 |
| | Walt Disney Co. | 7,353,478 | | | $481,799,879 |
| | Time Warner | 6,147,979 | | | $398,389,039 |
| | Verizon Communications | 6,866,471 | | | $330,345,920 |
| | Amazon | 966,455 | | | $306,414,558 |
| | Sprint Nextel | 44,300,700 | | | $284,410,494 |
| | Youku.com | 8,630,852 | | | $241,491,239 |
| | SBA Communications | 1,957,030 | | | $152,080,801 |
| | Yahoo | 4,586,200 | | | $142,218,062 |
| | CenturyLink | 4,350,902 | | | $141,186,770 |
| | Netflix | 423,860 | | | $132,960,643 |
| | National CineMedia | 6,774,978 | | | $127,369,586 |
| | International Game Technology | 5,926,755 | | | $124,639,658 |

| | | | |
|---|---|---|---|
| | Total Liberty Media Interactive | | $106,375,394 |
| | Ericsson ADR - Class B | 7,591,516 | $105,370,242 |
| | Total, including stakes <100M | | $13,886,970,046 |
| Goldman Sachs (US) | Google Inc | 2,259,482 | $1,989,181,000 |
| | Microsoft | 26,625,374 | $919,774,000 |
| | Amazon.com Inc | 3,020,663 | $915,955,641 |
| | AT&T | 22,932,972 | $811,828,000 |
| | American Tower Corp | 10,180,312 | $731,862,630 |
| | SBA Communications | 6,356,015 | $471,108,000 |
| | Verizon Communications | 9,203,347 | $463,296,000 |
| | Liberty Global Class A | 6,102,299 | $448,458,000 |
| | CBS Corp Class B | 7,938,225 | $428,584,768 |
| | Comcast Corporation – CMCSA | 7,736,065 | $345,183,220 |
| | Disney | 4,696,684 | $296,595,000 |
| | News Corp Class A | 8,722,711 | $284,186,000 |
| | Sprint | 39,743,799 | $279,001,000 |
| | Time Warner Cable | 1,891,981 | $212,810,000 |
| | The McGraw Hill Company | 3,989,696 | $212,213,000 |
| | Vodafone PLC | 7,270,680 | $208,996,000 |
| | Liberty Media Class A | 1,571,719 | $199,232,000 |
| | Liberty Media Interactive | 8,517,454 | $195,986,000 |
| | Scripps Networks Interactive | 2,842,279 | $189,751,000 |
| | Liberty Global Class K | 2,484,296 | $168,659,000 |
| | Accenture PLC | 2,263,641 | $160,922,239 |
| | Facebook | 6,226,215 | $154,909,000 |
| | Yahoo | 5,948,194 | $149,478,000 |
| | Time Warner | 2,464,158 | $142,477,000 |
| | CenturyLink | 4,406,304 | $137,829,189 |

*continued*

Table 36-4. *continued*

| OWNERS | COMPANIES | SHARES | % EQUITY | % VOTES | VALUE |
|---|---|---|---|---|---|
| Goldman Sachs (US) (*cont.*) | Discovery Communications Class A | 1,752,828 | | | $135,388,000 |
| | TW Telecom | 4,385,023 | | | $123,395,000 |
| | Crown Castle International | 1,673,909 | | | $117,608,846 |
| | Dish Network | 2,675,258 | | | $113,752,000 |
| | Total, including stakes <100M | | | | $12,348,883,670 |
| ClearBridge (US) | Total Comcast | | | | $1,966,715,320 |
| | Liberty Media (Malone) (US) A | 3,770,631 | 3.11 | 1.79 | $1,333,086,576 |
| | Total Liberty Media | | | | $911,294,698 |
| | Microsoft Corporation | 20,279,439 | | | $664,962,805 |
| | Walt Disney Co. | 9,582,478 | | | $622,956,895 |
| | AT&T | 14,844,063 | | | $509,299,802 |
| | Broadcom | 17,013,430 | | | $465,147,176 |
| | DirecTV | 6,582,971 | | | $417,426,191 |
| | Google | 458,367 | | | $413,955,821 |
| | Cablevision Systems | 22,886,094 | | | $403,252,976 |
| | Time Warner | 5,688,779 | | | $365,788,490 |
| | Facebook | 7,324,848 | | | $347,857,032 |
| | AMC Networks | 4,325,612 | | | $292,022,066 |
| | Amazon | 828,532 | | | $262,097,813 |
| | Akamai Technologies | 3,559,473 | | | $187,619,822 |
| | Madison Square Garden | 2,758,645 | | | $156,056,548 |
| | Total Liberty Global | | | | $136,753,228 |
| | Total Discovery Communication | | | | $133,805,753 |
| | DISH Network Corporation – Class A | 2,727,325 | | | $128,838,833 |
| | Time Warner Cable | 1,032,667 | | | $117,300,645 |

| | | | | | |
|---|---|---|---|---|---|
| | Total News Corp | | | | $116,695,930 |
| | Verizon Communication | 2,301,058 | | | $109,944,551 |
| | Starz – Class A | 4,065,448 | | | $109,563,824 |
| | Total, including stakes <100M | | | | $10,980,117,342 |
| Baillie Gifford & Co (UK) | Baidu (China) | 3,036,275 | 8.7 | 8.7 | $1,911,401,981 |
| | Google | 2,108,770 | | | $1,893,464,583 |
| | eBay | 28,607,817 | | | $1,556,265,245 |
| | Facebook | 23,690,777 | | | $1,082,905,417 |
| | LinkedIn | 2,517,917 | | | $624,493,774 |
| | Walt Disney Co. | 4,039,684 | | | $265,811,207 |
| | Yoku (China) | 5,406,675 | | | $155,874,440 |
| | Total, including stakes <100M | | | | $7,573,016,169 |
| Handsome Reward Limited(UK) | Baidu (China) | 5,490,000 | 15.7 | 15.7 | $7,411,500,000 |
| | Total, including stakes <100M | | | | $7,411,500,000 |
| Oppenheimer Funds (US) | Baidu (China) | 2,980,070 | 8.5 | 8.5 | $1,564,615,703 |
| | Google | 1,061,377 | | | $953,530,483 |
| | Microsoft | 21,341,582 | | | $717,930,818 |
| | The Walt Disney Company | 10,743,502 | | | $707,137,302 |
| | Smiths News Plc (NWS)Class B | 2,299,120 | | | $485,114,320 |
| | Facebook | 9,769,079 | | | $449,182,252 |
| | Grupo Televisa | 14,239,034 | | | $410,084,179 |
| | Time Warner | 3,674,571 | | | $236,642,372 |
| | Youku | 6,626,800 | | | $190,255,428 |
| | Amazon.com | 440,873 | | | $137,578,828 |
| | Time Warner Cable | 1,118,313 | | | $127,431,766 |
| | AT & T | 3,185,088 | | | $110,872,913 |
| | Verizon Communications | 2,285,204 | | | $110,855,246 |
| | Total, including stakes <100M | | | | $6,798,563,851 |

*continued*

Table 36-4. *continued*

| OWNERS | COMPANIES | SHARES | % EQUITY | % VOTES | VALUE |
|---|---|---|---|---|---|
| Thornburg Investment Management (US) | Baidu (China) | 1,206,805 | 3.5 | 3.5 | $1,458,860,000 |
| | Baidu (China) | 7,882,016 | | | $1,158,025,791 |
| | Liberty Global (Malone) (US) A | 10,816,837 | 4.21 | 4.45 | $996,670,881 |
| | Yandex NV | 19,250,139 | | | $696,085,026 |
| | Sina Corp | 4,888,348 | | | $398,791,430 |
| | Microsoft | 11,653,987 | | | $382,250,774 |
| | Telefonica Brasil | 14,874,181 | | | $337,346,425 |
| | Publicis Groupe | 7,784,360 | | | $159,268,006 |
| | AT&T | 3,830,000 | | | $131,215,800 |
| | Google | 137,055 | | | $123,327,571 |
| | Total, including stakes <100M | | | | $6,281,463,081 |
| Coatue Management (US) | CBS | 14,441,084 | | | $796,570,193 |
| | Time Warner Cable TWX | 11,874,982 | | | $776,267,573 |
| | Liberty Global LBTYA | 9,146,979 | | | $722,245,462 |
| | Google | 432,663 | | | $382,521,685 |
| | Charter Communications | 2,784,806 | | | $368,318,442 |
| | Netflix | 1,065,828 | | | $348,920,112 |
| | Broadcom Corp | 12,292,671 | | | $320,347,006 |
| | News Corp | 11,873,652 | | | $189,859,695 |
| | Amazon | 565,679 | | | $179,320,243 |
| | Sirius Xm | 42,914,664 | | | $169,083,776 |
| | Time Warner Cable TWC | 1,224,854 | | | $137,171,399 |
| | Yandex | 2,694,820 | | | $104,720,705 |
| | Total, including stakes <100M | | | | $4,612,871,021 |
| State Farm (US) | Disney (US) | 42,206,018 | 2.4 | 2.38 | $2,750,144,133 |
| | AT&T, (US) | 29,793,395 | | | $1,021,913,449 |
| | Verizon Communication (US) | 7,910,800 | | | $378,057,132 |
| | Vodafone (UK) | 9,574,678 | | | $322,187,915 |
| | Total, including stakes <100M | | | | $4,519,545,462 |

| | | | | |
|---|---|---|---|---|
| Berkshire Hathaway (Warren Buffet, US) | DirecTV (US) | 37,275,400 | | $2,297,655,000 |
| | The Washington Post Company (US) | 1,727,765 | | $835,840,000 |
| | Liberty Media Corporation (US) | 5,622,340 | | $712,688,000 |
| | Starz (US) | 5,622,340 | | $124,254,000 |
| | Total, including stakes <100M | | | $4,044,957,000 |
| Viking Global Investors (US) | Time Warner (US) | 26,511,363 | 67.73 | $1,795,614,616 |
| | Comcast Corp A (US) | 21,951,155 | 46.24 | $1,015,021,407 |
| | News Corp A (US) | 12,219,800 | 16.96 | $207,247,808 |
| | McGraw-Hill (US) | 2,516,600 | 67.44 | $169,719,504 |
| | Total, including stakes <100M | | | $3,253,723,819 |
| Lazard (US) | Turkcell | 49,062,750 | 15.85 | $777,644,588 |
| | Mobile TeleSystems | 18,245,184 | 23.22 | $423,653,172 |
| | PT Telekomunikasi Indonesia | 9,068,928 | 40.41 | $366,475,380 |
| | Philippine Long Distance Telephone | 4,943,403 | 66.19 | $327,203,845 |
| | China Mobile | 5,642,626 | 55.13 | $311,077,971 |
| | Baidu | 1,768,235 | 152.84 | $270,257,037 |
| | Grupo Televisa | 6,499,565 | 29.35 | $190,762,233 |
| | America Movil | 8,103,680 | 20.98 | $170,015,206 |
| | Vodacom Group | 12,471,812 | 11.53 | $143,799,992 |
| | Total, including stakes <100M | | | $3,093,586,281 |
| Horizon Asset Management (US) | Liberty Media | 4,813,479 | 149.73 | $720,722,211 |
| | Dreamworks Animation A | 11,708,672 | 28.17 | $329,833,290 |
| | Starz | 7,775,923 | 29.11 | $226,357,119 |
| | Dish Network A | 3,492,256 | 47.67 | $166,475,844 |
| | Viacom B | 1,713,184 | 82.76 | $141,783,108 |
| | Total, including stakes <100M | | | $1,933,223,461 |

*continued*

Table 36-4. *continued*

| OWNERS | COMPANIES | SHARES | % EQUITY | % VOTES | VALUE |
|---|---|---|---|---|---|
| AXA Financial (France) | Google | 263,254 | | | $233,464,177 |
| | Comcast | 5,127,672 | | | $224,027,990 |
| | Microsoft | 6,755,017 | | | $219,267,852 |
| | Yahoo | 5,277,147 | | | $165,280,244 |
| | General Electric | 5,726,044 | | | $139,257,390 |
| | AT&T | 3,164,448 | | | $107,876,032 |
| | Total, including stakes <100M | | | | $1,622,191,177 |
| Commerzbank (Germany) | Google A | 467,122 | | 882.69 | $412,323,918 |
| | SK Telecom | 10,061,729 | | 24.28 | $244,298,780 |
| | AT&T | 4,315,798 | | 33.79 | $145,830,814 |
| | Verizon Communications | 2,230,126 | | 46.5 | $103,700,859 |
| | Total, including stakes <100M | | | | $1,415,519,588 |
| Scout Capital Management (US) | IHS A | 2,840,000 | | 113.01 | $320,948,400 |
| | Google A | 265,000 | | 869.88 | $230,518,200 |
| | Charter Communications A | 1,600,000 | | 136.75 | $218,800,000 |
| | Comcast A | 2,600,000 | | 46.29 | $120,354,000 |
| | Scripps Networks Interact A | 1,400,000 | | 77.85 | $108,990,000 |
| | Total, including stakes <100M | | | | $1,071,880,600 |
| BNP Paribas (France) | Mobile Telesystems | 1,131,894 | | 265.12 | $300,087,737 |
| | Google | 140,946 | | 876.11 | $123,484,200 |
| | Verizon | 2,344,922 | | 46.81 | $109,765,799 |
| | Total, including stakes <100M | | | | $1,054,671,062 |
| Childrens Investment Fund Management (UK) | News Corp. | 25,378,867 | | 16.55 | $420,020,249 |
| | Walt Disney | 4,815,052 | | 66.96 | $322,415,882 |
| | McGraw | 1,691,970 | | 67.79 | $114,698,646 |
| | Total, including stakes <100M | | | | $857,134,777 |

| | | | | |
|---|---|---|---|---|
| Pointstate Capital (US) | CBS Corp Cl. B | 9,570,660 | 56.9 | $544,570,554 |
| | News Corp Cl. A | 8,982,940 | 16.54 | $148,577,828 |
| HSBC (UK) | Google | 335,960 | 898.03 | $301,702,159 |
| | AT&T | 8,239,422 | 34.2 | $281,788,232 |
| | Verizon | 3,051,481 | 47.25 | $144,182,477 |
| | Total, including stakes <100M | | | $636,655,980 |
| Cascade Investment | GRUPO TELEVISA | 20,680,400 | 29.5 | $610,071,800 |
| | Total, including stakes <100M | | | $610,071,800 |
| Deutsche Bank (Germany) | Amazon | 1,185,091 | | $329,076,000 |
| | Baidu | 1,757,710 | | $166,280,000 |
| | CBS B | 2,093,505 | | $102,303,000 |
| | Chunghwa Telecom | 3,473,617 | | $111,537,000 |
| | Comcast K | 5,938,660 | | $235,582,000 |
| | Comcast A | 7,481,322 | | $312,337,000 |
| | Walt Disney | 6,221,276 | | $392,865,000 |
| | Google | 1,404,602 | | $1,236,559,000 |
| | Total, including stakes <100M | | | $3,305,526,000 |

# Bibliography


"The World's 500 Largest Asset Managers." Towers Watson, 2013. <http://www.towerswatson.com/en/Insights/IC-Types/Survey-Research-Results/2013/11/The-Worlds-500-Largest-Asset-Managers-Year-end-2012>.

La Porta, Rafael, Florencio Lopez-de-Silanes, and Andrei Shleifer. "Corporate Ownership around the World." *Journal of Finance*, 54.2 (1999): 471–517.

Noam, Eli. *Telecommunications in Europe*. New York: Oxford University Press, 1992.

Noam, Eli. *Television in Europe*. New York: Oxford University Press, 1991.

Shleifer, Andrei, Caralee Mcliesh, Tatiana Nenova, and Simeon Djankov. "Who Owns the Media?" World Bank 2001. Also Published in the *Journal of Law and Economics*, 46 (October 2003).

Soloski, John, and Robert Picard. "The New Media Lords: Why Institutional Investors Call the Shots." *Columbia Journalism Review* (September 19, 1996): 11.

# 37

# Analysis of Media Concentration

IN THIS CHAPTER WE summarize, present, and analyze data from the preceding chapters. We cover the following:

- Convergence trends
- Factors for industry concentration
- Explanatory variables for intra-industry by variations in concentration
- Diversity and media voices
- Cross-ownership around the world
- The export of media by countries
- The share of "Hollywood" industry in the various countries' content media
- Ranking the world's largest companies
- Market power in news media, both nationally and worldwide.

## CONVERGENCE OF MEDIA INDUSTRIES

People have been talking about "media convergence" for a long time. Digital technology would lead formerly separate segments of the media sector—audiovisual, TV, telecom, text media, and online media—to overlap and assume similar characteristics. Can we observe this convergence also in the structure of media industries, as expressed in market concentration? The answer is yes. Graph 37.1 shows how the concentrations move toward each other.

Graph 37.1 shows global concentration for the major sectors of the media industry trends from 2000 to 2013. The highly concentrated

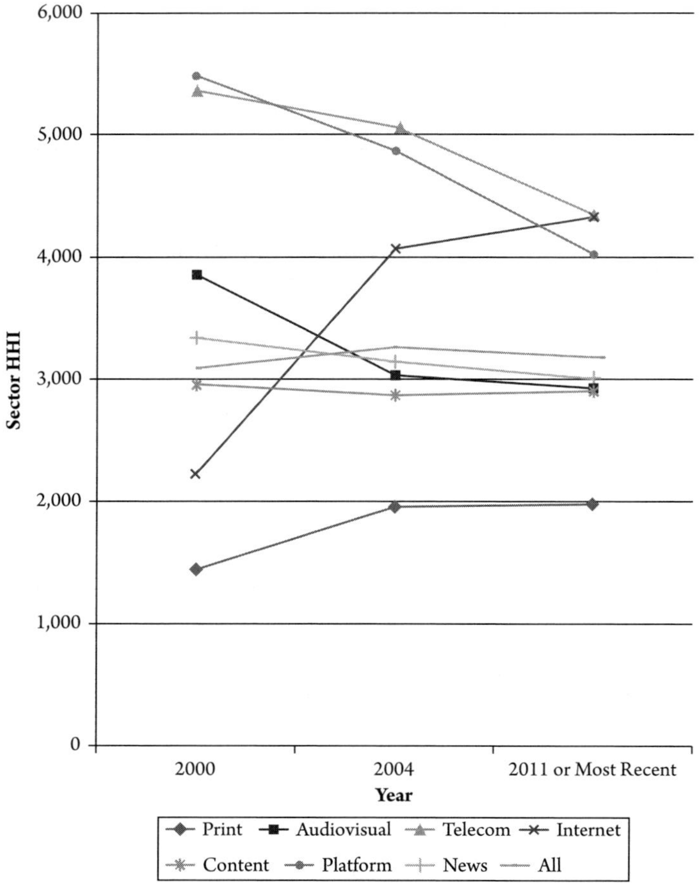


GRAPH 37.1 Convergence of Media Industries 2000–2012

Telecom and Audiovisual industries declined in concentration, whereas the low concentration print sector rose slightly. The Internet sector rose during that period (though much more slowly after 2004). The standard deviation—the measure of the industries' divergence from the average—in the concentration of print, audiovisual, telecom, and Internet sectors dropped from 1,512 in 2000 to 1,159 in 2004 to 1,002 in 2013.

Using greater aggregations we observe in Graph 37.1 the concentration of content media rising and of platform media declining, and then moving closer to each other. In 2000, the HHI for content was 2,962 while it was 5,486 for platforms, for a gap of 2,524. In 2004, the gap was 2,001 (2,871 for content and 4,872 for platforms). It narrowed to 1,120 in 2013 (2,906 content and 4,026 platforms).

For News Media, the trend of concentration (by time spent) has been a modest decline, from 3,338 in 2000, 3,146 in 2004, to 3,006 by 2013, due to the expansion of online news media and of new electronic media outlets in large market countries.

We can make a similar analysis for all 13 media industries without aggregating them into sub-sectors as the preceding graph did. Graph 37.2 shows the global trends. The high-concentration industries have been trending downward, while low-concentration industries have risen. The exception is the search engine industry. The standard deviation for the 13 industries dropped from 1,800 in 2000 to 1,713 in 2004 (rising slightly to 1,720 in 2013). If we omit search engines as an exception, the standard deviation for the 12 industries declined from 1,858 in

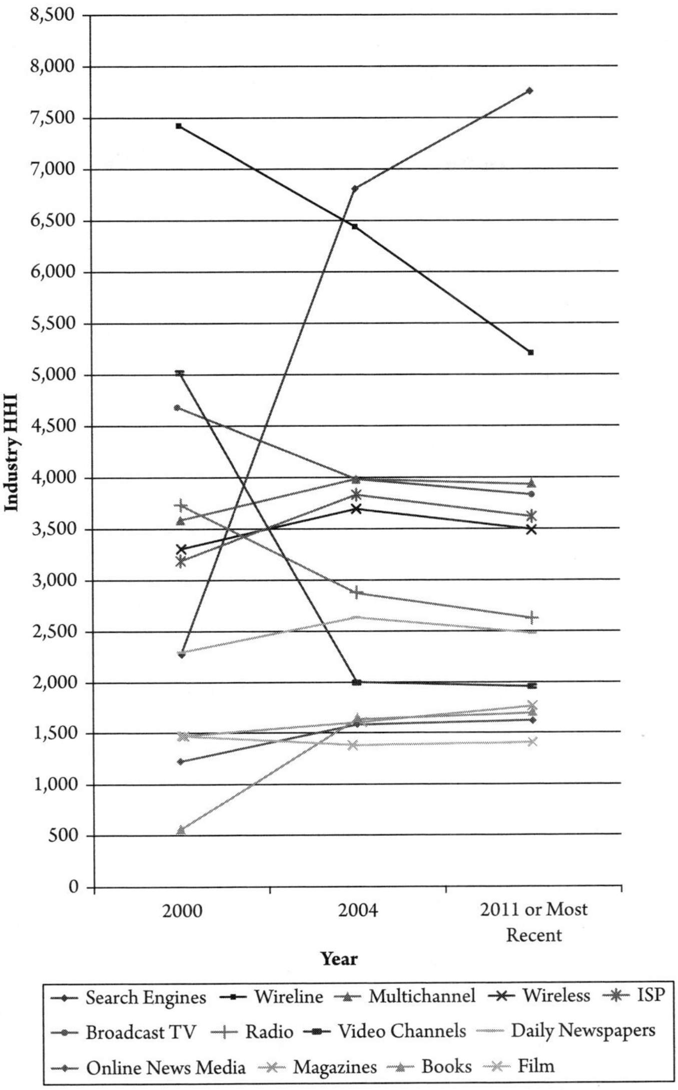


GRAPH 37.2 Convergence of 13 Media Industries

2000 to 1,430 in 2004 to 1,147 in 2013. Thus, we can observe a convergence across media.

A similar analysis can be done for the concentration trends geographically. We look at different regions of the world. These regions are Europe (EU and Switzerland); North America; Latin America (including Mexico); Mideast (Egypt, Israel, Turkey); Asia-Pacific; and BRICS (Brazil, Russia, India, China, and South Africa). For each region, average concentrations for content and for platform media were calculated. They are depicted in Graphs 37.3 and 37.4, below.

The graph shows a convergence of content media by region from 2000 to 2013. Regions with lower concentration in content media are rising in concentration and regions with higher concentration were declining. The world average was basically steady, with a small decline. The two extremes (North America and Europe) are low but increasing; the Mideast, the BRICS countries and Latin America are high but dropping. Asia-Pacific is rising slightly. The standard deviation of concentrations by region dropped from 1,210 in 2000 to 1,033 in 2004 to 764 in 2013. Thus, here, too, one can observe convergence.

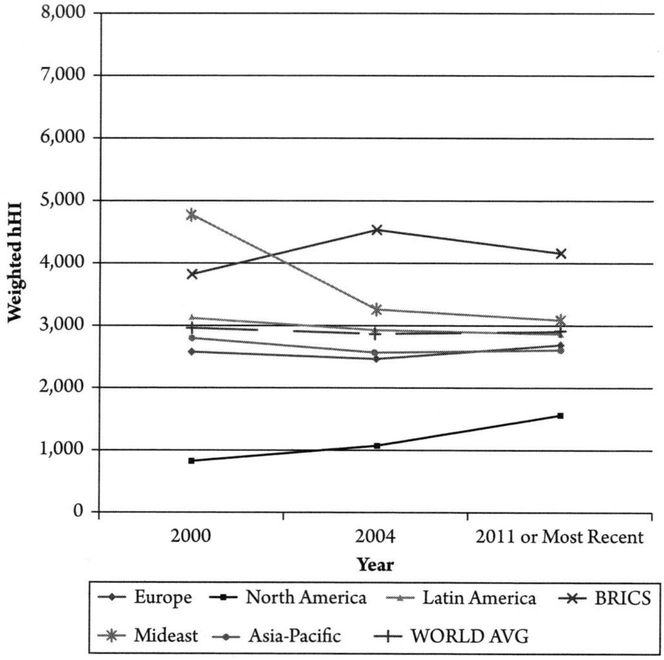


GRAPH 37.3 Convergence of Content Media by Region

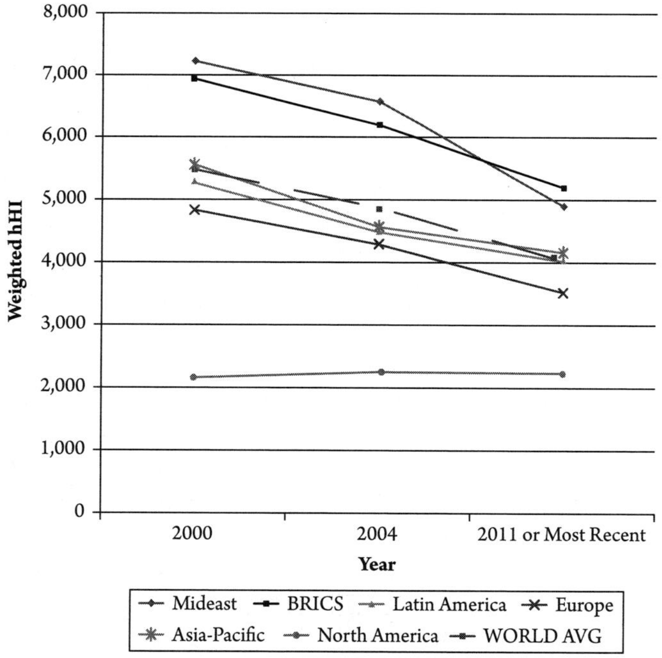


GRAPH 37.4 Convergence of Platform Media by Region

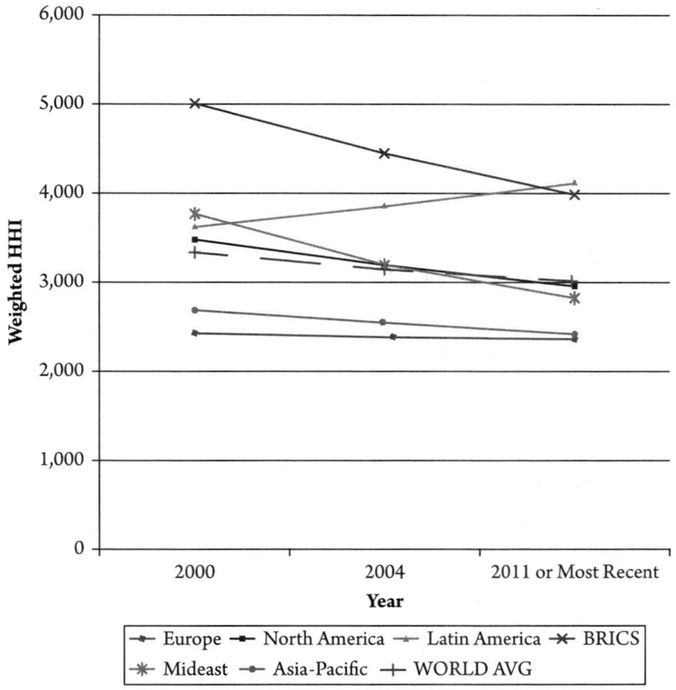


GRAPH 37.5 Regional Convergence of News Media

Similar concentration trends exist for platform media across the world's regions (Graph 37.4). The higher concentration regions (Mideast, Asia-Pacific, the BRICS countries, Latin America, and Europe) decreased in concentration while North America fell after 2000 (but rose very slightly again in 2013). The standard deviation declined from 1,660 in 2000 to 1,413 in 2004 to 966 in 2013.

We extend the analysis to another measure, that of "news media." These are the media that provide much of the information for the civil and political discourse. They are, in our investigation, defined as newspaper, magazines, radio, broadcast TV, multi-channel TV, and online news. These six industries do not carry equal weight as news sources, however. Newspapers and broadcast TV carry more news value than radio, for example. The relative "news weights" are based on the time people spend for each medium's news content (this is discussed near the end of this chapter, in the section on news media. That discussion also qualifies the results, given the limitations of the country-specific time data). The results of the calculation are depicted in Graph 37.5.

The findings show declining concentration in the world news concentration by time spent. Except Latin America, concentration is declining. News concentration was lowest in North America and Europe. Finally, we can observe here as well a convergence, with high concentration regions declining. The standard deviation fell from 835 in 2000 to 713 in 2004 to 698 in 2013.

## FACTORS FOR INDUSTRY CONCENTRATION

The concentration levels of different industries differ even, as we have seen, the differences are narrowing in a process of convergence. So the question is, why the differences? And what are the economic factors of convergence of industry structure? When consolidation in a single country is considered, the particularities of a national situation may obscure a more fundamental explanation. But once 30 countries' figures and trends are considered, one can look at the bigger picture with greater confidence.

What then are factors that can explain concentration levels of different industries? We ran regressions to test a number of variables. The maturity of an industry, for example, might have provided the time for industry consolidation to take place; or, the global export intensity of an industry might create scale effects that go beyond a country's borders; or, the size of an industry might give more firms an opportunity to flourish, etc. The economic relationship that both made sense as well as found statistical support was that market concentration is explained by capital intensity. Capital Intensity proved to be the best measure in terms of statistical correlation and economics. The more capital intense an industry, the more concentrated it is. The logic is that of economies of scale: the higher the fixed costs are relative to marginal cost, the greater these economies of scale. The high fixed costs get distributed over more users, and, given low marginal cost average cost per user, they decline. This gives cost advantage to large providers. Their price can be lower, or the production more elaborate, than that of small providers. On top of that, large providers also benefit from positive network effects, when users get utility from the presence of other users.

The regression analysis below specifies industry concentration (HHI) as the dependent variable, and capital intensity the independent variable.

Capital intensity is defined as an industries' average ratio of total assets to total revenue. Total assets and revenues are reported by companies in their annual reports. We selected firms that were "pure plays" (i.e., not involved in several industries), so as to provide measures relevant to a particular industry. The industry HHIs that are the dependent variable are the international averages for each industry that have been calculated across 30 countries in the chapter on "Countries."

Graph 37.6 below shows a fairly pronounced correlation: the higher the capital intensity, the higher the concentration. (One could add other explanatory variables, but with the number of

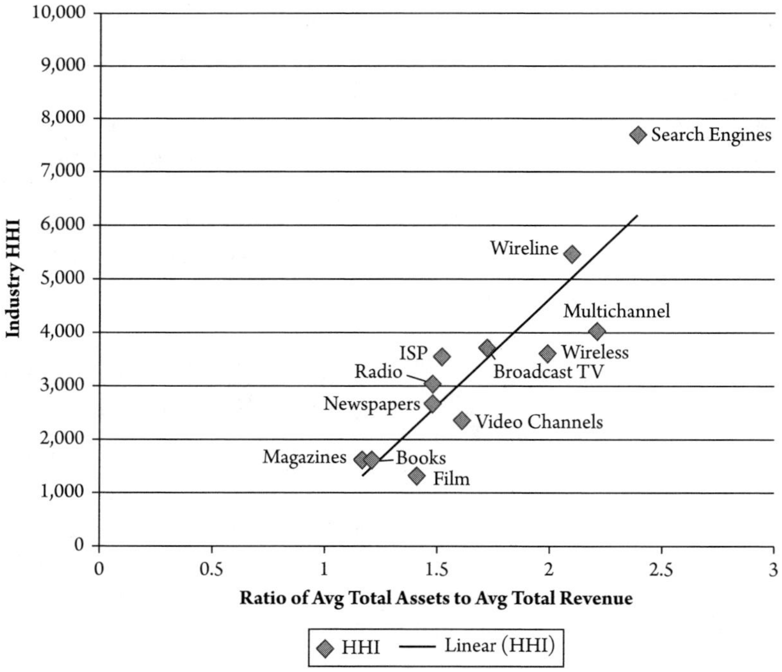


GRAPH 37.6 Media Concentration and Capital Intensity

observations (industries) small, not much is gained). The statistical details are provided in the OLS regression below.

The regression shows a fairly decent fit ($R^2 = 0.78$), and a statistically significant coefficient for capital intensity ($t = 5.879$). We thus conclude that the market concentration of a media industry is strongly associated with the capital intensity of that industry.

## VARIATION OF MEDIA CONCENTRATION ACROSS COUNTRIES

The preceding section analyzed how to explain the concentration levels of the various media industries. We found a strong correlation with capital intensity. The next question is how to explain different industry concentration within the same industry in different countries. Whereas before we analyzed across industries, we now analyze across countries for each industry.

Such an analysis would try to capture country-specific characteristics that affect concentration in that industry. A number of variables were hypothesized and tested:

- Population size
- Geographical size
- Income
- Education level
- Per capita spending
- Regulatory quality (for regulated media industries)
- Years as a democracy since 1900. (Of the 30 countries analyzed, only 9 were democracies 100 years ago, and of these only 6 for the entire period. Of the 193 UN member states, only 7 were continuous democracies for the past 100 years.[1])

We also tested for an impact of economic growth, R&D spending, and other variables. The results are provided in Table 37.1.

The results show that:

- Magnitude of population was a fairly decent-sized factor for the concentration in the newspaper and multi-channel TV industries, where it is associated with reduced concentration (negative correlation). Population was not a statistically significant factor in the other industries.
- Geographic size of a country was a moderately sized factor for newspapers (positive) and ISPs (negative). Conceivably (with moderate statistical significance), it was a factor for radio, TV, online news, and wireline. It is a contributing factor in the film industry, perhaps because larger countries are exporters of premium products with high-scale effect.
- Per capita income in a country was a negative factor for the concentration in mobile and wireless industries and a positive one for multi-channel
- Educational levels were a factor for newspapers and for multi-channel (both negative).
- Per-capita spending on a medium was a negative factor for newspapers and ISPs.
- The quality of regulation (from an index provided by the World Bank) made no statistical difference for any of the industries.
- Years as a democracy were, interestingly, a factor for platform industries (wireline, wireless, ISPs) but not for content media.
- We also looked at other variables: government spending (correlation with wireless), R&D spending (correlation with wireline, ISP), economic growth (correlation with online news, wireless, and radio). There was no gain in explanatory power.

---

1. This concept of democratization does not take into account the legal limitations on the voting franchise for female suffrage in France until 1945 and in Switzerland until 1990; and on black voters in the US South until after 1965.

Table 37-1. Factors for Media Industry Concentration

| INDUSTRY | CONSTANT | LN(POPULATION) | LN(GEOSIZE) | LN(INCOME) | LN(EDUCATION LEVEL) | LN(PER CAPITA SPENDING) | LN(REG QUALITY) | LN(YEARS AS DEMOCRACY) | R2 |
|---|---|---|---|---|---|---|---|---|---|
| Newspapers | 14.115 | −0.412 (−4.30) | 0.106 (1.83) | 0.47 (1.66) | −1.676 (−2.49) | −0.284 (−2.19) | 0.606 (0.17) | −0.192 (−1.59) | 0.6 |
| Books | 8.841 | 0.038 (0.10) | 0.043 (0.21) | 1.933 (1.21) | −3.008 (−1.32) | −0.213 (−0.34) | −3.239 (−1.27) | −0.323 (−0.61) | 0.4 |
| Magazines | 6.833 | −0.229 (−1.40) | 0.1 (0.94) | 0.836 (1.19) | −1.087 (−0.90) | −0.337 (−1.27) | −0.324 (−0.56) | −0.104 (−0.55) | 0.2 |
| Radio | 18.634 | −0.179 (−0.95) | −0.154 (−1.22) | −0.803 (−1.21) | 0.531 (0.39) | 0.372 (1.26) | 0.213 (0.31) | −0.219 (−1.06) | 0.3 |
| Broadcast TV | 14.134 | −0.432 (−0.38) | −0.916 (−1.15) | −0.795 (−1.48) | 1.131 (1.11) | 0.113 (0.05) | 0.279 (0.56) | −0.031 (−0.21) | 0.4 |
| Multichannel | 9.948 | −0.277 (−1.98) | 0.021 (0.27) | 0.769 (1.84) | −1.917 (−2.06) | −0.177 (−1.02) | 0.061 (0.13) | −0.062 (−0.25) | 0.4 |
| Online News | 13.48 | 0.126 (0.73) | −0.203 (−1.47) | −0.088 (−0.15) | −0.891 (−0.77) | 0.293 (1.16) | −0.624 (−0.71) | −0.220 (−1.25) | 0.4 |
| Wireline | 12.108 | −0.025 (−0.37) | −0.057 (−1.14) | −0.194 (−0.60) | 0.372 (0.63) | 0.045 (0.27) | −0.218 (−0.73) | −0.166 (−1.86) | 0.4 |
| Wireless | 6.845 | 0.021 (0.58) | −0.014 (−0.53) | 0.341 (2.42) | −0.340 (−1.11) | −0.115 (−0.95) | 0.011 (0.01) | −0.228 (−4.90) | 0.5 |
| ISP | 7.065 | −0.029 (−0.41) | −0.137 (−2.74) | 0.430 (1.23) | 0.001 (0.00) | −0.486 (−2.93) | 0.451 (1.36) | −0.291 (−2.97) | 0.6 |
| Film | 1.745 | 0.539 (2.37) | −0.122 (−0.81) | −0.556 (−0.62) | 2.311 (1.15) | −0.336 (−0.80) | 0.465 (0.46) | −0.147 (−0.55) | 0.6 |

## ANALYZING THE DIVERGENCE IN CONCENTRATION

The data that was collected from around the world subject to methodological uniformity permits an econometric cross-section analysis. Such an analysis was done in the preceding section. The results show which factor affects concentration levels of different media industries and which factors affect concentration within a media industry.

We now take this analysis a step further and try to extract information out of nonconformity. That is, how should one evaluate the fact that a country's industry deviates from the concentration level predicted by the econometric model? What the estimates of Table 37.1 do is to provide adjustment factors from mean concentration levels. It shows, for example, that size matters. A large population is associated with lower concentration, as Table 37.1 with its generally negative coefficients for population shows.

Similarly, a large geography is associated with lower concentration levels for many media industries. Income often plays a role, as it raises consumption of media services, as well as of products that are advertised, thereby attracting advertiser dollars to media. The same goes for education, which tends to raise demand for some media (and lower them for others). Overall, the models that were estimated provide a prediction: for a given industry and country, using its actual socio-demographic, economic, and geographic values, a prediction for concentration can be readily obtained, based on the explanatory power of the estimated coefficients. But, of course, the actually observed concentration levels will tend to deviate from predicted levels. What this means is that we can observe deviations of media from levels that can be explained by poverty, geography, population, and so on.[2] (Note that the earlier regression in Graph 37.6 also explained an industry's concentration level by its capital intensity. In the present set of estimations this factor helps explain the constants in the equations.)

The deviations of the predicted concentration from the one actually observed should be the measure of judging a country's concentration. The deviations of actual to predicted values are the "residuals" of the estimation. These residuals can be plotted as they are in the following graphs, in ascending order (the horizontal axis conveys no particular meaning for our purpose). We begin with a look at Graph 37.7, which shows the divergence—after adjustment for size, population, income, and so on—for newspaper concentration for the various countries.

China is an outlier, but so are Australia, Belgium, Mexico, and even France. These countries have, when it comes to the concentration of newspapers, a level that is much higher than can be explained from the socio-demographic, economic, and geographic circumstances. Such a divergence in a country needs to be explained by policy or business reasons, and if these are not persuasive, the question should be considered whether and how to bring the concentration level closer to the levels that are predicted from other countries.[3]

Those countries that clustered around the horizontal "zero-residual" line can take some comfort in that their concentration levels seem to be explainable by a variety of exogenous factors and thus likely to be fairly impervious to change. But their safety margin is not high. In contrast, those countries well below the line have more breathing space. For those countries, advocates of media pluralism and de-concentration should prioritize their advocacy and focus on those media industries in their country where divergences are largest, and have most impact.

---

2. In this estimation, we leave out the variables of "regulatory quality," "per capita spending," and "years as a democracy" as being possibly endogenous.

3. It should be noted that the divergence of high-concentration countries is actually understated. High-concentration countries already pull the regression coefficients in their direction. An estimation omitting them would shift the expected values further away and show their divergence as still larger.

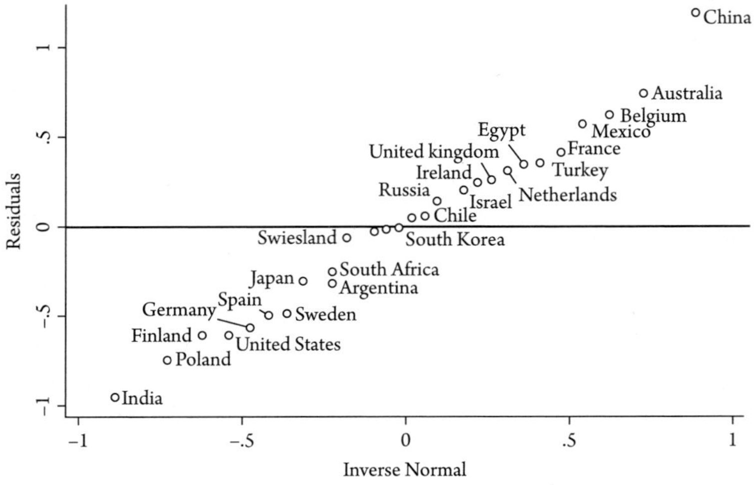


GRAPH 37.7 Divergence of Actual National Media Concentration over Predicted Concentration – Newspapers

Going beyond their own countries, they could also observe where elsewhere the deviations from media pluralism are the highest. Clearly, countries such as China or Egypt are at the high end of the concentration spectrum, for various reasons of politics and history. But the residual analysis also ferrets out unexpected observations. For Online News (Graph 37.8), based on the data supplied, the highest unexpected (and statistically unexplained) concentration is in Brazil, whereas the relatively lowest such concentration is in another Latin American country, Mexico.

Space does not permit the presentation of similar graphs for all of the 13 media industries. Here are some of the observations.

- In magazine publishing the divergence of actual above expected concentration is high for Russia, France, Turkey, and Australia.
- In radio, the outliers are China, Finland, Sweden, and the United Kingdom. In the latter three countries, alternatives are available but public service radio is popular and has a high market share
- In TV broadcasting, the high outliers are Ireland, Italy, France, India, and Taiwan.
- In multi-channel platforms, Italy, Spain, United Kingdom, Chile, and Poland are at the high end. At the low end are Taiwan and Sweden.
- For film, the most divergent are France (with Vivendi's Studio Canal and its "national champion" role) and Italy. In both cases, there is a very strong vertical integration into TV.
- In wireless, China shows a very high divergence in terms of actual vs. expected concentration. Also high are Mexico, Turkey, and Switzerland. India is at the other extreme.
- For wireline, it is Japan that is at the high end, together with Turkey, Australia, South Africa, Spain, and Mexico. At the low end are the United States, The Netherlands, Finland, and India.

We now move beyond the industry-specific analyses to sectoral aggregations. We start with "news media," which includes newspapers, magazines, radio, TV, online news, and multi-channel. We test the divergence of actual values from predicted ones, for several combinations of variables:

1. Average HHI concentration of the news media and the Power Index of news media are used as alternative dependent variables.

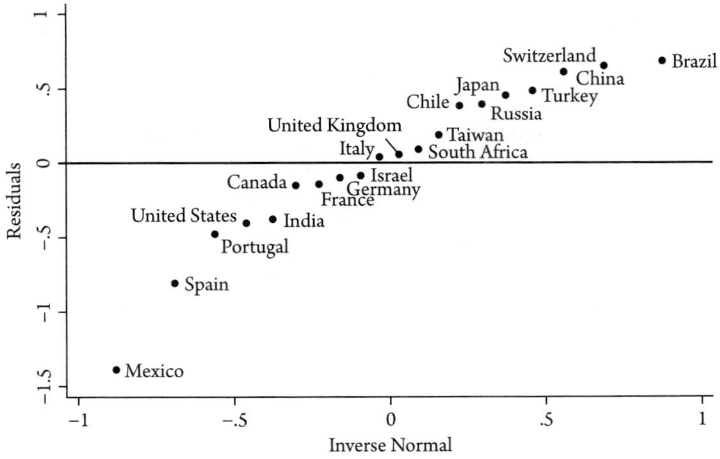


GRAPH 37.8 Divergence of Actual National Media Concentration over Predicted Concentration – Online News

2. We use, in the alternative, two kinds of averaging across news media: according to revenue and according to the attention factor (see explanation in this chapter, section 11 on news media).
3. We estimate for two years—2004 and 2011 or most recent.

For an example of the results, see Graph 37.9.

The findings show:

- China, by every measure, is a significant outlier, though this has slightly improved after 2005.
- Also high are Italy, Ireland, Mexico, Australia, Russia, France, and Egypt.
- The United States is low in comparison. Also low are Spain, Taiwan, India, and surprisingly Argentina.

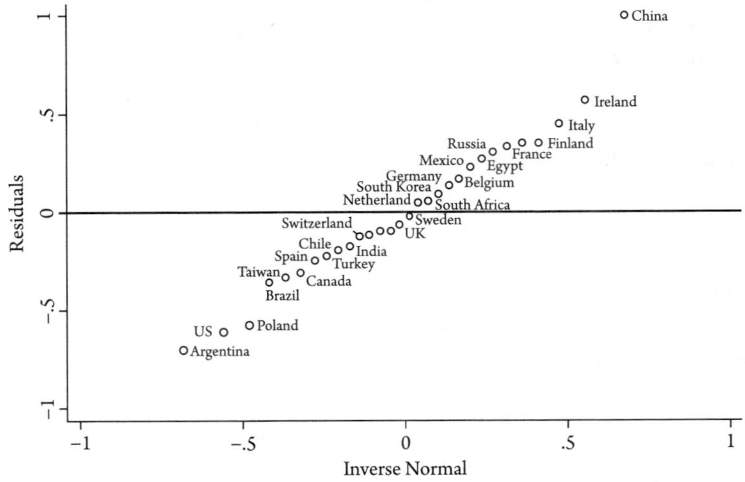


GRAPH 37.9 Divergence of Actual National Media Concentration over Predicted Concentration – News HHI by Time

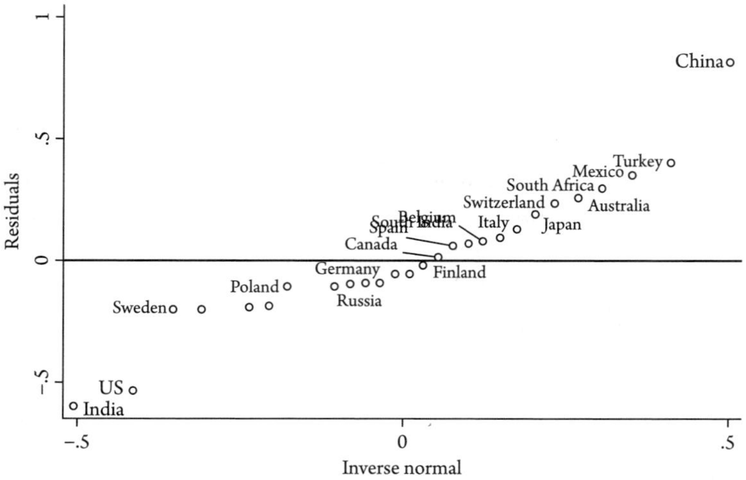


GRAPH 37.10 Divergence of Actual National Media Concentration over Predicted Concentration

- For most countries, the changes from 2004 to 2011 and newer have been minor. There seems to be a great stability.

Next, we analyze the residuals for all "content media," "platform media," and finally "all media" using the Power Index.

- China is the highest in divergence from the statistically explained concentration. For platforms, its divergence has increased over time. For content, the divergence is much higher but has declined somewhat.
- Also high for content are Ireland, South Korea, Italy, and Finland. All of them show concentration levels well above the predicted values.
- High for platforms are Turkey, South Africa, Mexico, Australia, Switzerland, and Japan.
- Low for platforms are India, United States, the Netherlands, Brazil, and Russia. Moreover, India has moved strongly in the direction of still lower than expected concentration, whereas the United States has moved in the opposite direction, though it is still low.

Finally, aggregating across all media, we can observe the following (Graph 37.10):

- China and India are the two extremes. Furthermore, they have moved in opposite directions, with China's media growing still more concentrated than predicted, while India's moving the other way, to the greatest non-concentration relative to prediction. India diverges now more toward de-concentration than the United States, factoring in its poverty, low average education, and so on.
- Other countries on the high end are Turkey, Mexico, and South Africa. These are relatively poor countries with lower educational averages, but that is already factored into the predictive model. Also on the high end are several rich countries: Australia, Switzerland, Japan, Italy, and Belgium.
- On the low end, though at considerable distance from the outliers India and United States, are Sweden, France, the Netherlands, Brazil, and Argentina.

## MEDIA VOICES

One way to look at the diversity of media is through market shares, and this is the approach taken in much of this book. However, there is also another perspective. That perspective says that pluralism is not defined by concentration but rather by the diversity of sources available. This was also discussed in the chapter on methodology. The argument made is that if there are 20 radio stations in the market, their actual audience share is immaterial; what counts is the number of options for a listener, not how popular they are. In this case, there would be 20 "voices." But this needs refinement. The same firm might own several of these stations and thus there would not be 20 truly different voices. One must therefore consolidate the voices supplied by the same media organization and calculate "net voices."

A second refinement is that of a limiting principle. Some voices are just too small to be considered an alternative option. As mentioned, is a college newspaper or a shopping magazine a "voice"? There has to be some threshold for a voice to be a meaningful participant. We chose, as such a threshold, for a firm to have at least a 1% share of that media industry's market.[4] To do so has reasons of principle and practicality. A media operation that is tiny does not, in most cases, meaningfully contribute to the pluralism of media choices facing a citizen, and to the shape of public opinion facing a politician. True, tiny publications have at times an impact, but typically only indirectly, as a catalyst to larger media operations that pick up the story. On a practical level, it is difficult to define, measure, and count tiny media. Picking a threshold of 1% of a market means applying a sliding scale that rises in absolute terms with the size of the market. Comparing countries' voice count on that basis is defensible on the level of pluralism, where voices are relative, not absolute in size. It is methodologically more difficult to do so for the already relative measure of voices-per-capita.

Counting different voices is one way to measure diversity for user choice. But it does not measure diversity in terms of political perspective or subject matter focus. To measure such differences would be highly subjective to define or measure. Instead, the underlying assumption is that when the number of options rises, different perspectives emerge. One example is the diversification of TV news perspectives in the United States. After several decades of three fairly centrist national TV news options, the emergence of multi-channel cable TV led to the creation of TV news channels with pronounced political perspectives, with Fox News Channel on the Right, and MSNBC on the Left. Thus, more options also meant a wider content diversity.

Industry-specific voices have been counted in the country reports. We now aggregate them for the content media industries: newspapers, magazines, book publishers, radio, broadcast TV, multi-channel TV, video channels, film, search engines, and online news media.[5] Duplicative voices of the same media organization are then consolidated.

Graph 37.11 shows the "net-n" number of voices (i.e., without double-counting the several media of the same companies). The United States had the highest number of net voices, with 59. (40 of its 99 "gross" voices had been duplicative, cross-owned by multi-voice firms.) Most countries had a net voice count of between 30 and 50. China had fewer than 20 net voices if one considers the state a single voice. This would be the lowest number for the 30-country world. If one segments Chinese media according to various media organizations, then China (segmented) reported 62 net voices, higher than any other country.

The absolute number of voices is generally higher in large countries. A larger population is likely to support a larger number of media. To estimate the diversity of the media supply side in a country, a good measure is therefore the number of media voices (net)-per-capita. Where such a value is high, it means that the country produces a wide variety of voices, given its size.

Graph 37.12 shows the number of voices divided by the population of the country. The distribution looks quite different than that of the absolute number, given in Graph 37.11. Ireland and Finland have the highest number, at

4. For radio and book publishing, we use a 2% national market share threshold because a large number of sub-national stations or specialty publishers would otherwise dominate country voice counts.

5. Although multi-channel is a platform industry, it overlaps with content so its voices are counted fully.

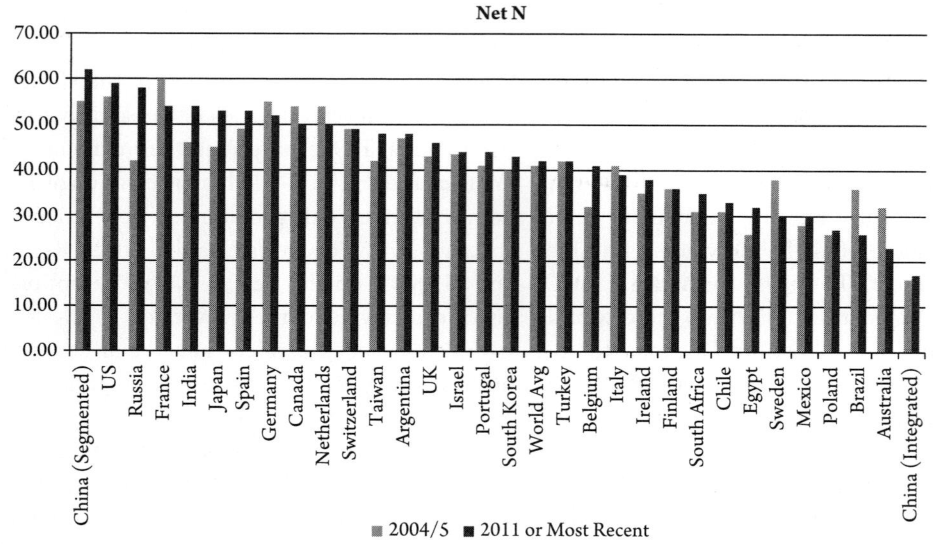


GRAPH 37.11 Countries' Number of Voices (Net)

(COMPANIES WITH >1% OF MARKET SHARE, 2011 OR MOST RECENT)

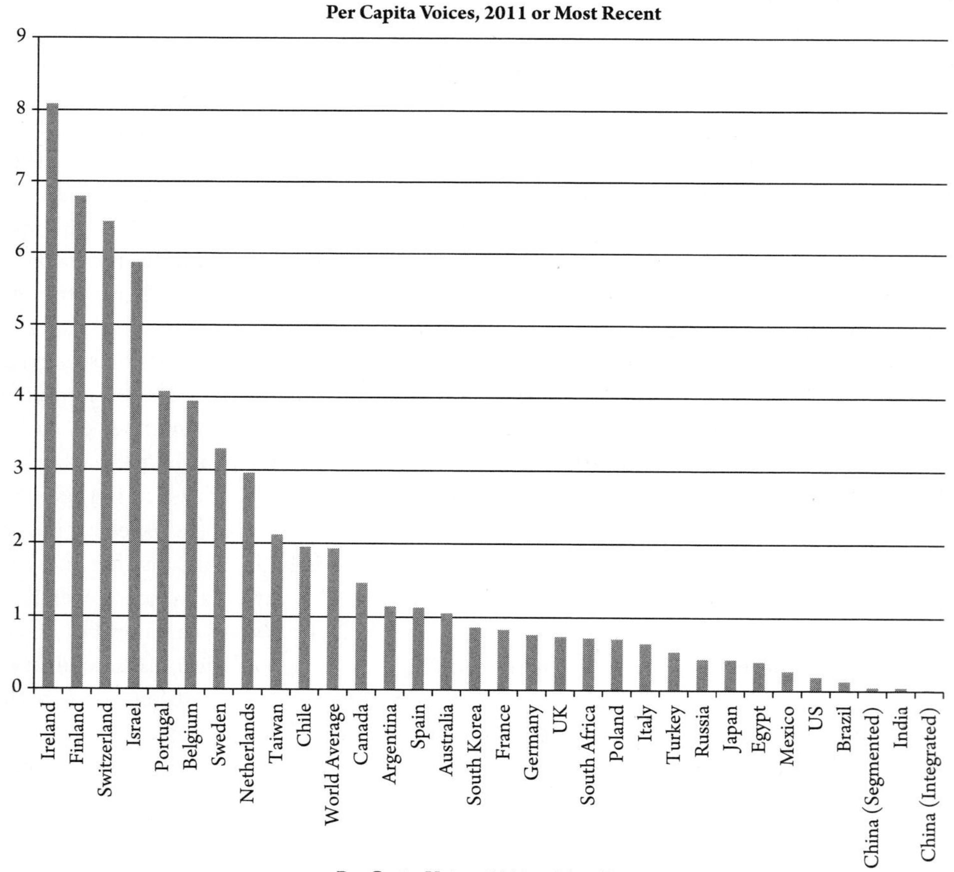


GRAPH 37.12 Countries' Per Capita Voices

(COMPANIES WITH >1% OF MARKET SHARE, 2011 OR MOST RECENT)

8.1 and 6.8 net voices per million population, respectively, followed by Israel (5.9), Switzerland (6.5), Belgium (3.9), and Sweden (3.3). Most countries have values below 1.0. Finland and Ireland have high per-capita voices, but one needs to consider that almost half of Ireland's media voices are spillovers from the United Kingdom. Similarly, about half of voices in Finland are from Sweden or Norway, and a large share of the voices for Belgium originate in the neighboring language partners France and the Netherlands. In Switzerland similarly, many of the voices are from Germany, Italy, France, and Austria. These are not foreign-owned operations inside these countries or aimed at them, but rather they are foreign broadcasts that spill in and/or are carried by cable systems. Given these qualifications, the highest per-capita voice count in Israel, whose tables do not include media from the adjoining countries. Countries with a high voice/capita measure are small (in population), wealthy, educated, with active politics, and often with more than one language community. The lowest values attained by the countries with the largest population, being China, India, Brazil, and United States. By this measure, the United States is not the most diverse environment at all, with less than one voice (0.2) per million population, much less than in Switzerland, Sweden, Belgium, Finland, or Israel.

The scattergram in Graph 37.13 shows how the size of a country's population affects its media structure. The regression analyses confirm the visual: voices-per-capita are highly correlated with population:

$$\ln\frac{\text{net voices}}{\text{cap}} = 3.5937 - 0.9361 \ln \text{population}.$$

The $R^2$ is a measure of how good the fit of the regression model is to the data. Its value of 0.7659 shows that there is a decent fit. The equation above can be raised out of the logarithmic form as:

$$\frac{\text{net voices}}{\text{pop}} = 43.929\ \text{pop}^{-0.9361},$$

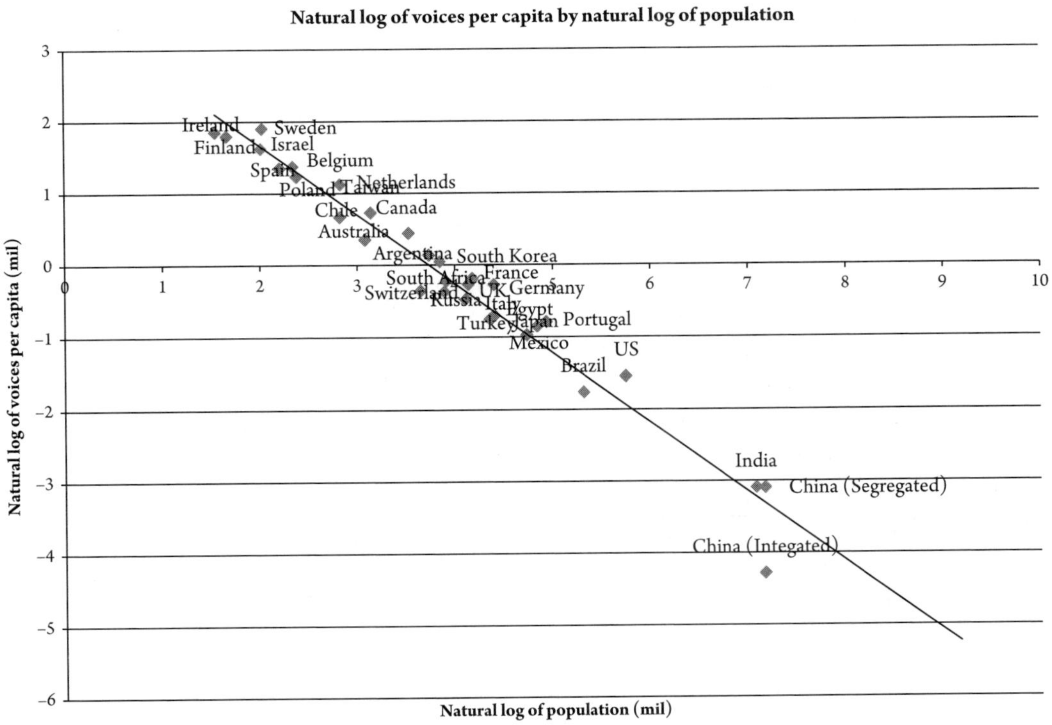


GRAPH 37.13 Regression of Net n Per Capita, 2011 or Most Recent

which can be rewritten as:

$$\text{net voices} = 43.93^{0.0639}.$$

This can be approximated as:

$$\text{NetVoices} \approx 44\,\text{pop}\,(\text{mil})^{0.064}.$$

What this equation tells us is that the number of voices rises with population, though at a slow pace. A doubling of population raises the number of voices by about 6.4%, or about 3 voices. This is not a very steep increase. However, it must also be remembered that the definition of a voice as above 1% is relative to national market size. To deal with the problem of a dynamic definition of voices, we look at the number of voices—holding that definition constant—that are added or subtracted when the population size is doubled (or halved), we find, looking at the 30 countries, that on average this adds (or subtracts) about 4.5 voices. Thus, taken together, the number of voices (holding the voice definition constant in terms of size) rises, for each doubling of population, by about 7.5 voices.

This result shows how media concentration, in terms of voices, declines with a country's population size, *ceteris paribus*. The number of voices rises. Yet on a per-capita basis, the ratio of voices declines, because a doubling of population increases the voice count by about 15%. Smaller countries have more voices-per-capita. These smaller countries are able to support voices at a scale that does not seem sustainable in larger countries. It must therefore be reasons of competitiveness, and factors such as economies of scale and network effects that make it harder for smaller voices to survive in large countries. In that sense, the larger countries under-perform in terms of sustaining voices. And this suggests that in an increasingly global media system, the number of voices-per-capita will decline.

## PREVALENCE OF NATIONAL CROSS-OWNERSHIP OF VOICES

In what countries is there much cross-ownership of media? How many media voices are owned by larger entities that combine several voices? This might be based on regulatory policy and enforcement and on economic and business circumstances. To answer that question, we calculate the percentage of media voices owned by a multi-voice firm, of the overall number of voices (Graph 37.14).

The world average voice cross-ownership ratio is 26.9% (27% in 2004/05). On average, then, about 26.9% of content media entities in a country's media market do share common owners. But, the number varies greatly across countries. The countries with the highest voice cross-ownership ratios are China (using the integrated definition of state ownership) at 73.9%; Sweden, 44.4%, Brazil, 40.9%; and the United States, 40.4%. In Canada (35.9%), Germany (35%), and Australia (34.3%) it is a combination of the strong position of the private companies and of the public broadcasters in a tight oligopoly. The US's high cross-ownership is the result of diverse activities of the top groups Murdoch, Redstone, Disney, and Time Warner. Canada, the United Kingdom (31.3%), and Italy (25%) also have relatively high voice cross-ownership ratios, due to a combination of very active public broadcasters (CBC, BBC, and RAI) as well as private companies (Bell Canada, Murdoch, Fininvest).

Countries with low cross-ownership include South Korea (10.4%), Belgium (8.9%) and Egypt (8.6%). In Argentina (7.7%) and Mexico (9.1%), even though the overall content media market is highly concentrated because the audiovisual and newspaper markets are oligopolies, the main cross-owning firm is one media firm (Clarín in Argentina, Televisa in Mexico) so the ratios are low.

Depending how one measures China's state media holdings—either as a single group (as we also do for Murdoch, Redstone, and Malone) or as a set of independent media—China has either the world's lowest share of voices independent of each other (11.4%) or the very highest (73.9%).

One can observe that voice cross-ownership is higher in well-developed media markets such as those of Sweden, the United States, Germany, Australia, Canada, the United Kingdom, Italy, Israel, Portugal, Japan, and France. It is least in evidence in the less developed media markets

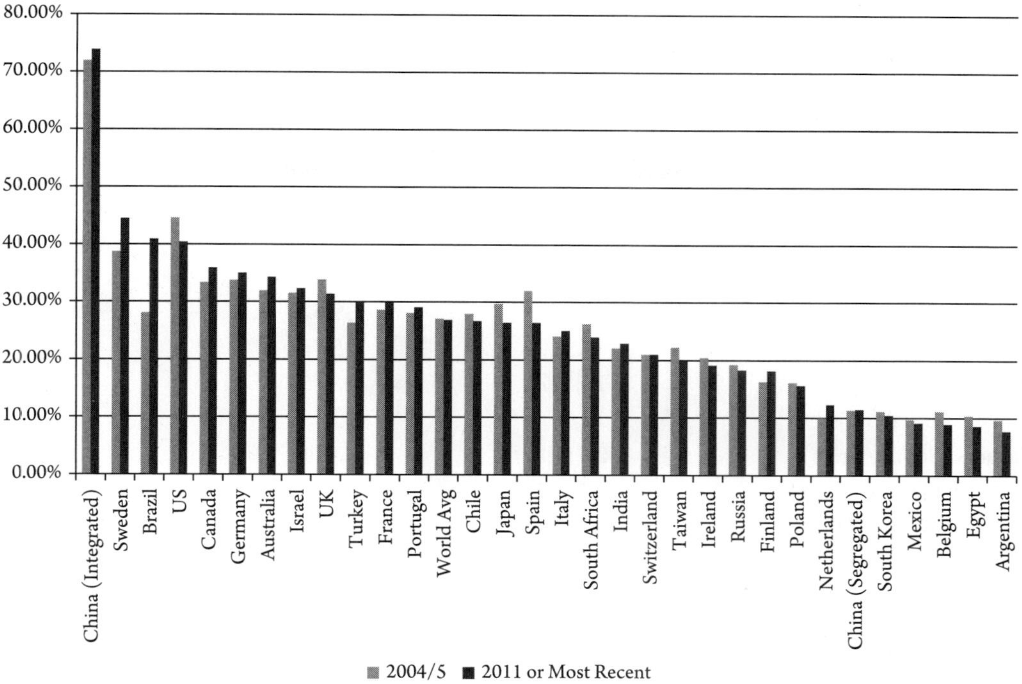


GRAPH 37.14 Cross-Ownership Ratios

of Mexico, Egypt, and Argentina. This suggests that the presence of media firms in multiple media industries rises with economic and media development. In consequence, as countries develop economically and their media grow in technological and business complexity, one should expect further trends to the cross-ownership of media voices.

## MEDIA EXPORTS

This section discusses the extent of the top media companies' out-of-home-country activity. We call this, somewhat imprecisely, "export" activities, even though many of them are produced by a company in foreign countries by its foreign subsidiaries. The term "international" might be better, but it has several meanings and is thus ambiguous.

Table 37.2 below shows the percentage of revenues made by companies outside of their home countries. Data sources are of two kinds. Some companies report their international revenues. For other companies, we identify domestic revenues, often from this book's country chapters, and allocate the remaining media revenues to foreign activities. In still other cases, the revenue of foreign operation has already been identified in the various country chapters and overarching summaries.

Of the top companies, the average share of exports as part of their revenues was 38.6% in 2013. The company with the greatest share of revenue outside of its home country is Vodafone, which makes 89% of its revenue outside of the United Kingdom.[6] Following Vodafone, the companies with the largest export profiles are Pearson (United Kingdom, 86.7%),

6. Vodafone operates in Albania, Australia, Czech Republic, Egypt, Germany, Ghana, Greece, Hungary, India, Ireland, Italy, Malta, The Netherlands, New Zealand, Portugal, Qatar, Romania, South Africa, Spain, Turkey, and the United Kingdom. Until 2013, it was also active in the United States by holding 45% in Verizon Communications.

Table 37-2. Top Media Companies by Export Percentage, 2010 or Most Recent (>10% of Total Revenue From Exports)

| COMPANY | TOTAL REVENUE (MIL $) | NON-DOMESTIC REVENUES (EXPORT) (MIL $) | EXPORT% |
|---|---|---|---|
| Vodafone (UK)[a] | 73,766 | 65,681 | 89.0 |
| Pearson (UK) | 9,043 | 7,841 | 86.7 |
| Google (US) | 50,175 | 39,829 | 79.4 |
| America Móvil (Mexico) | 59,300 | 45,731 | 77.1 |
| Liberty (US) | 12,197 | 8,775 | 72.0 |
| MTN (South Africa) | 11,667 | 8,213 | 70.4 |
| Naspers (South Africa) | 2,719 | 1,914 | 70.4 |
| Softbank (Japan) | 50,804 | 34,497 | 67.9 |
| Murdoch Group (US) | 64,424 | 41,792 | 64.8 |
| *-21st Century Fox* | 30,718 | 21,168 | 68.9 |
| *-News Corp.* | 33,706 | 20,561 | 61.0 |
| Bertelsmann (Germany) | 16,065 | 10,259 | 63.9 |
| Vimpelcom | 23,061 | 14,068 | 61.0 |
| TeliaSonera (Sweden/Finland)[b] | 16,204 | 9,704 | 60.0 |
| Bauer (Germany) | 2,842 | 1,697 | 59.7 |
| Lagardère (France) | 11,204 | 6,676 | 59.6 |
| Vivendi (France) | 38,657 | 22,523 | 58.3 |
| Telefónica + Telecom Italia | 122,471 | 70,482 | 57.6 |
| Orange (France) | 55,349 | 24,898 | 45.0 |
| Bonnier (Sweden) | 4,300 | 1,932 | 44.9 |
| Turkcell (Turkey) | 6,400 | 2,628 | 41.1 |
| Deutsche Telekom (Germany) | 77,556 | 30,494 | 39.3 |
| British Telecom (UK) | 29,400 | 11,466 | 39.0 |
| Redstone Group (US) | 28,396 | 10,047 | 35.4 |
| *-Viacom* | 13,887 | 8,316 | 59.9 |
| *-CBS* | 14,509 | 1,731 | 11.9 |
| Time Warner (US) | 27,950 | 9,865 | 35.3 |
| Disney (US) | 13,879 | 4,826 | 34.8 |
| PRISA (Spain) | 3,591 | 1,220 | 34.0 |
| SK Group (South Korea) | 13,600 | 4,073 | 29.9 |
| Swisscom (Switzerland) | 12,295 | 3,675 | 29.9 |
| Fininvest (Italy) | 6,854 | 1,836 | 26.8 |
| Globo Group (Brazil) | 10,398 | 2,667 | 25.6 |
| Telstra (Australia) | 25,500 | 6,299 | 24.7 |
| KPN (Netherlands) | 16,698 | 4,090 | 24.5 |
| China Film Group | 1,306 | 282 | 21.6 |
| Hearst (US) | 9,000 | 1,910 | 21.3 |
| Yahoo (US) | 5,607 | 1,160 | 20.7 |

Table 37-2. *continued*

| COMPANY | TOTAL REVENUE (MIL $) | NON-DOMESTIC REVENUES (EXPORT) (MIL $) | EXPORT% |
|---|---|---|---|
| ZDF (Germany) | 2,731 | 555 | 20.3 |
| Belgacom | 8,750 | 1,700 | 19.4 |
| Dish Network (US) | 14,270 | 2,633 | 18.5 |
| Televisa (Mexico) | 5,300 | 954 | 18.0 |
| BBC (UK) | 8,200 | 1,448 | 17.7 |
| MTS (Russia) | 12,400 | 1,861 | 15.0 |
| KDDI (Japan) | 35,605 | 1,780 | 15.0 |
| Telkom (South Africa) | 2,865 | 406 | 14.2 |
| Planeta (Spain) | 2,597 | 326 | 12.6 |
| ProSiebenSat.1 (Germany) | 3,150 | 387 | 12.3 |
| Verizon (US) | 115,846 | 13,176 | 11.4 |
| AT&T (US) | 129,434 | 14,095 | 10.9 |
| Bharti Enterprises (India) | 9,372 | 964 | 10.3 |
| Oi + Portugal Telecom | 32,195 | 3,244 | 10.1 |
| *-Telemar Participações (Oi) (Brazil)* | 27,197 | 1,808 | 6.6 |
| *-Portugal Telecom (Portugal)* | 4,998 | 1,436 | 28.7 |
| Korea Telecom (South Korea) | 18,338 | 16,504 | 10.0 |

[a] Incorporates the 2013 Verizon buyout of Vodafone's part ownership of Verizon Wireless.

[b] TeliaSonera's Sweden/Finland market is counted as a single domestic market since both countries' governments have stakes in the company's ownership.

Google (United States, 79.4%), America Móvil (Mexico, 77.1%), TeliaSonera (Sweden/Finland, 72.8%), and Liberty (United Kingdom, 72%). Two South African companies, the mobile telecom firm MTN (70.4%) and Naspers (70.4%) also derive much of their revenue from abroad, in MTN's case primarily in Middle Eastern and sub-Saharan mobile markets, and for Naspers primarily in Europe and sub-Saharan Africa. French firms also have large export percentages, primarily in Europe and the United States, with Lagardère (59.6%), Vivendi (58.3%), and the mobile provider Orange (45%). In contrast, primarily domestic companies are the telecom firms (mostly incumbents) Telus (Canada, 9.5%), Korea Telecom (10%), AT&T (United States, 10.9%), Verizon (United States, 88.6%), and Bharti Enterprises (India, 10.3%).

Is there a common denominator for the export-intensive companies? Several are based in rich and large countries where media firms expand globally. Others originate in less developed countries, some of whose companies seek expansion by moving beyond their borders. Companies also originate in very different industries. If one industry is singled out, it would be mobile telecom with 4 of the top 10 companies with highest export intensity.

Overall, the export (i.e., non-domestic) media activity is $575.1 billion. Of these, by far the largest share is that of telecom with 68.3%. This includes both wireline and wireless, but by far the greatest part is for mobile phone operations. The share of audiovisual (TV, multi-channel, cable channels, and film) is 16.5%, for print publishing (newspapers, books, magazines) 10.1%, and for search engines 5.1%.

The countries whose companies had the largest share of media exports in the world in 2013 (Table 37.4) were the United States with nearly one quarter (23.9%), distantly followed by the United Kingdom (14.8%), Spain (9.2%), France

### Table 37-3. Country Shares of World Media Exports, 2010 or Most Recent

| COUNTRY | COUNTRY'S SHARE OF WORLD MEDIA EXPORTS | COUNTRY'S SHARE OF WORLD MEDIA MARKET |
|---|---|---|
| US | 23.9% | 35.4% |
| UK | 14.8% | 5.7% |
| Spain | 9.5% | 3.9% |
| France | 9.2% | 4.5% |
| Mexico | 8.0% | 1.1% |
| Germany | 7.5% | 6.7% |
| Japan | 6.4% | 8.8% |
| South Korea | 3.5% | 3.9% |
| Italy | 3.2% | 3.0% |
| Russia | 2.7% | 2.1% |
| Sweden | 2.3% | 0.9% |
| South Africa | 1.8% | 0.9% |
| China | 1.7% | 6.7% |
| Australia | 1.2% | 2.0% |
| Brazil | 0.8% | 4.3% |
| Netherlands | 0.7% | 1.3% |
| Switzerland | 0.6% | 0.7% |
| Canada | 0.6% | 2.5% |
| Turkey | 0.4% | 1.0% |
| Belgium | 0.3% | 0.7% |
| Portugal | 0.2% | 0.7% |
| India | 0.2% | 2.0% |
| Finland | 0.2% | 0.4% |
| Egypt | 0.2% | 0.4% |
| Taiwan | 0.1% | 0.6% |
| Argentina | <0.1% | 0.4% |
| Chile | <0.1% | 0.3% |
| Ireland | 0.05% | 0.5% |
| Israel | <0.1% | 0.3% |
| Poland | <0.1% | 0.7% |

(9.5%), Mexico (8%), and Germany (7.5%). ("Exports," as explained, are defined as non-domestic revenues.) These export numbers are not shares in the world's non-domestic media presence. Those shares in the overall markets are provided in the second column of Table 37.3.

Observations:

- The United States is highest, at 23.9% in world share of non-domestic media activities. However, the US world export share is much smaller than its share in the overall world media market, which is 35.4%. That means that the domestic role of US media is much higher than the world average, and its export activity, while large in absolute terms, is lower than world average.
- Other export "under-performers" are Japan, South Korea, China (1.7% vs. 6.7%), Australia (counting Rupert Murdoch as a US

Table 37-4. Big Six Hollywood Studios' Share in National Content Market, 2012

| REVENUE IN MILLIONS $US | REVENUE OF 6 HOLLYWOOD DISTRIBUTORS (USD MILLIONS) | TOTAL MEDIA REVENUE | SHARE OF TOP 6 HOLLYWOOD DISTRIBUTORS IN TOTAL NATIONAL MEDIA REVENUE | TOTAL NATIONAL CONTENT REVENUE | SHARE OF TOP 6 HOLLYWOOD DISTRIBUTORS IN TOTAL NATIONAL CONTENT REVENUE |
|---|---|---|---|---|---|
| Argentina | 83 | 8,671 | 0.95% | 3,694 | 2.24% |
| Australia | 552 | 46,568 | 1.19% | 17,882 | 3.09% |
| Belgium | 134 | 15,184 | 0.88% | 5,557 | 2.41% |
| Brazil | 504 | 95,447 | 0.53% | 20,903 | 2.41% |
| Canada | 694 | 56,645 | 1.22% | 22,148 | 3.13% |
| Chile | 30 | 6,009 | 0.49% | 2,482 | 1.19% |
| China | 69 | 156,582 | 0.04% | 38,519 | 0.18% |
| Egypt | 25 | 7,791 | 0.33% | 1,776 | 1.43% |
| Finland | 42 | 10,767 | 0.39% | 4,781 | 0.88% |
| France | 614 | 100,408 | 0.61% | 31,528 | 1.95% |
| Germany | 838 | 150,354 | 0.56% | 56,383 | 1.49% |
| India | 108 | 45,102 | 0.24% | 10,734 | 1.00% |
| Ireland | 76 | 11,365 | 0.67% | 4,898 | 1.55% |
| Israel | 61 | 7,546 | 0.81% | 1,720 | 3.55% |
| Italy | 395 | 65,951 | 0.60% | 23,149 | 1.70% |
| Japan | 575 | 218,063 | 0.26% | 81,119 | 0.71% |
| Mexico | 295 | 24,922 | 1.18% | 5,362 | 5.24% |
| Netherlands | 115 | 30,203 | 0.38% | 8,880 | 1.29% |
| Portugal | 88 | 15,298 | 0.58% | 3,580 | 2.46% |
| Russia | 86 | 47,744 | 0.18% | 19,098 | 0.45% |
| South Africa | 215 | 21,744 | 0.99% | 3,544 | 6.06% |
| South Korea | 158 | 44,107 | 0.36% | 7,410 | 2.14% |
| Spain | 927 | 87,365 | 1.06% | 34,422 | 2.69% |
| Sweden | 52 | 19,231 | 0.27% | 8,673 | 0.60% |
| Switzerland | 95 | 15,367 | 0.62% | 7,883 | 1.20% |
| Taiwan | 130 | 14,209 | 0.91% | 4,746 | 2.74% |
| Turkey | 186 | 23,049 | 0.81% | 5,209 | 3.56% |
| UK | 871 | 127,407 | 0.68% | 24,080 | 3.62% |
| US | 8,579 | 795,890 | 1.08% | 335,070 | 2.56% |
| World Totals | 16,593 | 2,270,262 | 0.73% | 794,524 | 2.09% |

national), Brazil, Turkey, India, Taiwan, and Canada.

- Export "over-achievers" are the United Kingdom (14.8% vs. 5.7%), Spain, France, Mexico (8.0% vs. 1.1%, mostly due to Carlos Slim's America Móvil), Sweden, and South Africa.

## THE ROLE OF HOLLYWOOD

It is often argued that the Hollywood studios dominate content media worldwide. This may well be the case culturally. But in economic terms the case is harder to substantiate. In pure revenue terms, the combined share in content media industries' revenues of the six Hollywood "majors" box office revenues adds up to 2.09%. It is higher in Australia, South Africa, and Mexico, and lower in China, Japan, Russia, Sweden, and India.[7]

Quite likely, several income streams are not reported beyond those of theatrical box office. These include licensing to TV networks and cable channels, as well as home video sales. These additional income streams would raise the percentages, but even if they tripled it, it would merely rise to 6.27% of global content media revenues. One might also include the Hollywood firms' TV series production. In the United States, such revenues are 70.9% of film box office revenues (see the chapter on the United States). If one generously assumes the same percentage to hold worldwide (even though most TV series are not exported, and there is much more competition, both domestic and from third countries), then the Hollywood share of global content would rise to 7.75% for the six firms. Thus, the claim that Hollywood dominates the world's content creation and distribution is much exaggerated. It is accurate when applied to film specifically (see the Chapter "Media Industries," section "Film") but not to the wider content sector.

## MARKET SIZE

Table 37.5 shows the total platform and content revenue figures for every country. The revenue numbers in this section include advertising revenues as well as viewer fees, ear-marked taxes, and so on. They do not include in-kind income or voluntarist services such as bloggers.

Table 37-5. Content, Platform, and Overall Media Revenues (MIL$)

| | CONTENT MEDIA | | PLATFORM MEDIA | | ALL MEDIA | |
|---|---|---|---|---|---|---|
| | 2004/05 | 2011 OR MOST RECENT YEAR | 2004/05 | 2011 OR MOST RECENT YEAR | 2004/05 | 2011 OR MOST RECENT YEAR |
| Argentina | 1,675 | 3,229 | 4,416 | 5,789 | 6,091 | 9,018 |
| Australia | 10,284 | 16,998 | 21,624 | 33,132 | 31,908 | 50,130 |
| Belgium | 3,854 | 5,011 | 11,999 | 13,525 | 15,853 | 18,536 |
| Brazil | 10,662 | 23,354 | 65,361 | 103,697 | 76,023 | 127,050 |
| Canada | 15,217 | 24,089 | 27,855 | 49,644 | 43,072 | 73,733 |
| Chile | 1,719 | 2,129 | 2,382 | 3,528 | 4,102 | 5,657 |
| China | 8,598 | 35,272 | 50,300 | 173,430 | 58,898 | 208,702 |
| Egypt | 913 | 1,258 | 3,166 | 6,721 | 4,079 | 7,979 |
| Finland | 2,890 | 3,868 | 4,678 | 5,399 | 7,568 | 9,267 |
| France | 32,856 | 37,462 | 43,066 | 68,955 | 75,922 | 106,417 |
| Germany | 48,194 | 55,999 | 79,198 | 89,152 | 127,392 | 145,151 |
| India | 4,957 | 8,774 | 10,235 | 34,683 | 15,192 | 43,457 |
| Ireland | 3,455 | 3,788 | 6,180 | 6,469 | 9,635 | 10,257 |
| Israel | 1,436 | 1,961 | 4,939 | 5,468 | 6,375 | 7,429 |
| Italy | 17,453 | 22,873 | 46,787 | 42,450 | 64,240 | 65,323 |

7. These figures do not include the income of the domestic movie theaters' net of their license fee payments ("rentals" to the Hollywood distributors). The share of the theaters is not part of Hollywood firms' revenues.

Table 37-5. *continued*

| | CONTENT MEDIA | | PLATFORM MEDIA | | ALL MEDIA | |
|---|---|---|---|---|---|---|
| | 2004/05 | 2011 OR MOST RECENT YEAR | 2004/05 | 2011 OR MOST RECENT YEAR | 2004/05 | 2011 OR MOST RECENT YEAR |
| Japan | 58,770 | 63,861 | 115,104 | 171,942 | 173,874 | 235,803 |
| Mexico | 3,533 | 4,825 | 18,177 | 25,701 | 21,710 | 30,526 |
| Netherlands | 8,674 | 9,323 | 17,055 | 21,286 | 25,729 | 30,609 |
| Poland | 1,802 | 1,988 | 9,617 | 12,691 | 11,419 | 14,679 |
| Portugal | 4,893 | 6,450 | 7,489 | 8,772 | 12,381 | 15,222 |
| Russia | 2,674 | 9,122 | 18,117 | 36,492 | 20,792 | 45,614 |
| South Africa | 2,059 | 3,725 | 13,561 | 15,800 | 15,619 | 19,525 |
| South Korea | 5,846 | 7,827 | 23,087 | 32,730 | 28,933 | 40,558 |
| Spain | 28,254 | 34,088 | 44,559 | 52,893 | 72,813 | 86,981 |
| Sweden | 6,638 | 9,041 | 8,712 | 10,459 | 15,350 | 19,500 |
| Switzerland | 4,672 | 5,562 | 8,404 | 11,781 | 13,076 | 17,342 |
| Taiwan | 3,441 | 3,181 | 10,638 | 11,495 | 14,079 | 14,677 |
| Turkey | 1,675 | 3,465 | 9,523 | 16,053 | 11,198 | 19,518 |
| UK | 45,912 | 58,004 | 58,186 | 66,506 | 104,098 | 124,510 |
| US | 243,167 | 268,222 | 409,398 | 521,350 | 652,565 | 789,572 |
| Total | 586,173 | 734,749 | 1,153,813 | 1,657,993 | 1,739,986 | 2,392,742 |

The largest platform media markets are the United States ($521 billion, 31.7%), China ($173 billion, 10.5%), Japan ($172 billion, 10.5%), Brazil ($104 billion, 6.3%), Germany ($89.2 billion, 5.4%), and the United Kingdom ($66.5 billion, 4%). China and India's growth has been especially significant, with the market tripling in the period under review in both countries: in 2004, the Chinese platform industries accounted for 4.4% of the world at $50.3 billion, it was 10.5% in 2013. In India, it was $10.2 billion (0.9%) but grew to $34.7 billion (2.1%) by 2013.

Table 37.5 also shows the size of content markets. It is largest, by far, in the United States ($268 billion, 36.5%). This is followed by Japan ($63.1 billion, 8.7%), the United Kingdom ($58 billion, 7.9%), Germany ($56 billion, 7.6%), France ($37.5 billion, 5.1%), China ($35.3 billion, 4.8%), and Spain ($34.1 billion, 4.6%). China's market size has quadrupled since 2004. India's market size for content has almost doubled, from $5 billion to $8.8 billion (or 1.2% of the world market), since 2005. What these numbers show is that three countries (the United States, Japan, and the United Kingdom) account for over half (53.1%) of the world's (30-country) media market in content. Add Germany, France, and China, and these six countries account for 70.6% of the world's content media market. Add Canada (3.3%), Brazil (3.2%), Italy (3.1%), and Australia (2.3%), and the percentage of the top 10 media markets rises to 82.5%. A media exporter, by focusing just on these 10 countries, would therefore reach most of the world's media content market, at least by size of revenue (though not necessarily in terms of receptivity to imports).

## CONTENT REVENUES VS. PLATFORM REVENUES

In this section, we analyze the proportions of content revenues and platform revenues in the overall media sector. Graph 37.15 shows content revenues as a percentage of the national media industry.

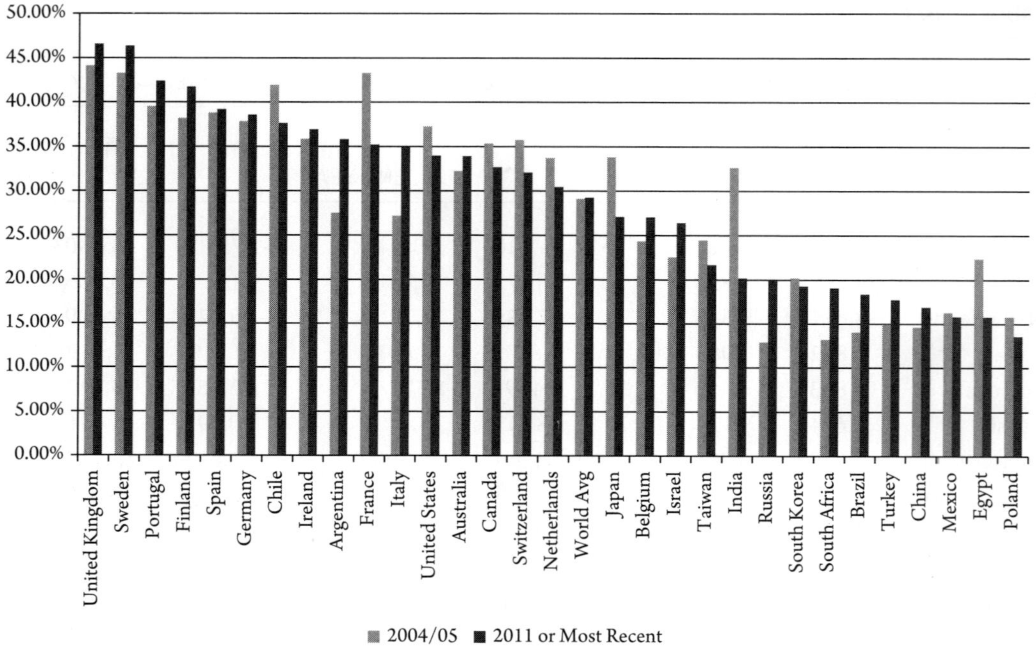


GRAPH 37.15 Content Revenue as a Percentage of National Media Industry

In most countries, platform media revenues are twice as high as content revenues. In no country does content's share of the national media revenues exceed 50%. The world average for content as a percentage of all revenues is 29.3%. The share of national content media revenue to total revenue is highest in the United Kingdom (46.6%), Sweden (46.4%), Portugal (42.4%), Finland (41.7%), and Spain (39.2%). It is lowest in Poland (13.5%), China (16.9%), Egypt (15.8%), and Turkey (17.8%). In the United States, it is 34%.

The share of platforms (telecom, ISPs, and cable/satellite)[8] is correspondingly high. Telecom and ISP platforms reach across society and economy, and demand for these services is fairly inelastic with respect to price and income. As a result, the world average share of platforms over all media was 70.8% in 2013. Shares are especially high in emerging markets: the largest shares are for China (83.1%), Poland (86.5%), South Africa (80.9%), Turkey (82.3%), Russia (80%), Egypt and Mexico (both 84.2%), Brazil (81.6%), and South Korea (80.7%). It is lower in richer countries (e.g., Sweden, with 53.6% or Finland, with 58.3%) or where content is oligopolistic and expensive, but widely consumed (Ireland, with 63.1% and Argentina, with 64.2%).

This means that of media spending in poor countries, most goes to platforms and connectivity, not to content. This leaves content relatively underfinanced, even in the context of these poor countries. The implications are then of two kinds: first, creation of lower quality (in budget terms) domestic content; and second, greater reliance on content imports. In contrast, platforms cannot be easily imported.[9]

## THE WORLD'S LARGEST MEDIA COMPANIES

In past chapters we identified the largest media companies in the 30 countries analyzed, and in the 13 industries. We are now ready to calculate the largest such firms for the world as a whole, and to do so—for platform, content, and overall media.

8. Only their platform concentration.
9. Satellite TV is one possibility, and, in the future, a reliance on offshore media clouds with their server infrastructure.

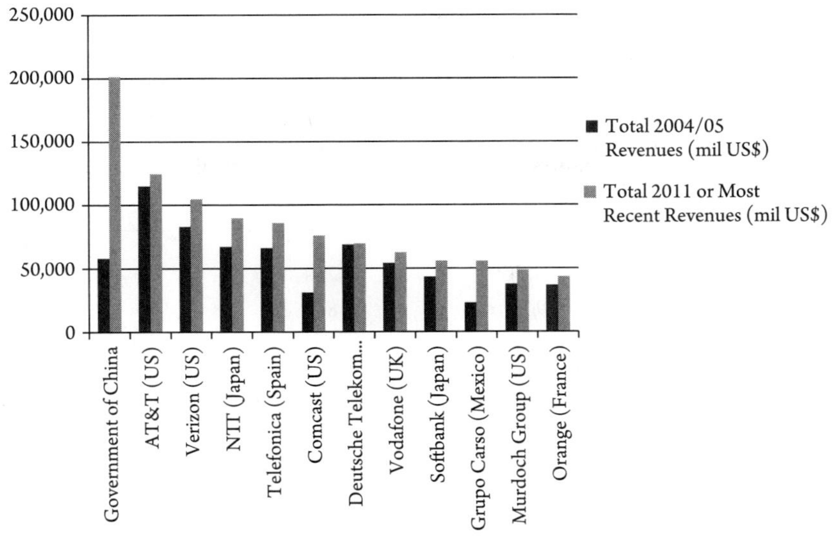


GRAPH 37.16 Top 12 Media Companies By Revenue

The rankings are by revenues, as well as by the "Power Index," described earlier. More details are provided in the chapter "Companies."

Graph 37.16 repeats the results from Chapter 35, Table 35.18, shows the top 12 global overall media companies by revenues. The revenues are for their activities across the 13 media industries (Table 37.6). Some of these firms are active in non-media industries too, or media not covered. These operations are not included.

As already mentioned, we find that all of the top 12 companies are platform telecommunications providers, of which three also have significant content media holdings. This is not surprising given the analysis of the preceding section. The top five ownership organizations are the Government of China (the combination of China Mobile, China Unicom, China Telecom, CCTV, and other media activities), AT&T (United States), Telefónica (Spain), NTT (Japan), and Verizon (United States). In the top 12, Comcast (a US platform cable company with extensive content media operations), Murdoch (some ISP and multichannel in addition to all other content media industries), Carso (multichannel platforms), Softbank (with mobile, ISP, search, and other online activities in Japan and the United States) are private-sector firms with a significant content presence, too. Of these, Murdoch and Comcast are closer to being content companies than Softbank, whose revenues are mostly from wireless, wireline, and ISP activities.

The Chinese companies experienced major growth. Even separately, the three main state telecom enterprises would be among the world's largest overall media organizations by revenues. (See Chapter 35, "Companies.")

We now turn to the platform sector and to the top 100 platform companies' revenues (this means separating the revenues of companies that are both platform and content providers). We also calculate and report the Power Index for the company. The Power Index, as described before, is the sum of market shares of the company, squared, across all countries and industries where it operates, and weighted by the revenue of that country and industry's overall media market. The data has been presented in Chapter 35 in Table 35.19.

The integrated entity "Government of China" was at the top by a wide margin with total revenues of US$173 billion dollars, and a Power Index of 1,046.

- The combined state holdings of China are largest in revenue terms, and it is far ahead

Table 37-6. Top Media Companies by Revenue

| TOP MEDIA COMPANIES | | | |
|---|---|---|---|
| COMPANY | TOTAL 2004/05 REVENUES (MIL US$) | TOTAL 2011 OR MOST RECENT REVENUES (MIL US$) | % CHANGE PER ANNUM (8 YRS) |
| Government of China[a] | 58,069 | 201,366 | 30.8% |
| *China Mobile* | *19,326* | *84,579* | *42.2%* |
| *China Telecom* | *14,685* | *44,243* | *25.2%* |
| *China Unicom* | *7,768* | *42,692* | *56.2%* |
| *CCTV* | *2,146* | *6,619* | *26.1%* |
| *Shanghai Media Group* | *732* | *1,493* | *13.0%* |
| *Zhejiang Publishing United Group* | *128* | *1,757* | *158.7%* |
| *Beijing Publishing Group* | *138* | *1,332* | *108.1%* |
| *Rayli* | *23* | *1,303* | *711.5%* |
| *Hunan Media Group* | *389* | *907* | *16.7%* |
| *Trend Media Group* | *66* | *889* | *156.6%* |
| *China Satcom* | *93* | *874* | *105.0%* |
| *China Railcom* | *114* | *842* | *79.8%* |
| *Jiangsu Media Group* | *60* | *706* | *135.7%* |
| AT&T (US) | 114,958 | 124,611 | 1.0% |
| Verizon (US) | 83,074 | 104,767 | 3.3% |
| NTT (Japan) | 67,093 | 89,716 | 4.2% |
| Telefónica (Spain) [b] | 66,054 | 85,895 | 3.8% |
| Comcast (US) | 30,815 | 75,954 | 18.3% |
| Deutsche Telekom (Germany) | 68,746 | 69,558 | 0.1% |
| Vodafone (UK) | 54,055 | 62,366 | 1.9% |
| Softbank (Japan) | 43,109 | 55,707 | 3.7% |
| Grupo Carso (Mexico) | 22,700 | 55,559 | 18.1% |
| Murdoch Group (US) | 37,320 | 48,425 | 3.7% |
| Orange (France) | 36,459 | 43,138 | 2.3% |

[a] Not all state enterprises are specified and listed here under GOC revenue totals.
[b] Includes the share of Telecom Italia (Italy) revenues for 2013.

of the second group of very substantial telecom firms: AT&T, Telefónica, Verizon, NTT, Deutsche Telekom, Vodafone, Grupo Carso, and Orange. A third tier adds somewhat different types of platform companies—Softbank, Comcast, TWC, Altice, the Murdoch Group, Bell Canada, and DirecTV with their significant TV and Internet media activities. Also in that size league are Telecom Italia, BT, Telstra, KT, SingTel, Vimpelcom, and Oi Telemar.

- There follows a fourth tier, comprised of several companies that are typically their countries' number 2 telecom incumbent (or for the United States, number 3) or the primary new entrant—KDDI, Century Link, MTN, the SK Group, Bharat Sanchar Nigam, MTS, Rogers, TeliaSonera, and Telus.
- By Power Index, China's number 1 position is greater still, given its population and the dominance of the triumvirate of state firms.

- As mentioned, several firms' Power Index ranking is stronger than their revenue ranking, which is the case where a company dominates a platform in a medium-sized or a poor country—BT (United Kingdom), KT (South Korea), Svyazinvest (Russia), Oger and Turkcell (Turkey), and Telkom (South Africa). Conversely, firms are lower ranked in Power Index terms than their revenues would suggest if they operate in large or rich markets as a medium-sized market share company—Century Link, Level 3, US Cellular, and Dish Network in the United States, Liberty in Europe, Bharti, Reliance, and BSN (India), and MTN (South Africa).

We next look at *content* companies. Revenues are those made in the industries of film, TV, multi-channel (one-third of multi-channel revenues), search engine, newspapers, book publishing, magazines, video channels, and online news. Graph 37.17 shows the top 15 content media companies by revenue in 2004/05 and 2011, or most recent year available.

In 2005, the content company with the highest revenues was Murdoch (News Corp. and 21st Century Fox) with $32.5 billion. Time Warner was a close second, at $22.8 billion. However, after spinning off Time Warner Cable and Time Inc.), Time Warner's revenues had dropped to $12.6 billion and eighth place position. In 2013, Murdoch's holdings retained their first place standing with $32.5 billion. Collectively, the Chinese government came in second with $27.9 billion, a major increase from $7.8 billion in 2004.

Google, with $7.4 billion in 2004/05, became the company with the second highest revenues in 2013, with $26.8 billion. The company with the highest growth rate was Google, with a growth of 33% per annum. With such a growth rate, Google is on track to become the world's largest content company. Most other companies have much lower growth rates, excluding the Chinese state enterprises.

The company with the highest content Power Index is Google with 278. Its market shares

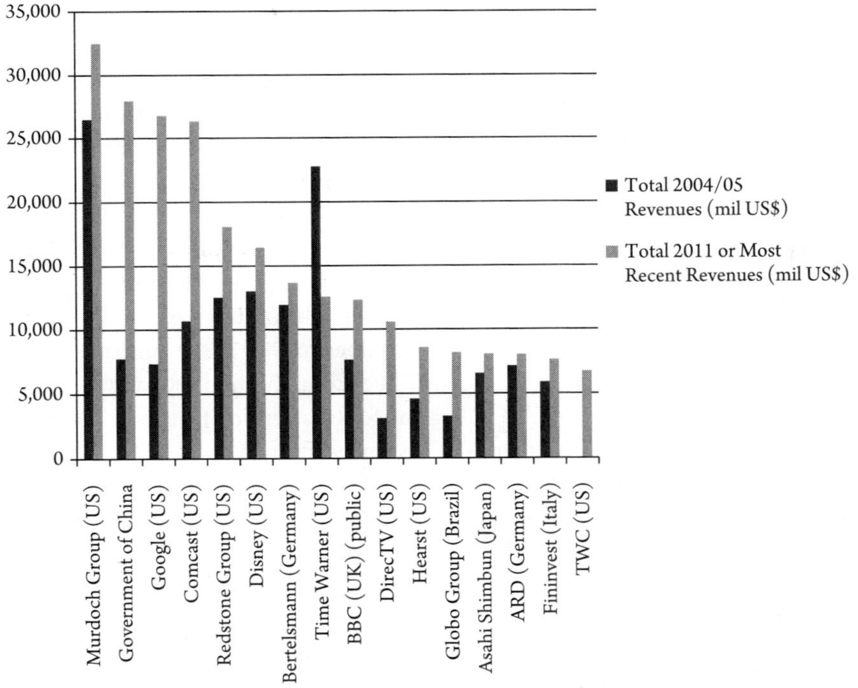


GRAPH 37.17 Top 15 Content Media Companies By Revenue[1]

[1]Time Warner Cable did not exist in its current configuration in 2004/05 – so there is no data for that year (TWC was part of Time Warner). It is shown here as "#16" in the world to visualize the impact of Time Warner's divestiture.

around the world are huge, even if its revenues are not at the top. The Government of China has the highest, at 359, but this is a combined number: the largest firm within this figure is CCTV, which has 53.2, putting it behind Murdoch, Google, Comcast (67.5), and the BBC (73.2) but ahead of any other one entity. The Murdoch Group is third with 129 and Comcast (67.8) is fourth. The graph also shows Time Warner Cable, since it ranked among the top fifteen in 2013, but did not exist in its present configuration for 2004/05, and might not exist independently after 2015. A more extensive list of the top 100 content media companies is provided in Chapter 35, Table 35.20.

- Several firms high in the revenue ranking rank much lower in Power Index—Redstone, Advance, Disney, Bertelsmann, DirecTV, Time Warner and Time Warner Cable, SkyPerfecTV, Fininvest, Fuji TV, ARD, BBC, and PRISA.
- These firms have typically a strong but non-dominant market share in medium or large markets.
- A third tier is medium-sized firms in medium-sized industry (often Print)—Advance, Lagardère, PRISA, Time Warner Cable, and a set of Japanese firms: Fuji TV, NHK, NTV, TBS, and Asahi.

## THE MARKET CONCENTRATION OF NEWS MEDIA

### National Concentrations of News Media

No debate over concentration in any industry—media or other sectors—is more impassioned than that about concentration in the news. The specter of control over what we think and how we vote haunts all who care for democracy. It is a fear that is shared across the political spectrum. The legitimate concern typically is colored by one's political perspective. People tend to attribute excessive power to those media—and their owners—with whom they disagree. To those on the Left, proprietors such as Berlusconi or Murdoch pull all the strings. To those on the Right, liberal "mainstream media" and public service TV organizations are setting the public agenda. The section that follows tries to develop and look dispassionately at the extent of dominance over news media.

It should be emphasized at the outset that the calculation of "news shares" is approximate only, relying on a number of simplifying assumptions. The reader should take the results as orders of magnitude rather than as precise metrics.

"News" is used here in the sense of information about current events. "News media" are organizations that collect, organize, and distribute such information. Among the industries we analyze, six are news media, at least in part: daily newspapers, magazines, radio, broadcast TV, non-broadcast TV (cable and satellite), and online news. Several of these media also provide other types of content, such as music and video entertainment, and in our calculations below, only their news activity will be considered. Other content media are not included, such as film or book publishing. Their products, in part, convey information and perspective. But they lack the immediacy of news sources.

From our analyses we know the market shares and the concentration indices for these six news media industries across countries. The question is how to add them up in order to get a company's share in the overall news media, both nationally and globally. This really is the question of how important the different news industries are in news terms. There are several proxies. The first measure is to use an industry's revenues to weight the various news industries. The second method is to assign weights according to the attention time they receive from users as a news source.

We first look at revenues. The results are shown in Table 37.7 below, which presents the weighted average HHI for the news industries for each country. The countries with high news media concentration by revenue are China (7,661), Egypt (4,199), Mexico (4,266), Italy (3,878), and Russia (3,853). The world average is a dismal 2,818. On the positive side, the world average has declined modestly since 2004/05 when it stood at 3,022. In almost no country is the number below the "highly concentrated"

Table 37-7. Weighted Average HHI – news Media (Weighted by Revenue)

| | 2004/5 | 2011 OR MOST RECENT |
|---|---|---|
| China (Integrated) | 8,676 | 7,661 |
| Mexico | 4,927 | 4,266 |
| Egypt | 5,339 | 4,199 |
| Italy | 3,503 | 3,878 |
| Russia | 4,549 | 3,853 |
| South Korea | 4,447 | 3,789 |
| South Africa | 4,033 | 3,724 |
| Ireland | 3,649 | 3,601 |
| Portugal | 3,402 | 3,276 |
| Chile | 3,408 | 3,175 |
| Australia | 2,749 | 3,110 |
| World Avg | 3,022 | 2,818 |
| Israel | 3,303 | 2,765 |
| UK | 2,389 | 2,730 |
| China (Segmented) | 3,546 | 2,663 |
| India | 4,083 | 2,492 |
| Switzerland | 2,171 | 2,489 |
| Netherlands | 2,383 | 2,488 |
| France | 2,656 | 2,459 |
| Turkey | 2,745 | 2,458 |
| Germany | 2,487 | 2,445 |
| Sweden | 1,852 | 2,442 |
| Brazil | 2,092 | 2,359 |
| Finland | 3,056 | 2,301 |
| Taiwan | 2,262 | 2,131 |
| Canada | 1,427 | 1,849 |
| Belgium | 1,700 | 1,685 |
| Poland | 1,724 | 1,591 |
| Argentina | 1,748 | 1,541 |
| Japan | 1,621 | 1,540 |
| Spain | 1,678 | 1,409 |
| US | 604 | 839 |

definition threshold of 1,800. It is only less than 1,800 in the United States (839), Spain (1,409), Poland (1,591), Argentina (1,541), and Japan (1,540).

Measuring news media by their overall revenues, however, does not seem quite satisfactory. Some media are greater producers of revenues due to a richer advertising market, for example. And some media combine news with entertainment content. An alternative methodology is therefore to use an "attention weight" based on time, the time people spent reading, listening to, or watching the medium's news content. Of course, some news attention has more of an impact, such as reading an editorial opinion column versus a sports report, and some media may leave more of an impact on a user per time unit. However, trying to gauge and measure subjective dimensions such as importance, value, or impact raises formidable conceptual and methodological barriers.

The "attention weight" of a news medium is calculated by the average time spent by an average user in that country with that medium, prorated to the share of that medium's news content in its overall content, and also adjusting for the penetration of that medium in a country's population, because the user population is likely to be smaller than the national population. The limitation we faced was that the data necessary for

Table 37-8. National Attention Weights of News Media (%), 2012

| | DAILY NEWSPAPERS | MAGAZINES | RADIO | BROADCAST TV | CABLE & SATELLITE TV | ONLINE NEWS MEDIA |
|---|---|---|---|---|---|---|
| Argentina | 19.6 | 5.8 | 6.6 | 27.0 | 29.0 | 12.2 |
| Australia | 24.1 | 5.8 | 6.6 | 41.0 | 15.0 | 7.7 |
| Belgium | 23.7 | 5.8 | 6.6 | 37.0 | 19.0 | 8.1 |
| Brazil | 25.5 | 5.8 | 6.6 | 38.0 | 18.0 | 6.3 |
| Canada | 14.8 | 5.8 | 6.6 | 29.0 | 27.0 | 17.0 |
| Chile | 24.9 | 5.8 | 6.6 | 27.0 | 29.0 | 6.9 |
| China | 28.0 | 5.8 | 6.6 | 43.0 | 13.0 | 3.8 |
| Egypt | 27.9 | 5.8 | 6.6 | 38.0 | 18.0 | 3.9 |
| Finland | 20.5 | 5.8 | 6.6 | 37.0 | 19.0 | 11.3 |
| France | 21.6 | 5.8 | 6.6 | 38.0 | 18.0 | 10.2 |
| Germany | 22.6 | 5.8 | 6.6 | 39.0 | 18.0 | 9.2 |
| India | 31.1 | 5.8 | 6.6 | 36.0 | 21.0 | 0.7 |
| Ireland | 25.0 | 5.8 | 6.6 | 36.0 | 20.0 | 6.8 |
| Israel | 21.5 | 5.8 | 6.6 | 31.0 | 25.0 | 10.3 |
| Italy | 27.0 | 5.8 | 6.6 | 39.0 | 18.0 | 4.8 |
| Japan | 25.1 | 5.8 | 6.6 | 49.0 | 8.0 | 6.7 |
| Mexico | 27.0 | 5.8 | 6.6 | 47.0 | 9.0 | 4.8 |
| Netherlands | 17.9 | 5.8 | 6.6 | 24.0 | 33.0 | 13.9 |
| Poland | 23.1 | 5.8 | 6.6 | 28.0 | 28.0 | 8.7 |
| Portugal | 23.2 | 5.8 | 6.6 | 50.0 | 6.0 | 8.6 |
| Russia | 25.1 | 5.8 | 6.6 | 54.0 | 3.0 | 6.7 |
| South Africa | 27.8 | 5.8 | 6.6 | 40.0 | 16.0 | 4.0 |
| South Korea | 19.5 | 5.8 | 6.6 | 22.0 | 34.0 | 12.3 |
| Spain | 24.2 | 5.8 | 6.6 | 36.0 | 20.0 | 7.6 |
| Sweden | 22.6 | 5.8 | 6.6 | 37.0 | 19.0 | 9.2 |
| Switzerland | 25.5 | 5.8 | 6.6 | 36.0 | 20.0 | 6.3 |
| Taiwan | 26.1 | 5.8 | 6.6 | 16.0 | 41.0 | 5.7 |
| Turkey | 25.2 | 5.8 | 6.6 | 29.0 | 27.0 | 6.6 |
| UK | 18.3 | 5.8 | 6.6 | 42.0 | 14.0 | 13.5 |
| US | 15.9 | 5.8 | 6.6 | 22.0 | 34.0 | 15.9 |

usage time were available for several countries only but not for others. Where they were not, we use the average of those that were available. This is a simplifying assumption, but, for those countries where the data were available, we did not observe huge differences across countries in the consumption-time among news users of a medium, given access to that medium. What differs more are penetrations (e.g., for newspaper readership or of online news access). Table 37.8 shows the attention weights for each country.[10]

For example, a 34% broadcast TV weight in the United States means of every hour of news consumption in that country, 20.4 minutes are spent in front of the TV watching news content.

These weights are then applied to the various news industries, companies, and countries.

To get a country's total attention-weighted HHI, we add up the sums of the news media industry HHIs multiplied by the attention weights in Table 37.8. Graph 37.18 and Table 37.9 show

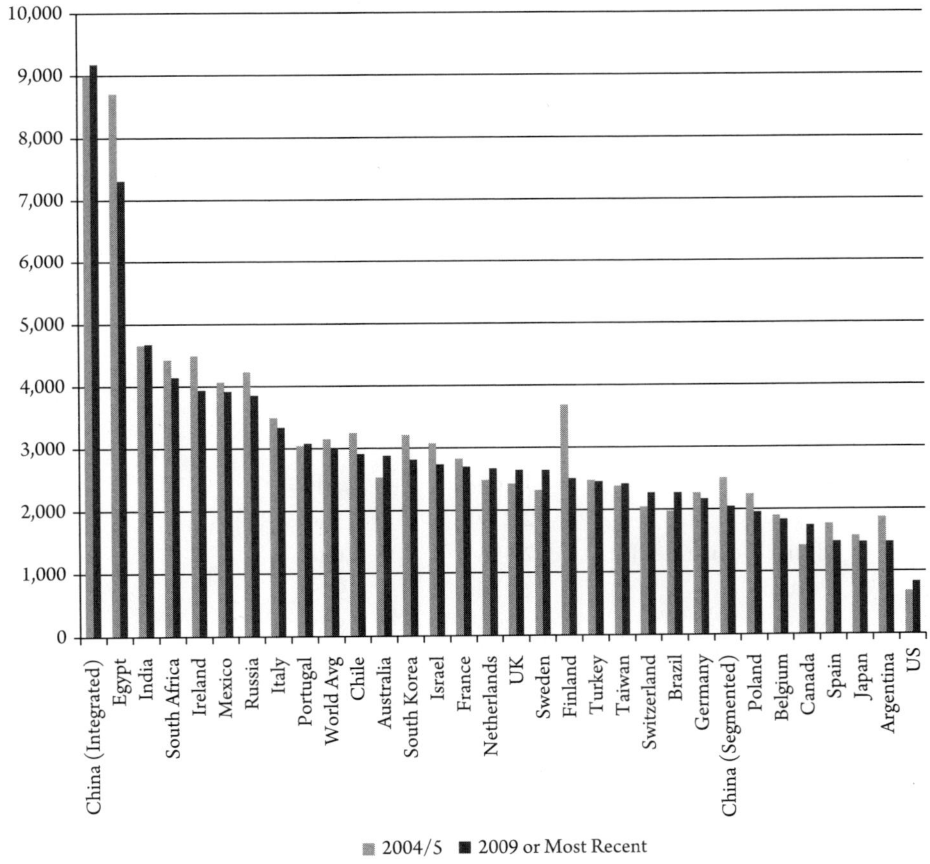


GRAPH 37.18 National News Media Concentration (HHI Weighted By Attention)

10. Coefficients calculated by Liwei Wang. US news usage shares from FCC/Nielsen survey. Europe shares from International Central Institute for Youth and Educational Television, and a UNESCO study. Attention weights for magazines and radio were assumed to be the same as their average overall weight. For broadcast TV and cable/satellite TC assumed to be the same, distributed according to national usage of cable channels. Overall weight for newspapers and online news assumed the same, distributed based on reported Internet usage. Federal Communication Commission, Report and Order and Notice of Proposed Reulemaking for Market Definitions and Ownership of Broadcast Radio and Newspaper Media, 18 F.C.C.R. 13,620, 327 (2003). vom Orde, Heike. "Digital natives, and their media worlds: Starting points for promoting MIL." International Central Institute for Youth and Educational Television, UNESCO (2012).

Table 37-9. News Media HHI (Weighted by Attention)

| | 2004/5 | 2011 OR MOST RECENT |
|---|---|---|
| China (Integrated) | 8,997 | 9,181 |
| Egypt | 8,708 | 7,310 |
| India | 4,657 | 4,677 |
| South Africa | 4,419 | 4,137 |
| Ireland | 4,487 | 3,934 |
| Mexico | 4,063 | 3,918 |
| Russia | 4,227 | 3,852 |
| Italy | 3,496 | 3,340 |
| Portugal | 3,042 | 3,075 |
| World Avg | 3,146 | 3,006 |
| Chile | 3,246 | 2,907 |
| Australia | 2,527 | 2,877 |
| South Korea | 3,212 | 2,810 |
| Israel | 3,073 | 2,732 |
| France | 2,818 | 2,688 |
| Netherlands | 2,472 | 2,658 |
| UK | 2,412 | 2,632 |
| Sweden | 2,309 | 2,629 |
| Finland | 3,686 | 2,495 |
| Turkey | 2,463 | 2,443 |
| Taiwan | 2,371 | 2,408 |
| Switzerland | 2,039 | 2,264 |
| Brazil | 1,966 | 2,264 |
| Germany | 2,259 | 2,164 |
| China (Segmented) | 2,493 | 2,038 |
| Poland | 2,236 | 1,955 |
| Belgium | 1,895 | 1,833 |
| Canada | 1,420 | 1,742 |
| Spain | 1,761 | 1,478 |
| Japan | 1,572 | 1,471 |
| Argentina | 1,867 | 1,468 |
| US | 679 | 828 |

the news media HHI (weighted by attention shares for the various countries). News media concentration on that basis is quite high in most countries, even higher than on a revenue basis. The world average high at 3,006, though down from 3,146 in 2005. Most countries' concentration was in the range of 3,000–2,500.

There are outliers. China (integrated) had the highest news concentration (9,181, and slightly rising), followed by Egypt with 7,310 (down from 8,708 in 2004/05), India (4,677), Ireland (3,934), South Africa (4,137), Russia (3,852), Mexico (3,918), and Italy (3,340).

In several countries the attention time-weighted news HHIs have been declining since 2004, especially in Russia, Ireland, Mexico, Egypt, South Korea, Chile, Japan, Finland, Israel, Poland, Argentina, and Spain.

Concentration has remained stable in Portugal, Turkey, and India. It rose in China, the United States, Canada, Sweden, the United Kingdom, Brazil, Switzerland, Australia, and the Netherlands.

The concentration of news attention is lowest in the United States (828), Argentina (1,468), Spain (1,478), Japan (1,471), and Canada (1,742). This also matches their position in terms of revenue concentration.

## Company Shares in National News Media Attention

In the next segment, we calculate most important news media companies and organizations in their countries, based on their shares of news attention.[11]

The number of people who might access a given news medium is called the potential audience size. It is the population of a country multiplied by the penetration rate of that news medium (Table 37.10). In that table, "print" encompasses both daily newspapers and magazines; "audiovisual" includes radio, cable & satellite, and broadcast TV operations. "Internet" includes online news. These three measures represent the share of people in each of the 30 countries that accesses one of the three types of news media, print, audiovisual, and Internet.

For the two print news media, penetration is given by the newspapers in circulation as a percentage of numbers of household in each country. For the three audiovisual news media, it is the percentage of households in each country with a TV set. And for Internet news, we use the Internet access rate in that country. In the United States, for example, Table 37.10 shows that household penetration (accessibility) is 44.2% for print, 98% for audiovisual, and 81% for Internet.

These coefficients are then applied to each of the six news media industries, along with the population, to calculate the news attention of companies, industries, and countries.

The potential audience market ($M$) for a news industry ($i$) in a country ($n$) is:

$$\mathrm{M}_{i,n} = \text{Population}_n \times \text{Penetration}_{i,n} \ .$$

The news attention ($A$) for a news media industry $i$ in a country $n$ is:

$$\mathrm{A}_i = \mathrm{M}_{i,n} * \mathrm{a}_i,$$

where $a_i$ is the news attention share of a medium $i$ in news.

Total news attention $A$ in a country $n$ for all news media $i$ is then:

$$A_n = \sum_n A_i$$

News attention for company $c$ in country $n$ for all of its activities in a given news industry is:

$$\mathrm{A}_{\hat{c},\hat{n},i} = \text{Population}_{\hat{n}} \times \text{Penetration}_{\hat{n}\hat{i}} \times \text{Attention Share}_{\hat{i}} \times \text{Market share}_{\hat{i}}$$

For all of company $c$'s news activities across industries $i$ in country $n$, news attention would be:

$$A_{\hat{c},\hat{n},i} = \sum_c A_{\hat{c},\hat{n},i}$$

The total news attention share of company $c$ in country $n$ is therefore:

$$S_{c,n} = \frac{A_{c\hat{n}}}{A_{\hat{n}}} \ .$$

These values, for the shares of a news company in overall news attention, can be calculated. Table 37.11 provides, for each country, the companies with the national top news attention. For space reasons, the short version is shown in Table 37.11A. A complete version is provided in the Appendix to this chapter, as Table 37.11B.

11. Included are consumer-level news outlets such as newspapers, TV networks, and so on, but not wholesalers such as news agencies.

### Table 37-10. Penetration Rates of News Media, 30 Countries (%), 2012[a]

| | PRINT PENETRATION | AUDIOVISUAL PENETRATION | INTERNET PENETRATION |
|---|---|---|---|
| Argentina | 48.1 | 97.0 | 55.8 |
| Australia | 42.9 | 96.0 | 82.4 |
| Belgium | 36.6 | 98.0 | 82.0 |
| Brazil | 21.7 | 91.0 | 49.9 |
| Canada | 57.7 | 99.0 | 86.8 |
| Chile | 17.1 | 87.0 | 61.4 |
| China | 34.3 | 76.0 | 34.0 |
| Egypt | 37.6 | 89.0 | 44.1 |
| Finland | 50.3 | 94.0 | 91.0 |
| France | 46.3 | 95.0 | 83.0 |
| Germany | 51.9 | 95.0 | 84.0 |
| India | 13.9 | 32.0 | 12.6 |
| Ireland | 70.4 | 95.0 | 79.0 |
| Israel | 72.4 | 93.0 | 73.4 |
| Italy | 28.7 | 96.0 | 58.0 |
| Japan | 43.3 | 99.0 | 79.1 |
| Mexico | 33.7 | 93.0 | 38.4 |
| Netherlands | 62.5 | 99.0 | 93.0 |
| Poland | 35.4 | 91.0 | 65.0 |
| Portugal | 19.5 | 99.0 | 64.0 |
| Russia | 19.8 | 98.0 | 53.3 |
| South Africa | 17.9 | 59.0 | 41.0 |
| South Korea | 33.2 | 98.0 | 84.1 |
| Spain | 30.8 | 100 | 72.0 |
| Sweden | 44.9 | 94.0 | 94.0 |
| Switzerland | 35.5 | 100 | 85.2 |
| Taiwan | 23.1 | 99.0 | 76.0 |
| Turkey | 42.6 | 92.0 | 45.1 |
| UK | 60.8 | 98.0 | 87.0 |
| US | 44.2 | 98.0 | 81.0 |

[a] For Newspapers: World Newspapers Association: <http://www.economist.com/blogs/graphicdetail/2013/06/daily-chart-1>

For Audiovisual: <http://www.nationmaster.com/graph/med_hou_wit_tel-media-households-with-television>

For Internet: International Telecommunications Union. <https://www.google.com/url?sa=t&rct=j&q=&esrc=s&source=web&cd=1&ved=0CCwQFjAA&url=http%3A%2F%2Fwww.itu.int%2Fen%2FITU-D%2FStatistics%2FDocuments%2Fstatistics%2F2013%2FIndividuals_Internet_2000-2012.xls&ei=ozW4UpriOK7gsATj7IK4Cg&usg=AFQjCNEblMHo-xiFshzx4wKHsI4WkPCosQ&sig2=KNl_qkmxdFN9UhAX1Mbk9w&bvm=bv.58187178,d.cWc>

Table 37-11A. C3 National News Companies in National News Attention Share, 2011 or Most Recent[1]

| COUNTRY | #1 | | #2 | | #3 | | C10 | HHI |
|---|---|---|---|---|---|---|---|---|
| Argentina | Grupo Clarín | 26.9% | América 2 | 9.2% | Canal 9 (Mexico) | 8.3% | 63.8% | 955 |
| Australia | Seven Network | 17.8% | Nine Entertainment | 14.7% | Murdoch Group (US) | 13.6% | 87.3% | 1,046 |
| Belgium | VRT (public) | 17.3% | Bertelsmann (Germany) | 11.6% | RTBF (public) | 9.4% | 74.7% | 767 |
| Brazil | Globo | 35.5% | Folhapar | 9.6% | Estado Group | 8.2% | 78.3% | 1,534 |
| Canada | Bell Canada/CTV | 17.4% | CBC (public) | 16.8% | Shaw | 15.3% | 76.0% | 975 |
| Chile | Universidad Catholica/ Luksic Group | 12.5% | TVN (public) | 12.1% | El Mercurio SAP | 8.1% | 73.6% | 696 |
| China | Government of China | 96.0% | Hearst (US) | 0.8% | Sina.com | 0.5% | 98.7% | 9,220 |
| Egypt | Govenrment of Egypt | 89.8% | Al Masry Al Youm | 3.7% | Nile Radio Productions | 3.0% | 98.2% | 8,080 |
| Finland | Yle (public) | 32.4% | Bonnier (Sweden) | 20.2% | Sanoma Oyj | 9.4% | 77.4% | 1,588 |
| France | France Télevisions (public) | 15.7% | Bouygues Group | 13.8% | Vivendi | 11.3% | 64.4% | 689 |
| Germany | ARD (public) | 17.4% | ProSiebenSat.1 | 16.2% | Bertelsmann | 15.9% | 73.9% | 949 |
| India | Prasar Bharati (public) | 65.4% | The Times of India Group (BCCL) | 6.9% | Zee | 3.3% | 82.2% | 4,342 |
| Ireland | RTE (public) | 40.0% | Independent News and Media | 10.9% | Murdoch Group (US) | 9.9% | 83.9% | 1,938 |
| Israel | Channel 2 | 22.6% | Yedioth Ahronoth Group | 10.9% | Channel 10 | 9.2% | 73.1% | 881 |
| Italy | RAI (public) | 34.4% | Fininvest | 22.4% | Murdoch Group (US) | 8.9% | 84.2% | 1,855 |
| Japan | Asahi Shimbun | 13.6% | Fujisankei (NBS) | 11.9% | NHK (public) | 11.7% | 76.1% | 772 |
| Mexico | Televisa | 44.4% | TV Azteca | 19.5% | OEM | 8.2% | 82.3% | 2431 |
| Netherlands | NPB (public) | 16.6% | Bertelsmann (Germany) | 10.5% | Redstone (US) | 9.1% | 75.8% | 721 |
| Poland | TP (public) | 21.1% | Polsat | 17.5% | Grupa TVN | 10.0% | 80.1% | 1,031 |

*continued*

Table 37-11A. *continued*

| COUNTRY | #1 | | #2 | | #3 | | C10 | HHI |
|---|---|---|---|---|---|---|---|---|
| Portugal | Media Capital (PRISA Group, Spain) | 27.8% | RTP (public) | 27.8% | Impresa | 22.6% | 91.6% | 2,097 |
| Russia | Government of Russia | 54.8% | CTC | 15.6% | Komsomolskaya Pravda | 3.5% | 83.5% | 3,271 |
| South Africa | SABC (public) | 38.5% | Naspers Group | 27.2% | Hosken | 17.3% | 97.8% | 2,563 |
| South Korea | Government of South Korea | 38.3% | SBS | 12.5% | Tbroad | 4.4% | 74.5% | 1,702 |
| Spain | Planeta | 14.7% | PRISA | 14.1% | Fininvest (Italy) | 13.8% | 66.8% | 752 |
| Sweden | Bonnier | 25.0% | Sveriges (public) | 24.7% | Investment AB Kinnevik | 16.9% | 86.4% | 1,609 |
| Switzerland | SRG SSR (public) | 25.3% | Tamedia | 9.5% | Liberty (US) | 6.5% | 62.2% | 854 |
| Taiwan | Government of Taiwan | 19.6% | China Times News Group (Want Want Ltd.) | 9.8% | Liberty Times News Group | 7.8% | 68.8% | 700 |
| Turkey | Dogan Group | 37.5% | Turkuvaz-Kalyone Group | 14.0% | Cukurova Group | 13.4% | 86.4% | 1,860 |
| UK | BBC (public) | 33.9% | Murdoch Group (US) | 14.3% | ITV | 11.3% | 77.8% | 1,549 |
| US | Comcast | 12.2% | Disney | 9.6% | Redstone Group | 9.3 | 55.8 (68.0)[2] | 476 |

[1]For China, Russia, Egypt, South Korea, and Taiwan, the aggregate government share is given as the top company, followed by the first and second largest privately-owned companies as firms #2 an #3. The component firms of the state attention share are listed in the appendix.

[2]C20 is given in parentheses for the United States.

The shorter table for each country shows the top three (C3) companies by share of news attention.

The right-most column lists the share, in news attention, of the top 10 firms (C10) and the concentration as expressed by an HHI.

- The C10 in news attention is above 90% in four countries (China, Egypt, Portugal, and South Africa). It is above 80% in nine countries (Australia, India, Ireland, Italy, Mexico, Poland, Russia, Sweden, and Turkey). It is nowhere below 50%. Only the United States has a C10 under 60%, at 55.8% (up from 51% in 2004).
- In most of the world, then, the top 10 firms command over 75% of news attention. And nowhere do they command less than 50%.
- In terms of HHI, concentrations are highest in China (9,220), Egypt (7,544), India (4,342), Russia (3,271), South Africa (2,563), Mexico (2,431), and Portugal (2,097).[12] The good news is that in most other countries the news concentration, in terms of attention, is either close to or less than an HHI of 1,800, the old threshold for high concentration.
- The share of the top firm (C1) in terms of news attention is, on average, a very high 32.2%.
- Very high HHIs exist where governments control media operations and hold a very high C1 share of the news attention:
  - Government of China (96%)
  - Government of Egypt (89.8%)
  - Government of Russia (54.8%)
- In many countries, the public service broadcasters are strong and at the number 1 spot:
  - India (Prasar Bharati, 65.4%)
  - Ireland (RTE, 40%)
  - Finland (Yle, 32.4%)
  - South Africa (SABC, 38.5%)
  - Italy (RAI, 34.4%)
  - Government of South Korea (38.3%)
  - Government of Taiwan (19.6%)
  - United Kingdom (BBC, 34%)
  - Poland (TP, 21.1%)
  - Switzerland (SRG-SSR, 25.4%)
  - Belgium (VRT, 17.3%)
  - France (France Televisions, 15.7%)
  - Germany (ARD, 17.4%)
  - The Netherlands (NPB, 16.7%)
- Private media firms are at the number 1 spot in:
  - the United States (Comcast, 12.2%)
  - Mexico (Televisa, 44.4%)
  - Brazil (Globo, 35.5%)
  - Luksic Group (Chile, 12.5%)
  - Portugal (PRISA, 27.8%)
  - Sweden (Bonnier, 25%)
  - Turkey (Dogan, 37.5%)
  - Argentina (Grupo Clarín, 26.9%)
  - Spain (Planeta, 14.7%)
  - Israel (Channel 2, 22.6%)
  - Japan (Asahi Shimbun, 13.4%)
  - Canada (Bell Canada, 17.4%)
  - Australia (Seven Network, 17.8%)
- Private media firms with more than 30% do exist, but rarely:
  - Mexico (Televisa, 44.4%)
  - Brazil (Globo, 35.5%)
  - Turkey (Dogan, 37.5%)
- In the more highly developed countries, the largest private-sector firms rarely reach shares above 20%.
- The high attention share of government-controlled media organizations gets higher still when one looks at major countries not part of this 30-country study, (which encompassed 64% of the world's population and 85% by GDP). The top 11 countries by population that have not been included account for 18% of the world's population.
  - Of these, five countries have a strong state dominance over both audiovisual (TV, radio) and print media—Vietnam, Ethiopia, Iran, Congo, Burma. These countries account for 5.3% of the world's population.
  - Three countries have a substantial governmental dominance over audiovisual but not over print: Thailand, Bangladesh, and Nigeria. They account for 5.5% of the world's population.

12. In Tables 37.11 and 37.11B, these are "pooled" HHIs, because they are the summed squares of the individual companies' attention shares of the national audiences in a country.

- Three countries have no state dominance over audiovisual and print: Indonesia, Pakistan, and the Philippines. They account for 7.3% of the world's population.
- In these countries, large media firms exist but the industries are not more concentrated than those of many countries in this study.
- In none of these 11 countries do foreign media firms play a significant content media role. There are several newspapers in Nigeria owned by UK firms (The Guardian, Trinity Mirror). Saban (US) holds 7.5% of Indonesia's leading MNC Group.
- In platform media, in 8 of these 11 countries, foreign ownership in parts of the mobile telecom industry exists. Of companies from the countries of the study, these are the South African MTN (Iran, Nigeria), the Indian firm Bharti Enterprises (Bangladesh, Democratic Republic of the Congo, Nigeria), Orange and Vodafone (both in the Democratic Republic of the Congo). None of these operations is large enough to change their companies' or countries' shares more than minimally.

## Company Shares in Global News Media Attention

We now extend this analysis to a company's attention share worldwide, by aggregating its country-specific attention shares, with population weights to account for different country population markets.

$$S_{cglobal} = \sum_{n} S_{cnational} \frac{Pop_n}{\sum_{n} Pop_n}$$

This way, we can calculate the global news media share of companies that are active in several countries, and we can compare the share in worldwide attention of the various media companies. For example, Fininvest has, in Italy, a news attention share of 22.4%. Italy has a weight of 1.45%, in population, of the 30-country world. This would give Fininvest, for its Italian news operations, a share of 0.33%. To that are added its news attention shares for Spain, where the company has 13.8% of an attention market that accounts for 1.11% globally (i.e., 0.15%). Fininvest has a global news attention market share of 0.32% + 0.15% = 0.48% in 2013 for the 30-country world, and 0.31% for the entire world.[13]

The results are shown in Table 37.12, which lists news media organizations with their share of world news attention in two ways: as a share of the 30-country world, and as a share of the world's entire population of which the 30 countries comprise 64%.[14]

These findings show:

- The government of China, through its several media organizations, accesses a truly vast share of global news attention. In the aggregate, it has 29.7% of the 30-country world news attention in 2013, and even more in previous years. If we include the population of other countries beyond the 30 that are covered, it would still be a huge share: 19% of global news attention. Even if we unbundle China's news organization, CCTV the government would still command 11.1% of the 30-country world's news attention (and 7.1% of the global news attention) and be the second largest news media company in the world. The explanations for these high shares are simple: a huge population (1.3 billion) and state control over most news media outside several online portals and print magazines.
- Other large shares in global news attention time are held by the governments of Russia (1.81%) and Egypt (1.75%); both countries governments prioritize broadcast TV control, and Egypt maintains significant state-owned print newspapers. India's public service broadcaster Prasar Bharati became autonomous

13. The 36% of population and 15% of GDP not covered by these 30 countries.

14. The cutoff for inclusion was if a firm held less than a 0.5% audience share in our 30-country world when all its markets were aggregated.

Table 37-12. Top News Media Companies by Attention Share of World Population, 2004/05 and 2012/13 (>0.1% Share, or #1 in their Country)

| COMPANY | ATTENTION SHARE (2004/05) – 30 COUNTRIES | ATTENTION SHARE (2012/13) – 30 COUNTRIES | ATTENTION SHARE (2012/13) – WORLD | COUNTRY OF ORIGIN |
|---|---|---|---|---|
| Government of China | 30.85% | 29.65% | 18.98% | China |
| *China Central Television (public)* | *12.17%* | *11.14%* | *7.13%* | *China* |
| *Shanghai Media Group (public)* | *4.28%* | *2.58%* | *1.65%* | *China* |
| *Hunan Media Group (public)* | *2.20%* | *1.53%* | *0.98%* | *China* |
| *Jiangsu Media Group (public)* | *0.34%* | *1.19%* | *0.76%* | *China* |
| *People's Daily (public)* | *0.29%* | *0.64%* | *0.41%* | *China* |
| *Guangdong Provincial Government (public)* | *0.88%* | *0.46%* | *0.29%* | *China* |
| *Reference News (Xinhua News Agency, public)* | *0.34%* | *0.41%* | *0.26%* | *China* |
| *Rayli (public)* | *0.07%* | *0.41%* | *0.26%* | *China* |
| *Southern Media Corporation (public)* | *0.33%* | *0.39%* | *0.25%* | *China* |
| *China Radio International (public)* | *0.52%* | *0.38%* | *0.24%* | *China* |
| *Trend Media Group (public)* | *0.20%* | *0.28%* | *0.18%* | *China* |
| Prasar Bharati (public) | 18.23% | 18.47% | 11.82% | India |
| BCCL (The Times of India Group) | 1.62% | 1.96% | 1.25% | India |
| Government of Russia | 1.93% | 1.81% | 1.16% | Russia |
| *Gazprom Media (public)* | *0.52%* | *0.85%* | *0.54%* | *Russia* |
| *First Channel (public)* | *0.77%* | *0.50%* | *0.32%* | *Russia* |
| *VGTRK (public)* | *0.63%* | *0.43%* | *0.28%* | *Russia* |
| *RIA Novosti (public)* | *0.01%* | *0.02%* | *0.01%* | *Russia* |
| Government of Egypt | 1.82% | 1.75% | 1.12% | Egypt |
| *ERTU (public)* | *1.30%* | *1.35%* | *0.86%* | *Egypt* |
| *Al Akhbar (public)* | *0.17%* | *0.13%* | *0.08%* | *Egypt* |
| *Al Ahram (public)* | *0.18%* | *0.11%* | *0.07%* | *Egypt* |
| *Al Gomhouria (public)* | *0.09%* | *0.09%* | *0.06%* | *Egypt* |
| *Nile Television Network (public)* | *0.08%* | *0.06%* | *0.04%* | *Egypt* |
| Globo | 1.46% | 1.71% | 1.10% | Brazil |
| Murdoch Group | 1.15% | 1.35% | 0.86% | US |
| Televisa | 1.26% | 1.21% | 0.77% | Mexico |
| Disney | 0.85% | 1.10% | 0.70% | US |

*continued*

Table 37-12. *continued*

| COMPANY | ATTENTION SHARE (2004/05) – 30 COUNTRIES | ATTENTION SHARE (2012/13) – 30 COUNTRIES | ATTENTION SHARE (2012/13) – WORLD | COUNTRY OF ORIGIN |
|---|---|---|---|---|
| Zee | 0.73% | 0.93% | 0.59% | India |
| Comcast | 0.30% | 0.91% | 0.58% | US |
| Dogan Group | 0.52% | 0.70% | 0.45% | Turkey |
| TV Azteca | 0.49% | 0.53% | 0.34% | Mexico |
| CTC | 0.78% | 0.51% | 0.33% | Russia |
| BBC (public) | 0.47% | 0.51% | 0.33% | UK |
| Bertelsmann | 0.54% | 0.51% | 0.33% | Germany |
| Redstone Group | 0.54% | 0.78% | 0.50% | US |
| GE | 0.43% | 0.00% | 0.00% | US |
| Sun (India) | 0.31% | 0.50% | 0.32% | India |
| RAI (public) | 0.50% | 0.50% | 0.32% | Italy |
| Time Warner | 0.80% | 0.50% | 0.32% | US |
| Fininvest | 0.46% | 0.48% | 0.31% | Italy |
| Folhapar | 0.53% | 0.46% | 0.30% | Brazil |
| Government of South Korea | 0.51% | 0.45% | 0.29% | South Korea |
| *KBS (public)* | *0.27%* | *0.21%* | *0.13%* | *South Korea* |
| *MBC (public)* | *0.15%* | *0.18%* | *0.12%* | *South Korea* |
| *EBS (public)* | *0.03%* | *0.02%* | *0.01%* | *South Korea* |
| SABC (public) | 0.47% | 0.45% | 0.29% | South Africa |
| Asahi Shimbun Company | 0.36% | 0.41% | 0.26% | Japan |
| Estado Group | 0.36% | 0.39% | 0.25% | Brazil |
| Hearst | 0.20% | 0.39% | 0.25% | US |
| Fujisankei (NBS) | 0.48% | 0.36% | 0.23% | Japan |
| NHK (public) | 0.44% | 0.35% | 0.23% | Japan |
| ProSiebenSat.1 | 0.29% | 0.34% | 0.22% | Germany |
| ARD (public) | 0.37% | 0.34% | 0.22% | Germany |
| NTV | 0.36% | 0.34% | 0.22% | Japan |
| Naspers Group | 0.29% | 0.32% | 0.20% | South Africa |
| Universal Church Group | 0.12% | 0.31% | 0.20% | Brazil |
| Tokyo Broadcasting System | 0.34% | 0.31% | 0.20% | Japan |
| Living Media Group | 0.12% | 0.28% | 0.18% | India |
| Grupo Clarín | 0.31% | 0.27% | 0.17% | Argentina |
| Turkuvaz-Kalyone Group | 0.19% | 0.26% | 0.17% | Turkey |
| France Televisions (public) | 0.31% | 0.26% | 0.17% | France |
| Abril Group | 0.33% | 0.26% | 0.16% | Brazil |
| Advance | 0.07% | 0.25% | 0.16% | US |
| Cukurova Group | 0.25% | 0.25% | 0.16% | Turkey |
| Bouygues Group | 0.27% | 0.25% | 0.16% | France |
| Liberty | 0.26% | 0.24% | 0.15% | US |
| PRISA | 0.28% | 0.24% | 0.15% | Spain |
| Gannett | 0.22% | 0.23% | 0.15% | US |
| OEM | 0.21% | 0.22% | 0.14% | Mexico |

Table 37-12. *continued*

| COMPANY | ATTENTION SHARE (2004/05) – 30 COUNTRIES | ATTENTION SHARE (2012/13) – 30 COUNTRIES | ATTENTION SHARE (2012/13) – WORLD | COUNTRY OF ORIGIN |
|---|---|---|---|---|
| Discovery Communications | 0.17% | 0.21% | 0.13% | US |
| Yahoo | 0.23% | 0.21% | 0.13% | US |
| DirecTV | 0.00% | 0.20% | 0.13% | US |
| Hathway | 0.14% | 0.20% | 0.13% | India |
| Hosken | 0.14% | 0.20% | 0.13% | South Africa |
| TP (public) | 0.24% | 0.19% | 0.12% | Poland |
| HT Media | 0.15% | 0.19% | 0.12% | India |
| Vivendi | 0.17% | 0.19% | 0.12% | France |
| Network18 Group | 0.06% | 0.18% | 0.11% | India |
| Bandeirantes Group | 0.07% | 0.17% | 0.11% | Brazil |
| JSAT(SkyPerfecTV) | 0.19% | 0.17% | 0.11% | Japan |
| ITV | 0.21% | 0.17% | 0.11% | UK |
| Sina.com | 0.19% | 0.17% | 0.11% | China |
| Planeta | 0.11% | 0.16% | 0.10% | Spain |
| ZDF (public) | 0.19% | 0.16% | 0.10% | Germany |
| Polsat | 0.15% | 0.16% | 0.10% | Poland |
| Nihon Keizai Shimbun (TV Tokyo) | 0.12% | 0.16% | 0.10% | Japan |
| SBS | 0.18% | 0.15% | 0.10% | South Korea |
| Axel Springer AG | 0.15% | 0.15% | 0.09% | Germany |
| Bhaskar Group | 0.00% | 0.15% | 0.09% | India |
| Bell Canada/CTV | 0.11% | 0.14% | 0.09% | Canada |
| Reliance | 0.00% | 0.14% | 0.09% | India |
| Tribune | 0.17% | 0.14% | 0.09% | US |
| Sohu.com | 0.15% | 0.14% | 0.09% | China |
| Canadian Broadcasting Corporation (public) | 0.13% | 0.14% | 0.09% | Canada |
| Deccan Chronicle Holdings Ltd. (DCHL) | 0.03% | 0.13% | 0.08% | India |
| Telefonica | 0.18% | 0.13% | 0.08% | Spain |
| Diários Associados Group | 0.11% | 0.13% | 0.08% | Brazil |
| Jagran Prakashan | 0.03% | 0.13% | 0.08% | India |
| New Delhi Television (NDTV) | 0.09% | 0.13% | 0.08% | India |
| DEN Networks | 0.04% | 0.12% | 0.08% | India |
| Shaw | 0.06% | 0.12% | 0.08% | Canada |
| RBS Group | 0.17% | 0.12% | 0.08% | Brazil |
| Komsomolskaya Pravda | 0.11% | 0.12% | 0.07% | Russia |
| Kasturi & Sons | 0.13% | 0.11% | 0.07% | India |
| Albavision | 0.08% | 0.11% | 0.07% | Mexico |
| Government of Taiwan | 0.11% | 0.11% | 0.07% | Taiwan |
| *Formosa TV (public)* | *0.03%* | *0.04%* | *0.03%* | *Taiwan* |

*continued*

Table 37-12. *continued*

| COMPANY | ATTENTION SHARE (2004/05) – 30 COUNTRIES | ATTENTION SHARE (2012/13) – 30 COUNTRIES | ATTENTION SHARE (2012/13) – WORLD | COUNTRY OF ORIGIN |
|---|---|---|---|---|
| *BCC (public)* | *0.03%* | *0.03%* | *0.02%* | *Taiwan* |
| *Taiwan TV Enterprise (public)* | *0.03%* | *0.02%* | *0.01%* | *Taiwan* |
| *China Television Corporation (Public)* | *0.02%* | *0.02%* | *0.01%* | *Taiwan* |
| RCS Media Group | 0.08% | 0.11% | 0.07% | Italy |
| Jornal do Brasil | 0.09% | 0.11% | 0.07% | Brazil |
| RTVE (public) | 0.14% | 0.11% | 0.07% | Spain |
| India Today Group (ITG) | 0.19% | 0.11% | 0.07% | India |
| Gruppo Espresso | 0.12% | 0.11% | 0.07% | Italy |
| Ongoing (Arca Group, Portugal) | 0.12% | 0.11% | 0.07% | Brazil |
| The Washington Post Company | 0.09% | 0.10% | 0.07% | US |
| Feza | 0.05% | 0.10% | 0.06% | Turkey |
| The New York Times | 0.10% | 0.10% | 0.06% | US |
| EMG | 0.11% | 0.10% | 0.06% | Russia |
| AOL | 0.01% | 0.10% | 0.06% | US |
| ABP Group | 0.09% | 0.09% | 0.06% | India |
| Yomiuri Shimbun | 0.09% | 0.09% | 0.06% | Japan |
| Seven Network Ltd | 0.08% | 0.09% | 0.06% | Australia |
| América 2 | 0.07% | 0.09% | 0.06% | Argentina |
| Grupa TVN S.A. | 0.08% | 0.09% | 0.06% | Poland |
| Clear Channel | 0.11% | 0.09% | 0.06% | US |
| Silvio Santos Group | 0.14% | 0.09% | 0.05% | Brazil |
| 163.com | 0.02% | 0.08% | 0.05% | China |
| Time Warner Cable | 0.00% | 0.08% | 0.05% | US |
| Bonnier | 0.07% | 0.08% | 0.05% | Sweden |
| Amaury Group | 0.07% | 0.08% | 0.05% | France |
| Channel 4 | 0.09% | 0.08% | 0.05% | UK |
| Nine Entertainment | 0.09% | 0.08% | 0.05% | Australia |
| Al Masry Al Youm | 0.00% | 0.07% | 0.05% | Egypt |
| RTP (public) | 0.07% | 0.07% | 0.05% | Portugal |
| McClatchy | 0.06% | 0.07% | 0.05% | US |
| El Universal | 0.05% | 0.07% | 0.05% | Mexico |
| EBC (public) | 0.07% | 0.07% | 0.04% | Brazil |
| Radio RMF FM | 0.07% | 0.07% | 0.04% | Poland |
| Rogers | 0.06% | 0.07% | 0.04% | Canada |
| Independent News & Media | 0.07% | 0.07% | 0.04% | Ireland |

Table 37-12. *continued*

| COMPANY | ATTENTION SHARE (2004/05) – 30 COUNTRIES | ATTENTION SHARE (2012/13) – 30 COUNTRIES | ATTENTION SHARE (2012/13) – WORLD | COUNTRY OF ORIGIN |
|---|---|---|---|---|
| Dogus | 0.06% | 0.07% | 0.04% | Turkey |
| Netherland Public Broadcasting (public) | 0.07% | 0.07% | 0.04% | Netherlands |
| TRT (public) | 0.12% | 0.06% | 0.04% | Turkey |
| Bharti Enterprises | 0.00% | 0.06% | 0.04% | India |
| Sanoma Oyj | 0.07% | 0.06% | 0.04% | Finland |
| Kabel Deutschland | 0.07% | 0.06% | 0.04% | Germany |
| Nile Radio Productions | 0.07% | 0.06% | 0.04% | Egypt |
| Australian Broadcasting Corporation (public) | 0.05% | 0.06% | 0.04% | Australia |
| Tata Group | 0.00% | 0.06% | 0.04% | India |
| Impresa | 0.06% | 0.06% | 0.04% | Portugal |
| Radio ZET | 0.06% | 0.06% | 0.04% | Poland |
| Polskie Radio (public) | 0.06% | 0.06% | 0.04% | Poland |
| Daily Mail | 0.05% | 0.06% | 0.04% | UK |
| Kanal 7 | 0.08% | 0.05% | 0.04% | Turkey |
| Sveriges (public) | 0.07% | 0.05% | 0.03% | Sweden |
| Quebecor (Videotron, Canada) | 0.05% | 0.05% | 0.03% | Canada |
| China Times News Group (Want Want Ltd.) | 0.06% | 0.05% | 0.03% | Taiwan |
| ProfMedia | 0.07% | 0.05% | 0.03% | Russia |
| Cablevision | 0.05% | 0.05% | 0.03% | US |
| Tbroad | 0.04% | 0.05% | 0.03% | South Korea |
| FORTA (public) | 0.08% | 0.05% | 0.03% | Spain |
| Sada Group | 0.03% | 0.05% | 0.03% | Brazil |
| Universidad Catholica/ Luksic Group | 0.05% | 0.05% | 0.03% | Chile |
| Socpresse Group | 0.05% | 0.05% | 0.03% | France |
| Cox | 0.05% | 0.05% | 0.03% | US |
| TVN (public) | 0.05% | 0.05% | 0.03% | Chile |
| EchoStar (DISH) | 0.06% | 0.05% | 0.03% | US |
| Altice | 0.02% | 0.05% | 0.03% | France |
| Demirören | 0.00% | 0.05% | 0.03% | Turkey |
| Ten Network | 0.05% | 0.05% | 0.03% | Australia |
| Bauer | 0.05% | 0.05% | 0.03% | Germany |
| SRG SSR (public) | 0.05% | 0.05% | 0.03% | Switzerland |
| CJ Group | 0.02% | 0.05% | 0.03% | South Korea |

[1]GE held a 49% stake in NBCuniversal in the year 2004/05 while Comcast held 51%. For the latest year, Comcast's numbers reflect full ownership of NBCuniversal, and the exit of GE from the news media sector.

[2]This is what GE's share of news media attention outside the 30-country world would have been in 2004/05.

from direct state control after 1997 and has a terrestrial broadcasting monopoly in a country with a population of 1.1 billion people, giving it 18.5% of world attention for 30 countries (11.8% for the whole world).

- India's BCCL is the largest privately owned media firm by news attention, with 1.25% of the global news attention, due to the country's large population. The second largest private media firm in the world, by attention time, is Globo in Brazil, with 1.71% (1.1% for the entire world). Brazil, too, has a large population (192 million).
- Rupert Murdoch's two companies combined are the third largest privately owned news providers by attention, and the largest US-headquartered news firm (1.35% for 30 countries, 0.83% for the entire globe).
- Of other US companies, Comcast, Disney, Redstone's CBS and Viacom, and Time Warner are also in the top 20.
- The combined global news share of those top five US companies (technically, seven companies) is 4.6% in the 30-country world, and 3% for the whole world. The combined share of all US media firms in Table 37.14 is 7.2% of the 30-country world and 4.6% globally.
- The combined news share of EU-headquartered media firms is 6.2% in the 30-country world, and 4% in the global market. As is the case for the United States, only the top EU companies, listed in Table 37.14, are covered. However, only 11 of the 27 EU member states (the large ones) are part of our study, and counting only them is therefore a lower bound. On the other hand, for the EU countries, sales to other EU member states are counted as exports, even though it stays within the EU bloc.
- Of individual owners, the largest controls over news attention are held by the Marinho family (Globo), Rupert Murdoch (News Corp and 21st Century Fox), Emilio Azcárraga (Televisa), the Berlusconi family (Fininvest), Aydın Doğan (Dogan Group), Brian Roberts (Comcast), Alexander Rodnyansky (CTC Media), Sumner Redstone (CBS and Viacom), Ricardo Salinas Pliego (TV Azteca), S. Narsing Rao (BCCL), and Subhash Chandra (Zee).
- Similarly, in the Philippines, the Lopez family (ABS-CBN) controls much news attention. It also owns the telephone company Bayan Telecommunications. Others are the Gozon, Duavit, and Jiminez families (GMA); and Manuel Pangilian (TV5 and PLDT, the largest wireline and mobile telecom company).
- In Indonesia, the leading media-owning families are the Hary Tanoesobijo (MNC Group), Tohir (Mahaka Media), Oetama and Adiprasetyo (Kompas Gramedia), Iskan (Jawa Pos Group), and Narada (Media Bali Post). MNC's owner Hary was a candidate for Vice President in the country's 2014 election.

## PUBLIC MEDIA SHARE OF NEWS ATTENTION

Based on Table 37.12, it is possible to calculate what percent of news audiences are watching content from non-commercial outlets (Graph 37.19). News attention to public media averages approximately 21.9%. This is a considerable share. It is slightly down from 23.8% in 2004/05, with public broadcasters losing audiences to an increasing number of private companies. The biggest decreases have been in Egypt, Finland, Israel, Sweden, and Spain, with public companies losing 5–10% of their 2004/05 attention share by 2013. The shares of state enterprises/independent public broadcasters have risen slightly in Russia, Canada, Portugal, the United Kingdom, China, and Switzerland. In the United States, that share is very low (0.3%). In China (93.2%) and Egypt (83.1%), it is very high, though Egypt's is decreasing.

## FOREIGN MEDIA SHARE OF NEWS ATTENTION

News attention to foreign-owned media averaged 12.8% in the 30 countries, and remained constant (Graph 37.20). India has relatively low foreign provision of news (4.6%), but its importation increased because of a relaxation on laws that banned foreigners from owning audiovisual content producers. Foreign ownership of news providers is high in the Netherlands (38.1%),

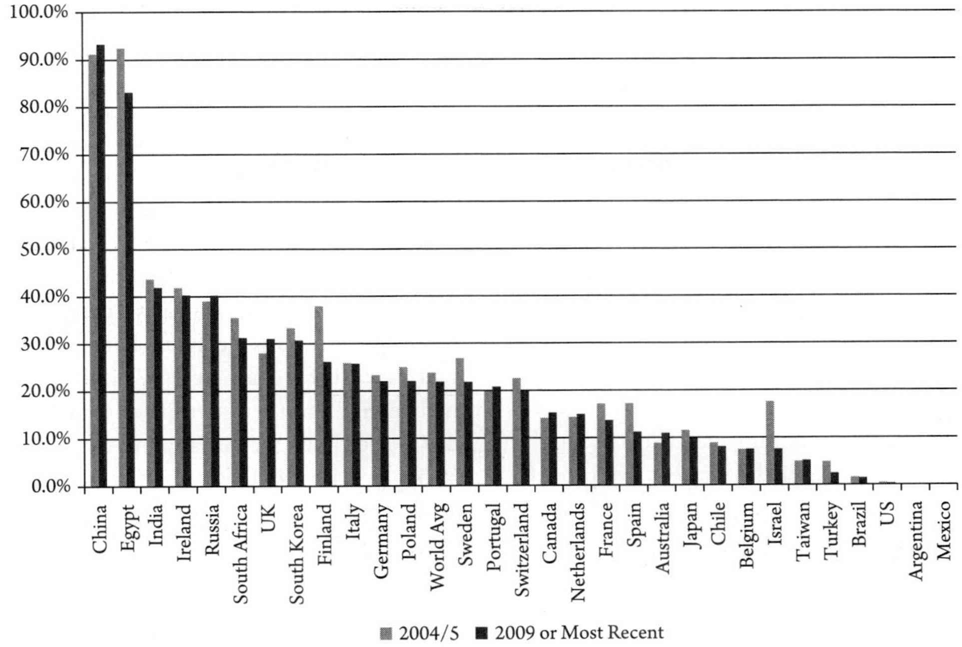


GRAPH 37.19 % Public Ownership of News Media (By News Attention)

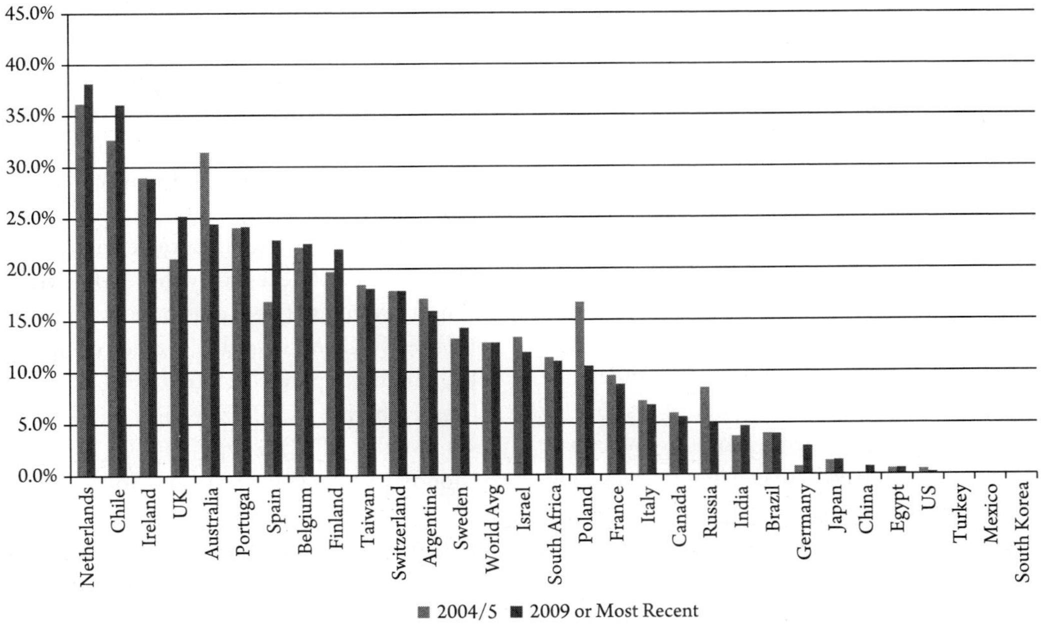


GRAPH 37.20 % Foreign Ownership of News Media (By News Attention)

Chile (36%), and Ireland (28.9%), but remains effectively nonexistent in Turkey, China, South Korea, and Mexico due to continuing restrictive laws. It is also very low in Japan (1.4%) and the United States (0.2%), where the major companies are domestic owned.

## NEWS CONCENTRATION, POVERTY, AND ECONOMIC DEVELOPMENT

The top company in news attention has, by definition, the greatest "mindshare" in its society. But there are degrees of dominance. What we find is that the dominance of the single news provider is especially strong in poorer countries. A simple regression analysis of the market share of the top firm (the "C1") with the average per-capita income in that country and without control variables

$$lnC1 = a - b\ln Y$$

shows a correlation

$$lnC1 = 7.24 - 0.4018 lnY$$

with an $R^2$ of 0.5801 and a $t$-value for the coefficient, of −6.22.

Thus, the poorer the country, the higher the dominance of the top firm. Media concentration in news is associated with lesser economic development, not with more.

There are several factors that contribute to this:

- **Resources.** The first factor is that poorer countries are economically smaller, and hence sustain fewer media firms. There is less of direct purchases of media products and consumption that drives advertising, and less government funding capacity.
- **Government control.** Poorer countries tend to be characterized by powerful governments whose TV and press dominate the media. Cause and effect in this relation are intertwined.
- **Crony capitalism.** Even where the state does not dominate directly, well-connected private media firms with close relations to the political establishment obtain privileges that lead to their dominance, again in a vicious circle. Many poor countries therefore exhibit astonishingly high market shares of such privileged media firms—such as Televisa in Mexico (44.4% in overall news and 61% in the TV market), and Globo in Brazil (35.5% of the news market and 52.4% of TV).
- In some instances, a strong governmental role in media is rationalized as a defense to offset powerful private media firms. This was the case, for example, in Venezuela, as it had been in France under President Charles de Gaulle. The reverse happens, too, when a strong private media firm is seen as an offset against powerful state media. An example is Italy, where Fininvest has a major share and emerged as an offset to the state RAI. Thus, one can observe a system of duopolies evolving where a strong state system and a strong private media firm derive justification from each other's power.
- Both also derive justification by positioning themselves as bulwarks against the erosion of national culture. The argument is that without strong domestic media organizations, domestic cultural productions would be rolled over by a tide of imports. Concentrated media provide that protection. In France, such a role is claimed by both the public France Télévisions and the private Vivendi/Canal Plus.
- The organizations with the greatest news attention are not necessarily those with the highest revenues. The media organizations of China (CCTV, Shanghai Media Group, etc., whether separate or integrated), of India (Prasar Bharati, BCCL, Zee), Russia (Gazprom Media, CTC), Egypt (ERTU), Brazil (Globo), Mexico (Televisa), and South Africa (SABC, Naspers) have a vastly greater hold on people's attention than on revenues. These are the media organizations of the emerging world—the *BRICS media*.
- In contrast, the leading media organizations of the developed world are a combination of traditional media conglomerates and successful startups. Their hold on attention, large as it is, is dwarfed by the BRICS media. But

when it comes to revenues, it is the other way around (see Graph 37.17). Here, Murdoch, Redstone, Google, Comcast, Disney, Bertelsmann, and Time Warner outpace most individual Chinese companies like CCTV (though not China as a whole). Vivendi, ARD, Hearst, the Asahi Shimbun Company, and Fininvest are nearly as large. Only one non-Chinese BRICS media organization is among the top 30 media groups (Globo, ranking as number 12).

- These media companies are headquartered in OECD countries[15] and among these countries the economically strongest account for 28 of the 30 largest media content firms by revenue: the United States (14), Japan (5), Spain (1), the United Kingdom (1), Italy (2), France (2), and Germany (2).
- Thus, the emerging media system of the world is one of BRICS vs. Top-OECD, of mindshare vs. marketshare.

## GLOBAL CONCENTRATION

The worldwide market shares of the major companies in a media industry enable us to determine measures of global concentration (See "Companies"). This is different from an average concentration across countries. Instead, it is a measure of how much the major firms in an industry control worldwide. Table 37.13 shows the results. As one could expect, the shares of a particular firm on a worldwide basis are much smaller than it is in the few countries where it is active. Thus, the combined share of the top few firms, the C4, is quite low for newspapers, at 17%. It is also relatively low for broadcast TV (18.5%), online news (19.6%), and radio (20%). Of content industries, books (33.6%) and non-broadcast video channels (33.5%) are high. The C4 measures are much higher for platforms and reach 26% for wireline telecom and 38.3% for wireless; ISPs are 34.5% and Multichannel-

Table 37-13. Worldwide Industry Concentration (using global market shares of major firms)

| | 2011 OR MOST RECENT (30 COUNTRIES) | | |
|---|---|---|---|
| | W-HHI | C4 (%) | INDUSTRY REVENUES (MIL US$) |
| Newspapers | 126 | 17.0% | 106,734 |
| Magazines | 189 | 24.0% | 79,590 |
| Books | 394 | 33.6% | 72,224 |
| Radio | 154 | 20.0% | 47,687 |
| Broadcast TV | 206 | 18.5% | 184,379 |
| Multichannel Platforms | 599 | 43.0% | 247,666 |
| Video Channels | 431 | 33.5% | 70,636 |
| Film | 1,057 | 47.0% | 30,710 |
| Wireline | 308 | 26.0% | 589,986 |
| Wireless | 655 | 38.3% | 740,038 |
| ISP | 443 | 34.5% | 162,033 |
| Search Engines | 4,339 | 84.8% | 41,300 |
| Online News | 193 | 19.6% | 19,760 |

15. Technically, Mexico is a member of the OECD.

Platforms are 43%. By far the highest concentrations are for search engines (84.4%) and film (47% for C4 and 76.4% for C6 concentrations).

We have come to the end of this chapter, in which we investigated and summarized media concentration around the world. The analysis included:

- Convergence trends
- Factors for industry concentration
- Explanatory variables for intra-industry by variations in concentration
- Diversity and media voices
- Cross-ownership around the world
- The export of media by countries
- The share of "Hollywood" industry in the various countries' content media
- Ranking the world's largest companies
- Market power in news media, both nationally and worldwide.

With the results in hand we are ready to reach our conclusions. This will be the subject of the next chapter.

# Appendix

Table 37-11B. Top 10 National News Companies in National News Attention Share (Full Version)

| COMPANY | 2004/05 | 2011 OR MOST RECENT |
|---|---|---|
| ARGENTINA | | |
| Grupo Clarín | 31.6% | 26.9% |
| América 2 | 7.4% | 9.2% |
| Canal 9 (Mexico) | 6.6% | 8.3% |
| Telefónica (Spain) | 12.1% | 5.9% |
| DirecTV (US) | 0.0% | 3.0% |
| Grupo Haddad | 6.0% | 2.7% |
| Editorial Atlátida SA | 1.2% | 2.3% |
| Grupo Monera | 0.0% | 2.1% |
| Del Plata | 0.0% | 1.7% |
| El Día | 1.3% | 1.6% |
| C10 | 66.2% | 63.8% |
| HHI | 1301 | 955 |
| AUSTRALIA | | |
| Seven Network | 15.0% | 17.8% |
| Nine Entertainment | 18.4% | 14.7% |
| Murdoch Group (US) | 10.4% | 13.6% |
| Australian Broadcasting Corporation (public) | 9.1% | 11.4% |
| Ten Network | 10.2% | 8.9% |
| Southern Cross Media Group | 5.2% | 7.5% |
| Fairfax | 3.4% | 4.5% |
| Telstra | 2.3% | 3.7% |
| Win Television | 2.7% | 2.6% |
| Prime Network | 2.6% | 2.6% |
| C10 | 79.2% | 87.3% |
| HHI | 937 | 1046 |
| BELGIUM | | |
| VRT (public) | 18.6% | 17.3% |
| Bertelsmann (Germany) | 11.3% | 11.6% |
| RTBF (public) | 9.3% | 9.4% |
| VMMA | 8.4% | 9.1% |
| Telenet (Liberty, US) | 6.6% | 6.6% |
| Bouygues Group (France) | 5.7% | 5.2% |
| France Télévisions (France) (public) | 4.8% | 4.8% |

*continued*

Table 37-11B. *continued*

| COMPANY | 2004/05 | 2011 CR MOST RECENT |
|---|---|---|
| Corelio/VUM | 4.0% | 3.9% |
| VOO & Coditel | 3.5% | 3.5% |
| Rossel | 3.1% | 3.4% |
| C10 | 75.3% | 74.7% |
| HHI | 798 | 767 |
| BRAZIL | | |
| Globo Group | 31.7% | 35.5% |
| Folhapar | 11.5% | 9.6% |
| Estado Group | 7.8% | 8.2% |
| Universal Church Group | 2.5% | 6.4% |
| Abril Group | 7.1% | 5.3% |
| Bandeirantes Group | 1.6% | 3.6% |
| Diários Associados Group | 2.3% | 2.7% |
| RBS Group | 3.7% | 2.5% |
| Jornal do Brasil | 2.1% | 2.3% |
| Ongoing (Arca Group, Portugal) | 2.6% | 2.2% |
| C10 | 72.9% | 78.3% |
| HHI | 1,301 | 1,534 |
| CANADA | | |
| Bell Canada Enterprise/CTV | 13.0% | 17.4% |
| Canadian Broadcasting Corporation (public) | 15.6% | 16.8% |
| Shaw | 7.9% | 15.3% |
| Rogers | 7.5% | 8.4% |
| Quebecor (Videotron, Canada) | 6.4% | 6.7% |
| Astral | 1.7% | 4.4% |
| Postmedia Network Canada | 2.7% | 2.8% |
| Bing (Microsoft, US) | 3.0% | 2.4% |
| Google (US) | 0.9% | 1.7% |
| Torstar | 1.8% | 1.3% |
| C10 | 58.9% | 76.0% |
| HHI | 659 | 975 |
| CHILE | | |
| Universidad Catholica/Luksic Group | 13.2% | 12.5% |
| TVN (public) | 13.3% | 12.1% |
| El Mercurio SAP | 8.5% | 8.1% |
| DirecTV (US) | 12.7% | 8.0% |
| Discovery Communications (US) | 4.4% | 7.4% |
| Albavisión (Mexico) | 4.4% | 7.3% |

Table 37-11B. *continued*

| COMPANY | 2004/05 | 2011 OR MOST RECENT |
|---|---|---|
| Bethia Group | 7.7% | 6.8% |
| Time Warner (US) | 7.6% | 6.2% |
| COPESA | 5.2% | 5.1% |
| Iberoamericana (Spain) | 4.3% | 4.5% |
| C10 | 77.0% | 73.6% |
| HHI | 796 | 696 |
| CHINA | | |
| Government of China | 94.8% | 96.0% |
| *China Central Television (public)* | *37.4%* | *36.1%* |
| *Shanghai Media Group (public)* | *13.2%* | *8.4%* |
| *Hunan Media Group (public)* | *6.8%* | *4.9%* |
| *Jiangsu Media Group (public)* | *1.0%* | *3.8%* |
| *People's Daily (public)* | *0.9%* | *2.1%* |
| *Guangdong Provincial Government (public)* | *2.7%* | *1.5%* |
| *Reference News (Xinhua News Agency, public)* | *1.0%* | *1.3%* |
| *Rayli (public)* | *0.2%* | *1.3%* |
| *Southern Media Corporation (public)* | *1.0%* | *1.3%* |
| *China Radio International (public)* | *1.6%* | *1.2%* |
| *Trend Media Group (public)* | *0.6%* | *0.9%* |
| Hearst (US) | 0.3% | 0.8% |
| Sina.com | 0.6% | 0.5% |
| Sohu.com | 0.5% | 0.4% |
| Conde Nast (US) | 0.0% | 0.4% |
| 163.com | 0.1% | 0.3% |
| TenCent Holdings | 0.2% | 0.1% |
| Caijing | 0.1% | 0.0% |
| C10 | 96.5% | 98.7% |
| HHI | 8,994 | 9,220 |
| EGYPT | | |
| Government of Egypt | 95.5% | 89.8% |
| *ERTU (public)* | *68.4%* | *69.1%* |
| *Al Akhbar (public)* | *8.7%* | *6.9%* |
| *Al Ahram (public)* | *9.2%* | *5.7%* |
| *Al Gomhouria (public)* | *4.8%* | *4.8%* |
| *Nile Television Network (public)* | *4.3%* | *3.3%* |
| Al Masry Al Youm | 0.1% | 3.7% |
| Nile Radio Productions | 3.7% | 3.0% |
| Hayat | 0.0% | 0.3% |

*continued*

Table 37-11B. *continued*

| COMPANY | 2004/05 | 2011 OR MOST RECENT |
|---|---|---|
| Panorama Action TV | 0.0% | 0.3% |
| Rotana (Saudi Arabia) | 0.1% | 0.3% |
| MBC (Saudi Arabia) | 0.3% | 0.2% |
| Dream TV | 0.4% | 0.2% |
| Al Baraheen (Saudi Arabia) | 0.0% | 0.2% |
| Melody Entertainment TV | 0.1% | 0.1% |
| C10 | 100% | 98.2% |
| HHI | 9,125 | 8,080 |
| FINLAND | | |
| Yle (public) | 46.9% | 32.4% |
| Bonnier (Sweden) | 18.5% | 20.2% |
| Sanoma Oyj | 9.2% | 9.4% |
| Alma Media Oyj | 4.9% | 4.9% |
| Otava Kuvalehdet OY | 2.1% | 2.5% |
| Vivendi (France) | 2.2% | 2.2% |
| ProSiebenSat1 (Germany) | 1.4% | 2.1% |
| Keskisuomalainen | 1.4% | 1.5% |
| A-Lehdet | 1.2% | 1.1% |
| TS-Yhtymä | 1.1% | 1.1% |
| C10 | 88.8% | 77.4% |
| HHI | 2,668 | 1,588 |
| FRANCE | | |
| France Télevisions (public) | 19.5% | 15.7% |
| Bouygues Group | 15.7% | 13.8% |
| Vivendi | 11.1% | 11.3% |
| Bertelsmann (Germany) | 8.0% | 8.1% |
| Amaury Group | 4.8% | 5.1% |
| Socpresse Group | 3.3% | 3.2% |
| Lagardère | 2.7% | 2.8% |
| Altice | 0.6% | 2.4% |
| NRJ Group | 1.7% | 2.1% |
| C10 | 67.4% | 64.4% |
| HHI | 871 | 689 |
| GERMANY | | |
| ARD (public) | 17.9% | 17.4% |
| ProSiebenSat.1 | 12.8% | 16.2% |
| Bertelsmann | 16.9% | 15.9% |
| ZDF (public) | 9.2% | 8.2% |

Table 37-11B. *continued*

| COMPANY | 2004/05 | 2011 OR MOST RECENT |
|---|---|---|
| Axel Springer | 5.2% | 5.9% |
| Kabel Deutschland | 3.2% | 3.1% |
| Unitymedia | 1.9% | 2.0% |
| Der Spiegel | 2.3% | 1.8% |
| Murdoch (US) | 0.3% | 1.8% |
| Burda | 1.9% | 1.7% |
| C10 | 71.6% | 73.9% |
| HHI | 910 | 949 |
| INDIA | | |
| Prasar Bharati (public) | 68.3% | 65.4% |
| The Times of India Group (BCCL) | 6.1% | 6.9% |
| Zee | 2.8% | 3.3% |
| Sun Group | 1.2% | 1.8% |
| Disney (US) | 1.1% | 1.2% |
| Living Media Group | 0.5% | 1.0% |
| Hathway | 0.5% | 0.7% |
| HT Media | 0.6% | 0.7% |
| Network18 Group | 0.2% | 0.6% |
| Murdoch (US) | 0.1% | 0.6% |
| C10 | 81.3% | 82.2% |
| HHI | 4,719 | 4,342 |
| IRELAND | | |
| RTE (public) | 45.8% | 40.0% |
| Independent News and Media | 10.6% | 10.9% |
| Murdoch (US) | 7.3% | 9.9% |
| TV3 (UK) | 7.9% | 8.3% |
| Liberty (US) | 5.3% | 4.1% |
| TG4 (public) | 0.0% | 4.0% |
| Daily Mail & General Trust (UK) | 1.8% | 2.0% |
| Thomas Crosbie Holdings | 1.2% | 1.7% |
| Communicorp | 1.7% | 1.5% |
| ITV (UK) | 2.1% | 1.4% |
| C10 | 83.6% | 83.9% |
| HHI | 2,370 | 1,938 |
| ISRAEL | | |
| Channel 2 | 18.9% | 22.6% |
| Yedioth Ahronoth Group | 12.5% | 10.9% |

*continued*

Table 37-11B. *continued*

| COMPANY | 2004/05 | 2011 OR MOST RECENT |
|---|---|---|
| Channel 10 | 6.2% | 9.2% |
| Bezeq | 5.5% | 6.4% |
| Altice-HOT (France) | 9.7% | 6.2% |
| Israel Broadcasting Authority (public) | 12.6% | 5.7% |
| Dori Media Group | 0.9% | 3.6% |
| RGE | 4.3% | 3.4% |
| Ha'aretz | 2.7% | 2.5% |
| IDF (Army Radio, public) | 2.3% | 2.5% |
| C10 | 75.6% | 73.1% |
| HHI | 926 | 881 |
| ITALY | | |
| RAI (public) | 34.6% | 34.4% |
| Fininvest | 23.8% | 22.4% |
| Murdoch (US) | 9.4% | 8.9% |
| Gruppo Espresso | 8.3% | 7.5% |
| RCS Media Group | 4.4% | 5.3% |
| Caltagirone Editori | 1.5% | 1.5% |
| Gruppo Finelco | 1.4% | 1.4% |
| Il Sole 24ore | 1.5% | 1.4% |
| Poligrafici Editoriale | 1.1% | 1.0% |
| Telecom Italia | 1.7% | 0.4% |
| C10 | 87.6% | 84.2% |
| HHI | 1,949 | 1,855 |
| JAPAN | | |
| Asahi Shimbun | 11.2% | 13.6% |
| Fujisankei (NBS) | 15.0% | 11.9% |
| NHK (public) | 13.7% | 11.7% |
| NTV | 11.3% | 11.3% |
| Tokyo Broadcasting System | 10.7% | 10.2% |
| JSAT (SkyPerfecTV) | 5.8% | 5.8% |
| Nihon Keizai Shimbun (TV Tokyo) | 3.8% | 5.2% |
| Yomiuri Shimbun | 3.0% | 3.1% |
| Softbank (Yahoo, Japan/US) | 2.0% | 2.1% |
| Mainichi Shimbun | 1.3% | 1.2% |
| C10 | 77.9% | 76.1% |
| HHI | 845 | 772 |

Table 37-11B. *continued*

| COMPANY | 2004/05 | 2011 OR MOST RECENT |
|---|---|---|
| MEXICO | | |
| Televisa | 47.7% | 44.4% |
| TV Azteca | 18.7% | 19.5% |
| OEM | 7.9% | 8.2% |
| El Universal | 1.8% | 2.6% |
| Megacable | 1.9% | 1.6% |
| Medios Masivos Mexicanos | 1.4% | 1.4% |
| Editorial Ovaciones | 1.4% | 1.4% |
| Grupo Reforma | 1.8% | 1.4% |
| MVS Comunicaciones | 0.6% | 1.2% |
| Grupo Carso | 0.5% | 0.8% |
| C10 | 83.6% | 82.3% |
| HHI | 2,706 | 2,431 |
| THE NETHERLANDS | | |
| NPB (public) | 16.0% | 16.6% |
| Bertelsmann (Germany) | 8.8% | 10.5% |
| Redstone (US) | 6.7% | 9.1% |
| Ziggo (Zesko Holding BV) | 0.7% | 7.8% |
| Telegraaf Media Groep | 4.8% | 7.7% |
| Sanoma Oyj (Finland) | 8.1% | 6.8% |
| Discovery Communications (US) | 4.1% | 5.2% |
| Liberty (US) | 5.1% | 4.8% |
| Disney (US) | 1.2% | 4.2% |
| PCM Uitgevers | 2.7% | 3.2% |
| C10 | 58.3% | 75.8% |
| HHI | 541 | 721 |
| POLAND | | |
| TP (public) | 24.8% | 21.1% |
| Polsat | 15.7% | 17.5% |
| Grupa TVN | 8.0% | 10.0% |
| Radio RMF FM | 7.1% | 7.5% |
| Radio ZET | 6.4% | 6.2% |
| Polskie Radio (public) | 6.3% | 6.2% |
| Wizja TV (Liberty, US) | 9.6% | 4.7% |
| Agora | 3.2% | 3.7% |
| Axel Springer (Germany) | 4.3% | 3.2% |

*continued*

Table 37-11B. *continued*

| COMPANY | 2004/05 | 2011 OR MOST RECENT |
|---|---|---|
| C9 | 85.4% | 80.1% |
| HHI | 1,174 | 1,031 |
| PORTUGAL | | |
| Media Capital (PRISA Group, Spain) | 28.8% | 27.8% |
| RTP (public) | 26.6% | 27.8% |
| Impresa | 24.9% | 22.6% |
| Cofina | 3.6% | 3.9% |
| Controlinveste | 2.4% | 3.2% |
| Group RR | 2.9% | 3.2% |
| Zon | 2.1% | 1.6% |
| Portugal Telecom (Oi-Telemar, Brazil/ Portugal) | 0.0% | 0.9% |
| Murdoch (US) | 0.1% | 0.7% |
| C10 | 91.3% | 91.6% |
| HHI | 2,188 | 2,097 |
| RUSSIA | | |
| Government of Russia | 53.6% | 54.8% |
| *Gazprom Media (public)* | *14.5%* | *25.8%* |
| *First Channel (public)* | *21.3%* | *15.1%* |
| *VGTRK (public)* | *17.5%* | *13.1%* |
| *RIA Novosti (public)* | *0.3%* | *0.7%* |
| CTC | 21.6% | 15.6% |
| Komsomolskaya Pravda | 2.9% | 3.5% |
| EMG | 3.2% | 2.9% |
| ProfMedia | 1.9% | 1.6% |
| Kommersant | 2.6% | 1.2% |
| NMG | 1.2% | 1.1% |
| RMG | 2.1% | 1.1% |
| RBC.ru | 1.3% | 0.9% |
| Sanoma Oyj (Finland) | 0.8% | 0.7% |
| C10 | 91.0% | 83.5% |
| HHI | 3,377 | 3,271 |
| SOUTH AFRICA | | |
| SABC (public) | 43.5% | 38.5% |
| Naspers Group | 26.7% | 27.2% |
| Hosken | 13.5% | 17.3% |
| Independent News and Media (Ireland) | 4.0% | 3.7% |

Table 37-11B. *continued*

| COMPANY | 2004/05 | 2011 OR MOST RECENT |
|---|---|---|
| Primedia | 2.2% | 2.9% |
| Avusa | 2.5% | 2.6% |
| Kagiso Media | 2.1% | 1.9% |
| Caxton | 1.8% | 1.8% |
| African Media Entertainment | 0.3% | 1.0% |
| MSN (Microsoft, US) | 0.6% | 0.9% |
| C10 | 97.3% | 97.8% |
| HHI | 2,824 | 2,563 |
| SOUTH KOREA | | |
| Government of South Korea | 41.9% | 38.3% |
| *KBS (public)* | *22.2%* | *17.9%* |
| *MBC (public)* | *12.6%* | *14.9%* |
| *EBS (public)* | *2.8%* | *1.4%* |
| SBS | 14.4% | 12.5% |
| Tbroad | 3.6% | 4.4% |
| CJ Group | 1.5% | 3.8% |
| Chosun Ilbo | 3.6% | 3.5% |
| Jmnet (JoongAng Ilbo) | 3.1% | 3.2% |
| C&M | 3.0% | 3.1% |
| Dong-A Ilbo | 2.7% | 2.7% |
| KT | 3.6% | 2.5% |
| OhMyNews | 0.6% | 0.6% |
| C10 | 78.0% | 74.5% |
| HHI | 2,032 | 1,702 |
| SPAIN | | |
| Planeta | 11.2% | 14.7% |
| PRISA | 18.3% | 14.1% |
| Fininvest (Italy) | 11.5% | 13.8% |
| RTVE (public) | 14.0% | 9.8% |
| FORTA (public) | 8.2% | 4.6% |
| RCS (Italy) | 1.4% | 3.0% |
| Vocento | 2.1% | 2.3% |
| Ono | 0.6% | 1.8% |
| Telefónica | 0.6% | 1.4% |
| Zeta | 1.3% | 1.3% |
| C10 | 69.1% | 66.8% |
| HHI | 871 | 752 |

*continued*

Table 37-11B. *continued*

| COMPANY | 2004/05 | 2011 OR MOST RECENT |
|---|---|---|
| SWEDEN | | |
| Bonnier | 18.5% | 25.0% |
| Sveriges (public) | 31.8% | 24.7% |
| Investment AB Kinnevik | 11.0% | 16.9% |
| proSiebenSat1 (Germany) | 4.9% | 7.3% |
| Com Hem (BC Partners, UK) | 2.4% | 3.9% |
| Stampen | 1.5% | 2.1% |
| Teracom (public) | 1.8% | 2.1% |
| Schibsted (Norway) | 2.5% | 1.7% |
| Telenor (Norway) | 1.6% | 1.3% |
| Boxer | 0.0% | 1.3% |
| C10 | 75.8% | 86.4% |
| HHI | 1,519 | 1,609 |
| SWITZERLAND | | |
| SRG SSR (public) | 28.2% | 25.3% |
| Tamedia | 7.0% | 9.5% |
| Liberty (US) | 6.5% | 6.5% |
| Ringier | 3.7% | 5.0% |
| ProSiebenSat.1 (Germany) | 4.0% | 4.0% |
| Edipresse | 2.8% | 0.0% |
| Neue Zürcher Zeitung (NZZ) | 2.1% | 2.5% |
| ARD (Germany) (public) | 2.5% | 2.2% |
| ZDF (Germany) (public) | 2.2% | 2.0% |
| Axel Springer Suisse (Germany) | 1.2% | 1.8% |
| C10 | 60.2% | 62.2% |
| HHI | 953 | 854 |
| TAIWAN | | |
| Government of Taiwan | 18.8% | 19.6% |
| *Formosa TV (public)* | *4.6%* | *4.8%* |
| *BCC (public)* | *3.8%* | *4.0%* |
| *Taiwan TV Enterprise (public)* | *4.2%* | *3.2%* |
| *China Television Corporation (Public)* | *2.2%* | *3.0%* |
| China Times News Group (Want Want Ltd.) | 9.7% | 9.8% |
| Liberty Times News Group | 5.0% | 7.8% |
| Next Media (Hong Kong) | 4.5% | 6.0% |
| United Daily News Group | 7.1% | 5.9% |
| SET | 5.5% | 4.2% |
| ET | 4.7% | 4.2% |

Table 37-11B. *continued*

| COMPANY | 2004/05 | 2011 OR MOST RECENT |
|---|---|---|
| TVBS (TVB, Hong Kong) | 4.1% | 3.9% |
| VL | 4.3% | 3.7% |
| Murdoch (US) | 2.5% | 3.6% |
| C10 | 66.3% | 68.8% |
| HHI | 664 | 700 |
| TURKEY | | |
| Dogan Group | 29.8% | 37.5% |
| Turkuvaz-Kalyone Group | 10.8% | 14.0% |
| Cukurova Group | 14.2% | 13.4% |
| Feza | 2.7% | 5.5% |
| Dogus | 3.5% | 3.6% |
| TRT (public) | 6.9% | 3.5% |
| Kanal 7 | 4.4% | 2.9% |
| Demirören | 0.0% | 2.6% |
| Power Group | 1.3% | 1.8% |
| Ciner Group | 8.5% | 1.7% |
| C10 | 82.1% | 86.4% |
| HHI | 1,373 | 1,860 |
| UK | | |
| BBC (public) | 30.5% | 33.9% |
| Murdoch Group (US) | 13.0% | 14.3% |
| ITV | 14.0% | 11.3% |
| Channel 4 | 6.0% | 5.1% |
| Daily Mail and General Trust | 3.1% | 3.5% |
| Viacom (US) | 1.1% | 2.3% |
| Trinity Mirror | 2.9% | 2.2% |
| Bauer (Germany) | 1.5% | 1.9% |
| Northern & Shell | 1.7% | 1.6% |
| The Guardian | 1.7% | 1.6% |
| C10 | 75.5% | 77.8% |
| HHI | 1,363 | 1,549 |
| USA | | |
| Comcast[1] | 4.1% | 12.2% |
| *GE* | *5.8%* | *0.0%* |
| Disney | 7.4% | 9.6% |
| Redstone Group | 6.5% | 9.3% |
| Murdoch Group | 8.2% | 8.5% |

*continued*

Table 37-11B. *continued*

| COMPANY | 2004/05 | 2011 OR MOST RECENT |
|---|---|---|
| Time Warner[a] | 9.8% | 6.1% |
| Gannett | 3.1% | 3.3% |
| Tribune | 2.4% | 1.9% |
| DirecTV | 0.0% | 1.6% |
| Discovery Communications | 1.1% | 1.7% |
| Yahoo | 2.3% | 1.7% |
| Liberty Media | 1.0% | 1.5% |
| Advance | 0.9% | 1.5% |
| The Washington Post Company | 1.2% | 1.4% |
| Hearst | 1.1% | 1.3% |
| AOL-Huffington Post[b] | 0.0% | 1.3% |
| The New York Times | 1.2% | 1.2% |
| Clear Channel | 1.5% | 1.2% |
| Time Warner Cable[c] | 0.0% | 1.1% |
| McClatchy | 0.8% | 1.0% |
| Cablevision | 0.7% | 0.7% |
| Univision | 1.10% | 0.20% |
| Dish Network | 0.80% | 0.70% |
| C4 | 31.9% | 39.6% |
| C10 | 51.0% | 55.8% |
| C20 | 61.0% | 68.0% |
| HHI | 345 | 476 |

[a] Acquisition by AT&T pending in 2015.
[b] Acquisition by Verizon pending in 2015.
[c] Merger with Charter/Liberty pending in 2015.

## Bibliography

For the sources mentioned in this chapter, see the general bibliography at the end of this book.

# 38

# Findings: The Questions Addressed, From A to Z

## OVERVIEW

There are few industries whose ownership and control are more important than those of media, given their central role in social, cultural, economic, and political life. The preceding chapters have analyzed media industries, companies, and countries. We have proceeded in a funnel-like fashion, developing the data for each country, and then aggregating and analyzing it along different dimensions. The findings across the world will now be summarized and interpreted, with some repetitions of previous sections.[1] This will answer the series of questions A-Z posed in the introductory chapter.

## A. How large is the media sector?

Content media[2] accounted for a combined $735 billion, about 1% of the world's GDP, and platform media[3] accounted for $1.7 trillion, 2.3% of world GDP. Together then, media spending—both direct consumer spending and media advertising—added up to 3.3% of world GDP. To that we can add other media activities or related businesses. Music sales added up to $16.5 billion in 2012[4] and video games to $67 billion. Consumer electronics devices used in the consumption of media accounted for the following revenues: mobile phone handsets, $210 billion; consumer microcomputers,

1. Our "world" consists of the 30 countries and 13 media industries. While these countries account for 82% of the world GDP and 64% of its population, it under-represents the less developed countries. We add a brief discussion of 11 of the largest countries that were not included, and which account for another 15% of the world's population.

2. "Content media" include newspapers, magazines, book publishing, film, radio, broadcast television, video channels, search engines, online news media, and one-third of the multi-channel industry's total revenues.

3. "Platform media" include wireless and wireline telecom, ISPs, and two-thirds of multi-channel industry revenues. The allocation for the multi-channel industry is based on its dual role or both platform and content aggregator, and approximates the cost shares.

4. According to the International Federation of the Phonographic Industry Report 2012, as reported by Billboard: <http://www.billboard.com/biz/articles/news/digital-and-mobile/1549915/ifpi-digital-music-report-2013-global-recorded-music>.

$160 billion; TV sets, $100 billion; and music players, $30 billion. If these are included, the consumption of media content, platforms, and devices is about $2.9 trillion, 5.9% of the 30-country gross world product of $61.2 trillion USD.

These are revenues on the platform/vendor/producer end. The price to end-users/consumers rises by the retail and wholesale markups for print media, film, music, games, and most consumer electronics. If we make a simplifying but fairly realistic assumption that the wholesale and retail distribution chain doubles the price of these content items and devices to consumers, the share in GDP rises to $3.6 trillion, or 7.3%. (There is, however, some double counting involved due to a vertical value chain, as is often the case in GDP and industry aggregations.)

This percentage still understates its actual weight. Much of household income is spent on necessities. In the United States, 30.43% of household income goes toward housing (excluding telecom), 12.81% on food, 6.92% on healthcare, 2.35% on education, and 17.5% on transportation, etc.[5] These necessities leave only 30% for discretionary categories of spending. Media spending, including telecom connectivity, thus accounts for almost 20% of discretionary spending, a substantial chunk.[6] It is even higher in terms of discretionary *time*. In the United States, average annual media consumption per person, as reported by the Census Bureau, has been measured to be an astonishing 3,545 hours per year for 2005–2009, not including time for e-mail and telephone calls, or 9.7 hours per day.[7] This number implies that media consumption—including background music, multi-tasking, and multimedia—would occupy or overlap with 60% of all non-sleep time, including work time, meals etc. Thus, media takes a major share of people's time and money.

## B. What are the world's largest national media markets?

Overall, the United States accounts for 33% of the 30-country world media markets, by revenues. Following are Japan (9.9%), China (8.7%), Germany (6.1%), the United Kingdom (5.2%), France (4.5%), Spain (3.6%), and Canada (3.1%).

The content market is largest, by far, in the United States, with $268 billion, or 36.5% by revenue, followed by Japan ($63.9 billion, 8.7%), Germany ($56 billion, 7.6%), the United Kingdom ($58 billion, 7.9%), France ($37.5 billion, 5.1%), China ($35.3 billion, 4.8%), and Spain ($34.1 billion, 4.6%). China's content revenue value has increased greatly from 2004. India's content revenues grew substantially between 2004 and 2013, rising to $8.8 billion, 1.2% of the world total. So did Brazil's, from $10.7 billion in 2005 to $23.4 billion in the latest year, 3.2% of the world total.

The largest platform media markets are the United States ($521 billion, 31.4%), China ($173 billion, 10.5%), Japan ($172 billion, 10.4%), Brazil ($104 billion, 6.3%), Germany ($89.2 billion, 5.4%), and the United Kingdom ($66.5 billion, 4%). China's and India's ($34.7 billion, 2.1%) growth have been especially high, with market sizes nearly tripling in the period under review.

## C. What are the largest media industries? Is content really "king"?

Is content king, as the cliché asserts? Actually, in no country do the top content companies have higher market power in their industries than platform companies do.[8] Similarly, in most countries, platform media revenues are on average three times larger than those of content media. In no country does content's revenue match that of platforms. It came closest in the United Kingdom and Sweden where content industries' share in overall media revenue is 46.6% and 46.4%, respectively. This ratio is lower in richer countries

5. US Bureau of Labor Statistics, *Consumer Expenditure Survey*, September 2013.

6. For many people telecommunications and Internet connectivity is increasingly becoming a necessity.

7. US Census Bureau, *Statistical Abstract of the United States: 2011*, Table 1130, attributed to Suhler Stevenson, New York, NY, *Communications Industry Forecast*, annual.

8. They are trivially higher in Finland, Russia, and the Netherlands.

with competitive mobile communications (e.g., Sweden or Finland). On average, the top four content companies have an especially low proportion of overall media revenues in emerging markets: China (16.9%), Poland (13.5%), South Africa (19.1%), Turkey (17.8%), Mexico (15.8%), Egypt (15.8%), and Brazil (18.4%).

The largest media industries, by far, are mobile telecom ($740 billion, 30.9%) and wireline telecom ($590 billion, 24.7%), followed by multi-channel platforms (10.4%). Of content media, broadcast TV is largest (7.7%), followed by ISPs (6.8%), newspapers (4.5%), magazines (3.3%), book publishing (3%), video channels (3%), radio (2%), search engines (1.7%), film (1.3%), and online news (0.8%).[9]

Not only are platform media revenue streams larger, they are also more stable and face less competition and potential entry. Analogously, content media are smaller, relatively more competitive, and have usually lower barriers to entry. It is hard to view them as holding superior bargaining power, with a few exceptions such as for unique sports events.

## GLOBAL CONCENTRATION TRENDS

### D. What are the levels and trends of media concentration in different countries and industries?

The world average concentration levels are now summarized. For each country, we calculated average national concentrations for "all media," "content media," "platform media," and "news media." The latter was presented in two ways, by "revenues," and by "news attention." There are three steps to these calculations:

1. The countries' various industry concentration measures are provided in the country reports by their respective authors.

2. These figures are then aggregated on the national level by a weighted averaging, with the weights being each country's industries' share in that country's media sector.

3. The national averaged figures are then aggregated internationally. Two methods of averaging are used, weighted and arithmetic. Arithmetic averaging gives equal weight to each country whether large or small, and thus provides measures that approximate those of a typical country. In contrast, the weighted averaging takes account of the size of a country's media market, and thus approximates the concentration prevailing in the global media sector taken together.

The world average weighted concentration measures across industries and across countries are reported in Table 38.1. The arithmetic averages are provided in Table 38.1A in the Appendix. Findings now follow.

The weighted world average HHI for the overall media sector is a very high 3,253, up from 3,125 in 2004/05.[10] The arithmetic average, which reduces the weight of large countries such as the United States, China, and India, is 3,729. These numbers are considerably above the threshold of a "highly concentrated" industry of antitrust enforcement standards (HHI > 2,500, weakened in 2010 from 1,800). In 2013, the five countries with the highest average media industry HHIs were China (9,700), South Africa (5,535), Mexico (5,315), Turkey (5,150), and Egypt (5,041). The lowest average concentrations measured were for the United States (1,517), Canada (2,506), France (2,694), Brazil (2,773), and the Netherlands (2,754).

For content media, the world average weighted HHI is much lower, at 2,219 (and 1,999 without the Internet media of search engines and online news). But it has also been rising considerably, up from 1,660 and 1,603, respectively.

9. If one looks only at content media, these figures would scale up by about three times, with broadcast TV the largest at 21%.

10. To check for robustness with respect to the high-population countries, we calculated the following: For "all media," the weighted HHI is 3,253. Taking out China—a highly concentrated market with a large population—changes the HHI to 2,637. Taking out both China and India results in an HHI of 2,631. When the United States, which makes up approximately one-third of all media revenues in the world, is also excluded, the HHI rises to 3,283. Thus, the results are fairly robust with respect to India and the United States. China has the biggest impact, and we therefore provide in the tables also measures without it.

Table 38-1. National Media Concentrations—Weighted Country Average, 2012

| | ALL MEDIA | PLATFORM MEDIA | CONTENT MEDIA (W/INTERNET) | CONTENT MEDIA (W/O INTERNET) | NEWS MEDIA (BY REVENUES) | NEWS MEDIA (BY ATTENTION) | NEWS MEDIA BY ATTENTION (W/O CHINA) |
|---|---|---|---|---|---|---|---|
| HHI | 3,253 | 3,711 | 2,219 | 1,999 | 1,987 | 5,194 | 3,089 |
| C1 | 42.7% | 47.7% | 32.2% | 30.0% | 30.0% | 60.2% | 40.7% |
| C4 | 81.2% | 87.7% | 66.5% | 65.0% | 66.7% | 80.4% | 65.8% |
| Power Index | 3,223 | 3,707 | 2,134 | 1,934 | 1,912 | 5,162 | 3,045 |
| Noam Index | | | 1,167 | 1,064 | 984 | 4,450 | 2,197 |
| Voices | | | 42 | 36 | 31 | 31 | 32 |
| Cross-ownership | | | 25.9% | 24.4% | 28.7% | 28.7% | 26.0% |
| Pooled HHI | 1,884 | 2,982 | 863 | 993 | 969 | 3,811 | 1,473 |
| Pooled C1 | 31.0% | 41.7% | 18.2% | 19.5% | 20.1% | 51.6% | 29.9% |
| Pooled C4 | 59.7% | 78.5% | 39.8% | 42.4% | 45.0% | 69.6% | 52.1% |

For platform media, country weighted HHI has been slightly decreasing from a very high 3,869 to 3,711. The arithmetic HHI averages were still higher, 2,907 for content and 4,026 for platforms, reflecting the lesser downward pull on the international average by the United States.

The top four companies in platform media accounted, by weighted-country average, for a dominant 81.2% of their particular market. For content media markets the figure was 66.5%, also high but lower than for platforms. The gap is still higher for the pooled C4 for the sector, where the top four firms account for 78.5% of platforms and 39.8% of content, a difference close to 40%.

When it comes to news media, we find that concentration is very high in most countries, with a world average HHI of 5,194 when measured by attention time.[11] Because population size heavily affects attention measures, we also calculate the average without China due to its unusual characteristics. The news media HHI concentration by attention is then measured at 3,089, a very high figure.

On average, the national pooled C4 (i.e., the top four companies in the overall combined market of the 13 industries) for the world is 59.7% weighted by country (up from 55.5% in 2004), which means that on average, four companies control over one half of each country's 13 national media industries, combined. This is an astonishingly high percentage and it is based on the large size of platform media and their high concentration. In the United States, that pooled C4 is 42.7%, with AT&T[12] holding 15.8%; Verizon,[13] 13.3%; Comcast[14] 9.5%; and Time Warner Cable,[15] with 4.2%. Google has 1.3%. For the United Kingdom, the pooled C4 is 42.7%; for Germany, 49.9%; France, 61.9%; India, 62.2%; Japan, 66.8%; and China, 98.5%.

The share of the top firm in each country's national media market, as measured by the pooled C1 ratio weighted by country, is 41.7% for platforms, while for content media it is 18.2%, and for news media, by revenue, it is 20.1%). Without China, the country weighted world average pooled C1 for attention is lower, at 29.9% (with China, it is over 50%).

An important reason for the high overall media numbers is the large size of the platform media—wireline, wireless, ISPs, multi-channel platforms—coupled with their generally high market shares held by two or three firms. Pooled platform C4 in international average weighted by countries is a very high 78.5%. In contrast, the pooled content C4 concentration weighted by countries is 39.8%.

If we look at the "unpooled" measures, that is, averaged across industries (and hence with a different set of companies for each industry), the C4 for all media is a huge 81.2%; for platforms, it is 87.7%; for content media, 66.5%; and for news media, 66.7% by revenues, 65.8% for attention excluding China, and 80.4% with China. The arithmetic averages are even higher.

These are all high numbers. They are those of oligopolistic markets. For the average industry in the average country, barely a handful of companies control most of the markets. For the United States, the pooled C4 for content is 24.7%, whereas it is 91% for China. In between are Italy (59.7%), Russia (55.3%), Poland (48.1%), Japan (38.2%), India (37.6%), Germany (40.7%), France (39.3%), and the United Kingdom (47.7%).

## E. Are the world's media becoming more concentrated?

As the preceding section shows, the concentration numbers are high in almost every country. Has this concentration been declining or rising still further? The changes in the various measures of concentration show the following annual change, averaged over the recent past nine years or so (depending on the available time series). The changes in average national media concentration, in terms of annual percent change, are given in Tables 38.2 and 38.3.

11. "Attention time" is explained below.

12. seeking to acquire DirecTV in 2015.

13. seeking to acquire AOL in 2015.

14. Rebuffed in attempt to acquire Time Warner Cable in 2015.

15. Seeking to merge with Charter and Bright House in 2015.

Table 38-2. Average Annual Change in National Media Concentrations—Weighted Country (2004/5–2011/2)

| | ALL MEDIA | PLATFORM MEDIA | CONTENT MEDIA W/INTERNET | CONTENT W/O INTERNET | NEWS MEDIA (BY REVENUES) | NEWS MEDIA (BY ATTENTION) | NEWS MEDIA BY ATTENTION (WITHOUT CHINA) |
|---|---|---|---|---|---|---|---|
| HHI | 0.51% | – 0.51% | 4.22% | 3.08% | 2.17% | –0.12% | 0.02% |
| C1 | 0.12% | –0.80% | 2.89% | 2.20% | 1.46% | –0.08% | 0.02% |
| C4 | 0.93% | –0.59% | 1.48% | 1.35% | 1.12% | 0.23% | 0.31% |
| Power Index | 0.60% | –0.53% | 4.88% | 3.71% | 2.68% | –0.19% | –0.14% |
| Noam Index | | | 6.35% | 5.26% | 3.57% | –0.29% | –0.06% |
| Voices | | | 0.39% | 0.30% | 0.17% | 0.17% | 0.17% |
| Pooled HHI | 2.44% | –0.08% | 5.03% | 6.57% | 4.11% | –0.13% | –0.36% |
| Pooled C1 | 0.01% | –1.11% | 3.26% | 4.07% | 3.04% | –0.17% | –0.08% |
| Pooled C4 | 0.93% | –0.20% | 1.46% | 1.93% | 1.57% | 0.10% | 0.20% |
| Cross-ownership | | | 0.13% | 0.20% | 1.00% | 1.00% | 0.88% |
| Trend Average | 0.79% | –0.32% | 3.32% | 3.27% | 2.31% | –0.05% | 0.00% |

Table 38-3. Average Annual Change in National Media Concentrations—Arithmetic (2004/5–2011/12)

| | ALL MEDIA | PLATFORM MEDIA | CONTENT MEDIA W/INTERNET | CONTENT W/O INTERNET | NEWS MEDIA (BY REVENUES) | NEWS MEDIA (BY ATTENTION) | NEWS MEDIA BY ATTENTION (WITHOUT CHINA) |
|---|---|---|---|---|---|---|---|
| HHI | – 2.23% | –2.89% | 0.20% | –0.75% | –1.13% | –0.74% | –0.86% |
| C1 | –1.61% | –2.13% | 0.15% | –0.49% | –0.76% | –0.46% | –0.52% |
| C4 | 0.07% | –0.03% | 0.18% | 0.04% | –0.05% | –0.15% | –0.18% |
| Power Index | –2.08% | –2.82% | 0.34% | –0.67% | –1.03% | –0.82% | –0.94% |
| Noam Index | | | 0.05% | –1.06% | –1.52% | –0.81% | –1.01% |
| Voices | | | 0.51% | 0.41% | 0.23% | 0.23% | 0.23% |
| Pooled HHI | –2.30% | –2.87% | –2.09% | –0.74% | –1.37% | –0.71% | –1.01% |
| Pooled C1 | –2.43% | –2.80% | –1.39% | –0.55% | –0.81% | –0.56% | –0.67% |
| Pooled C4 | –0.41% | –0.30% | –0.77% | 0.08% | –0.15% | –0.02% | –0.05% |
| Cross-ownership | | | 0.18% | 0.26% | 1.33% | 1.33% | 1.17% |
| Trend Average | –1.57% | –1.98% | –0.48% | –0.44% | –0.76% | –0.49% | –0.60% |

## CONTENT MEDIA CONCENTRATION TRENDS

- For content media, average national HHI industry concentration rose by an annual rate of 4.2% per year, a fairly strong rate.[16] That increase was mostly due to search engines. If we exclude the Internet sector, content concentration rose by a still strong 3.1% per year.[17]
- If we look at the C4 concentration measure, which does not have the exponential nature of the HHI that magnifies increases, the average annual increase for content media is a substantial 1.5% (and 1.4% without Internet media). In other words, the top four firms increased collectively their share by 1.5%. The top firm increased its share by twice that amount (2.9%).
- What drives these increases?
  1. A strong rise in content concentration in the United States;
  2. A strong rise in the weights of China and India (even as their concentration measures slightly decline);
  3. Rises in Turkey, Italy, Argentina, Canada, the Netherlands, Switzerland, and others. Only a few countries had offsetting reductions—Mexico, South Korea, Taiwan, Egypt, and Finland, mostly from very high levels to still high ones.
- The HHI of pooled content media rose by 5% per year, which suggests a strong increase in the overall position of major media conglomerates. The top companies increased their collective market shares, with pooled C1 and C4 rising by hefty 3.3% and 1.5% per year, respectively.
- The power index for content media grew by 4.9% per year (3.7% without Internet), a high number. In comparison, the power index for platform media was slightly declining (−0.5%). If we look at the arithmetic average (which gives a greater weight to small countries), a similar gap exists, but at a lower base—content power index growth is 0.3% per year, and for platform media it declines by –2.8%. HHI, C4, and C1 are rising at about 0.2%, i.e. at a slower rate, indicating that increases are driven, in particular, by the larger countries.
- The world average of net voices per country has slightly increased. On an arithmetic average, each country has 42 independent content media producers (up from 41) active in its national media market and possessing at least 1% in one of that country's media industries. (Several of these voices may be owned by foreign firms.) The number of voices, on average, rose 0.4% per year.

## PLATFORM MEDIA CONCENTRATION TRENDS

- For platform media, average weighted industry HHI concentration declined by −0.5% per year. On the basis of arithmetic averaging, it fell more strongly, by –2.9%.
- The share of the top four firms (C4) declined by 0.6% per year, and that of the top firm (C1) dropped by 0.8%.
- The dominance of the top firm (pooled C1) in a platform industry weighted by countries declined by −1.1% per year, and the share of the top four (pooled C4) dropped by a lower 0.2%. This suggests a shift from near-monopoly to oligopoly.

---

16. These are substantial growth rates and they must be properly interpreted. The exponential nature of the HHI means that a rise in absolute terms gets magnified in its squaring. The increase in the HHI due to an increase in the market of a company by one percent is $\frac{dHHI}{dS_i}=\frac{2*100}{S_i}$. For example, if a company's share rises from 30% to 31%, the HHI rises by 61 points, which is as a percentage change of the contribution to the HHI by that firm, is 61/900=6.77%. If there are three firms, each with a 30% share rising to 31%, the overall HHI would rise by the same percentage. In other words, the 1% increment in these firms' shares (for a C4 increase of 3%) increased the HHI contribution by 6.77%. This is about double the C4 increase in percent terms. The exact ratio will depend on the various market shares. But a rough estimate is the square root of the change in the HHI gives an order of magnitude for the increase in the market shares similar to the C4.

17. For all concentration measures averaged together, there was an increase of 0.5%. This number represents an average of the growth rates for HHI and pooled HHI, the National Power Index, and the pooled and weighted C1s and C4s in the United States.

## NEWS MEDIA CONCENTRATION TRENDS

- Average country-weighted national news media HHI industry concentration, based on attention time, and without China, rose very slightly by 0.02% per year. It rose much more (2.2% per year) for news media by revenues, driven by rises in the large countries, and is negative for the arithmetic average.

## F. What are the overall concentration trends of media industries?

- Overall, the average country-weighted national media industry rose in HHI concentration by 0.5% per year, driven by the increase in content concentration that was higher than the downward pull of the platform media's despite their greater size.
- The top firm in a pooled (all-industries) country market gained very little, (on average, 0.01% per year). But the top four firms, whether by a pooled or industry-averaged measure, gained over 0.9% per year.
- In conclusion: average HHI concentration for content media rose by an annual 4.2%, while it declined by an annual −0.5% for platforms. The overall trend towards increased world concentration for all media was 0.5% per year. The C4 measures rose 1.5% per year for content media and dropped 0.6% for platforms.
- The picture is more positive in an arithmetic averaging (Table 38.3). HHI concentration for platform media declined by a robust −2.9%. For content media, it rose by a moderate 0.2%. For news media, concentration declined (if measured by revenues) by −1.1%, and about −0.7% by attention. The top firm (pooled C1) lost −2.4% (−2.8% in platforms and −1.4% in content).

The trends can also be presented as absolute changes per year (rather than as rates of change, i.e., as a percent change of a percentage). See Tables 38.4A and 38.4B in the Appendix. On average (arithmetic), the top single platform company in a country lost −1.3 points of market share each year, while the combined share of the top four firms stayed almost flat (−0.03). In content media, the top firm gained 0.06 points per year; the top four firms gained 0.14. The increase in average voices was fairly small (0.2). For news media by revenues and attention, the top firm lost

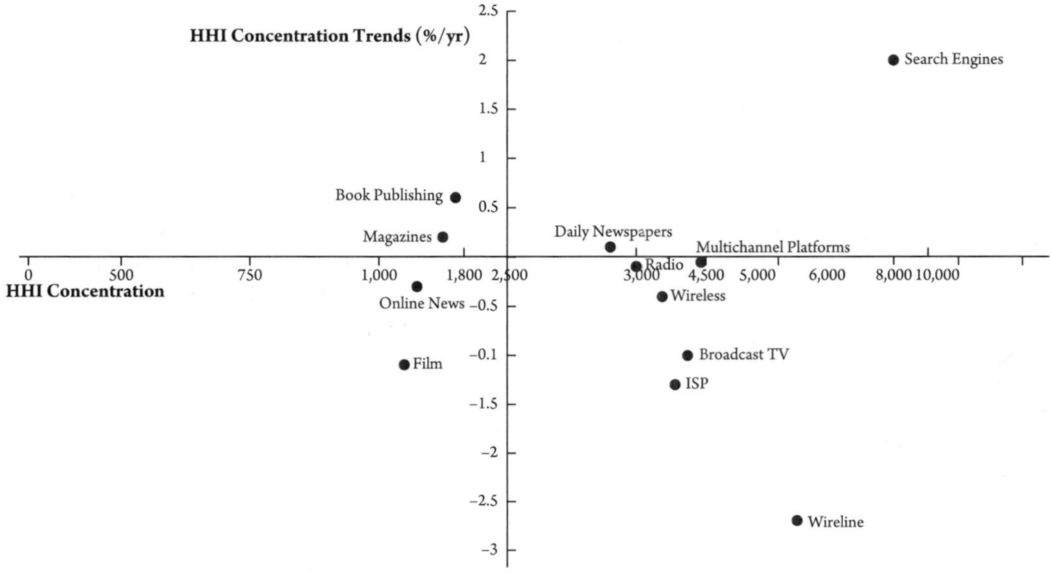


GRAPH 38.1 Media Industry HHI Concentrations and their Trends

each year about −0.2 points, respectively, the C4 decrease was of a lesser magnitude, and pooled C4 ownership increased by 0.4 points p.a. The top four companies in a content industry have, on average, raised their combined market shares by 0.14 points per year. Half of it is attributable to the largest firm.

We can depict the concentration trends and level of different industries.

Graph 38.1 shows the position and direction of concentration of media industries. The horizontal axis shows the existing concentration levels as represented by the HHI for the weighted world average of an industry. The vertical axis depicts the average annual change in that industry for the six-year period 2004/05 to 2011/12 or more recent. The graph is log-normal, since the HHI is an exponential measure.

The graph shows, first, that there is almost no media industry where concentration is both low and falling (the south-west quadrant). The exceptions seem to be film and online news. The figures for film, however, are somewhat misleading. It is true that on the country level typically no company dominates and a handful of companies form an oligopoly, each with about 10–20%. The problem is, however, that this small group is usually the same six companies everywhere, namely the Hollywood majors. When we look at the media industries globally, we find that the film industry is the second highest in worldwide concentration, with only search engines having a higher score. The top six companies hold 76.4% of the world's pooled film revenues. Hence, one cannot really consider the film industry to have a low concentration.

The other low/negative concentration growth industry is that of online news, but only barely so. It is similar for several industries where concentration is low but slightly rising—magazines and book publishing. In a larger number of industries, concentration is high but falling—wireline, wireless, broadcast TV, and ISP. These are intermediate situations where it is not clear, without additional facts and considerations, whether a long-term problem exists. Most problematic is the north-east sector: here, concentration is high and rising. This is an issue in just one industry, that of search engines.

## G. Have American media become more concentrated?

Table 38.5 shows the concentration trends in the United States.

### CONTENT MEDIA

- US content media industry concentration rose by 3.3% p.a., over the nine years 2004 to 2013, when all indices are averaged together. It grew by 6.9% per year for the HHI, a strong rise (but recall the exponential nature of the HHI). Without Internet media, the annual HHI increase was still a high 4.9%.
- The increase was driven by the audiovisual industries' growing concentrations: HHI rose in broadcast TV, video channels, and multi-channel platforms.
- The increase was also high for the content power index (6.2% p.a.), the pooled HHI (4.4% p.a.), and C4 (0.35% p.a.), indicating an increase in conglomeration.
- The combined market share of the top four content firms rose, on average per industry, by 2.7% p.a. The average trend across all 8 indices was a decline of 0.6% p.a.
- The share of the top single firm in the content market rose strongly, too, though not faster than that of firms number 2–4.
- The number of net voices grew at a pace of 0.6%.

### PLATFORM MEDIA

- Average industry HHI concentration fell by −0.2% per year. Most other indices fell as well, in particular, the indicators of cross-platform ownership (inter-industry pooled C1 (−1.9%) and the national power index (−0.6%)).
- While the top firms lost market share (C1 down by 1.4% p.a.), the aggregate for the top four rose by 1.3%.
- The information in chapter 18 (USA) shows that the platform concentration in the United States trended in as "S-shape," first dropping in the mid-1980s through regulation and entry,

Table 38-5. US Media Concentration Trends, Per Annum Changes (2004–2013)

| | ALL MEDIA | PLATFORM MEDIA | CONTENT MEDIA W/INTERNET | CONTENT W/O INTERNET | NEWS MEDIA (BY REVENUES) | NEWS MEDIA (BY ATTENTION) |
|---|---|---|---|---|---|---|
| HHI | 1.23% | – 0.24% | 6.89% | 4.94% | 4.32% | 2.43% |
| C1 | –0.37% | –1.37% | 2.41% | 1.46% | 0.57% | –0.15% |
| C4 | 1.79% | 1.26% | 2.68% | 2.67% | 2.37% | 1.36% |
| Power Index | 0.77% | –0.63% | 6.20% | 4.59% | 3.63% | 1.77% |
| Noam Index | | | 8.91% | 5.87% | 5.40% | 3.59% |
| Voices | | | 0.60% | –1.13% | –0.46% | –0.46% |
| Pooled HHI | 1.02% | –1.17% | 4.42% | 5.44% | 4.18% | 3.50% |
| Pooled C1 | –1.16% | –1.85% | 0.17% | 3.41% | 2.83% | 1.52% |
| Pooled C4 | 1.12% | 0.00% | 0.35% | 2.32% | 1.69% | 2.38% |
| Cross-ownership | | | –0.46% | –0.34% | 0.29% | 0.29% |
| Trend Average | 0.63% | –0.57% | 3.30% | 3.55% | 2.80% | 1.83% |

then rising again as incumbents reasserted themselves, merged, and as the fragmented mobile market consolidated, followed by a smaller U.

### A NEWS MEDIA

- News HHI concentration rose by 4.3% per year by revenues and 2.4% by attention. It was lower for the top firm's share. It was high for the pooled C1s and C4s as well as the national power index measures, indicating that cross-media ownership trends were strong (0.3% increase p.a.). The average of the eight indices was a rise of 1.8% (by attention) or 2.8% (by revenue).

### OVERALL US MEDIA

- Overall media concentration in the United States rose over the nine years, by an average rate of 1.2% per year for the HHI measure, and on average, for the eight indices, at 0.6%.
- The share of the top firm in an industry, on average, fell by −0.4% per year, but the combined share of the top four firms rose by 1.8% per year. (The absolute increase is discussed below).
- The MOCDI Index shows for content a high growth of 8.9% per year (5.9% without Internet) and of 5.4% for news media by revenues. The growth in voices in the United States was 0.6%, higher than it was worldwide (0.5%). But the US growth in content media HHI concentration was also higher than the international average, and as a result the MOCDI measure of pluralism rose in the United States faster than it did on average worldwide.
- The power index, as a measure for cross-industry conglomeration, rose in the United States by a hefty 6.2% for content media, much higher than the worldwide trends of 0.3% arithmetically.

We can also calculate the increases in the various shares and indices on an absolute basis (Table 38.6 in the Appendix). The C1 firm in a content industry gained in the United States, on average, market share of 0.4 points each year, and an added 1.6 points in the pooled share. The top four firms gained 1.3 points in their markets weighted, and 3.2 pooled. This was mostly due to consolidation of audiovisual media by the top domestic companies.

For platform media, the top firm lost about −0.5 of market share each year. For news media, the top four firms gained each year 0.7 points by attention and 1.1 points by revenues.

In conclusion, the concentration trend of platform media shows a decline by −0.6% per year, while content media rose by 3.3%. The absolute levels in the United States were still lower than the world average. But the difference has been getting smaller. (US weighted average C4 gained 1.8% per year, twice the world average of 0.9%).

For Europe, concentration trends were also investigated and interpreted. This is presented in chapter 33 of this volume.

## H. Does the transition to Internet-based media reduce concentration?

The conventional wisdom is that new media is less concentrated than legacy media. But that is not what the data show.

Graph 38.2 shows media concentrations for three types of media: (1) *traditional media,* such as newspapers, books, and magazines (print); (2) *twentieth-century audiovisual media,* such as film, radio, TV, cable TV; and (3) *Internet media,* such as online news, search, and ISPs. Graph 38.2 shows that each newer generation of media is more concentrated than the preceding ones. This is true in almost all of the world's regions. Traditional media on average has a national concentration index, worldwide, of about 1,100—which is almost unconcentrated. Twentieth-century audiovisual media has a concentration index of about 2,000, which is already highly concentrated (the threshold is 1,800). And Internet media has an index of over 3,000, which is very highly concentrated.

The ISP market is highly concentrated, worldwide, with the arithmetic average HHI standing at 3,616 (down from 3,829 in 2004/05) and ranked sixth out of all industries for concentration. The search engines industry

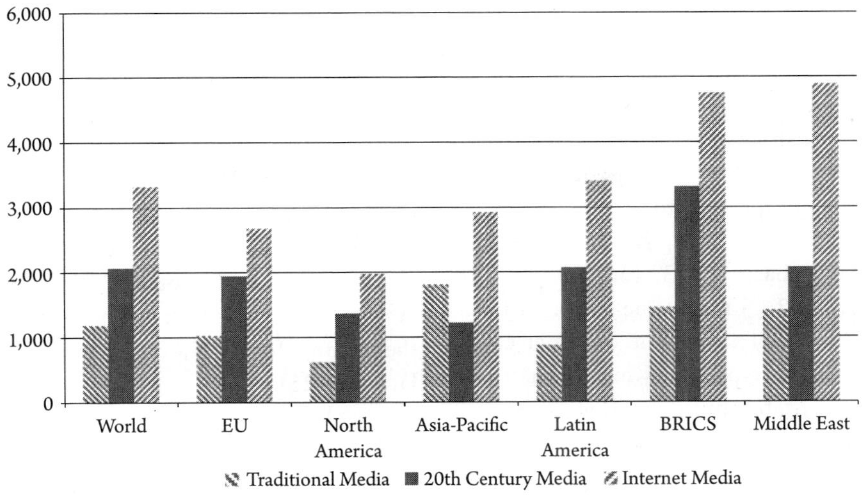


GRAPH 38.2 Old and New Media—Average Concentration by Region

is the most concentrated media industry of all. Its average HHI concentration was an extraordinarily high 7,760 in 2013, up from 6,809 in 2004/05.

In the handful of countries where Google is not the dominant search engine, the largest search engine by volume tends to be one based in that country, and they too, hold huge market shares. Baidu is the largest search engine in China, holding 78.6% of the Chinese market share. The top search engines in South Korea are those of NHN, with 68.8% of the market. In Russia, Yandex controlled 47.1% of the market, followed by Mail.ru (34.8%).

This observation about search engines raises an important question. Do the new Internet media make a difference on media industry concentration, in the way its enthusiasts believe? Internet media, after an early stage of a dynamically competitive market structure, often become highly concentrated. Various market segments have their dominant players—Amazon, EBay, Microsoft, Google, Facebook, Twitter, YouTube, Apple's App Store, and others. The Internet sector was believed to be wide open and competitive and would open things up for other industries, but it exhibits strong concentration trends. The underlying economics on the supply side are, high fixed cost and low marginal cost; and on the demand side, strong network effects. Because these factors will remain, this trend is likely to continue, especially if the pace of disruptive innovation in the sector slows down a bit.

Online news media are less concentrated than newspapers, though not in all countries. In the United States, the incumbency advantage of the major multimedia producers means that online news is more concentrated as the daily newspapers industry, though less concentrated than for broadcast TV news. Around the world, many major online news companies were successful due to their pre-existing strong brands in print news, coupled with economies of scope in extending traditional news operations to the online mode.

In many countries, online news media HHIs are actually higher than print dailies' HHIs. This is the case in the United States (569 for online news, 304 for newspapers), Italy (2,188 for online news, 1,863 for newspapers), Germany (1,521 versus 558), Japan (1,965 versus 762), Turkey (3,052 versus 2,637), India (1,996 versus 1,228), Spain (1,382 versus 1,090), South Africa (2,471 versus 2,450), and Brazil (2,916 versus 2,321). In fairness, the analysis deals with "commodity news" rather than with specialized online information sources. But such "long tail" information does not negate the fact that most online news attention by national audiences is focused on a few mass-audience outlets.

On the negative side of the ledger is that revenues for online news, on a per capita basis, are dismal in comparison to print news. In the United States, for example, the annual revenue on a per capita basis is $132 for print newspapers and only $26.5 for online news. In Japan with $191 the annual revenue on a per capita basis for print newspapers is higher than that in the United States, but for online news, with annual revenue per capita of $14.2, it is performing even lower. This also is true for Switzerland ($256 versus $13.9) and Spain ($167 versus $0.5). Similarly large ratios exist in Taiwan ($28.1 versus $2.7), the United Kingdom ($184 versus $24.4), Israel ($42.8 versus $4.7), Germany ($69.1 versus $4.8), Chile ($48.4 versus $2.7), South Africa ($26.5 versus $2.7) and Canada ($151 versus $2.9). Globally, for the reporting countries, it is $76.2 for newspapers versus $8.2 for online news. Only in China ($0.68 for online news, $0.48 for newspapers) and in Russia ($6.54 for online news and $3.58 for newspapers) are the ratios reversed. Given such low revenues, the likely market scenario is one of the larger providers surviving due to scale advantages and network effects, and because they could charge the users more easily for content, and the advertisers more for placement.

One must therefore conclude that the Internet does not overcome the problems of media concentration. To the contrary, it may accelerate it.

## NATIONAL MEDIA CONCENTRATIONS

### I. What countries have particularly high media concentrations?

As mentioned, China, Egypt, and South Africa have particularly high average overall concentration, due to considerable state ownership of media (Table 38.7). In 2013, the five countries with the highest average industry HHIs were China (9,700), South Africa (5,535), Mexico (5,315), Turkey (5,150), and Egypt (5,041). The lowest concentrations measured were for the United States (1,517), Canada (2,506), France (2,694), the Netherlands (2,754), and Brazil (2,773).

Looking at content media only, China has the highest average concentration at 8,224. Also high are South Africa (3,945), Mexico (3,778), Ireland (3,736), and South Korea (3,726). The lowest average content industry concentrations are in the United States (1,080), Spain (1,606), and Japan (1,725).

For platform media, the list of the highest concentration countries is similar: China (10,000), South Africa (5,910), Mexico (5,603), Turkey (5,600), and Egypt (5,328). The lowest concentrations are in the United States (1,742), the Netherlands (2,628), and Canada (2,729). The United States is the only country whose average platform concentration is below the antitrust standard of "highly concentrated," but barely so.

Graph 38.3 shows HHI concentration and the trend for platform media.

One can readily see how concentrated the platform sub-sector is—its industries are typically in the HHI range of 3,000–6,000. A second observation is that in almost all cases, this concentration is declining, often quite substantially. That reduction is low for the United States, but that country also has the lowest concentration as a base. Concentration is both high and rising for the United Kingdom, Canada, and Switzerland. Only the United States is in the southwest quadrant.

A similar presentation is made in Graph 38.4 for content media. The HHI concentrations for content media as well as its trends are shown.

- Only one country—Taiwan—is in the southwest quadrant, where concentration is low and falling.
- Under the previous definition of high concentration (HHI = 1,800), only one country (the United States) was in the low concentration sector, though with a substantial rise. Under the substantially more lenient recent definition (HHI = 2,500), 9 more countries are below that threshold.

Table 38-7. Weighted Country HHIs, 2011 or Most Recent

| | ALL MEDIA | PLATFORM MEDIA | CONTENT MEDIA | CONTENT MEDIA W/INTERNET | NEWS MEDIA (BY ATTENTION) | NEWS MEDIA (BY REVENUE) |
|---|---|---|---|---|---|---|
| Argentina | 3,151 | 3,840 | 1,914 | 1,540 | 1,468 | 1,541 |
| Australia | 4,238 | 4,796 | 3,150 | 2,827 | 2,877 | 3,110 |
| Belgium | 3,351 | 3,815 | 2,098 | 2,098 | 1,833 | 1,685 |
| Brazil | 2,773 | 2,846 | 2,453 | 2,255 | 2,264 | 2,359 |
| Canada | 2,506 | 2,729 | 2,046 | 1,826 | 1,742 | 1,849 |
| Chile | 3,607 | 3,781 | 3,318 | 3,169 | 2,907 | 3,175 |
| China | 9,700 | 10,000 | 8,224 | 8,585 | 9,181 | 7,661 |
| Egypt | 5,041 | 5,328 | 3,510 | 3,291 | 7,310 | 4,199 |
| Finland | 2,793 | 2,949 | 2,576 | 2,259 | 2,495 | 2,301 |
| France | 2,694 | 2,868 | 2,373 | 2,213 | 2,688 | 2,459 |
| Germany | 3,001 | 3,468 | 2,257 | 1,914 | 2,164 | 2,445 |
| India | 2,935 | 3,001 | 2,673 | 2,394 | 4,677 | 2,492 |
| Ireland | 3,956 | 4,084 | 3,736 | 3,520 | 3,934 | 3,601 |
| Israel | 3,517 | 3,759 | 2,842 | 2,672 | 2,732 | 2,765 |
| Italy | 3,667 | 3,957 | 3,129 | 3,200 | 3,340 | 3,878 |
| Japan | 3,328 | 3,924 | 1,725 | 1,534 | 1,471 | 1,540 |
| Mexico | 5,315 | 5,603 | 3,778 | 3,750 | 3,918 | 4,266 |
| Netherlands | 2,754 | 2,628 | 3,044 | 2,455 | 2,658 | 2,488 |
| Poland | 3,877 | 4,106 | 2,416 | 1,538 | 1,955 | 1,591 |
| Portugal | 3,550 | 3,584 | 3,504 | 3,226 | 3,075 | 3,276 |
| Russia | 4,091 | 4,240 | 3,497 | 3,771 | 3,852 | 3,853 |
| South Africa | 5,535 | 5,910 | 3,945 | 3,764 | 4,137 | 3,724 |

*continued*

Table 38-7. *continued*

| | ALL MEDIA | PLATFORM MEDIA | CONTENT MEDIA | CONTENT MEDIA W/INTERNET | NEWS MEDIA (BY ATTENTION) | NEWS MEDIA (BY REVENUE) |
|---|---|---|---|---|---|---|
| South Korea | 4,036 | 4,110 | 3,726 | 3,475 | 2,810 | 3,789 |
| Spain | 2,867 | 3,679 | 1,606 | 1,257 | 1,478 | 1,409 |
| Sweden | 2,881 | 3,246 | 2,459 | 2,233 | 2,629 | 2,442 |
| Switzerland | 3,807 | 4,341 | 2,677 | 2,453 | 2,264 | 2,489 |
| Taiwan | 3,375 | 3,802 | 1,830 | 1,807 | 2,408 | 2,131 |
| Turkey | 5,150 | 5,600 | 2,907 | 2,420 | 2,443 | 2,458 |
| UK | 2,871 | 3,043 | 2,674 | 2,474 | 2,632 | 2,730 |
| US | 1,517 | 1,742 | 1,080 | 830 | 828 | 839 |
| Arithmetic Average | 3,729 | 4,026 | 2,906 | 2,692 | 3,006 | 2,818 |
| Country Weighted Average | 3,254 | 3,711 | 2,219 | 1,999 | 5,194 | 1,987 |

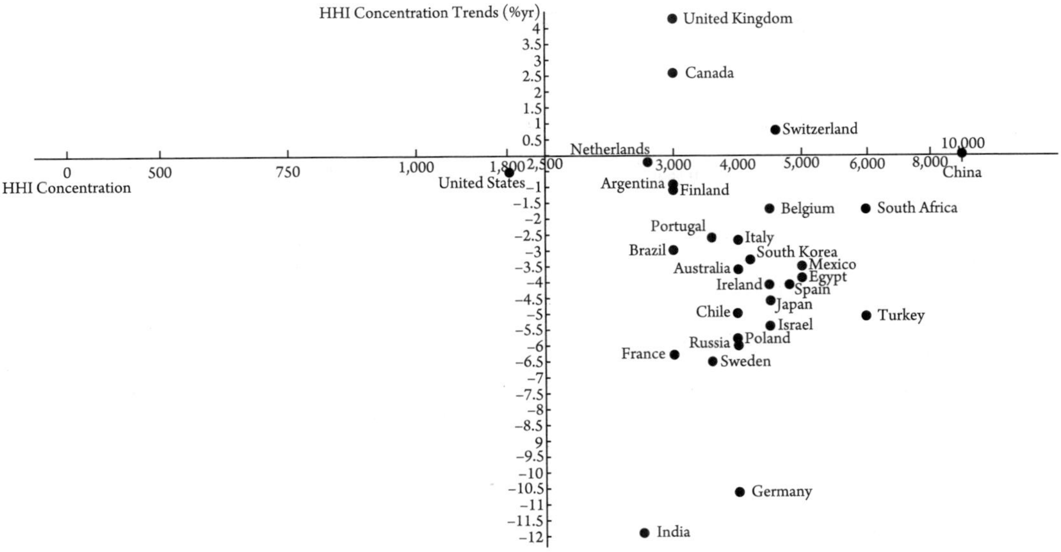


GRAPH 38.3 Platform Media HHI Concentration and its Trends

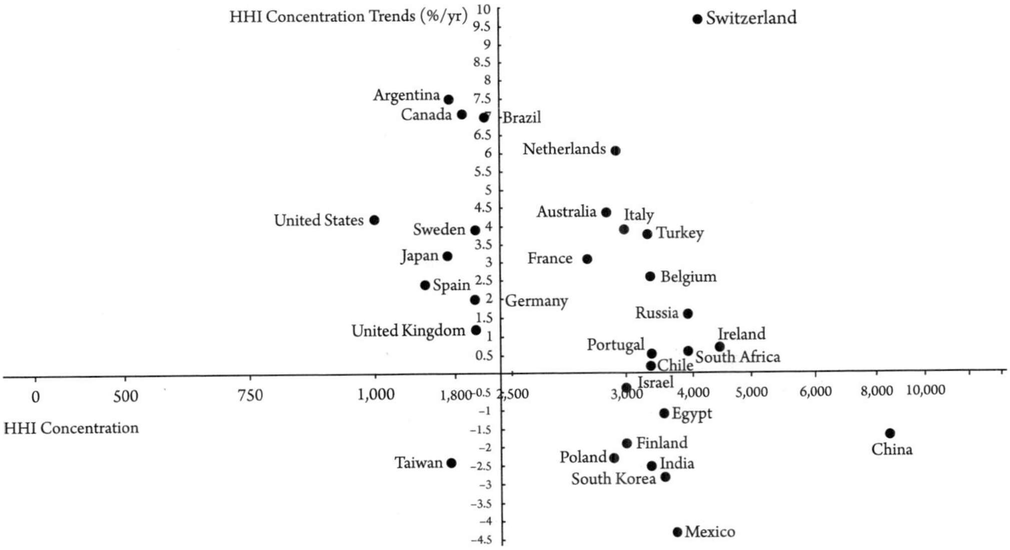


GRAPH 38.4 Content Media HHI Concentration and its Trends

- A number of countries are highly concentrated, but the trend is one of a reduction. These are China, Mexico, India, South Korea, Poland, Finland, and Egypt.
- Several countries are both highly concentrated (or nearly so) and rising. These problematic cases are Switzerland, the Netherlands, Australia, Italy, Turkey, France, and Russia. South Africa and Ireland are highly concentrated to the point that even with a low growth trend in their HHI there is a problem.

The national media power index, which incorporates cross-ownership media power by the same company or organization, is even relatively

Table 38-8. National Power Indices, 2011 or Most Recent

| | ALL MEDIA | PLATFORM MEDIA | CONTENT MEDIA | CONTENT MEDIA W/O INTERNET | NEWS MEDIA (BY ATTENTION) | NEWS MEDIA (BY REVENUE) |
|---|---|---|---|---|---|---|
| Argentina | 3,366 | 4,064 | 2,114 | 1,748 | 738 | 1,772 |
| Australia | 3,984 | 4,524 | 2,930 | 2,594 | 1,628 | 2,833 |
| Belgium | 3,355 | 4,004 | 2,113 | 2,113 | 835 | 1,701 |
| Brazil | 2,868 | 2,896 | 2,746 | 2,583 | 1,055 | 2,698 |
| Canada | 3,251 | 3,828 | 2,215 | 1,983 | 616 | 2,026 |
| Chile | 3,319 | 3,449 | 3,113 | 2,906 | 1,287 | 3,003 |
| China | 9,702 | 10,000 | 8,234 | 8,596 | 8,977 | 7,685 |
| Egypt | 4,774 | 5,015 | 3,488 | 3,269 | 5,092 | 4,170 |
| Finland | 2,861 | 3,095 | 2,533 | 2,215 | 1,069 | 2,257 |
| France | 2,524 | 2,888 | 1,852 | 1,673 | 1,228 | 1,926 |
| Germany | 2,603 | 3,098 | 1,814 | 1,444 | 697 | 1,818 |
| India | 2,922 | 3,020 | 2,673 | 2,402 | 4,145 | 2,496 |
| Ireland | 3,650 | 3,656 | 3,642 | 3,430 | 2,024 | 3,511 |
| Israel | 3,591 | 3,875 | 2,800 | 2,627 | 1,253 | 2,729 |
| Italy | 3,654 | 3,953 | 3,101 | 3,180 | 1,992 | 3,857 |
| Japan | 3,274 | 3,922 | 1,531 | 1,535 | 539 | 1,491 |
| Mexico | 5,298 | 5,407 | 3,699 | 3,672 | 2,017 | 4,171 |
| Netherlands | 2,726 | 2,676 | 2,842 | 2,232 | 1,134 | 2,234 |
| Poland | 3,869 | 4,010 | 2,965 | 2,154 | 868 | 2,271 |
| Portugal | 3,468 | 3,531 | 3,382 | 3,097 | 1,551 | 3,010 |
| Russia | 4,092 | 4,201 | 3,654 | 4,023 | 1,674 | 3,959 |
| South Africa | 5,536 | 5,914 | 3,932 | 3,759 | 2,650 | 3,710 |

| | | | | | | |
|---|---|---|---|---|---|---|
| South Korea | 3,997 | 4,074 | 3,677 | 3,417 | 1,235 | 3,764 |
| Spain | 2,844 | 3,675 | 1,564 | 1,212 | 599 | 1,383 |
| Sweden | 2,887 | 3,284 | 2,427 | 2,200 | 1,225 | 2,324 |
| Switzerland | 3,690 | 4,180 | 2,652 | 2,427 | 1,040 | 2,474 |
| Taiwan | 3,408 | 3,856 | 1,789 | 1,756 | 1,350 | 2,122 |
| Turkey | 5,143 | 5,599 | 3,033 | 2,400 | 849 | 2,613 |
| UK | 2,888 | 3,043 | 2,711 | 2,515 | 1,226 | 2,741 |
| US | 1,482 | 1,711 | 1,035 | 806 | 273 | 809 |
| Arithmetic Average | 3,701 | 4,015 | 2,875 | 2,666 | 1,696 | 2,785 |
| Arithmetic Average | 3,223 | 3,707 | 2,135 | 1,934 | 4,450 | 1,912 |

higher for China[18] (9,702), South Africa (5,536), Mexico (5,298), Turkey (5,143), Egypt (4,774), and South Korea (4,092) (Table 38.8). The countries with the lowest national media power index are Spain (2,844), the Netherlands (2,726), Germany (2,603), France (2,524), and the United States (1,482).

Of course, when comparing countries one needs to take account of various factors such as population, geographic size, income, education, and so on. When we account for those factors (see the preceding chapter of "Analysis of Media Concentration"), we can identify concentrations that are higher (and lower) than would be predicted by the country variables.

The findings show that China and India are the two extremes, with China the highest and India the lowest in divergence of actual concentration from the one predicted by socio-demographic and other factors. Furthermore, they have moved in opposite directions, with China's media growing somewhat more concentrated than predicted while India has been moving the other way, to the lowest concentration relative to prediction. India is less concentrated than the United States, factoring in its poverty, low average education, population size, and so on.

Looking at news media only, the countries with the highest such concentration, whether by revenue or attention time, are China, Russia, India, South Africa, Italy, Egypt, Mexico, Russia, and Ireland. The least concentrated are the United States, Spain, Argentina, Japan, and Canada. It is rising in the United States and Canada, but falling in Japan, Argentina, and Spain.

The Government of China, through its several media organizations, accounts for a truly vast share of the global news attention. In the aggregate, it has 29.7% of global news attention in 2013 for the 30-country world and about 19% of the entire world. Even if we unbundle China's news organizations, the state TV broadcaster CCTV alone would still command 11.1% (or 7.1%) of the world's news attention and be the second largest news media company in the world. The explanations for these high shares are simple: a huge population (1.3 billion), relatively good access to media (print, audiovisual, and Internet) compared to other emerging countries like India, and state control over most news media outside some online portals and print magazines. This share by the Government of China over the world's news attention, through its domestic dominance, is truly extraordinarily high.

## J. Where is there a high pluralism of voices, and why?

Most countries had a "net" voice count between 30 and 60. But the absolute number of net voices must also be seen in relation to a country's market size. The United States had the highest number of net voices, with 59 voices. (40 of its 99 overall voices are cross-owned).

But on a per capita basis the United States actually ranks quite low, due to its population size. The European countries' average of voices per capita per million people (3.1) is over twice as high as the North American (0.8). One explanation is that countries with a high count of voices per capita tend to be relatively small in population but with active politics and culture. Europe has many such countries, for example, Finland, Sweden, Ireland, and Portugal. The same factors hold true for Israel and Australia. Another explanation for

18. To recapitulate: The figures for China represent the primary measure we use for common control state media firms, that is, of "integrated" concentration, whether run by different ministries or levels of a government. We present a second measure for China as well, that of a "segmented" concentration measure, which treats each state enterprise as an independent entity. For platforms, the "segmented" average HHI for China is still a high 4,276 because all telecom companies are state-owned. The same integrated approach is also taken for state holdings in Egypt, Taiwan, and Russia. It is also used for the different media properties of the same company or proprietor, across industries, countries, and corporate entities if they share controlling owners, as in the case of Murdoch (21st Century Fox and News Corp.), Redstone (CBS and Viacom), and Malone (Liberty).

the often higher ratios of smaller countries is more methodological in nature: the definition of voices is relative to the size of the national market (1% of one of a country's media industries), and hence this threshold becomes higher for large countries. This is discussed in the chapter on "National Media Concentrations Compared."

On the other extreme are countries that are large, as well as poor and sometimes nondemocratic. The reported number of voices per capita is the smallest—in India, Brazil, and Mexico—and is amazingly low in China.

Ireland, Finland and Switzerland have the highest number of voices per capita (with the definition of voices relative to market size), followed by Belgium, Portugal, and Israel. Complicating this comparison is that many media voices in Ireland, Portugal, Belgium, and Switzerland are spillovers from the adjoining countries (the United Kingdom for Ireland; Spain for Portugal; France and the Netherlands for Belgium; and Germany, France, Austria, and Italy for Switzerland), rather than domestic in origin. If the spillover media from neighboring countries were counted for Israel too, it would have the highest number of voices per capita.

The absolute number of voices is generally higher in large countries. A larger population is likely to support a larger number of media. To estimate the diversity of the media supply side in a country, a good measure is therefore the number of media voices (net) per capita. Where such a value is high, it means that the country produces pluralism, given its size. The number of voices rises with population, though at a much slower pace. A doubling of population raises the number of voices by about 6.4%, or about 3–4 voices. This is not a very steep increase. However, as mentioned, the definition of voices is relative (1% of a national media industry). If we hold that definition constant in terms of size (and exclude China as an anomaly), we find that for each doubling of population, the number of voices rises by about 7.5.

Even with a static definition of a voice, smaller countries have more voices per capita. These smaller countries are able to support voices at a scale that does not seem sustainable in larger countries. It must therefore be not absolute size but relative size with its negative impact on competitiveness—factors such as economies of scale and network effects—that make it harder for smaller voices to survive in large countries. In that sense, the larger countries under-perform in terms of sustaining voices. And this suggests that in an increasingly global media system, the number of voices *per capita* will decline, and the aggregate number of voices will decline. On the positive side, users will have more options in a global media system than they had in a national market. But on average, the national production of media voices will decline as media markets become more global.

## K. What are the factors for high national media concentration?

One must distinguish between differences in concentration among the various media industries and the differences in concentration among countries within a given industry.

How can the market concentration of different industries be explained? We show further below that this concentration is correlated with the capital intensity of a media industry. The more capital-intense an industry, the higher the fixed costs, the lower the variable cost, and hence the higher the economies of scale are. This, in turn, favors large firms and thus industry concentration.

But what factors explain different national industry concentrations within the *same* industry? The following factors were found to be statistically correlated with concentration:

- The size of population was a fairly decent-sized negative (i.e., reducing) factor for concentration in the newspaper and multi-channel TV industries.
- Geographic size of a country was a moderately sized factor for newspapers (higher concentration) and ISPs (lower concentration).
- Per capita income in a country was a reducing factor for the concentration in mobile and wireless industries and a raising one for multichannel platforms.

- Educational levels were a factor for newspapers and for multi-channel platforms (both associated with lower concentration).
- The "quality of regulation" and a variety of other government performance metrics showed no statistically significant correlation for any of the industries. This is a disappointing result from a good-government perspective. There is some correlation for the outliers on the high end, but for most countries at the center of the distribution, no correlation could be observed. This should be subject for further research.

## L. What countries have particularly high cross-media ownership?

On average, about one-quarter (26.9%) of those content media entities that account for over 1% in a country's media markets (a "voice") share common owners (i.e., are cross-owned). The countries with the highest voice-count cross-ownership percentages are China (73.9%), Sweden (44.4%), Brazil (40.9%), the United States (40.4%), and Canada (35.9%). Those with the lowest are Mexico (9.1%), Argentina (7.7%), and Egypt (8.6%). Thus, voice cross-ownership is greater in highly developed media markets and less in poorer countries. This suggests that the presence of cross-media firms in multiple media industries rises with economic and media development. In consequence, as countries develop economically and their media grow in technological and business complexity, one should expect further trends to cross-ownership. Cross-ownership is also high in countries with a strong government presence, typically less developed countries. Cross-ownership thus trends with economic development in a U-shape—high at both the lower and the higher ends of economic development.

## M. What countries have a high foreign ownership of media?

Foreign ownership of media has been increasing in most countries, rising from a world average of 24.5% in 2005 to 28.2% in 2013 for all media industries. The growth has been driven primarily by internationalization in telecom, especially in mobile, and in the non-broadcast audiovisual media, enabled by lowering of laws favoring domestic ownership.

The countries with the highest percentage of foreign-ownership of media are Ireland (72.1%), Brazil (65.8%), Argentina (62.8%), Chile (62.4%), the United Kingdom (52.1%), and Egypt (50.3%).

The countries with the lowest foreign ownership are mostly countries with major regulatory restrictions: Canada (2.5%), Japan (1.8%), South Korea (0.9%), and China (0.5%). The United States has a foreign ownership percentage of 8.6%, up from 4%, primarily due to the acquisitions of Sprint by Softbank (Japan) and T-Mobile by Deutsche Telekom.

For content media, foreign ownership is highest in the Netherlands (55.2%) and Chile (41.6%), due to the heavy presence of German and French firms in the former, and of Spanish and Mexican firms in the latter. It is also high in Ireland (39.4%) due to the presence of British firms. Media firms in these countries are also active abroad (Reed-Elsevier of the Netherlands and Independent News and Media of Ireland). Foreign ownership of content media is lowest in the United States (3.1%), South Korea (4.9%), China (3%), and Japan (5.8%).

The world average of content foreign ownership, including film imports, is 21%, increased from 18.3% in 2004/05. In platform media, however, foreign ownership is much higher: averaging 31.5% worldwide, up from 26.6% in 2004/05. Regionally, it is highest in Latin America (64%) due to the dominant presence of companies such as America Móvil (Mexico) and Telefónica (Spain). These are high numbers indeed. It is lowest in North America (5.7%), partly due to historical restrictions on foreign ownership of infrastructure, and in the Asia-Pacific region (8.3%).

## N. What countries are high exporters of media?

For the top 50 media companies, the average share of exports (defined as non-domestic revenues) as part of their overall turnover was 38.6% in 2013. The company with the greatest share of revenue outside of its home country was Vodafone (89% of revenue outside of the United Kingdom). Other major exporters are Pearson (United Kingdom, 86.7%), Google (United States, 79.4%), America Móvil (Mexico, 77.1%), and Liberty (United States, 72%). Two South African companies, the mobile telecom firm MTN (70.4%) and Naspers (70.4%) also derive much of their revenue from abroad.

Is there a common denominator for these export-intensive companies? Several are based in rich and large countries where media firms expand globally from positions of success. Others originate in less developed countries, with companies with a secure national base seeking expansion by moving beyond their borders. If one industry stands out in export activity it is mobile telecom, with 4 of the top 10 companies with the highest export intensity.

The countries whose companies had the largest share of media exports in the world were the United States with nearly one-quarter of world media exports (23.9%), distantly followed by the United Kingdom (14.8%), Spain (9.2%), France (9.5%), Mexico (8%), and Germany (7.5%).

However, the US world export share is much smaller than its share in the overall world media market, which is 35.3%. That means that the domestic role of US media is much higher than the world average, but its export activity, while large in absolute terms, is below average in comparison to GDP, and relatively low in comparison to domestic media activity.

Other export "under-performers" are Japan, South Korea, China (1.7% vs. 6.7%), Australia (counting Rupert Murdoch as a US national), Brazil, Turkey, India, Taiwan, and Canada.

In contrast, export "over-achievers" are the United Kingdom (14.8% vs. 5.7%), Mexico (8.0% vs. 1.1%, mostly due to Carlos Slim's America Móvil), Spain, France, Sweden, and South Africa.

## O. Do American media companies dominate the world media? Does Hollywood dominate content?

The US share in the 30-country world in various media industries is very high (Table 38.9). For all media, the US share is 33%, while its population share is 7.4% and its GDP share is 27.8%. For several media, the US share is large but roughly corresponding to GDP share (Newspapers, 30%; broadcast TV, 23.4%; wireline, 35.6%; wireless, 21.6%; radio, 33.6%; and ISPs, 28%). In several media, however, the US share is much larger: online news, 42.3%; video channels, 40.4%; multi-channel platforms, 64%; film, 76.4% (with exports included; it is 30.9% without exports); and search engines, 74% (again, with exports included—and 35.8% without).

But revenue is only one way to measure top firms. A large and high-income market such as the United States will result in its firms being large, even if their American market shares are intermediate rather than dominant. This is the reason for the usefulness of the power index, which adds the dimension of market power as well as a cross-media and cross-national calculation. Using it, the position of the United States declines in most of the 13 industries, except for film and search engines.

The reduced position of US firms with the use of the power index is most noticeable for print media. In newspapers, Murdoch, whose newspapers are rooted in Australia and the UK, ranks first for both power index and revenue (power index = 249 and $8.4 billion), but the other top firms are non-American: Amaury of France (36.8), Independent News & Media of Ireland (62.1), the Government of China (46.4), the Yomiuri (68.2) and Asahi Shimbun (42.2) of Japan, the UK's Daily Mail Trust (40.2), Canada's Postmedia (32.6), Italy's RCS Media Group (40.4), Bonnier in Sweden (52.9), and Fairfax in Australia (32.8), all rank ahead of any other American companies (Gannet is the next largest US firm, with a power index of 27.7).

For book publishing, US firms rank relatively low. The Chinese Government (1,590), the German-Anglo venture Bertelsmann-Pearson (168), France's Lagardère (108), and

Table 38-9. US Presence in Global Media

| INDUSTRY | US PRESENCE IN WORLD |
|---|---|
| Newspapers | Largest firm: News Corp. |
| Book Publishing | 3 of the top 10 firms: News Corp. (Harper Collins, #4), McGraw-Hill (#5), Redstone (Simon & Schuster, #9) |
| Magazines | 3 of the largest firms: Time Inc., Advance Communications, and Hearst |
| Radio | Top 2nd, 3rd, and 4th largest radio firms by revenue: Liberty (Sirius XM), Clear Channel, and CBS |
| Multichannel Platforms | Top 5 firms: 21st Century Fox, Comcast, Time Warner Cable, DirecTV, Dish, and Liberty |
| Broadcast TV | 5 of the 10 largest broadcasters by revenue: CBS, 21st Century Fox, ABC, NBC |
| Cable Channels | 5 top firms: Time Warner, Disney, Redstone Group (Viacom), 21st Century Fox, Discovery, and Comcast |
| Film | 5 of the 6 largest film companies: Universal, Disney, Paramount, Warner Bros, 21st Century Fox[a] |
| Wireline | 2 top firms by revenue: AT&T and Verizon |
| Mobile Telecom | 2 of the top 5 firms: AT&T and Verizon. |
| ISP | 4 of the top 7 firms, by revenues: AT&T, Verizon, Comcast, and Time Warner Cable |
| Search Engines | Top 3: Google, Yahoo, and Microsoft |
| Online News | Top 4[b] by revenue: Yahoo, AOL, Comcast, and News Corp. |

[a] The 6th (Sony) is Japanese-owned, but US-based.
[b] Data is incomplete.

Italy's Fininvest (52.1) all score by power index above McGraw Hill (22.3) and Murdoch's HarperCollins (24). For magazines, the American majors (Hearst, Time Inc., and Advance) are about even with non-US majors (Lagardère, Bauer, Nine Entertainment) by power index.

In radio, Germany's public ARD[19] (240), the Chinese Government (223), and BBC (172) all rank in power index ahead of Sirius XM (79.7) and Clear Channel (76.7). In broadcast TV, the Chinese government leads with 607 (CCTV alone has a power index of 212), followed by the BBC (225), Globo (Brazil, 155), RAI (Italy, 127), Fininvest (Italy, 116), and the Governments of South Korea (86.9) and Russia (112). The largest US firms here are Comcast with 72.5 and Disney with 60.7. In multi-channel platforms, though, the largest firms by power index and revenue are American: 21st Century Fox/Sky, Liberty, DirecTV, Time Warner Cable, the Dish Network, and Comcast. The situation is similar for video channels (Time Warner, Disney, 21st Century Fox, Discovery, and Viacom)—though France's Vivendi has the largest company power index (247) and Time Warner is number 3 by power index (222).

In wireline telecom, Japan's NTT has the largest power index (567) due to its strong market share in Japan, ranking it above the regional provider AT&T in the United States (254): AT&T is the wireline leader in the share of world revenue, however (9.5%). Although Verizon has a large power index (123) and high revenues (6.6%), its

19. Aggregating the regional stations and networks.

PI is less than that of the Chinese state (448), Deutsche Telekom (278), Telefónica (245), Oi Telemar (165), and Telecom Italia (135). In mobile telephony, the Chinese state leads in terms of power index (1,709), and Verizon (244) and AT&T (234) are the second and third largest by PI, respectively.

The combined global news share of the top five United States companies (technically, seven companies) is 4.6% for the 30 countries, and 3% for all nations. The top ten American firms have 7.2% for the 30-country world news market (4.6% globally). Rupert Murdoch's two companies combined are the third largest privately owned news provider in the world by attention (after BCCL and Globo) and the largest US-headquartered news firm (1.35%). Of other US companies, Comcast, Disney, the Redstone companies of CBS and Viacom, and Time Warner are in the top 20 in news attention.

What this means is that the US share of global news attention by its top 10 (i.e., 12) firms is roughly equal to its population share of about 7%. (These top firms hold 55.8% of US-based news share, leaving almost half the US news attention to other firms). There is relatively little US ownership of news providers and distributors in other countries. They are primarily multichannel platforms (Liberty and Murdoch's Sky in Europe, DirecTV and Dish in Latin America), some magazines (Hearst, Advance, Time Inc.), and video channels (Time Warner, Disney, and Murdoch).

How does the news attention share of European companies compare? The aggregate shares of EU-based firms add up to 6.2% for the 30-country world and 4% globally. This is a lower bound figure, since our study covers only 11 of the 27 EU member countries, though they are mostly the larger ones in terms of population or GDP. The EU share in news attention is thus comparable in size to that of the United States. However, the European countries' numbers count exports to other EU countries. If we subtract those, US exports are larger in comparison.

The next question is that of Hollywood's dominance. In pure revenue terms, the box office revenues of the six Hollywood "majors" add up to just 2.1% of the combined content media industries' revenues. With generous assumptions on non-box-office revenue streams, Hollywood share of global content revenues would rise to 7.8% for the six firms. Thus, the claim that Hollywood dominates the world's content creation and distribution are exaggerated. It is accurate when applied narrowly to the film industry but not when this observation is widened to media content more generally.

National film market concentration is actually highest in those countries where there is a state-supported domestic industry, which favors the emergence of a very large domestic firm receiving protectionist policy support, as is the case in China, Russia, South Korea, and France.[20]

## P. Where is there a strong role of public (state) ownership in media concentration?

Public (i.e., state and state-corporate) ownership has been decreasing, combined for all media, from 18.6% in 2004 to 15.3% in 2013. Public ownership for content fell from 17.9% to 15.5%. For platforms, it fell from 17.7% to 13.7%. It is highest in countries where public incumbents, usually in platform industries, control the national infrastructure. China is the largest market, with public ownership of platform media at 100% (unchanged since 2004/05), and content at 72.9% (down from 90.4% in 2004/05). Public ownership for content media is also high in Egypt (35.5%), Russia (36.4%), and South Korea (31.6%). It has been declining in Egypt and South Korea, and slightly rising in Russia. It should be noted that many countries that are usually private-enterprise oriented have significant residual public holdings in their telecom/mobile incumbent (Germany, 38% and Japan, 33.7%). It is 37% in Sweden, and 27% in France. Some of the media are companies held, in some form

20. In France, this is accomplished through the favored licensing of Vivendi's Canal Plus, whose film production arm, Studio Canal, finances over 60% of French-made films.

of arm's length relationship, e.g., the public service broadcasters, which can hold a considerable share of radio and TV.

In radio, even after liberalization in the 1980s, the national market shares of public service radio broadcasters remained quite high (54.5% for the BBC in the United Kingdom, 36.5% for the ARD regional stations in Germany and 13.5% for ZDF, 48.2% for RAI in Italy, 30.6% for India's national public broadcaster, and 31.6% for SABC in South Africa. This is remarkable insofar as these shares are not based on a formal monopoly. Public media prevails in the face of private competition. Hence, market concentration is still fairly high in these countries, but lower than in the past.[21]

The countries with the lowest percentage of public-owned market share are Argentina, Brazil, Mexico, and the United States, all of which had 1% or less.

The high share of government-controlled media organizations gets still higher when we venture beyond the 30 countries of this study and look at other countries. The 11 largest countries by population that have not been included in the 30-country world (with its 64% share of the world's population), account for another 18% of the world's population. Of these, five countries have a strong state dominance over both audiovisual (TV, radio) and print media—Vietnam, Ethiopia, Iran, the Democratic Republic of Congo, and Burma/Myanmar. These countries, each with populations above 50 million, account for 5.3% of the world's population and populations. Three countries have a substantial governmental dominance over audiovisual but not over print: Thailand, Bangladesh, and Nigeria. They account for 5.5% of the world's population. Three countries have no state dominance over audiovisual and print: Indonesia, Pakistan, and the Philippines. They account for 7.3% of the world's population.

This matches others' findings. The World Bank study mentioned earlier (Shleifer et al. 2001; Djankov 2003) reviewed 97 countries and identified a state ownership, on average, of 29% of major newspapers, 60% of TV stations, and 72% of the top radio stations. Twenty-one percent of the countries had a government monopoly on newspapers and 44% on TV stations. In Africa, governments control 61% of the top five newspapers (by circulation) and 84% of TV audiences, with 71% of countries having a state monopoly. In the Middle East, and North Africa, in 2001 all countries but one (Israel) had a state monopoly over TV broadcasting and held 90% of newspapers by circulation. In Singapore, the dominant Lee family controls the Singapore Press Holdings, which publishes all five top newspapers. The family owns 47.23% through four companies, and possibly more indirectly.

## Q. Are there different market characteristics for media in the countries of the North versus those of the South?

We found that concentration in news media is associated with less economic development, not with more (and with a strong state role, as discussed in the preceding section). In many emerging markets, individual news media owners have an amazingly high share in news attention: the Marinho family (Globo, Brazil, 35.5%), the Azcárraga family (Televisa, Mexico, 44.4%), the Dogan family (DMG, Turkey, 37.5%), Alexander Rodnyansky (CTC Media, Russia, 15.6%), and the de Noble family (Grupo Clarín, Argentina, 26.9%). It is also very high in Italy, where Silvio Berlusconi's firms hold 22.4% of news attention, and in Sweden, where the Bonnier family holds 25%. The Murdoch Group controls mid-sized shares across the United States, United Kingdom, Ireland, and Australia.

Similarly, in the Philippines, large media owners are the Lopez family (ABS-CBN),

21. In the United States, the opposite trend occurred; the deregulation of ownership rules led to a consolidation of a previously atomistic industry. But this, in turn, was challenged by new platforms: satellite radio, cable-based music channels, and online music providers.

which also owns the telephone company Bayan Telecommunications, as well as the Gozon, the Duavit, and the Jiminez families (GMA); and Manuel Pangilian (TV5 and the PLDT, the largest wireline and mobile telecom company).

In Indonesia, the leading media-owning families are the Hary Tanoesoedibjo (MNK Group), Tohir (Mahaka Media), Oetama Adierasetya (Kompas Gramedia), Narada (Media Bali Post). MNC's owner Hary was a candidate for Vice President in the country's 2014 election.

Around the world, the organizations with the greatest news attention are not those with the highest revenues. The media organizations in BRICS countries, especially audiovisual media in China and India, have a vastly greater hold on people's attention than on revenues. These are the media organizations of the emerging world—the "*BRICS media*."

In contrast, the leading media organizations of the developed world—one might call them "*OECD media*"—consist of a combination of traditional media conglomerates and successful startups. Their hold on national attention, large as it is, is dwarfed by the BRICS media.

But when it comes to revenues, such ranking is quite the other way around. Now, the Murdoch Group, Google, Comcast, Disney, Bertelsmann, the Redstone Group, and Time Warner outpace Chinese and Indian firms. Vivendi and Fininvest are nearly as large. Aside from the Chinese and Indian entities, only one BRICS media organization is by revenues among the top 30 media groups: Globo at number 12.

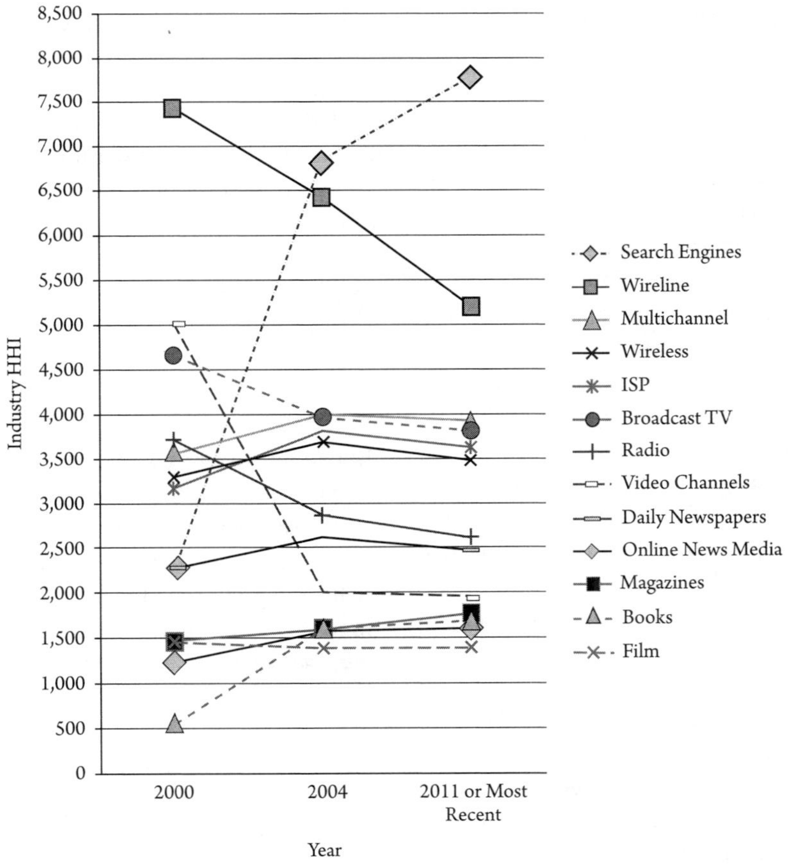


GRAPH 38.5 Convergence of 13 Media Industries

## R. Are there trends of convergence among countries in media concentration?

A convergence of concentration can be observed across geographic regions. Regions with lower concentration in content media have been rising in concentration, and regions with higher concentration showed a decline. North America and Europe were low but rising and Mideast, Asia-Pacific, and Latin America high but falling. A measure of the divergence—the standard deviation—of concentrations among the regions for content media dropped from 1,210 in 2000 to 1,033 in 2004 and to 764 in 2013, and for platform media from 1,660 in 2000 to 1,413 in 2004 to 966 in 2013. For news media, the standard deviation across regions fell from 835 in 2000 to 713 in 2004 to 698 in 2013.

# INDUSTRIES

## S. What are media industries with particularly high—and low—concentrations? What are the explanations?

We find that differences in the concentration levels of different media industries are correlated with the capital intensity of a media industry. The more capital-intensive an industry is, the relatively higher are fixed costs and the relatively lower are variable cost, and hence economies of scale. This, in turn, favors large firms and thus industry concentration.

The consequence is that as media becomes more capital-intensive in the future, its industry concentration will rise.

We can analyze the concentrations of the global average HHI of the 13 industries (Graph 38.5). Due to the worldwide market power of Google, the search engine industry is the most concentrated (7,760). It also experiences the highest increase in concentration (1.7% per annum during the observed period). In comparison, the worldwide average HHI of the telecom wireline industry, characterized by former or current public telecom monopolies, is the second most concentrated, but has been decreasing by –2.4% per annum. At the lower end of concentration are the print media industries. As discussed earlier, the film industry has the lowest average national HHI (1,403) of all media industries, due to the absence of a single dominating firm. It is an oligopoly of six firms, each with about 10–15%. That industry, however, is unusual insofar as the same six firms dominate in almost all countries where they are free to operate. Worldwide, of the 13 industries, the top six firms hold 76.4%.

## T. Is there a convergence in market structure among the various media industries?

We find a convergence in the concentration of media industries relative to each other. High-concentration industries have mostly been trending downward, while low-concentration industries have risen (the exception are search engines). The standard deviation—the measure of the industries' divergence from the average—in the concentration of the four media industry sectors dropped from 1,512 in 2000 to 1,159 in 2004 to 1,002 in 2013.

The concentration of content media has been rising and that of platform media declining, and they have thus been moving closer to each other.

# COMPANIES

## U. Who are the world's most dominant media companies?

The first way to rank companies is by revenues. Of the world's top 12 media companies by revenues, all are platform telecommunications providers. The top five organizations are the Government of China (with the combination of China Mobile, China Unicom, China Telecom, CCTV, and other media activities), AT&T (United States), Telefónica (Spain), NTT (Japan), and Verizon (United States). In that top 12 group, only Comcast (a US platform cable company), Murdoch, and Softbank (with mobile, ISP, search, and other online activities in Japan and the United States), as well as the Government of China have a major content presence, too.

The Chinese organizations experienced major growth. Even separately, the three firms of China Mobile, China Telecom, and China Unicom, would be among the world's largest overall media organizations by revenues while CCTV would be counted among top content providers.

In the content media, the Murdoch Group (News Corp and 21st Century Fox combined) is the private media organization with the highest revenues, at $32.5 billion. In 2004/05, it also held first place with $26.5 billion. In that period, Time Warner was second with $22.8 billion. However, after several spin-offs, Time Warner's revenues had dropped to $12.6 billion.

Google, with $7.4 billion in 2004/05, became the company with the second highest content revenues in 2010, with $26.8 billion. Its total revenues grew to $66 billion by 2014. With its enormous growth rate, it either already is or will be shortly the world's largest content company. Google already is the content company with the highest power index with a score of 278 for content.

The largest media companies by revenues are headquartered in OECD countries, and among these countries the economically largest account for 28 of the 30 largest media content firms by revenue: the United States (14), Japan (5), Spain (1), the United Kingdom (1), Italy (2), France (2), and Germany (2).

The second way to rank media organizations is by the power index. The privately owned content company with the highest power index was Google with a score of 278. Its market shares around the world are huge, and it effectively dominates the search engine industry.

The content producer with the highest global power index in the world is the Government of China, with a power index of 359. The Murdoch Group is third with 129. For more details, see Chapter 35 on "Companies."

## V. Which companies dominate the attention for news?

A third way to rank media companies is by news attention, by the aggregate time spent on news by their audiences. The combined share of firms that are top in news attention is enormous everywhere. The C10 in news attention in four countries is above 90%. It is between 80–90% in nine countries. It is nowhere below 50%, not even in the United States (55.8%). Thus, in most of the world, the top 10 firms command over 75% of their countries' news attention. The share of the top firm in terms of news attention is, on average, 32.2%.

High shares exist, in particular, where governments control media operations (i.e., China, Egypt, and Russia). As mentioned, the government of China, through its several media organizations, accesses a vast share of global news attention. In the aggregate, it has 29.7% of the 30-country world news attention in 2013, and 19% of the entire globe. Even if we unbundled China's news organizations, CCTV alone would still command 11.1% of the world's news attention (and 7.1% for the entire globe) and be the second largest news media company in the world. The explanations for these high shares are, as mentioned, the country's huge population (1.3 billion), relatively good accessibility to media (print, audiovisual, and Internet), and state control over most news media except several online portals and print magazines.

Other large shares in global news attention time are held by the governments of Russia (1.81%) and Egypt (1.75%): both countries' governments prioritize broadcast TV and radio control. Egypt also maintains significant state-owned print newspapers. Russia's government, in contrast, is looking to increase its ownership of online news media. India's public service broadcaster Prasar Bharati (18.5% in the 30-countries and 11.8% for the world) became autonomous from direct state control after 1997 and has a terrestrial broadcasting monopoly in a country with a population of 1.1 billion people.

India's BCCL is the largest privately owned media firm in the world, with 1.25% for the entire world of global news attention share in 2012. The second largest private media firm, by news attention time, is Globo in Brazil, with 1.1% for the entire world.

Rupert Murdoch's two companies combined are the second largest privately owned news providers, holding a global attention share of 1.35% of the 30-country world and 0.86% of the entire world. It is the largest US-headquartered news firm.

Most other large firms dominate large, single-country markets. Televisa of Mexico is the fourth largest private news firm in world news attention (1.21% for the 30 countries and 0.77% for the world). Other significant players such as BBC, Disney, the Redstone Group, Fininvest, PRISA, France Télévisions, the Bouygues Group, and ProSiebensat.1 all primarily operate within their home country's news markets. Bertelsmann is active in Europe more generally, and BBC and Time Warner also have an international presence.

As noted above, the combined global news share of the top five United States companies/groups is 4.6% for the 30 countries, and 3% for the whole world. The top ten firms have 7.2% for the 30-country world and 4.6% for the entire world.

## OWNERS

### W. Who are the media owners, and what do they own?

There are major differences between the ownership of content media and platform media.

- Many of the platform companies, even after the corporatizations and privatizations of the 1980s, have major ownership stakes by their governments. This includes, as mentioned, NTT in Japan (33%), Deutsche Telekom in Germany (38%), Orange in France (27%), the Chinese telecom majors, Svyazinvest in Russia (53%), and Telkom in South Africa (50.7%). These governmental ownership stakes are particularly high given the fragmentation of the private stock holdings in platform companies among numerous small investors who have traditionally considered telecom shares as safe "widows-and-orphans" stock.
- Few of the top 20 platform companies have major individual owners. The main exception is America Móvil with Carlos Slim holding 52%. Several other large platform firms have high individual ownership stakes but originated as a media or Internet company with a presence in platforms. Most of these companies have a dual stock structure. This is designed to allow an individual owner to maintain control while accessing outside capital investors. The other major platform companies are majority-owned by large institutional investors or governments and have no dual stock structure.
- US institutional investors have good-sized ownership stakes in many of the top 20 platform companies around the world, though rarely over 5%.
- Content firms tend to be privately owned. Most of the top content companies have major individual owners—Globo Group (Marinho family, Brazil); Lagardère (Lagardère family, France); Bertelsmann (Mohn family, Germany); Fininvest (former Prime Minister Silvio Berlusconi, Italy); Softbank (Masayoshi Son, Japan); Fuji TV (Shikanai family, Japan); TF1 (Bouygues family, France); Fox/News Corp (Murdoch family); Google (Larry Page and Sergey Brin); Comcast (Roberts family); Liberty (John Malone); CBS/Viacom (Sumner Redstone); Baidu (Robin Lee); Yomirui group (Shoriki family); and Disney (Steve Jobs family with 7.8%). The major exceptions are Time Warner/Time Warner Cable; Yahoo; and ProSiebenSat.1.[22]
- A World Bank study (Shleifer et al. 2001) of major media enterprises in 97 countries finds that only 4% of them were widely held by shareholders and 2% were owned by employees. In contrast, families control 57% of newspapers and 34% of TV stations. What are the reasons for such high levels of individual or

22. The latter is the only non US content media firm in the top 20 that is not controlled by an individual family, after having been owned by a group led by Haim Saban.

family ownership? Economists, going back to Demsetz (1989) concluded that the non-financial "amenity potential" of controlling media outlets, such as fame, influence, and favorable policy are high and therefore create incentives to acquire control.
- Family control is often assured by foregoing shareholder democracy. Great discrepancies exist for the equity percentage and voting percentage of many major content companies: Google; News Corp/21st Century Fox; Comcast; Lagardère; Bertelsmann; Liberty; Televisa; Bouygues; CBS; Viacom; etc. In each case, founders or their heirs protect their control through complex stock structures.

There are three major kinds of owners—individual, institutional, and state/public. There are 56 media billionaires as individuals or families.

- Of the billionaires, two-thirds substantially created their media properties themselves (37) rather than inheriting established firms (19).
- The "new media billionaires" were mostly active in Internet (12), mobile telecom (11), and information services (2).
- A significant share of the 56 media billionaires are US citizens (22); 4 each are from Japan and France; 3 from India; 2 each from Germany, Canada, Turkey, and China; 5 from Latin America; and 2 from Africa. Overall, there are 17 media billionaires in developing countries and the BRICS.
- The top 10 individual owners hold, in aggregate, $225 billion worth of media companies. The top 20 individual owners hold $313 billion. And the top 50 such owners had $422 billion.
- For 2013, the media sector's global aggregate (i.e., 30-countries) revenues were about $2.25 trillion dollars. If we estimate a multiple of enterprise value to revenue of 2.07, this would give a rough estimate for valuation of $4.7 trillion. The share of the top 10 individual owners in the world's media companies is then about 4.8% of the top 20 owners about 6.7%, and of the top 50 owners, 9.0%. A lower multiple, such as 1, would double these shares.

## X. Who are the Institutional Owners?

Generally, the stake of institutional investors' is larger than those of individuals. In 2013, the investment company State Street Corp. had $65 billion invested in major media companies, mostly on behalf of the clients of its investment funds. Rupert Murdoch, in comparison, had "only" $11.6 billion. Janus Capital, with over $11 billion, had more money tied up in media than Berlusconi, Malone, Redstone, and Lagardère combined.

The popular belief that technological convergence in the information sector, plus worldwide mergers, have resulted in a small group of media moguls is not an accurate one. Such individual owners do exist, of course, and the 56 media billionaires attest to that. Almost every country seems to have several such large-scale private media owners. But for the developed world, larger and growing faster is the ownership by fairly anonymous financial institutions that own, on behalf of their fund investors, narrow slices of a very big pie.

We identify at least 30 institutional owners with more than $1 billion in media assets.

- The total media assets of the top 10 institutional owners add up to $332.5 billion; the top 20 have $423.4 billion. And the top 30 have $449 billion.
- Of overall media value, estimated as $4.7 trillion (before including state-owned media), the top 10 institutional owners hold 6.1%, and the top 30 hold 9.6%.
- Of the headquarters of the top 30 asset management companies (in terms of media holdings), 73% are based in the United States (22 companies).

## Table 38-10. Top Media Owners Worldwide (as of Sept. 2013 )

| OWNER OR ASSET MANAGER | VALUE OF MEDIA HOLDINGS ($ BILLIONS) |
|---|---|
| Government of China | 317.2 |
| Government of Japan | 67.2 |
| State Street (US) | 64.8 |
| Vanguard (US) | 63.8 |
| Fidelity (US) | 46.5 |
| Capital Group (US) | 35.2 |
| Government of Germany | 29.9 |
| Carlos Slim (America Móvil, Mexico) | 29.2 |
| Larry Page (Google, US) | 26.7 |
| Government of France | 26.4 |
| T. Rowe Price Assoc. (US) | 26.1 |
| Sergey Brin (Google, US) | 26.0 |
| Government of Russia | 25.4 |
| BlackRock (US) | 24.3 |
| Cox family (Chambers, Kennedy, Parry-Sheden, Anthony) (Cox Communications (US)) | 24.0 |
| Michael Bloomberg (Bloomberg LP, US) | 24.0 |
| David Thompson family (Thompson Reuters, Canada) | 20.3 |
| Marinho family (Globo, Brazil) | 20.0 |
| Dodge & Cox (US) | 20.0 |
| Government of India | 19.8 |
| Mark Zuckerberg (Facebook, US) | 19.0 |
| Massachusetts Finance (US) | 18.7 |
| Brian Roberts family (Comcast, US) | 18.5 |
| JP Morgan Chase (US) | 17.9 |
| Newhouse family (Advance Publications, US) | 17.1 |
| Government of Norway[a] | 16.2 |
| Wellington Management (US) | 14.0 |
| Government of the UK | 13.9 |
| Janus Group (US) | 13.8 |
| Goldman Sachs (US) | 12.3 |
| Government of Taiwan | 12.1 |
| Sawiris Family (Orascom, Egypt) | 12.0 |
| Murdoch family (News Corp./21st Century Fox, US) | 11.6 |
| Government of South Africa | 11.6 |
| ClearBridge (US) | 10.9 |
| Charles Ergen (Dish Network, US) | 10.6 |
| Government of Switzerland | 9.1 |
| Jobs Family (Disney, US) | 8.7 |
| Masayoshi Son (Softbank, Japan) | 8.6 |

Table 38-10. *continued*

| OWNER OR ASSET MANAGER | VALUE OF MEDIA HOLDINGS ($ BILLIONS) |
|---|---|
| Government of Poland | 8.4 |
| Eric Schmidt (Google, US) | 8.2 |
| Legal & General (UK) | 7.7 |
| Edward Rogers Family (Rogers Cable, Canada) | 7.6 |
| Robin Yanhong Li (Baidu, China) | 7.53 |
| Baille Gifford (UK) | 7.5 |
| Government of Egypt | 7.5 |
| Government of Italy | 7.1 |
| Hearst Family (Hearst Publishing and TV, US) | 7.0 |
| Oppenheimer Funds (US) | 6.7 |
| Xavier Nile (France) | 6.6 |
| Government of Spain | 6.4 |
| Thornburg Invest. Mgmt. (US) | 6.2 |
| Ambani family (Reliance, India)[b] | 6.2 |
| Silvio Berlusconi (Fininvest, Italy)[b] | 6.2 |
| Government of Sweden | 6.0 |
| David Geffen (film and music, US) | 5.6 |
| Avande Krishnan (Moxis Mobile, Malaysia) | 5.85 |
| Government of Turkey | 5.8 |
| Government of the United States | 5.7 |
| John Malone (Liberty, US) | 5.6 |
| Denis O'Brien (Communicorp Radio, Ireland) | 5.2 |

[a] Includes the Government Pension Fund of Norway, which invests worldwide, with media assets of $15.2 billion.
[b] Media activities only.

## Y. What is the Overall Ownership of the Media Sector?

The Top 30 *individual* owners: 7.9–15.8%.[23]
The Top 30 *institutional* owners: 9.6–19.2%.
The 30 *government* holdings: 13.4%.

These 90 owners, then, account for about 30–50% of all media assets (Table 38.10). Institutional investors will likely grow their ownership share, while state ownerships will likely continue to decline in the developed world. Individual non-entrepreneurial ownership will also decline, though it is being kept around through various forms of economically inefficient multiple share classes.

Entrepreneurial ownership is likely to grow in periods of innovation but shift to institutional ownership in periods of consolidation. Industry concentration and institutional ownership are therefore closely associated, as is government ownership and concentration.

Thus, the actual ownership of firms has shown two seemingly contradictory trends. On the one hand, it has become more fragmented—a large number of institutional owners with stakes that are moderate in terms of control.

23. The reason for a range is to account for different multiples of value to revenues. The upper bound figure is based on a 2014 average for US media industries (2.07). The lower bound is a conservative multiple of 1.0.

On the other hand, some of these same institutional owners have stakes in many media and information firms, and they add up to huge amounts.

Public attention has centered on highly visible media moguls such as media Murdoch, Redstone, or Berlusconi. But that kind of personalized portrayal is dated given the prevalence of institutional investors, dot-com startup founders, and national governments.

# PROBLEMS

## Z. What are the priority problems?

We mentioned in the introductory chapter what this book is *not*: it is not a policy recommendation and it is not an analysis of societal impacts. We have created a fact base as the basis for public policy, media strategy, and further academic interpretation. Adding policy recommendations in this book would only detract from the analysis. In addition, given the wide diversity of authors in this project's country studies, such a recommendation would end up dividing an otherwise collegial group. This does not mean, however, that we would not endeavor to do so in the future. But this volume, long as it already is, is dedicated to data analysis and interpretation, not to a call for specific actions. We do, however, identify 11 key problems.

### 1. THE PROBLEM OF CONTENT MEDIA CONCENTRATION

A convergence of concentration can be observed across geographic regions, countries, and industries. Those with higher concentration in content media have been dropping in concentration, and this is positive in terms of competition and pluralism. But convergence has two sides—countries with low concentration have moved in the opposite direction, toward greater concentration.

There has been a high concentration in the average content media (HHI = 2,219 for country-weighted weighted; 2,906 for arithmetic average), and it has been strongly rising 3.1% per year by country-weighted HHI without the Internet, and 4.2% with it. (Internet content media are defined as search engines and online news.) The (arithmetic average) has 41%, and top four firms have 76.5%. In the overall (pooled) content sector, the top four firms account for 48% (40% weighted by countries).

The growth rate for content (weighted country averages) are, for the HHI, 4.2%; for the C1, 2.9%; for the C4, 1.5%; for the power index, indicating cross-industry expansion, 4.9%; for the Noam Index, indicating pluralism, 6.4%. These are high numbers. In the United States, content concentration is much lower (HHI = 1,080), but HHI was growing at a rate of 6.9% per year, 4.9% without the Internet. The average growth rate for all eight index definitions was 3.3% per year in the US.

The high and rising levels of concentration in content and news cannot be satisfactory for those who believe in pluralism in content and competition in platforms.

The rise in content media concentration was only partly offset by the decline in the platform industry concentration. And that sub-sector is still very highly concentrated.

The magnitude of these increases was higher than expected by the author (who was agnostic about the findings). This trend should be cause for concern. It may be a reflection of inherent market equilibria, which means that some of the richer and more developed countries will move to greater concentration than in the past, and that market forces are unlikely to be able to deal with this trend. Even regulatory policies will have a hard time being effective in the face of fundamental trends. This is therefore a serious challenge to pluralist media.

### 2. THE PROBLEM OF DIVERGENCE IN PLATFORM AND CONTENT CONCENTRATION

The concentration of platform media has been declining slowly and that of content media has increased more rapidly, and they have thus been moving closer to each other. However, the gap is still substantial. This means that in the vertical value chain, a relatively more competitive content sector is facing a more highly concentrated platform sector. For platforms, the pooled C4 is 78.5% weighted by country, which means a tight oligopoly, whereas it is 39.8% for content media. Even if the platform market structure declines gradually its basic structure is based on the economics of distribution technology that favor a large scale. Content providers, while not small, are significantly more competitive, and are facing a less competitive provider market in the essential service of distribution. This imbalance leads to rent extraction by the less competitive segment, which therefore points to several strategic options for content firms: (a) continuous battles before governments to adjust the imbalance in bargaining power through regulatory interventions; or, (b) a greater priority for the acquisition of content elements that are so unique and noncompetitive that they create advantage in dealing with platform providers; or, (c) incentives to further concentration in the content industries in order to match that of the platform industries.

Option (a) would lead to a more regulated media system. Options (b) and (c) mean that the highly concentrated platform market leads to greater concentration in the content market. Conversely, it would suggest (though not prove) that creating greater incentives to a more rapid de-concentration in platforms would have a multiplier in content media de-concentration, too.

## 3. THE CHINA PROBLEM

Much has been written about media in China.[24] That discussion, however, is usually about the editorial independence of journalists and publications. It is more about content and process than about ownership, market share, and international comparisons.

The share by the government of China in the world's media ownership and especially in news attention, through its domestic dominance, is truly extraordinary. As mentioned, the government of China, through its several media organizations, accesses a large share of global news attention. In the aggregate, it has 29.7% of the 30-country world news attention in 2013, and about 19% of the entire globe's. While the media holdings of the Chinese state dwarf those of all other governments that keep a grip over their media such as Russia and Egypt, even countries that are market-oriented, such as Japan and Germany, have substantial state ownership in the media sector. But these countries extend pluralism through transmission over content-neutral telecom, and in providing, over public service TV, content that is often not produced by the commercial sector. In China ownership is not only much more extensive but its explicit aim has been to lower pluralism. This aim has loosened somewhat but is still substantial. As China becomes an economic superpower, such centralized control might either decline as part of a more general liberalization, or remain as a major problem for the world's overall media diversity.

## 4. THE GOOGLE PROBLEM

The media organizations with the highest revenues are not necessarily the ones with the highest news attention or the highest market power globally. Google, with just $7.4 billion in 2004/05, became the company with the second highest content revenues, with $26.8 billion. Google's growth rate will make it the world's largest content company. Google is the private content company with the, by far, highest global power index, a score of 278. US and EU antitrust actions deal primarily with the manifestations of market power in vertical markets, rather than with the cause itself. Even aside from its brilliance, the company's sheer cause and entry barriers in a high fixed-cost, low marginal-cost industry create problems that are hard to overcome conceptually or practically.

24. Examples are Chan (1993); Chu (1994); Kennedy (2009); Lee (2006 and 2007); Liu (2006); Stockmann and Gallagher (2011); Qin et al. (2014, in progress); Winfield (2005); Wu (2000); Zhao (2000).

## 5. THE DEVELOPING COUNTRIES AND BRICS PROBLEM

Even where media companies are private, they might not be independent. In many emerging markets, the top individual news media owners have an amazingly high share in news attention, often well above 30%. These companies have often achieved their strong mindshare by close relations with a government in power that awarded preferential licenses. (This is not limited to the developing world: in France, President Mitterrand in the 1980s awarded the country's exclusive pay-TV license to Canal Plus, a company headed by his former chief-of-staff, golfing partner, and executor of his estate. In the United States, the FCC, under President Eisenhower, awarded no TV license to any newspapers company that had editorially endorsed his Democratic rival.) Once they achieve dominance, the media companies are hard to dislodge by subsequent governments or competitors. Indeed, given their influence over public opinion, they may receive additional benefits. In some cases, their principals become political players themselves, as in the case of Italy, Indonesia, or Thailand. The implicit quid-pro-quo of economic favors and media support eradicates the concept of media as "speaking truth to power." It would be comforting to believe that the role of governments in the media world declines and with it such leverage. But that is not so. On a large range of issues—telecom infrastructure, content production support and protectionism, spectrum licensing, intellectual property rights, merger approvals, access and interconnection rules and pricing, standards, and so on—the role of governments may well be increasing, and with it the potential for favoritism.

A great deal of attention has been given to the media power of those such as Murdoch, Redstone, or Berlusconi. One does not condone such individual media power in rich countries if one also flags the problem of media in developing and emerging countries. In a good number of these countries, the market shares and mindshares of domestic media firms are, within their societies, even higher than in richer countries, often by a substantial margin. The critique of media power cannot be asymmetric. Developing and emerging countries cannot get a free pass on anti-pluralism. Indeed, they need active media even more than well-developed democracies. In Mexico, three companies dominate their respective market niches, Carlos Slim in telecom (85% in wireline and 70.8% in wireless), the Azcárraga family in TV via Televisa (61%), and the Vázquez family OEM in newspapers (59.4%). Supporting a media protectionism that serves as a shield for major domestic companies under the guise of cultural autonomy does no favor to these societies. Of course, a delicate balance must be found of foreign and domestically grown content and distribution, but such balance is not found in a domestic dominance of politically well-connected companies.

The incentive to create national media empires in developing, low-income markets is still rising. For now, cross-ownership is higher in well-developed media markets. This suggests that the presence of media firms in multiple media industries rises with economic and media development. In consequence, as countries develop economically and their media grow in technological and business complexity, one should expect further trends to cross-ownership, without necessarily the counter-forces of competitive markets and public-interest regulation.

## 6. THE INDUSTRIAL MEDIA POLICY PROBLEM

With scale economies an important factor, one obvious strategic conclusion for media firms is to grow in scale, whether through internal expansion and value pricing, or through mergers and joint ventures. Governments, often at the behest of powerful constituencies, regularly take initiatives to consolidate a market structure. In theory they aim to make it more efficient, but in practice they reduce consumer choice. And it is questionable whether this approach will work. While there are some success stories, "national champions" fail once protectionist walls are lowered. First, there usually are other firms with even larger scale out there. And second, dominance behind protected walls leads in general to complacency and lower accountability.

## 7. THE INTERNET PROBLEM

The Internet has been the great hope for media pluralism. It is supposed to open established powerful media and create pluralism around the world, impervious to efforts by governments to control it. This optimistic scenario is unlikely to happen. First, because governments can control its use, as China has demonstrated. More fundamentally, the underlying economics of the Internet media is one of great economies of scale and of high network effects. Our findings, more generally, show that differences in the concentration levels of different media industries are correlated with the capital intensity of that media industry. The more capital-intense an industry, the higher fixed costs, lower variable cost, and hence the higher economies of scale are. This, in turn, favors large firms and thus industry concentration.

The consequence is that if media become more capital-intensive in the future, their industry concentration will rise.

The Internet will play a positive role in niche markets of the "long tail," where entry becomes easier, scale is low, and competition from large players is not a major factor. But off the long tail—at the center of media activities, with general news, mass entertainment, infrastructure, and central nodes such as clouds—the opposite is the case. The Internet raises capital intensity of key segments of media. Companies such as Netflix, Apple iStore, Amazon.com, Google, and its YouTube, have global market shares that far exceed those of conventional media. Successful Internet companies with high market share in some of their operations are also able to extend them into other media activities, as Apple or Google have demonstrated. This goes beyond online. The example of Jeff Bezos's (principal of Amazon.com) buying *The Washington Post* might be part of a broader pattern. Internet and concentrations are becoming high enough and large enough in key activities that they have been pulling up the average concentration of content industries. Thus, the Internet, far from being the solution to media power, will prove to be part of the problem, a force of concentration.

## 8. THE GLOBALIZATION PROBLEM

Smaller countries tend to have more voices per capita. These smaller countries are able to support voices at a scale that does not seem sustainable in larger countries. It must therefore be not absolute size but the relative size with its negative impact on competitiveness that makes it harder for smaller voices to survive in large countries. In that sense, the larger countries under-perform in terms of sustaining voices. And this suggests that in an increasingly global media system, the number of voices per capita will decline, and the aggregate number of voices will decrease. For example, instead of two countries having each 30 voices, a joint media market might have 45 voices.[25] On the positive side, users will have more options in that global media system than they had in a national market. But on average, the national production of media voices will decline as media markets become more global. The number of voices per capita will also decline. And the content, in order to be successful in the larger world market, will be less domestic in orientation.

## 9. THE PROBLEM OF PRIVATE MEDIA BARONS

In the 30 countries analyzed, there are 56 media billionaires as individuals or families. (About one-third are from developing countries or the BRICS.) Two-thirds of these moguls substantially created their media properties themselves rather than inheriting them. This suggests openness. In the aggregate, never has there been so much wealth created through the ownership of media. The financial stakes are much higher than they used to be, and this will raise the importance of the business side relative to the editorial side.

25. Hervas-Drane and Noam (2013).

On the business side, this wealth leads to various maneuvers to perpetuate control into the next generations, whose competence might not equal that of the founders. Various inefficient arrangements such as multiple classes of shares protect control and distort investment decisions. Inefficiency, in turn, creates the need for a shield against potential competitors, and this leads such firms to seek market power (concentration) or regulatory protection.

### 10. THE PROBLEM OF ABSENTEE INSTITUTIONAL OWNERSHIP

Public attention has centered on highly visible moguls. But the reality of media ownership is that of institutional investors that hold small to medium-sized pieces of many media companies. Institutional investors will likely grow their ownership share. There are several potential issues associated with such an ownership system: too much control; not enough control; and the absence of localism. The first issue is that an institutional owner or a small group of such owners would affect media behavior. They do not normally exercise a direct role in management in the way that personal owners do, but instead do so in an indirect way. Through their setting priorities on short-term stock performance and through their buy-and-sell decisions they set behavioral parameters for the actual managers. The second potential problem is the opposite, giving managers too much of a free hand, without the alleged vision and civic responsibilities of proprietors or their heirs. But this faith in the noblesse oblige of owners and their heirs romanticizes such quasi-feudal arrangement. And the third and most real type of problem is that of absentee ownership—a lack of sensitivity and concern by institutional owners for distant localities. This is exacerbated by the length of the ownership span: Of the headquarters of the top 30 asset management companies (in terms of media holdings), 73% are based in the United States.

### 11. THE PROBLEM OF LOW IMPACT REGULATION

Earlier in this book, in the statistical estimation of correlations with a country's level of media concentration, neither the variable "quality of regulation" nor several other "good government" metrics showed statistically significant association with concentration. We concluded that this was a disappointing result from a good-government perspective, and a challenge for further data analysis. It also points out a major problem of media control. It is generally undesirable to give governments an undue role in the structure or behavior of media, since such powers are regularly abused or corrupted. To that principled objection is added one of practicality. Media regulation was easier to accomplish before digital convergence and global information flows. But now, traditional tools such as licenses, ownership ceilings, cross-ownership rules, and so on, are becoming ineffectual. How would they deal with the market power of Google? With the quality issues of Skype? With protecting national culture? With a balance in electoral campaign presence?

In the end, the most effective tool for government to assure pluralism, outside of curbing clear abuses, is to help generate alternative media, and to protect their access and interconnection. It is important for academics, public-policy analysts, NGOs, companies, and governments to think creatively about new approaches to these issues, balancing the public interest, technological innovation, and financial investment in the emerging environment.

## Outlook

It is the nature of a fast-paced, short attention span, hyper-information society to seek a single and simple "bumper sticker" answer to complex and divergent set of facts and trends. These people will be disappointed. This book has tried to assemble and sift through large amounts of data, with many results, without seeking a single answer or conclusion. But there are some broader findings. They come in five parts:

1. The concentration problem is growing considerably in content media and declining in platforms (though it is quite high there).

2. Although rising considerably, one may observe–and maybe hope based on past decades–that in rich countries, a combination of market forces, technological innovation, fundamental economics, shareholder pressure, consumer preferences, regulatory interventions, and democratic politics will keep media monopolization more or less under control, in a system of loose oligopolies. This requires vigilance.

Beyond this challenge, there are two looming problem areas.

3. The developing and emerging nations, where a symbiotic relationship of large media organizations and governments leads to the market dominance by a very few media companies or by state-controlled media.

4. The rise of high-tech, capital-intensive media operations using or providing Internet and related instrumentalities. The proliferation of "long tail" content, as well as the dynamic technology, obscures that the core of Internet-based media content and of its distribution are becoming more concentrated and globalized. Far from being the solution of media diversity, the Internet with its fundamental economics is actually a major part of the problem in the near future.

5. For either kind of concentration problem, there are no easy policy tools and remedies. Hence, they will grow in significance.

To consider remedies is therefore the task for our next round of research.

We conclude with the counsel to keep a perspective. Media concentration is not a new phenomenon. On the contrary, it has a long and contentious history, often leading to governmental counter-efforts. And it exists as a problem around the world.

Of course, just because a problem is old or widespread does not make it less important. The opposite is true. But a broader perspective may moderate the rhetoric of a falling sky and a lost golden age that pervades many of today's debates. The issue of media concentration has a long past and will have an even longer future. It is one of those fundamental issues of distribution of power and wealth that every generation needs to resolve. Now, that task has reached the Internet generation. The data and its analysis presented in this volume, we hope, will help in that process.

## BIBLIOGRAPHY

For the sources mentioned in this chapter, see the general bibliography at the end of this book.

## APPENDIX

### Who Dominates Print Publishing?

Daily newspapers are a concentrated industry with a world average HHI of 2,848.[26] Yet despite those high numbers, newspapers were among the less concentrated media. Even so, once one looks beyond worldwide averages to several specific countries, one can observe several astonishingly high shares of the top firms. In Chile, El Mercurio, 53.9%; in Ireland, INM with 52%; in Mexico, OEM with 59.4%; in Turkey, Dogan with 46.4%; in Australia, News Corp. with 57.5%; in Portugal, Cofinae with 42.2%; in Russia, Komsomolskaya Pravda with 39.6%; in Switzerland, INM in Ireland with 52%; Tamedia with 44.3%; in the United Kingdom, Murdoch with 32.5%; and in France, the Amaury Group with 30.2%. The countries with the lowest concentrations are Germany and the United States, the latter with an unweighted industry HHI of 304 and

26. Weighted market-size average was significantly lower, at 1,328, suggesting that smaller and poorer countries have higher concentrations.

a top share, by Gannett, of 9.9%. The low concentration is in part due to the mostly local or regional nature of American and German newspapers. On the local level, most US newspapers are unchallenged by other local daily newspapers.

News Corp's US share is 8.3%, but its worldwide market share and global power score are the world's highest for newspapers. The company's revenue in this industry is $8.4 billion, giving it a 7.9% share of the world newspaper market by revenue.

Magazine world average concentration is moderate (1,761 unweighted and 1,031 weighted). Despite the seemingly low entry barriers and the large number of titles, the magazine industry is often dominated by a few firms in an oligopolistic market structure. The world's most significant magazine firms are all US publishers: Hearst (8.6%), TIME Inc. (6.4%), and Advance (5.1%).

For book publishing, the worldwide market concentration of the industry is low (1,690 unweighted and 2,160 weighted). Most of the tens of thousands of publishing houses are small or medium-sized operations that produce only a few dozen titles each year. There are, of course, several major multimedia, multinational conglomerates with a major international presence. The combined market share of the world's top 10 book publishers is 42.9%. The world's largest privately owned book publishers by power index are mostly European firms: the German/UK Penguin Random House collaboration of Bertelsmann and Pearson (9.2% share of the global book publishing market), the French company Lagardère (Hachette) (5.3%), Murdoch's Harper Collins (3.2%), McGraw-Hill (2.7%), Italy's Fininvest Mondadori (1.8%), and Simon and Schuster of the Redstone Group (1.1%).

Exceptions with rising concentration are Russia (2,545), partly a relic of its past, partly the result of the two major private publishers AST and Eksmo merging in 2012. France's high book industry concentration (1,748) is led by the Lagardère Group with its high share of 29.4%. The Chinese government is the largest book publisher in the world by power index (1,590) due to its control over book publishing in a large market. Its world market share in revenue is also large (15.9%).

The summaries for Europe (chapter 33 in this volume, by Badillo, Bourgeois, and Lesourd) show an average concentration for daily newspapers of around 2,000 (lower than the worldwide average). For books, it is 1,896 (slightly above the world average), and for magazines, 2,116 (above world average).

## What are the largest multi-channel platforms, and the largest providers of TV and video?

In multi-channel platforms, the United States' market dominates the world by volume: 64% of the world total by revenue, due to high penetrations, high prices, and an international presence of US firms such as Liberty, 21st Century Fox, DirecTV, and Dish. Per capita spending was highest in the United State ($502), reflective of the extent of subscribership, and also of high prices. In most other heavily cabled countries—Canada, Switzerland, the Netherlands, United Kingdom, Ireland, Finland, and Sweden—the spending was less than half of that.

By revenue, the top six companies in the world are American: Comcast ($36.8 billion 14.9%), DirecTV ($32 billion, 12.9%), Time Warner Cable ($20.5 billion, 8.3%), Murdoch's Fox/Sky ($17.8 billion, 7.2%), Dish Network ($13 billion, 5.2%), Cox ($7.6 billion, 3.1%), and Liberty ($7.6 billion, 3.1%). The top non-American firms are SkyPefecTV of Japan (1.8%), PRISA of Spain (1.1%), Globo in Brazil (1.5%), and Vivendi of France (1.2%).

Of the world's top 45 multi-channel platform firms, 5 are satellite-platform providers and 6 are IPTV providers, mostly incumbent telecom firms. The other 29 firms are cable TV platform providers. Of these, five are mixed cable/DBS or cable/IPTV providers.

The world's television broadcasters with the highest power indices are the Government of China (including its CCTV entity), Globo Group (Brazil); Prasar Bharati (India); and Fininvest (Italy). In terms of revenues, the BBC has 4.5%,

Fininvest (Mediaset) 3.3%, Globo controls 3%, Comcast (NBC) has 4.1%, Disney 3.8%, Redstone 3.4%, Murdoch 3.1%, and Bertelsmann (RTL) 2.9%, with networks in several European countries. The global share of the top four firms is only 18.5% of this $184 billion international industry, as most terrestrial broadcasters limit their activities to their home countries. Nationally, however, market shares are high.

Many of the largest television broadcasters are national public service organizations. Prasar Bharati has a legal monopoly over TV broadcast television in India. In China, CCTV, with its several channels in multiple languages, has 59.1% of the national market, and the largest television audience in the world measured by total viewers. If one adds the other governmental TV operators, the state's share is 100%. State broadcast TV operations account for 83.6% in Taiwan, 70.9% in Russia, 57.9% in South Korea, 67.6% in South Africa, 50.1% in the United Kingdom, and 48.3% in Italy.

For the combined TV content industry—broadcasting and video channels (worth $255 million)—the largest companies by power index, revenue, and market share are four American companies and one British: Disney (76.6, $12.8 billion, 5%), Redstone (75.4, $12.3 billion, 4.8%), Comcast (68.8, $11 billion, 4.3%), Murdoch (75.4, $11 billion, 4.3%). The British entity is the BBC is also quite large (163, $8.2 billion, 3.2%). The next five largest are Bertelsmann, Time Warner, Fininvest, NHK, and ProSiebenSat.1. China's CCTV ranks in the top 10 as well when counted alone with a PI of 153, revenues of $6.6 billion (2.6% of combined video industries). By revenue, however, Disney and Redstone eclipse the combined Chinese state enterprises, which have an aggregate PI of 439, revenues of $11.2 billion, and 2.6% of the world market. The HHI of the combined video industries is 199.

Of the top 20 content channel companies (broadcast and cable), six are based in the United States, five in Japan, two in Italy, and three in Germany. No firm has more than 5% globally. This reflects the decentralized nature of broadcast TV in many countries, where most TV content is produced by local broadcasters for local markets.

The European HHI averages (see chapter 33) are, for radio, 2,968; for broadcast TV and for multi-channel platforms, about 3,000; for film, about 1,166. These concentrations are lower than world averages, except for radio.

## Who dominates telecom and ISPs?

Wireline telecom providers have some of the highest revenues and power indices of any companies in this study. The top companies by power index, revenue, and market share of the wireline global market are quite huge and dominant: AT&T (254, $56.1 billion, 9.5%), NTT (567, $46.3 billion, 7.8%), Verizon (123, $39 billion, 6.6%), Deutsche Telekom (278, $29.1 billion, 4.9%), and Telefónica (245, $30.6 billion, 5.2%). By revenue, the largest are AT&T, Verizon, and NTT. The Chinese government has a power index of 448, revenues of $26.4 billion, and 4.5% global wireline market share: China Telecom is the state's largest component enterprise (154, $15.5 billion, 2.6%). India's top state firm Bharat Sanchar Nigam has a power index of 66.7, $4.9 billion in revenues, and a world revenue share of 0.8%.

National wireline concentration is often high, and often even higher than the numbers show, because the incumbents in several countries (such as the United States, Canada, and Argentina) operate on a regional basis and do not compete with each other in much of the wireline business. The wireline market is not as internationalized as the wireless market is. This is due to the advantages that many former (or current) state wireline operators enjoy, some by law and most by the advantages of incumbency. However, some of the larger incumbents, mostly European, leverage their scale, experience, and resources to buy into smaller markets. Internationalization of the wireline industry is common in many of the smaller markets, where older, more established, mostly European providers have been able to move in after the privatizations of the 1990s.

Mobile telecom is somewhat different, though many of the players are the same. However, they compete with each other. The typical market structure for mobile wireless is three to four national footprints. Among the national operators, the incumbent traditional wireline operators, formerly the monopoly providers, typically has the highest market share, for example, Orange in France, NTT in Japan, Deutsche Telekom in Germany, Telmex in Mexico, Telefónica in Spain and Argentina. The number 2–4 firms are frequently foreign companies, often the national incumbents of other countries, for example, Orange, Telefónica, Deutsche Telekom. These firms then can use their domestic experience, resources, and economies of scale (and the needs of developing markets) into increasingly global footprints. This process is still taking place. At present, there are over a dozen of such firms. Consolidation is likely, probably through international partnerships. In Europe, the EU Commission is actively encouraging such consolidations.

By revenue, the leading mobile operator is the Government of China (power index of 1,709; $127 billion in revenues, and 17.1% of worldwide revenues), comprising of the three state-owned firms: China Mobile (698; $80.8 billion; 10.9%), China Unicom (79, $27.2 billion, 3.7%), and China Telecom (35.4, $18.2 billion, 2.5%). The other top firms are Vodafone (6.8%), AT&T (7.1%), NTT (4.9%), Verizon (7.3%), Softbank (6.5%), Deutsche Telekom/T-Mobile (4.9%), France Telecom/Orange (3.5%), America Móvil/Claro (4.9%), and Telefónica (5.6%).

The markets of wireline telecom, mobile wireless, and ISPs are closely related. If we add up their major participants, we find the following: the top 10 firms by market share are the Government of China, 11.6%; AT&T, 7.8%; Verizon, 6.5%; NTT, 6%; Telefónica, 5.1%; Deutsche Telekom, 4.7%; Vodafone, 4.2%; Softbank, 3.27%; Grupo Carso, 3.7%; and Orange, 2.9% (see chapter 35). The C4 is 32% and the C10 is 56.2%. The three industries' combined global (W-HHI) concentration is 276.

Across Europe, the platform industry averages are 4,264 for wireline, 3,119 for wireless, and 3,056 for ISPs (chapter 33). In all cases, these concentrations are below the world averages.

Table 38-1A. National Media Concentrations—Arithmetic Average, 2012

| | ALL MEDIA | PLATFORM MEDIA | CONTENT MEDIA (W/INTERNET) | CONTENT MEDIA (W/O INTERNET) | NEWS MEDIA (BY REVENUES) | NEWS MEDIA (BY ATTENTION) | NEWS MEDIA BY ATTENTION (W/O VCHINA) |
|---|---|---|---|---|---|---|---|
| HHI | 3,729 | 4,026 | 2,906 | 2,692 | 2,818 | 3,006 | 2,793 |
| C1 | 48.3% | 52.9% | 40.9% | 38.8% | 40.5% | 41.8% | 40.0% |
| C4 | 88.6% | 93.3% | 76.5% | 75.6% | 78.0% | 78.3% | 77.6% |
| Power Index | 3,701 | 4,015 | 2,875 | 2,666 | 2,785 | 2,948 | 2,733 |
| Noam Index | | | 1,608 | 1,462 | 1,514 | 1,696 | 1,444 |
| Voices | | | 42 | 36 | 31 | 31 | 32 |
| Cross-ownership | | | 25.9% | 24.4% | 28.7% | 28.7% | 26.0% |
| Pooled HHI | 1,824 | 2,908 | 1,132 | 1,298 | 1,466 | 1,586 | 1,341 |
| Pooled C1 | 32.2% | 42.6% | 23.6% | 25.5% | 27.8% | 28.3% | 26.1% |
| Pooled C4 | 66.8% | 86.7% | 47.7% | 52.5% | 57.0% | 57.5% | 56.2% |

Table 38-4A. Absolute Change in National Media Concentrations—Weighted Country (2004/5–2011/12 )

| | ALL MEDIA | PLATFORM MEDIA | CONTENT MEDIA W/INTERNET | CONTENT W/O INTERNET | NEWS MEDIA (BY REVENUES) | NEWS MEDIA (BY ATTENTION) | NEWS MEDIA BY ATTENTION (WITHOUT CHINA) |
|---|---|---|---|---|---|---|---|
| HHI | 16.08 | – 19.70 | 69.98 | 49.41 | 36.76 | –6.49 | 0.71 |
| C1 (%) | 0.05 | –0.40 | 0.76 | 0.56 | 0.39 | –0.05 | 0.01 |
| C4 (%) | 0.70 | 0.49 | 0.88 | 0.79 | 0.69 | 0.18 | 0.20 |
| Power Index | 18.43 | –20.62 | 74.90 | 55.27 | 42.24 | –9.96 | –4.19 |
| Noam Index | | | 49.18 | 39.39 | 27.31 | –13.23 | –1.37 |
| Voices | | | 0.16 | 0.11 | 0.05 | 0.05 | 0.05 |
| Pooled HHI | 38.43 | –2.55 | 30.95 | 42.77 | 29.98 | –4.92 | –5.49 |
| Pooled C1 (%) | 0.00 | –0.51 | 0.47 | 0.60 | 0.49 | –0.09 | –0.02 |
| Pooled C4 (%) | 0.52 | 0.16 | 0.52 | 0.71 | 0.63 | 0.07 | 0.10 |
| Cross-ownership (%) | | | 0.03 | 0.05 | 0.26 | 0.26 | 0.21 |

Table 38-4B. Absolute Change in National Media Concentrations—Arithmetic (2004/5–2011/12 )

| | ALL MEDIA | PLATFORM MEDIA | CONTENT MEDIA W/INTERNET | CONTENT W/O INTERNET | NEWS MEDIA (BY REVENUES) | NEWS MEDIA (BY ATTENTION) | NEWS MEDIA BY ATTENTION (WITHOUT CHINA) |
|---|---|---|---|---|---|---|---|
| HHI | – 96.05 | –141.04 | 5.77 | –21.00 | –34.01 | –23.36 | –25.23 |
| C1 (%) | –0.86 | –1.29 | 0.06 | –0.20 | –0.32 | –0.20 | –0.21 |
| C4 (%) | 0.07 | –0.03 | 0.14 | 0.03 | –0.04 | –0.12 | –0.14 |
| Power Index | –88.04 | –136.14 | 9.65 | –18.58 | –30.44 | –25.38 | –27.35 |
| Noam Index | | | 0.75 | –16.54 | –25.38 | –14.52 | –15.47 |
| Voices | | | 0.21 | 0.14 | 0.07 | 0.07 | 0.07 |
| Pooled HHI | –48.66 | –100.82 | –27.11 | –10.06 | –21.87 | –11.81 | –14.34 |
| Pooled C1 (%) | –0.92 | –1.43 | –0.36 | –0.15 | –0.24 | –0.17 | –0.18 |
| Pooled C4 (%) | 0.28% | –0.26 | 0.39 | 0.04 | –0.09 | –0.01 | –0.03 |
| Cross-ownership (%) | | | 0.05 | 0.06 | 0.35 | 0.35 | 0.28 |

Table 38-6. US Media Concentration Trends, Absolute Changes (2004–2013)

| | ALL MEDIA | PLATFORM MEDIA | CONTENT MEDIA W/INTERNET | CONTENT W/O INTERNET | NEWS MEDIA (BY REVENUES) | NEWS MEDIA (BY ATTENTION) |
|---|---|---|---|---|---|---|
| HHI | 18.93 | – 4.86 | 51.67 | 31.91 | 29.38 | 18.60 |
| C1 (%) | –0.11 | –0.48 | 0.44 | 0.25 | 0.10 | –0.03 |
| C4 (%) | 1.16 | 0.95 | 1.25 | 1.19 | 1.08 | 0.68 |
| Power Index | 11.94 | –12.94 | 46.35 | 29.45 | 24.93 | 13.65 |
| Noam Index | | | 22.75 | 11.98 | 11.36 | 8.35 |
| Voices | | | 0.38 | –0.63 | –0.25 | –0.25 |
| Pooled HHI | 6.48 | –17.31 | 8.94 | 10.94 | 9.89 | 11.68 |
| Pooled C1 (%) | –0.23 | –0.58 | 1.54 | 0.29 | 0.27 | 0.16 |
| Pooled C4 (%) | 0.49 | 0.00 | 3.15 | 0.57 | 0.44 | 0.78 |
| Cross-ownership (%) | | | –4.15 | –0.15 | 0.13 | 0.13 |

# General References on Media Ownership and Concentration


Abramson, Bram Dov. "Interpreting Current Statistics: Internet Backbone Market Shares." *TelecomReform* 1, no. 2 (July 2001): 7.

Adelman, M. A. "Concept and Statistical Measurement of Vertical Integration." In Stigler, G. J., ed., *Business Concentration and Price Policy.* Princeton, NJ: Princeton University Press, 1955, 281–322.

Ader, Thorsten, Fueg, Oliver C., Kevin, Deirdre, Pertzinidou, E., and Schoenthal, Max. *Final Report of the Study on the Information of the Citizen in the EU: Obligations for the Media and the Institutions Concerning the Citizen's Right to Be Fully and Objectively Informed.* Düsseldorf: The European Institute for the Media, 2004.

Advisory Panel on Media Diversity. "Transnational Media Concentrations in Europe." Strasbourg: Council of Europe, November 2004.

Ahn, Hoekyun, and Litman, Barry R. "Vertical Integration and Consumer Welfare in the Cable Industry." *Journal of Broadcasting and Electronic Media* 41, no. 3 (1997): 453–477.

Akhavan-Majid, Roya, Gopinath, Sheila, and Rife, Anita. "Chain Ownership and Editorial Independence: A Case Study of Gannet Newspapers." *Journalism Quarterly* 68(1991): 59–67.

Albarran, Alan. *Media Economics: Understanding Markets, Industries and Concepts.* Ames: Iowa State University Press, 2002.

Albarran, Alan B., and Dimmick, J. "Economics of Multiformity and Concentration in the Communication Industries." *Journal of Media Economics* 9(1996): 41–49.

Albarran, Alan B., and Gormly, Karen. "Strategic Response or Strategic Blunder? An Examination of AOL Time Warner and Vivendi Universal." In Picard, Robert G., ed. *Strategic Responses to Media Market Changes.* Jönköping, Sweden: Jönköping International Business School Research Reports, 2004, 35–46.

Albarran, Alan B., and Mierzejewska, Bozena I. "Media Concentration in the U.S. and European Union: A Comparative Analysis," presented at

*6th World Media Economics Conference.* Centre d'Etudes Sure les Médias and Journal of Media Economics, HEC Montreal, Montreal, Canada, May 12–15, 2004.
Albornoz, L. A. and García Leiva, M. T. (2013). *A Internet e o Futuro da TV.* Lisboa: MediaXXI.
Alexander, Peter J., and Cunningham, Brendan M. "Diversity and Market Structure: Preliminary Evidence from Broadcast Television News." In *Media Diversity and Localism: Meaning, Metrics, and the Public Interest.* New York: The Donald McGannon Communication Research Center, Fordham University, Dec. 15–16, 2003, 3–4.
Allam, Rasha. "Media Landscape: Egypt." *European Journalism Centre.* Nov. 8, 2010. Last accessed on Dec. 29, 2011 at <http://www.ejc.net/media_landscape/article/egypt/>.
Allen, David S., Blanks, S. Elizabeth, and Glasser, Theodore L. "The Influence of Chain Ownership on New Play: A Case Study." *Journalism Quarterly* 66 (Autumn 1989): 607–615.
Almirón, Nuria. "Grupos privados propietarios de medios de comunicación en España: principales datos estructurales y financieros." *Comunicación y Sociedad* 22, no. 1 (2009): 243–273.
Amez-Droz, Philippe. "L'économie des médias en Suisse." *Les Médias en Suisse, Structure et audience.* Cornu, Daniel, ed. 7th ed. Lausanne: FCJ/CRFJ, 2012.
Amin, Hussein. *Strengthening the Rule of Law and Integrity in the Arab World: Report on the State of the Media in Egypt.* The Arab Center for the Development of the Rule of Law and Integrity, 2009, 6. Last accessed on June 10, 2012 at <http://www.arabruleoflaw.org/Files/PDF/Media/English/P2/Egypt_MediaReportP2_En.pdf>.
*ANATEL 2011.* Annual Report: National Agency of Telecommunications of Brazil. Feb. 28, 2013. <http://www.anatel.gov.br/Portal/verificaDocumentos/documento.asp?numeroPublicacao=278637&pub=original&filtro=1&documentoPath=278637.pdf>.
Anderson, Chris. "The Long Tail." *WIRED Magazine,* no. 12.10 (October 2004).
Anderson, S., Coate, S., "Market Provision of Broadcasting: A Welfare Analysis." *Review of Economic Studies* 72(2005): 947–972.
Anderson, S. P., and McLaren, J., "Media Mergers and Media Bias with Rational Consumers. " *Journal of the European Economic Association* 10(2012): 831–859.
Anderson, Simon, and Coate, Steve. "Market Provision of Public Goods: The Case for Broadcasting." NBER Working Paper No. W7513, University of Virginia and Cornell, 2001.
Anderson, S. P., de Palma, A., Thisse, J. F., *Discrete Choice Theory of Product Differentiation.* MIT Press, Cambridge, MA. 1992.
Angelmar, Reinhard. "Market Structure and Research Intensity in High-Technological-Opportunity Industries." *Journal of Industrial Economics* 34, no. 1 (1985): 69–79.
Antonelli, C. *The Diffusion of Advanced Telecommunications in Developing Countries.* Part 56. Paris: Organization for Economic Co-operation and Development (OECD) Development Centre, 1991.
Archer, Gleason. *Big Business and Radio.* New York: The American Historical Company, 1939, 8.
Ardıyok, Ş., and Oğuz, F. "Competition Law and Regulation in the Turkish Telecommunications Industry: Friends or Foes?" *Telecommunications Policy* 34, no. 4 (2010). Accessed on Apr. 8, 2013. <http://www.sciencedirect.com/science/article/pii/S030859610900113X>.
Argentesi, E., and Filistrucci, L. "Estimating Market Power in a Two-Sided Market: The Case of Newspapers." *Journal of Applied Econometrics* 22(2007): 1247–1266.
Armstrong, Mark, "Public Service Broadcasting." *Fiscal Studies* 26 (2005) 281–299.
Arrow, Kenneth J., "Vertical Integration and Communication." *The Bell Journal of Economics* 6, no. 1 (1975): 173–183.
*Associated Press v. United States,* 326 U.S. 1, 20 (1945).
Attalah, Lina, and Rizk, Nagla. "Egypt's Evolving Media Landscape: Access, Public Interest and Control." In *A New Frontier, an Old Landscape.* Report for the Ford Foundation, November 2011, 118–136. Last accessed on Jan. 16, 2012 at <http://participatorymedia.lab.asu.edu/files/NewFrontierOldLandscape.pdf>.
Aufderheide, Pat. "Too Much Media." *In These Times* 29, no. 12 (May 9, 2005).
Auletta, Ken. *Three Blind Mice: How the TV Networks Lost Their Way.* New York: Random House, 1991.
Baade, Robert C., Bantz, Charles R., and McCorkle, Suzanne. "The News Factory." *Communications Research* 7, no. 1 (1980): 45–68.
Bachen, Christine, Craft, Stephanie, Hammond, Allen, and Mason, Laurie. "Diversity of Programming in the Broadcast Spectrum: Is There a Link between Owner Race or Ethnicity and News and Public Affairs Programming?" Washington, DC: Federal Communications Commission, 1999.

Bachen, Christine M., Craft, Stephanie L., and Mason, Laurie. "Support for FCC Minority Ownership Policy: How Broadcast Station Owner Race or Ethnicity Affects News and Public Affairs Programming Diversity." *Communication Law & Policy* 6(2001): 37–73.

Badillo, Patrick-Yves, and Lesourd, Jean-Baptiste, eds. *The Media Industries and Their Markets: Quantitative Analyses*. London: Palgrave-MacMillan, 2010.

Badillo, Patrick-Yves. *L'écologie des médias*. Bruxelles: Editions Bruylant, 2008.

Badillo, Patrick-Yves, and Proulx, Serge. "Mondialisation de la Communication : à la recherche du sens perdu." In *Les essentiels d'Hermès "Sociétés de la Connaissance - Fractures et évolutions."* Paris: CNRS Éditions, 2009, 117–134.

Baer, Walter S. *Concentration of Mass Media Ownership: Assessing the State of Current Knowledge*. Santa Monica: Rand, 1974.

Bagdikian, Ben. *The New Media Monopoly*. Boston: Bacon Press, 2004.

Bagdikian, Ben. *The Media Monopoly*. Boston: Beacon, 1983, 1987, 1990, 1992, 1997.

Bahlmann, T., Brakman, I., and van Cuilenburg, J. "Mediaconcentratie in Beeld: Concentratie en pluriformiteit van de Nederlandse media 2005." Hilversum: Commissariaat voor de Media, 2006. Accessed on Jan. 31, 2013. <http://www.mediamonitor.nl/dsresource?objectid=320&type=org>.

Bainbridge, Jane. "Sector Insight: Fixed-Line Telecoms." *Brand Republic*, May 20, 2011. Accessed on July 14, 2012. <http://www.brandrepublic.com/analysis/1070132>.

Baker, C. Edwin. *Media Concentration and Democracy: Why Ownership Matters*. Cambridge University Press, Cambridge. 2007.

Balbi, Gabriele, and Prario, Benedetta. "The History of Fininvest/Mediaset's Media Strategy: 30 Years of Politics, the Market, Technology and Italian Society." *Media, Culture, and Society* 32, no. 3 (2010): 391–409. <http://mcs.sagepub.com/content/32/3/391.full.pdf>.

Balio, T. *The American Film Industry*. Madison: The University of Wisconsin Press, 1985.

Balle, Francis. *Médias et Sociétés*. 15th ed. Paris: Montchrestien, 2011.

Band, D. "Media Ownership and the Flow of Information." *Nieman Reports* 46, no. 4 (1992).

Barboza, David. "China Yearns to Form Its Own Media Empires." *The New York Times*, Oct. 5, 2010. Accessed on Feb. 28, 2013. <http://www.nytimes.com/2009/10/05/business/global/05yuan.html>.

Barker, Jolyon. "Technology, Media and Telecommunications Predictions." Deloitte, 2014. Last accessed on Jan. 29, 2014 at <http://www2.deloitte.com/content/dam/Deloitte/global/Documents/Technology-Media-Telecommunications/gx-tmt-predictions-2014.pdf>.

Barry, Paul. *The Rise and Rise of Kerry Packer Uncut*. Sydney: Bantam, 2007.

Bates, Benjamin J. "Concentration in Local Television Markets." *Journal of Media Economics* 6 (Fall 1993): 3–22.

Bates, Benjamin J. "Station Trafficking in Radio: The Impact of Deregulation." *Journal of Broadcasting and Electronic Media* 37, no. 1 (1993): 21–30.

Bauer, T. "A Democratized Market? Development of South Africa's daily newspapers 1990–2006." *Global Media Journal*–African *Edition* 3: 1 (2009): 1–17.

Baumgarten, Paul, and Farber, Donald. *Producing, Financing and Distributing Film*. New York: Drama Book Specialists/Publishers, 1973.

Baumol, William J., Panzar, John C., and Willig, Robert D. *Contestable Markets and the Theory of Industry Structure*. San Diego, CA: Harcourt Brace, 1982.

Baxter, William F. "Antitrust Policy." In Feldstein, Martin, ed. *American Economic Policy in the 1890s*. Chicago: University of Chicago Press, 1994.

Baynes, Leonard M. "WHITEOUT: The Absence and Stereotyping of People of Color by the Broadcast Networks in Prime Time Entertainment Programming," presented at *Media Diversity and Localism: Meaning, Metrics, and the Public Interest*. The Donald McGannon Communication Research Center, Fordham University, New York, Dec. 15–16, 2003, 9–12.

Beam, Randal A. "Size of Corporate Parent Drives Market Orientation." *Newspaper Research Journal* 23, no. 2/3 (Spring/Summer 2002): 46–63.

Becerra, Martin, et al. "La concentracion de las industrias culturales." *Industrias culturales: Mercado y politicas publicas en Argentina*. Buenos Aires: Ciccus, 2003.

Becerra, Martin, Marino, Santiago, and Mastrini, Guillermo. "Mapping Digital Media: Argentina." Open Society Foundation (OSI), Feb. 29, 2012. Accessed on July 16, 2012. <http://www.soros.org/sites/default/files/mapping-digital-media-argentina-20120404.pdf>.

Beebe, Jack H. "Institutional Structure and Program Choices in Television Markets." *The Quarterly Journal of Economics* 91, no. 1 (1977): 15–37.

Beebe, J. H., Manning, W. G., and Owen, Bruce M. *Television Economics*. Lexington, MA: D.C. Health, 1974.
Bek, M. G. "Tabloidization of News Media: An Analysis of Television News in Turkey." *European Journal of Communication* 19, no. 3 (2004). Accessed on Apr. 1, 2013. <http://ejc.sagepub.com/content/19/3/371.full.pdf+html>.
Benkler, Yochai, Faris, Rob, Gasser, Urs, Miyakawa, Laura, and Schultze, Stephen. *Next Generation Connectivity: A Review of Broadband Internet Transitions and Policy from Around the World*. Cambridge, MA: Berkman Center for Internet and Society, 2010. Available at: <http://cyber.law.harvard.edu/publications/2010/NextGenerationConnectivity>. Last visited on November 2013.
Benkler, Yochai. *The Wealth of Networks: How Social Production Transforms Markets and Freedom*. Yale University Press, 2006.
Benson, R., and Powers, M., *Public Media and Political Independence: Lessons for the Future of Journalism from Around the World*. Free Press, Washington, DC. 2011.
Berndt, Ernst R., Klein, Lisa R., and Silk, Alvin J. "Intermedia Substitutability and Market Demand by National Advertisers." *Review of Industrial Organization* 20 (June 2002): 323–348.
Bernstein, J. M., and Lacy, S. "Contextual Coverage of Government by Local Television News." *Journalism Quarterly* 69, no. 2 (1992): 329–341.
Bernstein, Peter W. "The Record Business: Rocking to the Big Money Beat." *Fortune* (April 23, 1979): 58.
Berry, Steven, and Waldfogel, Joel. "Do Mergers Increase Product Variety? Evidence from Radio Broadcasting." *The Quarterly Journal of Economics* 116, no. 3 (2001): 1009–10025.
Berry, Steven, and Waldfogel, Joel. "Mergers, Station Entry, and Programming Variety in Radio Broadcasting." Working Paper, National Bureau of Economic Research, April 1999.
Berry, Steven, and Waldfogel, Joel. "Public Radio in the United States: Does it Correct Market Failure or Cannibalize Commercial Stations?" *Journal of Public Economics* 71(1999): 189–211.
Besen, Stanley M., and Woodbury, John R. "Rate Regulation, Effective Competition, and the 1992 Cable Act." *Hastings Communications and Entertainment Law Journal* 17 (Fall 1994): 203–224.
Betzel, Marcel, Lauf, Edmund, Negenborn, Rini, and Vosselman, Jan B. "Mediaconcentratie in Beeld: Concentratie en pluriformiteit van de Nederlandse media 2007." Hilversum: Commissariaat voor de Media, 2008. Accessed on Jan. 31, 2013. <http://www.mediamonitor.nl/dsresource?objectid=8373&type=org>.
Betzel, M., van der Burg, M., Lauf, Edmund, and Negenborn, R. "Mediabedrijven en MediaMarkten 2001–2010." Hilversum: Commissariaat voor de Media, 2011. Accessed on Jan. 31, 2013. <http://www.mediamonitor.nl/dsresource?objectid=12148&type=org>.
Bezanson, Randall, Cranberg, Gilbert, and Soloski, John. *Taking Stock: Journalism and the Publicly Traded Newspaper*. Ames, IA: Iowa State University Press, 2001.
Bhagwat, Ashutosh. "Unnatural Competition?: Applying the New Antitrust Learning to Foster Competition in the Local Exchange." *Hastings Law Journal* 50 (1999): 1479.
Bhawan, Mahanagar Doorsanchar, and Nehru, Jawaharlal. *Consultation Paper on Issues Relating to Media Ownership*. Telecom Regulatory Authority of India, 2013.
Bigman, Stanley K. "Rivals in Conformity: A Study of Two Competing Dailies." *Journalism Quarterly* 25 (Spring 1948): 127–131.
Bishop, Ronal, and Hakanen, Ernest A. "In the Public Interest? The State of Local Television Programming Fifteen Years after Deregulation." *Journal of Communications Inquiry* 26, no. 3 (July 2002): 261–276.
Bittlingmayer, George, and Hazlett, Thomas W. "DOS Kapital: Has Antitrust Action against Microsoft Created Value in the Computer Industry?" *Journal of Financial Economics* 55, no. 3 (March 2000): 329–359.
Bittlingmayer, George, and Hazlett, Thomas W. "The Political Economy of Cable 'Open Access.'" *Stanford Tech. Law Review*, Fall 2001. Available at <http://www.manhattaninstitute.org/hazlett/working_01_06.pdf>.
Bizberge, Ana, Mastrini, Guillermo, and Becerra, Martin. "La Television Digital Terrestre en Argentina: entre la geopolitica regional y la iniciativa estatal." In Badillo, Angel, and Sierra, Francisco, eds., *La transicion a la television digital terrestre en Iberoamerica: diagnostic y prospective*. Quito: CIESPAL, 2011.
Blackstone, Erwin A., and Bowman, Gary W. "Vertical Integration in Motion Pictures." *Journal of Communication*, 49, no. 1 (Winter 1999): 123–139.
Blaiklock, A. Richard M., and Krotoszynski Jr., Ronald J. "Enhancing the Spectrum: Media Power, Democracy, and the Marketplace of Ideas." *University of Illinois Law Review* 3 (2000): 867.

Blankenburg, William B. "A Newspaper Chain's Pricing Behavior." *Journalism Quarterly* 60 (Summer 1983): 275–280.

Blankenburg, William B., and Ozanich, Gary W. "The Effects of Public Ownership on the Financial Performance of Newspaper Corporations." *Journalism Quarterly* 70 (Spring 1993): 68–75.

Blaskopf, Lawrence P. "Note, Defining the Relevant Product Market of the New Video Technologies." *Cardozo Arts & Entertainment Law Journal* 4, no. 1 (1985): 75.

Blau, Andrew, and Schwartzman, Andrew J. *What's Local About Local Broadcasting?* Washington, DC: Media Access Project and the Benton Foundation, 1998.

Blommenstein, H. J. "Impact of Institutional Investors on Financial Markets." In Blommenstein, H. J., and Funke, N., eds., *Institutional Investors in the New Financial Landscape*. Paris: OECD, 1998.

Blumer, Jay G., and Spicer, Carolyn Martin. "Prospects for Creativity in the New Television Marketplace: Evidence from Program-Makers." *Journal of Communications* 40 (1990): 78–101.

Boadwee, Harry. "Note, Product Market Definition for Video Programming." *Columbia Law Review* 86 (1986): 1210–1211.

Bogart, Leo. *Commercial Culture, the Media System, and the Public Interest*. New York: Oxford University Press, 1994.

Bohlin, E., Caves, K. W., and Eisenach, J. A. (2013). *Mobile Wireless Market Performance in Canada: Lessons from the EU and the US*. Report prepared for Telus. <http://www.navigant.com/~/media/WWW/Site/Insights/Economics/Navigant-Mobile-Wireless-Canada-FINAL.ashx >. Last visited on December 2013.

Bollinger Jr., Lee C. "Freedom of the Press and Public Access: Toward a Theory of Partial Regulation of the Mass Media." *Michigan Law Review* 75, no. 1 (1976): 30–38.

Bork, Robert. *Antitrust Paradox: A Policy at War with Itself*. New York: Free Press, 1993.

Braima, Mahmoud, Johnson, Timothy, and Sothirajah, Jayanthi. "Measure for Measure: The Relationship Between Different Broadcast Types, Formats, Measures and Political Behaviors and Cognitions." *Journal of Broadcasting & Electronic Media* 44 (2000): 43–61.

Brakman, I., van Cuilenburg, J., and van der Meulen, L. "A View on Media Concentration: Concentration and Diversity of the Dutch Media in 2001." Hilversum: Commissariaat voor de Media, 2002. Accessed on Jan. 31, 2013. <http://www.mediamonitor.nl/dsresource?objectid=423&type=org>.

Brancato, Carolyn Kay, and Rabimov, Stephan. "The 2008 Institutional Investment Report: Trends in Institutional Investor Assets and Equity Ownership of US Corporations." Conference Board. 2008.

Brendon, Piers. *The Life and Death of the Press Barons*. New York: Athenium, 1983, 100–101.

Bresnahan, Timothy F. "New Modes of Competition and the Future Structure of the Computer Industry." In Eisenach, J. A., and Lenard, T. M., eds., *Competition, Innovation, and the Microsoft Monopoly: Antitrust in the Digital Marketplace*. Washington, DC: The Progress and Freedom Foundation, Kluwer Press, 1999.

Brinkley, Joel. *Defining Vision: The Battle for the Future of Television*. San Diego, CA: Harcourt Brace, 1998.

The British Film Institute (BFI). *BFI Statistical Yearbook 2011*. London: BFI, 2011. Dec. 20, 2012. <http://www.bfi.org.uk/sites/bfi.org.uk/files/downloads/bfi-statistical-yearbook-2011.pdf.>

*Broadcasting & Cable Industry Annual*. Laura Mars, ed. Amenia, NY: Grey House Pub., 2013.

Broadcasting Authority of Ireland. *Ownership and Control Policy 2010*. Dublin: Broadcasting Authority of Ireland, 2010, 2, 11.

Brown, Keith, and Williams, George. *Consolidation and Advertising Prices in Local Radio Markets*. Media Bureau Staff Research Paper, Federal Communications Commission, September 2002.

Brown, Richard. "Early American Origins of the Information Age." In Chandler Jr., Alfred D., and Cartada, James W., eds., *A Nation Transformed by Info: How Information Has Shaped U.S. from Colonial Times to the Present*. New York: Oxford University Press, 2000.

Budde, Paul. "China's Cable TV Market in Transition." BuddeBlog.com, July 6, 2012. Accessed on Feb. 28, 2013. <http://www.buddeblog.com.au/frompaulsdesk/chinas-cable-tv-market-in-transition/>.

Bush, C. Anthony. *On the Substitutability of Local Newspaper, Radio and Television Advertising in Local Business Sales*. Media Bureau Staff Research Paper, Federal Communications Commission, September 2002.

Busterna, John C. "How Managerial Ownership Affects Profit Maximization in Newspaper Firms." *Journalism Quarterly* 66 (Summer 1989): 302–307.

Busterna, John C. "Television Station Ownership Effects of Programming and Idea Diversity:

Baseline Data." *Journal of Media Economics* 1, no. 2 (1988): 36–74.
Busterna, John C. "The Cross-Elasticity of Demand for National Newspaper Advertising." *Journalism Quarterly* 64, no. 2/3 (Summer/Fall 1987): 346–351.
Busterna, John C. "Trends in Daily Newspaper Ownership." *Journalism Quarterly* 65 (Winter 1988): 831–838.
Candeub, Adam. "Media Ownership Regulation, the First Amendment, and Democracy's Future." *UC Davis L. Rev.* 41 (2007): 1547.
Carroll, Raymond L., and Tuggle, C. A. "The World Outside: Local TV News Treatment of Imported News." *Journalism and Mass Communications Quarterly* 74 (Spring 1997): 123–133.
Carvajal, M., and Sanchez-Taberno, A. *Media Concentration in the European Market: New Trends and Challenges*. Pamplona: EUNSA, 2002.
Caspi, Dan. *Media Decentralization: The Case of Israel's Local Newspapers*. New Brunswick: Transaction Publishers, 1986.
Caspi, D., and Limor, Y. *The In/Outsiders: The Media in Israel*. Cresskill, NJ: Hampton Press. 1999.
Center for Media Pluralism and Media Freedom. *European Union Competencies in Respect of Media Pluralism and Media Freedom*. European Commission (2013).
Chan, J. M. "Commercialisation without Independence: Trends and Tensions of Media Development in China." In J. Y. S. Cheng and M. Brosseau, eds, *China Review 1993*. Hong Kong: Chinese University Press of Hong Kong, 1993, 25.1–25.21.
Chan, Joseph Man, and Lee, Chin-Chuan. *Mass Media and Political Transition: The Hong Kong Press in China's Orbit*. New York: Guilford, 1991.
Chan-Olmsted, Sylvia M. "A Structural Analysis of Market Competition in the U.S. TV Syndication Industry, 1981–1990." *Journal of Media Economics* 3 (Fall 1991): 9–28.
Chan-Olmsted, Sylvia M., and Frank, Stacy. "How Americans Get Their Political Information: Print versus Broadcast News." *The Annals of the American Academy of Political and Social Science* 546, no. 1 (1996): 48–58.
Chan-Olmsted, Sylvia M., and Litman, B. R. "Antitrust and Horizontal Mergers in the Cable Industry." *Journal of Media Economics* 1 (1988): 63–74.
Chan-Olmsted, Sylvia M., and Park, Jung Suk. "From On-Air to Online World: Examining the Content and Structures of Broadcast TV Stations' Web Sites." *Journalism & Mass Communication Quarterly* 77, no. 2 (2000): 321–339.
Chandler Jr., Alfred D., and Cortada, James W., eds. *A Nation Transformed by Information: How Information Has Shaped the United States from Colonial Times to the Present*. New York: Oxford University Press, 2000.
Chen, Ping-Hung. "Who Owns Cable Television? Media Ownership Concentration in Taiwan." *Journal of Media Economics*, 15, 1 (2002): 44–55.
Chenoweth, Neil. *Virtual Murdoch: Reality Wars on the Information Superhighway*. London: Vintage, 2002.
Chester, Jeffrey, and Larson, Gary O. "A 12-Step Program for Media Democracy." *The Nation*, July 23, 2002. <http://www.thenation.com/doc.mhtml?I=20020805&s=larson20020723>.
Children Now. *Big Media, Little Kids: Media Consolidation and Children's Television Programming*. Oakland, CA: Children Now, 2003.
Chipty, Tasneem. "Vertical Integration, Market Foreclosure, and Consumer Welfare in the Cable Television Industry." *American Economics Review* 91, no. 3 (June 2001): 428–453.
Chomsky, Noam. *Media Control: The Spectacular Achievements of Propaganda*. New York: Seven Stories Press, 2nd ed., 2002.
Chomsky, Noam, and Herman, Edward S. *Manufacturing Consent: The Political Economy of the Mass Media*. New York: Pantheon, 1988.
Christensen, Clayton M. *The Innovator's Dilemma: When New Technologies Cause Great Firms to Fail*. Boston: Harvard Business School Press, 1997.
Chu, L. L. "Continuity and Change in China's Media Reform." *Journal of Communication* 44, no. 3 (1994): 4–21.
Church, J. (2010). *Foreign Ownership Restrictions of Canadian Telecoms: An Analysis of Industry Canada's Proposals*. Prepared for Rogers Communications submission and comments on Industry Canada's Telecommunications Foreign Ownership Consultation Paper. <http://www.ic.gc.ca/eic/site/smt---gst.nsf/vwapj/Rogers.pdf/$file/Rogers.pdf>
Cochrane, Paul. "Saudi Arabia's Media Influence." *Arab Media and Society*. 2007. Last accessed on Jan. 17, 2012 at <http://www.arabmediasociety.com/?article=421> and <http://www.insightsmena.com/en/#!place=category&cat=Media+Usage>.
Cohen, A., Lemish, D., and Schejter, A. *The Wonder Phone in the Land of Miracles: Mobile Telephony in Israel*. Cresskill: Hampton Press, 2008.
Cole, Barry G., ed. *After the Break-Up: Assessing the New Post-AT&T Divestiture Era*. New York: Columbia University Press, 1991.
Commission of the European Union. *Notice on the Definition of Relevant Market for the Purposes of Community Competition Law*. Brussels: European Union. Dec. 9, 1997.

Commission on Freedom of the Press. *A Free And Responsible Press: A General Report on Mass Communication: Newspapers, Radio, Motion Pictures, Magazines, and Books* (The Hutchins Report). Chicago: University of Chicago Press, 1947.

Comreg. *Irish Communications Market Quarterly Key Data Report Q4 2011*. Dublin: Comreg, 2012, 68.

Compaine, Benjamin M. "The Impact of Ownership on Content: Does It Matter?" *Cardozo Arts & Entertainment Law Journal* 13 (1995): 755–775.

Compaine, B., and Gomery, D. *Who Owns the Media?: Competition and Concentration in the Mass Media Industry*. 3rd ed. Mahwah, NJ: Lawrence Erlbaum Associates, Inc., 2000.

Compaine, Benjamin, Guback, Thomas, Noble Jr., J. Kendrick, and Sterling, Christopher, eds. *Who Owns The Media?* White Plains, NY: Knowledge Industry, 1979, 1982.

Cooper, Mark. "Inequality in Digital Society." *Cardozo Journal on Media and the Arts* 73(2002): 383.

Cooper, Mark N. *Media Ownership and Democracy in the Digital Information Age: Promoting Diversity with First Amendment Principles and Market Structure Analysis*. Palo Alto, CA: Center for Internet & Society, Stanford Law School, 2003.

Copps, Michael. "Crunch Time at the FCC." *The Nation*, New York, February 3, 2003.

Cornu, Daniel, and Borruat, Régis, eds. "Structure des médias suisses." *Les Médias en Suisse, Structure et audience*. 7th ed. Lausanne: FCJ/CRFJ, 2012.

Cornu, Daniel, Borruat, Régis, Léchot, Pierre-André and Nerini, Marlène. "Inventaire des médias suisses." In Cornu, Daniel and Borruat, Régis, eds., *Les Médias en Suisse, Structure et audience*. 7th ed. Lausanne: FCJ/CRFJ, 2012.

Coulson, David, Hiromi, Cho, and Lacy, Stephen. "Competition for Readers among U.S. Metropolitan Daily, Non-metropolitan Daily and Weekly Newspapers." *Journal of Media Economics* 15 (January 2002): 21–40.

Council of Europe, "Synthesis Report on Media Concentrations in Europe." *Journal of Media Economics* 5.1 (1992):39–63.

Council of Europe. "Media Diversity in Europe." Strasbourg: Directorate General of Human Rights, Media Division, 2002.

Council of the European Communities. "Television without Frontiers Directive." Brussels, Belgium: CONSLEG System of the Office for Official Publications of the European Communities, March 7, 1997.

Cowhey, P., and Aronson, J. (2009). *Transforming Global Information and Communication Markets*. Cambridge, MA: MIT.

Crandall, Robert W. *After the Breakup: U.S. Telecommunications in a More Competitive Era*. Washington, DC: The Brookings Institution, 1991.

Cricelli, April, Livio, Gastaldi, Massimo, and Levialdi, Nathan . "Vertical Integration in International Telecommunication System." *Review of Industrial Organization* 14, no. 4 (June 1999): 337–353. 11, 2001.

CSA. "L'offre de médias et le pluralisme en Fédération Wallonie-Bruxelles." CSA, 2013. Accessed on Jan. 10, 2013. <http://www.csa.be/pluralisme>.

Cukier, Kenneth Neil, and Hindman, Mathew. "Measuring Media Concentration Online and Offline." Feb. 12, 2004. CUKIER.com. Last accessed on May 29, 2007 <http://www.cukier.com/writings/webmedia-jan04.htm>.

Curran, J., and Seaton, J. *Power Without Responsibility: The Press, Broadcasting and New Media in Britain*. 6th ed. London: Routledge, 2003.

Dale, Dennis F. "What Subscribers Think of Group Ownership of Newspapers." *Journalism Quarterly* 57 (Summer 1980): 314–316.

Daly, Charles P. "The Business of Magazine Publishing." In Daly, Charles, and Henry, Patrick, eds., *The Magazine Publishing Industry*. Boston: Allyn & Bacon, 1996.

Dantas, M. *A lógica do capital informação*, 2nd ed. Rio de Janeiro: Contraponto, 2002.

Darbishire, H., and Harrison, F. *Transparency of Media Ownership in Europe*. Open Society Media Program, Madrid: Access Info Europe (2012).

Datta, Rathin. *The Indian Entertainment and Media Industry: Unraveling the Potential*. PriceWaterhouseCoopers, 2006. Last accessed on Jan. 29, 2014 at <http://www.pwc.in/en_IN/in/assets/pdfs/ficci-pwc-indian-entertainment-and-media-industry.pdf>.

Davenport, Lucinda, and Lacy, Stephen. "Daily Newspaper Market Structure, Concentration, and Competition." *The Journal of Media Economics* 7, no. 3 (1994): 33–46.

Davidson, R., and Schejter, A. "'Their Deeds Are the Deeds of Zimri; but They Expect a Reward Like Phineas': Neoliberal and Multicultural Discourses in the Development of Israeli DTT Policy." *Communication, Culture and Critique* 4(2011): 1–22.

De Bens, E., et al. *Media Landscape: Belgium*. European Journalism Centre, Nov. 8, 2010. Accessed on Jan. 10, 2013. <http://www.ejc.net/media_landscape/article/belgium/>.

De Bens, E. "Belgium." In Kelly, M., Mazzoleni, G., and McQuail, D., eds. *The Media in Europe: The Euromedia Handbook*. 3rd ed. London: SAGE Publications, 2004.

Dejong, A. S., and Bates, B. J. "Channel Diversity in Cable Television." *Journal of Broadcasting and Electronic Media* 35, no. 2 (1991): 159–167.

Della Vigna, S., Kennedy, A. *Does Media Concentration Lead to Biased Coverage? Evidence from Movie Reviews*. Mimeo, University of California Berkeley, 2011.

Demers, David P. "Corporate Structure and Emphasis on Profits and Product Quality at U.S. Daily Newspapers." *Journalism Quarterly* 68 (1991): 15–26.

Demers, David P. *Global Media: Menace or Messiah?* Cresskill: Hampton Press, 1999.

Demers, David P. "Effect of Corporate Structure on Autonomy of Top Editors at U.S. Dailies." *Journalism Quarterly* 70(1993): 499–508.

Demers, David P. *The Menace of the Corporate Newspaper: Fact or Fiction?* Ames: Iowa University Press, 1996.

Demers, David K. "Structural Pluralism, Corporate Newspaper Structure, and News Source Perceptions: Another Test of the Editorial Vigor Hypothesis." *Journalism & Mass Communication Quarterly* 75 (Autumn 1998): 572–592.

De Miguel, Juan Carlos. *Los grupos multimedia: estructuras y estrategias de los medios–europeos*. Barcelona: Bosch, 1993.

Demsetz, Harold. "The Amenity Potential of Newspapers and the Reporting of Presidential Campaigns." In Demsetz, H., ed., *Efficiency, Competition and Policy*. London: Basil Blackwell, 1989.

Demsetz, Harold, and Lehn, Kenneth. "The Structure of Corporate Ownership: Causes and Consequences." *Journal of Political Economy* 93 (1985): 1155–1177.

U.S. Department of Justice and Federal Trade Commission. "Horizontal Merger Guidelines." Washington, DC: Government Printing Office, April 2, 1992, revised April 8, 1997.

Dertouzous, J. N., and Trautman, W. B. "Economic Effects of Media Concentration: Estimates from a Model of the Newspaper Firm." *The Journal of Industrial Economics* 39, no. 1 (September 1990): 1–14.

Dimmick, J. *Media Competition and Coexistence: The Theory of Niche*. Mahwah, NJ: Erlbaum Associates, 2003.

"Directive 2010/13/EU of the European Parliament and of the Council of 10 March 2010 on the coordination of certain provisions laid down by law, regulation or administrative action in Member States concerning the provision of audiovisual media services (Audiovisual Media Services Directive)." *Official Journal of the European Union*, Apr. 15, 2012. Jan. 10, 2013. <http://eur-lex.europa.eu/LexUriServ/LexUriServ.do?uri=OJ:L:2010:095:0001:0024:EN:PDF>

Djankov, Simon, et al. "Who Owns the Media?" *Journal of Law and Economics* 46, no. 2 (2003): 341–381.

Dodge, John. "Ten Years of Events That Shook Our World: Ten Major Events in the Computer Industry, 1984–1994." *PC Week* 11 (Feb. 28, 1994): 70.

"Does Ownership Matter in Local Television News: A Five-Year Study of Ownership and Quality." Washington, DC: Project for Excellence in Journalism, Feb. 17, 2003.

Donald, Stephanie, Keane, Michael, and Hong, Yin, eds. *Media in China: Consumption, Content, and Crisis*. New York: Routledge. 2002.

Doremus, Mark, Hakanen, Ernest A., and Slattery, Karen L. "The Expression of Localism: Local TV News Coverage in the New Video Marketplace." *Journal of Broadcasting and Electronic Media* 40(1996): 403–413.

Doshi, J. K., et al. "Diffusion of Television in India." *Media Asia* 15 (1988): 4. Accessed on Feb. 21, 2013. <http://utminers.utep.edu/asinghal/Articles%20and%20Chapters/Singhal-Rogers-Doshi-Rahman-1988-Diffusion%20of%20television.pdf>.

Downing, John D. H. "Media Ownership, Concentration, and Control: The Evolution of Debate." In Janet Wasko, Graham Murdock, and Helena Sousa, eds., *The Handbook of Political Economy of Communications*, Wiley-Blackwell, 2011, 140–168.

Doyle, Gillian. "The Economics of Monomedia and Cross Media Expansion: A Study of the Case Favouring Deregulation of TV and Newspaper Ownership in the UK." *Journal of Cultural Economics* 24(2000): 1–26.

Doyle, Gillian. *Media Ownership: The Economics and Politics of Convergence and Concentration in the UK and European Media*. London: SAGE, 2002.

Doyle, Gillian. "Regulation of Media Ownership and Pluralism in Europe: Can the European Union Take Us Forward." *Cardozo Arts & Ent. LJ* 16(1998): 451.

Doyle, Gillian. "Undermining Media Diversity: Inaction on Media Concentrations and Pluralism in the EU." *European Studies: A Journal of European Culture, History and Politics* 24, no. 1 (2007): 135–156. Print.

*Duncan's Radio Market Guide*. 2002 ed. Kalamazoo, MI: Duncan Media Enterprises, 2002.

Dugger, Ronald. "The Corporate Domination of Journalism." In Seriin, William, ed. *The Business of Journalism*. New York: New Press, 2000, 27–56.

Durante, Ruben, and Knight, Brian. "Partisan Control, Media Bias, and Viewer Responses: Evidence from Berlusconi's Italy." *Journal of the European Economic Association* 10, no. 3 (2012): 451–481.
Eco, Umberto. "The Multiplication of Media." In Eco, Umberto, ed. *Travels in Hyperreality: Essays.* San Diego, CA: Harcourt Brace Javanovich, 2001.
Edwin, Baker C. *Media Concentration and Democracy: Why Ownership Matters.* Boston: Cambridge University Press, 2007.
Eickelman, D. F., and Anderson, J. W. *New Media in the Muslim World: The Emerging Public Sphere,* 2nd ed. Bloomington, IN: Indiana University Press, 2003.
Einstein, M. *Prime Time Power and Politics: The Financial Interest and Syndication Rules and Their Impact on the Structure and Practices of the Television Industry.* Ph.D. dissertation, Department of Culture and Communications, New York University, 2000.
Einstein, Mara. "Program Diversity and the Program Selection Process on Broadcast Network Television." Working Paper, Federal Communications Commission, Media Ownership Working Group, September 2002.
Ekelund Jr., Robert B., Ford, George S., and Jackson, J. D. "Is Radio Advertising a Distinct Local Market? An Empirical Analysis." *Review of Industrial Organization* 14(1999): 239–256.
Ekclund Jr., Robert B., Ford, George S., and Koutsky, Thomas. "Market Power in Radio Markets: An Empirical Analysis of Local and National Concentration." *Journal of Law and Economics* 43, no. 1 (April 2000): 157–184.
Electronic Industries Alliance. *Electronic Market Data Book, 1996-2000.* Arlington, VA: Electronic Industries Alliance, 2001, 14.
Ellul, Jacques. *Propaganda: The Formation of Men's Attitudes.* New York: Random House Inc., 1962.
Emery, Edwin, and Emery, Michael. *The Press & America: An Interpretive History of the Mass Media.* Englewood Cliffs, NJ: Prentice Hall, 1984, 39.
Emord, Jonathan W. "The First Amendment Invalidity of FCC Ownership Regulations." *Catholic University Law Review* 38(1989): 401, 449.
Entman, R. M., & Paletz, David L. *Media, Power, Politics.* New York: Free Press, 1981.
EPRA. *Transparency of Media Ownership: What Role for Regulators?,* 2013. Last accessed on Jan. 29, 2014 at <http://www.epra.org/news_items/transparency-of-media-ownership-what-role-for-regulators>.
European Audiovisual Observatory. *Yearbook: Film and Home Video.* Vol. 3. Strasbourg: European Audiovisual Observatory, 2010.
European Commission. "Commission Opens Consultations on Media Freedom and Pluralism and Audiovisual Media Regulator Independence." *Digital Agenda for Europe.* A Europe 2020 Initiative, 2013. <http://europa.eu/rapid/press-release_IP-13-267_en.htm>.
European Commission. Green Paper on the Convergence of the Telecommunications, Media and Information Technology Sectors, and the Implications for Regulation, 2007. Available at <http://ec.europa.eu/green-papers/>. Accessed on May 14, 2012.
European Federation of Journalists. *European Media Ownership: Threats on the Landscape: A Survey of Who Owns What in Europe.* Brussels: European Federation of Journalists, 2003.
European Parliament. *Concentration and Pluralism in the Media in the European Union.* European Parliament resolution, 2008. <http://www.europarl.europa.eu/sides/getDoc.do?type=TA&reference=P6-TA-2008-0459&language=EN>.
Evans, David S., ed. *Breaking Up Bell.* New York: Elsevier Science Publishing Co. Inc., 1983.
Ezrahi, Y., Goshen, Z., and Leshem, S. *Cross Ownership–Control and Competition in the Israeli Media: Economic and Legal Aspects and Their Impact on the Israeli Democracy.* Jerusalem: Israel Democracy Institute, 2003. <http://en.idi.org.il/media/274944/%D7%91%D7%A2%D7%9C%D7%95%D7%AA%20%D7%A6%D7%95%D7%9C%D7%91%D7%AA%20-%20%D7%9E%D7%9C%D7%90.PDF>.
"False Freedom: Online Censorship in the Middle East and North Africa." *UNHCR: Human Rights Watch.* Nov. 15, 2005. Last accessed on Feb. 13, 2012 at <http://www.unhcr.org/refworld/docid/45cc445f2.html>.
Fandy, Mamoun. *(Un)Civil War of Words: Media and Politics in the Arab World.* New York: Praeger, 2007.
Faustino, Paulo. *Concentração, Pluralismo e Regulação dos Media.* Lisbon: MediaXXI, 2011.
Federal Communications Commission. 2001 Video Competition Report, 17 FCC Rcd. at 1282, 79.
Federal Communications Commission. 2002 Biennial Regulatory Review: Review of the Commission's Broadcast Ownership Rules and Other Rules Adopted Pursuant to Section 202 of the Telecommunications Act of 1996, 18 FCC Rcd 13620, 13711-47 (2003) ("2002 Biennial Review Order").
Federal Communications Commission. Annual Assessment of the Status of Competition in Markets for the Delivery of Video Programming, CC Docket No. 00-132, Fifteenth Report, in BIA

Financial. *Television Market Report: 2013.* Chantilly, VA: November 2013.
Federal Communications Commission. Newspaper/Radio Cross-Ownership, Notice of Proposed Rulemaking, 16 FCC Rcd. 17292 (2002).
Federal Communications Commission, Media Bureau. *Report on Ownership of Commercial Broadcast Stations.* Washington DC: Federal Communications Commission, 2012.
Federal Communications Commission. Rules and Policies Concerning Multiple Ownership of Radio Broadcast Stations in Local Markets, Notice of Proposed Rulemaking, MM Docket No. 00-244, 16 FCC Rcd 19861 (2001).
Federman, A. "Moscow's New Rules: Islands of Press Freedom in a Country of Control." *Columbia Journalism Review* Jan.-Feb., 2010. <http://www.cjr.org/feature/moscows_new_rules.php?page=all>.
Fenton, Natalie. "Deregulation or Democracy? New Media, News, Neoliberalism and the Public Interest. *Continuum* 25, no. 1 (2011): 63–72.
Fernández, F., Godoy, S., Grassau, D., Julio, P., Pellegrini, S., Puente, S. and Soto, J.: *Ventanas y Espejos; televisión local en red.* Santiago: El Mercurio/Aguilar, 2009.
Fetschering, M., and Knolmayer, G. "Business Models for Content Delivery: An Empirical Analysis of the Newspaper and Magazine Industry." *International Journal on Media Management* 6(2004): 4–11.
"Film Sector Monopoly Broken Up." *China Daily,* June 10, 2003. Accessed on Feb. 28, 2013.
Finlay, A. (Ed.). *Global Information Society Watch 2009.* Uruguay: APC/Hivos, 104–105.
Finn, A., Hoskins, C., and McFayden, S. *Global Television and Film: An Introduction to the Economics of the Business.* Oxford: Clarendon Press, 1997.
Flores-Roux, Ernesto, and Mariscal, Judith. *Política de Generación de Infraestructura de Telecomunicaciones en México: Una Crítica.* Mexico City: Centro de Investigación y Docencia Económicas, 2010. <http://telecomcide.org/docs/publicaciones/DTAP-246.pdf>.
"Forbes 400." *Forbes.com.* September 2013. Last accessed on Jan. 29, 2014 at <http://www.forbes.com/forbes-400/list>.
Fox, E. *Latin American Broadcasting, from tango to Telenovela.* Luton, Hertforshire: John Libbey/University of Luton, 1997.
Fox, Kraig G. "Paramount Revisited: The Resurgence of Vertical Integration in the Motion Picture Industry." *Hofstra Law Review* 21 (Winter 1992): 505–536.
Freedman, Des. "Metrics, Models and the Meaning of Media Ownership." *International Journal of Cultural Policy* 20, no. 2 (2014): 170–185.
Freedman, Des. *The Politics of Media Policy.* Cambridge: Polity Press, 2008.
Froeb, Luke M., and Werden, Gregory J. "The Effects of Mergers in Differentiated Products Industries: Logit Demand and Merger Policy." *Journal of Law, Economics and Organization* 10, no. 2 (October 1994): 407, 423–424.
Froeb, Luke M., and Werden, Gregory. "The Effects of Mergers in Differentiated Products Industries: Demand and Merger Policy." *The Journal of Law, Economics, and Organization* 10, no. 2 (1994): 407–416.
Fung, A.Y.H. "Political Economy of Hong Kong Media: Producing a Hegemonic Voice." *Asian Journal of Communication,* 17, no. 2 (2007): 159–171.
Gal, M. *Competition Policy for Small Market Economies.* Cambridge, MA: Harvard University Press, 2003.
Galambos, Louis, and Temin, Peter. *The Fall of the Bell System: A Study in Prices and Politics.* New York: Cambridge University Press, 1987.
Gandal, N. "Compatibility, Standardization, and Network Effects: Some Policy Implications." *Oxford Review of Economic Policy* 18 (2002): 80–91.
Garabova, Madina I. *Osobennosti funkcionirovaniya rossiiskih SMI v usloviyah ih monopolizacii I cincentracii (Particular Qualities of Russian Media Functioning in Terms of Monopolization and Concentration).* Diss. Moscow: RUDN, 2004.
George, Lisa. "What's Fit to Print: The Effect of Ownership Concentration on Product Variety in Daily Newspaper Markets." Working Paper, City University of New York, 2001.
George, Lisa, and Waldfogel, Joel. "Who Affects Whom in Daily Newspaper Markets?" *Journal of Political Economy* 11(2003): 765–785.
Gerbarg, Darcy, Groebel, Jo, and Noam, Eli M., eds. *Internet Television.* London: Lawrence Erlbaum Associates, 2004.
Germano, Fabrizio, and Meier, Martin. "Concentration and Self-Censorship in Commercial Media." *Journal of Public Economics* 97(2013): 117–130.
Getino, O. *Las industrias culturales en la Argentina: Dimensión económica y políticas públicas.* Buenos Aires: Colihue, 1995.
Ghiglione, Lauren, ed. *The Buying and Selling of American Newspapers.* Indianapolis: R.J. Berg, 1984.
Gitlin, Todd. *The Whole World Is Watching: Mass Media in the Making and Unmaking of the New Left.* Berkeley: University of California Press,. 1981/2003.

Godoy, S. *An Unlikely Mixture. An Evaluation of Chile's Market Oriented Model of Public Television*. Köln: Lambert Academic Publishing, 2009.

Godoy, S., and Gronemeyer, M. *Mapping Digital Media: Country Report—Chile*. A report by the Open Society Foundations, London, Oct. 23, 2012. <http://www.opensocietyfoundations.org/reports/mapping-digital-media-chile>

Goggin, G. *Global Mobile Media*. New York: Routledge, 2011.

Goldman, Merle. "The Role of the Press in Post-Mao Political Struggles." In Lee, Chin-Chuan, ed., *China's Media, Media's China*. Boulder, CO: Westview, 1994.

Gomery, Douglas. "Interpreting Media Ownership." In Compaine, B. and Gomery, D., eds., *Who Owns the Media?* 3rd ed. Mahwah, NJ: Lawrence Earlbaum Associates, 2000, 507–535.

Gómez, Rodrigo, Sosa, Gabriel, Bravo, Jorge, and Tellez, Primavera. *Mapping Digital Media: México*. London: Open Society Foundations–Media Program, 2011. June 13, 2011. <http://www.soros.org/initiatives/media/articles_publications/publications/mapping-digital-media-mexico-20110613>.

Goolsbee, Austan, and Petrin, Amil. "The Consumer Gains from Direct Broadcast Satellites and the Competition with Cable Television." Working Paper No. 8317, National Bureau of Economic Research, May 29, 2001, 4, 27–28, 32. Available at <http://gsbwww.uchicago.edu/fac/austan.goolsbee/research/satfin.pdf>.

Gopbetz, R. H., and Scott, D. K. "Hard News/Soft News Content of the National Broadcast Networks: 1972–1987." *Journalism Quarterly* 69, no. 2 (Summer 1992): 406–412.

Graham, Daniel A., and Vernon, John M. "Profitability of Monopolization by Vertical Integration." *Journal of Political Economy* 79, no. 4 (July/August 1971): 924–925.

Grant, A. E. "The Promise Fulfilled? An Empirical Analysis of Program Diversity on Television." *The Journal of Media Economics* 7, no. 1 (1994): 51–64.

Greco, Albert N. "The Impact of Horizontal Mergers and Acquisitions on Corporate Concentration in the U.S. Book Publishing Industry, 1989–1994." *Journal of Media Economics* 12, no. 3 (1999): 165–180.

Greco, Albert N. "Market Concentration Levels in the U.S. Consumer Book Industry: 1995–1996." *Journal of Cultural Economics* 24, no. 4 (November 2000): 321–336.

Groebel, J., Feldmann, V., and Noam, Eli M., eds. *Media Content and Services for Mobile Wireless Communications*. London: Lawrence Erlbaum Associates, 2006, 225–237.

Grönlund, Mikko, and Björkroth, Tom. "Newspaper Market Concentration, Competitive Pressure and Financial Performance: The Case of Finland." *Journal of Media Business Studies* 8, no. 3 (2011). Accessed on Jan. 10, 2013. <http://www.econis.eu/PPNSET?PPN=722188722>.

Guback, Thomas. "Theatrical Film." In Compaine, Benjamin, Sterling, Christopher, Guback, Thomas, and Noble Jr., J. Kendrick., eds., *Who Owns the Media?* White Plains, NY: Knowledge Industry Publications, 1982, 199–290.

"Guide to Media and Content Regulation in Asia Pacific" 2012. <http://www.commsalliance.com.au/_data/assets/pdf_file/0016/42136/Guide-to-Media-and-Content-Regulation-in-Asia-Pacific.pdf>.

Gustaffson, K. E. "The Circulation Spiral and the Principle of Household Coverage." *The Scandinavian Economic History Review* 26, no. 1 (1978): 1–14.

Gwiasda, Gregory W. "Network News Coverage of Campaign Advertisements: Media's Ability to Reinforce Campaign Messages." *American Politics Research* 29, no. 5 (2001): 461–482.

Hachmeister, Lutz, and Rager, Günther, eds. *Wer beherrscht die Medien? Die 50 größten Medienkonzerne der Welt; Jahrbuch*. Munich: C.H. Beck, 2005.

Häckner, Jonas, and Nyberg, Sten. "Advertising and Media Market Concentration." *Journal of Media Economics* 21, no. 2 (2008):79–96.

Haddock, David D., and Polsby, Daniel D. "Bright Lines, the Federal Communications Commission's Duopoly Rule, and the Diversity of Voices." *Federal Communications Law Journal* 42 (1990): 331–364.

Hahn, Tae-youl, and Nixon, Raymond B. "Concentration of Press Ownership: A Comparison of 32 Countries." *Journalism Quarterly* 47 (Spring 1971): 5–16.

Halimi, Serge. "United States: An Unfree Press." *Le Monde Diplomatique*, Paris, June 8, 2003.

Hallin, Daniel, and Mancini, Paolo. *Comparing Media Systems: Three Models of Media and Politics*. New York: Cambridge University Press, 2004.

Hallin, Daniel, and Papathanassopoulos, Styliano. "Political Clientelism and the Media: South Europe and Latin America in Comparative Perspective." *Media, Culture & Society* 24(2002): 175–95.

Hans-Bredow-Institut. *Medien von A bis Z*. Wiesbaden: VS Verlag für Sozialwissenschaften, 2006.

Harcourt, Alison. "The regulation of Media Markets in Selected EU Accession States in Central and

Eastern Europe." *European Law Journal* 9, no. 3 (2003): 316–340.

Harcourt, Alison J. "EU Media Ownership Regulation: Conflict over the Definition of Alternatives." *JCMS: Journal of Common Market Studies* 36, no. 3 (1998): 369–389.

Harding-Smith, Rob and Center for Policy Development. *Media Ownership and Regulation in Australia*. Center for Policy Development, 2011. Last accessed on Jan. 29, 2014 at <http://cpd.org.au/wp-content/uploads/2011/11/Centre_for_Policy_Development_Issue_Brief.pdf>.

Hargreaves Heap, Shaun, et al. *Consultation on the Assessment of Media Mergers under the Competition Act 2002: Final Report*. Dublin: Competition Authority, 2006, 8–11, 50.

Hartman, Barrie, and Rarick, Galen. "The Effects of Competition on One Daily Newspaper's Content." *Journalism Quarterly* 43 (Autumn 1966): 459–463.

Haryanto, Ignatius, Hill, D., and Sen, K. "Media Ownership and Its Implication for Journalists and Journalism in Indonesia." *Politics and the New Media in 21st Century Indonesia*. London: Routledge, 2010, 104–118.

He, Zhou. "Chinese Communist Party Press in a Tug of War: A Political Economy Analysis of the Shenzhen Special Zone Daily." In Lee, Chin-Chuan, ed., *Power, Money, and Media: Communication Patterns and Bureaucratic Control in Cultural China*. Evanston, IL: Northwestern University Press, 2000.

Hellman, Heikki, and Soramaki, Martti. "Competition and Content in the U.S. Video Market." *Journal of Media Economics* 7, no. 1 (1994): 29–49.

Hepp, Erica. "Barking up the Wrong Channel: An Analysis of the Communication Law Problems through the Lens of Media Concentration Rules." *Boston University Law Review* 85 (April 2004): 553–584.

Herrero, Monica, and Medina, Mercedes. "Concentration and Vertical Integration in the European Television Production Market," presented at 6th *World Media Economics Conference*. Centre d'Etudes Sure les Médias and Journal of Media Economics, HEC Montreal, Montreal, Canada, May 12–15, 2004.

Hervas-Drane, Andres, and Noam, Eli. "Online Sharing and Cultural Trade Policy." *Journal of Public Economies* (2013).

Hesmondhalgh, David. "Media and Cultural Policy as Public Policy: The Case of the British Labour Government." *International Journal of Cultural Policy* 11, no. 1 (2005): 95–109.

Heuvel, J. V., and Dennis, E.E. *The Unfolding Lotus: East Asia's Changing Media*. New York: Freedom Forum Media Studies Center, Columbia University, 1993.

Hickey, Neil. "Money Lust: How Pressure for Profits Is Perverting Journalism." *Columbia Journalism Review* 37, no. 2 (July/August, 1998): 28–36.

Hickey, Neil. "Unshackling Big Media." *Columbia Journalism Review* 40, no. 2 (July/August 2001): 30–33.

Hindman, Matthew, Johnson, Judy A., and Tsioutsiouliklis, Kostas. "Measuring Media Diversity Online and Offline: Evidence from Political Websites," presented at the *Proceedings of the 32nd Annual Telecommunications Policy Research Conference*. Telecommunications Policy Research Conference, Arlington, VA, Sept. 17, 2004.

Hitchens, L. P. "Media Ownership and Control: A European Approach." *The Modern Law Review* 57, no. 4 (1994): 585–601.

"Horizontal Merger Guidelines." April 8, 1997. US Department of Justice. Last accessed on May 24, 2007 at <http://www.usdoj.gov/atr/public/guidelines/horiz_book/hmg1.html>

Horgan, John. *Irish Media: A Critical History Since 1922*. London: Routledge, 2001.

Horwitz, Robert B., and Currie, W. "Another Instance Where Privatization Trumped Liberalization: The Politics of Telecommunications Reform in South Africa—A Ten-Year Retrospective." *Telecommunications Policy* 31, no. 8–9 (2007), 445–462.

Horwitz, R. "On Media Concentration and the Diversity Question." *The Information Society*, 21(2005): 181–204.

Howard, P. N. "Ownership Diversity in Muslim Media Systems." *World Information Access Project (WIA)*. Oct. 6, 2008. Last accessed on 16 Jan. 2012 at <http://www.wiaproject.org/index.php/68/ownership-diveristy-in-muslim-media-systems>

Huang, C. J. 'China's State-Run Tabloids: The Rise of "City Newspapers."' *Gazette* 63, no. 5 (2001): 435–450.

Huang, Chengju. "The Development of a Semi-Independent Press in Post-Mao China: An overview and a Case Study of Chengdu Business News." *Journalism Studies* 1, no. 4 (2000): 649–664.

Huang, Chengju. "Editorial: From Control to Negotiation: Chinese Media in the 2000s." *International Communication Gazette* 69(2007): 402–413.

Huang, Wanwu. "Market Economy and Transformation of Periodical Industry in China." *Chinese Journal of Scientific and Technical Periodicals* 4(1993): 3.

Huang, Y. "Peaceful Evolution: The Case of the Television Reform in Post-Mao China." *Media Culture and Society* 16(1994): 217–241.
Huber, P., Kellog, M., and Thorne, J. "*The Geodesic Network II: 1993 Report on Competition in the Telephone Industry*." Washington, DC: Geodesic Company, 1992.
Hull, Geoffrey. *The Recording Industry*. Boston: Allyn and Bacon, 1998.
Hunter, C. "The Ever-Increasing Cost of Media." *Journal of Marketing* Oct-Nov (2006): 42–43.
Institut De l'Audiovisuel et des Télécommunications en Europe (IDATE). <http://www.idate.org/en/Home/>. Jan. 4, 2013.
International Telecommunication Union (ITU). "Measuring the Information Society 2012." Geneva: ITU, 2012.
Iosifidis, P. "Cómo medir la concentración medial." *Cuadernos de Información* 12(1997): 94–111.
Iosifidis, P., Steemers, J., and Wheeler, M.: *European Television Industries*. London: BFI, 2005.
Iosifides, Petros. "Methods of Measuring Media Concentration." *Media Culture & Society* 19(1997): 643–663.
Iosifidis, Petros. *Public Television in the Digital Era: Technological Challenges and New Strategies for Europe*. 2nd sed. Basingstoke: Palgrave Macmillan, 2012.
Iyengar, Shanto, and Kinder, Donald R. *News That Matters: Television and American Opinion*. Chicago: University of Illinois Press, 1987.
Jagannathan, N. S. *Independence and the Indian Press: Heirs to a Great Tradition*. 1st ed. New Delhi: Konark Publishers, 1999.
Jha, Shilpi. "Need of Research Related to Indian Ownership Patterns in Media." *Mass Communicator: International Journal of Communication Studies* 7, no. 3 (2013): 28.
JiJi Eiga Tsushinsha. "Eiga Nenkan (Film Industry Yearbook)," Tokyo: Nihon Tosho Senta 1993, 1997, 2001, 2005, 2009.
José, María, and Serrano, Pérez. "Analysis of the right to information and pluralism in Spanish media holdings." *Global Media Journal: Mediterranean Edition* 7, no. 2 (2012): 52–60.
Just, Natascha. "Measuring Media Concentration and Diversity: New Approaches and Instruments in Europe and the US." *Media, Culture & Society* 31, no. 1 (2009): 97–117.
Just, N., and Latzer, M. "EU Competition Policy and Market Power Control in the Mediamatics Era." *Telecommunications Policy* 24, no. 5 (June 2000): 395–441.
Kachkaeva, Anna G. *Rossijskie sredstva massovoi informacii, vlast' i capital: k voprosu o koncentracii SMI v Rossii (Russian Mass Media, Power and Capital: On Concentration of Russian Media)*. Moscow: Centre Prava SMI, 2012.
Kaitatzi-Whitlock, Sophia. "Pluralism and Media Concentration in Europe Media Policy as Industrial Policy." *European Journal of Communication* 11, no. 4 (1996): 453–483.
Kalathil, S. "Chinese Media and the Information Revolution." *Harvard Asia Quarterly* Winter, 2002. <www.carnegie.endowment.org/publications/index.cfm?fa=view&id=924>. Accessed on Nov. 8, 2006.
Kapfer, Jack, Kurpius, David, Shano-Yeon Chern, David, and Voakes, Paul S. "Diversity in the News: A Conceptual and Methodological Framework." *Journalism and Mass Communications Quarterly* 73 (Autumn 1996): 582–593.
KatholiekeUniversiteit Leuven (ICRI), Jönköping International Business School (MMTC), Central European University (CMCS), and Ernst & Young Consultancy Belgium. "Independent Study on Indicators for Media Pluralism in the Member States: Towards a Risk-Based Approach." Leuven: KatholiekeUniversiteit Leuven, 2009. Accessed on June 19, 2012, <http://ec.europa.eu/information_society/media_taskforce/doc/pluralism/study/final_report_09.pdf>.
Kathuria, Raiat, Singh, H. V., and Soni, Anita. "Telecom Policy Reform in India." Washington, DC: The World Bank, 2005. Accessed on Feb. 21, 2013, <http://siteresources.worldbank.org/INTRANETTRADE/Resources/Singh.pdf>.
Katz, Michael L. "Old Rules and New Rivals: An Examination of Broadcast Television Regulation and Competition," submitted as an appendix to the Emergency Petition for Relief and Supplemental Comments of Fox Television Stations, Inc., in MM Docket No. 98-35, November 18, 1999.
Katz, Michael L., and Shapiro, Carl. "Systems Competition and Network Effects." *Journal of Economic Perspectives* 8(1994): 93, 110.
Kaya, Rasit. "A Fait Accompli: Transformation of Media Structures in Turkey." *Metu Studies in Development* 21(1994): 3.
Kaya, Rasit, and Cakmur, Baris. "Politics and the Mass Media in Turkey." *Turkish Studies* 11(2010): 4.
Kennedy, J. "Maintaining Popular Support for the Chinese Communist Party: The Influence of Education and the State-Controlled Media." *Political Studies* 57(2009): 517–536.
Kenny, Colum. "A Double-Edged Sword: Irish Media Merger Policy in Transition." *Journal of Media Business Studies* 6, no. 3 (2009): 93–108.

Khanna, T. and Palepu, K. "The Evolution of Concentrated Ownership in India: Broad Patterns and a History of the Indian Software Industry." Cambridge: National Bureau of Economic Research, 2004. Accessed on June 14, 2012. <http://www.nber.org/papers/w10613>.

Khlebnikov, Pavel Y. *Krestnyi otetz Kremlja Boris Berezovsky (The Kremlin Godfather Boris Berezovsky)*. Moscow: Detektiv-Press, 2001.

Klass, Michael W., and Salinger, Michael A. "Do New Theories of Vertical Foreclosure Provide Sound Guidance for Consent Agreements in Vertical Merger Cases?" *Antitrust Bulletin* 40(1995): 667.

Klieman, H. "Content Diversity and the FCC's Minority and Gender Licensing Policies." *Journal of Broadcasting and Electronic Media* 35(1991): 411–429.

Klimkiewicz, Beata. *Media Freedom and Pluralism Media Policy Challenges in the Enlarged Europe*. Budapest, Hungary; New York: Central European University Press, 2010.

Klinenberg, Eric. *Fighting for Air: The Battle to Control America's Media*. New York: Henry Holt & Co., 2007.

Kohli-Khandekar, Vanita. *The Indian Media Business*. 3rd ed. New Delhi: SAGE Publications, 2010.

Koltsova, Olessia. *News Media and Power in Russia*. Kentucky: Taylor & Francis, 2006.

Korea Association for ICT Promotion, *Statistics of Broadcasting and Telecommunications Industry*. 2011 < http://www.kait.or.kr/index.jsp>

Korea Communications Commission. *Annual Report on Broadcasting Industry*, 2003–2011. <http://www.kisdi.re.kr/kisdi/fp/kr/board/listSingleBoard.do?cmd=listSingleBoard&sBoardId=BCAST_DB2>

Korean Film Council. *Korean Cinema 2011*. Seoul: Dong Kwang Printing. 2011.

Korea Press Foundation. *Statistics of Newspaper Industry*. 2011 <http://www.kpf.or.kr/journal/mdata_result_view.jsp?ctg=%C1%B6%BB%E7%BA%D0%BC%AE%BC%AD&bd_seq=7372&pg=1&url=4>

Krasilovsky, M. William, and Shemel, Sidney. *This Business of Music*. 3rd ed. New York: Billboard, 1995.

Krasnow, Erwin G., Stern, Jill A., and Senkowski, R. Michael. "New Video Marketplace and the Search for a Coherent Regulatory Philosophy." *Catholic University Law Review* 32, no. 3 (Spring 1983): 529, 541–543.

Krattenmaker, Thomas G., and Powell Jr., Lucas A. *Regulating Broadcast Programming*. Cambridge, MA/London: MIT Press and Washington, DC: AEI Press, 1995.

Kunz, W. M. *Culture Conglomerates: Consolidation in the Motion Picture and Television Industries*. Lanham: Rowman & Littlefield, 2007.

Künzler, M. "Switzerland: Desire for Diversity without Regulation—A Paradoxical Case? *International Communication Gazette* 71, 2 (2009): 67–76.

La Porta, Rafael, Lopez-de-Silanes, Florencio, and Shleifer, Andrei. "Corporate Ownership Around the World." *Journal of Finance* 54, no. 2 (1999): 471–517.

Lacy, Stephen. "A Model of Demand of News: Understanding the Impact of Competition on Daily Newspaper Content." *Journalism Quarterly* 66, no. 1 (Spring 1989): 40–48, 128.

Lacy, Stephen. "Effects of Group Ownership on Daily Newspaper Content." *Journal of Media Economics* 4, no. 1 (Spring 1991): 35–47.

Lacy, Stephen. "The Effects of Intracity Competition on Daily Newspaper Content." *Journalism Quarterly* 64 (Summer/Autumn 1987): 282–290.

Lacy, Stephen. "The Financial Commitment Model of News Media Competitions." *Journal of Media Economics* 5, no. 2 (Summer 1992): 5–21.

Lacy, Stephen. "The Impact of Intercity Competition on Daily Newspaper Content." *Journalism Quarterly* 65 (Summer 1988): 399–406.

Lacy, Stephen, and Blanchard, Alan. "The Impact of Public Ownership, Profits, and Competition on Number of Newsroom Employees and Starting Salaries in Mid-Sized Daily Newspapers." *Journalism and Mass Communication Quarterly* 80, no. 4 (Winter 2003): 949–968.

Lacy, Stephen, and Fico, Fredrick. "Newspaper Content Quality and Circulation." *Newspaper Research Journal* 12 (Spring 1991): 46–57.

Lacy, Stephen, and Fico, Fredrick. "Newspaper Quality and Ownership: Rating the Groups." *Newspaper Research Journal* 11 (Spring 1990): 42–56.

Lacy, Stephen, Fico, Frederick, and Simon, Todd F. "The Relationship Among Economic Newsroom and Content Variables: A Path Model." *Journal of Media Economics* 2, no. 2 (1989): 51–66.

Lacy, Stephen, and Picard, Robert G. "Interactive Monopoly Power in the Daily Newspaper Industry." *Journal of Media Economics* 3, no. 3 (1990), 27–37.

Lacy, Stephen, and Shaver, Mary Alice. "The Impact of Intermedia and Newspaper Competition on Advertising Linage in Daily Newspapers." *Journalism & Mass Communication Quarterly* 76 (Winter 1999): 729–744.

Lacy, Stephen, Shaver, Mary Alice, and St. Cyr, Charles. "The Effects of Public Ownership and Newspaper Competition on the Financial

Performance of Newspaper Corporations: A Replication and Extension." *Journalism & Mass Communication Quarterly* 73 (Summer 1996): 332–341.
Lacy, Stephen, and Simon, Todd F. "Intercounty Group Ownership of Daily Newspapers and the Decline of Competition for Readers." *Journalism & Mass Communication Quarterly* 74 (Winter 1997): 814–825.
Lacy, Stephen, and Simon, Todd F. *The Economics and Regulation of United States Newspapers*. Norwood, NJ: Ablex, 1993.
Latham, Kevin. "Nothing But Truth: News Media, Power, and Hegemony in South China." *China Quarterly* 163 (2000):633–54.
Lawson, C. *Building the Fourth Estate: Democratization and the Rise of a Free Press in Mexico*. Berkeley, CA: University of California Press, 2002.
Leanza, Cheryl, and Feld, Harold. "More Than 'a Toaster with Pictures': Defending Media Ownership Limits." *Communications Lawyer* (Fall 2003): 12–22.
Lee, Chin-Chuan, He, Zhou, and Huang, Yu. "'Chinese Party Publicity Inc.' Conglomerated: The Case of the Shenzhen Press Group." *Media, Culture & Society* 28, no. 5 (2006): 581–602.
Lee, Chin-Chuan. "The Global and the National of the Chinese Media: Discourses, Market, Technology, and Ideology." In *Chinese Media, Global Contexts*, ed. Chin-Chuan Lee. London: RoutledgeCurzon, 2003.
Lee, Chin-Chuan. "Party-Market Corporatism, Clientelism, and Media in Shanghai." *The International Journal of Press/Politics* 12 (July 2007: 321–342.
Lee, Chin-Chuan. "State, Capital, and Media: The Case of Taiwan." In Curran, James and Park, Myunk-jin, eds., *De-westernizing Media Studies*. London: Routledge, 2000.
Lee, P. S. "Mass Communication and National Development in China: Media Roles Reconsidered." *Journal of Communication* 44, no. 3 (1994): 22–37.
Lehman-Wilzig, S., and Schejter, A. "Mass Media in Israel." In Kamalipour, Y., & Mowlana, H., eds. *Mass Media in the Middle East*. Westport, CT: Greenwood Press, 1994, 109–125.
Lessig, Lawrence. *Code and Other Laws of Cyberspace*. New York: Basic Books, 1999.
Leuven, K. U., and Valcke, Peggy. "Independent Study on Indicators for Media Pluralism in the Member States: Towards a Risk-Based Approach." Budapest: Ernst & Young and Central European University CMCS. Prepared for the European Commission Directorate-General Information Society and Media, 2007.
Levin, Harvey J. *Broadcast Regulation and Joint Ownership of Media*. New York: New York University Press, 1960.
Levin, Harvey. *Fact and Fancy in Television Regulation: An Economic Study of Policy Alternatives*. New York: Russell Sage, 1980.
Levy, Jonathan D., and Setzer, Florence O. *Measurement of Concentration in Home Video Markets*. Washington, DC: Office of Plans and Policy, Federal Communications Commission, Dec. 23, 1982, 51–53.
Li, Ziwei. "Media Transformation in China in the New Media Era." *Contemporary Manager* 23(2005): 6.
Li, Zhanjun. "A Look at the Radio Industry in the New Media Era." *China Broadcasts* 23(2011): 4.
Lichter, S. Robert, Lichter, Linda S., and Rothman, Stanley. *The Media Elite: America's New Powerbrokers*. New York: Hastings House, 1990.
Limor, Y. "Mass Media in Israel." *Trends in Israeli Society*. Ya'ar, E. and Shavit, Z., eds. Tel Aviv: Open University, 2003.
Lin, F. 'Dancing Beautifully, but with Hands Cuffed? A Historical Review of Journalism Formation during Media Commercialization in China." *Perspectives* 7(2): 79–98, 2006; <www.oycf.org/perspectives/33_06302006/Lin_Fen_DancingBeautifully.pdf>. Accessed on Nov. 9, 2006.
Litman, Barry. *The Motion Picture Mega-Industry*. Boston: Allyn & Bacon, 1988.
Litman, Barry R., & Bridges, J. "An Economic Analysis of Daily Newspaper Performance." *Newspaper Research Journal* 7, no. 3 (Spring 1986): 9–26.
Liu, Bonnie Rui. "New Strength of Competition and Innovation: China's Independent Television Production." *Matters of Communication*, 2010.
Liu, Y. L. (2008). "Broadband Policy, Market Competition, and User Adoption in Taiwan." In Dwivedi, Y. K., et al., eds., *Handbook of Research on Global Diffusion of Broadband Data Transmission*. New York: IGI Global, 47–59.
Llorens-Maluquer, Carles. *Concentración de empresas de comunicación y el pluralismo: la acción de la UE*. Doctoral thesis, Universidad Autónoma de Barcelona. Barcelona: TDX, 2001. (Publication No. B.24.148-2002). n.d. <http://www.tdx.cat/handle/10803/4095>.
Loveless, M. "Media Dependency: Mass Media as Sources of Information in the Democratizing Countries of Central and Eastern Europe." *Democratization*, 15(2008): 162–183.

Lv, Jie. "Rise of China's Local Radio Media." *China Newspaper Industry* 14(2011): 24.
Ma, E. K. "Rethinking Media Studies: The Case of China." In J. Curran & M. J. Park, eds., *De-Westernizing Media Studies*. London: Routledge, 2000, 21–34.
Ma, Shengrong. "A Discussion on Media Development and Strategy in China." *International Communications* 7(2011): 6.
Mahanagar, D. Bhawan, and Marg, J. Nehru. *Telecom Regulatory Authority of India: Consultation Paper on Issues relating to Media Ownership*. TRAI, New Delhi, 2013.
Malhotra, Dina N., ed. *60 Years of Book Publishing in India, 1947–2007*. New Delhi: Federation of Indian Publishers, 2006.
Marché du film. *FOCUS 2012: World Film Market Trends*. Cannes: Marché du film, 2011. Accessed on July 10, 2011, <http://www.obs.coe.int/oea_publ/market/focus.html>.
Mariscal, Judith. *The Mexican Telecommunications Reform: A Political Economy Approach*. Diss. The University of Texas at Austin, 1998. Austin: UT, 1998.
Mariscal, Judith, Bonina, Carla, and Luna, Julio. *Los nuevos escenarios del mercado en América Latina. Documento de Trabajo*. Lima: International Development Research Centre, 2007. <http://www.dirsi.net/sites/default/files/dirsi_07_PD03_es.pdf>.
Martin, Hugh. *A Study of How a Strategy Creating Clusters of Commonly Owned Newspapers Affects Prices, Quality and Profits*. Ph.D. dissertation, Michigan State University, 2003.
Masalha, Omar Amin. "Arabic Press in Israel." reader.co.il, 2009. Jan. 25, 2013. <http://www.reader.co.il/article/48627/תונותיעה-תיברעה-לארשיב>.
Mastrini, Guillermo, and Becerra, Martìn Alfredo. "Concentration in Media." In Donsbach, W., ed., *The International Encyclopedia of Communication*, vol. 3. Malden, MA: Wiley-Blackwell, 2008, 904–910.
Mastrini, Guillermo, ed. *Mucho Ruido y pocas leyes. Economia y politicas de comunicacion en la Argentina (1920–2004)*. 2nd ed. Buenos Aires: La Crujia, 2008.
Mastrini, Guillermo, and Becerra, Martin. *Periodistas y Magnates. Estructura y concentracion de las industrias culturales en America Latina*. Buenos Aires: Prometeo, 2006.
McCormick, Robert E. "The Strategic Use of Regulation: A Review of the Literature." In *The Political Economy of Regulation: Private Interests in the Regulatory Process*. Washington, DC: Federal Trade Commission, March 1984, 13, 18–25.
McEwen, M. "Media Ownership: Rules Regulations and Practices in Selected Countries and Their Potential Relevance to Canada." *Report for the Canadian Radio-Television and Telecommunications Commission* Toronto (2007).
McMurria, John, and Valdivia, Angharad N. "Citizenship and Media Ownership." *The International Encyclopedia of Media Studies*. Oxford: Blackwell Publishing Ltd., 2012.
McNair, Brian. "Power, Profit, Corruption, and Lies: The Russian Media in the 1990s." In Curran, James and Park, Myung-Jin, eds. *De-Westernizing Media Studies*. London: Routledge, 2000, 79–94.
MDDA. "Trends of Ownership and Control of Media in South Africa." Johannesburg: Media Development and Diversity Agency, 2009.
"Media Landscape: Germany." European Journalism Centre. 2012. Last accessed on Aug. 10, 2012 at <http://www.ejc.net/media_landscape/article/germany/>
Medina, Mercedes. "Algunas claves de la expansión de los grupos de comunicación españoles en Latinoamérica." *Comunicación y Sociedad* 16(2001): 71–99.
Meier, W. A., and Trappel, J. "Media Concentration and the Public Interest." In Mcquail, D., & Siune, K., eds., *Media Policy: Convergence, Concentration and Commerce*. London: Sage, 1998, 38–59.
Meyer, Philip, and Weardon, Stanley T. "The Effects of Public Ownership on Newspaper Companies: A Preliminary Inquiry." *Public Opinion Quarterly* 48(1984): 564–577.
Miller, Mark Crispin. "The Crushing Power of Big Publishing." *The Nation*, March 17, 1997, 11.
Miller, Mark Crispin. "Who Controls the Music?" *The Nation*, Aug. 25, 1997, 11.
Milgrom, Paul, and Roberts, John. *Economics, Organization and Management*. Englewood Cliffs, NJ: Prentice Hall, 1992, 552–569.
Millonzi, Joel C., and Noam, Eli M. *The International Market in Film and Television Programs*. Norwood, NJ: Ablex, 1993.
Ministry of Communications and Information Technology (MCIT). "Key Indicators." *Egypt ICT Indicators*. May 2010. Last accessed on Dec. 15, 2011 at <http://www.new.egyptictindicators.gov.eg/en/Indicators/_layouts/KeyIndicatorsViewer.aspx>
Ministry of Culture, Sports and Tourism of Korean Government. *Annual Report of Contents Industry*. 2008–2011. <http://www.mcst.go.kr/web/dataCourt/research/researchView.jsp>
Ministry of Internal Affairs and Communications, Japan. "2009 White Paper: Information and Communications in Japan." Ministry of Internal

Affairs and Communications, Japan. 2009. (Published quadri-annually) <http://www.soumu.go.jp/johotsusintokei/whitepaper/ja/h21/pdf/index.html>.

M-Net. "M-Net: History." M-Net, 2011. July 4, 2012. <http://www.mnetcorporate.co.za/ArticleDetail.aspx?Id=82>.

Monckeberg, M. O.: *Los Magnates de la Prensa en Chile. Concentración de los Medios de Comunicación en Chile*. Santiago: Random House Mondadori, 2009.

Moreira, Daniela. "Google inaugura novo centro de pesquisas em BH e busca engenheiros." *IDG Now!*, Nov. 10, 2006. Mar. 28, 2010. <http://idgnow.uol.com.br/mercado/2006/11/10/idgnoticia.2006-11-10.1445065867/>.

Moreira, Sonia Virgínia. "Geografias do público e do privado na configuração de um marco legal para as comunicações." *Intercom: Sociedade Brasileira de Estudos Interdisciplinares da Comunicação*, Sept. 2–6, 2011. Accessed on Dec. 26, 2012. <http://www.geografias.net.br/papers/2011/R6-0804-1.pdf>.

Moreira, S. V. "Para além dos clichês: o Brasil e o contexto internacional da radiodifusão digital." In Magnoni, A. F. e Carvalho, J. M. de. *O novo rádio, cenários da radiodifusão na era digital*. São Paulo: Editora SENAC SP, 2010, 177–201.

Morton, John. "Farewell to More Family Dynasties." *American Journalism Review* 17 (October 1995): 68.

Morton, John. "Wall Street Squeeze." *American Journalism Review* 27 (December/January 2006): 7.

Mott, Frank Luther. *Golden Multitudes: The Story of Best Sellers in the United States*. New York: Macmillan Co., 1947.

Myllylahti, Merja. *New Zealand Media Ownership Report 2013*. Centre for Journalism, Media and Democracy (JMAD), 2013.

Napoli, P. "The Marketplace of Ideas Metaphor in Communications Regulation." *Journal of Communication* 49(1999): 151–169.

Napoli, Philip M. "A Principal-Agent Approach to the Study of Media Organizations: Toward a Theory of the Media Firm." *Political Communication* 14 (April 1, 1997): 207–219.

Napoli, Philip M. "Audience Valuation and Minority Media: An Analysis of the Determinants of the Value of Radio Audiences." *Journal of Broadcasting and Electronic Media* 46(2002): 180–181.

Naor, M. *History of the Press in Israel: An Introduction*. Tel Aviv: Miksam, 1998.

Nasaw, David. *The Chief: The Life of William Randolph Hearst*. Boston: Houghton Mifflin, 2001.

National Telecom Regulatory Authority (NTRA). "Media Convergence Issues in Egypt and the Region and Prospects for the Future." *Scribd*. 2009. Last accessed on Nov. 27, 2011 at <www.scribd.com/doc/38482708/Doc9-NTRA>.

Navasky, Victor S. *A Matter of Opinion*. New York: Farrar Straus and Giroux, 2005.

Netanal, Neil. "Is the Commercial Mass Media Necessary, or Even Desirable, for Liberal Democracy?" Working paper, TPRC Conference on Information, Communications, and Internet Policy, October 2001, 20–24, Arlington, VA.

Nieto, Alfonso. *Time and the Information Market: The Case of Spain*. Pamplona: Media Market Monographs, 2000.

Nishuilleabhain, Aine, and Noam, Eli M., eds. *Private Networks and Public Objectives*. Amsterdam: Elsevier, 1996.

Nissen, C. *Public Service Broadcasting in the Information Society*. Strasbourg: Council of Europe, Directorate General of Human Rights, Media Division, 2006. Available at <http://www.coe.int/t/dghl/standardsetting/media/Doc/H-Inf(2006)003_en.pdf>. Accessed on Dec. 10, 2011.

Nixon, Raymond B. "Trends in U.S. Newspaper Ownership: Concentration with Competition." *Gazette* 14 (March 1968): 181–193.

Neumann, A. L. "The Great Firewall." CPJ (Committee to Protect Journalists) Press Freedom Report. Jan 1, 2001, at: <www.cpj.org/Briefings/2001/China_jan01/Great_Firewall.pdf>. Accessed on Mar. 3, 2003.

Noam, Eli. "Deregulation and Market Concentration: An Analysis of Post-1996 Consolidations." *Federal Communications Law Journal* 58, no. 3 (2006): 539–549. Accessed on Jan. 29, 2013. <http://www.law.indiana.edu/fclj/pubs/v58/no3/l-noam.pdf>..

Noam, Eli M. "The Internet: Still Wide Open and Competitive?" *Oxford Internet Institute, Internet Issue Brief* 1, August 2003.

Noam, Eli. *Media Ownership and Concentration in America*. Oxford: Oxford University Press, 2005.

Noam, Eli M. "The Next Frontier for Openness: Wireless Communications." In Noam, Eli M., & Steinbock, Dan, eds., *Competition for the Mobile Internet*. New York: Springer-Verlag, 2003, 21–38.

Noam, Eli M. "Opening the 'Walled Airwave,'" In Entman, R. ed. *Telecommunications Competition in a Consolidating Marketplace*. Queenstown, MD: The Aspen Institute, 2002, 33–55.

Noam, Eli M. *Privacy in Telecommunications: Markets, Rights, and Regulations*. Prospects, Cleveland: United Church of Christ, 1994.

Noam, Eli M. "Reconnecting Communications Studies With Communications Policy." *Journal of Communications* 43, no. 3 (Summer 1993): 199–206.
Noam, Eli M. *Telecommunications in Latin America*. New York: Oxford University Press, 1998.
Noam, Eli. *Telecommunications in Europe*. New York: Oxford University Press, 1992.
Noam, Eli.*Television in Europe*. New York: Oxford University Press, 1991.
Noam, Eli. "The Effect of Deregulation on Market Concentration: An Analysis of the Telecom Act of 1996 and the Industry Meltdown." *The Columbia Science and Technology Law Review* 4 (2003): 1–12. Last accessed on Jan. 4, 2013 at <http://www.citi.columbia.edu/elinoam/articles/Effect_of_Deregulation_on_MarketConcentration.pdf>
Noam, Eli M., ed. *Video Media Competition: Regulation, Economy, and Technology*. New York: Columbia University Press, 1985.
Noam, Eli M., and Wolfson, A. J., eds. *Globalism and Localism in Telecommunications*. Oxford: Elsevier Science Publishers, 1997.
Noble, J. Kendrick. "Book Publishing." In Compaine, Benjamin, Guback, Thomas, Noble Jr., J. Kendrick and Sterling, Christopher, eds., *Who Owns The Media?* White Plains, NY: Knowledge Industry Publications, 1982, 30.
Nowell-Smith, G., and Ricci, S. eds. *Hollywood and Europe: Economics, Culture, National Identity 1945-95*. 6th ed. London: BFI Publishing, 1998.
Odlyzko, Andrew. "The Economics of Electronic Journals." In Ekman, R., and Quandt, R., eds., *Technology and Scholarly Communication*. Berkeley, CA: University of California Press, 1998.
OECD. Communications Outlook. Geneva: OECD, 2013. <http://www.swisscom.ch/content/dam/swisscom/de/ghq/media/documents/OECE_Communications_Outlook_2013.pdf.dl.res/OECE_Communications_Outlook_2013.pdf>.
Ofcom. "Media Ownership Rules Review Consultation Document." London: Office of Communications, July 31, 2009, 35–40. Accessed on May 18, 2012 <http://stakeholders.ofcom.org.uk/binaries/research/cmr/cmr09.pdf>.
Organisation for Economic Co-operation and Development. *News in the Internet Age: New Trends in News Publishing*. Paris: OECD, 2010.
Organization of Economic Co-operation and Development (OECD). "Total Broadband Subscriptions in the OECD Area." *OECD Communication Outlook 2011*. Paris: OECD, 2011.
Orozco, G. *Historias de La televisión en América Latina*. Barcelona: Gedisa, 2002.
Owen, Bruce M. "Regulatory Reform: The Telecommunications Act of 1996 and the FCC Media Ownership Rules." Prepared by the *Law Review of Michigan State University–Detroit College of Law* for the Fourth Annual James H. Quello Communication Policy and Law Symposium, Washington DC, Feb. 27, 2003.
Owen, Bruce M., et al. *An Economic Analysis of the Broadcast Television National Ownership, Local Ownership and Cross-Ownership Rules*. Washington, DC: Economists Incorporated, 1995, 26.
Owen, Bruce M., and Wildman, Steven S. *Video Economics*. Cambridge, MA: Harvard University Press, 1992.
Ozanich, G. W., and Wirth, M. O. "Media Mergers and Acquisitions: A Communications Industry Overview." In Alexander A., Owers, J., and Carveth, R., eds., *Media Economics: Theory and Practice*. 2nd ed. Mahwah, NJ: Lawrence Erlbaum Associates, 1998, 95–107.
Paltridge, Sam. *Internet Infrastructure Indicators*. Paris: OECD, Working Party on Telecommunication and Information Service Policies, Directorate for Science, Technology, and Industry, Oct. 28, 1998.
Papandrea, Franco. "Media Diversity and Cross Media Regulation." *Prometheus* 24, no. 3 (2006): 301–322.
Papathanassopoulos, S. "The Mediterranean/Polarized Pluralist Media Model Countries." In Terzis, George, ed., *European Media Governance: National and Regional Dimensions*. Bristol, UK; Chicago: Intellect, 2007.
Park, Myung-jin, Kim, Chang-Nam, and Sohn, Byung-Woo. "Modernization, Globalization, and the Powerful State: The Korean Media." In Curran, James Curran, and Park, Myung-, eds., *De-westernizing Media Studies*. London: Routledge, 2000.
Paxman, A., and Saragoza, A. "Globalization and Latin Media Powers: The Case of Mexico's Televisa." In Mosco, V., and Schiller, D., eds. *Continental Order? Integrating North America for Cybercapitalism*. Lanham: Rowman & Littlefield, 2001.
Perlman, Allison, and Amaya, Hector. "Owning a Voice: Broadcasting Policy, Spanish Language Media, and Latina/o Speech Rights." *Communication, Culture & Critique* 6, no. 1 (2013): 142–160.
Peruško, Zrinjka, and Popović, Helena. "Media Concentration Trends in Central and Eastern Europe." In *Finding the Right Place on the Map: Central and Eastern European Media Change in a Global Perspective*. Chicago: Intellect Books/University of Chicago Press, 2008, 165–189.

Peruško, Zrinjka. "The Link that Matters: Media Concentration and Diversity of Content." In Klimkiewicz, Beata, ed., *Media Freedom and Pluralism: Media Policy Challenges in the Enlarged Europe*. Budapest: Central European University Press, 2010.

Petkovic, Brankica, and Basic Hrvatin, Sandra. "Media Ownership Regulation in South-East European and New EU Member Countries." *MulitMedia und Recht*, beck. de Zeitungen, MMR 2004, Heft 8, XXI. 2004.

Picard, Robert G. "Economics of the Daily Newspaper Industry." In Alexander, A., Owes, J., and Carveth, R., eds., *Media Economics: Theory and Practice*. New York: Lawrence Erlbaum Association, 1993, 181–204.

Picard, Robert G. "Measures of Concentration in the Daily Newspaper Industry." *Journal of Media Economics* 1 (1988): 61–74.

Picard, Robert G. *The Economics and Financing of Media Companies*.New York: Fordham University Press, 2002.

Picard, Robert, and Soloski, John. "The New Media Lords: Why Institutional Investors Call the Shots." *Columbia Journalism Review* 35, no. 3 (September/October 1996): 11–12.

Piramal, G. *Business Maharajas*. New Delhi: Penguin, 1996.

Posner, Richard A. "The Chicago School of Antitrust Analysis." *University of Pennsylvania Law Review* 127(1979): 925.

Powell, Michael K. "The Public Interest Standard: A New Regulator's Search for Enlightenment," presented at the *17th Annual Legal Forum on Communications Law*. Las Vegas, Apr. 5, 1998.

Powers, A. "Competition, Conduct, and Ratings in Local Television News: Applying the Industrial Organization Model." *Journal of Media Economics* 6, no. 2 (1993): 37–44.

Pozen, Robert. "Institutional Investors: The Reluctant Activists." *Harvard Business Review*, January/February 1994, 140.

Prario, Benedetta. *Le trasformazioni dell'impresa tv verso l'era digitale*. Bern: Peter Lang, 2005.

PricewaterhouseCoopers. *German Entertainment and Media Outlook: 2010–2014—Die Entwicklung des deutschen Unterhaltungs- und Medienmarktes*. Frankfurt am Main, 2010, 5–119.

Primeaux, W. J., Rice, E., and Simon, J. "The Price Effects of Monopoly Ownership in Newspapers." *Antitrust Bulletin* 31, no. 1 (Spring 1986): 113–131.

Pritchard, David. "A Tale of Three Cities: 'Diverse and Antagonistic' Information in Situations of Local Newspaper/Broadcast Cross-Ownership." *Federal Communications Law Journal* 54 (December 2001): 31–51.

Qin, Bei, Wu, Yanhui, and Strömberg, David. "The Determinants of Media Bias in China," unpublished paper in progress, University of Stockholm, 2013.

Raff, Daniel M. G. "Superstores and the Evolution of Firm Capabilities in American Bookselling." *Strategic Management Journal* 21, no. 10/11 (October/November 2000): 1043–1059.

Rantanen, Terhi. *Globalnoe I nacionalnoe. Mass media I communicacii v postcommunisticheskoi Rossii. (The Global and the Local. Mass Media and Communications in Postcommunist Russia)*. Moscow: Izdatelstvo MGU, 2005.

Rawnsley, G. D., and Rawnsley, M.-Y. T. "Regime Transition and the Media in Taiwan." In V. Randall (ed.), *Democratization and the Media* (pp. 106–124). London, UK: Frank Cass, 1998.

Ray, W. B. *FCC: The Ups and Downs of Radio–TV Regulation*. Iowa: Iowa State University Press, 1990.

Reed, John. "Debate: Will Publishers Soon Be Irrelevant?" *Publishing Talk*, Apr. 18, 2011. Dec. 20, 2012.

Regulation and Competition Disputes. *Media plurality (déjà vu): The UK Department for Culture Media & Sport (DCMS) consultation*. Competition Disputes Regulation, 2013. Last accessed on Jan. 29, 2014 at www.ceg-europe.com.

Reporters Without Borders: *Press Freedom Index 2010*, <http://en.rsf.org/press-freedom-index-2010,1034.html>. Accessed on Sept. 15, 2011.

Richeri, Giuseppe. "Broadcasting and the Market: The Case of Public Television." In Calabrese, Andrew and Sparks, Colin, eds., *Toward a Political Economy of Culture*. Maryland: Rowman & Littlefield Publishers, 2004.

Richter, Andrei. *A Landmark for Mass Media in Russia*. Strasburg: European Audiovisual Observatory, 2011.

Riordan, Michael H., and Salop, Steven C. "Evaluating Vertical Mergers: A Post-Chicago Approach." *Antitrust Law Journal* 63(1995): 513.

Rizk, Nagla. "Stories from Egypt's Music Industry: De Facto Commons as Alternative to Copyright." In Rizk, Nagla and Shaver, Lea, eds., *Access to Knowledge: New Research on Intellectual Property and Development*. Bloomsbury Academic, 2010, 103. Last accessed on May 29, 2012 at <http://www.aucegypt.edu/Business/A2K4D/Documents/A2KEgypt.pdf>

Roger, G. "Media Concentration with Free Entry." *Journal of Media Economics* 22, no. 3 (2009): 134–163.

Rolland, A. "The Norwegian Media Ownership Act and the Freedom of Expression." *International Journal of Media and Cultural Politics*, 4, no. 3 (2008): 313–330.

Rosse, James N. "The Evolution of One-Newspaper Cities," presented at the *Federal Trade Commission Symposium on Media Concentration*, Dec. 14–15, 1978, Washington, DC.:

Rothenbuhler, Eric W., and Dimmick, John W. "Popular Music: Concentration and Diversity in the Industry, 1974–1980," *Journal of Communication* 33, no. 1 (Winter 1982): 143–149.

Ruffolo, John, Gluckman, Gary, and Lee, A. Richard. *Directing the Future*. Deloitte Canada: Deloitte Technology Media and Telecommunications (TMT) Group, 2013.

Rugh, William. A. *Arab Mass Media: Newspapers, Radio & Television in Arab Politics*. New York: Praeger, 2004.

Rutenberg, Jim. "Fewer Media Owners, More Media Choices." *The New York Times*, Dec. 2, 2002, sec. C, col. 2, 1.

Safire, William. "Localism's Last Stand." *The New York Times*, July 17, 2003, 27.

Sakr, Naomi. "Satellite Television and Development in the Middle East." *Middle East Research and Information Project 2010*. Last accessed on May 29, 2012 at <http://www.merip.org/mer/mer210/satellite-television-development-middle-east>.

Sánchez-Ruiz, E. "Globalization, Cultural Industries and Free Trade: The Mexican Audiovisual Sector in the NAFTA Age." In Mosco, V., and Schiller, D., eds., *Continental Order? Integrating North America for Cybercapitalism*. Lanham: Rowman & Littlefield, 2001.

Sánchez-Tabernero, Alfonso. "Competition between Public Service and Commercial Television Broadcasting in the European Market," presented at 6th *World Media Economics Conference*. Centre d'Etudes Sure les Médias and Journal of Media Economics, HEC Montreal, Montreal, Canada, May 12–15, 2004.

Sánchez-Tabernero, Alfonso. *Media Concentration in Europe: Commercial Enterprise and the Public Interest*. Düsseldorf: European Institute for the Media, 1993.

Sánchez-Tabernero, Alfonso. "The Future of Media Companies: Strategies for an Unpredictable World." In Picard, R. G., ed., *Strategic Responses to Media Market Changes*. Jönköping, Sweden: Media Management and Transformation Centre, Jönköping International Business School, 2004, 19–34.

Sánchez-Tabernero, Alfonso, and Carvajal, Miguel. *Media Concentration in the European Market: New Trends and Challenges*. Pamplona: Universidad de Navarra, 2002.

Schejter, A. "'Rulers of Thousands, Rulers of Hundreds, Rulers of Fifties, and Rulers of Tens': Does Democracy Count?" *Beyond Broadband: Developing Data-Based Information Policy Strategies*. Taylor, R. and Schejter, A., eds. New York: Fordham University Press, 2013.

Schejter, A., and Han, S. "Regulating the Media: Four Perspectives." In Levi-Faur, D., ed., *Politics of Regulation Handbook*. London: Edward Elgar, 2011, 243–253.

Scherer, F. M. *Industrial Market Structure and Economic Performance*. Chicago: Rand McNally, 1971.

Schiffrin, Andre. *The Business of Books*. New York: Verso, 2000.

Schlesinger, Arthur. *Prelude to Independence: The Newspaper War on Britain, 1764–1776*. New York: Knopf, 1958.

Schudson, Michael. *Discovering the News: A Social History of American Newspapers*. New York: Basic Books, 1978.

Select Committee on Communications of the United Kingdom. *Inquiry into Media Convergence and Its Public Policy Impact*. London: House of Lords, United Kingdom, 2012.

Serrin, William, ed. *The Business of Journalism*. New York: New Press, 2000.

Shahine, Alaa. "New Media Flourish in Egypt." *UCLA Center for Communications and Community*. Apr. 27, 2004. Last accessed on 17 Jan. 2012 at <http://www.c3.ucla.edu/newsstand/global/new-media-flourish-in-egypt/>

Shapiro, Carl. "Exclusivity in Network Industries." *George Mason Law Review* 7, no. 3 (Spring 1999): 673, 678.

Shapiro, Carl. "Mergers with Differentiated Products." *Antitrust*, Spring 1996, 23, 28–29.

Shapiro, Carl, and Varian, Hal. *Information Rules*. Boston, MA: Harvard Business School Press, 1999.

Shih, Jun-ji, Liu, Kungchung, Shu Yeh, and Chou, Yuntsai. *The Market Survey of Cable TV and the Competition Policy Analysis*. Taipei: GIO Research Report, 2003.

Shirk, Susan L., ed. *Changing Media, Changing China*. Oxford University Press, 2011.

Shleifer, Andrei, Mcliesh, Caralee, Nenova, Tatiana, and Djankov, Simeon. "Who Owns the World's Media?" *World Bank Policy Research Working Papers*, June 2001, 58. <http://dx.doi.org/10.1596/1813-9450-2620>

Sidak, J. Gregory, and Spulber, Daniel F "Deregulation and Managed Competition in Network Industries." *Yale Journal of Regulation* 15, no. 1 (Winter 1998): 117, 122–125.

Snider, James H., and Page, Benjamin I. "Does Media Ownership Affect Media Stands? The Case of the Telecommunications Act of 1996," presented at the *55th MPSA Annual National Conference*. Midwest Political Science Association, Chicago, IL, April 1997.

Smirnov, Sergey S. *Koncentraciya sredstv massovoi informacii Rossii v usloviyah transformacii nacionalnoi mediasistemy (1991–2006). (Concentration of Russian Mass Media in the context of national media system transformation (1991–2006).* Diss. Moscow: Moscow State University, Faculty of Journalism, 2006.

Smirnov, Sergey S. "Koncentraciya v industrii SMI (Concentration in the Media Industry)." In Vartanova, Elena L., ed. *Osnovy mediabiznesa*. Moscow: Aspekt Press, 2009, 152–189.

Smith, Rachael Craufurd, and Tambini, Damian. "Measuring Media Plurality in the United Kingdom: Policy Choices and Regulatory Challenges." *Journal of Media Law* 4, no. 1 (2012): 35–63. Print.

Squire, Jason, ed. *The Movie Business Book*. New York, NY: Fireside, 1992, 283.

Starr, Paul. *The Creation of the Media: Political Origins of Modern Communications*. New York: Basic Books, 2004, 131.

The State Administration of Radio, Film and Television (SARFT). "China Radio, Film, and TV Development." Beijing: SARFT: 2009.

Steinbock, Dan. *The Nokia Revolution*. New York: American Management Association, 2001, 102.

Stockmann, Daniela. "Media Commercialization and Authoritarian Rule in China." *Communication, Society, and Politics*. Cambridge University Press. 2012.

Stockmann, Daniela, and Gallagher, Mary E. "Remote Control: How the Media Sustain Authoritarian Rule in China." *Comparative Political Studies*. Vol. 44, no. 4 (April 2011): 436–467.

Sturm, John F. "Time for Change on Media Cross-Ownership Regulation." *Federal Communications Law Journal* 57, no. 2 (March 2005): 201.

Sturke and Grunes, (2013). The AT&T/T–Mobile Merger: What Might Have Been? *Journal of European Competition Law & Practice*, 3(2). <http://jeclap.oxfordjournals.org/content/3/2/196.full.pdf?keytype=ref&ijkey=xQTu254rjEa8vMX>.

Sunstein, Cass. *Republic.com*. Princeton, NJ: Princeton University Press, 2001.

Sussman, G. Ownership in the Media. In Donsbach, W., *The International Encyclopedia of Communication,* vol. 3. Malden, MA: Wiley-Blackwell, 2008, 3480–3485.

Tebbel, John, and Zuckerman, Mary Ellen. *The Magazine in America, 1741–1990*. New York: Oxford University Press, 1991, 140–141.

Telecom Egypt. "Ownership Structure." *Telecom Egypt*. Last accessed on June 27, 2012 at <http://telecom.tedata.net/uipages/Ownership_Structure.aspx>.

*Television & Cable Fact Book*. Washington, DC: Warren Publishing, Annual editions.

The Cairo Institute for Human Rights Studies. *Media in North Africa: Obstacles and Challenges*. United Nations Economic Commission for Africa, 2007, 4. Last accessed on Jan. 16, 2012 at <http://www.uneca.org/africanmedia/documents/NorthAfrica-eng-report.pdf>.

Thorsen, Einar. *Three-D Issue 21: Tackling Ownership at a Snail's Place, the Government's Consultation on Plurality*. Available at: <http://www.meccsa.org.uk/news/three-d-21-tackling-ownership-at-a-snails-place-the-governments-consultation-on-plurality/>

Tiffen, Rodney. "From Technological Abundance to Commercial Monopoly in Australian Pay TV: Key Relationships in Institutionalising Subscription Television." In Kenyon, Andrew T. (ed.), *TV Futures: Digital Television Policy in Australia*. Melbourne: Melbourne University Press, 2007.

Tirole, Jean. *The Theory of Industrial Organization*. Cambridge, MA: MIT Press, 1988, 173–181.

Tomaselli, K. "Media Ownership and Democratization: The Political Economy of the South African Print Media." In Hyden G., Leslie M., and Ogundimu F. F., eds. *Media and Democracy in Africa*. London: Transaction Books, 2003.

Tong, Jingrong, and Sparks, Colin. "Investigative Journalism in China Today." *Journalism Studies* 10, No. 3 (2009): 337–352.

Trakhtenberg, Anna. "Transformation without Change. Russian Media as a Political Institution and Cultural Phenomenon." In Vartanova, Elena L., ed. *Media and Change*. Moscow: MediaMir, 2007.

Tunç, A. "Faustian Acts in Turkish Style: Structural Change in National Newspapers as an Obstacle to Quality Journalism in 1990–2003." In O. Spassov, O., ed. *Quality Press in Southeast Europe*. Sofia: SOEMZ/Sofia University, 2003.

Tunstall, Jeremy. "A Media Industry Perspective." In Anderson, James A., ed. *Communication Yearbook 14*. Beverly Hills: Sage, 1978, 163–186.

UK Department for Culture Media and Sport. *Media Ownership and Plurality*. London: UK Department for Culture Media & Sport, 2013.

UNESCO Institute for Statistics, *The Media Landscape in 28 Countries*. UNESCO, Paris, 2012.

United States Department of Justice and Federal Trade Commission (2010). *Horizontal Merger Guidelines*. Washington, DC: Authors. <http://www.justice.gov/atr/public/guidelines/hmg--2010.pdf>

Valckie, Peggy, and Lievens, Eva. *Media Law in Belgium*. Alphen aan den Rijn: Kluwer Law International BV, 2011.

Van Loon, Ad. *Europäische Kommission: Pluralismus and Medienkonzentration*. Europäische Audiovisuelle Informationstelle Brüssel (2013).

Vartanova, Elena L. "The Russian Media Models in the Context of Post-Soviet Dynamics." Hallin, Daniel and Mancini, Paolo, eds. *Comparing Media Systems beyond the Western World*. Cambridge: University Press, 2012.

Vartanova, Elena L. and Smirnov, Sergey S. "Contemporary Structure of the Russian Media Industry." In Rosenholm, Arja, Nordenstreng, Kaarle, and Trubina, Elena, eds., *Russian Mass Media and Changing Values*. London: Routledge, 2010, 21–40.

Veronis Suhler & Associates. *Communications Industry Forecast*. New York: Veronis Suhler & Associates, Annual Reports.

"Vertical Integration and Media Regulation in the New Economy." *Yale Journal on Regulation* 19, no. 1 (Winter 2002): 171.

Vīķe-Freiberga, Vaira u. a. *A Free and Pluralistic Media to Sustain European Democracy*. High Level Group on Media Freedom and Pluralism, European Commission, 2013.

Vīķe-Freiberga, Vaira, Herta Däubler-Gmelin, and Ben Hammersley. "High Level Group on Media Freedom and Pluralism." Brussels: Digital Agenda for Europe, 2013.

Vizcarrondo, Tom. "Measuring Concentration of Media Ownership: 1976–2009." *International Journal on Media Management* 15, no. 3 (2013): 177–195.

Vogel, Harold L. *Entertainment Industry Economics*. 7th ed. Cambridge: Cambridge University, 2013.

Waldfogel, Joel. "Consumer Substitution among Media," Media Bureau Staff Research Paper, Media Ownership Working Group, Federal Communications Commission, September 2002.

Waldman, Stephen, and the Working Group on Information of Communities. "The Information Needs of Communities: The Changing Media Landscape in a Broadband Age. Washington, DC: FCC, 2011. Jan. 29, 2013. <http://www.fcc.gov/info-needs-communities>.

Wall Communications. *Price Comparisons of Wireline, Wireless and Internet Services in Canada and with Foreign Jurisdictions* (Prepared for CRTC). Ottawa: Author, 2013. <http://www.wallcom.ca/pdfs/price---comp---report_2013 update.pdf>.

Wallsten, Scott. "Understanding International Broadband Comparisons." Washington, DC: Technology Policy Institute, 2009. Jan. 18, 2011. <http://www.techpolicyinstitute.org/files/understanding%20international%20broadband%20comparisons%202009%20update%207-9.pdf>.

Ward, David, Fueg, Oliver C., and D'amo, Alessandro. "A Mapping Study of Media Concentration and Ownership in Ten European Countries." Hilversum: Commissariaat voor de Media, 2004. Jan. 31, 2013. <http://www.mediamonitor.nl/dsresource?objectid=435&type=org>.

Werbach, Kevin. "A Layered Model for Internet Policy." *Journal of Telecommunications and High Technology Law* 1(2002): 58–64.

Werbowski, Michael. "'El Excelsior': The Rise and Fall of a Great Mexican Newspaper." *Worldpress.org*, Feb. 18, 2007. Accessed on Jan. 17, 2013. <http://www.worldpress.org/Americas/2678.cfm#down>.

Wiio, O. A. "Open and Closed Mass Media Systems and Problems of International Communication Policy." *Studies of Broadcasting* 13(1977): 67–90.

Williams, Dmitri. "Synergy Bias: Conglomerates and Promotion in the News." *Journal of Broadcasting and Electronic Media* 46 (September 2002): 453–472.

Williamson, Oliver E. *Markets and Hierarchies, Analysis and Antitrust Implications*. New York: Free Press, 1975, 29–30, 35–37.

Wimmer, Kurt A. "Deregulation and the Future of Pluralism in the Mass Media: The Prospects for Positive Policy Reform." *Mass Communications Review*, 1988.

Winfield, B. H., Mizuno, T., and Beaudoin, C.E. "Confucianism, Collectivism and Constitutions: Press Systems in China and Japan." *Communication Law and Policy* 5, no. 3 (2000): 323–347.

Winfield, Betty Houchin, and Peng, Zengjun. "Market or Party Controls? Chinese Media in Transition." *International Communication Gazette* 67, no. 3 (June 2005): 255–270.

Winseck, D. (forthcoming). "The Network Media Economy: Triumph of the Media Infrastructure Industries, or Crisis of Media?" In Hong, J. H. (eds.). *New Approaches to Media and Communication Studies*. Beijing: Xinhua University (forthcoming, in Chinese).

Winseck, Dwayne. "The State of Media Ownership and Media Markets: Competition or Concentration and Why Should We Care?" *Sociology Compass*, 2, no. 1 (2008): 34–47.
Winseck, D. (1998). *Reconvergence: Toward a Political Economy of Telecommunications in Canada*. Cresskill, NJ: Hampton.
Wolff, Michael. *The Man Who Owns the News: Inside the Secret World of Rupert Murdoch*. New York: Knopf, 2008.
Woodhull, N., and Snyder, R., eds. *Media Mergers*. New Brunswick: Transaction Publishers, 1998.
World Association of Newspapers and News Publishers (WAN-IFRA). *World Press Trends 2010*. WAN-IFRA/ZenithOptimedia, 2010. Dec. 28, 2012. <http://www.wan-ifra.org/reports/2010/07/23/world-press-trends-2010-edition>.
Wu, G. "One Head, Many Mouths: Diversifying Press Structures in Reform China." In C.-C. Lee, ed., *Power, Money, and Media: Communication Patterns*, Evanston, IL: Northwestern University Press 2000.
Wyly, Brendan. "Competition in Scholarly Publishing? What Publisher Profits Reveal." in *ARL: A Bimonthly Report on Research Library Issues and Actions from ARL, CNI, and SPARC*. Washington, DC: Association of Research Libraries, 1998.
Yang, Guobin, and Calhoun, Craig. "Media, Civil Society, and the Rise of a Green Public Sphere in China." *China Information* 21.2 (2007): 211–236.
Yoo, Christopher S. "Architectural Censorship and the FCC." *Southern California Law Review* 78 (Spring 2005): 669.
Yu, X. 'Professionalization without Guarantees: Changes of the Chinese Press in Post-1989 Years." *Gazette* 53(1–2): 23–41, 1994.
Zhao, J. "A Snapshot of Internet Regulation in Contemporary China. *China Media Research, 4*, no. 3 (2008): 37–42.
Zhao, Y. Z. *Media, Market and Democracy in China: Between the Party Line and the Bottom Line*. Urbana and Chicago: University of Illinois Press, 1998.
Zhao, Y. Z. "From Commercialisation to Conglomeration: The Transformation of the Chinese Press within Orbit of the Party State," *Journal of Communication* 50(2): 3–26, 2000.
Zhao, Y. "The State, Market, and Media Control in China." In P. Thomas & Z. Nain, eds., *Who Owns the Media: Global Trends and Local Resistance*). Penang, Malaysia: Southbound, 2005, 179–212.
Zheng, Baoyu. "Business Model for China's TV Broadcasting Industry." *Modern Audio-Video Arts* 13 (2009):1.
Zongyuan, Shi. "Chinese Media Official Interviewed on Central Control of the Internet." *BBC Monitoring Asia Pacific – Political*. London. Oct. 29, 2002.

# Index

Figures, graphs and tables are indicated by f, g and t following the page number.